Endotoxin in Health and Disease

Endotoxin in Health and Disease

edited by

Helmut Brade
Research Center Borstel
Borstel, Germany

Steven M. Opal
Brown University School of Medicine, Providence,
and Memorial Hospital of Rhode Island,
Pawtucket, Rhode Island

Stefanie N. Vogel
Uniformed Services University of the Health Sciences
Bethesda, Maryland

David C. Morrison
Saint Luke's Hospital and University of Missouri–Kansas City
Kansas City, Missouri

MARCEL DEKKER, INC.   NEW YORK • BASEL

Library of Congress Cataloging-in-Publication Data

Endotoxin in health and disease / edited by Helmut Brade . . . [et al.].
   p. cm.
   Includes bibliographical references and index.
   ISBN 0-8247-1944-1 (alk. paper)
   1. Endotoxins. I. Brade, Helmut.
QP632.E4E53   1999
615.9′5293—dc21   99-29165
                                                                                                      CIP

This book is printed on acid-free paper.

Headquarters
Marcel Dekker, Inc.
270 Madison Avenue, New York, NY 10016
tel: 212-696-9000; fax: 212-685-4540

Eastern Hemisphere Distribution
Marcel Dekker AG
Hutgasse 4, Postfach 812, CH-4001 Basel, Switzerland
tel: 41-61-261-8482; fax: 41-61-261-8896

World Wide Web
http://www.dekker.com

The publisher offers discounts on this book when ordering in bulk quantities. For more information, write to Special Sales/Professional Marketing at the headquarters address above.

Copyright © 1999 by Marcel Dekker, Inc. All Rights Reserved.

Neither this book nor any part may be reproduced or transmitted in any form or by any means, electronic or mechanical, including photocopying, microfilming, and recording, or by any information storage and retrieval system, without permission in writing from the publisher.

Current printing (last digit):
10 9 8 7 6 5 4 3 2 1

PRINTED IN THE UNITED STATES OF AMERICA

Preface

It was just about a century ago that endotoxin, a heat-stable product of bacteria, was recognized as a microbial constituent capable of causing deleterious effects when administered to experimental animals. Since that time, endotoxin has been repeatedly demonstrated by a multitude of investigative scientists to manifest biological activities of almost unprecedented diversity when tested in either in vitro or in vivo systems in virtually any field of science. Although this remarkable array of endotoxin-dependent responses would have ensured almost any investigator publication in an earlier era of scientific investigation, the apparent nonspecificity of this microbial toxin, in comparison to the relatively precise narrow spectrum of effects of other, more well-characterized bacterial toxins, hindered its acceptance by the infectious disease and bacteriological sciences disciplines as a "true" pathological factor of the virulent disease-causing microbial pathogen.

Adding to this problem of nonspecificity of host responses, many early studies led to diverse findings, even as to the chemical nature of the active constituent. A further confounding observation was that endotoxin could be isolated from both virulent and avirulent organisms, thus contrasting with exotoxins whose expression in microbes correlated well with manifestation of disease pathogenesis. Also of concern to investigators was the fact that exotoxins were microbial products that were actually synthesized and secreted by microbes, whereas it was a generally held belief that bacteriolysis was a prerequisite for liberation of significant quantities of endotoxic materials. Finally, it was also confusing that, in contrast to bacterial exotoxins, doses of endotoxin that were orders of magnitude greater were required in order to elicit detectable disease. Even when the active principle of endotoxin, lipopolysaccharide (LPS), and its biologically active subcomponent, lipid A, were identified, concerns persisted about the importance of endotoxin (LPS) to the pathophysiology of infection-related diseases.

This situation changed dramatically in the years following the discovery of lipid A and methods to isolate and purify LPS free of other microbial constituents. This chemical evolution coincided with significant advancement in techniques to purify and evaluate in in vitro cultures a variety of host immune and inflammatory systems in a reproducible and quantitative manner. What emerged from these parallel technological evolutionary pathways was a synthesis of investigative opportunities that quickly established endotoxin (LPS) as among the most potent of stimuli for activation of a diverse array of host cells and soluble cell-free biological fluid systems. In this respect, endotoxin became one of the prototype stimuli for in vitro activation of lymphocytes, monocytes and macrophages, fibroblasts, platelets, and polymorphonuclear leukocytes, as well as the plasma intrinsic coagulation pathway and both pathways for serum complement activation. Collectively these in vitro findings seemed to provide a compelling rationale for the earlier in vivo studies that reported such a wide array of effects of endotoxin.

Based on this explosion of information regarding the chemical and biological properties of endotoxin, there were strong expectations that this increased knowledge would, and should, be readily transferable to the gram-negative microbe-infected septic patient. Gram-negative sepsis accounted for almost half of all cases of severe sepsis seen by infectious disease physicians, and the implementation of appropriate antimicrobial chemotherapy and improved supportive care had not resulted in a dramatically decreased incidence of morbidity and mortality. It was therefore convincingly argued that the application of this new knowledge base regarding endotoxin would significantly amplify the overall effectiveness of treatment of this important disease. Enthusiasm for this concept derived from a number of very promising preclinical studies. These studies clearly established that in vivo immunotherapies that targeted either the endotoxin molecule itself or the proinflammatory mediators induced in inflammatory cells following their noncytotoxic interaction with endotoxin were effective in reducing mortality in a variety of experimental models of endotoxemia and gram-negative sepsis. Additional fuel for this therapeutic approach was provided by the very promising published clinical trials that seemed to support the preclinical results regarding protective efficacy of immunotherapeutic intervention strategies in the treatment of sepsis.

Regrettably, the extrapolation of these very promising preclinical studies to the septic patient has almost invariably led to the conclusion that no demonstrable efficacy could be reproducibly documented by such interventions. These conclusions were based on the results of almost two dozen carefully conducted double-blind, placebo-controlled studies on septic patients using a variety of antiendotoxin antibody, anticytokine, and cytokine receptor reagents and other soluble mediators of inflammation. Such findings have, in general, had a profound dampening effect on the enthusiasm of the pharmaceutical industry to support continued investment of resources into further novel treatment strategies in sepsis. However, they have also caused a number of infectious disease physicians and basic scientists to begin to question the very premise on which these expensive clinical trials were based. What is the precise role of endotoxin and the induction of the host inflammatory response in the overall pathogenic events that lead to multiple organ failure? The answer to this question is obviously complex and dependent upon many possibilities, all of which are likely to contribute to some degree in individual patient subpopulations.

What has emerged from these critical evaluative postclinical trial examinations is the not unanticipated conclusion that the existing knowledge base on which these early clinical trials were based was insufficient to accurately predict or assess the consequences of intervention by application of single-target therapeutic reagents. What was urgently needed, therefore, was increased understanding of the biology, chemistry, pharmacology, immunology, and pathology of endotoxin and its effects on host systems. Advocates of this position, of which there are many, will therefore not be disappointed by the contents of this volume, which provides information on the very latest advances in knowledge in almost all areas relating to endotoxin. Because there has been an explosion of new findings regarding endotoxin in virtually every scientific discipline that this molecule impacts to some extent, we found it impossible to compile these findings into a concise and easily manageable text. Rather, as can be readily concluded by a quick perusal of the Contents, this collection of contributions by the leading investigators in the field of endotoxin research requires more than 60 comprehensive and information-filled chapters.

Nevertheless, we feel confident that this comprehensive volume of *Endotoxin in Health and Disease* will provide the interested reader with an invaluable and up-to-date reference standard for the current status of endotoxin research as we move into the twenty-first century. It is inevitable that, given the current level of interest in and excitement about understanding the cellular and molecular bases for the immunological, pharmacological, and pathophysiological effects of endotoxin, new advances will continue to appear in the scientific and medical literature. These ever-expanding pathways of inquiry will result in the more rational design of truly effective therapeutic intervention strategies. The ultimate benefit, of course, will be to the septic patient and significantly improved health care treatments with increased survival and return to society. We envision that this projected scenario will not be long in coming.

Helmut Brade
Steven M. Opal
Stefanie N. Vogel
David C. Morrison

Contents

Preface — iii
Contributors — xi

1. Endotoxin: Historical Perspectives — 1
   Ernst T. Rietschel and Otto Westphal

2. Lipopolysaccharide and the Permeability of the Bacterial Outer Membrane — 31
   Martii Vaara

3. Lipopolysaccharide Phase Variation in Haemophilus and Neisseria — 39
   Derek W. Hood and E. Richard Moxon

4. Antigenic Mimicry in Neisseria Species — 55
   Peter C. Giardina, Andrew Preston, Brad Gibson, and Michael A. Apicella

5. Antibiotic-Induced Endotoxin Release: Important Parameters Dictating Responses — 67
   Jesse J. Jackson and Helmut Kropp

6. Complement-Mediated Lipopolysaccharide Release — 77
   Vernon L. Tesh

7. Chemical Structure of Lipid A: Recent Advances in Structural Analysis of Biologically Active Molecules — 93
   Ulrich Zähringer, Buko Lindner, and Ernst T. Rietschel

8. Chemical Structure of the Core Region of Lipopolysaccharides — 115
   Otto Holst

9. The Chemistry of O-Polysaccharide Chains in Bacterial Lipopolysaccharides — 155
   Per-Erik Jansson

10	The Chemistry and Biology of Lipooligosaccharides: The Endotoxins of Bacteria of the Respiratory and Genital Mucosae *J. McLeod Griffiss and Herman Schneider*	179
11	A Biophysical View on the Function and Activity of Endotoxins *Ulrich Seydel, Andre Wiese, Andra B. Schromm, and Klaus Brandenburg*	195
12	Lipopolysaccharide Preparations in Aqueous Media: Implications for Solution Versus Suspension *Pasupati Mukerjee, Manfred Kastowsky, Stefan Obst, and Kuni Takayama*	221
13	Chlamydial Lipopolysaccharide *Helmut Brade*	229
14	The Chemical Synthesis of Lipid A *Shoichi Kusumoto, Koichi Fukase, and Masato Oikawa*	243
15	Chemical Synthesis of Core Structures *Paul Kosma*	257
16	Microbial Pathways of Lipid A Biosynthesis *Paul D. Rick and Christian R. H. Raetz*	283
17	Biosynthesis and Genetics of Lipopolysaccharide Core *David E. Heinrichs, Miguel A. Valvano, and Chris Whitfield*	305
18	Genetics and Biosynthesis of Lipopolysaccharide O-Antigens *Wendy J. Keenleyside and Chris Whitfield*	331
19	Lipopolysaccharide-Binding Protein *Peter S. Tobias*	359
20	Bactericidal/Permeability-Increasing Protein, p15s and Phospholipases A_2, Endogenous Antibiotics in Host Defense Against Bacterial Infection *Peter Elsbach*	369
21	Interactions of Lipopolysaccharides and Lipoproteins *Sander J. H. van Deventer and Dasja Pajkrt*	379
22	Effects of Human Hemoglobin on Bacterial Endotoxin In Vitro and In Vivo *Robert J. Roth and Jack Levin*	389
23	LPS/Lipid A–Binding Synthetic Peptides *Massimo Porro*	403
24	The Interaction of Lipid A and Lipopolysaccharide with Human Serum Albumin *Sunil A. David*	413
25	Endothelial Cell Activation by Lipopolysaccharide: Role of Soluble CD14 *Moshe Arditi*	423
26	Scavenger Receptors and Lipopolysaccharide *Alexander Shnyra*	437

27	The Role of Platelet-Activating Factor in Endotoxin-Related Disease *Taco W. Kuijpers and Tom van der Poll*	449
28	CD14, An Innate Immune Receptor for Various Bacterial Cell Wall Components *Artur J. Ulmer, Volker T. El-Samalouti, Ernst T. Rietschel, Hans-Dieter Flad, and Roman Dziarski*	463
29	The Role of MAP Kinases, Phosphatidylinositol 3-Kinase, and Ceramide in LPS-Induced Signaling in Macrophages *Anthony L. DeFranco, Mary T. Crowley, Alexander J. Finn, Julie Hambleton, Mary Lee MacKichan, and Steven L. Weinstein*	473
30	Endotoxin Effects on Synthesis of Phosphatidic Acid and Phosphatidic Acid-Derived Diacylglyceride Species *Stuart L. Bursten*	483
31	Endotoxic Shock and the Sphingomyelin Pathway *Cecil K. Joseph and Richard N. Kolesnick*	497
32	Role of NF-κB in Macrophage Activation *Tsuneo Suzuki*	509
33	Internalization of Lipopolysaccharide by Phagocytes *Richard L. Kitchens and Robert S. Munford*	521
34	Multifunctional G Proteins: Implication in Endotoxin Cellular Responses *Lawrence P. Fernando, Michel A. Makhlouf, and J. A. Cook*	537
35	Immediate Cytokine Responses to Endotoxin: Tumor Necrosis Factor-α and the Interleukin-1 Family *Charles A. Dinarello*	549
36	Platelet-Activating Factor in Sepsis: An Update *Reuven Rabinovici, Fizan Abdullah, Guenther Mathiak, and Giora Feuerstein*	561
37	Interleukin-10 and Other Suppressive Cytokines *Arnaud Marchant, Michel Goldman, and Tom van der Poll*	581
38	Nitric Oxide as a Signaling Molecule in the Systemic Inflammatory Response to LPS *Michael J. Parmely*	591
39	Studies of the Primate Inflammatory Hemostatic Axis and Its Response to Inflammatory Mediators *Fletcher B. Taylor, Jr.*	605
40	Biological Functions of Lipopolysaccharide Antibodies *Matthew Pollack*	623
41	Specificity and Neutralizing Properties of Cross-Reactive Anti-Core LPS Monoclonal Antibodies *Franco E. Di Padova, Didier Heumann, Michel Pierre Glauser, and Ernst T. Rietschel*	633

42	Effects of Lipopolysaccharide on T Cells *Masayasu Nakano, Teruo Kirikae, and Toshimasa Nitta*	643
43	Immunological Properties of Microbial Outer Membrane Proteins and Their Effects as Modulators of LPS Immunobiology *Kathryn Nixdorff, Dagmar Schilling, and Waltraud Ruiner*	651
44	Opsonization of *Actinobacillus actinomycetemcomitans* by LPS-Directed IgG Antibodies in Sera of Juvenile Periodontitis Patients *Mark E. Wilson*	667
45	Apoptotic Cell Death in Response to Lipopolysaccharide *Takashi Yokochi*	679
46	Nontoxic RsDPLA As a Potent Antagonist of Toxic Lipopolysaccharide *Nilofer Qureshi, Bruce W. Jarvis, and Kuni Takayama*	687
47	Synthetic Endotoxin Antagonists *Daniel P. Rossignol, William J. Christ, Lynn D. Hawkins, Seiichi Kobayashi, Tsutomu Kawata, Melvyn Lynn, Isao Yamatsu, and Yoshito Kishi*	699
48	Bacteria-Induced Hypersensitivity to Endotoxin *Marina A. Freudenberg, Thomas Merlin, Andreas Sing, Reinaldo Salomao, and Chris Galanos*	719
49	Genetic Control of Endotoxin Responsiveness: The *Lps* Gene Revisited *Stefanie N. Vogel, Nayantara Bhat, Danielle Malo, and Salman T. Qureshi*	735
50	Endotoxin Tolerance *F. Ulrich Schade, Regina Flach, Sascha Flohé, Matthias Majetschak, Ernst Kreuzfelder, Emilio Domínguez-Fernández, Jochen Börgermann, Martin Reuter, and Udo Obertacke*	751
51	Glucocorticoid Control of Endotoxin Responses *Richard Silverstein, Donald C. Johnson, and Mari Norimatsu*	769
52	Gene Knockout Technology and the Host Response to Endotoxin: Role of CD14 and Other Inflammatory Mediators *Sanna M. Goyert*	781
53	Endotoxemia in Primate Models *Heinz Redl, Günther Schlag, and Soheyl Bahrami*	795
54	The Value of Animal Models in Endotoxin Research *Steven M. Opal*	809
55	Pathophysiological Responses to Endotoxin in Humans *Anthony F. Suffredini and Naomi P. O'Grady*	817
56	Endotoxin Detection in Body Fluids: Chemical Versus Bioassay Methodology *Thomas J. Novitsky*	831
57	The Relevance of Endotoxin Detection in Sepsis *James Hurley and Jack Levin*	841

Contents

58	The Role of Gut-Derived Endotoxin in the Pathogenesis of Multiple Organ Dysfunction *Mitchell P. Fink and Michael G. Mythen*	855
59	Therapeutic Approaches Targeting Endotoxin-Derived Mediators *Jean-Daniel Baumgartner, Didier Heumann, and Michel Pierre Glauser*	865
60	Human Responses to Endotoxin: Role of the Genetic Background *Frank Stüber*	877
61	Endotoxin, Antibiotics, and Inflammation in Gram-Negative Infections *Jan M. Prins*	887
62	Lipopolysaccharide from Oral Bacteria: Role in Innate Host Defense and Chronic Inflammatory Disease *Brian W. Bainbridge and Richard P. Darveau*	899
63	Endotoxin and Cancer *Minghuang Zhang and Kevin J. Tracey*	915
64	The Future of Endotoxin Research *Helmut Brade, Steven M. Opal, Stefanie N. Vogel, and David C. Morrison*	927
Index		929

Contributors

Fizan Abdullah, M.D., Ph.D. Department of Surgery, Yale University School of Medicine, New Haven, Connecticut

Michael A. Apicella, M.D. Department of Microbiology, University of Iowa, Iowa City, Iowa

Moshe Arditi, M.D. Department of Pediatrics, Cedars-Sinai Medical Center, UCLA School of Medicine, Los Angeles, California

Soheyl Bahrami, Ph.D. Ludwig Boltzmann Institute for Experimental and Clinical Traumatology, Vienna, Austria

Brian W. Bainbridge Department of Periodontics, University of Washington, School of Dentistry, Seattle, Washington

Jean-Daniel Baumgartner, M.D. Service of Internal Medicine, Hôpital de Morges, Morges, Switzerland

Nayantara Bhat, Ph.D. Department of Microbiology and Immunology, Uniformed Services University of the Health Sciences, Bethesda, Maryland

Jochen Börgermann Department of Thoracic Surgery, University Hospital of Essen, Essen, Germany

Helmut Brade, M.D. Department of Immunochemistry and Biochemical Microbiology, Research Center Borstel, Center for Medicine and Biosciences, Borstel, Germany

Klaus Brandenburg, Ph.D. Department of Immunochemistry and Biochemical Microbiology, Research Center Borstel, Center for Medicine and Biosciences, Borstel, Germany

Stuart L. Bursten, M.D. Department of Lipid Biology and Analytical Lipid Biochemistry, Cell Therapeutics, Inc., Seattle, Washington

William J. Christ, Ph.D. Department of Process Research and Development, Eisai Merrimack Valley Laboratories, Inc., Andover, Massachusetts

J. A. Cook, Ph.D.   Department of Physiology, Medical University of South Carolina, Charleston, South Carolina

Mary T. Crowley, Ph.D.   Department of Immunology, The Scripps Research Institute, La Jolla, California

Richard P. Darveau, Ph.D.   Department of Periodontics, University of Washington, School of Dentistry, Seattle, Washington

Sunil A. David, M.B.B.S., Ph.D.   Department of Microbiology, University of Kansas Medical Center, Kansas City, Kansas

Anthony L. DeFranco, Ph.D.   Department of Microbiology and Immunology, University of California, San Francisco, San Francisco, California

Charles A. Dinarello, M.D.   Department of Medicine, University of Colorado Health Science Center, Denver, Colorado

Franco E. Di Padova, M.D.   Department of Preclinical Research, Novartis Pharma, Ltd., Basel, Switzerland

Emilio Domínguez-Fernández, M.D.   Department of General Surgery, University Hospital of Essen, Essen, Germany

Roman Dziarski, Ph.D.   Department of Microbiology and Immunology, Northwest Center for Medical Education, Indiana University School of Medicine, Gary, Indiana

Volker T. El-Samalouti   Department of Immunology and Cell Biology, Research Center Borstel, Center for Medicine and Biosciences, Borstel, Germany

Peter Elsbach, M.D., Ph.D.   Departments of Medicine and Microbiology, New York University School of Medicine, New York, New York

Lawrence P. Fernando, Ph.D.   Department of Physiology, Medical University of South Carolina, Charleston, South Carolina

Giora Feuerstein, M.D.   Department of Cardiovascular Disease Research, DuPont Pharmaceuticals, Wilmington, Delaware

Mitchell P. Fink, M.D.   Department of Surgery, Beth Israel Deaconess Medical Center, and Harvard Medical School, Boston, Massachusetts

Alexander J. Finn   Department of Microbiology and Immunology, University of California, San Francisco, San Francisco, California

Regina Flach, Ph.D.   Clinical Research Group and Shock and Multiple Organ Failure, University Hospital of Essen, Essen, Germany

Hans-Dieter Flad, M.D.   Department of Immunology and Cell Biology, Research Center Borstel, Center for Medicine and Biosciences, Borstel, Germany

Sascha Flohé, M.D.   Clinical Research Group and Shock and Multiple Organ Failure, University Hospital of Essen, Essen, Germany

Marina A. Freudenberg, M.D.   Max Planck Institut für Immunbiologie, Freiburg im Breisgau, Germany

Contributors

Koichi Fukase, Ph.D. Department of Chemistry, Graduate School of Science, Osaka University, Osaka, Japan

Chris Galanos, Ph.D. Max Planck Institut für Immunbiologie, Freiburg im Breisgau, Germany

Peter C. Giardina, Ph.D. Department of Microbiology, University of Iowa, Iowa City, Iowa

Brad Gibson The University of California, San Francisco, San Francisco, California

Michel Pierre Glauser, M.D. Department of Internal Medicine, Division of Infectious Diseases, Centre Hospitalier Universitaire Vaudois, Lausanne, Switzerland

Michel Goldman, M.D., Ph.D. Department of Immunology, Erasme Hospital, Free University of Brussels, Brussels, Belgium

Sanna M. Goyert, Ph.D. Department of Medicine, Division of Molecular Medicine, North Shore University Hospital, Manhasset, New York

J. McLeod Griffiss, M.D. Department of Laboratory Medicine, University of California, San Francisco, San Francisco, California

Julie Hambleton, M.D. Department of Medicine, University of California, San Francisco, San Francisco, California

Lynn D. Hawkins, Ph.D. Department of Chemistry, Eisai Research Institute, Andover, Massachusetts

David E. Heinrichs, Ph.D. Department of Microbiology, University of Guelph, Guelph, Ontario, Canada

Didier Heumann, Ph.D. Department of Internal Medicine, Division of Infectious Diseases, Centre Hospitalier Universitaire Vaudois, Lausanne, Switzerland

Otto Holst, Ph.D. Department of Medical and Biochemical Microbiology, Research Center Borstel, Center for Medicine and Biosciences, Borstel, Germany

Derek W. Hood, Ph.D. Molecular Infectious Disease Group, University Department of Paediatrics, John Radcliffe Hospital, The University of Oxford, Headington, Oxford, England

James Hurley, Ph.D., F.R.A.C.P., M.B. Intensive Care Unit, Wimmera Health Care Group, Ballarat Base Hospital, Ballarat, Victoria, Australia

Jesse J. Jackson, B.S. Department of Antibiotics and Biochemical Research, Merck Research Laboratories, Rahway, New Jersey

Per-Erik Jansson, Ph.D. Clinical Research Center, Karolinska Institute, Huddinge Hospital, Huddinge, Sweden

Bruce W. Jarvis, Ph.D. Department of Research and Development, Promega Corporation, Madison, Wisconsin

Donald C. Johnson, Ph.D. Departments of Obstetrics and Gynecology, and Molecular and Integrative Physiology, University of Kansas Medical Center, Kansas City, Kansas

Cecil K. Joseph, Ph.D. Department of Pharmacology, Toxicology, and Medicinal Biochemistry, Arnold & Marie Schwartz College of Pharmacy & Health Sciences, Brooklyn, New York

Manfred Kastowsky, Ph.D. Free University of Berlin, Berlin, Germany, and William S. Middleton Memorial Veterans Hospital, Madison, Wisconsin

Tsutomu Kawata, Ph.D. Department of Exploratory Drug Research, Tsukuba Research Laboratories, Eisai Co., Ltd., Tsukuba, Ibaraki, Japan

Wendy J. Keenleyside, Ph.D. Department of Microbiology, University of Guelph, Guelph, Ontario, Canada

Teruo Kirikae, M.D., Ph.D. Jichi Medical School, Minamikawachi-machi, Tochigi-ken, Japan

Yoshito Kishi, Ph.D. Eisai Company, Ltd., Koishikawa, Bunkyo-ku, Tokyo, Japan

Richard L. Kitchens, Ph.D. Department of Internal Medicine, Division of Infectious Diseases, The University of Texas Southwestern Medical Center, Dallas, Texas

Seiichi Kobayashi, Ph.D. Biology Unit I, Tsukuba Research Laboratories, Eisai Co., Ltd., Tsukuba, Ibaraki, Japan

Richard N. Kolesnick, M.D. Department of Pharmacology, Sloan-Kettering Institute, Memorial Sloan-Kettering Cancer Center, New York, New York

Paul Kosma, Ph.D. Department of Chemistry, University of Agricultural Sciences, Vienna, Austria

Ernst Kreuzfelder, Ph.D. Institute of Immunology, University Hospital of Essen, Essen, Germany

Helmut Kropp, B.Sc. Department of Antibiotics and Biochemical Research, Merck Research Laboratories, Rahway, New Jersey

Taco W. Kuijpers, M.D., Ph.D. Department of Pediatrics, Emma Children's Hospital, Academic Medical Center, University of Amsterdam, Amsterdam, The Netherlands

Shoichi Kusomoto, Ph.D. Department of Chemistry, Graduate School of Science, Osaka University, Osaka, Japan

Jack Levin, M.D. Departments of Medicine and Laboratory Medicine, University of California School of Medicine, and the Department of Veterans Affairs Medical Center, San Francisco, California

Buko Linder, Ph.D. Department of Immunochemistry and Biochemical Microbiology, Division of Biophysics, Research Center Borstel, Center for Medicine and Biosciences, Borstel, Germany

Melvyn Lynn, Ph.D. Department of Clinical Research, Eisai Research Institute, Andover, Massachusetts

Mary Lee MacKichan, Ph.D. Chiron Corporation, Emeryville, California

Matthias Majetschak, M.D. Department of Trauma Surgery, University Hospital of Essen, Essen, Germany

Michel A. Makhlouf Department of Physiology, Medical University of South Carolina, Charleston, South Carolina

Danielle Malo, D.V.M., Ph.D. McGill Centre for the Study of Host Resistance, Montreal General Hospital, Montreal, Quebec, Canada

Arnaud Marchant, M.D., Ph.D. Department of Immunology, Erasme Hospital, Free University of Brussels, Brussels, Belgium

Contributors

Guenther Mathiak, M.D. Department of Surgery, University of Cologne, Cologne, Germany

Thomas Merlin, M.D. Max Planck Institut für Immunbiologie, Freiburg im Breisgau, Germany

David C. Morrison, Ph.D.* Saint Luke's Hospital and University of Missouri–Kansas City, Kansas City, Missouri

Richard Moxon Molecular Infectious Disease Group, University Department of Paediatrics, John Radcliffe Hospital, The University of Oxford, Headington, Oxford, England

Pasupati Mukerjee, Ph.D. School of Pharmacy, University of Wisconsin–Madison, Madison, Wisconsin

Robert S. Munford, M.D. Department of Internal Medicine, Division of Infectious Diseases, The University of Texas Southwestern Medical Center, Dallas, Texas

Michael G. Mythen Department of Anesthesia and Intensive Care, University College of London Hospital, London, England

Masayasu Nakano, M.D., Ph.D. Department of Microbiology, Jichi Medical School, Minamikawachi-machi, Tochigi-ken, Japan

Toshimasa Nitta, Ph.D. Ohu University School of Dentistry, Koriyama, Japan

Kathryn Nixdorff, Ph.D. Department of Microbiology and Genetics, Darmstadt University of Technology, Darmstadt, Germany

Mari Norimatsu, D.V.M., Ph.D. Departments of Microbiology, Molecular Genetics, and Immunology, University of Kansas Medical Center, Kansas City, Kansas

Thomas J. Novitsky, Ph.D. Associates of Cape Cod, Inc., Falmouth, Massachusetts

Udo Obertacke, M.D. Department of Trauma Surgery, University Hospital of Essen, Essen, Germany

Stefan Obst, Ph.D. Department of Chemistry, Institute of Crystallography, Free University of Berlin, Berlin, Germany

Naomi P. O'Grady Department of Critical Care Medicine, Warren G. Magnuson Clinical Center, National Institutes of Health, Bethesda, Maryland

Masato Oikawa, Ph.D. Department of Chemistry, Graduate School of Science, Osaka University, Osaka, Japan

Steven M. Opal, M.D. Infectious Disease Division, Brown University School of Medicine, Providence, and Memorial Hospital of Rhode Island, Pawtucket, Rhode Island

Dasja Pajkrt Department of Experimental Internal Medicine, Academic Medical Center, University of Amsterdam, Amsterdam, The Netherlands

Michael J. Parmely, Ph.D. Departments of Microbiology, Molecular Genetics, and Immunology, University of Kansas Medical Center, Kansas City, Kansas

---

*Formerly at University of Kansas Medical Center, Kansas City, Kansas.

Matthew Pollack, M.D. Department of Medicine, Uniformed Services University of the Health Sciences, F. Edward Hébert School of Medicine, Bethesda, Maryland

Massimo Porro, Ph.D. Department of Immunochemistry, BiosYnth Research Laboratories, Terme, Siena, Italy

Andrew Preston, Ph.D. Department of Clinical Veterinary Medicine, The University of Cambridge, Cambridge, England

Jan M. Prins, M.D., Ph.D. Department of Infectious Diseases, Tropical Medicine, and AIDS, Academic Medical Center, Amsterdam, The Netherlands

Nilofer Qureshi, Ph.D. Departments of Bacteriology, Animal Health, and Biomedical Sciences, William S. Middleton Memorial Veterans Hospital, and University of Wisconsin–Madison, Madison, Wisconsin

Salman T. Qureshi, M.D., F.R.C.P.C. McGill Centre for the Study of Host Resistance, Montreal General Hospital, Montreal, Quebec, Canada

Reuven Rabinovici, M.D. Department of Surgery, Yale University School of Medicine, New Haven, Connecticut

Christian R. H. Raetz, M.D., Ph.D. Department of Biochemistry, Duke University Medical Center, Durham, North Carolina

Heinz Redl, Ph.D. Ludwig Boltzmann Institute for Experimental and Clinical Traumatology, Vienna, Austria

Martin Reuter, M.D. Department of Trauma Surgery, University Hospital of Essen, Essen, Germany

Paul D. Rick, Ph.D. Department of Microbiology and Immunology, Uniformed Services University of the Health Sciences, Bethesda, Maryland

Ernst T. Rietschel Department of Immunology and Cell Biology, Research Center Borstel, Center for Medicine and Biosciences, Borstel, Germany

Daniel P. Rossignol, Ph.D. Eisai Research Institute, Andover, Massachusetts

Robert J. Roth, M.D., Ph.D. Department of Pathology, University of California School of Medicine, and the Department of Veterans Affairs Medical Center, San Francisco, California

Waltraud Ruiner Department of Microbiology and Genetics, Darmstadt University of Technology, Darmstadt, Germany

Reinaldo Salamao, M.D. Department of Infectious Diseases, Escola Paulista de Medicina, São Paulo, Brazil

F. Ulrich Schade, Ph.D. Clinical Research Group Shock and Multiple Organ Failure, University Hospital of Essen, Essen, Germany

Dagmar Schilling, Dip.-Biol. (M.S.) Department of Microbiology and Genetics, Darmstadt University of Technology, Darmstadt, Germany

Richard L. Schilsky, M.D. Cancer Research Center, University of Chicago, Chicago, Illinois

Günther Schlag, M.D. Ludwig Boltzmann Institute for Experimental and Clinical Traumatology, Vienna, Austria

Contributors

Herman Schneider, Ph.D.   Department of Bacterial Diseases, Walter Reed Army Institute of Research, Washington, D.C.

Andra B. Schromm, Ph.D.   Department of Immunochemistry and Biochemical Microbiology, Research Center Borstel, Center for Medicine and Biosciences, Borstel, Germany

Ulrich Seydel, Ph.D.   Department of Immunochemistry and Biochemical Microbiology, Research Center Borstel, Center for Medicine and Bioscience, Borstel, Germany

Alexander Shnyra, M.D., Ph.D.   Departments of Microbiology, Molecular Genetics, and Immunology, University of Kansas Medical Center, Kansas City, Kansas

Richard Silverstein, Ph.D.   Departments of Biochemistry and Molecular Biology, University of Kansas Medical Center, Kansas City, Kansas

Andreas Sing, M.D.   Max Planck Institut für Immunbiologie, Freiburg im Breisgau, Germany

Frank Stüber, M.D.   Department of Anesthesiology and Intensive Care Medicine, Bonn University, Bonn, Germany

Anthony F. Suffredini, M.D.   Department of Critical Care Medicine, Warren G. Magnuson Clinical Center, National Institutes of Health, Bethesda, Maryland

Tsuneo Suzuki, M.D., Ph.D.   Departments of Microbiology, Molecular Genetics, and Immunology, University of Kansas Medical Center, Kansas City, Kansas

Kuni Takayama, Ph.D.   Mycobacteriology Research Laboratory, William S. Middleton Memorial Veterans Hospital, Madison, Wisconsin

Fletcher B. Taylor, Jr., M.D.   Oklahoma Medical Research Foundation, Oklahoma City, Oklahoma

Vernon L. Tesh, Ph.D.   Department of Medical Microbiology and Immunology, Texas A&M University Health Science Center, College Station, Texas

Peter S. Tobias, Ph.D.   Department of Immunology, The Scripps Research Institute, San Diego, California

Kevin J. Tracey, M.D.   Department of Surgery, Division of Neurosurgery, North Shore University Hospital, and The Picower Institute for Medical Research, Manhasset, New York

Artur J. Ulmer, Ph.D.   Department of Immunology and Cell Biology, Research Center Borstel, Center for Medicine and Biosciences, Borstel, Germany

Martti Vaara, M.D., Ph.D.   Department of Bacteriology, National Public Health Institute and University of Helsinki Medical School, Helsinki, Finland

Miguel A. Valvano, M.D.   Department of Microbiology and Immunology, University of Western Ontario, London, Ontario, Canada

Tom van der Poll, M.D., Ph.D.   Department of Internal Medicine, Academic Medical Center, University of Amsterdam, Amsterdam, The Netherlands

Sander J. H. van Deventer, M.D., Ph.D. Laboratory for Experimental Internal Medicine, Academic Medical Center, University of Amsterdam, Amsterdam, The Netherlands

Stefanie N. Vogel, Ph.D. Department of Microbiology and Immunology, Uniformed Services University of the Health Sciences, Bethesda, Maryland

Steven L. Weinstein, Ph.D. Department of Biology, San Francisco State University, San Francisco, California

Otto Westphal, M.D., D.Sc. Clinique La Prairie Research Institute, Montreux, Switzerland

Chris Whitfield, Ph.D. Department of Microbiology, University of Guelph, Guelph, Ontario, Canada

Andre Wiese, Ph.D. Department of Immunochemistry and Biochemical Microbiology, Research Center Borstel, Center for Medicine and Biosciences, Borstel, Germany

Mark E. Wilson, Ph.D. Departments of Oral Pathology, Biology, and Diagnostic Sciences, University of Medicine and Dentistry of New Jersey, Newark, New Jersey

Isao Yamatsu Department of Joint Research Coordination, Eisai Company, Ltd., Koishikawa, Bunkyo-ku, Tokyo, Japan

Takashi Yokochi, M.D., Ph.D. Department of Microbiology and Immunology, Aichi Medical University, Nagakute, Aichi, Japan

Ulrich Zähringer, Ph.D. Division of Immunochemistry, Research Center Borstel, Center for Medicine and Biosciences, Borstel, Germany

Minghuang Zhang Department of Surgery, Division of Neurosurgery, North Shore University Hospital, and The Picower Institute for Medical Research, Manhasset, New York

Endotoxin in Health and Disease

1

Endotoxin: Historical Perspectives

Ernst T. Rietschel
Research Center Borstel, Borstel, Germany

Otto Westphal
Clinique La Prairie Research Institute, Montreux, Switzerland

> Bacteriological research cannot be sufficiently advanced without knowledge of its historical development. A scientist must know what has been achieved and what information has been accumulated in the field he works in. What self criticism would one or the other have expressed with regard to his own work, had he been aware of the historical developments in bacteriology.
>
> Maybe, numerous papers would have never been published.
>
> Friedrich Loeffler* (1852–1915)

INTRODUCTION

The very first international conference dedicated primarily to bacterial endotoxin took place in New York in the year 1952. It was organized by Florence Seibert (1897–1991), a pioneer in pyrogen research (Fig. 1), who introduced a sensitive test for pyrogenicity in rabbits (2). In her introductory remarks to the 1952 conference, she said: "The isolation of a pure and stable pyrogen with well defined chemical and biological properties which could be easily reproduced, would furnish a valuable standard for the evaluation of pyrogen use" (3). Obviously, at that time the exact nature of pyrogens was not yet known, but from one of her studies she concluded: "The pyrogen is a filterable product produced by a specific bacterium . . ." (4). In her attempts to identify the pyrogenic material that contaminated pharmaceutical drugs and infusion fluids, she always ended up with such small amounts of what may have been the pyrogenic substance that a characterization of the material was impossible. Florence Seibert called the elusive bioactive material her "little blue devil," showing both her affection towards and frustration with the material that would later turn out to be endotoxin. One of her students, Dennis W. Watson, working with her in 1939 in Philadelphia will play, as described later in this chapter, an important role in one of the most exciting chapters of the history of endotoxin research, i.e., the identification of the bioactive principle of endotoxin.

The second major endotoxin conference, and the first one in Europe, was organized by Otto Westphal and took place in Freiburg, Germany, in August 1958. The conference was planned to be held in the then most comfortable, fashionable, and recognized hotel in the center of the city of Freiburg, the Colombi. Two weeks before the opening ceremony of the meeting, the Saudi Arabian king Ibn Saud had come to Freiburg for a medical consultation with Professor Ludwig Heilmeyer, a

---

*Coworker of Robert Koch, discoverer of *Corynebacterium diphtheriae* and *Salmonella typhimurium*, Professor for Hygiene at the University of Greifswald, Director of the Robert Koch-Institute in Berlin, Germany (1).

Fig. 1 Florence Seibert (1897–1991) receiving the Achievement Award from Eleanore Roosevelt (right) at the White House (1944).

leading physician and head of the University Clinic for Internal Medicine. Heilmeyer had once saved the life of the most beloved daughter of Ibn Saud and, therefore, was highly esteemed by the king. Ibn Saud had not arrived unaccompanied: with him had come an enormous number of family members, friends, and servants. The royal group occupying the hotel was scheduled for the flight back home a few days before the start of the endotoxin conference, but one day before that date, Ibn Saud decided to extend his stay in Freiburg. For the organizers of the endotoxin conference this was a disastrous decision. As nobody dared to prompt the Saudis to leave the hotel as scheduled, within 2 days new rooms for the approximately 50 symposium guests had to be found—a formidable task since it was the main summer season and all major hotels were completely booked. But after the first shock, all went well: the meeting took place in the demonstration rooms of a local photoshop (Photo Stober) and the guests were dispersed in the many hotels of the nearby Black Forest area. Finally, the conference guests began to enjoy spending beautiful nights and mornings in the mountains, and the whole affair created great amusement, the sessions being animated by a particularly good spirit. One evening session was held in a hotel on top of the second highest mountain of the Black Forest, the Belchen. Just after its start (the subject was "Endotoxin and the Properdin System"), an enormous thunderstorm with heavy lightning blew out all electricity, so that the session had to be continued with candlelight and without projection of slides. Many participants still remember the great spirit of this early symposion with all its improvisations at the last minute. These pioneering endotoxin conferences were followed by many others with increasing frequency (at almost annual intervals) with an increasing number of scientists from all over the world attending.

Due to the initiative, enthusiasm, and endurance of the late L. Joe Berry (1910–1987), professor of microbiology at the University of Texas, Austin (Fig. 2), the International Endotoxin Society (IES) was founded in the year 1988 with the late Alois Nowotny (1922–1992), professor of immunology at the University of Pennsylvania, Philadelphia, being the first president (Fig. 3). This Society grew rapidly and today represents with its approximately 600 members one of the larger international scientific organizations. As another important step, in 1994 the *Journal of Endotoxin Research* (JER) was founded, with David C. Morrison, Kansas Masons Distinguished Professor of Cancer Research at the University of Kansas Medical Center, being the first editor-in-chief. Within its short life, the *Journal* has gained significant scientific recognition, which is expressed by its remarkably increasing ICI citation index. The many conferences, the creation of the IES, and the establishment of the JER are true witnesses to the particular and still growing academic and industrial interest in bacterial endotoxin.

The fascination associated with this bacterial product certainly relates to its manifold and diverse biological activities, its complex and unique chemical and physical structure, its serological properties, and its physiological function for bacteria (5). Endotoxins are formed by a certain group of bacteria, which differ from other microorganisms by the unique architecture of their cell wall and are termed gram-negative bacteria. Endotoxins are known to be integral components of the cell wall of these microorganisms and to be essential for bacterial growth and viability. Endotoxin has, therefore, attracted the interest of microbiologists and bacterial geneticists who wanted to understand its biosynthesis and the molecular basis of its vital func-

Fig. 2 L. Joe Berry (1910–1987), founder and father of the International Endotoxin Society (IES).

Fig. 3 Alois Nowotny (1922–1992), first president of the IES, together with Otto Westphal (honorary member of the IES) at the occasion of an endotoxin conference in New York (1966).

tion for bacteria. Endotoxins are potent toxins that, in higher organisms, can elicit a broad spectrum of harmful effects and that contribute to the pathogenic potential of bacteria. Endotoxins, therefore, intrigued clinical and biological researchers who set out to elucidate their mode of action and to devise strategies aiming at control of the detrimental endotoxin effects observed during bacterial infection and sepsis. Depending upon the amount and route of introduction, endotoxins are also capable of producing beneficial effects in higher organisms. Thus, endotoxins belong to the most active stimulators of the mammalian defense system. They induce the formation of antibodies not only to themselves but also to unrelated antigens, and they activate immune cells resulting in the body's enhanced capacity to cope with microbial infections as well as malignant tumors. Endotoxins have, therefore, become a major subject of immunological and immunochemical research. They even seem to be involved in the normal development of the host's defense system and in the physiological interplay between bacteria and mammalian organisms. Endotoxins have accompanied the phylogenesis of higher organisms as they appeared as constituents of blue-green algae 2 billion years ago—long before the ascent of man. On the other hand, endotoxins are often feared—like they were by Florence Seibert—by researchers in cellular immunology and related fields: small amounts of endotoxin are ever-present and may contaminate glassware, recombinant proteins, culture media as well as sera, hybridoma supernatants, and research tools or reagents in general. Endotoxin expresses bioactivity in extremely small doses, and a number of examples are known where a biological phenomenon ascribed to a certain compound turned out to be due to the activity of a contaminating endotoxin. Finally, endotoxins possessing a complex primary structure and an even more complex three-dimensional architecture have always fascinated chemists and physicists who wished to elucidate those structures that are responsible for endotoxin activity and function and to modify the structures in such a way that they loose their toxic properties while still displaying beneficial, e.g., immunostimulating, effects.

Many excellent volumes and reviews cover early and recent scientific progress in these fields of endotoxin research, which has led to the recognition that endotoxins not only represent powerful poisons, but also the O-antigens of gram-negative bacteria, and that they chemically constitute lipopolysaccharides (5–9). Only occasionally, however, has the story been described of those pioneering scientists and medical doctors who paved the way for the discovery of endotoxin and the elucidation of its chemistry and biology (10–12). In the present review, which may at certain sections unintentionally express a more personal view of events—the preendotoxin era is described as well as major developments that gradually led to the identification of endotoxin and the characterization of its chemical structure and toxic region.

INITIATION OF ENDOTOXIN RESEARCH

Through the work of Albrecht von Haller (1708–1777) and other physicians and scientists, it was recognized that intravenous injection of putrid fluids, e.g., obtained from decomposing organic matter such as putrid fish or meat, into experimental animals evoked fever and other manifestations of illness (13). As extracts from fresh fish or meat did not produce febrile reactions, it appeared that during decomposition and putrefaction a fever- and disease-producing toxic principle was formed. François Magendie (1783–1855), director of the medical service of the oldest Parisian hospital, L'Hôtel-Dieu, intelligently pointed to the frequent observation of physicians about the unhealthy influence of nonhygienic conditions like those in and around major harbors. Here, masses of material of plant and animal origin underwent deterioration and putrefaction, and here severe illnesses occurred, including plague and cholera. Magendie asked the fundamental question, whether there might exist a connection between putrefaction, toxin production, and the initiation of fever or disease (14).

Magendie performed the most revealing experiments with filtrates of putrid fish studying primarily the host response. He did not try to purify the bioactive principle of his toxic fluids. Exactly this problem, however, was addressed by the Dane Peter L. Panum (1820–1885), then associated professor of physiology and pathology at Kiel, Germany (Fig. 4). Panum attempted to characterize the putrid poison, work that now can be considered as the initiation of systematic and scientific endotoxin research. In 1874, he published his work dealing with his earlier attempts (around 1856) to purify the active principle of putrid poison, which may be summarized as follows (15):

Fig. 4 Peter L. Panum (1820–1885).

The putrid poison is not volatile and is not a known simple end product of putrefaction or fermentation. It can be differentiated from living microorganisms, which may be a source but not the cause.

The toxin resists heat and, thus, differs from typical enzymes (at that time called in German *Fermente*).

It is insoluble in pure alcohol but soluble in water.

The protein-like substances frequently present in putrid fluids are not toxic by themselves, but they absorb ("condense") the toxin on their surface when precipitated. The toxic principle can be, at least partially, eluted from the precipitates.

Injection of 12 mg of the concentrate suffices to produce high fever and to kill a dog.

Looking at Panum's results and applying our present knowledge on the physicochemical and biological properties of endotoxin, it becomes evident that Panum indeed was dealing with endotoxin, a term that had not yet been coined. Panum, therefore, may be regarded as the first researcher having systematically and scientifically addressed the problem of endotoxin.

The ideas of Panum were later followed by Ernst von Bergmann (1836–1906), professor of surgery in Würzburg and Berlin, Germany. Von Bergmann believed that a chemically defined substance was responsible for putrid intoxication, which he termed *sepsin*. In 1868, he published a first paper on sepsin, which he believed to be the poison of putrid substances (16). Subsequently, he attempted to isolate the postulated compound from putrid fluids that, in a purified form, would induce toxic manifestations in animals (17). His work on sepsin was unfortunately interrupted by the Franco-Prussian war (1870–71) and thereafter not continued.

During these years of emerging research on the molecule that would later turn out to be endotoxin and that at that time was often recognized by its fever-producing properties, the terms *pyrogenic material* or *pyrogen* were coined. With these terms a material was described, which, after intravenous injection into experimental animals such as rabbits, would cause fever. For the first use of the term pyrogenic substance, credit must be given to Theodor Billroth (1829–1894), professor of surgery in Zurich, Switzerland, and Vienna, Austria. In his 1862 publication "Observations on fever caused by wounds and accidental wound diseases," he stated: "The pyrogenic substances are equally present in dried putrid material and dried pus as well as in putrid fluids and in fresh pus" (18). Independently, the term "pyrogenic substance" was used in 1873 by Bernhard Naunyn (1839–1925), professor of internal medicine in Dorpat, Bern, Switzerland, Königsberg, Russia (now Kaliningrad), and Strasbourg, France, to designate a defined fever-producing principle (19). Finally, Sir John Burdon-Sanderson (1828–1905), an eminent British physiologist and pathologist, stated in a review of 1896: "In 1875 I prepared a substance from putrid extract of flesh, which I ventured to call pyrogen. This sterile product produced fever" (20). Burdon-Sanderson, in the tradition of Panum and von Bergmann, attempted and apparently partially succeeded in purifying his pyrogen from putrid material. Of course, it is not possible today to decide whether the preparations obtained were endotoxin or, perhaps, concentrates of endogenous mediators such as interleukin 1 and 6 or tumor necrosis factor α. In any case, it was also Burdon-Sanderson who reflected in 1876 on the question as to the origin of the pyrogenic substance (21). He wondered whether it was of endogenous origin, being present in blood or tissue fluid. Alternatively, he argued, the pyrogenic material could come from exogenous sources, such as microbes.

MICROBIAL ORIGIN OF DISEASES AND TOXINS

The years 1870–1905 comprise the period in which the concept that infections were due to microorganisms gained scientific ground. Until then, the endemically or epidemically occurring diseases, such as plague and cholera, were believed to be due to miasma, a nonvisible volatile material originating from the soil or areas below the soil. The disease-causing miasma was also called "bad air" (inale d'aria), hence the origin of the word *malaria*. It was thought that the spread of infections did not take place from person to person but rather through environmental factors—miasma. The famous painting by Arnold Böcklin (1827–1901) of plague shows a demon riding a flying dragon-like animal, which exhausts bad air, i.e., miasma (Fig. 5). On the other hand, it was obvious to certain physicians that some diseases, such as syphilis, were transferred from person to person and, thus, most likely not by bad air. The transferred material was termed *contagion*. Thus, two schools were created: one favoring the idea of miasma, the other following the concept of contagion. The existence of these two separate theories, even in 1885, is remarkable, because in 1840 Jakob Henle (1809–1885), professor for pathology and anatomy in Göttingen, Germany, and a teacher of Robert Koch, had pointed out that contagion and miasma may, in fact, be identical and represent organic living matter that possesses the ability to multiply and grow in the sick individual, thereby causing irritation and disease (22). Of course, Henle had no proof for his ingenious concept. It took almost 25 years for the idea of microbe-based fermentation and putrefaction resulting in disease to be accepted.

It was Louis Pasteur (1822–1895), in Paris, who convincingly showed that the class of diseases we now refer to as infectious diseases was brought about by small animate microorganisms, which, being associated with dust in the air, could be transported everywhere and could reach wounds, where they would multiply and cause infections. It was argued for some time that one microorganism, perhaps in various forms, might be responsible for many different diseases. Thus, Billroth

Fig. 5 Plague. Painting by Arnold Böcklin (1827–1901).

thought that a peculiar microorganism, *Coccobacteria septica*, caused several types of infections (23). He thought that these bacteria were present in normal and healthy tissue and were activated to multiply if the human organism was irritated by an inflammatory or septic "zymoid." Thus, Billroth remained in the tradition of the theory of endogenous putrefaction. In contrast, the German bacteriologist Edwin Klebs (1834–1913), whose name is remembered in the enterobacterial genus *Klebsiella*, clearly stated that severe infections were caused by microorganisms that originated from external sources and could invade healthy but irritated tissue. During the Franco-Prussian war, Klebs performed 115 autopsies between August 17 and October 17, 1870, and realized not only that 75% of the deaths were due to septicemia, but also that septicemia was caused by microorganisms, which he called *Microsporon septicum*. In one of his reports, he made the following remarkable statement: "During the development of *Microsporon septicum* a fever-producing substance is generated. . . . The continuous fever is caused by the continuous transport of this substance into the organism" (24). It is obvious that Klebs was an excellent clinician and scientific observer and that he described the action of bacterial products that would include endotoxin.

It was Robert Koch (1843–1910), at the end of his career as director of the Berlin Institute for Infectious Diseases, who characterized microorganisms as causative agents of specific diseases. Koch based his brilliant conclusions initially on animal experiments. The connection with human infectious diseases was first made by the Scottish professor for surgery, Sir Alexander Ogston (1844–1929). Ogston isolated pus from patients and stained it with anilinviolett. Under a normal microscope he saw what he described in the following way: "My delight may be conceived when there were revealed to me beautiful tangles, tufts, and chains of round organisms in great number which stood out clear and distinct among the pus cells and debris." Ogston found in all pus isolates bacteria, which, if injected into experimental animals, caused septicemia (25). It was, nevertheless, Koch who finally contributed greatly to the understanding of microbial participation in wound infections from 1880 on, drawing on experiences with thousands of wounded soldiers in wartime, profiting from expeditions to countries such as Egypt and India, which suffered from disastrous epidemics such as cholera, and by supporting his ideas with animal experiments using pure cultures of bacteria as infectious agents.

From the time of Pasteur, Koch, and Ogston on, the search for the origin of disease-promoting, poisonous, and pyrogenic principles concentrated on microorganisms. Bacterial strains could be characterized, grown to large amounts, and treated with defined procedures, which facilitated the enrichment and characterization of toxic molecules. Gustave Roussy (1874–1948), a French medical doctor who founded close to Paris the Institut du Cancer at Villejuif, isolated from bacteria two types of substances: one causing hypothermia (termed *frigorigenine*) and one causing fever (termed *pyretogenine*). He believed that these principles resided in bacteria and that they were released by the lytic action of body enzymes (26). Ludwig Brieger (1849–1919) analyzed typhoid bacilli, which had been discovered in 1880 by Joseph Eberth (1835–1926). Brieger demonstrated the generation of a toxic material, which he called *typhotoxin* (27,28). In fact, it was Brieger who for the first time introduced the term *toxin*. During investigations on cultures of other bacilli, a high molecular weight and toxic substance of protein-like nature could be enriched and prepared as a sterile, white powder, which the authors called *toxalbumin*. Toxalbumins were soon isolated from many bacterial culture fluids. (Today we may conclude that the toxalbumins were proteins being complexed with varying amounts of endotoxin). These toxalbumines or albumoses often proved to be pyrogenic and toxic in animal experiments, and Hans Buchner (1850–1902) pointed to the strong leukocytic response as an alarm reaction of the stimulated defense mechanisms after injection of toxalbumines (29,30). But Ludolf von Krehl (1861–1937), in extended and careful investigations, showed that the albumoses, such as those isolated from *Escherichia coli* cultures, were of significantly lower pyrogenicity than the starting material contained in whole bacteria (31). Therefore, von Krehl considered the methods for isolating the pyrogenic principle to be inadequate. "Who tells us," he wrote, "that during lysis of the bacterial bodies in vivo substances may not be generated which are pyrogenically different" (from the extracted albumoses) (31)?

CREATION OF THE TERM ENDOTOXIN

In the context of endotoxin research, the time around the year 1892 is of particular importance. During 1892 and the following years, dramatic cholera outbreaks threatened large harbor cities including Hamburg, Germany, and St. Petersburg, Russia, causing in Hamburg alone between August 15 and November 12 about 17,000 cases of cholera and 8,600 deaths. It may be mentioned in passing that Gustaf Mahler, who since 1891 had been conductor of the City Theatre Orchestra, escaped cholera by leaving Hamburg just before the main outbreak in 1892. Peter Iljitsch Tschaikowsky, however, swallowed cholera-contaminated water and died in 1893 in St. Petersburg.

In the year 1884, Robert Koch had identified *Vibrio cholerae* as the causative microbial agent of cholera, and active research was going on in his laboratories to characterize the factors rendering these bacteria pathogenic. Studies performed mainly in France, Germany, and the United States had revealed that many pathogenic bacteria produced toxic substances, which, in an isolated form, induced many of the fatal reactions seen during severe bacterial infections. Diphtheria toxin, which was isolated in 1888–90 from culture filtrates by Pierre Paul Emile Roux (1853–1933) and Alexandre Yersin (1863–1943) at the Pasteur Institute in Paris, proved to be of proteinaceous nature, and the concept that practically all bacterial toxins were proteins was accepted for quite some time. The characterization of other toxins such as the tetanus and botulinum toxins (also proteins), the discovery of antitoxins, and the demonstration of the toxin-neutralizing and curative properties of antitoxins provided impressive support for the important pathogenic role of toxins in bacterial disease. Therefore, cholera bacilli were assumed also to produce such toxins. Among Koch's best co-workers was Richard Pfeiffer (1858–1945). Figure 6 shows Koch and Pfeiffer in the year 1892, Pfeiffer standing behind Koch, who looks into a microscope. Pfeiffer, in fact, observed that cholera vibrios released a proteinaceous toxin into the culture medium, a heat-labile *exotoxin*. He noted, however, that cholera vibrios produced, in addition, a second, quite different toxin that appeared to be firmly attached to the bacterial cell. In his first publication on the matter (32), Pfeiffer states: "In very young aerobically grown cholera cultures, a specific toxic substance is contained which exerts extraordinarily intense toxic effects. This primary cholera toxin is closely attached to, and probably an integral part of, the bacterial body. By chloroform, thymol, or

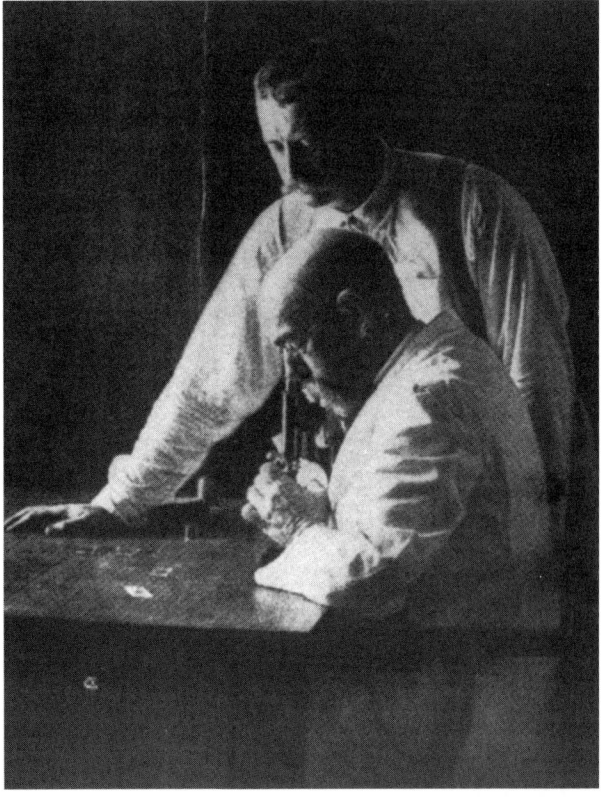

Fig. 6 Robert Koch (1843–1910) sitting in front of a microscope with Richard Pfeiffer (1858–1945), the discoverer of endotoxin, standing behind him.

by drying, the cholera vibrios can be killed without any detectable change of the toxin.''

It is obvious that Pfeiffer had detected endotoxin, and it was also he who introduced the term *endotoxin*. We can conclude this from a masterly review of Alfred Wolff, physician at the Medical University Clinic in Munich, Germany, summarizing the then pertinent concepts on bacterial toxins as well as immunity to toxins and disease-producing bacteria such as *V. cholerae* (33). In this 1904 review, Wolff writes: "Richard Pfeiffer gave a specific name to the bacterial substrate exerting a toxic action which is definitely different from the already known toxins. He called it endotoxin. The creator of this name based his concept on the correct opinion that his toxic principle is not excreted from the living bacteria, but firmly fixed to the body constituents. In contrast to typical toxins, these poisonous substances are not secreted into the culture medium, but they will be set free if the bacteria undergo lysis, called bacteriolysis.'' In the same article, Wolff gives a biological definition of endotoxin: "Toxin and endotoxin are poisonous substances, which in defined doses cause death of injected animals after a certain time of incubation'' (33).

In the available literature including the first 30 volumes of the then important *Zeitschrift für Hygiene* (1886–1899) and Volumes 1 to 10 (1883–1890) of the *Archiv für Hygiene* (34), there is no authentic publication in which Pfeiffer actually introduces the term endotoxin. Brieger (27,28) refers to the original 1892 paper of Pfeiffer in an 1895 publication in which he describes toxins of *Salmonella typhi* and *E. coli*, which appeared similar to that of *V. cholerae* without, however, using the term endotoxin (35). After 1904, one finds many more articles and reviews referring to Pfeiffer, but in none of the quoted publications is endotoxin really mentioned. Hans Raettig, Professor Emeritus of the Robert-Koch-Institute, Berlin, an expert in history of bacteriology of the Robert Koch era, concluded from extensive search through the literature that Pfeiffer's definition was probably laid down by him in the notes for one of his lectures, given in Königsberg (now Kaliningrad) around 1903/4. But copies of these notes are no longer available. In any case, there is no doubt that it was Richard Pfeiffer who introduced the term endotoxin.

FIRST STEPS IN THE CHARACTERIZATION OF ENDOTOXINS

At about the same time, an important step forward was made in the characterization of endotoxin by the Italian Eugenio Centanni (1863–1948), director of the Institute of General Pathology at the University of Bologna (Fig. 7). In 1894, he started to publish a series of papers, the first of which was entitled "Investigations on infectious fever—the fever toxin of bacteria" (36–38). With great enthusiasm he stated the following:

I have undertaken a general investigation about infectious fever using the newest achievements of the sciences. It comprises three sets of investigations:
- on the agents which induce infectious fever,
- on mechanisms by which they act on the organism, and
- on therapeutic questions with regard to fever.

In chapter 1 of *The Fever Toxin of Bacteria* (36), Centanni describes the preparation of what he called *pyrotoxina bacterica*. In principle, he used long-term (up to 10 days) autolysates of pure bacterial cultures, working up the sterile filtrate by alcoholic and other fractiona-

Fig. 7 Eugenio Centanni (1863–1948).

tion procedures, and ending up with a sterile white powder. He was able to produce more or less the same highly pyrogenic material from a large variety of bacteria including *E. coli*, *S. typhi*, and others. He finally stated: "Thus, we can conclude that the whole family of bacteria possess essentially the same toxin, a poison which is tied inseparably to their existence and upon which depends the typical picture of the general disturbances caused by bacterial infections." Centanni's statement that his pyrotoxina is not proteinaceous in nature and that it is quite stable to heat (in contrast to the toxalbumins) is of great importance (39). On the basis of his data on the pyrogenicity of his preparations, one might assume that he had reached a rather high degree of purification of endotoxin. Although, according to present knowledge, some of Centanni's claims appear somewhat too generalized, there is no question that he was the first to recognize the intimate relationship between the pyrogenic and the endotoxic principle of bacteria, which he found chemically inseparable, a fact expressed by the term *pyrotoxina*.

In 1898, the French bacteriologist and pathologist André Chantemesse (1851–1919) used filtrates of typhoid cultures to show that a material can be precipitated with $CaCl_2$, which was toxic to experimental animals (40). He also noted that horses were particularly sensitive to the material and that guinea pigs responded with hypothermia instead of fever, as rabbits did. It is obvious that Chantemesse worked with endotoxin, and his studies greatly contributed to our understanding of its chemical and, in particular, biological properties. It was the same Dr. Chantemesse who was consulted in Paris in April 1911 by Gustav Mahler, who suffered from ultimately fatal bacterial endocarditis and who died about 3 weeks later in Vienna.

Numerous researchers—too many to mention— worked with bacterial filtrates or extracts that must have contained endotoxin as the bioactive principle. Of these, E. E. Ecker refused to acknowledge that he was working with endotoxin, because his heat-stable bioactive material was present in filtrates of young cultures that did not show any sign of disintegration (41) and, therefore, seemingly differed from the preparation described by Pfeiffer. Further, in 1916 J. L. Jona isolated a nonprotein thermogenic factor from *Bacillus coli communis* (*E. coli*) and *Eberthella typhosa* (*S. typhi*). It was colloidal in form, thermostable in dry heat to 110°C, and resistant to boiling. Organic solvents were ineffective, and in combination with calcium the substance formed inert salts. In quantities of 4 μg it produced fever in rabbits (42). These properties indicate the material investigated to be endotoxin.

Pfeiffer, Centanni, Ecker, Jona, Chantemesse, and others had studied endotoxin present in autolysates of *V. cholerae*, *E. coli*, and *S. typhi*, but no report had appeared on an endotoxin-like material being present in heat-treated or lysed cultures of, e.g., *Mycobacteria*, *Treponema*, *Streptococcus*, or *Corynebacteria*. How then did *V. cholerae*, *S. typhi*, and *E. coli* differ from other bacteria, and did they have something in common? As we know today, endotoxin is produced by a group of bacteria called *gram-negative*. This term relates to the Danish Physician Hans-Christian Gram (1853–1939), who in 1884 had devised a staining procedure by which the world of bacteria could be divided into two major groups (43). Bacteria belonging to one group are, after staining with a blue dye, decolorized by alcohol, and were therefore called gram-negative. In contrast, gram-positive bacteria such as *Corynebacteria*, *Streptococcus*, and *Staphylococcus* retained the dye and appeared blue after staining. The large group of gram-negative bacteria comprises many important pathogens such as *V. cholerae*, *S. typhi*, *E. coli*, *Serratia marcescens*, and, in addition, *Yersinia pestic* (causing plague), *Neisseria meningitidis*, and *Hemophilus influenzae* (causing meningitis in newborns and children), *Helicobacter pylori* (causing gastritis and duodenal ulcers), *Shigella dysenteria* (causing gastrointestinal infections), *Klebsiella pneumoniae* (causing pneumoniae), *Pseudomonas aeruginosa* (causing pulmonary and wound infections), and *Bordetella pertussis* (causing whooping cough), as well as *Chlamydia pneumoniae* (today associated not only with pneumonia but also with arteriosclerosis and myocardial infarction). We know today that gram-negative bacteria in general— and only gram-negative bacteria—produce endotoxin, which is a common constituent of and, therefore, characteristic for this class of microorganisms.

ISOLATION AND FURTHER CHARACTERIZATION OF ENDOTOXIN

It is evident that Pfeiffer's endotoxin, Centanni's pyrotoxina, and Chantemesse's filtrates were only poorly defined products and certainly represented mixtures of several bacterial components. It took several decades before selective extraction procedures were developed allowing a more detailed examination of the toxin.

A great pioneer was the French microbiologist André Boivin (1895–1949) (Fig. 8), who, together with the Rumanian Lydia Mesrobeanu (1908–1978), whom he had met when working in Bukarest, Rumania, devised in 1932 for the first time a generally applicable

Fig. 8 Profile of André Boivin (1895–1949) on a Commemorative Medal of the 1963 Endotoxin Conference held at Rutgers State University, New Brunswick, New Jersey.

procedure for the extraction of endotoxin from many gram-negative bacteria: the trichloroacetic acid (TCA) method. On the basis of their analyses, Boivin and Mesrobeanu called the purified, antigenic, and toxic extracts *antigènes glycido-lipidiques* to indicate that the main components were polysaccharide and lipid in nature, with only small amounts of additional protein (44–46). Pennel and Huddeleson used TCA (0.25 N) for the preparation of toxic fractions from *Brucella* bacteria. They named the bioactive material that contained lipid and reducing substances after hydrolysis (i.e., sugars) *endoantigen* (47).

Using a similar approach, Walter T. J. Morgan (48–50) in London and Walther F. Goebel (1899–1994) in New York developed further extraction procedures using mixtures of organic solvents and water. Their purified substances, similar to Boivin and Mesrobeanu's products, were composed of polysaccharide, lipid, and protein. Morgan's diethylene glycol extracts from *Salmonella* and *Shigella* appeared as physicochemically homogeneous and uniform substances. In systematic studies, Morgan then described methods for the dissociation of the complex into structural subunits like undegraded polysaccharide or degraded polysaccharide, conjugated or simple protein, and loosely bound kephalin-like lipid.

In the 1940s, Walther F. Goebel (Fig. 9) in New York, working with *Shigella sonnei* and *Shigella flexneri*, showed together with Chloe Tal that bioactive preparations contained a firmly bound component, which he called *toxic component T* (51). By mild acid treatment he obtained haptenic O-specific polysaccharide (Morgan's degraded polysaccharide) and a toxic protein: *protein T* (Morgan's conjugated protein). In contrast, by alkaline treatment and alcoholic fractionation he was able to isolate an O-antigenic, toxic polysaccharide (polysaccharide T) and nontoxic protein (Morgan's simple protein). In the O-antigenic complex, therefore, the component T appeared to form the link between O-antigenic polysaccharide and protein, so that, depending on the dissociation procedure, Goebel obtained two different toxic substances in which T was either bound to polysaccharide or to protein. The chemical nature of compound T, however, remained obscure and turned out only much later to be identical with the bioactive principle of endotoxin, i.e., the lipid A component. Goebel and coworkers termed the bioactive material obtained from *S. flexneri* (52,53) *toxic carbohydrate* (CT) and that isolated from *S. sonnei lipocarbohydrate* (54). Another interesting aspect of these bacterial antigens was described in 1934 by Gough and Frank M. Burnet (55). These authors had isolated bacterial extracts consisting mainly of polysaccharide. They showed that this polysaccharide fraction was important not only for its serological properties but also—and this was new—for the phage susceptibility of bacteria. As the biological activity of the polysaccharide material was destroyed by alkali, it is certain that the authors worked with O-antigen or "somatic antigen" as they discussed in their revealing study.

Goebel and Burnet, like others, had used the term O-antigen or somatic antigen, and it is worthwhile to briefly reflect on their origin. For the discovery of O- and H-antigens, credit must be given to the great American microbiologist Theobald Smith (1859–1934). Smith, perhaps better known for his discovery of *Salmonella cholerae-suis* (hog cholera bacillus) and his work on Texas fever, was at the end of his career as director of the Department of Plant and Animal Pathology at the Rockefeller Institute for Medical Research at Princeton. Together with A. L. Reagh he discovered that antibacterial (*S. cholerae-suis*) agglutinins (antibodies) may be directed against two distinct microbial components, i.e., thermolabile and heat-resistant structures associated with the body (Greek, *soma*) of bacteria (56). The agglutinins reacting with the body of bacteria were one year later termed by H. G. Beyer and Reagh as *somatic agglutinins* (57). At the same time, a study of A. Joos working with *S. typhi* bacteria

Fig. 9 Picture taken at the occasion of the Fifth Conference on Polysaccharides in Biology (70) showing many scientists who have greatly contributed to endotoxin research. First row: Frank Fremont-Smith, Kenneth McQuillen, Anne-Marie Staub, Paul György, Dorothy LaGuardia, Ward Pigman, M. R. J. Stalton, Richard J. Winzler, Elvin A. Kabat. Second row: Jack L. Strominger, William E. Ehrich, Michael Heidelberger, James T. Park, Elizabeth F. Purcell, A. G. Norman, Guy T. Barry, Maclyn McCarty. Third row: Ernest C. Pollard, Albert Dorfman, Henry W. Scherp, Saul Roseman, Colin M. MacLeod, Georg F. Springer, Murray J. Shear, James V. Neel, Walther F. Goebel, and Friedrich Zilliken.

provided similar data. Joos termed the heat-stable bacterial antigen (which represents, as we know today, endotoxins) as β-*agglutinogens* (58). Unaware of these experiments, Edmund Weil (1879–1922) and Arthur Felix (1887–1956), working with *Proteus* bacteria, noticed that bacterial cultures can appear in two different forms: swarming and nonswarming. Cultures the bacteria of which swarmed (like a "breath," or, in German, *Hauch*) were characterized as containing the H-antigens which were shown by Braun and Schaeffer (59) as well as by K. W. Jötten (60) to be identical with flagella. Those that did not swarm (in German, *ohne Hauch*) were called the O-forms, which possess the O-antigens (also termed by Weil and Felix as *Körpersubstanz*, or body substance). As H. Sachs demonstrated (61), the flagellar (H)-antigen was heat-labile (to boiling for 2.5 hr), whereas the somatic O-antigen was thermoresistant. Thus, the term *O-antigen* was coined by Weil and Felix, who showed in 1920 that O-antigens are also present in *Salmonella* (62–64). Both, Weil and Felix were eminent microbiologists, who are perhaps better known for the discovery of the Weil-Felix reaction, pioneering work on rickettsiosis (Weil), the discovery of the Vi-antigen (Felix), and the introduction of phage typing (Felix). At the end of his short but remarkable scientific career, Edmund Weil (Fig. 10) worked at the German University in Prague on the etiopathology for typhus fever. In order to study the significance of both the louse and rickettsia for human disease, he went to Poland, where epidemic typhus represented a great health problem. During attempts to inject the intestinal content of infected lice into mice, part of the gut material squirted into Weil's eyes. He fell sick and died at the age of 43 years from typhus on June 15, 1922. Figure 11 shows a portrait of Arthur Felix at the time when he was working in Jerusalem (1921–1927), from where he went to London to join the Lister Institute where he would later become the director of the Central Enteric Reference Laboratory at the Medical Research Council.

Fig. 10 Edmund Weil (1879–1922).

Fig. 11 Arthur Felix (1887–1956).

In the early 1940s, Murray Shear and colleagues had taken up the studies of William Bradley Coley (1862–1936), a physician working in New York who used a pyrogenically highly active mixture of live and, later, killed *S. marcescens* (*Bacillus prodigiousus*) and *Streptococcus* (Coley's toxin) to treat patients suffering from carcinomas and sarcomas (65,66). In attempts to isolate the tumor-necrotizing agent from *S. marcescens*, the active principle of Coley's vaccine, a material was obtained by Shear, the analytical figures of which indicated the presence of mainly polysaccharide and lipoidic material besides small quantities of protein. To designate this compound, in 1943 Shear et al. introduced the term *lipopolysaccharide* (67,68).

In the late 1940s, Otto Westphal and Otto Lüderitz, first working at the Wander Forschungsinstitut (Säckingen, Germany) and later at the Max Planck Institut für Immunobiologie (Freiburg, Germany), started their systematic work on the fever-producing and toxic principle of *Salmonella enterica*, *E. coli*, and other Enterobacteriaceae. They showed that the pyrogenic activity resided mainly in the O-antigenic complex by testing all characteristic components, which were kindly supplied by Walter Morgan. Undegraded polysaccharide (Goebel's polysaccharide T) was highly pyrogenic, conjugated protein (Goebel's protein T) somewhat less, whereas degraded polysaccharide and simple protein were completely devoid of activity. Westphal and colleagues then developed a new and efficient procedure, which was applicable to a variety of gram-negative bacteria and by which biologically active material could be extracted in high yield and pure form: the *hot phenol-water extraction* method (69). This procedure celebrated in 1997 its forty-fifth anniversary. The development of this method proved to be a milestone in endotoxin research. Application of this now classical method yielded preparations that were essentially free of protein and consisted of only carbohydrate, phosphorus, and fatty acids. These preparations exhibited extreme biological activity, with an intravenous dose as small as 1 ng/kg causing fever in humans. To designate this bioactive material, Westphal and Lüderitz used the term lipopolysaccharide (LPS) throughout their work and thereby gave it the popularity it now enjoys.

In fact, the discussion on the occasion of the Second Conference on Polysaccharides in Biology on April 25–27, 1956 in Princeton (70) shows that the term lipopolysaccharide had yet to be accepted. In his talk Otto Westphal stated "... it was Dr. György's (Fig. 9) idea to have an extended discussion on special pyrogenic products that have been isolated from bacteria and which we and others have called 'lipopolysaccharides'. I realize that Dr. Pillemer [who attended this meeting and who had recently discovered properdin] does not approve of this expression, but I think at present it is reasonable to call this substance 'lipopolysaccharide.'" And later Westphal adds: "(Of) the purified lipopolysaccharides of *E. coli* 08 and *S. abortus equi* ... 95 to 90 percent of the building stones ... have been isolated and identified. Up to present we found only carbohydrate and lipid constituents. We feel, therefore, that we may be allowed to term these substances 'lipopolysaccharide'" (70,71). It is obvious that the term lipopolysaccharide was not generally accepted at that time. Other terms such as *endotoxic liposaccharide*, proposed later, did not receive the wide acceptance later enjoyed by the term *lipopolysaccharide* (72). The chemical, serological, and biological studies of the Westphal-Lüderitz group also established that endotoxin, pyrotoxina, O-antigen, and lipopolysaccharide (LPS) were one and the same bacterial component and that both pathogenic and nonpathogenic bacteria carried this molecule. The availability of purified LPS now finally allowed studies as to the definition of those regions responsible for the endotoxic, pyrogenic, and O-antigenic properties of LPS and the elucidation of their chemical structures.

CHEMISTRY AND SEROLOGY OF ENDOTOXIN

Initial studies on the chemical structure of endotoxin were performed on those members of the group of gram-negative bacteria that are either part of the normal flora of the intestine or that may cause gastrointestinal (enteric) diseases. These bacteria, which include the genera *Salmonella* and *Escherichia*, are therefore called the enterics or enterobacteria. The colony morphology of naturally occurring (wild-type) *Salmonella* and *Escherichia* bacteria appears entire and smooth (S) and their LPS was therefore denoted as *wild-type* or *S-form* LPS.

Based on qualitative—and later also quantitative—sugar analyses of a large variety of different S-form LPS, the concept emerged around 1960 that lipopolysaccharides (or O-antigens or endotoxins) of different enterobacterial origin are built up according to a common architecture. This concept was, in particular, promoted by Otto Lüderitz (Fig. 12), who postulated (73) that LPS consists of three regions: *O-specific side chain*, *basal core*, and *lipid A*. From the chemical analysis of about 100 different O-antigens derived from various *Salmonella* and *E. coli* serotypes, it became obvious that the O-antigen contained complex polysaccharides, which may be comprised of five to eight different monosaccharides (74–76). However, without exception, all LPS preparations contained five common sugars: the well-known hexoses D-glucose (Glc), D-galactose (Gal), and D-glucosamine (GlcN), and two unusual glycosyl components, 2-keto-3-deoxyoctonic acid, usually abbreviated as Kdo, correctly termed 3-

Fig. 12 Otto Lüderitz (honorary member of the IES) in 1977.

deoxy-D-manno-oct-2-ulosonic acid (dOclA), and L-glycero-D-manno-heptose (L,D-Hep). This heptose was first described in 1952 by Jesaitis and Goebel (77) as a component of *Shigella sonnei* LPS, with the correct stereochemistry being determined in 1953 by Slein and Schnell (78) and in 1955 by Weidel (79). Kdo was later discovered to be an *E. coli* LPS constituent in 1963 by Heath and Ghalambor (80) and in the same year in *Salmonella* LPS by Mary Jane Osborne (81). These sugars were also found in lipopolysaccharides derived from various R-form bacteria (see below), no matter how complex in composition the parent S-forms were. From these findings it was concluded that *Salmonella* S-form LPS contains a common region, which was called by Lüderitz the *common basal core* (or R-polysaccharide). To this common core are attached the specific chains composed of the sugars characteristic of a serotype. For the designation of this region, the term *O-specific side chain* was coined by Lüderitz, Staub, and Westphal in 1966 (73). It was also recognized that the side chains represented heteropolymers composed of several identical oligosaccharides, which were named by Phil Robbins et al. when working on lysogenic conversion as *repeating units* (82–85). It is of interest that Nikaido et al. introduced in 1966 the term *S-specific side chain*, S signifying "serotype" (86). Although this expression appears appropriate, it did not gain general acceptance. This is also true for the term *repeat unit*, which was occasionally used instead of repeating unit.

O-Specific Chain

Studies performed in several laboratories (Phil Robbins in Cambridge, MA, Anne-Marie Staub in Paris, Hiroshi Nikaido in Berkeley, and Otto Lüderitz in Freiburg, Germany) showed that the repeating oligosaccharide units of O-specific chains consisted of up to eight sugar residues. Disclosing the structure of different LPS, the fundamental discovery was made that the LPS of a given *Salmonella* serotype differed from that of others in the nature, ring size, type of linkage, and substitution of the individual sugar residues as well as their sequence within a repeating unit of the O-specific chain. Thus, the O-specific chain renders LPS a characteristic and unique feature of a given bacterium. Based on the structure of the O-specific chain, the genus *Salmonella* could, therefore, be subdivided into structural types called *chemotypes*, a term coined by Kauffmann Lüderitz and Westphal in 1960 (74).

It was long known that antisera raised against a defined *S. enterica* species were specific for this very species and did, in general, not react with others. This serological specificity prompted Fritz Kauffmann (1898–1978), a pioneer in bacteriological research and director at the Statens Seruminstitut in Copenhagen, Denmark (Fig. 13), to define *Salmonella* "serotypes" and to establish a serological agglutination scheme to classify isolates of the genus *Salmonella*, which is still used today in medical, veterinary, and plant microbiology (Kauffmann-White scheme). Lüderitz and Westphal, then cooperating with Kauffmann, and in particular with Anne-Marie Staub at the Institut Louis Pasteur in Paris, recognized that experimental animals inoculated with gram-negative bacteria produced specific antibodies against the O-specific chain of LPS. Taking the knowledge of O-chain structures into account, bacterial serotypes could now be correlated to LPS chemotypes. Thus, the structure recognized by the somatic agglutinins of Reagh, the β-agglutinogens of Joos, and the O-antigens of Weil and Felix turned out to be lipopolysaccharide. It became evident that LPS

Fig. 13 Fritz Kauffmann (1898–1978).

constituted surface antigens of gram-negative bacteria, the O-specific chain being responsible for the O-antigenic properties of LPS. Later the fine structure of those determinants within the O-specific chain reacting with serotype-specific antibodies were elucidated. These determinants were first called by Staub *motifs antigéniques*, but later became known as *O-factors* (73,87–90). The term *immunodominance* was coined in 1966 by Lüderitz and Staub after an extended discussion with Michael Heidelberger in New York in order to define the sugar to which an antibody exhibits the strongest affinity, i.e., the sugar that best inhibits (i.e., contributes most to) the binding of this antibody (73). It should be mentioned that in the characterization and the understanding of the creation of structural diversity of the O-specific chain, work on lysogenic phage conversion performed mainly by the groups of Staub and Robbins was essential (82,83,85,87,90). The basic finding was that certain phages, after infecting *Salmonella* bacteria, caused changes in the architecture of the serotype of the Kauffmann-White scheme, i.e., in the fine structure of the O-antigen. Lysogenic conversion also greatly helped to exactly define the size of serological sugar or oligosaccharide determinants and to pinpoint immunodominant regions. The first synthesis of an LPS-related immunodominant sugar (i.e., an O-factor), was performed in 1960 by Lüderitz et al. (91), who prepared a colitose-protein conjugate, which induced antibodies cross-reactive with colitose-containing bacteria or their LPS including *E. coli* 0111 and 05. As a culmination of this type of research, groups in Canada, Sweden, and Russia have, during the last two decades, chemically synthesized partial structures of repeating units of O-chains and shown that the antisera against such structures reacted with the corresponding LPS and the bacterial strain from which the LPS was isolated (92–94).

A number of gram-negative pathogens, including *Neisseria gonorrhoeae*, *N. meningitidis*, *Bordetella pertussis*, *Haemophilus influenzae*, *Campylobacter jejuni*, and *Vibrio parahaemolyticus*, produce LPS that lacks an O-specific chain. This was first recognized in 1975 by Stead and colleagues (95). In these cases, the serospecificity is mediated by the terminal sugars of the LPS glycosyl region. As these bacteria posssess, due to the absence of the O-specific chain, a shorter glycosyl chain, the term *lipooligosaccharide* (LOS) instead of LPS was proposed for their designation by Schneider in 1984 (96). Although a number of arguments were promoted against this new term (97), it is now often used by researchers in the field of human mucosal pathogens.

Core Oligosaccharide

In the structural characterization of the core oligosaccharide, LPS from enterobacterial rough mutants proved to be a most valuable tool. Rough mutants (R-mutants) had been known for a long time to microbiologists by the irregular and rough appearance of the bacterial colonies and their unique serological behavior (they could serologically not be typed by means of antisera to the O-specific chain) and investigated in pioneering studies by scientists like Arkwright (98), Kröger (99), and Schlosshardt (100). The elegant studies of Hiroshi Nikaido, Mary Jane Osborn, P. Helen Mäkela, Bruce A. D. Stocker, Klaus Jann, Stefan Stirm, Gunther Schmidt, Otto Lüderitz, and others revealed that *Salmonella* R-mutants have defects in genes that code for enzymes involved in the synthesis or transfer of LPS sugars or other constituents such as phosphate. As a result, these organisms synthesize a truncated LPS (called rough-type or R-form LPS), which contains, besides lipid A, a smaller carbohydrate portion than S-form LPS. LPS of all R-mutants lack the O-specific chain and, depending on the biosynthetic block, contain the complete core or only parts of this LPS region. In the case of *S. minnesota*, the mutant synthesizing a complete core was termed as *Ra*, that lacking the terminal core sugar as *Rb*, and so on, up to the most defective mutant, which was termed *Re* (101). For *S. minnesota*, *S. typhimurium*, and later *E. coli*, a complete series of such R-mutants became available and after Chris Galanos (Max-Planck-Institut für Immunbiologie, Freiburg) had developed an efficient extraction method, the phenol-chloroform-petroleum ether (PCP) procedure (102), R-form LPS could be studied both chemically and biologically. The first proposal of the structure of a core, i.e., that of *S. minnesota*, was made in 1967 by Lüderitz, Westphal, and colleagues. It was shown to consist of an O-chain–proximal *outer region* containing the hexoses Glc, Gal, and GlcN (this region was, therefore, also termed the *hexose region*) and a lipid A–proximal *inner core* containing Kdo, Hep, and phosphate. Subsequently, a great number of enterobacterial and other core structures were elucidated, and the core of certain bacteria still represent the object of great scientific interest (for review see Ref. 103 and Chapter XX, this volume).

Based on the reducing power and the behavior of LPS fractions upon Sephadex filtration, in 1964 the hypothesis was forwarded that heptose units of various core strands were interconnected by phosphate bridges forming a macromolecular polymer-like structure (104). In the late 1960s, mutants were characterized, the LPS of which lacked Hep-bound phosphate groups (P⁻-mutants). As LPS fractions of such P-lacking bacteria behaved similar to the P-containing LPS, the idea of phosphate bridges was abandoned and later experimentally disproven. Of interest was the demonstration that all LPS, independent of their bacterial origin, contained at least one Kdo residue, which, according to the original suggestion of Osborn (81), serves as a link between the polysaccharide and lipid A component. It took more than two decades for Brade (105) to recognize that a single Kdo residue (and not two as postulated previously) provides this link in enterobacterial LPS and LPS of other bacterial origins. Today we know that Kdo or a derivative represents a common and obligatory constituent of all LPS. With the exception of oligosaccharides from certain plants and algae, Kdo is only found in gram-negative bacteria, and thus appears to be characteristic for this class of microorganisms.

Two types of R-mutants deserve special mention: the Rc-chemotype M-mutants and the Re-chemotype mutants, the LPS of which have occasionally been termed *glycolipids* (106) or *glycolipid A*. As glycolipids by definition do not contain phosphoric acid, this term was later used only for LPS of P⁻-mutants (101). It appears, however, that these terms have not found a permanent place in the nomenclature of endotoxin research. In any case, the first mutant to be dealt with here in more detail is the M-mutant (from mutabile-type), which was originally discovered in 1959 by Murase (107) and biochemically characterized by Nikaido (85,108–110) to lack UDP-galactose-4-epimerase. Such an Rc-chemotype mutant derived from *E. coli* 0111 and termed J5 (111) was used by the groups of Braude (112) and Young (113,114) to engender anti-LPS human and murine monoclonal antibodies, which were expected to cross-react with LPS of different serotypes and to cross-protect from LPS effects believed to cause the toxic and disastrous manifestations of severe bacterial sepsis. Seemingly, the first clinical studies involving antibodies obtained by this approach were successful, a fact that made the J5 mutant quite famous and a popular tool in many follow-up studies. Repetition of clinical trials, however, revealed that the antibodies were ineffective in preventing lethality resulting from gram-negative sepsis (115,116). This demonstration had serious economic and scientific consequences. First, the two companies involved in the production of the J5-induced antibodies largely terminated their J5 program and dramatically reduced their scientific personnel. Second, governmental as well as industrial funding of work concerning monoclonal antibodies to LPS was reduced or completely abandoned. Third, in medical circles the failure of the clinical trials was interpreted to show that the concept of immunological control of LPS-induced toxicity was wrong. The more likely interpretation of the failure of clinical trials (i.e., that the concept was right but the antibodies used were inappropriate), is still only hesitantly accepted (117).

The second group of R-mutants to be mentioned, the Re-chemotype mutants, have also been proposed for the production of cross-reactive and cross-protective antibodies (118,119). After the experience with antibodies obtained by immunization of the J5-mutant, it appears, however, unlikely that antibodies generated by Re-mutants will ever be studied in large clinical trials. On the other hand, Re-mutants were important and helpful in connection with the search for the endotoxic principle. First described by Goebel and Jesaitis in 1952 (54) and by Weidel in 1954 (120) as phage-resistant strains, Lüderitz and coworkers chemically characterized LPS of Re-mutants to consist of only lipid A and Kdo (101). Thus, LPS of Re-mutant bacteria appeared to represent an ideal substrate to study the relative role that the LPS regions lipid A, Kdo, and—indirectly—the lacking polysaccharide portion may play in the expression of endotoxin activity.

Lipid A

The lipoidal LPS region, lipid A, is of particular interest. Lipid A had been postulated to represent the toxic center of LPS—a postulate that was for quite some time not universally accepted. The controversy concerning the significance of lipid A for LPS bioactivity, therefore, represents an important chapter in the history of endotoxin research. Further, the elucidation of its structure turned out to be considerably more difficult and, therefore, took much longer than that of the O-specific chain and also of the core. Thus, lipid A will be dealt with in more detail in the next section.

PRIMARY STRUCTURE OF LIPID A

Probably the first scientists to recognize a lipoidal precipitate on treatment of LPS (of *Brucella melitensis*) with mineral acid were Miles and Pirie in 1939 (121). Of course, Miles and Pirie did not call their starting material LPS—they used the term *native antigen*. Systematic studies, however, were after the pioneer studies of the groups of Morgan and Goebel only resumed more than a decade later by Westphal and Lüderitz (122–125), who stated in their often-cited review of

1954: "Lipid A is obtained, if undegraded polysaccharide (i.e., LPS) is treated for approximately 30 min at elevated temperature with N mineral acid. It is not soluble in diethyl ether and petroleum ether, but readily soluble in chloroform and pyridine" (125). Figure 14 shows a copy of page 506 of the laboratory notebook of Otto Lüderitz of October 9, 1951, where initial experiments involving the hydrolysis of *S. abortus equi* ("AE") and *E. coli* ("Coli") LPS and the formation of a lipoidal precipitate (lipid A) is described. The term *lipid A*, which was used for the first time in this review, had been adopted to distinguish the covalently bound lipid component of LPS from loosely associated and formamide-extractable lipid B, later identified as phosphatidyethanolamine (Morgan's kephalin-type phospholipid). Similar observations were made by other groups studying different gram-negative organisms. To cite one, Freeman and Anderson obtained on acid hydrolysis of the antigen of *B. typhosa* Ty2 a polysaccharide (50–60%), an insoluble polypeptide fraction (16%), a soluble nitrogenous material (10–20%), and a smaller amount of lipid (126).

If LPS is treated with acid, the linkage between Kdo and lipid A is cleaved, yielding a lipid A precipitate, the primary hydroxyl group of which at GlcN II is now free. In 1977 it was recognized for the first time that lipid A may carry positively charged groups (polar head groups) such as 4-amino-4-deoxy-L-arabinose (Ara4N), which is bound to phosphate groups through a highly acid-labile linkage (127). On treatment of LPS with mild acid—the procedure used to prepare lipid A—Ara4N was cleaved off from lipid A. It had, therefore, to be realized that lipid A as present in LPS differed structurally from isolated lipid A after its acid-catalyzed release from LPS, not only in the substitution of the primary hydroxyl group of GlcN II, but also in other structural features. To account for this fact, the term *free lipid A* was coined to distinguish liberated (i.e., polysaccharide-free lipid from bound lipid A) (128). Although in its endotoxic properties free lipid A is of comparable activity as bound lipid A, the two molecules behave, as shown by Brade et al. (129), very different serologically. Thus, monoclonal antibodies raised against free lipid A do not cross-react with LPS, i.e., bound lipid A. This finding suggests that the free primary hydroxyl group is part of the epitope(s) of anti–lipid A antibodies (130) and that, in fact, free lipid A represents a neoantigen em-

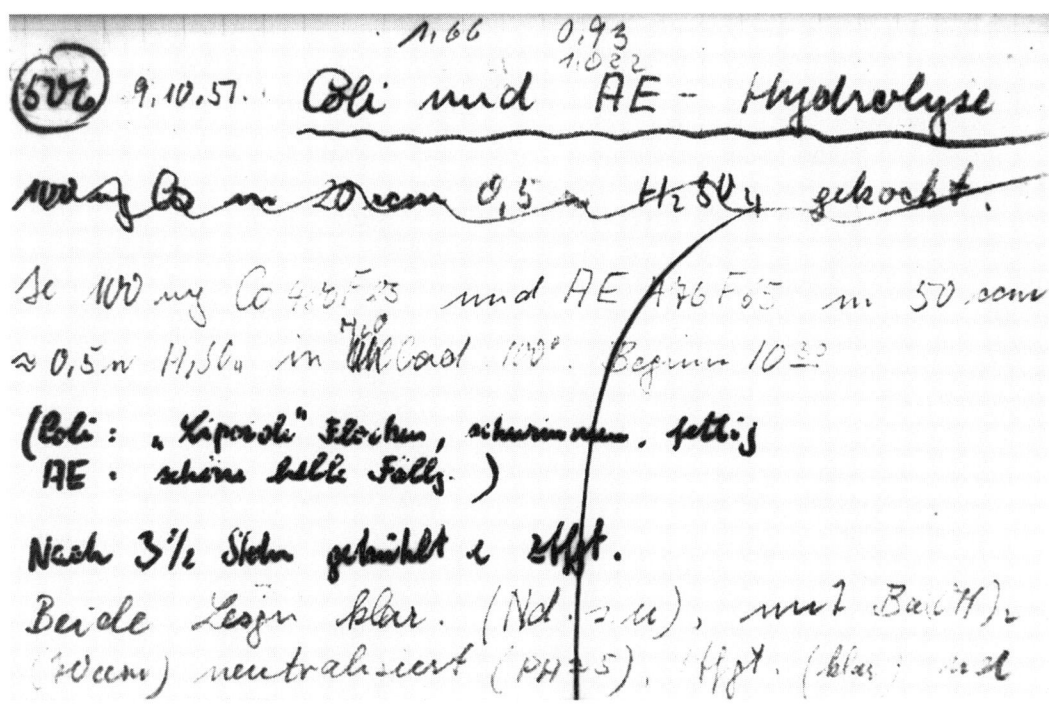

Fig. 14 Page 506 of the laboratory note book of Otto Lüderitz describing his initial experiments on the preparation of lipid A in 1951. Coincidentally the page number is identical with the designation number of the first synthetic *E. coli* lipid A (preparation 506).

phasizing structural differences between the two molecules.

The story of the elucidation of the chemical structure of lipid A would merit a separate historical review, and here only certain milestones will be briefly summarized. Pioneering work was performed in the laboratory of Carl Niemann in New York, who, in 1958, then working with tumor-necrotizing *E. coli* LPS, was the first to describe the main constituents of lipid A: D-GlcN, phosphorus, and (R)-3-hydroxymyristic acid (130,131). Lüderitz and Westphal confirmed these data and extended them to show that *Salmonella* lipid A contained the same constituents. How these constituents were linked together remained obscure for a long time. The first structural proposals were made in 1961 suggesting the presence of a polymeric phosphodiester-containing architecture (132) or of glycosidic bonds between D-GlcN residues of unknown position (133). It was only in 1969, when Jobst Gmeiner, then working at the Max Planck Institut in Freiburg and using LPS of a *S. minnesota* Re-mutant, showed for the first time that lipid A contained a D-GlcN disaccharide, which is β(1-6) interlinked and carries phosphate residues in positions 1 and 4′ (134). In extending this work, Sumihiro Hase and Rietschel, then also working at the Max Planck Institut, showed that the bisphosphorylated β(1-6)-linked D-GlcN disaccharide, which was termed the *lipid A backbone*, was also present in lipid A of *E. coli* and all other enterobacterial as well as nonenterobacterial LPS preparations studied (135).

Attention was then focused on the lipid A fatty acids bound to the hydroxyl and amino groups of the lipid A backbone, which confer the lipoidal properties to lipid A. In *E. coli* lipid A, a total of six fatty acid residues per lipid A backbone was found (i.e., one lauric, one myristic, and four 3-hydroxymyristic acids possessing the (R)-configuration) (136). As a characteristic feature of lipid A, the presence of a 3-acyloxyacyl residue was identified (137). By selective chemical degradation it was realized that each of the two amino groups of the backbone carried a 3-hydroxymyristic acid and that the other four fatty acids were ester-linked. It was then shown by our group in 1982 that two of the latter (i.e., lauric and myristic acid), were bound to the 3-hydroxyl group of one amide- and one esterlinked 3-hydroxymyristic acid, respectively (138,139). The attachment site of the two remaining ester-linked acyl groups was then determined in a collaborative study with Shoichi Kusumoto and Tetsuo Shiba (Osaka University), who applied two-dimensional nuclear magnetic resonance (140). It was found that the hydroxyl groups at carbons 4 and 6′ of the lipid A backbone were not substituted, suggesting that the two acyl residues were linked to the hydroxyl groups at carbons 3 and 3′, a hypothesis compatible with the then-emerging knowledge about the biosynthesis of lipid A. The exact location of fatty acids in lipid A was finally determined by application of fast atom bombardment–mass spectrometry (141–144) and laser desorption–mass spectrometry (145). The attachment site of Kdo, i.e., of the polysaccharide component, had been previously determined to be position 3′ of the lipid A backbone (146). Just after the elucidation of the location of acyl groups, evidence was provided by chemical and mass spectrometric studies (147), which were supported (148) and proven (149) by nuclear magnetic resonance, that in LPS of *P. mirabilis*, *E. coli*, and *S. typhimurium* Kdo was linked to position 6′ rather than to position 3′. Thus, the complete covalent structure of *E. coli* lipid A with a molecular weight of 1798 daltons was established in 1983. Independently, the complete structure of *S. typhimurium* lipid A had been determined (142), and *E. coli* and *S. typhimurium* lipid A proved to be structurally identical. (The phosphate groups of lipid A of *Salmonella* and other enterobacterial species or serotypes may carry in stoichiometric or nonstoichiometric amounts substituents such as 4-amino-4-deoxy-L-arabinose and ethanolamine, which, however, will not be considered here further; for literature, see Ref. 150).

The deduced lipid A structure was further supported by studies on the biosynthesis of lipid A by Chris R. H. Raetz, professor of biochemistry at Duke University, Durham, North Carolina, which had shown that UDP-GlcNAc represents a key precursor that is, after transfer of two 3-hydroxymyristic acids and deacetylation, condensed with 2,3-diacyl glucosamine 1-phosphate (a compound termed *lipid X*) to a tetraacyl disaccharide 1-phosphate (6). This molecule is then phosphorylated at position 4′ to yield the so-called precursor Ia lipid IV A according to Ruetz's nomenclature of lipid A biosynthesis. Earlier studies had demonstrated that Kdo is transferred to precursor Ia before the completion of lipid A by addition of lauric and myristic acid (151).

With the successful elucidation of the chemical and physical structure of LPS, an important chapter in the history of endotoxin research was completed (150). Many other problems, however, remained to be solved, among which was the nature of the bioactive region of LPS (i.e., the nature of the endotoxic principle).

SEARCH FOR THE ENDOTOXIC PRINCIPLE

In fact, a major driving force in endotoxin research had been the search for the very structure within the LPS macromolecule that was responsible for its endotoxic properties (i.e., its fever- and lethality-inducing capacity). In view of the fact that the apparent molecular weight of endotoxin preparations had been estimated to be in the order of 10^4-10^6 daltons, the definition of a toxophor group seemed to be a formidable, if not impossible task. The presence of a toxic component in LPS had been first postulated in 1950 by Tal and Goebel, who termed this component *factor T* (51). In 1954, Westphal and Lüderitz proposed that the lipid A component represented the endotoxic principle of LPS (125). They reasoned that the polysaccharide portion carrying the O-antigenic specificity differed so widely in composition and structure that theoretically this LPS region was unlikely to bear common endotoxic entities. Lipid A, however, appeared to be an ubiquitous component of the, at that time, investigated enterobacterial LPS. Further, complexing of LPS with inert proteins, like pyrogen-free serum albumin, followed by splitting with mild acid (Goebel's procedure) led to degraded polysaccharide and an artificial, soluble protein–lipid A complex (comparable to Morgan's conjugated protein), which was endotoxically active.

The view that lipid A represented the endotoxic principle was at that time, however, not shared by all researchers. Their arguments were based in part on the fact that free lipid A, either highly dispersed or complexed to albumin, expressed maximal pyrogenicity or toxicity in the order of only one fifth to one tenth of the original LPS. In addition, it had been observed that highly toxic LPS preparations obtained by the aqueous ether method contained only approximately 2% firmly bound lipid. These and other findings led Edgar Ribi (1920–1986) and colleagues to state: "On the basis of present evidence, lipid cannot, therefore, be visualized as a toxophore" (152). These researchers rather concluded from their data that LPS adopts a peculiar configuration—a quarternary complex—which is responsible for endotoxic activity (72,153). It was, however, postulated that the lipid component of LPS would play an important role in the organization of the toxic complex. The Swiss-born Ribi was a passionate pilot and tragically died when the airplane he was flying entered a snow blizzard and crashed in the Rocky Mountains on October 5, 1989. Figure 15 shows Ribi together with coworkers and guest scientists in 1965.

Another very different view was expressed by Chandler A. Stetson, who postulated that endotoxin was not a toxin in its own right at all (154). It was argued that all major effects of endotoxin such as fever and lethality could be reproduced by antigen-antibody complexes. Thus, the endotoxic properties of LPS appeared to be based on its quality as an antigen—in other words, endotoxicity was merely an immunological phenomenon. Although it is true that an Arthus reaction can be produced with complexes of S-form LPS and anti–O-chain antibodies (155), the theory of Stetson was soon disproven. In Minneapolis, Yoon Berm Kim and Dennis W. Watson at the Department of Medical Microbiology of the University of Minnesota (Fig. 16) showed that piglets obtained by hysterectomy and raised on an antigen-free diet were immunologically virgin and did not contain detectable antibodies in their serum. Yet such piglets were highly sensitive to the lethal effect of endotoxin (156). It was obvious that endotoxins were endowed with an intrinsic toxicity for which the term *primary toxicity* was coined by Watson and Kim.

The hypothesis that lipid A was the carrier of the toxic properties was strongly supported in 1968 when the same scientists showed that polysaccharide-deficient LPS, derived from Re-mutant bacteria and consisting of only Kdo and lipid A, was endotoxically fully active (157). In 1967, similar data were provided by Kasai and Nowotny (106) after the original report of Lüderitz et al. in 1966 on the toxic properties of LPS of Re-mutant bacteria (101). These results provided strong evidence that the O-specific chain and the larger part of the core (i.e., the polysaccharide component of LPS), were dispensable for endotoxicity. The finding that the Kdo region could be chemically modified without loss of bioactivity further emphasized the importance of lipid A (158). Final proof for the essential role of lipid A in endotoxicity was provided by Galanos, who demonstrated that completely polysaccharide-free lipid A could be rendered water soluble (as the triethylammonium salt after electrodialysis) without carrier, and that free lipid A exhibited in this form strong pyrogenic and toxic properties qualitatively and quantitatively comparable to the most active LPS (159).

Although it was evident that lipid A harbored the endotoxic principle, it was far from clear whether this principle was identical with or contained within the lipid A structure elucidated in the year 1983. Doubts were justified because of the strategy that had been employed for the structural elucidation of lipid A. Fragments of LPS and lipid A had been chemically prepared, their structural make-up elucidated, and this in-

Fig. 15 Edgar Ribi (1920–1989), standing, with guest scientists and coworkers at the Rocky Mountain Laboratories in 1965. Front row, from left: Jon A. Rudbach (at present Secretary of the IES), Kelsey C. Milner, W. Ted Haskins, Robert L. Anacker; second row, from left: Werner Brehmer (Berlin, Germany), Claes Weilbull (Lund, Sweden), William Wicht, Robert List, and William D. Bickle.

formation integrated into a formula of lipid A. This strategy, however, harbored the danger that during isolation and purification of one partial structure, another perhaps biologically very important one may have escaped detection; consequently, it would be lacking in the final structural proposal for lipid A. In addition, Alois Nowotny had demonstrated that lipid A expresses a certain degree of heterogeneity making the assignment of bioactivity to a defined structural entity difficult if not impossible (160). The doubts that lipopolysaccharide or lipid A per se are toxic were still expressed in 1974 by Milner et al.: "And the suspicion that their (i.e., LPS) potency may reside chiefly in some tiny component "X" remains with us" (72). It was, therefore, felt that for proving that the proposed lipid A structure contained the group(s) essential for endotoxic activity, chemically synthesized (i.e., structurally defined) and homogeneous lipid A was needed.

Several research groups including those of Hans Paulsen (Hamburg, Germany), Laurens Anderson (Madison, Wisconsin), Akira Hasegawa (Gifu, Japan), Ladislas Szabó (Paris, France), Frank M. Unger, Peter Stütz, and Paul Kosma (Vienna, Austria), and J. H. van Boom (Utrecht, The Netherlands) synthesized lipid A partial structures, which later proved to be of great value for the analysis of structure-activity relationships. But it was only the group of Tetsuo Shiba, Shoichi Kusumoto (Fig. 17), and their associates at the Faculty of Sciences of Osaka University at Toyonaka, Osaka, Japan, who undertook the total chemical synthesis of lipid A. Initial synthetic approaches were based on a lipid A structure, which was incorrect with respect to the attachment site of Kdo. Thus, the first synthetic products proved to be of disappointingly low biological activity. After the elucidation of the correct primary lipid A structure in 1983, Kusumoto and Shiba resumed their synthetic work and, in collegial contact with our groups, prepared tetraacyl precursor Ia (preparation 406). In murine assay systems, such as lethality in D-GalN-sensitized animals and B-cell mitogenicity, this compound proved to be endotoxically highly active (161).

Finally, as a culmination of the synthetic efforts, on June 18, 1984, the first total chemical synthesis of *E.*

Fig. 16 Dennis W. Watson (honorary member of the IES) in 1968.

Fig. 17 Tetsuo Shiba (right) (honorary member of the IES) and Shoichi Kusumoto (left) in 1997.

coli lipid A was completed (162). The structure of the synthetic compound (preparation 506 or LA-15-PP) corresponded to that identified previously analytically, i.e., of bacterial lipid A. The structural identity of the synthetic compound and the dominant fraction of bacterial free *E. coli* lipid A was demonstrated by proton nuclear magnetic resonance. Figure 18 shows the original ^{1}H-NMR spectrum of (dimethyl-4′-monophosphate) *E. coli* lipid A obtained by chemical synthesis on the one hand and from bacteria on the other.

The biological analysis of synthetic lipid A or preparation 506 in various endotoxin test systems in vivo and in vitro, in comparison with bacterial free *E. coli* lipid A, showed that in all test systems, including lethal toxicity in mice, pyrogenicity in rabbits, and activation of human monocytes, the synthetic *E. coli* lipid A had, on a weight basis, activity identical to that of its bacterial counterpart (163). Similar results were obtained independently by the groups of Shozo Kotani (Osaka, Japan), the late I. Yuzuru Homma (1919–1991) (Tokyo, Japan), and Shiro Kanegasaki (Tokyo). These findings unequivocally demonstrated that, indeed, lipid A with the structure elucidated harbored the endotoxic principle of LPS.

The successful total synthesis of *E. coli* lipid A and the demonstration of its chemical as well as biological identity with natural *E. coli* lipid A has certainly closed that chapter of endotoxin research that dealt with the search for the endotoxic principle and its structure. Endotoxic activity of the large LPS molecule had been reduced to a small active component—lipid A. Also the classical sequence of steps to be followed in the characterization of natural products had also now been verified for lipid A: isolation, purification, structural characterization, total synthesis, comparison of the chemical and biological properties of the natural and synthetic products, and, thus, demonstration of identity.

The availability of synthetic lipid A and partial structures and, very importantly, the detection by Hubert Mayer (Freiburg, Germany) of nontoxic LPS harboring a lipid A component with structural features differentiating it from *E. coli* lipid A (150,164) opened the way for studies concerning quantitative structure-activity relationship in vivo and in vitro, the immunogenicity of lipid A, the definition of serological lipid A epitopes (165,166), and, perhaps most importantly, the conformation of lipid A. It was only recently found by Ulrich Seydel and Klaus Brandenburg (Fig. 19) that lipid A assumes at physiological conditions (>90%, H_2O, 37°), a peculiar three-dimensional organization comprising mainly cubic phases (165,166). This result allowed the conclusion that the molecular shape of an individual lipid A molecule resembles a cone, the top of which

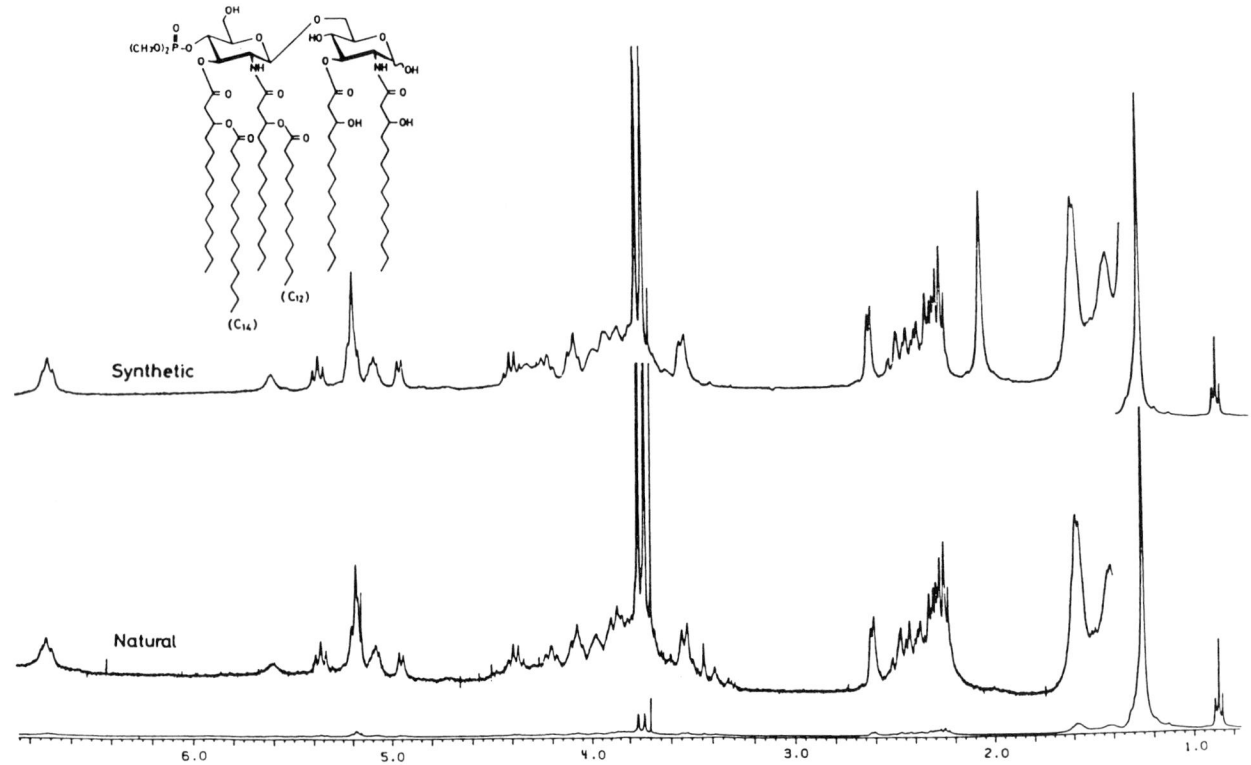

Fig. 18 Original ^{1}H-NMR spectrum of synthetic and bacterial dimethyl-4'-monophosphory *E. coli* lipid A recorded in 1983 by S. Kusumoto.

has been cut off (167,168). How the lipid A constituents are conformationally organized within this molecular shape is presently under investigation (169,170). In order to point out that endotoxin activity is not dependent on one single lipid A constituent per se, but rather is associated with a unique molecular shape that allows optimal interaction with humoral and cellular receptors, we have used the term *endotoxic conformation* (171), previously used in a different context (172). The recognition that lipid A possesses a unique structure and conformation suggested that a specific receptor should exist on endotoxin target cells such as macrophages and monocytes. It was around the time of the synthesis of lipid A that David C. Morrison revitalized the concept of a specific LPS or lipid A receptor, which had originally been forwarded by Erwin Neter, Georg F. Springer, James Watson, and Antonio Coutinho. It was only 7 years ago that Richard Ulevitch and Peter Tobias at the Scripps Research Institute and Samuel Wright at the Rockefeller Institute finally demonstrated how LPS is specifically recognized by serum factors and cellular binding sites (173–176) leading to cellular activation, signal transduction, and mediator production (i.e., the key events in endotoxin action). The mid-1980s were also the years during which the first proteins supposed to represent endogenous mediators of endotoxin action were purified, sequenced, and cloned, with human interleukin 1 (177) and human tumor necrosis factor α (178) being the prototypes of such mediators. Using pure proteins and the new animal gene knock-out technology, it was now possible to analyze the relative importance of such proteins in the mediation of endotoxin effects. It should be recalled that the first milestone experiments in this area were performed approximately only one decade ago—too short a time to be looked upon with the distance necessary for a historical review.

FINAL REMARKS

In the present retrospective we have covered about 150 years of endotoxin research. The progress described from ill-defined tissue extracts and bacterial filtrates to defined molecules and structures (i.e., from phenomenological studies to scientific understanding), was made

Fig. 19 Endotoxinologists working in the first author's (ETR) Department at the Research Center Borstel. Left to right: K. Brandenburg (Biophysics), A. J. Ulmer (Immunology), P. Zabel (Internal Medicine), S. Uhlig (Experimental Pneumology), O. Holst (Immunochemistry), Lore Brade (Serology), U. Seydel (Biophysics), H. Brade (Medical Microbiology, Chemistry and Molecular Biology), U. Zähringer (Chemistry), U. Mamat (Molecular Biology), and B. Lindner (Biophysics).

possible, of course, by the brilliant minds of clinicians and basic researchers, scientists who devoted their experimental and intellectual life to infection, microbes, and toxins. It was, however, also greatly facilitated by the advancement of methodology—in the case of endotoxin in many disciplines, in particular molecular biology, immunology, pharmacology, serology, chemistry, and physics. Limitations of space, unfortunately, do not allow reflections on the more recent history of many important and seminal endotoxin-related discoveries, including:

Cellular degradation and detoxification of endotoxin in vivo (179)
Neutralization of endotoxin by specific and cross-reactive antibodies (117,180)
Identification of nontoxic LPS (143,150,164)
Endotoxin antagonism by lipid A partial structures (181,182)
Endogenous pyrogens (177)
Endotoxin tolerance (183–185)
Limulus amebocyte lysate reactivity with LPS (186,187)
Antibiotic-induced release of endotoxin from bacteria (188)
Cytokine (endogenous mediators)-mediated endotoxic effects (177,178)
Discovery of macrophages/monocytes as endotoxin target cells (189,190)
Humoral and cellular endotoxin receptors (173,175)
Role of endotoxin in sepsis (191)
Role of LPS for microbial physiology (192,193)
Genetic determination and biosynthesis of LPS (6,194)
LPS binding and neutralizing proteins and peptides such as BPI, PXB, ENP, and sCD14 (195–197)

Each of these exciting topics would merit its own historical review. We are certain that these reviews will be written by competent scientists in the not too distant future, adding new pages to and thereby completing the book of the history of endotoxin.

Today, we know about the nature and structure of the material that Panum, Pfeiffer, Centanni, Chantemesse, and others worked with, and we can interpret the results of the experiments of Naunyn, von Bergmann, and Billroth on the cellular, molecular, and often the submolecular level. It appears, however, that the past decades of research have only laid the foundation for important studies and discoveries yet to be made. Thus, molecular details of the interaction of endotoxin with the host remain to be elucidated, including the exact nature of the membrane receptor complex and molecular details of the individual steps of the cascade transmitting the LPS signal from the cellular surface to the nucleus. Also, the importance of endotoxin in disease waits to be unequivocally identified. Endotoxin is only one of several types of toxins produced by microbes, and the disease-promoting interplay of exo- and endotoxins will certainly represent an important area of future active research.

Perhaps one day all problems relating to the toxicity phlogistic activity and disease-promoting properties of endotoxin will have been solved. In fact, the last 150 years of endotoxin research have been characterized by purification, isolation, fragmentation, dissection, and the definition within complex systems such as bacteria of single bioactive molecules and structures. The total synthesis of lipid A was a first step of a new view on the endotoxin molecule. The time has come to combine, to put together all the information gathered in order to understand the significance of endotoxin in an integrated framework. This will be a major task for the present and upcoming generation of endotoxinologists (Fig. 19). Thus, future research could concentrate on the role endotoxin may play in the maintenance of human health. There is no endotoxin-free life because endotoxin is omnipresent. In addition it was hitherto impossible to experimentally create microorganisms free of endotoxin. Each human individual carries approximately 10^{14} bacteria, which colonize the digestive tract, other mucosal surfaces, and the skin. Thus, our body is potentially confronted with gram amounts of endotoxin, approximately 3–10 pg/ml serum being continuously found in the human circulation. From the moment of birth on, we have to interact with gram-negative bacteria and their surface structures, including LPS, and there is evidence that this interaction is required for the physiological development of vital body systems such as the immune apparatus. It may be further speculated that during normal life endotoxin is constantly resorbed from the intestinal flora, thus acting like a vitamin and perhaps contributing to a physiological state of activation—a certain tonus—of the immune system. In this sense, perhaps 2 million patients die annually from endotoxin-based septic shock, but 6 billion people do in fact profit from the immunostimulating properties of LPS (198). Bacteria also profit from LPS, as otherwise they would not generate it (199). Despite numerous studies, we do not know at present which physical properties or structural/conformational features render LPS an obligatory constituent of the gram-negative bacterial outer membrane that is absolutely necessary for microbial viability. It is certain that bacteria produce LPS because they need it and not in order to induce proinflammatory responses in higher organisms. This follows from evolutionary considerations and also from the simple observation that not all LPS represent endotoxins, the LPS of *R. sphaeroides* being a good example (143,150,164).

Endotoxin represents an outstanding example of a molecule where biology and chemistry, where clinical and laboratory science, where basic and applied research have equally contributed. Therefore, endotoxin will continue to fascinate scientists in numerous disciplines, and many decades will pass before the ultimate review on the history of endotoxin research will be written. May the here presented look back to the past illustrate how rewarding work with endotoxin has been and show that it certainly will continue to be and, thus, stimulate young scientists to devote their attention to this fascinating microbial product.

ACKNOWLEDGMENTS

Part of the work performed in the first author's (ETR) laboratory was supported by the Deutsche Forschungsgemeinschaft (SFB 367, project B2, SFB 470, project B4), the BMBF ("Sepsis"-project), the German-Israeli Foundation (GIF, project "Mycoplasma"), and the Fonds der Chemischen Industrie. We thank Mrs. Frauke Richter, Renate Mohr, and Ingrid Stegelmann-Müller for typing this manuscript and for photographical work. We also thank Helmut Brade, Stefan Ehlers, Sheldon E. Greisman, Shoichi Kusumoto, David C. Morrison, Ulrich Seydel, and Ulrich Zähringer for reading this manuscript and for valuable suggestions. We are grateful to Shlomo Rottem, Hubert Mayer, Anni Nowotny, Jon Rudbach, Shoichi Kusumoto, Dennis W. Watson, Magdalena Lukacova, Jan Závada, and Werner Köhler for helping to trace important historical information and for providing unpublished photographs.

REFERENCES

1. Machmann H, Köhler W. Meilensteine der Bakteriologie, Edition Wötzel. Frankfurt, Germany, 1997.
2. Seibert FB. Fever-producing substance found in some distilled waters. Am J Physiol 1923; 67:90–104.

3. Seibert FB. Introduction. Symposium on bacterial pyrogens. NY Acad Sci 1952; 14:157–158.
4. Seibert FB. The cause of many febrile reactions following intravenous infections. I. Am J Physiol 1925; 71:621–652.
5. Rietschel ET, Brade H. Bacterial endotoxins. Sci Am 1992; 267:26–33.
6. Raetz CRH. Biochemistry of endotoxins. Annu Rev. Biochem 1990; 59:129–170.
7. Morrison DC, Ryan JL. Bacterial endotoxins and host immune responses. Adv Immunol 1979; 28:293–450.
8. Morrison DC, Ryan JL. Endotoxins and disease mechanisms. Annu Rev Med 1987; 38:417–432.
9. Rietschel ET, Brade H, Holst O, Brade L, Müller-Loennies S, Mamat U, Zähringer U, Beckmann F, Seydel U, Brandenburg K, Ulmer AJ, Mattern T, Heine H, Schletter J, Loppnow H, Schönbeck U, Flad H-D, Hauschildt S, Schade FU, Di Padova F, Kusumoto S, Schumann RR. Bacterial endotoxin: chemical constitution, biological recognition, host response, and immunological detoxification. Curr Top Microbiol Immunol 1996; 216:39–81.
10. Westphal O, Westphal U, Sommer T. The history of pyrogen research. In: Schlessinger D, ed. Microbiology 1977. Washington, DC: American Society of Microbiology, 1977:221–238.
11. Westphal O, Lüderitz O, Galanos C, Mayer H, Rietschel ET. The story of endotoxin. Proc 3rd Internat Conf Adv Immunopharmacol 1975:13–34.
12. Berry LJ. Retrospective and prospective view of endotoxin research. In: Szentivanyi A, Friedman H, Nowotny A, eds. Immunobiology and immunopharmacology of bacterial endotoxins. New York: Plenum Press, 1986:13–34.
13. von Haller A. Elementa physiologiae corporis humani. Vol III. p. 154.
14. Magendie F. Remarques sur la notice précédente (de Dupre), avec quelques expériences sur les effets des substances en putréfaction. J Physiol (Paris) 1823; 3: 81–88.
15. Panum PL. Das putride Gift, die Bakterien, die putride Infektion oder Intoxikation und die Septikämie. Arch Pathol Anat Physiol Klin Med (Virchow's Archiv) 1874; 60:301–352.
16. Bergmann von E. Schwefelsaures Sepsin. Centralbl Med Wissensch 1868; Nr. 32.
17. Bergmann von E. Zur Lehre von der putriden Intoxikation. Dtsch Z Chir 1872; 1:373–398.
18. Billroth, Th. Beobachtungsstudien über das Wundfieber und accidentelle Wundkrankheiten. Arch klin Chir 1862; 2:578–667.
19. Dubzanski V, Naunyn B. Beiträge zur Lehre von der fieberhaften (durch pyrogene Substanzen bewirkten) Temperaturerhöhung. Arch Pathol 1873; 1:1–32.
20. Burdon-Sanderson J. Aetiology of fever. In: Albutt TC, ed. A System of Medicine. New York: Macmillan and Co, 1896.
21. Burdon-Sanderson, J. On the process of fever, part III: pyrexia. Practitioner 1876; 417–431.
22. Henle J. Von den Miasmen und den miasmatisch-kontagiösen Krankheiten. In: Sudhoff K, ed. Klassiker der Medizin. Leipzig: Joh Ambrosius Barth, 1910.
23. Billroth Th. Untersuchungen über die Vegetationsformen von *Coccobacteria septica* und den Anteil, welchen Sie an der Entstehung und Verbreitung der accidentellen Wundkrankheiten haben. Arch Exp Pathol Pharmakol 1874; 2:206–209.
24. Klebs E. Beiträge zur pathologischen Anatomie der Schußwunden. Leipzig: Vogel, 1872.
25. Elik SD. *Staphylococcus pyogenes* and its relation to disease. Edinburgh: Livingstone, 1959.
26. Roussy G. Recherches experimentales. Substances calorigenes et frigorigenes d'origine microbienne; Pyretongenine et Frigorigenine. Gaz. d'Hopit. (Paris) 1989; 62:171.
27. Brieger L. Zur Erkenntnis der Fäulnisalkaloide. Berl Dtsch Chem Ges 1883; 16:1186–1191, 1405–1407.
28. Brieger L, Fraenkel C. Untersuchungen über Bakteriengifte. Berlin Klin Wochenschr 1890; 27:231–246, 268–271.
29. Buchner H. Über pyrogene Stoffe in den Bakterienzellen. Berlin Klin Wochenschr 1890; 27:673–677.
30. Buchner H. Die chemische Reizbarkeit der Leukozyten und deren Beziehung zur Entzündung und Eiterung. Berlin Klin Wochenschr 1890; 27:1084–1089.
31. von Krehl L. Versuche zur Erzeugung von Fieber bei Tieren. Arch Exp Pathol Pharmakol 1895; 35:222–268.
32. Pfeiffer R. Untersuchungen über das Choleragift. Z Hygiene 1892; 11:393–412.
33. Wolff M. Beiträge zur Immunitätslehre (Contributions to the teaching of immunity). Zentralbl Bakteriol Parasit Infekt Hyg I Orig 1904; 37:390–397.
34. Knoke M. Endotoxine in der Gastroenterologie—ein medizinhistorischer Abriss. Dtsch Z Verdau Stoffwechselkr 1984; 44:109–117.
35. Brieger L. Weitere Erfahrungen über Baktericngifte. Z Hyg Infektionskr 1897; 19:101–112.
36. Centanni, E. Über Infektionsfieber. Chem Zentr (4th Series) 1894; 6:597.
37. Centanni E. Untersuchungen über das Infektionsfieber—das Fiebergift der Bakterien. Dtsch Med Wochenschr 1894; 20:148.
38. Centanni E, Bruschettini A. Untersuchungen über das Infektionsfieber—das Antitoxin des Bakterienfiebers. Dtsch Med Wochenschr 1894; 20:270.
39. Centanni E. Weitere Beiträge zur Kenntnis des pyrogenen Wirkstoffes des Fiebers. Dtsch Med Wochenschr (Anal Series) 1940; 10:263–265.
40. Chantemesse A. Lösliches Typhustoxin und antitoxisches Serum des typhösen Fiebers. Wiener Medizinische Blätter. Wochenschr Gesamte Heilkunde 1898; 21:279–280.
41. Ecker EE. J Infect Dis 1917; 21:541.
42. Jona JL. A contribution to the experimental study of fever. J Hyg 1916; 15:169–194.
43. Gram Chr. Über die isolierte Färbung der Schizomyceten in Schnitt- und Trockenpräparaten. Fortschr Med 1884; 2:185–189.
44. Boivin A, Mesrobeanu J, Mesrobeanu L. Technique pour la préparation des polyosides microbiens spécifiques. Compt Rend Soc Biol 1933; 113:490–492.
45. Boivin A, Mesrobeanu L. Recherches sur les antigènes somatiques et sur les endotoxines des bactéries. I.

Considérations générales et exposé des technique utilisées. Rev Immunol 1935; 1:553–569.
46. Boivin A. Traveaux récents sur la constitution chimique et sur les propriétés biologiques des antigènes bactériens. Schweiz Z Pathol Bakteriol 1946; 9:505–541.
47. Penell CB, Huddleson IF. Mich. State Coll. Agr. Expt. Sta. Tech. Bull. No. 15, 1937.
48. Morgan WTJ, Patridge SM. Studies in immunochemistry—the fractionation and nature of antigenic material isolated from *Bact dysenteriae* (Shiga). Biochem J 1940; 34:169–191.
49. Morgan WTJ, Partridge SM. Studies in immunochemistry. 6. The use of phenol and of alkali in the degradation of antigenic material isolated from *Bact. dysenteriae* (Shiga). Biochem J 1941; 35:1140–1163.
50. Morgan WTJ, Partridge SM. An examination of the O antigen complex of *Bact typhosum*. Br J Exp Pathol 1942; 23:151–165.
51. Tal C, Goebel WF. On the nature of the toxic component of somatic antigen of *Shigella paradysenteriae* type Z (Flexner). J Exp Med 1950; 92:25.
52. Goebel WF, Binkley F, Perlman E. Studies on Flexner group of dysentery bacilli. I. Specific antigens of *Shigella paradysenteriae* (Flexner). J Exp Med 1945; 81:315–330.
53. Binkley F, Goebel WF, Perlman E. Studies on the Flexner group of dysentery bacilli. II. The chemical degradation of the specific antigen of type Z *Shigella paradysenteriae* (Flexner). J Exp Med 1945; 81:331–347.
54. Goebel WF, Jesaitis MA. The somatic antigen of a phage-resistant variant of phase II *Shigella sonnei*. J Exp Med 1952; 96:425–438.
55. Gough GAC, Burnet FM. The chemical nature of the phage-inactivating agent in bacterial extracts. J Path Bact 1934; 38:301–303.
56. Smith T, Reagh AL. The non-identity of agglutinins acting upon the flagella and upon the body of bacteria. J Med Res 1903; 10:89.
57. Beyer HG, Reagh AL. The further differentiation of flagellar and somatic agglutinins. J Med Res 1904; 12:313–328.
58. Joos A. Untersuchungen über die verschiedenen Agglutinine des Typhusserums. Zentralbl Bakteriol I Orig 1903; 33:762–782.
59. Braun H, Schaeffer H. Zur Biologie der Fleckfieber-Proteusbazillen. Berlin Klin Wochenschr 1919; 18:409–412.
60. Jötten KW. Vergleichende Untersuchungen über das kulturelle und serologische Verhalten gewöhnlicher und Fleckfieber-X-Proteusstämme mit besonderer Berücksichtigung ihrer Abspaltungsvarietäten. Berlin Klin Wschr 1919; 12:270–274.
61. Sachs H. Zur Kenntnis der Weil-Felixschen Reaktion. Serodiagnostik des Fleckfiebers II. Dtsch Med Wschr 1918; 17:459–462.
62. Weil W, Felix A. Weitere Untersuchungen über das Wesen der Fleckfieberagglutination. Wien Klin Wochenschr 1917; 30:1509–1511.
63. Weil E, Felix A, Mitzenmacher F. Über die Doppelnatur der Rezeptoren in der Typhus-Paratyphus-Gruppe. Wien Klin Wochenschr 1918; 31:1226–1228.
64. Weil E, Felix A. Über den Doppeltypus der Rezeptoren in der Typhus-Paratyphus-Gruppe. Zeitschr Immunitätsforsch 1920; 29:24–91.
65. Coley WB. The treatment of inoperable sarcoma with the mixed toxins of erysipelas and *Bacillus prodigiosus* and final results in 140 cases. J Am Med Assoc 1989; 31:389–456.
66. Coley-Nauts H, Swift WE, Coley BL. The treatment of malignant tumors by bacterial toxins as developed by the late William B. Coley, MD, revised in the light of modern research. Cancer Res 1946; 6:205–216.
67. Shear MJ, Turner FC. Chemical treatment to tumors. V. Isolation of the hemorrhage producing fraction from *Serratia marcescens* (*Bacillus prodigiosus*) culture filtrate. J Natl Cancer Inst 1943; 4:81–97.
68. Hartwell JL, Shear MJ, Adams JR Jr. Chemical treatment of tumors. VII. Nature of the hemorrhage-producing fraction from *Serratia marcescens* (*Bacillus prodigiosus*) culture filtrates. J Natl Cancer Inst 1943; 4:107–122.
69. Westphal O, Lüderitz O, Bister F. Über die Extraktion von Bakterien mit Phenol/Wasser. Z Naturforsch 1952; 7:148–155.
70. Springer GF, ed. Polysaccharides in Biology. Transactions of the Fifth Conference, Josiah Macy, Jr. Foundation, 1959, Princeton, NJ, 1959.
71. Westphal O, Lüderitz O. Chemische und biologische Analyse hochgereinigter Bakterienpolysaccharide. Deutsch Med Wochenschr 1953; 2:17–19.
72. Milner KC, Rudbach JA, Ribi E. Bacterial endotoxins: general characteristics. In: Weinbaum G, Kadis S, Ajl SJ, eds. Microbial Toxins. Vol. IV. Bacterial Endotoxins. New York and London: Academic Press, 1971, 1–65.
73. Lüderitz O, Staub AM, Westphal O. Immunochemistry of O and R antigens of *Salmonella* and related *Enterobacteriaceae*. Bacteriol Rev 1966; 30:192.
74. Kauffmann F, Lüderitz O, Stierlin H, Westphal O. Zur Immunchemie der O-Antigene der Enterobacteriaceen. I. Analyse der Zuckerbausteine von Salmonella-O-Antigenen. Zentralbl Bakt I Orig 1960; 178:442–458.
75. Kauffmann F, Braun OH, Lüderitz O, Stierlin H, Westphal O. Zur Immunchemie der O-Antigene von Enterobacteriaceae. IV. Analyse der Zuckerbausteine von *Escherichia*-O-Antigenen. Zentralbl Bakteriol Parasitenk Abt I Orig 1960; 180:180–188.
76. Kauffmann F, Krüger L, Lüderitz O, Westphal O. Zur Immunchemie der O-Antigene von Enterobacteriaceae. VI Vergleich der Zuckerbausteine von Polysacchariden aus *Salmonella*-S- und R-Formen. Zentralbl Bakteriol Parasitenk Abt I Orig 1961; 182:57–66.
77. Jesaitis MA, Goebel WF. The chemical and antiviral properties of the somatic antigen of phage II: *Shigella sonnei*. J Exp Med 1952; 96:409–424.
78. Slein MW, Schnell GW. The polysaccharide of *Shigella flexneri* type 3. J Biol Chem 1953; 203:837–848.
79. Weidel W. L-Gala-D-mannoheptose als Baustein von Bakterienzellwänden. Z Physiol Chem 1955; 299:253–257.
80. Heath EC, Ghalambor MA. 2-Keto-3-deoxyoctonate, a constituent of cell wall lipopolysaccharide prepara-

tions obtained from *E. coli*. Biochem Biophys Res Commun 1963; 10:340–345.
81. Osborn MJ. Studies on the gram-negative cell wall. I. Evidence for the role of 2-keto-3-deoxyoctonate in the lipopolysaccharide of *Salmonella typhimurium*. Proc Natl Acad Sci USA 1963; 50:499–506.
82. Robbins PW, Uchida T. Studies on the chemical basis of the phage conversion of O-antigens in the E group *Salmonella*. Biochemistry 1962; 1:323–335.
83. Robbins PW, Uchida T. Determinants of specificity in *Salmonella*: changes in antigenic structure mediated by bacteriophage. Fed Proc 1962; 21:702–710.
84. Uchida T, Robbins PW, Luria SE. Analysis of the serologic determinant groups of the *Salmonella* O antigen. Biochemistry 1963; 2:663–668.
85. Losick R, Robbins PW. Mechanism of ε^{15} conversion studied with a bacterial mutant. J Mol Biol 1967; 30:445–455.
86. Nikaido H, Naide Y, Mäkelä PH. Biosynthesis of O-antigenic polysaccharides in *Salmonella*. Ann NY Acad Sci 1966; 133:209–314.
87. Pon G, Staub AM. Etude chimique du polyside somatique typhique. Bull Soc Chim Biol 1952; 34:1132.
88. Staub AM, Tinelli R. Structure chimique de certains motifs antigéniques présents dans les antigènes 09 et 12 du tableau de Kauffmann-White. Bull Soc Chim Biol 1957; 39:65–83.
89. Staub AM. Constitution chimique de quelque sites présents à la surface des *Enterobacteriaceae* responsables de leur spécificité vis à vis des anticorps et des bactériophages. Pathol Microbiol 1961; 24:890–909.
90. Staub AM. The role of the polysaccharide moiety in determining the specificity and immunological activity of the O antigen complex of *Salmonellae*. In: Landy M, Braun W, eds. Bacterial Endotoxins. New Brunswick: Rutgers University Press, 1964:38–48.
91. Lüderitz O, Westphal O, Staub AM, Le Minor L. Preparation and immunological properties of an artificial antigen with colitose (3-deoxy-L-fucose) as determinant group. Nature 1960; 188:556–558.
92. Jörbeck HJ, Svenson SB, Lindberg AA. Artificial Salmonella vaccines: Salmonella typhimurium O-antigen-specific oligosaccharide-protein conjugates elicit opsonizing antobodies that enhance phagocytosis. Infect Immun 1981; 32:497–502.
93. Bundle DR, Josephson S. Artificial carbohydrate antigens: the synthesis of a tetrasaccharide hapten, a *Shigella flexneri* O-antigen repeating unit. Carbohydr Res 1980; 80:75–85.
94. Dimitriev BA, Nikolaev AV, Shashkov AS, Kochetkov NK. Block-synthesis of higher oligosaccharides: Synthesis of hexa- and nonasaccharide fragments of the O-antigenic polysaccharide of *Salmonella newington*. Carbohydr Res 1982; 100:195–206.
95. Stead A, Main S, Ward ME, Walt PJ. Studies on lipopolysaccharides isolated from strains of *Neisseria gonorrhoeae*. J Gen Microbiol 1975; 88:123.
96. Schneider H, Hale TL, Zollinger WD, Seid RC jr, Hammack CA, Griffiss JM. Heterogenicity of molecular size and antigenic expression within lipopolysaccharides of individual strains of *Neisseria meningitidis*. Infect Immun 1984; 45:544–549.
97. Hitchcock PJ, Leive L, Mäkelä PH, Rietschel ET, Strittmatter W, Morrison DC. Lipopolysaccharide nomenclature—past, present, and future. J Bacteriol 1986; 166:699–705.
98. Arkwright JA, Felix A. Typhus fever. In: A System of Bacteriology in Relation to Medicine. London: 1930: 393–432.
99. Kröger E. Experimentelle Untersuchungen zum morphologischen S/R- und antigenen O/o-Formenwechsel gram-negativer Darmbakterien. Z Naturforsch 1953; 8:133–141.
100. Schlosshardt J. Untersuchungen über die Entstehung von T-Antigenen im S-R-Formenwechsel bei Salmonellen. Zentralbl Bakteriol Parasit Abt I Orig 1960; 177:176–185.
101. Lüderitz O, Galanos C, Risse HJ, Ruschmann E, Schlecht S, Schmidt G, Schulte-Holthausen H, Wheat R, Westphal O, Schlosshardt J. Structural relationships of *Salmonella* O and R antigens. Ann NY Acad Sci 1966; 133:349–374.
102. Galanos C, Lüderitz O, Westphal O. A new method for the extraction of R lipopolysaccharides. Eur J Biochem 1969; 9:245–249.
103. Holst O, Brade H. Chemical structure of the core region of lipopolysaccharides. In: Morrison DC, Ryan JL, eds. Molecular Biochemistry and Cellular Biology. Boca Raton: CRC Press, 1992:135–170.
104. Kuriki Y, Kurahashi K. Polymannoheptose isolated from cell wall of uridine diphosphate glucose pyrophosphorylase less mutant of *Escherichia coli* K12. J Biol Chem (Tokyo) 1965; 58:308–311.
105. Brade H, Rietschel ET. α-2→4-Interlinked 3-deoxy-D-manno-octulosonic acid disaccharide. A common constituent of enterobacterial lipopolysaccharides. Eur J Biochem 1984; 145:231–236.
106. Kasai N, Nowotny A. Endotoxic glycolipid from a heptoseless mutant of *Salmonella minnesota*. J Bacteriol 1967; 94:1824–1836.
107. Murase W. Jpn J Bacteriol 1959; 440:975 (in Japanese).
108. Fukasawa T, Nikaido H. Galactose-sensitive mutants of *Salmonella*. Nature 1959; 184:1168–1169.
109. Nikaido H. Galactose sensitive mutants of *Salmonella*. I. Metabolism of galactose. Biochim Biophys Acta 1961; 48:460–469.
110. Nikaido H. Studies on the biosynthesis of cell wall polysaccharide in mutant strains of *Salmonella*. II. Proc Natl Acad Sci USA 1962; 48:1542–1548.
111. Elbein AD, Heath EC. The biosynthesis of cell wall lipopolysaccharide in *Escherichia coli*. I. The biochemical properties of a uridine diphosphate galactose 4-epimeraseless mutant. J Biol Chem 1965; 240:1919–1925.
112. Teng NNH, Kaplan HS, Herbert JM, Moore C, Douglas H, Wunerlich A, Braude H. Protection against gram-negative bacteremia and endotoxemia with human monoclonal IgM antibodies. Proc Natl Acad Sci USA 1985; 82:1790–1794.
113. Young LS, Stevens P. Cross-protective immunity to gram-negative bacilli: studies with core glycolipid of *Salmonella minnesota* and antigens of *Streptococcus pneumoniae*. I Infect Dis 1977; 136:174–180.

114. Young LS, Gascon R, Alam S, Bermudez LEM. Monoclonal antibodies for treatment of gram-negative infections. Rev Infect Dis 1989; 11:1564–1571.
115. Ziegler EJ, Fischer CJ, Sprung CI, Straube RC, Sadoff JC, Foulke GE, Wortel CH, Fink MP, Dellinger RP, Teng NNH, Allen IE, Berger HJ, Knatterud GI, LoBuglio AF, the HA-1A Sepsis Study Group. Treatment of gram-negative bacteremia and septic shock with HA-1A human monoclonal antibody against endotoxin. A randomized, double-blind, placebo-controlled trial. N Engl J Med 1991; 324:429–436.
116. Greenman RL, Schein RMH, Martin MA, Wenzel RP, McIntyre NR, Emmanuel G. A controlled clinical trial of E5 murine monoclonal antibody to endotoxin in the treatment of gram-negative sepsis. JAMA 1991; 266:1097–1102.
117. Greisman SE, Johnston CA. Evidence against the hypothesis that antibodies to the inner core of lipopolysaccharides in antisera raised by immunization with enterobacterial deep-rough mutants confer broad spectrum protection during gram-negative bacterial sepsis. J Endotox Res 1997; 4:123–153.
118. McCabe WR. Immunization with R mutants of *S. minnesota*. I. Protection against challenge with heterologous gram-negative bacilli. J Immunol 1972; 108:601–610.
119. Bruins SC, Stumacher R, Johns MA, McCabe WR. Immunization with R mutants of *Salmonella minnesota*. II. Comparison of the protective effect of immunization with lipid A and the Re mutant. Infect Immun 1977; 17:16–20.
120. Weidel W, Koch G, Lohss F. Über die Zellmembran von *E. coli* B II. Der Rezeptorkomplex für die Bakteriophagen T3, T4, T7. Vergleichende chemisch analytische Untersuchungen. Z Naturforsch 1954; 9b:398–406.
121. Miles AA, Pirie NW. Antigenic preparations from *Brucella melitensis*. Br J Exp Pathol 1939; 20:278–296.
122. Westphal O. Bakterienreizstoffe und ihre Wirkungsweise. Angew Chem 1952; 64:314.
123. Westphal O, Lüderitz O, Eichenberg E, Keiderling W. Über bakterielle Reizstoffe I. Mitteilung. Reindarstellung eines Polysaccharid-Pyrogens aus *Bacterium coli*. Ztschr Naturforsch 1952:536.
124. Lüderitz O, Westphal O. Über bakterielle Reizstoffe. II. Mitteilung. Qualitative und quantitative papierchromatographische Bestimmung der Zuckerbausteine eines hochgereinigten Polysaccharids aus Colibakterien. Z Naturforsch 1952; 7:548–55.
125. Westphal O, Lüderitz O. Chemische Erforschung von Lipopolysacchariden gram-negativer Bakterien. Angew Chemie 1954; 66:407–417.
126. Freeman GG, Anderson TH. The hydrolytic degradation of the antigenic complex of Bact. Typhosum Ty 2. Biochem J 1941; 35:564–577.
127. Hase S, Rietschel ET. The chemical structure of the lipid A component of lipolysaccharides from *Chromobacterium violaceum* NCTC 9694. Eur J Biochem 1977; 75:23–34.
128. Rietschel ET, Hase S, King M-T, Redmond J, Lehmann V. Chemical structure of lipid A. In: Schlessinger D, ed. Microbiology. Washington, DC: American Society for Microbiology, 1977:262–268.
129. Brade L, Engel R, Christ WJ, Rietschel ET. A non-substituted primary hydroxyl group in position 6' of free lipid A is required for binding of lipid A monoclonal antibodies. Infect Immun 1997; 65:3961–3965.
130. Ikawa M, Koepfli JB, Mudd SG, Niemann C. An agent from *E. coli* causing hemorrhage and regression of an experimental mouse tumor. III. The component fatty acids of the phospholipid moiety. J Am Chem Soc 1953; 75:1035–1038.
131. Ikawa M, Koepfli JB, Mudd SG, Niemann C. An agent from *E. coli* causing hemorrhage and regression of an experimental mouse tumor. IV. Some nitrogenous components of the phospholipid moiety. J Am Chem Soc 1953; 75:3439–3442.
132. Nowotny A. Chemical structure of a phosphomucolipid and its occurrence in some strains of *Salmonella*. J Am Chem Soc 1961; 83:501–503.
133. Burton HJ, Carter HE. Purification and characterization of the lipid A component of the lipopolysaccharide from *Escherichia coli*. Biochemistry 1964; 3:411–418.
134. Gmeiner J, Lüderitz O, Westphal O. Biochemical studies on lipopolysaccharides of *Salmonella* R mutants 6. Investigations on the structure of the lipid A component. Eur J Biochem 1969; 7:370–379.
135. Hase S, Rietschel ET. Isolation and analysis of the lipid A backbone: lipid A structure of lipopolysaccharides from various bacterial groups. Eur J Biochem 1976; 63:101–107.
136. Rietschel ET. Absolute configuration of 3-hydroxy fatty acids present in lipopolysaccharides from various bacterial groups. Eur J Biochem 1976; 64:423–428.
137. Rietschel ETh, Gottert H, Lüderitz O, Westphal O. Nature and linkages of the fatty acids present in the lipid A component of *Salmonella* lipopolysaccharides. Eur J Biochem 1972; 28:166–173.
138. Wollenweber H-W, Broady KW, Lüderitz O, Rietschel ET. The chemical structure of lipid A: demonstration of amide-linked 3-acyloxyacyl residues in *Salmonella minnesota* Re lipopolysaccharide. Eur J Biochem 1982; 124:191–198.
139. Wollenweber H-W, Seydel U, Lindner B, Lüderitz O, Rietschel ET. Nature and location of amide-bound (R)-3-acyloxyacyl groups in lipid A of lipopolysaccharides from various gram-negative bacteria. Eur J Biochem 1984; 145:262–272.
140. Imoto M, Shiba T, Naoki H, Iwashita T, Rietschel ET, Wollenweber H-W, Galanos C, Lüderitz O. Chemical structure of *E. coli* lipid A: linkage site of acyl groups in the disaccharide backbone. Tetrahedron Lett 1983; 24:4017–4020.
141. Takayama K, Qureshi N, Mascagni P. Complete structure of lipid A obtained from the lipopolysaccharide of the heptoseless mutant of *Salmonella typhimurium*. J Biol Chem 1983; 258:12801–12803.
142. Qureshi N, Takayama K, Heller DN, Fenselau C. Position of ester groups in the lipid A backbone of lipopolysaccharides obtained from *Salmonella typhimurium*. J Biol Chem 1983; 258:12947–12951.
143. Takayama K. In: Qureshi N. Chemical structure of lipid A. In: Morrison DC, Ryan JL, eds. Molecular Biochemistry and Cellular Biology. Boca Raton: CRC Press, 1992:43–60.

144. Qureshi N, Mascagni P, Ribi E, Takayama K. Monophosphoryl lipid A obtained from the lipopolysaccharides from *Salmonella minnesota* R595. Purification of the dimethyl derivative by high performance liquid chromatography and complete structural determination. J Biol Chem 1985; 260:5271–5278.
145. Seydel U, Lindner B, Wollenweber H-W, Rietschel ET. Structural studies on the lipid A component of enterobacterial lipopolysaccharides by laser desorption mass spectrometry. Location of acyl groups at the lipid A backbone. Eur J Biochem 1984; 145:505–509.
146. Gmeiner J, Simon M, Lüderitz O. The linkage of phosphate group and of 2-keto-3-deoxy-octonate to the lipid A component in a *Salmonella minnesota* lipopolysaccharide. Eur J Biochem 1971; 21:355–356.
147. Sidorczyk Z, Zähringer U, Rietschel ET. Chemical structure of the lipid A component of the lipopolysaccharide from a *Proteus mirabilis* Re-mutant. Eur J Biochem 1983; 137:15–22.
148. Strain SM, Fesik SW, Armitage IM. Characterization of lipopolysaccharide from a heptoseless mutant of *Escherichia coli* by carbon 13 nuclear magnetic resonance. J Biol Chem 1983; 258:2906–2910.
149. Zähringer U, Lindner B, Seydel U, Rietschel ET, Naoki H, Unger FM, Imoto M, Kusumoto S, Shiba T. Structure of de-O-acylated lipopolysaccharide from the *Escherichia coli* Re mutant strain F515. Tetrahedron Lett 1985; 26:6321–6324.
150. Zähringer U, Lindner B, Rietschel ET. Molecular structure of lipid A, the endotoxic center of bacterial lipopolysaccharides. Adv Carbohydr Chem Biochem 1994; 50:211–276.
151. Osborn MJ, Rick PD, Lehmann V, Rupprecht E, Singh M. Structure and biogenesis of the cell envelope of gram-negative bacteria. Ann NY Acad Sci 1974; 235:52–65.
152. Ribi E, Anacker RL, Fukushi K, Haskins WT, Landy M, Milner KC. Relationship of chemical composition to biological activity. In: Landy M, Braun W, eds. Bacterial Endotoxins. New Brunswick, NJ: Rutgers University Press, 1964:16–28.
153. Ribi E, Haskins WT, Milner KC, Anacker RL, Ritter DB, Goode G, Trapani R-J, Landy M. Physicochemical changes in endotoxin associated with loss of biological potency. J Bacteriol 1962; 84:803–814.
154. Stetson CA. Studies on the mechanism of the Shwartzman phenomenon. Similarities between reactions to endotoxins and certain reactions of bacterial allergy. J Exp Med 1955; 101:421.
155. Davis CE, Brown KR, Douglas H, Tate III WJ, Braude AI. Prevention of death from endotoxin with antisera. I. The risk of fatal anaphylaxis to endotoxin. J Immunol 1969; 102:563–572.
156. Kim YB, Watson DW. Role of antibodies in reactions to gram-negative bacterial endotoxins. Ann NY Acad 1966; 133:727–745.
157. Kim YB, Watson DW. Biologically active endotoxins from *Salmonella* mutants deficient in O- and R-polysaccharides and heptose. J Bacteriol 1967; 94:1320–1326.
158. Rietschel ET, Galanos Ch, Tanaka A, Ruschmann E, Lüderitz O, Westphal O. Biological activities of chemically modified endotoxins. Euro J Biochem 1971; 22:218–224.
159. Galanos C. Physical state and biological activity of lipopolysaccharides. Toxicity and immunogenicity of the lipid A component. Z Immunol Forsch 1975; 149:214.
160. Nowotny A. Heterogenity of endotoxins. In: Rietschel ET, ed. Handbook of Endotoxin. Vol. 1. Chemistry of Endotoxin. Amsterdam: Elsevier Science Publishers, 1984:308–338.
161. Galanos C, Lehmann V, Lüderitz O, Rietschel ET, Westphal O, Brade H, Brade L, Freudenberg MA, Hansen-Hagge T, Lüderitz T, McKenzie G, Schade U, Strittmatter W, Tanamoto K, Zähringer U, Imoto M, Yoshimura H, Yamamoto M, Shimamoto T, Kusumoto S, Shiba T. Endotoxic properties of chemically synthesized lipid A part structures: comparison of synthetic lipid A precursor and synthetic analogues with biosynthetic lipid A precursor and free lipid A. Eur J Biochem 1984; 140:221–227.
162. Imoto M, Yoshimura H, Kusumoto S, Shiba T. Total synthesis of lipid A, active principle of bacterial endotoxin. Proc Jpn Acad Sci 1984; 60:285–288.
163. Galanos C, Lüderitz O, Rietschel ET, Westphal O, Brade H, Brade L, Freudenberg MA, Schade FU, Imoto M, Yoshimura S, Kusumoto S, Siba T. Synthetic and natural *Escherichia coli* free lipid A express identical endotoxic activities. Eur J Biochem 1985; 148:1–5.
164. Mayer H, Campos-Portuguez SA, Busch M, Urbanik-Sypniewska T, Bhat RU. Lipid A variants—or, how constant are the constant regions in lipopolysaccharides? Excerpta Med Int Congr Ser 1990; 923:111–120.
165. Brade L, Brade H. Characterization of two different antibody specificities recognizing distinct antigenic determinants in free lipid A of *Escherichia coli*. Infect Immun 1985; 48:776–781.
166. Brade L, Brandenburg K, Kuhn H-M, Kusumoto S, Rietschel ET, Brade H. The immunogenicity and antigenicity of lipid A are influenced by its physicochemical state and environment. Infect Immun 1987b; 55:2636–2644.
167. Brandenburg K, Mayer H, Koch MHJ, Weckesser J, Rietschel ET, Seydel U. Influence of the supramolecular structure of free lipid A on its biological activity. Eur J Biochem 1993; 218:555–563.
168. Brandenburg K, Seydel U, Schromm AB, Loppnow H, Koch MHJ, Rietschel ET. Conformation of lipid A, the endotoxic center of bacterial lipopolysaccharide. J Endotox Res 1996; 3:173–178.
169. Wang Y, Hollingworth RI. An NMR spectrometry and molecular mechanics study of the molecular basis for the supramolecular structure of lipopolysaccharides. J Am Chem Soc 1996; 35:5647–5654.
170. Oikawa M, Shintaku T, Fukase K, Kusumoto S. Conformational Analysis of a Biosynthetic Precursor of Lipid A. Eurocarb9, Utrecht, The Netherlands July 6–11, 1997, p. 369.
171. Rietschel ETh, Brade H, Brade L, Brandenburg K, Schade FU, Seydel U, Zähringer U, Galanos C, Lüderitz O, Westphal O, Labischinski H, Kusumoto S,

Shiba T. Lipid A, the endotoxic center of bacterial lipopolysaccharides: relation of chemical structure to biological activity. Prog Clin Biol Res 1987; 231:25–53.
172. Tripodi D, Nowotny A. Relation of structures to function in bacterial O-antigens. V. Nature of active sites in endotoxic lipopolysaccharides of *Serratia marcescens*. Ann NY Acad Sci 1966; 133:604.
173. Schumann RR, Leong SR, Flaggs GW, Gray PW, Wright SD, Mathison JC, Tobias PS, Ulevitch RJ. Structure and function of lipopolysaccharide binding protein. Science 1990; 249:1429–1431.
174. Ulevitch RJ. Recognition of bacterial endotoxin by receptor dependent mechanisms. Adv Immunol 1993; 53:267–288.
175. Wright SD, Ramos RA, Tobias PS, Ulevitch RJ, Mathison JC. CD14, a receptor for complexes of lipopolysaccharide (LPS) and LPS binding protein. Science 1990; 249:1431–1433.
176. Ziegler-Heitbrock HWL, Ulevitch RJ. CD14: cell surface receptor and differentiation marker. Immunol Today 1993; 14:121–125.
177. Dinarello CA. The interleukin-1 family: 10 years of discovery. FASEB J 1994; 8:1314–1325.
178. Brouckaert P, Fiers W. Tumor necrosis factor and the systemic inflammatory response syndrome. Curr Top Microbiol Immunol 1996; 216:167–207.
179. Munford RS, Hall CL. Purification of acyloxyacyl hydrolase, a leukocyte enzyme that removes secondary acyl chains from bacterial lipopolysaccharides. J Biol Chem 1989; 264:15613–15619.
180. Di Padova FE, Brade H, Barclay R, Poxton IR, Liehl E, Schuetze E, Kocker HP, Ramsay G, Schreier MH, McClelland DBL, Rietschel ET. A broadly cross-protective monoclonal antibody binding to *Escherichia coli* and *Salmonella* lipopolysaccharides. Infect Immun 1993; 61:3863–3872.
181. Lynn WA, Golenbock DT. Lipopolysaccharide antagonists. Immunol Today 1992; 13:271–276.
182. Flad H-D, Loppnow H, Rietschel ET, Ulmer AJ. Agonists and antagonists for lipopolysaccharide-induced cytokines. Immunobiology 1993; 187:303–316.
183. Greisman SE, Young EJ, Carozza FA. Mechanisms of endotoxin tolerance. V. Specificity of early and late phases of pyrogenic tolerance. J Immunol 1969; 103:1223–1236.
184. Greisman SE, Young EJ, DuBuy B. Mechanisms of endotoxin tolerance. VIII. Specificity of serum transfer. J Immunol 1973; 111:1349–1360.
185. Johnston CA, Greismann SE. Mechanisms of endotoxin tolerance. In: Hinshaw LB, ed. Handbook of Endotoxin. Vol. 2: Pathophysiology of Endotoxin. Amsterdam: Elsevier, 1985:359–401.
186. Levin J, Bang FB. The role of endotoxin in the extracellular coagulation of Limulus blood. Bull Johns Hopkins Hosp 1964; 115:265–274.
187. Levin J, Bang F. In: Watson SW, Levin J, Novitsky TJ, eds. Endotoxins and Their Detection with the Limulus Amebocyte Lysate Test. New York: Alan R. Liss, 1982:xvii–xix.
188. Antibiotic mediated endotoxin release contribution to pathogenesis in gram-negative sepsis. J Endo Res 1996; 3:1–280.
189. Michalek SM, Moore RN, McGhee JR, Rosenstreich DL, Mergenhagen SE, The primary role of lymphoreticular cells in the mediation of host responses to bacterial endotoxins. J Infect Dis 1980; 141:55.
190. Rosenstreich DL, Vogel SN. Central role of macrophages in the host response to endotoxin. In: Schlesinger D, ed. Microbiology 1980. Washington, DC: American Society for Microbiology, 1980:314–320.
191. Martich GD, Boujoukos AJ, Suffredini AF. Response of man to endotoxin. Immunobiology 1993; 187:403–416.
192. Lehmann V, Rupprecht E, Osborn MJ. Isolation of mutants conditionally blocked in the biosynthesis of the 3-deoxy-D-manno-octulosonic-acid-lipid-A part of lipopolysaccharides de-rived from *Salmonella typhimurium*. Eur J Biochem 1977; 76:41–49.
193. Vaara M, Nikaido H. Molecular organization of bacterial outer membrane. In: Rietschel, E.Th, ed. Handbook of Endotoxin, Elsevier Science Publishers BV Chemistry of Endotoxin 1984; 1:1–45.
194. Mamat U, Seydel U, Grimmecke D, Holst O, Rietschel ET. Lipopolysaccharides. In: Pinto BM, ed. Comprehensive Natural Products Chemistry. In Press.
195. Weiss J, Elsbach P, Shu C, Castillo J, Horwitz A, Theofan G. Human bactericidal/permeability-increasing protein and a recombinant NH_2-terminal fragment causing killing of serum-resistant Gram-negative bacteria in whole blood and inhibiting tumor necrosis factor release induced by the bacteria. J Clin Invest 1992; 90:1122–1130.
196. Hoess A, Watson S, Siber GR, Liddington R. Crystal structure of an endotoxin neutralizing protein from horseshoe crab, Limulus anti-LPS factor, at 1.5 Å resolution. EMBO J 1993; 12:3351–3356.
197. Schütt C, Schilling T, Grünwald U, Schoenfeld W, Krüger C. Endotoxin-neutralizing capacity of soluble CD14. Res Immunol 1992; 143:71–78.
198. Bocci V. The neglected organ: bacterial flora has a curial immunostimulatory role. Perspect Biol Med 1992; 35:251–260.
199. Rietschel ETh, Kirikae T, Schade FU, Mamat U, Schmidt G, Loppnow H, Ulmer AJ, Zähringer U, Seydel U, Di Padova F, Schreier M, Brade H. Bacterial endotoxin: molecular relationships of structure to activity and function. FASEB J 1994; 218:217–225.

2

Lipopolysaccharide and the Permeability of the Bacterial Outer Membrane

Martti Vaara
National Public Health Institute, University of Helsinki Medical School, Helsinki, Finland

Lipopolysaccharide (LPS) is not only a potent toxin (endotoxin), but also the major constituent of the unique cell surface structure, the outer membrane (OM), of gram-negative bacteria (Fig. 1). It covers approximately 75% of the outer surface of the OM, while the rest is mainly covered by porin proteins that form small (cut-off ~600 daltons) channels for hydrophilic diffusion through the OM (1). All four domains of LPS—lipid A, the inner core oligosaccharide with two unusual sugars (heptose and KDO), the outer core oligosaccharide, as well as the O-antigenic polysaccharide that gives the colonies a smooth appearance—are essential for the virulence of such pathogens as *Salmonella* spp., whereas only lipid A and KDO are vital to the viability and growth in optimal laboratory conditions (1). There are approximately 3–4 million molecules of LPS per cell (1). LPS comprises 3% of the total cell dry weight in typical laboratory strains of *Escherichia coli* such as *E. coli* K-12 (2) and more than 10% of the total cell dry weight in smooth, wild-type *E. coli* isolated from clinical sources.

The carboxyl groups in KDO as well as the phosphate residues in lipid A and heptoses make LPS an anionic molecule (Fig. 2). Divalent cations (Mg^{2+}, Ca^{2+}) probably link the LPS molecules through ionic bridges to each other to form an highly ordered structure (see below). However, part of the phosphate residues carry cationic substituents that reduce the anionicity of the LPS (Fig. 2).

The OM acts as an effective permeability barrier against external noxious agents, and, as will be described below, LPS is the key molecule for this function of the OM. While the porin channels allow the diffusion of small hydrophilic compounds such as the mono- and disaccharide nutrients, larger compounds and all of the hydrophobic compounds are effectively retained by the OM of many clinically important bacteria such as the members of the Enterobacteriaceae family, *Pseudomonas aeruginosa*, and many others (1,3–7).

DIFFUSION OF HYDROPHOBIC COMPOUNDS THROUGH THE LPS LEAFLET OF THE OM

The characteristic features of the OM of gram-negative enteric bacteria include that the LPS molecules are located exclusively in the outer leaflet and the glycerophospholipids exclusively in the inner leaflet (1,3,8). While the glycerophospholipid monolayer is fluid at normal growth temperatures and closely resembles the thermal behavior of the glycerophospholipid bilayer in the cytoplasic membrane, both x-ray diffraction and Fourier transform infrared spectrometry studies have indicated that the LPS monolayer is in a highly ordered quasicrystalline structure with very low fluidity (1,8–10). Hydrophobic probe molecules have been shown to

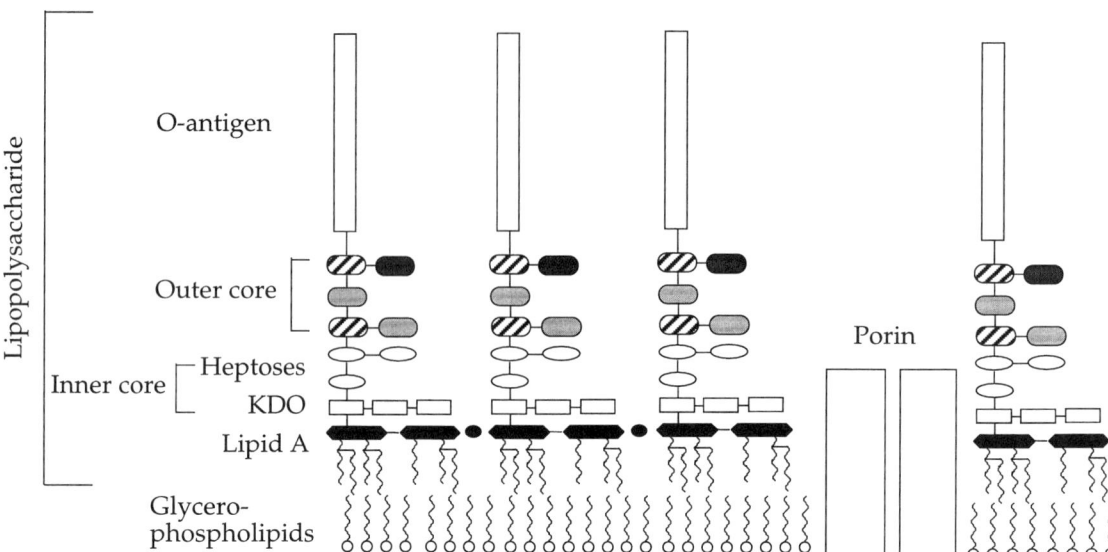

Fig. 1 Schematic representation of the structure of the outer membrane of gram-negative enteric bacteria such as *Escherichia coli* and *Salmonella typhimurium*. A notable feature is the asymmetry of the membrane; all LPS molecules are located in the outer leaflet and all of the glycerophospholipids in the inner leaflet. The small black dots denote divalent cations (Mg^{2+}, Ca^{2+}) that connect LPS molecules via ionic bridges to form a network. In this simplistic drawing, only the most dominant class of proteins, the porins, are presented. The molecule sizes are not drawn to scale, and the shapes of the molecules are imaginary. The length of the O-antigen varies.

partition very poorly into the hydrophobic interior portion of isolated LPS, which forms bilayers (8,11). The highly ordered structure of the LPS layer can be due to (1) tight lateral interaction of LPS molecules, mediated by divalent cations that bridge the anionic LPS molecules, (2) the rigidifying effect of the deep core oligosaccharide domain, as well as (3) the occurrence of a tightly packed set of fatty acid residues (six to seven per one LPS molecule, four linked to the nonreducing glucosamine residue) (8).

The highly ordered structure of the LPS layer of the OM probably explains why the OM is an effective permeability barrier to hydrophobic compounds (1,3,8). Plésiat and Nikaido (12) developed a very elegant assay to determine the OM permeability rates for hydrophobic steroid probes oxidized in the cytoplasmic membrane (CM) and showed that these probes permeate across the outer membrane bilayer at a rate approximately two orders of magnitude lower than the rate of penetration through the cytoplasmic membrane.

As a result of the poor penetration of hydrophobic compounds, gram-negative enteric bacteria are much less susceptible to hydrophobic antibiotics and other drugs than are the gram-positive bacteria. The pioneering studies performed by Nikaido, Kuo and Stocker, Schlecht and Westphal, Roantree, and their coworkers as well as subsequent studies have uniformly demonstrated that the minimum inhibitory concentrations (MICs) of a large number of hydrophobic inhibitors are 10- to more than 1000-fold higher for wild-type strains of *E. coli* and *Salmonella typhimurium* than for their OM-defective mutants (see below) or gram-positive bacteria (1,3,4,8,13; see also Table 1). This is also the reason why most of the novel antibiotics discovered from natural sources or invented in the laboratory are inhibitory for gram-positive bacteria but lack activity against wild-type *E. coli* (13).

E. coli AND *S. typhimurium* LPS MUTANTS WITH DEFECTIVE OM PERMEABILITY BARRIER TO HYDROPHOBIC ANTIBIOTICS

Mutants that are defective in lipid A biosynthesis are usually conditionally thermosensitive; some do not grow at elevated temperatures such as 42°C, some do not grow even at 37°C (13,15). A typical characteristic is that at growth-permitting temperatures they are very susceptible to hydrophobic antibiotics (Table 1), indicating that their outer membrane permeability barrier is defective (13,16). Other mutants that are supersusceptible include those that elaborate very truncated LPS

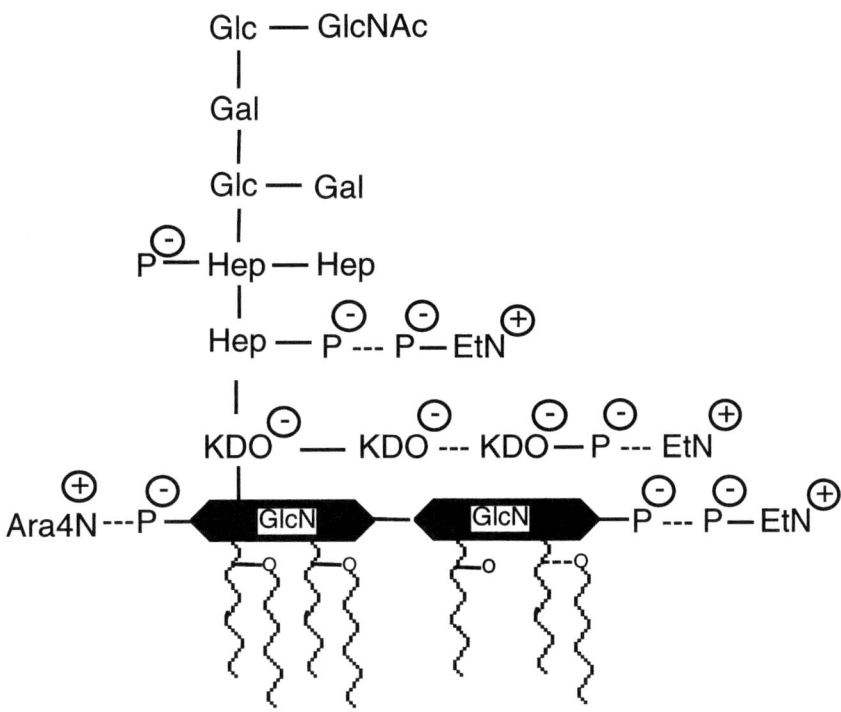

Fig. 2 Structure of the lipid A and core oligosaccharide parts of LPS of *Salmonella typhimurium*. Dotted lines represent incomplete substitution. The anionic and cationic residues are indicated. The wavy lines represent fatty acid residues. The details of the chemical structure of LPS are described elsewhere in this book. Gal, D-Galactose; Glc, D-glucose; GlcN, D-glucosamine; GlcNac, N-acetyl-D-glucosamine; Hep, L-glycero-D-mannoheptose; KDO, 2-keto-3-deoxy-octonic acid (3-deoxy-D-mannooctulosonic acid); EtN, 2-aminoethanol (ethanolamine); P, phosphate; Ara4N, 4-amino-4-deoxy-L-arabinose.

Table 1 Minimum Inhibitory Concentrations (MICs) of Various Hydrophobic Antibiotics and Large Hydrophilic Antibiotics (Vancomycin and Bacitracin) for Wild-type *E. coli* (IH3080), for the Lipid A Defective *lpxA* Mutant Strain SM101 of *E. coli*, for IH3080 in the Presence of the OM Permeabilizing Agent PMBN, and for the Gram-Positive Bacterium *Micrococcus luteus* ATCC9341

Antibiotic	MIC (μg/ml) for			
	E. coli IH3080 (wild-type)	*E. coli* SM101 (lpxA)	*E. coli* IH3080 with PMBN (3 μg/ml)	*Micrococcus luteus* ATCC 9341
Rifampin	10	0.03	0.1	0.03
Fusidic acid	300	1	10	0.25
Mupirocin	100	1	1	ND
Erythromycin	100	0.25	10	0.25
Novobiocin	30	4	3	0.125
Clindamycin	300	2	10	ND
Vancomycin	300	8	10	0.5
Bacitracin	≥1000	2	ND	1

PMBN, Polymyxin B nonapeptide; ND, not determined.
Source: Adapted from Refs. 13,14 and unpublished data by M. Vaara.

inner core (1,3,4,8,13). Comparative studies that have determined the MICs of a large set of hydrophobic antibiotics indicate that the lipid A mutants are more supersusceptible than the LPS inner core mutants. The OM permeability barrier properties of all these mutants have been reviewed in detail by Vaara (13).

Since the amount of lipid A synthesized by the lipid A mutants is low, the most probable explanation for the defectiveness of their permeability barrier to hydrophobic compounds is the simple lack of a continuous LPS layer in the outer leaflet of their OM and the resultant compensatory existence of glycerophospholipids in this leaflet (13). This creates glycerophospholipid bilayers or "patches" in the OM that allow diffusion of hydrophobic solutes. The lipid A mutants are also susceptible to the large peptide antibiotics vancomycin and bacitracin (Table 1, 13). Since these hydrophilic peptides cannot penetrate the OM of the lipid A mutants through the hydrophobic pathway, i.e., through the putative glycerophospholipid patches, they probably do it through transient ruptures (13). Such ruptures also allow the leakage of proteins from the periplasm of the lipid A mutants (13,16,17,18).

The lipid A mutants thus far extensively characterized (13,15–19) include those defective in *lpxA* (codes for the first acyltransferase of lipid A synthesis), *lpxC* (formerly known as *envA*, codes for the deacetylase), as well as *lpxD* (formerly *firA*, *omsA*, *ssc*, codes for the second acyltransferase), whereas the phenotypes and OM permeability barrier properties of the mutants defective in the late acyltransferase genes of lipid A synthesis, namely *htrB* (20) and *msbB* (21,22), have not yet been characterized in detail. The lipid A mutants have found use in pharmaceutical industry, since new important antibiotics against gram-negative bacteria could perhaps be discovered from natural sources by using in the screenings OM permeability barrier defective, maximally antibiotic supersusceptible target bacteria (13).

In contrast to the lipid A mutants, the deep core LPS mutants produce unchanged or even increased amounts of LPS (4,13). Since the defective LPS does not facilitate trimerization of the monomeric OmpF porin (1), the translocation of OmpF to the OM is inhibited, the porin content of the OM is low (1,3), and there is an observed compensatory appearance of glycerophospholipids in the outer leaflet of the OM, and, accordingly phospholipid patches in the OM (1,3,8,13), as suggested above for the lipid A mutants. The deep core LPS mutants have been known for more than 30 years, and they have been instrumental in the research on LPS structure and function as well as in endotoxin research.

EFFECT OF POLYCATIONS AND CHELATORS ON LPS AND ON THE PERMEABILITY BARRIER FUNCTION OF THE OM

The above-mentioned cation-binding sites of LPS are the Achilles' heel of the OM (23). The outer membranes of gram-negative enteric bacteria are susceptible to strongly cationic agents including the naturally occurring antibiotics of the polymyxin group, to cationic detergents, to synthetic cationic peptides, as well as to the antibacterial peptides present in neutrophil and macrophage phagosomes (23,24). Polymyxins complex avidly with the anionic groups of LPS and lipid A, probably reduce the highly ordered supramolecular structure of the LPS layer, and disorganize the whole OM (23). These actions are not their lethal action, but their means to permeate the OM to reach their final target, the cytoplasmic membrane. Some polycationic agents with a suitable structure practically lack the lethal action on the cytoplasmic membrane but are very effective permeabilizers of the OM, and hence sensitize the target bacteria to hydrophobic antibiotics (23). An example of this is shown in Table 1. The OM-disorganizing and -permeabilizing action of EDTA and other chelators that chelate Ca^{2+} and Mg^{2+} is also well known (1,23). The agents that damage the outer membrane and increase its permeability have been reviewed extensively in a previous review (23).

The study of polymyxin-resistant mutants has yielded important data on the structure-function relationships in the LPS molecule. The first *pmrA* mutants of *S. typhimurium* were reported by Mäkelä et al. in 1978, and thereafter, numerous other *pmrA* mutant alleles were made and studied by Vaara and coworkers (as reviewed in Refs. 1,3,4,23). The mutants were found to tolerate 20–100 times higher concentrations of polymyxin B and bind 4 times less of it than their parents. The isolated outer membranes and LPS from the *pmrA* mutants was found to bind remarkably less polymyxin than the outer membranes and LPS, respectively, from the parent strains. In contrast to their parents, polymyxin did not permeabilize their outer membrane to macromolecules such as lysozyme and the periplasmic β-lactamase and to the anionic detergent, deoxycholate.

The *pmrA* strains were also found to be resistant to the effects of EDTA as well as of polylysines, protamine, high concentrations of Tris, Na^+, and Mg^{2+} in the cold (1,3,4,23), bactericidal/permeability-increasing protein (BPI or CAP57) as well as azurocidin (CAP37) (23,25,26), but not to defensins (27), tachyplesins (28), cecropin A (29), and cationic detergents

(24,29,30). Since the *pmrA* mutants are also fully susceptible to octapeptin (30), an antibiotic resembling polymyxin but more hydrophobic, I have suggested that hydrophobic interaction plays a more predominant role in the interaction of the latter compounds with the outer membrane (29).

The LPS from four independent *pmrA* mutants was found to contain four- to sixfold larger amounts of 4-aminoarabinose and also larger amounts of ethanolamine than the LPS from their corresponding parent strains (31). These amino compounds esterify the phosphate groups in LPS and thus make the mutant LPS less acidic. The ester-linked phosphate at position 4' is extensively (80–90%) substituted with 4-aminoarabinose in the *pmrA* mutants (31). These findings have recently been verified by using the phosphorus NMR technique (32,33). The same mechanism, i.e., esterification of the 4' phosphate, has also been shown to underlie the polymyxin resistance in the polymyxin-resistant mutants of *E. coli* (34) and *Klebsiella pneumoniae* (35). As a result one could speculate that in the polymyxin-resistant mutants of enteric bacteria, intermolecular LPS-LPS bridging takes place, at least partially, by electrostatic interaction of the basic charge of 4-aminoarabinose, located in the 4' end of lipid A, with the anionic group(s) located elsewhere in the LPS (30).

The *pmrA* gene and its downstream flanking *pmrB* resemble structurally two-component regulatory systems previously described, and hence one could surmise that they regulate genes more directly involved in the expression of polymyxin resistance (36). Very intriguingly, recent works by Miller and coworkers (27) as well as by Groisman and coworkers (28) have shown that pmrA-pmrB is preceded by a gene (called *pagB* in Ref. 27 and *pmrC* in Ref. 37) that is activated by the two-component *Salmonella* virulence regulator PhoP-PhoQ. Expression of *pmrAB* activates a promoter 5' to *pagB* and increases the transcription of *pagB-pmrA-pmrB* (27). In wild-type strains, Mg^{2+} limitation induces the PhoP-PhoQ–dependent expression of the polymyxin-resistance operon (27,37). That Mg^{2+} limitation renders bacteria resistant to polymyxin has been known for 30 years (38). Another promoter of *prmA-pmrB* is located within the *pagB* sequence, is not dependent on PhoP-PhoQ, and is apparently activated by mild acidic conditions. In polymyxin-resistant *E. coli* mutants, probably very analogous genes are involved (27,37), since the resistant phenotype can be made to revert by plasmids carrying the wild-type *pagB-pmrA-pmrB* operon from *S. typhimurium* (K. Nummila, I. M. Helander, and M. Vaara, unpublished).

Accordingly, I conclude that even though many cationic agents as well as chelators damage the outer membrane, bacteria are able to modify their lipid A in vivo by using sophisticated regulator systems so that they acquire adaptive resistance to those agents. Furthermore, some species of gram-negative bacteria are inherently resistant to polycations. These include *Proteus mirabilis*, *Chromobacterium violaceum*, and *Pseudomonas cepacia*, all of which have a high content of phosphate-linked 4-aminoarabinose in their LPS, as reviewed previously (23).

OM PERMEABILITY BARRIER PROPERTIES AGAINST HYDROPHOBIC SOLUTES IN SELECTED OTHER GROUPS OF BACTERIA

In contrast to members of the Enterobacteriaceae family, *Pseudomonas aeruginosa*, and many other gram-negative bacteria, certain other gram-negatives show high intrinsic susceptibility to hydrophobic inhibitors. These include *Neisseria gonorrhoeae* (13), *Brucella* spp., (39), *Vibrio cholerae* (40), *Comamonas testosteroni* (12), *Pseudomonas acidovorans* (12), *Acinetobacter calcoaceticus* (12), as well as *Sphingomonas paucimobilis* (M. Vaara, unpublished observations). In an elegant series of assays, Plésiat and Nikaido determined the OM permeability rates for hydrophobic steroid probes in various gram-negatives and showed that the rates were significantly (up to more than 20-fold) higher in *C. testosteroni*, *P. acidovorans*, and *A. calcoaceticus* than in *P. aeruginosa*, *P. putida*, *E. coli*, and *S. typhimurium* (12). The molecular mechanism underlying this difference was not studied. It would be interesting to assay whether the permeable species expose phospholipids in the outer leaflet of the OM, as does *V. cholerae* (40). In any case, increased permeability of hydrophobic molecules could be advantageous for gram-negative bacteria in certain environmental habitats, both terrestrial and aquatic, in order to facilitate the uptake of hydrophobic nutrients.

The intracellular parasite *Brucella* sp. has a rather unique susceptibility pattern to inhibitors. It is very susceptible to hydrophobic antibiotics and notably permeable to the hydrophobic probe NPN but shares with the polymyxin-resistant *pmrA* mutants of *S. typhimurium* (see above) the property of being resistant to polymyxin, certain other cationic peptides, and Tris-EDTA (39,41,42). It would be most important to learn more about the structure-function relationships in the *Brucella* OM, since such findings could be more general and applicable to a larger group of intracellularly living gram-negatives. However, since the *pmrA* mutants of *Salmonella* are not supersusceptible to hydrophobic solutes (30), the *Brucella* OM is certainly not a direct analog to the OM of the *pmrA* mutants.

Most interestingly, *Sphingomonas paucimobilis* lacks LPS but contains charged glycosphingolipids, somewhat analogous to LPS, instead, as an integral part of the OM (43,44). Potential explanations for the increased susceptibility of *S. paucimobilis* to hydrophobic antibiotics include the possibility that the sphingolipid layer is more permeable to hydrophobic solutes than an LPS layer. Alternatively, it could be suggested that sphingolipids are perhaps not present in a sufficient amount to cover the entire outer leaflet of the *S. paucimobilis* OM; this results in the appearance of glycerophospholipid bilayer patches. *S. paucimobilis* is the first gram-negative bacterium shown to lack LPS, and even though this lack is probably a rare property among gram-negatives, at least the close relatives of *S. paucimobilis* could also lack LPS.

OTHER FACTORS IMPORTANT FOR THE RESISTANCE OF GRAM-NEGATIVE ENTERIC BACTERIA TO HYDROPHOBIC COMPOUNDS

Even though the intact OM with no genetic defects in LPS deep core and lipid A synthesis plays the key role in the resistance of enteric bacteria to hydrophobic inhibitors, it can ultimately only retard the diffusion of those drugs through the OM. Bacteria additionally need energy-dependent mechanisms to pump hydrophobic solutes, diffused into the cytoplasm, out from the cell. Such efflux mechanisms have recently been a subject of intensive research. The AcrAB and EmrAB efflux systems have been discovered and characterized (45–48). The *acrA* mutants were reported in 1968 as supersusceptible to acriflavine and thereafter intensively studied, but the underlying defect in them was only identified 25 years later. It is probable that also some other previously partially characterized, antibiotic supersusceptible enterobacterial mutants such as the SS-B mutant of *S. typhimurium* (49) will eventually prove to be efflux mutants.

In addition, mutants seriously defective in the synthesis of multiple OM proteins, such as the mutants defective in both OmpA and the murein lipoprotein, are known to possess unstable OM that is partially defective in its barrier function to hydrophobic solutes (3,4). Other mutants supersusceptible to hydrophobic drugs include the *pss* mutant (defective in phosphatidylserine synthetase) (50) as well as the *E. coli* mutant LH530 that synthesizes reduced amounts of lipid A but probably has the primary defect in a more general step of lipid biosynthesis, since the defect can be compensated for by the gene that encodes phosphopantetheinyl transferase (51–53).

CONCLUDING REMARKS

LPS is an integral part of the outer leaflet of the OM of gram-negative bacteria, and the intact apparatus for the biosynthesis of its lipid A-KDO part appears to be essential for the viability of such bacteria as *E. coli* and *Salmonella typhimurium* (1,15). Mutations that affect the synthesis of this part are lethal. Furthermore, LPS is responsible for the relative impermeability of the OM to noxious hydrophobic agents, including many of those antibiotics effective against gram-positives (1,3,13). Disorganizing the LPS layer on the OM results in a drastic increase in susceptibility to hydrophobic antibiotics (23). It has also been shown that bacteria that have disorganized LPS are susceptible to the complement-mediated bactericidal activity of normal fresh serum (23,54,55).

Accordingly, inhibition of LPS synthesis and/or disorganizing the LPS structure would be expected to be an ideal way to treat gram-negative infections. Drugs that appear to be reasonably effective lipid A synthesis inhibitors in vivo and during infection have recently been developed (56); they are metalloprotease inhibitors that inhibit the deacetylase encoded by envA (see above). The future will show whether these drugs and their derivatives will prove clinically useful (57). Furthermore, it can be anticipated that some lipid A inhibitors can be found in microorganisms in environmental habitats, provided that sufficiently susceptible target organisms such as *E. coli lpxA*, which has a very defective OM, are used in the screenings (13,16). Subsequent chemical modification, through combinatorial chemistry, would then potentially widen the spectra of those inhibitors to include clinically relevant gram-negative bacteria as well.

REFERENCES

1. Nikaido H. Outer membrane. In: Neidhardt FC, Curtiss III R, Ingraham JL, Lin ECC, Low Jr KB, Magasanik B, Reznikoff WS, Riley M, Schaechter M, Umbarger HE, eds. Escherichia coli and Salmonella typhimurium, Cellular and Molecular Biology. 2d ed. Washington, DC: American Society for Microbiology, 1996:29–47.

2. Neidhardt FC. Chemical composition of Escherichia coli. In: Neidhardt FC, Ingraham JL, Low Jr KB, Magasanik B, Schaechter M, Umbarger HE, eds. Escherichia coli and Salmonella typhimurium, Cellular and Molecular Biology. 1st ed. Washington, DC: American Society for Microbiology, 1987:3–6.
3. Nikaido H, Vaara M. Molecular basis of bacterial outer membrane permeability. Microbiol Rev 1985; 49:1–32.
4. Vaara M, Nikaido H. Molecular organization of bacterial outer membrane. In: Rietschel ET, ed. Handbook of Endotoxin, Vol. 1: Chemistry of Endotoxin. Amsterdam: Elsevier Science Publishers B.V., 1984:1–45.
5. Nikaido H. Role of the outer membrane of Gram-negative bacteria in antimicrobial resistance. In: Bryan LE, ed. Handbook of Experimental Pharmacology. Vol. 91. Berlin: Springer-Verlag, 1989:1–34.
6. Nikaido H, Hancock REW. Outer membrane permeability of Pseudomonas aeruginosa. In: Sokatch JR, ed. The Bacteria, a Treatise on Structure and Function. Vol. X. Orlando: Academic Press, Inc., 1986:145–193.
7. Trias J, Nikaido H. Diffusion of antibiotics via specific pathways across the outer membrane of Pseudomonas aeruginosa. In: Silver S, ed. Pseudomonas: Biotransformations, Pathogenesis, and Evolving Biotechnology. Washington, DC: American Society for Microbiology, 1990:319–327.
8. Nikaido H. Permeability of the lipid domains of bacterial membranes. In: Aloia RC, Curtain CC, Gordon LM, eds. Membrane Transport and Information Storage. Advances in Membrane Fluidity. Vol. 4. New York: Alan Riss, 1990:165–190.
9. Brandenburg K, Seydel U. Investigation into the fluidity of lipopolysaccharide and free lipid A membrane systems by Fourier-transform infrared spectroscopy and differential scanning calorimetry. Eur J Biochem 1990; 191:229–236.
10. Labischinski H, Naumann D, Schulz C, Kusumoto S, Shiba T, Rietschel ET, Giesbrecht P. Comparative X-ray and Fourier-transform-infrared investigations of conformational properties of bacterial and synthetic lipid A of Escherichia coli and Salmonella minnesota as well as partial structures and analogues thereof. Eur J Biochem 1989; 179:659–665.
11. Vaara M, Plachy WZ, Nikaido H. Partitioning of hydrophobic probes into lipopolysaccharide bilayers. Biochim Biophys Acta 1990; 1024:152–158.
12. Plésiat P, Nikaido H. Outer membranes of gram-negative bacteria are permeable to steroid probes. Mol Microbiol 1992; 6:1323–1333.
13. Vaara M. Antibiotic-supersusceptible mutants of Escherichia coli and Salmonella typhimurium. Antimicrob Agents Chemother 1993; 37:2255–2260.
14. Viljanen P, Matsunaga H, Kimura Y, Vaara M. The outer membrane permeability-increasing action of deacylpolymyxins. J Antibiot 1991; 44:517–523.
15. Raetz CRH. Lipopolysaccharide biosynthesis. In: Neidhardt FC, Curtiss III R, Ingraham JL, Lin ECC, Low Jr KB, Magasanik B, Reznikoff WS, Riley M, Schaechter M, Umbarger HE, eds. Escherichia coli and Salmonella typhimurium, Cellular and Molecular Biology. 2d ed. Washington, DC: American Society for Microbiology, 1996:1035–1063.
16. Vuorio R, Vaara M. The lipid A biosynthesis mutation lpxA2 of Escherichia coli results in drastic antibiotic supersusceptibility. Antimicrob Agents Chemother 1992; 36:826–829.
17. Vuorio R, Vaara M. Comparison of the phenotypes of the lpxA and lpxD mutants of Escherichia coli. FEMS Microbiol Lett 1995; 134:227–232.
18. Young K, Silver LL. Leakage of periplasmic enzymes from envA1 strains of Escherichia coli. J Bacteriol 1991; 173:3609–3614.
19. Young K, Silver LL, Bramhill D, Cameron P, Eveland SS, Raetz CR, Hyland SA, Anderson MS. The envA permeability/cell division gene of Escherichia coli encodes the second enzyme of lipid A biosynthesis, UDP-3-O-(R-3-hydroxymyristoyl)-N-acetylglucosamine deacetylase. J Biol Chem 1995; 270:30384–30391.
20. Clementz T, Bednarski JJ, Raetz CR. Function of the htrB high temperature requirement gene of Escherichia coli in the acylation of lipid A: HtrB catalyzed incorporation of laurate. J Biol Chem 1996; 271:12095–12102.
21. Somerville JE Jr, Cassiano L, Bainbridge B, Cunningham MD, Darveau RP. A novel Escherichia coli lipid A mutant that produces an antiinflammatory liposaccharide. J Clin Invest 1996; 97:359–365.
22. Clementz T, Zhou Z, Raetz CRH. Function of the Escherichia coli msbB gene, a multicopy suppressor of htrB knockouts, in the acylation of lipid A. J Biol Chem 1997; 272:10353–10360.
23. Vaara M. Agents that increase the permeability of the outer membrane. Microbiol Rev 1992; 56:395–411.
24. Vaara M, Porro M. Group of peptides that act synergistically with hydrophobic antibiotics against gram-negative enteric bacteria. Antimicrob Agents Chemother 1996; 40:1801–1805.
25. Shafer WM, Martin LE, Spitznagel JK. Cationic antimicrobial proteins isolated from human neutrophil granulocytes in the presence of diisopropyl fluorophosphate. Infect Immun 1984; 45:29–35.
26. Farley MM, Shafer WM, Spitznagel JK. Lipopolysaccharide structure determines ionic and hydrophobic binding of a cationic antimicrobial neutrophil granule protein. Infect Immun 1988; 56:1589–1592.
27. Gunn JS, Miller SI. PhoP-PhoQ activates transcription of pmrAB, encoding a two-component regulatory system involved in Salmonella typhimurium antimicrobial peptide resistance. J Bacteriol 1996; 178:6857–6864.
28. Ohta M, Ito H, Masuda K, Tanaka S, Arakawa Y, Wacharotayankun R, Kato N. Mechanisms of antibacterial action of tacgyplesins and polyphemysins, a group of antimicrobial peptides isolated from horseshoe crab hemocytes. Antimicrob Agents Chemother 1992; 36:1460–1465.
29. Vaara M, Vaara T. Ability of cecropin to penetrate the enterobacterial outer membrane. Antimicrob Agents Chemother 1994; 38:2498–2501.
30. Vaara M. Increased outer membrane resistance to ethylenediaminetetraacetate and cations in novel lipid A mutants. J Bacteriol 1981; 149:523–528.
31. Vaara M, Vaara T, Jensen M, Helander I, Nurminen M, Rietschel ET, Mäkelä PH. Characterization of the lipopolysaccharide from the polymyxin-resistant pmrA

mutants of Salmonella typhimurium. FEBS Lett 1981; 129:145–149.
32. Helander IM, Kilpeläinen I, Vaara M. Increased substitution of phosphate groups in lipopolysaccharides and lipid A of the polymyxin-resistant pmrA mutants of Salmonella typhimurium: a 31P-NMR study. Mol Microbiol 1994; 11:481–487.
33. Guo L, Lim KB, Gunn JS, Bainbridge B, Darveau RP, Hackett M, Miller SI. Regulation of lipid A modifications by Salmonella typhimurium virulence genes phoP-phoQ. Science 1997; 276:250–253.
34. Nummila K, Kilpeläinen I, Zähringer U, Vaara M, Helander IM. Lipopolysaccharides of polymyxin B-resistant mutants of Escherichia coli are extensively substituted by 2-aminoethyl pyrophosphate and contain aminoarabinose in lipid A. Mol Microbiol 1995; 16:271–278.
35. Helander IM, Kato Y, Kilpeläinen I, Kostiainen R, Lindner B, Nummila K, Sugiyama T, Yokochi T. Characterization of lipopolysaccharides of polymyxin-resistant and polymyxin-sensitive Klebsiella pneumoniae O3. Eur J Biochem 1996; 237:272–278.
36. Roland KL, Martin LE, Esther CR, Spitznagel JK. Spontaneous pmrA mutants of Salmonella typhimurium LT2 define a new two-component regulatory system with a possible role in virulence. J Bacteriol 1993; 175:4154–4164.
37. Soncini FC, Groisman EA. Two-component regulatory systems can interact to process multiple environmental signals. J Bacteriol 1996; 178:6796–6801.
38. Brown MRW, Melling J. Role of divalent cations in the action of polymyxin B and EDTA on Pseudomonas aeruginosa. J Gen Microbiol 1969; 59:263–274.
39. Martínez de Tejada G, Moriyón I. The outer membranes of Brucella spp. are not barriers to hydrophobic permeants. J Bacteriol 1993; 175:5273–5275.
40. Paul S, Chaudhuri K, Chatterjee AN, Das J. Presence of exposed phospholipids in the outer membrane of Vibrio cholerae. J Gen Microbiol 1992; 138:755–761.
41. Martínez de Tejada G, Pizarro-Cerdá J, Moreno E, Moriyón I. The outer membranes of Brucella spp are resistant to bactericidal cationic peptides. Infect Immun 1995; 63:3054–3061.
42. Freer E, Moreno E, Moriyón I, Pizarro-Cerdá J, Weintraub A, Gorvel J-P. Brucella-Salmonella lipopolysaccharide chimeras are less permeable to hydrophobic probes and more sensitive to cationic peptides and EDTA than are their native Brucella sp. counterparts. J Bacteriol 1996; 178:5867–5876.
43. Kawahara K, Seydel U, Matsuura M, Danbara H, Rietschel ET, Zähringer U. Chemical structure of glycosphingolipids isolated from Sphingomonas paucimobilis. FEBS Lett 1991; 292:107–110.
44. Kawasaki S, Moriguchi R, Sekiya K, Nakai T, Ono E, Kume K, Kawahara K. The cell envelope structure of the lipopolysaccharide-lacking gram-negative bacterium Sphingomonas paucimobilis. J Bacteriol 1994; 176:284–290.
45. Ma D, Cook DN, Alberti M, Pon NG, Nikaido H, Hearst JE. Genes acrA and acrB encode a stress-induced efflux system of Escherichia coli. Mol Microbiol 1995; 16:45–55.
46. Nikaido H. Prevention of drug access to bacterial targets: permeability barriers and active efflux. Science 1994; 264:382–388.
47. Ma D, Cook D, Hearst JE, Nikaido H. Efflux pumps and drug resistance in gram-negative bacteria. Trends Microbiol 1994; 2:489–493.
48. Thanassi DG, Cheng LW, Nikaido H. Active efflux of bile salts by Escherichia coli. J Bacteriol 1997; 179:2512–2518.
49. Vaara M. Antimicrobial susceptibility of Salmonella typhimurium carrying the outer membrane permeability mutation SS-B. Antimicrob Agents Chemother 1990; 34:853–857.
50. Raetz CRH, Foulds J. Envelope composition and antibiotic hypersensitivity of Escherichia coli mutants defective in phopshatidylserine synthetase. J Biol Chem 1977; 252:5911–5915.
51. Hirvas L, Nurminen M, Helander IM, Vuorio R, Vaara M. The lipid A biosynthesis deficiency of the Escherichia coli antibiotic-supersensitive mutant LH530 is suppressed by a novel locus, ORF195. Microbiology 1997; 143:73–81.
52. Nurminen M, Hirvas L, Vaara M. The outer membrane of lipid A-deficient Escherichia coli mutant LH530 has reduced levels of OmpF and leaks periplasmic enzymes. Microbiology 1997; 143:1533–1537.
53. Lambalot RH, Gehring AM, Flugel RS, Zuber P, LaCelle M, Marahiel MA, Reid R, Khosla C, Walsh CT. A new enzyme superfamily—the phosphopantetheinyl transferases. Chem Biol 1996; 3:923–936.
54. Vaara M, Vaara T. Sensitization of gram-negative bacteria to antibiotics and complement by a nontoxic oligopeptide. Nature (Lond) 1983; 303:526–528.
55. Vaara M, Viljanen P, Vaara T, Mäkelä PH. An outer membrane-disorganizing peptide PMBN sensitizes E. coli strains to serum bactericidal action. J Immunol 1984; 132:2582–2589.
56. Onishi HR, Pelak BA, Gerckens LS, Silver LL, Kahan FM, Chen MH, Patchett AA, Galloway SM, Hyland SA, Anderson MS, Raetz CR. Antibacterial agents that inhibit lipid A biosynthesis. Science 1996; 274:980–982.
57. Vaara M. Lipid A, target for antibacterial drugs. Science 1996; 274:939–940.

3

Lipopolysaccharide Phase Variation in *Haemophilus* and *Neisseria*

Derek W. Hood and E. Richard Moxon
John Radcliffe Hospital, The University of Oxford, Headington, Oxford, England

Lipopolysaccharide (LPS) is a critical, if not essential, component of the cell wall of gram-negative bacteria. Through lipid A functioning as a bacterial endotoxin and the saccharide portion contacting host cell surfaces, LPS is responsible for many aspects of host-parasite interaction. Lipid A acts as a membrane anchor and is linked to a heterogeneous surface exposed core oligosaccharide composed mainly of neutral sugars. In many genera of bacteria there is a third hydrophilic O-antigen polysaccharide region, distal to the core, that varies greatly both in composition and the number of repeated units from strain to strain (see Chapter 9).

Because of the relationships between bacterial virulence and LPS, the biology of LPS is most studied in bacteria that are pathogens, being best understood for members of the Enterobacteriaceae (especially *Escherichia coli* and *Salmonella* species). However, in recent years major advances have been made in the study of LPS of other pathogenic bacteria, especially the mucosal pathogens, e.g., *Haemophilus influenzae*, *Neisseria meningitidis*, *Neisseria gonorrhoeae*, and *Bordetella pertussis*.

The key role of LPS structure in membrane architecture and function makes it no surprise that LPS contributes to many aspects of host-pathogen interactions. In enteric bacteria, the polymerized sugar units of the O-antigen side chains impart to the cell surface relatively hydrophilic properties that facilitate resistance to bile salts and intestinal enzymes. But loss of O-antigen also results in severely decreased virulence through increased susceptibility to complement-mediated killing.

Several pathogens that reside in the respiratory tract have LPS molecules that lack the O-antigen extensions, and therefore their outer surfaces are relatively hydrophobic as well as more susceptible to cell surface–directed antibodies, salts, and some antibiotics such as bactericidal cationic peptides. These more truncated LPS structures have been designated lipooligosaccharides by some workers. Despite the apparently greater simplicity of the LPS of mucosal pathogens, these structures maintain a significant degree of heterogeneity and will be our focus for discussion in this chapter.

The exact role of LPS during each stage of infection by any given bacterium has not been defined, but the LPS structure of *H. influenzae*, for example, has been shown to be important at each stage in the pathogenesis of systemic infections in the infant rat model of bacteremia and meningitis (1). Defined *H. influenzae* mutants have implicated particular LPS structures in colonization, attachment, and invasion (2,3), evasion of host clearance (4), and induction of host response (5). Progressive reduction in the size and complexity of the LPS results in increasing clearance of the bacterial strains from the blood in an animal model (6). The adherence of *N. meningitidis* and *N. gonorrhoeae* to cultured mucosal epithelial cells has been shown to be dependent on LPS phenotype variants, while all other known invasion-associated variants are constant (7,8). LPS structure alters colonization of infant rats by *N. meningitidis* (9), and in *N. gonorrhoeae* LPS that carries a particular epitope promotes the ingestion of bacteria by host cells through binding cellular receptors involved in the removal of N-acetylglucosamine containing human glycoproteins (10).

A characteristic of the LPS from mucosal pathogens such as *H. influenzae* and *N. meningitidis* is that the variable oligosaccharide moieties can mimic host structures, generally glycosphingolipids. Frequently isolated forms of *Neisseria* LPS contain either lacto-*N*-neotetraose (Galβ1-4 GlcNAcβ1-3Galβ1-4Glc) or digalactoside (αGal1-4βGal) at their terminal ends (11). The latter is also found as a terminal structure on *H. influenzae* LPS (12,13). These terminal LPS structures are immunologically related to the paraglobiside glycolipids and the Pi and Pk antigen, respectively, which are found on host tissue (see Chapter 4). This molecular mimicry may result in poor immunogenicity, can contribute to evasion of host immune responses, and allows a role for LPS as a ligand for attachment. Variation of terminal LPS structures can therefore alter the properties and function of the LPS molecule and creates a potential biological advantage for pathogenic bacteria in being able to vary the structures expressed. Heterogeneity of LPS structure could allow some bacteria within a population to have an advantage at any given time within specific compartments or different microenvironments within the host. Variation of LPS structure could allow changes in cell surface properties, expose or protect other cell surface components, and maximize antigenic diversity to help attachment and evade the host immune response. For example, in *Neisseria* the addition of sialic acid to galactose residues masks some previously exposed epitopes and blocks the formation of the membrane attack complex (C5-C9) of complement (14). A highly sialylated LPS phenotype allows *Neisseria* cells to adhere to epithelial cells, but they are invasion deficient (7); thus the ability to vary LPS structures offers a potential advantage to the organisms in terms of host survival and the potential for causing disease. Some LPS variation will be a function of the biosynthetic processes that result in the complex tertiary structure of the molecule. Biosynthesis involves numerous functionally diverse proteins, which are required to synthesize and activate component lipids and sugars, to attach the sugars in stereospecific linkages, and to assemble, transport, and anchor the mature glycolipid molecules on the outer surface of the bacterial cell wall. Additional LPS heterogeneity will result when considering the degree of substitution of the LPS by modifications such as phosphate (PO_4), phosphoethanolamine (PEA), pyrophosphoethanolamine (PPEA), O-acetyl (OAc), or phosphorylcholine (PC) groups. Heterogeneity of LPS structure may be thought to arise in the following ways

1. Microheterogeneity that is a function of whether the biosynthetic processes go to completion; all possible sugar units and substituents may be non-stoichiometric in the absence of any known proofreading function. A particularly striking heterogeneity is the ladder pattern seen when enteric bacterial LPS is purified and resolved by gel electrophoresis. Consecutive bands are considered to be LPS with loss or gain of an O-antigen repeat unit.
2. Microheterogeneity may result from competition, either from different enzymes involved in alternative LPS biosynthetic pathways or precursor supply between LPS biosynthesis and other metabolic processes within the cell.
3. Microheterogeneity may reflect genetic programming, e.g., gene expression, in response to altered environmental stimuli. The LPS repertoire can respond, within defined limits, to changes in the growth phase, growth conditions, and to specific environmental signals.
4. Microheterogeneity through specific genetic mechanisms that promote antigenic or phase variation through so-called contingency genes (15). This type of heterogeneity is a characteristic of mucosal pathogens such as *Haemophilus* and *Neisseria* and significantly enhances variation of LPS structural epitopes over that seen in other bacteria under uniform growth conditions. Phase variation can be defined as a high-frequency (typically 1 in 100–1000 bacterial cells per generation) on-off switching leading to the reversible loss or gain of specific epitopes. The functional importance of this is that it is rapid, occurs randomly with respect to time, and is reversible. Phase variation results in multiple LPS structures from a population of cells (Fig. 1) and has confounded efforts both to study the biology of the molecule and to develop its potential as a common antigen for vaccine development.

Monoclonal antibodies (mAbs) specific to LPS provide a necessary tool for detecting variations in LPS structure and have been used to characterize the extensive variability and cross-reactivity of LPS structural epitopes from different organisms. Kimura and Hansen (16) described spontaneous high-frequency loss and acquisition of reactivity of *H. influenzae* with oligosaccharide-specific mAbs. These LPS changes correlated with altered serum resistance and virulence in the infant rat model of bacteraemia and meningitis. All serotype and nontypable (NT) *H. influenzae* strains

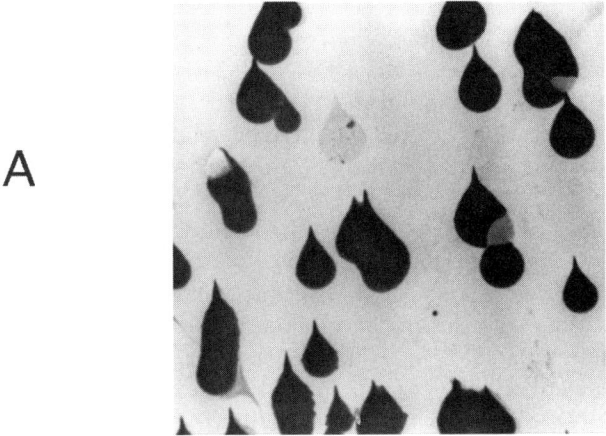

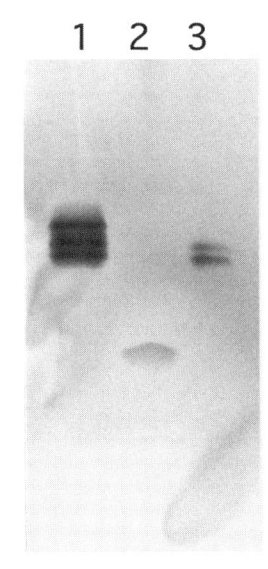

Fig. 1 Phase variation of *H. influenzae* strain RM7004 LPS. (A) Immunoblot of colonies of RM7004 grown on an agar plate then reacted with mAb 4C4. Colonies are generally reactive, with an identifiable nonreactive colony just above the middle. To the right are two colonies containing nonreactive sectors. (B) Fractionation profiles of purified RM7004 LPS after gel electrophoresis and staining with silver. Lane 1, RM7004 wild-type; lane 2, RM7004 deep rough LPS mutant; lane 3, RM7004*lgtC* mutant. This LPS is isolated from a strain mutant in a phase-variable gene and shows a simplified banding pattern with loss of higher molecular weight glycoforms.

show extensive cross-reactivity of LPS epitopes and the potential for LPS antigenic, or phase, variation. This variability produced a basis for grouping different strains by their reactivities with the mAbs (17); the same mAbs have been used to study phase variation of LPS epitopes in individual strains of *H. influenzae* (Fig. 1). For example, a component of the LPS structure(s) recognized by one mAb (4C4) (17) is the digalactoside αGal(1-4)βGal. Significantly, the digalactoside and the lacto-*N*-neotetraose mimic host structures and are the most variable moieties in the LPS of *H. influenzae* and *Neisseria* spp., respectively. Several of the *H. influenzae* mAbs show cross-reactivity to LPS structures produced by *Neisseria* species (18). For *N. meningitidis*, the reactivity with LPS specific mAbs has utility as a typing system for strains. Despite the variable nature of the target antigen, 12 distinct LPS types have been identified. The serum resistance (19) and relative virulence in the mouse model of infection (9) are altered for strains displaying different LPS structures.

The basis for the mechanism by which most characterized LPS phase variation is exerted is the presence of short nucleotide repeats situated within biosynthetic genes. All reiterated short nucleotide repeats are presumed to vary through a process related to slipped-strand mispairing (20). Polymerase slippage during nucleic acid replication is *recA* independent and results in loss or gain of a repeat unit. This phase-variation mechanism is distinct from other types of genetic antigenic variation that occur in pathogenic bacteria: the pilin genes of *Neisseria* vary by recombination between a functional and many silent copies of the major structural gene (21); surface proteins of gram-positive cocci contain large repetitive tracts within the genes, which are altered by recombinational events (22). In mucosal pathogens, the phase-variable expression of LPS is known to be promoted by tetranucleotide repeats in *H. influenzae* and by mononucleotide repeats in *N. meningitidis* and *N. gonorrhoeae*. Reports of other organisms displaying LPS capable of phase variation include *Franciscella*, *Chlamydia*, *Coxiella*, *Helicobacter*, and *Bordetella* species, but the nature and mechanisms of these variations are not fully understood.

H. influenzae lic LOCI

The first report of the molecular basis of LPS phase variation was by Weiser and his colleagues in 1989 working on *H. influenzae*. A genomic library from a type b strain, RM7004, was screened for clones that would allow a recipient strain, Rd, to express a novel

LPS epitope recognized by mAb 4C4 (23). Strain Rd does not naturally react with mAbs 4C4, 5G8, 6A2, and 12D9, whereas strain RM7004 reacts in a phase-variable manner with each (24). A clone was identified that conferred expression of the epitopes recognized by the mAbs 6A2 and 12D9 and allowed mAb 4C4 expression but only at extremely reduced frequency. This locus was designated *lic1* (*l*ipopolysaccharide *c*ore 1), and it is now evident that the major genes associated with mAb 4C4 epitope expression are located at another chromosomal locus, *lic2*, as well as elsewhere on the genome. A total of three *lic* loci have been characterized for *H. influenzae*: *lic1*, *lic2*, and *lic3*. A common feature of the *lic* loci is that just within the first open reading frame (orf) of each are a number of repeats of the tetranucleotide 5'-CAAT-3'. Loss or gain of copies of this repeat by slipped-strand mispairing would cause frame shifting of the reading frame with respect to the start codons, resulting in variable gene expression (Fig. 2) (25,26). Gene function associations have now been made between the *lic1* and *lic2* loci and LPS biosynthesis, but the function of *lic3* remains unclear. The *lic1* locus will now be described in some detail to illustrate the molecular analysis and the mechanism of LPS phase variation in *H. influenzae*, and then some comparisons will be made with the other phase-variable genes to show similarities and the complexity of the process.

The nucleotide sequence of *lic1* comprises four orfs encompassing 3.4 kb of DNA. The orfs are designated *lic1A* (339-amino-acid (aa) product), *lic1B* (305-aa product), *lic1C* (198-aa product) and *lic1D* (628-aa product) and are transcribed in the same direction (27) (Fig. 2). A series of isogenic mutants in strain RM7004 with specific deletions in *lic1* confirmed that the expression of two mAb-specific LPS epitopes, 6A2 and 12D9, were associated with this locus (27). Transformation with subclones of this region revealed that a determinant for mAb 6A2 epitope expression was located within *lic1C* and mAb 12D9 expression within *lic1D*. Studies have indicated that the mAb 12D9 reactive epitope is required in LPS biosynthesis before the mAb 6A2 epitope can be added. Recent studies have shown that *lic1* is involved with the addition to the LPS of phosphorylcholine (PC) (28). The *lic1A* and *lic1B* gene products have regions of homology to proteins involved with choline metabolism in other organisms, mAb 12D9 reactivity to colonies mirrors that of

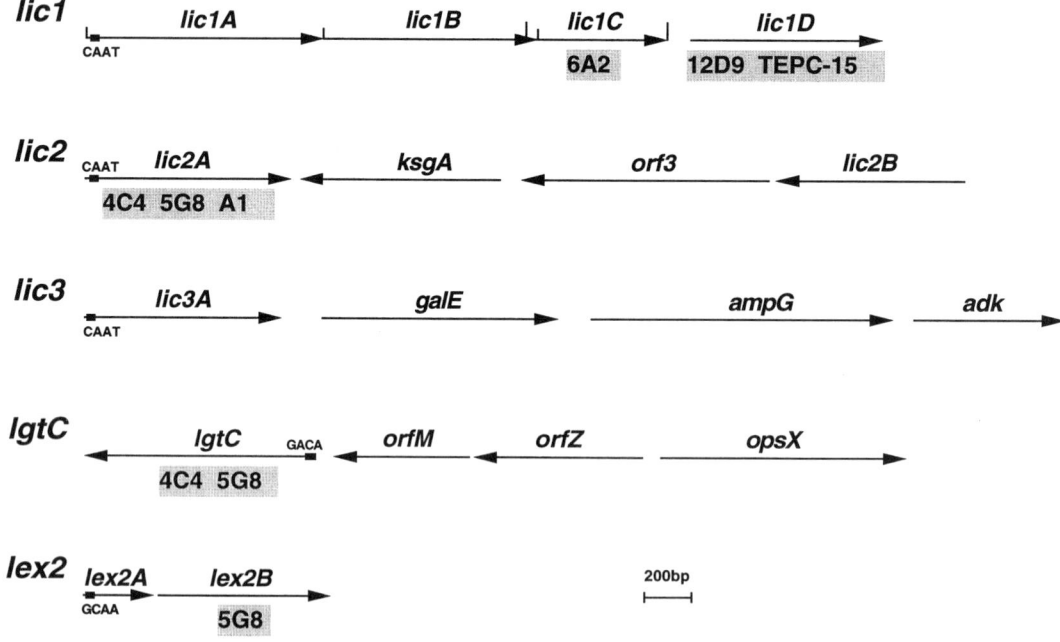

Fig. 2 Schematic representation of the tetranucleotide containing loci associated with phase variation of *H. influenzae* LPS. The direction of transcription is shown by the arrows and the gene designations are as detailed in the text. Illustrated below the genes are monoclonal antibody reactivities dependent upon particular orfs. Regions of repeats are shown by thick lines and the relevant tetranucleotide motifs are given below. The vertical lines for *lic1* indicate the extent of the contiguous and overlapping reading frames present in this locus.

a choline-specific mAb, TEPC-15. The first gene of this locus, *lic1A*, was shown to mediate phase variation. Two of the identified start codons would initiate translation in one frame, while the third would allow translation in a second frame (Fig. 3). Three levels of expression of the phase-variable mAb 6A2 and 12D9 epitopes were reported (27) upon colony immunoblotting of strain RM7004: strong (++++), weak (+), and undetectable (−). These were present in the ratio of weak 97% to strong and undetectable 3%. Similar phenotypes and frequencies have been observed for the PC-specific mAb, TEPC-15. The polymerase chain reaction (PCR) and oligonucleotides flanking the tetranucleotide region were used to amplify a portion of *lic1A* from chromosomal DNA of colonies exhibiting each of the three phenotypes. The most prevalent size of repeat DNA found in mAb 6A2 nonreactive colonies was 184 bp or 29 repeats. In mAb 6A2++++ variants, the size of the repeat was 192 bp, 31 repeats, and the predominant mAb 6A2+ colonies had 188 bp or 30 (5′-CAAT-3′) repeats. These results correlate with the alternative reading frames: 6A2+ variants (30 repeats) have a single ATG in frame, 6A2++++ variants (31 repeats) had either of two closely sited ATGs in frame and 6A2− variants (29 repeats) would be out of frame (Fig. 3). Thirty repeats is apparently favored, and loss or gain of repeats results in strong or undetectable mAb binding, respectively. This implication, that the number of 5′-CAAT-3′ repeats and translation of *lic1A* correlates with altered expression of LPS epitopes, was supported by constructing an insertion mutation within the *licA* orf. This single-site mutation in RM7004 allowed constitutive low-level expression of the mAbs 6A2 and 12D9 specific LPS epitopes (27). Thus, *lic1A* mediates phase variation of the mAb 6A2 and 12D9 specific epitopes, but the mechanism by which this is achieved remains unclear. *lic1A* is the first of four contiguous genes, the third and fourth of which have been shown to mediate mAb 6A2 and 12D9 reactivity. Either *lic1C* or *lic1D* expression has a dependence on the Lic1A protein, or some form of translation coupling regulates the expression of the latter gene products. Transcript analysis by Northern blotting and primer extension has shown a 3 kb transcript for *lic1*

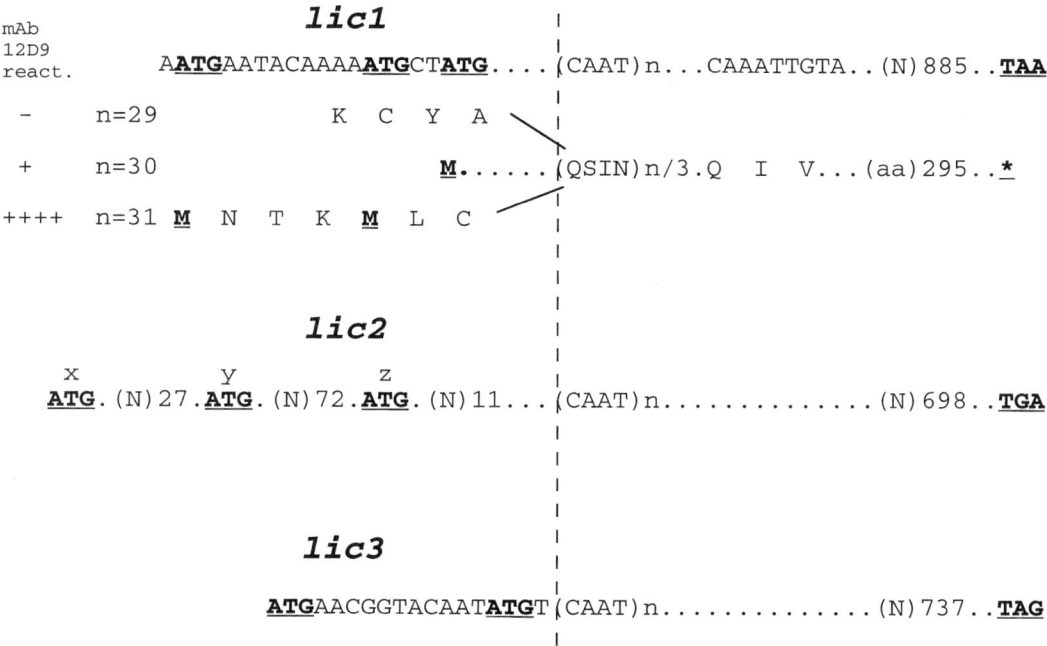

Fig. 3 Multiple initiation codons in the *lic* genes of *H. influenzae*. Initiation (ATG) and termination (TAA) codons are given in bold and are underlined. Shown in greater detail is the relationship between the number of repeats of the tetranucleotide 5′-CAAT-3′ and the reading frame of the *lic1A* gene. The nucleotide and the amino acid sequence of the correct reading frame are shown to the right of the vertical dashed line. The upstream sequences in each frame relevant when 29, 30, or 31 copies of the repeat are present are shown to the left of the line. Alternative initiation codons are available for correct translation of *lic1A* with both 30 and 31 but not 29 copies of the repeat. On the left are shown the corresponding reactivities observed with mAb 12D9.

(unpublished results), which is presumed to be rapidly processed to individual transcripts for the *lic1A*, *lic1B*, and *lic1C* genes. *lic1D* transcription may initiate within the *lic1C* gene (unpublished results). To allow further analysis of *lic1* gene expression, a *lacZ* reporter fusion was constructed in frame with *lic1D* on the *H. influenzae* genome (29). The strain showed reversible high-frequency variation between β-galactosidase positive (++++ or +) and negative (−) colony phenotypes. These correlated with ++++ = 31, + = 29, and − = 29 or 31 copies of 5′-CAAT-3′ in *lic1A*, results not entirely consistent with mAb 12D9 reactivity described above. Thus, the number of repeats in *lic1A* influences, but is not the only factor involved with, the expression of *lic1D*. Introduction of an omega cassette containing multiple stops into *lic1C*, upstream of the β-galactosidase fusion, removed expression and would support the above observation that interruption of *lic1C* could have a polar effect on *lic1D*.

Additional *lic* loci, *lic2* and *lic3*, were discovered when digested chromosomal DNA from strain RM7004 was hybridized using a (5′-CAAT-3′)$_5$ oligonucleotide probe (25). The relevant loci were identified and isolated from a lambda library of strain RM7004 genomic DNA. *lic2* was isolated on a 3.7 kb *Bgl*II fragment and when analyzed appeared to contain only one gene of relevance to LPS biosynthesis. *lic3* was found on a 5.5 kb *Bgl*II DNA fragment and when sequenced was found to comprise 4 orfs.

Deletion and insertion mutagenesis of the 3.7 kb of DNA around the *lic2* locus indicated that only one of 4 orfs, *lic2A*, had function in LPS biosynthesis (Fig. 2) (26). *lic2A* contained multiple copies of 5′-CAAT-3′ within its 5′ end (26) and when first identified had no homology to any known data base sequence. Mutation of *lic2A* resulted in the loss of expression of oligosaccharide epitopes recognized by the mAbs 4C4, 5G8, and A1-12E-5G-8E(A1) (25). An identical gene, *lex1*, has been described in another *H. influenzae* type b strain, DL42 (30). The mAbs 4C4, 5G8, and A1 have an element of their binding specificity dependent on the αGal(1-4)βGal epitope. This structure is found on the LPS of other bacteria, is a component of the globoseries glycolipids of some human epithelial cells, and can act as receptor for P fimbriae expressed by *E. coli* and the B subunit of shiga toxin. This makes the αGal(1-4)βGal structure a strong candidate for molecular mimicry, and as a terminal structure and virulence determinant, it represents an intuitively likely candidate for phase variation. *lic2A* contains two possible start codons, ATGx and ATGy, upstream of the repeats in one frame and a third potential start codon, ATGz, closer to the repeats in another frame (Fig. 3). By PCR amplification and direct sequencing methods (31), some correlation was again found between the number of 5′-CAAT-3′ repeats in *lic2A* and mAb (4C4) specific LPS epitope expression (26). Three levels of mAb reactivity could be detected upon colony immunoblotting of the type b strains RM7004 and RM153 (Eagan): on, intermediate, and off. These occurred at different relative frequencies in the two strains, 60% on, 30% intermediate, and 10% off for RM7004, whereas RM153 showed equal numbers of each. Direct sequencing of DNA amplified by PCR from immunostained colonies showed that each phenotype had 15, 16, or 17 copies of the 5′-CAAT-3′ repeat within the *lic2A* gene. It appeared that 16 copies of 5′-CAAT-3′, presumably allowing translation initiation from ATGx or ATGy, were required for strong expression of mAb 4C4 reactive epitopes in strain RM7004. No initiation was ever seen from ATGz, requiring 17 copies of 5′-CAAT-3′, in strain RM7004. In strain RM153, colonies expressing the on phenotype all contained 17 copies of the repeat (ATGz in frame), those expressing intermediate reactivity to mAb 4C4 had 16 copies of 5′-CAAT-3′ (ATGx/ATGy in frame), and strains expressing the off phenotype had 16 or 15 copies of the repeat (ATGx/ATGy or no translation, respectively). The equivalent gene, *lex1*, sequenced from strain DL42, had 19 copies of the 5′-CAAT-3′ repeat, but no analysis of phase variation of LPS structure was made. In no strain can an absolute correlation of colony reactivity to mAb 4C4 and the number of repeats in *lic2A* be made.

lic3 was identified as a third locus that contained the 5′-CAAT-3′ repeats in *H. influenzae* type b strains. DNA sequence analysis of 5.0 kb of the *lic3* locus identified four orfs of potential relevance to the 5′-CAAT-3′ repeats, designated *lic3A*, *lic3B*, *lic3C*, and *lic3D* (32) (Fig. 2). *lic3A* has no known function or homology with data base sequence but contains 5′-CAAT-3′ repeats just within its 5′ end. Two possible start codons are located 1 and 15 bp upstream of the repeats (Fig. 3). Mutation of *lic3A* has no detectable effect on LPS structure, and so the gene has no correlate to a phase-varying phenotype. The second orf, *lic3B*, codes for a protein that has 56% amino acid identity with UDP-galactose-4-epimerase and is thus designated *galE*. Mutation of the *galE* gene perturbs the balance of activated galactose and its availability for incorporation for LPS synthesis. *galE* maps to a separate chromosomal location to other galactose biosynthetic genes (*galK*, *galT*, and *galR*) and is therefore unlike the operon found in *E. coli*. A *galEgalK* mutant of strain RM7004 elaborates LPS with no galactose included. The third orf of *lic3*, *lic3C*, produces a protein with homology to AmpG of the Enterobacteriaceae. This protein has a function in signal transduction, but mutants in this gene in *H. influenzae* have no obvious phenotype. The fourth orf, *lic3D*, has high homology with the *adk* gene of *E. coli* whose product (adenylate kinase) is a key enzyme in cAMP biosynthesis and cell energy level control. It has been postulated that this operon may provide a regulatory mechanism by which LPS biosynthesis, which is a high energy consumption process, can be linked to energy production within the cell. Only circumstantial evidence of gene position and some possible cotranscription of the *lic3* genes (unpublished results) support this hypothesis. When a *lacZ* reporter fusion was constructed downstream of the repeats in *lic3A*, three levels of β-galactosidase activity were noted from colonies grown on plates: blue (++++), bull's eye (+), and white (−) (33). PCR amplification and direct sequencing of the repeat region DNA from individual colonies showed some correlation between *lacZ* expression and the number of 5′-CAAT-3′ repeats. High-level expression (++++) correlated with 21 or 24 repeats of 5′-CAAT-3′ making the ATG adjacent to the repeats in frame with the remainder of the orf. Medium expression (+) arose from 21, 19, or 22 copies of the repeat, corresponding to use of either start codon. Low or nonexpressing colonies contained 20, 21, 22, or 25 copies of 5′-CAAT-3′, making it possible that all reading frames were utilized. A further construct was made in *lic3A* by deleting the repeats and fusing *lacZ* to the ATG start codon (33). Medium and low-level expression were still observed, but at much reduced frequencies. It was concluded from this that other independent mechanisms must be operating to influence the level of *lic3A* expression; a likely mechanism would be transcription. Northern blot experiments have shown comparable levels of *lic3A* transcript in both high and low/nonproducing colonies (unpublished results). The exact nature of this unknown control may not be tractable until a function has been assigned to the Lic3A protein.

Two further genes associated with LPS biosynthesis and containing tetranucleotide repeats have been identified in *H. influenzae*: *lex2* and *lgtC*.

lex2

A genomic library was constructed from the *H. influenzae* type b strain DL42, which reacts with mAb 5G8, and was used to transform another type b strain, DL180, which is nonreactive with mAb 5G8 (34). A 5.5 kb clone was identified, which conferred mAb 5G8 reactivity, and 1.9 kb of this was sequenced revealing two contiguous orfs (Fig. 2). The first orf had the tetranucleotide 5′-GCAA-3′ repeated 18 times within the 5′ end (34) and would produce a small protein of 101 amino acids. The second orf, separated by 13 bp, would produce a product of 247 amino acids, and transposon insertion mutagenesis indicated that this second orf is required for expression of the mAb 5G8 reactive LPS epitope. Eighteen copies of the 5′-GCAA-3′ repeat put the first orf in frame with an initiation codon 66 bp upstream of the repeats, and DL180, which was unreactive, had 17 copies of the repeat. There is a second potential ATG codon in an alternative frame and closer to the repeats. A detailed study of mAb 5G8 reactive LPS epitope expression and repeat numbers has not been carried out. Similar to *lic1*, the mechanism by which changes in the number of repeats affect expression of the product from an adjacent orf needs to be explained. Jarosik and Hansen showed that the *lex2* locus was present in all type b strains tested (34), but the recently completed genome sequence of the type d–derived strain, Rd, revealed no equivalent locus. A homolog of *lex2* has been reported in a related species, *Haemophilus somnus*, but associated with 5′-CAAT-3′ repeats (U94833 GenBank). A search of the *H. influenzae* genome sequence for other nucleotide repeat motifs that may be associated with LPS biosynthetic genes did, however, reveal a further locus of interest, *lgtC*, described in more detail below.

lgtC

Analysis of the complete genome sequence of *H. influenzae* strain Rd (35) revealed that the bacterium contained all three of the *lic* loci but no sequence for the *lex2* locus described above. One of the nine novel loci identified with multiple tandem tetranucleotide repeats in the genome sequence was a homolog of a glycosyl transferase, *lgtC*, implicated with LPS biosynthesis in *Neisseria* (4; see below). The *lgtC* homolog was one of 25 novel loci identified from the genome sequence (6) with potential functions related to LPS biosynthesis. *lgtC* is the third of three contiguous orfs and has 22 copies of the tetranucleotide 5′-GACA-3′ located just within the 5′ end (Fig. 2). Unlike the *lic* loci, only one translation start codon immediately upstream of the repeats would appear to be relevant to gene expression. The function of the *lgtC* gene was investigated by insertional mutagenesis, and the relevant mutation in strain RM7004 had a very similar phenotype to that of a RM7004 *lic2A* mutant. RM7004*lgtC* showed loss of reactivity to mAbs 4C4 and 5G8, and isolated LPS had an altered electrophoretic profile, showing a reduced number of bands staining with silver than the wild-type (Fig. 1). This indicated that extension of the LPS to form the highest molecular weight species is blocked in the mutant. Mass spectrometric analysis of the LPS from an RM153*lgtC* (Eagan*lgtC*) mutant indicated that the terminal αGal(1-4)βGal epitope was absent (Fig. 4) (6). This correlates with the loss of reactivity and phase variation to the mAbs 4C4 and 5G8. Reactivity and phase variation to other mAbs, 6A2 and 12D9, were retained in this mutant, adding evidence that these mAbs recognize alternative epitopes of the LPS which are synthesized largely independent of the digalactoside. PCR amplification and direct sequencing of the 5′-GACA-3′ repeat region showed a correlation between the number of repeats in *lgtC* and mAb 4C4 reactive and nonreactive RM7004 colonies (4).

Our current interpretation of the genetics of LPS phase variation in *H. influenzae* in terms of its biology are as follows. First, some observations can be made from our knowledge of the structure of *H. influenzae* LPS. To understand the complex process of LPS phase variation, we require some knowledge of the LPS epitopes involved. The detailed LPS structure of Hi type b strain RM153 has been elucidated by mass spectrometric and NMR analyses (Fig. 4) (13). LPS from strain RM7004 has the same basal structure (36) but with further extension of sugars from heptoseI equivalent to the extension from heptoseII in strain RM153 (unpublished results). Structural analysis of LPS derived from RM153 *lgtC* and *lic2A* mutants has revealed that both contain no digalactoside as the distal extension from heptoseII (6; unpublished results). This αGal(1-4)βGal structure is known to direct reactivity with mAb 4C4, which is lost in RM153 *lgtC* and *lic2A* mutants. A mutation in *lic1* results in lack of incorporation of phosphorylcholine as a substituent in the hexose sugars (unpublished results). The digalactoside epitope and PC substitution can be considered, at least to a degree, to be independent, as mutant strains lacking each still phase variably express the alternative epitope. Where structural correlates with phase variable genes are known, the epitopes involved are peripheral in the LPS molecule. This is in agreement with our earlier statement that the distal epitopes will be responsible

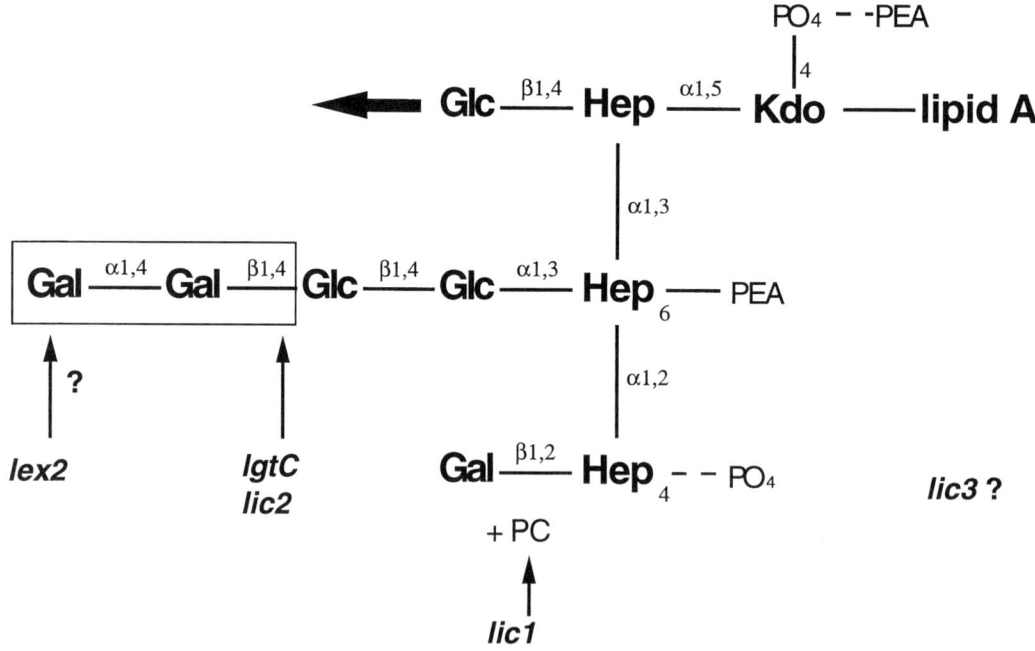

Fig. 4 A schematic representation of the structure of LPS from *H. influenzae* type b strains determined after mass spectrometric and NMR analyses. The unshaded structure is the highest glycoform found for strain RM153; the phase-varying terminal digalactoside epitope is indicated in the box. The thick arrow represents the direction of further extension from heptose I found in the higher glycoforms from strain RM7004. The vertical arrows indicate the proposed point of action of phase variable genes in LPS biosynthesis. Represented in the LPS structure: Kdo, 2-keto-3-deoxyoctulosonic acid; Hep, L-glycero-D-manno-heptose; Glc, D-glucose; Gal, D-galactose; PEA, phosphoethanolamine; PO_4, phosphate. For the heptose residues, listed top to bottom are heptose I, heptose II, and heptose III.

for initial interactions with host structures and would be those of greatest potential benefit to the bacterium if varied. Some evidence exists to support these observations in terms of the host-pathogen interaction (see below).

The importance of the digalactoside epitope to LPS biology is demonstrated by the evidence that two apparently independent phase variable genes, *lgtC* and *lic2A*, contribute to its biosynthesis. By experiment, the mAb 5G8 reactive epitope of *H. influenzae* LPS has been inferred to be, at least partly, the same as the mAb 4C4 reactive epitope. Its co-phase varies with mAb 4C4 reactivity and has again been correlated with virulence in some strains (16). Thus, there is a potential influence of a third phase-variable gene, *lex2*, on the digalactoside expression. Studies to investigate numbers of repeats and mAb 4C4 reactivity typically have considered only a single locus, the apparent contradictions observed in each set of results (as described above for *lic2*) are likely to be due to the contribution of interdependent phase-variable genes. This can likely be extended to explain contradictions found in repeat number and phenotype observations at other phase-variable loci. More recently, a mutant of *lic2A*, RM7004lic2AΔCAAT, with the 5′-CAAT-3′ repeats removed, has been constructed in strain RM7004 (37). The LPS from this strain was identical to wild-type as assessed by mAb reactivity with colonies upon immunoblotting and by gel fractionation of purified LPS. Deletion of the 5′-CAAT-3′ repeats reduced the rate of phase variation to mAb 4C4 but did not abolish it, supporting the notion of multiple genes being involved. Direct sequencing of strain RM7004lic2AΔCAAT mAb 4C4 reactive and nonreactive colonies showed a correlation with the number of 5′-GACA-3′ repeats in *lgtC*. Colonies that were nonreactive with mAb 4C4 had 32, 33, or 35 repeats of 5′-GACA-3′, which would place *lgtC* out of frame with the putative initiation codon. Colonies that bound mAb 4C4 had 34 or 35 copies of the repeat, placing *lgtC* either in or out of frame with its initiation codon. Thus a correlation between *lgtC* expression and mAb 4C4 reactivity can be drawn, with LgtC being off in nonreactive colonies. These studies indicate a potential correlation in the expression of the phase variable gene functions, but a comprehensive study of the relative contributions of *lic2A* and *lgtC* (and perhaps *lex2*) to the expression of the αGal1-4βGal structure needs to be undertaken. Assessing the numbers of repeats at these loci, in wild-type and repeat-depleted strains, should establish their interdependence and any hierarchy of the genes required for synthesis of the epitope. Many aspects of the incorporation of the αGal(1-4)βGal epitope into *H. influenzae* LPS remain to be investigated. The exact nature and biochemical function of all the gene products remain to be established. It is also unclear to what extent other LPS epitopes and substitutions, themselves potentially under phase-variable control, constitute levels of heterogeneity that may influence the frequency of digalactoside incorporation into the LPS.

Second, some evidence has shown the importance of these peripheral phase-variable LPS epitopes to biological function of the LPS and to host-parasite interactions of *H. influenzae*. Organisms contained in the CSF of newly diagnosed cases of meningitis showed a majority (>99%) of organisms binding mAbs 4C4 and 12D9 compared to only <0.1% when organisms were cultured in vitro (27). This would imply that the mAb 4C4 and 12D9 reactive LPS epitopes are selected in vivo; the occasional bacteria that did not bind the mAbs were consistent with the occurrence of phase variation in vivo. The role of LPS in the virulence of *H. influenzae* has been studied mainly through experimental infection in infant rats. LPS has been shown to influence all stages of the disease process in the infection model, i.e. persistence in the nasopharynx, invasion across cellular barriers, intravascular survival, and microbial damage to tissues (1,38). The digalactoside epitope of *H. influenzae* LPS, which phase varies and directs reactivity with mAb 4C4, has been shown to have a role in pathogenesis. Chemical mutagenesis produced mAb 4C4 nonreactive strains, which differed only from the wild-type in the ability to cause invasive disease after intranasal challenge (39), but the structure of the LPS elaborated by these strains has not been defined. A strain with a defined genetic lesion in the *lgtC* gene (4) has LPS that lacks the digalactoside structure and has a 10-fold increase in clearance from the bloodstream. An impairment of virulence has also been noted for a *lic1*/*lic2* double mutant strain, which constitutively lacks the digalactoside epitope, when compared to the wild-type strain. This mutant is comparable with the wild-type in its ability to colonize the nasopharynx and survive in the bloodstream but shows a reduced incidence and magnitude of bacteremia after intranasal colonization (3). The combined results from these experiments would indicate that the digalactoside structure of *H. influenzae* LPS has a role in translocation from the respiratory tract to the bloodstream and subsequent survival in the blood. The contribution of the digalactoside epitope to virulence is likely to reflect the contribution of other phase-variable epitopes to the infection process, and more recently Weiser reported that *lic1*-dependent incorporation of PC into *H. influenzae*

LPS alters the bacterium's susceptibility to the bactericidal activity of human serum (28).

So far we have discussed three tetranucleotide repeats, 5'-CAAT-3', 5'-GACA-3' and 5'-GCAA-3', associated with the phase variation of LPS in *H. influenzae*. Why are these particular repeats found in their respective genes, and are these likely to be the only candidates found in other strains? Weiser (25) found 5'-CAAT-3' repeats in all capsular and NT strains tested, and more recent extensive studies have found between two and five 5'-CAAT-3' loci in all capsular and NT strains upon hybridization with a (5'-CAAT-3')$_5$ oligonucleotide probe (unpublished results). Are the further loci duplicates or derivatives of *lic1*, *lic2*, or *lic3*, or are other genes present in some strains? The gene *lex2* contains 5'-GCAA-3' repeats in *H. influenzae* type b strains, but the equivalent gene in a related species, *H. somnus*, contains 5'-CAAT-3' repeats. Also, for the methyltransferase gene associated with the tetranucleotide 5'-AGTC-3' in *H. influenzae* strain Rd, an alternative tetranucleotide of 5'-AGCC-3' was found at the equivalent locus in capsular types a, b, c, and e and two nontypable strains (4). This raises the distinct possibility that related species and strains exhibiting LPS phase variation may contain other loci associated with alternative (even absent) repeats. The availability of the complete genome sequence of *H. influenzae* has allowed us to conclude that, in strain Rd at least, there are no obvious further candidate LPS genes with associated nucleotide repeats. So far we have limited our discussion to the genetics of LPS phase variation in *H. influenzae*, but meningococcal and gonococcal terminal LPS structures also undergo high-frequency reversible switching of expression at a rate similar to that found for *H. influenzae*.

Neisseria *lgt* LOCUS

LPS changes in *Neisseria* are a combination of intrinsic changes in LPS biosynthesis and modification of the molecules with sialic acid. Although the repertoire of LPS molecules produced by a single strain is often simpler than is seen for *H. influenzae*, between one and six structures can be produced (40). Again, changes in these terminal LPS structures reflect altered bacterial behavior and virulence; in particular, interaction with human cells and evasion of the host immune defenses. Monoclonal antibodies have again been used to assess the LPS phase variation. For example, in *N. gonorrhoeae* the conversion between cells making mAb IB2 reactive and mAb 2-1-48 reactive LPS occurs at a rate of 10^{-3} per cell per generation. This change in mAb reactivity occurs through loss of lactosamine from the LPS alpha side chain (Fig. 5, see later) (41). In *N. meningitidis*, switching between the L3 and L8 immunotypes (Fig. 5) occurs at about the same frequency and can be detected by changes in reactivity with the mAbs Mn4A8-B2 and Mn8D6A, respectively (42,43).

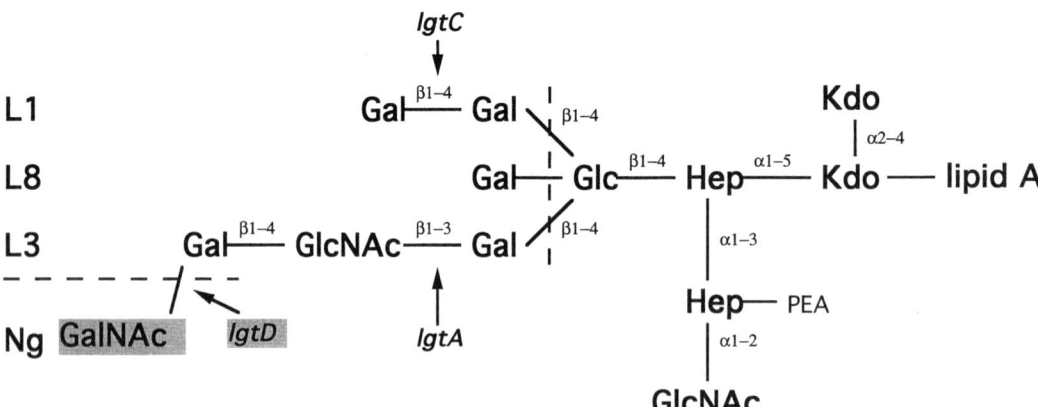

Fig. 5 A schematic representation of the saccharide structure of the primary immunotypes of *N. meningitidis* LPS, L1, L3, and L8. The variable portion of each immunotype is shown to the left of the vertical dotted line. The presumed functions of the phase-variable genes in LPS biosynthesis are indicated by the arrows. *lgtD* functions only in *N. gonorrhoeae*, and its point of action is shown below the horizontal dotted line. Represented in the LPS structure: Kdo, 2-keto-3-deoxyoctulosonic acid; Hep, L-glycero-D-manno-heptose; Glc, D-glucose; Gal, D-galactose; GalNAc, *N*-acetylgalactosamine; GlcNAc, *N*-acetylglucosamine; PEA, phosphoethanolamine.

The molecular bases for the LPS variation were investigated in both *N. meningitidis* and *N. gonorrhoeae* and found to be similar. Jennings and coworkers (43), searched for genes involved with LPS biosynthesis in *N. meningitidis* by hybridization using *H. influenzae* genes as heterologous probes. The *lic2A* gene was used to identify a homolog and a specific 2.8 kb *Sau*3A clone from an enriched library of the *N. meningitidis* type b strain, MC58. Nucleotide sequence analysis of an extended 3.6 kb chromosomal segment revealed three orfs, *lgtA, lgtB,* and *lgtE* (Fig. 6), putatively involved in LPS biosynthesis. The first orf, *lgtA*, encoded a product of 333 amino acids, the second orf, *lgtB*, a product of 275 amino acids and the third orf, *lgtE,* a product of 276 amino acids. The latter two genes both had homology to the *lic2A* gene of *H. influenzae* and form part of a group of genes whose products have predicted functions as glycosyl transferases. At about the same time, a similar locus involved in LPS biosynthesis was reported for *N. gonorrhoeae* (44), which was larger and contained two further genes, *lgtC* and *lgtD* (Fig. 6). The *N. meningitidis* genes are found in the same orientation and order as in *N. gonorrhoeae*. *lgtA* of *N. gonorrhoeae* is also described as *lsi-2* (45). In *N. gonorrhoeae*, internal homology was found between the *lgtA* and *lgtD* and the *lgtB* and *lgtE* gene pairs. A notable feature of the *N. meningitidis* *lgtA* orf was a homopolymeric tract of 14 guanosine (G) residues within the 5′ end of the coding sequence. These tracts were presumed to promote phase variation of *Neisseria* LPS, making the repeats associated with the process fundamentally different from those in *H. influenzae* but in an equivalent position. Homopolymeric tracts of 17, 10, and 11 G residues were found within the *lgtA, lgtC,* and *lgtE* genes of *N. gonorrhoeae* strain F62, respectively (44), and 13 G residues in the equivalent gene, *lsi-2*, in its nonexpressed form and 12 G residues in its expressed form (45).

Insertional mutagenesis of the *lgtABE* cluster of *N. meningitidis*, construction of mutant strains, and analysis of their LPS showed a contribution for each of these genes to LPS biosynthesis. Mutants showed altered reactivity to LPS-directed mAbs. Fractionated LPS showed reductions of one, two, and at least three sugars from mutants in the *lgtB, lgtA,* and *lgtE* genes, respectively. This would assign putative functions to the genes of *lgtB*–galactosyl transferase, *lgtA*–*N*-acetyl glucosamine transferase, and *lgtE*–galactosyl transferase, all involved in the sequential steps in the synthesis of the terminal lacto-*N*-neotetraose epitope of the L3 immunotype (Fig. 5). Biochemical functions have now been confirmed for the products of these genes (46). This type of genetic analysis and gene function prediction is somewhat simpler than is found for *H. influenzae* as described above. In *N. gonorrhoeae*, the *lgtA, lgtB,* and *lgtE* genes had the same predicted function in the assembly of the lacto-*N*-neotetraose, with *lgtC* having predicted function as a galactosyl transferase and *lgtD* as an *N*-acetyl galactose transferase for further side chain elaboration of the LPS structure (Fig. 5) (44). The modification of LPS encoded by *lgtD* has not been reported in *N. meningitidis*, consistent with the absence of the gene in the *N. meningitidis* locus. A survey of *N. meningitidis* strains elaborating different LPS immunotypes by Jennings using *lgt* gene probes revealed that the genotype reflected the observed LPS phenotype; most strains expressed either a terminal lacto-*N*-neotetraose (L3) or digalactoside (L1) structure (47). The contribution of the homopolymeric tract of G residues in the *lgt* genes to LPS phase variation has only been studied for the *lgtA* gene of *N. meningitidis* (43) and *N. gonorrhoeae* (48). The proposed function of the LgtA protein, and *N*-acetyl glucosamine transferase, would provide an obvious switch point between the L8 and the fully extended L3 immunotypes (Fig. 5). Experiments similar to those described previously

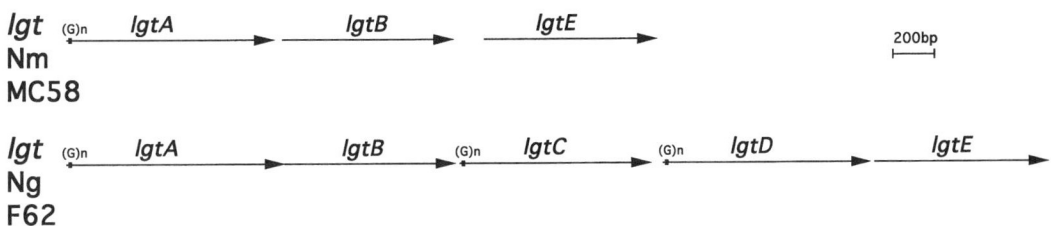

Fig. 6 Schematic representations of the *lgt* loci of *Neisseria meningitidis* strain MC58 (NmMC58) and *Neisseria gonorrhoeae* strain F62 (NgF62). The direction of transcription is given by the arrows, and the gene designations are as detailed in the text. Regions of the phase-variable genes containing homopolymeric tracts are indicated by the thick lines, and the relevant base is indicated above.

for the *lic* genes of *H. influenzae*, using PCR and direct sequencing of immunostained colonies, failed to yield meaningful results. The limitation was that the DNA replication during PCR and sequencing produced mixed populations of products by polymerase slippage. Homopolymeric tracts appear to be hypermutable under such conditions. A method was developed whereby chromosomal DNA was digested using a restriction enzyme with a frequent recognition sequence, run on a denaturing polyacrylamide sequencing gel and then blotted to a nylon membrane (43). Hybridization using a radiolabeled oligonucleotide, complementary to sequence adjacent to the homopolymeric tract, allowed identification and sizing of the specific DNA fragment containing the G tract. 14 G residues, a number for the correct reading frame of *lgtA* from the single initiation codon, were found in strain *N. meningitidis* c3, producing the lacto-*N*-neotetraose and the L3 immunotype. A spontaneously derived mutant, c2, expressing the L8 immunotype was found to have 15 G residues in the homopolymeric tract, a number placing the upstream initiation codon out of frame. Burch (48) sequenced the *lgtA* equivalent genes from *N. gonorrhoeae* strains FA91 and 1291 and compared them with the sequence from strain F62 (44). The only sequence variation that could explain the differential activity of the *lgtA* gene was through changes in the homopolymeric tract. Although LPS phase variation is routinely reported as 10^{-3} per cell per generation, LPS expression is determined from the majority of cells in a population and could be an oversimplification of LPS expression and phase variation in single cells. It has been shown that individual *N. gonorrhoeae* cells coexpressed LPS epitopes recognized by both mAbs 2-1-L8 and IB2 (48). This corresponds to a loss or gain of two sugar residues (Gal-GlcNAc) controlled by the *lgtA* gene and its homopolymeric tract.

As with *Haemophilus*, observations can be made from LPS structural and virulence data to help understand the biology of LPS phase variation in *Neisseria* spp. The variable LPS structure controlled by these homopolymeric tracts is primarily the lacto-*N*-neotetraose, which structural analysis has confirmed as the major distal extension in the highly related LPS of both *N. meningitidis* and *N. gonorrhoeae*. This structure contributes to much of the specific host-microbe interactions and can be sialylated endogenously or exogenously by a bacterial sialyl transferase. When present, the sialylated structure reduces the efficiency of *N. meningitidis* entry into host mucosal cells and of bactericidal killing by antibodies and complement (7,49).

Thus in *Neisseria*, LPS heterogeneity, although reduced in the laboratory when compared to *H. influenzae*, is no less important to the biology of the LPS and to bacterial virulence.

Many questions remain to be answered about features of the DNA repeat regions, the amino acids that they encode, and the mechanism by which they vary in each organism. These regions of DNA are prone to slippage, yet variations in repeat numbers tend to be constrained for any given locus in a particular bacterium. High (37) reported that for 26 *H. influenzae* isolates, including representatives of each capsular serotype and nontypable strains, the number of 5'-CAAT-3' repeats in *lic2A* ranged between 7 and 22, with the majority of strains having 16 copies. In contrast, the number of 5'-CAAT-3' repeats in the *lic1A* locus in the same isolates was between 5 and 57 copies, with no distinct bias in favor of any particular number of repeats. Thus, it would appear that a substantial range of size of DNA repeats can be stably maintained. There was no trend for isolates to have a bias towards higher or lower numbers of repeats at different loci (N. High, personal communication). The distribution of repeats was maintained (+/− 4 bp) when strains were subcultured whether they had high or low numbers of repeats.

The question of the contribution of the deduced amino acid sequence in a repeated DNA motif to gene product function is important for tetranucleotide sequences and has been addressed by High et al. (37). Different numbers of 5'-CAAT-3' repeats in *lic2A* can maintain an apparently functional Lic2A protein in each case. *lic2A* containing 22 copies of the repeat predicts a product in which the amino acids SINQ would constitute 10% of the total protein. With *lic1A*, 57 copies of the repeat would constitute approximately 20% of the gene product. Such large peptide sequences may be tolerated in the proteins simply because they occur at the N termini and might suggest that the resultant amino acid repeats (SINQ) play no crucial role in protein function. Consistent with this, removal of the 5'-CAAT-3' repeats had no obvious effect on Lic2A function (37). The SINQ tetrapeptide has been predicted to form a random coiled structure (37) with high flexibility and thus be less likely to interfere with catalytic activity. A similar flexibility for the 5'-GCAA-3' tetranucleotide repeat encoding the ASKQ peptide of Lex2 and the 5'-GACA-3' encoded TDRQ peptide of LgtC were shown. Such structural constraints on peptide flexibility would limit the number of tetranucleotide motifs capable of having neutral effect on protein function. For homopolymeric tracts, the parameters that affect phase variation will be different, and less is known about the numbers of repeats across a range of *Neisseria* strains. A maximum of six amino acids in the gene product is likely to be encoded by the homopolymeric tract. The contribution of the homopolymeric tract in *lgtA* was investigated in *N. gonorrhoeae* (48) by altering every third base in the tract by PCR mutagenesis. Transformants containing this modified tract correctly expressed LgtA and displayed only one LPS type reactive with the mAb 2-1-L8 on single cells and colonies showing that the phenotype had been fixed. In each case investigated, phase variation results from the insertion or deletion of only one tetranucleotide unit in *H. influenzae* and one base in homopolymeric tracts in *Neisseria*, not by multiples.

Some common features of the DNA around the repeats can be found in *H. influenzae*. First, many have multiple potential independent start codons, although the use of any predicted initiation codon with particular repeat numbers remains to be proven. In the case of *lic2A*, for example, there are (at least) three predicted start codons. When a number of different strains were investigated, repeats were 7, 10, 13, 16, 19, or 22, i.e., a periodicity of three. These would maintain translation of *lic2A* but might imply that only one initiation codon is preferred or functional. DNA sequence immediately upstream of the tetranucleotide repeat regions is relatively AT rich compared to the average content (AT 69%: GC 31%) of the *H. influenzae* genome. For the 120 bp immediately upstream of loci the GC contents are 22.5% for *lic1*, 20.0% for *lic2*, 20.8% for *lic3*, 25.8% for *lex2*, and 22.5% for *lgtC* (although only 18.4% for the intergenic 76 bp). This might favor denaturation and help initiate mispairing in the downstream tetranucleotide repeat regions, which are themselves either 75% or minimally 50% AT rich. No particular sequence features upstream or downstream of the repeat, e.g., base composition, were found to be associated with homopolymeric tract function in *Neisseria*. The use of guanosine or cytosine bases to make homopolymeric tracts known to function in phase variation might indicate that, unlike tetranucleotide repeats in *H. influenzae*, an appropriate stability to maintain these repeats requires the stronger G = C hydrogen bonding.

Phase-variable expression of peripheral moieties, which mimic host structures, is a feature of the LPS of mucosal pathogens. This process maintains a repertoire of functionally variant organisms within populations, and particular forms of these glycolipids are of benefit to the bacterium in different host compartments or microenvironments. Phase variation of surface structures plays a role in many aspects of host pathogen interactions. This type of variation, controlled by so-called contingency genes, is one of the two main ways that bacteria can adapt to a changing environment and is a recurring strategy in pathogenic bacteria. *H. influenzae* strain Rd contains 12 potentially phase-variable genes associated with tetranucleotide repeats, most encoding likely virulence-related determinants (4). Also, an example of dinucleotide repeats located in the promoter region of fimbriae biosynthetic genes has been reported to control fimbrial variation in type b strains in the same organism (50). Homopolymeric tracts have been shown to mediate phase variation of pilin (51), a surface-exposed protein Opc (52), and capsule expression (53) in *Neisseria* spp. Further, in *Neisseria* a pentanucleotide, 5′-CTCTT-3′, mediates phase variation of surface-exposed proteins (54), and some tetranucleotide repeats have been reported but no correlation with a phase-variable phenotype has been established (55). Other pathogens, notably *Bordetella pertussis* (56), *Helicobacter pylori* (57), and *Mycoplasma hominis* (58), express phase-variable surface structures from genes containing homopolymeric tracts. Indeed, *H. pylori* has five potentially phase-variable loci associated with LPS biosynthesis identified from the recent genome sequence. Phase variation by stochastic change in the numbers of short tandem repeats controlling expression of important surface structures is a method by which the cells can avoid the complex sensing and response pathways controlling the fine-tuned gene expression seen in other bacteria such as *E. coli*. Indeed, the genome sequence of *H. influenzae* strain Rd revealed only 2 two component regulatory systems in contrast to the 40 or so reported for *E. coli* (59), an organism that has no documented phase variation of the type described in this manuscript. Bacteria such as *Haemophilus* may have sacrificed the genetic load of regulatory systems either because of, or leading to, their host dependence and restricted environmental changes. Obviously these alternative regulatory strategies cannot operate completely independently, and other environmental and regulatory factors must affect phase variation via short nucleotide repeats as, for example, levels of supercoiling, methylation and repressor/activator proteins change within the cells. The availability of bacterial genome sequences will help us to compose an ordered list of DNA interacting proteins and then test them in the laboratory for their effects on DNA repeats and phase variation.

REFERENCES

1. Smith AL, Smith DH, Averill DR, Moxon ER. Production of *Haemophilus influenzae* b meningitis in infant rats by intraperitoneal inoculation. Infect Immun 1973; 8:278–290.
2. Zwahlen A, Rubin LG, Connelly CJ, Inzana TJ, Moxon ER. Alteration of the cell wall of *Haemophilus influenzae* type b by transformation with cloned DNA: association with attenuated virulence. J Infect Dis 1985; 152:485–492.
3. Weiser JN, Williams A, Moxon ER. Phase-variable lipopolysaccharide structures enhance the invasive capacity of *Haemophilus influenzae*. Infect Immun 1990; 58:3455–3457.
4. Hood DW, Deadman ME, Jennings MP, Bisceric M, Fleischmann RD, Venter JC, Moxon ER. DNA repeats identify novel virulence genes in *Haemophilus influenzae*. Proc Natl Acad Sci USA 1996; 93:11121–11125.
5. Patrick D, Betts J, Frey EA, Prameya R, Dorovini ZK, Finlay BB. *Haemophilus influenzae* lipopolysaccharide disrupts confluent monolayers of bovine endothelial cells via a serum-dependent cytotoxic pathway. J Infect Dis 1992; 165:865–872.
6. Hood DW, Deadman ME, Allen T, Masoud H, Martin A, Brisson JR, Fleischmann R, Venter JC, Richards JC, Moxon ER. Use of the complete genome sequence information of *Haemophilus influenzae* strain Rd to investigate lipopolysaccharide biosynthesis. Mol Microbiol 1996; 22:951–965.
7. Van Putten JPM. Phase variation of LPS directs interconversion of invasive and immuno-resistant phenotypes of *Neisseria gonorrhoeae*. EMBO J 1993; 12: 4043–4051.
8. Virji M, Makepeace K, Peak IRA, Ferguson DJP, Jennings MP, Moxon ER. Opc- and pilus-dependent interactions of meningococci with human endothelial cells: molecular mechanism and modulation by surface polysaccharides. Mol Microbiol 1995; 18:741–754.
9. Jones DM, Borrow R, Fox AJ, Gray S, Cartwright KA, Poolman JT. The lipooligosaccharide immunotype as a virulence determinant in *Neisseria meningitidis*. Microbiol Pathogen 1992; 13:219–224.
10. Mandrell RE, Apicella MA, Lindstedt R, Leffler H. Meth Enzymol 1994; 236:231–254.
11. Jennings HJ, Johnson KG, Kenne L. The structure of an R-type oligosaccharide core obtained from some lipopolysaccharides of *Neisseria meningitidis*. Carbohydr Res 1983; 233:233–241.
12. Mandrell RE, Mclaughlin R, Kwaik YA, Lesse A, Yamosaki R, Gibson B, Spinola SM, Apicella MA. Lipooligosaccharides of some *Haemophilus* species mimic host glycosphingolipids and are sialylated. Infect Immun 1992; 60:1322–1328.
13. Masoud H, Moxon ER, Martin A, Krajcarski D, Richards JC. Structure of the variable and conserved lipopolysaccharide oligosaccharide epitopes expressed by *Haemophilus influenzae* serotype b strain Eagan. Biochemistry 1997; 36:2091–2103.
14. Elkins C, Carbonetti NH, Varela VA, Stirewalt D, Klapper DG, Sparling PF. Antibodies to N-terminal peptides of gonococcal porin are bactericidal when gonococcal lipopolysaccharide is not sialylated. Mol Microbiol 1992; 6:2617–2628.
15. Moxon ER, Rainey PB, Nowak MA, Lenski RE. Adaptive evolution of highly mutable loci in pathogenic bacteria. Curr Biol 1994; 4:24–33.
16. Kimura A, Hansen EJ. Antigenic and phenotypic variants of *Haemophilus influenzae* type b lipopolysaccharide and their relationship in virulence. Infect Immun 1986; 51:60–79.
17. Gulig PA, Patrick CC, Hermanstorfer L, McCracken Jr. GH, Hansen EJ. Conservation of epitopes in the oligosaccharide portion of the lipooligosaccharide of *Haemophilus influenzae* type b. Infect Immun 1987; 55: 513–520.
18. Virji M, Weiser JN, Lindberg AA, Moxon ER. Antigenic similarities in lipopolysaccharides of *Haemophilus* and *Neisseria* and expression of a digalactoside structure also present on human cells. Microb Pathogen 1990; 9:441–450.
19. Moran EE, Brandt BL, Zollinger WD. Expression of the L8 lipopolysaccharide determinant increases the sensitivity of *Neisseria meningitidis* to serum bactericidal activity. Infect Immun 1994; 62:5290–5295.
20. Levinson G, Gutman GA. Slipped strand mispairing: a major mechanism for DNA sequence evolution. Mol Biol Evol 1987; 4:203–221.
21. Seifert HS. Questions about gonococcal pilus phase- and antigenic variation. Mol Microbiol 1996; 21:433–440.
22. Gravekamp C, Horensky DS, Michel JL, Modoff LC. Variation in repeat number within the alpha C protein of group B streptococci alters antigenicity and protective epitopes. Infect Immun 1996; 64:3576–3583.
23. Weiser JN, Lindberg AA, Manning EJ, Hansen EJ, Moxon ER. Identification of a chromosomal locus for expression of lipopolysaccharide epitopes in *Haemophilus influenzae*. Infect Immun 1989; 57:3045–3052.
24. Patrick CC, Pelzel SE, Miller EE, Haanes-Fritz E, Ruolf JD, Gulig PA, McCracken GH, Hansen EJ. Antigenic evidence for simultaneous expression of two different lipooligosaccharides by some strains of *Haemophilus influenzae* type b. Infect Immun 1989; 57: 1971–1978.
25. Weiser JN, Maskell DJ, Butler PD, Lindberg AA, Moxon ER. Characterisation of repetitive sequences controlling phase variation of *Haemophilus influenzae* lipopolysaccharide. J Bacteriol 1990; 172:3304–3309.
26. High NJ, Deadman ME, Moxon ER. The role of the repetitive DNA motif (5′-CAAT-3′) in the variable expression of the *Haemophilus influenzae* lipopolysaccharide epitope Galα(1-4)βGal. Mol Microbiol 1993; 9:1275–1282.
27. Weiser JN, Love JM, Moxon ER. The molecular mechanism of *Haemophilus influenzae* lipopolysaccharide epitopes. Cell 1989; 59:657–665.
28. Weiser JN, Shchepetov M, Chong STH. Decoration of lipopolysaccharide with phosphorylcholine: a phase variable characteristic of *Haemophilus influenzae*. Infect Immun 1997; 65:943–950.
29. Moxon ER, Maskell DJ. *Haemophilus influenzae* lipopolysaccharide: the biochemistry and biology of a vir-

ulence factor. In: Hormaeche CE, Penn CW, Smythe CJ, eds. Molecular Biology of Bacterial Infection, Current Status and Future Perspectives. Society for General Microbiology Symposium 49. Cambridge: Cambridge University Press, 1992:75–96.
30. Cope LD, Yogev R, Mertsola J, Latimer LJ, Hanson MS, McCracken GH, Hansen EJ. Molecular cloning of a gene involved in lipooligosaccharide biosynthesis and virulence expression by *Haemophilus influenzae* type b. Mol Microbiol 1991; 5:1113–1124.
31. Roche RJ, High NJ, Moxon ER. Phase variation of *Haemophilus influenzae* lipopolysaccharide: characterisation of lipopolysaccharide from individual colonies. FEMS Microbiol Lett 1994; 120:279–284.
32. Maskell DJ, Szabo MJ, Butler PD, Williams AE, Moxon ER. Molecular analysis of a complex locus from *Haemophilus influenzae* involved in phase-variable lipopolysaccharide biosynthesis. Mol Microbiol 1991; 5:1013–1022.
33. Szabo M, Maskell D, Butler P, Love J, Moxon ER. Use of chromosomal gene fusions to investigate the role of repetitive DNA in regulation of genes used in lipopolysaccharide biosynthesis in *Haemophilus influenzae*. J Bacteriol 1992; 174:7245–7252.
34. Jarosik GP, Hansen EJ. Identification of a new locus involved in the expression of *Haemophilus influenzae* type b lipooligosaccharide. Infect Immun 1994; 62:4861–4867.
35. Fleischmann RD, Adams MD, White O, Clayton RA, Kirkness EF, Kerlavage AR, Butt CJ, Tomb J-F, Dougherty BA, Merrick JM, McKenney K, Sutton G, FitsHugh W, Fields C, Gocayne JD, Scott J, Shirley R, Liu L-I, Glodek A, Kelley JM, Weidman JF, Phillips CA, Spriggs T, Hedblom E, Cotton MD, Utterback TR, Hanna MC, Nguyen DT, Saudek DM, Brandon RC, Fine LD, Fritchman JL, Fuhrmann JL, Geoghagen NSM, Gnehm CL, McDonald LA, Small KV, Fraser CM, Smith HO, Venter JC. The genome of *Haemophilus influenzae* Rd. Science 1995; 269:496–512.
36. Schweda EKM, Hegedus OE, Borrelli S, Lindberg AA, Weiser JN, Maskell, DJ, Moxon ER. Structural studies of the saccharide portion of cell envelope lipopolysaccharide from *Haemophilus influenzae* strain AH1-3 (lic3$^+$). Carbohydr Res 1993; 246:319–330.
37. High NJ, Jennings MP, Moxon ER. Tandem repeats of the tetramer 5′-CAAT-3′ present in *lic2A* are required for phase variation but not LPS biosynthesis in *Haemophilus influenzae* Mol Microbiol 1996; 20:165–174.
38. Moxon ER, Smith AL, Averill DR, Smith DH. *Haemophilus influenzae* meningitis in infant rats after intraperitoneal inoculation. J Infect Dis 1974; 129:154–162.
39. Cope LD, Yogev R, Mertsola J, Argyle JC, McCracken GH, Hansen EJ. Effect of mutation in lipopolysaccharide biosynthetic genes on virulence of *Haemophilus influenzae* type b. Infect Immun 1990; 58:2343–2351.
40. Schneider H, Hale TL, Zollinger WD, Seid RC, Hammack CA, Griffis JM. Heterogeneity of molecular size and antigen expression within lipooligosaccharides of *Neisseria gonorrhoeae* and *Neisseria meningitidis*. Infect Immun 1984; 45:544–549.
41. Schneider H, Hammack CA, Apicella MA, Griffis JM. Instability of expression of lipooligosaccharide and their epitopes in *Neisseria gonorrhoeae*. Infect Immun 1988; 56:942–946.
42. Kim JJ, Mandrell RE, Zhen H, Westerbrink MAJ, Poolman JT, Griffiss JM. Electrophoretic characterisation and description of conserved epitopes of the lipooligosaccharides of group A *Neisseria meningitidis*. Infect Immun 1988; 56:2631–2638.
43. Jennings MP, Hood DW, Peak IRA, Virji M, Moxon ER. Molecular analysis of a locus for the biosynthesis and phase-variable expression of the lacto-N-neotetraose terminal lipopolysaccharide structure in *Neisseria meningitidis*. Mol Microbiol 1995; 18:729–740.
44. Gotschlich EC. Genetic locus for the biosynthesis of the variable portion of *Neisseria gonorrhoeae* lipooligosaccharide. J Exp Med 1994; 180:2181–2190.
45. Danaher RJ, Levin JC, Arking D, Burch CL, Sandlin R, Stein DC. Genetic basis of *Neisseria gonorrhoeae* lipooligosaccharide antigenic variation. J Bacteriol 1995; 177:7275–7279.
46. Wakarchuk W, Martin A, Jennings MP, Moxon ER, Richards JC. Functional relationships of the genetic locus encoding the glycosyl transferase enzymes involved in the expression of the lacto-N-neotetraose terminal lipopolysaccharide structure in *Neisseria meningitidis*. J Biol Chem 1996; 271:19166–19173.
47. Preston A, Mandrell RE, Gibson BW, Apicella MA. The lipooligosaccharides of pathogenic G-ve bacteria. Crit Rev Microbiol 1996; 22:139–180.
48. Burch CL, Danaher RJ, Stein DC. Antigen variation in *Neisseria gonorrhoeae*: production of multiple lipooligosaccharides. J Bacteriol 1997; 179:982–986.
49. Van Putten JPM, Robertson BD. Molecular mechanism and implications for infection of lipopolysaccharide variation in *Neisseria*. Mol Microbiol 1995; 16:847–853.
50. van Ham SM, van Alphen L, Mool FR, van Putten JPM. Phase variation of *H. influenzae* fimbriae: transcriptional control of two divergent genes through a variable combined promoter region. Cell 1993; 73:1187–1196.
51. Jonsson AB, Pfeiffer J, Normark S. *Neisseria gonorrhoeae* PilC expression provides a selective mechanism for structural diversity of gonococcal pili. Proc Natl Acad Sci USA 1992; 89:3204–3208.
52. Sarkari J, Pantid N, Moxon ER, Achtman M. Variable expression of the Opc outer membrane protein in *Neisseria meningitidis* is caused by size variation of a promoter containing poly-cytidine. Mol Microbiol 1994; 13:207–217.
53. Hammerschmidt S, Muller A, Sillmann H, Muhlenhoff M, Borrow R, Fox A, van Putten J, Zollinger WD, Gerardy SR, Frosch M. Capsule variation in *Neisseria meningitidis* group B by slipped strand mispairing in the polysialyltransferase gene (*siaD*); correlation with bacterial invasion and the outbreak of meningococcal disease. Mol Microbiol 1996; 20:1211–1220.
54. Stern A, Brown M, Nickel P, Meyer TF. Opacity genes in *Neisseria gonorrhoeae*: control of phase and antigenic variation. Cell 1986; 47:61–67.
55. Peak IRA, Jennings MP, Hood DW, Bisercic M, Moxon ER. Tetrameric repeat units associated with virulence

factor phase variation in *Haemophilus* also occur in *Neisseria* spp. and *Moraxella catarrhalis*. FEMS Microb Lett 1996; 137:109–114.
56. Stibitz S, Aaronson W, Monack D, Falkow S. Phase variation in *Bordetella pertussis* by frameshift mutation in a gene for a novel two-component system. Nature 1989; 338:3204–3208.
57. Tomb JF, et al. The complete genome sequence of the gastric pathogen *Helicobacter pylori*. Nature 1997; 388:539–547.
58. Theiss P, Wise KS. Localised frameshift mutation generates selective high frequency phase variation of a surface lipoprotein encoded by a *Mycoplasma* ABC transporter system. J Bacteriol 1997; 179:4013–4022.
59. Blattner FR, Plunkett G, Bloch CA, Perna NT, Burland V, Riley M, Collado-Vides J, Glasner JD, Rode CK, Mayhew GF, Gregor J, Davis NW, Kirkpatrick HA, Goeden MA, Rose DJ, Mau B, Shau Y. The complete genome sequence of *Escherichia coli* K12. Science 1997; 277:1453–1462.

4

Antigenic Mimicry in *Neisseria* Species

Peter C. Giardina and Michael A. Apicella
University of Iowa, Iowa City, Iowa

Brad Gibson
University of California San Francisco, San Francisco, California

Andrew Preston
University of Cambridge, Cambridge, England

INTRODUCTION

The nonenteric gram-negative mucosal bacteria from the genera *Neisseria, Moraxella, Haemophilus, Helicobacter,* and *Bordetella* have evolved a unique set of glycolipids expressed on the outer leaf of the outer membrane, termed lipooligosaccharides (LOS). Members of these genera are strict human pathogens. LOS is analogous to lipopolysaccharides (LPS) in other gram-negative bacteria. LPS and LOS are similar in structure and function with respect to lipid A, but differ significantly in their carbohydrate moieties; LOS are discernible by their lack of repeating terminal glycan (O-antigen) (1,2), a predominant feature of LPS. So-called rough (R-type) LPS variants that lack repeating O-antigen have been well studied in enteric bacteria, and it should be noted that LOS is structurally and functionally distinct from R-type LPS, contrary to a previous report (1).

The majority of LOS studies have been performed on *Neisseria gonorrhoeae* and *N. meningitidis* (pathogenic *Neisseria*), enhancing our understanding of the structure, biosynthesis, and functions of these glycolipids. Little was known about the LOS antigenic or chemical structure until the late 1960s, when Maeland and coworkers began to define the antigenicity of gonococcal endotoxin (3–5). Their studies demonstrated that multiple LOS antigenic specificities were expressed by different gonococcal strains. Thus, gonococcal LOS were segregated into six LOS serotypes; subsequently 12 LOS immunotypes of the meningococcus were identified (6,7). Perry and coworkers demonstrated that gonococcal LOS lacked O-antigens (1). Since that time, the basic chemical structure of the pathogenic *Neisseria* LOS have been resolved, the LOS biosynthesis genes are being elucidated, and some of the functions of the LOS in pathogenesis are being defined.

Among the most intriguing discoveries with regard to gonococcal LOS composition was that the carbohydrate moiety resembled glycolipid structures expressed on the surface of human cells. The glycan, lacto-*N*-neotetraose [Gal(β1$\rightarrow$4)GlcNAc(β1$\rightarrow$3)Gal(β1$\rightarrow$4)Glc], was shown to be a component of gonococcal LOS and immunochemically identical to the terminal tetrasaccharide on human paragloboside, precursor to the human I and i blood group antigens (8). Monoclonal antibody (mAb) 3F11, raised in a mouse to gonococcal LOS and specific for the terminal lacto-*N*-neotetraose, agglutinated 47 out of 47 human erythrocyte samples but failed to agglutinate mouse and rabbit erythrocytes (9). Conversely, mAb 1B2, raised in a mouse to human erythrocyte glycolipid, recognized the same 4.5 kDa band as did mAb 3F11 in *Neisseria* LOS Western blot analysis.

Mandrell speculated on the potential evolutionary implications and possible benefit(s) of mimicking these paraglobside carbohydrate structures and proposed three possible selective advantages with regard to virulence and survival. First, antigenic mimicry may be important for evasion of the host immune response, as normal human serum does not contain antibodies to these self antigens (10–12). LOS has also been shown to inhibit complement-mediated killing and opsonization by normal human serum (13). Second, LOS may be involved in tissue specificity and invasion, based on the observation that human cells express N-acetyllactosamine [Gal(β1→4)GlcNAc]-binding lectins, which are involved in intercellular communication and are part of the normal scavenging uptake system (14). Third, the action of bacterial glycosyltransferases released into tissues may alter host cell glycan-containing structures triggering a tissue-damaging immune response. These three hypotheses are not mutually exclusive and do not preclude other redundant virulence factors that may be expressed.

Bacterial LOS structures have been implicated in sequelae associated with *Campylobacter jejuni* infection of the human intestine and in *Helicobacter pylori* infection associated with peptic ulcer disease (15,16). These organisms, along with other gram-negative mucosal pathogens (17,18), also express LOS carbohydrate structures that mimic human glycosphingolipid epitopes similar to the pathogenic *Neisseria*.

We will evaluate data collected and published by several laboratories over the past 10 years to better understand the role that the LOS terminal glycan plays in the pathogenesis of *Neisseria* spp. First we will consider the LOS terminal glycan structure. Second we will discuss the function of the LOS oligosaccharide pertaining to pathogenesis. In this regard we will focus on evasion of host humoral and cell-mediated immunity as it pertains to the LOS glycan. Finally, we will explore the role of LOS in tissue tropism and describe mechanisms by which pathogenic *Neisseria* may invade host tissues.

NEISSERIA LOS STRUCTURE AND ANTIGENIC MIMICRY: CARBOHYDRATE MOIETIES

LOS Structure

The terminal tetrasaccharides of the LOS from pathogenic *Neisseria* (also *H. influenzae*, *H. ducreyi*, *C. jejuni*, and *H. pylori*) contain an antigen that is immunochemically and structurally identical to paragloboside. mAb 3F11 produced by Apicella et al. (19) has been particularly useful because it binds to the LOS terminal N-acetyllactosamine [Galβ(1→4)GlcNAc] of the lacto-N-tetraose component and to human paragloboside (20). This epitope is present on most gonococcal strains and on many meningococcal strains (19). Most clinical gonococcal isolates bind mAb 3F11, but laboratory-derived truncated LOS mutants and population variant strains have been isolated that lack the 4.5 kDa component and therefore do not bind mAb 3F11 (10,19,21,22).

Mandrell showed that mAb 3F11 and a second mAb of a similar specificity, 6B4, could agglutinate human erythrocytes and that purified glycosphingolipids from human cells could bind to these mAbs (23). Both mAbs could bind to a series of neutral glycosphingolipids that contain terminal Galβ(1→4)GlcNAc, but terminal sialic acid, galactose, or fucose at the non-reducing end of the glycolipids blocked the binding of both mAbs (23). These data support the hypothesis that *Neisseria* LOS share a N-acetyllactosamine–containing structure expressed on the surface of human cells.

The first partial structure reported for a gonococcal LOS was by Gibson for the pyocin-resistant mutant strain JW31R (24). The lack of mAb 3F11 binding to this LOS was consistent with the absence of the putative Galβ(1→4)GlcNAc disaccharide epitope in JW31R LOS. Subsequent structural analysis of LOS from other gonococcal strains has confirmed that lacto-N-neotetraose is present in gonococcal LOS that bind the antilactosamine mAbs (20,25) and that it is a constituent of the epitope for mAbs 3F11 and 6B4.

LOS Sialylation

The paragloboside tetrasaccharide nonreducing end is a substrate for sialyltransferase, and in mammals N-acetyllactosamine-containing glycolipids are sialylated in vivo. The presence of the paragloboside epitope on *Neisseria* LOS led investigators to speculate whether this epitope was sialylated. Analysis of LOS isolated from *N. gonorrhoeae* grown in vitro failed to provide evidence for sialylation. This did not exclude the possibility that the lacto-N-neotetraose was sialylated in vivo by host enzymes. Gonococci in urethral exudates possess a virulence factor that is lost upon subculture to an artificial medium (26), suggesting that host factors can modify either directly or indirectly the bacteria during infection. This transient factor contributed to serum resistance in bactericidal assays. An explanation for some of the differences between in vitro and in vivo grown gonococci has been provided through a series of detailed studies by Smith and colleagues (27–29). Modification of gonococci in human fluids and secretions resulted in the conversion of serum-sensitive gonococci to serum resistance in vitro (30,31). Subsequently, it was confirmed that a change in gonococcal pyocin sensitivity (32) correlated with the induction of serum resistance and a difference in the LOS structure (33–35). A factor that could induce serum-susceptible gonococci to serum resistance was isolated from guinea pig serum, human red blood cells, and human white blood cells. Moreover, a 100- to 400-fold higher concentration of inducing activity was present in lymphocytes compared to whole blood (36), indicating that the resistance-inducing factor was concentrated in an environment relevant to gonococcal infection (30).

Attempts to purify the inducing factor resulted in detection of both a high and a low Mr factor in extracts from human red blood cells. The low Mr factor was <5000 daltons, very heat and acid labile (37), and contained carbohydrate, N-acetyl groups, and possibly a pyrimidine nucleotide (38). The low Mr activity was discovered to be the nucleotide sugar for sialic acid, cytidine 5′-monophospho-N-acetylneuraminic acid (CMP-NANA). Purified CMP-NANA duplicated all of the effects of the low Mr factor in gonococcal cell culture, including a major alteration in the LOS structure (28). Furthermore, a 1-hour incubation of gonococci with neutrophils decreased the binding of an anti-4.5 kDa LOS mAb to bacteria, and this effect could be reversed by treating the bacteria with neuraminidase (39).

A number of recent studies have provided definitive information regarding the relevance of LOS sialylation in natural gonococcal infection. Apicella and coworkers used immunoelectron microscopy to analyze the expression of LOS during natural infection of males (40). Six of seven gonococcal strains isolated from male exudates expressed the mAb-defined 4.5 kDa LOS. However, mAb binding to gonococci present in five of the exudates was minimal unless the exudate preparations were first pretreated with neuraminidase, indicating in vivo sialylation of the 4.5 kDa LOS component.

In a related set of experiments, gonococci isolated from male patient exudates were resistant to the bacterial activity of fresh human serum but became serum-sensitive when the exudates were treated with neuraminidase. Schneider reported that two male volunteers infected with a gonococcus expressing only a single 3.6 kDa LOS in the absence of the 4.5 kDa species developed urethritis containing predominately variant gonococci expressing 4.5 kDa or larger LOS (11). The bacteria isolated from the exudates appeared to be the same as variants isolated subsequently from the inoculating strain in vitro (frequency of 1 in 103). These studies together suggested that gonococcal LOS is sialylated during uncomplicated infections and that the presence of the 4.5 kDa component is associated with the virulence of gonococci.

Endogenous and Exogenous Sialylation

LOS mimicry of human glycolipid precursors suggested that host enzymes might modify bacterial LOS in vivo. Mandrell and coworkers identified two mammalian sialyltransferases, rat liver CMP-NANA: Galβ(1$\rightarrow$4)GlcNAcα(2$\rightarrow$6) sialyltransferase and porcine submaxillary gland CMP-NANA:Galα(2$\rightarrow$3) sialyltransferase, that incorporated sialic acid into LOS, however, sialic acid was incorporated into other bacterial molecules not sialylated by the putative bacterial sialyltransferase (39). These results suggested that exogenous LOS sialylation by host-derived sialyltransferase(s) might occur if the substrates CMP-NANA and bacterial LOS co-localize with host-derived sialyltransferase during infection.

The 4.5 kDa lactoneoseries LOS is also present in meningococcal serogroup B and C strains (41). However, the pattern of binding of mAb 3F11 to these meningococcal LOS was similar to the pattern of mAb binding to sialylated and nonsialylated gonococci. Since serogroup B and C meningococci synthesize a sialic acid capsule, and thus CMP-NANA, it seemed possible that endogenous CMP-NANA might also be used for sialylating LOS. The presence of sialic acid on meningococcal LOS was confirmed by a set of experiments similar to those used for studies of gonococcal LOS (39). The LOS of meningococci grown in the absence of CMP-NANA contained both endogenously sialylated LOS and nonsialylated LOS (42); bacteria grown in the presence of CMP-NANA appeared to further sialylate the 4.5 kDa acceptor LOS and, therefore, had mostly sialylated LOS. Analysis of partially deacylated LOS of a serogroup B meningococcus by liquid secondary ion mass spectrometry revealed that a fragmentation pattern characteristic of a sialylated molecule was present, confirming the terminal location of sialic acid on a tetrasaccharide consistent with lacto-N-neotetraose.

The detection of endogenously sialylated LOS in meningococci-stimulated population studies that led to the identification of sialylated LOS in other Neisseria strains (42). Most strains of meningococci that synthesize a sialic acid capsule (serogroups B, C, W, and Y) and a 4.5 kDa LOS endogenously sialylate their LOS. However, endogenously sialylated LOS are absent in meningococci that cannot synthesize sialic acid (e.g., serogroups A, 29E, X) and in nonpathogenic *Neisseria*, even those nonpathogenic strains that express the 4.5 kDa LOS acceptor for sialic acid. The development of an assay to measure the LOS-specific sialyltransferase activity in cell-free extracts (42), however, revealed that all strains of pathogenic *Neisseria* (>100 tested) and most strains of *N. lactamica* have an LOS-specific sialyltransferase (42); nonpathogenic *Neisseria* are negative for endogenous sialyltransferase activity. Larger amounts of sialic acid are incorporated into LOS of pathogenic *Neisseria* compared to *N. lactamica* when grown in the presence of CMP-NANA (unpublished observation). This result correlates with virulence, but does not correlate well with sialyltransferase expression.

LOS AND EVASION OF THE HOST IMMUNE RESPONSE

Antigenic Phase Variation

Neisseria LOS has been observed to antigenically phase vary in culture at a rate of 10^{-3}, and this phase variation is defined by fluctuating expression of sugar residues in the LOS terminal oligosaccharide (43). Therefore, sodium dodecyl sulfate–polyacrylamide gel electrophoresis (SDS-PAGE) analysis of cultured *Neisseria* LOS generally reveals multiple bands representing terminal oligosaccharides of varying length. Experimental *N. gonorrhoeae* infections of male volunteers revealed a shift from low Mr (3.6 kDa) LOS to a high Mr LOS (4.7 kDa) species, indicating that expression of a truncated LOS oligosaccharide may be contrary to survival in vivo (11). In this study, disease progression from dysuria to leukorrhea directly correlated with the LOS antigenic variation. The high molecular weight LOS species contained the mAb 3F11 epitope, *N*-acetyllactosamine, which mimics the human paragloboside terminal glycan. More than 90% of urethral exudates from naturally infected men showed gonococcal LOS expressing this 3F11 epitope suggesting a selective advantage in vivo.

The host is not able to generate protective antibodies to these self antigens, as they would prove detrimental to the host tissues. In fact, the basis of serum resistance in *N. gonorrhoeae* is believed to be the absence of bactericidal antibodies to the 4.7 kDa LOS terminal glycan (44). Serum-sensitive gonococci do not express the high Mr lactosamine containing LOS. Free LOS from these serum-sensitive bacteria blocks normal human sera bactericidal activity, unlike *N*-acetyllactosamine–containing free LOS from serum-resistant organisms. Furthermore, antibodies specific for the gonococcal LOS *N*-acetyllactosamine moiety will promote complement-mediated killing in vitro, suggesting that the protective nature of the gonococcal LOS stems from the inability of the host to generate protective antibodies to this major bacterial structure.

Schneider and coworkers assessed variation in the expression of LOS components and their epitopes within populations of a strain of *N. gonorrhoeae* by using the mAbs 6B4 and 3F11 and immunoenzymatic, immuno-colloidal gold electron microscopic, and SDS-PAGE procedures (10). Wild-type organisms varied in binding of both mAbs. The intensity of immunoenzymatic colony blot color allowed the identification of four binding variants for each mAb: red (R), pink (P), and colorless (nonreactive [N]) and an N back to R (N-R) revertant. R to P to R and R to N to R variations occurred at frequencies of 0.2% and 0.02%, respectively. The electrophoretic LOS profiles and mAb immunoblot patterns of the R, P, and N-R variants were the same as those of the wild-type. LOS of the N variants, in contrast, were of lower Mr, bound neither 3F11 nor 6B4 mAb, and contained as their major component the 3.6 kDa LOS that bears the L8 LOS epitope of *N. meningitidis*. Results of immunoelectron microscopic studies were consistent with LOS binding patterns. Large numbers of colloidal gold particles were deposited about both R and P variants, distally from R organisms, but proximally from P organisms. N variant organisms, like their LOS, bound neither of the mAbs. N-R variant organisms were like the wild-type in that they showed much variation in the amounts of mAb they bound. It is not known if the *N. gonorrhoeae* LOS phase varies during natural infection. It is known that the LOS phenotype of gonococci can change during experimental human urethral infection (11), but whether this is due to phase variation or selection of different phenotypes is unclear.

To investigate whether all members of clonally selected populations of *N. gonorrhoeae* express antigenically similar LOS, gonococcal strains 4505 and 220 were studied with mAbs 6B4 and 3F11, which have specificity for different oligosaccharide epitopes on the same or co-migrating LOS unit(s) on SDS-PAGE (22). Fluorescent antibody and immunoelectron microscopy studies indicated that all members of the clonally selected populations were not homogeneous for the 6B4 and 3F11 epitopes. Fluorescence-activated cell sorting studies of 3F11-coated gonococcal strain 220 indicated that the density of epitope expression was a function of time of growth. The population could be separated into two broad groups corresponding to organisms staining strongly or weakly for the 3F11 epitope, and the epitope density decreased during the late-log and stationary phases of growth.

Sequentially staining organisms on Formvar-coated grids with 6B4 and 3F11, followed by staining with either 5- or 15-nm colloidal gold spheres conjugated to goat antimouse immunoglobulin M demonstrated the following populations of cells among organisms derived from a single clone: (1) organisms that stained for both 6B4 and 3F11 epitopes, (2) organisms that stained for 6B4 epitopes alone, and (3) organisms that stained for 3F11 epitopes alone. Immunofluorescence microscopy studies with rhodamine and fluorescein goat antimouse immunoglobulin M conjugates sequentially staining organisms on Formvar-coated grids with 3F11 and 6B4 also demonstrated these three populations. Analysis of LOS preparations made over the last 5 years indicated no change in serotype antigen concentration or in SDS-PAGE migration pattern. These studies indicate that while clonally selected strains of *N. gonorrhoeae* undergo phenotypic variation at the epitope level, the impact of this variation on the total LOS of the population has little overall effect on its antigenic or physicochemical properties.

Mechanisms of LOS Variation

Phase variation of LOS, which is the reversible loss and gain of epitopes corresponding to synthesis of altered LOS molecules, is believed to occur primarily through differential expression of LOS biosynthetic genes. Several forms of DNA rearrangement affect phase variation of bacterial components, e.g., inversion of DNA segments for phase variation of *Salmonella* flagellin and *E. coli* type I fimbriae (45,46), intragenic recombination for phase variation of gonococcal pili (47,48), and addition or deletion of repetitive DNA elements. This last type of rearrangement is responsible for phase variation of *B. pertussis* fimbriae (49), pili of *H. influenzae* (50), gonococcal opacity proteins (51), and also LOS phase variation.

The number of repeats determines which, if any, initiation codon is in frame with the repeats and the downstream coding frame. Thus the repeat number determines the presence or absence of a complete orf and, in the case of a complete orf being present, the amino acid sequence of the reading frame upstream of the repeats. The *lgt* locus of *N. gonorrhoeae* strain F62 contains five genes (*lgtABCDE*) encoding glycosyl transferases, which assemble the terminal lacto-*N*-neotetraose and add a galactose residue of an alternative structure (52). *lgtA*, -*C*, and -*D* contain poly-G tracts in their coding regions and as such are potential sites of slip-strand mispairing. *lgtA* in *N. meningitidis* strain MC58 also contains a poly-G tract, and variation in the number of residues in this tract correlates with phase variation of lacto-*N*-neotraose expression (53).

Thus slip-strand mispairing generates variation in gene expression of a large number of LOS genes. However, it is unknown if this is a random process that generates a population with a heterologous LOS repertoire or if there are factors controlling slip-strand mispairing. Whichever the case, slip-strand mispairing is unlikely to be the only process governing phase variation.

Additional levels of control may operate through regulation of metabolic genes that affect LOS biosynthesis. *H. influenzae* and *N. gonorrhoeae* LOS show variation in composition and amount produced depending on the growth rate of the culture (54). *N. gonorrhoeae* LOS varies according to the pH of culture conditions (55). The level of aeration of cultures affects the LOS phenotype of *N. meningitidis* (56). The mechanism by which these changes are effected is unknown. However, the variation in LOS that occurs through changes in central metabolism may mean that LOS biosynthesis is regulated by factors involved in regulation of central metabolism, such as signaling through cAMP/CRP and gene regulation through regulators such as LRP. In vitro experiments are often performed with bacterial cultures grown in rich laboratory media. It is likely that in vivo nutrients are far more limited. It is thus possible that in vivo organisms do not produce the LOS structures seen in laboratory experiments but instead produce LOS characteristic of those structures seen under conditions of limitation.

Regulation of LOS biosynthesis may operate at various levels, including random phase variation of individual genes producing a large LOS repertoire and at a more global level in response to metabolic status or specific environmental signals. Much of the existing study of LOS genetics has identified the array of genes involved in biosynthesis. As the list of genes involved in this process becomes complete, increasing attention must be paid to understanding the regulation of these genes in order to appreciate the mechanisms by which organisms produce the LOS structures that we observe and to understand how regulation of LOS biosynthesis affects the pathogenicity of the organism.

Interference with the Host Immune Response

The majority of the inflammatory effects associated with the inflammatory response to *Neisseria* are related to the release of LOS. The elegant studies of Norwegian investigators in studies of sepsis associated with fulminent meningococcal infection have shown that the release of cytokines, the activation of the complement system, and the effects on plasminogen activation induced by meningococcal LOS are primary factors in the etiology and prolongation of the shock state. Since the inflammatory state does not, in general, serve these organisms well since it stimulates host defenses, it would seem that LOS must play other roles in pathogenesis that directly benefit the microorganism. The immunochemical similarity of LOS epitopes with human glycosphingolipid antigens may be an example of host evasion by molecular mimicry. The sialylation of the LOS in *Neisseria* spp. and *Haemophilus* spp. may provide protection through the biologic masking provided by the addition of the sialic acid (57). There is evidence to support this contention. The initial observation that led Smith and coworkers to identify CMP-NANA as a substrate that could modify the LOS in vivo was the fact that freshly isolated genital gonococcal strains were resistant to killing by normal human serum (58). This resistance was lost on a single passage in vitro. Many gonococcal strains have been identified that do not require CMP-NANA for their resistance to killing by human serum. Strains isolated from patients with disseminated infection remain serum resistant in in vitro assays without the addition of CMP-NANA (57,58), and this includes strains that do not express the LOS acceptor for sialic acid (39). In addition, there are serum-sensitive strains that are unaffected by growth in CMP-NANA even though their LOS are sialylated (59).

It is possible that the primary role of LOS sialylation is something different than protecting the bacteria from the lytic effects of serum antibody and complement. For example, recent studies have shown that LOS sialylation also prevents, or at least decreases, the phagocytosis of gonococcal or meningococcal strains by human neutrophils (59–61). Three separate studies of the effect of LOS sialylation on the interaction of gonococci with human neutrophils have yielded somewhat different results. One study described a dramatically increased survival of a sialylated versus nonsialylated gonococcal strain by complement-dependent opsonophagocytosis (62), whereas in a different study with a different strain only a moderate increase in the survival of the sialylated versus the nonsialylated strain was measured (63). The results of the second study were consistent with those of a third study showing a limited increase (30%) in the survival of a sialylated strain under the same conditions (60). In addition, it was shown that sialylated gonococci have a decreased ability to stimulate an oxidative burst in neutrophils and to adhere to neutrophils in the absence of complement and antibody compared to the nonsialylated gonococci (63). The effects of sialylation on phagocytosis suggest a potential survival mechanism for pathogenic *Neisseria* in mucosal environments, although other experiments will be necessary to determine exactly how these various factors are involved.

The protective effects of sialylation of *Neisseria* do not appear to be limited to inhibition of phagocytosis. Elkins and coworkers demonstrated that affinity-purified antisera against both PI.A and PI.B N-terminal peptides were bactericidal for homologous gonococci and many heterologous PI serovars. However, sialylation of gonococcal LPS by growth of gonococci in the presence of CMP-NANA abrogated the bactericidal activity of these antisera (64). Binding of anti-PI monoclonal antibodies to whole gonococci was reduced two-to fourfold by sialylation of LPS, suggesting that sialylation may inhibit bactericidal activity by masking porin epitopes. However, binding of anti-PII (Opa) mAbs was not inhibited. Binding of complement components C3 and C9 was inhibited in the presence of

Table 1 Glycosphingolipid-like Structures Found on *Neisseria* LOS

LOS glycosphingolipid-like structure[a]	Common name[b]
Galβ(1→4)Glcβ1-	LacCer
GalNAcβ(1→4)Galβ(1→4)Glcβ1-	Asialo-GM2
GalNAcβ(1→3)Galβ(1→4)GlcNAcβ(1→3)Galβ(1→4)Glc β1-	Asialo-G_3
Galα(1→4)Galβ(1→4)Glcβ1-	p^k
Galβ(1→4)GlcNAcβ(1→3)Galβ(1→4)Glcβ1-	Paragloboside
NeuNAcα(2→3)Galβ(1→4)GlcNAcβ(1→3)Galβ(1→4)Glcβ1-	Sialylparaglogoside
Galα(1→4)Galβ(1→4)GlcNAcβ(1→3)Galβ(1→4)Glcβ1-	P1

[a]LOS structures are terminal branches of heptose-kdo-lipid A core region.
[b]Glycosphingolipid structures are branches from ceramide.

either anti-PI or anti-PII monoclonals when gonococci were grown in the presence of CMP-NANA. Thus, sialylation appeared to inhibit both anti-PI antibody binding and complement deposition, with a resultant decrease in bactericidal activity.

Wetzler and coworkers also demonstrated that sialylated gonococci also become resistant to the bactericidal effect of immune sera containing antibodies that recognize exposed components of the outer membrane besides LOS (62). Prevention of antibody binding to the organism was not the cause, since the same percentage of bactericidal antibodies to the major outer membrane protein, Protein I, can be absorbed with sialylated organisms as with wild-type organisms. In addition, gonococcal sialylation prevents opsonophagocytosis by antigonococcal antisera. They concluded that the negative effect of sialic acid on the complement pathway might be the reason for the findings in this study. In contrast, others have shown that it should be possible to induce complement-mediated killing by directing the immune response to those surface-exposed epitopes that are least susceptible to the potential inhibitory effect of LOS sialylation.

NEISSERIA TISSUE TROPISM AND INVASION

Neisseria Are Intracellular Pathogens

Neisseria gonorrhoeae requires exogenous CMP-NANA in order to generate sialo-LOS. CMP-NANA is found intracellularly in the host. Therefore, these organisms must invade host cells to acquire CMP-NANA. It has been well accepted that *N. gonorrhoeae* are able to invade polymorphonuclear leukocytes (PMN). Infected PMN are readily observed in patient exudates by confocal and electron microscopy. Also, in these exudates, infected epithelial cells have been observed. Several labs have shown that *N. gonorrhoeae* successfully invade the apical surface of genital epithelial cells, survive intracellularly, and can traverse the cytoplasm to emerge at the cells basolateral surface in vitro (65,66).

The Role of LOS as a Ligand

Previous evidence has suggested that the tight intercellular adhesions between the outer membrane of gonococci displaying the opacity colony phenotype occurred because Opa proteins expressed on one gonococcus adhered to the LOS of the opposing bacterium. Blade and coworkers, employing a noncompetitive inhibition assay previously used to determine the carbohydrate structures recognized by the major hepatic asialoglycoprotein receptor, demonstrated that gonococcal LOS structures bind Opa proteins (67). The LOS carbohydrate used in these assays were LOS structures purified from pyocin-resistant LOS mutants of *N. gonorrhoeae* strain 1291 (21). These data suggest that the gonococcal Opa proteins had the highest affinity for the Gal β(1→4)GlcNAc residue present on the gonococcal lactoneoseries LOS. This affinity was comparable to that reported for the binding of the major hepatic asialoglycoprotein receptor to glycoconjugates containing terminal galactose and *N*-acetylgalactosamine. After sialylation of the lactoneoseries LOS, presumably on the terminal galactose residue, the interaction with the Opa proteins was ablated.

The Role of LOS in Adherence and Invasion

It has been known for many years that the gonococcus could adhere to and invade a wide variety of tissue culture cells. Recent studies have demonstrated that gonococci can invade urethral epithelial cells during urethral infection in males (68). As gonococci invade these cells, membrane fusion occurs, suggesting that a very tight ligand receptor–based interaction is occurring between the gonococci and the host epithelial cell. Recent studies in experimental models suggest that the gonococcal LOS may be participating in the adhesion and invasion events as a ligand. van Putten and coworkers have investigated the function of the *lsi-1* gene of *N. gonorrhoeae*, which had been previously implicated in LOS inner core biosynthesis. Disruption of the gene in gonococcal strain MS11 resulted in the production of LOS that migrated faster than that from an isogenic *galE* mutant, typical for a mutation that influences the inner core region (69). Complementation of a panel of *Salmonella typhimurium* mutants with defined defects in the *rfa* loci demonstrated conclusively that the *lsi-1* gene of MS11 is functionally homologous to the *rfaF* gene, which encodes heptosyltransferase II in both *E. coli* and *S. typhimurium*. Immunochemical analysis of the LPS using monoclonal antibodies directed against chemically defined inner-core glycoconjugates revealed that the gonococcal and *Salmonella* Rd2-chemotypes were antigenically similar, further extending the genetic and functional homology. Infection experiments in vitro demonstrated that the *lsi-1* mutant could not invade human Chang epithelial cells despite expression of a genetically defined invasion-promoting gonococcal opacity protein. These data imply that the LPS phenotype is a critical factor for gonococcal invasiveness.

The human equivalent of this structure based on paragloboside can act as the ligand for asialoglycoprotein receptors (ASGP-R) contained on numerous human macrophages, sperm cells, and hepatocytes. The most well studied of this large family of receptors is that expressed on the surface of hepatic cells. In a model cell system, using the hepatoma tissue culture cell line HepG2, Porat and coworkers (70,71) analyzed the role of the asialoglycoprotein receptor in the adherence and/or invasion of gonococci expressing the lacto-N-neotetraose structure. Using well-established assays for the utilization of the ASGP-R, these investigators found that incubation of HepG2 cells with gonococci expressing the terminal Galβ(1→4)GlcNAc asialo-LOS carbohydrate structure competitively inhibited the ASGP-R from binding to one of its well-known ligands, asialo-α-acid-1-glycoprotein. The inhibition was specific to the ASGP-R, since binding of two other ligands to their specific receptors in the same model cell system was not affected. Immunoblot analysis for the ASGP-R suggested that the gonococci seemed to stimulate the HepG2 cells to increase the expression of the major (46 kDa) receptor species. This observation was confirmed both by functional analysis, which showed that the concentration of total receptor molecules, as well as surface receptors, was about 60% higher after incubation with gonococci than in control cells, and by Northern (RNA) blot analysis using a cDNA probe of the major human H1 subunit. Poly(A) RNA purified from control and HepG2 cells exposed to gonococci indicated the presence of increased amounts of mRNA coding for the ASGP-R after incubation with gonococci. This result supports the idea that the molecular mechanism controlling the receptor level after gonococcal exposure is under transcriptional regulation. This same group also demonstrated that gonococcal LOS could bind to a 70 kDa receptor on the HepG2 cell and that this interaction was similar to the ligand receptor interaction between LOS and the Opa protein (70,71).

SUMMARY

Pathogenic *Neisseria* spp. as well as several other gram-negative mucosal pathogens express a unique phase-variable glycolipid, LOS, on the bacterial outer membrane. LOS is structurally and functionally distinct from the well-studied glycolipid LPS expressed by *E. coli*, *Salmonella* spp., and other gram-negative enteric bacteria.

The LOS outer core glycan, lacto-N-neotetraose, is structurally and immunochemically identical to the terminal tetrasaccharide expressed on human paragloboside, precursor to the I and i blood group antigens. Paragloboside-based glycolipids are sialylated in vivo, and it was found that mammalian sialyltransferases could exogenously sialylate N-acetyllactosamine–containing LOS. While both *N. gonorrhoeae* and *N. meningitidis* express sialyltransferase, only meningococci produce CMP-NANA and thus endogenously sialylate LOS in vitro.

LOS antigenic mimicry is important for pathogenesis with respect to evasion of the immune response. The host humoral immune response to these self antigens is poor, as production of such antibodies would prove detrimental. Moreover, sialylation of the LOS N-acetyllactosamine moiety provides protection against complement-mediated killing and cell-mediated immunity through the action of sialic acid as a so-called biological mask.

Recent studies suggest that N-acetyllactosamine–containing asialo-LOS is involved in host cell invasion. *Neisseria* LOS is a potential ligand for the host asialoglycoprotein receptor, which is expressed on several tissues. Thus, LOS may contribute to host tissue specificity as well as invasion.

REFERENCES

1. Perry MG, Daoust V. The lipopolysaccharides of *Neisseria gonorrhoeae* colony type 1 and 4. Can J Biochem 1975; 53:623–629.
2. Griffiss JM, Schneider H, Mandrell RE, et al. Lipooligosaccharides: the principal glycolipids of the neisserial outer membrane. Rev Infect Dis 1988; 10:S287–S295.
3. Maeland JA. Antigenic properties of various preparations of *Neisseria gonorrhoeae* endotoxin. Acta Pathol Microbiol Scand 1968; 73:413–422.
4. Maeland JA. Serological cross-reactions of aqueous ether extracted endotoxon from *Neisseria gonorrhoeae* strains. Acta Pathol Microbiol Scand 1969; 77:505–517.
5. Maeland JA, Kristoffersen T, Hofstad T. Immunochemical investigations of *Neisseria gonorrhoeae* endotoxin. Acta Pathol Microbiol Scand 1971; 79:233–236.
6. Mandrell RE, Zollinger WD. Lipopolysaccharide serotyping of *Neisseria meningitidis* by hemagglutination inhibition. Infect Immun 1977; 16:471–475.
7. Zollinger WD, Mandrell RE. Type-specific antigens of group A *Neisseria meningitidis*: lipopolysaccharide and heat-modifiable outer membrane proteins. Infect Immun 1980; 28:451–458.
8. Jennings HJ, Johnson KG. The structure of an R-type oligosaccharide core obtained from some lipopolysaccharides of *Neisseria meningitidis*. Carbohydr Res 1983; 121:233–241.

9. Mandrell RE, Griffiss JM, Macher BA. Lipooligosaccharides (LOS) if *Neisseria gonorrhoeae* and *Neisseria meningitidis* have components that are immunochemically similar to precursors of human blood group antigens. Carbohydrate sequence specificity of the mouse monoclonal antibodies that recognize crossreacting antigens on LOS and human erythrocytes. J Exp Med 1988; 168:107–126.
10. Schneider H, Hammack CA, Apicella MA, Griffiss JM. Instability of expression of lipooligosaccharides and their epitopes in *Neisseria gonorrhoeae*. Infect Immun 1988; 56:942–946.
11. Schneider H, Griffiss JM, Boslego JW, Hitchcock PJ, Zahos KM, Apicella MA. Expression of paragloboside like lipooligosaccharides may be a necessary component of gonococcal pathogenesis in men. J Exp Med 1991; 174:1601–1605.
12. Ward ME, Glynn AA. Human antibody response to lipopolysaccharides from *Neisseria gonorrhoeae*. J Clin Pathol 1972; 25:56–59.
13. Rice PA, Blake MS, Joiner KA. Mechanisms of stable serum resistance of *Neisseria gonorrhoeae*. Antonie Van Leeuwenhoek 1987; 53:565–574.
14. Ashwell G, Morell AG. The role of surface carbohydrates in the hepatic recognition and transport of circulating glycoproteins. Adv Enzymol 1974; 41:49–128.
15. Appelmelk B, Negrini R, Moran A, Kuipers E. Molecular mimicry between *Helicobacter pylori* and the host. Trends Microbiol 1997; 5:70–73.
16. Aspinall GO, McDonald AG, Pang H, Kurjanczyk LA, Penner JL. Lipopolysaccharides of *Campylobacter jejuni* serotype O:19: structures of core oligasaccharide regions from the serostrain and two bacterial isolates from patients with the Guillain-Barré syndrome. Biochemistry 1994; 33:241–249.
17. Campagnari AA, Gupta MR, Dudas KC, Murphy TF, Apicella MA. Antigenic diversity of lipooligosaccharides of nontypable *Haemophilus influenzae*. Infect Immun 1987; 55:882–887.
18. Campagnari AA, Spinola SM, Lesse AJ, Abu Kwaik Y, Mandrell RE, Apicella M. Lipooligosaccharide epitopes shared among Gram-negative non-enteric mucosal pathogens. Microb Pathogen 1990; 8:353–362.
19. Apicella MA, Bennett KM, Hermerath CA, Roberts DE. Monoclonal antibody analysis of lipopolysaccharide from *Neisseria gonorrhoeae* and *Neisseria meningitidis*. Infect Immun 1981; 34:751–756.
20. Yamasaki R, Nasholds W, Schneider H, Apicella MA. Epitope expression and partial structural characterization of F62 lipooligosaccharide (LOS) of *Neisseria gonorrhoeae*: IgM monoclonal antibodies (3F11 and 1-1-M) recognize non-reducing termini of the LOS components. Mol Immunol 1991; 28:1233–1242.
21. Dudas KC, Apicella MA. Selection and immunochemical analysis of lipooligosaccharide mutants of *Neisseria gonorrhoeae*. Infect Immun 1988; 56:499–504.
22. Apicella MA, Shero M, Jarvis GA, Griffiss JM, Mandrell RE, Schneider H. Phenotypic variation in epitope expression of the *Neisseria gonorrhoeae* lipooligosaccharide. Infect Immun 1987; 55:1755–1761.
23. Mandrell R, Schneider H, Apicella M, Zollinger W, Rice PA, Griffiss JM. Antigenic and physical diversity of *Neisseria gonorrhoeae* lipooligosaccharides. Infect Immun 1986; 54:63–69.
24. Gibson BW, Webb JW, Yamasaki R, et al. Structure and heterogeneity of the oligosaccharides from the lipopolysaccharides of a pyocin-resistant *Neisseria gonorrhoeae*. Proc Natl Acad Sci USA 1989; 86:17–21.
25. Yamasaki R, Bacon BE, Nasholds W, Schneider H, Griffiss JM. Structural determination of oligosaccharides derived from lipooligosaccharide of *Neisseria gonorrhoeae* F62 by chemical, enzymatic, and two dimensional NMR methods. Biochemistry 1991; 30:10566–10575.
26. Ward ME, Watt PJ, Glynn AA. Gonococci in urethral exudates possess a virulence factor lost on subculture. Nature 1970; 227:382–384.
27. Fox AJ, Curry A, Rowland PL, et al. A surface polysaccharide forms when gonococci are converted to serum resistance by cytidine 5'-monophospho-N-acetyl neuraminic acid. FEMSLETT 1989; 75–80.
28. Parsons NJ, Patel PV, Tan EL, et al. Cytidine 5'-monophospho-N-acetyl neuraminic acid and a low molecular weight factor from human blood cells induce lipopolysaccharide alteration in gonococci when conferring resistance to killing by human serum. Microb Pathogen 1988; 5:303–309.
29. Parsons NJ, Cole JA, Smith H. Resistance to human serum of gonococci in urethral exudates is reduced by neuraminidase. Proc R Soc Lond B 1990; 241:3–5.
30. Patel PV, Martin PM, Tan EL, et al. Protein changes associated with induced resistance of *Neisseria gonorrhoeae* to killing by human serum are relatively minor. J Gen Microbiol 1988; 134:499–507.
31. Patel PV, Parsons NJ, Andrade JRC, et al. White blood cells including polymorphonuclear phagocytes contain a factor which induces gonococcal resistance to complement-mediated serum killing. FEMS Microbiol Lett 1988; 50:173–184.
32. Winstanley FP, Blackwell CC, Tan EL, et al. Alteration of pyocin-sensitivity pattern of *Neisseria gonorrhoeae* is associated with induced resistance to killing by human serum. J Gen Microbiol 1984; 130:1303–1306.
33. Tan EL, Patel PV, Parsons NJ, Martin PMV, Smith H. Lipopolysaccharide alteration is associated with induced resistance of *Neisseria gonorrhoeae* to killing by human serum. J Gen Microbiol 1986; 132:1407–1413.
34. Demarco de HR, Bundell C, Chong H, Taylor DW, Wildy P. Definition of a virulence-related antigen of *Neisseria gonorrhoeae* with monoclonal antibodies and lectins. J Infect Dis 1986; 153:535–546.
35. Demarco De Hormaeche R, Thornley MJ, Holmes A. Surface antigens of gonococci: correlation with virulence and serum resistance. J Gen Microbiol 1983; 129:1559–1567.
36. Patel PV, Veale R, Fox JE, Martin V, Parson NJ, Smith H. Fractionation of guinea pig serum for an inducer of gonococcal resistance to killing by human serum: active fractions containing glucopeptides similar to those from human red blood cells. J Gen Microbiol 1984; 130:2757–2769.
37. Martin PMV, Patel PV, Parsons NJ, Smith H. Induction of serum resistance in recent isolates of *Neisseria gonorrhoeae* by a low-molecular-weight fraction of guinea pig serum. J Infect Dis 1983; 148:334.

38. Nairn CA, Cole JA, Patel PV, Parsons NJ, Fox JE, Smith H. Cytidine 5′-monophospho-N-acetylneuraminic acid or a related compound is the low Mr factor from human red blood cells which induces gonococcal resistance to killing by human serum. J Gen Microbiol 1988; 134:3295–3306.
39. Mandrell RE, Lesse AJ, Sugai JV, et al. In vitro and in vivo modification of Neisseria gonorrhoeae lipooligosaccharide epitope structure by sialylation. J Exp Med 1990; 171:1649–1664.
40. Apicella MA, Mandrell RE, Shero M, et al. Modification of sialic acid of Neisseria gonorrhoeae lipooligosaccharide epitope expression in human urethral exudates: an immunoelectron microscopic analysis. J Infect Dis 1990; 162:506–512.
41. Kim JJ, Mandrell RE, Griffiss JM. Neisseria lactamica and Neisseria meningitidis share lipooligosaccharide epitopes but lack common capsular and class 1, 2 and 3 protein epitopes. Infect Immun 1989; 57:602–608.
42. Mandrell RE, Griffiss JM, Smith H, Cole JA. Distribution of a lipooligosaccharide-specific sialyltransferase in pathogenic and nonpathogenic Neisseria. Microb Pathogen 1993; 14:315–327.
43. Schneider H, Hale TL, Zollinger WD, Seid RC, Hammack CA, Griffiss JM. Heterogeneity of molecular size and antigenic expression within lipooligosaccharides of individual strains of Neisseria gonorrhoeae and Neisseria meningitidis. Infect Immun 1984; 45:544–549.
44. Schneider H, Griffiss JM, Williams GD, Pier GB. Immunological basis of serum resistance of Neisseria gonorrhoeae. J Gen Microbiol 1982; 128:13–22.
45. Klemm P. Two regulatory fim genes, fimB and fimE control the phase variation of type 1 fimbria in Escherichia cola. FEMS Microbiol Lett 1986; 72:43.
46. Zieg J, Silverman M, Simon M. Recombinational switch for gene expression. Science 1977; 196:170.
47. Haas R, Meyer TF. Molecular principles of antigenic variation in Neisseria gonorrhoeae. Antonie Van Leeuwenhoek 1987; 53:431–434.
48. Swanson J, Bergstrom S, Robbins K, Barrera O, Corwin D, Koomey JM. Gene conversion involving the pilin structural gene correlates with pilus+ in equilibrium with pilus− changes in Neisseria gonorrhoeae. Cell 1986; 47:267–276.
49. Willems R, Paul A, van der Heide H, ter Avest A, Mooi F. Fimbrial phase variation in Bordetella pertussis: a novel mechanism for transcription regulation. EMBO J 1990; 9:2803–2806.
50. van Alphen L, Van den Broek LG, Van Ham M. In vivo and in vitro expression of outer membrane components of Haemophilus influenzae. Microb Pathogen 1990; 8:279–288.
51. Murphy GL, Connell TD, Barritt DS, Koomey M, Cannon JG. Phase variation of gonococcal protein II: regulation of gene expression by slipped strand mispairing of a repetitive DNA sequence. Cell 1989; 56:539–547.
52. Gotschlich EC. Genetic locus for the biosynthesis of the variable portion of Neisseria gonorrhoeae lipooligosaccharide. J Exp Med 1994; 180:2181–2190.
53. Jennings MP, Hood DW, Peak RA, Virji M, Moxon ER. Molecular analysis of a locus for the biosynthesis and phase-variable expression of the lacto-N-neotetraose terminal lipopolysaccharide structure in Neisseria meningitidis. Mol Microbiol 1995; 18:729–740.
54. Morse SA, Mintz CS, Sarafian SK, Barenstein L, Bertram B, Apicella MA. Effect of dilution rate on lipopolysaccharide and serum resistance of Neisseria gonorrhoeae grown in continuous culture. Infect Immun 1983; 41:74–82.
55. Pettit RK, Martin ES, Wagner SM, Bertolino VJ. Phenotypic modulation of gonococcal lipooligosaccharide in acidic and alkaline culture. Infect Immun 1995; 63:2773–2775.
56. Tsai CM, Boykins R, Frasch CE. Heterogeneity and variation among Neisseria meningitidis lipopolysaccharides. J Bacteriol 1983; 155:498–504.
57. Schauer R. Sialic acids and their role as biological masks. TIBS 1985; 357–360.
58. Smith H, Cole JA, Parsons NJ. The sialylation of gonococcal lipopolysaccharide by host factors: a major impact on pathogenicity. FEMS Microbiol Lett 1992; 100:287–292.
59. Estabrook MM, Christopher NN, Griffiss JM, Baker CJ, Mandrell RE. Sialylation and human neutrophil killing of group C Neisseria meningitidis. J Infect Dis 1992; 166:1079–1085.
60. Frangipane JV, Rest RF. Anaerobic growth and cytidine 5′-monophospho-n-acetylneuraminic acid act synergistically to induce high-level serum resistance in Neisseria gonorrhoeae. Infect Immun 1993; 61:1657–1666.
61. Kim JJ, Zhou D, Mandrell RE, Griffiss JM. Effect of exogenous sialylation of the lipooligosaccharide of Neisseria gonorrhoeae on opsonophagocytosis. Infect Immun 1992; 60:4439–4442.
62. Wetzler LM, Barry K, Blake MS, Gotschlich EC. Gonococcal lipooligosaccharide sialylation prevents complement-dependent killing by immune sera. Infect Immun 1992; 60:39–43.
63. Rest RF, Frangipane JV. Growth of Neisseria gonorrhoeae in CMP-N-acetylneuraminic acid inhibits nonopsonic (opacity-associated outer membrane protein-mediated) interactions with human neutrophils. Infect Immun 1992; 60:989–997.
64. Elkins C, Carbonetti NH, Varela VA, Stirewalt D, Klapper DG, Sparling PF. Antibodies to N-terminal peptides of gonococcal porin are bactericidal when gonococcal lipopolysaccharide is not sialylated. Mol Microbiol 1992; 6:2617–2628.
65. McGee ZA, Johnson AP, Taylor-Robinson D. Pathogenic mechanisms of Neisseria gonorrhoeae: observations on damage to human fallopian tubes in organ culture by gonococci of colony type 1 or type 4. J Infect Dis 1981; 143:413–422.
66. Harvey HA, Ketterer MR, Preston A, Lubaroff D, Williams R, Apicella MA. Ultrastructural analysis of primary human urethral epithelial cell cultures infected with Neisseria gonorrheae. Infect Immun 1997; 65:2420–2427.
67. Blake MS, Blake CM, Apicella MA, Mandrell RE. Gonococcal opacity: lectin-like interactions between Opa proteins and lipooligosaccharide. Infect Immun 1995; 63:1434–1439.
68. Apicella MA, Ketterer M, Lee FKN, Zhou D, Rice PA, Blake MS. The pathogenesis of gonococcal urethritis in

men: confocal and immunoelectron microscopic analysis of urethral exudates from men infected with *Neisseria gonorrhoeae*. J Infect Dis 1996; 173:636–646.
69. Schwan ET, Robertson BD, Brade H, van Putten JPM. Gonococcal rfaF mutants express Rd_2 chemotype LPS and do not enter epithelial host cells. Mol Microbiol 1995; 15:267–275.
70. Porat N, Apicella MA, Blake MS. A lipooligosaccharide binding site on HepG2 cells similar to the gonococcal opacity-associated surface protein Opa. Infect Immun 1995; 63:2164–2172.
71. Porat N, Apicella MA, Blake MS. *Neisseria gonorrhoeae* utilizes and enhances the biosynthesis of the asialoglycoprotein receptor expressed on the surface of the hepatic HepG2 cell line. Infect Immun 1995; 63:1498–1506.

5

Antibiotic-Induced Endotoxin Release: Important Parameters Dictating Responses

Jesse J. Jackson and Helmut Kropp
Merck Research Laboratories, Rahway, New Jersey

INTRODUCTION

The pathophysiology of septic shock is complex and involves a bacterial-induced web of pro-inflammatory and anti-inflammatory mediators that may ultimately lead to death (1). Free endotoxin (lipopolysaccharide, LPS) released from gram-negative bacteria, one of several bioreactive subcellular cell wall components, is among the most potent microbial initiators of inflammatory cytokines. The magnitude of the inflammation is dependent upon the quantity, bioreactivity, and duration of LPS stimulation. When antibiotics are used to kill bacteria, endotoxin release from the microbe may become intensified. Normally, bacteria that are highly sensitive to antibiotics are rapidly destroyed by most antibiotics, with only a minimal release of endotoxin. However, these organisms can become more pathogenic when the host is compromised or antibiotic therapy is delayed or inappropriate. Delayed or inappropriate antibiotic therapy permits an increase in bacterial cell mass along with its accompanying endotoxin production. When bacterial biomass or LPS levels reach potentially harmful or threshold amounts that cannot be rapidly cleared by normal host defenses, antibiotic choice, its concentration, timing, and duration of treatment become more critical.

Despite optimistic results from animal studies, to date therapeutic strategies designed to eliminate deleterious levels of free LPS from the circulation of more severe gram-negative infections have failed to develop into reliable methods of treatment (2). These unsuccessful strategies have brought into question the role that antibiotic-released LPS itself plays in the survival rate of septic patients. Adding to this uncertainty are cases that show a lack of correlation between the presence of gram-negative bacteria and endotoxin/cytokine levels detected in the plasma of patients. In theory, at least two factors (with respect to antibiotic treatment) may have contributed to the above anti-LPS/anticytokine therapeutic failures. The first factor is associated with data interpretation and involves the lack of a consensus as to just what the predictive harmful or threshold endotoxin/cytokine levels or endpoints should actually be with respect to patient outcome. The second factor more directly influences poor patient outcome and is due to the lack of prospective considerations (in sepsis trials) for the relative endotoxin-liberating potential of the specific antibiotics that are also administered to patients selected for novel anti-LPS/anticytokine therapies.

In support of these concepts, it would be relevant to revisit the effects that three major components of the infectious process may have had on these failed strategies through regulating the detectable levels of antibiotic-liberated endotoxin. These components are the infecting organism, the specific choice of antibiotic, and the endotoxin sensitivity of the infected host/animal. Such a review may allow future in vitro/in vivo studies regarding antibiotic-induced endotoxin release to be more carefully designed, and this may ultimately help to establish guidelines that will simplify interpretation of data as well as aid in future clinical evaluations.

BACTERIAL SPECIES–RELATED DIFFERENTIAL REGULATION OF DETECTABLE ENDOTOXIN LEVELS

Results from experimental data suggest that unique qualities of the bacterial species itself may differentially influence or regulate the detectable levels of endotoxin found in the plasma of infected patients. The mechanism in this instance would be strictly related to the production and release of endotoxin. Three bacteria-related factors that would contribute to this are the relative ability of the organism to liberate endotoxin, the types or physiochemical properties (bioreactivity) of endotoxins that are released, and the actual quantities of endotoxins that are produced and liberated.

Gram-negative bacteria, in general, may differ both quantitatively and qualitatively among and within species in the production and spontaneous release of endotoxin. In fact, some bacteria are even non–LPS liberators, and these organisms are usually less pathogenic than their LPS-liberating variants (3). However, these less virulent organisms may become considerably more infective in an immune-compromised host. It may, therefore, be important to determine the relative spontaneous LPS-releasing potential of those gram-negative organisms isolated from patients who do not show detectable levels of endotoxin/cytokine and compare it to the LPS-liberating potential of bacterial isolates from patients in which circulating endotoxin is observed.

When bacteria are LPS-liberating, the type or physiochemical property (i.e., serotype, immunotype, smooth, rough) of endotoxin released, as well as the amount produced, can be expected to influence its ability to modulate the pro-inflammatory and anti-inflammatory responses (bioreactivity) important in natural host defense. Although the quantity of LPS spontaneously released may vary among isolates of a given bacterial species, each species may also differ in the relative amount of endotoxin produced. For example, LPS-liberating *Escherichia coli*, for the most part, are generally considered to produce and spontaneously liberate more endotoxin than LPS-liberating *Pseudomonas aeruginosa* and perhaps *Klebsiella pneumoniae* as well (4). The corresponding cytokine (e.g., tumor necrosis factor, TNF) response from LPS-stimulated white cells (e.g., macrophages) would, therefore, at least in part, simply reflect the different amounts of bacterial endotoxin released from each of the organisms (5).

It would thus follow that treatment of these same organisms with the same LPS-liberating antibiotic under identical conditions (time, concentration, etc.) would subsequently amplify the corresponding differential release of even higher amounts of free endotoxin and its corresponding cytokine production from each of the above bacterial species. For the above hypothesis to be true, all of the characteristics of the bacterial isolates from each bacterial species would obviously have to be identical in every respect, and this, of course, is not the case. In fact, any additional or alternative variations or factors may render antibiotic-induced endotoxin release dramatically more complex than in the circumstances considered above, while ultimately resulting in a multitude of possible influences on the final detectable endotoxin levels among and within bacterial species. Such modifying factors may include variations in host sensitivity to LPS, immune competence of the host, bioreactivity of the LPS, change of organism or mixed bacterial cultures and antibiotics with other modes of action, and their use at different concentrations. The remainder of this chapter will concern itself with a number of these modulating factors that will ultimately determine the final amount of detectable endotoxin and its influence on the pathophysiologic status of the host.

NON-SPECIES-RELATED DIFFERENTIAL INFLUENCE ON DETECTABLE ENDOTOXIN LEVELS

A number of bacteria-related factors may also influence detectable amounts of endotoxin by regulating the in vivo bacterial growth rate along with its accompanying endotoxin production without regard to its species. These cell mass–related factors include differential bacterial virulence (infectivity/invasiveness), serum sensitivity (levels of complement, specific antibodies, etc.), sensitivity to antibiotics (e.g., MIC, MBC), availability to antibiotics (i.e., intracellular vs. extracellular), and the inoculum size (6). These latter two factors are associated with the ability of antibiotics to kill bacteria. Failure to consider the above factors may lead to misleading results. For example, data resulting from studies with intracellular organisms (e.g., *Salmonella typhimurium*) may yield observations of little value, because most antibiotics used for human gram-negative sepsis would be of limited use due to the lack of ability to penetrate and accumulate inside macrophages at effective antibiotic concentrations. On the other hand, high initial bacterial inoculum size ($>10^8$ CFU) may differentially reduce the effective antimicrobial actions of certain antibiotics in response to increases in bacterial biomass (6).

Bacterial cultures evaluated under static growth conditions (high, non-log phase, concentrations $\geq 10^8$ CFU) may produce larger than usual amounts of pencillin-binding proteins (PBP), enzymes that would of necessity require an increase in the effective antibiotic concentration for complete inhibition of their activity. Thus, initial bacterial concentrations of 10^4–10^6 CFU would be preferable in antibiotic-induced LPS release studies, because bacteria at these lower concentrations would permit a growth period for those antibiotics with delayed killing actions. Static growth conditions (10^8 CFU), on the other hand, would reduce the possibility of normal antibiotic-induced differential bacterial growth and may actually mask the potential for differential endotoxin release. In addition, when high bacterial numbers are exposed to highly bactericidal antibiotic concentrations, cells would be rapidly lysed, releasing all the LPS present on a fixed cell population. This would dramatically reduce the potential for differential LPS release regardless of antibiotic class or subclass. Besides, with the exception of imipenem, most β-lactam antibiotics are rather inactive against slow or nongrowing bacteria (7) that would normally exist in high initial inoculum studies ($>10^8$ CFU). Finally, bacteria highly sensitive to serum may be readily lysed by complement, which also makes it difficult to distinguish differential antibiotic-induced LPS release, whereas bacteria highly sensitive to antibiotics may liberate only minimal endotoxin when exposed to most antibiotics. In the latter case, such low amounts of free LPS (although it may actually be released in differential amounts) would have little or no impact on host survival provided these low/harmless levels are below the threshold LPS sensitivity of the host.

ANTIBIOTIC CLASS AND SUBCLASS DIFFERENTIALLY REGULATE ENDOTOXIN RELEASE

To date, all classes of antibiotics studied, including aminoglycosides, quinolones, macrolides, other bacteriostatic antibiotics, and especially the cell wall active β-lactams, have been shown to induce release of at least some endotoxin. Antibiotic-induced endotoxin release is, therefore, simply a matter of magnitude. Although β-lactams, in general, may indeed induce release of more endotoxin than the other classes, disparity in the ability to release endotoxin from gram-negative bacteria exists among and within the β-lactam subclasses as well, including the penicillins, cephalosporins, monobactams, and carbapenems. The antimicrobial cell wall actions of all β-lactams are mediated through their binding affinities for PBPs important in outer cell wall (peptidoglycan) biosynthesis. Subtle differences in their relative specificities for the few critical gram-negative associated PBPs (PBP-1, -2, and -3) important for growth, shape, and lysis have been shown to be responsible for the ability of β-lactams to differentially induce release of endotoxin from gram-negative bacteria (8–16).

PBP BINDING AFFINITY AFFECTS BACTERIAL MORPHOLOGY AND ENDOTOXIN RELEASE

The above-cited β-lactam antibiotic subclasses differentially induce various PBP-specific morphologies that have been associated with the potential to release endotoxin (8–17). For example, penicillins, cephalosporins, and monobactams, in general, all primarily inhibit the PBP-3 enzyme, which is responsible for rod septation. As a result of this inhibition, bacterial rods continue to grow individually and produce endotoxin as elongated cells or filaments prior to lysis, as shown in Figures 1C and 2B–D, respectively, for *P. aeruginosa* and *E. coli*. While only a minimal increase in CFU (if any) may occur during filamentation, the observed continued individual rod growth (increased cell mass) may occur over a wide range of concentrations that ultimately release substantially larger amounts of free LPS at low and high antibiotic concentrations, due to the high affinity of these antibiotics for PBP-3 and the relatively low affinity for PBP-1 (responsible for bacterial lysis). Carbapenems (e.g., imipenem and meropenem), on the other hand, may bind more readily to the PBP-2 enzyme, thereby converting rod-shaped bacteria into round cells. These oval cells/spheroplasts are unable to divide or to continue producing endotoxin. PBP-2–specific imipenem induces spheroplast formation exclusively with both *P. aeruginosa* (Fig. 1B) and *E. coli* (Fig. 3F and I) over a wide range of concentrations, which are subsequently rapidly destroyed due to the additional high binding affinity of imipenem for PBP-1. Another carbapenem, meropenem, unlike imipenem, possesses high binding affinities for PBP-2 and -3 of *P. aeruginosa* and may induce a combination spheroplast and filament morphology on the same bacterial rod (Figs. 1D and 4B–D). The primary affinity of meropenem in *P. aeruginosa* is, however, for PBP-3 as

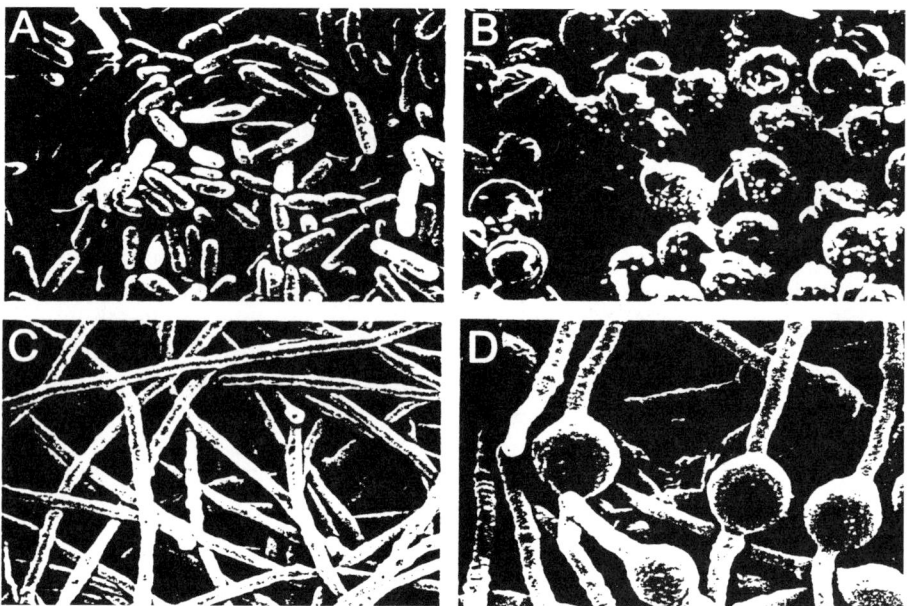

Fig. 1 Scanning electromicrographs of *P. aeruginosa* MB3286. (A) No antibiotic; (B) spheroplasts induced by imipenem at 2 μg/ml (2 × MIC) for 3 hours; (C) filaments induced by ceftazidime at 4 μg/ml (2 × MIC) for 7 hours; and (D) combination filaments containing spheroid-like segments on the same rods induced by meropenem at 2 μg/ml (2 × MIC) for 6 hours. (Magnification 10,000×, bar = 1 μm.)

illustrated by filament induction at low sub-MIC concentrations in Figure 2A. At much higher meropenem inhibitory concentrations, the PBP-2 enzyme of *P. aeruginosa* is saturated, and spheroplasts are formed reminiscent of those shown for imipenem treatment (Fig. 4E and F).

INFLUENCE OF DIFFERENT BACTERIAL SPECIES ON THE SPECIFICITY/AFFINITY OF PBP BINDING OF THE SAME ANTIBIOTIC

Antibiotic Choice

Unlike many representative antibiotics of the other β-lactam subclasses (including other carbapenems, (e.g., imipenem), the primary PBP-binding affinity of meropenem may change in accordance with the bacterial species. For example, *E. coli* exposed to meropenem (even at sub-MICs) do not form filaments (as shown with *P. aeruginosa*) at any antibiotic concentration evaluated but form somewhat imipenem-like spheroplasts at very low and very high concentrations (Fig. 3D, E, G, and H). Thus, the primary PBP-binding affinity of meropenem against *E. coli* is for PBP-2 rather than for PBP-3 (as with *P. aeruginosa*). Although meropenem still demonstrates high dual PBP-binding affinity with *E. coli*, the binding capacity of meropenem for PBP-2 and -3 appears to be closer in binding affinity than that observed with *P. aeruginosa*. These closer binding affinities for PBP-2 and -3 are demonstrated by the somewhat elongated "obese sea lion" bacterial shapes of *E. coli* that are observed at intermediate meropenem concentrations (Fig. 3E and G). These unusual *E. coli* morphologies are quite large in comparison to spheroplasts associated with imipenem (Fig. 3F). Unpublished LPS release data show higher release of endotoxin from these larger meropenem-treated cells than amounts released by smaller (cherry-shaped) imipenem-induced spheroplasts (18). The level of antibiotic-liberated endotoxin is thus dependent upon the choice of antimicrobial agent among and within the β-lactam subclasses even for different bacterial species (e.g., *P. aeruginosa* and *E. coli*).

Antibiotic Dose/Concentration

Morphological variations and LPS release are associated with changes in antibiotic concentrations—especially at subinhibitory antibiotic levels where high amounts of endotoxin are also released (8–16). Pharmacokinetic studies in humans suggest that concentrations of important β-lactam antibiotics (i.e., imipenem, meropenem, and ceftazidime) exist in vivo below the

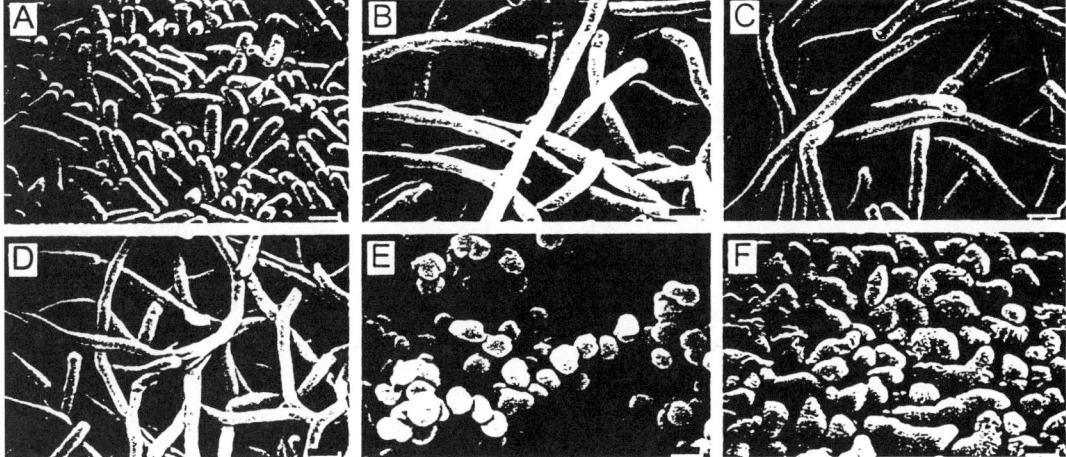

Fig. 2 Scanning electromicrograph of *E. coli* KN126. (A) No antibiotic; filaments induced by ceftriaxone at (B) high (4 × MIC, 8 hours) and (C) low (0.25 × MIC, 8 hours) concentrations; (D) filaments induced by ceftazidime at 0.25 × MIC, 8 hours; (E) spheroplasts induced by imipenem 0.25 × MIC, 8 hours, and (F) large obese cells induced with meropenem at 0.25 MIC, 8 hours. (Magnification 10,000×, bar = 1 μm.)

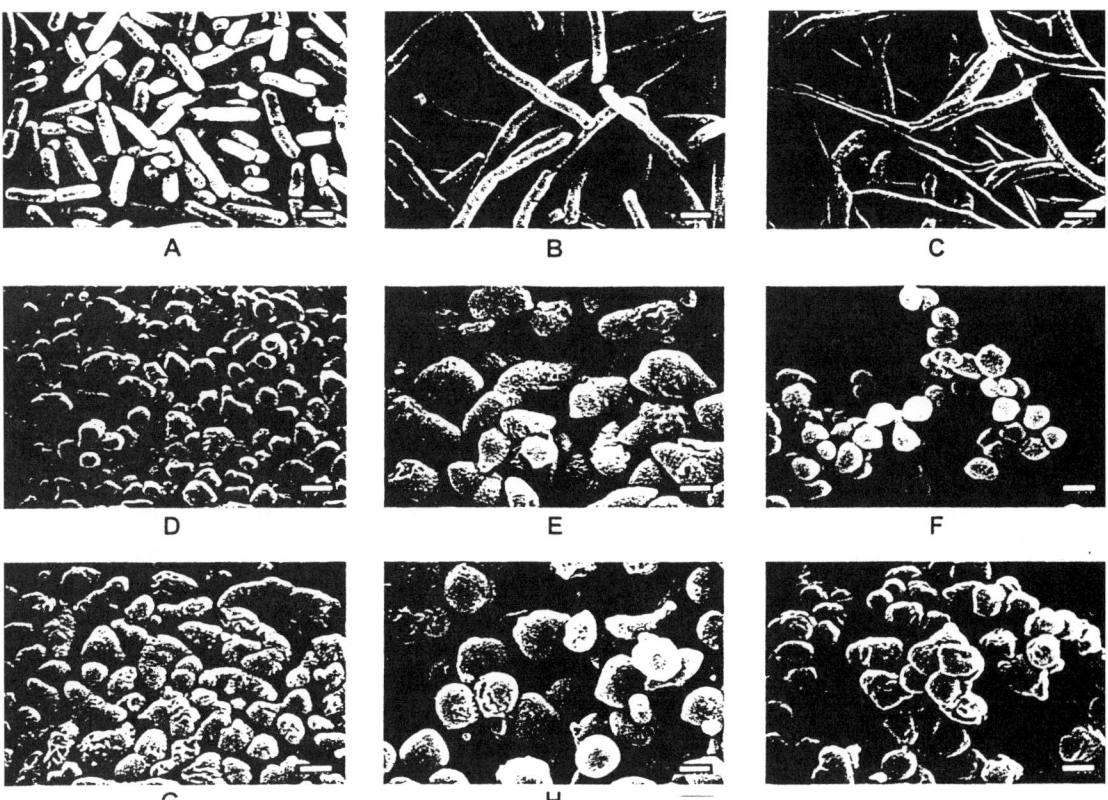

Fig. 3 Scanning electromicrographs of *E. coli* KN126. (A) No antibiotic; ceftazidime at (B) 0.125 and (C) 0.25 μg/ml; meropenem at (D) 0.0625 μg/ml, 0.125 × MIC, (G) 0.125 μg/ml, 0.25 × MIC, (E) 0.25 μg/ml, 0.5 × MIC and (H) 0.5 μg/ml, 1 × MIC; and imipenem at (F) 0.125 μg/ml, 0.25 × MIC and (I) 0.25 μg/ml, 0.5 × MIC. (Magnification 10,000×, bar = 1 μm.)

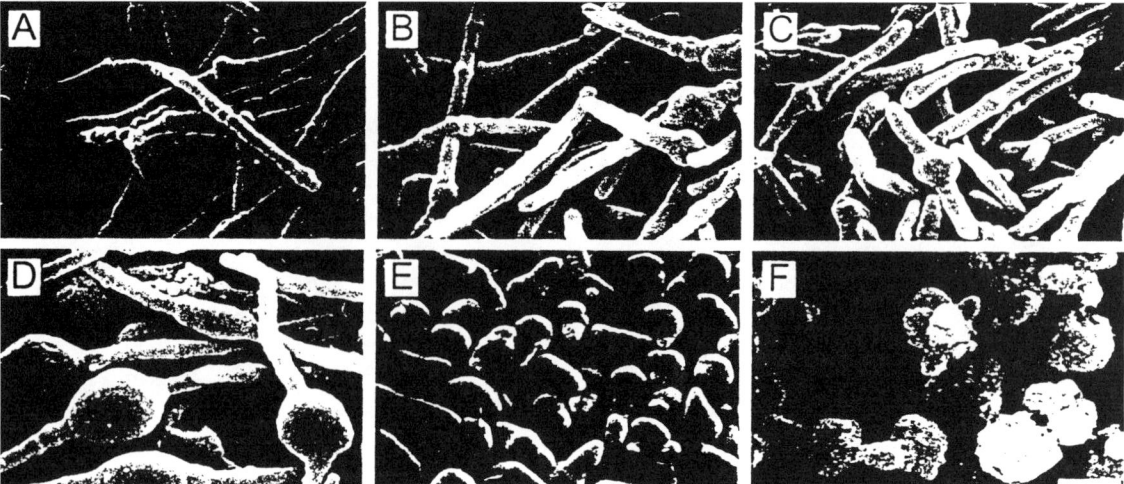

Fig. 4 Scanning electromicrographs of *P. aeruginosa* MB3286 expressed to meropenem at (A) 0.5 × MIC (6 hours); (B) 1 × MIC (4 hours); (C) 2 × MIC (2 hours); (D) 2 × MIC (6 hours); (E) 4 × MIC (2 hours); (F) 10 × MIC (2 hours). (Magnification 10,000×, bar = 1 μm.)

effective inhibitory concentrations from one third to one half of the suggested dosing schedules (8,19–21).

Therefore, studies investigating the relative potential of antibiotics to release endotoxin should be conducted at both subinhibitory and inhibitory antibiotic concentrations, as long as evaluations are conducted at clinically achievable levels. For humans, the recommended clinically achievable single doses of β-lactam antibiotics range from 0.5 to 2.0 g/dose (∼7.5–30 mg/kg) (19–21) and not the very high antibiotic amounts (100–400 mg/kg = 7–28 g/dose) reported in some publications. Although the primary objective, indeed, is to kill the invading bacteria as rapidly as possible to minimize increased cell mass and endotoxin release, antibiotic levels should not exceed acceptable standard levels of safety. Use of nonsafe antibiotic doses as high as ≥100 mg/kg may give the illusion that all antibiotic treatments (regardless of class) induce release of less endotoxin than is spontaneously liberated by non–antibiotic-treated controls (regardless of the differential amounts released). On the other hand, published data do show that certain highly active antibiotics, such as aminoglycosides and certain carbapenems (both used at clinically achievable doses), may indeed induce less endotoxin release than is spontaneously released (5,8,10). Ironically, while it is, indeed, most important to use (safe) clinically achievable antibiotic doses, even the misuse of high antibiotic concentrations (≥100 mg/kg) in experimental animal studies inadvertently confirmed the idea that high amounts (although inappropriate) of antibiotic may be more appropriate than much reduced amounts in limiting the release of endotoxin from more resistant organisms. It is simply a matter of choosing the correct antibiotic at the appropriate concentration.

In addition, investigators should also consider that antibiotics not only have affinity for bacteria but may bind irreversibly to plasma proteins and other tissues. Therefore, the actual amount of bioavailable (free) drug necessary for efficacy against the infecting pathogen may be quite different from the amount administered. The percent of free β-lactam antibiotic may range from as high as 95–100% to as low as 3–5% and should be taken into account (19–21).

Other Antibiotic Factors that may Influence Detectable Endotoxin/Cytokine Levels

Several other factors associated with cell mass and detectable levels of endotoxin released by antibiotic treatment of bacteria include the number of antibiotic treatments (single vs. multiple doses), route of administration (local vs. systemic), method of administration (bolus vs. continuous infusion), timing of treatment (early vs. late intervention), and length/duration of intervals (+6, +8, or +24 hr) between antibiotic treatments.

Local drug administration may result in local release of higher LPS/cytokine levels due to compartmentalization. Systemic drug administration may, on the other hand, result in the release of higher plasma endotoxin/cytokine levels. This latter statement may not necessarily be true since some evidence suggests that plasma cytokine levels may be just the tip of the iceberg and that membrane-associated cytokine levels may be far more predictive of the actual amount of LPS released. This factor alone may account for a lack of predictable plasma cytokine levels in studies with established infections. Continuous infusion of antibiotics leads to higher sustained levels of antibiotics above that of bolus treatments (22), while higher antibiotic concentrations may result in a more rapid reduction in bacterial biomass, which would leave fewer organisms from which to release LPS. Early antibiotic intervention would manifest as an effect similar to that observed with continuous infusion (i.e., maintaining higher antibiotic concentrations). The length or duration of antibiotic treatment not only affects cell mass, but also sustains LPS release, which in turn promotes production of higher cytokine levels.

The number of antibiotic treatments not only affects reductions in cell mass, but may also regulate detection of LPS and cytokines in plasma (especially in small laboratory animals). Pharmacokinetics is a valuable tool for deciding the frequency and dose of antibiotic treatment, which may differ dramatically among the animal species and with respect to the particular antibiotic. Pharmacokinetic data, however, are not available for a large number of commonly used antibiotics in many animal species that are used as animal models of infection, and thus investigators are left without specific guidelines regarding the number and frequency of appropriate antibiotic treatments. Reports from single and multiple antibiotic treatments of laboratory animals with the same antibiotic nonetheless show that multiple treatments with some antibiotics may protect infected animals better than single (1×) treatment with the same antibiotic through prolonging higher antibiotic concentrations. Multiple antibiotic treatments may, therefore, be ideal for rapidly inhibiting bacterial growth in host-survival studies. Multiple dosing, on the other hand, complicates the interpretation of the relative effects antibiotics may have upon LPS release, bacterial growth/morphology, cytokine production, and even the relative antibiotic efficacy that would normally follow single antibiotic treatment prior to the next antibiotic dose. Single dosing is, therefore, recommended for LPS-release studies, or, in the case of multiple antibiotic treatments in larger animals, samples for endotoxin and cytokine detection should be taken prior to the next antibiotic treatment (8).

DIFFERENCES IN SENSITIVITY TO ENDOTOXIN AND CYTOKINE PRODUCTION

Different animal species may vary widely in their sensitivity to endotoxin and perhaps also in the production of cytokines in response to LPS. Humans are among the most LPS-sensitive species, while rats appear to be among the most LPS-resistant. The chronology of a variety of animal species, listed in a decreasing order of sensitivity to LPS, is suggested to be as follows: human, horse, *rabbit, dog, swine, guinea pig, hamster, Rhesus monkey,* mouse, and rat. The chronology of the species in italic type is the result of speculation, because no comparative data that delineate the hierarchy of LPS sensitivity of these species are available. However, more detailed studies conducted with a variety of mouse strains show a wide range of sensitivity to endotoxin. Mouse strains also listed in decreasing order of sensitivity to LPS and TNF production are as follows: DDY, ICR, CD1, DBA/2, DDD, C57Bl/6J, A/J, C3H/HeN Crj, BALB/c, and C3H/HeJ (23).

While rats are indeed more resistant to the lethal effects of LPS and even produce less TNF in response to LPS than do mice, disparity in LPS sensitivity also exists even within the more resistant rat species. For example, Wistar rats appear to be far more resistant to LPS than Sprague-Dawley rats (23). The relevance of host sensitivity to endotoxin and differential antibiotic-induced endotoxin levels may well be determined in studies utilizing a variety of LPS-sensitive and -resistant animals of the same species (e.g., rats or mice). It is of interest, therefore to consider studies conducted in separate laboratories under very similar experimental conditions to evaluate the relative antibiotic-induced endotoxin-liberating potential of various antibiotic classes and subclasses in different rat strains infected with nearly identical numbers of *E. coli*. Surprisingly, results in one laboratory showed a 1000-fold difference in the absolute amount of detectable LPS (ng/ml) in the plasma of Sprague-Dawley rats (11) compared to endotoxin levels (pg/ml) detected in a second laboratory utilizing Wistar rats (24), although similar rates of reduction in *E. coli* CFU were observed following similar antibiotic treatments.

Despite the wide disparity in detectable levels of LPS (ng vs. pg), both studies independently documented that PBP-3–specific cephalosporins liberated more free endotoxin than PBP-2–specific carbapenems. It would, however, have been of interest to compare the relative differential effects (if any) of ng and pg amounts of endotoxin released from the respective

*E. coli*s upon the survival of the different rat strains. Unfortunately, both studies were terminated after only 6 hours, and no lethality data were reported. The lack of data also made it impossible to determine the potential significance of differential antibiotic-induced endotoxin released by treatment with various subclasses of antibiotics using the same rat strains. In the absence of animal survival data, one can only speculate that differential LPS release may only become relevant when the amount of free endotoxin reaches a quantitative or bioreactive level above threshold amounts tolerated by the individual host regardless of animal or bacterial strains/species. To avoid speculation, it would be of value to carry out similar studies in a single laboratory in which both rat strains are challenged with a variety of bacterial doses (including lethal infection levels) of the same isolate followed by a variety of identical antibiotic therapeutic treatments (peak, break point, and sub-MIC amounts of antibiotics). In addition to monitoring LPS, TNF, and CFU levels and survival of infected rats over at least a 24-hour period, comparative endotoxin lethality and its plasma clearance rates should be determined in both rat strains following injection with various amounts of free endotoxin.

Animal variations also exist among and within species with respect to susceptibility to bacterial infections. In fact, the age (young vs. mature) of the animal also influences its susceptibility to infection but not in a predictable manner. For example, rabbits become more sensitive with age, while guinea pigs and chickens become more resistant with age. Even the presence or absence of gut flora may influence sensitivity to LPS. Mice, for example, that harbor gram-negative bacteria are more sensitive to LPS than germ-free mice (25).

The immune status of the host is perhaps among the most critical variables with respect to bacterial infections and sensitivity to endotoxin. Immunosuppression of animals may be due to genetics, or it may be induced via chemicals (cyclophosphamide, D-galactosamine, actinomycin D, anti-WBCs, etc.) or through stress, both physical (surgery, burns, starvation, travel, etc.) or mental (loud or sustained noise, death in family, divorce, marriage, buying/selling home, etc.). Thus, both the sensitivity of the host to endotoxin and bacterial infection may influence the effects that antibiotic treatment has upon survival. Unfortunately, the diversity in sensitivity to endotoxin illustrated by the various animal species exists among individual humans as well.

CONCLUSION

In summary, the infecting organism, the choice of antibiotic, and the sensitivity of infected host may all influence the detectable plasma levels of endotoxin and cytokines and subsequently influence host outcome. Similar heterogeneity in LPS-induced cytokine production exists between humans and animals, and thus harmful LPS/cytokine endpoints predictive of patient outcome generally would not be readily anticipated. The relevance of differential antibiotic-induced endotoxin release thus depends on the magnitude and bioreactivity of endotoxin as well as on the individual LPS sensitivity of the host. In order to prevent the release of possible harmful LPS levels into the circulation of the host, empirical antibiotic therapeutic strategy should include early intervention with fast-killing antibiotics that release minimal endotoxin and administered at maximum (but safe) concentrations as frequently as is safe. Conjunctive therapy with an "excess" of anti-LPS/anticytokine agent(s) to reduce/neutralize endotoxin should also be administered with a carefully selected antibiotic in cases of severe infection.

ACKNOWLEDGMENTS

We thank Charles J. Gill, Jon Sundelof, and Sandra L. Hobbis, respectively, for in vitro, in vivo studies, pharmacokinetic evaluations, and slide/graphic preparations, respectively, and Solomon Scott, Douglas W. Kawka, and Irwin I. Singer for electron micrographic preparations and analysis.

REFERENCES

1. Brandtzaeg P, Kierulf P, Gaustad P, et al. Plasma endotoxin as a predictor of multiple organ failure and death in systemic meningococcal disease. J Infect Dis 1989; 159:195–204.
2. Opal SM. Lessons learned from clinical trials of sepsis. J Endotoxin Res 1995; 2:221–226.
3. Anderson BM, Solgberg O. Release of endotoxin from *Neisseria meningitidis*: a short survey with a preliminary report on virulence in mice. Natl Inst Public Health Ann 1980; 3:49–55.
4. Eng RH, Smith SM, Fan-Harvard P, Ogbara R. Effects of antibiotics on endotoxin release from gram-negative bacteria. Diag Microbiol Infect Dis 1993; 16:185–189.
5. Simon DM, Koenig G, Trenholme GM. Differences in release of tumor necrosis factor from THP-1 cells stimulated by filtrates of antibiotic killed *Escherichia coli*. J Infect Dis 1991; 164:800–802.

6. Stevens DL, Yan S, Bryant AE. Penicillin-binding protein expression at different growth stages determines penicillin efficacy in vitro and in vivo: an explanation for the inoculum effect. J Infect Dis 1993; 167:1401–1405.
7. Tuomanen E, Tomasz A. Induction of autolysis in nongrowing *Escherichia coli*. J Bacteriol 1986; 167:1077–1080.
8. Jackson JJ, Kropp H. Differences in mode of action of β-lactam antibiotics influence morphology, LPS release and in vivo antibiotic efficacy. J Endotoxin Res 1996; 3:201–218.
9. Morrison DC, Bucklin SE, Leeson MC, Norimatsu M. Contribution of soluble endotoxin released from gram-negative bacteria by antibiotics to the pathogenesis of experimental sepsis in mice. J Endotoxin Res 1996; 3:237–244.
10. Dofferhoff ASM, Buys J. The influence of antibiotic-induced filament formation on the release of endotoxin from gram-negative bacteria. J Endotoxin Res 1996; 3:187–194.
11. Opal SM, Horn DL, Palardy JE, Jhung J, Bhattacharjee A, Young LD. The in vivo significance of antibiotic-induced endotoxin release in experimental gram-negative sepsis. J Endotoxin Res 1996; 3:245–252.
12. Nakano M, Kirikae T. Biological characterization of *Pseudomonas aeruginosa* endotoxin released by antibiotic treatment in vitro. J Endotoxin Res 1996; 3:195–200.
13. Arditi M, Zhou J, Huang SH. Antibiotic induced bacterial killing activates vascular endothelial cells and whole blood cells. Role of free lipopolysaccharide and soluble CD14. J Entodoxin Res 1996; 3:179–186.
14. Rokke O, Sveen G, Lehmann AK, Digranes A, Halstensen A. Endotoxin and cytokines in plasma and clinical manifestations during treatment of septicemia with imipenem or cefuroxime/metronidazole. Immunol Infect Dis 1996; 6:159–166.
15. Mock CN, Jurkovich GJ, Dries D, Maier RV. The clinical significance of endotoxin released by antibiotics; what is the evidence. J Endotoxin Res 1996; 3:253–260.
16. Yokochi T, Kusumi A, Kido N, Kato Y, Sugiyama T, Koide N, Jiang G-Z, Narita K, Takahashi K. Differential release of smooth-type lipopolysaccharide from *Pseudomonas aeruginosa* treated with carbapenem antibiotics and its relation to production of tumor necrosis factor alpha and nitric oxide. Antimicrob Agents Chemother 1996; 40:2410–2412.
17. Whelen AC. *E. coli* damaged by antibiotics. ASM News 1993; 59:225.
18. Norimatsu M, Morrison DC. Correlation of antibiotic-induced endotoxin release and cytokine production in *Escherichia coli*-inoculated mouse whole blood ex vivo. J Infect Dis. In press.
19. Drusano GI, Standiford HC, Bustamante C, et al. Multiple dose pharmacokinetics of imepenem-cilastatin. Antimicrob Agents Chemother 1984; 26:715–721.
20. Bax RP, Bastain W, Featherstone A, Wilkinson DM, Hutchinson M, Haworth SJ. The pharmacokinetics of meropenem in volunteers. J Antimicrob Chemother 1989; 24(suppl A):311–320.
21. Harding SM, Monro AJ, Thornton JE, Ayrton J, Hogg MIJ. The comparative pharmacokinetics of ceftazidime and cefotaxime in healthy volunteers. J Antimicrob Chemother 1991; 8(suppl B):263–272.
22. Dofferhoff ASM, Buys J. The influence of antibiotic-induced filament formation on the release of endotoxin from gram-negative bacteria. J Endotoxin Res 1996; 3:187–194.
23. Haranaka K, Satomi N, Sakurai A. Differences in tumor necrosis factor productivity ability among rodents. Br J Cancer 1984; 50:471–478.
24. Nitsche D, Schulze C, Oesser S, Dalhoff A, Sack M. Impact of different classes of antimicrobial agents on plasma endotoxin activity. Arch Surg 1996; 131:192–199.
25. Milner KC, Rudbach JA, Ribi E. General characteristics. In: Weinbaum G, Kadis S, Ajl SJ, eds. Microbiol Toxins, Vol IV. Bacterial Endotoxins. New York: Academic Press, 1971:1–65.

6

Complement-Mediated Lipopolysaccharide Release

Vernon L. Tesh
Texas A&M University Health Science Center, College Station, Texas

INTRODUCTION

Activation and Activating Surfaces

The complement system is comprised of a series of glycoproteins found circulating in the blood in inactive precursor forms. Activation of complement involves a sequential cascade of enzymatic (primarily proteolytic) reactions that produces the active forms of the complement proteins. The general formula for complement activation is X → Xa + Xb, where a is the smaller molecular weight fragment. Initiation of the complement cascade requires the presence of an activating surface that allows the formation of stable enzyme complexes called C3 and C5 convertases. The stable deposition of these multicomponent enzyme complexes on bacterial surfaces may then trigger the assembly of the membrane attack complex (MAC) composed of terminal complement components C5b through C9 (1,2). The MAC is capable of killing some bacteria targeted by complement activation. Killing mechanisms may include the formation of membrane-spanning pores and a detergent-like membrane perturbation effect (3,4). Activating surfaces may require prior coating of the surface with antibody molecules in order to activate complement or may directly "fix" or cross-link complement proteins to the surface, as is the case for some lipopolysaccharide (LPS) molecules found in gram-negative bacterial outer membranes.

Complement activation proceeds through two distinct mechanisms (Fig. 1). The classical pathway of complement activation utilizes complement proteins: C1 complex ($C1qr_2s_2$), C2, C3, and C4. Antibody-dependent classical pathway activation requires the C1q subunit of the C1 complex to recognize and bind multiple Fc regions of antigen-bound IgG molecules or a single antigen-bound pentameric IgM molecule. C1q also binds directly to gram-negative bacteria in the absence of antibodies, particularly to bacteria expressing deep rough LPS. C1q binding may result in the activation of the two C1 complex serine proteases, C1r and C1s. These proteases cleave the serum proteins C4 and C2. A fragment of C4, designated C4b, covalently binds to the activating surface, and C2a, a split product of C2 that is also a serine protease, binds to C4b. The surface-bound C4b2a macromolecular complex is a stable C3 convertase, catalytically cleaving C3 molecules and focusing C3b binding onto the activating surface. C4b2a is an efficient enzyme; millions of molecules of C3b may be covalently bound to activating surfaces following classical pathway activation (1,2). As will be discussed below, the ability of the host to "pepper" microorganisms with covalently bound C3b molecules has important biological implications.

The alternative pathway of complement activation involves the interaction of C3 with Factor B, Factor D, and properdin (Fig. 1). C3 is the most abundant complement protein in the blood, and at any given time a fraction of the C3 molecules undergoes metabolic turnover by a spontaneous conformational change that exposes an intrachain thiolester bond ($-S-C=O$) between cysteine and glutamyl residues (5,6). The thiolester bond is extremely labile, its half-life mea-

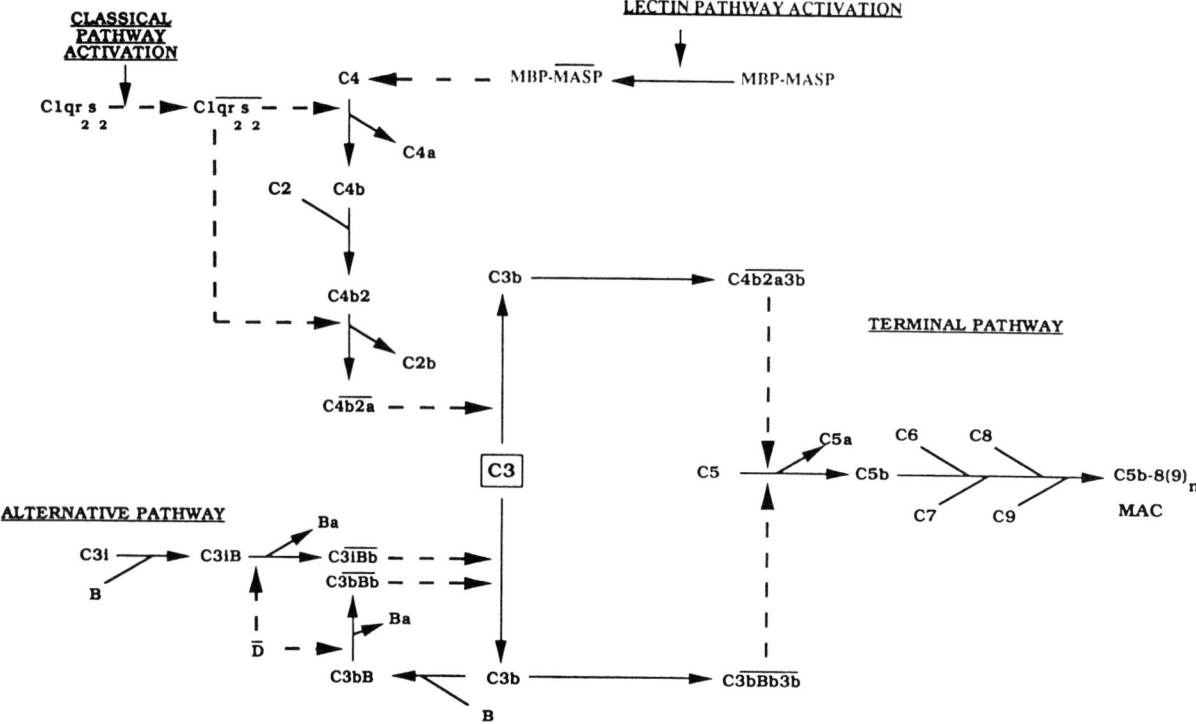

Fig. 1 Activation steps in the classical and alternative pathways of complement. (From Ref. 2.)

sured in μsec, and is subjected to nucleophilic attack by water to form $C3(H_2O)$ or C3bi. Factor B may bind to $C3(H_2O)$, and after cleavage of Factor B by Factor D, the $C3(H_2O)$Bb complex may serve as a fluid-phase C3 convertase (Fig. 2). However, this complex is susceptible to inactivation by Factors H and I. Factor H accelerates the disassembly of the $C3(H_2O)$Bb complex and acts as a cofactor for the serine protease Factor I, which cleaves C3b and renders it inactive in terms of complement activation. In the presence of an activating surface, such as LPS, $C3(H_2O)$Bb-derived C3b molecules may form amide or ester linkages with amino or hydroxyl side groups found on the activating surface. The covalent cross-linking of C3b to a surface favors the interaction of C3b with Factor B and properdin (a C3bB stabilizer) and prohibits the binding of Factor H. The subsequent cleavage of C3b-bound Factor B by Factor D produces the C3bBb complex, a stable membrane-bound C3 convertase (half-life measured in min) that will catalytically cleave C3 and produce millions of membrane-bound C3b molecules (1,2,5). Antibody-dependent activation of the alternative pathway has been characterized and probably involves the binding of C3b to carbohydrates on the Fab portion of certain IgG subclasses.

At this stage in the complement cascade, the classical and alternative pathways merge in that both C4b2a3b and C3bBbC3b complexes possess C5 convertase activity (Fig. 1). Surface-bound C3b and C4b act as ligands for C5. Bound C5 is cleaved and the membrane-bound C5b fragment serves as the initiator for the self-assembly of the MAC (7). Thus, the ability to covalently link C3b to gram-negative outer membranes is a crucial step in the complement activation pathways in that C3b "targets" the bacteria for killing by MAC or by phagocytic cells (see below).

Hepatocytes are the primary source of most serum complement proteins, although other cell types may be involved in the biosynthesis of complement (8). The cellular source of C1q remains controversial, with epithelial cells, fibroblasts, and monocyte/macrophages implicated in production. Monocytes also produce C1r, C1s, properdin, and C7. The primary source of Factor D is the adipocyte. Extrahepatic production of complement proteins, particularly proteins involved in the regulation of complement activation (e.g., Factor H) may

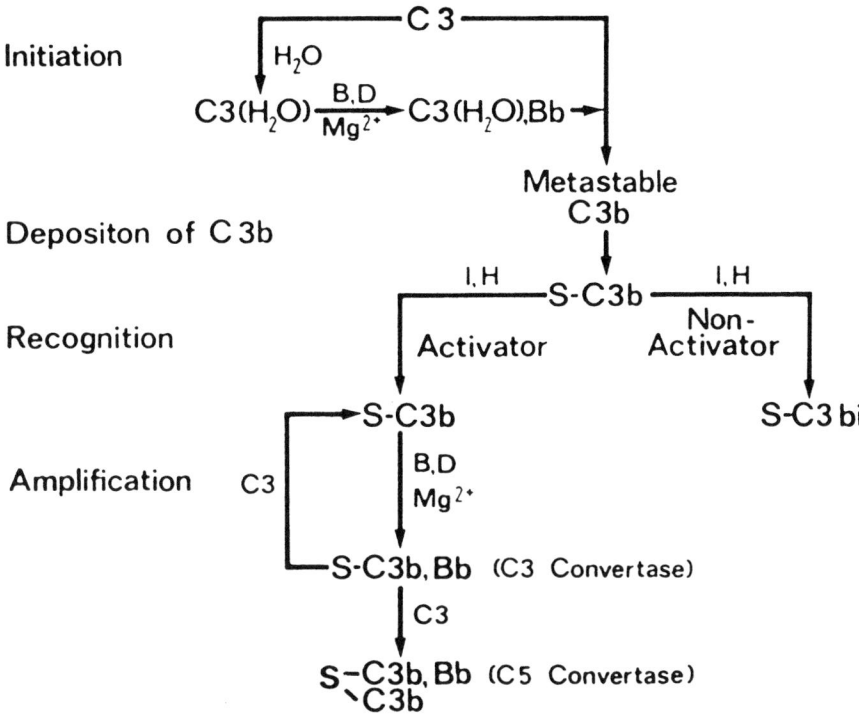

Fig. 2 Initiation of the alternative pathway of complement. (From Ref. 6.)

be critical in controlling localized damage to host tissues during an inflammatory response.

Biological Consequences of Complement Activation

Activation of the complement cascade sets in motion a series of events that (1) facilitate the extravasation of mononuclear phagocytes into areas of infection, (2) destroy invading microorganisms by opsonization or by direct MAC-mediated membrane damage, and (3) clear the blood of antigens such as bacteria and bacterial debris by routing C3b-bound antigens to the liver and spleen for ingestion by hepatic and splenic macrophages. Anaphylatoxins, the low molecular weight split products C3a, C4a, and C5a, act on mast cells and basophils to release histamine and other inflammatory mediators, increasing vasopermeability and vasodilation. C5a is also a potent chemotactic agent, facilitating the extravasation of neutrophils into inflamed tissue. Mononuclear phagocytes possess receptors for C3b and C4b and will readily bind, internalize, and degrade microorganisms coated with these complement proteins (9). Human erythrocytes also possess receptors for C3b, and erythrocyte binding to C3b-coated particulate antigens or immune complexes is important in clearing antigen from the bloodstream (1). Finally, as discussed above, under the appropriate circumstances, the MAC mediates direct bacteriolysis. Thus, although the biological properites of activated complement components are legion (Table 1), the proteins are designed to recognize and eliminate infectious agents, foreign particles, or altered self molecules.

INTERACTION OF COMPLEMENT PROTEINS WITH LPS

C1 Complex and MBP-MASP

Early investigation into complement fixation by gram-negative bacteria demonstrated that C1 will bind to lipid A and LPS isolated from rough mutants in an antibody-independent manner (10,11). C1 binding to smooth LPS was undetectable. Tenner et al. (12) demonstrated that low levels of C1q bound to smooth strains of *E. coli*, but binding did not result in classical pathway activation since C1q bound to smooth LPS was acted on by the complement regulatory protein C1-Inhibitor. In contrast, C1 binding to rough strains of *E. coli* activated the classical pathway and led to the formation of MAC and bacterial killing (13,14). C1 binding and antibody-independent classical pathway acti-

Table 1 Biological Consequences of Complement Activation

Biological activity	Complement proteins mediating effect
Degranulation of mast cells, basophils, and eosinophils	C3a, C4a, C5a
Neutrophil extravasation and chemotaxis	C5a
Platelet aggregation	C3a, C5a
Opsonization of coated particles	C3b, C4b, iC3b, C1q, MBP
Solubilization and clearance of LPS/immune complexes	C3b, iC3b
Phagocyte activation to a bactericidal state	C1q, C3b (+Ab), MAC
LPS release	MAC, C5b-8
Export of host proteins lacking signal sequences	MAC
Bacteriolysis and virus neutralization	MAC

vation usually involves the interaction of the globular heads of the C1q component with the activating surface (Fig. 3). C1q is a highly basic protein, possessing an isoelectric point of pH 9.2. Therefore, C1q will bind to polyanionic molecules via charge-dependent interactions. Zohair et al. (15) showed that purified C1q bound to the LPS of the deep rough (heptose-less) mutant *E. coli* D31m4, but binding to smooth strains of *E. coli* was minimal. Interestingly, the C1q collagen-like fragment (C1qCLF) inhibited C1q binding while purified globular heads did not affect C1q binding to the bacteria, suggesting that rough LPS may interact with multiple sites on the C1q macromolecule. Treatment of C1q and C1qCLF with phenylglyoxal, a reagent that modifies arginine residues and interrupts C1q binding to immune complexes, did not affect C1q binding to *E. coli* D31m4. In contrast, reagents that modify histidine residues blocked C1q and C1qCLF binding to the bacteria and to immune complexes. When KDO was removed from *E. coli* D31m4 LPS by acid hydrolysis, binding by C1qCLF was lost, suggesting that the interaction of cationic histidines with anionic KDO was necessary to mediate C1q binding to deep rough strains. In addition to direct C1q-LPS binding, a number of studies have shown that C1q will bind cationic outer membrane proteins from gram-negative bacteria in an antibody-independent manner (16,17).

Several investigators noted that the human mannose-binding protein (MBP) and lung surfactant proteins are structurally similar to the C1q molecule (reviewed in Ref. 18). This led to the speculation that proteins other than C1q may participate in complement activation. Human MBP directly binds to the C1 complex serine proteases C1r and C1s and can mediate activation of the classical pathway of complement. However, when isolated from serum, MBP is usually associated with a 100 kDa serine protease. This dimeric complex was originally designated Ra-reactive factor because it was shown to specifically bind rough LPS of the Ra chemotype (19). The MBP-associated serine protease (MASP) (1) shares approximately 39% amino acid homology with human C1r and C1s, (2) contains a histidine loop structure, which is highly conserved among serine proteases, and (3) can cleave C4 and C2, or may directly bind to and hydrolyze C3, to produce functional classical and alternative pathway C3 convertases (18). Thus, the MBP-MASP complex may bind *N*-acetylglucosamine-, *N*-acetylmannosamine-, and mannose-containing lipopolysaccharides synthesized by *E. coli*

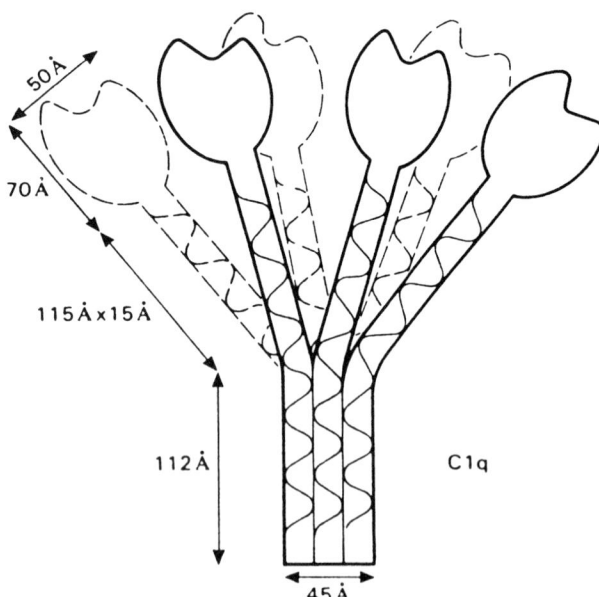

Fig. 3 Structure of C1q. The macromolecule is composed of 18 polypeptide chains (6 A-chains, 6 B-chains, and 6 C-chains). Each globular head is attached to a triple-helical collagen-like domain (wavy lines). (From Ref. 2.)

and *Salmonella* spp. and directly activate the classical and alternative pathways of complement in an antibody- and C1q-independent manner.

C3

Over 40 years ago, Pillemer and colleagues established that gram-negative bacteria elicit a serum bactericidal response in the absence of antibody (20). Subsequently, many studies were carried out confirming the initial characterization of alternative pathway activation by gram-negative bacteria (reviewed in Ref. 21). Definitive proof of antibody-independent alternative pathway activation by gram-negative bacteria was provided by Schreiber and colleagues (22), who mediated the killing of *E. coli* K12 with 11 purified alternative pathway proteins in the absence of antibody (Fig. 4). As stated above, C3b covalently binds activating surfaces via amide or ester linkages. What makes a bacterial membrane an efficient alternative pathway activator remains to be precisely explained but involves, in part, the ability of polyanionic surfaces to inhibit the interaction of C3b with the regulatory component Factor H (23). Thus, pathogenic bacteria frequently possess sialylated LPS or capsules that are poor alternative pathway activators.

MAC

Following the covalent association of the classical and alternative pathway C5 convertases on bacterial surfaces, C5 is split and C5b associates with C3b in a noncovalent manner. This surface association may induce a conformational change in the C5b molecule, revealing a hydrophobic binding domain for two closely related proteins, C6 and C7. The C5b-7 complex dissociates from the C5 convertase and undergoes a hydrophilic to amphiphilic transition, allowing the complex to integrate into the bacterial outer membrane. After C8 binding, the C5b-8 complex serves as a nidus for C9 binding. The MAC may contain 1–18 C9 molecules inserted into the lipid bilayer via the hydrophobic carboxy-terminal regions of the proteins. When more than six C9 molecules are found in the MAC, characteristic "ring-like" lesions are formed on the bacterial membrane, whereas the presence of fewer C9 proteins results in the deposition of aggregates on the surface (1,2,7). However, membrane damage may take place in the absence of formation of polymeric C9 cylinders (24). Although poly-C9 is deposited in the outer membrane of gram-negative bacteria, bacterial killing is thought to require dissipation of the membrane potential of the inner membrane (7). The precise mechanism(s) of MAC-mediated bacterial killing is incompletely understood, although the formation of poly-C9 pores allows periplasmic access to low molecular weight compounds such as [^{3}H]-proline or [^{86}Rb]. Tomlinson et al. (25) showed that the transfer of preformed MAC to the outer membranes of *Salmonella minnesota* Re595 or *E. coli* did not result in loss of bacterial viability, while exposure of the bacteria to lysozyme-free human serum killed 99% of the organisms. Recent studies suggest that the deposition of MAC on homologous host cells may also allow for the transient export of biologically active growth factors and cytokines (26). These data suggest that biochemical and membrane signaling events may occur during the C9 polymerization reaction on bacterial and host membranes that contribute to bacterial killing.

Considerable insight into the mechanism of complement-mediated bactericidal activity has come from

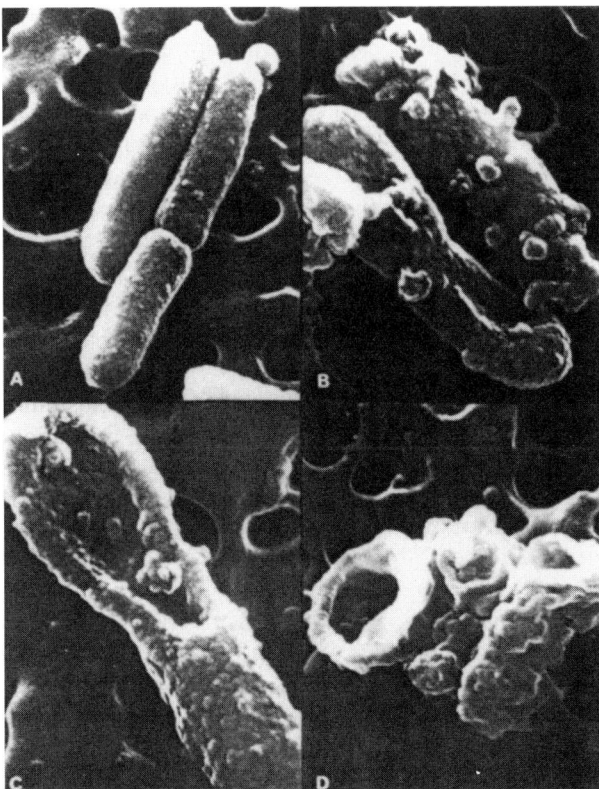

Fig. 4 Scanning electron micrographs of *E. coli* K12 treated with (A) heat-inactivated alternative pathway proteins; (B,C) purified alternative pathway proteins; and (D) alternative pathway proteins + lysozyme. Note the prominent membrane blebs apparent on complement-treated cells. (From Ref. 22.)

studies on the virulence determinants expressed by serum-resistant gram-negative bacteria. Serum-resistant Enterobacteriaceae frequently express O-antigen–rich LPS molecules, causing the MAC to form at sites distal to the outer membrane bilayer leaflet and resulting in the release of MAC (reviewed in Ref. 27). Rck (*r*esistance to *c*omplement-mediated *k*illing) is a 17 kDa outer membrane protein encoded on the virulence plasmid of *Salmonella typhimurium*. Transformation of plasmids containing the *rck* gene into rough or smooth strains of *E. coli* or *Salmonella* conferred striking increases in resistance to complement-mediated killing, with Rck$^+$ transformants showing a >6 log increase in survival in 50% normal human serum compared to non–Rck-expressing wild-type strains. The introduction of Rck did not alter the LPS chemotypes of the recipient strains. Rck neither affected the amount of C3b or C5b bound to bacteria nor altered C5 convertase activity. However, C9 was bound to Rck-expressing bacteria in a trypsin-sensitive form, and MAC released from Rck$^+$ bacteria contained C9 that was readily dissociated in SDS, properties of C9 subunits that fail to form polymeric "tubular" C9 complexes (28). If the polymerization of C9 is essential for the rapid and efficient killing of Enterobacteriaceae (29), then the expression of a bacterial outer membrane protein that prevents C9 polymerization would represent an effective virulence determinant.

LPS Structural Requirements for Complement Activation

The exquisite structural specificity of antibody-independent classical pathway activation by LPS was described by Vukajlovich (30). The classical pathway activation potential of an extensive panel of *S. minnesota* R chemotype LPS preparations was examined using Factor D–depleted human sera. Furthermore, to ensure that the classical pathway activation being measured was not mediated by immune complexes, Vukajlovich absorbed Factor D–depleted sera with formalinized rabbit erythrocytes, a treatment that removes LPS reactive "natural" antibodies from the sera. Only lipid A and the Re chemotype LPS were capable of antibody-independent classical pathway activation. The addition of a single L-glycero-D-mannoheptose residue (Rd$_2$ chemotype) inhibited lipid A–associated classical pathway activation. In contrast to classical pathway activation, Rd$_2$ chemotype LPS and LPS with more extensive core polysaccharides and O-antigen were proficient activators of the alternative pathway (31). The presence of O-antigen polysaccharides facilitates alternative pathway activation by restricting the interaction of LPS-bound C3b with Factors H and I (32).

Qualitative differences in O-antigen polysaccharide constituents also affect the anticomplementary activities of LPS. Leive and Jimenez-Lucho (33) used *Salmonella* mutants that differed solely in the expression of O-antigen polysaccharides to show an inverse relationship between lethality and phagocyte ingestion/C3b deposition; that is, the most virulent organisms were internalized more slowly by murine macrophages and fixed C3 at a slower rate. Paeng et al. (34) transformed *E. coli* K-12 strains with plasmids containing the entire rfb regions of *E. coli* 08 or 09, conferring on the rough recipient strains the ability to synthesize O-antigen consisting of homopolymeric mannose residues. The recombinant strains were approximately 50-fold more effective in activitating serum complement compared to *E. coli* 0111, a strain synthesizing O-antigen composed of heterogeneous saccharide subunits. Subtle differences in the O-antigen moiety of LPS may affect the relative affinities of Factors H and B for C3b and alter degradation rates of C3b to iC3b.

Finally, the physical conformation of purified LPS was shown to affect complement activation. Galanos and Lüderitz (35) compared the anticomplementary activity of purified smooth LPS preparations, which form high molecular weight aggregates in aqueous solution, versus the more soluble triethylamine salt forms. Using an antibody-coated sheep erythrocyte hemolytic assay, they found that high molecular weight LPS aggregates possessed the greatest anticomplementary activity. Wilson and Morrison (36) utilized phenol-water–extracted high molecular weight *Serratia marcescens* LPS preparations (sedimentation coefficients = 22S − 89S) and the respective triethylamine-solubilized LPS preparations (sedimentation coefficients = 3S − 7S) with antibody-coated sheep erythrocytes (classical pathway activation) or rabbit erythrocytes in Mg^{2+}-EGTA chelated serum (alternative pathway activation) to demonstrate that the alternative pathway of complement activation was markedly less dependent on the physical state of LPS aggregation. Thus, the physical state of LPS in solution differentially influences activation of the classical pathway rather than the alternative pathway of complement. Following the deposition of complement proteins onto gram-negative bacterial membranes, it is unlikely that liberated LPS exists in a homogeneous, high molecular weight micellar form, as is the case for chemically derived purified LPS preparations. Rather, a number of investigators have shown that LPS in serum will rapidly bind to a complex array of proteins and lipoproteins (reviewed in Ref. 37; see also Chapters 18–24, this volume). The ability of released LPS to rapidly bind serum proteins may up- or downregulate the host inflammatory response. The effect of LPS binding to specific serum proteins on complement activation, however, remains to be fully explored.

COMPLEMENT-RELEASED LPS

Role of Complement in LPS Release

Biochemical and electron microscopy studies have shown that during growth in enriched media, enteric bacteria release LPS-, phospholipid-, and membrane protein–containing vesicles into the media. However, LPS release may be markedly increased when bacteria are incubated with human serum. When viable *E. coli* J5 in which the LPS had been specifically radiolabeled by growth of the bacteria in medium containing [^{3}H]-galactose were exposed to normal human serum, the bacteria were rapidly killed, and in the process a fraction of the [^{3}H]-LPS consistently representing about 30% of the total labeled LPS was released into the serum (38). Both the classical and alternative pathways contributed to the bacteria-killing and LPS release phenomena. Neither killing nor LPS release was detected when *E. coli* were treated with C7-deficient serum. C9-deficient serum, however, mediated LPS release in the absence of bacterial killing, demonstrating that the two events could be dissociated. Optimal complement-mediated LPS release occurred during the mid-logarithmic phase of growth of the bacteria. Interestingly, EDTA treatment of the bacteria released comparable levels of [^{3}H]-LPS from *E. coli* J5, and the "releasable" fractions appeared to overlap in that sequential treatment of bacteria with normal human serum and EDTA, or vice versa, did not mediate the release of additional [^{3}H]-LPS (39,40). Although the precise mechanism(s) of LPS release remains to be elucidated, these data suggest that divalent cation chelation may be a common mechanism for both serum- and EDTA-mediated LPS release from intact *E. coli*. Finally, LPS released from bacteria by the action of serum complement or EDTA did not appear to represent a selective fraction of LPS molecules; rather, both R- and S-forms of LPS molecules were detected in serum (40,41).

Falk et al. (42) showed that shortly after mixing *E. coli* 0111:B4 LPS with rat serum, a precipitate containing LPS and serum proteins formed. While the protein content of the precipitates remained constant for 60 minutes, LPS levels declined over time. The precipitates were resolved by SDS-PAGE and bands excised for microsequencing of amino-termini. Albumin, C3b and iC3b, and IgM and IgG composed 90% of the serum proteins bound to LPS. Minor LPS-binding serum proteins included fibronectin, C1 complex, C2, C4, C5, C6, C8, Factors H and B, transferrin, properdin, mannose-binding protein, and high-density lipoproteins. Interestingly, the only major differences in LPS–serum protein interactions in normal rat serum versus serum from LPS-tolerized animals were increased binding of LPS to C3 and immunoglobulins in the tolerized rats. In contrast to naive rats, both the LPS and protein content of precipitate formed in sera from animals preexposed to LPS declined at roughly equivalent rates. The authors speculated that the rapid reversible solubilization of LPS-serum protein precipitates in tolerized animals may be related to the rapid clearance rates of aggregated LPS from the circulation. The fact that LPS released from intact bacteria by the action of serum complement may complex with other bacterial membrane constituents and rapidly associate with serum proteins may alter the biological activities of LPS when compared to purified LPS (41,43,44). Furthermore, serum proteins may prevent the aggregation of LPS into protein-free, high molecular weight micelles, a form of LPS known to activate many biological cascades.

The clinical relevance of serum complement–mediated LPS release in the development of disease is perhaps best highlighted by the pathogenesis of meningococcal meningitis. The risk of recurrent meningococcal meningitis is approximately 5000 times greater in patients with terminal complement component deficiencies compared to immunocompetent individuals. Yet, paradoxically, mortality rates from meningitis are lower in complement-deficient patients (45). Platonov et al. (46) carried out an extensive comparative evaluation of the immunologic status of normal individuals versus patients with terminal complement component deficiencies and recurrent meningitis. They found no significant differences between the two groups in absolute numbers or percentages of total lymphocytes in $CD4^+$ and $CD8^+$ T-cell, B-cell, or NK-cell populations. The concentrations of C4 and circulating immune complexes were also similar. Patients with terminal complement deficiencies displayed a significant reduction in serum IgM, IgG, and IgA titers compared to normal controls, and neutrophils isolated from complement-deficient patients expressed a marked diminution in the capacity to reduce nitroblue tetrazolium in vitro. The authors speculated that MAC may signal B lymphocytes and neutrophils, possibly through triggering an influx of Ca^{2+}, to maximize humoral immunity and phagocyte oxidative activity. A recent case report of a

C6-deficient patient with meningococcal meningitis demonstrated that following the administration of fresh frozen plasma, the patient's plasma endotoxin levels increased ~7.5-fold and plasma C6 levels increased ~50-fold. The patient's serum was incapable of releasing [^{3}H]-LPS from *E. coli* J5 and reconstitution of the serum with 25% normal human plasma or purified C6 restored LPS-releasing ability (47). Thus, patients with deficiencies in terminal complement components are susceptible to infection with *Neisseria* spp., may respond to infection with suboptimal immune responses, and their mortality rates are decreased in comparison to immunocompetent patients, perhaps because of limited "self-inflicted" tissue damage. The ability of serum complement to mediate LPS release may be an important determinant in the development of overwhelming sepsis and shock associated with meningococcal meningitis.

Role of Complement in LPS Clearance

The injection of antigens into a variety of animal species has been employed to assess the role of complement in antigen clearance. Brown et al. (48) mixed washed [^{51}Cr]-labeled rabbit erythrocytes with human IgM and C6-deficient rabbit serum, or rabbit serum depleted of C3 by treatment with cobra venom factor, to form radiolabeled EAC34 and EAC4 complexes, respectively. When injected into C6-deficient rabbits, EAC34 complexes rapidly ($t_{1/2}$ = 2–4 min), disappeared from the circulation, and were found associated with macrophages in the liver, lungs, and spleen. In contrast, approximately 90% of [^{51}Cr]-labeled EAC4 complexes were circulating in the bloodstream of C3-depleted rabbits 300 minutes after injection. Michael Frank and colleagues (49) showed that C4-deficient guinea pigs manifested normal humoral and cellular immunity, and leukocytes readily extravasated in response to foreign bodies implanted in the abdominal wall. However, the immune clearance of antibody-coated autologous erythrocytes was impaired in C4-deficient guinea pigs, and the intravenous administration of normal guinea pig serum as a source of C4 resulted in accelerated reticuloendothelial clearance.

The injection of purified LPS preparations into animals has been used to monitor the in vivo clearance of LPS from the circulation, and collectively these studies showed that LPS preparations were rapidly cleared from the bloodstream and primarily localized in hepatocytes, Kupffer cells, and splenic macrophages. LPS injection into animals was also used to examine the role of complement activation in specific pathophysiologic responses characteristic of endotoxic shock. Studies using C3-depleted murine serum (50) or splenocytes derived from C5-deficient mice (51) showed that LPS-mediated B-cell mitogenic activity occurred independent of complement activation. Curry and Morrison (52) compared serum C3 consumption and lethality in LPS-responsive (C3H/St) and LPS-hyporesponsive (C3H/HeJ) mice injected intravenously with (1) *S. minnesota* R595 LPS (primarily activates the classical pathway), (2) a high molecular weight, polysaccharide-rich form of *E. coli* O111:B4 LPS (primarily activates the alternative pathway), or (3) a form of *E. coli* O111:B4 LPS that contains relatively fewer polysaccharides and is a poor complement activator. There was no clear correlation between C3 consumption and lethality (Table 2). In both mouse strains, R595 was the most potent complement activator, yet only LPS-responsive mice were killed by LPS injection. Therefore, LPS-mediated activation of the classical and alternative pathways, resulting in extensive C3

Table 2 Lack of Correlation Between Complement Activation and Lethality

LPS	LPS-responsive mice		LPS-hyporesponsive mice	
	% C3 decrease	% lethality	% C3 decrease	% lethality
S. minnesota R595	42	100	36	0
E. coli O111:B4 (polysaccharide-rich)	27	100	20	0
E. coli O111:B4 (polysaccharide-poor)	5	100	0	0

Source: Ref. 52.

consumption, was not sufficient to induce lethality in mice.

C5-deficient mice did not develop hemorrhagic necrosis of the skin when injected subcutaneously with TNF and LPS (53). Administration of normal mouse plasma, but not heat inactivated (56°C for 30 min) plasma, restored the ability of TNF + LPS to mediate skin necrosis. Furthermore, plasma fractions containing proteins in the molecular weight range of C5a reconstituted the necrotic effect, suggesting that the generation of C5a was a critical determinant in the localized Shwartzman reaction in the mouse. Mice given TNF and LPS intravenously develop a generalized Shwartzman reaction characterized by bowel necrosis and shock. Mice congenitally deficient in C5 are protected from low dose TNF + LPS–induced shock, displaying reduced hypotension and bowel necrosis (54).

In contrast to mice, C6-deficient rabbits were shown to be more susceptible to lethality following the intravenous injection of *E. coli* LPS (55). Infusion of *Salmonella* LPS into C6-deficient rabbits resulted in the development of disseminated intravascular coagulation (DIC) and the formation of intraglomerular microthrombi, suggesting that terminal complement components may not be necessary in the elicitation of the generalized Shwartzman reaction in the rabbit (56). Ulevitch et al. (57) injected LPS into normal, C3-depleted, and C6-deficient rabbits and measured changes in hemodynamic and coagulation parameters. Regardless of complement status, all of the animals displayed roughly equivalent hypotensive changes and developed neutropenia and DIC. Compared to normocomplementemic animals, C3 depletion did not alter the rate of clearance or tissue distribution [^{125}I]-labeled *S. minnesota* R595 LPS in the rabbit (58). A fraction of the injected purified LPS associated with platelets and a fraction bound to rabbit serum lipoproteins. The rate of clearance of lipoprotein-LPS complexes from the blood was prolonged.

When Quezado et al. (59) administered LPS to normal dogs and dogs with inherited C3 deficiency, the C3-deficient dogs experienced increased endotoxinemia, more pronounced acute hypotension and diffuse vascular leak syndrome, and increased hepatic and pulmonary tissue damage compared to normal animals. Normal dogs receiving LPS displayed reductions in serum C5 levels, whereas levels of C5 in C3-deficient animals were elevated. However, C3-deficient dogs did manifest lower febrile responses to LPS, emphasizing the importance of the generation of anaphylatoxins and prostaglandins in fever induction. The clinical studies of McCabe showed no significant differences in serum C3 levels between control patients and patients with uncomplicated gram-negative bacteremia but significant differences in C3 consumption in controls versus patients in gram-negative septic shock (60). In humans, C3 consumption is a prognostic indicator of the occurrence of shock and fatal outcome. Collectively, these data suggest that, in humans and the canine shock model, an intact complement system is essential for protection against LPS-induced shock and multiple organ failure.

Recent studies utilizing intact gram-negative bacteria highlight an important caveat in the interpretation of these in vivo studies. Shaw Warren and colleagues (61) used confocal microscopy and immunofluorescent staining techniques to demonstrate that the tissue distribution of LPS following the injection of rats with live *E. coli* O111:B4 or purified *E. coli* O111:B4 LPS was markedly different. Purified LPS localized to hepatic sinusoidal Kupffer cells and hepatocytes, whereas LPS derived from intact bacteria primarily localized to tissue macrophages. Freudenberg et al. (41) used biotinylated viable *Salmonella* to show that ~15% of the total bacterial LPS was consistently released following incubation in normal human serum. A fraction of the released LPS was associated with bacterial outer membrane proteins, and this fraction was not found to associate with serum lipoproteins. These findings suggest that the results of studies utilizing purified LPS preparations to study LPS clearance and biological activities may not fully reproduce the biological properties of LPS released from bacteria in vivo.

A number of investigators have shown that treatment of serum-resistant, smooth *E. coli* with compounds that inhibit LPS biosynthesis renders the bacteria more susceptible to complement-mediated serum killing, increases the rate of neutrophil uptake of the bacteria, and increases the rate of bloodstream clearance of bacteria after injection into mice (62,63). Ohno et al. (64) used isogenic pairs of *Pseudomonas aeruginosa* and *S. typhimurium*, which differed only in their ability to synthesize rough or smooth LPS, to examine the effect of O-antigen on bacterial clearance in mice. The rough strains were rapidly cleared from the bloodstream of mice with a >2 log reduction in CFU in the blood within 10 minutes after injection. The clearance rate of the smooth strains was much longer. Furthermore, when phagocytic indices of Kupffer cells from the mice were compared, the numbers of intracellular rough bacteria were 10^2- to 10^4-fold higher than intracellular smooth bacteria. Thus, the ability of bacteria to synthesize high molecular weight O-antigen polysaccharides not only renders the organisms less sensitive to the bactericidal effects of serum, but also decreases the rate of blood clearance and decreases trapping of the bacteria by the hepatic reticuloendothelial system.

The complement system may play an important role in the reduction in mortality documented in septic patients given anti-LPS antisera. Krieger et al. (65) showed that in the presence of normal human serum, an antiendotoxin monoclonal antibody mediated the adherence of radioiodinated *E. coli* LPS to human erythrocytes and neutrophils in a time- and dose-dependent manner. LPS adherence to the cells was inhibited by heat inactivation, EDTA treatment, or Mg^{2+}/EGTA treatment of serum, suggesting that the classical pathway of complement was required for LPS-immune complex binding to the cells. Treatment of human cells with antibody to the C3b receptor CR1 (CD35) blocked the increased LPS-immune complex adherence. Thus, antiendotoxin antibodies may reduce the availability of circulating LPS by activating the classical pathway and directing LPS to Fc- and C3b-receptor–bearing cells. Pollack et al. (66) reported that monocytes in whole blood did not readily internalize fluoresceinated *E. coli* O111:B4 LPS, and treatment of whole blood with anti-CR1 antibodies did not alter monocyte uptake of LPS. However, monocyte uptake of LPS was inhibited by an anti–O-antigen monoclonal antibody, presumably by blocking the ability of LPS to bind to soluble serum proteins that facilitate LPS binding to monocyte membranes (see Chapters 18–24 of this volume). The ability of an anti–O-antigen antibody to block monocyte uptake of LPS ex vivo was, however, reduced by first lysing erythrocytes, suggesting that red blood cells may represent the major cellular source of C3b-binding receptors, and in the presence of LPS-specific antibodies and complement, erythrocytes may "outcompete" monocytes for LPS liberated from bacteria in the course of bacteriolysis. The ability of erythrocytes to route LPS to macrophages within the liver sinusoids and the spleen may be a critical determinant in regulating peripheral blood monocyte activation and the concomitant systemic production of proinflammatory cytokines.

Role of Complement in Opsonization and Phagocyte Activation

The phagocytosis of gram-negative bacteria is greatly facilitated when bacteria are coated, or opsonized, with immunoglobulin and complement proteins. This facilitated bacterial uptake is mediated by the engagement of opsonins with specific phagocyte receptors for the Fc portion of immunoglobulins and for membrane-bound C1q, C4b, C3b, and C3b degradation products iC3b and C3d,g. Animal studies using the depletion of specific complement components and clinical studies of patients with complement deficiencies have demonstrated the importance of complement in the ingestion of bacteria by phagocytic cells. For example, Noel et al. (67) showed that normal and congeneic C5-deficient mice rapidly cleared an encapsulated, serum-resistant strain of *Haemophilus influenzae* from the blood. However, mice made C3 deficient by treatment with cobra venom factor experienced prolonged bacteremia. MBP also acts as an opsonin. Kuhlman et al. (68) showed the MBP binds to *Salmonella montevideo* expressing a mannose-rich LPS and the bacteria are efficiently ingested and killed by human granulocytes.

The early studies of Tenner et al. (12) suggested that C1q binding to smooth strains of gram-negative bacteria did not necessarily result in classical pathway activation. Macrophages synthesize C1q and may incorporate endogenous C1q into the membrane. Loos et al. (69) identified membrane-bound C1q as a specific murine macrophage receptor for *Salmonella* expressing rough LPS chemotypes. Attachment of the rough strains led to endocytosis of the organisms, the generation of the oxidative burst, and release of lysosomal enzymes, interleukin-1 and prostaglandin E_2. Bacterial attachment and macrophage activation was inhibited by pretreatment of the bacteria with purified C1q or pretreatment of murine macrophages with a monoclonal anti-C1q $F(ab')_2$. Recent studies have suggested that C1q binding to smooth strains may also be a sufficient determinant for triggering the phagocyte oxidative burst (70). The oxidative burst generated by differentiated U937 cells following incubation with smooth strains of *S. typhimurium* required C3, but not C5, excluding a role for C5a-C5a receptor interaction in triggering the oxidative reaction. Heating normal human serum to 50°C for 15 minutes (altering Factor B and disrupting the alternative pathway,) did not affect *Salmonella*-induced oxidative burst. In contrast, C1q depletion of serum resulted in a marked diminution in chemiluminescence. Absorption of serum with protein A-Sepharose or smooth *Salmonella* to remove cross-reactive antibodies did not inhibit the oxidative response. Although lipid A has been shown to be involved in antibody-independent classical pathway activation, O-antigen polysaccharides were necessary for the elicitation of enhanced chemiluminescence by *Salmonella*-treated monocytic cells. Monoclonal antibodies directed against the complement receptor CR3 (CD11b/CD18) partially inhibited the induction of oxidative burst. Collectively, these data suggest that while antibody-independent binding of C1q to smooth gram-negative bacteria may not activate the complement cascade, it is sufficient to trigger the oxygen-dependent bactericidal machinery of phagocytes.

While the deposition of the MAC within the outer lipid bilayer of gram-negative bacteria may directly kill microorganisms, there is also evidence that the MAC can alter the intracellular fate of phagocytosed bacteria. Treatment of *E. coli* with nonlethal doses of nonimmune human sera increased the rate and extent of intracellular bacterial killing by rabbit neutrophils (71). Serum-mediated facilitation of neutrophil bacteriolysis required terminal complement components, as treatment of bacteria with C7- or C9-depleted sera resulted in prolonged viability of the microorganisms within the neutrophils. There appears to be a major quantitative difference between MAC lytic effects versus MAC-associated promotion of intracellular killing, with fewer than 200 MAC per bacterium sufficient to activate neutrophil bactericidal activity (72).

Although it is clear that the lysis of bacteria and release of outer membrane constituents into the bloodstream can mediate deleterious effects by the dysregulated activation of numerous biological pathways, including cytokine, coagulation, and arachidonic acid oxidation cascades, one must not overlook the putative beneficial effects of LPS liberated from the surface of bacteria in an "appropriate" fashion, i.e., a finite amount of "releasable" LPS coupled to antibodies, C3b, or serum proteins involved in host cell transmembrane signaling. The protective role of covalently binding complement proteins to LPS may be more complex than focusing the assembly of the MAC on the bacterial outer membrane, facilitating mononuclear phagocyte infiltration into infected sites, and cleaning up bacterial debris after the lysis of serum-sensitive bacteria. For example, bacteria that produce both long-chain O-antigen polysaccharides and antiphagocytic capsules are particularly virulent. Approximately 20% of all bacteremic strains of *E. coli* possess K and O antigens. Cross et al. (73) undertook a comparative analysis of the innate immune response of LPS-responsive (C3H/HeN) and LPS-hyporesponsive (C3H/HeJ) mice to encapsulated smooth *E. coli* isolates. Both responsive and hyporesponsive mice cleared the bacteria from the bloodstream within 4 hours following intravenous administration of the bacteria. Electron microscopic analyses of liver tissue from responsive mice showed that 48 hours postinjection bacteria within hepatic macrophages were undergoing degradation. In contrast, replicating bacteria were evident within the liver macrophages of the LPS-hyporesponsive mice. The inability of the LPS-hyporesponsive macrophage to carry out bacteriolysis was restored by the treatment of animals with BCG, a well-characterized activator of macrophage microbicidal activity. These data suggest, therefore, that LPS, directed to mononuclear phagocytes by opsonic antibodies and C3b, trigger signaling pathways that are crucial in establishing an activated state for killing ingested encapsulated smooth gram-negative bacteria. The mechanisms of LPS-induced phagocyte activation are complex but may involve, in part, alterations of complement receptors. Treatment of U937 cells with *Salmonella enteritidis* LPS increased the phagocytic activity of the cells for the bacteria without altering the expression of complement receptors on the cell surface (74). The LPS-induced increase in phagocytic indices was inhibited by an anti-CR3 (anti-CD11b) monoclonal antibody. The authors of this study speculated that LPS may induce a conformational change in the CD11b molecule to a phagocytosis-competent form.

Some facultative intracellular pathogens not only evade complement-mediated killing but have also evolved the capacity to utilize complement receptors to gain entry into host cells. The intracellular fate of some opsonized intracellular bacterial pathogens, such as salmonellae, may be dependent on the complement receptor engaged by the organism. CR1 (CD35)-mediated binding and internalization of *Salmonella* is associated with intracellular survival, while CR3 (CD11b/CD18)-mediated binding leads to bacteriolysis (75). The mechanism for the differential fates of CR1-bound versus CR3-bound bacteria is not known. However, earlier studies suggested that bacteria-phagocyte receptor interactions influence the granule constituents ultimately found in the neutrophil phagosome. When *S. typhimurium* were coated with IgG or C3b, radioiodinated azurophilic granule proteins were abundant in the phagosome, as were labeled FcR and C3b receptor molecules. In contrast, phagosomes containing *Salmonella* internalized by IgG-FcR interactions contained specific granule proteins, while bacteria opsonized with C3b failed to trigger the release of specific granule proteins into the phagosome (76).

Effect of LPS on Complement Production

The bactericidal effects of serum complement and the potential for complement to release biologically active LPS into the circulation have been appreciated for some time. Only more recently has it become clear that complement-released LPS may have a "feedback" effect on the synthesis of complement proteins, and this effect may be mediated, in large part, by LPS-induced cytokines. LPS is a potent inducer of TNF-α, IL-1, and IL-6 expression, and these cytokines are major activators of the hepatic acute phase response. Thus, the presence of circulating LPS may initiate a cytokine cascade that results in the transcriptional activation of genes encoding IL-6, IL-1, C-reactive protein, serum amyloid A, fibrinogen, and mannose-binding protein. Serum levels of complement proteins C3, C4, Factor B, and Factor H are also increased during the acute phase response (77). IFN-γ, a cytokine produced by lymphocytes in response to LPS-stimulated monocytes, appears to be an important regulator of complement gene expression, inducing the expression of complement proteins by some cell types and inhibiting production by others (77,78). The IFN-γ–inducing effects are augmented by TNF-α, a proinflammatory monokine synthesized in response to LPS. There may also be significant differences in the capability of macrophages derived from different tissues to be stimulated to produce complement proteins. Finally, there is increasing evidence to support the concept that the multiple isoforms of transforming growth factor-β (TGF-β) are essential regulators in the control of inflammation (79). Drouin et al. (80) recently showed that TGF-$β_2$ transcriptionally activated the C3 gene in human monocytes but suppressed C3 gene expression in cytokine-treated astrocytes. Thus, in addition to the previously characterized pleiotropic effects of LPS-induced cytokines, one can now add the tissue-specific regulation of complement protein production.

SUMMARY

The complement system is a critical component of both innate host defense and acquired immunity against pathogenic microorganisms. Complement proteins initiate inflammation, kill bacteria (and other pathogenic microorganisms), mediate the release of gram-negative outer membrane constituents from intact bacteria, and facilitate clearance of bacterial debris from the blood. Despite impressive advancements in the genetics and biochemistry of complement, the precise mechanisms of serum complement–mediated bacterial killing and LPS release remain to be clarified. Complement proteins also influence the biological activities of phagocytic cells, determine the intracellular fate of internalized bacteria, may be crucial in the successful resolution of infection, and protect against the development of septic shock and multiple organ failure. Finally, the components of the complement system have been expanded by recent studies demonstrating that lectin-binding proteins with structural similarities to C1q can also bind gram-negative bacteria, activate the complement cascade, and serve as opsonins.

ACKNOWLEDGMENT

I thank Dr. David C. Morrison for his careful reading of this chapter and his advice during the preparation of the chapter.

REFERENCES

1. Morgan BP. Complement. Clinical Aspects and Relevance to Disease. London: Academic Press, 1990.
2. Reid KBM. The complement system. In: Hames BD, Glover DM, eds. Molecular Immunology. 2d ed. Oxford: Oxford University Press, 1996:326–381.
3. Esser AF. Big MAC attack: complement proteins cause leaky patches. Immunol Today 1991; 12:316–318.
4. Bhakdi S, Tranum-Jensen J. Complement lysis: a hole is a hole. Immunol Today 1991; 12:318–320.
5. Lambris JD, ed. The Third Component of Complement. Chemistry and Biology. Berlin: Springer-Verlag, 1990.
6. Pangburn MK, Schreiber RD, Müller-Eberhard HJ. Formation of the initial C3 convertase of the alternative complement pathway. Acquisition of C3b-like activities by spontaneous hydrolysis of the putative thiolester in native C3. J Exp Med 1981; 154:856–867.
7. Esser AF. The membrane attack complex of complement. Assembly, structure and cytotoxic activity. Toxicology 1994; 87:229–247.
8. Morgan BP, Gasque P. Extrahepatic complement biosynthesis: where, when and why. Clin Exp Immunol 1997; 107:1–7.
9. Frank MM, Fries LF. The role of complement in inflammation and phagocytosis. Immunol Today 1991; 12:322–326.
10. Loos M, Bitter-Suermann D, Dierich M. Interaction of the first (C1), the second (C2) and the fourth (C4) components of complement with different preparations of bacterial lipopolysaccharides and with lipid A. J Immunol 1974; 112:935–940.
11. Morrison DC, Kline LF. Activation of the classical and properdin pathways of complement by bacterial lipopolysaccharides (LPS). J Immunol 1977; 118:362–368.
12. Tenner AJ, Ziccardi RJ, Cooper NR. Antibody-independent C1 activation by E. coli. J Immunol 1984; 133:886–891.
13. Cooper NR, Morrison DC. Binding and activation of the first component of human complement by the lipid A region of lipopolysaccharides. J Immunol 1978; 120:1864–1868.
14. Betz SJ, Isliker H. Antibody-independent interactions between Escherichia coli J5 and human complement components. J Immunol 1981; 127:1748–1754.

15. Zohair A, Chesne S, Wade RH, et al. Interaction between complement subcomponent C1q and bacterial lipopolysaccharides. Biochem J 1989; 257:865–873.
16. Mintz CS, Arnold PI, Johnson W, et al. Antibody-independent binding of complement component C1q by Legionella pneumophila. Infect immun 1995; 63:4939–4943.
17. Alberti S, Marqués G, Hernández-Allés, et al. Interaction between complement subcomponent C1q and the Klebsiella pneumoniae porin OmpK36. Infect Immun 1996; 64:4719–4725.
18. Epstein J, Eichbaum Q, Sheriff S, et al. The collectins in innate immunity. Curr Opin Immunol 1996; 8:29–35.
19. Kawakami M., Ihara I, Suzuki A, et al. Properties of a new complement-dependent bactericidal factor specific for Ra chemotype Salmonella in sera of conventional and germ-free mice. J Immunol 1982; 129:2198–2201.
20. Pillemer L, Blum L, Lepow IH, et al. The properdin system and immunity. I. Demonstration and isolation of a new serum protein, properdin, and its role in immune phenomena. Science 1954; 120:279–285.
21. Taylor PW. Non-immunoglobulin activators of the complement system. In: Sim RB, ed. Activators and Inhibitors of Complement. Dordrecht, The Netherlands: Kluwer Academic Publishers, 1993:37–68.
22. Schreiber RD, Morrison DC, Podack ER, et al. Bactericidal activity of the alternative complement pathway generated from 11 isolated plasma proteins. J Exp Med 1979; 149:870–882.
23. Meri S, Pangburn MK. Discrimination between activators and nonactivators of the alternative pathway of complement regulation by a sialic acid/polyanion binding site on factor H. Proc Natl Acad Sci USA 1990; 87:3982–3986.
24. Dankert Jr, Esser AF. Proteolytic modification of human complement protein C9. Loss of poly(C9) and circular lesion formation without impairment of function. Proc Natl Acad Sci USA 1985; 82:2128–2132.
25. Tomlinson S, Taylor PW, Luzio JP. Transfer of preformed terminal C5b-9 complement complexes into the outer membrane of viable gram-negative bacteria: effect on viability and integrity. Biochemistry 1990; 29:1852–1860.
26. Acosta JA, Benzaquen LR, Goldstein DJ, et al. The transient pore formed by homologous terminal complement complexes functions as a bidirectional route for the transport of autocrine and paracrine signals across human cell membranes. Mol Med 1996; 2:755–765.
27. Joiner KA. Complement evasion by bacteria and parasites. Annu Rev Microbiol 1988; 42:201–230.
28. Heffernan EJ, Reed S, Hackett J, et al. Mechanism of resistance to complement-mediated killing of bacteria encoded by Salmonella typhimurium virulence plasmid gene rck. J Clin Invest 1992; 90:953–964.
29. Bloch EF, Schmetz MA, Foulds J, et al. Multimeric C9 within C5b-9 is required for inner membrane damage to Escherichia coli J5 during complement killing. J Immunol 1987; 138:842–848.
30. Vukajlovich SW. Antibody-independent activation of the classical pathway of human serum complement by lipid A is restricted to Re-chemotype lipopolysaccharides and purified lipid A. Infect Immun 1986; 53:480–485.
31. Vukajlovich SW, Hoffman J, Morrison DC. Activation of human serum complement by bacterial lipopolysaccharides: structural requirements for antibody independent activation of the classical and alternative pathways. Mol Immunol 1987; 24:319–331.
32. Pangburn MK, Morrison DC, Schreiber RD, et al. Activation of the alternative complement pathway: recognition of surface structures on activators by bound C3b. J Immunol 1980; 124:977–982.
33. Leive LL, Jimenez-Lucho VE. Lipopolysaccharide O-antigen structure controls alternative pathway activation of complement: effects on phagocytosis and virulence of Salmonellae. In: Leive LL, ed. Microbiology—1986. Washington, DC: American Society of Microbiology, 1986:14–17.
34. Paeng N, Kido N, Schmidt G, et al. Augmented immunological activities of recombinant lipopolysaccharide possessing the mannose homopolymer as the O-specific polysaccharide. Infect Immun 1996; 64:305–309.
35. Galanos C, Lüderitz O. The role of the physical state of lipopolysaccharides in their interaction with complement. High molecular weight as prerequisite for the expression of anti-complementary activity. Eur J Biochem 1976; 65:403–408.
36. Wilson ME, Morrison DC. Evidence for different requirements in physical state for the interaction of lipopolysaccharides with classical and alternative pathways of complement. Eur J Biochem 1982; 128:137–141.
37. Tesh VL, Vukajlovich SW, Morrison DC. Endotoxin interactions with serum proteins: relationship to biological activity. Prog Clin Biol Res 1988; 272:47–62.
38. Tesh VL, Duncan RL, Morrison DC. Interaction of Escherichia coli with normal human serum: kinetics of serum-mediated lipopolysaccharide release and its dissociation from bacterial killing. J Immunol 1986; 137:1329–1335.
39. Leive L. Release of lipopolysaccharide by EDTA treatment of E. coli. Biochem Biophys Res Commun 1965; 21:290–296.
40. Tesh VL, Morrison DC. The interaction of Escherichia coli with normal human serum: factors affecting the capacity of serum to mediate lipopolysaccharide release. Microb Pathog 1988; 4:175–187.
41. Freudenberg MA, Meier-Dieter U, Staehelin T, et al. Analysis of LPS released from Salmonella abortus equi in human serum. Microb Pathog 1991; 10:93–104.
42. Falk MC, Fletcher MA, Williams TJ, et al. Aggregation of serum proteins with lipopolysaccharide (LPS): characterization of the precipitable LPS-protein complex. J Endotoxin Res 1996; 3:129–142.
43. Tesh VL, Morrison DC. The physical-chemical characterization and biologic activity of serum released lipopolysaccharides. J Immunol 1988; 141:3523–3531.
44. Mattsby-Baltzer I, Lindgren K, Lindholm B, et al. Endotoxin shedding by enterobacteria: free and cell-bound endotoxin differ in Limulus activity. Infect Immun 1991; 59:689–695.

45. Figueroa JE, Densen P. Infectious diseases associated with complement deficiencies. Clin Microbiol Rev 1991; 4:359–395.
46. Platonov AE, Beloborodov VB, Gabrilovitch DI, et al. Immunological evaluation of late complement component-deficient individuals. Clin Immunol Immunopathol 1992; 64:98–105.
47. Lehner PJ, Davies KA, Walport MJ, et al. Meningococcal septicaemia in a C6-deficient patient and effects of plasma transfusion on lipopolysaccharide release. Lancet 1992; 340:1379–1381.
48. Brown DL, Lachmann PJ, Dacie JV. The in vivo behavior of complement-coated red cells: studies in C6-deficient, C3-depleted and normal rabbits. Clin Exp Immunol 1970; 7:401–422.
49. Ellman L, Green I, Judge F, et al. In vivo studies in C4-deficient guinea pigs. J Exp Med 1971; 134:162–175.
50. Janossy G, Humphrey JH, Pepys MB, et al. Complement independence of stimulation of mouse splenic B-lymphocytes by mitogens. Nature (New Biol) 1973; 245:108–112.
51. Moatamed F, Karnovsky MJ, Unanue ER. Early cellular responses to mitogens and adjuvants in the mouse spleen. Lab Invest 1975; 32:303–312.
52. Curry BJ, Morrison DC. Role of complement in endotoxin initiated lethality in mice. Immunopharmacology 1979; 1:125–135.
53. Rothstein JL, Lint TF, Schreiber H. Tumor necrosis factor/cachectin induction of hemorrhagic necrosis in normal tissue requires the fifth component of complement C5. J Exp Med 1988; 168:2007–2021.
54. Hsueh W, Sun X, Rioja LN, et al. The role of the complement system in shock and tissue injury induced by tumor necrosis factor and endotoxin. Immunology 1990; 70:309–314.
55. Johnson KJ, Ward PA. Protective function of C6 in rabbits treated with bacterial endotoxin. J Immunol 1971; 106:1125–1127.
56. Müller-Berghaus G, Lohmann E. The role of complement in endotoxin-induced disseminated intravascular coagulation: studies in congenitally C6-deficient rabbits. Br J Haematol 1974; 28:403–418.
57. Ulevitch RJ, Cochrane CG, Henson PM, et al. Mediation systems in bacterial lipopolysaccharide-induced hypotension and disseminated intravascular coagulation. I. The role of complement. J Exp Med 1975; 142:1570–1590.
58. Mathison JC, Ulevitch RJ, Fletcher JR, et al. The distribution of lipopolysaccharide in normocomplementemic and C3-depleted rabbits and rhesus monkeys. Am J Pathol 1980; 101:245–263.
59. Quezado ZMN, Hoffman WD, Winkelstein JA, et al. The third component of complement protects against Escherichia coli endotoxin-induced shock and multiple organ failure. J Exp Med 1994; 179:569–578.
60. McCabe WR. Serum complement levels in bacteremia due to gram-negative organisms. New Engl J Med 1973; 288:21–23.
61. Ge Y, Ezzell RM, Tompkins RG, et al. Cellular distribution of endotoxin after injection of chemically purified lipopolysaccharide differs from that after injection of live bacteria. J Infect Dis 1994; 169:95–104.
62. Lam C, Turnowsky F, Högenauer G, et al. Effect of a diazaborine derivative (Sa 84.474) on the virulence of Escherichia coli. J Antimicrob Chemother 1987; 20:37–45.
63. Hammond SM. Inhibitors of lipopolysaccharide biosynthesis impair the virulence potential of Escherichia coli. FEMS Microbiol Lett 1992; 100:293–298.
64. Ohno A, Isii Y, Tateda K, et al. Role of LPS length in clearance rate of bacteria from the bloodstream of mice. Microbiology 1995; 141:2749–2756.
65. Krieger JI, Fletcher RC, Siegel SA, et al. Human anti-endotoxin antibody HA-1A mediates complement-dependent binding of Escherichia coli J5 lipopolysaccharide to complement receptor type 1 of human erythrocytes and neutrophils. J Infect Dis 1993; 167:865–875.
66. Pollack M, Espinoza AM, Guelde G, et al. Lipopolysaccharide (LPS)-specific monoclonal antibodies regulate LPS uptake and LPS-induced tumor necrosis factor-α responses by human monocytes. J Infect Dis 1995; 172:794–804.
67. Noel GJ, Katz S, Edelson PJ. Complement-mediated early clearance of Haemophilus influenzae type b from blood is independent of serum lytic activity. J Infect Dis 1988; 157:85–90.
68. Kuhlman M, Joiner K, Ezekowitz RAB. The human mannose-binding protein functions as an opsonin. J Exp Med 1989; 169:1733–1745.
69. Loos M, Euteneuer B, Clas F. Interaction of bacterial endotoxin (LPS) with fluid phase and macrophage membrane associated C1q, the Fc-recognizing component of the complement system. Adv Exp Med Biol 1990; 256:301–317.
70. Château M-T, Caravano R. The oxidative burst triggered by Salmonella typhimurium in differentiated U937 cells requires complement and a complete bacterial lipopolysaccharide. FEMS Immunol Med Microbiol 1997; 17:57–66.
71. Mannion BA, Weiss J, Elsbach P. Separation of sublethal and lethal effects of polymorphonuclear leukocytes on Escherichia coli. J Clin Invest 1990; 86:631–641.
72. Madsen LM, Inada M, Weiss J. Determinants of activation by complement of group II phospholipase A2 acting against Escherichia coli. Infect Immun 1996; 64:2425–2430.
73. Cross A, Asher L, Seguin, et al. The importance of a lipopolysaccharide-initiated, cytokine-mediated host defense mechanism in mice against extraintestinally invasive Escherichia coli. J Clin Invest 1995; 96:676–686.
74. Ikewaki N, Tamauchi H, Inoko H. Modulation of cell surface antigens and regulation of phagocytic activity mediated by CD11b in the monocyte-like cell line U937 in response to lipopolysaccharide. Tissue Antigens 1993; 42:125–132.
75. Ishibashi Y, Arai T. A possible mechanism for host-specific pathogenesis of Salmonella serovars. Microb Pathog 1996; 21:435–446.
76. Joiner KA, Ganz T, Albert J, et al. The opsonizing ligand on Salmonella typhimurium influences incorporation of specific, but not azurophil, granule constituents into neutrophil phagosomes. J Cell Biol 1989; 109:2771–2782.

77. Volanakis JE. Transcriptional regulation of complement genes. Annu Rev Immunol 1995; 13:277–305.
78. Williams SA, Vik DP. Characterization of the 5′ flanking region of the human complement Factor H gene. Scand J Immunol 1997; 45:7–15.
79. McCartney-Francis NL, Wahl SM. Transforming growth factor-β: a matter of life and death. J Leukocyte Biol 1994; 55:401–409.
80. Drouin SM, Carlino JA, Barnum SR. Transforming growth factor-β2-mediated regulation of C3 gene expression in monocytes. Mol Immunol 1996; 33:1025–1034.

7

Chemical Structure of Lipid A: Recent Advances in Structural Analysis of Biologically Active Molecules

Ulrich Zähringer, Buko Lindner, and Ernst T. Rietschel
Research Center Borstel, Borstel, Germany

INTRODUCTION

Lipopolysaccharides (LPS) of gram-negative bacteria are composed of three genetically and structurally distinct regions: the O-antigenic polysaccharide (O-specific chain); the core oligosaccharide; and a lipophilic portion, termed lipid A, which anchors the LPS molecule to the bacterial outer membrane (1,2). Lipid A was first described in 1954 by Westphal and Lüderitz as a hydrophobic water-insoluble precipitate obtained on acid treatment of LPS (3). The term lipid A was introduced in order to distinguish it from a further LPS-associated lipid B investigated at that time by Morgan and Partridge (4), which was later identified as phosphatidylethanolamine (5).

The preparation of largely intact (i.e., biologically active) lipid A from Enterobacteriaceae was possible because of the acid-labile ketosidic linkage between the core oligosaccharide and lipid A. This linkage is between the lipid A proximal 3-deoxy-D-*manno*-oct-2-ulosonic acid (Kdo) residue and the distal glucosamine [GlcN (II)] of the lipid A backbone [α-Kdo-(2→6)-D-GlcN (II)]. This ketosidic linkage is one of the most acid-labile glycosidic bonds found in nature. As early as 1954 it was postulated that lipid A constituted the endotoxically active region of the LPS molecule, expressing all the pathophysiological effects known to be induced by the complete LPS molecule (3). Today it is well established that lipid A bears the endotoxic principle of LPS (6). It has been assumed that parts of the core oligosaccharide may also contribute to endotoxic activities such as interleukin-1 (IL-1) release by human monocytes (7), and in fact it is now generally accepted that the core oligosaccharide may modulate the biological activity of lipid A (2,8).

From the very beginning of structural research on endotoxin, a number of research groups performed studies on lipid A obtained from various bacterial sources with the goal of defining the structural elements carrying or modifying the biological activity. These efforts were based on the idea that knowledge of the lipid A structure is an absolute prerequisite for an understanding of the molecular mechanisms operative during endotoxic effects.

During the last 5 years analytical techniques have provided unequivocal and complete sets of data for the structural analysis of lipid A and have reached new levels in understanding lipid A chemistry, especially with respect to the elucidation of the primary structure of mature lipid A, its molecular conformation and its three-dimensional organization. Here we summarize the present knowledge of all lipid A structures presently described, thus updating earlier reviews (6,9).

GENERAL STRUCTURAL FEATURES OF LIPID A

Intensive analytical investigations between the 1950s and 1970s yielded some characteristic structural ele-

---

This article is dedicated to Professor Dr. Dr. med. h.c. Otto Westphal on the occasion of his 85th birthday (February 1, 1998).

ments of the lipid A molecule, such as β-hydroxylated fatty acids, the β-(1'→6)-interlinked D-glucosamine (GlcpN) disaccharide backbone, and the partial structures of phosphorylated GlcpN (10,11). However, the correct structure of lipid A remained an enigma over three decades. One reason for this was the lack of synthetic homogeneous and well-defined lipid A preparations and the inability to prepare pure lipid A from natural sources. A second reason concerned the lack of physicochemical techniques allowing structural analyses of complex molecules like lipid A. With the advanced development of more sophisticated tools in modern analytical chemistry, the structure of lipid A from *Escherichia coli* and *Salmonella typhimurium* (later described as *Salmonella enterica* sv. typhimurium) was correctly established in 1983 (12,13). Based on these analytical data, *E. coli* type lipid A was successfully synthesized one year later (14). The finding that the synthetic material was quantitatively and qualitatively identical to bacterial lipid A by various in vitro and in vivo endotoxic activities (fever, Shwartzman, lethality) (15) finally proved the early hypothesis of Westphal and Lüderitz to be correct, thus closing the chapter of the search for the primary structure of lipid A in 1984.

The structure of *E. coli* lipid A is shown in Figure 1 and consists of a β-(1'→6)-interlinked GlcpN-disaccharide to which two phosphate residues are bound, one in ester linkage at position 4' of the distal GlcN (II) and the other being α-glycosidically linked to the proximal GlcN (I) residue, forming the lipid A backbone [4'-P-β-D-GlcpN-(1'→6)-α-D-GlcpN-(1→P]. A characteristic feature of this backbone is that two hydroxyl groups are not substituted, one in position 4 and the other in position 6', the last serving as the attachment site for the core oligosaccharide. Four (R)-3-hydroxytetradecanoic acids [14:0(3-OH)] are bound by either amide (positions 2' and 2) or ester linkage (positions 3 and 3) to the lipid A backbone. These acyl groups are termed primary fatty acids (16), distinguishing them from those ester-linked to the β-hydroxyl groups of 14:0(3-OH), which are termed secondary fatty acids. These latter are exemplified by tetradecanoic (14:0) acid being ester-linked to the 14:0(3-OH) in position 3' and dodecanoic acid (12:0) being linked to the hydroxyl group of the 14:0(3-OH) in amide linkage to position 2'. Thus, these two pairs of fatty acids form 3-acyloxyacyl acid residues, one being tetradecanoyloxytetradecanoyl [14:0(3-O(14:0))] ester-linked to position 3', the other being dodecanoyloxytetradecanoyl [14:0(3-O(12:0))] amide-linked to position 2'. *E. coli* lipid A, therefore, represents a hexaacyl lipid A, having an asymmetric fatty acid distribution (4 + 2).

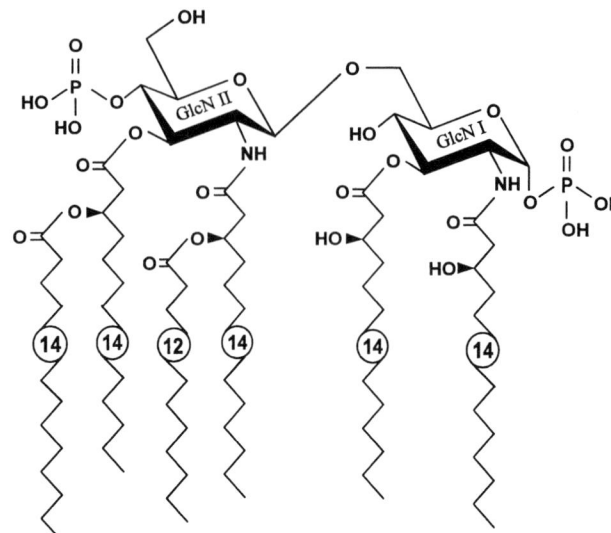

Fig. 1 Chemical structure of lipid A of the *Escherichia coli* Re mutant strain F515. The hydroxyl group at position 6' constitutes the attachment site of Kdo. The numbers in circles indicate the number of carbon atoms present in the fatty acyl chains. The 14:0(3-OH) residues possess the (R)-configuration. The glycosidic phosphate group may be substituted by a further phosphate group in non-stoichiometric amount. (See Table 1.)

The structural elements observed in *E. coli* are also present in the lipid A from other bacterial sources. Generally, the lipid A backbone disaccharide contains two *gluco*-configurated and pyranosidic D-hexosamine residues (2-amino-2-deoxy-D-glucose, GlcpN, or 2,3-diamino-2,3-dideoxy-D-glucose, GlcpN3N [also termed DAG (17,18)], which are present as a β-(1'→6)-linked disaccharide. An unusual backbone structure has been encountered in *Rhizobium leguminosarum* bv. *phaseoli*, namely a lipid A containing a β-D-GlcpN (II) and a 2-amino-2-deoxy-gluconic acid (GlcN-onic) instead of the proximal GlcN (I) in the classical lipid A backbone to yield a β-D-GlcpN-(1'→6)-D-GlcN-onic disaccharide backbone (19). It should be noted that the early steps of lipid A backbone biosynthesis in *Rhizobium* spp. appear to be comparable to those identified in Enterobacteriae (20) and that additional enzymes such as phosphatases, which are not active or not present in enterics, contribute to the observed structural differences (21). As the same structural principle (with modifications) is found in all gram-negative bacteria, it is concluded that the lipid A backbone constitutes a biosynthetically and structurally highly conserved element.

It has long been assumed that the β-D-GlcpN-(1'→6)-α-D-GlcpN disaccharide represents a unique structural feature not found in other biomolecules such as polysaccharides, glycolipids, or glycoproteins (6). Recently, however, a linear β-1,6-linked glucosaminoglycan from the biofilm-producing gram-positive bacterium *Staphylococcus epidermidis* was isolated and structurally characterized (22).

GlcpN disaccharides with a β-glycosidic linkage are also present in other natural molecules such as glycoproteins (23), bacterial murein (peptidoglycan), and chitin (24). In these cases, however, the disaccharide linkage is β-(1′→4). A lipid A partial structure expressing the β-(1′→4)-interlinked GlcpN disaccharide as lipid A "backbone" has been recently synthesized (25). These lipid A–related compounds do not contain phosphate, but exhibit some biological activity including mitogenicity and NO-inducing capacity (25).

The lipid A backbone carries two phosphate groups (at positions 4′ and 1), both of which are often substituted by nonacylated charged glycosyl or nonglycosyl residues. These polar head groups are, in general, not present in stoichiometric amounts, thus yielding a certain intrinsic heterogeneity of lipid A. Some of the substituents have been structurally elucidated, and their nature and location are summarized in Tables 1 and 2. It is noteworthy that most of the phosphate substituents such as 2-aminoethanol (Etn), 4-amino-4-deoxy-L-arabinopyranose (L-Arap4N) and GlcpN carry, at neutral and acidic pH, a positively charged amino group. Their presence in the neighborhood of phosphate residues can be considered as a regulating element, affecting the electrostatic interaction of negatively charged residues (phosphates) with bivalent cations, such as Ca^{2+} or Mg^{2+}, of the medium. In particular lipid A (*Rhizobium* or *Rhodospirillum*), these phosphate groups may be replaced by D-galacturonic acid (D-GalpA) (see Sec. III.B).

As mentioned above, the acyl groups of lipid A are present as primary and secondary fatty acids. Whereas the primary acyl groups are, with only one exception [*Chlamydia* (26)], always identical with (R)-3-hydroxy or 3-oxo fatty acids, the secondary fatty acids comprise acyl groups of various nature including unsaturated fatty acids, long-chain fatty acids with a chain length of up to 28 carbon atoms, and 2-hydroxylated long-chain acyl groups (27). In addition to 3-oxo-fatty acids (28,29) also (n-1)-hydroxylated (30) and (n-1)-oxo fatty acids (31) were demonstrated to be lipid A constituents. In a few cases, 3-hydroxy fatty acids were characterized, which contain additional functional groups, such as unsaturated bonds, e.g., (R)-3-hydroxydodec-5-Z-enoic acid [Δ^5-12:1(3-OH)] (32). In lipid A of *Legionella pneumophila*, 2,3-dihydroxy-12-methyltridecanoic acid [*i*14:0(2,3-diOH)] and 2,3-dihydroxytetradecanoic acid [14:0(2,3-diOH)] were identified in amide-linkage (33). Despite the considerable variability of the acylation pattern, certain rules are obeyed and the known acyl groups exist according to type and position of linkage, as summarized in Table 3. For a more systematic review of lipid A fatty acids, the reader is referred to a previous review (6).

NEWER DEVELOPMENTS IN METHODS OF ANALYSIS OF LIPID A STRUCTURE AND CONFORMATION

Mass Spectrometry

Application of Gas Chromatography–Mass Spectrometry to the Stereochemical Analysis of Lipid A Fatty Acids

The determination of the double bond positions of various lipid A fatty acids has previously been carried out by gas chromatography–mass spectrometry (GC-MS) analysis of their picolinylesters (34,35). Mayer and co-workers recently introduced the dimethyldisulfide (DMDS) reagent to label the position of the double bonds and also to determine their *cis/trans*-isomerism by retention time in comparison to authentic standards. By this procedure, octadec-11-Z-enoic acid ($cis\Delta^{11}$-18:1) and a hitherto unknown biosynthetically related 11,12-methylene-octadecanoic acid (c19:0) containing a cyclopropane ring could be unequivocally identified for the first time in lipid A (36). With the help of a chemical derivatization method introduced by Wollenweber et al. (37,38), these rare fatty acids were identified as secondary fatty acids 18:0[3-O($cis\Delta^{11}$-18:1) and 14:0[3-O(c19:0)] in the lipid A of *Rhodospirillum salinarum* 40 (36). It deserves special mention that the DMDS procedure is not helpful for determination of olefinic bonds in fatty acids containing additional functional groups (39). Examples of such fatty acids are (R)-3-hydroxydodec-5-Z-enoic acid [Δ^5-12:1(3-OH)] in *Phenylobacterium immobile* (32) or unsaturated 3-hydroxy fatty acids in *R. trifolii* (40) or *R. meliloti* (41).

In a few cases, 3-hydroxy fatty acids were shown to contain additional functional groups. In *L. pneumophila*, 2,3-di-hydroxy-12-methyltridecanoic acid [*i*14:0(2,3-diOH)] and 2,3-dihydroxy-tetradecanoic acid [14:0(2,3-diOH)] were identified to be present in amide-linkage (42). We recently described these fatty acids as primary fatty acids attached to positions 3′ and 3 of the backbone disaccharide 4′-P-β-D-GlcpN3N-(1′→6)-α-D-GlcpN3N-(1→P backbone disaccharide (33). (R)-3-Hydroxylated fatty acids were long assumed to be of considerable chemotaxonomic value (1,5), thus rendering lipid A–associated 3-hydroxylated

Table 1 Phosphate-Linked Substituents of the Lipid A Backbone [4'-P-β-D-GlcpN(3N)-(1'→6)-α-D-GlcpN(3N)-(1→P]

Organism	Nature of substituent of backbone phosphate group at		Ref.
	C-4'	C-1	
Bordetella pertussis	—	—	141
Brucella abortus[a]	—	—	130
Burkholderia (Pseudomonas) cepacia	L-Arap4N	R[b]	142
Campylobacter jejuni[c]	Etn-P[d]	Etn	76
Chromobacterium violaceum	L-Arap4N	D-GlcpN	143
Comamonas testeroni	—	—	116
Erwinia carotovora	—	—	107
Escherichia coli	—	P[d]	144,145
Haemophilus influenzae	—	—	146
Haemophilus ducreyi	—	—	100
Klebsiella pneumoniae	L-*Ara*p4N[d]	L-Arap4N[d]	81
Legionella pneumophila[a]	—	—	33
Moraxella catarrhalis	—	P-Etn[d]	97,120
Neisseria meningitidis	Etn-P[d]	P-Etn[d]	79
Pectinatus cerevisiiphilus	L-Arap4N	—	83
Pectinatus frisingensis	L-Arap4N	—	83
Porphyromonas gingivalis[e]	—	Etn[d]	114,121
Proteus mirabilis	L-Arap4N	—	1
Providencia rettgeri	L-Arap4N	—	147
Pseudomonas aeruginosa	L-Arap4N[d]	—	133
Rhodobacter capsulatus	Etn[d]	P-Etn[d]	51
Rhodocyclus gelatinosus	—	—	77
Rhodopseudomonas gelatinosa	Etn[d]	Etn[d]	148
Rhodospirillum fulvum	L,D-Hep	D-GalpA	82
Rhodospirillum salinarum[c]	—	—	36
Rhodospirillum tenue	L-Arap4N	D-Ara*f*	149
Salmonella enterica sv. Minnesota	L-Arap4N[d]	P-Etn[d]	117
Salmonela enterica sv. Typhimurium	Etn-P[d]	L-Arap4N[d]	126
Shigella sonnei	—	—	150
Sphaerotilus natans	Etn-P	—	80
Vibrio cholerae	—	P-Etn	151
Yersinia pestis	L-Arap4N	D-Ara*f*[d]	152

[a] β-D-GlcpN3N-(1'→6)-α-D-GlcpN3N lipid A backbone.
[b] Unidentified residue R.
[c] hybrid lipid A backbone (β-D-GlcpN3N-(1'→6)-α-D-GlcpN).
[d] Nonstoichiometric.
[e] Nonstoichiometric 4'-P.
P, Phosphate; Etn, 2-aminoethanol; L-Arap4N, 4-amino-4-deoxy-L-arabionpyranose; D-Ara*f*, D-arabionfuranose; L,D-Hep, L-glycero-D-manno-heptopyranose; GalpA, D-galacturonic acid.

fatty acids suitable markers for the qualitative and quantitative determination of endotoxins in biological samples (43,44). However, (*R*)-3-hydroxylated fatty acids have also been identified in other lipids such as ornithine-containing lipids of various gram-negative bacteria (45), rhamnolipids of *Pseudomonas* spp. (46,47), and in the fish poison pahutoxin (48). Finally, (*R*)-hydroxylated long-chain fatty acids have been identified in a variety of glycosylated peptidolipids both in gram-negative bacteria such as herbicolin A and

Table 2 Substituents of Phosphateless Lipid A Backbone

Organism	Nature of substituent of β-D-GlcpN-(1'→6)-D-GlcpNX[a] backbone		Ref.
	C-4'	C-1	
Porphyromonas (Bacteroides) fragilis	—	P	78
Rhizobium leguminosarum[b]	α-D-GalpA		19
Rhodospirillum fulvum	α-L,D-Hepp	D-GalpA	82
Rhodomicrobium vaniellii	β-D-Manp[c]	n.d.	34

[a]GlcpNX represents either GlcpN or GlcN-onic.
[b]Contains GlcN-onic instead of GlcpN (I).
[c]Nonstoichiometric.
GlcpN, 2-Amino-2-deoxy-D-glucose; GlcN-onic, 2-amino-2-deoxy-D-gluconic acid; L,D-Hep, L-glycero-D-manno-heptopyranose; GalpA, D-galactopyranuronic acid; n.d., not determined.

B from *Erwinia herbicola* (49) and in gram-positive bacteria such as gordonin [20:0(3-OH) and 22:1(3-OH)] in *Gordona hydrophobica* (50) and others (45).

In addition to modern analytical strategies using nuclear magnetic resonance (NMR) and various techniques of mass spectrometry (MS), chemical compositional analysis is still often required in order to corroborate the interpretation of NMR spectra. Today, chemical analysis is still useful, and, moreover, it can sometimes provide unequivocal data to elucidate certain structural features in the lipid A molecule that cannot be achieved by either NMR or modern MS methods. The following example may illustrate this point. Despite various NMR and fast atom bombardment mass spectrometry (FAB-MS) analyses of *Rhodobacter capsulatus* lipid A, the position of the double bond as well as the Z,E(cis/trans)-configuration of the dodecenoic acid residue (12:1) could not be deduced (51). ^{1}H-NMR analysis of lipid A (and derivatives) is not an appropriate method to reveal the position of the double bond as well as the Z,E-configuration unambiguously. This limitation can be explained by the fact that the double bond of an unsaturated fatty acid is separated by a number of methylene protons (—CH$_2$—) from the well-distinguished methylene protons at C2 (≈2.2 ppm) and C3 (≈1.4 ppm), and a clear-cut interpretation for their interconnectivities in a ^{1}H,^{1}H-COSY experiment is therefore not possible.

A similar restriction is found for the investigation of the Z,E-configuration. In this case, a clear-cut interpretation of the coupling constants of the multiplet raised from the two olefinic protons (—CH=CH—; ≈5.25 ppm) to reveal their Z- or E-configuration (J$_Z$ ≈ 10 Hz and J$_E$ ≈ 15 Hz) is hampered by the fact that these protons are coupling with four neighboring methylene protons, therefore raising to nonresolved multiplets of higher order. Since these coupling constants are also dependent on the conformation of the fatty acid chains, a clear-cut assignment of the coupling constants cannot be deduced unambiguously, although some characteristical differences for the chemical shift and shape of Z- and E-olefinic protons are found in pure and isolated fatty acid methyl esters (52,53).

However, with the help of simple chemical derivatization and in combination with combined GC-MS analysis, this problem could be solved (39). A similar problem is the assignment of the Z,E-configuration of 14:1 present in *Rhodobacter sphaeroides* lipid A (28,54–56), which, despite several NMR investigations, is still in dispute (57,58).

Analysis of the stereochemistry of 3- and 2-hydroxylated fatty acids present in lipid A either as secondary or primary fatty acids, revealed, without exception, the (R)-configuration for 3-hydroxylated fatty acids and the (S)-configuration for 2-hydroxy fatty acids (59,60). It is noteworthy that in the lipid A of *Chlamydia trachomatis* 14:0, 15:0, and 16:0 were found to be linked as primary fatty acids to positions 3' and 3 of the backbone. Their chemical character (iso-, anteiso-, or straight-chain fatty acid) has, however, not yet been determined (26).

Without exception all unsaturated fatty acids investigated so far are cis-configured with isolated double bonds located at C5 in dodecenoic acid (12:1) (51), at C7 in tetradecenoic acid (14:1) (57), and C11 in octadecenoic acid residues (18:1) (36). Therefore, unsaturated fatty acids in lipid A carry a double bond generally located at the n-7 position (sometimes also designated as ω-7). This generalization suggests a similar biosynthetic pathway for these fatty acids as well as a particular enzyme specificity for the late acyl transferases (61).

Table 3 Distribution of Acyl and 3-Acyloxyacyl Residues over Hydroxyl and Amino Groups of the Lipid A Backbone [β-D-GlcpN(3N)(II)-(1′→6)-D-GlcpN(3N)(I)] of Various Bacteria in the Main Lipid A Fraction

Bacteria	Nature and linkage type of acyl substituents of the lipid A backbone bound to					Ref.
	GlcpN(3N)(II) at position		GlcpN(3N)(I) at position			
	3′	2′	3	2		
Heptaacyl lipid A						
Acyl distribution (4+3)						
Enterobacter agglomerans	14:0[3-O(14:0)][a] 14:0[3-O(14:0(2-OH))][a,b]	14:0[3-O(12:0)]	14:0(3-OH)	14:0[3-O(16:0)]		94,95,153
Pantoea agglomerans	14:0[3-O(14:0)]	14:0[3-O(12:0)]	14:0(3-OH)	14:0[3-O(16:0)]		119
Acyl distribution 3+4						
Moraxella catarrhalis	12:0(3-OH)	12:0[3-O(10:0)]	12:0[3-O(10:0)]	12:0[3-O(12:0)]		97,120
Hexaacyl lipid A						
Acyl distribution (4+2)						
Erwinia carotovora [4+2(3)]	14:0[3-O(12:0)]	14:0[3-O(12:0)]	14:0(3-OH)	14:0(3-OH) 14:0[3-O(16:0)][c]		66
Salmonella enterica [4+2(3)] sv. Minnesota	14:0[3-O(14:0)][a] 14:0[3-O(14:0(2-OH))][a]	14:[3-O(12:0)]	14:0(3-OH)	14:0[3-O(16:0)][c]		38,71
Actinobacillus actinomycetemcomitans	14:0[3-O(14:0)]	14:0[3-O(14:0)]	14:0(3-OH)	14:0[3-O(16:0)][c]		92
Campylobacter jejuni[d]	14:0[3-O(16:0)]	14:0[3-O(16:0)]	14:0(3-OH)[a]	14:0(3-OH)		76
Escherichia coli	14:0[3-O(14:0)]	14:0[3-O(12:0)]	14:0(3-OH)	14:0(3-OH)		1
Haemophilus influenzae	14:0[3-O(14:0)]	14:0[3-O(14:0)]	14:0(3-OH)	14:0(3-OH)		99,146
Haemophilus ducreyi	14:0[3-O(14:0)]	14:0[3-O(14:0)]	14:0(3-OH)	14:0(3-OH)		100
Helicobacter pylori	16:0(3-OH)[c]	18:0[3-O(18:0)] 16:0[3-O(12:0)][a,c] 16:0[3-O(14:0)][a,c]	16:0(3-OH)]	18:0(3-OH)		84
Klebsiella pneumoniae [4+2(3)]	14:0[3-O(14:0)]	14:0[3-O(14:0)]	14:0(3-OH)	14:0(3-OH) 14:0[3-O(16:0)][c]		81
Legionella pneumophila	i14:0[3-O(28:0(27-oxo))][a,e] i14:[2-OH,3-O(28:0(27-oxo))][a,e]	18:0[3-O(i16:0)][a] 20:0[3-O(i16:0)][a] 22:0[3-O(i16:0)][a]	i14:0(3-OH)[a] i14:0(2,3-diOH)[a]	18:0(3-OH)[a] 20:0(3-OH)[a] 22:0(3-OH)[a]		33
Pectinatus cerevisiiphilus	13:0[3-O(11:0)]	13:0[3-O(11:0)][a] 13:0[3-O(13:0)][a]	13:0(3-OH)	13:0(3-OH)		83
Pectinatus frisingensis	13:0[3-O(11:0)]	13:0[3-O(11:0)][a] 13:0[3-O(13:0)][a]	13:0(3-OH)	13:0(3-OH)		83
Providencia rettgeri	14:0[3-O(14:0)]	14:0[3-O(14:0)]	14:0(3-OH)	14:0(3-OH) 14:0[3-O(16:0)][c]		147
Proteus mirabilis [4+2(3)]	14:0[3-O(14:0)]	14:3-O(14:0)]	14:0(3-OH)	14:0(3-OH) 14:0[3-O(16:0)][c]		154
Rhodospirillum fulvum	14:0[3-O(12:0)]	16:0[3-O(16:0)]	14:0(3-OH)	16:0(3-OH)		82
Salmonella enterica sv. Typhimurium	14:0[3-O(14:0)]	14:0[3-O(12:0)]	14:0(3-OH)	14:0(3-OH)		12,62

Chemical Structure of Lipid A

Acyl Distribution (3+3)					
Chromobacterium violaceum	10:0(3-OH)	12:0[3-O(12:0)][a] 12:0[3-O(12:0(2-OH))][a]	10:0(3-OH)	12:0[3-O(12:0)]	37
Comamonas testosteroni	10:0(3-OH)	10:[3-O(12:0)]	10:0(3-OH)	10:0[3-O(14:0)]	116
Neisseria gonorrhoeae [3+3(2)]	12:0(3-OH)	14:0[3-O(12:0)]	12:0(3-OH)	14:0[3-OH)	64
				14:0[3-O(12:0)][c]	
Neisseria meningitidis	12:0(3-OH)	14:0[3-O(12:0)]	12:0(3-OH)	14:0[3-O(12:0)]	79
Plesiomonas shigelloides	14:0(3-OH)	14:0[3-O(12:0)]	14:0(3-OH)	14:0[3-O(14:0)]	155
Rhodocyclus gelatinosus	10:0(3-OH)	10:0[3-O(12:0)]	10:0(3-OH)	10:0[3-O(12:0)][a]	77
				10:0[3-O(14:0)][a]	
Rhodopseudomonas gelatinosa	10:0(3-OH)	10:0[3-O(12:0)]	10:0(3-OH)	10:0[3-O(14:0)]	148
Shigella sonnei	14:0(3-OH)	14:0[3-O(14:0)]	14:0(3-OH)	14:0[3-O(12:0)]	150
Acyl Distribution (3+2)					
Bordetella pertussis	14:0(3-OH)[c]	14:0[3-O(14:0)]	14:0(3-OH)	14:0(3-OH)	141
Burkholderia caryophylli [3+2(1)]	14:0(3-OH)	16:0[3-O(14:0)]	14:0(3-OH)[c]	16:0(3-OH)	O. Holst, personal communication
Chlamydia trachomatis	14:0[a]	20:0[3-O(18:0)][a]	14:0[a]	20:0(3-OH)	26
	15:0[a]	20:0[3-O(19:0)][a]	15:0[a]		
	16:0[a]	20:0[3-O(20:0)][a]			
Escherichia coli msbB mutant	14:0(3-OH)	14:0[3-O(12:0)]	14:0(3-OH)	14:0(3-OH)	125
Porphyromonas (Bacteroides) fragilis	16:0(3-OH)[a]	i17:0[3-O(i15:0)][a]	15:0(3-OH)[a]	i17:0(3-OH)[a]	78
	15:0(3-OH)[a]	16:0[3-O(i15:0)][a]	16:0(3-OH)[a]	16:0(3-OH)[a]	
Porphyromonas (Bacteroides) gingivalis	i15:0(3-OH)[f]	i17:0[3-O(16:0)]	16:0(3-OH)[f]	i17:0(3-OH)	114
Pseudomonas aeruginosa [3+2(3)]	10:0(3-OH)	12:0[3-O(12:0)][a]	10:0(3-OH)[c]	12:0[3-O(12:0)][a]	68
		12:0[3-O(12:0(2-OH))][a]		12:0[3-O(12:0(2-OH))][a]	
Pseudomonas fluorescens [2(3)+2(3)]	10:0(3-OH)	12:0[3-O(12:0)][a]	10:0(3-OH)[c]	12:0[3-O(12:0)][a]	156
		12:0[3-O(12:0(2-OH))][a]		12:0[3-O(12:0(2-OH))][a]	
Rhodobacter sphaeroides	10:0(3-OH)	14:0[3-O(cisΔ⁷-14:1)][a]	10:0(3-OH)	14:0(3-oxo)	28,29,39
		14:0[3-O(14:0)][a]		14:0(3-OH)	
Rhodobacter capsulatus	14:0[3-O(cisΔ⁵-12:1)]	14:0(3-oxo)	10:0(3-OH)	14:0(3-oxo)	39,51
Shigella flexneri [3(4)+2]	14:0(3-OH)	14:0[3-O(12:0)]	14:0(3-OH)	14:0(3-OH)	96
	14:0[3-O(14:0)][c]				
Sphaerotilus natans	10:0(3-OH)	10:0[3-O(12:0)][g]	10:0(3-OH)	10:0[3-O(12:0)][g]	80
Acyl Distribution (2+2)					
Rhizobium meliloti	28:0(27-OH)	18:0(3-OH)	14:0(3-OH)	18:0(3-OH)	41

[a]Fatty acids in one column add up to 1 mol equivalent of lipid A.
[b]Position of the hydroxy group (2-OH or 3-OH) was not determined.
[c]Present in nonstoichiometric amounts.
[d]Contains mainly a hybrid lipid A backbone [4'-P-β-D-GlcpN3N-(1'→6)-α-D-GlcNp-P].
[e]Intrinsic heterogeneity, ester-linked 28:0(27-oxo) can be replaced by 27:0-dioic acid.
[f]Fatty acids within this line may have to be reversed.
[g]Intrinsic heterogeneity, 12:0 may be replaced by 10:0, 10:1, 12:1, 14:0, and 14:1.

Soft Ionization Mass Spectrometry Analysis

Since the development of soft ionization methods for the mass spectrometric analysis of complex biomolecules in the early 1980s, these methods have been used for the structural determination of LPS and of its partial structures. However, the intrinsic biological variability of LPS (number, type and distribution of fatty acids, and phosphate substitution), its thermolability and nonvolatility have hindered straightforward analyses. Sample preparation techniques for the different desorption techniques developed for the analysis of other classes of macromolecules (peptides, proteins, neutral oligosaccharides) were not applicable to lipid A due to its high polarity and its amphiphilic character leading to aggregate formation. Thus, for FAB-MS [respectively liquid secondary ion mass spectrometry (LSI-MS)] and for laser desorption mass spectrometry (LD-MS), lipid A samples are normally chemically modified either by blocking negative charges via derivatization by methylation with etheral diazomethane or by removal of one or all phosphate groups [mild acid hydrolysis and hydrogen fluoride (HF) treatment to obtain monophosphoryl and dephosphorylated lipid A, respectively]. It is worth mentioning that these chemical modifications are essential even with the use of modern desorption/ionization techniques such as MALDI-MS and electrospray ionization mass spectrometry (ESI-MS) to induce specific diagnostic fragments normally generated only in the positive ion mode (e.g., oxonium ion, see below).

In what follows, examples of lipid A structural analysis are discussed with special reference to the advantages and disadvantages of these methods.

Fast Atom Bombardment or Liquid Secondary Ion Mass Spectrometry The terms fast atom bombardment mass spectrometry (FAB-MS) and liquid secondary ion mass spectrometry (LSI-MS) are used synonymously. The nature of the liquid matrix in which the sample is dispersed rather than the charge of the incident particle is of importance for the generation of ions. The first FAB-MS investigations by Takayama and coworkers were performed on lipid A monophosphate preparations of *S. enterica* sv. Typhimurium, purified by high-performance liquid chromatography (HPLC). FAB mass spectra were obtained in the positive and negative ion mode, and both modes gave quasimolecular ions allowing the exact determination of the molecular masses (62). The HPLC elution profile allowed an analysis of the heterogeneity of the lipid A preparation. In addition to the molecular mass peaks, FAB-MS analysis provided the diagnostic fragment originating from cleavage at the C-1'-O-C-6 glycosidic bond (m/z 1087) and permitting the conclusion that GlcN (II) carried one dimethyl phosphate, two 14:0(3-OH) residues, one 14:0, and one 12:0 residue (63). Fragments derived from the reducing region of lipid A [GlcN (I)] were not observed in the positive ion FAB-MS but could be calculated from the difference of the molecular mass and the fragment of the oxonium ion (m/z 1087). In this case, GlcN (I) carried two 14:0(3-OH) fatty acyl groups, which, however, could not be assigned to specific positions 2, 3, or 4 of GlcN (I) or to one of the hydroxyl groups of the two 14:0(3-OH) residues. Nevertheless, the FAB-MS analysis not only confirmed the presence of acyloxyacyl residues, as earlier suggested by chemical analysis (38), but also allowed for the first time the assignment of the two acyloxyacyl residues to the distal GlcN (II). It was thus realized that in *S. enterica* sv. *typhimurium* lipid A, GlcN (II) carried four, and GlcN (I) only two acyl groups, i.e., that the fatty acid distribution over the two GlcpN residues was asymmetric (see Table 3). In quite a similar way, the symmetric distribution of fatty acids (3 + 3) was demonstrated in *Neisseria gonorrhoeae* lipid A (64), and an acylation pattern first recognized in *Chromobacterium violaceum*, however, in this first case by chemical analysis (37).

FAB-MS has also been applied to lipid A of the *S. enterica* sv. Minnesota Re mutant, strain R595 (65), *Erwinia carotovora* (66), *Chlamydia trachomatis* (26), and *Pseudomonas aeruginosa* (67), the latter study using FAB-MS and NMR confirming results previously obtained in our laboratory (68). LSI-MS of heterogeneous lipid A samples followed by collision-induced decomposition (CID) experiments performed on selected molecular ions have been performed by Kaltashov et al. (69) to confirm the presence of two isoforms in dimethyl-diphosphoryl lipid A isolated from *R. sphaeroides* (69). Recently, MS/MS analyses were also performed with a pentaacyl monophosphoryl lipid A of *E. carotovora* in the positive and negative ion mode, giving evidence that different fatty acids are cleaved in the two modes (70).

Laser Desorption Mass Spectrometry Laser desorption mass spectrometry (LD-MS) of dephosphorylated and nonpurified lipid A was first introduced by Seydel and Lindner for studies on the acylation pattern of lipid A of *S. enterica* sv. Minnesota, *E. coli*, and *Proteus mirabilis* (71,72). Although these earlier studies did not establish whether some mass peaks originated from laser-induced fragmentation or from smaller intact molecular species (monomeric subunits formed during acid hydrolysis), the first LD-MS study in combination with results of chemical analysis (37,38) revealed the location of the individual fatty acids in these bacteria. An advantage of the LD-MS method resides in its ability to control the degree of fragmentation by the amount and kind of alkali salts (NaI, CsI) added to the sample for cationization. In this way, both the fragmentation pattern and the heterogeneity of a preparation can be studied in detail (73). Using the same approach, even extremely heterogeneous structures, such as a highly complex lipid A of *Legionella pneumophila*, could be successfully elucidated (33). Unfortunately, the method failed to give information on bisphosphorylated lipid A species (72). LD-MS analysis of dephosphorylated lipid A preparations was also used to determine the distribution of the acyl and acyloxyacyl acid residues over the lipid A backbone. This type of analysis is based on the observation by Cotter et al. (74) that LD-MS in the positive ion mode generates fragmentation of the proximal GlcN (I) pyranoside leading to diagnostic fragment ions from the distal GlcN (II) [$^{0,4}A_2$ and $^{1,3}A_2$, according to the fragmentation nomenclature introduced by Domon and Costello (75)]. Using this technology, the general acylation pattern in the lipid A backbone of the distal GlcN (II) and the proximal GlcN (I) can be determined.

LD-MS contributed to the structural elucidation of a variety of lipid A structures; these analyses have been performed on dephosphorylated or methylated monophosphoryl lipid A of *R. sphaeroides* (28), *N. gonorrhoeae* (64), *Campylobacter jejuni* (76), *Rhodocyclus gelatinosus* (77), *Rhodobacter capsulatus* (51), *Porphyromonas (Bacteroides) fragilis* (78), *Rhizobium meliloti* (41), *Neisseria meningitidis* (79), *Sphaerotilus natans* (80), *P. aeruginosa* (68), *Rhodospirillum salinarum* (36), *Klebsiella pneumoniae* (81), *Rhodospirillum fulvum* (82), *Pectinatus cerevisiiphilus* (83), and *Helicobacter pylori* (84). The results of those and other studies are summarized in Table 3.

Plasma Desorption Mass Spectrometry Plasma desorption mass spectrometry (PD-MS) is especially suitable for the study of larger molecules and has been successfully used to analyze native, underivatized LPS isolated from rough mutants of *S. enterica* sv. Minnesota and *E. coli* strains (85). The negative ion mass spectra comprise the major molecular species of the heterogeneous LPS samples with respect to the number of fatty acids in lipid A as well as the phosphorylation and composition of the core oligosaccharide. Also, the heterogeneity of underivatized lipid A preparations isolated from a variety of enterobacterial and nonenterobacterial strains was determined (86,87). Thus, purified and derivatized (methylated) mono and bisphosphoryl lipid A of *E. coli*, *R. sphaeroides*, and *R. capsulatus* were analyzed by Cotter and Takayama (88,89). In the positive ion mode, besides the quasimolecular ions, ions corresponding to the loss of ester-linked fatty acyl, glycosidic phosphate groups, and oxonium ions, the latter formed by the cleavage of the distal GlcN(I), were observed. PD-MS of the phosphomethylated and purified LPS of the *E. coli* Re mutant (strain D31m4) showed, in addition to the hexaacyl (>90%), a pentaacyl (>10%) and a tetraacyl (traces) LPS species. However, after acid hydrolysis (0.1 M HCl, 100°C, 0.5 hr) the molar ratios of the hexa-, penta-, and tetraacyl structures were determined to be approximately 2:1:1, indicating that the preparation of lipid A had introduced additional heterogeneity with respect to the acylation pattern (88).

Karibian et al. investigated lipid A isolated from LPS of various bacterial strains (*E. coli*, *Escherichia hermanii*, *S. enterica* svs. Typhimurium, Typhosa, Minnesota, Milwaukee, Haifa, Enteritidis (90). Some of these LPS preparations (*E. coli* and *S. enterica* sv. Minnesota) were available either as S- or R-form LPS and were fractionated according to the length of the O-specific chain (86). They found in *E. coli* O97 the degree of acylation in LPS with long O-chain (smooth-type) to be higher (hexaacyl lipid A dominating) than in short-chain LPS (rough-type), the latter containing predominantly tetraacyl lipid A (86). This finding has been recently confirmed in *H. pylori* LPS (84), but is in contrast to previous findings on endotoxin preparations of *S. enterica* sv. *abortus equi*, where it was shown by GC analysis that LPS with long O-specific chain had a significantly reduced degree of lipid A acylation (91).

PD-MS analysis has so far been successfully performed on lipid A of *Actinobacillus actinomycetemcomitans* (92), *E. coli*, *E. hermannii*, *Shigella flexneri*, *Salmonella enterica* svs. Minnesota and Typhimurium, *K. pneumoniae*, *Serratia marcescens*, *Yersinia enterolytica*, *Aeromonas salmonicida*, *Haemophilus influenzae*, *Haemophilus pleuropneumoniae*, *N. meningitidis*, *N. gonorrhoeae*, *P. aeruginosa*, *Pseudomonas maltophila* (87), *R. sphaeroides* (89,93), *R. capsulatus* (89), and *Enterobacter agglomerans* (94) (see also Table 3).

Electrospray Ionization Mass Spectrometry Electrospray ionization mass spectrometry (ESI-MS) has proven to be a "very soft" and sensitive ionization method, where the surrounding liquid of the droplets has a moderating effect on the internal and translational energies of desorbed multiple-charged ions. ESI-MS was first applied by Harrata et al. to lipid A of *E. agglomerans*, which is believed to be the causative agent of a pulmonary affliction of textile mill workers (95). These authors showed that the acidity of the phosphate group of crude, underivatized lipid A provides improved ionization efficiency in the negative ion mode yielding only intact molecular ions. In a similar way, Chan and Reinhold demonstrated that the molecular mass profiling by ESI-MS in combination with CID tandem mass spectrometry provides an effective and sensitive technique for clarifying acyl components and related structural details of monophosphoryl lipid A (96). ESI-MS seems to be an efficient analytical method when combined with a separation system such as capillary electrophoresis (CE) prior to mass analysis. This combination (CE–ESI-MS) has been successfully used for *O*-deacylated LPS of *Moraxella catarrhalis* (97). It has to be pointed out in this context that lipid A could not be separated and analyzed by this method due to its insolubility in aqueous buffer systems.

ESI-MS has been especially employed for the study of LPS biosynthesis, especially for the late acylation steps (98). Thus, the characterization of an *htrB* mutant of *H. influenzae* LPS was facilitated by analyzing the acylation pattern of the mutant by ESI-MS (99). The data obtained with LPS of the *htrB* mutant in comparison to *O*-deacylated LPS of the wild-type strain revealed that this mutant not only lost one (or two) of the myristic acid residues (14:0) [present in acyloxyacyl residues attached to the GlcN (II)], but also had undergone changes in the core structure, in particular with respect to the substitution by 2-aminoethyl phosphate (Etn-*P*) (100).

Furthermore, the ESI-MS method has been used to determine for the first time the degree of substitution of Etn-*P* or 2-aminoethyl pyrophosphate (Etn-*PP*) residues in *H. influenzae*, *N. meningitidis*, and *S. enterica* sv. Typhimurium (101). However, the procedure harbors the disadvantage that, during the alkali treatment to prepare *O*-deacylated LPS, part of the phosphate groups (including pyrophosphate and even triphosphate) (102,103) as well as Etn-*P* or Etn-*PP* residues may be cleaved off, thus introducing additional heterogeneity not expressed in the native LPS.

Matrix-Assisted Laser Desorption/Ionization Mass Spectrometry Matrix-assisted laser desorption/ionization mass spectrometry (MALDI-MS) with a time-of-flight (TOF) analyzer has attracted special interest in protein analysis due to its sensitivity and capability to generate the intact molecular ion from molecules with even very high molecular masses (>100 kDa). In the meantime, MALDI-MS has proven to also represent an excellent tool for the analysis of glycopeptides, oligonucleotides, and carbohydrates. Despite methodical progress, sample preparation with respect to the selection of the appropriate matrix and solvent, desalting of the sample, and the technique of sample deposition on the target remains of utmost importance, as these seemingly trivial procedures may dramatically influence the quality of the spectra (sensitivity, fragmentation, and mass resolution) (104).

MALDI-MS spectra of complete S-form LPS showed a series of broad ion peaks up to $m/z \sim 15,000$, the mass difference between the main components indicating the size of the repeating sugar unit (105,106). At present it is impossible to obtain significant information on the lipid A composition from the complex spectra of complete S-form LPS. However, we have recently analyzed the native R-type LPS of *E. carotovora* FERM P-7576 without further purification (107). These spectra offered information on the heterogeneity of the complete LPS structure in both the lipid A and core oligosaccharide regions. In combination with the analysis of the *O*-deacylated LPS derivative and taking into account the results of previous analyses of the lipid A component by chemical methods and LSI-MS (66) and of the core oligosaccharide by NMR spectroscopy, the complete LPS structure of *E. carotovora* could be unequivocally established (107).

Gibson et al. (108), who analyzed intact and *O*-deacylated rough-type LPS, demonstrated the advantages of delayed ion extraction. Whereas under continuous ion extraction only a broad peak of low intensity was detected, delayed ion extraction yielded well-resolved quasimolecular ion peaks, which significantly improved both mass resolution and stability of molecular ions against elimination of phosphate groups.

Postsource decay (PSD), which allows TOF analyzers equipped with a reflectron to acquire mass spectra of metastable fragment ions of a selected parent molecular ion, can provide important structural information on the phosphate and Etn-*P* substitution pattern of the lipid A and oligosaccharide portion (108). We have recently shown that PSD of selected molecular ions from heterogeneous dephosphorylated lipid A samples can yield detailed information about the fatty acid composition and substitution. PD spectra of selected lipid A preparations (*E. coli* F515, *S. enterica* sv. Minnesota R595) comprised, besides expected fragment ions originating from cleavage of primary and secondary ester-linked fatty acids, diagnostic fragment ions from fatty acids in amide-linkage (109).

MALDI-MS of HPLC purified methylated hexaacyl bisphosphoryl Re LPS from *E. coli* strain D31m4 (110) and of HPLC purified monophosphorylated lipid A from *Chlamydia trachomatis* (26) was performed by Cotter and Takayama. It is noteworthy that the latter lipid A represents the first example where the primary fatty acids attached to positions 3 and 3′ of GlcpN are not β-hydroxylated. Unknown is whether these structural features influence the biological activities of *Chlamydia* lipid A (see Chapter 13).

Nuclear Magnetic Resonance Spectroscopy in Lipid A Analysis

Of all modern structural methods discussed so far, nuclear magnetic resonance (NMR) spectroscopy yields the most complete picture with respect to the structure of lipid A (LPS) as well as to its conformational behavior in solution. With the aid of ^{1}H-, ^{13}C-, and ^{31}P-NMR analysis, characteristics of lipid A structures can be revealed which are otherwise difficult or impossible to elucidate. These include (1) the anomeric configuration of the GlcpN residues in lipid A, (2) the linkage between the GlcpN residues in the disaccharide [β-(1′→6)], (3) the presence of primary and/or secondary fatty acids (but not their exact positions, see below), (4) the degree of acylation, (5) the position of the phosphate residue(s), (6) the position of unsubstituted hydroxyl groups in the lipid A backbone, and (7) the nature, degree, and position of in general positively charged head groups linked to the phosphates of the lipid A backbone (Table 1). Finally, NMR is also suitable for determination of lipid A conformational states in solution, i.e., the relative position and arrangement of the GlcN pyranosides in the lipid A backbone as well as their position relative to the fatty acids or the spatial arrangement of the phosphate residues relative to the β-(1′→6)-interlinked GlcpN disaccharide. Of particular importance is the fact that NMR spectroscopy represents a nondestructive method allowing samples to be still available for further chemical, biochemical, and, especially, biological investigations.

Some fundamental restrictions limit the application of NMR analysis of lipids in general, particularly of lipid A. These molecules are amphiphilic and amphipathic. In aqueous solutions they form aggregates that complicate NMR analysis in that they lead to barely resolved signals and line broadening (111). In the past, these difficulties could be partly overcome by using lipid A derivatives such as 4′-dimethyl-monophosphoryl lipid A, which enabled the recording of satisfactory NMR spectra in organic solvents. Other lipid A partial structures, e.g., obtained by *O*-deacylation or by dephosphorylation, were also recorded with well-resolved signals. However, in these cases not all the advantages of the NMR method could be made useful, because derivatization or destruction of samples limited its availability for further biological analysis.

In order to investigate native and underivatized lipid A preparations by NMR spectroscopy, purification procedures and special solvent systems (112) as well as suitable salt forms reducing the tendency of lipid A to form aggregates in organic solution have been introduced. Conditions were found that were suitable to obtain significantly improved signal resolution, whereby in some lipid A preparations individual signal assignments, coupling constants, and even intraresidual nuclear Overhauser effects could be determined (113).

Before dealing with specific applications of NMR methods to lipid A, certain disadvantages limiting its application should be mentioned. In contrast to the MS methods described above, ^{1}H-NMR is able to furnish information about the presence of unsubstituted and/or substituted (acyloxyacyl) fatty acids, but not on their chain length because of overlapping multiple methylene (—CH_2—) proton signals. It has to be pointed out in this context that integration of the methylene protons, a routine procedure to determine the number of protons in a molecule, could not be used so far to give reliable data for the exact determination of the number of methylene protons, and, therefore, the chain length of the fatty acids in lipid A cannot be determined in this way. For unequivocal allocation of the fatty acids to individual hydroxyl and amino groups of the lipid A backbone, the NMR data must be supplemented by chemical and by mass spectrometric analysis. As already discussed in more detail, NMR analysis is incapable of assigning unambiguously either the position or the *Z,E*-isomerism of an olefinic fatty acid in lipid A. Despite these restrictions, NMR analysis of various derivatized and nonderivatized lipid A samples and LPS partial structures has been performed in recent years with considerable success, and some important data obtained by these studies will be summarized in the following chapter.

One-Dimensional and Two-Dimensional ^{1}H-NMR Spectroscopy

Stereochemistry of β-Hydroxylated Fatty Acids
Recently, a new ^{1}H-NMR method has been introduced to determine the absolute configuration of (*R*)-3-hydroxylated (3-OH) fatty acids (114). This is achieved by investigation of the carboxymethyl protons of the

fatty acid methyl ester (COOC$\underline{\text{H}}_3$) in a ^{1}H-NMR spectrum recorded in CDCl$_3$ usually resonating around 3.7 ppm. This signal shifts in a dose-dependent manner to a lower field in the presence of Tris-[3-(heptafluoroprolyl-hydroxymethylene)-(+)-camphorato] europium(III) as complexing reagent (114). Depending on the shift induced by this reagent and with the help of synthetic reference compounds, the absolute stereochemistry of the hydroxylated fatty acid can be determined. The disadvantages of this procedure as compared to a previous method employing (R)-phenylethylamine and GLC analysis (115) are that the methyl ester has to be isolated and purified to homogeneity prior to NMR investigations and that several measurements have to be done for each hydroxy fatty acid being investigated. This is not required in case of the GLC procedure, where 2-hydroxylated fatty acids can be analyzed in the same run. Nevertheless, NMR has been used in order to determine the stereochemistry of 3-hydroxylated fatty acids of *Porphyromonas gingivalis* (114) and *Comamonas testeroni* (116), which was shown to be (R) in both cases.

One-Dimensional and Two-Dimensional ^{1}H-NMR Spectroscopy of Lipid A and Derivatives As mentioned above, natural and underivatized lipid A dispersed in aqueous or organic solvents yields poorly resolved spectra because of its tendency to form aggregates (112,117). Although the solutions may be seemingly clear, lipid A aggregates are present that limit the investigation of lipid A by ^{1}H-NMR spectroscopy. The first NMR experiments could in part overcome this problem by removing the glycosidic phosphate, blocking of the ionic phosphate group in position 4′ by methylation with diazomethane (12,13) and by substituting the free hydroxyl groups of the lipid A backbone and of the β-hydroxy fatty acids with trimethylsilyl residues (12).

Recently, Wang and Hollingsworth reported on a complex solvent system for high-resolution ^{1}H-NMR spectroscopy, which appeared suitable also for underivatized membrane lipids such as phosphatidylethanolamine (PE), phosphatidylserine (PS), disialogangliosides as well as mono- and bisphosphorylated lipid A. This solvent system is composed of a complex mixture of pyridine-d_5, deuterium chloride (DCl) in deuterium oxide (D$_2$O) (37 wt%), methanol-d_4, and chloroform-d in a volume ratio of 1:1:2:10 (112). However, in this case nonpurified lipid A preparations from commercial sources were used and, due to the heterogeneity in the preparation, neither the complete assignment of signals nor the coupling constants of the individual protons, which are necessary for complete NMR analysis, could be obtained. Despite the heterogeneity of the preparation, the proton-phosphorous heteronuclear multiple-quantum coherence (^{1}H, ^{31}P-HMQC) spectroscopy was found useful for the assignment of the phosphate resonances in mono- and bisphosphoryl lipid A (112).

The dimethyl-monophosphoryl lipid A derivatives turned out to be suitable for analytical NMR spectroscopy, because these can be separated by preparative HPLC or layer chromatography (pLC) prior to NMR analysis. In using this protocol, lipid A structures of the following bacteria were analyzed: *S. enterica* svs. Typhimurium (12) and Minnesota (118), *E. coli* (13), *E. carotovora* (66), *Pantoea agglomerans* (119), *Comamonas testosteroni* (116), and *Porphyromonas gingivalis* (114). Despite the different origins of these lipid A and despite the differences in the location and chemical nature of the various fatty acids, the lipid A backbone in all these studies was found to consist of a β-(1′→6)-interlinked Glc$p$N disaccharide with unsubstituted hydroxyl groups at C-6′ and C-4, as was shown for the first time by this protocol for HPLC-purified dimethyl-monophosphoryl lipid A preparations of *S. enterica* sv. Typhimurium (12,62).

The determination of the nature and linkage type of the phosphate residue to GlcN (I) as well as the nature and location of polar head groups was of great interest, and NMR studies have provided useful information. In several studies, however, O-deacylated LPS (LPS-OH) or O-deacylated lipid A (lipid A-OH) partial structures were used (67). As for mass spectrometric analysis, the advantage of lipid A-OH preparations is their lower degree of heterogeneity with respect to fatty acid substitution as well as their reduced tendency to form aggregates in aqueous solutions. In addition, O-deacylated lipid A can be measured in D$_2$O. In this way, lipid A-OH of *M. catarrhalis* was investigated (120). A similar protocol was used for *P. aeruginosa* (67), whereby the lipid A structure established previously by chemical methods was confirmed (68). Dephosphorylated lipid A of *P. gingivalis* has also been investigated by one- and two-dimensional ^{1}H-NMR spectroscopy, the analysis having been performed in CDCl$_3$, whereby a satisfactory resolution of signals, allowing the determination of coupling constants, was obtained (121). In this study, an unequivocal interpretation of the individual proton signals was possible, which could not have been achieved with phosphorylated lipid A.

In order to investigate purified and homogeneous native and underivatized lipid A structures suitable for

NMR spectroscopy, Baltzer and Mattsby-Baltzer introduced preparative layer chromatography (pLC) using chloroform-methanol-water (130:45:7) as the solvent system (122). We have recently established a similar purification protocol and obtained monophosphoryl hexaacyl, monophosphoryl pentaacyl, bisphosphoryl hexaacyl, and bisphosphorylated pentaacyl lipid A preparations in an overall satisfactory yield ($\approx$37%) (123) and high purity (>95%). Their well-resolved NMR spectra could be obtained in a mixture of chloroform-d–methanol-d_4 85:15 (by volume). In combination with chemical and MALDI-TOF analysis, we were able to show that both the mono- and bisphosphorylated pentaacyl lipid A were devoid of the tetradecanoyl fatty acid (14:0) (123). Therefore, this lipid A partial structure is assumed to represent the lipid A precursor of a late (defective) acylation step (124). This finding is in good agreement with other reports showing the presence of pentaacyl lipid A in *E. coli* strains (13,122) and studies on the lipid A structure obtained from an *msbB* mutant of *E. coli* which is known to lack the myristoyl-transferase (125).

One-Dimensional ^{13}C- and Two-Dimensional Heteronuclear Correlated ^{13}C,^{1}H-NMR Spectroscopy

^{13}C-NMR spectroscopy was used for LPS and lipid A analysis prior to ^{1}H-NMR because of the lower susceptibility of the carbon signals to line broadening (126–129). Mattsby-Baltzer and Baltzer were the first to study lipid A by ^{13}C-NMR spectroscopy in 1986 (122). Although the lipid A preparations were of high purity and although the solvent system used for NMR analysis was optimized for each of the different lipid A preparations, the ^{13}C-NMR assignments could not be done unambiguously, since neither heteronuclear correlated spectroscopy (^{13}C,^{1}H-COSY) nor proton-carbon heteronuclear multiple-quantum coherence (^{1}H,^{13}C-HMQC) NMR spectroscopy was routinely available at that time. Assignment of signals could only be performed by comparison with synthetic reference partial structures. Therefore, the structural proposal of lipid A made in this early ^{13}C-NMR lipid A study differed from other *E. coli* lipid A analyses, as only pentaacyl lipid A was identified. Today, with the help of ^{13}C,^{1}H-NMR and ^{1}H,^{13}C-HMQC spectroscopy, the assignment of ^{13}C-NMR spectra can be easily achieved. This was demonstrated for the structural determination of *Brucella abortus* lipid A (130) and for investigations of the supramolecular structure of LPS (131).

Phosphorus (^{31}P)-NMR Analysis of Polar Head Groups

The disaccharide backbone of lipid A carries, in general, two phosphate groups. Of these, one is α-linked to the glycosidic hydroxyl group of GlcN (I) and the other is ester-bound to the hydroxyl group at C-4' of GlcN (II). Both the glycosidic and the nonglycosidic phosphate residue may be replaced in general, by nonacylated charged glycosyl or nonglycosyl residues. These polar head groups are often not present in stoichiometric amounts, and their amount appears to be dependent on growth conditions yielding a certain intrinsic variability of the lipid A backbone. Some of the phosphate substituents have been structurally elucidated, and their nature and locations are summarized in Tables 1 and 2.

It should be emphasized that routine one-dimensional ^{31}P-NMR analysis carried out in the ^{1}H-decoupled mode is suitable for the determination of the degree of substitution on the phosphates [monophosphate (*P*), pyrophosphates (*PP*, 2-aminoethylphosphate (Etn-*P*), and β-L-Ara*p*4N-(1→P] as well as to distinguish between glycosidic (O-1) and the ester-bound (O-4') phosphates, but the anomeric configuration or even the position of the phosphate residue in the lipid A backbone cannot be determined by simple one-dimensional ^{31}P-NMR experiments.

Using one-dimensional ^{31}P-NMR spectroscopy, phosphates and polar phosphate substituents at the lipid A backbone were investigated by Mühlradt et al. to determine the occurrence of L-Ara*p*4N at the 4'-phosphate in *S. enterica* sv. Minnesota (132). Other investigators used ^{31}P-NMR spectroscopy for the analysis of substituents on the phosphate groups in *S. enterica* sv. Minnesota (117), *P. aeruginosa* (133), *N. meningitidis* (79), *S. natans* (80), *Pectinatus frisingensis*, and *P. cerevisiiphilus* (83). Determination of the position of polar head groups (attached to either the glycosidic or the ester-bound phosphate), however, requires additional NMR experiments, wherein the proton-coupled spin systems are used (126). ^{1}H,^{31}P-HMQC experiments have been applied to lipid A of *E. coli* and *S. enterica* sv. Minnesota R595 (112,131).

Studying a lipid A precursor isolated from a temperature-sensitive mutant of *S. enterica* sv. Typhimurium (STi50), Strain et al., by applying proton-coupled ^{31}P-NMR, found that L-Ara*p*4N was linked to the glycosidic phosphate at C-1 and 2-aminoethylphosphate (Etn-*P*) to the phosphate at position 4' of GlcN (II), thus forming a nonglycosidic pyrophosphate linkage

(126). Therefore, the polar head groups in the lipid A precursor produced by this mutant are at opposite locations as found for lipid A of *S. enterica* sv. Minnesota (117,132). It is presently not known whether this discrepancy is due to differences in the two *Salmonella* strains investigated (svs. Typhimurium vs. Minnesota), or if this is an unique structural feature of this temperature-sensitive mutant, or whether the substitution pattern of the polar head groups in the biosynthetic precursor(s) is different from that of mature lipid A.

The ester-bound phosphate group usually present in lipid A may be lacking in certain cases. Thus, in *Porphyromonas* (*Bacteroides*) *fragilis* (78) and *Porphyromonas gingivalis* (114,121) the hydroxyl group in position 4' is free, and in *Rhodomicrobium vannielii* (34) this phosphate group is positionally replaced by a β-linked D-mannopyranose (Table 2). In the phototrophic bacterium *R. fulvum* it is replaced by an α-glycosidically linked L-*glycero*-D-*manno*-heptopyranose (L,D-Hep$p$) (82), and in *R. leguminosarum* (19) the 4'-position of GlcN (II) is substituted by α-glycosidically linked D-galacturonic acid (Gal$p$A). In *R. fulvum*, Gal$p$A is bound to C1, most likely in α-anomeric linkage (82).

These examples indicate that the polar components at the β-(1'→6)-Glc$p$N-disaccharide at positions C-4' and C-1 are not necessarily phosphates, as is the case in the "classic" lipid A backbone [4'-P-β-D-Glc$p$N(3N)-(1'→6)-α-D-Glc$p$N(3N)-(1→P]. Moreover, it seems that these substituents at C-4' and C-1, in general, carry negative charges.

Conformational Analysis

With the introduction of complex solvent systems, native and underivatized lipids such as PE, PS, and gangliosides and also underivatized lipid A became accessible to NMR spectroscopy (112,131). With such solvents it was possible to analyze mono- and bisphosphorylated lipid A without further derivatization, leaving the possibility of recording even long-range heteronuclear couplings, which is useful not only for the study of primary lipid A structures, but also for conformational analysis. Nuclear Overhauser effect (NOE) measurements by two-dimensional spectroscopy (NOESY) have gained special interest with respect to conformational analysis. They became available for underivatized lipid A due to the quality of signal resolution in the above-mentioned solvent system. In this way, data on the conformation(s) of lipid A molecules in solution could be obtained (see below) (131).

The first conformational analysis of lipid A using NMR data combined with molecular mechanics calculations was recently achieved (131). In the lipid A backbone the only interresidual NOE to be expected is observed between the (β-anomeric) proton H-1' of distal GlcN (II) and H-4, H-6R, and H-6S of GlcN (I). In accordance with the amphiphilic nature of lipid A, no rotation around the interglycosidic bond was observed, and various NOE signals indicated the planes of the two glycosyl rings to be orthogonal. Based on the Haasnoot-DeLeeuw-Altona relationship, the value of the coupling constant for the two vicinal coupling constants as a function of the torsion angle ω [dihedral angle of the bond between C-5 and C-6 of GlcN (I)] could be calculated as $-53°$, which was in good agreement with the coupling constants determined (~4 Hz). The two protons of the alkoxymethyl group (H-6R and H-6S) were found to adopt the *gauche-gauche*(gg)-conformation, which is the most stable conformation for the hydroxyl group of gluco-configurated hexopyranosides. These data are in agreement with an atomic model of *E. coli* lipid A previously presented by Kastowsky et al. using energy-minimization calculations (134).

A similar approach for achieving suitable ^{1}H-NMR spectra of underivatized lipid A samples has been carried out by Kusumoto and coworkers (113) using DMSO-d_6 as solvent and the diisopropylethylammonium [H$^+$N(^{i}Pr)$_2$Et] salt form of synthetic lipid A (compound 506). This hydrophobic ammonium ion salt reduces the polar interaction with the solvent and gives rise to well-resolved ^{1}H-NMR spectra. Based on NOE signals and on optimization by semiempirical MO calculations, it was shown that the two Glc$p$N residues of the backbone express mainly two conformations. In one of these, the two Glc$p$N rings are orthogonally orientated (GL form), a finding that is in agreement with the data of Wang and Hollingsworth (131). In the second conformation (GR form), the planes of the two Glc$p$N-rings are twisted 180° to each other. It should be noted that the solvent system used for these NMR studies was selected to reduce intermolecular interactions between lipid A molecules. The conformation(s) in this solvent system may thus differ from that adopted by lipid A in an aqueous system.

CURRENT VIEW OF LIPID A STRUCTURE

Since the appearance of a recent overview on the primary structure of lipid A (6), a considerable improvement in lipid A analytical methods has occurred. This enables researchers to study in great detail the chemical and physical structure of lipid A, thus enlarging our knowledge of various aspects of lipid A. In particular, modern MS and NMR analysis has deepened our understanding of the primary structure, the heterogeneity, and also the conformation of lipid A. The results of these scientific efforts are, in part, summarized in Tables 1 through 3.

Backbone

Lipid A is considered to be the structurally most conserved element of the LPS molecule (6), and all lipid A structural data accumulated in the recent years have further confirmed and strengthened this view. The glycosyl backbone of lipid A, without exception, is always composed of β-(1′→6)-interlinked D-gluco-configurated hexosamine disaccharide. Another element of structural constancy is represented by the two phosphate residues, usually attached to positions 4′ and 1, respectively. There are only a few examples known where the lipid A backbone is different from the highly conserved [4′-P-β-D-GlcpN(3N)-(1′→6)-α-D-GlcpN(3N)-(1→P] structure, and these are shown in Table 2. But these exceptional structures share an architecture principally similar to the classical structure. In addition, the substitution pattern at the lipid A backbone expresses peculiar conserved elements. In all lipid A structures thoroughly investigated, position 4 of GlcN (I) was found to be unsubstituted like position 6′ of GlcN (II), which serves as the attachment site for the core oligosaccharide. It should be kept in mind, however, that within one lipid A preparation of a single bacterial strain, there exists a moderate natural heterogeneity with respect to the degree of substitution of phosphate groups by polar head groups with regard to the degree of acylation, this heterogeneity being generated by incomplete biosynthesis and/or degradation processes during the preparation of lipid A.

Despite the similar chemical make-up of the backbone of lipid A of different origin, a considerable variability exists with respect to (1) the nature of polar head groups (Tables 1 and 2) and (2) the nature, chain length, and distribution of acyl groups substituting the lipid A backbone (Table 3).

Polar Head Groups

The various types of polar head groups present in lipid A so far investigated are summarized in Tables 1 and 2. This list was considerably enlarged during the last 3 years (6) due to investigations based on the newer methods described in this review, in particular modern soft-ionization MS and ^{31}P-NMR techniques.

As a general feature, the polar substituents of the lipid A backbone carry negative charges (phosphate, pyrophosphate, galacturonic acid), positive charges (4-amino-4-deoxy-L-arabinopyranose, ethanolamine, glucosamine), or zwitterionic groups (2-aminoethyl phosphate). In addition, in a few cases neutral substituents were identified such as D-arabinofuranose, L-*glycero*-D-*manno*-heptopyranose, and β-D-mannopyranose. There are no experimental data suggesting that head groups would contribute to the endotoxic properties of lipid A. However, they may be of importance for the organization and permeability of the outer membrane as well as the resistance of bacteria to certain antibiotics, including polymyxin B (135–138).

Fatty Acids

The acylation pattern of those lipid A that were structurally completely characterized are shown in Table 3. In the table a subdivision of lipid A structures is made according to the number (hepta-, hexa-, penta-, and tetra-acyl) of fatty acids present. Furthermore, this table is subdivided with regard to the distribution of the fatty acids over the backbone. For example, hexaacyl lipid A can have either an asymmetric, i.e., GlcN (II) having four and GlcN (I) containing two fatty acids (4 + 2), or symmetric acyl distribution (3 + 3). As lipid A isolated from different bacteria represent generally a mixture of several fractions varying in the number and position of the acyl groups, the classification given in Table 3 reverse only to the major fraction.

As a rule, the primary fatty acids of lipid A are β-hydroxylated, and only a few exceptions to this are known. Such exceptions involve the presence of 3-oxo fatty acids [14:0(3-oxo) in *R. sphaeroides* (position 2 of GlcN (I)] and *R. capsulatus* (positions 2 and 2′), respectively. Another exception is seen in *C. trachomatics* (26), where nonhydroxylated fatty acids are bound to the backbone.

The presence of olefinic fatty acids deserves special mention. These fatty acids appear exclusively as secondary fatty acids, thus forming acyloxyacyl residues. The configuration of dodecenoic acid being attached as a secondary fatty acid to 10:0(3-OH) in *R. capsulatus* (51) has recently been determined as *cis* (39). Thus, the structure of *R. capsulatus* is now fully elucidated. A synthetic analog of *R. sphaeroides*, which is known to be a potent lipid A antagonist, has been chemically synthesized (57,58). However, the structure of the natural counterpart is, in part, still unknown. This uncertainty concerns the *cis/trans* configuration of the tetradecenoic acid (14:1), which had never been investigated but was always presented in structural formulas as being *trans* (28,54–56). This fact became a matter of controversy, since small but significant differences between the ^{1}H-NMR spectra of natural and synthetic lipid A of *R. sphaeroides* were observed, but the reason for this discrepancy could not be assigned to a defined structure. This problem awaits further analysis (28,58).

Recently, mutants have been characterized that express LPS harboring a lipid A with an incomplete acylation pattern. Somerville et al. have shown that the *msbB* mutant of *E. coli* lacks myristoyl-transferase, i.e., a late acylation step of lipid A biosynthesis (98,124). Because of this defect, this mutant produces a pentaacyl lipid A with dodecanoyloxytetradecanoic acid [14: 0(3-*O*(12:0))] as the only acyloxyacyl residue in amide linkage (125). Interestingly, this mutation was conditionally not lethal for growth, thus indicating that outer membrane function and stability is maintained by LPS containing a pentaacylated lipid A. We have demonstrated pentaacyl lipid A to be present in lipid A preparations isolated from *E. coli* Re LPS (122,123,139). It represents a minor fraction not only in lipid A of *E. coli* LPS, but also in *S. enterica* sv. Typhimurium (S-form) (62,140), *S. enterica* sv. Minnesota (Re-mutant) (65), and *N. gonorrhoeae* (64). Recently, a *waaN*-mutant of *Salmonella enterica* sv. Typhimurium was constructed having intact S-form LPS but expressing a pentaacyl lipid A with diminished toxicity (157). In a mouse typhoid model this mutant provided to be nontoxic despite excellent growth in vivo. This study, therefore, provided for the first time evidence that death is directly dependent on the toxicity of lipid A (i.e., on its acylation pattern).

FINAL REMARKS

Lipid A structures of different bacterial origin are built up according to a conserved and common architectural principle. The backbone of lipid A, without exception, is always composed of a β-(1'$\rightarrow$6)-interlinked-D-gluco-configured hexosamine disaccharide. Lipid A, therefore, represents the most conserved structural element of the LPS molecule as compared to the core region or the O-specific chains, which express moderate to enormous structural variability. Modern lipid A analysis, however, has shown that with respect to the nature and number of polar head groups and fatty acids, a considerable variability is observed among lipid A of different origin. The term lipid A, therefore, does not define a single molecular entity possessing one defined structure, but rather describes the lipoid region of LPS, which represents the toxic component of LPS and which is important for the organization and function of the bacterial outer membrane and, thus, for microbial viability.

ACKNOWLEDGMENTS

The financial support of the German Research Ministry for Research and Development, "Septischer Schock, molekulare Pathophysiologie und therapeutische Ansätze" (01 KI 9471, UZ), the German-Israeli Foundation for Research & Development (Nr. I 373-169.13/94, UZ and ETR), the Deutsche Forschungs Gemeinschaft (DFG) ZA 149/3-1 (UZ), DFG/RFFI, project Nr. 436 RUS 113/314/0 (R/S) (UZ), Sonderforschungsbereich (SFB) 367, project B2 (ETR), and SFB 470, projects B4 (ETR) and B5 (UZ), the Graduiertenkolleg (GRK 288/1-96 project A2, UZ), and Fonds der Chemie (ETR) is greatly appreciated.

We are particularly grateful to our friends and colleagues H. Brade, K. Brandenburg, O. Holst, Y. A. Knirel, R. L. Pardy, and U. Seydel for valuable comments and suggestions.

REFERENCES

1. Rietschel ET, Galanos C, Lüderitz O, Westphal O. The chemistry and biology of lipopolysaccharides and their lipid A component. In: Webb DR, ed. Immunopharmacology and the Regulation of Leukocyte Function. New York: Marcel Dekker, 1982:183–229.
2. Rietschel ET, Kirikae T, Schade FU, Mamat U, Schmidt G, Loppnow H, Ulmer AJ, Zähringer U, Seydel U, Di Padova FE, Schreier M, Brade H. Bacterial endotoxin: molecular relationships of structure to activity and function. FASEB J 1994; 8:217–225.
3. Westphal O, Lüderitz O. Chemische Erforschung von Lipopolysacchariden Gram-negativer Bakterien. Angew Chem 1954; 66:407–417.
4. Morgan WTJ, Partridge SM. The use of phenol and of alkali in the degradation of antigenic material isolated from *Bact. dyseneteriae* (Shiga). Biochem J 1941; 35:1140–1163.
5. Lüderitz O, Jann K, Wheat R. Somatic and capsular antigens of gram-negative bacteria. In: Florkin M, Stotz EH, eds. Comprehensive Biochemistry. Amsterdam: Elsevier Publishing Company, 1968:105–228.
6. Zähringer U, Lindner B, Rietschel ET. Molecular structure of lipid A, the endotoxic center of bacterial lipopolysaccharides. Adv Carbohydr Chem Biochem 1994; 50:211–276.
7. Haeffner-Cavaillon N, Cavaillon J-M, Moreau M, Szabó L. Interleukin 1 secretion by human monocytes stimulated by the isolated polysaccharide region of the

Bordetella pertussis endotoxin. Mol Immunol 1984; 21:389–395.
8. Holst O, Brade H. Chemical structure of the core region of lipopolysaccharides. In: Morrison DC, Ryan JL, eds. Bacterial Endotoxic Lipopolysaccharides. Boca Raton, FL: CRC Press, 1992:135–170.
9. Takayama K, Qureshi N. Chemical structure of lipid A. In: Morrison DC, Ryan JL, eds. Bacterial Endotoxic Lipopolysaccharides. Vol. I: Molecular Biochemistry and Cellular Biology. Boca Raton, FL: CRC Press, 1992:43–65.
10. Burton AJ, Carter HE. Purification and characterization of the lipid A component of the lipopolysaccharides from *Escherichia coli*. Biochemistry 1964; 3:411–418.
11. Gmeiner J, Lüderitz O, Westphal O. Biochemical studies on lipopolysaccharides of *Salmonella* R mutants 6. Investigations on the structure of the lipid A component. Eur J Biochem 1969; 7:370–379.
12. Takayama K, Qureshi N, Mascagni P. Complete structure of lipid A obtained from the lipopolysaccharide of the heptoseless mutant of *Salmonella typhimurium*. J Biol Chem 1983; 258:12801–12803.
13. Imoto M, Shiba T, Naoki H, Iwashita T, Rietschel ET, Wollenweber H-W, Galanos C, Lüderitz O. Chemical structure of *E. coli* lipid A: linkage site of acyl groups in the disaccharide backbone. Tetrahedron Lett 1983; 24:4017–4020.
14. Imoto M, Yoshimura H, Sakaguchi N, Kusumoto S, Shiba T. Total synthesis of *Escherichia coli* lipid A. Tetrahedron Lett 1985; 26:1545–1548.
15. Galanos C, Lehmann V, Lüderitz O, Rietschel ET, Westphal O, Brade H, Brade L, Freudenberg MA, Hansen-Hagge T, Lüderitz T, McKenzie G, Schade U, Strittmatter W, Tanamoto K-I, Zähringer U, Imoto M, Yamamoto M, Shimamoto T, Kusumoto s, Shiba T. Endotoxic properties of chemically synthesized lipid A part structures: Comparison of synthetic lipid A precursor and synthetic analogues with biosynthetic lipid A precursor and free lipid A. Eur J Biochem 1984; 140:221–227.
16. Erwin AL, Munford RS, Jr. Deacylation of structurally diverse lipopolysaccharides by human acyloxyacyl hydrolase. J Biol Chem 1990; 265:16444–16449.
17. Weckesser J, Mayer H. Different lipid A types in lipopolysaccharides of phototrophic and related non-phototrophic bacteria. FEMS Microbiol Lett 1988; 54:143–154.
18. Mayer H, Krauss JH, Yokota A, Weckesser J. Natural variants of lipid A. In: Friedman H, Klein TW, Nakano M, Nowotny A, eds. Endotoxin. New York: Plenum Publishing Corporation, 1990:45–70.
19. Bhat UR, Forsberg LS, Carlson RW. Structure of lipid A component of *Rhizobium leguminosarum* bv. *phaseoli* lipopolysaccharide. Unique nonphosphorylated lipid A containing 2-amino-2-deoxygluconate, galacturonate, and glucosamine. J Biol Chem 1994; 269:14402–14410.
20. Price NPJ, Kelly TM, Raetz CRH, Carlson RW. Biosynthesis of a structurally novel lipid A in *Rhizobium leguminosarum*: identification and characterization of six metabolic steps leading from UDP-GlcNAc to 3-deoxy-D-*manno*-2-octulosonic acid$_2$-lipid IV$_A$. J Bacteriol 1994; 176:4646–4655.
21. Price NPJ, Jeyaretnam B, Carlson RW, Kadrmas JL, Raetz CRH, Brozek KA. Lipid A biosynthesis in *Rhizobium leguminosarum*: role of a 2-keto-3-deoxyoctulosonate-activated 4′ phosphatase. Proc Natl Acad Sci USA 1995; 92:7352–7356.
22. Mack D, Fischer W, Krokotsch A, Leopold K, Hartmann R, Egge H, Laufs R. The intercellular adhesin involved in biofilm accumulation of *Staphylococcus epidermidis* is a linear β-1,6-linked glucosaminoglycan: purification and structural analysis. J Bacteriol 1996; 178:175–183.
23. Montreuil J. Primary structure of glycoprotein glycans. Basis for the molecular biology of glycoproteins. Adv Carbohydr Chem Biochem 1980; 37:157–223.
24. Seidl PH, Schleifer KH, eds. Biological Properties of Peptidoglycan. Berlin: Walter de Gruyter, 1986.
25. Ikeda K, Miyajima K, Achiwa K. Lipid A and related compounds. 30. Synthesis of biologically active N,N'-diacyl chitobiose derivatives structurally related to lipid A. Chem Pharm Bull (Tokyo) 1996; 44:1958–1961.
26. Qureshi N, Kaltashov I, Walker K, Doroshenko V, Cotter RJ, Takayama K, Sievert TR, Rice PA, Lin J-SL, Golenbock DT. Structure of the monophosphoryl lipid A moiety obtained from the lipopolysaccharide of *Chlamydia trachomatis*. J Biol Chem 1997; 272:10594–10600.
27. Sonesson A, Moll H, Jantzen E, Zähringer U. Long-chain α-hydroxy-(ω-1)-oxo fatty acids and α-hydroxyl-1,ω-dioic fatty acids are cell wall constituents of *Legionella* (*L. jordanis*, *L. maceachernii* and *L. micdadei*). FEMS Microbiol Lett 1993; 106:315–320.
28. Qureshi N, Honovich J, Hara H, Cotter RJ, Takayama K. Location of fatty acids in lipid A obtained from lipopolysaccharide of *Rhodopseudomonas sphaeroides* ATCC 17023. J Biol Chem 1988; 263:5502–5504.
29. Salimath PV, Weckesser J, Strittmatter W, Mayer H. Structural studies on the non-toxic lipid A from *Rhodopseudomonas sphaeroides* ATCC 17023. Eur J Biochem 1983; 136:195–200.
30. Hollingsworth RI, Carlson RW. 27-Hydroxyoctacosanoic acid is a major structural fatty acyl component of the lipopolysaccharide of *Rhizobium trifolii* ANU 843. J Biol Chem 1989; 264:9300–9303.
31. Moll H, Sonesson A, Jantzen E, Marre R, Zähringer U. Identification of 27-oxo-octacosanoic acid and heptacosane-1,27-dioic acid in *Legionella pneumophila*. FEMS Microbiol Lett 1992; 97:1–6.
32. Weisshaar R, Lingens F. The lipopolysaccharide of a chloridazon-degrading bacterium. Eur J Biochem 1983; 137:155–161.
33. Zähringer U, Knirel YA, Lindner B, Helbig JH, Sonesson A, Marre R, Rietschel ET. The lipopolysaccharide of *Legionella pneumophila* serogroup 1 (strain Philadalphia 1): chemical structure and biological significance. Prog Clin Biol Res 1995; 392:113–139.
34. Holst O, Borowiak D, Weckesser J, Mayer H. Structural studies on the phosphate-free lipid A of *Rhodomicrobium vannielii* ATCC 17100. Eur J Biochem 1983; 137:325–332.

35. Strittmatter W, Weckesser J, Salimath PV, Galanos C. Nontoxic lipopolysaccharide from *Rhodopseudomonas sphaeroides* ATCC 17023. J Bacteriol 1983; 155:153–158.
36. Rau H, Seydel U, Freudenberg MA, Weckesser J, Mayer H. Lipopolysaccharide of *Rhodospirillum salinarum* 40: structural studies on the core and lipid A region. Arch Microbiol 1995; 164:280–289.
37. Wollenweber H-W, Seydel U, Lindner B, Lüderitz O, Rietschel ET. Nature and location of amide-bound (R)-3-acyloxyacyl groups in lipid A of lipopolysaccharides from various gram-negative bacteria. Eur J Biochem 1984; 145:265–272.
38. Wollenweber H-W, Broady KW, Lüderitz O, Rietschel ET. The chemical structure of lipid A: demonstration of amide-linked 3-acyloxyacyl residues in *Salmonella minnesota* Re lipopolysaccharide. Eur J biochem 1982; 124:191–198.
39. Mayer H, Merkofer T, Warth C, Weckesser J. Position and configuration of double bonds of lipid A-associated monounsaturated fatty acids of *Proteobacteria* and *Rhodobacter capsulatus* 37b4. J Endotoxin Res 1996; 3:345–352.
40. Russa R, Lüderitz O, Rietschel ET. Structural analyses of lipid A from lipopolysaccharides of nodulating and non-nodulating *Rhizobium trifolii*. Arch Microbiol 1985; 141:284–289.
41. Urbanik-Sypniewska T, Seydel U, Greck M, Weckesser J, Mayer H. Chemical studies on the lipopolysaccharide of *Rhizobium meliloti* 10406 and its lipid A region. Arch Microbiol 1989; 152:527–532.
42. Mayberry WR. Dihydroxy and monohydroxy fatty acids in *Legionella pneumophila*. J Bacteriol 1981; 147:373–381.
43. Maitra SK, Schotz MC, Yoshikawa TT, Guze LB. Determination of lipid A and endotoxin in serum by mass spectroscopy. Proc Natl Acad Sci USA 1978; 75:3993–3997.
44. Brandtzaeg P, Bryn K, Kierulf P, Ovstebo R, Namork E, Aase B, Jantzen E. Meningococcal endotoxin in lethal septic shock plasma studied by gas chromatography, mass-spectrometry, ultracentrifugation, and electron microscopy. J Clin Invest 1992; 89:816–823.
45. Wilkinson SG. Gram-negative bacteria. In: Ratledge C, Wilkinson SG, eds. Microbial Lipids. London: Academic Press, 1988:299–488.
46. Choe B-Y, Krishna NR, Pritchard DG. Proton NMR study on rhamnolipids produced by *Pseudomonas aeruginosa*. Magn Reson Chem 1992; 30:1025–1029.
47. de Koster CG, Vos B, Versluis C, Heerma W, Haverkamp J. High-performance thin-layer chromatography/fast atom bombardment (tandem) mass spectrometry of *Pseudomonas* rhamnolipids. Biol Mass Spectrom 1994; 23:179–185.
48. Boylan DB, Scheuer PJ. Pahutoxin: a fish poison. Science 1967; 155:52–56.
49. Aydin M, Lucht N, König WA, Lupp R, Jung G, Winkelmann G. Structure elucidation of the peptide antibiotics herbicolin A and B. Liebigs Ann Chem 1985: 2285–2300.
50. Moormann M, Zähringer U, Moll H, Kaufmann R, Schmidt R, Altendorf K. A new glycosylated lipopeptide incorporated into the cell wall of a smooth variant of *Gordona hydrophobica*. J Biol Chem 1997; 272:10729–10738.
51. Krauss JH, Seydel U, Weckesser J, Mayer H. Structural analysis of the nontoxic lipid A of *Rhodobacter capsulatus* 37b4. Eur J Biochem 1989; 180:519–526.
52. Purcell JM, Morris SG, Susi H. Proton magnetic resonance spectra of unsaturated fatty acids. Anal Chem 1966; 38:588–592.
53. Frost DJ, Gunstone FD. The PMR analysis of non-conjugated alkenoic and alkynoic acids and esters. Chem Phys Lipids 1975; 15:53–85.
54. Takayama K, Qureshi N, Beutler B, Kirkland TN. Diphosphoryl lipid A from *Rhodopseudomonas sphaeroides* ATCC 17023 blocks induction of cachectin in macrophages by lipopolysaccharide. Infect Immun 1989; 57:1336–1338.
55. Qureshi N, Takayama K, Kurtz R. Diphosphoryl lipid A obtained from the nontoxic lipopolysaccharide of *Rhodopseudomonas sphaeroides* is an endotoxin antagonist in mice. Infect Immun 1991; 59:441–444.
56. Kirkland TN, Qureshi N, Takayama K. Diphosphoryl lipid A derived from lipopolysaccharide (LPS) of *Rhodopseudomonas sphaeroides* inhibits activation of 70Z/3 cells by LPS. Infect Immun 1991; 59:131–136.
57. Rose JR, Christ WJ, Bristol JR, Kawata T, Rossignol DP. Agonistic and antagonistic activities of bacterially derived *Rhodobacter sphaeroides* lipid A: comparison with activities of synthetic material of the proposed structure and analogs. Infect Immun 1995; 63:833–839.
58. Christ WJ, McGuiness PD, Asano O, Wang Y, Mullarkey MA, Perez M, Hawkins LD, Blythe TA, Dubuc GR, Robidoux AL. Total synthesis of the proposed structure of *Rhodobacter sphaeroides* lipid A resulting in the synthesis of new potent lipopolysaccharide antagonists. J Am Chem Soc 1994; 116:3637–3638.
59. Rietschel ET, Gottert H, Lüderitz O, Westphal O. Nature and likages of the fatty acids present in the lipid-A component of *Salmonella lipopolysaccharides*. Eur J Biochem 1972; 28:166–173.
60. Rietschel ET. Absolute configuration of 3-hydroxy fatty acids present in lipopolysaccharides from various bacterial groups. Eur J Biochem 1976; 64:423–428.
61. Clementz T, Bednarski JJ, Raetz CRH. Function of the htrB high temperature requirement gene of *Escherichia coli* in the acylation of lipid A. J Biol Chem 1996; 271:12095–12102.
62. Qureshi N, Takayama K, Ribi EE. Purification and structural determination of nontoxic lipid A obtained from the lipopolysaccharide of *Salmonella typhimurium*. J Biol Chem 1982; 257:11808–11815.
63. Qureshi N, Takayama K, Heller DN, Fenselau C. Position of ester groups in the lipid A backbone of lipopolysaccharides obtained from *Salmonella typhimurium*. J Biol Chem 1983; 258:12947–12951.
64. Takayama K, Qureshi N, Hyver K, Honovich J, Cotter RJ, Mascagni P, Schneider H. Characterization of a structural series of lipid A obtained from the lipopolysaccharides of *Neisseria gonorrhoeae*. J Biol Chem 1986; 261:10624–10631.
65. Johnson RS, Her G-R, Grabarek J, Hawiger J, Reinhold VN. Structural characterization of monophos-

phoryl lipid A homologs obtained from *Salmonella minnesota* Re595 lipopolysaccharide. J Biol Chem 1990; 265:8108–8116.
66. Fukuoka S, Kamishima H, Nagawa Y, Nakanishi H, Ishikawa K, Niwa Y, Tamiya E, Karube I. Structural characterization of lipid A component of *Erwinia carotovora* lipopolysaccharide. Arch Microbiol 1992; 157:311–318.
67. Karunaratne DN, Richards JC, Hancock REW. Characterization of lipid A from *Pseudomonas aeruginosa* O-antigenic B band lipopolysaccharide by 1D and 2D NMR and mass spectral analysis. Arch Biochem Biophys 1992; 299:368–376.
68. Kulshin VA, Zähringer U, Lindner B, Jäger K-E, Dmitriev BA, Rietschel ET. Structural characterization of the lipid A component of *Pseudomonas aeruginosa* wild-type and rough mutant lipopolysaccharides. Eur J Biochem 1991; 198:697–704.
69. Kaltashov IA, Walker K, Doroshenko V, Cotter RJ, Qureshi N. Development of strategies for lipid A characterization by methods of mass spectrometry. Proceedings of the 44th ASMS Conference on Mass Spectrometry and Allied topics. Orlando, FL, 1996; p. 584.
70. Fukuoka S, Kakita H, Obika H, Ishikawa K. Structural characterization of lipid A using fast atom bombardment mass spectrometry based on ionization method dependent fragmentation. Proceedings of Advances in Mass Spectrometry. Vol. 14. Amsterdam: Elsevier Publishing Company (in press).
71. Seydel U, Lindner B, Wollenweber H-W, Rietschel ET. Structural studies on the lipid A component of enterobacterial lipopolysaccharides by laser desorption mass spectrometry. Location of acyl groups at the lipid A backbone. Eur J Biochem 1984; 145:505–509.
72. Seydel U, Lindner B, Zähringer U, Rietschel ET. Laser desorption mass spectrometry of synthetic lipid A-like compounds. Biomed Mass Spectrom 1984; 11: 132–141.
73. Lindner B, Zähringer U, Rietschel ET, Seydel U. Structural elucidation of lipopolysaccharides and their lipid A component: application of soft ionization mass spectrometry. In: Fox A, Morgan SL, Larsson L, Odham G, eds. Analytical Microbiology Methods. New York: Plenum Press, 1990:149–161.
74. Cotter RJ, Honovich J, Qureshi N, Takayama K. Structural determination of lipid A from Gram negative bacteria using laser desorption mass spectrometry. Biomed Environ Mass Spectrom 1987; 14:591–598.
75. Domon B, Costello CE. Oligosaccharide fragmentation nomenclature. Glycoconjugate J 1988; 5:397–409.
76. Moran AP, Zähringer U, Seydel U, Scholz D, Stütz P, Rietschel ET. Structural analysis of the lipid A component of *Campylobacter jejuni* CCUG 10936 (serotype O:2) lipopolysaccharide. Description of a lipid A containing a hybrid backbone of 2-amino-2-deoxy-D-glucose and 2,3-diamino-2,3-dideoxy-D-glucose. Eur J Biochem 1991; 198:459–469.
77. Masoud H, Lindner B, Weckesser J, Mayer H. The structure of the lipid A component of *Rhodocyclus gelatinosus* Dr₂ lipopolysaccharide. System Appl Microbiol 1990; 13:227–233.
78. Weintraub A, Zähringer U, Wollenweber H-W, Seydel U, Rietschel ET. Structural characterization of the lipid A component of *Bacteroides fragilis* strain NCTC 9343 lipopolysaccharide. Eur J Biochem 1989; 183: 425–431.
79. Kulshin VA, Zähringer U, Lindner B, Frasch CE, Tsai C-M, Dmitriev BA, Rietschel ET. Structural characterization of the lipid A component of pathogenic *Neisseria meningitidis*. J Bacteriol 1992; 174:1793–1800.
80. Masoud H, Urbanik-Sypniewska T, Lindner B, Weckesser J, Mayer H. The structure of the lipid A component of *Sphaerotilus natans*. Arch Microbiol 1991; 156:167–175.
81. Helander IM, Kato Y, Kilpeläinen I, Kostiainen R, Lindner B, Nummila K, Sugiyama T, Yokochi T. Characterization of lipopolysaccharides of polymyxin-resistant and polymyxin-sensitive *Klebsiella pneumoniae* 03. Eur J Biochem 1996; 237:272–278.
82. Rau H, Seydel U, Freudenberg MA, Weckesser J, Mayer H. Lipopolysaccharide of the phototrophic bacterium *Rhodospirillum fulvum*. System Appl Microbiol 1995; 18:154–163.
83. Helander IM, Kilpeläinen I, Vaara M, Moran AP, Lindner B, Seydel U. Chemical structure of the lipid A component of lipopolysaccharides of the genus *Pectinatus*. Eur J Biochem 1994; 224:63–70.
84. Moran AP, Lindner B, Walsh EJ. Structural characterization of the lipid A component of *Helicobacter pylori* rough- and smooth-form lipopolysaccharides. J Bacteriol 1997; 179:6453–6463.
85. Caroff M, Deprun C, Karibian D. ^{252}Cf plasma desorption mass spectrometry applied to the analysis of underivatized rough-type endotoxin preparations. J Biol Chem 1993; 268:12321–12324.
86. Lebbar S, Karibian D, Deprun C, Caroff M. Distribution of lipid A species between long and short chain lipopolysaccharides isolated from *Salmonella*, *Yersinia*, and *Escherichia* as seen by ^{252}Cf plasma desorption mass spectrometry. J Biol Chem 1994; 269: 31881–31884.
87. Karibian D, Deprun C, Szabó L, Le Beyec Y, Caroff M. ^{252}Cf-plasma desorption mass spectrometry applied to the analysis of endotoxin lipid A preparations. Int J Mass Spectr 1991; 111:273–286.
88. Qureshi N, Takayama K, Mascagni P, Honovich J, Wong R, Cotter RJ. Complete structural determination of lipopolysaccharide obtained from deep rough mutant of *Escherichia coli*. J Biol Chem 1988; 263: 11971–11976.
89. Wang R, Chen L, Cotter RJ, Qureshi N, Takayama K. Fragmentation of lipopolysaccharide anchors in plasma desorption mass spectrometry. J Microbiol Methods 1992; 15:151–166.
90. Karibian D, Deprun C, Caroff M. Comparison of lipids A of several *Salmonella* and *Escherichia* strains by ^{252}Cf plasma desorption mass spectrometry. J Bacteriol 1993; 175:2988–2993.
91. Jiao B, Freudenberg MA, Galanos C. Characterization of the lipid A component of genuine smooth-form lipopolysaccharide. Eur J Biochem 1989; 180:515–518.
92. Masoud H, Weintraub ST, Wang R, Cotter RJ, Holt SC. Investigation of the structure of lipid A from *Ac-*

tinobacillus actinomycetemcomitans strain Y4 and human clinical isolate PO 1021-7. Eur J Biochem 1991; 200:775–779.
93. Qureshi N, Takayama K, Meyer KC, Kirkland TN, Bush CA, Chen L, Wang R, Cotter RJ. Chemical reduction of 3-oxo and unsaturated groups in fatty acids of diphosphoryl lipid A from the lipopolysaccharide of *Rhodopseudomonas sphaeroides*. J Biol Chem 1991; 266:6532–6538.
94. Cole RB, Domelsmith LN, David CM, Laine RA, DeLucca AJ. ^{252}Cf plasma-desorption mass spectrometry of lipid A from *Enterobacter agglomerans*. Rapid Commun Mass Spetrom 1992; 6:616–622.
95. Harrata AK, Domelsmith LN, Cole RB. Electrospray mass spectrometry for characterization of lipid A from *Enterobacter agglomerans*. Biol Mass Spec 1993; 22:59–67.
96. Chan S, Reinhold VN. Detailed structural characterization of lipid A: electrospray ionization coupled with tandem mass spectrometry. Anal Biochem 1994; 218:63–73.
97. Kelly J, Masoud H, Perry MB, Richards JC, Thibault P. Separation and characterization of O-deacylated lipooligosaccharides and glycans derived from *Moraxella catarrhalis* using capillary electrophoresis electrospray mass spectrometry and tandem mass spectrometry. Anal Biochem 1996; 233:15–30.
98. Raetz CRH. Bacterial endotoxins: extraordinary lipids that activate eucaryotic signal transduction. J Bacteriol 1993; 175:5745–5753.
99. Lee NG, Sunshine MG, Engstrom JJ, Gibson BW, Apicella MA. Mutation of the *htrB* locus of *Haemophilus influenzae* nontypable strain 2019 is associated with modifications of lipid A and phosphorylation of the lipo-oligosaccharide. J Biol Chem 1995; 270:27151–27159.
100. Melaugh W, Phillips NJ, Campagnari AA, Karalus R, Gibson BW. Partial characterization of the major lipooligosaccharide from a strain of *Haemophilus ducreyi*, the causative agent of chancroid, a genital ulcer disease. J Biol Chem 1992; 267:13434–13439.
101. Gibson BW, Melaugh W, Phillips NJ, Apicella MA, Campagnari AA, Griffiss JM. Investigation of the structural heterogeneity of lipooligosaccharides from pathogenic *Haemophilus* and *Neisseria* species and of R-type lipopolysaccharides from *Salmonella typhimurium* by electrospray mass spectrometry. J Bacteriol 1993; 175:2702–2712.
102. Horton D, Riley DA. ^{31}P nuclear magnetic resonance spectroscopy of lipopolysaccharides from *Pseudomonas aeruginosa*. Biochim Biophys Acta 1981; 640:727–733.
103. Knirel YA, Helbig JH, Zähringer U. Structure of a decasaccharide isolated by mild acid degradation and dephosphorylation of the lipopolysaccharide of *Pseudomonas fluorescens* strain ATCC 49271. Carbohydr Res 1996; 283:129–139.
104. Karas M, Bahr U, Strupat K, Hillenkamp F. Matrix dependence of metastable fragmentation of glycoproteins in MALDI TOF mass spectrometry. Anal Chem 1995; 67:675–679.
105. Niedziela T, Petersson C, Helander A, Jachymek W, Kenne L, Lugowski C. Structural studies of the O-specific polysaccharide of *Hafnia alvei* strain 1209 lipopolysaccharide. Eur J Biochem 1996; 237:635–641.
106. Jachymek W, Petersson C, Helander A, Kenne L, Niedziela T, Lugowski C. Structural studies of the O-specific chain of *Hafnia alvei* strain 32 lipopolysaccharide. Carbohydr Res 1996; 292:117–128.
107. Fukuoka S, Knirel YA, Lindner B, Moll H, Seydel U, Zähringer U. Elucidation of the structure of the core region and the complete structure of the R-type lipopolysaccharide of *Erwinia carotovora* FERM P-7576. Eur J Biochem 1997; 250:55–62.
108. Gibson BW, Engstrom JJ, John CM, Hines W, Falick AM. Characterization of bacterial lipopolysaccharides by delayed extraction matrix-assisted laser desorption time-of-flight mass spectrometry. J Am Soc Mass Spectrom 1997; 8:645–648.
109. Lindner B, PSD-MALDI-MS of lipid A: a tool for fatty acid analysis. Proceedings of Advances in Mass Spectrometry, Vol. 14: Amsterdam: Elsevier Publishing Company (in press).
110. Qureshi N, Takayama K, Sievert TR, Manthey CL, Vogel S, Hronowski X, Cotter RJ. Novel method for the purification and characterization of lipopolysaccharide from *Escherichia coli* D31m3. Prog Clin Biol Res 1995; 392:151–159.
111. Agrawal PK, Bush CA, Qureshi N, Takayama K. Structural analysis of lipid A and Re-lipopolysaccharide by NMR spectroscopic methods. Adv Biophys Chem 1994; 4:179–236.
112. Wang Y, Hollingsworth RI. A Solvent system for the high-resolution proton nuclear magnetic resonance spectroscopy of membrane lipids. Anal Biochem 1995; 225:242–251.
113. Oikawa M, Shintaku T, Fukase K, Kusumoto S. Conformational analysis of a biosynthetic precursor of lipid A. Eurocarb$_9$, Utrecht, The Netherlands July 6-11, 1997, p. 369.
114. Kumada H, Haishima Y, Umemoto T, Tanamoto K-I. Structural study on the free lipid A isolated from lipopolysaccharide of *Porphyromonas gingivalis*. J Bacteriol 1995; 177:2098–2106.
115. Bryn K, Rietschel ET. L-2-Hydroxytetradecanoic acid as a constituent of *Salmonella* lipopolysaccharides (lipid A). Eur J Biochem 1978; 86:311–315.
116. Iida T, Haishima Y, Tanaka A, Nishiyama K, Saito S, Tanamoto K-I. Chemical structure of lipid A isolated from *Comamonas testosteroni* lipopolysaccharide. Eur J Biochem 1996; 237:468–475.
117. Batley M, Packer NH, Redmond JW. Configurations of glycosidic phosphates of lipopolysaccharides from *Salmonella minnesota* R595. Biochemistry 1982; 21:6580–6586.
118. Qureshi N, Mascagni P, Ribi E, Takayama K. Monophosphoryl lipid A obtained from the lipopolysaccharides of *Salmonella minnesota* R595. Purification of the dimethyl derivative by high performance liquid chromatography and complete structural determination. J Biol Chem 1985; 260:5271–5278.
119. Tsukioka D, Nishizawa T, Miyashita T, Achiwa K, Suda T, Soma G-I, Mizuno D-I. Structural characterization of lipid A obtained from *Pantoea agglomerans* lipopolysaccharide. FEMS Microbiol Lett 1997; 149:239–244.

120. Masoud H, Perry MB, Richards JC. Characterization of the lipopolysaccharide of *Moraxella catarrhalis*. Structural analysis of the lipid A from *M. catarrhalis* serotype A lipopolysaccharide. Eur J Biochem 1994; 220:209–216.

121. Ogawa T. Chemical structure of lipid A from *Porphyromonas (Bacteroides) gingivalis* lipopolysaccharide. FEBS Lett 1993; 332:197–201.

122. Baltzer LH, Mattsby-Baltzer I. Heterogeneity of lipid A: structural determination by ^{13}C and ^{31}P NMR of lipid A fractions from lipopolysaccharide of *Escherichia coli*. Biochemistry 1986 25:3570–3575.

123. Zähringer U, Salvetzki R, Ulmer AJ, Rietschel ET. Isolation, chemical analyses and biological investigations of natural lipid A and lipid A-antagonists. J Endotoxin Res 1996; 3 (suppl):33.

124. Brozek KA, Raetz CRH. Biosynthesis of lipid A in *Escherichia coli*. J Biol Chem 1990; 265:15410–15417.

125. Somerville JE, Jr., Cassiano L, Bainbridge B, Cunningham MD, Darveau RP. A novel *Escherichia coli* lipid A mutant that produces an antiinflammatory lipopolysaccharide. J Clin Invest 1996; 97:359–365.

126. Strain SM, Armitage IM, Anderson L, Takayama K, Qureshi N, Raetz CRH. Location of polar substituents and fatty acyl chains on lipid A precursor from a 3-deoxy-D-manno-octulosonic acid-deficient mutant of *Salmonella typhimurium*. J Biol Chem 1985; 260:16089–16098.

127. Strain SM, Fesik SW, Armitage IM. Structure and metal-binding properties of lipopolysaccharides from heptoseless mutants of *Escherichia coli* studied by ^{13}C and ^{31}P nuclear magnetic resonance. J Biol Chem 1983; 258:13466–13477.

128. Strain SM, Armitage IM. Selective detection of 3-deoxymannooctulosonic acid in intact lipopolysaccharides by spin-echo ^{13}C NMR. J Biol Chem 1985; 260:12974–12977.

129. Strain SM, Fesik SW, Armitage IM. Characterization of lipopolyaccharide from a heptoseless mutant of *Escherichia coli* by carbon 13 nuclear magnetic resonance. J Biol Chem 1983; 258:2906–2910.

130. Qureshi N, Takayama K, Seydel U, Wang R, Cotter RJ, Agrawal PK, Bush CA, Kurtz R, Berman DT. Structural analysis of the lipid A derived from the lipopolysaccharide of *Brucella abortus*. J Endotoxin Res 1994; 1:137–148.

131. Wang Y, Hollingsworth RI. An NMR spectroscopy and molecular mechanics study of the molecular basis for the supramolecular structure of lipopolysaccharides. Am Chem Soc 1996; 35:5647–5654.

132. Mühlradt PF, Wray V, Lehmann V. A ^{31}P-nuclear-magnetic-resonance study of the phosphate groups in lipopolysaccharide and lipid A from *Salmonella*. Eur J Biochem 1977; 81:193–203.

133. Bhat UR, Marx A, Galanos C, Conrad RS. Structural studies of lipid A from *Pseudomonas aeruginosa* PAO1: occurrence of 4-amino-4-deoxyarabinose. J Bacteriol 1990; 172:6631–6636.

134. Kastowsky M, Sabisch A, Gutberlet T, Bradaczek H. Molecular modelling of bacterial deep rough mutant lipopolysaccharide of *Escherichia coli*. Eur J Biochem 1991; 197:707–716.

135. Vaara M, Vaara T, Jensen M, Helander IM, Nurminen M, Rietschel ET, Mäkelä PH. Characterization of the lipopolysaccharide from the polymyxin-resistant *pmrA* mutants of *Salmonella typhimurium*. FEBS Lett 1981; 129:145–149.

136. Helander IM, Kilpeläinen I, Vaara M. Increased substitution of phosphate groups in lipopolysaccharides and lipid A of the polymyxin-resistant *pmrA* mutants of *Salmonella typhimurium*: a ^{31}P-MNR study. Mol Microbiol 1994; 11:481–487.

137. Nummila K, Kilpeläinen I, Zähringer U, Vaara M, Helander IM. Lipopolysaccharides of polymyxin B-resistant mutants of *Escherichia coli* are extensively substituted by 2-aminoethyl pyrophosphate and contain 4-amino-4-deoxy-L-arabinose. Mol Microbiol 1995; 16:271–278.

138. Seltmann G, Lindner B, Holst O. Resistance of *Serratia marcescens* to polymyxin B: a comparative investigation of two S-form lipopolysaccharides obtained from a sensitive and a resistant variant of strain 111. J Endotoxin Res 1996; 3:497–504.

139. Mattsby-Baltzer I, Gemski P, Alving CR. Heterogeneity of lipid A: comparison of lipid A types from different Gram-negative bacteria. J Bacteriol 1984; 159:900–904.

140. Raetz CRH, Purcell S, Meyer MV, Qureshi N, Takayama K. Isolation and characterization of eight lipid A precursors from a 3-deoxy-D-manno-octulosonic acid-deficient mutant of *Salmonella typhimurium*. J Biol Chem 1985; 260:16080–16088.

141. Caroff M, Deprun C, Richards JC, Karibian D. Structural characterization of the lipid A of *Bordetella pertussis* 1414 endotoxin. J Bacteriol 1994; 176:5156–5159.

142. Cox AD, Wilkinson SG. Ionizing groups in lipopolysaccharides of *Pseudomonas cepacia* in relation to antibiotic resistance. Mol Microbiol 1991; 5:641–646.

143. Hase S, Rietschel ET. The chemical structure of the lipid A component of lipopolysaccharides from *Chromobacterium violaceum* NCTC 9694. Eur J Biochem 1977; 75:23–34.

144. Rosner MR, Khorana HG, Satterthwait AC. The structure of lipopolysaccharide from a heptose-less mutant of *Escherichia coli* K-12. II. The application of ^{31}P NMR spectroscopy. J Biol Chem 1979; 254:5918–5925.

145. Rosner MR, Tang J, Barzilay I, Khorana HG. Structure of the lipopolysaccharide from an *Escherichia coli* heptose-less mutant. I. Chemical degradation and identification of products. J Biol Chem 1979; 254:5906–5017.

146. Helander IM, Lindner B, Brade H, Altmann K, Lindberg AA, Rietschel ET, Zähringer U. Chemical structure of the lipopolysaccharide of *Haemophilus influenzae* strain I-69 Rd$^-$/b$^+$: Description of a novel deep-rough chemotype. Eur J Biochem 1988; 177:483–492.

147. Basu S, Radziejewska-Lebrecht J, Mayer H. Lipopolysaccharide of *Providencia rettgeri*. Chemical studies and taxonomical implications. Arch Microbiol 1986; 144:213–218.

148. Tharanathan RN, Salimath PV, Weckesser J, Mayer H. The structure of lipid A from the lipopolysaccharide

149. Tharanathan RN, Weckesser J, Mayer H. Structural studies on the D-arabinose-containing lipid A from *Rhodospirillum tenue* 2761. Eur J Biochem 1978; 84:385–394.
150. Bhat UR, Kontrohr T, Mayer H. Structure of *Shigella sonnei* lipid A. FEMS Microbiol Lett 1987; 40:189–192.
151. Broady KW, Rietschel ET, Lüderitz O. The chemical structure of the lipid A component of lipopolysaccharides from *Vibrio cholerae*. Eur J Biochem 1981; 115:463–468.
152. Dalla Venezia N, Minka S, Bruneteau M, Mayer H, Michel G. Lipopolysaccharides from *Yersinia pestis* studies on lipid A of lipopolysaccharides I and II. Eur J Biochem 1985; 151:399–404.
153. Wang Y, Cole RB. Acid and base hydrolysis of lipid A from *Enterobacter agglomerans* as monitored by electrospray ionization mass spectrometry: pertinence to detoxification mechanisms. J Mass Spectrom 1996; 31:138–149.
154. Sidorczyk Z, Zähringer U, Rietschel ET. Chemical structure of the lipid A component of the lipopolysaccharide from a *Proteus mirabilis* Re-mutant. Eur J Biochem 1983; 137:15–22.
155. Basu S, Tharanathan RN, Kontrohr T, Mayer H. Chemical structure of the lipid A component of *Plesiomonas shigelloides* and its taxonomical significance. FEMS Microbiol Lett 1985; 28:7–10.
156. Knirel YA, Helbig JH, Großkurth H, Zähringer U. Structures of decasaccharide and tridecasaccharide tetraphosphates isolated by strong alkaline degradation of *O*-deacylated lipopolysaccharide of *Pseudomonas fluorescens* strain ACTC 49271. Carbohydr Res 1995; 279:215–226.
157. Khan SA, Everest P, Servos S, Foxwell N, Zähringer U, Brade H, Rietschel ET, Dougan G,, Charles IG, Maskell DJ. A lethal role for Lipid A in *Salmonella* infections. Mol Microbiol 1998; 29:571–579.

8

Chemical Structure of the Core Region of Lipopolysaccharides

Otto Holst
Research Center Borstel, Borstel, Germany

INTRODUCTION

Depending on the size of the saccharide portion, there exist two types of lipopolysaccharides (LPS), i.e., smooth- and rough-form (S- and R-form) LPS (1–3). Both consist of lipid A and, covalently linked to it, a saccharide portion composed of up to 15 sugars—the core region (3–5). In S-form LPS, this core region is replaced by the O-specific polysaccharide. Both LPS forms are found in wild-type gram-negative bacteria: the S-form, for example, in *Escherichia coli*, *Klebsiella pneumoniae*, *Acinetobacter*, or *Vibrio cholerae*, and the R-form in *Neisseria meningitidis*, *N. gonorrhoeae*, *Haemophilus influenzae*, and *Bordetella pertussis*. Mutants that are not able to synthesize a minimal core structure are not viable. Thus, the core region and lipid A represent a common structural unit occurring in all LPS, suggesting these components to be important for viability and membrane function. Herein, the chemical structure of the core regions of LPS is reviewed. Although part of the work was summarized recently (3–5), this review contains all structures presently known and the respective references.

GENERAL STRUCTURAL FEATURES

A comparison of known structures of core regions of different bacterial origin shows that chemical variation is more limited than in the structures of O-antigenic chains. However, the only structural element that is present in *all* core regions is the 3-deoxy-D-*manno*-oct-2-ulopyranosonic acid [Kdo, Fig. 1 (6)] residue that links the core to the lipid A. A majority of core regions is characterized by the additional presence of L-*glycero*-D-*manno*-heptopyranose (L,D-Hep, Fig. 1) and of the oligosaccharide L-α-D-Hep-(1→3)-L-α-D-Hep-(1→5)-[α-Kdo-(2→4)]-α-Kdo. In this chapter, for the residues of this structure the short forms Hep II, Hep I, Kdo II, and Kdo I, respectively, are used. Variations in these core structures occur by different substitutions. Despite other sugar substituents, most of the investigated core regions also carry phosphate residues. In addition to L,D-Hep, some LPS contain D-*glycero*-D-*manno*-heptopyranose (D,D-Hep, Fig. 1), which is the biosynthetic precursor of L,D-Hep (7–9). Some LPS contain only D,D-Hep (10) or even lack heptose (11–13). D,D-Hep may be condensed to a heptoglycan, which was identified in strains of *Klebsiella pneumoniae* (α-(1→2)-linked; see Table 1) and *Helicobacter pylori* (α-(1→3)-linked; see Table 6). In the last case, the heptoglycan builds up an intervening region between core and O-specific polysaccharide. Nothing is known about the biosynthesis of such heptoglycans and their biological functions. Either Kdo I [in *Acinetobacter* (11–13)] or Kdo II [in *Burkholderia cepacia* (14,15)] may be replaced by another, stereochemically similar sugar, D-*glycero*-D-*talo*-oct-2-ulopyranosonic acid [Ko, Fig. 1 (16)]. The biosynthesis of Ko and the regulation of the exchange between Kdo and Ko are so far unknown.

The structural analysis of the core region was hampered for a long time by the difficult chemistry of Kdo and the phosphate substitution of core sugars. While the first problem was solved in the mid-1980s, the last

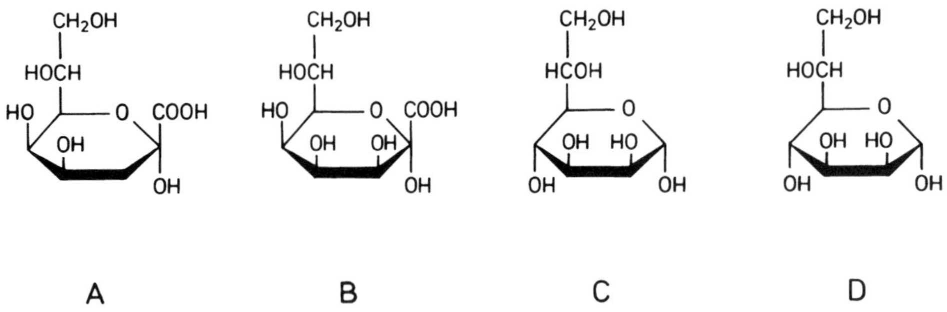

Fig. 1 Characteristic sugars of the core region: (A) 3-deoxy-D-*manno*-oct-2-ulosonic acid (Kdo); (B) D-*glycero*-D-*talo*-oct-2-ulosonic acid (Ko); (C) D-*glycero*-D-*manno*-heptopyranose; (C) D-*glycero*-D-*manno*-heptopyranose.

has still not been fully overcome. We have developed a degradation pathway to obtain phosphorylated oligosaccharides from LPS (17,18) using successive mild hydrazinolysis and strong alkali treatment and isolation of pure compounds by high-performance anion-exchange chromatography (HPAE). The application of this methodology has led to the characterization of various core regions from bacteria of remote genetic origin (for review see Ref. 19). However, phosphodiester, diphosphate, diphosphodiester, acetyl, and carbamoyl groups, which are present in many LPS, are cleaved under strong alkaline conditions, and their positions can thus not be identified. Also, the alkaline treatment gives rise to phosphate migration in case of substitution by 2-aminoethyl phosphate [PE (20–22)] and to β-elimination of 4-substituted galacturonic acid (GalA) residues (23,24).

If not otherwise identified in the following presentation, sugars are present as α-D-pyranosides.

THE CORE STRUCTURES

Heptose-Containing Core Regions

Enterobacteriaceae

General Structural Features of Salmonella enterica–*Type Core Oligosaccharides* Core structures belonging to this type (Table 1) contain a common structural element of L,D-Hep-(1→7)-L,D-Hep-(1→3)-L,D-Hep-(1→5)-α-Kdo, which is replaced at O-3 of Hep II by Glc. Heptose residues are phosphorylated, and position O-4 of Hep I is not replaced by a saccharide.

Salmonella enterica Until recently, all LPS of *S. enterica* were thought to possess only one core structure, which was elucidated mainly using LPS from rough mutants of serovar (sv.) *minnesota*. This structure (25–39), together with those proposed for the core region of LPS from *E. coli* and *Shigella* (see below), gave rise to a further subdivision of the core region into the inner and the outer core regions, which could be distinguished by their sugar composition: the inner core consists of heptopyranoses and Kdo, and the outer core of hexopyranoses. Although it has been shown in the meantime that hexopyranoses may be present in the inner and heptopyranoses in the outer core (see below), this subdivision is still useful for descriptive reasons. The *Salmonella* core structure was reported to be substituted by a phosphate residue at Hep II (at unknown position) and by a 2-aminoethyl diphosphate (PPE) residue at O-4 of Hep I. In addition, Kdo 7-(2-aminoethyl phosphate) was identified in various rough mutants of *S. enterica* sv. *minnesota*, and it was shown that this substituent is localized at Kdo II (36).

It is now proven that a second core type exists in *Salmonella* LPS (40–43). Using a monoclonal antibody directed against the terminal α-(1→2)-linked GlcNAc residue of the Ra core of *S. enterica* sv. *minnesota* (compare Table 1), a variety of wild-type isolates and laboratory strains were serologically screened. It was found that this antibody failed to react with strains of *S. enterica* sv. *adelaide*, *djakarta*, and *arizonae*. Structural analyses of rough mutant LPS of *S. enterica* sv. *djakarta* indicated that the terminal GlcNAc residue was replaced by Glc (42). Recently, this new core type was fully established for S-form LPS of *S. enterica* sv. *arizonae* (43). Both the terminal GlcNAc and Glc residues are present in nonstoichiometric amounts in the S-form LPS of *S. enterica* sv. *minnesota* and *arizonae*, respectively (42,43). It is not known whether this indicates the attachment of an O-specific polysaccharide to the Rb$_1$ core structure (25) followed by the introduction of the terminal GlcNAc or Glc residue later in biosynthesis or the elimination of these residues in late S-form LPS biosynthesis.

Escherichia coli Five core types have been identified in LPS of *E. coli* [R1–R4, K-12 (44)], for which partial structures had been published (26,45–51). So far, the core types R1–R3 and K-12 have been further investigated by our group (4,52–61; E. V. Vinogradov et al., unpublished; D. Grimmecke, personal communication), and their structures are given in Table 1. They all contain the oligosaccharide L,D-Hep-(1→7)-[Glc-(1→3)-]-L,D-Hep-(1→3)-L,D-Hep-(1→5)-[Kdo-(2→4)]-Kdo, and substitution of the Glc residue differs in all five core types. In the R1 and R3 core types, Hep III is additionally substituted in nonstoichiometric amounts at O-7 by GlcN (which is partially *N*-acetylated in R3), and in R2, Gal was identified in nonstoichiometric amounts at O-7 of Kdo II in strain EH100, one example of a hexose residue in the *inner core* region. However, there are other strains that lack this residue. The structure of the R4 core type has not yet been fully established. The nonreducing terminus of the K-12 core region is characterized by the trisaccharide β-GlcNAc-(1→7)-L,D-Hep-(1→6)-Glc, which represents a unique partial structure in all core regions investigated so far. Thus, and as shown below, heptopyranose residues may be present in the *outer core* region. Three monoclonal antibodies specific for K-12 LPS were obtained and serologically characterized (59). They reacted with 17 K-12 strains and are thus useful tools to identify *E. coli* K-12 bacteria.

The phosphate substitution of the R1, R2, and K-12 core types and of *E. coli* J-5, which is an Rc-mutant strain of the R3 core type (60), was investigated in more detail. In these cases, Hep I and Hep II were substituted by phosphoryl residues at position O-4, Hep II in nonstoichiometric amounts (61; E. V. Vinogradov et al., unpublished; D. Grimmecke, personal communication). However, it remains unclear which of these positions carry phosphodiester and diphosphate/diphosphodiester groups. Former investigations proposed for R1 LPS the presence of PPE at O-4 of Hep I and a phosphate residue at O-4 of Hep II (35) and for R4 LPS PPE at O-4 of both heptose residues (48). In *E. coli* K-12, PE was identified at O-7 of Kdo II (36).

Shigella In *Shigella sonnei*, LPS of the *E. coli* R1 core type and in *S. flexneri* 6, LPS similar to this core type were detected [Table 1 (62–64; A. Gamian, personal communication)]. In the last case, however, the unusual β-configuration of the Glc residue linked to O-3 of Hep II was proposed, solely owing to the fact that alkali-treated rough LPS (from strain R6 288) gave no reaction with concanavalin A. Thus, the anomeric configurations of all sugar residues should be investigated. The LPS of *S. flexneri* 4b possessed an outer core region of the *E. coli* R3 type (31). In LPS of *S. sonnei* and *S. flexneri* 6, O-4 of Hep I was substituted by PPE, as it was in some R mutants O-4 of Hep II.

Hafnia alvei Only one carbohydrate backbone has been identified in the core region of several strains of *Hafnia alvei* (65–69). The linkage point of the O-antigen of LPS from strain 2 was identified at O-6 of the first Glc residue. In most strains, the core region carries PPE at O-4 of Hep I and phosphate residues at O-4 of Hep II. In strain 1211, Hep II is not substituted at O-4 but at O-6 by PE. Finally, it should be mentioned that from strains 32, 1192 (70), 2, and 1211 (71) the trisaccharide L,D-Hep-(1→4)-[Gal(1→7)-]-Kdo was isolated, which was in the last two strains acetylated at O-6 of the Gal residue. From another strain, the linear trisaccharide Gal-(1→2)-L,D-Hep-(1→4)-Kdo was obtained (72).

Citrobacter freundii The core region of strain 1487 possessed a structure similar to that of *E. coli* R3: instead of a branching GlcNAc in the last, a GalNAc residue was present in the *outer core region*, and Hep III was substituted at O-7 by PPE instead of GlcNAc (73–75). The LPS of strains O4 and O36 contained core structures similar to each other [the terminal Glc residue in the O4 core is replaced by a Gal residue in O36 (76)]. Both possessed at the nonreducing terminus a trisaccharide built up from *N*-acetylated amino sugars. The phosphate substitution (75,79) comprises in all core oligosaccharides a PPE residue linked to O-4 of Hep I and, except in strain O23, a second PPE residue linked to O-7 of Hep II. The last is replaced in strain O23 by β-GalA (80).

Erwinia carotovora One partial and one complete structure have been described to date, obtained from LPS of *Erwinia carotovora* strains B374 (81) and FERM P-7576 (82), respectively. Whereas in the first case there was no information on phosphate substituents, the core oligosaccharide of strain FERM P-7576 possessed a phosphate residue at O-4 of Hep I. A characteristic structural feature of this core region was the substitution of O-7 of Hep III by Gal. Both residues were present in nonstoichiometric amounts.

General Structural Features of Core Oligosaccharides Different from the Salmonella enterica *Type*
In such cores, the common partial structure L,D-Hep-(1→7)-L,D-Hep-(1→3)-L,D-Hep-(1→5)-Kdo is not substituted at O-3 of Hep II by Glc. Heptose residues

Table 1 Structures of Core Regions of Enterobacterial LPS

Organism	Structure	Ref.
Salmonella enterica	$$\begin{array}{c} R^{1a} \\ \downarrow \\ 4 \end{array}$$ O-chain-(1→4)-Glc-(1→2)-Gal-(1→3)-Glc(1→3)-L,D-Hep-(1→3)-L,D-Hep-(1→5)-Kdo^b $$\begin{array}{ccc} 2 & & 4 \\ \uparrow & & \uparrow \\ 1 & & 2 \end{array}$$ R^{2c,d} Gal L,D-Hep Kdo^e-(2→4)-Kdo7←PE^f	25–43
Escherichia coli		
R1	$$\begin{array}{c} P \\ \downarrow \\ 4 \end{array}$$ Gal-(1→2)-Gal-(1→2)-Glc-(1→3)-Glc-(1→3)-L,D-Hep-(1→3)-L,D-Hep-(1→5)-Kdo^b $$\begin{array}{ccc} 3 & 7 & 4 \\ \uparrow & \uparrow & \uparrow \\ 1 & 1 & 2 \end{array}$$ β-Glc GlcN-(1→7)^d-L,D-Hep Kdo	26^{AA}
R2	$$\begin{array}{c} P^g \\ \downarrow \\ 4 \end{array}$$ GlcNAc-(1→2)-Glc-(1→2)-Glc-(1→3)-Glc-(1→3)-L,D-Hep-(1→3)-L,D-Hep-(1→5)-αKdo $$\begin{array}{ccc} 4 & 6 & 4 \\ \uparrow & \uparrow & \uparrow \\ 1 & 1 & 2 \end{array}$$ β-Gal^d D-Gal αKdo $$\begin{array}{c} \uparrow \\ \end{array}$$ L,D-Hep Kdo-(2→4) or Gal-(1→7)^h	26, 46, 54^{AA}

Chemical Structure of LPS Core Region

R3 (26, 47, 55, 56, 60, 61)

```
                        P^{d,i}            P^j
                         ↓                  ↓
                         4                  4
Glc-(1→2)-Glc-(1→2)-Gal-(1→3)-Glc-(1→3)-L,D-Hep-(1→3)-L,D-Hep-(1→5)-Kdo^b
          3                                                  7            ↑
          ↑                                                  ↑            2
          1                                                  1       Kdo-(2→4)-Kdo
       GlcNAc^k                                           L,D-Hep
                                                             7
                                                             ↑
                                                             1
                                                          GlcNAc^k
```
⟵ Strain J-5 ⟶

R4 (26, 48)

```
Gal-(1→2)-Gal-(1→2)-Glc -(1→3)-Glc-(1→3)-inner core^l
                     4
                     ↑
                     1
                   β-Gal
```

K-12 (26, 36, 50–53, 57–59^{BB})

```
                                        P^g               P^g
                                         ↓                 ↓
                                         4                 4
L,D-Hep^{d,g}-(1→6)-Glc-((1→2)-Glc-((1→3)-Glc-(1→3)-L,D-Hep-(1→3)-L,D-Hep-(1→5)-Kdo^b
     7                             6                       7                       4
     ↑                             ↑                       ↑                       ↑
     1                             1                       1                       2
β-GlcNAc^{d,g}                    Gal                   L,D-Hep                   Kdo
                                                                                   ↑
                                                                    Kdo-(2→4)- or L-Rha-(1→5)^m
```

Table 1 Continued

Organism	Structure	Ref.
S. sonnei	Gal-(1→2)-Gal-(1→2)-Glc-(1→3)-Glc-(1→3)-L,D-Hep-(1→3)-L,D-Hep-(1→5)-Kdo with R[3n] →4, R[4o]-(1→3)-β-Glc at 3, R[5p]-(1→7)-L,D-Hep at 7, PPE →4	62, 63[cc]
R-form	Glc[d]-(1→3)-Glc-(1→3)-L,D-Hep-(1→3)-L,D-Hep-(1→5)-Kdo with R[5d,p]-7L,D-Hep at 3, R[5p]-(1→7)-L,D-Hep, PPE →4	64
S. flexneri Strain 6	Gal-(1→2)-Gal-(1→2)-Glc[r]-(1→3)-β-Glc-(1→3)-L,D-Hep-(1→3)-L,D-Hep-(1→4/5)-Kdo with β-Glc at 3, R[5p]-(1→7)-L,D-Hep at 7, PPE[q] →4, PPE →4 Strain R 6 488	
Strain 4b	Glc-(1→2)-Glc-(1→2)-Gal-(1→3)-β-Glc-(1→)-inner core[s] with GlcNAc at 3, R[6t] →4, PPE	
Hafnia alvei	Glc-(1→3)-Glc-(1→3)-L,D-Hep-(1→3)-L,D-Hep-(1→5)-Kdo with X[u] at 7, L,D-Hep at 1	66–72

Chemical Structure of LPS Core Region

Citrobacter freundii

O4 and O36

β-GalNAc-(1→4)-GalNAc-(1→3)-β-GlcNAc-(1→4)-Glc-(1→2)-Glc-(1→3)-L,D-Hep-(1→3)-L,D-Hep-(1→5)-Kdo
$$4$$
$$\uparrow$$
$$\text{PPE}$$
$$71$$
$$\uparrow$$
$$\text{L,D-Hep7}\leftarrow\text{PPE}$$
$$R^{?v}\text{-}(1\rightarrow 6)\text{-}\beta\text{-Glc}$$

←— O4 and O36 incomplete cores —→

76, 79

O27 and PCM 1487

Glc-((1→2)-Glc-(1→2)-Gal-(1→3)-Glc-(1→3)-L,D-Hep-(1→3)-L,D-Hep-(1→5)-Kdo
$$374$$
$$\uparrow\uparrow\uparrow$$
$$11\text{PPE}$$
$$\text{GalNAc}\text{L,D-Hep7}\leftarrow\text{PPE}$$

73–75, 77, 79

O23

Glc-(1→2)-Gal-(1→2)-Glc-(1→4)-Glc-(1→3)-L,D-Hep-(1→3)-L,D-Hep-(1→5)-Kdo
$$7$$
$$\uparrow$$
$$1$$
$$\beta\text{-GalA-}(1\rightarrow 7)\text{-L,D-Hep}$$

79, 80

Erwinia carotovora

B374[w]

β-Glc-(1→6)-Glc-(1→3)-Hep-(1→3)-Hep
$$7$$
$$\uparrow$$
$$1$$
$$\text{Hep}$$

81

FERM P-7576

Gal-(1→6)-Glc-(1→3)-L,D-Hep-(1→3)-L,D-Hep-(1→5)-Kdo[b]
$$74$$
$$\uparrow\uparrow$$
$$1P^g$$
$$\beta\text{-Gal}^d\text{-}(1\rightarrow 7)\text{-L,D-Hep}^d2$$
$$\text{Kdo}$$

82

K. pneumoniae[x]

O1/R20

D,D-Hep[d]-(1→6)-GlcN-(1→4)-GalA-(1→3)-L,D-Hep-(1→3)-L,D-Hep-(1→5)-Kdo[b]
$$274$$
$$\uparrow\uparrow\uparrow$$
$$11\text{Kdo}$$
$$\text{D,D-Hep-}(1\rightarrow 2)\text{-D,D-Hep}\text{L,D-Hep}4$$
$$2\uparrow$$
$$\uparrow6\beta\text{-Glc}$$
$$1\uparrow$$
$$\text{D,D-Hep}^d12$$
$$\beta\text{-GalA}^d\text{Kdo}^d$$

24, 83, 85

Table 1 Continued

Organism	Structure	Ref.

O8/RFK11	Glc-(1→4)-GalA-(1→3)-L,D-Hep-(1→3)-L,D-Hep-(1→5)-Kdo^b
	                7 4 2
	                ↑ ↑ Kdo
	                1 1
	            L,D-Hep β-Glc
	                                        6
	                                      ↑
	                                        1
	                                β-GalA
		84

O1/R29 and O8/RFK9	L,D-Hep-(1→3)-L,D-Hep-(1→5)-Kdo^b
	                   4 4
	                   ↑ ↑
	                   1 2
	              β-Glc Kdo
		84, 85

P. mirabilis R110/1959	PE^d
	                                                ↓
	                                                6
	D,D-Hep-(1→6)-β-GlcN-(1→4)-GalA-(1→3)-L,D-Hep-(1→3)-L,D-Hep-(1→5)-Kdo
	              2 7 4
	              ↑ ↑ ↑
	              1 1 1
	            Glc^d L-D-Hep β-Glc
	                                            7 6
	                                          ↑ ↑
	                                          1 1
	                                        β-GalA^d Glc^d
		88^{AA,DD}

R45/1959	β-L-Ara4N^d-(1→8)-Kdo
	                    4
	                    ↑
	                    2
	                  Kdo
		86, 92

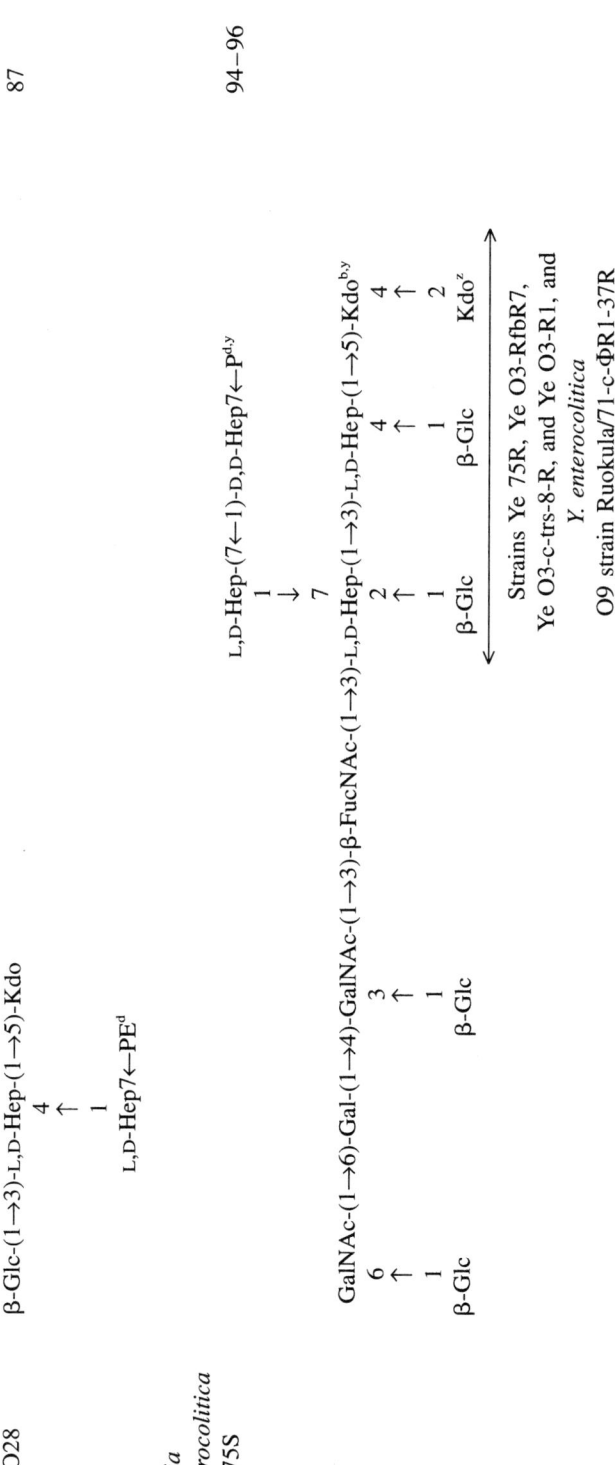

Unless stated otherwise, sugars are α-D-pyranosides. Kdo III is always present in nonstoichiometric amounts. [a]R_1 = PPE in serovars Typhimurium, minnesota, and arizonae (43). [b]This residue was identified to link the core region to GlcN II of lipid A. [c]R^2 = GlcNAc in svs. Typhimurium and minnesota and Glc in sv. arizonae (43). [d]Nonstoichiometric substitution. [e]Not identified in sv. arizonae. [f]Nonstoichiometric substitution in sv. minnesota. [g]Product obtained after de-O-acylation. Further substituents are not known. [h]Nonstoichiometric substitution in strain EH 100. [i]Product obtained after de-O-acylation of LPS from strain J-5. Further substituents are not known. [j]Partially N-acetylated in the complete core. [k]The Hep/Kdo region has not been fully determined. [l]In strains W3100 and 3110, and not in strain AB1133. [m]R^3 = H in LPS from S. sonnei Phase I and P in LPS from Phase II. [n]R^4 = linkage point of the repeating unit(s) in PhI, and PhI$_2$ core oligosaccharides from Phase I LPS, and R^4 = H in Phase II LPS and in unsubstituted core oligosaccharide Ph$_3$ from Phase II. [o]R^5 = GlcN in S. sonnei Phase I and in the R-form, and in S. flexneri strains S 6 standard, S 6 975, R V b 975, R V 6 551, and R V 6 288, and H in S. sonnei Phase II and S. flexneri strain R 6 488. [p]Glc II was not substituted by β-Glc in S. flexneri strains R V 6 975 and R 6 551 LPS. [q]Structure not characterized. [r]R^6 = P in LPS of the strains ATCC 13337, 1187, 2, 1191, 1196, 1220, and 481L, and = PE in that of strain 1211. [s]x = 4 in LPS of the strains ATCC 13337, 1187, 2, 1191, 1196, 1220, and 481L, and = 6 in that of strain 1211. [t]R^7 = Glc in O4 and Gal in O36. [u]Absolute configurations not determined. [v]Strain R20 is a rough mutant of K. pneumoniae spp. pneumoniae O1:K20, and strains RFK 11 and RFK9 of K. pneumoniae spp. ozaenae O8:K. [w]Only in strains Ye 75R, Ye O3-c-trs-8-R, and Ye O3-Rfb-R7. [x]The position of Kdo II was only determined for the core region of Y. enterocolitica O9 strain Ruokula/71-c-ΦR1-37R, as was the linkage of Kdo I to GlcN II of lipid A. [AA]Vinogradov, E., et al. Unpublished. [BB]Grimmecke, D. Personal communication. [CC]Gamian, A. Personal communication. [DD]Radziejewska-Lebrecht, J. Personal communication.

may not be phosphorylated, and position O-4 of Hep I is substituted by a hexose residue or oligosaccharide.

Klebsiella pneumoniae The core regions of serotypes O1 (24,83) and O8 (84) have recently been investigated. Both structures contained as partial structure the oligosaccharide L,D-Hep-(1→7)-L,D-Hep-(1→3)-L,D-Hep-(1→5)-[Kdo-(2→4)]-Kdo and, most interestingly, were not phosphorylated, which is very unusual within enterobacterial LPS. Hep I carries in both core regions at O-4 the disaccharide β-GalA-(1→6)-β-Glc. Hep II is substituted at O-3 by GalA, which is in turn substituted at O-4 either by GlcN (in serotype O1) or Glc (in serotype O8). The most striking structural feature is the presence of an oligosaccharide furnished of up to four α-(1→2)-linked D,D-Hep residues, which substitutes the GlcN in serotype O1 at O-6. This heptoglycan structure is so far unique in enterobacterial core regions, but similar structures have been found and characterized in LPS of *Helicobacter pylori* (see below and Table 6). From both serotypes, the structures of the core regions of Rc-analog chemotype LPS were characterized [strain R29 in case of O1 (85) and strain RFK9 in case of O8 (84)] and found to be identical.

Proteus Most of the structural investigations on the core region were performed with LPS from *P. mirabilis* (86–93). Studies on core regions of other *Proteeae* showed that GalA is a characteristic constituent (89). The core of LPS from *P. mirabilis* strain R110/1959 [Table 1 (88, E. Vinogradov, personal communication, J Radziejewska-Lebrecht, personal communication)] contained the oligosaccharide L,D-Hep-(1→7)-L,D-Hep-(1→3)-L,D-Hep-(1→5)-Kdo. The substitution pattern of this oligosaccharide shares some similarities with that of the core of *K. pneumoniae* O1. Hep I is substituted at O-4 by β-Glc (which carries Glc at O-6), and Hep II carries β-GlcN-(1→4)-GalA at O-3, which is also substituted at O-6 by D,D-Hep. This terminal trisaccharide is identical to that of the core of *K. pneumoniae* serotype O1 (see above), however, only one D,D-Hep residue was identified in the core of *P. mirabilis* strain R110/1959. Furthermore, Hep III is substituted at O-7 by β-GalA, and Hep II at O-6 by PE (nonstoichiometric). A second PE residue was supposed to be linked to Kdo. Investigations on the cores of serotypes O27 and O28 indicated basically the same structures as in R110/1959, but GlcN was exchanged by GalN in O27 and D,D-Hep by L,D-Hep in serotype O28 (90,91). The structural elucidation of the complete carbohydrate backbone of an Re-mutant of *P. mirabilis* 1959 (strain R45/1959) proved the presence of an Kdo-(2→4)-Kdo disaccharide, which is α-(2→6)-linked to lipid A and in which Kdo I is partially substituted at O-8 by 4-amino-4-deoxy-β-L-arabinose [Ara4N (86,92)].

The structure of the core region of an Rc-type mutant from serotype O28 (strain R4/O28) is unusual since it comprises the oligosaccharide L,D-Hep-(1→4)-L,D-Hep-(1→5)-Kdo substituted at O-3 of Hep I by β-Glc (87). Hep II is further substituted at O-7 by PE in nonstoichiometric amounts. However, this structure was deduced from results of methylation analysis by which the positions of Hep II and Glc cannot be distinguished. Since in other cases (including the core of *P. mirabilis* R110/1959, see above) the oligosaccharide L,D-Hep-(1→3)-L,D-Hep-(1→5)-Kdo is substituted at O-4 of Hep I by β-Glc, the proposed carbohydrate structure of strain R4/O28 should be confirmed. A similar but more incomplete structure has been proposed for the core region of LPS from strain 17301, a deep rough clinical isolate (93).

Yersinia In the complete core structure of *Y. enterocolitica* O3 strain Ye 75S (94,95), Hep I (at O-4) and Hep II (at O-2) are substituted by β-Glc, Hep III at O-7 by D,D-Hep, which in turn is phosphorylated in nonstoichiometric amounts at O-7, and Hep II at O-3 by a branched hexasaccharide. In this hexasaccharide, the FucNAc residue carries at O-4 a phosphate residue. In strain O3-R1, the same carbohydrate structure was identified, however, here the FucNAc residue carries at O-4 an ethyl phosphate group (J. Radziejewska-Lebrecht, personal communication), which to our knowledge was identified there for the first time in LPS. LPS from strains Ye 75R, O3-RfbR7, and O3-c-trs-8-R are of the Rc chemotype (J. Radziejewska-Lebrecht, personal communication) and possess the core structure published for strain Ye 75R in which the branched hexasaccharide (at O-3 of Hep II) is missing and the D,D-Hep (phosphate) and Glc (at O-2 of Hep II) residues are present in nonstoichiometric amounts. We have investigated the carbohydrate backbone of LPS (core oligosaccharide and lipid A backbone) from an Rc-type mutant strain of *Y. enterocolitica* O9 (96). The structure of the core region is essentially the same as that of the O3 Rc-type LPS, but all substitutions are stoichiometric and the D,D-Hep residue is not phosphorylated at al.

Some structural investigations have been performed on the core region of LPS from *Y. pseudotuberculosis* (97). The *inner core* region also comprises four heptose residues of unknown configurations. The *outer core* region consists of a branched pentasaccharide of three Glc, one Gal, and one GlcNAc residues in so far unknown linkages.

Neisseria

LPS of *Neisseria* are all of the rough type, and core structures (Table 2) consist of the trisaccharide L,D-Hep-(1→3)-L,D-Hep-(1→x)-Kdo (in which x is either 5 or, in some cases, unknown), which is substituted at O-4 of Hep I by β-Glc and at O-2 of Hep II by GlcNAc. Structural variations occur by different substituents at O-4 of the first residue and at Hep II (98–117).

The resistance of *N. gonorrhoeae* (causative agent of gonorrhoeae) in patients and when grown in serum containing media against the bactericidal effects of complement may be caused by sialylation (attachment of *N*-acetyl-neuraminic acid, Neu5Ac) of the LPS (118–124). As indicated by investigations using monoclonal antibodies, sialylation occurs at a terminal Gal-residue β-(1→4)-linked to GlcNAc (119). It was proposed that complement-fixing IgM antibodies are not able to bind to this structure. However, no complete structures of sialylated gonococcal LPS have been published to date. A further interesting observation is that phase variation of LPS caused by a different degree of sialylation controls both the entry of *N. gonorrhoeae* into human mucosa cells (low degree of sialylation) and the resistance to complement-mediated killing [high degree of sialylation (125)].

N. meningitidis is a causative agent of severe human diseases, such as sepsis and meningitis. At present, it can be differentiated into 12 LPS serotypes (126). Of these, 8 possess lacto-*N*-neotetraose [β-Gal-(1→4)-β-GlcNAc-(1→3)-β-Gal-(1→4)-β-Glc] as a common partial structure. This was shown by structural analyses and reaction of LPS with a monoclonal anti-lacto-*N*-neotetraose antibody (serotypes L2–L5, L7, and L9) or indicated by reaction of LPS with this antibody alone [serotypes L8 and L10 (126)]. Since this tetrasaccharide is identical to the glycosyl moiety of lactoneotetraglycosylceramide, which is present on the membrane of human erythrocytes, it is thought to represent a mimicry antigen that helps to evade the host defense. Sialylation also occurs in some meningococcal LPS, and Neu5Ac is α-(2→3)-linked to the terminal Gal residue of lacto-*N*-neotetraose. Whereas sialylated LPS and the sialyltransferase appear to be important virulence factors of gonococci, they are thought to be less important in meningococci, which possess a capsule built up of α-(2→8)-linked Neu5Ac that acts similarly (123).

Vibrionaceae

Vibrio *Vibrio cholerae* is an important causative agent of severe diarrhea (127). It is serologically and structurally divided into the O1 and the non-O1 serogroups. Bacteria of the first (which is further subdivided into the Ogawa and Inaba types) are known as cholera vibrios and were thought to be solely responsible for this disease for a long time. However, this view was revised when an epidemic in India and Bangladesh in 1992 was shown to be caused by a non-O1 *Vibrio* [serotype O139 (128,129)].

Several structures of LPS core regions from *V. cholerae* (Table 3) have been investigated to date (130–137). A common feature is the presence of one Kdo residue, which is phosphorylated at O-4 and substituted at O-5 by the tetrasaccharide GlcNAc-(1→7)-L,D-Hep-(1→2)-L,D-Hep-(1→3)-L,D-Hep, which carries β-Glc at O-4 and Glc at O-6 of Hep I. Variations in core structures occur by different substituents at O-6 of the two Glc residues and of Hep II and at O-2 or O-4 of Hep III. Interestingly, all characterized structures possess a β-linked ketofuranose at O-6 of the β-linked Glc, either fructose or sedoheptulose (D-altro-heptulose) which was identified in LPS for the first time (133). In strain H11 (non-O1), a second residue of this sugar was shown to function as a link between the O-antigen and the core region (133).

In addition to earlier publications on core structures from *V. parahaemolyticus* O12 (138) and *V. ordalii* (142), a partial structure from the core region of *V. parahaemolyticus* O2 has been reported (139). Finally, the unique core structure of *V. salmonicida* NCMB 2262 (140) comprises a rather rarely occurring nonulosonic acid, which was identified here in a core region for the first time, and a D,D-Hep-(1→5)-Kdo disaccharide, which was also identified in the core region of *Coxiella burnetii* (see below).

Aeromonas Bacteria of the genus *Aeromonas* are mainly associated with diseases in freshwater fish. To date, core structures of LPS from the species *A. hydrophila* and *A. salmonicida* have been investigated (Table 3) (143–147). However, that of *A. salmonicida salmonicida* strain SJ-15 has been revised (147) with respect to the terminal trisaccharide, which is Gal-(1→4)-GalNAc-(1→6)-Glc rather than Gal-(1→6)-GalNAc-(1→4)-Glc. Most interestingly, the core regions of *A.*

Table 2 Structures of Core Regions of LPS from *Neisseriaceae*

General structure:

$$R^1\text{-}(1\to 4)\text{-}\beta\text{-Glc-}(1\to 4)\text{-Hep-}(1\to x)\text{-Kdo-}R^2$$
$$\overset{3}{\underset{1}{\uparrow}}$$
$$\text{GlcNAc}^a\text{-}(1\to 2)\text{-Hep-}R^3$$

Strain/serotype	R^1	R^2	R^3	x	Ref.
N. gonorrhoeae					
F62	β-GalNAc^b-(1→3)-β-Gal-(1→4)-β-GlcNAc-(1→3)-β-Gal-	n.d.	H	n.d.	98, 102
MS11mk	β-Gal-	n.d.	H	n.d.	103
1291wt	β-Gal-(1→4)-β-GlcNAc-(1→3)-β-Gal-	n.d.	PE→3	5	104
1291$_b$	Gal-(1→4)-β-Gal-	n.d.	PE→3	5	104
15253	β-Gal-	n.d.	β-Gal-(1→3)-β-Glc-(1→3)-	n.d.	101
N. meningitidis					
L1^c	Gal-(1→4)-β-Gal	n.d.	PEA→3	5	105
L2	β-Gal-(1→4)-β-GlcNAc-(1→3)-β-Gal	n.d.	Glc-(1→3), PEA→6/7^{b,d}	5	108, 113
L3	Neu5Ac(2→3)-β-Gal-(1→4)-β-GlcNAc-(1→3)-β-Gal	(4←2)-Kdo	PEA→3	5	22, 106, 108, 111
L4	Neu5Ac-(2→3)-β-Gal-(1→4)-β-GlcNAc-(1→3)-β-Gal	n.d.	PEA→6	5	107, 117
L5	β-Gal-(1→4)-β-GlcNAc-(1→3)-β-Gal-(1→4)-β-Glc	n.d.	Glc-(1→3)	5	107, 108, 112
L6^c	β-GlcNAc-(1→3)-β-Gal	n.d.	PEA→7	5	105
L7	β-Gal-(1→4)-β-GlcNAc-(1→3)-β-Gal	n.d.	PEA→3	5	106, 117
L9	β-Gal-(1→4)-β-GlcNAc-(1→3)-β-Gal	n.d.	PEA→x^e	5	106
L10	β-Gal-(1→4)-β-Glc	n.d.	PEA n.d.	5	116
6275	Neu5Ac^a-(2→3)-β-Gal-(1→4)-β-GlcNAc-(1→3)-β-Gal	n.d.	2 PEA→x,y^{e,f}	5	109, 110
NMB	Neu5Ac^a-(2→3)-β-Gal-(1→4)-β-GlcNAc-(1→3)-β-Gal	n.d.	Glc-(1→3), PEA→x^e	5	114, 115

Unless stated otherwise, sugars are α-pyranosides (in case of *N. gonorrhoeae*) or α-D-pyranosides (*N. meningitidis*). n.d., Not determined. ^{a}Partially *O*-acetylated at unknown position(s) in *N. gonorrhoeae* strains 1291wt and 1291$_b$, at GlcNAc in *Neisseria meningitidis* serotypes L2 and L5, and at O-3 in serotype L4. ^{b}Present in nonstoichiometric amounts. ^{c}*O*-Acetylated at unknown position(s). ^{d}Probably at O6 and at O7. ^{e}x,y unknown positions. ^{f}Two moieties at unknown positions.

Table 3 Structures of Core Regions of LPS from *Vibrionaceae*

General structure of *Vibrio cholerae*:

$$R^1\text{-Glc}$$
$$1$$
$$\downarrow$$
$$6$$
$$R^2\text{-}\beta\text{-Glc-}(1\rightarrow 4)\text{-L,D-Hep-}(1\rightarrow 5)\text{-Kdo}^a\text{-R}^3$$
$$3$$
$$\uparrow$$
$$1$$
$$\text{GlcNAc-}(1\rightarrow 7)\text{-L,D-Hep-}(1\rightarrow 2(\text{-L,D-Hep-R}^4$$
$$|$$
$$R^5$$

with P→4 on Kdo.

Organism	R^1	R^2	R^3	R^4	R^5	Ref.
Vibrio cholerae						
O1 569B Inaba 95R Ogawa	H	β-Fruf^b-(2→6)-	H	L,D-Hep^c-(1→6)-	H	135
O139 MO10-T4^d	H	β-Fru$f$-(2→6)-	7←PE	L,D-Hep-(1→6)-	O-chain-(1→3)-	21
NRCC#4740^e	Glc-(1→6)-	β-Fru$f$-(2→6)-	H	L,D-Hep-(1→6)-	O-chain-(1→2)-	136, 137
H11	H	β-Sedf^f-(2→6)-	H	H	O-chain^c-(1→4)-β-Sedf^c-(2→3)-β-Gal-(1→3)-	133, 134
V. parahaemolyticus						
O12						138

V. parahaemolyticus O12:

β-3,6-dd-Glc3NAc^g-(1→3)-β-GalNAc-(1→4)-β-Glc-(1→4)-L,D-Hep-(1→5)-Kdo

with β-GlcA(1→2), P→4 on Kdo,
and β-Gal-(1→2)→L,D-Hep linked at position 3.

Table 3 Continued

Organism		Ref.
O2[h]	GlcA-(1→2)-L,D-Hep-(1→3)-D,D-Hep	139
V. salmonicida	$\begin{array}{cc}(PE)_2 & P\\ \downarrow & \downarrow\\ 2,7 & 4\end{array}$ Fuc4NBA[i]-(1→7)-NonA¹-(2→6)-β-Glc-(1→4)-D,D-Hep-(1→5)-Kdo $\begin{array}{c}3\\ \uparrow\\ 1\end{array}$ L-Rha-(1→4)-Glc-(1→2)-L,D-Hep	140
V. ordalii	$\begin{array}{c}\text{L,D-Hep}\\ 1\\ \downarrow\\ 6\end{array}$ β-GlcN-(1→7)-L,D-Hep-(1→4)-L,D-Hep $\begin{array}{c}3\\ \uparrow\\ 1\end{array}$ Glc-(1→3)-L,D-Hep-(6←1)-Glc	142
Aeromonas hydrophila Chemotype II	$\begin{array}{c}\text{L,D-Hep}\\ 1\\ \downarrow\\ 6\end{array}$ β-Glc-(1→4)-L,D-Hep-(1→2)-L,D-Hep $\begin{array}{c}3\\ \uparrow\\ 1\end{array}$ GlcN-(1→7)-L,D-Hep	143

Chemical Structure of LPS Core Region

Chemotype III

```
                    Glc
                     1
                     ↓
                     6
     Glc-(1→4)-L,D-Hep-(1→2)-L,D-Hep
                     3
                     ↑
                     1
          β-3,6-dd-Glc3NAc^g-(1→3)-Gal
```
144

A6
```
                              L,D-Hep
                                 1
                                 ↓
                                 6
     β-Gal-(1→4)-D,D-Hep-(1→6)-β-Glc-(1→4)-L,D-Hep-(1→2)-L,D-Hep
                    D,D-Hep                    3
                       1                       ↑
                       ↓                       1
                       6                  GlcN-(1→7)-L,D-Hep
```
145

A. salmonicida   L,D-Hep-(1→2)-L,D-Hep-(1→3)-L,D-Hep-(1→6)-Kdof^l 146

A. salmonicida SJ-15
```
                                                            L,D-Hep
                                                               1
                                                               ↓
                                                               6
     Gal-(1→4)-GalNAc-(1→6)-Glc-(1→4)-L,D-Hep-(1→2)-L,D-Hep-(1→3)-L,D-Hep-(1→6)-Kdof^l
                                                               4
                                                               ↑
                                                               1
                                              GlcN-(1→7)-L-β-D-Hep
```
147

Where not stated otherwise, sugars are α-D-pyranosides. [a]This residue was identified to link the core region to GlcN II of lipid A. [b]Fru*f*, D-fructofuranose. [c]Nonstoichiometric substitution. [d]SR-type strain. In addition, the core region possesses one *O*-acetyl group and one D-Fru*f* residue, both at unknown positions. [e]SR-type strain. [f]Sed*f*, Sedoheptulofuranose (D-*altro*-heptulofuranose). [g]3,6-dd-Glc3NAc, 3-acetamido-3,6-dideoxy-D-glucopyranose. [h]Partial structure. [i]Fuc4NBA, 4-[(*R*)-3-hydroxybutanoyl]amido-4,6-dideoxy-D-galactopyranose. [k]NonA, 5-acetamidino-7-acetamido-3,5,7,9-tetradeoxy-D-glycero-D-galacto-non-2-ulosonic acid. [l]Kdo*f*, 3-deoxy-D-*manno*-octulofuranosonic acid.

Table 4 Structures of Core Regions of LPS from *Pasteurellaceae*

$$R^1\text{-}\beta\text{-Glc-}(1\to4)\text{-L,D-Hep-}(1\to x)\text{-Kdo-}4\leftarrow R^2$$
$$3$$
$$\uparrow$$
$$1$$
$$R^3\to2)\text{-L,D-Hep-}(1\to2)\text{-L,D-Hep-}6\leftarrow R^6$$
$$\begin{matrix}4 & & 3\\ \uparrow & & \uparrow\\ R^4 & & R^5\end{matrix}$$

General structure of *Haemophilus influenzae*

Organism	R^1	R^2	R^3	R^4	R^5	R^6	x	Ref.
Haemophilus influenzae								
NtHi 2019[a]	β-Gal-(1→4)-	P	H	PE[b]	H	PE[b]	n.d.	155, 167
AH1-3 (*lic3* +)	β-Glc-(1→4)-	H	β-Gal-(1-	H	H	PE	n.d.	157
A2[a]	Glc-(1→4)-	P	H	H	Glc-(1→4)-Glc-(1	PE[b]	n.d.	158
galEgalK (Rm 7004)	β-Glc-(1→4)-[c]	H	H	H	β-Glc-(1→4)-Glc-(1[c]	PE	5	159, 162
281.25	β-Glc-(1→4)-	P	H	H	β-Glc-(1→4)-Glc-(1	PE	n.d.	163
RM.118	PCho[d]→6)	H	β-GalNAc[c]-(1→3)-Gal[c]-(1→4)-β-Gal-(1→4)-β-Glc-(1	H	H	PE	n.d.	168
RM.118-28	H	H	Pcho[d]→6)-β-Glc-(1	H	H	PE	n.d.	165
Eagan[e]	H	PPE[c]	β-Gal-(1	P[c]	Gal[c]-(1→4)-β-Gal[c]-(1→4)-β-Glc-(1→4)-Glc-(1	PE	5	169

H. ducreyi

$$R^1\text{-}\beta\text{-Glc-}(1\to4)\text{-L,D-Hep-}(1\to x^f)\text{-Kdo}^g$$
$$3$$
$$\uparrow$$
$$1$$
$$\text{L,D-Hep-}(1\to2)\text{-L,D-Hep}$$

General structure

Chemical Structure of LPS Core Region

	R^1	
ITM 2665[h]	β-Gal[c]-(1→4)-β-GlcNAc-(1→3)-β-Gal-(1→4)-D,D-Hep-(1→6)-	160
ITM 4747	β-Gal-(1→4)-	160
ITM 5535	β-Gal-(1→4)-β-GlcNAc-(1→3)-β-Gal-(1→4)-D,D-Hep-(1→6)- Neu5Ac[c,i]-(2→3)-β-Gal-(1→4)-β-GlcNAc-(1→3)-β-Gal-(1→4)-D,D-Hep-(1→6)-	161
ITM 3147, ACY1	β-Gal-(1→4)-β-GlcNAc-(1→3)-β-Gal-(1→4)-β-GlcNAc-(1→3)-β-Gal-(1→4)-D,D-Hep-(1→6)- β-Gal-(1→4)-β-GlcNAc-(1→3)-β-Gal-(1→4)-D,D-Hep-(1→6)- β-Gal-(1→4)-β-GlcNAc-(1→3)-β-Gal-(1→4)-β-GlcNAc-(1→3)-β-Gal-(1→4)-D,D-Hep-(1→6)-	161
Pasteurella haemolytica		174

```
                                               Glc
                                                1
                                                ↓
                                                6
OAg[k]-β-Gal-(1→7)-Hep^I-(1→6)-Hep^I-(1→6)-β-Glc-(1→4)-Hep^I-(1→5)-Kdo
                                                3      4
                                                ↑      ↑
                                                1      P
                                              Hep^I-(1→2)-Hep^I
```

Where not stated otherwise, sugars are α-D-pyranosides. [a]The absolute configurations of sugars are unknown. [b]Position unknown. [c]Nonstoichiometric substitution. [d]PCho, phosphocholine. [e]Structures of a number of mutants of this strain are published in Ref. 98. [f]x, not determined. [g]Phosphorylated at unknown position in strains ITM 5535, ITM 3147, and ACY1. [h]Proposed also for strain 35000 (109). [i]Neu5Ac, 5-*N*-acetyl-neuraminic acid. [k]O-Ag, O-antigen. [l]Relative molar ratio of L,D-Hep:D,D-Hep is 2.2:2.0 (175).

salmonicida possess a furanosidic Kdo, which is substituted at O-6 by L,D-Hep.

Pasteurellaceae

Haemophilus All LPS of *Haemophilus* (Table 4) so far investigated are of the rough type (109,148–173). The identified structures of core regions from LPS of various strains of *H. influenzae* and *H. ducreyi* comprise as a common partial structure the tetrasaccharide L,D-Hep-(1→2)-L,D-Hep(1→3)-L,D-Hep-(1→5)-Kdo, in which position 5 was in some cases not determined. Hep I is substituted at O-4 by β-Glc, which in turn may carry other substituents in different strains. Only one Kdo residue is present that links the core region to lipid A, which is in several cases phosphorylated.

Haemophilus influenzae is the causative agent of invasive and bacteremic infections such as meningitis in young children. In addition to the common structural features outlined above, the tetrasaccharide may be further substituted at O-3 and O-6 of Hep II, and O-2 and O-4 of Hep III. Some substituents are of great interest. First, phosphorylation by phosphorylcholine of the outer core region occurs in Hib strains RM.118 and Rm.118-28, which is a so far unique structural feature of *Haemophilus* LPS (165,166,168). Choline is taken up from the growth medium, and this process together with the expression of the resulting carbohydrate epitope undergo high-frequency phase variation (166). The biological function of this decoration remains unclear, however, it is not considered as a tool to evade host defense mechanisms. Second, other phase-varying epitopes were identified by using monoclonal antibodies that mimic the human blood group antigen P^k [Gal-(1→4)-β-Gal(1→4)-Glc] and the trisaccharide β-Gal-(1→4)-β-GlcNAc-(1→3)-Gal, which is present in human glycosphingolipids and in LPS of *N. meningitidis* and *N. gonorrhoeae* (see Table 2) and *H. ducreyi* (see below) (154,170–172). Also, like in human glycolipids and neisserial LPS, some Hib LPS were found to be sialylated (109,154), however, their complete structures are presently unknown. Finally, the structure of the LPS from the deep-rough mutant I-69 Rd^-/b^+, which possesses as core only one Kdo residue that is phosphorylated either at O-4 or O-5 (152,153,173), was shown to be the minimal LPS structure for bacterial survival (mutants that cannot synthesize any core structure are not viable).

Sialylated LPS was also identified in *H. ducreyi* (109,161), and, in strain ITM 5535 (161), Neu5Ac was identified as the terminal residue α-(2→3)-linked to the trisaccharide β-Gal-(1→4)-β-GlcNAc-(1→3)-Gal. This trisaccharide is also present in the cores of strains ITM 2665, ITM3147, and ACY1, and in all cases connected via a D,D-Hep residue to O-6 of β-Glc.

Pasteurella *Pasteurella haemolytica* causes respiratory infections in cattle and sheep. At present, one core structure of LPS from the Al serotype (Table 4) has been reported (174). Similar to *Haemophilus* cores, it possesses the tetrasaccharide Hep-(1→2)-Hep-(1→3)-Hep-(1→x)-Kdo in which Kdo is also phosphorylated at O-4 and carries Hep I at O-5. To Hep I, a second tetrasaccharide β-Gal-(1→7)-Hep-(1→6)-Hep-(1→6)-β-Glc is linked at O-4. The relative molar ratio of L,D-Hep:D,D-Hep was determined as 2.2:2.0 (175), but the distribution of these two sugars in the core is unknown. With respect to *Haemophilus* and other D,D-Hep–containing core structures, it may be expected that L,D-Hep is exclusively present in the first and D,D-Hep in the second tetrasaccharide.

Pseudomonadaceae

Pseudomonas All characterized core regions from LPS of *Pseudomonas aeruginosa* and *P. fluorescens* (Table 5) (176–182; F. Beckmann and U. Zähringer, personal communication; J. C. Richards, personal communication) possess a GalN residue that is amidated by alanin and which substitutes (O-3 of) Hep II of the trisaccharide L,D-Hep-(1→3)-L,D-Hep-(1→x)-Kdo, where x was in most cases determined as 5. In *P. aeruginosa* strain PAC605 (serotype Habs O3) and several strains of serotype IATS O6, and in *P. fluorescens* strain ATCC 49271, Hep II is further substituted at O-7 by a carbamoyl residue, which was identified for the first time in carbohydrates (182). Screening of several LPS for this residue revealed that it is only present in LPS from bacteria of RNA group I ("authentic" *Pseudomonas*) and may thus be of help for taxonomical classification of bacteria of the genus *Pseudomonas* (F. Beckmann and U. Zähringer, personal communication). It should be noted that in the core region of *P. aeruginosa* strain PAC605 14 phosphate residues were detected, two of which have so far been located at O-2 and O-4 of Hep I. To date, this represents the highest number of phosphate residues in LPS. In the core region of *P. fluorescens* ATCC49271, di- and, rather unusual, triphosphate residues were identified (179,181). Whether the high number of negatively charged groups determines a distinct function in the outer membrane of *Pseudomonas* is currently investigated.

Burkholderia At present, only the core region from LPS of *B. cepacia* strain GIFU 645 is structurally characterized (Table 5) (14). Its most prominent feature is the presence of Ko, a sugar that was first identified in LPS of *Acinetobacter* (see below and Table 8). However, one significant difference from the core structures of the last genus (which may also be of taxonomical importance) is that Ko in the *B. cepacia* core is present as a branch, substituting Kdo at O-4. In strain GIFU 645, Ko is partially substituted at O-8 by β-L-Ara4N. A similar disaccharide [β-L-Ara4N-(1→8)-Kdo] is present in the core of *Proteus mirabilis* strain R45 (compare Table 1).

Campylobacter and Helicobacter

Campylobacter Of the genus *Campylobacter*, the species *C. jejuni* and *C. coli* are the leading causative agents of human enteritis, and, furthermore. *C. jejuni* is involved in neurologic complications, e.g., meningitis and Guillain-Barré syndrome and its variants (183–186). All core structures of *C. jejuni*, *C. coli*, and *C. lari* comprise L,D-Hep-(1→3)-L,D-Hep-(1→5)-Kdo, which is substituted at O-4 of Hep I by β-Glc (Table 6) (183–194). The cores of *C. jejuni* vary in different strains by further substitution of this tetrasaccharide at O-6 or O-7 of Hep I and O-2 and O-3 of Hep II. As its most striking feature, core regions of *C. jejuni* possess an impressive variety of structures in mimicry of those of human gangliosides, which is expected to play an important role in evading from the immune defense of the host. The terminal branched trisaccharide of the O:1, O:23, and O:36 outer core regions (compare R_1 in Table 6) is identical to that of the ganglioside GM_2 (187,188). The outer core regions of serotypes O:4 and O:19 comprise partial structures identical to those of GD_{1a} (complete R_1) and of GM_1 [R_1 without the terminal Neu5Ac residue (185–188)], which is also present in strain OH 4384, a clinical isolate serotyped as O:19 (184,185). The last core region comprises further partial structures, which are identical to those of GM_2 (see above), GM_1 (GM_2 substituted by β-Gal), GT_{1a} (complete R_1), and GD_3 (terminal Neu5Ac-Neu5Ac-Gal trisaccharide), the last of which is also present in the outer core regions of serotype O:10 (186,189) and strain OH 4382, another clinical isolate serotyped as O:19 (184,185). Interestingly, a structural variation of the carbohydrate moiety of ganglioside GM_2 is present in serotype O:3 in which the terminal disaccharide β-GalNAc-(1→4)-β-Gal (present in GM_2 and the O:1, O:23, and O:36 outer core regions) is changed to α-GalNAc-(1→4)-β-GalNAc. Furthermore, the first disaccharide is in GM_2 (1→4)-linked on β-Glc, however, in serotypes O23 and O36 this is varied to a (1→3)-linkage on β-Glc and in serotype O:1 to a (1→3)-linkage on β-Gal. The second disaccharide is in serotype O:3 (1→3)-linked on α-Gal. These examples show how slight structural variations can lead to a large number of serologically distinct strains.

The core region of *C. coli* O:30 possesses as unusual features an additional substitution of the β-Glc by a third L,D-Hep residue and at the nonreducing terminus two quinovosamine residues, both of which are N-acylated by either (R)-3-hydroxybutanoyl or 3-hydroxy-2,3-dimethyl-5-oxoprolyl (190). In *C. lari*, the core of strain ATCC 35221 possesses a Kdo residue substituted at O-4 by PE and linked α-(2→5) on a second Kdo residue (194). Such an unusual Kdo disaccharide has so far only been identified here and in *Acinetobacter haemolyticus* (see below and Table 8).

Helicobacter *Helicobacter pylori* causes gastritis and gastric and duodenal ulcers and is associated with gastric carcinoma in humans. The core regions of LPS from various strains of *H. pylori* share a partial structure that comprises hexoses and two residues each of D,D- and L,D-Hep (Table 6). A most prominent feature is the presence of a heptoglycan of α-(1→3)-linked D,D-Hep residues in serotypes O:3 and O:6 and strain MO19 (195,196), which builds up an intervening region between the core region and the O-antigen. Also, 2- and 2/7-linked D,D-Hep residues were identified in LPS of *H. felis* (197; M. A. Monteiro, personal communication), however, no complete structures have been reported to date. It is not clear whether such heptoglycans belong to the core region or represent an optional fourth LPS region (in addition to lipid A, core region, and O-antigen). A similar heptoglycan was identified in *Klebsiella pneumoniae* ssp. *pneumoniae* strain R20 (compare Table 1). A recent investigation of the outer core of *H. mustelae* (causes gastritis in ferrets) proved the expression of the monofucosyl A Type 1 blood group determinant in mimicry of animal cell-surface glycolipids and glycoproteins (199). Interestingly, and in contrast to the above-described core regions of *Helicobacter*, this core does not contain any D,D-Hep. Its complete structure has not yet been published.

Miscellaneous

An incomplete structure of the core region of LPS from *C. burnetii* Phase II has been published (200,201), which possesses the unusual trisaccharide D,D-Hep-D,D-Hep-Kdo and no L,D-Hep residues (Table 7). Phase

Table 5 Structures of Core Regions of LPS from *Pseudomonadaceae*

Organism	Structure	Ref.
Pseudomonas aeruginosa Habs 2B/NCTC 1999	$\begin{array}{c} \text{Ala}^a \\ | \\ \beta\text{-Glc-}(1\to2)\text{-L-Rha-}(1\to6)\text{-Glc-GalN}^b\text{-}(1\to3)\text{-L,D-Hep-}(1\to3)\text{-L,D-}(1\to4/5)\text{-Kdo} \\ \uparrow \\ \text{Glc-}(1\to6)\text{-Glc} \end{array}$	176
Habs O3/PAC1R[b]	$\begin{array}{c} \text{OAg}^c \quad\quad\quad \text{Glc}^g \\ 1 \quad\quad\quad\quad 1 \\ \downarrow \quad\quad \text{Ala} \downarrow \\ 3 \quad | \quad 7 \\ \text{Glc}^d\text{-}(1\to6)\text{-}\beta\text{-Glc}^e\text{-}(1\to3)\text{-GalN}^b\text{-Hep-}(1\to4/5)\text{-Kdo} \\ 6 \\ \uparrow \\ 1 \\ \text{Rha}^f \end{array}$	177
Habs O3/PAC605[h]	$\begin{array}{c} \text{CONH}_2 \\ | \\ \text{Ala} \quad 7 \\ | \\ \beta\text{-Glc}^a\text{-}(1\to3)\text{-GalN-}(1\to3)\text{-L,D-Hep-}(1\to3)\text{-L,D-Hep-}(1\to5)\text{-Kdo}^i \\ 4 \\ \uparrow \\ 2 \\ \text{Kdo} \end{array}$	182[o]
IATS O6	$\begin{array}{c} \text{CONH}_2 \quad\quad \text{P} \\ | \quad\quad\quad \downarrow \\ \text{Ala} \quad 7 \quad\quad\quad 4 \\ | \\ R^1\text{-}3)\text{-GalN-}(1\to3)\text{-L,D-Hep-}(1\to3)\text{-L,D-Hep-}(1\to5)\text{-Kdo}^i \\ 4 \quad\quad\quad 2 \quad 4 \\ \uparrow \quad\quad\quad \uparrow \quad \uparrow \\ R^2 \quad\quad\quad \text{P} \quad 2 \\ \quad\quad\quad\quad\quad\quad\quad \text{Kdo} \\ \text{General structure} \end{array}$	

Chemical Structure of LPS Core Region

	R^1	R^2	
A28	β-Glc-(1 ↑ 6 Glc	Glc-(1 ↑ 1 L-Rha	180^p
R5	H	Glc-(1	178
H4	H	H	180

P. fluorescens ATCC 49271

$$\text{Ochain}^k \rightarrow 4)\text{-L-FucNAc}^m\text{-}(1\rightarrow 3)\text{-QuiNAc}^n\text{-}(1\rightarrow 3)\text{-}\beta\text{-Glc-}(1\rightarrow 3)\text{-GalN-}(1\rightarrow 3)\text{-L,D-Hep-}(1\rightarrow 3)\text{-L,D-Hep-}(1\rightarrow 5)\text{-Kdo}^i$$

$$\begin{array}{c} \text{Ala} \\ | \\ 7 \end{array} \quad \begin{array}{c} 2 \\ \uparrow \\ 1 \end{array} \quad \begin{array}{c} 4 \\ \uparrow \\ 1 \end{array} \quad \begin{array}{c} \text{CONH}_2 \\ | \\ 7 \end{array} \quad \begin{array}{c} P \\ \downarrow \\ 4 \end{array}$$

$$\begin{array}{c} 3 \\ \uparrow \\ \text{Ac}^l \end{array} \quad \begin{array}{c} 2 \\ \uparrow \\ P \end{array} \quad \begin{array}{c} 4 \\ \uparrow \\ 2 \end{array}$$

β-GlcNAc   Glc-(6—Ac^l)   Kdo

179, 181

Burkholderia cepacia GIFU 645

$$\text{L-Rha-}(1\rightarrow 2)\text{-L,D-Hep-}(1\rightarrow 3)\text{-L,D-Hep-}(1\rightarrow 5)\text{-Kdo}^i$$

$$\begin{array}{c} 7 \\ \uparrow \\ 1 \end{array} \quad \begin{array}{c} 4 \\ \uparrow \\ 2 \end{array}$$

L,D-Hep   β-GlcKo-(8←1)- β-L-Ara4N^a

14, 15

Where not stated otherwise, sugars are α-D-pyranosides. ^{a}Nonstoichiometric substitution. ^{b}The absolute configurations were not determined. ^{c}OAg, O-antigen, not present in mutant strains PAC608 and PAC557. ^{d}Not present in mutant strains PAC611 and PAC605. ^{e}Not present in mutant strain PAC605. ^{f}Not present in mutant strains PAC556, PAC611, and PAC605. ^{g}Not present in mutant strains PAC609, PAC611, and PAC605. ^{h}Dephosphorylated core region. ^{i}This residue was identified to link the core region to lipid A. ^{k}The O-chain was characterized as homopolymer of α-(2→4)-linked 5-acetamidino-7-acetamido-3,5,7,9-tetradeoxy-D-glycero-L-galacto-nonulosonic acid (ca. 75% O-acetylated at position 8). In LPS of strain ATCC 49271, one residue of this sugar is α-(2→4)-linked to L-FucNAc. Ac, acetyl group, present in nonstoichiometric amounts. ^{m}L-FucNAc, 2-acetamido-2,6-dideoxy-L-galactopyranose. ^{n}QuiNAc, 2-acetamido-2,6-dideoxy-D-glucopyranose. ^{o}Richards, JC. Personal communication. ^{p}Beckman, F. and Zähringer, U. Personal communication.

Table 6 Structures of Core Regions of LPS from the Genera *Campylobacter* and *Helicobacter*

Campylobacter jejuni

General structure:

$$R^1 \rightarrow x)\text{-L-}\alpha\text{-D-Hep-}(1\rightarrow 3)\text{-L-}\alpha\text{-D-Hep-}(1\rightarrow 5)\text{-Kdo}$$

$$\begin{array}{c}\beta\text{-Glc}\\1\\\downarrow\\4\end{array}$$

with substituents at position 2 (R²) and position y (R³).

Organism	R¹	R²	R³	x	y	Ref.
C. jejuni						
O:1/ATCC 43429	β-GalNAc-(1→4)-β-Gal-(1→3)-β-Gal-(1→ 　　　　　　　　　　3 　　　　　　　　　　↑ 　　　　　　　　　　2 　　　　　α-Neu5Ac　α-Gal 　　　　　　　　　　1	β-Glc	PE	3	6	187, 188
O:2/CCUG 10936	β-Gal-(1→3)-β-Gal-(1→ 　　　　3 　　　　↑ 　　　　2 　α-Neu5Ac　α-Gal 　　　　1	β-Glc	PE	3	6	191
O:3/ATCC 43431	GalNAc-(1→4)-β-GalNAc-(1→3)-α-Gal-(1→ 　　　　3　　　　　　　　2 　　　　↑　　　　　　　　↑ 　　　P^a　　　　　　　　1 　　　　　β-QuiNAc　　　β-GlcNAc	β-Glc	P	3	7	192
O:4/ATCC 43422 O:19/ATCC 43446	β-Gal-(1→3)-β-GalNAc-(1→4)-β-Gal-(1→ 　　　　　　　　　3 　　　　　　　　　↑ 　　　　　　　　　2 　　　　　　α-Neu5Ac^b 　　　　　　　　　1 　　　　　α-Neu5Ac	H	PE	3	6	185, 187, 188

Chemical Structure of LPS Core Region

Strain	Structure			Ref.
OH 4384	β-Gal-(1→3)-β-GalNAc-(1→4)-β-Gal-(1→ 3 ↑ 2 α-Neu5Ac-(8←2)-α-Neu5Ac α-Neu5Ac	H	P	3	6/7	184, 185
OH 4382	α-Neu5Ac-(2→8)-α-Neu5Ac-(2→3)-β-Gal-(1→	H	PE	3	6/7	184, 185
O:10/PG 836	β-Gal-(1→3)-β-GalNAc-(1→4)-β-Gal-(1→ 3 ↑ 2 α-Neu5Ac-(8←2)-α-Neu5Ac	β-Glc	P	3	7	186, 189
O:23/ATCC 43449 O:36/ATCC 43456	β-GalNAc-(1→4)-β-Gal-(1→3)-β-Glc-(1→ 3 ↑ 2 α-Neu5Ac	L-R¹	PE	2	6	187, 188

C. coli
O:30/

$$\beta\text{-Qui3NAcyl}^E\text{-}(1\to2)\text{-}\beta\text{-Qui3NAcyl}^E\text{-}(1\to3)\text{-}\beta\text{-GlcNAc-}(1\to3)\text{-}\beta\text{-Gal-}(1\to3)\text{-L-}\alpha\text{-D-Hep-}(1\to3)\text{-L-}\alpha\text{-D-Hep-}(1\to5)\text{-Kdo}$$

with branches:
- L-α-D-Hep-(1→3)-β-Glc at position 4
- α-Gal at position 2 (→1)
- α-Gal at position 2 (→1)

190

C. lari
PC 637

$$\alpha\text{-Glc-}(1\to3)\text{-}\beta\text{-GalNAc-}(1\to3)\text{-}\beta\text{-GalNAc-}(1\to3)\text{-}\beta\text{-Gal-}(1\to3)\text{-L-}\alpha\text{-D-Hep-}(1\to3)\text{-L-}\alpha\text{-D-Hep-}(1\to5)\text{-Kdo}$$

with branches:
- β-Glc at position 4
- α-Gal at position 2
- α-Gal at position 2
- αGal at position 2

193

Table 6 Continued

Organism		Ref.
ATCC 35221		194

$$\text{α-Glc-(1→3)-α-GalNAc-(1→3)-β-Gal-(1→3)-L-α-D-Hep-(1→3)-L-α-D-Hep-(1→5)-Kdo-(2→5)-Kdo}$$

with β-Glc at position 4 (PE→4), α-Glc at position 2 (1→), α-Gal at position 2 (1→).

Helicobacter pylori

$$\text{Glc-(1→3)-Glc-(1→4)-β-Gal-(1→7)-D,D-Hep-(1→2)-L,D-Hep-(1→3)-L,D-Hep-(1→5)-Kdo}$$

with P→7, D,D-Hep-(y←1)-R² at position 2, and R¹ at position x.

General structure

	R¹	R²	x	y	Ref.
O:3	O-chain-(1→3)-[→3)-D,D-Hep-(1]₅→6)-D,D-Hep-(1→2)-D,D-Hep-	Glc-(1[→6)-Glc-(1]₁₋₂→6)-Glc-	7	2	195
O:6A	Le^y-(1→3)-{-Gal-(1→[→3)-D,D-Hep-(1]ₙ→6)-D,D-Hep-(1→2)-D,D-Hep-	Glc-(1[→6)-Glc-(1]₁₋₂→6)-Glc-	7	2	195
O:6B	D,D-Hep-(1→[→3)-D,D-Hep-(1]ₙ→6)-D,D-Hep-(1→2)-D,D-Hep-	Glc-(1[→6)-Glc-(1]₁₋₂→6)-Glc-	7	2	195
P466	O-chain	H	7	2	196
MO19	Le^y-(1→3)-β-Gal-(1[→3)-D,D-Hep-(1]₅→6)-D,D-Hep-(1→2)-D,D-Hep-	Glc-(1→6)-Glc-(1→6)-Glc-	n.d.	n.d.	196
NCTC 11637	O-chain	Glc-(1→6)-Glc-(1→6)-Glc-^d	7	2	198

H. mustelae
ATCC 43772: GalNAc-(1→3)-β-Gal-(1→3)-β-GlcNAc-(1→inner core, with L-Fuc at position 2. Ref. 199

To show the slight structural variations in the outer core of LPS from *Campylobacter*, all anomeric configurations are depicted. Unless stated otherwise, sugars are α-D-pyranosides. For *Helicobacter*, where not stated otherwise, sugars are α-D-pyranosides. n.d., Not determined. ᵃNonstoichiometric substitution at unknown position. ᵇNonstoichiometric substitution in O:19. ᶜNAcyl, either (R)-3-hydroxybutanoyl or 3-hydroxy-2,3-dimethyl-5-oxoprolyl. ᵈOnly present in the rough-type fraction of the LPS.

Chemical Structure of LPS Core Region

Table 7 Structures of Core Regions of LPS from *Coxiella burnetii*, *Bordetella pertussis*, *Rhodocyclus tenuis*, *R. gelatinosus*, and *Phenylobacterium immobile*

Coxiella burnetii Phase II

$$R^a{\rightarrow}4)\text{-}\beta\text{-Man-}(1{\rightarrow}2)\text{-}_{D,D}\text{-Hep}^b\text{-}(1{\rightarrow}3/4)\text{-}_{D,D}\text{-Hep-}(1{\rightarrow}5)\text{-Kdo}$$

positions 3 and 4 substituents:
- 3 ← R^a
- 3/4 ← β-Man^b (1)
- 4 ← Kdo-(4←2)-Kdo (2)

200, 201

Bordetella pertussis

$$\text{GlcNAc}^f\text{-}(1{\rightarrow}4)\text{-}2,3\text{-}d\text{-Man-}2,3\text{-NAcA}^{c,g}\text{-}(1{\rightarrow}3)\text{-FucNAcMe}^{c,h}\text{-}(1{\rightarrow}6)\text{-GlcN}^f\text{-}(1{\rightarrow}4)\text{-}\beta\text{-Glc-}(1{\rightarrow}4)\text{-}_{L,D}\text{-Hep-}(1{\rightarrow}5)\text{-Kdo}$$

With substituents:
- GalNA^{c,d} (1→6) on the FucNAcMe
- Hep^c (1→4) and P^c (→4) on the Kdo/Hep region
- 3 ← 1

205–208

Rhodocyclus tenuis

General structure:
$$_{L,D}\text{-Hep-}(1{\rightarrow}3)\text{-}_{L,D}\text{-Hep-}(1{\rightarrow}x^i)\text{-Kdo}$$

with 7←1 $_{L,D}$-Hep and 4←R substituents

- EU1: R = $_{L,D}$-Hep
- 2761: R = $_{D,D}$-Hep-(1→2)-$_{L,D}$-Hep
- 3661: R = $_{D,D}$-Hep-(1→2)-$_{D,D}$-Hep
- G FUy: R = y^k-(1→2)-$_{D}$-Man

GlcA-(1→2)-$_{L,D}$-Hep-(7←1)-GlcN

202

Rhodocyclus gelatinosus Dr2

$$\text{GalA-}(1{\rightarrow}4)\text{-GalA-}(1{\rightarrow}4)\text{-Kdo}$$

with 5←1 $_{D,D}$-Hep

10

Phenylobacterium immobile

$$\beta\text{-Glc-}(1{\rightarrow}3)\text{-}_{L,D}\text{-Hep-}(1{\rightarrow}3)\text{-}_{L,D}\text{-Hep-}(1{\rightarrow}3)\text{-}_{L,D}\text{-Hep-}(1{\rightarrow}4/5)\text{-Kdo}$$

with 7←1 2-d-Hep^{f,j}

209

Where not stated otherwise, sugars are α-D-pyranosides. ^{a}R, positions of Phase I sugars. ^{b}Position not unequivocally identified. ^{c}Anomeric and absolute configurations unknown. ^{d}GalNA, galactosaminuronic acid. ^{e}Not present in all LPS-types. ^{f}Anomeric configurations unknown. ^{g}2,3-d-Man-2,3-NAcA, 2,3-diacetamido-2,3-dideoxy-mannosaminuronic acid. ^{h}FucNAcMe, methylacetamidofucosamin. ^{i}x, not determined. ^{j}2-d-Hep, 2-deoxyheptose. ^{k}Unknown substituent.

I-specific sugars are linked to Hep II and to a Man residue.

The structures of the core regions of *R. tenuis* (202), *R. gelatinosus* (10), and, basically, of *B. pertussis* LPS (203–207) are as reviewed earlier, however, in the latter the absolute configuration of most sugars of the inner core region has been determined (208).

Unique features of the core region of LPS from *P. immobile* (209) are the tetrasaccharide L,D-Hep-(1→3)-L,D-Hep-(1→3)-L,D-Hep-(1→4/5)-Kdo and its substitution by a 2-deoxyheptose at O-7 of Hep I.

Heptose-Deficient Core Regions

Moraxellaceae

Acinetobacter *Acinetobacter* represents a causative agent of nosocomial infections of increasing prevalence. Its taxonomical description has undergone many revisions, and today at least 19 DNA groups can be distinguished, several of which can be classified as nomenspecies (210,211). At present, four different core structures are known (Table 8) (11–13,212–214; E. Vinogradov et al., unpublished), two of which (strain ATCC 17905 and *A. haemolyticus* strain ATCC 17906) possess one Ko residue linking the core region to lipid A (12,13). However, this residue may be structurally replaced to Kdo in nonstoichiometric amounts. The biosynthesis of Ko is to date not known, thus, it is unclear whether this sugar is transferred by a specific Ko-transferase to lipid A (or its precursor) or if it originates from a transferred Kdo residue by virtue of oxidation at C-3. The core region of strain ATCC 17906 contains two residues of 3-deoxy-D-*lyxo*-hept-2-ulosaric acid at its nonreducing terminus, a sugar that has also been found in LPS of *Rhizobium* and *Agrobacterium tumefaciens* and in polysaccharides from plants and green algae (for references, see Ref. 13). The other two core regions (*A. baumannii* strains NCTC 10303 and ATCC 19606) possess as a unique feature a branched Kdo trisaccharide (E. Vinogradov et al., unpublished) with the structure proposed for the Kdo region of LPS from *S. enterica* and *E. coli* until the early 1980s but proved false in 1985. Most prominent of these two core regions is the structure of strain NCTC 10303 in which the second Kdo of the main chain is substituted at O-7 by a fourth Kdo residue, which in turn carries at O-8 a rhamnan. This is only the second case of an LPS where a cluster of four Kdo residues was identified (see *Chlamydia*, below, and Table 10).

Moraxella The core regions of LPS of serotypes A, B, and C from *M. catarrhalis*, which causes otitis media and sinusitis in children, have been investigated (Table 8) (215–218). They all consist of a branched neutral oligosaccharide of Glc, Gal, and GlcNAc residues, which is α-(1→5)-linked to Kdo. Differences in the core structures occur by various substituents at O-2 of the β-Glc-(1→4)-Glc moiety.

Rhizobiaceae

Bacteria of the genera *Rhizobium* and *Bradyrhizobium* live in symbiosis with legume plants and participate in the process of nitrogen assimilation. Several partial core structures of LPS from *R. leguminisarum*, *R. meliloti*, and *B. japonicum* have been published (Table 9) (219–226). The core structure of *R. etli* CE3 comprises two oligosaccharides, which have also been isolated from *R. leguminisarum* bv. trifolii strains ANU843 and 24.1, a branched tetrasaccharide consisting of Gal, Man, GalA, and Kdo, which is substituted at O-4 of Kdo by a branched trisaccharide built up from two GalA and one Kdo residues (226,227). In both species, the O-antigen is linked to O-6 of the Gal residue of the tetrasaccharide unit, and in the core of *R. etli* this is furnished via a third Kdo residue. The anomeric configuration of the Kdo residues are not published.

Chlamydia

For the structure and function of *Chlamydia* LPS, the reader is referred to Chapter 13 of this book, which includes all aspects of chlamydia core regions. Briefly, we established the structures of the core regions of *C. trachomatis* and *C. psittaci* by analysis of LPS of recombinant deep-rough *E. coli* bacteria, which express the respective chlamydia Kdo transferase (Table 10) (17,18,228–234). Chlamydial Kdo transferases are multifunctional enzymes that transfer three Kdo residues in *C. trachomatis* and *C. pneumoniae* (235) and at least four in *C. psittaci*. In the last case, the characterized branched tetrasaccharide represents the first oligosaccharide consisting of a cluster of four Kdo residues (see *Acinetobacter*, Table 8).

Miscellaneous

The structures of *B. fragilis* (236), *P. maltophila* (237), and *R. sphaeroides* (238) are as reviewed earlier (4,5), and the oligosaccharide of the core region of *R. capsulatus* (239) represents a partial structure (Table 10). The structure of the core region of LPS from *L. pneumophila* (240–242) possesses a highly lipophilic character because of the presence of deoxy sugars (Rha, QuiN) and *N*- and *O*-acetyl groups. For the first time,

Chemical Structure of LPS Core Region

Table 8 Structures of Core Regions of LPS from *Moraxellaceae*

Organism	Structure	Ref.
Acinetobacter ATCC 17905	Glc-(1→6)-β-Glc-(1→4)-Glc-(1→5)-Sug^{a,b} 4 ↑ 2 Glc-(1→6)-β-Glc-(1→3)-β-Gal-(1→7)-Kdo	12
A. haemolyticus ATCC 17906	Dha^c-(1→6)-β-Glc-(1→4)-Glc-(2→6)-β-Glc-(1→6)-β-Glc-(1→4)-Glc-(1→5)-Sug^{a,b} 6 4 ↑ ↑ P 2 Kdo	13
A. baumannii NCTC 10303	GlcNAc-(1→4)-GlcNA^d-(1→4)-Kdo-(2→5)-Kdo^b 5 4 ↑ ↑ 2 2 L-Rha-(1[→3)-L-Rha-(1→]$_n$8)-Kdo Kdo n = 1–4, mainly 2	†
ATCC 19606	Glc^e-(1→2)-β-Glc^e-(1→4)-β-Glc^e-(1→4)-β-Glc^e-(1→3)-β-GlcNAcA^f-(1→4)-Kdo-(2→5)-Kdo^b 4 4 7 4 ↑ ↑ ↑ ↑ 1 1 1 2 GalNAc β-GlcN Kdo	†
Moraxella	β-Glc 1 ↓ 3 Gal^g-(1→4)-β-Gal^b-(1→4)-Glc-(1→2)-β-Glc-(1→6)-Glc-(1→5)-Kdo^{b,h} 4 4 ↑ ↑ 1 2 R^1→2)-β-Glc Kdo^h General structure R^1	
M. catarrhalis A ATCC 25238	GlcNAc-(1	215, 216
M. catarrhalis B CCUG 3292	β-Gal-(1→4)-Glc-(1	217
M. catarrhalis C RS26 RS10	Gal^j-(1→4)-β-Gal^k-(1→4)-GlcNAc-(1	218

Where not stated otherwise, sugars are α-D-pyranosides. ^{a}Sug, Kdo (in a major portion of LPS of strain ATCC 17905 and in a minor of that of strain ATCC 17906) or D-*glycero*-D-*talo*-octulopyranosonic acid (Ko, in a major portion of LPS of strain ATCC 17906 and in a minor of that of strain ATCC 17905). ^{b}This residue was identified to link the core region to lipid A. ^{c}Dha, 3-deoxy-D-*lyxo*-heptulosaric acid. ^{d}GlcNA, 2-amino-2-deoxy-D-glucopyranosuronic acid. ^{e}Nonstoichiometric substitution. GlcNAcA, 2-acetamido-2-deoxy-D-*gluco*-pyranosuronic acid. ^{g}Nonstoichiometric substitution in strains CCUG and RS 26. ^{h}Determined for strains ATCC 25238, RS26, and RS10. ^{i}Nonstoichiometric substitutions in strain RS 26. ^{j}Nonstoichiometric substitution in strains RS 10 and RS 26. ^{k}Vinogradov, E. et al. Unpublished.

Table 9 Structures of Core Regions of LPS from *Rhizobiaceae*

Species/Strain	Structure	Ref.
Rhizobium etli CE3	O-chain-(1→4)-Kdo^a-(1→6)-Gal-(1→6)-Man-(1→5)-Kdo^{a,b} 　　　　　　　　　　　4　　　　　　　　　　4 　　　　　　　　　　↑　　　　　　　　　　↑ 　　　　　　　　　　1　　　　　　　　　　2 　　　　　　　　　GalA　　　　Kdo^a-(4←1)-GalA 　　　　　　　　　　　　　　　　　　　5 　　　　　　　　　　　　　　　　　　　↑ 　　　　　　　　　　　　　　　　　　　1 　　　　　　　　　　　　　　　　　　GalA	226, 227
R. leguminosarum 　bv. *trifolii* ANU843^d 　bv. *trifolii* 24.1^d 　bv. *phaseoli*^{d} 　bv. *viciae* VF39	Gal^c-(1→6)-Man-(1→5)-Kdo　　　　Gal-(1→5)-Kdo 　　　　　　4　　　　　　　　　　　　　4 　　　　　　↑　　　　　　　　　　　　　↑ 　　　　　　1　　　　　　　　　　　　　1 　　　　　GalA　　　　　　　　　　　　GalA	219–222a
R. leguminosarum 　bv. *viciae* VF39^d	Man-(1→5)-Kdo 　　　7 　　　↑ 　　　1 　　　Gal	223
R. meliloti 102F51^d	R$_1$^{e}-(1→6)-β-GlcN-(1→4)-Kdo　　　GalA-(1→3)-β-Glc-(1→4)-Kdo 　　　　　　　　　　　　　　　　　　　　6　　　　　　　5 　　　　　　　　　　　　　　　　　　　　↑　　　　　　↑ 　　　　　　　　　　　　　　　　　　　　R$_2$^{e}　　　　R$_3$^{e}	224
Bradyrhizobium japonicum 61A101c^d	Man-(1→4)-β-Glc-(1→4)-Kdo	225
JS314^d	Man-(1→4)-β-Glc-(1→4)-Kdo　　　　4-O-Me^f-Man-(1→5)-Kdo	

Where not stated otherwise, sugars are α-D-pyranosides. ^{a}Anomeric configuration unknown. ^{b}This residue was identified to link the core region to lipid A. ^{c}Substituted at O-6 by the O-antigen. ^{d}Partial structures. ^{e}R$_1$–R$_3$, unknown substituents. ^{f}Me, methyl.

a substitution of Kdo II at O-8 by Man was identified. The O-antigen is attached at O-3 of Rha at the nonreducing terminus. The core region of *O. anthropi* possesses a symmetrically branched oligosaccharide in which both Kdo residues are substituted at O-5 by a disaccharide (243).

CONCLUDING REMARKS

In the past 6 years, there has been considerable progress in the structural elucidation of core regions of LPS obtained from a broad variety of bacterial species. A comparison of all structures known to date shows that there is considerable structural variability in the core region. However, most core regions comprise as a partial structure the trisaccharide L,D-Hep-(1→3)-L,D-Hep-(1→5)-Kdo, and if core structures from one bacterial family (e.g., *Enterobacteriaceae*) are compared, a common structural theme is often identified (e.g., the oligosaccharide L,D-Hep-(1→7)-L,D-Hep-(1→3)-L,D-Hep-(1→5)-[Kdo-(2→4)]-Kdo in *Enterobacteriaceae*), which varies in different species. The most general structure of rough-type LPS (Fig. 2) is thus only an abstraction, but it still displays essential properties of LPS. First, the common principle of LPS is represented by the core oligosaccharide and lipid A, as proven by the findings that (1) mutants that cannot synthesize a core are not viable and (2) wild-type LPS may be S-form or R-form, thus, the expression of an O-specific polysaccharide is not a prerequisite for bacterial survival. The minimal structure of an LPS was identified as lipid A substituted by one phosphorylated Kdo residue (present in *Haemophilus influenzae* I-69 Rd$^-$/b$^+$).

Table 10 Structures of Core Regions of LPS from *Bacteroides fragilis*, *Chlamydia*, *Legionella pneumophila*, *Pseudomonas maltophila*, *Rhodobacter*, and *Ochrobacterium anthropi*

Species	Structure	Ref.
Bacteroides fragilis NCTC 9343	PS I β-Gal-(1→6)-β-Gal-(1→6)-β-Gal-(1→6)-β-Gal^a-(1→4/6)-Glc-(1→2)-L-Rha $\qquad\qquad\qquad\qquad\qquad\qquad\qquad\qquad\qquad\qquad$ 4/6 $\qquad\qquad\qquad\qquad\qquad\qquad\qquad\qquad\qquad\qquad$ ↑ $\qquad\qquad\qquad\qquad\qquad\qquad\qquad\qquad\qquad\qquad$ 1 $\qquad\qquad\qquad\qquad\qquad\qquad\qquad\qquad\qquad\qquad$ β-Gal^a	236
	PS II β-Gal-(1→4)-Glc-(1→2)-L-Rha $\qquad\qquad\qquad\qquad$ 6 $\qquad\qquad\qquad\qquad$ ↑ $\qquad\qquad\qquad\qquad$ 1 $\qquad\qquad\qquad\qquad$ β-Gal	
Chlamydia trachomatis	Kdo-(2→8)-Kdo(2→4)-Kdo^b	17, 18, 229–231
C. psittaci	Kdo-(2→8)-Kdo-(2→4)-Kdo^b $\qquad\qquad\qquad\quad$ 4 $\qquad\qquad\qquad\quad$ ↑ $\qquad\qquad\qquad\quad$ 2 $\qquad\qquad\qquad\quad$ Kdo	232
Legionella pneumophila	O-chain $\qquad\qquad\qquad\qquad\qquad\qquad\qquad$ GlcNAc 1 $\qquad\qquad\qquad\qquad\qquad\qquad\qquad\qquad\quad$ 1 ↓ $\qquad\qquad\qquad\qquad\qquad\qquad\qquad\qquad\quad$ ↓ 3 $\qquad\qquad\qquad\qquad\qquad\qquad\qquad\qquad\quad$ 6 L-Rha-(1→3)-L-Rha-(1→3)-β-QuiNAc-(1→4)-β-GlcNAc-(1→4)-Man-(1→5)-Kdo^b $\quad$ 2 $\qquad\qquad$ 2 $\qquad\qquad\quad$ 4 $\qquad\qquad\qquad$ 3 $\qquad\qquad\qquad\qquad$ 4 $\quad$	$\qquad\qquad$	$\qquad\qquad\quad$	$\qquad\qquad\qquad$	$\qquad\qquad\qquad\qquad$ ↑ $\quad$ Ac $\qquad\quad$ Ac $\qquad\qquad$ Ac $\qquad\qquad\quad$ Ac $\qquad\qquad\qquad$ 2 $\qquad\qquad\qquad\qquad\qquad\qquad\qquad\qquad\qquad\qquad\quad$ Man-(1→8)-Kdo	240–242
P. maltophila^c	Glc-(1→4)-Glc-(1→4)-Man-(1→5)-Kdo $\qquad\qquad\qquad\qquad\qquad$ 6 $\qquad\qquad\qquad\qquad\qquad$ ↑ $\qquad\qquad\qquad\qquad\qquad$ 1 $\qquad\qquad\qquad\qquad\qquad$ GalN	237
Rhodobacter sphaeroides ATCC 17023	GlcA-(1→4)-GlcA-(1→4)-GlcA-(1→4)-Kdo $\qquad\qquad\qquad\qquad\qquad\qquad\qquad$ 6 $\qquad\qquad\qquad\qquad\qquad\qquad\qquad$ ↑ $\qquad\qquad\qquad\qquad\qquad\qquad\qquad$ 1 $\qquad\qquad\qquad\qquad\qquad\qquad\qquad$ Thr^d	238
R. capsulatus 37b4^e	Gal-(1→6)-Glc-(1→7)-Neu5Ac	239
Ochrobacterium anthropi	β-GlcN-(1→3)-Man-(1→5)-Kdo^b $\qquad\qquad\qquad\qquad\quad$ 4 $\qquad\qquad\qquad\qquad\quad$ ↑ $\qquad\qquad\qquad\qquad\quad$ 2 Glc^f-(1→4)-GalA-(1→5)-Kdo	243

Where not stated otherwise, sugars are α-D-pyranosides. ^{a}Position not unequivocally identified. ^{b}This residue was identified to link the core region to lipid A. ^{c}The anomeric configurations were not determined. ^{d}Thr, threonine. ^{e}Partial structure. ^{f}Nonstoichiometric substitution.

Second, the binding of the core region to lipid A always occurs via a Kdo residue [except in two strains of *Acinetobacter* (see Table 8), where this Kdo is replaced in nonstoichiometric amounts by Ko]. According to present knowledge, this Kdo residue (and also the Ko residue) is substituted at O-4 either by a Kdo disaccharide [Kdo-(2→4)-Kdo, e.g., in *S. enterica* or *E. coli*, or Kdo-(2→8)-Kdo in *Chlamydia*], or a Kdo monosaccharide (e.g., in *S. enterica* or *E. coli*, which may be further substituted by PE or monosaccharidic residues), or phosphate (e.g., in *Vibrio cholerae* or *Haemophilus influenzae*). Very interesting and presently exceptional

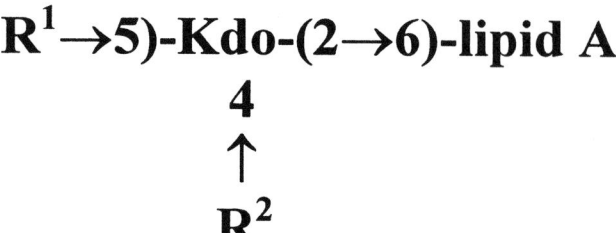

Fig. 2 Most general structure of the core-lipid A region of LPS. Kdo may exceptionally be nonstoichiometrically replaced by Ko in some *Acinetobacter* strains. The core elongation R^1 is in most cases provided by a *manno*-configurated sugar, e.g., L,D-Hep. When determined, R^2 was shown to represent a negatively charged residue, i.e., Kdo, Ko, or phosphate.

are the attachment at this position of the branched trisaccharide Kdo-(2→4)-[Kdo-(2→8)-]-Kdo (in *Chlamydia psittaci*) and the presence of the disaccharide Kdo-(2→5)-Kdo [in *Acinetobacter haemolyticus* NCTC 10303 and ATCC 19606 (see Table 8) and *Campylobacter lari* ATCC 35221 (see Table 6)]. In *Acinetobacter*, this disaccharide was shown to be linked to lipid A and to be substituted at O-4 of the first Kdo by a third Kdo residue and at O-7 of the second Kdo by either a fourth Kdo (in strain NCTC 10303, which is further substituted by a short rhamnoglycan) or β-GlcN (in strain ATCC 19606). Thus, there are two core regions known to date that possess a cluster of four Kdo residues in proximity to lipid A. In *E. coli* and *Chlamydia* it was shown that only one Kdo transferase is present, which transfers either two (in *E. coli*) or three (in *C. trachomatis* and *C. pneumoniae*) or four (in *C. psittaci*) Kdo residues. It is not known to date how many Kdo transferases are present in *Acinetobacter* or how Ko is synthesized. As a third general feature of LPS, it should be noted that the core region is always negatively charged, which is provided by phosphoryl substituents and/or sugar acids (e.g., Kdo, uronic acids) and which is thought to contribute to the rigidity of the gram-negative cell wall through intermolecular cross-links via bivalent cations. The absence of such intermolecular cross-links in mutants that are not able to synthesize a core region could be a reason for their lethality.

Core structures may be classified into two types: with and without heptoses. In the first type, L,D-Hep or D,D-Hep alone or both may be present in a particular core structure. If present, D,D-Hep is in most cases transferred in later steps of biosynthesis and either decorates the *inner core* region (e.g., in *Y. enterocolitica*) or is attached to more remote parts of the carbohydrate chain. The regulation of the distribution of L,D-Hep and D,D-Hep in the core region as well as that of the synthesis of D,D-heptoglycans (as in *K. pneumoniae* O1 or *Helicobacter*) are not understood, and it is not known whether L,D-Hep and D,D-Hep are transferred by different transferases. However, the identification of one strain of *S. enterica* sv. *typhimurium* in which the gene encoding for ADP-D,D-Hep-6-epimerase (*rfaD*) was defective and which used D,D-Hep for biosynthesis of the inner core region, indicates that heptosyltransferases may accept both L,D-Hep and D,D-Hep. A certain D,D-Hep residue may eventually be replaced by L,D-Hep [as shown for *P. mirabilis* O28 (see Table 1)].

NOTE ADDED IN PROOF

The following structures or partial structures of core oligosaccharides have been published since October 1998.

Vibrio cholerae 022 (2344,245), *Haemophilus sommus* 738 (246), *Pseudomonas aeruginosa* 05 (247), *Neisseria meningitidis* L1 (248), *N. meningitidis* NMB (249), and *Burkholderia pseudomallei* GIFU 12046 (250).

ACKNOWLEDGMENTS

I am indebted to the following colleagues for sharing data prior to publication and/or critical reading of the typescript (in alphabetical order): Gerald O. Aspinall (York University, North York, Canada), Klaus Bock (Carlsberg Laboratory, Valby, Denmark), Helmut Brade (Research Center Borstel, Borstel, Germany), Russell W. Carlson (Complex Carbohydrate Research Center, Athens, Georgia), Martine Caroff (Université de Paris-Sud, Orsay, France), Andrzej Gamian (Ludwik Hirszfeld Institute of Immunology and Experimental Therapy, Wroclaw, Poland), Dieter Grimmecke (Research Center Borstel, Borstel, Germany), Per-Erik Jansson (Karolinska Institute, Huddinge Hospital, Huddinge, Sweden), Yuriy Knirel (N. D. Zelinsky Institute of Organic Chemistry, Moscow, Russia), Mario Monteiro (Institute for Biological Sciences, NRCC, Ottawa, Canada), Joanna Radziejewska-Lebrecht (Institute of Microbiology, Silesian University, Katowice, Poland), James C. Richards (Institute for Biological Sciences, NRCC, Ottawa, Canada), Elke Schweda (Karolinska Institute, Huddinge Hospital, Huddinge, Sweden), Zygmunt Sidorzyk (Institute of Microbiology and Immunology, University of Lodz, Lodz, Poland), Jane E. Thomas-Oates (Michael Barber Centre for Mass Spectrometry, Manchester, England), Evgeny Vinogradov (Carlsberg Laboratory, Valby, Denmark), and Ulrich Zähringer (Research Center Borstel, Borstel, Germany). The research of our group is supported by the Deutsche Forschungsgemeinschaft (grant SFB 470-B1), the Deutscher Akademischer Austauschdienst (grants 314-VIGONI and ppp-313), and the Bundesministerium für Bildung und Forschung (grant POL-150).

REFERENCES

1. Rietschel ET, Brade H, Holst O, Brade L, Müller-Loennies S, Mamat U, Zähringer U, Beckmann F, Seydel U, Brandenburg K, Ulmer AJ, Mattern T, Heine H, Schletter J, Loppnow H, Schönbeck U, Flad H-D, Hauschildt S, Schade UF, Di Padova F, Kusumoto S, Schumann RR. Bacterial endotoxin: chemical constitution, biological recognition, host response, and immunological detoxification. Curr Top Microbiol Immunol 1996; 216:39–81.
2. Mamat U, Seydel U, Grimmecke D, Holst O, Rietschel ET. Lipopolysaccharides. In: Pinto M, ed. Comprehensive Natural Products Chemistry. Amsterdam: Elsevier Science, in press.
3. Wilkinson SG, Bacterial lipopolysaccharides—themes and variations. Prog Lipid Res 1996; 35:283–343.
4. Holst O, Brade H. Chemical structure of the core region of lipopolysaccharides. In: Morrison DC, Ryan JL, eds. Bacterial Endotoxic Lipopolysaccharides. Boca Raton, FL: CRC Press, 1992;135–170.
5. Knirel YA, Kochetkov NK. The structure of lipopolysaccharides of gram-negative bacteria. II. The structure of the core region (a review). Biokhimya 1993; 58:182–201.
6. Unger FM. The chemistry and biological significance of 3-deoxy-D-manno-2-octulosonic acid (Kdo). Adv Carbohydr Chem Biochem 1983; 348:323–387.
7. Coleman Jr. WG. The rfaD gene codes for ADP-L-glycero-D-manno-heptose-6-epimerase: an enzyme required for lipopolysaccharide core biosynthesis. J Biol Chem 1983; 258:1985–1990.
8. Kocsis B, Kontrohr, T. Isolation of adenosine 5'-diphosphate-L-glycero-D-manno-heptose, the assumed substrate for heptose transferase(s) from Salmonella minnesota R595 and Shigella sonnei Re mutants. J Biol Chem 1984; 259:11858–11860.
9. Ding L, Seto BL, Ahmed SA, Coleman WG. Purification and properties of the Escherichia coli K-12 NAD-dependant nucleotide diphosphosugar epimerase, ADP-L-glycero-L-mannoheptose-6-epimerase. J Biol Chem 1994; 269:24384–24390.
10. Masoud H, Mayer H, Kontrohr T, Holst O, Weckesser J. The structure of the core region of the lipopolysaccharide from Rhodocyclus gelatinosus Dr2. Syst Arch Microbiol 1991; 12:222–227.
11. Kawahara K, Brade H, Rietschel ET, Zähringer U. Studies on the chemical structure of the core-lipid A region of the lipopolysaccharide of Acinetobacter calcoaceticus NCTC 10305. Detection of a new 2-octulosonic acid interlinking the core oligosaccharide and lipid A component. Eur J Biochem 1987; 163:489–495.
12. Vinogradov EV, Bock K, Petersen BO, Holst O, Brade H. The structure of the carbohydrate backbone of the lipopolysaccharide from Acinetobacter strain ATCC 17905. Eur J Biochem 1997; 243:122–127.
13. Vinogradov EV, Müller-Loennies S, Petersen BO, Meshkov S, Thomas-Oates JE, Holst O, Brade H. Structural investigation of the lipopolysaccharide from Acinetobacter strain NCTC 10305 (ATCC 17906, DNA group 4). Eur J Biochem 1997; 247:82–90.
14. Kawahara K, Isshiki Y, Ezaki T, Moll H, Kosma P, Zähringer U. Chemical characterization of the inner core-lipid A region of the lipopolysaccharide isolated from Pseudomonas (Burkholderia) cepacia (abstr). J Endotoxin Res 1994; 1(suppl 1):52.
15. Isshiki Y, Zähringer U, Kawahara K, Ezaki T. Chemical structure of the core region of the lipopolysaccharide from Burkholderia cepacia and B. pseudomallei. Research Center Borstel, Annual Report 1996, p. 168.
16. Gass J, Strobl M, Loibner A, Kosma P, Zähringer U. Synthesis of allyl O-[sodium (α-D-glycero-D-talo-2-octulopyranosyl)onate]-(2→6)-2-acetamido-2-deoxy-β-D-glucopyranoside, a core constituent of the lipopolysaccharide from Acinetobacter calcoaceticus NCTC 10305. Carbohydr Res 1993; 244:69.
17. Holst O, Broer W, Thomas-Oates JE, Mamat U, Brade H. Structural analysis of two oligosaccharide bisphosphates from the lipopolysaccharide of a recombinant strain of Escherichia coli F515 (Re chemotype) expressing the genus-specific epitope of Chlamydia lipopolysaccharide. Eur J Biochem 1993; 214:703–710.
18. Holst O, Thomas-Oates JE, Brade H. Preparation and structural analysis of oligosaccharide monophosphates obtained from the lipopolysaccharide of recombinant strains of Salmonella minnesota and Escherichia coli expressing the genus-specific epitope of Chlamydia lipopolysaccharide. Eur J Biochem 1994; 222:183–194.
19. Holst O, Ulmer AJ, Brade H, Flad H-D, Rietschel ET. Biochemistry and cell biology of bacterial endotoxins. FEMS Immunol Med Microbiol 1996; 16:83–104.
20. Brabetz W, Müller-Loennies S, Holst O, Brade H. Deletion of the heptosyltransferase genes rfaC and rfaF in Escherichia coli K-12 results in an Re-type lipopolysaccharide with a high degree of 2-aminoethanol phosphate substitution. Eur J Biochem 1997; 247:716–724.
21. Knirel YA, Widmalm G, Senchenkova SN, Jansson P-E, Weintraub A. Structural studies on the short-chain lipopolysaccharide of Vibrio cholerae O139 Bengal. Eur J Biochem 1997; 247:402–410.
22. Wakarchuk W, Martin A, Jennings MP, Moxon ER, Richards JC. Functional relationships of the genetic locus encoding the glycosyltransferase enzymes involved in expression of the lacto-N-neotetraose ter-

minal lipopolysaccharide structure in *Neisseria meningitidis*. J Biol Chem 1996; 271:19166–19173.
23. Lindberg B, Lönngren J, Svensson S. Specific degradation of polysaccharides. Adv Carbohydr Chem Biochem 1975; 31:185–240.
24. Süsskind M, Müller-Loennies S, Nimmich W, Brade H, Holst O. Structural investigation on the carbohydrate backbone of the lipopolysaccharide from *Klebsiella pneumoniae* rough mutant R20/O1. Carbohydr Res 1995; 269:C1–C7.
25. Galanos C, Lüderitz O, Rietschel ET, Westphal O. Newer aspects on the chemistry and biology of bacterial lipopolysaccharides, with special reference to their lipid A component. In: Goodwin TW, ed. International Review of Biochemistry, Biochemistry of Lipids II, Vol. 14. Baltimore: University Park Press, 1997:239–335.
26. Jansson P-E, Lindberg AA, Lindberg B, Wollin R. Structural studies on the hexose region of the core in lipopolysaccharides from Enterobacteriaceae. Eur J Biochem 1981; 115:571–577.
27. Volk WA, Salomonsky NL, Hunt D. *Xanthomonas sinensis* cell wall lipopolysaccharide. I. Isolation of 4,7-anhydro- and 4,8-anhydro-3-deoxy-octulosonic acid following acid hydrolysis of *Xanthomonas sinensis* lipopolysaccharide. J Biol Chem 1972; 247:3881–3887.
28. Brade H, Brade L, Rietschel ET. Structure-activity relationships of bacterial lipopolysaccharides (endotoxins). Zentralbl Bakteriol Hyg 1988; A268:151–179.
29. Brade H, Zähringer U, Rietschel ET, Christian R, Schulz G, Unger FM. Spectroscopic analysis of a 3-deoxy-D-manno-2-octulosonic acid(KDO)-disaccharide from the lipopolysaccharide of a *Salmonella godesberg* Re mutant. Carbohydr Res 1984; 134:157–166.
30. Brade H, Rietschel ET. α-2→4-Interlinked 3-deoxy-D-manno-2-octulosonic acid-disaccharide. A common constituent of enterobacterial lipopolysaccharides. Eur J Biochem 1984; 145:231–236.
31. Tacken A, Brade H, Unger FM, Charon D. GLC-MS of partially methylated and acetylated derivatives of 3-deoxyoctitols. Carbohydr Res 1986; 149:263–277.
32. Brade H, Moll H, Rietschel ET. Structural investigations on the inner core region of lipopolysaccharides from *Salmonella minnesota* rough mutants. Biomed Mass Spectrom 1985; 12:602–609.
33. Tacken A, Rietschel ET, Brade H. Methylation analysis of the heptose/3-deoxy-D-*manno*-2-octulosonic acid-region (inner core) of the lipopolysaccharide from *Salmonella minnesota* rough mutants. Carbohydr Res 1986; 149:279–291.
34. Holst O, Brade H, Dziewiszek K, Zamojski A. GLC-MS of partially methylated and acetylated derivatives of L-*glycero*-D-*manno*- and D-*glycero*-D-*manno*-heptopyranoses and -heptitols. Carbohydr Res 1990; 204:1–9.
35. Rietschel ET, Galanos C, Lüderitz O, Westphal O. Chemical structure, physiological function and biological activity of lipopolysaccharides and their lipid A component. In: Webb D, ed. Immunopharmacology and Regulation of Leukocyte Function. New York: Marcel Dekker, 1982:183–229.
36. Holst O, Röhrscheidt-Andrzejewski E, Brade H, Charon D. Isolation and characterisation of 3-deoxy-D-manno-2-octulopyranosonate 7-(2-aminoethyl phosphate) from the inner core region of *Escherichia coli* K-12 and *Salmonella minnesota* lipopolysaccharides. Carbohydr Res 1990; 204:93–102.
37. Holst O, Brade H. Structural studies of the core region of the lipopolysaccharide from *Salmonella minnesota* strain R7 (rough mutant chemotype Rd$_1$). Carbohydr Res 1991; 219:247–251.
38. Holst O, Brade H. Isolation and structural analysis of the tetrasaccharide 3-deoxy-5-*O*-[3-*O*-(3-*O*-α-D-glucopyranosyl-L-*glycero*-D-*manno*-heptosyl)-L-*glycero*-D-*manno*-heptosyl]-D-*manno*-octulosonic acid from the core region of the lipopolysaccharide from *Salmonella minnesota* strain R5 (rough mutant chemotype RcP$^-$). Carbohydr Res 1993; 245:159–163.
39. Zähringer U, Sinnwell V, Peter-Katalinic J, Rietschel ETh, Galanos C. Isolation and characterization of the tetrasaccharide (bis)phosphate from the glycosyl backbone of *Salmonella minnesota* and *Escherichia coli* Re-mutant lipopolysaccharides. Tetrahedron 1993; 49:4193–4200.
40. Tsang RSW, Schlecht S. Smooth lipopolysaccharide of *Salmonella adelaide* has an atypical *Salmonella* Ra core. Res Microbiol 1990; 141:671–678.
41. Tsang RSW, Schlecht S, Aleksic S, Chan KH, Chau PY. Lack of the α-1,2-linked *N*-acetyl-D-glucosamine epitope in the outer core structures of lipopolysaccharides from certain O serogroups and subspecies of *Salmonella enterica*. Res Microbiol 1991; 142:521–533.
42. Tsang RSW, Schlecht S, Mayer H. Structural differences in the outer core region of lipopolysaccharides derived from members of the genus *Salmonella*. Zbl Bakt Hyg I Abt Orig A 1992; 276:330–339.
43. Olsthoorn MMA, Petersen BO, Schlecht S, Haverkamp J, Bock K, Thomas-Oates JE, Holst O. Identification of a novel core type in *Salmonella* lipopolysaccharide. Complete structural analysis of the core region of the lipopolysaccharide from *Salmonella enterica* sv. Arizonae. J Biol Chem 1998; 273:3817–3829.
44. Schmidt G, Jann B, Jann K. Immunochemistry of R lipopolysaccharides of *Escherichia coli*. Studies on R mutants with an incomplete core, derived from *E. coli* O8:K27. Eur J Biochem 1970; 16:382–392.
45. Feige U, Stirm S. On the structures of the *Escherichia coli* C cell wall lipopolysaccharide core and on its X174 region. Biochem Biophys Res Commun 1976; 71:566–573.
46. Hämmerling G, Lüderitz O, Westphal O, Mäkelä PH. Structural investigations on the core polysaccharide of *Escherichia coli* O100. Eur J Biochem 1971; 22:331–344.
47. Jansson P-E, Lindberg AA, Lindberg B, Wollin R. Structural studies on the hexose region of the Enterobacteriaceae type R3 core polysaccharide. Carbohydr Res 1979; 68:385–389.
48. Feige U, Jann B, Jann K, Schmidt G, Stirm S. On the primary structure of the *Escherichia coli* R4 cell wall lipopolysaccharide core. Biochem Biophys Res Commun 1977; 79:88–95.

49. Jansson P-E, Lindberg B, Bruse G, Lindberg AA, Wollin R. Structural studies on the hexose region of the lipopolysaccharide from *Escherichia coli* C. Carbohydr Res 1977; 54:261–267.
50. Prehm P, Stirm S, Jann B, Jann K, Boman HG. Cell-wall lipopolysaccharide of ampicillin-resistant mutants of *Escherichia coli* K-12. Eur J Biochem 1976; 66:369–377.
51. Prehm P, Schmidt G, Jann B, Jann K. The cell-wall lipopolysaccharide of *Escherichia coli* K-12: structure and acceptor site for O-antigen and other substituents. Eur J Biochem 1976; 70:171–177.
52. Holst O, Zähringer U, Brade H, Zamojski A. Structural analysis of the heptose/hexose region of the lipopolysaccharide from *Escherichia coli* K-12. Carbohydr Res 1991; 215:323–335.
53. Pakulski Z, Zamojski A, Holst O, Zähringer U. The synthesis and characterisation of 6-O-L-*glycero*-D-*manno*-heptopyranosyl-D-gluco-pyranose. Carbohydr Res 1991; 215:337–344.
54. Holst O, Röhrscheidt-Andrzejewski E, Cordes H-P, Brade H. Isolation and identification of 3-deoxy-7-O-α-D-galactopyranosyl-D-*manno*-2-octulopyranosonate from the inner core region of the lipopolysaccharide of *Escherichia coli* E 100. Carbohydr Res 1989; 188:212–218.
55. Haishima Y, Holst O, Brade H. Structural investigation on the lipopolysaccharide of *Escherichia coli* rough mutant F653 representing the R3 core type. Eur J Biochem 1992; 203:127–134.
56. Kaca W, de Jongh-Leuvenink J, Zähringer U, Rietschel ETh, Brade H, Verhoef J, Sinnwell V. Isolation and chemical analysis of 7-O-(2-amino-2-deoxy-glucopyranosyl)-L-*glycero*-D-*manno*-heptose as a constituent of the lipopolysaccharide of the UDP-galactose epimerase-less mutant J-5 of *Escherichia coli* and *Vibrio cholerae*. Carbohydr Res 1988; 179:289–99.
57. Holst O, Brade H. Isolation and identification of 3-deoxy-5-O-α-L-rhamnopyranosyl-D-*manno*-2-octulopyranosonate from the inner core region of the lipopolysaccharide of *Escherichia coli* K-12. Carbohydr Res 1990; 207:327–331.
58. Nepogodiev SA, Backinowsky LV, Grzeszczyk B, Zamojski A. Synthesis of linear oligosaccharides: L-*glycero*-D-*manno*-heptopyranosyl derivatives of allyl α-glycosides of D-glucose, kojibiose, and 3-O-α-kojibiosyl-D-glucose, substrates for synthetic antigens. Carbohydr Res 1994; 254:43–60.
59. Brade L, Grimmecke H-D, Holst O, Brabetz W, Zamojski A, Brade H. Specificity of monoclonal antibodies against *Escherichia coli* K-12 lipopolysaccharide. J Endotox Res 1996; 3:39–47.
60. Müller-Loennies S, Holst O, Brade H. Chemical structure of the core region of *Escherichia coli* J-5 lipopolysaccharide. Eur J Biochem 1994; 224:751–760.
61. Müller-Loennies S, Holst O, Lindner B, Brade H. Isolation and structural analysis of phosphorylated oligosaccharides obtained from *Escherichia coli* J-5 lipopolysaccharide. Eur J Biochem 1998. In press.
62. Gamian A, Romanowska E. The core structure of *Shigella sonnei* lipopolysaccharide and the linkage between O-specific polysaccharide and the core region. Eur J Biochem 1982; 129:105–109.
63. Gamian A, Romanowska E. Heterogeneity of *Shigella sonnei* core oligosaccharides. 7th International Symposium on Glycoconjugates, Lund-Ronneby, Sweden, Jul 17–23, 1983.
64. Katzenellenbogen E, Romanowska E. Structural studies on *Shigella flexneri* serotype 6 core region. Eur J Biochem 1980; 113:205–211.
65. Lugowski C, Kulakowska M, Romanowska E. Characterisation and diagnostic application of a lipopolysaccharide core oligosaccharide-protein conjugate. J Immunol Meth 1986; 95:187–194.
66. Gamian A, Romanowska E, Dabrowski U, Dabrowski, J. Structure of the O-specific, sialic acid containing polysaccharide chain and its linkage to the core region in lipopolysaccharide from *Hafnia alvei* strain 2 as elucidated by chemical methods, gas-liquid chromatography/mass spectrometry, and ^{1}H NMR spectroscopy. Biochemistry 1991; 30:5032–5038.
67. Gamian A, Katzenellenbogen E, Romanowska E, Dabrowski U, Dabrowski J. Lipopolysaccharide core region of *Hafnia alvei*: structure elucidation using chemical methods, gas chromatography-mass spectrometry, and NMR spectroscopy. Carbohydr Res 1995; 266:221–228.
68. Jachymek W, Petersson C, Helander A, Kenne L, Lugowski C, Niedziela T. Structural studies of the O-specific chain and a core hexasaccharide of *Hafnia alvei* strain 1192 lipopolysaccharide. Carbohydr Res 1995; 269:125–138.
69. Lugowski C, Jachymek W, Niedziela T, Romanowska A, Witkowska D, Romanowska E. Lipopolysaccharide core region of *Hafnia alvei*: serological characterisation. FEMS Immunol Med Microbiol 1995; 10:119–124.
70. Jachymek W, Lugowski C, Romanowska E, Witkowska D, Petersson C, Kenne L. The structure of a core oligosaccharide component from *Hafnia alvei* strain 32 and 1192 lipopolysaccharides. Carbohydr Res 1995; 251:327–330.
71. Katzenellenbogen E, Gamian A, Romanowska E, Dabrowski U, Dabrowski J. 3-Deoxy-octulosonic acid-containing trisaccharide fragment of an unusual core type of some *Hafnia alvei* lipopolysaccharides. Biochem Biophys Res Comm 1993; 194:1058–1064.
72. Romanowska E, Katzenellenbogen E, Gamian A, Dabrowski U, Dabrowski J, Lugowski C, Jachymek W, Niedziela T. Core structure of *Hafnia alvei* LPS (abstr). J Endotoxin Res 1996; 3(suppl 1):35.
73. Gamian A, Romanowska E, Romanowska A, Lugowski C, Dabrowski J, Trauner K. *Citrobacter* lipopolysaccharides: structure elucidation of the O-specific polysaccharide from strain PCM 1487 by mass-spectrometry, one-dimensional and two-dimensional ^{1}H-NMR spectroscopy and methylation analysis. Eur J Biochem 1985; 146:641–647.
74. Romanowska E, Gamian A, Dabrowski J. Core region of *Citrobacter* lipopolysaccharide from strain PCM 1487: structure elucidation by two dimensional ^{1}H-NMR spectroscopy at 500 MHz and methylation analysis/mass spectrometry. Eur J Biochem 1986; 161:557–564.
75. Dabrowski J, Hauck M, Romanowska E, Gamian A. Structure elucidation of the core octasaccharide from

Citrobacter PCM 1487 with the aid of 500-MHz, two-dimensional phase-sensitive correlated, relayed-coherence transfer, double-quantum, triple-quantum filtered, and n.o.e. ^{1}H-NMR spectra. Carbohydr Res 1988; 180: 163–174.

76. Romanowska E, Gamian A. Lugowski C, Romanowska A, Dabrowski J, Hauck M, Opferkuch HJ, and von der Lieth C-W. Structure elucidation of the core regions from *Citrobacter* O4 and O36 lipopolysaccharides by chemical and enzymatic methods, gas chromatography/mass spectrometry, and NMR spectroscopy at 500 MHz. Biochemistry 1988; 27:4153–4161.

77. Romanowska E, Romanowska A, Dabrowski J, Trauner K. The structure of the lipopolysaccharide core region of *Citrobacter* O27. FEMS Microbiol Lett 1989; 58:107–110.

78. Gamian A, Romanowska E. Structure of the heptose-3-deoxyoctulosonic acid region of *Citrobacter* lipopolysaccharide core. Carbohydr Res 1990; 198:381–383.

79. Romanowska E, Gamian A, Katzenellenbogen E, Romanowska A, Lugowski C, Kulakowska M, Dabrowski J, Dabrowski U. Lipopolysaccharide core regions of *Citrobacter*: structure and serology. In: Nowotny A, Spitzer JJ, Ziegler E, eds. Cellular and Molecular Aspects of Endotoxin Reactions. Amsterdam: Excepta Medica, 1990:103–111.

80. Katzenellenbogen E, Gamian A, Romanowska E, Dabrowski U, Dabrowski J. Core region of *Citrobacter* O23 lipopolysaccharide. Structure elucidation by chemical methods, gas chromatography/mass spectrometry and NMR spectroscopy at 500 MHz. Eur J Biochem 1991; 196:197–201.

81. Sandulache R, Prehm P. Structure of the core oligosaccharide from lipopolysaccharide of *Erwinia carotovora*. J. Bacteriol 1985; 161:1226–1227.

82. Fukuoka S, Knirel YA, Lindner B, Moll H, Seydel U, Zähringer U. Elucidation of the structure of the core region and the complete structure of the R-type lipopolysaccharide of *Erwinia carotovora* FERM P-7576. Eur J Biochem 1997; 250:55–62.

83. Süsskind M, Brade L, Brade H, Holst O. Identification of a novel heptoglycan of α-(1→2)-linked D-*glycero*-D-*manno*-heptopyranose. Chemical and antigenic structure of lipopolysaccharides from *Klebsiella pneumoniae* ssp. *pneumoniae* rough strain R20 (O1$^-$:K20$^-$). J Biol Chem 1998; 273:7006–7017.

84. Severn WB, Kelly RF, Richards JC, Whitfield C. Structure of the core oligosaccharide in the serotype O8 lipopolysaccharide from *Klebsiella pneumoniae*. J Bacteriol 1996; 178:1731–1741.

85. Süsskind M. Chemische Strukturen der Lipopolysaccharide aus *Klebsiella* R20 (O1$^-$:K20$^-$) und *Klebsiella oxytoca* R29 (O1$^-$:K29$^-$). Ph.D. dissertation. Lübeck: Medizinische Universität zu Lübeck, 1997.

86. Sidorczyk Z, Kaca W, Brade H, Rietschel ET, Sinnwell V, Zähringer U. Isolation and structural characterisation of an 8-O-(4-amino-4-deoxy-β-L-arabinosyl)-3-deoxy-D-manno-octulosonic acid disaccharide in the lipopolysaccharide of a *Proteus mirabilis* deep rough mutant. Eur J Biochem 1987; 168:269–273.

87. Radziejewska-Lebrecht J, Bhat UR, Brade H, Mayer H. Structural studies on the core and lipid A region of a 4-amino-L-arabinose-lacking Rc-type mutant of *Proteus mirabilis*. Eur J Biochem 1988; 172:535–541.

88. Radziejewska-Lebrecht J, Mayer H. The core region of *Proteus mirabilis* R110/1959 lipopolysaccharide. Eur J Biochem 1989; 183:573–581.

89. Radziejewska-Lebrecht J, Krajewska-Pietrasik D, Mayer H. Terminal and chain-linked residues of D-galacturonic acid: characteristic constituents of the R-core region of *Proteeae* and *Serratia marcescens*. Syst Appl Microbiol 1990; 13:214–219.

90. Krajewska D, Gromska W. Heterogeneity of the lipopolysaccharide from *Proteus mirabilis* O27. Arch Immunol Ther Exp (Wroclaw) 1981; 29:581–587.

91. Radziejewska-Lebrecht J, Kotelko K, Krajewska D, Mayer H. The R-core region of *Proteus mirabilis* lipopolysaccharides. EOS Immunol Immunopharm 1986; 6:167.

92. Vinogradov EV, Thomas-Oates JE, Brade H, Holst O. Structural investigation of the lipopolysaccharide from *Proteus mirabilis* R45 (Re-chemotype). J Endotox Res 1994; 1:199–206.

93. Krajewska-Pietrasik D, Rozalski A, Bartodziejska B, Radziejewska-Lebrecht J, Mayer H, Kotelko K. Properties of a deep *Proteus* R mutant isolated from clinical material. APMIS 1991; 99:499–506.

94. Radziejewska-Lebrecht J, Shashkov AS, Stroobant V, Wartenberg K, Warth C, Mayer H. The inner core region of *Yersinia enterocolitica* Ye75R (O:3) lipopolysaccharide. Eur J Biochem 1994; 221:343–351.

95. Shashkov AS, Radziejewska-Lebrecht J, Kochanowski H, Mayer H. The chemical structure of the outer core region of the *Yersinia enterocolitica* O:3 lipopolysaccharide. EUROCARB VIII, 8th European Carbohydrate Symposium, Seville, Spain, July 2–7, 1995.

96. Müller-Loennies S, Rund S, Ervelä E, Skurnik M, Holst O. Structural analysis of the core-lipid A region of the lipopolysaccharide from a rough mutant of *Yersinia enterocolitica* O:9. EUROCARB IX, 9th European Carbohydrate Symposium, Utrecht, The Netherlands, July 6–11, 1997.

97. Ovodov YS, Gorshkova RP, Tomshich SV, Komandrova NA, Zubkov VA, Kalmykova EN, Isakov VV. Chemical and immunochemical studies on lipopolysaccharides of some *Yersinia* species—a review of some recent investigations. J Carbohydr Chem 1992; 11:21–35.

98. Yamasaki R, Bacon BE, Nasholds W, Schneider H, Griffiss JM. Structural determination of oligosaccharides derived from lipooligosaccharide of *Neisseria gonorrhoeae* F62 by chemical, enzymatic, and two-dimensional NMR methods. Biochemistry 1991; 30: 10566–10575.

99. Griffiss JM, Schneider H, Mandrell RE, Yamasaki R, Jarvis GA, Kim JJ, Gibson BW, Hamadeh R, Apicella MA. Lipooligosaccharides: the principal glycolipids of the neisserial outer membrane. Rev Inf Dis 1988; 10: S287–S295.

100. Gibson BW, Webb JW, Yamasaki R, Fisher SJ, Burlingame AL, Mandrell RE, Schneider H, Griffiss JM. Structure and heterogeneity of the oligosaccharides

from the lipopolysaccharides of a pyocin-resistant *Neisseria gonorrhoeae*. Proc Natl Acad Sci USA 1989; 86:17–21.
101. Yamasaki R, Kernwood DE, Schneider H, Quinn KP, McLeod Griffiss J, Mandrell RE. The structure of lipooligosaccharide produced by *Neisseria gonorrhoeae*, strain 15253, isolated from a patient with disseminated infection. Evidence for a new glycosylation pathway of the gonococcal lipooligosaccharide. J Biol Chem 1994; 269:30345–30351.
102. Yamasaki R, Bacon BE, Nasholds W, Schneider H, Griffis JM. Structural determination of oligosaccharides derived from lipooligosaccharide of *Neisseria gonorrhoeae* F62 by chemical, enzymatic, and two-dimensional NMR methods. Biochemistry 1991; 30: 10566–10575.
103. Kerwood DE, Schneider H, Yamasaki R. Structural analysis of lipooligosaccharide produced by *Neisseria gonorrhoeae*, strain MS11mk (variant A): A precursor for a gonococcal lipooligosaccharide associated with virulence. Biochemistry 1992; 31:12760–12768.
104. John CM, Griffiss JM, Apicella MA, Mandrell RE, Gibson BW. The structural basis for pyocin resistance in *Neisseria gonorrhoeae* lipooligosaccharides. J Biol Chem 1991; 266:19303–19311.
105. Di Fabio JL, Michon F, Brisson J-R, Jennings HJ. Structure of the L1 and L6 core oligosaccharide epitopes of *Neisseria meningitidis*. Can J Chem 1990; 68: 1029–1034.
106. Jennings HJ, Johnson JG, Kenne L. The structure of an R-type oligosaccharide core obtained from some lipopolysaccharides of *Neisseria meningitidis*. Carbohydr Res 1983; 121:233–241.
107. Dell A, Azadi P, Tiller P, Thomas-Oates J, Jennings HJ, Beurret M, Michon F. Analysis of oligosaccharide epitopes of meningococcal lipopolysaccharides by fast-atom-bombardment mass spectrometry. Carbohydr Res 1990; 200:59–76.
108. Jennings HJ, Beurret M, Gamian A, Michon F. Structure and immunochemistry of meningococcal lipopolysaccharides. Antonie van Leeuwenhoek; J Microbiol Serol 1987; 53:519–522.
109. Gibson BW, Melaugh W, Phillips NJ, Apicella MA, Campagnari AA, McLeod Griffiss J. Investigation of the structural heterogeneity of lipooligosaccharides from pathogenic *Haemophilus* and *Neisseria* species and of R-type lipooligosaccharides from *Salmonella typhimurium* by electrospray mass spectrometry. J Bacteriol 1993; 175:2702–2712.
110. Yamasaki R, Griffiss JM, Quinn KP, Mandrell RE. Neuraminic acid is α2→3 linked in the lipooligosaccharide of *Neisseria meningitidis* serogroup B strain 6275. J Bacteriol 1993; 175:4565–4568.
111. Pavliak V, Brisson J-R, Michon F, Uhrin D, Jennings HJ. Structure of sialylated L3 lipopolysaccharide of *Neisseria meningitidis*. J Biol Chem 1993; 268: 14146–14152.
112. Michon F, Beurret M, Gamian A, Brisson J-R, Jennings HJ. Structure of the L5 lipopolysaccharide core oligosaccharides of *Neisseria meningitidis*. J Biol Chem 1990; 265:7243–7247.
113. Gamian A, Beurret M, Michon F, Brisson J-R, Jennings HJ. Structure of the L2 lipopolysaccharide core oligosaccharides of *Neisseria meningitidis*. J Biol Chem 1992; 267:922–925.
114. Lee FKN, Stephens DS, Gibson BW, Engstrom JJ, Zhou D, Apicella MA. Microheterogeneity of *Neisseria* lipooligosaccharide: Analysis of a UDP-glucose 4-epimerase mutant of *Neisseria meningitidis* NMB. Infect Immun 1995; 63:2508–2515.
115. Kahler CM, Carlson RW, Rahman MM, Martin LE, Stephens DS. Inner core biosynthesis of lipooligosaccharide (LOS) in *Neisseria meningitidis* serogroup B: Identification and role in LOS assembly of the α1,2-N-acetylglucosamine transferase (RfaK). J Bacteriol 1996; 178:1265–1273.
116. Jennings HJ, Beurret M, Gamian A, Michon F. Structure and immunochemistry of menigococcal lipopolysaccharides. In: Gonococci and Meningococci. Dordrecht: Kluwer Academic Publishers, 1988:553–556.
117. Kogan G, Uhrin D, Brisson J-R, Jennings HJ. Structural basis of the *Neisseria meningitidis* immunotypes including the L4 and L7 immunotypes. Carbohydr Res 1997; 298:191–199.
118. Parsons NJ, Andrade JRC, Patel PV, Cole JA, Smith H. Sialylation of lipopolysaccharide and loss of absorption of bactericidal antibody during conversion of gonococci to serum resistance by cytidine 5′-monophospho-N-acetyl neuraminic acid. Microb Pathogen 1989; 7:63–72.
119. Mandrell RE, Lesse AJ, Sugai JV, Shero M, McLeod Griffiss J, Cole JA, Parsons NJ, Smith H, Morse SA, Apicella MA. In vitro and in vivo modification of *Neisseria gonorrhoeae* lipooligosaccharide epitope structure by sialylation. J. Exp Med 1990; 171:1649–1664.
120. Demarco de Hormaeche R, van Crevel R, Hormaeche CE. *Neisseria gonorrhoeae* LPS variation, serum resistance and its induction by cytidine 5′-monophospho-N-acetylneuraminic acid. Microbial Pathogenesis 1991; 10:323–332.
121. Smith H, Cole JA, Parsons NJ. The sialylation of gonococcal lipopolysaccharide by host factors: a major impact on pathogenicity. FEMS Microbiol Lett 1991; 100:287–292.
122. Mandrell RE, McLeod Griffiss J, Smith H, Cole JA. Distribution of a lipooligosaccharide-specific sialyltransferase in pathogenic and non-pathogenic *Neisseria*. Microb Pathogen 1993; 14:315–327.
123. Smith H, Parsons NJ, Cole JA. Sialylation of neisserial lipopolysaccharide: a major influence on pathogenicity. Microb Pathogen 1995; 19:365–377.
124. Rest RF, Mandrell RE. Neisseria sialyltransferases and their role in pathogenesis. Microb Pathogen 1995; 19: 379–390.
125. van Putten JPM. Phase variation of lipopolysaccharide directs interconversion of invasive and immuno-resistant phenotypes of *Neisseria gonorrhoeae*. EMBO J 1993; 12:4043–4051.
126. Tsai C-M, Civin CI. Eight lipooligosaccharides of *Neisseria meningitidis* react with a monoclonal antibody which binds lacto-N-neotetraose (Galβ1-4GlcNAcβ1-3Galβ1-4Glc). Infect Immun 1991; 59: 3604–3609.
127. Finkelstein RA. Cholera. Crit Rev Microbiol 1973; 2: 553–623.

128. Swerdlow DL, Ries AA, *Vibrio cholerae* non-O1—the eighth pandemic? Lancet 1993; 342:382–383.
129. Cholera Working Group. Large epidemic of cholera-like disease in Bangladesh caused by *Vibrio cholerae* O139 synonym Bengal. Lancet 1993; 342:387–390.
130. Brade H. Occurrence of 2-keto-3-deoxyoctonic acid 5-phosphate in lipopolysaccharides of *Vibrio cholerae* Ogawa and Inaba. J Bacteriol 1985; 161:795–798.
131. Kondo S, Haishima Y, Hisatsune K. Analysis of the 2-keto-3-deoxyoctonate (KDO) region of lipopolysaccharides isolated from non-O1 *Vibrio cholerae* O5R. FEMS Microbiol Lett 1990; 68:155–158.
132. Vinogradov EV, Holst O, Thomas-Oates JE, Broady KW, Brade H. The structure of the O-antigenic polysaccharide from lipopolysaccharide of *Vibrio cholerae* strain H11 (non-O1). Eur J Biochem 1992; 210:491–498.
133. Bock K, Vinogradov EV, Holst O, Brade H. Isolation and structural analysis of oligosaccharide phosphates containing the complete carbohydrate chain of the lipopolysaccharide from *Vibrio cholerae* strain H11 (non-O1). Eur J Biochem 1994; 225:1029–1039.
134. Vinogradov EV, Stuike-Prill R, Bock K, Holst O, Brade H. The structure of the carbohydrate backbone of the core-lipid A region of the lipopolysaccharide from *Vibrio cholerae* strain H11 (non-O1). Eur J Biochem 1993; 218:543–554.
135. Vinogradov EV, Bock K, Holst O, Brade H. The structure of the lipid A-core region of the lipopolysaccharides from *Vibrio cholerae* O1 smooth strain 569B (Inaba) and rough mutant strain 95R (Ogawa). Eur J Biochem 1995; 233:152–158.
136. Cox AD, Brisson J-R, Varma V, Perry MB. Structural analysis of the lipopolysaccharide from *Vibrio cholerae* O139. Carbohydr Res 1996; 290:43–58.
137. Cox AD, Perry MB. Structural analysis of the O-antigen-core region of the lipopolysaccharide from *Vibrio cholerae* O139. Carbohydr Res 1996; 290:59–65.
138. Kondo S, Zähringer U, Seydel U, Sinnwell V, Hisatsune K, Rietschel ETh. Chemical structure of the carbohydrate backbone of *Vibrio parahaemolyticus* serotype O12 lipopolysaccharide. Eur J Biochem 1991; 200:689–698.
139. Kondo S, Watabe T, Haishima Y, Hisatsune K. Identification of oligosaccharides consisting of D-glucuronic acid and L-*glycero*-D-*manno*- and D-*glycero*-D-*manno*-heptose isolated from *Vibrio parahaemolyticus* O2 lipopolysaccharide. Carbohydr Res 1993; 245:353–359.
140. Edebrink P, Jansson P-E, Bogwald J, Hoffman J. Structural studies of the *Vibrio salmonicida* lipopolysaccharide. Carbohydr Res 1996; 287:225–245.
141. Banoub JH, Michon F, Shaw DH, Roy R. E.i. and c.i. mass-spectral identification of some derivatives of 7-O-(2-amino-2-deoxy-α-D-glucopyranosyl)-L-*glycero*-D-*manno*-heptose, obtained from lipopolysaccharide representative of the *Vibrionaceae* family. Carbohydr Res 1984; 128:203–216.
142. Banoub JH, Hodder HJ. Structural investigation of the lipopolysaccharide core isolated from a virulent strain of *Vibrio ordalii*. Can J Biochem Cell Biol 1985; 63:1199–1205.
143. Banoub JH, Choy Y-M, Michon F, Shaw DH. Structural investigations on the core oligosaccharides of *Aeromonas hydrophila* (chemotype II) lipopolysaccharide. Carbohydr Res 1983; 114:267–276.
144. Banoub JH, Shaw DH. Structural investigations on the core oligosaccharide of *Aeromonas hydrophila* (chemotype III) lipopolysaccharide. Carbohydr Res 1981; 98:93–103.
145. Michon F, Shaw DH, Banoub JH. Structure of the lipopolysaccharide core isolated from a human strain of *Aeromonas hydrophila*. Eur J Biochem 1984; 145:107–114.
146. Shaw DH, Squires MJ, Ishiguro EE, Trust TJ. The structure of the heptose-3-deoxy-D-*manno*-octulosonic-acid region in a mutant form of *Aeromonas salmonicida* lipopolysaccharide. Eur J Biochem 1986; 145:309–313.
147. Shaw D, Hart MJ, Lüderitz O. Structure of the core oligosaccharide in the lipopolysaccharide isolated from *Aeromonas salmonicida* ssp. *salmonicida*. Carbohydr Res 1992; 231:83–91.
148. Flesher AR, Insel RA. Characterization of lipopolysaccharide of *Haemophilus influenzae*. J Infect Dis 1978; 138:719–730.
149. Inzana TJ. Electrophoresis heterogeneity and interstrain variation of the lipopolysaccharide of *Haemophilus influenzae*. J Infect Dis 1983; 148:492–499.
150. Zamze SE, Moxon ER. Composition of the lipopolysaccharide from different capsular serotype strains of *Haemophilus influenzae*. J Gen Microbiol 1987; 133:1443–1451.
151. Inzana TJ, Seifert WE Jr, Williams RP. Composition and antigenic activity of the oligosaccharide moiety of *Haemophilus influenzae* type b lipopolysaccharide. Infect Immun 1985; 48:324–330.
152. Zamze SE, Ferguson MAJ, Moxon ER, Dwek RA, Rademacher TW. Identification of phosphorylated 3-deoxy-*manno*-octulosonic acid as a component of *Haemophilus influenzae* lipopolysaccharide. Biochem J 1987; 245:583–587.
153. Helander IM, Lindner B, Brade H, Altmann K, Lindberg AA, Rietschel ETh, Zähringer U. Chemical structure of the lipopolysaccharide of *Haemophilus influenzae* strain I-69 Rd−/b+: description of a novel deep-rough chemotype. Eur J Biochem 1988; 177:483–492.
154. Mandrell RE, McLaughlin R, Kwaik YA, Lesse A, Yamasaki R, Gibson B, Spinola SM, Apicella MA. Lipooligosaccharides (LOS) of some *Haemophilus* species mimic human glycosphingolipids, and some LPS are sialylated. Infect Immun 1992; 60:1322–1328.
155. Phillips NJ, Apicella MA, Griffiss JM, Gibson BW. Structural characterization of the cell surface lipooligosaccharides from a non-typable strain of *Haemophilus influenzae*. Biochemistry 1992; 31:4515–4526.
156. Melaugh W, Phillips NJ, Campagnari AA, Karalus R, Gibson BW. Partial characterization of the major lipooligosaccharide from a strain of *Haemophilus ducreyi*, the causative agent of chancroid, a genital ulcer disease. J Biol Chem 1992; 267:13434–13439.
157. Schweda EKH, Hegedus O, Borrelli S, Lindberg AA, Jansson P-E, Weiser JN, Maskell DJ, Moxon ER.

Structural studies of the saccharide part of the cell envelope lipopolysaccharide from *Haemophilus influenzae* strain AH1-3 (*lic3+*). Carbohydr Res 1993; 246:319–330.
158. Phillips NJ, Apicella MA, Griffiss JM, Gibson BW. Structural studies of the lipooligosaccharides from *Haemophilus influenzae* type b strain A2. Biochemistry 1993; 32:2003–2012.
159. Masoud H, Martin A, Moxon ER, Richards JC. Structural analysis of the oligosaccharide components of the lipopolysaccharides expressed by the *Gal* mutants of *Haemophilus influenzae* type b. XVIIth International Carbohydrate Symposium, Ottawa, Canada, July 17–22, 1994.
160. Schweda EKH, Sundström ACh, Eriksson LM, Jonasson JA, Lindberg AA. Structural studies of the cell envelope lipopolysaccharides of *Haemophilus ducreyi* strains ITM 2665 and ITM 4747. J Biol Chem 1994; 269:12040–12048.
161. Schweda EKH, Jonasson JA, Jansson P-E. Structural studies of lipooligosaccharides from *Haemophilus ducreyi* strains ITM 5535, 3147 and a fresh clinical isolate ACY1. Evidence for intrastrain heterogeneity with the production of mutually exclusive sialylated or elongated glycoforms. J Bacteriol 1995; 177:5316–5321.
162. Schweda EKH, Jansson P-E, Moxon ER, Lindberg AA. Structural studies of the saccharide part of the cell envelope lipooligosaccharide from *Haemophilus influenzae* strain galEgalK. Carbohydr Res 1995; 272: 213–224.
163. Phillips NJ, McLaughlin R, Miller TJ, Apicella MA, Gibson BW. Characterization of two transposon mutants from *Haemophilus influenzae* type b with altered lipooligosaccharide biosynthesis. Biochemistry 1996; 35:5937–5947.
164. Hood DW, Deadman ME, Allen T, Masoud H, Martin A, Brisson JR, Fleischmann R, Venter JC, Richards JC, Moxon ER. Use of the complete genome sequence information of *Haemophilus influenzae* strain Rd to investigate lipopolysaccharide biosynthesis. Mol Microbiol 1996; 22:951–965.
165. Risberg A, Schweda EKH, Jansson P-E. Structural studies of the cell-envelope oligosaccharide from the lipopolysaccharide of *Haemophilus influenzae* strain RM.118-28. Eur J Biochem 1997; 243:701–707.
166. Weiser JN, Shchepetov M, Chong STH. Decoration of lipopolysaccharide with phosphorylcholine: a phase-variable characteristic of *Haemophilus influenzae*. Infect Immun 1997; 65:943–950.
167. Nichols WA, Gibson BW, Melaugh W, Lee N-G, Sunshine M, Apicella MA. Identification of the ADP-L-*glycero*-D-*manno*-heptose-6-epimerase (*rfaD*) and heptosyltransferase II (*rfaF*) biosynthesis genes from nontypable *Haemophilus influenzae* 2019. Infect Immun 1997; 65:1377–1386.
168. Risberg A, Masoud H, Martin A, Richards JC, Jansson P-E, Schweda EKH. Structural studies of the cell envelope oligosaccharides from the lipopolysaccharides of *Haemophilus influenzae* strain RM.118. Royal Society of Chemistry Carbohydrate Group Spring Meeting, Galway, Ireland, March 24–27, 1997.
169. Masoud H, Moxon ER, Martin A, Krajcarski, Richards JC. Structure of the variable and conserved lipopolysaccharide oligosaccharide epitopes expressed by *Haemophilus influenzae* strain Eagan. Biochemistry 1997; 36:2091–2103.
170. Virji M, Weiser JN, Lindberg AA, Moxon ER. Antigenic similarities in lipopolysaccharides of *Haemophilus* and *Neisseria* and expression of a digalactoside structure also present on human cells. Microb Pathogen 1990; 9:441–450.
171. Borrelli S, Altmann K, Jansson P-E, Lindberg AA. Binding specificity for four monoclonal antibodies recognizing terminal Galα1→4Gal residues in *Haemophilus influenzae*. Microb Pathogen 1995; 19:139–157.
172. Borrelli S, Roggen EL, Hendriksen D, Jonasson J, Ahmed HJ, Piot P, Jansson P-E, Lindberg AA. Monoclonal antibodies against *Haemophilus* lipopolysaccharides: Clone DP8 specific for *Haemophilus ducreyi* and clone DH24 binding to lacto-*N*-neotetraose. Infect Immun 1995; 63:2665–2673.
173. Rozalski A, Brade L, Kosma P, Moxon R, Kusumoto S, Brade H. Characterization of monoclonal antibodies recognizing three distinct, phosphorylated carbohydrate epitopes in the lipopolysaccharide of the deep rough mutant I-69 Rd$^-$/b$^+$ of *Haemophilus influenzae*. Mol Microbiol 1997; 23:569–577.
174. Severn WB, Johnston RAZ, Kelly RF, Richards JC. The structural analysis of short chain lipopolysaccharide. XVIIth International Carbohydrate Symposium, Ottawa, Canada, July 17–22, 1994.
175. Lacroix RP, Duncan JR, Jenkins RP, Leitch RA, Perry JA, Richards JC. Structural and serological specificities of *Pasteurella haemolytica* lipopolysaccharides. Infect Immun 1993; 61:170–181.
176. Drewry DT, Symes KC, Gray GW, Wilkinson SG. Studies of polysaccharide fractions from the lipopolysaccharide of *Pseudomonas aeruginosa* NCTC 1999. Biochem J 1975; 149:93–106.
177. Rowe PSN, Meadow PM. Structure of the core oligosaccharide from the lipopolysaccharide of *Pseudomonas aeruginosa* PAC1R and its defective mutants. Eur J Biochem 1983; 132:329–337.
178. Masoud H, Altman E, Richards JC, Lam JS. General strategy for structural analysis of the oligosaccharide region of lipooligosaccharides. Structure of the oligosaccharide component of *Pseudomonas aeruginosa* IATS serotype O6 mutant R5 rough-type lipopolysaccharide. Biochemistry 1994; 33:10568–10578.
179. Knirel YA, Grosskurth H, Helbig JH, Zähringer U. Structures of decasaccharide and tridecasaccharide tetraphosphates isolated by strong alkaline degradation of *O*-deacylated lipopolysaccharide of *Pseudomonas fluorescens* strain ATCC 49271. Carbohydr Res 1995; 279:215–226.
180. Masoud H, Sadovskaya I, De Kievit T, Altman E, Richards JC, Lam JS. Structural elucidation of the lipopolysaccharide core region of the O-chain-deficient mutant strain A28 from *Pseudomonas aeruginosa* serotype O6 (International Antigenic Typing System). J Bacteriol 1995; 177:6718–6726.
181. Knirel YA, Helbig JH, Zähringer U. Structure of a decasaccharide isolated by mild acid degradation of

and dephosphorylation of the lipopolysaccharide of *Pseudomonas fluorescens* strain ATCC 49271. Carbohydr Res 1996; 283:129–139.
182. Beckmann F, Moll H, Jäger K-E, Zähringer U. 7-*O*-Carbamoyl-L-*glycero*-D-*manno*-heptose: a new core constituent in the lipopolysaccharide of *Pseudomonas aeruginosa*. Carbohydr Res 1995; 267:C3–C7.
183. Yuki N, Taki T, Inagaki F, Kasama T, Takahashi M, Saito K, Handa S, Miyatake T. A bacterium lipopolysaccharide that elicits Guillain-Barré syndrome has a GM1 ganglioside-like structure. J Exp Med 1993; 178:1771–1775.
184. Aspinall GO, Fujimoto S, McDonald AG, Pang H, Kurjanczyk L, Penner JL. Lipopolysaccharides from *Campylobacter jejuni* associated with Guillain-Barré syndrome patients mimic human gangliosides in structure. Infect Immun 1994; 62:2122–2125.
185. Aspinall GO, McDonald AG, Pang H, Kurjanczyk L, Penner JL. Lipopolysaccharides of *Campylobacter jejuni* serotype O:19: Structures of core oligosaccharide regions from the serostrain and two bacterial isolates from patients with the Guillain-Barré syndrome. Biochemistry 1994; 33:241–249.
186. Salloway S, Mermel LA, Seamans M, Aspinall GO, Nam Shin JE, Kurjanczyk L, Penner JL. Miller-Fisher syndrome associated with *Campylobacter jejuni* bearing lipopolysaccharide molecules that mimic human ganglioside GD_3. Infect Immun 1996; 64:2945–2949.
187. Aspinall GO, McDonald AG, Raju TS, Pang H, Mills SD, Kurjanczyk L, Penner JL. Serological diversity and chemical structures of *Campylobacter jejuni* low-molecular-weight lipopolysaccharides. J Bacteriol 1992; 174:1324–1332.
188. Aspinall GO, McDonald AG, Raju TS, Pang H, Moran AP, Kurjanczyk L, Penner JL. Chemical structures of the core regions of *Campylobacter jejuni* serotypes O:1, O:4, O:23, and O:36 lipopolysaccharides. Eur J Biochem 1993; 213:1017–1027.
189. Nam Shin JE, Ackloo S, Mainkar AS, Monteiro M, Pang H, Penner JL, Aspinall GO. Lipooligosaccharides of *Campylobacter jejuni* serotype O:10. Structures of core oligosaccharide regions from a bacterial isolate from a patient with the Miller-Fisher syndrome and from the serostrain. Carbohydr Res 1998; 305:223–232.
190. Aspinall GO, McDonald AG, Pang H, Kurjanczyk L, Penner JL. Lipopolysaccharide of *Campylobacter coli* serotype O:30. Fractionation and structure of liberated core oligosaccharide. J Biol Chem 1993; 268:6263–6268.
191. Aspinall GO, McDonald AG, Raju TS, Pang H, Kurjanczyk L, Penner JL, Moran AP. Chemical structure of the core regions of *Campylobacter jejuni* serotype O:2 lipopolysaccharide. Eur J Biochem 1993; 213:1029–1037.
192. Aspinall GO, Lynch CM, Pang H, Shaver RT, Moran AP. Chemical structures of the core region of *Campylobacter jejuni* O:3 lipopolysaccharide and an associated polysaccharide. Eur J Biochem 1995; 231:570–578.
193. Aspinall GO, Monteiro MA, Pang H, Kurjanczyk L, Penner JL. Lipo-oligosaccharide of *Campylobacter lari* strain PC 637. Structure of the liberated oligosaccharide and an associated extracellular polysaccharide. Carbohydr Res 1995; 279:227–244.
194. Aspinall GO, Monteiro MA, Pang H. Lipo-oligosaccharide of the *Campylobacter lari* type strain ATCC 35221. Structure of the liberated oligosaccharide and an associated extracellular polysaccharide. Carbohydr Res 1995; 279:245–264.
195. Aspinall GO, Monteiro MA, Shaver RT, Kurjanczyk L, Penner JL. Lipopolysaccharides of *Helicobacter pylori* serogroups O:3 and O:6: Structures of a class of lipopolysaccharides with reference to the location of oligomeric units of D-*glycero*-α-D-*manno*-heptose residues. Eur J Biochem 1997; 248:592–601.
196. Aspinall GO, Monteiro MA. Lipopolysaccharides of *Helicobacter pylori* strains P466 and MO19: Structures of the O antigen and core oligosaccharide regions. Biochemistry 1996; 35:2498–2504.
197. Monteiro MA, Perry MB, Moran AP, Lee A. Preliminary structural studies of lipopolysaccharides from *Helicobacter mustelae* and *Helicobacter felis* (abstr). Ir J Med Sci 1997; 166 (suppl 3):32.
198. Aspinall GO, Monteiro M, Pang H, Walsh EJ, Moran AP. Lipopolysaccharide of the *Helicobacter pylori* type strain NCTC 11637 (ATCC 43504): structure of the O antigen chain and core oligosaccharide regions. Biochemistry 1996; 35:2489–2497.
199. Monteiro MA, Zheng PY, Appelmelk BJ, Perry MB. The lipopolysaccharide of *Helicobacter mustelae* type strain ATCC 43772 expresses the monofucosyl A type 1 histo-blood group epitope. Microbiol Lett 1997; 154:103–109.
200. Mayer H, Radziejewska-Lebrecht J, Schramek S. Chemical and immunochemical studies on lipopolysaccharides of *Coxiella burnetii* phase I and phase II. In: Wu AM, Adams LG, eds. The Molecular Immunology of Complex Carbohydrates. New York: Plenum Press, 1988:577–591.
201. Toman R, Škultéty L. Structural study on a lipopolysaccharide from *Coxiella burnetii* strain Nine Mile in avirulent phase II. Carbohydr Res 1996; 283:175–185.
202. Radziejewska-Lebrecht J, Feige U, Mayer H, Weckesser J. Structure of the heptose region of lipopolysaccharides from *Rhodospirillum tenue*. J Bacteriol 1981; 145:138–144.
203. Le Dur A, Chaby R, Szabo L. Isolation of two protein-free and chemically different lipopolysaccharides from *Bordetella pertussis* phenol-extracted endotoxin. J Bacteriol 1980; 143:78–88.
204. Caroff M, Lebbar S, Szabo L. Detection of 3-deoxy-2-octulosonic acid in thiobarbiturate-negative endotoxins. Carbohydr Res 1987; 161:C4–C7.
205. Caroff M, Chaby R, Karibian D, Perry J, Deprun C, Szabo L. Variations in the carbohydrate regions of *Bordetella pertussis* lipopolysaccharides: electrophoretic, serological, and structural features. J Bacteriol 1990; 172;1121–1128.
206. Chaby R, Szabo L. 7-*O*-(2-Amino-2-deoxy-α-D-glucopyranosyl)-L-*glycero*-D-*manno*-heptose: a constituent in the endotoxin of *Bordetella pertussis*. Eur J Biochem 1976; 70:115–122.

207. Chaby R, Szabo L. 3-Deoxy-2-octulosonic acid 5-phosphate: a component of the endotoxin of *Bordetella pertussis*. Eur J Biochem 1975; 59:277–280.
208. Lebbar S, Caroff M, Szabó L, Mérienne C, Szilógyi L. Structure of a hexasaccharide proximal to the hydrophobic region of lipopolysaccharides present in *Bordetella pertussis* endotoxin preparations. Carbohydr Res 1994; 259:257–275.
209. Bellmann W, Lingens F. Structural studies on the core oligosaccharide of *Phenylobacterium immobile* strain K_2 lipopolysaccharide. Chemical synthesis of 3-hydroxy-5c-dodecenoic acid. Biol Chem Hoppe-Seyler 1985; 366:567–575.
210. Bergogne-Bérézin E, Towner KJ. *Acinetobacter* spp. as nosocomial pathogens: microbiological, clinical, and epidemiological features. Clin Microbiol Rev 1996; 9:148–165.
211. Dijkshoorn L. Acinetobacter—microbiology. In: Bergogne-Bérézin E, Joly-Guillou ML, Towner KJ, eds. *Acinetobacter*. Microbiology, Epidemiology, Infections, Management. Boca Raton, FL: CRC Press, 1996:37–69.
212. Brade H, Galanos C. Isolation, purification, and chemical analysis of the lipopolysaccharide and lipid A of *Acinetobacter calcoaceticus*. Eur J Biochem 1982; 122:233–237.
213. Brade H, Rietschel ETh. Identification of a 2-keto-3-deoxy-1,7-dicarboxyheptonic acid as a constituent of the lipopolysaccharide of NCTC 10305. Eur J Biochem 1985; 153:249–254.
214. Brade H, Tacken A, Christian R. Isolation and identification of a rhamnosyl-rhamnosyl-3-deoxy-D-manno-octulosonic acid trisaccharide from the lipopolysaccharide of *Acinetobacter calcoaceticus* (10303 NCTC London). Carbohydr Res 1987; 167:295–300.
215. Edebrink P, Jansson P-E, Rahman MM, Widmalm G, Holme T, Rahman M, Weintraub A. Structural studies of the O-polysaccharide from the lipopolysaccharide of *Moraxella (Branhamella) catarrhalis* serotype A (strain ATCC 25238). Carbohydr Res 1994; 257:269–284.
216. Masoud H, Perry MB, Brisson J-R, Uhrin D, Richards JC. Structural elucidation of the backbone oligosaccharide from the lipopolysaccharide of *Moraxella catarrhalis* serotype A. Can J Chem 1994; 72:1466–1477.
217. Edebrink P, Jansson P-E, Widmalm G, Holme T, Rahman M. The structures of oligosaccharides isolated from *Moraxella catarrhalis* serotype B, strain CCUG 3292. Carbohydr Res 1996; 295:127–146.
218. Edebrink P, Jansson P-E, Rahman MM, Widmalm G, Holme T, Rahman M. Structural studies of the O-polysaccharides from two strains of *Moraxella catarrhalis* serotype C. Carbohydr Res 1995; 266:237–261.
219. Carlson RW, Hollingsworth RI, Dazzo FB. A core oligosaccharide component from the lipopolysaccharide of *Rhizobium trifolii* ANU 843. Carbohydr Res 1988; 176:127–135.
220. Hollingsworth RI, Carlson RW, Garcia F, Gage DA. A new core tetrasaccharide component from the lipopolysaccharide of *Rhizobium trifolii* ANU 843. J Biol Chem 1990; 265:12752.
221. Carlson RW, Garcia F, Noel D, Hollingsworth RI. The structures of the lipopolysaccharide core components from *Rhizobium leguminosarum* BIOVAR phaseoli CE3 and two of its symbiotic mutants, CE109 and CE309. Carbohydr Res 1989; 195:101–110.
222. Bhat UR, Krishnaiah BS, Carlson RW. Re-examination of the structures of the lipopolysaccharide core oligosaccharides from *Rhizobium leguminosarum* biovar phaseoli. Carbohydr Res 1991; 220:219–227.
222a. Zhang Y, Hollingsworth RJ, Priefer UB. Characterization of structural defects in the lipopolysaccharides of symbiotically impaired *Rhizobium leguminosarum* biovar VF-39 mutants. Carbohydr Res 1992; 237:261–271.
223. Hollingsworth RI, Zhang Y, Priefer UB. Structure of the unusual trisaccharide lipopolysaccharide component produced by a symbiotically defective mutant of *Rhizobium leguminosarum* biovar *viciae*. Carbohydr Res 1994; 264:271–280.
224. Russa R, Bruneteau M, Shashkov AS, Urbanik-Sypniewska T, Mayer H. Characterization of the lipopolysaccharides from *Rhizobium meliloti* strain 102F51 and its nonnodulating mutant WL113. Arch Microbiol 1996; 165:26–33.
225. Carlson RW, Krishnaiah BS. Structures of the oligosaccharides obtained from the core regions of the lipopolysaccharides of *Bradyrhizobium japonicum* 61A101c and its symbiotically defective lipopolysaccharide mutant, JS314. Carbohydr Res 1992; 231:205–219.
226. Kannenberg EL, Reuhs B, Forsberg LS, Carlson RW. Lipopolysaccharides and K-antigens: their structures, biosynthesis, and functions in *Rhizobium*-legume interactions. In: Spaink HH, Kondrosi A, Hooykaas PJJ, eds. The *Rhizobiaceae*. Amsterdam, The Netherlands: Kluwer, 1998; 119–154.
227. Carlson RW, Reuhs B, Chen T-B, Bhat UR, Noel KD. Lipopolysaccharide core structures in *Rhizobium etli* and mutants deficient in O-antigen. J Biol Chem 1995; 270:11783–11788.
228. Brade L, Nano FE, Schlecht S, Schramek S, Brade H. Antigenic and immunogenic properties of recombinants from *Salmonella typhimurium* and *Salmonella minnesota* rough mutants expressing in the lipopolysaccharides a genus-specific chlamydial epitope. Infect Immun 1987; 55:482–486.
229. Brade H, Brade L, Nano FE. Chemical and serological investigations on the genus-specific lipopolysaccharide epitope of *Chlamydia*. Proc Natl Acad Sci USA 1987; 84:2508–2512.
230. Kosma P, Schulz G, Brade H. Synthesis of a trisaccharide of 3-deoxy-D-*manno*-2-octulo-pyranosylonic acid (KDO) residues related to the genus-specific lipopolysaccharide epitope of *Chlamydia*. Carbohydr Res 1988; 183:183–199.
231. Holst O, Brade L, Kosma P, Brade H. Structure, serological specificity, and synthesis of the genus-specific lipopolysaccharide epitope of *Chlamydia*. J Bacteriol 1991; 173:1862–1866.
232. Holst O, Bock K, Brade L, Brade H. The structures of oligosaccharide bisphosphates isolated from the lipopolysaccharide of a recombinant *Escherichia coli*

strain expressing the gene *gseA* [3-deoxy-D-*manno*-octulopyranosonic acid (Kdo) transferase] of *Chlamydia psittaci* 6BC. Eur J Biochem 1995; 229:194–200.

233. Brade H, Brabetz W, Brade L, Holst O, Löbau S, Lukacova M, Mamat U, Rozalski A, Zych K, Kosma P. Chlamydial lipopolysaccharide: chemical and antigenic structure, biosynthesis and biomedical application. Pure Appl Chem 1995; 67:1617–1626.

234. Brade H, Brabetz W, Brade L, Holst O, Löbau S, Lukacova M, Mamat U, Rozalski A, Zych K, Kosma P. Chlamydial lipopolysaccharide. J Endotoxin Res 1997; 4:67–84.

235. Löbau S, Mamat U, Brabetz W, Brade H. Molecular cloning, sequence analysis, and funcitonal characterization of the lipopolysaccharide biosynthetic gene *kdtA* encoding 3-deoxy-α-D-*manno*-octulosonic acid transferase of *Chlamydia pneumoniae* strain TW-183. Mol Microbiol 1995; 18:391–399.

236. Weintraub A, Zähringer U, Lindberg AA. Structural studies on the polysaccharide part of the cell wall lipopolysaccharide from *Bacteroides fragilis* NCTC 9343. Eur J Biochem 1985; 151:657–661.

237. Neal DJ, Wilkinson SG. Lipopolysaccharides from *Pseudomonas maltophila*. Structural studies of the side-chain, core, and the lipid A regions of the lipopolysaccharide from strain NCTC 10257. Eur J Biochem 1982; 128:143–149.

238. Salimath PV, Tharanathan RN, Weckesser J, Mayer H. The structure of the polysaccharide moiety of *Rhodopseudomonas sphaeroides* ATCC 17023 lipopolysaccharide. Eur J Biochem 1984; 144:227–232.

239. Krauss JH, Himmelspach K, Reuter G, Schauer R, Mayer H. Structural analysis of a novel sialic-acid-containing trisaccharide from *Rhodobacter capsulatus* 37b4 lipopolysaccharide. Eur J Biochem 1992; 217–223.

240. Knirel YA, Moll H, Zähringer U. Structural study of a highly *O*-acylated core of *Legionella pneumophila* serogroup 1 lipopolysaccharide. Carbohydr Res 1996; 293:223–234.

241. Moll H, Knirel YA, Helbig JH, Zähringer U. Identification of an α-D-Man*p*-(1→8)-Kdo disaccharide in the inner core region and the structure of the complete core region of *Legionella pneumophila* serogroup 1 lipopolysaccharide. Carbohydr Res 1997; 304:91–95.

242. Zähringer U, Knirel YA, Lindner B, Helbig JH, Sonesson A, Marre R, Rietschel ET. The lipopolysaccharide of *Legionella pneumophila* serogroup 1 (strain Philadelphia 1): chemical structure and biological significance. In: Levin J, Alving CR, Munford RS, Redl H, eds. Bacterial Endotoxins: From Genes to Therapy. New York: John Wiley & Sons, Inc., 1995:113–139.

243. Velasco J, Moll H, Knirel YA, Sinnwell V, Moriyón I, Zähringer U. Structural studies on the lipopolysaccharide from a rough strain of *Ochrobactrum anthropi* containing a 2,3-diamino-2,3-dideoxy-D-glucose disaccharide lipid A backbone. Carbohydr Res 1998; 306:283–290.

244. Cox AD, Brisson J-R, Thibault P, Perry MB. Structural analysis of the lipopolysaccharide from *Vibrio cholerae* O22. Carbohydr Res 1997; 304:191–208.

245. Knirel YA, Senchenkova SN, Jansson P-E, Weintraub A. More on the structure of *Vibrio cholerae* O22 lipopolysaccharide. Carbohydr Res 1998; 310:117–119.

246. Cox AD, Howard MD, Brisson J-R, Van Der Zwan M, Thibault P, Perry MB, Inzana TI. Structural analysis of the phase-variable lipooligosaccharide from *Haemophilus sommus* strain 738. Eur J Biochem 1998; 253:507–516.

247. Sadovskaya I, Brisson J-R, Lam JS, Richards JC, Altman E. Structural elucidation of the lipopolysaccharide core regions of the wild-type strain PAO1 and O-chain-deficient mutant strains AK1401 and AK1O12 from *Pseudomonas aeruginosa* serotype O5. Eur J Biochem 1998; 255:673–684.

248. Wakarchuk WW, Gilbert M, Martin A, Wu Y, Brisson J-R, Thibault P, Richards JC. Structure of an α-2,6-sialylated lipooligosaccharide from *Neisseria meningitidis* immunotype L1. Eur J Biochem 1998; 254:626–633.

249. Rahman MM, Stephens DS, Kahler CM, Glushka J, Carlson RW. The lipooligosaccharide (LOS) of *Neisseria meningitidis* serogroup B strain NMB contains L2, L3, and a novel oligosaccharides, and lacks the lipid-A 4′-phosphate substituent. Carbohydr Res 1998; 307:311–324.

250. Kawahara K, Isshiki Y, Dejsirilert S, Ezaki T, Zähringer U. Structural analysis of the core region of the lipopolysaccharide isolated from *Burkholderia pseudomallei*. The Fifth Conference of the International Endotoxin Society, Santa Fe, New Mexico, September 12–15, 1998.

9

The Chemistry of O-Polysaccharide Chains in Bacterial Lipopolysaccharides

Per-Erik Jansson
Karolinska Institute, Huddinge Hospital, Huddinge, Sweden

INTRODUCTION

Lipopolysaccharides (LPS) constitute one of the main components of the outer surface of gram-negative bacteria. The O-antigen polysaccharide (O-polysaccharide, O-specific side chain, O-chain), which is the polysaccharide part of LPS, generally has a regular structure and determines part of the immunospecificity of the cell. Apart from O-polysaccharide, the LPS contains a core oligosaccharide and lipid A. The structures of core oligosaccharides, which bridge the O-chain to lipid A, are reviewed in Chapter 10. Chapters 17 and 18 deal with the biosynthesis of O-antigens. Two reviews have dealt with the structure of O-polysaccharides, one by Kenne and Lindberg (1), and one by Knirel and Kochetkov (2). A significant number of the determined structures are collected in the Complex Carbohydrate Structural Database/CarbBank (CCSD/CarbBank), a computer database and search program, accessible via the internet (3). Although not complete, CCSD/CarbBank offers a possibility to search for structural elements in polysaccharides and may prove most efficient in future similarity studies. The structures given in this chapter were for the most part published after the last review (2), i.e., from 1994 to 1997.

FEATURES OF O-POLYSACCHARIDES

The reason why O-antigenic chains usually have regular structures is their mode of biosynthesis, in which preformed oligosaccharide blocks are transferred to a growing polymer chain. The length of the resulting chain varies considerably, and the distribution and the approximate length can be determined, e.g., by polyacrylamide gel electrophoresis with sodium dodecyl sulfate dispersing the lipopolysaccharide (SDS/PAGE). In some cases the biosynthetic route differs from the usual one, and monosaccharides are added one by one to the chain. This has, however, only been demonstrated in a few cases. For a limited number of O-chains the structure of the biological repeating unit, i.e., the repeating unit transferred in the biosynthesis, has been demonstrated. It is generally only the chemical repeating unit that is determined. The reducing end has in some cases a structurally different sugar attached, mostly a mono-O-methyl sugar. It has been speculated that this is a chain terminator. The diversity of sugars present in the O-chain is large, and new sugars are added to the list constantly. Data on monosaccharides may be found in Refs. 2 and 4.

Generally only one lipopolysaccharide (and, if any, only one capsular polysaccharide) is found in each strain. These two are usually different, but in a few examples the O-antigen and the capsular polysaccharide are identical. In some instances LPS preparations contain two different polysaccharide chains, and it is difficult to determine whether these are separate units or joined together. Both alternatives have been observed, and each possibility must be taken into consideration.

The core type of a certain species may be investigated chemically if at least part of the material contains few O-chains or none. A serological approach to the problem has been attempted by some groups (5,6).

STRUCTURES

Enterobacteriaceae

Escherichia coli

E. coli comprises more than 170 serotypes, and some 40% of the structures of the O-polysaccharides have been determined. Most polysaccharides are neutral, but acidic groups are present in approximately 10 serotypes, containing Neu5Ac, uronic acids, or ether-bound lactic acid. Furanosidic D-Rib and D-Gal and O-acetyl groups are fairly frequent. The structures from recent investigations are shown in Table 1, where all sugars, unless otherwise stated, are D and pyranosidic (as in all other tables).

Diarrheagenic E. coli can be divided into a number of groups with different pathogenicity. Several of those have attracted attention recently. Thus, the enterotoxigenic (ETEC) O153 (7), the enteroinvasive (EIEC) O28 (8) and O143 (9), the enteropathogenic (EPEC) O125 (10), O126 (11), and O142 (12), the enterohemorrhagic (EHEC) O26 (13), and the enteroaggregative (EAggEC) O3 (14) and O44 (15) have been investigated.

Two different O6 strains have been found, each with a different LPS (16). The different strains were consequently subgrouped, as serotypes O1, O4, and O18 had been. Among the more unusual sugars found in E. coli are L-RhaNAc in O3 (14) and 6-deoxy-L-talose (L-6dTal) in O45, O45-rel (O45-related), O66 (17) and O88 (18). Type O16 (19) cross-reacts with O4, i.e., a serological reaction is obtained between O16 and antibodies against O4, but they have a completely different O-polysaccharides. Type O45 cross-reacts with O45rel and O66, which is expected as they have the common sequence →2)-β-D-Glcp-(1→3)-α-L-6d-Talp2Ac-(1→. The O88, also with a L-6dTal residue in its O-chain, does not cross-react with O45 and O66, probably because L-6dTal is lateral in O88 and not joined to a Glc residue. Consensus has now been reached concerning the structures of O24 and O56 (20,21), which are similar, after careful reinvestigation using nuclear magnetic resonance (NMR) and mass spectrometry (MS). The O28 O-polysaccharide (8) is a glycerol teichoic acid polymer, linked through positions 1 and 2 of the glycerol. The chirality of the glycerol residue was determined to be R, by oxidizing the unsubstituted hydroxymethyl group in the glycerol to form glyceric acid, which, as its ester with chiral 2-butanol, was analyzed on gas-liquid chromatography (GLC).

Blood group specificity has previously been observed, e.g., for O86 (B), and O127 (H) and now also for O90 [H, (22)]. This could be explained for all three groups by the presence of the element α-L-Fuc-(1→2)-β-D-Gal-(1→3)-α-D-GalpNAc-(1→3)-D-GalpNAc. Despite the fact that the L-Fuc residue is acetylated in O90, it nevertheless has blood group specificity.

The structure of O101 was determined by NMR (23) and confirmed by a computerized analysis of the NMR data using the CASPER program (24). The O104 cross-reacts with E. coli K9, which, except for the O-acetyl groups, has identical repeating units (25). The O143 (11) has 2-amino-1,3-propanediol linked as an amide to the GalA residue and the same backbone as that of Shigella boydii type 8, with which it cross-reacts. The O167 (26) is unique in that different strains have different virulence mechanisms, something not observed for other serotypes. Thus, out of 29 studied isolates 19 were enterotoxigenic and 3 enteroinvasive. The O167 cross-reacts with Shigella boydii type 3, the structure of which is not yet known. The polysaccharide is degraded by base, which may be attributed to the fact that the uronic acid is amidated. E. coli K-12, the widely used laboratory strain, is a rough strain that does not produce an O-antigen due to mutations in the rfb gene cluster but which could be complemented with the aid of a plasmid (27). The O-polysaccharide has the same backbone as that from E. coli O16 but with an additional lateral α-D-Glc group. K-12 strains 30 and 64 were used to produce recombinant strains with Vibrio cosmids to produce typical Vibrio cholerae O1 antigens (28).

The core structures have only been determined for a few LPS. Thus, E. coli O56 (20,21) and O104 (25) have the R2 core. The repeating unit of O56 was shown to be linked to the subterminal α-D-Glc residue of the core. The core type of O24 (20,21) could not be established; it was not R2 despite the similarities with O56 with respect to the O-chains. In this context it is important to mention the possibility of probing core structures with antibodies (5,6).

Citrobacter

Strains of the heterologeneous species Citrobacter freundii are subdivided into some 45 serogroups, almost a dozen of which have been studied, nine of those only recently. Thus, O8a,O8b (29) contains D-Rha and

Table 1 Structures of O-Polysaccharides from *Escherichia coli*

Serogroup or strain	Structure of repeating unit	Ref.
O3, 17-2	-3)Gal(α1-3)GlcNAc(β1-3)GlcNAc(β1- 　　　　Glc(α1-4)⌐　　　　LRhaNAc(β1-4)⌐	14
O6:K2, O6:K13	-4)Man(β1-3)GlcNAc(α1-4)GalNAc(α1-3)Man(β1- 　　　　　　　　　　　Glc(β1-2)⌐	16
O6:K54	-4)Man(β1-3)GlcNAc(α1-4)GalNAc(α1-3)Man(β1- 　　　　　　　　　　GlcNAc(β1-2)⌐	16
O16	-6)Glc(α1-3)LRha2Ac(α1-3)GlcNAc(α1-2)Gal$f$(β1-	19
O17	-2)Man(β1-3)GlcNAc(α1-6)Man(α1-2)Man(α1- 　　　　Glc(α1-6)⌐	110
O22	-6)Glc(α1-4)GlcA(β1-4)GalNAc3Ac(β1-3)Gal(α1-3)GalNAc(β1-	111
O24	-7)Neu5Ac(α2-3)Glc(β1-3)GalNAc(β1- 　　　Glc(α1-2)⌐	20, 21
O26	-3)LRha(α1-4)LFucNAc(α1-3)GlcNAc(β1-	13
O28	4)GlcNAc(β1-3)Gal$f$2Ac(β1-3)GlcNAc(α1-2)RGro1-P-(O-	8
O44	-2Man(α1-2)Man(β1-3)GlcNAc(α1-6)Man(α1- 　　　　　　Glc(α1-4)⌐	15
O45	-3)L6dTal2Ac(α1-3)FucNAc(α1-2)Glc(β1-	17
O45rel	-3)L6dTal2Ac(α1-3)GlcNAc(β1-2)Glc(β1-	17
O56	-7)Neu5Ac(α2-3)Glc(β1-3)GlcNAc(β1- 　　　Gal(α1-2)⌐	20, 21
O66	-3)L6dTal(α1-3)GlcNAc(α1-2)Man(β1-3)GlcNAc(α1-2)Glc3Ac(β1-	17
O83	-6)Glc(α1-4)GlcA(β1-6)Gal(β1-4)Gal(β1-4)GlcNAc(β1-	112
O88	-3)Man(α1-3)GlcNAc(β1-4)Man(α1- 　　　　L6dTal(α1-3)⌐	18
O90	-4)LFuc2/3Ac(α1-2)Gal(β1-3)GalNAc(α1-3)GalNAc(β1-	22
O98[a]	-3)LQuiNAc(α1-4)GalNAcA(α1-3)LQuiNAc(α1-3)GlcNAc(β1-	113
O101	-6)GlcNAc(α1-4)GalNAc(α1-	23
O104	-4)Gal(α1-4)Neu5,7,9Ac$_3$(α2-3)Gal(β1-3)GalNAc(β1-	25
O113	-4)GalA(α1-3)Gal(α1-3)GlcNAc(β1-4)GalNAc(α1- 　　　　　　　　　　Gal(β1-3)⌐	114
O125	-2)Man(α1-3)LFuc(α1-3)GalNAc(α1-4)GalNAc(β1- 　Glc(α1-3)⌐　　　　　　Gal(β1-3)⌐	10
O126	-2)Man(β1-3)Gal(α1-3)GlcNAc(β1- 　　LFuc(β1-2)⌐	11
O128	-3)GalNAc(β1-4)Gal(α1-3)GalNAc(α1-6)Gal(β1- 　　　　　　　　　　LFuc(α1-2)⌐	115
O138	-4)GalNAcA(α1-3)GlcNAc(β1-2)LRha(α1-3)LRha(α1-	116

Table 1 Continued

Serogroup or strain	Structure of repeating unit	Ref.
O142	-6)GalNAcA(α1-4)GalNAc(α1-3)GalNAc(α1-2)LRha(α1- ⌐GlcNAc(β1-3)⌐	12
O143[b]	-3)GalNAc(α1-4)GlcA(β1-3)GlcNAc(β1-2)GalA3,4Ac$_2$6NHCH(CH$_2$OH)$_2$(β1-	9
O153	-4)Gal(β1-4)GlcNAc(α1-4)Gal(β1-3)GlcNAc(α1-2)Rib$f$(β1-	7
O167	-3)GlcNAc(β1-2)GalA6Ala(β1-3)GlcNAc(α1-2)Gal$f$(β1-5)Gal$f$(β1-	26
K12	-2)Gal$f$(β1-6)Glc(α1-3)LRha2Ac(α1-3)GlcNAc(α1- ⌐Glc(α1-6)⌐	27
K12/30[c] K12/64	-2)Rha4NAcyl(α1-	28

[a]QuiNAc = 2-Acetamido-2,6-dideoxyglucose.
[b]OAc not stoichiometric.
[c]Acyl = 3-Deoxy-L-*glycero*-tetronyl.

furanosidic D-Xyl, the latter found for the first time in an LPS. Xylose has been reported as a component of three species of *C. freundii* (Table 2). The O16 (30) contains a terminal glycerol phosphate, but the absolute configuration of the glycerol moiety has not been determined. During dephosphorylation the glycosidic linkage of one of the β-D-GalNAc residues was cleaved to a large extent. The same reaction has been observed for other β-HexNAc–containing polysaccharides. The repeating unit, →3)-α-L-Rha$p$-(1→3)-α-L-Rha$p$-(1→2)-β-D-Rib$f$-(1→, of O28,1c (31), comprises a partial structure of the *Klebsiella* O7 O-polysaccharide. Type O32 (32) is a homopolymer of 3,6-dideoxy-3-glyceroylamino-D-galactose. A different structure was reported previously for O32, but it actually belonged to a contaminating bacterium. The O32 polymer is not a polysaccharide in the chemical sense since it contains alternating glycosidic and amidic linkages The same polysaccharide backbone structure was previously observed for the gram-positive *Eubacterium sabbureum* L13 cell wall polysaccharide, but in addition it had a D-fructofuranose side chain. Another example of amidic linkages in the main chain is found for the O-polysaccharide of *Pseudomonas aeruginosa* O9, which the contains a glycosidic linkage to a 3-hydroxybutyric acid linked to a diamino-3-deoxy-nonulosonic acid residue.

Since *Citrobacter* is closely related to *Salmonella*, many cross-reactions occur. Thus, O35 (33) cross-reacts with *Salmonella enterica* sv. *arizonae* O59 (new nomenclature), and in fact they have identical O-polysaccharides. Type O38 (33) cross-reacts with *S. enterica* sv. Kentucky, and the two differ only by the O-acetyl substitution in the former. O41 (34) also contains a 3,6-dideoxy-3-amino-D-galactose, but the acid to which it is amidated is (*R*)-3-hydroxybutyric acid. A weak cross-reaction of O41 (34) is observed with *Hafnia alvei* 1211, which has a similar backbone but a side chain that differs in anomerity and site of attachment.

Hafnia

Hafnia alvei is the only species within this genus and has been divided into 39 O-serotypes. *H. alvei* sometimes causes nosocomial infections and is rarely capsulated. The O-polysaccharides are often acidic and contain Neu5Ac, hexuronic acids, and phosphate groups (Table 3).

The molecular mass of the repeating unit could be established inter alia for strain 32 (35) by MALDI-TOF mass spectrometry. The presence of 0.7 moles of acetate was established from a series of MS peaks 29 atomic mass units higher than observed for the native material. It was shown also by MALDI-TOF mass spectrometry that the O-antigen contained up to 16 repeating units. Strain PCM1185 was found to be unrelated to any of the existing serotypes A–C and therefore designated serotype D. It contains (*R*)-3-hydroxybutyric acid, as does strain 1216 (36), in both cases amidically linked to 3-amino-3,6-dideoxy-D-glucose.

Table 2 Structures of Polysaccharides from *Citrobacter freundii*

Serogroup or strain	Structure of repeating unit	Ref.

O8a, 8b	-3)Rha(α1-3)Rha(α1-2)Rha(β1-
	     ⌐
	  Xyl*f*(α1-2)┘
		29

O16	Gro1-*P*-(O-3)⌐ Gal(α1-6)⌐ -6)Gal(β1-4)GalNAc(β1-4)Glc(β1-3)GalNAc(β1- Glc(α1-2)┘	30
O28, 1c	-3)LRha(α1-3)LRha(α1-2)Rib*f*(β1-	31
O29a	-3)ManNAc(β1-4)Glc(β1-	117

O29b	-3)ManNac(β1-4)Glc(β1-
	     ⌐
	  Glc(α1-2)┘
		117

O32[a]	-2)GroA(1-3)Fuc3N2Ac(α1-	32
O35	-3)LFucNAc(α1-3)GlcNAc(β1-2)Gal(β1-	33
O38[b]	Abe4Ac(α1-3)⌐ -2)Man(α1-2)Man(α1-3)Gal(β1-4)LRha(β1- Glc(α1-2)┘	33
O41[c]	-2)Glc(β1-2)Fuc3N*R*3HOBu(β1-6)GlcNAc(α1-4)Gal(β1-3)GalNAc(β1- Glc(α1-2)┘	34

[a] 75% OAc, GroA = glyceric acid.
[b] Abe = 3,6-Dideoxy-D-*xylo*-hexose.
[c] 3HOBu = 3-Hydroxybutyric acid.

This substituent has also been found linked to other sugars in *H. alvei*. MALDI-TOF mass spectrometry was also used in the study of strains PCM1192 (37) and 1209 (38) in which the average number of repeating units was approximately 15 and 10, respectively. For PCM1192 the core structure was also determined. The O-chain of strain 1220 (39) is a teichoic acid polymer, with a glycerol phosphate in the main chain. This type of polymer has previously been demonstrated for strains PCM1191 and 1205. Strain PCM1190 (40) contains two side chains and thereby differs from most other *Hafnia* strains. The PCM1206 strain contains an unusual amino acid, D-allothreonine (41).

The carbohydrate backbone of the *Hafnia alvei* strain Y166/91 LPS has been characterized (42). The polysaccharide had a block structure with two distinct regions, consisting of the blocks Gal*p*-Gal*p* and Gal*p*-Gal*f*, which are connected. The proportion of the blocks was approximately 2:1. It was not determined which block was linked to the core. The structure is identical to that suggested for the O-specific polysaccharide chain of *Klebsiella pneumoniae* O1:K2.

Proteus

Proteus has three different clinically important species: *P. mirabilis* and *P. vulgaris*, the latter with 60 serogroups, and *P. penneri*, which was formerly called *P. vulgaris* biogroup 1. Most of their O-polysaccharides contain acidic groups such as uronic acids, phosphate groups, and, less frequently, pyruvate ketals and lactic acid (Table 4). Partial O-acetylation is a common feature in many of these polysaccharides. Amino acids, amidically linked through their α-amino group, are also frequent. It has been suggested for bacteria involved in infections in the bladder and the kidney that the acidic nature of the O-polysaccharides may contribute to the formation of bladder and kidney stones.

The O-polysaccharide from *P. mirabilis* type O10 (43,44) contains altruronic acid, which has the L-configuration, as demonstrated by the use of ^{13}C-NMR glycosylation shifts. The chemical shift of the C-1 signal of the uronic acid was only compatible with a different absolute configuration than that of the D-GalNAc, to which it was linked. Unlike altrose it was found in a

Table 3 Structures of Polysaccharides from *Hafnia alvei*

Strain	Structure of repeating unit	Ref.
32	-4)GalA2,3Ac$_2$(α1-2)LRha(α1-4)Gal(β1-3)GalNAc(β1-4)GlcNAc(α1-	35
744[a]	-6)GlcNR3HOBu(α1-4)GalNAc(α1-3)GalNAc(β1-2)Glc(α1-P-O-	118
PCM1194	Glc(α1-6)⎦	
PCM1185[a]	-4)GlcA(β1-3)GlcNAc(α1-2)Qui3NR3HOBu(β1-6)Glc(α1-	119, 120
	Glc(α1-4)⎦	
1188	-4)Gal(β1-3)GlcNAc(β1-4)GlcA(α1-2)Man(α1-	120
PCM1190[b]	Glc(α1-5)⎤ -4)Gal(α1-3)GlcNAc(β1-3)LRha(α1-2)Rib$f$(β1- Gal$f$(α1-2)LRha(α1-2)⎦	40
PCM1192	-4LRha(α1-3)GlcNAc(β1-3)LRha(α1-3)LRha(β1- Rib$f$(β1-4)GlcA(α1-2)⎦	37
1204[c]	-3)Man(α1-2)Man(α1-3)GlcNAc(β1-2)Qui3NFo(β1-3)GalNAc(α1-4)GlcA(α1-	121
PCM1206[d]	-4)GalA6alloThr(α1-2)LRha(α1-2)Rib$f$(β1-4)Gal(β1-3)GalNAc(β1-	41
1209	-3)GalNAc(β1-3)Gal(β1-4)Glc(α1-4)GlcA(β1- LRha(β1-4)⎦	38
PCM1210	-6)Gal(α1-4)Gal(β1-3)GlcNAc(β1-3)GlcNAc(α1- LRha(β1-4)⎦	118
1216	-4)Qui3NR3HOBu(α1-4)Gal6Ac(β1-4)GlcNAc(β1-4)GlcA(β1-3)GlcNAc(β1-	36
1220	Glc(α1-6)⎤ -6)Glc(β1-4)LFucNAc(α1-3)GlcNAc(β1-1)Gro3-P-(O- Glc(α1-6)Gal(α1-3)⎦	39
PCM1222[e]	-3)GlcNAc(α1-2)LRha(α1-2)LRha3PEtN(α1-2)Rib$f$(β1-4)Gal(α1- Gal$f$(β1-3)⎦	122
Y166/91	[-3)Gal(β1-3)Gal(α1]$_n$[-3)Gal(α1-3)Gal$f$(β1-]$_n$	42

[a]3HOBu = 3-Hydroxybutyric acid, Qui3N = 3-amino-3,6-dideoxyglucose.
[b]Glc(α1-5) is partial.
[c]Fo = Formyl.
[d]alloThr = Allothreonine.
[e]PEtN = phosphorylethanolamine.

conformation close to that of $^4C_1$. The conformation also differed from that of the altruronic acid found in the capsular polysaccharide of *Aerococcus viridans* var. *homari*. The acid was suggested to derive biosynthetically from D-galacturonic acid (D-GalA). Types O26 (45,46) and O28 (47) have closely related structures with galacturonic acids amidated with the amino acids lysine (O26) or serine and lysine (O28). The O28 O-chain has amides on both GalA residues. The importance of GalA-Lys as an immunodominant factor was demonstrated using synthetic glycopolymers made from 2-acrylamidoethyl glycosides of amides of GalA (47). *P. mirabilis* O43 (48) cross-reacts with O10, and both contain a GalA residue. Periodate oxidation of O43, which destroys two of the four residues, results in loss of serological activity as determined by precipitation and inhibition assays. It was also noted that the level of negative charges was high: two charges per four sugar residues.

P. mirabilis O30 (45) is a linear hexosaminoglycuronan. As an aid in the structural determination computer-assisted analysis of the ^{13}C-NMR chemical shifts

Table 4 Structures of Polysaccharides from *Proteus*

Serogroup or strain	Structure of repeating unit	Ref.
P. mirabilis		
O3	-3)GalNAc(β1-6)GalNAc(β1-4)GlcA(β1- 　　　　　　GalA6Lys(α1-4)⁻ˈ　　Glc(α1-2)⁻ˈ	123, 124
O10	-3)GlcNAc(α1-4)GalNAc(α1-3)Gal(α1- 　　　　　　　　　LAltA(α1-3)⁻ˈ	43, 44
O13[a]	-3)GlcNAc(β1-3)Gal(α1- GalA6AlaLys(α1-4)⁻ˈ	125
O23, 56[b]	-4)Gal(α1-3)GlcNAc(β1-2)Gal(β1-3)GalNAc(α1-	49
O26	-4)GalA6Lys(α1-4)Gal(α1-3)GalA4Ac(α1-3)GlcNAc(β1-	45
O28	-4)GalA6Lys(α1-4)Gal(α1-3)GalA4Ac6Ser(α1-3)GlcNAc(β1-	47
O30[c]	-4)GlcA(β1-6)GalNAc(α1-6)GlcNAc(β1-3)GlcNAc4Ac(β1-	45, 126
O43	-4)Gal(α1-3)Gal(α1-3)GlcNAc(α1-4)Glc(α1-	48
O57	-6)Gal(β1-3)GalNAc(β1-4)GalNAc(β1-3)Gal(α1- 　　　　　Glc(α1-6)⁻ˈ　Gro1-*P*-(O-3)⁻ˈ	127
R14/1959	-2)Glc(β1-4)Glc(β1-3)GlcNAc(β1-4)Gal(α1-	50
P. penneri		
12	-3)LRha(β1-4)GlcNAc6Ac(β1-4)GalNAc(β1- 　　　　　　　GalA3Ac6Thr(α1-3)⁻ˈ	51
14	-2)Rib*f*(β1-4)Gal(β1-3)GlcNAc(β1-2)Qui3N(β1-4)GalA6Ala(α1- 　　　　　　　　　　　　　　　　　　　AcDAla⁻ˈ	128
19[d] 35	-4)GlcNAc3SLac(β1-3)Gal(α1-3)GlcNAc(β1-	52
25[e]	-4)GlcA(β1-3)GlcNAc6Ac(β1-6)GlcNAla3Ac(α1- 　　　　　　　GlcA3/4Ac(α1-4)⁻ˈ	129
42	-2)Glc(β1-4)Glc(β1-3)GlcNAc(β1-4)GalA(α1-	130
52	-4)Gal(β1-3)GlcNAc(β1-4)GalNAc(β1-3)GalNAc(α1- 　　　　　　　　GlcNAc(β1-3)⁻ˈ	131
P. vulgaris		
O1	-4)LQuiNAc(α1-3)GlcNAc(β1-4)GalNAc(α1-4)Galα1-*P*-(O- 　　　　　　　　LQuiNAc(α1-3)⁻ˈ	53, 123
O2	-2)Glc(β1-6)GlcNAc(α1-3)LQuiNAc(α1-3)GlcNAc6Ac(β1-	123, 132
O25[d]	-4)GalNAc(β1-3)GlcNAc(β1-2)LRha(α1-2)Rib*f*(β1- 　　　　　　　Glc3RLac(α1-3)⁻ˈ	54

[a] AlaLys = *N*ᵉ-(1-carboxyethyl)lysine.
[b] OAc at random locations.
[c] 70% OAc.
[d] Lac = 1-Carboxyethyl, lactyl.
[e] OAc not stoichiometric.

was made, and the NMR spectrum was compatible with only one linear sequence. The O-chain of *P. mirabilis* serotype O23,56 [strain 7570 (49)] contains two GalA residues, thus resembling O26 and O28 but none of them with an amide. The O-antigen of the mutant R14/1959, called the T-antigen (50), is identical to that of strain *P. penneri* strain 42. More correctly it should be called T-like antigen as T-forms do not carry any normal O-antigen.

P. penneri strain 12 (51) was shown to have its uronamide residue as the immunodominant group, but this was not true for strain 14, which also has a similar uronamide group and a 3-amino-3,6-dideoxy-D-glucose (D-Qui3N) residue to which an amino acid is linked. *E. coli* O114 also contains this residue, but only a weak cross-reaction is observed. The *P. penneri* strains 19 and 35 (52), which are identical, have the same trisaccharide repeating unit in their O-polysaccharide as *P. penneri* strain 62, but they are polymerized differently. They all contain a 2-amino-3-[(S)-carboxyethyl]-2-deoxy-D-glucose (iso-muramic acid) residue. On delipidation the carboxyethyl group in strains 19 and 35 catalyzes depolymerization via its carboxyl group. This, however, is not true for strain 62.

The *P. vulgaris* OX19 O-antigen (53) chain is not a true polysaccharide as its main chain contains galactosyl phosphate due to the phosphodiester. The polysaccharide is acid labile at pH 4.5 and depolymerizes readily. Serotype O25 contains an (R)-lactic acid, etherbound to position 3 of a lateral Glc residue (54). Both forms of lactyl ethers have been found in nature.

Klebsiella, Providencia, Salmonella, Serratia, Shigella, and Yersinia

A strain (22535) from *Klebsiella pneumoniae* (55), a bacterium normally found in small amounts in the intestinal flora, has pathogenic potential and cross-reacts with *Shigella flexneri* serotype 6 (Table 5). The cross-reaction is assumed to arise from the presence of the common structural element α-L-Rha$p$-(1→2)-α-L-Rha$p$.

The O23 antigen of *Providencia alcalifaciens* (56) also contains an amidated uronic acid, D-GlcA, amidated with N^ε-(1-carboxyethyl)-L-Lys (Table 5). This same unusual amino acid was found in the *Proteus mirabilis* O13 LPS.

The O-antigen polysaccharide from *Salmonella enterica* sv. *arizonae* O21 (57,58) contains Neu5Ac and L-fucosamine *N*-acylated with an acetimidoyl group. It is related to *S. enterica* sv. *arizonae* O61, which has a similar repeat but another C-9 ulosonic acid, namely, 5,7-diamino-3,5,7,9-tetradeoxy-L-glycero-D-galacto-nonulosonic acid (Table 5). This nonulosonic acid was first assigned the D-glycero-L-galacto configuration, but is actually the enantiomer (59). It is worth noting that the nonulosonic acid is designated α despite the fact that it carries an equatorial hydroxyl group at the anomeric carbon. This is due to the IUPAC nomenclature rules, which use the chirality at C-7 for comparison.

In common with most other *S. enterica* sv. *arizonae* O-antigens, that of O62 (60) is acidic and has a D-GalNAcA residue in its repeating unit (Table 5). The repeating unit also contains D-GlcNAc and three L-Rha residues, in analogy with the *S. flexneri* Y O-antigen, and most of the sequence is identical, but it is unclear whether this is immunologically significant. The O-antigen from *S. enterica* sv. *arizonae* (61) has a structure that includes a fragment, α-Col$p$-(1→2)-β-D-Gal$p$-(1→3)-β-D-Glc$p$NAc (Col is 3,6-dideoxy-L-xylo-hexose), that is Lewis d (Le^d) blood group antigen-like. The *S. enterica* sv. *borreze* has an O-polysaccharide that differs from others of *Salmonella*, as it contains D-ManNAc only. The biosynthesis of O54 O-polysaccharide does not require the host chromosomal *rfb* functions (62) but is plasmid encoded. The LPS of *S. enterica* sv. *enteritidis* has been implicated as a virulence factor. Therefore, the LPS from a stable virulent isolate was compared with that from an avirulent isolate. The high molecular mass LPS O-antigen polysaccharides from both isolates are comprised of two different repeating units, I and II, where I is the polysaccharide devoid of D-Glc and II that with a D-Glc residue. In the virulent isolate, the high molecular mass LPS contains structure I and II in the ratio 1:1, but in the avirulent isolate this ratio is 7:1 (63).

Two newly proposed serotypes of *Serratia marcescens*, O25 and O26, have been studied (64). The former was reported to lack both LPS and CPS, and the latter had two O-antigen polysaccharides found in other serotypes, and they are consequently not new serotypes (Table 5). Strain S111 is immunologically distinct from the other *Serratia* and is therefore suggested as serotype O29 (65). The O-polysaccharide is similar to that of *S. marcescens* O18.

The structures of the O-chains of several *Shigella boydii* serotypes have been determined to date (Table 5). The structure of the O-chain of *S. boydii* 5 (66,67) has been reported from two groups, but with slightly different results. The reinvestigated structure with additional data is the correct one (67).

The structure of *Yersinia enterocolitica* O8 O-chain is only partial and contains an unusual sugar, 6-deoxy-D-gulose (Table 5). The structures of the O-polysac-

Table 5 Structures of Polysaccharides from *Klebsiella*, *Providencia*, *Salmonella*, *Serratia*, *Shigella*, and *Yersinia*

Serogroup/Serovar or strain	Structure of repeating unit	Ref.
Klebsiella pneumoniae		
22535	-3)LRha(α1-3)LRha(α1-2)LRha(α1-2)LRha(α1-2)LRha(α1-	55
Providencia alcalifaciens		
O23[a]	-6)Gal(β1-6)Glc(β1-3)GalNAc(β1-4)GlcA6ALys(β1-	56
Salmonella enterica sv. *arizonae*		
O21[b]	-3)LFucNAm(α1-3)GlcNAc6Ac(β1-7)Neu5Ac(α2-	57, 58
O50[c]	-3)Gal(α1-3)GlcNAc(1-6)GlcNAc(β1- Col(α1-2)Gal(β1-3)⌐	61
O62	-3)GlcNAc(β1-3)LRha(α1-2)LRha(α1-3)LRha(α1-2)LRha(α1- GalANAc(α1-2)⌐	60
S. enterica sv. *borreze*		
O54	-4)ManNAc(β1-3)ManNAc(β1-	62
S. enterica sv. *enteritidis*[d]		
	-2)Man([1-4)LRha(α1-3)Gal(α1- Tyv(α1-3)⌐ [Glc(α1-4)]⌐	63
Serratia marcescens		
O26	-3)GalNAc(β1-4)LRha(α1- (a) major -3)GalNAc(β1-3)LRha(α1- (b) minor	64
O27	-3)LRha(α1-4)Glc(α1-	133
O28	-3)Man(β1-2)Man(α1-2)Man(α1-	133
O29	-6)Glc(α1-2)LRha(α1-2)LRha(α1-2)LRha(α1-	65
Shigella boydii		
O5	-3)Man(β1-4)GlcA(β1-3)GlcNAc(α1-2)Gal(β1-4)Man6Ac(β1- LRha(α1-3)⌐	66, 67
Yersinia bercovieri		
O10[e]	-3)Rha(α1-3)Rha(α1- Yer(α1-2)⌐	68
Y. enterocolitica		
O8	6dGul(1-3)⌐ -3)GalNAc(α1-4)Man(1-3)Gal(1- LFuc(1-2)⌐	134
O10	-3)Rha(α1- LXul$f$(β2-2)⌐	69
O:11,23	-3)LQuiNAc(α1-4)GalNAcA3Ac(α1-3)LQuiNAc(α1-3)GlcNAc(β1-	113
O:11,24	-3)LQuiNAc(α1-4)GalNAcA(α1-3)LQuiNAc(α1-3)GlcNAc(β1-	113
Y. ruckeri		
O1[f]	-3)LFucNAm(α1-3)GlcNAc(α1-8)Sug(α2- GlcNAc(β1-4)⌐	70

[a] AlaLys = $N^{\varepsilon-}$ (1-carboxyethyl)lysine.
[b] Am = CH_3C=NH.
[c] Col = 3,6-Dideoxy-L-*xylo*-hexose.
[d] Tyv = 3,6-Dideoxy-D-*arabino*-hexose, substitution with Glc is not stoichiometric.
[e] Yer = 3,6-Dideoxy-4-*C*-(L-*glycero*-1-hydroxy-ethyl)-D-*xylo*-hexose.
[f] Sug = 7-Acetamido-5-(4-hydroxybutyramido)-3,5,7,9-tetradeoxy-L-*glycero*-D-*galacto*-nonulosonic acid.

charides from *Y. bercovieri* O10 (68) and *Y. enterocolitica* O10 (69) have been determined, and both have a 3-linked D-rhamnan backbone to which 3,6-dideoxy-4-C-(L-*glycero*-1-hydroxyethyl)-D-*xylo*-hexopyranose (Yersiniose A, YerA) and β-L-xylulofuranose, respectively, are linked to the 2-position of a Rha residue. The structure of the LPS from *Y. ruckeri* O1 (70) was found to contain inter alia the two sugars 2-acetamidino-2-deoxy-L-fucose and 7-acetamido-3,5,7,9-tetradeoxy-5-(4-hydroxybutyramido)-L-*glycero*-D-*galacto*-nonulosonic acid. The absolute configuration of the nonulosonic acid was reported incorrectly (59).

Vibrionaceae

The LPS of one of the common causes of vibriosis, *V. anguillarum* O2 (71), has been studied. It is one of 10 distinct serotypes of this species (Table 6). Its O-antigen polysaccharide has several unusual features, such as three different 2,3-diaminohexuronic acids (with D-*gluco*-, D-*manno*, and L-*galacto* configuration) and 2,4-diamino-2,4,6-trideoxy-D-glucose (bacillosamine). *Vibrio cholerae* is divided into O1 and non-O1, where O1 is the major causative agent for cholera. Unusual structures were found for the *V. cholerae* O76 (72) and

Table 6 Structures of Polysaccharides from Vibrionaceae

Serogroup or strain	Structure of repeating unit	Ref.
Vibrio anguillarum		
O2[a]	-4)GlcNAc3NAN(β1-4)ManNAc3NAm(β1-4)LGalNAc3NAcA(α1-3)QuiNAc4NAc(β1- FoAla⌐	71
V. cholerae		
O10	-3)ManNAc(α1-4)GlcA(β1-3)Gal(β1-3)GlcNAc(β1-	135

O22	αCol(α1-2)GalA3,4Ac(β1-3)⌐
	                    GlcNAc(α1-4)GalA(α1-3)QuiNAc(β1-
	       α-Col(α1-4)⌐
		76

O76[b]	-2)LRha4N*S*2HOPr(α1-	72

O139	αCol(α1-2)4,6*P*Gal(β1-3)⌐
	                    GlcNAc(β1-4)GalA(α1-3)QuiNAc(β1-
	       α-Col(α1-4)⌐
		75

O144[b]	-2)LRha4N*R*2HOPr(α1-	73
Aeromonas trota		
1354[c]	Col(α1-4)⌐ -3)Gal(β1-3)GlcNAc(β1-4)LRha(α1-3)GalNAc(α1- Col(α1-2)⌐	74
A. caviae		
11212	LRha(α1-3)⌐ -3)GalNAc(β1-6)ManNAc(β1-4)GlcA(β1- Gal(β1-4)⌐	136
Plesiomonas shigelloides		
22074 12254	-4)Gal(α1-3)GlcNAc(α1-3)LRha(α1-2)LRha(α1-2)LRha(α1-	77

[a] Am = CH₃C=NH, Fo = formyl.
[b] 2HOPr = 2-Hydroxypropionic acid.
[c] Col = 3,6-Dideoxy-L-*xylo*-hexose.

O144 (73) antigens. Thus, they are homopolymers built up of 4-amino-4,6-dideoxy-L-mannose L-perosamine residues, in contrast to O1 which is a homopolymer of D-perosamine. The acid forming the amide in O1 is 3-deoxy-L-*glycero*-tetronic acid (or (S)-2,4-dihydroxybutanoic acid), but for O76 it was (S)-2-hydroxypropionic acid, and in O144 (R)-(2-hydroxyl)-propionic acid. As expected, no serological cross-reaction was observed between O1 and the two other species.

Aeromonas trota strain 1354 (74) is cross-reactive with *V. cholerae* O139 Bengal (75), which is a second causative agent of epidemic cholerae. This could be rationalized from the presence in their LPS of a common tetrasaccharide, containing inter alia the two 3,6-dideoxy-L-xylo-hexose (colitose, Col) residues. Another cross-reacting species is *V. cholerae* O22, which differs in the polysaccharide from O139 by two sugars (76).

Two strains from *Plesiomonas shigelloides* (77) have the same pentasaccharide repeating units and cross-react with *Shigella flexneri* 6 and Y and *S. dysenteriae* 1. The disaccharide elements α-L-Rha*p*-(1→2)-α-L-Rha*p* and α-D-Glc*p*NAc-(1→3)-α-L-Rha*p*- are present in their O-antigen repeats of the latter and should be responsible for the serology.

Campylobacter

Campylobacter bacteria are associated with several diseases such as human enteritis. The LPS of *C. jejunii* have highly variable structures accounting for the classification into different serotypes (Table 7). *C. jejunii* serotype O19 (78) is associated with the Guillian-Barré syndrome, a neuropathy, and has an O-antigen with a hyaluronic acid backbone (→4)-β-D-Glc*p*A-(1→3)-β-D-Glc*p*NAc-(1→) in which the D-GlcA residues are amidated with 2-amino-1,3-propanediol. This amide is also present in *E. coli* O143 and some *Shigella* and *Vibrio* LPS. Both serotypes of *C. fetus*, A (79) and B (80), have been studied. The type A O-antigen is a

Table 7 Structures of Polysaccharides from *Campylobacter* and *Helicobacter*

Serogroup or strain	Structure of repeating unit	Ref.
Campylobacter fetus		
A	-3)Man2Ac(α1-	79
B	Rha3Me(α1[-3)Rha(β1-2)Rha(α1-]$_n$	80
C. jejuni		
O19	-4)GlcA6NH(CH$_2$OH)$_2$(β1-3)GlcNAc(β1-	78
Helicobacter pylori		
O1[a]	Gal(β1-4)GlcNAc(β1-[3)Gal(β1-4)GlcNAc(β1]$_n$- LFuc(α1-3)⌐ [LFuc(α1-3)]⌐	137

O3[a]	[LFuc(α1-2)]⌐
	       Gal(β1-4)GlcNAc(β1-[3)gal(β1-4)GlcNAc(β1]$_n$-
	       LFuc(α1-3)⌐ [LFuc(α1-3)]⌐
		82

O6	LFuc(α1-2)⌐
	       Gal(β1-4)GlcNAc(β1-
	       LFuc(α1-3)⌐
		82

P466[a]	LFuc(α1-2)⌐
	       Gal(β1-4)GlcNAc(β1-[3)Gal(β1-4)GlcNAc(β1]$_n$-
	       LFuc(α1-3)⌐ [LFuc(α1-3)]⌐
		81

MO19[a]	LFuc(α1-2)⌐
	       Gal(β1-4)GlcNAc(β1-
	       LFuc(α1-3)⌐
		81

[a] L-Fuc in brackets not stoichiometric.

D-mannan and that of B a D-rhamnan. The B polysaccharide is terminated by a 3-*O*-methyl-D-Rha residue. If the polysaccharide is elongated from the nonreducing end in the biosynthesis, this sugar then stops further elongation.

Helicobacter

Helicobacter pylori is implicated as a causative agent for gastritis, gastric and duodenal ulcers, and gastric carcinoma. Studies of *H. pylori* O-antigens indicate complex structures mimicking blood group substances, which may explain the low endotoxicity and low immunochemical response. Thus, Lewis x and y (Le^x and Le^y) determinants are present in their LPS and in some an intervening heptan is found. In P466 (81) there is a terminal Le^y determinant but also internal Le^x residues. Similar heterogeneity is found for MO19 (81) in which there is a single Le^y epitope terminally and an internal chain of 3-substituted D-*glycero*-α-D-*manno*-heptose residues. However, the inner cores are identical. In type O6 two populations of related molecules are present with chains of 3-substituted D-*glycero*-D-*manno*-heptose similar to those in the MO19 strain, one with and one without a single terminal Le^y epitope. In contrast, in the O3 LPS (82) Le^x and Le^y epitopes terminate a partially fucosylated *N*-acetyllactosaminoglycan, but a heptan chain similar to that in the O6 LPS (82) was shown to connect the outer chains to the inner core.

It is evident that the complex structures found in *H. pylori* require an elaborate biosynthetic system and further studies are needed to determine whether the intervening heptoglycan region belongs to the core, the O-polysaccharide, or should be considered as a new, fourth region in the LPS.

Pseudomonadaceae

A large number of structures of O-polysaccharides from *Pseudomonas* have been determined (Table 8), many from *P. aeruginosa*, a species which inter alia causes nosocomial infections. Amino-, aminodeoxy-, and diaminodideoxysugars and aminouronic, diaminouronic and diaminodideoxynonulosonic acids are common constituents of these O-antigens. O-Antigens from other *Pseudomonas* species differ significantly from those present in *P. aeruginosa* and have common components and small repeating units.

P. caryophylli (now *Burkholderia*) is a phytopathogenic bacterium responsible for the wilting of carnations. Two homopolysaccharides were found in the LPS of this bacterium. Two novel sugars with 10 and 9 carbons were discovered in two different O-chains (83). The chains were homopolymers of the sugars 3,6,10-trideoxy-4-C-(D-*glycero*-1-hydroxyethyl)-D-*erythro*-D-*gulo*-decose (caryophyllose) and 4,8-cyclo-3,9-dideoxy-L-*erythro*-D-*ido*-nonose (caryose), respectively (Fig. 1). The structure of the polymers have also been determined (Fig. 1). The structure of caryose has been confirmed by synthesis (84).

An L-rhamnan backbone is present in the O-polysaccharides from *P. fluorescens* IMV4125 (85), *P. syringae* pv. *coriandricola* W-43 (86), and three strains of *P. syringae* ssp. *savastanoi* (87). The O-chains found for *P. pseudomallei* are similar to those found previously for other *pseudomallei* strains.

The genus *Burkholderia* has been removed from the genus *Pseudomonas* and comprises organisms actually or potentially pathogenic to plants and humans. *B. cepacia* can cause inter alia pulmonary infections in patients with cystic fibrosis. Several typing schemes have been proposed for *B. cepacia*. The type I and O1 (88) O-antigens both contain two different chains. This has proved to be a common feature in *B. cepacia* LPS. Serotype E [O2, (89)] has a D-galacto-D-mannan trisaccharide repeat in the O-chain and also a D-mannan chain as a minor component. Serotype C [O4, (88,90)] also produces two chains. The individual chains have been demonstrated for other serotypes, namely Canadian type A, and in Japanese serotype C. The O-polysaccharide from *B. pickettii* (91) has a backbone that is similar to that of *Shigella flexneri* Y, differing only in the anomeric configuration of one L-Rha residue. In *B. plantarii* (92) two polysaccharides were found, one of which, a 6-deoxy-α-D-talan, →3)-α-D-6dTal*p*-(1→2)-α-D-6dTal*p*-(1→2)-α-D-6dTal*p*-(1→, could be extracted with 2-propanol, leaving the O-polysaccharide in the remaining biomass. *B. solanacearum* is a harmful phytopathogen and is most dangerous for *Solanaceae* species. Five strains from three biovars were investigated (93,94); the structures of the O-chains all contain a backbone with D-glucosamine and L-rhamnose (Table 8). The polysaccharides from ICMP 7945 and ICMP 8093 are xylosylated forms of the polysaccharides from ICMP 750, ICMP 8115, and ICMP 7864. The structure from ICMP 750 and ICMP 8115 has been found in the O-antigen from *Serratia marcescens* O22 and some *B. solanacearum* serotypes.

The only species of the *Stenotrophomonas* (formerly *Pseudomonas* and subsequently *Xanthomonas*) genus is *maltophilia*, an opportunistic pathogen. Five serotypes have been investigated: O2 (95), O3 (96), O6 (96), O10 (97), and O20 (98).

Table 8 Structures of Polysaccharides from Pseudomonadaceae

Serogroup, serovar, or strain	Structure of repeating unit		Ref.
Pseudomonas caryophylli	-7Sug1^a(α1-	(a)	83
	-7)Sug2^b(β1-	(b)	83
P. fluorescens			
A(IMV472)	-3)LRha2Ac(β1-4)LRha(α1-3)Fuc(α1- GlcNAc(β1-2)⎦		138
IMV4125^c ATCC13525	[Fuc3NAc(α1-2)]⏋ -3)LRha(α1-3)LRha(α1-2)LRha(α1- Fuc3NAc(α1-2)⎦		85
ATCC49271^d	-4)(Sug3)8Ac(α2-		139
P. syringae v. *coriandricola* W-43	-3)LRha(α1-2)LRha(α1-2)LRha(α1-3)LRha(α1- Fuc3NAc(α1-2)⎦		86
P. syringae ss. *savastanoi* ITM519 ITM317 PVF5	-3)LRha(α1-2)LRha(α1-3)LRha(α1- Fuc3NAc(α1-3)⎦		87
Burkholderia cepacia			
A	-3)GalNAc(α1-3)GalNAc(β1-4)LRha(α-1-		140
C (O4)	-3)Gal(α1-3)Gal(β1-3)GalNAc(β1-	(a) major	88, 90
	-3)GalNAc(α1-3)GalNAc(β1-4)LRha(α1-	(b) minor	90
E (O2)	-2)Man(α1-2)Man(α1-4)Gal(β1-	(a) major	89
	-2)Man(α1-2)Man(α1-3)Man(β1-	(b) minor	
I	-3)Fuc(α1-4)GalNAc(β1-	(a)	88
	-3)Fuc(α1-2)LRha(α1-	(b)	
O1	-4)Glc(α1-3)LGlcNAc(α1-	(a)	88
	-4)Glc(α1-3)LRha(α1-	(b)	
O9	-4)Glc(α1-3)LRha(α1-		141
B. gladioli pv. *gladioli* NCPPB1891	-3)Gal(α1-3)Man2Ac(β1-4)LRha(α1-		142
B. pickettii NCTC 11149	-3GlcNAc(β1-2)LRha(α1-2)LRha(β1-3)LRha2Ac(α1-		91
B. plantarii DSM 6535	-4)LRha(α1-3)ManNAc(β1-		92
B. pseudomallei			
304b^e	-3)6dD*man*Hep(β1-	(a)	143
	-3)Glc(β1-3)6dTal4Ac2Me(α1-	(b)	
824a^e	-3)6d*man*Hep(β1-	(a)	143

Table 8 Continued

Serogroup, serovar, or strain	Structure of repeating unit	Ref.
B. solanacearum		
8089	-3)Rha(α1-3)Rha(α1-4)GalNAc(α1-	94
ICMP 750 8115	-3)GlcNAc(α1-2)LRha(α1-2)LRha(α1-3)LRha(α1-	144
ICMP 7864	-3)GlcNAc(β1-3)LRha(α1-2)LRha(α1-3)LRha(α1-	144
ICMP 7945	-3)GlcNAc(α1-2)LRha(α1-2)LRha(α1-3)LRha(α1- LXyl(β1-4)⏋	144
ICMP 8093	-3)GlcNAc(β1-3)LRha(α1-2)LRha(α1-3)LRha(α1- LXyl(β1-4)⏋	144
biotype II	-6)Glc(α1-3)LRha(α1-3)LRha(α1-	145
B. vietnamiensis		
LMG 6998	-3)LRha(α1-3)Man(β1-4)Man(α1-	146
Stenotrophomonas maltophilia		
O2	-4)Man(α1-3)LRha(α1- LXyl(β1-2)⏋	95
O3	-3)GlcNAc(β1-3)Fuc(α1- Fuc4NAc(α1-4)⏋	96
O6	-3)LRha(α1-3)GlcNAc(β1- Xyl(β1-4)⏋	147
O10	-2)LRha(β1-2)LRha(α1-2)LRha(α1- LXyl(β1-3)⏋	97
O20	-2)Man(α1-3)Rha(β1-2)Rha(α1-2)Rha(α1-	98

[a]Sug1 = 3,6,10-Trideoxy-4-*C*-(D-*glycero*-l-hydroxyethyl)-D-*erythro*-D-*gulo*-decose.
[b]Sug2 = 4,8-Cyclo-3,9-dideoxy-L-*erythro*-D-*ido*-nonose.
[c]Fuc3NAc(α1-2) in brackets is not stoichiometric.
[d]Sug3 = 5-Acetamidino-7-acetamido-3,5,7,9-tetradeoxy-L-*glycero*-D-*galacto*-nonulosonic acid, OAc is partial.
[e]6dDmanHep = 6-deoxy-D-*manno*-heptose, Me is partial.

Miscellaneous: *Acetobacter*, *Acinetobacter*, *Actinobacillus*, *Alcaligenes*, *Chromobacterium*, *Legionella*, *Ochrobactrum*, *Pectinatus*, *Rhizobium*, and *Thiobacillus*

In *Acetobacter* MB 58 (99) the capsular polysaccharide (CPS) and the O-antigen have the same structure (99). Identical CPS and O-antigen was also reported for the MB70 (100) and MB 135 strains (101). The core–lipid A linkage showed unusually high resistance to acid hydrolysis for several of these species. Polysaccharide I of *A. methanolicus* MB70 has been found in strain MB58/4 (Table 9).

Acinetobacter, a species now being recognized as an etiological agent for nosocomial infections, has been divided into some 18 different species from DNA-DNA hybridization studies. Seven of these groups are named: *A. calcoaceticus* (DNA group 1), *A. baumannii* (DNA group 2), *A. haemolyticus* (DNA group 4), *A. junii* (DNA group 5), *A. johnsonii* (DNA group 7), *A. lwoffii* (DNA group 8), and *A. radioresistens* (DNA group 12). O-serotyping schemes have also been devised with up to 34 distinct groups. The O-polysaccharides have conventional structures, and amino sugars are abundant (Table 9). *Acinetobacter* often occurs as R-forms, containing the complete or partial core but no O-antigen,

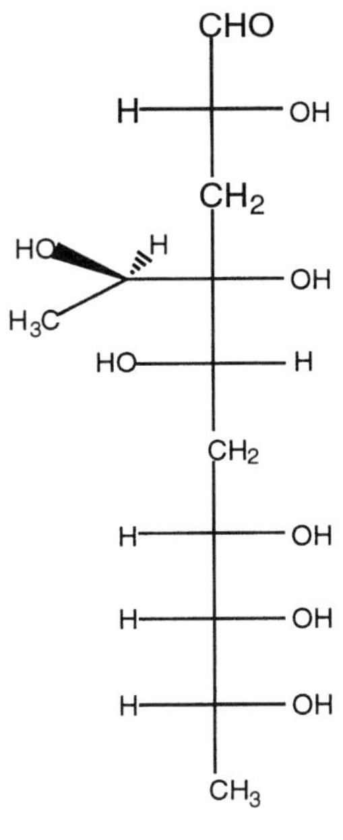

Fig. 1 The structures of caryphyllose (**1**), caryose (**2**), and the corresponding polysaccharides (**3** and **4**, respectively).

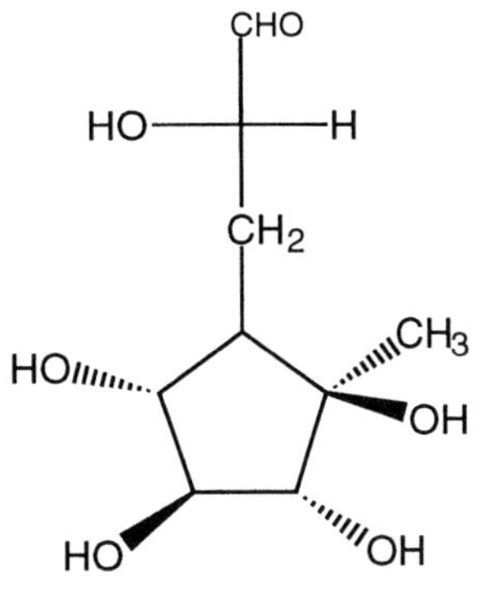

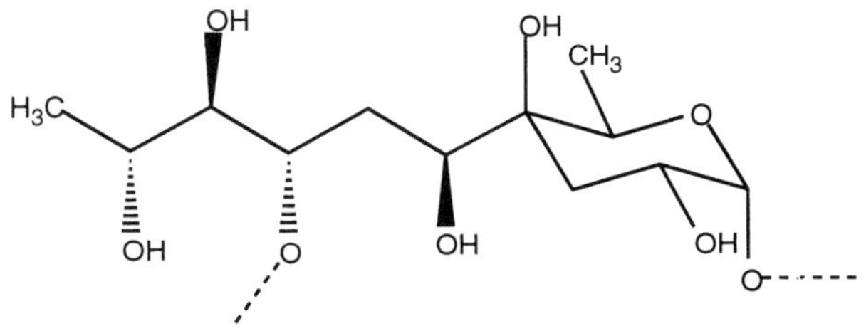

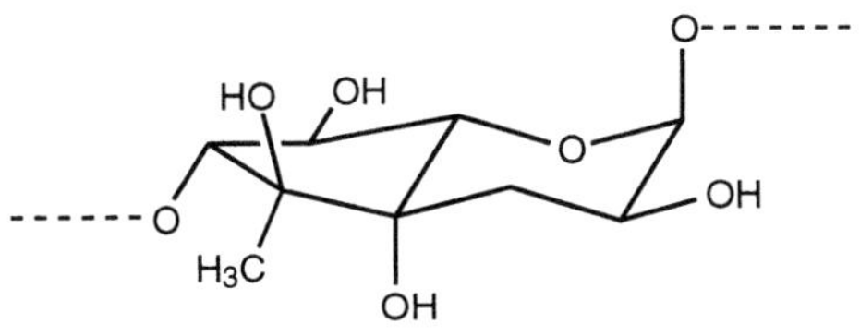

but also with O-chains, sometimes with unusual features. One unusual feature is that for many species, the silver stain does not work in the SDS-PAGE analysis. Some species seem to contain more than one polymer. Thus, the *A. baumanni* O16 polymer may be a minor component in the O11 strain. Pyruvic acid, an unusual component in LPS, is present in DNA group 1.

All five serotypes (a–e) of *Actinobacillus actinomycetemcomitans*, a prominent member of the subgingival microflora, have been studied (102,103). They all have structurally distinct O-polysaccharides and show no cross-reactions. The presence of both L- and D-6dTal (in homopolymers) in serotypes a and c is noteworthy (Table 9).

Chromobacterium violaceum is generally considered as nonpathogenic but can cause infections in humans and in animals. In the NCTC 9694 O-antigen (104) two residues of D-*glycero*-D-*galacto*-heptose are present in the repeat, of which one is α- and the other β-linked.

Heptoses are typical core constituents but have also been found in some O-polysaccharides. Thus L-*glycero*-D-*manno*-heptose has been found in *Pseudomonas cepacia*, D-*glycero*-D-*manno*-heptose in *Vibrio cholerae* O3 and O21, and D-*glycero*-D-*altro*-heptose in *Campylobacter jejuni* O23 and O36. D-*Glycero*-D-*galacto*-heptose has only been found among gram-negative species in *Chromobacterium violaceum* (Table 9). It is, however, a typical component of the gram-positive bacterium *Eubacterium sabbureum*.

Legionella pneumophila, the etiological agent of legionellosis, causes severe respiratory tract infections in humans, which may be lethal. The O-chain of type 1 (105) is a homopolymer of 5,7-diamino-3,5,7,9-tetradeoxy-*glycero*-*galacto*-nonulosonic acid (Table 9). The absolute configuration was not determined but is most probably L-*glycero*-D-*galacto*, which is the configuration demonstrated for several other O-antigens (59).

Table 9 Structures of Polysaccharides from *Acetobacter*, *Acinetobacter*, *Actinobacillus*, *Alcaligenes*, *Chromobacterium*, *Legionella*, *Ochrobactrum*, *Pectinatus*, *Rhizobium*, and *Thiobacillus*

Serogroup or strain	Structure of repeating unit	Ref.
Acetobacter diazotrophicus		

Pal 5	-2)Rib$f$(β1-3)LRha(α1-3)LRha(α1-2)LRha(α1-
	              ┘
	   Glc(β1-2)
		148

A. methanolicus		
MB 58	-6)Glc(α1-2)Gal(α1-6)Gal(α1-	99
MB 70	-2)Gal$f$(β1-3)Gal(β1- (a) -2)Glc(α1-6)Glc(α1- (b)	100
MB 135	-4)Man(α1-2)Man(α1-2)Man(α1-2)Man(α1-2)Man(α1-	101
Acinetobacter spp.		
34 (DNA group 2)	-3)GlcNAc(β1-3)GalNAcA(α1-3)LFucNAc(α1- ┘ LFucNAc(α1-4)	149
90[a] (DNA group 10)	-3)Gal(α1-3)GlcNAc(β1-3)Gal(α1-4)GalNAc(β1- ┘ Fuc4NR3HOBu(α1-4)	150
94[b]	-3)GalNAc(β1-3)Gal(α1- ┘ Fuc3NAcyl(α1-4)GalNAc(β1-4)	151
108[a]	-6)Gal(β1-3)GalNAc(β1-4)Gal(α1- ┘ Fuc3NR3HOBu(β1-3)GalNAc(α1-3)GalNAc(β1-3)	149
A. baumanni		
214	-3)Glc(β1-3)GalNAc(β1- ┘ Gal(α1-6)	152
O2[a]	-6)Gal(β1-3)GalNAc(β1-4)Gal(α1- ┘ Fuc3NR3HOBu(β1-3)GalNAcA(α1-3)GalNAc(β1-3)	153
O5	-3)GlcNAc(β1-3)GalNAcA(α1-3)LFucNAc(α1- ┘ LFucNAc(α1-4)	154
O10	-3)GlcNAc(α1-2)LRha(α1-2)LRha(α1-3)LRha(α1- ┘ ManNAc(α1-3)	155

O11	-3)GalNAc(α1-4)GalNAc(β1-3)Gal(α1-6)Gal(β1-
	            ┘
	  Glc(β1-6)
		156

O12 O23[a]	-3)GlcNAc(β1-3)GalNAc(β1-3)Gal(α1- ┘ QuiN3R3HOBu(β1-6)GlcNAc(α1-4)	157
O12 O16	-6)GlcNAc(α1-4)GalNAc(α1-3)GlcNAc(α1- ┘ Glc(β1-3)	157 158
O18	-3)GalNAc(β1-3)Gal(β1- ┘ ManNAc(β1-3)Gal(α1-4)	159
O24[c]	-6)GlcNAc(α1-3)L-FucNAc(α1-3)Gal(α1-3)GlcNAc(α1-4)Sug1(β1-	160
A. calcoaceticus		
7[d] (DNA group 1)	-2)Gal3Ac4,6RPy(β1-3)GlcNAc(β1-4)GlcA(β1-3)GalNAc(β1-	161
A. haemolyticus		
ATCC17906	-4)GalNAcA(α1-3)QuiNAc4NAc(β1-4)GalNAcA6DAla(α1-	162
A. junii		
65	-2)LRha(α1-3)LRha(α1-2)LRha(α1-3)LRha(α1-3)Gal(β1-	163

Table 9 Continued

Serogroup or strain	Structure of repeating unit	Ref.
Actinobacillus actinomycetemcomitans		
(a)	-3)6dTal2Ac(α1-2)6dTal(α1-	102
(b)	-3)Fuc(α1-2)LRha(α1- GalNAc(β1-3)⏌	103
(c)	-3)L6dTal4Ac(α1-2)L6dTal(α1-	102
(d)	-3)Glc(β1-4)Man(β1-4)Man(α1- LRha(α1-3)⏌	102
(e)	-4)GlcNAc(α1-3)LRha(α1-	102
Alcaligenes latus		
B-16	-2)Man(α1-3)LFuc(α1-	164
Chromobacterium violaceum		
NCTC 9694[e]	-2LRha(α1-4)DD*gal*Hep(β1-3)GlcNAc(α1-4)DD*gal*Hep(α1-	104
Legionella pneumophila		
O1[f]	-4)(Sug2)8Ac(α2-	105
Ochrobactrum anthropi		
LMG 3331	-3(GlcNAc(α1-2)LRha(α1-	165
Pectinatus cerevisiiphilus	-2Fuc*f*(β1-2)Glc(α1-	166
P. frisingensis	-2)L6dAlt*f*(β1-3)L6dAlt*f*(β1-2)L6dAlt*f*(α1- L6dAlt*f*(α1-2)⏌	166
Rhizobium leguminosarium bv. *trifolii*		
24[g]	-3)L6dTal(α1-2)LRha(α1-5)Sug3(2-	106
R. loti		
NZP2213	-3)L6dTal2Ac(α1-	167, 168
R. trifolii		
4s	-4)GlcNAc(β1-3)LRha(α1-3)LRha(α1-3)LRha(α1- ManNAc(α1-2)⏌	169
R. tropici		
CIAT 899	-4)Glc(β1-3)6dTal2Ac(α1-3)LFuc(α1-	107
Thiobacillus		
IFO 14570	-3)GlcNAc4NAcA(β1-3)QuiNAm4NAc(α1-4)GlcNAc3NAcA(β1-	170
T. ferrooxidans		
IFO 14262	-3)Rha(α1-3)LRha(1-3)Glc(1-3)Glc(α1- LRha3Me(α1-4)⏌	171

[a]3HOBu = 3-Hydroxybutyric acid.
[b]Acyl = 2-Acetoxypropionyl or 2-hydroxypropionyl.
[c]Sug1 = 5-Acylamino-7-acetamido-3,5,7,9-tetradeoxy-L-*glycero*-D-*galacto*-nonulosonic acid.
[d]Py = Pyruvic acid.
[e]DD*gal*-Hep = D-*glycero*-D-*galacto*-heptose.
[f]Sug2 = 5-Acetamidino-7-acetamido-3,5,7,9-tetradeoxy-L-*glycero*-D-*galacto*-nonulosonic acid.
[g]Sug3 = 3-Deoxy-*lyxo*-heptulosaric acid.

Ochrobactrum anthropi, a newly defined species, is related to *Brucella* and other bacteria belonging to the *Proteobacteria*. The strain investigated, LMG 3331, has a simple disaccharide repeat (Table 9).

O-antigens from two species of *Pectinatus* have been investigated. The most striking feature observed in the presence of furanosidic 6-deoxyaltrose residues in *P. frisingensis* (Table 9).

Rhizobial bacteria have the unique property of nitrogen fixation, through a complex multistep interaction. The O-specificity is changed when they convert to nitrogen-fixing bacteria, and the LPS has therefore attracted attention. The *R. leguminosarium* bv. *trifolii* strain 24 LPS (106) contains a sugar previously only found in plants, 3-deoxy-*lyxo*-heptulosaric acid (absolute configuration not determined) and in addition 6-deoxy-L-talose. In *R. loti* the O-polysaccharide was found in the phenol phase after phenol-water extraction, due to the hydrophobicity of the chain (Table 9). The enrichment of O-antigen in the phenol phase was also observed for *R. tropici* strain CIAT899 (107). Other phenol-soluble S-type LPS has been observed from *Pseudomonas aeruginosa* O7 (108), *Legionella pneumophila* O1 (105), and *Citrobacter freundii* (109).

Thiobacillus, like *Ochrobactrum*, belongs to the *Proteobacteria*. The bacterium can be used for leaching, and the attachment to copper and iron sulfide ores has been attributed to the hydrophobicity of its surface structures. For the same reason the LPS is enriched in the phenol phase upon extraction.

ACKNOWLEDGMENTS

I would like to thank Bengt Lindberg, Göran Widmalm (Stockholm University, Stockholm), Yuriy Knirel (Zelinski Institute, Moscow), Otto Holst (Forschungszentrum Borstel, Borstel, Germany), Elke Schweda, Andrej Weintraub (Karolinska Institute, Stockholm), and Ralfh Wollin (Swedish Institute for Infectious Diseases, Stockholm) for proofreading this manuscript. Special thanks to Yuriy Knirel and Otto Holst for indepth comments and to Bengt Lindberg for continuous, interesting, and stimulating polysaccharide discussions throughout the years. Financial support from the Swedish Research Council for Engineering Sciences and the Swedish Natural Science Research Council is gratefully acknowledged.

REFERENCES

1. Kenne L, Lindberg B. In: The Polysaccharides. Orlando: Academic Press, 1983:287–363.
2. Knirel Y, Kochetkov NK. The structure of lipopolysaccharides of gram-negative bacteria. III. The structure of O-antigens: a review. Biochemistry (Moscow) 1994; 59:1325–1383.
3. CARBBANK/Complex Carbohydrate Structural Database. Windows: ftp://ncbi.nlm.nih.gov/repository/carbbank/. Webversion: http://www.ccrc.uga.edu/ .
4. Lindberg B. Components of bacterial polysaccharides. Adv Carbohydr Chem Biochem 1990; 48:281–318.
5. Gibb AP, Barclay GR, Poxton IR, Di Padova F. Frequencies of lipopolysaccharide core types among clinical isolates of *Escherichia coli* defined with monoclonal antibodies. J Infect Dis 1992; 166:1051–1057.
6. Appelmelk BJ, An YQ, Hekker TAM, Thijs LG, MacLaren DM, De Graaf J. Frequencies of lipopolysaccharide core types in *Escherichia coli* strains from bacteraemic patients. Microbiology 1994; 140:1119–1124.
7. Ratnayake S, Weintraub A, Widmalm G. Structural studies of the enterotoxigenic *Escherichia coli* (ETEC) O153 O-antigenic polysaccharide. Carbohydr Res 1994; 265:113–120.
8. Rundlöf T, Weintraub A, Widmalm G. Structural studies of the enteroinvasive *Escherichia coli* (EIEC) O28 O-antigen polysaccharide. Carbohydr Res 1996; 291:127–139.
9. Landersjö C, Weintraub A, Widmalm G. Structure determination of the O-antigen polysaccharide from the enteroinvasive *Escherichia coli* (EIEC) O143 by component analysis and NMR spectroscopy. Carbohydr Res 1996; 291:209–216.
10. Kjellberg A, Urbina F, Weintraub A, Widmalm G. Structural analysis of the O-antigenic polysaccharide from the enteropathogenic *Escherichia coli* O125. Eur J Biochem 1996; 239:532–538.
11. Bhattacharyya T, Basu S. Structure of the O-specific side chain of the lipopolysaccharide from *E. coli* O126. Carbohydr Res 1994; 254:221–228.
12. Landersjö C, Weintraub A, Widmalm G. Structural analysis of the O-antigenic polysaccharide from the enteropathogenic *Escherichia coli* O142. Eur J Biochem 1997; 244:449–453.
13. Manca MC, Weintraub A, Widmalm G. Structural studies of the *Escherichia coli* O26 O-antigen polysaccharide. Carbohydr Res 1996; 281:155–160.
14. Medina EC, Widmalm G, Weintraub A, Vial PA, Levine MM, Lindberg AA. Structural studies of the O-antigenic polysaccharides of *Escherichia coli* O3 and the enteroaggregative *Escherichia coli* strain 17-2. Eur J Biochem 1994; 224:191–196.
15. Staaf M, Widmalm G, Weintraub A, Nataro JP. Structural elucidation of the O-antigenic polysaccharide from *Escherichia coli* O44:H18. Eur J Biochem 1995; 233:473–477.
16. Jann B, Shashkov AS, Kochanowski H, Jann K. Structural comparison of the O6 specific polysaccharides from *E. coli* O6:K2:H1, *E. coli* O6:K13:H1, and *E. coli* O6:K54:H10. Carbohydr Res 1994; 263:217–225.
17. Jann B, Shashkov A, Torgov V, Kochanowski H, Seltmann G, Jann K. NMR investigation of the 6-deoxy-L-talose-containing O45, O45-related (O45rel), and

O66 polysaccharides of *Escherichia coli*. Carbohydr Res 1995; 278:155–165.
18. Torgov VI, Shashkov AS, Kochanowski H, Jann B, Jann K. NMR analysis of the structure of the O88 polysaccharide (O88 antigen) of *Escherichia coli* O88:K⁻:H25. Carbohydr Res 1996; 283:223–227.
19. Jann B, Shashkov AS, Kochanowski H, Jann K. Structure of the O16 polysaccharide from *Escherichia coli* O16:K1: an NMR investigation. Carbohydr Res 1994; 264:305–311.
20. Torgov VI, Shashkov AS, Jann B, Jann K. NMR reinvestigation of two N-acetylneuraminic acid-containing O-specific polysaccharides (O56 and O24) of *Escherichia coli*. Carbohydr Res 1995; 272:73–90.
21. Gamian A, Kenne L, Mieszala M, Ulrich J, Defaye J. Structure of the *Escherichia coli* O24 and O56 O-specific sialic-acid-containing polysaccharides and linkage of these structures to the core region in lipopolysaccharides. Eur J Biochem 1994; 225:1211–1220.
22. Ratnayake S, Widmalm G, Weintraub A, Medina EC. Structural studies of the *Escherichia coli* O90 O-antigen polysaccharide. Carbohydr Res 1994; 263:209–215.
23. Staaf M, Urbina F, Weintraub A, Widmalm G. Structure determination of the O-antigenic polysaccharide from the enterotoxigenic *Escherichia coli* (ETEC) 0101. Carbohydr Res 1997; 297:297–299.
24. Jansson PE, Kenne L, Widmalm G. Computer-assisted structural analysis of polysaccharides with an extended version of CASPER using ^{1}H and ^{13}C-n.m.r. data. Carbohydr Res 1989; 188:169–191.
25. Gamian A, Romanowska E, Ulrich J, Defaye J. The structure of the sialic acid-containing *Escherichia coli* O104 O-specific polysaccharide and its linkage to the core region in lipopolysaccharide. Carbohydr Res 1992; 236:195–208.
26. Linnerborg M, Wollin R, Widmalm G. Structural studies of the O-antigenic polysaccharide from *Escherichia coli* O167. Eur J Biochem 1997; 246:565–573.
27. Stevenson G, Neal B, Liu D, Hobbs M, Packer NH, Batley M, Redmond JW, Lindquist L, Reeves P. Structure of the O antigen of *Escherichia coli* K-12 and the sequence of its *rfb* gene cluster. J Bacteriol 1994; 176:4144–4156.
28. Hisatsune K, Kondo S, Iguchi T, Ito T, Hiramatsu K. Lipopolysaccharides of *Escherichia coli* K12 strains that express cloned genes for the Ogawa and Inaba antigens of *Vibrio cholerae* O1—identification of O-antigenic factors. Microbiol Immunol 1996; 40:621–626.
29. Kocharova NA, Knirel YA, Shashkov AS, Kochetkov NK, Kholodkova EV, Stanislavsky ES. The structure of the *Citrobacter freundii* O8a,8b O-specific polysaccharide containing D-xylofuranose. Carbohydr Res 1994; 263:327–331.
30. Kocharova NA, Thomas-Oates JE, Knirel YA, Shashkov AS, Dabrowski U, Kochetkov NK, Stanislavsky ES, Kholodkova EV. The structure of the O-specific polysaccharide of *Citrobacter* O16 containing glycerol phosphate. Eur J Biochem 1994; 219:653–661.
31. Kocharova NA, Knirel YA, Kholodkova EV, Stanislavsky ES. Structure of the O-specific polysaccharide chain of *Citrobacter freundii* O28, 1c lipopolysaccharide. Carbohydr Res 1995; 279:327–330.
32. Kocharova NA, Knirel YA, Shashkov AS, Kochetkov NK, Kholodkova EV, Stanislavsky ES. The structure of the O-specific polysaccharide of *Citrobacter freundii* O32, a partially O-acetylated homopolymer of 3,6-dideoxy-3-(L-glyceroylamino)-α-D-galactopyranose. Carbohydr Res 1994; 264:123–128.
33. Kocharova NA, Knirel YA, Stanislavsky ES, Kholodkova EV, Lugowski C, Jachymek W, Romanowska E. Structural and serological studies of lipopolysaccharides of *Citrobacter* O35 and O38 antigenically related to *Salmonella*. FEMS Immun Med Microbiol 1996; 13:1–8.
34. Ravenscroft N, Dabrowski J, Romanowska E. Structural elucidation of the biological repeating unit of O-specific polysaccharide from *Citrobacter* serotype O41. Eur J Biochem 1995; 229:299–307.
35. Jachymek W, Petersson C, Helander A, Kenne L, Niedziela T, Lugowski C. Structural studies of the O-specific chain of *Hafnia alvei* strain 32 lipopolysaccharide. Carbohydr Res 1996; 292:117–128.
36. Katzenellenbogen E, Romanowska E, Shashkov AS, Kocharova NA, Knirel YA, Kochetkov NK. The structure of the O-specific polysaccharide of *Hafnia alvei* strain 1216. Carbohydr Res 1994; 259:67–76.
37. Jachymek W, Petersson C, Helander A, Kenne L. Lugowski C, Niedziela T. Structural studies of the O-specific chain and a core hexasaccharide of *Hafnia alvei* strain 1192 lipopolysaccharide. Carbohydr Res 1995; 269:125–138.
38. Niedziela T, Petersson C, Helander A, Jachymek W, Kenne L, Lugowski C. Structural studies of the O-specific polysaccharide of *Hafnia alvei* strain 1209 lipopolysaccharide. Eur J Biochem 1996; 237:635–641.
39. Dabrowski U, Dabrowski J, Katzenellenbogen E, Bogulska M, Romanowska E. Structure of the O-specific polysaccharide, containing a glycerol phosphate substituent, of *Hafnia alvei* strain 1220 lipopolysaccharide. Carbohydr Res 1996; 287:91–100.
40. Petersson C, Jachymek W, Kenne L, Niedziela T, Lugowski C. Structural studies of the O-specific chain of *Hafnia alvei* strain PCM 1190 lipopolysaccharide. Carbohydr Res 1997; 298:219–227.
41. Petersson C, Niedziela T, Jachymek W, Kenne L, Zarzecki P, Lugowski C. Structural studies of the O-specific polysaccharide of *Hafnia alvei* strain PCM 1206 lipopolysaccharide containing D-allothreonine. Eur J Biochem 1997; 244:580–586.
42. Karlsson C, Jansson PE, Wollin R. Structure of the O-polysaccharide from the LPS of a *Hafnia alvei* strain isolated from a patient with suspect yersinosis. Carbohydr Res 1997; 300:191–197.
43. Swierzko A, Shashkov AS, Senchenkova SN, Toukach FV, Ziolkowski A, Cedzynski M, Paramonov NA, Kaca W, Knirel YA. Structural and serological studies of the O-specific polysaccharide of the bacterium *Proteus mirabilis* O10 containing L-altruronic acid, a new component of O-antigens. FEBS Lett 1996; 398:297–302.
44. Shaskov AS, Senchenkova SN, Toukach FV, Ziolkowski A, Paramonov NA, Kaca W, Knirel YA, Ko-

chetkov NK. Structure of the O-specific polysaccharide of the bacterium *Proteus mirabilis* O10 containing L-altruronic acid, a new component of O-antigens. Biochemistry (Moscow) 1996; 61:1554–1562.
45. Shashkov AS, Toukach FV, Paramonov NA, Ziolkowski A, Senchenkova SN, Kaca W, Knirel YA. Structures of new acidic O-specific polysaccharides of the bacterium *Proteus mirabilis* serogroups O26 and O30. FEBS Lett 1996; 386:247–251.
46. Shashkov AS, Toukach FV, Paramonov NA, Senchenkova SN, Kaca W, Knirel YA. Structure of a new lysine-containing O-specific polysaccharide of the bacterium *Proteus mirabilis* O26. Biochemistry (Moscow) 1996; 61:15–22.
47. Radziejewska-Lebrecht J, Shashkov AS, Grosskurth H, Bartodziejska B, Knirel YA, Vinogradov EV, Rozalski A, Kaca W, Kononov LO, Chernyak AY, Mayer H, Kochetkov NK. Structure and epitope characteristic of O-specific polysaccharide of *Proteus mirabilis* O28 containing amides of D-galacturonic acid with L-serine and L-lysine. Eur J Biochem 1995; 230:705–712.
48. Cedzynski M, Knirel YA, Rozalski A, Shashkov AS, Vinogradov EV, Kaca W. The structure and serological specificity of *Proteus mirabilis* O43 O-antigen. Eur J Biochem 1995; 232:558–562.
49. Uhrin D, Chandan V, Altman E. Structural characterization of the O-chain polysaccharide from *Proteus mirabilis* strain 7570. Can J Chem 1995; 73:1600–1604.
50. Bartodziejska B, Radziejewska-Lebrecht J, Lipinska M, Knirel YA, Kononov LO, Chernyak AY, Mayer H, Rozalski A. Structural and immunochemical studies on the lipopolysaccharide of the 'T-antigen'-containing mutant *Proteus mirabilis* R14/1959. FEMS Immunol Med Microbiol 1995; 13:113–121.
51. Sidorczyk Z, Swierzko A, Knirel YA, Vinogradov EV, Chernyak AY, Kononov LO, Cedzynski M, Rozalski A, Kaca W, Shashkov AS, Kochetkov NK. Structure and epitope specificity of the O-specific polysaccharide of *Proteus penneri* strain 12 (ATCC 33519) containing the amide of D-galacturonic acid with L-threonine. Eur J Biochem 1995; 230:713–721.
52. Senchenkova SN, Shashkov AS, Knirel YA, Kochetkov NK, Zych K, Sidorczyk Z. Structure of a new N-acetylisomuramic acid-containing O-specific polysaccharide of *Proteus penneri* strains 19 and 35. Carbohydr Res 1996; 293:71–78.
53. Senchenkova SN, Shashkov AS, Toukach FV, Ziolkowski A, Swierzko A, Amano K-I Kaca W, Knirel YA, Kochetkov NK. Structure of the acid-labile galactosyl phosphate-containing O-antigen of *Proteus vulgaris* OX19 (serogroup O1) used in Weil-Felix test. Biochemistry (Moscow) 1997; 62:461–468.
54. Knirel YA, Paramonov NA, Vinogradov EV, Shashkov AS, Kochetkov NK, Kaca W, Cedzynski M, Ziolkowski A, Rozalski A. Structure of the O-specific polysaccharide of *Proteus vulgaris* O25 containing 3-O-[(R)-1-carboxyethyl]-D-glucose. Eur J Biochem 1997; 247:951–954.
55. Ansaruzzaman M, Albert MJ, Holme T, Jansson P, Rahman MM, Widmalm G. A *Klebsiella pneumoniae* strain that shares a type-specific antigen with *Shigella flexneri* serotype 6. Characterization of the strain and structural studies of the O-antigen polysaccharide. Eur J Biochem 1996; 237:786–791.
56. Kocharova NA, Shcherbakova OV, Shashkov AS, Knirel YA, Kochetkov NK, Kholodkova EV, Stanislavsky ES. Structure of the O-specific polysaccharide of the bacterium *Providencia alcalifaciens* serogroup O23 containing a novel amide of D-glucuronic acid with N^ε-(1-carboxyethyl)lysine. Biochemistry (Moscow) 1997; 62:501–508.
57. Vinogradov EV, Paramonov NA, Knirel YA, Shashkov AS, Kochetkov NK. Structural study of a new sialic acid-containing O-specific polysaccharide of *Salmonella arizonae* O21—formation of anhydro derivatives of neuraminic acid upon treatment with anhydrous hydrogen fluoride. Carbohydr Res 1993; 242:C11–C14.
58. Vinogradov EV, Knirel YA, Shashkov AS, Paramonov NA, Kochetkov NK, Stanislavsky ES, Kholodkova EV. The structure of the O-specific polysaccharide of *Salmonella arizonae* O21 (Arizona 22) containing N-acetylneuraminic acid. Carbohydr Res 1994; 259:59–65.
59. Edebrink P, Jansson PE, Bøgwald J, Hoffman J. Structural studies of the *Vibrio salmonicida* lipopolysaccharide. Carbohydr Res 1996; 287:225–245.
60. Vinogradov EV, Knirel YA, Kochetkov NK, Schlecht S, Mayer H. The structure of the O-specific polysaccharide of *Salmonella arizonae* O62. Carbohydr Res 1994; 253:101–110.
61. Senchenkova SN, Shashkov AS, Knirel YA, Schwarzmüller E, Mayer H. Structure of the O-specific polysaccharide of *Salmonella enterica* spp *arizonae* O50 (Arizona O9a,9b). Carbohydr Res 1997; 301:61–67.
62. Keenleyside WJ, Perry M, MacLean L, Poppe C, Whitfield C. A plasmid-encoded rfbO:54 gene cluster is required for biosynthesis of the O:54 antigen in *Salmonella enterica* serovar Borreze. Mol Microbiol 1994; 11:437–448.
63. Rahman MM, Guard-Petter J, Carlson RW. A virulent isolate of *Salmonella enteritidis* produces a *Salmonella typhi*-like lipopolysaccharide. J Bact 1997; 179:2126–2131.
64. Aucken HM, Merkouroglou M, Miller AW, Galbraith L, Wilkinson SG. Structural and serological studies of lipopolysaccharides from proposed new serotypes (O25 and O26) of *Serratia marcescens*. FEMS Microbiol Lett 1995; 130:267–272.
65. Holst O, Aucken HM, Seltmann G. Structural and serological characterisation of the O-specific polysaccharide of the lipopolysaccharide from proposed new serotype O29 of *Serratia marcescens*. J Endotoxin Res 1997; 4:215–220.
66. Albert MJ, Holme T, Lindberg B, Lindberg J, Mosihuzzaman M, Quadri F, Rahman MM. Structural studies of the *Shigella boydii* type 5 O-antigen polysaccharide. Carbohydr Res 1994; 265:121–127.
67. L'vov VL, Shashkov AS, Knirel YA, Arifulina AE, Senchenkova SN, Yakovlev AV, Dmitriev BA. Structure of the O-specific polysaccharide chain of *Shigella boydii* type 5 lipopolysaccharide—a repeated study. Carbohydr Res 1995; 279:183–192.

68. Gorshkova RP, Isakov VV, Zubkov VA, Ovodov YS. Structure of O-specific polysaccharide of lipopolysaccharide from *Yersinia bercovieri* 0:10. Bioorg Khim 1994; 20:1231–1235.
69. Gorshkova RP, Isakov VV, Kalmykova EN, Ovodov YS. Structural studies of O-specific polysaccharide chains of the lipopolysaccharide from *Yersinia enterocolitica* serovar 0:10. Carbohydr Res 1995; 268:249–255.
70. Beynon LM, Richards JC, Perry MB. The structure of the lipopolysaccharide O-antigen from *Yersinia ruckeri* serotype O1. Carbohydr Res 1994; 256:303–317.
71. Sadovskaya I, Brisson JR, Altman E, Mutharia LM. Structural studies of the lipopolysaccharide O-antigen and capsular polysaccharide of *Vibrio anguillarum* serotype 0:2. Carbohydr Res 1996; 283:111–127.
72. Kondo S, Sano Y, Isshiki Y, Hisatsune K. The O polysaccharide chain of the lipopolysaccharide from *Vibrio cholerae* O76 is a homopolymer of N-[(S)-(+)-2-hydroxypropionyl]-α-L-perosamine. Microbiology 1996; 142:2879–2885.
73. Sano Y, Kondo S, Isshiki Y, Shimada T, Hisatsune K. An N-[(R)-(-)-2-hydroxypropionyl]-α-L-perosamine homopolymer constitutes the O polysaccharide chain of the lipopolysaccharide from Vibrio cholerae O144 which has antigenic factor(s) in common with *V. cholerae* O76. Microbiol Immunol 1996; 40:735–741.
74. Knirel YA, Senchenkova SN, Jansson PE, Weintraub A, Albert MJ. Structure of the O-specific polysaccharide of *Aeromonas trota* strain cross-reactive with *Vibrio cholerae* Bengal. Eur J Biochem 1996; 238:160–165.
75. Knirel YA, Widmalm G, Senchenkova SN, Jansson PE, Weintraub A. Structural studies on the short-chain lipopolysaccharide of *Vibrio cholerae* O139 Bengal. Eur J Biochem 1997; 247:402–410.
76. Cox AD, Brisson JR, Thibault P, Perry MB. Structural analysis of the lipopolysaccharide from *Vibrio cholerae* serotype O22. Carbohydr Res 1997; 304:191–208.
77. Linnerborg M, Widmalm G, Weintraub A, Albert MJ. Structural elucidation of the O-antigen lipopolysaccharide from two strains of *Plesiomonas shigelloides* that share a type-specific antigen with *Shigella flexneri* 6, and the common group 1 antigen with *Shigella flexneri* spp and *Shigella dysenteriae*. Eur J Biochem 1995; 231:839–844.
78. Aspinall GO, McDonald AG, Pang H. Lipopolysaccharides of *Campylobacter jejuni* serotype O19: structures of O antigen chains from the serostrain and two bacterial isolates from patients with the Guillain-Barré syndrome. Biochemistry 1994; 33:250–255.
79. Senchenkova SN, Shashkov AS, Knirel YA, McGovern JJ, Moran AP. The O-specific polysaccharide chain of *Campylobacter fetus* serotype A lipopolysaccharide is a partially O-acetylated 1,3-linked X-D-mannan. Eur J Biochem 1997; 245:637–641.
80. Senchenkova SN, Knirel YA, Shashkov AS, McGovern JJ, Moran AP. The O-specific polysaccharide chain of *Campylobacter fetus* serotype B lipopolysaccharide is a linear D-rhamnan terminated with 3-O-methyl-D-rhamnose (D-acofriose). Eur J Biochem 1996; 239:434–438.
81. Aspinall O, Monteiro MA. Lipopolysaccharides of *Helicobacter pylori* strains P466 and MO19—structures of the O antigen and core oligosaccharide regions. Biochemistry 1996; 35:2498–2504.
82. Aspinall GO, Monteiro MA, Shaver R, Kurjanczyk L, Penner J. Lipopolysaccharides of *Helicobacter pylori* serogroups O:3 and O:6: Structures of a new class of lipopolysaccharides with reference to the location of oligomeric units of D-glycero-α-D-manno-heptose residues. Eur J Biochem 1997; 248:592–601.
83. Adinolfi M, Corsaro MM, De Castro C, Evidente A, Lanzetta R, Lavermicocca P, Parrilli M. Analysis of the polysaccharide components of the lipopolysaccharide fraction of *Pseudomonas caryophylli*. Carbohydr Res 1996; 284:119–133.
84. Adinolfi M, Barone G, Iadonisi A, Mangoni L, Manna R. Synthesis of caryose, the carbocyclic monosaccharide component of the lipopolysaccharide from *Pseudomonas caryophylli*. Tetrahedron 1997; 53:11767–11780.
85. Knirel YA, Zdorovenko GM, Paramonov NA, Veremeychenko SN, Toukach FV, Shashkov AS. Somatic antigens of pseudomonads: structure of the O-specific polysaccharide of the reference strain for *Pseudomonas fluorescens* (IMV 4125, ATCC 13525, biovar A). Carbohydr Res 1996; 291:217–224.
86. Das S, Ramm M, Kochanowski H, Basu S. Structural studies of the O-side chain of outer membrane lipopolysaccharide from *Pseudomonas syringae* pv. *coriandricola* W-43. J Bacteriol 1994; 176:6550–6557.
87. Adinolfi M, Corsaro MM, Lanzetta R, Marciano CE, Parrilli M, Evidente A, Surico G. Structure of the O-chain polysaccharide of three strains of *Pseudomonas syringae* ssp. *savastanoi*. Can J Chem 1994; 72:1839–1843.
88. Paramonov NA, Shashkov AS, Knirel YA, Soldatkina MA, Zakharova IY. Antigenic polysaccharides of bacteria. 39. Structure of the O-specific polysaccharides of *Pseudomonas cepacia* serogroups C, I, O1, and O4. Bioorg Khim 1994; 20:984–993.
89. Beynon LM, Cox AD, Taylor CJ, Wilkinson SG, Perry MB. Characterization of a lipopolysaccharide containing two different trisaccharide repeating units from *Burkholderia cepacia* serotype E (O2). Carbohydr Res 1995; 272:231–239.
90. Cox AD, Taylor CJ, Anderson AJ, Perry MB, Wilkinson SG. Structures of the two polymers present in the lipopolysaccharide of *Burkholderia (Pseudomonas) cepacia* serogroup O4. Eur J Biochem 1995; 231:784–789.
91. Galbraith L, George R, Wyklicky J, Wilkinson SG. Structure of the O-specific polysaccharide from *Burkholderia pickettii* strain NCTC 11149. Carbohydr Res 1996; 282:263–269.
92. Zähringer U, Rettenmaier H, Moll H, Senchenkova SN, Knirel YA. Structure of a new 6-deoxy-α-L-talan from *Burkholderia (Pseudomonas) plantarii* strain DSM 6535 which is different from the O-chain of the lipopolysaccharide. Carbohydr Res 1997; 300:143–151.
93. Varbanets LD, Kocharova NA, Knirel YA, Moskalenko NV. Structure of the O-specific polysaccharide

chains of the lipopolysaccharides of *Burkholderia (Pseudomonas) solanacearum* ICMP 750, 8093, 8115, 7864, and 7945. Biochemistry (Moscow) 1996; 61: 807–814.

94. Kocharova NA, Shashkov AS, Knirel YA, Varbanets LD, Moskalenko NV. The structure of the O-specific polysaccharide of *Pseudomonas solanacearum* strain 8089. Carbohydr Res 1994; 259:153–157.

95. Winn AM, Wilkinson SG. Structure of the O2 antigen of *Stenotrophomonas (Xanthomonas or Pseudomonas) maltophilia*. Carbohydr Res 1997; 298:213–217.

96. Winn AM, Miles CT, Wilkinson SG. Structure of the O3 antigen of *Stenotrophomonas (Xanthomonas or Pseudomonas) maltophilia*. Carbohydr Res 1996; 282: 149–156.

97. Winn AM, Miller AM, Wilkinson SG. Structure of the O10 antigen of *Stenotrophomonas (Xanthomonas) maltophilia*. Carbohydr Res 1995; 267:127–133.

98. Winn AM, Wilkinson SG. Structure of the O20 antigen of *Stenotrophomonas (Xanthomonas or Pseudomonas) maltophilia*. Carbohydr Res 1996; 294:109–115.

99. Grimmecke HD, Mamat U, Kuhn I, Shashkov AS, Knirel YA. Structure of the capsular polysaccharide and the O-Sidechain of the lipopolysaccharide from *Acetobacter methanolicus* MB 58 (IMET B346). Carbohydr Res 1994; 252:309–316.

100. Grimmecke HD, Knirel YA, Shashkov AS, Kiesel B, Lauk W, Voges M. Structure of the capsular polysaccharide and the O-side-chain of the lipopolysaccharide from *Acetobacter methanolicus* MB 70, and of oligosaccharides resulting from their degradation by the bacteriophage Acm6. Carbohydr Res 1994; 253:277–282.

101. Grimmecke HD, Voges M, Knirel YA, Shashkov AS, Lauk W, Kiesel B. Structure of the capsular polysaccharide and the O-side-chain of the lipopolysaccharide from *Acetobacter methanolicus* MB 135 (IMET 11402). Carbohydr Res 1994; 253:283–286.

102. Perry MB, Maclean LM, Brisson JR, Wilson ME. Structures of the antigenic O-polysaccharides of lipopolysaccharides produced by *Actinobacillus actinomycetemcomitans* serotypes a, c, d and e. Eur J Biochem 1996; 242:682–688.

103. Perry MB, Maclean LL, Gmur R, Wilson ME. Characterization of the O-polysaccharide structure of lipopolysaccharide from *Actinobacillus actinomycetemcomitans* serotype B. Infect Immun 1996; 64:1215–1219.

104. Vinogradov EV, Brade H, Holst O. The structure of the O-specific polysaccharide of the lipopolysaccharide from *Chromobacterium violaceum* NCTC 9694. Carbohydr Res 1994; 264:313–317.

105. Knirel YA, Rietschel ET, Marre R, Zähringer U. The structure of the O-specific chain of *Legionella pneumophila* serogroup 1 lipopolysaccharide. Eur J Biochem 1994; 221:239–245.

106. Russa R, Urbaniksypniewska T, Shashkov A, Banaszek A, Zamojski AHM. Partial structure of lipopolysaccharides isolated from *Rhizobium leguminosarum* by trifolii 24 and its GalA-negative exo⁻ mutant AR20. System Appl Microbiol 1996; 19:1–8.

107. Gilserrano AM, González-Jiménez I, Mateo PT, Bernabé M, Jimenez-Barbero J, Megias M, Romero-Vázquez MJ. Structural analysis of the O-antigen of the lipopolysaccharide of *Rhizobium tropici* CIAT899. Carbohydr Res 1995; 275:285–294.

108. Dmitriev BA, Knirel YA, Kocharova NA, Kochetkov NK, Stanislavsky ES, Mashilova GM. Somatic antigens of *Pseudomonas aeruginosa*. The structure of the polysaccharide chain of *Ps.aeruginosa* O-serogroup 7 (Lanyi) lipopolysaccharide. Eur J Biochem 1980; 106: 643–651.

109. Raff RA, Wheat RW. Carbohydrate composition of the phenol-soluble lipopolysaccharide of *Citrobacter freundii*. J Bacteriol 1968; 95:2035–2043.

110. Masoud H, Perry MB. Structural characterization of the O-antigenic polysaccharide of *Escherichia coli* serotype O17 lipopolysaccharide. Biochem Cell Biol 1996; 74:241–248.

111. Bartelt M, Shashkov AS, Kochanowski H, Jann B, Jann K. Structure of the O-specific polysaccharide of the O22-antigen (LPS) from *Escherichia coli* O22: K13. Carbohydr Res 1994; 254:203–212.

112. Jann B, Shashkov AS, Hahne M, Kochanowski H, Jann K. Structure of the O83-specific polysaccharide of *Escherichia coli* O83:K24:H31. Carbohydr Res 1994; 261:215–222.

113. Marsden BJ, Bundle DR, Perry MB. Serological and structural relationships between *Escherichia coli* O:98 and *Yersinia enterocolitica* O:11,23 and O:11,24 lipopolysaccharide O-antigens. Biochem Cell Biol 1994; 72:163–168.

114. Parolis H, Parolis LAS. The structure of the O-specific polysaccharide from *Escherichia coli* O113 lipopolysaccharide. Carbohydr Res 1995; 267:263–269.

115. Sengupta P, Bhattacharyya T, Shashkov AS, Kochanowski H, Basu S. Structure of the O-specific side chain of the *Escherichia coli* O128 lipopolysaccharide. Carbohydr Res 1995; 277:283–290.

116. Linnerborg M, Weintraub A, Widmalm G. Structural studies of the O-antigen polysaccharide from *Escherichia coli* O138. Eur J Biochem 1997; 247:567–571.

117. Kocharova NA, Bystrova OV, Borisova SA, Shashkov AS, Knirel YA, Kholodkova EV, Stanislavsky ES. Structures of two O-specific polysaccharides of *Citrobacter* O29. Carbohydr Lett 1997; 2:287–292.

118. Petersson C, Jachymek W, Klonowska A, Lugowski C, Niedziela T, Kenne L. Structural studies of the O-specific chains of *Hafnia alvei* strains 744, PCM 1194 and PCM 1210 lipopolysaccharides. Eur J Biochem 1997; 245:668–675.

119. Katzenellenbogen E, Kubler J, Gamian A, Romanowska E, Shashkov AS, Kocharova NA, Knirel YA, Kochetkov NK. Structural study and serological characterisation of the O-specific polysaccharide of *Hafnia alvei* PCM 1185, another *Hafnia* O-antigen that contains 3,6-dideoxy-3-[(R)-3-hydroxybutyramido]D-glucose. Carbohydr Res 1996; 293:61–70.

120. Gamian A, Katzenellenbogen E, Romanowska E, Fernandez JMG, Pedersen C, Ulrich J, Defaye J. Structure of the *Hafnia alvei* strain PCM 1188 O-specific polysaccharide. Carbohydr Res 1995; 277:245–255.

121. Katzenellenbogen E, Romanowska E, Kocharova NA, Shashkov AS, Knirel YA, Kochetkov NK. Structure of

the O-specific polysaccharide of *Hafnia alvei* 1204 containing 3,6-dideoxy-3-formamido-D-glucose. Carbohydr Res 1995; 273:187–195.
122. Toukach FV, Shashkov AS, Katzenellenbogen E, Kocharova NA, Czarny A, Knirel YA, Romanowska E, Kochetkov NK. Structure of the O-specific polysaccharide of *Hafnia alvei* strain 1222 containing 2-aminoethyl phosphate. Carbohydr Res 1996; 295:117–126.
123. Ziolkowski A, Senchenkova SN, Swierzko A, Shashkov AS, Toukach FV, Cedzynski M, Amano KI, Kaca W, Knirel YA, Kochetkov NK. Structures of the O-antigens of *Proteus* group OX strains used in Weil-Felix test. FEBS Lett 1997; 411:221–224.
124. Swierzko A, Cedzynski M, Knirel YA, Senchenkova SN, Kocharova NA, Shashkov AS, Amano KI, Kyohno K, Kaca W. Structural and serological studies of the O-antigen of *Proteus* OXK (*Proteus mirabilis* O3) used in Weil-Felix text. Biochemistry (Moscow) 1997; 62:21–27.
125. Shashkov AS, Toukach FV, Senchenkova SN, Ziolkowski A, Paramonov NA, Kaca W, Knirel YA, Kochetkov NK. Structure of the O-specific polysaccharide of the bacterium *Proteus mirabilis* O13 containing a novel amide of D-galacturonic acid with N$^\varepsilon$-(1-carboxyethyl)lysine. Biochemistry (Moscow) 1997; 62: 509–513.
126. Shashkov AS, Toukach FV, Ziolkowski A, Paramonov NA, Senchenkova SN, Kaca W, Knirel YA. Structure of the O-specific polysaccharide of the bacterium *Proteus mirabilis* O30. Biochemistry (Moscow) 1996; 61: 800–806.
127. Uhrin D, Brisson JR, MacLean LL, Richards JC, Perry MB. Application of 1D and 2D NMR techniques to the structure elucidation of the O-polysaccharide from *Proteus mirabilis* O:57. J Biomolec NMR 1994; 4: 615–630.
128. Sidorczyk Z, Swierzko A, Vinogradov EV, Knirel YA, Shashkov AS. Structural and immunochemical studies on the O-specific polysaccharide of *Proteus penneri* strain 14. Arch Immun Ther Exp 1994; 42:209–215.
129. Arbatsky NP, Shashkov AS, Widmalm G, Knirel YA, Zych K, Sidorczyk Z. Structure of the O-specific polysaccharide of *Proteus penneri* strain 25 containing N-(L-alanyl) and multiple O-acetyl groups in a tetrasaccharide repeating unit. Carbohydr Res 1997; 298: 229–235.
130. Knirel YA, Shashkov AS, Vinogradov EV, Kochetkov NK, Swierzko A, Sidorczyk Z. The structure of the O-specific polysaccharide chain of *Proteus penneri* strain 42 lipopolysaccharide. Carbohydr Res 1995; 275:201–206.
131. Sidorczyk Z, Zych K, Swierzko A, Vinogradov EV, Knirel YA. The structure of the O-specific polysaccharide of *Proteus penneri* 52. Eur J Biochem 1996; 240;245–251.
132. Cedzynski M, Knirel YA, Amano KI, Swierzko A, Paramonov NA, Senchenkova SN, Kaca W. Structural and serological studies of the O-antigen of *Proteus* OX2 (*Proteus vulgaris* O2) used in Weil-Felix test. Biochemistry (Moscow) 1997; 62:15–20.
133. Aucken HM, Wilkinson SG, Pitt TL. Immunochemical characterisation of two new serotypes of *Serratia marcescens* (O27 and O28). FEMS Microbiol Lett 1996; 138:77–82.
134. Zhang LJ, Radziejewska-Lebrecht J, Krajewska-Pietrasik D, Toivanen P, Skurnik M. Molecular and chemical characterization of the lipopolysaccharide O-antigen and its role in the virulence of *Yersinia enterocolitica* serotype O:8. Mol Microbiol 1997; 23: 63–76.
135. Kjellberg A, Weintraub A, Albert MJ, Widmalm G. Structural analysis of the O-antigenic polysaccharide from *Vibrio cholerae* O10. Eur J Biochem 1997; 249: 758–761.
136. Linnerborg M, Widmalm G, Rahman MM, Jansson PE, Holme T, Qadri F, Albert MJ. Structural studies of the O-antigenic polysaccharide from an *Aeromonas caviae* strain. Carbohydr Res 1996; 291:165–174.
137. Aspinall GO, Monteiro MA, Pang H, Walsh EJ, Moran AP. Lipopolysaccharide of the *Helicobacter pylori* type strain NCTC 11637 (ATCC 43504)—structure of the O antigen chain and core oligosaccharide regions. Biochemistry 1996; 35:2489–2497.
138. Knirel YA, Veremeychenko SN, Zdorovenko GM, Shashkov AS, Paramonov NA, Zakharova IY, Kochetkov NK. Somatic antigens of pseudomonads: Structure of the O-specific polysaccharide of *Pseudomonas fluorescens* biovar A strain IMV 472. Carbohydr Res 1994; 259:147–151.
139. Knirel YA, Grosskurth H, Helbig JH, Zähringer U. Structures of decasaccharide and tridecasaccharide tetraphosphates isolated by strong alkaline degradation of O-deacylated lipopolysaccharide of *Pseudomonas fluorescens*. Carbohydr Res 1995; 279:215–226.
140. Beynon LM, Perry MB. Structure of the lipopolysaccharide O-antigen of *Pseudomonas cepacia* serotype A. Biochem Cell Biol 1993; 71:417–420.
141. Taylor CJ, Anderson AJ, Wilkinson SG. Structure of the O9 antigen from *Burkholderia (Pseudomonas) cepacia*. FEMS Microbiol Lett 1994; 115:201–204.
142. Galbraith L, Wilkinson SG. Structural studies of the O-specific side chain of lipopolysaccharide from *Burkholderia gladioli* pv. *gladioli* NNCPPB 1891. Carbohydr Res 1997; 303:245–249.
143. Perry MB, MacLean LL, Schollaardt T, Bryan LE, Ho M. Structural characterization of the lipopolysaccharide O antigens of *Burkholderia pseudomallei*. Infect Immun 1995; 63:3348–3352.
144. Varbanets LD, Kocharova NA, Knirel YA, Moskalenko NV. Structure of O-specific polysaccharide chains of lipopolysaccharides of *Burkholderia (Pseudomonas) Solanacearum* ICMP 750, 8093, 8115, 7864, and 7945. Biochemistry (Moscow) 1996; 61: 580–585.
145. Bhattacharyya T, Basu S. Primary structure of the polysaccharide chain of virulent *Pseudomonas solanacearum* biotype II lipopolysaccharide. Carbohydr Res 1993; 250:335–337.
146. Gaur D, Wilkinson SG. Structure of the O-specific polysaccharide from *Burkholderia vietnamiensis* strain LMG 6998. Carbohydr Res 1996; 295:179–184.
147. Winn AM, Wilkinson SG. Structure of the O6 antigen of *Stenotrophomonas (Xanthomonas* or *Pseudomonas) maltophilia*. Carbohydr Res 1995; 272:225–230.

148. Previato JO, Jones C, Stephan MP, Almeida LPA, Mendonça-Previato L. Structure of the repeating oligosaccharide from the lipopolysaccharide of the nitrogen-fixing bacterium *Acetobacter diazotrophicus* strain Pal 5. Carbohydr Res 1997; 298:311–318.
149. Vinogradov EV, Pantophlet R, Dijkshoorn L, Brade L, Holst O, Brade H. Structural and serological characterisation of two O-specific polysaccharides of *Acinetobacter*. Eur J Biochem 1996; 239:602–610.
150. Haseley SR, Holst O, Brade H. Structural and serological characterisation of the O-antigenic polysaccharide of the lipopolysaccharide from *Acinetobacter* strain 90 belonging to DNA group 10. Eur J Biochem 1997; 245:470–476.
151. Haseley SR, Holst O, Brade H. Structural studies of the O-antigenic polysaccharide of the lipolysaccharide from *Acinetobacter* (DNA group 11) strain 94 containing 3-amino-3,6-dideoxy-D-galactose substituted by the previously unknown amidelinked L-2-acetoxypropionic acid or L-2-hydroxypropionic acid. Eur J Biochem 1997; 247:815–819.
152. Haseley SR, Galbraith L, Wilkinson SG. Structure of a surface polysaccharide from *Acinetobacter baumannii* strain 214. Carbohydr Res 1994; 258:199–206.
153. Haseley SR, Wilkinson SG. Structural studies of the putative O-specific polysaccharide of *Acinetobacter baumannii* O2 containing 3,6-dideoxy-3-N-(D-3-hydroxybutyryl)amino-D-galactose. Eur J Biochem 1995; 233:899–906.
154. Haseley SR, Wilkinson SG. Structure of the O-specific polysaccharide of *Acinetobacter baumannii* O5 containing 2-acetamido-2-deoxy-D-galacturonic acid. Eur J Biochem 1996; 237:229–233.
155. Haseley SR, Wilkinson SG. Structure of the putative O10 antigen from *Acinetobacter baumannii*. Carbohydr Res 1994; 264:73–81.
156. Haseley SR, Wilkinson SG. Structural studies of the putative O-specific polysaccharide of *Acinetobacter baumannii* O11. Eur J Biochem 1996; 237:266–271.
157. Haseley SR, Traub WH, Wilkinson SG. Structures of polymeric products isolated from the lipopolysaccharides of reference strains for *Acinetobacter baumannii* O23 and O12. Eur J Biochem 1997; 244:147–154.
158. Haseley SR, Diggle HJ, Wilkinson SG. Structure of a surface polysaccharide from *Acinetobacter baumannii* O16. Carbohydr Res 1996; 293:259–265.
159. Haseley SR, Wilkinson SG. Structure of the O-18 antigen from *Acinetobacter baumannii*. Carbohydr Res., 1997; 301:187–192.
160. Haseley SR, Wilkinson SG. Structural studies of the putative O-specific polysaccharide of *Acinetobacter baumannii* O24 containing 5,7-diamino-3,5,7,9-tetradeoxy-L-glycero-D-galacto-nonulosonic acid. Eur J Biochem 1997; 250:617–623.
161. Vinogradov EV, Pantophlet R, Haseley SR, Brade L, Holst O, Brade H. Structural and serological characterisation of the O-specific polysaccharide from lipopolysaccharide of *Acinetobacter calcoaceticus* strain 7 (DNA group 1). Eur J Biochem 1997; 243:167–173.
162. Haseley SR, Holst O, Brade H. Structural and serological characterisation of the O-antigenic polysaccharide of the lipopolysaccharide from *Acinetobacter haemolyticus* strain ATCC 17906. Eur J Biochem 1997; 244:761–766.
163. Haseley SR, Pantophlet R, Brade L, Holst O, Brade H. Structural and serological characterisation of the O-antigenic polysaccharide of the lipopolysaccharide from *Acinetobacter junii* strain 65. Eur J Biochem 1997; 245:477–481.
164. Nohata Y, Azuma JI, Kurane R. Structural studies of a neutral polysaccharide produced by *Alcaligenes latus*. Carbohydr Res 1996; 293:213–223.
165. Velasco J, Moll H, Vinogradov EV, Moriyon I, Zähringer U. Determination of the O-specific polysaccharide structure in the lipopolysaccharide of *Ochrobactrum anthropi* LMG 3331. Carbohydr Res 1996; 287:123–126.
166. Senchenkova SN, Shashkov AS, Moran AP, Helander I, Knirel YA. Structure of the polysaccharide chains of *Pectinatus cerevisiiphilus* and *Pectinatus frisingensis* lipopolysaccharides. Eur J Biochem 1995; 232:552–557.
167. Russa R, Urbanik-Sypniewska T, Lindström K, Mayer H. Chemical characterization of two lipopolysaccharide species isolated from *Rhizobium loti* NZP2213. Arch Microbiol 1995; 163:345–351.
168. Russa R, Urbanik-Sypniewska T, Shashkov AS, Kochanowski H, Mayer H. The structure of the homopolymeric O-specific chain from the phenol soluble LPS of the *Rhizobium loti* type strain NZP2213. Carbohydr Polym 1995; 27:299–303.
169. Wang Y, Hollingsworth RI. The structure of the O-antigenic chain of the lipopolysaccharide of *Rhizobium trifolii* 4s. Carbohydr Res 1994; 260:305–317.
170. Shashkov AS, Campos-Portuguez S, Kochanowski H, Yokota A, Mayer H. The structure of the O-specific polysaccharide from *Thiobacillus* sp. IFO 14570, with three different diaminopyranoses forming the repeating unit. Carbohydr Res 1995; 269:157–166.
171. Vinogradov EV, Campos-Portuguez S, Yokota A, Mayer H. The structure of the O-specific polysaccharide from *Thiobacillus ferrooxidans* IFO 14262. Carbohydr Res 1994; 261:103–109.

10

The Chemistry and Biology of Lipooligosaccharides: The Endotoxins of Bacteria of the Respiratory and Genital Mucosae

J. McLeod Griffiss
University of California San Francisco, San Francisco, California

Herman Schneider
Walter Reed Army Institute of Research, Washington, D.C.

Lipooligosaccharides (LOS) are made by gram-negative bacteria that colonize mucosal surfaces other than those of the gut (1). The structures of these surface-exposed outer membrane organelles more closely resemble human cell membrane glycosphingolipids (GSL), with which they share glycose moieties, than the analogous lipopolysaccharides (LPS) of enteric gram-negative bacteria (1–3).

LOS glycose synthesis is flexible. Mucosal bacteria vary LOS glycose moieties so as to mimic different GSL (3). LOS phase variation is quite rapid (5×10^{-2}) (4,5); this enables the organisms to evade immune recognition (6–8) and survive in different molecular environments, e.g., the mucosae of venereal consorts of different blood types.

The shortest LOS glycose structure that is surface-expressed terminates in lactose (Lac; Galβ1→4Glc) and mimics lacto series GSL (3,9). The terminal galactose (Gal) of this lactosyl moiety is the first biosynthetic "toggle switch" or decision point, as the glycose linked to it determines the structure of the mature LOS. If *N*-acetylglucosamine (GlcNAc) is added β1→3 to the lactosyl Gal, then a second Gal residue is added β1→4 to the GlcNAc residue to form the *N*-acetyllactosamine (LacNAc) terminal disaccharide of lacto-*N*-neotetraose (LNnT; LacNAcβ1→3Lac) (2,10–12), the glycose moiety of paraglobosideseries GSL (13). Alternatively, a second Gal residue may be linked α1→4 to the lactosyl Gal to form the P^k globo series structure (11).

The terminal lactosaminyl Gal of LNnT is a second biosynthetic decision point. It may remain unsubstituted (paraglobosyl) (2) or be substituted β1→3 with *N*-acetylgalactosamine (GalNAc) to form the asialo-G_3 gangliosyl glycose moiety (12,14), or α1→4 with Gal to form the P_1 globo series structure (3).

The unsubstituted terminal Gal of LNnT provides a binding site for C3, the third component of the complement system, and facilitates lysis of organisms through both complement pathways (immune lysis) (15). Unsubstituted terminal Gal residues also facilitate invasion of epithelial cells (16), probably by providing ligands for human lectins (17).

Unsubstituted globosyl and paraglobosyl LOS may be sialylated (18,19), which stops further chain extension. Sialylation of LNnT structures prevents immune lysis (15,20) and retards killing by polymorphonuclear leukocytes (PMNs) (21–23).

Neisseria meningitidis sialylate (19,24), but do not galactosaminylate (25), LNnT structures. *Neisseria gonorrhoeae* and *Neisseria lactamica* cannot make sialic acid (26), but all three *Neisseria* and many *Haemophilus* species can transfer sialic acid during infection from exogenous sources to their paraglobosyl LOS (18,19,27–29).

Galactosaminylation of LNnT structures, which also stops further chain extension, creates an antigenic determinant that binds bactericidal IgM molecules that are ubiquitous in human sera (30,31). These antibodies initiate immune lysis of organisms that enter the blood stream and thereby restrict colonization to the mucosa. Gonococcal strains that cause local disease galactosaminylate their LNnT LOS structures; those that cause disseminated disease overexpress lactosyl LOS molecules that can be neither sialylated nor galactosaminylated (6,32,33).

INTRODUCTION

Gram-negative bacteria invariably have glycosylated lipid components in the outermost leaflet of their cell envelopes that enable them to interact with aqueous environments (34). Enteric bacteria that must cope with environments rich in bile acids have outer membrane glycolipids that bear long glycose moieties created by the sequential addition of an oligosaccharide (O-antigen), which is reduplicated to various degrees on each glycolipid molecule. The resulting long, relatively hydrophilic and neutral polysaccharide prevents dispersion of the bacteria's lipid membrane by bile acids. Such glycolipids are called lipopolysaccharides (LPS).

Gram-negative bacteria that colonize respiratory and genital mucosal surfaces are not exposed to bile acids. Their membrane glycolipids serve different functions than LPS, and their structures reflect this (1). The different biologies of the two sorts of glycolipids led us to propose, in 1984, that those of mucosal bacteria be called lipooligosaccharides (LOS), rather than lipopolysaccharides (35). Use of this chemically correct term serves to remind us of the unique roles that these molecules play in the lives of the bacteria that make them (36).

LOS have lipoidal moieties that are structurally analogous to the lipids A of enteric LPS and that, like them, anchor the glycolipid in the bacterial membrane and are responsible for the endotoxic properties of the molecule (37). It is in their surface-exposed carbohydrate structures that LOS differ in such important ways from LPS.

The glycose moieties of LOS usually contain no more than 10 sugar residues arranged as three antennae that arise from two or three basal heptose (Hep) residues (1). These antennae are composed, for the most part, of just four sugars: glucose (Glc) and galactose (Gal) and their respective N-acetylated amino derivatives, GlcNAc and GalNAc. The length of each antenna varies among species and strains and with growth conditions. Although LOS antennae may be elongated by $\beta1\rightarrow3$ duplication of lactosamine (Gal$\beta1\rightarrow$4GlcNAc; LacNAc) residues, LOS do not have O-antigen repeats in the same meaning as those of LPS (38).

LOS glycose synthesis is highly flexible, and despite their relatively small size, LOS exhibit considerable structural and antigenic diversity (30,35,39–41). Individual bacteria usually can make several different LOS molecules—each with a different antigenic structure (4,42)—and LOS phase variation is quite rapid (5 × 10^{-2}) (4,5). The ability to make different LOS structures—and to surface-express more than one simultaneously—is thought to assist in the transmission of these organisms from person to person.

As they pass between hosts, mucosal bacteria encounter very different molecular environments—e.g., those of the male urethra and the female cervix—and find themselves on epithelial cells that have different membrane glycoconjugates and are covered by mucus with oligosaccharide structures that are changing continuously, in the case of the cervix, because of hormonal influences (43). The effects of the LOS triantennary structural motif, the expression of multiple molecular species of LOS by a single bacterium, rapid LOS phase variation, and the faithful mimicry of different human glycoconjugate glycose structures by the different LOS antennae—even including the sialylation of some—ensure that at least some bacteria of the infecting clone will have on their surface during passage spatially regulated clusters of the carbohydrate structures they will need to evade immune recognition by the new host (6–8), direct their internalization by epithelial cells (5,17,44,45), survive within phagocytic cells (21–23), and promote or defeat complement-mediated killing (15,18,20,30) as needed for the survival of both bacteria and host. If enough bacteria mimic enough of the new host's glycoconjugates, the clone as a whole will survive the passage and can be passed to yet another, and different, host. It is not surprising, then, that LOS synthesis is subject to such rapid phase variation (4,5).

The past decade has seen an explosion in our understanding of the structure and function of bacterial LOS, particularly those of the genus *Neisseria*. This has come from an appreciation of their physical heterogeneity, determination of their structural chemistry, the development of monoclonal antibodies (mAb) that can serve as surrogates for chemical structures, and the development of model systems with which to study pathogenesis. Most recently, the identification of LOS biosynthetic genes has led to a partial understanding of how LOS molecular heterogeneity is regulated (25,46,47).

In this chapter we will summarize our current understanding of this important bacterial organelle. We will concentrate on the LOS of *Neisseria gonorrhoeae*, as they have served as our model and we understand them best, but we will also point out similarities and differences between gonococcal LOS and those made by *Neisseria meningitidis* and between the LOS of *Neisseria* and those of *Haemophilus* species.

LOS STRUCTURES AND THEIR GLYCOSPHINGOLIPID EQUIVALENTS

LOS are triantennary glycolipids of bacterial membranes that structurally resemble the glycosphingolipids (GSL) of human membranes (1,2,18). They range in mass from 3229, the M_r of the smallest trans-membrane species, to slightly more than 5000. They have a highly conserved lipoidal moiety that is integrated into the outermost leaflet of the bacterial outer membrane; a proximal basal region that is polar and highly variable (Table 1), and three short, somewhat conserved, distal oligosaccharide chains, termed α, β, and γ (1), following the systematic nomenclature of Domon and Costello (48).

Lipoidal Moiety

LOS have diphosphoryl lipoidal moieties that are structurally similar to those of enteric LPS except for a symmetric acylation pattern and *O*-acylation with C_{12}, rather than C_{14}, fatty acids (49,50). Deacylation of different LOS yields a homogeneous lipid of M_r 953 that consists of a 1,4′-bisphosphorylated β1′→6 D-glucosamine (GlcN) disaccharide backbone and two amide-linked 3-OH-C_{14} fatty acids (51). Heterogeneity among lipoidal moieties is accounted for by different numbers of ester-linked 3-OH-C_{12} and C_{12} fatty acids (32), but a hexaacyl lipoidal moiety that has two amide-linked 3-OH-C_{14} fatty acids, two ester-linked 3-OH-C_{12} fatty acids, and two C_{12} fatty acids acyloxyacyl-linked to the C_{14} fatty acids is highly conserved (49,50). The M_r of this lipoidal moiety is 1713.

Kulshin et al. found that both lipoidal phosphate groups on a meningococcal LOS were substituted by *O*-phosphorylethanolamine (PEA) residues (50), but this has not been found for other neisserial LOS, either gonococcal or meningococcal. The failure to find lipoidal PEA substitutions may reflect degradation of pyrophosphoryl linkages during *O*-deacylation prior to mass spectrometric analyses (49–51).

Basal Region

LOS share a trisaccharidic core that is composed of two L-glycero-D-manno-heptose residues and a 3-deoxy-D-manno-oct-2-ulosonic acid (Kdo) residue, linked Hep-2α1→3Hep-1α1→5Kdo (Fig. 1) (1,10–12,51–54). The Kdo residue is linked α2→6′ to one of the GlcN residues of the lipoidal moiety backbone.

The highly conserved core structure is modified by the variable addition of polar residues (51). PEA residues may be added to heptose residues (51,55), and C4 of the Kdo residue is substituted with additional acidic functions—a second Kdo residue by *Neisseria* (51,55) and a phosphate group by *Haemophilus* (56–58).

PEA residues provide primary amino functions that are free to form hydrogen bonds with Kdo carboxylate

Table 1 LOS Basal Region Substitutions

Polar adornments	*Neisseria*	(PEA)$_{0-1}$	Kdo
	Haemophilus		PO$_3$H$_2$
		↓	↓
		?	4
α Chain	*Haemophilus* and *Neisseria*	Glcβ1→4Hep-1α1→5Kdo	
		3	
		↑	
		1	
	N. meningitidis	α 6←	
β Chain	*Neisseria* and *Haemophilus*	Glcα1→3Hep-23←(PEA)$_{0-1}$	
	N. meningitidis	GlcNAcβ 2 7←	
		↑	
γ Chain	*Neisseria*	GlcNAc(*O*Ac)α1	
	Haemophilus	Hepα1?←(PEA)$_{0-1}$	

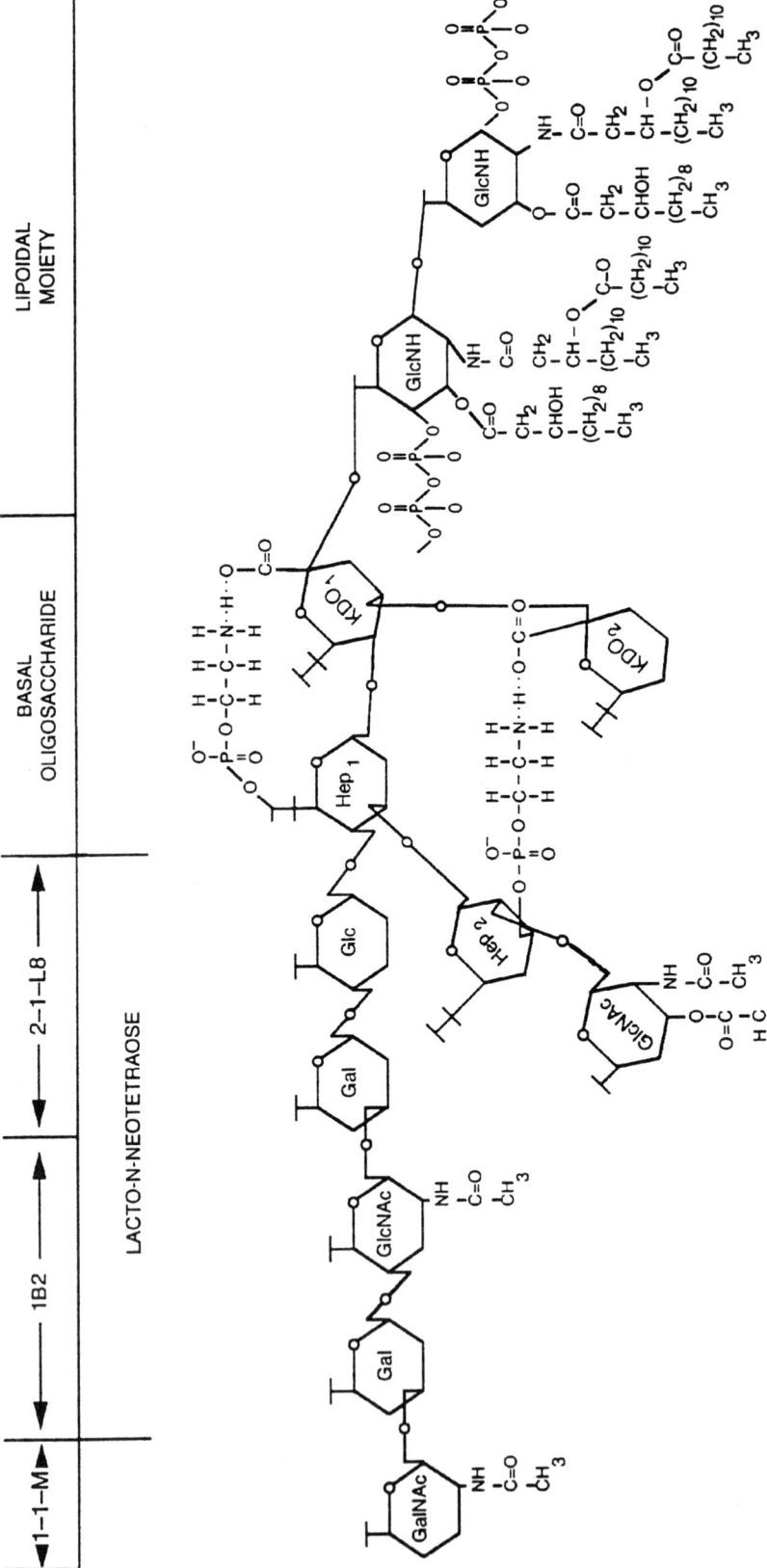

Fig. 1 Schematic of the highest M_r (gangliosyl) LOS made by *N. gonorrhoeae* strain F62$_{hs}$ (12,51). The hexaacyl, diphosphoryl lipoidal moiety is shown with pyrophosphates, as phosphoethanolaminylation of neisserial LOS diphosphoryl lipoidal moieties has been reported (50); the conservation of this function remains in question. Basal region oligosaccharide phosphoethanolamine substituents are shown forming electrostatic bonds with Kdo carboxyl groups. Although *N. gonorrhoeae* strain F62$_{hs}$ LOS have two basal region PEA substituents (51), the location of the second, shown here on C7 of Hep-1, has not been determined. The binding specificities of three mAb that have been extensively used in LOS studies are shown above the gangliosyl α chain of this M_r 4040 LOS. MAb 2-1-L8 requires the internal lactose disaccharide of lacto-*N*-neotetraose to be unsubstituted (9); it also requires one, but not two, PEA substituents (94). mAb 1B2 requires the complete and unsubstituted LNnT structure (80), an intact lipoidal moiety (81), and at least one PEA substituent. mAb 1-1-M requires the terminal GalNAc of the gangliosyl LOS (80) and at least one PEA substituent. *N. meningitidis* do not galactosaminylate C3 of the LNnT terminal Gal; instead, group B and C strains endogenously sialylate it (24). *N. gonorrhoeae* strain F62$_{hs}$ does not make an LOS β chain, as C3 of Hep-2 is phosphoethanolaminylated. The *O*-acetylated α-GlcNAc substitution at C2 of Hep-2 initiates the γ chain.

and/or phosphate groups (Fig. 1). Acidic and amino functionalities may be balanced or unbalanced. Most commonly they are unbalanced, with none or one PEA residues and two acidic functions—either the two Kdo carboxylates of neisserial LOS or the carboxylate and phosphate moieties of the Kdo phosphate of *Haemophilus* LOS (9,12,51,55–59). The number of PEA substitutions (one, two, or none) has little effect on the electrophoretic mobility of an LOS molecule in SDS-PAGE.

Both *Neisseria* and *Haemophilus* add Glc β1→4 to Hep-1 to initiate the α chain (1), and both often add Glc α1→3 to Hep-2 to initiate the β chain. *Neisseria* add GlcNAc α1→2 to Hep-2 to create the γ chain (1,52,53,55), whereas *Haemophilus* add a third Hep α1→2 to Hep-2 to form their γ chain (56–58). The γ chain α-GlcNAc of neisserial LOS usually is *O*-acetylated (Fig. 1) (1,9,12,52); Hep-3 of *Haemophilus* LOS may accept a second PEA substitution.

Whereas α and γ chains are always present and conserved in their initiation within genera, the presence and composition of β chains is highly variable, even among LOS molecules made by a single strain. PEA substitution at C3 of Hep-2 prevents α1→3Glc initiation of the β chain. Since many neisserial LOS molecules are phosphoethanolaminylated at C3 of Hep-2, they do not have β chains (9,12,59). LOS molecules that have β chains phosphoethanolaminylate C6 or C7 of Hep-2 (60). Neither the site of second PEA substitutions nor the linkage of basal region β-GlcNAc residues on some *N. meningitidis* LOS have been established.

Each LOS molecule has a "selection" of basal region adornments; one set predominates, whereas others are at lower abundances (55). Each basal region, regardless of its abundance, may be adorned with the same α, β, and γ chains. The biological importance of differences in LOS basal regions remains to be investigated.

α Chains Mimic GSL Oligosaccharides

When organisms are cultured ex vivo, they tend to synthesize an extended α chain with minimal, if any, synthesis of the other two oligosaccharide chains (Fig. 1). LOS α chain oligosaccharides share sequence and linkage with oligosaccharides of several human GSL series (Table 2) (1–3,11,28). Reference to their analogous GSL series provides a way to distinguish LOS structures, a convention that also serves to emphasize the pathogenetic importance of their molecular mimicry (3). Table 2 shows the relationships of known LOS α chain structures to one another and to GSL structures.

The addition of Gal β1→4 to the Hep-1-linked Glc that invariably initiates the α chain forms the disaccharide, lactose (Lac; Galβ1→4Glc), and creates an α chain that mimics lactosylceramide GSL. Neisserial lactosyl LOS were estimated to have a mass of 3.6 kDa by reference to the mass of rough mutant LPS of *Salmonella* with which they were co-electrophoresed (6,35), and this denominator remains current even though it overestimates their mass by a few hundred daltons (55). Lactosyl LOS are made by all *Neisseria* and many *Haemophilus* (56), although the latter genus also makes a diglucoside α chain (57).

Lactosyl structures are found on the bacterial surface. Their terminal Gal residues act as "toggle switches," or biosynthetic decision points, as they may be substituted with either Gal α1→4 or GlcNAc β1→3. Addition of Gal α1→4 to lactosyl LOS creates

Table 2 LOS and Equivalent GSL Structures, Structure-Specific mAb, and Meningococcal L-Types

GSL series	mAb	L-type	Structure
Globo	P	None	GalNAcβ1→3Galα1→4Galβ1→4Glcβ1→4Hep1→Kdo
Sialoglobo	None	None	NeuNAcα2→?Galα1→4Galβ1→4Glcβ1→4Hep1→Kdo
Globo	P^k	L1	Galα1→4Galβ1→4Glcβ1→4Hep1→Kdo
Lactosyl	2-1-L8	L8	Galβ1→4Glcβ1→4Hep1→Kdo
	4C4	L11	
Lacto	L6	L6	GlcNAcβ1→3Galβ1→4Glcβ1→4Hep1→Kdo
Paraglobo	1B2	L2, 4, 5, 7, 9	Galβ1→4GlcNAcβ1→3Galβ1→4Glcβ1→4Hep1→Kdo
Paraglobo	1B2	None	[Galβ1→4GlcNAcβ1]$_n$→3Galβ1→4Glcβ1→4Hep1→Kdo
Sialoparaglobo	None	L3	NeuNAcα2→3Galβ1→4GlcNAcβ1→3Galβ1→4Glcβ1→4Hep1→Kdo
Ganglio	1-1-M	None	GalNAcβ1→4GlcNAcβ1→3Galβ1→4Glcβ1→4Hep1→Kdo
Globo	P$_1$	None	Galα1→4Galβ1→4GlcNAcβ1→3Galβ1→4Glcβ1→4Hep1→Kdo

globotriaose, the glycose of human P^k GSL, and prevents further chain extension (3,11,55). L1 strains of *N. meningitidis* and many *N. gonorrhoeae* and *Neisseria lactamica* strains make LOS with this structure (3); they run as 4.0 kDa molecules in SDS-PAGE. *Haemophilus influenzae*, including biogroup *aegyptius* strains, also make this globosyl LOS, and it is commonly expressed by *H. influenzae* type b (28).

If GlcNAc is added β1→3 to the lactosyl Gal, then a second Gal residue usually is added β1→4 to the GlcNAc residue to form Galβ1→4GlcNAcβ1→3Galβ1→4Glc (2,10–12). This LacNAcβ1→3Lac tetrasaccharide—the most fully studied α chain oligosaccharide—is lacto-*N*-neotetraose (LNnT), the glycose of paraglobside series GSL (13,61,62) and the H precursor of human ABO blood type antigens (2). *H. influenzae, Haemophilus ducreyi, N. gonorrhoeae, N. lactamica*, group B, C, Y, and W *N. meningitidis*, and L9 strains of group A *N. meningitidis* make paraglobosyl LOS, often preferentially (2,3,53,56–58).

The terminal Gal of LNnT is a second biosynthetic decision point. It may remain unsubstituted [paraglobosyl (2)], or be substituted α1→4 with Gal or β1→3 with either GlcNAc or GalNAc (*N. gonorrhoeae*) (Fig. 1). Substitutions other than by β1→3 GlcNAc prevent further chain extension.

Gonococci, but neither meningococci nor *Haemophilus*, usually substitute 30–60% of their LNnT structures β1→3 with GalNAc (3,5,11,12). This creates V³-(β-*N*-acetylgalactosaminyl)LNnT (Fig. 1) (12), the pentaose of X₂ and G₃ gangliosyl GSL (14,63). A few *N. gonorrhoeae*, including the often used laboratory strain, F62, also substitute some LNnT structures α1→4 with Gal to create globopentaose, the glycose of human P₁ GSL (3). The P₁ globosyl oligosaccharide does not appear to be made by other *Neisseria* or *Haemophilus* species (3). Terminal galactosylation or galactosaminylation of LNnT stops further chain extension.

α Chain Extension by Polylactosaminylation

N. gonorrhoeae often substitute LNnT β1→3 with GlcNAc. This leads to further chain extension by recreating an acceptor for the β-galactosyl transferase that adds Gal β1→4 to GlcNAc and that then forms a second LacNAc disaccharide (38). C3 of the terminal Gal of this second LacNAc may, in turn, accept either GalNAc, which caps the chain, or yet another GlcNAc, which continues the process of linear chain extension by β1→3-polylactosaminylation until the chain finally is capped by substitution with GalNAc at C3 of a LacNAc Gal. Linear polylactosamine (pLacNAc) structures frequently are made by freshly isolated gonococci, but lost on subculture (5), so their presence has been underappreciated. To date we have not extracted LOS with more than three LacNAc repeats (lacto-*N*-neooctaose, the oligosaccharide of nLc₈Ceramide) from cultured gonococci (5).

β Chain Extension

Because α oligosaccharide synthesis by ex vivo cultured organisms so predominates, we know little about β chain synthesis or how often the β chain is extended beyond a single glycose (Table 1). The first LOS structures to be solved partially were made by JW31R, a pyocin-selected gonococcus (53). They have an extended β chain with a unique GalNAc→Gal→Gal→Glc→Hep-2 structure [Table 3 (53)], but this variant could be using an LOS biosynthetic pathway that is not much used by wild strains.

Gonococci that cause disseminated infections (DGI), as well as many of those cultured from patients with uncomplicated disease, do add Gal β1→4 to the β chain α-Glc on some of their LOS molecules (64). This creates a second lactosyl structure that is parallel to the α chain lactoside. Although we have not revisited JW31R LOS, nor completely solved its structure, its P^k-like Gal→Gal α chain and parallel GalNAc→Gal→Gal β chain suggest a globotetraose β chain that mimics that of the Gb₄Ceramide P blood group antigen. Synthesis of such a globosyl structure would need only those glycosyl transferases known to be used by *N. gonorrhoeae* for LOS synthesis (46). *H. influenzae* also make parallel α and β chains, but they are diglucosyl, rather than lactosyl, disaccharides (57).

We expect to find other β chain extensions as more complex *N. meningitidis* and *N. lactamica* LOS struc-

Table 3 LOS of *N. gonorrhoeae* Strain JW31R

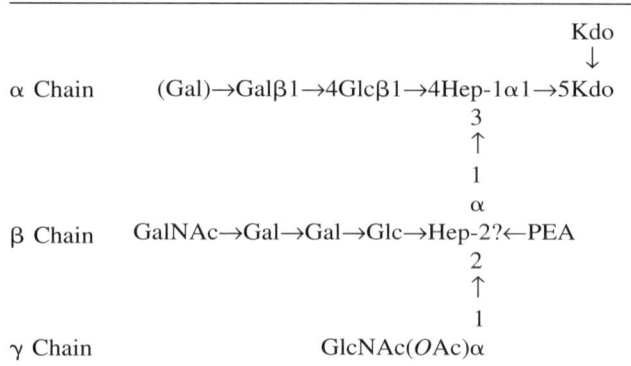

tures and higher M_r N. gonorrhoeae LOS structures are solved. For example, lactosyl moieties could be extended to form parallel α and β chain paraglobosyl structures, each or both of which could in turn be either galactosaminylated or extended by pLacNAc formation.

γ Chain Extension

Some N. meningitidis extend their LOS γ chains, but these structures have not been solved (55). We as yet have no evidence that N. gonorrhoeae or Haemophilus species extend their LOS γ chains.

Higher M_r LOS

N. gonorrhoeae, but not N. lactamica or N. meningitidis, often make LOS that migrate more slowly in SDS-PAGE than gangliosyl LOS (35, 65). Many, but not all, of these higher M_r LOS bind the same mAb as lower M_r LOS (5). The increased mass of those that do often can be accounted for by pLacNAc structures; β chain extensions with the same, or different, glycose sequences as those of the α antennae would account for those higher M_r LOS molecules that do not bind LacNAc or gangliosyl mAb (1,5).

SIALYLATION OF LOS

The terminal Gal residues of human glycoconjugates often are substituted with sialic acid (also called N-acetylneuraminic acid or NANA). Sialylation affects many biological functions. Sialylation of cell membrane glycoconjugates aborts assemblage of the complement membrane attack complex and prevents its lysis of cells (66–68). It also protects cells from recognition and clearance by antibody and complement-independent immune mechanisms, interferes with ligand binding to receptors on phagocytic cells, and prevents clearance of circulating glycoconjugates by hepatocytes.

Terminally unsubstituted globosyl and paraglobosyl LOS also may be sialylated by two different mechanisms (18,19,24,55). N. meningitidis of serogroups B, C, Y, and W synthesize sialic acid, incorporate it into their capsules, and usually sialylate a variable number of their globosyl (55,69) and paraglobosyl (15,19,23,24) LOS terminal Gal residues. This mechanism has been termed endogenous sialylation (19). It is not known whether the same sialyl transferase (STase) endogenously sialylates both globosyl and paraglobosyl LOS. Endogenous sialylation stops further chain extension.

Group A N. meningitidis, N. gonorrhoeae, and N. lactamica cannot make sialic acid (26) and do not endogenously sialylate their LOS when grown on artificial media, but nearly all strains of all three species make at least one STase that can transfer sialic acid from CMP-NANA of human origin to their paraglobosyl LOS in vitro and in vivo (18–20,24,27,29). This mechanism has been termed exogenous sialylation (18). It is not known whether globosyl or polylactosaminyl LOS can be exogenously sialylated.

Haemophilus paraglobosyl LOS are sialylated when grown on artificial media that contain blood to support their growth and, probably, low levels of CMP-NANA (28,56–58).

The endogenous STases made by N. meningitidis may not be the same as the exogenous STases made by them and other Neisseria. The exogenous STase is anchored in the cell membrane and active on the cell surface (70), whereas endogenous STases must be cytoplasmic or periplasmic enzymes. Meningococci that have endogenously sialylated their LOS can be sialylated further by the exogenous STase, and the effects on complement-dependent immune clearance mechanisms of exogenous LOS sialylation are different from those of endogenous LOS sialylation (15,20,22,23,69,71).

One meningococcal endogenous STase sialylates the same lacto-N-neotetraose terminal Gal C3 that is alternatively galactosaminylated or glucosaminylated by N. gonorrhoeae (24). Thus, the terminal paraglobosyl Gal is the toggle for sialylation as well as hexosaminylation. Meningococci do not galactosaminylate their paraglobosyl LOS (25), but endogenous sialylation prevents glucosaminylation and pLacNAc formation. Gonococcal paraglobosyl LOS are either galactosaminylated or sialylated when the organisms are grown in the presence of CMP-NANA. The effect of partial (approximately 30–60%) galactosaminylation of paraglobosyl LOS by N. gonorrhoeae is to leave paraglobosyl molecules unsubstituted and therefore available for sialylation during in vivo growth and passage between individuals. The proportion of paraglobosyl (or polylactosaminyl) molecules that remain unsubstituted, sialylated, or galactosaminylated in vivo is not known.

LOS ANTIBODIES AND IMMUNE CLEARANCE

LOS molecules that mimic human GSL are immunogenic in rabbits (LOS serotyping) and mice (mAb), but poorly so in humans (6–8). Nonetheless, LOS antibodies that can initiate clearance of organisms through complement-mediated (immune) lysis are ubiquitous in human serum (6,30–32,72–75). The protracted search for their exact specificities, however, has been frustrated by our minimal understanding of the tertiary conformation of LOS molecules and its influence on epitope formation.

Conformation and Epitope Formation

LOS epitope structures are conformationally constrained by basal region hydrogen bonds (55,76), by the steric influences of the various cations that form salts within basal region electronegative boxes (77), by interactions between LOS and adjacent membrane constituents (78,79), and by hydrophobic interactions between glycose and lipoidal moieties (39,76,80,81). Monoclonal antibodies have provided some useful insights into how various LOS glycose structures are conformed into epitopes, particularly how the basal region influences the formation of α chain epitopes.

Basal Region Adornments

Basal region anions affect epitope conformation in two ways: by forming hydrogen bonds with amino functions (Fig. 1) and by forming electronegative cavities, or boxes. The number of basal region acidic groups is constant, since phosphorylation of Kdo by *Haemophilus* is invariate, and most, if not all, *Neisseria* LOS molecules have two Kdo residues. Therefore, hydrogen bonding within the basal region is determined by phosphoethanolaminylation. Several LOS epitopes, including the meningococcal serotype determinants, L1, L8, L10, and L11, require the presence of one or more PEA substituents, even though the epitopes themselves reside in glycose moieties (55,76). Other epitopes either are unaffected by PEA adornments (76,80) or are lost after phosphoethanolaminylation (55). Clearly the hydrogen bond must change the conformation and, hence, epitope presentation of the oligosaccharide.

Acidic functions that are in excess of primary amino functions are free to form salts with available cations. The valency and atomic weight of such cations also affects epitope conformation (77). Divalent cations can bridge, or cross-link, any open electronegative boxes that are formed by basal region acidic functions (e.g., Kdo-Kdo carboxylates). The distance across the dipole of a cation that fills a basal region electronegative box will determine the spatial arrangement of the basal region and influence the molecule's conformation. Since different cations have different dipolar moments, they will have different effects (77).

Effect of the Lipoidal Moiety

Distal α chain epitopes, such as those conformed of lacto-*N*-neotetraose or its galactosaminylated derivative, are not affected by basal region adornments (77,80). Rather, they require the presence of a complete lipoidal moiety (81). Presumably, the lipoidal moiety determines the thermodynamics of the interaction between these relatively long glycose moieties and the bacterial membrane. Similar thermodynamic interactions between human GSL ceramide fatty acids and globosyl oligosaccharides have been found (82).

Interactions between LOS and adjacent outer membrane proteins and pili also influence glycose moiety epitope expression (39,78,79).

mAb as Surrogates for LOS Structures

A library of mouse mAb, each binding to a different LOS molecule, has been developed over many years (2,3,12,35,39,80) and used to denote the molecules to which they bind (5). Their usefulness as surrogates for LOS structures is only as good as the precision with which their binding specificities have been determined. The most useful are given in Table 2.

Meningococcal L-Types

N. meningitidis strains have been divided into 11 LOS serotypes, or L-types, based on rabbit immune responses to them (40,41). L-typing provides relatively little useful epidemiologic information (83), because most strains make more than one L-type (55), but L-typing has helped structural work. L-type epitopes are formed from LOS α chains, but basal region substitutions account for much of their diversity. L-type α chains are noted in Table 2.

Lactosyl LOS bear the L8 and L11 meningococcal LOS serotype determinants (6,76,84). L8 requires the presence of one PEA substituent (55); L11 requires two (76).

L1 and L10 add a Hex residue to the lactosyl-diheptaside structure. The identity and position of the L10 +Hex are not established; L10 LOS may have one or two PEA substituents (76). L1 is the P^k globosyl structure (55,85); it requires a PEA residue to discriminate it from L8 and can be endogenously sialylated (55).

β1→3 substitution of lactosyl LOS with GlcNAc creates the L6 LOS (85). It is phosphoethanolaminylated on C7 of Hep-2, but does not have a β chain.

L2-5, L7, and L9 LOS have paraglobosyl α chains (59,60,86–88). L3, L4, L7, and L9 LOS do not have β chains; all except L4, which is phosphoethanolaminylated on C6 of Hep-2 (86), are phosphoethanolaminylated on C3 of Hep-2 (59,60,86). L2 and L5 LOS have Glcα1→3Hep-2 β chains (87,88).

L3 and L7 usually are co-expressed (55). They are made by group B and C strains that endogenously sialylate some of their LNnT structures; it has been proposed that L3 is the sialylated structure and L7 the unsialylated one (59,86). L9 is made by group A strains that cannot endogenously sialylate LOS; a difference in phosphoethanolaminylation may distinguish L9 LOS from L3,7 LOS (51,59,60).

L5, an extremely rare L-type (55), has an additional β1→4Glc residue interposed between Hep-1 and the lacto-N-neotetraose structure (88).

Human LOS Antibodies in Immune Lysis

Bactericidal antibodies that initiate immune lysis prevent *H. influenzae* (89), *N. meningitidis* (90), and *N. gonorrhoeae* (31,32,91) from disseminating from mucosal sites of colonization. Humans seldom circulate antibodies that bind GSL structures (6–8), but they do circulate bactericidal LOS antibodies. Human LOS antibodies that initiate lysis of *Neisseria* have been partially characterized (30–32,54,72–74); less is known about *Haemophilus* LOS antibodies.

Gangliosyl (GalNAcβ1→3LNnT; Fig. 1) LOS bind IgM molecules that are ubiquitous in human sera but are not found on mucosal surfaces. These antibodies prevent *N. gonorrhoeae* from disseminating by activating the classical complement pathway (CP) on the surface of any organisms that enter the blood stream and lysing them (31). Therefore, *N. gonorrhoeae* that make gangliosyl LOS are serum sensitive (ser^s) (6).

Properdin (P), a regulatory protein of the alternative complement pathway (ACP) that stabilizes the ACP C3 convertase and thereby augments the density of C3b deposited following CP activation, binds to a gonococcal outer membrane protein that is variously surface exposed (31). Strains of *N. gonorrhoeae* bind various amounts of P, depending on their expression of the P-binding protein, and therefore variously support ACP augmentation of the C3b that is deposited following binding of gangliosyl IgM (31). Ser^s strains that bind equivalent amounts of gangliosyl IgM are therefore lysed at different titers (31).

Lactosyl LOS lack the LNnT acceptor for GalNAc (Fig. 1), cannot be galactosaminylated, and do not bind lytic IgM. Increased expression of lactosyl LOS by gonococci decreases the maximum density of bound IgM and, therefore, of C3b. Enhanced expression of lactosyl LOS make *N. gonorrhoeae* serum resistant (ser^r) unless they bind enough P to support lysis through the ACP (6,33).

The gangliosyl LOS epitope (Fig. 1) has been denoted as lytic locus 1; gonococci also have a second, IgG-dependent, LOS lytic locus (32); it is thought to be dependent on the expression of an extended β chain (47,73).

Meningococci do not make gangliosyl LOS and do not bind lytic gangliosyl IgM, but by 6 years of age most humans circulate LOS IgG molecules that are bactericidal for *N. meningitidis* (72,74,75). These IgG molecules bind to low M_r LOS that are not sialylated (74) and that have basal regions that are not phosphoethanolaminylated (92–94). They probably are not made by *N. gonorrhoeae*. Epidemiological evidence suggests that paraglobosyl α chains occlude the meningococcal low M_r LOS lytic locus, as meningococci that make predominantly lactosyl or lactosyl + Hex LOS (L1, L8, and L10) are less likely to disseminate than those that make paraglobosyl LOS (L3,7,9) (69,95). This observation also could be explained by the effects of endogenous sialylation of paraglobosyl LOS molecules (69,96).

Antibody-Independent Immune Clearance

Even though LNnT antibodies are not usually found in human sera (6–8), paraglobosyl α chains can support immune lysis in their absence (15). The activated thioester binding site of metastable C3b shows a strong preference for the hydroxyl groups of terminal glycose residues, particularly galactose (97). C6 of a terminal glycose provides an hydroxyl group that is not constrained by ring formation and that is free from steric interference from adjacent hydroxyl groups; a terminal Gal C6 hydroxyl group is considerably more reactive with metastable C3b than is the average glycose hydroxyl group. The unsubstituted terminal Gal of LNnT provides such a privileged C3b binding site and facilitates lysis of organisms through both complement pathways (15).

LOS SIALYLATION AND IMMUNE CLEARANCE

LOS sialylation was discovered because growth of *N. gonorrhoeae* in the presence of human blood that contains CMP-NANA converts ser^s to ser^r (18,20,27). Sialylation of LNnT structures would prevent them from being galactosaminylated to gangliosyl LOS with its lytic IgM epitope(s), but *N. gonorrhoeae* cannot synthesize sialic acid (26), and the exogenous STase is anchored in the cell membrane and active on the cell surface, not in the cytoplasm (70). As a result, ser^s gonococci galactosaminylate 30–60% of LOS LNnT structures even when grown in the presence of CMP-NANA or mammalian fluids that contain CMP-NANA (5,18). Gonococci observed in vivo by electron microscopy also are not sialylated fully (98). Thus, sialylation of LOS must interfere with immune clearance through additional mechanisms that also may explain the different effects that LOS sialylation has on the sensitivity to immune lysis of different pathogenic *Neisseria* (20,69,71,96).

Conformational Effects

Sialylation provides an additional, and distal, carboxylate. If the molecule's minimum energy requirements are better met by a hydrogen bond between the sialic acid carboxylate and PEA than by one between Kdo and PEA, as seems likely, the former would replace the latter. Similarly, the sialic acid carboxylate is free to form an electronegative box, either with an adjacent NANA residue or with a Kdo acidic function. A NANA-Kdo carboxylate cavity would be expected to preferentially attract different cations than those that are attracted by Kdo-Kdo carboxylates. Gonococci that have been exogenously sialylated no longer can absorb lytic IgM from human sera even though they continue to have gangliosyl LOS epitopes (20). Thus, sialylation can alter epitope expression without masking terminal glycose moieties.

Masking Terminal Galactose Moieties

The sensitivity of group C meningococci to complement-mediated lysis by nonimmune human sera is correlated directly with the availability of unsubstituted LNnT terminal Gal residues (15). Substitution of LNnT terminal Gal residues with a bulky and highly polar sialic acid residue would interfere with the binding of metastable C3b to its privileged C6 site, and sialylation appears to interfere with immune lysis by masking, or "subtracting," LNnT terminal Gal residues from the immunochemical equation through either steric or electrostatic constraints, and by redirecting C3b deposition to less favorable sites where it can be degraded by factors H and I.

Complement Activation Regulation

The effect of sialic acid on the ACP is opposite to that of properdin. Whereas P stabilizes the surface-phase ACP C3 convertase, sialic acid prevents its formation by attracting factor H, which binds to C3b, prevents the formation of C3bBb, and increases the degradation of C3b by factor I (15,67). Sialylation of the bacterial surface, then, would prevent ACP augmentation of any C3b that is bound by glycose hydroxyl groups or deposited through the CP by the binding of potentially lytic antibodies to gangliosyl or low M_r LOS epitopes (30,31,72,74,75).

Opsono-Phagocytosis and Sialylation

Exogenous LOS sialylation has relatively little effect on PMN uptake and killing of gonococci (21,22) or of endemic group C meningococci (23), but can make otherwise sensitive epidemic C:2b:P1.2:L3,7 meningococcal isolates resist neutrophil killing (23).

Among endemic group C *N. meningitidis*, polysialyl encapsulation and endogenous LOS sialylation are associated with each other and with resistance to killing by neutrophils (23). Epidemic strains, in contrast, intrinsically resist neutrophil killing, and this resistance is not related to endogenous sialylation of their LOS or to polysialyl capsule production (23). This suggests that endogenous sialylation evolved in meningococci as an adaptive mechanism that enabled them to evade clearance by PMNs and thereby maintain colonization of the pharyngeal mucosa, rather than as a means of avoiding immune lysis (69,71,96), which would not subserve their evolution in that dissemination through the bloodstream would result in the death of their only known host (36).

LOS AS ADHESINS

Neisseria and *Haemophilus* are facultative intracellular parasites (99) that must attach to and invade epithelial cells (100–103). Mammalian cell membranes contain lectins that are specific for LOS glycose sequences (104–108), and unsubstituted terminal LNnT Gal residues provide ligands that are necessary for invasion of epithelial cells (16,17,45).

Mammalian C- and S-type lectins are highly sensitive to the spatial arrangement of terminal Gal residues. Elongation of β chains with parallel LNnT structures would provide the bacterium with clusters of terminal Gal ligands and ensure high-affinity binding to host

LOS AND THE PATHOGENESIS OF URETHRAL GONORRHEA IN MEN

Gonococcal variants that make only lactosyl LOS do not cause urethritis in men (5,109). Gonorrhea in men is characterized by the onset after a variable incubation of dysuria followed by urethral leukorrhea (110). During the period between exposure and the onset of dysuria, organisms can not be recovered from urine (5) and are thought to be hidden within urethral epithelial cells from which they emerge to be ingested by phagocytes (102,111) that are evoked by the inflammatory mediators released by the infected epithelial cells (112). Organisms recovered as dysuria commences make paraglobosyl LOS; those associated with the onset of discharge make gangliosyl (Fig. 1), but not lactosyl, molecules (5).

Not all exposed men develop leukorrhea (110). An inoculum of only 250 gonococcal variants that made paraglobosyl and gangliosyl, but not lactosyl, LOS caused urethritis in three of seven volunteers who were challenged with them (109)—a 40% attack rate that is somewhat higher than the 22% risk of naturally acquired infection following sexual contact with an infected female (110). This suggests either that the number of organisms shed from the cervix of infected women is quite low (113) or that many shed organisms are not of the appropriate paraglobosyl/gangliosyl LOS phenotype (5,109). Surprisingly, sialylated gonococci are not able to infect men (114), so some LNnT LOS structures must remain unsialylated during cervical infection and transmission.

As leukorrhea develops in men, gonococcal cells are found packed closely together within PMNs in what appear to be parasitophorous vesicles, rather than a phagolysosomal vacuole; some bacteria can be seen to be actively dividing (102). Just how gonococci survive and multiply within PMNs remains unclear. LOS lacto-N-neotetraose structures are sialylated in vivo within phagocytes (98), and sialylation of their LOS is an obvious candidate mechanism for ensuring survival within PMNs. Rest and Frangipane reported that sialylation of LOS interferes with nonopsonic (Opa-mediated) binding to PMNs, quenches the phagocytes' oxidative burst and delays, but does not abolish, killing (21). We, too, found that LOS sialylation delayed, but did not abolish, PMN killing of N. gonorrhoeae (22). It is not clear whether sialylation occurs before or after internalization.

CONCLUSION

It should be clear from this review that LOS are key components of the biology of mucosal bacteria and of the pathogenesis of the diseases they cause. Although we have made enormous progress in understanding the structure and biology of these glycolipids, some structures remain unsolved; we still have little information about how LOS phase variation occurs in vivo, and sorting out the pathogenesis of cervical infections in women will be much harder than doing so for urethral infections in men. Our understanding of N. gonorrhoeae LOS is better integrated than that of N. meningitidis or Haemophilus LOS, but we do not know precisely how any LOS functions as an adhesin on any mucosal surface, or how LOS variations and substitutions affect survival within PMNs.

The study of LOS has had distinct phases: appreciation of their binding of protective antibodies and of their distinctness from LPS, the development of mAb and cataloguing of their specificities, considerable structural work, and the use of mAb of known specificity to study the biology of disease. We are nearing the end of the structural chemistry phase and at the beginning of the genetic phase (25,46,47). The availability of mutants with defined genetic defects should enable us to solve the remaining biological problems.

ACKNOWLEDGMENTS

The Centre for Immunochemistry of the University of California, San Francisco, is supported by The Public Health Service of the United States of America (Grants AI 21620, AI 21171) and the Research Service of the U.S. Department of Veterans Affairs. It is administered by Ms. May Fong. This is paper No. 90 from the Centre for Immunochemistry of the University of California, San Francisco.

REFERENCES

1. Griffiss JM, Schneider H, Mandrell RE, Yamasaki R, Jarvis GA, Kim JJ, Gibson B, Hamadeh R, Apicella MA. Lipooligosaccharides: the principal glycolipids of the neisserial outer membrane. Rev Infect Dis 1988; 10:S287–S295.

2. Mandrell RE, Griffiss JM, Macher BA. Lipooligosaccharides (LOS) of *Neisseria gonorrhoeae* and *Neisseria meningitidis* have components that are immunochemically similar to precursors of human blood group antigens. Carbohydrate sequence specificity of the mouse monoclonal antibodies that recognize cross-reacting antigens on LOS and human erythrocytes. J Exp Med 1988; 168:107–126.

3. Mandrell RE. Further antigenic similarities of *Neisseria gonorrhoeae* lipooligosaccharides and human glycosphingolipids. Infect Immun 1992; 60:3017–3020.

4. Schneider H, Hammack CA, Apicella MA, Griffiss JM. Instability of expression of lipooligosaccharides and their epitopes in *Neisseria gonorrhoeae*. Infect Immun 1988; 56:942–946.

5. Schneider H, Griffiss JM, Boslego JW, Hitchcock PJ, Zahos KM, Apicella MA. Expression of paraglobo-side-like lipooligosaccharides may be a necessary component of gonococcal pathogenesis in men. J Exp Med 1991; 174:1601–1605.

6. Schneider H, Griffiss JM, Mandrell RE, Jarvis GA. Elaboration of a 3.6 kDa lipooligosaccharide, antibody against which is absent from human sera, is associated with serum resistance of *Neisseria gonorrhoeae*. Infect Immun 1985; 50:672–677.

7. Mandrell RE, Apicella MA, Boslego J, Chung R, Rice PA, Griffiss JM. Human immune response to monoclonal antibody-defined epitopes on lipooligosaccharides of *Neisseria gonorrhoeae*. In: Poolman JT, Zanen H, Mayer T, Heckels J, Mäkelä PH, Smith H, Beuvery C, eds. Gonococci and Meningococci. Dordrecht, The Netherlands: Martinus Nijhoff, 1988:569–574.

8. Zahos KM, Kind P, Griffiss JM, Schneider H. Induction of human paraglobo side antibodies by gonococcal LOS during urethritis. Abstracts, 32nd Intersci. Conf. Antimicrob. Agents Chemother. Anaheim, CA: American Society for Microbiology, 1992:325.

9. Kerwood DE, Schneider H, Yamasaki R. Structural analysis of lipooligosaccharide produced by *Neisseria gonorrhoeae* strain MS11MK (variant A). Biochemistry 1992; 32:12760–12768.

10. Phillips NJ, John CM, Reinders LG, Gibson BW, Apicella MA, Griffiss JM. Structural models for the cell surface lipooligosaccharide (LOS) of *Neisseria gonorrhoeae* and *Haemophilus influenzae*. Biomed Environ Mass Spectrom 1990; 19:731–745.

11. John CM, Griffiss JM, Apicella MA, Mandrell RE, Gibson BW. The structural basis for pyocin-resistance in *Neisseria gonorrhoeae* lipooligosaccharides. J Biol Chem 1991; 266:19303–19311.

12. Yamasaki R, Bacon BE, Schneider H, Griffiss JM. Structural determination of oligosaccharides derived from lipooligosaccharide of *Neisseria gonorrhoeae* F62 by chemical, enzymatic and two-dimensional NMR methods. Biochemistry 1991; 30:10566–10575.

13. Niemann H, Watanabe K, Hakomori SI, Childs RA, Feizi T. Blood group i and I activities of "Lacto-N-norhexaosylceramide" and its analogues: the structural requirements for i-specificities. Biochem Biophys Res Commun 1978; 81:1286–1293.

14. Watanabe K, Hakomori SI. Gangliosides of human erythrocytes. A novel ganglioside with a unique N-acetylneuraminosyl-(2→3)-N-acetylgalactosamine structure. Biochemistry 18 1979:5502–5504.

15. Estabrook MM, Jarvis GA, Griffiss JM. Sialylation of *Neisseria meningitidis* lipooligosaccharide inhibits serum bactericidal activity by masking lacto-N-neotetraose. Infect Immun 1997; 65:4436–4444.

16. Griffiss JM, Wang J, Zhao S, Schneider H. *Neisseria gonorrhoeae* must express ganglioside-like lipooligosaccharides (LOS) to invade cervical epithelial cells. Clin Res 1994; 42:84A.

17. Porat N, Apicella MA, Blake MS. *Neisseria gonorrhoeae* utilizes and enhances the biosynthesis of the asialoglycoprotein receptor expressed on the surface of the hepatic HepG2 cell line. Infect Immun 1995; 63:1498–1506.

18. Mandrell RE, Lesse AJ, Sugai JV, Shero M, Griffiss JM, Cole JA, Parsons NJ, Smith H, Morse SA, Apicella MA. *In vitro* and *in vivo* modification of *Neisseria gonorrhoeae* lipooligosaccharide epitope structure by sialylation. J Exp Med 1990; 171:1649–1664.

19. Mandrell RE, Kim JJ, John CM, Gibson BW, Sugai JV, Apicella MA, Griffiss JM, Yamasaki R. Endogenous sialylation of the lipooligosaccharide of *Neisseria meningitidis*. J Bacteriol 1991; 173:2823–2832.

20. Parsons NJ, Andrade JRC, Patel PV, Cole JA, Smith H. Sialylation of lipopolysaccharide and loss of absorption of bactericidal antibody during conversion of gonococci to serum resistance by cytidine 5′-monophospho-N-acetyl neuraminic acid. Microb Pathogen 1989; 7:63–72.

21. Rest RF, Frangipane JV. Growth of *Neisseria gonorrhoeae* in CMP-N-acetylneuraminic acid inhibits nonopsonic (opacity-associated outer membrane protein-mediated) interactions with human neutrophils. Infect Immun 1992; 60:989–997.

22. Kim JJ, Zhou D, Mandrell RE, Griffiss JM. Effect of exogenous sialylation of the lipooligosaccharide of *Neisseria gonorrhoeae* on opsonophagocytosis. Infect Immun 1992; 60:4439–4442.

23. Estabrook MM, Christopher NC, Griffiss JM, Baker CJ, Mandrell RE. Sialylation and human neutrophil killing of group C *Neisseria meningitidis*. J Infect Dis 1992; 166:1079–1088.

24. Yamasaki R, Griffiss JM, Quinn KP, Mandrell RE. Neuraminic acid is α2→3 linked in the lipooligosaccharide of *Neisseria meningitidis* serogroup B strain 6275. J Bacteriol 1993; 175:4565–4568.

25. Jennings MP, Hood DW, Peak IRA, Virji M, Moxon ER. Molecular analysis of a locus for the biosynthesis and phase-variable expression of the lacto-N-neotetraose terminal lipooligosaccharide structure in *Neisseria meningitidis*. Mol Microbiol 1995; 18:729–740.

26. Frosch M, Weisgerber C, Meyer TF. Molar characterization and expression in *Escherichia coli* of the gene complex encoding the polysaccharide capsule of *Neisseria meningitidis* group B. Proc Natl Acad Sci USA 1989; 86:1669–1673.

27. Nairn CA, Cole JA, Patel PV, Parsons NJ, Fox JE, Smith H. Cytidine 5′-monophospho-N-acetylneuraminic acid or a related compound is the low M_r factor from human red blood cells which induces gonococcal resistance to killing by human serum. J Gen Microbiol 1988; 134:3295–3306.

28. Mandrell RE, McLaughlin R, Kwaik YA, Lesse AJ, Yamasaki R, Gibson BW, Spinola SM, Apicella MA. Lipooligosaccharides (LOS) of some *Haemophilus* species mimic human glycosphingolipids, and some LOS are sialylated. Infect Immun 1992; 60:1322–1328.
29. Mandrell RE, Griffiss JM, Smith H, Cole JA. Distribution of a lipooligosaccharide-specific sialytransferase in pathogenic and non-pathogenic *Neisseria*. Microb Pathogen 1993; 14:315–327.
30. Griffiss JM, O'Brien JP, Yamasaki R, Williams GD, Rice PA, Schneider H. Physical heterogeneity of neisserial lipooligosaccharides reflects oligosaccharides that differ in apparent molecular weight, chemical composition, and antigenic expression. Infect Immun 1987; 55:1792–1800.
31. Griffiss JM, Jarvis GA, Schneider H, O'Brien JP, Eads MM. Lysis of *Neisseria gonorrhoeae* initiated by binding of normal human IgM to a hexosamine-containing lipooligosaccharide epitope is augmented by strain-specific, properdin binding-dependent alternative complement pathway activation. J Immunol 1991; 147:298–305.
32. Schneider H, Griffiss JM, Williams GD, Pier GB. Immunological basis of serum resistance of *Neisseria gonorrhoeae*. J Gen Microbiol 1982; 128:13–22.
33. Stein DC, Petricoin EF, Griffiss JM, Schneider H. Use of transformation to construct *Neisseria gonorrhoeae* strains with altered lipooligosaccharides. Infect Immun 1988; 56:762–765.
34. Nikaido H, Vaara M. Molecular basis of bacterial outer membrane permeability. Microbiol Rev 1985; 49:1–32.
35. Schneider H, Hale TL, Zollinger W, Seid Jr. RC, Hammack CA, Griffiss JM. Heterogeneity of molecular size and antigenic expression within the lipooligosaccharides of individual strains of *Neisseria gonorrhoeae* and *Neisseria meningitidis*. Infect Immun 1984; 45:544–549.
36. Griffiss JM, Artenstein MS. The ecology of the genus *Neisseria*. Mt Sinai J Med 1976; 43:746–761.
37. Roth RI, Yamasaki R, Mandrell RE, Griffiss JM. Ability of gonococcal and meningococcal lipooligosaccharides to clot limulus amebocyte lysate. Infect Immun 1992; 60:762–767.
38. John CM, Schneider H, Griffiss JM. *Neisseria gonorrhoeae* that infect men have lipooligosaccharides with terminal *N*-acetyl lactosamine repeats. J Biol Chem 1999, In Press (January).
39. Mandrell RE, Schneider H, Apicella MA, Zollinger W, Rice PA, Griffiss JM. Antigenic and physical diversity of *Neisseria gonorrhoeae* lipooligosaccharides. Infect Immun 1986; 54:63–69.
40. Zollinger WD, Mandrell RE. Outer membrane protein and lipopolysaccharide serotyping of *Neisseria meningitidis* by inhibition of a solid phase radioimmunoassay. Infect Immun 1977; 18:424–433.
41. Zollinger WD, Mandrell RE. Type-specific antigens of group A *Neisseria meningitidis*: lipopolysaccharide and heat-modifiable outer membrane proteins. Infect Immun 1980; 28:451–458.
42. Apicella MA, Shero M, Jarvis GA, Griffiss JM, Mandrell RE, Schneider H. Phenotypic variation in epitope expression of the *Neisseria gonorrhoeae* lipooligosaccharide. Infect Immun 1987; 55:1755–1761.
43. Chantler E, Debruyne E. Factors regulating the changes in cervical mucous in different hormonal states. In: Elstein M, ed. Mucus in Health and Disease. New York: Plenum Press, 1977:131–141.
44. Wang J, Griffiss JM, Schneider H. *Neisseria gonorrhoeae* must express the paraglobosyl LOS in order to invade human genitourinary epithelial cells. In: Zollinger WD, Frasch CE, Deal CD, ed. Proc. 10th Internatl. Pathogenic Neisseria Conference. Baltimore 1996:112–113.
45. Apicella MA, Zhou D, Lee F, Ketterer M, Porat N, Blake M, Gibson BW, Stephens D. The biology of the lipooligosaccharides of the pathogenic *Neisseria*. In: Evans JS, Yost SE, Maiden MCJ, Feavers IM, ed. Neisseria 94. Winchester, England: S.C.C., 1994:10–11.
46. Gotschlich EC. Genetic locus for the biosynthesis of the variable portion of *Neisseria gonorrhoeae* lipooligosaccharide. J Exp Med 1994; 180:2181–2190.
47. Erwin AL, Haynes PA, Rice PA, Gotschlich EC. Conservation of the lipooligosaccharide synthesis locus *lgt* among strains of *Neisseria gonorrhoeae*: requirement for *lgtE* in synthesis of the 2C7 epitope and of the β-chain of strain 15253. J Exp Med 1996; 184:1233–1241.
48. Domon B, Costello CE. A systematic nomenclature for carbohydrate fragmentations in FAB-MS/MS spectra of glycoconjugates. Glycoconj J 1988; 5:397–409.
49. Takayama K, Qureshi N, Hyver K, Honovich J, Cotter RJ, Mascagni P, Schneider H. Characterization of a structural series of lipid A obtained from the lipopolysaccharides of *Neisseria gonorrhoeae*. J Biol Chem 1986; 261:10624–10631.
50. Kulshin VA, Zähringer U, Lindner B, Frasch CE, Tsai CM, Dmitriev BA, Rietschel ET. Structural characterization of the lipid A component of pathogenic *Neisseria meningitidis*. J Bacteriol 1992; 174:1793–1800.
51. Gibson BW, Melaugh W, Phillips NJ, Apicella MA, Campagnari AA, Griffiss JM. Investigation of the structural heterogeneity of lipooligosaccharides from pathogenic *Haemophilus* and *Neisseria* species and R-type lipopolysaccharides from *Salmonella typhimurium* by electrospray mass spectrometry. J Bacteriol 1993; 175:2702–2712.
52. Jennings HJ, Lugowski C, Ashton FE. The structure of an R-type of oligosaccharide core obtained from some lipopolysaccharides of *Neisseria meningitidis*. Carbohydr Res 1983; 121:233–241.
53. Gibson BW, Webb JW, Yamasaki R, Fisher SJ, Burlingame AL, Mandrell RE, Schneider H, Griffiss JM. Structure and heterogeneity of the oligosaccharides from lipopolysaccharides of a pyocin-resistant *Neisseria gonorrhoeae*. Proc Natl Acad Sci USA 1989; 86:17–21.
54. Gibson BW, Schneider H, John CM, Mandrell RE, Griffiss JM. Relationship of lacto-*N*-neotetraose to the gonococcal LOS receptor for lytic serum IgM. Abstracts, 89th Annual Meet. Am. Soc. Microbiol. Washington, DC: American Society for Microbiology, 1989:34.

55. Griffiss JM, Brandt BL, Engstrom JJ, Schneider H, Zollinger WD, Gibson BW. Structural relationships and sialylation among meningococcal lipooligosaccharide (LOS) serotypes. In: Evans JS, Yost SE, Maiden MCJ, Feavers IM, ed. Neisseria 94. Winchester, England: S.C.C., 1994:12.

56. Phillips NJ, Apicella MA, Griffiss JM, Gibson BW. Structural characterization of the cell surface lipooligosaccharides from a nontypable strain of *Haemophilus influenzae*. Biochemistry 1992; 31:4515–4526.

57. Phillips NJ, Apicella MA, Griffiss JM, Gibson BW. Structural studies of the lipooligosaccharides from *Haemophilus influenzae* type b strain A2. Biochemistry 1993; 32:2003–2012.

58. Melaugh W, Phillips NJ, Campagnari AA, Karalus R, Gibson BW. Partial characterization of the major lipooligosaccharide from a strain of *Haemophilus ducreyi*, the causative agent of chancroid, a genital ulcer disease. J Biol Chem 1992; 267:13434–13439.

59. Pavliak V, Brisson JR, Michon F, Uhrin D, Jennings HJ. Structure of the sialylated L3 lipopolysaccharide of *Neisseria meningitidis*. J Biol Chem 1993; 268:14146–14152.

60. Jennings HJ, Gamian A, Michon F, Beurret M. Structure and immunochemistry of meningococcal lipopolysaccharides. Antonie van Leeuwenhoek J Microbiol 1987; 53:519–522.

61. Young Jr. WW, Portoukalian J, Hakomori SI. Two monoclonal anticarbohydrate antibodies directed to glycosphingolipids with a lacto-N-glycosyl type II chain. J Biol Chem 1981; 256:10967–10972.

62. Siddiqui B, Hakomori SI. A ceramide tetrasaccharide of human erythrocyte membrane reacting with anti-type XIV pneumococcal polysaccharide antiserum. Biochim Biophys Acta 1973; 330:147–155.

63. Kannagi R, Fukuda MN, Hakomori SI. A new glycolipid antigen isolated from human erythrocyte membranes reacting with antibodies directed to globo-N-tetraosylceramide (globoside). J Biol Chem 1982; 257:4438–4442.

64. Yamasaki R, Kerwood DE, Quinn KP, Griffiss JM, Schneider H, Mandrell RE. The structure of lipooligosaccharide produced by *Neisseria gonorrhoeae*, strain 15253, isolated from a patient with disseminated infection: evidence for a new glycosylation pathway of the gonococcal lipooligosaccharide. J Biol Chem 1994; 269:30345–30351.

65. Kim JJ, Mandrell RE, Griffiss JM. *Neisseria lactamica* and *Neisseria meningitidis* share lipooligosaccharide epitopes, but lack common capsular and class 1, 2, and 3 protein epitopes. Infect Immun 1989; 57:602–608.

66. Fearon DT, Austen KF. The alternative pathway of complement—a system for host resistance to microbial infection. N Engl J Med 1980; 303:259–263.

67. Pangburn MK, Müller-Eberhard HJ. Complement C3 convertase: cell surface restriction of β1H. control and generation of restriction on neuraminidase-treated cells. Proc Natl Acad Sci USA 1978; 75:2416–2420.

68. Hostetter MK, Gordon DL. Biochemistry of C3 and related thiolester proteins in infection and inflammation. Rev Infect Dis 1987; 9:97–109.

69. Mackinnon FG, Borrow R, Fox AJ, Robinson A, Jones DM. Effects of capsule and lipooligosaccharide sialylation on the serum resistance of *Neisseria meningitidis*. In: Evans JS, Yost SE, Maiden MCJ, Feavers IM, ed. Neisseria 94. Winchester, England: S.C.C., 1994:38–40.

70. Mandrell RE, Smith H, Jarvis GA, Griffiss JM, Cole JA. Detection and some properties of the sialytransferase implicated in the sialylation of lipopolysaccharide of *Neisseria gonorrhoeae*. Microb Pathogen 1993; 14:307–313.

71. Fox AJ, Jones DM, Scotland SM, Rowe B, Smith A, Brown MRW, Fitzgeorge RG, Baskerville A, Parsons NJ, Cole JA, et al. Serum killing of meningococci and several other gram-negative bacterial species is not decreased by incubating them with cytidine-5'-monophospho-N-acetyl neuraminic acid. Microb Pathogen 1989; 7:317–318.

72. Griffiss JM, Brandt BL, Broud DD, Goroff DK, Baker CJ. Immune response of infants and children to disseminated infections with *Neisseria meningitidis*. J Infect Dis 1984; 150:71–79.

73. Gulati S, McQuillen DP, Mandrell RE, Jani DB, Rice PA. Immunogenicity of *Neisseris gonorrhoeae* lipooligosaccharide epitope 2C7, widely expressed in vivo with no immunochemical similarity to human glycosphingolipids. J Infect Dis 1996; 174:1223–1237.

74. Estabrook MM, Baker CJ, Griffiss JM. The immune response of children to meningococcal lipooligosaccharides during disseminated disease is directed primarily against two monoclonal antibody-defined epitopes. J Infect Dis 1993; 167:966–970.

75. Estabrook MM, Mandrell RE, Apicella MA, Griffiss JM. Measurement of the human immune response to meningococcal lipooligosaccharide antigens by using serum to inhibit monoclonal antibody binding to purified lipooligosaccharide. Infect Immun 1990; 58:2204–2213.

76. Kim JJ, Phillips NJ, Gibson BW, Griffiss JM, Yamasaki R. Meningococcal group A lipooligosaccharides (LOS): preliminary structural studies and characterization of serotype-associated and conserved LOS epitopes. Infect Immun 1994; 62:1566–1575.

77. Mandrell RE, Yamasaki R, Apicella MA, Schneider H, Griffiss JM. Analysis of gonococcal lipooligosaccharides with mouse monoclonal antibodies: vanishing and re-emerging epitopes due to NaOH, EDTA, and divalent cation treatment. In: Schoolnik GK, Brooks GK, Falkow S, Frasch CE, Knapp JS, McCutchan JA, Morse SA, ed. The Pathogenic Neisseria. Washington, DC: American Society for Microbiology, 1985:385–389.

78. Blake MS, Blake CM, Apicella MA, Mandrell RE. Gonococcal opacity: lectin-like interactions between Opa proteins and lipooligosaccharide. Infect Immun 1995; 63:1434–1439.

79. Judd RC, Shafer WM. Topographical alterations in proteins I of *Neisseria gonorrhoeae* correlated with lipooligosaccharide variation. Mol Microbiol 1989; 3:637–643.

80. Yamasaki R, Nasholds W, Schneider H, Apicella MA. Epitope expression and partial structural character-

ization of F62 lipooligosaccharide (LOS) of *Neisseria gonorrhoeae*. IgM monoclonal antibodies (3F11 and 1-1-M) recognize non-reducing terminus of F62 LOS components. Mol Immunol 1991; 28:1233–1242.
81. Yamasaki R, Schneider H, Griffiss JM, Mandrell RE. Epitope expression of gonococcal lipooligosaccharide (LOS): importance of the lipoidal moiety for expression of an epitope that exists in the oligosaccharide moiety of LOS. Mol Immunol 1988; 25:799–809.
82. Jones DH, Lingwood CA, Barber KR, Grant CW. Globoside as a membrane receptor: a consideration of oligosaccharide communication with the hydrophobic domain. Biochemistry 1997; 36:8539–8547.
83. Griffiss JM. Epidemiological value of lipopolysaccharide and heat-modifiable outer-membrane protein serotyping of group-A strains of *Neisseria meningitidis*. J Med Microbiol 1982; 15:327–330.
84. Kim JJ, Mandrell RE, Hu Z, Apicella MA, Poolman JT, Griffiss JM. Electromorphic characterization and description of conserved epitopes of the lipooligosaccharides of group A *Neisseria meningitidis*. Infect Immun 1988; 56:2631–2638.
85. Di Fabio JL, Michon F, Brisson JR, Jennings HJ. Structure of the L1 and L6 core oligosaccharide epitopes of *Neisseria meningitidis*. Can J Chem 1990; 68:1029–1034.
86. Kogan G, Uhrin D, Brisson JR, Jennings HJ. Structural basis of the *Neisseria meningitidis* immunotypes including L4 and L7 immunotypes. Carbohydr Res 1997; 298:191–199.
87. Michon F, Beurret M, Gamian A, Brisson JR, Jennings HJ. Structure of the L5 lipopolysaccharide core oligosacchrides of *Neisseria meningitidis*. J Biol Chem 1990; 265:7243–7247.
88. Gamian A, Beurret M, Michon F, Brisson JR, Jennings HJ. Structure of the L2 lipopolysaccharide core oligosaccharides of *Neisseria meningitidis*. J Biol Chem 1992; 267:922–925.
89. Fothergill LD, Wright J. Influenzal meningitis: the relation of age incidence to the bactericidal power of blood against the causal organism. J Immunol 1933; 24:273–284.
90. Goldschneider I, Gotschlich EC, Artenstein MS. Human immunity to the meningococcus. I. The role of humoral antibodies. J Exp Med 1969; 129:1307–1326.
91. Schoolnik K, Buchanan TM, Holmes KK. Gonococci causing disseminated gonococcal infections are resistant to the bactericidal action of normal human sera. J Clin Invest 1976; 58:1163–1173.
92. Hamadeh RM, Zhou P, Griffiss JM. Human IgG that kill group B and C *Neisseria meningitidis* bind non-sialylated lipooligosaccharides (LOS). Clin Res 1993; 41:250A.
93. Zhao S, Griffiss JM, Kim JJ, Jarvis GA. Human antibodies that kill group A *Neisseria meningitidis* are directed at α lactosyl determinants on lipooligosaccharides of M_r <4,500. Abstracts, 93rd Annual Meet. Am. Soc. Microbiol. Washington, DC: American Society for Microbiology, 1993:156.
94. Griffiss JM, Brandt BL, Engstrom JJ, Schneider H, Zollinger W, Gibson BW. Meningococcal lipooligosaccharide serotypes are structurally related. Abstracts, 93rd Annual Meet. Am. Soc. Microbiol. Washington, DC: American Society for Microbiology, 1993:156.
95. Jones DM, Borrow R, Fox AJ, Gray S, Cartwright KA, Poolman JT. The lipooligosaccharide immunotype as a virulence determinant in *Neisseria meningitidis*. Microb Pathogen 1992; 13:219–224.
96. Borrow R, Fox AJ, Jones DM. The relationship between encapsulation and sialylation of meningococcal lipooligosaccharide. In: Evans JS, Yost SE, Maiden MCJ, Feavers IM, ed. Neisseria 94. Winchester, England: S.C.C., 1994:43–45.
97. Sahu A, Kozel TR, Pangburn MK. Specificity of the thioester-containing reactive site of human C3 and its significance to complement activation. Biochem J 1994; 302:429–436.
98. Apicella MA, Mandrell RE, Shero M, Wilson M, Griffiss JM, Brooks GF, Fenner C, Breen JF, Rice PA. Modification by sialic acid of *Neisseria gonorrhoeae* lipooligosaccharide epitope expression in human urethral exudates: An immunoelectron microscope analysis. J Infect Dis 1990; 162:506–512.
99. Chen JRC, Bavoil P, Clark VL. Enhancement of the invasive ability of *Neisseria gonorrhoeae* by contact with HecIB, an adenocarcinoma endometrial cell line. Molec Microbiol 1991; 5:1531–1538.
100. Ward ME, Watt PJ. Adherence of *Neisseria gonorrhoeae* to urethral mucosal cells: an electron-microscopic study of human gonorrhea. J Infect Dis 1972; 126:601–605.
101. Ward ME, Watt PJ, Robertson JN. The human fallopian tube: a laboratory model for gonococcal infection. J Infect Dis 1974; 129:650–659.
102. Novotny P, Short JA, Walker PD. An electron-microscopic study of naturally occurring and cultured cells of *Neisseria gonorrhoeae*. J Med Microbiol 1975; 8:413–427.
103. McGee ZA, Stephens DS, Hoffman LH, Schlech III WF, Horn RG. Mechanisms of mucosal invasion by pathogenic *Neisseria*. Rev Infect Dis 1983; 5:S708–S714.
104. Hirabayashi J, Kasai KI. Human placenta β-galactoside-binding lectin. Purification and some properties. Biochem Biophys Res Commun 1984; 122:938–944.
105. Kinane DF, Weir DM, Blackwell CC, Winstanley FP. Binding of *Neisseria gonorrhoeae* by lectin-like receptors on human phagocytes. J Clin Lab Immunol 1984; 13:107–110.
106. Sparrow CP, Leffler H, Barondes SH. Multiple soluble β-galactoside-binding lectins from human lung. J Biol Chem 1987; 262:7383–7390.
107. Cornil I, Kerbel RS, Dennis JW. Tumor cell surface β1-4-linked galactose binds to lectin(s) on microvascular endothelial cells and contributes to organ colonization. J Cell Biol 1990; 111:773–781.
108. Drickamer K. Two distinct classes of carbohydrate-recognition domains in animal lectins. J Biol Chem 1988; 263:9557–9560.
109. Schneider H, Cross AS, Kuschner RA, Taylor DN, Sadoff JC, Boslego JW, Deal CD. Experimental hu-

man gonococcal urethritis: 250 *Neisseria gonorrhoeae* MS11mkC are infective. J Infect Dis 1995; 172:180–185.
110. Holmes KK, Johnson DW, Trostle HJ. An estimate of the risk of men acquiring gonorrhea by sexual contact with infected females. Am J Epidemiol 1970; 91:170–174.
111. Harkness AH. The pathology of gonorrhoea. Br J Vener Dis 1948; 24:137–147.
112. Ramsey KH, Schneider H, Cross AS, Boslego JW, Hoover DL, Staley TL, Kuschner RA, Deal CD. Inflammatory cytokines produced in response to experimental human gonorrhea. J Infect Dis 1995; 172:186–191.
113. Johnson DW, Holmes KK, Kvale PA, Halverson CW, Hirsch WP. An evaluation of gonorrhea case finding in the chronically infected female. Am J Epidemiol 1969; 90:438–448.
114. Schneider H, Schmidt KA, Skillman DR, van de Verg L, Warren RL, Wylie HJ, Sadoff JC, Deal CD, Cross AS. Sialylation lessens the infectivity of *Neisseria gonorrhoeae* MS11mkC. J Infect Dis 1996; 173:1422–1427.

11

A Biophysical View on the Function and Activity of Endotoxins

Ulrich Seydel, Andre Wiese, Andra B. Schromm, and Klaus Brandenburg
Research Center Borstel, Borstel, Germany

INTRODUCTION

Lipopolysaccharides (LPS) are known to constitute amphiphilic macromolecules located on the surface of gram-negative bacteria (1,2). They participate in the physiological membrane functions of the bacterial organism and are essential for its growth and survival (3). LPS are, at the same time, the primary target for interaction with antibacterial drugs and components of the immune system of the host.

Released from the bacterial surface or in isolated form, LPS evoke an overwhelming spectrum of biological activities when administered to animals or humans or in vitro. They play an important role in the pathogenesis and manifestation of gram-negative infection in general and of septic shock in particular. Thus, lipopolysaccharides are also termed endotoxins (4) and are among the most potent agents capable of inducing local or generalized inflammatory reactions in both humans and experimental animals.

In gram-negative bacteria, the composition of the lipid matrix of the outer membrane is extremely asymmetric with respect to chemical structure and charge of the lipids. Thus, the outer leaflet is built up from LPS and the inner from a mixture of phospholipids (5). Chemically, LPS consist of a hydrophilic heteropolysaccharide, which is covalently linked to a hydrophobic lipid portion, termed lipid A, which anchors the molecule in the outer membrane. In wild-type strains, the polysaccharide portion consists of an O-specific chain and the core oligosaccharide (S-form LPS). Rough mutant strains do not express the O side chain but retain core oligosaccharides of varying length. The LPS of the various rough mutants are characterized by chemotypes in the sequence of decreasing length of the core sugar as Ra (complete core), Rb, Rc, Rd, and Re, the latter representing the minimal structure of LPS consisting only of lipid A and two 2-keto-3-deoxyoctonate (Kdo) monosaccharides. Lipid A is composed of a β-glucosaminyl-(1→6)-α-D-glucosamine disaccharide backbone, which is phosphorylated in positions 1 and 4' of the disaccharide backbone and carries, for Enterobacteriaceae, six or seven (hydroxy) fatty acid residues, which are either ester- or amide-linked (for review see Ref. 6). Thus, the main characteristics that distinguish endotoxins from other membrane lipids are the presence of three hydroxyl groups in the apolar moiety and of unsubstituted phosphate groups at the lipid A backbone and, in some cases, also in the sugar region.

As free lipid A expresses all the characteristic in vivo activities of LPS, such as pyrogenicity and lethality, it represents the "endotoxic principle" of LPS (6). It was found that in nonenterobacterial species, lipid A variants exist that deviate in chemical structure, e.g., in *Rhodobacter capsulatus* with only five partially unsaturated acyl chains of, on the average, only 12 C-atoms (7) (enterobacterial lipid A normally carry at least six saturated acyl chains of, on the average, more than 12 C-atoms). Many of these variants exhibit—despite a backbone identical to that of enterobacterial lipid A—significantly reduced biological activity, nevertheless such LPS may continue to act as antagonists to biologically active LPS/lipid A (8,9).

In 1991, Kawahara et al. (10) succeeded in the elucidation of the complete chemical structure of a glycolipid that substitutes for LPS in the outer membrane of the strictly aerobic gram-negative rod *Sphingomonas paucimobilis* [formerly named *Flavobacterium devorans* (11) and *Pseudomonas paucimobilis* (12)]. This glycolipid was found to have unexpected and unusual structural features and was determined to be a glycosphingolipid (GSL): the hydrophobic portion was found to be heterogeneous with respect to the dihydrosphingosine residue but was, in any case, quantitatively substituted by a (S)-2-hydroxymyristic acid in amide linkage. The oligosaccharide portion of the two main fractions, GSL-4A and GSL-1, consisted of a Man-Gal-GlcN-GlcA tetrasaccharide and a GlcA monosaccharide, respectively. Because of the rather disparate structural relationship of this glycolipid to that of the LPS, in this review only the function of GSL as a constituent of the bacterial outer membrane, not its biological activity, will be discussed.

LPS and lipid A are amphiphilic molecules and therefore form aggregates in aqueous environments above a critical concentration (critical aggregate concentration, CAC), which depends on their hydrophobicity (i.e., on the particular primary chemical structure) (13). The actual structure of these aggregates depends on the conformation (shape) of the contributing molecules, which is again determined by their primary chemical structure and is influenced by ambient conditions like temperature, pH, water content, and concentration of mono- and divalent cations.

The molecular shape of a given amphiphilic molecule like LPS or lipid A within a supramolecular aggregate is not a constant but will depend, at least in part, on the fluidity (inversely correlated to the state of order) of its acyl chains, which can assume two phase states: a gel (β) phase and a liquid-crystalline (α) phase. Between these two phase states, a reversible transition takes place at a characteristic phase-transition temperature, T_c, which depends on both the length and the degree of saturation of the acyl chains as well as on the conformation and the charge density and its distribution within the headgroup region (14). From the variability in the acylation patterns and in the chemical structure and composition of the sugar moiety, a complex phase behavior and structural polymorphism are to be expected for free lipid A and for LPS. Their importance for the biological action of endotoxins is at present not fully understood.

Both the function and activity of endotoxins may be discussed in terms of membrane aspects: the function of endotoxin is related to its presence as a major constituent of the outer membrane of the bacterial organism; in contrast, the activity of endotoxin relates primarily to its interaction with host cells and inflammatory mediator cascade pathways.

Membranes, in general, constitute the boundary between a cell or a cell compartment and its environment. They are composed of (glyco)lipids and proteins, function as permeability barriers, maintain constant ion gradients across the membrane, and guarantee a controlled steady state of fluxes in the cell. Furthermore, the vast majority of cell membranes carry recognition sites for components of the immune system and for interaction/communication with other cells.

For these functions to work properly, a particular lipid composition on each side and distribution between both sides of the lipid bilayer are required. Thus, a membrane is built up from a large variety of lipids, differing in charge and fatty acid substitution (length and degree of saturation), and these lipids are in a delicate equilibrium providing a suitable environment for protein function and membrane permeability. By a complex interaction of passive and active transport processes (diffusion through the lipid matrix or protein-aligned transmembrane channels and by energy-dependent ion pumps and transport proteins, respectively) ion gradients are built up that contribute, together with the charge distribution of the lipids on the two leaflets of the membrane, to a transmembrane potential.

The cellular biological activity of endotoxins may be discussed within the context of the interaction of LPS molecules with the membranes of host cells, because the intercalation of endotoxin molecules is considered to be an important step in the activation cascade (15,16).

As already pointed out, a prerequisite for normal cell functioning is the maintenance of a particular composition of the lipid matrix at given ambient conditions (17). Disturbances of this composition (e.g., by uptake of extraneous lipids) that differ in their chemical structure (e.g., acylation pattern, headgroup conformation, net electrical charge) from that of the normal constituents of the cell matrix, may lead to (1) alterations of membrane fluidity and/or permeability, (2) phase separation and domain formation, (3) disturbance of the lamellar membrane architecture, and (4) even internalization of the extraneous lipids. In many cases, the cell may be able to compensate for those changes by altering the composition of the lipid matrix by "homoviscous adaptation" (18). If this is not possible—certainly not on a time scale of minutes—any one of these membrane alterations may cause severe dysfunctions of the cell. Such dysfunctions may manifest themselves, for example, in transient or permanent alterations in the normal functioning of transmembrane proteins that might be involved in signal transduction. It would be predicted that membrane alterations and their influence on cell functioning would vary in severity in direct relationship to differences in the chemical structure of the constituents and the interacting lipids.

Like other amphiphilic molecules, endotoxins should, from a biophysical perspective, interact with cell membranes unspecifically via hydrophobic interaction, i.e., minimization of Gibbs free energy. This interaction has been postulated in various investigations (15,16,19,20) and should consist of a direct intercalation of small endotoxin aggregates down to monomers into the cell membrane. Another potential mechanism of interaction is proposed to proceed via specific coupling directly to a membrane-bound receptor protein (mCD14) (21,22) or indirectly to the lipopolysaccharide-binding protein (LBP), an acute phase protein, which then transfers LPS to mCD14 (8,23). In the case of CD14-negative endothelial and epithelial cells, it has been shown that a soluble form of CD14 (sCD14) can mediate LPS binding (24). Because CD14 is anchored in the membrane by glycosylphosphatidylinositol and, therefore, lacks a transmembrane domain, it may be hypothesized that none of these pathways leads directly to cell activation, but that further steps are necessarily involved in the transmembrane signal transduction. One early step could be binding (directly or CD14-mediated) to a transmembrane protein, another the internalization of endotoxin (25), and both processes may subsequently be involved in triggering an intracellular signaling pathway. Independent of the detailed mechanism of endotoxin-cell interaction, however, it may be expected that both unspecific intercalation as well as specific receptor binding of endotoxin should ultimately depend on the LPS physical parameters such as CAC, phase state of the acyl chains, and the size and conformation of the hydrophobic moiety of LPS, i.e., of lipid A.

In the first part of this chapter, we will review the function of LPS as a constituent of the outer membrane. The results from investigations with a planar reconstitution model of the outer membrane, the asymmetric LPS/phospholipid matrix, will be reviewed, focusing on studies of the function of porins and on the interaction with the complement system and with polycationic oligo- and polypeptides. In the second part of the chapter, more recent data on biophysical investigations into the phase behavior and structural polymorphism of LPS and free lipid A will be reviewed, and their possible relevance will be discussed in light of experimental findings and theoretical considerations of the biological activity of endotoxins, in particular, their interaction with target cell membranes.

THE FUNCTION OF ENDOTOXINS AS CONSTITUENTS OF THE BACTERIAL OUTER MEMBRANE

A detailed characterization of the various functions of the bacterial outer membrane is complicated by its complexity. Therefore, the reconstitution of simpler model systems (e.g., as a first step that of the unmodified lipid matrix being composed on one side of LPS and on the other side of phospholipids) is a feasible approach to studying the role of single amphiphilic components of the outer membrane, in particular that of LPS. This refers in the first place to the role of LPS in its potential interaction with membrane-active substances like drugs, detergents, and components of the immune system. The influence of LPS on the function of transmembrane proteins may, in subsequent steps, be studied by reconstitution of the proteins into the bilayer.

The influence of externally applied proteins or drugs on the membrane can be manifold. Their adsorption to the membrane leaflet facing the side of their addition may influence the membrane profile due to changes in the electrostatic environment of the lipid bilayer. Furthermore, interaction with lipids may influence the state of order of the acyl chains of membrane lipids, resulting in their fluidization or rigidification depending, among other parameters, on the depth of intercalation of the molecules into the membrane and on the functional groups involved in the interaction. These interactions may, in turn, result in membrane permeabilization either by disturbance of the lamellar structure of the bilayer or by the formation of transmembrane pores.

In this review, we will focus on events influencing the electrical parameters of the reconstituted membranes, thus being accessible to electrophysiological methods. Our review will not be a comprehensive survey of all of the work done with planar bilayers, but rather will focus on investigations taking into account the particular asymmetric architecture of the outer membrane of gram-negative bacteria.

Membrane Potential

Electrostatic properties of membranes are, in addition to their hydrophobic characteristics, of particular importance for interaction with biomolecules and may play a crucial role in many membrane-associated biological effects. In lipid bilayers, several distinct electrostatic potentials arise from different sources and combine in a characteristic intrinsic potential profile. These various potentials are briefly summarized before discussing the particular contribution of LPS as a constituent of the outer membrane or as a target for drug and protein interaction.

Living cells actively maintain specific intracellular cation concentrations differing from the respective extracellular concentrations, which results in a net negative charge inside the cell. The charge separation across the membrane causes a potential difference of up to -100 mV. This transmembrane potential (resting membrane potential), which can be measured by microelectrodes, is important for the regulation of voltage-sensitive ion channels (26).

Charged components of lipid molecules (as phosphate, carboxyl, or amino groups) expressed at the membrane surface generate a surface potential that, according to Gouy and Chapman (for review see Ref. 27), decreases exponentially within a distance of a few nanometers from the membrane surface. According to the specific lipid composition of the membrane, surface potentials of biomembranes are typically negative and of the order of a few tens of mV. Nevertheless, they play an important role in controlling biological processes within the immediate vicinity of both bacterial and host cell membranes [e.g., repulsion of anionic or attraction and binding of polycationic compounds causing, in turn, alterations of the potential (28)].

The largest contribution to the intrinsic membrane potential arises from the dipole potential (U_D) inside the lipid bilayer (≤ 300 mV). Experimental data as well as electrostatic calculations allow the conclusion that this potential is caused by oriented dipoles of bound water molecules, in particular in the interface region (29). Unlike the surface and the transmembrane potentials, the internal dipole potential is independent of the ionic strength of the electrolyte solutions on both sides of the membrane. The dipole potential is responsible for observed differences in the passive permeability of cationic and anionic molecules through the membrane and for conformational changes in proteins when they are incorporated into the membrane (30,31). An additional energy barrier arises from the Born self-energy (Born potential, U_B), which is the energy necessary to transfer a charge from a medium with a high dielectric constant ε (water) to one with a low ε (membrane) (32).

The transmembrane potential is the sum of the intrinsic membrane potential and the external potentials resulting from an ion gradient between both sides on the membrane and an externally applied clamp voltage (for review on membrane electrostatics see also Ref. 27).

Lipid asymmetry may provoke an additional potential difference between the two surfaces of the bilayer membrane. An extreme asymmetry in charge densities as well as in headgroup conformation occurs for asymmetric LPS/phospholipid bilayers such as those found in the lipid matrix of the outer membrane of gram-negative bacteria. The inner leaflet consists of a phospholipid (PL) mixture of phosphatidylethanolamine (PE), phosphatidylglycerol (PG), and diphosphatidylglycerol (DPG) in a molar ratio of PE:PG:DPG = 81:17:2. At neutral pH, only the PG molecules carry negative charges (one per molecule). The LPS component of the membrane at its simplest consists of Re chemotype–like molecules. In this case, each molecule carries four negative charges (LPS Re from *E. coli* F515). For the calculation of the surface charge densities of each leaflet, i.e., the number of charges per unit area, the molecular area occupied by the respective lipid molecules must be considered, which is 1.23 nm^2 for LPS Re and 0.55 nm^2 for phospholipids by a factor of four. Thus, the surface charge density of the LPS Re leaflet is higher by a factor of 10 than that of the PL leaflet. From these surface charge densities, the surface potential, the height of which is one determinant for cell function as well as for the interaction of drugs with the membrane, can be calculated according to the Gouy equation (33). Further determinants are the height and the profile (innermembrane potential difference $\Delta\Phi$) of the potential wall, which can be determined experimentally (32).

Planar Lipid Bilayer Techniques

To allow a better understanding of the discussion of the results obtained with different membrane systems, a brief introduction to the most frequently used techniques is given. In 1962, Mueller et al. (34) described a method for forming planar lipid bilayer membranes separating two aqueous phases: a dispersion of phospholipids in a nonpolar solvent such as *n*-decane is spread beneath an aqueous phase over an aperture up to several millimeters in diameter separating two plastic compartments. The lipid thins out in the center of the aperture until it forms a bilayer that is optically black when viewed in incident light (black lipid membrane, BLM). The membrane is essentially impermeable to ions, and thus, application of a voltage across the pure lipid bilayer does not result in any detectable electrical current. Induced membrane disturbances can, therefore, be monitored via current measurements (for review see Ref. 35).

This method of reconstructing biological membranes is not suitable for the reconstitution of the outer membrane of gram-negative bacteria, because the researcher has no influence on the arrangement of the different lipids in regard to the specific (asymmetric) composition of the outer membrane. This difficulty can be overcome using techniques introduced by Montal and Mueller (36) and Schindler (37). In both methods, bilayers are formed over a small aperture (diameter ≤200 μm) from two lipid monolayers on top of bathing solutions in two compartments separated by the aperture. In the Montal-Mueller technique, lipid solutions in a highly volatile solvent (e.g., chloroform) are spread on the air/water interface, whereas in the case of the Schindler technique the monolayers are built from vesicles added to the bathing solution. When monolayer formation is completed—either after solvent evaporation or after vesicle fusion at the air/water interface— the monolayers are successively raised over the aperture to form the bilayer membrane. Asymmetric membranes can thus be obtained if different lipids or lipid mixtures are used on the two sides of the aperture (for review see Ref. 38). Recently, solid supported bilayers were introduced, e.g., as biosensors (39), to monitor ligand binding (40), or to study the charge transport by ion-translocating membrane proteins (41).

In unsupported planar bilayer systems, the membranes are voltage-clamped via a pair of Ag/AgCl electrodes, and the transmembrane current is then monitored. Such measurements allow the determination of (1) the membrane capacitance (32,36,42,43) and with that also the membrane thickness, (2) the formation of static or transient transmembrane pores or lesions (44–47), (3) gating voltages of transmembrane channels, i.e., voltages required for switching between different conductance states (48–52), and (4) the inner-membrane potential profile (32,52–54).

For obvious reasons, the Montal-Mueller and Schindler techniques are most suitable for studying bacterial outer membrane function, the Schindler technique having the advantage of being entirely solvent-free but the disadvantage of the presence of lipids in the subphase. We will focus here on work based on these techniques and refer to respective reviews for work done with BLM.

Investigations into the Role of Endotoxins in Various Outer Membrane Functions

In the 1970s, a surprisingly low permeability of the membrane of gram-negative bacteria towards hydrophobic antibiotics was observed (55). Those experiments showed that the outer membrane of *Escherichia coli* or *Salmonella typhimurium* can serve as a very effective barrier against the penetration of hydrophobic molecules. At the same time, an asymmetric composition of the outer membrane, with LPS on the outer and a phospholipid mixture on the inner leaflet, was found (56,57), suggesting an important contribution of the LPS component to this unusual behavior. Since that time, the function of the outer membrane has been studied intensively and reviews on different topics such as membrane permeability (5,58) and membrane biogenesis (59,60) have appeared. In this review, we will focus on the role of LPS in the proper function of bacterial outer membrane proteins, interaction with the complement system, and with polycationic substances, e.g., antibiotics such as polymyxin B or antibacterial proteins of polymorphonuclear neutrophils (PMN), such as the bactericidal/permeability-increasing protein (BPI).

Porins

The outer membrane of gram-negative bacteria serves as a molecular sieve, allowing small hydrophilic molecules (≤600 daltons) to permeate through particular pores that are formed by special outer membrane proteins (Omp) termed porins (61). The porins, in general, have characteristic molecular weights between 30 and 50 kDa and normally form trimers within the outer membrane (62,63).

Although porins have been studied intensively for many years (for review see Refs. 64–66), there still exist controversial aspects of some fundamental questions. These include, among others, the influence of LPS on protein biosynthesis and on functional properties such as channel size, gating behavior, and the orientation of porins in the membrane.

Planar bilayers have been used in many studies to approach these questions. However, even though the importance of LPS for the function of porins has often been pointed out (67–69) and reviewed (70,71), in most studies symmetric planar phospholipid membranes have been utilized (for reviews on porin function mostly based on BLM studies see Refs. 64,65,72). Schindler and Rosenbusch (73) used on one side of their apparatus outer membrane vesicles containing LPS and porins and on the other side phospholipid vesicles to form asymmetric lipid bilayers. Only a few patch-clamp studies on porins using spheroplasts or giant proteoliposomes containing outer membrane fragments were carried out (74,75).

Contradictory results on the degree of voltage sensitivity of bacterial porins (voltage gating) have been reported even for the same porin but investigated in different systems. This holds in particular for OmpF (76,77) and OmpC (78,79) of *E. coli*. Potential reasons for these discrepancies might be found in the different membrane systems used for reconstitution, salt concentrations, pH, and, perhaps most importantly, the presence or absence of LPS. For example, Lakey and Pattus (76) compared the voltage-dependent gating of OmpF from *E. coli* reconstituted into symmetric lipid membranes prepared as BLM or according to the Montal-Mueller and Schindler techniques, respectively. Independent of the chosen method, the authors observed channel closing of the incorporated pores, however, the absolute height of the gating voltage differed for the different techniques. Furthermore, for Schindler-type membranes, the pores had to be activated by an initially applied external potential (see also Ref. 73) but were then independent of the sign of the voltage.

In 1989, our laboratory, applying the Montal-Mueller technique, for the first time successfully reconstituted the lipid matrix of the outer membrane as an asymmetric bilayer composed of LPS on one side and the natural phospholipid mixture on the other side (43). Using this reconstitution model, we were then able to answer questions on the influence of the lipid environment, in particular that of LPS, on the function of porin channels. For these experiments, we used LPS-free porins from *Paracoccus denitrificans*, which we incorporated into different symmetric and asymmetric phospholipid and LPS/PL planar bilayers (51).

In accordance with earlier published studies of Parr et al. (80), we were able to show that the presence of LPS is not required for manifestations of channel activity. The effective channel diameter of the porin trimer was calculated to be 1.5 nm in all lipid matrices. This value is in good agreement with that determined by Zalman and Nikaido (81) for the porin of *P. denitrificans* with the liposome-swelling method and to that of the closely related porin from *Rhodobacter capsulatus* (82) as determined in BLM studies (83).

LPS does, however, influence other parameters of porin function such as their incorporation rate into the reconstituted membrane and their gating behavior, i.e., channel closing in dependence on the transmembrane potential. The incorporation rates into LPS-free phospholipid bilayer systems were more than a factor of 10 lower than into asymmetric membranes with an LPS leaflet on the opposite side. The differences in the incorporation rates for the different membrane systems cannot be accounted for by differences in the surface charge densities between the two membrane leaflets of each system, because no differences in the incorporation rates into PG/PL and PL/PL bilayers were observed. This implies the existence of specific molecular interactions pulling the porin molecules into the bilayer when they reach the LPS leaflet. This observation is, therefore, a strong indication that specific interactions between the porin molecules and the LPS take place within the LPS leaflet and is in accordance with the model of Weiss et al. (84,85), who proposed that the very polar zones on the outer surfaces of the barrel-shaped porin monomers with their numerous negatively charged residues are directed towards the extracellular medium. The carboxylate groups of the porin are, according to these authors, likely to participate in the strong and tight network of divalent cations, stabilizing the outer leaflet, and LPS carboxylate groups in such a way that the interface between LPS and porin would become as tight as the LPS layer itself. The complete prevention of porin incorporation from the LPS side may be explained by the rigidity of the corresponding membrane leaflets or the steric hindrance caused by the sugar moieties of the glycolipids. This pathway would, in any case, not be physiological.

Using this established experimental model, we were able to study gating effects with respect to bilayer asymmetry, in particular with respect to the resulting charge distribution and the arising potential gradient in the case of LPS/PL membranes (51). We have observed voltage-dependent closing of porin channels in symmetric as well as in asymmetric bilayers, however, this occurred at different voltages depending on the lipid composition of the membranes. For asymmetric membrane systems, the absolute value of the gating voltage was lower when the applied potential was negative on the side of the more negatively charged lipids. Taking into account the surface charges and Gouy-Chapman potentials derived therefrom, a correlation between the height of the surface potential, which is most negative in the case of LPS, on the one hand and the gating voltage on the other could be stated. These data emphasize the strong influence of LPS on the voltage-dependent gating of porin channels.

Besides its influence on the transmembrane potential of the membrane, LPS is a dominant factor in membrane fluidity (5). LPS from *Salmonella minnesota* R595 undergoes a gel↔liquid ($\beta \leftrightarrow \alpha$) phase transition of the hydrocarbon chains at $T_c \approx 32°C$ (86), and a similar phase behavior has been observed for GSL-1, the glycosphingolipid with the shortest sugar moiety (52). When the LPS/PL bilayer—with the incorporated porin pores—was cooled down below T_c, we readily observed an influence of the phase state of the lipid matrix on the gating behavior of the porin channels (51).

Our observations emphasize the importance of the lipid matrix for studying porin functions and could help to explain the reported differences in the gating behavior of porins investigated in different membrane systems.

Activation of the Complement System by LPS and GSL Surfaces

The complement system is an important part of the host defense against invading bacteria. To study the mechanisms of complement activation by invading gram-negative bacteria, in particular with the aim of defining the functional groups of the surface glycolipids directly involved in the activation process, the utilization of asymmetric glycolipid/phospholipid membranes should be a method of choice, because complement acts primarily on the bacterial surface, which is, in the case of gram-negative bacteria, the outer membrane with its outermost LPS or GSL leaflet. In a planar bilayer membrane, the reconstituted surface glycolipids are presented in the natural orientation independent of their primary chemical structure and salt form. In this way any influence of the state of lipopolysaccharide aggregation on complement activation, which has already been described by Wilson and Morrison (87), can be ruled out. In the literature, very few investigations into conductance changes induced by complement as measured in symmetric planar phospholipid bilayers (88–92) as well as in patch-clamped cells (93) have been reported, and no work except that of the authors of this chapter using asymmetric LPS/PL membranes has been reported. The results of these studies concerning pore size, pore stability, and the requirements of various complement components for the induction of conductance changes differ considerably. In many early investigations, the complement system was activated by addition of C5b-6 (reactive lysis) (90–92), after the addition of antigen and antibody (88,94) or the adsorption of antibodies onto the membranes (93). In all cases, an increase in membrane conductivity was observed. From the increase in conductivity in single steps and bursts or from the amplitude of current fluctuations, estimates for the diameter of the induced membrane pores/lesions, based on the assumption of cylindrical water-filled channels, have been derived. These calculations were based on two different pore lengths: a pore length of 6 nm according to the membrane thickness or a length of about 15 nm based on electron microscopic observations (95,96). Assuming a pore length of 15 nm, two different pore sizes have been found: small pores about 1–2 nm in diameter with short lifetimes and leading to current fluctuations (91) and larger ones about 8–10 nm in diameter, which were more stable (90). Whereas the smaller fluctuations were suggested to represent transient precursors of the final membrane attack complex (93), the larger ones were correlated with the insertion of poly C9 complexes. In none of these studies was the influence of LPS on complement activation, especially on the activation pathway and the role of the lipid matrix in pore size, able to be accurately determined.

Concerning complement activation by isolated LPS and lipid A as it may depend on the length of the polysaccharide chain, fundamental work has been published by Vukajlovich et al. (97–99). These authors found that both LPS and lipid A led to an antibody-independent activation of the complement cascade, but the pathway of activation depended on the length of the polysaccharide chain: lipid A and deep rough mutant LPS (LPS Re) activated the complement system via the classical pathway, whereas lipopolysaccharides with a longer polysaccharide chain (including R-mutant chemotype LPS beyond LPS Re) activated the system via the alternative pathway. Mey et al. (100) showed that the acylation of the lipid A of a *Klebsiella pneumoniae* LPS was important for the activation of the alternative pathway of the complement system, and Clas et al. (101) pointed out the importance of the length of the sugar chains for binding of Clq. The disadvantage of all these studies is the fact that LPS and lipid A were presented as aggregates in an aqueous environment rather than in their natural structural environment—i.e., as constituents of a lipid bilayer.

Our reconstitution model of the lipid matrix of the outer membrane (43) has allowed us to determine the activation pathway utilized by LPS within a microbial-like environment as well as to investigate the influence of the lipid matrix on the pore sizes generated (46,52). Whereas no changes in conductance were detected when serum was added to the PL side of the membrane, serum addition to the LPS Re side led to a significant increase in membrane conductivity (Fig. 1). At the beginning of the conductance trace, the incorporation of C9 monomers was reflected by a stepwise increase of the conductance—about nine steps—until the formation of one individual pore was completed and the assembly of the next pore began. The small current steps either had identical amplitudes or exhibited a quadratic increase in the amplitudes. From these current traces, a model for the formation of complement pores could

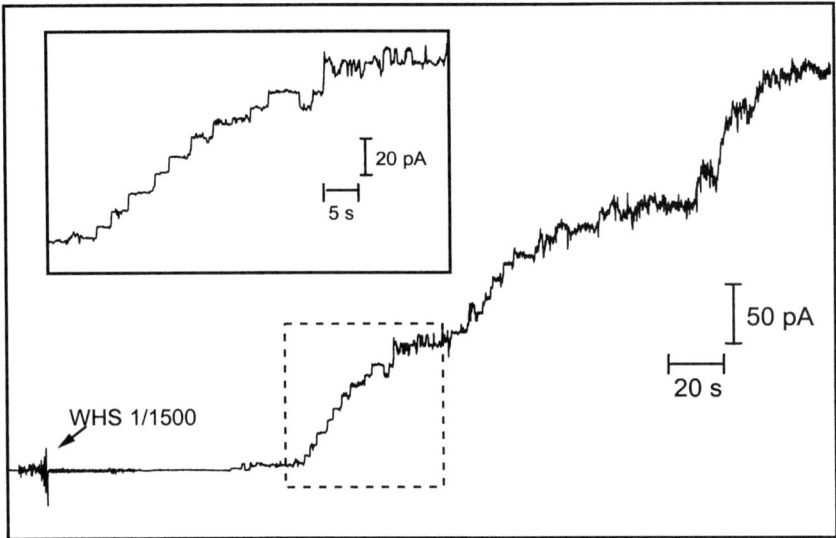

Fig. 1 Macroscopic current response of an asymmetric LPS/PL-bilayer upon addition (marked by an arrow) of 10:1 of whole human serum to the LPS side (final dilution of serum 1:1500). Each burst corresponds to the formation of a complement pore. Inset shows sequence of microscopic current steps within a burst. Each step corresponds to the insertion of one C9-monomer. (From Ref. 46.)

be derived, which is based on the subsequent incorporation of about nine C9 monomers to form either a cylindrical pore (corresponding to the quadratical increase in conductance) or a so-called leaky patch, which is aligned only on one side by C9 monomers (corresponding to current steps of same amplitude). From the amplitude of the current increase related to the formation of one complement pore, a pore diameter of about 8 nm was determined (46). Interestingly, experiments with LPS differing in the polysaccharide moiety (Re, Rdl, and Rc) and GSL-1 (52) led to very similar results.

In studies aiming at the elucidation of the pathways of complement activation, we found that neither the Ca^{2+} chelator EGTA nor the depletion of Clq had an influence on the activation by GSL-1, whereas in the case of LPS Re a significant deactivation was observed. From these results we confirmed that LPS Re activates complement predominantly along the classical pathway without the participation of antibodies, whereas GSL-1 activates the alternative pathway (46,52).

Polycationic Drugs and Proteins

The asymmetric LPS/PL planar reconstitution model also represents a suitable system for studying the interaction of bactericidal substances with the bacterial outer membrane. Here we will briefly review data on antibiotics such as polymyxin B or antibacterial proteins of PMN such as BPI, which has recently been discussed as a possible therapeutic agent (102).

For the polycationic decapeptide polymyxin B (PMB), which possesses antibacterial activity against most genera of gram-negative bacteria, including *Escherichia*, *Salmonella*, or *Pseudomonas*, the outer membrane is known to be the primary target (103,104). A rapid increase in membrane permeability with respect to charged or polar molecules of low molecular weight (105) and visible morphological alterations have been observed upon the addition of PMB (106). In the following decades, numerous studies on the action of PMB were reported, including lipid monolayers (107,108), liposomal membranes (109–111), black lipid membranes (112,113), and red blood cell membranes (114). For these studies, a variety of physical techniques such as electron spin resonance, nuclear resonance, fluorescence polarization, calorimetric methods, and x-ray diffraction were applied. Data obtained from these investigations mainly concerned structural information and binding properties of PMB to lipids. As would be expected from the polycationic character of the antibiotic, strong interactions have been observed with negatively charged amphiphiles including phospholipids like phosphatidic acid (115) or PG (116) and, particularly, LPS (106,117). Measurements of the electrical conductivity of planar lipid bilayers yielded contradictory results. Whereas Antonov et al. (112) proposed a carrier mechanism for the action of PMB, Miller et al. (113) found that PMB caused no conductance increase in BLM made from negatively charged phosphatidylserine or a mixture of PE and PG, but that it led to an unspecific destabilization of the membranes. In these studies, however, LPS was not used as a membrane component.

Using asymmetric LPS/PL membranes, we were able to show that the addition of PMB at concentrations corresponding to the minimal inhibitory concentration of PMB-sensitive gram-negative bacteria to the LPS side caused transient current fluctuations, which increased in frequency with time and finally combined with a stationary current (47). From a single-channel analysis, we found that the membrane lesions were large enough to allow the passage of PMB molecules through the membrane. This was not the case when LPS was replaced by GSL-1 (Fig. 2). This different behavior might explain the PMB resistance of *S. paucimobilis*. Furthermore, from the current traces and the critical concentration dependence of the observed effects, a model for PMB-induced pore formation was developed, which is based on the detergent-like character of the antibiotic (47,47a).

A great potential disadvantage of many antibiotics is the induction of LPS release from the bacterial surface leading to an augmentation of the inflammatory response. A new and promising therapeutic approach towards the treatment of gram-negative infections avoiding this disadvantage aims at the exploitation of endogenous peptide antibiotics, in particular BPI, having the capability of killing bacteria and at the same time detoxifying LPS.

Although various studies concerning antimicrobial activity (118–121), LPS binding (121,122), LPS detoxification (120,123), animal experiments (124–127), and human trials (128) with BPI and its fragments have been performed, the mechanism of action of BPI with the LPS surface of gram-negative bacteria is still poorly understood. Until now, the permeability-increasing effect of BPI has been determined very indirectly from the enhancement of the activity of actinomycin D, an antibiotic that normally cannot pass through the outer membrane (129). In our experiments with reconstituted asymmetric LPS/PL membranes, we used the recombinant N-terminal fragment $rBPI_{21}$ (130) instead of the holo protein. The addition of $rBPI_{21}$ to the LPS Re side led to membrane rupture. Following $rBPI_{21}$ addition, the inner membrane potential difference showed only a slight increase from 0 to 5 mV for symmetric DOPG membranes, but changed from -36 to $+8$ mV for asymmetric LPS/PL membranes (Fig. 3). In both cases, the addition of $rBPI_{21}$ led to an increase in membrane current (131). Furthermore, $rBPI_{21}$ reduced the absolute value of the surface potential of LPS Re and PG ag-

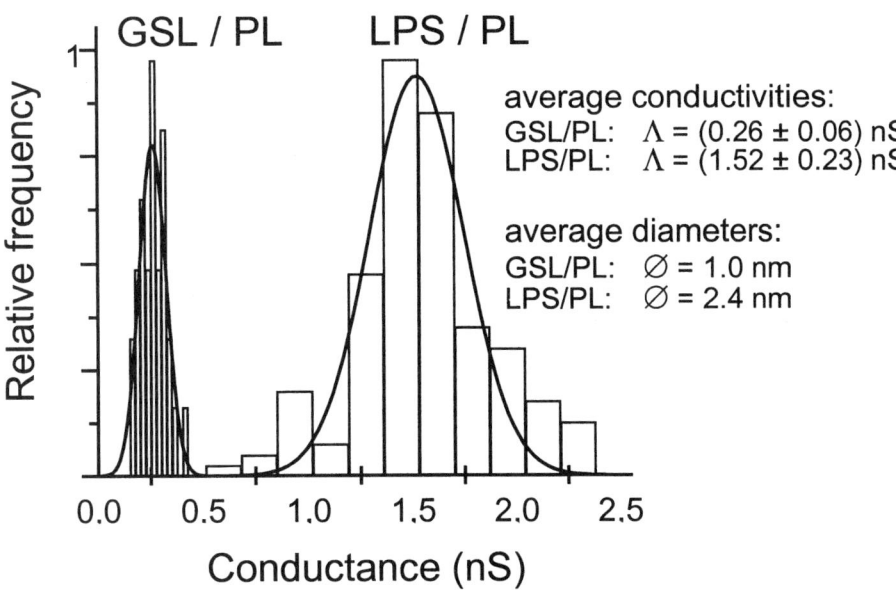

Fig. 2 Comparison of the distributions of the conductances λ of transient membrane lesions induced by PMB in asymmetric planar GSL/PL (left) and LPS/PL bilayers (right) and the respective calculated average diameters φ of the lesions. (From Ref. 47a.)

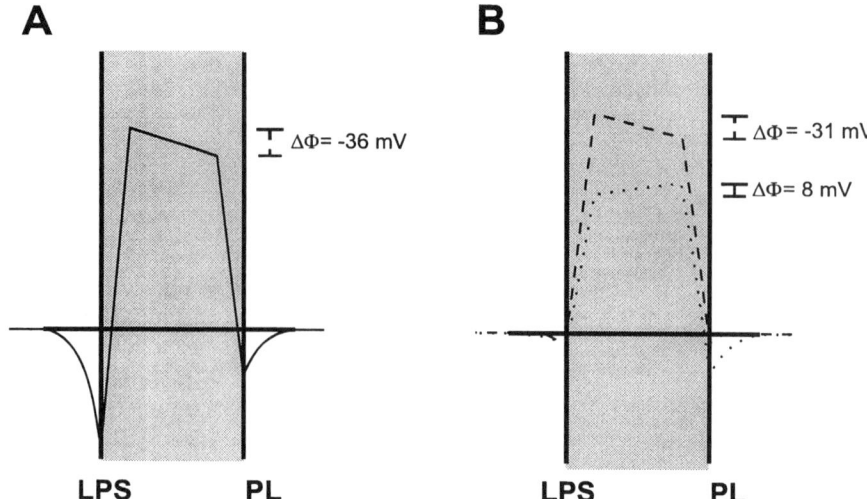

Fig. 3 Intrinsic membrane potential profiles of nonactin-doped LPS/PL membranes: (A) In the absence of rBPI$_{21}$ and MgCl$_2$ and (B) after the addition of rBPI$_{21}$ to the LPS side in the absence (dotted lines) and presence (dashed lines) of 40 mM MgCl$_2$ in the bathing solution. (From Ref. 131.)

gregates and caused the rigidification of the lipid acyl chains in the gel as well as in the liquid-crystalline phase and led to a drastic immobilization of the phosphate groups. Also, rBPI$_{21}$ could be shown to incorporate into liposomes made from or containing negatively charged phospholipids. In all systems, high concentrations of Mg^{2+} ions had a protective effect (132). Thus, in the presence of Mg^{2+}, the rBPI$_{21}$-induced potential shift was significantly reduced (from -36 to -31 mV; see Fig. 3). On the basis of these results, we discussed a model for the action of rBPI$_{21}$ on the outer membrane of gram-negative bacteria that includes the following steps: rBPI$_{21}$ binds to the negatively charged lipid surface, followed by its insertion, which causes rigidification of the acyl chains of the LPS layer and its destabilization, eventually provoking membrane rupture. The binding and insertion of the protein leads to significant changes of the membrane current, influencing membrane permeability for hydrophobic molecules, and of the transmembrane potential, which may be assumed to influence channel gating and with that cause membrane dysfunction (131).

BIOLOGICAL ACTIVITY OF ENDOTOXIN

In the 1960s, the physicochemical parameters of LPS were considered to be important for an understanding of its bioactivity. However, at that time and in the following decade, the available techniques and the lack of clearly defined endotoxin structures allowed the establishment of only qualitative structure-activity relationships. Thus, in 1976 Galanos and Lüderitz (133) found that the physical state of lipopolysaccharide, i.e., its "state of aggregation," played a decisive role in the interaction with complement, but a detailed characterization of LPS and lipid A aggregate structures could not be performed.

The importance of LPS physicochemical parameters involves the critical aggregate concentration (CAC), the phase state (fluidity of the endotoxin acyl chains), which is a property of the bulk lipid, and the molecular conformation (shape), which is a property of the individual molecules but is deduced from the type of supramolecular aggregate structure built up from a large number of identical endotoxin molecules. As the lipid A part of LPS is known to represent its "endotoxic principle" and is the molecular entity that primarily interacts with the host cell membrane, determination of the conformation of free lipid A should be of the utmost importance for an understanding of biological action. The role of the sugar moiety of LPS should, with this underlying concept in mind, be restricted to a modulation of lipid A bioactivity, mainly due to its influence on the hydrophobicity of the endotoxin molecules and therefore on their CAC but also on the fluidity of the lipid A acyl chains.

From these considerations, a more detailed physicochemical characterization of the above-mentioned parameters mainly of lipid A, but also of LPS, remains an important issue to address in order to understand why some endotoxin structures are biologically highly active and others are inactive.

Phase States

For amphiphilic compounds like lipid A and LPS, an endothermic transition between a highly ordered gel phase (β phase), in which the acyl chains are in the *all-trans* (zig-zag) configuration, and a less ordered liquid-crystalline phase (α phase), which is characterized by an increasing number of *gauche* conformers—caused by rotations by 120° chain segments around the acyl chain long axis—(134) takes place. The phase behavior and the phase-transition temperature T_c can be monitored by various spectroscopic methods like electron spin resonance (ESR), nuclear magnetic resonance (NMR), and fluorescence and infrared spectroscopy and calorimetry, allowing in addition the determination of the enthalpy change ΔH_c. The transition temperature T_c depends on properties of the hydrophilic headgroup (charge, size, conformation, water-binding capacity) as well as of the hydrophobic moiety (number, length, and degree of saturation of the acyl chains). A transition between the different phase states is usually accompanied by changes in the geometry of the involved molecules, i.e., of the geometrical cross sections of the molecules, and may, therefore, have an impact on the structure of the supramolecular assemblies.

Emmerling et al. (135) were the first to observe a β↔α phase transition of endotoxin, labeled with the fluorophore *N*-phenylnaphthylamine. Van Alphen et al. (136), applying light scattering spectroscopy, extended these measurements to LPS from bacteria grown at 12°C and Coughlin et al. (137), using ESR spin probes, to native and electrodialyzed LPS. All endotoxins were from *E. coli* K12. Peterson et al. (138) also compared LPS with different sugar chain lengths and explained the finding that the mobility of the spin probe CAT_{12} (4-dodecyl dimethylammonium-1-oxyl-2,2,6,6-tetramethylpiperidine bromide) labeling the headgroup region was greater in the long-chain than in the short-chain fraction by the tightly packed sugar portion of the latter. Furthermore, the addition of cations led to a stronger displacement of the spin probe in the case of the short-chain fraction. These results indicate that the aggregate shape and reactivity of LPS are affected by O-antigen length.

We have found a very characteristic dependence of T_c on the chemical structure of the sugar moiety of LPS by applying fluorescence and Fourier-transform infrared spectroscopy (FTIR) as well as differential scanning calorimetry (DSC) (86,139–142). Thus, for enterobacterial strains (natural salt forms), T_c were highest for free lipid A (around 45°C), lowest for deep rough mutant LPS (around 30°C), and, with increasing length of the polysaccharide portion toward completion of the O-chain (wild-type LPS), increased again up to 37–40°C. Naumann et al. (143), using FTIR, arrived at similar results. Moreover, for synthetic lipid A of *E. coli* and *S. minnesota* (compounds 506 and 516, respectively) T_c values in close agreement with the natural compounds were found (144). The phase-transition temperatures of various nonenterobacterial LPS and lipid A were found to be significantly lower than those of Enterobacteriaceae (139,145). This observation correlates with the fact that the former often contain shorter acyl chains with a higher degree of unsaturation (146). In an infrared spectroscopic study, the T_c values and corresponding states of order of various lipid A analogs and partial structures were determined (147). A characteristic dependence of T_c on the number of acyl chains was observed, with T_c lying significantly below 0°C for the diacyl analog 606, between 15 and 20°C for the tetraacyl compound 406, corresponding to the lipid A precursor Ia or IVa, and at 43°C for the hexaacyl compound 506. For a given LPS, the values of T_c seem to be strongly dependent on its salt form, being lowest for the Na salt and highest for the K salt form (139).

The addition of divalent cations or a lowering of pH caused a significant rigidification of the acyl chains of free lipid A and LPS preparations and also sometimes led to an increase in T_c. At basic pH, a lowering of T_c was observed. Concomitant with a decrease in T_c, fluidization of the LPS and lipid A acyl chains occurred (139,148,149). However, the values of the shift of T_c were found in different studies to vary, possibly due to problems in the pH determination of endotoxic dispersions. Excess concentrations of Mg^{2+} led to an increase in T_c for LPS from *S. minnesota* (139). A similar effect was observed for LPS from *Erwinia carotovora* at [endotoxin]:[cation] = 1:1 M (Mg^{2+} or Ca^{2+} (149). A study of the influence of divalent cations on the phase behavior of LPS Re showed a cation-specific increase of T_c in the sequence $Mg^{2+} < Ni^{2+} < Co^{2+} < Zn^{2+}$ (150). These effects were similar in buffer and in serum.

Furthermore, free lipid A and LPS exhibit extremely strong lyotropic behavior, i.e., a strong dependence of the expression of the β↔α chain-melting transition and of its enthalpy on the water content (139,151–153). Thus, for example, for LPS Re the phase transition appears distinctly only at water concentrations of >50%. It seems reasonable to assume that the strong lyotropic behavior of LPS and lipid A is one reason for partly contradictory statements in the literature about the phase behavior as well as the aggregate structure of endotoxin. This observation may explain why Labischinski et al. (154), in an earlier investigation with x-ray diffraction, did not detect any phase transition up to 50°C with dry or 'hydrated' lipid A and LPS samples from *S. minnesota*.

Ramos-Sanchez et al. (155) and Rodrigues-Torres et al. (156) investigated the phase behavior of endotoxin preparations of *Brucella* and other gram-negative bacteria. They describe endotherms in biologically irrelevant temperature regions. Thus, endotherms between 108 and 200°C were assigned to transitions in the polysaccharide moiety due to depolymerization and an endotherm between −13 and −36°C ("cooling phase transition") was attributed to transitions between structural forms differing in hydrogen bonding within the polysaccharide moiety of LPS. These authors did not report on the β↔α transition probably because they investigated dry LPS samples (see above).

It is well known that bacteria adapt the fluidity of their cytoplasmic membranes to a given growth temperature by the synthesis of fatty acids of proper length and degree of saturation (157). In this way, T_c of the phospholipid matrix is shifted by approximately the same value as the growth temperature. For LPS, the situation is very different: T_c of LPS from *Proteus mirabilis* R45 and of its free lipid A, for example, are shifted only from 38 to 32°C and from 52 to 48°C, respectively, when the growth temperature is shifted from 37 to 15°C (own unpublished data in collaboration with H. Mayer). These data are in good overall agreement with those published by van Alphen et al. (136). It must be pointed out that even at the growth temperature of 37°C the fluidity of the LPS leaflet of the outer membrane is considerably lower than that of the inner phospholipid leaflet.

Critical Aggregate Concentration, Supramolecular Structure, and Molecular Conformation

Similar to other amphiphilic molecules like phospholipids, endotoxins aggregate to supramolecular structures above the CAC. The three-dimensional supramolecular structure of the aggregates depends on the molecular conformation (shape) of the contributing individual molecules, which is determined by their primary chemical structure and ambient conditions. Thus, the molecular conformation can be derived from the aggregate structure.

As a first approximation, the aggregate structure of the amphiphilic molecules can be predicted from a simple geometric model, which relates the resulting structure to the ratio of the effective cross-sectional areas, a_o, of the hydrophilic polar and, a_h, of the hydrophobic apolar regions, respectively. Israelachvili et al. (158) introduced a dimensionless shape parameter S, which is defined as $S = v/(a_o Al_c) = a_h/a_o$, where v is volume per molecule of the hydrophobic moiety and l_c is length of the fully extended hydrophobic portion. From the absolute values of the shape parameter, which may be estimated from energy minimization calculations (see next section), a structure of the supramolecular aggregates can be deduced.

The correlation between different molecular shapes and the corresponding supramolecular structures is schematically outlined in Figure 4 (for a more detailed description, see Ref. 158). In this figure, only those three-dimensional structures relevant for an understanding of the structural polymorphism of lipid A, the "endotoxic center" of LPS, are considered. Lipid A with $1/2 \leq S \leq 1$ assumes lamellar structures (Fig. 4A). Lipid A with a prominent axis of the headgroup and $S > 1$ adopts inverted hexagonal H_{II} structures (Fig. 4C). In the latter case, the water component is assumed to form idealized circular rods lined with the hydrated lipid headgroups and with the remaining volume filled by the fluid hydrocarbon chains. More recent investigations have shown the importance of other nonlamellar, so-called cubic structures (Fig. 4B), which may occur as stable intermediate phases between lamellar L and inverted hexagonal H_{II} (159–162). Other than the L and H_{II} phases, the cubic Q phases cannot be described by a unique structure. They rather exist as various structures, which may be assigned to different crystallographic space groups (161). In Figure 4, only the most important cubic structures relevant for lipid A are shown.

Since the molecular shape varies as a function of ambient conditions, alterations of the latter may lead to modifications of spatial requirements of the molecules and, with that, of the shape parameter S. Thus, for instance, an increase in a_h or a decrease in a_o may induce a transition from a lamellar into a hexagonal H_{II} phase. An increase in a_h may be caused by an increase in temperature; a decrease in a_o may be provoked, particularly for the negatively charged free lipid A, by higher concentrations of mono- or divalent cations or by a decrease in pH. The reversible lamellar-to-hexagonal H_{II} transition occurs mainly in the liquid-crystalline phase.

So far, no exact value for the CAC of lipid A has been determined due to extreme experimental difficulties in the low concentration range (e.g., detection limits, loss of material due to absorption to walls, etc.). A rough estimation can be given from the limited number

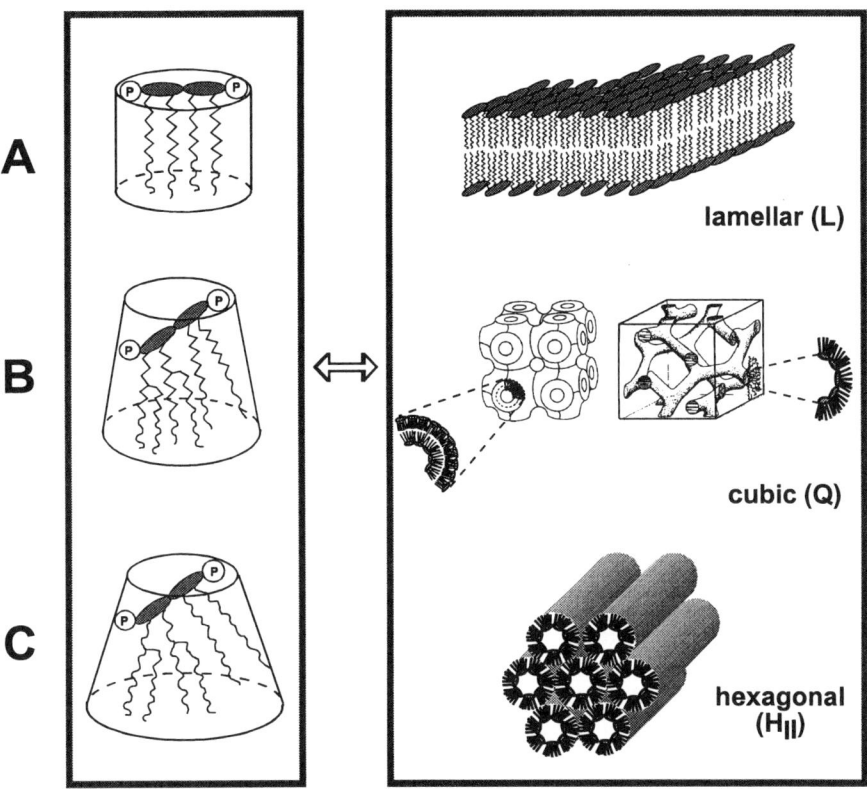

Fig. 4 Molecular conformations of lipid A (left) in relation to their supramolecular structures (right).

of available data for other lipids. Thus, from the change in CAC from 5×10^{-10} M for dipalmitoylphosphatidylcholine to 7×10^{-6} for lysopalmitoylphosphatidylcholine (13), which have the same headgroup but differ in the number of acyl chains by a factor of two, for hexaacyl lipid A a value of well below 10^{-10} M may be assumed on the basis of the approximated CAC $< 10^{-7}$ M for the lipid A precursor IVa, a tetraacyl lipid A (163). This approximation is backed by data on the solubility of LPS Re determined experimentally applying equilibrium dialysis by Takayama et al. (164,165). The authors found saturation values of the solubility of 3.3×10^{-8} M at 22°C and 2.8×10^{-8} M at 37°C. As the monomer concentration decreases with increasing aggregate concentration (13), these measurements would imply even higher values for the CAC than the values listed above. As LPS Re should be less hydrophobic than lipid A, the CAC of the latter should be lower.

For the determination of the long-range order (supramolecular structure) and of the short-range order (arrangement of the acyl chains), small-angle and wide-angle diffraction, respectively, with x-rays or neutrons can be used. Emmerling et al. (135) have published x-ray diffraction data on LPS Ra from *E. coli* K12 and found one ordered lamellar phase in the temperature range 0–50°C for an LPS K12 sample between 0 and 30% water content. From the observation of a wide-range reflection at 0.433 nm, the authors concluded that the ordered acyl chain conformation is considerably less ordered than the hexagonally packed acyl chains in biological membranes. Wawra et al. (166) arrived at similar results with dried samples of wild-type LPS and LPS Re from *S. minnesota* R595; however, in contrast to Emmerling et al., they observed a wide-angle reflection at 0.41 nm, which they interpreted as arising from the hexagonal dense packing of the acyl chains. Labischinski et al. (154) found no evidence for other than lamellar structures at temperatures up to 50°C for dry or hydrated wild-type form and LPS Re and lipid A from *Salmonella* spp. Experiments with synthetic lipid A analogs and partial structures in the dried state—differing in the number of the hydrocarbon chains attached to an identical bisphosphoryl diglucosamine backbone—led in each case to diffraction patterns typical for lamellar structures (167). These and other synthetic compounds were analyzed at water concentrations of >90% mainly applying infrared spectroscopy and, for calibration, in some experiments also x-ray diffraction (147). From these experiments it was concluded that di- and triacyl lipid A analogs adopt micellar, tetra- to pentaacyl samples lamellar, and other analogs cubic or H_{II} structures. The possible importance of these finding for biological activity will be discussed later.

Labischinski et al. (168) applied small-angle neutron scattering to monophosphoryl lipid A from LPS Re of *S. minnesota* and found again only lamellar structures between 10 and 50°C at a water content of 99%. Hayter et al. (169) applied the same technique for the structural analysis of LPS Re from *E. coli* D21f2 and best fit their data to a randomly coiled tubular micelle structures. In contrast, the x-ray diffraction measurements of Naumann et al. (143) on LPS Re from *S. minnesota* R595 as dried films and as hydrated samples (suspensions or centrifuged pellets) in the temperature range 10–60°C gave only lamellar structures. Indirect evidence for the existence of nonlamellar structures for lipid A and LPS Re was obtained from a pH titration study applying ESR by Coughlin et al. (148). These authors proposed that at high pH values LPS should assume H_I or tubular micellar structures, at low pH inverted micellar or hexagonal H_{II} structures, and at normal pH lamellar structures.

The results presented so far seem to reflect the different experimental conditions under which the samples were investigated, in particular (as emphasized earlier) the strong lyotropism of endotoxins. Expressed in terms of bound water molecules, at 75% water content one LPS Re molecule binds approximately 400 water molecules, and one LPS Ra 730.

Therefore, to completely account for structural diversity, phase diagrams for lipid A from *S. minnesota* and *E. coli* were established over a wide range of water content (20–95%) and Mg^{2+} concentrations (molar ratio [lipid]:[Mg^{2+}] from 1:0 to 1:1) and dependence on temperature (140,141,151,170). These parameter ranges included those under approximate physiological conditions. From this, the conformation of the individual lipid A molecules can be approximated as described above. Similar measurements have also been performed with LPS from rough mutants Re to Ra (152,153). Although it may be assumed that the conformation of the lipid A component is only marginally changed in the presence of additional sugar groups, the three-dimensional structural preference of the different rough mutant LPS may nevertheless be important for an understanding of the aggregate stability and the interaction with serum proteins. The phase transition behavior was investigated with FTIR spectroscopy and DSC, and the supramolecular structure with small-angle x-ray diffraction utilizing the high brilliance of synchrotron radiation.

Briefly, the results can be summarized as follows: in pure lipid/water systems, free lipid A aggregates into lamellar structures at water contents below approximately 60% and into nonlamellar cubic structures at higher water concentrations already below T_c. In other words, a lyotropic structural transition takes place around 60% water content. This observation is also true for deep rough mutant LPS Re, but for other rough mutant LPS the tendency to assume only lamellar structures at $T < T_c$ increases. In the presence of divalent cations (e.g., Mg^{2+}) in a lipid-to-cation molar ratio of ≤ 1, nonlamellar structures are suppressed below T_c. At the same time, T_c is shifted to higher values (see previous section). Once the acyl chain melting process begins, free lipid A assumes nonlamellar cubic structures over the whole range of water contents. With the completion of chain melting, the cubic structures change into inverted hexagonal H_{II} structures. For deep rough mutant LPS, the situation is similar in that with the chain melting a transition into nonlamellar cubic structures takes place, but the transition into H_{II} occurs at considerably higher temperatures (>70°C). The tendency for other rough mutant LPS to adopt nonlamellar structures at and above T_c decreases with increasing completeness of the core region. The presence of divalent cations at concentrations as described above suppresses the formation of nonlamellar cubic structures.

The transitions between different cubic (Q_1, Q_2), lamellar (L) to cubic, and cubic or lamellar to H_{II} structures have been shown to have impact on infrared-active functional groups within the lipid A moiety (171). Thus, for LPS Re a distinct shift of the low-frequency ester carbonyl stretching vibration C=O (similarly also for the amide II vibration) is observed for the $Q_1 \leftrightarrow Q_2$, $Q_2 \leftrightarrow H_{II}$, or $L \leftrightarrow H_{II}$ transitions corresponding to different hydrational states within each structure. This finding indicates the possibility to monitor structural changes via IR spectroscopy without the use of synchrotron radiation x-ray diffraction, which may be of importance in light of the fact that the latter technique is available only to a limited degree and requires more than one order of magnitude more material than required by IR experiments.

In further studies, measurements of the aggregate structure were extended to other lipid A samples, i.e., enterobacterial LPS and lipid A in different salt forms as well as monophosphoryl lipid A and lipid A from nonenterobacterial sources like those of *Rhodobacter capsulatus*, *Rhodopseudomonas viridis*, *Rhodocyclus gelatinosus, Rhodospirillum fulvum, Campylobacter jejuni,* and *Chromobacterium violaceum* (145,172,173). The measurements were carried out exclusively under near physiological conditions with the purpose of directly correlating the results to data from biological test systems. It was found that the kind of counterions present (endotoxins in different salt forms) significantly influenced the aggregate structure of LPS and that different nonenterobacterial lipid A samples showed a variety of aggregate structures ranging from H_{II} (lipid A from *R. gelatinosus*) over mixed cubic/lamellar (monophosphoryl lipid A from *S. minnesota* and lipid A from *C. jejuni*) to pure lamellar structures (lipid A from *C. violaceum, R. capsulatus, R. viridis,* and *R. fulvum*) (145,172,173). It was later shown that these different structural preferences are important for the expression of biological activity (see below).

Based on the primary chemical structure and by application of molecular modeling techniques, it has been possible to calculate the potential conformations accessible to a given molecule. Labischinski et al. (154) have published results from theoretical calculations on the conformation of the heptaacyl lipid A component of *S. minnesota*. The most striking result of these calculations suggest that an angle of approximately 45° exists between the bisphosphoryl disaccharide backbone and the fatty acids chains. The fatty acids occupy positions lying almost exactly on a hexagonal lattice, and the terminal methyl groups do not form a plane parallel to the membrane surface. The authors arrive at a length of 2.6 nm and a cross section of 0.6–0.8 nm for the smaller and 1.2–1.6 nm for the longer side of the rectangular cross section of the acyl chains. Similar results were obtained for the hexaacyl LPS Re from *E. coli*. For the latter, however, Kastowsky et al. (174) found a number of slightly different conformations. Using these models, anisotropic lateral dimensions of 1.0–1.1 nm and 1.7–2.0 nm for the longer and shorter side, respectively, were determined. The Kdo moiety was found to be centered on top of the molecule preferring two orientations, each of them stabilized by hydrogen bonds involving only one phosphate group of the lipid A moiety at a time. The precise orientation of the Kdo was found, not unexpectedly, to be sensitive to the charge state of the molecule. Some free hydroxyl groups of the R-(3)-hydroxymyristoyl chains were shown to cause acyl chain packing perturbations within the hydrophobic domain. In this model, which gives a hexagonal dense packing of the acyl chains, the tilt of the diglucosamine was determined to (53 ± 7°) with respect to the membrane normal with the reducing side of the diglucosamine, i.e., the 1-phosphate emerged in the hydrophobic moiety of the neighbouring molecule, while the 4′-phosphate sticks out into the aqueous phase. We could, however, verify via IR spectroscopy that the 1-phosphate is surrounded by water and the 4′-phosphate is buried in the backbone-close hydrophobic region, probably facing the 3-hydroxyl groups of the neighboring molecules (147). Evidence in support of the existence of this more unfavorable packing of the acyl chains was provided by the following observations: (1) several lipid A analogs and partial structures do not have a hexagonal dense packing and (2) lipid A and LPS have strongly reduced ΔH_c-values as compared to saturated phospholipids with the same acyl chain length. From this, a comparison on the basis of the number of methylene groups indicates a reduced packing density of the former.

Naumann et al. (143) compared their computer modeling data from LPS Re with those for phosphatidylethanolamine, which is known to form inverted H_{II} structures at higher temperatures (175). Based on the similarity of these data as well as the fact that their model calculations were performed on an LPS Re in the β phase, the authors concluded that LPS Re could also form inverted phases at elevated temperatures. However, since they did not find experimental evidence for the H_{II} structure, they deduced that a purely geometric, static interpretation of such data had inherent limitations, and that the effective polar headgroup area of charged lipids should be much larger than that of noncharged lipids. In this respect, the authors state that LPS Re should be rather comparable with phosphatidic acid, which they assumed to adopt a lamellar phase at physiologically relevant conditions, although their model calculations for this compound interestingly suggest a preference for the H_{II} phase. We could show, however, that dioleoylphosphatidic acid in the presence of divalent cations at room temperature (151) as well as natural phosphatidic acid at 37°C (U. Seydel and K. Brandenburg, unpublished results) adopt H_{II} structures. Thus, these model calculations do not exclude the formation of inverted structures by LPS Re and, even less, by free lipid A.

Recently, the three-dimensional structure and conformational flexibility of a complete wild-type LPS was modeled (176) (Fig. 5). As anticipated, the LPS-core polysaccharide (LPS Ra) moiety should have an approximately cylindrical shape, whereas the O-specific chain was found to be the most flexible portion within the molecule preferring bent conformations and partially lying flat on top of the headgroup of neighboring molecules. When the dimensions of the calculated model were compared to x-ray diffraction data of sev-

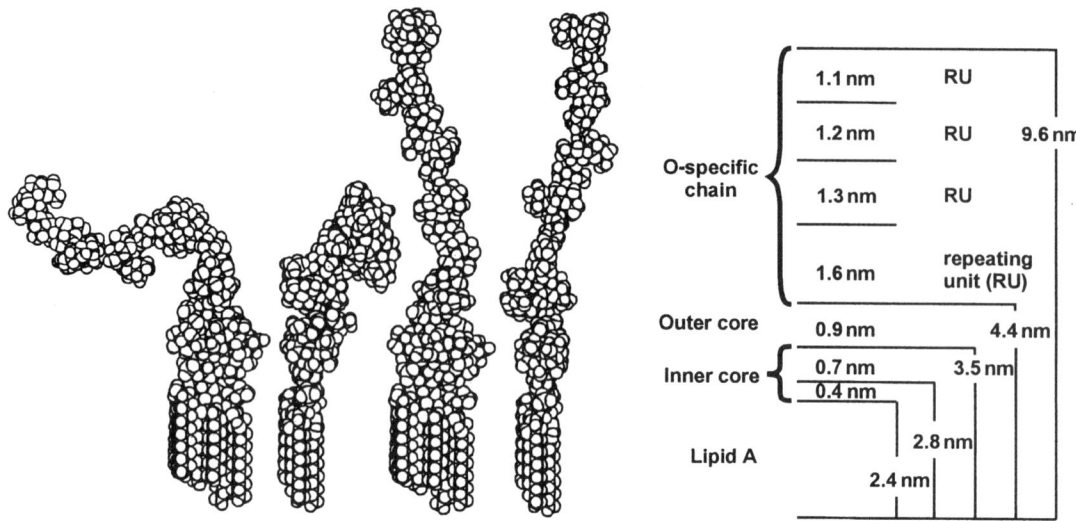

Fig. 5 Dimensions of the calculated LPS S-form model and of partial structures. The dimensions of partial LPS structures up to LPS Ra (complete core) fit well with x-ray powder diffraction data. (Adapted from Ref. 176.)

eral dried LPS structures, a reasonable agreement between the two was found.

By combining molecular modeling with data from NMR spectroscopic measurements, Jung et al. and Wang and Hollingworth (177,178) investigated the three-dimensional structure of the LPS lipid A moiety on dried LPS Re films. When looking at the membrane from above, they found that the calculated structures can easily form extended hexagonal structures assuming important contributions of van der Waals interactions between the fatty acid chains and electrostatic interactions by Mg- or Ca-phosphate bridges. These investigators arrived at a triangular cross section of the lipid A part as the most efficient packing of the acyl chains with six lipid A molecules forming a microdomain. These microdomains are associated with each other by cation-phosphate bridges. Under these conditions, however, large gaps between the single microdomains must exist, and the acyl chains would, therefore, have to be surrounded completely by the aqueous medium, a thermodynamically most unfavorable configuration. Thus, the physiological relevance of these findings remains unclear.

A different approach for the determination of the molecular geometry was chosen by Kato et al. (179–181). These authors successfully crystallized synthetic lipid A and rough mutant LPS Re to Ra, but the crystallization of natural lipid A and wild-type LPS failed. When crystals were analyzed by electron diffraction, the anisotropic cross section of the acyl chains was determined to be 1.67 nm × 0.924 nm for hexaacyl LPS Re and Ra, and the length of the lipid A part 2.96 nm. For the various rough mutant LPS, different morphological shapes were observed, including squares, rectangular plates, lozenge plates, discoids, and truncated hexagonal pyramid forms. The acyl chains of synthetic *E. coli*-type lipid A—viewed perpendicular to the membrane surface—were found to form a hexagonal lattice with a lattice constant of 0.462 nm. The longitudinal axis was determined to 4.93 nm corresponding to the bilayer thickness. The monolayer thickness derived from this value (2.47 nm) is thus slightly lower but still in quite reasonable agreement with the length of the lipid A molecule given by Labischinski et al. (154).

One restriction of the molecular modeling studies and the experiments with crystallized endotoxins results from the necessity to neglect the potential influence of water molecules, thus not allowing an extrapolation to physiological conditions. In a very recent paper, Obst et al. (182) carried out molecular dynamics simulation of fully hydrated partial structures of LPS Re from *E. coli* in the presence and absence of Ca^{2+}. They found the cations to be located between the carboxylate and phosphate groups leading to a rigidification of the headgroup and an alteration of the conformation of the backbone, thus influencing both the structure and flexibility of the hydrophobic region as well. Aggregation of these LPS Re monomers would be predicted to lead to lamellar structures. However, because intermolecular binding between neighboring LPS Re molecules was not considered in these calculations, the authors stated that a direct comparison to experimental results derived from data on aggregated lipids is not possible.

A comparison of the data from various investigations of the molecular geometry of lipid A collectively suggest very similar values for the bilayer thickness in the absence of water or at low water content ranging from 4.8 to 5.2 nm and increasing to a value of 5.5 nm at high water content (151). The respective values for LPS Re are 5.7–6.4 nm (152), for LPS Ra 7.8–8.8 nm, and the other rough mutant LPS lying somewhere in between (153). It should be noted that the values at high water content for lipid A and rough mutant LPS were obtained in the presence of Mg^{2+}, which was necessary because of the tendency of these endotoxin samples to adopt nonlamellar structures (see above). Very interesting are the values of the bilayer length for smooth form LPS. These are reported to lie in the range of 7–12 nm at low water content (135,154,166) and 24–36 nm at high water content (Brandenburg and Seydel, unpublished data) indicating a dramatic swelling of the bilayer periodicity due to the hydration of the O-chain. Thus, under these conditions the O-antigen constituent of the LPS is far from being "heavily coiled" as found at low water content (154).

Relation of Physicochemical Data to Biological Activity

Assuming that the phase behavior and the three-dimensional structure determined under near physiological conditions and outlined here reliably describe the structural polymorphism of these substances, there are some striking correlations between the phase-transition temperature of the endotoxins as well as of their supramolecular structures and various biological effects. However, since the reversible chain melting transition frequently accompanies a transition between different structures, it is in many cases difficult to establish whether the phase states of the acyl chains or the structural behavior of the entire lipid assembly, or a combination of both, are the primary processes dictating the resultant biological activity. Nevertheless, various biological effects can be correlated with the value of T_c or with the state of order of the acyl chains at 37°C. Here, only a very recent example is given for the induction of monokine secretion by LPS Re. Wellinghausen et al. (150) correlated a strong Zn^{2+}-induced decrease of fluidity of LPS Re at 37°C with its capability to enhance monokine secretion (TNF-α and IL-1β) from human monocytes. These authors found that the increasing order of the hydrocarbon chains induced by Zn^{2+} facilitated a stronger bond between LPS and LBP, thus enhancing the transport of LPS to the target membrane and subsequent release of monokines [for a more complete presentation of the influence of the acyl chain fluidity, the reader is referred to an earlier review by Seydel and Brandenburg (141)].

A most striking correlation was found between the biological activity of lipid A from different species and their ability to adopt particular supramolecular structures: lipid A samples adopting lamellar structures (having a cylindrical conformation) such as those from *R. capsulatus* and *C. violaceum* were completely inactive, those assuming mixed lamellar/cubic structures (partly conical conformation) had intermediate activity, and those samples preferring pure nonlamellar (Q, H_{II}) structures (conical conformation) were highly active (Fig. 6). A quite different correlation between structure and activity was found by Brade et al. (183) for im-

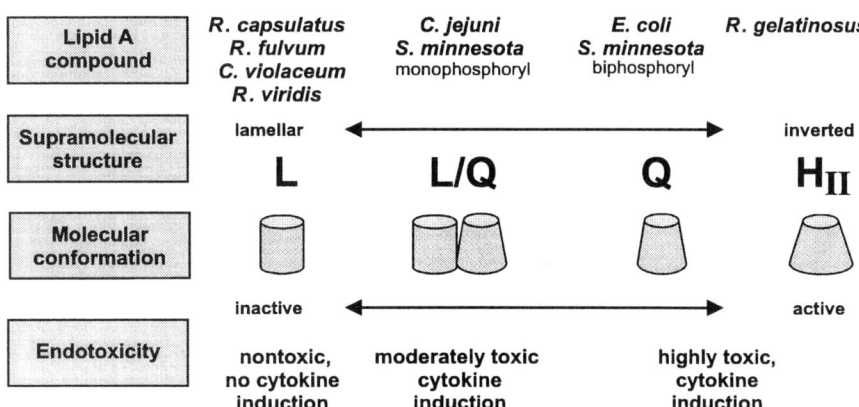

Fig. 6 Correlation between supramolecular structure/conformation and biological activity for various lipid A at 37°C. (Adapted from Ref. 145.)

munogenicity and antigenicity, which was comparably low for isolated lipid A, but considerably higher when lipid A was incorporated into the lipid matrix of phospholipid liposomes. This behavior can readily be explained by the existence of an inverted state in the former, but of a lamellar phase in the latter case, then exposing critical epitopes necessary for the full manifestation of these immunological activities.

In this context, the consequences of the structural characteristics of synthetic lipid A analogs and partial structures (see above) for their bioactivity can also be appreciated in this respect, all samples with fewer than six acyl chains are nearly completely inactive (184), which is again apparently correlated with their preferential adoption of lamellar structures.

A very interesting aspect of the induction of biological activity was raised by Tahri-Jouti and Chaby (185). They found that LPS (from *Bordetella pertussis*, *E. coli* J5, and *S. minnesota* R595) bound to mouse peritoneal macrophages by both specific and nonspecific interactions. They hypothesize that the observed nonspecific interactions most probably occurred as a result of the insertion of LPS into the lipid bilayer of the cellular membrane. These investigations have suggested that LPS-LPS associations may also contribute to the nonspecific binding and that the nonspecific binding may be less temperature-dependent than specific binding. Above 22°C (and particularly at 37°C) specific binding was completely obscured. The authors concluded that the latter observation could be explained by increased nonspecific interactions with the lipid layer, resulting from a modification of the fluidity of the cellular membrane at this temperature. Of course, the fluidity of the LPS aggregates is also temperature-dependent in a characteristic way (see above).

Interaction Mechanisms

Presently the detailed mechanisms of the interaction of endotoxins with host cell membranes finally leading to their activation are far from being understood. Because the membrane-bound mCD14, the proposed cellular receptor for endotoxin, lacks a transmembrane domain, this protein is not suitable for mediating a signal. Therefore, it is reasonable to assume that, in addition to receptor-mediated activation, the insertion of endotoxin molecules into the lipid matrix could also be a prerequisite for the activation of host cells (see, e.g., Refs. 186,187). This intercalation can be achieved in different ways: (1) via direct intercalation by hydrophobic interaction of a priori existing endotoxin monomers, which are present in sufficient numbers at least for those chemotypes with longer sugar moieties (188), or (2) via the intercalation of endotoxin monomers, which are produced by the disaggregating properties of LBP and transported into the membrane either by LBP alone (189) or via a complex interplay of soluble and membrane-bound CD14 (sCD14 and mCD14) and LBP (190). The mere intercalation of endotoxin molecules into the lipid matrix would, however, most likely not be sufficient for activation. As pointed out in the introductory remarks to this review, two principally different fates for intercalated endotoxins are conceivable. One possible mechanism is the internalization of endotoxin, which is described in several investigations (191–193) and could be either associated with cell activation or with clearance and detoxification. The other mechanism could be the direct or CD14-mediated interaction of endotoxin with other membrane proteins, as, for example, the 80 kDa protein described by Schletter et al. (194) or transmembrane ion channels (195). The correlation between the conformation of endotoxin molecules and their bioactivity is in favor of the latter mode, the existence of a transmembrane signal transducing protein which is triggered by binding of endotoxin molecules. The triggering signal requires a particular conformation of the lipid A moiety of the endotoxin molecule (Fig. 7). Thus, the only endotoxins that will activate are those that possess a lipid A part leading, in the isolated form, to nonlamellar aggregate structures. We further postulate that for binding to the putative signaling protein, the existence of a sufficient number of hydroxy fatty acids in the endotoxin molecules is necessary, allowing the formation of hydrogen bonds. This conformational concept would readily explain the antagonistic action of biologically inactive endotoxins. In these cases, the binding sites of the transmembrane protein would be occupied by the inactive molecules, thus inhibiting the binding of the active structures. If the transmembrane protein was an ion channel, the interaction of endotoxin with the channel protein could lead to mismatches in intracellular ion concentrations, provoking the activation of the succeeding signaling cascade.

The above-mentioned types of endotoxin intercalation should express different efficiencies. Thus, the direct intercalation of monomers and the LBP-mediated process will lead to an intercalation somewhere in the lipid matrix, whereas the mCD14-mediated process will bind the endotoxin directly to the signaling protein, assuming that mCD14 is located in the direct vicinity of the signal transducer. This assumption is backed by the observation that mCD14-mediated activation can be

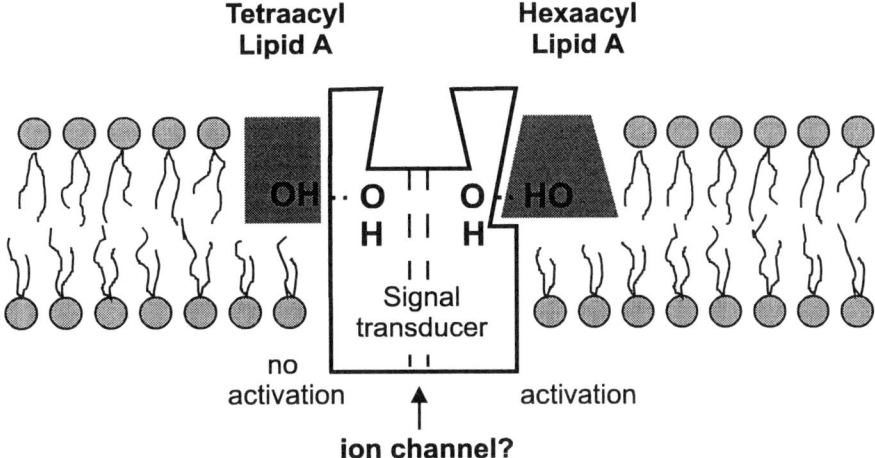

Fig. 7 Proposed model of cell activation by endotoxin. The endotoxin molecules bind with their lipid A moiety to a transmembrane signal transducing molecule, probably an ion channel. Binding is facilitated via hydrogen bonding. This requires the existence of hydroxy fatty acids in the lipid A part. A further prerequisite for activation is a particular conformation of lipid A.

blocked by anti-CD14 antibodies. At high endotoxin concentration the blockade by anti-CD14 antibodies can be overcome (196), and obviously in that case the CD14-independent activation pathway is initiated. One further point deserves attention. LBP, initially defined as lipopolysaccharide-binding protein, turns out not to be LPS specific but rather to interact with and transport other negatively charged lipids. Thus, LBP seems to be a lipid transfer protein in a more general sense (189).

ACKNOWLEDGMENTS

The financial support of the Deutsche Forschungsgemeinschaft (SFB 367, project B8; SFB 470, project B5; GRK 288/1-92, project A3) and of the German Federal Minister for Education, Science, Research, and Technology (BMBF-grant 01KI9471, project B6) is gratefully acknowledged.

REFERENCES

1. Raetz C. Biochemistry of endotoxins. Ann Rev Biochem 1990; 59:129–170.
2. Rietschel ET, Brade H, Holst O, Brade L, Müller-Loennies S, Mamat U, et al. Bacterial endotoxin: chemical constitution, biological recognition, host response, and immunological detoxification. In: Rietschel ET, Wagner H, eds. Pathology of Septic Shock. Berlin: Springer-Verlag, 1996:39–81.
3. Rietschel ET, Brade L, Lindner B, Zähringer U. Biochemistry of lipopolysaccharides. In: Morrison DC, Ryan JL, eds. Bacterial Endotoxic Lipopolysaccharides, Vol. I: Molecular Biochemistry and Cellular Biology. Boca Raton, FL: CRC Press, 1992:3–41.
4. Westphal O, Westphal U, Sommer T. The history of pyrogen research. In: Schlessinger D, ed. Microbiology. Washington, DC: American Society of Microbiology, 1977:221–238.
5. Nikaido H, Vaara M. Molecular basis of bacterial outer membrane permeability. Microbiol Rev 1985; 49:1–32.
6. Zähringer U, Lindner B, Rietschel ET. Molecular structure of lipid A, the endotoxic center of bacterial lipopolysaccharides. Adv Carbohydr Chem Biochem 1994; 50:211–276.
7. Krauss JH, Seydel U, Weckesser J, Mayer H. Structural analysis of the nontoxic lipid A of *Rhodobacter capsulatus* 37b4. Eur J Biochem 1989; 180:519–526.
8. Loppnow H, Libby P, Freudenberg MA, Kraus JH, Weckesser J, Mayer H. Cytokine induction by lipopolysaccharide (LPS) corresponds to the lethal toxicity and is inhibited by nontoxic *Rhodobacter capsulatus* LPS. Infect Immun 1990; 58:3743–3750.
9. Ulmer AJ, Feist W, Kirikae T, Kirikae F, Kusumoto S, Kusama T, et al. Modulation of endotoxin-induced monokine release in human monocytes by lipid A partial structures inhibiting the binding of ^{125}I-LPS. Infect Immun 1992; 60:5145–5152.
10. Kawahara K, Seydel U, Matsuura M, Danbara H, Rietschel ET, Zähringer U. Chemical structure of glycosphingolipids isolated from *Sphingomonas paucimobilis*. FEBS Lett 1991; 292:107–110.
11. Yamamoto A, Yano I, Masui M, Yabuuchi E. Isolation of a novel sphingoglycolipid containing glucuronic acid and 2-hydroxy fatty acid from *Flavobacterium devorans* ATCC 10829. J Biochem 1978; 83:1213–1216.
12. Kawahara K, Uchida K, Aida K. Isolation of an unusual 'lipid A' type glycolipid from *Pseudomonas paucimobilis*. Biochim Biophys Acta 1982; 712:571–575.

13. Israelachvili JN. Intermolecular and Surface Forces. 2d ed. London: Academic Press, 1991.
14. Aloia RC. Membrane Fluidity in Biology, Vol. I: Concept of Membrane Structure. New York: Academic Press, 1983.
15. Morrison DC, Rudbach JA. Endotoxin-cell-membrane interactions leading to transmembrane signaling. Contemp Top Mol Immunol 1981; 8:187–218.
16. Jackson SK, James PE, Rowlands CC, Mile B. Binding of endotoxin to macrophages: interactions of spin-labelled saccharide residues. Biochim Biophys Acta 1992; 1135:165–170.
17. Shinitzky M. Membrane fluidity and cellular functions. In: Shinitzky M, ed. Physiology of Membrane Fluidity. Vol I Boca Raton, FL: CRC Press, 1984:1–51.
18. Cossins AR, Sinensky M. Adaptation of membranes to temperature, pressure, and exogenous lipids. In: Shinitzky M, ed. Physiology of Membrane Fluidity. Vol II Boca Raton, FL: CRC Press, 1984:1–20.
19. Portoles MT, Pagani R, Diaz-Laviada I, Municio AM. Effect of *Escherichia coli* lipopolysaccharide on the microviscosity of liver plasma membranes and hepatocytes suspensions and monolayers. Cell Biochem Function 1987; 5:55–61.
20. Jacobs DM, Yeh H, Price RM. Fluorescent detection of lipopolysaccharide interactions with model membranes. Adv Exp Med Biol 1990; 25b:233–245.
21. Wright SD, Ramos RA, Tobias PS, Ulevitch RJ, Mathison JC. CD14, a receptor for complexes of lipopolysaccharide (LPS) and LPS binding protein. Science 1990; 249:1431–1433.
22. Golenbock DT, Liu Y, Millham FH, Freeman MW, Zoeller RA. Surface expression of human CD14 in Chinese hamster ovary fibroblasts imparts macrophage-like responsiveness to bacterial endotoxin. J Biol Chem 1993; 268;22055–22059.
23. Hailman E, Lichenstein HS, Wurfel MM, Miller DS, Johnson DA, Kelley M, et al. Lipopolysaccharide (LPS)-binding protein accelerates the binding of LPS to CD14. J Exp Med 1994; 179:269–277.
24. Pugin J, Schürer-Maly CC, Leturcq D, Moriarty A, Ulevitch RJ, Tobias PS. Lipopolysaccharide activation of human endothelial and epithelial cells is mediated by lipopolysaccharide-binding protein and soluble CD14. Proc Natl Acad Sci USA 1993; 90:2744–2748.
25. Gegner JA, Ulevitch RJ, Tobias PS. Lipopolysaccharide (LPS) signal transduction and clearance. J Biol Chem 1995; 270:5320–5325.
26. Gallin EK, McKinney LC. Monovalent ion transport and membrane potential changes during activation in phagocytic leukocytes. In: Grinstein S, Rotstein OD, eds. Current Topics in Membranes and Transport. New York: Academic Press, 1990:127–152.
27. Cevc G. Membrane electrostatics. Biochim Biophys Acta 1990; 1031:311–382.
28. Beschiaschvili G, Seelig J. Peptide binding to lipid bilayers. Binding isotherms and z-potential of a cyclic somatostatin analogue. Biochemistry 1990; 29:10995–11000.
29. Zheng C, Vanderkooi G. Molecular origin of the internal dipole potential in lipid bilayers: calculation of the electrostatic potential. Biophys J 1992; 63:935–941.
30. Franklin JC, Cafiso DS, Internal electrostatic potentials in bilayers: measuring and controlling dipole potentials in lipid vesicles. Biophys J 1993; 65:289–299.
31. Cafiso D. Lipid bilayers: membrane-protein electrostatic interactions. Curr Opin Struct Biol 1991; 1: 185–190.
32. Schoch P, Sargent DF, Schwyzer R. Capacitance and conductance as tools for the measurement of asymmetric surface potentials and energy barriers of lipid bilayer membranes. J Membr Biol 1979; 46:71–89.
33. McLaughlin S. The electrostatic properties of membranes. Annu Rev Biophys Biophys Chem 1989; 18: 113–136.
34. Mueller P, Rudin DO, Tien HT. Reconstitution of cell membrane structure in vitro and its transformation into an excitable system. Nature 1962; 194:979–981.
35. Tien HT. Bilayer Lipid Membranes (BLM): Theory and Practice. New York: Marcel Dekker, Inc., 1974.
36. Montal M, Mueller P. Formation of bimolecular membranes from lipid monolayers and a study of their electrical properties. Proc Natl Acad Sci USA 1972; 69: 3561–3566.
37. Schindler H. Formation of planar bilayers from artificial or native membrane vesicles. FEBS Lett 1980; 122:77–79.
38. White SH. The physical nature of planar bilayer membranes. In: Miller C, ed. Ion Channel Reconstitution. New York: Plenum Press, 1986;3–35.
39. Lu X, Ottova AL, Tien HT. Biophysical aspects of agar-gel supported bilayer lipid membranes: a new method for forming and studying planar bilayer lipid membranes. Bioelectrochem Bioenerg 1996; 39:285–289.
40. Stelze M, Miehlich R, Sackmann E. Two-dimensional microelectrophoresis in supported lipid bilayers. Biophys J 1992; 63:1346–1354.
41. Seifert K, Fendler K, Bamberg E. Charge transport by ion translocating membrane proteins on solid supported membranes. Biophys J 1993; 64:384–391.
42. Benz R, Janko K. Voltage-induced capacitance relaxation of lipid bilayer membranes. Effects of membrane composition. Biochim Biophys Acta 1976; 455:721–738.
43. Seydel U, Schröder G, Brandenburg K. Reconstitution of the lipid matrix of the outer membrane of gram-negative bacteria as asymmetric planar bilayer. J Membr Biol 1989; 109:95–103.
44. Hanke W, Schlue W-R. Planar Lipid Bilayers. New York: Academic Press. 1993.
45. Haydon DA, Hladky SB. Ion transport across thin lipid membranes: a critical discussion of mechanisms in selected systems. Q Rev Biophys 1972; 5:187–282.
46. Schröder G, Brandenburg K, Brade L, Seydel U. Pore formation by complement in the outer membrane of gram-negative bacteria studied with asymmetric planar lipopolysaccharide/phospholipid bilayers. J Membr Biol 1990; 118:161–170.
47. Schröder G, Brandenburg K, Seydel U. Polymyxin B induces transient permeability fluctuations in asymmetric planar lipopolysaccharide/phospholipid bilayers. Biochemistry 1992; 31:631–638.

47a. Wiese A, Münstermann M, Gutsmann T, Lindner B, Kawahara K, Zähringer U, and Seydel U. Molecular mechanisms of polymyxin B-membrane interactions: direct correlation between surface charge density and self-promoted transport. J Membrane Biol 1998; 162: 127–138.
48. Haydon DA. Functions of the lipid in bilayer ion permeability. Ann NY Acad Sci 1975; 264:2–16.
49. Buehler LK, Rosenbusch JP. Single channel behaviour of matrix porin of *Escherichia coli*. Biochem Biophys Res Commun 1993; 190:624–629.
50. Brunen M, Engelhardt H. Asymmetry of orientation and voltage gating of the *Acidovorax delafieldii* porin Omp34 in lipid bilayers. Eur J Biochem 1993; 212:129–135.
51. Wiese A, Schröder G, Brandenburg K, Hirsch A, Welte W, Seydel U. Influence of the lipid matrix on incorporation and function of LPS-free porin from *Paracoccus denitrificans*. Biochim Biophys Acta 1994; 1190:231–242.
52. Wiese A, Reiners JO, Brandenburg K, Kawahara K, Zähringer U, Seydel U. Planar asymmetric lipid bilayers of glycosphingolipid or lipopolysaccharide on one side and phospholipids on the other: membrane potential, porin function, and complement activation. Biophys J 1996; 70:321–329.
53. Hall JE, Latorre R. Nonactin-K^+ complex as a probe for membrane asymmetry. Biophys J 1976; 15:99–103.
54. Seydel U, Eberstein W, Schröder G, Brandenburg K. Electrostatic potential barrier in asymmetric planar lipopolysaccharide/phospholipid bilayers probed with the Valinomycin-K^+ complex. Z Naturforsch 1992; 47c:757–761.
55. Nikaido H. Outer membrane of *Salmonella typhimurium*. Transmembrane diffusion of some hydrophobic substances. Biochim Biophys Acta 1976; 433:118–132.
56. Smit J, Kamio Y, Nikaido H. Outer membrane of *Salmonella typhimurium*: chemical analysis and freeze-fracture studies with lipopolysaccharide mutants. J Bacteriol 1975; 124:942–958.
57. Kamio Y, Nikaido H. Outer membrane of *Salmonella typhimurium*: accessibility of phospholipid head groups to phospholipase C and cyanogen bromide activated dextran in the external medium. Biochemistry 1976; 15:2561–2570.
58. Nikaido H. Prevention of drug access to bacterial targets: permeability barriers and active efflux. Science 1994; 264:382–388.
59. Mizushima S. Structure, assembly, and biogenesis of the outer membrane. In: Nanninga N, ed. Molecular Cytology of *Escherichia coli*. London: Academic Press, 1985:39–75.
60. Tommassen J. Biogenesis and Membrane Topology of Outer Membrane Proteins in Escherichia coli, NATO ASI Series Vol. H16 (Membrane Biogenesis). Berlin-Heidelberg: Springer-Verlag, 1988.
61. Nakae T. Identification of the outer membrane protein of *E. coli* that produces transmembrane channels in reconstituted vesicle membranes. Biochem Biophys Res Commun 1976; 71:877–884.
62. Benz R, Schmid A, Hancock REW. Ion selectivity of gram-negative bacterial porins. J Bacteriol 1985; 162:722–727.
63. Mauro A, Blake M, Labarca P. Voltage gating of conductance in lipid bilayers induced by porin from outer membrane of *Neisseria gonorrhoeae*. Proc Natl Acad Sci USA 1988; 85:1071–1075.
64. Jap BK, Walian PJ. Biophysics of the structure and function of porins. Q Rev Biophys 1990; 23:367–403.
65. Benz R, Bauer K. Permeation of hydrophilic molecules through the outer membrane of gram-negative bacteria. Review on bacterial porins. Eur J Biochem 1988; 176:1–19.
66. Rosenbusch JP. Structural and functional properties of porin channels in *E. coli* outer membranes. Experientia 1990; 46:167–173.
67. Rocque WJ, McGroarty EJ. Structure and function of an OmpC deletion mutant porin from *Escherichia coli* K-12. Biochemistry 1990; 29:5344–5351.
68. Brunen M, Engelhardt H, Schmid A, Benz R. The major outer membrane protein of *Acidovorax delafieldii* is an anion-selective porin. J Bacteriol 1991; 173:4182–4187.
69. Ishii J, Nakae T. Lipopolysaccharide promoted opening of the porin channel. FEBS Lett 1993; 320:251–255.
70. Nikaido H, Reid J. Biogenesis of prokaryotic pores. Experientia 1990; 46:174–180.
71. Nikaido H. Porins and specific channels of bacterial outer membranes. Mol Microbiol 1992;6:435–442.
72. Benz R. Structure and selectivity of porin channels. Curr Top Membr Transport 1984; 21:199–217.
73. Schindler H, Rosenbusch JP. Matrix protein in planar membranes: clusters of channels in a native environment and their functional reassembly. Proc Natl Acad Sci USA 1981; 78:2302–2306.
74. Berrier C, Coulombe A, Houssin C, Ghazi A. Voltage-dependent cationic channel of *Escherichia coli*. J Membr Biol 1993; 133:119–127.
75. delaVega AL, Delcour AH. Cadaverine induces closing of *E. coli* porins. EMBO J 1995; 14:6058–6065.
76. Lakey JH, Pattus F. The voltage-dependent activity of *Escherichia coli* porins in different planar bilayer reconstitutions. Eur J Biochem 1989; 186:303–308.
77. Morgan H, Lonsdale J, Alder G. Polarity-dependent voltage-gated porin channels from *Escherichia coli* in lipid bilayer membranes. Biochim Biophys Acta 1990; 1021:175–181.
78. Lakey JH, Lea EJA, Pattus F. OmpC mutants which allow growth on maltodextrins show increased channel size and greater voltage sensitivity. FEBS Lett 1991; 278:31–34.
79. Buehler LK, Kusumoto S, Zhang H, Rosenbusch JP. Plasticity of *Escherichia coli* porin channels. Dependence of their conductance on strain and lipid environment. J Biol Chem 1991; 266:24446–24450.
80. Parr TR, Poole K, Crockford GWK, Hancock REW. Lipopolysaccharide-free *Escherichia coli* OmpF and *Pseudomonas aeruginosa* protein P porins are functionally active in lipid bilayer membranes. J Bacteriol 1986; 165:523–526.
81. Zalman LS, Nikaido H. Dimeric porin from *Paracoccus denitrificans*. J Bacteriol 1985; 162:430–433.

82. Fox GE, Stackebrandt E, Hespell RB, Gibson J, Maniloff J, Dyer TA, et al. The phylogeny of prokaryotes. Science 1980; 209:457–463.
83. Woitzik D, Weckesser J, Benz R, Stevanovic S, Jung G, Rosenbusch JP. Porin of *Rhodobacter capsulatus*: biochemical and functional characterization. Z Naturforsch 1990; 45c:576–582.
84. Weiss MS, Schulz GE. Structure of porin refined at 1.8 Å resolution. J Mol Biol 1992; 227:493–509.
85. Weiss MS, Abele U, Weckesser J, Welte W, Schiltz E, Schulz GE. Molecular architecture and electrostatic properties of a bacterial porin. Science 1991; 254: 1627–1630.
86. Brandenburg K, Seydel U. Physical aspects of structure and function of membranes made from lipopolysaccharides and free lipid A. Biochim Biophys Acta 1984; 775:225–238.
87. Wilson ME, Morrison DC. Evidence for different requirements in physical state for the interaction of lipopolysaccharides with the classical and alternative pathways of complement. Eur J Biochem 1982; 128: 137–141.
88. Wobschall D, McKeon C. Step conductance increases in bilayer membranes induced by antibody-antigen-complement action. Biochim Biophys Acta 1975; 413: 317–321.
89. Michaels DW, Abramowitz AS, Hammer CH, Mayer MM. Increased ion permeability of planar lipid bilayer membranes after treatment with the C5b-9 cytolytic attack mechanism of complement. Proc Natl Acad Sci USA 1976; 73:2852–2856.
90. Benz R, Schmid A, Wiedmer T, Sims PJ. Single-channel analysis of the conductance fluctuations induced in lipid bilayer membranes by complement proteins C5b-9. J Membr Biol 1986; 94:37–45.
91. Young JD-E, Young TM. Channel fluctuations induced by membrane attack complex C5b-9. Mol Immunol 1990; 27:1001–1007.
92. Shiver JW, Dankert JR, Esser AF. Formation of ion-conducting channels by the membrane attack complex proteins of complement. Biophys J 1991; 60:761–769.
93. Jackson MB, Stephens CL, Lecar H. Single channel currents induced by complement in antibody-coated cell membranes. Proc Natl Acad Sci USA 1981; 78: 6421–6425.
94. Mountz JD, Tien HT. Bilayer lipid membranes (BLM): study of antigen-antibody interactions. J Bioenerg Biomembr 1978; 10:139–151.
95. Tranum-Jensen J, Bhakdi S, Bhakdi-Lehnen B, Bjerrum OJ, Speth V. Complement lysis: the ultrastructure and orientation of the C5b-9 complex on target sheep erythrocyte membranes. Scand J Immunol 1978; 7:45–56.
96. Tranum-Jensen J, Bhakdi S. Freeze-fracture ultrastructural analysis of the complement lesion. J Cell Biol 1983; 97:618–626.
97. Vukajlovich SW. Antibody-independent activation of the classical pathway of human serum complement by lipid A is restricted to Re-chemotype lipopolysaccharide and purified lipid A. Infect Immun 1986; 53: 480–485.
98. Vukajlovich SW, Sinoway P, Morrison DC. Activation of human serum complement by bacterial lipopolysaccharides. EOS J Immunol Immunopharmacol 1986; 6(Suppl 3):73–75.
99. Vukajlovich SW, Hoffman J, Morrison DC. Activation of human serum complement by bacterial lipopolysaccharides: structural requirements for antibody independent activation of the classical and alternative pathway. Mol Immunol 1987; 24:319–331.
100. Mey A, Ponard D, Colomb M, Normier G, Binz H, Revillard J-P. Acylation of the lipid A region of *Klebsiella pneumoniae* LPS controls the alternative pathway activation of human complement. Mol Immunol 1994; 31:1239–1246.
101. Clas F, Schmidt G, Loos M. The role of the classical pathway for the bactericidal effect of normal sera against gram-negative bacteria. Curr Top Microbiol Immunol 1985; 121:19–72.
102. Weiss J, Elsbach P, Gazzano-Santoro H, Parent JB, Grinna L, Horwitz A, et al. Bactericidal and endotoxin-neutralizing activities of the bactericidal/permeability-increasing protein and its bioactive N-terminal fragment. In: Levin J, Alving CR, Munford RS, Stütz PL, eds. Bacterial Endotoxin: Recognition and Effector Mechanisms. 2d ed. Amsterdam: Elsevier, 1993: 103–111.
103. Neter E, Gorzynski EA, Westphal O, Lüderitz O. The effects of antibiotics on enterobacterial lipopolysaccharides (endotoxins), hemagglutination and hemolysis. J Immunol 1958: 80:66–72.
104. Vaara M, Vaara T. Polycations as outer membrane disorganizing agents. Antimicrob Agents Chemother 1983; 24:114–122.
105. Schindler PRG, Teuber M. Ultrastructural study of *Salmonella typhimurium* treated with membrane active agents: specific reaction of dansylchloride with cell envelope components. J Bacteriol 1978; 135:198–206.
106. Schindler PRG, Teuber M. Action of polymyxin B on bacterial membranes: morphological changes in the cytoplasm and in the outer membrane of *Salmonella typhimurium* and *Escherichia coli* B. Antimicrob Agents Chemother 1975; 8:95–104.
107. Theretz A, Theissie J, Tocanne JF. A study of the structure and dynamics of complexes between polymyxin B and phosphatidylglycerol in monolayers by fluorescence. Eur J Biochem 1984; 142:113–119.
108. Beurer G, Warncke F, Galla H-J. Interaction of polymyxin B_1 and polymyxin B_1 nonapeptide with phosphatidic acid monolayer and bilayer membranes. Chem Phys Lipids 1988; 47:155–163.
109. Imai M, Inoue K, Nojima S. Effect of polymyxin B on liposomal membranes derived from *Escherichia coli* lipids. Biochim Biophys Acta 1975; 375:130–137.
110. Mushayakarara E, Levin IW. Effects of polypeptide-phospholipid interactions on bilayer reorganizations. Raman spectroscopic study of the binding of polymyxin B to dimyristoylphosphatidic acid and dimyristoylphosphatidylcholine dispersions. Biochim Biophys Acta 1984; 769:585–595.
111. Feingold DS, HsuChen CC, Sud IJ. Basis for the se-

lectivity of action of the polymyxin antibiotics on cell membranes. Ann NY Acad Sci 1974; 235:480–492.
112. Antonov VF, Korepanova EA, Vladimirov YA. Bilayer membranes charged by detergents as model to study the role of the surface charge in ionic permeability. Studia Biophys 1976; 58:87–101.
113. Miller IR, Bach D, Teuber M. Effect of polymyxin B on the structure and the stability of lipid layers. J Membr Biol 1978; 39:49–56.
114. Carr C, Jr., Morrison DC. Mechanism of polymyxin B-mediated lysis of lipopolysaccharide-treated erythrocytes. Infect Immun 1985; 49:84–89.
115. Kubesch P, Boggs J, Luciano L, Maass G, Tümmler B. Interaction of polymyxin B nonapeptide with anionic phospholipids. Biochemistry 1987; 26:2139–2149.
116. Boggs JM, Rangaraj G. Phase transitions and fatty acid spin label behavior in interdigitated lipid phases induced by glycerol and polymyxin. Biochim Biophys Acta 1985; 816:221–233.
117. Peterson A, Hancock REW, McGroarty EJ. Binding of polycationic antibiotics and polyamines to lipopolysaccharides of *Pseudomonas aeruginosa*. J Bacteriol 1985; 164:1256–1261.
118. Weiss J, Franson RC, Beckerdite S, Schmeidler K, Elsbach P. Partial characterization and purification of a rabbit granulocyte factor that increases permeability of *Escherichia coli*. J Clin Invest 1975; 55:33–42.
119. Weiss J, Elsbach P, Olsson I, Odeberg H. Purification and characterization of a potent bactericidal and membrane active protein from the granules of human polymorphonuclear leukocytes. J Biol Chem 1978; 253:2664–2672.
120. Elsbach P, Weiss J. The bactericidal/permeability-increasing protein (BPI), a potent element in host-defense against gram-negative bacteria and lipopolysaccharide. Immunobiol 1993; 187:417–429.
121. Capodici C, Chen S, Sidorczyk Z, Elsbach P, Weiss J. Effect of lipopolysaccharide (LPS) chain length on interactions of bactericidal/permeability-increasing protein and its bioactive 23-kilodalton NH_2-terminal-fragment with isolated LPS and intact *Proteus mirabilis* and *Escherichia coli*. Infect Immun 1994; 62:259–265.
122. Gazzano-Santoro H, Parent JB, Conlon PJ, Kasler H, Tsai C-M, Lill-Elghanian DA, et al. Characterization of the structural elements in lipid A required for binding of a recombinant fragment of bactericidal/permeability-increasing protein $rBPI_{23}$. Infect Immun 1995; 63:2201–2205.
123. Ooi CE, Weiss J, Doerfler ME, Elsbach P. Endotoxin-neutralizing properties of the 25 kD N-terminal fragment of the 55-60 kD bactericidal/permeability-increasing protein of human neutrophils. J Exp Med 1991; 174:649–655.
124. Kartalija M, Kim Y, White ML, Nau R, Tureen JH, Täuber MG. Effect of a recombinant N-terminal fragment of bactericidal/permeability increasing protein ($rBPI_{23}$) on cerebrospinal fluid inflammation induced by endotoxin. J Infect Dis 1995; 171:948–953.
125. Koyama S, Shibamoto T, Ammons WS, Saeki Y. $rBPI_{23}$ attenuates endotoxin-induced cardiovascular depression in awake rabbits. Shock 1995; 4:74–78.

126. Lechner AJ, Lamprech KE, Johanns CA, Matuschak GM. The recombinant 23-kDa N-terminal fragment of bactericidal/permeability-increasing protein ($rBPI_{23}$) decreases *Escherichia coli*-induced mortality and organ injury during immunosuppression-related neutropenia. Shock 1995; 4:298–306.
127. Hansbrough J, Tenenhaus M, Wikström T, Braide M, Rennekampff OH, Kiessig V, et al. Effects of recombinant bactericidal/permeability-increasing protein ($rBPI_{23}$) on neutrophil activity in burned rats. J Trauma 1996; 40:886–892.
128. de Winter RJ, von der Möhlen M, van Lieshout H, Wedel N, Nelson B, Friedmann N, et al. Recombinant endotoxin binding protein ($rBPI_{23}$) attenuates endotoxin-induced circulation changes in humans. J Inflamm 1995; 45:193–206.
129. Elsbach P, Weiss J, Franson RC, Beckerdite-Quagliata S, Schneider A, Harris L. Separation and purification of a potent bactericidal/permeability-increasing protein and a closely associated phospholipase A_2 from rabbit polymorphonuclear leukocytes. J Biol Chem 1979; 245:11000–11009.
130. Huang K, Fishwild DM, Wu H-M, Dedrick RL. Lipopolysaccharide-induced E-selectin expression requires continuous presence of LPS and is inhibited by bactericidal/permeability-increasing protein. Inflammation 1995; 19:389–404.
131. Wiese A, Brandenburg K, Carroll SF, Rietschel ET, Seydel U. Mechanisms of action of the bactericidal/permeability-increasing protein BPI on reconstituted outer membranes of gram-negative bacteria. Biochemistry 1997; 36:10311–10319.
132. Wiese A, Brandenburg K, Lindner B, Schromm AB, Carroll SF, Rietschel ET, et al. Mechanisms of action of the bactericidal/permeability-increasing protein BPI on endotoxin and phospholipid monolayers and aggregates. Biochemistry 1997; 36:10301–10310.
133. Galanos C, Lüderitz O. The role of the physical state of lipopolysaccharides in the interaction with complement. Eur J Biochem 1976; 65:403–408.
134. Jain MK. Nonrandom lateral organization in bilayers and biomembranes. In: Aloia RC, ed. Membrane Fluidity in Biology, Vol I: Concepts of Membrane Structure. New York: Academic Press, 1983:1–37.
135. Emmerling G, Henning U, Gulik-Krzywicki T. Order-disorder conformational transition of hydrocarbon chains in lipopolysaccharides. Eur J Biochem 1977; 78:503–509.
136. Van Alphen L, Lugtenberg B, Rietschel ET, Mombers C. Architecture of the outer membrane of *Escherichia coli* K12. Eur J Biochem 1979; 101:571–579.
137. Coughlin RT, Haug A, McGroarty EJ. Physical properties of defined lipopolysaccharide salts. Biochemistry 1983; 22:2007–2013.
138. Peterson AA, Haug A, McGroarty EJ. Physical properties of short- and long-O-antigen-containing fractions of lipopolysaccharide from *Escherichia coli* O111:B4. J Bacteriol 1986; 165:116–122.
139. Brandenburg K, Seydel U. Investigations into the fluidity of lipopolysaccharide and free lipid A membrane systems by Fourier-transform infrared spectroscopy and differential scanning calorimetry. Eur J Biochem 1990; 191:229–236.

140. Seydel U, Brandenburg K. Conformations of endotoxin and their relationship to biological activity. In: Novotny A, Spitzer JJ, Zeigler EJ, eds. Cellular and Molecular Aspects of Endotoxin Reactions. Amsterdam: Elsevier, 1990:61–71.
141. Seydel U, Brandenburg K. Supramolecular structure of lipopolysaccharides and lipid A. In: Morrison DC, Ryan J, eds. Bacterial Endotoxic Lipopolysaccharides. Boca Raton, FL: CRC Press, 1992:225–250.
142. Seydel U, Labischinski H, Kastowsky M, Brandenburg K. Phase behaviour, supramolecular structure, and molecular conformation of lipopolysaccharide. Immunobiology 1993; 187:191–211.
143. Naumann D, Schultz C, Sabisch A, Kastowsky M, Labischinski H. New insights into the phase behaviour of a complex anionic amphiphile: architecture and dynamics of bacterial deep rough mutant lipopolysaccharide membranes as seen by FTIR, X-ray, and molecular modelling techniques. J Mol Struct 1989; 214:213–246.
144. Naumann D, Schultz C, Born J, Labischinski H, Brandenburg K, von Busse G, et al. Investigations into the polymorphism of lipid A from lipopolysaccharides of *Escherichia coli* and *Salmonella minnesota* by Fourier-transform infrared spectroscopy. Eur J Biochem 1987; 164:159–169.
145. Brandenburg K, Mayer H, Koch MHJ, Weckesser J, Rietschel ET, Seydel U. Influence of the supramolecular structure of free lipid A on its biological activity. Eur J Biochem 1993; 218:555–563.
146. Mayer H, Krauss JH, Yokota A, Weckesser J. Natural variants of lipid A. In: Friedman H, Klein TW, Nakano M, Nowotny A, eds. Endotoxin. New York: Plenum Press, 1990:45–70.
147. Brandenburg K, Kusumoto S, Seydel U. Conformational studies of synthetic lipid A analogues and partial structures by infrared spectroscopy. Biochim Biophys Acta 1997; 1329:183–207.
148. Coughlin RT, Peterson AA, Haug A, Pownall J, McCroarty EJ. A pH titration study on the ion binding within lipopolysaccharide aggregates. Biochim Biophys Acta 1992; 821:404–412.
149. Fukuoka S, Kodama M, Karube I. Thermotropic phase behavior of membranes made from *Erwinia carotovora* rough form lipopolysaccharide. Thermochim Acta 1995; 257:93–102.
150. Wellinghausen N, Schromm AB, Seydel U, Brandenburg K, Luhm J, Kirchner H, et al. Zinc enhances lipopolysaccharide-induced monokine secretion by alteration of fluidity state of lipopolysaccharide. J Immunol 1996; 157:3139–3145.
151. Brandenburg K, Koch MHJ, Seydel U. Phase diagram of lipid A from *Salmonella minnesota* and *Escherichia coli* rough mutant lipopolysaccharide. J Struct Biol 1990; 105:11–21.
152. Brandenburg K, Koch MHJ, Seydel U. Phase diagram of deep rough mutant lipopolysaccharide from *Salmonella minnesota* R595. J Struct Biol 1992; 108:93–106.
153. Seydel U, Koch MHJ, Brandenburg K. Structural polymorphisms of rough mutant lipopolysaccharides Rd to Ra from *Salmonella minnesota*. J Struct Biol 1993; 110:232–243.
154. Labischinski H, Barnickel G, Bradaczek H, Naumann D, Rietschel ET, Giesbrecht P. High state of order of isolated bacterial lipopolysaccharide and its possible contribution to the permeation barrier property of the outer membrane. J Bacteriol 1985; 162:9–20.
155. Ramos-Sanchez MC, Orduna-Domingo A, Rodriguez-Torres A, Martin-Gil FJ, Martin-Gil J. Thermal analysis of lipopolysaccharides from *Brucella* and other gram-negative bacteria. Thermochim Acta 1991; 191:299–309.
156. Rodriguez-Torres A, Ramos-Sanchez MC, Orduna-Domingo A, Martin-Gil FJ, Martin-Gil J. Differential scanning calorimetry investigations of LPS and free lipids A of the bacterial cell wall. Res Microbiol 1993; 144:729–740.
157. Lenaz G, Parenti Castelli G. Membrane fluidity, molecular basis and physiological significance. In: Benga G, ed. Structure and Properties of Cell Membranes. Boca Raton, FL: CRC Press, 1985:93–136.
158. Israelachvili J, Marcelja S, Horm RG. Physical principles of membrane organisation. Q Rev Biophys 1980; 13:121–200.
159. Luzzati V, Gulik A, Gulik-Krzywicki T, Tardieu A. Lipid polymorphism revisited: Structural aspects and biological implications. In: Op den Kamp JAF, Roelofsen B, Wirtz KWA, eds. Lipids and Membranes: Past, Present, and Future. Amsterdam: Elsevier, 1986:137–151.
160. Luzzati V, Mariani P, Gulik-Krzywicki T. The cubic phases of liquid-containing systems: physical structure and biological implications. In: Mennier J, Langevin D, Boccara N, eds. Physics of Amphiphilic Layers. Vol. 21. Berlin: Springer-Verlag, 1987:131–137.
161. Mariani P, Luzzati V, Delacroix H. Cubic phases of lipid containing systems. Structure analysis and biological implications. J Mol Biol 1993; 204:165–189.
162. Seddon JM. Structure of the inverted hexagonal (HII) phase, and non-lamellar phase transitions of lipids. Biochim Biophys Acta 1990; 1031:1–69.
163. Hofer M, Hampton RY, Raetz CRH, Yu H. Aggregation behavior of lipid IV_A in aqueous solutions at physiological pH. 1: Simple buffer solutions. Chem Phys Lipids 1991; 59:167–181.
164. Takayama K, Din ZZ, Mukerjee P, Cooke PH, Kirkland TN. Physicochemical properties of the lipopolysaccharide unit that activates B lymphocytes. J Biol Chem 1990; 265:14023–14029.
165. Din ZZ, Mukerjee P, Kastowsky M, Takayama K. Effect of pH on solubility and ionic state of lipopolysaccharide obtained from the deep rough mutant of *Escherichia coli*. Biochemistry 1993; 32:4579–4586.
166. Wawra H, Buschmann H, Formanek H, Formanek S. Strukturuntersuchung mit Röntgenbeugungsmethoden an Lipopolysacchariden von *Salmonella minnesota* Mutanten S SF1111 und R SF 1167. Z Naturforsch 1979; 34C:171–178.
167. Labischinski H, Naumann D, Schultz C, Kusumoto S, Shiba T, Rietschel ET, et al. Comparative X-ray and Fourier-transform-infrared investigations of conformational properties of bacterial and synthetic lipid A of *Escherichia coli* and *Salmonella minnesota* as well as partial structures and analogues thereof. Eur J Biochem 1989; 179:659–665.

168. Labischinski H, Vorgel E, Uebach W, May RP, Bradaczek H. Architecture of bacterial lipid A in solution. A neutron small-angle scattering study. Eur J Biochem 1990; 190:359–363.
169. Hayter JB, Rivera M, McGroarty EJ. Neutron scattering analysis of bacterial lipopolysaccharide phase structure. J Biol Chem 1987; 262:5100–5105.
170. Seydel U, Brandenburg K, Koch MHJ, Rietschel ET. Supramolecular structure of lipopolysaccharide and free lipid A under physiological conditions as determined by synchrotron small-angle X-ray diffraction. Eur J Biochem 1989; 186:325–332.
171. Brandenburg K. Fourier transform infrared spectroscopy characterization of the lamellar and nonlamellar structures of free lipid A and Re lipopolysaccharides from *Salmonella minnesota* and *Escherichia coli*. Biophys J 1993; 64:1215–1231.
172. Brandenburg K, Schromm AB, Koch MHJ, Seydel U. Conformation and fluidity of endotoxins as determinants of biological activity. In: Levin J, Alving CR, Munford RS, Redl H, eds. Bacterial Endotoxins: Lipopolysaccharides from Genes to Therapy. New York: John Wiley & Sons, 1995:167–182.
173. Seydel U, Brandenburg K, Rietschel ET. A case for an endotoxic conformation. Prog Clin Biol Res 1994; 388:17–30.
174. Kastowsky M, Sabisch A, Gutberlet T, Bradaczek H. Molecular modelling of bacterial deep rough mutant lipopolysaccharide of *Escherichia coli*. Eur J Biochem 1991; 197:707–716.
175. Cullis PR, De Kruijff B. The polymorphic phase behaviour of phosphatidylethanolamines of natural and synthetic origin. Biochim Biophys Acta 1978; 513:31–42.
176. Kastowsky M, Gutberlet T, Bradaczek H. Molecular modelling of the three-dimensional structure and conformational flexibility of bacterial lipopolysaccharide. J Bacteriol 1992; 174:4798–4806.
177. Jung S, Min D, Hollingworth RI. A metropolis Monte Carlo method for analysing the energetic and dynamics of lipopolysaccharide supramolecular structure and organization. J Comput Chem 1996; 17:238–249.
178. Wang Y, Hollingsworth RI. An NMR spectroscopy and molecular mechanics study of the molecular basis for the supramolecular structure of lipopolysaccharides. Biochemistry 1996; 35:5647–5654.
179. Kato N, Ohta M, Kido N, Arakawa Y, Sugiyama T, Naito S, et al. Polymorphism of crystals of *Salmonella minnesota* Re and Ra lipopolysaccharides. Microbiol Immunol 1993; 37:549–555.
180. Kato N, Arakawa Y, Sugiyama T, Ito H, Naito S, Kido N, et al. Crystallization and analyses of crystals of various chemotypes of R-form lipopolysaccharides from *Salmonella* spp. Microbiol Immunol 1994; 38:629–637.
181. Kato N, Naito S, Arakawa Y, Sugiyama T, Ito H, Ohta M, et al. Crystallization of synthetic *Escherichia coli*-type lipid A. Microbiol Immunol 1996; 40:33–38.
182. Obst S, Kastowsky M, Bradaczek H. Molecular dynamics simulations of six different fully hydrated monomeric conformers of *Escherichia coli* Re-lipopolysaccharide in the presence and absence of Ca^{2+}. Biophys J 1997; 72:1031–1046.
183. Brade L, Brandenburg K, Kuhn H-M, Kusumoto S, Macher I, Rietschel ET, et al. The immunogenicity and antigenicity of lipid A are influenced by its physicochemical state and environment. Infect Immun 1987; 55:2636–2644.
184. Rietschel ET, Brade L, Schade U, Seydel U, Zähringer U, Loppnow H, et al. Bacterial endotoxins: relationship between chemical structure and biological activity. In: Gregoridias G, Allison AC, Poste G, eds. Immunological Adjuvants and Vaccines. New York: Plenum Press, 1989:61–74.
185. Tahri-Jouti M-A, Chaby R. Specific binding of lipopolysaccharide to mouse macrophages. I. Characteristics of the interaction and inefficiency of the polysaccharide region. Mol Immunol 1990; 27:751–761.
186. Morrison DC. Nonspecific interactions of bacterial lipopolysaccharides with membranes and membrane components. In: Berry LJ, ed. Handbook of Endotoxin. Vol. 3: Cellular Biology of Endotoxins. Amsterdam: Elsevier, 1985: 25–55.
187. Morrison, DC, Rudbach JA. Endotoxin-cell-membrane interactions leading to transmembrane signaling. Contemp Top Mol Immunol 1981; 8:187–218
188. Schromm AB, Brandenburg K, Rietschel ET, Seydel U. Do endotoxin aggregates intercalate into phospholipid membranes in a nonspecific, hydrophobic manner? J Endotox Res 1995; 2:313–323.
189. Schromm AB, Brandenburg K, Rietschel ET, Flad HD, Carroll SF, Seydel U. Lipopolysaccharide-binding protein mediates CD14-independent intercalation of lipopolysaccharide into phospholipid membranes. FEBS Lett 1996; 399:267–271.
190. Yu B, Hailman E, Wright SD. Lipopolysaccharide binding protein and soluble CD14 catalyze exchange of phospholipids. J Clin Invest 1997; 99:315–324.
191. Kang Y-H, Dwivedi RS, Lee C-H. Ultrastructural and immunocytochemical study of the uptake and distribution of bacterial lipopolysaccharide in human monocytes. J Leuk Biol 1990; 316–332.
192. Kriegsmann J, Gay S, Bräuer R. Endocytosis of lipopolysaccharide in mouse macrophages. Cell Mol Biol 1993; 39:791–800.
193. Gegner JA, Ulevitch RJ, Tobias PS. Lipopolysaccharide (LPS) signal transduction and clearance. Dual roles for LPS binding protein and membrane CD14. J Biol Chem 1995; 270:5320–5326.
194. Schletter J, Brade H, Brade L, Krüger C, Loppnow H, Kusumoto S, et al. Binding of lipopolysaccharide (LPS) to an 80-kilodalton membrane protein of human cells is mediated by soluble CD14 and LPS-binding protein. Infect Immun 1995; 63:2576–2580.
195. Maruyama N, Yasunori K, Yamauchi K, Aizawa T, Ohrui T, Nara M, et al. Quinine inhibits production of tumor necrosis factor-α from human alveolar macrophages. Am J Respir Cell Mol Biol 1994; 10:514–520.
196. Lynn WA, Liu Y, Golenbock DT. Neither CD14 nor serum is absolutely necessary for activation of mononuclear phagocytes by bacterial lipopolysaccharide. Infect Immun 1993; 61:4452–4461.

12

Lipopolysaccharide Preparations in Aqueous Media: Implications for Solution Versus Suspension

Pasupati Mukerjee
University of Wisconsin, Madison, Wisconsin

Manfred Kastowsky
Free University of Berlin, Berlin, Germany, and William S. Middleton Memorial Veterans Hospital, Madison, Wisconsin

Stefan Obst
Free University of Berlin, Berlin, Germany

Kuni Takayama
William S. Middleton Memorial Veterans Hospital, Madison, Wisconsin

INTRODUCTION AND BACKGROUND

The lipopolysaccharide (LPS, also called endotoxin) from the outer leaflet of the outer membrane of gram-negative bacteria (1,2) is a highly acidic and amphipathic macromolecule (3–6). As shown in Figure 1, the LPS of Enterobacteriaceae consists of three general parts: a variable region called the O-specific chain (repeating oligosaccharides, O-antigen), a relatively conserved core region (heterooligosaccharide), and a conserved lipid A (highly acylated and phosphorylated glucosamine disaccharide) (5,8,9). The amphipathic nature of LPS results from the fact that one part of the molecule (lipid A) is composed of numerous hydrophobic groups (5–7 fatty acyl groups) and the other parts consist of many polar substituents (O-antigen, core, and glucosamine backbone of lipid A) and negative charges (inner core and glucosamine moiety of lipid A). Anions can be present in the O-antigen of certain LPS. LPS is the compound responsible for the numerous pathophysiological effects expressed in gram-negative septic shock (10). The effects (including lethal toxicity, pyrogenicity, activation of B cells, and induction of cytokines) can all be attributed directly to the lipid A moiety of the LPS (11). The reader is directed to a recent review of the general aspects of bacterial LPS by Wilkinson (12).

This built-in asymmetry of LPS is characteristic of much simpler lipids, soaps, detergents, and cationic surfactants, which have been studied extensively (13–18). The basic asymmetry allows lipid monomers to be interfacially active. They pack in a bewildering variety of organized and self-assembled structures such as micelles, vesicles, and a large number of mesophases (15,17,18). To give an example, the phase diagram of a simple two-component system of sodium palmitate and water has been determined at various compositions and temperatures (18,19). It showed more than a dozen different two-phase regions along with a single-phase liquid region. If a simple compound like sodium pal-

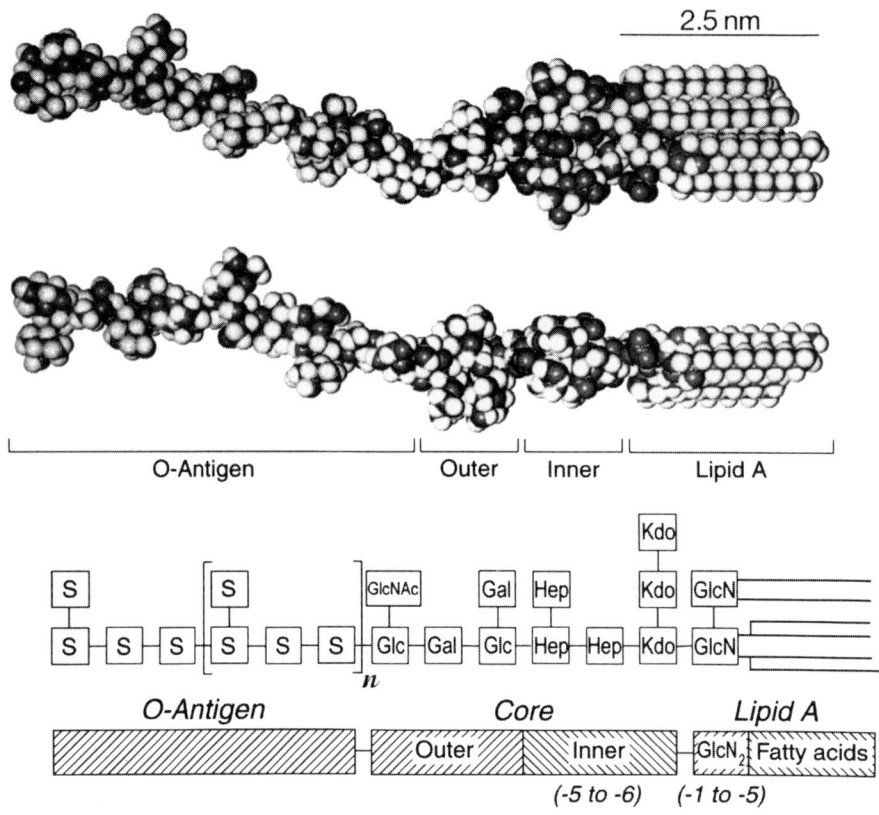

Fig. 1 Schematic representation of the general structure of smooth LPS from *S. typhimurium* showing the three parts (O-antigen, core, and lipid A) and the charge distribution. The numbers in parentheses are the range of net charges in the inner core region [due to the presence of phosphate, pyrophosphorylethanolamine (these two are not shown), and two to three Kdo units] and the disaccharide of lipid A [due to the presence of phosphate, pyrophosphate, aminoarabinose, and ethanolamine (not shown)]. The abbreviations are as follows: GlcN, glucosamine; Kdo, 3-deoxy-D-manno-octulosonic acid; Hep, L-glycero-D-mannno-heptose; Glc, glucose; Gal, galactose; GlcNAc, *N*-acetyl-glucosamine; *n*, number of repeating sugar units within the O-antigen; S, sugar. The space-filling molecular models (above this schematic representation) show front and side views of a smooth LPS bearing four O-antigen repeating units ($n = 3$). Hydrogen atoms are light; all other atom types have darker grey tones. The O-antigen chain is shown in its most extended conformation. For further details, see Ref. 7. The deep rough chemotype LPS (ReLPS) used in laboratory experiments would be the highly truncated LPS represented by lipid A containing only two Kdo units.

mitate can yield numerous phases, we can expect an even greater variety of aggregate structures to exist in the more complex LPS-water two-phase system.

In order to understand the activity of LPS molecules in pathophysiological situations and in laboratory research with isolated LPS, we must determine the nature of the molecular species in solution. We again turn to sodium palmitate as an example. The phase diagram of this compound indicates that at low concentrations of the solute as found in biological fluids, a single-phase solution is the thermodynamically stable phase and the only relevant one. The solution phase of sodium palmitate and many other simple lipids and surfactants may contain micelles in addition to monomers. These micelles are usually in rapid and dynamic association-dissociation equilibrium with monomers. They form spontaneously, and the monomer-to-micelle equilibria are governed by the laws of mass action (13,20). The micellar systems are single-phase systems, even when more than 99% of the molecules are micellized (13,20).

For LPS some of the most important information needed include the extent to which free molecules exists in an aqueous fluid in biological and laboratory systems along with self-aggregates such as micelles and membrane- or protein-bound molecules. Distinctions must be made between single fluid phase solutions, with or without micelles, and systems where some self-aggregates are present as one or more solid-like phases, however finely dispersed. These latter systems are best described as suspensions (16). The concentration of free monomers is needed to understand the binding of LPS to proteins (as carriers, receptors, and enzymes) on a quantitative basis. The present discussion is restricted to dilute systems of LPS of less than 0.1% by weight (less than 1 mg/mL). LPS is known to be active in many in vivo and in vitro experiments as well as in pathophysiological situations at extremely low concentrations (21,22). Our recent work in this field (21,23–25) and the present review are based on the application of some fundamental principles developed over many years in surfactant science and colloidal systems (13–17).

SUSPENSION AND SOLUTION OF LPS

For experimental use, dried samples of isolated LPS or lipid A need to be suspended in an aqueous medium, usually at about 1 mg/ml. This preparation must be stable as a dispersion or a suspension. Such suspensions are usually clear to opalescent in physical appearance. If a suitable LPS preparation cannot be obtained by simply mixing dry sample with water or buffer, the mixture is treated in one or more of the following ways: [1] heating to about 50°C, [2] sonicating, and [3] removal of the divalent cations from the LPS by electrodialysis or the use of ion exchange resin and conversion to the monovalent cation salts, e.g., the triethylamine salt (26,27). The effectiveness of such a treatment is illustrated by the fact that the triethylamine salt of ReLPS yields a clear suspension at 1 mg/ml in buffer. Unlike the smooth LPS, the isolated ReLPS is normally very difficult to suspend in water.

A clear suspension of LPS made at the usual concentration is not a true solution. It is a two-phase system. When the solution content is relatively high (>1 mg/mL), experimental methods are available to study the properties of these self-aggregates (28–31). Thus, several types of self-aggregates have been described for LPS (31). However, they constitute phases that are distinctly separate from the solution phase.

When the solute content is low, it is not always easy to distinguish between solution and suspension by visual inspection, particularly when the suspended phase is finely divided. The use of spectral changes in dilute solutions of cationic dyes produced by anionic surfactants such as sodium dodecyl sulfate (SDS) to determine its critical micelle concentration (CMC) may lead to erroneous results (32). Dilute solution of SDS well below CMC changes the color of pinacyanol. This is caused by the formation of an insoluble adduct of two oppositely charged hydrophobic species of pinacyanol and SDS ions (32). Because of the low concentration of dye used, the resulting suspension appears to be a solution. It was shown to be a suspension using high-speed centrifugation to sediment the colored adduct and demonstrating lack of permeation through cellophane membranes in dialysis experiments. In such spectral change methods, the interactions of the probe dyes with the monomers produces serious systematic errors in the CMC values (14,32).

For lipids in general it is reasonable to accept the dispersion as a single-phase system if the dispersion of a solute is spontaneous and does not need the input of energy through sonication and other techniques. Sonnino et al. (33) have shown that some gangliosides produce spontaneous micellar solutions even when the CMC values were calculated to be as low as 1.1×10^{-8} M. In contrast to ordinary surfactants such as SDS or gangliosides (discussed above), the highly purified deep rough chemotype LPS (ReLPS) from *Escherichia coli* D31m4 that we have investigated needs sonication and the addition of triethylamine to produce a clear dispersion at concentrations as low as 0.1% by weight. ReLPS was used as the model in laboratory experiments. In order to determine what fraction of the solute in such a preparation (suspension) is in the monomeric form, dialysis equilibrium experiments were performed (23,24). The membranes in these experiments do not permit large aggregates to traverse the membrane. This procedure had previously been shown to be suitable for measuring solubilities in water and micellar solutions (34). From an initial concentration of 82.5 μM ReLPS in 50 mM Tris-HCl, pH 7.5, 100 mM KCl on the donor side, the equilibrium concentration on the acceptor side was found to be only 3.3×10^{-8} M or about 0.04% of the donor side concentration. The conclusion was that 3.3×10^{-8} M is the solubility of ReLPS in this buffer system so that 99.96% of the initial donor side was present as a suspension phase. Thus, at a dilute concentration of ReLPS of 10^{-7} M, only 33% of the molecules would be in solution at equilibrium.

CRITICAL MICELLE CONCENTRATION AND SOLUBILITY

The formation of micelles is associated with the existence of CMC (13,14,20,35). The proper definition of CMC and appropriate evidence for its existence have been analyzed and presented (35) as part of a project for the International Union of Pure and Applied Chemistry. A large number of experimental methods and published values were evaluated many years ago (14). We are not aware of any careful CMC measurement for any LPS or lipid A. A related system of lipid X was investigated by Lipka et al. (36). Their primary method for determining the CMC was based on the spectral change of a cationic dye, as previously mentioned. This method produces significant errors in the CMC even when the values are high (13,14). The method is particularly unreliable when the CMC values are as low as in the case of ReLPS, as shown in some studies on palmitoyl coenzyme A (37).

A question of some importance is the extent to which the model ReLPS may self-associate at or below saturation (3.3×10^{-8} M). Evidence suggests that these solutions might not be micellar. Micellar solutions above the CMC level typically show very pronounced effects of increasing temperature on solubility (38,39). The solubility of sodium dodecyl sulfonate, for example, increases 60-fold as the temperature is raised from 23 to 37°C (39). In contrast, the solubility of ReLPS was reduced by 15% when the temperature changed from 22 to 37°C (23). The small effect observed by changing the ionic strength is also consistent with the absence of micelles. Some preliminary data suggested that undersaturated solutions of ReLPS do not have small oligomers to any significant extent and are thus essential monomeric (25). However, further study is needed to determine if LPS micelles really exist.

EFFECT OF pH ON SOLUBILITY AND IONIC STATE OF ReLPS

When equilibrium dialysis experiments with ReLPS were performed at different pH values over a pH range of 6.50–8.20, a rather small change in solubility from $2.91 \pm 0.01 \times 10^{-8}$ M to $4.55 \pm 0.07 \times 10^{-8}$ M was observed (24). The increase in solubility above pH 7.0 was attributed to a pK_5 of 8.58, corresponding to the formation of the pentaanion from the tetraanion of ReLPS. This pK should reflect primarily the second dissociation of the 1-phosphate group. The estimated pK_5 of 8.58 is much higher than the pK of 6.1 shown by simple model compounds (monosaccharide monophosphates) for the second dissociation. The interpretation of the above pK_5 and the estimated value of pK_4 of 5.5 ± 0.3 requires a detailed consideration of the field effects in the multiple charged ReLPS monomer as well as the nature of the microenvironments, including electrostatic potentials, at the aqueous interfaces of the suspended particles (24). Investigations on simpler model systems (40–44) have provided much of the conceptual framework needed for these interpretations. Physicochemical studies of this kind are likely to be of great importance in understanding the ionic character, charge distributions, and different populations of differently charged forms (e.g., pentaanions vs. tetraanions) of monomeric LPS and also bound LPS at the membrane interfaces. This is significant in view of the possibility that electrostatic interaction plays a major role in the binding of LPS with LBP, CD14, and other proteins.

ACTIVITY OF MONOMERIC ReLPS

To examine the activity of monomeric ReLPS, aqueous suspensions of 82.5 µM ReLPS, containing less than 0.1% in monomeric form, were compared with monomeric dialysates at 3.4×10^{-8} M (21). These two different preparations were compared at various dilutions using two selected assays of Limulus amebocyte lysate (LAL) and induction of Egr-1 mRNA in macrophages. Both of these assays require relatively short incubation periods. The monomeric ReLPS prepared from the dialysate was 179- and 1000-fold more active than the suspension in the LAL assay and the induction of Egr-1 mRNA by thioglycolate-elicited murine peritoneal macrophages, respectively. The LAL assay required a 60-minute incubation, whereas the macrophages were exposed to ReLPS for only 3 minutes. The detection of a clotting response in the LAL assay for the monomeric preparation at 0.043 pg/ml (1.9×10^{-14} M) and the 4.1-fold increase in the induction of Egr-1 mRNA over control at 1.0 pg/ml shows the remarkable potency of the monomeric preparation. Since the suspension must contain some monomers, the activity shown by the suspended preparations can be ascribed to some extent to monomers present in the suspension. The aggregates present in the suspension may thus be inactive.

IMPLICATIONS FOR LABORATORY STUDIES AND PATHOGENESIS

Much more study is needed for a clearer understanding of all the factors that control the activity of LPS in laboratory research with isolated LPS and in infection. A few brief remarks about the implications of the recent physicochemical studies with ReLPS may be of interest. The solubility studies demonstrate that the concentration of total LPS in even a fairly dilute suspension of isolated LPS may be above the solubility limit. Much more important for laboratory studies is the observation that suspensions are much less active than monomeric preparations from dialysate even when the concentrations are well below solubility (21). This points towards an important problem arising from the slow rate of dissolution (monomerization) of suspended particles. Thus, when a suspension is diluted, the actual monomeric concentration may be far lower than the formal concentration because of both equilibrium considerations (concentrations above the solubility limit) and kinetic considerations (slow rate of dissolution of suspended particles). Quantitative descriptions of the action of LPS (where the incubation time in the assay is short) using the formal concentrations may thus contain errors because only a very small percentage of the LPS used is in an active (i.e., monomeric) state due to overestimation of monomer concentrations in studies conducted with isolated LPS. These considerations need to be modified when lipid transfer proteins such as LBP are present because they can greatly increase the rates of dissociation of LPS from suspended particles (45,46).

In infection, smooth LPS originates from the outer leaflets of the infecting gram-negative bacteria. As compared to the model ReLPS used in laboratory experiments, the structures of the smooth LPS preparations are complex and very heterogeneous (47,48). Because of the presence of the polar O-antigen and core oligosaccharide as well as additional charges in the inner core (Fig. 1), their solubility and how they interact with proteins can be quite different from ReLPS. Presently it is not possible to prepare highly purified smooth LPS to study its physicochemical properties. Thus we must extrapolate the results from the study of ReLPS. The catalytic roles of LBP and CD14 and other such molecules in mobilizing and delivering bacterial membrane-bound LPS to active sites of immune cells (macrophages, monocytes, neutrophils, and endothelial cells) for signal transduction are major areas of investigation. Because of the presence of such mediating proteins, the expected slow release of LPS molecules from bacterial membrane fragments to the aqueous fluids through noncatalyzed processes based on thermal agitation may be of secondary importance. In infection, the concentration of free monomers is likely to remain far below saturation or the CMC levels so that the formation of self-assembled large aggregates of such monomers, e.g., micelles, vesicles, or mesophases, is unlikely to be of any significance. On the other hand, the extraordinary potency of monomeric ReLPS from dialysate (21) does suggest that if free monomers are available near the active sites, their potency may be high. It is thus possible to formulate two alternate hypotheses for the recognition of LPS at the active sites for signal transduction: (1) only LPS monomers attached to proteins, e.g., LBP or CD14, are active and (2) LPS monomers are active by themselves, the function of LBP and CD14 being effective delivery agents of the LPS monomers. Much more study will be needed to evaluate such hypotheses.

MONOMERIC LPS WOULD BE REQUIRED FOR STRUCTURAL SPECIFICITY

The binding of LPS to a biologically responding unit in immune cells probably leads to either signaling by a hypothetical membrane receptor(s), as suggested by several groups (49–51), or direct activation of an enzyme(s), i.e., the ceramide-activated protein kinase (52). In either case, the responding unit must recognize differences in the fine structure. For example, it differentiates small structural differences between highly toxic ReLPS from *E. coli* (53), the nontoxic sDPLA derived from the LPS of *Rhodobacter sphaeroides* (54,55), and lipid IV$_A$ (tetraacyl bisphosphoryl lipid A) from the temperature-sensitive mutant of *Salmonella typhimurium* (56,57). In these cases, the biological system actually discriminates lipid A with four fatty acyl groups (not active) from that with six fatty acyl groups (highly active). It also discriminates lipid A containing hydroxytetradecanoate (OHC$_{14}$) (highly active) from that containing hydroxydecanoate (OHC$_{10}$) (not active). Molecular models of these lipid As are shown in Figure 2. Other examples are given in a review by Rietschel et al. (59). Ultimately, the differentiation of these structures can only be achieved on a single molecule of LPS and not an aggregate. Thus, it seems clear that LPS action in infection involves monomers either in free solution or bound to suitable proteins.

DIRECTION OF FUTURE INQUIRY

Because of its amphipathic nature, LPS can exist in numerous aggregated forms in aqueous preparations. It also exists in the "true" solution state as monomers. In the biological system this monomeric LPS, in the free form or bound to suitable proteins, is the likely active species. It is clear that low concentrations of LPS (in the pg to ng/ml range) are involved in the activation of immune cells. This activation may occur via the so-

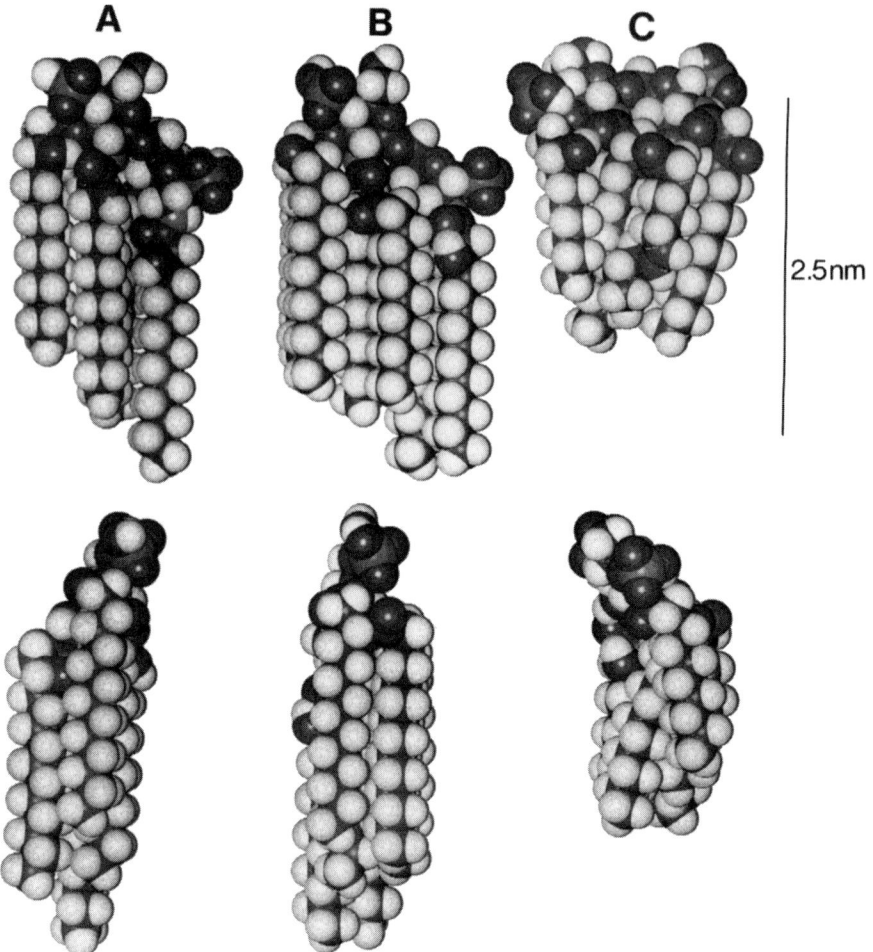

Fig. 2 Energy-minimized molecular models of (A) tetraacyl (lipid IV$_A$), (B) hexaacyl (*E. coli*), and (C) pentaacyl (*R. sphaeroides*) bisphosphoryl lipid A. The top row shows the broad side of the molecules with the 4'-phosphate groups (upper left) placed at the same horizontal plot level. The corresponding side views (with 4'-phosphates pointed at the viewer) are given in the lower row. Hydrogen atoms are light; all other atom types are shaded in darker grey tones. Note the markedly reduced chain length and differences in headgroup arrangement in (C) as compared to the others. The differences between (A) and (B) are much more subtle. Whereas the head group is almost identically arranged in (A) and (B), the packing of acyl chains and hydrogen bonding of the hydroxy fatty acids is different. (From Ref. 58.)

called CD14-dependent pathway of signal transduction. Thus, the nature of interaction of the monomeric LPS with the physiological signaling protein(s) would be an important area of inquiry in the future. This would involve [1] the identification of the hypothetical transmembrane signaling protein(s), [2] a study of the binding of monomeric LPS to these protein(s), and [3] a determination of how this binding leads to signaling.

The structure-to-activity relationships of different LPS species should be studied at the level of physicochemical interactions involved in solubilities, state of ionization, intramolecular interactions in solution, and binding to carrier proteins (especially LBP and CD14).

Molecular insights into all these relevant interactions can only come from a proper understanding of the roles of hydrophobic and polar (including electrostatic) interactions in solution. Based on such insights, we might find answers to several outstanding questions, i.e., why C3H/HeJ and similar mutant mice are hyposensitive to toxic LPS. Such studies may also lead to new approaches in developing drugs against gram-negative septic shock.

ACKNOWLEDGMENTS

This review was supported by the Medical Research Service of the Veterans Administration. We thank Kathyrn Kleckner for preparing the illustrations.

REFERENCES

1. Mühlradt PF, Golecki JR. Asymmetrical distribution and artificial reorientation of lipopolysaccharide in the outer membrane bilayer of *Salmonella typhimurium*. Eur J Biochem 1975; 51:343–352.
2. Funahara Y, Nikaido H. Asymmetrical localization of lipopolysaccharides on the outer membrane of *Salmonella typhimurium*. J Bacteriol 1980; 141:1463–1465.
3. Shands JW Jr, Graham JA, Nath K. The morphologic structure of isolated bacterial lipopolysaccharide. J Mol Biol 1967; 25:15–21.
4. Emmerling G, Henning U, Gulik-Krzywicki T. Order-disorder conformational transition of hydrocarbon chains in lipopolysaccharide from *Escherichia coli*. Eur J Biochem 1977; 78:503–509.
5. Galanos C, Lüderitz O, Rietschel ET, et al. Newer aspects of the chemistry and biology of bacterial lipopolysaccharides, with special reference to their lipid A component. In: Goodwin TW, ed. International Review of Biochemistry, Biochemistry of Lipids II. Baltimore: University Park Press, 1977:239–335.
6. Galanos C, Lüderitz O. Lipopolysaccharide: properties of an amphipathic molecule. In: Rietschel ET, ed. Handbook of Endotoxin. Vol. 1. Chemistry of Endotoxin. Amsterdam: Elsevier Science Publishers BV, 1984:46–58.
7. Kastowsky M, Gutberlet T, Bradaczek H. Molecular modelling of the three-dimensional structure and conformational flexibility of bacterial lipopolysaccharide. J Bacteriol 1992; 174:4798–4806.
8. Jann K, Westphal O. Microbial polysaccharides. In: Sela M, ed. The Antigen. New York: Academic Press, 1975:1–125.
9. Jansson P-E, Lindberg AA, Lindberg B, et al. Structural studies on the hexose region of the core lipopolysaccharides from enterobacteriaceae. Eur J Biochem 1981; 115:571–577.
10. Morrison DC, Ryan JL. Endotoxins and disease mechanisms. Ann Rev Med 1987; 38:417–432.
11. Morrison DC, Ryan JL. Bacterial endotoxins and host immune response. Adv Immunol 1979; 28:293–450.
12. Wilkinson SG. Bacterial lipopolysaccharides—themes and variations. Prog Lipid Res 1997; 35:283–343.
13. Mukerjee P. The nature of the association equilibria and hydrophobic bonding in aqueous solutions of association colloids. Adv Colloid Interf Sci 1967; 1:241–275.
14. Mukerjee P, Mysels KJ. Critical Micelle Concentrations of Aqueous Surfactant Systems. NSRDS-NBS-36. Washington, DC: U.S. Government Printing Office, 1971.
15. Attwood D, Florence AT. Surfactant Systems. London: Chapman and Hall, 1983.
16. Vold RD, Vold MJ. Colloid and Interface Chemistry. Reading, MA: Addison-Wesley, 1983.
17. Mittal KL, Mukerjee P. The wide world of micelles. In: Mittal KL, ed. Micellization, Solubilization, and Microemulsions. New York: Plenum Press, 1977:1–21.
18. Laughlin RG. The Aqueous Phase Behavior of Surfactants. New York: Academic Press, 1994.
19. Vold RD, Ferguson RH. A phase study of the system sodium palmitate-sodium chloride-water. J Am Chem Soc 1938; 60:2066–2076.
20. Mukerjee P. Micellar properties of drugs: micellar and non-micellar patterns of self-association of hydrophobic solutes of different molecular structures. Monomer fraction, availability and misuses of micellar hypothesis. J Pharm Sci 1974; 63:972–981.
21. Takayama K, Mitchell DH, Din ZZ, et al. Monomeric Re lipopolysaccharide from *Escherichia coli* is more active than the aggregated form in the *Limulus* amebocyte lysate assay and in inducing Egr-1 mRNA in murine peritoneal macrophages. J Biol Chem 1994; 269:2241–2244.
22. Michie HR, Manogue KR, Spriggs DR, et al. Detection of circulating tumor necrosis factor after endotoxin administration. N Engl J Med 1988; 318:1481–1486.
23. Takayama K, Din ZZ, Mukerjee P, et al. Physicochemical properties of the lipopolysaccharide unit that activates B lymphocytes. J Biol Chem 1990; 265:14023–14029.
24. Din ZZ, Mukerjee P, Kastowsky M, et al. Effect of pH on solubility and ionic state of lipopolysaccharide obtained from the deep rough mutant of *Escherichia coli*. Biochemistry 1993; 32:4579–4586.
25. Takayama K, Din ZZ, Mukerjee P. Physical state of biologically active lipopolysaccharide. In: Levin J, Alving CR, Munford RS, Redl H, eds. Bacterial Endotoxins: Lipopolysaccharides from Genes to Therapy. New York: Wiley-Liss, Inc, 1995:183–193.
26. Galanos C, Lüderitz O. Electrodialysis of lipopolysaccharides and their conversion to uniform salt form. Eur J Biochem 1975; 54:603–610.
27. Qureshi N, Mascagni P, Ribi E, et al. Monophosphoryl lipid A obtained from lipopolysaccharides of *Salmonella minnesota* R595. Purification of dimethyl derivative by high performance liquid chromatography and complete structural determination. J Biol Chem 1985; 260:5271–5278.
28. Brogden KA, Phillips M. The ultrastructural morphology of endotoxins and lipopolysaccharides. Electron Microsc Rev 1988; 1:261–278.
29. Hayter JB, Rivera M, McGroarty EJ. Neutron scattering analysis of bacterial lipopolysaccharide phase structure. Changes at high pH. J Biol Chem 1987; 262:5100–5105.
30. Brandenburg K, Koch MHJ, Seydel U. Phase diagram of deep rough mutant lipopolysaccharide from *Salmonella minnesota* R595. J Structural Biol 1992; 108:93–106.
31. Seydel U, Labischinski H, Kastowsky M, et al. Phase behavior, supramolecular structure, and molecular conformation of lipopolysaccharide. Immunobiology 1993; 187:191–211.
32. Mukerjee P, Mysels KJ. A re-evaluation of the spectral change method of determining critical micelle concentrations. J Am Chem Soc 1955; 77:2937–2943.

33. Sonnino S, Cantu L, Corti M, et al. Aggregative properties of gangliosides in solution. Chem Phys Lipids 1994; 71:21–45.
34. Gumkowski MJ. Surfactant micelles; solubilization and related phenomenon. Ph.D. dissertation, University of Wisconsin, Madison, Wisconsin, 1986.
35. Mysels KJ, Mukerjee P. Reporting experimental data dealing with critical micellization concentrations (c.m.c.'s) of aqueous surfactant system. Pure Appl Chem 1979; 51:1083–1089.
36. Lipka G, Demel RA, Hauser H. Phase behavior of lipid X. Chem Phys Lipids 1988; 48:267–280.
37. Constantinides PP, Stein JM. Critical micelle concentration and micellar size and shape. J Biol Chem 1985; 260:7573–7580.
38. Eggenberger DN, Harwood HJ. Conductometric studies of solubility and micelle formation. J Am Chem Soc 1951; 73:3353–3355.
39. Tartar HV, Wright KA. Studies of sulfonates. III. Solubilities, micelle formation and hydrates of the sodium salts of the higher alkyl sulfonates. J Am Chem Soc 1939; 61:539–544.
40. Kirkwood JG, Westheimer FH. The electrostatic influence of substituents on the dissociation constants of organic acids. J Chem Phys 1938; 6:506–512.
41. Mukerjee P, Ray A. Charge transfer interactions and the polarity at the surface of micelles of long-chain pyridinium iodides. J Phys Chem 1966; 70:2144–2149.
42. Mukerjee P, Banerjee K. A study of the surface pH of micelles using solubilized indicator dyes. J Phys Chem 1964; 68:3567–3574.
43. Mukerjee P, Cardinal JR, Desai NR. The nature of the local microenvironments in aqueous micellar systems. In: Mittal KL, ed. Micellization, Solubilization, and Microemulsions. New York: Plenum Press, 1977:241–261.
44. Ramachandran C, Pyter RA, Mukerjee P. Microenvironmental effects on energies of visible bands of nitroxides in electrolyte solutions and when solubilized in micelles of different charge types. Significance of effective polarity estimates. Implications for spectroscopic probe studies in lipid assemblies. J Phys Chem 1982; 86:3198–3205.
45. Schumann R, Leong SR, Flaggs GW, et al. Structure and function of lipopolysaccharide binding protein. Science 1990; 249:1429–1431.
46. Hailman E, Lichenstein HS, Wurfel MM, et al. Lipopolysaccharide (LPS)-binding proteins accelerates the binding of LPS to CD14. J Exp Med 1994; 179:269–277.
47. Jann B, Reske K, Jann K. Heterogeneity of lipopolysaccharide. Analysis of polysaccharide chain length by sodium dodecyl sulfate-polyacrylamide gel electrophoresis. Eur J Biochem 1975; 60:239–246.
48. Hitchcock PJ, Brown TM. Morphological heterogeneity among Salmonella lipopolysaccharide chemotypes in silver-stained polyacrylamide gels. J Bacteriol 1983; 154:269–277.
49. Labeta MO, Durieux J-J, Fernandez N, et al. Release from a human monocyte-like cell line of two different soluble forms of lipopolysaccharide receptor, CD14. Eur J Immunol 1993; 23:2144–2151.
50. Frey EA, Miller DS, Jahr TG, et al. Soluble CD14 participates in the response of cells to lipopolysaccharide. J Exp Med 1992; 176:1665–1671.
51. Lee JD, Kravchenko V, Kirkland TN, et al. Glycosylphosphatidylinositol-anchored or integral membrane forms of CD14 mediate identical cellular responses to endotoxin. Proc Natl Acad Sci USA 1993; 90:9930–9934.
52. Joseph CK, Wright SD, Bornmann WG, et al. Bacterial lipopolysaccharide has structural similarity to ceramide and stimulates ceramide-activated protein kinase in myeloid cells. J Biol Chem 1994; 269:17606–17610.
53. Qureshi N, Takayama K, Mascagni P, et al. Complete structural determination of lipopolysaccharide obtained from deep rough mutant of Escherichia coli. Purification by high performance liquid chromatography and direct analysis by plasma desorption mass spectrometry. J Biol Chem 1988; 263:11971–11976.
54. Qureshi N, Honovich JP, Hara H, et al. Location of fatty acids in lipid A obtained from lipopolysaccharide of Rhodopseudomonas sphaeroides ATCC 17023. J Biol Chem 1988; 263:5502–5504.
55. Takayama K, Qureshi N, Beutler B, et al. Diphosphoryl lipid A from Rhodopseudomonas sphaeroides ATCC 17023 blocks induction of cachectin in macrophages by lipopolysaccharide. Infect Immun 1989; 57:1336–1338.
56. Strain SM, Armitage IM, Anderson L, et al. Location of polar substituents and fatty acyl chains on lipid A precursors from a 3-deoxy-D-manno-octulosonic acid-deficient mutant of Salmonella typhimurium. Studies by ^{1}H, ^{13}C, and ^{13}P nuclear magnetic resonance. J Biol Chem 1985; 260:16089–16098.
57. Golenbock DT, Hampton RY, Qureshi N, et al. Lipid A-like molecules that antagonize the effects of endotoxins on human monocytes. J Biol Chem 1991; 266:19490–19498.
58. Kastowsky M, Sabisch A, Gutberlet T, Bradaczek H. Molecular modelling of bacterial deep rough mutant lipopolysaccharide of Escherichia coli. Eur J Biochem 1991; 197:707–716.
59. Rietschel ET, Kirikae T, Schade FU, et al. Bacterial endotoxin: molecular relationships of structure to activity and function. FASEB J 1994; 8:217–225.

13

Chlamydial Lipopolysaccharide

Helmut Brade
Research Center Borstel, Borstel, Germany

INTRODUCTION

As shown in many chapters of this book, the diversity of structures and functions of lipopolysaccharide (LPS) is tremendous. The first publications of so far unknown structures or functions usually point to the uniqueness of the discovery. Most often, however, it is understood sooner or later how this specific observation follows a general principle. At present, our knowledge of the structure of chlamydial LPS is still preliminary, and even less is known about the physiological function of this unique microorganism and its role in chlamydial infections.

The unique structural features of chlamydial LPS and lipid A described in the following are presently attracting our interest, e.g., the unique Kdo region or the description of nonhydroxylated fatty acids directly bound to the lipid A backbone. It may turn out that these features are indeed unique and that a review of chlamydial LPS will appear in extended form in a future edition, but I would also not be surprised to see chlamydial LPS as the subject of a short section of a general LPS review.

Nevertheless, the following review may be equally helpful for juniors and seniors in the field to discover new aspects of LPS biology whether they are unique or following a general line.

CHLAMYDIA AND *CHLAMYDIA*-INDUCED DISEASE
Description of the Genus *Chlamydia*

Chlamydia psittaci, *C. trachomatis*, *C. pneumoniae*, and *C. pecorum* are bacterial species of the monogeneric family of Chlamydiaceae. These bacteria are pathogenic, obligatory phagosomal intracellular parasites, which cause acute and chronic diseases in animals and humans (1–3).

The natural reservoir of *C. psittaci* and *C. pecorum* are animals, where they cause organ-specific reactions such as arthritis, pneumonia, enteritis, encephalitis, or abortion. Avian strains, particularly those from paroquets and related birds, may be transmitted to humans causing severe atypical pneumonia (ornithosis, psittacosis).

C. trachomatis serovars A, B, Ba, and C cause endemic trachoma in North Africa and Asia, the major cause of secondary blindness. Serovars L1 to L3 cause lymphogranuloma venerum, and serovars D through K cause sexually transmitted diseases in men and women. Chronic salpingitis in women deserves particular attention since it is a major cause of female infertility. These diseases have been known for a long time, but nevertheless the underlying pathogenic mechanisms have been only poorly defined. For further details, the reader is referred to textbooks on microbiology and infectious diseases.

Specific Chlamydial Diseases

The species *C. pneumonia* has been known for only a decade, and despite the fact that our knowledge about this microorganism is still rudimentary (4), we will discuss this species in more detail since it has been hypothesized that it is associated with arteriosclerosis in general and myocardial infarction in particular. If this hypothesis proves to be correct, it may become the most significant discovery of this century in infectious disease biology.

Although the microorganism had already been isolated in the 1960s from trachoma cases and in 1983 from the respiratory tract (5), it was recognized as a separate species only in 1987. At this time it became evident that *C. pneumoniae* caused mild respiratory tract infections. In the following years, the microorganism was also isolated in other geographic areas of North America and in Europe. The typical *C. pneumoniae* infection is mild or even asymptomatic, however, severe pneumonias are also observed, and *C. pneumoniae* accounts for 10% of community acquired pneumonias (6,7). Seroepidemiological investigations have shown that the prevalence of *C. pneumoniae* antibodies in the adult population is about 50% all over the world.

In 1988, Saikku et al. (8) reported on a statistically significant higher frequency of antibodies against *C. pneumoniae* in patients with coronary heart disease or myocardial infarction than in healthy people. The authors speculated that *C. pneumoniae* may be associated with these diseases. The report initiated an extremely controversial discussion, which is still going on. Nevertheless, the following two facts are now generally accepted:

1. The seroprevalence of *C. pneumoniae* antibodies is higher in patients with arteriosclerotic disease than in healthy people (9–12).
2. *C. pneumoniae* DNA or antigens can be detected in atheromas of arteries at various anatomic sites by PCR and immunohistology, respectively (13–18).

In addition, there are three independent reports (19–21) on the isolation of living *C. pneumoniae* from arteriosclerotic plaques. Whatever the conclusions, even the slightest possibility that *C. pneumoniae* represents one factor in the complex pathogenesis of arteriosclerosis demands further attention, since safe and effective antibiotics against chlamydiae are available. Ongoing clinical trials are presently exploring the potential benefits of antibiotic treatment of patients with myocardial infarction, the preliminary analyses of which show a significant improvement of those patients receiving antibiotics (22).

Such trials will very soon give a definite answer as to whether chlamydiae play a clinically relevant role in arteriosclerosis. However, how chlamydiae act in the different stages of the disease and whether they are the primary or a concomitant cause will require many more years of basic research. Some of the most crucial questions yet to be answered include:

1. Are *C. pneumoniae* strains isolated from arteriosclerotic plaques as are those isolated from the lung? We currently hypothesize that the lung is the source of these bacteria and, though this seems very plausible, no experimental evidence exists for this assumption.
2. Are the bacteria transported by monocytes or other vehicles? The ability of *C. pneumoniae* to persist in monocytes makes this an interesting hypothesis, which should be verified by culture and PCR.
3. Is the adherence of chlamydiae-laden monocytes the same as that of monocytes free of bacteria or infected with other obligatory or facultative intracellular bacteria? Do infected monocytes adhere to normal endothelial cells, or are fatty streaks a prerequisite?
4. Which chlamydial factors are important in maintaining the chronic inflammation in the plaque?
5. Which of the known risk factors for arteriosclerosis could be directly connected with the infection? For example, smoking is certainly a factor influencing the susceptibility to respiratory tract infections, and it has been shown that HDL cholesterol is lower in *C. pneumoniae* than in viral pneumonias (23).
6. Are there known or unknown pathogenicity factors that render certain strains atherogenic?
7. Are the animal infection models described relevant for the pathogenesis of arteriosclerosis in man?

One should keep in mind that the function of only a very few genes of *Chlamydia* are known. The ongoing genome sequencing project of a *C. trachomatis* serotype B (24) will be helpful in this respect.

Special attention should be paid to the heat-shock proteins and the LPS. The 57 kDa heat-shock protein of *C. trachomatis* is a hypersensitivity antigen, which induces a delayed-type hypersensitivity in the conjunctiva of experimental animals after primary infection of the eye or the genital tract (25–27). The histopathological findings resemble those found in chronic trachoma. This hypersensitivity may also be the main immunopathological mechanism from which occluding salpingitis results after repeated or persistent infection of the female genital tract. Whether a primary respiratory tract infection with *C. pneumoniae* can sensitize for a challenge by a genital tract infection with *C. trachomatis* is open to discussion. However, there are already some reports supporting the hypothesis that antibodies to heat-shock proteins are a risk factor for the development of pelvic inflammatory disease (28–31).

The LPS of *Chlamydia* has been reported to be of low endotoxic activity in *C. psittaci* (32) and *C. trachomatis* (33); no data are available on *C. pneumoniae*, and one should also consider the possibility that the biological activity of LPS from tissue culture–grown chlamydiae differs from that synthesized in the natural host. Independent of what endoxic activity is exhibited by chlamydial LPS in a natural infection, this molecule is in the center of our investigations since LPS is the most potent inducer of proinflammatory cytokines and able to induce a response in nearly all cell types (34,35).

Another disease considered to be associated with chlamydial infections is genital tract infections caused by *C. trachomatis*. Reactive arthritis is preferably observed after enteric or genital tract infections with gram-negative bacteria such as *Yersinia enterocolitica, Neisseria gonorrhoeae, Shigella*, and *Salmonella* (36). However, *C. trachomatis* seems to be by far the most important agent connected with this as well as with Reiter's disease (37,38). It is evident that heat-shock proteins play a significant role in chronic inflammatory processes on the one hand; on the other hand, the aforementioned bacteria are gram-negative and thus contain in their outer membrane LPS as a common constituent. Since LPS is the most potent activator for macrophages and obviously able to persist in different organs after infection, it is interesting to speculate on the combined action of heat-shock proteins and LPS in reactive arthritis. It is presently investigated whether *C. pneumoniae* infections may also be followed by reactive arthritis.

THE MICROORGANISM

Chlamydiae undergo a unique developmental cycle inside the infected host cell which is so characteristic and unique that they were taxonomically set apart from other bacteria into their own order Chlamydiales (38). Although the species share a DNA homology of only 10% as determined by DNA-DNA hybridization (40), their common origin is deduced from the homology of their 16S rRNA of more than 95% (41,42).

The infection is initiated by the contact of the so-called elementary body with the host cell (Fig. 1a and 1b). The elementary body measures only 200–300 nm in diameter and has by electron microscopy a three-layered outer membrane (Fig. 1c and 1d) (43). It is

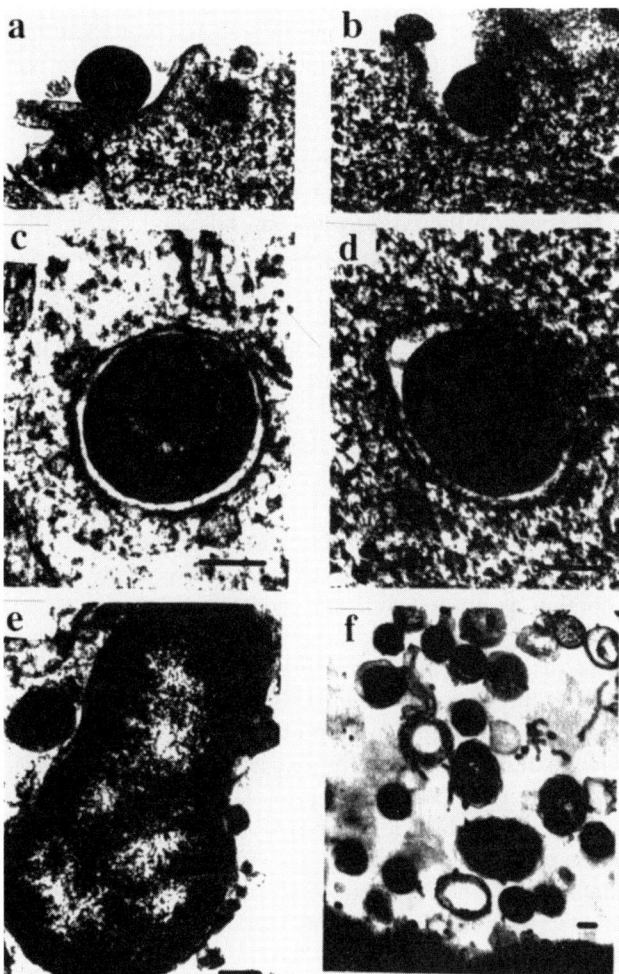

Fig. 1 The developmental cycle of *Chlamydia* as monitored by transmission electron microscopy. Mouse fibroblasts (L929) were infected with *C. trachomatis* serotype L2. The elementary body attaches to the host cell membrane (a) followed by invagination (b) and internalization (c and d). Note the changes in electron density; the trilamellar outer membrane of the microorganism and the phagosomal membrane are well resolved (c and d). Also shown is a dividing reticulate body and an outer membrane bleb (e) and a section of a mature inclusion with elementary and reticulate bodies and intermediate forms (f). Bar represents 100 nm. (From Ref. 62.)

extremely resistant against chemical and physical forces, metabolically inactive, and thus particularly suited to survive outside the host cell. Elementary bodies do not contain a peptidoglycan layer, and their physical resistance is attributed to the presence of highly cross-linked OMP1 (44,45). Typical adhesins are not known for chlamydiae, and a cellular receptor for chlamydiae has been postulated but not identified unequivocally. Glycosaminoglycans have been reported to be involved in chlamydial adhesion (46). Whereas one group has postulated that a heparan sulfate–like ligand is synthesized by chlamydiae functioning as a bridging molecule between an acceptor and a receptor on the chlamydial and host cell surface, respectively (47), others proposed that the heparan sulfate–like ligand is part of a host cell proteoglycan, which binds to chlamydial MOMP (48). Data of the latter group were obtained with a recombinant MOMP expressed as a fusion protein with maltose-binding protein of *E. coli*. This seems to contradict the results of a third group suggesting that the glycosyl moiety of chlamydial MOMP, which is of the N-linked high-mannose type, is the adhesive principle (49). However, as discussed in detail elsewhere (48), a special conformation of MOMP seems to be a prerequisite for its function as an adhesin. This conformation may be realized only when the protein is glycosylated or when it is fused to maltose-binding protein. Because adhesion is the first step in infection, clarification of these discrepant data will be important for all types of chlamydial infections. Attachment is followed by the uptake through a phagocytic process, which seems to be a receptor-mediated phagocytosis. Inside the phagosome the elementary body differentiates into the reticulate body. The reticulate body measures 1 μm in diameter, is noninfectious but metabolically active and able to multiply by binary fission (Fig. 1e), however, since chlamydiae are unable to generate energy-rich phosphates, they make use of the substrate and energy pool of the host. Therefore, chlamydiae have been also called energy parasites (39). The reticulate body is extremely fragile and adapted to the safe intracellular environment. During the whole developmental cycle the developing microcolony stays inside the phagosome, which does not fuse with lysosomes. Phagosome-lysosome fusion is inhibited by live chlamydiae with an intact protein synthesis but the molecular basis is not understood. Late in the infectious cycle, reticulate bodies differentiate into infectious elementary bodies, which are released from the lysed cell and may infect surrounding cells (Fig. 1f).

A specific property of chlamydiae is their ability to persist in cells without a productive infectious cycle and, thus, possibly cause latent chronic infections (50–52). Although the molecular basis for this phenomenon is also unknown, another observation may be related to it. Chlamydiae may develop morphologically aberrant forms (giant reticulate bodies), which do not divide by binary fission normally nor differentiate into elementary bodies (53). They may, however, regain their infectious abilities by unknown mechanisms. Giant reticulate bodies can be induced experimentally with γ-IFN (54) or α-TNF (55). The latter two cytokines act on chlamydiae by inducing the host cell enzyme 3-indolamine oxygenase (catalyzing the metabolization of tryptophan to formylkynurenin), and thus, cause a decrease of intracellular tryptophan to levels limiting the multiplication of chlamydiae, which are unable to synthesize this amino acid. Consequently, this effect can be counteracted by the addition of tryptophane to the growth medium. Although it has not been shown that this mechanism is also relevant under in vivo conditions, it could be one of the explanations for prolonged or repeated chlamydial infections as they are clinically observed. Since cytokine production by foam cells (the transformed macrophage type found in atheromatous plaques) certainly plays a role in the development of arteriosclerosis, it could at the same time be a mechanism to sustain persistent chlamydial infection.

Similar effects on the development of giant bodies are observed after treatment of chlamydiae in tissue culture with β-lactam antibiotics. Although chlamydiae have no peptidiglycan layer (45), they possess penicillin-binding proteins of unknown physiological function. Most interesting are preliminary data on sequence homologies of *C. trachomatis* DNA sequences to those of genes in *E. coli* known to be involved in peptidoglycan biosynthesis (56).

GENERAL PROBLEMS IN *CHLAMYDIA* RESEARCH

Although the biology of these unique microorganisms opens so many questions relevant to modern pathogenicity research, such as attachment to and entry into host cells, intracellular survival and persistence, and differentiation back and forth from metabolically active into inactive forms (in some aspects similar to spore-forming bacteria), very little is known about it. The reasons for this are manifold such as the complex growth requirements and purification protocols usually known for viruses combined with the compositional complexity of bacteria and the fact that some of the chlamydial strains are class III pathogens. However, most prominent seems the lack of defined systems of recombination. Although a bacteriophage has been described for a *C. psittaci* strain (57) and although there exists a common plasmid for *C. trachomatis* (58), these vectors have not been successfully used to introduce recombinant DNA into the microorganism. Therefore, characterization of chlamydial components has been limited to conventional biochemical characterizations and to those protein components that could be readily expressed in appropriate recipients transformed with cloned chlamydial genes. For these reasons, a genome-sequencing project was started in the United States on *C. trachomatis* serotype B (24), from which many genes will be identified by homology searches. For secondary gene products such as LPS, however, this approach needs the aid of the tools of biochemistry. Using this combined approach, our group has focused on the LPS; the results of this work are summarized and discussed in the following section.

THE LIPOPOLYSACCHARIDE (GROUP-SPECIFIC GLYCOLIPID) OF *CHLAMYDIA*

It has been known since the early beginnings of chlamydial research that these bacteria possess a genus-specific antigen against which antibodies are raised by the infected host. This group-specific antigen was then shown (59) to be a glycolipid containing as an immunodominant group a 3-deoxy-2-ulosonic acid, which was similar but not identical to 3-deoxy-D-manno-oct-2-ulosonic acid (Kdo). This observation led to the hypothesis that the chlamydial glycolipid may be related to the LPS of gram-negative bacteria (60), which was further supported by a report of Nurminen et al. on the extraction of a glycolipid from *C. trachomatis* serotype L_2, which was composed of D-glucosamine, phosphate, and long-chain fatty acids, the latter including β-hydroxy fatty acids as typical LPS markers (61). The same components were found later in the LPS of a ewe abortion strain of *C. psittaci* (32). With these results it became generally accepted that the chlamydial genus-specific antigen reported so far as the "group-specific glycolipid antigen" was an LPS with a chemical composition similar to that of well-characterized enterobacterial core-deficient Re mutants.

Chemical and Antigenic Structure

Details of the approaches leading to today's structural knowledge of chlamydial LPS have been reviewed recently (62). The first milestone of these investigations was the description of a chlamydial gene conferring *Chlamydia* specificity to the LPS of Re-type bacteria (63,64); the LPS of these recombinants contained the unique structure of an α-Kdo-(2→8)-α-Kdo-(2→4)-α-Kdo trisaccharide (65). The gene was later identified as a Kdo-transferase (KdtA) (66,67). At the end of this work, the complete structure of the phosphorylated carbohydrate backbone (Fig. 2) as well as the monophosphoryl derivatives of LPS from recombinant *E. coli*

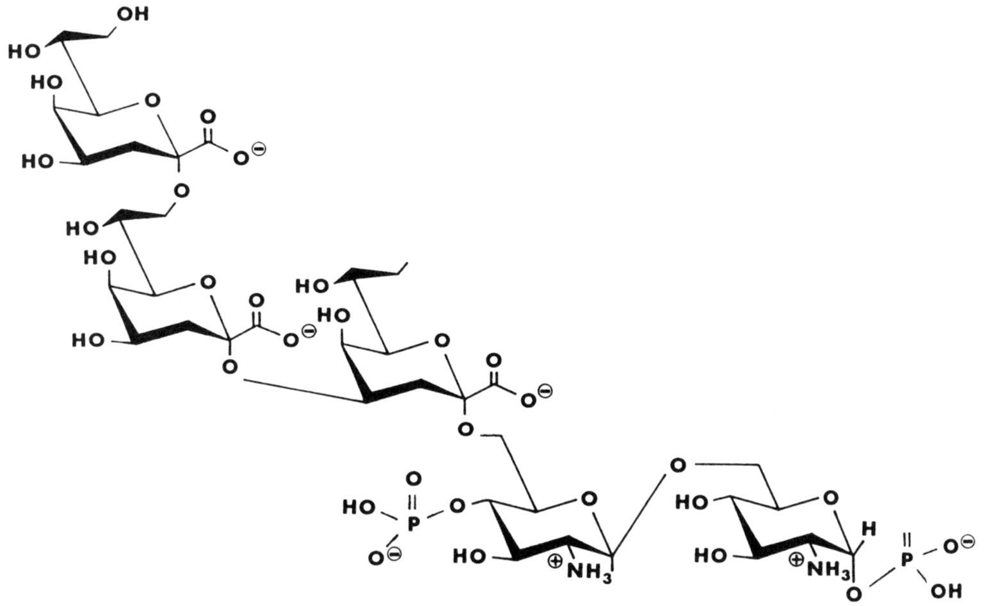

Fig. 2 Chemical structure of the phosphorylated carbohydrate backbone of the LPS from *E. coli* recombinants expressing the Kdo-transferase gene *kdtA* from *C. pneumoniae* TW-183.

F515, carrying the *KdtA* gene on a plasmid, was determined unequivocally by ^{1}H-, ^{13}C-, and ^{31}P-NMR spectroscopy and by fast-atom-bombardment mass spectrometry (68,69). In addition, we isolated from a recombinat *E. coli* strain expressing the Kdo transferase of *C. psittaci* 6BC two oligosaccharide bisphosphates containing the linear Kdo trisaccharide α-Kdo-(2→4)-α-Kdo-(2→4)-α-Kdo or the branched tetrasaccharide α-Kdo-(2→8)-[α-Kdo-(2→4)]-α-Kdo-(2→4)-α-Kdo (Fig. 6) (70). Most of the structures have been confirmed by chemical synthesis.

The availability of synthetic structures and artificial glycoconjugate antigens obtained after conjugation to proteins enabled us to determine the epitope specificities of monoclonal antibodies against the Kdo region, obtained after immunization with chlamydiae or Re-type bacteria (Fig. 3). The immunogenic properties of the artificial glycoconjugates were also investigated in mice (71). Relevant monoclonal antibodies were selected on the basis of binding to [1] the immunizing antigen, [2] LPS attached to plastic surfaces, [3] LPS imbedded in natural (sheep erythrocytes) or artificial (liposomal) membranes, [4] isolated chlamydial elementary bodies, or [5] chlamydial inclusions of infected monolayers. The results indicated that these monoclonal antibodies were 100-fold higher in affinity than those obtained after immunization with chlamydial elementary bodies from either *C. psittaci* or *C. trachomatis* (Fig. 4). Thus, the immunogenic and antigenic properties of the synthetic compounds are not only similar but superior to their natural counterparts.

Based on these data, a solid-phase ELISA was designed, which allows the determination of antibodies in body fluids against chlamydial LPS (72) and which is used in clinical laboratories in Europe. Since this is so far the most sensitive assay for the detection of a humoral immune response against chlamydiae at the genus level and is easy to perform on automatic equipment, it may become the method of choice as a screening parameter for those at risk for developing arteriosclerosis.

The binding of our monoclonal antibodies to their respective ligands has been further investigated to understand the principles of protein-carbohydrate interaction in general (73). After cyrstallization of the α2→8 linked Kdo disaccharide (74), we are now trying in cooperation with S. Evans (Ottawa, Canada) and P. Kosma (Vienna, Austria), to crystallize the Fab fragments of monoclonal antibody S25-2 (71). These crystal data together with NMR data on the antigen-antibody complex in solution will let us understand the three-dimensional interaction of the two molecules.

Biosynthesis

Readers who are not familiar with the principles of LPS biosynthesis are referred to Chapter 17 in this volume. Core biosynthesis starts with the transfer of Kdo to a lipid A precursor called precursor Ia (75), composed of the 1,4′-bisphosphorylated β1→6-linked glucosamine disaccharide and four mole equivalents of 3-hydroxytetradecanoic acid linked to positions 2, 3, 2′, and 3′. Kdo is synthesized from arabinose-5-P and phosphoenolpyruvate by the enzyme Kdo-8-P synthase encoded by the gene *kdsA* (76), which has been cloned, sequenced, and functionally characterized in *C. psittaci* (77). After dephosphorylation of Kdo-8-P, Kdo is converted into the activated sugar nucleotide CMP-Kdo by the enzyme CMP-Kdo synthase, which is encoded by the gene *kdsB*. *kdsB* has been cloned and sequenced first from *E. coli* (78) and later also from *C. trachomatis* (79). The activated sugar is transferred by a Kdo transferase encoded by the gene *kdtA*. The enzyme KdtA was shown to be bifunctional in *E. coli* (80), catalyzing two different glycosylation steps, namely the transfer of Kdo to position 6′ of precursor Ia and the transfer of a second Kdo residue to position 4 of the former. The gene *kdtA* was also cloned and sequenced from *C. psittaci* (81), *C. trachomatis* (66,82), and *C. pneumoniae* (67); the latter was expressed in the gram-positive bacterium *Corynebacterium glutamicum* and functionally characterized in vitro (67). The results obtained indicated that this enzyme is trifunctional (Fig. 5), which is in contrast to the dogma of glycobiology "one enzyme—one glycosidic bond." When the *kdtA* gene of *C. psittaci* strain 6BC was expressed in the *E. coli* Re-mutant F515, the resultant recombinant synthesized an LPS which contained, in addition to the trisaccharide α-Kdo-(2→8)-α-Kdo-(2→4)-α-Kdo, the trisaccharide α-Kdo-(2→4)-α-Kdo-(2→4)-α-Kdo and the tetrasaccharide α-Kdo-(2→8)-[α-Kdo-2→4]-α-Kdo-(2→4)-α-Kdo (70). These differences in function may be the reason why the homology among chlamydial *kdtA* genes is only 69%, which is low for a comparison at the genus level.

Up to now we have no evidence for a functional difference between the Kdo-transferases of *C. trachomatis* and *C. pneumoniae*, although the homology and identity with values of 70% and 51%, respectively, between them is still low. There is, however, one piece of evidence that LPS of *C. pneumoniae* is different from that of *C. trachomatis*. Peterson et al. have reported on a monoclonal antibody named CP33, obtained after immunization with elementary bodies of *C. pneumoniae* strain TW183, which exhibited cross-reac-

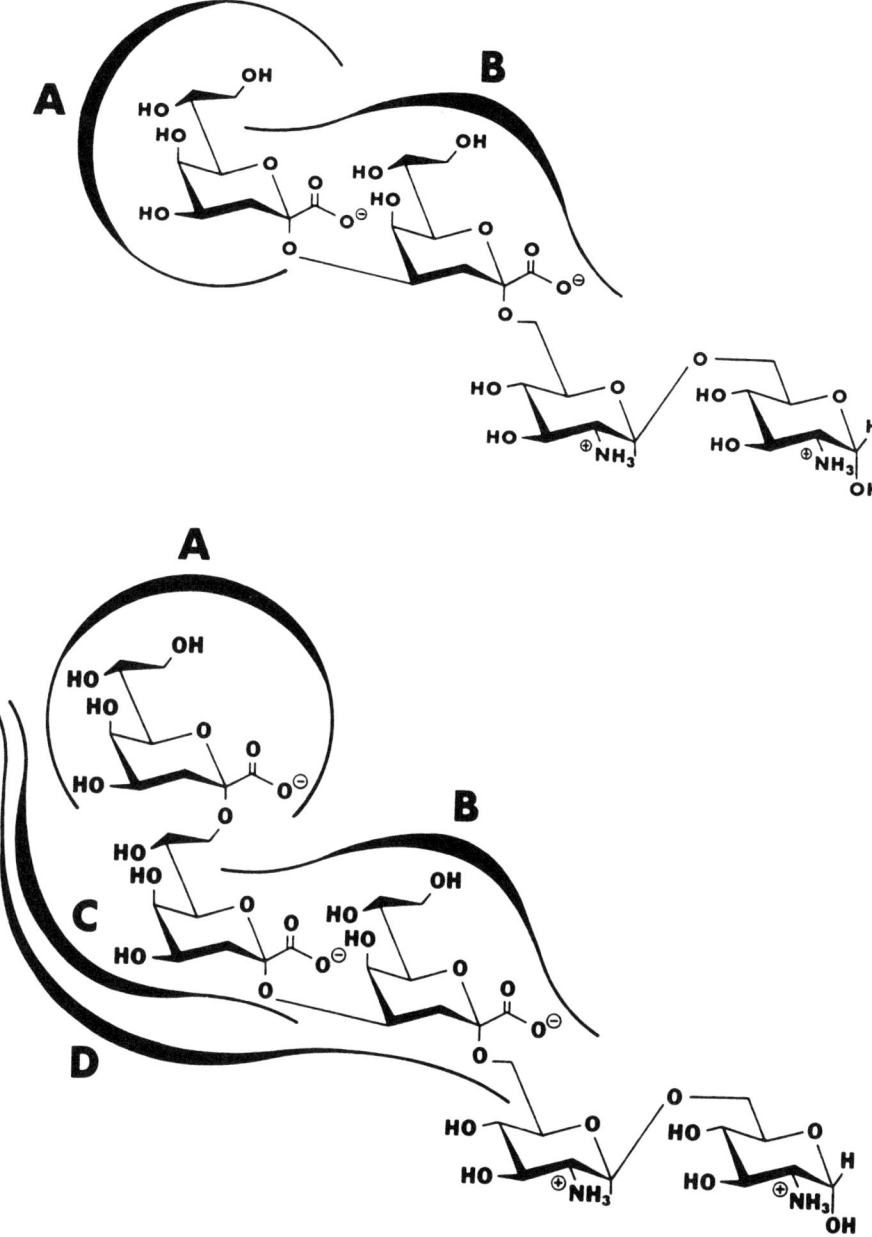

Fig. 3 Schematic representation of the carbohydrate backbone of LPS from *E. coli* Re (top) and *Chlamydia* (bottom) and the epitope specificities of monoclonal antibodies against a single Kdo residue (A), the α2→4- (B) and α2→8-linked (C) Kdo disaccharide, and the trisaccharide α-Kdo-(2→8)-α-Kdo-(2→4)-α-Kdo (D). Whereas type A and B antibodies react with Re and chlamydial LPS, type C and D antibodies are chlamydiae-specific. (From Ref. 62.)

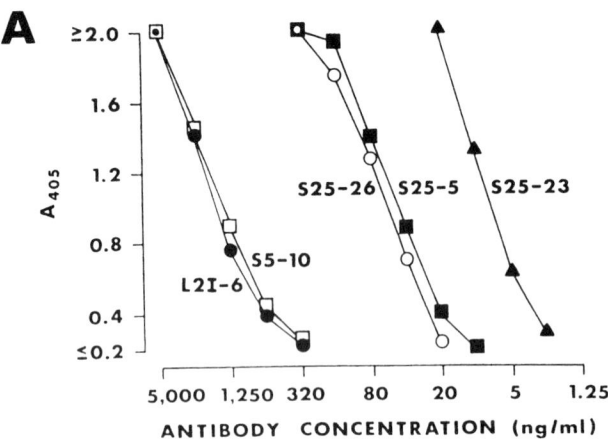

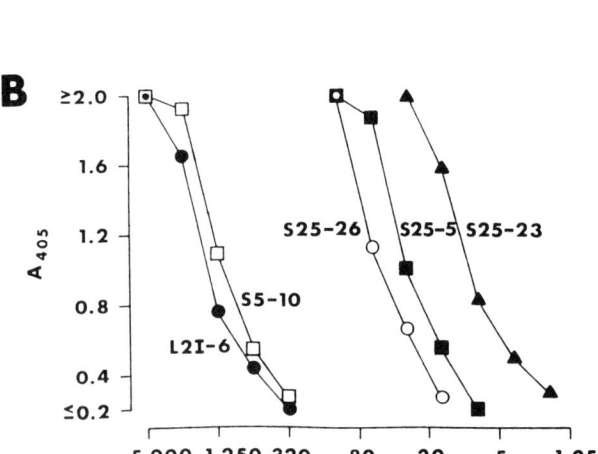

Fig. 4 Binding of monoclonal antibodies to artificial neoglyconjugate antigens composed of bovine serum albumin to which the oligosaccharide ligands α-Kdo-(2→8)-α-Kdo-(2→4)-α-Kdo (A) or α-Kdo-(2→8)-α-Kdo-(2→4)-α-Kdo-β-GlcNac (B) were covalently linked. Monoclonal antibodies L2I-6 (95) and S5-10 (96) were obtained after immunization with elementary bodies of *C. trachomatis* and *C. psittaci*, respectively; the other antibodies were obtained after immunization with an artificial glycoconjugate containing as a carbohydrate ligand the tetrasaccharide α-Kdo(2→8)-α-Kdo-(2→4)-Kdo-(2→6)-β-GlcNac (71).

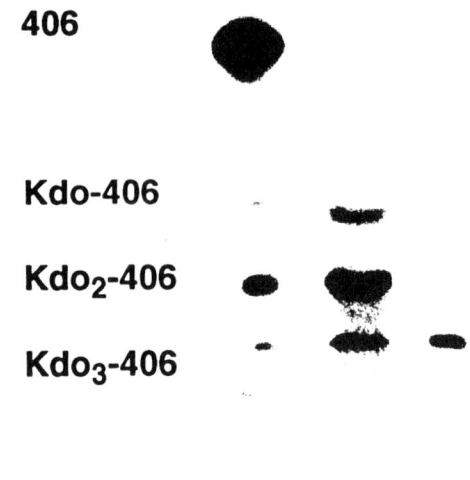

Fig. 5 In vitro activity of the Kdo-transferase from *C. pneumoniae* TW-183. The cloned Kdo-transferase gene of *C. pneumoniae* TW-183 was expressed in the gram-positive bacterium *Corynebacterium glutamicum*, cell-free extracts of which were reacted with lipid A precursor Ia and a CMP-Kdo–generating system. The in vitro products were separated by thin layer chromatography and developed by autoradiography (lane 1) or with monoclonal antibodies recognizing a single Kdo (lane 2) or the *Chlamydia*-specific trisaccharide α-Kdo-(2→)-α-Kdo-(2→4)-α-Kdo (lane 3). (From Ref. 67.)

tivity at the genus level as determined by immunofluorescence but was strain-specific in neutralization experiments (83). Detailed characterization of the epitope specificity of CP33 (83a) showed that it required the presence of the genus-specific Kdo-trisaccharide. However, binding was significantly enhanced when additional components were present. So far, it cannot be decided whether these additional components are part of the epitope or influence the conformation of the trisaccharide epitope.

Chlamydial Lipid A

Whereas the molecular cloning of the chlamydial Kdo transferases allowed structural and immunochemical analyses of the Kdo region, this approach could not be used to investigate chlamydial lipid A. It is noted again that all LPS obtained from recombinant bacteria contained the lipid A moiety from *E. coli* or *Salmonella enterica*. Structural knowledge on chlamydial lipid A is required to investigate the endotoxic and other biological activities of chlamydial LPS in higher organisms and to study its physiological role for the microorganism. Of specific interest are also the early biosynthetic steps of LPS assembly, i.e., the lipid A biosynthesis.

Early investigations using analytical methods for compositional analyses indicated that lipid A of *C. trachomatis* (61) contained approximately five fatty acids per two moles of glucosamine, whereby only two

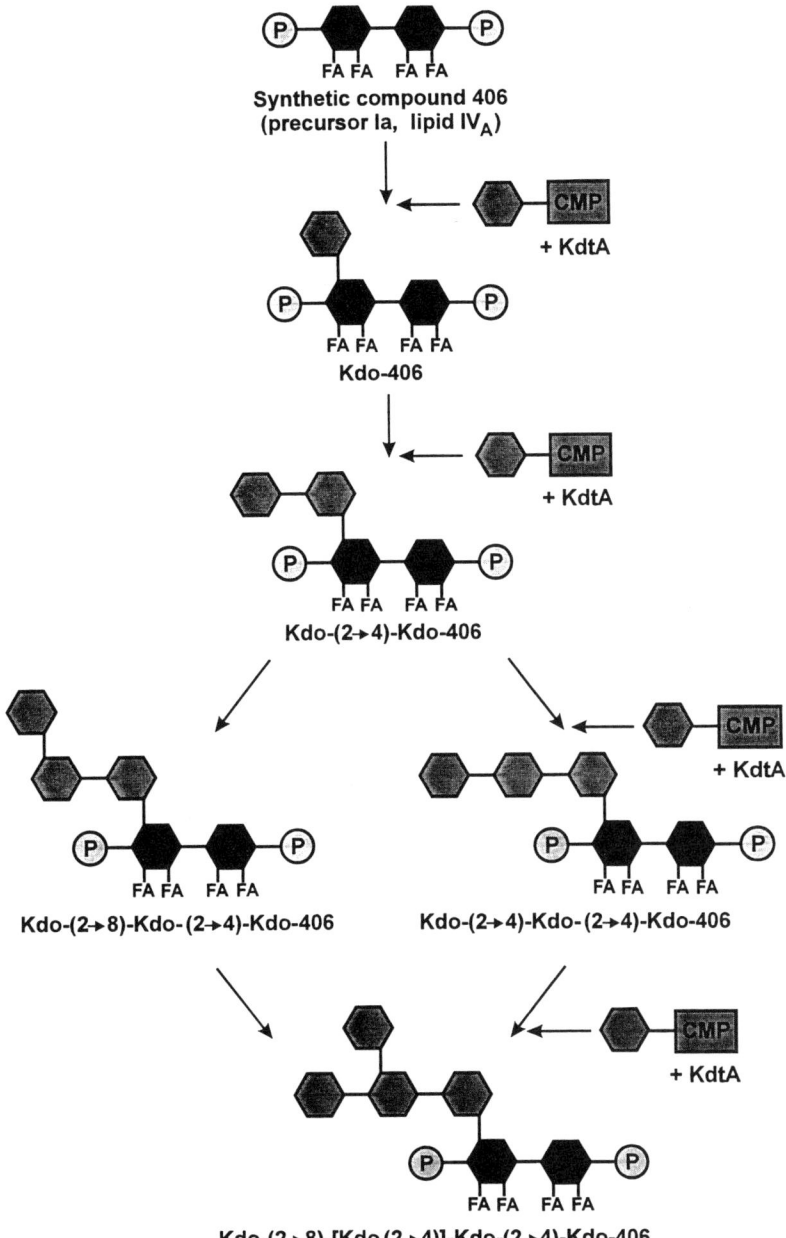

Fig. 6 LPS biosynthesis in *Chlamydia*. In *C. trachomatis* and *C. pneumoniae*, synthesis up to the α-Kdo-(2→8)-α-Kdo-(2→4)α-Kdo-406 product was followed in vitro. The synthesis of the other two structures has been determined in LPS from *E. coli* recombinants expressing the Kdo-transferase of *C. psittaci* 6BC.

moles of 3-hydroxy fatty acids were found. A similar compositional analysis was obtained with a ewe-abortion strain of *C. psittaci* (32), however, an additional constituent was galactosamine, and six instead of five fatty acids were detected. Also in this case, only two moles of 3-hydroxy fatty acids were found per molecule. These data suggested that, unlike in all other lipid A structures known, lipid A in *Chlamydia* contained only amide-linked 3-hydroxy fatty acids. Also unusual was the nature of these fatty acids, with up to 22 carbon atoms. These data were confirmed and extended by an elegant study by Qureshi et al., who investigated the structure of lipid A in *C. trachomatis* serotype F by matrix-assisted laser desorption and liquid secondary ion mass spectrometry of 4-monophosphoryl lipid A fractions separated by HPLC (84). The five most prominent fractions were analyzed to be a 4'-monophosphoryl-glucosamine disaccharide with two amide-linked 3-hydroxy eicosanoic acids, an ester-linked tetradecanoic acid at position 3, and a C14 to C16 fatty acid at 3' of the glucosamine backbone. The amide-linked 3-hydroxy eicosanoic acid at position 2' was esterified at its hydroxyl group with a C18 to C21 fatty acid. Due to the acid hydrolysis step used to liberate the lipid A moiety from LPS, no information could be given on the presence of substituents at the reducing end or on substituents of phosphate groups. These data, together with those obtained on the Kdo region, may motivate organic chemists to synthesize chlamydial lipid A to allow detailed biological and enzymatic investigations.

So far, nothing is known about the substitution of the phosphate groups as it has been found in many LPS with ethanolamine, 4-aminoarabinose, or both. The binding characteristics of monoclonal antibody CP33 may indicate that the polar head groups of lipid A are important for binding. It was shown that the antibody bound to LPS from *E. coli* recombinants but not to LPS from *Salmonella* recombinants, both expressing the chlamydial Kdo-transferase. However, dephosphorylation or de-O-acylation of the *Salmonella*-derived LPS allowed binding of the antibody. Since both degradations also affect the polar head groups, they may either be part of the epitope itself or influence the conformation of the epitope (83a).

Biology of Chlamydial LPS

Since LPS are the endotoxins of gram-negative bacteria acting on a variety of cells, the endotoxic potential of chlamydial LPS is of great interest, particularly to understand its role in infection. However, our knowledge on this aspect is rudimentary. Whereas it has been shown that chlamydial LPS is mitogenic for mouse spleenocytes and induces prostaglandin E_2 (32) in mouse peritoneal macrophages and is active in the limulus amoebocyte lysate assay (60), its ability to induce α-TNF is much less than that of enterobacterial LPS (33). Although little is known on this topic, we are still tempted to speculate on its role in infection. The low endotoxic activity may be attributed to the presence of unusual long-chain fatty acids, which increase the hydrophobicity of the molecule. The low endotoxic activity would be advantageous to the microorganism since it prevents the formation of α-TNF, γ-IFN, and the induction of NO synthase, mechanisms that certainly are relevant to the elimination of intracellular bacteria. This field deserves more attention in the future, and the structural basis provided by Qureshi et al. (84) is a first step into this direction.

The developmental cycle of chlamydiae requires a reorganization of the bacterial membrane, which is different in elementary and reticulate bodies. Since LPS is a major membrane constituent in both forms, its fate during the differentiation process is of great interest. Again, few data are available on this aspect of chlamydial physiology, which will be reviewed in the following section. It is an interesting speculation that phagosome-lysosome fusion is inhibited by the release of LPS from dividing reticulate bodies into the phagosome and subsequent association with the phagosomal membrane. The incorporation of the hydrophobic LPS into the phagosomal membrane would certainly decrease its fluidity. Thus, inhibition of the fusion with the phospholipid membrane of the lysosome could be simply due to physical incompatibility of the two membranes. This hypothesis is supported by a report in the literature showing that chlamydia-infected cultures, superinfected with vesicular stomatitis virus (an enveloped virus), release virus particles that contain LPS in their envelope (85). In addition, they provided evidence that LPS is also incorporated into the host cell membrane.

Another interesting observation was made when we studied LPS from chlamydiae grown either in an embryonated egg or in tissue culture. For *C. psittaci* 6BC we showed that they made, in addition to the well-known rough LPS, a typical smooth LPS when grown in embryonated eggs (86). This result is so far not generally accepted among chlamydiologists (the reasons have been discussed recently), but one new piece of evidence has come up, reporting on the chemical composition of the polysaccharide of S-LPS from a ewe-abortion strain (87). Nevertheless, the data on S-LPS need to be confirmed by an independent approach, ideally by the cloning and expression of those genes that encode O-chain biosynthesis. We constructed a number of cosmid banks, which were expressed in *E. coli* K-12 and screened with the anti-smooth LPS antibodies, however, so far without success. Also the attempts to complement a number of enterobacterial mutants with defects in the *rfa* locus failed. Either the genetic organization of chlamydial LPS is different from what we know from other bacterial genera, or the genes are not expressed in *E. coli* K-12, or we have isolated after all, and despite all the precautions and arguments reported, a contamination. We do not believe the latter possibility and conclude that the biosynthesis of LPS in *Chlamydia* depends on the host system in which the bacteria multiply. LPS phase variation as a mechanism of changing bacterial pathogenicity has been reported for other intracellular pathogens such as *Coxiella burnetii* (88) and *Francisella tularensis* (56), which may allow here some speculation. It is well established that the O-antigen in S-LPS protects bacteria from host defense mechanisms, e.g., attack by complement or by phagocytes. For a facultative intracellular pathogen or an obligatory intracellular pathogen with extracellular developmental forms, it would be highly advantageous to synthesize inside the host cell, where the bacterium is in a protected environment, a small R-LPS, the biosynthesis of which requires less host nutrients than does that of S-LPS, whereas outside the protection by O-chains would be desirable. Similar mechanisms have been reported for *Neisseria* (89) and *Haemophilus influenzae* (90), facultative intracellular mucosal pathogens, which change their LPS structures (not usually containing O-chains) and are able to substitute the terminal outer core region by sialic acid residues (91,92). Since evidence has been given that sialylation in these bacteria is related to virulence and regulated by host factors (93,94), one may allow us to also speculate that these bacteria may be able to synthesize O-chains under certain environmental conditions. This hypothesis is supported by a recent publication reporting that *B. pertussis*, also reported to have a rough-type LPS, possesses a gene with similarity to that of *rfbP* of *Salmonella enterica*, where it is involved in the early steps of O-chain biosynthesis (94).

FUTURE ASPECTS

The fascinating nature of the unique microorganisms of the genus *Chlamydia*, which so far have attracted only a small group of researchers, will spread rapidly into many areas of the biomedical sciences due to the hypothetical association with the pathogenesis of arteriosclerosis. More clinicians, biochemists, cell and molecular biologists, and immunologists will become engaged in this area. In addition, pharmaceutical companies and granting institutions will hopefully redistribute their funding resources.

ACKNOWLEDGMENTS

I am grateful to my coworkers W. Brabetz, L. Brade, O. Holst, S. Löbau, and S. Müller-Loennies for their work during the last years; U. Agge, D. Brötzmann, S. Cohrs, A. Denzin, R. Engel, P. Hellerung, S. Ruttkowski, V. Susott, and M. Willen for their expert technical assistance; H. Moll and H.-P. Cordes for recording the mass and NMR spectra, respectively; I. Bouchain, F. Richter, and M. Lohs for illustrations; B. Köhler and G. Stegelmann-Müller for photographs; and M. Kohlmorgen for typing the manuscript. I also acknowledge the cooperation with P. Kosma, Vienna, Austria, and S. Evans, Ottawa, Canada. This work was supported by the Deutsche Forschungsgemeinschaft (SFB 367 and SFB 470).

REFERENCES

1. Schachter J, Caldwell HD. Chlamydiae. Annu Rev Microbiol 1980; 34:285–309.
2. Storz J. Chlamydia and Chlamydia-Induced Diseases. Springfield, IL: Charles C. Thomas, 1971.
3. Fukushi H, Hirai K. *Chlamydia pecorum*—the fourth species of genus *Chlamydia* Microbiol Immunol 1993; 37:515–522.
4. Kuo CC, Jackson LA, Campbell LA, Grayston JT. *Chlamydia pneumoniae* (TWAR). Clin Microbiol Rev 1995; 8:451–461.
5. Grayston JT, Kuo C-C, Wang S-P, Altman J. A new *Chlamydia psittaci* strain, TWAR, isolated in acute respiratory tract infections. N Engl J Med 1986; 315:161–168.
6. Grayston JT, Aldous MB, Easton A, Wang S, Kuo C, Campbell LA, et al. Evidence that *Chlamydia pneumoniae* causes pneumonia and bronchitis. J Infect Dis 1993; 168:1231–1235.
7. Dalhoff K, Maass M. *Chlamydia pneumoniae* pneumonia in hospitalized patients—clinical characteristics and diagnostic value of polymerase chain reaction detection in BAL. Chest 1996; 110:351–356.
8. Saikku P, Mattila K, Nieminen MS, Mäkelä PH, Huttunen JK, Valtonen V. Serological evidence of an association of a novel *Chlamydia*, TWAR, with chronic coronary heart disease. Lancet 1988; 2:983–986.
9. Linnanmäki E, Leinonen M, Mattila K, Nieminen MS, Valtonen V, Saikku P. *Chlamydia pneumoniae*-specific circulating immune complexes in patients with chronic coronary heart disease. Circulation 1993; 87:1130–1134.
10. Saikku P, Leinonen M, Tenkanen L, Linnanmäki E, Ekman M-R, Manninen V, et al. Chronic *Chlamydia pneumoniae* infection as a risk factor for coronary heart disease in the Helsinki Heart Study. Ann Intern Med 1992; 116:273–278.
11. Thom DH, Wang S-P, Grayston JT, Siscovick DS, Stewart DK, Kronmal RA, et al. *Chlamydia pneumoniae* strain TWAR antibody and angiographically demonstrated coronary artery disease. Arterioscler Thromb 1991; 11:547–551.
12. Maass M, Gieffers J. Prominent serological response to *Chlamydia pneumoniae* in cardiovascular disease. Immunol Infect Dis 1996; 6:65–70.
13. Kuo C, Gown AM, Benditt EP, Grayston JT. Detection of *Chlamydia pneumoniae* in aortic lesions of ather-

osclerosis by immunocytochemical stain. Arterioscler Thromb 1993; 13:1501–1504.
14. Kuo C, Shor A, Campbell LA, Fukushi H, Patton DL, Grayston JT. Demonstration of *Chlamydia pneumoniae* in atherosclerotic lesions of coronary arteries. J Infect Dis 1993; 167:841–849.
15. Kuo C-C, Grayston JT, Campbell LA, Goo YA, Wissler RW, Benditt EP. *Chlamydia pneumoniae* (TWAR) in coronary arteries of young adults (15–34 years old). Proc Natl Acad Sci USA 1995; 92:6911–6914.
16. Campbell LA, O'Brien ER, Cappuccio AL, Kuo C, Wang S, Stewart D, et al. Detection of *Chlamydia pneumoniae* TWAR in human coronary atherectomy tissues. J Infect Dis 1995; 172:585–588.
17. Shor A, Kuo CC, Patton DL. Detection of *Chlamydia pneumoniae* in coronary arterial fatty streaks and atheromatous plaques. S Afr J Med 1992; 82:158–161.
18. Muhlestein JB, Hammond EH, Carlquist JF, Radicke E, Thomson MJ, Karagounis LA, et al. Increased Incidence of *Chlamydia* species within the coronary arteries of patients with symptomatic atherosclerotic versus other forms of cardiovascular disease. J Am Coll Cardiol 1996; 27:1555–1561.
19. Ramirez JA, Ahkee S, Summersgill JT, Ganzel BL, Ogden LL, Quinn TC, et al. Isolation of *Chlamydia pneumoniae* from the coronary artery of a patient with coronary atherosclerosis. Ann Intern Med 1996; 125:979–982.
20. Jackson LA, Campbell LA, Kuo CC, Rodriguez DI, Lee A, Grayston JT. Isolation of *Chlamydia pneumoniae* from a carotid endarterectomy specimen. J Infect Dis 1997; 176:292–295.
21. Maass M, Krause E, Krüger S, Engel PM, Bartels C. Coronary arteries harbour viable *Chlamydia pneumoniae*. Clin Microbiol Infect 1997; 3:136.
22. Gupta S, Camm AJ. Chlamydia pneumoniae and coronary heart disease. Br Med J 1997; 314:1778–1779.
23. Grayston JT, Kuo CC, Campbell LA, Benditt EP. *Chlamydia pneumoniae*, strain TWAR and atherosclerosis. Eur Heart J 1993; 14:66–71.
24. Stephens RS, Stary A, editor. Chlamydial genomic sequencing and manipulation. 1996; 7–10. Bologna, Italy: Societa Editrice Esculapio.
25. Watkins NG, Hadlow WJ, Moss AB, Caldwell HD. Ocular delayed hypersensitivity: a pathogenic mechanism of chlamydial conjunctivitis in guinea pigs. Proc Natl Acad Sci USA 1986; 83:7480–7484.
26. Morrison RP, Belland RJ, Lyng K, Caldwell HD. Chlamydial disease pathogenesis. The 57-kD chlamydial hypersensitivity antigen is a stress response protein. J Exp Med 1989; 170:1271–1283.
27. Morrison RP, Lyng K, Caldwell HD. Chlamydial disease pathogenesis. Ocular hypersensitivity elicited by a genus-specific 57-kD protein. J Exp Med 1989; 169:663–675.
28. Toye B, Laferrière C, Claman P, Jessamine P, Peeling R. Association between antibody to the chlamydial heat-shock protein and tubal infertility. J Infect Dis 1993; 168:1236–1240.
29. Claman P, Honey L, Peeling RW, Jessamine P, Toye B. The presence of serum antibody to the chlamydial heat shock protein (CHSP60) as a diagnostic test for tubal factor infertility. Fertil Steril 1997; 67:501–504.
30. Money DM, Hawes SE, Eschenbach DA, Peeling RW, Brunham R, Wölner-Hanssen P, et al. Antibodies to the chlamydial 60 kd heat-shock protein are associated with laparoscopically confirmed perihepatitis. Am J Obstet Gynecol 1997; 176:870–877.
31. Peeling RW, Kimani J, Plummer F, Maclean I, Cheang M, Bwayo J, et al. Antibody to Chlamydial hsp60 predicts an increased risk for chlamydial pelvic inflammatory disease. J Infect Dis 1997; 175:1153–1158.
32. Brade L, Schramek S, Schade U, Brade H. Chemical, biological, and immunochemical properties of the *Chlamydia psittaci* lipopolysaccharide. Infect Immun 1986; 54:568–574.
33. Ingalls RR, Rice PA, Qureshi N, Takayama K, Lin JS, Golenbock DT. The inflammatory cytokine response to *Chlamydia trachomatis* infection is endotoxin mediated. Infect Immun 1995; 63:3125–3130.
34. Ulevitch RJ, Tobias PS. Receptor-dependent mechanisms of cell stimulation by bacterial endotoxin. Annu Rev Immunol 1995; 13:437–457.
35. Rietschel ET, Kirikae T, Schade FU, Mamat U, Schmidt G, Loppnow H, et al. Bacterial endotoxin: molecular relationships of structure to activity and function. FASEB J 1994; 8:217–225.
36. Sieper J, Braun J, Brandt J, Miksits K, Heesemann J, Laitko S, et al. Pathogenetic role of *Chlamydia*, *Yersinia* and *Borrelia* in undifferentiated oligoarthritis. J Rheumatol 1992; 19:1236–1242.
37. Keat A, Thomas BJ, Dixey J, Osborn MJ, Taylor-Robinson D. *Chlamydia trachomatis* and reactive arthritis: the missing link. Lancet 1983; 1:72–74.
38. Taylor-Robinson D, Gilroy CB, Thomas BJ, Keat ACS. Detection of *Chlamydia trachomatis* DNA in joints of reactive arthritis patients by polymerase chain reaction. Lancet 1992; 340:81–82.
39. Moulder JW. Interaction of Chlamydiae and host cells in vitro. Microbiol Rev 1991; 55:143–190.
40. Kingsbury DT, Weiss E. Lack of deoxyribonucleic acid homology between species of the genus *Chlamydia*. J Bacteriol 1968; 96:1421–1423.
41. Gaydos CA, Palmer L, Quinn TC, Falkow S, Eiden JJ. Phylogenetic relationship of *Chlamydia pneumoniae* to *Chlamydia psittaci* and *Chlamydia trachomatis* as determined by analysis of 16S ribosomal DNA sequences. Int J Syst Bacteriol 1993; 43:610–612.
42. Sheehy N, Markey B, Quinn PJ. Analysis of partial 16S rRNA nucleotide sequences of *Chlamydia pecorum* and *C-psittaci*. FEMS Immunol Med Microbiol 1997; 17:201–205.
43. Raulston JE. Chlamydial envelope components and pathogen-host cell interactions. Mol Microbiol 1995; 15:607–616.
44. Hatch TP. Disulfide cross-linked envelope proteins: the functional equivalent of peptidoglycan in chlamydiae? J Bacteriol 1996; 178:1–5.
45. Barbour AG, Amato KI, Hackstadt T, Perry L, Caldwell HD. *Chlamydia trachomatis* has penicillin-binding proteins but not detectable muramic acid. J Bacteriol 1982; 151:420–428.
46. Zhang JP, Stephens RS. Mechanism of C. trachomatis attachment to eukaryotic host cells. Cell 1992; 69:861–869.

47. Chen JCR, Zhang JP, Stephens RS. Structural requirements of heparin binding to *Chlamydia trachomatis*. J Biol Chem 1996; 271:11134–11140.
48. Su H, Raymond L, Rockey DD, Fischer E, Hackstadt T, Caldwell HD. A recombinant *Chlamydia* trachomatis major outer membrane protein binds to heparan sulfate receptors on epithelial cells. Proc Natl Acad Sci USA 1996; 93:11143–11148.
49. Kuo CC, Takahashi N, Swanson AF, Ozeki Y, Hakomori SI. An N-linked high-mannose type oligosaccharide, expressed at the major outer membrane protein of *Chlamydia trachomatis*, mediates attachment and infectivity of the microorganism to HeLa cells. J Clin Invest 1996; 98:2813–2818.
50. Beatty WL, Belanger TA, Desai AA, Morrison RP, Byrne GI. Tryptophan depletion as a mechanism of gamma interferon-mediated chlamydial persistence. Infect Immun 1994; 62:3705–3711.
51. Malinverni R, Kuo C-C, Campbell LA, Lee A, Grayston JT. Effects of two antibiotic regimens on course and persistance of experimental *Chlamydia pneumoniae* TWAR pneumonitis. Antimicrob Agents Chemother 1995; 39:45–49.
52. Koehler L, Nettelnbreker E, Hudson AP, Ott N, Gerard HC, Branigan PJ, et al. Ultrastructural and molecular analyses of the persistance of *Chlamydia trachomatis* (serovar K) in human monocytes. Microb Pathog 1997; 22:133–142.
53. Matsumoto A, Manire GP. Electron microscopic observations on the effect of penicillin on the morphology of *Chlamydia psittaci*. J Bacteriol 1970; 101:278–285.
54. Byrne GI, Lehmann LK, Landry GJ. Induction of tryptophane catabolism is the mechanism for gamma-interferon-mediated inhibition of intracellular *Chlamydia psittaci* replication in T24 cells. Infect Immun 1986; 53:347–351.
55. Shemer-Avni Y, Wallach D, Sarov I. Inhibition of *Chlamydia trachomatis* growth by recombinant tumor necrosis factor. Infect Immun 1988; 56:2503–2506.
56. Cowley SC, Myltseva SV, Nano FE. Phase variation in *Francisella tularensis* affecting intracellular growth, lipopolysaccharide antigenicity and nitric oxide production. Mol Microbiol 1996; 20:867–874.
57. Richmond SJ, Stirling P, Ashley CR. Virus infecting the reticulate bodies of an avian strain of *Chlamydia psittaci*. FEMS Microbiol Lett 1982; 14:31–36.
58. Palmer L, Falkow S. A common plasmid of *Chlamydia trachomatis*. Plasmid 1986; 16:62.
59. Dhir SP, Hakomori S, Kenny GE, Grayston JT. Immunochemical studies on chlamydial group antigen (presence of a 2-keto-3-deoxycarbohydrate as immunodominant group). J Immunol 1972; 109:116–122.
60. Nurminen M, Leinonen M, Saikku P, Mäkelä PH. The genus-specific antigen of *Chlamydia*: resemblance to the lipopolysaccharide of enteric bacteria. Science 1983; 220:1279–1281.
61. Nurminen M, Rietschel ET, Brade H. Chemical characterization of *Chlamydia trachomatis* lipopolysaccharide. Infect Immun 1985; 48:573–575.
62. Brade H, Brabetz W, Brade L, Holst O, Löbau S, Lucakova M, et al. Chlamydial lipopolysaccharide. J Endotoxin Res 1997; 4:67–84.
63. Nano FE, Caldwell HD. Expression of the chlamydial genus-specific lipopolysaccharide epitope in *Escherichia coli*. Science 1985; 228:742–744.
64. Brade L, Nano FE, Schlecht S, Schramek S, Brade H. Antigenic and immunogenic properties of recombinants from *Salmonella typhimurium* and *Salmonella minnesota* rough mutants expressing in their lipopolysaccharide a genus-specific chlamydial epitope. Infect Immun 1987; 55:482–486.
65. Brade H, Brade L, Nano FE. Chemical and serological investigations on the genus-specific lipopolysaccharide epitope of *Chlamydia*. Proc Natl Acad Sci USA 1987; 84:2508–2512.
66. Belunis CJ, Mdluli KE, Raetz CRH, Nano FE. A novel 3-deoxy-D-manno-octulosonic acid transferase from *Chlamydia trachomatis* required for expression of the genus-specific epitope. J Biol Chem 1992; 267:18702–18707.
67. Löbau S, Mamat U, Brabetz W, Brade H. Molecular cloning, sequence analysis, and functional characterization of the lipopolysaccharide biosynthetic gene *kdt*A encoding 3-deoxy-a-D-manno-octulosonic acid transferase of *Chlamydia pneumoniae* strain TW-183. Mol Microbiol 1995; 18:391–399.
68. Holst O, Broer W, Thomas-Oates JE, Mamat U, Brade H. Structural analysis of two oligosaccharide biphosphates isolated from the lipopolysaccharide of a recombinant strain of *Escherichia coli* F515 (Re chemotype) expressing the genus-specific epitope of *Chlamydia* lipopolysaccharide. Eur J Biochem 1993; 214:703–710.
69. Holst O, Thomas-Oates JE, Brade H. Preparation and structural analysis of oligosaccharide monophosphates obtained from the lipopolysaccharide of recombinant strains of *Salmonella minnesota* and *Escherichia coli* expressing the genus-specific epitope of *Chlamydia* lipopolysaccharide. Eur J Biochem 1994; 222:183–194.
70. Holst O, Bock K, Brade L, Brade H. The structures of oligosaccharide bisphosphates isolated from the lipopolysaccharide of a recombinant *Escherichia coli* strain expressing the gene gseA [3-deoxy-D-manno-octulopyranosonic acid (Kdo) transferase] of *Chlamydia psittaci* 6BC. Eur J Biochem 1995; 229:194–200.
71. Fu Y, Baumann M, Kosma P, Brade L, Brade H. A synthetic glycoconjugate representing the genus-specific epitope of chlamydial lipopolysaccharide exhibits the same specificity as its natural counterpart. Infect Immun 1992; 60:1314–1321.
72. Brade L, Brunnemann H, Ernst M, Fu Y, Holst O, Kosma P, et al. Occurrence of antibodies against chlamydial lipopolysaccharide in human sera as measured by ELISA using an artificial glycoconjugate antigen. FEMS Immunol Med Microbiol 1994; 8:27–42.
73. Bundle DR, Young NM. Carbohydrate-protein interactions in antibodies and lectins. Curr Opin Struct Biol 1992; 2:666–673.
74. Mikol V, Kosma P, Brade H. Crystal and molecular structure of allyl *O*-(sodium 3-deoxy-a-D-manno-2-octulopyranosylonate)-(2->8)-*O*-(sodium 3-deoxy-a-D-manno-2-octulopyranosidonate)-monohydrate. Carbohydr Res 1994; 263:35–42.

75. Lehmann V, Rupprecht E, Osborn MJ. Isolation of mutants conditionally blocked in the biosynthesis of the 3-deoxy-D-*manno*-octulosonic-acid-lipid A part of lipopolysaccharides derived from *Salmonella typhimurium*. Eur J Biochem 1977; 76:41–49.
76. Woisetschläger M, Högenauer G. Cloning and characterization of the gene encoding 3-deoxy-D-*manno*-octulosonic acid 8-phosphate synthetase from *Escherichia coli*. J Bacteriol 1986; 168:437–439.
77. Brabetz W, Brade H. Molecular cloning, sequence analysis and functional characterization of the gene *kdsA*, encoding 3-deoxy-D-manno-2-octulosonate-8-phosphate synthase of *Chlamydia psittaci* 6BC. Eur J Biochem 1997; 244:66–73.
78. Goldman RC, Kohlbrenner WE. Molecular cloning of the structural gene coding for CTP:CMP-3-deoxy-manno-octulosonate cytidyltransferase from *Escherichia coli* K-12. J Bacteriol 1985; 163:256–261.
79. Wylie JL, Iliffe ER, Wang LL, McClarty G. Identification, characterization, and developmental regulation of *Chlamydia trachomatis* 3-deoxy-D-*manno*-octulosonate (KDO)-8-phosphate synthetase and CMP-KDO synthetase. Infect Immun 1997; 65:1527–1530.
80. Clementz T, Raetz CRH. A gene coding for 3-deoxy-D-*manno*-octulosonic-acid transferase in *Escherichia coli*. Identification, mapping, cloning, and sequencing. J Biol Chem 1991; 266:9687–9696.
81. Mamat U, Baumann M, Schmidt G, Brade H. The genus-specific lipopolysaccharide epitope of *Chlamydia* is assembled in *C. psittaci* and *C. trachomatis* by glycosyltransferases of low homology. Mol Microbiol 1993; 10:935–941.
82. Mamat U, Löbau S, Persson K, Brade H. Nucleotide sequence variations within the lipopolysaccharide biosynthesis gene *gseA* (Kdo transferase) among the *Chlamydia trachomatis* serovars. Microb Pathog 1994; 17:87–97.
83. Qu Z, Cheng X, De la Maza LM, Peterson EM. Stary A, ed. Identification of a Neutralizing Monoclonal Antibody to the Lipopolysaccharide of *Chlamydia pneumoniae*. Bologna, Italy: Societa Editrice Esculapio, 1996:375.
83a. Peterson EM, De la Maza LM, Brade L, and Brade H. Characterization of a neutralizing monoclonal antibody directed at the lipopolysaccharide of *Chlamydia pneumoniae*. Infect Immun 1998; 66:3848–3855.
84. Qureshi N, Kaltashov I, Walker K, Doroshenko V, Cotter RJ, Takayama K, et al. Structure of the monophosphoryl lipid a moiety obtained from the lipopolysaccharide of *Chlamydia trachomatis*. J Biol Chem 1997; 272:10594–10600.
85. Karimi ST, Schloemer RH, Wilde III CE. Accumulation of chlamydial lipopolysaccharide antigen in the plasma membranes of infected cells. Infect Immun 1989; 57:1780–1785.
86. Lukácová M, Baumann M, Brade L, Mamat U, Brade H. Lipopolysaccharide smooth-rough phase variation in bacteria of the genus *Chlamydia*. Infect Immun 1994; 62:2270–2276.
87. Toman R, Skultéty L, Kovácová E, Pätoprsty V. Some characteristics of a smooth type lipopolysaccharide of *Chlamydia psittaci*. Acta Virol (Praha) 1997; 41:55–56.
88. Lukácová M, Kazár J, Gajdosová E, Vavreková M. Phase variation of lipopolysaccharide of *Coxiella burnetii*, strain Priscilla during chick embryo yolk sac passaging. FEMS Microbiol Lett 1993; 113:285–290.
89. Van Putten JPM, Robertson BD. Molecular mechanisms and implications for infection of lipopolysaccharide variation in *Neisseria*. Mol Microbiol 1995; 16:847–853.
90. Roche RJ, High NJ, Moxon ER. Phase variation of *Haemophilus influenzae* lipopolysaccharide: characterization of lipopolysaccharide from individual colonies. FEMS Microbiol Lett 1994; 120:279–284.
91. Smith H, Parsons NJ, Cole JA. Sialylation of neisserial lipopolysaccharide: a major influence on pathogenicity. Microb Pathog 1995; 19:365–377.
92. Mandrell RE, McLaughlin R, Kwaik YA, Lesse A, Yamasaki R, Gibson B, et al. Lipooligosaccharides (LOS) of some *Haemophilus* species mimic human glycosphingolipids, and some LOS are sialylated. Infect Immun 1992; 60:1322–1328.
93. Mandrell RE, Lesse A, Sugai JV, Shero M, McLeod Griffiss J, Cole JA, et al. In vitro and in vivo modification of *Neisseria gonorrhoeae* lipooligosaccharide epitope structure by sialylation. J Exp Med 1990; 171:1649–1664.
94. Allen A, Maskell D. The identification, cloning and mutagenesis of a genetic locus required for lipopolysaccharide biosynthesis in *Bordetella pertussis*. Mol Microbiol 1996; 19:37–52.
95. Caldwell HD, Hitchcock PJ. Monoclonal antibody against a genus-specific antigen of *Chlamydia* species: location of the epitope on chlamydial lipopolysaccharide. Infect Immun 1984; 44:306–314.
96. Brade L, Holst O, Kosma P, Zhang Y, Paulsen H, Krausse R, et al. Characterization of murine monoclonal and murine, rabbit, and human polyclonal antibodies against chlamydial lipopolysaccharide. Infect Immun 1990; 58:205–213.

14

The Chemical Synthesis of Lipid A

Shoichi Kusumoto, Koichi Fukase, and Masato Oikawa
Osaka University, Osaka, Japan

INTRODUCTION

Chemical synthesis has played a role in the modern field of lipopolysaccharide (LPS) research, particularly in relation to the structure and function of lipid A, which is the active entity of LPS. The success of the first chemical synthesis of lipid A (1,2) made an epoch-making contribution to the field: a synthetic specimen (Fig. 1, **1**) prepared according to the structure proposed for lipid A of *Escherichia coli* F515 (3) showed identical full endotoxic activity observed with the corresponding bacterial preparation (4,5). This result gave the final, decisive evidence for the concept that lipid A is the active principle of endotoxin. Simultaneously, we obtained an entirely new ability to investigate endotoxic action by the use of fully defined specific molecules, which may be rationally designed at will. It has also become possible to precisely study the relationship between chemical structure and biological function. In fact, antagonistic activity against LPS and lipid A was first observed by the aid of a chemically synthesized specimen of a biosynthetic precursor of lipid A (Fig. 1, **2**) (6–9), which was designated precursor Ia (10) or lipid IV_A (11). This compound shares the same hydrophilic backbone structure of bisphosphorylated glucosamine β(1-6)disaccharide with mature lipid A (Fig. 1, **1**) but contains two fewer fatty acyl groups than the latter. Such a line of research may lead to the discovery of new, possibly artificial compounds with clinical values.

In this chapter recent advances in the chemical synthesis of lipid A and closely related compounds will be described. The focus will be on those topics that appeared after a previous review by the present author (12).

SYNTHESIS OF LIPID A AND ITS FATTY ACYL ANALOGS

Preparation of Optically Pure 3-Hydroxy and 3-Acyloxy Acids

After the first synthesis of *E. coli* lipid A (Fig. 1, **1**) and its precursor (Fig. 1, **2**) (1,2,6,7), many structural analogs of lipid A have been synthesized. Since 3-hydroxy (β-hydroxy) fatty acids are one of the characteristic and important components of lipid A, several efficient preparative methods have been described for optically pure 3-hydroxy acids of various chain lengths and configurations. The major contributions that appeared after those cited in the previous review (12) are described below.

The first practical synthesis of optically pure 3-hydroxytetradecanoic acid was achieved by asymmetric hydrogenation of the corresponding 3-keto ester with a modified Raney nickel catalyst (13). The most generally applicable method presently available is based on the same asymmetric hydrogenation of the 3-keto ester but by the use of a homogeneous binap ruthenium catalyst (Fig. 2) (14). Both enantiomers of the hydroxy acid of desired chain length can be obtained in high yield and high enantiomeric purity by employing the correct enantiomer of the commercial binap ligand for the preparation of the catalyst. Optically pure 3-hy-

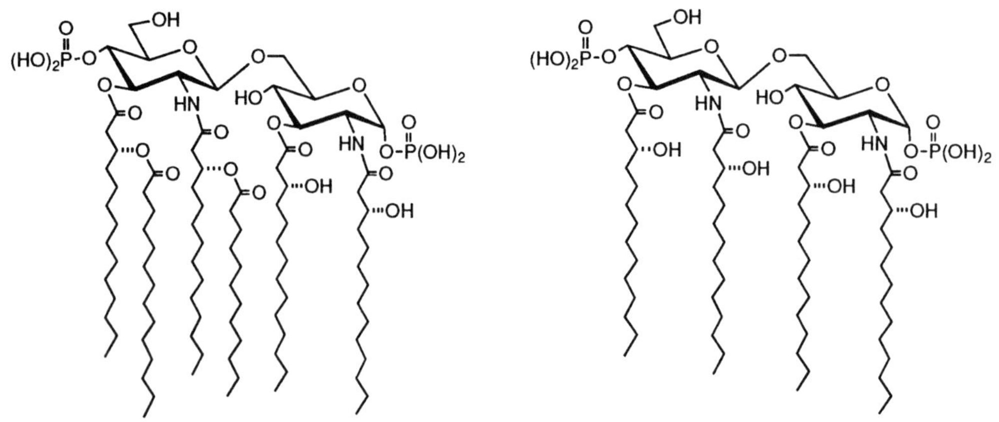

Fig. 1 The chemical structures of lipid A of *E. coli* and one of its biosynthetic precursors.

droxy fatty acids of 10 and 14 carbons prepared by this procedure were used in the recent syntheses of lipid A analogs (15,16).

An improved process of the same asymmetric hydrogenation of 3-keto acids, which employs a commercially available binap ruthenium catalyst at low pressure, was recently reported. A new method for the determination of the optical purity of the products was also presented in the same paper (17).

Another synthesis was based on the asymmetric dihydroxylation of α,β-unsaturated acid *t*-butyl ester. This procedure presents a direct access to optically pure 3-acyloxy fatty acids occurring in many natural lipid As (18).

Optical resolution of racemic hydroxy acids was achieved both by enzymatic and conventional procedures. Enzymatic kinetic resolution was achieved by the use of lipase. When the methyl ester of racemic 3-hydroxy acid was treated with vinyl acetate in the presence of lipase, the (*S*)-isomer was preferentially converted into the corresponding 3-*O*-acetate, whereas the (*R*)-isomers remained unchanged. After chromatographic separation of the free and 3-*O*-acetylated esters and their alkaline hydrolysis, both (*R*)- and (*S*)-acids of approximately 80% optical purities were obtained. Recrystallization as suitable amine salts gave the corresponding optical pure acids (19). The lipase-catalyzed resolution was also applied to free 3-hydroxytetradecanoic acid: in this case free (*R*)-acid was obtained by direct crystallization after the enzymatic reaction (20). Achiwa et al. (21) and Kiso et al. (22) employed resolution of the racemic hydroxy acid as its ephedrine and dehydroabietylamine salt, respectively, for their preparation of lipid A analogs.

The 3-hydroxy group in the fatty acid was normally protected as its benzyl ether. A quite efficient method was described for reductive *O*-benzylation of 3-hydroxy fatty acid esters with the aid of benzaldehyde

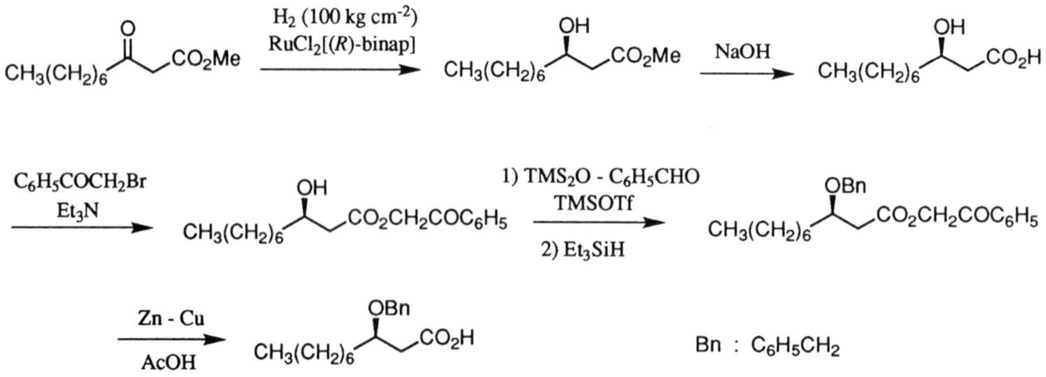

Fig. 2 Catalytic asymmetric synthesis of (*R*)-3-hydroxydecanoic acid and protection of its hydroxy group.

and triethylsilane in the presence of trimethylsilyl triflate and hexamethydisiloxane (Fig. 2) (16).

Improved Route to the Synthesis of Lipid A Analogs

The successful synthetic work during 1982–1987 was of great value in giving the final structural proof for the active entity and opening an entirely new era of endotoxin research (12). The structure proposed for lipid A was chemically constructed de novo as shown in Figure 3 (1,2). It thus became theoretically possible to obtain any related structures as desired. But the synthetic route contained many steps of chemical transformation, and there still remained several important points to be improved in order to make the chemical synthesis a truly efficient and useful tool. The major areas to be improved were (1) selection of protecting groups and their efficient and regioselective introduction onto the carbohydrate or fatty acyl moieties (creation of new protecting procedures was also important), (2) elaboration of efficient and reproducible reaction conditions for glycosyl phosphorylation and the workup procedure after that step, (3) establishing a method for the purification of the final deprotected products. During the past decade, most of these problems have been greatly resolved by continuous efforts of synthetic chemists. Chemical synthesis has now become a valuable tool in the field.

Figure 4 shows the synthetic route to a lipid A analog (**4**) not occurring in nature. This compound corresponds to *E. coli*–type lipid A containing the 3-hydroxy acids of *S*-configuration instead of the (*R*)-acid present in natural lipid A (19). This work and another new synthesis (15) of the biosynthetic precursor (**2**) include most of the typical recent advances in chemical construction of this type of glycoconjugates. The 4′-phosphate residues on the distal glucosamine were protected with a new cyclic benzyl-type ester, i.e., *o*-xylidene ester, first described by Watanabe et al. (23). This protected phosphate group, which is introduced by the use of the corresponding phosphoroamidite followed by mild oxidation, is more stable than the conventional dibenzyl phosphate and hence survives throughout the multistep transformations leading to the final deprotection step. In the first synthesis of lipid A (Fig. 3) this particular 4′-phosphate was protected as the diphenyl ester, which required an additional step of

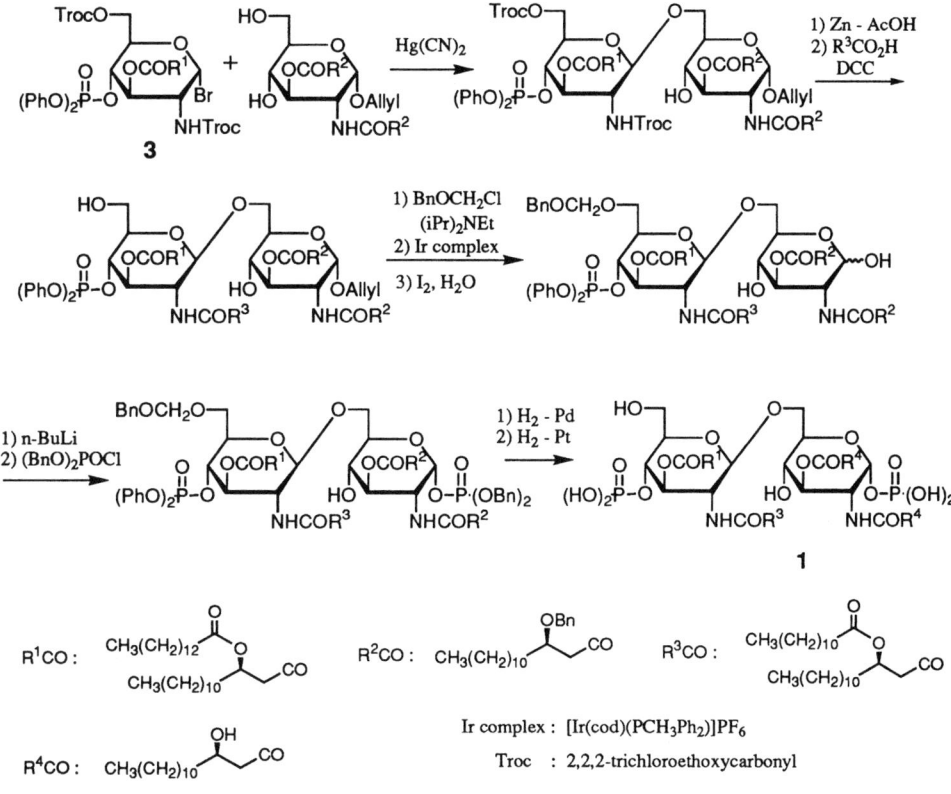

Fig. 3 The first total synthesis of lipid A of *E. coli*.

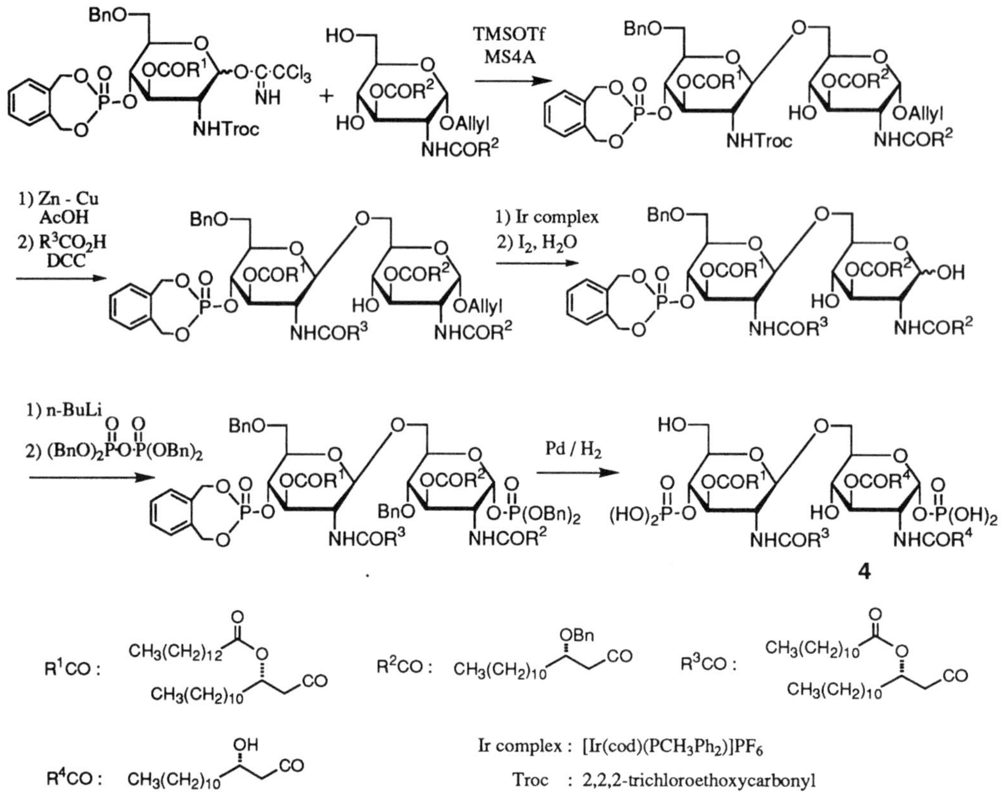

Fig. 4 Recent synthesis of an analog of lipid A–containing (S)-hydroxytetradecanoic acids.

hydrogenolysis with a platinum catalyst after removal of all the benzyl-type protections with a palladium catalyst (1,2). By contrast, deprotection of the phosphates and hydroxy functions are completed in one step in the recent synthesis (Fig. 4) and the undesired hydrolytic cleavage of the unstable glycosyl phosphate during the deprotection step was minimized.

In addition to the cyclic ester protection of the phosphate, the protection strategy for the glucosamine residues was also changed in the following several points. The 6-hydroxy group in the glycosyl donor (Fig. 4) was protected as the benzyl ether via a known regioselective opening of the 4,6-O-benzylidene ring. The use of the benzyl ether, which was retained until the final step, was more advantageous than the trichloroethoxycarbonyl (Troc) group in the previous synthesis. The latter was cleaved at a later synthetic step and the resultant free 6′-hydroxy group had to be protected again before glycosyl phosphorylation (Fig. 3).

The coupling of two glucosamine residues was effected by means of the imidate method to give, without the use of a mercury reagent, the desired β(1-6)disaccharide in high yield. After introduction of one more acyl group at the 2′-amino function and cleavage of the 1-O-allyl protection, the product was subjected to phosphorylation. The same reagents, butyllithium and tetrabenzyl diphosphate, were employed for this reaction as in the earlier synthesis, but neutralization of the reaction mixture followed by rapid extraction of the phosphorylated product with an organic solvent improved the yield of the protected phosphate (15,19). The use of lithium bis(trimethylsilyl)amide in place of butyllithium was then shown to enable simpler operation: the former reagent is much more easy to handle even on a small scale (16).

The method of purification of the final synthetic product was also improved. The product after complete deprotection was purified without loss of the sensitive glycosyl phosphate by the centrifugal countercurrent chromatography (CPC), which operates on the principle of liquid-liquid partition. The E. coli–type lipid A with (S)-3-hydroxytetradecanoic acids (Fig. 4, 4) was, for example, purified with a solvent system of butanol/ THF/water/triethylamine = 45:35:100:22 (19). Purification was more conveniently achieved by the same principle but not with the expensive CPC apparatus. Partition chromatography with a column of Sephadex LH-20 was found to be even more effective and con-

The Chemical Synthesis of Lipid A

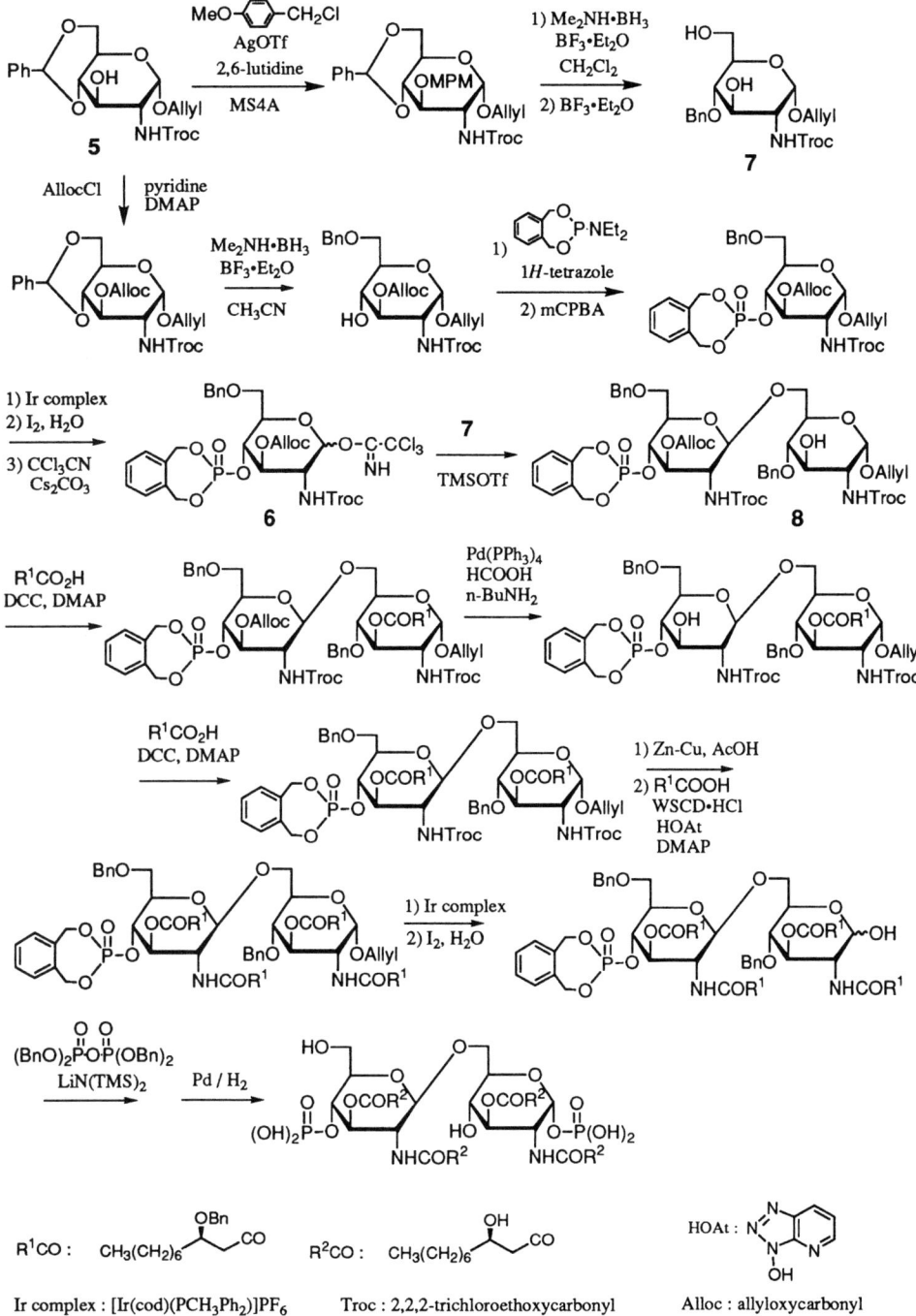

Fig. 5 Synthesis of an analog of lipid A with shorter acyl groups via a divergent route.

venient than CPC with similar solvent systems. In a typical example, synthetic **2** (Fig. 1) was purified with a solvent system of chloroform/methanol/isopropanol/water/triethylamine = 20:20:2.5:22.5:0.01. This column chromatographic purification is readily applicable to a hundred milligram scale (15).

The synthetic scheme shown in Figure 4 was further modified in order to facilitate efficient preparation of lipid A analogs having various combination of acyl groups (16). As illustrated in Figure 5, the 3,3′-hydroxy and two amino groups are differently protected in the key intermediate **8** so that any fatty acyl groups can be introduced at individual positions. The key disaccharide intermediate was formed by coupling of the two monosaccharide precursors, which were prepared from the same 4,6-O-benzylidene derivative **5**: the benzylidene ring was opened selectively in both directions to give either the 6- or the 4-benzyl ether separately (24) and converted to the glycosyl donor **6** or acceptor **7**, respectively (16). The route has been so far utilized only for the preparation of a limited number of analogs but opens a route to make up a lipid A library containing a large number of analogs.

Another research group has recently reported synthesis of a differentially protected glucosamine β(1-6)disaccharide, which may be useful for synthesizing a variety of analogs of lipid A with different acylation patterns (25). Although the synthesis of the entire lipid A structure has not yet been completed, several new useful ideas for selective protection are described.

In the scheme in Figures 4 and 5, knowledge accumulated up to now concerning the chemical synthesis of lipid A is concentrated to represent a simplified route to this complex bacterial glycoconjugate and its analogs. The total yield of **4** (Fig. 4), for example, amounted to 6.9% in 13 steps starting from N-Troc-glucosamine. Further effort is still being made in the authors' laboratory to elaborate a more divergent route through which the two hydroxy and two amino groups can be differentiated for acylation. Such a route will be soon available to construct a completely divergent family of acyl analogs on the common hydrophilic backbone of lipid A.

New Route to Lipid A Analogs Containing Unsaturated Fatty Acyl Groups

Most lipid A isolated from bacterial cells contains only saturated fatty acyl groups, although the majority of them are 3-hydroxylated. The synthetic routes so far described are all designed for the chemical construction of such saturated-type lipid A analogs. In all cases, the final step of the synthesis was hydrogenolytic deprotection to give completed free lipid A. This strategy has so far been quite satisfactory since catalytic cleavage of benzyl, xylidene, or phenyl protecting groups results in the formation of only volatile byproducts such as toluene and xylene, which can be readily removed to give the final product. Furthermore, the mild reaction conditions for the hydrogenolysis are advantageous to retain the unstable glycosidic phosphate intact.

The strategy of benzyl protection becomes basically incompetent when the final product contains a double bond. This situation happened in the synthesis of *Rhodobacter sphaeroides* lipid A (Fig. 6, **9**). According to the reported structural studies (26,27), this particular lipid A shares the same hydrophilic backbone of the bisphosphorylated glucosamine β(1-6) disaccharide but contains unusual amide-bound 3-keto acid in place of the 3-hydroxy acid commonly found in other lipid A. Furthermore, an unsaturated fatty acid was also found on the other amino group.

In view of the potent antagonistic activity of *R. sphaeroides* lipid A to suppress the endotoxic action of other lipid A and LPS, Christ et al. elaborated a new synthetic route where hydrogenolysis is avoided by employing a system of allyl protection, as shown in Figure 7 (28). The hydroxy groups on the glucosamine residues and those on the 3-hydroxy acids were protected by allyloxycarbonylation, while the phosphates were diallyl esters.

A 2-azido sugar imidate (Fig. 7, **10**) was prepared as the glycosyl donor for the synthesis of the disaccharide structure. As in the case of the previous synthesis of lipid A shown in Figures 4 and 5, the donor **10** already contains the 3-O-acyl group and the 4-phosphate protected as the diallyl ester. The glycosyl acceptor **11** (Fig. 7) already had both fatty acids, i.e., an amide-bound keto acid and an allyloxycarbonylated 3-hydroxy fatty acid at the 3-position. The glycosidic position was temporarily protected with a silyl group, which can be selectively removed before the later glycosyl phosphorylation. Coupling of **10** and **11** gave the desired β(1-6) disaccharide, **12** (Fig. 7). After reductive conversion of the azide into the amino group, the latter was acylated with the unsaturated fatty acid. Cleavage of the silyl group was carried out with aqueous hydrogen fluoride, and the resultant free glycosidic position was phosphorylated. This reaction was effected by the phosphoroamidite method with bis(allyloxy)diisopropylaminophosphine to give the desired α-phosphate exclusively in high yield.

All the allyl-type protecting groups were then cleaved by treating with a palladium(0) catalyst in good yield, leaving the double bond, 4'- and glycosidic phosphates intact. The final product was purified by DEAE column chromatography.

Unexpectedly, however, neither of the synthetic products (*cis* and *trans* isomers of the double bond, synthesized separately) were found to be identical with natural *R. sphaeroides* lipid A. The reason for this discrepancy has not yet been explained. Nevertheless the synthesis brought about an interesting fact that the synthetic compound **9** showed potent in vivo suppression of TNF-α generation induced by LPS in human mon-

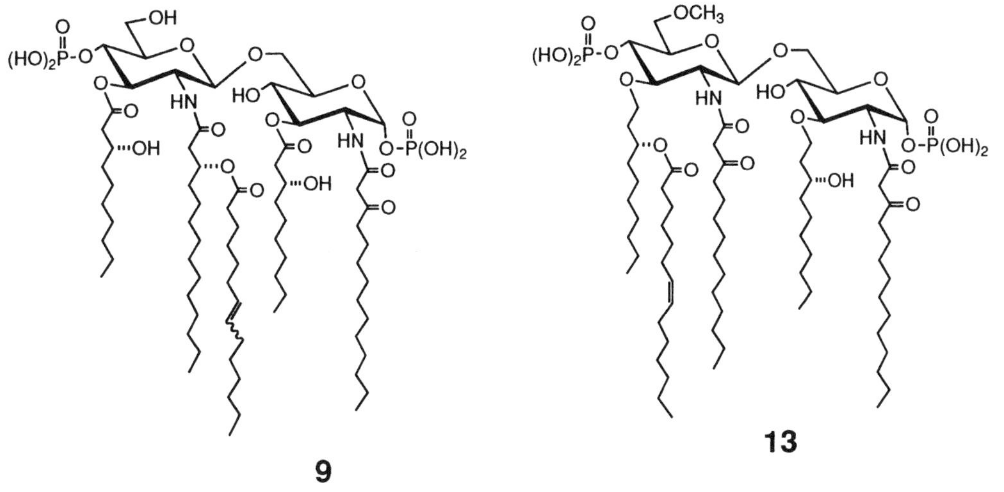

The Chemical Synthesis of Lipid A

Fig. 6 The structure proposed for lipid A of *R. sphaeroides* and an artificial structural analog derived from it.

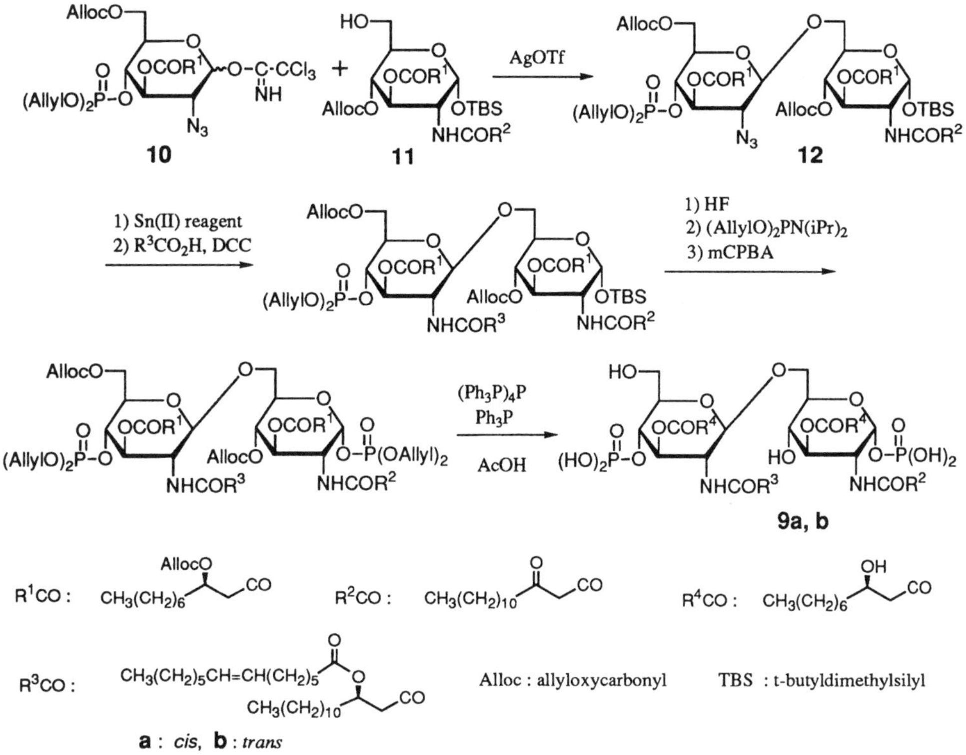

Fig. 7 Synthesis of the proposed structure for *R. sphaeroides* lipid A.

ocytes. This observation led to the design of an artificial molecule **13** (Fig. 6) based on the structure of **9**. Compound **13** was synthesized by similar strategy and proved to act as an antagonist, which may find clinical application in future (29).

SYNTHESIS OF UNNATURAL STRUCTURES BASED ON LIPID A

Synthesis of Monosaccharide Derivatives

After the discovery of lipid X (Fig. 8, **14**) as an early biosynthetic precursor of lipid A (11,30), N,O-diacylated glucosamine 1- and 4-phosphates corresponding to both halves of lipid A were synthesized. Among them some 4-monophosphates such as **15** (Fig. 8) showed definite lipid A–like activity. In view of the much shorter synthetic steps required for the chemical construction of monosaccharide derivatives than those required for disaccharides, many structural modifications were attempted to obtain clinically useful novel compounds. The major part of the works in this line was already described in the previous review (12), but some additional papers appeared recently (31–33).

Further synthetic modifications of acylated monosaccharide structure mimicking lipid A have also been reported. A typical example is a series of 2,3-diacyl glucosamine 4′-monophosphates whose 1-hydroxy group is acylated by a linear dicarboxylic acids of several chain lengths (Fig. 8, **16**) (34). 2,3-Diacylated glucosamines containing β-glycosidically bound alkyl phosphates or amino acid serine were also synthesized (35,36). Some authors describe such compounds as acyclic analogs of lipid A. But these are acylated monosaccharide derivatives, some of which even lack phosphate substituents. Although these compounds were created after extensive modification of the structure of natural lipid A, the term "lipid A analogs" should be used carefully enough to avoid confusion.

A series of 3-O-phosphorylated β-D-glucosyl amine derivatives which are acylated at the amino and 2-hydroxy groups were synthesized (37). They correspond to positional isomers of **15**, which is already an artificial analog of the nonreducing-side monosaccharide of lipid A. Synthesis was reported of a new artificial antagonist, which has two acidic moieties, one phosphono and one carboxyl group, on a 2,3-diacylglucosamine and its ester-linked dimer (38).

Synthesis of Glucosamine Disaccharide Derivatives

Synthetic substitution of the glycosyl phosphate with other acidic groups to form artificial structural analogs was attempted by several research groups in their attempt to find clinically applicable compounds that maintain the beneficial biological activity but hopefully lack the toxic activity of natural lipid A. The presence of the glycosyl phosphate causes many difficulties in chemical synthesis and purification of lipid A, mainly because of its chemical instability. It is thus understandable that the initial efforts of structural modification were focused on this phosphate part of the molecule. Although some of the biological function of lipid A was observed for several monosaccharide partial structures (see Sec. III.A), relative activity of the corresponding disaccharide derivatives is generally much higher. Therefore, the substitution of the glycosyl phos-

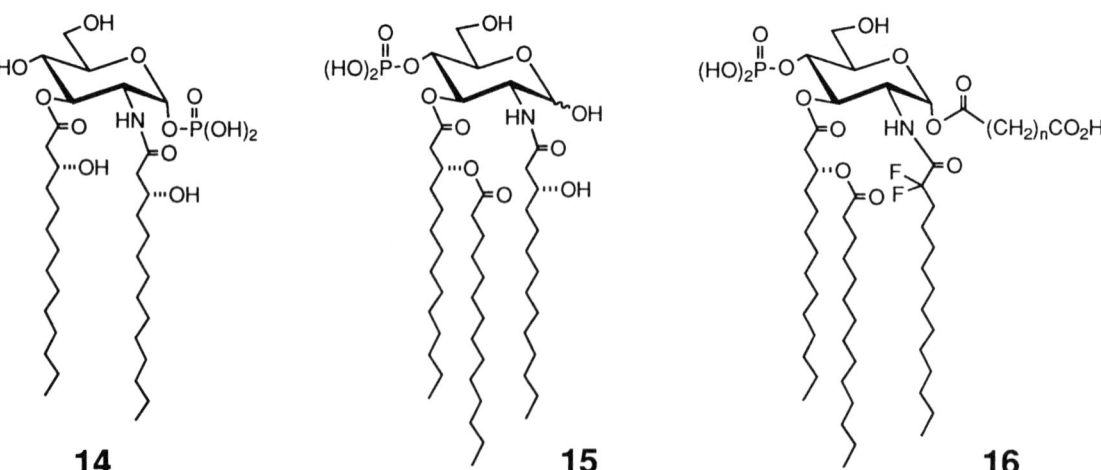

Fig. 8 The chemical structures of lipid X and acylated glucosamine 4-phosphate derivatives.

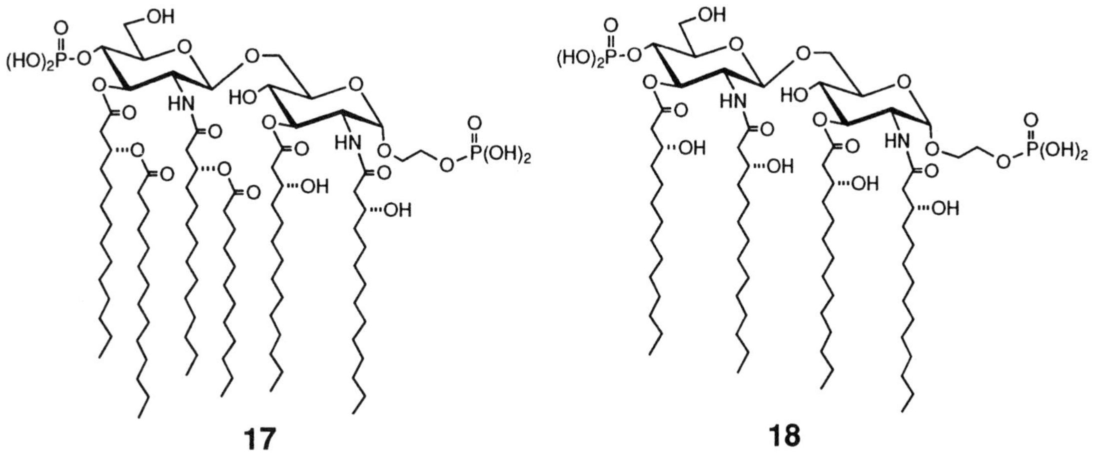

Fig. 9 The chemical structures of the phosphonooxyethyl analogs of lipid A and a biosynthetic precursor.

phate was attempted mainly for the β(1-6) disaccharide component of lipid A.

Synthesis of Phosphonooxyethyl Analog of Lipid A and Its Radiolabeled Derivative

The first reported example of this type of substitution was a phosphonooxyethyl (PE) analog (Fig. 9, **17**) of lipid A (38). In this structure the phosphate group is bound not directly but via an α-glycosidically linked ethylene glycol unit to the disaccharide. Ethylene glycol is the shortest linker available if the original O-glycosidic linkage and the phosphodiester structure are to be retained. But in this PE-type analog, three additional atoms are inserted between the glucosamine and the phosphate.

The PE analogs of *E. coli*–type and the biosynthetic precursor–type acylation patterns (**17** and **18**, respectively) were synthesized by a strategy basically similar to that employed in the early synthesis of lipid A (1,2), except for the timing of phosphorylation (39,40). Thus, in the case of **17**, a glucosamine derivative (Fig. 10, **19**), which contains two acyl groups, and protected phosphonooxyethyl group was prepared and coupled with the same glycosyl donor (Fig. 3, **3**) used previously to form the β(1-6) disaccharide backbone. The

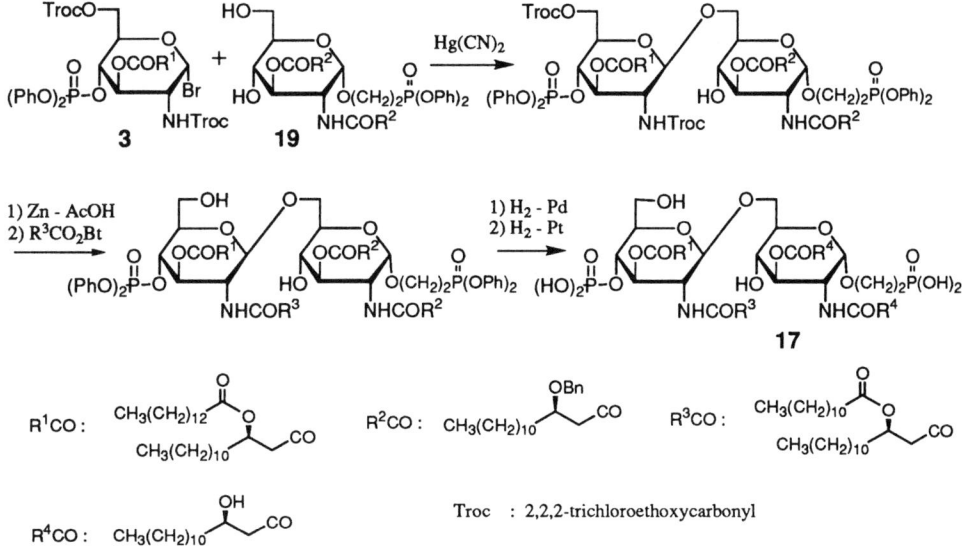

Fig. 10 Synthesis of the phosphonooxyethyl analog of lipid A.

reaction sequence leading to the final product was straightforward (Fig. 10): removal of the N-Troc function followed by N-acylation and deprotection readily gave **17**, which was purified by simple ion-exchange chromatography.

The PE analog **17** was shown to have the same activity, both toxic and beneficial, as the original *E. coli* lipid A, whereas the PE derivative **18** showed the same antagonistic activity as the biosynthetic precursor **2** (Fig. 1) (39,41). Interestingly, a stereoisomer of **17** which has β-glycosidically linked ethylene glycol lacks the activity (39). These observations, together with the previously accumulated information, demonstrate that the glycosyl phosphate is important but its strict location is not required for the expression of the biological activity. In spite of this allowance for its distance from the disaccharide, the change of the anomeric configuration seems to result in a loss of the activity owing to a not negligible structural change around the phosphate moiety.

In view of its biological functions being indistinguishable from those of lipid A (**1**), the PE analog **17** is assumed to act via a mechanism identical to that of **1**, and the same might apply to the biosynthetic precursor **2** and its PE analog **18**. The chemically stable PE analog was regarded as an ideal substrate to prepare a radiolabeled derivative for the study of the action mechanism of lipid A.

Chemical synthesis was achieved for **17** tritium labeled at the ethylene glycol moiety (Fig. 11). In this synthesis a disaccharide intermediate **20** was prepared in which an ethylene glycol is α-linked to the disaccharide. Oxidation of its primary hydroxy group into aldehyde **21** followed by reduction with NaB^{3}H$_4$ afforded a tritium-labeled intermediate **22**. Phosphorylation with a phosphoroamidite reagent followed by ox-

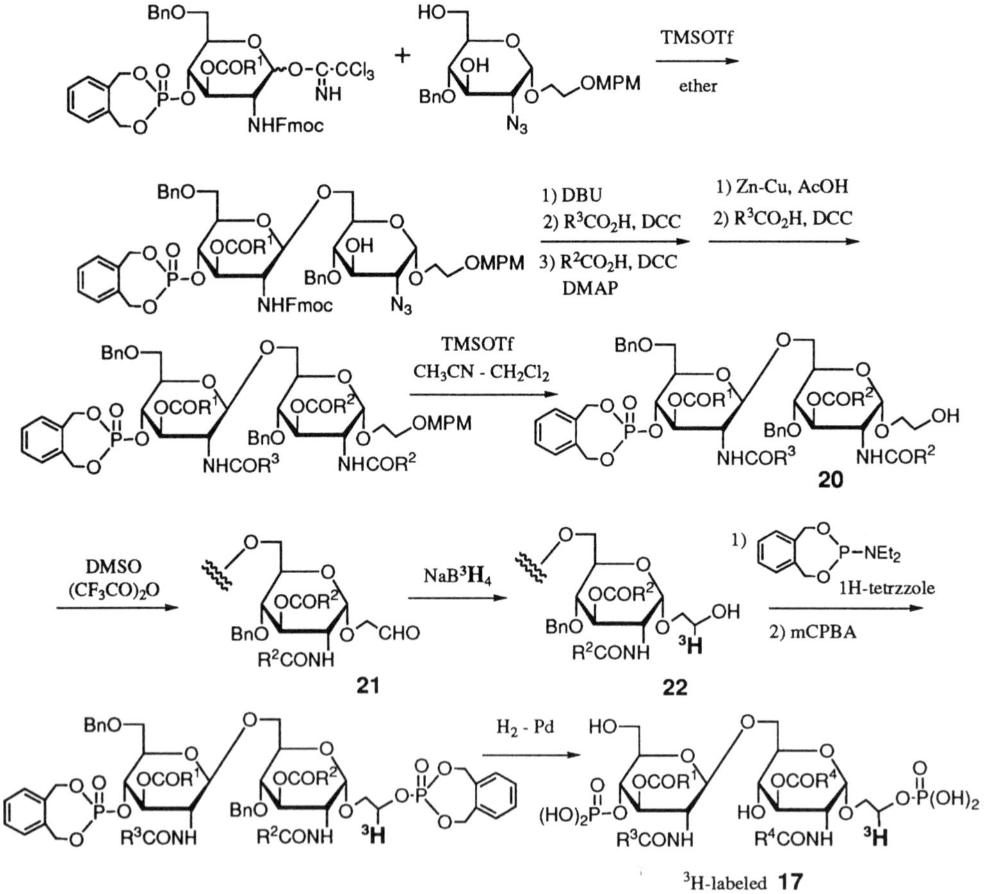

Fig. 11 Chemical synthesis of a tritium-labeled phosphonooxyethyl analog of lipid A.

R^{1}CO to R^{4}CO : the same as in FIGURE 10

Fmoc : 9-fluorenylmethoxycarbonyl   Troc : 2,2,2-trichloroethoxycarbonyl   Alloc : allyloxycarbonyl

idation and hydrogenolysis gave the labeled compound **17**.

Synthesis of Other Disaccharide Analogs of Lipid A

Several other disaccharide analogs were synthesized in which the glycosyl phosphate was replaced with carboxylic acid(s). One of the analogs has the structure **23** (Fig. 12), which contains two carboxylic acids mimicking the divalent nature of the phosphate, whereas the other (**24**) has only one carboxyl group. These compounds were synthesized by the same basic strategy employed for the PE analog **17**. Thus, a glycosyl bromide with the same acylation and protection pattern as **3** was coupled with 2,3-diacylated glycosyl acceptors having benzyl-protected carboxylic acid group(s) to give the corresponding disaccharides. Cleavage of the Troc group, followed by N-acylation and deprotection, afforded **23** and **24** (43).

The anomeric α-configuration is retained in both of them. A malonic acid–type structure [-CH(CO$_2$H)$_2$] where two carboxyl groups are located on the same carbon atom is not stable and readily loses one of them as carbon dioxide; the -CH(CH$_2$CO$_2$H)$_2$ structure in **23** is the minimum one containing two equivalent carboxyl groups. The distance between acidic moieties and the glucosamine residues are not the same in **23** and **24**. Furthermore, fatty acyl groups in these compounds are also different from the natural ones. Therefore, no direct effects of the substitution of the phosphate into carboxylic acid can be discussed. Nevertheless, it was again confirmed that the glycosyl phosphate can be replaced with other acidic group without loss of activity. It should be noted here that oxidation of the aldehyde function in **21** into a carboxyl

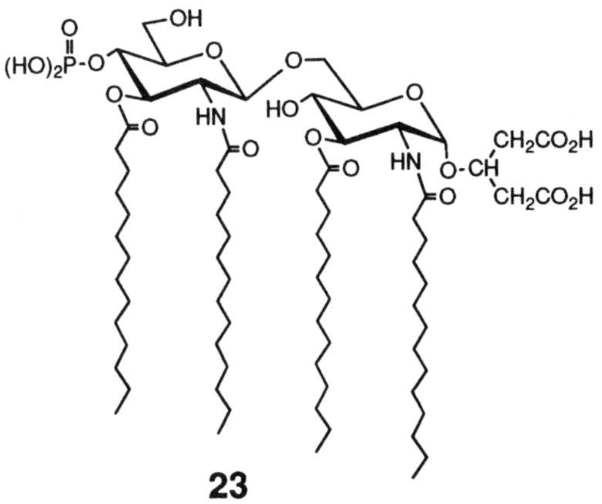

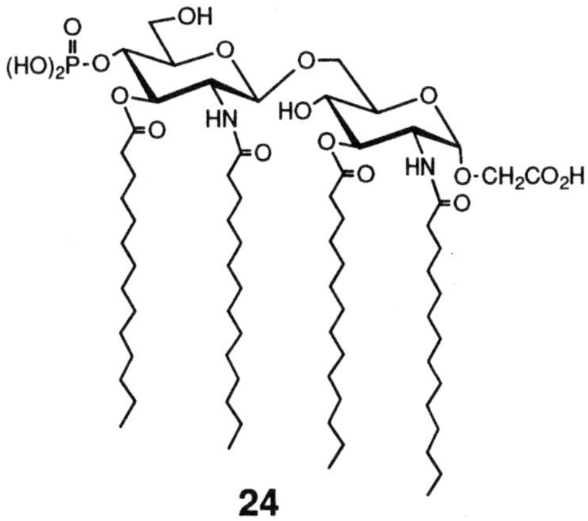

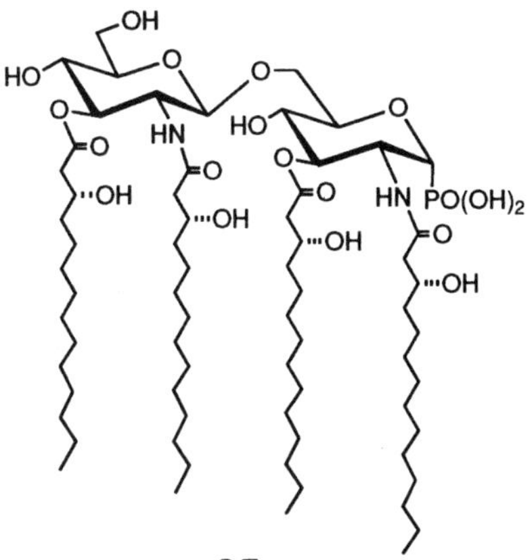

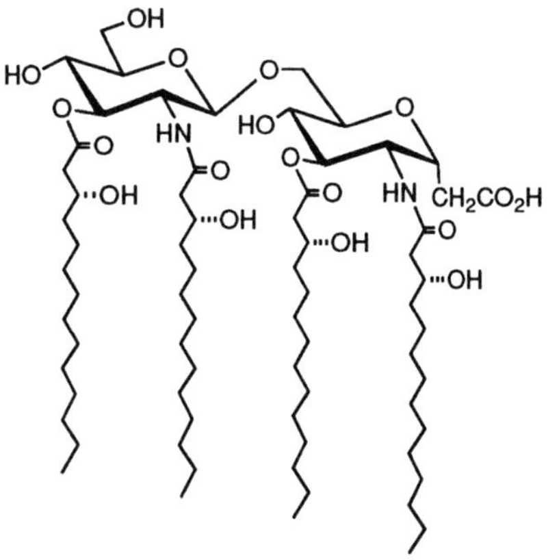

Fig. 12 Chemical structures of disaccharide structural analogs of lipid A.

group gives easy and direct access to carboxymethyl analogs like **24**.

Another type of artificial analog was prepared by the use of lipid A synthase, which is an enzyme acting in the biosynthesis of lipid A in bacterial cells. The original reaction catalyzed by this enzyme is the condensation of UDP-lipid X and another molecule of lipid X (**14**) to give a disaccharide 1-monophosphate as one of the biosynthetic precursors of lipid A (**11**). Because of its low specificity, the enzyme accepts glucosamine derivatives other than lipid X to form unnatural analogs of the polyacylated disaccharide 1-phosphate structure. Synthesis of disaccharide 1-phosphonate (**25**) and C-glycoside–type carboxymethyl analog (**26**) and some of their biological activities were reported (44,45). Lipid A analogs that contain 2-deoxy-2-fluoro glucose were also prepared by the same enzymatic method (46). Lipid A–like molecules carrying a 4′-phosphate can, however, not be obtained by this strategy since the enzyme is not able to accept any UDP–lipid X derivative carrying an additional 4-phosphate group.

CONCLUDING REMARKS

As summarized in this review, synthetic work on lipid A and related compounds has achieved remarkable development during the last decade making up the successful second era in this field of endotoxin research. Creation of novel protecting groups as well as methods for their regioselective introduction and chemoselective cleavage have made the synthetic procedure of these complex glycoconjugates quite efficient. Natural-type lipid A and many of their modified derivatives have been synthesized and studied for their biological functions. Several synthetic compounds were shown to have antitumor activity, and others exhibit potent antagonistic activity to abolish the toxic effect of endotoxin. Although none of these lipid A–based compounds are yet used in the actual clinical treatment of diseases, some might find practical application in a near future.

Recent improved synthetic methods and purification procedures have enabled the synthesis of highly homogeneous compounds in a wide range of scales from a few milligrams up to several grams. In fact, not only standard lipid A but also isotope-labeled lipid A analogs have been prepared. Besides the tritium-labeled derivative described above, ^{13}C-labeled lipid A has also been synthesized recently in the author's laboratory. A variety of synthetic compounds are expected to open unlimited routes to wide aspects of researches ranging from biology to physics of lipid A and related compounds. The relationship between the chemical structure and biological function is still an important issue but the more precise discussion will be possible hereafter on the reasons for the different biological activities such as the efficiency of transportation, molecular conformations and supramolecular structures of individual compounds. Chemical synthesis is expected to play increasing role in these respects.

REFERENCES

1. Imoto M, Yoshimura H, Sakaguchi N, Kusumoto S, Shiba T. Total synthesis of Escherichia coli lipid A. Tetrahedron Lett 1985; 26:1545–1548.
2. Imoto M, Yoshimura H, Sakaguchi N, Shimamoto T, Kusumoto S, Shiba T. Total synthesis of Escherichia coli lipid A, the endotoxically active principle of cell-surface lipopolysaccharide. Bull Chem Soc Jpn 1987; 60:2205–2214.
3. Imoto M, Kusumoto S, Shiba T, Rietschel ETh, Galanos C, Lüderitz O. Chemical structure of Escherichia coli lipid A. Tetrahedron Lett 1985; 26:907–908.
4. Galanos C, Lüderitz O, Rietschel ETh, Westphal O, Brade H, Brade L, Freudenberg M, Schade U, Imoto M, Yoshimura H, Kusumoto S, Shiba T. Synthetic and natural Escherichia coli free lipid A express identical endotoxic activities. Eur J Biochem 1985; 148:1–5.
5. Kotani S, Takada H, Tsujimoto M, Ogawa T, Takahashi I, Ikeda T, Otsuka K, Shimauchi H, Kasai N, Mashimo J, Nagao N, Tanaka A, Harada K, Nagai K, Kitamura H, Shiba T, Kusumoto S, Imoto M, Yoshimura H. Synthetic lipid A with endotoxic and related biological activities comparable to those of a natural lipid A from an Escherichia coli Re-mutant. Infect Immun 1985; 49:225–237.
6. Imoto M, Yoshimura H, Yamamoto M, Shimamoto T, Kusumoto S, Shiba T. Chemical synthesis of phosphorylated tetraacyl disaccharide corresponding to a biosynthetic precursor of lipid A. Tetrahedron Lett 1984; 25:2667–2670.
7. Imoto M, Yoshimura H, Yamamoto M, Shimamoto T, Kusumoto S, Shiba T. Chemical synthesis of a biosynthetic precursor of lipid A with a phosphorylated tetraacyl disaccharide structure. Bull Chem Soc Jpn 1987; 60:2197–2204.
8. Wang M-H, Feist W, Herzbeck H, Brade H, Kusumoto S, Rietschel ETh, Flad H-D, Ulmer AJ. Suppressive effect of lipid A partial structures on lipopolysaccharide or lipid A-induced release of interleukin 1 by human monocytes. FEMS Microbiol Immunol 1990; 64:179–186.
9. Wang M-H, Flad H-D, Feist W, Musehold J, Kusumoto S, Brade H, Gerdes J, Rietschel ETh, Ulmer AJ. Inhibition of endotoxin or lipid A-induced tumor necrosis factor production by synthetic lipid A partial structures in human peripheral blood mononuclear cells. Lymphokine Cytokine Res 1992; 11:23–31.

10. Lehmann V. Isolation, purification and properties of an intermediate in 3-deoxy-D-manno-octulosonic acid–lipid A biosynthesis. Eur J Biochem 1977; 75:257–266.
11. Raetz CRH. Biosynthesis of lipid A. In: Morrison DC, Ryan JL, eds. Bacterial Endotoxic Lipopolysaccharides. Vol. 1. Molecular Biochemistry and Cellular Biology. Boca Raton, FL: CRC Press, 1992:67–80.
12. Kusumoto S. Chemical Synthesis of Lipid A. In: Morrison DC, Ryan JL, eds. Bacterial Endotoxic Lipopolysaccharides. Vol. 1. Molecular Biochemistry and Cellular Biology. Boca Raton, FL: CRC Press, 1992: 81–105.
13. Tai A, Nakahata M, Harada T, Izumi Y, Kusumoto S, Inage M, Shiba T. A facile method for preparation of the optically pure 3-hydroxytetradecanoic acid by an application of asymmetrically modified nickel catalyst. Chem Lett 1980; 1125–1126.
14. Noyori R, Ohkuma T, Kitamura M, Takaya H, Sayo N, Kumobayashi H, Akutagawa S. Asymmetric hydrogenation of β-keto carboxylic esters. A practical, purely chemical access to β-hydroxy esters in high enantiomeric purity. J Am Chem Soc 1987; 109:5856–5858.
15. Oikawa M, Wada A, Yoshizaki H, Fukase K, Kusumoto S. New efficient synthesis of a biosynthetic precursor of lipid A. Bull Chem Soc Jpn 1997; 70:1435–1440.
16. Fukase K, Fukase Y, Oikawa M, Liu W-C, Suda Y, Kusumoto S. Divergent synthesis of lipid A analogues of shorter acyl chains. Tetrahedron. 1998; 54:4033–4050.
17. Keegan DS, Hagen SR, Johnson DA. Efficient asymmetric synthesis of (R)-3-hydroxy- and alkanoyloxytetradecanoic acids and method for the determination of enantiomeric purity. Tetrahedron: Asymmetry 1996; 7: 3559–3564.
18. Oikawa M, Kusumoto S. On a practical synthesis of β-hydroxy fatty acid derivatives. Tetrahedron: Asymmetry 1995; 6:961–966.
19. Liu W-C, Oikawa M, Fukase K, Suda Y, Winarno H, Mori S, Hashimoto M, Kusumoto, S. Enzymatic preparation of (S)-3-hydroxytetradecanoic acid and synthesis of unnatural analogues of lipid A containing the (S)-acid. Bull Chem Soc Jpn 1997; 70:1441–1450.
20. Sugai T, Ritzén H, Wong C-H. Towards the chemoenzymatic synthesis of lipid A. Tetrahedron: Asymmetry 1993; 4:1051–1058.
21. Shimizu T, Akiyama S, Masuzawa T, Yanagihara Y, Nakamoto S-I, Takahashi T, Ikeda K, Achiwa K. Comparison of biological activities of chemically synthesized monosaccharide analogues of reducing and non-reducing sugar moieties of lipid A. Chem Pharm Bull 1986; 34:5169–5175.
22. Kiso M, Tanaka S-I, Tanahashi M, Fujishima Y, Ogawa Y, Hasegawa A. Synthesis of 2-deoxy-4-phosphono-3-O-tetradecanoyl-2-[(3R)- and (3S)-3-tetradecanoyloxytetradenamido]-D-glucose: a diastereoisomeric pair of 4-O-phosphono-D-glucosamine derivatives (GLA-27) related to bacterial lipid A. Carbohydr Res 1986; 148: 221–234.
23. Watanabe Y, Kodama Y, Ebisuya K, Ozaki S. An efficient phosphorylation method using a new phosphitylating agent, 2-diethylamino-1,3,2-benzodioxaphophepane. Tetrahedron Lett 1990; 31:255–256.
24. Oikawa M, Liu W-C, Nakai Y, Koshida S, Fukase K, Kusumoto S. Regioselective reductive opening of 4,6-O-benzylidene acetals of glucose or glucosamine derivatives by $BH_3 \cdot ME_2NH-BF_3 \cdot OEt_2$. Synlett 1996; 1179–1180.
25. Griffiths SL, Madsen R, Fraser-Reid B. Studies toward lipid A: synthesis of differentially protected disaccharide fragments. J Org Chem 1997; 62:3654–3658.
26. Salimath PV, Weckesser J, Strittmatter W, Mayer H. Structural studies on the non-toxic lipid A from Rhodopseudomonas sphaeroides ATCC 17023. Eur J Biochem 1983; 136:195–200.
27. Qureshi N, Honovich JP, Hara H, Cotter RJ, Takayama K. Location of fatty acids in lipid A obtained from lipopolysaccharide of Rhodopseudomonas sphaeroides ATCC 17023. J Biol Chem 1988; 263:5502–5504.
28. Christ WJ, McGuinness PD, Asano O, Wang Y, Mullarkey MA, Perez M, Hawkins LD, Blythe TA, Dubuc GR, Robidoux AL. Total synthesis of the proposed structure of Rhodobacter sphaeroides lipid A resulting in the synthesis of new potent lipopolysaccharide antagonists. J Am Chem Soc 1994; 116:3637–3638.
29. Christ WJ, Asano O, Robidoux ALC, Perez M, Wang Y, Dubuc GR, Gavin WE, Hawkins LD, McGuinness PD, Mullarkey MA, Lewis MD, Kishi Y, Kawata T, Bristol JR, Rose JR, Rossignol DP, Kobayashi S, Hashinuma I, Kimura A, Asakawa N, Katayama K, Yamatsu I. E5531, a pure endotoxin antagonist of high potency. Science 1995; 268:80–83.
30. Nishijima M, Raetz CRH. Characterization of two membrane-associated glycolipids from an Escherichia coli mutant deficient in phosphatidylglycerol. J Biol Chem 1981; 256:10690–10696.
31. Ogawa Y, Wakida M, Ishida H, Kiso M, Hasegawa A. Synthesis of 3-O-(alkyl-branched acyl)-2-deoxy-2-[(3R)-3-hydroxytetradecanamido]-4-O-phosphono-D-glucose derivatives related to bacterial lipid A. Carbohydr Res 1992; 242:303–309.
32. Ogawa Y, Wakida M, Ishida H, Kiso M, Hasegawa A. Synthesis of novel nonreducing-sugar subunit analogs of lipid A carrying 2-acyloxytetradecanoyl and 2-hydroxyacyl groups of different carbon chain length. J Carbohydr Chem 1994; 13:433–446.
33. Ishida H, Fujishima Y, Ogawa Y, Kumazawa Y, Kiso M, Hasegawa A. Synthesis and mitogenic activity of nonreducing-sugar subunit analogs of bacterial lipid A composed only of 3-hydroxytetradecanoic acid and its homologs. Biosci Biotechnol Biochem 1995; 59:1790–1792.
34. Shiozaki M, Miyazaki H, Arai M, Hiraoka T, Kurakata S-I, Tatsuta T, Ogawa J, Nishijima M, Akamatsu Y. Synthesis of 1-O-[5-(carboxy)pentanoyl]-2-deoxy-2-(2,2-difuluorotetradecanamido)-3-O-[(R)-3-(tetradecanoyloxy)tetradecanoyl]-4-O-phosphono-α-D-glucopyranose and its analogues. Biosci Biotechnol Biochem 1995; 59:501–506.
35. Eustache J, Grob A, Retscher H. New acyclic analogues of lipid A: synthesis of 4-phosphonoxybutyl and 3-phosphonoxypropyl glycosides of 2-amino-2-deoxy-D-glucose. Carbohydr Res 1994; 251:251–267.
36. Ikeda K, Asahara T, Achiwa K. Synthesis of biologically active N-acylated L-serine-containing D-glucosa-

mine-4-phosphate derivatives of lipid A. Chem Pharm Bull 1993; 41:1879–1881.
37. Shiozaki M, Arai M, Macindoe WM, Mochizuki T, Wakabayashi T, Kurakata S-I, Tatsuta T, Maeda H, Nishijima M. The first syntheses of GLA-60 positional isomers and their biological activities. Bull Chem Soc Jpn 1997; 70:1149–1161.
38. Shiozaki M, Mochizuki T, Wakabayasi T, Kurakata S-I, Tatsuta T, Nishijima M. Synthesis of 2,6-anhydro-3-deoxy-5-O-phosphono-3-tetradecanamido-4-O-[(R)-3-(tetradecanoyloxy)tetradecanoyl]-D-glycero-D-ido-heptanoic acid as a new potent endotoxin antagonist and its dimeric analogue. Tetrahedron Lett 1996; 37:7271–7274.
39. Kusama T, Soga T, Shioya E, Nakayama K, Nakajima H, Osada Y, Ono Y, Kusumoto S, Shiba T. Synthesis and antitumor activity of lipid A analogs having a phosphonooxyethyl group with α- and β-configuration at position 1. Chem Pharm Bull 1990; 38:3366–3372.
40. Kusama T, Soga T, Ono Y, Kumazawa E, Shioya E, Osada Y, Kusumoto S, Shiba T. Synthesis and biological activities of analogs of a lipid A biosynthetic precursor: 1-O-phosphonooxyethyl-4′-O-phosphono-disaccharides with (R)-3-hydroxytetradecanoyl or tetradecanoyl groups at positions 2, 3, 2′ and 3′. Chem Pharm Bull 1991; 39:1994–1999.
41. Heine H, Brade H, Kusumoto S, Kusama T, Rietschel ETh, Flad H-D, Ulmer AJ. Inhibition of LPS binding on human monocytes by phosphonooxyethyl analogs of lipid A. J Endotoxin Res 1994; 1:14–20.
42. Fukase K, Kinoshita I, Suda Y, Aoki Y, Liu W-C, Oikawa M, Kurosawa M, Zähringer U, Seydel U, Rietschel ETh, Kusumoto S. Synthetic study of a bioactive ^{3}H-labeled analogue of lipid A. Synlett 1996:252–254.
43. Kusama T, Soga T, Ono Y, Kumazawa E, Shioya E, Nakayama K, Uoto K, Osada Y. Synthesis and biological activities of lipid A analogs: modification of a glycosidically bound group with chemically stable polar acidic groups and lipophilic groups on the disaccharide backbone with tetradecanoyl and N-dodecanoylglycyl groups. Chem Pharm Bull 1991; 39:3244–3253.
44. Scholz D, Bednarik K, Ehn G, Neruda W, Janzek E, Loibner H, Briner K, Vasella A. Enzymatic synthesis and comparative biological evaluation of a phosphonate analogue of the lipid A precursor. J Med Chem 1992; 35:2070–2074.
45. Bulusu M, Hildebrandt J, Lam C, Liehl E, Loibner H, Macher I, Scholz D, Schütze E, Stütz P, Vypel H, Unger F. Enzymatic synthesis of analogs of bacterial lipid A and design of biologically active LPS-antagonists and -mimetics. Pure Appl Chem 1994; 66:2171–2174.
46. Vypel H, Scholz D, Loibner H, Kern M, Bednarik K, Schaller H. Synthesis of fluorinated analogues of lipid A. Tetrahedron Lett 1992; 33:1261–1264.

15

Chemical Synthesis of Core Structures

Paul Kosma
University of Agricultural Sciences, Vienna, Austria

INTRODUCTION

This chapter covers significant contributions to the chemical synthesis of bacterial lipopolysaccharide (LPS) core structures that have been achieved within the last decade. A related review on the synthesis of core structures appeared in 1992 (1). The material presented does not strictly follow the somewhat arbitrary division of the core region into Kdo (3-deoxy-D-*manno*-oct-2-ulosonic acid)-, heptose- and outer core region, since many oligosaccharides of impressive size have been made comprising almost the whole region from outer core to the lipid A domain.

Starting in the early 1980s, chemical synthesis of LPS structures attracted organic chemists as one of the paramount challenges in the chemistry of natural products. This fact may be ascribed to the complexity of protecting-group strategies needed, which have to take into account the compatibility of numerous blocking groups, the presence of ester- and amide-linked fatty acid residues, glycosidic and ester-bound phosphate groups and the acid-sensitivity of the ketosidic bonds of Kdo residues. Furthermore, efficient stereoselective methods, both chemical and enzyme catalyzed, had to be developed for the preparation of sufficient quantities of the two complex monosaccharide constituents of the core, L-*glycero*-D-*manno*-heptose (Hep) and Kdo. The selective formation of α-anomeric glycosidic linkages for both sugars was another problem to be solved, and increasing the moderate glycosidation yields encountered for Kdo units could not be regarded as a trivial task, either. Synthetic compounds were previously needed as authentic model samples for comparison with natural compounds or degradation products obtained from LPS. These endeavors culminated in the successful chemical synthesis of *Escherichia coli* lipid A by Shiba's group, proving for the first time the chemical identity of synthetic lipid A with its natural counterpart (2). Furthermore, structural elucidation performed on oligosaccharides isolated from deacylated and dephosphorylated deep rough mutant LPS of *Salmonella enterica* serovar *minnesota* Re 595 and subsequent chemical synthesis provided unambiguous evidence for the α-(2→4) linkage between the Kdo units and for the attachment site of the Kdo disaccharide to O-6 of the distal glucosamine unit of lipid A (3–5).

Meanwhile, enormous progress had been made regarding the structure determination of LPS core oligosaccharides by high-field NMR spectroscopy or mass spectrometry. Furthermore, the purification efficiency for natural LPS isolates has been greatly improved. Thus, the objectives for organic chemistry in the field of endotoxin research have been shifted towards the synthesis of compounds to be used in the diagnosis and therapy of bacterial infections and for studies of structure-activity relationships. For immunochemical applications, synthetic oligosaccharides have been designed as multivalent haptens employed in enzyme-linked immunoassays and in serological studies with poly- and monoclonal antibodies directed against core determinants. In addition, compounds have been covalently linked to affinity-chromatography matrices as carbohydrate ligands and to carrier proteins for immunization experiments and vaccine development.

For a list of synthetic derivatives related to LPS core structures, see Table 1.

Table 1 Synthetic Derivatives Related to LPS Core Structures[a]

Structure	Aglycon	Ref.
α-Kdo-(2→6)-D-GlcN	—	15
α-Kdo-(2→6)-β-D-GlcNAc	Allyl	27
α-Ko-(2→6)-β-D-GlcNAc[b]	Allyl	73
α-Kdo-(2→6)-β-D-GlcNhm-(1→6)-D-GlcNhm[c]	—	13,16
α-Kdo-(2→6)-β-D-GlcNAc-(1→6)-α-D-GlcNAc	Allyl	28
α-Kdo-(2→6)-β-D-GlcNAc-(1→6)-β-D-GlcNAc	Allyl	28
α-Kdo-(2→4)-α-Kdo	Allyl, Me	8,74
α-Kdo-(2→4)-α-Kdo-(2→6)-D-GlcN	—	15
α-Kdo-(2→4)-α-Kdo-(2→6)-β-D-GlcNAc	Allyl	27
α-Kdo-(2→4)-α-Kdo-(2→6)-β-D-GlcN-(1→6)-D-GlcN	—	5
α-Kdo-(2→4)-α-Kdo-(2→6)-β-D-GlcNhm-(1→6)-D-GlcNhm	—	13
α-Kdo-(2→4)-α-Kdo-(2→6)-β-D-GlcNAc-(1→6)-α-D-GlcNAc	Allyl	28
α-Kdo-(2→4)-α-Kdo-(2→6)-β-D-GlcNAc-(1→6)-β-D-GlcNAc	Allyl	28
α-Kdo-(2→4)-α-Kdo-(2→4)-α-Kdo	Allyl	17
α-Kdo-(2→4)-α-Kdo-(2→4)-α-Kdo-(2→6)-β-D-GlcNhm-(1→6)-D-GlcNhm	—	12
α-Kdo-(2→8)-α-Kdo	Allyl	23
α-Kdo-(2→8)-α-Kdo-(2→4)-α-Kdo	Allyl	23
α-Kdo-(2→8)-α-Kdo-(2→4)-α-Kdo-(2→6)-β-D-GlcNAc	Allyl	27
α-Kdo-(2→8)-α-Kdo-(2→4)-α-Kdo-(2→6)-β-D-GlcNAc-(1→6)-α-D-GlcNAc	Allyl	28
α-Kdo-(2→8)-α-Kdo-(2→4)-α-Kdo-(2→6)-β-D-GlcNAc-(1→6)-β-D-GlcNAc	Allyl	28
α-Kdo-(2→6)-β-D-GlcNhm 4-phosphate	Allyl	42
α-Kdo-(2→6)-D-GlcR^{1}NR^2 4-phosphate		
R^1 = C$_{14}$OC$_{14}$, R^2 = C$_{14}$OC$_{12}$	—	18,41
R^1 = R^2 = C$_{14}$OC$_{14}$	—	41
R^1 = R^2 = C$_{14}$OC$_{16}$	—	41
α-Kdo-(2→6)-β-D-GlcR^{1}NR^2 4-phosphate-(1→6)-D-GlcR^{3}NR^3		
R^1 = C$_{14}$OC$_{14}$, R^2 = C$_{14}$OC$_{12}$, R^3 = C$_{14}$—OH	—	18
R^1 = R^2 = R^3 = C$_{14}$—H	—	18
α-Kdo-(2→4)-α-Kdo-(2→6)-β-D-GlcR^{1}NR^2 4-phosphate-(1→6)-D-GlcR^{3}NR^3		
R^1 = C$_{14}$OC$_{14}$, R^2 = C$_{14}$OC$_{12}$, R^3 = C$_{14}$—OH	—	18
L-α-D-Hep-(1→5)-α-Kdo-(2→6)-β-D-GlcRNR 4-phosphate		
R = C$_{14}$OC$_{14}$	—	50
α-Kdo 2-phosphate		75
β-Kdo 2-phosphate		75
α-Kdo 5-phosphate		76
α-Kdo 8-phosphate	—	77,78
α-Kdo 4-phosphate	Allyl, Me	30,32
α-Kdo 5-phosphate	Allyl, Me	32
α-Kdo 7-phosphate	Me	33
α-Kdo 4-phosphate-(2→6)-β-D-GlcNAc	Allyl	34
α-Kdo 5-phosphate-(2→6)-β-D-GlcNAc	Allyl	34
L-α-D-Hep-(1→4)-α-Kdo	Allyl	37
α-D-Man-(1→5)-α-Kdo	Allyl	36
L-α-D-Hep-(1→5)-α-Kdo	Allyl, (CH$_2$)$_3$NH$_2$	46,51
L-α-D-Hep-(1→5)-α-Kdo	—	44
α-D-Man-(1→5)-α-Kdo 4-phosphate	Allyl	36
L-α-Hep-(1→5)-α-Kdo 4-phosphate	Allyl	37
L-α-D-Hep-(1→5)-α-Kdo 4-phosphate-(2→6)-β-D-GlcNAc	Allyl	37
L-α-D-Hep-(1→5)-α-Kdo-(2→6)-β-D-GlcNhm-(1→6)-D-GlcNhm	—	45

L-α-D-Hep-(1→5)-Kdo
     4
     ↑
  α-Kdo-2
	—	44

Table 1 Continued

Structure	Aglycon	Ref.

L-α-D-Hep-(1→5)-Kdo
         4
         ↑
     α-Kdo-2
	Allyl	46

L-α-D-Hep-(1→5)-α-Kdo-(2→6)-β-D-GlcNhm-(1→6)-D-GlcNhm
         4
         ↑
     α-Kdo-2
	—	45

L-α-D-Hep-(1→3)-L-α-D-Hep-(1→5)-α-Kdo	Allyl	47
L-α-D-Hep-(1→3)-L-α-D-Hep-(1→5)-α-Kdo-(2→6)-β-D-GlcNhm-(1→6)-D-GlcNhm	—	48
β-D-Gal-(1→4)-β-D-Glc-(1→4)-L-α-D-Hep-(1→5)-α-Kdo	$(CH_2)_3NH_2$	51

β-D-Glc-1
          ↓
          6
β-D-Glc-(1→4)-α-D-Glc-(1→5)-α-Kdo
         3
         ↑
     β-D-Glc-1
	$(CH_2)_2C_6H_4NHCOCF_3$	72

D-α-D-Hep 1-phosphate		79
D-α-D-Hep	ADP	79
L,D-Hep 2-phosphate	—	61
L,D-Hep 3-phosphate	—	61
L,D-Hep 4-phosphate	—	61
L,D-Hep 6-phosphate	—	61
L,D-Hep 7-phosphate	—	61
L-α-D-Hep-(1→3)-L-α-D-Hep 4-phosphate	Me, $(CH_2)_2C_6H_4NHCOCF_3$	60
L-α-D-Hep 4-phosphate-(1→3)-L-α-D-Hep	Me, $(CH_2)_2C_6H_4NHCOCF_3$	60
L-α-D-Hep 4-phosphate-(1→3)-L-α-D-Hep 4-phosphate	Me, $(CH_2)_2C_6H_4NHCOCF_3$	60
L-α-D-Hep-(1→3)-L-α-D-Hep	Allyl	46
L-α-D-Hep-(1→3)-L-α-D-Hep	—	43,56
L-α-D-Hep-(1→6)-L-α-D-Hep	Me	64
L-α-D-Hep-(1→7)-L-α-D-Hep	Allyl, Me	46,58
L-α-D-Hep-(1→7)-L-α-D-Hep	—	43,56
L-α-D-Hep-(1→2)-L-α-D-Hep-(1→3)-L-α-D-Hep	$(CH_2)_2C_6H_4NHCOCF_3$	62
L-α-D-Hep-(1→7)-L-α-D-Hep-(1→3)-L-α-D-Hep	—	56
L-α-D-Hep-(1→7)-L-α-D-Hep-(1→3)-L-α-D-Hep	Allyl, $(CH_2)_2C_6H_4NHCOCF_3$	43,56,59
α-D-Glc-(1→3)-L-α-D-Hep	Me	58
β-D-Glc-(1→4)-L-α-D-Hep	Me, $(CH_2)_3NH_2$	52
α-D-GlcN-(1→7)-L-α-D-Hep	—	63
α-D-GlcN-(1→7)-L-α-D-Hep	Me	64
β-D-GlcN-(1→7)-L-α-D-Hep	—	63
α-D-GlcNAc-(1→2)-L-α-D-Hep-(1→3)-L-α-D-Hep	$(CH_2)_3NH_2$	53

α-D-Glc-(1→3)-L-α-D-Hep
         7
         ↑
    L-α-D-Hep-1
	Me	58

β-D-Glc-(1→4)-β-D-Glc-(1→4)-L-α-D-Hep	$(CH_2)_2C_6H_4NHCOCF_3$	62
β-D-Gal-(1→4)-β-D-Glc-(1→4)-L-α-D-Hep	$(CH_2)_2C_6H_4NHCOCF_3$	62
β-D-Gal-(1→2)-L-α-D-Hep-(1→2)-L-α-D-Hep-(1→3)-L-α-D-Hep	$(CH_2)_2C_6H_4NHCOCF_3$	62
β-D-Gal-(1→2)-L-α-D-Hep-(1→2)-L-α-D-Hep-(1→3)-L-α-D-Hep	$(CH_2)_2C_6H_4NHCOCF_3$	62

Table 1 Continued

Structure	Aglycon	Ref.
L-α-D-Hep-(1→6)-D-Glc	—	65
L-α-D-Hep-(1→6)-α-D-Glc	Allyl	67
L-α-D-Hep-(1→6)-α-D-Glc-(1→2)-D-Glc	—	66
L-β-D-Hep-(1→6)-α-D-Glc-(1→2)-D-Glc	—	66
L-α-D-Hep-(1→6)-α-D-Glc-(1→2)-α-D-Glc	Allyl	67
L-α-D-Hep-(1→6)-α-D-Glc-(1→2)-α-D-Glc-(1→3)-α-D-Glc	Allyl	67

α-D-Gal-(1→6)-α-D-Glc
       3
       ↑
α-D-Gal-1
	Me, $(CH_2)_2C_6H_4NHCOCF_3$, $(CH_2)_7CH_3$	69

α-D-GlcNAc-(1→2)-α-D-Glc-(1→2)-α-D-Gal	Me	68

α-D-Glc-(1→2)-α-D-Gal-(1→3)-α-D-Glc
                   6
                   ↑
            α-D-Gal-1
	Me	68

β-D-Glc-(1→3)-β-D-Glc-(1→4)-α-D-Glc-(1→1)-α-D-Glc		81
β-D-3-OMe-Glc-(1→3)-β-D-Glc-(1→4)-β-D-Glc-(1→6)-α-D-Glc-(1→1)-α-D-Glc		82

α-D-Glc-(1→3)-L-α-D-Hep-(1→3)-L-α-D-Hep
                   7
                   ↑
L-α-D-Hep-1
	$(CH_2)_2C_6H_4NHCOCF_3$	70

α-D-Gal-(1→3)-α-D-Glc-(1→3)-L-α-D-Hep-(1→3)-L-α-D-Hep
                                      7
                                     ↑
                            L-α-D-Hep-1
	$(CH_2)_2C_6H_4NHCOCF_3$	70

α-D-Gal-(1→3)-α-D-Glc-(1→3)-L-α-D-Hep-(1→3)-L-α-D-Hep
              6   7
              ↑   ↑
α-D-Gal-1   L-α-D-Hep-1
	$(CH_2)_2C_6H_4NHCOCF_3$	70

α-D-Glc-(1→2)-β-D-Glc-1
                                   ↓
                                   6
α-D-GlcNAc-(1→2)-β-D-Glc-(1→4)-α-Glc
                            3
                            ↑
                       β-D-Glc-1
	$(CH_2)_2C_6H_4NHCOCF_3$	71

β-D-Gal-(1→4)-α-D-Glc-(1→2)-β-D-Glc-1
                                      ↓
                                     6
α-D-GlcNAc-(1→2)-β-D-Glc-(1→4)-α-D-Glc
                            3
                            ↑
                       β-D-Glc-1
	$(CH_2)_2C_6H_4NHCOCF_3$	71

α-D-Gal-(1→4)-β-D-Gal-(1→4)-α-D-Glc-(1→2)-β-D-Glc-1
                                                            ↓
                                                         6
                α-D-GlcNAc-(1→2)-β-D-Glc-(1→4)-α-D-Glc
                                            3
                                            ↑
                                    β-D-Glc-1
	$(CH_2)_2C_6H_4NHCOCF_3$	71

[a]Nonreducing units occur as pyranosides. Mono- and oligosaccharide analogs have not been included.
[b]Ko corresponds to an octulosonic acid with D-*glycero*-D-*talo* configuration.
[c]hm = (R)-3-Hydroxytetradecanoyl.

SYNTHESIS OF Kdo-CONTAINING INNER CORE OLIGOSACCHARIDES

Synthesis of Kdo Oligosaccharides Corresponding to Re LPS

Numerous syntheses of oligosaccharide structures related to the enterobacterial Kdo region but also comprising Kdo units attached to the neighboring heptose region and linked to parts of the lipid A entity have been performed since 1980. As one of the major accomplishments, the synthesis of the reducing tetrasaccharide α-Kdo-(2→4)-α-Kdo-(2→6)-β-D-GlcN-(1→6)-D-GlcN confirmed its structural identity with a product isolated from degraded LPS of the mutant strain *Salmonella enterica* serovar *minnesota* R*e* 595 (5). In addition, synthetic studies were undertaken with the objective of characterizing epitope specificities of monoclonal antibodies directed against the core and lipid A region. For this purpose, the aglyconic part of oligosaccharide derivatives was designed to allow further transformation into multivalent haptens or neoglycoconjugates, respectively (Fig. 1). In general, synthetic strategies have relied upon the use of a terminal amino group (Fig. 1, **A** and **B**) introduced in protected form at the outset of the synthetic route or exploiting the reactivity of the allyl group (at the expense of using benzyl ethers as protecting groups) to introduce a spacer arm at the deprotected stage (**C**). Addition of cysteamine hydrochloride to the allyl glycoside **C** proceeds efficiently under UV irradiation to yield 3-(2-amino-ethyl)thiopropyl glycosides (**D**) (6,7). Covalent attachment of the amino-spacer derivatives **A**, **B**, or **D** to ε-amino groups of lysine residues of a protein may be accomplished by various activation procedures, e.g., with glutardialdehyde or thiophosgene, respectively. Since the reaction and the subsequent coupling step to the protein are performed at neutral or slightly alkaline pH, hydrolysis of the acid-sensitive ketosidic linkages of Kdo and the formation of intramolecular lactones of Kdo moieties may be avoided. Moreover, the allyl group itself or a terminal acrylamide function—attached to the ω-amino group of the spacer derivatives **A**, **B**, or **D** via reaction with acryloyl chloride—may be copolymerized with acrylamide, yielding high molecular weight, water-soluble glycopolymers of type **F** and **G**, respectively; these polymers are useful in various immunoassays (7,8). As an example, by using these synthetic antigens, antibody specificities directed against α-Kdo mono- and α-(2→4)–linked Kdo disaccharide epitopes, and against those comprising parts of the glucosamine region, have been described (9–11).

Alternatively, core oligosaccharides were coupled to lipid A derivatives containing two amide-linked (*R*)-3-hydroxytetradecanoyl residues, which may be incorporated into membranes of sheep erythrocytes or liposomes to serve as antigens or inhibitors in passive hemolysis experiments. To illustrate the chemistry involved, the synthesis of the pentasaccharide derivative α-Kdo-(2→4)-α-Kdo-(2→4)-α-Kdo-(2→6)-β-D-GlcNhm-(1→6)-D-GlcNhm **13** [hm = (*R*)-3-hydroxytetradecanoyl] (Fig. 2) corresponding to a part structure occurring in LPS from *S. enterica* serovar *minnesota* R 345 will be described (12).

First, the α-benzyl glycoside **2**, containing one amide-linked (*R*)-3-benzyloxytetradecanoyl residue, was subjected to regioselective glycosylation with the *N*-trichloroethoxycarbonyl (Troc) protected 2-amino-2-deoxy-glucopyranosyl bromide **1** to give the β-(1→6)–linked disaccharide **3** in 94% yield and excellent stereoselectivity (13). After reductive removal of the *N*-trichloroethoxycarbonyl group with zinc in acetic acid, the second fatty acid residue was coupled to the distal glucosamine in the presence of a carbodiimide reagent (EEDQ) to furnish **4** in 85% yield. Subsequent introduction of a 1,1,3,3-tetraisopropyldisiloxane-1,3-diyl (TIPS) group at O-4′ and O-6′ followed by acid-catalyzed rearrangement (13) of the silyl ether to O-3′ and O-4′ of the distal glucosamine unit provided the activated glycosyl acceptor **5**. For the glycosidic attachment of Kdo residues, the Kdo bromide derivative **6** (14) has frequently been employed as a glycosyl donor in carefully optimized variants of the Helferich procedure. This has been demonstrated for the assembly of α-(2→4)–linked Kdo disaccharide units (15) and for the linkage of Kdo to the 6′-position of the nonreducing glucosamine moiety of lipid A (16). Thus, condensation of disaccharide **5** with the Kdo methyl ester bromide derivative **6** in nitromethane-toluene and in the presence of HgBr$_2$ afforded the α-(2→6)–linked trisaccharide **8** in 40% yield with only trace amounts of the β-isomer and together with the glycal ester derivative **7** resulting from the donor **6** via elimination of HBr. Further transformation of **8** into the 7″,8″-O-TIPS–protected trisaccharide derivative **9**—ready for glycosylation at the reactive 4″-OH group of Kdo—was effected in 61% yield, again exploiting the versatile bifunctional silylether protecting group. On the other hand, the *O*-acetylated α-Kdo-(2→4)-α-Kdo disaccharide bromide derivative **11** was elaborated in several steps from the readily available β-benzyl derivative **10** (12) or β-allyl glycoside (17) of Kdo, respectively. Glycosylation of the TIPS-protected acceptor **9** with 2 equivalents of the disaccharide donor **11** fur-

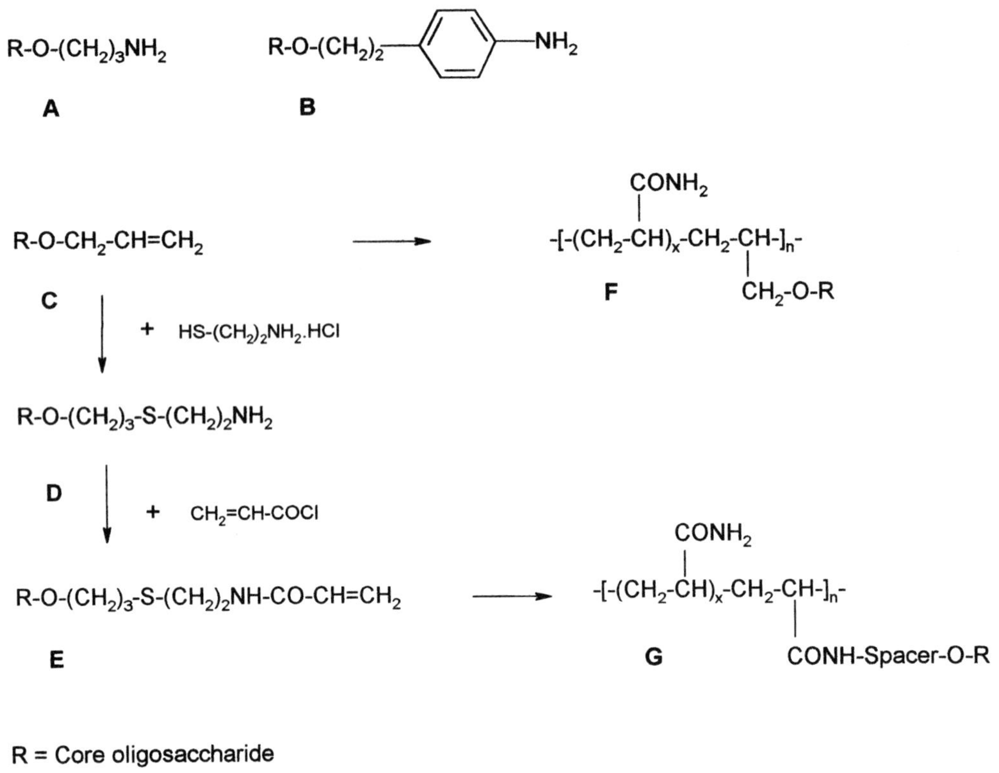

Fig. 1 Haptens and glycopolymers derived from synthetic core oligosaccharides.

nished in a regioselective fashion the α-(2→4)–linked pentasaccharide derivative **12** in 21% yield without formation of the corresponding β-anomer. Sequential removal of the protecting groups by treatment with Bu$_4$NF, sodium methoxide, and aqueous NaOH followed by quantitative hydrogenolysis of the benzyl groups afforded the pentasaccharide derivative **13**.

Besides Kdo methyl ester bromide derivatives, other glycosyl donors of Kdo residues such as fluorides (18), thioglycosides (see page 270), and phenyl selenyl triflates (19) have been employed for efficient glycosylation reactions, whereas the anomeric O-alkylation procedure developed by Schmidt (20) and the stereo-controlled approach based on iodocyclization of enol ethers have not yet found general use (21). The properties of Kdo phosphites as glycosyl donors remain to be studied.

Synthesis of Kdo Oligosaccharides Corresponding to *Chlamydia* LPS

The genus *Chlamydia* comprises four species (*C. trachomatis*, *C. psittaci*, *C. pneumoniae*, and *C. pecorum*), which harbor a genus-specific carbohydrate antigen resembling the LPS of core-deficient enterobacterial R*e* mutants (22). The structure of the genus-specific epitope was established as a Kdo-trisaccharide of the sequence α-Kdo-(2→8)-α-Kdo-(2→4)-α-Kdo, which, in nature, is assembled by the action of a single multifunctional Kdo transferase (see Chapter 13). Syntheses of the Kdo trisaccharide and of oligosaccharides containing in addition the backbone residues of the adjoining lipid A region have been accomplished in the form of the respective allyl glycosides which are amenable to further elongation with a spacer group and subsequent covalent attachment to a protein carrier (Fig. 3).

The synthetic strategy is based on a straightforward, blockwise assembly utilizing the α-(2→8)–linked disaccharide bromide donor **17**. Regioselective silylation of the O-allyl α-glycoside **14** at the 8-OH position, followed by O-acetylation and subsequent HF treatment produced **15**, which was coupled with the Kdo bromide derivative **6** in MeNO$_2$ in the presence of Hg(CN)$_2$. Thus, a 4:1 mixture of α- and β-(2→8)–linked disaccharides was obtained in 90% yield, from which the α-isomer **16** was isolated by fractional crystallization. Deprotection of **16** with NaOMe and alkaline hydrolysis of the methyl ester groups afforded the

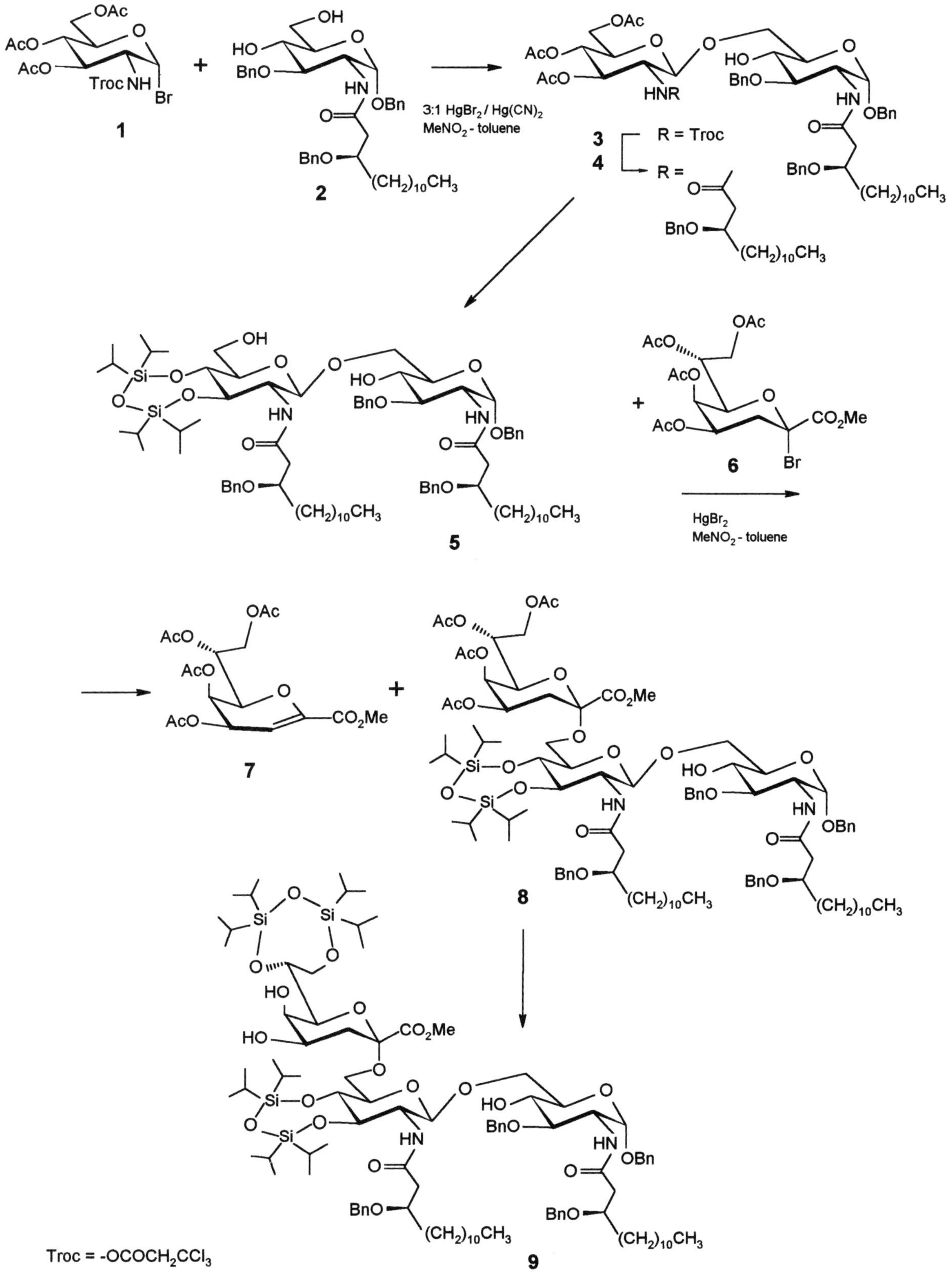

Fig. 2 Chemical synthesis of a pentasaccharide related to the inner core of *Salmonella enterica* serovar *minnesota* R345. (From Ref. 13.)

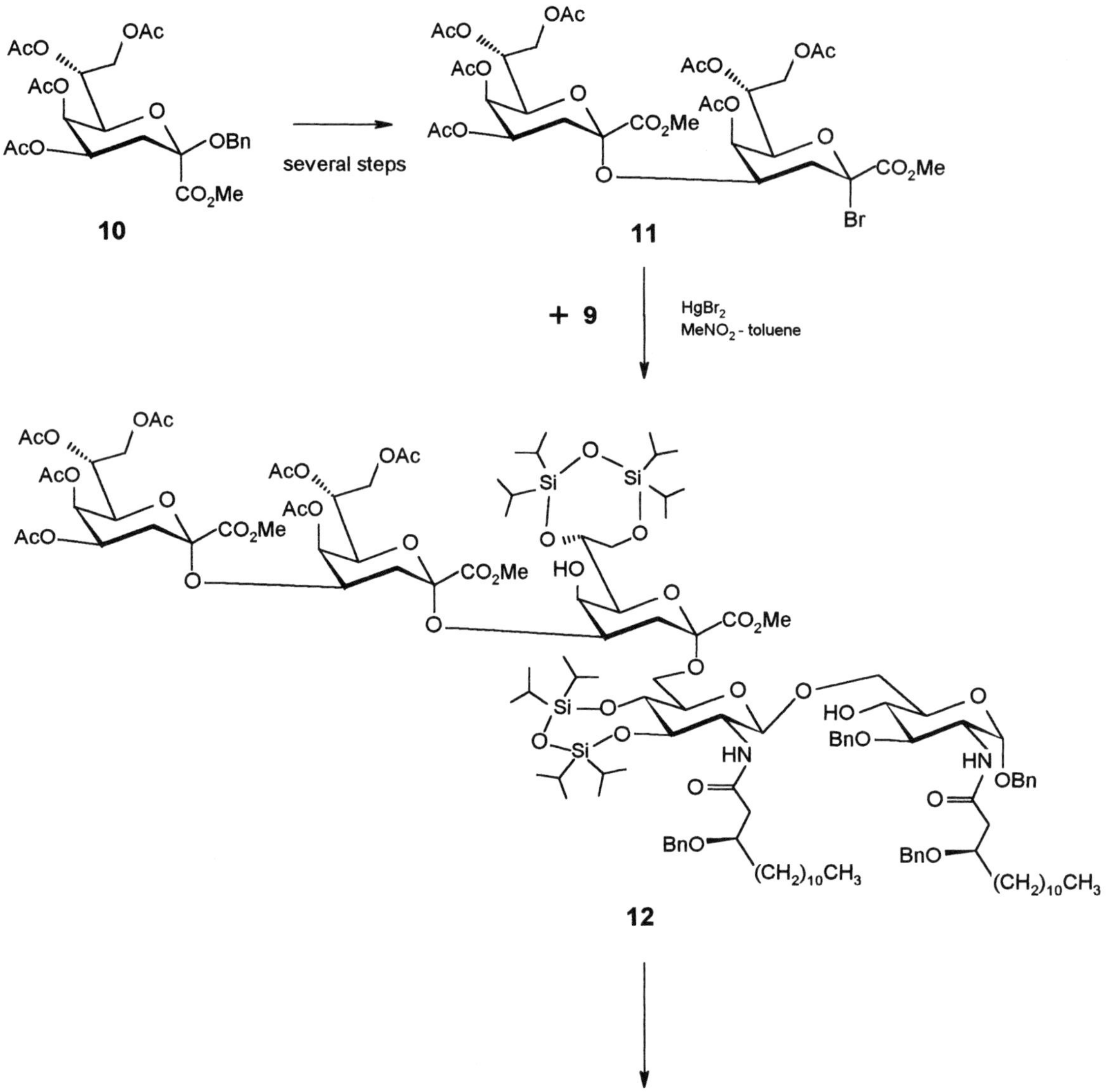

α-Kdo-(2-4)-α-Kdo-(2-4)-α-Kdo-(2-6)-β-GlcNhm-(1-6)-GlcNhm

13

hm = (R)-3-hydroxytetradecanoyl

Fig. 2 Continued

disaccharide α-Kdo-(2→8)-α-Kdo-(2→OAll) **18** as a crystalline disodium hemihydrate (23). The crystal structure determination (24) of **18** revealed the presence of an interresidue hydrogen bond between the terminal carboxyl group and 7-OH of the reducing Kdo unit. Furthermore, the disaccharide bromide **17** was made available in four steps from **16** via selective cleavage of the glycosidic allyl group; alternative synthetic pathways for **17** from the readily accessible β-benzyl ketoside and from the 2-*O*-acetyl derivative of Kdo have been reported (25,26). Condensation of **17** with the equatorially oriented, reactive 4-OH group of the 7,8-*O*-carbonyl protected Kdo unit in mono-, di-, and trisaccharide glycosyl acceptor derivatives **19a–c**

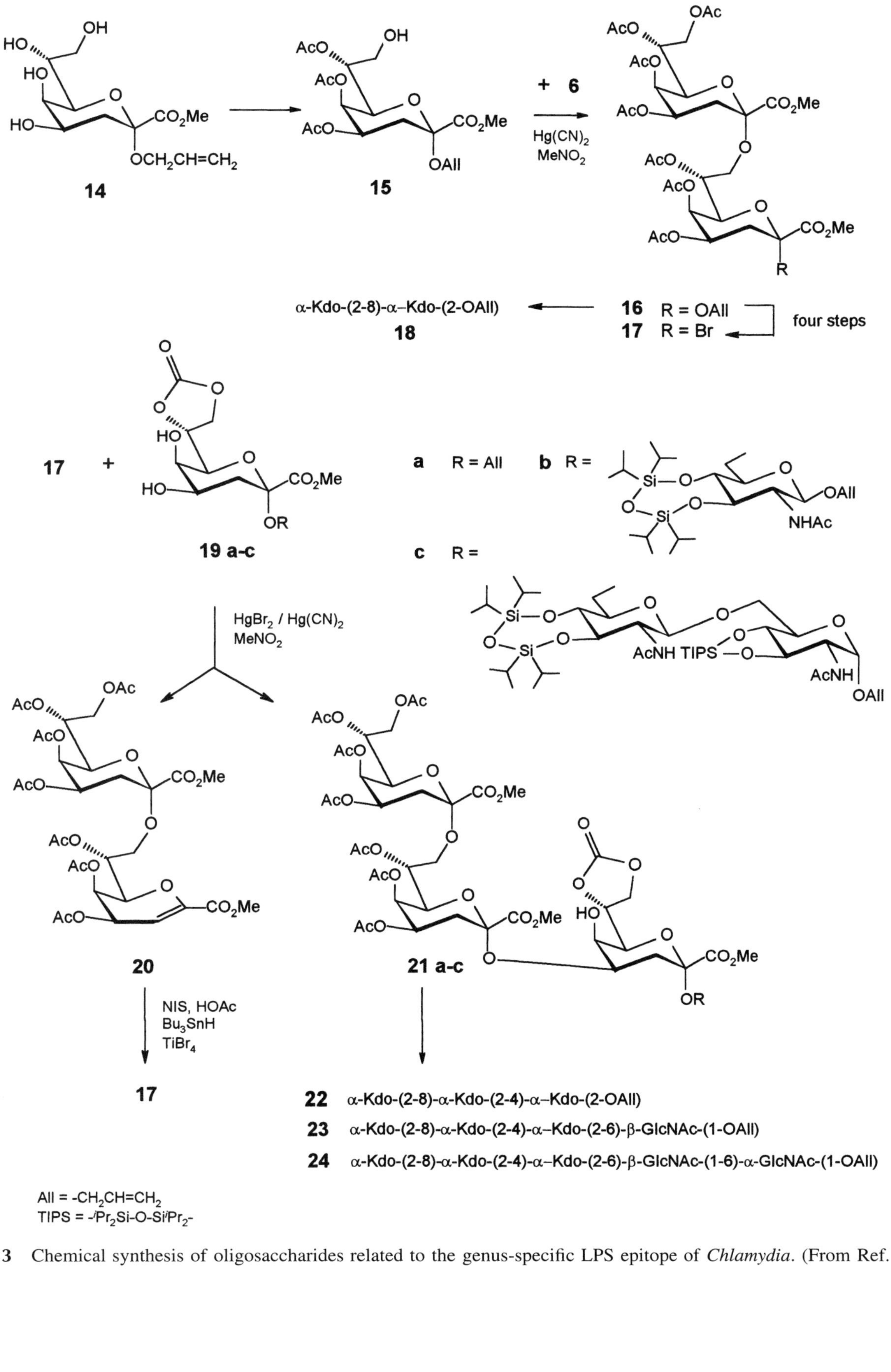

Fig. 3 Chemical synthesis of oligosaccharides related to the genus-specific LPS epitope of *Chlamydia*. (From Ref. 23.)

was effected by variants of the Helferich procedure in fair to modest yields and stereoselectivity to give, after separation of small amounts of β-isomers and subsequent deprotection of **21a–c**, the tri-, tetra-, and pentasaccharide derivatives **22** (23), **23** (27), and **24** (28). An improvement in the overall conversion has been accomplished, since the disaccharide glycal ester **20** formed by the competing elimination reaction during the glycosylation steps may be recycled in good overall yields into the disaccharide donor **17** via acetoxyiodination, reduction with Bu$_3$SnH, and treatment with TiBr$_4$ (29). Moreover, a series of Kdo di- and trisaccharide analogs containing carboxyl-reduced Kdo moieties was prepared employing octulopyranosyl fluorides as highly α-selective glycosyl donors in the presence of borontrifluoride etherate (25,26). Immunochemical investigations with neoglycoproteins derived from the oligosaccharides confirmed the trisaccharide and the α-(2→8)-linked disaccharide as the major immunodominant epitopes (22). In addition, a carboxyl-reduced trisaccharide analog containing the carboxyl-reduced moiety at the central Kdo unit also displayed *Chlamydia* reactivity with monoclonal antibodies (see Chapter 13).

Synthesis of Oligosaccharides Containing Kdo Phosphate Groups

Phosphorylated Kdo residues have been detected as constituents in numerous lipopolysaccharides from *Bordetella, Haemophilus, Vibrio, Aeromonas*, and *Bacteroides* strains. Chemical syntheses of Kdo monosaccharide phosphates have consequently been performed which include substitution at O-4, O-5, and O-7 of Kdo (30–33). Moreover, anomeric allyl glycosides of Kdo 4- and 5-phosphate, as well as disaccharides containing the Kdo phosphate entities α-(2→6)–linked to GlcNAc residues, were elaborated without affecting the allyl group upon deblocking of the phosphate protecting groups (e.g., by using 2-cyanoethyl functions). Thus, phosphorylated polyacrylamide copolymers and neoglycoproteins, respectively, were obtained (34). Thereafter, monoclonal antibodies have been produced recognizing specifically Kdo 4- and 5-phosphate epitopes (35), thereby confirming the previous finding of the presence of both moieties in LPS of the deep rough mutant of *Haemophilus influenzae* I-69 Rd$^-$/b$^+$. Furthermore, attachment of a mannopyranosyl residue α-(1→5)–linked to the α-allyl glycoside of Kdo 4-phosphate has been accomplished. Condensation of benzobromomannose with a suitably protected Kdo acceptor derivative in 1,2-dichloroethane as solvent under promotion with Ag-triflate and tetramethyl urea afforded the protected disaccharide α-Man*p*-(1→5)-α-Kdo-(2-OAll) in good yield (70%). Selective removal of a silylether protecting group at O-4 of the Kdo unit followed by subsequent treatment with bis(trichloroethyl)phosphochloridate and *N*-methylimidazole produced the corresponding phosphotriester in 90% yield. The protected compound was deblocked giving α-D-Man*p*-(1→5)-α-Kdo-(2-OAll) 4-phosphate, which was subsequently converted into a glycopolymer by copolymerization of the aglyconic allyl group with acrylamide (36). Meanwhile, the synthesis of a neoglycoconjugate containing the trisaccharide determinant L-α-D-Hep*p*-1(1→5)-α-Kdo 4-phosphate-(2→6)-β-D-GlcNAc has been achieved. First, a protected heptose-(1→5)-Kdo disaccharide bromide donor was prepared, then coupled to an appropriate glucosamine acceptor, and after selective deblocking of O-4 of the Kdo residue, the resulting trisaccharide acceptor was subjected to efficient phosphoramidite-mediated phosphorylation. The deprotected allyl glycoside was finally converted into a neoglycoprotein with bovine serum albumin (37).

Total Synthesis of R*e* Lipopolysaccharide Structures and of Kdo Units Linked to Glucosamine 4-Phosphate

The total synthesis of lipopolysaccharide structures comprising the Kdo-region linked to *N,O*-acylated, 1,4′-bisphosphorylated lipid A components required not only sophisticated chemical transformations and protecting group strategies, but also refined methods for the purification of final products which display amphiphilic properties due to the presence of fatty acid residues. The approach toward the total synthesis of R*e* LPS elaborated by Kusumoto et al. (18,38) was based on the progress made in the chemistry of lipid A synthesis and on the synthesis of the isopropylidene-protected Kdo benzyl ester fluoride donor **29**, prepared via a novel approach from D-mannose, as a key intermediate (Fig. 4). Coupling of the α-allyl glycoside acceptor **26** containing two (R)-3-benzyloxytetradecanoyl residues with the 4-*O*-diphenylphosphono glucosamine bromide derivative **25** promoted by Hg(CN)$_2$ afforded in a regioselective manner the β-(1→6)–linked glucosamine disaccharide compound **27** in 42% yield. After removal of the amide- and ester-linked Troc-protecting groups of the distal glucosamine group, the second amide-linked fatty acid residue was introduced via carbodiimide-activated condensation, giving **28**, which in turn was subjected to glycosylation with the

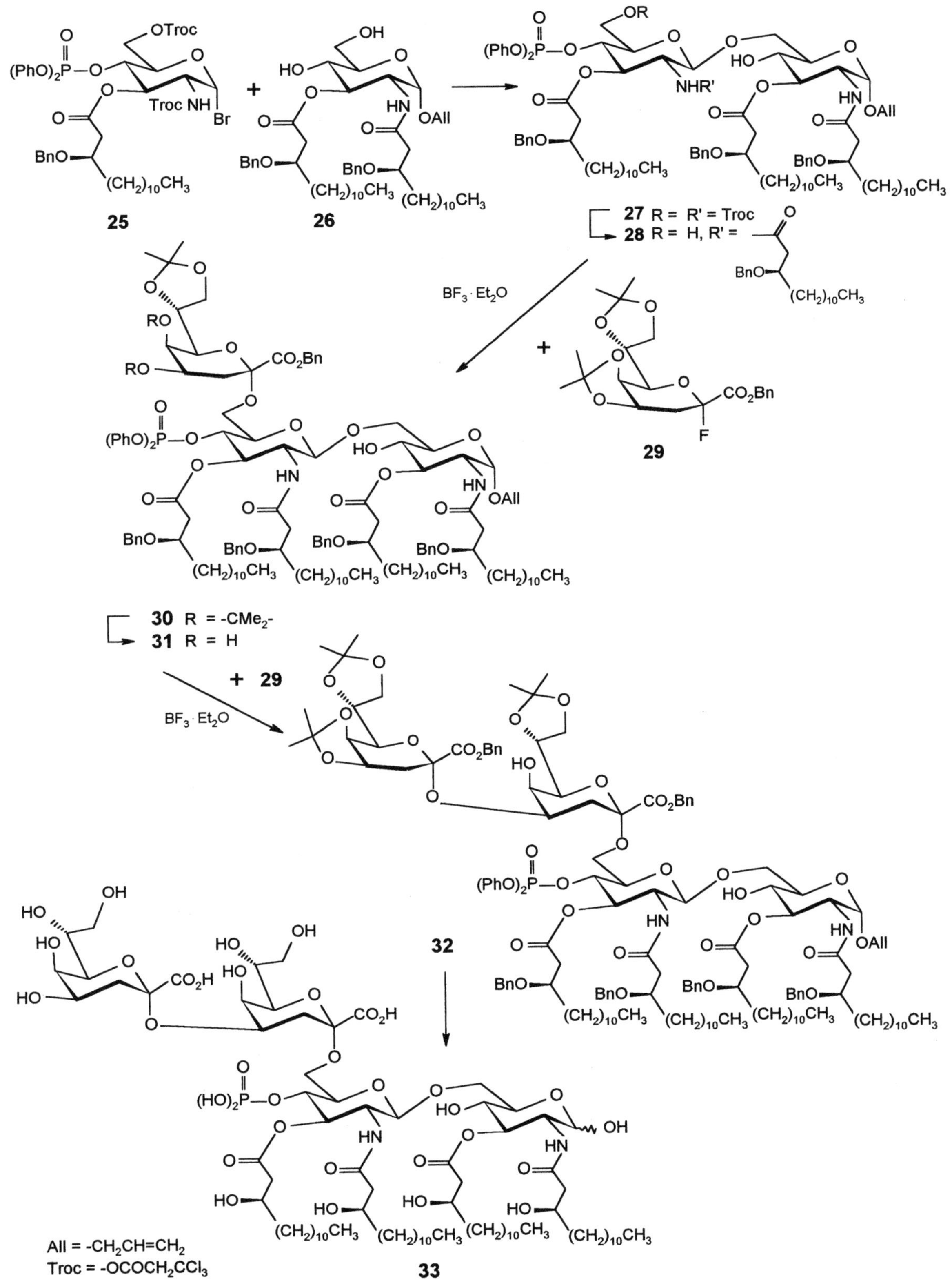

Fig. 4 Chemical synthesis of 1-dephospho derivative of *Escherichia coli* Re LPS. (From Ref. 18.)

Kdo benzyl ester fluoride **29** in the presence of BF$_3$-etherate and *N,N*-ethyldiisopropylamine to furnish the α-(2→6)–linked trisaccharide **30** in good yield. Acidic cleavage of acetal protecting groups with trifluoroacetic acid (TFA) and selective formation of the 7″,8″-*O*-isopropylidene acetal provided the acceptor derivative **31**, which was again condensed with the fluoride donor **29** to furnish the tetrasaccharide **32** in 31% yield and α-anomeric stereocontrol. Selective cleavage of the glycosidic allyl group followed by sequential hydrogenolysis with Palladium black and PtO$_2$ afforded the reducing tetrasaccharide 4′-phosphate **33**, which was purified by centrifugal partition chromatography (38). The compound corresponds to the 1-dephospho derivative of *E. coli* Re LPS and provides unambiguous proof of structure for the material obtained by selective cleavage of the glycosidic phosphate from natural Re-type LPS.

Additional oligosaccharide derivatives were prepared in partially protected form, since the protecting groups involved did not allow for complete deblocking. This holds true for the attachment of Kdo monosaccharide (**39**) as well as α-(2→4)–linked Kdo disaccharide units (**40**) to lipid A subunit derivatives such as GLA-60 containing two (*R*)-3-tetradecanoyloxytetradecanoyl residues and a 4-*O*-diphenylphosphotriester group at the glucosamine moiety. A subunit derivative, the disaccharide α-Kdo-(2→6)-D-GlcN 4-phosphate substituted with different ester- and amide-linked diacyloxyacyl residues was prepared using a levulinoyl-protected Kdo benzylester bromide derivative for the glycosylation step (41). Moreover, the preparation of a polyacrylamide copolymer containing the α-Kdo-(2→6)-GlcN 4-phosphate determinant acylated with a single amide-linked (*R*)-3-hydroxytetradecanoyl residue was performed (42). Phosphorylation was achieved in good yield employing bis(trichloroethyl)phosphochloridate/*N*-methylimidazole as reagent followed by Zn-mediated removal of the phosphate protecting groups.

SYNTHESIS OF INNER CORE OLIGOSACCHARIDES RELATED TO THE HEPTOSE-Kdo REGION

Synthesis of Oligosaccharides Corresponding to the Inner Core of *Enterobacteriaceae*

In most bacterial inner core structures investigated thus far, the Kdo region is further extended by sugars attached to the axially oriented, and thus unreactive, O-5 group of Kdo. Considerable synthetic efforts were therefore devoted to the development of efficient procedures for the synthesis of L-*glycero*-α-D-*manno*-heptopyranosyl residues α-(1→5)–linked to Kdo—the most abundant structural entity found in the inner core of enterobacterial LPS. The synthesis of the disaccharide methyl glycoside L-α-D-Hep*p*-(1→5)-β-Kdo was first achieved by Paulsen and Heitmann (43) using appropriate activation of the Kdo-acceptor molecule via electron-donating protecting groups in positions O-4, O-7, and O-8 and by employing the reactive heptopyranosyl chloride **34** bearing an ester group as stereocontrolling auxiliary at O-2 (Fig. 5). Reaction conditions needed to be chosen so as to minimize the competing formation of an orthoester derivative from **34**. Further synthetic routes were designed for the subsequent introduction of the lateral, α-(2→4)–linked Kdo residue obtained in good to excellent yields using the donor **6** and HgBr$_2$ as promoter in MeNO$_2$ (44). Moreover, protected di- and trisaccharide derivatives were converted into *O*-acetylated heptopyranosyl-(1→5)-Kdo bromide derivatives, which in turn were attached to the *N*-acylated glucosamine disaccharide derivative **5** (along similar lines as described for **13**) to afford the corresponding tetra- and pentasaccharide derivatives in good anomeric selectivity (45). Due to the concomitant formation of elimination products from the halide donors, the yields of glycosylation products did not exceed 23–35% even under carefully optimized reaction conditions. Deprotection gave the tetrasaccharide L-α-D-Hep*p*-(1→5)-α-Kdo-(2→6)-β-D-GlcNhm-(1→6)-D-GlcNhm and the branched pentasaccharide L-α-D-Hep*p*-(1→5)-[α-Kdo-(2→4)]-α-Kdo-(2→6)-β-D-GlcNhm-(1→6)-D-GlcNhm. The products were incorporated into membranes of sheep erythrocytes and used in passive hemolysis assays. Exploiting a similar synthetic strategy, structurally related inner core units—lacking the glucosamine residue—were prepared as allyl glycosides and finally copolymerized with acrylamide to give glycopolymers with molecular masses of 20–30 kDa (46).

Improvements with respect to the transfer of the heptose units were accomplished by utilizing the trichloroacetimidate procedure under promotion with trimethylsilyl triflate (47) (Fig. 5). First, the acetylated L-α-D-Hep*p*-(1→3)-L-α-D-Hep*p* trichloroacetimidate **36** was generated in several steps from the α-benzyl heptofuranoside **35**. TMS-triflate promoted coupling of **36** with either α- or β-configured allyl 4-*O*-*p*-methoxybenzyl Kdo derivatives **37a** and **37b**, respectively, proceeded in good yields giving the corresponding α-(1→5)–linked trisaccharide derivatives **38a** and **38b**.

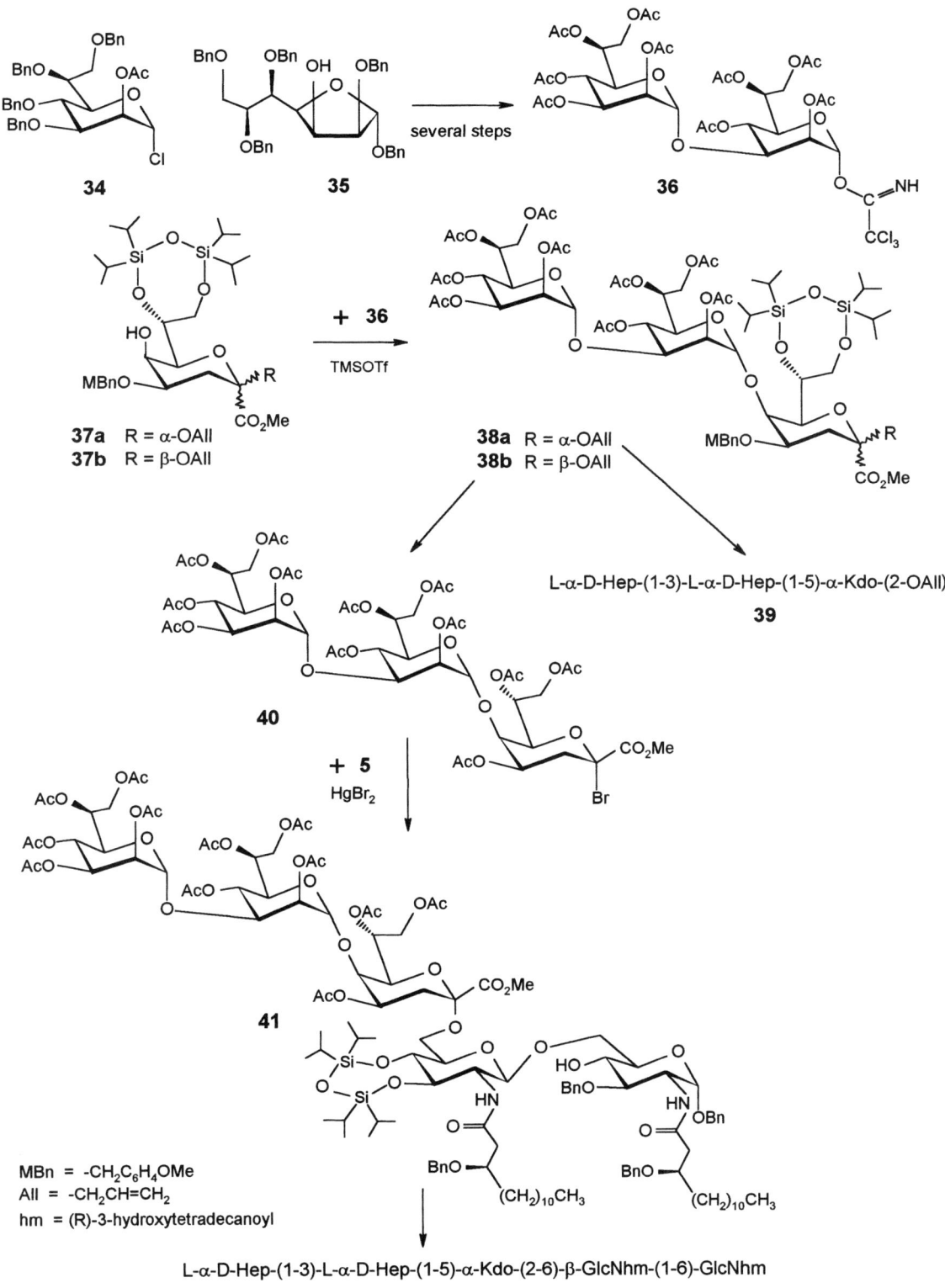

Fig. 5 Chemical synthesis of a pentasaccharide related to the enterobacterial Heptose-Kdo-Lipid A region. (From Ref. 48.)

Conventional deprotection of the α-allyl glycoside **38a** furnished the trisaccharide allyl glycoside L-α-D-Hepp-(1→3)-L-α-D-Hepp-(1→5)-α-Kdo **39**, which was incorporated into polyacrylamide copolymers. Furthermore, ether-type protecting groups of the β-allyl compound **38b** were replaced by acetates, and the allyl group was selectively removed via isomerization using [Ir(COD)(PMePh$_2$)]PF$_6$ and subsequent hydrolysis of the resulting propenyl group with HgO/HgCl$_2$. O-Acetylation of the anomeric OH group followed by reaction with TiBr$_4$ provided the trisaccharide donor **40**, which was coupled with the glucosamine acceptor **5** to furnish **41** in 37% yield and with excellent stereoselectivity. Successive removal of the protecting groups with tetrabutylammonium fluoride, sodium methoxide, hydrogenolysis of the benzyl groups with reactive palladium hydroxide as catalyst, and hydrolysis of the methyl ester of the Kdo moiety with aqueous NaOH gave the reducing pentasaccharide L-α-D-Hepp-(1→3)-L-α-D-Hepp-(1→5)-α-Kdo-(2→6)-β-D-GlcNhm-(1→6)-D-GlcNhm **42** (48). In combination with natural compounds, the synthetic antigens were of value in a study of the specificity of rabbit antisera against *Salmonella enterica* serovar *minnesota* R4 and R7 strains (49). The use of a benzyl-protected heptopyranosyl trichloroacetimidate donor has also been reported for the synthesis of the trisaccharide L-α-D-Hepp-(1→5)-α-Kdo-(2→6)-β-D-GlcN 4-phosphate (50).

Synthesis of Oligosaccharides Corresponding to the Inner Core of *Neisseria meningitidis*

A different approach to the synthesis of the heptose-Kdo region related to important epitopes of lipooligosaccharides from *Neisseria meningitidis* was developed by van Boom's group on the basis of iodonium ion–promoted glycosidations of thioglycosides (51–53). A thioglycoside group may be maintained as a temporary protection for the anomeric center through multistep reaction sequences, yet it may be selectively activated to serve as a versatile anomeric leaving group for glycosidation (Fig. 6). Thus, an aglyconic spacer-arm unit was introduced via glycosylation of N-benzyloxycarbonyl (Z) protected 3-aminopropanol with methyl or ethyl 2-thioglycosides of Kdo in the presence of N-iodosuccinimide (NIS) and catalytic amounts of triflic acid (TfOH) in yields of 50–80%, albeit with low anomeric selectivity not exceeding a ratio of 3:1 in favor of the α-anomer (51). Transformation of the protecting groups gave the 4-O-benzyl-7,8-O-isopropylidene–protected Kdo acceptor **44**, which was reacted with the ethyl 1-thioheptopyranosyl donor **43** under promotion with NIS/TfOH to furnish the α-(1→5)-linked disaccharide **45** in excellent yield (87%) and anomeric stereocontrol. Deprotection of **45** was effected in three steps, namely deacetonation in acetic acid, debenzoylation, and catalytic hydrogenolysis on palladium-carbon, which gave, after Sephadex S-100 chromatography, the homogeneous disaccharide L-α-D-Hepp-(1→5)-α-Kdo-2-O-(CH$_2$)$_3$NH$_2$ **46**. Proceeding toward epitopes containing also outer core determinants, the 2,3-O-isopropylidene–protected heptosyl donor **47** allowing for selective chain extension at O-4 was obtained from the ethyl 1-thio heptopyranosyl donor **43** in a few steps. Subsequent trimethylsilyl triflate–mediated glycosylation of **47** with the benzoyl-protected lactosyl trichloroacetimidate **48** afforded the trisaccharide ethyl 1-thioglycoside **49** in 91% yield. Exchange of the 2,3-O-isopropylidene group for benzoates afforded the trisaccharide donor **50**, which was coupled in the presence of the thiophilic promoter system NIS/TfOH with the Kdo acceptor **44** to give the α-(1→5)–linked tetrasaccharide derivative **51** in excellent yield (55%). Removal of the blocking groups and final purification of the product on Sephadex S-100 furnished the tetrasaccharide β-D-Galp-(1→4)-β-D-Glcp-(1→4)-L-α-D-Hepp-(1→5)-α-Kdo-2-O-(CH$_2$)$_3$NH$_2$ **52** corresponding to inner core determinant from *N. meningitidis* immunotypes L1–L9, which was subsequently used for the preparation of a synthetic vaccine.

A different donor for heptopyranosyl moieties, 7-dimethyl(phenyl)silyl heptopyranosyl chloride **53**, was obtained in few steps via Grignard reaction from allyl 2,3,4-tri-O-benzyl-α-D-*manno*-hexodialdo-1,5-pyranoside. The advantage of the 7-dimethyl(phenyl)silyl group as a masked form of the hydroxy function was convincingly demonstrated for the assembly of the trisaccharide α-D-Glcp NAc-(1→2)-L-α-D-Hepp-(1→3)-L-α-D-Hepp-1-O-(CH$_2$)$_3$NH$_2$ corresponding to a part structure of the dephosphorylated inner core region of *N. meningitidis* (53). In addition, **53** was employed for the synthesis of the disaccharide L-α-D-Hepp-(1→3)-L-α-D-Hepp-1-O-(CH$_2$)$_3$NH$_2$ **55**, which corresponds to the core of *N. meningitidis* immunotype 6 containing a 7'-O-(2-aminoethyl)phosphate group. Following Ag-triflate–mediated coupling of **53** with an appropriate acceptor to give an α-(1→3)–linked heptobioside intermediate, the 2-aminoethylphosphate moiety at the hydroxymethyl group of the distal heptose unit was obtained in two steps. Introduction of an intermediate 7'-O-phosphoramidite group and in situ condensation with 2-trifluoroacetamidoethanol followed by oxidation of the resulting phosphite triester furnished, after deprotection, the disaccharide phosphodiester **54**. The N-tri-

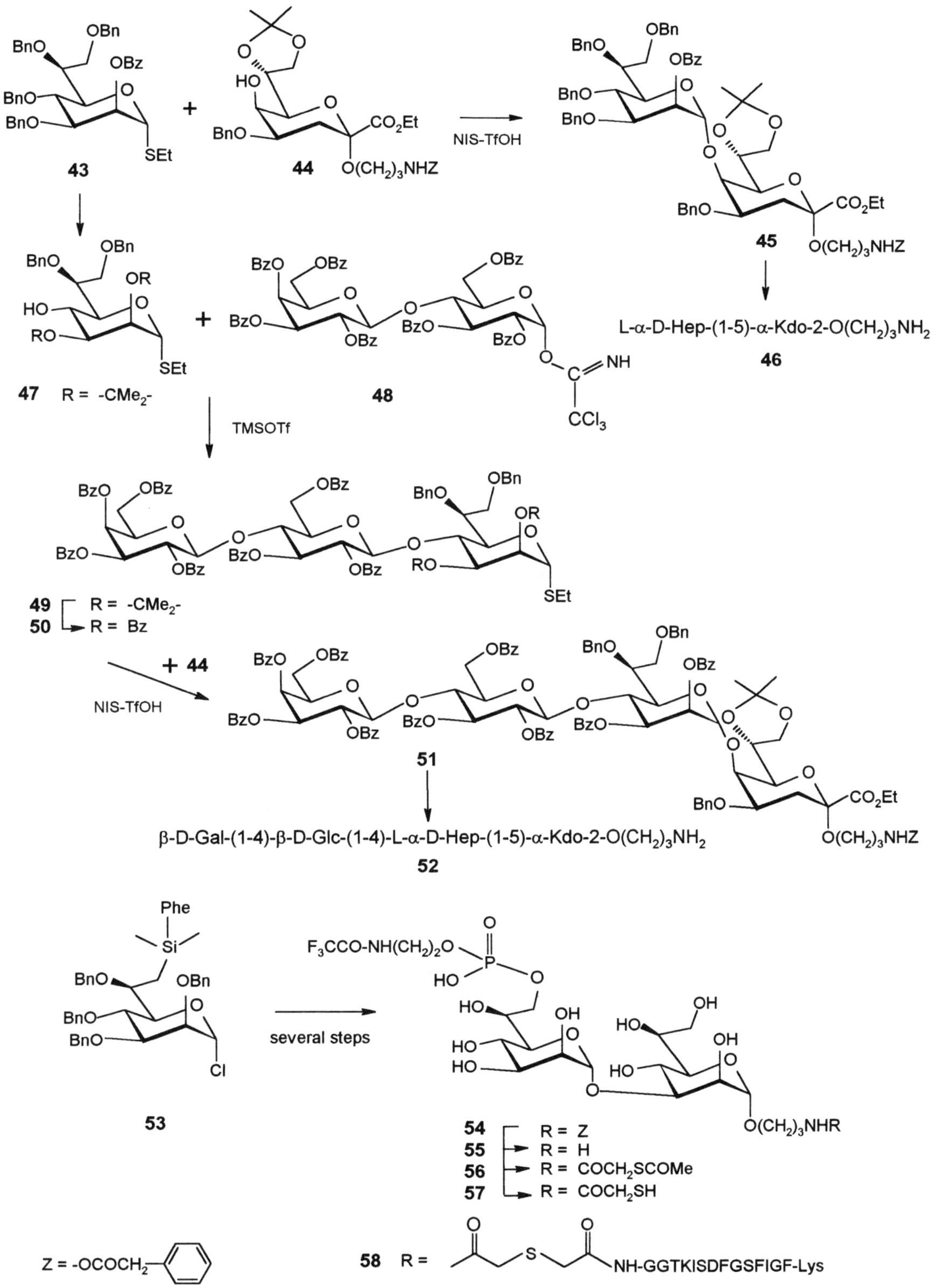

Fig. 6 Chemical synthesis of a tetrasaccharide and a disaccharide-peptide conjugate related to inner core determinants from *Neisseria meningitidis*. (From Ref. 54.)

fluoroacetamido group was kept until the very last step of the synthesis, which comprised hydrogenolysis of **54** to afford **55**. Further elongation of the aminopropyl spacer with a mercaptoacetamide unit and subsequent cleavage of the thioester gave the ω-thiol derivative **57**, which in turn was coupled with a synthetic bromoacetyl peptide. The peptide sequence Gly-Gly-TKISDFGSFIGF-Lys corresponds to a part of a meningococcal outer membrane protein. Ammonolysis of the trifluoroacetamide **58** followed by HPLC purification afforded the sugar-peptide conjugate containing T-helper cell epitopes for the generation of well-defined vaccines (54).

A semisynthetic approach was applied for the preparation of a novel type of artificial antigen containing lipooligosaccharides of *N. meningitidis* group B. The 2-aminoethylphosphate group of isolated oligosaccharides from *N. meningitidis* was used for the conversion into glycopolymers following introduction of a terminal acrylamide unit; in addition, the amino group was coupled to a synthetic lipopeptide derived from *E. coli* lipoprotein to give a glycopeptidolipid conjugate (55).

SYNTHESIS OF OLIGOSACCHARIDE FRAGMENTS OF THE HEPTOSE REGION

Synthesis of Oligosaccharides Related to *Salmonella* Ra LPS

Chemical syntheses of oligosaccharides comprising α-(1→3)- and α-(1→7)-linked disaccharides of L-*glycero*-D-*manno*-heptopyranosyl residues (L,D-Hep$p$), which are common subunits of enterobacterial LPS, were first accomplished using a heptopyranosyl trichloroacetimidate donor activated by benzyl ether–protecting groups (56). Under catalysis with anhydrous toluene sulfonic acid, the α-(1→3)–linked disaccharide was obtained in 50% yield, whereas the α-(1→7)–linked isomer was prepared in 91% yield as a 3.5:1 α-to-β mixture. Further extension of the synthetic scheme under similar conditions gave, after deblocking, the trisaccharide L-α-D-Hep$p$-(1→7)-L-α-D-Hep$p$-(1→3)-L,D-Hep. In an alternative preparation, the disaccharide structures were elaborated using 2-*O*-acetyl-3,4,6,7-tetra-*O*-benzyl-heptopyranosyl chloride **34** as a donor and silver triflate as promoter to give the α-(1→3)- and the α-(1→7)–linked disaccharide derivatives in good yields (43). The heptopyranosyl chloride **34** was also employed for the assembly of trisaccharides occurring in the inner core of *Citrobacter* PCM 1487 LPS (57). Simultaneous formation of two glycosidic linkages by silver triflate promoted condensation of **34** with methyl 2,4,6-tri-*O*-benzyl L-*glycero*-α-D-*manno*-heptopyranoside as acceptor afforded the protected branched trisaccharide methyl glycoside L-α-D-Hep$p$-(1→3)-[L-α-D-Hep$p$-(1→7)]-L-α-D-Hep$p$-*O*Me in excellent yield (71%). Furthermore, a heptopyranoside monosaccharide acceptor containing the versatile dimethyl(phenyl)silyl function as a hydroxy-masking group at C-7 was compatible with a coupling reaction using tetra-*O*-benzyl-glucopyranosyl chloride as donor in the presence of silver triflate giving the α-(1→3)–linked disaccharide in 90% yield without formation of the β anomer. Subsequent transformation of the carbon-silicon bond afforded a free 7-OH group, which in turn was glycosylated with the heptopyranosyl chloride **34** in 89% yield. Conventional deblocking afforded the branched trisaccharide methyl glycoside α-D-Glc$p$-(1→3)-[L-α-D-Hep$p$-(1→7)]-L-α-D-Hep$p$-*O*Me (57). The same trisaccharide had previously been obtained in comparable overall yield using acetylated heptopyranosyl bromide as the glycosyl donor in the presence of silver triflate and collidine (58).

Oligosaccharides of similar structure corresponding to core determinants being present in LPS of *Salmonella* Ra core were obtained as 2-(4-trifluoroacetamidophenyl)ethyl glycosides and coupled to protein carriers via the terminal amino function after hydrolytic removal of the trifluoroacetamido group (59). The compounds were used for studying the receptor structure of the phage G13 and the characterization of epitope specificities of monoclonal antibodies. The oligosaccharides were assembled in a sequential manner starting with the *N*-iodosuccinimide (NIS)/silver triflate-catalyzed conversion of the ethyl 1-thioheptopyranoside donor **59** (Fig. 7) into an anomeric mixture of the 2-(4-trifluoroacetamidophenyl)ethyl spacer derivatives, which was isomerized in the presence of borontrifluoride etherate into the pure α-anomer. Elegant transformation of the protecting groups via 2,3-*O*-orthoacetate formation, followed by 4-*O*-chloroacetylation and acid-catalyzed ring opening of the orthoester afforded **61** containing a free 3-OH group. Condensation of **61** with the 7-*O*-Si^{t}BuMe$_2$-protected ethyl 1-thioglycoside donor **60** under catalysis with dimethylthio(methylthio)sulfonium triflate (DMTST) gave the α-(1→3)–linked disaccharide **62** in 73% yield. Deblocking of the 7'-*O*-silyl ether group by treatment with acetic acid afforded **63** ready for the attachment of the α-(1→7)–linked heptopyranosyl moiety. Subsequent coupling was performed with benzoheptopyranosyl bromide **65** in the presence of silver triflate to furnish the protected trisaccharide **66** in 68% yield. Conventional deprotection gave the trisaccharide L-α-D-Hep$p$-(1→7)-L-α-D-Hep$p$-(1→3)-L-α-D-Hep$p$-O(CH$_2$)$_2$PhNHCOCF$_3$ **67**.

Chemical Synthesis of Core Structures

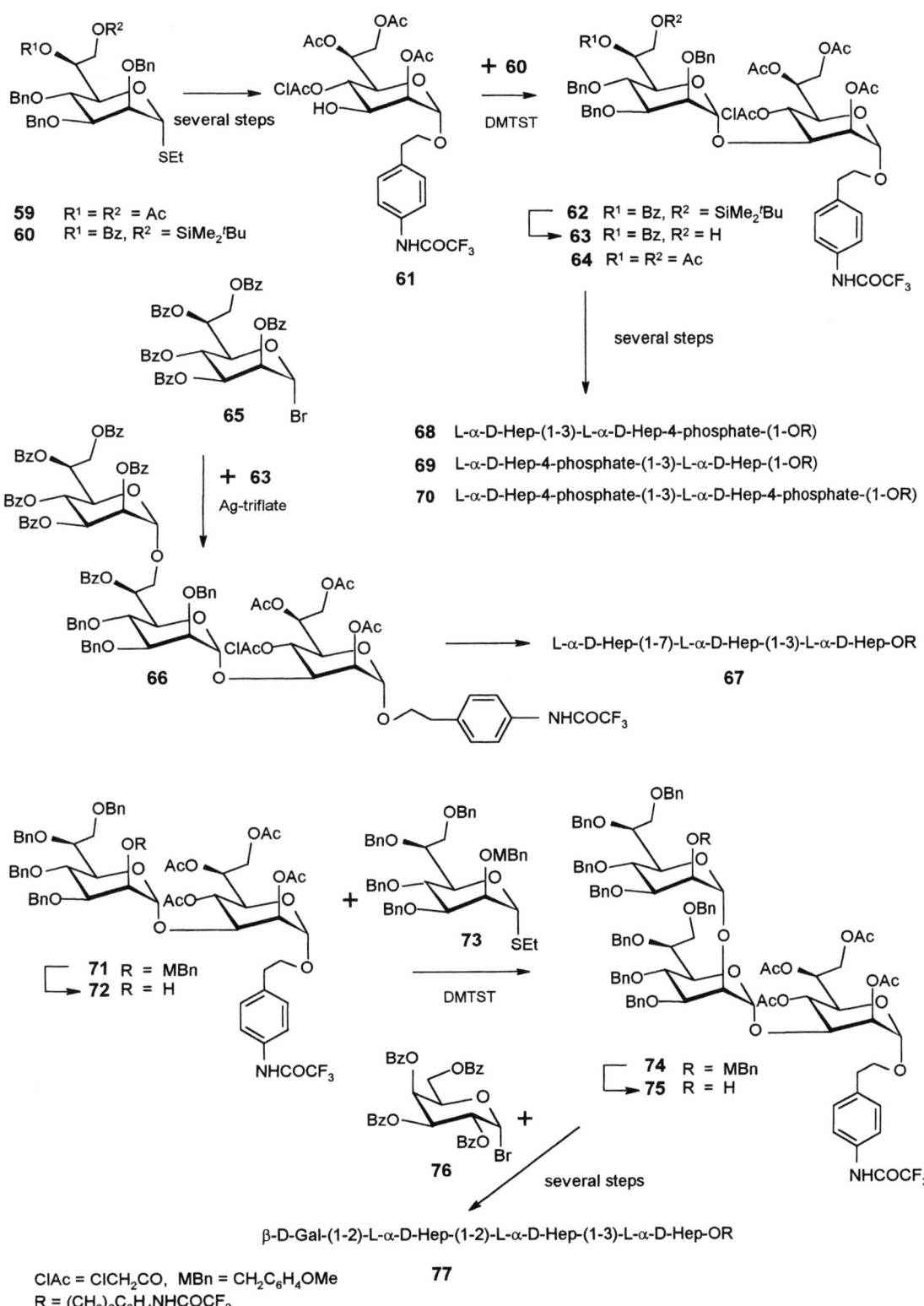

Fig. 7 Chemical synthesis of trisaccharides from the heptose region of *Salmonella* and *Haemophilus influenzae*. (From Ref. 59.)

Furthermore, α-(1→3)–linked heptobioside derivatives phosphorylated at the 4- and 4′-positions, respectively, which correspond to fragments proposed to occur in the *Salmonella* core were obtained as spacer glycosides. The phosphorylation steps on both heptosyl units of the α-(1→3)-linked disaccharide derivative **64** were performed in good overall yields—after selective liberation of the O-4 and O-4′ blocking groups—using phosphorus trichloride/imidazole, followed by addition of benzyl alcohol and in situ oxidation with *m*-chloroperbenzoic acid. Deprotection comprised deacetylation and catalytic hydrogenolysis, and the final purification of the products was achieved by FPLC in pyridine acetate buffer to afford L-α-D-Hep*p*-(1→3)-L-α-D-Hep*p*-4-phosphate-O(CH$_2$)$_2$PhNHCOCF$_3$ (**68**), L-α-D-Hep*p*-4-phosphate-(1→3)-L-α-D-Hep*p*-O(CH$_2$)$_2$-PhNHCOCF$_3$ (**69**), and L-α-D-Hep*p*-4-phosphate-(1→3)-L-α-D-Hep*p*-4-phosphate-O(CH$_2$)$_2$PhNHCOCF$_3$ (**70**) as sodium salts (60).

Meanwhile five regioisomeric L-glycero-D-*manno*-heptose monophosphates have been prepared as cyclohexylammonium salts using 2-dimethylamino-5,6-dibenzo-1,3,2-dioxaphosphepane (DMABDP) as a phosphorylating agent (61). The monosaccharide phosphates may be used as model compounds for NMR spectroscopic investigations and methylation analysis of phosphorylated substituents occurring in the heptose region.

Synthesis of Oligosaccharides Occurring in the Heptose Region of *Haemophilus influenzae*

The efficiency of heptopyranosyl thioglycoside donors has also been demonstrated for the assembly of a tetrasaccharide as well as three trisaccharide derivatives corresponding to important epitopes of dephosphorylated lipooligosaccharides of *Haemophilus influenzae* which lack the repeating units of the O-antigenic chain. The oligosaccharide structures contain α-(1→2)–linked heptopyranosyl units, which are absent in the *Salmonella* core; therefore, a different protecting scheme was elaborated (Fig. 7). The assembly of the heptotrioside backbone was achieved via DMTST-catalyzed condensation of the activated thioglycoside donor **73** with the α-(1→3)–linked spacer-arm disaccharide **72** obtained from **71** by oxidative removal of the 2-O-*p*-methoxybenzyl ether. Despite the presence of a nonparticipating *p*-methoxybenzyl group at O-2 of the donor **73**, exclusive formation of an α-1,2-*trans* configured glycosidic linkage was observed in the formation of the trisaccharide **74**. Following cleavage of the 2″-O-methoxybenzyl group to give **75**, the final glycosylation step at the 2″-OH group of the terminal heptose was performed with benzobromo galactose **76** under promotion with silver triflate to furnish the protected tetrasaccharide in 74% yield. Conventional deblocking afforded β-D-Gal*p*-(1→2)-L-α-D-Hep*p*-(1→2)-L-α-D-Hep*p*-(1→3)-L-α-D-Hep*p*-O(CH$_2$)$_2$-PhNHCOCF$_3$ (**77**) (62).

Synthesis of Hexose-Heptose Units

The disaccharides α- and β-D-Glc*p*N-(1→7)-L,D-Hep corresponding to core structures described for the LPS of *Aeromonas hydrophila*, *Bordetella pertussis*, *Vibrio ordalii*, *E. coli*, *Shigella sonnei*, and *Shigella flexneri* serotype 6 were prepared using acetylated 2-azido-2-deoxy-glucopyranosyl bromide as glycosyl donor (63). Under conditions of in situ anomerization, the α-(1→7)–linked disaccharide derivative was obtained in the presence of silver triflate at −60°C in 67% yield (together with 9% of β-isomer), whereas heterogeneous catalysis by silver silicate afforded the β-(1→7)–linked compound as the major product in 87% yield with concomitant formation of a small amount of α-isomer. An alternative synthesis for the α-(1→7)–linked disaccharide methyl glycoside and the synthesis of the disaccharide L-α-D-Hep*p*-(1→6)-L-α-D-Hep*p*-OMe occurring in *Aeromonas* LPS has been described (64).

SYNTHESIS OF OUTER CORE STRUCTURES

Synthesis of *Escherichia coli* K-12 Outer Core Fragments

For the unambiguous structural proof of the revised structure of the outer core of *Escherichia coli* K-12, containing one heptopyranosyl unit attached to a glucose residue of the outer core, the synthesis of the disaccharide L-α-D-Hep*p*-(1→6)-D-Glc was performed employing the trichloroacetimidate procedure (65). NMR and mass spectral data of the synthetic compound were identical with those of the natural derivative isolated from *E. coli* K-12 LPS. The synthetic scheme was further extended to include the neighboring glucose units (66). For the preparation of neoglycoconjugates, the heptopyranosyl derivatives of 3-O-kojibiosyl-glucose, i.e., L-α-D-Hep*p*-(1→6)-α-D-Glc*p*-(1→2)-α-D-Glc*p*-(1→3)-α-D-Glc*p* and part structures thereof were prepared as the respective allyl glycosides (67).

Synthesis of Outer Core Fragments from *Salmonella*

Outer core determinants related to *Salmonella* LPS have been prepared by a sequence of reactions with *gluco-* or *galacto-*configured glycosyl bromide donors carrying nonparticipating groups in the 2-position. Thus, glycosylation reactions with appropriate thioglycosides under halide-ion assistance, silver triflate or DMTST-promotion were used for connecting the glycosyl units followed by deprotection to afford the branched tetrasaccharide methyl glycoside α-D-Glc*p*-(1→2)-α-D-Gal*p*-(1→3)-[α-D-Gal*p*-(1→6)]-α-D-Glc*p*-OMe (**68**). Furthermore, synthesis of the 2-(4-trifluoroacetamidophenyl)ethyl spacer glycoside of the trisaccharide part structure α-D-Gal*p*-(1→3)-[α-D-Gal*p*-(1→6)]-α-D-Glc*p* has been described (69).

Recently, the synthesis of a hexasaccharide and subunits thereof related to the hexose-heptose region of the *Salmonella* Ra core has been achieved (70) (Fig. 8). First, the spacer-arm heptotrioside derivative **78** was prepared along similar lines as described for **66**. Subsequent condensation of **78** was then attempted with the α-(1→6)–linked methyl 1-thiodisaccharide donor **79** under the agency of dimethyl(methylthio)sulfonium triflate (DMTST). Since a tetrasaccharide derivative and the 1,6-anhydro compound **80** were formed instead of the expected pentasaccharide, the sequence of glycosidation steps was changed. Extension of **78** was thus accomplished using the protected α-Gal*p*-(1→3)-Glc*p* disaccharide thioglycoside donor **81** in the presence of NIS/silver triflate to give the pentasaccharide **82** in 72% yield. Finally, treatment with acetic acid released the benzylidene protecting group, and the α-(1→6)–linked terminal galactopyranosyl residue was finally introduced in 90% yield in a regio- and stereoselective reaction using 2,3,4,6-tetra-*O*-benzyl-α-D-galactopyranosyl bromide under halide-assisted conditions in the presence of Bu₄NBr. Final deprotection of the protected hexasaccharide **84** under conventional conditions afforded α-D-Gal*p*-(1→3)-[α-D-Gal*p*-(1→6)]-α-D-Glc*p*-(1→3)-[L-α-D-Hep*p*-(1→7)]-L-α-D-Hep*p*-(1→3)-L-α-D-Hep*p*-O(CH₂)₂PhNHCOCF₃ (**85**).

Synthesis of Outer Core and Inner Core Units from *Moraxella catarrhalis*

Recently, chemical synthesis of a complex octasaccharide and partial structures thereof, representing the outer part of the lipooligosaccharide from *Moraxella catarrhalis* serotype A, has been accomplished (71) (Fig. 9). Starting from a central β-(1→3)–linked ethyl 1-thiodisaccharide **86**, the α-linked 2-(4-trifluoroacetamidophenyl)ethyl spacer group was introduced, followed by further attachment of the β-(1→4)–connected glucopyranosyl residue to furnish the trisaccharide **87**. Removal of the 2-*O*-benzoate afforded **88**, which underwent a series of glycosylations employing the thioglycoside donors **89** and **92**—by in situ conversion into the corresponding bromides—to produce the highly branched pentasaccharide intermediate **93**. Selective cleavage of the 2-*O*-chloroacetyl–protecting group by treatment with hydrazine dithiocarbonate (HDTC) furnished the glycosyl acceptor **94**, which was elongated with various glycosyl donors. Whereas donors such as thioglycosides and trichloroacetimidates proved not to be efficient, the use of ether as solvent and bromide as the anomeric leaving group in the presence of silver triflate was found to give excellent product yields. Thus, coupling of **94** with the globotriosyl methyl 1-thioglycoside **95** provided the protected octasaccharide **96**. Deprotection of **96** and reductive conversion of the azido group into the acetamido group afforded the target compound α-D-Glc*p*NAc-(1→2)-β-D-Glc*p*-(1→4)-[α-D-Gal*p*-(1→4)-β-D-Gal*p*-(1→4)-α-D-Glc*p*-(1→2)-β-D-Glc*p*-(1→6)]{β-D-Glc*p*-(1→3)}-α-D-Glc*p* (**97**).

Furthermore, the authors also succeeded in the synthesis of *M. catarrhalis* inner core structures containing the highly branched α-glucopyranosyl residue in (1→5)-linkage to Kdo. To this end, the acetylated Kdo glycal methyl ester derivative **7** was elaborated into the 2-(4-trifluoroacetamidophenyl)ethyl spacer derivative of Kdo using the method of Achiwa via phenyl selenyl triflate–promoted addition, followed by triphenyltin hydride–mediated reduction (19). After appropriate protection of the 4-, 7- and 8-OH functions of Kdo, DMTST-catalyzed reaction with ethyl 1-thio-3,4,6-tri-*O*-acetyl-2-benzyl glucopyranoside afforded the protected α-Glc*p*-(1→5)-α-Kdo disaccharide in 84% yield. The further introduction of the remaining three, β-connected glucosyl residues was performed in a single step in 38% yield using benzobromo glucose in the presence of Ag-triflate followed by HPLC separation of the reaction mixture. Deprotection afforded the highly branched pentasaccharide β-D-Glc*p*-(1→3)-[β-D-Glc*p*-(1→4)]-{β-D-Glc*p*-(1→6)}-α-D-Glc*p*-(1→5)-α-Kdo-2-O(CH₂)₂PhNHCOCF₃ (**98**) (72).

In *Acinetobacter* and *Burkholderia* strains, the Kdo-isosteric octulosonic acid D-*glycero*-D-*talo*-oct-2-ulosonic acid (Ko) provides an acid-stable linkage of the core region to lipid A. Ko has been obtained by conversion of the glycal ester **7** via an α-2,3-*O*-anhydro derivative (29) to give a Ko bromide donor, which was

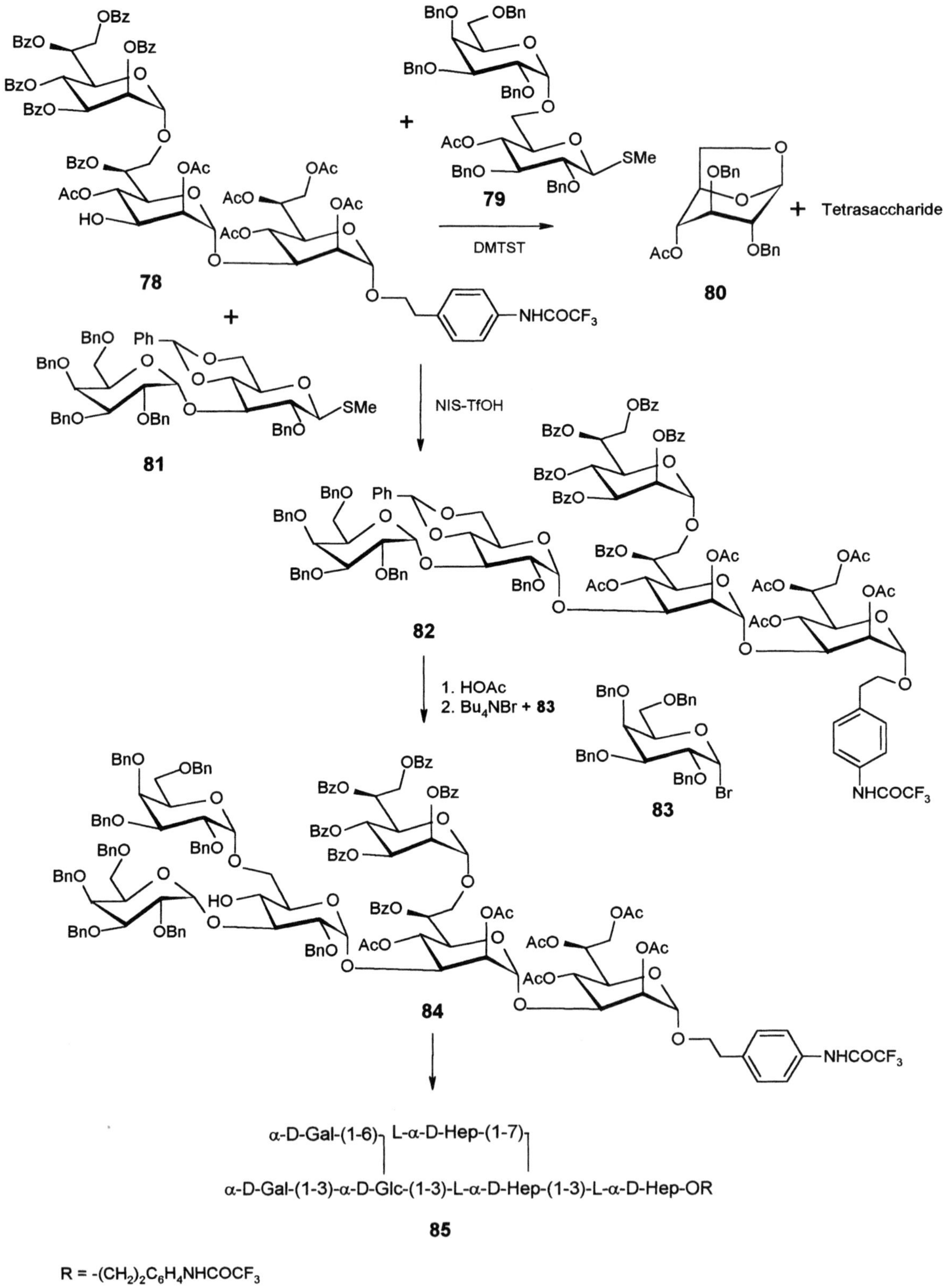

Fig. 8 Chemical synthesis of a hexasaccharide related to the core region of *Salmonella*. (From Ref. 70.)

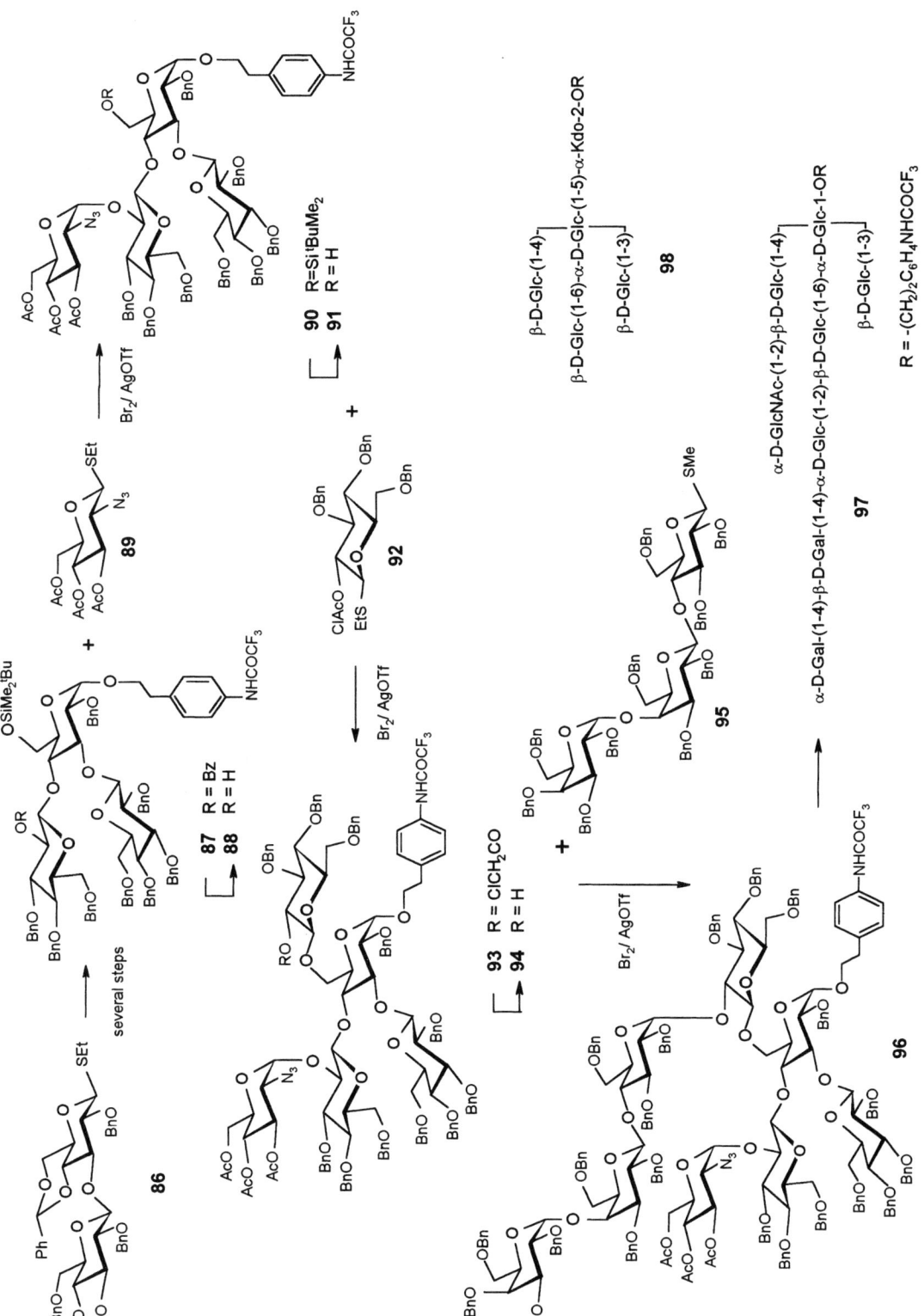

Fig. 9 Chemical synthesis of an octasaccharide related to the core region of *Moraxella catarrhalis*. (From Ref. 71.)

coupled in α-(2→6)-linkage to N-acetylglucosamine (73).

CLOSING REMARKS

Increasing evidence for the involvement of core structures in the interaction of gram-negative bacteria of the rough-mutant type with host defense systems has led to multidisciplinary efforts to study the molecular basis for these phenomena. Moreover, progress made in the field of oligosaccharide synthesis by advanced glycosylation procedures has had its impact on the synthesis of core determinants, where highly complex determinants including phosphorylated ligands of the heptose and Kdo region have been made available in defined haptenic and antigenic forms.

Future applications may be foreseen in the increasing demand for synthetic inhibitors and labeled substrates needed for studying the enzymatic glycosyl transfer steps involved in the biosynthesis of the core region. Furthermore, progress made in the crystallization of Fab fragments of core-directed monoclonal antibodies as well as crystallization of glycosyl transferases will provide plentiful applications of synthetic chemistry once the three-dimensional structures of the proteins have been sufficiently resolved. Given the powerful tools available in molecular biology today, the use of cloned transferases in the chemoenzymatic synthesis of core structures may soon become reality, adding to the repertoire required for the synthesis of LPS core determinants.

ACKNOWLEDGMENTS

Financial support by the Austrian Science Foundation (FWF-grants P8203, P9884, and P11449) of research performed in the author's laboratory is gratefully acknowledged. The author also thanks F. M. Unger and R. Müller for critically reading this manuscript and helpful suggestions.

REFERENCES

1. Stütz PL, Unger FM. Chemical synthesis of core structures. In: Morrison DC, Ryan JL, eds. Bacterial Endotoxic Lipopolysaccharides. Vol. 1. Boca Raton, FL: CRC Press, 1992:171–204.
2. Kusumoto S. Chemical synthesis of lipid A. In: Morrison DC, Ryan JL, eds. Bacterial Endotoxic Lipopolysaccharides. Vol. 1. Boca Raton, FL: CRC Press, 1992: 81–105.
3. Brade H, Rietschel ET. α-2→4-Interlinked 3-deoxy-D-manno-octulosonic acid disaccharide. A common constituent of enterobacterial lipopolysaccharides. Eur J Biochem 1984; 145:231–236.
4. Christian R, Schulz G, Waldstätten P, et al. Zur Struktur der 3-Desoxyoctulosonsäure-(KDO)-Region des Lipopolysaccharides von Salmonella minnesota Re 595. Tetrahedron Lett 1984; 25:3433–3436.
5. Paulsen H, Stiem M, Unger FM. Synthese eines 3-Desoxy-D-manno-2-octulosonsäure (KDO)-haltigen Tetrasaccharides und dessen Strukturvergleich mit einem Abbauprodukt aus Bakterien-Lipopolysacchariden. Tetrahedron Lett 1986; 27:1135–1138.
6. Lee RT, Lee YC. Synthesis of 3-(2-aminoethylthio)propyl glycosides. Carbohydr Res 1974; 37:193–201.
7. Roy R, Laferrière CA. Michael addition as the key step in the syntheses of sialyl oligosaccharide-protein conjugates from N-acryloylated glycopyranosylamines. J Chem Soc Chem Commun 1990; 1709–1711.
8. Kosma P, Gass J, Schulz G, et al. Artificial antigens: synthesis of polyacrylamide copolymers containing 3-deoxy-D-manno-2-octulopyranosylonic acid (Kdo) residues. Carbohydr Res 1987; 167:39–54.
9. Brade L, Kosma P, Appelmelk BJ, et al. Use of synthetic antigens to determine the epitope specificities of monoclonal antibodies against the 3-deoxy-D-manno-2-octulosonate region of bacterial lipopolysaccharide. Infect Immun 1987; 55:462–466.
10. Rozalski A, Brade L, Kosma P, et al. Epitope specificities of murine monoclonal and rabbit polyclonal antibodies against enterobacterial lipopolysaccharides of the Re chemotype. Infect Immun 1989; 57:2645–2652.
11. Rozalski A, Brade L, Kuhn HM, et al. Determination of the epitope specificity of monoclonal antibodies against the inner core region of bacterial lipopolysaccharides by use of 3-deoxy-D-manno-octulosonate-containing synthetic antigens. Carbohydr Res 1989; 193: 257–270.
12. Paulsen H, Krogmann C. Synthese einer KDO-haltigen Pentasaccharidsequenz der inneren Core- und Lipid A Region von Lipopolysacchariden. Carbohydr Res 1990; 205:31–44.
13. Paulsen H, Krogmann C. Synthese von KDO-haltigen Tri- und Tetrasaccharid-Sequenzen der inneren Core- und Lipoid A-Region von Lipopolysacchariden. Liebigs Ann Chem 1989, 1203–1213.
14. Paulsen H, Hayauchi Y, Unger FM. Synthese von Disacchariden der 3-Desoxy-D-manno-2-octulosonsäure (KDO) and D-Glucosamin. Liebigs Ann Chem 1984; 1270–1287.
15. Paulsen H, Stiem M, Unger FM. Synthese der Sequenz α-KDO-(2→4)-α-KDO-(2→6)-D-GlcN der inneren Core-Struktur von Lipopolysacchariden. Liebigs Ann Chem 1987; 273–281.
16. Paulsen H, Schüller M. Synthese von KDO-haltigen Lipoid-A-Analoga. Liebigs Ann Chem 1987; 249–258.
17. Kosma P, Schulz G, Unger FM, et al. Synthesis of trisaccharides containing 3-deoxy-D-manno-2-octulosonic acid residues related to the KDO-region of enterobacterial lipopolysaccharides. Carbohydr Res 1989; 190:191–201.

18. Imoto M, Kusunose N, Kusumoto S, et al. Synthetic approach to bacterial lipopolysaccharide. Preparation of trisaccharide part structures containing KDO and 1-dephospho lipid A. Tetrahedron Lett 1988; 29:2227–2230.
19. Ikeda K, Akamatsu S, Achiwa K. A new and stereospecific approach to Kdo-containing disaccharides using phenylselenyl triflate. Carbohydr Res 1989; 189: C1–C4.
20. Esswein A, Rembold H, Schmidt RR. O-Alkylation at the anomeric centre for the stereoselective synthesis of Kdo-α-glycosides. Carbohydr Res 1990; 200:287–305.
21. Haudrechy A, Sinaÿ P. Cyclization of hydroxy enol ethers: a stereocontrolled approach to 3-deoxy-D-manno-2-octulosonic acid containing disaccharides. J Org Chem 1992; 57:4142–4151.
22. Brade H, Brabetz W, Brade L, et al. Chlamydial lipopolysaccharide: chemical and antigenic structure, biosynthesis and biomedical application. Pure Appl Chem 1995; 67:1617–1626.
23. Kosma P, Schulz G, Brade H. Synthesis of a trisaccharide of 3-deoxy-D-manno-2-octulopyranosylonic acid (Kdo) residues related to the genus-specific lipopolysaccharide epitope of Chlamydia. Carbohydr Res 1988; 183:183–199.
24. Mikol V, Kosma P, Brade H. Crystal and molecular structure of allyl O-(sodium 3-deoxy-α-D-manno-2-octulopyranosylonate)-(2→8)-O-(sodium-3-deoxy-α-D-manno-2-octulopyranosidonate)-monohydrate. Carbohydr Res 1994; 263:35–42.
25. Kosma P, d'Souza FW, Brade H. Synthesis of Kdo-trisaccharide derivatives of chlamydial and enterobacterial LPS containing carboxyl-reduced or β-configurated Kdo residues. J Endotoxin Res 1995; 2:63–76.
26. D'Souza FW, Kosma P, Brade H. Synthesis of carboxyl-reduced analogues related to the Chlamydia-specific Kdo trisaccharide epitope. Carbohydr Res 1994; 262:223–244.
27. Kosma P, Bahnmüller R, Schulz G, et al. Synthesis of a tetrasaccharide of the genus-specific lipopolysaccharide epitope of Chlamydia. Carbohydr Res 1990; 208: 37–50.
28. Kosma P, Strobl M, Allmaier G, et al. Synthesis of pentasaccharide core structures corresponding to the genus-specific lipopolysaccharide epitope of Chlamydia. Carbohydr Res 1994; 254:105–132.
29. Kosma P, Sekljic H, Balint G. Addition reactions of glycal esters: Access to glycosyl donors of Kdo, D-glycero-D-talo- and D-glycero-D-galacto-2-octulosonic acid residues. J Carbohydr Chem 1996; 15:701–714.
30. Sarfati SR, Le Dur A, Szabó L. Phosphorylated sugars. Part 24. Methyl 3-deoxy-α-D-manno-oct-2-ulopyranosidonic acid 4-(dihydrogen phosphate): synthesis, stability in acidic medium, and colorimetric estimation. J Chem Soc Perkin Trans I 1988; 707–709.
31. Auzanneau FI, Charon D, Szabó L. Phosphorylated sugars. Part 27. Synthesis and reactions, in acid medium, of 5-O-substituted methyl 3-deoxy-α-D-manno-oct-2-ulopyranosidonic acid 4-phosphates. J Chem Soc Perkin Trans I 1991; 509–517.
32. Fukase K, Kamikawa T, Iwai Y, et al. Synthesis of allyl 3-deoxy-D-manno-2-octulopyranosidic acid 4- and 5-phosphates. Bull Chem Soc Jpn 1991; 64:3267–3273.
33. Auzanneau FI, Charon D, Szabó L. Phosphorylated sugars. Part 26. Synthesis of 3-deoxy-D-manno-oct-2-ulosonic acid 7-phosphate. J Chem Soc Perkin Trans I 1990; 2831–2834.
34. Sekljic H, Kosma P, Bartek J, et al. Synthesis of neoglycoproteins containing 5-O-phosphorylated Kdo-monosaccharide, 4-O- and 5-O-phosphorylated α-Kdo-(2→6)-2-acetamido-2-deoxy-β-glucopyranosyl disaccharide residues. J Endotoxin Res 1996; 3:151–164.
35. Rozalski A, Brade L, Kosma P, et al. Characterization of monoclonal antibodies recognizing three distinct, phosphorylated carbohydrate epitopes in the lipopolysaccharide of the deep rough mutant I-69 Rd$^-$/b$^+$ of Haemophilus influenzae. Mol Microbiol 1997; 23: 569–577.
36. Auzanneau FI, Charon D, Szilágyi L, et al. Chemistry of bacterial endotoxins. Part 6. Synthesis of allyl 5-O-(α-D-mannopyranosyl)-(3-deoxy-α-D-manno-oct-2-ulopyranosid)onic acid and of allyl 5-O-(α-D-mannopyranosyl)-(3-deoxy-α-D-manno-oct-2-ulopyranosid)onic acid 4-phosphate and their copolymers with acrylamide. J Chem Soc Perkin Trans I 1991; 803–809.
37. Sekljic H, Wimmer N, Hofinger A, et al. Synthesis of neoglycoproteins containing L-glycero-α-D-manno-heptopyranosyl-(1→4)- and (1→5)-linked 3-deoxy-α-D-manno-2-octulopyranosylonic acid (Kdo) phosphate determinants. J Chem Soc Perkin Trans I 1997; 1973–1982.
38. Kusumoto S, Kusunose N, Kamikawa T, et al. Chemical synthesis of 1-dephospho derivative of Escherichia coli Re lipopolysaccharide. Tetrahedron Lett 1988; 29: 6325–6326.
39. Nakamoto SI, Achiwa K. Lipid A and related compounds. XVI. Synthesis of biologically active tetraacetyl-3-deoxy-D-manno-2-octulosonic acid (Kdo)-(α2→6)-D-glucosamine-4-phosphates, novel analogs of the nonreducing sugar moiety of lipid A. Chem Pharm Bull 1987; 35:4537–4543.
40. Kiso M, Fujita M, Ogawa Y, et al. Synthesis of α-Kdo-(2→4)-Kdo disaccharide derivatives and their conjugation with a protected form of GLA-60, a biologically active analog of a lipid A subunit. Carbohydr Res 1990; 196:59–73.
41. Nakamoto SI, Achiwa K. Lipid A and related compounds. XVII. Synthesis of 3-deoxy-D-manno-2-octulosonic acid (Kdo)-(α2→6)-D-glucosamine-4-phosphates, analogs of the biologically active moiety of lipopolysaccharides from Escherichia coli Re mutant. Chem Pharm Bull 1988; 36:202–208.
42. Auzanneau FI, Mondange M, Charon D, et al. Synthesis of allyl 6-O-(3-deoxy-α- and -β-D-manno-oct-2-ulopyranosylonic acid)-(2→6)-2-deoxy-2-[(3R)-3-hydroxytetradecanamido]-β-D-glucopyranoside 4-phosphate and of the copolymer of the α anomer with acrylamide. Carbohydr Res 1992; 228:37–45.
43. Paulsen H, Heitmann AC. Synthese von Strukturen der inneren Core-Region von Lipopolysacchariden. Liebigs Ann Chem 1988; 1061–1071.
44. Paulsen H, Heitmann AC. Synthese von Trisaccharid-Einheiten der inneren Core-Region von Lipopolysacchariden. Liebigs Ann Chem 1989; 655–663.

45. Paulsen H, Brenken M. Synthese von Oligosacchariden der inneren Core- und Lipoid-A-Region von Lipopolysacchariden. Liebigs Ann Chem 1991; 1113–1126.
46. Paulsen H, Wulff A, Brenken M. Darstellung von synthetischen Antigenen der inneren Core-Region von Lipopolysacchariden durch Copolymerisation mit Acrylamid. Liebigs Ann Chem 1991; 1127–1145.
47. Paulsen H, Höffgen E. Darstellung synthetischer Antigene von LD-Heptose-haltigen Trisacchariden der Core-region von Lipopolysacchariden durch Copolymerisation mit Acrylamid. Liebigs Ann Chem 1993; 543–550.
48. Paulsen H, Höffgen E. Synthese des Pentasaccharids L-α-D-Hep-(1→3)-L-α-D-Hep-(1→5)-α-Kdo-(2→6)-β-D-GlcNhm-(1→6)-D-GlcNhm der linearen Core- und Lipoid-A Struktur von Lipopolysacchariden. Liebigs Ann Chem 1993; 531–541.
49. Swierzko A, Brade L, Höffgen E, et al. Specificity of rabbit antisera against the rough lipopolysaccharide of Salmonella minnesota strain R7 (Chemotype Rd1P-). FEMS Immun Med Microbiol 1993; 7:265–271.
50. Akamatsu S, Ikeda K, Achiwa K. Synthesis of the sequence Hept-(α1→5)-Kdo-(α2→6)-D-glucosamine-4-phosphate of lipopolysaccharide. Chem Pharm Bull 1991; 39:518–520.
51. Boons GJPH, van Delft FL, van der Klein PAM, et al. Synthesis of LD-Hepp and Kdo containing di- and tetrasaccharide derivatives of Neisseria meningitidis inner-core region via iodonium ion promoted glycosidations. Tetrahedron 1992; 48:885–904.
52. Boons GJPH, van der Marel GA, Poolman JT, et al. Synthesis of L-glycero-α-D-manno-heptopyranose-containing disaccharide of the Neisseria meningitidis dephosphorylated inner-core region. Rec Trav Chim Pays-Bas 1989; 108:339–343.
53. Boons GJPH, Overhand M, van der Marel GA, et al. Application of the dimethyl(phenyl)silyl group as a masked form of the hydroxy group in the synthesis of an L-glycero-α-D-manno-heptopyranoside-containing trisaccharide from the dephosphorylated inner core region of Neisseria meningitidis. Angew Chem Int Ed 1989; 28:1504–1506.
54. Boons GJPH, Hoogerhout P, Poolman JT, et al. Preparation of a well-defined sugar-peptide conjugate: a possible approach to a synthetic vaccine against Neisseria meningitidis. Bioorg Med Chem Lett 1991; 1:303–308.
55. Dmitriev BA, Ovchinnikov MV, Lapina EB, et al. Glycopeptidolipids—a new class of artificial antigens with carbohydrate determinants. Synthesis of artificial antigen with type-specific oligosaccharide hapten from Neisseria meningitidis group B. Glycoconjugate J 1992; 9:168–173.
56. Dziewiszek K, Banaszek A, Zamojski A. The synthesis of the heptose region of the gram-negative bacterial core oligosaccharides. Tetrahedron Lett 1987; 28:1569–1572.
57. Boons GJPH, Overhand M, van der Marel GA, et al. Synthesis of two branched heptopyranoside-(L,D-Hepp)-containing trisaccharides of the inner-core region of Citrobacter PCM 1487. Recl Trav Chim Pays-Bas 1992; 111:144–148.
58. Garegg PJ, Oscarson S, Szönyi M. Synthesis of methyl 3-O-(α-D-glucopyranosyl)-7-O-(L-glycero-α-D-manno-heptopyranosyl)-L-glycero-α-D-manno-heptopyranoside. Carbohydr Res 1990; 205:125–132.
59. Garegg PJ, Oscarson S, Ritzén H, et al. Synthesis of 2-(4-trifluoroacetamidophenyl)ethyl O-(L-glycero-α-D-manno-heptopyranosyl)-(1→7)-O-(L-glycero-α-D-manno-heptopyranosyl)-(1→3)-L-glycero-α-D-manno-heptopyranoside, corresponding to the heptose region of the Salmonella Ra core structure. Carbohydr Res 1992; 228:121–128.
60. Ekelöf K, Oscarson S. Syntheses of 4- and/or 4′-Phosphate derivatives of methyl 3-O-L-glycero-α-D-manno-heptopyranosyl-L-glycero-α-D-manno-heptopyranoside and their 2-(4-trifluoroacetamidophenyl)ethyl glycoside analogues. J Carbohydr Chem 1995; 14:299–315.
61. Grzeszczyk B, Holst O, Zamojski A. The synthesis of five L-glycero-D-manno-heptose monophosphates. Carbohydr Res 1996; 290:1–15.
62. Bernlind C, Oscarson S. Synthesis of L-glycero-D-manno-heptopyranose-containing oligosaccharide structures found in lipopolysaccharides from Haemophilus influenzae. Carbohydr Res 1997; 297:251–260.
63. Paulsen H, Wulff A, Heitmann AC. Synthese von Disacchariden aus L-glycero-D-manno-Heptose und 2-Amino-2-desoxy-D-glucose. Liebigs Ann Chem 1988; 1073–1078.
64. Boons GJPH, Steyger R, Overhand M, et al. Synthesis of naturally occurring LD-Hepp containing disaccharides. J Carbohydr Chem 1991; 10:995–1007.
65. Pakulski Z, Zamojski A, Holst O, et al. The synthesis and characterization of 6-O-L-glycero-α-D-manno-heptopyranosyl-D-glucopyranose. Carbohydr Res 1991; 215:337–344.
66. Nepogodev SA, Pakulski Z, Zamojski A, et al. The synthesis and characterisation of 2-O-(6-O-L-glycero-α,β-D-manno-heptopyranosyl-α-D-glucopyranosyl)-α,β-D-glucopyranose. Carbohydr Res 1992; 232:33–45.
67. Nepogodev SA, Backinowsky LV, Grzeszczyk B, et al. Synthesis of linear oligosaccharides: L-glycero-α-D-manno-heptopyranosyl derivatives of allyl α-glycosides of D-glucose, kojibiose, and 3-O-α-kojibiosyl-D-glucose, substrates for synthetic antigens. Carbohydr Res 1994; 254:43–60.
68. Garegg PJ, Helland AC. Synthesis of methyl O-(2-acetamido-2-deoxy-α-D-glucopyranosyl)-(1→2)-O-α-glucopyranosyl-(1→2)-α-D-galactopyranoside and of methyl O-α-D-glucopyranosyl-(1→2)-O-α-D-galactopyranosyl-(1→3)-O-[α-D-galactopyranosyl-(1→6)]-α-D-glucopyranoside, corresponding to parts of the outer core of the Salmonella cell wall lipopolysaccharide. J Carbohydr Chem 1993; 12:105–117.
69. Norberg T, Walding M, Westman E. Synthesis of the methyl, 1-octyl, and 4-trifluoro-acetamidophenylethyl α-glycosides of 3,6-di-O-(α-D-galactopyranosyl)-D-glucopyranose and an acyclic analogue thereof. J Carbohydr Chem 1988; 7:283–292.
70. Oscarson S, Ritzén H. Synthesis of a hexasaccharide corresponding to part of the heptose-hexose region of the Salmonella Ra core, and a penta- and a tetrasaccharide that compose parts of this structure. Carbohydr Res 1994; 254:81–90.

71. Ekelöf K, Oscarson S. Synthesis of oligosaccharide structures from the lipopolysaccharide of Moraxella catarrhalis. J Org Chem 1996; 61:7711–7718.
72. Ekelöf K, Oscarson S. Synthesis of 2-(4-aminophenyl)ethyl 3-deoxy-5-O-(3,4,6-tri-O-β-D-glucopyranosyl-α-D-glucopyranosyl)-α-manno-oct-2-ulopyranosidonic acid, a highly branched pentasaccharide corresponding to structures found in lipopolysaccharides from Moraxella catarrhalis. Carbohydr Res 1995; 278:289–300.
73. Gass J, Strobl M, Loibner A, et al. Synthesis of allyl O-[sodium (α-D-glycero-D-talo-2-octulopyranosyl)onate]-(2→6)-2-acetamido-2-deoxy-β-D-glucopyranoside, a core constituent of the lipopolysaccharide from Acinetobacter calcoaceticus NCTC 10305. Carbohydr Res 1993; 244:69–84.
74. Brade H, Zähringer U, Rietschel ET, et al. Spectroscopic analysis of a 3-deoxy-D-manno-2-octulosonic acid (KDO) disaccharide from the lipopolysaccharide of a Salmonella godesberg Re mutant. Carbohydr Res 1984; 134:157–166.
75. Kohen A, Belakhov V, Baasov T. Towards the synthesis of the putative reaction intermediate in the Kdo8P Synthase-catalyzed reaction. Synthesis and evaluation of 3-deoxy-D-manno-2-octulosonate-2-phosphate. Tetrahedron Lett 1994; 35:3179–3182.
76. Sarfati R, Mondange M, Szabó L. Phosphorylated sugars. Part 21. Synthesis of 3-deoxy-D-manno- and 3-deoxy-D-gluco-oct-2-ulosonic acid 5-(dihydrogen phosphates). J Chem Soc Perkin Trans I 1977; 2074–2077.
77. Charon D, Szabó L. Phosphorylated sugars. Part XX. Synthesis of 3-deoxy-D-manno-octulosonic acid 8-(dihydrogen phosphates). J Chem Soc Perkin Trans I 1976; 1628–1632.
78. Baasov T, Kohen A. Synthesis, inhibition, and acid-catalyzed hydrolysis studies of model compounds of the proposed intermediate in the Kdo8P-Synthase-catalyzed reaction. J Am Chem Soc 1995; 17:6165–6174.
79. Paulsen H, Pries M, Lorentzen JP, Synthese von DD-Heptosephosphaten als Substrate oder potentielle Inhibitoren für die Heptose-Synthetase. Liebigs Ann Chem 1994; 389–397.
80. Van Straten NCR, Kriek NMAJ, Timmers CM, et al. Synthesis of a trisaccharide fragment corresponding to the lipopolysaccharide region of Vibrio parahaemolyticus. J Carbohydr Chem 1997; 16:947–966.
81. Lipták A, Kerékgyártó J, Szurmai Z, et al. Synthesis of the tetrasaccharide core region of antigenic lipo-oligosaccharides characteristic of Mycobacterium kansasii. Carbohydr Res 1988; 175:241–248.
82. Szurmai Z, Kerékgyártó J, Harangi J, et al. glycosylated trehalose. Synthesis of the oligosaccharides of the glycolipid-type antigens from Mycobacterium smegmatis. Carbohydr Res 1987; 164:313–325.

16

Microbial Pathways of Lipid A Biosynthesis

Paul D. Rick
The Uniformed Services University of the Health Sciences, Bethesda, Maryland

Christian R. H. Raetz
Duke University Medical Center, Durham, North Carolina

INTRODUCTION

The outer membrane of *Escherichia coli* and most other gram-negative bacteria contains the endotoxin or lipopolysaccharide (LPS) molecule that is characteristic of these organisms. *E. coli* contains approximately 2×10^7 phospholipid molecules per cell, whereas each cell contains approximately 2×10^6 LPS molecules (1,2). Accordingly, the LPS molecule is a major component of the cell envelope, and it is responsible in large part for the "barrier function" of the outer membrane, which renders gram-negative bacteria resistant to a variety of detergents, dyes, and other agents. Many of the structural features of LPS molecules obtained from a wide variety of organisms have been determined, and comparison of these molecules has revealed that they consist of three covalently linked regions: a distal O-antigen region, a core region, and a proximal lipid A region (Fig. 1). The O-antigen region is the immunodominant portion of LPS, and it is comprised of oligosaccharide repeat units. O-antigen structures vary considerably among gram-negative bacteria, and indeed the structural diversity and immunoreactivity of this region is the basis for the classification of these organisms into O-serogroups. Compared with the O-antigen region, the structure of the core region is relatively invariant; nevertheless, important structural differences in the core regions of various organisms have been observed. Details concerning the structure and biosynthesis of the O-antigen and core regions are presented elsewhere in this volume (Chapters 8, 9, 17, and 18) as well as in several previous reviews (2–6). The specific structural features of lipid A are described in detail in Chapter 7 of this volume; however, a brief review of the structure of lipid A is presented in the following section of this chapter. Lipid A can generally be described as a phosphorylated β-1',6-linked glucosamine disaccharide that contains both *N*- and *O*-linked fatty acyl chains. The hydrophobic character of lipid A is responsible for anchoring the LPS molecule in the outer leaflet of the outer membrane. Although important structural differences have been determined for the lipid A regions of various gram-negative bacteria, the structure of this region of LPS, like that of the core region, is relatively conserved.

It is now well recognized that the lipid A region of LPS is responsible for the majority of the biological properties attributed to endotoxin. Accordingly, lipid A is critically involved in the pathophysiology of septic shock resulting from gram-negative bacteremia, and the morbidity and mortality resulting from sepsis and shock due to gram-negative bacteremia constitute an increasingly important medical problem (7). Our current knowledge of the specific roles of lipid A in the manifestation of septic shock are reviewed in detail throughout this volume. Attempts to develop specific therapies to combat sepsis have primarily followed two general approaches. One approach involves attempts to

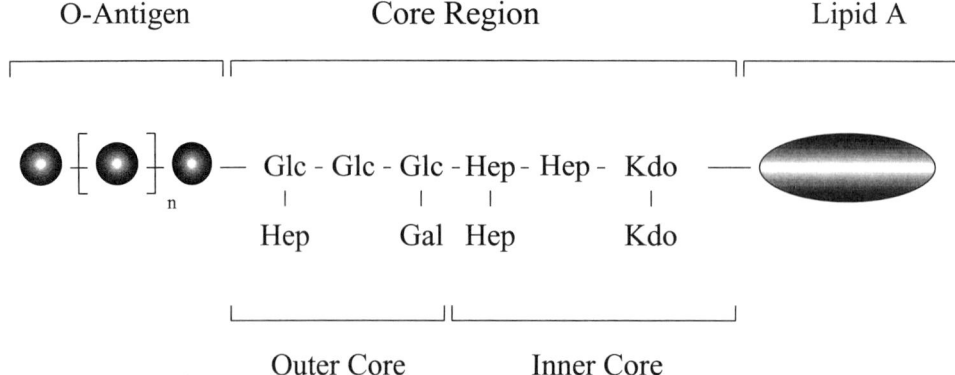

Fig. 1 The three structural regions of bacterial lipopolysaccharides. The structure of the *E. coli* K-12 core region is shown here as a representative example. *E. coli* K-12 does not possess an O-antigen due to two independent *wbb* (formerly *rfb*) mutations in different lineages of this organism (146). However, complementation studies have revealed that the *wbb* genes of *E. coli* K-12 encode the enzymes required for synthesis of an O16 O-antigen (147). Details concerning the structures and biosynthesis of various core and O-antigen regions are presented in Chapters 8, 9, 17, and 18 of this volume. The structure of *E. coli* lipid A is summarized in this review, and additional details are presented in Chapter 7 of this volume.

nullify the adverse biological effects of endotoxin through the development of strategies that result in altered host responses to lipid A. An alternative approach is based on the rational design of drugs that interfere with the synthesis of endotoxin in bacterial cells. The latter approach stems from the results of several studies that have established that the ability of gram-negative bacteria to synthesize a complete lipid A is critical for normal cell growth and physiology (8–10). Thus, it is quite likely that the development of clinically useful drugs that selectively target essential steps in lipid A synthesis will be extremely important for the effective management of sepsis. However, the development of such drugs is contingent upon an in-depth knowledge of the enzymes that catalyze the individual reactions involved in the pathway of lipid A biosynthesis. Considerable efforts to define this pathway in *E. coli* have resulted in a wealth of information that now makes this approach feasible. It is the aim of this chapter to review the latest information regarding lipid A biosynthesis. The reader is also referred to previous reviews on this subject (6,11–13) in order to obtain additional information on certain aspects of this topic.

LIPID A STRUCTURE

Although variations in the structure of the lipid A regions of *E. coli* and *Salmonella* strains have been reported (14), the fundamental structural features of the lipid A molecules of these organisms are essentially identical (15). The lipid A backbone consists of a β-1′,6-linked glucosamine disaccharide, which is phosphorylated at positions 1 and 4′ (Fig. 2) (16). However, the nature of the substituent in the 1-position appears to be somewhat variable; approximately half of the lipid A molecules of *E. coli* K-12 contain phosphate in the 1-position, whereas the remainder contain pyrophosphate (17,18). The disaccharide is primarily acylated with four molecules of (*R*)-3-hydroxymyristate residues; *N*-acylation occurs at positions 2 and 2′, whereas *O*-acylation occurs at positions 3 and 3′ (16,19–24). In addition, the 3-hydroxy groups of the *N*- and *O*-linked (*R*)-3-hydroxymyristoyl groups at positions 2′ and 3′ are acylated with lauric acid and myristic acid, respectively, to yield acyloxyacyl substituents (20,24). It is important to note that the lipid A molecule shown in Figure 2 has been chemically synthesized (25) and that it possesses all of the biological properties of free lipid A isolated from *E. coli* (26).

Structural analysis of the lipopolysaccharide obtained from a "deep-rough" or heptoseless mutant of *E. coli* revealed that lipid A is also substituted with a disaccharide of Kdo (24). The Kdo residues are 2,4-linked, and the 2-position of the proximal Kdo is linked to the primary hydroxyl group located at the 6′-position of the nonreducing terminal glucosamine residue. As we shall discuss later, the biosynthesis of a complete lipid A in *E. coli* is dependent on the incorporation of Kdo. However, Kdo is not considered to be part of the lipid A molecule; rather, it functions as the structural link between the LPS core region and lipid A.

Many of the general structural features of *E. coli* and *S. enterica* serovar *typhimurium* lipid A are representative of those found in the lipid A molecules of

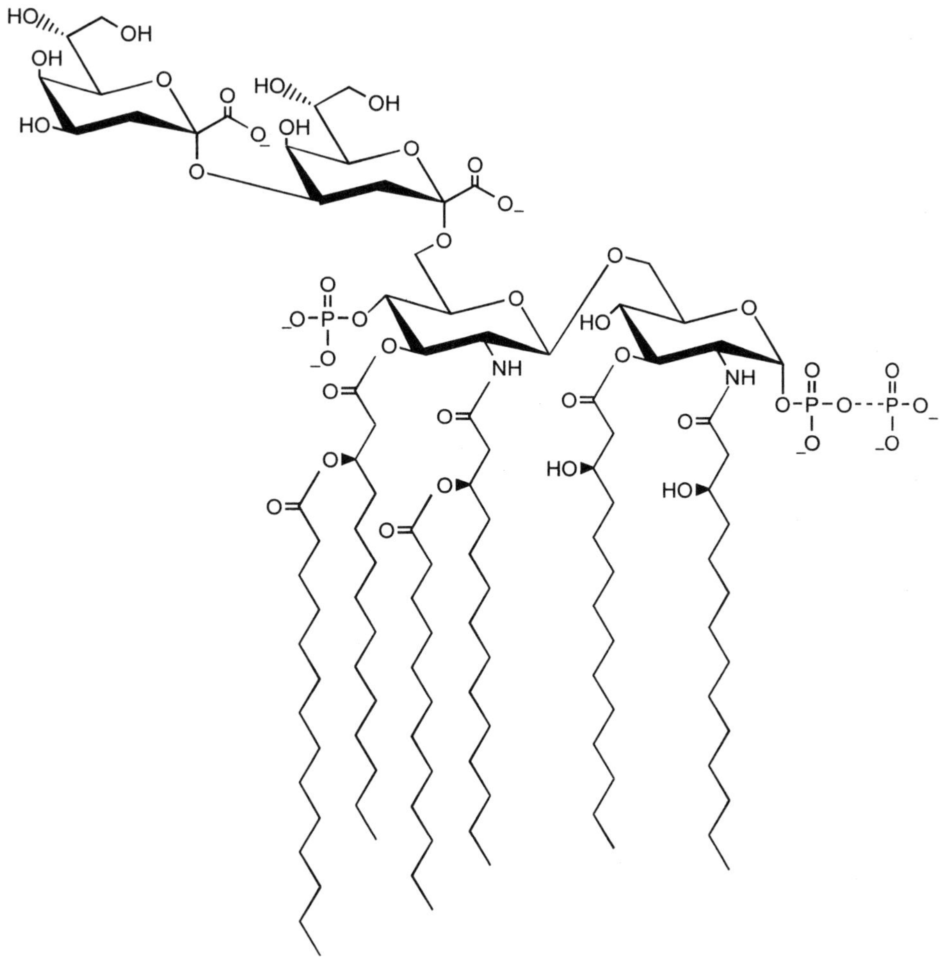

(Kdo)₂-Lipid A

(Re Endotoxin)

Fig. 2 Structure of the $(Kdo)_2$-lipid A region of *E. coli* lipopolysaccharide. Approximately 50% of the molecules contain pyrophosphate linked to the 1-position of the reducing terminal glucosamine residue (17,18). Additional polar substituents are described in this chapter, and a detailed description of the chemical structure of lipid A is presented in Chapter 7 of this volume.

most other gram-negative bacteria. Nevertheless, several gram-negative bacteria possess lipid A molecules with structural features that are quite different from those of *E. coli* lipid A, and in most cases these differences appear to render the molecules less toxic than *E. coli* lipid A. These differences include the degree to which the molecule is phosphorylated (27–29), the structure of the fatty acyl chains and the extent of fatty acylation (13,27–35), the possible absence of acyloxyacyl substituents (27), the occurrence of amino sugars other than glucosamine as a component of the backbone (35,36), and the presence of additional polar constituents (13,27).

LIPID A BIOSYNTHESIS

Early Observations and Historical Perspective

Early investigations to identify the steps involved in the biosynthesis of LPS were greatly facilitated by the use of phage-selection procedures. These procedures resulted in the isolation of a variety of mutants defective in the synthesis of the O-antigen and core regions, including the incorporation of the heptosyl residues (37); however, the genetic defects did not affect the viability of the respective mutants. In contrast, repeated use of these selection procedures failed to result in the isolation of mutants defective in the synthesis of Kdo, the incorporation of Kdo into lipid A, or the biosynthesis of lipid A per se. Accordingly, these observations led to the conclusion that synthesis of the Kdo-lipid A region is essential for normal cell growth and function, and attempts were made to isolate mutants of *S. enterica* serovar *typhimurium* that were conditionally defective in the synthesis of Kdo. These attempts resulted in the successful isolation of the first mutants defective in the biosynthesis of lipid A. Two independently isolated mutants were obtained that were conditionally defective in the synthesis of Kdo due to mutations in the structural gene for Kdo-8-phosphate synthetase (*kdsA*) (9,38). The characterizaton of these mutants revealed that the inability to synthesize Kdo at nonpermissive temperatures resulted in the accumulation of incomplete and underacylated lipid A precursors as well as growth inhibition (8,9). The major lipid A precursor was an acidic glucosamine disaccharide, which lacked Kdo as well as laurate and myristate; however, it was acylated with three or four residues of (R)-3-hydroxymyristic acid, and it also was substituted with phosphate residues at positions 1 and 4'. Although the complete structure of this molecule was not conclusively determined at the time, indirect evidence suggested that it contained both *O*-linked and *N*-linked (R)-3-hydroxymyristic acid residues. Munson et al. (39) subsequently showed that the lipid A precursor served as an in vitro acceptor of Kdo residues; accordingly, these investigators demonstrated the transfer of 2 Kdo residues from CMP-Kdo into the lipid A precursor using a cell-free extract obtained from *S. enterica* serovar *typhimurium*. It is interesting to note that the latter studies also provided the first data suggesting that a single enzyme is responsible for catalyzing the incorporation of both Kdo residues. The complete covalent structure of the "acidic" lipid A precursor (lipid IV$_A$) was later determined using fast atom bombardment mass spectrometry (FAB/MS)(40) as well as both ^{13}C- and ^{1}H-nuclear magnetic resonance spectrometry (NMR) (41), and it was indeed shown to be a β-1',6–linked glucosamine disaccharide acylated with (R)-3-hydroxymyristic acid residues in positions 2, 2', 3, and 3' and substituted with phosphates in positions 1 and 4' (see Fig. 4).

The investigations cited above are important from the standpoint that they provided the first data in support of the conclusion that synthesis of a complete Kdo–lipid A is indeed essential for normal cell growth and physiology. They also provided the first evidence that Kdo is incorporated into lipid A prior to the complete incorporation of fatty acyl substituents. Nevertheless, these studies provided very little information about the individual steps involved in lipid A synthesis. However, shortly after the initial description of the acidic lipid A precursor, Nishijima and Raetz (42,43) succeeded in identifying a key early intermediate in lipid A synthesis as an unexpected consequence of studies designed to isolate mutants of *E. coli* that were defective in phosphatidylglycerol synthesis. Accordingly, these studies resulted in the isolation of two glucosamine-containing glycolipids, designated lipids X and Y. The structures of lipids X and Y were later shown to be 2-deoxy-2-[(R)-3-hydroxytetradecamido]-3-*O*-[(R)-3-hydroxytetradecanoyl]-α-D-glucopyranose 1-phosphate (2,3-diacylglucosamine 1-phosphate) and 2-deoxy-2-[(R)-3-hexadecanoyloxytetradecanamido]-3-*O*-[(R)-3-hydroxytetradecanoyl]-α-D-glucopyranose 1-phosphate (2,β-2,3-triacylglucosamine 1-phosphate), respectively (23,44) (Fig. 4). As discussed below, lipid X is an early intermediate in the pathway of lipid A biosynthesis, and the isolation and characterization of lipid X proved to be of pivotal importance in the eventual elucidation of the steps involved in this pathway.

Sequence of Reactions Involved in the Pathway

The synthesis of (Kdo)$_2$-lipid A in *E. coli* involves nine enzymatically catalyzed reactions. The sequence of these reactions and the specific reaction catalyzed by each enzyme are shown in Figures 3–6. The following discussion summarizes our current understanding of the reactions involved in lipid A synthesis in *E. coli*. Additional information regarding how specific reactions in this pathway differ from those that occur in organisms other than *E. coli* is also included. For the sake of clarity, the reactions will be discussed in the same order that they occur in the pathway.

The synthesis of several important molecules located in the cell envelope of gram-negative bacteria is initiated by reactions involving UDP-GlcNAc. These include lipid A, peptidoglycan (45), enterobacterial common antigen (ECA) (46), and a variety of *wecA* (formerly *rfe*)-dependent O-antigens (46). ECA and *wecA*-dependent O-antigens are assembled from lipid carrier–linked intermediates, and the initial reaction in the synthesis of these molecules involves the transfer of GlcNAc-1-phosphate from UDP-GlcNAc to the polyisoprenoid lipid carrier, undecaprenylphosphate, to yield GlcNAc-pyrophosphorylundecaprenol (46,47). In contrast, the syntheses of lipid A (48) and peptidoglycan (45) are initiated by reactions that result in the covalent modification of the *N*-acetyl-D-glucosamine

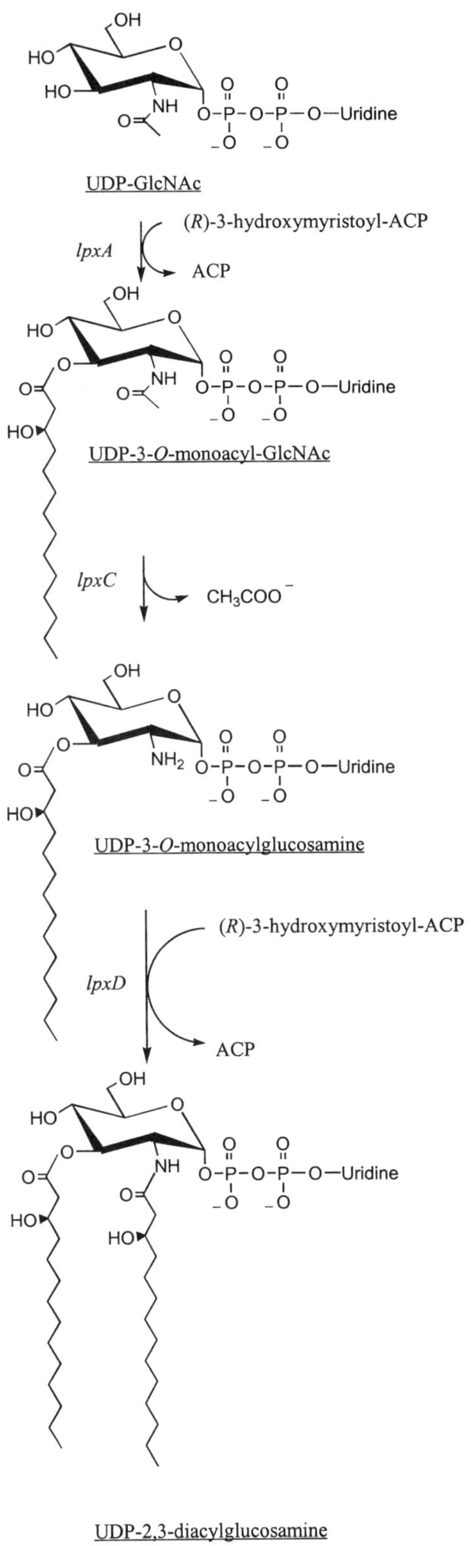

Fig. 3 Enzymatic reactions involved in the synthesis of UDP-2,3-diacylglucosamine from UDP-N-acetyl-D-glucosamine (UDP-GlcNAc).

moiety of UDP-GlcNAc. The role of UDP-GlcNAc as a precursor of all of these molecules suggests that the intracellular pool of this nucleotide sugar must be maintained at sufficient levels to meet this biosynthetic demand. In addition, it seems likely that the availability of UDP-GlcNAc for the synthesis of these various molecules is regulated in some manner. Indeed, it is also possible that a regulatory mechanism exists for the purpose of partitioning GlcNAc-pyrophosphorylundecaprenol for the synthesis of ECA and *wecA*-dependent O-antigens. However, essentially nothing is known about how the biosynthesis of any of these cell envelope components is regulated.

Synthesis of UDP-3-O-[(R)3-Hydroxymyristoyl]-N-Acetyl-D-Glucosamine

Lipid A synthesis is initiated by the transfer of the (*R*)-3-hydroxymyristate moiety from (*R*)-3-hydroxymyristoyl-acyl-carrier-protein (ACP) to the 3′-hydroxyl of UDP-GlcNAc to yield UDP-3-*O*-[(*R*)-3-hydroxymyristoyl]-*N*-acetyl-D-glucosamine (*UDP-3-O-monoacyl-N-acetylglucosamine*) (48). This reaction is catalyzed by the cytosolic enzyme UDP-*N*-acetylglucosamine acyltransferase (LpxA) (48,49). The *lpxA* gene has been cloned and sequenced, and it is located clockwise of *dnaE* at min 4 on the *E. coli* chromosome as part of a complex operon that includes three other genes involved in lipid A synthesis (*lpxB, fabZ*, and *firA[lpxD]*) as well as several additional open reading frames of unknown function (49–52). The deduced amino acid sequence of LpxA revealed it to be a 28 kDa protein comprised of 262 amino acids (50), and the enzyme has been purified to near homogeneity (53). Indeed, the x-ray crystal structure of the enzyme has been determined to 2.6 Å resolution, and an atomic model of LpxA based on the crystallographic data revealed the enzyme to be comprised of three identical subunits and a new kind of helical protein fold (54). The active site appears to be situated between the subunits. LpxA is the only acyltransferase for which a crystal structure is available.

Early studies using cell-free extracts demonstrated that the acyl donor in the reaction catalyzed by *E. coli* LpxA is ACP, and both (*R*)-3-hydroxymyristoyl-CoA and free (*R*)-3-hydroxymyristic acid were inactive as substrates for the enzyme (55). Subsequent studies using partially purified enzyme demonstrated a strict requirement for (*R*)-3-hydroxymyristoyl-ACP, whereas (*R*)-3-hydroxymyristoyl-CoA, myristoyl-ACP, (*R,S*)-3-hydroxypalmitoyl-ACP, palmitoyl-ACP, and (*R,S*)-3-hydroxylauryl-ACP were inactive (48). The apparent

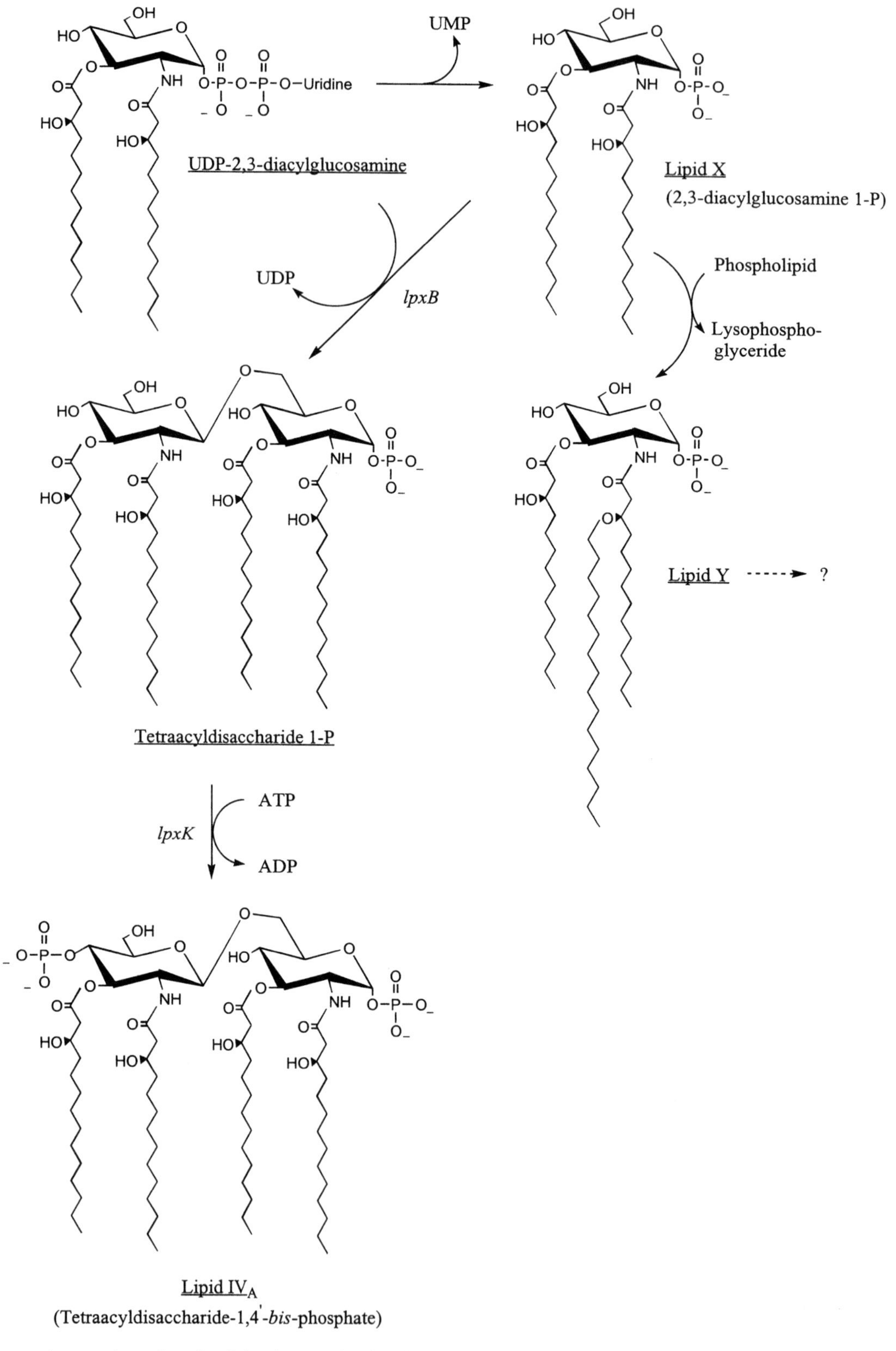

Fig. 4 Enzymatic reactions involved in the synthesis of the precursor, lipid IV$_A$ (tetraacyldisaccharide-1,4'-bis-phosphate), from UDP-2,3-diacylglucosamine. The synthesis of lipid Y from lipid X is also shown.

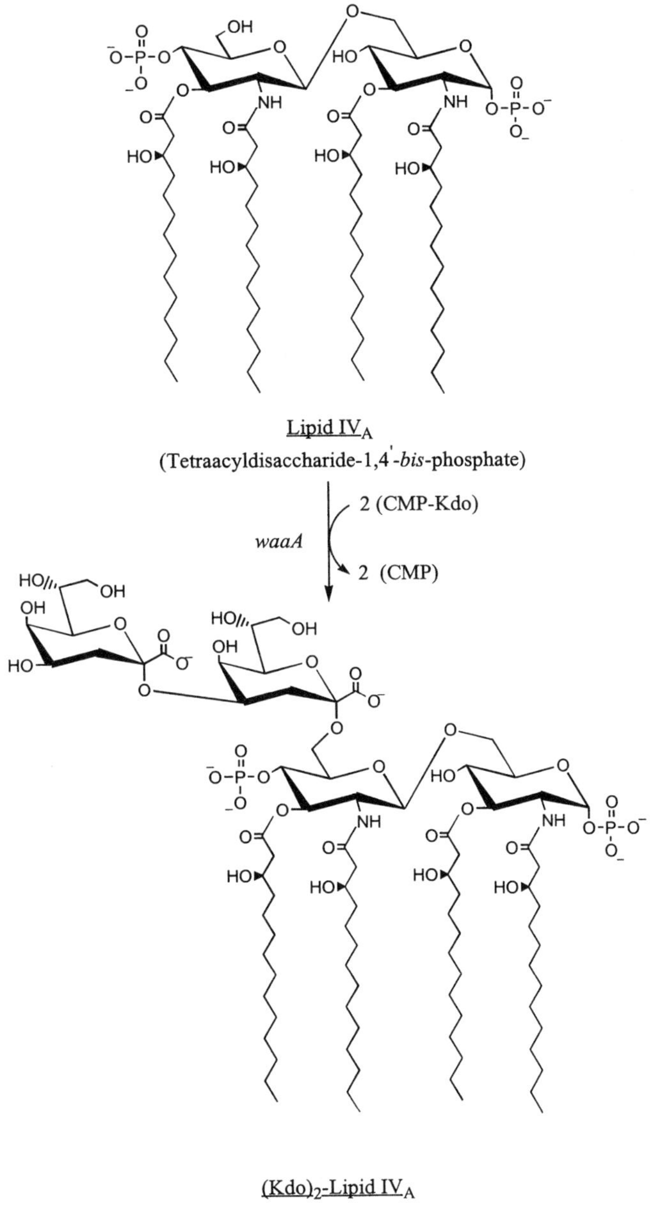

Fig. 5 Transfer of Kdo residues from CMP-Kdo to lipid IV$_A$ to yield (Kdo)$_2$-lipid IV$_A$. The bifunctional Kdo-transferase is encoded by the *waaA* (formerly *kdtA*) gene, and the enzyme catalyzes the successive transfer of Kdo residues from CMP-Kdo to lipid IV$_A$ and (Kdo)$_1$-lipid IV$_A$.

Km of *E. coli* LpxA for UDP-GlcNAc and (R)-3-hydroxymyristoyl-ACP are 99 and 1.6 μM, respectively (53). Quite surprisingly, the equilibrium constant of the reaction was found to be approximately 0.01, thus indicating that synthesis of the 3-*O*-(R)-3-hydroxymyristoyl derivative of UDP-GlcNAc is very unfavorable (53). Both GDP-GlcNAc and CDP-GlcNAc are inactive as acceptors of (R)-3-hydroxymyristate, whereas TDP-GlcNAc was found to be about 20% as active as UDP-GlcNAc (48). In addition, relatively early studies demonstrated that UDP-glucosamine was unable to serve as an acceptor of acyl chains (55), thus providing the first evidence that the initial acylation reaction catalyzed by LpxA occurs prior to the deacetylation of the *N*-acetylglucosamine moiety.

A temperature-sensitive *lpxA* mutant (*lpxA2*) of *E. coli* has been isolated by Galloway and Raetz (1). A shift of the mutant from permissive temperature (30°C)

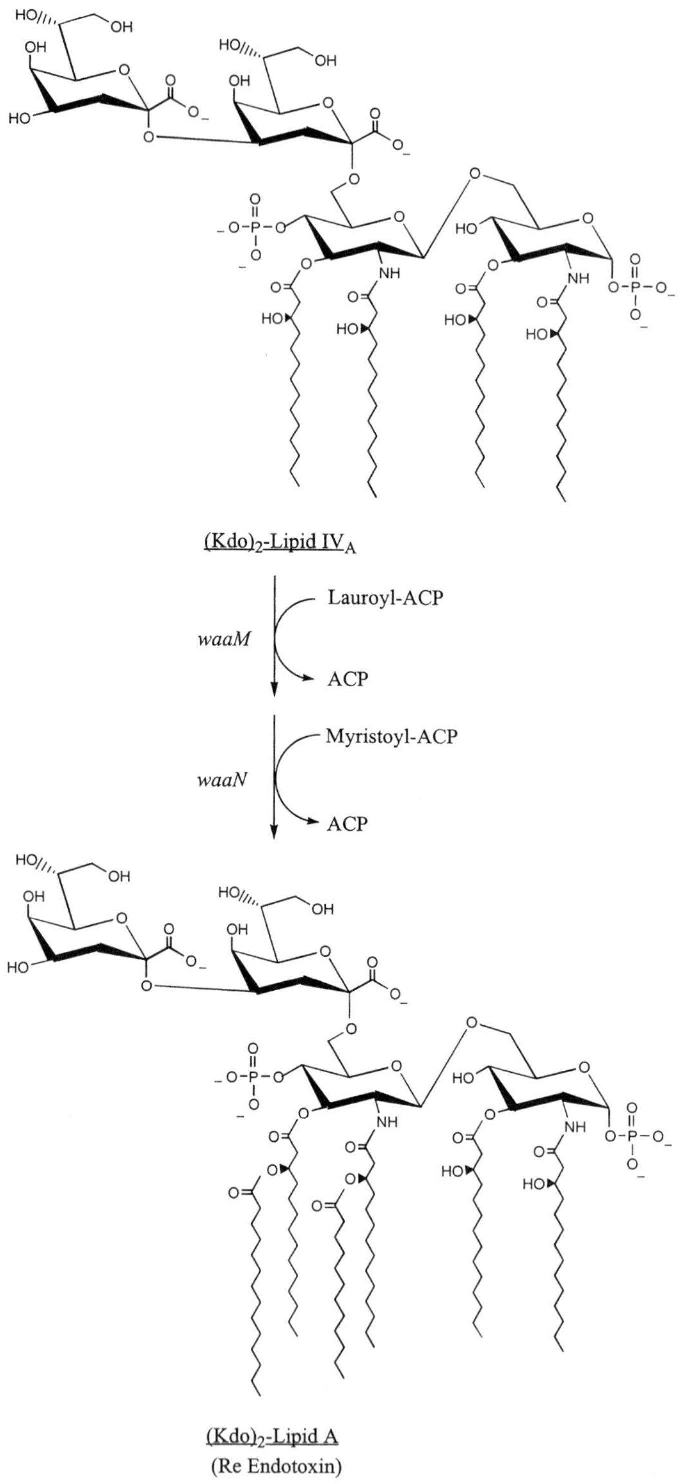

Fig. 6 Enzymatic synthesis of $(Kdo)_2$-lipid A from $(Kdo)_2$-lipid IV_A. The final steps in the assembly of lipid A involve the successive synthesis of the laurate- and myristate-containing acyloxyacyl substituents located at the 2'- and 3'-positions of the nonreducing terminal glucosamine residue, respectively. It should be noted that approximately 50% of naturally occurring Re endotoxin molecules contain pyrophosphate in the 1-position (see Fig. 2); however, the mechanism responsible for synthesis of the pyrophosphate moiety has not yet been determined.

to nonpermissive temperature (42°C) resulted in the complete cessation of lipid A synthesis, and this was accompanied by cell death. Since the inhibition of the initial step of lipid A synthesis would not be expected to result in the accumulation of lipid A precursors that are deleterious to the cell, the lethal expression of the *lpxA2* allele provides compelling evidence in support of the conclusion that lipid A plays an essential role in growth and replication. It is interesting to note that at permissive temperature the *lpxA2* mutant is hypersensitive to a variety of hydrophobic antibiotics, and it is also sensitive to hypoosmotic conditions (56). This phenotype suggests that the function of LpxA may be somewhat compromised at 30°C; indeed, the amount of lipid A synthesized at 30°C appears to be about 30% below normal (1).

Attempts to isolate temperature-resistant revertants of *E. coli* mutants possessing the *lpxA2* allele resulted in the isolation of pseudorevertants with a mutation in an open reading frame designated *orf17* located immediately upstream of *lpxA* (52). Subsequent experiments revealed that *orf17* encodes an (R)-3-hydroxymyristoyl-ACP dehydrase, and this locus was renamed *fabZ*. The specific activity of the dehydrase in extracts of the pseudorevertants was two- to fivefold less than that of the parental strain. Accordingly, the decreased level of dehydrase activity in the *fabZ* mutants appears to result in an increase in the pool of (R)-3-hydroxymyristoyl-ACP sufficient to support lipid A synthesis in *lpxA* mutants at elevated temperature.

The fatty acyl chain length specificities of the UDP-GlcNAc-*O*-acyltransferases and UDP-GlcNAc-*N*-acyltransferases of many gram-negative bacteria differ from one another, and these specificities are directly reflected by the 3,3′ (*O*-linked) and 2,2′ (*N*-linked) (R)-3-hydroxy fatty acyl components of their respective lipid A molecules (57). For example, the lipid A of *Neisseria meningitidis* contains primarily (R)-3-hydroxylaurate at both the 3- and 3′-positions (58,59); however, (R)-3-hydroxydecanoate has also been detected at these positions in some strains of *Neisseria gonorrhoeae* (60). Accordingly, the LpxA of *Neisseria meningitidis* selectively utilizes (R)-3-hydroxylauroyl-ACP and (R)-3-hydroxydecanoyl-ACP as acyl donors. Similarly, the 3- and 3′-hydroxyls of the lipid A of *Pseudomonas aeruginosa* are acylated with (R)-3-hydroxydecanoate, and this is in agreement with the observation that the LpxA of this organism is highly selective for (R)-3-hydroxydecanoyl-ACP as an acyl donor (57). Both growth and synthesis of wild-type levels of lipid A were rescued at 42°C following the introduction of the *lpxA* allele of either *P. aeruginosa* or *N. meningitidis*

into the *E. coli lpxA2* mutant (61,62). In each case complementation of the mutant phenotype was accompanied by synthesis of lipid A molecules whose 3 and 3′-*O*-fatty acyl components reflected the fatty acyl–ACP substrate specificity of the respective LpxA. The above observations also suggest that enzymes involved in subsequent steps of lipid A synthesis in *E. coli* can utilize substrates possessing 3- and 3′-linked fatty acyl groups other than (R)-3-hydroxymyristate.

A single amino acid residue accounts for the remarkable acyl chain length selectivity of *E. coli* and *Pseudomonas* LpxA. The *E. coli lpxA* mutation G173M switches the chain length selectivity from 14 carbons to 10 carbons, whereas the reciprocal mutation in the *Pseudomonas lpxA* (M169G) switches the *Pseudomonas* LpxA selectivity from 10 carbons to 14 carbons (63).

Deacetylation of UDP-3-*O*-[(*R*)-3-Hydroxymyristoyl]-*N*-Acetyl-D-Glucosamine

Anderson et al. (55) revealed that incubation of UDP-GlcNAc with extracts of *E. coli* in the presence of (R)-3-hydroxymyristoyl-ACP resulted in the formation of UDP-2,3-diacylGlcN and 2,3-diacylGlcN 1-P. Subsequent investigations demonstrated that incubation of UDP-3-*O*-[(R)-3-hydroxymyristoyl]GlcNAc with membrane-free extracts of *E. coli* in the absence of (R)-3-hydroxymyristoyl-ACP resulted in the accumulation of only UDP-3-*O*-[(R)-3-hydroxymyristoyl]GlcN (64). However, the addition of (R)-3-hydroxymyristoyl-ACP to these reaction mixtures resulted in the formation UDP-2,3-diacylGlcN, 2,3-diacylGlcN 1-P, as well as tetraacyldisaccharide 1-P, the immediate precursor of compound IV$_A$ lacking the 4′-phosphate (64). Accordingly, these observations clearly support the conclusion that synthesis of UDP-3-*O*-[(R)-3-hydroxymyristoyl]GlcNAc is followed by its deacetylation to yield UDP-3-*O*-[(R)-3-hydroxymyristoyl]GlcN, which serves as a precursor of UDP-2,3-diacylGlcN as well as other intermediates in the pathway (Fig. 3). Indeed, experiments utilizing UDP-3-*O*-[(R)-3-hydroxymyristoyl]-*N*-[³H]acetyl-D-glucosamine revealed that the deacetylase releases acetate and not a derivative of acetate (64).

The *E. coli* deacetylase is encoded by the *lpxC*(*envA*) gene (65). The *lpxC* gene was initially identified as a mutation (*envA1*) that resulted in increased permeability to a variety of antibiotics and other agents (66–68). The phenotype of the *envA1* mutant also included decreased amounts of LPS in the outer membrane (69) and a defect in cell septation (66,67). The *lpxC* gene was mapped to 2 min on the

E. coli genetic map in a region containing several other genes involved in cell division and cell envelope assembly (70). The nucleotide sequence of *lpxC* established LpxC to be a 34 kDa protein containing 305 amino acids; these studies also defined the alteration in the original *envA1* mutant allele as a missense mutation (71). It now seems apparent that the *envA1* gene product maintained residual activity since null mutations in *lpxC* have revealed that this gene is essential (71).

Although acylation of the 3-position of the glucosamine moiety of UDP-GlcNAc with (R)-3-hydroxymyristate is the initial step in lipid A synthesis, as stated earlier, this reaction is reversible and thermodynamically unfavorable (53). In contrast, hydrolysis of the acetamido linkage is considered to be essentially irreversible. Accordingly, the deacetylase reaction is the first committed step in lipid A synthesis, and hydrolysis of the acetamido linkage most likely serves to promote the formation of product in the reaction catalyzed by LpxA. The committed step in many biosynthetic pathways is frequently the subject of regulatory control, and the available data support the conclusion that LpxC is regulated in some manner. Thus, the in vivo inhibition of LpxA or the in vivo inhibition of the deactylase with a specific inhibitor resulted a 5- to 10-fold increase in the specific activity of LpxC in extracts prepared from such cells (72,73). Other enzymes involved in lipid A synthesis were not affected. The increase in specific activity of the deacetylase does not appear to be due to a change in the Km of the enzyme for UDP-3-O-[(R)-3-hydroxymyristoyl]GlcNAc or to increased transcription of *lpxC*. Rather, the increase in specific activity of the deacetylase is due to increased levels of the enzyme; however, it is not known whether this is a result of increased translation of mRNA or to decreased turnover of the enzyme.

Synthesis of UDP-2,3-Diacylglucosamine

The synthesis of UDP-N^2,O^3-bis[(R)-3-hydroxymyristoyl]-α-D-glucosamine (UDP-2,3-diacylglucosamine) results from the transfer of the (R)-3-hydroxymyristate moiety of (R)-3-hydroxymyristoyl-ACP to the amino group of UDP-3-O[(R)-3-hydroxymyristoyl]glucosamine (UDP-3-monoacylglucosamine) (Fig. 3). This reaction is catalyzed by the enzyme UDP-3-O-monoacylglucosamine N-acyltransferase, which is encoded by the *lpxD* gene (74); the *ssc* gene of *S. enterica* serovar *typhimurium* is the homolog of *lpxD* (74). The *lpxD* gene encodes a 36 kDa protein, and it is located in the same complex operon as *lpxA* and *lpxB* at 4 min on the E. coli chromosome (51,74,75). The *lpxD* gene was originally designated *firA*, and it was discovered based on the observation that mutations in *firA* resulted in the suppression of rifampin resistance in mutants of E. coli that possessed defects in *rpoB*, the gene encoding the β-subunit of RNA polymerase (76). One such mutant allele, the *firA200*(Ts) allele, not only reversed the rifampin resistance of *rpoB* mutants, it also caused temperature-sensitive growth (76–78). Thus, it was postulated that FirA interacted with the β-subunit of RNA polymerase as part of the machinery involved in transcription (76). However, a later comparison of the nucleotide sequences of *firA* and *lpxA* suggested that the products of these genes were significantly homologous (50,51,73,76,79). Indeed, Kelly et al. (74) subsequently demonstrated that lipid A synthesis was temperature sensitive in a mutant of E. coli possessing the *firA200*(Ts) allele. These investigators also provided additional data that established *firA* as the structural gene for the N-acyltransferase, and they renamed the gene *lpxD*.

Although the N-acyltransferase is highly specific for (R)-3-hydroxymyristoyl-ACP as the acyl donor (74), the enzyme will also utilize palmitoyl-ACP; however, it does so at a substantially slower rate (64). Indeed, the utilization of palmitate by LxpD is proabably not of significance since palmitate is not a component of lipid A in E. coli. In addition, the enzyme appears to lack O-acyltransferase activity since mutations in the *lpxA* gene cannot be complemented by *lpxD* (80).

Disaccharide 1-P Synthesis

Both lipid X (2,3-diacylglucosamine 1-P) and UDP-2,3-diacylglucosamine were found to accumulate in phosphatidylglycerol-deficient mutants of E. coli that possessed a mutation in the structural gene for phosphatidylglycerol synthetase (*pgsA*) as well as a second mutation in a gene called *pgsB* (42,43,81,82). Although the accumulation of these compounds was dependent on the defect in *pgsB*, the function of *pgsB* was not immediately recognized. Nevertheless, the structural similarity of these compounds to lipid A suggested that they functioned as early intermediates in lipid A synthesis (81). Subsequent studies demonstrated the conversion of UDP-GlcNAc to UDP-2,3-diacylglucosamine and 2,3-diacylglucosamine 1-P in cell-free extracts of E. coli (55). These studies also demonstrated a precursor-product relationship between UDP-2,3-diacylglucosamine and 2,3-diacylgluucosamine 1-P. Thus, 2,3-diacyglucosamine 1-P results from cleavage between P^1 and P^2 of the pyrophosphate linkage of UDP-2,3-diacylglucosamine (Fig. 4). The physiologically relevant pyrophosphatase that cleaves UDP-2,3-diacylglucosamine has recently been identified as the product of the *lpxH* gene (83). The foregoing observations also suggested that *pgsB* encodes the disaccharide synthase responsible for the synthesis of 2′,3′diacylglucosaminyl-β-(1→6)-2,3-diacylucosamine 1-P (tetraacyldisaccharide 1-P). Indeed, a considerable body of data has been obtained that clearly supports this conclusion. Thus, an in vitro assay for the disaccharide synthase was developed, and the use of this assay revealed that synthesis of tetraacyldisaccharide 1-P occurs as a result of the transfer of 2,3-diacylglucosamine from UDP-2,3-diacylglucosamine to 2,3-diacylglucosamine 1-P (84). In addition, extracts obtained from *pgsB* mutants were found to lack disaccharide synthase activity, whereas enzyme activity was only slightly decreased in *pgsA* mutants (84). Finally, the *pgsB* gene was mapped to 4 min on the *E. coli* chromosome, and it was found to be located immediately downstream of the *lpxA* gene (49,81,85). This locus was given the new designation, *lpxB*, and subcloning and expression of *lpxB* resulted in overexpression of disaccharide synthase activity in strains possessing the cloned fragment (49).

Earlier studies revealed the disaccharide synthase to be a cytosolic enzyme (84). Accordingly, LpxB was subsequently purified from membrane-free extracts of *E. coli*, and analyses of the purified protein indicated that it was a dimer with a subunit molecular mass of 42 kDa (86). In vitro assays of the enzyme using crude cell-free extracts revealed that ADP-, CDP-, and GDP-2,3-diacylglucosamine are not active as substrates (84).

The relationship between mutations in *lpxB* and a further decrease of PG synthesis in *pgsA* mutants is unclear. It has been proposed that perhaps UDP-2,3-diacylglucosamine inhibits PG synthesis by interacting with the product of the mutant *pgsA* allele (84). In contrast, an in vitro association between LpxB and glycerol-3-phosphate dehydrogenase (aerobic) has been demonstrated, and it has been suggested that this association may also occur in vivo, thus providing a link between the synthesis of lipid A and phospholipid (87).

Phosphorylation of Tetraacyldisaccharide 1-P: The 4′-Kinase

A wealth of data have been obtained that clearly demonstrate that the synthesis of tetraacyldisaccharide 1-P is followed by phosphorylation in the 4′-position to yield lipid IV$_A$ (Fig. 4) and that this reaction occurs prior to the incorporation of Kdo. Accordingly, the inability of cells to either synthesize Kdo or to incorporate Kdo into lipid A results in the accumulation of lipid IV$_A$ as the major product (8,9,40,41,88,89).

The 4′-kinase is a membrane-bound enzyme, which requires Mg^{+2} for activity; enzyme activity is also stimulated by phospholipids (particularly cardiolipin) (90). Although the nucleotide triphosphates CTP, GTP, UTP, and ATP are able to function as phosphate donors for the 4′-kinase, maximal rates of phosphorylation occur when ATP is as the phosphate donor (90). In addition, the enzyme can utilize de-*O*-acylated 1-monophosphoryl IV$_A$ as a substrate in vitro; however, it seems likely that tetraacyldisaccharide 1-P is utilized as the substrate in vivo since lipid IV$_A$ accumulates in Kdo-deficient mutants or in cells where the incorporation of Kdo into lipid A has been precluded by the addition to cultures of specific inhibitors.

Recent investigations have resulted in the identification of the structural gene for the 4′-kinase (*lpxK*) in *E. coli* (91). The *lpxK* gene is located immediately downstream of the *msbA* gene in the region of min 21 on the *E. coli* chromosome. Indeed, the *msbA* gene is believed to play a role in LPS transport, and *msbA* together with *lpxK* form an operon. Analyses of the deduced amino acid sequence of LpxK indicate that it shares no obvious homology with other kinases. Membranes from LpxK overproducing strains are extremely useful for the efficient synthesis of [4′-^{32}P]-lipid IV$_A$, a valuable probe for examining the late steps of lipid A biosynthesis and for detection of endotoxin-binding proteins.

Biosynthesis and Incorporation of Kdo

The synthesis of Kdo occurs as the result of three sequential reactions (92–94); these reactions are as follows:

a. D-Ribulose-5-phosphate ⇌ D-Arabinose-5-phosphate
b. D-Arabinose-5-phosphate + Phosphoenolpyruvate → Kdo-8-phosphate
c. Kdo-8-phosphate → Kdo + Pi

Reactions (a), (b), and (c) are catalyzed by the enzymes D-ribulose-5-phosphate isomerase, Kdo-8-phosphate synthetase, and Kdo-8-phosphate phosphatase, respectively. Free Kdo is then converted to CMP-Kdo, the nucleotide-sugar donor of Kdo residues for LPS synthesis, by the enzyme CTP:CMP-3-deoxy-D-*manno*-octulosonate cytidylyltransferase (CMP-Kdo synthetase) (95). The synthesis of CMP-Kdo is somewhat unusual since prior activation of free Kdo by phosphorylation is not required, and the overall reaction is as follows:

$$CTP + Kdo \rightleftharpoons CMP\text{-}Kdo + PPi$$

The β-pyranose form of free Kdo is the preferred substrate for the CMP-Kdo synthetase, and the reaction proceeds by a mechanism that results in retention of the β-configuration in CMP-Kdo (96). The structural gene for CMP-Kdo synthetase (*kdsB*) is located at min 20.5 on the *E. coli* chromosome (97), and analysis of the nucleotide sequence of *kdsB* predicts that the enzyme consists of 248 amino acids and has a molecular weight of 27,486 daltons (98,99).

An interesting aspect of LPS synthesis is the fact that the synthesis of the Kdo-containing region of the molecule occurs prior to the completion of lipid A synthesis (3). Specifically, the synthesis of the Kdo region of the inner core occurs prior to the complete incorporation of fatty acyl substituents into lipid A, and it has been well established that the underacylated lipid A precursor, lipid IV_A, is the initial acceptor of Kdo residues (Fig. 5). Indeed, the in vivo accumulation of lipid IV_A under conditions where the incorporation of Kdo has been precluded by mutations or by specific inhibitors is in agreement with this conclusion.

Munson et al. (39) were the first to demonstrate the in vitro transfer of Kdo from CMP-Kdo into lipid IV_A. These investigators used detergent-solubilized cell envelope preparations of *S. enterica* serovar *typhimurium* as a source of Kdo transferase activity. Although the mature lipopolysaccharide from some serovars of *S. enterica* has been reported to contain three Kdo residues (100–102), the product of the in vitro reaction contained only two Kdo residues, and no data to explain the lack of incorporation of the third Kdo residue were obtained. In addition, this investigation did not establish if the incorporation of the Kdo residues occurred as the result of the successive incorporation of two Kdo residues—catalyzed either by two separate enzymes or by a single bifunctional enzyme—or alternatively, if the product resulted from synthesis of a Kdo disaccharide followed by the *en bloc* transfer of the disaccharide to lipid IV_A. It should be noted that $(Kdo)_2$-IV_A has also been generated in vivo, and evidence has been obtained that indicates that this acyl-deficient molecule is capable of functioning in vivo as an acceptor of sugars for core synthesis (103,104).

The structural gene for the Kdo-transferase (*waaA*, formerly *kdtA*) encodes a 43 kDa protein, which has been purified to near homogeneity (89,105). Studies utilizing the purified transferase clearly demonstrated that the enzyme is bifunctional, and it catalyzes the successive transfer of two Kdo residues from CMP-Kdo to lipid IV_A to yield $(Kdo)_2$-IV_A (Fig. 5) (105).

Indeed, the enzyme was shown to catalyze the incorporation of a single Kdo residue at low concentrations of CMP-Kdo, and conversion of $(Kdo)_1$-IV_A to $(Kdo)_2$-IV_A was observed following reincubation of the isolated $(Kdo)_1$-IV_A with enzyme and CMP-Kdo. In addition, the LPS of *Haemophilus influenzae* contains a single Kdo linked to the 6′-position of lipid A, and the Kdo-transferase of this organism is monofunctional (106). Thus, the in vitro synthesis of $(Kdo)_1$-IV_A has been demonstrated using extracts of *H. influenzae* as the source of Kdo-transferase, and this product was found to be an excellent substrate for the addition of a second Kdo catalyzed by the bifunctional transferase of *E. coli* (106). The presence of the 4′-phosphate of lipid IV_A was shown to be an absolute requirement for the incorporation of Kdo catalyzed by the *E. coli* transferase (107); however, the substrate specificity of the enzyme with respect to the degree of acylation of the acceptor appears to be somewhat less stringent since the enzyme will recognize substrates containing a fifth or sixth fatty acyl substituent (105). Nevertheless, maximal activity occurs when lipid IV_A, which is the naturally occurring substrate, is used as the acceptor of Kdo residues. In this regard, the early steps of lipid A synthesis in *P. aeruginosa* differ from those that have been demonstrated in *E. coli* or *S. enterica* serovar *typhimurium* since the incorporation of a full complement of fatty acids into *P. aeruginosa* lipid A appears to occur prior to the incorporation of Kdo (108).

The LPS of *Chlamydia trachomatis* is structurally analagous to the Re LPS that occurs in heptoseless mutants of *E. coli* and *S. enterica* serovar *typhimurium* with the exception that the *C. trachomatis* LPS contains three or more Kdo residues (109,110), whereas the heptoseless mutants contain only two Kdo residues (18,24,101,102,111). The *gseA* gene of *C. trachomatis* is the genetic homolog of the *waaA* gene, and the deduced amino acid sequence of GseA reveals a 23% identity and 66% similarity with *E. coli* WaaA (112). The enzyme encoded by *gseA* consists of a single subunit that functions as a trifunctional Kdo-transferase (112). Thus, GseA has been demonstrated to catalyze the in vitro incorporation of a third Kdo from CMP-Kdo into $(Kdo)_2$-IV_A, and expression of *gseA* in an *E. coli* mutant possessing a temperature sensitive Kdo-transferase enables extracts of the mutant to incorporate all three Kdo residues into lipid IV_A. Furthermore, expression of *gseA* in *E. coli* or *S. enterica* serovar *typhimurium* in vivo appears to result in the synthesis of $(Kdo)_3$-lipid A as well as native LPS (109,113). The incorporation of the third Kdo residue apparently precludes transfer of heptose to the 5-position of the innermost Kdo residue. It is also important to note that multifunctional Kdo-transferases have been demonstrated to occur in other members of the genus *Chlamydia*. Indeed, *C. pneumoniae* possesses a trifunctional *gseA* gene product (144), whereas the Kdo-transferase of *C. psittaci* appears capable of catalyzing the synthesis of a Kdo tetrasaccharide (145).

Mutants of *E. coli* conditionally defective in Kdo-transferase activity accumulate lipid IV$_A$ under nonpermissive conditions (10,89). As expected, the phenotype of these mutants is the same as that of mutants conditionally defective in KdsA and KdsB, and expression of the defect in *waaA* results in growth stasis. These observations support the conclusion that the incorporation of Kdo into lipid A is essential for normal cell growth and function. Indeed, this conclusion has been further supported by the results of experiments obtained with a *waaA::kan* insertion mutant of *E. coli* (10). The ability of the insertion mutant to grow was dependent on expression of the wild-type *waaA* allele located on a temperature-sensitive plasmid, and growth cessation and accumulation of lipid IV$_A$ resulted if replication of the plasmid was precluded by a shift to nonpermissive temperature. The mutant could also be complemented with the *gseA* gene of *C. trachomatis*, and expression of *gseA* resulted in the synthesis of LPS bearing the Kdo-trisaccharide characteristic of the *Chlamydia*.

The monofunctional Kdo transferase encoded by the *H. influenzae waaA* gene can support the growth of *E. coli* strains in which the endogenous, bifunctional transferase is inactivated. However, such constructs grow very slowly, and they are temperature sensitive (114).

Synthesis of Acyloxyacyl Substituents

Lipid IV$_A$ is the acceptor of Kdo residues for LPS synthesis in *E. coli* and *S. enterica* serovar *typhimurium*. Accordingly, synthesis of the Kdo region of the inner core occurs prior to the incorporation of acyloxyacyl-linked lauroyl and myristoyl substituents. The in vitro incorporation of laurate and myristate into (Kdo)$_2$-lipid IV$_A$ by extracts of *E. coli* was first demonstrated by Brozek et al. (Fig. 6) (115). These investigators established that the donors of laurate and myristate were lauroyl- and myristoyl-ACP; thioesters of coenzyme A are not able to function as donors. In addition, the acyltransferases are not able to utilize lipid IV$_A$ as an acceptor. The structural gene for the *E. coli* lauroyl transferase (*waaM*, formerly *htrB*) is located at min 23.4 on the *E. coli* chromosome (116), whereas the structural gene for the myristoyl transferase (*waaN*, formerly *msbB*) is located at min 40.5 (117). The deduced amino acid sequence of the *waaM* and *waaN* gene products indicate that they are membrane proteins with molecular masses of 35,407 daltons (116,118) and 37,410 daltons (117), respectively. Analyses of purified WaaM verified that the enzyme catalyzes the transfer of lauroyl residues from lauroyl-ACP to (Kdo)$_2$-lipid IV$_A$ (Fig. 6) (119). These studies also showed that (Kdo)$_1$-lipid IV$_A$ and lipid IV$_A$ were not acceptors and that the rate of incorporation of laurate was considerably faster than was the rate of incorporation of myristate or decanoate (119). The incorporation of lauroyl residues into lipid A appears to precede the incorporation of myristoyl residues. Accordingly, partially purified WaaN was demonstrated to catalyze the in vitro incorporation of myristoyl residues into (Kdo)$_2$-lauroyl-lipid IV$_A$ at a rate 100 times faster than was the case when (Kdo)$_2$-lipid IV$_A$ was used as an acceptor (120). In addition, the lipid A synthesized by *waaN*-insertion mutants possess an acyloxyacyl-linked lauroyl group (121–123). The WaaN transferase appears capable of utilizing either lauroyl- or myristoyl-ACP as the acyl donor; however, decanoyl-, palmitoyl-, palmitoleoyl-, and (*R*)-3-hydroxymyristoyl-ACP are very poor donors. The locations of the lauroyl and myristoyl residues of in vitro synthesized products have not been firmly established; however, based on the structures of *E. coli* and *S. enterica* serovar *typhimurium* lipid A, it seems likely that they are linked to the hydroxyl substituents of the (*R*)-3-hydroxymyristoyl residues located at the 2'- and 3'-positions, respectively (20,24).

A functional *waaM* gene product is essential for cell viability above 32°C, and it appears that WaaM is somehow involved in coupling phospholipid synthesis, LPS synthesis, and growth rate. Accordingly, the LPS molecules of cells possessing an insertionally inactivated *waaM* gene are depleted in both laurate and myristate at both 30 and 42°C (121). However, such mutants lose viability when grown in rich media at temperatures greater than 33°C, and cell death is accompanied by filamentation and bulging at potential septa sites or at the ends of the cell (116,118). In contrast, WaaM appears to be dispensable at slow growth rates regardless of the temperature. Thus, when cells possessing an insertionally inactivated *waaM* gene are grown at 42°C in defined media containing a poor carbon source, generation times of greater than 70 minutes result without attendant morphological aberrations or a loss in viability (116,118). When *waaM* null mutants are shifted from permissive conditions (33°C, rich media) to nonpermissive conditions (42°C, rich media), they continue to grow at the same rate, but they synthesize phospholipids at the rate required for growth at 42°C, resulting in a greatly increased phospholipid:protein ratio (121). Spontaneous extragenic suppressor mutations have been isolated in *accA* and *accB* genes, which allow *waaM* null mutants to grow at elevated temperature in rich media (121); these genes encode the biotin carboxyl carrier protein and biotin carboxylase subunits of acetyl coenzyeme A carboxylase complex, respectively. The mutations in *accA* and *accB* result in a decreased rate of phospholipid synthesis, which apparently restores the balance between phospholipid synthesis and growth rate at elevated temperature in rich media.

Insertions in *waaN* result in myristate-deficient lipid A, but unlike null mutations in *waaM*, these mutations are not conditionally lethal (121–123). In addition, overexpression of *waaN* can also suppress the temperature-sensitive phenotype of *waaM* mutants (117,120). The overexpression of *waaN* apparently compensates for the slow rate of acylation of $(Kdo)_2$–lipid IV_A by this enzyme, and considerable synthesis of penta-acylated lipid A has been detected following overproduction of WaaN in *waaM* insertion mutants (120). Although it has been assumed that the acyloxyacyl substituent present in the penta-acylated lipid A contains a myristate, the fatty acyl composition of the penta-acylated lipid A has not been determined. In any event, the available data suggest that lipid A must possess at least one acyloxyacyl substituent in order for rapid growth to occur at temperatures above 33°C.

The temperature-sensitive phenotype of *waaM* null mutants can also be suppressed by multiple copies of the *msbA* gene (124). The *msbA* gene maps to 20.5 min on the *E. coli* chromosome, and it appears to encode an integral membrane protein of 64,460 daltons. The primary amino acid sequence of MsbA is very similar to a number of ATP-dependent translocator proteins, especially the Mdr proteins of animal cells; however, the function of MsbA is not yet fully established (124). The *msbA* gene and a downstream gene, *orfE*, appear to constitute an operon that is essential for bacterial viability (124). Accordingly, *msbA*::Ωcam insertions can only be introduced into cells that are diploid for both *msbA* and *orfE*; the insertion in *msbA* presumably exerts a polar effect on *orfE* gene expression. Indeed, *orfE* is now known to be the same as *lpxK*, the structural gene for the 4′-kinase (91). Recent experiments have demonstrated that lipopolysaccharide accumulates in the inner membranes of *waaM* mutants, and partial restoration of LPS transport has been observed following introduction of *msbA* (125,126). Thus, it is tempting to speculate that the *msbA* gene product—and possibly the *lpxK* gene product—function in the translocation of LPS from the cytoplasmic membrane to the outer membrane; however, the function of these gene products in transport remains to be validated by more direct biochemical assays.

The structure of the lipid A of *P. aeruginosa* differs from that of *E. coli* and other gram-negative enteric organisms (13). Accordingly, *P. aeruginosa* lipid A contains *N*-linked (*R*)-3-hydroxylaurate at positions 2 and 2′ and *O*-linked (*R*)-3-hydroxydecanoate at positions 3 and 3′. In addition, the acyloxyacyl substituents of this *P. aeruginosa* lipid A are located at the 2- and 2′-positions, and the predominant acyloxy substituent is lauric acid. The sequence of reactions involved in lipid A synthesis in *P. aeruginosa* also appears to differ from the sequence of reactions demonstrated in *E. coli*. Thus, the complete fatty acylation of lipid A in *P. aeruginosa* occurs prior to the incorporation of Kdo residues (108). Indeed, Mohan and Raetz (127) demonstrated the in vitro transfer of laurate from lauroyl-ACP into lipid IV_A in extracts of *P. aeruginosa*, but acylation of $(Kdo)_2$-lipid IV_A was also possible. In addition, only one laurate was incorporated, and although the structure of the reaction product was not determined, the laurate was presumed to be incorporated as an acyloxyacyl substituent. In this regard it should be noted that *E. coli* lipid IV_A was used as the acceptor in these studies, and it is possible that incorporation of a second laurate acyloxy substituent would have occurred if the *P. aeruginosa* equivalent of *E. coli* lipid IV_A had been used as an acceptor.

The accumulation of lipid X in *lpxB* mutants of *E. coli* is accompanied by the accumulation of a related compound, lipid Y, containing palmitate as an acyloxy substituent linked to the *N*-linked (*R*)-3-hydroxymyristoyl group (Fig. 4) (42,43,45). A related lipid, lipid IV_B, has also been isolated from mutants of *S. enterica* serovar *typhimurium* conditionally defective in the synthesis of Kdo (40,41). Lipid IV_B is a structural analog of lipid IV_A, it contains all of the structural features of lipid IV_A plus a palmitoyl residue linked to the (*R*)-3-hydroxymyristoyl group located at the 2-position of the reducing terminal glucosamine moiety. The available data clearly indicate that lipid Y is derived from lipid X. Accordingly, Brozek et al. (128) demonstrated the in vitro synthesis of lipid Y utilizing lipid X as an acceptor. Unlike the synthesis of lauroyl- and myristoyl-containing acyloxyacyl substituents, the acyl donor for the synthesis of lipid Y was found to be glycerophospholipid. The membrane-bound acyl transferase does not preferentially use a particular class of phospholipid as an acyl donor; however, it appears to specifically transfer palmitoyl residues that are located in the S_N1 position. The enzyme is also capable of utilizing lipid IV_A as an acceptor of palmitoyl residues, but it will not catalyze the incorporation of palmitate into the de-O-acylated derivative of lipid X. Accordingly, the presence of an (R)-3-hydroxymyristoyl moiety in the 3-position appears to be essential. In this regard, it is not known if the synthesis of lipid IV_B results from the transfer of palmitate from glycerophospholipid into lipid IV_A or, alternatively, if palmitate is incorporated at the level of lipid Y. Furthermore, the functions of lipids Y and IV_B remain to be established. The mature lipid A molecules isolated from cultures of *E. coli* and *S. enterica* serovar *typhimurium* grown under standard laboratory conditions are essentially devoid of palmitate. Thus, as suggested by Brozek et al. (128), these molecules may be the products of side reactions that occur in conditional mutants following a shift to nonpermissive conditions. Alternatively, the compositions of lipid A molecules obtained from cells grown under a variety of environmental conditions not normally encountered in the laboratory have not been determined. Thus, the palmitoyltransferase may become activated when bacteria are present in human or animal hosts or when they encounter stressful conditions.

Polar Substituents Attached to Lipid A

As shown in Figure 2, a portion of the lipid A moieties found in wild-type *E. coli* K-12 may be substituted with a pyrophosphate residue at position 1 rather than with the more abundant monophosphate (111). The origin of this pyrophosphate is unknown. Recent studies in mutants with reduced levels of the MsbA protein (125) (an ABC transporter involved in lipid A transfer across the inner membrane) suggest that the formation of the pyrophosphate occurs in conjunction with transport (126). When MsbA levels are low, hexa-acylated lipid A moieties on nascent LPS accumulate in the inner membrane, but only the monophosphate form is present (126). If the extra phosphate is indeed added on the periplasmic surface of the inner membrane or in the outer membrane, then ATP is unlikely to be the phosphate donor. Other high-energy compounds, such as bactoprenyl pyrophosphate (3) or diacylglycerol pyrophosphate (129), would have to be considered, since they are more likely to gain access to the outer layers of the cell envelope.

In strains of *Salmonella* and many other types of gram-negative organisms, lipid A is modified with substituents that reduce its net negative charge (13,130,131). L-4-Deoxy-4-aminoarabinose (4-aminoarabinose) and phosphoethanolamine are especially common (Fig. 7) (13,130,131). 4-Aminoarabinose is attached by means of a phosphodiester linkage to its anomeric carbon, and phosphoethanolamine is usually attached as a pyrophosphate (13,130,131). High levels of modification with these substituents are thought to confer resistance to polymyxin, a basic peptide with a high affinity for lipid A (132–134).

Recent work with *Salmonella enterica* serovar *typhimurium* suggests that modification with 4-aminoarabinose may also be necessary for intracellular survival in the presence of basic antibacterial proteins and for virulence (135). In this organism, 4-aminoarabinose modification is under the control of the *phoP* regulatory system (135). Additional modifications of *S. enterica* serovar *typhimurium* lipid A with palmitate and with 2-hydroxymyristate (Fig. 7) are similarly elevated in *phoP* constitutive mutants and, like 4-aminoarabinose, are absent in *phoP*-defective strains (135).

The biosynthetic origins of the 4-aminoarabinose and the ethanolamine phosphate groups are unknown. The phosphoethanolamine might be derived from phosphatidylethanolamine (136). The 4-aminoarabinose might be synthesized from UDP-glucuronic acid, assuming that the pathways for the biosynthesis of UDP-xylose and UDP-L-arabinose in plants are a relevant precedent (137,138). No one has actually investigated the ability of UDP-glucuronic acid–deficient (ugd^-) bacteria (139) to make 4-aminoarabinose.

Most investigators suggest that ethanolamine pyrophosphate, when present on lipid A, is attached to the 1-position and that 4-aminoarabinose is linked to the 4′-position, as shown in Figure 7 (130,131,133,134). However, there is some uncertainty about the generality of this claim. In Kdo-deficient strains of *S. enterica* serovar *typhimurium*, the lipid IV_A that accumulates under nonpermissive conditions may be extensively decorated with 4-aminoarabinose and/or phosphoethanolamine (40,41). In this case, which is one of the best characterized, all the 4-aminoarabinose is attached to position 1 and all the phosphoethanolamine is on position 4′ (40,41). This interesting discrepancy needs to be resolved, especially as the more generally accepted pattern of substitution (Fig. 7) has been confirmed in the case of *Salmonella enterica* serovar *minnesota* R595 LPS using the same techniques employed with the Kdo-deficient strains of *S. enterica* serovar *typhimurium* (Z. Zhou and C. R. H. Raetz, unpublished). One explanation is that the 4-aminoarabinose and the phosphoethanolamine switch places after Kdo transfer and late acylation are completed. Another scenario is

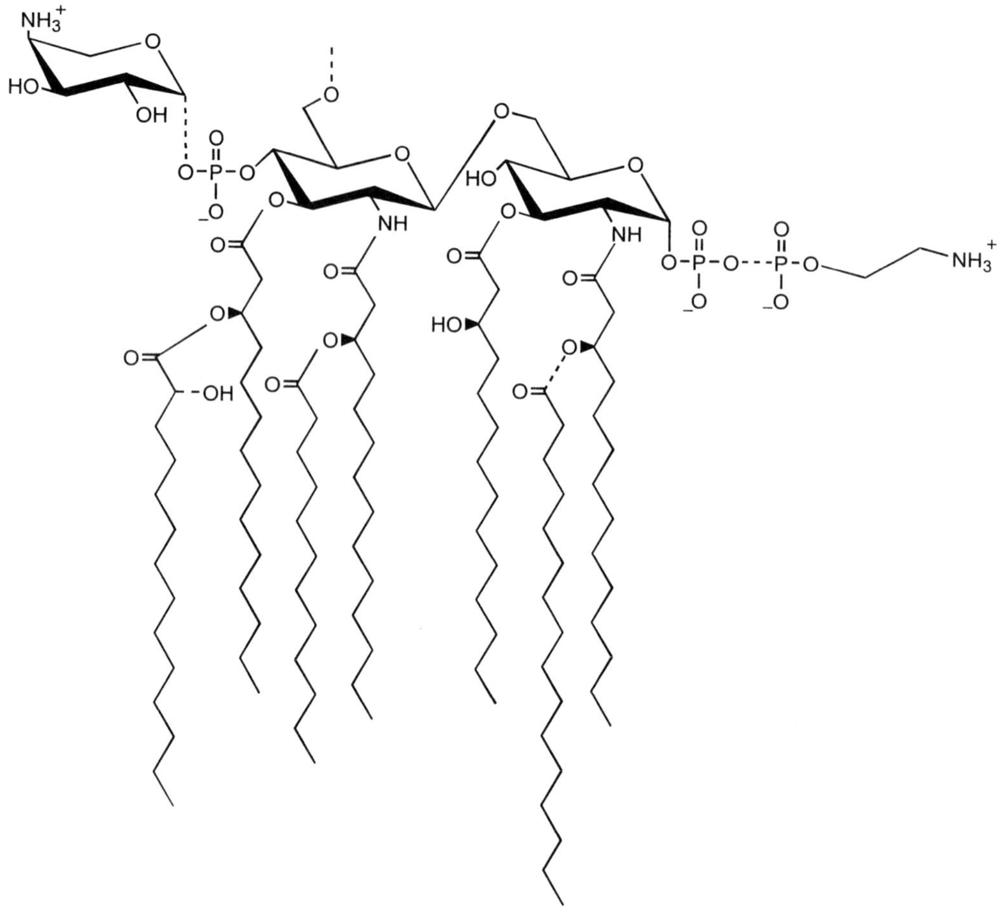

Fig. 7 Partial substitutions of *Salmonella tyhimurium* lipid A with polar headgroups and other moieties. The partial substitutions with L-4-aminoarabinose, 2-hydroxymyristate, and palmitate seem to be under *phoP* control (135). The regulation of ethanolamine pyrophosphate formation has not yet been characterized. In Kdo-deficient mutants, the lipid IV$_A$ derivatives that accumulate under nonpermissive conditions may also be decorated with L-4-aminoarabinose and phosphoethanolamine, but in that case the locations (1 vs. 4' phosphates) are reversed (41).

that *S. enterica* serovar *typhimurium* has different enzymes for attaching these substituents to its lipid A than does *S. enterica* serovar *minnesota*. These remaining uncertainties regarding the structure and biosynthesis of lipid A are worthy of clarification, given the importance of 4-aminoarabinose substitution in mediating polymyxin resistance and virulence (133–135).

CONCLUSIONS AND FUTURE DIRECTIONS

Although most of the key reactions of lipid A biosynthesis have been elucidated in *E. coli* (6), there remain many interesting variations on the general theme of lipid A structure and assembly that need to be studied in other organisms. Progress with the biochemistry of diverse gram-negative systems will be greatly facilitated by genome sequencing. Opportunities to generate bacteria with hybrid lipid A structures, as illustrated by the studies of *E. coli lpxA* mutants complemented with *Pseudomonas* (62) or *Neisseria lpxA* (61) will be facilitated. By modifying the structure of lipid A in living cells, the function of lipid A in outer membrane assembly and in pathogenesis may finally be explained. For instance, *S. enterica* serovar *typhimurium msbB* mutants with only five acyl chains on their lipid A are orders of magnitude less lethal to infected mice than are wild-type bacteria, even though these mutants replicate in mice as rapidly as the wild-type organism (140).

The regulation of lipid A biosynthesis and the mechanisms of LPS secretion and assembly into outer membranes deserve considerable further study. If MsbA indeed functions in the initial step of core lipid A transport across the inner membrane (125), it may be possible to find other components of the transport machinery that interact with MsbA by using biochemical and genetic approaches. With regard to regulation, the *phoP* connection (135) and the effects of cold shock (which causes palmitoleate to be incorporated into lipid A in place of laurate) (141–143) are especially intriguing.

The recently determined genomic sequence of the hyperthermophilic bacterium, *Aquifex aeolicus*, is especially provacative from an evolutionary perspective (148). *A. aeolicus*, an organism isolated from hot springs, grows at 95°C under pressure on a mixture of H_2, CO_2, and O_2. Ribosomal RNA sequencing indicates that *A. aeolicus* diverged very early during the evolution of bacteria from the Archaea and Eukaryotae. Although no actual biochemical studies of LPS have been done using the *Aquifex* system, the fact that many of the genes encoding the enzymes of lipid A, Kdo, heptose, and outer core biosynthesis are well conserved suggests that lipid A appeared very early during membrane evolution. Presumably, outer membranes helped early bacterial species survive in hostile environments. Whatever the explanation, the availability of thermostabile variants of the enzymes of lipid A biosynthesis should greatly facilitate the preparation of lipid A analogs that are not readily accessible by chemical synthesis. The complete characterization of lipid A biosynthesis in diverse bacterial systems, including its underlying molecular and structural biology, will also enable the preparation of novel, radioactive endotoxinlike molecules with which to probe for lipid A receptors in animal systems.

ACKNOWLEDGMENTS

This work was supported by Uniformed Services University of the Health Sciences Grant R07382 and by U.S. Public Health Service Grant GM52882 from the National Institutes of Health to P.D.R. and by U. S. Public Health Service Grants GM51310 and GM51796 from the National Institutes of Health to C.R.H.R.

REFERENCES

1. Galloway SM, Raetz CRH. A mutant of *Escherichia coli* defective in the first step of endotoxin biosynthesis. J Biol Chem 1990; 265:6394–6402.
2. Raetz CRH. Molecular genetics of membrane phospholipid synthesis. Annu Rev Genet 1986; 20:253–295.
3. Rick PD. Lipopolysaccharide synthesis. In: Neidhardt FC, ed. *Escherichia coli* and *Salmonella typhimurium*. Cellular and Molecular Biology. Vol. 1. Washington, DC: ASM Publications, 1987:648–662.
4. Schnaitman CA, Klena JD. Genetics of lipopolysaccharide biosynthesis in enteric bacteria. Microbiol Rev 1993; 57:655–682.
5. Reeves P. Biosynthesis and assembly of lipopolysaccharide. In: Neuberger A, van Deenen LLM, eds. Bacterial Cell Wall: New Comprehensive Biochemistry. Vol. 27. New York: Elsevier Science, 1994:281–314.
6. Raetz CRH. Bacterial lipopolysaccharides: a remarkable family of bioactive macroamphiphiles. In: Neidhardt FC, ed. *Escherichia coli* and *Salmonella typhimurium*. Cellular and Molecular Biology. Vol. 1. Washington, DC: ASM Publications, 1996:1035–1063.
7. Morrison DC, Dinarello CA, Munford RS, Natanson C, Danner R, Pollack M, Spitzer JJ, Ulevitch RJ, Vogel SN, McSweegen E. Current status of bacterial endotoxins. ASM News 1994; 60:479–484.
8. Rick PD, Osborn MJ. Lipid A mutants of *Salmonella typhimurium*. Characterization of a conditional lethal mutant in 3-deoxy-D-mannooctulosonate-8-phosphate synthetase. J Biol Chem 1977; 252:4895–4903.
9. Lehmann V, Rupprecht E, Osborn MJ. Isolation of mutants conditionally blocked in the biosynthesis of the 3-deoxy-D-mannooctulosonic-acid part of lipopolysaccharide derived from *Salmonella typhimurium*. J Biol Chem 1977; 76:41–49.
10. Belunis CJ, Clementz T, Carty SM, Raetz CRH. Inhibition of lipopolysaccharide biosynthesis and cell growth following inactivation of the *kdtA* gene of *Escherichia coli*. J Biol Chem 1995; 270:27646–27652.
11. Raetz CRH. Structure and biosynthesis of lipid A. In: Neidhardt FC, ed. *Escherichia coli* and *Salmonella typhimurium*. Cellular and Molecular Biology. Vol. 1. Washington, DC: ASM Publications, 1987:498–503.
12. Raetz CRH. Bacterial endotoxins: extraordinary lipids that activate eucaryotic signal transduction. J Bacteriol 1993; 175:5745–5753.
13. Raetz CRH. Biochemistry of endotoxins. Annu Rev Biochem 1990; 59:129—170.
14. Karibian D, Deprun C, Caroff M. Comparison of lipids A of several *Salmonella* and *Escherichia* strains by ^{252}Cf plasma desorption mass spectrometry. J Bacteriol 1993; 175:2988–2993.
15. Imoto M, Kusumoto S, Shiba T, Rietschel ET, Galanos C, Lüderitz O. Tetrahedron Lett 1985; 26:907–908.
16. Rietschel Et, ed. Handbook of Endotoxin. Vol. 1: Chemistry of Endotoxin. Amsterdam: Elsevier Biomedical Press, 1984.
17. Rosner MR, Khorana HG, Satterhwait AG. The structure of lipopolysaccharide from a heptose-less mutant of *Escherichia coli* K-12. II. The application of ^{31}P-NMR spectroscopy. J Biol Chem 1979; 254:5918–5925.
18. Strain SM, Fesik SW, Armitage IM. Characterization of lipopolysaccharide from a heptoseless mutant of *Escherichia coli* by carbon 13 nuclear magnetic resonance. J Biol Chem 1983; 258:2905–2910.
19. Imoto M, Kusumoto S, Shiba T, Naoki H, Iwashita T, Rietschel ET, Wollenweber H-W, Galanos, C, Lüderitz

O. Chemical structure of *E. coli* lipid A: linkage site of acyl groups in the disaccharide backbone. Tetrahedron Lett 19983; 24:4017–4020.
20. Qureshi N, Takayama K, Heller D, and Fenselau C. Position of ester groups in the lipid A backbone of lipopolysaccharide obtained from *Salmonella typhimurium*. J Biol Chem 1983; 258:12947–12951.
21. Rietschel ET, Wollenweber H-W, Sidorczyk Z, Zähringer U. Lüderitz O. Analysis of the primary structure of lipid A. In: Anderson L, Unger FM, eds. Bacterial Lipopolysaccharides. Structure, Synthesis, and Biological Activities. Washington, DC: American Chemical Society, 1983:214.
22. Takayama K, Qureshi N, Mascagni P. Complete structure of lipid A obtained from the lipopolysaccharides of the heptoseless mutant of *Salmonella typhimurium*. J Biol Chem 1983; 258:12801–12803.
23. Takayama K. Qureshi N, Mascagni P, Nashed MA, Anderson L, Raetz CRH. Fatty acyl derivatives of glucosamine 1-phosphate in *Escherichia coli* and their relation to lipid A. Complete structure of a diacyl GlcN-1-P found in a phosphatidylglycerol-deficient mutant. J Biol Chem 1983; 258:7379–7385.
24. Qureshi N, Takayama K, Mascagni P, Honovich J, Wong R, Cotter RJ. Complete structural determination of lipopolysaccharide obtained from deep rough mutant of *Escherichia coli*. Purification by high performance liquid chromatography and direct analysis by plasma desorption mass spectrometry. J Biol Chem 1988; 263:11971–11976.
25. Imoto M, Yoshimura H, Kusumoto S, Shiba T. Total synthesis of lipid A, active principle of bacterial endotoxin. Proc Jpn Acad 1984; 60B:285–288.
26. Galanos C, Lüderitz O, Rietschel E Th, Westphal O, Brade H. Brade L, Freudenberg M, Schade U, Imoto M, Yoshimura H, Kusumoto S, Shiba T. Synthetic and natural *Escherichia coli* free lipid A express identical endotoxic activities. Eur J Biochem 1985; 148:1–5.
27. Bhatt UR, Forsberg LS, Carlson RW. Structure of lipid A component of *Rhisobium luguminosarum* bv. *phaseoli* lipopolysaccharide. Unique nonphosphorylated lipid A containing 2-amino-2-deoxygluconate, galacturonate, and glucosamine. J Biol Chem 1994; 269:14402–14410.
28. Ogawa T. Chemical structure of lipid A from *Porphyromonas* (*Bacteroides*) *gingivalis* lipopolysaccharide. FEBS Lett 1993; 332:197–201.
29. Weintraub A, Zähringer U, Wollenweber H-W, Seydel U, Rietschel ET. Structural characterization of the lipid A component of *Bacteriodes fragilis* strain NCTC 9343 lipopolysaccharide. Eur J Biochem 1989; 183:425–431.
30. Karunaratne DN, Richards JC, Hancock REW. Characterization of lipid A from *Pseudomonsa aeruginosa* O-antigenic B band lipopolysaccharide by 1D and 2D NMR and mass spectral analysis. Arch Biochem Biophys 1992; 299:368–376.
31. Kulshin VA, Zähringer U, Lindner B, Jäger K.-E, Dmitriev BA, Rietschel ET. Structural characterization of the lipid A component of *Pseudomonas aeruginosa* wild-type and rough mutant lipopolysaccharide. Eur J Biochem 1991; 198:697–704.
32. Qureshi N, Honovich JP, Hara H, Cotter RJ, and Takayama. Location of fatty acids in lipid A obtained from lipopolysaccharide of *Rhodopseudomonas sphaeroides* ATCC 17023. J Biol Chem 1988; 263:5502–5504.
33. Salimath PV, Tharanathan RN, Weckesser J, Mayer H. Structural studies on the non-toxic lipid A from *Rhodopseudomonas sphaeroides* ATCC 17023. Eur J Biochem 1983; 136:195–200.
34. Salimath PV, Tharanathan RN, Weckesser J, and Mayer H. The Structure of the polysaccharide moiety of *Rhodopseudomonas sphaeroides* ATCC 17023 lipopolysaccharide. Eur J Biochem 1984; 144:227–232.
35. Mayer H, Salimath PV, Holst O, Weckesser J. Unusual lipid A types in phototropic bacteria and related species. Rev Infect Dis 1984; 6:542–545.
36. Mayer H, Bock E, Weckesser J. 2,3-Diamino-2,3-dideoxyglucose containing lipid A in the *Nitrobacter* strain X_{14}. FEMS Microbiol Lett 1983; 17:93–96.
37. Eidels L, Osborn MJ. Lipopolysaccharide and aldoheptose biosynthesis in transketolase mutants of *Salmonella typhimurium*. Proc Natl Acad Sci USA 1971; 68:1673–1677.
38. Rick PD, Osborn MJ. Isolation of a mutant of *Salmonella typhimurium* dependent on D-arabinose-5-phosphate for growth and synthesis of 3-deoxy-D-mannooctulosonate (ketodeoxyoctonate). Proc Natl Acad Sci USA 1972; 69:3756–3760.
39. Munson RS Jr, Rasmussen NS, Osborn MJ. Biosynthesis of lipid A. Enzymatic incorporatiion of 3-deoxy-D-mannooctulosonate into a precursor of lipid A in *Salmonella typhimurium*. J Biol Chem 1978; 253:1503–1511.
40. Raetz CRH, Purcell S, Meyer MV, Qureshi N, Takayama K. Isolation and characterization of eight lipid A precursors from a 3-deoxy-D-*manno*-octulosonic acid-deficient mutant of *Salmonella typhimurium*. J Biol Chem 1985; 260:16080–16088.
41. Strain SM, Armitage IM, Anderson L, Takayama K. Qureshi N, Raetz CRH. Location of polar substituents and fatty acyl chains on lipid A precursors from a 3-deoxy-D-*manno*-octulosonic acid-deficient mutant of *Salmonella typhimurium*. Studies by ^{1}H, ^{13}C, and ^{31}P nuclear magnetic resonance. J Biol Chem 1985; 260:16089–16098.
42. Nishijima M, Raetz CRH. Membrane lipid biogenesis in *Escherichia coli*: identification of genetic loci for phosphatidylglycerophosphate synthetase and construction of mutants lacking phosphatidylglycerol. J Biol Chem 1979; 254:7837–7844.
43. Nishijima M, Raetz CR. Characterization of two membrane-associated glycolipids from an *Escherichia coli* mutant deficient in phosphatidylglycerol. J Biol Chem 1981; 256:10690–10696.
44. Takayama K. Qureshi N, Mascagni P, Anderson L, Raetz CRH. Glucosamine-derived phospholipids in *Escherichia coli*. Structure and chemical modification of a triacyl glucosamine 1-phosphate found in a phosphatidylglycerol-deficient mutant. J Biol Chem 1983; 258:14245–14252.
45. Park JT. Murein synthesis. In: Neidhardt FC, ed. *Escherichia coli* and *Salmonella typhimurium*. Cellular and

Molecular Biology. Vol. 1. 2d ed.Washington, DC: ASM Publications, 1987:663–671.
46. Rick PD, Silver RP, Enterobacterial common antigen and capsular polysaccharides. In: Neidhardt FC, ed. *Escherichia coli* and *Salmonella typhimurium*. Cellular and Molecular Biology. Vol. 1. 2nd ed. Washington, DC: ASM Publications, 1996:104–122.
47. Barr K, Rick PD. Biosynthesis of enterobacterial common antigen in *Escherichia coli*. In vitro synthesis of lipid-linked intermediates. J Biol Chem 1987; 262:7142–7150.
48. Anderson MS, Raetz CRH. Biosynthesis of lipid A precursors in *Escherichia coli*. A cytoplasmic acyltransferase that converts UDP-*N*-acetylglucosamine to UDP-3-*O*-(*R*-3-hydroxymyristoyl)-*N*-acetyl-glucosamine. J Biol Chem 1987; 262:5159–5169.
49. Crowell DN, Anderson MS, Raetz CRH. Molecular cloning of the genes for lipid A disaccharide synthase and UDP-*N*-acetylglucosamine acyltransferase in *Escherichia coli*. J Bacteriol 1986; 168:152–159.
50. Coleman J. Raetz CRH. First committed step of lipid A biosynthesis in *Escherichia coli*: sequence of the *lpxA* gene. J Bacteriol 1988; 170:1268–1274.
51. Dicker IB, Seetharam S. Cloning and nucleotide sequence of the *firA* gene and the *firA200*(Ts) allele from *Escherichia coli*. J Bacteriol 1991; 173:334–344.
52. Mohan S, Kelly TM, Eveland SS, Raetz CRH, Anderson M. An *Escherichia coli* gene (*fabZ*) encoding (3*R*)-hydroxymyristoyl acyl carrier protein dehydrogenase. J Biol Chem 1994; 269:32896–32903.
53. Anderson MS, Bull HG, Galloway SM, Kelly TM, Mohan S, Radika K, Raetz CRH. UDP-*N*-acetylglucosamine acyltransferase of *Escherichia coli*. The first committed step of endotoxin biosynthesis is thermodynamically unfavorable. J Biol Chem 1993; 268:19858–19865.
54. Raetz CRH, Roderick SL. A left-handed parallel β helix in the structure of UDP-*N*-acetylglucosamine acyltransferase. Science 1995; 270:997–1000.
55. Anderson MS, Bulawa CE, Raetz CRH. The biosynthesis of gram-negative endotoxin. Formation of lipid A precursors from UDP-GlcNAc in extracts of *Escherichia coli*. J Biol Chem 1985; 260:15536–15541.
56. Vuorio R, Vaara M. The lipid A biosynthesis mutation *lpxA2* of *Escherichia coli* results in drastic antibiotic supersusceptibility. Antimicrob Agents Chemother 1992; 36:826–829.
57. Williamson JM, Anderson MS, Raetz CRH. Acyl-acyl carrier protein specificity of UDP-GlcNAc acyltransferases from gram-negative bacteria: Relationship to lipid A structure. J Bacteriol 1991; 273:3591–3596.
58. Kulshin VA, Zähringer U, Lindner B, Frasch CE, Tsai C-M, Dmitriev BA, Rietschel ET. Structural characterization of the lipid A component of pathogenic *Neisseria meningitidis*. J Bacteriol 1992; 174:1793–1800.
59. Karibian D, Deprun C, Szabó L, Le Beyee Y, Caroff M. ^{252}Cf plasma desorption mass spectrometry applied to the analysis of endotoxin lipid A preparations. Int J Mass Spectrom Ion Proc 1991; 111:273–286.
60. Stead A, Sylvia Main J, Ward ME, Watt PJ. Studies on lipopolysaccharides isolated from strains of *Neisseria gonorrhoeae*. J Gen Microbiol 1975; 88:123–131.
61. Odegaard TJ, Kaltoshov IG, Cotter RJ, Steeghs L, van der Ley P, Khan S, Maskell DJ, Raetz CRH. Shortened hydroxyacyl chains on lipid A of *Escherichia coli* cells expressing a foreign UDP-*N*-acetylglucosamine *O*-acyltransferase. J Biol Chem 1997; 272:19688–19696.
62. Dotson GD, Kaltashov IG, Cotter RJ, Raetz CRH. Expression cloning of a *Pseudomonas* gene encoding a hydroxydecanoyl-acyl carrier protein dependent UDP-GlcNAc acyltransferase. J Bacteriol 1998; 180:330–337.
63. Wyckoff TJO, Raetz CRH. Altering the acyl chain specificity of UDP-GlcNAc *O*-acyltransferases from *E. coli* and *P. aeruginosa*. FASEB J. 1998; 12:L53.
64. Anderson MS, Robertson AD, Macher I, Raetz CRH. Biosynthesis of lipid A in *Escherichia coli*: identification of UDP-3-*O*-[(*R*)-3-hydroxymyristoyl]-α-D-glucosamine as a precursor of UDP-*N*^{2},*O*^{3}-bis[(*R*)-3-hydrosymyristoyl]-α-D-glucosamine. Biochemistry 1988; 27:1908–1917.
65. Young K. Silver LL, Bramhill D, Cameron P, Eveland SS, Raetz CRH, Hyland SA, Anderson MS. The *envA* permeability/cell division gene of *Escherichia coli* encodes the second enzyme of lipid A biosynthesis. UDP-3-*O*-(*R*-3-hydroxymyristoyl)-*N*-acetyl-glucosamine deacetylase. J Biol Chem 1995; 270:30384–30391.
66. Normark S. Genetics of a chain-forming mutant of *Escherichia coli*. Transduction and dominance of the *envA* gene mediating increased penetration to some antibacterial agents. Genet Res 1970; 16:63–78.
67. Normark S, Boman HG, Matsson E. Mutant of *Escherichia coli* with anomalous cell division and ability to decrease episomally and chromosomally mediated resistance to ampicillin and several other antibiotics. J Bacteriol 1969; 97:1334–1342.
68. Young K, Silver L. Leakage of periplasmic enzymes from *envA1* strains of *Escherichia coli*. J Bacteriol 1991; 173:3609–3614.
69. Gundstrom T, Normark S, Magnusson K-E. Overproduction of outer membrane protein suppresses *envA*-induced hyperpermeability. J Bacteriol 1980; 144:884–890.
70. Sullivan NF, Donachie WD. Transcriptional organization within an *Escherichia coli* cell division gene cluster: direction of transcription of the cell separation gene *envA*. J Bacteriol 1984; 160:724–732.
71. Beall B, Lutkenhaus J. Sequence analysis, transcriptional organization, and insertional mutagenesis of the *envA* gene of *Escherichia coli*. J Bacteriol 1987; 169:5408–5415.
72. Sorensen PG, Lutkenhaus J, Young K, Eveland SS, Anderson MS, Raetz CRH. Regulation of UDP-3-*O*-[*R*-3-hydroxymyristoyl]-*N*-acetylglucosamine deacetylase in *Escherichia coli*. The second enzymatic step of lipid A biosynthesis. J Biol Chem 1996; 271:25898–25905.
73. Onishi HR, Pelak BA, Gerckens LS, Silver LL, Kahan FM, Chen M-H, Patchett AA, Galloway SM, Hyland SA, Anderson MS, Raetz CRH. Antibacterial agents

that inhibit lipid A biosynthesis. Science 1996; 274: 980–982.
74. Kelly TM, Stachula SA, Raetz CRH, Anderson MS. The *firA* gene of *Escherichia coli* encodes UDP-3-O-(*R*-3-hydroxymyristoyl)-glucosamine *N*-acyltransferase. The third step of endotoxin biosynthesis. J Biol Chem 1993; 268:19866–19874.
75. Lathe R. Fine-structure mapping of the *firA* gene, a locus involved in the phenotypic expression of rifampicin resistance in *Escherichia coli*. J Bacteriol 1977; 131:1033–1036.
76. Dicker IB, Seetharam S. What is known about the structure and function of the *Escherichia coli* protein FirA? Mol Microbiol 1992; 6:817–823.
77. Lathe R, Lecocq J-P. The *firA* gene, a locus involved in the expression of rifampicin resistance in *Escherichia coli*. I. Mol Gen Genet 1977; 154:43–51.
78. Lathe R, Lecocq J-P. The *firA* gene, a locus involved in the expression of rifampicin resistance in *Escherichia coli*. II. Mol Gen Genet 1977; 154:53–60.
79. Vuorio R, Hirvas L, Vaara M. The Ssc protein of enteric bacteria has significant homology to the acyltransferase of LpxA of lipid A biosynthesis, and to three acetyltransferases. FEBS Lett 1991; 292:90–94.
80. Helander IM, Lindner B, Seydel U, Vaara M. Defective biosynthesis of the lipid A component of temperature-sensitive *firA* (*omsA*) mutant of *Escherichia coli*. Eur J Biochem 1993; 212:363–369.
81. Nishijima M, Bulawa CE, Raetz CRH. Two interacting mutations causing temperature-sensitive phosphatidylglycerol synthesis in *Escherichia coli* membranes. J Bacteriol 1981; 145:113–121.
82. Bulawa CE, Raetz CRH. The biosynthesis of gram-negative endotoxin. Identification and function of UDP-2,3-diacylglucosamine in *Escherichia coli*. J Biol Chem 1984; 259:4846–4851.
83. Babinski KJ, Raetz CRH. Identification of a gene encoding a novel *Escherichia coli* UDP-2,3-diacylglucosamine hydrolase. FASEB. 1998; 12:L63.
84. Ray BL, Painter G, Raetz CRH. The biosynthesis of gram-negative endotoxin. Formation of lipid A disaccharides from monosaccharide precursors in extracts of *Escherichia coli*. J Biol Chem 1984; 259:4852–4859.
85. Crowell DN, Reznikoff WS, Raetz CRH. Nucleotide sequence of the *Escherichia coli* gene for lipid A disaccharide synthase. J Bacteriol 1987; 269:5727–5734.
86. Radika K, Raetz CRH. Purification and properties of lipid A disaccharide synthase of *Escherichia coli*. J Biol Chem 1988; 263:14859–14867.
87. Milla ME, Raetz CRH. Interaction of lipid A disaccharide synthase (LpxB) with the aerobic glycerol-3-phosphate dehydrogenase (GlpD) (abstr 318) FASEB J 1995; 9:A1310.
88. Goldman RC, Doran CC, Capobianco JO. Analysis of lipopolysaccharide biosynthesis in *Salmonella typhimurium* and *Escherichia coli* by using agents which specifically block incorporation of 3-deoxy-D-*manno*-octulosonate. J Bacteriol 1988; 170:2185–2191.
89. Clementz T, Raetz CRH. A gene coding for 3-deoxy-D-*manno*-octulosonic-acid transferase in *Escherichia coli*. Identification, mapping, cloning, and sequencing. J Biol Chem 1991; 266:9687–9696.
90. Ray BL, Raetz CRH. The biosynthesis of gram-negative endotoxin. A novel kinase in *Escherichia coli* membranes that incorporates the 4′-phosphate of lipid A. J Biol Chem 1987; 262:1122–1128.
91. Garrett TA, Kdrmas JL, Raetz CRH. Identification of the gene encoding the *Escherichia coli* lipid A 4′-kinase. Facile phosphorylation of endotoxin analogs with recombinant LpxK. J Biol Chem 1997; 272: 21855–21864.
92. Ghalambor MA, Levine EM, Heath EC. The biosynthesis of cell wall lipopolysaccharide in *Escherichia coli*. III. Isolation and characterization of 3-deoxyoctulosonic acid. J Biol Chem 1966; 241:3207–3215.
93. Levine DH, Racker E. Condensation of arabinose-5-phosphate and phosphorylenolpyruvate by 2-keto-3-deoxy-8-phosphooctonic acid synthetase. J Biol Chem 1959; 234:2532–2539.
94. Lim R, Cohen SS. D-Phosphorarabinoisomerase and D-ribulokinase in *Escherichia coli*. J Biol Chem 1966; 241:4303–4315.
95. Ghalambor MA, Heath EC. The biosynthesis of cell wall lipopolysaccharide in *Escherichia coli*. J Biol Chem 1966; 241:3222–3227.
96. Kohlbrenner WE, Fesik SW. Determination of the anomeric specificity of the *Escherichia coli* CTP:CMP-3-deoxy-D-*manno*-octulosonate cytidylyltransferase by ^{13}C NMR spectroscopy. J Biol Chem 1985; 260: 14695–14700.
97. Berlyn MKB, Low KB, Rudd KE. Linkage map of *Escherichia coli* K-12, Edition 9. In: Neidhardt FC, ed. *Escherichia coli* and *Salmonella typhimurium*. Cellular and Molecular Biology. Vol. 2. 2d ed. Washington, DC: ASM Publications, 1996:1715–1902.
98. Goldman RC, Bolling TJ, Kohlbrenner WE, Kim Y, Fox JL. Primary structure of CTP:CMP-3-deoxy-D-*manno*-octulosonate cytidylyltransferase (CMP-KDO synthetase) from *Escherichia coli*. J Biol Chem 1986; 261:15831–15835.
99. Goldman RC, Kohlbrenner WE. Molecular cloning of the structural gene coding for CTP:CMP-3-deoxy-*manno*-octulosonate cytidylyltransferase from *Escherichia coli* K-12. J Bacteriol 1985; 163:256–261.
100. Brade H, Galanos C. Common lipopolysaccharide specificity: new type of antigen residing in the inner core region of S- and R-form lipopolysaccharides from different families of gram-negative bacteria. Infect Immun 1983; 42:250–256.
101. Brade H, Galanos C, Lüderitz O. Isolation of a 3-deoxy-D-*manno*-octulosonic acid disaccharide from *Salmonella minnesota* rough-form lipopolysaccharides. Eur J Biochem 1983; 131:201–203.
102. Brade H, Rietschel E-T. α-2→4-Interlinked 3-deoxy-D-manno-octulosonic acid disaccharide. A common constituent of enterobacterial lipopolysaccharides. Eur J Biochem 1984; 145:231–236.
103. Walenga RW, Osborn MJ. Biosynthesis of lipid A. In vivo formation of an intermediate containing 3-deoxy-D-manno-octulosonate in a mutant of *Salmonella typhimurium*. J Biol Chem 1980; 255:4252–4256.
104. Walenga RW, Osborn MJ. Biosynthesis of lipid A. Formation of acyl-deficient lipopolysacharides in *Sal-*

monella typhimurium and Escherichia coli. J Biol Chem 1980; 255:4257–4263.
105. Belunis CJ, Raetz CRH. Biosynthesis of endotoxin. Purification and catalytic properties of 3-deoxy-D-manno-octulosonic acid transferase from Escherichia coli. J Biol Chem 1992; 267:9988–9997.
106. White, KA, Kaltashov IA, Cotter RJ, Raetz CRH. A mono-functional 3-deoxy-D-manno-octulosonic acid (Kdo) transferase and a Kdo kinase in extracts of Haemophilus influenzae. J Biol Chem 1997; 272:16555–16563.
107. Brozek KA, Hosaka K, Robertson AD, Raetz CRH. Biosynthesis of lipopolysaccharide in Escherichia coli. Cytoplasmic enzymes that attach 3-deoxy-D-manno-octulosonic acid to lipid A. J Biol Chem 1989; 264:6956–6966.
108. Goldman RC, Doran CC, Kadam SK, Capobianco JO. Lipid A precursor from Pseudomonas aeruginosa is completely acylated prior to addition of 3-deoxy-D-manno-octulosonate. J Biol Chem 1988; 263:5217–5223.
109. Brade L, Nano RE, Schlecht S, Schramek S, Brade H. Antigenic and immunogenic properties of recombinants from Salmonella typhimurium and Salmonella minnesota rough mutants expressing in their lipopolysaccharide a genus-specific chlamydial epitope. Infect Immun 1987; 55:482–486.
110. Brade H, Brade L, Nano FE. Chemical and serological investigations on the genus-specific lipopolysaccharide epitope of Chlamydia. Proc Natl Acad Sci USA 1987; 84:2508–2512.
111. Strain SM, Fesik SW, Armitage IM. Structure and metal-binding properties of lipopolysaccharides from heptoseless mutants of E. coli studied by ^{13}C and ^{31}P nuclear magnetic resonance. J Biol Chem 1983; 258:13466–13477.
112. Belunis CJ, Mdluli KE, Raetz CRH, Nano FE. A novel 3-deoxy-D-manno-octulosonic acid transferase from Chlamydia trachomatis required for expression of the genus-specific epitope. J Biol Chem 1992; 267:18702–18707.
113. Nano FE, Caldwell HD. Expression of the chlamydial genus-specific lipopolysaccharide epitope in Escherichia coli. Science 1985; 228:742–744.
114. White KA, Raetz CRH. Complementation of a genomic disruption of the E. coli bifunctional KDO transferase by the H. influenzae mono-functional KDO transferase. FASEB J. 1998; 12:L44.
115. Brozek KA, Raetz CRH. Biosynthesis of lipid A in Escherichia coli. Acyl carrier protein-dependent incorporation of laurate and myristate. J Biol Chem 1990; 265:15410–15417.
116. Karow M, Fayet O, Cegeilska A, Ziegelhoffer T, Georgopoulos C. Isolation and characterization of the Escherichia coli htrB gene, whose product is essential for bacterial viability above 33°C in rich media. J Bacteriol 1991; 173:741–750.
117. Karow M. Georgopoulos C. Isolation and characterization of the Escherichia coli msbB gene, a multicopy suppressor of null mutations in the high-temperature requirement gene htrB. J Bacteriol 1992; 174:702–710.
118. Karow M. Georgopoulos C. Sequencing, mutational analysis, and transcriptional regulation of the Escherichia coli htrB gene. Mol Microbiol 1991; 5:2285–2292.
119. Clementz T, Bednarski JJ, Raetz CRH. Function of the htrB high temperature requirement gene of Escherichia coli in the acylation of lipid A. J Biol Chem 1996; 271:12095–12102.
120. Clementz T, Zhou Z, Raetz CRH. Function of the Escherichia coli msbB gene, a multicopy suppressor of htrB knockouts, in the acylation of lipid A. J Biol Chem 1997; 272:10353–10360.
121. Karow M, Fayet O, Georgopoulos C. The lethal phenotype caused by null mutations in the Escherichia coli htrB gene is suppressed by mutations in the accBC operon, encoding two subunits of acetyl coenzyme A carboxylase. J Bacteriol 1992; 174:7407–7418.
122. Clementz T, Bednarski J, Raetz CRH. Escherichia coli genes encoding Kdo dependent acyltransferases that incorporate laurate and myristate into lipid A. FASEB J 1995; 9:A1311.
123. Somerville JE Jr, Cassiano L, Bainbridge B, Cunningham MD, Darveau RP. A novel Escherichia coli lipid A mutant that produces an antiinflammatory lipopolysaccharide. J Clin Invest 1996; 97:359–365.
124. Karow M. Georgopoulos C. The essential Escherichia coli msbA gene, a multicopy suppressor of null mutations in the htrB gene, is related to the universally conserved family of ATP-dependent translocators. Mol Microbiol 1993; 7:69–79.
125. Polissi A, Georgopoulos C. Mutational analysis and properties of the msbA gene of Escherichia coli, coding for an essential ABC family transporter. Mol Microbiol 1996; 20:1221–1233.
126. Zhou Z, White KA, Polissi A, Georgopolous C, Raetz CRH. Function of Escherichia coli MsbA, and essential ABC family transporter, in lipid A and phospholipid biosynthesis. J Biol Chem. 1998; 273:12466–12475.
127. Mohan S, Raetz CRH. Endotoxin biosynthesis in Pseudomonan aeruginosa: enzymatic incorporation of laurate before 3-deoxy-D-manno-octulosonate. J Bacteriol 1994; 176:6944–6951.
128. Brozek KA, Bulawa CE, Raetz CRH. Biosynthesis of lipid A precursors in Escherichia coli. A membrane-bound enzyme that transfers a palmitoyl residue from a glycerophospholipid to lipid X. J Biol Chem 1987; 262:5170–5179.
129. Dillon DA, Wu WI, Riedel B, Wissing JB, Dowhan W, Carman GM. The Escherichia coli pgpB gene encodes a diacylglycerol pyrophosphate phosphatase activity. J Biol Chem 1996; 271:30548–30553.
130. Rietschel ET, Brade L, Lindner B, Zähringer U. Biochemistry of lipopolysaccharides. In: Morrison DC, Ryan JL, eds. Bacterial Endotoxic Lipopolysaccharides: Molecular Biochemistry and Cellular Biology. Vol. I. Boca Raton, FL: CRC Press, 1992:3–41.
131. Rietschel ET, Kirikae T, Schade FU, Mamat U, Schmidt G, Loppnow H, Ulmer AJ, Zähringer U, Seydel U, Di Padova F, Schreier M, Brade H. Bacterial endotoxin: molecular relationships of structure to activity and function. FASEB J 1994; 8:217–225.

132. Boll M, Radziejewska-Lebrecht J, Warth C, Krajewska-Pietrasik D, Mayer H. 4-Amino-4-deoxy-L-arabinose in LPS of enterobacterial R-mutants and its possible role for their polymyxin reactivity. FEMS Immunol Med Microbiol 1994; 8:329–341.
133. Helander IM, Kilpeläinen I, Vaara M. Increased substitution of phosphate groups in lipopolysaccharides and lipid A of the polymyxin-resistant *pmrA* mutants of *Salmonella typhimurium*: a ^{31}P-NMR study. Mol Microbiol 1994; 11:481–487.
134. Nummila K. Kilpelainen I, Zähringer U, Vaara M, Helander IM. Lipopolysaccharides of polymyxin B-resistant mutants of *Escherichia coli* are extensively substituted by 2-aminoethylpyrophosphate and contain aminoarabinose in lipid A. Mol Microbiol 1995; 16:271–278.
135. Guo L, Lim KB, Gunn JS, Bainbridge B, Darveau RP, Hackett M, Miller SI. Regulation of lipid A modifications by *Salmonella typhimurium* virulence genes *phoP-phoQ*. Science 1997; 276:250–253.
136. Raetz CRH, Dowhan W. Biosynthesis and function of phospholipids in *Escherichia coli*. J Biol Chem 1990; 265:1235–1238.
137. Dalessandro G, Northcote DH. Changes in enzymic activities of nucleoside diphosphate sugar interconversions during differentiation of cambium to xylem in sycamore and poplar. Biochem J. 1977; 162:267–279.
138. Gebb C, Baron D, Grisebach H. Spectroscopic evidence for the formation of a 4-keto intermediate in the UDP-apiose/UDP-xylose synthase reaction. Eur J Biochem 1975; 54:493–498.
139. Blattner FR, Plunkett GR, Bloch CA, Perna NT, Burland V, Riley M, Collado-Vides J, Glasner JD, Rode CK, Mayhew GF, Gregor J, Davis NW, Kirkpatrick HA, Goeden MA, Rose DJ, Mau B, Shao Y. The complete genome sequence of *Escherichia coli* K-12. Science 1997; 277:1453–1474.
140. Kahn SA, Everest P, Servos S, Foxwell N, Zähringer U, Brade H, Rietschel ET, Dougan G, Charles IG, Maskell DJ. A lethal role for lipid A in *Salmonella* infections. Mol Microbiol 1998; 29:571–579.
141. Carty SM, Sreekumar K. Raetz CRH. A cold-induced, palmitoleoyl transferase involved in lipid A biosynthesis in *Escherichia coli*. FASEB J 1997; 11:A1423.
142. Van Alphen L, Lugtenberg B, Rietschel ET, Mombers C. Architecture of the outer membrane of *Escherichia coli*: phase transitions of the bacteriophage K3 receptor complex. Eur J Biochem 1979; 101:571–579.
143. Wollenweber H-W, Schlecht S, Lüderitz O, Rietschel ET. Fatty acid in lipopolysaccharides of *Salmonella* species grown at low temperatures. Eur J Biochem 1983; 130:167–171.
144. Löbau S, Mamat U, Brabetz W, Brade H. Molecular cloning, sequence analysis, and functional characterization of the lipopolysaccharide biosynthetic gene *kdtA* encoding 3-deoxy-α-D-*manno*-octulosonic acid transferase of *Chlamydia pneumoniae* strain TW-183. Mol Microbiol 1995; 18:391–399.
145. Holst O, Bock K, Brade L, Brade H. The structures of oligosaccharide bisphosphates isolated from the lipopolysaccharide of a recombinant *Escherichia coli* strain expressing the gene *gseA* [3-deoxy-D-*manno*-octulopyranosonic acid (Kdo) transferase] of *Chlamydia psittaci* 6BC. Eur J Biochem 1995; 229:194–200.
146. Liu D, Reeves PR. *Escherichia coli* regains its O antigen. Microbiology 1994; 140:49–57.
147. Stevenson G, Neal B, Liu D, Hobbs M, Packer NH, Batley M, Redmond JW, Lindquist L, Reeves P. Structure of the O antigen of *Escherichia coli* K-12 and the sequence of its *rfb* gene cluster. J Bacteriol 1994; 176:4144–4156.
148. Deckert G, Warren PV, Gaasterland T, Young WG, Lenox AL, Graham DE, Overbeek R, Snead MA, Keller M, Aujay M, Huber R, Feldman RA, Short JM, Olsen GJ, Swanson RV. The complete genome of the hyperthermophilic bacterium *Aquifex aeolicus*. Nature 1998; 392:353–358.

17

Biosynthesis and Genetics of Lipopolysaccharide Core

David E. Heinrichs and Chris Whitfield
University of Guelph, Guelph, Ontario, Canada

Miguel A. Valvano
University of Western Ontario, London, Ontario, Canada

OVERVIEW

The typical core oligosaccharide (core OS) of lipopolysaccharide (LPS) is a phosphorylated heterooligosaccharide of less than 15 sugars, whose structure is described in detail elsewhere in this volume. In some gram-negative bacteria, LPS can terminate with the core OS to form so-called rough LPS (R-LPS). In others, the core OS can be capped by an O-polysaccharide (O-PS) to form smooth LPS (S-LPS). The "smooth" and "rough" terminology originates from observations that colonies of O-PS–containing strains often have a smooth appearance on solid media, whereas colonies of mutants that do not produce an O-PS tend to have a rough surface. Production of an O-PS is implicated in resistance to host defense mechanisms, such as complement-mediated serum killing, and has been shown to increase the pathogenicity of many gram-negative species. However, some pathogenic organisms including members of the genera *Neisseria, Haemophilus, Bordetella, Branhamella,* and *Campylobacter*, among others, lack O-PS but produce lipooligosaccharide (LOS) instead (see Ref. 1). Therefore, while the importance of the O-PS should certainly not be overlooked, it is clearly not an indispensable component of the pathogenicity and survival of some organisms.

The basic structure of the LPS core OS of *Escherichia coli* and *Salmonella enterica* serovar Typhimurium (hereafter referred to simply as *Salmonella**) was elucidated in the 1960s and early 1970s. As structural and analytical methods have advanced, the fine structural details of these core OS have been established. The structure of the core OS is made up of a backbone of 3-deoxy-D-*manno*-oct-2-ulosonic acid (Kdo), heptose, and hexose sugars (Fig. 1). The core OS structure is divided into inner (Kdo and heptose) and outer (hexose) core regions. The structure of the inner core is well conserved among members of the family Enterobacteriaceae, while the outer core can vary between strains. Until recently, only one core OS type was known for *Salmonella* spp. A variant structure is found in serovar Arizonae IIIa (serotype 062) (2). In *E. coli*, however, there are five distinct core structures, termed K-12, R1, R2, R3, and R4. Further advancements in analytical chemical/biochemical techniques will undoubtedly enhance our understanding of these structures, especially with respect to nonstoichiometric substitutions (e.g., phosphate) of the core.

The term R-LPS has historically referred to LPS species that simply lack an O-PS, owing to mutations

---

*Since the majority of genetic and biochemical data in this area are confined to *Salmonella enterica* serovar Typhimurium, this bacterium will simply be referred to as *Salmonella* in this chapter.

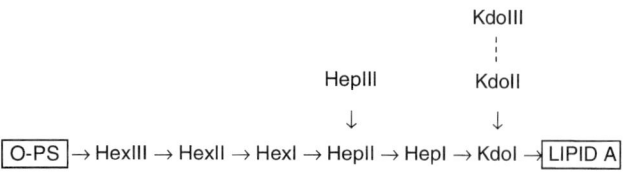

Fig. 1 Generalized structure of the carbohydrate backbone of the LPS core OS. The terminology for the glycosyl residues is taken from Ref. 66. The dotted line for the KdoIII linkage indicates that this is a partial (nonstoichiometric) substituent.

in genes involved with (1) O-PS synthesis, (2) ligation of O-PS to lipid A-core, or (3) core biosynthesis. With the possible exception of terminal branch and phosphate substitutions (discussed below), mutants blocked at a particular step in core OS biosynthesis fail to attach an O-PS to the distal end of the core OS. The synthesis of O-PS occurs independent of core biosynthesis and is detectable in cell lysates of core mutants. In this case, the O-PS remains attached to the lipid carrier undecaprenol diphosphate (3), in a form termed O-hapten. Many bacteriophages utilize either O-PS or core OS as their principal site of adsorption. Bacteria resistant to these phages lack the appropriate receptor, and the resulting R-LPS mutants provide a crucial source of material to facilitate structural elucidation of the core molecule of *E. coli* K-12 and *Salmonella*, as well as allowing initial insight into the assembly pathway. The data arising from many of these early mutants is discussed in a classic 1984 review by Mäkelä and Stocker (3).

Synthesis of the carbohydrate backbone of the core OS requires sugar nucleotide precursors. Assembly of the LPS core occurs using a lipid A–based derivative as an acceptor. In fact, in some bacteria the early steps in core assembly (Kdo addition) occur prior to full acylation and completion of lipid A (4). Assembly of the core then proceeds in a processive manner with sequential sugar addition and core extension. A model for the assembly process was first proposed by Rothfield and colleagues (reviewed in Ref. 5) and was revisited more recently by Rick (6). Briefly, the model proposes that sequential addition of sugar residues to the growing LPS core is catalyzed by a series of membrane-associated glycosyltransferases. More recent sequence data predict that most of these enzymes are peripheral membrane proteins. The glycosyltransferases are proposed to be arranged in the proper sequence and located in close proximity to one another, perhaps forming a multienzyme complex. Based on the peripheral membrane-association of the proteins and the cytoplasmic location of the precursors, the assembly of core OS likely occurs at the interface of the cytoplasm and plasma membrane. Growing LPS molecules, situated in the inner leaflet of the plasma membrane with their growing OS chains projecting into the cytoplasm, are suggested to be capable of movement within the fluid membrane. As they pass across the glycosyltransferases, the core OS is extended in a sequential fashion.

Many of the strains that contain truncated LPS cores harbor mutations in genes in the *rfa* (core OS biosynthesis) region of the chromosome. The core OS biosynthesis region is best characterized in *E. coli* K-12 and *Salmonella*, where it maps to between 81 and 82 min (between *cysE* and *pyrE*) on the chromosome (7,8). The core OS biosynthesis region consists of three operons that contain most, but not all, of the genes required for the synthesis of LPS core. Some of the unlinked genes exert a specific function in LPS core biosynthesis, including heptose-precursor biosynthesis. Others are "housekeeping genes," whose products are involved in metabolism or the formation of common biosynthetic precursors (e.g., UDP-galactose, UDP-glucose, UDP–*N*-acetylglucosamine); these will not be covered in this chapter. Recently a proposal has been made to rename the core OS biosynthesis genes within the framework of a new nomenclature system for bacterial polysaccharide genes (1). In this system, genes encoding glycosyltransferases are identified by the prefix *waa*, and genes for enzymes involved in precursor formation are named after the precursor involved. Details of this system are available on the World Wide Web (http://www.microbio.su.oz.au/BPGD/default.html). For clarity, the old name is also indicated in the text that follows and in Table 1.

Current knowledge of the biosynthesis and genetics of core OS leans heavily on information from *E. coli* and *Salmonella*, and for the most part biochemical data are confined to these organisms. Consequently this review will focus primarily on these systems. We will discuss the basis for gene assignments and, where appropriate, the limitations of these assignments. Homologs of core OS biosynthetic genes are now being identified in other organisms, and these will be discussed using the enteric bacteria as a framework for comparison. However, it should be recognized that the structurally different core OSs of other bacteria will result in the identification of novel enzymes with no homologs in enteric bacteria. It can be predicted that the advent of whole genome sequencing will result in a rapid expansion of data in this area. Core OS biosynthesis has been covered in a comprehensive manner by Schnaitman and Klena (10), and historical aspects are described by

Biosynthesis and Genetics of LPS Core

Table 1 Genes Involved in the Production and Modification of Core OS of *E. coli* and *Salmonella*

New gene name	Former gene name	Known or proposed product function for core OS synthesis
kdsA	*kdsA*	Kdo-8-phosphate synthase
kdsB	*kdsB*	CTP: CMP-Kdo synthase
waaA	*kdtA*	Bifunctional (or trifunctional) Kdo transferase
waaM	*htrB*	Lipid A–associated Kdo-dependent lauroyltransferase
waaN	*msbB*	Lipid A–associated late acyltransferase
gmhA	*lpcA*	Sedoheptulose-7-phosphate isomerase
gmhB/C	*rfaE*	Phosphoheptose kinase-ADP-heptose synthase fusion protein
gmhD	*rfaD*	ADP-L-glycero-D-manno-heptose epimerase
waaC	*rfaC*	HepI transferase
waaF	*rfaF*	HepII transferase
waaQ	*rfaQ*	HepIII transferase
rfaH	*rfaH*	*rfa* operon antiterminator
waaG	*rfaG*	HexI transfer: UDP-glucose: (heptosyl) LPS α-1,3-glucosyltransferase
waaI	*rfaI*	HexII transfer: UDP-galactose: (glucosyl) LPS α-1,3-galactosyltransferase of *Salmonella* and *E. coli* R3
waaO	*rfaI*	HexII transfer: UDP-glucose: (glucosyl) LPS α-1,3-glucosyltransferase of *E. coli* K-12, R1, R2, and R4
waaJ	*rfaJ*	HexIII transfer: UDP-glucose: (galactosyl) LPS α-1,2-glucosyltransferase of *Salmonella* and *E. coli* R3
waaR	*rfaJ*	HexIII transfer: UDP-glucose: (glucosyl) LPS α-1,2-glucosyltransferase of *E. coli* K-12 and R2
waaT		HexIII transfer: UDP-galactose: (glucosyl) LPS α-1,2-galactosyltransferase of *E. coli* R1 and R4
waaB	*rfaB*	HexI substitution: UDP-galactose: (glucosyl) LPS α-1,6-galactosyltransferase of *Salmonella*, *E. coli* K-12 and R2
waaK	*rfaK*	HexIII substitution: UDP-*N*-acetylglucosamine: (glucosyl) LPS α-1,2-*N*-acetylglucosaminyltransferase of *Salmonella* and *E. coli* R2
waaU	*rfak*	HexIII substitution: HepIV or inner core GlcNAc transferase of *E. coli* K-12
waaX		HexII substitution: β-1,4-galactosyltransferase of *E. coli* R4
waaW		HexIII substitution: UDP-galactose: (galactosyl) LPS α-1,2-galactosyltransferase of *E. coli* R1 and R4
waaV		HexII substitution: β-1,3-glucosyltransferase of *E. coli* R1
waaD		HexIII or HexII substitution: possible UDP-glucose: (glucosyl) LPS α-1,2-glucosyltransferase or UDP-*N*-acetylglucosamine: (galactosyl) LPS α-1,3-*N*-acetylglucosaminyltransferase of *E. coli* R3
waaL	*rfaL*	Lipid A core:surface polymer ligase
waaP	*rfaP*	Involved in the phosphorylation of HepI
waaS	*rfaS*	Unknown; possibly involved in the formation of α-Rha-1→5-Kdo in *E. coli* K-12
waaY	*rfaY*	Involved in the phosphorylation of HepII
waaZ	*rfaZ*	Unknown; possible involvement with production of an "LOS form" of LPS of *E. coli* K-12
wabA	*rfaS*	Unknown; possibly involved in the formation of α-Gal-1→7-Kdo in *E. coli* R2

Mäkelä and Stocker (3). Here we will focus on more recent developments (147) that add to or clarify the issues discussed in these earlier reviews. Since the inner core regions of core oligosaccharides tend to show structural similarities, the enzymes and reactions involved in their syntheses often exhibit a high degree of conservation. Structural diversity increases in the distal, outer core regions, and this is reflected in the enzymology. It is therefore convenient to separate the inner and outer core regions for the purposes of discussion. Addition of nonstoichiometric substituents to the inner core oligosaccharide will be considered separately.

BIOSYNTHESIS AND GENETICS OF THE INNER CORE

Introduction

The inner core of most gram-negative bacteria is comprised of Kdo and heptose, and its structure tends to be highly conserved among many different organisms. The structure of this region of the core OS of *E. coli* K-12 and *Salmonella* is shown in Figure 2, together with the defects arising from specific mutations. In contract, the inner cores of, for example, *Acinetobacter*, *Legionella*, and *Rhizobium* strains are quite different, since they lack heptose and contain unusual sugars (11–13). Very little is known at the present time about the biosynthesis and genetics of these different inner core structures.

In the laboratory, the minimal LPS structure required for *E. coli* viability consists of lipid A glycosylated with two Kdo residues (4,14–16). This minimal structure is termed Re LPS. It is likely that structures smaller than Re LPS (i.e., lipid A precursors) fail to translocate to the outer membrane, and the lack of a minimal LPS component would create a nonviable outer membrane. Indeed, only conditional mutants of lipid A or Kdo synthesis have been isolated in *E. coli*. Interestingly, the I-69 mutant of *Haemophilus influenzae* survives with a truncated LPS, whose structure consists of lipid A and a single phosphorylated Kdo residue (17,18). The differences between *H. influenzae* and the Enterobacteriaceae likely reflects altered specificity between the Kdo transferase enzymes (see below) rather than altered LPS structural requirements for viability.

Mutants with defects in the backbone heptose residues are stable under laboratory conditions but give rise to the pleiotropic phenotype known as "deep-rough." However, the precise involvement of heptosyl residues in the deep-rough phenotype is obscured because these residues provide the site of modification by phosphoryl groups and phosphoryl derivatives (Fig. 2). The deep-rough phenotype has been discussed at length in a number of recent reviews (4,10,19,147), and we refer the reader to these reviews for a more detailed description and pertinent references. Briefly, deep-rough LPS mutants characteristically have changes in surface hydrophilicity, resulting in hypersensitivity to hydrophobic dyes, detergents, hydrophobic antibiotics, fatty acids, phenols, and polycyclic hydrocarbons. The wild-type outer membrane normally has limited permeability to these compounds, and they are toxic only in high concentrations. The increased outer membrane permeability of these mutants appears to reflect major compositional and structural changes in the outer membrane, resulting from the defect in LPS biosynthesis. LPS interacts with divalent cations that help maintain the structural integrity of the outer membrane. Mutants with deep-rough LPS typically release significant amounts of periplasmic enzymes into the culture media. The outer membrane is stabilized by growth in media supplemented with high concentrations of Mg^{2+}, and the release of enzymes is reduced. The outer membrane of some deep-rough mutants has also been reported to have a decreased protein content with a concomitant increase in phospholipid. The reduction in protein content may be a consequence of reduced trimerization of porin monomers. The compromised outer membrane integrity of the deep-rough LPS mutants increases their susceptibility to attack by lysosomal fractions of polymorphonuclear leukocytes and phagocytosis by macrophages (20). In addition to the above observations, deep-rough mutants of *E. coli* K-12 are associated with increased expression of colanic acid

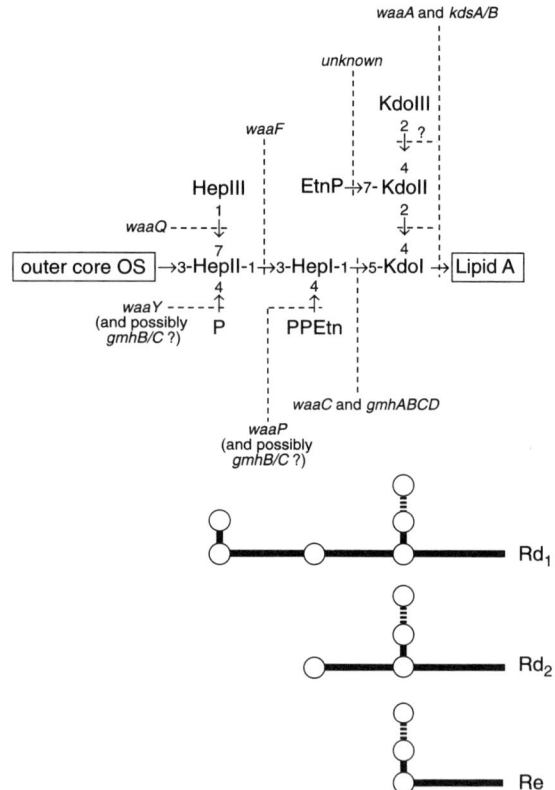

Fig. 2 Structure of the inner core OS from the LPS of *E. coli* and *Salmonella* and the genetic determinants involved in its biosynthesis. Note that KdoII can be partially modified by the addition of KdoIII and by addition of either PEtn (as shown) or phosphate. In *E. coli* K-12, KdoII is the site of rhamnose addition and in the R2 core type, it carries an α-D-Gal*p* residue. In R3, and some other core types, HepIII can be modified by an α-D-Glc*p*NAc group. The basis for the assignment of specific genes is described in the text. The Re, Rd$_1$, and Rd$_2$ chemotype designations were originally given to truncated *Salmonella* LPS structures.

capsular polysaccharide (21) and the loss of pili and flagella in deep-rough mutants of *E. coli* K-12 and *Salmonella* (21,22). It is possible that these latter phenotypes are a general response to outer membrane perturbation.

Synthesis and Transfer of Kdo

The study of the biosynthesis of Kdo was only made possible by the isolation of conditional mutants (16,23). Biochemical studies with these mutants were very important in providing direct evidence that LPS containing Kdo is essential for cell growth. The biosynthetic pathway of Kdo is shown in Figure 3 (24–26) and involves sequential reactions catalyzed by three enzymes: D-phosphoarabinose isomerase, Kdo-8-phosphate synthetase, and Kdo-8-phosphate phosphatase. Kdo is then convertd by a CMP-Kdo synthetase into cytidine-5′-monophosphate-Kdo (CMP-Kdo). The genes involved in Kdo biosynthesis in *E. coli* and *Salmonella* are not clustered. *kdsA*, encoding the Kdo-8-phosphate synthetase, maps at 27 min, and *kdsB*, encoding the CMP-Kdo synthetase, maps at 85 min. The gene for the phosphatase activity has not been identified. *kdsA* is located within a putative operon comprised of six open reading frames; the operon includes other genes not associated with LPS biosynthesis (27). *kdsA* is the last gene of this cluster and is transcribed from its own promoter. Experiments involving Northern blot analysis and operon fusions have shown that *kdsA* is subject to growth phase–dependent regulation at the transcriptional level with maximal expression achieved during early log phase (27). Protein sequence alignments of known Kdo-8-phosphate synthases with bacterial and fungal 3-deoxy-D-*arabino*-2-heptulosonate-7-phosphate synthases suggest that both classes of enzymes are structurally related and may belong to a family of 2-keto-3-deoxy-aldonic acid synthases (28). Intriguingly, a *kdsA* homolog has recently been cloned and sequenced from plant origin (W. Brabetz, personal communication). The plant homolog can functionally complement the KdsA^{ts} mutant of *Salmonella*, suggesting that Kdo-8-phosphate is synthesized in plants by an enzyme with structural and functional similarities to KdsA.

The gene *kdsB* is also transcriptionally regulated by growth rate in a similar manner to *kdsA* (27). In some *E. coli* strains, a homolog of this gene has been identified in region I of the *kps* gene cluster, where it is involved with group II capsular polysaccharide biosynthesis (29). Effective competitive inhibitors of the CMP-Kdo synthase that can act as antimicrobial agents have been synthesized (30). The clinical use of these inhibitors has been hampered by their low permeability across the bacterial envelope, although derivatives conjugated with a dipeptide can be taken up via the oligopeptide permease system (30,31). These issues have recently been reviewed elsewhere (4).

The transfer of Kdo to lipid IV$_A$ is catalyzed by the bifunctional (and possibly trifunctional) transferase WaaA (formerly KdtA) (32). This enzyme has a high specificity for the recognition of lipid A disaccharide biphosphate substrates, however, it shares no homology with mammalian proteins that recognize and bind to lipid A (33). The transfer of Kdo to lipid IV$_A$ is necessary for the late acylation and subsequent completion of the lipid A molecule. Of interest, the Kdo-transferase of *Chlamydia trachomatis* is at least trifunctional (34), while the *Haemophilus* Kdo-transferase appears simply to be monofunctional (4). Further and more detailed aspects of Kdo transfer in the LPS biosynthetic process have been subject to a recent review (4).

Synthesis of ADP-L-glycero-D-manno-heptose

Pioneering work by Eidels and Osborn (35) first demonstrated that transketolase-negative mutants of *Salmonella* do not synthesize sedoheptulose-7-phosphate (an intermediate in the nonoxidative portion of the pentose phosphate pathway) and have an incomplete

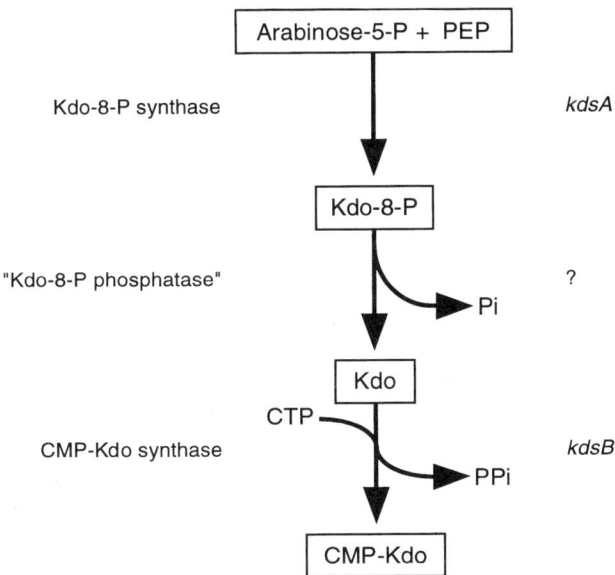

Fig. 3 Pathway for the biosynthesis of the precursors of Kdo residues in the inner core OS. Note that the enzyme and corresponding gene for the second step in this pathway has not been identified. (Adapted from Ref. 4.)

heptose-deficient LPS. Addition of exogenous sedoheptulose-7-phosphate to growing cultures of the transketolase mutants rescued the LPS defect, implicating sedoheptulose-7-phosphate as an obligatory precursor of L-*glycero*-D-*manno*-heptose (35). Based on the above observations, Eidels and Osborn proposed a pathway for the biosynthesis of L-*glycero*-D-*manno*-heptose (Fig. 4A) involving the following four enzymatic reactions: (1) conversion of sedoheptulose-7-phosphate to D-*glycero*-D-*manno*-heptose-7-phosphate by a phosphoheptose isomerase; (2) conversion of D-*glycero*-D-*manno*-heptose-7-phosphate to D-*glycero*-D-*manno*-heptose-1-phosphate by a phosphoheptose mutase; (3) conversion of D-*glycero*-D-*manno*-heptose-1-phosphate with ATP to ADP-D-*glycero*-D-*manno*-heptose and PP_i by an ADP-heptose synthase; and (4) racemization by an epimerase of ADP-D-*glycero*-D-*manno*-heptose to the L isomeric form.

Biochemical evidence supporting this pathway has only been obtained for the isomerase and the epimerase activities, while the mutase and the ADP-heptose synthase have not yet been characterized. One of the major difficulties in studying these enzymes is that the corresponding genes are dispersed throughout the chromosome, and more importantly, all give the same heptose-deficient LPS phenotype. Another complication is the lack of commercially available sugar phosphate precursors to investigate the enzymology of these intermediate steps. The latter could be overcome by the chemical synthesis of the 1-phosphate and the 7-phosphate forms of D-*glycero*-D-*manno*-heptose, followed by the study of their interconversion using enzyme extracts from wild-type and mutant strains. However, with the rapid appearance of bacterial genome sequences, it may be straightforward to characterize heptose-deficient mutants by now standard molecular genetic approaches. A number of mutants of *E. coli* and *Salmonella* are known to produce a heptose-deficient LPS (see below). Some of the mutated genes are located in the core OS biosynthesis cluster, while others are dispersed throughout the chromosome. These other loci have received a variety of designations depending on the various groups discovering them. For instance, the same locus has been called *lpcA* (36), *tfrA* (37), and *con* (38) in *E. coli* and *isn69* in *H. influenzae* (39). We have recently proposed replacing all of these designa-

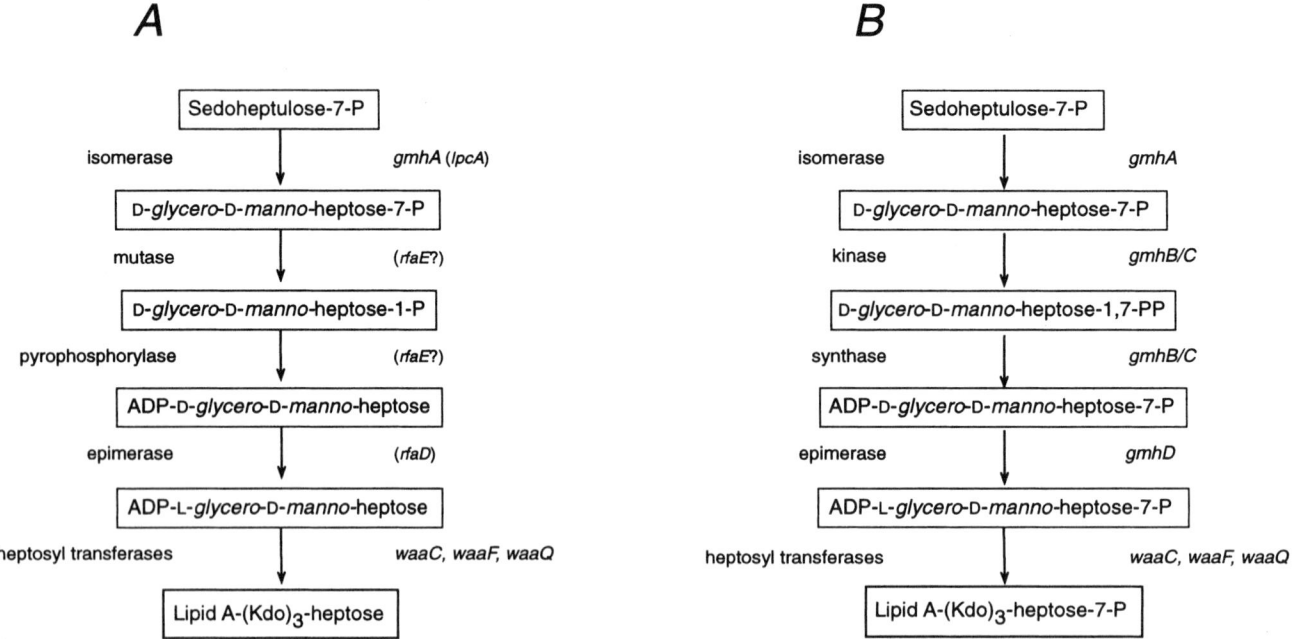

Fig. 4 Pathways for the biosynthesis of the precursors of heptosyl residues in the inner core OS and their incorporation into the core OS. (A) The pathway for ADP-L-*glycero*-D-*manno*-heptose synthesis, essentially as proposed by Eidels and Osborn (35). (B) A new hypothesis in which the novel precursor ADP-L-*glycero*-D-*manno*-heptose-7-P provides phosphoryl groups found on heptosyl residues in the inner core OS (51). The relevant genes are shown, and their assignments in the new pathway are described in the text.

tions with *gmh* (for glyceromannoheptose synthesis) (9). The fourth letter identifies a specific enzyme in the biosynthesis pathway.

Phosphoheptose Isomerase

Eidels and Osborn used radiolabeled sedoheptulose-7-phosphate to show specific labeling of the L-*glycero*-D-*manno*-heptose residues of LPS in cultures of the transketolase mutants of *Salmonella* and reported the identification and partial purification of the phosphoheptose isomerase as well as the isolation and characterization of its product as D-*glycero*-D-*manno*-heptose-7-phosphate. More recently, Brooke and Valvano (40,41) identified and isolated the phosphoheptose isomerase gene, *gmhA*, from both *E. coli* and *H. influenzae* and confirmed its biochemical role. *gmhA* is a monocistronic gene encoding a 29 kDa polypeptide containing a region of amino acid similarity shared by proteins of the glucosamine-6-phosphate transaminase family. Interestingly, the conserved sequence identifies a region proposed to be involved in the isomerization reaction and not in the transfer of amino groups (40). One of the GmhA homologs includes LmbN, a protein that appears to be involved in an isomerization step in the synthesis of the 8-carbon sugar-containing antibiotic lincomycin (42; W. Piepersberg, personal communication). As might be predicted given the structural conservation of heptose residues in the inner core OS, Southern hybridization shows the *gmhA* gene to be widely conserved in many enteric and nonenteric bacteria including *Bordetella pertussis*, *Actinobacillus pleuropneumoniae*, and *Neisseria* spp.

Epimerase

The second known enzyme for heptose synthesis is ADP-L-*glycero*-D-*manno*-heptose-6-epimerase. This enzyme catalyzes the last step of the biosynthesis of L-*glycero*-D-*manno*-heptose and is encoded by the *gmhD* gene (previously *rfaD*). The function of the *E. coli* GmhD enzyme was determined by Coleman (43) using chromatographic and mass-spectroscopic methods to examine the heptose components of LPS extracted from the *gmhD* mutant (44). These experiments indicated that the *gmhD* mutant LPS contained more D-*glycero*-D-*manno*-heptose than L-*glycero*-D-*manno*-heptose, consistent with the observation that the strain accumulated ADP-D-*glycero*-D-*manno*-heptose. The *gmhD* gene is part of the core OS biosynthesis gene cluster of both *E. coli* K-12 and *Salmonella* (45–47). Homologs of *gmhD* are found in *H. influenzae* (48) and *N. gonorrhoeae* (49), but core OS assembly genes are dispersed among various gene clusters in these organisms.

Intermediate Steps

The genes and corresponding enzymes involved in the intermediate steps of the Eidels and Osborn pathway are not assigned, although one candidate gene is *rfaE*. This is a designation for a locus in *Salmonella* not physically linked to the major core OS biosynthesis gene cluster. A *Salmonella rfaE* mutant produces a heptose-deficient LPS, and the locus was suggested to encode the ADP-heptose synthase (50). This was based on rather indirect evidence showing that a cell-free extract from the *rfaE* mutant transfers heptose residues to $[4'-^{32}P](Kdo)_2$-lipid IV$_A$ in vitro, provided that preformed ADP-heptose is added. Consequently, RfaE is not a heptosyltransferase.

Recently, Valvano et al. (51) cloned the *rfaE* homolog from *E. coli*. This gene is part of an operon that includes *glnE* and *orfXE* of unknown function (52). *glnE* encodes the adenylyltransferase involved in the postranscriptional regulation of glutamine synthase, a central enzyme for nitrogen assimilation, although *glnE* itself is not regulated by nitrogen levels (52). It is intriguing but unclear why genes involved in LPS synthesis and nitrogen assimilation form an operon. *rfaE* encodes a 55-kDa protein of which the C-terminal 190 amino acids have strong similarities to the cytidylyltransferase superfamily, including the enzyme glycerol-3-phosphate cytidylyltransferase from *Bacillus subtilis* (TagD) and other enzymes with ADP transferase activity (53). The most conserved region in this superfamily resembles the ATP-binding HiGH (His-Ile-Gly-His) motif of class I aminoacyl-tRNA synthases. A BLASTP search using the *E. coli* RfaE C-terminal domain reveals strong similarities with Aut, a small protein identified in *Ralstonia* (*Alcaligenes*) *eutrophus* (54). An *aut*::Tn5 mutant is defective in autotrophic growth, but the phenotype includes morphological changes in the colony appearance. Valvano et al (51) showed that the *aut*::Tn5 mutant, unlike the wild-type strain, cannot grow in the presence of novobiocin and produces an LPS with an electrophoretic mobility in SDS-PAGE typical of heptose-deficient LPS. More importantly, the *aut* phenotypes are rescued with the *E. coli rfaE*. Collectively, these results strongly suggest that the *aut* homologs encode ADP-heptose synthases.

The 343 N-terminal amino acids of the *E. coli* RfaE lack characteristics expected of sugar pyrophosphorylases but share strong conservation with members of the ribokinase family. Given these characteristics, it is possible that the pathway of synthesis of heptose involves a phosphotransfer step rather than a mutase reaction. In the Eidels and Osborn pathway, a mutase reaction transfers the phosphate group from carbon 7 to carbon 1 (Fig. 4A). In the alternative pathway proposed by Valvano et al. (51) (Fig. 4B), a new phosphate group would be added to carbon 1 and used later in the formation of the phosphodiester bond with ADP. If this is the case, the product of the second reaction should be D-*glycero*-D-*manno*-heptose-1,7-diphosphate. This alternative pathway will result in a heptose that will be phosphorylated at carbon 7. Interestingly, a phosphate group is found in at least two of the three heptoses isolated from *E. coli* LPS samples (see Fig. 2). It is therefore proposed that the *rfaE* gene product encodes a bifunctional protein acting in the intermediate steps of heptose biosynthesis.

rfaE gene homologs have recently been identified and cloned from *H. influenzae* and *Neisseria gonorrhoeae* by functional complementation of the *Salmonella rfaE* mutant (55,56). However, in *N. gonorrhoeae* and *N. meningitidis*, the ADP transferase domain of RfaE is encoded by a separate gene like in the case of *R. eutrophus* (51,56). In contrast, the two domains are part of the same gene in *E. coli*, *H. influenzae*, *Vibrio cholerae*, and *Helicobacter pylori* (51). Since *E. coli* and *Salmonella* are genetically related, it is likely that the *Salmonella rfaE* also encodes the two domains, and the *rfaE* mutation is nonpolar. In light of these findings, the *rfaE* gene has tentatively been assigned as *gmhB/C* to reflect the bifunctional nature of its product.

Heptosyltransferases

Heptosyl I Transfer

The *waaC* (formerly *rfaC*) mutation of *Salmonella* yields heptose-deficient LPS and results in a deep-rough phenotype as described above (57). The *waaC* gene of *Salmonella* has been cloned and sequenced, and plasmids containing the *waaC* gene complement the *waaC* phenotype (50). The identification of WaaC as the HepI transferase was made using an indirect thin layer chromatography (TLC) assay and was based on the inability of *waaC* mutants to transfer heptose residues from ADP-D-*glycero*-D-*manno*-heptose to [4'-^{32}P](Kdo)$_2$-lipid IV$_A$; the cloned *waaC* gene complemented the defect. As expected if WaaC is the HepI transferase, addition of ADP-D-*glycero*-D-*manno*-heptose to extracts of *gmhC* and *gmhD* mutants restores heptosyl transfer in vitro (50).

The *E. coli* K-12 *waaC* mutation, originally termed *rfa-2*, exhibits a similar phenotype to *Salmonella waaC*, resulting in heptose-deficient LPS (58). The *waaC* gene of *E. coli* K-12 was cloned and sequenced (59), and its predicted product shares 90% total similarity with the WaaC protein of *Salmonella* (46). Not surprisingly, the *E. coli* K-12 *waaC* gene is capable of complementing the *Salmonella waaC* mutant to restore wild-type LPS (59,60).

Homologs of WaaC have now been identified in *Pseudomonas aeruginosa* (61), *B. pertussis* (62), *Campylobacter hyoilei* (accession number X91082), *N. gonorrhoeae* (63), and *N. meningitidis* (accession numbers U35454 and U40862).

Heptosyl II Transfer

The product of the *waaF* gene is implicated in the transfer of the second heptosyl (HepII) residue onto the LPS core OS molecule. This conclusion is based upon the following results: (1) examination of *Salmonella waaF* mutants indicates that they produce LPS containing only one heptose unit (3,57,64); and (2) LPS from *waaF* mutants migrates slightly slower on SDS-PAGE than completely heptose-deficient LPS from a *waaC* mutant (10). The *Salmonella* and *E. coli* K-12 *waaF* genes have both been cloned and sequenced (45–47), and the deduced polypeptide products share 95% similarity (46). The *waaF* gene is situated immediately downstream of *waaD* and upstream of *waaC* in both organisms. The deduced WaaF protein is approximately 49% similar to the deduced WaaC protein (46), which is suggestive of a similar enzymatic function, presumably involving an interaction with ADP-heptose. Homologs of WaaF have been identified in *H. influenzae* (48), *P. aeruginosa* (61), and *N. meningitidis* (65).

Heptosyl III Transfer

The core OS structure of both *E. coli* and *Salmonella* has been shown to be substituted at HepII by a third heptose residue in an α-1,7 linkage (66,67). Recent analyses of a precisely defined, nonpolar *waaQ* mutant in an *E. coli* R1 strain yield LPS completely devoid of the HepIII residue (148). This, together with data indicating that the deduced products of the *waaQ*, *waaC*, and *waaF* genes share regions of similarity, identifies WaaQ as the HepIII transferase. The *waaQ* gene is the first gene of the long, central operon of the core OS biosynthesis gene cluster (10,45,47,68). This gene has now been completely sequenced from *Salmonella* LT2 and *E. coli* R1, R2, R3, and R4 reference strains (147). Alignments of the deduced WaaQ homologs indicate that this protein is highly conserved among the six strains.

The HepIII residue is substituted with GlcNAc in the cores of *E. coli* R3, *Vibrio cholerae*, and *B. pertussis*, but not in *Salmonella* (66). Currently, no genetic determinants have been assigned to this substitution.

Phosphate Substitution in the Inner Core OS

Where present, phosphate substitution occurs in a number of forms and at a number of sites in the inner core region of LPS (see Fig. 2). HepI can be substituted at position 4 with either phosphate or pyrophosphorylethanolamine (2-aminoethyl pyrophosphate; PPEtn), while substitution of HepII with phosphate occurs at an unknown position. Further, the KdoII residue can also be substituted at the 7 position by phosphoethanolamine (PEtn) (69). All of the phosphate substituents have been found in nonstoichiometric amounts, and due to technical difficulties in structural determination of these charged groups on the core OS, the substitutions as illustrated in Figure 2 have been regarded as tentative (66). Phosphate substitution at the HepI residue is important for the integrity and/or biogenesis of the outer membrane (21,70,148). It is suggested that the lack of phosphate substitution at HepI (resulting in the loss of a negatively charged group) gives rise to an outer membrane in which adjacent LPS molecules cannot be cross-linked by divalent cations or polyamines, nor can the LPS interact with positively charged groups on proteins (21). Loss of activity in the secreted form of *E. coli* hemolysin (71) also occurs in mutants with nonphosphorylated LPS. A possible explanation is that an altered outer membrane affects the normal secretion mechanism of the toxin, which may be key to its activation. It is interesting to note here that there is increasing evidence to indicate that there are complex interactions between a variety of cell surface–associated components and that many of these components appear to be coordinately regulated.

In the literature, many of the phosphoryl modifications of both heptosyl and, perhaps, Kdo residues have been attributed to the activity of the *waaP* (formerly *rfaP*) gene product. The same enzyme has also been implicated in transfer of HepIII to HepII (21), although WaaQ has since been identified as the HepIII transferase (148). These diverse phenotypes can be explained by (1) a requirement for a phosphorylated acceptor for HepIII, or by (2) disruption of critical protein-protein interactions within a concerted complex resulting from the *waaP* mutation. Using ^{31}P–NMR, coupled with methylation–linkage and fast–atom bombardment spectrometry analyses, Yethon et al. have recently implicated the *waaP* gene product in the phosphorylation of HepI, and the *waaY* gene product in the phosphorylation of HepII (148). In agreement with previous studies, the *waaP* mutant lacks all phosphoryl substituents in the heptose region of the core as well as HepIII. However, the *waaY* mutant, while retaining phosphoryl substituents at HepI, lacks phosphate substitution of HepII. Interestingly, the predicted products of *waaP* and *waaY* share homology with kinase proteins in PSI-BLAST searches of the databases.

In *Salmonella*, *waaP* mutations certainly result in the lack of phosphate substitution on core OS, but they also result in a truncated core OS since only small amounts of O-PS are detectable (72,73). Such mutants exhibit the deep-rough phenotype, and most of the core OS molecules are not extended beyond the proximal hexose (GlcI) residue (73). These mutants are thought to be "leaky," in that minor quantities of phosphorylated core OS molecules and O-PS are detectable (72). The same is not true for *E. coli* K-12, where *waaP* mutants appear to have a fully extended core OS (21). Insertional inactivation of the *E. coli* R1 *waaP* gene with a nonpolar cassette also results in a deep-rough phenotype, although the core OS is completed and is capable of being "capped" with an O-PS (J. A. Yethon et al., unpublished). Based on these results, it is conceivable that the *Salmonella waaP* mutants carry additional, and unrecognized, defects in core assembly, which results in their failure to fully extend the core OS structure.

Mutants of *S. enterica* serovar Minnesota with an LPS chemotype termed RcP$^-$ have been isolated that contain GlcI and lack the third heptose as well as phosphate substitution in the heptose region (74). Unfortunately, it is still unclear whether these mutants carry a defect in the same locus (i.e., *waaP*) as those in the Typhimurium derivatives described in the preceding section. However, soluble enzyme fractions from wild-type strains of serovars Typhimurium or Minnesota both catalyze the transfer of [γ-^{32}P]-ATP into RcP$^-$ acceptor LPS (75). Interestingly, similar experiments indicate that the amount of phosphate incorporated into Rd$_1$P$^-$ LPS is significantly less than the amount incorporated into RcP$^-$ LPS (76). This would seem to indicate that addition of GlcI to the heptose region occurs prior to phosphorylation and provides the preferred substrate for modification. Moreover, incorporation of GlcI seems to be a requirement for subsequent incorporation of HepIII and phosphorylation of HepII (67).

The heterogeneity of phosphoryl substituents (PPEtn or P) found at HepI is intriguing, and two potential processes have been hypothesized. There exist partial and indirect data in support of either mechanism but no experimental evidence to directly confirm the involvement of WaaP or WaaY in either alternative. In one scenario, the membrane phospholipid phosphatidylethanolamine provides the donor for transfer of the PEtn head group to an already phosphorylated HepI residue to form the PPEtn substituent. As discussed above, however, addition of the phosphate group may occur during the synthesis of ADP-L-*glycero*-D-*manno*-heptose and therefore prior to its incorporation into the core. Preliminary pulse-labeling studies support the involvement of phosphatidylethanolamine as the PPEtn donor (77). *E. coli* K-12 Δ*pss* mutants lack phosphatidylserine synthase, which is essential for the synthesis of phosphatidylethanolamine. These mutants produce virtually no phosphatidylethanolamine (78) and normal levels of LPS, although the LPS has aberrant electrophoretic mobility suggestive of less mass and/or an increase in net negative charge (10). Although not determined unequivocally, it is likely that the LPS does not contain ethanolamine substituents (82). Interestingly, the outer membrane protein content appears not to be altered in Δ*pss* mutants, indicating that the deep-rough phenotype does not result directly from the lack of ethanolamine substituents on the LPS molecule (10). In an alternative proposal, Parker et al. (21) suggested that during the process of LPS assembly and/or export, PEtn could be enzymatically cleaved from PPEtn-substituted HepI to leave phosphate. If a processing reaction is involved, a processing enzyme must be invoked, but there is no evidence to its identity.

It is interesting to note here that not all core OS structures contain a phosphorylated inner core region. In *K. pneumoniae*, negative charges in the inner core OS are provided by galacturonic acid and Kdo residues, and this has a potential impact on the outer membrane stability in this organism (79). The core OS biosynthesis gene cluster from *K. pneumoniae* has not yet been analyzed, but it will be interesting to determine whether it contains a *waaP* homolog. In summary, the role of WaaP and the biosynthetic processes leading to the modification of the inner core region provide a number of interesting questions for further investigation.

BIOSYNTHESIS AND GENETICS OF THE OUTER CORE

Introduction

The structures of the five outer core OS types from *E. coli* and the single structure from *Salmonella* are shown in Figure 5. Mutants blocked in the synthesis of specific sugar nucleotide precursors gave early indications of the sequence in outer core OS assembly. For example, in the absence of exogenous galactose in the growth medium, *Salmonella* mutants with a defect in the UDP-galactose-4-epimerase gene (*galE*) contain incomplete LPS lacking all sugars beyond the proximal glucose (Hexose I) (80,81). Mutants unable to synthesize UDP-glucose (e.g., *pgi, galU*) produce hexose-deficient LPS. The presence of glycosyltransferases that are capable of elongating the core OS in such mutants was subsequently demonstrated through assays for glycosyltransferase activity. These assays identify incorporation of radiolabeled sugars (from nucleotide sugar precursors) into mutant LPS acceptor in the presence of crude cell extracts and membranes. Of note, purified LPS is incapable of serving as an acceptor in combination with crude preparations of soluble glycosyltransferases (82). Membrane phospholipid, or phosphatidylethanolamine, is a requisite for enzymatic activity (5,82–86). Membrane reconstitution involving LPS, phospholipid, and glycosyltransferases was successfully used for the study of galactosyltransferase activity (86,87,149) and led to the tentative model for core OS assembly by Rothfield and coworkers (5). In the following sections we attempt to summarize the current status of biochemical and genetic data pertaining to outer core OS biosynthesis.

LPS is a heterogeneous molecule, and the exact structures of many mutant LPS species have not been confirmed. The same is true for products generated in vitro using glycosyltransferase assays. Many of the specific glycosyltransferase assignments involved in core OS biosynthesis should therefore be taken as tentative. A further complication is that much of the enzymology has been performed using *Salmonella* and many of the *E. coli* K-12 gene assignments have been made by reference to *Salmonella* based on similarities in structure and following complementation of *Salmonella* mutants. Despite this biochemical limitation, there remains compelling evidence for the function of a number of the outer core OS glycosyltransferases, and this lays the foundation for more detailed future studies. Moreover, the sequencing of the core OS bio-

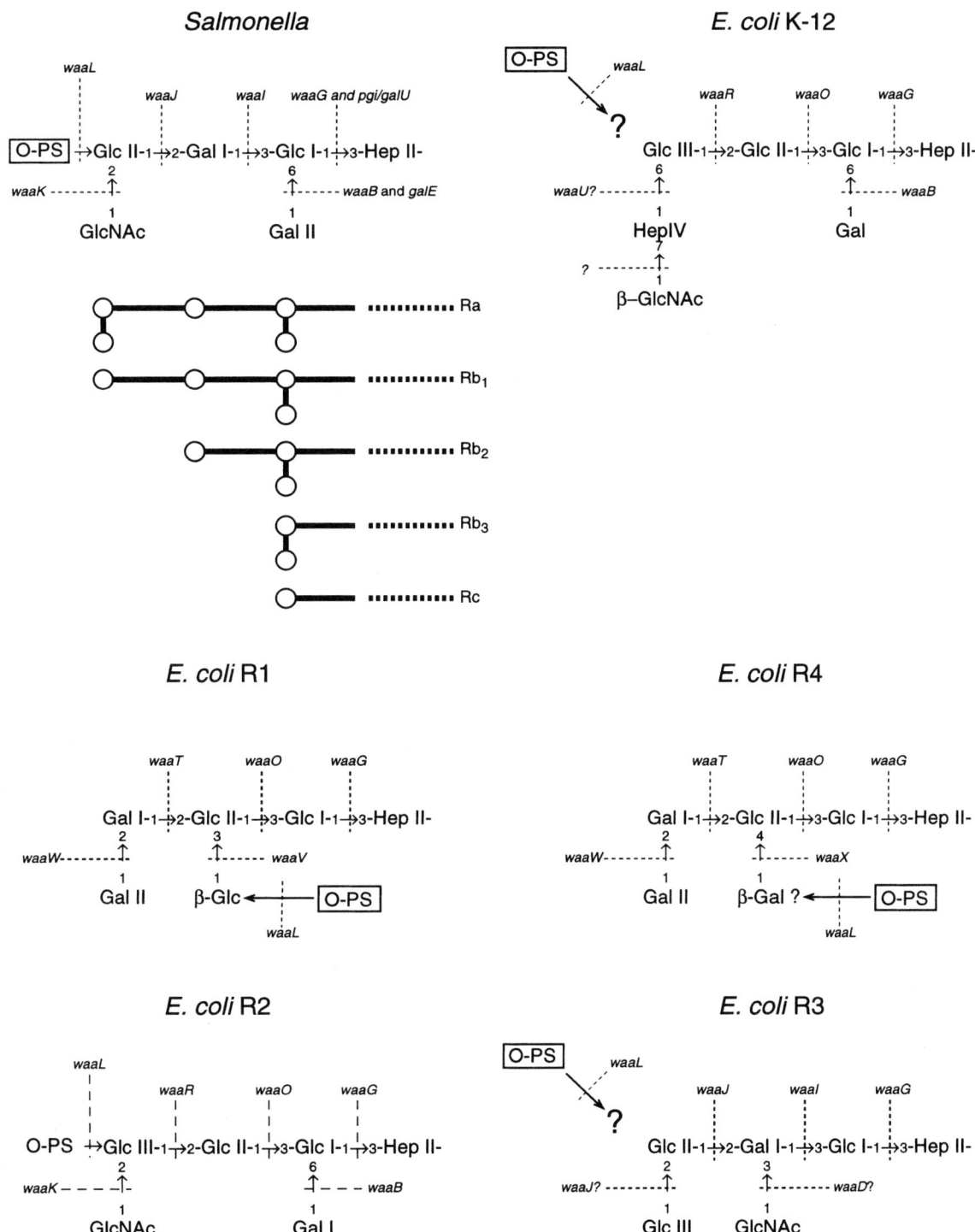

Fig. 5 Structures of the outer core OSs from the LPS of *E. coli* and *Salmonella* and the genetic determinants involved in their biosynthesis. The structures are taken from Ref. 66, and the basis for assignment of the individual genes is described in the text. The chemotype designations were originally given to truncated *Salmonella* LPS structures. The ligation site for linking O-PS to the R4 core is predicted by data from R1 (147). Ligation sites in the K-12 and R3 core OSs are currently unknown.

synthesis gene clusters from *E. coli* R1-R4 and the additional sequencing in this region of *Salmonella* (147,149) (Fig. 6), combined with the known core OS structures, will aid in the tentative assignment of transferase activities. However, in order to ascribe specific and definitive functions to these proteins, precise acceptor molecules will need to be chemically synthesized.

Hexose I (GlcI) Transfer

E. coli and *Salmonella* both contain a glucose-α1,3-heptose disaccharide at the junction of the inner and outer core (Fig. 5). *Salmonella pgi* mutants are deficient in the production of glucose and therefore lack the proximal glucose residue (GlcI) of the core OS (Rd_1-chemotype LPS). Extracts of these mutants show UDP-glucose:(heptosyl) lipopolysaccharide α-1,3-glucosyltransferase activity by incorporating glucose into Rd_1 acceptor LPS (82). The α-1,3-glucosyltransferase I protein has been purified and characterized and shows an absolute requirement for phospholipid (84). The corresponding genetic locus is designated *waaG* (formerly *rfaG*).

In the first molecular-genetic studies on LPS core OS assembly, a *Salmonella waaG* mutant was functionally complemented with plasmids containing *E. coli* K-12 DNA from the Clarke and Carbon collection. The Ffm phage-sensitivity pattern (R-LPS) of the mutant was restored to the Ffm phage-resistant pattern of a smooth strain (89). Enzyme activity from the product of the cloned gene was confirmed as UDP-glucose:(heptosyl) lipopolysaccharide α-1,3-glucosyltransferase. Heterologous complementation of this *Salmonella* mutation is not surprising given the identical structures in this region of the core OS. The *Salmonella waaG* gene was subsequently cloned using a similar complementation strategy (90) and has recently been sequenced (149). Pairwise alignments of the deduced amino acid sequences of WaaG proteins from *Salmonella* and *E. coli* K-12, R1, R2, R3, and R4 strains yield values with 85% similarity in all cases (147), correlating with the conserved glucose-α-1,3-heptose structure. A homolog of WaaG has been identified in *P. aeruginosa* (91,92), although its precise function is unclear since the corresponding part of the *P. aeruginosa* core OS has a galactosaminyl-heptose structure.

It has been suggested that the lack of flagella and pili associated with the deep-rough phenotype is due specifically to the loss of WaaG (21). This is based on the observation that the flagellar and pili defects in an *E. coli* K-12 Δ*waaGPSBO* strain (which has a deep-rough phenotype) are complemented by addition of single-copy *waaG* via *waaG*-containing λ prophage. This is in agreement with earlier studies showing that *galU* (UDP-glucose–deficient) mutants of *E. coli* also show a loss of flagella and pili (93).

Hexose I Substitution

The first hexose (GlcI) of the core OS of *E. coli* K-12 and R2 and *Salmonella* is substituted with α-1,6-galactose. Historically, this α-1,6-galactose substitution has been referred to as GalII in *Salmonella* and simply as Gal in *E. coli* K-12.

Salmonella mutants have been isolated with LPS that lack the HexII (α-1,3-galactose) residue but contain α-1,6-galactose linked to GlcI (94), indicating that addition of the side branch precedes further elongation of the core OS during assembly. Mutants deficient in α-1,6-galactose attachment are designated *waaB* (formerly *rfaB*). Although WaaB activity was predicted for some time based on the structure of *Salmonella* core OS, *waaB* mutants were not identified until 1978 (95) and not characterized in any detail until 1983 (96). Extracts from *Salmonella waaB* mutants show no UDP-galactose:(glucosyl)lipopolysaccharide α-1,6-galactosyltransferase activity, and LPS from these strains can act as acceptors for α-1,6-galactosyltransferase activity from extracts of WaaB+ cells (96). A complementing DNA fragment carrying the *waaB* gene of *Salmonella* has been isolated (90), and the gene has been sequenced (149). The sequence of the *E. coli* K-12 *waaB* gene is also known, and the gene complements a *Salmonella waaB* mutant (97). Whereas the *waaB* gene product of *E. coli* R2 shows greater than 90% identity with the predicted WaaB protein of *E. coli* K-12, the *Salmonella* WaaB protein is slightly less conserved (149). It shares approximately 75% similarity with WaaB proteins from *E. coli* K-12 and R2. Core OS structures of *E. coli* R1, R3, and R4 types do not contain the α-1,6-galactose substitution (Fig. 5), and a *waaB* homolog is absent from the corresponding core OS biosynthesis regions (147) (Fig. 6).

Hexose II Transfer

It is at the HexII position of the outer core that the structures of the core OS backbones of *E. coli* K-12 and *Salmonella* diverge. In *E. coli* K-12 HexII is α-1,3-glucose, whereas in *Salmonella*, it is α-1,3-galactose (see Fig. 5). As a consequence, the core OS structures in a *galE* mutant and in a strain lacking the HexII

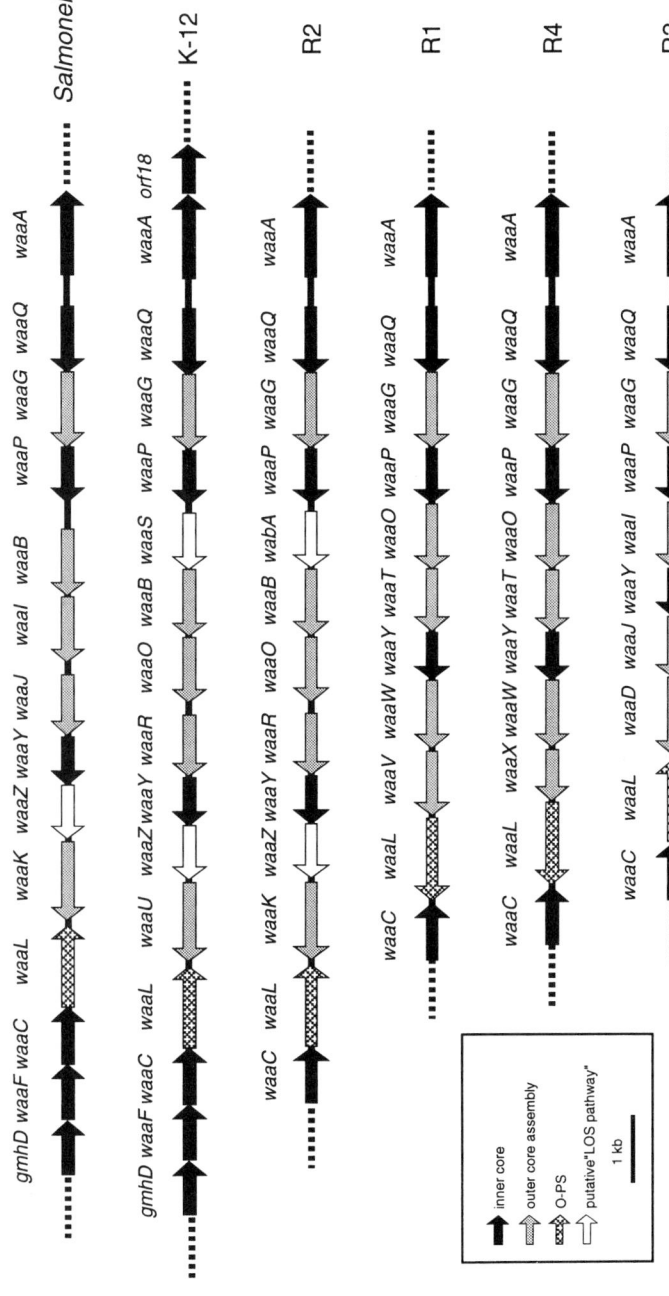

Fig. 6 Organization of the major core OS biosynthesis regions on the chromosomes of *E. coli* and *Salmonella*. Data are available only for the *waaC* to *waaA* region of the *E. coli* R1, R2, R3, and R4 reference strains (10a). The gene assignments are described in the text.

transferase (chemotype Rb$_3$) are identical in *Salmonella* (98).

There exists some confusion in the literature concerning the nomenclature of HexII transferase mutants. In *Salmonella*, they were initially termed *rfaH* (99) but were subsequently renamed *rfaI* (now *waaI*). RfaH is the more recent designation for the protein whose function regulates transcription through the core OS biosynthesis operon. The designation *rfaM* was given to *E. coli* K-12 mutants deficient in UDP-glucose:(glucosyl) lipopolysaccharide α-1,3-glucosyltransferase (89) in order to distinguish the activity from the distinct enzyme specificity in *Salmonella*. However, Schnaitman et al. suggested that both should be designated *rfaI* based on the similarity of these enzymes in structure, function, and location, despite the fact that they have different sugar specificities (45). Under the new nomenclature scheme we have given a separate designation to all genes whose products exhibit different acceptor and substrate specificities in all six core OS biosynthesis clusters (see Figs. 5 and 6). The *Salmonella* standard has been retained as *waaI*, but the gene responsible for UDP-glucose:(glucosyl) lipopolysaccharide α-1,3-glucosyltransferase activity in *E. coli* K-12 is now designated *waaO* (see Figs. 5 and 6 and Table 1).

In *Salmonella*, the GlcI-deficient acceptor LPS (Rd$_1$) does not support incorporation of galactose (98). The corresponding UDP-galactose:(glucosyl) lipopolysaccharide α-1,3-galactosyltransferase of *Salmonella* (WaaI) has been purified and partially characterized (100). Although much evidence has been obtained to suggest that the WaaI protein is the α-1,3-galactosyltransferase, definitive data showing galactose transfer to precise acceptors are not available.

The deduced protein encoded by the *waaO* gene of *E. coli* K-12 (97) and *waaI* of *Salmonella* (101) were shown to be highly similar, although the deduced sequence of WaaO has an additional 198 amino-terminal residues. Questions were therefore raised about whether this simply reflected a sequencing error (46), and this is indeed the case. The WaaI gene of *Salmonella* has been resequenced (149). The WaaI protein is actually 338 amino acids in length, whereas the *E. coli* K-12 WaaO protein is 339 amino acids long. Both WaaI and WaaO proteins are basic and contain no transmembrane domains. Genes whose products have a high degree of similarity to WaaI and WaaO are found in all of the *E. coli* R-type strains. Three of these core types (R1, R2, and R4) contain a glucose residue in the HexII position (and contain a *waaO* homolog). The R3 core OS resembles *Salmonella* in containing a galactose residue at HexII (Fig. 5) and, as predicted, conservation in the WaaI homologs of *Salmonella* and R3 is higher than that between WaaI and WaaO homologs (D. E. Heinrichs and C. Whitfield, unpublished). The features of these highly conserved enzymes, which might dictate the specificity of glucosyl- or galactosyl transfer, remains an intriguing but unanswered question.

Hexose II Substitution

The HexII position of *Salmonella* and *E. coli* K-12 core OS structures is not substituted, but the equivalent position of *E. coli* core types R1, R3, and R4 carry side-branch residues of β-1,3-glucose, α-1,3-*N*-acetylglucosamine (α-1,3-GlcNAc), and β-1,4-galactose, respectively (Fig. 5). These substitutions are known to be provided by the activities of the WaaV protein in *E. coli* R1 and the WaaX protein in *E. coli* R4 (150). In agreement, the predicted WaaV and WaaX proteins share similarities with other β-glycosyltransferases in the databases. These substitutions complete the outer core assembly (see Hexose III Substitution below). The gene responsible for the α-1,3-GlcNAc substitution in R3 may be *waaD*, whose predicted product has a significant amount of similarity to WaaK of *Salmonella*. This remains speculative at this point in time, and functional investigation of this gene product is underway in our laboratories.

Hexose III Transfer

The third hexose in both *E. coli* K-12 and *Salmonella* is an α-1,2–linked glucose residue. Selection for resistance of *Salmonella* to phage Felix O often results in the isolation of mutants containing LPS lacking the distal glucose residue (chemotype Rb$_2$). Initially, these mutants were designated as R-res-1 (3), but they are now termed *waaJ* (formerly *rfaJ*).

In *E. coli* K-12, the *waaR* (formerly *rfaJ*) gene product is involved in UDP-glucose:(glucosyl) lipopolysaccharide α-1,2-glucosyltransferase activity, as opposed to UDP-glucose:(galactosyl) lipopolysaccharide α-1,2-glucosyltransferase activity in *Salmonella*. The *waaJ* and *waaR* genes have been cloned and sequenced, and the deduced proteins share a high degree of similarity (90,97,101). As with WaaI and WaaO, the deduced polypeptide sequence of WaaJ was found to be significantly shorter than WaaR. Again, this reflects errors in the published sequence of the *Salmonella waaJ* gene. The *Salmonella waaJ* gene has now been resequenced and is predicted to encode a protein of 336 amino acids in length compared to the 338 amino acids predicted for the *E. coli* WaaR protein (149). The *Salmonella* WaaJ and *E. coli* K-12 WaaR proteins share approximately 69% total similarity.

The third hexose is either glucose or galactose in the *E. coli* R1-R4 strains. A gene whose product shows homology to both WaaJ and WaaR is present in the core OS biosynthesis cluster of all of these strains, and the sizes of the orfs are consistent with those of the *waaJ* gene of *Salmonella* and the *waaR* gene of *E. coli* K-12 (147). The protein involved in the transfer of this third hexose is designated WaaT in R1 and R4 strains, WaaR in R2, and WaaJ in R3, based on activities predicted by the various core OS structures. Interestingly, the position of the *waaJ* homolog from R3 differs in comparison to the other core OS biosynthesis gene clusters; its position, relative to the conserved *waaY* gene, is reversed (Fig. 6). The significance of this is unknown. Interestingly, the WaaI, WaaJ, WaaO, WaaR, and WaaT proteins all share a significant amount of similarity. This may reflect the similar acceptor structures and donor molecules with which they must interact. Although significant degrees of similarity are seen between the WaaR protein of *E. coli* K-12 and WaaJ of *Salmonella*, plasmids carrying the *E. coli* K-12 *waaR* gene only complement *Salmonella waaJ* mutations at low efficiency (97). The altered efficiency presumably reflects different substrate and acceptor specificities of these proteins due to variation in core OS structure. The structure of the core OS that is assembled in the complemented strain has not been established.

Hexose III Substitution

The HexIII position of the *Salmonella* core OS is substituted with an α-1,2-linked GlcNAc residue (3,102). However, literature on the HexIII substitution in *E. coli* K-12 is rather confusing, in that original studies reported a partial substitution with β-1,6-GlcNAc (103), while later and more detailed analyses determined this substitution to be α-1,6-heptose (commonly referred to as HepIV) (104). It is now known that the HexIII position in *E. Coli* K-12 is partially substituted with the disaccharide β-D-GlcNAc-(1→7)-L-α-D-Hep (151, Fig. 5).

Early studies showed that incorporation of GlcNAc into the cell wall of *Salmonella* requires the prior incorporation of both glucose and galactose into Rc chemotype LPS and that the GlcNAc transferase is membrane-associated (105). *Salmonella waaK* (previously *rfaK*) mutants produce LPS that lacks the terminal GlcNAc residue (Rb_1 LPS) and the frequency of "capping" of the core OS with O-PS is reduced (3). It is possible that the GlcNAc-substituted core OS is the preferred, but not absolute, recognition site for attachment of O-PS. Alternatively, the *waaK* mutation is leaky. More recently, Schnaitman and Klena (10) proposed that the presence of GlcNAc-substituted core OS may direct attachment of O-PS to particular lipid A-core molecules. They further speculated that the GlcNAc residue could be cleaved off during attachment of O-PS in a mechanism analogous to oligosaccharide processing pathways in eukaryotic glycoproteins. This is an interesting proposal, which can now be tested directly using an appropriate combination of chemical and molecular genetic-approaches.

The predicted *Salmonella waaK* gene product is a peripheral membrane-associated or soluble protein (106), and homologs have been identified in the core OS biosynthesis gene clusters of both *E. coli* R2 and R3 (88a). Due to structural similarities between the WaaK-determined core OS structure of *Salmonella* and the *E. coli* R2 OS core structures, the R2 *waaK* homolog is also designated *waaK* while the R3 homolog is termed *waaD* to indicate its different function. Plasmids carrying the *E. coli* R2 *waaK* gene complement a *Salmonella waaK* mutant (107). It is currently unknown whether *waaD* is involved in the transfer of the branch α-1,2–linked GlcIII residue or the HexII-linked α-1,3-GlcNAc in the formation of the R3 core OS (indicated in Figs 5 and 6).

Most of the genes of the core OS biosynthesis gene cluster of *E. coli* K-12 have a significant degree of similarity (>70% deduced amino acid sequence similarity) to homologs in *Salmonella*. This relationship does not hold true for the "*waaK*" genes (108). Although the two genes are present in identical positions within the core OS biosynthesis gene cluster, the respective gene products share very little similarity (<20% identity, 45% similarity). Due to the structural differences in the core OS structures of *Salmonella* and *E. coli* K-12, the *E. coli* K-12 gene has been renamed *waaU*. As might be expected, the deduced products of the *waaK* (R2) and *waaD* (R3) genes show striking similarities to the *waaK* gene of *Salmonella* but not to WaaU from *E. coli* K-12 (149). Like WaaK, WaaU contains limited regions of similarity with other core-associated glycosyltransferases (10,97). It has previously been suggested that WaaU might be involved in the transfer of GlcNAc to different (nonterminal) parts of the core OS of *E. coli* K-12 (10,60,109). However, BLASTP searches identify regions of local similarity shared by WaaU and a variety of known and predicted heptosyltransferases. Such results suggest that WaaU may be involved in the addition of the HepIV substitution in *E. coli* K-12 and would explain the difference in the primary structures of the WaaK and WaaU proteins. If this is correct, a most surprising observation is that plasmids carrying the *E. coli* K-12 *waaU* gene show partial complementation of the *Salmonella waaK* phenotype (108). The basis for this observation is difficult to understand given the differences in core OS structure (specifically HexIII substitution; Fig. 5), and to interpret this result the structure of the resulting LPS will need to be determined.

A WaaK homolog has been identified in *N. meningitidis*, where it is involved in the production of LOS (110). The deduced amino acid sequence of this protein shows 21% identity and 47% total similarity to the WaaK protein of *Salmonella*. However, the neisserial WaaK homolog shows very little similarity to the *E. coli* K-12 WaaU protein. Unfortunately, it is currently not known whether the neisserial *waaK* gene is capable of complementing either the *waaK* phenotype of *Salmonella* or the *waaU* phenotype of *E. coli* K-12. Sequence from the *H. influenzae* genome project (accession numbers U32734 and L42023) revealed an open reading frame whose deduced amino acid sequence indicates similarity, albeit over a small region, to the WaaU protein of *E. coli* K-12. Whether this open reading frame codes for a protein, and what its possible substrates are are unknown at this time.

The HexIII substitution in *E. coli* R1 and R4 is an α-1,2–linked galactose residue, attached in both cases to a terminal galactose. The product of *waaW*, as designated in Figures 5 and 6, is involved in this substitution (150). The WaaW protein shows greater than 90% similarity between the two organisms and shares significant similarity with WaaI, WaaJ, WaaO, and WaaR proteins, as might be expected given their related activities (150). Heinrichs et al. have shown, using specific mutants of *E. coli* R1, that a *waaW* mutant contains only minor quantities of the β3-linked glucose residue at HexII (150). This result suggests that the β-linked substitution occurs as the final step in core OS assembly and this activity is much higher in the presence of WaaW or its reaction product.

It is interesting to note that the core OS biosynthesis gene cluster from *E. coli* R3 does not contain the necessary number of genes predicted by the structure of the core OS (see Fig. 6). Based on the putative functions of the homologs that it does contain, there appears to be no gene whose product is dedicated to the substitution of HexIII. It may be that one of the genes of the R3 core OS biosynthesis region is bifunctional or that one of the core OS biosynthesis genes is unlinked with the major core OS biosynthesis gene cluster of this organism.

COMPLETION OF THE LPS MOLECULE: EXPORT OF LIPID A CORE AND LIGATION TO O-PS

We have covered the various steps in the assembly of the lipid A core molecule and will now discuss the steps involved in its assembly into completed LPS. These steps include (1) the export of lipid A core to the periplasmic face of the plasma membrane and (2) the ligation of lipid A core to O-PS. The translocation of completed LPS to the outer membrane is discussed elsewhere in this book.

Data showing that undecaprenol-linked O-PS is detectable at the periplasmic face of the plasma membrane indicates that the O-PS and lipid A core components of LPS, which are synthesized separately at the cytoplasmic face of the plasma membrane, are exported in an energy-dependent manner to the periplasmic face of the plasma membrane before being joined in a ligation step (111,112). The processes involved in O-PS export are described in detail elsewhere in this book and will therefore not be discussed here. Whereas the biosynthesis of lipid A core molecules does not require a membrane potential in the presence of excess ATP, export of lipid A core from the inner to the outer face of the plasma membrane requires both the proton-motive force and ATP (113–115). The ATP requirement is consistent with the possibility that export of lipid A core occurs via the recently identified MsbA protein, an inner membrane protein belonging to the ABC family of transporters/exporters (116,117).

Early studies concerning the ligation of O-PS to lipid A core molecules in *Salmonella* identified two loci—*waaL* (formerly *rfaL*) and *rfbT*—whose functions were thought to be involved in this ligation step. Mutations in these genes were found to lead to the accumulation of immunologically detectable undecaprenol pyrophosphate-linked O-hapten (although direct chemical confirmation of this lipid:polymer linkage is not available). The LPS in these strains contains a complete core OS. *rfbT* mutants are now known to be defective in export of O-antigen and are not involved in ligation per se. Thus the *waaL* gene product is the only remaining candidate that is involved in the ligation of lipid A core to O-PS. Although the periplasmic location of O-PS attachment has been defined, the mechanism of ligation remains elusive. An interesting aspect of O-

PS ligation is that the reaction is not highly specific with respect to the O-PS that can be ligated to a given lipid A core molecule. Indeed, *E. coli* and *Salmonella* can ligate and express many foreign O-PS structures to their lipid A core structure when provided with O-PS biosynthesis genes of a heterologous organism. Also, *E. coli* WaaL is required for the attachment of group I K antigen and one form of enterobacterial common antigen (ECA) to the core OS (see Ref. 118 for further discussion).

The *waaL* gene of *Salmonella* was identified by complementation of *waaL* mutants (106). The *E. coli waaL* designation was assigned to an open reading frame in the *E. coli* K-12 sequence because its gene product is similar to that of the *Salmonella* homolog in terms of size and hydropathy profile. The *waaL* homologs occupy similar locations within their respective cluster (108). Primary sequences of the predicted WaaL homologs of *E. coli* K-12 and *Salmonella* share almost no similarity. However, both proteins are predicted to be integral membrane proteins with 8 or more membrane-spanning domains. Furthermore, hydropathy profiles of the two proteins are virtually identical; they share hydrophobic domains of similar length and distribution. WaaL homologs have been identified within the core OS biosynthesis gene clusters of *E. coli* R1-R4 strains (147). As with *E. coli* K-12, identification of these *waaL* homologs was based upon the deduced size and hydropathy profile of the protein and the location of the gene within the core OS biosynthesis gene cluster. Interestingly, the putative ligase gene from *E. coli* R1 and R4 strains is oriented opposite to its position in *Salmonella*, *E. coli* K-12, R2, and R3. Moreover, in *E. coli* R1 and R4 it forms the last gene of the long, central operon of the core OS biosynthesis region and not the last gene of the *waaC*-containing operon as seen in the other strains. Ligation-defective phenotypes have been confirmed for *waaL* mutants in *E. coli* R1 and R2 (149,150). Pairwise alignments of the various WaaL homologs give total similarity scores above 50% only for *E. coli* R2, R3, and *Salmonella* WaaL homologs. WaaL proteins of R1 and R4 core-type strains resemble one another but show little primary sequence relatedness to other WaaL homologs. In spite of the limited similarities between the putative ligase proteins and the structural differences between the core OS, the lipid A core molecules of *E. coli* K-12, R1, R2, R3, and R4 strains are all capable of being "capped" by the O-PS D-galactan I of *K. pneumoniae* (149,150). Heinrichs et al. have shown that the β-3-linked glucose residue of the *E. coli* R1 core serves as the attachment site for O-PS to an R1-type core (150).

This explains the sequence dissimilarity between the R1 WaaL protein and the *Salmonella* and *E. coli* R2 proteins. Similarly, given the higher degree of similarity between the R1 and R4 WaaL proteins, it is reasonable to assume that O-PS is attached to the β-4-linked galactose residue of the R4 core OS (Fig. 5).

The ligase enzyme, or enzyme complex, is effectively a glycosyltransferase with a more complex substrate requirement than most, but the WaaL homologs contain none of the currently known glycosyltransferase features. The WaaL protein likely functions as part of a complex that involves highly specific interactions with O-PS, lipid A core, and one or more O-PS biosynthesis proteins (10,108). As part of this complex, WaaL proteins would be involved in specific protein-protein and protein-carbohydrate interactions, and this may explain the observed specificity of WaaL for a particular acceptor structure. For example, restriction fragments carrying the *waaL* gene of *E. coli* K-12 do not complement *Salmonella waaL* mutants (60). This presumably reflects the differences in the terminal regions of the core OS acceptor. Plasmids carrying the *waaL* gene of *E. coli* R2, however, are capable of complementing a *Salmonella waaL* mutant (107,149). This likely reflects the similar nature of the core OS structures between *Salmonella* and *E. coli* R2. Cross-complementation studies involving the six known *waaL* genes are currently underway in the Whitfield laboratory in an attempt to clearly define their core OS acceptor specificity.

REGULATION OF CORE BIOSYNTHESIS

The core OS biosynthesis gene cluster consists of three operons (Fig. 6), which we will refer to as the *gmhD-*, *waaQ-*, and *waaA-*containing operons, an assignment based on the first gene of the transcriptional unit. Below, we describe the factors involved in the regulation of these operons and discuss implications for coordinated regulation of cell-surface components.

Regulation of the *gmhD* Operon

The *gmhD* gene is the first of a block of four genes (*gmhD-waaFCL*) that form an operon at one end of the core OS biosynthesis region (Fig. 6) (10,45,47). The *gmhD*, *waaF*, and *waaC* genes encode proteins involved in the biosynthesis and transfer of heptose. The region of DNA immediately upstream of *gmhD* from *E. coli* K-12 contains three putative promoters, designated P1 (*gmhD* proximal), P2, and P3 (*gmhD* distal, 130–160 bp upstream of the *gmhD* start codon) (119). P1 has a consensus σ^{70} promoter sequence, P2 is an *rpoN*-dependent promoter, and P3 is a heat-shock promoter. As confirmation of the physiological relevance of the P3 promoter, the *htrM* gene of *E. coli* K-12 was identified as a heat-shock–responsive gene required for growth above 43°C; *htrM* and *gmhD* are the same gene (119). The fact that the *gmhD* gene is heat-shock–responsive has been interpreted as indicating that the heptose domain of LPS, at least in *E. coli* K-12, is required for growth at elevated temperatures (4,119). Interestingly, the region upstream of *gmhD* in *Salmonella* contains only P1, suggesting fundamental differences in *gmhD*-operon regulation, specifically the absence of heat-shock regulation (46). *E. coli* K-12 has an extra 1.2 kb of DNA containing P2 and P3, which is entirely absent in *Salmonella* (46).

Regulation of the *waaQ* and *waaA* Operons

There are approximately 400–500 bp of intervening DNA between the divergently transcribed *waaQ* and *waaA* operons (Fig. 6). The *waaQ* operon is the long, central operon of the core OS biosynthesis gene cluster, which contains genes necessary for the biosynthesis of the outer core and for core modification and/or decoration. The *waaA* transcript contains *waaA* and a gene (designated 18K) of unknown function.

It has been known for some time that the *rfaH* locus is necessary for the production of full-length core OS molecules. The *rfaH* designation has not been changed because it has well-documented functions beyond its involvement in core OS biosynthesis (see below). The *rfaH* gene maps at 86 min, outside of the core OS biosynthesis region. Defects in *rfaH* result in the production of LPS chemotypes ranging from Ra to Rc (120). Originally, it was assumed that *rfaH* was the UDP-galactose:(glucosyl) lipopolysaccharide α-1,3-galactosyltransferase since the phenotype of these mutants was similar to that of *galE* mutants in *Salmonella*. However, in *E. coli* K-12, *rfaH* mutants were also found to have core defects, despite the different core OS structure, suggesting a different role for *rfaH* (121).

The *rfaH* locus is required for expression of a number of systems in *E. coli* and *Salmonella*. These include hemolysin (122), F-factor (123,124), LPS (120,121,124,125), and group II capsules (126,127). There is now overwhelming data to indicate that RfaH is a transcriptional antiterminator (153). Transcription of the *traYZ* operon (part of the F factor) is prematurely terminated in *rfaH* mutants compared with *rfaH*$^+$ cells (128,129), and suppressors of *rfaH* map to *rho* and *rpoBC* indicating that RfaH operates at *rho*-dependent termination sites (130). Whether or not RfaH acts at *rho*-independent termination sites is still in question. Reporter fusions throughout the *waaQ* operon indicate that expression of most, if not all of the genes of the operon are regulated by *rfaH* (125,131). The effect of *rfaH* mutations is more dramatic on promoter-distal gene fusions (125). Although initial findings indicated the possible existence of secondary promoters within the *waaQ*-operon (132), a more recent study using polar mutations has more or less ruled this out (47). Further, Brazas et al. have reported the presence of multiple termination sequences throughout the *waaQ* operon (131). Thus it would appear that there is a requirement for an antitermination mechanism that would allow transcription of distal genes in the operon.

The intervening DNA between *waaQ* and *waaA* appears to be a complex regulatory region. Two promoters, oriented opposite one another, are identified in this region, and the respective 5' mRNA start sites are mapped (33,133). The close proximity of these transcriptional start sites may facilitate use of common regulatory elements to coordinate their expression. Regions that resemble −10 and −35 consensus sequences are not identifiable between *waaA* and *waaQ* (133,134), which may not be surprising given that the transcription of these operons is likely highly regulated.

A 39 bp sequence, which is designated JUMPStart (for just upstream of many polysaccharide-associated gene starts), is found in the noncoding region upstream of *waaQ* (134). Part or all of this sequence has been identified in regulatory regions for operons that include *rfb*, group II *kps, hly,* and *tra*, suggesting a common regulatory mode (134,135). An 8 bp region designated *ops* (for operon polarity suppressor) is found within the highly conserved JUMPStart sequence and is required for operon polarity suppression through the *hly* operon (135,153). It is now known that JUMPStart sequences and RfaH protein are not required for *hly* transcription initiation (136). However, recent data suggest that the JUMPStart sequence may cause RNA polymerase to pause, and it is suggested that this may allow time for the transcription complex to be modified (possibly by RfaH) to a more processive form (136,152). Due to similarities with NusG, Bailey et al. have suggested that RfaH may function in an analogous manner by interacting with Rho and RNA polymerase at the transcribed *ops* element and forming a complex capable of extending transcription through certain termination elements (137). This process occurs in N-mediated antitermination of phage λ (138). However, it cannot be ruled out that RfaH recognizes the DNA form of *ops* (or JUMPStart) before interacting with RNA polymerase to form a termination-resistant complex, similar to the Q-mediated antitermination mechanism of phage λ (137,138). It appears that the JUMPStart sequences are oriented in such a manner as to regulate expression of genes in the long, *waaQ* operon. Moreover, given the small size of *waaA* operon (two genes), it seems that there is no need for the actions of antiterminators such as RfaH. Data concerning promoter activity for the *waaA* operon are limited.

It is interesting to note that RfaH is involved in the coordinate regulation of a variety of cell surface–associated components, although it is perhaps not surprising that both O-PS biosynthesis and core OS biosynthesis gene clusters are regulated by a common mechanism since the production of O-PS and lipid A core are intimately dependent on one another.

UNRESOLVED ISSUES

Introduction

In *Salmonella* and *E. coli* K-12, putative functions have been assigned to many of the core biosynthesis genes, based on their ability to complement known core defects or by homology to other genes of known function. A few genes, however, have been identified through sequencing whose function remains unresolved. These genes, which include *waaS*, *wabA*, and *waaZ*, have been given names that link them to core OS biosynthesis (and reflect their location in the major core OS assembly operon), but they are not currently associated with defined defects in the core OS structure. Similarly there are some structural features of LPS molecules for which the enzymes and relevant genes are unknown. A good example is the rhamnose substituent found in *E. coli* K-12 core OS.

Rhamnose Substituents in Inner Core OS

Rhamnose has been detected in the *E. coli* K-12 inner core region, attached to the KdoII residue as a partial substituent (66,139). The biosynthetic precursor TDP-rhamnose is encoded by the *rmlABCD* (formerly *rfbABCD*) genes located in the O-PS biosynthesis region of *E. coli* K-12 although enzymes for part of the TDP-rhamnose pathway are duplicated in the *wec* (formerly *rff*) (enterobacterial common antigen biosynthesis) gene cluster (140). The original O-PS (serotype O16) of *E. coli* K-12 contained rhamnose residues (141), but *E. coli* K-12 strains of the Y10 lineage harbor a mutation in the *rmlD* gene and are therefore unable to make TDP-rhamnose, which results in the typical R-LPS phenotype of these strains (141,142). The *rmlD* mutation also alters the LPS banding pattern in SDS-PAGE, and it has been proposed that this observation reflects the loss of the core OS rhamnose substituents (60,143). However, in the absence of chemical analysis of LPS, such conclusions should be viewed as tentative. The identity of the gene encoding rhamnosyl transferase activity is unknown.

Does *E. coli* K-12 Have an Alternative LOS Pathway?

The functions of the products of *waaQ*, *-S*, and *-Z* have been studied in recent years. It has been suggested that these three genes play an integral role in the production of two forms of LPS in *E. coli* K-12: an "LOS form," which is not ligated to O-PS but different from R-LPS, and the classical LPS form, which contains O-PS. This conclusion has been based primarily on the banding patterns of *E. coli* K-12 R-LPS derivatives in SDS-PAGE analysis, but firm conclusions have been hampered by the lack of chemical and enzymatic data. We now know that *waaQ* encodes the HepIII transferase (148).

Nonpolar *waaQ* mutants in an *E. coli* K-12 *rmlD* background show a significant reduction in one of the lipid A core bands in SDS-PAGE. The same *waaQ* derivatives were also unable to ligate significant amounts of a heterologous O antigen to the *E. coli* K-12 core OS. The O-PS was provided by a plasmid carrying the cloned O-PS biosynthesis region from *Shigella dysenteriae* type 1; this contains functional *rmlABCD* genes (60, 143). Based on SDS-PAGE examination of a number of constructs in this hybrid system, Klena et al. interpret their results as indicating that WaaQ adds a substituent required for the transfer of all or part of the O-PS to the core OS (60,143). This substituent would subsequently be removed (or processed) during O-PS ligation by a WaaL-dependent mechanism, leading to the heterogeneous SDS-PAGE profiles (10,60). Based upon these data, it was suggested that *E. coli* K-12 contains a unique form of R-LPS, which is destined to terminate without O-PS addition. This species of LPS has been termed LOS based upon similarities to LOS of *Haemophilus*, *Neisseria*, *Chlamydia*, and *Bordetella* spp., none of which contain O-PS–capped lipid A core molecules. The LOS band in SDS-PAGE appears to carry the substitution added by WaaQ, but this substitution is not required for LOS synthesis (60).

Sequencing of *E. coli* K-12 DNA downstream of *waaP* revealed open reading frames designated *waaS* and *waaZ*. The corresponding gene products are predicted to be soluble or peripheral membrane-associated proteins (97,108) with no similarity to known glycosyltransferases. Nonpolar mutations in the *waaS* or *waaZ* genes do not appear to affect the core OS banding pattern of *rmlD* derivatives in SDS-PAGE, but slight differences are detected in the lipid A core bands in the presence of an O-PS. These observations are interpreted as the products of *waaS* and *waaZ* (but not *waaL*) affecting the production of an unsubstituted form of LPS in *E. coli* K-12 to give the LOS form (60,143). In SDS-PAGE profiles, *waaS* mutations in an $rmlD^+$ background seem to mimic the *rmlD* defect (143). It is conceivable that *waaS* encodes a rhamnosyltransferase. Alternatively, and in the absence of any similarity to known glycosyltransferases, WaaS may be required for the function of an as yet unidentified rhamnosyltransferase enzyme. The presence of rhamnose on the core OS might then be a signal that acts to switch from the synthesis of S-LPS molecules to the LOS pathway (143).

Based on SDS-PAGE profiles, it has been suggested that *Salmonella* lacks the equivalent LOS pathway (10). Consistent with this, the core OS biosynthesis region of *Salmonella* LT2 lacks *waaS*, although remnants of DNA from the appropriate region of the cluster can be translated by computer to give WaaS-related peptides (144). It is possible that this region of the *Salmonella* LT2 core OS biosynthesis gene cluster contains the remnants of an ancestral *waaS* gene. However, a *waaZ* homolog was identified in *Salmonella* whose predicted product shows greater than 80% similarity with the *E. coli* K-12 WaaZ protein.

One would expect that an alternative system for the production of an LOS form of LPS would not be confined to *E. coli* K-12. As already indicated, all of the *E. coli* core types have a highly conserved WaaQ homolog. However, R2 is the only other *E. coli* core type strain that possesses *waaS* and *waaZ* homologs. The predicted product of the *E. coli* R2 *waaS* homolog has only limited similarity to its *E. coli* K-12 counterpart (149). In contrast, WaaZ homologs from *Salmonella* and *E. coli* K-12 and R2 are highly conserved with the greatest similarity (approximately 90%) shared by the *E. coli* proteins. A close relationship to the *Salmonella* gene content is certainly expected, given their similar core OS structures. Interestingly, the core OS biosynthesis gene clusters of *E. coli* R1, R3, and R4 do not contain any intervening DNA between *waaP* and *waaI* or *waaO* (Fig. 6) and lack both *waaS* and *waaZ* homologs. There is no clear evidence of an LOS form in SDS-PAGE profiles of R1, R2, R3, and R4 LPS (D. E. Heinrichs and C. Whitfield, unpublished).

To date, any LOS-processing pathway seems to be confined to *E. coli* K-12. It is noteworthy to mention that strains with the most clinically prevalent core types (i.e., *Salmonella* and *E. coli* R1) (145) appear to lack the LOS pathway and are missing one or more of the requisite genes. Although the existence of homologs of these genes outside of the core OS biosynthesis gene cluster cannot be ruled out, this raises important questions concerning the biological significance of the alternate pathway. Nevertheless, the LOS pathway remains an intriguing hypothesis and warrants further investigation. Future studies should concentrate on a complete structural elucidation of the LPS in nonpolar chromosomal *waaS*, *wabA*, and *waaZ* mutants. To allow unambiguous interpretation, these systems would be best studied in a background retaining the ability to synthesize and ligate the "natural" O-PS from chromosomal (rather than plasmid-encoded) genetic determinants.

CONCLUDING REMARKS

In recent years, the sequencing of the *E. coli* K-12 and most of the *Salmonella* LT2 core OS biosynthesis gene clusters has resulted in great advances in our knowledge of the genetics and biosynthesis of the core OS. Completion of the genomes of various bacteria will undoubtedly lead to a rapid advancement in this field. Indeed, the application of genome sequencing, directed chromosomal mutagenesis, and careful chemical characterization to studies of *H. influenzae* LPS provides a classic example (146). As an increasing number of core OS biosynthesis genes are sequenced, conserved features will be established, and as studies of the various core types from *E. coli* have shown, important differences will be revealed. However, it must be recognized that homologies identified in GenBank searches are useful but are not definitively predictive of function. Many currently assigned functions are based upon assumptions or putative functions ascribed to other homologs. The limitation in resolution of the various core OS assembly systems will be the availability of accurate chemical structures to allow initial interpretations of gene functions. Biochemical approaches will still be essential in order to move the data from tentative to unequivocal interpretations.

ACKNOWLEDGMENTS

D.E.H. is a recipient of a Natural Sciences and Engineering Research Council of Canada (NSERC) Postdoctoral Fellowship. Research in the authors' laboratories is generously supported by funding from NSERC and the Medical Research Council (MRC) of Canada (M.A.V.) and from NSERC, MRC, and the Canadian Bacterial Diseases Network (C.W.).

REFERENCES

1. Preston A, Mandrell RE, Gibson BW, Apicella MA. The lipooligosaccharides of pathogenic gram-negative bacteria. Crit Rev Microbiol 1996; 22:139–180.
2. Olsthoorn MMA, et al. Identification of a novel core type in *Salmonella* lipopolysaccharide. Complete structural analysis of the core region of the lipopolysaccharide from *Salmonella enterica* sv Arizonae 062. J Biol Chem 1998; 273:3817–3829.
3. Mäkelä PH, Stocker BAD. Genetics of lipopolysaccharide. In: Rietschel ET, ed. Handbook of Endotoxin. Vol. I: Chemistry of Endotoxin. Amsterdam: Elsevier Science Publishers, B.V., 1984:59–137.
4. Raetz CRH. Bacterial lipopolysaccharides: a remarkable family of bioactive macroamphiphiles. In: Neidhardt FC, Curtiss III R, Ingraham JL, et al., eds. *Escherichia coli* and *Salmonella*. Cellular and Molecular Biology. Vol. 1. Washington, DC: ASM Press, 1996: 1035–1036.
5. Rothfield L, Romeo D. Role of lipids in the biosynthesis of the bacterial cell envelope. Bacteriol Rev 1971; 35:14–38.
6. Rick PD. Lipopolysaccharide biosynthesis. In: Neidhardt FC, Ingraham JL, Low KB, Magasanik B, Schaechter M, Umbarger HE, eds. *Escherichia coli* and *Salmonella typhimurium*: cellular and molecular biology. Vol. 1. Washington, DC: American Society for Microbiology, 1987:648–662.
7. Berlyn MKB, Low KB, Rudd KE. Linkage map of *Escherichia coli* K-12, Edition 9. In: Neidhardt FC, Curtiss III R, Ingraham JL, et al., eds. *Escherichia coli* and *Salmonella*: Cellular and Molecular Biology. Vol. 2. Washington, DC: ASM Press, 1996:1715–1902.
8. Sanderson KE, Hessel A, Liu S-L, Rudd KE. The genetic map of *Salmonella typhimurium*, Edition VIII. In: Neidhardt FC, Curtiss III R, Ingraham JL, et al., eds. *Escherichia coli* and *Salmonella*: Cellular and Molecular Biology. Vol. 2. Washington, DC: ASM Press, 1996:1903–1999.
9. Reeves PR, Hobbs M, Valvano MA, et al. Bacterial polysaccharide synthesis and gene nomenclature. Trends Microbiol 1996; 4:495–503.
10. Schnaitman CA, Klena JD. Genetics of lipopolysaccharide biosynthesis in enteric bacteria. Microbiol Rev 1993; 57:655–682.
11. Kawahara K, Brade H, Rietschel ET, Zähringer U. Studies on the chemical structure of the core-lipid A region of the lipopolysaccharide of *Acinetobacter calcoaceticus* NCTC 10305. Eur J Biochem 1987; 163: 489–495.
12. Knirel YA, Moll H, Zähringer U. Structural study of a highly O-acetylated core of *Legionella pneumophila* serogroup 1 lipopolysaccharide. Carbohydr Res 1996; 293:223–234.
13. Kadrmas JL, Brozek KA, Raetz CRH. Lipopolysaccharide core glycosylation in *Rhizobium leguminosarum*. An unusual mannosyl transferase resembling the heptosyl transferase I of *Escherichia coli*. J Biol Chem 1996; 271:32119–32125.
14. Galloway SM, Raetz CRH. A mutant of *Escherichia coli* defective in the first step of endotoxin biosynthesis. J Biol Chem 1990; 265:6394–6402.
15. Rick PD, Fung LW-M, Ho C, Osborn MJ. Lipid A mutants of *Salmonella typhimurium*. Purification and characterization of a lipid A precursor produced by a mutant in 3-deoxy-D-mannooctulosonate-8-phosphate synthetase. J Biol Chem 1977; 252:4904–4912.
16. Rick PD, Osborn MJ. Lipid A mutants of *Salmonella typhimurium*. Characterization of a conditional lethal mutant in 3-deoxy-D-mannooctulosonate-8-phosphate synthetase. J Biol Chem 1977; 252:4895–4903.
17. Helander IM, Lindner B, Brade H, et al. Chemical structure of the lipopolysaccharide of *Haemophilus influenzae* strain I-69 Rd$^-$/b$^+$: description of a novel deep-rough chemotype. Eur J Biochem 1988; 177: 483–492.
18. Zamze SE, Ferguson MAJ, Moxon ER, Dwek RA, Rademacher TW. Identification of phosphorylated 3-deoxy-*manno*-octulosonic acid as a component of *Haemophilus influenzae* lipopolysaccharide. Biochem J 1987; 245:583–587.
19. Nikaido H. Outer membrane. In: Neidhardt FC, Curtiss III R, Ingraham JL, et al., eds. *Escherichia coli* and *Salmonella*. Cellular and Molecular Biology. Vol. 1. Washington, DC: ASM Press, 1996:29–47.
20. Hammond SM, Lambert PA, Rycroft AN. The cell envelope in bacterial disease. In: Hammond SM, Lambert PA, Rycroft AN, eds. The Bacterial Cell Surface. Washington, DC: Kapitan Szabo, 1984:147–193.
21. Parker CT, Kloser AW, Schnaitman CA, Stein MA, Gottesman S, Gibson BW. Role of the *rfaG* and *rfaP* genes in determining the lipopolysaccharide core structure and cell surface properties of *Escherichia coli* K-12. J Bacteriol 1992; 174:2525–2538.
22. Ames GF-L, Spudich EN, Nikaido H. Protein composition of the outer membrane of *Salmonella typhimurium*: effect of lipopolysaccharide mutations. J Bacteriol 1974; 117:406–416.
23. Lehmann V, Rupprecht E, Osborn MJ. Isolation of mutants conditionally blocked in the biosynthesis of the 3-deoxy-D-mannooctulosonic acid-lipid A part of lipopolysaccharides derived from *Salmonella typhimurium*. Eur J Biochem 1977; 76:41–49.
24. Ghalambor MA, Levine EM, Heath EC. The biosynthesis of cell wall lipopolysaccharide in *Escherichia coli*. III. Isolation and characterization of 3-deoxyoctulosonic acid. J Biol Chem 1966; 241:3207–3215.

25. Levine DH, Racker E. Condensation of arabinose-5-phosphate and phosphorylenolpyruvate by 2-keto-3-deoxy-8-phosphooctonic acid synthetase. J Biol Chem 1959; 234:2532–2539.

26. Lim R, Cohen SS. D-Phosphoarabinoisomerase and D-ribulokinase in *Escherichia coli*. J Biol Chem 1966; 241:4304–4315.

27. Strohmaier H, Remler P, Renner W, Högenauer G. Expression of genes *kdsA* and *kdsB* involved in 3-deoxy-D-*manno*-octulosonic acid metabolism and biosynthesis of enterobacterial lipopolysaccharide is growth phase regulated primarily at the transcriptional level in *Escherichia coli* K-12. J Bacteriol 1995; 177:4488–4500.

28. Brabetz W, Brade H. Molecular cloning, sequence analysis and functional characterization of the gene *kdsA*, encoding 3-deoxy-D-*manno*-2-octulosonate-8-phosphate synthase of *Chlamydia psittaci* 6BC. Eur J Biochem 1997; 244:66–73.

29. Pazzani C, Rosenow C, Boulnois GJ, Bronner D, Jann K, Roberts IS. Molecular analysis of region 1 of *Escherichia coli* K5 antigen gene cluster: a region encoding proteins involved in cell surface expression of capsular polysaccharide. J Bacteriol 1993; 175:5978–5983.

30. Goldman RC, Kohlbrenner WE, Lartey P, Pernet A. Antibacterial agents specifically inhibiting lipopolysaccharide synthesis. Nature 1987; 329:162–164.

31. Hammond SM, Claesson A, Jansson AM, et al. A new class of synthetic antimicrobials acting on lipopolysaccharide biosynthesis. Nature 1987; 327:730–732.

32. Hammond SM, Lambert PA, Rycroft AN. The envelope of gram-negative bacteria. In: Hammond SM, Lambert PA, Rycroft AN, eds. The Bacterial Cell Surface. Washington, DC: Kapitan Szabo, 1984:57–118.

33. Clementz T, Raetz CRH. A gene coding for 3-deoxy-D-*manno*-octulosonic-acid transferase in *Escherichia coli*. Identification, mapping, cloning, and sequencing. J Biol Chem 1991; 266:9687–9696.

34. Belunis CJ, Mdluli KE, Raetz CRH, Nano FE. A novel 3-deoxy-D-*manno*-octulosonic acid transferase from *Chlamydia trachomatis* required for expression of the genus-specific epitope. J Biol Chem 1992; 267:18702–18707.

35. Eidels L, Osborn MJ. Lipopolysaccharide and aldoheptose biosynthesis in transketolase mutants of *Salmonella typhimurium*. Proc Natl Acad Sci USA 1971; 68:1673–1677.

36. Tamaki S, Sato T, Matsuhashi M. Role of lipopolysaccharides in antibiotic resistance and bacteriophage adsorption of *Escherichia coli* K-12. J Bacteriol 1971; 105:968–975.

37. Curtiss III R, Charamella LJ, Stallions DR, Mays JA. Parental functions during conjugation in *Escherichia coli* K-12. Bacteriol Rev 1968; 32:320–348.

38. Havekes LM, Lugtenberg BJJ, Hoekstra WPM. Conjugation deficient *Escherichia coli* K-12 F$^-$ mutants with heptose-less lipopolysaccharide. Mol Gen Genet 1976; 146:43–50.

39. Preston A, Maskell D, Johnson A, Moxon ER. Altered lipopolysaccharide characteristic of the I69 phenotype in *Haemophilus influenzae* results from mutations in a novel gene, *isn*. J Bacteriol 1996; 178:396–402.

40. Brooke JS, Valvano MA. Biosynthesis of inner core lipopolysaccharide in enteric bacteria identification and characterization of a conserved phosphoheptose isomerase. J Biol Chem 1996; 271:3608–3614.

41. Brooke JS, Valvano MA. Molecular cloning of the *Haemophilus influenzae gmhA* (*lpcA*) gene encoding a phosphoheptose isomerase required for lipooligosaccharide biosynthesis. J Bacteriol 1996; 178:3339–3341.

42. Peschke U, Schmidt H, Zhang H-Z, Piepersberg W. Molecular characterization of the lincomycin-production gene cluster of *Streptomyces lincolnensis* 78–11. Mol Microbiol 1995; 16:1137–1156.

43. Coleman Jr, WG. The *rfaD* gene codes for ADP-L-*glycero*-D-mannoheptose-6-epimerase. J Biol Chem 1983; 258:1985–1990.

44. Coleman Jr, WG, Leive L. Two mutations which affect the barrier function of the *Escherichia coli* K-12 outer membrane. J Bacteriol 1979; 139:899–910.

45. Schnaitman CA, Parker CT, Klena JD, et al. Physical maps of the *rfa* loci of *Escherichia coli* K-12 and *Salmonella typhimurium*. J Bacteriol 1991; 173:7410–7411.

46. Sirisena DM, MacLachlan PR, Liu S-L, Hessel A, Sanderson KE. Molecular analysis of the *rfaD* gene, for heptose synthesis, and the *rfaF* gene, for heptose transfer, in lipopolysaccharide synthesis in *Salmonella typhimurium*. J Bacteriol 1994; 176:2379–2385.

47. Roncero C, Casadaban MJ. Genetic analysis of the genes involved in synthesis of the lipopolysaccharide core in *Escherichia coli* K-12: three operons in the *rfa* locus. J Bacteriol 1992; 174:3250–3260.

48. Nichols WA, Gibson BW, Melaugh W, Lee N-G, Sunshine M, Apicella MA. Identification of the ADP-L-*glycero*-D-*manno*-heptose-6-epimerase. (*rfaD*) and heptosyltransferase II (*rfaF*) biosynthesis genes from nontypeable *Haemophilus influenzae* 2019. Infect Immun 1997; 65:1377–1386.

49. Drazek ES, Stein DC, Deal CD. A mutation in the *Neisseria gonorrhoeae rfaD* homolog results in altered lipooligosaccharide expression. J Bacteriol 1995; 177:2321–2327.

50. Sirisena DM, Brozek KA, MacLachlan PR, Sanderson KE, Raetz CRH. The *rfaC* gene of *Salmonella typhimurium*. Cloning, sequencing, and enzymatic function in heptose transfer to lipopolysaccharide. J Biol Chem 1992; 267:18874–18884.

51. Valvano MA, et al. Molecular Characterization of an *Escherichia coli* gene encoding a bifunctional protein involved in the biosynthesis of the lipopolysaccharide core precursor ADP-L-*glycero*-D-*manno*-heptose. (submitted).

52. van Heeswijk WC, Rabenberg M, Westerhoff HV, Kahn D. The genes of the glutamine synthetase adenylylation cascade are not regulated by nitrogen in *Escherichia coli*. Mol Microbiol 1993; 9:443–457.

53. Bork P, Koonin EV, Holm L, Sander C. The cytidylyltransferase superfamily: identification of the nucleotide-binding site and fold prediction. Proteins 1995; 22:259–266.

54. Freter A, Bowien B. Identification of a novel gene, *aut*, involved in autotrophic growth of *Alcaligenes eutrophus*. J Bacteriol 1994; 176:5401–5408.

55. Lee N-G, Sunshine MG, Apicella MA. Molecular cloning and characterization of the nontypeable *Haemophilus influenzae* 2019 *rfaE* gene required for lipopolysaccharide biosynthesis. Infect Immun 1995; 63:818–824.
56. Levin JC, Stein DC. Cloning, complementation, and characterization of an *rfaE* homolog from *Neisseria gonorrhoeae*. J Bacteriol 1996; 178:4571–4575.
57. Sanderson KE, Van Wyngaarden J, Lüderitz O, Stocker BAD. Rough mutants of *Salmonella typhimurium* with defects in the heptose region of the lipopolysaccharide core. Can J Microbiol 1974; 20:1127–1134.
58. Coleman Jr, WG, Deshpande KS. New *cysE-pyrE*-linked *rfa* mutation in *Escherichia coli* K-12 that results in a heptoseless lipopolysaccharide. J Bacteriol 1985; 161:1209–1214.
59. Chen L, Coleman Jr, WG. Cloning and characterization of the *Escherichia coli* K-12 *rfa-2* (*rfaC*) gene, a gene required for lipopolysaccharide inner core synthesis. J Bacteriol 1993; 175:2534–2540.
60. Klena JD, Ashford II RS, Schnaitman CA. Role of *Escherichia coli* K-12 *rfa* genes and the *rfp* gene of *Shigella dysenteriae* 1 in generation of lipopolysaccharide core heterogeneity and attachment of O antigen. J Bacteriol 1992; 174:7297–7307.
61. de Kievit TR, Lam JS. Isolation and characterization of two genes, *waaC* (*rfaC*) and *waaF* (*rfaF*), involved in *Pseudomonas aeruginosa* serotype O5 inner-core biosynthesis. J Bacteriol 1997; 179:3451–3457.
62. Allen A, Maskell D. The identification, cloning and mutagenesis of a genetic locus required for lipopolysaccharide biosynthesis in *Bordetella pertussis*. Mol Microbiol 1996; 19:37–52.
63. Zhou D, Lee N-G, Apicella MA. Lipooligosaccharide biosynthesis in *Neisseria gonorrhoeae*: cloning, identification and characterization of the α-1,5-heptosyltransferase I gene (*rfaC*). Mol Microbiol 1994; 14:609–618.
64. Wilkinson RG, Gemski P, Stocker BAD. Non-smooth mutants of *Salmonella typhimurium*: differentiation by phage sensitivity and genetic mapping. J Gen Microbiol 1972; 70:527–554.
65. Jennings MP, Bisercic M, Dunn KL, et al. Cloning and molecular analysis of the Isi1 (*rfaF*) gene of *Neisseria meningitidis* which encodes a heptosyl-2-transferase involved in LPS biosynthesis: evaluation of surface exposed carbohydrates in LPS mediated toxicity for human endothelial cells. Microb Pathog 1995; 19:391–407.
66. Holst O, Brade H. Chemical structure of the core region of lipopolysaccharides. In: Morrison DC, Ryan JL, eds. Bacterial Endotoxic Lipopolysaccharides. Vol. I. Boca Raton, FL: CRC Press, 1992:134–170.
67. Hämmerling G, Lehmann V, Lüdertiz O. Structural studies on the heptose region of *Salmonella* lipopolysaccharides. Eur J Biochem 1973; 38:453–458.
68. Parker CT, Pradel E, Schnaitman CA. Identification and sequences of the lipopolysaccharide core biosynthetic genes *rfaQ*, *rfaP*, and *rfaG* of *Escherichia coli* K-12. J Bacteriol 1992; 174:930–934.
69. Brabetz W, Müller-Loennies S, Holst O, Brade H. Deletion of the heptosyltransferase genes *rfaC* and *rfaF* in *Escherichia coli* K-12 results in an Re-type lipopolysaccharide with a high degree of 2-aminoethanol phosphate substitution. Eur J Biochem 1997; 247:716–724.
70. Vaara M. Antibiotic-supersusceptible mutants of *Escherichia coli* and *Salmonella typhimurium*. Antimicrob Agents Chemother 1993; 37:2255–2260.
71. Stanley PLD, Diaz P, Bailey MJA, Gygi D, Juarez A, Hughes C. Loss of activity in the secreted form of *Escherichia coli* haemolysin caused by an *rfaP* lesion in core lipopolysaccharide assembly. Mol Microbiol 1993; 10:781–787.
72. Helander IM, Kilpeläinen I, Vaara M. Phosphate groups in lipopolysaccharides of *Salmonella typhimurium rfaP* mutants. FEBS Lett 1997; 409:457–460.
73. Helander IM, Vaara M, Sukupolvi S, et al. *rfaP* mutants of *Salmonella typhimurium*. Eur J Biochem 1989; 185:541–546.
74. Mühlradt P, Risse HJ, Lüderitz O, Westphal O. Biochemical studies on lipopolysaccharides of *Salmonella* R mutants. 5. Evidence for a phosphorylating enzyme in lipopolysaccharide biosynthesis. Eur J Biochem 1968; 4:139–145.
75. Mühlradt P. Biosynthesis of *Salmonella* lipopolysaccharide. The *in vitro* transfer of phosphate to the heptose moiety of the core. Eur J Biochem 1969; 11:241–248.
76. Mühlradt PF. Biosynthesis of *Salmonella* lipopolysaccharide. Studies on the transfer of glucose, galactose, and phosphate to the core in a cell free system. Eur J Biochem 1971; 18:20–27.
77. Hasin M, Kennedy EP. Role of phosphatidylethanolamine in the biosynthesis of pyrophosphoethanolamine residues in the lipopolysaccharide of *Escherichia coli*. J Biol Chem 1982; 257:12475–12477.
78. DeChavigny A, Heacock PN, Dowhan W. Sequence and inactivation of the *pss* gene of *Escherichia coli*. Phosphatidylethanolamine may not be essential for cell viability. J Biol Chem 1991; 266:5323–5332.
79. Severn WB, Kelly RF, Richards JC, Whitfield C. Structure of the core oligosaccharide in the serotype O8 lipopolysaccharide from *Klebsiella pneumoniae*. J Bacteriol 1996; 178:1731–1741.
80. Nikaido H. Studies on the biosynthesis of cell-wall polysaccharide in mutant strains of *Salmonella*, I. Proc Natl Acad Sci USA 1962; 48:1337–1341.
81. Nikaido H. Studies on the biosynthesis of cell wall polysaccharide in mutant strains of *Salmonella*, II. Proc Natl Acad Sci USA 1962; 48:1542–1548.
82. Rothfield L, Horecker BL. The role of cell-wall lipid in the biosynthesis of bacterial lipopolysaccharide. Proc Natl Acad Sci USA 1964; 52:939–946.
83. Rothfield L, Pearlman M. The role of cell envelope phospholipid in the enzymatic synthesis of bacterial lipopolysaccharide. J Biol Chem 1966; 241:1386–1392.
84. Müller E, Hinckley A, Rothfield L. Studies of phospholipid-requiring bacterial enzymes. III. Purification and properties of uridine diphosphate glucose: lipopolysaccharide glucosyltransferase I. J Biol Chem 1972; 247:2614–2622.
85. Hinckley A, Müller E, Rothfield L. Reassembly of a membrane-bound multienzyme system. I. Formation

85. of a particle containing phosphatidylethanolamine, lipopolysaccharide, and two glycosyltransferase enzymes. J Biol Chem 1972; 247:2623–2628.
86. Rothfield L, Takeshita M. The role of cell envelope phospholipid in the enzymatic synthesis of bacterial lipopolysaccharide: binding of transferase enzymes to a lipopolysaccharide-lipid complexz. Biochem Biophys Res Comm 1965; 20:521–527.
87. Romeo D, Girard A, Rothfield L. Reconstitution of a functional membrane enzyme system in a monomolecular film. I. Formation of a mixed monolayer of lipopolysaccharide and phospholipid. J Mol Biol 1970; 53:475–490.
88. Romeo D, Hinckley A, Rothfield L. Reconstitution of a functional membrane enzyme system in a monomolecular film. II. Formation of a functional ternary film of lipopolysaccharide, phospholipid and transferase enzyme. J Mol Biol 1970; 53:491–501.
89. Creeger ES, Rothfield LI. Cloning of genes for bacterial glycosyltransferases. I. Selection of hybrid plasmids carrying genes for two glucosyltransferases. J Biol Chem 1979; 254:804–810.
90. Kadam SK, Rehemtulla A, Sanderson KE. Cloning of *rfaG, B, I,* and *J* genes for glycosyltransferase enzymes for synthesis of the lipopolysaccharide core of *Salmonella typhimurium.* J Bacteriol 1985; 161:277–284.
91. Walsh AG, Matewish M, Burrows LL, Lam JS. *rfa* genes of *Pseudomonas aeruginosa* Abstract #B-211, 97th American Society for Microbiology General Meeting, Miami Beach, FL, 1997.
92. Coyne Jr. MJ, Goldberg JB. Cloning and sequence analysis of lipopolysaccharide (LPS)-core genes of *Pseudomonas aeruginosa.* Abstract #D-44, 97th American Society for Microbiology General Meeting, Miami Beach, FL, 1997.
93. Komeda Y, Icho T, Iino T. Effects of *galU* mutation on flagellar formation in *Escherichia coli.* J Bacteriol 1977; 129:908–915.
94. Osborn MJ. Biochemical characterization of mutants of *Salmonella typhimurium* lacking glucosyl or galactosyl lipopolysaccharide transferases. Nature 1968; 217:957–960.
95. Hudson H, Lindberg AA, Stocker BAD. Lipopolysaccharide core defects in *Salmonella typhimurium* which are resistant to Felix O phage but retain smooth character. J Gen Microbiol 1978; 109:97–112.
96. Wollin R, Creeger ES, Rothfield LI, Stocker BAD, Lindberg AA. *Salmonella typhimurium* mutants defective in UDP-D-galactose:lipopolysaccharide α-1,6-D-galactosyltransferase. Structural, immunochemical, and enzymologic studies of *rfaB* mutants. J Biol Chem 1983; 258:3769–3774.
97. Pradel E, Parker CT, Schnaitman CA. Structures of the *rfaB, rfaI, rfaJ,* and *rfaS* genes of *Escherichia coli* K-12 and their roles in assembly of the lipopolysaccharide core. J Bacteriol 1992; 174:4736–4745.
98. Rothfield L, Osborn MJ, Horecker BL. Biosynthesis of bacterial lipopolysaccharide. II. Incorporation of glucose and galactose catalyzed by particulate and soluble enzymes in *Salmonella.* J Biol Chem 1964; 239:2788–2795.
99. Wilkinson RG, Stocker BAD. Genetics and cultural properties of mutants of *Salmonella typhimurium* lacking glucosyl or galactosyl lipopolysaccharide transferases. Nature 1968; 217:955–957.
100. Endo A, Rothfield L. Studies of a phospholipid-requiring bacterial enzyme. I. Purification and properties of uridine diphosphate galactose:lipopolysaccharide α-3-galactosyl transferase. Biochemistry 1969; 8:3500–3507.
101. Carstenius P, Flock J-I, Lindberg A. Nucleotide sequence of *rfaI* and *rfaJ* genes encoding lipopolysaccharide glycosyl transferases from *Salmonella typhimurium.* Nucl Acids Res 1990; 18:6128.
102. Galanos C, Lüderitz O, Rietschel ET, Westphal O. Newer aspects of the chemistry and biology of bacterial lipopolysaccharides, with special reference to their lipid A component. In: Goodwin TW, ed. International Review of Biochemistry, Biochemistry of Lipids II. Vol. 14. Baltimore: University Park Press, 1977:239–335.
103. Jansson P-E, Lindberg AA, Lindberg B, Wollin R. Structural studies on the hexose region of the core in lipopolysaccharides from Enterobacteriaceae. Eur J Biochem 1981; 115:571–577.
104. Holst O. Zähringer U, Brade H, Zamojski A. Structural analysis of the heptose/hexose region of the lipopolysaccharide from *Escherichia coli* K-12 strain W3100. Carbohydr Res 1991; 215:323–335.
105. Osborn MJ, D'Ari L. Enzymatic incorporation of *N*-acetylglucosamine into cell wall lipopolysaccharide in a mutant strain of *Salmonella typhimurium.* Biochem Biophys Res Comm 1964; 16:568–575.
106. MacLachlan PR, Kadam SK, Sanderson KE. Cloning, characterization, and DNA sequence of the *rfaLK* region for lipopolysaccharide synthesis in *Salmonella typhimurium* LT2. J Bacteriol 1991; 173:7151–7163.
107. Heinrichs DE, Whitfield C. Cloning and characterization of *rfaL* and *rfaK* of *Escherichia coli* F632. Abstract #B-204, 97th American Society for Microbiology, Miami Beach, FL, 1997.
108. Klena JD, Pradel E, Schnaitman CA. Comparison of lipopolysaccharide biosynthesis genes *rfaK, rfaL, rfaY,* and *rfaZ* of *Escherichia coli* K-12 and *Salmonella typhimurium.* J Bacteriol 1992; 174:4746–4752.
109. Rapin AMC, Kalckar HM. The relation of bacteriophage attachment to lipopolysaccharide structure. In: Weinbaum G, Kadid S, Ajl SJ, eds. Microbial Toxins. Vol. IV. New York: Academic Press, 1971:267–307.
110. Kahler CM, Carlson RW, Rahman MM, Martin LE, Stephens DS. Inner core biosynthesis of lipooligosaccharide (LOS) in *Neisseria meningitidis* serogroup B: identification and role in LOS assembly of the α-1,2-*N*-acetylglucosamine transferase (RfaK). J Bacteriol 1996; 178:1265–1273.
111. Mulford CA, Osborn MJ. An intermediate step in translocation of lipopolysaccharide to the outer membrane of *Salmonella typhimurium.* Proc Natl Acad Sci USA 1983; 80:1159–1163.
112. McGrath BC, Osborn MJ. Localization of the terminal steps of O-antigen synthesis in *Salmonella typhimurium.* J Bacteriol 1991; 173:649–654.

113. Marino PA, Phan KA, Osborn MJ. Energy dependence of lipopolysaccharide translocation in *Salmonella typhimurium*. J Biol Chem 1985; 260:14965–14970.
114. Marino PA, McGrath BC, Osborn MJ. Engery dependence of O-antigen synthesis in *Salmonella typhimurium*. J Bacteriol 1991; 173:3128–3133.
115. McGrath BC, Osborn MJ. Evidence for energy-dependent transposition of core lipopolysaccharide across the inner membrane of *Salmonella typhimurium*. J Bacteriol 1991; 173:3134–3137.
116. Karow M, Georgopoulos C. The essential *Escherichia coli msbA* gene, a multicopy suppressor of null mutations in the *htrB* gene, is related to the universally conserved family of ATP-dependent translocators. Mol Microbiol 1993; 7:69–79.
117. Polissi A, Georgopoulos C. Mutational analysis and properties of the *msbA* gene of *Escherichia coli*, coding for an essential ABC family transporter. Mol Microbiol 1996; 20:1221–1233.
118. Whitfield C, Amor PA, Köplin R. Modulation of the surface architecture of gram-negative bacteria by the action of surface polymer:lipid A-core ligase and by determinants of polymer chain length. Mol Microbiol 1997; 23:629–638.
119. Raina S, Georgopoulos C. The *htrM* gene, whose product is essential for *Escherichia coli* viability only at elevated temperatures, is identical to the *rfaD* gene. Nucl Acids Res 1991; 19:3811–3819.
120. Lindberg AA, Hellerqvist C-G. Rough mutants of *Salmonella typhimurium*: immunochemical and structural analysis of lipopolysaccharides from *rfaH* mutants. J Gen Microbiol 1980; 116:25–32.
121. Creeger ES, Schulte T, Rothfield LI. Regulation of membrane glycosyltransferases by the *sfrB* and *rfaH* genes of *Escherichia coli* and *Salmonella typhimurium*. J Biol Chem 1984; 259:3064–3069.
122. Bailey MJA, Koronakis V, Schmoll T, Hughes C. *Escherichia coli* HlyT protein, a transcriptional activator of haemolysin synthesis and secretion, is encoded by the *rfaH* (*sfrB*) locus required for expression of sex factor and lipopolysaccharide genes. Mol Microbiol 1992; 6:1003–1012.
123. Beutin L, Achtman M. Two *Escherichia coli* chromosomal cistrons, *sfrA* and *sfrB*, which are needed for expression of F factor *tra* functions. J Bacteriol 1979; 139:730–737.
124. Sanderson KE, Stocker BAD. Gene *rfaH*, which affects lipopolysaccharide core structure in *Salmonella typhimurium*, is required also for expression of F-factor functions. J Bacteriol 1981; 146:535–541.
125. Pradel E, Schnaitman CA. Effect of *rfaH* (*sfrB*) and temperature on expression of *rfa* genes of *Escherichia coli* K-12. J Bacteriol 1991; 173:6428–6431.
126. Stevens MP, Hänfling P, Jann B, Jann K, Roberts IS. Regulation of *Escherichia coli* K5 capsular polysaccharide expression: evidence for involvement of RfaH in the expression of group II capsules. FEMS Microbiol Lett 1994; 124:93–98.
127. Stevens MP, Clarke BR, Roberts IS. Regulation of the *Escherichia coli* K5 capsule gene cluster by transcription antitermination. Mol Microbiol 1997; 24:1001–1012.
128. Beutin L, Manning PA, Achtman M, Willetts N. *sfrA* and *sfrB* products of *Escherichia coli* are transcriptional control factors. J Bacteriol 1981; 145:840–844.
129. Gaffney D, Skurray R, Willets N. Regulation of the F conjugation genes studied by hybridization and *tra-lacZ* fusion. J Mol Biol 1983; 168:103–122.
130. Farewell A, Brazas R, Davie E, Mason J, Rothfield LI. Suppression of the abnormal phenotype of *Salmonella typhimurium rfaH* mutants by mutations in the gene for transcription termination factor Rho. J Bacteriol 1991; 173:5188–5193.
131. Brazas R, Davie E, Farewell A, Rothfield LI. Transcriptional organization of the *rfaGBIJ* locus of *Salmonella typhimurium*. J Bacteriol 1991; 173:6168–6173.
132. Austin EA, Graves JF, Hite LA, Parker CT, Schnaitman CA. Genetic analysis of lipopolysaccharide core biosynthesis by *Escherichia coli* K-12: insertion mutagenesis of the *rfa* locus. J Bacteriol 1990; 172:5312–5325.
133. Clementz T. The gene coding for 3-deoxy-*manno*-octulosonic acid transferase and the *rfaQ* gene are transcribed from divergently arranged promoters in *Escherichia coli*. J Bacteriol 1992; 174:7750–7756.
134. Hobbs M, Reeves PR. The JUMPstart sequence: a 39 bp element common to several polysaccharide gene clusters. Mol Microbiol 1994; 12:855–856.
135. Nieto JM, Bailey MJA, Hughes C, Koronakis V. Suppression of transcription polarity in the *Escherichia coli* haemolysin operon by a short upstream element shared by polysaccharide and DNA transfer determinants. Mol Microbiol 1996; 19:705–713.
136. Leeds JA, Welch RA. Enhancing transcription through the *Escherichia coli* hemolysin operon, *hlyCABD*: RfaH and upstream JUMPStart DNA sequences function together via a postinitiation mechanism. J Bacteriol 1997; 179:3519–3527.
137. Bailey MJA, Hughes C, Koronakis V. Increased distal gene transcription by the elongation factor RfaH, a specialized homologue of NusG. Mol Microbiol 1996; 22:729–737.
138. Greenblatt J, Nodwell JR, Mason SW. Transcription antitermination. Nature 1993; 364:401–406.
139. Holst O, Brade H. Isolation and identification of 3-deoxy-5-O-α-L-rhamnopyranosyl-D-*manno*-2-octulopyranosate from the inner core region of the lipopolysaccharide of *Escherichia coli* K-12. Carbohydr Res 1990; 207:327–331.
140. Marolda CL, Valvano MA. Genetic analysis of the dTDP-rhamnose biosynthesis region of *Escherichia coli* VW187 (O7:K1) *rfb* gene cluster: identification of functional homologs of *rfbB* and *rfbA* in the *rff* cluster and correct location of the *rffE* gene. J Bacteriol 1995; 177:5539–5546.
141. Stevenson G, Neal B, Liu D, et al. Structure of the O antigen of *Escherichia coli* K-12 and the sequence of its *rfb* gene cluster. J Bacteriol 1994; 176:4144–4156.
142. Liu D, Reeves PR. *Escherichia coli* K-12 regains its O antigen. Microbiology 1994; 140:49–57.
143. Klena JD, Schnaitman CA. Genes for TDP-rhamnose synthesis affect the pattern of lipopolysaccharide het-

erogeneity in *Escherichia coli* K-12. J Bacteriol 1994; 176:4003–4010.
144. Klena JD, Pradel E, Schnaitman CA. The *rfaS* gene, which is involved in production of a rough form of lipopolysaccharide core in *Escherichia coli* K-12, is not present in the *rfa* cluster of *Salmonella typhimurium*. J Bacteriol 1993; 175:1524–1527.
145. Gibb AP, Barclay GR, Poxton IR, di Padova F. Frequencies of lipopolysaccharide core types among clinical isolates of *Escherichia coli* defined with monoclonal antibodies. J Infect Dis 1992; 166:1051–1057.
146. Hood DW, Deadman ME, Allen T, et al. Use of the complete genome sequence information of *Haemophilus influenzae* strain Rd to investigate lipopolysaccharide biosynthesis. Mol Microbiol 1996; 22:951–965.
147. Heinrichs DE, et al. Molecular basis for structural diversity in the core regions of the lipopolysaccharides of *Escherichia coli* and *Salmonella enterica*. Mol Microbiol 1998; 30:221–232.
148. Yethon JA, et al. Involvement of *waaY*, *waaQ*, and *waaP* in the modification of *Escherichia coli* lipopolysaccharides, and their role in the formation of a stable outer membrane. J Biol Chem 1998; 273:26310–26316.
149. Heinrichs DE, et al. The assembly system for the lipopolysaccharide R2 core-type of *Escherichia coli* is a hybrid of those found in *Escherichia coli* K-12 and *Salmonella enterica*. Structure and function of WaaK and WaaL homologs. J Biol Chem 1998; 273:8849–8859.
150. Heinrichs DE, et al. The assembly system for the outer core portion of R1 and R4-type lipopolysaccharides of *Escherichia coli*. The R1 core-specific β-glucosyltransferase provides a novel attachment site for O polysaccharides. J Biol Chem 1998; 273:29497–29505.
151. Brade L, et al. Specificity of monoclonal antibodies against *Escherichia coli* K-12 lipopolysaccharide. J Endotox Res 1996; 3:39–47.
152. Marolda CL, Valvano M. Promoter region of the *Escherichia coli* 07-specific lipopolysaccharide gene cluster: structural and functional characterization of an upstream untranslated mRNA sequence. J Bacteriol 1998; 180:3070–3079.
153. Bailey MJ, Hughes C, Koronakis V. RfaH and the *ops* element components of a novel system controlling bacterial transcription elongation. Mol Microbiol 1997; 26:845–851.

18

Genetics and Biosynthesis of Lipopolysaccharide O-Antigens

Wendy J. Keenleyside and Chris Whitfield
University of Guelph, Guelph, Ontario, Canada

INTRODUCTION

In most gram-negative pathogens, and particularly in most wild-type members of the families Enterobacteriaceae, Pseudomonadaceae, Pasteurellaceae, and Vibrionaceae, the LPS core is capped by an O-polysaccharide (O-PS) side chain to form smooth LPS (S-LPS). Structural diversity in the O-PSs defines the O-antigen specificity used in serological classifications. Although the fine structure of a particular O-antigen determines the serospecificity of a given strain, there are examples where taxonomically distant species may express similar or identical polysaccharides. Serotype specificity is therefore important for classification and/or epidemiology within a species, and its genetic basis is of particular interest in terms of the evolution of O-antigenic diversity.

Much of the early interest in the chemistry, biosynthesis, and genetics of O-PSs originated from their roles as essential virulence determinants and their potential application in the development of vaccines. The presence of O-PS at the periphery of the cell gives rise to a hydrophilic surface layer, which, based on location and structure, plays a critical role in the interactions between the bacterium and its environment. Traditionally, most functional analyses of S-LPSs have been confined to animal pathogens. More recently, researchers have become interested in their roles in plant-associated bacteria. The function of S-LPS in free-living bacteria has received less attention. The primary role(s) of the O-PSs appear to be protective. In animal pathogens, O-PS structural diversity may contribute to bacterial evasion of host immune responses, particularly the alternative complement cascade. Assembly of the membrane attack complex (MAC) is affected by the chemistry of O-PS, the length of the O-polysaccharide, and the relative amounts of long-chain O-polysaccharide–substituted LPS (reviewed in Ref. 1). More recently, the presence and chain length of O-PS in *Shigella flexneri* has been implicated in directing the invasion protein (IcsA) to its correct location at one pole of the cell. This is essential for invasiveness and subsequent inter- and intracellular spread (2). This very brief consideration demonstrates that the role played by O-PS in different bacteria is quite variable and the precise structural details of the S-LPS molecules can have a profound influence on the biology of a given bacterium. These features are dictated by the processes involved in their biosynthesis. However, it must be recognized that O-PS is not essential for virulence, as many important pathogenic bacteria produce only lipooligosaccharides (LOSs). The involvement of LOS in virulence is covered in other chapters of this volume.

The chemistry, biosynthesis, and genetics of bacterial O-PSs has been most intensively studied in *Salmonella* and *Escherichia coli*. However, there is a growing body of information from comparative studies with other enteric organisms as well as from the Pseudomonadaceae and the Vibrionaceae. Despite significant differences in the biology of these microorganisms and the variation in O-PS structures, common themes are becoming evident. In this review we will summarize what is currently known about the genetics and biosynthetic processes for the assembly and attachment of the O-PS to lipid A core.

CHEMISTRY AND DIVERSITY OF LIPID A CORE–LINKED POLYSACCHARIDES

The structural diversity of O-PSs is tremendous (Table 1). A review of the compositional data reveals greater than 60 monosaccharides that may be present as well as more than 30 different noncarbohydrate substituents (3). The O-PSs possess repeating unit structures and the O units vary among different serotypes based on the nature of the sugar constituents, the position and anomeric configuration (α or β) of the O-glycosidic linkages, the ring form of the substituent sugars (pyranose or furanose), as well as the presence or absence of noncarbohydrate "decorations" and the types of linkages by which these substituents are attached. The O units may comprise anywhere from one to eight monosaccharides, may be linear or branched, and may be homopolymers (i.e., a single monosaccharide component) or, more frequently, heteropolymers. In some cases, the precise O unit may be masked by nonstoichiometric modifications (e.g., O-acetylation or glycosylation).

The LPS molecules extracted from any smooth strain are heterogeneous in size. They include at least some rough LPS (R-LPS) devoid of O-PS and, in some cases, truncated R-LPS molecules with an incomplete core (4–8). The heterogeneity in the degree of polymerization is readily demonstrated by SDS-PAGE, where LPS gives a characteristic ladder pattern in which each incremental "rung" of the ladder represents a lipid A core molecule substituted with one additional O unit. The spacing between the "rungs" provides an indication of the size of the O unit. The S-LPS from most isolates exhibits a preferred size distribution pattern, which is strain-specific. As will become apparent, the mechanism of size selectivity is not well understood but results from particular aspects of the biosynthesis pathway. Within a given strain, variations in growth conditions may also result in subtle changes to the SDS-PAGE pattern of LPS due to alterations in capping frequency, O-PS chain length, or chemical modifications. For the most part, the molecular mechanism(s) underlying these finer controls has not been elucidated.

Although the role of lipid A core as the anchor for O antigens has been well established for decades, the distinction of what may be considered an O antigen has become complicated by more recent studies. In *E. coli*, for example, lipid A core can act as an anchor for several different polysaccharides. Enterobacterial common antigen (ECA) is expressed by virtually every member of the Enterobacteriaceae. Although normally anchored by phospholipid, ECA can be found attached to lipid A core in strains that do not synthesize a typical O antigen (9). Similarly, the group I capsular K antigens of *E. coli* exist in both a lipid A core–bound form (known as K_{LPS}) as well as a high molecular weight LPS-free capsular form (10–14). These K antigens are co-expressed with an O antigen, and the designation of each polymer as the O or K antigen is operationally (serology) based (see Ref. 10). This versatility of lipid A core as an anchor is not confined to *E. coli*. In *Salmonella* strains, lipid A core may serve as an acceptor for the classical O antigen as well as the minor antigens T1 and T2 (15), and among members of the O:54 *Salmonella* serogroup, two structurally distinct O polysaccharides compete for the same lipid A core acceptor (see below). *Serratia marcescens* serotype O16 expresses two structurally distinct lipid A core–bound polymers: the serotype O16 O antigen and a riban homopolymer, which has the same structure as the T1 antigen of *Salmonella* (16). Among nonenteric bacteria, *Pseudomonas aeruginosa* also expresses two lipid A core–linked polymers. One is a conserved polyrhamnose homopolymer known as A-band LPS (Table 1), which is analogous to lipid A core–linked ECA. The other is the serotype-specific heteropolymer known as B-band LPS (reviewed in Ref. 17). Interestingly, while both O-polysaccharides are attached to lipid A core, preliminary evidence suggests that the outer core structures of the two LPS are different (18,19).

Structural and genetic studies with *Klebsiella pneumoniae* serotypes O1, O2(2a,2c), and O8 provide examples of O-PSs not characterized by a consistent O-unit structure. In these strains, the O-PS contains different domains, each possessing a distinct repeating unit structure (20–23). While it has not yet been confirmed, a similar situation may exist with *Serratia plymuthica*, which synthesizes the same two lipid A core–bound polymers as *K. pneumoniae* O1 (24). Variations in the extent of structural modifications can also result in the presence of chemically distinct LPS populations within a bacterial culture. The molecular basis for this variation, sometimes referred to as "form variation," has not been characterized, however, chemical analyses performed with differentially fractionated LPS from *S. enterica* sv. *typhimurium* cultures (25) and more recently with sv. *enteritidis* (26) identified two chemically different populations of S-LPS, differing only in the presence or absence of a glucose modification.

Genetics and Biosynthesis of LPS O-Antigens

Table 1 Selected O-Polysaccharide Repeating Unit Structures

Organism	Structure of repeating unit
Salmonella enterica	
serogroup A	α-Par*p* α-D-Glc*p* 1 1 ↓ ↓ 3 4 →2)-α-D-Man*p*-(1→4)-α-L-Rha*p*-(1→3)-α-D-Gal*p*-(1→
serogroup B	α-Abe*p*(2-OAc) α-D-Glc*p* 1 1 ↓ ↓ 3 4/6 →2)-α-D-Man*p*-(1→4)-α-L-Rha*p*-(1→3)-α-D-Gal*p*-(1→
serogroup D1	α-Tyv*p* α-D-Glc*p* 1 1 ↓ ↓ 3 4 →2)-α-D-Man*p*-(1→4)-α-L-Rha*p*-(1→3)-α-D-Gal*p*-(1→
serogroup D2	α-Tyv*p* α-D-Glc*p* 1 1 ↓ ↓ 3 4 →6)-β-D-Man*p*-(1→4)-α-L-Rha*p*-(1→3)-α-D-Gal*p*-(1→
serogroup E1	→2)-α-D-Man*p*-(1→4)-α-L-Rha*p*-(1→3)-α-D-Gal*p*-(1→ 6 ↑ OAc
Escherichia coli	
O8	→3)-β-D-Man*p*-(1→2)-α-D-Man*p*-(1→2)-α-D-Man*p*-(1→
O9	→3)-α-D-Man*p*-(1→3)-α-D-Man*p*-(1→2)-α-D-Man*p*-(1→2)-α-D-Man*p*-(1→2)-α-D-Man*p*-(1→
Klebsiella pneumoniae	
O1	→3)-β-D-Gal*f*-(1→3)-α-D-Gal*p*-(1→ (D-Galactan I)
	→3)-β-D-Gal*p*-(1→3)-β-D-Gal*p*-(1→ (D-Galactan II)
O8	→3)-β-D-Gal*f*-(1→3)-α-D-Gal*p*-(1→ (D-Galactan I-*O*Ac) 2/6 ↑ *O*-acetyl
	→3)-β-D-Gal*p*-(1→3)-β-D-Gal*p*-(1→ (D-Galactan II)
Serratia marcescens O16	→3)-β-D-Gal*f*-(1→3)-α-D-Gal*p*-(1→ (D-Galactan I)
Pseudomonas aeruginosa	
O5	→4)-β-D-Man(2NAc3N)A-(1→4)-β-D-Man(2NAc3NAc)A-(1→3)-α-D-FucNAc-(1→ 3 ↑ CH$_3$C=NH
A-band LPS	→2)-α-D-Rha*p*-(1→3)-α-D-Rha*p*-(1→3)-α-D-Rha*p*-(1→

Source: Adapted from Refs. 3 and 187.

OVERVIEW OF O-POLYSACCHARIDE BIOSYNTHESIS

The biosynthesis of O-PSs is a complex process. Like other cell-surface polysaccharides, O-PSs are assembled from sugar nucleotides, which are synthesized in the cytoplasm (27). Sugar monomers are transferred to a lipid carrier (antigen-carrier lipid, ACL; glycosyl-carrier lipid, GCL) present in the plasma membrane. This membrane-bound carrier is undecaprenol-phosphate (Und-P), the same C_{55}-isoprenoid alcohol derivative used for synthesis of peptidoglycan as well as other cell-surface polysaccharides (reviewed in Ref. 28). It remains unclear how the common pool of Und-P is distributed among the various biosynthesis systems to ensure that cellular needs are met. Specific glycosyl transferases act sequentially to generate a repeating unit structure. Structural fidelity is entirely maintained in the O units during synthesis by the specificities of enzymes (transferases or polymerases), which catalyze the formation of particular linkages and which display very specific binding properties for acceptor and donor molecules. Remarkably little is known about the mechanisms of specificity.

Assembly of O-PS is a trans–plasma membrane process. Activated precursors are available in the cytoplasm, and assembly of lipid-linked O units occurs at the cytoplasm/plasma membrane interface. Polymerization of O-PS is terminated by its transfer from the carrier lipid and its ligation to lipid A core, a process thought to occur at the periplasmic face of the plasma membrane. As a result, either lipid-linked O-PS or lipid-linked O units must be transferred across the plasma membrane. As will become apparent, the initiation of O-PS synthesis and the termination step (ligation) are conserved, but the precise steps that occur between can vary. In particular, the components involved in the trans–plasma membrane export processes define the three known pathways for O-PS assembly (29,30). Models for the three pathways are illustrated in Figures 1, 2, and 3. The little known about O-PS ligation to lipid A core is discussed in Chapter 9 in this volume. After ligation, the completed LPS molecule is translocated to the surface of the outer membrane, and there are virtually no data for the components and processes involved in these important later steps.

The classic O-PS assembly system is the Wzy-dependent (formerly Rfc-dependent) pathway first described in *Salmonella*. This involves growth of the polymer at the reducing terminus by polymerization of individual repeat units en bloc. Polymerization occurs at the periplasmic face of the plasma membrane and follows export of the newly synthesized lipid-linked O units across the plasma membrane. In contrast, the ATP-binding cassette (ABC) transporter–dependent pathway (formerly the Rfe-dependent pathway) is distinguished by growth at the nonreducing terminus through the processive polymerization of nascent O-polysaccharide at the cytoplasmic face of the plasma membrane. After polymerization, nascent O-PS is exported across the plasma membrane by a process involving an ABC transporter. It is becoming increasingly apparent that the Wzy-dependent and ABC transporter–dependent pathways are widespread in bacteria. Similar systems have been reported for capsular and extracellular polysaccharides. To date, the newly described third pathway, the synthase-dependent pathway, has only been characterized for the O:54 O-PS of *Salmonella*, but it resembles the synthetic process for bacterial cellulose, chitin, and possibly certain streptococcal capsular polysaccharides. The critical feature of this pathway is a unique glycosyl transferase, which is proposed to catalyze a vectorial polymerization, sequentially adding monosaccharide substituents while simultaneously extruding the nascent polymer across the plasma membrane.

While there is a growing body of genetic information for many O-PS biosynthesis gene clusters, biochemical investigations are relatively limited, and the individual enzymatic mechanisms are often a matter of conjecture and extrapolation based on sequence homologies. The following sections describe what is currently known about the pathways for assembly of O-PS. With few exceptions (see below), the enzymes involved in O-PS assembly are encoded by the gene cluster historically referred to as the *rfb* locus. Individual genes within it are designated *rfbA*, *rfbB*, etc. The *rf** designation was originally assigned to gene loci implicated in LPS biosynthesis (*rfa*, *rfb*, and *rfc*) based on the rough phenotype of strains possessing mutations within these loci. More recently, a new nomenclature has been proposed to accommodate the polymorphic nature of the genetic loci and simplify interspecies comparative studies by providing a universal means of designating genes based on conserved biological function (31). The new nomenclature will be used throughout this chapter, but to provide a link to earlier literature the previous names will be indicated in parentheses. Details of this new system are available online (http://www.angis.su.oz.au/BacPolGenes).

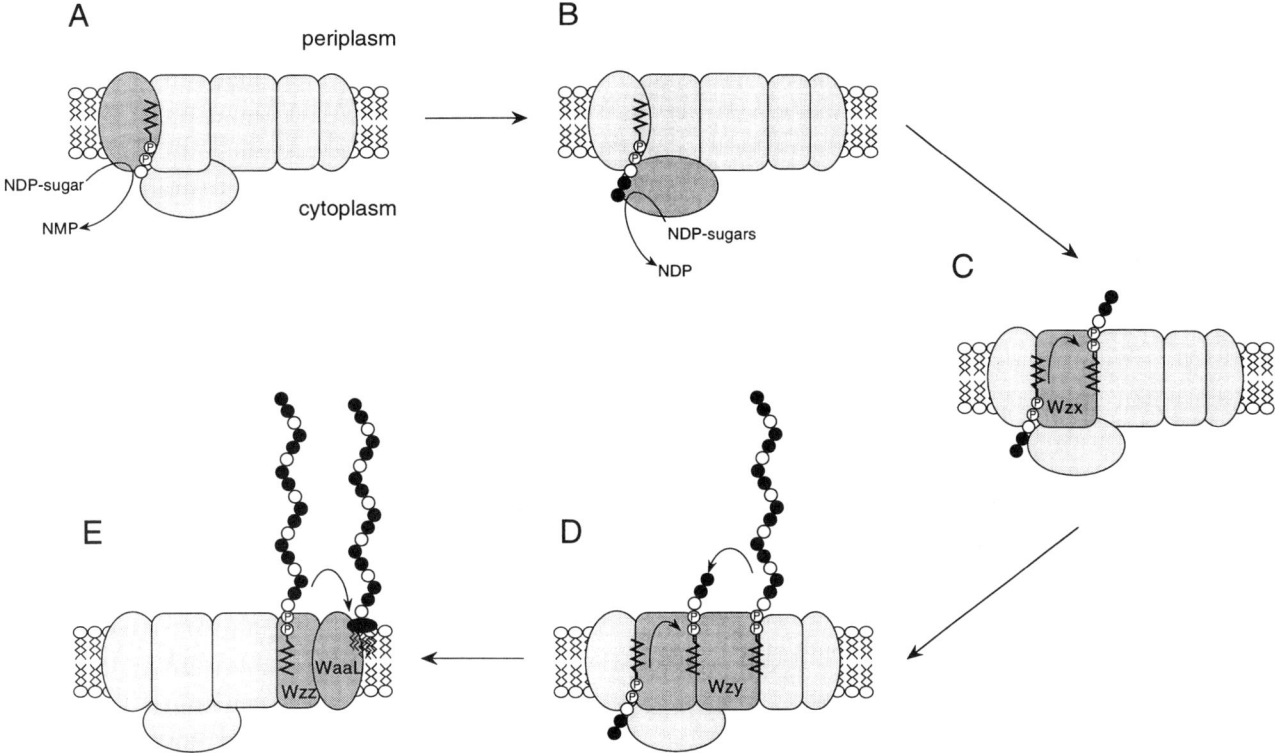

Fig. 1 Model for the assembly of an O-PS synthesized by a Wzy-dependent pathway. Synthesis is initiated at the cytoplasmic face of the plasma membrane by transfer of a sugar-1-P to Und-P, catalyzed by a protein with multiple membrane-spanning domains (A). Peripheral membrane proteins with specific glycosyl transferase activities then complete formation of the Und-PP–linked O unit, depicted as a trisaccharide (B). Individual Und-PP–linked O units are then transferred to the periplasmic face of the membrane by Wzx (C). These then serve as substrates for the Wzy O-PS polymerase in a blockwise process: the polymerized chain is transferred from its Und-PP carrier to the new Und-PP–linked O unit with the result that the chain grows at the reducing terminus (D). The final step involves transfer of the nascent O-PS to the separately assembled lipid A core acceptor. The process involves the O-PS: lipid A core ligase (WaaL) and the modality of O-PS chain length is dictated by the Wzz protein (E). (Adapted from Ref. 29.)

INITIATION REACTIONS

Despite the tremendous diversity of structures among the O antigens, only two initiation reactions have been unambiguously identified for O-PS biosynthesis, and as indicated above, the general features of initiation reactions are conserved in different O-PS assembly pathways. The known reactions are catalyzed by the enzymes WbaP (formerly RfbP) from *S. enterica* (32) and WecA (formerly Rfe) from *E. coli* (33). The initiating enzymes catalyze transfer of a sugar-1-P residue to Und-P. The energy of the sugar-phosphate linkage in the donor molecule is therefore conserved in the resulting Und-PP–linked intermediate and is available to drive the postpolymerization lipid A core ligation reaction. In contrast to the initiation step, subsequent transferase reactions in O-unit formation involve transfer of glycosyl monomers only. The glycosyl transferases that initiate O-PS assembly are therefore distinguished from postinitiation glycosyl transferases by both their binding specificity and catalytic activity: the acceptor they recognize is the hydrophobic lipid carrier, Und-P, and the bond catalyzed is a pyrophosphate linkage. In contrast, noninitiating glycosyl transferases recognize and bind sugar acceptors and catalyze the formation of an *O*-glycosidic linkage. Given the chemical nature of the acceptors for noninitiating transferases and the location of the nucleotide sugar donors, it is not surprising that the majority of these transferases are soluble proteins. Both WecA and WbaP are predicted to be integral membrane proteins (33,34), and their hydrophobicity profiles are quite similar (35). This prop-

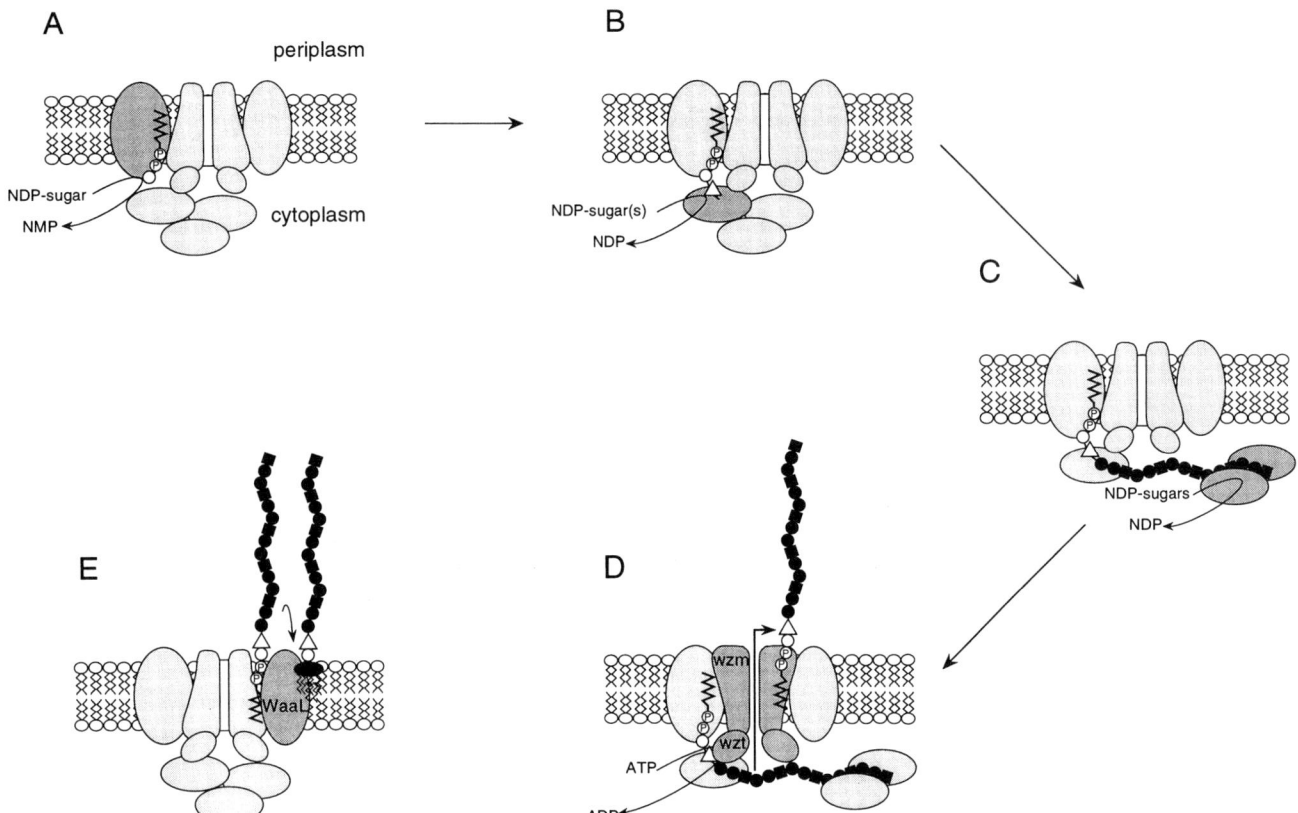

Fig. 2 Model for assembly of an O-PS by an ABC transporter–dependent pathway. Synthesis is initiated at the cytoplasmic face of the plasma membrane by transfer of a sugar-1-P to Und-P, catalyzed by a protein with multiple membrane-spanning domains (A). A specific glycosyl transferase enzyme then adds one or more sugars to commit the common primer to this pathway (B). The repeating unit structure is then formed by the processive transfer of glycosyl residues to the nonreducing terminus of the Und-PP–linked O-PS at the cytoplasmic face of the plasma membrane. O-repeat units are identified by the order of glycosyl residues or by the linkage sequences; in some cases a single enzyme catalyzes formation of multiple linkages (C). The nascent O-PS is then transferred to the periplasmic face by a process involving a dedicated ABC transporter (D). Although the exported O-PS is shown retaining its Und-PP carrier, this has not been demonstrated experimentally. The final step involves transfer of the nascent O-PS to the separately assembled lipid A core acceptor, catalyzed by the O-PS:lipid A core ligase (WaaL) (E). (Adapted from Ref. 29.)

erty may reflect the hydrophobic nature of the acceptor molecule (Und-P). However, it is also possible that the initiating enzyme forms a membrane-associated "scaffold" for assembly of a coordinated biosynthetic enzyme complex, and structural similarities may result from these functional requirements. Interestingly, there are a number of structural homologs of WbaP in systems involved in capsular and exopolysaccharide synthesis in various bacteria (36).

The biochemistry of O-PS synthesis was first characterized in *S. enterica* serovars *typhimurium* and *anatum* (serogroups B and E, respectively). These two strains synthesize heteropolymeric O antigens using the Wzy-dependent pathway (see below), and in both cases, assembly is initiated by WbaP catalyzing the reversible transfer of galactose-1-P (Gal-1-P) from uridine diphosphogalactose (UDP-Gal) to Und-P (32). The correlation between the *wbaP* open reading frame and galactosyl-1-P transferase activity has been clearly demonstrated (36). Galactose is the first sugar of each O unit.

WecA (Rfe) is a GlcNAc-1-phosphate transferase with a much wider role in the assembly of cell-surface polysaccharides. WecA is encoded by the ECA biosynthesis cluster and not surprisingly initiates ECA formation in a manner resembling Wzy-dependent O-PSs (33). This enzyme is the only known initiating enzyme for *E. coli* O-PS, and it is involved in representatives

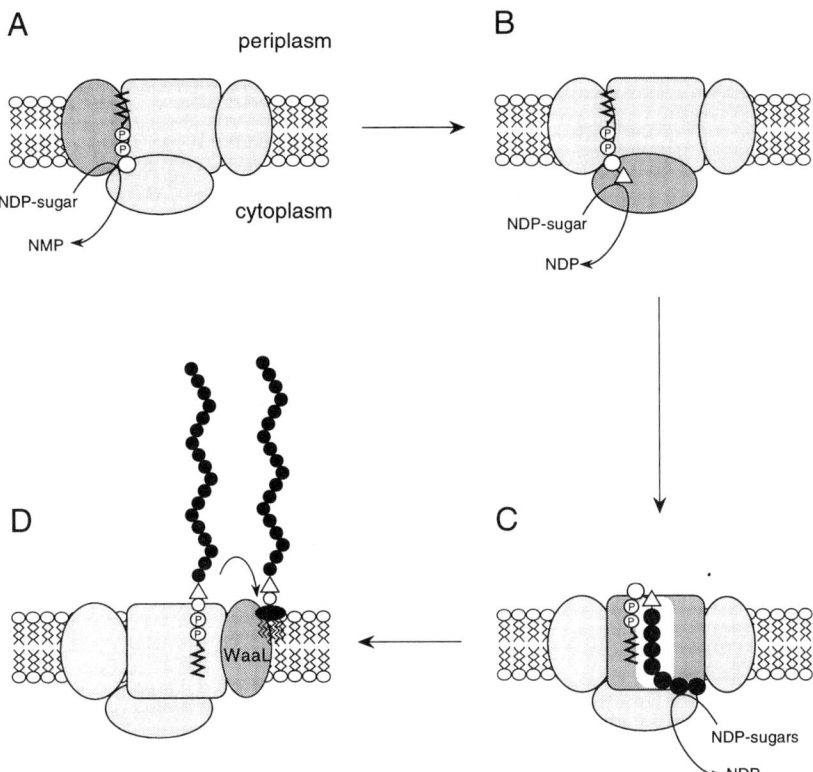

Fig. 3 Model for assembly of an O-PS by the synthase-dependent pathway. Synthesis is initiated at the cytoplasmic face of the plasma membrane by transfer of a sugar-1-P to Und-P, catalyzed by a protein with multiple membrane-spanning domains (A). A specific glycosyl transferase enzyme then adds one or more sugars to commit the common primer to this pathway (B). The repeating unit structure is then formed by the processive transfer of glycosyl residues to the nonreducing terminus of the Und-PP–linked O-PS at the cytoplasmic face of the plasma membrane. A single enzyme, the synthase, is responsible for this step (C). This pathway has no additional components for transfer of nascent O-PS across the plasma membrane. By analogy to chitin synthases, it has been proposed that the O-PS catalyzes vectorial polymerization, simultaneously elongating the polymer and extruding it across the plasma membrane. The final step involves transfer of the nascent O-PS to the separately assembled lipid A core acceptor, catalyzed by the O-PS: lipid A core ligase (WaaL) (D).

from each of the O-PS synthesis pathways. In *E. coli* (37), *Shigella dysenteriae* type 1 (35), and *S. flexneri* (38), WecA transfers the GlcNAc residue, which often forms the first sugar in each O unit. The same reaction is required for initiation of the reconstructed "O16 antigen" in *E. coli* K-12 (39). Recently, Zhang et al. demonstrated that expression of the *Yersinia enterocolitica* O:8 antigen in *E. coli* K-12 requires WecA, despite the fact that the O-PS contains GalNAc rather than GlcNAc (40). This observation was confirmed with the finding that some GalNAc-containing *E. coli* K_{LPS} are also initiated by WecA (10).

WecA is not confined to Wzy-dependent O-PSs, and in the other pathways the enzyme forms an Und-PP-GlcNAc primer on which the O-PS is assembled. In this case, the GlcNAc residue is retained at the terminus but does not form part of the O-unit structure. A number of ABC transporter–dependent O-PSs are known to be initiated by WecA, and for this reason this pathway was originally designated Rfe-dependent. This term is clearly now inaccurate. The best characterized examples of this pathway are in the mannose-containing O-PSs of *E. coli* O8 (41) and O9 (42) and D-galactan I in *K. pneumoniae* (43) (see Table 1 for structures). A WecA-mediated primer is also required for the *Salmonella* O:54 antigen formed by the synthase-dependent pathway (44,45). In the *E. coli* examples the actual identity of the WecA-transferred initiating sugar has been the subject of some controversy. Transfer of GlcNAc-1-P has been unequivocally identified as the initial reaction for *E. coli* O8 (41); this correlates with the demonstrated activity of the enzyme in ECA synthesis. However, in the case of *E. coli* O9, in vitro analyses identified Und-P-P-Glc as the product of the

WecA reaction (46). More recently, in vitro analyses have shown that Und-P-P-GlcNAc can also serve as a mannose acceptor for the first O9-specific mannosyl transferase (42). These data, together with the recent finding of WecA-mediated GalNAc-1-P transfer, suggest that WecA has a broad specificity. Since some of these bacteria should contain one or both of the precursors UDP-GalNAc and UDP-Glc, together with UDP-GlcNAc for peptidoglycan synthesis, it is not clear how fidelity of the O-unit structure is maintained.

No initiating enzymes or reactions for O-PS synthesis have yet been biochemically characterized among nonenteric bacteria, however, the successful expression of heterologous O antigens in *E. Coli* K-12 hosts indicates that either the O-PS biosynthetic clusters encode the necessary activity or, alternatively, that the host is able to complement the missing activity. For example, transposon mutagenesis studies with *V. cholerae* O1 (47) identified at least three O1 polysaccharide biosynthesis genes required for expression in *V. cholerae* but complemented in *E. coli* K-12 (48). Surprisingly, one of these genes, $wbaP_{VcO1}$ is predicted to encode a sugar transferase with homology to the initiating enzyme $WbaP_{Se}$ (RfbP) from *S. enterica*. Inactivation of $wbaP_{VcO1}$ in *V. cholerae* results in loss of O1 expression, suggesting that this gene may encode an initiating galactosyl transferase. $WbaP_{VcO1}$ is also predicted to share significant homology (68.7%) with the putative galactosyl transferase encoded by *wbfU (orf7)* of *V. cholerae* (O139 (49,50). Surprisingly, compositional analysis of the O139 polysaccharide indicates that this polymer does not contain galactose (51), and the identity of the initiating enzyme for O139 synthesis remains undetermined.

Characterization of the O-PS gene cluster from *P. aeruginosa* O5 identified one predicted gene product, WbpL, with primary and secondary structures resembling WecA (52). WbpL also shares sequence similarity with other transferases that recognize and bind undecaprenol. Based on these observations, WbpL is proposed to be the initiating transferase for O5 synthesis. Although the transferase activity of this protein has not been determined, structural analysis indicates that *N*-acetylfucosamine (Fuc2NAc) is the first sugar of the biological O repeat unit, suggesting that WbpL may transfer Fuc2NAc-1-P to Und-P (53). Interestingly, synthesis of the *P. aeruginosa* A-band LPS (an ABC transporter–dependent O-PS) is also initiated by WbpL, but in this pathway the enzyme appears to transfer a GlcNAc residue (53). This variable specificity is reminiscent of WecA.

Sequence analysis of a DNA fragment from *Anabaena* sp. has identified a predicted protein with significant similarity to $WbaP_{Se}$, and insertional mutagenesis of the corresponding $wbaP_A$ gene eliminates synthesis of O-PS (54). These observations represent the first published data for an O-PS biosynthesis gene in a cyanobacterium and suggest the possibility that, at least in this strain, O-PS assembly could be initiated by transfer of Gal-1-P to Und-P.

Wzy-DEPENDENT (Rfc-DEPENDENT) PATHWAY OF O-POLYSACCHARIDE ASSEMBLY

Synthesis of O-Repeat Units

The biochemistry of O-PS biosynthesis has been extensively studied in *S. enterica* serogroups B and E1, the prototype strains for the Wzy-dependent pathway. Serogroups B and E1 synthesize O units with identical Man-Rha-Gal trisaccharide backbones and differ only in side branch substitutions (Table 1). Classical work by M. J. Osborn, H. Nikaido, A. Wright, P. Robbins, and others established the sequence of glycosyl transfer and the required sugar nucleotide precursors (reviewed in Ref. 28). Although the genes/proteins have now been identified, knowledge of the mechanisms involved and the organization of functional complexes has advanced little beyond this pioneering work. In this pathway, assembly occurs at the cytoplasmic face of the plasma membrane. Postinitiation O-unit glycosyltransferases are peripheral membrane proteins, which, given their sequential action and specificity, are assumed to form part of a biosynthesis complex surrounding the growing Und-PP–linked O unit (55) (Fig. 1B). Following initiation, each glycosyl transfer is an irreversible step requiring the product of the previous reaction (28).

Polysaccharide modification by glycosyl and noncarbohydrate substituents may be stoichiometric or nonstoichiometric, and the difference most likely lies in the stage of biosynthesis at which the substitution reaction occurs. In the case of *Salmonella* serogroups E and B, in vitro and in vivo studies have both shown that incorporation of the stoichiometric side-branch sugars occurs prior to completion of the O unit and the subsequent polymerization reactions. Studies with serogroup B strains unable to produce the side chain precursor, CDP-abequose, fail to produce O-PS (56). The donor for nonstoichiometric glucosyl substituents in *S. enterica* O-PSs is an unusual lipid intermediate, α-glucosylmonophosphorylundecaprenol (57–59). In serogroups B and D, glucosylation occurs postpolymerization and is assumed to occur at the level of the O-hapten (Und-PP–linked O-PS) (58–61). In contrast, in serogroups C1 and C2, glucosylation occurs before polymerization, at the level of individual Und-PP–linked O units (62). O-Acetylation has been studied in S. enterica serogroup E1 both in vivo (63) and in vitro (64). The donor for the acetyl group is acetyl-CoA, and the substrate for the acetyltransferase reaction is the single Und-PP–linked O unit.

Wzx and Its Role in Transmembrane Assembly

After the individual Und-PP–linked O units are assembled, they are transported to the site of polymerization at the periplasmic face of the plasma membrane (65). Trans–plasma membrane export has received the most attention in Salmonella, where the Wzx (formerly RfbX) protein is currently the only component implicated in the process (Fig. 1C). Earlier reports of a role for WbaP in the export process (66) likely have a different explanation (36,37). A Wzx homolog is encoded by all known Wzy-dependent O-PS biosynthesis gene clusters (Table 2). However, detailed analyses and mechanistic studies are hampered by difficulties in cloning and overexpressing the wzx gene or its mutated derivatives (67,68). Recent in vivo labeling studies have utilized an S. enterica serogroup B–S. dysenteriae type I hybrid system where the S. enterica host strain lacks its own O-PS biosynthesis cluster and a wzx-deficient S. dysenteriae O1 biosynthesis cluster is provided on a plasmid (67). This hybrid strain accumulated Und-PP–linked O units, which appeared to be located at the cytoplasmic face of the plasma membrane, providing the basis for the proposal that Wzy is the O-unit transporter ("flippase"). It is interesting to note that predicted Wzx proteins are all highly hydrophobic proteins with 12 potential transmembrane domains, and while they share little primary sequence similarity, they do share structural features with bacterial permeases (68). However, it should be acknowledged that no enzymatic activity has been ascribed to the Wzx protein, and the mechanism by which such large molecules (Und-PP–linked O units) might be flipped across the lipid bilayer remains obscure.

Wzy-Dependent Polymerization

Once the Und-PP–linked O units have been translocated to the external face of the plasma membrane, they are polymerized en bloc by transfer of nascent polymer to the nonreducing and (i.e., the distal end) of the new Und-PP–linked O repeat (Fig. 1D) (69,70). Polymerization is catalyzed by the O-PS polymerase, Wzy, and mutants affected in the wzy gene produce LPS consisting of lipid A core capped with a single O unit (SR-LPS) (71) (Fig. 4A). The released Und-PP must be recycled to the active monophosphoryl form. The dephosphorylation reaction is inhibited by the antibiotic bacitracin, and sensitivity to this antibiotic is therefore a characteristic of this biosynthetic pathway (72). O-PSs synthesized using the Wzy-dependent pathway appear to be exclusively heteropolysaccharides, often with side-branch residues. The Wzy-dependent mechanism of assembly provides a relatively efficient process for maintaining the fidelity of synthesis for such complex structures. Similar strategies involving a repeat unit polymerase have also been implicated in synthesis of other cell-surface heteropolysaccharides (reviewed in Ref. 28), such as ECA (73) and a variety of exopolysaccharides such as colanic acid (74).

Bacteria for which the wzy determinant has been characterized are listed in Table 2. The respective gene products are all predicted to be integral membrane proteins with 11–13 transmembrane domains, however, like the Wzx proteins, they exhibit little or no primary sequence homology (73). The absence of such conserved features has complicated the identification of catalytic and binding residues, and consequently the mechanism of Wzy activity has not been resolved. Given the lack of sequence conservation among different polymerases, it is not surprising that individual polymerases are specific for the cognate O unit or for closely related structures (reviewed in Refs. 28, 75). To date, attempts to overexpress and detect a Wzy protein have been unsuccessful (71,73,76), and this is thought to be due to the fact that Wzys are characterized by a relatively high percentage of rare codons (73,77).

Wzz-Dependent Polymer Chain Length Determination

Most wild-type bacteria that express S-LPS exhibit a strain-specific, sometimes multimodal distribution of O-PS chain lengths. This size distribution is more complex than would be predicted by a simple competition between Wzy and the O-PS:lipid A-core ligase (78) and is a function of the chain-length determinant Wzz (formerly Rol or Cld; Fig. 1E). In the absence of Wzz, modality is lost and the amount of a given O-PS chain length is inversely proportionally to the chain length (Fig. 4B). Wzz proteins have been identified in all of the strains known to have Wzy-dependent O-PSs (Table 2). Regions of Wzz primary sequences are relatively well conserved among the various homologs, as

Table 2 Putative or Known Genes Characteristic of WZY-Dependent O-Polysaccharide Biosynthesis Clusters

Gene	Putative or known activity	Organism	Accession number	Ref.
wzy	Polymerase	Salmonella enterica		
		serogroup A[a]		15
		serogroup B[a]	M60066	71
		serogroup C1	M84642	188
		serogroup C2	X61917 (orf16.7)	189
		serogroup D1[a]		15,190
		serogroup D2	U04165	176
		serogroup E1	X60666 (orf9.6)	175,176
		Shigella dysenteriae 1	L07293	35
		Shigella flexneri Y	X71970	73
		Escherichia coli		
		O4	U39042	192
		O16 (K-12)	U09876	157
			U03041	38
		Yersinia enterocolitica O:8	U18674	77
		Pseudomonas aeruginosa O5	U17294	52,193,194
			U26685	76
		Vibrio cholerae O139	Y07786; Y07787 (orf41x8)	50,117
			U47057	49
wzz	Chain-length determinant	Escherichia coli		
		O8[b]	U78086	10
		O8[c]	U39305	12
		O9[d]	U39306	12
		O16 (K-12)	U09876	157
			Y07559	198
		O75	M89934	112
		O111	Z17241	79
		Yersinia		
		Y. enterocolitica O:8	U43708	199
		Y. pseudotuberculosis IIA	U13685	113
		Pseudomonas aeruginosa O5	U50397	114
		Salmonella enterica		
		serogroup B	Z17278	79
			M89933	112
		serogroup D$_2$		176
		Shigella		
		S. dysenteriae 1	Y07560	198
		S. flexneri Y	M25995 (cld$_{pHS-2}$)	113
			X71970	80
wzx	O-unit transporter	Yersinia		
		Y. enterocolitica O:8	U46859	40
		Y. pseudotuberculosis IIA	L01777	197
		Pseudomonas aeruginosa O5	U50397 (wbpF)	114
		Salmonella enterica		
		serogroup C1	M84642	159
		serogroup C2	X61917 (orf12.6)	177
		serogroup D1	M65054	195
		serogroup D2		173
		serogroup E1	X60665 (orf7.9)	175
		Shigella flexneri Y	X71970	68
		Escherichia coli		
		O16 (K-12)	U09876	157
			U03041	38
		O111	U13629	196
		Salmonella enterica		
		serogroup A	M65054 (orf12.8)	195
		serogroup B	X56793; M29713	158

[a]These genes map outside of the O-polysaccharide biosynthesis cluster, at 32 min on the Salmonella chromosome (15,191).
[b]This gene is involved in expression of the type IB K40 K$_{LPS}$.
[c]This gene regulates chain length of the type IB K87 K$_{LPS}$.
[d]This gene regulates chain length of the type IB K9$_{LPS}$.

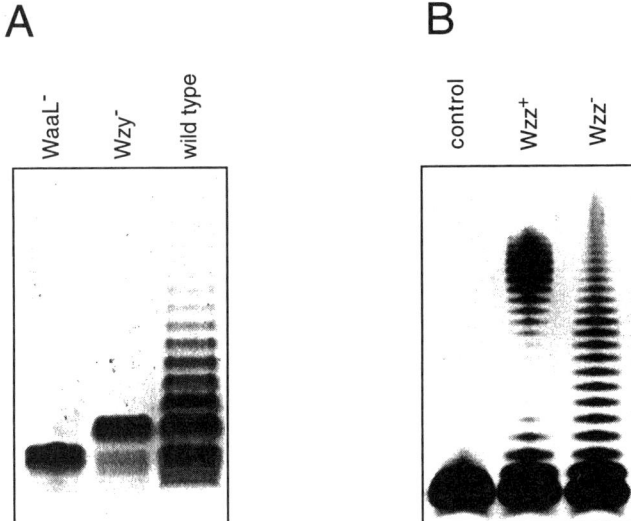

Fig. 4 SDS-PAGE profiles of LPS molecules, demonstrating the activities of the characteristic Wzy and Wzz proteins in Wzy-dependent O-PS assembly. (A) S-LPS from *S. enterica* serovar *typhimurium* (wild type) together with R-LPS with a complete core resulting from a defect in the O-PS ligase (WaaL⁻) and the SR-LPS comprising lipid A-core and a single O unit, which results from a defect in the O-PS polymerase (Wzy⁻). (B) The influence of Wzz in an *E. coli* K-12 strain on expression of the Wzy-dependent *S. dysenteriae* type 1 O-PS. The host K-12 strain (control) produces R-LPS. In a Wzz⁺ chromosomal background, the type 1 O-PS is polymerized and ligated to the lipid A core with a preference for a particular range of O-chain lengths, giving a modal distribution. The modal distribution is lost in the Wzz⁻ background, and the unregulated LPS contains more molecules with lower molecular weight.

are the predicted secondary structures (reviewed in Ref. 30). All share highly conserved transmembrane domains at the C- and N-termini. The ability of Wzz to somehow modulate the kinetics of the polymerization and/or ligation reactions suggests that the protein must recognize and form a complex with either Wzy, the ligase enzyme (WaaL; formerly RfaL), or both.

The biochemical basis for Wzz function has not yet been determined, but two models have been proposed (reviewed in Ref. 30). After quantifying the distribution of chain lengths in the presence and absence of a functional Wzz determinant, Bastin et al. (79) concluded that Wzz confers a preference for chain length on either the polymerase or the ligase or both. In the absence of Wzz, the distribution of chain lengths was shown to be independent of chain length, correlating instead with a constant probability of transfer to lipid A core. These authors therefore proposed a model where Wzz acts as a timing "clock," interacting with the Wzy polymerase. In this model, Wzy exists in two functional states, similar to the ribosome and fatty acid synthetase: the E-state favors extension or polymerization and the T-state favors transfer to the O-PS ligase. They propose that the switch from one state to the other is triggered by Wzz, with different modalities arising from the different predetermined "times" for the various Wzz proteins; multimodal distributions are suggested to arise from a "retesting" of the clock, allowing Wzy to enter a fresh E-state before ligation can proceed. The molecular basis for this timing mechanism is unknown.

Morona et al. have presented a different hypothesis for the function of Wzz (80) in which the protein acts as a molecular chaperone to assemble a complex consisting of Wzy, the O-PS ligase, and Und-PP–linked O-PS chain. The specific modality is then a result of the Wzz-dependent ratio of Wzy to WaaL, with the ratio determining the kinetics of the ligation reaction. The absence of a functional Wzz is suggested to result in a loss of modality due to the random interaction between the polymerase and ligase, as the two proteins diffuse through the two-dimensional plane of the membrane.

Recent studies with *E. coli* group IB K antigens indicate that Wzz regulates the modality of K_{LPS} but does not influence the chain length of LPS-independent capsular K antigen (10). Since both are polymerized by the same Wzy-dependent pathway, the O-PS ligase must play a central role in the establishment of modality.

ABC TRANSPORTER–DEPENDENT PATHWAY FOR O-POLYSACCHARIDE ASSEMBLY

Polymerization by Growth at the Nonreducing Terminus

In contrast to the Wzy-dependent biosynthetic pathway, O-PSs synthesized using an ABC transport–dependent pathway are polymerized in the cytoplasm by the processive addition of glycosyl residues to the nonreducing terminus of the growing chain. The system is found in enteric as well as nonenteric bacteria (Table 3). To date this pathway appears to be restricted to relatively simple O-antigen structures. This may reflect the limited number of examples examined to date because the synthetic pathway is similar to that used for assembly of group II capsular K antigens, many of which are branched heteropolysaccharides (81).

Table 3 Putative or Known O-Polysaccharide ABC-Transporter Genes

Gene	Putative or known activity	Organism	Accession number	Ref.
wzm	Membrane component of ABC transporter	*Escherichia coli* O9	D43637 (*orf*261)	42
		Klebsiella pneumoniae		
		O1	L31775 (*rfbA*)	83
		O8	L41518 (*rfbA*)	91
		O9		91
		Serratia marcescens O16	L34166 (*rfbA*)	92
		Yersinia enterocolitica O:3	Z18920 (*rfbD*)	82
		Vibrio cholerae O1	X59554 (*rfbH*)	200
		Myxococcus xanthus	U36795 (*rfbA*)	201
		Pseudomonas aeruginosa[a]	U63722	85
wzt	ATP-binding component of ABC transporter	*Escherichia coli* O9	D43637 (*orf*431)	42
		Klebsiella pneumoniae		
		O1	L31775 (*rfbB*)	83
		O8	L41518 (*rfbB*)	91
		O9		93
		Serratia marcescens O16	L34166 (*rfbB*)	92
		Yersinia enterocolitica O:3	Z18920 (*rfbE*)	82
		Vibrio cholerae O1	X59554 (*rfbI*)	200
		Myxococcus xanthus	U36795 (*rfbB*)	201
		Pseudomonas aeruginosa[a]	U63723	85
		Aeromonas salmonicida	L11870 (*abcA*[b])	86

[a]Involved in expression of A-band LPS.
[b]This protein is structurally and genetically unique among the previously characterized O- and capsular polysaccharide ABC transporters; see text for details.

After WecA-dependent formation of the Und-PP-GlcNAc primer, a specific glycosyl transferase modifies the primer and commits it to the O-PS polymerization pathway. In *E. coli* O9 (42) and *K. pneumoniae* O1 (43), this step is catalyzed by WbdC (MtfC) or WbbO (RfbF), respectively (Fig. 2B). These enzymes are not required for subsequent polymerization, and additional glycosyltransferases perform this role. Subsequent residues are rapidly added to the nonreducing end of the acceptor (46). There is no need for a polymerase per se since the glycosyl transferases act in a processive fashion (Fig. 2C). Such mechanistic simplicity might pose inherent difficulties for the maintenance of fidelity of O-repeat unit structure. In most of the O-PSs assembled by this pathway, the repeat unit is defined by the sequence of linkages (see the *E. coli* O8 and O9 structures in Table 1). In the case of the O9 polymer, fidelity is maintained by transferases, each of which catalyzes the simultaneous formation of two or more glycosidic linkages (42).

ATP-Dependent Transport of Nascent O Polymer

Studies with *Yersinia enterocolitica* serotype O:3 first showed intracellular O-PS accumulation in this pathway (82). Electron microscopy subsequently confirmed the cytoplasmic location of O-PS in *K. pneumoniae* O1 mutants (83). In both cases, surface assembly requires a plasma membrane ABC transporter encoded by the O-PS biosynthesis cluster (Fig. 2D) (82,83). There is therefore no requirement for a Wzx homolog in this pathway. The genes for the ABC transporters were identified based on predicted protein sequence homologies with the ABC-2 subfamily of transporters (84). ABC-2 transporters consist of an integral membrane protein with an average of six membrane-spanning domains and a hydrophilic protein containing an ATP-binding motif or Walker box. The genes for these two components have now been identified in a number of O-PS biosynthesis clusters (Table 2). As with other

ABC transporters involved in transmembrane export, the membrane-spanning (Wzm) homologs for O-PS biosynthesis generally exhibit little primary sequence identity but have nearly identical hydropathy plots (83,85). In contrast, primary sequences of the ATP-binding (Wzt) homologs are much more highly conserved, with the highest degree of sequence homology in the nucleotide-binding region (83,85).

Expression of S-LPS in *Aeromonas salmonicida* also requires an ABC transporter, AbcA, however, this transporter is unique among cell-surface polysaccharide ABC transporters characterized to date. The N-terminal part of the protein possesses an ATP-binding domain and exhibits sequence homology with the ATP-binding component of other ABC transporters, however, the C-terminal end possesses a potential DNA-binding domain; mutations within the protein lead to the production of R-LPS but no intracellular accumulation of O-PS (86). The genetic organization of this transport system also differs, as the *abcA* gene is not flanked by a gene for an integral membrane component, as in the case with the other carbohydrate ABC transporters (86).

Little is known about the organization and mode of action of ABC transporters for O-PSs. By analogy with better characterized ABC transporters, O-PS export is believed to involve a Wzm/Wzt complex of homodimers with the membrane component forming a pore through which polymers are transported. The only functional studies have been performed with the exporters for group II capsular polysaccharides in *E. coli*, and Bliss and Silver recently proposed a working model for the system (87,88). ABC transporters are believed to move substances across membranes using the energy derived from ATP hydrolysis, and an ATP requirement has been demonstrated for K1 capsule export (89). Polymerized K antigen associates with the ATP-binding component of the transporter (KpsT), which then undergoes an ATP binding–induced conformational change, resulting in its insertion into the membrane via an interaction with the integral membrane component (KpsM). In this manner, the polymer is inserted into the channel. Hydrolysis of ATP then returns KpsT to its initial conformation, promoting its deinsertion and release of polymer. The model also proposes that polymerization and transport of the capsular polysaccharide occur concurrently and that this lack of temporal separation results from the assembly of the components of the biosynthetic pathway into an oligomeric complex, of which the ABC transporter is the central component (87).

The exporters for structurally distinct *E. coli* group II K antigens are functionally interchangeable (90).

ABC transporters for identical O-PS structures from different bacteria are interchangeable (91,92), but unequivocal data are lacking for systems with diverse O-PS structures. The *E. coli* K1 exporter cannot substitute for the ABC-2 transporter of *K. pneumoniae* O1 in export of D-galactan I (83). One possible difference in the K antigen and O-PS synthetic systems is the nature of the carrier lipid on which these polymers are assembled. While most evidence points to polymerization of the O-PS on Und-PP (42), no conclusive evidence has been obtained for involvement of Und-PP–linked K polysaccharide (93). Group II K antigens do terminate in *sn*-α-glycerophosphate, and this lipid may be essential for export (94). It is possible that common lipid carriers provide the unifying concepts for the different types of ABC transporters. Since the ligation reaction is independent of the polymerization pathway and occurs at the periplasmic face of the plasma membrane, we assume that in each pathway the nascent O-PS presented for ligation is in the same (Und-PP–linked) form. That said, it is difficult to envisage a mechanism by which the hydrophobic Und-PP could be transported along with the hydrophilic O-PS through the ABC transporter.

Regulation of O-Antigen Chain Length

Although the ABC transport–dependent pathway does not involve a Wzz chain-length determinant, most smooth LPSs from these strains do exhibit simple modal distributions, which are strain-specific (30). In the extreme case of *Aeromonas salmonicida*, the O-PSs have a uniform chain length (95). This suggests that the modal distribution is a function of one or more of the components encoded by the biosynthesis cluster. By analogy with the group II capsule biosynthesis system, synthesis of ABC transport–dependent O-PSs is believed to be highly coordinated, with the transferases assembled into a biosynthetic complex with the transporter, allowing the polymerization and export processes to proceed concurrently. It is possible that the modal distribution of these O-PSs is simply a function of a strain-specific size preference of the transporter (30). Given the fact that repeat unit fidelity and delineation is maintained despite the presumed continuity of polymerization and export, one must conclude that the glycosyl transferases that form part of these complexes have highly complex substrate and acceptor-binding specificities. These observations suggest that a preferred chain-length distribution may in fact be a function of the transferase specificities and/or a length-dependent discontinuity in the polymerization and export processes (42). A role for the transferases and transporter in the determination of chain-length distribution is further suggested by the observation that altering the stoichiometry of the various components by selectively increasing the copy number of the transporter components results in an altered chain-length distribution (83).

SYNTHASE-DEPENDENT PATHWAY FOR O-POLYSACCHARIDE ASSEMBLY

Polymerization and Export

Recent studies have revealed a third pathway for O-PS assembly in *S. enterica* serovar *borreze*, a member of the unusual O:54 serogroup (45). Synthesis of the poly-N-acetylmannosamine O:54 polysaccharide is initiated by WecA to form the primer for subsequent polymerization of the ManNAc homopolymer (Fig. 3). The first ManNAc residue that commits the Und-PP-GlcNAc primer to O:54 biosynthesis is transferred by the ManNAc transferase WbbE (RfbA). WbbE is a member of the ExoU family of nonprocessive β-glycosyltransferases (45,96). These enzymes are characterized by a common structural domain, which correlates with a β-glycosyltransferase activity and possesses features predicted for the catalytic domain of inverting glycosyltransferases.

The first two steps in the synthase pathway resemble those in the ABC transporter–dependent process. The distinguishing feature of this pathway centers around the second transferase encoded by the O:54 biosynthesis cluster, WbbF (RfbB). This enzyme is a processive ManNAc transferase, which belongs to the HasA family of processive β-glycosyltransferases or "synthases" (45,96). In addition to HasA, the synthase for the hyaluronic acid capsule of *Streptococcus pyogenes*, this family includes Cps3S, the synthase for the type 3 capsule of *S. pneumoniae* and the chitin and cellulose synthases. These proteins all share two conserved structural regions, which are believed to represent catalytic domains. The presence of two catalytic domains is predicted to facilitate the processive incorporation of the characteristic disaccharide repeats of these cell-surface polysaccharides. Members of the HasA family of transferases are all integral membrane proteins with similar topology and are believed to catalyze a vectorial polymerization reaction, simultaneously extending the polysaccharide chain and extruding the nascent polymer across the plasma membrane (45). Because of the central role of these transferases in the synthesis and export of cell surface polysaccharides, the O:54 pathway has been termed the "synthase-dependent" pathway. As with the ABC transporter–dependent pathway, the product of the synthase pathway is ligated to lipid A core by the *E. coli* O-PS ligase (44), and it is assumed that the donor for the ligase reaction is Und-PP-linked polymer. How the carrier lipid might be translocated across the plasma membrane in this system remains an intriguing but unresolved question.

Chain-Length Determination in the Synthase-Dependent Pathway

SDS-PAGE analysis of the O:54 LPS reveals a pattern typical of a nonmodal chain-length distribution (30). The integral role of WbbF in both the polymerization and export processes suggests that chain length might be determined by substrate availability or altered kinetics of polymerization versus export as chain length increases. Interestingly, the SDS-PAGE profile of O:54 LPS suggests a lack of molecules substituted with only one or two O repeats. This chain-length bias is accommodated by the biosynthesis model, where a minimum number of sugars would be polymerized by WbbF before the nascent polymer attained a length sufficient to traverse the plasma membrane and be delivered to the site of ligation.

EXPORT TO THE CELL SURFACE

The method of translocation of LPS to the cell surface from its site of completion at the periplasmic face of the plasma membrane represents the least well-understood process in LPS biosynthesis. The same is true for other cell-surface polysaccharides. This lack of information is partly due to the fact that LPS translocation is essential for the viability of gram-negative bacteria, necessitating the construction of conditional lethal mutants to dissect the components and reactions. Studies by Osborn et al. (97) revealed that R-LPS with complete core is rapidly translocated to the cell surface, and LPS with a core defect in the Kdo region is also translocated, albeit at reduced rates. It was therefore concluded that the translocation machinery must recognize features of lipid A and inner core but is not specific for the O-PS portion of the molecule. This is supported by the observation that the *E. coli* K-12 system efficiently translocates a variety of hybrid LPS molecules containing O-PSs from diverse bacteria to its cell surface. The translocation machinery does not discriminate based on the structure of the O-PS or the assembly pathway used in its synthesis.

Unlike outer membrane proteins, there are no obvious structural features in the LPS molecule that would target it to the cell surface. LPS is known to be tightly associated with newly synthesized outer membrane proteins, but this does not necessarily mean that such an association is a requirement for translocation, since LPS is efficiently exported in the absence of protein synthesis (98). Targeting of LPS therefore appears to be a function of a dedicated export pathway minimally traversing the periplasm but potentially also containing outer membrane components. The components involved most likely form part of an LPS biosynthetic complex. In early studies, Mühlradt et al. reported that nascent LPS synthesized by a conditional LPS mutant of *S. enterica* sv. *typhimurium* first appeared on the cell surface at approximately 220 discrete sites (99). Subsequent studies implicated membrane-adhesion sites (i.e., Bayer junctions (100,101) in the export of nascent LPS, although the number of sites observed per cell is approximately 10-fold lower (102,103). These structures are only visualized by specific electron microscopy techniques and have been quite controversial (for a discussion see Ref. 28), but they have now been implicated in a variety of processes involving the outer membrane. These processes include transport of LPS, CPS, some outer membrane proteins and phospholipids, assembly of peptidoglycan and bacteriophages, as well as bacteriophage adsorption and DNA injection.

It is a conceivable and attractive hypothesis that electron microscopy identifies a scaffold-like translocation system exporting nascent LPS (or other) molecules to the cell surface. Although no LPS-specific translocation candidates have been identified, studies on group II capsule synthesis in *E. coli* may provide some insight into a possible mechanism. The model recently proposed for synthesis and export of the group II capsular polysaccharides (87) also implicates zones of adhesion in polysaccharide export. In this model, proteins encoded by the *kps* cluster (capsular polysaccharide biosynthesis) form a multicomponent synthetic/translocation complex consisting of the cytoplasmic transferases, the ABC transporter, plus (minimally) two periplasmic proteins and a porin. Interestingly, one of the periplasmic proteins, KpsE, shares similarities with membrane-fusion proteins (87). The other periplasmic protein, KpsD, is postulated to span the periplasm, where it recruits an outer membrane porin, as well as interacting with the nascent polysaccharide as it emerges from the ABC transporter. Given the flexibility of the *E. coli* K-12 system for translocating heterologous O-PSs, it seems unlikely that any "fusion" protein such as KpsE interacts specifically with the plasma membrane components of the O-PS synthetic complex.

Recent studies by Gaspar and Valvano (104) have implicated a component of the Tol import system in export of the O7 LPS of *E. coli*. The Tol proteins are required for import of bacteriocins and bacteriophages across the outer membrane, and the system consists of four proteins: TolA, TolB, TolQ, and TolR. TolQ and TolR are believed to be embedded in the plasma membrane, and the C-terminus of TolQ is predicted to be in the periplasm (105–107). Characterization of various *tol* mutants led Gaspar and Valvano to conclude that TolA was involved in surface expression of O7 LPS. This finding may provide an entry point for research into this important aspect of cell-surface assembly.

Bacteria that express the same polysaccharide structure in a traditional LPS-linked form (S-LPS) and in an LPS-free capsular form (traditionally called O-antigen capsules) are becoming increasingly prevalent with expanded investigations. This phenomenon has been described in some *E. coli* strains, in *Proteus*, and in *Vibrio cholerae* O139 (reviewed in Ref. 30). The factors that distinguish expression of traditional LPS molecules from O-antigen capsules remain unresolved. In the case of the *E. coli* group IB capsules, there is nothing evident within the gene clusters that could account for an LPS-independent expression (10). The gene clusters are indicative of a typical Wzy-dependent O-PS assembly pathway, and there is no evidence of any additional export components, for example, homologs of capsular polysaccharide export machinery. The analogous system in *V. cholerae* O139 has not been characterized biochemically, but the genetic locus required for expression of the O139 SR-LPS and the O139 capsule encodes some homologs of both O-PS and capsular polysaccharide export proteins. The *wbfF (otnB)* product showed homology with the periplasmic KpsD protein involved in *E. coli* K1 export (see above) and this may have significance in the expression of the O-antigen capsule. The *otnA* gene product is predicted to encode a Wzz homolog (50,108), although any role for a Wzz protein in formation of an S-LPS remains unclear.

ORGANIZATION OF O-POLYSACCHARIDE BIOSYNTHESIS GENES

Chromosomal Determinants

With only a few exceptions, the genes for O-PS biosynthesis are chromosomal. Since these polymers are assembled separately from lipid A core, then transferred en bloc from the lipid carrier, it is not surprising that the majority of biosynthesis genes are located in a dedicated O-PS biosynthesis gene cluster. In the Enterobacteriaceae, this cluster maps near the *his* locus and encodes O-PS–specific sugar nucleotide synthetases and glycosyl transferases. In some *E. coli* strains, this region contains two distinct loci required for synthesis on O-PSs synthesized by a Wzy-dependent and an ABC transporter–dependent pathway, respectively (10). Some activated precursors that may be required for O-PS biosynthesis are also required for other metabolic and/or biosynthetic functions. These are synthesized by products of "housekeeping" genes, which map outside of the O-PS biosynthesis cluster. In the Enterobacteriaceae, the central enzyme, WecA, maps to the ECA biosynthesis locus and not to an O-PS biosynthesis gene cluster (109,110). However, the potential WecA homologue of *P. aeruginosa* O5, WbpL, is encoded within the O-PS biosynthesis gene cluster.

The gene for the Wzy polymerase was first identified in *S. enterica* serovars A, B, and D1 (15) and mapped to a chromosomal locus separate from the O-PS biosynthesis cluster. Subsequent studies have revealed that, with the possible exception of one strain of *S. enterica* sv. *dublin*, where a wzy-like function is associated with a virulence plasmid (111), *wzy* homologs in other bacteria (Table 2) form part of the chromosomal O-PS biosynthesis cluster. Interestingly, a Wzy analog has been identified in the ECA biosynthesis locus (73; P. Rick, personal communication), an observation that is perhaps not surprising given the existence of a lipid A core–linked ECA fraction in some bacteria. In the enteric bacteria, *wzz* has been mapped near the O-PS biosynthesis cluster, between *his* and *gnd* (10,73,79,112). Some *S. flexneri* strains contain a chromosomally encoded Wzz with a distinct modality as well as an additional plasmid-encoded gene, which encodes a protein with Wzz-like activity and structure (113). *wzz* homologs have recently been identified in *P. aeruginosa* O5 and O16, where the gene product regulates chain length of the B-band (serotype-specific) O antigen (114). In these strains, *wzz* maps immediately upstream of the O polysaccharide biosynthesis locus.

Among nonenteric bacteria, the majority of O-PS biosynthesis genes are also found in a dedicated cluster. Studies with the two major serotypes of *Vibrio cholerae* O1, Inaba and Ogawa, have localized the O-PS biosynthesis region to the same location of the chromosome, adjacent to *gmhD (rfaD)* (47,48,115). The biosynthesis genes are organized in two separate, convergent operons (47). The two serotypes are distinguished by an O-PS modification directed by the *rfbT* gene (116). The genes required for synthesis of the *V. cholerae* O139 O-PS/capsule map to the same region of the chromosome as the O1 biosynthesis cluster (117). The entire cluster has now been sequenced and the organization summarized (50,117). The O-PS biosynthesis clusters for both A-band and B-band LPS of *P. aeruginosa* have been cloned from serotypes O5 and O11 (see Tables 2 and 3). Chromosomal mapping with serotype O5 (strain PAO1) localized the two clusters to 11–13 min and 37 min, respectively (118).

Although information on O-PS biosynthesis operon structure is limited, genes within the clusters can be organized in one or more transcriptional units. Many of these clusters have been cloned and expressed in *E. coli* K-12. This strain possesses its own defective O-PS biosynthesis cluster (39) but supports expression of heterologous O antigens attached to the K-12 core. In this context, it should be noted that proteins encoded by the defective K-12 O-PS biosynthesis cluster may in some cases interact with the products of cloned O-PS biosynthesis genes, complicating interpretation of the results (38,119). For this reason, much of the cloning is now done in a K-12 strain such as *E. coli* SØ874 (120), where the entire O-PS biosynthesis region has been deleted.

Plasmid-Encoded Biosynthesis Determinants

To date, there are several reports of plasmid-encoded O-PS biosynthesis clusters. Probably the most curious example is *E. coli* O111, where the cluster occurs on a 54 MDa EPEC plasmid in one strain (121) but in the normal *his*-linked locus in another (122). The evolutionary origin of the plasmid-borne sequences has not been investigated. Plasmids play a variable role in expression of O-PSs in *Shigella*. *S. sonnei* comprises only one serotype, *S. sonnei* form I, and the form I O-antigen biosynthesis cluster is plasmid-encoded (123–125). In contrast, the biosynthesis cluster is located on the chromosome in *S. flexneri* (126–128) and *S. boydii* (129), while synthesis of the serotype 1 O polysaccharide of *S. dysenteriae* requires the cooperation of a chromosomal biosynthesis cluster and a plasmid-encoded galactosyl transferase gene, *rfpA*. The product of this enzyme catalyzes the transfer of the second glycosyl residue in an Und-PP–linked O unit, which is polymerized by a Wzy-dependent mechanism (130). Loss of the plasmid results in a rough phenotype.

Expression of the O:54 O antigen by the synthase-dependent pathway in *S. enterica* O:54 serovars is directed by a small, ColE1-type mobilizable plasmid

Genetics and Biosynthesis of LPS O-Antigens

(44,131). The heterogeneous O:54 serogroup comprises, to date, 13 strains. Based on the reported O-serotypes of plasmid-cured strains (132) as well as SDS-PAGE analysis of LPS from the different O:54 serovars (44), each member of this serogroup also possesses a functional and distinct chromosomal O-PS biosynthesis cluster.

Genes That Direct Modification Reactions

In many bacteria, O-antigen variation results from modifications to a common O-repeat structure. These modifications are generally nonstoichiometric and nonessential for O side-chain expression. To date, the only genetic data for these modifications involves Wzy-dependent O-PSs. Most involve O-acetylation, glucosylation, or alterations in specificity of the Wzy-directed linkage between O units. With the exception of the O-acetyltransferase of *E. coli* K-12 (38), the enzymes responsible for these modifications are usually encoded by genes outside of the O polysaccharide gene cluster. While the genetic basis for these "decorations" has not been well characterized, in at least a few cases antigenic variation has been shown to result from lysogenic phage conversion (15,28). This type of phage conversion is thought to occur after infection with bacteriophages that recognize and adsorb to O side chains and is a mechanism for the prevention of superinfection by related bacteriophage(s). This phenomenon was first identified in *S. enterica*. This species expresses more than 60 distinct O antigens, and the chemical structures of more than half of these have been determined (reviewed in Ref. 3). Examination of these structures reveals that most of the serovars within a given serogroup express common O-repeat unit backbones with subtle differences arising from side-branch substitutions, O-acetylation, or changes in the glycosyl linkages. Within the E serogroup, subgroups E2 and E3 result from lysogenic phage conversion of subgroup E1. Lysogenization by phage ε^{15} causes a change in the configuration of the Wzy-catalyzed glycosidic linkage of the trisaccharide repeat to give the O antigen of E2 (133,134); further lysogenization by ε^{34} results in the addition of a glucose side branch to yield the E3 O antigen (135).

Phage conversion is also the basis for much of the antigenic variation seen among the 13 O-serotpyes of *S. flexneri* (136). Ten of these serotypes share a common tetrasaccharide O-repeat unit backbone whose synthesis is directed by the *his*-linked O-PS biosynthesis cluster. O-acetylation and glucosylation modifications of this structure are directed by prophage integrated near the *pro-lac* region of the chromosome (137). In one well-documented example, *S. flexneri* serotype Y strains express S-LPS with the unmodified tetrasaccharide O-repeat unit, and this S-LPS functions as a receptor for *Shigella* bacteriophage SF6 (138). After infection, the SF6-directed O-acetylation of the O-PS converts the group 3,4 antigen (characterizing the unmodified O-repeat unit) to the group 6 antigen of serotype 3b (139–141). The gene responsible for this acetylation, *oac*, has been cloned and sequenced and the predicted protein found to share homology with a putative acetylase, NodX, from *Rhizobium leguminosarum* (142). Recently, Slauch and coworkers identified the gene for an O-acetylase which, when expressed, confers O factor 5 on *S. enterica* sv. *typhimurium* (143). This gene, *oafA*, maps to 48.5 min on the chromosome, away from the O-PS biosynthesis locus, and encodes a protein with homology to the SF6-encoded Oac and the NodX acetylase (144). Given the map location of *oafA* and the fact that factor 5 is not expressed in all *typhimurium* isolates, expression of this gene is also expected to result from phage conversion. Bacteriophage-mediated modifications have also been reported in *P. aeruginosa* (145).

Lysogenization of *S. enterica* sv. *choleraesuis* by phage 14 converts the O-PS from serotype O-6,7 to O-6,7,14, and the appearance of O factor 14 appears to result from a phage-induced glycosylation of the O-repeat (146). Lysogenic conversion by this phage is somewhat more complex than serotyping results would indicate however, since the O-PS from a lysogenic strain has been shown to be, on average, six times longer than that from a nonlysogenic strain (146). Given the relationship between O-chain length and serum resistance, it is not surprising that the lysogenic strain is serum resistant, in contrast to the serum-sensitive nonlysogenic strains, and is more virulent in mice (146).

These O-PS modifications are not confined to O-PSs synthesized by Wzy-dependent pathways. A number of *Klebsiella* O serotypes express galactose-containing O-PSs synthesized by an ABC transporter–dependent pathway. These share a common D-galactan I backbone but differ in modifications to this structure (Table 1). The O-PS biosynthesis clusters of serotypes O1 and O8 have been cloned and sequenced and shown to direct synthesis of D-galactan I only (91,147); the locus responsible for the D-galactan II modification is undetermined. The O antigen of serotype O8 differs from O1 in the presence of O-acetyl groups on the D-galactan I backbone. Characterization of the O8 D-galactan I biosynthesis cluster indicates that the O-acetyl transferase is not encoded by the biosynthesis cluster (91). It is conceivable that prophages are also responsible for the serotype-specific modifications of D-galactan I.

REGULATION

Although lipid A core is a required component of the gram-negative cell wall, the O-PS chain is not essential for survival in the laboratory. There is considerable evidence that, while on/off switching of O-PS biosynthesis does not occur, specific alterations in growth conditions will affect the pattern of S-LPS in SDS-PAGE (reviewed in Ref. 28). Serial passage of *S. enterica* sv. *anatum* in the laboratory has been found to lead to the synthesis of shorter O side chains, although the basis for this shift is not known (148). In contrast, selection for serum resistance in *E. coli* O111 results in an increase in the average O-chain length and a higher frequency of capping of R-LPS with O-PS (149). Again, the basis for these changes is not understood. In both cases, the bacteria appear to have some means of sensing the changes in environmental conditions. Given what we know about O-PS synthesis, it would appear that in these two cases the bacteria are able to respond by altering the kinetics of polymerization and ligation of nascent O chains while still maintaining a modal distribution.

Transcriptional regulation of the O-PS biosynthesis cluster has only been demonstrated with one species: *Y. enterocolitica* O:3 (150). In this strain, as well as in serotypes O:1 (151) and O:8 (152), raising the growth temperature to 37°C results in less capping of R-LPS with O-PS (153) as well as a decrease in the average preferred chain length (150). In serotype O:3, these changes result from repression of transcription of the O-PS biosynthesis cluster and is mediated by a repressor encoded outside of the cluster. There is no equivalent repressor gene in *E. coli*, and as a result, expression of the O:3 biosynthesis cluster in *E. coli* K-12 is not temperature-regulated (154).

In 1994, Hobbs and Reeves described a 39 bp sequence located upstream of a number of cell-surface polysaccharide gene clusters in a variety of different gram-negative organisms (155). They called this sequence JUMPstart (for just upstream of many polysaccharide-associated gene starts). Within the second half of JUMPstart is a conserved 8 bp element (GGCGGTAG), which has been termed *ops* (for operon polarity suppressor) (156). To date, JUMPstart sequences have been identified upstream of most of the O-PS biosynthesis gene clusters that have been examined, although a concerted search for this sequence has not been made with all of the clusters so far characterized. The sequence has been identified upstream of the O-PS biosynthesis gene clusters of *E. coli* K-12 (O16) (157), *E. coli* O9 (42), *K. pneumoniae* O1 (84) *K. pneumoniae* O8 (D. Heinrichs and C. Whitfield, unpublished), *S. enterica* B (158), *S. enterica* C1 (159), *V. cholerae* O1 (116), *V. cholerae* O139 (160), *Y. pseudotuberculosis* IIA (82), *Y. enterocolitica* O:3 (155), and *S. flexneri* (73,161). Interestingly, while the sequence has been identified upstream of the O-PS biosynthesis gene clusters in *S. enterica* serogroups B and C1, it is not present upstream of the plasmid-borne O:54 cluster of *S. enterica* sv. *borreze* (131). The significance of this is currently unclear.

The JUMPstart sequence is always found in the same orientation to the direction of transcription, an observation that, combined with its position upstream of transcription start sites, suggests a possible role in transcription. The *ops* subsequence is also found in the regulatory regions of a number of gram-negative operons involved in toxin production and conjugal transfer of DNA. Studies on the involvement of *ops* in expression of the *E. coli* hemolysin operon have provided an insight into a possible mechanism of transcriptional regulation and have implicated the product of the *rfaH* gene in this process. By examining transcription of promoter-distal genes and mRNA transcript lengths in various *ops* and *rfaH* mutants, two groups (162,163) reported evidence to support the idea that the *E. coli* RfaH protein interacts with *ops* to suppress transcription polarity. Bailey et al. (163) also reported a structural relationship between RfaH and the essential transcription/termination cofactor NusG. Based on their experimental data and this homology, these workers have proposed that RfaH regulates expression of the *ops*-containing operons by interacting with the *cis*-acting *ops* element and the RNA polymerase, suppressing polarity and enabling transcription to continue through the operon. More recently, Leeds and Welch (164) speculated that RfaH might interact with the RNA polymerase to help create a termination-resistant complex in a manner similar to λ N-mediated antitermination at the λ*nut* site (165).

The *ops* sequence is also found upstream of the LPS core biosynthesis locus (166), suggesting the possibility of coregulation of synthesis of the core and O-PSs. The role of this sequence in regulation of the LPS core oligosaccharide is described elsewhere in this volume. This additional role complicates a study of O-PS regulation by *ops*-RfaH since formation of S-LPS is not a suitable measure of O-PS regulation. Transcriptional studies comparable to those reported for the *E. coli* hemolysin operon have not yet been reported with the relevant O-PS biosynthesis operons, so the role of JUMPstart in RfaH-mediated transcription elongation of these operons remains to be confirmed. However, the conservation of the *ops* sequence within the JUMPstart sequences provides strong support for an antitermination process in O-PS regulation.

There are several unexplained phenomena in the literature which potentially reflect regulatory processes. For example, lysogeny of *Acetobacter methanoliticus* MB58/4 by bacteriophage Acm1 results in conversion of S-LPS to the rough LPS phenotype (167). Studies with *E. coli* K-12 recombinants expressing other *E. coli* O-PSs has implicated an antisense RNA in this phenomenon (167). However, the target for this RNA molecule has not yet been identified, nor has this mechanism of regulation been confirmed in the endogenous host. Because phage Acm1 appears to be restricted to strain MB 58/4, it is not clear whether this mechanism of regulation is peculiar to this bacteriophage-host interaction. Also, *Coxiella burnetii*, the bacterium that causes Q fever, undergoes smooth-to-rough LPS phase variation, with a concomitant loss of virulence (168). This loss of virulence reflects the role of the O-PS in conferring serum resistance and conversion to R-LPS results from growth of the organism in the absence of immune selection (169). The molecular events underlying this switch have not been elucidated, however, given that the change appears to be irreversible, one possibility is the existence of a mutational "hot spot" involving sequences required for O-PS expression.

GENERATION OF ANTIGENIC DIVERSITY

The serological diversity seen among O antigens is a reflection of the molecular diversity of the different biosynthesis clusters. Given the relative stability of the surrounding DNA sequences, it is apparent that there are forces specifically driving the (evolutionary) changes at these loci. In the case of gram-negative pathogens, it is generally accepted that these changes are largely a response to selective pressure imposed by the host immune response (170). Some insight into potential mechanisms for the generation of the diversity seen among O antigens is now possible with the growing number of O-PS biosynthesis clusters that have been cloned and sequenced.

Gene mutation is probably the simplest mechanism for the generation of antigenic diversity. This has been demonstrated in the seroconversion of *V. cholerae* serotype *ogawa* to serotype *inaba*. The *ogawa* serotype has been shown to switch to the *inaba* serotype in vivo, and seroconversion correlates with the host immune response, implying that the selective pressure for the conversion comes from the immune response. The structural differences between the two O serotypes has not yet been established, however, the genetic basis has been identified as the inactivation of the *rfbT* gene of the *ogawa* O-PS biosynthesis cluster (116). The experimental strain examined by Stroeher et al. was found to have suffered a single base change giving rise to a truncated protein product, however, the nature of the mutation was reported to vary among isolates (116). The precise function of the RfbT protein has yet to be determined.

Gene mutation has also been implicated in the genesis of *S. enterica* serogroup A from serogroup D1 (171). These two strains synthesize an O-PS with the same Man-Rha-Gal backbone (Table 1), but differ in the identity of the 3,6-dideoxyhexose side branch: paratose in A and tyvelose in D1. Synthesis of these two sugars follows the same pathway, with CDP-tyvelose resulting from the epimerization of CDP-paratose. This epimerization reaction is catalyzed by the product of *tyv* (previously *rfbE*), but the gene, while present in serogroup A, is nonfunctional due to a frameshift mutation (171).

The major source of antigenic diversity is believed to be repeated lateral gene transfer and recombination events involving the O-PS biosynthesis region, followed by adaptation to a new niche or environment (172,173). It has been suggested that one indication of "captured DNA" in a given O-PS biosynthesis gene cluster is the common observation that the G+C content of these clusters, or blocks of genes within the clusters, is lower than the average for the species (173). The same is often true for genes involved in LPS core oligosaccharide synthesis. It has also been argued that the lower G+C of some of these genes may instead reflect atypical codon use as a means of translational regulation (28,73,174). However, there is now supporting evidence for the involvement of lateral gene transfer in the generation of O-PS diversity. Much of this information has come from the Reeves laboratory, where the O-PS biosynthesis clusters of *S. enterica* have been characterized and compared to generate a model for molecular evolution. Their studies have demonstrated a mosaic structure of clusters synthesizing structurally related O-PSs, which contrasts with the high degree of sequence conservation in the regions flanking the O-PS biosynthesis gene clusters (175–177). In several members of the Enterobacteriaceae, the O-PS biosynthesis cluster is located adjacent to *gnd*

(78,80,147), and the surprisingly high degree of sequence polymorphism seen in *gnd* alleles is speculated to reflect its cotransfer with flanking O-PS biosynthesis genes (178). Sequence analysis of the O-PS biosynthesis gene cluster from serogroup D2 suggests that this cluster is a hybrid of the D1 and E1 clusters. An H-repeat (H-rpt) element has been identified in the region between the D1- and E1-like sequences. Based on its position in the junction between these two regions and the resemblance of the H-rpt to IS elements, serogroup D2 is speculated to have evolved by intraspecific gene transfer between a serogroup D1 recipient and a serogroup E1 donor.

Intraspecific recombination and IS elements have also been implicated in the generation of the novel *V. cholerae* epidemic strain O139. Based on the available evidence, this strain appears to have resulted from the capture of DNA sequences for the synthesis of a novel O-PS and capsular polysaccharide by an O1 El Tor strain (reviewed in Ref. 179). The LPS of O139 is semi-rough (SR-LPS) and possesses a different sugar composition from the O1 serotypes (51). Interestingly, while synthesis of the O139 antigen is directed by genes that map to the same chromosomal region as O1, evidence indicates that the O139-specific DNA was acquired through lateral gene transfer, possibly involving an IS element associated with a number of the O-PS biosynthesis loci in this species (50,179–181). *V. cholerae* O139 also differs from O1 in that the O139 strain produces an O-antigen capsule (51), and sequence analysis of the O139-specific DNA identified genes with homology to genes involved in synthesis of both O- and capsular polysaccharides (49,50).

The association of the *V. cholerae* DNA sequences with a mobile element, previously termed *rfbQRS* (116), and since renamed IS*1358* (116,179), has led to the idea that this element played a role in the genesis of the new strain. Interestingly, the H-repeat of *Salmonella* exhibits homology to IS*1358* (182). The idea that IS*1358* was involved in the genetic changes that gave rise to O139 is supported by the observation that a related element, IS*AS1*, in *Aeromonas salmonicida* is transposable (183). In *A. salmonicida* this element is also implicated in mediating changes within the O-PS biosynthesis cluster. Chu and coworkers analyzed a spontaneous rough mutant of *A. salmonicida* and found that the O-PS–deficient phenotype was due to the insertional inactivation of an O-polysaccharide ABC transport gene by the element (86). Is-*1* elements have also been implicated in gene duplication events involving the *manB* (formerly *rfbK*) and *manC* (*rfbM*) genes from the O-PS biosynthesis cluster of *E. coli* O9 (184).

Sequence analysis revealed that one of the *manB-manC* copies is flanked by IS*1* elements, forming a compound transposon and providing a mechanism for the gene duplication.

The observation that different organisms, some closely related and some distantly related, may produce similar or identical O-polysaccharides provided the first indications for lateral gene transfer of O-PS biosynthesis genes. Further evidence for this phenomenon has come from the examination of the relationships between the biosynthesis genes directing the synthesis of related O antigens. The most extensively characterized are the different serovars of the species *S. enterica*. These studies, which come from Peter Reeve's laboratory (173), have revealed that genes for similar functions are similar if not identical, conserved regions of the biosynthesis cluster exhibit remarkably little restriction site polymorphism, and the biosynthesis clusters within a given serogroup are highly conserved (185). There are also a number of examples of structurally related O-PSs in different species and even different genera; DNA homology studies of the relevant O-PS biosynthesis gene clusters have provided further evidence for lateral gene transfer. For example, *E. coli* O7 and *Shigella boydii* type 12 express structurally similar O-PSs, and this is reflected in high levels of homology between the respective O-PS biosynthesis loci (129).

There are a number of reports of bacteria that synthesize related or identical O-PSs but fall into different clonal groups based on hybridization and sequence comparison of the O-PS biosynthesis clusters. Probably the best example of this is seen with *K. pneumoniae* O1 and O8 and *S. marcescens* O16. All three synthesize D-galactan I O-PS (Table 1). Sequence analysis has revealed that the genes in the three D-galactan biosynthesis gene clusters are arranged in the same order, but the homology at the nucleotide sequence level is relatively low (91,92). Despite this overall low homology, the protein components of the ABC transporters encoded by the three clusters were found to be much more conserved and were functionally interchangeable (91,92). It has been speculated that the relatively low sequence homology seen among the three clusters may indicate that the clusters were not recently exchanged. Over a long period of time, adaptation of the clusters to their chromosomal environment would result in the low sequence homology (91). The authors have also suggested that the glycosyl transferases may be able to accommodate a greater degree of sequence variation than the transporters, without affecting function. The generally low protein homology seen among related glycosyl transferases would support this hypothesis (96,186).

Although there has been much speculation regarding the role of lateral gene transfer in the generation of antigenic diversity, one area about which there is very little information is the mechanism of DNA transfer. Logically, DNA sequences could be mobilized either through plasmid transfer, transduction by bacteriophage, or by DNA transformation with free DNA. Integration into the recipient cluster could then be mediated by homologous recombination or possibly transposition. The relative involvement of each of these processes is unknown. As previously discussed, lysogenic phage conversion is one source of O-PS diversity, however the phage-encoded changes characterized to date all map outside of the O-PS biosynthesis gene cluster. As discussed above, plasmids are known to be involved to varying extents in O-PS expression, but lateral transfer has only been shown for the O:54 antigen of *S. enterica* serovar *borreze* (131). The ColE1-type plasmid carrying the O:54 biosynthesis gene cluster can be mobilized to confer O:54 serospecificity on recipient strains. However, while the various members of the O:54 serogroup are heterogeneous in terms of their endogenous chromosomal O polysaccharide biosynthesis clusters (44,132), the unique structure of the O:54 antigen compared to other *Salmonella* (or indeed enteric) O antigens (44) and the absence of associated IS or IS-like elements suggest that this plasmid is not a likely source of homologous recombination with chromosomal O-PS biosynthesis sequences.

CONCLUDING REMARKS

The synthesis and surface assembly of LPS O-PSs is accomplished by the highly coordinated expression of a variety of different gene products, encoded within a number of different operons. These proteins function with outstanding accuracy and specificity to give repeat unit fidelity and delineation, as well as a strain-specific length distribution. The last decade has witnessed a tremendous increase in our understanding of the synthesis of LPS O-PSs due to the application of a combination of genetic and biochemical analyses of the multicomponent systems responsible for their expression. Despite these advances, however, many of the questions that were unresolved at the time of the Mäkelä and Stocker review of 1984 are still unanswered today. These questions comprise a black box representing the various aspects of export and regulation:

1. How are the newly synthesized LPS molecules exported from the site of ligation to the cell surface, and what are the steps involved? Given the essential nature of LPS in the assembly of the gram-negative outer membrane, the identification of individual components involved in this process would provide valuable targets for novel antimicrobial therapies.
2. The involvement of Und-P in the synthesis of a variety of distinct cell surface polysaccharides, both essential and nonessential, suggests the bacteria have a method of controlling the relative availability of this molecule to these different pathways. Such a control mechanism has not yet been identified.
3. How does the cell regulate expression of O-PSs?
4. What are the features that determine the specificity of glycosyl transferases and O-PS polymerases? What are the precise mechanisms of action of glycosyl transferases and the Wzy polymerases? The identification of common features among these enzymes might also provide valuable information for the design of specific complex carbohydrates or antimicrobial targets.

The relatively recent identification of the genes involved in O chain-length regulation and plasma membrane export has also raised a number of mechanistic questions, which have yet to be elucidated:

5. How do the Wzx O unit transporters, O-PS ABC transporters, and the O:54 synthase effect the trans–plasma membrane export of their substrate molecules? Does the carrier lipid remain attached during export?
6. What is the mechanism of action of the Wzz O chain-length determinant?

ACKNOWLEDGMENTS

Research in CW's laboratory has been generously supported by funding from the Natural Sciences and Engineering Research Council of Canada and the Medical Research Council of Canada.

REFERENCES

1. Joiner KA. Studies on the mechanism of bacterial resistance to complement-mediated killing and on the mechanism of action of bactericidal antibody. Curr Top Microbiol Immunol 1985; 121:49–133.

2. Sandlin RC, Goldberg MB, Maurelli AT. Effect of O side-chain length and composition on the virulence of *Shigella flexneri* 2a. Mol Microbiol 1996; 22:63–73.
3. Knirel YA, Kochetkov NK. The structure of lipopolysaccharides of gram-negative bacteria. III. The structure of O-antigens: a review. Biochemistry (Moscow) 1994; 59:1325–1383.
4. Goldman RC, Leive L. Heterogeneity of antigenic-side-chain length in lipopolysaccharide from *Escherichia coli* O111 and *Salmonella typhimurium* LT2. Eur J Biochem 1980; 107:145–153.
5. Hitchcock PJ, Brown TM. Morphological heterogeneity among *Salmonella* lipopolysaccharide chemotypes in silver-stained polyacrylamide gels. J Bacteriol 1983; 154:269–277.
6. Jann B, Reske K, Jann K. Heterogeneity of lipopolysaccharides. Analysis of polysaccharide chain lengths by sodium dodecylsulfate-polyacrylamide gel electrophoresis. Eur J Biochem 1975; 60:239–246.
7. Palva L, Mäkelä PH. Lipopolysaccharide heterogeneity in *Salmonella typhimurium* analyzed by sodium dodecyl sulfate/polycarylamide gel electrophoresis. Eur J Biochem 1980; 107:137–143.
8. Peterson AA, McGroarty EJ. High-molecular-weight components in lipopolysaccharides of *Salmonella typhimurium, Salmonella minnesota*, and *Escherichia coli*. J Bacteriol 1985; 162:738–745.
9. Kuhn H-M, Meier-Dieter U, Mayer H. ECA, the enterobacterial common antigen. FEMS Microbiol Rev 1988; 54:195–222.
10. Amor PA, Whitfield C. Molecular and functional analysis of genes required for expression of group IB K antigens in *Escherichia coli*: characterization of the *his*-region containing gene clusters for multiple cell-surface polysaccharides. Mol Microbiol 1997; 26:145–161.
11. Dodgson C, Amor P, Whitfield C. Distribution of the *rol* gene encoding the regulator of lipopolysaccharide O-chain length in *Escherichia coli* and its influence on the expression of group I capsular K antigens. J Bacteriol 1996; 178:1895–1902.
12. Franco AV, Liu D, Reeves PR. A Wzz (Cld) protein determines the chain length of K lipopolysaccharide in *Escherichia coli* O8 and O9 strains. J Bacteriol 1996; 178:1903–1907.
13. MacLachlan PR, Keenleyside WJ, Dodgson C, Whitfield C. Formation of the K30 (group I) capsule in *Escherichia coli* O9:K30 does not require attachment to lipopolysaccharide lipid A-core. J Bacteriol 1993; 175:7515–7522.
14. Jann K, Dengler T, Jann B. Core-lipid A on the K40 polysaccharide of *Escherichia coli* O8:K40:H9, a representative of group I capsular polysaccharides. Zentrabl Bakteriol 1992; 276:196–204.
15. Mäkelä PH, Stocker BAD. Genetics of lipopolysaccharide. In: Reitschel ET, ed. Handbook of Endotoxin. Vol. I. Chemistry of Endotoxin. Amsterdam: Elsevier, 1984:59–137.
16. Oxley D, Wilkinson SG. Structures of neutral glycans isolated from the lipopolysaccharides of reference strains for *Serratia marcescens* serogroups O16 and O20. Carbohydr Res 1989; 193:241–248.
17. Lam JS, de Kievit TR, Currie HL. Genes involved in the biosynthesis of *Pseudomonas aeruginosa* lipopolysaccharide. In: Nakazawa T, et al., eds. Molecular Biology of Pseudomonads. Washington, DC: ASM Press, 1996:451–461.
18. Rivera M, Chivers TR, Lam JS, McGroarty EJ. Common antigen lipopolysaccharide from *Pseudomonas aeruginosa* AK1401 as a receptor for bacteriophage A7. J Bacteriol 1992; 174:2407–2411.
19. Yokota S, Kaya S, Sawada S, Kawamura T, Araki Y, Ito E. Characterization of a polysaccharide component of lipopolysaccharide from *Pseudomonas aeruginosa* IID 1008 (ATCC 27584) as D-rhamnan. Eur J Biochem 1987; 167:203–209.
20. Kol O, Wieruszeski J-M, Strecker G, Fournet B, Zalisz R, Smets P. Structure of the O-specific polysaccharide chain of *Klebsiella pneumoniae* O1:K2 (NCTC 5055) lipopolysaccharide. A complementary elucidation. Carbohydr Res 1992; 236:339–344.
21. Kol O, Wieruszeski J-M, Strecker G, Montreuil J, Fournet B. Structure of the O-specific polysaccharide chain from *Klebsiella pneumoniae* O1:K2 (NCTC 5055) lipopolysaccharide. Carbohydr Res 1991; 217:117–125.
22. Kelly RF, Severn WB, Richards JC, et al. Structural variation in the O-specific polysaccharides of *Klebsiella pneumoniae* serotype O1 and O8 lipopolysaccharide: evidence for clonal diversity in *rfb* genes. Mol Microbiol 1993; 10:615–625.
23. Whitfield C, Richards JC, Perry MB, Clarke BR, MacLean LL. Expression of two structurally distinct D-galactan O-antigens in the lipopolysaccharide of *Klebsiella pneumoniae* serotype O1. J Bacteriol 1991; 173:1420–1431.
24. Auken HM, Oxley D, Wilkinson SG. Structural and serological characterisation of an O-specific polysaccharide from *Serratia plymuthica*. FEMS Microbiol Lett 1993; 111:295–300.
25. Helander IM, Moran AP, Mäkelä PH. Separation of two lipopolysaccharide populations with different contents of O-antigen factor 12_2 in *Salmonella enterica* serovar Typhimurium. Mol Microbiol 1992; 6:2857–2862.
26. Rahman MM, Guard-Petter J, Carlson RW. A virulent isolate of *Salmonella enteritidis* produces a *Salmonella typhi*-like lipopolysaccharide. J Bacteriol 1997; 179:2126–2131.
27. Shibaev VL. Biosynthesis of bacterial polysaccharide chains composed of repeating units. Adv Carbohydr Chem Biochem 1986; 44:277–339.
28. Whitfield C, Valvano MA. Biosynthesis and expression of cell surface polysaccharides in gram-negative bacteria. Adv Micro Physiol 1993; 35:135–246.
29. Whitfield C. Biosynthesis of lipopolysaccharide O antigens. Trends Microbiol 1995; 3:178–185.
30. Whitfield C, Amor PA, Köplin R. Modulation of the surface architecture of Gram-negative bacteria by the action of surface polymer:lipid A-core ligase and by determinants of polymer chain length. Mol Microbiol 1997; 23:629–638.
31. Reeves PR, Hobbs M, et al. Bacterial polysaccharide synthesis and gene nomenclature. Trends Microbiol 1996; 4:495–503.

32. Osborn MJ, Gander JE, Parisi E. Mechanisms of assembly of the outer membrane of *Salmonella enterica*: site of synthesis of lipopolysaccharide. J Biol Chem 1972; 247:3973–3986.
33. Meier-Dieter U, Barr K, Starman R, Hatch L, Rick PD. Nucleotide sequence of the *Escherichia coli rfe* gene involved in the synthesis of enterobacterial common antigen. J Biol Chem 1992; 267:746–753.
34. Ohta M, Ina K, Kusuzaki K, Kido N, Arakawa Y, Kato N. Cloning and expression of the *rfe-rff* gene cluster of *Escherichia coli*. Mol Microbiol 1991; 5:1853–1862.
35. Klena JD, Schnaitman CA. Function of the *rfb* gene cluster and the *rfe* gene in the synthesis of O antigen by *Shigella dysenteriae* 1. Mol Microbiol 1993; 9:393–402.
36. Wang L, Liu D, Reeves PR. C-terminal half of *Salmonella enterica* WbaP (RfbP) is the galactosyl-1-phosphate transferase domain catalyzing the first step of O-antigen synthesis. J Bacteriol 1996; 178:2598–2604.
37. Alexander DC, Valvano MA. Role of the *rfe* gene in the biosynthesis of the *Escherichia coli* O7-specific lipopolysaccharide and other O-specific polysaccharides containing *N*-acetylglucosamine. J Bacteriol 1994; 176:7079–7084.
38. Yao Z, Valvano MA. Genetic analysis of the O-specific lipopolysaccharide biosynthesis region (*rfb*) of *Escherichia coli* K-12 W3110: identification of genes that confer group 6 specificity to *Shigella flexneri* serotypes Y and 4a. J Bacteriol 1994; 176:4133–4143.
39. Liu D, Reeves PR. *Escherichia coli* K12 regains its O antigen. Microbiology 1994; 140:49–57.
40. Zhang L, Radziejewska-Lebrecht J, Krajewska-Pietrasik D, Toivanen P, Skurnik M. Molecular and chemical characterization of the lipopolysaccharide O-antigen and its role in the virulence of *Yersinia enterocolitica* serotype O:8. Mol Microbiol 1997; 23:63–76.
41. Rick PD, Hubbard GL, Barr K. Role of the *rfe* gene in the synthesis of the O8 antigen in *Escherichia coli* K-12. J Bacteriol 1994; 176:2877–2884.
42. Kido N, Torgov VI, Sugiyama T, et al. Expression of the O9 polysaccharide of *Escherichia coli*: sequencing of the *E. coli* O9 *rfb* gene cluster, characterization of mannosyltransferases, and evidence for an ATP-binding cassette transport system. J Bacteriol 1995; 177:2178–2187.
43. Clarke BR, Bronner D, Keenleyside WJ, Severn WB, Richards JC, Whitfield C. Role of Rfe and RfbF in the initiation of biosynthesis of D-galactan I, the lipopolysaccharide O antigen from *Klebsiella pneumoniae* serotype O1. J Bacteriol 1995; 177:5411–5418.
44. Keenleyside WJ, Perry M, MacLean L, Poppe C, Whitfield C. A plasmid-encoded $rfb_{O:54}$ gene cluster is required for biosynthesis of the O:54 antigen in *Salmonella enterica* serovar Borreze. Mol Microbiol 1994; 11(3):437–448.
45. Keenleyside WJ, Whitfield C. A novel pathway for O-polysaccharide biosynthesis in *Salmonella enterica* serovar Borreze. J Biol Chem 1996; 271:28581–28592.
46. Weisgerber C, Jann K. Glucosyldiphosphoundecaprenol, the mannose acceptor in the synthesis of the O9 antigen of *Escherichia coli*. Biosynthesis and characterization. Eur J Biochem 1982; 127:165–168.
47. Fallarino A, Mavrangelos C, Stroeher UH, Manning PA. Identification of additional genes required for O-antigen biosynthesis in *Vibrio cholerae* O1. J Bacteriol 1997; 179:2147–2153.
48. Manning PA, Heuzenroeder MW, Yeadon J, Leavesley DI, Reeves PR, Rowley D. Molecular cloning and expression in *Escherichia coli* K-12 of the O-antigens of the Inaba and Ogawa serotypes of the *Vibrio cholerae* O1 lipopolysaccharides and their potential for vaccine development. Infect Immun 1986; 53:272–277.
49. Comstock LE, Johnson JA, Michalski JM. Morris Jr. JG, Kaper JB. Cloning and sequence of a region encoding a surface polysaccharide of *Vibrio cholerae* O139 and characterization of the insertion site in the chromosome of *Vibrio cholerae* O1. Mol Microbiol 1996; 19:815–826.
50. Stroeher UW, Manning PA. *Vibrio cholerae* serotype 139: swapping genes for surface polysaccharide biosynthesis. Trends Microbiol 1997; 5:178–180.
51. Cox AD, Perry MB. Structural analysis of the O-antigen-core region of the lipopolysaccharide from *Vibrio cholerae* O139. Carbohydr Res 1996; 290;59–65.
52. Burrows LL, Charter DF, Lam JS. Molecular characterization of the *Pseudomonas aeruginosa* serotype O5 (PAO1) B-band lipopolysaccharide gene cluster. Mol Microbiol 1996; 22:481–495.
53. Rocchetta L, Burrows LL, Pacan JC, Lam JS. Three rhamnosyl transferases responsible for assembly of the A-band D-rhamnan polysaccharide in *Pseudomonas aeruginosa*: a fourth transferase, WbpL, is required for the initiation of both A-band and B-band lipopolysaccharide synthesis. Mol Microbiol 1998; 28:1103–1120.
54. Xu X, Khudyakov I, Wolk CP. Lipopolysaccharide dependence of cyanophage sensitivity and aerobic nitrogen fixation in *Anabaena* sp. strain PCC 7120. J Bacteriol 1997; 179:2884–2891.
55. Anderson RG, Hussey H, Baddiley J. The mechanism of wall synthesis in bacteria: the organisation of enzymes and isoprenoid phosphates in the membrane. Biochem J 1972; 127:11–25.
56. Yuasa R, Levinthal M, Nikaido H. Biosynthesis of cell wall lipopolysaccharide in mutants of *Salmonella*. V. A mutant of *Salmonella typhimurium* defective in the synthesis of cytidine diphosphoabequose. J Bacteriol 1969; 100:433–444.
57. Mäkelä PH. Glucosylation of lipopolysaccharide in *Salmonella*: mutants negative for O antigen factor 12_2. J Bacteriol 1973; 116:847–856.
58. Nikaido K, Nikaido H. Glucosylation of lipopolysaccharide in *Salmonella*: biosynthesis of O antigen factor 12_2. II. Structure of the lipid intermediate. J Biol Chem 1971; 246:3912–3919.
59. Takashita M, Mäkelä PH. Glucosylation of lipopolysaccharide in *Salmonella*: biosynthesis of O antigen factor 12_2. III. The presence of 12_2 determinants in haptenic polysaccharides. J Biol Chem 1971; 246:3920–3927.

60. Sasaki T, Uchida T, Kurahashi K. Glucosylation of O-antigen in *Salmonella* carrying epsilon 15 and epsilon 34 phages. J Biol Chem 1974; 249:761–772.
61. Wright A, Kanegasaki S. Molecular aspects of lipopolysaccharides. Physiol Rev 1971; 51:748.
62. Shibaev VN, Druzhinina RN, Popova AN, Rozhinova SS, Kilesso VA. Mechanism of O-specific polysaccharide biosynthesis in *Salmonella* serogroups C2 and C3. Eur J Biochem 1979; 101:309–316.
63. Mäkelä PH. Glucosylation of lipopolysaccharide in *Salmonella*: mutants negative for O antigen factor 12_2. J Bacteriol 1996; 116:847.
64. Keller JM. Ph.D. Thesis, Massachusetts Institute of Technology, Cambridge, MA, 1966.
65. McGrath BC, Osborn MJ. Localization of terminal steps of O-antigen synthesis in *Salmonella typhimurium*. J Bacteriol 1991; 173:649–654.
66. Wang L, Reeves PR. Involvement of the galactosyl-1-phosphate transferase encoded by the *Salmonella enterica rfbP* gene in O-antigen subunit processing. J Bacteriol 1994; 176:4348–4356.
67. Liu D, Cole RA, Reeves PR. An O-antigen processing function for Wzx (RfbX): a promising candidate for O-unit flippase. J Bacteriol 1996; 178:2102–2107.
68. MacPherson DF, Manning PA, Morona R. Genetic analysis of the *rfbX* gene of *Shigella flexneri*. Gene 1995; 155:9–17.
69. Bray D, Robbins PW. The direction of chain growth in *Salmonella anatum* O-antigen biosynthesis. Biochem Biophys Res Commun 1967; 28:334–339.
70. Robbins PW, Bray D, Dankert M, Wright A. Direction of chain growth in polysaccharide synthesis. Science 1967; 158:1536–1542.
71. Collins LV, Hackett J. Molecular cloning, characterization, and nucleotide sequence of the *rfc* gene, which encodes an O-antigen polymerase of *Salmonella typhimurium*. J Bacteriol 1991; 173:2521–2529.
72. Jann K, Jann B. Structure and biosynthesis of O-antigens. In: Reitschel ET, ed. Handbook of Endotoxin. Vol I. Chemistry of Endotoxin. Amsterdam: Elsevier, 1984:138–186.
73. Morona R, Mavris M, Fallarino A, Manning PA. Characterization of the *rfc* region of *Shigella flexneri*. J Bacteriol 1994; 176:733–747.
74. Stevenson G, Andrianopoulos K, Hobbs M, Reeves PR. Organization of the *Escherichia coli* K-12 cluster responsible for production of the extracellular polysaccharide colanic acid. J Bacteriol 1996; 178:4885–4893.
75. Schnaitman CA, Klena JD. Genetics of lipopolysaccharide biosynthesis in enteric bacteria. Microbiol Rev 1993; 57:655–682.
76. Coyne Jr. MJ, Goldberg JB. Cloning and characterization of the gene (*rfc*) encoding O-antigen polymerase of *Pseudomonas aeruginosa* PAO1. Gene 1995; 167:81–86.
77. Zhang L, Toivanen P, Skurnik M. The gene cluster directing O-antigen biosynthesis in *Yersinia enterocolitica* serotype O:8: identification of the genes for mannose and galactose biosynthesis and the gene for the O-antigen polymerase. Microbiology 1996; 142:277–288.
78. Goldman RC, Hunt F. Mechanism of O-antigen distribution in lipopolysaccharide. J Bacteriol 1990; 172:5352–5359.
79. Bastin DA, Stevenson G, Brown PK, Haase A, Reeves PR. Repeat unit polysaccharides of bacteria: a model for polymerization resembling that of ribosomes and fatty acid synthetase, with a novel mechanism for determining chain length. Mol Microbiol 1993; 7:725–734.
80. Morona R, Van Den Bosch L, Manning PA. Molecular, genetic, and topological characterization of O-antigen chain length regulation in *Shigella flexneri*. J Bacteriol 1995; 177:1059–1068.
81. Roberts IS. The biochemistry and genetics of capsular polysaccharide production in bacteria. Ann Rev Microbiol 1996; 50:285–315.
82. Zhang L, Al-Hendy A, Toivanen P, Skurnik M. Genetic organization and sequence of the *rfb* gene cluster of *Yersinia enterocolitica* serotype O:3: similarities to the dTDP-L-rhamnose biosynthesis pathway of *Salmonella* and to the bacterial polysaccharide transport systems. Mol Microbiol 1993; 9:309–321.
83. Bronner D, Clarke BR, Whitfield C. Identification of an ATP-binding cassette tranpsort system required for translocation of lipopolysaccharide O-antigen side chains across the cytoplasmic membrane of *Klebsiella pneumoniae* serotype O1. Mol Microbiol 1994; 14:505–519.
84. Reizer J, Reizer A, Saier MH. A new subfamily of bacterial ABC-type transport systems catalyzing export of drugs and carbohydrates. Prot Sci 1992; 1:1326–1332.
85. Rocchetta HL, Lam JS. Identification and functional characterization of an ABC transport system involved in polysaccharide export of A-band LPS in *Pseudomonas aeruginosa*. J Bacteriol 1997; 179:4713–4724.
86. Chu S, Noonan B, Cavaignac S, Trust TJ. Endogenous mutagenesis by an insertion sequence element identifies *Aeromonas salmonicida* AbcA as an ATP-binding cassette transport protein required for biogenesis of smooth lipopolysaccharide. Proc Natl Acad Sci USA 1995; 92:5754–5758.
87. Bliss JM, Silver RP. Coating the surface: a model for expression of capsular polysialic acid in *Escherichia coli*. Mol Microbiol 1996; 21:221–231.
88. Pigeon RP, Silver RP. Topological and mutational analysis of KpsM, the hydrophobic component of the ABC-transporter involved in the export of polysialic acid in *Escherichia coli* K1. Mol Microbiol 1994; 14:871–881.
89. Pavelka JMS, Wright LF, Sivler RP. Characterization of KpsT, the ATP-binding component of the ABC-transporter involved with the export of capsular polysialic acid in *Escherichia coli* K-1. J Biol Chem 1994; 269:20149–20158.
90. Roberts IS, Mountford R, Hodge R, Jann K, Boulnois GJ. Common organization of gene clusters for production of different capsular polysaccharide (K antigens) in *Escherichia coli*. J Bacteriol 1988; 170:1305–1310.
91. Kelly RF, MacLean LL, Perry MB, Whitfield C. Clonally diverse *rfb* gene clusters are involved in expres-

sion of a family of related D-galactan O antigens in *Klebsiella* species. J Bacteriol 1996; 178:5205–5214.
92. Szabo M, Bronner D, Whitfield C. Relationships between *rfb* gene clusters required for biosynthesis of identical D-galactose-containing O antigens in *Klebsiella pneumoniae* serotype O1 and *Serratia marcescens* serotype O16. J Bacteriol 1995; 177:1544–1553.
93. Finke A, Bronner D, Nikolaev AV, Jann B, Jann K. Biosynthesis of the *Escherichia coli* K5 polysaccharide, a representative of group II capsular polysaccharides: polymerization in vitro and characterization of the product. J Bacteriol 1991; 173:4088–4094.
94. Frosch M, Müller A. Phospholipid substitution of capsular polysaccharides and mechanisms of capsule formation in *Neisseria meningitidis*. Mol Microbiol 1993; 8:483–493.
95. Chart H, Shaw DH, Ishiguro EE, Trust TJ. Structural and immunochemical homogeneity of *Aeromonas salmonicida* lipopolysaccharide. J Bacteriol 1984; 158:16–22.
96. Saxena IM, Brown Jr. RM, Fevre M, Geremia RA, Henrissat B. Multidomain architecture of. J Bacteriol 1995; 177:1419–1424.
97. Osborn MJ, Rick PD, Rasmussen NS. Mechanism of assembly of the outer membrane of *Salmonella typhimurium*. J Biol Chem 1980; 255:4246–4251.
98. Osborn MJ. Biosynthesis and assembly of lipopolysaccharide of the outer membrane. In: Inouye M, ed. Bacterial Outer Membranes: Biogenesis and Functions. New York: Wiley, 1979:15–34.
99. Mühlradt PF, Menzel J, Golecki JR, Speth V. Outer membrane of *Salmonella*. Sites of export of newly synthesized lipopolysaccharide on the bacterial surface. Eur J Biochem 1973; 35:471–481.
100. Bayer ME. Areas of adhesion between wall and membrane of *Escherichia coli*. J Gen Microbiol 1968; 53:395–403.
101. Bayer ME. Zones of membrane adhesion in the cryofixed envelope of *Escherichia coli*. J Struct Biol 1991; 107:268–280.
102. Kupla CF, Jr., Leive L. Mode of insertion of lipopolysaccharide into the outer membrane of *Escherichia coli*. J Bacteriol 1976; 126:467–477.
103. Bayer ME. The fusion sites between outer membrane and cytoplasmic membrane in bacteria: their role in membrane assembly and virion infection. In: Inouye M, ed. Bacterial Outer Membranes: Biogenesis and Function. New York: Wiley, 1979:167–202.
104. Gaspar JA, Valvano MA. The TolA Protein is Required for Surface Expression of O7-Specific Lipopolysaccharide in *Escherichia coli*. Miami: American Society for Microbiology, 1997:Abstract K-101.
105. Kampfenkel K, Braun V. Membrane topologies of the TolQ and TolR proteins of *Escherichia coli*: inactivation of TolQ by a missense mutation in the proposed first transmembrane segment. J Bacteriol 1993; 175:4485–4491.
106. Muller MM, Vianney A, Lazzaroni J-C, Portalier R, Webster R. Membrane topology of the *Escherichia coli* TolR protein required for cell envelope integrity. J Bacteriol 1993; 175:6059–6061.
107. Vianney A, Lewin TM, Beyer WFJ, Lazzaroni J-C, Portalier R, Webster RE. Membrane topology and mutational analysis of the TolQ protein of *Escherichia coli* required for the uptake of macromolecules and cell envelope integrity. J Bacteriol 1994; 176:822–829.
108. Bik EM, Bunschoten AE, Gouw RD, Mooi FR. Genesis of the novel epidemic *Vibrio cholerae* O139 strain: evidence for horizontal transfer of genes involved in polysaccharide synthesis. EMBO J 1995; 14:209–216.
109. Lew HD, Nikaido H, Mäkelä PH. Biosynthesis of uridine diphosphate *N*-acetyl-D-mannosaminuronic acid in *rff* mutants of *Salmonella typhimurium*. J Bacteriol 1978; 136:227–233.
110. Meier U, Mayer H. Genetic location of genes encoding enterobacterial common antigen. J Bacteriol 1985; 163:756–762.
111. Kawahara K, Hamaoka T, Suzuki S, et al. Lipopolysaccharide alteration mediated by the virulence plasmid of *Salmonella*. Microb Pathog 1989; 7:195–202.
112. Batchelor RA, Alifano P, Biffali E, Hull SI, Hull RA. Nucleotide sequences of the genes regulating O-polysaccharide chain length (*rol*) from *Escherichia coli* and *Salmonella typhimurium*: protein homology and functional complementation. J Bacteriol 1992; 174:5228–5236.
113. Stevenson G, Kessler A, Reeves PR. A plasmid-borne O-antigen chain length determinant and its relationship to other chain length determinants. FEMS Microbiol Lett 1995; 125:23–30.
114. Burrows LL, Chow D, Lam JS. *Pseudomonas aeruginosa* B-band O-antigen chain length is modulated by Wzz (Rol). J Bacteriol 1997; 179:1482–1489.
115. Ward HM, Morelli G, Kamke M, et al. A physical map of the chromosomal region determining O-antigen biosynthesis in *Vibrio cholerae* O1. Gene 1987; 55:197–204.
116. Stroeher UH, Karageorgos LE, Morona R, Manning PA. Serotype conversion in *Vibrio cholerae* O1. Proc Natl Acad Sci USA 1992; 89:2566–2570.
117. Stroeher UH, Parasivam G, Dredge BK, Manning PA. Novel *Vibrio cholerae* O139 genes involved in lipopolysaccharide biosynthesis. J Bacteriol 1997; 179:2740–2747.
118. Lightfoot JL, Lam JS. Chromosomal mapping, expression and synthesis of lipopolysaccharide in *Pseudomonas aeruginosa*; a role for guanosine diphospho (GDP)-D-mannose. Mol Microbiol 1993; 8:771–782.
119. Kogan G, Haraguchi G, Hull SI, et al. Structural analysis of O4-reactive polysaccharides from recombinant *Escherichia coli*. Changes in the O-specific polysaccharide induced by cloning of the *rfb* genes. Eur J Biochem 1993; 214:259–265.
120. Neuhard J, Thomassen E. Altered deoxynucleotide pools in P2 eductants of *Escherichia coli* K-12 due to the deletion of the *dcd* gene. J Bacteriol 1976; 126:999–1001.
121. Riley LW, Junio L, Libaek LB, Schoolnik GK. Plasmid-encoded expression of lipopolysaccharide O-antigenic polysaccharide in enteropathogenic *Escherichia coli*. Infect Immun 1987; 55:2052–2056.
122. Bastin DA, Romana LK, Reeves PR. Molecular cloning and expression in *Escherichia coli* K-12 of the *rfb*

gene cluster determining the O antigen of an *E. coli* O111 strain. Mol Microbiol 1991; 5:2223–2231.
123. Kopecko DJ, Washington O, Formal SB. Genetic and physical evidence for plasmid control of *Shigella sonnei* form I cell surface antigen. Infect Immun 1980; 29:207–214.
124. Sansonetti PJ, Hale TL, Dammin GJ, Kapfer C, Collins HHJ, Formal SB. Alterations in pathogenicity of *Escherichia coli* K-12 after transfer of plasmid and chromosomal genes from *Shigella flexneri*. Infect Immun 1983; 39:1392–1402.
125. Yoshida Y, Okamura N, Kaot J, Watanabe H. Molecular cloning and characterization of form I antigen genes of *Shigella sonnei*. J Gen Microbiol 1991; 137:867–874.
126. MacPherson DF, Morona R, Beger DW, Cheah K-C, Manning PA. Genetic analysis of the *rfb* region of *Shigella flexneri* encoding the Y serotype O-antigen specificity. Mol Microbiol 1991; 5:1491–1499.
127. Cheah K-C, Beger DW, Manning PA. Molecular cloning and genetic analysis of the *rfb* region from *Shigella flexneri* type 6 in *Escherichia coli* K-12. FEMS Microbiol Lett 1991; 83:213–218.
128. Yao Z, Liu H, Valvano MA. Acetylation of O-specific lipopolysaccharides from *Shigella flexneri* 3a and 2a occurs in *Escherichia coli* K-12 carrying cloned *S. flexneri* 3a and 2a *rfb* genes. J Bacteriol 1992; 174:7500–7508.
129. Valvano MA, Marolda CL. Relatedness of O-specific lipopolysaccharide side chain genes from strains of *Shigella boydii* type 12 belonging to two clonal groups and from *Escherichia coli* O7:K1. Infect Immun 1991; 59:3917–3923.
130. Fält IC, Schweda EKH, Weintraub H, Sturm S, Timmis KN, Lindberg AA. Expression of the *Shigella dysenteriae* type-1 lipopolysaccharide repeating unit in *Escherichia coli* K-12/*Shigella dysenteriae* type-1 hybrids. Eur J Biochem 1993; 213:573–581.
131. Keenleyside WJ, Whitfield C. Lateral transfer of *rfb* genes: a mobilizable ColE1-type plasmid carries the *rfb* (O:54 antigen biosynthesis) gene cluster from *Salmonella enterica* serovar Borreze. J Bacteriol 1995; 177:5247–5253.
132. Popoff MY, Le Minor L. Expression of antigenic factor O:54 is associated with the presence of a plasmid in *Salmonella*. Annal Inst Pasteur/Microbiol 1985; 136B:169–179.
133. Robbins PW, Uchida T. Chemical and macromolecular structure of O-antigens from *Salmonella anatum* strains carrying mutants of bacteriophage ϵ^{15}. J Biol Chem 1965; 240:375–383.
134. Robbins PW, Keller JM, Wright A, Bernstein RL. Enzymatic and kinetic studies on the mechanism of O-antigen conversion by bacteriophage ϵ^{15}. J Biol Chem 1965; 240:384–390.
135. Wright A. Mechanism of conversion of the *Salmonella* O antigen by bacteriophage ϵ^{34}. J Bacteriol 1971; 105:927–936.
136. Simmons DAR, Romanowska E. Structure and biology of *Shigella flexneri* O antigens. J Med Microbiol 1987; 23:289–302.
137. Petrovska VG, Licheva TA. A provisional chromosomal map of *Shigella* and the regions related to pathogenicity. Acta Microb Acad Sci Hung 1982; 29:41–53.
138. Lindberg AA, Wollin R, Gemski P, Wohlheiter JA. Interaction between bacteriophage SF6 and *Shigella flexneri*. J Virol 1978; 27:38–44.
139. Gemski PJ, Koeltzow DE, Formal SB. Phage conversion of *Shigella flexneri* group antigens. Infect Immun 1975; 11:685–691.
140. Lindberg B, Lonngren J, Romanowska E, Ruden U. Location of O-acetyl groups in *Shigella flexneri* types 3c and 4b lipopolysaccharides. Acta Chem Scand 1972; 26:3808–3810.
141. Verma NK, Brandt JB, Verma DJ, Lindberg AA. Molecular characterization of the O-acetyl transfersase gene of converting bacteriophage Sf6 that adds group antigen 6 to *Shigella flexneri*. Mol Microbiol 1991; 5:71–75.
142. Fisher RF, Long SR. *Rhizobium*-plant signal exchange. Nature (London) 1992; 357:655–660.
143. Slauch JM, Mahan MJ, Michetti P, Neutra MR, Mekalanos JJ. Acetylation (O-factor 5) affects the structural and immunological properties of *Salmonella typhimurium* lipopolysaccharide O antigen. Infect Immun 1995; 63:437–441.
144. Slauch JM, Lee AA, Mahan MJ, Mekalanos JJ. Molecular characterization of the oafA locus responsible for acetylation of *Salmonella typhimurium* O-antigen: OafA is a member of a family of integral membrane trans-acylases. J Bacteriol 1996; 178:5904–5909.
145. Kuzio J, Kropinski AM. O-antigen conversion in *Pseudomonas aeruginosa* PAO1 by bacteriophage D3. J Bacteriol 1983; 155:203–212.
146. Nnalue NA, Newton S, Stocker BAD. Lysogenization of *Salmonella choleraesuis* by phage 14 increases average length of O-antigen chains, serum resistance and intraperitoneal mouse virulence. Microb Pathog 1990; 8:393–402.
147. Clarke BR, Whitfield C. Molecular cloning of the *rfb* region of *Klebsiella pneumoniae* serotype O1:K20: the *rfb* gene cluster is responsible for synthesis of the D-galactan I O polysaccharide. J Bacteriol 1992; 174:4614–4621.
148. McConnell M, Schoelz JE. Evidence for shorter average O-polysaccharide chain length in the lipopolysaccharide of a bacteriophage Felix O1-sensitive variant of *Salmonella anatum* A1. J Gen Microbiol 1983; 129:3177–3184.
149. Goldman RC, Joiner K, Leive L. Serum-resistant mutants of *Escherichia coli* O111 contain increased lipopolysaccharide, lack an O antigen-containing capsule, and cover more of their lipid A-core with O antigen. J Bacteriol 1984; 159:877–882.
150. Al-Hendy A, Toivanen P, Skurnik M. The effect of growth temperature on the biosynthesis of *Yersinia enterocolitica* O:3 lipopolysaccharide: temperature regulates the transcription of the *rfb* but not the *rfa* region. Microb Pathog 1991; 10:81–86.
151. Lüderitz O, Westphal O, Staub AM, Nikaido H. Isolation and chemical and immunological characterization of bacterial lipopolysaccharides. In: Weinbaum G, Kadis S, Ajl SJ, eds. Microbial Toxins. Vol. 4. London: Academic Press, 1971:145–233.

152. Portnoy DA, Martinez RJ. Role of a plasmid in the pathogenicity of *Yersinia* species. Curr Top Microbiol Immunol 1985; 118:29–51.
153. Wartenberg K, Knapp W, Ahamed NM, Widemann C, Mayer H. Temperature-dependent changes in the sugar and fatty acid composition of lipopolysaccharides from *Yersinia enterocolitica* strains. Zbl Bakt Hyg I Abt Orig A 1983; 253:523–530.
154. Al-Hendy A, Toivanen P, Skurnik M. Expression cloning of *Yersinia enterocolitica* O:3 *rfb* gene cluster in *Escherichia coli* K12. Microb Pathog 1991; 10:11–13.
155. Hobbs M, Reeves PR. The JUMPstart sequence: a 39 bp element common to several polysaccharide gene clusters. Mol Microbiol 1994; 12:855–856.
156. Nieto JM, Bailey MJA, Hughes C, Koronakis V. Suppression of transcription polarity in the *Escherichia coli* haemolysin operon by a short upstream element shared by polysaccharide and DNA transfer determinants. Mol Microbiol 1996; 19:705–713.
157. Stevenson G, Neal B, Liu D, et al. Structure of the O antigen of *Escherichia coli* K-12 and the sequence of its *rfb* cluster. J Bacteriol 1994; 176:4144–4156.
158. Jiang X-M, Neal B, Santiago F, Lee SJ, Romana LK, Reeves PR. Structure and sequence of the *rfb* (O antigen) gene cluster of *Salmonella* serovar typhimurium (strain LT2). Mol Microbiol 1991; 5:695–713.
159. Lee SJ, Romana LK, Reeves PR. Sequence and structural analysis of the *rfb* (O antigen) gene cluster from a group C1 *Salmonella enterica* strain. J Gen Microbiol 1992; 138:1843–1855.
160. Bik EM, Bunschoten AE, Willems RJL, Chang ACY, Mooi FR. Genetic organization and functional analysis of the *otn* DNA essential for cell-wall polysaccharide synthesis in *Vibrio cholerae* O139. Mol Microbiol 1996; 20:799–811.
161. Rajakumar K, Jost BH, Sasakawa C, Okada N, Yoshikawa M, Adler B. Nucleotide sequence of the rhamnose biosynthetic operon of *Shigella flexneri* 2a and role of lipopolysaccharide in virulence. J Bacteriol 1994; 176:2362–2373.
162. Leeds JA, Welch RA. RfaH enhances elongation of *Escherichia coli* hlyCABD mRNA. J Bacteriol 1996; 178:1850–1857.
163. Bailey MJA, Hughes C, Koronakis V. Increased distal gene transcription by the elongation factor RfaH, a specialized homologue of NusG. Mol Microbiol 1996; 22:729–737.
164. Leeds JA, Welch RA. Enhancing transcription through the *Escherichia coli* hemolysin operon, *hlyCABD*: RfaH and upstream JUMPstart DNA sequences function together via a postinitiation mechanism. J Bacteriol 1997; 179:3519–3527.
165. Das A. How the phage lambda N gene product suppresses transcription termination: communication of RNA polymerase with regulatory proteins mediated by signals in nascent RNA. J Bacteriol 1992; 174:6711–6716.
166. Parker CT, Pradel E, Schnaitman CA. Identification and sequences of the lipopolysaccharide core biosynthetic genes *rfaQ, rfaP*, and *rfaG* of *Escherichia coli* K-12. J Bacteriol 1992; 174:930–934.
167. Mamat U, Rietschel ET, Schmidt G. Repression of lipopolysaccharide biosynthesis in *Escherichia coli* by an antisense RNA of *Acetobacter methanoliticus* phage Acm1. Mol Microbiol 1995; 15:1115–1125.
168. Vishwanath S, Hackstadt T. Lipopolysaccharide phase variation determines the complement-mediated serum susceptibility of *Coxiella burnetii*. Infect Immun 1988; 56:40–44.
169. Lukácová M, Kazár J, Gajdosová E, Vavreková M. Phase variation of lipopolysaccharide of *Coxiella burnetii*, a strain Priscilla during chick embryo yolk sac passaging. FEMS Microbiol Lett 1993; 113:285–290.
170. Reeves PR. Evolution of *Salmonella* O antigen variation by interspecific gene transfer on a large scale. Trends Genet 1993; 9:17–22.
171. Verma N, Reeves PR. Identification and sequence of *rfbS* and *rfbE*, which determine antigenic specificity of group A and group D salmonellae. J Bacteriol 1989; 171:5694–5701.
172. Reeves PR. Variation in O-antigens, niche-specific selection and bacterial populations. FEMS Microbiol Lett 1992; 100:509–516.
173. Reeves PR. Biosynthesis and assembly of lipopolysaccharide. In: Ghuysen J-M, Hakenbeck R, eds. Bacterial Cell Wall. New York: Elsevier, 1994:281–317.
174. Sirisena DM, MacLachlan PM, Liu S-L, Hessel A, Sanderson KE. Molecular analysis of the *rfaD* gene, for heptose synthesis, and the *rfaF* gene, for heptose transfer, in lipopolyhsaccharide synthesis in *Salmonella typhimurium*. J Bacteriol 1994; 176(8):2379–2385.
175. Wang L, Romana LK, Reeves PR. Molecular analysis of a *Salmonella enterica* group E1 *rfb* gene cluster: O antigen and the genetic basis of the major polymorphism. Genetics 1992; 130:429–443.
176. Xiang S-H, Hobbs M, Reeves PR. Molecular analysis of the *rfb* gene cluster of a D2 *Salmonella enterica* strain: evidence for its origin from an insertion sequence-mediated recombination event between group E and D1 strains. J Bacteriol 1994; 176:4357–4365.
177. Brown PK, Romana LK, Reeves PR. Cloning of the *rfb* gene cluster of a group C2 *Salmonella* strain: comparison with the *rfb* regions of groups B and D. Mol Microbiol 1991; 5:1873–1881.
178. Nelson K, Selander RK. Intergeneric transfer and recombination of the 6-phosphogluconate dehydrogenase gene (*gnd*) in enteric bacteria. Proc Natl Acad Sci USA 1994; 91:10227–10231.
179. Mooi FR, Bik EM. The evolution of epidemic *Vibrio cholerae* strains. Trends Microbiol 1997; 5:161–165.
180. Stroeher UW, Jedani KE, Dredge BK, et al. Genetic rearrangements in the *rfb* regions of *Vibrio cholerae* O1 and O139. Proc Natl Acad Sci USA 1995; 92:10374–10378.
181. Dumontier S, Trieu-Cuot P, Berche P. Distribution and Characterization of the *rfbQRS* Element in Various Serogroups of *Vibrio cholerae*. Miami: American Society for Microbiology, 1997:Abstract H-71.
182. Zhao S, Sandt CH, Feuler G, Vlazny DA, Gray JA, Hill CW. Rhs elements of *Escherichia coli* K-12: complex composites of shared and unique components that have different evolutionary histories. J Bacteriol 1993; 175:2799–2808.

183. Gustafson C, Chu S, Trust TJ. Mutagenesis of the paracrystalline surface protein array of *Aeromonas salmonicida* by endogenous insertion elements. J Mol Biol 1994; 237:452–463.
184. Drummelsmith J, Amor PA, Whitfield C. Polymorphism, duplication and IS*1*-mediated rearrangement in the chromosomal *his-rfb-gnd* region of *Escherichia coli* strains with group IA capsular K antigens. J Bacteriol 1997; 179:3232–3238.
185. Xiang S-H, Haase AM, Reeves PR. Variation of the *rfb* gene clusters in *Salmonella enterica*. J Bacteriol 1993; 175:4877–4884.
186. Geremia RA, Petroni EA, Ielpi L, Henrissat B. Towards a classification of glycosyltransferases based on amino acid sequence similarities: prokaryotic α-mannosyltransferases. Biochem J 1996; 318:133–138.
187. Arsenault TL, Huges DW, MacLean DB, Szarek WA, Kropinski AMB, Lam JS. Structural studies on the polysaccharide portion of "A-band" lipopolysaccharide from a mutant (AK1401) of *Pseudomonas aeruginosa* strain PAO1. Can J Chem 1991; 69:1273–1280.
188. Lee SJ, Romana LK, Reeves PR. Cloning and structure of group C1 O antigen (*rfb* gene cluster) from *Salmonella enterica* serovar montevideo. J Gen Microbiol 1992; 138:305–312.
189. Brown PK, Romana LK, Reeves PR. Molecular analysis of the *rfb* gene cluster of *Salmonella* serovar muenchen (strain M67): the genetic basis of the polymorphism between groups C2 and B. Mol Microbiol 1992; 6:1385–1394.
190. Nurminen M, Hellerqvist CE, Valtonen VV, Mäkelä PM. The smooth lipopolysaccharide character of 1,4,(5),12 and 1,9,12 transductants formed as hybrids between groups B and D of *Salmonella*. Eur J Biochem 1971; 22:500–505.
191. Sanderson KE, Roth JR. Linkage map of *Salmonella typhimurium*. Microbiol Rev 1989; 52:485–532.
192. Lukomski S, Hull RA, Hull SI. Identification of the O antigen polymerase (*rfc*) gene in *Escherichia coli* O4 by insertional mutagensis using a nonpolar chloramphenicol reistance cassette. J Bacteriol 1996; 178:240–247.
193. Dasgupta T, Lam JS. Identification of *rfbA*, involed in B-band lipopolysaccharide biosynthesis in *Pseudomonas aeruginosa* serotype O5. Infect Immun 1995; 63:1674–1680.
194. de Kievit TR, Dasgupta T, Schweizer H, Lam JS. Molecular cloning and characterization of the *rfc* gene of *Pseudomonas aeruginosa* (serotype O5). Mol Microbiol 1995; 16:565–574.
195. Liu D, Verma NK, Romana LK, Reeves PR. Relationships among the *rfb* regions of *Salmonella* serovars A,B, and D. J Bacteriol 1991; 173:4814–4819.
196. Bastin DA, Reeves PR. Sequence analysis of the O antigen gene (*rfb*) cluster of *Escherichia coli* O111. Gene 1995; 164:17–23.
197. Kessler AC, Haase A, Reeves PR. Molecular analysis of the 3,6-dideoxyhexose pathway genes of *Yersinia tuberculosis* serogroup IIA. J Bacteriol 1993; 175:1412–1422.
198. Klee SR, Tzschaschel BD, Timmis KN, Guzmán CA. Influence of different *rol* gene products on the chain length of *Shigella dysenteriae* type 1 lipopolysaccharide O antigen expressed by *Shigella flexneri* carrier strains. J Bacteriol 1997; 179:2421–2425.
199. Pierson DE, Carlson S. Identification of the *galE* gene and a *galE* homolog and characterization of their roles in the biosynthesis of lipopolysaccharide in a serotype O:8 strain of *Yersinia enterocolitica*. J Bacteriol 1996; 178:5916–5924.
200. Manning PA, Stroeher UH, Karageorgos LE, Morona R. Putative O-antigen transport genes within the *rfb* region of *Vibrio cholerae* O1 are homologous to those for capsule transport Gene 1995; 158:1–7.
201. Guo D, Gabriela Bowden M, Pershad R, Kaplan HB. The *Myxococcus xanthus rfbABC* operon encodes an ATP-binding cassette transporter homolog required for O-antigen biosynthesis and multicellular development. J Bacteriol 1996; 178:1631–1639.

19

Lipopolysaccharide-Binding Protein

Peter S. Tobias
The Scripps Research Institute, San Diego, California

INTRODUCTION

The innate immune system is an ancient evolutionary response by multicellular organisms to infectious pathogens. The receptors that activate the innate immune response must recognize a broad variety of "coats of arms" heralding attack. The molecules accomplishing this function have been dubbed pattern recognition receptors by Janeway (1) because they have evolved not to recognize any one pathogen with exquisite sensitivity but to recognize families of pathogens by virtue of the pathogen's conserved patterns of structure. In the case of recognizing attack by gram-negative organisms as mediated by the pattern recognition receptor CD14 (2), mammals have evolved in addition a mechanism for enhancing the ability of CD14 to recognize lipopolysaccharide (LPS): plasma LPS-binding protein (LBP). Although it is unlikely that we will ever really know why or how LBP evolved, it is likely that the lipophilic properties of LPS figure heavily. As a strongly amphipathic molecule, it is no simple matter to bring LPS out of an aggregate to interact with a water-soluble protein (3). LBP has a unique ability to accomplish this process. Doubtless, this ability of LBP is related to its unique ability to sensitively opsonize LPS coated particles for recognition by leukocytes (4). Multiple facets of the role of LBP in facilitating LPS interaction with LPS "receptors" are presented schematically in Figure 1.

In the late 1970s Ulevitch et al. showed that LPS in serum or plasma would bind to lipoproteins and become less endotoxic (5). We then postulated that host responses that alter the structure of lipoproteins, such as an acute phase response that elevates levels of the apolipoprotein serum amyloid A, should alter the reactivity of LPS with high-density lipoprotein (HDL). However, the elevated level of serum amyloid A in acute phase HDL did not affect the binding of LPS. We did nevertheless observe that LPS binding to HDL in acute phase sera was considerably slowed (6,7), and we used this phenomenon to isolate LBP (8). LBP is a strong acute phase reactant, rising 50 fold or more in rabbits and humans subjected to an inflammatory stimulus (9,10). To our surprise LBP did not appear to be an apolipoprotein. While we did anticipate that LBP would assist in protecting the host against the effects of LPS, we had no idea that it would do so by dramatically enhancing host sensitivity to LPS. Many examples of LBP mediated enhancement of responses to LPS have now been documented, e.g. (11–15). LBP also contributes to removal or deactivation of LPS, for example in the delivery of LPS to non-activating cellular uptake pathways (16–18) or in the enhancement of LPS binding to lipoproteins (19). However, the ability of antibodies to LBP to protect animals from the effects of LPS (11,20) and the recent description of two strains of LBP "knockout" mice (21,22) make it clear that a principal function of LBP is to enhance host responses to LPS. In this review I will try to summarize the salient features of LBP and indicate some areas needing attention.

PHENOMENA: LBP IN VIVO

Four studies are relevant to the function of LBP in vivo. Two of these used antibodies to inactivate or de-

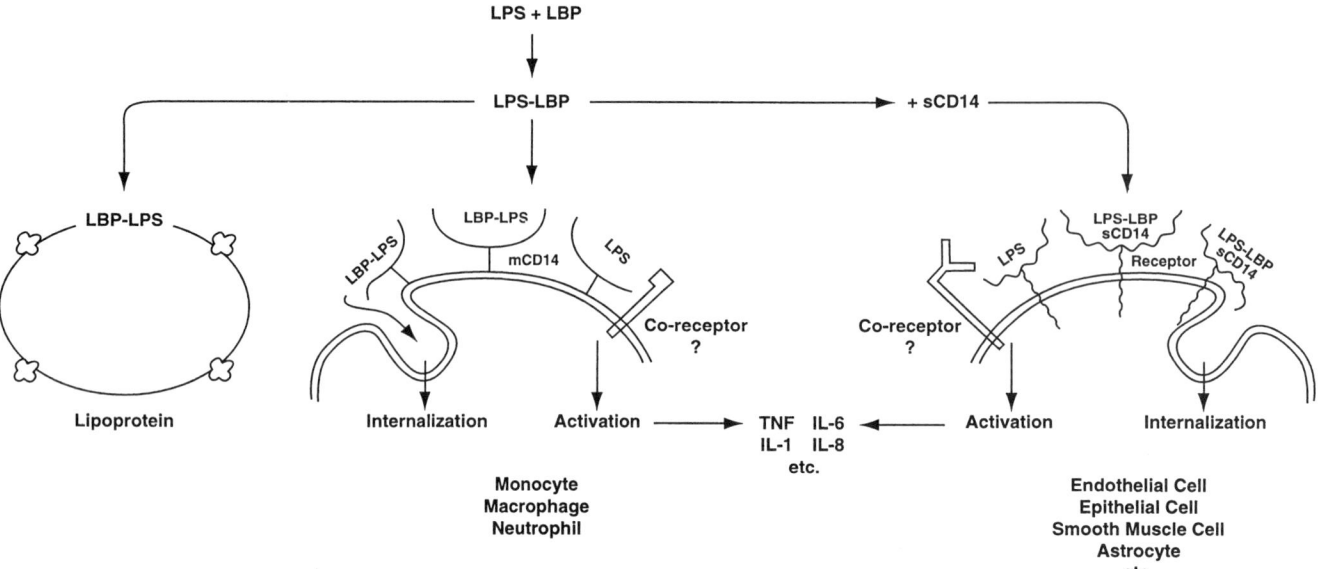

Fig. 1 The involvement of LBP in endotoxin interactions with cells and lipoproteins. The figure shows the interaction of LPS with lipoproteins and two types of LPS-responsive cells-mCD14 expressing (center, e.g., monocytes) and not mCD14 expressing (right, e.g., endothelial cells). The interaction with lipoproteins results in lowered endotoxic activity of the LPS. For each type of cell, there is an activating interaction, indicated as LPS interacting with a co-receptor and an arrow pointing to "Activation," as well as a nonactivating interaction, indicated as LPS in complex with protein(s) entering an invagination and an arrow pointing to "Internalization." The molecular details of the activation and internalization processes are not known.

plete LBP, and two used homologous recombination to create an LBP-free environment.

The earliest study is that of Gallay et al. (20). These workers isolated murine LBP and used it to prepare a polyclonal IgG, which would block LPS-initiated tumor necrosis factor (TNF) production in whole blood ex vivo. In vivo, the IgG led to depletion of LBP from the circulation by mechanisms that were explored later (23). They then slowed, in an LPS lethality model using D-galactosamine–treated mice, that functional LBP in the circulation was required for LPS-dependent lethality as well as TNF expression. Although this murine endotoxemia model has been widely used, it does suffer from the difficulty that the murine liver is temporarily put out of action by D-galactosamine in order to enhance the sensitivity of the mice to TNF. A somewhat different approach was taken by Mathison and coworkers (unpublished results) using a rabbit model of endotoxemia. Since rabbits are much more sensitive to the lethal effects of LPS than mice, they do not require any sensitization as is accomplished with D-galactosamine in the murine model. J. Mathison et al. isolated rabbit LBP, prepared murine monoclonal antibodies, and selected one that blocked LPS-dependent TNF production in whole rabbit blood ex vivo. They found that this antibody would block TNF production and prevent lethality as in the mouse model.

Jack et al. (21) and Wurfel et al. (22) have both disrupted the LBP gene in mice by homologous recombination. Both groups convincingly show a lack of LBP functionality in whole blood taken from the $LBP^{-/-}$ mice, resulting in ablation of a TNF response to administered LPS. However, somewhat different results were obtained in their in vivo studies. Jack et al. used the D-galactosamine model and found that the $LBP^{-/-}$ mice neither made TNF nor died in response to administered LPS. Wurfel et al. used $LBP^{-/-}$ mice untreated with D-galactosamine and did not observe a deficit in TNF production when the mice were challenged with LPS. Wurfel et al. did not report results using a lethality endpoint. To people with experience using models of endotoxemia, it is no surprise that different models yield different answers. However, it is a surprise that the $LBP^{-/-}$ mice of Wurfel et al. made TNF in amounts indistinguishable from their $LBP^{-/-}$ littermates in response to LPS. This finding prompted the suggestion that in vivo an LBP-independent pathway for LPS-dependent cellular activation exists (22). Obviously, the first step that should be taken is for the two groups to use each other's experimental protocols and determine if there really are phenotypic differences between the two knockouts or if the different results derive from the different models used. It would be fascinating if an LBP-independent, D-galactosamine–sensitive mechanism existed for LPS responses in vivo that is not seen in whole blood. Alternatively, it seems unlikely but not inconceivable that a truncated, cell-associated form of LBP might have been expressed. Jack et al. (21) also studied the effects of LBP loss on an *Salmonella typhimurium* infection. They found that the LBP$^{-/-}$ mice were unable to check multiplication of the bacteria and succumbed to the infection, whereas the LBP$^{-/-}$ littermates were much more resistant.

N. Lamping et al. (personal communication, 1997) have studied the role of LBP in vivo in the opposite direction, administering murine LBP to mice to reach acute phase levels of LBP and measuring the subsequent response to LPS and infection. Using the D-galactosamine–treated mouse model they observed significant diminution of cytokine production and liver damage in mice to whom LBP had been administered. They also found that administering acute phase levels of LBP to mice enhanced their survival to a subsequent challenge with *E. coli* 0111:B4 bacteria.

In a review that will hopefully be read by persons outside the field, it will be puzzling if no mention of the term "Septin" is made, since it is common enough in the literature cited. Previous studies have suggested that a substance referred to as Septin was actually accomplishing the functions ascribed to LBP (24). This reviewer's comments on septin have been recorded elsewhere (25,26), and the originator of the term has declared it "unnecessary" (27).

MECHANISMS: LBP IN VITRO

Even if we cannot define precisely the entire role of LBP, certainly we may conclude that LBP does have an important role in cellular responses to LPS. How is this accomplished? In many ways, the two most important observations are that LBP will opsonize LPS-bearing particles for interaction with CD14 and that LBP enables cells to respond to lower levels of LPS than they otherwise could (11,28); the rest has been filling in an understanding of the mechanics of these phenomena.

SOME BASIC BIOCHEMISTRY

Rabbit LBP, the first LBP to be isolated, is a 456-amino-acid plasma glycoprotein of about 60 kDa (8) typically present in plasma at several micrograms or less per milliliter. However, in inflammatory states that lead to elevated levels of plasma LBP, LBP may also be present in extravascular spaces (15,29). The sequences of human (Fig. 2), mouse, and rat LBP are also known, and all exhibit a high degree of similarity (GenBank: LBP_RAT, Q3313; LBP_MOUSE, Q61805; LBP_RABIT, P17454, LBP_HUMAN, P18428). Elucidation of the LBP sequence eventually led to the recognition that LBP was one member of a protein family (30); in addition to LBP the members are bactericidal/permeability increasing protein (BPI) (31), cholesterol ester transport protein (CETP) (32), and phospholipid transport protein (PLTP) (33). All are capable of binding LPS (34,35), but only LBP fosters interaction with CD14. CETP and PLTP have no defined roles in host interactions with LPS. The alignment studies that we have carried out have not indicated any feature clearly associable with LBP that would suggest a structural basis for its unique ability to transfer LPS to CD14. Several proteins possibly involved in olfactory reception may have sequence similarity to LBP (36), but none of these have been tested for LPS-binding activity. LBP expression has been observed in several tissues in the rat, but it appears to be principally synthesized in hepatocytes (37,38). Studies of the control of its expression have been carried out in several species (39–42), which generally detail its dependence on inflammatory agents and cytokines for elevated expression during an acute phase response. The exon/intron organization of the LBP and bactericidal permeability increasing protein (BPI) genes have been determined and were found to be very similar (43).

An x-ray crystallographic study of the LBP homolog BPI has been carried out (44). BPI binds LPS with an even higher affinity than LBP (45). There are some

```
  1  ANPGLVARIT DKGLQYAAQE GLLALQSELL RITLPDFTGD LRIPHVGRGR
 51  YEFHSLNIHS CELLHSALRP VPGQGLSLSI SDSSIRVQGR WKVRKSFFKL
101  QGSFDVSVKG ISISVNLLLG SESSGRPTGY CLSCSSDIAD VEVDMSGDSG
151  WLLNLFHNQI ESKFQKVLES RICEMIQKSV SSDLQPYLQT LPVTTEIDSF
201  ADIDYSLVEA PRATAQMLEV MFKGEIFHRN HRSPVTLLAA VMSLPEEHNK
251  MVYFAISDYV FNTASLVYHE EGYLNFSITD DMIPPDSNIR LTTKSFRPFV
301  PRLARLYPNM NLELQGSVPS APLLNFSPGN LSVDPYMEID AFVLLPSSSK
351  EPVFRLSVAT NVSATLTFNT SKITGFLKPG KVKVELKESK VGLFNAELLE
401  ALLNYYILNT LYPKFNDKLA EGFPLPLLKR VQLYDLGLQI HKDFLFLGAN
451  VQYMRV
```

Fig. 2 The amino acid sequence of human LBP.

puzzling features of the structural results. One is the finding that BPI crystals contain molecules of phospholipid. At first glance these phospholipid-binding sites would seem to indicate potential LPS-binding sites. However, it is not clear that either site could accommodate a molecule as large as lipid A. In addition, the phospholipid-binding sites are distant from residues in the sequence between 90 and 100 that have been implicated as important in LPS binding by LBP (46). However, the model does show how these residues are exposed at the end of the molecule to make a contact with LPS. There is another peculiarity to the BPI structure with regard to its relevance to LBP. Human LBP contains four cysteines. Our titration experiments with dithionitrobenzoic acid suggest that there are no free sulfhydryl groups (unpublished results). The crystal structure of BPI clearly shows a disulfide bond between BPI cysteines 136 and 177, which are also cysteine in human LBP. However, BPI residues 61 and 133, both of which are also cysteine in human LBP, are much too far apart to form a disulfide bond.

ASSAYS

In terms of defining the functional properties of LBP, many assays have proven useful. Opsonization by LBP has been demonstrated by counting binding of LPS-coated particles to cells (4) or by measuring cell-associated, fluoresceinated gram-negative bacteria (47). There are many assays of the ability of LBP to enhance cellular activation, but in essence they are all similar to that of Mathison et al. (12) in which monocytic cells are challenged with LPS under conditions of limiting LBP, LPS, or both and some response such as TNF-α secretion is measured. Perhaps the simplest is a binding assay in which LPS, immobilized by adsorption in a microtiter plate, is used to capture LBP, which is then detected immunologically or by previous radioactive labeling. This assay has been used, for example, to determine that LBP binds to the lipid A moiety of LPS (48), to estimate the relative affinities of LBP and BPI for LPS (45), and to study the effects of site-specific mutations in LBP (46). The ability of LBP to facilitate the binding of fluorescein-labeled LPS (FITC-LPS) to CD14-bearing cells has been useful in many tests of LBP activity (e.g., Ref. 46) as well as in estimating the concentration of LBP in serum or plasma (49). Direct observation of LPS-LBP complex formation in solution has been accomplished with four techniques-sucrose density gradients, fluorescence enhancement of FITC-LPS (50), native PAGE (51), and sensitivity to proteolysis (see below). Cellular association of LBP with cells has been measured using biosynthetic labeling of the protein (18). By the time this chapter appears in print, high specific activity [^{32}P] LBP should have been obtained using a mutant of LBP that incorporates a heterologous protein kinase A site (17).

STRUCTURE-FUNCTION STUDIES

The ability of LBP to interact with both LPS and CD14 suggests binding sites for both these molecules, and it turns out that these two sites are encoded in separate regions of the LBP molecule. Two groups have studied the properties of amino-terminal fragments of LBP with essentially identical results (52,53). The LBP$^{1-197}$ fragment binds LPS with very nearly the same affinity as the intact protein but has no ability to transfer LPS to CD14. The other fragment, comprising amino acids 198–456 of LBP, has not been studied except in the context of a composite with an amino-terminal fragment of BPI (45). This composite was able to deliver LPS to CD14 and promote cellular activation. Whether the LBP fragment 198–456 has any ability to interact with CD14 in the absence of LPS remains to be seen. Separate roles for the amino and carboxy terminal domains of BPI have also been observed (54).

Several studies have begun to map the functional activities of LBP to the level of specific amino acids (46,55,56). Plasmin or trypsin cleavage of the 99–100 bond in rabbit LBP destroys its LPS-binding activity, and LPS blocks the proteolytic cleavage (57; P. Tobias et al., unpublished data). Taylor et al. (56) showed that peptides encompassing the LBP sequence between 91 and 108 could bind to LPS and inhibit LPS-dependent cellular activation in the presence of LBP (56). This type of analysis has also been pursued by Lamping et al. (46), who observed that peptides covering the LBP sequence from 31–50, 81–110, 281–315, and 396–425 inhibited LBP binding to LPS immobilized in a microtiter plate. The latter study (46) then focused on the region 81–110 and showed by alanine substitution that point mutations at positions Arg-94, Lys-95, and Lys-99 inhibited LBP binding to LPS. After studying transfer of fluorescent LPS to CD14 as well as neutrophil activation, Lamping et al. (46) concluded that Lys-95 was the single most important residue in this group but that complete abrogation of activity required loss of both Arg-94 and Lys-95.

We have conducted some experiments in the area of LBP structure-function relationships that are as yet unpublished. They confirm the importance of the region around residues 90–100 in that a peptide having the rabbit LBP sequence 92–106 inhibited LBP binding to LPS, as did a polyclonal antibody to this same peptide. We have also studied a longer peptide covering the region 134–172. This peptide had the ability to bind to LPS, as assessed by fluorescence enhancement (50). When the peptide was subdivided into three segments, the sequence 158–172 had the highest LPS-binding activity, binding to LPS with a Kd of 30 μM. A peptide of the same composition but scrambled in sequence did not bind LPS. This peptide has three positively charged residues, which may account for the activity. However, the sequence 158–172 is homologous to part of an amphipathic helix in the structure of BPI (44). Our definition of the interaction of LBP with lipid A indicated important interactions with the fatty acid tails (48). It could be the hydrophobic part of the 158–172 peptide that is actually important for LPS binding. Additionally we have observed (unpublished data) that acylation, using acetyl N-hydroxysuccinimide, causes a loss of LPS-binding activity with rabbit mouse LBP but not with human LBP. Comparing sequences and looking for acylation-susceptible (i.e., lysine) residues in rabbit and mouse LBP, which are not lysine in human LBP and which are within the first 197 residues, reveals that human Pro-44 and Arg-171 are lysine in both rabbit and mouse LBP. Arg-171 is of course within the sequence 158–172 discussed above as capable of binding to LPS. LBP point mutants are being constructed to test the roles of residues 44 and 171. Thus, some of the determinants of LPS binding to LBP have been defined, and more are being pursued within the first 200 residues of the LBP sequence. None of the determinants for interaction of LBP with CD14 have been identified beyond their localization in the carboxy-terminal part of the molecule.

CD14 AS THE LPS ACCEPTOR

What is the mechanism by which LBP fosters the interaction of LPS with CD14? Before dealing with this question of mechanism or pathway, it is useful first to consider what sorts of LPS, LBP, and CD14 complexes are formed under various circumstances. The simplest situation is in solution with purified components. When LPS is mixed with LBP, complexes of LBP with either multiple LPS molecules or with only one or two LPS molecules form depending on the initial molar ratio of the two components. This is readily seen using sucrose density gradient analysis of mixtures of LPS with LBP (50). Complexes of LBP with multiple LPS molecules may also have been observed by native PAGE (51). LPS mixed with soluble CD14 (sCD14) alone makes complexes containing only one or two molecules of LPS (51). When LPS, LBP, and sCD14 are mixed, no ternary complex is observed (50,51). However, when LPS, LBP, and sCD14 are mixed at appropriate concentrations, it is clear that LBP catalyzes the formation of LPS-sCD14 complexes and these complexes contain one to two LPS per sCD14 (50,51,58,59).

At present there are two principal unresolved questions with respect to these reactions. The first involves the kinetic mechanism by which the catalysis occurs. One position is that the reaction occurs during a ternary encounter of an LPS aggregate, LBP and sCD14 (59). In favor of this position is an analysis of the rates of appearance of LPS-sCD14 complexes. The other position is that LBP extracts an LPS molecule from an LPS aggregate and then donates this LPS to sCD14 (50). In favor of this position is the demonstrated existence of low-stoichiometry LPS:LBP complexes, the observation that the maximal rate of appearance of LPS-sCD14 complexes is approximately equal to the maximal rate of appearance of LPS-LBP complexes, and the observation of a significant solvent effect, such that high ionic strength slows the formation of LPS-sCD14 complexes. There are several possible explanations, none of which can be favored at present. One is that both mechanisms are possible under different conditions. The data of Tobias et al. (50) were obtained at significantly lower LPS concentrations than the data of Yu and Wright (59). Another is that although the reagents used by the two groups seem to be identical, for one reason or another they may differ significantly, perhaps by virtue of different expression systems or derivatization or purification procedures. The other unresolved question concerns the relative affinities of LBP and sCD14 for LPS. We have estimated dissociation constants for LPS-LBP and LPS-sCD14 complexes to be 3.5×10^{-9} M^{-1} and 29×10^{-9} M^{-1}, respectively. A similar ratio of affinities was found using [^{125}I]ASD-LPS to photolabel mixtures of LBP and sCD14 (50). However, Yu and Wright (59) interpret their data to indicate that sCD14 has a higher affinity for LPS than LBP. While they have not specifically quantified either affinity, their research suggests a Kd for LPS-LBP near 2×10^{-9} M^{-1}, similar to the value we and others have determined. Again, differences in methods may account for these variations in results. However, especially with respect to these measurements of affinity and interpreting them with respect to reactions in vivo, it should be kept in mind that the milieu in which one would like to know the relative affinities of LBP and sCD14 for LPS is blood—a much more complex environment. Finally, LPS in blood rapidly reacts to form complexes with lipoproteins (5), and it is possible that the binding of LPS to LBP and sCD14 never really reaches equilibrium before most of the LPS binds to lipoprotein. This is of concern especially if LBP is an apolipoprotein, as discussed below.

ALTERNATE LPS ACCEPTORS

Two laboratories have observed LBP-mediated transfer of LPS and other lipids to acceptors other than CD14. Wurfel et al. (60,61) have found that LBP is able to catalyze the transfer of LPS to lipoproteins and phospholipid vesicles. Although it is clear that these reactions take place, some of the details are difficult to assess because most of the measurements were made using an indirect assay of LPS association with the lipids. As noted earlier, LPS association with lipoproteins leads to a loss of endotoxic activity; this phenomenon is termed neutralization. One way of measuring LPS neutralization is to determine the ability of an LPS preparation exposed to lipoproteins to stimulate neutrophil adhesion to fibrinogen-coated plastic surfaces. While this assay does show changes in the endotoxic activity of lipid-complexed LPS, not all lipids lead to the same degree of neutralization. For example, dihexanoyl phosphatidyl choline leads to little change in endotoxic activity, but didecanoyl phosphatidyl choline inhibits 80% of the endotoxic activity of 10 ng/ml LPS. Apparently paradoxically, dioctanoyl phosphatidyl choline leads to a 73% enhancement of endotoxic activity. Interpreting these activity data in terms of the ability of LBP to use various lipids as LPS acceptors is simply not possible because the complexes have not been biochemically characterized. Their stoichiometric compositions are unknown, their physical states are unknown, and the specific effects of the various lipids on neutrophil adhesion are unknown. Furthermore, what does it mean to observe the "neutralization" of LPS by ceramide (61) if one also hypothesizes that LPS may activate cells through ceramide activated protein kinase as a ceramide mimetic (62)? However, the existence of alternative acceptors for LPS from LPS-LBP complexes does offer useful tools for the further investigation of the mechanism by which LBP catalyzes the transfer of LPS from aggregates to acceptors. The work of Schromm et al. (54) shows more directly the LBP-catalyzed association of LPS with phospholipid vesicles. These workers used quenching of resonance energy transfer in a donor-acceptor pair incorporated in the phospholipid vesicles to observe LPS binding to the vesicles; LPS binding dilutes the donor and acceptor concentrations and leads to less energy transfer. They found that LBP was required for LPS insertion and that phosphatidyl serine vesicles were much better acceptors of LPS than were phosphatidyl choline vesicles—if one can assume that resonance energy transfer in the two types of vesicles would be similarly quenched by insertion of equal amounts of additional lipid. These observations highlight the point that LBP is capable of acting as a lipid transfer protein in a more general sense than simply as an LPS transfer protein, in agreement with the amino acid sequence similarities between LBP, PLTP, and CETP. As yet there is no evidence that LBP functions as a lipid transfer protein in biological systems and, if it does, what the functional consequences are.

LBP IN PLASMA

In addition to transferring LPS to lipoproteins, Wurfel et al. observed that LBP in plasma appeared to be bound directly to plasma lipoprotein (60). We have confirmed that LBP can bind to lipoprotein and have shown that specific epitopes of apo A-1 mediate the interaction (63). This observation, together with the observation that LBP accelerated the binding of LPS to HDL (60), is quite surprising given that the assay used to purify LBP was based on inhibition of the binding of LPS to lipoproteins. One explanation of this may be that in acute phase serum, from which LBP was purified, the level of LBP is significantly raised. There may simply be more LBP in acute phase serum than can be accommodated on HDL. The excess LBP in acute phase serum could then act as a competitive binder of LPS, slowing its delivery to the lipoprotein and potentially altering the acceptors to which the LPS ultimately goes.

These observations clearly indicate a critical need for further studies of the time-dependent distribution of LPS between LBP, sCD14, and the other acceptors present in plasma and whole blood.

OPSONIZATION AND PHAGOCYTOSIS

Host cellular recognition of gram-negative bacteria is an important function of a host defense system (1). It is thus not surprising that LBP should be involved in opsonization and phagocytosis. In fact, opsonization leading to phagocytosis by macrophages was one of the activities of LBP that led to the discovery of CD14 as an LPS receptor (28). More recently, two groups have reexamined the role of LBP in phagocytosis and confirmed that this is a CD14-dependent event in both monocytes and neutrophils (47,64). However, signaling via CD14, as assessed by NFκB activation, is not required for this phagocytic event and is independent of CR3 and Fc receptors. It seems likely that loss of the opsonizing capacity of LBP is related to the inability of LBP$^{-/-}$ mice to check a bacterial infection (21). In addition to delivering LPS-bearing particles such as gram-negative bacteria to cells, LBP also delivers free LPS, which may have been shed from gram-negative bacteria, to cells. LBP facilitates this process for cells expressing membrane-bound CD14 (mCD14) such as monocytes, macrophages, and neutrophils or transfected CHO cells (18), as well as for cells that do not bear CD14 but respond to LPS in the context of sCD14, such as endothelial or epithelial cells (17). As with phagocytosis, LBP-dependent uptake of the bound LPS is independent of CD14-mediated cellular activation, although CD14 is essential for cellular association of the LPS.

PERSPECTIVE

In the last decade or so there have been major advances in understanding how LPS interacts with and activates cells. One of the principal components for these events is LBP, which functions to make LPS more accessible to the innate immune system receptor, CD14, of the host. The progress that has been made enables us to study the details of the LPS, LBP, CD14 system. However, that progress also leaves us with three important new foci for study. First, what is the biochemical mechanism by which LPS bound to CD14 activates cells? Our present knowledge shows how LPS binds to CD14, but transmembrane signaling is just a black box at present, despite much effort by many investigators. Second, we know that cell activation is not critical to cellular internalization of LPS, implying an alternative pathway for internalization. The biochemistry of internalization is unknown. And finally, if it is indeed true that LPS administered to an LBP$^{-/-}$ mouse is still completely able to activate the immune system to produce TNF, how does this occur? There is still plenty of big game to be found in the wilds of LBP, not to mention LPS, biochemistry.

ACKNOWLEDGMENTS

I have been privileged to collaborate with many excellent scientists in the course of exploring LPS-binding protein. Their names appear in the cited references. Special thanks go to Katrin Soldau, for unflagging labors in the lab, and to Richard Ulevitch, for providing the opportunity. And, let's not forget the taxpayers, who paid for it via NIH grants GM37696, AI32021, and HL23584. This is publication 11431-IMM from The Scripps Research Institute.

REFERENCES

1. Janeway CA, Jr. The immune system evolved to discriminate infectious nonself from noninfectious self. Immunol Today 1992; 13:11–16.
2. Pugin J, Heumann D, Tomasz A, Kravchenko VK, Akamatsu Y, Nishijima M, et al. CD14 is a pattern recognition receptor. Immunity 1995; 1:509–516.
3. Din ZZ, Mukerjee P, Kastowsky M, Takayama K. Effect of pH on solubility and ionic state of lipopolysaccharide obtained from the deep rough mutant of *Escherichia coli*. Biochemistry 1993; 32:4579–4586.
4. Wright SD, Tobias PS, Ulevitch RJ, Ramos RA. Lipopolysaccharide binding protein opsonizes LPS-bearing particles for recognition by a novel receptor on macrophages. J Exp Med 1989; 170:1231–1241.
5. Ulevitch RJ, Johnston AR, Weinstein DB. New function for high density lipoproteins. Their participation in intravascular reactions of bacterial lipopolysaccharides. J Clin Invest 1979; 64:1516–1524.
6. Tobias PS, McAdam KP, Soldau K, Ulevitch RJ. Control of lipopolysaccharide-high-density lipoprotein interactions by an acute-phase reactant in human serum. Infect Immun 1985; 50:73–76.
7. Tobias PS, Ulevitch RJ. Control of lipopolysaccharide-high density lipoprotein binding by acute phase protein(s). J Immunol 1983; 131:1913–1916.
8. Tobias PS, Soldau K, Ulevitch RJ. Isolation of a lipopolysaccharide-binding acute phase reactant from rabbit serum. J Exp Med 1986; 164:777–793.
9. Calvano SE, Thompson WA, Marra MN, Coyle SM, de Riesthal HF, Trousdale RK, et al. Changes in polymorphonuclear leukocyte surface and plasma bactericidal/permeability-increasing protein and plasma lipopolysaccharide binding protein during endotoxemia or sepsis. Arch Surg 1994; 129:220–226.
10. Tobias PS, Soldau K, Hatlen LE, Schumann RR, Einhorn G, Mathison JC, et al. Lipopolysaccharide Binding Protein. J Cell Biochem 1992; 16C:151.
11. Schumann RR, Leong SR, Flaggs GW, Gray PW, Wright SD, Mathison JC, et al. Structure and function of lipopolysaccharide (LPS) binding protein; a plasma protein that controls the response of macrophages to LPS. Science 1990; 249:1429–1431.
12. Mathison JC, Tobias PS, Wolfson E, Ulevitch RJ. Plasma lipopolysaccharide (LPS)-binding protein: a key component in macrophage recognition of Gram-negative LPS. J Immunol 1992; 149:200–206.
13. Camussi G, Mariano F, Biancone L, De Martino A, Bussolati B, Montrucchio G, et al. Lipopolysaccharide binding protein and CD14 modulate the synthesis of

platelet-activating factor by human monocytes and mesangial and endothelial cells stimulated with lipopolysaccharide. J Immunol 1995; 155:316–324.
14. Goldblum SE, Brann TW, Ding X, Pugin J, Tobias PS. Lipopolysaccharide (LPS)-binding protein and soluble CD14 function as accessory molecules for LPS-induced changes in endothelial barrier function, in vitro. J Clin Invest 1994; 93:692–702.
15. Martin TR, Mathison JC, Tobias PS, Leturcq DJ, Moriarty AM, Maunder RJ, et al. Lipopolysaccharide binding protein enhances the responsiveness of alveolar macrophages to bacterial lipopolysaccharide. Implications for cytokine production in normal and injured lungs. J Clin Invest 1992; 90:2209–2219.
16. Kitchens RL, Ulevitch RJ, Munford RS. Lipopolysaccharide (LPS) partial structures inhibit responses to LPS in a human macrophage cell line without inhibiting LPS uptake by a CD14-mediated pathway. J Exp Med 1992; 176:485–494.
17. Tapping RI, Tobias PS. Cellular binding of soluble CD14 requires lipopolysaccharide (LPS) and LPS-binding protein. J Biol Chem 1997; 272:23157–23164.
18. Gegner JA, Ulevitch RJ, Tobias PS. LPS signal transduction and clearance. Dual roles for LBP and mCD14. J Biol Chem 1995; 270:5320–5325.
19. Wurfel MM, Kunitake ST, Lichenstein H, Kane JP, Wright SD. Lipopolysaccharide (LPS)-binding protein is carried on lipoproteins and acts as a cofactor in the neutralization of LPS. J Exp Med 1994; 180:1025–1035.
20. Gallay P, Heumann D, Le Roy D, Barras C, Glauser MP. Lipopolysaccharide-binding protein as a major plasma protein responsible for endotoxemic shock. Proc Natl Acad Sci USA 1993; 90:9935–9938.
21. Jack RS, Fan X, Bernheiden M, Rune G, Ehlers M, Weber A, et al. Lipopolysaccharide-binding protein is required to combat a murine gram-negative bacterial infection. Nature 1997; 389:742–745.
22. Wurfel MM, Monks BG, Ingalls RR, Dedrick RL, Delude R, Zhou D, et al. Targeted deletion of the lipopolysaccharide (LPS)-binding protein gene leads to profound suppression of LPS responses Ex vivo, whereas In vivo responses remain intact. J Exp Med 1997; 186:2051–2056.
23. Gallay P, Heumann D, Le Roy D, Barras C, Glauser M-P. Mode of action of anti-lipopolysaccharide-binding protein antibodies for prevention of endotoxemic shock in mice. Proc Natl Acad Sci USA 1994; 91:7922–7926.
24. Wright SD, Ramos RA, Patel M, Miller DS. Septin: A factor in plasma that opsonizes lipopolysaccharide-bearing particles for recognition by CD14 on phagocytes. J Exp Med 1992; 176:719–727.
25. Tobais PS, Mathison JC, Lee J-D, Kravchenko V, Mintz D, Pugin J, et al. LPS binding protein, LPS, and CD14 mediated activation of myeloid cells. In: Levin J, Alving CR, Munford RS, Stutz PL, eds. Bacterial Endotoxin: Recognition and Effector Mechanisms. Elsevier Science Publishers, 1993:135–137.
26. Tobias PS, Gegner JA, Han J, Kirkland T, Kravchenko V, Leturcq D, et al. LPS binding protein and CD14 in the LPS dependent activation of cells. In: Levin J, van Deventer SJH, van der Poll T, Sturk A, eds. Bacterial Endotoxins. Basic Science to Anti-sepsis Strategies. New York: Wiley-Liss, 1993:31–39.
27. Park CT, Wright SD. Plasma lipopolysaccharide-binding protein is found associated with a particle containing apolipoprotein A-I, phospholipid, and factor H-related proteins. J Biol Chem 1996; 271:18054–18060.
28. Wright SD, Ramos RA, Tobias PS, Ulevitch RJ, Mathison JC. CD14 serves as the cellular receptor for complexes of lipopolysaccharide with lipopolysaccharide binding protein. Science 1990; 249:1431–1433.
29. Dubin W, Martin TR, Swoveland P, Leturcq DJ, Moriarty A, Tobias PS, et al. Asthma and endotoxin: lipopolysaccharide-binding protein and soluble CD14 in bronchoalveolar compartment. Am J Physiol 1995; 270:L736–734.
30. Tobias PS, Mathison JC, Ulevitch RJ. A family of lipopolysaccharide binding proteins involved in responses to gram-negative sepsis. J Biol Chem 1988; 263:13479–13481.
31. Gray PW, Flaggs G, Leong SR, Gumina RJ, Weiss J, Ooi CE, et al. Cloning of the cDNA of a human neutrophil bactericidal protein structural and functional correlations. J Biol Chem 1989; 264:9505–9509.
32. Drayna D, Jarnagin AS, McLean J, Henzel W, Kohr W, Fielding C, et al. Cloning and sequencing of human cholesteryl ester transfer protein cDNA. Nature 1987; 327:632–634.
33. Day JR, Albers JJ, Lofton-Day CE, Gilbert TL, Ching AFT, Grant FJ, et al. Complete cDNA encoding human phospholipid transfer protein from human endothelial cells. J Biol Chem 1994; 269:9388–9391.
34. Hailman E, Albers JJ, Wolfbauer G, Tu A-y, Wright SD. Neutralization and transfer of lipopolysaccharide by phospholipid transport protein. J Biol Chem 1996; 271:12172–12178.
35. Elsbach P, Weiss J. Prospects for use of recombinant BPI in the treatment of gram-negative bacterial infections [review]. Infect Agents Dis 1995; 4:102–109.
36. Dear TN, Boehm T, Keverne EB, Rabbitts TH. Novel genes for potential ligand-binding proteins in subregions of the olfactory mucosa. EMBO J 1991; 10:2813–2819.
37. Su GL, Freeswick PD, Geller DA, Wang Q, Shapiro RA, Wan Y-H, et al. Molecular cloning, characterization, and tissue distribution of rat lipopolysaccharide binding protein. J Immunol 1994; 153:743–752.
38. Ramadori G, Meyer zum Buschenfelde K-H, Tobias PS, Mathison JC, Ulevitch RJ. Biosynthesis of lipopolysaccharide binding protein in rabbit hepatocytes. Pathobiol 1990; 58:89–94.
39. Geller DA, Kispert PH, Su GL, Wang SC, di Silvio M, Tweardy DJ, et al. Induction of hepatocyte lipopolysaccharide binding protein in models of sepsis and the acute-phase response. Arch Surg 1993; 128:22–28.
40. Wong HR, Pitt BR, Su GL, Rossignol DP, Steve AR, Billiar TR, et al. Induction of lipopolysaccharide-binding protein gene expression in cultured rat pulmonary artery smooth muscle cells by interleukin 1beta. Am J Respir Cell Mol Biol 1995; 12:449–454.
41. Grube BJ, Cochane CG, Ye RD, Green CE, McPhail ME, Ulevitch RJ, et al. Lipopolysaccharide binding protein expression in primary human hepatocytes and

HepG2 hepatoma cells. J Biol Chem 1994; 269:8477–8482.
42. Kirschning C, Unbehaun A, Lamping N, Pfeil D, Herrmann F, Schumann RR. Control of transcriptional activation of the lipopolysaccharide binding protein (LBP) gene by proinflammatory cytokines. Cytokines Cell Mol Ther 1997; 3:59–62.
43. Hubacek JA, Buchler C, Aslanidis C, Schmitz G. The genomic organization of the genes for human lipopolysaccharide binding protein (LBP) and bactericidal permeability increasing protein (BPI) is highly conserved. Biochemical & Biophysical Research Communications 1997; 236:427–430.
44. Beamer LJ, Carroll SF, Eisenberg D. Crystal structure of human BPI and two bound phospholipids at 2.4 angstrom resolution. Science 1997; 276:1861–1864.
45. Abrahamson SL, Wu H-M, Williams RE, Der K, Ottah N, Little R, et al. Biochemical characterization of recombinant fusions of lipopolysaccharide binding protein and bactericidal/permeability-increasing protein. J Biol Chem 1997; 272:2149–2155.
46. Lamping N, Hoess A, Yu B, Park TC, Kirschning CJ, Pfeil D, et al. Effects of site-directed mutagenesis of basic residues (Arg 94, Lys 95, Lys 99) of lipopolysaccharide (LPS)-binding protein on binding and transfer of LPS and subsequent immune cell activation. J Immunol 1996; 157:4648–4656.
47. Schiff DE, Kline L, Soldau K, Lee JD, Pugin J, Tobias PS, et al. Phagocytosis of gram-negative bacteria by a unique CD14-dependent mechanism. J Leukoc Biol 1997; 62:786–794.
48. Tobias PS, Soldau K, Ulevitch RJ. Identification of a lipid A binding site in the acute phase reactant lipopolysaccharide binding protein. J Biol Chem 1989; 264:10867–10871.
49. Gallay P, Barras C, Tobias PS, Calandra T, Glauser MP, Heumann D. LPS-binding protein in human serum determines the TNF responses to of monocytes to lipopolysaccharide. J Infect Dis 1994; 170:1319–1322.
50. Tobias PS, Soldau K, Gegner JA, Mintz D, Ulevitch RJ. Lipopolysaccharide binding protein-mediated complexation of lipopolysaccharide with soluble CD14. J Biol Chem 1995; 270:10482–10488.
51. Hailman E, Lichenstein HS, Wurfel MM, Miller DS, Johnson DA, Kelley M, et al. Lipopolysaccharide (LPS)-binding protein accelerates the binding of LPS to CD14. J Exp Med 1994; 179:269–277.
52. Han J, Mathison J, Ulevitch R, Tobias P. Lipopolysaccharide (LPS) binding protein, truncated at Ile-197, binds LPS but does not transfer LPS to CD14. J Biol Chem 1994; 269:8172–8175.
53. Theofan G, Horwitz A, Williams R, Liu P, Chan I, Birr C, et al. An amino-terminal fragment of human lipopolysaccharide-binding protein retains lipid A binding but not CD14-stimulatory activity. J Immunol 1994; 152:3623–3629.
54. Schromm AB, Brandenburg K, Rietschel ET, Flad HD, Carroll SF, Seydel U. Lipopolysaccharide-binding protein mediates CD14-independent intercalation of lipopolysaccharide into phospholipid membranes. FEBS Lett 1996; 399:267–271.
55. Battafaraono RJ, Dahlberg PS, Ratz CA, Johnston JW, Gray BH, Haseman JR, et al. Peptide derivatives of three distinct lipopolysaccharide binding proteins inhibit lipopolysaccharide-induced tumor necrosis factor-alpha secretion in vitro. Surgery 1995; 118:318–324.
56. Taylor AH, Heavner G, Nedelman M, Sherris D, Brunt E, Knight D, et al. Lipopolysaccharide (LPS) neutralizing peptides reveal a lipid A binding site of LPS binding protein. J Biol Chem 1995; 270:17934–17938.
57. Tobias PS, Mathison J, Mintz D, Lee J, Kravchenko V, Kato K, et al. Participation of lipopolysaccharide binding protein in lipopolysaccharide dependent macrophage activation. Am J Respir Cell Mol Biol 1992; 7:239–245.
58. Pugin J, Schurer-Maly C-C, Leturcq D, Moriarty A, Ulevitch RJ, Tobias PS. Lipopolysaccharide activation of human endothelial and epithelial cells is mediated by lipopolysaccharide-binding protein and soluble CD14. Proc Natl Acad Sci USA 1993; 90:2744–2748.
59. Yu B, Wright SD. Catalytic properties of lipopolysaccharide (LPS) binding protein. Transfer of LPS to soluble CD14. J Biol Chem 1996; 271:4100–4105.
60. Wurfel MM, Hailman E, Wright SD. Soluble CD14 acts as a shuttle in the neutralization of lipopolysaccharide (LPS) by LPS-binding protein and reconstituted high density lipoprotein. J Exp Med 1995; 181:1743–1754.
61. Wurfel MM, Wright SD. Lipopolysaccharide-binding protein and soluble CD14 transfer lipopolysaccharide to phospholipid bilayers: preferential interaction with particular classes of lipid. J Immunol 1997; 158:3925–3934.
62. Wright SD, Kolesnick RN. Does endotoxin stimulate cells by mimicking ceramide? Immunol Today 1995; 16:297–302.
63. Massamiri T, Tobias PS, Curtis LK. Structural determinants for the binding of lipopolysaccharide binding protein to purified human high density lipoproteins: role of apolipoprotein A-1, J Lipid Res 1997; 38:516–525.
64. Grunwald U, Fan X, Jack RS, Workalemahu G, Kallies A, Stelter F, et al. Monocytes can phagocytose gram-negative bacteria by a CD 14-dependent mechanism. J Immunol 1996; 157:4119–4125.

20

Bactericidal/Permeability-Increasing Protein, p15s and Phospholipases A_2, Endogenous Antibiotics in Host Defense Against Bacterial Infections

Peter Elsbach
New York University School of Medicine, New York, New York

INTRODUCTION

The effects of endotoxin on the animal host provide a paradigm for microbe-host interactions in general. Thus, the ability of the lipopolysaccharides (LPS) isolated from the gram-negative bacterial envelope to trigger a broad spectrum of cellular and extracellular responses in the intact host as well as in vitro has served as a model for the dissection of antimicrobial defense systems that constitute innate immunity (1).

The extraordinary efficiency and effectiveness of these systems is well illustrated by the fact that we are generally unaware of the daily penetration of our epithelial barriers by microbial inhabitants of our internal and external environment and their products, including cell-free LPS. This reflects the integration of mechanisms of recognition, sequestration, and disposal of the invaders by the host. However, in more severe infections and in a weakened host, these normally tightly regulated and beneficial responses may become unbalanced, manifest as excessive output of proinflammatory mediators causing cardiovascular instability, impaired hemostasis, and tissue damage.

Three proteins that are components of mammalian antimicrobial host defenses have been isolated and characterized in this laboratory (Table 1). Because the principal characteristics of each of these three proteins have been summarized in several recent reviews (2–10), only a brief synopsis of these features will follow, with emphasis in this chapter on the role that these proteins play in pathophysiological settings.

BACTERICIDAL/PERMEABILITY-INCREASING PROTEIN

Bactericidal/Permeability Increasing Protein (BPI), a 456-residue lysine-rich protein (11), is expressed solely by myeloid precursors of polymorphonuclear leukocytes (PMN) (12) and is stored in the primary granules of the mature PMN (12) but is also present on the external surface of the plasma membrane of these cells (13). The origin of BPI reported to be detectable on the surface of monocytes that do not synthesize the protein is not clear (14). BPI has been isolated from the PMN of humans, rabbits, and cows and represents 0.5–1% of total cell protein (15,16). The known biological properties of BPI include potent antibacterial activity directed selectively at a broad spectrum of gram-negative bacterial species and strains. BPI in laboratory media or physiological fluids is nontoxic for gram-positive bacteria or eukaryotic cells (2). This cell-specific toxicity of BPI is attributable to its strong attraction for LPS. BPI forms complexes with LPS via the highly conserved lipid A region of LPS, accounting for the fact that the high affinity of BPI (apparent Kd 2–5 nM) (17) is the same for all forms of isolated LPS tested, regardless of bacterial origin and the length of

Table 1 Three Protein Components of Mammalian Antimicrobial Host-Defenses

	BPI	p15s	PLA2
Source	Human PMN	Rabbit PMN	Human[a]
	Rabbit PMN		Rabbit[a]
	Cow PMN		
M_r (kDa)	50–55	15	14–18
Target	GNB	GNB	GNB; GPB
	LPS	LPS	
Potency	nM	µM	nM

BPI = Bactericidal/Permeability-increasing protein; GNB = gram-negative bacteria; GPB = gram-positive bacteria.
[a]PLA2 with antibacterial activities is present in both human and rabbit PMN, but other sources are likely to account for the high levels in extracellular fluids under inflammatory conditions.

the polysaccharide chain. In contrast, the length of the polysaccharide chain of LPS within the bacterial envelope is an important determinant of the antibacterial potency of BPI (18,19). This may be explained at least in part by the denser packing of the LPS molecules in the outer membrane of the gram-negative bacterial envelope than in aggregates of isolated LPS (1). As a consequence, in the intact bacterial envelope, the longer polysaccharide chains restrict access to lipid A and higher concentrations of BPI are required for its antibacterial actions against gram-negative bacteria carrying O-chain extensions (19). In contrast to the effect of the length of the polysaccharide chain on the relative sensitivity of a range of gram-negative bacterial species to BPI, the presence of capsular antigens does not alter bacterial sensitivity (20). The divalent cations Ca^{2+} and Mg^{2+} that occupy the anionic sites on lipid A of envelope LPS also influence the presentation and accessibility of LPS (21,22). Complexing by BPI of LPS, whether free (3,4,6,23–25) or bacteria-associated, inhibits LPS-induced host responses to both whole bacteria and isolated LPS in vitro (26–28) as well as in vivo (7). The insertion of BPI into the outer leaflet (LPS layer) of the outer membrane of the gram-negative bacterial envelope disrupts its characteristic permeability barrier for and resulting resistance to noxious hydrophobic substances (including antibiotics) (29). This may account for the synergy that has been observed between BPI and certain antibiotics against multiple antibiotic–resistant gram-negative bacterial species (26,30,31).

While the binding of BPI to the outer membrane coincides with immediate growth arrest, this action and the accompanying envelope alterations are all reversible (bacteriostasis) (32,33). Irreversible growth arrest (killing) occurs later and coincides with impairment of biochemical processes that are linked to the inner membrane (33,34). Various conditions and host factors can accelerate or retard the transition from reversible to irreversible growth inhibition (26,33,35) (for details, see earlier reviews) (5,6). *E. coli* exposed to isolated BPI or ingested by PMN either in room air or with oxygen removed undergo closely similar fates (33,36,37). Thus, O_2-independent antibacterial systems in the intact PMN suffice against these bacteria, and among these BPI is apparently the principal agent. A prominent role of BPI in the whole PMN is supported by the ability of BPI-neutralizing antiserum to abolish the growth-inhibitory activity of whole PMN lysates against gram-negative bacteria (38).

The antibacterial and anti-LPS activities of BPI are fully expressed by the N-terminal half of the molecule that can be obtained by limited proteolysis (39–41) or as a recombinant fragment (rBPI-21; residues 1–193) (17,25–27,42,43). Expression of truncated forms of BPI by in vitro transcription/translation indicates that antibacterial and LPS-binding activities require a mostly intact N-terminal fragment (43). If the known functions of BPI are met by one half of the molecule, what is the role of the other half?

That the BPI molecule is composed of two structurally distinct domains with interdependent functions is supported by what has been learned about the structural and functional properties of another member of the family of LPS-interactive proteins, the LPS-binding protein (LBP) (44). LBP is an acute phase plasma protein produced in the liver that serves a primary role in transmitting the LPS signal to responsive host cells (44). LBP is also a 456-residue protein and shares ~45% amino acid sequence identity with BPI (45). LBP is a bifunctional protein in which the N-terminal half contains the determinants of LPS binding and the C-terminal half mediates delivery of LPS to a CD14 LPS receptor, thereby markedly amplifying host cell responses to very low concentrations of LPS. CD14 exists either as a plasma membrane–anchored protein of myeloid cells or as a soluble protein in the body fluids (46). Thus, while LBP and BPI represent opposing vectors in LPS signaling (47–49) [in ways that are related to the different physical interactions of the proteins with LPS (50,51)], the former promoting and the latter preventing host cell activation, the two proteins have many structural properties in common. To explore the possibility that the C-terminal half of BPI also participates in transferring LPS (recognized on intact bacteria) the ability of five recombinant proteins (52)—holo-BPI, BPI-21, LBP, and two chimeric proteins (BPI/LBP and LBP/BPI)—to stimulate the ingestion by PMN of encapsulated (phagocytosis-resistant) *E. coli* was compared. For this strain of *E. coli*, only the bacteria coated by full-length BPI were ingested by PMN (53). This opsonic effect of BPI equalled that of nonlethal (5%) normal serum. As expected, the serum-mediated opsonic activity was blocked by anti-CR3 antibodies, but neither anti-CR3 nor anti-CD14 antibodies inhibited BPI-mediated phagocytosis (53). By what route BPI promotes phagocytosis remains to be determined. Whatever pathway of internalization BPI employs, these findings implicate both the N-terminal and the C-terminal domains of BPI in opsonophagocytosis, the former providing high-affinity binding to the bacteria, the latter serving to deliver the BPI-coated bacteria to the PMN. Thus, the opposing actions of BPI and LBP reside in similar but clearly divergent two-domain structures. Finally, the recently determined three-dimensional structure of human BPI further establishes that BPI (and therefore LBP) is organized in two distinct domains (54).

Both added purified or recombinant holo-protein or the N-terminal fragment (rBPI-21) are fully active in whole blood ex vivo (10,26,55). In inflammatory settings a small portion of the cellular holo-BPI is secreted by the PMN into the extracellular environment, where its antibacterial activity is readily demonstrable (56). The actions of BPI, stored in the cytoplasmic granules of the PMN, are therefore not limited to the intraphagosomal environment but extend into the extracellular space. The study of an in vivo–ex vivo rabbit model of sterile peritoneal inflammation has also revealed, not surprisingly, that in complex biological settings the many components of the host-defense systems do not act alone (56). Neutralizing BPI antiserum blocks all antibacterial activity against gram-negative bacteria in the cell-free inflammatory fluid, indicating the strict dependence of this activity on BPI. However, the BPI content of the fluid can account only for a portion of the total activity, reflecting synergy with other components, most prominently the p15s (56) (see below).

Among all known antibacterial and anti-LPS constituents of mammalian host-defense systems, BPI stands out because of its target cell selectivity and potency. These features of BPI plus the availability of recombinant human BPI and its biologically active N-terminal fragments have provided the background for testing of human recombinant (r) BPI in vivo (57–62).

In experiments with a range of experimental animals, rBPI-21 has been shown to protect against the lethal and other toxic effects of administered isolated LPS (57–62) (Table 2). In combination with antibiotics, BPI also protects against lethal inocula or gram-negative bacteria in rabbits and in some animal models without added antibiotics (58) (Table 2).

These preclinical studies have cleared the way for trials in humans. Phase I studies in human volunteers have established that intravenously administered BPI is nontoxic and nonimmunogenic and also inhibits LPS-induced cytokine release, cardiovascular responses, and changes in adhesive properties of PMN and in levels of coagulation factors in endotoxin challenge studies (63–65). Ongoing Phase II trials have included patients with meningococcemia, hemorrhagic trauma, partial hepatectomy, and complicated abdominal infections (unpublished observations). In none of these studies involving hundreds of patients, have problems of safety arisen (Table 3). In an open-label multicenter trial, 26 patients with severe meningococcemia were treated with BPI and compared with a historical control group of 56 patients of similar severity, recently cared for in the same clinical centers. An uncontrolled trial because of the exceedingly high morbidity and mortality in this disease, especially in children. There was one death in the BPI-treated group (4%) and 12 deaths in the control group (20%) (94). These results have led to the initiation of a Phase III double-blind, placebo-controlled, meningococcemia trial that is being conducted in the United States, Canada, and Europe.

Table 2 Protective Effects in Animals of Administered rBPI PBI-21 Against Isolated LPS and Live Gram-Negative Bacteria (57–62)

Challenge	Protective effects
LPS	
Mice	Increased survival
Rats	Reduced cytokine responses
Rabbits	and pulmonary,
Pigs	cardiovascular, and
	metabolic alterations
Live GNB (i.v.; intraperitoneal; intratracheal) ± antibiotics	
Mice	Increased survival
Rats	Reduced systemic responses
Rabbits	Enhanced bacterial clearance

Table 3 rBPI-21 Is Apparently Safe and Nonimmunogenic in Man (Summary of Completed and Ongoing Trials)

Phases I and II (>700 individuals):
  Healthy volunteers: study completed (n = 71)
  Clinical trials:
    Placebo-controlled:
      hemorrhagic trauma (n = 400)—study completed
      partial hepatectomy (n = 72)—ongoing
    Open, escalating doses:
      complicated abdominal infections
      severe pediatric meningoccemia (n = 26)—study completed
Phase III
  Placebo-controlled:
    severe pediatric meningoccemia
    hemorrhagic trauma

These studies were carried out by the XOMA Corporation, Berkeley, California.

15 kDa PROTEINS OF RABBIT PMN (p15s)

Several isoforms of this cationic (pI > 11.0) 15 kDa protein from the secondary granules of rabbit PMN have been isolated and cloned (66,67). The structural features of the p15s have revealed that they belong to a recently recognized family of proteins found in pig, mouse, cow, rabbit, and human PMN that have been named cathelicidins (10,68,69). Common features include a conserved N-terminal proprotein and a highly variable C-terminal segment with antibacterial and LPS-binding activities (10,28,55,67,68,70). All members of this family contain four cysteines in the same positions of the pro-region. The two disulfide arrangements of the four cysteines in one of the cathelicidins have been identified (71). None of the holoproteins of the family members so far identified, except the p15s, possess antibacterial or other known bioactivity (10). Expression of the antibacterial activity of the other cathelicidin family members requires release of the C-terminal portion by proteolytic cleavage. In contrast, the full-length p15s are active without proteolysis as a condition for activity. Thus, the p15s apparently represent more distant members of the cathelicidin family. So far homologs of rabbit p15s have only been reported in mouse PMN (cDNA accession). The antibacterial activity of the p15s appears to be limited to gram-negative bacteria, probably reflecting, as in the case of BPI, specific interaction with the LPS component of the bacterial envelope. The p15s also inhibit the induction of host cell responses by LPS (55).

In contrast to the indiscriminate inhibition by BPI of the activities of isolated LPS regardless of the length of the polysaccharide chains, the inhibition by p15s of long-chain LPS is less effective than of rough LPS (55). The potency of both the antibacterial and LPS-inhibitory activities of the p15s is at least 100-fold less than of BPI (28,55). However, under conditions where both BPI and p15s [the latter at concentrations (0.5 μM) that by themselves are devoid of activity] are present either as added agents (to whole blood) (55) or as natural constituents released into inflammatory (cell-free) fluids, the two proteins show antibacterial synergy (56). The location of the p15s in the secondary granules is consistent with much greater extracellular accumulation of p15s than of BPI, a resident of the primary granules, which do not as readily release their contents (70). Thus, the antibacterial synergy between BPI and p15s in the inflammatory fluid might benefit economical extracellular deployment of BPI and preservation of intracellular stores (70). The most potent rabbit defensins added to whole blood at subbioactive concentrations (0.5 μM) also exhibit synergy with BPI (55). However, in contrast to the p15s, defensins released by PMN extracellularly in inflammatory fluid do not participate in antibacterial activity because they form inactive complexes with other proteins in the fluid (56,72–74).

The secondary granules of rabbit as well as human PMN also contain another cationic member of the cathelicidin family, the CAP18, of which the C-terminal proteolytic cleavage product or a synthetic analog possesses roughly the same antibacterial potency and LPS-inhibitory activity as do the p15s (74–76).

While the function of the cathelicidins as antimicrobial agents has received the most experimental attention, additional roles should be considered as well. Thus, one member of this family (PR-39) has been shown to stimulate proteoglycan production by fibroblasts (75), raising the possibility that this peptide contributes to tissue repair. The fact that p15s, CAP18, and probably other cathelicidins are stored in the secondary granules that readily release their contents, thereby generating substantial extracellular accumulation of these proteins, is further reason to search for other (mediator?) functions. Further support for more functions of PR-39 and other cathelicidins comes from reports of chemoattractant activity and modulatory effects on oxidative metabolism (76,77).

INFLAMMATORY 14 kDa GROUP II PHOSPHOLIPASES A$_2$

These phospholipases A$_2$ (PLA2s) belong to a large family of more than a 100 isolated and sequenced 14 kDa PLA2s that are widely distributed throughout nature (78). All members of the 14 kDa PLA2s share many common structural features, including the compact three-dimensional organization of the proteins that contain ≥ 6 disulfides and a strictly conserved catalytic pocket and Ca^{2+} binding site (78). However, surface-exposed regions of these PLA2s show a high level of diversity. Growing evidence indicates that these variable regions are related to distinct functional roles of these enzymes in biology (8,79,80). This conclusion is well documented by comparative studies of the ability of different 14 kDa PLA2s to contribute to the destruction of bacteria (81,82). Most PLA2s studied, including pancreatic (group I) PLA2s from various animal species, lack this ability. Comparison of a wide range of 14 kDa PLA2s (79), chemical modification of specific residues (83), and site-specific mutagenesis (81,82,84) have implicated certain basic residues in the N-terminal region of the enzyme as determinants of binding to the envelope of bacteria and hydrolysis of their phospholipids (8). This activity directed toward gram-negative bacteria (*E. coli*) is only evident in combination with other antimicrobial proteins with surface-perturbing properties, notably BPI, p15s, and the membrane attack complex of complement (35,55,85). Among the PLA2s active in this setting are the so-called inflammatory PLA2s that accumulate in the circulation during inflammatory events in animals and humans (86) and locally at sites of inflammation (87–89). In contrast, these same PLA2s also attack gram-positive bacteria without dependence on other antimicrobial agents or systems (89). Both in the cell-free fluid of a sterile inflammatory exudate elicited in the peritoneal cavity of the rabbit (89) and in the circulation of baboons that had received a lethal inoculum of *E. coli*, PLA2 levels rise exponentially after this challenge in parallel with antistaphylococcal activity (90). This activity can be eliminated by neutralizing antiserum against the enzyme (89,90). The antibacterial activity is also abolished by interference with catalytic activity (89).

Comparison of many PLA2s with and without antibacterial activity has shown that the structural determinants of the activities against gram-negative or gram-positive bacteria overlap but are not identical (89), presumably reflecting different structural requirements for interaction (binding) with the different microbial envelopes.

The potent bactericidal action of these PLA2s against bacteria such as *Staphylococcus aureus* (including antibiotic-resistant strains) and other gram-positive species as well as their contribution to the destruction of gram-negative bacteria suggest that they may be important elements in antimicrobial host defense. These findings have raised questions about the common belief that the primary role of these PLA2s is a potentially harmful pro-inflammatory one by contributing to release of arachidonic acid and hence production and release of inflammatory prostaglandins and prostacyclins (91). Which of these two activities predominates in inflammatory conditions remains to be determined.

CONCLUSION

The vast range and complexity of antimicrobial host defenses is evident throughout evolution and must reflect the need of a broad spectrum of antimicrobial devices in order to match the enormous diversity of potential microbial pathogens. The accelerating pace of discovery of new agents, especially polypeptides, with antimicrobial activities that are being isolated and characterized from plants, insects, and higher animals (92,93) is much aided by methodological advances of cloning and isolation. The functional characterization of many of these novel "endogenous antibiotics" most often has been limited to assays of antimicrobial activity under laboratory conditions that do not necessarily reveal their actual pathophysiological contribution to host defense. In the study of each of the three proteins discussed in this review, we have searched for evidence of a role in their complex natural environment. In the case of the PMN proteins BPI and p15s, no reliable methods are available to establish unequivocally what their intracellular role is. However, with neutralizing antibodies and other tools at our disposal, it has been possible to demonstrate that BPI and p15s secreted by PMN into inflammatory fluid in vivo are major contributors to antibacterial activity against gram-negative bacteria. The function of the secretory Group II 14 kDa PLA2s is thought to be mainly extracellular. Our findings show that the high levels of this PLA2 in body fluids during infection and inflammation are indeed sufficient to account for potent bactericidal action against major gram-positive pathogens. In combination with BPI, p15s, complement, and, very likely other host defense elements, these PLA2s, both within the PMN and extracellularly, also participate in the destruction of gram-negative bacteria.

Finally, the view that the antibacterial and LPS-neutralizing activities of BPI probably are important in the daily maintenance of antimicrobial control by the healthy host may be supported by the demonstration of protection by administered BPI of infected or LPS-challenged animals and humans.

ACKNOWLEDGMENT

The work in this laboratory is supported by US PHS grants DK05472 and AI32021 and a grant from the U-XOMA Corporation.

REFERENCES

1. Rietschel ET, Brade H, Holst O, et al. Bacterial endotoxin: chemical constitution, biological recognition, host response, and immunological detoxification. Curr Top Microbiol Immunol 1996; 216:39–81.
2. Elsbach P, Weiss J. Oxygen-independent antimicrobial systems of phagocytes. In: Gallin JI, Goldstein IM, Snyderman R, eds. Inflammation: Basic Principles and Clinical Correlates. New York: Raven Press, 1992: 603–636.
3. Elsbach P, Weiss J. Role of the Bactericidal/Permeability increasing protein in host defence. Curr Opin Immunol 1998; 10:45–49.
4. Elsbach P, Weiss J. The bactericidal/permeability-increasing protein (BPI), a potent element in host-defense against gram-negative bacteria and lipopolysaccharide. Immunobiology 1993; 187:417–429.
5. Elsbach P, Weiss J, Levy O. Integration of antimicrobial host defenses: role of the bactericidal/permeability-increasing protein. Trends Microbiol 1994; 2:324–328.
6. Elsbach P. Bactericidal permeability-increasing protein in host defense against gram-negative bacteria and endotoxin. In: Marsh J. Goode JA, eds. Antimicrobial Peptides. Chichester: John Wiley & Sons Ltd, 1994: 176–189.
7. Elsbach P, Weiss J, Prospects for the use of recombinant BPI in the treatment of gram-negative bacterial infections. Infect Agents Dis 1995; 4:102–109.
8. Elsbach P. Determinants of the antimicrobial action of 14-kD phospholipases A2. In: Uhl W, Nevalainen TJ, Buchler MW, eds. Phospholipase A2. Basic and Clinical Aspects in Inflammatory Diseases. Vol. 24. Basel: Karger, 1997:17–22.
9. Weiss J. Leukocyte-derived antimicrobial proteins. Curr Opin Hematol 1994; 1:78–84.
10. Levy O. Antibiotic proteins of polymorphonuclear leukocytes. Eur J Haematol 1996; 56:263–277.
11. Gray PW, Flaggs G, Leong SR, et al. Cloning of the cDNA of a human neutrophil bactericidal protein. Structural and functional correlations. J Biol Chem 1989; 264:9505–9509.
12. Weiss J, Olsson I. Cellular and subcellular localization of the bactericidal/permeability-increasing protein of neutrophils. Blood 1987; 69:652–659.
13. Weersink AJ, van Kessel KP, van den Tol ME, et al. Human granulocytes express a 55-kDa lipopolysaccharide-binding protein on the cell surface that is identical to the bactericidal/permeability-increasing protein. J Immunol 1993; 150:253–263.
14. Dentener MA, Francot GJ, Buurman WA. Bactericidal/permeability-increasing protein, a lipopolysaccharide-specific protein on the surface of human peripheral blood monocytes. J Infect Dis 1996; 173:252–255.
15. Weiss J, Elsbach P, Olsson I, Odeberg H. Purification and characterization of a potent bactericidal and membrane active protein from the granules of human polymorphonuclear leukocytes. J Biol Chem 1978; 253: 2664–2672.
16. Elsbach P, Weiss J, Franson RC, Beckerdite-Quagliata S, Schneider A, Harris L. Separation and purification of a potent bactericidal/permeability-increasing protein and a closely associated phospholipase A2 from rabbit polymorphonuclear leukocytes. Observations on their relationship. J Biol Chem 1979; 254:11000–11009.
17. Gazzano-Santoro H, Parent JB, Grinna L, et al. High-affinity binding of the bactericidal/permeability-increasing protein and a recombinant amino-germinal fragment to the lipid A region of lipopolysaccharide. Infect Immun 1992; 60:4754–4761.
18. Weiss J, Beckerdite-Quagliata S, Elsbach P. Resistance of gram-negative bacteria to purified bactericidal leukocyte proteins: relation to binding and bacterial lipopolysaccharide structure. J Clin Invest 1980; 65: 619–628.
19. Capodici C, Chen S, Sidorczyk Z, Elsbach P, Weiss J. Effect of lipopolysaccharide (LPS) chain length on interactions of bactericidal/permeability-increasing protein and its bioactive 23-kilodalton NH2-terminal fragment with isolated LPS and intact *Proteus mirabilis* and *Escherichia coli*. Infect Immun 1994; 62:259–265.
20. Weiss J, Victor M, Cross AS, Elsbach P. Sensitivity of K1-encapsulated *Escherichia coli* to killing by the bactericidal/permeability-increasing protein of rabbit and human neutrophils. Infect Immun 1982; 38: 1149–1153.
21. Weiss J, Victor M, Elsbach P. Role of charge and hydrophobic interactions in the action of the bactericidal/permeability-increasing protein of neutrophils on gram-negative bacteria. J Clin Invest 1983; 71:540–549.
22. Elsbach P, Weiss J, Kao L. The role of intramembrane Ca^{2+} in the hydrolysis of the phospholipids of *Escherichia coli* by Ca^{2+}-dependent phospholipases. J Biol Chem 1985; 260:1618–1622.
23. Marra MN, Wilde CG, Collins MS, Snable JL, Thornton MB, Scott RW. The role of bactericidal/permeability-increasing protein as a natural inhibitor of bacterial endotoxin. J Immunol 1992; 148:532–537.
24. Betz-Corradin S, Heumann D, Gallay P, Smith J, Mauel J, Glauser MP. Bactericidal/permeability increasing protein inhibits induction of macrophage ni-

tric oxide production by lipopolysaccharide. J Infect Dis 1994; 169:105–111.
25. Huang K, Conlon PJ, Fishwild DM. A recombinant aminoterminal fragment of bactericidal/permeability-increasing protein (rBPI23) inhibits CD14-mediated LPS-induced endothelial adherence of human neutrophils. Shock 1993; 1:81–86.
26. Weiss J, Elsbach P, Shu C, et al. Human bactericidal/permeability-increasing protein and a recombinant NH2-terminal fragment cause killing of serum-resistant gram-negative bacteria in whole blood and inhibit tumor necrosis factor release induced by the bacteria. J Clin Invest 1992; 90:1122–1130.
27. Meszaros K, Parent JB, Gazzano SH, et al. A recombinant amino terminal fragment of bactericidal/permeability-increasing protein inhibits the induction of leukocyte responses by LPS. J Leuk Biol 1993; 54:558–563.
28. Levy O, Ooi CE, Elsbach P, Doerfler ME, Lehrer RI, Weiss J. Antibacterial proteins of granulocytes differ in interaction with endotoxin: comparison of bactericidal/permeability-increasing protein p15s, and defensins. J Immunol 1995; 154:5403–5410.
29. Vaara M. Lipid A: target for antibacterial drugs. Science 1996; 274:939–940.
30. Reed D. Synergy between endogenous anitmicrobials and clinical antibiotics against multidrug resistant gram-negative bacteria. Dissertation, New York University School of Medicine, New York, 1997.
31. Lin Y, Leach WJ, Ammons WS. Synergistic effect of a recombinant N-terminal fragment of bactericidal/permeability-increasing protein and cefamandole in treatment of rabbit gram-negative sepsis. Antimicrob Agents Chemother 1996; 40:65–69.
32. Weiss J, Muello K, Victor M, Elsbach P. The role of lipopolysaccharides in the action of the bactericidal/permeability-increasing neutrophil protein on the bacterial envelope. J Immunol 1984; 132:3109–3115.
33. Mannion BA, Weiss J, Elsbach P. Separation of sublethal and lethal effects of the bactericidal/permeability increasing protein on *Escherichia coli*. J Clin Invest 1990; 85:853–860.
34. in't Veld G, Mannion B, Weiss J, Elsbach P. Effects of the bactericidal/permeability-increasing protein of polymorphonuclear leukocytes on isolated bacterial cytoplasmic membrane vesicles. Infect Immun 1988; 56:1203–1208.
35. Wright GC, Weiss J, Kim KS, Verheij H, Elsbach P. Bacterial phospholipid hydrolysis enhances the destruction of *Escherichia coli* ingested by rabbit neutrophils. Role of cellular and extracellular phospholipases. J Clin Invest 1990; 85:1925–1935.
36. Weiss J, Stendahl O, Elsbach P. O_2-independent killing of gram-negative bacteria by intact granulocytes. The role of a potent bactericidal membrane-perturbing protein. Adv Exp Med Biol 1982; 141:129–137.
37. Mannion BA, Weiss J, Elsbach P. Separation of sublethal and lethal effects of polymorphonuclear leukocytes on *Escherichia coli*. J Clin Invest 1990; 86:631–641.
38. Weiss J, Victor M, Stendhal O, Elsbach P. Killing of gram-negative bacteria by polymorphonuclear leukocytes: role of an O_2-independent bactericidal system. J Clin Invest 1982; 69:959–970.
39. Ooi CE, Weiss J, Elsbach P, Frangione B, Mannion B. A 25-kDa NH_2-terminal fragment carries all the antibacterial activities of the human neutrophil 60-kDa bactericidal/permeability-increasing protein. J Biol Chem 1987; 262:14891–14894.
40. Ooi CE, Weiss J, Elsbach P. Structural and functional organization of the human neutrophil 60 kDa bactericidal/permeability-increasing protein. Agents Actions 1991; 34:274–277.
41. Ooi CE, Weiss J, Doerfler ME, Elsbach P. Endotoxin-neutralizing properties of the 25 kD N-terminal fragment and a newly isolated 30 kD C-terminal fragment of the 55-60 kD bactericidal/permeability-increasing protein of human neutrophils. J Exp Med 1991; 174:649–655.
42. Meszaros K, Aberle S, Dedrick R, et al. Monocyte tissue factor induction by lipopolysaccharide (LPS): dependence on LPS-binding protein and CD14, and inhibition by a recombinant fragment of bactericidal/permeability-increasing protein. Blood 1994; 83:2516–2525.
43. Capodici C, Weiss J. Both N- and C-terminal regions of the bioactive N-terminal fragment of the neutrophil granule bactericidal/permeability-increasing protein are required for stability and function. J Immunol 1996; 156:4789–4796.
44. Ulevitch RJ, Tobias PS. Receptor-dependent mechanisms of cell stimulation by bacterial endotoxin. Ann Rev Immunol 1995; 13:437–457.
45. Schumann RR, Leong SR, Flaggs GW, et al. Structure and function of lipopolysaccharide binding protein. Science 1990; 249:1429–1431.
46. Ulevitch RJ, Dunn DL, Fink MP, Taylor CE. Endotoxin-related intracellular pathways: implications for therapeutic intervention. Shock 1996; 6:1–2.
47. Dentener MA, Von Asmuth E, Francot GJ, Marra MN, Buurman WA. Antagonistic effects of lipopolysaccharide binding protein and bactericidal/permeability-increasing protein on lipopolysaccharide-induced cytokine release by mononuclear phagocytes. Competition for binding to lipopolysaccharide. J Immunol 1993; 151:4258–4265.
48. Gazzano-Santoro H, Meszaros K, Birr C, et al. Competition between rBPI23, a recombinant fragment of bactericidal/permeability-increasing protein, and lipopolysaccharide (LPS)-binding protein for binding to LPS and gram-negative bacteria. Infect Immun 1994; 62:1185–1191.
49. Opal SM, Palardy JE, Marra MN, Fisher CJ, McKelligon BM, Scott RW. Relative concentrations of endotoxin-binding proteins in body fluids during infection. Lancet 1994; 344:429–431.
50. Tobias PS, Soldau K, Gegner JA, Mintz D, Ulevitch RJ. Lipopolysaccharide binding protein-mediated complexation of lipopolysaccharide with soluble CD14. J Biol Chem 1995; 270:10482–10488.
51. Tobias PS, Soldau K, Iovine N, Elsbach P, Weiss J. Lipopolysaccharide (LPS) binding proteins BPI and LBP form different complexes with LPS. J Biol Chem 1997; 272:18682–18685.

52. Abrahamson SL, Wu HM, Williams RE, et al. Biochemical characterization of recombinant fusions of lipopolysaccharide binding protein and bactericidal/permeability-increasing protein. Implications in biological activity. J Biol Chem 1997; 272:2149–2155.
53. Iovine N, Elsbach P, Weiss J. An opsonic function of the neutrophil bactericidal/permeability-increasing protein depends on both its N- and C-terminal domains. Proc Natl Acad Sci USA 1997; 94:10973–10978.
54. Beamer LJ, Carroll SF, Eisenberg D. Crystal structure of human BPI and two bound phospholipids at 2.4 angstrom resolution. Science 1997; 276:1881–1884.
55. Levy O, Ooi CE, Weiss J, Lehrer RI, Elsbach P. Individual and synergistic effects of rabbit granulocyte proteins on *Escherichia coli*. J Clin Invest 1994; 94:672–682.
56. Weinrauch Y, Foreman A, Shu C, et al. Extracellular accumulation of potently microbicidal bactericidal/permeability-increasing protein and p15s in an evolving sterile rabbit peritoneal inflammatory exudate. J Clin Invest 1995; 95:1916–1924.
57. Ammons WS, Kung AH. Recombinant amino terminal fragment of bactericidal/permeability-increasing protein prevents hemodynamic responses to endotoxin. Circ Shock 1993; 41:176–184.
58. Ammons WS, Kohn FR, Kung AHC. Protective effects of an N-terminal fragment of bactericidal/permeability-increasing protein in rodent models of gram-negative sepsis: role of bactericidal properties. J Infect Dis 1994; 170:1473–1482.
59. Kohn FR, Ammons WS, Horwitz A, et al. Protective effect of a recombinant amino-terminal fragment of bactericidal/permeability-increasing protein in experimental endotoxemia. J Infect Dis 1993; 168:1307–1310.
60. Kung AH, Ammons WS, Lin Y, Kohn FR. Efficacy of a recombinant amino terminal fragment of bactericidal/permeability-increasing protein in rodents challenged with LPS or *E. coli* bacteria. Prog Clin Biol Res 1994; 338:255–263.
61. Lin Y, Ammons WS, Leach WJ, Kung AH. Protective effects of a recombinant N-terminal fragment of bactericidal/permeability-increasing protein on endotoxic shock in conscious rabbits. Shock 1994; 2:324–331.
62. Lin Y, Kohn FR, Kung AH, Ammons WS. Protective effect of a recombinant fragment of bactericidal/permeability increasing protein against carbohydrate dyshomeostasis and tumor necrosis factor-alpha elevation in rat endotoxemia. Biochem Pharmacol 1994; 47:1553–1559.
63. von der Mohlen MA, van Deventer SJ, Levi M, et al. Inhibition of endotoxin-induced activation of the coagulation and fibrinolytic pathways using a recombinant endotoxin-binding protein (rBPI23). Blood 1995; 85:3437–3443.
64. von der Mohlen MA, Kimmings AN, Wedel NI, et al. Inhibition of endotoxin-induced cytokine release and neutrophil activation in humans by use of recombinant bactericidal/permeability-increasing protein. J Infect Dis 1995; 172:144–151.
65. de Winter RJ, von der Mohlen MA, van Lieshout H, et al. Recombinant endotoxin-binding protein (rBPI23) attenuates endotoxin-induced circulatory changes in humans. J Inflamm 1995; 45:193–206.
65a. Giroir BP, Quint PA, Barton P, Kirsch EA, Kitchen L, Goldstein B, Margraf D, Sastry S, Orlowski JP, Bradley JS, Nelson BJ, Wedel NI, White ML, Bauer RJ, Carroll SF, Scannon PJ. Evaluation of RBPI21 (Recombinant amino terminal fragment of Human Bactericidal/Permeability-Increasing Protein) in children with severe meningococcal disease. Lancet 1997; 350:1439–1443.
66. Ooi CE, Weiss J, Levy O, Elsbach P. Isolation of two isoforms of a novel 15-kDa protein from rabbit polymorphonuclear leukocytes that modulate the antibacterial actions of other leukocyte proteins. J Biol Chem 1990; 265:15956–15962.
67. Levy O, Weiss J, Zarember K, Ooi CE, Elsbach P. Antibacterial 15-kDa protein isoforms (p15s) are members of a novel family of leukocyte proteins. J Biol Chem 1993; 268:6058–6063.
68. Zanetti M, Gennaro R, Romeo D. Cathelicidins: a novel protein family with a common proregion and a variable C-terminal antimicrobial domain. FEBS Lett 1995; 374:1–5.
69. Boman HG. Peptide antibiotics and their role in innate immunity. Annu Rev Immunol 1995; 13:61–92.
70. Zarember K, Elsbach P, Shin-Kim K, Weiss J. p15s (15-kD antimicrobial proteins) are stored in secondary granules of rabbit granulocytes: implications for antibacterial synergy with the bactericidal/permeability-increasing protein in inflammatory fluids. Blood 1997; 89:672–679.
71. Storici P, Tossi A, Lenarcic B, Romeo D. Purification and structural characterization of bovine cathelicidins, precursors of antimicrobial peptides. Eur J Biochem 1996; 238:769–776.
72. Panyutich A, Ganz T. Activated α-2-macroglobulin is a principal defensin-binding protein. Am J Respir Cell Mol Biol 1991; 5:101–106.
73. Panyutich AV, Baturevich EA, Kolesnikova TS, Ganz T. The effect of biotinylation on the antigenic specificity of anti-defensin monoclonal antibodies. J Immunol Methods 1993; 158:237–242.
74. Panyutich AV, Szold O, Poon PH, Tseng Y, Ganz T. Identification of defensin binding to C1 complement. FEBS Lett 1994; 356:169–173.
75. Gallo RL, Ono M, Povsic T, et al. Syndecans, cell surface heparan sulfate proteoglycans, are induced by a proline-rich antimicrobial peptide from wounds. Proc Natl Acad Sci USA 1994; 91:11035–11039.
76. Huang H, Ross CR, Blecha F. Chemoattractant properties of PR-39, a neutrophil antibacterial peptide. J Leuk Biol 1997; 61:624–629.
77. Shi J, Ross CR, Leto TL, Blecha F. PR-39, a proline-rich antibacterial peptide that inhibits phagocyte NADPH oxidase activity by binding to Src homology 3 domains of p47*phox*. Proc Natl Acad Sci USA 1996; 93:6014–6018.
78. Dennis EA. Diversity of group types, regulation, and function of phospholipase A2. J of Biol Chem 1994; 269:13057–13060.
79. Kini RM, Evans HJ. Structure-function relationships of phospholipases. The anticoagulant region of phospholipases A2. J Biol Chem 1987; 262:14402–14407.

80. Kini RM, Evans HJ. A model to explain the pharmacological effects of snake venom phospholipases A2. Toxicon 1989; 27:613–635.
81. Weiss J, Bekkers ACAPA, van den Bergh CJ, Verheij HM. Conversion of pig pancreas phospholipase A2 by protein engineering into enzyme active against *Escherichia coli* treated with the bactericidal/permeability-increasing protein. J Biol Chem 1991; 266:4162–4167.
82. Weiss J, Inada M, Elsbach P, Crowl RM. Structural determinants of the action against *Escherichia coli* of a human inflammatory fluid phospholipase A2 in concert with polymorphonuclear leukocytes. J Biol Chem 1994; 269:26331–26337.
83. Forst S, Weiss J, Elsbach P. The role of phospholipase A2 lysines in phospholipolysis of *Escherichia coli* killed by a membrane-active neutrophil protein. J Biol Chem 1982; 257:14055–14057.
84. Inada M, Crowl RM, Bekkers AC, Verheij H, Weiss J. Determinants of the inhibitory action of purified 14-kDa phospholipases A2 on cell-free prothrombinase complex. J Biol Chem 1994; 269:26338–26343.
85. Madsen LM, Inada M, Weiss J. Determinants of activation by complement of group II phospholipase A2 acting against *Escherichia coli*. Infect Immun 1996; 64:2425–2430.
86. Vadas P, Browning J, Edelson J, Pruzanski W. Extracellular phospholipase A2 expression and inflammation: the relationship with associated disease states. J Lipid Mediat 1993; 8:1–30.
87. Franson R, Dobrow R, Weiss J, Elsbach P, Weglicki WB. Isolation and characterization of a phospholipase A2 from an inflammatory exudate. J Lipid Res 1978; 19:18–23.
88. Forst S, Weiss J, Elsbach P, Maraganore JM, Reardon I, Heinrikson RL. Structural and functional properties of a phospholipase A2 purified from an inflammatory exudate. Biochemistry 1986; 25:8381–8385.
89. Weinrauch Y, Elsbach P, Madsen LM, Foreman A, Weiss J. The potent anti-Staphylococcus aureus activity of a sterile rabbit inflammatory fluid is due to a 14-kD phospholipase A2. J Clin Invest 1996; 97:250–257.
90. Weinrauch Y, Abad C, Liang NS, Lowry SF, Weiss J. Mobilization of potent plasma bactericidal activity during systemic bacterial challenge. Role of group 11a phospholipase a2. J Clin Invest 1998; 102:633–638.
91. Pruzanski W, Vadas P. Secretory nonpancreatic phospholipase A2 (sPLA2) as a mediator of inflammation. In: Buchler MW et al., eds. Phospholipase A2. Basic and Clinical Aspects of Inflammatory Disease. Vol. 24. Basel: Karger, 1997:38–42.
92. Broekaert WF, Terras FR, Cammue BP, Osborn RW. Plant defensins: novel antimicrobial peptides as components of the host defense system. Plant Physiol 1995; 108:1353–1358.
93. Hoffmann JA. Innate immunity of insects. Curr Opin Immunol 1995; 7:4–10.

21

Interactions of Lipopolysaccharides and Lipoproteins

Sander J. H. van Deventer and Dasja Pajkrt
University of Amsterdam, Amsterdam, The Netherlands

INTRODUCTION

It has been long known that bacterial lipopolysaccharides (LPS) induce important changes in lipid metabolism, and more recently the notion has emerged that lipoproteins (LP) may alter the susceptibility to gram-negative endotoxin. The precise mechanisms of LP-LPS interactions and the role of various lipid-binding proteins have only been elucidated in the past few years. In addition, several enzymes that are involved in lipid metabolism, but do not directly interact with LPS, influence the ability of LP to interact with LPS. These findings have led to the general concept that the human body handles LPS and (phospho)lipids (PL) in a similar fashion, and this opened the prospect of therapeutically modulating the sensitivity to LPS by administration of LP particles.

LIPOPROTEIN METABOLISM

LP are water-soluble complexes composed of a neutral lipid core, surrounded by a phospholipid layer that contains cholesterol and one or more apolipoproteins. These proteins serve as ligands for cell membrane receptors, as cofactors for enzymes, and can dock certain LPS-binding proteins. According to their density, LP are categorized as very-low-density lipoprotein (VLDL, d < 1.006 g/ml), low-density lipoprotein (LDL, d = 1.006–1.063 g/ml), and high-density lipoprotein (HDL, d = 1.063–1.21 g/ml). The primary function of LP is to transport lipid, phospholipid, cholesterol, and cholesteryl esters in blood and lymph, and a schematic overview of lipid transport pathways is provided in Figure 1. Dietary lipids are absorbed in the intestine and transported by chylomicrons that deliver free fatty acids (FFA) to the peripheral tissues following the activity of lipoprotein lipase. In addition, VLDL transports FFA synthesized by hepatocytes to the peripheral tissues and are hence converted in VLDL remnants. Both chylomicrons and VLDL contain apolipoprotein E (apoE), which is important for chylomicron and VLDL remnant removal by the liver, through interaction with various receptors including the LDL receptor, and the LDL receptor–related protein. VLDL remnants are lipolytically processed to form LDL. The liver and peripheral tissues catabolize LDL following binding of apolipoprotein B100 (apoB100) to the LDL receptor, and this constitutes a major pathway of cholesterol removal. Both liver and intestine secrete HDL, which contains either apolipoprotein AI (apoAI) and apoAII (liver-derived) or apo AI and apo AIV (intestinal HDL). A major function of HDL is reverse cholesterol transport (RCT), the transport of cholesterol from peripheral tissues via plasma to the liver. RCT is necessary because the liver and the adrenals are the only tissues capable of cholesterol catabolism. During the process of RCT, the HDL particle, which consists of approximately 40% lipid and 50% protein, changes in size through exchange of the lipid fraction (Fig. 2). The HDL that is secreted by the liver and the intestine are small spherical or disc-shaped, lipid-poor particles, and the lipid constituents for these particles are in part derived by lipolysis of triglyceride-rich particles by lipo-

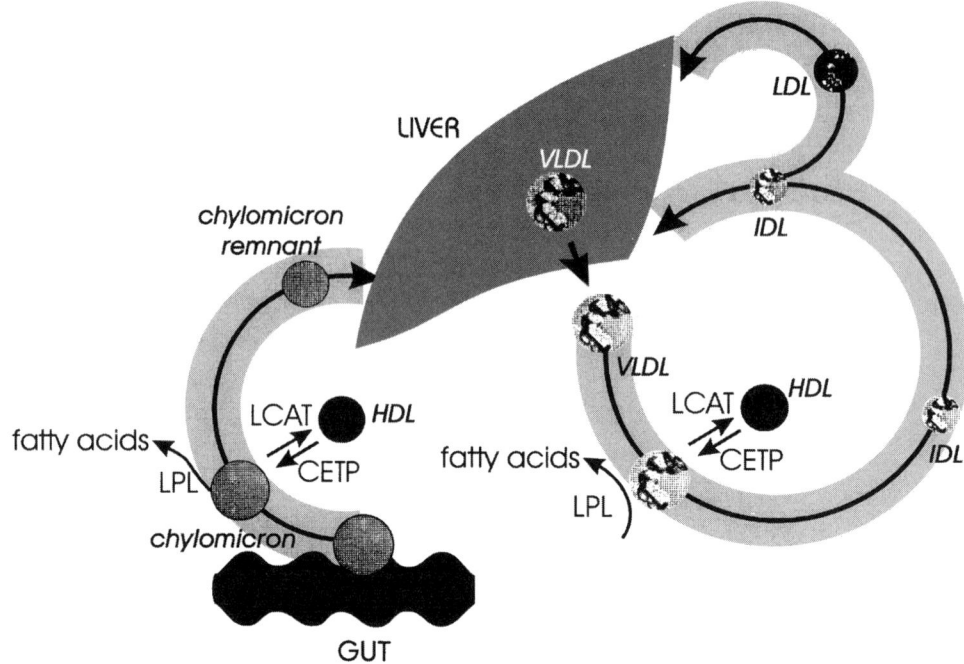

Fig. 1 Schematic overview of lipid transport pathways: Dietary lipids are absorbed by the small intestinal mucosa and transported by chylomicrons. Free fatty acids are delivered to the peripheral tissues through the action of the lipoprotein lipase. VLDL transports free fatty acids that are synthesized in the liver. HDL can exchange lipids (including cholesterol) with chylomicrons and VLDL through the activities of LCAT and CETP.

protein lipase. Mainly through the activity of the enzyme lecithin-cholesterol acyltransferase (LCAT), which is secreted by the liver and carried on HDL particles through an interaction with apoA1, HDL disks are converted into large spherical particles. LCAT catalyzes the conversion of free cholesterol at the surface of LP (mainly HDL) into cholesteryl esters, which can subsequently be incorporated in the LP core, thereby maintaining a concentration gradient that promotes uptake of cholesterol from the tissues into the LP. The remodeling of HDL is further dependent on two other proteins, phospholipid transfer protein (PLTP) and cholesteryl ester transfer protein (CETP). The latter protein transfers cholesteryl esters from HDL to VLDL and LDL in exchange for triglyceride, thus enabling hepatic uptake of HDL-derived cholesteryl esters through LDL and VLDL. The activity of CETP yields triglyceride-enriched HDL, which can serve as a substrate for hepatic lipase, which hydrolyzes HDL triglycerides and phospholipids, resulting in formation of small HDL particles that can serve as cholesterol acceptors, and thereby completing the HDL cycle (Fig. 2). PLTP does not exchange neutral lipids, but transfers phospholipids between LP and from membranes into LP.

THE FAMILY OF LIPID TRANSPORTERS

The main function of the lipid transport proteins CETP and PLTP is to facilitate effective transport of plasma lipids between LP (1). These proteins belong to a family that includes the endotoxin-binding proteins lipopolysaccharide-binding protein (LBP) and bactericidal permeability–increasing protein (BPI) (2). About 20% of the amino acids in CETP and PLTP are identical, whereas CETP and LBP share 23% amino acid sequence homology (3). The recent resolution of the three-dimensional structure of BPI has led to the notion that the lipid transporters are structurally even more related than predicted by their amino acid homology. BPI, resolved at 2.4 Å, was demonstrated to resemble a boomerang-like molecule with two domains of similar size being connected by a proline-rich linker (4). Despite a lack of sequence homology, the two domains

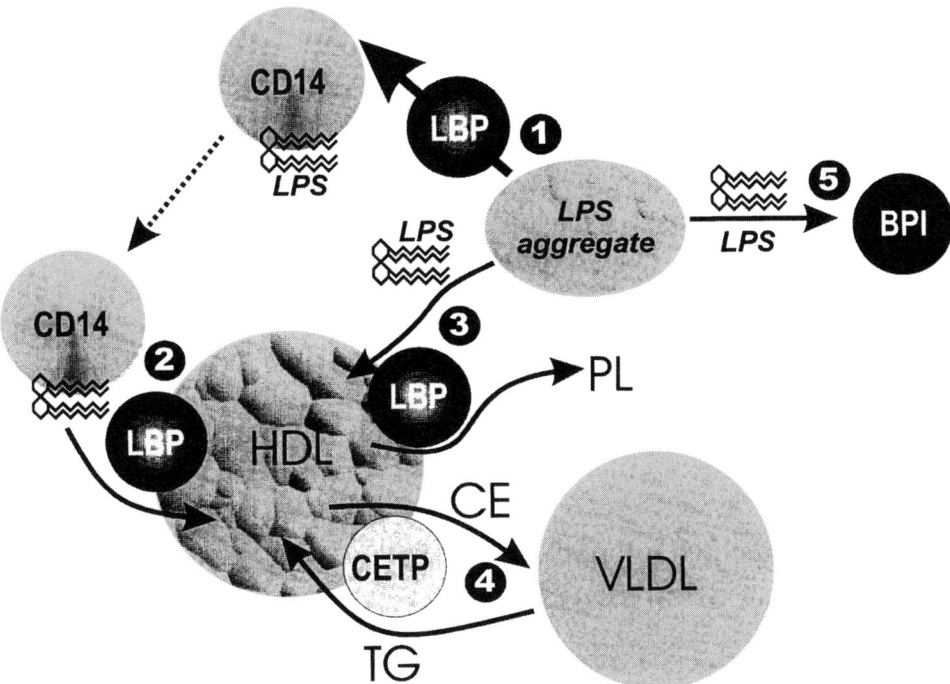

Fig. 2 LPS that is released from bacteria forms large aggregates. LBP momomerizes LPS, and serves as a transport protein to present LPS to membrane-bound CD14, causing cell activation (not shown). LBP also transfers LPS to sCD14 (1), that serves as an endotoxin "sink." SCD14-bound LPS can be transferred to a subfraction of HDL particles, and this process is catalyzed by LBP (2). Alternatively, HDL-bound LBP can directly transfer LPS to HDL, in exchange for phospholipids (3). CETP influences the composition of HDL, which is important for the ability to accept LPS, by exchanging cholesterol esters for triglycerides (4). Another member of the LPS-binding protein family was identified by its capacity to neutralize LPS, but was recently found to bind phospholipids (5).

are structurally very similar, and both domains contain a phospholipid-binding pocket. In fact, the crystallized CHO-expressed BPI molecule was unexpectedly found to contain two phospholipid molecules within the lipid-binding pockets. These lipid-binding pockets are apolar and are thought to interact with the acyl chains of phospholipids and the lipid A moiety of LPS. The protein sequence homology of BPI, LBP, CETP, and PLTP demonstrated conservation of structurally important residues and predicted that all proteins have the same two-domain structure. These observations strongly suggest that, in human plasma, LPS and phospholipids can be recognized and transported by the same proteins. This notion is further supported by the observation that essentially all LBP in human plasma is associated with a complex containing apoA-I, factor H–related proteins 1 and 2, and lipids (5). This LP particle resembles very high-density lipoprotein, but contains only a small amount of phospholipids (0.9%) and constitutes only a minor fraction of all plasma LP (6). Recently, the LBP-binding domains of apoA1 have been mapped to three regions (residues 1–31, 95–164, and 178–200) using apoA-I–specific monoclonal antibodies and synthetic peptides (7).

BINDING AND NEUTRALIZATION OF LPS BY LIPOPROTEINS

It has long been known that serum can inactivate bacterial LP, and the interaction of endotoxin with a plasma LP was first demonstrated in the late 1960s (8). In 1979 Ulevitch et al. reported that the buoyant density of purified radiolabeled LPS markedly changed in normal human, rabbit, and murine serum, and this phenomenon was not observed in delipidated serum preparations (9). When the serum was reconstituted with VLDL, LDL, or HDL, only the latter LP affected the buoyant density of added O111:B4 and R595 LPS, whereas LDL partially reconstituted the activity of R595, but not O111:B4 LPS. Interestingly, these early experiments already suggested that factors present in the nonlipid serum fraction affected the interaction of LPS with LP, because the combination of HDL and delipidated serum, but not the individual components, antagonized the capacity of LPS to induce a febrile response and to cause neutropenia in rabbits. It was subsequently demonstrated that the interaction of LPS with HLD in rat plasma was temperature-dependent and that the physical state of the LPS was important: calcium chloride, which causes LPS aggregation, inhibited the LPS-HDL interaction, whereas sodium deoxycholate, which deaggregates LPS, accelerated it (10). Following intravenous endotoxin injection, the capacity of rabbit serum to neutralize endotoxin remarkably increases, and the rabbits become tolerant to the effects of systemic administration of endotoxin. Warren et al. showed that in "tolerant" rabbit serum LPS was more rapidly shuttled to the LP fraction and suggested that the main difference between "tolerant" and normal rabbit sera was the larger capacity of the former to deaggregate LPS (11). These experiments were at odds with the findings of Tobias et al., who reported a *reduction* of LPS-LP binding in acute phase rabbit serum due to formation of an intermediate complex with a glycoprotein subsequently characterized as LBP (12). Although the endotoxin dose as well as the preparation used (smooth versus rough) differed in the two experiments, it is unlikely that these factors are the cause for this striking difference, which remains unexplained. The fate of LP-bound endotoxin is still a matter of debate, but in rats LPS that is not bound to LP is rapidly distributed to the liver, spleen, and other organs, whereas HDL-bound LPS is more slowly taken up by the liver and adrenal glands. Early studies in mice indicated that endotoxin induced a loss of hepatocyte-associated HDL immunoreactivity and the appearance of HDL-immunoreactive particles in the bile caniculi, suggesting HDL secretion into the bile (13). Specific anti-LPS antibodies (recognizing the polysaccharide chain) were shown to interfere with LPS-HDL complex formation ex vivo and shifted the tissue uptake of radiolabeled LPS towards the liver and spleen (14). Hence, the effects of anti-LPS antibodies in vivo extend beyond simple neutralization and include a redistribution of tissue uptake secondary to the formation of antibody-LPS complexes and inhibition of the LPS-LP interaction. In these experiments, dexamethasone pretreatment augmented the uptake of radiolabeled LPS by the adrenal gland, spleen, and lung and decreased uptake by the liver.

LP do not only alter the tissue distribution of intravenously injected LPS, they also decrease endotoxicity. LPS-LP complexes, generated by incubation of LPS in serum and subsequent density centrifugation, induced significantly less production of IL-1, TNF-α, and IL-6 than noncomplexed LPS. Because the LPS-LP complexes were not able to compete with excess radiolabeled *Neisseria meningitides* LPS for binding to murine macrophages, it was concluded that the reduction of LPS-induced cytokine production by LP was a consequence of inactivation rather than reduced binding to the target cells (15). However, it is now known that the binding assay used in these experiments provides a rather inaccurate estimate of the amount of LPS that binds to receptors that are involved in intracellular signaling. However, experiments using other assay systems have also provided evidence that LP inactivate endotoxin. For example, HDL-complexed LPS had a greatly reduced capacity to prime polymorphonuclear granulocytes and to induce cytokine synthesis by mononuclear cells (16,17). Using Limulus amebocyte activation as a readout, it was demonstrated that HDL, LDL, apoA1, and apoB reduced the "recovery" of endotoxin from serum, but the triglyceride-rich intralipid interfered with the ability of apoB (but not apoA1) to decrease the recovery of endotoxin from serum (18).

Following injection into rats, a substantial fraction (>60%) of LPS associates with plasma LP (19). Treatment with 4-aminopyrolo-(3,4-D)pyrimide (4APP), a reagent that prevents secretion of LP by the liver, resulted in a substantial decrease of serum cholesterol and triglyceride levels, and in these circumstances only a minor fraction (16.9%) of LPS is associated with LP. The hypolipemic animals had much higher peak TNF-α levels when challenged with LPS, and this was associated with a significant increase in mortality, but when exogenous LP (prepared by density centrifugation of normal human plasma) were administered to these hypolipemic animals, this increase of mortality was completely prevented. Similar results were obtained when serum lipids were lowered using estradiol (19). These data indicate that LP, at physiological serum concentrations, bind and inactivate endotoxin and protect against the deleterious effects of systemic endotoxemia. It should be noted that the LP composition rather than the total serum lipid level determines survival in endotoxemia. In fact, dietary induction of hypercholesterolemia in rabbits increased endotoxin-induced TNF-α production and mortality (20). Unfortunately, in the latter experiment no data on the changes of LP profiles were reported.

LIPID-BINDING PROTEINS AND LPS TRANSPORT

The biological activity of LPS is dramatically increased in the presence of plasma proteins. Although LPS in blood interacts with many different plasma proteins, lipopolysaccharide-binding protein has been shown to be necessary and sufficient for effective presentation of LPS to both membrane-associated and soluble CD14, the LPS receptor (21). In addition, in a series of elegant experiments, LBP was also demonstrated to increase the uptake by R-HDL particles of LPS derived from either LPS micelles or LPS·sCD14 complexes, and in this process LPS molecules are exchanged for phospholipids (22). The LPS-binding site on LBP has been mapped to a region spanning amino acids 91–108, and synthetic peptides that overlap this region interfere with LBP/LPS binding (23). Mutations in LBP amino acids 94/95 caused a greatly reduced capacity (up to 40% of native LBP) to bind immobilized LPS, both mutations being additive, and although retained in several single amino acid mutants, the capacity to shuttle LPS from aggregates to sCD14 was lost in the 94/95 double mutant (23). In addition, the 94/95 double mutant was unable to effectively increase transport of LPS to R-HDL particles. Because apoAI-containing LP isolated from plasma carry LBP, these particles, in contrast to R-HDL, do not require exogenous LBP in able to neutralize LPS (5). However, the ability of R-HDL to neutralize LPS in vitro (when serum or LBP are present) as well as in vivo indicates that R-HDL is able to rapidly interact with LBP present in serum.

Additional experiments by Wright's group demonstrated a role for both sCD14 and LBP in the transport of lipids (22). LBP catalyzed the binding of phosphatidyl inositol to sCD14, facilitated exchange of phospholipids into and out of sCD14, and, in combination, LBP and sCD14 remarkably increased the speed of lipid uptake by R-HDL. Because human blood contains considerable amounts of both LBP and sCD14, it is possible that these proteins normally function as fat transport proteins, the interaction with LPS being a secondary activity. What is the functional implication of LPS-sCD14 binding, when LBP can also directly present LPS to HDL? Although the precise answer to this question remains unknown, it is now known that septic humans have very low (sometimes undetectable) plasma HDL levels, and in these conditions sCD14 would be the major endotoxin "sink." Furthermore, it has been demonstrated that sCD14 increased the uptake of LPS in HDL particles and that the exit of LPS from LPS·sCD14 complexes was the rate-limiting step in this process (22). Hence, even when sufficient amounts of HDL are present, sCD14 may be the preferred LPS-removal pathway. Because uptake of LPS into LP particles occurs in exchange for (phospho)lipid molecules and the lipid portion, but not the apoA-I, is responsible for the endotoxin-neutralizing effect of HDL (5), Wurfel et al. studied the ability of various phospholipids to neutralize endotoxin (24). It was shown that particles consisting of only PC and cholesterol, in the absence of apoA-I, were able to neutralize LPS and that this effect was augmented by the addition of LBP. LBP also increased the transfer of LPS from LPS micelles into phospholipid vesicles, and this was accelerated by addition of sCD14. These findings led to the question of whether the lipid composition of the acceptor vesicle (or bilayer) would affect the efficiency of LPS transfer, and, indeed, it was shown that various classes of phospholipids and glycosphingolipids differed extensively in the ability to accept LBP-presented LPS. Interestingly, sCD14 increased the transfer of LPS to vesicles composed of some, but not all, of the tested lipids, and this effect was related to the length and composition (saturation) of the acyl chains, which determine membrane thickness and fluidity. The relevance of these findings for LPS transfer to native LP particles remains uncertain, because sCD14 did not alter the LBP-mediated uptake of LPS by LP purified from human plasma. Hence, in the presence of LBP, human LP accept LPS equally well from LPS micelles as from LPS·sCD14 complexes. Therefore, the prime function of sCD14 seems to be that of a temporary LPS sink, allowing transport of LPS to sites where LPS-accepting lipid layers (cell membranes, LP) are available.

LIPID METABOLISM IN SEPSIS

Infection and inflammation cause significant changes in lipid metabolism, which are in large part mediated by systemic release of cytokines (for review see Ref. 25). Low-dose endotoxin, TNF-α, and IL-1 increase serum triglycerides by augmenting hepatic fatty acid production and VLDL secretion and stimulating lipolysis (26–28). High-dose endotoxin and TNF-α decrease LPL activity, but this cannot be the only cause of TNF-α–induced hypertriglyceridemia, because the acute hypertriglyceridemia following TNF-α administration precedes the decrease of LPL activity in vivo (29). Recently, it has been reported that endotoxin and the combination of TNF-α and IL-1 importantly decreased the mRNA levels of apoA1 and other apolipoproteins in various extrahepatic tissues, including the intestine (30). In Syrian hamsters, endotoxin, TNF-α, and IL-1 increase the expression of leptin mRNA in adipose tissues (31). Leptin is a major regulator of food intake and body weight, and the endotoxin-induced increase of leptin mRNA and protein levels was strongly correlated to food intake. Hence, the endotoxin-induced anorexia is mediated at least in part by leptin, and decreased dietary intake contributes to changes in serum lipids. In nonprimate animals endotoxin, TNF-α, and IL-1 increase hepatic cholesterol synthesis, presumably by an effect on HMG-CoA reductase, but decrease HDL levels (32), and these effects can be partially blocked by administration of antibodies that neutralize TNF-α. The mechanism of the endotoxin-induced decrease of serum HDL levels is not yet clear and may differ between species. In primates, endotoxin, TNF-α, IL-2, but not IL-1 decrease serum cholesterol, LDL, HDL, apoB, and apoA1 (33,34). Endotoxin also causes a change in the size of circulating HDL particles, with an increase of the total mass of the smallest particles and a loss of the intermediate-size subfractions. It has been suggested that these changes are secondary to an endotoxin-related reduction of the hepatic activity of LCAT (35), but it is likely that other factors also affect the composition of HDL in sepsis. In a nonprimate model, endotoxin reduced the extrahepatic mRNA and protein levels of CETP, and because CETP activity is inversely correlated to HDL cholesterol levels, this mechanism has been proposed to preserve HDL cholesterol in sepsis (36). However, it remains to be seen whether the HDL particles that result from CETP deficiency effectively neutralize endotoxin, because the lipid composition of HDL, rather than the total HDL concentration, is likely to be more important in this respect. As has been discussed, LPS uptake by R-HDL in several models is linked to exchange for phospholipids, and CETP is known to be responsible for 50% of all PL transport in plasma. Moreover, CETP preferentially binds to the small discoid HDL particles (37) and converts the larger spherical HDL particles into the small discoid, pre-beta migrating (38), which fraction contains the most effective endotoxin acceptors.

In critically ill surgical patients the total, LDL, and HDL cholesterol levels were found to be dramatically decreased, and patients with bacterial infections had even lower HDL levels than noninfected patients (39). Addition of R-HDL dramatically reduced endotoxin-induced TNF-α production ex vivo in whole blood obtained from patients as well as normal volunteers. The plasma concentrations of various endotoxin-binding proteins also change in sepsis. In seriously ill patients with gram-negative sepsis, serum BPI and LBP levels were both significantly increased, but no difference was detected between survivors and nonsurvivors (40). In contrast, the soluble sCD14 levels were not elevated in septic patients when compared to normals.

ENDOTOXIN-NEUTRALIZING EFFECTS OF RECONSTITUTED HDL PARTICLES

Investigators at the Rogosin institute in New York and the Swiss Red Cross in Bern, Switzerland, have pursued the idea of therapeutic use of reconstituted HDL particles in gram-negative sepsis. Most of the work has been performed using HDL-like particles containing apoA-I, phospholipids, and, in some instances, cholate. The R-HDL was prepared by HDL precipitation from human plasma, various protein extraction procedures yielding "apoHDL," and subsequent reconstitution with phosphatidyl choline (PC) by cholate dialysis. Thus, prepared R-HDL particles very effectively and dose-dependently neutralized both smooth and rough LPS, as indicated by a reduction of TNF-α production in human whole blood (41). R-HDL particles were found to be more effective than native HDL, and the efficacy of the preparation was strongly related to the concentration of added PC. Likewise, a similar, but differently prepared, R-HDL reduced TNF-α production in human whole blood by several orders of magnitude and, after infusion into rabbits, diminished TNF-α production following ex vivo stimulation of blood by LPS (42). In this model, the efficacy of R-HDL in reducing LPS-induced cytokine release was directly related to the amount of PC in the reconstituted particle (43). The cytokine response in experimental endotoxemia in rabbits was significantly blunted by R-HDL when prophylactically administered, and this was associated with prevention of LPS-induced hypotension. In a canine model of gram-negative sepsis, the systemic levels of endotoxin and TNF-α were reduced by R-HDL treatment, but the animals treated with control lipids or albumin had a significantly better survival (44). Subsequent studies demonstrated that hepato- and neurotoxicity likely caused the adverse effects of human R-HDL in canine sepsis. Transient increases of liver transaminase levels have been observed following infusion of R-HDL in rats, rabbits, and pigs, but not in human volunteers, and seem to be related to dose and infusion rate. It should be noted that the R-HDL dose used in the canine model (500 mg/kg) was much higher than that used in other studies (<75 mg/kg). The therapeutic potential of R-HDL was also tested in gram-negative (*Escherichia coli*) bacteremia in rabbits (45).

Prophylactic R-HDL administration had no effect on the number of circulating bacteria, but reduced the amount of circulating endotoxin, as well as the systemic TNF-α release, and prevented metabolic acidosis and hypotension. R-HDL therapy, when initiated one hour after induction of bacteremia, did not alter the TNF-α peak level, but caused a more rapid reduction of blood LPS levels and prevented metabolic acidosis. A nonsignificant trend towards higher mean arterial blood pressures was demonstrated in R-HDL–treated rabbits, but no data on survival where reported. In this model, R-HDL had no effect on sepsis induced by *Staphylococcus aureus*. The tolerability and pharmacokinetics of R-HDL were investigated in a model of hemorrhagic shock with resuscitation in pigs (46). Reconstituted HDL was well tolerated and did not (adversely) change the physiological responses to shock. The calculated volume of distribution for apoA-I in pigs (1.39 ± 0.08 L) indicated that R-HDL remained confined to the vascular space, and the half-life was 24.5 ± 5.3 hours. Finally, the anti-inflammatory effects of R-HDL were investigated in a model of low-dose endotoxemia in volunteers (47). The study was designed so that the volunteers were challenged on two occasions by intravenous injection of *E. coli* endotoxin (4 ng/kg). On one of these occasions the volunteers were pretreated with R-HDL (total dose 40 mg/kg, infused over a 4-hour period starting 3.5 hours before the endotoxin injection), and on the other occasion placebo was administered. R-HDL pretreatment increased the amount of Limulus-reactive LPS in blood, but potently reduced the release of pro-inflammatory cytokines, prevented upregulation of CD11b/CD18 by granulocytes, and attenuated LPS-induced intravascular coagulation (47,48). These data indicate that in humans, R-HDL bound and subsequently neutralized endotoxin within the vascular space. In these experiments the half-life of infused HDL (as measured by HDL cholesterol or the increase of apoA-I) was much longer than the increase of serum phospholipids (hours) and far exceeded the period that circulating LPS was detected using the Limulus test (D. Pajkrt and S. Van Deventer, unpublished data). These data indicate that PL within the infused R-HDL was extensively exchanged and suggest that the fate of the subfraction of LPS-containing R-HDL particles may significantly differ from the bulk of the infused R-HDL particles. It remains uncertain whether the altered half-life is a consequence of the interaction of LPS with R-HDL per se or whether the R-HDL preparation used was heterogeneous, only a minor fraction being important for the neutralizing capacity. Alternatively, endotoxin may become undetectable in the Limulus test after incorporation within the lipid fraction of the HDL particle.

Apart from neutralizing endotoxin, HDL may exert anti-inflammatory effects through other mechanisms. For example, HDL isolated from human plasma by density centrifugation, as well as R-HDL, reduced the TNF-α and IL-1–induced expression of VCAM-1, E-selectin, and ICAM-1 by endothelial cells by reducing the transcription of these genes (49). R-HDL also reduced the capability of peripheral blood mononuclear cells to respond to LPS, by a reduction of CD14 expression (47). The mechanism of R-HDL–induced CD14 expression remains unknown, but it is tempting to speculate that CD14 was internalized following binding of R-HDL–delivered phospholipids.

THERAPEUTIC EFFECTS OF TRIGLYCERIDE-RICH PARTICLES

Triglyceride-rich particles can interact with LPS and have been reported to protect against the lethality of experimental endotoxemia, but the mechanism of protection seems to differ from that of R-HDL. Preincubation with chylomicrons increased the clearance of radiolabeled LPS in rats and doubled LPS uptake by the liver (50). Interestingly, after infusion of chylomicrons, the distribution of the LPS within the liver shifted from the hepatic macrophages towards the hepatocytes. Pretreatment with chylomicrons reduced the LPS-induced TNF-α serum levels in these experiments and diminished mortality. In subsequent studies chylomicrons were shown to prevent death from polymicrobial peritonitis induced by cecal ligation and puncture and to improve survival even when administered up to 30 minutes after LPS infusion in rats (51,52). Recently, "recombinant" chylomicrons, prepared from apoE and a lipid emulsion consisting of mainly triolein and PC, were tested in experimental endotoxemia in rats (53). These particles bound LPS in an apoE-dependent manner and *reduced* the uptake of radiolabeled LPS in the liver. It is possible that the apparent difference in the effect of native and recombinant chylomicrons on the rate of endotoxin uptake by the liver may have been a consequence of the endotoxin used in the experiments (smooth versus rough) rather than LP. Nonetheless, a significant amount of LPS was still taken up by the liver after treatment with recombinant chylomicrons, and, analogous to the effect of native chylomicrons, a shift of LPS uptake away from Kupffer cells towards parenchymal cells was observed. Hence, triglyceride-rich apoE-containing particles alter the fate of endotoxin in vivo, by shifting the uptake from Kupffer cells towards hepatic parenchymal cells, thereby reducing the release of pro-inflammatory cytokines.

CONCLUSIONS

The biological effects of endotoxin are importantly altered by interactions with lipid-binding proteins and LP. LBP and BPI, first identified as endotoxin-binding proteins, can also bind phospholipids, and conversely, LBP may primarily function as a lipid transporter. Soluble CD14 is able to bind both phospholipids and endotoxin, and it may serve to shuttle endotoxin towards membranes (LP, specialized cells) that are able to exchange LPS for the appropriate (phospho)lipids. Several other enzymes and lipid-binding proteins, including lipoprotein lipase, LCAT, and CETP, do not interact with endotoxin directly, but importantly alter the composition of (high-density) LP. These alterations, in particular when altering the lipid composition of HDL particles, importantly influence endotoxin uptake and may also alter the subsequent LP-dependent tissue distribution of endotoxin. Several observations indicate that endotoxin preferentially binds to a small discoid HDL particle, and it is tempting to speculate that the same particle also carries LBP. However, the precise nature of the endotoxin-binding particle within the HDL fraction has not yet been characterized. It remains unknown to what extent the changes in lipid metabolism that occur in patients with severe sepsis will affect the efficacy of therapeutic administration of R-HDL, and a more detailed knowledge of the fate of endotoxin-carrying LP is required. For example, the activities of both CETP and LCAT, which are altered in sepsis, importantly change the lipid composition of HDL. Therefore, successful development of LP administration as a therapeutic tool in sepsis will require detailed knowledge of the importance of lipid and protein composition of the LPS-binding fraction and of the metabolic route of LP particles following uptake of endotoxin.

REFERENCES

1. Tall A. Plasma lipid transfer proteins. Ann Rev Biochem 1995; 64:235–257.
2. Tobias PS, Mathison JC, Ulevitch RJ. A family of lipopolysaccharide binding proteins involved in responses to gram-negative sepsis. J Biol Chem 1988; 263:13479–13481.
3. Day JR, Albers JJ, Lofton-Day CE, et al. Complete cDNA encoding human phospholipid transfer protein from human endothelial cells. J Biol Chem 1994; 269: 9388–9391.
4. Beamer LJ, Carroll SF, Eisenberg D. Crystal structure of human Bpi and two bound phospholipids at 2.4 angstrom resolution. Science 1997; 276:1861–1864.
5. Wurfel MM, Kunitake ST, Lichenstein H, Kane JP, Wright SD. Lipopolysaccharide (LPS)-binding protein is carried on lipoproteins and acts as a cofactor in the neutralization of LPS. J Exp Med 1994; 180:1025–1035.
6. Park CT, Wright SD. Plasma lipopolysaccharide-binding protein is found associated with a particle containing apoliproprotein A-I, phospholipid, and factor H-related proteins. J Biol Chem 1996; 271:18054–18060.
7. Massamiri T, Tobias PS, Curtiss LK. Structural determinants for the interaction of lipopolysaccharide binding protein with purified high density lipoproteins—role of apolipoprotein a-I. J Lipid Res 1997; 38:516–525.
8. Skarnes RC. In vivo interaction of endotoxin with a plasma lipoprotein having esterase activity. J Bacteriol 1968; 95:2031–2034.
9. Ulevitch RJ, Johnston AR, Weinstein DB. New function for high density lipoproteins. Their participation in intravascular reactions of bacterial lipopolysaccharides. J Clin Invest 1979; 64:1516–1524.
10. Munford RS, Hall CL, Dietschy JM. Binding of Salmonella typhimurium lipopolysaccharides to rat high-density lipoproteins. Infect Immun 1981; 34:835–843.
11. Warren HS, Knights CV, Siber GR. Neutralization and lipoprotein binding of lipopolysaccharides in tolerant rabbit serum. J Infect Dis 1986; 154:784–791.
12. Tobias PS, McAdam KP, Ulevitch RJ. Interactions of bacterial lipopolysaccharide with acute-phase rabbit serum and isolation of two forms of rabbit serum amyloid A. J Immunol 1982; 128:1420–1427.
13. Konig V, Hopf U, Moller B, et al. The significance of high-density lipoproteins (HDL) in the clearance of intravenously administered bacterial lipopolysaccharides (LPS) in mice. Hepato-Gastroenterology 1988; 35: 111–115.
14. Munford RS, Dietschy JM. Effects of specific antibodies, hormones, and lipoproteins on bacterial lipopolysaccharides injected into the rat. J Infect Dis 1985; 152: 177–184.
15. Cavaillon JM, Fitting C, Haeffner-Cavaillon N, Kirsch SJ, Warren HS. Cytokine response by monocytes and macrophages to free and lipoprotein-bound lipopolysaccharide. Infect Immun 1990; 58:2375–2382.
16. Baumberger C, Ulevitch RJ, Dayer JM. Modulation of endotoxic activity of lipopolysaccharide by high-density lipoprotein. Pathobiology 1991; 59:378–383.
17. Vosbeck K, Tobias P, Mueller H, et al. Priming of polymorphonuclear granulocytes by lipopolysaccharides and its complexes with lipopolysaccharide binding protein and high density lipoprotein. J Leuk Biol 1990; 47: 97–104.
18. Emancipator K, Csako G, Elin RJ. In vitro inactivation of bacterial endotoxin by human lipoproteins and apolipoproteins. Infect Immun 1992; 60:596–601.

19. Feingold KR, Funk JL, Moser AH, Shigenaga JK, Rapp JH, Grunfeld C. Role for circulating lipoproteins in protection from endotoxin toxicity. Infect Immun 1995; 63:2041–2046.
20. Brito BE, Romano EL, Grunfeld C. Increased lipopolysaccharide-induced tumour necrosis factor levels and death in hypercholesterolaemic rabbits. Clin Exp Immunol 1995; 101:357–361.
21. Hailman E, Lichenstein HS, Wurfel MM, et al. Lipopolysaccharide (LPS)-binding protein accelerates the binding of LPS to CD14. J Exp Med 1994; 179:269–277.
22. Yu B, Hailman E, Wright SD. Lipopolysaccharide binding protein and soluble CD14 catalyze exchange of phospholipids. J Clin Invest 1997; 99:315–324.
23. Lamping N, Hoess A, Yu B, et al. Effects of site-directed mutagenesis of basic residues (Arg 94, Lys 95, Lys 99) of lipopolysaccharide (LPS)-binding protein on binding and transfer of LPS and subsequent immune cell activation. J Immunol 1996; 157:4648–4656.
24. Wurfel MM, Wright SD. Lipopolysaccharide-binding protein and soluble cd14 transfer lipopolysaccharide to phospholipid bilayers—preferential interaction with particular classes of lipid. J Immunol 1997; 158:3925–3934.
25. Hardardottir I, Grunfeld C, Feingold KR. Effects of endotoxin and cytokines on lipid metabolism [see comments]. Curr Opin Lipidol 1994; 5:207–215.
26. Chajek-Shaul T, Friedman G, Stein O, Shiloni E, Etienne J, Stein Y. Mechanism of the hypertriglyceridemia induced by tumor necrosis factor administration to rats. Biochim Biophys Acta 1989; 1001:316–324.
27. Feingold KR, Serio MK, Adi S, Moser AH, Grunfeld C. Tumor necrosis factor stimulates hepatic lipid synthesis and secretion. Endocrinology 1989; 124:2336–2342.
28. Feingold KR, Adi S, Staprans I, et al. Diet affects the mechanisms by which TNF stimulates hepatic triglyceride production. Am J Physiol 1990; 259:E177–E184.
29. Semb H, Peterson J, Tavernier J, Olivecrona T. Multiple effects of tumor necrosis factor on lipoprotein lipase in vivo. J Biol Chem 1987; 262:8390–8394.
30. Hardardottir I, Sipe J, Moser AH, Fielding CJ, Feingold KR, Grunfeld C. Lps and cytokines regulate extra hepatic mrna levels of apolipoproteins during the acute phase response in syrian hamsters. Biochim Biophys Acta Lipids Lipid Metab 1997; 1344:210–220.
31. Grunfeld C, Zhao C, Fuller J, et al. Endotoxin and cytokines induce expression of leptin, the ob gene product, in hamsters. J Clin Invest 1996; 97:2152–2157.
32. Feingold KR, Hardardottir I, Memon R, et al. Effect of endotoxin on cholesterol biosynthesis and distribution in serum lipoproteins in Syrian hamsters. J Lipid Res 1993; 34:2147–2158.
33. Malmendier CL, Lontie JF, Sculier JP, Dubois DY. Modifications of plasma lipids, lipoproteins and apolipoproteins in advanced cancer patients treated with recombinant interleukin-2 and autologous lymphokine-activated killer cells. Atherosclerosis 1988; 73:173–180.
34. Ettinger WH, Miller LA, Smith TK, Parks JS. Effect of interleukin-1 alpha on lipoprotein lipids in cynomolgus monkeys: comparison to tumor necrosis factor. Biochim Biophys Acta 1992; 1128:186–192.
35. Auerbach BJ, Parks JS. Lipoprotein abnormalities associated with lipopolysaccharide-induced lecithin: cholesterol acyltransferase and lipase deficiency. J Biol Chem 1989; 264:10264–10270.
36. Hardardottir I, Moser AH, Fuller J, Fielding C, Feingold K, Grunfeld C. Endotoxin and cytokines decrease serum levels and extra hepatic protein and mRNA levels of cholesteryl ester transfer protein in syrian hamsters. J Clin Invest 1996; 97:2585–2592.
37. Bruce C, Davidson WS, Kussie P, et al. Molecular determinants of plasma cholesteryl ester transfer protein binding to high density lipoproteins. J Biol Chem 1995; 270:11532–11542.
38. Kunitake ST, Mendel CM, Hennessy LK. Interconversion between apolipoprotein A-I-containing lipoproteins of pre-beta and alpha electrophoretic mobilities. J Lipid Res 1992; 33:1807–1816.
39. Gordon BR, Parker TS, Levine DM, et al. Low lipid concentrations in critical illness: implications for preventing and treating endotoxemia. Crit Care Med 1996; 24:584–589.
40. Von der Mohlen M, Wortel C, Obradov D, et al. Endotoxin-binding proteins in patients with severe gram-negative sepsis. In: Faist E, ed. Immune Consequences of Trauma, Shock and Sepsis: Mechanisms and Therapeutic Approaches. Vol. 1. Heidelberg: Springer-Verlag, 1990.
41. Parker TS, Levine DM, Chang JC, Laxer J, Coffin CC, Rubin AL. Reconstituted high-density lipoprotein neutralizes gram-negative bacterial lipopolysaccharides in human whole blood. Infect Immun 1995; 63:253–258.
42. Casas AT, Hubsch AP, Rogers BC, Doran JE. Reconstituted high-density lipoprotein reduces LPS-stimulated TNF alpha. J Surg Res 1995; 59:544–552.
43. Cue JI, DiPiro JT, Brunner LJ, et al. Reconstituted high density lipoprotein inhibits physiologic and tumor necrosis factor alpha responses to lipopolysaccharide in rabbits. Arch Surg 1994; 129:193–197.
44. Quezado ZM, Natanson C, Banks SM, et al. Therapeutic trial of reconstituted human high-density lipoprotein in a canine model of gram-negative septic shock. J Pharmacol Exp Ther 1995; 272:604–611.
45. Hubsch AP, Casas AT, Doran JE. Protective effects of reconstituted high-density lipoprotein in rabbit gram-negative bacteremia models. J Lab Clin Med 1995; 126:548–558.
46. DiPiro JT, Cue JI, Richards CS, Hawkins ML, Doran JE, Mansberger AR. Pharmacokinetics of reconstituted human high-density lipoprotein in pigs after hemorrhagic shock with resuscitation. Crit Care Med 1996; 24:440–444.
47. Pajkrt D, Doran JE, Koster F, et al. Antiinflammatory effects of reconstituted high-density lipoprotein during human endotoxemia. J Exp Med 1996; 184:1601–1608.
48. Pajkrt D, Lerch PG, van der Poll T, et al. Differential effects of reconstituted high-density lipoprotein on co-

agulation, fibrinolysis and platelet activation during human endotoxemia. Thrombosis Haemostasis 1997; 77: 303–307.

49. Cockerill GW, Rye KA, Gamble JR, Vadas MA, Barter PJ. High-density lipoproteins inhibit cytokine-induced expression of endothelial cell adhesion molecules. Arteriosclerosis Thromb Vasc Biol 1995; 15:1987–1994.

50. Harris HW, Grunfeld C, Feingold KR, et al. Chylomicrons alter the fate of endotoxin, decreasing tumor necrosis factor release and preventing death. J Clin Invest 1993; 91:1028–1034.

51. Read TE, Grunfeld C, Kumwenda ZL, et al. Triglyceride-rich lipoproteins prevent septic death in rats. J Exp Med 1995; 182:267–272.

52. Read TE, Grunfeld C, Kumwenda Z, et al. Triglyceride-rich lipoproteins improve survival when given after endotoxin in rats. Surgery 1995; 117:62–67.

53. Rensen PC, Oosten M, Bilt E, Eck M, Kuiper J, Berkel TJ. Human recombinant apolipoprotein E redirects lipopolysaccharide from Kupffer cells to liver parenchymal cells in rats In vivo. J Clin Invest 1997; 99:2438–2445.

22

Effects of Human Hemoglobin on Bacterial Endotoxin In Vitro and In Vivo

Robert J. Roth and Jack Levin
University of California School of Medicine, and the Department of Veterans Affairs Medical Center, San Francisco, California

INTRODUCTION

Toxicities of hemoglobin (Hb) solutions, described in numerous animal resuscitation models, have prominently included fever, hypertension, thrombocytopenia, activation of the complement and coagulation cascades, disseminated intravascular coagulation with parenchymal organ damage, vasculitis with resultant hemorrhagic lesions, reduced tolerance to sepsis, susceptibility to bacterial infections, reticuloendothelial cell blockade, and lethal toxicity (1–13). Many of these effects (most of which were reported prior to 1991) may have been due to impure or bacteriologically contaminated preparations of Hb. Recent clinical trials of cross-linked Hb have not demonstrated most of these toxicities, but have been associated with production of hypertension and gastrointestinal dysmotility.

It was recently demonstrated that injection of nonlethal doses of gram-negative bacteria into mice produced 50% and 100% mortality when the animals had been preinfused with either native or cross-linked preparations of cell-free Hb, respectively (14). Mortality was lessened when sepsis was not produced until 3 hours after administration of either Hb. However, these results were not confirmed in apparently similar models performed with either mice (15) or rats (16). In vitro, Hb has been shown to stimulate tissue factor production by mononuclear cells (17), to cause endothelial cell injury (18), and to activate complement (17). These in vivo and in vitro effects are characteristic of bacterial endotoxins (lipopolysaccharide, LPS). Investigations of the possibility that LPS may contribute to the observed side effects of Hb infusions have been a major focus of our laboratory during the past several years, and a significant role for LPS in Hb toxicity has been suggested by our studies.

One of the most critical aspects of LPS toxicity is the high in vivo potency of LPS, even at very low concentrations (pg/ml). LPS is a potentially ubiquitous contaminant during the preparation of Hb-based resuscitation fluids, and even low levels of LPS contamination become a major clinical concern when large volumes of resuscitation solutions are required for infusion. In addition, physiologically significant levels of LPS are present in the circulating blood in a variety of clinical conditions, including sepsis, hepatic injury, hypotension, and damage to the gastrointestinal tract. Because many clinical circumstances for which Hb-based resuscitation fluids would be administered are likely to be associated with shock and hypoxia (pathological states that lead to deterioration of mucosal barriers and hepatic function), significant concentrations of endotoxin would be expected to be potentially present in the circulation of many patients. Since there is increasing evidence that cell-free Hb and LPS synergistically produce toxicities, the infusion of Hb-based resuscitation fluids may potentiate the toxicity of preexisting endotoxemia (or of endotoxemia that subsequently occurs when Hb remains present in the plasma), thus compounding the problem of the high intrinsic biological potency of LPS. The co-infusion of LPS and Hb into rabbits activated blood coagulation and produced a marked increase in mortality (50–100%) compared to the toxicity of LPS or cell-free Hb alone (only 0–10%) (19). A similar increase in mortality was produced in rabbits by the intravenous administration of LPS and αα cross-linked Hb (20). We have shown (following sections) that LPS clearance in vivo is retarded in the presence of hemoglobinemia. LPS biological effects in vitro, such as activation of coagulation mechanisms (both the direct activation of coagulation cascades and the production of monocyte- and endothelial cell–derived procoagulant activity), can be enhanced up to 100-fold by cell-free Hb. Furthermore, rates of Hb oxidation to methemoglobin and hemichromes are dramatically increased in the presence of LPS. Thus, the binding of cell-free Hb to LPS produces complexes that result in both enhancement of the biological activities of LPS and degradation of Hb.

Investigations in our laboratory with Hb solutions, including both native human HbA_0 and cross-linked Hb (human Hb, αα cross-linked using bis(3,5-dibromosalicyl) fumarate [DBBF]; and bovine Hb, fumaryl ββ cross-linked), have led to an understanding of the complex contributions of LPS to the observed toxicities of Hb solutions. Initial experiments suggested the possibility that Hb was a previously unrecognized LPS-binding protein. Subsequently, detailed experiments documented the formation of Hb-LPS complexes, characterized the complexes, and identified consequences of the LPS-Hb interaction that might contribute to toxicity.

DEMONSTRATION THAT Hb IS AN LPS-BINDING PROTEIN

An extensive series of experimental approaches have been utilized to document that mixtures of LPS and Hb produce stable complexes (21). In all experiments, equivalent results were obtained using either purified native, unmodified human HbA_0 or cross-linked Hb prepared as a potential red cell substitute. Direct evidence of saturable binding of LPS to immobilized Hb was obtained (Fig. 1). The calculated K_d (4.7×10^{-4} g/liter [3.1×10^{-8} M, assuming a monomer molecular mass of 1.5×10^4 for *E. coli* LPS] based on the microliter plate binding assay and 6.3×10^{-4} g/liter based on a sucrose centrifugation assay) indicated that the interaction Hb and LPS is of moderate affinity. Complex formation was demonstrated by affinity-labeling of Hb with a photoactivatable form of LPS (Fig. 2). Using density gradient centrifugation, co-migration of LPS with Hb was shown, and it was demonstrated that

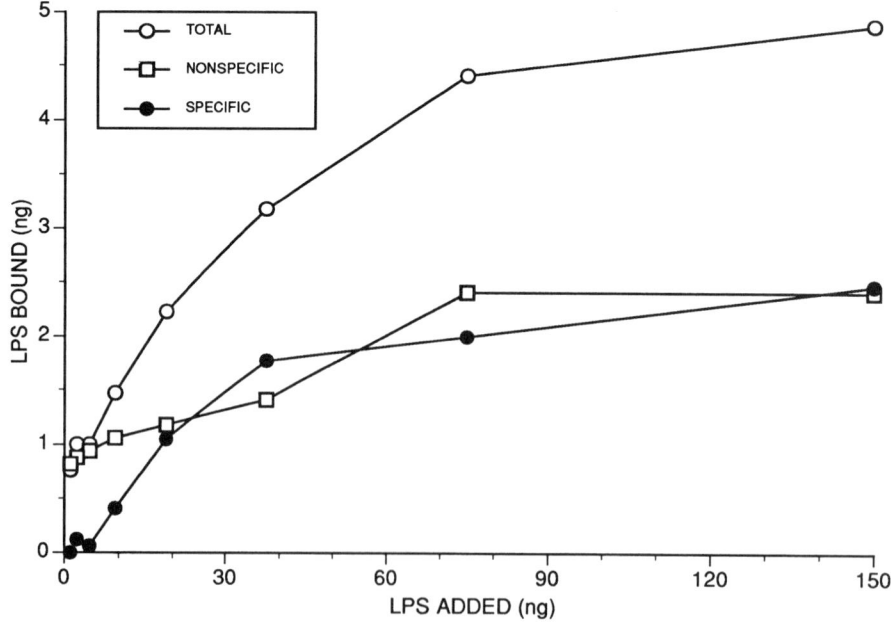

Fig. 1 Binding of LPS to immobilized Hb. ααHb (1 mg/well) was immobilized in microtiter plate wells, and ^{125}I-LPS was added. Bound LPS was determined by gamma counting, and specific binding was calculated by subtracting bound ^{125}I-LPS in wells without Hb. (From Ref. 21.)

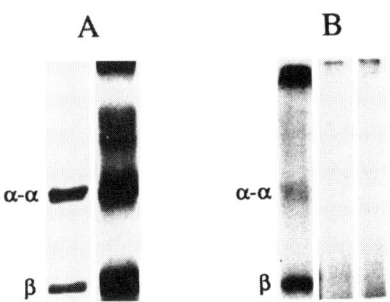

Fig. 2 Photoaffinity labeling of Hb with ^{125}I-LPS-ASD. ^{125}I-LPS-ASD was incubated with ααHb, photolyzed with UV light, and electrophoresed in SDS and 2-mercaptoethanol. Following electrophoresis, the gel was stained with Coomassie blue (A, left lane), dried, and subjected to autoradiography (A, right lane). Another photoaffinity-labeled ααHb preparation from a separate experiment is shown (B, left lane), along with controls consisting of an incubation mixture containing 100-fold excess unlabeled LPS as a blocking agent to demonstrate inhibition of specific binding (B, middle lane) and ^{125}I-LPS-ASD alone (B, right lane). LPS-ASD is *S. minnesota* Re595 LPS-(*p*-azidosalicylamido)-1,3′-dithioproprionamide. (From Ref. 21.)

the sedimentation velocity of LPS was decreased in the presence of Hb preparations. This indicated that there had been disaggregation of LPS and formation of lower-density Hb-LPS complexes. Additional evidence of LPS dissociation was obtained by nondenaturing polyacrylamide gel electrophoresis, which demonstrated that LPS, when complexed with Hb, entered the gel and co-migrated with Hb, whereas LPS alone remained within the stacking gel. Ultrafiltration experiments demonstrated that LPS, which alone in aqueous solutions has a very high molecular weight (typically $\geq 10^6$ daltons), co-filtered with Hb through 300 kDa and 100 kDa membranes. Whereas only 10–16% of LPS alone was filterable through the 300 kDa membrane and LPS alone was not filterable at all through the 100 kDa membrane, in the presence of Hb, 87–97% of LPS was filtered through the 300 kDa membrane and 64–72% through the 100 kDa membrane. Thus, these data provide further evidence that Hb greatly decreased the aggregate molecular weight of LPS.

Independent evidence that Hb is an LPS-binding protein has been provided by recent investigations that demonstrated binding of porcine Hb to the LPS of *Actinobacillus pleuropneumoniae* as well as binding to the surface of intact bacteria of this species (22). In addition, the LPS was shown to bind to both the α and β chains of porcine Hb (22), confirming our observations that LPS binds to both the α and β chains of human Hb (21).

As described above, Hb alters several characteristics of LPS. Conversely, LPS can produce Hb denaturation, with production of methemoglobin and hemichromes (Fig. 3) (23). Degradation of Hb by LPS is time dependent, with maximal production of methemoglobin and hemichromes (each approximately 30%, with 40% residual oxyHb) after 2 hours, and also is LPS-concentration dependent. There also are structural changes indicative of Hb oxidation as demonstrated by circular dichroic analysis between 210 and 600 nm. However, there is no demonstrable change in the overall tertiary structure of the globin molecule (23).

The oxygen affinity of Hb was measured in the absence and presence of LPS in order to evaluate the possible influence of LPS binding on Hb function (23). These measurements were made after a 2-hour incubation period, a time sufficient to result in Hb-LPS complex formation (24), but in these experiments prior to the formation of substantial quantities of oxidized Hb species unable to bind oxygen. One mg/ml Hb (16 μM) and 1 mg/ml of LPS were utilized because the two components of Hb-LPS complexes are of approximately equal concentration by weight and little unbound Hb is calculated to be present. P_{50} values for ααHb (26.6 mmHg) and HbA_0 (9.6 mmHg) were only slightly decreased by smooth LPS to 25.1 and 8.7 mmHg), respectively, and by rough LPS to 25.6 and 7.3 mmHg, respectively. Non–cross-linked cell-free HbA_0, which exhibited high oxygen affinity (P_{50} = 9.6 mmHg) similar to that measured with lysed whole blood (P_{50} = 10.0 mmHg), best demonstrated the small trend toward higher oxygen affinity in the presence of LPS (P_{50} = 7.3 mmHg in the presence of OH37 LPS). Overall, there is little change in oxygen affinity of Hb when complexed to LPS.

EFFECT OF Hb ON LPS CLEARANCE IN VIVO

LPS clearance in rabbits was shown to be delayed in the presence of Hb (free Hb levels were 2 g/dl, which produced a 15% increase in total circulating Hb) compared to LPS clearance in animals given equivalent doses of human serum albumin (HSA) or NaCl (25). The intravascular retention of injected ^{125}I-LPS during the 30-minute period analyzed was significantly longer in the LPS + Hb group than in the LPS + NaCl or LPS + HSA group, especially during the initial 10 minutes. The intravascular half-life ($T_{1/2}$) of LPS in the

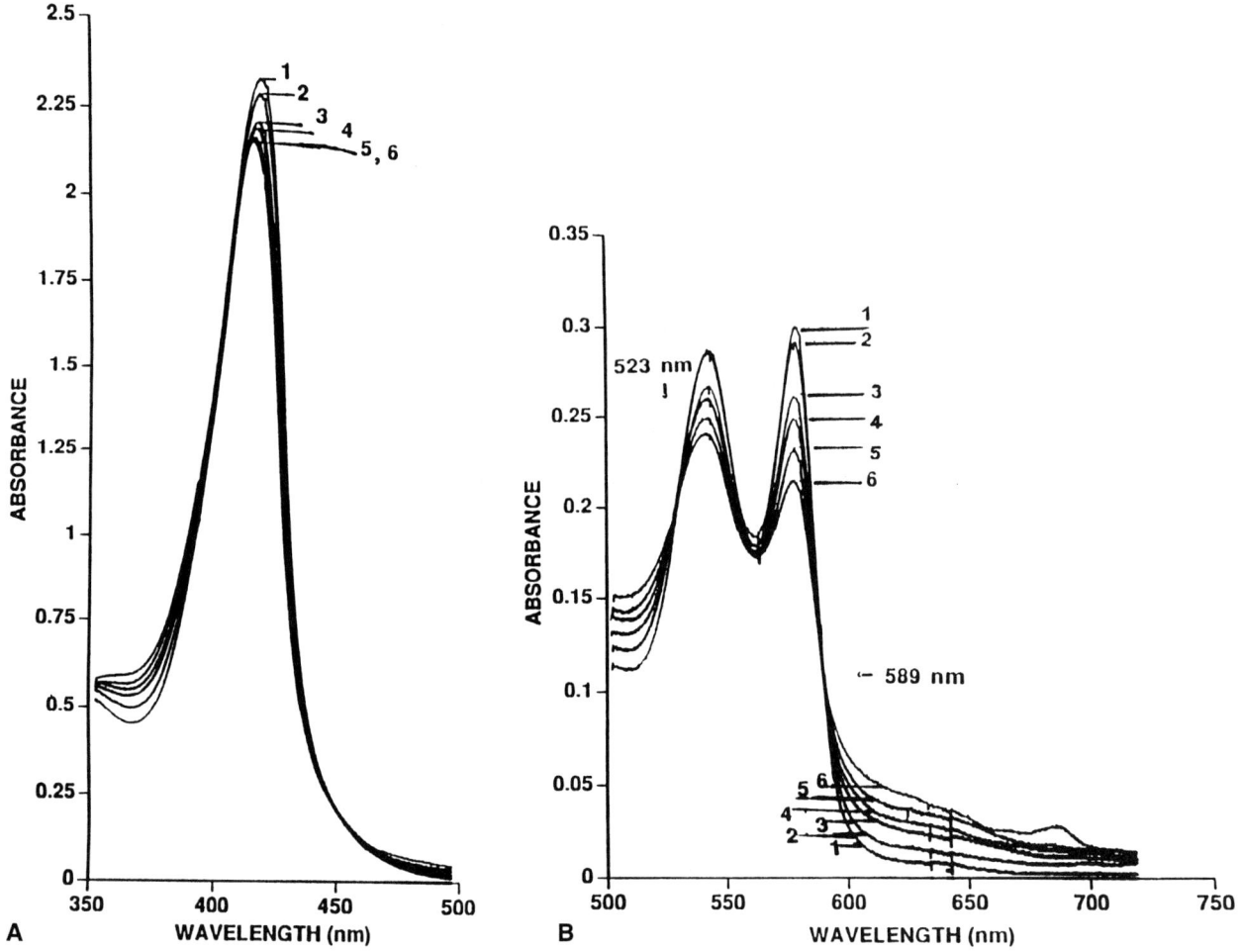

Fig. 3 Time course of changes in the hemoglobin absorption spectrum in the presence of LPS. ααHb (21 μM in PBS, pH 7.4) was incubated at 37°C in the presence of 0.3 mg/ml *S. minnesota* 595 OH37 LPS, and absorbance spectra in the Soret (A) and visible (B) regions of the Hb spectrum were obtained at various times of incubation. 1, Initial spectrum of ααHb alone; 2, 10 min; 3, 20 min; 4, 40 min; 5, 90 min; 6, 120 min. The sample cuvette contained Hb in PBS with or without LPS, and the reference cuvette contained PBS (for ααHb spectra alone) or LPS alone (0.3 mg/ml in PBS) (for ααHb-LPS mixture spectra). Spectral changes indicative of oxidation of Hb to methemoglobin and hemichrome include a decrease in the Soret peak at 414 nm, decreases in the major visible peaks at 541 nm and 577 nm, and appearance of a visible absorbance peak at 630 nm. Arrows indicate the apparent isosbestic points. (From Ref. 23.)

LPS + NaCl control, LPS + HSA control, and LPS + Hb groups was 2.8, 4.0, and 4.9 minutes; the area under the curve was 1369 ± 483, 1594 ± 360, and 1731 ± 481 (ng/ml × min, mean ± SD); and the total body clearance was 24.7 ± 9.2, 20.1 ± 5.4, and 18.9 ± 6.0 (ml/min, mean ± SD), respectively. The proportion of LPS associated with blood cells was very small at the initial 1-minute time period (3–4%) and decreased even further during the 30-minute period analyzed. Over 96% of injected LPS was associated with the cell-free plasma, with 51–54% of LPS in the apoprotein fraction at the initial time point and 35–37% in the high-density lipoprotein (HDL) fraction. The proportion of LPS increased significantly in the HDL fraction and decreased significantly in apoproteins during the period analyzed. However, there were no differences between the three groups (25), consistent with previous in vitro studies (26). Of the six organs evaluated (liver, kidney, lung, spleen, adrenal, and heart), the liver was the main distribution site (74% of injected LPS (25). In the Hb group, the accumulation of ^{125}I-LPS in the spleen was significantly lower than in the HSA group. The synergism between LPS and Hb in the reported production of in vivo toxicity may be due, in part, to the decreased rate of intravascular clearance of endotoxin.

DEMONSTRATION THAT Hb ENHANCES THE BIOLOGICAL ACTIVITY OF LPS

Limulus Amebocyte Lysate Activation

The effect of Hb on the biological activity of LPS was initially investigated using Limulus amebocyte lysate (LAL), the most sensitive in vitro assay for LPS. LAL, a preparation from the blood cells of the horseshoe crab *Limulus polyphemus*, contains an LPS-activated coagulation cascade that is sensitive to pg/ml concentrations of LPS. Each of a variety of LPS preparations obtained from *Proteus mirabilis* and spiked into solutions of LPS-free Hb demonstrated greatly increased activation of LAL (8- to 27-fold), using a chromogenic endpoint, in comparison to identical concentrations of LPS assayed in saline (Fig. 4) (24). Similar results were obtained with endotoxin obtained from *Salmonella minnesota* 595 and with purified lipid A (24). Furthermore, all three Hb preparations tested ($\alpha\alpha$Hb, HbA$_0$, and $\alpha\alpha$HbCO) produced enhanced activation of LAL over a wide range of LPS concentrations (1–1000 µg/ml). The enhancement by each Hb of LAL activation also was demonstrated with the gelation LAL test and was shown to be dependent on protein concentration, with increasing enhancement between 0.1 and 5 mg/ml Hb. LPS biological activity was enhanced >1000-fold at the concentrations of Hb that would be achieved in vivo for purposes of resuscitation. Pertinently, similar Hb concentrations have been detected in plasma following hemolysis associated with experimental endotoxemia (27). These results are of great interest because LAL activation is an excellent model for the intravascular coagulation commonly seen in humans during endotoxemia and which has been described during infusion of hemoglobin solutions in animals.

In order to further establish the generalized nature of the Hb-enhancement effect, we studied the effect of $\alpha\alpha$Hb on biological activities of several other LPSs, including LPSs from different bacterial species (24). Prominent and similar extents of enhancement by both $\alpha\alpha$Hb and $\alpha\alpha$HbCO in the LAL assay were shown with three defined salts of *E. coli* 026:B6 (smooth LPS), i.e., the calcium, sodium, and triethylamine forms, suggesting that the specific cations bound to LPS did not influence the Hb-enhancement process. Enhancement of LPS biological activity also was demonstrated with a smooth *Salmonella* LPS (*S. abortus equi*) and a rough *E. coli* LPS (Re F515), but was not

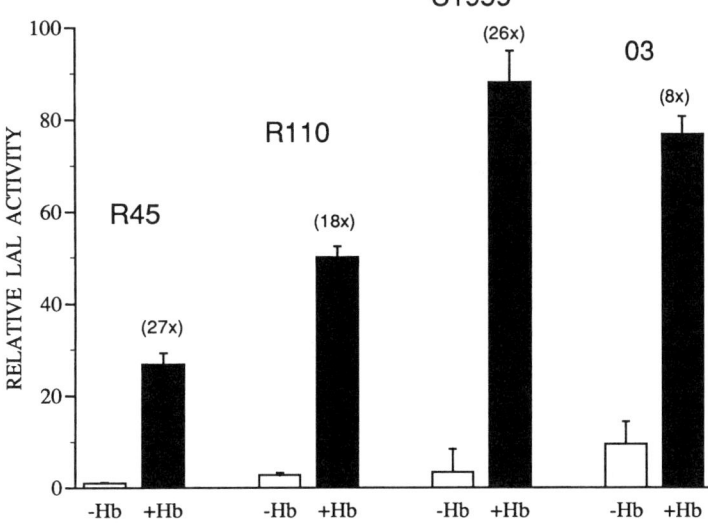

Fig. 4 Enhancement by hemoglobin of the activation of Limulus amebocyte lysate (LAL) by *Proteus* LPS. LAL reactivities of LPS (500 ng/ml) in the absence (−Hb) or presence (+Hb) of $\alpha\alpha$ cross-linked hemoglobin (1 mg/ml) were determined with the chromogenic LAL assay. To determine relative LAL activities, a standard curve of *P. mirabilis* R45 LPS was prepared, which related absorbance to LPS concentration. Using this standard curve, the absorbance for each sample (LPS alone or LPS-Hb) was converted into the equivalent R45 LPS concentration. 500 ng/ml R45 LPS was assigned a relative LAL activity of 1. The increase in LAL activity of each LPS, induced by Hb, is indicated in parentheses. Samples were assayed with eight replicates, and results are expressed as the mean ± 1 S.D. (From Ref. 24.)

observed with nontoxic *Rhodobacter spheroides*, *Rhodobacter capsulatus*, and *Rhodopseudomonas viridis* LPSs.

Many of the LPS preparations studied had poor aqueous solubility and were visually turbid (especially *S. minnesota* 595 LPS, lipid A and monophosphoryl lipid A, and *P. mirabilis* R110). Hb enhancement of LPS biological activity was a prominent feature of some of these LPSs and their partial structures, suggesting that a possible mechanism for the Hb-enhancement effect was via increased LPS solubility. Therefore, we compared turbidity and the LAL biological activity of these LPSs in the absence and presence of Hb (24). With increasing concentrations of ααHb, *P. mirabilis* R110 and *S. minnesota* 595 LPS each demonstrated a concomitant progressive decrease in turbidity and increase in LAL biological activity (Fig. 5).

To further demonstrate the effect of Hb on the physical state of LPS, electron microscopic studies were performed. In the absence of Hb, *S. minnesota* (Re) 595 LPS was highly aggregated and consisted of variable ribbon-like, mesh-like, and/or membrane-like structures, with the largest dimensions being >1 μm (Fig. 6). However, in the presence of HbA_0, marked disaggregation of all of the highly aggregated LPS structures was demonstrated, with production of discoidal 5- to 20-nm particles (Fig. 6) (28). Similar results were shown with LPS from *E. coli* (Re) F515. Recent investigations with *Actinobacillus pleuropneumoniae* LPS and porcine Hb have confirmed that Hb causes disaggregation (and decreased density) of LPS (29).

Tissue Factor Production

To further investigate the ability of Hb to modify LPS-activated coagulation, we evaluated the effect of Hb on LPS stimulation of peripheral blood mononuclear cell procoagulant activity (i.e., tissue factor, TF). This is another coagulation-based assay for LPS activity, which is quantitative (as is the LAL assay) and which is known to correlate well with LPS activity as determined by LAL. A Hb concentration–dependent enhancement of LPS-stimulated procoagulant activity in mononuclear cells was observed (Fig. 7) (30).

Since Hb has the ability to increase the production of TF by mononuclear cells, we reasoned that vascular endothelium might demonstrate a similar response. Cultured human umbilical vein endothelial cell (EC) monolayers were incubated with LPS, in the presence and absence of Hb, and the generation of EC procoagulant activity (TF) was determined. LPS alone (0.01–10 μg/ml) caused a concentration-dependent increase in production of EC TF activity, compared to the TF produced by unstimulated cells. Hb resulted in aug-

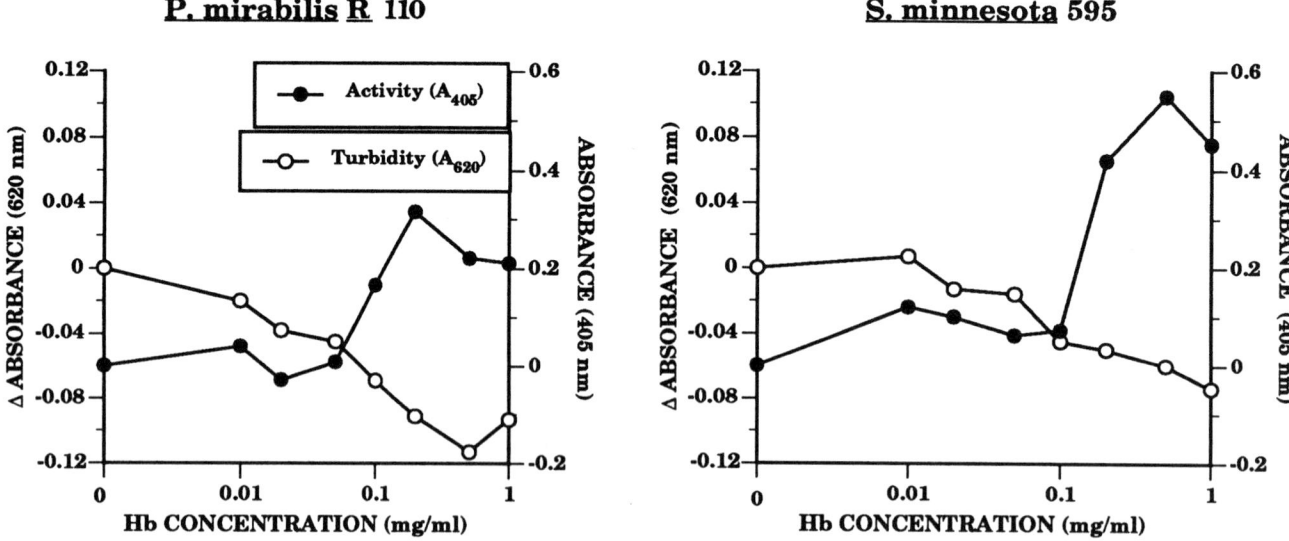

Fig. 5 Turbidity and biological activity of LPS in the absence and presence of Hb. Various concentrations of αα cross-linked Hb (from 0.01 to 1.0 mg/ml) were added to LPS (final concentration 1 mg/ml) in microtiter plate wells and absorbances were measured at 620 nm. The turbidity of each LPS (absorbance at 620 nm) in the absence of Hb has been designated as 0, and the change in absorbance induced by Hb is shown. Absorbances due to Hb have been subtracted. Actual baseline LPS absorbances were as follows: *P. mirabilis* R110, 0.21; *S. minnesota* R595, 0.12. LAL was then added to each well and chromogenic activities determined at 405 nm. (From Ref. 24.)

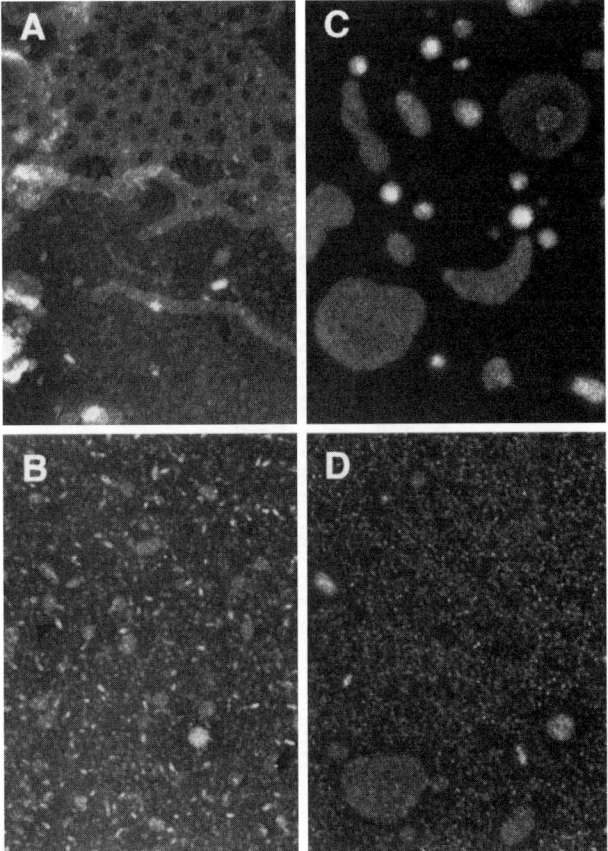

Fig. 6 Disaggregation of LPS in the presence of Hb. Highly aggregated rough *S. minnesota* (Re) 595 LPS (A) demonstrated primarily ribbon-like (arrows) and mesh-like (open arrows) structures. Following incubation at 37°C for 18 hours with HbA$_0$, there was complete disaggregation of LPS into small disc (arrow) and lens-shaped (arrow head) particles of 5–20 nm (B). Heterogeneous particles (10–100 nm) of smooth *E. coli* 055:B5 LPS (C) showed loss of most of the largest particles after incubation at 37°C for 18 hours with HbA$_0$ (D). All electron micrographs are ×180,000. (From Ref. 28.)

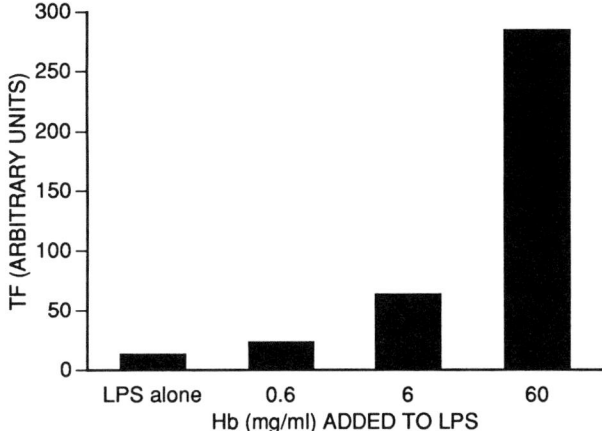

Fig. 7 Tissue factor (TF) production by human mononuclear cells. Human mixed mononuclear cells were incubated with LPS in the presence of various concentrations of endotoxin-free Hb (0.6–60 mg/ml). TF generated by LPS alone and the Hb-LPS mixtures was determined following addition of citrated plasma and calcium (plasma recalcification assay). The contribution of the Hb alone (at each concentration, respectively) to the total TF generated by the mononuclear cells was subtracted from the measured total.

mented production of TF in response to LPS (Fig. 8) (31). Enhancement was not produced by IgG or human serum albumin (32). Enhancement was demonstrated with both native HbA$_0$ and cross-linked Hbs and was shown to be concentration-dependent between 0.1 and 100 mg/ml Hb. In both the presence and absence of Hb, the production of TF activity by LPS was completely inhibited by actinomycin D or cycloheximide, indicating a requirement for new protein synthesis. Elevated levels of TF protein in response to Hb-LPS, as assessed by an ELISA assay, were also demonstrated. Inhibition of nitric oxide synthesis, using *N*-monomethyl-L-arginine (NMMA), resulted in attenuated TF production (10–80% decrease of TF) by the EC in response to both LPS alone and Hb-LPS.

A possible mechanism for the enhancement by Hb of the stimulation of LPS-induced TF production was suggested by the demonstration that Hb increased the binding of LPS to EC (up to 11-fold) (32). The increase in binding was directly related to concentrations of Hb between 0.001 and 20 mg/ml. Furthermore, the increase in binding of LPS was produced only when LPS and Hb had been incubated prior to addition to the EC culture. Increased binding was demonstrable both in serum-containing and serum-free medium, as well as in plasma. This indicated that soluble CD14 was not necessary for the binding of LPS under the conditions of these experiments. However, in the absence of serum, LPS binding to EC did not produce the biological response characterized by synthesis of TF.

Platelet Adherence to Endothelial Cells

Because of the critical role of the vascular endothelium in promoting pathological hemostatic responses to LPS in vivo (LPS transforms the endothelium from an anticoagulant surface to a procoagulant surface), we also examined whether Hb modified LPS-induced platelet adherence to EC. Cultured human EC monolayers were incubated with LPS in the presence and absence of

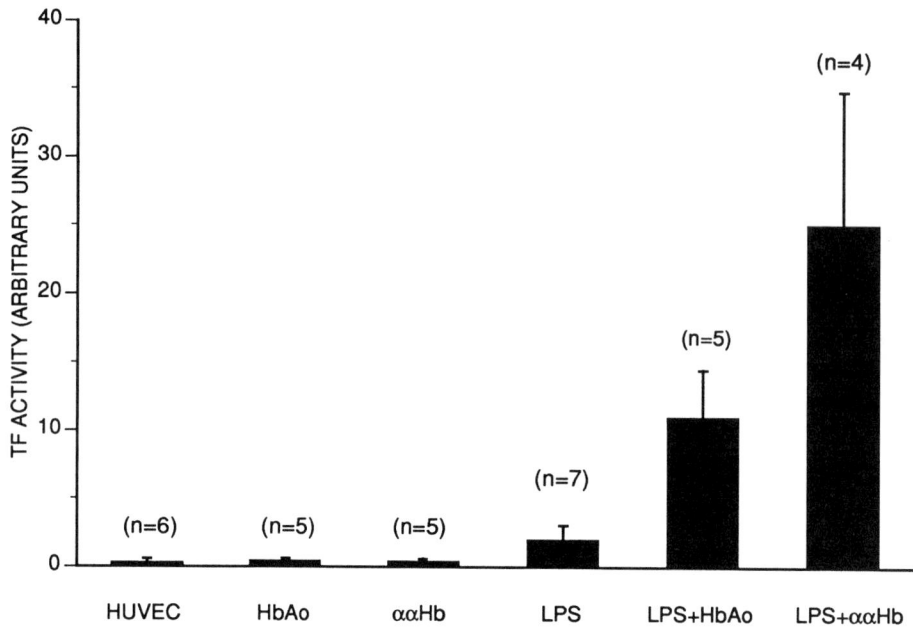

Fig. 8 Effect of Hb on the production of human umbilical vein endothelial cell (EC) tissue factor (TF) in response to LPS. Cultured human endothelial cells (EC) were incubated with HbA_0 alone, $\alpha\alpha Hb$ alone, LPS alone, or LPS in the presence of Hb. After 4 hours, the EC were washed, freeze-thawed, and sonicated, and TF activities were then determined with the plasma recalcification assay. Mean ± 1 SD TF activities from replicate wells are presented. The number of replicate wells is indicated in parentheses. (From Ref. 31.)

HbA_0, and the binding of radiolabeled human platelets was examined. LPS alone resulted in slightly increased binding of human platelets to EC in culture (20% increase compared to platelet binding in the absence of LPS), and Hb-LPS complexes further increased platelet binding to EC (35% increase compared to platelet binding in the control without LPS or Hb). Incubation of the EC with Hb alone resulted in a slight decrease in platelet binding.

Complement Activation

Enhancement by Hb of the biological activity of LPS in the activation of a proteolytic coagulation cascade in LAL suggested that there may be an impact of Hb on the ability of LPS to activate other protease cascades. We studied whether formation of Hb-LPS complexes altered the ability of LPS to activate and fix complement (a process thought to contribute to the in vivo toxicity of Hb in some animal studies). Addition of Hb had little or no effect on the intrinsic complement-fixing abilities of eight smooth endotoxins, rough endotoxins, or endotoxin partial structures (33). At concentrations of >0.2 mg/ml, Hb by itself also was capable of fixing complement, in the absence of LPS, via the classical pathway of complement activation.

Lethality in Mice

Because the extensive in vitro data we had obtained demonstrated the ability of Hb to enhance the biological activity of LPS, in vivo experiments were initiated to determine whether LPS-induced mortality was affected by the presence of hemoglobinemia. Mice were injected i.p. with *E. coli* LPS (doses ranged from 0.1 to 1 mg LPS/mouse) and 8 hours later received an i.v. infusion of Hb (60 mg) sufficient to raise the blood Hb level by 4–5.5 g/dl. At each LPS dose, LPS-induced mortality was increased by Hb infusion (Fig. 9) (34). Mortality in the Hb-treated mice was also noted many hours earlier than in mice that had received only LPS. At a given endotoxin dose, enhancement of mortality was dependent on the dose of Hb administered. In the presence of endotoxemia, Hb doses of 45 and 60 mg resulted in increased mortality, whereas doses of 6, 11, or 22 mg had no effect. Hb itself caused no mortality, and mice that received Hb alone appeared completely normal throughout the study. Furthermore, Hb increased endotoxin-related mortality in mice whether it was infused intravenously 12 hours prior to, coincident with, or 8–10 hours subsequent to intraperitoneal endotoxin injection. Increased mortality in mice that had received LPS was observed for all preparations of Hb

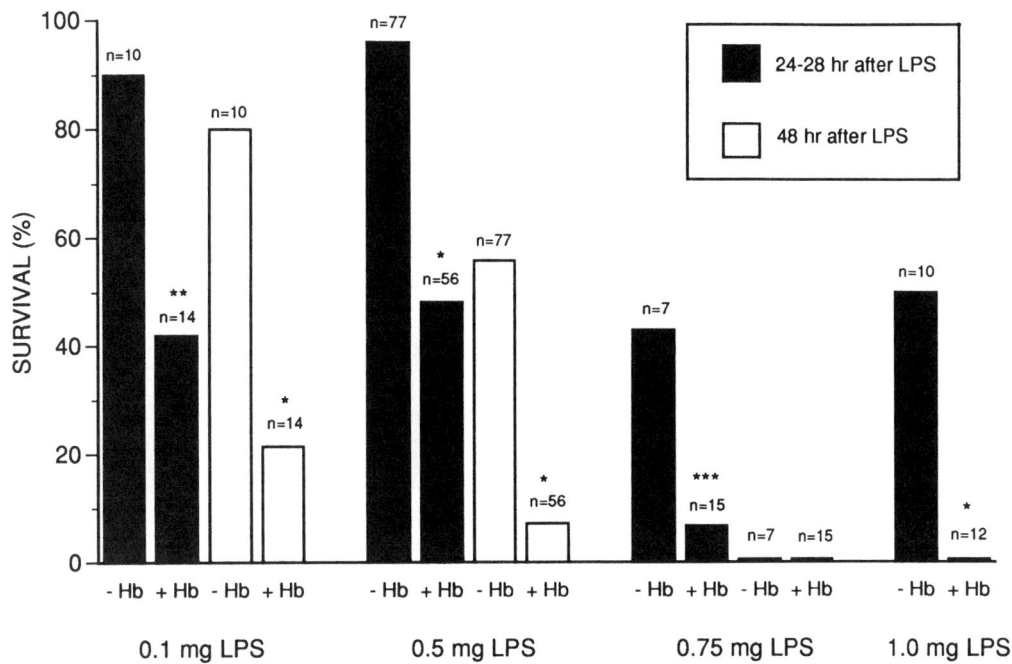

Fig. 9 Effect of Hb on mortality from LPS. Swiss Webster female mice (28–32 g) were injected intraperitoneally with 0.1, 0.5, 0.75, or 1.0 mg LPS (*E. coli* 055:B5 LPS, in sterile, pyrogen-free saline). Eight to ten hours after LPS injection, mice were infused by tail vein with either 0.6–0.8 ml saline or 0.6–0.8 ml ααHb in Ringer's acetate, pH 7.4 (60 mg/mouse). Survival was monitored at 24–28 and 48 hours after LPS injection. n = Number of mice in each group. $*p < 0.01$, +Hb vs. −Hb; $**p = 0.3$, +Hb vs. −Hb; $***p = 0.07$, +Hb vs. −Hb. (Fisher's exact p-value). (From Ref. 34.)

tested, i.e., ααHb, HbA$_0$, ββHb, ββ/ααHb, and polymerized Hb. This established that the effect of Hb on mortality was not limited to a single preparation of Hb nor was uniquely produced by the nature of the specific chemical cross-link.

A series of experiments was conducted to identify potential mechanisms for the observed Hb enhancement of LPS mortality (34). Intraperitoneal LPS generated plasma LPS levels that rose over 4–10 hours, plateaued, and then gradually decreased during the following 2 days. Plasma LPS concentrations were not altered by the subsequent intravenous administration of Hb. LPS-induced hypoglycemia (to approximately 40% of normal) also was not altered by Hb infusion. However, Hb clearance was slower ($t_{1/2} = 7.2$ hr) in animals that received intraperitoneal LPS than in control animals that did not receive LPS ($t_{1/2} = 3.7$ hr), suggesting that reticuloendothelial cell function may have been depressed by LPS. Direct assessment of reticuloendothelial cell function was accomplished by measurement of particulate carbon clearance. The combination of intraperitoneal LPS and intravenous Hb resulted in slower carbon particle clearance compared to the effects of either Hb or LPS alone, further indicating an alteration in the reticuloendothelial cell system. Our observations are consistent with the previous reports indicating that the presence of free Hb in the circulation can compromise reticuloendothelial system function and increase susceptibility to bacterial infection (13,35–40).

Tumor Necrosis Factor Production

Based on the above data that clearance properties of the reticuloendothelial cell system were altered by the combination of LPS and Hb, we considered the possibility that the responsiveness of the cytokine-producing cells of this system to LPS was modified by the infusion of Hb. Increased production of cytokines by these cells in response to LPS could potentially contribute to enhancement by Hb of LPS-induced mortality. The intravenous infusion of Hb (60 mg/mouse) either prior to or coincident with LPS resulted in peak tumor necrosis factor (TNF) concentrations in plasma approximately twice those of animals that received only LPS (Fig. 10) (40a). This suggested that Hb infusion may have primed TNF-producing cells for subsequent stimulation by LPS. Supporting this hypothesis was the finding that Kupffer cells and peripheral blood mononuclear cells, obtained from Hb-treated mice and then placed in culture, demonstrated increased sensitivity to LPS ex vivo compared to cultured cells from control, untreated mice (Fig. 11) (40a). We concluded that LPS-responsive cells became hypersensitive to stimulation by LPS as a result of Hb infusion and that such a mechanism might contribute to Hb enhancement of mortality in mice. However, the importance of the hypersensitivity of LPS-responsive cells for lethality is not entirely clear, and we (40a) and others (41,42) have noted that peak plasma TNF levels do not correlate with survival. It has been proposed that localized cytokine production may be of greater importance than circulating plasma cytokines in the pathogenesis of inflammatory disorders such as sepsis (43,44).

SUMMARY AND CONCLUSIONS

Our data strongly support the conclusion that hemoglobin is an endotoxin-binding protein and that, as a result, LPS and Hb form complexes. The interaction between LPS and Hb alters each of the components of the Hb-LPS complex. Importantly, the biological effects of LPS are enhanced and the UV spectrum of Hb is changed, consistent with methemoglobin formation and denaturation of the Hb molecule. Our observations indicate that the interaction between LPS and Hb results in marked disaggregation of the LPS macromolecule into smaller units that may approximate LPS monomers. The association between disaggregation and an increase in the biological activity of LPS is consistent with recent studies that have emphasized the relationship between the physical state of the LPS and biological activity (45–49). The spatial conformation of lipid A aggregates (e.g., lamellar vs. nonlamellar) may play an important role in increasing the biological activity of LPS in aqueous biological systems. In addition, the relative concentrations of monomeric versus aggregated forms of LPS may also influence biological activity. However, this remains a controversial issue and may depend, in part, upon the concentration of LPS, its solubility, and the biological system utilized (29,50–52).

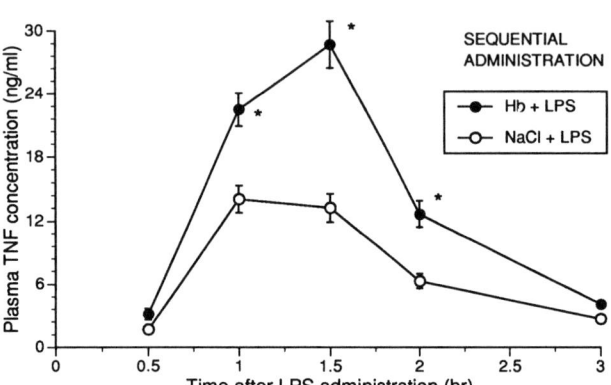

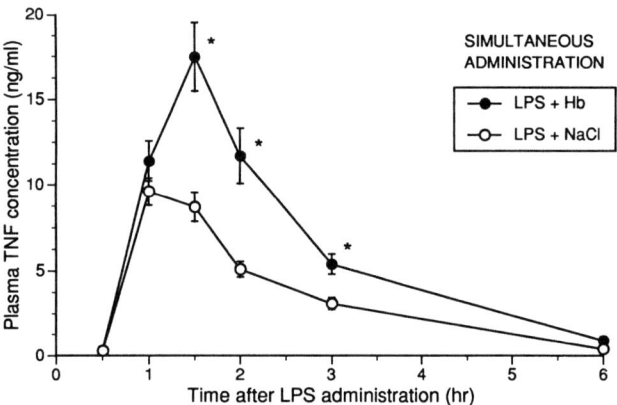

Fig. 10 Effect of Hb on induction of plasma tumor necrosis factor (TNF) by LPS. (Top) Mice were injected intravenously with Hb (60 mg/mouse) or NaCl, and 10 hours later were injected intraperitoneally with an LD_{50} dose of E. coli LPS (500 μg). Plasma TNF concentrations following LPS administration were determined by ELISA. TNF levels (mean ± SE) of 35 mice (Hb + LPS) and 20 mice (NaCl + LPS), respectively, are shown. (Bottom) Mice were injected intraperitoneally with an LD_{50} dose of LPS (500 μg), and then immediately were injected intravenously with Hb (60 mg/mouse) or NaCl. Plasma TNF concentrations following LPS administration were determined by ELISA. TNF levels (mean ± SE) of 16 mice (LPS + Hb) and 15 mice (LPS + NaCl), respectively, are shown. *$p < 0.05$ (Mann-Whitney U test). (From Ref. 40a.)

The interaction between Hb and LPS occurs with native HbA_0, ααHb, or ααHbCO, the three forms of Hb that we have investigated in vitro. Furthermore, a wide variety of LPSs have been shown to interact with Hb in the systems we have examined. Importantly, we have demonstrated that the administration of Hb also significantly increases the biological activity of LPS in vivo, as manifested by a marked increase in the mortality of mice that received both LPS and either native or cross-linked Hb. Therefore, our observations have potential relevance for the utilization of hemoglobin solutions as substitutes for red blood cells.

Our data suggest that hemoglobin-based blood substitutes, which are currently undergoing clinical trials (53,54), may intensify the potentially fatal effects of

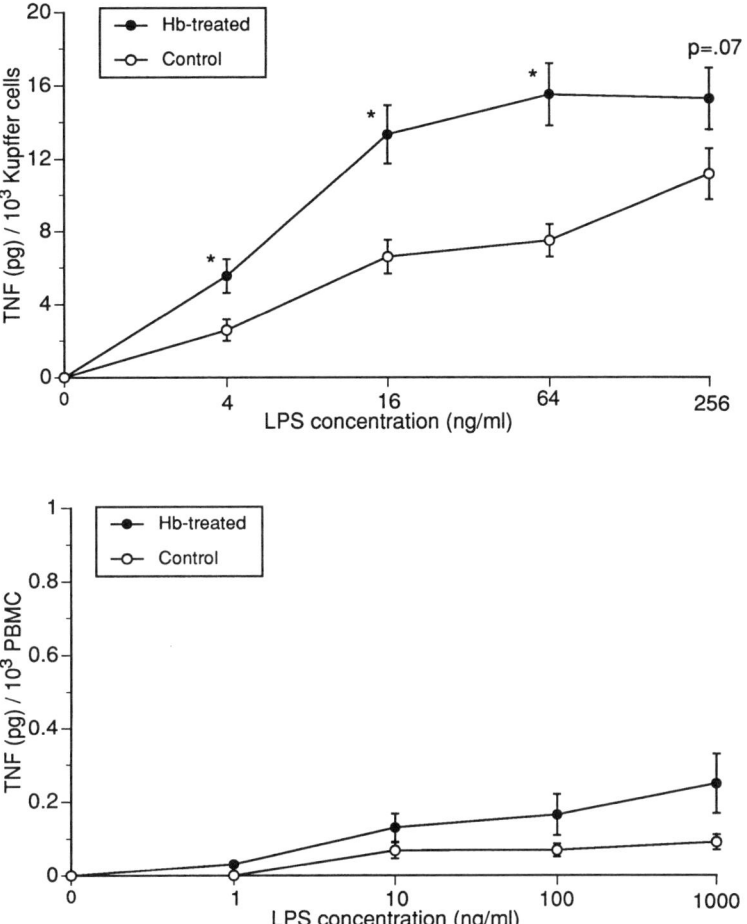

Fig. 11 Tumor necrosis factor (TNF) production by LPS-stimulated Kupffer cells and peripheral blood mononuclear cells (PBMC) obtained from Hb-infused or control mice. (Top) Mice were injected intravenously with Hb (60 mg/mouse) or NaCl. Ten hours later, pronase was injected, the liver was excised and digested with DNase and pronase, and the digested liver preparations were centrifuged on Accudenz to isolate Kupffer cells. The Kupffer cell–enriched preparations were placed in culture and stimulated with LPS for 5 hours; TNF in the culture medium was measured by ELISA. Six independent experiments were performed and the results pooled. TNF concentrations are shown (mean ± SE of 17–23 measurements at each LPS concentration). *$p < 0.05$ (Mann-Whitney U test). The Hb-treated and control groups also were significantly different by repeated measures ANOVA ($p < 0.05$). (Bottom) Mice were injected intravenously with Hb (60 mg/mouse) or NaCl. Ten hours later, PBMC were isolated, placed in culture, and stimulated with LPS for 5 hours; TNF in the culture medium was measured by ELISA. Monocytes comprised 3–7% of total PBMC. Four independent experiments were performed and the results pooled. TNF concentrations are shown (mean ± SE of eight measurements at each LPS concentration). PBMC from Hb-treated mice appeared to generate more TNF in response to LPS ex vivo than control mice, but the differences were not statistically significant. (From Ref. 40a.)

the sepsis syndrome in patients with trauma, infection, or hypotension who receive hemoglobin for red blood cell replacement. Others have also recently expressed concern about the potential danger of administration of hemoglobin-based red blood cell substitutes to patients with sepsis, ischemia, or shock (the latter two clinical conditions can predispose to the development of endotoxemia, even if endotoxin is not the precipitating cause of ischemia or hypotension) (13,55,56). Therefore, Hb should be administered to such patients with caution and thorough serial physiological observations performed in order to detect any worsening of signs or symptoms that may be attributable to endotoxemia and the sepsis syndrome.

ACKNOWLEDGMENTS

These studies were supported in part by U.S. Army Medical Research, Development, Acquisition and Logistics Command Research Contracts MIPR Nos. 90-MM0535 and 94-MM4585; Research Grant No. DK 43102 from the NIDDKD, National Institutes of Health; Grant No. 95-26 from the National Blood Foundation; and the Veterans Administration. Opinions, interpretations, conclusions, and recommendations are those of the authors and are not necessarily endorsed by the U.S. Army. In conducting research using animals, the investigators adhered to the *Guide for the Care and Use of Laboratory Animals* prepared by the Committee on Care and Use of Laboratory Animals of the Institute of Laboratory Animal Resources, National Research Council (NIH Publication No. 86-23, revised 1985).

REFERENCES

1. Brandt JL, Frank NR, Lichtman HC. The effects of hemoglobin solutions on renal function in man. Blood 1951; 6:1152–1158.
2. Bornside GH, Bouis PI, Cohn I. Enhancement of Escherichia coli infection and endotoxic activity by hemoglobin and ferric ammonium citrate. Surg 1970; 68:350–355.
3. Savitsky JP, Doczi J, Black J, Arnold JD. A clinical safety trial of stroma-free hemoglobin. Clin Pharmacol Ther 1978; 23:73–78.
4. Bolin R, Smith D, Moore G, Boswell G, Devenuto F. Hematologic effects of hemoglobin solutions in animals. In: Bolin RB, Geyer RP, Nemo GJ, eds. Advances in Blood Substitute Research. New York: Alan R. Liss, 1983:117–126.
5. White CT, Murray AJ, Greene JR, Smith DJ, Medina F, Makovec GT, Martin EJ, Bolin RB. Toxicity of human hemoglobin solution infused into rabbits. J Lab Clin Med 1986; 108:121–131.
6. Feola M, Simoni J, Canizaro PC, Tran R, Raschbaum G. Toxicity of polymerized hemoglobin solutions. Surg Gyn Obstet 1988; 166:211–222.
7. Feola M, Simoni J, Dobke M, Canizaro PC. Complement activation and toxicity of stroma-free hemoglobin solutions in primates. Circ Shock 1988; 25:275–290.
8. Marks DH, Cooper T, Makovec T, Okerberg C, Lollini LO. Effect of polymyxin B on hemoglobin-mediated hepatoxicity. Mil Med 1989; 154:180–184.
9. Smith DC, Schuschereba ST, Hess JR, McKinney L, Bunch D, Bowman PD. Liver and kidney injury after administration of hemoglobin cross-linked with bis(3,5-dibromosalicyl) fumarate. Biomat Art Cells Art Org 1990; 18:251–261.
10. Feola M, Simoni J, Tran R, Canizaro PC. Nephrotoxicity of hemoglobin solutions. Biomat Art Cells Art Org 1990; 18:233–249.
11. Winslow RM. The toxicity of hemoglobin. In: Winslow RM, ed. Hemoglobin-Based Red Cell Substitutes. Baltimore: The Johns Hopkins University Press, 1992:136–163.
12. Bleeker W, Agterberg J, La Hey E, Rigter G, Zappaij L, Bakker J. Hemorrhagic disorders after administration of glutaraldehyde-polymerized hemoglobin. In: Winslow RM, Vandegriff KD, Intaglietta M, eds. Blood Substitutes. New Challenges. Boston: Birkhäuser, 1996:112–131.
13. Everse J, Hsia N. The toxicities of native and modified hemoglobins. Free Rad Biol Med 1997; 22:1075–1099.
14. Griffiths E, Cortes A, Gilbert N, Stevenson P, MacDonald S, Pepper D. Haemoglobin-based blood substitutes and sepsis. Lancet 1995; 345:158–160.
15. Langermans JAM, van Vuren-van der Hulst MEB, Bleeker WK. Safety evaluation of a polymerized hemoglobin solution in a murine infection model. J Lab Clin Med 1996; 127:428–434.
16. Mourelatos MG, Enzer N, Ferguson JL, Rypins EB, Burhop KE, Law WR. The effects of diaspirin cross-linked hemoglobin in sepsis. Shock 1996; 5:141–148.
17. Smith DJ, Winslow RM. Effects of extraerythrocytic hemoglobin and its components on mononuclear cell procoagulant activity. J Lab Clin Med 1992; 119:176–182.
18. Feola M, Simoni J, Fishman D, Tran R, Canizaro PC. Compatibility of hemoglobin solutions. I. Reactions of vascular endothelial cells to pure and impure hemoglobins. Artif Organs 1989; 13:209–215.
19. White CT, Murray AJ, Smith DJ, Greene JR, Bolin RB. Synergistic toxicity of endotoxin and hemoglobin. J Lab Clin Med 1986; 108:132–137.
20. Krishnamurti C, Carter AJ, Maglasang P, Hess JR, Cutting MA, Alving BM. Cardiovascular toxicity of human cross-linked hemoglobin in a rabbit endotoxemia model. Crit Care Med 1997; 25:1874–1880.
21. Kaca W, Roth RI, Levin J. Hemoglobin: a newly recognized lipopolysaccharide (LPS) binding protein which enhances LPS biological activity. J Biol Chem 1994; 269:25078–25084.
22. Bélanger M, Bégin C, Jacques M. Lipopolysaccharides of *Actinobacillus pleuropneumoniae* bind pig hemoglobin. Infect Immun 1995; 63:656–662.
23. Kaca W, Roth RI, Vandegriff KD, Chen GC, Kuypers FA, Winslow RM, Levin J. Effects of bacterial endotoxin on human cross-linked and native hemoglobins. Biochemistry 1995; 34:11176–11185.
24. Kaca W, Roth RI, Ziolkowski A, Levin J. Human hemoglobin increases the biological activity of bacterial lipopolysaccharides in activation of *Limulus* amebocyte lysate and stimulation of tissue factor production by endothelial cells. J Endotoxin Res 1994; 1:243–252.
25. Yoshida M, Roth RI, Levin J. The effect of cell-free hemoglobin on intravascular clearance and cellular, plasma, and organ distribution of bacterial endotoxin in rabbits. J Lab Clin Med 1995; 126:151–160.

26. Roth RI, Levin FC, Levin J. Distribution of bacterial endotoxin in blood. Infect Immun 1993; 61:3209–3215.
27. Brain MC, Hourihane DO'B. Microangiopathic haemolytic anaemia: the occurrence of haemolysis in experimentally produced vascular disease. Br J Haematol 1967; 13:135–142.
28. Roth RI, Wong JS, Hamilton RL. Ultrastructural changes in bacterial lipopolysaccharide induced by human hemoglobin. J Endotoxin Res 1996; 3:361–366.
29. Archambault M, Olivier M, Foiry B, Diarra MS, Paradis SE, Jacques M. Effects of pig hemoglobin binding on some physical and biological properties of *Actinobacillus pleuropneumoniae* lipopolysaccharides. J Endotoxin Res 1997; 4:53–65.
30. Roth RI, Levin J, Chapman KW, Schmeizl M, Rickles FR. Production of modified crosslinked cell-free hemoglobin for human use: the role of quantitative determination of endotoxin contamination. Transfusion 1993; 33:919–924.
31. Roth RI. Hemoglobin enhances the production of tissue factor by endothelial cells in response to bacterial endotoxin. Blood 1994; 83:2860–2865.
32. Roth RI. Hemoglobin enhances the binding of bacterial endotoxin to human endothelial cells. Thrombos Haemost 1996; 76:258–262.
33. Kaca W, Roth RI. Activation of complement by human hemoglobin and by mixtures of hemoglobin and bacterial endotoxin. Biochim Biophys Acta 1995; 1245:49–56.
34. Su D, Roth RI, Yoshida M, Levin J. Hemoglobin enhances mortality from bacterial endotoxin. Infect Immun 1997; 65:1258–1266.
35. Kaye D, Hook WE. The influence of hemolysis or blood loss on susceptibility to infections. J Immunol 1963; 91:65–75.
36. Kaye D, Hook EW. The influence of hemolysis on susceptibility to Salmonella infection: additional observations. J Immunol 1963; 91:518–527.
37. Litwin MS, Walter CW, Ejarque P, Reynolds ES. Synergistic toxicity of gram-negative bacteria and free colloidal hemoglobin. Ann Surg 1963; 157:485–493.
38. Kaye D, Gill FA, Hook EW. Factors influencing host resistance to Salmonella infections: the effects of hemolysis and erythrophagocytosis. Am J Med Sci 1967; 254:205–215.
39. Schneidkraut ML, Loegering DJ. Reticuloendothelial system depression with hemolyzed blood. Adv Shock Res 1980; 3:272–282.
40. Eaton JW, Brandt P, Mahoney JR. Haptoglobin: a natural bacteriostat. Science 1982; 215:691–693.
40a. Su D, Roth RJ, Levin J. Hemoglobin infusion augments the tumor necrosis factor response to bacterial endotoxin (lipopolysaccharide) in mice. Crit Care Med 1999 (in press).
41. Casey LC, Balk RA, Bone RC. Plasma cytokine and endotoxin levels correlate with survival in patients with the sepsis syndrome. Ann Intern Med 1993; 119:771–783.
42. Dofferhoff ASM, Bom VJJ, de Vries-Hospers HG, van Ingen J, vd Meer J, Hazenberg BPC, Mulder POM, Weits J. Patterns of cytokines, plasma endotoxin, plasminogen activator inhibitor, and acute-phase proteins during the treatment of severe sepsis in humans. Crit Care Med 1992; 20:185–192.
43. Waage A, Halstensen A, Shalaby R, Brandtzaeg P, Kierulf P, Espevik T. Local production of tumor necrosis factor α, interleukin 1, and interleukin 6 in meningococcal meningitis. Relation to the inflammatory response. J Exp Med 1989; 170:1859–1867.
44. Sekut L, Menius JA, Brackeen MF, Connolly KM. Evaluation of the significance of elevated levels of systemic and localized tumor necrosis factor in different animal models of inflammation. J Lab Clin Med 1994; 124:813–820.
45. Falk MC, Fletcher MA, Williams TJ, Ching WM, Wu Y, Morrison TK. Aggregation of serum proteins with lipopolysaccharide (LPS): characterization of the precipitable LPS-protein complex. J Endotoxin Res 1996; 3:129–142.
46. Schromm AB, Brandenburg K, Rietschel ET, Seydel U. Do endotoxin aggregates intercalate into phospholipid membranes in a nonspecific, hydrophobic manner? J Endotoxin Res 1995; 2:313–323.
47. Takayama K, Din ZZ, Mukerjee P, Cooke PH, Kirkland TN. Physicochemical properties of the lipopolysaccharide unit that activates B lymphocytes. J Biol Chem 1990; 265:14023–14029.
48. Takayama K, Mitchell DH, Din ZZ, Mukerjee P, Li C, Coleman DL. Monomeric Re lipopolysaccharide from *Escherichia coli* is more active than the aggregated form in the *Limulus* amebocyte lysate assay and in inducing Egr-1 mRNA in murine peritoneal macrophages. J Biol Chem 1994; 269:2241–2244.
49. Seydel U, Brandenburg K, Rietschel ET. A case for an endotoxic conformation. In: Levin J, van Deventer SJH, van der Poll T, Sturk A, eds. Bacterial Endotoxins. Basic Science to Anti-Sepsis Strategies. New York: Wiley-Liss, 1994:17–30.
50. Galanos C, Lüderitz O. Lipopolysaccharide: properties of an amphipathic molecule. In: Rietschel ET, ed. Handbook of Endotoxin, Vol. 1: Chemistry of Endotoxin. Amsterdam: Elsevier, 1984:46–58.
51. Shnyra A, Hultenby K, Lindberg AA. Role of the physical state of *Salmonella* lipopolysaccharide in expression of biological and endotoxic properties. Infect Immun 1993; 61:5351–5360.
52. Brandenburg K, Schromm AB, Koch MHJ, Seydel U. Conformation and fluidity of endotoxins as determinants of biological activity. In: Levin J, Alving CR, Munford RS, Redl H, eds. Bacterial Endotoxins. Lipopolysaccharides from Genes to Therapy. New York: Wiley-Liss, 1995:167–182.
53. Ogden JE, Mac Donald SL. Haemoglobin-based red cell substitutes: current status. Vox Sang 1995; 69:302–308.
54. Dietz NM, Joyner MJ, Warner MA. Blood substitutes: fluids, drugs, or miracle solutions? Anesth Analg 1996; 82:390–405.
55. Lieberthal W. Stroma-free hemoglobin: a potential blood substitute (editorial). J Lab Clin Med 1995; 126:231–232.
56. Eaton JW. Hemoglobin-based blood substitutes: a dream-like trade of blood and guile (editorial)? J Lab Clin Med 1996; 127:416–417.

23

LPS/Lipid A–Binding Synthetic Peptides

Massimo Porro
BioSynth Research Laboratories, Siena, Italy

THE PUTATIVE LIPID A–BINDING SITE AND ITS RECOGNITION BY PEPTIDE STRUCTURES

The amphipathic structure of bacterial endotoxins (lipopolysaccharides, LPS) relates to the supramolecular architecture of lipid A, the biologically active moiety of LPS (1). Lipid A is structurally conserved among LPS from different species of gram-negative bacteria and results from the association of several glycolipid monomers of general structure N,O-acyl-β-1,6-D-glucosamine disaccharide 1,4′-bisphosphate. In an aqueous environment, the glycolipid monomers form micelles according to the critical micellar concentration (CMC) of the system (1,2). The biological activity of LPS is expressed through the interaction of the lipid A component with a variety of cell- and serum-binding proteins/receptors inducing the immune system to react by secretion of proinflammatory cytokines, a process that often leads to very severe biological effects (3,4). The interaction of LPS, via lipid A, with mammalian receptor proteins appears to be a remarkably efficient process, since it is today well recognized that LPS interacts with a variety of serum proteins, with specialized cells of the immune system, and with tissues of different key organs like liver, lungs, and spleen.

During the past few years, our laboratories have undertaken studies designed to determine the thermodynamic efficiency of the interaction between the lipid A moiety of R(rough)- and S(smooth)-chemotype LPS with well-defined peptide molecules. Our primary objectives have been to characterize in detail the physicochemical characteristics of the putative binding site expressed by lipid A, which, in turn, would be of value in defining the corresponding features required by the amino acid sequences of mammalian LPS-binding/receptor proteins with which LPS interacts. The focus of our study has been based on two fundamental observations reported several years ago and today widely recognized:

1. The capability of the peptide antibiotic polymyxin B (PMXB) to bind with high affinity to LPS and to reduce its toxicity and improve survival of mice challenged systemically with purified LPS (5)
2. The molecular explanation for the protective activity of PMXB, based on the stoichiometric interaction of the peptide antibiotic with the lipid A moiety of LPS (6), demonstrating for the first time that lipid A contains a saturable binding site.

FEATURES OF PEPTIDE STRUCTURES RECOGNIZED BY LIPID A

The antibiotic PMXB is a cyclic cationic peptide, which is composed of 10 amino acids and a hydrophobic tail, a heptanoyl/octanoyl chain, at the N-terminus of the molecule. PMXB contains in its structure several uncommon amino acids not normally found in mammalian cells, like D-phenylalanine and α,γ-L-diaminobutyric acid (DAB, six residues accounting for about 50% of the molecular mass of PMXB). DAB is a homolog of L-lysine, and the available evidence would suggest that it is responsible for the secondary rearrangement of this cyclic peptide through the internal condensation of a DAB residue with L-threonine. It is also thought to be responsible for the proteolytic stability of PMXB to serine proteases, such as trypsin and chymotrypsin. Also of interest from the toxicological point of view, as a free amino acid it can function to replace L-lysine in the protein synthesis of mammalian cells (7). This lack of biodegradability results in the accumulation of PMXB in target organs (mainly kidneys and tissues of the nervous system) following its administration to experimental animals leading to a significant toxicity, which often ends with organ failure.

In order to investigate the contribution of the different amino acid residues in PMXB to the physical-chemical characteristics of the cyclic peptide for binding and detoxifying the lipid A moiety of LPS, we have synthesized a series of synthetic peptides containing natural amino acids, each of which replaces one of the residues found in the original structure of PMXB (8). For the primary structure of synthetic peptides, L-lysine was used for replacement of DAB residues in PMXB, L-phenylalanine for replacement of D-phenylalanine, and L-isoleucine for replacement of the heptanoyl/octanoyl chain at the N-terminal side of the molecule. Recreation of the secondary rearrangement of the peptides was achieved by insertion of two L-cysteine residues in the primary structure, followed by oxidation of the two sulfhydryl groups to form an internal disulfide bridge, thereby generating the cystine disulfide. In this case, a cystine residue was replacing the DAB-L-threonine internal bridge of PMXB, providing a natural rearrangement for the synthetic peptide (Fig. 1) in a way similar to that by which proteins naturally develop secondary structure. The primary structure of PMXB and one of its synthetic analogs are indicated in Figure 1.

Synthetic peptides prepared according to this strategy have been of considerable value in allowing us to elucidate the molecular constraints on amino acid sequences required for efficient thermodynamically stable binding of such peptides to lipid A. Analysis of these peptides binding to lipid A has also led to the definition of the size of its binding site. Further, this approach has provided important information on some of the molecular characteristics that mammalian receptor proteins should have for potential recognition of lipid A, a crucial step in the design of peptide-based antagonists of LPS (8). Accordingly, the biological activity of synthetic antiendotoxin peptides (SAEP), as assessed by their ability to inhibit the toxic properties of a variety of R- and S-chemotype LPS, has been determined in both in vitro and in vivo studies by measuring the competitive (dose-related) inhibition of LPS-induced clotting of Limulus amebocyte lysate (LAL), local and systemic TNF and IL-6 production in mice, hemorrhagic dermonecrosis in the rabbit (local Schwartzman reaction), and lethality in mice (8,9). The results of these studies are reviewed in the following paragraphs.

GENERAL STRUCTURE OF LIPID A–BINDING PEPTIDES

Analysis of the primary as well as secondary structural features of SAEP in relation to PMXB has revealed that the characteristics of peptide-based structures required for thermodynamically efficient binding and detoxification of the lipid A moiety of LPS reside primarily in their cationicity and amphipathicity, with a relatively less, but nevertheless significant impact of the secondary rearrangement of the sequences for improving the overall affinity of binding of peptides to lipid A. In aqueous solutions, binding to lipid A appears to involve a dual-step process, through a preliminary interaction of the cationic amino acids (lysine, arginine) with the anionic groups of lipid A (phosphates), followed by the stabilization of the resulting molecular complex through hydrophobic interactions involving the fatty acid residues of lipid A, the hydrophobic amino acids (e.g., phenylalanine and leucine), and most likely the alkyl chain of lysine or arginine residues as well. These conclusions derive from experimental observations showing that anionic amino acids cannot replace cationic ones in the sequences of SAEP and still manifest high-affinity lipid A binding, that the complex between SAEP and LPS is stable over a broad range of pH

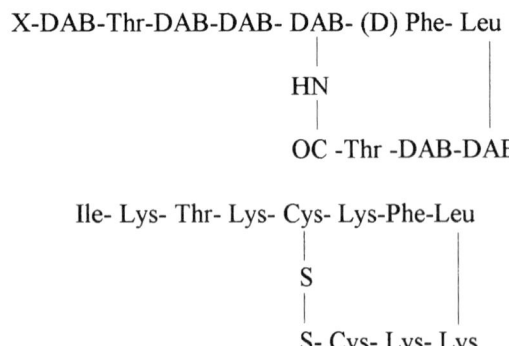

Fig. 1 Structure of polymyxin B and one of its peptide analogs. Amino acids are in three-letter code. X = 6-Methyl heptanoyl/octanoyl; DAB = α,γ-diaminobutyric acid. (From Ref. 8.)

(pH 2–11) and ionic strength (0.10–0.5 M), that the primary amino groups of lysine residues are still accessible for reaction with specific reagents following complex formation, and that the complex with LPS can be readily dissociated by 1% (w/v) SDS treatment. The length of peptide-based structures that would optimally possess these features was found to be about 10 amino acids. No detectable lipid A–binding activity was observed for sequences having less than 6–7 amino acids (8). Studies of molecular mechanics based on NMR spectra of SAEP (10) and of an analog peptide in complex with LPS have confirmed the conclusions reported above concerning the binding process in aqueous/polar solvents (11).

On the basis of these fundamental observations of peptide-LPS–binding interactions, an algorithm has been developed in order to search for potential amino acid sequences within the primary structure of well-documented LPS-binding proteins present in nature. The algorithm predicts that any cyclic or linear sequence encompassing a minimum of 6–7 amino acid residues, respectively, containing a minimum of three aliphatic cationic amino acids with solvent parameter values ≥ 1.5 kcal/mole (lysine and arginine) and hydrophobic amino acids with solvent parameter values ≥ -1.5 kcal/mole (tryptophan, phenylalanine, tyrosine, leucine, isoleucine, and valine), and characterized by a value of the ratio of number of cationic to number of hydrophobic amino acids (Rc/h) ≥ 1 has a significantly increased probability of binding with high affinity to the lipid A–binding site of LPS (12).

To test the hypothesis, we have analyzed the primary amino acid structure of well-established LPS-binding proteins to attempt to identify amino acid sequences fulfilling the three requirements of the algorithm defined above. In situations where the candidate anti-LPS sequences have been identified, the corresponding peptides were synthesized and employed in a variety of assays designed to assess binding to lipid A and neutralization of the toxic effects of LPS both in vitro and in vivo. The first set of LPS-binding proteins that were analyzed include phylogenetically different proteins reported to be involved in modulating different biological functions of LPS (Table 1). These include CD14 (the 55 kDa LPS receptor expressed on the cell membrane of some immunocompetent cells), LBP (a 60 kDa LPS-binding protein present in serum), BPI (a 55 kDa bactericidal permeability-increasing protein first identified in polymorphonuclear cell granules), LALF (a 15 kDa factor of Limulus amebocyte), and LEBP-PI (the 12 kDa limulus endotoxin-binding protein-protease inhibitor). In each protein we have identified at least one candidate amino acid sequence fulfilling all requirements of the algorithm defined above, and the corresponding synthetic peptides have shown the capability to bind and detoxify LPS in various assays (12–14). Comparable results have been reported for some of these protein sequences in independent studies carried out in other laboratories (15–19). In Table 1, selectivity expresses the ratio between the affinity constant value of the considered peptide for *E. coli* B5 LPS and that of polymyxin B, as detected in the competition assay described in Ref. 8. LAL inhibition indicates the minimal ratio of peptide to LPS (w/w) capable of inhibiting the clotting induced by 0.125 EU/ml (0.04 ng/ml) of *E. coli* B5 LPS, in the Limulus amebocyte lysate assay.

SEARCH FOR LIPID A–BINDING PEPTIDES IN THE PRIMARY STRUCTURE OF NATURAL LPS-BINDING POLYPEPTIDES

Several natural cationic polypeptides of the animal kingdom are known to have antibiotic properties against gram-negative bacteria, at least in part by increasing the permeability of the outer membrane (20). Among these are defensins and defensin-related peptides recently found in human skin, magainins, cecropins, melittins, tachyplesins, and rabbit cationic proteins (20–24). The algorithm has, therefore, been applied to these natural polypeptides in order to identify specific sequences as potential binding site for LPS and, if so, their potential use as synthetic anti-LPS antagonist molecules. Indeed, several amino acid sequences with cationic amphipathic characteristics were predicted in the primary sequence of these polypeptide antibiotics (25), and several of these are now under investigation in our laboratories for possible LPS-neutralizing capability. Since natural cationic antibiotic polypeptides seem to contain amino acid sequences with characteristics comparable to those required for binding and neutralizing lipid A, our attention has been also directed toward the investigation of associated complementary activities of these anti-LPS synthetic peptides; that is, the potential antibiotic activity of synthetic peptides designed de novo according to the sequences predicted by the algorithm on gram-negative bacteria (Tables 2 and 3). We have found that several of these synthetic peptides manifest direct antibiotic activity in vitro, as well as synergistic activity with hydrophobic antibiotics like rifampin and erythromycin (26).

Table 1 Characteristics of Synthetic Peptides Predicted in the Primary Amino Acid Sequences of LPS-Binding Natural Polypeptides

Peptide sequences	Rc/h	Selectivity	LAL inhibition (w/w)
Predicted			
CD14$_{67-75}$ Human VKALRVRRL	1.00	0.02	100
CD14$_{68-78}$ Mouse KSLSLKRLTVR	1.00	0.03	200
LBP$_{92-100}$ Human KVRKSFFKL	1.00	0.07	100
LBP$_{376-384}$ Human FLKPGKVKV	0.75	0.04	100
BPI$_{27-34}$ Human KELKRIKI	1.33	0.04	100
BPI$_{90-99}$ Human KWKAQKRFLK	1.67	0.20	10
LALF$_{41-51}$ Crab KRLKWKYKGKF	1.50	1.83	1
LEBP-PI$_{5-17}$ Crab CQSWKSSEIRCGK [S — S]	1.50	0.05	100
Control			
LEBP-PI$_{86-98}$ Crab CRQHGTYINCLHV [S — S]	0.25	0.00	>1000
BPI$_{153-160}$ Human IQLPHKKI	0.67	0.00	>1000
LALF$_{33-41}$ Crab HYRINPTVK	0.67	0.00	>1000

Amino acids appear in one-letter code. Rc/h value defines the ratio between aliphatic cationic (at physiological pH) and hydrophobic amino acids present in a selected sequence.

Another protein that is of potential relevant interest in the elucidation of the molecular basis of the interaction of proteins/peptides with the lipid A moiety of LPS is tubulin, the globular protein responsible for microtubule formation in the cytoskeleton. This somewhat ubiquitous protein has been also reported to bind LPS (27). In the primary structure of tubulin, the algorithm predicts seven amino acid sequences fulfilling all the requirements (25), specifically at residues 156–166, 287–299, 335–345, 362–373, 381–390, 383–401, and 408–415. The implications deriving from the interaction of tubulin with LPS may be of particular relevance, especially if one considers the capability of LPS to target various tissues in mammalian hosts and be responsible for the failure of key organs like the liver.

Another category of natural polypeptides that would be of potential interest as a source of molecular information useful for generating novel new synthetic peptides specific for the lipid A moiety of LPS are opioid-related peptides. In fact, lipid A is a pyrogenic molecule with reported somnogenic activity (28), and LPS is known for its ability to induce, through lipid A, an increase in the permeability of the blood-brain barrier

Table 2 Synergism of Synthetic Peptides Designed De Novo with Rifampin and Erythromycin Against *E. coli* IH3080 Evaluated by Minimal Inhibitory Concentration (MIC)

Peptide	MIC (μg/ml) at indicated peptide concentration (μg/ml)											
	Rifampin						Erythromycin					
	0.3	1	3	10	30	100	0.3	1	3	10	30	100
None	10						30					
Deacylpolymyxin B	10	0.1	0.01	—[a]	—	—	10	3	1	—	—	—
KFFKFFKFF	10	1	0.03	0.03	0.01	—	30	30	1	3	1	—
IKFLKFLKFL	10	0.3	0.01	0.01	—	—	30	10	1	1	—	—
CKFKFKFKFC [S—S]	10	10	3	1	0.1	—	30	30	30	30	1	—
IKTKCKFLKKC [S—S]	10	10	10	3	3	1	30	30	30	30	30	30
IRTRCRFLRRC [S—S]	3	3	3	1	1	0.1	30	30	30	10	1	1

[a]The peptide alone inhibited growth.

(29) and the release of proinflammatory cytokines in brain tissue (30) and to reproduce effects mimicking those of opioid peptides (31).

Inflammation is a process involving several mediators, and it is now recognized that LPS-induced inflammation in the brain of mice, following an intravenous challenge, results in the upregulation of neurotactin, a newly discovered membrane-anchored small protein of the chemokine families, shortly after LPS challenge (32). With this in mind, it could be important to note that the sensation of pain follows the activation of nociceptors, primary sensory neurons located on cells, which are targets for opioids and are specialized to detect tissue damage in the periphery and in the central nervous system (33). Nociceptin or Orphanin FQ is a new heptadecapeptide discovered in brain tissue of mammals, which resembles the opioid peptide dynorphin A and acts as an endogenous agonist of the opioid receptor-like 1 (ORL1), a new G-protein coupled receptor (34,35). ORL1 is an orphan receptor, resembling opioid receptors, whose cDNA has been detected in brain tissue of both humans and mice. Nociceptin increases reactivity to pain by stimulating the ORL1 receptor, a process that can therefore be defined as no-

Table 3 Synergism of Synthetic Peptides Designed De Novo with Rifampin Against *Klebsiella pneumoniae* and *E. cloacae* Evaluated by Minimal Inhibitory Concentration (MIC)

Peptide	MIC (μg/ml) of rifampin at indicated peptide concentration (μg/ml)							
	K. pneumoniae KY12854				*E. cloacae* KY12645			
	3	10	30	100	3	10	30	100
None	10				10			
Deacylpolymyxin B	0.3	0.1	NT	NT	1	1	NT	NT
KFFKFFKFF	0.3	0.1	0.03	0.01	0.1	0.1	0.03	0.01
IKFLKFLKFL	0.3	0.03	0.01	—[a]	0.03	0.03	0.01	0.01
CKFKFKFKFC [S—S]	10	10	1	0.3	10	3	1	0.3
IKTKCKFLKKC [S—S]	10	10	10	10	10	10	10	3
IRTRCRFLRRC [S—S]	10	10	3	1	10	10	1	1

NT, Not tested.
[a]The peptide alone inhibited growth.

ciceptive stimulation of a nociceptor. Of considerable potential interest is the fact that nociceptin appears to be unique within the family of dynorphin-related peptides, which fulfills all the criteria required by the algorithm for a peptide sequence efficiently binding lipid A and potentially resulting in its detoxification. This neurologic peptide has therefore been evaluated in a variety of in vitro and in vivo assays for its ability to neutralize the biological effects of LPS and has shown a remarkable ability to bind and neutralize LPS in all assays performed (25). Thus, there is a strong suggestion that the opioid-related peptide nociceptin might to date well serve as a potential target of LPS in the central and peripheral nervous system of mammalians. It is our hypothesis that it might function as an important recognition system, serving to alert the host's defenses on the basis of an imbalance in the nociceptin/nociceptor system. This reciprocal imbalance could in turn serve in mammals as a physiological detector triggering the early biological effects of an LPS insult (e.g., inflammation, fever) that follow gram-negative bacterial infections.

DETOXIFICATION MECHANISM(S) OF SAEP

The biological activity of SAEP and PMXB in different animal models of endotoxin-mediated cytokine release and lethality is dose-dependent in experiments in which R- and S-chemotype LPS has been used for either local or systemic challenge (8,9,13) despite the fact that in in vitro studies the binding of SAEP and PMXB to lipid A is stoichiometric (6,8). This experimental observation is most likely explained by the likely antagonistic effects exerted by SAEP and PMXB on the lipid A moiety of LPS, through an active competition with receptor proteins on cells and tissues of the mammalian host (9). The fact, as discussed earlier, that serum LBP and soluble or cell-associated CD14 contain amino acid sequences with characteristics comparable to SAEP would support this conclusion.

However, other experimental observations merit discussion. For instance, binding of SAEP and PMXB to lipid A in undiluted serum does not occur at the stoichiometry observed in aqueous solutions when measured by two independent methods [e.g., surface plasmon resonance (SPR) and equilibrium molecular dialysis (9)]. The presence of serum therefore, appears to interfere with the efficiency of binding, and a significant dilution of serum is required in order to restore full binding activity to the level observed in aqueous solutions. Since neither ionic strength nor pH of the medium offers reasonable explanations for this discrepancy, as binding to lipid A by SAEP and PMXB is not affected by these two parameters in a broad range of values (8), the observed interference may be due to either the presence of LPS-binding components or intrinsic physical-chemical features of the serum environment. Despite these in vitro observations, SAEP are very active in the neutralization of the toxic effects associated with LPS when given prophylactically and therapeutically in vivo (8,9,13,17). These effects are manifest even within a time frame of treatment broader than the experimentally observed half-life time for SAEP in the bloodstream (9). For instance, LPS-dependent production of tumor necrosis factor (TNF), the proinflammatory cytokine generally recognized as responsible for many of the toxic effects induced by LPS, is significantly inhibited by SAEP in the organs, tissues, and sera of animals challenged by LPS, either intravenously, intraperitoneally, or intradermally (9). On these bases, it is reasonable to postulate that the biological manifestations of SAEP activity as an LPS neutralization event occur at the tissue level rather than in the serum.

There are, however, recent findings concerning novel properties of cationic amphipathic peptides that suggest an additional molecular mechanism leading to the inhibition of LPS-induced effects. It has been reported that, at least in vitro, PMXB can serve as a selective and potent antagonist of calmodulin through the formation of a stable molecular complex (36). Also, D-amino-acid analogs of the peptide antibiotic melittin bind to calmodulin, which is remarkably tolerant sterically, and the resulting complexes have been shown to be highly stabilized by van der Waals forces (37). Since calmodulin is a multifunctional protein often involved in the cellular signaling pathways of mammalian cells, where it plays the role of a phosphokinase activator, the capability of cationic amphipathic peptides to interact with calmodulin could result in the inhibition of the LPS-induced signal pathway by a pathway independent of its ability to interact with LPS. This hypothesis would be supported by the observation that calmodulin is a well-characterized subunit of the inducible nitric oxide synthase (NOS) from macrophages (38) and human hepatocytes (39), the enzyme responsible for the synthesis of the biological mediator nitric oxide, a potent hypotensive agent, which can be overproduced in the vasculature after exposure to LPS (40). In this respect, our laboratory has investigated the capability of SAEP to inhibit in vitro the enhancing effect of calmodulin on the phosphodiesterase activity, which leads to the degradation of the cyclic-AMP to 5'-AMP.

Our results are comparable to those previously mentioned for PMXB, that is, several SAEP sequences were able to antagonize the enhancing effect of calmodulin on phosphodiesterase activity, in a dose-dependent fashion (unpublished observations). Thus, parallel mechanisms of LPS detoxification might exist for SAEP in vivo, and more studies are clearly necessary to elucidate this issue at the molecular and cellular level.

POTENTIAL CLINICAL USES OF SAEP

Since the structure of lipid A is quite constant among different LPS (1) and since detoxification of different LPS by SAEP has been demonstrated by a variety of different assays, a general approach for prophylaxis and therapy of LPS-mediated diseases by SAEP would appear to merit serious consideration. Efficacy of synthetic peptides in their prophylactic/therapeutic use as anti-LPS drugs in animal models of endotoxemia and sepsis has been well documented (8,9,17,18,23). The human use of peptide-based drugs should, however, also address the question of their safety, since the clinical experience with PMXB has shown significant toxicity for this antibiotic associated with its lack of biodegradability and its accumulation in tissues as consequence of diffusion (41). With this in mind, SAEP have been designed according to the principle of biodegradable sequences (8) in order to allow for complete degradation of the peptide structures (ideally to the level of amino acid) in a time frame that would be sufficient to perform one or more cycles of treatment. Also, complete biodegradability of SAEP would avoid accumulation phenomena, as they diffuse to tissues where they actively inhibit LPS-induced TNF production (9). Another important issue to consider in the clinical use of peptide-based drugs is the potential immunogenicity that could derive from repeated therapeutic treatments. However, at least in three animal models involving different species (mice, rabbits, and dogs), pharmacokinetic, toxicological, and immunological studies with a specific SAEP sequence have shown that these theoretical possibilities are not supported by experimental observations (9,13), in that no detectable cellular and humoral immunity has developed even following prolonged exposure of experimental animals to SAEP.

Detoxification of LPS by SAEP may also represent a novel new strategy by which to effectively target LPS to the immune system of mammals in the form of molecular complexes. Such an approach would avoid prior chemical hydrolysis of the lipid A moiety, which is usually performed to eliminate toxicity (42) but which also results in destruction of the supramolecular architecture of LPS (2). This strategy has been applied to LPS of *Neisseria meningitidis* group A and B for inducing a biologically functional immune response in mice (43). This approach is currently under investigation for other R- and S-LPS micelles complexed with SAEP in order to expand the concept of relative prophylactic potential of this novel strategy for delivering LPS-based vaccines.

Removal of LPS from blood or plasma of endotoxemic patients by extracorporeal treatment with solid phase matrixes supporting ligands specific for the lipid A moiety is another potential clinical use of SAEP. This strategy has been recently introduced through the use of PMXB-immobilized fibers (44) with encouraging preliminary results, at least on the basis of the effects related to the endotoxin levels measured. However, given the fact that the formation of a complex between PMXB and LPS is a process that requires significant solvation for high thermodynamic efficiency and stability (9), specific demonstration of reduction of LPS levels seems necessary in studies involving a larger number of patients. The use of SAEP in this clinical approach would have the advantage of safety in the event of accidental release of ligand during the extracorporeal treatment of the patient, provided that peptides with a sufficient degree of proteolytic stability are employed during the course of treatment.

CONCLUSIONS

Synthetic peptides originally designed to mimic the primary and secondary structures of the peptide drug antibiotic PMXB have provided significant new information about the definition of the molecular characteristics of the binding site expressed by the lipid A structure of LPS. This information has also allowed the identification of candidate amino acid sequences involved in the recognition process of LPS by phylogenetically different polypeptides having different biological functions. A common homogeneous recognition system for LPS seems to be employed. If so, this system is based on the efficient binding to lipid A, the structurally conserved region of LPS, of amino acid sequences that can be predicted with reasonable certainty on the basis of an algorithm whose criteria have been postulated from physical-chemical considerations on the structure of PMXB-related peptides in various conformations. Although the algorithm may be predictive only of lipid A–specific sequences present in the primary structure of natural polypeptides, the variety of peptide structures evaluated in various studies suggest that the physical-chemical characteristics of SAEP might also represent the primary target for lipid A in the conformationally defined structure of proteins. Also, as in the case of the opioid-like peptide nociceptin, the developed algorithm might help the discovery of novel LPS-binding polypeptides whose binding activity for LPS is still unknown or not yet related to their biological function in nature. The size of peptide sequences optimally determined to bind to lipid A in aqueous solvents (8,13) is comparable to the size reported for epitopes recognized by monoclonal antibodies (45,46) as well as for interaction with the major histocompatibility complex class I molecules (47). This consideration would allow the interesting speculation that the mechanism of LPS recognition by natural polypeptides may well bear similarities to the immunorecognition system. Finally, several different clinical uses of anti-LPS synthetic peptides can be envisioned that, if successful in the near future, would allow innovative approaches in clinical medicine for prophylaxis and treatment of diseases caused by gram-negative bacteria and the related effects due to the release of LPS during the infectious process, such as endotoxic shock.

REFERENCES

1. Rietschel ET, Brade H, Brade L, et al. Lipid A, the endotoxic center of bacterial lipopolysaccharides: relation of chemical structure to biological activity. Prog Clin Biol Res 1987; 231:25–53.
2. Seydel U, Brandenburg K, Koch MHJ, et al. Supramolecular structure of lipopolysaccharide and free lipid A under physiological conditions as determined by synchrotron small angle X-ray diffraction. Eur J Biochem 1989; 325–332.
3. Rietschel ET, Brade H, Holst O, et al. Bacterial endotoxin: chemical constitution, biological recognition, host response and immunological detoxification. In: Rietschel ET, Wagner H, eds. Pathology of Septic Shock. Berlin: Springer-Verlag, 1996:39–81.
4. Morrison DC, Lei MG, Kirikae T, et al. Endotoxin receptors on mammalian cells. Immunobiology 1993; 187:212–226.
5. Rifkind D. Prevention by polymyxin B of endotoxin lethality in mice. J Bacteriol 1967; 93:1463–1464.
6. Morrison DC, Jacobs DM. Binding of polymyxin B to the lipid A portion of bacterial lipopolysaccharide. Immunochemistry 1976; 13:813–818.
7. Christensen HN, Liang M. Transport of diamino acids into the Erlich cell. J Biol Chem 1966; 241:5542–5551.
8. Rustici A, Velucchi M, Faggioni R, et al. Molecular mapping and detoxification of the lipid A binding site by synthetic peptides. Science 1993; 259:361–365.
9. Demitri MT, Velucchi M, Bracci L, et al. Inhibition of LPS-induced systemic and local TNF production by a synthetic anti-endotoxin peptide (SAEP-2). J Endotox Res 1996; 3(6):445–454.
10. Liao SY, Ong GT, Wang KT et al. Conformation of polymyxin B analogs in DMSO from NMR spectra and molecular modeling. Biochim Biophys Acta 1995; 1252(2):312–320.
11. Bhattacharjya S, David SA, Mathan VI, et al. Polymyxin B nonapeptide: conformations in water and in the lipopolysaccharide-bound state determined by two dimensional NMR and molecular dynamics. Biopolymers 1997; 41(3):251–265.
12. Porro M. Structural basis of endotoxin recognition by natural polypeptides. Trends Microbiol 1994; 2:65–66.
13. Velucchi M, Rustici A, Porro M. Molecular requirements of peptide structures binding to the lipid A region of bacterial endotoxins. In: Norrby E, Brown F, Chanock RM, Ginsberg HS, eds. Vaccines '94. New York: Cold Spring Laboratory Press, 1994:141–146.
14. Porro M. Cyclic or linear conformations of sequences binding lipid A: Does it really matter? Trends Microbiol 1994; 2:338.
15. Little R, Kelner DN, Lim E, et al. Functional domains of recombinant bactericidal/permeability increasing protein (rBPI23). J Biol Chem 1994; 269:1865–1872.
16. Battafarano RJ, Dahlberg PS, Ratz CA, et al. Peptide derivatives of three distinct lipopolysaccharide binding proteins inhibit lipopolysaccharide-induced tumor necrosis factor-alpha secretion in vitro. Surgery 1995; 118:318–324.
17. Kloczewiak M, Black KM, Loiselle P, et al. Synthetic peptides that mimic the binding site of horseshoe crab antilipopolysaccharide factor. J Infect Dis 1994; 170:1490–1497.
18. Larrick JW, Hirata M, Zheng H, et al. A novel granulocyte-derived peptide with lipopolysaccharide-neutralizing activity. J Immunol 1994; 152:231–240.
19. Hoess A, Watson S, Siber GR, et al. Crystal structure of an endotoxin-neutralizing protein from the horseshoe crab, limulus anti-LPS factor, at 1.5 Å resolution. EMBO J 1993; 12:3351–3356.
20. Vaara M. Agents that increase the permeability of the outer membrane. Microbiol Rev 1992; 56:395–411.
21. Hirata M, Shimomura Y, Yoshida M, et al. Characterization of a rabbit cationic protein (CAP18) with lipopolysaccharide-inhibitory activity. Infect Immun 1994; 62:1421–1426.
22. Harder J, Bartels J, Chistophers E, et al. A peptide antibiotic from human skin. Nature 1997; 387:861.
23. Gough M, Hancock REW, Kelly NM. Antiendotoxin activity of cationic peptide antimicrobial agents. Infect Immun 1996; 64:4922–4927.
24. Iwanaga S, Muta T, Shigenaga T, et al. Structure-function relationships of tachyplesins and their analogues. In: Marsh J, Gooden JA, eds. Antimicrobial Peptides (Series Ciba Foundation Symposia). Chichester, UK: John Wiley & Sons, 1994:160–174.

25. Porro M, Rustici A, Velucchi M, et al. Natural and synthetic polypeptides that recognize the conserved lipid A binding site of lipopolysaccharides. In: Levin J, Pollack M, Yokichi T, Nakano M, eds. Endotoxin and Sepsis: Molecular Mechanisms of Pathogenesis, Host Resistance and Therapy. New York: John Wiley & Sons, 1998:315–323.
26. Vaara M, Porro M. Group of peptides that act synergistically with hydrophobic antibiotics against gram-negative enteric bacteria. Antimicrob Agents Chemother 1996; 40:1801–1805.
27. Ding A, Sanchez E, Tancino M, et al. Interactions of bacterial lipopolysaccharide with microtubule proteins. J Immunol 1992; 148:2853–2858.
28. Cady AB, Kotani S, Shiba T, et al. Somnogenic activities of synthetic lipid A. Infect Immun 1989; 57:396–403.
29. Quagliarello V, Scheld VM. Bacterial meningitis: pathogenesis, pathophysiology and progress. N Engl J Med 1992; 327:864–872.
30. Faggioni R, Benigni F, Ghezzi P. Proinflammatory cytokines as pathogenic mediators in the central nervous system: brain-periphery connections. Neuroimmunomodulation 1995; 2:2–15.
31. Yirmiya R, Rosen H, Donchin O, et al. Behavioural effects of lipopolysaccharide in rats: involvement of endogenous opioids. Brain Res 1994; 648:80–86.
32. Pan Y, Lloyd C, Zhou H, et al. Neurotactin, a membrane-anchored chemokine upregulated in brain inflammation. Nature 1997; 387:611–617.
33. Taddese A, Nah SY, McCleskey EW. Selective opioid inhibition of small nociceptive neurons. Science 1995; 270:1366–1369.
34. Meunier JC, Mollerau C, Toll L, et al. Isolation and structure of the endogenous agonist of opioid receptor-like ORL1 receptor. Nature 1995; 377:532–535.
35. Reinscheid RK, Nothaker H, Bourson A, et al. Orphanin FQ: neuropeptide that activates the opioidlike G protein-coupled receptor. Science 1995; 270:792–794.
36. Hegemann L, Vanrooijen LAA, Traber J, et al. Polymyxin B is a selective and potent antagonist of calmodulin. Eur J Pharm 1991; 207:17–22.
37. Fisher PJ, Prendergast FG, Ehrhardt MR, et al. Calmodulin interacts with amphiphilic peptides composed of all D-amino acids. Nature 1994; 368:651–653.
38. Cho HJ, Xie Q, Calaycay J, et al. Calmodulin is a subunit of nitric oxide synthase from macrophages. J Exp Med 1992; 176:599–604.
39. Geller DA, Lowenstein CJ, Shapiro RA, et al. Molecular cloning and expression of inducible nitric oxide synthase from human hepatocytes. Proc Natl Acad Sci USA 1993; 90:3491–3495.
40. Moncada S, Palmer RMJ, Higgs EA. Nitric oxide physiology, pathophysiology and pharmacology. Pharmacol Rev 1991; 43:109–142.
41. Craig WA, Turner JH, Kunin CM. Prevention of the generalized Shwartzman reaction and endotoxin lethality by polymyxin B localized in tissue. Infect Immun 1974; 10:287–292.
42. Gu XX, Tsai CM. Preparation, characterization and immunogenicity of meningococcal lipooligosaccharide-derived oligosaccharide-protein conjugates. Infect Immun 1993; 61:1873–1879.
43. Velucchi M, Rustici A, Meazza C, et al. A model of Neisseria meningitidis vaccine based on LPS micelles detoxified by synthetic anti-endotoxin peptides. J Endotox Res 1997; 4(4):261–272.
44. Aoki H, Kodama M, Tani T, et al. Treatment of sepsis by extracorporeal elimination of endotoxin using polymyxin B-immobilized fiber. Am J Surg 1994; 167:412–417.
45. Cygler M, Rose DR, Bundle DL. Recognition of cell-surface oligosaccharide of pathogenic Salmonella by an antibody Fab fragment. Science 1991, 253:442–445.
46. Nnalue NA, Lind SM, Lindberg AA. The disaccharide L-α-D-heptose 1→7-L-α-D-heptose 1→ of the inner core domain of Salmonella lipopolysaccharide is accessible to antibody and is the epitope of a broadly reactive monoclonal antibody. J Immunol 1992; 149:2722–2728.
47. Parker KC, Bednarek MA, Hull LK, et al. Sequence motifs important for peptide binding to the human MHC class I molecule, HLA-A2. J Immunol 1992; 149:3580–3587.

24

The Interaction of Lipid A and Lipopolysaccharide with Human Serum Albumin

Sunil A. David
University of Kansas Medical Center, Kansas City, Kansas

INTRODUCTION

Endotoxins, or lipopolysaccharides, are important structural constituents of the outer membrane of gram-negative bacteria (1). These amphiphilic macromolecules elicit a variety of biological effects in susceptible hosts (2) and have been implicated in the pathogenesis of septic shock, an important cause of mortality worldwide (3). Knowledge of the nature of interactions of endotoxin with serum components would help address the questions of how these potentially deleterious molecules are processed and distributed in vivo and how they ultimately reach their target tissues and mediate their effects. Because of their exterior location on the bacterial membrane, they are readily accessible for interaction with serum constituents. Indeed, several normal serum components other than lipopolysaccharide (LPS)-specific antibodies such as high-density lipoproteins (4,5), complement components (6,7), lysozyme (8,9), albumin (10), and elicited plasma proteins (important among which is lipopolysaccharide-binding protein) (11–13) are known to bind LPS.

Albumin was among the first serum proteins shown to bind and solubilize lipid A, the toxic glycolipid moiety of LPS (14,15); these early studies showed that complexes of albumin and the otherwise insoluble lipid A elicited almost the entire spectrum of endotoxic activities, which was pivotal in establishing that the toxic activities of LPS were ascribable to the lipid A component. Later studies showed that albumin, both in solution (10,16) and in the cell-bound form (17), bound native LPS. As the most abundant circulatory protein with typical concentrations of 5 g/100 ml ($\approx$0.75 mM), albumin is preeminent among the serum transport proteins, long recognized as a carrier for an extraordinarily diverse range of ligands including fatty acids, hormones, bilirubin and bile salts, metals, and drugs (18). More recent studies have established that nitric oxide (19) and bacterial cell surface constituents such as peptidoglycan (20,21) and lipoteichoic acid (22) are also part of the repertoire of ligands recognized by albumin. An examination of the physical properties of these ligands indicate that they are generally hydrophobic and anionic. Such molecules as fatty acids, bilirubin, hemin, and some steroidal hormones are so hydrophobic that they are practically insoluble in their noncomplexed forms. Many of the water-soluble negatively charged ligands are small aromatic compounds such as aspirin, warfarin, ibuprofen, and azidothymidine (23–26), although large polyanions such as sulfated heparinoids (22) and phospholipids (27) also bind to albumin.

The primary structure of albumin, highly conserved among mammals, is composed of three homologous domains of about 190 amino acids (I, II, and III), each domain consisting of two predominantly α-helical subdomains (IA, IB, etc.), bearing five (domain I) or six (domains II and III) disulfide bonds. Domain I has a single free cysteine at position 34. Human serum albumin (HSA) has a lone tryptophan at position 214, at the interface of domains II and III (28). The presence of the single free sulfhydryl group and indole moiety, as will be seen later, provide useful internal probes for monitoring ligand binding. The recent determination of the high-resolution crystal structure of HSA has not only allowed the precise delineation of the binding sites on the molecule, but has also permitted rationalization in terms of molecular structure, employing the wealth of information on the binding chemistry of albumin that has accumulated in the literature over the years (28,29). Most ligands bind to cavities in subdomains IIA and IIIA, the latter being the predominant site. These sites are distinctly amphipathic, characterized by the presence of a hydrophobic surface on the side of the cavity and a positively charged surface on the other, a well-recognized structural determinant of endotoxin-binding molecules (30–37). Long-chain fatty acids, however, appear to localize at or near the surface of the B subdomains (28).

The anionic and amphipathic nature of lipid A and LPS and the earlier observations that albumin could indeed act as a solubilizer for lipid A prompted a detailed characterization of the interactions of endotoxin with HSA in order to determine the binding parameters and site(s) of interaction. Since the structural heterogeneity of LPS (38) is a potential impediment to detailed analyses of such interactions, lipid A, which is structurally well characterized and considerably less heterogeneous than native LPS, was used. The availability of site-specific fluorescent probes and the presence of the single tryptophan and free cysteine thiol group in HSA serving as unique spectroscopic "handles" rendered fluorescent spectroscopic techniques an excellent means of characterizing the interactions (39).

DETERMINATION OF STOICHIOMETRY AND AFFINITY OF BINDING

The lone tryptophan residue at position 214 could be employed as a convenient intrinsic fluorescent probe to monitor the interaction with lipid A. The addition of lipid A to HSA resulted in blue-shifting of the maximal emission wavelength when excited at 275 nm, saturating at a lipid A:HSA molar ratio of 2:1 (Fig. 1). The emission intensity of Trp-214, however, was distinctly biphasic (Fig. 1), manifesting initially in quenching, which is maximal at 1 equivalent of lipid A and from 1 to 2 equivalents of lipid A in emission enhancement. These results are indicative of a lipid A:HSA stoichiometry of 2:1, and the biphasic behavior of Trp-214 intensity is suggestive of sequential occupancy. Dissociation constants of lipid A for HSA obtained by Scatchard treatment of the emission intensity data were 1.998 μM and 6.389 μM, respectively, for the two sites (Table 1); these values are intermediary between the high-affinity (fatty acids and bilirubin, 10^{-7} M) and low-affinity (bile salts and steroid hormones, 10^{-3}–10^{-4} M) interactions. The steady-state emission polarization value of Trp-214 in the absence of lipid A is 0.154 ± 0.003 and remained essentially unchanged upon lipid A addition, signifying that Trp-214 does not experience hindered rotational mobility in the presence of lipid A. These results, taken together, suggested that the intensity and emission wavelength changes were an indirect effect, rather than a consequence of lipid A binding in the vicinity of Trp-214 (domain II).

DETERMINATION OF LIPID A–BINDING SITES

Fluorescent probes specific for subdomains IA, IIA, and IIIA were used in order to identify the potential

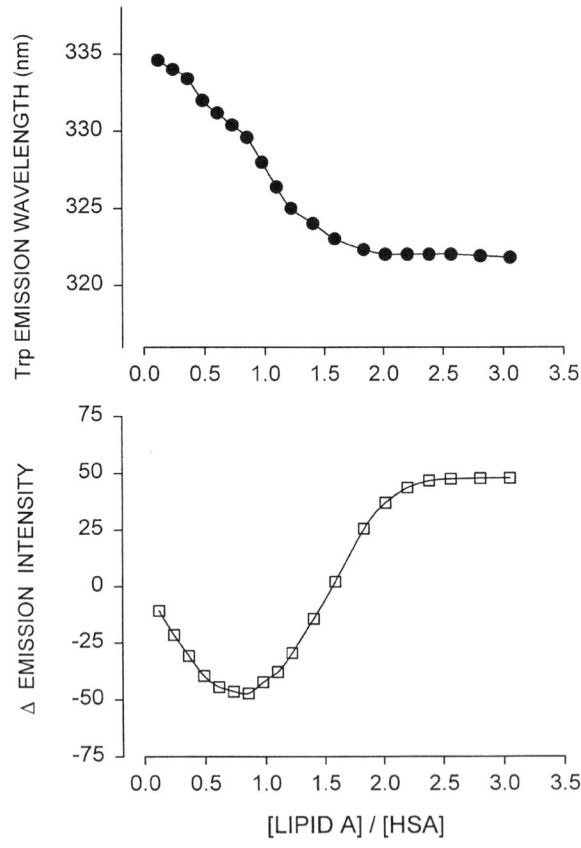

Fig. 1 (*Top*) Blue shift of tryptophanyl emission of HSA (●) plotted as a function of lipid A:HSA molar ratio. (*Bottom*) Biphasic intensity changes plotted as change in intensity from uncomplexed, free HSA (□). These, and all other titration experiments depicted in the figures to follow were carried out in 10 mM Tris-HCl buffer, pH 7.4, at 25°C. HSA concentration: 4.43 μM. Excitation: 275 nm.

Table 1 Binding Parameters of Lipid A–HSA Interactions

Parameter	$K_d(1)$ (μM)	$K_d(2)$ (μM)	Probe (μM)	Method	Ref.
Tryptophanyl emission intensity	1.998	6.389	—	Scatchard	71
Tryptophanyl emission wavelength	—	6.238	—	Scatchard	71
Dansylsarcosine displacement	0.954	—	6.11	LIGAND	42
Dansylsarcosine displacement	1.276	—	7.16	Horovitz-Levitzki	43

sites of interaction of lipid A with HSA. The free thiol of Cys-34 in domain IA was covalently coupled to a dansyl group using a sulfhydryl-specific reagent. The addition of lipid A to dansylated HSA was without effect on dansyl emission intensity, steady-state dansyl emission polarization, and efficiency of energy transfer from Trp-214 to dansyl, indicating that lipid A does not bind to this region. Although domain I is structurally homologous to domains II and III and bears an invariant group of basic amino acids (28), the packing geometry of the helical segments of this domain is such that it effectively eliminates the binding pocket in domain IA (28,29).

Dansylsarcosine and warfarin were used as fluorescent probes for subdomains IIIA and IIA, respectively (23–26,40,41). The addition of HSA to either dansylsarcosine or warfarin resulted in enhancements in emission intensities of the probes as expected, consistent with binding of the probes to HSA (23). The addition of lipid A to HSA-bound dansylsarcosine resulted in quenching of fluorescence (Fig. 2), while lipid A by itself was without effect on free dansylsarcosine, and the addition of dansylsarcosine to HSA preincubated with a 10-molar excess of lipid A did not result in enhancement of the probe fluorescence. These results are indicative of the occupancy of the probe and lipid A at the same site and that dansylsarcosine is competitively displaced from HSA by lipid A. Dissociation constants for lipid A were obtained by analyzing the displacement data by fitting a one-site competitive model using LIGAND (42) and by Horovitz-Levitzki (43) analysis. Both methods allow the dissociation constants of the probe as well as the displacing ligand to be estimated. The binding constants for lipid A were 0.95 and 1.27 μM and that of the probe were 6.11 and 7.16 μM, as computed by the two methods, respectively (Table 1). The latter values are in close agreement to those reported in the literature (6 μM) (23).

Titration of lipid A with HSA-bound warfarin showed a biphasic response, with fluorescence enhancement saturating at approximately 2 equivalents of lipid A, followed by quenching at higher lipid A concentrations (Fig. 2, bottom). Control experiments as described above verified that lipid A was without effect on free probe, and lipid A sequestered by polymyxin B did not alter the fluorescence of HSA-bound probe. The enhancement of warfarin emission in the initial phase could represent increased probe binding (24), or may be a consequence of lipid A–induced local conformational changes in HSA resulting in altered microenvironment and therefore of the quantum yield of the bound probe. The quenching occurring at higher lipid A concentrations may likewise mirror conformational perturbations or may be indicative of probe displacement. In order to further examine the possibilities, these experiments were also performed with indirect excitation of the bound probes via resonance energy transfer from Trp-214 of HSA. In these experiments, Trp-214 was excited at 275 nm, and the emission of the probe was followed as a function of lipid A added. Both warfarin and dansylsarcosine were examined, the latter being a control for warfarin. The relative efficiency of energy transfer from Trp-214 of HSA to bound dansylsarcosine decreased with increasing lipid A concentration (Fig. 3), which, taken together with the fluorescence quenching observed by direct dansylsarcosine excitation as mentioned above, means that results would be consistent with probe displacement. The emission intensity changes of bound warfarin, however, are only in the range of $\pm 1.5\%$ (Fig. 3), ruling out displacement. Therefore, although the tryptophan-214 data presented above may at first be suggestive of the involvement of subdomain IIA, the lack of changes in emission polarization and the inability of lipid A to displace HSA-bound warfarin indicate that this site does not contribute to lipid A binding. It is recognized that ligands that bind to domain III can induce conformational changes in domain II because of a shared interface region between the two regions, as is apparent in the crystal structure of HSA (29). Furthermore, Trp-214 also participates in the formation of the interface between the two domains (28,29). In light of these structural considerations, the changes in Trp-214 and warfarin fluorescence are interpreted to be the result of

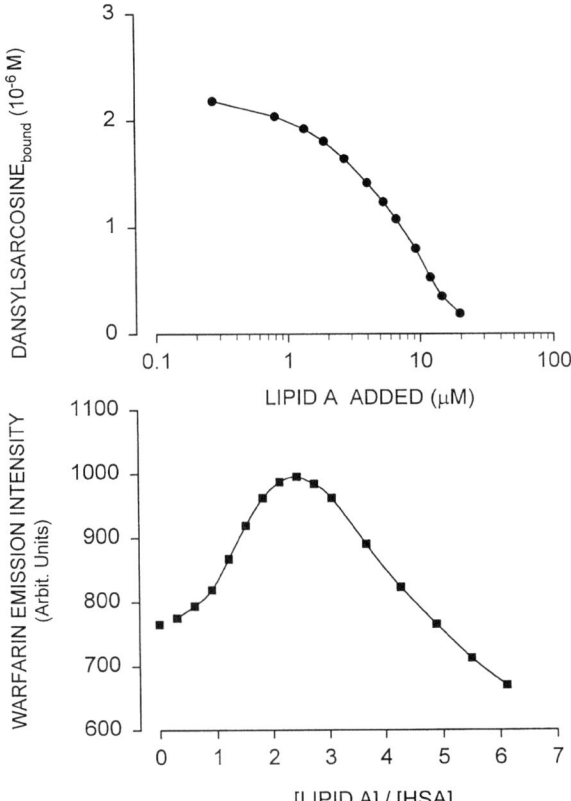

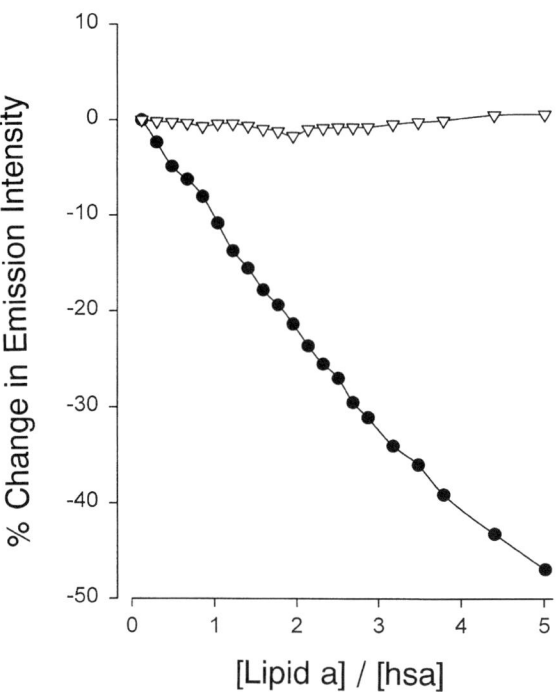

Fig. 2 Effect of lipid A on fluorescent probes bound to HSA. Dansylsarcosine displacement (●) and effect on warfarin emission (■) as probed by direct excitation of probe are shown. Bound dansylsarcosine concentrations (top) were computed from values of saturating fluorescence (maximal emission when the probe is completely bound to HSA); these values were derived from double-reciprocal plots constructed for the probe by titrating HSA against a fixed concentration of probe (71). HSA concentration was 4.43 μM in both experiments. Dansylsarcosine concentration: 8.35 μM. Warfarin concentration: 8.3 μM.

Fig. 3 Lipid A–induced changes in emission intensities of warfarin (●) and dansylsarcosine (△) bound to HSA as measured by indirect excitation and energy transfer from Trp-214. Excitation was at 280 nm; warfarin and dansylsarcosine emission intensities were recorded at 385 nm and 475 nm, respectively, when lipid A was titrated against a fixed concentration of HSA:probe complex. HSA concentration: 4.43 μM. Warfarin concentration: 8.5 μM. Dansylsarcosine concentration: 8.5 μM.

local conformational changes in domain II as a consequence of lipid A binding to domain III. This interpretation would be consistent with altered ligand binding to domain IIA upon domain IIIA occupancy (24) and the dependency of warfarin binding on the conformational state of albumin (44).

Since lipid A interacts with neither domain I nor II, it is to be expected that domain III represents the site of interaction. It is evident from the dansylsarcosine displacement data that subdomain IIIA is at least one of the sites, but does not resolve whether one or both sites are represented by this region. Based on the above data, the estimated dimensions of lipid A (45), and the fact that the domains A and B share a common structural motif (29), it is proposed that regions III-A and III-B represent the high and the low affinity sites of lipid A binding, respectively. Atomic coordinates of HSA, when made accessible in the Protein Data Bank, should enable this proposition to be examined by molecular modeling.

INTERACTION OF LIPOPOLYSACCHARIDE WITH HUMAN SERUM ALBUMIN

The addition of LPS labeled with fluorescein isothiocyanate (FITC) to HSA dansylated at Cys-34 resulted in incremental quenching of dansyl emission, which is coupled with fluorescein intensity enhancement indicative of spatial proximity of the donor and acceptor fluorophores on HSA and LPS, respectively. These changes are reversed upon adding either excess unlabeled LPS or lipid A, signifying that whole LPS also binds to HSA and that the lipid A moiety of LPS represents the interaction site on the toxin molecule. LPS displaces dansylsarcosine from HSA, but at rather high concentrations (ED_{50} = approx. 250 μg/ml) and, unlike lipid A, does not enhance HSA-bound warfarin fluorescence at low concentrations, but quenches it in a mass-dependent manner. These data are not plotted since the heterogeneity of LPS precludes the computation of meaningful ligand:protein molar ratios. Thus, although these results yield neither stoichiometry nor binding constants, it is apparent that subdomain IIIA is an LPS-binding site. Differential interactions, which are apparently dependent on the chemotype structure of LPS, have been noted with other circulatory proteins of mammalian origin such as lysozyme (46), which are thought to be a consequence of the different supramolecular states of LPS chemotypes. Smooth-form LPS containing O-specific polysaccharide units tend to exist as closed lamellar structures, which allow a limited exposure of the hydrophobic domains at the surface of the molecular aggregates, while lipid A can exist in several nonlamellar forms (47,48) under experimental conditions similar to those employed in this study. Steric differences between lipid A and LPS may also be of importance, the latter possessing a rigid and bulky anionic core oligosaccharide region (49,50) as well as an extensive branched structure comprising of repeating polysaccharide units. Furthermore, recent theoretical calculations have shown that ionization states at physiological pH of free lipid A and LPS (rough-mutant form) are considerably different (51), which is likely to influence their solubilities. This is of importance since HSA can be expected only to interact with soluble lipid A or LPS molecules below their critical micellar concentrations. Although on theoretical grounds, one would predict the solubility of whole LPS to be higher than that of free lipid A, most preparations of LPS such as the one used in the present study contain substantial amounts of divalent (Ca^{2+}, Mg^{2+}) and organic (pyridinium) cations, which greatly limit its solubility. Indeed, LPS yields a highly turbid suspension in its concentrated form, while lipid A in the presence of triethylamine is an optically clear solution. Further studies using a series of LPS chemotypes obtained in salt forms of uniform aqueous solubilities are essential to address this issue.

ALBUMIN AND POLYMYXIN B RECOGNIZE DIFFERENT SITES ON LPS

A wide variety of proteins have been reported to bind to LPS or lipid A, giving rise to complexes with either unaltered or attenuated endotoxic activity (see Table 2 for a partial list). Anti–lipid A monoclonal antibodies such as E-5 and HA-1A are known to bind to different epitopes on the lipid A molecule (52), and several nonantibody proteins such as pertussis toxin (53) and lysozyme (9,54,55) appear to bind different regions of lipid A. Adequate and detailed structural and/or binding data, while not available for the majority of these LPS-binding molecules, would be valuable in providing a structural basis for why molecules such as LBP and BPI, although they bind LPS with comparable affinities, differentially modulate LPS activity. It was of interest, therefore, to determine if polymyxin B and HSA recognize the same regions on lipid A or LPS.

In order to unambiguously examine the possibility of the formation of HSA:lipid A/LPS:polymyxin B ternary complexes using fluorescence spectroscopic methods, it was convenient to use resonance energy transfer from the intrinsic donor tryptophan fluorescence of HSA successively to acceptor fluorophores on lipid A or LPS and on polymyxin B. The lack of derivatizable groups on lipid A limited the application of this technique to LPS. Polymyxin B was dansylated on one of its primary amino groups, which served as the primary acceptor fluorophore, and the fluorescein label on LPS as the secondary acceptor. Spectral overlaps of the Trp-214/dansyl and dansyl/fluorescein pairs enables energy transfer from Trp to FITC via the dansyl fluorophore if the three interacting components are simultaneously within the Förster energy transfer–limiting distance (≈ 60 Å). This is manifested as a decrease in donor emission intensity paralleled by a concomitant increase in acceptor emission. Although the formalism of the technique, in principle, allows the estimation of interfluorophore distances, attempts were not made to derive such data from the experimental results since the site(s) and stoichiometry of FITC labeling of LPS are unknown, and the stoichiometry of the LPS:HSA complex itself is indeterminable due to heterogeneity inherent in LPS.

Dansyl-polymyxin B forms ternary complexes with FITC-LPS bound to HSA as manifested by progressive quenching of Trp-214 fluorescence of HSA accompanied by intensity enhancements of both dansyl and FITC fluorescence (Fig. 4) upon addition of dansyl-polymyxin B. These changes are indicative of energy transfer from Trp-214 to FITC via dansyl, since in control experiments, dansyl-polymyxin B quenches FITC-LPS in the absence of HSA. These data show that HSA and polymyxin B recognize different sites on LPS; because it is clear that it is the lipid A moiety that interacts with both polymyxin B and HSA, it would

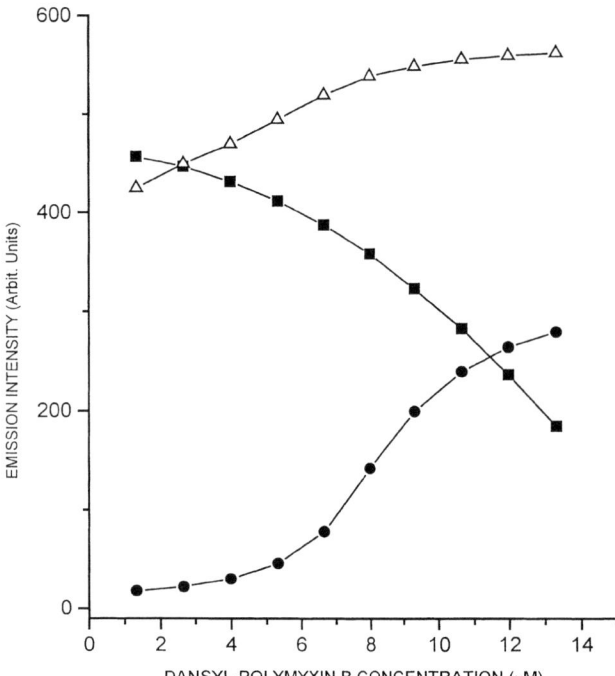

Fig. 4 Resonance energy transfer from *Trp*-214 (HSA) → *Dansyl*-polymyxin B → *FITC*-LPS demonstrating noncompetitive binding of dansylated polymyxin B to FITC-LPS complexed to HSA. HSA (10 μM) was preincubated with FITC-LPS (50 μg) in buffer. Small aliquots of dansylpolymyxin were added, equilibrated for 5 minutes, and spectra recorded with excitation at 275 nm. Intensities of Trp-214 (■), dansyl-polymyxin B (●), and FITC-LPS (△) were recorded at 325, 470, and 525 nm, respectively.

be reasonable to assume that these substances bind to different regions on lipid A with respect to polymyxin B.

GENERALIZED CONCEPTS OF ALBUMIN-LPS INTERACTIONS: RELATIONSHIP TO OTHER LPS-PROTEIN INTERACTIONS

Experimental data have been obtained that demonstrate that two molecules of lipid A bind to HSA, with apparent dissociation constants for the two sites corresponding to 1 and 6 μM, respectively. The considerable affinity toward lipid A and mass action effects by virtue of HSA being the most abundant serum protein would imply a major role as a carrier molecule for endotoxins and may therefore be contributory, in the clinical setting, to the generalized systemic response seen even in highly localized gram-negative infections. In this regard, it may further be noted that albumin, being the key factor in maintaining colloid osmotic pressure in the extracellular space and abundantly present even in transudates, seems a probable candidate to aid in the systemic dissemination of LPS released from otherwise highly localized foci of infection. Given the observation that LPS localizes on the cell membrane predominantly as cell-bound albumin (17), it is tempting to speculate that albumin may also serve as a conduit in the delivery of LPS to specific receptors on the cell membrane.

One of the sites on HSA is subdomain IIIA, while the other presumably, is IIIB. At neutral pH, only domain III is zwitterionic with a net charge of 0, while domains I and II have net charges of -10 and -8, respectively (18). Given the anionic nature of LPS and lipid A and the observation that cationicity is a necessary requisite for LPS-binding substances (32,33,37,56), it would appear that domain III would present electrostatically favorable sites for LPS binding. It is to be noted that this region of HSA has several highly basic arginine residues (57). A more precise understanding of the nature of endotoxin binding could potentially come from studying other members of the albumin family; α-fetoprotein, an albumin-related transport protein that shares a 39% homology with HSA lacks the double disulfide bridge in subdomain IIA, and vitamin D–binding protein, yet another member of the albumin family, lacks subdomain IIIB almost entirely (28). A novel 87 kDa protein with a domain structure similar to that of the albumins called afamin has recently be identified (58). A comparative study of LPS binding with these proteins may help better conceptualize the structural determinants that govern LPS binding in these proteins.

Although native LPS also binds to HSA through its lipid A moiety, there are differences between free lipid A and LPS with respect to their interactions with HSA. This is probably attributable to differences in the physical states of lipid A and LPS. It has also become apparent that HSA and polymyxin B bind to different sites on the lipid A portion of LPS. Complexes of lipid A or LPS with albumin exhibit the full spectrum of biological activities (14,15), being in this respect similar to other serum factors that behave as LPS opsonins, such as the septins (59) and lipopolysaccharide-binding protein (13,60–64), while other serum proteins such as lysozyme (9,54,55) or bactericidal/permeability increasing protein (65–68) can either partially or completely neutralize endotoxin activity. This suggests that these LPS-binding proteins may recognize disparate regions on the endotoxin and, consequentially, present parts of the toxin molecule which are either toxically active or quiescent to target cells. A unique "toxo-

Table 2 Correlation of LPS-Antagonistic Activity and Basicity in Some LPS-Binding Proteins

Protein	Activity	pK[a]	Ref.
LPS-binding protein (LBP), human	N	6.24	12, 13, 72, 73
Septin	N	?	59
Albumin, human	N	5.67	14, 17, 39, 74
Hemolymph LPS-binding protein, *Periplaneta americana*	N	5.11	75, 76
Lipophorin	N	5.58	77
BPI, human	I	9.44	66, 67, 78
Acyloxyacyl hydrolase, human	I	8.87, 7.63 (13 & 48 kDa subunits)	79–83
Lactoferrin, human	I	8.51	84, 85
CAP18 protein, rabbit	I	9.43	35, 36, 86–88
Anti-LPS factor, *Limulus polyphemus*	I	10.09	69, 89

N = Noninhibitory; I = inhibitory.
[a] pK computed by the SWISS-PROT databank (90).

phore'' on the lipid A/LPS molecule has not yet been identified, and it is of relevance in this regard to note that the ability of polymyxin B to form ternary complexes with LPS bound to macromolecular ligands correlates with the retention of toxicity in the LPS-protein complexes. Both HSA and lysozyme (9,54,55), which, respectively, do not or only partially modify the activity of complexed LPS, permit noncompetitive interactions with polymyxin B and the consequent formation of ternary complexes, whereas proteins that neutralize endotoxin activity such as the *Limulus* anti-LPS factor (69) or the HA-1A monoclonal antibody to lipid A (70), bind to LPS in a competitive manner. The elucidation of the structural basis of this phenomenon would be of value in structure-toxicity relationships in LPS and in designing specific endotoxin-sequestering agents. Upon a closer examination of the physical properties of such endotoxin-binding proteins, a rather intriguing trend became apparent: several proteins that only bind LPS and do not inhibit endotoxic activity are generally negatively charged at neutral pH with a pK of less than 7.0, while proteins that "sequester" LPS, inhibiting its toxicity, are basic with a much higher pK (Table 2). This observation, although preliminary and deserving of more careful and thorough scrutiny, might be a valuable tool in identifying potential molecular targets of lipopolysaccharide.

ACKNOWLEDGMENTS

This work was supported in part by grants-in-aid from the Indian Council of Medical Research, New Delhi, the Wellcome Trust, London, and the Rockefeller Foundation, New York. The authors are grateful to Dr. P. J. Munson for the LIGAND programs.

REFERENCES

1. Nikaido H, Vaara M. The gram-negative bacterial outer membrane. In: Neidhardt F, ed. *Escherichia coli* and *Salmonella typhimurium*. Washington, DC: ASM Publications, 1987:7–44.
2. Morrison DC, Ulevitch RJ. The effects of bacterial endotoxins on host mediation systems. Am J Pathol 1978; 93:527–617.
3. Young LS. Gram-negative sepsis. In: Mandell GL, et al., eds. Principles and Practice of Infectious Diseases. New York: Wiley, 1985:452.
4. Freudenberg MA, Bog-Hansen TC, Back U, Galanos C. Interaction of lipopolysaccharides with plasma high density lipoproteins in rats. Infect Immun 1980; 28:373–380.
5. Ulevitch RJ, Johnston AR, Weinstein DB. New function for high density lipoproteins. Isolation and characterization of a bacterial lipopolysaccharide-high density lipoprotein complex formed in rabbit plasma. J Clin Invest 1981; 67:827–837.
6. Cooper NR, Morrison DC. Binding and activation of the first component of human complement by the lipid A region of lipopolysaccharide. J Immunol 1978; 120:1862–1868.
7. Vukajlovich SW. Antibody-independent activation of the classical pathway of human serum complement by lipid A is restricted to Re-chemotype lipopolysaccharide and purified lipid A. Infect Immun 1986; 53:480–485.
8. Ohno N, Morrison DC. Lipopolysaccharide interaction with lysozyme. Binding of lipopolysaccharide to lysozyme and inhibition of lysozyme enzymatic activity. J Biol Chem 1989; 264:4434–4441.
9. Ohno N, Tanida N, Yadomae T. Characterization of complex formation between lipopolysaccharide and lysozyme. Carbohydr Res 1991; 214:115–130.

10. Tesh VL, Vukajlovich SW, Morrison DC. Endotoxin interactions with serum proteins: relationship to biological activity. In: Levin J, et al., eds. Bacterial Endotoxins: Pathophysiologic Effects, Clinical Significance, and Pharmacologic Control. New York: Alan R. Liss, 1988: 47–62.
11. Tobias PS, Soldau K, Ulevitch RJ. Isolation of a lipopolysaccharide-binding acute phase reactant from rabbit serum. J Exp Med 1986; 164:778–793.
12. Tobias PS, Soldau K, Ulevitch RJ. Identification of a lipid A binding site in the acute phase reactant lipopolysaccharide binding protein. J Biol Chem 1989; 264:10867–10871.
13. Mathison JC, Tobias PS, Wolfson E, Ulevitch RJ. Plasma lipopolysaccharide (LPS)-binding protein. A key component in macrophage recognition of gram-negative LPS. J Immunol 1992; 149:200–206.
14. Rietschel ET, Kim YB, Watson DW, Galanos C, Lüderitz O, Westphal O. Pyrogenicity and immunogenicity of lipid A complexed with bovine serum albumin or human serum albumin. Infect Immun 1973; 8:173–177.
15. Galanos C, Rietschel ET, Lüderitz O, Westphal O. Biological activities of lipid A complexed with bovine serum albumin. Eur J Biochem 1972; 31:230–233.
16. Wollenweber H-W, Morrison DC. Synthesis and biochemical characterization of a photoactivatable, iodinatable, cleavable bacterial lipopolysaccharide derivative. J Biol Chem 1985; 260:15068–15074.
17. Dziarski R. Cell-bound albumin is the 70-kDa peptidoglycan-, lipopolysaccharide-, and lipteichoic acid-binding protein on lymphocytes and macrophages. J Biol Chem 1994; 269:20431–20436.
18. Peters T, Jr. Albumin: An Overview and Bibliography. 2d ed. Kankakee, IL: Miles Inc. Diagnostic Division, 1992.
19. Stamler JS, Singel DJ, Loscalzo J. Biochemistry of nitric oxide and its redox-activated forms. Science 1992; 258:1898–1902.
20. Dziarski R. Peptidoglycan and lipopolysaccharide bind to the same binding site on lymphocytes. J Biol Chem 1991; 266:4719–4725.
21. Dziarski R. Demonstration of peptidoglycan binding sites on lymphocytes and macrophages by photoaffinity cross-linking. J Biol Chem 1991; 266:4713–4718.
22. Dziarski R, Gupta D. Heparin, sulfated heparinoids, and lipoteichoic acids bind to the 70-kDa peptidoglycan/lipopolysaccharide receptor protein on lymphocytes. J Biol Chem 1994; 269:2100–2110.
23. Sudlow G, Birkett DJ, Wade DN. The characterization of two specific drug binding sites on human serum albumin. Mol Pharmacol 1975; 11:824–832.
24. Sjoholm I, Ekman B, Kober A, Ljungstedt-Pahlman I, Sieving B, Sjodin T. Binding of drugs to human serum albumin: XI. The specificity of three binding sites as studied with albumin immobilized in microparticles. Mol Pharmacol 1979; 16:767–777.
25. Sudlow G, Birkett DJ, Wade DN. Further characterization of specific drug binding sites on human serum albumin. Mol Pharmacol 1976; 12:1052–1061.
26. Sollenne NO, Means GE. Characterization of a specific drug binding site on human serum albumin. Mol Pharmacol 1979; 15:754–757.
27. Jonas A. Interaction of phosphatidylcholine with bovine serum albumin: specificity and properties of the complexes. Biochim Biophys Acta 1976; 427:325–336.
28. Carter DC, Ho JX. Structure of serum albumin. Adv Prot Chem 1994; 45:153–203.
29. He XM, Carter DC. Atomic structure and chemistry of human serum albumin. Nature 1992; 358:209–215.
30. David SA, Balaram P, Mathan VI. Interaction of basic amphiphilic polypeptide antimicrobials, Gramicidin S, Tyrocidin, and Efrapeptin, with endotoxic lipid A. Med Microbiol Lett 1993; 2:42–47.
31. David SA, Mathan VI, Balaram P. Interaction of melittin with endotoxic lipid A. Biochim Biophys Acta 1992; 1123:269–274.
32. David SA, Bechtel B, Annaiah C, Mathan VI, Balaram P. Interaction of cationic amphiphilic drugs with lipid A: implications for development of endotoxin antagonists. Biochim Biophys Acta Lipids Lipid Metab 1994; 1212:167–175.
33. David SA, Mathan VI, Balaram P. Interactions of linear dicationic molecules with lipid A: structural features that correspond to optimal binding affinity. J Endotoxin Res 1995; 2:325–336.
34. David SA, Awasthi SK, Wiese A, Ulmer AJ, Lindner B, Brandenburg K, Seydel U, Rietschel ET, Sonesson A, Balaram P. Characterization of the interactions of a polycationic, amphiphilic, terminally branched oligopeptide with lipid A and lipopolysaccharide from the deep rough mutant of *Salmonella minnesota*. J Endotoxin Res 1996; 3:369–379.
35. Larrick JW, Hirata M, Balint RF, Lee J, Zhong J, Wright SC. Human CAP18: a novel antimicrobial lipopolysaccharide-binding protein. Infect Immun 1995; 63:1291–1297.
36. Hirata M, Shimomura Y, Yoshida M, Wright SC, Larrick JW. Endotoxin-binding synthetic peptides with endotoxin-neutralizing, antibacterial and anticoagulant activities. Prog Clin Biol Res 1994; 388:147–159.
37. Porro M. Structural basis of endotoxin recognition by natural polypeptides. Trends Microbiol 1994; 2:65–66.
38. Jann B, Reske K, Jann K. Heterogeneity of lipopolysaccharides. Analysis of polysaccharide chain lengths by sodium dodecylsulfate-polyacrylamide gel electrophoresis. Eur J Biochem 1975; 60:239–246.
39. David SA, Balaram P, Mathan VI. Characterization of the interaction of lipid A and lipopolysaccharide with human serum albumin: implications for an endotoxin carrier function for albumin. J Endotoxin Res 1995; 2:99–106.
40. Berde CB, Hudson BS, Simoni RD, Sklar LA. Human serum albumin: spectroscopic studies of binding and proximity relationships for fatty acids and bilirubin. J Biol Chem 1979; 254:391–400.
41. Fehske KJ, Schlafer U, Wollert U, Muller WE. Characterization of an important drug binding area on human serum albumin including the high-affinity binding sites for warfarin and azapropazone. Mol Pharmacol 1982; 21:387–393.
42. Munson PJ, Rodbard D. LIGAND: A versatile computerized approach for characterization of ligand-binding systems. Anal Biochem 1980; 107:220–239.
43. Horovitz A, Levitzki L. An accurate method for determination of receptor-ligand and enzyme-inhibitor dis-

sociation constants from displacement curves. Proc Natl Acad Sci USA 1987; 84:6654–6658.
44. Wilting J, van der Giesen WF, Janssen LHM, Weideman MM, Otagiri M. The effect of albumin conformation on the binding of warfarin to human serum albumin. J Biol Chem 1980; 255:3032–3037.
45. Kastowsky M, Sabisch A, Gutberlet T, Bradaczek H. Molecular modelling of bacterial deep rough mutant lipopolysaccharide of *Escherichia coli*. Eur J Biochem 1991; 197:707–716.
46. Ohno N, Morrison DC. Effects of lipopolysaccharide chemotype structure on binding and inactivation of hen egg lysozyme (abstr). Eur J Biochem 1989; 186:621–627.
47. Seydel U, Brandenburg K, Koch MHJ, Rietschel ET. Supramolecular structure of lipopolysaccharide and free lipid A under physiological conditions as determined by synchroton small-angle x-ray diffraction. Eur J Biochem 1989; 186:325–332.
48. Brandenburg K, Mayer H, Koch MHJ, Weckesser J, Rietschel ET, Seydel U. Influence of the supramolecular structure of free lipid A on its biological activity. Eur J Biochem 1993; 218:555–563.
49. Kastowsky M, Gutberlet T, Bradaczek H. Molecular modelling of the three-dimensional structure and conformational flexibility of bacterial lipopolysaccharide. J Bacteriol 1992; 174:4798–4806.
50. Kastowsky M. Modelling the three-dimensional structure of lipopolysaccharide. In: Levin J, Alving CR, Munford RS, Stuetz PL, eds. Bacterial Endotoxin: Recognition and Effector Mechanisms. Amsterdam:Elsevier Science Publishers B.V., 1995:61–70.
51. Din ZZ, Mukerjee P, Kastowsky M, Takayama K. Effect of pH on solubility and ionic state of lipopolysaccharide obtained from the deep rough mutant of *Escherichia coli*. Biochemistry 1993; 32:4579–4586.
52. Wisniewski MA, Kazemi M, Fang IS, Knight LS, Huntenburg CC, Bubbers JE, Schneidkraut MJ. Comparison of binding specificity and the function of two human IgM anti-lipid A monoclonal antibodies. Circ Shock 1994; 44:230–237.
53. Lei MG, Morrison DC. Lipopolysaccharide interaction with S2 subunit of pertussis toxin. J Biol Chem 1993; 268:1488–1493.
54. Takada K, Ohno N, Yadomae T. Detoxification of lipopolysaccharide (LPS) by egg white lysozyme. FEMS Immunol Med Microbiol 1994; 9:255–264.
55. Takada K, Ohno N, Yadomae T. Binding of lysozyme to lipopolysaccharide suppresses tumor necrosis factor production in vivo. Infect Immun 1994; 62:1171–1175.
56. Rustici A, Velucchi M, Faggioni R, Sironi M, Ghezzi P, Quataert S, Green B, Porro M. Molecular mapping and detoxification of the lipid A binding site by synthetic peptides. Science 1993; 259:361–365.
57. Jonas A, Weber G. Presence of arginine residues at the strong, hydrophobic anion binding sites of bovine serum albumin. Biochemistry 1971; 10:1335–1339.
58. Lichenstein HS, Lyons DE, Wurfel MM, Johnson DA, McGinley MD, Leidli JC, Trollinger DB, Mayer JP, Wright SD, Zukowski MM. Afamin is a new member of the albumin, alpha-fetoprotein, and vitamin D-binding protein gene family. J Biol Chem 1994; 269:18149–18154.
59. Wright SD, Ramos RA, Patel M, Miller DS. Septin: A factor in plasma that opsonizes lipopolysaccharide-bearing particles for recognition by CD14 on phagocytes. J Exp Med 1992; 176:719–727.
60. Hailman E, Lichenstein HS, Wurfel MM, Miller DS, Johnson DA, Kelley M, Busse LA, Zukowski MM, Wright SD. Lipopolysaccharide (LPS)-binding protein accelerates the binding of LPS to CD14. J Exp Med 1994; 179:269–277.
61. Mathison J, Wolfson E, Steinemann S, Tobias P, Ulevitch R. Lipopolysaccharide (LPS) recognition in macrophages. Participation of LPS-binding protein and CD14 in LPS-induced adaptation in rabbit peritoneal exudate macrophages. J Clin Invest 1993; 92:2053–2059.
62. Pugin J, Schürer-Maly C-C, Leturcq D, Moriarty A, Ulevitch RJ, Tobias PS. Lipopolysaccharide activation of human endothelial and epithelial cells is mediated by lipopolysaccharide-binding protein and soluble CD14. Proc Natl Acad Sci USA 1993; 90:2744–2748.
63. Betz Corradin S, Mauël J, Gallay P, Heumann D, Ulevitch RJ, Tobias PS. Enhancement of murine macrophage binding of and response to bacterial lipopolysaccharide (LPS) by LPS-binding protein. J Leuk Biol 1992; 52:363–368.
64. Schumann RR. Function of lipopolysaccharide (LPS)-binding protein (LBP) and CD14, the receptor for LPS/LBP complexes: A short review. Res Immunol 1992; 143:11–15.
65. Ooi CE, Weiss J, Doerfler ME, Elsbach P. Endotoxin-neutralizing properties of the 25 kD N-terminal fragment and a newly isolated 30 kD C-terminal fragment of the 55-60 kD bactericidal/permeability-increasing protein of human neutrophils. J Exp Med 1991; 174:649–655.
66. Gazzano-Santoro H, Parent JB, Grinna L, Horwitz A, Parsons T, Theofan G, Elsbach P, Weiss J, Conlon PJ. High-affinity binding of the bactericidal/permeability-increasing protein and a recombinant amino-terminal fragment to the lipid A region of lipopolysaccharide. Infect Immun 1992; 60:4754–4761.
67. Wilde CG, Seilhamer JJ, McGrogan M, Ashton N, Snable JL, Lane JC, Leong SR, Thornton MB, Miller KL, Scott RW, Marra MN. Bactericidal/permeability-increasing protein and lipopolysaccharide (LPS)-binding protein. LPS binding properties and effects on LPS-mediated cell activation. J Biol Chem 1994; 269:17411–17416.
68. Kohn FR, Kung AHC. Role of endotoxin in acute inflammation induced by gram-negative bacteria:specific inhibition of lipopolysaccharide-mediated responses with an amino-terminal fragment of bactericidal/permeability-increasing protein. Infect Immun 1995; 63:333–339.
69. Hoess A, Watson S, Siber GR, Liddington R. Crystal structure of an endotoxin-neutralizing protein from the horseshoe crab, *Limulus* anti-LPS factor, at 1.5 Å resolution. EMBO J 1993; 12:3351–3356.
70. Bogard WC, Jr., Siegel SA, Leone AO, Damiano E, Shealy DJ, Ely TM, Frederick B, Mascelli MA, Siegel RC, Machielse B, Naveh D, Kaplan PM, Daddona PE. Human monoclonal antibody HA-1A binds to endo-

71. David SA, Balasubramanian KA, Mathan VI, Balaram P. Analysis of the binding of polymyxin B to endotoxic lipid A and core glycolipid using a fluorescent displacement probe. Biochim Biophys Acta 1992; 1165:147–152.
72. Ishii Y, Wang Y, Haziot A, Del Vecchio PJ, Goyert SM, Malik AB. Lipopolysaccharide binding protein and CD14 interaction induces tumor necrosis factor-a generation and neutrophil sequestration in lungs after intratracheal endotoxin. Circ Res 1993; 73:15–23.
73. Gallay P, Barras C, Tobias PS, Calandra T, Glauser MP, Heumann D. Lipopolysaccharide (LPS)-binding protein in human serum determines the tumor necrosis factor response of monocytes to LPS. J Infect Dis 1994; 170:1319–1322.
74. Falk MC, Fletcher MA, Williams TJ, Ching WM, Wu Y, Morrison TK. Aggregation of serum proteins with lipopolysaccharide (LPS): characterization of the precipitable LPS-protein complex. J Endotoxin Res 1996; 3:129–142.
75. Jomori T, Natori S. Function of the lipopolysaccharide-binding protein of *Perplaneta americana* as an opsonin. FEBS Lett 1992; 296:283–286.
76. Kawasaki K, Kubo T, Natori S. A novel role of *Periplaneta* lectin as an opsonin to recognize 2-keto-3-deoxy octonate residues of bacterial lipopolysaccharides. Comp Biochem Physiol [B] 1993; 106B:675–680.
77. Kato Y, Motoi Y, Taniai K, Kadono-Okuda K, Yamamoto M, Higashino Y, Shimabukuro M, Chowdhury S, Xu J, Sugiyama M, Hiramatsu M, Yamakawa M. Lipopolysaccharide-lipophorin complex formation in insect hemolymph: a common pathway of lipopolysaccharide detoxification both in insects and in mammals. Insect Biochem Mol Biol 1994; 24:547–555.
78. Weersink AJL, Van Kessel KPM, Van den Tol ME, Van Strijp JAG, Torensma R, Verhoef J, Elsbach P, Weiss J. Human granulocytes express a 55-kDa lipopolysaccharide-binding protein on the cell surface that is identical to the bactericidal/permeability-increasing protein. J Immunol 1993; 150:253–263.
79. Erwin AL, Munford RS, Jr. Deacylation of structurally diverse lipopolysaccharides by human acyloxyacyl hydrolase. J Biol Chem 1990; 265:16444–16449.
80. Hagen FS, Grant FJ, Kuijper JL, Slaughter CA, Moomaw CR, Orth K, O'Hara PJ, Munford RS. Expression and characterization of recombinant human acyloxyacyl hydrolase, a leukocyte enzyme that deacylates bacterial lipopolysaccharides. Biochemistry 1991; 30:8415–8423.
81. Munford RS, Erwin AL. Eukaryotic lipopolysaccharide deacylating enzyme. Methods Enzymol 1992; 209:485–492.
82. Munford RS, Hunter JP. Acyloxyacyl hydrolase, a leukocyte enzyme that deacylates bacterial lipopolysaccharides, has phospholipase, lysophospholipase, diacylglycerollipase, and acyltransferase activities in vitro. J Biol Chem 1992; 267:10116–10121.
83. Munford RS, Hall CL. Purification of acyloxyacyl hydrolase, a leukocyte enzyme that removes secondary acyl chains from bacterial lipopolysaccharides. J Biol Chem 1989; 264:15613–15619.
84. Appelmelk BJ, An Y-Q, Geerts M, Thijs BG, De Boer HA, MacLaren DM, De Graaff J, Nuijens JH. Lactoferrin is a lipid A-binding protein. Infect Immun 1994; 62:2628–2632.
85. Ohno N. LPS binding proteins in granulocyte lysosomes. In: Morrison DC, Ryan JL, eds. Bacterial Endotoxic Lipopolysaccharides. Vol. I. Molecular Biochemistry and Cellular Biology. Boca Raton, FL: CRC Press, 1992:387–404.
86. Hirata M, Shimomura Y, Yoshida M, Morgan JG, Palings I, Wilson D, Yen MH, Wright SC, Larrick JW. Characterization of a rabbit cationic protein (CAP18) with lipopolysaccharide-inhibitory activity. Infect Immun 1994; 62:1421–1426.
87. Chen C, Brock R, Luh F, Chou PJ, Larrick JW, Huang RF, Huang TH. The solution structure of the active domain of CAP18—a lipopolysaccharide binding protein from rabbit leukocytes. FEBS Lett 1995; 370:46–52.
88. Larrick JW, Hirata M, Zheng H, Zhong J, Bolin D, Cavaillon JM, Warren HS, Wright SC. A novel granulocyte-derived peptide with lipopolysaccharide-neutralizing activity. J Immunol 1994; 152:231–240.
89. Muta T, Miyata T, Tokunaga F, Nakamura T, Iwanaga S. Primary structure of anti-lipopolysaccharide factor from American horseshoe crab, *Limulus polyphemus*. J Biochem (Tokyo) 1987; 101:1321–1330.
90. Bairoch A, Apweiler R. The SWISS-PROT protein sequence data bank and its new supplement TrEMBL. Nucleic Acids Res 1996; 24:21–25.

25

Endothelial Cell Activation by Lipopolysaccharide: Role of Soluble CD14

Moshe Arditi
Cedars-Sinai Medical Center, UCLA School of Medicine, Los Angeles, California

INTRODUCTION

Lipopolysaccharide (LPS, endotoxin) is a very potent stimulator of the cells of the immune system, including monocyte/macrophages, B cells, polymorphonucleocytes (PMNs), and vascular endothelial cells (EC). It is the ability of host cells to respond to LPS by producing a wide variety of immunological and pharmacological mediators that results in gram-negative septic shock.

While LPS-mediated macrophage activation mechanisms and signaling events have been intensely investigated for the past 10–15 years, the precise mechanisms involved in LPS-EC interactions and signaling have only been intensively studied for the last several years. The endothelium, perhaps by virtue of its strategic position between blood and tissue, is involved in a variety of critical responses to both physiologic and pathophysiologic stimuli. Although initially envisioned as a passive inert vascular lining cell, it has become clear that the endothelium plays important roles in the regulation of vascular tone, coagulation and fibrinolysis, cellular growth and differentiation, and immune and inflammatory responses.

The endothelium is now recognized as a critical target for LPS and cytokine actions (1–3). The vascular endothelium plays a major role in the pathogenesis of endotoxic shock and of blood-brain barrier damage in gram-negative meningitis (4–6). Vascular complications of septic shock, including hypotension and disseminated intravascular coagulation, are related to endothelial activation and injury induced by LPS (4–6).

Although a clear distinction exists between injury and activation, the functions displayed by an activated endothelial cell may lead to endothelial cell injury through indirect mechanisms (7). Once activated, the endothelial cells release inflammatory cytokines (IL-1, IL-6, IL-8, monocyte chemotactic protein-1, and PAF) and express surface adhesion molecules (E-selectin, P-selectin, ICAM, and VCAM) that are involved in recruitment and activation of leukocytes (8–10). In addition, activated EC express tissue factor in concert with the down-regulation of thrombomodulin expression and increased expression of tissue plasminogen activator inhibitor-1 (PAI-1), culminating in a prothrombotic endothelial cell surface (11).

Several earlier studies suggested that EC responds to LPS in a dose-dependent manner and that serum significantly potentiates these effects, decreasing the required concentration of LPS by more than 100- to 1000-fold (12–15). Recent studies have confirmed that LPS-induced EC responses are serum dependent (16–20).

SOLUBLE CD14

CD14, a 55 kDa glycosylphosphatidylinositol-anchored glycoprotein expressed on the surface of myeloid cells, has been shown to be an acceptor for LPS-LBP complexes (21,22). CD14 is found not only on the surface of myeloid cells, but also in the plasma; soluble forms of CD14 (sCD14) are present in sera of normal individuals (2–6 μg/ml) (23,24). While the LBP-LPS–independent pathway is closely linked to the initiation of cellular responses to LPS in CD14-bearing cells (22,25), cells that do no express CD14, such as endothelial, epithelial, dendritic cells, and astrocytes, still respond vigorously to low concentrations of LPS, but only in the presence of serum (17–19,26). Endothelial and epithelial cells respond to low concentrations of LPS and a variety of additional microbial constituents through the sCD14-dependent pathway. Once activated, these cells secrete inflammatory cytokines, oxygen and nitrogen radicals, modulate coagulation, and direct leukocyte trafficking into surrounding tissues.

We (18) and others (16,17,19) have shown that the serum factor required for LPS-mediated responses in endothelial cells is the soluble form of the CD14 molecule. These recent investigations have strongly implicated sCD14 as the serum component necessary for LPS-mediated endothelial cell responses, including LPS-mediated bovine EC cytotoxicity, induction of adhesion molecules or cytokine release (17–19), NF-κB activation, and tissue factor mRNA expression (20). Cells lacking CD14 respond to LPS-sCD14 complexes, and these responses do not require LBP, although LBP can accelerate the responses (17–20,43).

While a CD14- and LBP-independent pathway of LPS stimulation is observed in myeloid cells in the presence of high LPS concentration (>10 ng/ml), even μg concentrations of LPS are unable to activate the endothelium in the absence of sCD14 (18). Recent studies have also shown that cells lacking mCD14 but known to respond to LPS in a sCD14-dependent manner can also be activated by gram-positive bacterial cell walls in a sCD14-dependent mechanism (27,28). CD14 thus functions as a recognition molecule for a wide variety of bacterial molecules such as LPS, mycobacterial lipoarabinomannan, and component(s) of gram-positive cell walls; CD14 has been proposed to represent a pattern recognition receptor with multiple microbial ligand-binding specificities (28,29). Therefore, CD14 represents the prototype of a receptor for nonadaptive nonspecific early immune response to pathogenic microorganisms.

Brazil and Strominger have shown that CD14 may be released from the surface of monocytes and can be found in the plasma of healthy adults (23). Soluble CD14 is released in two forms (49 and 55 kDa) from monocytes, both lacking the GPI anchor (30). The lower molecular weight form is derived from the membrane by proteolytic cleavage and therefore has a shortened C-terminus. The higher molecular weight form is directly released from an intracellular source, without prior attachment of a GPI tail. Although monocytes can shed sCD14 from their membranes (23,24), it is not known whether this is the only source of circulating sCD14. Biosynthetic labeling experiments have revealed that sCD14 has a lower molecular mass (48 kDa) than the form released from membranes by PI phospholipase C (53 kDa) and does not contain ethanolamine, the first constituent of the PI anchoring system (31). At least two possible mechanisms may result in the production of sCD14. In one, all of the circulating sCD14 is derived from the membrane by shedding, produced in response to an endogenous enzymatic activity. Alternatively, at least some of the sCD14 detected in plasma could result from direct secretion. A lack of the ability to detect alternative splicing of CD14 mRNA (32) suggests that classic secretion is not occurring; however, it is possible that a population of CD14 molecules escapes the PI glycan anchor and that such molecules are then directly secreted. In addition, patients with paroxysmal nocturnal hemoglobinuria (PNH) lack CD14 on the surfaces of their monocytes; however, PNH monocytes synthesize and secrete an sCD14 molecule that is indistinguishable in size from the form shed by normal monocytes (31).

In one study, healthy volunteers had only the 49 kDa form of sCD14 in serum, whereas both molecular constituents were identified in patients in gram-negative septic shock, provided that the total sCD14 level was <3.5 μg/ml; for patients with >3.5 μg/ml of sCD14, the sera contained only the 55 kDa form (33). Other investigators observed both isoforms in healthy controls (34). Increased levels of sCD14 were found in patients with gram-negative septic shock (35,36), which could be due to a reduced sCD14 clearance or to enhanced shedding of sCD14 from monocytes. While the pathophysiologic significance of increased sCD14 levels in patients with gram-negative septic shock is unknown, it appears to correlate with increased mortality (35,36). Since sCD14 mediates LPS binding to endothelial cells and leads to the activation of these cells together with other CD14-negative cells, it has been suggested that the ensuing activation of these cells may be detrimental and thus it may partially explain the worse outcome in patients with increased sCD14 levels.

Soluble CD14 is abundant in plasma, accounting for at least 99% of the total CD14 content of blood, the remaining 1% being found on the surface of monocytic cells. Exogenous recombinant soluble CD14 at very high concentrations (70 μg/ml) can inhibit LPS-induced TNF-α production in whole blood (37,38) and has been shown to have a protective effect in mouse model of sepsis (57). These observations led some investigators to suggest that sCD14 may represent a new therapeutic approach for endotoxic shock and that the deleterious effects of LPS on the endothelium may not be as critical, at least in the mouse model of sepsis (57). However, the plasma concentration of sCD14 in healthy adults [2–6 μg/ml ($\cong 10^{-7}$ M)] is more than 1000 times greater than the peak concentration of LPS [0.1 ng/ml ($\cong 2 \times 10^{-11}$ M)] observed in human serum during sepsis (39). Thus, humans can respond to LPS briskly and fatally even in the presence of huge molar excess of sCD14 (39). This observation, together with the presence of the sCD14-dependent LPS-induced activation pathway in several critical target cells such as endothelial and epithelial cells, would seriously challenge the concept that sCD14 will serve as a useful therapeutic reagent in therapy of gram-negative sepsis in humans.

In conclusion, soluble CD14 may represent a large reservoir for binding LPS in the circulation, and the distribution of LPS into this and other molecular host reservoirs may play a critical role in modulating the host response in gram-negative sepsis. A better understanding of the differential effects of sCD14 in induction or inhibition of LPS effects in various cell types is necessary.

LPS-INDUCED ACTIVATION OF ENDOTHELIAL CELLS

Patrick et al. (16) were the first to postulate that the mechanism by which bovine brain EC are killed by LPS includes the interaction of LPS with a serum component and subsequent damage via a CD14-dependent pathway. The concept that soluble CD14 might be involved in LPS-dependent activation of cells was also initially suggested by experiments reported by these investigators, who showed that antibodies against human antiCD14 will block LPS-induced cytotoxicity in bovine cells in the presence of human serum, and by Schutt et al. (37), who suggested that sCD14 might negatively modulate LPS-dependent activation of monocytes by competing with mCD14 for LPS. Recently a role for sCD14 in endothelial cells and astrocytes was shown by Frey et al. (17). Using affinity chromatography on an anti-CD14-Sepharose column, these investigators showed for the first time that CD14-depleted serum failed to support the cytotoxic response of bovine pulmonary aortic EC to LPS and that the addition of the material eluted from the affinity column restored the response of the EC to LPS (17). However, the importance of other serum factors involved in this response could not be ruled out by these experiments, since it was not clear whether the sCD14 preparations (the eluted material) might have been contaminated with septin, LBP, or other proteins. More recently two groups of investigators, Arditi et al. (18) and Pugin et al. (19), investigated directly the role of sCD14 and LBP in LPS-induced EC responses using highly purified proteins in serum-free experiments. Both groups have independently shown that sCD14 was required for a direct activation of EC by LPS, independent of LBP. Pugin et al. have also shown evidence for specific LPS-sCD14 binding in epithelial cells (19). The initial reports by these investigators provided a strong experimental foundation for subsequent investigators, who have studied the involvement of sCD14 in endothelial and epithelial cells; several of these studies have subsequently confirmed a dominant role for sCD14 in LPS-induced EC and epithelial cell activation (20,40–42).

The observation by Frey et al. (17) that native sCD14 isolated from human serum was able to restore the ability of CD14-depleted serum to induce EC responses to LPS has suggested a potentially important physiologic role of sCD14. The recombinant sCD14 used in the studies by Arditi et al. (18) has allowed exclusion of the possibility that the results obtained might be due to contamination by one or more other serum proteins, such as septin. These studies suggested that, in contrast to myeloid cells, where mCD14 is a receptor, a soluble form of CD14 found in serum may be involved in the LPS-induced responses of human endothelial cells.

LPS-Induced Cytotoxicity in Bovine EC and IL-6 Release in Human EC

To investigate LPS-mediated responses in endothelial cells, Arditi et al. (18) have employed both LPS-induced cytotoxicity (LDH release) in bovine brain EC (BBEC) and IL-6 release from human brain microvessel EC (HBMEC) or human umbilical vein EC (HUVEC). In their studies it was observed that serum is absolutely required for these responses of EC to LPS, indicating a requirement for soluble protein(s) (18). It was also shown that both rough and smooth LPS as well as a purified lipid A induced a dose-dependent lethal injury (by apoptosis) in bovine EC with concentrations of LPS as low as 10 ng/ml (18, and unpublished observations). Rough and smooth LPS and lipid A also induced a dose-dependent release of IL-6 from HBMEC and HUVEC. Serum was once again absolutely required for the LPS-mediated cytotoxicity in BBEC and IL-6 release from HUVEC or HBMEC; serum concentrations as low as 0.5% were sufficient to support these LPS-induced EC responses (18). The inactive lipid A analog *Rhodopseudomonas sphaeroides* diphosphoryl lipid A (RSDPLA) was unable to induce cytotoxicity or IL-6 release. Moreover, active LPS-induced EC responses were blocked when cells were preincubated with 100-fold excess RSDPLA prior to LPS stimulation (M. Arditi et al., unpublished). Further supporting the serum requirement, LPS concentrations as high as 5 μg/ml were unable to induce IL-6 release from HUVEC and HBMEC in the absence of serum.

Soluble CD14 in LPS-Induced Endothelial Cell Responses

To define further the mechanism(s) of sCD14 involvement in LPS-mediated EC responses, BBEC monolayers were preincubated in 10% FBS or 6% pooled normal human serum with antihuman CD14 mAbs, including MY4, 28C5, and S39, all of which have been shown to be function-inhibiting antibodies. All three anti-CD14 mAbs, at concentrations of 2.5 μg/ml, provided complete protection against LPS-induced cytotoxicity in BBEC and blocked LPS-induced IL-6 release from HUVEC and HBMEC in the presence of human serum but not FBS (18). These experiments suggested that sCD14 plays an important role in the interaction of LPS with EC. Unlike the results for the interaction of monocytes with LPS, increasing the LPS concentrations up to 1 μg/ml did not overcome the blocking effect of these anti-CD14 mAbs. As an additional approach, EC monolayers were pretreated with 5 μg/ml of anti-CD14 mAb (MY4), the cells were then washed and subsequently incubated with LPS in the presence of serum-containing media; cytotoxicity in BBEC and IL-6 release from HBMEC were not affected (18). These findings, together with the observation that the blocking effect of the anti-human CD14 mAbs was specific for human serum, suggested that the anti-CD14 Abs were acting by inhibiting the ability of a serum factor(s) to function in LPS-EC interactions rather than by blocking a cryptic form of membrane CD14 at the surface of EC.

To determine the serum factor that is absolutely required for LPS-mediated responses in EC, confluent monolayers of BBEC, HUVEC, and HBEC were washed with RPMI 1640 and incubated in serum-free medium containing LPS (10–100 ng/ml), both with and without exogenous recombinant sCD14. The addition of recombinant sCD14 restored the ability of LPS to cause cytotoxicity in BBEC and IL-6 release from HUVEC or HBMEC in the absence of serum (18; Table 1). In dose-response experiments, sCD14 concentrations as low as 0.1 μg/ml were sufficient to restore LPS effects (18). These observations indicate that sCD14 is the serum factor required for LPS-mediated EC responses and suggest that sCD14 is a novel agonist that is required for mediating serum-dependent LPS responses in EC.

LBP and LPS-Induced EC Responses Under Serum-Free Conditions

Since LBP has been demonstrated to play an important role in LPS-dependent activation of human monocytes, it was of interest to evaluate its potential contribution to LPS-EC serum interactions. Confluent monolayers of BBEC or HBMEC were washed with RPMI 1640 and incubated in serum-free medium with LPS (10–20 ng/ml) in the presence or absence of rabbit LBP (bioactive across species). When LPS was preincubated with LBP at various concentrations for 30 minutes and the

Table 1 Effects of Serum and sCD14 on LPS-Induced IL-6 Secretion from HUVEC

	IL-6 release (pg/ml)[a] in the presence of:			
LPS (ng/ml)	6% PNHS	No serum	No serum + sCD14	6% PNHS + My4
10	254 ± 25	35 ± 7	240 ± 20	44 ± 7
20	312 ± 32	38 ± 8	285 ± 32	62 ± 5
100	408 ± 57	40 ± 10	350 ± 44	77 ± 8

[a]HUVEC were incubated with various concentrations of *E. coli* LPS for 6 hours in the presence or absence of PNHS, with or without an anti-CD14 mAb (My4; 2.5 μ/ml), and with recombinant human sCD14 (1 μg/ml) as indicated, and IL-6 levels were measured in the supernatants. Values are averages for four samples ± standard deviations.

mixture was then added to these EC monolayers in serum-free medium, it was found that the addition of LPS-LBP complexes alone even at a ratio of 1:100 (i.e., 20 ng/ml LPS and 2 μg/ml LBP) did not induce cytotoxicity in BBEC or IL-6 release from HBMEC (18). In addition, when a polyclonal goat antirabbit LBP Ab was used to immunodeplete LBP from rabbit serum and BBEC monolayers were incubated with LPS (100 ng/ml) in 10% normal rabbit serum or in 10% LBP-depleted rabbit serum, cytotoxicity was observed for both normal rabbit serum and LBP-depleted rabbit serum (18). These observations indicate that LBP is most likely not participating in the initiation of LPS-mediated EC responses. More recently several additional studies have provided support for these observations that LBP is not required in LPS and sCD14-induced EC responses (17,19,20), although collectively the data would support the conclusion that LBP can significantly enhance the binding of LPS to sCD14 and thus enhance the LPS effects (19). These observations and other studies suggest that the role of LBP is limited to accelerating movement of LPS into CD14 and that it may not play a further role (at least in signaling and cell-activation events). The only possible exception to these observations is a published study in which investigators observed that LPS-induced changes in EC barrier function in bovine pulmonary artery endothelial cells were also serum-dependent; however, in contrast to most other studies where LBP alone was unable to restore the serum effect and sCD14 was absolutely required, these investigators reported that LBP alone was able to restore some of the LPS effects without sCD14 (41). The reason for these discrepancies in observations is currently unclear.

sCD14 and LPS in EC Responses

In further experiments designed to investigate how sCD14 participates in LPS-mediated injury (BBEC and LDH release) or activation (HUVEC and IL-6 release), it was shown that preincubation of BBEC or HUVEC monolayers with recombinant sCD14 (up to 5 μg/ml for 60 minutes at 37, 22, or 4°C), washing of monolayers, and subsequent incubation with LPS under serum-free conditions did not induce either cytotoxicity in BBEC (18) (Fig. 1) or IL-6 release from HUVEC (M. Arditi et al., unpublished). Similarly, preincubation of BBEC with LPS (100 gn/ml) in serum-free medium

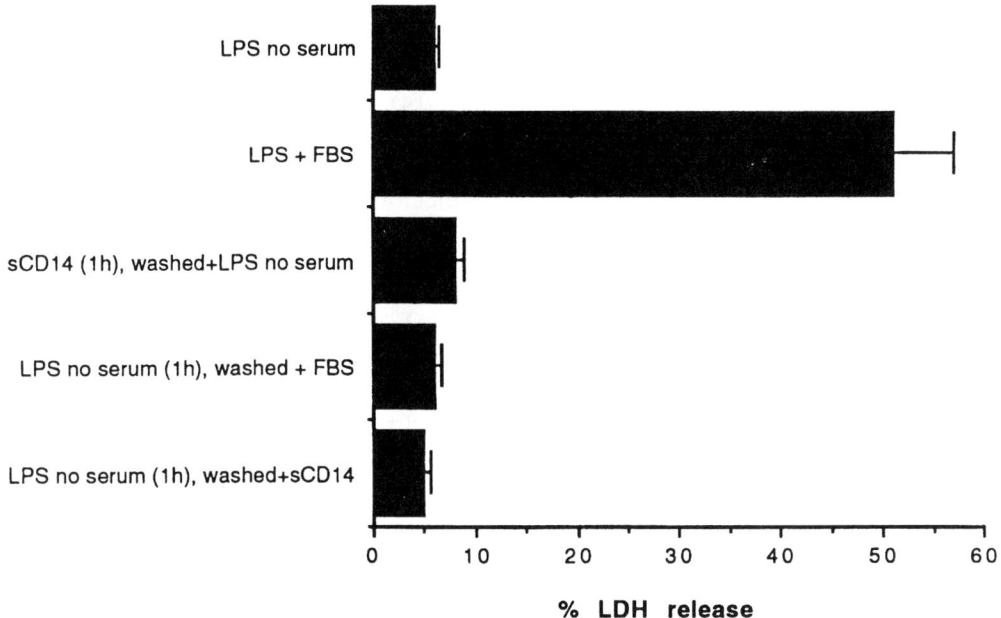

Fig. 1 EC injury mediated by LPS requires LPS and sCD14 to be present simultaneously, BBEC were incubated with sCD14 (1 μg/ml) in serum-free medium for 1 hour at 37°C, and monolayers were washed with RPMI 1640 and then incubated with E. coli O18 LPS (100 ng/ml) in serum-free medium. In separate experiments, monolayers were incubated with LPS (100 ng/ml) in serum-free medium for 1 hour at 37°C, washed with RPMI 1640, and then incubated with medium containing 10% FBS or recombinant human sCD14 (1 μg/ml). The specific percent LDH release was measured following overnight incubation. Values are averages for three samples ± standard deviation.

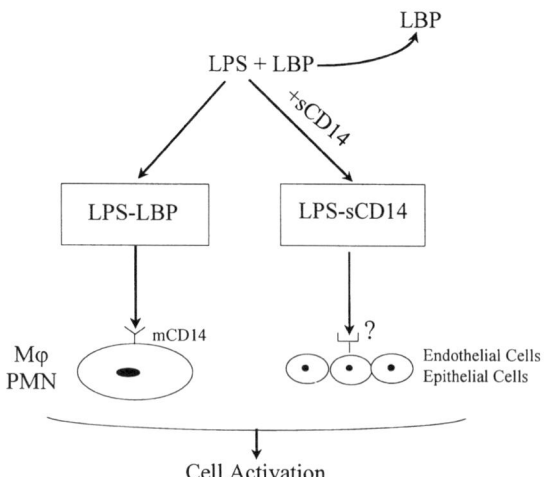

Fig. 2 LPS-induced activation of myeloid cells bearing membrane CD14 receptors versus activation of cells lacking mCD14, such as endothelial and epithelial cells.

for 60 minutes at 37°C, washing of the monolayers and subsequent incubation with medium containing 10% FBS or recombinant sCD14 (1 μg/ml) did not induce these EC responses (Fig. 1). These experiments indicate that sCD14 and LPS must be present simultaneously, presumably to form complexes, for the initiation of EC responses. However, the potential issue of cellular uptake of sCD14 or LPS was not addressed in these experiments. Of relevance to these findings are parallel observations showing that soluble CD14 will enable both normal monocytes and CD14-deficient PNH monocytes to respond to LPS in serum-free conditions (42). Collectively these studies allow the conclusion that sCD14 plays a critical role in LPS-mediated responses in cells lacking mCD14 and that a surface anchor is not needed for the function of CD14. They further indicate that sCD14 most likely binds to some additional membrane constituent on the cell surface in order to associate with the cell and transduce a signal (Fig. 2).

LPS-sCD14 INTERACTIONS

Structure-function analyses of sCD14 have shown that the ability of this protein to stimulate nonmyeloid cells in the presence of LPS clearly resides in the amino-terminal half of the molecule (49). Deletion mutagenesis studies have demonstrated that the N-terminal 152 amino acids of sCD14 are sufficient for mediating inflammatory responses induced by LPS, suggesting that the LPS-binding and cell-signaling domains of sCD14 are located within the first 152 amino acids (49). Each sCD14 molecule binds one or two LPS molecules (43) at a site mapped to residues 57–64 near the N-terminus (50,51). Deletion of this region leads to a mutant protein that neither binds LPS nor supports responses of cells to LPS (51). In contrast, the carboxy-terminal two thirds of sCD14 can be deleted without affecting either LPS binding or the ability to support cellular responses (49). Furthermore, by a series of site-directed alanine substitution mutants in sCD14, investigators have shown that the region between amino acids 7–10, $sCD14_{(7-10)A}$, was capable of binding LPS but was impaired in its ability to mediate cellular responses to LPS (52). It is thus possible that residues 7–10 are essential for interaction of sCD14 with the putative transmembrane receptor protein that interacts with LPS-sCD14 (52).

From the published findings summarized above, it is now clear that sCD14 functions as an agonist and enables cells that lack CD14 to respond to low concentrations of LPS and that sCD14 binds to LPS in either the absence or the presence of LBP. The resulting sCD14-LPS complexes then are able to stimulate not only CD14-negative cells, but also mCD14-bearing myeloid cells (43–46). Studies by Hailman et al. (43) have provided direct evidence that (1) recombinant sCD14 binds LPS in the absence of LBP or other proteins, (2) the binding of LPS to sCD14 is stable and of low stoichiometry (one or two molecules of LPS per sCD14), and (3) recombinant LBP does not form detectable ternary complexes with sCD14 and LPS, but it does accelerate the binding of LPS to sCD14 at substoichiometric concentrations. The latter findings suggest that LBP functions catalytically as a lipid transfer protein. Binding of LPS to sCD14 is dramatically accelerated by the lipid transfer protein LBP, resulting in enhanced responses to LPS (19,43); each LBP molecule has been implicated as capable of transferring literally hundreds of sCD14 molecules in an hour (43). Therefore, while LBP was initially thought to form a critical part of a complex with LPS and CD14 (22), the presence of LBP is now known not to be necessary for either CD14 to bind LPS or for CD14 to initiate responses to LPS. LPS binds to sCD14 in the complete absence of LBP (43) and the resulting LPS-sCD14 complexes stimulate endothelial cells (43), neutrophils (43,47), and monocytes (48).

In a very recent study, investigators have examined LPS-dependent binding events on the surface of nonmyeloid cells using a biologically active radiolabeled sCD14 molecule as ligand (53). Through this approach, these investigators have described a novel sCD14-binding phenomenon, exhibited by a variety of non–mCD14-bearing cells (i.e., epithelial, endothelial, and smooth muscle cells) that is dependent on both LPS and LBP; they have also shown that sCD14, LPS, and LBP will form a ternary molecular complex at 37°C and that, once bound, sCD14 is actively internalized by the cell (53). Furthermore, they observed that in contrast to the requirement of both LPS and LBP for binding of labeled sCD14 to cells, the sCD14-mediated stimulation of these nonmyeloid cells by LPS does not require LBP (53). A panel of monoclonal anti-CD14 and anti-LBP antibodies were added to the preincubation reaction containing labeled sCD14, LBP, and LPS, and the resulting complexes were tested for their ability to bind and stimulate IL-8 production using SW620 epithelial cells. The anti-CD14 mAb 63D3 had no effect on labeled sCD14 binding or cell activation. In contrast, the anti-CD14 mAb 28C5 blocked both binding and activation of the cells. However, while the anti-CD14 mAb 18E12 inhibited cell activation, it had no effect on the binding of sCD14 to the cells. Conversely, while the sCD14 binding to the cells was blocked by two anti-LBP Abs, these Abs had no effect on cellular activation (53). Thus, LBP- and LPS-dependent association of sCD14 does not appear to be directly involved in events leading to cellular activation. Collectively, these observations suggest that binding of sCD14 may not be required for cell activation and that this binding phenomenon may represent a general cellular mechanism for clearance of LPS.

Another group of investigators who used biologically active, radiolabeled sCD14 have shown that there is a specific and saturable binding of sCD14 in endothelial and epithelial cells, astrocytes, as well as human monocytes in the presence of LPS but in the complete absence of LBP (54). The differences between the results of these two studies may be related to the different methods used to radiolabel the sCD14 molecule and to the different experimental conditions used in each study.

INTERACTION OF GRAM-NEGATIVE BACTERIA, sCD14 AND ENDOTHELIAL CELLS

In intact gram-negative bacteria, the lipid A moiety is associated with the bacterial outer membrane and may not be accessible to CD14; however, Jack et al. (56) have shown that cells expressing CD14 on the surface can bind FITC-labeled *E. coli* in a serum-dependent manner, as this is inhibited by anti-CD14 Ab as well as sCD14. In the presence of serum, both mCD14 and sCD14 can bind to gram-negative bacteria, suggesting that CD14 may play a role in the detection and elimination of intact bacteria in vivo (56). Several investigators have shown that LPS, as an integral constituent of gram-negative bacterial outer membrane, can bind specifically to human sCD14 (55,56). Thus, not only LPS but also gram-negative bacteria may be a target for sCD14, and in addition to LPS, which requires sCD14 to activate endothelial cells, live or killed gram-negative bacteria also need sCD14 to activate these cells (55). These data suggest that sCD14 may play an important role in the responses of endothelial, epithelial, and monocytic cells to LPS and to gram-negative bacteria.

BINDING PROTEINS OF LPS-sCD14 COMPLEXES IN ENDOTHELIAL CELLS

The precise biochemical and molecular nature of cellular receptor for the LPS-sCD14 complexes on endothelial and epithelial cells is of considerable interest. The presence of a "specific receptor" for LPS on these cells is certainly strongly implicated by data showing specific binding of ^{3}H-LPS to epithelial cells (19) and that LPS-induced activation of EC could be inhibited by an LPS receptor antagonist (58) as well as by the inactive lipid A analog RSDPLA (M. Arditi et al., unpublished). Of interest, binding of ^{3}H-LPS to epithelial cells was minimal in the absence of serum; significant binding was, however, readily detectable in the presence of serum. Anti-CD14 Ab blocked the binding of ^{3}H-LPS, suggesting further that the presence of sCD14 is required for the binding of LPS to the cells, and binding of radio-labeled LPS to sCD14 was accelerated by LBP (19).

How then does sCD14 participate in signal transduction? The mechanism by which sCD14 is involved in LPS signaling remains unclear. As pointed out above, in the absence of serum, sCD14 alone does not stimulate EC; neither does LPS alone stimulate EC (even at microgram levels). Pretreatment of EC with LPS alone under serum-free conditions and subsequent incubation with recombinant sCD14 or pretreatment of EC with recombinant sCD14 alone and subsequent incubation with LPS under serum-free conditions also did not alter the responses of these cells. Taken together, these observations suggest that EC responses to LPS require that LPS and sCD14 be present simultaneously at the cell surface. We have postulated that such a requirement would support the concept that LPS and sCD14 interact to form a complex and that this complex formation is a *prerequisite* for LPS signal transduction in EC (18). Several studies have shown that sCD14 binds to LPS in either the absence or the presence of LBP, and the resulting sCD14-LPS complexes can stimulate endothelial or epithelial cells. Based on these observations, several investigators (17,29,43,54) have postulated the existence of a transmembrane receptor structure that interacts with LPS-sCD14 complexes to generate an activating signal. The identification and characterization of this structure represents an important area of current interest and is under intense investigation in a number of laboratories. It is interesting to speculate whether or not the postulated transmembrane signaling unit of the putative multimeric receptor in mCD14-bearing myeloid cells and the proposed sCD14-LPS receptor in CD14-negative cells such as endothelial cells are either the same or closely related structures.

The mechanism of sCD14-dependent activation of EC by LPS may involve the bridging of the LPS-receptor complex with the sCD14-receptor complex via the LPS-sCD14 interactions, which triggers the initiation of signaling (Fig. 3). Since sCD14 is not normally associated with EC, it is also possible that sCD14 only interacts with this receptor unit after ligation of sCD14 by LPS. Precedent for such a mechanism of activation is provided by analogy with the IL-6 receptor. The ligand-binding 80 kDa subunit of the IL-6 receptor interacts with a signal-transducing 130 kDa subunit only after binding of IL-6 (59).

Several LPS-specific binding proteins have been identified on various cell types (60). Recently Schletter et al. (61) investigated LPS binding to separated membrane proteins from human monocytes or endothelial cells and observed an 80 kDa protein that preferentially binds LPS in the presence of serum. These investigators reported that neither sCD14 nor LBP alone substituted for serum, and the binding required the presence of both sCD14 and LBP (61). How then does one reconcile the observations that purified soluble CD14 alone allows LPS to activate these cells in the absence of LBP, yet the binding of LPS requires both sCD14 and LPS? Once again, these findings support the concept that binding may not be required for cell activation. Vita et al. (54) used affinity cross-linking of ^{125}I-sCD14 and LPS complexes to several cell types and have shown that a 216 kDa binding protein for LPS-sCD14 complexes is expressed in these cells. While the binding of sCD14-LPS complexes to a specific cell surface structure and the concomitant induction of IL-6 suggest that this sCD14-LPS binding structure (216 kDa) may be a functional receptor with intrinsic signaling ability or may be linked to a signaling subunit, currently there are no convincing data to prove this. Using ^{125}I-ASD-LPS cross-linking studies, Arditi et al. (unpublished) have obtained data showing the presence of a 73 kDa LPS-binding protein on human brain microvessel endothelial cells (HBMEC) and BBMEC (Fig. 4). The binding was shown to be specific, since it was inhibited by excess underivatized LPS or lipid A, and ^{125}I-ASD alone failed to bind to the cells. Furthermore, the binding significantly increased in the presence of serum or sCD14 in the absence of LBP, and anti-CD14 antibodies blocked the binding, suggesting an important role for sCD14 in this LPS binding to HBMEC. Therefore, this 73 kDa LPS-binding protein in HBMEC is also a possible receptor structure, which may be responsible for initiating transmembrane signaling in LPS activation of endothelial cells.

ROLE OF sCD14 IN LPS-INDUCED SIGNALING IN ENDOTHELIAL CELLS

Arditi et al. (62) have shown that smooth and rough LPS as well as lipid A rapidly induces the tyrosine phosphorylation of several proteins, including ERK1, ERK2, and p38 MAP kinase in both bovine and human brain endothelial cells. Pretreatment of EC with the tyrosine kinase inhibitor herbimycin A was found to inhibit LPS-induced protein tyrosine phosphorylation and LPS-induced IL-6 release from HBMEC in a dose-dependent manner, suggesting that a herbimycin-sensitive step, presumably a tyrosine kinase, is most likely involved in mediating LPS-induced activation

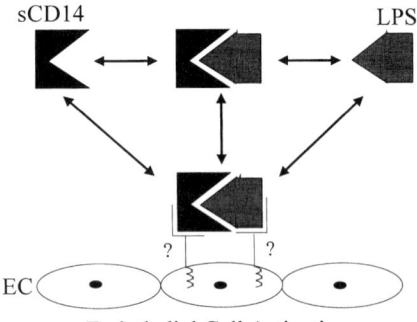

Fig. 3 A hypothetical scheme for endothelial cell activation by the LPS-sCD14 complex. LPS alone or sCD14 alone does not activate endothelial cells. Endothelial cells become activated only when LPS and sCD14 are present together and presumably simultaneously bind to their receptors.

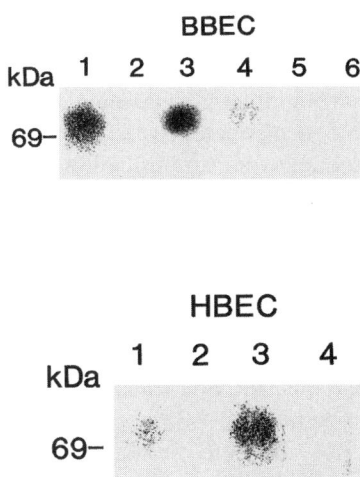

Fig. 4 Photo affinity cross-linking experiments. Bovine brain EC (BBEC) or human brain EC (HBEC) were washed five times to remove serum proteins and were incubated with ^{125}I-ASD-LPS (Re 595 LPS, 5 μg/ml). Following UV cross-linking, cell lysates were electrophoresed and binding of photo–cross-linked LPS was assessed by autoradiography. (BBEC) Lane 1 = BBEC incubated with labeled LPS in serum-free media; lane 2 = BBEC preincubated with 50-fold excess underivatized Re 595 LPS, followed by incubation with labeled LPS; 3 = BBEC incubated with labeled LPS in serum-free media plus exogenous recombinant sCD14 (2 μg/ml); 4 = same as lane 3 but also with anti-CD14 mAb (MY4, 5 μg/ml); 5 = same as lane 3 but pretreated with 50-fold excess underivatized LPS; 6 = same as lane 3 but pretreated with 50-fold excess underivatized lipid A. (HBEC) Lane 1 = HBEC incubated with labeled LPS in serum-free medium; 2 = HBEC pretreated with 50-fold excess underivatized Re 595 LPS, followed by labeled LPS; 3 = HBEC incubated with labeled LPS in serum-free media plus exogenous sCD14 (2 μg/ml); 4 = same as lane 3 but pretreated with 50-fold underivatized LPS.

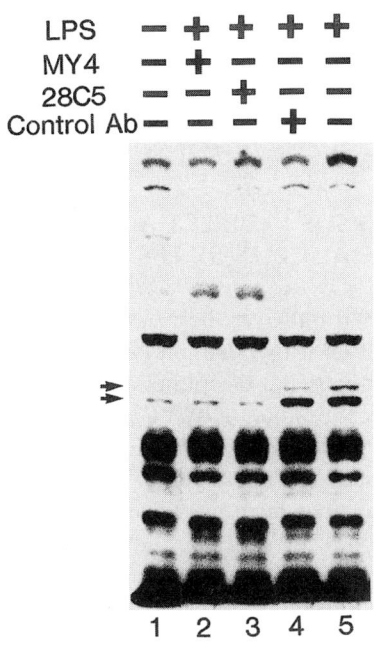

Fig. 5 LPS-induced tyrosine phosphorylation in human brain EC and effect of anti-CD14 antibodies. Antiphosphotyrosine immunoblot of HBEC grown in the presence of 5% heat-inactivated pooled human serum, stimulated for 30 minutes with 100 ng/ml *E. coli* O18 LPS in the presence or absence of antihuman CD14 mAb [MY4(2.5 μg/ml), 28C5 (2.5 μg/ml)] or an isotype matched control antibody (20 μg/ml). Lane 1 = resting cells; lane 2 = HBEC pretreated with MY4 for 30 minutes and stimulated with LPS; lane 3 = HBEC pretreated with 28C5 for 30 minutes and stimulated with LPS; lane 4 = HBEC pretreated with control antibody for 30 minutes and stimulated with LPS; lane 5 = HBEC stimulated with LPS for 30 minutes. Approximate molecular masses of the induced phosphoproteins indicated by the arrows are 44 and 42 kDa (ERK1 and ERK2). Note that the third prominent phosphoprotein induced by LPS (p38 MAP kinase), which migrate around 41 kDa, did not always clearly separate from 42 kDa phosphoprotein.

(IL-6 release) of human EC (62). To investigate the potential sCD14 dependence of LPS-induced tyrosine phosphorylation in HBMEC, Arditi et al. incubated monolayers in 5% human serum with or without antihuman CD14 mAbs (MY4 and 28C5) and then stimulated the cells with LPS. The induction of tyrosine phosphorylation by LPS was found to be inhibited by both Abs at concentrations ≥2.5 μg/ml (Fig. 5), while an isotype-matched control Ab at concentrations as high as 20 μg/ml had no effect on the LPS-induced increase tyrosine phosphorylation (62). The inhibition was observed over a broad range of LPS concentrations (1–100 ng/ml). These findings differ from observations made using human macrophages, in which inhibition of LPS-induced tyrosine phosphorylation by anti-CD14 Ab was only observed with LPS concentrations of <10 ng/ml (63). Since antibody-mediated cross-linking of FcR, at least in monocytes and macrophages, can under certain circumstances induce protein tyrosine phosphorylation, these investigators prepared Fab fragments of the anti-CD14 mAB MY4. The Fab fragments alone did not stimulate tyrosine phosphorylation, while they did clearly inhibit LPS-induced responses and tyrosine phosphorylation. Furthermore, neither the inactive lipid A analog RSDPLA nor LPS in the absence of serum or sCD14 induced tyrosine phosphorylation of ERK1, ERK2, and p38 MAP kinase in EC (62). Together these results suggest that binding of LPS to sCD14 is required for LPS-induced protein tyrosine phosphorylation in endothelial cells, a finding that is certainly consistent with earlier discussed conclusions regarding the role of these factors in EC activation by LPS.

DIRECT VERSUS INDIRECT ACTIVATION OF ENDOTHELIAL CELLS BY LPS

Studies carried out over the past decade have strongly implicated at least two distinct pathways for LPS stimulation of endothelial or epithelial cells. These include either a direct stimulation by LPS-sCD14 complexes or an indirect stimulation via cytokines released from LPS-stimulated macrophages (Fig. 6). Evidence for both pathways has been provided from several in vitro studies (19,64,65). While the direct pathway of EC activation is dependent on the presence of plasma, the ECs are in constant contact with blood cells and not only plasma. The synergistic action between LPS and various cytokines in various tissues is well documented. For example, when EC were incubated with diluted whole blood instead of plasma, they become sensitive to picomolar rather than nanomolar quantities of LPS (64,66). These picomolar concentrations of LPS are consistent with the levels measured in plasma from endotoxemic or septic patients (66). Blood cell fractionation experiments, experiments with blood from PNH patients, and use of anti-CD14 Abs have suggested that monocytes in whole blood may be the critical blood cells responsible for the observed amplification of LPS effects on EC (64).

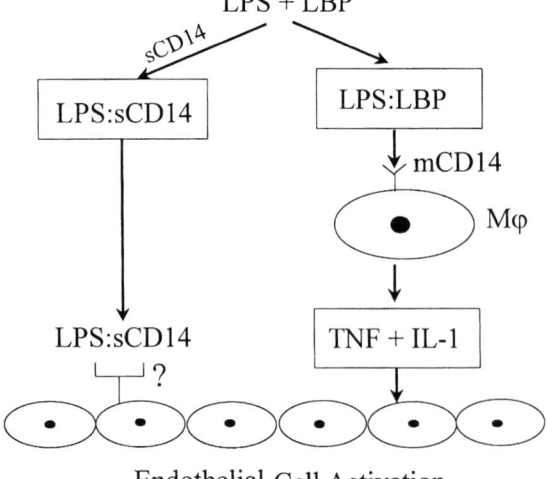

Fig. 6 The direct and indirect pathways of endothelial cell activation by LPS.

Perhaps not surprisingly, both direct and indirect activation pathways are inhibited by anti-CD14 mAb. The fact that conditioned plasma (plasma obtained from whole blood incubated with picomolar quantities of LPS) reproduced the amplification effect indicates that monocyte-EC contact may not be required and that the observed effects could be carried by a soluble mediator(s) (65,67). Of interest, it has been shown that inhibitors of TNF-α and IL-1β partially blocked LPS-induced activation of EC when added separately to whole blood, but blocked EC responses when added together (67). Thus, the amplification effect seen with whole blood begins with LPS activation of monocytes via interactions with cell membrane localized CD14 and is followed by further activation of EC by the effects of TNF-α and IL-1β. This latter cytokine-mediated EC activation reflects, of course, the indirect pathway. Supporting this concept, Arditi et al. (68) have shown in a transwell system of antibiotic-induced LPS release from live bacteria that EC activation (IL-6 release) by antibiotic-induced LPS release was significantly amplified when EC were co-cultivated with whole blood (68).

It is not clear which of these two pathways of LPS-induced EC activation would be more dominant in vivo. The relative importance of the two pathways has been investigated in vitro by experiments using blood or plasma for LPS-mediated EC activation (64). These studies have shown that HUVEC co-incubated with diluted whole blood required 1000 times less LPS for full activation than when only plasma was present (64). These in vitro observations suggest that, at very low concentrations of LPS, the direct sCD14-dependent pathway may have a lesser role in activating EC and that the indirect pathway may be a more efficient pathway for EC activation (64). These findings raise the important question from a very practical point of view of the very existence of such a direct pathway of EC activation with sCD14, since the indirect pathway seems to require three orders of magnitude less LPS to achieve the same level of EC activation. One possible answer may be the phenomenon of adaptation of blood cells to LPS. The characteristics of this phenomenon are that myeloid cells, in prolonged contact with LPS, become hyporesponsive, or refractory, to LPS stimulation (i.e., desensitized) both in vitro (69) and in vivo (70); thus, with time, the direct pathway might become more important for maintaining EC activation. In contrast to myeloid cells, EC do not show the phenomenon of adaptive desensitization to LPS activation, at least as assessed by IL-8 (64) or IL-6 (M. Arditi et al., unpublished) secretion as indicators of EC activation.

Moreover, the direct pathway might also play an important role in the extravascular space, where myeloid cells are not usually found and the LPS-sCD14 complexes are activators of epithelial cells. Finally, in a disorder such as PNH where the indirect pathway is nonfunctional, the direct pathway of EC activation would obviously be critical for an inflammatory response to LPS.

In summary, lipopolysaccharides interact with numerous blood components, including erythrocytes, monocytes and macrophages, neutrophils, platelets, and various plasma proteins, such as lipoproteins, LDH, LBP, and sCD14, and the relative affinities of LPS for these elements and the distribution of LPS among them are not well established; thus, the relevance of these in vitro findings and the relative importance of one pathway versus the other in various organ systems and various pathophysiologic conditions need to be carefully addressed in relevant in vivo systems. Elucidation and better understanding of these pathways will be critical in order to effectively treat gram-negative sepsis.

ACKNOWLEDGMENTS

I graciously acknowledge the essential contributions of those individuals who have been so generous with their advice and reagents. This work was supported by the NIH AI40275 and by the AHA Clinician Scientist Award 1019-CS.

REFERENCES

1. Cybulsky MI, Chan MKW, Movat HZ. Acute inflammation and microthrombosis induced by endotoxin, interleukin-1 and tumor necrosis factor and their implication in gram-negative infection. Lab Invest 1988; 58:365–378.
2. Morrison DC, Ulevitch RJ. The effects of bacterial endotoxins on host mediation systems. A review. Am J Pathol 1978; 93:526–617.
3. Pober JS, Cotran RS. Cytokines and endothelial cell biology. Physiol Rev 1990; 70:427–451.
4. Morrison DC, Ryan JL. Endotoxins and disease mechanisms. Annu Rev Med 1987; 38:417–432.
5. Bone R. Sepsis syndrome. New insights into its pathogenesis and treatment. Infect Dis Clin North Am 1991; 5:793–805.
6. Wispelwey B, Lesse AJ, Hansen EJ, Scheld WM. Haemophilus influenza lipopolysaccharide-induce blood brain barrier permeability during experimental meningitis in the rat. J Clin Invest 1988; 82:1339–1346.
7. Royall JA, Berkow RL, Beckman JS, et al. Tumor necrosis factor and interleukin 1 alpha increase vascular endothelial permeability. Am J Physiol 1989; 257:L399–L410.
8. Pober JS, Cotran RS. The role of endothelial cells in inflammation. Transplantation 1990; 50:537–544.
9. Jirik FR, Pober TJ, Hirano T, Kishimoto T, Loskutoff DJ, Carson DA, Lotz M. Bacterial lipopolysaccharide and inflammatory mediators augment IL-6 secretion by human endothelial cells. J Immunol 1989; 142:144–147.
10. Bevilacque MP, Pober JS, Mendrick DL, Cotran RS, Gimbrone MA. Identification of an inducible endothelial-leukocyte adhesion molecule. Proc Natl Acad Sci USA 1987; 84:9238.
11. Pohlman TH, Harlan JM. Endotoxin-endothelial cell interactions. In Morrison DC, Ryan J, eds. Bacterial Endotoxic Lipopolysaccharides. Boca Raton, FL: CRC Press, 1992:459–472.
12. Harlan JM, Harker LA, Reidy MA, Gajdusek CM, Schwartz SM, Striker GE. Lipopolysaccharide-mediated bovine endothelial cell injury in vitro. Lab Invest 1983; 48:269–274.
13. Meyrick B, Ryan US, Brigham KL. Direct effects of E. coli endotoxin on structure and permeability of pulmonary endothelial monolayers and the endothelial layer of intimal explants. Am J Pathol 1986; 122:140–151.
14. Harlan JM, Harker LA, Striker GE, Weaver J. Effects of lipopolysaccharide on human endothelial cells in culture. Thromb Res 1983; 29:15–26.
15. Meyrick B. Endotoxin-mediated pulmonary endothelial cell injury. Fed Proc 1986; 45:19–24.
16. Patrick D, Betts J, Frey EA, Prameya R, Dorovini-Zis K, Finley BB. Haemophilus influenzae lipopolysaccharide disrupts confluent monolayers of bovine brain endothelial cells via a serum-dependent cytotoxic pathway. J Infect Dis 1992; 165:865–872.
17. Frey EA, Miller DS, Jahr TG, Sundan A, Bazil V, Espevik T, Finlay BB, Wright SD. Soluble CD14 participates in the response of cells to lipopolysaccharide. J Exp Med 1992; 176:1665–1671.
18. Arditi M, Zhou J, Dorio R, Rong GW, Goyert SM, Kim KS. Endotoxin-mediated endothelial cell injury and activation: role of soluble CD14. Infect Immun 1993; 61:3149–3156.
19. Pugin J, Schurer-Maly CC, Leturcq D, Moriarty A, Ulevitch RJ, Tobias PS. Lipopolysaccharide activation of human endothelial and epithelial cells is mediated by lipopolysaccharide-binding protein and soluble CD14. Proc Natl Acad Sci USA 1993; 90:2744.
20. Read M, Cordle SR, Veach RA, Carlisle CD, Hawiger J. Cell-free pool of CD14 mediates activation of transcription factor NF-kB by lipopolysaccharide in human endothelial cells. Proc Natl Acad Sci USA 1993; 90:9887–9891.
21. Wright SD, Tobias PS, Ulevitch RJ, Ramos RA. Lipopolysaccharide binding protein opsonizes LPS-bearing particles for recognition by a novel receptor on macrophages. J Exp Med 1989; 170:1231–1241.
22. Wright SD, Ramos RA, Tobias PS, Ulevitch RJ, Mathison JC. CD14, a receptor for complexes of lipopolysaccharide and LPS-binding protein. Science (Washington, D.C.) 1990; 249:1431–1433.
23. Bazil V, Strominger JL. Shedding as a mechanism of down-modulating of CD14 on stimulated human monocytes. J Immunol 1991; 147:1567–1574.

24. Bazil V, Baudys M, Hilgert I, Stefanova I, Low MG, Zbrozek J, Horejsi V. Structural relationship between the soluble and membrane-bound forms of human monocyte surface glycoprotein CD14. Mol Immunol 1989; 26:657–662.
25. Tobias PS, Mathison J, Mintz D, Lee JD, Kravchenko V, Kato K, Pugin J, Ulevitch RJ. Participation of lipopolysaccharide-binding protein in lipopolysaccharide-dependent macrophage activation. Am J Respir Cell Mol Biol 1992; 7:239.
26. Verhasselt V, Buelens C, Willems F, DeGroote D, Haeffner-Cavaillon N, Goldman M. Bacterial lipopolysaccharide stimulates the production of cytokines and the expression of costimulatory molecules by human peripheral blood dendritic cells. Evidence for soluble CD14-dependent pathway. J Immunol 1997; 158: 2919–2925.
27. Kusunoki T, Hailman E, Juan TSC, Lichenstein HS, Wright SD. Molecules from *Staphylococcus aureus* that bind CD14 and stimulate innate immune responses. J Exp Med 1995; 182:1673–1682.
28. Wright SD. CD14 and innate recognition of bacteria. J Immunol. 1995; 155:6–8.
29. Pugin J, Heumann D, Tomasz A, Kraychenko VV, Akamatsu Y, Nishijima M, Glauser MP, Tobias PS, Ulevitch RJ. CD14 is a pattern recognition receptor. Immunity 1994; 1:509–516.
30. Labeta MO, Durieux JJ, Fernandez N, et al. Release from a human monocyte-like cell line of two different soluble forms of the LPS receptor, CD14. Eur J Immunol 1993; 23:2144–2151.
31. Haziot A, Chen S, Ferrero E, Low MG, Silber R, Goyert SM. The monocyte differentiation antigen, CD14, is anchored to the cell membrane by a phosphatidylinositol linkage. J Immunol 1988; 141:547–552.
32. Goyert SM, Ferrero E, Rettig WJ, Yenamandra AK, Obata F, LeBeau M. The CD14 monocyte differentiation antigen maps to a region encoding growth factors and receptors. Science (Washington, D.C.) 1988; 239: 497–500.
33. Landmann R, Zimmerli W, Sansano S, Link S, Hahn A, Glauser MP, Calandra T. Increased circulating soluble CD14 is associated with high mortality in gram-negative septic shock. J Infect Dis 1995; 171:639–644.
34. Durieux ZZ, Vita N, Popescu O, et al. The two soluble forms of the lipopolysaccharide receptor, CD14; characterization and release by normal human monocytes. Eur J Immunol 1994; 24:2006–2012.
35. Landmann R, Reber AM, Sansano S, Zimmerli W. Function of soluble CD14 in serum from patients with septic shock. J Infect Dis 1996; 173:661–668.
36. Landmann R, Calandra T, Zimmerli W. Soluble CD14 in septic shock. In: Bacterial Endotoxins: Lipopolysaccharides from Genes to Therapy. New York: Wiley-Liss, Inc., 1995:375–380.
37. Schutt C, Schilling T, Grunwald U, Schonfeld W, Kruger C. Endotoxin-neutralizing capacity of soluble CD14. Res Immunol 1992; 143:71.
38. Haziot A, Rong G-W, Bazil V, Silver J, Goyert SM. Recombinant soluble CD14 inhibits LPS-induced tumor necrosis factor alpha production by cells in whole blood. J Immunol 1994; 152:5868–5876.
39. Wright SD. Response. CD14 and immune response to lipopolysaccharide. Science (Washington, DC) 1991; 252:1321–1322.
40. Schumann RR, Pfeil D, Lamping N, et al. Lipopolysaccharide induces the rapid tyrosine phosphorylation of the mitogen-activated protein kinases erk1 and p38 in cultured human vascular endothelial cells requiring the presence of soluble CD14. Blood 1996; 87:2805–2814.
41. Goldblum SE, Brann TW, Ding X, Pugin J, Tobias PS. Lipopolysaccharide-binding protein and soluble CD14 function as accessory molecules for LPS-induced changes in endothelial barrier function, in vitro. J Clin Invest 1994; 93:692–702.
42. Golenbock DT, Bach RR, Lichenstein H, Juan TS-C, Tadavarthy A, Moldow CF. Soluble CD14 promotes LPS activation of CD14-deficient PNH monocytes and endothelial cells. J Lab Clin Med 1995; 125:662–671.
43. Hailman E, Lichenstein HS, Wurfel MM, Miller DS, Johnson DA, Kelley M, Busse LA, Zukowski MM, Wright SD. Lipopolysaccharide-binding protein accelerates the binding of LPS to CD14. J Exp Med 1994; 179:269–277.
44. Hailman E, Vasselon T, Kelley M, Busse LA, Hu MC, Lichenstein HS, Detmers PA, Wright SD. Stimulation of macrophages and neutrophils by complexes of lipopolysaccharide and soluble CD14. J Immunol 1996; 156:4384–4390.
45. Yu B, Hailman E, Wright SD, Lipopolysaccharide binding protein and soluble CD14 catalyze exchange of phospholipids. J Clin Invest 1997; 99:315–324.
46. Wurfel M, Hailman E, Wright SD. Soluble CD14 acts as a shuttle in the neutralization of lipopolysaccharide by LPS-binding protein and reconstituted high density lipoprotein. J Exp Med 1995; 181:1743–1754.
47. Detmers RA, Zhou D, Powell DE. Different signaling pathways for CD18-mediated adhesion and Fc-mediated phagocytosis. Response of neutrophils to LPS. J Immunol 1994; 153:2137–2145.
48. Sundan A, Gullstein-Jahr T, Otterlei M, Ryan L, Bazil V, Wright SD, Espevik T. Soluble CD14 from urine copurifies with a potent inducer of cytokines. Eur J Immunol 1994; 24:1779–1784.
49. Juan TS, Kelley MJ, Johnson DA, et al. Soluble CD14 truncated at amino acid 152 binds lipopolysaccharide and enables cellular response to LPS. J Biol Chem 1995; 270:1382–1387.
50. McGinley MD, Narhi LO, Kelley MJ, Davy E, Robinson J, Rohde MF, Wright SD, Lichenstein HS. CD14: physical properties and identification of an exposed site that is protected by lipopolysaccharide. J Biol Chem 1995; 270:5213–5218.
51. Juan TS, Hailman E, Kelley MJ, Busse LA, et al. Identification of lipopolysaccharide binding domain in CD14 between amino acids 57 and 64. J Biol Chem 1995; 270:5219–5224.
52. Todd S-C, Hailman E, Kelley MJ, Wright SD, Lichenstein HS. Identification of a domain in soluble CD14 essential for LPS signaling but not LPS binding. J Biol Chem 1995; 270:17237–17242.
53. Tapping RI, Tobias PS. Cellular binding of soluble CD14 requires lipopolysaccharide and LPS-binding protein. J Biol Chem 1997; 272:23157–23164.

54. Vita N, Lefort S, Sozzani P, Reeb R, Richards S, Borysiewicz LK, Ferrara P, Labeta MO. Detection and biochemical characteristics of the receptor for complexes of soluble CD14 and bacterial lipopolysaccharide. J Immunol 1997; 158:3457–3462.
55. Noel Jr RF, Sato TT, Mendez C, Johnson MC, Pohlman TH. Activation of human endothelial cells by viable or heat-killed gram-negative bacteria requires soluble CD14. Infect Immun 1995; 63:4046–4053.
56. Jack R, Grunwald U, Stelter F, Workalemahu G, Schutt C. Both membrane bound and soluble forms of CD14 bind to gram-negative bacteria. Eur J Immunol 1995; 25:1436–1441.
57. Haziot A, Rong CW, Lin X-Y, Silver J, Goyert SM. Recombinant soluble CD14 prevents mortality in mice treated with endotoxin. J Immunol 1995; 154:6529.
58. Somerville Jr JE, Cassiano L, Brainbridge B, Cunningham M, Darveau RP. A novel *Escherichia coli* lipid A mutant that produces an antiinflammatory lipopolysaccharide. J Clin Invest 1996; 97:359–365.
59. Taga T, Hibi M, Hirata Y, Yamasaki K, Yasukawa K, Matsuda T, Hirano T, Kishimoto T. Interleukin-6 triggers the association of its receptor with a possible signal transducer, gp 130. Cell 1989;58:573.
60. Morrison DC, Kirikae T, Kirikae F, Lei M-G, Chen T, Vukajlovich SW. The receptor(s) for endotoxin on mammalian cells. In: Bacterial Endotoxins. Basic Science to Anti-Sepsis Strategies. New York: Wiley-Liss Inc., 1994:3–15.
61. Schletter J, Brade H, Brade L, et al. Binding of lipopolysaccharide to an 80-kilodalton membrane protein of human cells is mediated by soluble CD14 and LPS-binding protein. Infect Immun 1995; 63:2576–2580.
62. Arditi M, Zhou J, Torres M, Durden DL, Stins M, Kim KS. Lipopolysaccharide stimulates the tyrosine phosphorylation of mitogen-activated protein kinases p44, p42, and p41 in vascular endothelial cells in a soluble CD14-dependent manner. J Immunol 1995; 155:3994–4003.
63. Weinstein SL, June CH, DeFranco AL. Lipopolysaccharide-induced protein tyrosine phosphorylation in human macrophages is mediated by CD14. J Immunol 1993; 151:3829.
64. Pugin J, Ulevitch RJ, Tobias PS. A critical role for monocytes and CD14 in endotoxin-induced endothelial cell activation. J Exp Med 1993; 178:2193–2200.
65. Montovani A, Bussolino F, Dejano E. Cytokine regulation of endothelial cell function. Fed Am Soc Exp Biol J 1992; 6:2591.
66. van Deventer SJH, Buller HR, tenCate JW, et al. Endotoxemia; an early predictor of speticemia in febrile patients. Lancet 1988; 1:605.
67. Pugin J, Ulevitch RJ, Tobias PS. Tumor necrosis factor and interleukin-1 mediate endothelial cell activation in blood at low concentrations. J. Inflamm 1995; 45:49–55.
68. Arditi M, Zhou J. Differential antibiotic-induced endotoxin release and IL-6 production by human umbilical vein endothelial cells (HUVEC): amplification of the response by coincubation of HUVEC with blood cells. J Infect Dis 1997;175:1255–1258.
69. Mathison JC, Virca GD, Wolfson E, Tobias PS, Glaser K, Ulevitch RJ. Adaptation to bacterial lipopolysaccharide controls lipopolysaccharide-induced tumor necrosis factor production in rabbit macrophages. J Clin Invest 1990; 85:1108.
70. McCall CE, Grosso-Wilmoth LM, LaRue K, Guzman RN, Cousart SL. Tolerance to endotoxin-induced expression of the interleukin-1 beta gene in blood neutrophils of humans with the sepsis syndrome. J Clin Invest 1993; 91:853.

26

Scavenger Receptors and Lipopolysaccharide

Alexander Shnyra
University of Kansas Medical Center, Kansas City, Kansas

INTRODUCTION

Scavenger receptors of macrophages represent a trimeric integral membrane glycoproteins implicated in the pathogenesis of atherosclerosis (1,2), apoptosis or programmed cell death (3,4), and host defense against microbial pathogens or their toxic products (5–8). Analysis of tissue distribution has revealed the presence of scavenger receptor(s) in a variety of organs and tissues and identified tissue macrophages as the primary cells expressing scavenger receptor(s) (9). A unique property of scavenger receptor(s) is that it exhibit a broad binding specificity towards a variety of negatively charged ligands including chemically modified proteins, polyribonucleotides, and anionic polysaccharides (2,10). Since initial studies were primarily related to evaluation of the role of scavenger receptors in the pathogenesis of atherosclerosis, scavenger receptors were defined by their ability to bind modified low-density lipoproteins (LDL), e.g., acetylated LDL (Ac-LDL) and oxidized LDL (ox-LDL). Based on the ligand-binding characteristics and amino acid sequence analysis, there have been described at least three independent classes of scavenger receptors, namely class A, B, and C scavenger receptors (5,11–14).

The cloning and characterization of molecular structure of scavenger receptor type I and type II class A showed that it consists of six specific domains (5,11), which are well conserved among different species (15–19). However, despite structural similarity of the scavenger receptors, there is both receptor-dependent and species-dependent diversity in binding characteristics of various ligands to scavenger receptors (17,20).

Several studies have suggested that scavenger receptors may well participate in macrophage-mediated clearance of microbial pathogens and/or neutralization of bacterial products including lipopolysaccharide (LPS) of gram-negative bacteria and lipoteichoic acid of gram-positive bacteria (5–7,21–23). However, only scavenger receptor class A was found to date to interact specifically with microbial components of both gram-positive and gram-negative bacteria. Recently accumulated experimental evidence suggest that the scavenger receptors may represent an evolutionary ancient receptor mechanism(s), which confers phagocytes with capacities to recognize nonself structural determinants normally not expressed on the plasma membrane of the host cells (24).

In this review we summarize the results of recent studies focused upon the characterization of LPS-dependent activation of macrophages mediated by a scavenger receptor pathway(s) and the potential role of the intestinal commensal microflora in modulation of scavenger receptor mechanism(s) in hepatic macrophages or Kupffer cells.

DOMAIN STRUCTURE OF SCAVENGER RECEPTOR A

Characterization of the scavenger receptors was greatly facilitated by purification of bovine scavenger receptor, which was found to be a trimeric glycoprotein comprised of 77 kDa monomers (25). The two types of bovine scavenger receptor, scavenger receptor type I and type II, were initially cloned (5,11). The subsequent analysis of the receptors' amino acid sequences suggested that type I and type II scavenger receptors A are the alternatively spliced products of a single gene. The cDNAs for bovine type I and type II scavenger receptor were used to clone human, rabbit, and murine cDNA homologs (17,16,19).

Extensive studies have revealed the domain structure of scavenger receptors type I and type II (Table 1), which were found to be comprised of six specific domains including N-terminal cytoplasmic domain, transmembrane domain, and four extracellular domains: spacer, α-helical coiled-coil, collagenous, and C-terminal domain (6,16,17,26). The only difference between the two types of receptor is that the type I C-terminal domain, which contains six conserved cysteines and is called the scavenger receptor cysteine-rich (SRCR) domain, is substituted with a truncated six-residue C terminus in the type II receptor. It was also found that the SRCR domain is a highly conservative structure among various proteins involved in the functioning of innate immunity (15). Despite this structural difference, both types of scavenger receptor A revealed similar binding characteristics, indicating that the C-terminal domain may not be essential for ligand-receptor interaction.

In contrast, the collagen-like domain V of scavenger receptor consists of a number of glycine-X-Y triplet repeats frequently containing prolines and lysines in positions X and Y, which may well be positively charged at physiological pH and be involved in specific interaction of scavenger receptor with polyanionic ligands. The primary role of domain V in ligand-receptor interaction was further confirmed in experiments, which demonstrated that truncation of the collagenous domain results in a decrease of ligand binding to the scavenger receptor (27).

Amino acid sequence of domain IV predicts that α-helical coiled-coil structure mediates the assembly of functionally active trimeric scavenger receptor via interaction of the interhelical hydrophobic residues. Point mutagenesis studies provided the evidence that domain IV is essential for dissociation of the ligands from scavenger receptor induced in lysosomes by conformational changes at low pH (28). Recent experimental evidence indicates that the α-helical coiled-coil domain also mediates a cation-independent adhesion of macrophages (29). The N-terminal cytoplasmic domain is believed to be essential for endocytosis and recycling of scavenger receptors (30). The amino acid structure of the cytoplasmic domain also suggested the potential substrate sites for protein kinase activity.

BINDING PROPERTIES OF SCAVENGER RECEPTORS

One of the unusual characteristics that suggests the pattern-recognition nature of the receptors is that scavenger receptors express a high affinity towards a broad variety of ligands (Table 2). The ligands for scavenger receptor include chemically modified LDL (2,10,12,14,31,34–36,39) in which the lysine residues

Table 1 Schematic Domain Structure of Macrophage Scavenger Receptor Class A and Proposed Functions for Different Domains

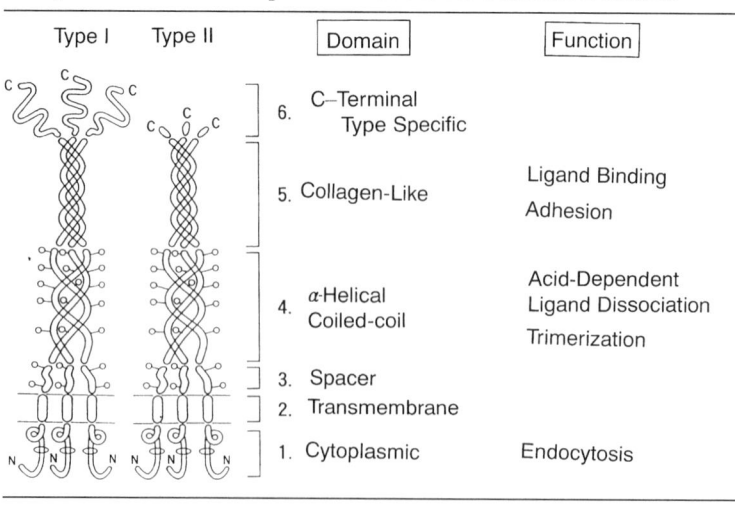

Reproduced with permission of [18].

Table 2 Broad Specificity and Similarity in Binding Characteristics of Different Classes of Macrophage Scavenger Receptors

Scavenger receptor	Specific ligands	Ref.
SR Class A Type I/II (bovine, human, mouse)	*Modified proteins*: acetyl-LDL; maleylated LDL; oxidized LDL maleylated HDL; maleylated albumin	2,31
	Polyribonucleic acids (four-stranded): polyvinyl sulfate; polyinosinic acid, poly(I) polyguanylic acid, poly(G); poly G:I (1:1) polyxanthinylic acid	2,32
	Polysaccharides: dextran sulfate; fucoidan; carrageenan	2
	Bacterial components: lipopolysaccharide, lipoteichoic acid	21–23
	Phospholipids: phosphatidylserine	33
MARCO SR Class A (mouse)	*Bacteria*: E. coli; S. aureus	13,40
SR Class B Type I/CD36 (hamster, human)	*Modified proteins*: oxidized LDL; maleylated albumin; collagen	12,34–36
	Native lipoproteins: HDL; LDL	12,37
	Phospholipids: phosphatidylserine; phosphatidylinositol; phosphatidic acid	38
SR Class C Type I (Drosophila melanogaster)	*Modified proteins*: acetyl-LDL; maleylated albumin	14,39
	Polyribonucleic acids (four-stranded): poly I; poly G; d(A_5G_{37})	14,39
	Polysaccharides: dextran sulfate; fucoidan laminarin (D-glucose polymer)	14,39

had been acetylated (Ac-LDL) or oxidized (ox-LDL) in order to increase the anionic property of LDL molecule; maleylated and formaldehyde-treated serum albumin (2,12,14); four-stranded polynucleotides such as polyinosinic (poly I) and polyguanylic (poly G) acid (2,10,14,32,39); polysaccharides (2,10,14,39); and phospholipids (33,38). Of particular importance, scavenger receptor A can mediate interaction of the host cells with either intact gram-negative and gram-positive bacteria or the membrane integral components of these bacteria including LPS and lipoteichoic acid (13,21–23,40,41). The common feature of these ligands for scavenger receptors is that all are polyanionic molecules except for native LDL and HDL, which were found to be specific ligands for scavenger receptor BI/CD36 (13,37). Analysis of amino acid sequences of the scavenger receptors and point mutagenesis experiments have confirmed the primary role of the positively charged collagen-like domain in specific binding of polyanionic ligands (27). Specifically, a cluster of four lysine residues within the carboxy-terminal end of the collagenous domain is believed to be essential for specific binding and subsequent internalization of the ligand-receptor complexes. Furthermore, computer-based analysis has proposed that the ligand-binding structure, containing a lysine cluster, forms a positively charged groove on the surface of scavenger receptor (27). This three-dimensional structure exhibits high affinity towards a broad array of biologically relevant molecules including microbial products—the characteristic feature of a pattern-recognition receptor system (24). Although it appears that this positively charged 22-amino-acid region in the extracellular collagenous domain is a highly conservative structure among bovine, rabbit, human, and mouse scavenger receptors, specific ligands are differentially recognized by the scavenger receptors (17,18). For instance, polyinosinic acid or poly (I) effectively inhibits ^{125}I-Ac-LDL binding, while polycytidylic acid or poly (C) does not complete for the binding. These data are strong evidence that the net negative charge is necessary but not sufficient for ligand binding to the scavenger receptors and that conformational properties and/or density of negative charge within the ligands are important for electrostatic interactions controlling the affinity of ligand receptor binding.

PROPERTIES OF LIPOPOLYSACCHARIDE AS A NATURALLY OCCURRING LIGAND FOR SCAVENGER RECEPTORS

LPS is a major integral component of the outer membrane of gram-negative bacteria, which has been implicated into the pathogenesis of gram-negative sepsis and septic shock (42). Chemically, LPS represents a complex glycolipid comprised of three structurally and functionally distinct domains: lipid A, a core polysaccharide, and O-antigen (43). Although the diversity of O-polysaccharide chain provides the molecular basis for the structural variety and serological specificity of LPS, similarity of the endotoxic reactions developed in response to LPS isolated from different gram-negative bacteria implies that LPS molecules may possess a common endotoxic structure(s) involved in the activation of LPS-responsive cells. Structural analysis has revealed that lipid A and the inner core region represent the most conservative parts of the LPS molecule and contain the unique bacterial sugars 3-deoxy-D-manno-octulosonic acid (Kdo) and heptose. In addition to structural conservation, this region is a negatively charged part of LPS since lipid A is biphosphorylated at positions 1 and 4′ of the D-glucosamine backbone and the inner core contains carboxyl groups of Kdo and usually phosphorylated heptasyl residues (43).

The net negative charge of the proximal moiety of LPS, provided by phosphate groups of lipid A and acidic Kdo residues, is considered to be another conserved feature of LPS (44). Structural and compositional analysis of LPS isolated from different bacteria has identified the minimal molecular requirements for LPS that are crucial for survival and growth of gram-negative microorganisms. Thus, experimental evidence indicates that, even in bacterial mutants with the most reduced LPS structure characterized to date, the inner core must retain at least two neighboring negative charges provided by either two carboxyl groups of a Kdo-disaccharide or one Kdo residue with an acidic substitution at the 4-position. Together with phosphoryl residues of lipid A, anionic groups of Kdo are believed to be essential for the integrity of gram-negative microorganisms, providing the binding sites for divalent cations which stabilize the outer membrane of the bacteria with abundant low-energy bonds of a chelating nature (45,46).

The concept of a pattern-recognition receptor pathway has been recently proposed to explain the unique ability of the innate immune system to recognize different microorganisms (47). The author has hypothesized that the innate immune response must employ a common, receptor-like mechanism(s) that recognizes structurally related determinants displayed on bacteria and, perhaps, on other pathogens, which are not expressed on the host cells. From this perspective, the existence of the polyanionic conservative structure in the proximal part of LPS may well modulate the development of specific receptor pathway(s) aimed at recognition of both intact bacteria and cell-free LPS by phagocytic cells of mammals. Broad specificity of the scavenger receptors towards polyanionic ligands (2,11), their specific distribution among tissue macrophages (9), and the capacity of scavenger receptors to mediate binding and internalization of LPS (21,41) strongly support the hypothesis that a scavenger receptor–dependent mechanism(s) may have evolved as an ancient pattern-recognition pathway of the innate immunity designed to control the harmonious symbiosis of the host with the acquired commensal gut microflora and participate in the host defense against pathogens in the course of infection.

ROLE OF SCAVENGER RECEPTORS IN LPS-DEPENDENT ACTIVATION OF KUPFFER CELLS

Kupffer cells, the resident hepatic macrophages that account for more than 90% of all tissue macrophages in the body, contribute to the scavenger function of the liver in the clearance of various waste products and toxins via receptor-mediated phagocytosis and endocytosis. Although the large amounts of endotoxin produced by normal gram-negative microflora in the intestinal tract (48) may penetrate the mucosa and appear in the portal circulation under certain pathological conditions (49,50), the liver is able to retain the traces of gut-derived endotoxin. Considerable experimental evidence indicates that the liver is the principal organ in the clearance of the portal or systemic LPS injections and that Kupffer cells are among the first cells of reticuloendothelial system involved in this process (51,52). In addition, recent data have identified the liver as the main source of circulating TNF-α induced by endotoxin administration in both human volunteers and experimental animals (53,54). Evidently, Kupffer cells are the most likely candidates responsible for hepatic TNF-α release, although the precise molecular mechanisms of LPS-induced activation in Kupffer cells have yet to be fully defined.

Recent studies have confirmed that Kupffer cells express the scavenger receptors and are able to accumulate chemically modified proteins such as Ac-LDL and ox-LDL (55,56). In another study using scavenger receptor–transfected CHO cells, it was demonstrated that Ac-LDL and other scavenger receptor ligands compete in binding and metabolism with lipid IVa, the biologically active precursor of lipid A (21). Taken together, these findings prompted the studies for evaluation of the physiological role of the naturally occurring scavenger receptors on Kupffer cells in recognition and clearance/neutralization of endotoxin (41).

To express biological activity, LPS must bind to membrane-localized LPS-binding constituents displayed on a variety of LPS-responsive cells. Although it has yet to be fully evaluated, experimental evidence suggests the existence of multiple LPS-binding sites expressed on different LPS-responsive cells or even on an individual cell. CD14, a 55 kDa glycoprotein, expressed on monocytes, macrophages, and activated neutrophils (57), a 73 kDa LPS-binding protein described on different LPS-responsive cells (58,59), CD11/18 (60), and scavenger receptors (21,41) are the candidates for a putative LPS receptor involved in selective interaction and stimulation of the target cells. The most studied CD14 receptor exists both as a membrane glycosyl phosphatidylinositol–anchored molecule (mCD14) and as a soluble serum protein (sCD14). LPS is transferred to CD14 by a serum acute-phase protein termed LPS-binding protein (LBP). Formation of a LPS-CD14 complex initiates LPS-induced activation of the cells via CD14-dependent pathway (57,61). However, it is becoming evident that LPS signal transduction and clearance or neutralization of LPS by the cells of reticuloendothelial system may represent separate events mediated by different receptor pathways (21,62,63). In addition, while enhanced cytokine responses of mCD14-positive cells to low LPS concentrations is mediated by a CD14-dependent pathway (64,65,66), neutralization of LPS (21), LPS-dependent induction of acute-phase response (67), and modulation of nitric oxide production (68,69) in macrophages may well be mediated by a scavenger receptor mechanism(s).

Transfection of LPS-unresponsive cells with either CD14 (65,66) or scavenger receptors (21) represents a unique approach to individual characterization of each receptor pathway in LPS signal transduction and clearance/detoxification of LPS. However, the potential existence of more than one LPS receptor mechanism in macrophages and other naturally LPS-responsive cells dictates the development of alternative experimental techniques for evaluation of specific contribution of each pathway in LPS-induced cellular responses. Because activation via CD14-dependent mechanisms requires the presence of either purified LBP or serum as a source of LBP (57), the in vitro stimulation of the cells by LPS under strictly serum-free conditions would be expected to bypass CD14 and reflect a CD14-independent pathway of LPS signaling. Using the experimental approach we have experimentally attempted to characterize the potential role of scavenger receptor mechanism(s) in LPS binding to and activation of rat Kupffer cells.

Since receptor-mediated signal is exclusively attributed to the ligand-receptor interaction that precedes internalization of ligand-receptor complexes and their delivery and degradation in lysosomes, our initial efforts have been focused on the identification and characterization of LPS-binding constituents on Kupffer cells. Interestingly, in these in vitro studies specific and saturable binding of ^{125}I-LPS to rat Kupffer cells was demonstrated only under serum-free conditions, while addition of as little as 1% homologous rat serum or fetal bovine serum (FBS) completely inhibited specific binding of radiolabeled LPS to the cells (41,70), suggesting that a CD14-dependent mechanism may not be essential to LPS interaction with rat Kupffer cells. To find out whether the negative charge of the LPS molecule modulates interaction with specific binding sites on Kupffer cells, the effect of different biologically relevant polyanions on ^{125}I-ReLPS binding to Kupffer cells was studied (Fig. 1). The data presented in Figure 1 show that the capacity of polyanions to inhibit specific LPS binding correlates with the net negative charge of these compounds derived from comparative analysis of their molecular structure. The inhibitory effects of the polyanions tested underscore the role of electrostatic interactions in LPS binding to the cells and indicate the potential involvement of scavenger receptor pathway in this process.

Further support for the concept of scavenger receptor–mediated LPS binding to Kupffer cells was obtained from inhibitory experiments in which scavenger receptor–dependent binding of ^{125}I-Ac-LDL to the cells was inhibited by ReLPS. The data summarized in Figure 2 show that ReLPS is able to block ^{125}I-Ac-LDL binding to Kupffer cells with an efficacy similar to that of unlabeled Ac-LDL. Collectively, these data support the hypothesis that a scavenger receptor mechanism(s) may well mediate LPS interaction with rat Kupffer cells.

In an attempt to examine the primary role of serum-independent scavenger receptor mechanism(s) in LPS binding to Kupffer cells, CD14 expression on isolated rat Kupffer cells and resident peritoneal macrophages was compared (Fig. 3). Data shown in Figure 3A indicate that CD14 is readily detectable on isolated rat peritoneal macrophages stained with anti-CD14 monoclonal antibody (My4) and analyzed by flow cytometry. In contrast, under identical experimental conditions, Kupffer cells highly purified by a centrifugal elutriation and stained with MY4 antibody revealed only a slight right shift in the profile of fluorescence intensity in comparison with isotype-matching control antibody (Fig. 3B). The observed low reactivity of anti-CD14 mAb with isolated Kupffer cells suggests low expression of CD14 on Kupffer cells.

The earlier findings that LPS-induced cytokine responses in CD14-positive human cells are 100- to 1000-fold higher in the presence of serum or purified

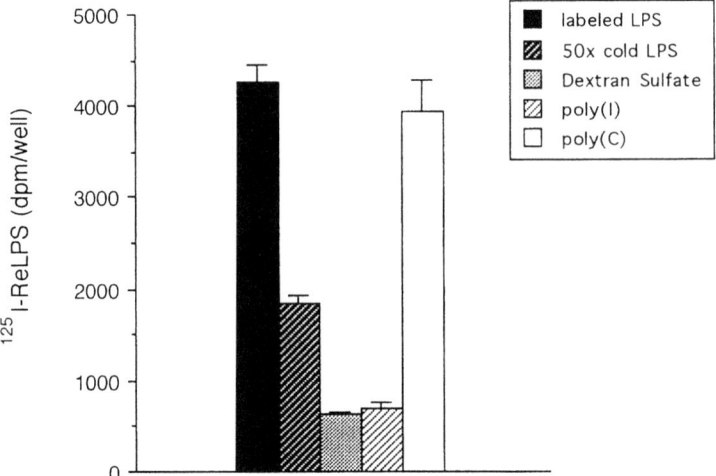

Fig. 1 Inhibitory effects of biologically relevant polyanions on ^{125}I-ReLPS binding to Kupffer cells. Kupffer cells were isolated from conventional Sprague-Dawley rats by a two-step collagenase perfusion technique (41). Isolated cells were placed into a 24-well plate at a density of 0.5×10^6 cells per well in order to obtain a monolayer cell culture. Cells were incubated with 2.5 μg of ^{125}I-ReLPS (*S. minnesota* R595) in the absence or presence of a 50-fold excess of unlabeled ligand or 5 μg/ml of different biological polyanions at 4°C for 4 hours. After incubation, free ligand was extensively washed with PBS and cells were solubilized with 0.5% Triton X-100 prior to counting cell-associated radioactivity. The data shown are means ± standard deviations for quadruplicate cultures of three independent experiments.

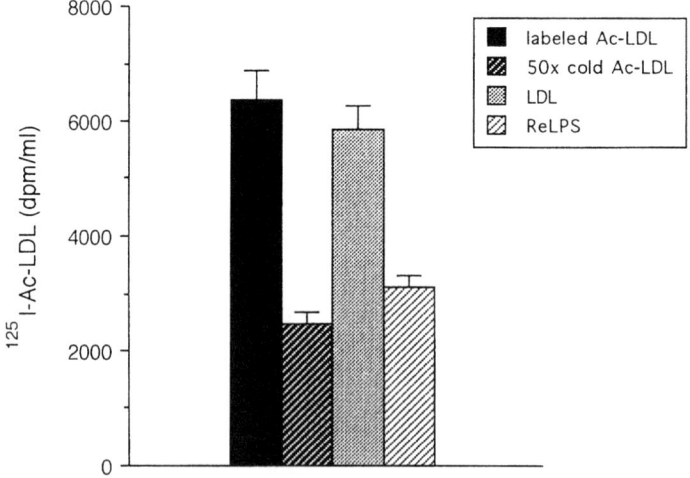

Fig. 2 Inhibition of specific binding of ^{125}I-Ac-LDL to Kupffer cells by Ac-LDL and ReLPS. Rat Kupffer cells were isolated from conventional animals as described for Figure 1. After monolayer cell culture was established, Kupffer cells were incubated with 2.5 μg/ml of ^{125}I-Ac-LDL, which was acetylated with acetic anhydride and then labeled with Na^{125}I using the IODO-GEN reagent (41). The binding assay was performed at 4°C for 4 hours in the absence of presence of 50-fold excess of unlabeled Ac-LDL, native LDL, or ReLPS from *S. minnesota* R595. The cells were washed with PBS, lysed with 0.5% Triton, and counted for cell-bound radioactivity. Data are means ± SD for triplicate cultures of one representative experiment of four performed.

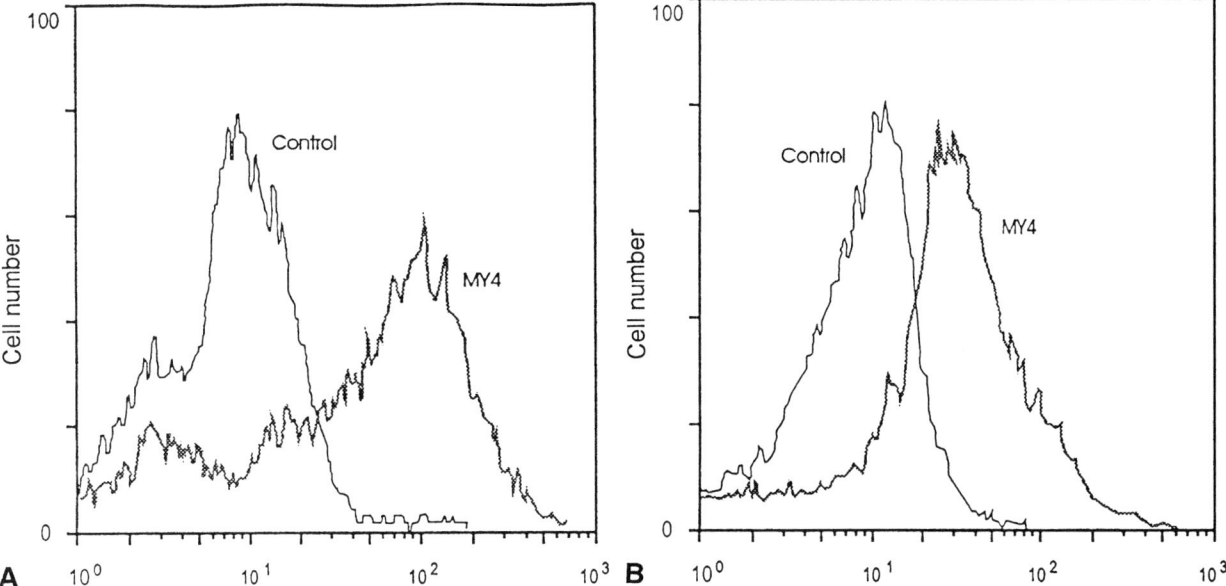

Fig. 3 CD14 antigen expression on resident peritoneal macrophages and Kupffer cells isolated from conventional Sprague-Dawley rats. Isolated from the same animal, resident peritoneal macrophages (A) and Kupffer cells (B) were stained with a fluorescein isothiocyanate (FITC)–conjugated monoclonal antibody (MY4 mAb, Coulter) or isotype-matching control mAb for 30 minutes at 4°C. The cells were fixed with buffered 1% paraformaldehyde (PBS) and then analyzed for fluorescence on a FACSort flow cytometer (Becton-Dickinson). The results are from one representative experiment of three independently performed.

LBP than under serum-free conditions (57) were utilized in our next study to confirm functionally the observed CD14 independence of LPS-induced activation in Kupffer cells. In these experiments isolated Kupffer cells were stimulated with different amounts of LPS in the absence or presence of fetal bovine serum (FBS), as a source of LBP, and the levels of tumor necrosis factor (TNF-α) in 24-hour culture supernatants were measured using a cytotoxicity assay on a TNF-α–sensitive mouse fibrosarcoma cell line (L929) (Fig. 4A). In accordance with our flow cytometry and LPS-binding data, LPS-induced TNF-α responses in Kupffer cells were found to be serum independent in the sense that similar cytokine profiles were produced by the cells stimulated with various LPS concentrations in the presence or absence of FBS (Fig. 4A). In spite of this, LPS-induced cytokine production by isolated rat resident peritoneal macrophages was potentiated in the presence of 10% of FBS, indicating the primary role of CD14-dependent pathway(s) in cell activation (Fig. 4B). These findings imply strongly that while LPS-induced activation of rat peritoneal macrophages is controlled by a CD14-dependent mechanism, CD14 may not be required for the activation of Kupffer cells.

Because of the potential presence of endotoxin in the portal circulation, it was hypothesized that exposure of hepatic cells to traces of gut-derived endotoxin might specifically modulate the observed CD14-independent LPS-dependent responses in Kupffer cells. To address this hypothesis we studied LPS-induced TNF-α production by Kupffer cells isolated from germ-free AUGUS rats, which were originally derived from Sprague-Dawley rats as a germ-free strain in 1948. The data summarized in Figure 5A indicate that, unlike CD14-independent activation in Kupffer cells from conventional rats (Fig. 4A), LPS-induced TNF-α response in Kupffer cells isolated from germ-free rats was only marginal under serum-free conditions. Furthermore, cell activation in the presence of FBS resulted in an effective TNF-α response, indicating strongly the involvement of accessory serum proteins and, perhaps, CD14 in LPS-induced activation in Kupffer cells isolated from germ-free rats. The observed serum dependence of cytokine production in Kupffer cells from germ-free animals also closely resembles LPS-induced activation in CD14-positive cells that manifest the requirements for serum or purified LBP (64,65).

Accumulated experimental evidence suggests the existence of complex interplay between the host and both commensal microflora and pathogenic bacteria (47,71). The symbiosis bewteen the host and microor-

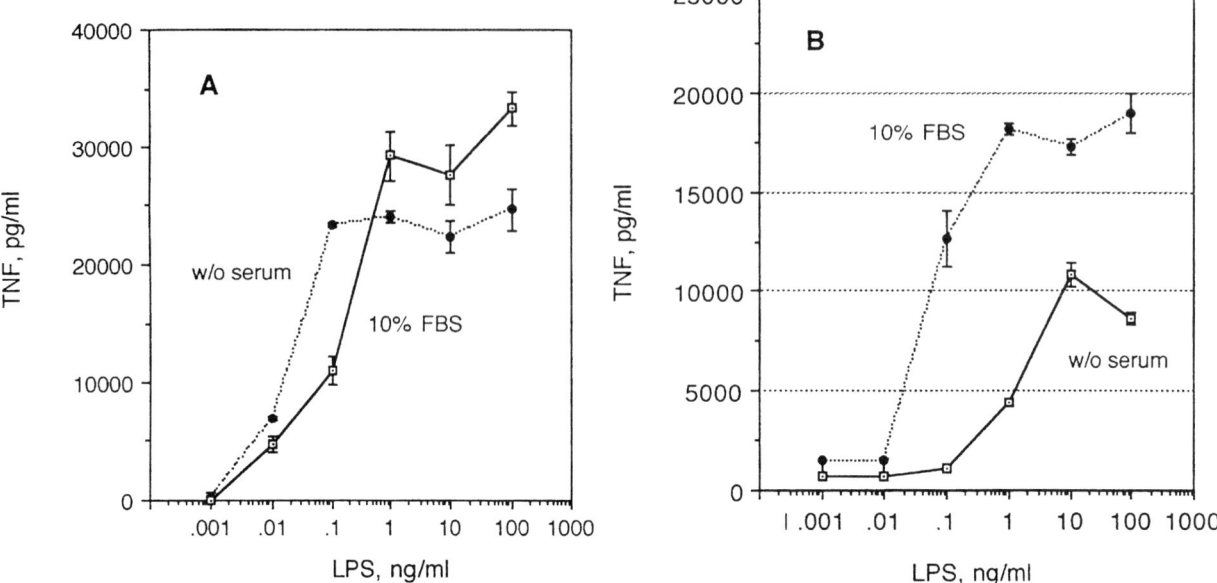

Fig. 4 LPS-induced TNF-α secretion by Kupffer cells and resident peritoneal macrophages isolated from conventional Sprague-Dawley rats. Kupffer cells (A) and resident peritoneal macrophages (B) were plated into a 24-well cluster at approximately density of 0.5×10^6 cells per well in RPMI-1640 culture medium without serum. After monolayer cultures were obtained, cells were stimulated with different amounts of LPS (*S. minnesota* R60, List Biological Laboratories) in the absence or presence of 10% FBS. TNF activity in 24-hour culture supernatants was determined using a cytotoxicity assay on L292 cells. The data are means ± SD of triplicate determinations from one representative experiment of five separately performed.

ganisms primarily depends on the interactive signaling mediated by a cytokine network of multicellular organisms and immunomodulatory properties of bacteria and their products. From this perspective, the following experiments were designed to investigate whether the germ-free status of experimental animals is associated with an insufficient portal supply of the regulatory mediators/signals derived from the commensal intestinal microflora and whether these mediators control cell differentiation and modulation of CD14-independent mechanism of LPS-induced activation in Kupffer cells. Therefore, germ-free rats were removed from the sterile environment and maintained for 30 days under conventional conditions in order to establish normal intestinal microflora. Isolated Kupffer cells from these rats were then used for characterization of LPS-induced TNF-α responses in the presence or absence of serum. The results summarized in Figure 5B show dose-response curves of TNF-α production by LPS-stimulated Kupffer cells isolated from the germ-free rats with acquired commensal intestinal microflora, which was confirmed by bacteriological analysis (data not shown). Unlike serum-dependent cytokine responses manifested by the cells from germ-free animals, a potent LPS-induced production of TNF-α in the absence of serum was observed in Kupffer cells after acquisition of the intestinal microflora. These findings suggest that colonization of the gastrointestinal tract by normal microflora results in acquisition of a unique CD14-independent mechanism of LPS-induced activation in Kupffer cells.

To explore further the concept of a regulatory relationship between the microorganisms constituting normal intestine microflora and cellular components of the immune system (47,71), we evaluated the potential role of gram-negative and gram-positive bacteria and their products in the modulation of CD14-independent LPS-dependent responses in Kupffer cells. Under rigorously controlled sterile conditions and routine bacteriological analysis of fecal microflora, either rat-derived *E. coli* ×7 or *Lactobacillus plantarum* was introduced into the environment of germ-free rats in order to obtain animals selectively associated with gram-negative or gram-positive bacteria, correspondingly. After the bacterial counts in fecal samples reached the steady-state levels within 7–10 days, LPS-induced TNF-α production by Kupffer cells isolated from the animals with selectively established gram-negative or gram-positive microflora was evaluated in the absence or presence of FBS. The results summarized in Figure 5C indicate that

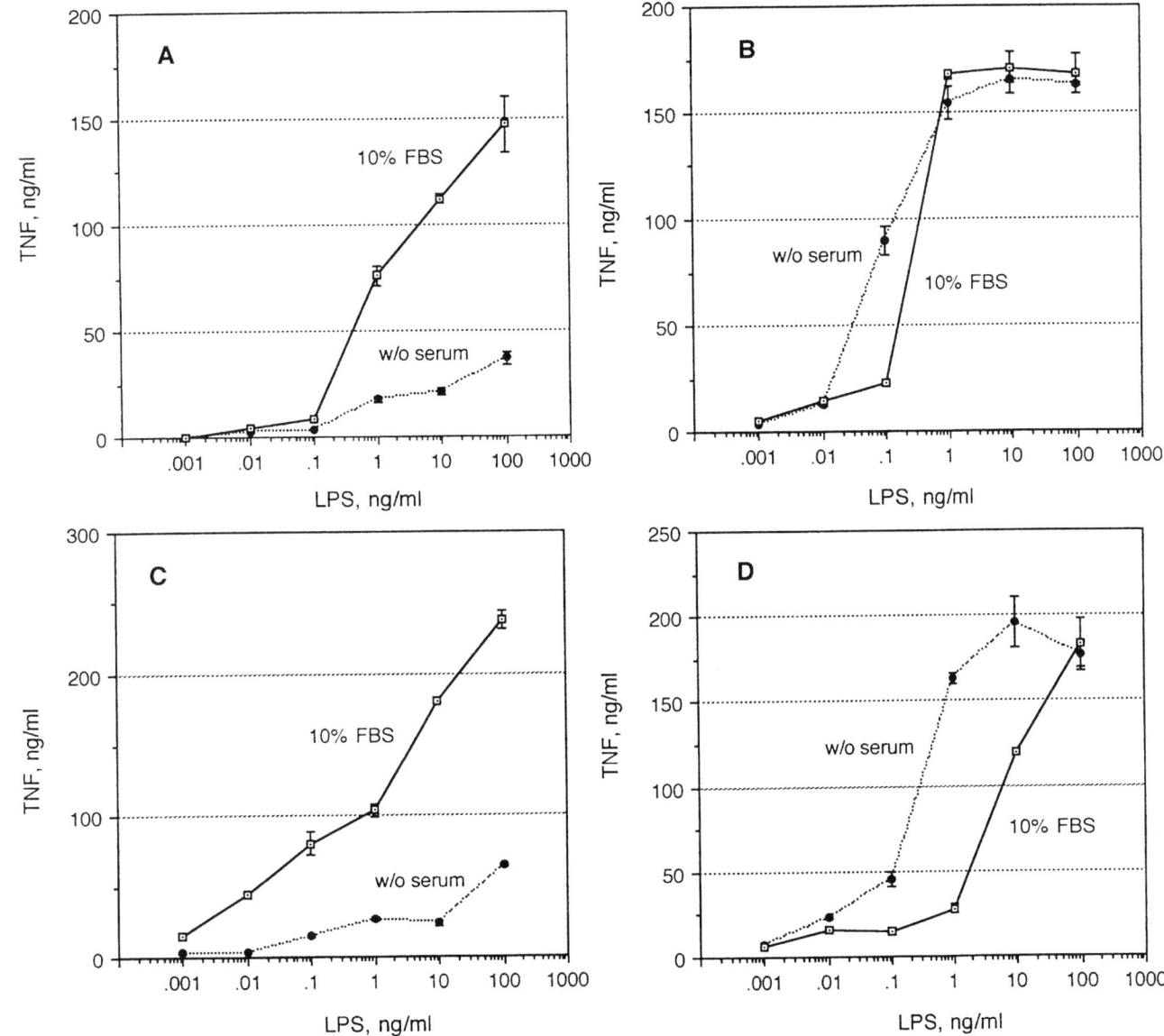

Fig. 5 The effects of acquisition of commensal, gram-positive or gram-negative microflora by germ-free rats on LPS-responsiveness of Kupffer cells. LPS-induced TNF-α production by Kupffer cells isolated from germ-free rats (A) and germ-free rats with either acquired commensal intestinal microflora (B) or selectively associated with *L. plantarum* (C) or *E. coli* ×7 (D) was studied. Kupffer cells were plated into a 24-well cluster at an approximate density of 0.5×10^6 cells per well in RPMI-1640 culture medium without serum. Then cells were extensively washed with PBS prior to be stimulated with different amount of LPS (*S. minnesota* R60) in the absence or presence of 10% FBS. TNF activity in 24-hour culture supernatants was determined using a cytotoxicity assay on L292 cells. The data are means ± SD of triplicate determinations from one representative experiment of four separately performed.

acquisition of gram-positive intestinal microflora (*L. plantarum*) did not affect the CD14-dependent phenotype of LPS-induced cytokine responses originally manifested by Kupffer cells from germ-free animals and characterized by low LPS response in the absence of FBS and potent production of TNF-α in the presence of FBS. In contrast, Kupffer cells isolated from the germ-free rats monoassociated with gram-negative bacterium (*E. coli* ×7) displayed substantial changes in LPS responsiveness (Fig. 5D). Kupffer cells isolated from these animals were able to respond effectively to LPS in the presence or absence of FBS displaying the modulation of CD14-independent mechanism of LPS responsiveness observed in conventional rats. Although the intestinal microflora consists of several hundred different microbial species, our data are direct evidence that monoassociation of germ-free animals with gram-negative (*E. coli* ×7) but not with gram-positive bacteria (*L. plantarum*) modulates a CD14-independent scavenger receptor pathway of LPS signaling in Kupffer cells.

CONCLUSION

Kupffer cells represent more than 90% of tissue macrophages in the body and, among other hepatic cells, are considered to be major cellular source of circulating TNF-α after endotoxin administration. Under serum-free conditions LPS interaction with rat Kupffer cells manifests the characteristic properties of scavenger receptors in the sense that various polyanions inhibit ^{125}I-LPS binding to Kupffer cells. Further, LPS effectively inhibited binding of ^{125}I-Ac-LDL to Kupffer cells mediated by scavenger receptor. The primary role of the scavenger receptor pathway in LPS interaction with Kupffer cells was further confirmed by flow cytometry analysis, which revealed a relatively low expression of CD14 on Kupffer cells as compared to rat peritoneal macrophages. Moreover, LPS-induced cytokine responses in Kupffer cells were found to be serum independent, thus confirming functionally the hypothesis that a unique CD14-independent scavenger receptor mechanism governs activation in Kupffer cells. Studies performed on germ-free rats have revealed the potential role of intestinal microflora and, specifically, gram-negative bacteria in modulation of CD14-independent LPS-dependent scavenger receptor pathway(s), which manifests the characteristic properties of a pattern-recognition receptor system. Further understanding and characterization of LPS-dependent activation in Kupffer cells will provide new insights into the patho-physiological processes associated with gram-negative sepsis and septic shock.

REFERENCES

1. Goldstein JL, Ho YK, Basu SK, Brown MS. Binding site on macrophages that mediates uptake and degradation of acetylated low density lipoprotein, producing massive cholesterol deposition. Proc Natl Acad Sci USA 1979; 76:333–337.
2. Brown MS, Goldstein JL. Lipoprotein metabolism in the macrophage: implications for cholesterol deposition in atherosclerosis. Annu Rev Biochem 1983; 52:223–261.
3. Ren Y, Silverstein RL, Allen J, Savill J. CD36 gene transfer confers capacity for phagocytosis of cells undergoing apoptosis. J Exp Med 1995; 181:1857–1862.
4. Platt N, Suzuki H, Kurihara Y, Kodama T, Gordon S. Role for the class A macrophage scavenger receptor in the phagocytosis of apoptotic thymocytes in vitro. Proc Natl Acad Sci USA 1996; 93:12456–12460.
5. Kodama T, Freeman M, Rohrer L, Zabrecky J, Matsudaira P, Krieger M. Type I macrophage scavenger receptor contains alpha-helical and collagen-like coiled coils. Nature 1990; 343:531–535.
6. Krieger M, Acton S, Ashkenas J, Pearson A, Penman M, Resnick D. Molecular flypaper, host defense, and atherosclerosis. Structure, binding properties, and functions of macrophage scavenger receptors. J Biol Chem 1993; 268:4569–4572.
7. Suzuki H, Kurihara Y, Takeya M, Kamada N, Kataoka M, Jishage K, Ueda O, Sakaguchi H, Higashi T, Suzuki T, Takashima Y, Kawabe Y, Cynshi O, Wada Y, Honda M, Kurihara H, Aburatani H, Doi T, Matsumoto A, Azuma S, Noda T, Toyoda Y, Itakura H, Yazaki Y, Kodama T, et al. A role for macrophage scavenger receptors in atherosclerosis and susceptibility to infection. Nature 1997; 386:292–296.
8. Haworth R, Platt N, Keshav S, Hughes D, Darley E, Suzuki H, Kurihara Y, Kodama T, Gordon S. The macrophage scavenger receptor type A is expressed by activated macrophages and protects the host against lethal endotoxic shock. J Exp Med 1997; 186:1431–1439.
9. Naito M, Kodama T, Matsumoto A, Doi T, Takahashi K. Tissue distribution, intracellular localization, and in vitro expression of bovine macrophage scavenger receptors. Am J Pathol 1991; 139:1411–1423.
10. Brown MS, Basu SK, Falck JR, Ho YK, Goldstein JL. The scavenger cell pathway for lipoprotein degradation: specificity of the binding sites that mediates the uptake of negatively-charged LDL by macrophages. J Supramol Struct 1980; 13:67–81.
11. Rohrer L, Freeman M, Kodama T, Penman M, Krieger M. Coiled-coil fibrous domains mediate ligand binding by macrophage scavenger receptor type II. Nature 1990; 343:570–572.
12. Acton SL, Scherer PE, Lodish HF, Krieger M. Expression cloning of SR-BI, a CD36-related class B scavenger receptor. J Biol Chem 1994; 269:21003–21009.
13. Elomaa O, Kangas M, Sahlberg C, Tuukkanen J, Sormunen R, Liakka A, Thesleff I, Kraal G, Tryggvason K. Cloning of a novel bacteria-binding receptor structurally related to scavenger receptors and expressed in a subset of macrophages. Cell 1995; 80:603–609.
14. Pearson A, Lux A, Krieger M. Expression cloning of dSR-CI, a class C macrophage-specific scavenger receptor from Drosophila melanogaster. Proc Natl Acad Sci USA 1995; 92:4056–4060.
15. Freeman M, Ashkenas J, Rees DJ, Kingsley DM, Copeland NG, Jenkins NA, Krieger M. An ancient, highly conserved family of cysteine-rich protein domains revealed by cloning type I and type II murine macrophage scavenger receptors. Proc Natl Acad Sci USA 1990; 87:8810–8814.

16. Matsumoto A, Natio M, Itakura H, Ikemoto S, Asaoka H, Hayakawa I, Kanamori H, Aburatani H, Takaku F, Suzuki H, Kobari Y, Myai T, Takahashi K, Cohen EH, Wydro R, Housman DE, Kodama T. Human macrophage scavenger receptors: primary structure, expression, and localization in atherosclerotic lesions. Proc Natl Acad Sci USA 1990; 87:9133–9137.

17. Ashkenas J, Penman M, Vasile E, Acton S, Freeman M, Krieger M. Structures and high and low affinity ligand binding properties of murine type I and type II macrophage scavenger receptors. J Lipid Res 1993; 34:983–1000.

18. Wada Y, Doi T, Matsumoto A, Asaoka H, Honda M, Hatano H, Emi M, Naito M, Mori T, Takahashi K, Nakamura H, Itakura H, Yazaki Y, and Kodama T. Structure and function of macrophage scavenger receptors. Ann NY Acad Sci 1995; 748:226–238.

19. Bickel PE, Freeman MW. Rabbit aortic smooth muscle cells express inducible macrophage scavenger receptor messenger RNA that is absent from endothelial cells. J Clin Invest 1992; 90:1450–1457.

20. Arai H, Kita T, Yokode M, Narumiya S, Kawai C. Multiple receptors for modified low density lipoproteins in mouse peritoneal macrophages: different uptake mechanisms for acetylated and oxidized low density lipoproteins. Biochem Biophys Res Commun 1989; 159:1375–1382.

21. Hampton RY, Golenbock DT, Penman M, Krieger M, Raetz CR. Recognition and plasma clearance of endotoxin by scavenger receptors. Nature 1991; 352:342–344.

22. Dunne DW, Resnick D, Greenberg J, Krieger M, Joiner KA. The type I macrophage scavenger receptor binds to gram-positive bacteria and recognizes lipoteichoic acid. Proc Natl Acad Sci USA 1994; 91:1863–1867.

23. Greenberg JW, Fischer W, Joiner KA. Influence of lipoteichoic acid structure on recognition by the macrophage scavenger receptor. Infect Immun 1996; 64:3318–3325.

24. Pearson AM. Scavenger receptors in innate immunity. Curr Opin Immunol 1996; 8:20–28.

25. Kodama T, Reddy P, Kishimoto C, Krieger M. Purification and characterization of a bovine acetyl low density lipoprotein receptor. Proc Natl Acad Sci USA 1988; 85:9238–9242.

26. Wada Y, Doi T, Matsumoto A, Asaoka H, Honda M, Hatano H, Emi M, Naito M, Mori T, Takahashi K, Nakamura H, Itakura H, Yazaki Y, Kodama T. Structure and function of macrophage scavenger receptors. Ann NY Acad Sci 1995; 748:226–238.

27. Doi T, Higashino K, Kurihara Y, Wada Y, Miyazaki T, Nakamura H, Uesugi S, Imanishi T, Kawabe Y, Itakura H, Yazaki Y, Matsumoto A, Kodama T. Charged collagen structure mediates the recognition of negatively charged macromolecues by macrophage scavenger receptors. J Biol Chem 1993; 268:2126–2133.

28. Doi T, Kurasawa M, Higashino K, Imanishi T, Mori T, Naito M, Takahashi K, Kawabe Y, Wada Y, Matsumoto A, Kodama T. The histidine interruption of an alpha-helical coiled coil allosterically mediates a pH-dependent ligand dissociation from macrophage scavenger receptors. J Biol Chem 1994; 269:25598–25604.

29. Hughes DA, Fraser IP, Gordon S. Murine macrophage scavenger receptor: in vivo expression and function as receptor for macrophage adhesion in lymphoid and non-lymphoid organs. Eur J Immunol 1995; 25:466–473.

30. Liao HS, Doi T, Wada Y, Matsumoto A, Kodama T. Multiple function of macrophage scavenger receptors mediated by fibrous coiled coil domains. Gerontology 1996; 42:37–47.

31. Freeman M, Ekkel Y, Rohrer L, Penman M, Freedman NJ, Chisolm GM, Krieger M. Expression of type I and type II bovine scavenger receptors in Chinese hamster ovary cells: lipid droplet accumulation and nonreciprocal cross competition by acetylated and oxidized low density lipoprotein. Proc Natl Acad Sci USA 1991; 88:4931–4935.

32. Pearson AM, Rich A, Krieger M. Polynucleotide binding to macrophage scavenger receptors depends on the formation of base-quartet-stabilized four-stranded helices. J Biol Chem 1993; 268:3546–3554.

33. Krieger M, Herz J. Structures and functions of multiligand lipoprotein receptors: macrophage scavenger receptors and LDL receptor-related protein (LRP). Annu Rev Biochem 1994; 63:601–637.

34. Endemann G, Stanton LW, Madden KS, Bryant CM, White RT, Protter AA. CD36 is a receptor for oxidized low density lipoprotein. J Biol Chem 1993; 268:11811–11816.

35. Nicholson AC, Frieda S, Pearce A, Silverstein RL. Oxidized LDL binds to CD36 on human monocyte-derived macrophages and transfected cell lines. Evidence implicating the lipid moiety of the lipoprotein as the binding site. Arterioscler Thromb Vasc Biol 1995; 15:269–275.

36. Tandon NN, Lipsky RH, Burgess WH, Jamieson GA. Isolation and characterization of platelet glycoprotein IV (CD36). J Biol Chem 1989; 264:7570–7575.

37. Acton S, Rigotti A, Landschulz KT, Xu S, Hobbs HH, Krieger M. Identification of scavenger receptor SR-BI as a high density lipoprotein receptor. Science 1996; 271:518–520.

38. Rigotti A, Acton SL, Krieger M. The class B scavenger receptors SR-BI and CD36 are receptors for anionic phospholipids. J Biol Chem 1995; 270:16221–16224.

39. Abrams JM, Lux A, Steller H, Krieger M. Macrophages in Drosophila embryos and L2 cells exhibit scavenger receptor-mediated endocytosis. Proc Natl Acad Sci USA 1992; 89:10375–10379.

40. van der Laan LJ, Kangas M, Dopp EA, Broug Holub E, Elomaa O, Tryggvason K, Kraal G. Macrophage scavenger receptor MARCO: in vitro and in vivo regulation and involvement in the anti-bacterial host defense. Immunol Lett 1997; 57:203–208.

41. Shnyra A, Lindberg AA. Scavenger receptor pathway for lipopolysaccharide binding to Kupffer and endothelial liver cells in vitro. Infect Immun 1995; 63:865–873.

42. Glauser MP, Zanetti G, Baumgartner J-D, Cohen J. Septic shock pathogenesis. Lancet 1991; 338:732–739.

43. Brade H, Brade L, Schade U, Zahringer U, Holst O, Kuhn H-M, Rozalski A, Rohrscheidt E, Rietschel ET. Structure, endotoxicity, immunogenicity and antigenic-

ity of bacterial lipopolysaccharides (endotoxins, O-antigens). In: Levin J, ten Cate JW, Buller HR, van Deventer SJH, Sturk A, eds. Bacterial Endotoxins: Pathophysiological Effects, Clinical Significance, and Pharmacological Control. New York: Alan R. Liss, 1988:17–45.
44. Mayer H, Campos-Portuguez SA, Busch M, Urabanik-Sypniewska T, Bhat UR. Lipid A variants—or, how constant are the constant regions in lipopolysaccharides? In: Nowotny A, Spitzer JJ, Ziegler EJ, eds. Cellular and Molecular Aspects of Endotoxin Reactions. Amsterdam: Elsevier Science Publishers BV, 1990: 111–120.
45. Schindler M, Osborn MJ. Interaction of divalent cations and polymyxin B with lipopolysaccharide. Biochemistry 1979; 18:4425–4430.
46. Hancock REW, Karuneratne DN. LPS integration into outer membrane structures. In: Nowotny A, Spitzer JJ, Zieler EJ, eds. Cellular and Molecular Aspects of Endotoxin Reactions. Amsterdam: Elsevier Science Publisher BV, 1990:185–191.
47. Janeway CA. The immune system evolved to discriminate infectious non-self from non-infectious self. Immunol Today 1992; 13:11–16.
48. van Saene JJM, Stoutenbeek CP, van Saene NKF. Faecal endotoxin in human volunteers: normal values. Microb Ecol Health Dis 1992; 5:179–184.
49. Jacob AI, Goldberg BS, Bloom N, Degenshein GA, Kozinn PJ. Endotoxin and bacteria in portal blood. Gastroenterology 1977; 72:1268–1270.
50. Schlag G, Redl H, Dinges HP, Davies J. Sources of endotoxin in the posttraumatic setting. Prog Clin Biol Res 1991; 367:121–134.
51. Mathison JC, Ulevitch RJ. The clearance, tissue distribution and cellular localization of intravenously injected lipopolysaccharide in rabbits. J Immunol 1979; 123:2133–2143.
52. Ruiter DJ, van der Meulen J, Brouwer A, Hummer MJR, Mauw BJ, van der Ploeg JCM, Wisse E. Uptake by liver cells of endotoxin following its intravenous injection. Lab Invest 1981; 45:38–45.
53. Kumins NH, Hunt J, Gamelli RL, Filkins JP. Partial hepatectomy reduces the endotoxin-induced peak circulating level of tumor necrosis factor in rats. Shock 1996; 5:385–388.
54. McGuinness OP, Lacy DB, Ejiofor J, Bagby GG. Hepatic release of tumor necrosis factor in the endotoxin-treated conscious dog. Shock 1996; 5:344–348.
55. De-Rijke YB, Biessen EA, Vogelezang CJ, van Berkel TJ. Binding characteristics of scavenger receptors on liver endothelial and Kupffer cells for modified low-density lipoproteins. Biochem J 1994; 304:69–73.
56. de Rijke YB, van Berkel TJ. Rat liver Kupffer and endothelial cells express different binding proteins for modified low density lipoproteins. Kupffer cells express a 95-kDa membrane protein as a specific binding site for oxidized low density lipoproteins. J Biol Chem 1994; 269:824–827.
57. Wright SD, Ramos RA, Tobias PS, Ulevitch RJ, Mathison JC. CD14, a receptor for complexes of lipopolysaccharide (LPS) and LPS binding protein. Science 1990; 249:1431–1433.
58. Lei M-G, Morrison DC. Specific endotoxic lipopolysaccharide-binding proteins on murine splenocytes. I. Detection of lipopolysaccharide-binding sites on splenocytes and splenocyte subpopulations. J Immunol 1988; 141:996–1005.
59. Lei M-G, Morrison DC. Specific endotoxic lipopolysaccharide-binding proteins on murine splenocytes. II. Membrane localization and binding characteristics. J Immunol 1988; 141:1006–1011.
60. Ingalls RR, Golenbock DT. CD11c/CD18, a transmembrane signaling receptor for lipopolysaccharide. J Exp Med 1995; 181:1473–1479.
61. Ulevitch RJ, Tobias PS. Receptor-dependent mechanisms of cell stimulation by bacterial endotoxin. Annu Rev Immunol 1995; 13:437–457.
62. Kitchens RL, Ulevitch RJ, Munford RS. Lipopolysaccharide (LPS) partial structures inhibit responses to LPS in a human macrophage cell line without inhibiting LPS uptake by a CD14-mediated pathway. J Exp Med 1992; 176:485–494.
63. Delude RL, Fenton MJ, Savedra R Jr, Perera P-Y, Vogel SN, Thieringer R, Golenbock DT. CD14-mediated translocation of NF-kB induced by lipopolysaccharide does not require tyrosine kinase activity. J Biol Chem 1994; 269:22253–22260.
64. Lee JD, Kato K, Tobias PS, Kirkland TN, Ulevitch RJ. Transfection of CD14 into 70Z/3 cells dramatically enhances the sensitivity to complexes of lipopolysaccharide (LPS) and LPS binding protein. J Exp Med 1992; 175:1697–1705.
65. Golenbock DT, Liu Y, Millham FH, Freeman MW, Zoeller RA. Surface expression of human CD14 in Chinese hamster ovary fibroblasts imparts macrophage-like responsiveness to bacterial endotoxin. J Biol Chem 1993; 268:22055–22059.
66. Delude RL, Savedra R, Zhao H, Thieringer R, Yamamoto S, Fenton MJ, Golenbock DT. CD14 enhances cellular resoinses to endotoxin without imparting ligand-specific recognition. Proc Natl Acad Sci USA 1995; 92:9288–9292.
67. Haziot A, Lin XY, Zhang F, Goyert SM. The induction of acute phase proteins by lipopolysaccharide uses a novel pathway that is CD14-independent. J Immunol 1998; 160:2570–2572.
68. Matsuna R, Aramaki Y, Arima H, Tsuchiya S. Scavenger receptors may regulate nitric oxide production from macrophages stimulated by LPS. Biochem Biophys Res Commun 1997; 237:601–605.
69. Kirikae T, Kodama T, Kirikae F, Suzuki H, Nakano M. The role of scavenger receptors in LPS-induced macrophage activation. J Endotoxin Res 1996; 3:S9.
70. Shnyra AA, Hirohachi N, Lei M-G, Ware J, Morrison DC. Lipopolysaccharide-dependent, CD14-independent scavenger receptor pathway for proinflammatory cytokine responses in rat Kupffer cells and endothelial liver cells. In: Faist E, Baue AE, Schildberg FW, eds. The Immune Consequences of Trauma, Shock and Sepsis: Mechanisms and Therapeutic Approaches. Lengerich, Germany: Pabst Science Publisher, 1996:839–846.
71. Henderson B, Poole S, Wilson M. Micobial/host interactions in health and disease: Who controls the cytokine network? Immunopharmacology 1996; 35:1–21.

27

The Role of Platelet-Activating Factor in Endotoxin-Related Disease

Taco W. Kuijpers and Tom van der Poll
University of Amsterdam, Amsterdam, The Netherlands

INTRODUCTION

Biochemistry and Biosynthesis of Platelet-Activating Factor

Platelet-activating factor (PAF) is a phospholipid implicated in various pathophysiological conditions. Its mediator role was originally identified in anaphylaxis and allergic diseases such as asthma. Recent studies suggest a broader role for PAF in inflammatory reactions, such as sepsis, acute respiratory distress syndrome (ARDS), disseminated intravascular coagulation (DIC), necrotizing pancreatitis, hemolytic uremic syndrome, and IL-2 immunotherapy (1–11). The potent bioactivity of PAF towards cells depends on its structural features (Fig. 1A), with an ether linkage at the sn-1 position and an acetate at the sn-2 position (1-O-alkyl-2-acetyl-sn-glyceryl-3-phosphocholine) (1,2,12). The precursor lipid for PAF is 1-O-alkyl-2-acyl-glyceryl-3-phosphocholine (GPC), an alkoxyether phospholipid present in the plasma membrane of most cell types. The key enzymes involved in the biosynthesis of GPC are localized in peroxisomes. Once formed, GPC is transferred to the plasma membrane. During inflammation, the rapid formation of PAF from cellular membranes takes place by the enzymatic activity of cytoplasmic phospholipase A_2 (cPLA$_2$) and acetyltransferase, both activated upon cellular stimulation (1,2,13). PAF can be produced in response to injury by various cells, including endothelial cells, macrophages, granulocytes, and platelets (1,2).

The de novo route does not contribute to PAF formation under these conditions. cPLA$_2$ liberates arachidonic acid (AA) from the sn-2 position of GPC to generate 2-lyso-PAF, which is rapidly acetylated to PAF (Fig. 1B). When AA has been liberated from lipid precursors, e.g., from the sn-2 position of GPC by the aforementioned activation of intracellular PLA$_2$, a wide range of eicosanoid metabolites consisting of leukotrienes, thromboxanes, and prostaglandins can be concomitantly generated through the activity of 5-lipoxygenase (5-LO) or the isoforms cyclooxygenase-1 (COX-1) and COX-2 (induced upon inflammation), depending on the tissue or cell type involved and the predominating enzymatic activity present (14). Such activation of parallel mediator cascades may have synergistic effects amplifying an inflammatory response once initiated, as will be discussed separately.

Plasma Membrane Exposition and Releasability

Even though PAF can be produced by various cells (1,2), significant differences exist between the cell types in their capacity to release PAF. Most of the PAF synthesized by tissue cells remains cell bound, therefore perfectly able to initiate and localize inflammatory reactions. In contrast, phagocytes release most of the PAF produced, most likely contributing more predominantly to systemic effects of PAF.

A convincing mechanistic view explaining these phenomena concerning releasability is not available. The multidrug resistance–associated protein (MRP), a 190 kDa product of an ATP-binding cassette transporter gene, has most recently been suggested to have a role

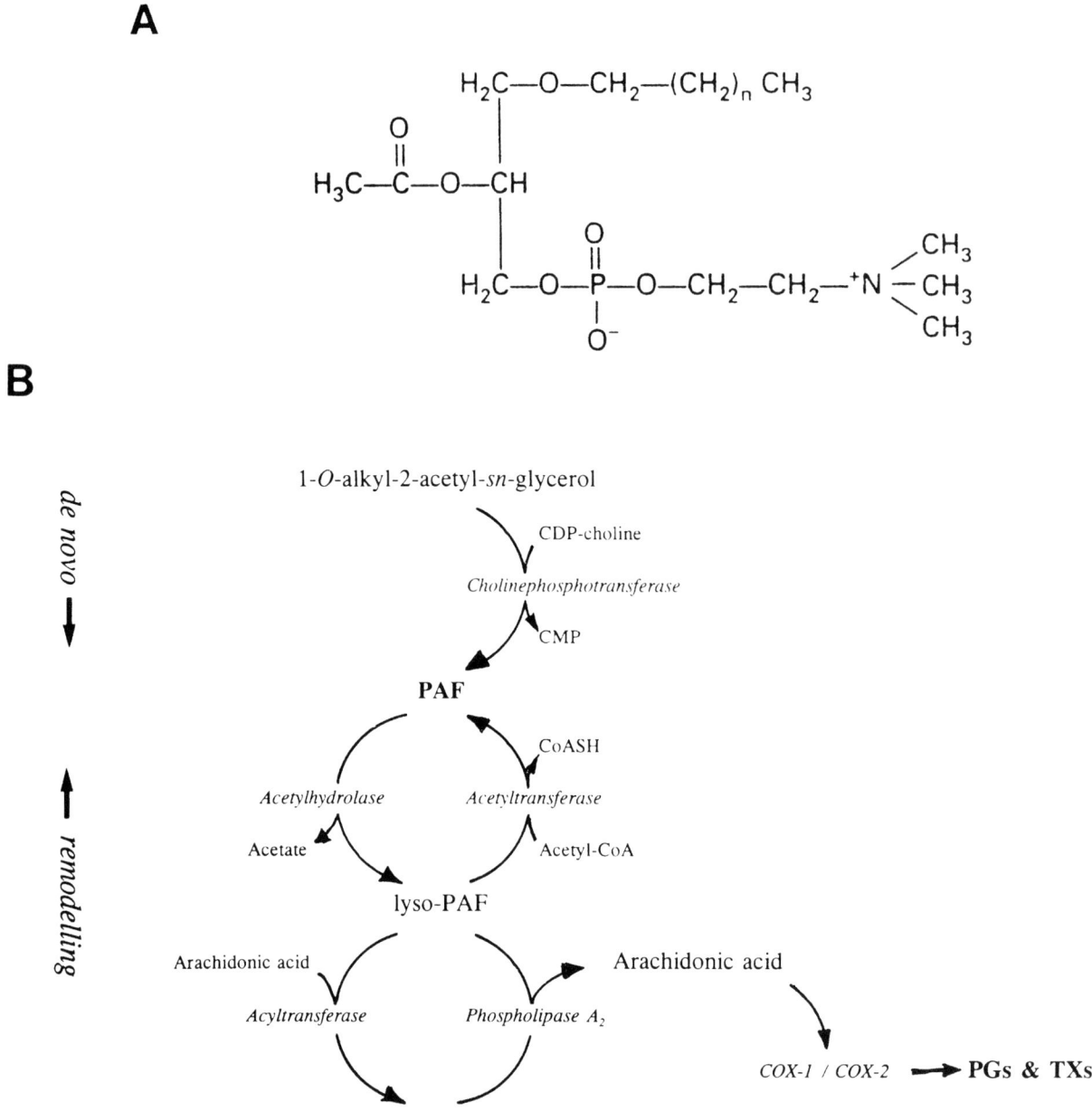

Fig. 1 (A) The chemical structure of PAF ($n = 14-18$). (B) The biosynthetic routes of PAF synthesis and metabolism. The so-called de novo pathway is not rapidly inducible (upper part), in contrast to the *remodeling* pathway (lower part). The latter pathway is therefore of major importance in the inflammatory responses as, for instance, during infections. The biochemical structure of PAF is known as 1-*O*-alkyl-2-acetyl-*sn*-glycerol-3-phosphocholine, whereas lyso-PAF stands for 1-*O*-alkyl-2-lyso-*sn*-glycerol-3-phosphocholine. Glycerol-3-phosphocholine has been abbreviated to GPC, the major source for the lipid mediator PAF, as well as arachidonic acid. Arachidonic acid can be metabolized by enzymatic activities of cyclooxygenase (COX-1 and the inducible COX-2) and 5-lipoxygenase (5-LO) to pro-inflammatory metabolites such as prostaglandins (PGs), thromboxanes (TXs), and leukotrienes (LTs), respectively. The pro-inflammatory lipid mediators are shown in bold.

in the release of leukotrienes from cells. The endogenous glutathione conjugate of LTC4 had the highest affinity for this transporter. A series of experiments (e.g., photoaffinity labeling and transfection studies) provided evidence that the MRP gene genuinely encodes an ATP-dependent export pump for conjugates of lipophilic compounds with glutathione and several other anionic residues (15). Whether similar "export pumps" are involved in PAF release remains to be clarified.

The Pro-Inflammatory Role of PAF

The action of PAF is pro-inflammatory. The biological effects of PAF are very diverse. For instance, PAF acts as a spasmogenic on smooth muscle cells, increases mucus secretion in bronchial tissue cultures, and, with respect to inflammatory cells, induces granulocytes and monocytes/macrophages to migrate, degranulate, and release cytokines and toxic oxygen metabolites. The in vivo effect of PAF depends in part on the mode of administration. Intradermal injection induces a rapid but transient influx of predominantly granulocytes, with a concomitant local reaction of edema and vasodilatation (2). Upon inhalation of PAF aerosol, PAF provokes an influx of inflammatory cells into the interstitium and bronchoalveolar lumen, together with a reaction of bronchoconstriction that resembles asthma, ventilation-perfusion mismatches, changes in pulmonary microcirculation, and vascular leakage (6,9). Finally, when administered intravenously, PAF is able to cause shock and multiple organ failure, characterized by a state of depressed cardiac output, vasoconstriction, and vascular leakage (16).

A SINGLE G-PROTEIN–COUPLED RECEPTOR FOR PAF

Cloning of the PAF Receptor

The diversity in vivo effects can be explained by the ubiquitous expression of the receptor for PAF (PAF-R). Cloned DNA for human PAF-R conferred high-affinity binding sites for PAF when transfected into COS-7 cells showing binding and desensitization properties similar to the human leukocyte receptor. Seven threonine and three serine residues conserved in the cytoplasmic loops of both receptors may function as phosphate acceptors and may be involved in the homologous desensitization of the receptor. Southern analysis using this cDNA indicates that the PAF receptor gene is present as a single copy in the human genome. The deduced protein sequence predicts seven hydrophobic regions for the PAF receptor (Fig. 2) characteristic of the rhodopsin gene family of G-protein–coupled receptors (GPCR) and is 83% identical to the deduced protein sequence of the corresponding guinea pig molecule (17–20). The gene for PAF-R consists of an open-reading frame without introns resulting in a single transcript. Alternative splicing events only occurs at the untranslated 5′-end site. Analysis of somatic cell hybrids suggests that the PAF receptor is encoded by a single gene on human chromosome 1 (21).

In general, GPCR members are involved in a wide spectrum of functional responses ranging from hormonal adaptations to a balanced regulation of cardiovascular function, blood pressure, and electrolyte levels. The nature of the ligands for the various GPCR members are diverse: lipid mediators, proteins, or electrolytes (Table 1).

Within the GPCR superfamily, inflammation can be mediated through lipid metabolites other than PAF, e.g., the aforementioned leukotrienes, thromboxanes, and prostaglandins (14). For instance, the leukotrienes are biologically active 5-lipoxygenase products of arachidonic acid metabolism involved in the mediation of many of the inflammatory disorders in which PAF is believed to act. Leukotriene B4 is generated by many cell types and is a chemoattractant for granulocytes. The sulfidopeptide leukotrienes C4, D4, and E4 have been incriminated in allergic reactions and increase vascular permeability and constrict smooth muscle. Leukotrienes also exert their biological actions through specific GPCR interactions (see Table 1).

The small proteins of the chemokine superfamily should also be mentioned briefly. This family of leukocyte activators consists of more than 30 chemotactic molecules. Most chemokines can be classified into two groups—α (CXC) and β (CC)—distinguished by the presence or absence of a single amino acid between the first two of four conserved cysteines. The γ (single C) and δ (CX_3C) classes of chemokines have only recently been coined, having one (lymphotactin; fractalkine or neurotactin) member each, respectively. As exemplified by the first CXC member described, IL-8 (22), most CXC chemokines activate neutrophils, whereas the CC chemokines act upon different sets of lymphocyte subsets, monocytes, as well as eosinophilic and basophilic granulocytes (23) (Table 1). The overwhelming interest in chemokine receptors stems from the recent findings that CXCR4 and CCR2b, −3, and −5 function as important coreceptors for virus internalization and disease upon infection with HIV-1 and HIV-2 (23).

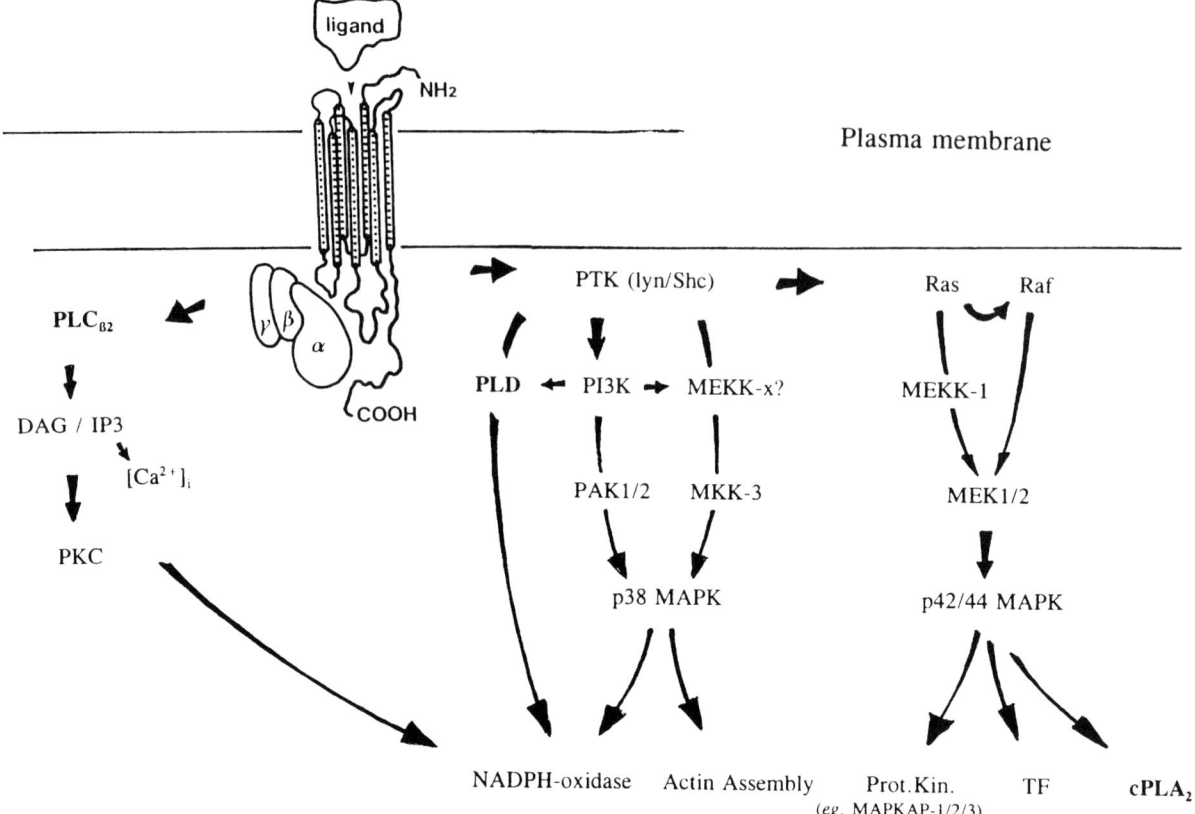

Fig. 2 G-protein–coupled receptor and signal transduction. After ligand binding, a series of signaling events takes place. First, the activated G-protein falls apart in the active βγ units, thus inducing phospholipase-C (PLC)–dependent cleavage of phosphoinositol (PI)-4,5-biphosphate into the diacylglycerol (DAG) and inositoltrisphosphate (IP3). DAG, alone or combined with increased levels of intracellular Ca^{2+}, stimulates protein kinase-C (PKC) activity, which is involved in multiple cellular responses. In addition, the liberated Gα-units are involved in protein tyrosine kinase (PTK) activation and subsequent signaling cascades, among which are phospholipase D (PLD), the proto-oncogenic Ras protein, and Raf kinase activities. Various mitogen-activated protein kinase (MAPKs) become activated, "translating" signaling events into specific cellular responses as has been indicated for neutrophils (e.g., toxic oxygen metabolites by NADPH-oxidase activity; adhesion and migration by actin assembly; other protein kinases, transcription factors (TF), and phospholipase A_2 involved in PAF generation).

PAF-R and Signal Transduction

Signal transduction through GPCR is thought to be imposed by changes in conformation upon ligand binding to the extracellular domain(s) (24). The subsequent process of G-protein–catalyzed GTP hydrolysis at the intracellular plane of the plasma membrane results in the activation of protein kinase activity and the generation of second messengers involved in intracellular signaling.

In contrast to PAF-R, some GPCR are subject to alternative splicing (25–28). G-protein coupling and downstream signaling may differ accordingly. Although the receptors have similar affinity for their respective ligand, coupling to different G proteins has clear implications for the induction of certain cellular functions (28). Differences in desensitization, receptor turnover, and prevalence in alternative splice-variant expression may explain part of the well-differentiated cellular responsiveness observed. However, most GPCR are not subject to alternative splicing.

Although certain schemes of downstream signaling pathways are currently used to explain the intracellular and nuclear changes observed after ligand binding to a receptor (Fig. 2), the coupling events of signaling to specific cellular functions become increasingly elucidated (29). Some important differences have also been

Table 1 G-Protein–Coupled Receptors: An Ever-Growing Family

Receptor	Ligand	Expression
Rhodopsins/Odorants		
Rhodopsin	Light	Retinal rod cells
Opsin	Spectral light	Retinal cone cells
OR	Odorants	Mucosa/Nervus I
Hormones, hormonal peptides, and transmitter(-like) substances		
1. Endocrine function		
ACTHR	ACTH	Adrenal cortex, leukocytes
MSHR	MSH	Melanocytes
TSHR	Thyrotropin	Thyroid
LHR	Lutropin	Gonads
GFR	GRHR	Pituitary
GLR	Glucagon	Hepatocytes
VIPR	Vasoactive intestinal peptide	Vasculature, neuroectoderm
AVPR2	Vasopressin	Kidney
2. Neurotransmitter/Cardiovascular function		
SPR	Substance P	Neurons, SMC, leukocytes
NKR	Neuromedin K	Nerve tissue, lung
5HT1/2R	Serotonin	Brain
ADRα1,2a/b	Adrenaline	Ubiquitous
ADRβ1,2	Adrenaline	Ubiquitous
D2/3/4R	Dopamine	Ubiquitous
mACR(1-4)	Acetylcholine	Ubiquitous
AT_2R1	Angiotensin-II	VSMC, adrenals, kidney
AT_2R2	Angiotensin-II	VSMC, adrenals, kidney
ET_AR	ET-1 (ET-2)	VSMC, heart
ET_BR	ET-1/ET-3	EC, brain, neuroectoderm
3. Calcium homeostatis		
CTR	Calcitonin	Kidney, osteoclasts
PTH/PTHrPR	PTH(1-36)/PTHrP(1-37)	Kidney, bone, leukocytes
CaR	Ca^{2+}	Parathyroid, kidney
Pro-inflammatory mediators and chemotactic cytokines		
1. Lipid mediators		
PAFR	PAF	Ubiquitous
TXRα/β	TxA_2	Placenta/HEC, platelets
TXR2	TxA_2	Kidney
BLTR	LTB4	Leukocytes, thymus, spleen
EP1,-2,-3	PGs	Ubiquitous
2. Peptidic mediators		
FMLPR	fMLP	Myeloid cells, hepatocytes
C5aR	C5a	Ubiquitous (e.g., myeloid cells, HEC, VSMC, BEpC, hepatocytes)
3. Protease-activated receptors (PAR)		
PAR1	Thrombin	Platelets, BM
PAR2	Trypsin?	Kidney, small intestine, stomach, eye
PAR3	Thrombin	BM, lymph node, heart, GI tract
4. Chemokines		
CXCR1	IL-8	Granulocytes
CXCR2	IL-8, NAP-2, MGSA/groα, ENA-78	Granulocytes
CXCR3	IP-10, Mig	Granulocytes, lymphocytes
CXCR4	SDF	Lymphocytes (macrophages)
CCR1	MIP-1α/β, RANTES, MCP-3	Lymphocytes (Th1), monocytes

Table 1 Continued

Receptor	Ligand	Expression
CCR2a/b	MCP-1/3/4	Lymphocytes (Th2), monocytes, basophils
CCR3	Eotaxin > RANTES, MCP-2/3/4	Eosinophils, macrophages
CCR4	TARC, MDC	Thymus, lymphocytes, macrophages, eosinophils, platelets
CCR5	MIP-1α/β, RANTES	Macrophages, lymphocytes
CCR6	LARC	Lymphocytes
CCR7	ELC	Lymphocytes
CCR8	I-309	Thymus, spleen, lymph nodes, monocytes, lymphocytes, eosinophils
DARC	IL-8, MGSA/groα, RANTES, MCP-1	Erythrocytes, HEC, Purkinje cells, lymphocytes

ET: Endothelin; IL: interleukin; ENA-78: 78-amino-acid epithelial cell–derived neutrophil activator; IP-10: interferon-inducible protein of 10 kDa; Mig: monokine induced by gamma-interferon; MCP: monocyte-chemotactic protein; MIP: macrophage inflammatory protein; RANTES: regulated on activation normal T cell expressed and secreted; SDF: stromal cell–derived factor; TARC: thymus and activation-regulated chemokine; LARC: liver and activation-regulated chemokine; ELC: EBV-induced gene-1 (EBI-1) ligand chemokine; DARC: Duffy antigen receptor complex; (V)SMC: (vascular) smooth muscle cell; HEC: human endothelial cell; BEpC: bronchial epithelial cell; BM: bone marrow; MDC: macrophage-derived chemokine.

noted among chemoattractants. For instance, PAF is unable to induce NADPH-oxidase activity in neutrophils but only primes for enhanced activity upon a second triggering event. FMLP (N-formyl-Methionine-Leucine-Phenylalamine), another strong chemoattractant for neutrophils through binding to a single GPCR member (Table 1), induces all of the responses that PAF does, but, in contrast, is also able to activate the respiratory burst for superoxide production (30). A plausible "upstream" explanation consists of the fact that a single GPCR can bind to several G-proteins of different classes (and pertussis toxin sensitivity). The range of signaling events of a single-gene GPCR member could then depend on the number of different G-proteins. This possibility has been most thoroughly investigated in case of the FMLP receptor (31,32). Like PAF-R, FMLP-R is encoded by a single-copy gene (on chromosome 19q13) with an intronless open-reading frame and no alternative splicing events occurring during translation (33). Although no data exist concerning this upstream possibility for PAF-R, co-transfection studies in various cell types may clarify this issue.

Regulation of PAF-R Expression

Many of the GPCR proteins are expressed ubiquitously (Table 1), being regulated at the level of expression by cellular activation or differentiation. The same holds true for the PAF-R. Whereas Northern blot analysis, flow cytometry, and immunoblotting with anti-hPAF-R antibody indicated that monocytes, neutrophils, and B-lymphocytic cell lines all shared a similar hPAF-R species, resting T cells and natural killer cells failed to express detectable levels of either hPAF-R protein or mRNA (34). Nonetheless, regulation of expression holds true for the PAF-R as well. For instance, macrophages possess PAF-R that can be increased significantly upon exposure to lipopolysaccharide (LPS) and interferon-gamma (IFN-γ) (35,36), while transcription of the *PAFR* gene can be downregulated by increased levels of intracellular cAMP (37). Lymphocytes not expressing PAF-Rs in their resting state already exhibited high-affinity PAF receptors after being activated for 1 day, with a maximum by days 4–6. The subsequent decline in expression is a consequence of a decreased transcription rate, which could, however, be prevented by further IL-2 addition (38).

Although some studies have indicated that a receptor-independent mechanism of internalization of PAF exists (39), transfection studies have indicated that after binding of PAF to its receptor, essentially all of the cell-associated PAF is internalized to become incorporated into a phospholipid pool (40). The fate of internalized PAF-R is less clear, albeit they are likely to be degraded upon internalization.

THE ROLE OF PAF IN LOCAL AND SYSTEMIC INFECTIOUS DISEASE

PAF Synthesis: Mechanism of LPS-Mediated Induction

The interactions of LPS with host cells can result in PAF synthesis, both in vitro as well as in vivo, during gram-negative infections. The molecular mechanism that controls cell recognition and response to LPS has been established to depend largely on the expression of CD14. In human phagocytes, cellular responses, e.g., synthesis of PAF, involves two proteins—plasma LPS-binding protein (LBP) and cell membrane CD14 (mCD14)—recognizing the natural ligand LBP-LPS complex (41–43). On the other hand, tissue cells, such as endothelial cells (EC), which do not express mCD14, use the soluble CD14 molecule from plasma for LPS-induced effector functions, among which is PAF synthesis (43). In contradiction to the specificity of CD14 for LPS as once proclaimed, capsular components of gram-positive bacteria may evoke innate inflammatory responses through interactions with CD14 as well, exemplifying the existence of a more universal "pattern recognition" not delineated or equaled by a simple gram-staining procedure (44,45). Although sCD14 and LBP are able to catalyze the effects of LPS and, in addition, the exchange of phospholipids in general, the effects of the phospholipid mediator PAF are induced independently, as demonstrated in CD14 knockout mice (46).

Membrane-Bound PAF: Localized Inflammation

In acute localized inflammatory reactions, granulocytes rapidly respond by adhesion to the vascular lining of endothelial cells and rapid extravasation before exerting toxic activities in the tissues. These early responses of neutrophil extravasation are mediated by PAF synthesized by endothelial cells, as we, among others, have shown (47,48). The adhesion molecules dominating the picture of leukocyte traffic consist of members of the selectin and integrin families, as well as their respective ligands (49,50) (see Table 2).

Membrane-bound endothelial PAF can act in a so-called "juxtacrine" fashion in order to localize the early immune response of leukocyte adhesion and extravasation. In this way, only the leukocytes rolling along the locally induced vascular lining via selectin-mediated interaction become activated as indicated by high-affinity interactions of the leukocyte integrin receptors and subsequent diapedesis (49).

The biosynthesis of endothelial PAF in this early phase can be induced by bacterial products, by histamine released by mast cells close to vessels, by local thrombin generation, by inflammatory cytokines such

Table 2 Adhesion Molecules in Leukocyte Extravasation

Leukocyte	Endothelial cell	Extracellular matrix or cell-associated proteins
Selectin/Ligand	**Selectin/Ligand**	
L-selectin (CD62L)	Undefined	
PSGL-1, other	P-selectin (CD62P)	
ESL-1, other	E-selectin (CD62E)	
β_2 Integrins	**Ig-supergene family**	**Coagulant/Complement**
CD11a/CD18 ($\alpha L\beta 2$; LFA-1)	ICAM-1, ICAM-2	
CD11b/CD18 ($\alpha M\beta 2$; CR3)	ICAM-1, undefined	Fibrin(ogen), C3bi
CD11c/CD18 ($\alpha X\beta 2$)	Undefined	Fibrin(ogen), C3bi
β_1 Integrins	**Ig-supergene family**	**ECM proteins/Mucins**
CD49b/CD29 ($\alpha 2\beta 1$)		Collagen
CD49d/CD29 ($\alpha 4\beta 1$)	VCAM-1 (variants)	FN(CS1)
CD49d/CD107 ($\alpha 4\beta 7$)	VCAM-1, MadCAM-1	GlyCAM-1
CD49e/CD29 ($\alpha 5\beta 1$)		FN(RGD)
CD49f/CD29 ($\alpha 6\beta 1$)		Laminin

ICAM, Intercellular adhesion molecule; VCAM, vascular cellular adhesion molecule; FN, fibronectin (CS1- or RGD-containing isoforms).

as TNF and IL-1 derived from macrophages or lymphocytes, or after hypoxic conditions (51,52).

The Systemic Inflammatory Response Syndrome

Under normal conditions, the localized synthesis of PAF helps to attract inflammatory cells required to eliminate the cause(s) of inflammation. Under severe conditions of inflammation or hypoxic stress, the tissue reaction will, however, activate infiltrating granulocytes excessively. As a consequence, these tissues may show major damage and further impairment of function, as observed in ischemia-reperfusion injury. Such aggravated reactions may become life-threatening due to spillover of bacteria or inflammatory products into the circulation and devastating systemic effects.

The systemic inflammatory response syndrome is a serious clinical condition with a grave prognosis, involving the release of cytokines and excessive activation of coagulation, fibrinolysis, and complement. The role of cytokines has been extensively studied in the pathogenesis of septic shock. PAF is also detectable in sera of patients with septic shock (53) or animals injected with LPS, but despite some challenging reports on PAF inhibition after the symptoms of shock were induced (54–56), most investigators have tested PAF inhibitors in a preventive manner in several infection-related shock models (using either gram-negative or gram-positive bacterial strains (57–59).

There is one report on human volunteers in which the prophylactic use of PAF antagonists was tested with respect to LPS-induced phenomena. Fewer symptoms including rigors and myalgia were scored after administration of LPS in concert with diminished cortisol levels and epinephrine secretion (60). On the other hand, plasma levels of TNF, IL-1β, and IL-6 remained unaffected in these human volunteers (60) [or rabbits (56)], in contrast to studies in chimpanzees (61) or mice (58,59). In conclusion, these studies support a role for PAF antagonists as an adjunct to cytokine blockade in the treatment of gram-negative sepsis.

PAF Breakdown and Natural Antagonism

The half-life of membrane-exposed PAF is in the order of minutes, similar to plasma PAF, with a half-life of about 2–3 minutes at 37°C (2,53; T. W. Kuijpers, unpublished). The enzymes responsible for the rapid degradation of PAF consist of a cytoplasmic and an extracellular PAF–acetyl hydrolase (PAF-AH) (62). Plasma PAF-AH is associated with high-density lipoproteins (HDL) and low-density lipoproteins (LDL), affecting its activity. The major source contributing to PAF-AH in plasma consists of the macrophage population and to some extent hepatocytes (62–64). Both the intracellular and plasma forms of PAF-AH are able to cleave the acetyl group from the sn-2 position of PAF, thus forming inactive 2-lyso-PAF.

Plasma levels and function of PAF-AH are subject to variation. PAF-AH levels are reduced in patients suffering from severe sepsis (53). Moreover, oxidative modifications of PAF-AH negatively affect its activity (65). The recent findings that lipoproteins are protective toward LPS-induced morbidity (66,67) as well as the fact that infusion of lipoprotein emulsions can prevent septic shock in animals (68) may relate not only to rapid neutralization of LPS but also to the amounts of PAF-AH infused concomitantly.

SYNERGISM: THE MAGIC WORD IN PATHOPHYSIOLOGY

Local Amplification of Inflammation: The Role of PAF

Adhesion molecules and PAF are components of an adhesion and activation cascade. The facilitated interaction of target cells with membrane-bound chemotactic factors such as PAF induce high-affinity integrin-receptor binding (47–49). Other responses can be initiated by the early "juxtacrine" interactions of *low avidity*. Only in combination are P-selectin and PAF able to induce monocytes to make MCP-1 or TNF-α (69) (Fig. 3A).

Adhesion via *high-avidity* interactions of the β_2-integrins on leukocytes can by itself induce many different responses in phagocytes, among which is PAF synthesis (70,71). In this way, PAF enables further signaling (Fig. 3B), e.g., by upregulated expression and further avidity regulation of the β_2-integrin receptors or by the recruitment of additional signal transducing elements to the integrin receptors. Interestingly, in animals in which the adhesion molecules have been "knocked-out" or functionally blocked by monoclonal antibodies (72–75), LPS-induced shock or ischemia-reperfusion injury is virtually completely prevented. In addition to reduced leukocyte-endothelium interactions of leukocytes, the subsequent generation of inflammatory mediators may be abolished as well (72). The latter observation is, however, not consistently made in all models (75). Thus, one cannot easily tell from these data whether protection from injury depends on the lack of neutrophil infiltration, the reduced (adhesion-

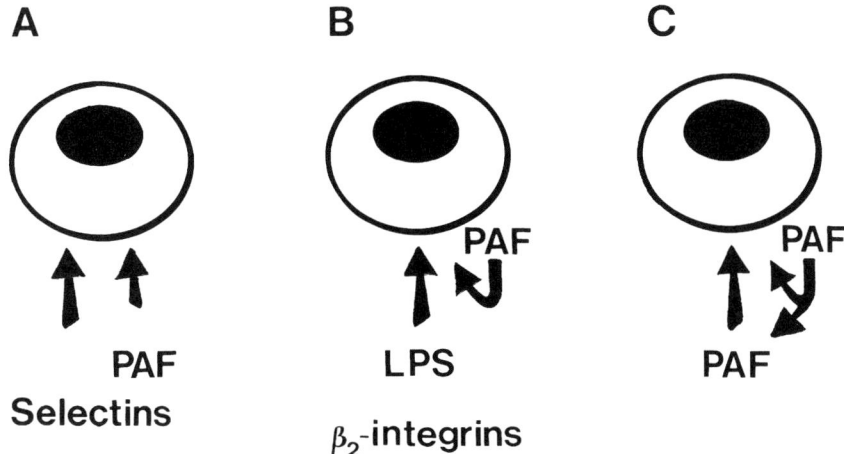

Fig. 3 (A) Early "juxtacrine" interactions of selectin interactions of low-avidity in conjunction with PAF can induce responses that either one alone is not able to. (B) High-avidity adhesion via integrin receptors or LPS-LBP complex binding to CD14 can induced PAF synthesis with paracrine as well as autocrine signaling functions. (C) PAF from paracrine sources is able to perpetuate its own production (self-amplification) in an autocrine fashion as shown for various cell types.

dependent) generation of inflammatory mediators, or both.

With respect to LPS-induced cell activation, some of the responses are clearly dependent on the intermediary or synergistic actions of PAF (Fig. 3B), as has for instance been shown for the LPS-induced synthesis of TNF and generation of toxic oxygen metabolites (76). Whether this form of heterologous amplification of LPS signaling via PAF depends on CD14/β_2-integrin receptor association, as has been suggested recently (77), deserves further study in β_2 knockout models. There is ample evidence for the simultaneous upregulation of CD14 and β_2-integrins upon stimulation of neutrophils and monocytes. Therefore, if the "cooperative signal-transduction devices" really exist and function as such, the PAF-mediated regulation of integrin avidity and LPS-mediated signaling events are directly tied together.

There is a third mechanism by which PAF may operate at the cellular level of activation. Once induced by an early stimulus, PAF is able to perpetuate its own production (self-amplification) as shown for various cell types, such as monocytes, endothelial cells, and mesangial cells (78–80) (Fig. 3C).

LPS-Induced Shock: The Main Source of Inflammatory Mediators

Injection of LPS into experimental animals induces shock associated with a rapid, almost sequential, release of pro-inflammatory cytokines such as TNF, IL-1β, IL-6, IFN-γ, and IL-8, as well as the counteracting IL-10. Most early cytokine mRNA expression (TNF, IL-1β, IL-6, IL-10, followed by IFN-γ and GM-CSF) is observed in macrophages of the liver and spleen, i.e., the reticuloendothelial system (RES) (81) and only a small fraction in the distal ileum (most likely the Peyer's patches) (82).

In cases of shock and ischemia, macrophages may be thought of as the main contributors to systemic PAF as concluded from studies with LPS-resistant mice (C3H/HeJ strain) and macrophage-depleted animals (16,83). In C3H/HeJ mice, resistance to LPS-induced shock resides in and can be transferred by their macrophages. PAF seems to have an intermediary role in LPS-induced shock, as indicated by experiments in these mice: these mice are also resistant to PAF-induced shock (16). One explanation for this finding would consist of an impaired responsiveness of the C3H/HeJ mice to *endogenous LPS* leaking from the gut into the circulation during PAF-induced intestinal hypoperfusion (i.e., PAF "primes" cells, such as macrophages, for activation by LPS yet fails to respond). Although elegant and straightforward, this explanation cannot be reconciled with findings in the CD14 knockout mice in which the LPS receptor has been eliminated: these mice may still succumb to PAF (46). The recent finding of a constitutive overexpression of secretory leukocyte protease inhibitor (SLPI) in C3H/HeJ macrophage cell lines (and suppression of LPS-induced activation of macrophages from LPS-sensitive strains after transfection of SLPI) (84) has not brought us any further in our understanding of either the primary defect or the actual phenotype of C3H/HeJ mice.

In C3H/HeJ mice, LPS- as well as PAF-induced biosynthesis of secondarily generated "endogenous" PAF was strongly reduced in bowel segments, as analyzed for PLA_2 activity and levels of PAF (16), indicative of a disease-promoting role for local PAF generation as well. The amount of "endogenous" PAF produced by (self-)amplification in bowel segments after intravenous LPS or PAF correlated with the presence and extent of intestinal injury and necrosis. Such a scenario of a primary plasma factor that secondarily induces a wide range of tissue-based inflammatory mediators can be envisaged to operate in other organs as well, resulting in multiple organ failure.

Systemic Amplification in LPS-Defined Organ Failure and the Impact of PAF

The clinical picture of septic shock is defined by cardiovascular depression with sinus tachycardia, lowered cardiac output, and decreased blood pressure, resulting in metabolic (tissue) acidosis and multiple organ failure. The inherent sensitivity of any organ to such dramatic changes in perfusion as well as the presence of superfused inflammatory mediators may depend on both local as well as systemic factors.

Regarding the cardiovascular situation, the cardiac myocytes show depressed contractibility in response to hypoxia and the cytokines TNF and IL-1 (85). The role of PAF herein seems to be modest, as indicated by the lack of detectable PAF-R in human heart (86) as well as the failure of PAF-R antagonists to influence the contractile function in heart muscle isolated from endotoxin-treated guinea pigs (87). On the other hand, PAF exerts an intermediary role in the increased iNOS expression and NO-dependent decrease in vascular pressor responses to norepinephrine (88).

In the lung, pulmonary hypertension during sepsis coincides with pulmonary leukostasis. This clinical picture may eventually culminate in severe lung injury or ARDS. Although pretreatment with different PAF-R antagonists has been shown to prevent symptoms of LPS-induced shock, these agents failed to block LPS-induced increase in levels of myeloperoxidase in the lung (89). The role of PAF in lung injury (under conditions of normotension) resides in priming the neutrophil for more vigorous adherence and activation, as well as dramatic increase of pulmonary artery pressure (PAP), even moreso in the presence of TNF (90,91).

Although neutrophil influx may not be greatly influenced, PAF antagonism results in less intrapulmonary shunting and edema in concert with improvements in gas exchange (54). In conclusion, PAF may promote lung damage and edema formation, while aggravating the cytokine-mediated cardiac depression by decreased arterial oxygen tension and increased PAP during sepsis.

The RES of liver and spleen seems to be the major source of inflammatory mediators during septic shock or ischemia/reperfusion injury (81,82,93,94). Hepatic TNF and PAF are almost simultaneously induced (93), making the rise in plasma levels of cytokines and PAF part of the acute-phase response (95). Injury to the gastrointestinal tract during shock complicates the clinical picture, tilting the balance to high mortality. In several experimental models of bowel injury, LPS- or TNF-induced generation of local intestinal PAF closely correlates with mucosal and vascular leakage, hypotension, metabolic acidosis, and intestinal necrosis. PAF antagonists are able to attenuate these pathological changes significantly (16). The cellular source(s) of PAF biosynthesis in the gastrointestinal tract has not yet been defined under these conditions. As soon as the intestinal mucosa loses its intact barrier function, bacterial products (such as LPS) from the normal gut flora will leak into the portal circulation draining to the liver. The Kupffer cells in the sinusoids of the portal triads will be overwhelmed by activating substances (96). The increased synthesis or spill-over of inflammatory mediators into the systemic circulation may aggravate (neutrophil-dependent) pulmonary injury (97), resulting in ARDS and death (Fig. 4). The fact that severe septic shock can be prevented by decontamination of the digestive tract further supports the view of gut-derived disease-promoting factors, possibly confined to some extent to a clearly diminished synthesis and activity of PAF (94,98,99).

CONCLUDING REMARKS

The complex mechanisms through which PAF may operate have not yet been unraveled completely. Our belief is that systemic PAF contributes to the initiation of septic and anaphylactic shock. Tissue-derived PAF, either secondarily induced by ischemia, self-amplification, or concurrent inflammatory mediators such as LPS and TNF, functions as a positive feedback system for the prolongation of systemic effects as well as local effects on microcirculation, infiltration and activation of neutrophils, and resulting tissue damage/organ fail-

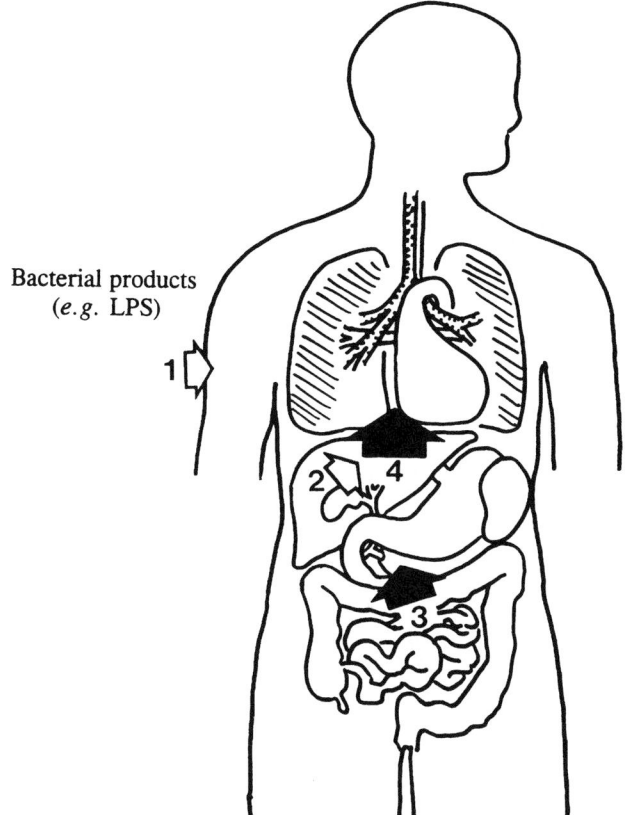

Fig. 4 Cascade of events due to bacteremia or bacterial products. Local PAF generation is involved in local tissue damage (1). If bacterial products or huge amounts of local inflammatory mediators are released into the blood, most of the cytokine mediators are subsequently synthesized in the liver (2). In addition to bacterial products and PAF, the systemic presence of cytokines can affect systemic as well as local perfusion dramatically. In case of hypoperfusion of the gut (3), lipid mediators such as PAF and bacterial products will leak from the intestinal space into the splanchnic and portal drainage circuit, hence further increasing the production of inflammatory mediators in the liver. The perpetuation of inflammation aggravates the clinical situation by ARDS and a progressive decline in cardiac function (4), resulting in high mortality.

ure. Blockade of the PAF-R by PAF-RA or an increased breakdown of PAF by the administration of rhPAF-AH can therefore help to counteract shock, even some time after its onset. Moreover, endogenous PAF generation, either associated with shock-like states or as a consequence of localized inflammation (not necessarily reflected by circulating levels of PAF), is a potential target for intervention by either of the two aforementioned inhibitory modalities to protect the organ(s) from ensuing failure.

REFERENCES

1. Snyder F. Biochemistry of platelet-activating factor: a unique class of biologically active phospholipids. Fed Proc (FASEB J) 1989; 190:125–135.
2. Chao W, Olson MS. Platelet-activating factor: receptors and signal transduction. Biochem J 1993; 292:617–629.
3. Koltai M, Hosford D, Braquet P. Role of PAF and cytokines in microvascular tissue injury. J Lab Clin Med 1992; 119:461–466.
4. Kingsnorth AN, Galloway SW, Formela LJ. Randomized, double-blind phase II trial of Lexipafant, a platelet-activating factor antagonist, in human acute pancreatitis. Br J Surg 1995; 82:1414–1420.
5. Hsieh KH, Ng CK. Increased plasma platelet-activating factor in children with acute asthmatic attacks and decreased in vivo and in vitro production of platelet-activating factor after immunotherapy. J All Clin Immunol 1993; 91:650–657.
6. O'Connor BJ, Uden S, Carty TJ, Eskra JD, Barnes PJ, Chung KF. Inhibitory effect of UK, 74505, a potent and specific oral platelet activating factor (PAF) receptor antagonist, on airway and systemic responses to inhaled PAF in humans. Am J Respir Crit Care Med 1994; 150:35–40.
7. Hozawa S, Haruta Y, Ishioka S, Yamakido M. Effects of a PAF antagonist, Y-24180, on bronchial hyperresponsiveness in patients with asthma. Am J Respir Crit Care Med 1995; 152:1198–1202.
8. Kuitert LM, Angus RM, Barnes NC, Barnes PJ, Bone MF, Chung KF, Fairfax AJ, Higenbotham TW, O'Connor BJ, Piotrowska B. Effect of a novel potent platelet-activating factor antagonist, modipafant, in clinical asthma. Am J Respir Crit Care Med 1995; 151:1331–1335.
9. Rodriguez-Roisin R, Felez MA, Chung KF, Barbera JA, Wagner PD, Cobos A, Barnes PJ, Roca J. Platelet-activating factor causes ventilation-perfusion mismatch in humans. J Clin Ivest 1994; 93:188–194.
10. Benigni A, Boccardo P, Noris M, Remuzzi G, Siegler RL. Urinary excretion of platelet-activating factor in haemolytic uraemic syndrome. Lancet 1992; 339:835–836.
11. Rabinovici R, Sofronski MD, Renz JF, Hillegas LM, Esser KM, Vernick J, Feuerstein G. Platelet activating factor mediates interleukin-2-induced lung injury in the rat. J Clin Invest 1992; 89:1669–1673.
12. Pinckard RN, Showell HJ, Castillo R, Lear C, Breslow R, McManus LM, Woodard DS, Ludwig JC. Differential responsiveness of human neutrophils to the autocrine actions of 1-O-alkyl-homologs and 1-acyl analogs of platelet-activating factor. J Immunol 1992; 148:3528–3535.
13. Dennis EA. Diversity of group types, regulation and function of phospholipase A2. J Biol Chem 1994; 269:13057–13060.
14. Henderson WR Jr. The role of leukotrienes in inflammation. Ann Intern Med 1994; 121:684–697.
15. Leier I, Jedlitschky G, Buchholz U, Cole SP, Deeley RG, Keppler D. The MRP gene encodes an ATP-de-

pendent export pump for leukotriene C4 and structurally related conjugates. J Biol Chem 1994; 269:27807–27810.
16. Sun X, Caplan MS, Liu Y, Hsueh W. Endotoxin-resistant mice are protected from PAF-induced bowel injury and death. Role of TNF, complement activation, and endogenous PAF production. Dig Dis Sci 1995; 40:495–502.
17. Honda Z, Nakamura M, Miki I, Minami M, Watanabe T, Seyama Y, Okado H, Toh H, Ito K, Miyamoto T, Shimizu T. Cloning by functional expression of platelet-activating factor receptor from guinea-pig lung. Nature 1991; 349:342–344.
18. Nakamura M, Honda Z, Izumi T, Skanaka C, Mutoh H, Minami M, Bito H, Seyama Y, Matsumoto T, Noma M, Shimizu T. Molecular cloning and expression of platelet-activating factor receptor from human leukocytes. J Biol Chem 1991; 266:20400–20405.
19. Ye RD, Prossnitz ER, Zou A, Cochrane CG. Characterization of a human cDNA that encodes a functional receptor for platelet activating factor. Biochem Biophys Res Commun 1991; 15:105–111.
20. Kunz D, Gerard NP, Gerard C. The human leukocyte platelet-activating factor receptor. cDNA cloning, cell surface expression, and construction of a novel epitope-bearing analog. J Biol Chem 1992; 267:9101–9106.
21. Seyfried CE, Schweickart VL, Godiska R, Gray PW. The human platelet-activating factor receptor gene (PAFR) contains no introns and maps to chromosome 1. Genomics 1992; 13:832–834.
22. Baggiolini M, Walz A, Kunkel SL. Neutrophil-activating peptide/interleukin-8: a novel cytokine that activates neutrophils. J Clin Invest 1989; 84:1045–1049.
23. Murphy PM. The molecular biology of leukocyte chemoattractant receptors. Annu Rev Immunol 1994; 12:593–633.
24. Spiegel AM, Weinstein LS, Shenker A. Abnormalities in G-protein-coupled signal transduction pathways in human disease. J Clin Invest 1993; 92:1119–1125.
25. Charo IF, Myers SJ, Herman A, Franci C, Connolly AJ, Coughlin SR. Molecular cloning and functional expression of two monocyte chemoattractant protein-1 receptors reveals alternative splicing of the carboxy-terminal tail. Proc Natl Acad Sci USA 1994; 91:2752–2756.
26. Namba T, Sugimoto Y, Negishi M, Irie A, Ushikubi F, Kakizuka A, Ito S, Ichikawa A, Narumiya S. Alternative splicing of C-terminal tail of the prostaglandin E receptor subtype EP3 determines G-protein specificity. Nature 1993; 365:166–170.
27. Monsma FJ, McVittie LD, Gerfan CR, Mahan LC, Sibley DR. Multiple D2 dopamine receptors produced by alternative splicing. Nature 1989; 342:926–929.
28. Hirata T, Ushikubi F, Kakizuka A, Okuma M, Narumiya S. Two TxA2 receptor isoforms in human platelets. J Clin Invest 1996; 97:949–956.
29. Bokoch GM. Chemoattractant signalling and leukocyte activation. Blood 1995; 86:1649–1660.
30. Nick JA, Avdi NJ, Young SK, Knall C, Gerwins P, Johnson GL, Worthen GS. Common and distinct intracellular signaling pathways in humen neutrophils utilized by platelet activating factor and FMLP. J Clin Invest 1997; 99:975–986.
31. Gierschik P, Sidiropoulos D, Jakobs KH. Two distinct G_i-proteins mediate formyl peptide receptor signal transduction in human leukemia (HL-60) cells. J Biol Chem 1989; 264:21470–21473.
32. Polakis PG, Evans T, Snyderman R. Multiple chromatographic forms of the formylpeptide chemoattractant receptor and their relationship to GTP-binding proteins. Biochem Biophys Res Commun 1989; 161:276–283.
33. Haviland DL, Borel AC, Fleisher DT, Haviland JC, Wetsel RA. Structure, 5′-flanking sequence, and chromosome location of the human N-formyl peptide receptor gene. A single copy gene comprised of two exons on chromosome 19q13.3 that yields two transcripts by alternative polyadenylation. Biochemistry 1993; 32:4168–4174.
34. Muller E, Dagenais P, Alami N, Rola-Pleszczynski M. Identification and functional characterization of platelet-activating factor receptors in human leukocyte populations using polyclonal anti-peptide antibody. Proc Natl Acad Sci USA 1993; 90:5818–5822.
35. Liu H, Chao W, Olson MS. Regulation of the surface expression of the platelet-activating factor receptor in IC-21 peritoneal macrophages. Effects of lipopolysaccharide. J Biol Chem 1992; 267:20811–20819.
36. Ouellet S, Muller E, Rola-Pleszczynski M. IFN-gamma up-regulates platelet-activating factor receptor gene expression in human monocytes. J Immunol 1994; 152:5092–5099.
37. Thivierge M, Alami N, Muller E, de Brum-Fernandes AJ, Rola-Pleszczynski M. Transcriptional modulation of platelet-activating factor receptor gene expression by cyclic AMP. J Biol Chem 1993; 268:17457–17466.
38. Calabresse C, Nguer MC, Pellegrini O, Benveniste J, Richard Y, Thomas Y. Induction of high-affinity paf receptor expression during T cell activation. Eur J Immunol 1992; 22:1349–1355.
39. Bratton DL, Dreyer E, Kailey JM, Fadok VA, Clay KL, Henson PM. The mechanism of internalization of platelet-activating factor in activated human neutrophils. Enhanced transbilayer movement across the plasma membrane. J Immunol 1992; 148:514–523.
40. Gerard NP, Gerard C. Receptor-dependent internalization of platelet-activating factor. J Immunol 1994; 152:793–800.
41. Schumann RR, Leong SR, Flaggs GW, Gray PW, Wright SD, Mathison JC, Tobias PS, Ulevitch RJ. Structure and function of lipopolysaccharide binding protein. Science 1990; 249:1429–1431.
42. Wright SD. CD14, a receptor for complexes of lipopolysaccharide (LPS) and LPS binding protein. Science 1990; 249:1231–1233.
43. Camussi G, Mariano F, Biancone L, De Martino A, Bussolati B, Montrucchio G, Tobias PS. Lipopolysaccharide binding protein and CD14 modulate the synthesis of platelet-activating factor by human monocytes and mesangial and endothelial cells stimulated with lipopolysaccharide. J Immunol 1995; 155:316–324.
44. Pugin J, Heumann D, Thomzasz A, Kravchenko V, Akamatsu Y, Nishijima M, Glauser MP, Tobias PS, Ulevitch RJ. CD14 is a pattern recognition receptor. Immunity 1994; 1:509–516.

45. Kusunoki T, Hailman E, Juan TSC, Lichenstein HS, Wright SD. Molecules from Staphylococcus aureus that bind CD14 and stimulate innate immune responses. J Exp Med 1995; 182:1673–1682.
46. Haziot A, Ferrero E, Kontgen F, Hijiya N, Yamamoto S, Silver J, Stewart CL, Goyert SM. Resistance to endotoxin shock and reduced dissemination of gram-negative bacteria in CD14-deficient mice. Immunity 1996; 4:407–414.
47. Lorant DE, Patel KD, McIntyre TM, McEver RP, Prescott SM, Zimmerman GA. Coexpression of GMP-140 and PAF by endothelium stimulated by histamine or thrombin: a juxtacrine system for adhesion and activation of neutrophils. J Cell Biol 1991; 115:223–234.
48. Kuijpers TW, Hakkert BC, Hart MHL, Roos D. Neutrophil migration across monolayers of cytokine-prestimulated endothelial cells: a role for platelet-activating factor and IL-8. J Cell Biol 1992; 117:565–572.
49. Springer TA. Traffic signals on endothelium from lymphocyte recirculation and leukocyte emeigration. Annu Rev Physiol 1995; 57:827–872.
50. Varki A. Selectin ligands: will the real ones please stand up. J Clin Invest 1997; 99:158–162.
51. Gaboury JP, Johnston B, Niu XF, Kubes P. Mechanisms underlying acute mast cell-induced leukocyte rolling and adhesion in vivo. J Immunol 1995; 154:804–813.
52. Nourshargh S, Larkin SW, Das A, Williams TJ. Interleukin-1-induced leukocyte extravasation across rat mesenteric microvessels is mediated by platelet-activating factor. Blood 1995; 85:2553–2558.
53. Graham RM, Stephens CJ, Silvester W, Leong LL, Sturm MJ, Taylor RR. Plasma degradation of platelet-activating factor in severely ill patients with clinical sepsis. Crit Care Med 1994; 22:204–212.
54. Siebeck M, Kohl J, Endres S, Spannagl M, Machleidt W. Delayed treatment with platelet activating factor receptor antagonist WEB2086 attenuates pulmonary dysfunction in porcine endotoxin shock. J Trauma 1994; 37:745–751.
55. Ou MS, Kambayashi, Kawasaki T, Uemara Y, Shinozaki K, Shiba E, Sakon M, Yukawa M, Mori T. Potential etiologic role for PAF in two major septic complications: disseminated intravascular coagulation and multiple organ failure. Thromb Res 1994; 73:227–238.
56. Yue TL, Farhat M, Rabinovici R, Perera PY, Vogel SN, Feuerstein G. Protective effect of BN 50739, a new PAF antagonist, in endotoxin-treated rabbits. J Pharm Exp Med 1990; 254:976–981.
57. DeJoy SQ, Jeyaseelan R Sr, Torley LW, Pickett WC, Wissner A, Wick MM, Oronsky AL, Kerwar SS. Effect of CL 184,005, a platelet-activating factor antagonist in a murine model of Staphylococcus aureus-induced gram-positive sepsis. J Inf Dis 1994; 169:150–156.
58. Ruggiero V, Chiappino C, Manganello S, Pacello L, Foresta P, Matelli EA. Beneficial effects of a novel PAF receptor antagonist, ST 899, on endotoxin-induced shock in mice. Shock 1994; 2:275–280.
59. Ogata M, Matsumoto T, Koga K, Kamochi M, Sata T, Yoshida S, Shigematsu A. An antagonist of PAF suppresses endotoxin-induced tumor necrosis factor and mortality in mice pretreated with carrageenan. Infect Immun 1993; 61:699–704.
60. Thompson WA, Coyle S, Van Zee K, Oldenburg H, Trousdale R, Rogy M, Felsen D, Moldawer L, Lowry SF. The metabolic effects of platelet-activating factor antagonism in endotoxemic man. Arch Surg 1994; 129:72–79.
61. Kuipers B, van der Poll T, Levi M, van Deventer SJ, ten Cate H, Imai Y, Hack EC, ten Cate JW. Platelet-activating factor antagonist TCV-309 attenuates the induction of the cytokine network in experimental endotoxemia in chimpanzees. J Immunol 1994; 152:2438–2446.
62. Tjoelker LW, Wilder C, Eberhardt C, Stafforini DM, Dietch G, Schimpf B, Hooper S, Trong HL, Cousens LS, Zimmerman GA, Yamada Y, McIntyre TM, Prescott SM, Gray PW. Anti-inflammatory properties of a platelet-activating factor acetylhydrolase. Nature 1994; 374:549–553.
63. Karasawa K, Kato H, Setaka M, Nojima S. Accumulation of PAF-acetylhydrolase in the peritoneal cavity of guinea pig after endotoxin shock. J Biochem 1994; 116:368–373.
64. Satoh K, Imaizumi T, Yoshida H, Takamatsu S. High-density lipoprotein inhibits the production of platelet activating factor acetylhydrolase by HepG2 cells. J Lab Clin Med 1994; 123:225–231.
65. Ambrosio G, Oriente A, Napoli C, Palumbo G, Chiariello P, Marone C, Condorelli M, Chiariello M, Triggiani M. Oxygen radicals inhibit human plasma acetylhydrolase, the enzyme that catabolizes platelet-activating factor. J Clin Invest 1994; 93:2408–2416.
66. Levine DM, Parker TS, Donelly TM, Walsh A, Rubin AL. In vivo protection against endotoxin by plasma high density lipoprotein. Proc Natl Acad Sci USA 1993; 90:12040–12044.
67. Netea MG, Demacker PNM, Kullberg BJ, Boerman OC, Verschueren I, Stalenhoef AFH, van der Meer JWM. Low-density lipoprotein receptor-deficient mice are protected against lethal endotoxemia and severe gram-negative infections. J Clin Invest 1996; 97:1366–1372.
68. Read TE, Grunfeld C, Kumwenda ZL, Calhoun MC, Kane JP, Feingold KR, Rapp JH. Triglyceride-rich lipoproteins prevent septic death in rats. J Exp Med 1995; 182:267–272.
69. Weyrich AS, McIntyre TM, McEver RP, Prescott SM, Zimmerman GA. Monocyte tethering by P-selectin regulates monocyte chemotactic protein-1 and tumor necrosis factor-α secretion. J Clin Invest 1995; 95:2297–2303.
70. Tool AT, Koenderman L, Kok PT, Blom M, Roos D, Verhoeven AJ. Release of PAF is important for the respiratory burst induced in human eosinophils by opsonized particles. Blood 1992; 79:2729–2732.
71. Au BT, Williams TJ, Collins PD. Zymosan-induced IL-8 release from human neutrophils involves activation via the CD11b/CD18 receptor and endogenous platelet-activating factor as an autocrine modulator. J Immunol 1994; 152:5411–5419.
72. Watanabe S, Mukaida N, Ikeda N, Akiyama M, Harada A, Nakanishi I, Watanabe Y, Matsushima K. Prevention of endotoxin shock by an antibody against leukocyte integrin β_2 through inhibiting production and action of TNF. Int Immunol 1995; 7:1037–1046.

73. Tedder TF, Steeber DA, Pizcueta P. L-selectin-deficient mice have impaired leukocyte recruitment to inflammatory sites. J Exp Med 1995; 184:2259–2264.
74. Horie Y, Wolf R, Anderson DC, Granger DN. Hepatic leukostasis and hypoxic stress in adhesion-molecule-deficient mice after gut ischemia/reperfusion. J Clin Invest 1997; 99:781–788.
75. Gutierrez-Ramos JC, Bluethmann H. Molecules and mechanisms operating in septic shock: lessons from knockout mice. Immunol Today 1997; 18:329–334.
76. Zhang F, Decker K. Platelet-activating factor antagonists suppress the generation of tumour necrosis factor-alpha and superoxide induced by lipopolysaccharide or phorbol-ester in rat liver macrophages. Eur Cytokine Netw 1994; 5:311–317.
77. Petty HR, Todd RF 3d. Integrins as promiscuous signal transduction devices. Immunol Today 1996; 17:209–212.
78. Valone H, Epstein LB. Biphasic PAF synthesis by human monocytes stimulated with IL-1β, TNF or IFNγ. J Immunol 1988; 141:3945–3950.
79. Camussi G, Biancone L, Iorio EL, Silvestro L, Da col R, Capasso C, Rossano F, Servbillo L, Balestrieri C, Tufano MA. Porins and lipopolysaccharide stimulate platelet activating factor synthesis by human mesangial cells. Kidney Int 1992; 42:1309–1318.
80. Heller R, Bussolino F, Ghigo D, Garbarino G, Pescarmona G, Till U, Bosia A. Human endothelial cells are target for platelet-activating factor. II. Platelet-activating factor induces platelet-activating factor synthesis in human umbilical vein endothelial cells. J Immunol 1992; 149:3682–3688.
81. Salkowski CA, Neta R, Wynn TA, Strassmann G, van Rooijen N, Vogel SN. Effect of liposome-mediated macrophage depletion on LPS-induced cytokine gene expression and radioprotection. J Immunol 1995; 155:3168–3179.
82. Huang L, Tan X, Crawford SE, Hsueh W. PAF and endotoxin induce tumour necrosis factor in rat intestine and liver. Immunology 1994; 83:65–69.
83. Tschaikowsky K, Brain JD. Effects of liposome-encapsulated dichloromethylene diphosphonate on macrophage function and endotoxin-induced mortality. Biochem Biophys Acta 1994; 1222:323–330.
84. Jin F, Nathan C, Radzioch D, Ding A. Secretory leukocyte protease inhibitor: a macrophage product induced by and antagonistic to bacterial lipopolysaccharide. Cell 1997; 88:417–426.
85. Kumar A, Thota V, Dee L, Olson J, Uretz E, Parrillo JE. Tumor necrosis factor α and interleukin 1β are responsible for in vitro myocardial cell depression induced by human septic shock serum. J Exp Med 1995; 182:949–958.
86. Schwinger RH, Bohm M, La Rosee K, Erdmann E. Existence of PAF receptors in human platelets and human lung tissue but not in the human myocardium. Am Heart J 1992; 124:320–330.
87. Heard SO, Toth I, Perkins M. The role of platelet-activating factor in LPS-induced myocardial depression in guinea-pigs. J Crit Care 1995; 10:7–14.
88. Szabo C, Wu CC, Gross SS, Thiemermann C, Vane JR. PAF contributes to the induction of nitric oxide by bacterial lipopolysaccharide. Clin Res 1993; 73:991–999.
89. Chang SW. Endotoxin-induced pulmonary leukostasis in the rat: role of platelet-activating factor and tumor necrosis factor. J Lab Clin Med 1994; 123:65–72.
90. Abu-Zidan FM, Walther S, Lennquist S. Modulation of lung injury by PAF antagonism in nonhypotensive porcine endotoxinemia. Circ Shock 1994; 44:148–153.
91. Chang SW. TNF potentiates PAF-induced pulmonary vasoconstriction in the rat: role of neutrophils and thromboxane A2. J Appl Physiol 1994; 77:2817–2826.
92. Rabinovici R, Bugelski PJ, Esser KM, Hillegass LM, Vernick J, Feuerstein G. ARDS-like lung injury produced by endotoxin in platelet-activating factor-primed rats. J Appl Physiol 1993; 74:1791–1802.
93. Wang KS, Monden M, Kanai T, Gotoh M, Umeshita K, Ukei T, Mori T. Protective effect of platelet-activating factor antagonist on ischemia-induced liver injury in rats. Surgery 1993; 113:76–83.
94. Zhou W, Chao W, Levine BA, Olson MS. Role of platelet-activating factor in hepatic responses after bile duct ligation in rats. Am J Physiol 1992; 263:G587–592.
95. Klosterhalfen B, Horstmann-Jungemann K, Vogel P, Flohe S, Offner F, Kirkpatrick CJ, Heinrich PC. Time course of various inflammatory mediators during recurrent endotoxemia. Biochem Pharmacol 1992; 43:2103–2109.
96. Haglind E, Xia G, Rylander R. Effects of anti-oxidants and PAF-receptor antagonist in intestinal shock in the rat. Circ Shock 1994; 42:83–91.
97. Colletti LM, Kunkel SL, Walz A, Burdick MD, Kunkel RG, Wilke CA, Strieter R. Chemokine expression during hepatic ischemia/reperfusion-induced lung injury in the rat. J Clin Invest 1995; 95:134–141.
98. Sun XM, MacKendrick W, Tien J, Huang W, Caplan MS, Hsueh W. Endogenous bacterial toxins are required for the injurious action of platelet-activating factor in rats. Gastroenterology 1995; 109:83–88.
99. Schiffrin EJ, Trop M, Schroeder S, Carter EA. Platelet-activating factor induces intestinal necrosis but not septic shock, in germ-free and specific-pathogen-free rodents. Burns 1991; 17:276–278.

28

CD14, An Innate Immune Receptor for Various Bacterial Cell Wall Components

Artur J. Ulmer, Volker T. El-Samalouti, Ernst T. Rietschel, and Hans-Dieter Flad
Research Center Borstel, Borstel, Germany

Roman Dziarski
Indiana University School of Medicine, Gary, Indiana

INTRODUCTION

During infection with bacteria the host is confronted with various toxins, which are either released by the bacteria or which are constituents of the microbe and only set free when bacteria die. An example of the first group are exotoxins such as α-toxin, hemolysin, and the superantigens (1,2). The latter group of toxins constitute cell wall components of bacteria. Lipopolysacharide (LPS), also termed endotoxin, of the outer leaflet of gram-negative bacteria and peptidoglycan (PG) represent such cell wall–associated compounds, which are of great biomedical importance in the mediation of pathophysiological reactions of an infected host (3). Minute amounts of LPS are able to induce pathophysiological reactions, which can be observed during severe infection, i.e., sepsis. Therefore, this prominent cell wall component has been the object of comprehensive investigations to elucidate its chemical structure and biological activity. An enormous knowledge on LPS has been accumulated during the last several decades. Consequently, LPS constitutes an exceptional focus of scientific interest regarding bacterial cell wall components.

The discovery that CD14 represents a cellular binding site for LPS expressed by monocytes/macrophages was the result of investigations aimed at the elucidation of the specific binding of LPS by cells associated with their activation (4). It is now well established that binding of LPS to membrane-bound CD14 (mCD14), which is facilitated by a serum protein [LPS-binding protein (LBP) (5)], is a prerequisite for activation and signal transduction of these cells: Monoclonal antibodies (mAb) directed against CD14 inhibit the binding of LPS to monocytes/macrophages as well as the LPS-mediated activation. In addition, CD14-transfected cell lines are able to bind LPS and can become activated in response to LPS (6,7). Furthermore, purified CD14 interacts physically with LPS, e.g., in a gel shift assay (8).

In addition to mCD14, a soluble form of CD14 (sCD14) also exists (9). SCD14 is detectable in normal sera and has been reported to be responsible for LPS activation of cells that lack mCD14 (e.g., endothelial cells, smooth muscle cells) (10–12). The mode, however, by which mCD14 as well as sCD14 mediate induction of transmembrane signals still remains to be clarified.

INTERACTION OF CD14 WITH PEPTIDOGLYCAN

Peptidoglycan (PG) was the first bacterial product demonstrated to bind to cell membranes via a 70 kDa LPS-binding protein (13,14). This led to subsequent studies in which it was investigated whether or not CD14, the thus far only established functional LPS binding site, was also involved in the interaction of cells with PG. Experiments showed an inhibitory activity of anti-CD14 mAb with regard to the stimulation of monocytes by PG or cell wall preparations of gram-positive bacteria, indicating that CD14 was involved not only in the activation of monocytes by LPS but also by other cell wall components (15–17). These results were confirmed by the use of LPS antagonists, in particular the synthetic lipid A analog compound 406 (compound 406 is structurally identical to the lipid A precursor Ia, also called lipid IV$_A$). It was shown that compound 406 inhibited not only LPS-induced monocyte activation (18–21) but also cytokine release by monocytes activated with soluble peptidoglycan (sPG) (15). This finding indicated a contribution of the endotoxin-receptor CD14 to the induction of cytokines in monocytes by sPG.

These results, performed with normal human monocytes, were confirmed with a CD14-transfected 70Z/3 murine cell line (22). The CD14-transfected 70Z/3 cells, but not the cells that were transfected with the control vector, were found to express surface IgM, to activate NF-κB, and to degrade I-κB-α in response to PG and sPG. This cell response is identical to that found with LPS. However, further experiments with CD14-deletion mutants did not show a response profile identical to endotoxin and PG. Although the N-terminal 151 amino acids of CD14 are sufficient for activation of the transfected 70Z/3 cells by both LPS and PG, various mutations within the N-terminal 65 amino acids had different effects on the response to PG and LPS (22). This finding suggested that LPS and PG may interact with the same molecule but not the same epitopes of CD14 during activation of the cells.

Recently the intracellular signaling involved in the activation of monocytes/macrophages by PG and LPS were compared (24). PG, like LPS, was found to induce rapid tyrosine phosphorylation of several cellular proteins involved in signaling, like *lyn* and extracellular signal-regulated kinases (ERK), and activation of ERK and *rsk* signal-transducing molecules. Furthermore, the activation of mitogen-activated protein (MAP) kinases was studied: ERK1 and ERK2 were strongly activated by sPG, whereas c-Jun NH$_2$ terminal kinase (JNK) was only moderately and p38 MAP kinase only weakly activated by sPG. In contrast, LPS always strongly activated these kinases. Not only was early phosphorylation of signal-transducing elements observed, but a late (4-hour) phosphorylation of cytosolic proteins p36/p38 was detected after stimulation with LPS (25) as well as with sPG (26).

A study of the inhibitory effect of various drugs affecting intracellular signaling in human monocytes suggested that calmodulin-dependent protein kinase, protein tyrosine kinase, and cholera toxin–sensitive G-proteins are involved in PG- as well as in LPS-induced TNF production by human monocytes (27). However, differences between the signal-transducing pathways of PG and LPS were described: The tyrosine kinase inhibitor tryphostin AG 126 blocked PG- but not LPS-induced TNF release. It should be noted, however, that, in contrast to human monocytes, the LPS-induced TNF release can be blocked in murine macrophages by AG 126 (28).

Taken together, the findings discussed suggest that PG as well as LPS is recognized via CD14 by monocytes/macrophages and that both inflammatory compounds use similar although not identical signal-transducing elements for cellular activation.

Regarding the recognition of CD14, however, it should be noted that the bioassays presented so far have yielded only indirect evidence for an interaction of PG with CD14, and the mechanism of cell activation remained unclear. This may not include the binding of PG to CD14. During recent years, therefore, efforts have been made to establish that indeed CD14 is the recognition receptor for PG essential for activation of cells.

Binding assays with FITC-labeled sPG (FITC-sPG) were appliedd to prove the specific binding of sPG to normal human monocytes (29). This specific binding was sensitive not only to unlabeled sPG, but also to the minimal bioactive structure of PG, i.e., the muramyl dipeptide (MDP). Competition of FITC-sPG binding by LPS and compound 406 revealed the contribution of LPS receptor(s), and inhibition by anti-CD14 mAb demonstrated the relevance of CD14 not only to activation but also to binding of sPG to monocytes (29).

Direct evidence of a physical interaction of sPG with CD14 was offered by a gel-shift assay in which the electrophoretic mobility of sCD14 in a native PAGE was examined. It was found that addition of sPG resulted in an increase in the electrophoretic mobility of CD14 (29). This enhanced electrophoretic mobility of CD14 in the presence of sPG is obviously a result of the direct interaction of the two molecules, i.e., the formation of a sPG-CD14 complex.

PG possesses a polymeric organization consisting of ß1,4-linked *N*-acetylmuramic acid-*N*-acetyl-D-glucosamine disaccharide. The units of this glycosyl polymer are cross-linked by short peptides and form a three-dimensional network, so-called murein or the sacculus of the bacteria (30). The sPG is released by bacteria upon treatment with ß-lactam antibiotics, which blocks enzymatic cross-linking (31,32). In this way a stranded polymeric structure is built with pentapeptide side chains. Lipid A, the biologically active principle of LPS (33), is composed of a 1,4'-bisphosphorylated β1,6-linked D-glucosamine disaccharide, which is acylated by up to six fatty acids (3,34). Its polymeric behavior is due to the formation of three-dimensional aggregates in aqueous solution (35). From a chemical point of view, PG and lipid A both possess a D-glucosamine disaccharide backbone. However, PG represents a glycopeptide, whereas lipid A is a glycolipid. Therefore, the only structural similarity between PG and lipid A is the carbohydrate backbone. This leads to the assumption that the D-gluco-configurated saccharide structure is relevant for the binding to CD14. Because other polysaccharide structures (possessing the D-manno configuration) bind to CD14, it is tempting to postulate that CD14 constitutes a lectin, i.e., a sugar-binding protein, which recognizes carbohydrate domains of LPS as well as of PG (29,36).

RECOGNITION OF NATURAL AND SYNTHETIC GLYCOSYLATED COMPOUNDS

LPS and PG are not the only bacterial cell wall compounds that express inflammatory properties. Lipoarabinomannan, lipoteichoic acid, lipopeptides, sphingolipids, and various bacterial cell wall preparations also induce the production of cytokines in monocytes/macrophages (16,37–44) and, therefore, may contribute to an inflammatory reaction in an infected host. In addition, nonbacterial and synthetic polymers are known to mediate inflammatory reactions at least in vitro (36,45).

The prominent role of CD14 in the binding of LPS to monocytes and their activation led to experiments in which the contribution of CD14 to the activation of monocytes/macrophages and binding by these bacterial and nonbacterial compounds was elucidated. Various inflammatory compounds that stimulate mononuclear phagocytes have indeed been described to be active in a CD14-dependent manner (see Tables 1 and 2). It should be noted, however, that CD14 is not involved in the stimulation of every cell wall component. Thus, structures have been detected (e.g., a synthetic lipopeptide resembling the cell wall lipopeptide of gram-negative bacteria as well as a glycosphingolipid from *Sphingiomonas paucibilis*) that induce cytokine production in human monocytes in a CD14-independent manner (43,46).

Table 1 Inflammatory Bacterial Cell Wall Components

Cell wall compound	CD14-dependent	Ref.
Gram-negative bacteria		
Lipopolysaccharide	Yes	4
β1,4-D-Mannuronic acid (poly M)	Yes	47
Glycosphingolipids	No	43
Lipopeptide	No	46
Gram-positive bacteria		
Peptidoglycan, soluble	Yes	15
Peptidoglycan, native	Yes	22
MDP	No	23
Lipoteichoc acid	Yes	57
Cell walls	Yes	16
Cell wall "active fraction"	Yes	17
Rhamnose-glucose polymers	Yes	77
Lipoarabinomannan	Yes	16, 53, 54
Group B streptococcal cell walls	Yes	44

The first report on the involvement of CD14 in the bioactivity of an inflammatory compound other than LPS concerned mannose-uronic polymers (47). It was shown that the binding of uronic acid polymers as well as the activation of human monocytes by these structures were CD14 dependent. However, cells of the astrocytoma cell line U373 were unable to respond to the polyuronic acid even in the presence of serum containing sCD14. These cells are known to produce IL-6 in response to LPS in the presence of sCD14. Therefore, it was concluded that these uronic acid polymers interact with membrane-bound CD14 but are unable to bind to the up-to-now unidentified receptor for LPS-sCD14 complexes present on CD14-negative cells (like endo-

Table 2 CD14-Dependent Nonbacterial Compounds

Compound	Source	CD14-dependent	Ref.
High M alginate	*Ascophyllum nodosum*	Yes	47
Chitosans	Arthropods	Yes	78
WI-1 (cell wall Ag)	*Blastomyces dermatitidis*	Yes	79
Fucoidan	*Fucus vesiculosus*	Yes	36
β1,4-Glucuronic acid	Synthetic	Yes	47
Phospholipids	Synthetic	Yes	76

thelial cells or cells of the astrocytoma cell line U373). Two candidates for this receptor have been described: a 80 kDa membrane protein, which is able to bind LPS in presence of sCD14 and LBP in a ligand blotting assay (48), and a 216 kDa protein, which binds sCD14 when LPS is present (49).

INTERACTION OF CD14 WITH LIPOARABINOMANNAN

Lipoarabinomannan (LAM) is a major antigen of the mycobacterial cell wall. This lipoglycan has been reported to stimulate cytokine production in human as well as in murine monocytes/macrophages (37,38). In addition, it has been demonstrated that this molecule has broad-spectrum reactivity to target cells including T and B lymphocytes (37,50–52). The involvement of membrane-bound CD14 (mCD14) in the induction of IL-1 and TNF-α was first recognized in 1993 (53). It was found that anti-CD14 mAb inhibited cytokine release in human THP-1 cell line induced by LAM. This response of the THP-1 cells can also be blocked by lipid IV_A (structurally identical to compound 406), providing further evidence for the contribution of LPS-binding sites to the recognition of LAM by mononuclear phagocytic cells. Evidence for a direct interaction of LAM with CD14 was later provided by the finding that the changes in fluorescence intensity of FITC-LPS mediated by binding of sCD14 was inhibited by LAM (16). These experiments directly demonstrated an interaction of CD14 with LAM. However, like the bioactivity of uronic acid polymers, this interaction of CD14 with LAM appears to be essential but not sufficient for the activation of cells. Cells of nonhemopoietic origin (U373 cells and even CD14-transfected CHO and CD14-HT1080 fibrosarcoma cells) could not be activated by LAM in the presence or absence of additional sCD14 and LBP, although they were responsive to LPS (54). Therefore, additional receptors or signal-transducing pathways in addition to CD14 are postulated to be required for cellular activation by LAM. These receptors/signal-transducing pathways seem to be present in 70Z/3 cells, because CD14-transfected 70Z/3 cells were able to respond to LAM (22).

INTERACTION OF CD14 WITH LIPOTEICHOIC ACID AND RELATED COMPOUNDS

Lipoteichoic acids (LTA) are glycolipids that are present in the cell wall of gram-positive but not gram-negative bacteria. They constitute, like LPS, negatively charged amphiphilic structures, which are hydrophobically anchored to the cell wall (55,56). Like other cell wall components, such as LPS and PG, LTA are also released from bacteria, in particular during treatment with cell wall–active antibiotics (56). It has, therefore, been discussed whether LTA may contribute to the inflammatory reactions observed in an infected host. Indeed, it has been shown by various groups that LTA preparations are able to induce the production of inflammatory cytokines (TNF, IL-18, IL-6, IL-8, and IL-12) as well as inducible nitric oxide synthase (iNOS) (57–60), both in vitro and in vivo. The activation of phosphatidylcholine-phospholipase C (PC-PLC), phosphorylation of tyrosine kinases, as well as the transcription factor NF-κB was also demonstrated to be involved in the bioactivity of LTA (61). Despite the release of inflammatory cytokines, involved in the septic shock reaction, lethality or organ failure was not observed at relevant concentrations of LTA (62,63). More recently it was shown that LTA are able to act in synergy with PG (64). The synergistic action of these two gram-positive cell wall components was demonstrated in vitro after stimulation of murine J774.2 macrophages as well as in vivo after treatment of rats, which reacted with the production of TNF and IFN-γ. Furthermore, a synergism of LTA and PG during induction of iNOS as well as multiple organ dysfunction syndrome (MODS) was described. These findings provide strong evidence that LTA as well as PG may be involved in the fatal response after severe bacterial infections.

Because LTA, like LPS, represents a negatively charged amphiphilic structure, it is possible that LPS receptors are also involved in the recognition of LTA by monocytes/macrophages. It was found that LTA shares with LPS the same macrophage scavenger receptor (65). The contribution of CD14 to the activation of the promonocytic human cell line THP-1 by LTA has been demonstrated (57). It was shown that LTA induces IL-12 and that this reaction could be inhibited by anti-CD14 mAb My4 but not by LeuM3. This response could also be suppressed by *Rhodobacter sphaeroides* LPS. *R. spaeroides* LPS is known to contain a penta-acylated lipid A and to lack endotoxic activity in murine and human cells. It rather blocks the activity of endotoxic LPS and, in this way, antagonizes endotoxic activity of LPS at an early stage (66–68). These results indicate that PG and LTA may use the same receptors for activation, namely LPS receptors. However, because of a synergistic activity of PG and LTA, the manner in which these cell wall compounds induce activation of cells must be different, at least in part.

It should be noted that the concept of biologically active LTA has been challenged (63). Chemically synthesized structures resembling LTA (LTA-1 and LTA-2) of *Enterococcus hirae* (69) ar not active in inducing cytokines and antitumor activity. It appears, therefore, that the biological activity of purified natural LTA is not due to the main constituent of the LTA preparations but to other, so far unknown, compounds (70).

The bioactive part of cell wall preparations of *Staphylococcus aureus* is also distinct from LTA (17,71). The cytokine-inducing factor fractionated on a reverse-phase column was distinct from LTA, and purified LTA failed to stimulate the IL-6 release in U373 cells or in human monocytes. Interestingly, the purified LTA fraction blocks the sCD14 dependent activation of U373 cells by LPS (17). Furthermore, it was demonstrated that LTA causes a shift in the electrophoretic mobility of sCD14 in a native PAGE showing a direct biochemical interaction between LTA and sCD14 (17). This indicates the possibility of a competitive antagonistic activity of LTA during stimulation of inflammatory cells with LPS.

The bioactive fraction isolated from cell wall preparation of *S. aureus* has been characterized to represent a mixture of heterogeneous amphiphilic molecules, which forms tight micelles in aqueous solution (17,71). It was suggested that these molecules have functional and structural similarities to LPS, and indeed the activation of cells by this active fraction of *S. aureus* cell wall preparation is CD14-dependent. Furthermore, the compound in this fraction causes a shift of the mobility of sCD14 in a PAGE.

FINAL REMARKS

The aim of this chapter was to summarize the results of investigations on the reactivity of CD14 with structures other than LPS. We have focused our review mostly on sPG, LAM, and LTA because these are bacterial cell wall compounds demonstrating direct binding to CD14.

The mode of recognition by CD14 of so many different structures is unknown. The best studied structure that is able to interact with CD14 is LPS. CD14-dependent stimulation of monocytes/macrophages by LPS can be observed using concentrations of 1 ng/ml or less. In contrast, when using other cell wall compounds, usually concentrations of more than 1 µg/ml are necessary. Furthermore, assays in which the binding of these structures were investigated show that higher concentrations of these non-LPS compounds are necessary to observe binding to cells. Thus, the binding of FITC-LPS to human monocytes can already be observed at concentrations of 10–30 ng/ml. In contrast, concentrations of about 1 µg/ml of FITC-sPG are required for the determination of binding to cells (29). Whether this higher reactivity of LPS in comparison to other cell wall components is simply due to the supporting activity of LBP [which is lacking for sPG (15)] remains to be studied. The low activation and binding capacity may reflect a rather low-affinity binding of the cell wall components. Besides this assumed low affinity, however, it should be noted that, especially for sPG, a distinct specificity of the binding and response, as determined by different inhibitors, was demonstrated (15,29). It is reasonable, therefore, to postulate a specific binding of low affinity of cell wall components like sPG, LAM, and LTA to CD14. However, measurements of affinity or interaction of CD14 with these cell wall compounds are still missing.

Also still unresolved is how cells handle the compounds recognized and how signal-transducing elements are activated after binding of microbial structures to CD14. For sPG as well as for LPS, it is known that they are internalized into the cells within a few minutes (29,72–74). However, it appears that internalization of LPS is not necessary for the activation of cells (75). This problem has not been studied in the case of sPG. More detailed examinations of signal-transducing elements, however, indicate that although the same receptor (CD14) is used, the resulting signaling pathways that are induced differ at least in part between LPS and PG (24–27).

Taken together, the available data show that with regard to the function of CD14 in binding and functional recognition, there is no exclusive specificity for LPS as other bacterial and nonbacterial compounds are recognized (see Tables 1 and 2). Nonclonal, polyspecific receptors with multiple distinct ligand-binding sites or sites binding a range of distinct ligands have been defined as "pattern recognition receptors" (80). CD14, therefore, has been classified as a "pattern-recognition receptor" (16). It is evident that all bacterial and nonbacterial compounds that are able to activate monocytes or macrophages via CD14 contain glycosyl residues. Therefore, it is reasonable to suggest that CD14 represents a lectin-like receptor (29,36). However, very recently a direct interaction of phosphatidylinositol, phosphatidylcholine, and phosphatidylethanolamine with sCD14 was demonstrated (76). As these phospholipids lack carbohydrate residues, the hypothesis that CD14 is a lectin-like molecule is not tenable. It, therefore, appears that CD14, like the other LPS-binding proteins (LBP and BPI), recognizes glyco-configured structures as well as lipids via a distinct pattern of ionic charges. However, this hypothesis needs to be confirmed by further investigations.

Note: The most recent and surprising finding, published just after this manuscript was finished, is the interaction of interleukin-2, a cytokine produced by T lymphocytes, with CD14 during activation of human monocytes (81).

ACKNOWLEDGMENT

This work was supported in part by a grant of the BMBF (grant No. 01KI9471 to AJU and EThR), by the DFG (SFB 367, project B2 [EThR] C5 [AJU], SFB 470, project B4 [EThR], and the Fonds der Chemischen Industrie (HDF and EThR).

REFERENCES

1. Bhakdi S, Grimminger F, Suttorp N, Walmrath D, Seeger W. Proteinaceous bacterial toxins and pathogenesis of sepsis syndrom and septic shock: The unknown connection. Med Microbiol Immunol 1994; 183:119–144.
2. Seidl PH, Schleifer KH. Biological Properties of Peptidoglycan. Berlin: Walter de Gruyter, 1986.
3. Rietschel ET, Kirikae T, Schade FU, Mamat U, Schmidt G, Loppnow H, Ulmer AJ, Zähringer U, Seydel U, Di Padov F, Schreier M, Brade H. Bacterial endotoxin: molecular relationships of structure to activity and function. FASEB J 1994; 8:217–225.
4. Wright SD, Ramos RA, Tobias PS, Ulevitch RJ, Mathison JC. CD14, a receptor for complexes of lipopolysaccharide (LPS) and LPS binding protein. Science 1990; 249:1431–1433.
5. Schumann RR, Leong SR, Flaggs GW, Gray PW, Wright SD, Mathison JC, Tobias PS, Ulevitch RJ. Structure and function of lipopolysaccharide binding protein. Science 1990; 249:1429–1431.
6. Golenbock DT, Liu Y, Millham FH, Freeman MW, Zoeller RA. Surface expression of human CD14 in Chinese hamster ovary fibroblasts imparts macrophage-like responsiveness to bacterial endotoxin. J Biol Chem 1993; 268:22055–22059.
7. Lee JD, Kato K, Tobias PS, Kirkland TN, Ulevitch RJ. Transfection of CD14 into 70Z/3 cells dramatically enhances the sensitivity to complexes of lipopolysaccharide (LPS) and LPS binding protein. J Exp Med 1992; 175:1697–1705.
8. Wright SD, Ramos RA, Tobias PS, Ulevitch RJ, Mathison JC. CD14, a receptor for complexes of lipopolysaccharide (LPS) and LPS binding protein. Science 1990; 249:1431–1433.
9. Bazil V, Horejsi V, Baudys M, Kristofova H, Strominger JL, Kostka W, Hilgert I. Biochemical characterization of a soluble form of the 53-kDa monocyte surface antigen. Eur J Immunol 1986; 16:1583–1589.
10. Hailman E, Lichenstein HS, Wurfel MM, Miller DS, Johnson DA, Kelley M, Busse LA, Zukowski MM, Wright SD. Lipopolysaccharide (LPS)-binding protein accelerates the binding of LPS to CD14. J Exp Med 1994; 179:269–277.
11. Frey EA, Miller DS, Jahr TG, Sundan A, Bazil V, Espevik T, Finlay BB, Wright SD. Soluble CD14 participates in the response of cells to lipopolisaccharide. J Exp Med 1992; 176:1665–1671.
12. Loppnow H, Stelter F, Schonbeck U, Schluter C, Ernst M, Schutt C, Flad HD. Endotoxin activates human vascular smooth muscle cells despite lack of expression of CD14 mRNA or endogenous membrane CD14. Infect Immun 1995; 63:1020–1026.
13. Dziarski R. Peptidoglycan and lipopolysaccharide bind to the same binding site on lymphocytes. J Biol Chem 1991; 266:4719–4725.
14. Dziarski R. Cell-bound albumin is the 70-kDa peptidoglycan-, lipopolysaccharide-, and lipoteichoic acid-binding protein on lymphocytes and macrophages. J Biol Chem 1994; 269:20431–20436.
15. Weidemann B, Brade H, Rietschel ETh, Dziarsi R, Bazil V, Kusumoto S, Flad HD, Ulmer AJ. Soluble peptidoglycan-induced monokine production can be blocked by anti-CD14 monoclonal antibodies and by lipid A partial structures. Infect Immun 1994; 62:4709–4715.
16. Pugin J, Heumann D, Tomasz A, Kravchenko V, Akamatzu Y, Nishijima M, Glauder M, Tobias P, Ulevitch R. CD14 is a pattern recognition receptor. Immunology 1994; 1:509–516.
17. Kusunoki T, Hailman E, Juan TSC, Lichenstein HS, Wright SD. Molecules from Staphylococcus aureus that bind CD14 and stimulate innate immune responses. J Exp Med 1995; 182:1673–1682.
18. Loppnow H, Brade H, Dürrbaum I, Dinarello CA, Kusumoto S, Rietschel ETh, Flad HD. Interleukin 1 induction capacity of defined lipopolysaccharide partial structures. J Immunol 1989; 142:3229–3238.
19. Kovach NL, Yee E, Munford RS, Raetz CRH, Harlan JM. Lipid IVa inhibits synthesis and release of tumor necrosis factor induced by lipopolysaccharide in human whole blood ex vivo. J Exp Med 1990; 172:77–84.
20. Lynn WA, Golenbock DT. Lipopolysaccharide antagonists. Immunol Today 1992; 13:271–276.
21. Flad HD, Loppnow H, Rietschel ETh, Ulmer AJ. Agonists and antagonists for LPS-induced cytokines. Immunobiology 1993; 187:303–316.
22. Gupta D, Kirkland TN, Viriyakosol S, Dziarski R. CD14 is a cell activating receptor for bacterial peptidoglycan. J Biol Chem 1996; 271:23310–23316.
23. Dziarski R, Tapping RI, Tobias PS. Binding of bacterial peptidoglycan to CD14. J Biol Chem 1998; 273:8680–8690.
24. Dziarski R, Jin YP, Gupta D. Differential activation of extracellular signal-regulated kinase (ERK)1, ERK2, p38, and c-Jun NH_2-terminal kinase mitogen activated

protein kinases by bacterial peptidoglycan. J Infect Dis 1996; 174:777–785.
25. Heine H, Uler AJ, Flad HD, Hauschildt S. LPS-induced change of phosphorylation of two cytosolic proteins in human monocytes is prevented by inhibitors of ADP-ribosylation. J Immunol 1995; 155:4899–4908.
26. Heine H, Flad H-D, Ulmer AJ, Hauschildt S. Phosphorylation of cytosolic proteins and ADP-ribosylation is involved in monocyte activation by LPS. Eur J Cell Biol 1996; 69(suppl. 42):133.
27. Mattsson E, Van Dijk H, Van Kessel K, Verhoef J, Fleer A, Rollof J. Intracellular pathways involved in tumor necrosis factor-α release by human monocytes on stimulation with lipopolysaccharide or Staphylococcal peptidoglycan are partly similar. J Infect Dis 1996; 173:212–218.
28. Novogrodsky A, Vanichkin A, Patya M. Gazit A, Osherov N, Levitski A. Prevention of lipopolysaccharide-induced lethal toxicity by tyrosine kinase inhibitors. Science 1994; 264:1319–1322.
29. Weidemann B, Schletter J, Dziarski R, Kusumoto S, Stelter F, Rietschel ETh, Flad HD, Ulmer AJ. Specific binding of soluble peptidoglycan and muramyldipeptide to CD14 on human monocytes. Infect Immun 1997; 65:858–864.
30. Heymer B, Seidl PH, Schleifer KH. Immunochemistry and biological activity of peptidoglycan. In: Stewart-Tull DES, Davies M, eds. Immunology of the Bacterial Cell Envelope. New York: Wiley 1985:11–46.
31. Mirelman D, Bracha R, Sharon N. Penicillin-induced secretion of a soluble, uncross-linked peptidoglycan by *Micrococcus luteus* cells. Biochemistry 1974; 13:5045–5053.
32. Zeiger AR, Wong W, Chatterjee AN, Young FE, Tuazon Cu. Evidence for the secretion of soluble peptidoglycans by clinical isolates of *Staphylococcus aureus*. Infect Immunol 1982; 37:1112–1118.
33. Imoto M, Yoshimura H, Shimamoto T, Sakaguchi N, Kusumoto S, Shiba T. Total synthesis of *Escherichia coli* lipid A. The endotoxically active principle of cell-surface lipopolysaccharide. Bull Chem Soc Jpn 1987; 60:2205–2214.
34. Rietschel ET, Brade H. Bacterial endotoxins. Sci Am 1992; 267:2–54.
35. Rietschel ET, Brade H, Holst O, Brade L, Müller-Loennies S, Mamat U, Zähringer U, Beckmann F, Seydel U, Brandenburg K, Ulmer AJ, Mattern T, Heine H, Schletter J, Hauschildt S, Loppnow H, Schönbeck U, Flad HD, Schade UF, Di Padova F, Kusumoto S, Schumann RR. Bacterial endotoxin: chemical constitution, biological recognition, host response and immunological detoxification. Curr Topics Microbiol Immunol 1996; 216:39–81.
36. Cavaillon JM, Marie C, Caroff M, Ledur A, Godard I, Poulain D, Fitting C, Haeffner-Cavaillon N. CD14/LPS receptor exhibits lectin-like properties. J Endotox Res 1996; 3:471–480.
37. Barnes PF, Chatterjee D, Adams JS, Lu S, Wang E, Yamamura M, Brennan PJ, Modlin RL. Cytokine production induced by *Mycobacterium tuberculosis* lipoarabinomannan. J Immunol 1992; 149:541–547.
38. Moreno C, Taverne J, Melhert A, Bate CAW, Bearley RJ, Meager A, Rook GAW, Playfair JHL. Lipoarabinamannan from *Mycobacterium tuberculosis* induces the production of tumor necrosis factor from human and murine macrophages. Clin Exp Immunol 1989; 76:240–245.
39. Bhakdi S, Klonisch T, Nuber P, Fischer W. Stimulation of monokine production by lipoteichoic acids. Infect Immun 1991; 59:4614–4620.
40. Keller R, Fischer W, Keist R, Bassetti S. Macrophage response to bacteria: induction of marked secretory and cellular activities by lipoteichoic acids. Infect Immun 1992; 60:3664–3672.
41. Riesenfeld-Orn I, Wolpe S, Garcia-Bustos JF, Hoffman MK, Tuomanen E. Production of interleukin-1 but not tumor necrosis factor by human monocytes stimulated with pneumococcal cell surface components. Infect Immun 1989; 57:1890–1893.
42. Hauschildt S, Hoffmann P, Beuscher HU, Dufhues G, Heinrich P, Wiesmüller KH, Jung G, Bessler WG. Activation of bone marrow-derived mouse macrophages by bacterial lipopeptide: cytokine production, phagocytosis and Ia expression. Eur J Immunol 1990; 20:63–68.
43. Krziwon C, Zähringer U, Kawahara K, Weidemann B, Kusumoto S, Rietschel ET, Flad HD, Ulmer AJ. Glycosphingolipids from Sphingomonas paucimobilis induce monokine production in human mononuclear cells. Infect Immun 1995; 63:2899–2905.
44. Medvedev AE, Flo T, Ingalls RR, Golenbock DT, Tetti G, Vogel SN, Espevik T. Involvement of CD14 and complement receptors CR3 and CR4 in NF-κB activation and TNF production induced by lipopolysaccharide and group B streptococcal cell walls. J Immunol 1998; 160:4535–4542.
45. Otterlei M, Sundan A, Skjåk-Bræk G, Ryan L, Smidsrød O, Espevik T. Similar mechanisms of action of defined polysaccharides and lipopolysaccharides: characterization of binding and tumor necrosis factor alpha induction. Infect Immun 1993; 61:1917–1925.
46. Kreuz M, et al.: Comparative analysis of cytokine production and tolerance inhibition by bacterial lipopeptides, LPS, and S. aureus in human monocytes. J Immunol 1997; 92:396–406.
47. Espevik T, Otterlei M, Skjåk-Bræk G, Ryan L, Wright SD, Sundan A. The involvement of CD14 in stimulation of cytokine production by uronic acid polymers. Eur J Immunol 1993; 23:255–261.
48. Schletter J, Brade H, Brade L, Krüger C, Loppnow H, Kusumoto S, Rietschel ET, Flad HD, Ulmer AJ. Binding of lipopolysaccharide (LPS) to an 80 kD membrane protein of human cells is mediated by soluble CD14 and LPS-binding protein. Infect Immun 1995; 63:2576–2580.
49. Vita N, Lefort S, Sozzani P, Reeb R, Richards R, Borysiewicz LK, Ferrara P, Labéta MO. Detection and biochemical characteristics of the receptor for complexes of soluble CD14 and bacterial lipopolysaccharide. J Immunol 1997; 158:3457–3462.
50. Kaplan G, Gandhi RR, Weinstein DE, Lewis WR, Patarroya ME, Brennan PJ, Cohn ZA. Mycobacterium

leprae antigen-induced suppression of T-cell proliferation in vitro. J Immunol 1987; 138:3028–3034.
51. Moreno C, Melhert A, Lamb J. The inhibitory effect of mycobacterial lipoarabinomannan and polysaccharides upon polyclonal and monoclonal human T cell proliferation. Clin Exp Immunol 1988; 74:206–210.
52. Molloy A, Gaudernack G, Lewis WR, Cohn ZA, Kaplan G. Suppression of T-cell proliferation by *Mycobacterium leprae* and its products: the role of lipopolysaccharide. Proc Natl Acad Sci USA 1990; 87:973–977.
53. Zhang Y, Doerfler M, Lee TC, Guillemin B, Rom WN. Mechanisms of stimulation of interleukin-1 beta and tumor necrosis factor-alpha by *Mycobycterium tuberculosis* components. J Clin Invest 1993; 91:2076–2083.
54. Savedra R, DeLude RL, Ingalls RR, Fenton MJ, Golenbock DT. Mycobacterial lipoarabinomannan recognition requires a receptor that shares components of the endotoxin signaling system. J Immunol 1996; 157:2549–2554.
55. Fischer W. Bacterial phospholipids and lipoteichoic acids. In: Kates M, ed. Glycolipids, Phosphoglycolipids, and Sulfoglycolipids. New York: Plenum Press, 1990:123–234.
56. Fischer W. Lipoteichoic acids and lipoglycans. In: Ghuysen JM, Hackenbeck R, eds. Bacterial Cell Wall. Amsterdam: Elsevier Science Publishing, 1994:199–215.
57. Cleveland MG, Gorham JD, Murphy TL, Tuomanen E, Murphy KM. Lipoteichoic acid preparations of gram-positive bacteria induce interleukin-12 through a CD14-dependent pathway. Infect Immun 1996; 64:1906–1912.
58. Bhakdi S, Klonisch T, Nuber P, Fisher W. Stimulation of monokine production by lipoteichoic acids. Infect Immun 1991; 59:4614–4620.
59. Mattsson E, Verhage L, Rollof J, Fleer A, Verhoef J, van Dijk H 49. Peptidoglycan and teichoic acid from *Staphylococcus epidermis* stimulate human monocytes to release tumour necrosis factor-alpha, interleukin-1 beta and interleukin-6. FEMS Immunol Med Microbiol 1993; 7:281–287.
60. Standiford TJ, Arenberg DA, Danforth JM, Kunkel SL, Van Otteren GM, Strieter RM. Lipoteichoic acid induces secretion and interleukin-8 from human blood monocytes: a cellular and molecular analysis. Infect Immun 1994; 62:119–125.
61. Kengatharan M, De Kimpe SJ, Thiemermann C. Analysis of the signal transduction in the induction of nitric oxide synthase by lipoteichoic acid in macrophages. Br J Pharmacol 1996; 117:1163–1170.
62. Himanen JP, Pyhala L, Olander RM, Merimskaya O, Kuzina T, Lysyuk O, Pronin A, Sanin A, Helander IM, Sarvas M. Biological activities of lipoteichoic acid and peptidoglycan-teichoic acid of *Bacillus subtilis* 168 (Marburg). J Gen Microbiol 1993; 139(Pt 11):2659–2665.
63. Takada H, Kawabata Y, Arakaki R, Kusumoto S, Fukase K, Suda Y, Yoshimura T, Kokeguchi S, Kato K, Komuro T, Tanaka N, Saito M, Yoshida T, Sato M, Kotani S. Molecular and structural requirements of a lipoteichoic acid from *Enterococcus hirae* ATCC 9790 for cytokine-inducing, antitumor, and antigenic activities. Infect Immun 1995; 63:57–65.
64. De Kimpe SJ, Kengatharan M, Thiemermann C, Vane JR. The cell wall components peptidoglycan and lipoteichoic acid from *Staphylococcus aureus* act in synergy to cause shock and multiple organ failure. Proc Natl Acad Sci USA 1995; 92:10359–10363.
65. Greenberg JW, Fischer W, Joiner KA. Influence of lipoteichoic acid structure on recognition by the macrophage scavenger receptor. Infect-Immun 1996; 64:3318–3325.
66. Takayama K, Quereshi N, Beutler B, Kirkland ThN. Diphosphoryl lipid A from *Rhodopseudomons sphaeroides* ATCC 17023 blocks induction of cachectin in macrophages by lipopolysaccharide. Infect Immun 1989; 57:1336–1338.
67. Quereshi N, Takayama K, Kurtz R. Diphosphoryl lipid A obtained from the nontoxic lipopolysaccharide of *Rhodopseudomonas sphaeroides* is an endotoxin antagonist in mice. Infect Immun 1991; 59:441–444.
68. Loppnow H, Lippy P, Freudenberg M, Krauss JH, Weckesser J, Mayer H. Cytokine induction by lipopolysaccharide (LPS) corresponds to lethal toxicity and is inhibited by nontoxic *Rhodobacter capsulatus* LPS. Infect Immun 1990; 58:3743–3750.
69. Fischer W. Bacterial phosphoglycolipids and lipoteichoic acids. Handb Lipid Res 1990; 6:123–234.
70. Suda Y, Tochio H, Kawano K, Takada H, Yoshida T, Kotani S, Kusumoto S. Cytokine-inducing glycolipids in the lipoteichoic acid fraction from *Enterococcus hirae* ATCC 9790. FEMS Immunol Med Microbiol 1995; 12:97–112.
71. Kusunoki T, Wright SD. Chemical characteristics of *Staphylococcus aureus* molecules that have CD14-dependent cell-stimulating activity. J Immunol 1996; 157:5112–5117.
72. Kang YH, Dwivedi RS, Lee CH. Ultrastructural and immunocytochemical study of the uptake and distribution of bacterial lipopolysaccharide in human monocytes. J Leukocyte Biol 1990; 48:316–332.
73. Kitchens RL, Ulevitch RJ, Munford RS. Lipopolysaccharide (LPS) partial structures inhibit responses to LPS in a human macrophage cell line without inhibiting LPS uptake by a CD14-mediated pathway. J Exp Med 1992; 176:485–494.
74. Luchi M, Munford RS. Binding, internalization, and deacylation of bacterial lipopolysaccharide by human neutrophils. J Immunol 1993; 151:959–969.
75. Gegner JA, Ulevitch RJ, Tobias PS. Lipopolysaccharide (LPS) signal transduction and clearance. Dual roles for LPS binding protein and membrane CD14. J Biol Chem 1995; 270:5320–5325.
76. Yu B, Hailman B, Wright SD. Lipopolysaccharide binding protein and soluble CD14 catalyse exchange of phospholipids. J Clin Invest 1997; 99:315–324.
77. Soell M, Lett E, Holveck F, Scholler M, Wachsmann D, Klein JP. Activation of human monocytes by streptococcal rhamnos glucose polymers is mediated by CD14 antigen, and mannan binding protein inhibits TNF-alpha release. J Immunol 1995; 154:851–860.
78. Otterlei M, Varum KM, Ryan L, Espevik T. Characterization of binding and TNF-alpha-inducing ability of

chitosans on monocytes: the involvement of CD14. Vaccine 1994; 12:825–832.

79. Newman SL, Chaturvedi S, Klein BS. The WI-1 antigen of *Blastomyces dermatitidis* yeast mediates binding to human macrophage CD11/CD18 (CR3) and CD14. J Immunol 1995; 154:753–761.

80. Janeway CA Jr. The immune system evolved to discriminate infectious nonself from noninfectious self. Immunol Today 1992; 13:11–16.

81. Bosco MC, Espinoza-Delgado I, Rowe TK, Malabarba MG, Longo DL, Varesio L. Functional role for the myeloid differentiation antigen CD14 in the activation of human monocytes by IL-2. J Immunol 1997; 159:2922–2931.

29

The Role of MAP Kinases, Phosphatidylinositol 3-Kinase, and Ceramide in LPS-Induced Signaling in Macrophages

Anthony L. DeFranco, Alexander J. Finn, and Julie Hambleton
University of California, San Francisco, San Francisco, California

Mary T. Crowley
The Scripps Research Institute, La Jolla, California

Mary Lee MacKichan
Chiron Corporation, Emeryville, California

Steven L. Weinstein
San Francisco State University, San Francisco, California

INTRODUCTION

Macrophages play a particularly important role in the immune defense against bacterial infection by recognizing and responding to a number of highly conserved components of bacterial cell walls. The best understood of these components is lipopolysaccharide (LPS) from the outer membrane of gram-negative bacteria. Macrophages recognizing LPS secrete a variety of proinflammatory mediators, including the cytokines tumor necrosis factor (TNF-α), interleukins 1, 6, and 12 (IL-1, IL-6, and IL-12), and interferons (IFN) α and β, the chemokines IL-8, MIP-1α, MIP-1β, and MIP-2, and lipid mediators such as platelet-activating factor and various prostaglandins (1–4). In addition, LPS induces the expression of a number of important cell surface proteins, such as tissue factor (5), which initiates the clotting cascade, and class II MHC and B7 molecules (6,7), which promote the ability of the stimulated macrophage to serve as an antigen-presenting cell for CD4$^+$ T cells. In combination with IFN-γ, LPS stimulates macrophages to transcribe the gene encoding inducible nitric oxide synthase (iNOS), leading to the generation of the toxic oxidant nitric oxide. Macrophages stimulated with IFN-γ and LPS also assume a highly activated phenotype characterized by strong bactericidal and tumoricidal capabilities (1). Thus, LPS stimulates macrophages to make a complex array of responses leading to local immune responses to gram-negative bacteria and, through IL-1, IL-6, and TNF-α, a variety of distant and systemic responses, such as the acute phase response, fever, and mobilization of energy stores that aid in fighting the infection.

Our understanding of how macrophages respond to LPS is still fragmentary. Virtually all of the cellular responses listed above depend upon recognition of the most conserved part of the LPS molecule, lipid A. Macrophage responses to physiological concentrations of LPS (e.g., probably levels in the low ng/ml range) typically involve the cell surface protein CD14 (8). First, LPS binds avidly to an acute phase protein from serum, LPS-binding protein (LBP), which acts as a catalyst to speed the binding of LPS monomers to CD14. The importance of CD14 for many macrophage responses to LPS has been established by antibody blocking experiments (8,9) and confirmed recently by studies with macrophages from mice in which the gene for CD14 was ablated by gene targeting (10).

Because CD14 is a glycosylphosphatidylinositol-linked protein, it is unclear how it can convey information to the cell's interior. An attractive hypothesis is that CD14 acts primarily as a binding subunit required for high-affinity binding and not as a signal-transducing component per se (8). According to this view, another currently unidentified polypeptide is responsible for transmitting signals to the cell's interior. As macrophages can respond to higher concentrations of LPS (>10 ng/ml) in the absence of CD14, a low-affinity receptor also presumably exists. It may be that at low LPS concentrations, CD14 cooperates with endogenous lower-affinity receptors, which then transmit the signal of LPS binding. Despite this lack of understanding of how LPS recognition by CD14 is converted into intracellular signals that mediate the varied responses of the macrophage to LPS, there is now a considerable amount known about the nature of those intracellular signals.

SIGNAL TRANSDUCTION IN LPS-STIMULATED MACROPHAGES

LPS-Induced Protein Tyrosine Phosphorylation

Among the earliest signaling events detected in LPS-stimulated macrophages is tyrosine phosphorylation of a number of proteins (11). LPS-induced protein tyrosine phosphorylation has been observed in many different macrophage populations by a number of groups and thus appears to be a widespread feature of the macrophage response to LPS (12–15). Activation of tyrosine kinases is used by many receptors to initiate downstream signal-transduction reactions. This may be the case with LPS-induced signal transduction as well, because several structurally diverse tyrosine kinase inhibitors are able to block many biological responses to LPS, including release of arachidonic acid metabolites (11) and production of TNF-α (12,14,16), IL-1 (13,14), IL-6 (12), and nitric oxide (16). A tyrosine kinase inhibitor also blocks LPS-induced endotoxic shock in mice (17). Thus, protein tyrosine phosphorylation appears to be a critical initiating event in macrophage responses to LPS.

Several proteins that become rapidly phosphorylated on tyrosine in macrophages stimulated with LPS have been identified. These include the adapter protein Shc, which is involved in the activation of Ras (18), the proto-oncogene Vav (19), the inositol phosphatase SHIP (18), and all three of the major classes of mammalian MAP kinases (20–22). The first three proteins are bona fide tyrosine kinase substrates, suggesting strongly that LPS does, indeed, activate one or more tyrosine kinase.

The tyrosine kinases that play key roles in mediating LPS-induced tyrosine phosphorylation are presently unidentified. Although the Src family tyrosine kinases Hck, Lyn, and Fgr associate with CD14 and/or become activated in cells treated with LPS (12,23), macrophages from mice genetically deficient in all three of these tyrosine kinases respond normally to LPS (24). Thus, the Src family tyrosine kinases Hck, Lyn, and Fgr do not play critical roles in LPS-induced signal transduction although it should be noted that macrophages do appear to express a low level of c-Src. Another intracellular protein tyrosine kinase, Syk, also becomes tyrosine phosphorylated in LPS-stimulated macrophages (18). Tyrosine phosphorylation of Syk is typically correlated with its activation, suggesting that Syk is also activated in response to LPS. Therefore, we tested macrophages from Syk-deficient mice for their responses to LPS. These macrophages were found to respond to LPS identically to wild-type macrophages with regard to several early signaling events and with regard to production of TNF-α, IL-1, IL-6, and IL-12, indicating that Syk is not required for LPS signaling (25). These genetic experiments indicate that the activations of Syk and the Src family tyrosine kinases are secondary events that do not play critical roles in mediating many of the responses of macrophages to LPS. Presumably, another currently unidentified tyrosine kinase has a central role in initiating signaling events following LPS binding to CD14.

Activation of Three MAP Kinase Modules in Response to LPS

Mammalian cells have three major MAP kinase pathways that act as relays between extracellular signals and transcription factors (26,27). These include Erk1 and Erk2 (the classical p44 and p42 MAP kinases), the c-Jun N-terminal protein kinases 1 and 2 (JNK1/2), and the p38 subfamilies of MAP kinases. LPS activates all three of these MAP kinase types strongly (20–22,28–30), and, as described in the next section, there is evidence that they play important roles in mediating the effects of LPS.

Although these three signaling pathways are incompletely understood, a consensus view of the nature of these pathways is shown in Figure 1. Typically, the Erk pathway is activated by Ras and is important for activating transcription from serum-response elements by

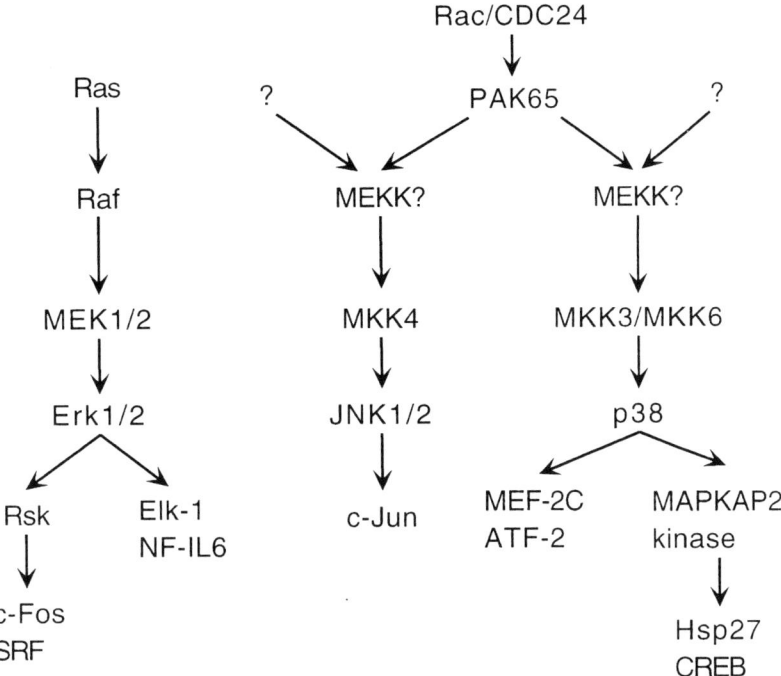

Fig. 1 Mammalian MAP kinase pathways. Mammalian MAP kinases fall into three major subgroups, Erk1/2, JNK1/2, and p38, each controlled by a separate upstream pathway of activating protein kinases. The MAP kinases are phosphorylated and activated by the MEK/MKK protein kinases, which in turn are activated by Raf or the MEKKs. Raf is activated primarily by Ras, although this may not be the only route to activation of the Erk pathway. There are multiple related MEKKs, which are regulated by Rac or Cdc42 GTP-binding proteins, at least in some circumstances. Rac and Cdc42 can activate protein kinases of the PAK65 type. Many elements of these pathways remain to be firmly established. MAP kinases can phosphorylate and activate transcription factors or other protein kinases, which in turn can regulate transcription or possibly other processes.

phosphorylating the Ets family ternary complex factors such as Elk-1. The JNK and p38 pathways may be downstream of another subgroup of low molecular weight GTP-binding proteins, which includes the Rac and CDC42 proteins. JNK and p38 MAP kinase pathways are often activated in cells subjected to stressful conditions such as osmotic shock and hence have also been called stress-activated protein kinases (SAPKs). A major target for JNK is, as the name implies, the bZIP family transcription factor c-Jun. Phosphorylation by JNK of serines 63 and 73 within c-Jun greatly increases its ability to stimulate transcription without affecting its binding to DNA. This promotes gene transcription at AP-1 sites, as can be observed with a reporter gene assay in macrophages stimulated by LPS (22). The p38 MAP kinase can also promote the activity of transcription factors, notably the myocyte enhancer factor MEF2C (31), which may regulate the *c-jun* promoter. In addition to their ability to act directly on transcription factors, MAP kinases can act on other protein kinases to activate them. For example, in many cell types, Erk1 and Erk2 activate $p90^{RSK}$ (27) and p38 activates MAPKAP kinase-2 (32).

How LPS activates these MAP kinase modules is still poorly understood. LPS has been shown to activate MEK1 (MAP kinase/ERK kinase; also called MAP kinase kinase), which is the immediate upstream activator of the Erk1 and Erk2 MAP kinases (30,33,34). In most systems, MEK1 is phosphorylated and consequently activated by Raf (35). Raf in turn is usually activated by the active form of Ras, Ras(GTP), perhaps acting in concert with other, presently undefined, inputs.

The fact that LPS strongly activates Erk1 and Erk2 in macrophages suggests that it may do so via Ras activation. Indirect evidence for this possibility comes from two directions. First, LPS induces rapid tyrosine phosphorylation of the adapter protein Shc, which is important for activation of Ras in numerous systems (18). Second, a dominant negative mutant form of Ras was found to inhibit TNF-α promoter-driven transcription in transiently transfected RAW264.7 macrophage cells (33). The simplest interpretation of the latter result is that LPS activates Ras and that this is directly important for TNF-α transcription, although alternative explanations can be imagined. The most definitive experiment is to directly assess the amount of Ras(GTP) in the cell and see if LPS causes an increase. Indeed, Geng et al. observed a substantial increase in Ras(GTP) in human monocytes stimulated with LPS (19). In contrast, Buscher et al. saw no increase in Ras(GTP) in the BAC-1.2F5 CSF-1–dependent macrophage cell line, which in contrast showed a clear increase in Ras(GTP) in response to CSF-1 (34). Similarly, we saw no clear increase in Ras(GTP) in LPS-stimulated RAW264.7 macrophages (S. L. Weinstein and A. L. DeFranco, unpublished results). Thus, although there is some direct and indirect evidence supporting the idea that LPS induces an increase in the amount of Ras(GTP), it is unclear whether this is a general phenomenon in macrophages or whether this is restricted to monocytes.

Activated Ras binds to Raf and promotes its activation by a mechanism that is not fully understood (36). The activation of Raf can be assessed by immunoprecipitating Raf and measuring its ability to phosphorylate MEK1 in vitro. We examined Raf-1 activation in macrophages and found that LPS stimulated a small increase in Raf-1 activity in RAW264.7 cells and a more impressive increase in cultured bone marrow–derived macrophages (S. L. Weinstein and A. L. DeFranco, unpublished observations). Interestingly, Buscher et al. did see activation of Raf-1 in BAC-1.2F5 cells, even though they failed to detect an increase in Ras(GTP) in response to LPS (34). They concluded that Raf-1 was being activated by LPS in a Ras-independent manner. Although it would be unusual, there is a precedent for this hypothesis (37). Alternatively, changes in Raf may be more readily detected than changes in Ras(GTP); thus, changes in Ras(GTP) induced by LPS may have occurred but gone undetected. At this point, this issue is not fully resolved.

Importance of the Erk1/Erk2 MAP Kinases for Macrophage Responses to LPS

Two strategies have been used to assess the role that the Erk1/2 MAP kinase pathway plays in mediating macrophage responses to LPS. The first strategy was to activate Erk1/2 independently of LPS in order to test whether the Erk1/2 MAP kinases are sufficient to activate downstream events. The second strategy was to use a pharmacological inhibitor of MEK1 and MEK2 to block specifically the activation of Erk1/2 following LPS addition. This strategy tests whether the MEK/Erk1/2 pathway is required for LPS to elicit the responses examined.

A form of the upstream component Raf that can be rapidly turned on was used to activate the Erk1/2 MAP kinases independently of LPS stimulation. Dr. Martin McMahon (UCSF Cancer Center, San Francisco, CA) developed a regulated form of Raf-1 by fusing the kinase domain of Raf-1 to the ligand-binding domain of the estrogen receptor. This chimeric protein (ΔRaf-1/ER) was rapidly activated in cells by addition of estradiol. Strong activation of Erk1/2 MAP kinases occurred within 15 minutes of the addition of estradiol (38). In collaborative experiments with Dr. McMahon we expressed the ΔRaf-1/ER fusion protein stably in RAW264.7 cells and tested the acute effects of its activation. As in fibroblasts, addition of estradiol to ΔRaf-1/ER-expressing RAW264.7 cells led to rapid and full activation of the Erk1 and Erk2 MAP kinases. Interestingly, activation of these MAP kinases by Raf was stronger at later times (e.g., 4–48 hr) than was the activation seen in response to LPS (39). LPS-induced signals upstream of Erk1/2 may decrease with prolonged stimulation or, alternatively, LPS may induce a downregulatory mechanism. In support of the latter possibility, we found that LPS induces the synthesis of the *mkp-1* gene (39), which encodes a MAP kinase phosphatase (40), whereas activation of Raf and Erk1/2 MAP kinases by the ΔRaf-1/ER chimera does not. The presence of the *mkp-1* gene product could decrease Erk1/2 phosphorylation, resulting in reduced enzymatic activity at later times following LPS stimulation.

The activation of the Erk1/2 pathway by the ΔRaf-1/ER chimeric protein is highly specific, as addition of estradiol to RAW264.7 cells expressing this chimera does not activate the JNK pathway (22), nor does it activate the transcription factor NF-κB detectably (39). In contrast, LPS activates JNK and NF-κB strongly. Interestingly, Raf activation led to a small increase in TNF-α mRNA accumulation and TNF-α production, amounting to about 5% of that seen with LPS. The murine TNF-α promoter has an NF-κB site that appears to be important for promoter activity (41,42), so it is likely that the ability of LPS to fully stimulate production of TNF-α depends on activation of NF-κB as well as activation of the Erk1/2 MAP kinases. These two signaling events may be in some way integrated at the TNF-α promoter to give a full response.

The hypothesis that the Erk1/2 MAP kinases play an important role in the LPS induction of TNF-α expression has been supported by studies with a highly specific inhibitor of MEK1 and MEK2, PD098059 (43). This compound appears to act by binding to MEK1 and MEK2 and preventing their activation by

Raf. No detectable inhibition of the p38 or JNK MAP kinases or NF-κB was seen under the conditions used, indicating the specificity of the inhibitor. In practice, this inhibitor often incompletely blocks Erk1/2 activation, especially at higher levels of receptor stimulation. This may be because the few MEK1/2 molecules that do become active can phosphorylate multiple Erk1/2 molecules and amplify any leakiness of the inhibitor. Consistent with this previous experience, we found that PD098059 blocks LPS activation of Erk1/2 MAP kinases at low doses of LPS (e.g., ≤500 pg/ml) but not at doses of LPS above 10 ng/ml (J. Hambleton et al., unpublished). Good production of TNF-α was seen at the lower doses of LPS, and this production was decreased by 90% in the presence of the MEK1/2 inhibitor. Similarly, PD098059 also inhibited production of IL-6 substantially. Thus, MEK1/2 and downstream components, presumably Erk1 and Erk2, are required for LPS-induced production of the cytokines TNF-α and IL-6, at least at low LPS concentrations.

The other two MAP kinase pathways also contribute to TNF-α expression, although they do so at the translational rather than the transcriptional level. A sequence in the 3′-untranslated region of the TNF-α mRNA represses translation of this mRNA in unstimulated macrophages, but not in LPS-stimulated macrophages (44). Although the mechanism by which LPS relieves this repression is not understood, it appears to involve the actions of both p38 and JNK. Inhibition of the p38 MAP kinases with the specific inhibitor SB203580 reduces TNF-α expression due to posttranscriptional regulation of TNF-α expression (45). More recently, it has been found that JNK also contributes to this regulatory event (46) as a kinase-defective mutant form of JNK blocked translation of TNF-α mRNA, presumably by competing with the endogenous JNK for upstream activators, thereby decreasing its activation. Thus, both p38 and JNK contribute to TNF-α expression at the translation level, although, as mentioned above, they also can promote the activation of transcription factors and therefore are likely to contribute to transcriptional regulation of other LPS-induced genes. A model for the mechanism of LPS-induced activation of TNF-α gene expression is shown in Figure 2.

Most cytokine gene expression also requires the activity of the transcription factor NF-κB (47). Evidence

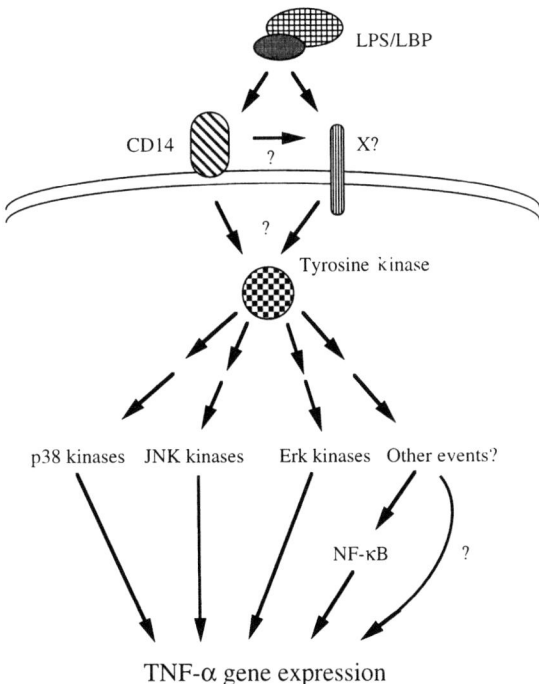

Fig. 2 Regulation of TNF-α gene expression in LPS-stimulated macrophages. LPS binds to CD14 and to a putative low-affinity receptor (X), which activate one or more intracellular protein tyrosine kinase. Through a series of poorly characterized events (see Fig. 1), this tyrosine kinase(s) promotes the activation of the three types of mammalian MAP kinases (Erk1/2, JNK, p38) and also the activation of NF-κB. Erk MAP kinases and NF-κB activate TNF-α gene transcription, whereas JNK and p38 contribute to increased expression of TNF-α by enhancing translation of the TNF-α mRNA. These signaling pathways likely contribute to expression of other LPS-induced cytokines as well.

to date indicates that activation of this transcription factor is not a direct consequence of the three MAP kinase pathways described above. Little is known about how LPS activates NF-κB, but recently there has been considerable progress in understanding how TNF-α receptors activate this transcription factor (48). Briefly, TNF receptor I activates a MEKK-like protein kinase called NIK, which in turn activates a large protein kinase complex referred to as the IκB kinase. At least two different subunits of the IκB kinase have protein kinase activity. IκB kinase phosphorylates two serine residues of IκB (serines 32 and 36 in IκBα), thereby targeting it for proteasome-mediated destruction. This degradation releases NF-κB, which then goes to the nucleus, binds to κB sites, and in conjunction with other transcription factors activates transcription of target genes such as those encoding TNF-α and IL-6.

Role of Ceramide in Mediating LPS Actions

Based on some limited structural similarity of LPS and the lipid second messenger ceramide, it has been proposed that LPS may act primarily by interacting with ceramide-responsive enzymes such as the ceramide-activated protein kinase (49). According to this hypothesis, CD14 and other cell surface molecules acting with lower affinity, such as the β2-integrins (50,51), would primarily serve as a way of getting LPS into the plasma membrane, where it would then act like ceramide. In support of this idea, LPS treatment of macrophages results in activation of a ceramide-activated protein kinase (49). A further prediction of this model is that addition of ceramide analogs to cells should result in biological responses similar to that seen with LPS. Addition of the water-soluble ceramide analog N-acetyl sphingosine (C2 ceramide) did induce some responses that are elicited by LPS, such as the activation of the c-Jun N-terminal kinase (JNK), growth arrest of RAW264.7 macrophage cells (51a), and the induction of several LPS-responsive genes (52). However, it did not induce other important intracellular events, such as activation of the Erk1/2 MAP kinases or activation of the transcription factor NF-κB (51a).

The failure of ceramide to induce important LPS-induced intracellular events does not support the ceramide-mimic hypothesis for LPS action. Rather, these observations suggest the alternative that ceramide is one of several second messengers elicited in macrophages responding to LPS and therefore may mediate some but not all LPS responses. In support of this latter hypothesis, we found that LPS caused a rapid, although modest, rise in cellular ceramide levels (51a), in contrast to a previous report (49). As ceramide is both a precursor for synthesis of sphingolipids and a second messenger derived from sphingomyelin, moderate changes in ceramide levels could be due to large changes in the second messenger pool of ceramide along with no or little change in lipid precursor ceramide levels. In any case, these latter observations suggest that LPS induces production of ceramide, which then activates ceramide-activated protein kinase or ceramide-activated protein phosphatase. This may be one of a handful of signaling reactions stimulated in macrophages by LPS and may make unique contributions to mediating the effects of LPS.

LPS Signaling Pathway Leading to Activation of p70 S6 Kinase

Another signaling pathway activated by LPS includes the lipid kinase phosphatidylinositol 3-kinase (PI 3-kinase) and two downstream serine/threonine protein kinases Akt/PKB and p70^{S6kinase}. In various cell types, this signaling pathway is thought to regulate translation of certain mRNAs and also to regulate apoptotic pathways (53,54). PI 3-kinase phosphorylates various inositol-containing phospholipids, generating second messengers that may work primarily by serving as ligands for pleckstrin homology (PH) domains in signaling proteins. The interaction between PH domains and the products of PI 3-kinase often serves to recruit signaling components to the plasma membrane in an inducible fashion (55,56). LPS has recently been shown to activate PI 3-kinase in monocytes (57) and in the RAW264.7 macrophage cell line (S. L. Weinstein et al., unpublished). We have also found that LPS activates Akt/PKB (A. J. Finn and A. L. DeFranco, unpublished data) and p70^{S6kinase} (S. L. Weinstein et al., unpublished). LPS induced approximately a threefold increase in p70^{S6kinase} activity in an in vitro kinase assay. Two inhibitors of PI 3-kinase, wortmannin and the chemically distinct inhibitor LY294002, blocked activation of p70^{S6kinase} in response to LPS. In addition, the activation of p70^{S6kinase} was inhibited by the immunosuppressant rapamycin, indicating a role for the mTOR protein, the target of rapamycin, in the activation of p70^{S6kinase}.

Akt/PKB activation in response to LPS stimulation was also blocked by the PI 3-kinase inhibitors but not by rapamycin, as has been seen in other cell types. Thus, Akt/PKB activation is dependent upon PI 3-kinase but not upon mTOR, whereas p70^{S6kinase} activity is dependent upon both. Current understanding of this signaling pathway is depicted in Figure 3.

Recent evidence indicates that Akt/PKB is directly activated by PI 3,4-P$_2$ (57a,57b). The origin of this sec-

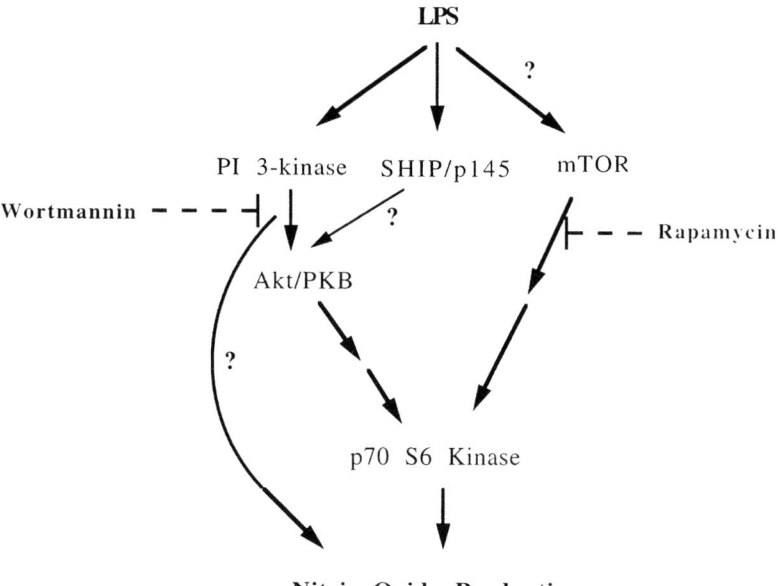

Fig. 3 The PI 3-kinase/Akt/p70^{S6kinase} signaling pathway. LPS activates a signaling pathway starting with PI 3-kinase and including two downstream serine/threonine protein kinases, Akt/PKB and p70^{S6kinase}. The latter kinase also requires the action of mTOR (mammalian target of rapamycin) for its activation. Activation of Akt/PKB is dependent upon the lipid second messenger phosphatidylinositol 3,4-bisphosphate, which can be created by the combined actions of PI 3-kinase and the inositol 5-phosphatase SHIP. The latter signaling molecule is tyrosine phosphorylated in response to LPS, suggesting that it does participate here. This signaling pathway plays an important role in mediating LPS-induced nitric oxide production. Based on the effects of PI 3-kinase inhibitors and rapamycin, PI 3-kinase appears to participate in nitric oxide production in two ways, one likely involving p70^{S6kinase} and the other independent of it.

ond messenger is not entirely clear, but it can be generated by the sequential actions of PI 3-kinase and SHIP (58–60). Interesting in this regard, LPS stimulation of macrophages activates PI 3-kinase and induces the tyrosine phosphorylation of SHIP, as well as the association of SHIP with the signaling protein Shc (18). Thus, a reasonable hypothesis is that PI 3,4-P$_2$ is being generated in response to LPS and is activating Akt/PKB, which in turn promotes the activation of p70^{S6kinase}.

The function of this signaling pathway in mediating the action of LPS is not yet known. Cytokine production was not inhibited substantially by PI 3-kinase inhibitors or by rapamycin, indicating that this functional response is not clearly dependent upon the PI 3-kinase/p70^{S6kinase} signaling pathway. In contrast, the PI 3-kinase inhibitor LY294002 almost completely blocked the production of nitric oxide in response to low concentrations of LPS, suggesting that at least one function of this pathway is to contribute to iNOS expression or activity (S. L. Weinstein et al., unpublished). Rapamycin also blocked nitric oxide production, although to a lesser extent. One possible interpretation of these results is that PI 3-kinase participates in nitric oxide production in two ways: one involving p70^{S6kinase} and hence also sensitive to inhibition by rapamycin, the other not involving p70^{S6kinase} (Fig. 3). When the macrophages were stimulated with a higher concentration of LPS, both the PI 3-kinase inhibitor and rapamycin inhibited less well, indicating that other signal transduction events stimulated by high doses of LPS can partially replace the requirement for the PI 3-kinase and mTOR-dependent signaling events.

CONCLUSIONS

Macrophages bind LPS through the high-affinity receptor CD14 and also via an as-yet-unidentified low-affinity receptor. Because CD14 lacks transmembrane or cytoplasmic domains, it is unlikely to transmit information of ligand binding to the cytoplasm of the cell by itself. Thus, an unidentified transmembrane signal-transducing component is postulated to perform this function. An early and critical event in LPS signaling appears to be the activation of one or more protein tyrosine kinases, since inhibitors of these enzymes block almost all LPS responses. Therefore, it was initially exciting when LPS recognition was found to lead to the rapid activation of Syk and the three Src-family members in macrophages: Hck, Fgr, and Lyn. Macrophages from mice with targeted mutations in the genes encoding these kinases all exhibit normal LPS responses, however, indicating that the key tyrosine kinase(s) mediating the response to LPS is not yet known. We hypothesize that LPS activates another, so-far-unidentified, intracellular or transmembrane protein tyrosine kinase that is important for mediating subsequent signaling events. The lack of a strong defect in these tyrosine kinase–deficient macrophages is surprising, and it is certainly possible that some LPS response not yet examined will depend on one or more of these kinases. Nevertheless, the lack of an important role for Syk or Src family tyrosine kinases is a reminder that the activation of a signaling component in response to LPS does not necessarily mean that that component is an important mediator of responses to LPS.

Among the targets of LPS-induced tyrosine phosphorylation are Shc and SHIP, signaling components believed to be important for Ras activation and activation of Akt/PKB, respectively, in other systems. LPS also induces the elevation of intracellular levels of the lipid second messenger ceramide and activates PI 3-kinase, which causes production of additional lipid second messengers. Ceramide may participate in the activation of JNK 1/2, whereas PI 3-kinase stimulates a signaling pathway involving the serine/threonine protein kinases Akt/PKB and $p70^{S6kinase}$. Signaling coming from PI 3-kinase appears to play two distinct roles in LPS-induced nitric oxide production. One role may involve $p70^{S6kinase}$, whereas the other appears to be independent of it.

Finally, LPS also induces the activation of all three of the major mammalian MAP kinase modules: Erk1/2, JNK, and p38 MAP kinases. How LPS activates these MAP kinases is not yet well understood, although there is some evidence supporting the idea that Ras and Raf are involved in the activation of the Erk MAP kinases and ceramide may be involved in the activation of JNK. All three types of MAP kinases participate in mediating LPS-induced TNF-α expression. Erk1/2 contribute at the transcriptional level, whereas the other two MAP kinases contribute at the translational level. None of these MAP kinases appears to be critical for activation of NF-κB, therefore signaling events leading to this important transcription factor are also likely to be important for LPS-induced expression of TNF-α and other cytokines. It is likely that the MAP kinases also participate in mediating many other of the responses of macrophages to LPS. For example, the MEK1/2 inhibitor PD098059 blocked production of IL-6 as well as TNF-α.

Clearly, a great deal has been learned in the past few years about the intracellular signaling events occurring after cell surface recognition of LPS by macrophages. Equally clearly, there are still many important gaps in our knowledge, starting with a poor understanding of how the information of LPS binding by CD14 is transmitted across the membrane and of the identity of the intracellular protein tyrosine kinase that is postulated to be critical for initiating many of the signaling cascades now known to be activated in response to LPS. In addition, the mechanisms by which downstream signaling events become activated and an understanding of how they contribute to important responses such as cytokine production and bactericidal activity remain areas of current interest and progress.

ACKNOWLEDGMENTS

The authors' studies were supported by grants AI-33442 and K11AI01164 from the U.S. National Institutes of Health. We also appreciate the contributions of our collaborators to the work described above, especially C. Lowell, F. Meng, V. Tybulewicz, M. McMahon, S. Pelech, J. Sanghera, and C. June.

REFERENCES

1. Adams DO, Hamilton TA. The cell biology of macrophage activation. Ann Rev Immunol 1984; 2:283–318.
2. Morrison DC, Ryan JL. Endotoxins and disease mechanisms. Ann Rev Med 1987; 38:417–432.
3. Davatelis G, Tekamp-Olson P, Wolpe SD, Hermsen K, Luedke C, Gallegos C, Coit D, Merryweather J, Cerami A. Cloning and characterization of a cDNA for murine macrophage inflammatory protein (MIP), a novel monokine with inflammatory and chemokinetic properties. J Exp Med 1988; 167:1939–1944.
4. Tekamp-Olson P, Gallegos C, Bauer D, McClain J, Sherry B, Fabre M, van Deventer S, Cerami A. Cloning and characterization of cDNAs for murine macrophage inflammatory protein 2 and its human homologues. J Exp Med 1990; 172:911–919.
5. Mackman N, Brand K, Edgington TS. Lipopolysaccharide-mediated transcriptional activation of the human tissue factor gene in THP-1 monocytic cells requires both activator protein 1 and nuclear factor κB binding sites. J Exp Med 1991; 174:1517–1526.
6. Marshall NE, Ziegler HK. Role of lipopolysaccharide in induction of Ia expression during infection with gram-negative bacteria. Infect Immun 1989; 57:1556–1560.

7. Ding L, Linsley PS, Huang LY, Germain RN, Shevach EM. IL-10 inhibits macrophage costimulatory activity by selectively inhibiting the up-regulation of B7 expression. J Immunol 1993; 151:1224–1234.
8. Ulevitch RJ, Tobias PS. Receptor-dependent mechanisms of cell stimulation by bacterial endotoxin. Annu Rev Immunol 1995; 13:437–457.
9. Wright SD, Ramos RA, Tobias PS, Ulevitch RJ, Mathison JC. CD14, a receptor for complexes of LPS and LPS binding protein. Science 1990; 249:1431–1433.
10. Haziot A, Ferrero E, Kontgen F, Hijiya N, Yamamoto S, Silver J, Stewart CL, Goyert SM. Resistance to endotoxin shock and reduced dissemination of gram-negative bacteria in CD14-deficient mice. Immunity 1996; 4:407–414.
11. Weinstein SL, Gold MR, DeFranco AL. Bacterial lipopolysaccharide stimulates protein tyrosine phosphorylation in macrophages. Proc Natl Acad Sci USA 1991; 88:4148–4152.
12. Beaty CD, Franklin TL, Uehara Y, Wilson CB. Lipopolysaccharide-induced cytokine production in human monocytes: role of tyrosine phosphorylation in transmembrane signal transduction. Eur J Immunol 1994; 24:1278–1284.
13. Geng Y, Zhang B, Lotz M. Protein tyrosine kinase activation is required for lipopolysaccharide induction of cytokines in human blood monocytes. J Immunol 1993; 151:6692–6700.
14. Shapira L, Takashiba S, Champagne C, Amar S, Van Dyke TE. Involvement of protein kinase C and protein tyrosine kinase in lipopolysaccharide-induced TNF-α and IL-1β production by human monocytes. J Immunol 1994; 153:1818–1824.
15. Weinstein SL, June CH, DeFranco AL. Lipopolysaccharide-induced protein tyrosine phosphorylation in human macrophages is mediated by CD14. J Immunol 1993; 151:3829–3838.
16. Dong Z, Qi X, Xie K, Fidler IJ. Protein tyrosine kinase inhibitors decrease induction of nitric oxide synthase activity in lipopolysaccharide-responsive and lipopolysaccharide-nonresponsive murine macrophages. J Immunol 1993; 151:2717–2724.
17. Novogrodsky A, Vanichkin A, Patya M, Gazit A, Osherov N, Levitzki A. Prevention of lipopolysaccharide-induced lethal toxicity by tyrosine kinase inhibitors. Science 1994; 264:1319–1322.
18. Crowley MT, Harmer SL, DeFranco AL. Activation-induced association of a 145kDa tyrosine-phosphorylated protein with Shc and Syk in B lymphocytes and macrophages. J Biol Chem 1996; 271:1145–1152.
19. Geng Y, Gulbins E, Altman A, Lotz M. Monocyte deactivation by interleukin 10 via inhibition of tyrosine kinase activity and the ras signaling pathway. Proc Natl Acad Sci USA 1994; 91:8602–8606.
20. Weinstein SL, Sanghera JS, Lemke K, DeFranco AL, Pelech SL. Bacterial lipopolysaccharide induces tyrosine phosphorylation and activation of mitogen-activated protein kinases in macrophages. J Biol Chem 1992; 267:14955–14962.
21. Han J, Lee JD, Bibbs L, Ulevitch RJ. A MAP kinase targeted by endotoxin and hyperosmolarity in mammalian cells. Science 1994; 265:808–811.
22. Hambleton J, Weinstein SL, Lem L, DeFranco AL. Activation of c-Jun N-terminal kinase in bacterial lipopolysaccharide-stimulated macrophages. Proc Natl Acad Sci USA 1996; 93:2774–2778.
23. Stephanova I, Corcoran ML, Horak EM, Wahl LM, Bolen JB, Horak ID. Lipopolysaccharide induces activation of CD14-associated protein tyrosine kinase p53/56^{lyn}. J Biol Chem 1993; 268:20725–20728.
24. Meng F, Lowell C. Lipopolysaccharide (LPS)-induced macrophage activation and signal transduction in the absence of Src-family kinases Hck, Fgr and Lyn. J Exp Med 1997; 185:1661–1670.
25. Crowley MT, Costello P, Turner M, Meng F, Lowell C, Tybulewicz V, DeFranco AL. A critical role for Syk in signal transduction and phagocytosis mediated by Fcγ receptors on macrophages. J Exp Med 1997; 186:1027–1039.
26. Karin M, Hunter T. Transcriptional control by protein phosphorylation: signal transmission from the cell surface to the nucleus. Curr Biol 1995; 5:747–757.
27. Treisman R. Regulation of transcription by MAP kinase cascades. Curr Opin Cell Biol 1996; 8:205–215.
28. Ding A, Sanchez E, Nathan CF. Taxol shares the ability of bacterial lipopolysaccharide to induce tyrosine phosphorylation of microtubule-associated protein kinase. J Immunol 1993; 151:5596–5602.
29. Dong Z, Qi X, Fidler IJ. Tyrosine phosphorylation of mitogen-activated protein kinases is necessary for activation of murine macrophages by natural and synthetic bacterial products. J Exp Med 1993; 177:1071–1077.
30. Sanghera JS, Weinstein SL, Aluwalia M, Girn J, Pelech SL. Activation of multiple proline-directed kinases by bacterial lipopolysaccharide in murine macrophages. J Immunol 1996; 156:4457–4465.
31. Han J, Jiang Y, Li Z, Kravchenko VV, Ulevitch RJ. Activation of the transcription factor MEF2C by the MAP kinase p38 in inflammation. Nature 1997; 386:296–299.
32. Rouse J, Cohen P, Trigon S, Morange M, Alonso-Llamazares A, Zamanillo D, Hunt T, Nebreda AR. A novel kinase cascade triggered by stress and heat shock that stimulates MAPKAP kinase-2 and phosphorylation of the small heat shock proteins. Cell 1994; 78:1027–1037.
33. Geppert TD, Whitehurst CE, Thompson P, Beutler B. Lipopolysaccharide signals activation of tumor necrosis factor biosynthesis through the ras/raf-1/MEK/MAPK pathway. Mol Med 1994; 1:93–103.
34. Buscher D, Hipskind RA, Krautwald S, Reimann T, Baccarini M. Ras-dependent and -independent pathways target the mitogen-activated protein kinase network in macrophages. Mol Cell Biol 1995; 15:466–475.
35. Blumer KJ, Johnson GL. Diversity in function and regulation of MAP kinase pathways. TIBS 1994; 19:236–239.
36. Morrison DK, Cutler RE. The complexity of Raf-1 regulation. Curr Opin Cell Biol 1997; 9:174–179.
37. Hou XS, Chou T-B, Melnick MB, Perrimon N. The torso receptor tyrosine kinase can activate Raf in a Ras-independent pathway. Cell 1995; 81:63–71.

38. Samuels ML, Weber MJ, Bishop JM, McMahon M. Conditional transformation of cells and rapid activation of the mitogen-activated protein kinase cascade by an estradiol-dependent human Raf-1 protein kinase. Mol Cell Biol 1993; 13:6241–6252.
39. Hambleton J, McMahon M, DeFranco A. Activation of raf-1 and mitogen-activated protein kinase in murine macrophages partially mimics lipopolysaccharide-induced signaling events. J Exp Med 1995; 182:147–154.
40. Sun H, Charles CH, Lau LF, Tonks NK. MKP-1 (3CH134), an immediate early gene product, is a dual specificity phosphatase that dephosphorylates MAP kinase in vivo. Cell 1993; 75:487–493.
41. Collart MA, Baeuerle P, Vassalli P. Regulation of tumor necrosis factor alpha transcription in macrophages: involvement of four kappa B-like motifs and of constitutive and inducible forms of NF-kappa B. Mol Cell Biol 1990; 10:1498–1506.
42. Shakhov AN, Collart MA, Vassalli P, Nedospasov SA, Jongeneel CV. Kappa B-type enhancers are involved in lipopolysaccharide-mediated transcriptional activation of the tumor necrosis factor alpha gene in primary macrophages. J Exp Med 1990; 171:35–47.
43. Alessi SR, Cuenda A, Cohen P, Dudley DT, Saltiel AR. PD098059 is a specific inhibitor of the activation of mitogen-activated protein kinase kinase in vitro and in vivo. J Biol Chem 1995; 270:27489–27494.
44. Han J, Brown T, Beutler B. Endotoxin-responsive sequences control cachectin/tumor necrosis factor biosynthesis at the translational level. J Exp Med 1990; 171:465–475.
45. Lee JC, Laydon JT, McDonnell PC, Gallagher TF, Kumar S, Green D, McNulty D, Blumenthal MJ, Heys JR, Landvatter SW, Strickler JE, McLaughlin MM, Slemens IR, Fisher SM, Livi GP, White JR, Adams JL, Young PR. A protein kinase involved in the regulation of inflammatory cytokine biosynthesis. Nature 1994; 372:739–746.
46. Swantek JL, Cobb MH, Geppert TD. Jun N-terminal kinase/stress-activated protein kinase (JNK/SAPK) is required for lipopolysaccharide stimulation of tumor necrosis factor alpha (TNF-α) translation: glucocorticoids inhibit TNF-α translation by blocking JNK/SAPK. Mol Cell Biol 1997; 17:6274–6282.
47. Sweet MJ, Hume DA. Endotoxin signal transduction in macrophages. J Leuk Biol 1996; 60:8–26.
48. Stancovski I, Baltimore D. NF-κB activation: The IκB kinase revealed? Cell 1997; 91:299–302.
49. Joseph CK, Wright SD, Bornmann WG, Randolph JT, Kumar ER, Bittman R, Liu J, Kolesnick RN. Bacterial lipopolysaccharide has structural similarity to ceramide and stimulates ceramide-activated protein kinase in myeloid cells. J Biol Chem 1994; 269:17606–17610.
50. Wright SD, Jong MTC. Adhesion-promoting receptors on human macrophages recognize *Escherichia coli* by binding to lipopolysaccharide. J Exp Med 1986; 164:1876–1888.
51. Ingalls RR, Golenbock DT. CD11c/CD18, a transmembrane signaling receptor for lipopolysaccharide. J Exp Med 1995; 181:1473–1479.
51a. MacKichan ML, DeFranco AL. J Biol Chem 1999; 274:1767–1775.
52. Barber SA, Detore G, McNally R, Vogel SN. Stimulation of the ceramide pathway partially mimics lipopolysaccharide-induced responses in murine peritoneal macrophages. Infect Immun 1996; 64:3397–3400.
53. Proud CG. p70 S6 kinase: an enigma with variations. Trends Biochem Sci 1996; 21:181–185.
54. Franke TF, Kaplan DR, Cantley LC. PI3K: downstream AKTion blocks apoptosis. Cell 1997; 88:435–437.
55. Hemmings BA. Akt signaling: linking membrane events to life and death decisions. Science 1997; 275:628–630.
56. Grammer TC, Cheatham L, Chou MM, Blenis J. The p70S6K signalling pathway: a novel signalling system involved in growth regulation. Cancer Surv 1996; 27:271–292.
57. Herrera-Velit P, Reiner NE. Bacterial lipopolysaccharide induces the association and coordinate activation of p53/56lyn and phosphatidylinositol 3-kinase in human monocytes. J Immunol 1996; 156:1157–1165.
57a. Franke TF, Kaplan DR, Cantley LC, Toker A. Science 1997; 275:665–668.
57b. Klippel A, Kavanaugh WM, Pot D, Williams LT. Mol Cell Biol 1997; 17:338–344.
58. Damen JE, Liu L, Rosten P, Humphries RK, Jefferson AB, Majerus PW, Krystal G. The 145-kDa protein induced to associate with Shc by multiple cytokines is an inositol tetraphosphate and phosphatidylinositol 3,4,5-triphosphate 5-phosphatase. Proc Natl Acad Sci USA 1996; 93:1689–1693.
59. Kavanaugh WM, Pot DA, Chin SM, Deuter-Reinhard J, Jefferson AB, Norris FA, Masiarz FR, Cousens LS, Majerus PW, Williams LT. Multiple forms of an inositol polyphosphate 5-phosphatase form signaling complexes with Shc and Grb2. Curr Biol 1996; 6:438–445.
60. Lioubin MN, Algate PA, Tsai S, Carlberg K, Aebersold RA, Rohrschneider LR. p150Ship, a signal transduction molecule with inositol polyphosphate-5-phosphatase activity. Genes Dev 1996; 10:1084–1095.

30

Endotoxin Effects on Synthesis of Phosphatidic Acid and Phosphatidic Acid–Derived Diacylglyceride Species

Stuart L. Bursten
Cell Therapeutics, Inc., Seattle, Washington

INTRODUCTION

The role of lipids, such as phosphatidylinositides (PI), diacylglycerides (DG), and phosphatidic acid (PA) species in both intracellular homeostatic and inflammatory signaling has been well established for over a decade (reviewed in Refs. 1,2), although initial evidence for the significance of PA-related signaling can be found in research from the period 1950–1955. Given that the recent reviews cited have extensively discussed mechanisms and dynamics of generalized lipid signaling in the context of PA, this chapter will center on aspects of PA and PA-related DG synthesis commensurate with inflammatory signaling that has been associated with endotoxin. The initial discussion will briefly summarize current understanding of PA and connected DG signaling, with analysis of their complex interconnections, and will include a section on the membrane effects of these lipids, which are related to some of their numerous biological activities.

PHOSPHATIDIC ACID SIGNALING AND RELATED FUNCTIONS

Calcium Currents, Small G Proteins, and Auto-Amplification

PA is an anionic phospholipid present in eukaryotic cells at concentrations ranging from 0.5 to 3% of total phospholipid (PL) mass. PA, along with DG, is a fundamental precursor of the storage lipid triacylglyceride (TG) and of all phospholipids, and hence both its de novo and phospholipid-based (i.e., PC-PLD–determined) syntheses are tightly regulated and its normal biological half-life is short (varying by cell type, but averaging less than 5 min). Both endogenous increases in PA mass and exogenous stimulation by physiological concentrations of PA (≤ 10–20 μM), which thus reflect probable induction of inflammatory or related signaling pathways, have been associated in many cell types with stimulation of calcium currents, probably through (1) PA activation of PI metabolism [i.e., release of inositol trisphosphate (IP_3) by phospholipase C (PLC); see below], (2) direct PA interaction with plasma membrane calcium channels, and the possible conversion of PA to lyso-PA (both intracellularly and on the external leaflet of the plasma membrane) by secretory type IIA phospholipase A_2 (PLA_2), and (3) conversion of PA to DG by phosphatidate phosphatase (PAP) (1–11). Lyso-PA in turn acts through a G-protein–linked seven-transmembrane receptor to affect both calcium and chloride currents (2,8,12). In some cell types, particularly oocytes, some evidence is available for PA receptors that also modulate or stimulate external calcium currents, although cloning efforts for these receptors have thus far been unsuccessful (1,2,13). Resulting increases in calcium-associated signaling through such transduction agents as protein kinase C (PKC), calmodulin, and calmodulin-associated kinases have thus also been associated with increases in PA mass or external PA administration. Conversion of PA to DG by PAP, in parallel with DG production by PLC isoforms β and γ, also has been associated with activation of PKC isoforms, particularly the atypical PKC isoforms (1,2).

Both lyso-PA and exogenous PA also act to stimulate endogenous cellular PA levels, possibly through small G-protein stimulation of phosphatidylcholine (PC)-directed and phosphatidylethanolamine (PE)-directed phospholipases D (PLD). Small *ras*-related G-proteins linked to lyso-PA receptor stimulation include rho, arf1, arf6, and possibly cdc42, which have been implicated to different degrees in activation of intracellular PC-PLD (2,14–19). In addition, administration of PA to microsomal systems from rat (RMC) and human mesangial cells (HMC) results in stimulation of lyso-PA acyl-CoA:acyltransferase (LPAAT), with resulting increases in specific *sn*-2–unsaturated PA species; exogenous administration of PA in these systems to intact cells also results in stimulation of linoleate and arachidonate uptake into cellular PA, with a corresponding increase in PA mass suggestive of intracellular activation of LPAAT (20–23). The amplification effect of exogenous PA is paralleled by exogenous administration of lipid A to RMC and HMC, which results in increases in PA and apparent stimulation of LPAAT activity (discussed in detail later). Tandem activation of LPAAT and PLD has also been described (24), suggesting generation of PA from (most probably) PC and possibly PE, followed by *sn*-2 remodeling through the sequential activity of PLA2 and LPAAT. Thus, it is probable that exogenous PA acts to modulate calcium currents through both native PA receptors and conversion to lyso-PA and possibly DG; in addition, PA acts as an auto-stimulant or amplification factor to its own intracellular synthesis both through small G-protein stimulation of PLDs and through stimulation of LPAAT, the modulation of which will be discussed below, as well as apparent interactions of LPAAT with endotoxin. Little to no evidence exists to support the original idea of PA as a direct calcium ionophore (1,2).

PA Interactions with Cellular Signaling Proteins

Endogenous PA levels have pronounced effects upon known intracellular signaling proteins over and above their effects on cellular calcium currents. Some of these effects may be relatively nonspecific in that they are associated with anionic membrane PL such as phosphatidylserine (PS) and PI as well as PA. For example, the binding of annexins, particularly annexin II, to a membrane is associated with high concentrations of anionic PL, but there seems to be no specific requirement for PA, as opposed to PS (25). Similarly, binding of src family kinases to the inner leaflet of the plasma membrane is associated with both the requirement for N-terminal myristoylation and a high concentration of anionic PL required to bind the lysine and arginine residues located near the N-terminus (26,27). Although PA is a possible contributor in these interactions, it is not exclusively required.

In contrast, it now seems apparent that PA specifically enhances the activation of some PLC isoforms, particularly PLCγ associated with tyrosine-kinase receptor signaling, through changes in PLC conformation, possibly guaranteeing exposure of src homology type 2 (SH2) and pleckstrin homology sites for complex protein binding and hence activity recruitment, SH2 binding to the receptor, or signaling (1,2,28). Thus, membrane synthesis of PA from PC may have immediate stimulatory or permissive consequences in terms of signal transduction related to PI signaling (this should also be viewed as another possible locus in which PA may affect calcium currents, i.e., by regulating PLCγ-related synthesis of inositol trisphosphate [IP3], which in turn regulates IP3-sensitive Ca^{2+} channels in endoplasmic reticulum). Similarly, evidence has accumulated recently, suggesting the existence of a PL-activated protein kinase that is most sensitive to PA (1,29). Whether this represents a direct signal transduction kinase activated by PA or simply a membrane-associated kinase that has PL requirements remains unclear.

At the next level of specificity of protein-lipid interaction, a distinct binding site for PA has been identified on the raf kinase, which itself interacts with the small Gp ras in activation of the MAP kinase pathway (30). Previous data had implicated PA and/or DG both upstream and downstream of the small Gp ras in the MAP kinase signaling pathway (31,32). Taken together, these data imply that a number of specific intracellular signaling pathways involved in mitogenesis and inflammation have specific checkpoints such as PLCγ and raf (and, more questionably, ras itself), where PA may be necessary when membrane association is required of a signaling protein. This suggests that observations implicating PA as a necessary but not sufficient factor in both mitogenesis and inflammation may be correct (1,2,33–35).

Of similar interest and import are the requirements of PA synthesis for assembly and activation of the NADPH–mixed function oxidase (NADPH-ox) in the respiratory burst of the neutrophil (1–3,36–40). In this complex system, evidence has again accumulated that

PA is a necessary but not sufficient component for activation of a family of membrane-proximal protein machinery; in addition, a small Gp is again involved in protein recruitment to the membrane: in this case, rac. Rac promotes the association of the two cytoplasmic components of the NADPH-ox, p47^{phox} and p67^{phox}, with the membrane component cytochrome b_{558}, leading to activation of the NADPH-ox enzyme; the stability of this association of the components with each other and the neutrophil membrane is apparently enhanced and stabilized by the synthesis of PA by hydrolysis of PC by PLD, synthesis of diglycerides, most probably by activation of PLCβ, and release of arachidonic acid (AA) by PLA2. AA also enhances the proton pump function of the membrane component of NADPH-ox, leading to augmented activity of the respiratory burst (36–41). It is noteworthy that several investigators have identified the necessity for formation of peroxidized lipids, as opposed to H_2O_2, in mitogenic signaling and in inflammatory signaling related to interleukin-1β (IL-1β) and tumor necrosis factor-α (TNF-α); this has been associated with the generation of reactive oxygen species by mixed-function oxidases (42). These data suggest a further interplay between generation of PA and oxidation of polyunsaturated acyl chains, such as linoleate (C18:2;ω-6) and arachidonate (C20:4;ω-6), in the plasma membrane. Recently, evidence has accumulated for the presence of a separate PA-dependent protein kinase involved in activation as well as translocation of NADPH-ox p47-67^{phox} subunits (43).

In summary, it is apparent that PA plays an important role in cooperativity of conformational shifts and stabilization of membrane function of enzymes involved in signaling through association with tyrosine kinase receptors, SH2 regions, and translocation to the membrane. A general rule appears to be stabilization by PA of multiple protein complexes contiguous to the plasma membrane inner leaflet related to generation of reactive oxygen species (NADPH-ox) and mitogenesis (raf). A specific interaction between small G proteins in the ras family and PA may have evolved in each of these protein complexes to promote membrane localization analogous to the functions of SH2 domains and protein lattices in tyrosine kinase receptors. PA may further modulate mitogenesis through activation of mixed-function oxidase enzymes and generation of peroxidized lipids from unsaturated intramembrane acyl chains. These peroxidized lipids may have effects on membrane proteins, including ion channels, or may be released by phospholipases A_2 and have their own signaling effects (1,44).

Biophysical Effects of PA on Membrane Changes

A noteworthy effect of PA on membrane properties is its induction of increased fusogenicity. The transformation of conical phospholipids (e.g., lyso-PA) to cylindrical phospholipids, such as PA (e.g., by LPAAT), may result in increased susceptibility of adjacent membranes to fusion. These observations have led to the postulated importance of PLD in promoting the evolution of the Golgi apparatus and the production of exocytotic granules. The finding of differently activated and inhibited PLD isoforms has further supported the concept of partial regulation of membrane fusion by PA (1,2,45). In addition, polyglycerophosphate lipids related to PA in structure, such as lyso-bis-phosphatidic acid (LBPA), have also been implicated in the formation and maintenance of the complex internal structure of endosomes (46).

It is therefore important to note that different acyl side-chain compositions of PA subspecies may also make important contributions to biophysical membrane effects, such as fusion, as well as activation of specific signaling proteins such as PLCγ and raf. The degree of membrane anisotropy induced by particular unsaturated side chains, such as linoleate (C18:2;ω-6), α-linolenate (C18:3;ω-3), and arachidonate (C20:4ω-6), may be highly variable and critical to functions of local intrinsic membrane proteins and association of extrinsic proteins with the membrane. For example, both *sn*-2–incorporated linoleate and α-linolenate cause conspicuous decreases in membrane-order parameters due to leaflet disruption as determined by both fluorescent and electron spin paramagnetic probes, whereas arachidonate, despite similarities to α-linolenate, may actually cause an increase in order parameters. The imposition of isotropy and transition to the gel phase from the liquid crystalline phase in local membrane substructures by arachidonate or by saturated acyl side chains such as palmitate (C16:0) or myristate (C14:0) may in turn cause dysfunction and even failure in membrane-associated proteins. Thus, modifications in PA species through *sn*-2 remodeling by PLA2 and LPAAT working in tandem may serve to restore membrane structure and hence enzyme integrity (20–23).

A signal difference between PA and DG in membrane biophysical functions is the relative quickness with which the uncharged and nonpolar DG "flip-flops" between membrane leaflets compared to the highly point-charged PA molecule. In addition, evidence exists suggesting that DG molecules may be rapidly rephosphorylated to PA by DG kinase isoforms on the inner leaflet of the membrane (1,2). Thus, the rapid interconversions of DG and PA on opposite leaflets of the plasma membrane may serve to stabilize charge density on the inner leaflet (where DG kinase is located and concentrations of PA are higher), while preventing accumulation of PA on the outer leaflet, where it may serve as a signaling molecule with specific PA receptors or be converted to lyso-PA by PLA2 and activate lyso-PA receptors.

The biochemical economy of PA synthesis and catabolism may thus be related to both the external signaling and the internal membrane biophysical dynamics of the PA molecule. In addition, acyl chain economy as determined by the acyl chain composition of the plasma membrane and the free fatty acid/acyl-CoA/acyl-CoA:acyl-CoA binding protein ratios of the cell may also determine the biochemical dynamics (by acting as regulatory elements to LPAAT and hence de novo and Lands cycle PA synthesis) and biophysical dynamics of PA signaling and biophysical interaction with the plasma membrane. The alteration of these normally balanced dynamic equilibria by the insertion of lipid A and/or LPS into the plasma membrane may thus be viewed as a point at which dynamic changes in membrane equilibria and function may occur (see Fig. 1). In concrete terms, reacylation of lyso-PA by lipid A–stimulated LPAAT activity utilizing unsaturated acyl-CoA molecules, which will significantly depress the order parameter (e.g., C18:2;ω-6 and C18:3;ω-3), may have a dramatic impact on intrinsic membrane protein function; production of PA on the inner leaflet of the plasma membrane through stimulation of PLD may alter charge dynamics of membrane-associated proteins as described above and cause activation of such proteins.

PHOSPHATIDIC ACID SIGNALING INDUCED BY LIPID A

Mechanisms by which bacterial lipid A moieties alter the functional properties of cells remain incompletely understood. Lipid A readily inserts into cellular membranes due to hydrophobic interactions between the myristoyl (C14:0), lauroyl (C12:0), and palmitoyl (C16:0) residues attached to the sugar groups, and this event may induce alterations in the physicochemical properties of membranes to initiate cellular activation (47). Evidence also exists for specific LPS-receptor proteins in splenic lymphocytes to which lipid A binds (48,49). In addition, a serum LPS/lipid A–binding protein that attaches to and activates macrophages (poten-

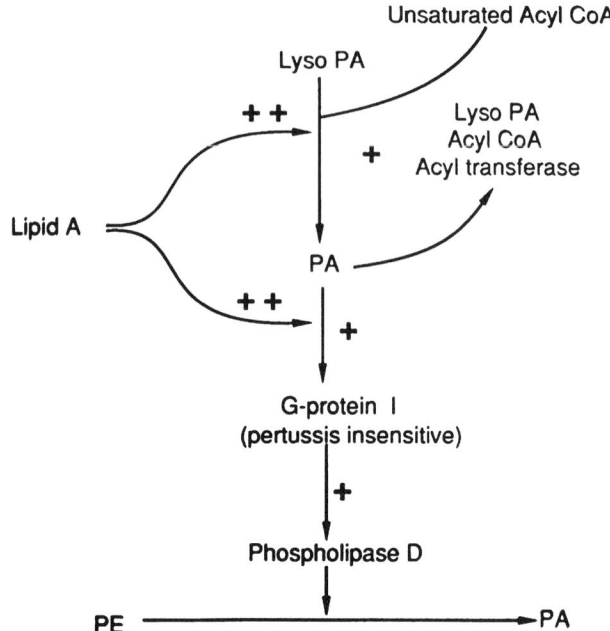

Fig. 1 Schematic diagram of possible effects of lipid A on PA metabolism in GMC and other selected cell types. Exogenous administration of lipid A to whole cells and to isolated plasma membrane–enriched microsomes results in enhanced conversion of lyso-PA to PA by lyso-PA:acyl-CoA acyltransferase (LPAAT) activity. As is characteristic of lysophosphatide:acyl-CoA acyltransferase enzymes in general, the production of the end product (in this case, PA) results in further enhancement of LPAAT activity. Production of PA in the presence of lipid A/LPS is further enhanced by apparent PA stimulation of a small *ras*-related G protein (evidence at this time indicates that this is probably *rho* and other *rho* family members such as *cdc42*, which induce filopodia and lamellipodia formation). This small G protein induces activation of PE-directed PLD in GMC and smooth muscle cells and induces activation of conventional hPLD1 (PC-directed PLD1) in other cells, including HL60 cells (1–23).

tiating TNF synthesis and release) through the CD14 molecule is well described (50–52) and is discussed elsewhere in this volume. Considering the broad range of biochemical responses induced by LPS or lipid A, it is increasingly clear that each membrane event previously described plays a separate but key role in total biological response and cellular activation.

The monosaccharyl form of lipid A (lipid X: 2,3-diacylglucosamine 1-phosphate) has previously been shown to bear a close formal structural resemblance to PA (20–23,53–56). Given the structural similarities, we postulated that lipid A might induce cellular activation through one set of mechanisms by functioning as a biological mimetic of PA. The structural and functional mimicry of biologically active lipids by lipid A may partially account for the diverse range of cellular responses ascribed to this pathophysiologically important substance. It is important to note that lipid A resembles a number of other putative lipid mediators or structural moieties in addition to PA, including glycosylated phosphatidylinositol (PI), particularly species of the latter containing acylated myoinositol. Energy-minimized three-dimensional representations of lipid A demonstrate how it may mimic both PA molecules containing saturated acyl chains and myristoyl-containing glycosylated PI (53,55,56). The forward portion of the lipid A molecule containing one glucosamine ring with a phosphate and a myristoyl/beta-hydroxy-myristoyl linkage demonstrates the similar orientation of the carbonyl oxygen atoms to the phosphate oxygen atoms (20,53,55,56). The mirror-image orientation of the glycosylated PI molecule to lipid A shows a similar orientation of glucosamine nitrogens and the terminal nonglucosaminyl acyl chain of the lipid A to the glucosamine and the 2'-inosityl acyl chain of the glycosylated PI. The effects of lipid A on membranes may be viewed as a summation of direct mimicry of glycosylated PI and PA due to acyl chain orientation and point-charge orientation, respectively; the impact of saturated acyl chains on the order parameter of the outer leaflet may also be a significant contributor to the effects of lipid A reviewed below.

Fluorescence and Mass Spectrometric Studies of Lipid A Effects on Glomerular Mesangial Cells

Glomerular mesangial cell (GMC) response to lipid A was first studied using uptake of cis-parinaric acid (9,11,13,15-all-cis-octadecatetraenoic acid, or cis-PnA) in whole cells. Cis-PnA is a naturally fluorescent fatty acid, which is used as a substrate by acyltransferases and transacylases in isolated microsomal and whole-cell plasma membranes (20–23). As transacylation of the acyl chain results in transference to an environment that is much less polar, fluorescence is enhanced. In addition, chain translational constraint of cis-PnA results in an increase in polarization of fluorescence (20–23). Incubation of 4 μM cis-PnA with GMC microsomes enriched in plasma membranes results in a steady increase in polarization of fluorescence over the first 3–4 minutes of incubation, followed by stabilization of the rate, reflecting a saturable system (20–23). Removal of cofactors for CoA ligase or acyltransferase activities, including Mg^{2+}, coenzyme CoA, or ATP, or heating at 60°C for 30 minutes results in abrogation of the increase in polarization of fluorescence, reflecting dependence of uptake on active, enzyme-mediated processes. These processes are stimulated significantly by addition of 50–100 ng/ml lipid A. Following addition of lipid A, the initial velocity of the reaction, as well as the total cis-PnA uptake into the membrane, is stimulated. The observation that the initial velocity of the reaction is accelerated by lipid A is reinforced by closer analysis of initial rates of uptake, which demonstrates stimulation of cis-PnA uptake as early as 5 seconds after addition of lipid A and an overall rate increase of >50%.

The target of acylation in the cell membrane was studied using high-performance liquid chromatography (HPLC) separation of lipids extracted from whole GMC incubated with lipid A and cis-PnA for 30 seconds with fluorescence analysis in series to detect covalently labeled fluorescent (cis-PnA–labeled) lipids. We found that several separable PA peaks were highly labeled, with little label entering other lipid fractions at rapidly determined time points (5–30 sec). In addition, certain polyglycerophosphate congeners, synthesized directly from PA through cytosine-diphosphate diacylglycerol synthase (CDP-DG synthase), such as phosphatidylglycerol (PG), were also labeled with cis-PnA. DG fractions with acyl content similar to that of the cis-PnA–labeled PA appeared later, around 2–5 minutes after stimulation of GMC with lipid A. The DG fractions were shown by exhaustive mass spectrometric analysis to originate from the labeled PA (20–23).

The PA peak isolated from GMC whole-cell membranes was collected and analyzed by fast-atom bombardment–mass spectrometry (FAB-MS). FAB-MS negative ion spectrometry (FAB-NI) verified that this peak was PA. The peak was subsequently shown to have two separable components. The first, which appeared within 5 seconds of lipid A stimulation of GMC and oscillated in concentration during the first 5 minutes after stimulation with a period of 45 seconds to 1 minute, was derived from the activity of LPAAT, as suggested from the cis-PnA data. This PA was characterized by masses (FAB NI $[M-H]^-/z$) in the 695–701 range. These LPAAT-derived PA species were shown by radioactive labeling, linked scan (tandem mass spectrometry: MS/MS), and gas-liquid chromatographic (GLC) analysis of acyl chains to be highly enriched in the fatty acid linoleate (C18:2;ω-6) in both the sn-1 and sn-2 positions (20–23), a somewhat unusual finding. Further verification for the provenance of this PA species was provided by fatty acyl competition experiments, in which incubation with a variety of unsaturated fatty acids, particularly linoleate and linolenate (C18:3;ω-3), inhibited both fluorescence uptake and formation of this specific PA species, strongly supporting the observation that this PA resulted from the activity of LPAAT [order of inhibitory strength: C18:2 ω-6 ≃ C18:3 ω-3 > C20:3 ω-3 = C20:5 ω-3 >> C20:4 (arachidonate) > C18:1 (oleate) >>>> palmitate (C16:0) or stearate (C18:0) (20-23)]. In analyzing the acyl content of membrane species, evidence was provided that the DG species appearing rapidly after the formation of linoleoyl PA species was synthesized directly from the LPAAT-dependent PA (i.e., was equally enriched in linoleate and linolenate) rather than from PI species (which are enriched in C20:4, arachidonate, and C20:3, eicosatrienoate). Mass spectrometric studies confirmed fluorescence and labeling data by showing a predominance of predicted dilinoleoyl, 1-oleoyl 2-linoleoyl, and dioleoyl DG generated in whole GMC and GMC plasma membrane–enriched microsomes (21). This indicated that lipid A also could be activating the enzyme phosphatidate phosphohydrolase (PAP) (recently considered to be a lipid phosphate phosphatase, addressing sphingosine 1-phosphate as well as PA), which was subsequently demonstrated by kinetic studies in GMC microsomes. It remains unclear as to whether PAP activation by lipid A occurs (1) due to direct interaction with the membrane, resulting in G-protein–mediated stimulation of the PAP, (2) due to direct interaction with PAP by lipid A, or (3) due to the increased synthesis of PA providing increased substrate for PAP.

The second PA species observed in this peak was synthesized later in GMC whole cell membranes or microsomal membranes after stimulation with lipid A (>60–90 sec). It was derived from the activity of a phospholipase D (PLD) apparently directed against phosphatidylethanolamine (PE) and was characterized by masses (FAB NI [M-H]-/z) in the 705–737 range, enriched in 1-alkyl/alkenyl (sn-1 ether and vinyl ether) species, and had significant C20 and C22 unsaturated species content (21). The provenance of this PA species was verified by the formation of phosphatidylethanol when GMC were stimulated with lipid A in the presence of ethanol (the transphosphatidylation reaction characteristic of PLD in the presence of C2-C4 primary alcohols (see Refs. 20–23). These PA species were subsequently also converted to diradylglycerols, indicating that PAP activation was persistent for at least the observed time course. In summary, lipid A stimulation of GMC resulted in the activation of several mechanisms for the synthesis of PA species, including the enzymes LPAAT and PE-directed PLD, to which the primary response, occurring very rapidly, was activation of LPAAT. Preliminary evidence indicates that the unsaturated PA produced by LPAAT may act as an amplification factor by stimulating PE-PLD but that this may also occur through direct stimulation of PE-PLD by lipid A. PA species were subsequently converted into DG and/or diradylglycerols by the action of PAP. It remains unclear in biological systems whether these diradylglycerol species have specific biological effects or if this represents simply catabolism (or deactivation) of PA. These effects of lipid A are summarized in Figure 1.

Structural Considerations in the Similarity Between Lipid A and Phosphatidic Acid

Lipid X resembles PA in that both are phosphomonoesters and can assume planar conformations facilitating membrane insertion. The respective 2-acyl carbonyl groups bear similar relationships to the charged phosphate groups, and the constraints imposed on acyl carbonyl movement would be expected to be similar (20–23). Lyso-phosphatide acyltransferase enzymes demonstrate end-product stimulation of activity (20–23,57,58), which is dependent upon incorporation of unsaturated fatty acids into acceptor lyso-phospholipids and may be the result of membrane fluidity changes affecting acyltransferase activity, as discussed above (20–23,57,58). Specific structural requirements for feedback stimulation of LPAAT were found to be present in lipid X or PA species. Removal of the phosphate moiety from either of these species resulted in loss of acyltransferase stimulation (20). Other researchers have reported that structural similarities between lipid A and PA may be responsible for monocyte activation (59,60). In monocytes, metabolic dephosphorylation of lipid A results in loss of biological activity. Nishijima and colleagues (59) demonstrated that lipid A dephosphorylation is specifically associated with loss of effects on glycerolipid metabolism. These studies further suggest that lipid A has a profound effect on membrane metabolism and structure and can rapidly effect phenotypic changes in cells independent of specific receptor binding or gene induction. The similarities in effects of lipid A and PA on membrane-active enzymes may be explained by phosphate–acyl chain interactions with specific proteins, which, in the case of cells such as GMC, are not specific receptor–lipid A interactions (20).

In view of the discussion above, in which increasing evidence suggests that PA may have membrane fusogenic and fluidity-modulating effects and, in addition, may effect perimembrane enzymes through possible point charge effects (e.g., NADPH oxidase, raf), a working hypothesis may be generated as follows: lipid A insertion into membranes, energetically favored by the dependent acyl chains in a planar conformation, may cause an increase in order parameter of the external leaflet of the plasma membrane due to the saturated nature of the dependent acyl chains. Due either to the increase in order parameter, which would act to decrease or abrogate the activity of intramembrane proteins and ion channels, or its similarity to the structure of PA, or both, insertion into the membrane causes activation of LPAAT. LPAAT may be predicted (and was observed) to produce PA in an oscillating fashion, as it uses up available lyso-PA substrate, which is regenerated through the action of membrane-associated external (secretory) PLA2, resulting in subsequent waves of PA synthesis. Production of PA in turn causes end-product stimulation of LPAAT, resulting in a positive feedback generation of more unsaturated PA. Catabolism of PA to DG by PAP, also apparently activated by lipid A, results in the possible flip-flop of unsaturated DG and either its rephosphorylation to PA (which may also account for wavelike and alternating formation of PA and DG) or its use in complexes with membrane-associated proteins. The net synthesis of PA on the external leaflet (and possibly its transport and reformation on the internal leaflet) results in stimulation of phospholipase D, with further synthesis of PA and diradylglycerol. Thus, amplification of PA synthesis induced by lipid A in membrane systems containing LPAAT and PAP may be expected and is consistent with current observations on PA effects on PLD (1,2). LPAAT stimulation by lipid A may be viewed as a natural subcellular response to membrane perturbation or, additionally, given more recent data, a component of the inflammatory response itself (see next section). Thus, formation of unsaturated PA may not only serve to restore membrane fluidity, and thus restore membrane protein function, but its reformation on the inner membrane may result in membrane association and activation of mixed-function oxidases, production of reactive oxygen species (ROS) and lipid peroxides, and resulting local inflammation. In addition, synthesis of both ROS and lipid peroxides results in signal transduction and gene induction of cytokines, including interleukin-1β (IL-1β) and TNF-α. Amplification of PA synthesis through PE-PLD may have a similar effect. In either case, these observed effects on membrane function should be considered in an analysis of lipid A effects on cellular pathways.

CLONING OF LPAAT AND TRANSFECTION INTO LPAAT-DEFICIENT *E. coli* AND MAMMALIAN CELLS

Cloning of Human LPAAT Enzymes and Insertion into LPAAT-Deficient *E. coli*

Sequence comparison of cDNA clones suggested the presence of four potential human LPAAT isoforms, two of which were cloned and designated as LPAAT-α and LPAAT-β (61). West et al. were able to clone both enzymes from a λzap human brain cDNA library and a pCMV.SPORT human leukocyte cDNA library, respectively. To verify that these enzymes possessed LPAAT activity, they were transfected into an *E. coli* strain, designated JC201, deficient in bacterial LPAAT activity due to a mutation in the plsC locus. This mutation leads to a temperature-sensitive phenotype that causes JC201 to grow slowly at 37°C and not at all at 44°C.

JC201 *E. coli* separately transfected with LPAAT-α and LPAAT-β vectors were observed to grow at both 42 and 44°C, whereas JC201 transfected with control vectors generated no colonies at these temperatures. To verify that specific phospholipid acyltransferase activity was present in transfected JC201 strains when appropriate vectors were present, all strains were labeled with ^{32}P, and total phospholipids were extracted from JC200, control JC201, and LPAAT-α– and -β–transformed JC201 *E. coli*. The results are seen in Figure 2, where the TLC-separated lipids are identified on the left side of the figure by intensity of ^{32}P and the quantitation by phosphorimager for the major lipid components on the right side of the figure. JC200, as expected, demonstrated a small amount of total labeled lipid in slow-migrating PA, which is relatively saturated (>80% myristate, C14:0, with the balance palmitate, C16:0; this is represented on the figure as dimyristoyl PA, C14:0/C14:0), with an even smaller amount in fast-migrating, unsaturated PA (>80% oleate, linoleate, and arachidonate) and an equally small amount of cellular lipid in lyso-PA. In contrast, JC201 strains deficient in LPAAT demonstrated a 65% reduction in both types of PA and a doubling in lyso-PA, as would be expected in an LPAAT-deficient strain. Transfection of both LPAAT-α and LPAAT-β into JC201 cells resulted in reversal of the normal JC201 lipid pattern, with some marked alterations in contrast to wild type. Slow-migrating, saturated PA species were reduced even further compared to nontransfected JC201, whereas there was a 13- to 14-fold increase in fast-migrating unsaturated PA species, with an expected reduction of 80–85% in lyso-PA. This strongly suggested

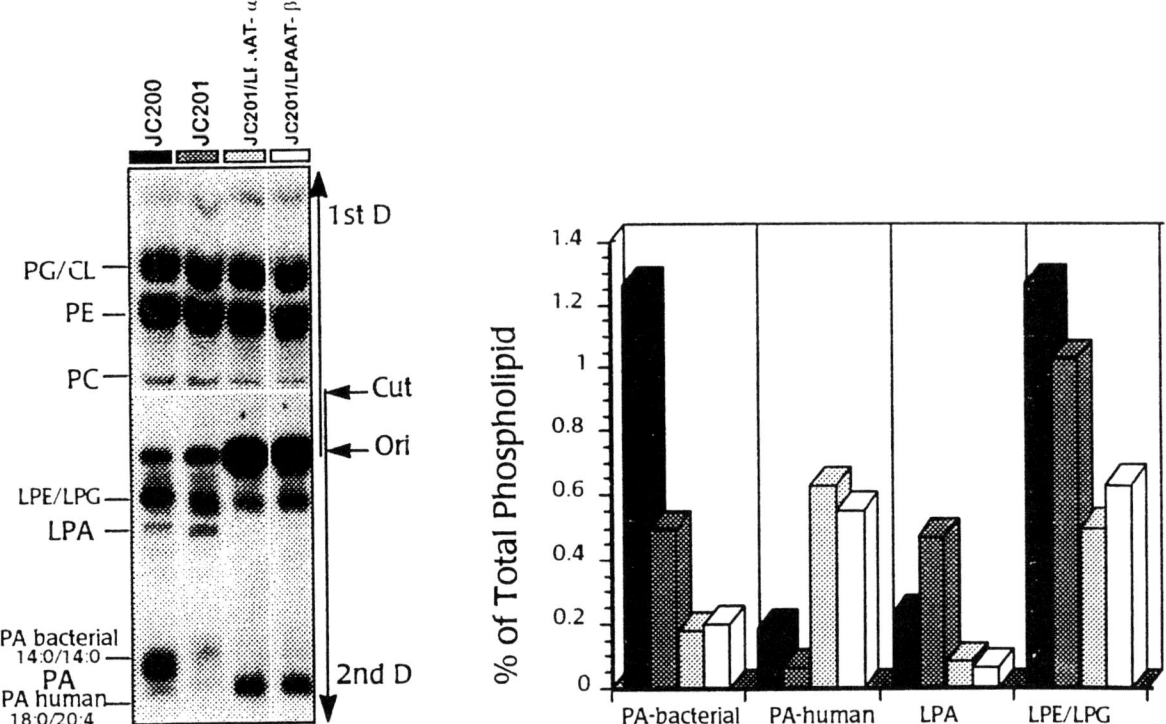

Fig. 2 Two-dimensional ("flip-flop") HPTLC of ^{32}P-labeled lipids from control, LPAAT-deficient, and LPAAT-α– and -β– transfected *E. coli*, with quantitation of lipids of interest (PA, lyso-PA, lyso-PE, lyso-PG) using phosphorimager. Lipids are run in the indicated first dimension with chloroform:methanol:NH$_4$OH as solvents (see Refs. 61,62 for details) to separate major phospholipids; the plate is then cut at the indicated location, and the lipids on the lower side of the plate, which migrated poorly or not at all in the ammonium hydroxide–containing gradient (lyso-phospholipids, including lyso-PA, PA, and LPS, which remains at the origin) are separated in a second, 180° dimension using chloroform:methanol:acetic acid (61,62). This gradient in turn not only separates each of the aforementioned phospholipids, but separates saturated and unsaturated subspecies of PA (as indicated here by PA bacterial, C14:0/C14:0, and PA human, C18:0/20:4). Changes in the major lipid species of concern are indicated in each type of *E. coli* by black/white/gray scale coding at the top of the HPTLC (JC201 black, JC201 dark gray, JC201/LPAAT-α light gray, and JC201/LPAAT-β white) and quantitation by phosphorimager as percent of total phospholipids shown in the histogram on the right side of the figure. In addition, note changes in the labeled phospholipid remaining at the origin, identified by FAB-NI and electrospray mass spectrometry as LPS. Quantitation of this lipid band normalized to JC200 is found in Table 1.

that transfection of both human LPAAT isoforms into JC201 mutant *E. coli*, while restoring PA synthesis and reducing lyso-PA accumulation consistent with predicted LPAAT activity, also changed the species acyl content of PA due to unsaturated acyl substrate specificity of both human isoforms. This further reinforced the observation that nonendogenous LPAAT activity was present after transfection in the JC201 strains (61,62).

Effect of Transfection on LPS Content of JC201 Strains

Labeling of total strain phospholipid content with ^{32}P resulted in the ability to analyze the effect of transfection on lipopolysaccharide in various strains. LPS was found to remain at the origin in the multi-one-dimensional HPTLC (63) as determined by electrospray mass spectrometry of both smooth and rough LPS standards from *Salmonella* and *E. coli*. In addition, the lipid bands remaining at the origin were also analyzed by FAB-MS and found to be identical to previously analyzed standard LPS (20–23). These bands were then analyzed using a phosphorimager as were the other lipids above. These results are seen in Table 1.

It is apparent that transfection with both available human isoforms of LPAAT induced a four- to fivefold increase in ^{32}P-labeled LPS mass. That this reflects genuine mass increase in LPS was verified by acyl chain analysis of the LPS band at the origin, using HPLC

Table 1 Relative *E. coli* Strain LPS Content (as determined by ^{32}P content) Normalized to Wild-Type (JC200) LPS

	JC200, wild type	JC201, nontransfected	JC201, LPAAT-α transfected	JC201, LPAAT-β transfected
LPS content	1.0	1.2	4.2	4.7

Relative concentration of LPS was confirmed by acyl chain hydrolysis, followed by quantitation of hydrolyzed free fatty acids by HPLC.

separation of hydrolyzed LPS acyl chains from the lipid A moiety (62,64). These findings showed a three- to fivefold increase in acyl chain mass, which correlated with the data from the ^{32}P-labeling of the lipid. These data gain suggest a complex interaction between the lipid A moiety and human LPAAT; it is difficult to attribute the increase in mass to direct acylation of lipid A by human LPAAT, especially in view of the evidence that the human LPAAT has a distinct substrate preference for polyunsaturated acyl chains with ≥18 carbons. In addition, specific bacterial enzymes exist for the synthesis of LPS that have definite preferences for short-chain unsaturated acyl moieties such as myristate and would be expected to address glucosamine residues with much greater efficiency than human LPAAT isoforms. It is thus more likely that a change in membrane properties occurring due to incorporation of unsaturated acyl chains (see Figure 2) has a tandem effect upon LPS-synthesizing enzymes or that the shift in acyl economy due to the transfection of mammalian LPAAT enzymes changes regulation of LPS synthesis. The processes involved may be viewed as the converse of LPS effects on the mammalian plasma membrane as monitored by LPAAT activity.

EFFECT OF LPAAT TRANSFECTION INTO MAMMALIAN CELLS

LPAAT Transfection into Lung Epithelial Cell Carcinoma A549 and ECV 304

A mammalian expression vector, pCE9, was used for the expression of LPAAT cDNA in mammalian cells. Stable transfections of pCE9.LPAAT-α or pCE9.LPAAT-β into lung epithelial adenocarcinoma A549 cells or endothelial ECV304 cells were achieved by digesting pCE9 vectors with BspH I before electroporating into these lines (61,65). In addition to Northern blot analysis to identify mRNA expression, A549 and ECV304 cells were recharacterized by a fluorescent TLC assay (61,66) using a lyso-PA substrate labeled either with the BODIPY fluorphore at the end of the *sn*-1 alkyl chain of lyso-PA or with nitrobenzoxadiazole (NBD) at the end of an *sn*-1 acyl chain (61). Virtually all cells stably transfected and expressing mRNA for LPAAT isoforms demonstrated 20- to 40-fold increases in fluorescent PA synthesis from either BODIPY-lyso-PA or NBD-lyso-PA compared to control cell lines transfected only with pCE9 vector. In addition, these cells produced little to no fluorescent monoacyl- or monoalkylglycerol (MAG/MA1kG), and when given either exogenous fluorescent MAG or MA1kG did not synthesize significant fluorescent DG (i.e., the cells had little or no MG acyltransferase or MRG acyltransferase). Finally, LPAAT-transfected cells given BODIPY-lyso-PC or NBD-lyso-PC in the reaction mix did not synthesize fluorescent PC. These latter experiments demonstrated both the stable transfection of an active LPAAT and that activity of the mammalian LPAAT is highly specific for lyso-PA and does not possess lyso-PC or MG acyltransferase activities.

LPAAT Overexpression and IL-1β–Induced Transcription of TNF-α and IL-6 mRNA in ECV304 Cells

Because LPAAT stimulation by LPS with subsequent synthesis of unsaturated PA was associated with cell-activation parameters such as shape change in GMC, we decided to examine whether LPAAT overexpression in ECV cells would alter a readily measurable activation parameter, cytokine transcription (61). IL-1β stimulation of endothelial cells is known to induce transcription of both TNF-α and IL-6. RNA was extracted from ECV304 cells transfected either with control pCE9 vector or with LPAAT-α expression plasmid at various times after IL-1β treatment, and Northern blot analyses were performed using ^{32}P-probes derived from TNF, IL-6, and EF-1α (elongation factor control) cDNAs. Results are seen in Table 2: EF-1α did not change under these conditions.

Table 2 TNF-α and IL-6 mRNA Transcription in Control and LPAAT-α–Transfected ECV304 Cells After IL-1β Stimulation

Cytokines	Vector					LPAAT-α, transfected				
	0	1°	4°	8°	24°	0	1°	4°	8°	24°
TNF-α	1.0	8.7	12	6.5	7.5	2.5	105	97	84	3.5
IL-6	1.0	22.5	15	7	9	3.5	195	225	270	245

Fold-increase in message detected by Northern blot normalized to time zero.

There was an 8- to 14-fold increase in TNF-α mRNA transcription in LPAAT-α–transfected ECV cells 1–4 hours after stimulation with IL-1 as compared to vector-transfected cells. This effect was greatly diminished 24 hours after stimulation. In contrast, IL-6 mRNA transcription, which was stimulated by a factor of 9–15 in LPAAT-transfected cells, was persistent after 24 hours (61). Also of interest was the observation that baseline mRNA levels (time zero) were increased in transfected cells (2.5- to 3.5-fold), suggesting that overexpression of LPAAT and the resulting PA synthesis may in and of itself promote either transcription or message stabilization. Recent studies have suggested that mRNA stability may be affected by lipids (67).

Cytokine Release in A549 and ECV304 Cells Transfected with LPAAT Expression Vectors

As well as increasing baseline and stimulated IL-6 and TNF-α mRNA transcription, LPAAT transfection increased cytokine release. Transfected A549 cells were incubated 16 hours in the absence or the presence of human IL-1β and murine TNF (Table 3). Transfected ECV304 cells were incubated in the absence or presence of IL-1β (Table 4). In contrast to induction of mRNA transcription at baseline in LPAAT-transfected cells, there was no enhancement of cytokine release. However, as with cytokine transcription, there was a >18-fold increase in IL-6 release and a >4-fold increase in TNF-α release from A549 cells.

There was again no baseline enhancement of cytokine release from ECV304 cells (Table 4). There was a >25-fold increase in IL-6 release in LPAAT-α–transfected ECV304 cells and a 15-fold increase in LPAAT-β–transfected ECV304 cells. Similarly, there was a 90- to 100-fold increase in TNF-α release in LPAAT-α–transfected ECV304 cells and a 30- to 35-fold increase in TNF-α release in LPAAT-β–transfected cells. We have concluded that LPAAT transfection increases baseline cytokine transcription and greatly enhances stimulated cytokine transcription; LPAAT transfection also greatly enhances stimulated cytokine release. These findings suggest that there may be some baseline control by lipids over either transcriptional regulation or, more likely, mRNA stability (61,67). In contrast, lipid membrane composition may be a necessary but not sufficient component controlling cytokine release; it is difficult from data obtained thus far to say whether this is directly related to PA composition of membranes or rather to changes in general lipid species composition dictated by a shift in general lipid acyl chain metabolism through LPAAT expression.

CONCLUSION

Exogenous PA and either LPS or lipid A have very similar effects on a variety of cell types, inducing transcription and release of IL-1β, eicosanoids, and TNFα and transducing such phenotypic changes as cell pro-

Table 3 IL-6 and TNF-α in the Supernatant of A549 (human lung epithelial adenocarcinoma) Cells Stably Transfected with LPAAT-α, Then Stimulated with IL-1β and Murine TNF-α (expressed in pg/ml cytokine in supernatant)

Stimulation conditions	Vector		LPAAT-α, transfected	
	IL-6	TNF-α	IL-6	TNF-α
No cytokines	<10	<10	<10	<10
IL-1β + murine TNF-α	225	22	4125	93

Table 4 IL-6 and TNF in the Supernatant of Human Endothelial ECV304 Cells Stably Transfected with LPAAT-α and LPAAT-β, Then Stimulated with IL-1β (expressed in pg/ml cytokine in supernatant)

Stimulation conditions	Vector		LPAAT-α, transfected		LPAAT-β, transfected	
	IL-6	TNF-α	IL-6	TNF-α	IL-6	TNF-α
No cytokines	<10	<10	<10	<10	<10	<10
IL-1β	725	<10	19250	925	13500	325

duction of lamellipodia and filopodia (20–23). PA and lipid A also cause amplification of PA synthesis through stimulation (either by translocation, alteration of membrane biophysical properties, or stimulation of small G-proteins, and possibly via all three pathways) of LPAAT and PLD (both PC-directed and PE-directed). Despite point charge density common to both lipids, PA and LPS are capable of inducing large biophysical changes in hydrophobic bilayers such as plasma membranes, and conversion of PA directly to DG or lyso-PA also results in profound changes, as well as possible stimulation of (external leaflet) membrane receptors by PA and lyso-PA. PA has now been demonstrated to be necessary and/or sufficient for translocation of perimembrane proteins important in mitogenesis such as raf kinase and membrane positioning of perimembrane proteins important in the inflammatory response and production of reactive oxygen species such as NADPH mixed function oxidases. The ability of LPS or lipid A to stimulate PA production through both LPAAT and PLD, either through direct intercalation into the membrane or through attachment to LPS binding proteins, followed by binding of the complex to CD14 or other related mechanisms may explain at least part of the intracellular and local inflammatory responses observed following lipid A administration.

Human LPAAT isoforms have been cloned and transfected into *E. coli* and mammalian cells, confirming their preferences for lyso-PA and polyunsaturated acyl chains and hence their possible influence on biophysical parameters of plasma membranes such as the order parameter. In addition, LPAAT transfection has been confirmed as having profound effects on cytokine transcription and release; this, however, cannot be concluded as providing proof that LPS or lipid A acts definitively through PA induction in even these selected cell types, and it remains to be proven that PA provides a definitive link in the mechanism of inflammation by LPS. Nonetheless, the currently available studies on mechanisms of PA activity, stimulation in LPAAT activity by lipid A and LPS, and the effect of LPAAT on cytokines suggest that this will be an area of fruitful study in the next decade.

REFERENCES

1. English D, Cui Y, Siddiqui RA. Messenger functions of phosphatidic acid. Chem Phys Lipids 1996; 80(1-2):117–132.
2. English D. Phosphatidic acid: a lipid messenger involved in intracellular and extracellular signaling. Cell Signal 1996; 8:341–347.
3. Imai A, Ishizuka Y, Kawai K, Nozawa Y. Evidence for coupling of phosphatidic acid formation and calcium influx in thrombin-activated human platelets. Biochem Biophys Res Commun 1982; 108:752–759.
4. Moolenaar WH, Kruijer W, Tilly BC, Verlaan I, Bierman AJ, deLaat SW, Growth factor-like action of phosphatidic acid. Nature 1986; 323:171–173.
5. Kawase T, Suzuki A. Phosphatidic acid-induced calcium mobilization in osteoblasts. J Biochem 1988; 103:581–582.
6. Kurz T, Wolf RA, Corr PB. Phosphatidic acid stimulates inositol 1,4,5-trisphosphate production in adult cardiac myocytes. Circ Res 1993; 72:701–706.
7. Harris RA, Schmidt J, Hitzemann BA, Hitzemann RJ. Phosphatidate as a molecular link between depolarization and neurotransmitter release in the brain. Science 1981; 212:1290–1291.
8. Jalink K, Hordijk PL, Moolenaar WH. Growth factor-like effects of lysophosphatidic acid, a novel lipid mediator. Biochim Biophys Acta 1994; 1198:185–196.
9. Murayama T, Ui M. Phosphatidic acid may stimulate membrane receptors mediating adenylate cyclase and phospholipid breakdown in 3T3 fibroblasts. J Biol Chem 1987; 262:5522–5529.
10. Billah MM, Eckel S, Mullmann TJ, Egan RW, Siegel MI. Phosphatidylcholine hydrolysis by phospholipase D determines phosphatidate and diglyceride levels in chemotactic peptide-stimulated human neutrophils. Involvement of phosphatidate phosphohydrolase in signal transduction. J Biol Chem 1989; 264:17069–17077.
11. English D, Taylor G, Garcia JG. Diacylglycerol generation in fluoride-treated neutrophils: involvement of phospholipase D. Blood 1991; 77:2746–2756.
12. van Corven EJ, Groenink A, Jalink K, Eichholtz T, Moolenaar WH. Lysophosphatidate-induced cell prolif-

eration: identification and dissection of signaling pathways mediated by G proteins. Cell 1989; 59:45–54.
13. Ferguson J, Hanley MR. Phosphatidic acid and lysophosphatidic acid stimulate receptor-regulated membrane currents in *Xenopus laevis* oocytes. Arch Biochem Biophys 1992; 297:388–392.
14. Bocckino SB, Blackmore PF, Wilson PB, Exton JH. Phosphatidate accumulation in hormone-treated hepatocytes via a phospholipase D mechanism. J Biol Chem 1987; 262:15309–15315.
15. Reinhold SL, Prescott SM, Zimmerman GA, McIntyre TM. Activation of human neutrophil phospholipase D by three separable mechanisms. FASEB J 1990; 4:208–214.
16. Lambeth JD, Kwak JY, Bowman EP, Perry D, Uhlinger DJ, Lopez I. ADP-ribosylation factor functions synergistically with a 50-kDa cytosolic factor in cell-free activation of human neutrophil phospholipase D. J Biol Chem 1995; 270:2431–2434.
17. Bowman EP, Uhlinger DJ, Lambeth JD. Neutrophil phospholipase D is activated by a membrane-associated Rho family small molecular weight GTP-binding protein. J Biol Chem 1993; 268:21509–21512.
18. Bauldry SA, Bass DA, Cousart SL, McCall CE. Tumor necrosis factor alpha priming of phospholipase D in human neutrophils. Correlation between phosphatidic acid production and superoxide generation. J Biol Chem 1991; 266:4173–4179.
19. Bourgoin S, Plante E, Gaudry M, Naccache PH, Borgeat P, Poubelle PE. Involvement of a phospholipase D in the mechanism of action of granulocyte-macrophage colony-stimulating factor (GM-CSF): priming of human neutrophils in vitro with GM-CSF is associated with accumulation of phosphatidic acid and diradylglycerol. J Exp Med 1990; 172:767–777.
20. Bursten SL, Harris WE. Rapid activation of phosphatidate phosphohydrolase in mesangial cells by Lipid A. Biochemistry 1991; 30:6195–6203.
21. Bursten SL, Harris WE, Resch K, Lovett D. Lipid A activation of glomerular mesangial cells: mimicry of the bioactive lipid, phosphatidic acid. Am J Physiol (Cell Physiol 31) 1992; 262:C328–C338.
22. Harris WE, Bursten SL. Lipid A stimulates phospholipase D activity in rat mesangial cells via a G-protein. Biochem J 1992; 281:675–682.
23. Bursten SL, Harris WE, Bomsztyk K, Lovett D. Interleukin-1 rapidly stimulates lysophosphatidate acyltransferase and phosphatidate phosphohydrolase activities in human mesangial cells. J Biol Chem 1991; 266:20732–20743.
24. Bursten SL, Harris WE. Interleukin-1 stimulates phosphatidic acid-mediated phospholipase D activity in human mesangial cells. Am J Physiol 1994; 266:C1093–1104.
25. van Klompenburg W, Nilsson I, von Heijne G, de Kruijff B. Anionic phospholipids are determinants of membrane protein topology. EMBO J 1997; 16:4261–4266.
26. Resh MD. Myristylation and palmitylation of Src family members: the fats of the matter. Cell 1994; 76:411–413.
27. Silverman L, Resh MD. J Cell Biol 1992; 119:415–425.
28. Jones GA, Carpenter G. The regulation of phospholipase C-gamma 1 by phosphatidic acid. Assessment of kinetic parameters. J Biol Chem 1993; 268:26206–26211.
29. Khan WA, Blobe GC, Richards AL, Hannun YA. Identification, partial purification, and characterization of a novel phospholipid-dependent and fatty acid activated protein kinase from human platelets. J Biol Chem 1994; 268:9729–9735.
30. Ghosh S, Strum JC, Sciorra VA, Daniel L, Bell RM. Raf-1 kinase possesses distinct binding domains for phosphatidylserine and phosphatidic acid. Phosphatidic acid regulates the translocation of Raf-1 in 12-O-tetradecanoylphorbol-13-acetate-stimulated Madin-Darby Canine kidney cells. J Biol Chem 1966; 271:8472–8480.
31. Tsai MH, Yu CL, Stacey DW. A cytoplasmic protein inhibits the GTPase activity of H-Ras in a phospholipid-dependent manner. Science 1990; 250:982–985.
32. Tsai MH, Yu CL, Wei FS, Stacey DW. The effect of GTPase activating protein upon ras is inhibited by mitogenically responsive lipids. Science 1989; 243:522–526.
33. Krabak MJ, Hui SW. The mitogenic activities of phosphatidate are acyl chain-length-dependent and calcium independent in C3H/1OT 1/2 cells. Cell Regul 1991; 2:57–64.
34. Kaszkin M, Richards J, Kinzel V. Proposed role of phosphatidic acid in the extracellular control of the transition from G2 phase to mitosis exerted by epidermal growth factor in A431 cells. Cancer Res 1992; 52:5627–5634.
35. Fukami K, Takenawa T. Phosphatidic acid that accumulates in platelet-derived growth factor-stimulated Balb/C 3T3 cells is a potential mitogenic signal. J Biol Chem 1992; 267:10988–10993.
36. Bellavite P, Corso F, Dusi S, Grzeskowiak M, Della-Bianca V, Rossi F. Activation of NADPH-dependent superoxide production in plasma membrane extracts of pig neutrophils by phosphatidic acid. J Biol Chem 1988; 263:8210–8214.
37. Agwu DE, McPhail LC, Sozzani S, Bass DA, McCall CE. Phosphatidic acid as a second messenger in human polymorphonuclear leukocytes. Effects on activation of NADPH oxidase. J Clin Invest 1991; 88:531–539.
38. Qualliotine-Mann D, Agwu DE, Ellenburg MD, McCall CE, McPhail LC. Phosphatidic acid and diacylglycerol synergize in a cell-free system for activation of NADPH oxidase from human neutrophils. J Biol Chem 1993; 268:23843–23849.
39. Knaus UG, Heyowrth PG, Evans T, Curnutte JT, Bokoch GM. Regulation of phagocyte oxygen radical production by the GTP binding protein Rac 2. Science 1991; 254:1512–1515.
40. Chuang TH, Bohl BP, Bokoch GM. Biologically active lipids are regulators of Rac.GDI complexation. J Biol Chem 1993; 268:26206–26211.
41. Fernandez B, Balboa MA, Solis-Herruzo JA, Balsinde J. Phosphatidate-induced arachidonic acid mobilization in mouse peritoneal macrophages. J Biol Chem 1994; 269:26711–26716.
42. Lee Z-W, Kweon S-M, Kim B-C, Leem S-H, Shin I, Kim J-H, Ha K-S. Phosphatidic acid-induced elevation

of intracellular Ca^{+2} is mediated by RhoA and H_2O_2 in rat-2 fibroblasts. J Biol Chem 1998; 273:12710–12715.
43. Waite KA, Wallin R, Qualliotine-Mann D, McPhail LC. Phosphatidic acid-mediated phosphorylation of the NADPH oxidase component p45-phox: evidence that phosphatidic acid may activate a novel protein kinase. J Biol Chem 1997; 272:15569–15578.
44. DeLeo FR, Renee J, McCormick S, Nakamura M, Apicella M, Weiss JP, Nauseef WM. Neutrophils exposed to bacterial lipopolysaccharide upregulate NADPH oxidase assembly. J Clin Invest 1998; 101:455–463.
45. Olson SC, Lambeth JD. Biochemistry and cell biology of phospholipase D in human neutrophils. Chem Phys Lipids 1996; 80:3–19.
46. Kobayashi T, Stang E, Fang KS, de Moerloose P, Parton RG, Gruenberg J. A lipid associated with the antiphospholipid syndrome regulates endosome structure and function. Nature 1998; 392:193–195.
47. Carr C, Morrison DC. Lipopolysaccharide interaction with rabbit erythrocyte membranes. Infect Immun 1984; 43:600–606.
48. Lei M-G, Morrison DC. Specific endotoxic LPS-binding protein on murine splenocytes. I. Detection of LPS-binding sites on splenocytes and splenocyte subpopulations. J Immunol 1988; 141:996–1005.
49. Lei M-G, Morrison DC. Specific endotoxic LPS-binding proteins on murine splenocytes. II. Membrane localization and binding characteristics. J Immunol 1988; 141:1006–1011.
50. Mathison JC, Wolfson E, Ulevitch RJ. participation of tumor necrosis factor in the mediation of gram negative bacterial lipopolysaccharide-induced injury in rabbits. J Clin Invest 1988; 81:1925–1937.
51. Schumann RR, Leong SR, Flaggs GW, Tobias PS, Mathison JC, Ulevitch RJ. Structure and function of lipopolysaccharide binding protein. Science 1990; 249:1429–1431.
52. Wright SD, Ramos RA, Tobias PS, Ulevitch RJ, Mathison JC. CD14, a receptor for complexes of LPS and LPS binding protein. Science 1990; 249:1431–1433.
53. Bulawa CE, Raetz CR. The biosynthesis of gram negative endotoxin. Identification and function of UDP-2,3-diacylglucosamine in Escherichia coli. J Biol Chem 1984; 259:4846–4851.
54. Ray BL, Painter G, Raetz CR. The biosynthesis of gram negative endoxotin. Formation of lipid A disaccharides from monosaccharide precursors in extracts of *Escherichia coli*. J Biol Chem 1984; 259:4852–4859.
55. Zoeller RA, Wightman PD, Anderson MS, Raetz CRH. Accumulation of lysophosphatidylinositol in RAW 264.7 macrophage tumor cells stimulated by lipid A precursors. J Biol Chem 1987; 262:17212–17219.
56. Bursten SL, Stevenson F, Torrano F, Lovett DH. Mesangial cell activation by bacterial endotoxin: induction of rapid cytoskeletal reorganization and gene expression. Am J Pathol 1991; 139:371–382.
57. Szamel M, Resch K. Modulation of enzyme activities in isolated lymphocyte plasma membranes by enzymatic modification of phospholipid fatty acids. J Biol Chem 1981; 256:11618–11623.
58. Lovett DH, Martin M, Bursten SL, Szamel M, Gemsa D, Resch K. Interleukin 1 and the glomerular mesangium. III. Il-1-dependent stimulation of mesangial cell protein kinase activity. Kidney Int 1988; 3:26–35.
59. Nishijima M, Amano F, Akamatsu Y, Akagawa K, Tokunaga T, Raetz CRH. Macrophage activation by monosaccharide precursors of E. coli lipid A. Proc Natl Acad Sci USA 1985; 82:282–286.
60. Prpic V, Weiel JE, Somers SD, et al. Effects of bacterial lipopolysaccharide on the hydrolysis of phosphatidylinositol-4,5-bisphosphate in murine peritoneal macrophages. J Immunol 1987; 239:526–533.
61. West J, Tompkins CK, Balantac N, Nudelman E, Meengs B, White T, Bursten SL, Coleman J, Kumar A, Singer JW, Leung DW. Cloning and expression of two human lysophosphatidic acid acyltransferase (LPAAT) cDNAs that enhance cytokine-induced signaling responses in cells. DNA Cell Biol 1997; 16:691–701.
62. White T, Bursten SL, Federighi D, Lewis RA, Nudelman E. High-resolution separation and quantification of neutral lipid and phospholipid species in mammalian cells and sera by multi-one-dimensional thin-layer chromatography. Anal Biochem 1998; 258:109–117.
63. Bursten SL. Interaction of lipopolysaccharide with a mammalian (human) lyso-phosphatidate acyltransferase (LPAAT) transfected into E. coli, and effect of lisofylline on LPAAT transfected into mammalian cells. In: Pollack M, ed. Proceedings of the Fourth Conference of the International Endotoxin Society. New York: John Wiley & Sons, Inc. In press.
64. Bursten SL, Federighi DA, Parsons PE, Harris WE, Abraham E, Moore EE, Moore FA, Bianco JA, Singer JW, Repine JE. An increase in serum C18 unsaturated free fatty acids as a predictor of the development of acute respiratory distress syndrome. Crit Care Med 1996; 24:1129–1136.
65. Cachianes G, Ho C, Weber RF, Williams SR, Goeddel DV, Leung DW. Epstein-Barr virus-derived vectors for transient and stable expression of recombinant proteins. BioTechniques 1993; 15:255–259.
66. Ella KM, Meier GP, Bradshaw CD, Huffman KM, Spivey EC, Meier KE. A fluorescent assay for agonist-activated phospholipase D in mammalian cell extracts. Anal Biochem 1994; 218:136–142.
67. Gonzalez CI, Martin CE. Fatty acid-responsive control of mRNA stability: unsaturated fatty acid-induced degradation of the Saccharomyces OLE1 transcript. J Biol Chem 1996; 271:25801–25809.

31

Endotoxic Shock and the Sphingomyelin Pathway

Cecil K. Joseph
Arnold & Marie Schwartz College of Pharmacy & Health Sciences, Brooklyn, New York

Richard N. Kolesnick
Memorial Sloan-Kettering Cancer Center, New York, New York

INTRODUCTION

Sepsis and septic shock are heterogeneous clinical syndromes that can be triggered by many microorganisms, including gram-negative bacteria, gram-positive bacteria, and fungi (1–3). Sepsis is generally defined as a systemic response to infection manifested by tachycardia, tachypnea, change in temperature, and leukopenia or leukocytosis. Septic shock is severe sepsis accompanied by hypotension (4).

The toxin responsible for the induction of septic shock by gram-negative bacteria is the glycolipid lipopolysaccharide (LPS). Major events in the pathogenesis of sepsis include neutrophil, monocyte, and macrophage inflammatory responses, intravascular coagulopathy resulting from activation of plasma complement and clotting cascades, endothelial cell damage, and hypotension. Thus, LPS exerts pleiotropic effects on many tissues and organs, resulting in multiple organ damage, circulatory collapse, and death.

The complex sequence of events that result in septic shock begins when local host defenses respond to microbial infection. In instances of overwhelming infection, a severe pro-inflammatory reaction ensues, resulting in the secretion of cytokines [tumor necrosis factor (TNF), interleukin (IL)-1, and IL-6] to combat infection, destroy damaged tissue, and promote wound repair (5). To control this inflammation and restore homeostasis, an anti-inflammation response is mounted with the release of anti-inflammatory agents such as IL-4, IL-10, IL-11, IL-13, soluble TNF receptors, IL-1 receptor antagonists, and transforming growth factor (TGF)-β (6). Failure to restore homeostasis results in shock, multiple organ failure, and death.

Of all these mediators, the central agent responsible for the lethal effects of LPS appears to be TNF-α. In LPS-treated macrophages, TNF-α is one of the first and most abundant cytokines produced, comprising approximately 1–2% of the total activated secretory product (7). Injection of purified TNF-α in animals evokes most of the effects of LPS including fever, shock, and death (8). Furthermore, therapies directed at inactivating TNF-α, such as neutralizing antibodies (7–9), chimeric inhibitors comprised of the extracellular domain of the TNF receptor fused with an immunoglobulin heavy chain fragment or as a polyethylene glycol-linked dimer (TNF-bp), or a TNF convertase metalloproteinase inhibitor, block experimental endotoxic shock in multiple models of sepsis (10). Thus, TNF-α represents a key intermediate in signal transduction initiated by LPS.

One effect of TNF-α is activation of the sphingomyelin pathway. TNF-α activates enzymes termed sphingomyelinases that hydrolyze the phosphodiester bond of sphingomyelin, generating the second messenger ceramide and the free headgroup phosphorylcholine. Ceramide has multiple effects on many cellular processes in different cell types. In this regard, monocytes, macrophages, neutrophils, and endothelium display significant responses to elevation of cellular ceramide levels. Of relevance to sepsis, ceramide induces both pro-inflammatory and apoptotic responses in cells critically involved in the septic shock response. In this chapter, we review the involvement of the sphingomyelin (SM) pathway in LPS-induced septic shock. Recent investigations suggest that ceramide is a critical molecule in the development of a disseminated form of microvascular endothelial cell apoptosis that may be the pathogenic lesion in endotoxic shock. Thus, ceramide and other intermediates of the sphingomeylin pathway may be novel targets for therapeutic intervention in the treatment of septic shock syndrome.

THE SPHINGOMYELIN PATHWAY AND CERAMIDE GENERATION

The sphingomyelin pathway appears to be evolutionarily conserved and functions in nearly all mammalian cells. This pathway is often initiated when a ligand binds to a specific cell surface receptor, leading to the activation of one or more sphingomyelinases (SMases), sphingomyelin-specific forms of phospholipase C, that hydrolyze the phosphodiester bond of sphingomyelin yielding ceramide and phosphorylcholine. Alternatively, the pathway can be activated in a receptor-independent fashion by environmental stresses such as ionizing radiation, heat, oxidative stress, and ultraviolet light. Several isoforms of SMase, distinguished by their pH optima, exist in human and murine cells. Human and murine acid (A)-SMases (pH optimum 4.5–5.0) were originally localized to lysosomal and endosomal compartments. However, A-SMase activity has been measured in the caveolae of cells treated with IL-1 and nerve growth factor (NGF) (11) and recently as a secretory product of endothelial cells (12,13). Patients suffering from the inherited disorder Nieman-Pick disease (NPD) are deficient in A-SMase activity due to a variety of known point mutations in the A-SMase gene (14).

Neutral (N)-SMases have not been characterized at the molecular level. Evidence indicates that they are products of a distinct gene or genes (15). One isoform (pH and optimum 7.4) appears to be membrane-bound and Mg^{2+} dependent. A cation-independent form has been observed in cytosolic fractions, and an alkaline SMase activity that appears to be involved in digestion and mucosal cell proliferation has been detected in the intestinal mucosa and bile (16,17). SMases are activated by diverse exogenous stimuli leading to increased levels of ceramide in a time frame of seconds to minutes. Therefore, SMases are considered to be principal pathways for the production of ceramide in early signal transduction. Receptors as distinct as CD28, CD95, TNF-α, IL-1β, and p75-NGF, to list a few, all signal via the SM pathway as a consequence of ligand binding (18).

Signaling via the p55-TNFR represents a model for the role of SMases and ceramide in many cellular functions. Engagement of the TNFR leads to activation of both neutral and acid SMases. Experiments with deletion mutants of the p55-TNFR receptor led to the identification of an 11-amino-acid region in the cytoplasmic domain termed the neutral sphingomyelinase domain (NSD). This motif binds a protein, factor activating neutral sphingomyelinase (FAN), which appears to be required for neutral SMase activation (19). A-SMase can also be activated by the p55-TNFR through a different region, the death domain (20). The death domain of CD95 has also been linked through mutational analysis to activation of A-SMase. It has been suggested that N-SMase may signal downstream to the extracellular regulated kinase (ERK) cascade and pro-inflammatory responses, while A-SMase is involved in TNF-mediated apoptosis (21). Additional investigations are required to substantiate this hypothesis. The involvement of SMases and ceramide in apoptosis is presented below.

Ceramide may also be generated from anabolic reactions. De novo synthesis of ceramide involves the condensation of serine and palmitoyl-CoA to form ketosphinganine (22). The latter is reduced to dihydrosphingosine, acylated to dihydroceramide by the enzyme ceramide synthase, and finally oxidized to ceramide by dihydroceramide reductase. Treatment of P388 and U937 cells with the chemotherapeutic agent daunorubicin results in elevated levels of ceramide (23). Ceramide elevation in this instance did not result from activation of a sphingomyelinase but rather from stimulation of ceramide synthase. The fungal toxin fumonisin B1, a specific inhibitor of ceramide synthase, was shown to block both daunorubicin-induced ceramide generation and apoptosis, suggesting that ceramide synthase activation was obligatory for the death response.

Once generated, ceramide may be subject to extensive modifications: it may be phosphorylated to ceramide-1-phosphate, deacylated to sphingosine, glycosylated, and converted back to SM. Ceramide, sphingosine, and its phosphorylated form sphingosine 1-phosphate all have been shown to have second messenger function (18). Thus, the multiplicity of potential second messengers may contribute to the wide array of effector functions of ceramide, which include proliferative, differentiating, pro-inflammatory, and apoptotic responses (24).

CERAMIDE TARGETS

Identification of direct targets of ceramide action provides a clearer understanding of the cell type–specific nature of ceramide signaling (reviewed in Ref. 18). One key direct target of ceramide action is ceramide-activated protein kinase (CAPK), a membrane-associated proline-directed serine/threonine kinase that recognizes substrates with the motif X-thr-leu-pro-X (25). CAPK has been identified as the mammalian homolog of kinase suppressor of Ras (KSR), a protein found in *C. elegans* and *Drosophila* (26). TNF-α and ceramide analogs induce autophosphorylation of KSR and enhance its ability to activate Raf-1, initiating signaling down the MAP kinase cascade. ERK activation may lead to phosphorylation and activation of cPLA$_2$ and the release of arachidonic acid from phospholipid. Thus, KSR may be an important mediator of the pro-inflammatory actions of TNF-α. It should be noted that this finding has recently been disputed, and additional studies will be required before it can be stated with certainty that CAPK is KSR (27).

Other direct targets of ceramide include a ceramide-activated protein phosphatase (CAPP) (28), and protein kinase (PKC)-ζ. CAPP is a heterotrimeric serine/threonine protein phosphatase (PP2A) that may be involved in downregulation of c-myc expression and apoptosis (29). Subunits A and B of the phosphatase are regulatory, and subunit C is catalytic. The B subunit is required for activation by ceramide. Elegant studies have demonstrated that CAPP mediates an antiproliferative response to ceramide in the yeast *Saccharomyces cerevisiae* by causing cells to arrest in the G1 phase of the cell cycle (30). The genes TPD3 and CDC55 encode the regulatory subunits and SIT4 the catalytic subunit. Evidence has also been provided that CAPP may also play a role in mammalian systems in the dephosphorylation and inactivation of c-jun (31).

PKC-ζ is a diacylglycerol (DAG)– and phorbol ester–insensitive PKC isoform that shows increased activity in response to treatment of cells with TNF-α, SMases, or ceramide analogs (32,33). Radiolabeled ceramide bound to and activated PKC-ζ and a dominant-negative, kinase defective PKC-ζ inhibited activation of NF-κB by SMase in NIH-3T3 cells, consistent with a role for PKC-ζ in the TNF-induced activation of NF-κB in some systems. Activation of NF-κB and its translocation to the nucleus involves phosphorylation and degradation of the inhibitor protein IκB, which is complexed with NF-κB in the cytosol (34). Studies by Machleidt and coworkers (35) show that SMase or ceramide can, in some cells, induce degradation of IκB and NF-κB activation. This may serve as an alternative mechanism to that involving TRAFs for TNF-induced NF-κB activation. Both neutral and acid SMase have been associated with activation of NF-κB by TNF (36,37).

Ceramide also interacts indirectly with the Jun N-terminal kinase (JNK)/stress-activated protein kinase (SAPK) cascade. Treatment of bovine aortic endothelial cells and U937 cells with TNF, UV and ionizing radiation, heat shock, and H$_2$O$_2$ results in sphingomyelinase activation, ceramide generation, stimulation of JNK/SAPK pathway, and apoptosis (38). At least for ionizing radiation, this event appears to require A-SMase, as cells from patients with Nieman-Pick disease (NPD), an inherited deficiency of A-SMase activity, fail to generate this response. Ceramide may activate this pathway by signaling through either a TGF-β–activated kinase (TAK1) or the small G-protein rac1 (39,40). The use of a kinase-negative TAK1 or dominant negative N17rac1 completely inhibited JNK/SAPK activation in response to ceramide treatment or stimulation with C95/Apo1/Fas.

CERAMIDE AND INFLAMMATION

Neutrophils must coordinate many biological processes while killing invading microorganisms. TNF-α is a primary signal for pro-inflammatory responses of neutrophils, augmenting chemotaxis, adhesion, priming, phagocytosis, inflammatory mediator release, and superoxide generation (24). In order to prevent the inadvertent destruction of normal tissue elements, neutrophils activated within the circulation must migrate to the site of inflammation prior to release of inflammatory mediators. While some evidence supports a role for ceramide in TNF-induced superoxide generation (41), ceramide appears not to be involved in stimulation of other pro-inflammatory processes. However, ceramide analogs mimic the delay in the respiratory burst and oxidant release induced by TNF, suggesting that ceramide may serve as a phase coordinator of the neutrophil response to soluble agonists (42).

Fibroblasts, chondrocytes, epithelial cells, and endothelial cells represent some of the nonleukocytic cells that function in immune and inflammatory responses. These cells produce cytokines, prostaglandins, and other mediators that modulate immune and inflammatory responses. Pioneering studies by Ballou and coworkers (43) demonstrated a link between cytokine signaling, inflammation, and the SM pathway. In these experiments, treatment of human foreskin fibroblasts with IL-1β resulted in SM hydrolysis and ceramide accumulation. Although elevation of ceramide alone induced only minimal effects, ceramide or SMase synergistically enhanced IL-1β– or TNF-α–induced production of the pro-inflammatory eicosanoid prostaglandin E_2 (PGE_2) production (44). The effect of ceramide elevation appeared mediated, at least in part, by activation of cyclo-oxygenase (Cox). However, the relative contributions of Cox-1 and Cox-2 to this effect were not established. These studies suggested that ceramide might serve as a co-signal for cytokine action in fibroblasts.

Recent studies by Modur et al. (45) delineate a role for ceramide in the inflammatory response of endothelium. In human umbilical vein endothelial cells (HUVEC), TNF induced rapid ceramide generation. Further, treatment of these cells with either ceramide analogs or exogenous SMase resulted in expression of E-selectin and binding to polymorphonuclear leukocytes, activities recognized as initial steps in physiological inflammation. Exogenous SMase and ceramide also induced endothelial cell expression of VCAM-1, ICAM-1, and secretion of the pro-inflammatory cytokines IL-6 and IL-8. While ceramide was particularly effective in activating Raf-1 and signaling down the MAP kinase cascade, it less effectively stimulated p38 or JNK-1 activation or NF-κB translocation. Thus, for some intracellular pathways involved in endothelial activation, ceramide appears to be a primary signal, whereas for other events it may serve only as a cosignal. Tabas and coworkers (12,13) have recently elucidated a novel mechanism involved in signaling through the sphingomyelin pathway in endothelium. This group showed that a Zn^{2+}-stimulable form of A-SMase is secreted at a high level by primary cultures of endothelium and that secretion of this SMase isoform is increased by pro-inflammatory cytokines such as IL-1β, IFN-β, IFN-γ, and IL-4. Hence, endothelium may serve to coordinate responses in a variety of pro-inflammatory cells by release of A-SMase locally or into the circulation.

Finally, studies by Balsinde and coworkers (46) showed that LPS and platelet-activating factor (PAF) stimulated a transient accumulation of ceramide in P388D1 macrophages. The accumulation of ceramide was likely due to increased de novo synthesis since fumonisin B1 inhibited this LPS/PAF-induced response.

CERAMIDE AND LPS

Studies from many laboratories have established that TNF, LPS, and IL-1 appear to use a common set of kinases, transcription factors, and promoter elements to provoke cellular activities associated with the inflammatory response (47–49). Activities as diverse as MAP kinase activation, NF-κB translocation, AP-1 stimulation, TNF gene expression, and phospholipase A_2 activation are common early events initiated by TNF, IL-1, and LPS. Because accumulating evidence supported the notion that ceramide was a second messenger for TNF and IL-1 action, we considered the possibility that LPS also acted through ceramide (50). However, LPS failed to elevate cellular ceramide levels. This prompted a further inspection of the lipid A moiety of LPS.

A comparison of the structures of lipid A and ceramide is shown in Figure 1. Carbons 1–3 of the reducing glucosamine of lipid A closely resemble Carbons 1–3 of ceramide (Fig. 1, boxed). This region is conserved in all biologically active LPS and ceramide analogs, and nearly all other portions of the molecules can be deleted or altered without destroying bioactivity. Experiments using molecular modeling and conformational dynamics to generate energy minimized structures of the reducing glucosamine of lipid A (GlcN-1, dephosphorylated form) and ceramide showed that positions C-1, C-2, and C-3 and their functional groups were nearly superimposable. A similar result was obtained when the 1-phosphorylated forms of each lipid were compared. In contrast, molecular modeling of 1,2-diacylglycerol generated showed far less similarity.

The discovery of the gram-negative bacterium *Sphingomonas* provided further support for a physical similarity between LPS and sphingolipids. In contrast to all other gram-negative species that require LPS as an essential structural component, this bacterium expresses glycosphingolipids rather than LPS (51), suggesting that the physical properties of these two lipids may be interchangeable. Further, degradation of LPS and sphingolipids may require similar enzymatic activities. The enzyme acyloxyacyl hydrolase (AOAH), which degrades LPS, displays significant homology to sphingolipid activator proteins (saposins), accessory molecules that function in sphingolipid degradation (52).

To determine whether the structural similarity between LPS and ceramide translated into function, we examined the effect of both on CAPK activation. Stimulation of HL-60 cells with LPS or lipid A, like ceramide, resulted in increased activity of CAPK within

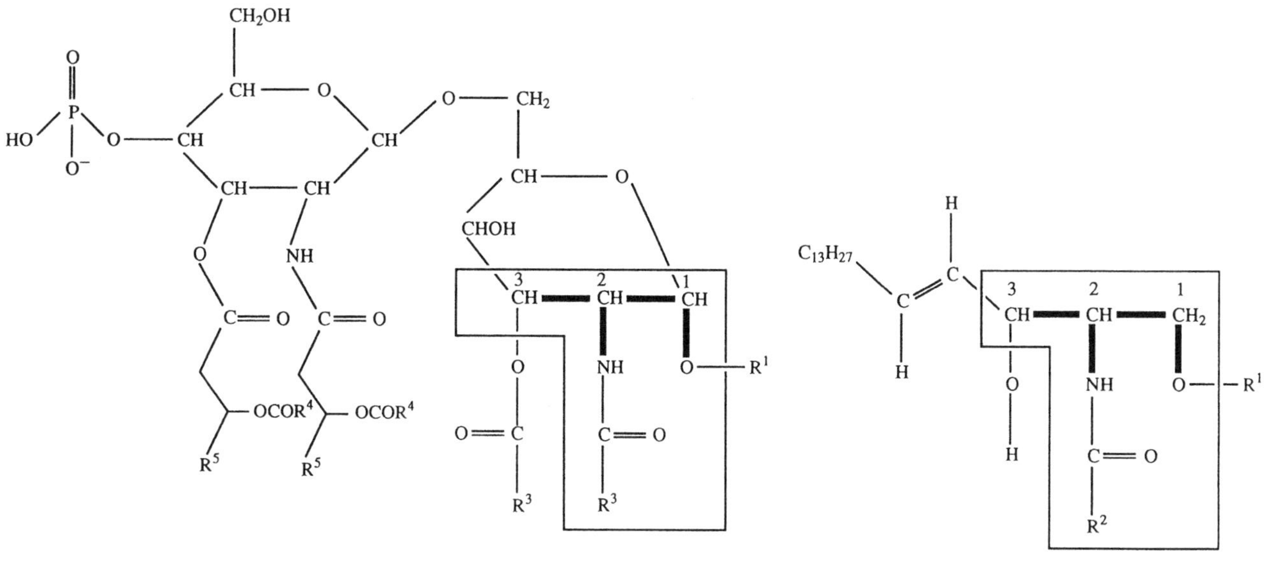

Lipid A **Ceramide**

Fig. 1 Chemical structures of lipid A and ceramide: Schematic representation of the primary chemical structures of lipid A (R^1 = phosphate, R^3 = $CH_2CH(OH)C_{11}H_{23}$, R^4 = $C_{14}H_{29}$, R^5 = $C_{11}H_{23}$) and of ceramide and ceramide-1-phosphate (R^1 = H and phosphate, respectively; R^2 = long hydrocarbon chain). Carbon atoms 2 and 3 are asymmetrical in both LPS and ceramide, with the absolute configuration identical at carbon 2 and opposite at carbon 3. The configurations at carbon 3 are considered opposite because the oxygen at carbon 3 of LPS is positioned opposite from the oxygen in ceramide. However, the long chains attached to carbon 3 are identically placed on LPS and ceramide. The boxes indicate regions of similarity in these schematic diagrams. (From Ref. 50.)

seconds. Finally, LBP enhanced LPS activation of CAPK, and an antibody to the cell surface receptor for LPS, CD14, blocked the effect of LPS, confirming the specificity of LPS stimulation. Thus, LPS may induce some of its effects by co-opting the second messenger function of ceramide and directly activating ceramide targets. This function would enable LPS to bypass and simultaneously mimic activation of cytokine receptors. In order for this to occur, LPS would have to be targeted to the same intracellular sites as ceramide. Wright and colleagues have proposed, based on studies with model membrane systems, that LBP might serve as a transfer protein unloading LPS from CD14 into membrane (53). Once in the membrane, LPS and ceramide appear to be subject to a similar mode of vesicular trafficking (54). These latter studies utilized a fluoroprobe attached to both lipids and followed their uptake and transfer to intracellular sites in macrophages derived from Lps-responsive (Lps^n) and LPS-hyporesponsive (Lps^d) mice. In Lps^n mice, both LPS and ceramide appeared to be transported to the Golgi apparatus. In contrast, Lps^d macrophages exhibited defective vesicular transport of both LPS and ceramide.

Studies comparing the biological effects of LPS and ceramide have also documented similarities in a number of systems. Using macrophages from (Lps^n) and (Lps^d) mice, Barber and coworkers (55) analyzed the expression of the LPS-inducible genes, TNF-α, IL-1β, interferon regulatory factor (IRF)-1, and TNFR-2, following stimulation with LPS, ceramide, and SMase. In contrast to Lps^n macrophages, Lps^d mice failed to respond to LPS, cell-permeable analogs of ceramide (C2, C6, C16), or SMase, suggesting that the product of the Lps gene played a key role in both LPS and ceramide signaling. An extension of these studies illustrated that a Ser/Thr phosphatase activity was required to induce secretion of TNF-α in response to LPS or SMase (56). Further, macrophages "tolerized" to LPS secreted very low levels of TNF in response to treatment with either LPS or SMase. These findings are strengthened by recent studies (discussed above) illustrating that macrophages from Lps^d mice exhibit a defect in intracellular transport of LPS and ceramide (54). It appears, therefore, that the Lps gene may encode a single protein that regulates macrophage responses to LPS and ceramide. This protein may play an important role in signaling by LPS and ceramide. It is important to note, however, that studies by Barber and coworkers have shown that activation of the sphingomyelin pathway results in stimulation of only a subset of LPS-inducible genes (56). Treatment of macrophages with ceramide analogs did not lead to IFN production, or induction of interferon-inducible protein 10 (IP-10) or interferon consensus sequence-binding protein (ICSBP) mRNA to levels comparable to those induced by LPS. Thus LPS stimulation of macrophages does not only mimic the second messenger function of ceramide.

Finally, a recent study (57) identified a novel gene, DIF-2, which is expressed in freshly isolated monocytes and downregulated during differentiation. LPS and ceramide both strongly increased DIF-2 transcription, while lysophosphatidylcholine treatment was minimally effective. The role of this gene in signaling or inflammation has not been established, but a homologous sequence identified in fibroblasts is a serum-inducible immediate early gene.

These studies support the notion that LPS may mimic the second messenger function of ceramide and suggest that in so doing LPS may integrate into a feed-forward pathway specifically utilized by cytokines to initiate inflammatory responses.

CERAMIDE AND APOPTOSIS

Cell death by necrosis results in the release of toxic enzymes into surrounding tissue, causing a local inflammatory response and tissue damage (61). Apoptosis, however, is a controlled form of cell death in which cells are removed without tissue damage. Apoptosis is absolutely essential during the processes of wound healing, embryogenesis, immune response, and normal maintenance of homeostasis (58–61). It is usually divided into three phases: initiation, execution, and degradation. The initiation phase often involves the stimulation of receptors such as CD95/Apo-1/Fas and p55-TNFR. These receptors contain death domains that assemble a signaling complex of proteins to convert the initiating signals into an execution phase, which signals activation of cysteine proteases (caspases) and in some cases disruption of mitochondrial function. End-stage apoptosis is the degradation stage involving cleavage of key cellular proteins and destruction of subcellular organelles.

An increasing body of evidence implicates ceramide as an important mediator in the process of apoptosis (21). Cytokines like TNF and CD95/Fas/Apo1, ionizing radiation, heat shock, ultraviolet-C, and oxidative stress induce ceramide generation as an early step in the development of apoptosis. Additional confirmation comes from studies on A-SMase knockout mice and patients with NPD. A-SMase mice are resistant to endothelial cell apoptosis in response to radiation, and lymphoblasts derived from patients with NPD show defects in the apoptotic response. As stated above, A-SMase links to the death domain of p55-TNFR (20) and CD95 (62). Alternatively, neutral sphingomyelinase may play a role in the development of the apoptotic response under conditions where glutathione is low (63).

Figure 2 shows some interactions of apoptotic components with intermediates of the SM pathway. The 55 kDa TNF receptor initiates apoptosis via formation of a death domain–adaptor protein complex consisting of the TNF receptor 1–associated death domain protein (TRADD), Fas-associated death domain protein (FADD), and caspase-8 (FLICE/MACH-1) or caspase 10. Investigations by Dixit and coworkers suggest that A-SMase activation and ceramide generation may function downstream of these protein complexes. In their studies, overexpression of a dominant-negative FADD blocked ligand-induced ceramide generation and apoptosis, and treatment with ceramide analogs was shown to bypass the antiapoptotic effect of dominant-negative FADD and restore apoptosis (64). Similarly, Hannun and coworkers showed that the viral caspase inhibitor crmA (cytokine response modifier A) prevented CD95-induced ceramide generation and apoptosis but not apoptosis in response to ceramide analogs, whereas overexpression of Bcl-2 blocked ceramide analog–induced apoptosis but not CD95-induced ceramide generation (65). Recently, Schwandner and Kronke (unpublished observations) provided direct evidence that A-SMase is downstream of the death domain adaptor proteins by showing that overexpression of TRADD and FADD enhances TNF-induced activation of A-SMase.

Although the precise role of ceramide in apoptosis remains to be defined, evidence suggests two independent mechanisms of action. Ceramide may signal transcriptionally, perhaps through c-Jun kinase, upregulation of pro-apoptotic proteins such as Fas ligand (66), TNF (67), or TRAIL (I. Herr et al., private communication). Alternatively, ceramide may initiate apoptosis nontranscriptionally by signaling mitochondrial membrane permeability transition (MPT) and/or generation of mitochondrial reactive oxygen species (ROS), key steps in one form of the apoptotic process (68). Ceramide-initiated MPT may signal the release of an apoptosis initiating factor (AIF) or activation of a CPP32-

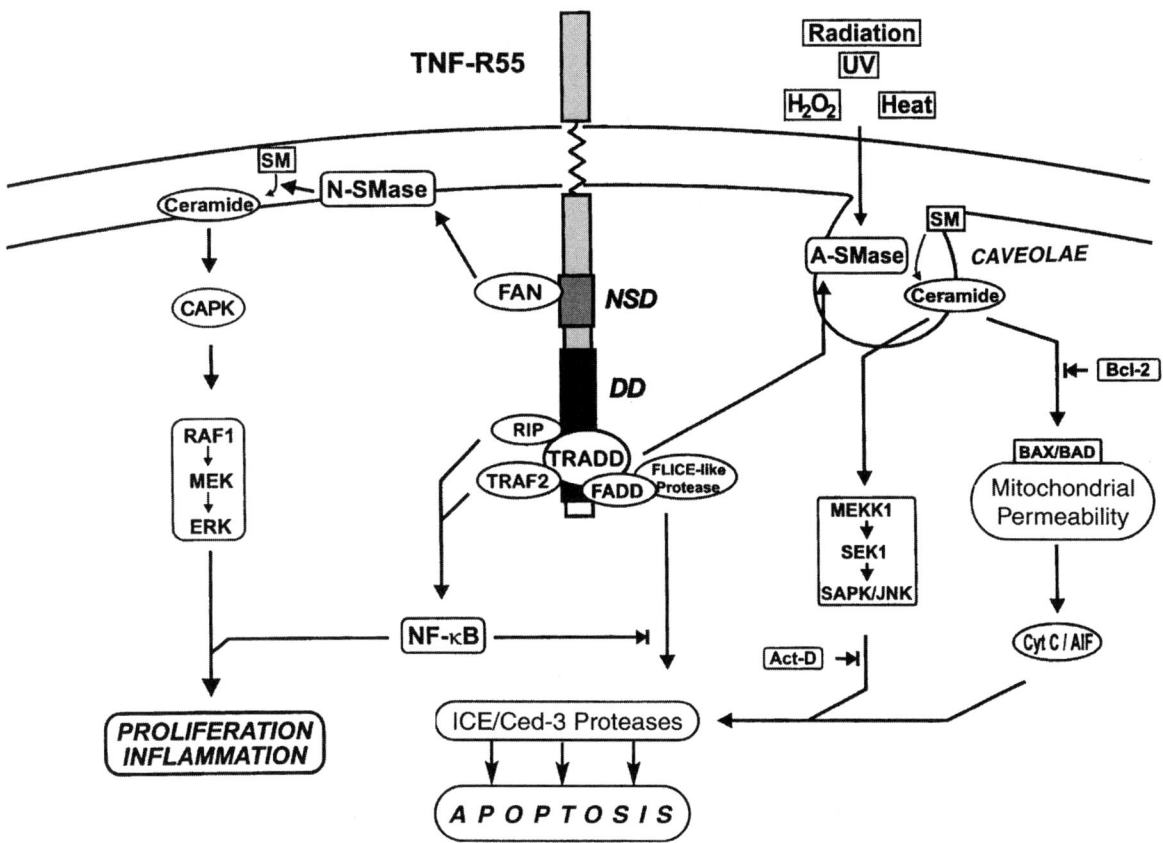

Fig. 2 Proposed mechanism for TNF-induced apoptosis via the sphingomyelin pathway: The 55 kDa TNF receptor initiates apoptosis via formation of a death domain–adaptor protein complex that links downstream to acid sphingomyelinase, SAPK/JNK, and the ICE/Ced-3 systems to signal apoptosis. Proliferative/pro-inflammatory effects of TNF are signaled through a different region of the 55 kDa TNF receptor. A membrane proximal region of the cytoplasmic domain links the adaptor protein FAN to neutral sphingomyelinase, ceramide generation, stimulation of ceramide-activated protein kinase (CAPK), Raf-1, and the ERK cascade. In addition, TRADD may link the TNF receptor to NF-κB activation via RIP and/or TRAFs. (From Ref. 21.)

like protease (Caspase 3). It has also been suggested that inhibition of mitochondrial carnitine palmitoyltransferase I (CPTI) activity enhances de novo synthesis of ceramide, which eventually leads to programmed cell death (21).

CERAMIDE AND ENDOTOXIC SHOCK

The mechanism of microvascular injury and its relevance to the evolution of septic shock have been a subject of substantial debate (69–75). The events that precede vascular collapse include disseminated intravascular thrombosis, endothelial necrosis concomitant with tissue necrosis, and generalized humoral dysfunction. These events have been reported to occur concomitantly with other aspects of inflammation and tissue destruction. Hence, it has been difficult to define the exact role of endothelial dysfunction in the progression of the septic shock response.

Support for the involvement of ceramide in endotoxicity is provided by recent investigations by Haimovitz-Friedman and coworkers (76). These studies identified generalized microvascular endothelial apoptosis as a primary factor in the development of endotoxic shock. Within one hour after intraperitoneal injection of *S. typhimurium* or *E. coli* LPS or TNF-α into C57BL/6 mice, ceramide levels increased in lung and small intestine. For the small intestine studies, the villous layer was dissected from the muscularis to provide an enriched specimen. Apoptosis of the microvascular but not the macrovascular endothelium of intestine, lung, fat, and thymus ensued after 6 hours. At this time the nonendothelium cellular components of these tissues were not damaged. Damage to nonendothelial cells of these tissues was not detected until 8 hours after LPS in our model, except in fat tissue, where even at 6 hours a small amount of fat cell apoptosis was observed. An inflammatory cell infiltrate, most easily seen in the lung, was not observed until 8–10 hours after the LPS challenge. These studies indicate that LPS induces a disseminated form of microvascular endothelial apoptosis that preceded nonendothelial tissue damage.

TNF-binding protein, which protects against LPS-induced death in various animal models (77), blocked LPS-induced ceramide generation and endothelial apoptosis, indicating that tissue ceramide was downstream of TNF action. A role for ceramide in this process was confirmed using the A-SMase knockout mice. These animals displayed a normal increase in serum TNF in response to LPS yet were defective in tissue ceramide generation. At an LD_{90}, they also manifested a marked reduction in disseminated endothelial apoptosis and were protected against death. Increasing the dose of LPS overcame the deficit in endothelial apoptosis in A-SMase knockout mice and restored the death response. These studies indicate that ceramide is a co-signal for the optimal death response. Hence, other signals such as nitric oxide, caspases, or PAF are likely capable of signaling endothelial death in the absence of ceramide.

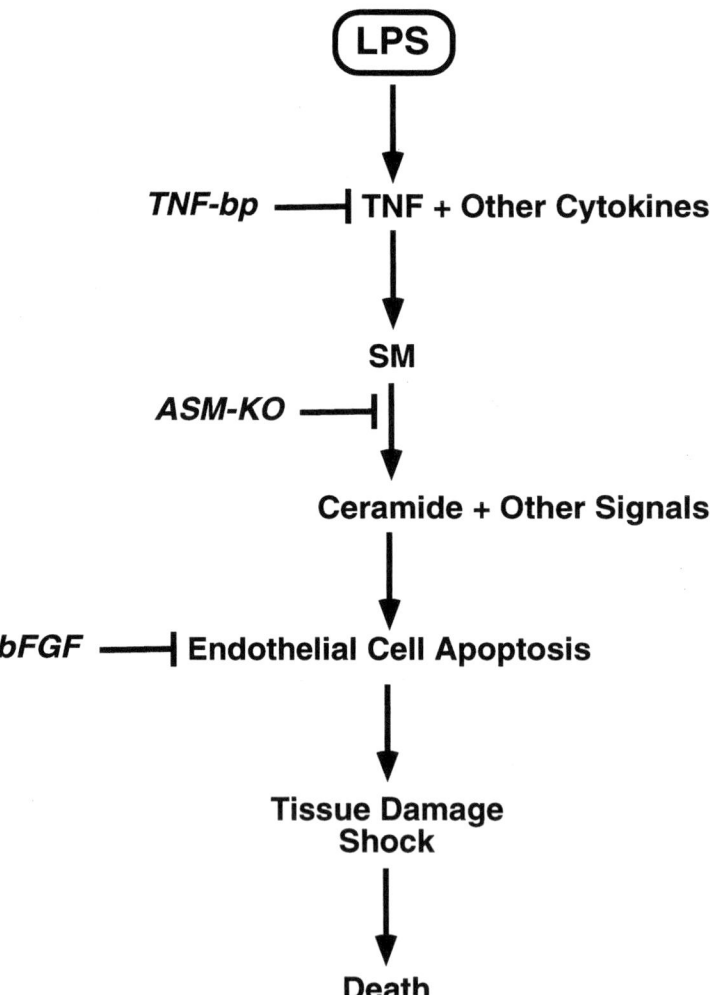

Fig. 3 Proposed schema for progression of the endotoxic response: LPS, released by gram-negative bacteria, interacts with inflammatory cells leading to generation of TNF-α and other cytokines. TNF-α, acting upon endothelium, stimulates sphingomyelin hydrolysis to ceramide, presumably via an ASMase, which then serves as a co-signal for apoptosis. Apoptosis of the endothelium ensues, which can be blocked by bFGF via inhibition of ceramide generation. We further propose that endothelial apoptosis results in generalized microvascular dysfunction sufficient to compromise the circulation to major organs, leading to nonendothelial tissue damage, circulatory collapse, and death. (From Ref. 76.)

These studies, which identified microvascular endothelial apoptosis as the pathogenic lesion in endotoxic shock, suggested that endothelial survival factors such as basic fibroblast growth factor (bFGF) might protect from septic shock (78). Indeed, bFGF injected intravenously just prior to LPS blocked LPS-induced ceramide elevation, endothelial apoptosis, and animal death. However, as anticipated, bFGF had no effect on TNF secretion into the circulation. It should be noted that the tissue inflammatory response was blocked in both the A-SMase knockout mouse and the wild-type mouse when pretreated with bFGF, suggesting that this even may be downstream, in part, of endothelial dysfunction.

A hypothetical ordering of events is shown in Figure 3. A cascade of events is initiated with LPS generation of TNF, binding to the TNFR, and assembly of death domain complexes. These studies suggest that pharmacological intervention at a step distal to cytokine secretion may be a feasible alternative to anticytokine therapies, which to date have for the most part failed (7). In this regard, preliminary studies suggest that bFGF can be added even after LPS and some protection attained (A. Haimovitz-Friedman and R. N. Kolesnick, unpublished observation). However, once the tissue ceramide level has peaked, bFGF no longer appears to provide protection. Whether other endothelial survival factors, antisphingomyelinase therapy, or agents that strengthen Bcl-2 function or antagonize caspase function will similarly protect against endotoxic shock in vivo is the subject of ongoing investigation.

CONCLUSION

The second messenger ceramide appears to play a pivotal role in the development of LPS-induced septic shock. Not only does LPS bear structural similarity to ceramide and possibly mimic some of its second messenger functions, but many of the key cytokines like TNF and IL-1 that mediate responses to LPS also activate components of the sphingomyelin pathway leading to increased levels of ceramide. Ceramide, once generated, may engage in cross-talk with many other biochemical pathways to modulate both pro- and antiapoptotic factors leading to cell death. Thus, it is possible to begin a molecular ordering of signals that eventually lead to the induction of endothelial apoptosis by LPS (Fig. 3). This scenario points to new targets for pharmacological manipulation of endotoxic shock with potential for clinical application.

REFERENCES

1. Parker MM, Shelhamer JH, Bacharach, SL, Green MV, Natanson C, Frederick TM, et al. Profound but reversible myocardial depression in patients with septic shock. Ann Intern Med 1984; 100:483–490.
2. Wiles JB, Cerra FB, Siegel JH, Border JR. The systemic septic response: does the organism matter? Crit Care Med 1980; 8:55–60.
3. Natanson C, Danner RL, Elin RJ, Hosseini JM, Peart KW, Banks SM, et al. Role of endotoxemia in cardiovascular dysfunction and mortality. *Escherichia coli* and *Staphylococcus aureus* challenges in a canine model of human septic shock. J Clin Invest 1989; 83:243–251.
4. Bone RC, Balk RA, Cerra FB, Dellinger RP, Fein AM, Knaus WA, et al. Definitions for sepsis and organ failure and guidelines for the use of innovative therapies in sepsis. The ACCP/SCCM Consensus Conference Committee. Chest 1992; 101:1644–1655.
5. Fukushima R, Alexander JW, Gianotti L, Ogle CK. Isolated pulmonary infection as a source of systemic tumor necrosis factor. Crit Care Med 1994; 22:114–120.
6. Bone RC. Why sepsis trials fail. JAMA 1996; 276:565–566.
7. Beutler B, Milsark IW, Cerami AC. Passive immunization against cachetin/tumor necrosis factor protects mice from lethal effect of endotoxin. Science 1985; 229:869–871.
8. Tracey KJ, Beutler SF, Lowry J, Merryweather S, Wolpe IW, Milsark RJ, et al. Shock and tissue injury induced by recombinant human cachetin. Science 1986; 234:470–474.
9. Mathison JC, Wolfson E, Ulevitch RJ. Participation of tumor necrosis factor in the mediation of gram negative bacterial lipopolysaccharide-induced injury in rabbits. J Clin Invest 1988; 81:1925–1937.
10. Solorzano CC, Ksontini R, Pruitt JH, Hess PJ, Edwards PD, Kaibara A, et al. Involvement of 26-kDa cell-associated TNF-alpha in experimental hepatitis and exacerbation of liver injury with a matrix metalloproteinase inhibitor. J Immunol 1997; 158:414–419.
11. Liu P, Anderson RG. Compartmentalized production of ceramide at the cell surface. J Biol Chem 1995; 270:27179–27185.
12. Schissel SL, Schuchman EH, Jon Williams K, Tabas I. Zn^{2+}-stimulated sphingomyelinase is secreted by many cell types and is a product of the acid sphingomyelinase gene. J Biol Chem 1996; 271:18431–18436.
13. Marathe S, Schissel SL, Yellin MJ, Beatini N, Mintzer R, Jon Williams K, Tabas I. Human vascular endothelial cells are a rich and regulatable source of secretory sphingomyelinase. J Biol Chem 1998; 273:4081–4088.
14. Schuchman EH. Two new mutations in the acid sphingomyelinase gene causing type A Niemann-Pick disease: N389T and R441X. Human Mutat 1995; 6:352–354.
15. Horinouchi K, Erlich S, Perl DP, Ferlinz K, Bisgaier CL, et al. Acid sphingomyelinase deficient mice: a model of types A and B Niemann-Pick disease. Nat Genet 1995; 10:288–293.

16. Nyberg L, Duan RD, Axelson J, Nilsson A. Identification of an alkaline sphingomyelinase activity in human bile. Biochim Biophys Acta 1996; 1300:42–48.
17. Duan RD, Hertervig E, Nyberg L, Hauge T, Sternby B, Lillienau J, Farooqi A, Nilsson A. Distribution of alkaline sphingomyelinase activity in human beings and animals. Tissue and species differences. Dig Dis Sci 1996; 41:1801–1806.
18. Spiegel S, Foster DA, Kolesnick RN. Signal transduction through lipid second messengers. Curr Opin Cell Biol 1996; 8:159–167.
19. Adam-Klages S, Adam D, Wiegmann K, Struve S, Kolanus W, et al. FAN, a novel WD-repeat protein, couples the p55 TNF-receptor to neutral sphingomyelinase. Cell 1996; 86:937–957.
20. Wiegmann K, Schutze S, Machleidt T, Witte D, Kronke M. Functional dichotomy of neutral and acidic sphingomyelinases in tumor necrosis factor signaling. Cell 1994; 78:1005–1015.
21. Kronke M, Kolesnick RN. Regulation of ceramide production and apoptosis. Ann Rev Physiol 1998; 60:643–665.
22. Merrill AH Jr, Jones DD. An update of the enzymology and regulation of sphingomyelin metabolism. Biochim Biophys Acta 1990; 1044:1–12.
23. Bose R, Verheij M, Haimovitz-Friedman A, Scotto K, Fuks Z, et al. Ceramide synthase mediates dunorubicin-induced apoptosis: an alternative mechanism for generating death signals. Cell 1995; 82:405–414.
24. Ballou LR, Lauledekind SJF, Rosloniec EF, Raghow R. Ceramide signalling and the immune response. Biochim Biophys Acta 1996; 1301:273–287.
25. Mathias S, Dressler KA, Kolesnick RN. Characterization of a ceramide-activated protein kinase: stimulation by tumor necrosis factor α. Proc Natl Acad Sci USA 1991; 88:18994–18999.
26. Zhang Y, Yao B, Delikat S, Bayoumy S, Lin XH, et al. Kinase suppressor of Ras is ceramide-activated protein kinase. Cell 1997; 89:63–72.
27. Michaud NR, Therrien M, Cacace A, Edsall LC, Spiegel S, Rubin GM, Morrison DK. KSR stimulates Raf-1 activity in a kinase-independent manner. Proc Natl Acad Sci USA 1997; 94:12792–12796.
28. Fishbein JD, Dobrowsky RT, Bielawaska A, Garrett S, Hannun YA. Ceramide-mediated growth inhibition and CAPP are conserved in Saccharomyces cerevisiae. J Biol Chem 1993; 268:9255–9261.
29. Dobrowsky RT, Kamibayashi C, Mumby MC, Hannun YA. Ceramide activates heterotrimeric protein phosphatase 2A. J Biol Chem 1993; 268:15523–15530.
30. Nickels JT, Broach JR. A ceramide-activated protein phosphatase mediates ceramide-induced G1 arrest of *Saccharomyces cerevisiae*. Genes Dev 1996; 10:382–394.
31. Pushkavera M, Obeid LM, Hannun YA. Ceramide: an endogenous regulator of apoptosis and growth suppression. Immunol Today 1995; 16:294–297.
32. Lozano J, Berra E, Municio MM, Diaz Meco MT, Dominguez I, et al. Protein kinase C ζ isoform is critical for κB-dependent promoter activation by sphingomyelinase. J Biol Chem 1994; 269:19200–19202.
33. Muller G, Ayoub M, Storz P, Rennecke J, Fabbro D, et al. PKC ζ is a molecular switch in signal transduction of TNF-α, bifunctionally regulated by ceramide and arachidonic acid. EMBO J 1995; 14:1961–1969.
34. Henkel T, Machleidt T, Alkalay I, Kronke M, Ben NY, Baeuerle PA. Rapid proteolysis of IκB-α is necessary for activation of transcription factor NF-κB. Nature 1993; 365:182–185.
35. Machleidt T, Wiegmann K, Henkel T, Schutze S, Baeuerle P, Kronke M. Sphingomyelinase activates proteolytic I kappa B-alpha degradation in a cell-free system. J Biol Chem 1994; 269:13760–13765.
36. Schutze S, Potthoff K, Machleidt T, Berkovic D, Wiegmann K, et al. TNF activates NF-κB by phosphatidylcholine-specific phospholipase C-induced "acidic" sphingomyelin breakdown. Cell 1992; 17:765–776.
37. Yang Z, Costanzo M, Golde DW, Kolesnick RN. Tumor necrosis factor activation of the sphingomyelin pathway signal nuclear factor kappa B translocation in intact HL-60 cells. J Biol Chem 1993; 268:20520–20523.
38. Verheij M, Bose R, Lin XH, Yao B, Jarvis WD, Grant S, Birrer MJ, et al. Requirement for ceramide-initiated SAPK/JNK signalling in stress-induced apoptosis. Nature 1996; 380:75–79.
39. Shirakabe K, Yamaguchi K, Shibuya H, Irie K, Matsuda S, Moriguchi T, et al. TAK1 mediates the ceramide signaling to stress-activated protein kinase/c-Jun N-terminal kinase. J Biol Chem 1997; 272:8141–8144.
40. Gulbins E, Coggeshall KM, Brenner B, Schlottmann K, Linderkamp O, Lang F. Fas-induced apoptosis of a Ras and Rac protein-regulated signaling pathway. J Biol Chem 1996; 271:26389–26394.
41. Yanaga F, Watson SP. Ceramide does not mediate the effect of tumor necrosis factor alpha on superoxide generation in human neutrophils. Biochem J 1994; 298:733–738.
42. Fuortes M, Wen-wen J, Nathan C. Ceramide selectively inhibits early events in the response of human neutrophils to tumor necrosis factor. J Leukoc Biol 1996; 59:451–460.
43. Ballou LR, Chao CP, Holness MA, Barker SC, Raghow R. Interleukin-1-mediated PGE_2 production and sphingomyelin metabolism. J Biol Chem 1992; 267:20044–20050.
44. Candela M, Barker SC, Ballou LR. Sphingosine synergistically stimulates tumor necrosis factor α-induced prostaglandin E_2 production in human fibroblasts. J Exp Med 1991; 174:1363–1369.
45. Modur V, Zimmerman GA, Prescott SM, McIntyre TM. Endothelial cell inflammatory responses to tumor necrosis factor α. J Biol Chem 1996; 271:13094–13102.
46. Balsinde J, Balboa MA, Dennis EA. Inflammatory activation of arachidonic acid signaling in murine P388D1 macrophages via sphingomyelin synthesis. J Biol Chem 1997; 272:20373–20377.
47. Bevilacqua MP. Endothelial-leukocyte adhesion molecules in human disease. Annu Rev Med 1994; 45:361–378.
48. Tracey KJ, Lowry SF. The role of cytokine mediators in septic shock. Adv Surg 1990; 23:21–56.
49. Wright SD, Ramos RA, Hermanowski-Vosatka A, Rockwell P, Detmers PA. Activation of the adhesive capacity of CR3 on neutrophils by endotoxin: dependence on lipopolysaccharide binding protein and CD14. J Exp Med 1991; 173:1281–1286.

50. Joseph CK, Wright SD, Bornmann WG, Randolph JT, Kumar ER, Bittman R, et al. Bacterial lipopolysaccharide has structural similarity to ceramide and stimulates ceramide-activated protein kinase in myeloid cells. J Biol Chem 1994; 269:17606–17610.
51. Kawasaki S, Moriguchi R, Sekiya K, Nakai T, Ono E, Kume K, Kawahara K. The cell envelope structure of the lipopolysaccharide lacking gram-negative bacterium *Sphingomonas paucimobilis*. J Bacteriol 1994; 176:284–290.
52. Hagen FS, Grant FJ, Kuijper JL, Slaughter CA, Mormaw CR, Orth K, et al. Expression and characterization of recombinant human acyloxyacyl hydrolase, a leukocyte enzyme that deacylates bacterial lipopolysaccharides. Biochemistry 1991; 30:8415–8423.
53. Wurfel MM, Wright SD. Lipopolysaccharide-binding protein and soluble CD14 transfer lipopolysaccharide to phospholipid bilayers: preferential interaction with particular classes of lipid. J Immunol 1997; 158:3925–3934.
54. Thieblemont N, Wright SD. Mice genetically hyporesponsive to lipopolysaccharide (LPS) exhibit a defect in endocytic uptake of LPS and ceramide. J Exp Med 1997; 185:2095–2100.
55. Barber SA, Perera P, Vogel SN. Defective ceramide response in C3H/HeJ (Lps^d) macrophages. J Immunol 1995; 155:2303–2305.
56. Barber SA, Defore G, McNally R, Vogel SN. Stimulation of the ceramide pathway partially mimics lipopolysaccharide-induced responses in murine peritoneal macrophages. Infect Immun 1996; 64:3397–3400.
57. Pietzsch A, Buchler C, Aslanidis C, Schmitz G. Identification and characterization of a novel monocyte/macrophage differentiation-dependent gene that is responsive to lipopolysaccharide, ceramide, and lysophosphatidylcholine. Biochem Biophys Res Commun 1997; 235:4–9.
58. Raff MC, Barres BA, Burne JF, Coles HS, Ishizaki Y, et al. Programmed cell death and the control of cell survival: lessons from the nervous system. Science 1993; 262:695–700.
59. Yuan J, Shaham S, Ledoux S, Ellis HM, Horvitz HR. The C. elegans cell death gene ced-3 encodes a protein similar to mammalian interleukin-1 beta converting enzyme. Cell 1993; 75:641–652.
60. Ellis HM, Horvitz HR. Genetic control of programmed cell death in nematode *C. elegans*. Cell 1986; 44:817–829.
61. Wyllie AH. Glucocorticoid-induced thymocyte apoptosis is associated with endonuclease activation. Nature 1997; 284:554–556.
62. Testi R. Sphingomyelin breakdown and cell fate. Trends Biochem Sci 1996; 12:468–471.
63. Liu B, Hannun RA. Inhibition of the neutral magnesium-dependent sphingomyelinase by glutathione. J Biol Chem 1997; 272:16281–16287.
64. Chinnaiyan AM, Tepper CG, Seldin MF, O'Rourke K, Kischkel FC, et al. FADD/MORT1 is a common mediator of CD95 (Fas/APO-1) and tumor necrosis factor receptor-induced apoptosis. J Biol Chem 1996; 271:4961–4965.
65. Dbaibo GS, Perry DK, Gamard CJ, Platt R, Poirier GG, Obeid LM, Hannun YA. Cytokine response modifier A (CrmA) inhibits ceramide formation in response to tumor necrosis factor (TNF)-alpha: CrmA and Bcl-2 target distinct components in the apoptotic pathway. J Exp Med 1997; 185:481–490.
66. Herr I, Wilhelm D, Bohler T, Angel P, Debatin KM. Activation of CD95 (APO-1/Fas) signaling by ceramide mediates cancer therapy-induced apoptosis. EMBO J 1997; 16:6200–6208.
67. Rivas CI, Golde DW, Vera JC, Kolesnick RN. Involvement of the sphingomyelin pathway in autocrine tumor necrosis factor signaling for human immunodeficiency virus production in chronically infected HL-60 cells. Blood 1994; 83:2191–2197.
68. Kroemer G, Zamzami N, Susin SA. Mitochondrial control of apoptosis. Immunol Today 1997; 18:44–51.
69. Morrison DC, Ryan JL. Endotoxins and disease mechanisms. Ann Rev Med 1987; 38:417–432.
70. Takei Y, Kawano S, Nishimura Y, Goto M, Nagai H, Chen SS, et al. Apoptosis: a new mechanism of endothelial and Kupffer cell killing. J Gastroent Hepat 1995; 10:S65–S67.
71. Eissner G, Kohlhuber F, Grell M, Ueffing M, Scheurich P, Hieke A, et al. Critical involvement of transmembrane tumor necrosis factor-alpha in endothelial programmed cell death mediated by ionizing radiation and bacterial endotoxin. Blood 1995; 86:4184–4193.
72. Xu DZ, Lu Q, Swank GM, Deitch EA. Effect of heat shock and endotoxin stress on enterocyte viability apoptosis and function varies based on whether the cells are exposed to heat shock or endotoxin first. Arch Surg 1996; 131:1222–1228.
73. Buchman TG, Abello PA, Smith EH, Bulkley GB. Induction of heat shock response leads to apoptosis in endothelial cells previously exposed to endotoxin. Am J Physiol 1993; 265:165–170.
74. Hoyt DG, Mannix RJ, Rusnak JM, Pitt BR, Lazo JS. Collagen is a survival factor against LPS-induced apoptosis in cultured sheep pulmonary artery endothelial cells. Am J Physiol 1995; 269:171–177.
75. Hoyt DG, Mannix RJ, Gerritsen ME, Watkins SC, Lazo JS, Pitt BR. Integrins inhibit LPS-induced DNA strand breakage in cultured lung endothelial cells. Am J Physiol 1996; 270:689–694.
76. Haimovitz-Friedman A, Cordon-Cardo C, Bayoumy S, Garzotto M, McLoughlin M, Gallily R, et al. Lipopolysaccharide induces disseminated endothelial apoptosis requiring ceramide generation. J Exp Med 1997; 186:1831–1841.
77. Colagiovanni DB, Evans RJ, McCabe JM, Bendele AM, Edwards CK. Protection of fas mutant MRL-lpr/lpr mice to LPS or staphylococcal enterotoxin B-induced septic shock by tumor necrosis factor binding protein (TNF-bp). In press.
78. Fuks Z, Persaud RS, Alfieri A, McLoughlin M, Ehleiter D, Schwartz JL, et al. Basic fibroblast growth factor protects endothelial cells against radiation-induced programmed cell death in vitro and in vivo. Cancer Res 1994; 54:2582–2590.

32

Role of NF-κB in Macrophage Activation

Tsuneo Suzuki
University of Kansas Medical Center, Kansas City, Kansas

PROPERTIES OF NF-κB IN MACROPHAGES

NF-κB is a ubiquitous transcription factor, which is a heterodimer consisting of proteins of the Rel/NF-κB family. It exists in the cytoplasm as an inactive form due to its association with a specific inhibitor protein family, designated as IκB, which inhibits the nuclear translocation of NF-κB by binding to nuclear localizing signal peptides of NF-κB (reviewed in Refs 1,2). Treatment of cells with a wide variety of stimulants including LPS has been shown to free NF-κB from IκB, most likely as a result of phosphorylation and/or proteolysis of IκB protein (1–8). Free NF-κB then translocates to the nucleus and binds to the κB consensus motif (GGGRNNYYCC), which exists in the enhancer/promoter regions of genes coding for various cytokines such as IFN-β (9), GM-CSF (10), G-CSF (11), IL-6 (12), or TNF-α (13), for cell surface proteins such as MHC class I (14) or II (15) or Fcγ receptor (16), and for cellular enzyme such as inducible nitric oxide synthase (iNOS) (17,18).

A mouse macrophage cell line, J774, responds to LPS (>10 ng/ml) by the activation of various cytokine genes and the *iNOS* gene, produce biologically active cytokines and NO, and develop bactericidal/tumoricidal activity (19). Immunoprecipitation studies show that J774 cells express, in cytoplasm, four members of the mammalian NF-κB/Rel family—p50, p52, RelA (p65), and c-Rel—and three forms of the IκB family —IκBα, p105, and p100. Electrophoretic mobility shift assay (EMSA) using a radiolabeled oligonucleotide containing HIV-κB motif shows that J774 cells constitutively express low levels of NF-κB proteins in the nucleus and respond to LPS (>0.1 ng/ml) treatment by a marked increase in the long-lasting expression of three types of NF-κB–binding nuclear proteins. These are designated as NF-κB1, NF-κB2, and NF-κB3, according to their electrophoretic mobilities (fast, intermediate, and slow, respectively) (20,21). Immunological and UV cross-linking studies showed that NF-κB1 consists of only p50 subunit, whereas both NF-κB2 and κB3 contain the p65 subunit and c-Rel. In addition, NF-κB2 was found to contain the p50 subunit of NF-κB. NF-κB2 and κB3 complexes appear in nuclei as early as 10 minutes after stimulation of J774 cells with LPS and remain substantially elevated for a least 8 hours. On the other hand, the expression of NF-κB1 (p50 dimer) in nuclei is delayed until between 60 and 120 minutes after LPS stimulation and continues to markedly increase up to the end of the observation period (8 hr). LPS-induced activation of NF-κB was partially inhibited by pretreatment with a protein kinase A inhibitor (H89) but not at all by inhibitors of other protein kinases (H-7, H-8, HA1004, tyrphostin, W-7, ML-7, or staurosporine) (20).

Composition of NF-κB Proteins and Binding Affinity

Since the three types of NF-κB proteins of J774 cells are composed of different subunits, their fine DNA-binding specificities are expected to differ. Indeed, the comparison of the order of IC_{50} values of different NF-κB motifs for NF-κB1, NF-κB2, and NF-κB3 revealed that p50 subunit–containing NF-κB1 and NF-κB2 bind H2-K[b] NF-κB motif (GGGGATTCCC) better than the

IFN-β motif (GGGAAATTCC), whereas the c-Rel and p65 subunit–containing NF-κB3, which lacks the p50 subunit, preferentially binds to the IFN-β motif (21). Similar, but not identical, difference in binding affinity depending on NF-κB subunit composition was also reported by several laboratories. Baldwin and Sharp (14) and Kieran et al. (22) separately noted that a p50 homodimer exhibits a high affinity for an NF-κB motif of the MHC class I gene promoter but shows a weaker binding to an NF-κB motif of the HIV enhancer than p50/p65 heterodimer. Fujita et al. (23) used recombinant p50 and p65 subunits to show that p50 homodimer binds to the NF-κB motif almost 10-fold more than a p65 homodimer, whereas the binding efficiency of a p50/65 heterodimer is intermediate. Nakayama et al. (24) presented evidence that the c-Rel homodimer has a higher affinity for NF-κB sites of IL-6 (GGGATTTTCC) or IFN-β than c-Rel/p50 dimer. However, the question of whether or not various combinations of NF-κB proteins differentially activate genes with different NF-κB motifs requires further studies.

NF-κB–Binding Affinity and 3′ Half-Site NF-κB Nucleotide Sequence

Competitive binding assays using various synthetic NF-κB motifs showed that these motifs could be classified into three groups based on their inhibitory capacities (21). Thus, the specific binding of NF-κB proteins to the HIV NF-κB motif could be effectively inhibited with IC_{50} of 3.4–27 nM by decameric NF-κB motifs present in the enhancer regions of HIV, IL-6, IFN-β, H2-K^b, I-$Eα^d$, and TNF-α2, whose 3′ half-site nucleotide sequences are T/A-T-T/C-CC. This group of NF-κB motifs will be referred to as high affinity. The inhibitory capacities of oligonucleotides containing NF-κB motifs with 3′ half-site sequences of TGCCC (TNF-α3), ATCTC (G-CSF), TATTC (FcγR), or TCCTT (TNF-α1) were found to be 10–100 times lower than those with T/A-T-T/C-CC sequence. This group of NF-κB motifs will be referred to as low affinity. The inhibitory capacity of GM-CSF NF-κB motif with 3′ half-site of CTACC was between the two groups and was designated as intermediate. Comparison of nucleotide sequences of three groups revealed that neither substitution of A to G at position 4 nor substitution of C to A at position 5 contributes to affinity change because these substitutions occur in high-, intermediate-, or low-affinity NF-κB motifs. The T to A substitution at position 6 appears to be tolerated by NF-κB proteins because both high-affinity (IFN-β) and low-affinity (G-CSF) motifs have this change.

However, T to C substitution at this position may weaken the affinity as seen in GM-CSF. The substitution of T with G, A, or C at position 7 appears to significantly contribute to the affinity, since all high-affinity NF-κB motifs have T at this position, whereas all low-affinity NF-κB motifs except G-CSF have G, A, or C. T at position 8 could be substituted by C without affecting the affinity, because both high- and low-affinity NF-κB motifs have either T or C at this position. However, the T to A substitution at this position may affect the affinity, because the intermediate affinity GM-CSF possesses such a substitution. The substitution of C to T or G at position 9 appears to be critical for NF-κB–binding affinity, because all of the high-affinity NF-κB motifs have C at this position, which is substituted by T or G in all of the low-affinity NF-κB motifs except that of TNF-α3. The nucleotide C at position 10 does not appear to influence the NF-κB–binding affinity, because all NF-κB motifs except that for TNF-α1 have C at this position regardless of their inhibitory capacities. The difference in the affinity of the NF-κB protein towards various NF-κB motifs was also reported by Zabel et al. (25). Their data showed that the binding of NF-κB proteins purified from human placenta to κB motif (5′GGGACTTTCC-3′) is most effectively inhibited by a homologous NF-κB motif and is also inhibited, in order of decreasing efficiencies, by heterologous NF-κB motifs of IL-2 receptor (5′-GGGAATCTCC-3′), IFN-β (5′-GGGAAATTCC-3′), IL-2 (5′-GGGATTTCAC-3′), GM-CSF (-107/-98) (5′-GAGATTCCAC-3′), and mutated IL-2 receptor (5′-GGGAATCTAA-3′). The comparison of these nucleotide sequences again reveals the importance of positions 7 and 9 in influencing the affinity of NF-κB, because strong inhibitors, such as the NF-κB motifs of HIV or IFN-β, have T and C at these positions, whereas the weak inhibitors, such as the NF-κB motifs of GM-CSF (-107/-98) or mutated IL-2 receptor, have substitutions at position 7 and 9 from T to C and C to A, respectively. Of 11 different NF-κB motifs examined in our study and 6 additional NF-κB motifs reported by Zabel et al. (25), the only exception to this rule is the NF-κB motif of the IL-2 receptor, which has a substitution at position 7 from T to C but was shown to effectively compete against a κB motif.

MECHANISMS OF NF-κB ACTIVATION

Role of IκBα Protein Kinases

The activation of NF-κB is defined as the dissociation of NF-κB from IκB and subsequent translocation of

NF-κB and Rel dimers to the nucleus (reviewed in Refs. 1,2). The prerequisite for the activation of NF-κB and its subsequent translocation to the nucleus is thought to involve the phosphorylation and/or degradation of IκB. IκBα, the best studied member of the IκB family, has been shown to consist of three domains, an N-terminal regulatory, a central Rel-interactive, and a highly acidic C-terminal domain (reviewed in Ref. 2). Serines at positions 32 and 36 (26) and tyrosine at position 42 (27) within the N-terminal regulatory domain have been identified as the phosphorylation sites of IκBα. Although phosphorylation of serines at 32 and 36 does not lead to dissociation of IκBα from NF-κB, mutation of the serines at these positions to alanine was found to prevent inducible phosphorylation and degradation of IκBα. The C-terminal region of IκBα contains a PEST domain, which is characterized by richness in proline, glutamic acid, serine, and threonine residues and has been shown to be the site of constitutive phosphorylation (28). Because phosphorylation of IκBα appears to be required for the subsequent degradation of IκBα and therefore the activation of NF-κB, extensive efforts are being made by a large number of different laboratories to identify the nature of the IκBα protein kinases that specifically phosphorylate IκBα. Results of these studies, which have utilized a wide variety of cell types and diverse stimuli, implicate the involvement of various protein kinases such as protein kinase C (PKC), protein kinase A (PKA), casein kinase II (CKII), raf-1 kinase, protein tyrosine kinase (PTK), mitogen-activated protein kinase/extracellular signal-regulated kinase kinase (MEKK), and double-stranded RNA activated kinase (PKR).

Role of Protein Kinase C

Because the addition of purified PKC and ATP to cytosol (29) or to partially purified NF-κB–IκB complexes (3) can activate NF-κB, and because PMA, a potent activator of PKC, strongly activates NF-κB in 70Z/3 pre-B cells (30,31), PKC has been considered to mediate phosphorylation of IκB. Based on studies using the PKC inhibitor calphostin C, Lindholm et al. (32) showed the involvement of PKC in human T-cell lymphotropic virus type 1 Tax1-mediated activation of NF-κB in Tax$_1$-transformed C81 cells and Tax$_1$-stimulated murine pre-B cells. More recently, based on results of dominant negative experiments and direct expression experiments, PKCζ was shown to induce phosphorylation and inactivation of IκBα in vitro (33–36). The potential role of PKCε in NF-κB activation was suggested by PKC overexpression experiments with rat fibroblast (3Y1) (37) and human T-cell line (JH6.2) (38). However, Janosch et al. (39) showed that none of the recombinant, purified PKC isozymes α, β, γ, δ, ε, η, and ζ were capable of phosphorylating recombinant IκBα in vitro. Thus, whether phosphorylation of IκBα by PKC is solely responsible for the activation of NF-κB and what PKC isoenzymes are responsible for IκBα phosphorylation remain controversial.

Role of Protein Kinase A

LPS-triggered NF-κB activation in J774 cells is, at least in part, mediated by type I PKA, but not by PKC (20). Evidence for this notion includes (1) lack of activation of NF-κB in response to PMA, a potent PKC activator; (2) LPS-triggered activation of NF-κB in J774 cells whose cellular PKC was depleted from both cytosol and particulate compartments by PMA pretreatment; (3) specific partial blockage of LPS-triggered activation of NF-κB by pretreatment of J774 cells with a PKA inhibitor, H-89, but not by a PKC inhibitor, H-7; and (4) activation of NF-κB in response to treatment with the cAMP-elevating agents, dibutyric cAMP, forskolin, prostaglandin E$_2$, or cholera toxin. However, a definitive role of PKA in the activation of NF-κB remains controversial, as discussed below. Shirakawa and Mizel (29,40) implicated PKA and PKC in the IL-1–induced activation of NF-κB in 70Z/3 pre-B cells and the human NK-like cell line, YT. Their evidence included (1) the inhibition of IL-1–induced activation of NF-κB by PKA inhibitor H8; (2) the activation of NF-κB by preincubation of cells with 8-bromo-cAMP, forskolin, or PMA; and (3) in vitro activation of NF-κB by PKA as well as PKC. More recently, Jeon et al. (41) showed a marked inhibition of LPS-induced NF-κB activation, *iNOS* gene transcription, and NO production in RAW 264.7 cells treated with delta 9-tetrahydro-cannabinol (delta 9-THC), which negatively regulates adenylate cyclase. Because RAW 264.7 cells respond to LPS or forskolin by the activation of cAMP response element as well as NF-κB, which could be inhibited by delta 9-THC, these data suggest the involvement of PKA in LPS-induced activation of NF-κB. On the other hand, Bomsztyk et al. (42) showed that neither PGE$_2$ nor dibutyric cAMP itself activates NF-κB in 70Z/3 cells, although the degree of the IL-1–induced activation of NF-κB could be modified by cAMP-elevating agents such as PGE$_2$ or dibutyric cAMP. Evidence was also presented that PKA in HL60 cells plays no role in the activation of NF-κB by TNF-α or TNF-β (lymphotoxin-α) (43).

Role of Non-PKC, Non-PKA Kinase

Meichle et al. (44) proposed an alternative pathway leading to the activation of NF-κB in two human leukemic cell lines, K562 and Jurkat. These cells responded to TNF or PMA by the activation of PKC, which was followed by the activation of NF-κB. However, pretreatment of Jurkat cells with H7 or staurosporine, which inhibits TNF- or PMA-mediated activation of PKC, inhibited PMA- but not TNF-induced activation of NF-κB. Furthermore, since PKC depletion by PMA pretreatment of Jurkat cells did not block the TNF-induced activation of NF-κB, they suggested that NF-κB activation by TNF may be mediated in a PKC-independent manner. A potential involvement of PKA in their experimental system was eliminated by the failure to activate NF-κB with dibutyric cAMP and the inability to block NF-κB activation with H8. Bomsztyk et al. (45) also suggested that physiological activators such as IL-1 and LPS activate NF-κB by pathways independent of PKC in both 70Z/3 and EL-4 6.1 C10 cells. Similarly, Vinceti et al. (46) concluded that NF-κB is activated in RAW264.7 cells by LPS via a pathway independent of PKC, because LPS-mediated activation was not inhibited by pretreatment of the cells with PMA or H7. Their results also showed that RAW264.7 cells, unlike J774 cells, do not respond to exogenous 8-bromo cAMP by activation of NF-κB. In none of these studies was an effort made to identify IκBα kinase.

Role of Casein Kinase II

Janosch et al. (39) showed that the IκBα kinase present in porcine spleen is casein kinase II (CKII), which efficiently phosphorylates the C-terminal portion of IκBα, but not raf-1 kinase or PKC isoenzymes. Phosphorylation of the IκBα PEST domain by CKII appears, however, to regulate basal rather than induced turnover of IκBα. Bennett et al. (47) identified two discrete kinases (36 and 41 kDa) in the cytoplasm of human umbilical vein endothelial cells that specifically bind to and phosphorylate IκBα. The activity of these kinases is upregulated in response to stimuli (TNF-α, IL-1β, or LPS) and precedes activation of either mitogen-activated kinase or jun kinase. Based on deletion mutagenesis of IκBα, these authors showed that these kinases bind in or around the ankyrin repeat domains and phosphorylate residues within the C-terminus of IκBα. Their data showed that these kinases are not identical to CKII and possess a pharmacological profile distinct from other known kinases.

Role of Other Protein Kinases

Raf-1 kinase is another candidate for an IκBα kinase (48–50). However, since the activation of raf-1 kinase appears not to lead to the nuclear translocation of NF-κB, raf-1 kinase is proposed to be involved in stimulation of the transcriptional activation domain of constitutive nuclear NF-κB (2). Several laboratories have investigated the role of protein tyrosine kinase (PTK) in NF-κB activation. Imbert et al. (27) showed that Jurkat cells respond to pervanadate (a protein tyrosine phosphatase inhibitor) by NF-κB activation through tyrosine phosphorylation but not degradation of IκBα. This process was dependent of the expression of the T-cell PTK p56[lck]. On the other hand, Delude et al. (51) showed that PTK activity is not involved in the LPS-mediated nuclear localization of NF-κB. This was confirmed by Yoza et al. (52) in THP-1 cells. Their data showed, however, that PTK is required for NF-κB–dependent activation of IL-1β gene transcription. Mitogen-activated protein kinase/extracellular signal–regulated kinase kinase (MEKK) kinase was also suggested to be involved in TNF-α–induced activation of NF-κB (53). Chen et al. (54) recently identified an IκBα kinase in unstimulated HeLa cell cytoplasmic extract that can phosphorylate IκBα at Ser-32 and Ser-36. This kinase activity was shown to be activated in vitro by ubiquitination. Subsequent studies by Maniatis group (55) showed that this IκBα kinase activity is inducible by TNF-α and can be directly activated in vitro by mitogen-activated protein kinase/ERK kinase kinase-1 (MEKK1), which induces the site-specific phosphorylation of IκBα in vivo. A role of double-stranded RNA-dependent protein kinase (PKR) in NF-κB activation was suggested by Maran et al. (56) using antisense oligonucleotide to ablate the mRNA encoding the PKR.

The nature of protein kinases involved in the activation of NF-κB thus remains highly controversial. Of particular difficulty to reconcile is the disparity of the data reported using the same cell type and the same ligand, as cited above.

Proteolysis of IκBα and Activation of NF-κB: Effect of Serine and Cysteine Protease Inhibitors on NF-κB Activation

NF-κB–activating stimuli have recently been shown to lead not only to phosphorylation but also to a rapid degradation of IκBα protein (5–7,57–66). As shown in Figure 1, Western analysis shows that IκBα present in the cytosol of J774 cells becomes undetectable by 15 minutes after stimulation of the cells with LPS (10 ng/ml), reappears by 30 minutes, and returns to the prestimulation level by 3 hours. The loss of IκBα from cytosol coincides with the appearance of NF-κB proteins in nuclear fraction (see lower panel). However, NF-κB remains activated for the entire period of observation (8 hr), despite the reappearance of IκBα in the cytosol 30 minutes after LPS treatment. Pretreatment of J774 cells with either phenylalanine–chloromethyl ketone (PCK) or tosylphenylalanine chloromethyl ketone (TPCK) effectively blocked LPS-triggered IκBα degradation and NF-κB activation, although in contrast to the other cell types, N^α-p-tosyl-L-lysine chloromethyl ketone (TLCK) pretreatment did not lead to a substantial inhibition (64). Other types of protease inhibitors such as E-64, PMSF, 4-amidinophenyl-p-(6-amidino-2-indolyl)phenyl ether (APMSF), 4-(2-aminoethyl)benzenesulfonyl fluoride (AEBSF) or their structural analogs [p-toluenesulfonamide (TSDA) or p-toluene-sulfonylfluoride (TSF)] or a chymotrypsin/trypsin substrate [N-benzoyl-L-tyrosine ethyl ester (BTEE) or its structural analogs (N-benzoyl-D, L-phenylalanine β-naphthyl ester (BPNE) or N-benzoyl-L-phenylalanine (BP)] had no demonstrable effect on LPS-triggered NF-κB activation in J774 cells (64). A comparison of the structures of these protease inhibitors or their structural analogs suggests the importance of a benzene ring and a chloromethyl ketone group within the phenylalanine backbone of TPCK and PCK for manifestation of the inhibitory action. The two most effective inhibitors, PCK and TPCK, possess these chemical groups, whereas TLCK, which differs from TPCK by the substitution of a benzene ring by an amino group, loses substantial inhibitory activity. A tosyl group in TPCK does not appear to be involved in the inhibition, because PCK, which lacks this group, was a very effective inhibitor and none of its structural analogs such as AEBSF, TSAD, TLCK, TSF, PMSF, or APMSF was capable of inhibiting LPS-triggered NF-κB activation.

Henkel et al. (6) originally reported that pretreatment of cells with each of six different serine protease inhibitors, including TPCK, TLCK, BTEE, benzyloxycarbonyl-leu-tyr-chloromethylketone, 3,4-dichloroisocoumarin, or N-acetyl-DL-Phe-β-naphtylester, substantially inhibited NF-κB activation in Jurkat cells stimulated with TNF-α or in 70Z/3 cells treated with PMA. Finco et al. (58) showed that pretreatment of HeLa or Jurkat cells with TPCK, TLCK, BTEE, or AEBSF blocked TNF-α–triggered NF-κB activation, whereas pretreatment of the cells with chymostatin, leupeptin, or N^α-benzoyl-L-arginyl ethyl ester did not. Mackman (60) and Kim et al. (61) also reported that LPS-induced NF-κB activation in human monocytes and THP-1 cells or in mouse peritoneal macrophages could be effectively inhibited by pretreatment of cells with either TPCK or TLCK. Although Henkel et al. (6) and Finco et al. (58) reported effective inhibition by pretreatment of cells with either BTEE or AEBSF, these reagents did not block LPS-triggered NF-κB activation in J774 cells. The reason for this discrepancy is uncertain but could be due to differences in the cell types studied.

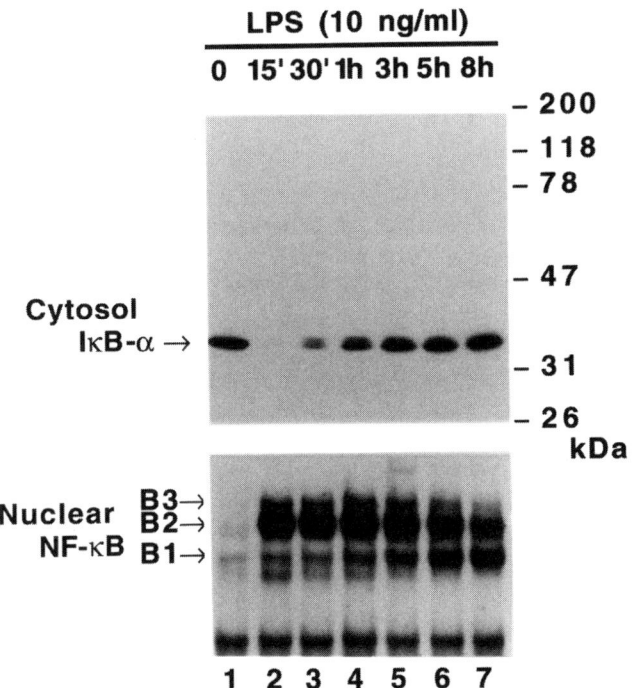

Fig. 1 LPS stimulation of J774 cells causes transient loss of IκBα from the cytosol (top panel) and persistent translocation of NF-κB proteins to the nucleus (bottom panel). J774 cells were stimulated with LPS (10 ng/ml) for the period indicated in the figure. At the end of each incubation period, the cytosol fraction and nuclear extracts were prepared from the homogenates of cells. IκBα proteins in the cytosol fraction were detected by Western analysis using antiserum against mouse IκBα following SDS-PAGE of the samples. Nuclear extracts were analyzed for NF-κB activity by electrophoretic mobility shift assay using radiolabeled HIV NF-κB probe as described in Refs. 20, 21, and 64. B1 is a p50 homodimer. B2 consists of p50, p65, and c-Rel. B3 is a heterodimer consisting of p65 and c-Rel.

Protease Inhibitors Directly Modify DNA-Binding of NF-κB Proteins

The in vitro modification of DNA-binding activity by protease inhibitors was reported by Finco et al. (58), although other investigators reported that TPCK has no effect on in vitro DNA-binding capacity of NF-κB (59). This discrepancy could be due to the presence or absence of DTT during nuclear extraction, as suggested by Finco et al. (58). Indeed, elimination of DTT from the buffers used during nuclear extraction led us to confirm that either TPCK or PCK directly inhibits DNA-binding activity of NF-κB extracted from J774 cells (64).

Finco et al. (58) showed that pretreatment of HeLa cells with TPCK or TLCK inhibited the DNA-binding activity of not only NF-κB but also AP-1. However, it did not inhibit Oct-1 or serum response element–binding protein. Mackman reported that TPCK or TLCK pretreatment of THP-1 cells blocked LPS-stimulated activation of NF-κB, but did not inhibit PMA-induced SP-1 or EGR-1 (60). Miyamoto et al. reported that TPCK pretreatment of WEHI231 cells inhibited DNA-binding of constitutively expressed NF-κB but did not inhibit that of AP-1 (66). Pretreatment of J774 cells with either PCK or TPCK suppressed both constitutive or LPS-stimulated levels of not only NF-κB, but also AP-1, CREB, and Oct-1 (64). The reason why TPCK or TLCK differentially affects the DNA-binding activities of different transcription factors is unclear at present and may be due to differences in the cell types and/or the properties of transcription factors studied.

Nature of IκBα-Degrading Protease

Since pretreatment of cells with TPCK or TLCK has been shown to prevent the stimuli-triggered degradation of IκBα and activation of NF-κB, LPS-mediated activation of a cellular chymotrypsin-like protease has been first proposed to play an important role in the NF-κB activation signal transduction (58–60,65). However, Palombella et al. (62) and Traenckner et al. (63) suggested, based on the use of specific inhibitors, a potential involvement of the ubiquitin-proteasome pathway in the degradation of IκBα and NF-κB activation induced by TNF-α in human MG-63 or HeLa cells. The 26S proteasome consists of a 20S multicatalytic protease complex associated with additional regulatory subunits required for the recognition and degradation of multiubiquitinated proteins (reviewed in Ref. 67). The inhibitors used by Palombella et al. include carbobenzoxyl-leucinyl-leucinyl-norvalinal-H (also called Z-LLnV) or carbobenzoxyl-leucinyl-leucinyl-leucinyl-H (also called Z-LLL) (62), whereas that used by Traenckner et al. was Cbz-Ile-Glu(O-t-Bu)-ala-leucinal (63). The latter inhibitor was also used to demonstrate effective inhibition of LPS-induced activation of NF-κB and *iNOS* gene (68,69). Chen et al. (70) provided evidence that phosphorylation of serine residues at 32 and 36 of IκBα targets the protein to the ubiquitin-proteasome pathway. They demonstrated the occurrence of in vivo and in vitro ubiquitination of IκBα following phosphorylation. Mutations that abolish phosphorylation and degradation of IκBα in vivo prevent ubiquitination in vitro. Furthermore, their data showed that ubiquitinated IκBα remains associated with NF-κB and that the bound IκBα is degraded by the 26S proteasome. In contrast, Roff et al. (71) showed that TNF-α treatment of HeLa cells results in phosphorylation-dependent ubiquitination and release of NF-κB from IκBα. Lin et al. (72) also concluded, based on the experiments using calpain inhibitor I and II, that signal-induced proteolysis, but not phosphorylation, of IκBα is an obligatory step for NF-κB activation, although they did not directly characterize any IκBα-degrading proteases. The question remains, however, why only a small portion of the IκBα accumulates in the ubiquitinated form in the cytosol. Using IκBα mutant proteins, in which a series of lysine-to-arginine mutations were introduced, the major sites of ubiquitin ligation that target the protein for rapid degradation by proteasome have been identified as lysine residues at positions 21 and 22 (73). Li et al. (74) showed, using the S100 fraction as a source of IκBα-degrading proteasome, apparent cleavage of recombinant IκBα in vitro in the presence of bovine ubiquitin and ATP. However, a question of whether or not their findings represent actual events in living cells remains, because (1) TPCK and TLCK, which are good inhibitors of signal-induced in vivo IκBα degradation, did not inhibit the in vitro degradation of recombinant IκBα, and (2) leupeptin and PMSF, which are poor inhibitors of in vivo degradation were good inhibitors of in vitro IκBα degradation. Thus, the exact nature of the cellular proteases involved in the NF-κB–activation process remains controversial.

Autoregulation of IκBα

The promoter region of the *IκBα* gene has a functionally important κB site (75). Activation of NF-κB by PMA treatment of 70Z/3 cells results in *IκBα* gene transcription reach maximum levels by 1 hour followed by a steady decline (65,76). Newly synthesized IκBα may be transported to the nucleus and inhibit NF-κB

DNA binding activity (77). This was suggested to be the mechanism to recycle the NF-κB/IκBα complex back to cytoplasm and to make cells capable to respond to further activation signals. Indeed, J774 cells do re-synthesize IκBα 30 minutes after LPS treatment. Cytosolic levels of IκBα appear to recover to prestimulation levels within 1 hour. However, the newly synthesized IκBα appears to remain in cytosol, as nuclear NF-κB proteins remain highly elevated (see Fig. 1). The IκBα protein appears to be inherently unstable and has been shown to have a short half-life (6,7,66). This rapid turnover of IκBα is likely to be mediated by the proteasome system. However, unlike signal-induced degradation of IκBα, basal turnover appears to require neither ubiquitination of amino-terminal signal response element or carboxy-terminal PEST sequence (78).

BIOLOGICAL SIGNIFICANCE OF NF-κB ACTIVATION IN MACROPHAGE ACTIVATION

Macrophages develop the ability to kill bacteria or tumor cells in response to IFN-γ and/or LPS. Full development of bactericidal/tumoricidal activity requires at least several hours and is dependent on the activation of various genes and subsequent production of biologically active cytokines and cytotoxic substances such as nitric oxide (NO). Some genes such as c-*jun*, *junB*, c-*fos*, c-*myc*, *JE*, *KC* and *TNF-α* are rapidly activated following LPS treatment of macrophages, whereas other genes such as *IL-1α* and *IL-1β*, *IL-6*, *IFN-α* and *IFN-β*, *M-CSF*, and *iNOS* respond more slowly. Activation of all of these genes is probably involved in the full development of cytocidal activity, although the exact relationships among these various gene products is still unknown. NF-κB plays a major role in the transcriptional activation of many of these genes, which possess NF-κB motifs in their promote/enhancer regions. Therefore, the inhibition of LPS-triggered NF-κB activation can be expected to suppress macrophage activation. Indeed, PCK pretreatment of J774 mouse macrophages, which inhibits LPS-induced IκBα degradation and NF-κB activation, causes a marked inhibition of expression of LPS-inducible genes specific for *TNF-α*, *IL-1α* and *IL-1β*, *IL-6*, *iNOS*, and *IκBα* and results in inhibition of LPS-triggered production of NO and tumoricidal activity (64). Similarly, pretreatment of human monocytes or monocytic THP-1 cells with either TPCK or TLCK, which inhibit LPS-induced IκBα degradation and nuclear translocation of Rel/RelA heterodimers, leads to inhibition of expression of tissue factor–specific mRNA and tissue factor protein by THP-1 cells (60).

Mouse macrophages express the *iNOS* gene in response to LPS and produce NO, which kills NO-sensitive tumor cells and microorganisms (19,79–81). Endogenous IFN-β, but not TNF-α, secreted from macrophages contributes to LPS-triggered activation of the *iNOS* gene (19). LPS-mediated NO production is markedly increased when macrophages are co-stimulated with IFN-γ (79). These findings suggest that mouse *iNOS* gene expression is transcriptionally regulated at least at two different enhancer/promoter regions. Indeed, deletional analysis of the 5′-flanking region of the *iNOS* gene cloned independently by two laboratories (17,18) revealed two regions, referred to as I (-209 to -48) and II (-1029 to -913), that are necessary for full gene induction. Region I, which contains Oct-1, NF-κB, NF-IL-6, and TNF-α–responsive elements but lacks an IFN-responsive element, plays an important role in mediating the *iNOS*-inducing effect of LPS. Region II, which contains the IFN-responsive elements ISRE and PIE, in addition to an NF-κB motif and a TNF-α–responsive element, mediates IFN-related responses (17,18). NF-κB motifs in Regions I and II (denoted as κBI and κBII, respectively) are similar but not identical. The κBI sequence of GGGGAATCTCCC is almost identical to the κB site of the *IL2Rα* gene (GGGGACTCTCCC), whereas the κBII sequence of GGGACTTTCC relates well to the κB site of the *IL-6* gene (GGGATTTCC). The transfection analysis of luciferase reporter gene constructs containing *iNOS* κBI, κBII elements, or their mutants showed that both κBI and κBII sites significantly but differently participate in LPS-induced promoter activity (79). Mutation of κBII substantially reduces LPS-induced promoter activity, whereas mutation of the κBI site has less severe effects. In vivo footprint analysis showed that both κB elements are bound by protein complexes when macrophages are stimulated with LPS with or without IFN-γ (80). Such protein complexes are most likely NF-κB proteins, since electrophoretic mobility shift assay using radiolabeled κBI or κBII motif as probes clearly showed a marked increase of NF-κB proteins in nuclear extracts of LPS-stimulated macrophages (80). Maximum activation of NF-κB was reached within 0.5 hour after stimulation with LPS and then decreased in intensity but remained activated substantially above the constitutive level for 3 hours. IFN-γ treatment of macrophages did not affect either the LPS-induced rate of activation of NF-κB proteins or their subsequent loss from the nuclear fraction. Supershift analysis using various antibodies showed that the NF-κB proteins bound to κBI and κBII oligonucleotides were not identical. Both κBI and κBII binding proteins consisted of a p50 homodimer, p50/p52 heterodimer, and a complex formed with p50, p65, and c-Rel. κBII oligonucleotide binds an additional complex consisting of p65 and c-Rel. Competition experiments showed that (1) the p50 homodimer and p50/52 heterodimer bind to both κBI and κBII with approximately equal affinity, (2) the p50/p65 heterodimer has higher affinity for κBII than for κBI, and (3) the p65/c-Rel heterodimer specifically interacts with κBII. A potential contribution of different NF-κB complex that binds to the κBI or κBII sites to transactivation of the iNOS gene requires further studies, however. Goldring et al. (82) found that LPS treatment of RAW 264.7 cells activates p50 homodimer and p50/p65 heterodimer, which bind to both κBI and κBII sites in the mouse iNOS promoter. Based on depletion studies of NF-κB/Rel activity through the use of a phosphorothioate-modified oligonucleotide containing three copies of κB consensus sequence, they suggested that NF-κB activators or repressors other than LPS might affect macrophage NO response. The importance of NF-κB in the activation of iNOS in RAW 264.7 cells transfected with plasmid harboring PKCε has been also suggested (83). In this study, transfection of macrophages with plasmids containing PKCα, β1, or δ did not activate the iNOS gene. Since PKCε-mediated iNOS gene activation was blocked by pyrrolidine dithiocarbamate, the effect of PKCε gene overexpression may be mediated through the reactive oxgen intermediate–triggered activation of NF-κB.

DOWNREGULATION OF NF-κB–MEDIATED TRANSCRIPTION

Role of NF-κB in LPS Tolerance

Monocytes/Macrophages activated by the initial exposure to LPS respond to the secondary exposure to LPS by the minimal activation of various genes. This phenomenon is called LPS tolerance or LPS-induced desensitization and has been the subject of extensive studies (see Chapter XX). Löms Ziegler-Heitbrock et al. (84) showed that human monocytic cell line Mono Mac 6 responds to LPS (20 ng/ml) by rapid, transient expression of mRNAs specific for TNF, IL-1, and IL-6 and production of biologically active cytokines. Pretreatment of the cells with LPS (20 ng/ml) for 2 days significantly reduced both the expression and the production of these mRNAs and cytokines in response to subsequent LPS (1 μg/ml) stimulation. LPS tolerance apparently increased CD14 expression on the cell surface and only slightly decreased the affinity of the LPS-CD14 interaction. Both naive and LPS-tolerant cells responded to LPS by translocation of NF-κB to nuclei. However, NF-κB proteins activated in native Mono Mac 6 cells consist mainly of p50-p65 heterodimer, whereas the predominant NF-κB protein activated in tolerance cells was the p50 homodimer. They furthermore showed, by reporter gene analysis using NF-κB–dependent luciferase constructs, that NF-κB proteins activated in LPS-tolerance cells are functionally inactive. Based on these data, they suggested that LPS tolerance is caused by postreceptor mechanisms that involve an altered composition of the NF-κB complex.

Inhibition of NF-κB DNA Binding by Nitric Oxide

NO produced by macrophages as a consequence of LPS-induced activation of iNOS gene via NF-κB has been shown to inactivate several proteins by modification of reactive thiol groups (85). All members of the NF-κB/Rel family possess a reactive cysteine residue towards the N-terminus of their Rel homology regions. NO produced by macrophages therefore potentially participates in downregulation of NF-κB. Because N-monomethyl arginine significantly enhances LPS-induced NF-κB activation in a mouse macrophage cell line, Chen et al. (81) suggested that endogenous NO may modify the cysteine at position 62 of p52 and downregulate NF-κB activity. The data of Matthews et al. (86) partially support this concept by showing the correlation between S-nitrosylation of cys 62 of p50 protein and in vitro inhibition of DNA binding of p50. However, the concentrations of NO donors used in this study were much higher than physiological concentrations.

Other Inhibitors of NF-κB–Mediated Transcription

Ollivier et al. (87) reported that the elevation of cAMP level or overexpression of the catalytic subunit of PKA inhibits the induction of NF-κB–dependent gene expression in transiently transfected human monocytic THP-1 cells and human umbilical vein endothelial cells. Because elevated levels of cAMP does not prevent nuclear translocation of p50/p65 or p65/c-Rel heterodimers, the noted inhibitory effect could be at the level of transactivation. Acetylsalicylic acid (aspirin) and sodium salicylate inhibit LPS-induced synthesis of tissue factor, TNF-α, and PGE$_2$ by human monocytes as a result of inhibition of NF-κB activation (88,89). Glucocorticoids may inhibit NF-κB by (1) binding of glucocorticoid-activated receptors to NF-κB subunits (90,91) or (2) transcriptionally activating IκBα (92,93).

SUMMARY

The activation of NF-κB is a critical step in LPS-induced macrophage activation, which depends on the activation of various genes for cytokines, cell surface molecules, and cellular enzymes. Extensive research by numerous laboratories in recent years rapidly advanced our knowledge on the structure and function of proteins of the NF-κB and IκB families. A large body of evidence clearly indicates that phosphorylation and degradation of IκBα in response to LPS are the essential events for NF-κB activation. Several important questions regarding the mechanism of LPS-triggered NF-κB activation still remain to be examined, however. Which of protein kinases so far implicated are indeed involved in LPS-triggered phosphorylation of IκBα? How are such protein kinase activities regulated by LPS? Is the proteasome system solely responsible for IκBα degradation? How is the activity of such a proteasome system regulated by LPS?

ACKNOWLEDGMENTS

Work from the author's laboratory was supported in part by grants from the National Cancer Institute (RO1 CA35977 and PO1 CA54474) and a grant from the Kansas Health Foundation. The author wishes to acknowledge the significant contributions to these studies by Drs. M. Muroi, M. Fujihara, N. R. Rice, K. Yamamoto, Y. Watanabe, W. J. Murphy, J. L. Pace, S. W. Russell, and C. X. Zhang. Special thanks are due to N. Ito and Y. Muroi for their excellent technical contributions.

REFERENCES

1. Baeurle PA, Henkel T. Function and activation of NF-κB in the immune system. Annu Rev Immunol 1994; 12:141–179.
2. Baldwin AS, Jr. The NF-κB and IκB proteins: new discoveries and insights. Annu Rev Immunol 1996; 14:649–681.
3. Ghosh S, Baltimore D. Activation in vitro of NF-κB by phosphorylation of its inhibitor IκB. Nature 1990; 344:678–682.
4. Beg AA, Baldwin AS, Jr. The IκB proteins: multifunctional regulators of Rel/NF-κB transcription factors. Genes Dev 1993; 78:2064–2070.
5. Cordle SR, Donald R, Read MA, Hawiger J. Lipopolysaccharide induces phosphorylation of MAD3 and activation of Rel and related NF-κB proteins in human monocytic THP-1 cells. J Biol Chem 1993; 268:11803–11810.
6. Henkel T, Machleidt T, Alkalay I, Krönke M, Ben-Neriah Y, Baeuerle PA. Rapid proteolysis of IκB-α is necessary for activation of transcription factor NF-κB. Nature 1993; 365:182–185.
7. Rice NR, Ernst MK. In vivo control of NF-κB activation by IκBα. EMBO J 1993; 12:4685–4695.
8. Brown K, Park S, Kanno T, Franzoso G, Siebenlist U. Mutual regulation of the transcriptional activator NF-κB and its inhibitor, IκBα. Proc Natl Acad Sci USA 1993; 90:2532–2536.
9. Dirks W, Mittnacht S, Rentrop M, Hauser H. Isolation and functional characterization of the murine interferon-β1 promoter. J Interferon Res 1989; 9:125–133.
10. Schreck R, Baeuerle PA. NF-κB as inducible transcriptional activator of the granulocyte-macrophage colony stimulating factor gene. Mol Cell Biol 1990; 10:1281–1286.
11. Nishizawa M, Nagata S. Regulatory elements responsible for inducible expression of the granulocyte colony-stimulating factor gene in macrophages. Mol Cell Biol 1990; 10:2002–2011.
12. Shimizu H, Mitomo K, Watanabe T, Okamoto S, Yamamoto K. Involvement of a NF-κB-like transcription factor in the activation of the interleukin-6 gene by inflammatory lymphokines. Mol Cell Biol 1990; 10:561–568.
13. Drouet C, Shakhov AN, Jongeneel CV. Enhancers and transcription factors controlling the inducibility of the tumor necrosis factor-α promoter in primary macrophages. J Immunol 1991; 147:1694–1700.
14. Baldwin AS, Sharp PA. Two transcription factors, NF-κB and H2TF1, interact with a single regulatory sequence in the class I major histocompatibility complex promoter. Proc Natl Acad Sci USA 1988; 85:723–727.
15. Blanar MA, Burkly LC, Flavell RA. NF-κB binds within a region required for B-cell-specific expression of the major histocompatibility complex class II gene Eα^d. Mol Cell Biol 1989; 9:844–846.
16. Hogarth PM, Witort E, Hulet MD, Bonnerot C, Even J, Fridman WH, McKenzie IC. Structure of the mouse Fcγ receptor II gene. J Immunol 1991; 146:369–376.
17. Xie Q-W, Whisnant R, Nathan C. Promoter of the mouse gene encoding calcium-independent nitric oxide synthase confers inducibility by interferon gamma and bacterial lipopolysaccharide. J Exp Med 1993; 177:1779–1784.
18. Lowenstein CJ, Alley EW, Raval P, Snowman AM, Snyder SH, Russell SW, Murphy WJ. Macrophage nitric oxide synthase gene: two upstream regions mediate induction by interferon γ and lipopolysaccharide. Proc Natl Acad Sci USA 1993; 90:9730–9734.
19. Fujihara M, Ito N, Pace JL, Watanabe Y, Russell SW, Suzuki T. Role of endogenous interferon-β in lipopolysaccharide-triggered activation of the inducible nitric-

oxide synthase gene in a mouse macrophage cell line, J774. J Biol Chem 1994; 269:12773–12778.
20. Muroi M, Suzuki T. Involvement of protein kinase A in LPS-induced activation of NF-κB proteins of a mouse macrophage-like cell line, J774. Cell Signal 1992; 5:289–298.
21. Muroi M, Muroi Y, Yamamoto K, Suzuki T. Influence of 3′ half-site sequence of NF-κB motifs on the binding of lipopolysaccharide-activatable macrophage NF-κB proteins. J Biol Chem 1993; 268:19534–19539.
22. Kieran M, Blank V, Logeat F, Vandekerckhove J, Lottspeich F, Bail OL, Urban MB, Kourilsky P, Baeuerle PA, Israel A. The DNA binding subunit of NF-κB is identical to factor KBF1 and homologous to the rel oncogene product. Cell 1990; 62:1007–1018.
23. Fujita T, Nolan GP, Ghosh S, Baltimore D. Independent modes of transcriptional activation by the p50 and p65 subunits of NF-κB. Genes Dev 1992; 6:775–787.
24. Nakayama K, Shimizu H, Mitomo K, Watanabe T, Okamoto S, Yamamoto K. A lymphoid cell-specific nuclear factor containing c-Rel-like proteins preferentially interacts with interleukin-6 κB-related motifs whose activities are repressed in lymphoid cells. Mol Cell Biol 1992; 12:1736–1746.
25. Zabel U, Schreck R, Baeuerle PA. DNA binding of purified transcription factor NF-κB. J Biol Chem 1991; 266:252–260.
26. Traenckner EB-M, Pahl HL, Henkel T, Schmidt K, Wilk S, Baeuerle PA. Phosphorylation of human IκB-α on serines 32 and 36 controls IκB-α proteolysis and NF-κB activation in response to diverse stimuli. EMBO J 1995; 14:2876–2883.
27. Imbert V, Rupec RA, Livolsi A, Pahl HL, Traenckner EB-M, Mueller-Dieckmann C, Farahifar D, Rossi B, Auberger P, Baeuerle PA, Peyron J-F. Tyrosine phosphorylation of IκB-α activates NF-κB without proteolytic degradation of IκB-α. Cell 1996; 86:787–798.
28. Haskill S, Beg AA, Thompkins SM, Morris JS, Yurochko AD, Sampson-Johannes A, Mondal K, Ralph P, Baldwin AS Jr. Characterization of an immediate-early gene induced in adherent monocytes that encodes IκB-like activity. Cell 1991; 65:1281–1289.
29. Shirakawa F, Mizel SB. In vitro activation and nuclear translocation of NF-κB catalyzed by cyclic AMP-dependent protein kinase and protein kinase C. Mol Cell Biol 1989; 9:2424–2430.
30. Sen R, Baltimore D. Inducibility of κ immunoglobulin enhancer-binding protein NF-κB by a posttranslational mechanism. Cell1 1986; 47:921–928.
31. Baeuerle PA, Baltimore D. Activation of DNA-binding activity in an apparently cytoplasmic precursor of the NF-κB transcription factor. Cell 1988; 53:211–217.
32. Lindholm PF, Tamami M, Makowski J, Brady JN. Human T-cell lymphotropic virus type 1 Tax_1 activation of NF-κB: involvement of the protein kinase C pathway. J Virol 1996; 70:2525–2532.
33. Diaz-Meco MT, Dominguez I, Sanz L, Dent P, Lozano J, Municio M, Berra E, Hay R, Sturgill T, Moscat J. PKCζ induces phosphorylation and inactivation of IκBα in vitro. EMBO J 1994; 13:2842–2848.
34. Dominguez I, Sanz L, Arenzana-Seisdedos F, Diaz-Meco MT, Virelizier JL, Moscat J. Inhibition of protein kinase Cζ subspecies blocks the activation of an NF-κB-like activity in Xenopus laevis oocytes. Mol Cell Biol 1993; 13:1290–1295.
35. Diaz-Meco MT, Berra E, Municio MM, Sanz L, Lozano J, Dominiguez I, Diaz-Golpe V, Lain de Lera MT, Arenzana F, Paya CV, Virelizier J, Moscat J. A dominant negative protein kinase Cζ subspecies blocks NF-κB activation. Mol Cell Biol 1993; 13:4770–4775.
36. Flogueira L, McElhinny JA, Bren GD, MacMorran WS, Diaz-Meco MT, Moscat J, Paya CV. Protein kinase C-zeta mediates NF-κB activation in human immunodeficiency virus-infected monocytes. J Virol 1996; 70: 223–231.
37. Hirano M, Hirai S-I, Mizuno K, Osada S-I, Hosaka M, Ohno S. A protein kinase C isozyme, nPKCε, is involved in the activation of NF-κB by 12-O-tetradecanoylphorbol-13-acetate (TPA) in rat 3Y1 fibroblasts. Biochem Biophys Res Commun 1995; 206:429–436.
38. Genot EM, Parker PJ, Cantrell DA. Analysis of the role of protein kinase C-α, -ε, and -ζ in T cell activation. J Biol Chem 1995; 270:9833–9839.
39. Janosch P, Schellerer M, Seitz T, Reim P, Eulitz M, Brielmeier M, Kölch W, Sedivy JM, Mischak H. Characterization of IκB kinases. J Biol Chem 1996; 271: 13868–13874.
40. Shirakawa F, Chedid M, Suttles J, Pollok BA, Mizel SB. Interleukin 1 and cyclic AMP induce κ immunoglobulin light chain expression via activation of an NF-κB-like DNA-binding protein. Mol Cell Biol 1989; 9: 959–964.
41. Jeon YJ, Yang KH, Pulaski JT, Kaminski NE. Attenuation of inducible nitric oxide synthase gene expression by delta 9-tetrahydrocannabinol is mediated through the inhibition of nuclear factor kappa B/Rel activation. Mol Pharmacol 1996; 50:334–341.
42. Bomsztyk K, Toivola B, Emery DW, Rooney JW, Dower SK, Rachie NA, Sibley CH. Role of cAMP in interleukin-1-induced kappa light chain gene expression in murine B cell line. J Biol Chem 1990; 265:9413–9417.
43. Hohmann H-P, Kolbeck R, Remy R, van Loon APGM. Cyclic AMP-independent activation of transcription factor NF-κB in HL60 cells by tumor necrosis factor α and β. Mol Cell Biol 1991; 11:1121–1133.
44. Meichle A, Schütze S, Hensel G, Brunsing D, Krönke M. Protein kinase C-independent activation of nuclear factor κB by tumor necrosis factor. J Biol Chem. 1990; 265:8339–8343.
45. Bomsztyk K, Rooney JW, Iwasaki T, Rachie NA, Dower SK, Sibley CH. Evidence that interleukin-1 and phorbol esters activate NF-κB by different pathways: role of protein kinase C. Cell Regul 1991; 2:329–335.
46. Vinceti MP, Burrell TA, Taffet SM. Regulation of NF-κB activity in murine macrophages: effect of bacterial lipopolysaccharide and phorbol ester. J Cell Physiol 1992; 150:204–213.
47. Bennett BL, Lacso RG, Chen CC, Cruz R, Wheeler JS, Kletzien RF, Tomasselli AG, Heinrikson RL, Manning AM. Identification of signal-induced IκBα kinases in human endothelial cells. J Biol Chem. 1996; 271: 19680–19688.
48. Li S, Sedivy J. Raf-1 protein kinase activates NF-κB transcription factor by dissociating the cytoplasmic NF-

κB-IκB complex. Proc Natl Acad Sci USA 1993; 90: 9247–9251.
49. Finco TS, Baldwin AS, Jr. κB site-dependent induction of gene expression by diverse inducers of nuclear factor κB requires Raf-1. J Biol Chem 1993; 268:17676–17679.
50. Devary Y, Rosette C, DiDonato JA, Karin M. NF-κB activation by ultraviolet light not dependent on a nuclear signal. Science 1993; 261:1442–1445.
51. Delude RL, Fenton MJ, Savedra R, Jr., Perera P-Y, Vogel SN, Thieringer R, Golenbock DT. CD14-mediated translocation of nuclear factor-kappa B induced by lipopolysaccharide does not require tyrosine kinase activity. J Biol Chem 1994; 269:22253–22260.
52. Yoza BK, Hu JYQ, McCall CE. Protein tyrosine kinase activation is required for lipopolysaccharide induction of interleukin 1β and NF-κB activation, but not NF-κB nuclear translocation. J Biol Chem 1996; 271:18306–18309.
53. Hirano M, Osada S, Aoki T, Hirai S, Hosaska M, Inoue J, Ohno S. MEK kinase is involved in tumor necrosis factor α-induced NF-κB activation and degradation of IκBα. J Biol Chem 1996; 271:13234–13238.
54. Chen ZJ, Parent L, Maniatis T. Site-specific phosphorylation of IκB-α by a novel ubiquitination-dependent protein kinase activity. Cell 1996; 84:853–862.
55. Lee FS, Hagler J, Chen ZJ, Maniatis T. Activation of the IκBα kinase complex by MEKK1, a kinase of the JNK pathway. Cell 1997; 88:213–222.
56. Maran A, Maitra RK, Kumar A, Dong B, Xiao W, Li G, Williams BRG, Torrence PF, Siverman RH. Blockage of NF-κB signaling by selective ablation of an mRNA target by 2-5 antisense chimeras. Science. 1994; 265:789–792.
57. Beg AA, Finco TS, Nantermet PV, Baldwin AS, Jr. Tumor necrosis factor and interleukin-1 lead to phosphorylation and loss of IκB: a mechanism for NF-κB activation. Mol Cell Biol 1993; 13:3301–3310.
58. Finco TS, Beg AA, Baldwin AS, Jr. Inducible phosphorylation of IκBα is not sufficient for its dissociation from NF-κB and is inhibited by protease inhibitors. Proc Natl Acad Sci USA 1994; 91:11884–11888.
59. Mellitis KH, Hay RT, Goodbourn S. Proteolytic degradation of MAD3 (IκBα) and enhanced processing of the NF-κB precursor p105 are obligatory steps in the activation of NF-κB. Nucl Acids Res 1993; 21:5059–5066.
60. Mackman N. Protease inhibitors block lipopolysaccharide induction of tissue factor gene expression in human monocytic cells by preventing activation of c-Rel/p65 heterodimers. J Biol Chem 1994; 269:26363–26367.
61. Kim H, Lee HS, Chang KT, Ko TH, Baek KJ, Kwon NS. Chloromethyl ketones block induction of nitric oxide synthase in murine macrophages by preventing activation of nuclear factor-κB. J Immunol 1995; 154: 4741–4748.
62. Palombella VJ, Rando OJ, Goldberg AL, Maniatis T. The ubiquitin-proteasome pathway is required for processing the NF-κB1 precursor protein and the activation of NF-κB. Cell 1994; 78:773–785.
63. Traenckner EB, Wilk S, Baeuerle PA. A proteasome inhibitor prevents activation of NF-κB and stabilizes a newly phosphorylated form of IκB-α that is still bound to NF-κB. EMBO J 1994; 13:5433–5441.
64. Muroi M, Muroi Y, Ito N, Rice NR, Suzuki T. Effects of protease inhibitors on LPS-mediated activation of a mouse macrophage cell line (J774). J Endotoxin Res 1995; 2:337–347.
65. Chiao PJ, Miyamoto S, Verma IM. Autoregulation of IκBα activity. Proc Natl Acad Sci USA 1994; 91:28–32.
66. Miyamoto S, Chiao P, Verma IM. Enhanced IκBα degradation is responsible for constitutive NF-κB activity in murine B-cell line. Mol Cell Biol 1994; 14:3276–3282.
67. Rechsteiner M, Hoffman L, Dubiel W. The multicatalytic and 26S proteases. J Biol Chem 1993; 268:6065–6068.
68. Griscavage JM, Wilk S, Ignarro LJ. Inhibitors of the proteasome pathway interfere with induction of nitric oxide synthase in macrophages by blocking activation of transcription factor NF-κB. Proc Natl Acd Sci USA 1996; 93:3308–3312.
69. Griscavage JM, Wilk S, Ignarro LJ. Serine and cysteine proteinase inhibitors prevent nitric oxide production by activated macrophages by interfering with transcription of the inducible NO synthase gene. Biochem Biophys Res Commun 1995; 215:721–729.
70. Chen Z, Hagler J, Palombella VJ, Melandri F, Scherer D, Ballard D, Maniatis T. Signal-induced site-specific phosphorylation targets IκBα to the ubiquitin-proteasome pathway. Genes Dev 1995; 9:1586–1597.
71. Roff M, Thompson J, Rodriguez MS, Jacque J-M, Baleux F, Arenzana-Seisdedos F, Hay RT. Role of IκBα ubiquitination in signal-induced activation of NF-κB in vivo. J Biol Chem 1966; 271:7844–7850.
72. Lin Y-C, Brown K, Siebenlist U. Activation of NF-κB requires proteolysis of the inhibitor IκB-α: Signal-induced phosphorylation of IκB-α alone does not release active NF-κB. Proc Natl Acad Sci USA 1995; 92:552–556.
73. Rodriguez MS, Write J, Thompson J, Thomas D, Baleux F, Virelizier J-L, Hay RT, Arenzana-Seisdedos F. Identification of lysine residues required for signal-induced ubiquitination and degradation of IκB-α in vivo. Oncogene 1996; 12:2425–2435.
74. Li C-H, Dai R-M, Longo DL. Inactivation of NF-κB inhibitor IκBα: Ubiquitin-dependent proteolysis and its degradation product. Biochem Biophys Res Commun 1995; 215:292–301.
75. LeBail O, Schmidt-Ullrich R, Israël A. Promoter analysis of the gene encoding the IκB-α/MAD-3 inhibitor of NF-κB: positive regulation by members of the rel/NF-κB family. EMBO J 1993; 12:5043–5049.
76. Sun S-C, Ganchi PA, Ballard DW, Green WC. NF-κB controls expression of inhibitor IκBα: evidence for an inducible autoregulatory pathway. Science 1993; 259: 1912–1915.
77. Arenzana-Seisdedos F, Thomson JA, Rodriguez MS, Bachelerie F, Thomas D, Hay RT. Inducible nuclear expression of newly synthesized IκBα negatively regulates DNA-binding and transcriptional activities of NF-κB. Mol Cell Biol 1995; 15:2689–2696.
78. Rappmann D, Wulczyn FG, Scheidereit C. Different mechanisms control signal-induced degradation and

basal turnover of the NF-κB inhibitor IκBα in vivo. EMBO J 1996; 15:6716–6726.
79. Lorsbach RB, Murphy WJ, Lowenstein CJ, Snyder SH, Russell SW. Expression of the nitric oxide synthase gene in mouse macrophages activated for tumor cell killing. Molecular basis for the synergy between interferon-γ and lipopolysaccharide. J Biol Chem 1993; 268:1908–1913.
80. Murphy EJ, Muroi M, Zang CX, Suzuki T, and Russell SW. Both basal and enhancer kB elements are required for full induction of the mouse inducible nitric oxide synthase gene. J Endotoxin Res 1996; 3:381–393.
81. Chen F, Kuhn DC, Sun SC, Gaydos LJ, Demers LM. Dependence and reversal of nitric oxide production on NF-κB in silica and lipopolysaccharide-induced macrophages. Biochem Biophys Res Commun 1995; 214: 839–846.
82. Goldring CE, Narayanan R, Lagadec P, Jeannin JF. Transcriptional inhibition of the inducible nitric oxide synthase gene by competitive binding of NF-κB/Rel proteins. Biochem Biophys Res Commun 1995; 209: 73–79.
83. Diaz-Guerra MJM, Bodelón OG, Velasco M, Whelan R, Parker PJ, Boscá L. Up-regulation of protein kinase C-ε promotes the expression of cytokine-inducible nitric oxide synthase in RAW 264.7 cells. J Biol Chem. 1996; 271:32028–32033.
84. Löms Ziegler-Heitbrock HW, Wedel A, Schraut W, Ströbel M, Wedelgass P, Sternsdorf T, Bäuerle PA, Haas JG, Riethmüller G. Tolerance to lipopolysaccharide involves mobilization of nuclear factor kB with predominance of p50 homodimers. J Biol Chem 1964; 269: 17001–17004.
85. Molina y Vedia L, McDonald B, Reep B, Brüne B, Di Silvio M, Billiar TR, Lapetina EG. Nitric oxide induced S-nitrosylation of glyceraldehyde-3-phosphate dehydrogenase inhibits enzymatic activity and increase endogenous ADP-ribosylation. J Biol Chem 1992; 267: 24929–24932.
86. Matthews JR, Botting CH, Panico M, Morris HR, Hay RT. Inhibition of NF-κB DNA binding by nitric oxide. Nucl Acids Res 1996; 24:2236–2242.
87. Ollivier V, Parry GCN, Cobb RR, Prost D, Mackman N. Elevated cyclic AMP inhibits NF-κB-mediated transcription in human monocytic cells and endothelial cells. J Biol Chem 1996; 271:20828–20835.
88. Kopp E, Ghosh S. Inhibition of NF-κB by sodium salicylate and aspirin. Science 1994; 265:956–959.
89. Osnes LTN, Foss KB, Joø GB, Okkenhaug C, Westvik Å-B, Øvstbø R, Kierulf P. Acetylsalicylic acid and sodium salicylate inhibit LPS-induced NF-κB/C-Rel nuclear translocation, and synthesis of tissue factor (TF) and tumor necrosis factor alpha (TNF-α) in human monocytes. Thrombosis Haemostasis 1996; 76:970–976.
90. Ray A, Prefontaine K. Physical association and functional antagonism between the p65 subunit of transcriptional factor NF-κB and the glucocorticoid receptor. Proc Natl Acad Sci USA 1994; 91:752–756.
91. Mukaida N, Morita M, Ishikawa Y, Rice N, Okamoto S, Kasahara T, Matsushima K. Novel mechanism of glucocorticoid-mediated gene repression: NF-κB is a target for glucocorticoid mediated IL-8 gene repression. J Biol Chem 1994; 269:13289–13295.
92. Scheinman R, Cogswell P, Lofquist A, Baldwin A. Role of transcriptional activation of IκBα in mediation of immunosuppression by glucocorticoids. Science 1995; 270:283–286.
93. Auphan N, DiDonato J, Rosette C, Helmberg A, Karin M. Molecular basis for immunosuppression by glucocorticoids: inhibition of NF-κB activity through induction of IκB synthesis. Science 1995; 270:286–290.

33

Internalization of Lipopolysaccharide by Phagocytes

Richard L. Kitchens and Robert S. Munford
The University of Texas Southwestern Medical Center, Dallas, Texas

INTRODUCTION

As the interactions between LPS and its receptors on eukaryotic cells have become more completely understood, scientific attention has turned to the mechanisms by which LPS moves to the cell interior and travels intracellularly. There is general agreement that the professional phagocytes (neutrophils, monocytes, macrophages) probably internalize LPS to clear it from the extracellular milieu—a disposal function that should limit its interactions with other host cells. Little is known about the actual mechanism(s) involved in LPS internalization and intracellular movement, however, and controversy has developed regarding the role that internalized LPS may play in signaling. Detmers and colleagues described evidence that LPS internalization and vesicle fusion are required for cell adhesion responses in neutrophils (1). Moreover, Thiéblemont and Wright subsequently found that, in macrophages from LPS-hyporesponsive mice, LPS is unable to traffic normally to the interior of the cell (2). These results suggest that signals may be transduced after LPS is present in one or more intracellular compartment(s). On the other hand, Tobias and colleagues presented evidence that internalized LPS does not signal. Their conclusion is based largely on the ability of a unique monoclonal antibody (mAb) to LPS-binding protein (LBP) to block most of the LPS binding to membrane CD14 without interfering with signaling. Signal transduction is thought to occur while LPS is on or in the plasma membrane and to involve relatively few molecules of cell-associated LPS (3). Signaling from the cell surface is also suggested by the ability of *Limulus* endotoxin–neutralizing protein to antagonize LPS that has been added to cells many minutes earlier (4). Clarification of these issues should be fostered by a more complete understanding of the intracellular fate of LPS, starting with a clearer definition of the membrane structures and biochemical mechanisms that underlie its internalization and intracellular movement.

LPS (or, in some instances, lipid A partial structures) can bind to several cell-surface molecules on eukaryotic cells, including CD14 (5,6), scavenger receptor(s) (7,8), CD18 (5), cell-associated albumin (9), and galectin (10). If LPS remained attached to the receptor to which it initially bound, its subsequent fate would likely be determined by the movement of that receptor. Such prolonged ligand-receptor association may not be assumed, however, since LPS may move from protein carriers such as soluble CD14 to lipid vesicles (11) and presumably to the lipid bilayers of biological membranes as well (11,12). Although most studies of LPS internalization have followed the movement of LPS per se without concern for the molecule(s) to which it is bound, the overwhelming evidence that CD14 is important for both LPS binding and signaling encourages dissection of the role(s) played by this protein in LPS internalization and intracellular trafficking. In this review we shall therefore focus on the internalization of LPS that is taken up by CD14-dependent pathways.

An important consideration in all studies of LPS-cell interactions is the possibility that the fate of the LPS followed by one or more techniques may be different from that of a much smaller population of LPS molecules that has important or different biological consequences. This distinction is at the heart of the current discussion over the role of internalized LPS in signaling: it is entirely possible that the "bulk" of cell-associated LPS followed by virtue of its radioactivity or visual tag does not include a small population of molecules that triggers cellular responses. Another possibility, given the pleiotropic nature of responses to LPS, is that different responses are initiated by the LPS that finds its way into different surface domains or intracellular compartments. Combining morphological, functional, and genetic studies is most likely to sort out these possibilities, but it must be admitted that the challenge is formidable.

DEFINITIONS

Many workers have used the term *uptake* to refer to the overall process by which a ligand associates with (cannot be washed off) a cell, whether the ligand is on the cell surface or inside the cell. In this review we distinguish *binding* (which occurs on the cell surface), *internalization* (passage from the cell surface into the cell), and *intracellular traffic* (movement among intracellular vesicles in one or more pathways). An extracellular ligand is usually considered to become internal when it reaches a membrane-bound vesicle that has no opening to the outside. In some cases, however, ligands may be internalized by an active cellular process into compartments that remain connected to the cell surface by small openings that exclude the entry of large molecules (13,14). Operationally, it is convenient to identify internalized LPS by its inaccessibility to extracellular molecules that can either remove it (e.g., proteases that release the protein to which the LPS is bound) or modify it (e.g., quenching the fluoresence of fluorescein-conjugated LPS with an antifluorescein antibody or trypan blue) when it is on the cell surface. Thus, LPS that has been internalized cannot be stripped or quenched by molecules that do not pass through the plasma membrane. This operational definition is somewhat unsatisfactory for LPS, however, since LPS may enter plasma membrane invaginations (tubules) that partially or intermittently close off from the surface. The ability of an extracellular reagent to reach LPS in these invaginations appears to be size dependent: a small molecule like trypan blue ($M_r = 961$) penetrates more deeply into the invaginations than can the larger proteinase K ($M_r = 29,900$) or IgG ($M_r = 146,000$). An intracellular location may also be inferred when LPS is found in compartments that also contain known intracellular molecules.

METHODS USED TO STUDY LPS INTERNALIZATION AND INTRACELLULAR MOVEMENT

Table 1 summarizes several methods used to study LPS internalization and intracellular traffic.

CD14-DEPENDENT BINDING AND INTERNALIZATION

Binding

CD14-dependent binding may be defined operationally as LPS-cell binding that (a) can be inhibited by specific anti-CD14 antibodies, (b) occurs in cells transfected to express recombinant membrane CD14 (rmCD14) but not in control cells transfected with an empty vector, (c) is substantially augmented in the presence of LPS binding protein (LBP), which transfers LPS to CD14, or (d) requires the presence of soluble CD14 (sCD14). In most systems studied, mCD14-dependent binding triggers cellular responses to low concentrations of LPS, while mCD14-independent interactions occur at higher ambient LPS concentrations (15,16). LPS that is bound to sCD14 may bind and stimulate some cells that do not express mCD14 (17–20), or LPS-sCD14 complexes may stimulate cells through mCD14 without a requirement for LBP (21,22).

LBP can function as an opsonin that binds LPS-bearing particles (23,24), LPS aggregates (25), or bacterial membranes (26) and facilitates their rapid attachment to mCD14, or LBP can function as a catalyst that results in the rapid binding of LPS monomers to CD14 (21,27,28). After LPS aggregates are exposed for prolonged periods to a molar excess of sCD14, monomeric LPS-sCD14 complexes are formed with an approximately 1:1 stoichiometry (29). These LPS monomers can be rapidly transferred to unoccupied molecules of sCD14 (21) or mCD14 (21,22) in the absence of LBP. Thus, cells expressing mCD14 can bind and internalize LPS monomers (1,2,22) or aggregates (3,22,25) by receptor-mediated internalization (see below), whereas larger LPS-bearing particles (e.g., whole bacteria) may be internalized by phagocytosis (26).

LPS Internalization

Electron microscope images of THP-1 cells and human monocytes show that their plasma membranes tend to be highly ruffled, with thin membrane protrusions imparting a large surface area (Fig. 1). Thin sections of these cells stained to identify mCD14 (Fig. 1) or mCD14-bound LPS show that these molecules are distributed along the surface of the plasma membrane, abundantly decorating the protruding membrane ruffles. In contrast, the invaginations that carry out endocytosis are found almost entirely on unruffled regions of the plasma membrane (R. L. Kitchens and R. S. Munford, unpublished), suggesting that the LPS that binds to mCD14 on membrane ruffles must move to the body of the cell before internalization can occur.

Recent experiments in our laboratory (30) indicate that the majority of the LPS molecules that bind to wild-type mCD14 on THP-1 cells are internalized by non–clathrin-coated membrane invaginations (NCCI) (Fig. 2), whereas a smaller number of molecules enter clathrin-coated pits. These conclusions were initially suggested by measurements of [^{3}H]LPS or FITC-LPS internalization by THP-1 cells or peripheral blood monocytes that had been exposed to hypertonic medium (0.45 M sucrose), a treatment that selectively disrupts coated pits. Whereas treatment with hypertonic sucrose almost completely blocked internalization of ^{125}I-transferrin, which enters cells via clathrin-coated pits, this maneuver only partially inhibited internalization of LPS. To discriminate between the two endocytic pathways, THP-1 cells were transfected with either wild-type GPI-anchored CD14 (CD14-GPI) or with a CD14-LDL receptor fusion protein (CD14-LDLR) in which the GPI-anchor was replaced by the transmembrane and cytoplasmic domains (including the coated pit targeting sequence) of the low-density lipoprotein receptor (LDLR). Internalization of [^{3}H]LPS via coated pits (CD14-LDLR) proceeded much more rapidly and completely than internalization via the GPI-mediated pathway(s). In cells transfected with CD14-LDLR and studied using electron microscopy, LPS was frequently found in coated pits, coated vesicles, and tubular invaginations (which often contained coated pits), whereas in cells that expressed CD14-GPI it was found mainly in noncoated membrane invaginations, in tubular structures that were frequently found to be connected to the surface, and in noncoated vesicles.

Entry of LPS into THP-1 cells by a nonclathrin pathway is reminiscent of findings described in CHO cells transfected with CD4-GPI and CD4-transmembrane anchor (31); only cells transfected with CD4-GPI internalized CD4-bound ligand via the slower, non–clathrin-mediated route. Similarly, interleukin-2 receptors were internalized by a T-lymphoma cell line in a clathrin-independent fashion (32). Ricin, the plant toxin, binds to terminal galactose residues on membrane glycolipids and glycoproteins and is internalized both by clathrin- and non–clathrin-mediated pathways. In Vero cells, ricin enters large membrane invaginations (150–300 nm) that are connected to the surface by a narrow neck (33). The appearance of these structures is very similar to that of some of the NCCI that internalize LPS in THP-1 cells (30).

When studied using electron microscopy of semi-thick (300 nm) sections, many of the plasma membrane invaginations that accumulate LPS in CD14-GPI cells resemble the tubular pinosomes described in rabbit alveolar macrophages by Nichols (34) and the tubular structures ("STEMs") described in murine macrophages by Myers et al. (13,35). The latter authors also noted that β-VLDL, which they found in these invaginations, could not be bound by antibodies applied to the cell surface, yet the lipoproteins could be reached by trypan blue, a much smaller extracellular molecule. Large molecules in the extracellular milieu may thus have limited access to ligands (in particular, large aggregates) that have moved into STEMs. When viewed in this way, LPS "internalization" measured by the protease protection and fluorescence quenching assays may represent, at least in part, passage of LPS into tubular invaginations of the plasma membrane rather than movement into vesicles that are closed to the surface. In keeping with this notion, trypan blue can quench 20–30% of the BODIPY-LPS that remains on THP-1 cells after proteinase K treatment (30). Internalization of LPS into a proteinase-inaccessible, trypan blue–accessible compartment (which presumably includes surface-connected tubular invaginations) is an active process that requires elevated temperature, is strongly inhibited by agents that deplete cellular ATP (22), and is partially inhibited by cytochalasins (30). The extent to which LPS moves into this compartment, how long it stays there, and its subsequent fate(s) are under investigation.

Two-color laser confocal microscopy was also used to study early vesicular traffic of LPS (3–20 minutes after internalization). In THP-1 cells that expressed wild-type CD14-GPI, Texas Red-LPS was usually found in intracellular vesicles that could be distinguished from those containing BODIPY fluorescein-transferrin (although there was some overlap), whereas in cells that expressed CD14-LDLR, the LPS was found almost exclusively in early endosomes containing transferrin. Taken together, these results suggest that CD14's GPI anchor directs LPS to an endocytic pathway that is distinct from the clathrin-mediated pathway taken by transferrin.

Table 1 Methods Used to Study the Internalization of LPS by Cells

Method	Step studied	Assumptions	Description	Caveats	Ref.
Electron microscopy	Various stages in cell surface binding and internalization	Fixation and labeling artifacts are not important. The detection method does not influence LPS entry or intracellular movement.	See Table 2	Tubular lysosomes may be disrupted by fixation or temperature changes (92). Antigenic epitopes may be obscured by fixation. Bulky gold particles may alter LPS intracellular traffic. When anti-LPS antibodies are used to locate LPS in or on living cells, care must be taken to exclude Fc receptor-mediated LPS internalization.	See Table 2
Density gradient sedimentation	Movement of LPS from plasma membrane to intracellular compartments that have different densities	LPS does not move between compartments during cell lysis and centrifugation. Density fractionation successfully separates intracellular organelles from each other and from plasma membrane.	Cells are exposed to labeled LPS, lysed using nitrogen cavitation, and fractionated on Percoll gradients. LPS location is correlated with that of various organelles, detected using biochemical or microscopic techniques.	LPS that "floats" with the plasma membrane might be in early endosomes that have the density of plasma membrane. Co-sedimentation of LPS with a specific organelle does not indicate that the LPS is necessarily contained within that organelle.	58
Protease protection	Disappearance from the cell surface into the cell or into inaccessible plasma membrane invaginations	Cell-surface LPS is protein-bound and is susceptible to protease cleavage.	Labeled LPS is removed from the cell surface by incubating cells with protease at low temperature.	LPS may lose protease susceptibility if it interacts with, or is transferred to, a protease-resistant structure, or is sequestered in a protease-inaccessible membrane invagination.	3, 22

Fluorescence quenching	Disappearance from the cell surface into the cell or into inaccessible plasma membrane invaginations	The linkage between fluorescent modifier and LPS is stable. The fluorescence of cell-surface FITC-LPS can be quenched by antifluorescein Ab or trypan blue.	Cells exposed to FITC-LPS or BODIPY-fluorescein LPS are incubated with Texas red-conjugated antifluorescein Ab (for FITC-LPS) or trypan blue, which quench extracellular fluorescence. Total and internal LPS are quantitated by fluorescence activated cell sorting.	The LPS conjugation site is not known; contaminants may be labeled. Antibodies may not reach LPS inside invaginations of the plasma membrane (13). Intracellular fluorescein may be quenched by low pH (a problem with FITC but not BODIPY). Trypan blue may have better access to LPS inside plasma membrane invaginations that are partially or intermittently open to the surface.	22, 30
Fluorescence microscopy	Cell-surface LPS binding; movement of LPS in intracellular compartments that are well-separated from the plasma membrane	LPS conjugates are stable. Intracellular quenching of fluorescence is not significant.	Cells exposed to fluorescent LPS (conjugated to FITC, BODIPY, Texas red, etc.) are studied using fluorescence or laser confocal microscopy. May be combined with protease treatment to strip cell-surface LPS. Confocal microscopy allows viewing optical sections of cells.	The conjugation chemistry is poorly understood. Cell-surface and intracellular (peripheral) LPS may be difficult to distinguish, especially in leukocytes that have large nuclei. In contrast to fluorescein, pH-insensitive derivatives such as BODIPY-fluorescein and Texas red are not quenched in acidic intracellular vesicles. All fluorescent molecules may be quenched by aggregation, however.	1, 2, 22, 30, 59

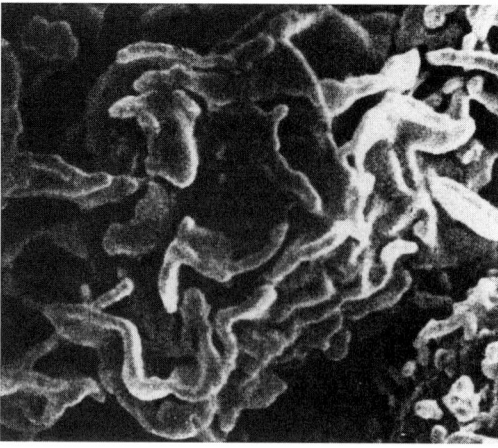
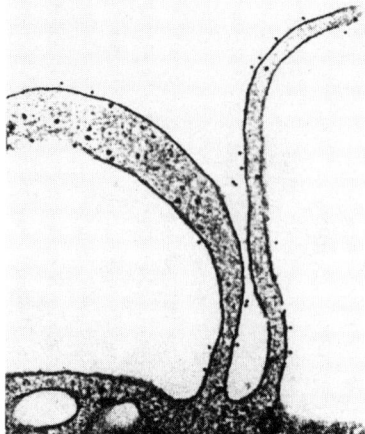

Fig. 1 (Left) Scanning electron micrograph of a peripheral blood monocyte with its prominent surface membrane ruffles. (Right) A transmission electron micrograph of a THP-1 cell. Gold particles indicate the locations of cell-surface CD14.

Biochemical Nature of Noncoated Invaginations That Internalize LPS

Wang et al. (36) reported that LPS binds to CD14 in low-density plasma membrane regions (fragments), which have biochemical similarities to the structures known as caveolae (37). Caveolae are small, often flask-shaped, non–clathrin-coated membrane invaginations found in many cell types (38–40). They (or the membrane regions in which they reside) are enriched in sphingomyelin and glycolipids, facilitating their isolation on various density gradient media (37,41). Caveolae typically are organized around one or more structural proteins called caveolins (VIP21) (42,43), and they may function as signaling centers for various extracellular ligands (44). In addition, caveolae may be portals for the internalization of ligands such as folate (potocytosis) (45) or the transcytosis of molecules such as albumin (46). Many GPI-anchored proteins seem to concentrate in (38), or near (44), caveolae. Like caveolae, the CD14-enriched membrane microdomains of THP-1 cells are also relatively enriched in several potential signaling molecules but depleted in β-adaptin, a coated pit protein (36). Moreover, some of the non–clathrin-coated invaginations (NCCI) of the monocyte plasma membrane morphologically resemble caveolae (30), although the NCCI are more often tubular than flask-shaped. Although caveolin has not been found in these cells, there is thus indirect evidence from cell fractionation studies as well as direct observations using electron microscopy (discussed above) that CD14-dependent LPS uptake can proceed via caveolae-like membrane invaginations. Structures that resemble caveolae but lack caveolin have also been described in human lymphocytes (47); these NCCI internalize another GPI-anchored protein, CD59, after it is crosslinked with antibodies.

Are monocyte-macrophage NCCI caveolae or their equivalent? Although these structures appear to be similar to caveolae in many respects, doubt has arisen principally from the apparent absence of caveolin in resting monocytes or THP-1 cells (or in other hematopoietic cells) (48). These cells also lack plasma membrane regions that contain numerous caveolae, as are usually observed in cells that normally express caveolin (e.g., fibroblasts) or cells that are transfected with caveolin cDNA (e.g., lymphocytes) (49). The NCCI found in THP-1 cells are more tubular in structure than are classical caveolae, which are usually flask-shaped, and they may be connected to the surface by a narrow neck. Similar structures have been found in the Vero cell membrane invaginations that internalize ricin (33) and in alveolar macrophage plasma membrane invagina-

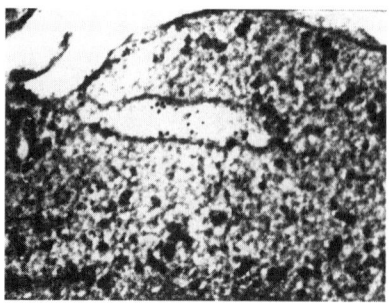

Fig. 2 A surface-connected tubular invagination in a THP-1 cell expressing GPI-anchored CD14. Gold particles indicate the location of dinitrophenol-conjugated LPS.

tions that contain acid-phosphatase (34). As noted above, the structures also resemble STEMs, the large tubular plasma membrane invaginations noted by Myers et al. (13) in murine macrophages.

In contrast to these findings in resting monocyte-macrophages, Kiss and Geuze found caveolin-1 outlining omega-shaped caveolae in the plasma membranes of elicited peritoneal macrophages (50). Neither caveolae nor caveolin was detected in nonelicited (nonactivated) cells. Although these authors suggested that macrophage caveolae are endocytic structures that pinch off from the plasma membrane and move inside the cell (50), whether or not caveolae detach from the plasma membrane in most cells is still controversial (40), and there is little precedent for the notion that caveolae deliver extracellular molecules to the degradative pathway (to late endosomes or lysosomes). Rather, their putative functions in other cells are potocytosis, transcytosis, and signal transduction (51).

Is LPS Internalized in Macropinosomes or Phagosomes?

Another potential LPS internalization mechanism is macropinocytosis, the process by which plasma membrane ruffles fuse to form vacuoles that contain large amounts of plasma membrane and extracellular fluid (52). Indeed, virulent gram-negative bacteria have been reported to enter epithelial cells by inducing macropinocytosis (53). The formation of macropinosomes may be stimulated by a variety of agonists, including EGF (54), M-CSF (55), and PMA (52). Macropinocytosis increases the uptake of fluid-phase markers such as horseradish peroxidase (HRP) (54) and lucifer yellow (52); it may be blocked by amiloride (54), dimethylamiloride (56), and nocodazole (55), as well as by inhibitors of the small GTPase rac1 (57). In the case of LPS, pinocytic (fluid phase) uptake per se has been shown to be quantitatively trivial compared to receptor (mCD14)–mediated uptake. For example, we (22) found that when LPS was prevented from binding to mCD14, fluid phase (or non-CD14) internalization of [³H]LPS by THP-1 cells or monocytes was typically less than 5% of levels measured for mCD14-mediated LPS internalization. However, this finding does not exclude the possibility that mCD14-bound LPS may be internalized by triggering membrane ruffling and fusion in a process that resembles macropinocytosis. We also found that treating THP-1 cells with nocodazole or dimethylamiloride blocked PMA-stimulated HRP uptake but had no effect on their ability to internalize LPS and that LPS stimulation of the cells did not trigger macropinocytosis (30). These data suggest that in these cells, macropinocytosis is not a major mechanism for LPS internalization.

Since large LPS aggregates can bind mCD14, another potential LPS internalization mechanism is phagocytosis. We (30) found that the internalization of [³H]LPS aggregates by THP-1 cells was only slightly (25–35%) inhibited by concentrations of cytochalasin D or cytochalasin H that completely inhibited mCD14-mediated phagocytosis of BODIPY-labeled *E. coli*. Phagocytosis is thus not required for internalization of LPS aggregates that bind to mCD14. Interestingly, the relatively weak inhibition by cytochalasins also suggests that the mechanism of LPS internalization mediated by GPI-anchored CD14 differs from that of GPI-anchored CD59, which is internalized by a cytochalasin-inhibitable, non–clathrin-dependent pathway in T lymphocytes (47). Internalization of cross-linked CD59 appears to involve an actin-dependent capping mechanism that does not occur during LPS internalization. On the other hand, the fact that cytochalasins partially inhibit LPS internalization points to some role for the actin cytoskeleton in this process.

Kinetics of LPS Internalization

In human blood monocytes and THP-1 cells, the rate at which LPS is internalized depends strongly upon the initial aggregation state of the LPS (22). Large LPS aggregates are internalized almost as quickly as they bind cell-surface CD14, smaller aggregates are internalized more slowly, and LPS monomers, which are internalized extremely slowly, remain on the cell surface for much longer periods of time. Regardless of the aggregation state, however, internalization proceeds in two phases: a rapid internalization phase, lasting less than 10–15 minutes, is followed by a phase in which internalization proceeds much more slowly. Remarkably, much CD14-bound LPS remains on the surface of the cell, particularly when small LPS aggregates or monomers are used for the experiment. This LPS can be released by treatment with proteinase K, indicating that a large fraction of the LPS that binds CD14 and stays on the cell surface remains protein-bound. CD14-mediated LPS internalization by THP-1 cells or murine macrophages appears to be a constitutive process in which the internalization kinetics are not influenced by cell responses to LPS. We (22) found that macrophages from LPS-responsive C3H/HeN and LPS-hyporesponsive C3H/HeJ mice internalized partially disaggregated [³H]LPS with identical kinetics, and LPS internalization kinetics in THP-1 cells were not altered by pretreating the cells with LPS or with the LPS-specific inhibitor, LA-14-PP. In contrast, intracellular traffic and processing of LPS may influence (or be influenced by) LPS-induced signals (see below). When considering these observations, it should be kept in mind that they were made using the protease protection and fluorescence quenching assays, which may not completely distinguish intracellular LPS from LPS that has moved into surface invaginations that are relatively inaccessible to these extracellular probes (see above). "Internalization," when assessed using these methods, may actually represent, at least in part, movement into such invaginations.

Summary

LPS may enter THP-1 cells and human monocytes via two CD14-dependent pathways (Fig. 3). The dominant route is via non–clathrin-coated invaginations of the plasma membrane. The LPS that enters cells via these invaginations follows an intracellular path that largely bypasses the early endosomes that contain transferrin, a protein that enters via coated pits. In addition, a smaller but substantial fraction of CD14-bound LPS enters via clathrin-coated pits and can be found in early endosomes that contain transferrin. The initial kinetics (i.e., the rate and extent of movement from the cell surface into protease-inaccessible compartments) are strongly influenced by LPS aggregation but not by LPS-induced cellular responses. In contrast, subsequent intracellular traffic and processing of LPS may influence (or be influenced by) LPS-induced signals (see below). It should be noted that the influence of monocyte maturation on these processes has not been studied and that the endocytic structure(s) that internalize LPS, as well as its intracellular pathway, may differ in the various cell types that express CD14.

INTRACELLULAR MOVEMENT

Movement of LPS from the plasma membrane to intracellular compartments has been studied by several investigators. Early studies examined cells that had been exposed to high concentrations of LPS prior to analysis using electron microscopy (Table 2). These studies found LPS in numerous intracellular compartments, including cytosol, mitochondria, vacuolar structures, and the nucleus. Although the cells studied include macrophages and monocytes, which may have taken up LPS via CD14-dependent pathways, it seems likely that much of the LPS in these studies reached intracellular sites via CD14-independent routes. This is particularly likely when high LPS concentrations (≥ 1 µg/ml) were used or a source of LBP or sCD14 was not present.

Luchi and Munford (58) used Percoll gradients to separate the intracellular organelles of human neutrophils that had been incubated with [^{3}H]LPS prior to disruption by nitrogen cavitation. They found that approximately 1% of the cell-associated LPS translocated from the plasma membrane to an intracellular location

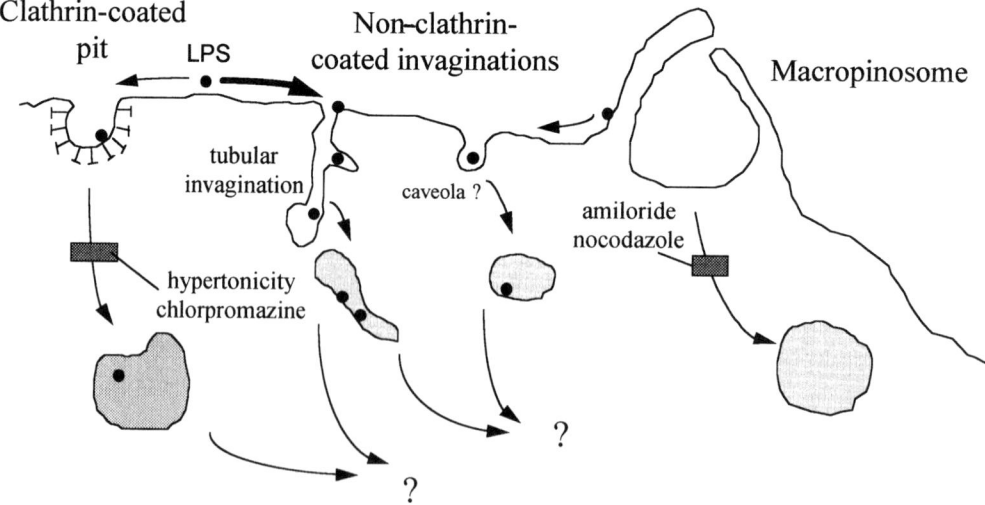

Fig. 3 Schematic showing potential LPS entry pathways. Of the three major internalization mechanisms, entry via non–clathrin-coated invaginations (tubular invaginations or caveolae) predominates. Although CD14-bound LPS can also enter via coated pits, macropinocytosis does not seem to be an important entry mechanism.

Table 2 EM Studies of LPS Internalization

Cell type	Species	LPS, concentration	Method	Results	Ref.
Macrophage	Guinea pig	^{14}C- or uranyl acetate labeled *S. enteritidis*, 10 µg/ml or "150 mg"	Also studied transfer of ^{14}C-LPS from macrophages to lymphocytes	"After 15 min it penetrates into the cytoplasm through invaginations or small channels." No loss of bioactivity even when LPS is in lysosomes.	93
Hepatocytes	Rat	*E. coli* O111:B4, 10 µg/ml or higher	^{14}C-LPS binding; immunocytochemistry using anti-LPS and gold protein A; serum-free.	Serum-free binding was greater at 4°C; gold was found on "microvilli" and dispersed in cytoplasm, reaching mitochondria. Non–receptor-mediated uptake.	94
Alveolar macrophage	Rat	*E. coli* O111:B4, 500 µg/ml	LPS or LPS from serum-treated bugs; fracture-flip, immunolabel	Serum-treated LPS showed more dispersed distribution on surface than commercial LPS.	95
Type II pneumocyte, alveolar macrophage	Rat	Same	Same as above; transmission EM too	Preferential attachment of LPS to microvilli; LPS aggregates move within cytoplasm and enter nucleus and mitochondria. LPS co-localizes with tubulin, disrupts microtubular network.	96
Platelets, monocytes	Human	Biotinylated *E. coli* O111:B4, 100 µg/ml	Avidin-biotin-HRP or streptavidin-gold; serum-free.	LPS "bilayers" found in vesicles. Disruption of cell membranes.	97
Monocytes	Human	*E. coli* J5 or biotin-O111:B4, or FITC-O55:B5; 100 µg/ml		LPS bilayers found in large membrane-bound vacuoles, small vesicles; the nucleus.	98
Macrophage	Mouse resident peritoneal macrophages	Gold-LPS conjugate, 50 ng/ml	No specificity controls; serum-free.	Gold-LPS in "deep labrinthic surface invaginations, coated pits (Fig. 4), tubular structures, "lysosomes."	99
Alveolar macrophage	Hamster	Lipid A–coated *E. agglomerans*	ICC–anti-lipid A antiserum; aerosolized bacteria	Authors counted gold particles in BAL cells. No EM photography shown.	100

per minute. Unexpectedly, LPS did not migrate to the well-characterized primary (azurophilic) or secondary neutrophil granules. Although the precise nature of the LPS-containing endocytic vesicles was not determined, the intracellular LPS "band" on the gradient was rather tight, implying uniform density, and it closely overlapped the distribution of the LPS-deacylating enzyme, acyloxyacyl hydrolase (AOAH). LPS translocation occurred by both CD14-dependent and -independent mechanisms (as conferred by the presence or absence of autologous plasma). One limitation of these studies was the inability of this technique to discriminate between LPS that was on the exterior and interior faces of the plasma membrane or between LPS on the plasma membrane and LPS in early endosomes.

More recently, several groups have used laser confocal microscopy to follow the intracellular movement of LPS tagged with fluorescent adjuncts such as FITC or BODIPY (1,2,22,30,59). These studies have confirmed that monocytes internalize LPS aggregates rapidly, so that discrete pockets of bright fluorescence become discernible within 5 minutes after LPS is added (22,30,59). BODIPY-LPS that was presented to cells as monomeric LPS-sCD14 complexes was internalized slowly (consistent with the slow internalization of monomeric [³H]LPS by THP-1 cells (22), but it also accumulated in discrete intracellular locations (1,2). The nature of the intracellular vesicles has not been determined.

INTERNALIZATION AND SIGNALING

Thiéblemont and Wright (2) followed BODIPY-LPS in murine macrophages as it moved from the plasma membrane into peripherally located, small intracellular vesicles, then concentrated in one or two large vesicles in the perinuclear region of the cell. Macrophages from LPS-hyporesponsive mice (C3H/HeJ) were unable to concentrate the LPS centrally, although they appeared to internalize it normally. They were similarly unable to concentrate substimulatory doses of BODIPY-ceramide (which has been shown to accumulate in the Golgi). Since cells from LPS-hyporesponsive mice are also unable to respond normally to ceramide, these results are consistent with the conclusion that their inability to translocate LPS-containing vesicles to a central location (possibly the Golgi) is somehow related to their hyporesponsiveness. The authors acknowledged the alternative possibility that, because of their hyporesponsiveness, the cells are unable to translocate either LPS or ceramide.

In addition to these morphological observations using fluorescent LPS derivatives, there are functional data that support a role for signaling by intracellular LPS. Lichtman et al. (60) found that the ability of LPS to stimulate TNF-α release by rat Kupffer cells was blocked by several compounds that should inhibit LPS internalization and/or prevent the acidification of endocytic compartments. In addition, Shinji et al. (61) found that treating murine macrophages for 5 minutes with cytochalasin D inhibited several responses to LPS, including TNF release, and Detmers et al. found evidence that LPS-induced activation of human neutrophils to express adhesion molecules could be blocked by cytochalasin D and by wortmannin, a PI-3 kinase inhibitor that blocks vesicle fusion. The LPS-induced responses studied in all of these experiments (TNF-α release by Kupffer cells, TNF release by macrophages, adhesion by neutrophils) are relatively slow responses to LPS, however, requiring from 20 minutes to 2 hours. Moreover, none of these studies actually quantitated LPS internalization or its inhibition, and in human monocytes, wortmannin also evidently does not interfere with LPS signaling (see Ref. 1). Nevertheless, all of these experiments suggest a possible signaling role for internalized LPS.

In contrast to these studies, Tobias's group has suggested that the large majority of the LPS that binds to the cell has no role in signaling. Their published data are based on the properties of two mAb. One (18E12) blocks signaling but minimally decreases the amount of LPS bound to the cell, while the other (18G4) does not block signaling but prevents more than 95% of the LPS from binding to the cell (3). They prefer a model in which LPS bound to CD14 interacts with a surface-associated signaling apparatus, while the LPS that enters the cell is destined for disposal or degradation and plays no role in signaling per se.

LPS DEGRADATION

Unlike primitive phagocytes, which often use bacteria as a foodstuff, the phagocytic cells found in animals tend to retain much of what they engulf (62). Gram-negative bacteria are typically killed within phagolysosomes, but their signature molecules, such as LPS, can be detected for days to months thereafter (63–65), and there is now evidence that the LPS in the bacterial carcass undergoes very slow deacylation inside phagocytes (see below). The ultimate intracellular fate of chemically isolated LPS, or even of the LPS that is presented in bacterial membrane fragments (blebs), is also uncertain, but it seems likely that it also can persist inside phagocytes for long periods of time. Indeed, the cellular machinery for degrading LPS appears to be limited to enzymes that remove phosphates or acyl chains—there is little evidence that animals have enzymes that degrade either the core oligosaccharides or the many diverse O-antigenic polymers found in nature. More remarkably, the bacteria themselves do not have such degradative enzymes—in violation of the dictum that microorganisms are usually able to break down the molecules they build.

Chemically isolated LPS is known to undergo deacylation and dephosphorylation inside phagocytes, although these degradative steps occur at relatively slow rates even in optimized in vitro settings. For a more comprehensive discussion of LPS processing in vivo and by phagocytes in vitro, the reader is referred to the review by Erwin and Munford (65).

Deacylation

Neutrophils and monocyte-macrophages have an unusual lipase that acts on the lipid A moiety of LPSs, removing the secondary ("piggyback") acyl chains that are substituted to the hydroxyl functions of the glucosamine-linked hydroxy-fatty acids. Discovered in 1983 (66), this enzyme ("acyloxyacyl hydrolase" or AOAH) has now been purified, cloned, and expressed in eukaryotic cells. It has several features that distinguish it from the other known lipases, including its heterodimeric structure, the presence of a saposin-like domain in one of its subunits, its ability to cleave acyl chains from the *sn*-1 backbone position of both glycerolipids and phosphoglycerolipids, and its ability to be selectively activated, by proteolytic cleavage, to act on LPS (67). The similarity to saposins is particularly intriguing. Saposins (sphingolipid activator proteins) are soluble peptide cofactors for specific sphingolipid hydrolases (68). Their signature structure—specifically, the locations of six cysteines, a site for N-linked glycosylation, and several hydrophobic regions—is thought to form a scaffold upon which various functions have been built. Other proteins that have a saposin-like motif include acid sphingomyelinase, surfactant protein B, and the pore-forming toxins of *Entamoeba histolytica* and NK cells (68). Since an intact saposin-like domain is required for maximal AOAH enzymatic activity, presumably this domain functions as the equivalent of a cofactor for the enzyme. In any case, it is intriguing that one component of a phagocytic cell enzyme that attacks LPS has structural similarity to cofactors for enzymes that hydrolyze sphingolipids, the eukaryotic counterpart of bacterial lipopolysaccharides.

AOAH activity has been found in lysates of human neutrophils and monocytes (66), murine peritoneal macrophages (69), and in lapine (70) and bovine (71) peripheral blood leukocytes. Blood leukocytes from horse and chicken can also deacylate LPS, as can homogenates of murine liver, kidney, spleen, and lung (72). Although human AOAH has not been detected in extracellular fluids, plasma AOAH activity increases manyfold in mice or rabbits challenged with LPS, and LPS stimulates the release of AOAH from rabbit leukocytes in vitro (70). LPS stimulation of murine macrophages also increases AOAH mRNA expression and enzymatic activity (73). The murine, lapine, and human AOAHs show striking sequence conservation, with preservation of all of the unique structural features of the protein (S. Fosmire, J. Staab, A. Varley, and R. Munford, unpublished).

Lipid A analogs that lack secondary acyl chains are unable to stimulate human cells (74,75). The observation that treatment with AOAH diminishes the biopotency of many different LPSs (76,77) suggests strongly that the acyloxyacyl configuration is a major feature of the signal information in LPS. Remarkably, LPS that has been treated with AOAH (dLPS) in vitro is also a potent LPS-specific inhibitor, able to block the ability of LPS to stimulate human endothelial cells (78,79), neutrophils (80), and monocytes (15). dLPS binds to CD14 in an LBP-dependent fashion (15), yet it can inhibit cellular responses to LPS without preventing LPS-CD14 binding (81). In THP-1 cells, low concentrations of dLPS can inhibit signaling by three- to five-fold higher concentrations of LPS (81). dLPS and LPS may compete for a site in the signal pathway that is downstream of CD14. Alternatively, dLPS may initiate a signal that interferes in some way with the LPS signaling pathway.

The actual biological role of LPS deacylation is uncertain. Deacylation of purified LPS occurs very slowly in both mononuclear phagocytes and neutrophils. Although very little is known about the ability of phagocytes to degrade the LPS that resides in native bacterial membranes, recent experiments found that rabbit peritoneal exudate macrophages deacylated the LPS in phagocytosed *E. coli* almost as rapidly as they cleaved the bacterial phospholipids, with removal of half of the nonhydroxylated (piggyback) fatty acids from the bacterial LPS within 4 hours. In contrast, the neutrophils in the exudate deacylated much less of the LPS in the *E. coli* they internalized (S. Katz, Y. Weinrauch, P. Elsbach, R. Munford, and J. Weiss, unpublished). These results point to a dominant role for macrophages in deacylating the LPS in bacterial membranes in vivo. Interestingly, even when macrophages were allowed to degrade phagocytosed bacteria for prolonged periods (20 hours), much of the LPS in the bacterial carcass remained at least partially acylated.

One plausible hypothesis is that partial deacylation prepares LPS for presentation as an antigen. Interesting recent evidence suggests that other lipoglycan molecules may be presented, bound to one or more isoforms of CD1, on the surfaces of antigen-presenting cells. CD1 has a deep hydrophobic groove (82) into which the lipid moiety of lipoarabinomannan, mycolic acid, and other lipoglycans may insert (83). Internalized ligand and CD1 evidently meet in multivesicular bodies (MIIC), where CD1 undergoes an acid-dependent conformational change that allows insertion of the lipid moiety of the ligand into its hydrophobic groove (84). The ligand-CD1 complex then moves to the cell surface, where it can be recognized by specific T cells. Partial deacylation of the lipid moiety of LPS might be required to trim the bulky, rigid lipid A structure to a size that would fit into the presentation groove of CD1 or another protein.

Another hypothetical role for LPS deacylation was suggested by the experiments of Matsuura et al. (85). These workers found that a lipid A analog with an alkyl (ether-linked) secondary fatty chain was a less potent adjuvant than an analog in which the secondary chain was joined to the primary fatty acid in the usual (ester-linked) fashion. The alkyl analog was also somewhat more toxic, as measured by its lethality for mice. They proposed that AOAH degrades lipid A to a less toxic structure that is able to function as an adjuvant. The notion that AOAH may degrade LPS to an adjuvant form is consistent with early speculation (86) that the enzyme may play a role analogous to that of lysozyme, which breaks peptidoglycan into muramylpeptides that can be potent adjuvants.

It is hoped that further insight into the biological role of LPS deacylation will come soon from the analysis of AOAH-null (-/-) mice.

Dephosphorylation

Mononuclear phagocytes also dephosphorylate lipid A, as shown by Peterson and Munford (87) and by Hampton et al. (88). An acidic intracellular compartment has again been implicated for this degradative reaction (87,88), although the responsible enzyme(s) have not been identified. It is also not certain whether dephosphorylation precedes or follows deacylation, and the stimulatory potency of LPS that has been both partially deacylated and dephosphorylated has not been studied.

A biological role for dephosphorylation is suggested by the fact that monophosphoryl lipid A, which retains one or more secondary acyl chains, is a nontoxic adjuvant (89). Similarly, removing the 1' phosphate from lipid IVA reduced its potency (90).

Regulation of LPS Dephosphorylation Deacylation

Peritoneal macrophages from LPS-hyporesponsive and LPS-responsive mice remove acyl chains from the lipid A moiety of LPS with equal facility (69). There is also no apparent difference in their ability to dephosphorylate lipid A (87). Lipid A may decrease its own dephosphorylation in certain cells (90), while LPS can induce increases in both AOAH mRNA and LPS deacylation in murine macrophages studied in vitro (73). Treating adherent human monocytes in vitro with GM-CSF and interleukin-4 (91) decreases intracellular AOAH activity as the cells differentiate into dendritic cells, while treatment with interferon-gamma has a transient stimulatory effect on AOAH activity (R. Munford, unpublished preliminary observations).

SUMMARY

The study of LPS internalization by phagocytes is in its infancy. As suggested by the literature discussed above, there is much yet to be learned about how LPS enters and is transported within the cell, whether it recycles to the plasma membrane, whether it is released from the cell, and where it resides during long-term "storage" within the phagocyte. Questions regarding the potential biological significance of these processes include:

Does LPS initially signal phagocytes from the cell surface, or is it internalized before it initiates signal transduction?

How long, and in which cellular compartments, does LPS remain bioactive (able to trigger mediator release by cells)? If LPS is exocytosed, is it bioactive or inactive (perhaps able to block cellular responses to LPS)?

Does the LPS that is present on the surfaces of phagocytes play a role in antigen presentation (to T or B cells, leading to cell-mediated immunity and anti-LPS antibody synthesis)? Is cell-surface LPS recycled to the cell surface from ingested bacteria or membrane fragments, and if so, how is this process regulated? What are the cofactors? Does CD1 play a role?

How does LPS become an adjuvant? Is this mediated entirely by LPS-stimulated cytokines, or does the LPS molecule play an independent role? Must the lipid A moiety be processed enzymatically before LPS can be an effective adjuvant?

The current evidence indicates that the processes underlying LPS internalization are very complex and that understanding them is likely to be difficult. Although the experimental challenge is great, the questions are important, and answering them is likely to yield significant biological and medical insights.

REFERENCES

1. Detmers PA, Thiéblemont N, Vasselon T, Pironkova R, Miller DS, Wright SD. Potential role of membrane internalization and vesicle fusion in adhesion of neutrophils in response to lipopolysaccharide and TNF. J Immunol 1996; 157:5589–5596.
2. Thiéblemont N, Wright SD. Mice genetically hyporesponsive to lipopolysaccharide (LPS) exhibit a defect in endocytic uptake of LPS and ceramide. J Exp Med 1997; 185:1–6.
3. Gegner JA, Ulevitch RJ, Tobias PS. Lipopolysaccharide (LPS) signal transduction and clearance. Dual roles for LPS binding protein and membrane CD14. J Biol Chem 1995; 270:5320–5326.
4. Kurt-Jones EA, Sattler F, Thompson C, Fleisher G, Novitsky T, Siber G. Endotoxin neutralizing protein, a recombinant Limulus anti-LPS factor, prevents monocyte activation by LPS (abstr). Conference on the Molecular Basis of Sepsis, Woods Hole, MA 1996.
5. Wright SD. CD14 and innate recognition of bacteria. J Immunol 1995; 155:6–8.
6. Ulevitch RJ, Tobias PS. Receptor-dependent mechanisms of cell stimulation by bacterial endotoxin. Annu Rev Immunol 1995; 13:437–457.
7. Hampton RY, Golenbock DT, Penman M, Krieger M, Raetz CRH. Recognition and plasma clearance of endotoxin by scavenger receptors. Nature 1991; 352:342–344.
8. Shnyra A, Lindberg AA. Scavenger receptor pathway for lipopolysaccharide binding to Kupffer and endothelial liver cells in vitro. Infect Immunol 1995; 63:865–873.
9. Dziarski R. Cell-bound albumin is the 70-kDa peptidoglycan, lipopolysaccharide, and lipoteichoic acid-binding protein on lymphocytes and macrophages. J Biol Chem 1994; 269:20431–20436.
10. Mey A, Leffler H, Hmama Z, Normier G, Revillard J-P. The animal lectin galectin-3 interacts with bacterial lipopolysaccharides via two independent sites. J Immunol 1996; 156:1572–1577.
11. Wurfel MM, Wright SD. Lipopolysaccharide-binding protein and soluble CD14 transfer lipopolysaccharide to phospholipid bilayers—preferential interaction with particular classes of lipid. J Immunol 1997; 158:3925–3934.
12. Schromm AB, Brandenburg K, Rietschel ET, Flad HD, Carroll SF, Seydel U. Lipopolysaccharide-binding protein mediates CD14-independent intercalation of lipopolysaccharide into phospholipid membranes. FEBS Lett 1996; 399:267–271.
13. Myers JN, Tabas I, Jones NL, Maxfield FR. β-Very low density lipoprotein is sequestered in surface-connected tubules in mouse peritoneal macrophages. J Cell Biol 1993; 123:1389–1402.
14. Kruth HS, Skarlatos SI, Lilly K, Chang J, Ifrim I. Sequestration of acetylated LDL and cholesterol crystals by human monocyte-derived macrophages. J Cell Biol 1995; 129:133–145.
15. Kitchens RL, Ulevitch RJ, Munford RS. Lipopolysaccharide (LPS) partial structures inhibit responses to LPS in a human macrophage cell line without inhibiting LPS uptake by a CD14-mediated pathway. J Exp Med 1992; 1760:485–494.
16. Perera PY, Vogel SN, Detore GR, Haziot A, Goyert SM. CD14-dependent and CD14-independent signaling pathways in murine macrophages from normal and CD14 knockout mice stimulated with lipopolysaccharide or taxol. J Immunol 1997; 158:4422–4429.
17. Frey EA, Miller DS, Jahr TG, et al. Soluble CD14 participates in the response of cells to lipopolysaccharide. J Exp Med 1992; 176:1665–1671.
18. Pugin J, Schürer-Maly C-C, Leturcq D, Moriarty A, Ulevitch RJ, Tobias PS. Lipopolysaccharide activation of human endothelial and epithelial cells is mediated by lipopolysaccharide-binding protein and soluble CD14. Proc Natl Acad Sci USA 1993; 90:2744–2748.
19. Goldblum SE, Brann TW, Ding X, Pugin J, Tobias PS. Lipopolysaccharide (LPS)-binding protein and soluble CD14 function as accessory molecules for LPS-induced changes in endothelial barrier function in vitro. J Clin Invest 1994; 93:692–702.
20. Troelstra A, Giepmans BNG, Van Kessel KPM, Lichenstein HS, Verhoef J, Van Strijp JAG. Dual effects of soluble CD14 on LPS priming of neutrophils. J Leukocyte Biol 1997; 61:173–178.
21. Hailman E, Vasselon T, Kelley M, et al. Stimulation of macrophages and neutrophils by complexes of lipopolysaccharide and soluble CD14. J Immunol 1996; 156:4384–4390.
22. Kitchens RL, Munford RS. CD14-dependent internalization of lipopolysaccharide (LPS) is strongly influenced by LPS aggregation but not by cellular responses to LPS. J Immunol 1998; 160:1920–1928.
23. Schumann RR, Leong SR, Flaggs GW, et al. Structure and function of lipopolysaccharide-binding protein. Science 1990; 249:1429–1431.
24. Wright SD, Ramos RA, Tobias PS, Ulevitch RJ, Mathison JC. CD14, a receptor for complexes of lipopolysaccharide (LPS) and LPS binding protein. Science 1990; 249:1431–1433.
25. Kirkland TN, Finley F, Leturcq D, et al. Analysis of lipopolysaccharide binding by CD14. J Biol Chem 1993; 268:24818–24823.
26. Grunwald U, Fan XL, Jack RS, et al. Monocytes can phagocytose gram-negative bacteria by a CD14-

dependent mechanism. J Immunol 1996; 157:4119–4125.
27. Hailman E, Lichenstein HS, Wurfel MM, et al. Lipopolysaccharide (LPS)-binding protein accelerates the binding of LPS to CD14. J Exp Med 1994; 179:269–277.
28. Tobias PS, Soldau K, Gegner JA, Mintz D, Ulevitch RJ. Lipopolysaccharide binding protein-mediated complexation of lipopolysaccharide with soluble CD14. J Biol Chem 1995; 270:10482–10488.
29. Wurfel MM, Hailman E, Wright SD. Soluble CD14 acts as a shuttle in the neutralization of lipopolysaccharide (LPS) by LPS-binding protein and reconstituted high density lipoprotein. J Exp Med 1995; 181:1743–1754.
30. Kitchens RL, Wang P-Y, Munford RS. Bacterial lipopolysaccharide can enter human monocytes via two CD14-dependent pathways. J Immunol 1998; 161:5534–5545.
31. Keller GA, Siegel MW, Caras IW. Endocytosis of glycophospholipid-anchored and transmembrane forms of CD4 by different endocytic pathways. Eur Mol Biol Org J 1992; 11:863–874.
32. Subtil A, Hémar A, Dautry-Varsat A. Rapid endocytosis of interleukin 2 receptors when clathrin-coated pit endocytosis is inhibited. J Cell Science 1994; 107:3461–3468.
33. van Deurs B, Petersen OW, Olsnes S, Sandvig K. The ways of endocytosis. Int Rev Cytol 1989; 117:131–177.
34. Nichols BA. Uptake and digestion of horseradish peroxidase in rabbit alveolar macrophages. Lab Invest 1982; 47:235–246.
35. Mukherjee S, Ghosh RN, Maxfield FR. Endocytosis. Physiol Rev 1997; 77:759–803.
36. Wang Ping-yuan, Kitchens RL, Munford RS. Bacterial lipopolysaccharide binds to low-density domains of the monocyte-macrophage plasma membrane. J Inflamm 1996; 47:126–137.
37. Smart EJ, Ying Y-S, Mineo C, Anderson RGW. A detergent-free method for purifying caveolae membrane from tissue culture cells. Proc Natl Acad Sci USA 1995; 92:10104–10108.
38. Ying Y-S, Anderson RGW, Rothberg KG. Each caveola contains multiple glycosyl-phosphatidylinositol-anchored membrane proteins. Cold Spring Harbor Symp Quant Biol 1992; 57:593–604.
39. Anderson RGW. Plasmalemmal caveolae and GPI-anchored membrane proteins. Curr Opin Cell Biol 1993; 5:647–652.
40. Parton RG. Caveolae and caveolins. Curr Opin Cell Biol 1996; 8:542–548.
41. Parolini I, Sargiacomo M, Lisanti MP, Peschle C. Signal transduction and glycophosphatidylinositol-linked proteins (LYN, LCK, CD4, CD45, G proteins, and CD55) selectively localize in Triton-insoluble plasma membrane domains of human leukemic cell lines and normal granulocytes. Blood 1996; 87:3783–3794.
42. Rothberg KG, Heuser JE, Donzell WC, Ying YS, Glenny JR, Anderson RC. Caveolin, a protein component of caveolae membrane coats. Cell 1992; 68:673–682.
43. Li SW, Couet J, Lisanti MP. Src tyrosine kinases, G_α subunits, and H-Ras share a common membrane-anchored scaffolding protein, caveolin—caveolin binding negatively regulates the auto-activation of Src tyrosine kinases. J Biol Chem 1996; 271:29182–29190.
44. Liu J, Oh P, Horner T, Rogers RA, Schnitzer JE. Organized endothelial cell surface signal transduction in caveolae distinct from glycosylphosphatidylinositol-anchored protein microdomains. J Biol Chem 1997; 272:7211–7222.
45. Anderson RG, Kamen BA, Rothberg KG, Lacey SW. Potocytosis: Sequestration and transport of small molecules by caveolae. Science 1992; 255:410–411.
46. Schnitzer JE, Oh P, Pinney E, Allard J. Filipin-sensitive caveolae-mediated transport in endothelium: reduced transcytosis, scavenger endocytosis, and capillary permeability of select macromolecules. J Cell Biol 1994; 127:1217–1232.
47. Deckert M, Ticchioni M, Bernard A. Endocytosis of GPI-anchored proteins in human lymphocytes: role of glycolipid-based domains, actin cytoskeleton, and protein kinases. J Cell Biol 1996; 133:791–799.
48. Parton RG, Simons K. Digging into caveolae. Science 1995; 269:1398–1399.
49. Fra AM, Williamson E, Simons K, Parton RG. De novo formation of caveolae in lymphocytes by expression of VIP21-caveolin. Proc Natl Acad Sci USA 1995; 92:8655–8659.
50. Kiss AL, Geuze HJ. Caveolae can be alternative endocytotic structures in elicited macrophages. Eur J Cell Biol 1997; 73:19–27.
51. Anderson RGW. Caveolae: where incoming and outgoing messengers meet. Proc Natl Acad Sci USA 1993; 90:10909–10913.
52. Swanson JA, Watts C. Macropinocytosis. Trends Cell Biol 1995; 5:424–428.
53. Francis CL, Ryan TA, Jones BD, Smith SJ, Falkow S. Ruffles induced by Salmonella and other stimuli direct macropinocytosis of bacteria. Nature 1997; 364:639–642.
54. West MA, Bretscher MS, Watts C. Distinct endocytotic pathways in epidermal growth factor-stimulated human carcinoma A431 cells. J Cell Biol 1989; 109:2731–2739.
55. Racoosin EL, Swanson JA. M-CSF-induced macropinocytosis increases solute endocytosis but not receptor-mediated endocytosis in mouse macrophages. J Cell Sci 1992; 102:867–880.
56. Sallusto F, Cella M, Danieli C, Lanzavecchia A. Dendritic cells use macropinocytosis and the mannose receptor to concentrate macromolecules in the major histocompatibility complex class II compartment: downregulation by cytokines and bacterial products. J Exp Med 1995; 182:389–400.
57. Ridley AJ, Paterson HF, Johnston CL, Diekmann D, Hall A. The small GTP-binding protein rac regulates growth factor-induced membrane ruffling. Cell 1992; 70:401–410.
58. Luchi M, Munford RS. Binding, internalization, and deacylation of bacterial lipopolysaccharides by human neutrophils. J Immunol 1993; 151:959–969.

59. Gallay P, Jongeneel CV, Barras C, et al. Short time exposure to lipopolysaccharide is sufficient to activate human monocytes. J Immunol 1993; 150:5086–5093.
60. Lichtman SN, Wang J, Zhang C, Lemasters JJ. Endocytosis and Ca^{2+} are required for endotoxin-stimulated TNF-α release by rat Kupffer cells. Am J Physiol Gastrointest Liver Physiol 1996; 271:G920–G928.
61. Shinji H, Akagawa KS, Yoshida T. Cytochalasin D inhibits lipopolysaccharide-induced tumor necrosis factor production in macrophages. J Leukoc Biol 1993; 54:336–342.
62. Holtzman E. Lysosomes. New York: Plenum Press, 1989:32.
63. Granfors K, Jalkanen S, Von Essen R, et al. *Yersinia* antigens in synovial-fluid cells from patients with reactive arthritis. N Engl J Med 1989; 320:216–221.
64. Granfors K, Jalkanen S, Toivanen P, Koski J, Lindberg AA. Bacterial lipopolysaccharide in synovial fluid cells in *Shigella* triggered reactive arthritis. J Rheumatol 1992; 19:500.
65. Erwin AL, Munford RS. Processing of LPS by phagocytes. In: Morrison DC, Ryan JL, eds. Bacterial Endotoxic Lipopolysaccharides. Molecular Biochemistry and Cellular Biology. Boca Raton, FL: CRC Press, 1992:405–434.
66. Hall CL, Munford RS. Enzymatic deacylation of the lipid A moiety of *Salmonella typhimurium* lipopolysaccharides by human neutrophils. Proc Natl Acad Sci USA 1983; 80:6671–6675.
67. Staab JF, Ginkel DL, Rosenberg GB, Munford RS. A saposin-like domain influences the intracellular localization, stability, and catalytic activity of human acyloxyacyl hydrolase. J Biol Chem 1994; 269:23736–23742.
68. Munford RS, Sheppard PO, O'Hara PJ. Saposin-like proteins (SAPLIP) carry out diverse functions on a common backbone structure. J Lipid Res 1995; 36:1653–1663.
69. Munford RS, Hall CL. Uptake and deacylation of bacterial lipopolysaccharides by macrophages from normal and endotoxin-hyporesponsive mice. Infect Immun 1985; 48:464–473.
70. Erwin AL, Munford RS. Plasma lipopolysaccharide-deacylating activity (acyloxyacyl hydrolase) increases following lipopolysaccharide administration to rabbits. Lab Invest 1991; 65:138–144.
71. McDermott CM, Cullor JS, Fenwick BW. Intracellular and extracellular enzymatic deacylation of bacterial endotoxin during localized inflammation induced by *Escherichia coli*. Infect Immun 1991; 59:478–485.
72. Coulthard MG, Swindle J, Munford RS, Gerard RD, Meidell RS. Adenovirus-mediated transfer of a gene encoding acyloxyacyl hydrolase (AOAH) into mice increases tissue and plasma AOAH activity. Infect Immun 1996; 64:1510–1515.
73. Cody MJ, Salkowski CA, Henricson BE, Detore GR, Munford RS, Vogel SN. Effect of inflammatory and anti-inflammatory stimuli on acyloxyacyl hydrolase gene expression and enzymatic activity in murine macrophages. J Endotoxin Res 1998.
74. Takada H, Kotani S. Structural requirements of lipid A for endotoxicity and other biological activities. CRC Crit Rev Microbiol 1989; 16:477–523.
75. Takada H, Kotani S. Structure-function relationships of lipid A. In: Morrison DC, Ryan JL, eds. Bacterial Endotoxic Lipopolysaccharides. Vol. I. Boca Raton, FL: CRC Press, 1992:107–135.
76. Erwin AL, Munford RS. Deacylation of structurally diverse lipopolysaccharides by human acyloxyacyl hydrolase. J Biol Chem 1990; 265:16444–16449.
77. Erwin AL, Mandrell RE, Munford RS. Enzymatically deacylated *Neisseria* LPS inhibits murine splenocyte mitogenesis induced by LPS. Infect Immun 1991; 59:1881–1887.
78. Pohlman TH, Munford RS, Harlan JM. Deacylated lipopolysaccharide inhibits neutrophil adherence to endothelium induced by lipopolysaccharide in vitro. J Exp Med 1987; 165:1393–1402.
79. Riedo FX, Munford RS, Campbell WB, Reisch JS, Chien KR, Gerard RD. Deacylated lipopolysaccharide inhibits plasminogen activator inhibitor-1, prostacyclin, and prostaglandin E2 induction by lipopolysaccharide but not by tumor necrosis factor-alpha. J Immunol 1990; 144:3506–3512.
80. Dal Nogare AR, Yarbrough WC, Jr. A comparison of the effects of intact and deacylated lipopolysaccharide on human polymorphonuclear leukocytes. J Immunol 1990; 144:1404–1410.
81. Kitchens RL, Munford RS. Enzymatically deacylated lipopolysaccharide (LPS) can antagonize LPS at multiple sites in the LPS recognition pathway. J Biol Chem 1995; 270:9904–9910.
82. Zeng Z-H, Castano AR, Segelke BW, Stura EA, Peterson PA, Wilson IA. Crystal structure of mouse CD1: an MHC-like fold with a large hydrophobic binding groove. Science 1997; 277:339–345.
83. Sieling PA, Chatterjee D, Porcelli SA, et al. CD1-restricted T cell recognition of microbial lipoglycan antigens. Science 1995; 269:227–230.
84. Jullien D, Stenger S, Ernst WA, Modlin RL. CD1 presentation of microbial nonpeptide antigens to T cells. J Clin Invest 1997; 99:2071–2074.
85. Matsuura M, Shimada S-I, Kiso M, Hasegawa A, Nakano M. Expression of endotoxic activities by synthetic monosaccharide lipid A analogs with alkyl-branched acyl substituents. Infect Immun 1995; 63:1446–1451.
86. Munford RS, Hall CL. Detoxification of bacterial lipopolysaccharides (endotoxins) by a human neutrophil enzyme. Science 1986; 234:203–205.
87. Peterson AA, Munford RS. Dephosphorylation of the lipid A moiety of *Escherichia coli* lipopolysaccharide by mouse macrophages. Infect Immun 1987; 55:974–978.
88. Hampton RY, Golenbock DT, Raetz CRH. Lipopolysaccharide stimulation regulates lipid A metabolism in macrophage cell lines. FASEB J 1990; 4:A1908.
89. Baker PJ, Hiernaux JR, Fauntleroy MB, et al. Ability of monophosphoryl lipid A to augment the antibody response of young mice. Infect Immun 1988; 56:3064–3066.
90. Hampton RY, Raetz CRH. Macrophage catabolism of lipid A is regulated by endotoxin stimulation. J Biol Chem 1991; 266:19499–19509.
91. Kiertscher SM, Roth MD. Human $CD14^+$ leukocytes acquire the phenotype and function of antigen-pre-

senting dendritic cells when cultured in GM-CSF and IL-4. J Leukocyte Biol 1996; 59:208–218.
92. Swanson J, Bushnell A, Silverstein SC. Tubular lysosome morphology and distribution within macrophages depend on the integrity of cytoplasmic microtubules. Proc Natl Acad Sci USA 1987; 84:1921–1925.
93. Bona CA. Fate of endotoxin in macrophages: biological and ultrastructural aspects. J Infect Dis 1973; 128: S74–S81.
94. Diaz-Laviada I, Ainaga J, Portolés MT, Carrascosa JL, Muncio AM, Pagani R. Binding studies and localization of *Escherichia coli* lipopolysaccharide in cultured hepatocytes by an immunocolloidal-gold technique. Histochem J 1991; 23:221–228.
95. Risco C, Pinto da Silva P. Binding of bacterial endotoxins to the macrophage surface: visualization by fracture-flip and immunocytochemistry. J Histochem Cytochem 1993; 41:601–608.
96. Risco C, Da Silva PP. Cellular functions during activation and damage by pathogens: Immunogold studies of the interaction of bacterial endotoxins with target cells. Microsc Res Tech 1995; 31:141–158.
97. Odeyale CO, Kang Y-H. Biotinylation of bacterial lipopolysaccharide and its applications to electron microscopy. J Histochem Cytochem 1988; 36:1131–1137.
98. Kang Y-H, Dwivedi RS, Lee C-H. Ultrastructural and immunocytochemical study of the uptake and distribution of bacterial lipopolysaccharide in human monocytes. J Leukocyte Biol 1990; 48:316–332.
99. Kriegsmann J, Gay S, Bräuer R. Endocytosis of lipopolysaccharide in mouse macrophages. Cell Mol Biol 1993; 39:791–800.
100. Keller GE, III, Dey RD, Burrell R. Immunocytochemical determination of the role of alveolar macrophages in endotoxin processing in vitro and in vivo. Int Arch Allergy Appl Immunol 1991; 96:149–155.

34

Multifunctional G Proteins: Implication in Endotoxin Cellular Responses

Lawrence P. Fernando, Michel A. Makhlouf, J. A. Cook
Medical University of South Carolina, Charleston, South Carolina

HISTORICAL PERSPECTIVE

Binding of a cell surface receptor by an external ligand triggers a cascade of cellular events via transmembrane signaling mechanisms. The observation that a nucleotide, guanosine triphosphate (GTP), is required for a hormone to activate an enzyme through adenylyl cyclase led to the identification of the significance of G proteins in cellular signaling (1). It was found that this factor was not GTP, but rather a protein bound specifically to GTP. This complex then leads to the selective activation of adenylyl cyclase (2). The use of cholera toxin (3,4), pertussis toxin (PTx) (5), and studies on light-sensitive cyclic GMP phosphodiesterase (6) distinguished the existence of two classes of adenylyl cyclase–associated G proteins—namely a stimulatory protein, Gs, and an inhibitory protein, Gi. It was later discovered that Gs and Gi each consist of three different subunits, which were termed α, β, and γ. The GTP-binding site is located in the α subunit. The intrinsic GTPase activity of α subunit regulates the association and dissociation of the trimeric state among α, β, γ and thereby regulates receptor effector coupling (Fig. 1). The β, γ subunits exist as a dimeric complex. Currently at least 23 α isoforms, 6 β isoforms, and 11 γ isoforms have been isolated (7). In a single class of subunits different isoforms arise as a result of different gene products or as a result of differential splicing from the same gene product. Comparison of G protein sequences from different species show that they are highly evolutionary conserved, suggesting that they have an essential role (8). Theoretically, any isoform of α, β, and γ can combine to give an active trimeric complex, which can function as a network of signal-transducing molecules. The experimental data suggest that the type of α, β, and γ combination in a trimeric complex determines the signaling event through G proteins (9). These trimeric complexes have different affinities for the effecter receptors and to downstream members of the signal cascade bringing diversity and multi functionality to the signaling cascade (10). Since the initial identification of G protein-mediated signaling associated with the β-adrenergic receptor effector system, numerous other examples of G protein–mediated signaling have been observed (9).

METHODS USED TO STUDY G PROTEINS

Most of the G-protein subunits, cDNAs and genes, have been cloned or the proteins purified in the native state (9). Specific antibodies have been raised against most of the subunits. These antibodies and DNA have been successfully used to analyze the relative abundance of G proteins at the protein and mRNA levels by quantitative Western (11,12) and Northern blotting (13). These antibodies have been used successfully to immunoprecipitate specific G proteins in the presence of a nonhydrolyzable [α-^{32}P] GTP azidoanilide, which chemically cross-links to activated Gα subunit (14). Agents such as fluoride or GTPγS, which nonspecifically activate G proteins, or cholera toxin, which activates the Gs class of proteins by ADP ribosylating Gαs

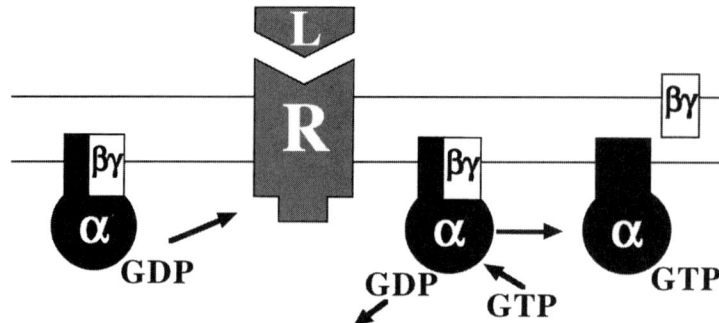

Fig. 1 A potential scheme for G-protein–mediated signaling. External ligands (L) interacts with receptors (R) sitting on the plasma membrane. These interactions (L/R) activate the heterotrimeric membrane G proteins to associate with the receptor-ligand complexes. As a result a guanine nucleotide GTP from the cytosol exchanges with the GDP on the α subunit of the heterotimeric complex leading to the dissociation of the trimer into α-GTP and a βγ dimer. Each component by interaction with a potential effector molecule can bring about an array of cellular events. The intrinsic GTPase activity of the α subunit hydrolyzes the GTP to GDP, resulting in association with the βγ dimer to enter into another cycle of signaling.

(3), have been successfully used to study G-protein–mediated signaling. Membrane GTPase activity has been used as a nonspecific indicator of the membrane G-protein function and content. In other studies, C-terminal specific antibodies to Gα subunits have been used to compete for receptor binding of G proteins (15–18). Another approach has been to reconstitute the receptor effector and intermediate G proteins in artificial membranes (19–22) or in intact cells pretreated with PTx to neutralize endogenous G-protein activity (23). PTx ADP ribosylates the α subunit of the Gi class of proteins and thereby uncouples the α subunit from the receptor to inhibit the function of Gi proteins (24). A novel and specific direct approach is to express mutant forms of G-protein subunit (10,25,26) isoforms by DNA transfection or to transfect an antisense oligonucleotide for a specific G-protein subunit isoform (7).

ROLE OF G PROTEINS IN LPS-MEDIATED CELLULAR RESPONSES

Lipopolysaccharide (LPS), a cell wall component of gram-negative bacteria, produces a plethora of effects in vitro and in vivo (27–31). The most affected are cells of myeloid origin, whose metabolic state is dramatically altered, leading to a release of a spectrum of potent immune modulators, cytokines, arachidonic acid metabolites, and reactive oxygen species (28–31). The signaling events during the development of cellular activation are poorly understood. There is evidence that LPS interacts via a glycosylphosphatidylinositol-anchored membrane glycoprotein, CD14 (31,32). Interaction of LPS with CD14 is mediated via LPS binding protein present in the serum (31). The relationship between CD14 and G proteins is not clear. However, there is evidence to show an association of G-protein α subunits with other glycosylphosphatidylinositol-anchored receptor proteins (33). Whether activated directly, in response to specific LPS receptor(s), or indirectly, via autocrine cytokines, G proteins can potentially modulate LPS signaling pathways. For example, nonspecific G protein activators (e.g., the nonhydrolyzable GTP analog GTPγS or NaF), like LPS, stimulate arachidonic acid metabolism in rat peritoneal macrophages (MØ)

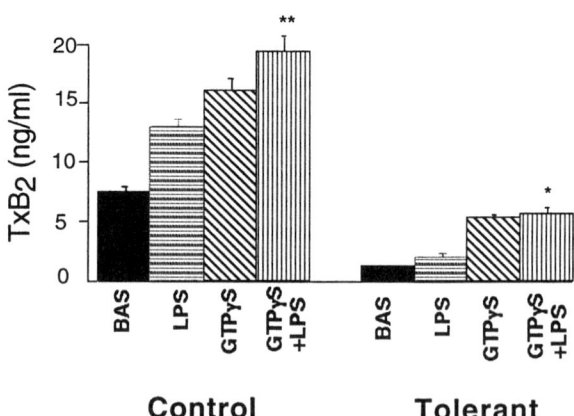

Fig. 2 Effect of GTPγS and LPS on TxB_2 production in rat peritoneal macrophages. Macrophages isolated from control and tolerant rats were treated with LPS (50 μg/ml), GTPγS (100 μM), or LPS plus GTPγS. The TxB_2 concentration of the medium was assayed by radioimmunoassay. The TxB_2 concentration is shown as mean ± SEM for at least for five different experiments. ** and * represent $p < 0.05$ compared to control BAS and tolerant BAS respectively.

(34,35). Figure 2 shows the relative levels of TxB$_2$, an arachidonic acid metabolite, in the culture media of rat MØ treated with indicated stimulators. Both LPS and GTPγS significantly increased TxB$_2$ levels over the unstimulated basal levels in both control and LPS-tolerant MØ.

Many observations also suggest that the mitogen-activated protein kinase (MAPK) cascade is essential for LPS stimulation of specific mediators (36). Interaction of LPS with cell surface receptors leads to activation of a cascade of kinase activity (37–40) that induces nuclear transcription factors. The transcription factors can interact with cis-acting elements of the promoter regions of the genes, leading to induction or repression of gene expression (41). In different cell types, Gi protein–coupled receptors can activate MAPK primarily through G βγ subunits (42–45). Furthermore, it has been shown that tyrosine kinases modulate Gi-coupled MAPK activation (46,47). Because tyrosine kinases are important in LPS-stimulated MØ activation (38), LPS activation of ERK may occur through a Gi-coupled pathway. The ability of PTx to inhibit the Gi class of proteins has been used to dissect out the LPS signaling pathways. Indeed, we have shown in the human monocytic leukemia cell line THP-1 that LPS activation of ERK is PTx sensitive (48,49).

We have studied the role of Gi proteins in LPS-induced mediator production in rat peritoneal MØ. Levels of TxB$_2$ (35), 6-keto-PGF$_{1α}$, nitric oxide (NO), or IL-6 (50) in the MØ culture media were quantitated as indicators of LPS activation. In rat peritoneal MØ, TxB$_2$ production is increased markedly in response to *Salmonella enteritidis* LPS (Fig. 3). Preincubation of MØ with PTx at concentrations of 0.1 or 1.0 ng/ml reduced TxB$_2$ relative to LPS-induced control levels. PTx had a similar trend on LPS induced 6 keto PGF$_{1α}$ levels (50). The results suggest that selective inhibition of the Gi class of proteins by PTx inhibits signaling events leading to arachidonic acid metabolism. However, LPS-stimulated induction of IL-6 and NO in rat peritoneal MØ is not inhibited by PTx pretreatment, suggesting that IL-6 and NO synthesis is independent of signaling through Gi proteins (35).

PTx also blocks LPS activation in human cell lines. LPS-induced TxB$_2$ synthesis in THP-1 cells is inhibited by PTx (48). Incubation of the THP-1 cells with the PTx B protomer, which lacks the ADP ribosylating factor, had no effect on the mediator production (48), suggesting that this inhibition was mediated by inactivating Gi proteins and is not due to a nonspecific effect of PTx (Fig. 4). It is of interest that PTx can block LPS induction of TxB$_2$ in undifferentiated human THP-1 cells that show a low expression of CD-14 receptors (48).

Many others have documented PTx-mediated inhibition of LPS-induced mediators. Jakway and Defranco

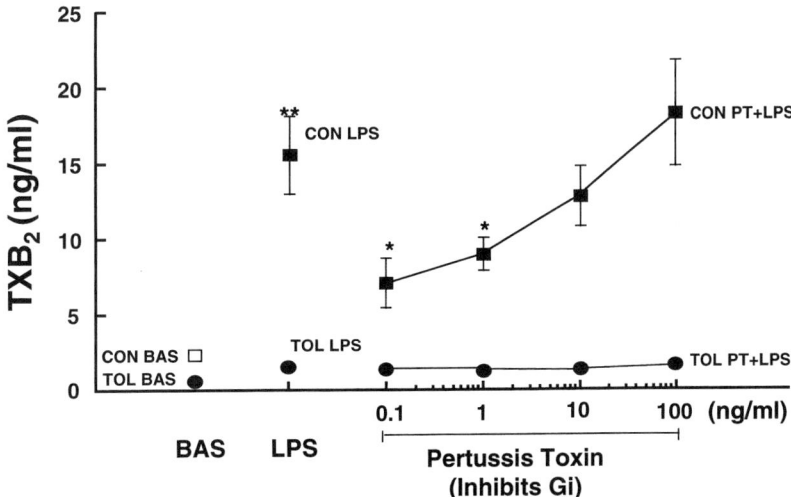

Fig. 3 Effect of pertussis toxin (PT) on LPS-mediated TxB$_2$ production in rat peritoneal macrophages. Peritoneal macrophages were isolated from control rats that were intraperitoneally injected with 5% dextrose solution. Cells were either stimulated directly with LPS (CON LPS) or incubated with PT at the indicated dose for 4 hours and then stimulated for 24 hours with LPS (50 μg/ml). The culture media was collected and assayed for TxB$_2$ content by radioimmunoassay. CON BAS represents the amount of TxB$_2$ from the culture media of unstimulated macrophages. The TxB$_2$ levels are presented as means ± SEM for five different experiments. ** and * represent $p < 0.05$ compared to CON BAS and CON LPS and CON LPS versus CON PT LPS, respectively.

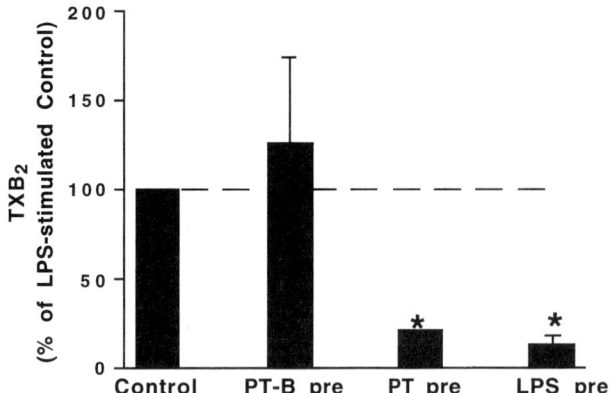

Fig. 4 The effect of PTx or PTx-B protomer on LPS simulated TxB_2 production in THP-1 cells. The cells were pretreated with PTx 10 ng/ml (PT pre) or an equimolar concentration of PTx-B protomer (PT-B pre) or media alone (control) for 4 hours or LPS (1 μg/ml) for 18 hours (LPS pre). Each group of cells were then treated with LPS (10 μg/ml) for 24 hours. The media was collected for TxB_2 assay, and data are presented as the mean ± SEM for four different experiments expressed as a ratio of tolerant over controls. * denotes $p < 0.05$ compared to control LPS-stimulated cells.

demonstrated that LPS-mediated responses are inhibited by PTx in the MØ cell line P388D1 and lymphoma cell line, suggesting that LPS activation of these cells occurs through a Gi-like receptor-effector coupling protein (51). Wang et al. (52) observed that LPS induced the synthesis of prostaglandin E_2 (PGE_2) in cultured rat mesangial cells via a PTx-sensitive G-protein–coupled activation of phospholipase A_2. In a phorbol myristate acetate primed promonocytic cell line, U937, LPS-induced IL-1β transcript synthesis is associated with the de novo synthesis of $Gi_{\alpha 2}$ protein (53). Further, this LPS-induced synthesis of IL-1β is PTx sensitive, suggesting the participation of Gi class of proteins (53). Other studies with LPS-stimulated peritoneal MØ from tumor-bearing rats suggest that a PTx-sensitive G protein regulates the formation of PGE_2, which in turn controls TNF-α synthesis (54). Zhang and Morrison studied the effect of PTx on LPS-mediated NO and TNF-α production in thioglycolate-elicited C3HeB/FeJ mouse peritoneal MØ (55). They observed that PTx differentially regulates mediator production. Pretreatment for 30 minutes or longer with PTx had a profound affect on E. coli O111:B4 S-LPS–induced mediator production. Over a range of increasing concentrations, PTx inhibited LPS-stimulated NO production, whereas TNF-α production increased. The data suggest that NO and TNF-α are controlled in a reciprocal fashion in the murine system by a PTx-sensitive G-protein pathway (55).

Other experimental approaches have implicated a role for G proteins in LPS signaling. Yasui et al. (56) observed that LPS primes human neutrophils for O_2 production in response to stimulation by chemotactic factor f Met-Leu-Phe. LPS induced an increase in the membrane $Gi_{\alpha 2}$ content, suggesting a redistribution of $Gi_{\alpha 2}$ in activated cells. Kugi et al. (25) observed a twofold increase in $Gi_{\alpha 2}$ protein when a murine macrophage cell line $P388D_1$ was treated with low doses of LPS. Further they expressed a mutated activated form of $Gi_{\alpha 2}$ in $P388D_1$ by DNA transfection. There was an increase in PAF induced Ca^{2+} influx across the cell membrane and arachidonic acid release in the activated $Gi_{\alpha 2}$ transfected cells similar to the effects of low concentrations of LPS (25). These composite observations demonstrate that LPS can activate numerous biochemical pathways, some of which may be mediated by signaling through Gi proteins.

ROLE OF G PROTEINS IN LPS TOLERANCE

Another line of evidence for G protein involvement in LPS cellular responses arises from studies of the LPS tolerance phenomenon. The response of cells or animals changes dramatically when preexposed to low concentrations or doses of LPS before stimulation in vitro with higher concentrations or in vivo with lethal LPS doses. The altered cellular responses to LPS or improved survival to LPS shock is referred to as LPS tolerance or desensitization (27). It has been shown that LPS tolerance is not due to downregulation of LPS-binding sites. In fact, in response to a single dose of LPS, CD14 receptors on human monocytes are upregulated (57).

We have studied the effects of LPS tolerance in rat peritoneal macrophages. The survival of rats of LPS shock for 72 hours after a tolerizing dose of LPS or TNF-α given at indicated time periods is shown in Figure 5. Control rats develop severe shock within 24 hours in response to a lethal dose of LPS. However, if a sublethal amount of LPS was introduced 24 hours prior to the lethal dose of LPS, survival increased dramatically and 80% of the rats survived (35). A similar improvement of survival can be seen if rats are pretreated with TNF-α, demonstrating a cross-tolerance between TNF and LPS (58). This observation of cross-tolerance with LPS and other noxious stimuli (e.g., TNF-α) thus further suggests a change in signaling pathways rather than specific receptors. In vitro

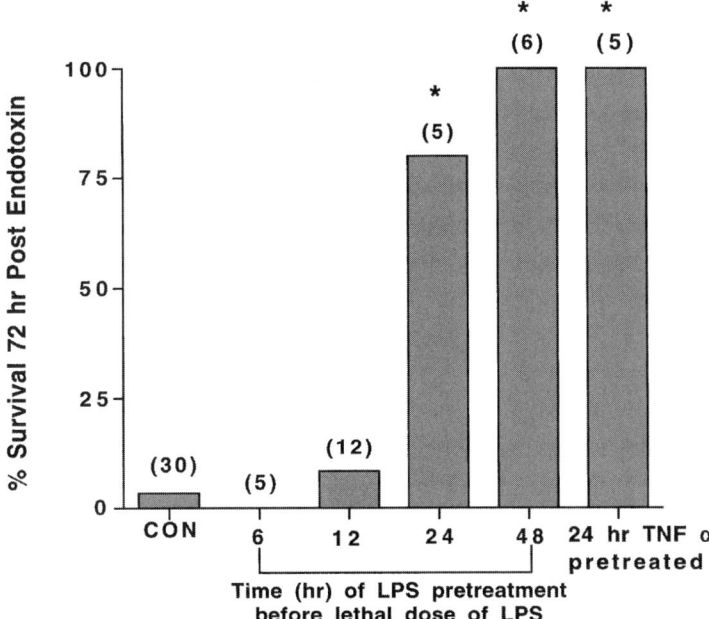

Fig. 5 Effect of LPS or TNF-α tolerance on survival of rats after the administration of a lethal dose of LPS. Rats were given a primary dose of LPS 100 μg/kg or TNF-α (10 μg/kg) and then were challenged with a lethal dose of LPS (15 mg/kg, iv) at 6, 12, 24, and 48 hours after the tolerizing dose of the respective agent. Survival was monitored for 72 hours (( ) = n, *$p < 0.05$).

changes in LPS-stimulated TxB_2 production of peritoneal MØ isolated from rats preinjected with LPS precedes the improved survival in tolerance. MØ from rats receiving LPS pretreatment for 6 hours exhibit markedly reduced LPS-stimulated TxB_2 levels, which persists for several days (Fig. 6). The decreased responsiveness to LPS cannot be attributed to a depletion of arachidonic acid substrate, since tolerant MØ respond to calcium ionophore stimulation with TxB_2 production (59). The latter observations suggest a potential defect in signal transduction during LPS activation prior to activation of phospholipase. MØ membrane GTPase activity was assessed at the same time intervals as were TxB_2 measurements (Fig. 7). The membrane GTPase activity also was markedly reduced at 6 hours and persists for 72 hours following in vivo LPS pretreatment. Thus, decreases in GTPase activity temporally correlate with reduction of TxB_2. The development of resistance to lethal LPS challenge in vivo that occurs with 24 hours of LPS preexposure is thus preceded by earlier changes in cellular mediator production and membrane GTPase activity (51).

Table 1 shows the effect of LPS on the production of several mediators in rat peritoneal MØ after tolerance induced by LPS or TNF-α. The levels of 6-keto $PGF_{1\alpha}$ and TxB_2 are significantly reduced in both LPS- and TNF-α–tolerant MØ populations, whereas NO levels were significantly increased in the same cell populations. Since we have shown that NO is not PTx sensitive, this is consistent with the notion of selective decreases in Gi-coupled signaling during LPS tolerance (50).

These experimental data suggest that tolerance induction is preceded by a reduction of G-protein function as measured by GTPase activities. Additionally, tolerant MØ are affected minimally by nonspecific G-protein agonists such as GTPγS (Fig. 2) and NaF, whereas these agents induce TxB_2, 6-keto-$PGF_{1\alpha}$ in control populations (35). These findings prompted the investigation of whether there is altered substrate binding for GTP in LPS-tolerance MØ membranes. Equilibrium binding data of GTPγS to membranes isolated from control and tolerant MØ show that the Kd values were not significantly different. However, maximum binding capacity was greatly reduced in tolerant MØ populations (approximately 70%) (Fig. 8) (60). Further studies were initiated to assess whether the substrate binding was due to a generalized effect or depletion of G proteins. MØ membrane G-protein content was determined by Western blotting with specific antisera to $Gi_{\alpha 1}$, $Gi_{\alpha 2}$, $Gi_{\alpha 3}$, $G\alpha s$, and the β subunit of G proteins (60). Scanning densitometric analysis demonstrated

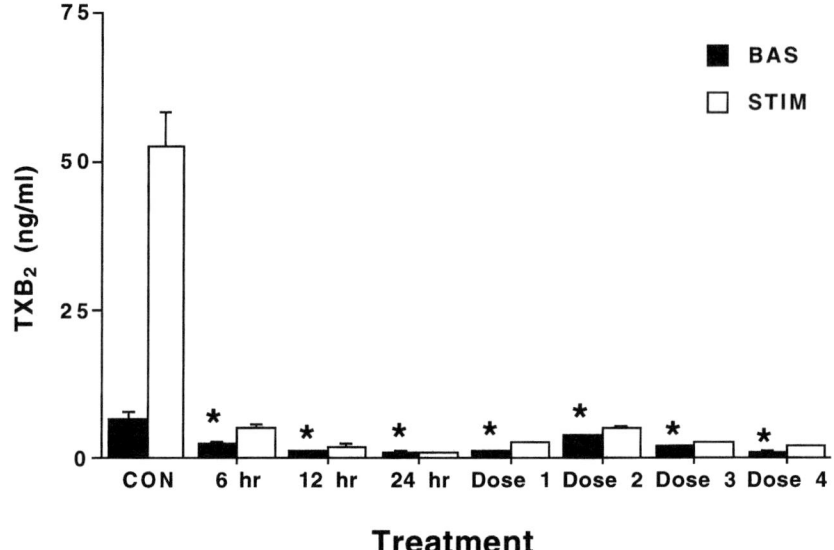

Fig. 6 In vitro effect of endotoxin on TxB_2 production by macrophages isolated from control and rats injected with LPS (100 μg/kg) at designated time intervals. Basal and endotoxin-stimulated TxB_2 were measured from macrophages harvested at 6, 12, and 24 hours after a single injection of endotoxin. Dose 1 represents basal and endotoxin-stimulated TxB_2 measured in macrophages at intervals up to 72 hours after first injection (100 μg/kg) of endotoxin; Dose 2 is a second injection of endotoxin, (500 μg/kg); Dose 3 is a third injection of endotoxin (1 mg/kg); and Dose 4 is a fourth injection of endotoxin (5 mg/kg)—given on four consecutive days. ND, Nondetectable (<250 pg/ml). Data are expressed as means ± SE ($n = 5$ per group, *$p < 0.05$ vs. control).

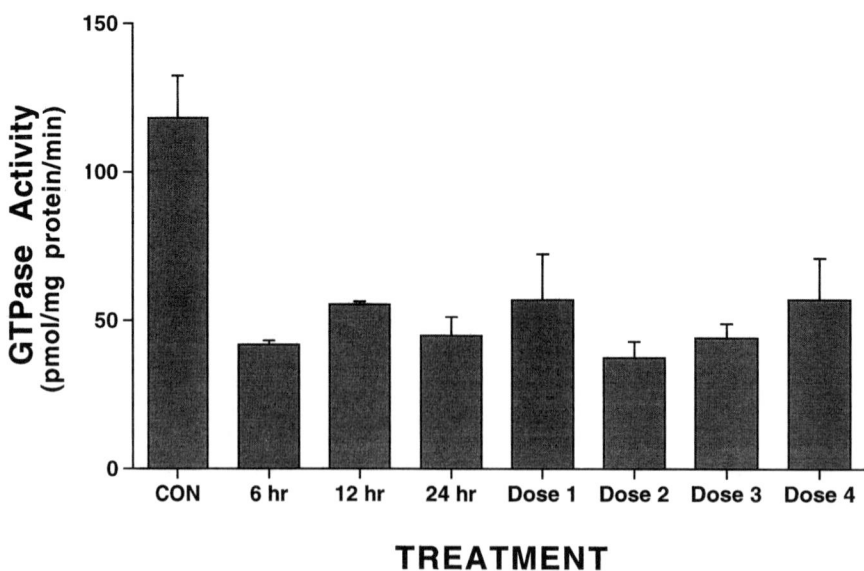

Fig. 7 Effect of tolerance induction on GTPase activity. GTPase activity was measured in rat peritoneal macrophage membranes harvested at 6, 12, and 24 hours after a single injection of endotoxin. Doses 1 to 4 represent GTPase activity measured in MØ membranes in the same treatment regime as in Figure 6. Data are expressed as means ± SE ($n = 5$ per group). All treatment groups $p < 0.05$ vs. control.

Table 1 Effect of LPS and TNF-α Tolerance on MØ Mediator Production

Mediator	CON	LPS-TOL	TNF-TOL
TXB$_2$ (ng/ml)	16.5 ± 3.0	2.0 ± 0.1*	ND
6-Keto-PGF$_{1\alpha}$ (ng/ml)	13.2 ± 2.7	1.7 ± 0.3*	6.0 ± 1.5*
NO (nmol/ml)	9.0 ± 1.0	13.0 ± 1.5*	26.0 ± 4*

Peritoneal MØ were harvested from control (CON) and LPS (LPS-TOL) or TNF-α–tolerant (TNF-TOL) rats. The cells were stimulated with LPS for 24 hours, and TXB$_2$, 6-keto-PGF$_{1\alpha}$, and NO (as nitrite) levels were determined. Each value represents mean ± SE of at least four different experiments. The TXB$_2$ and 6-keto-PGF$_{1\alpha}$ levels were significantly lower in tolerant rats, whereas NO levels were elevated.
*$p < 0.05$ vs. control.
ND = Not determined.

differential decreases in tolerant MØ membrane G proteins. Of the G proteins investigated, Gi$_{\alpha 3}$ showed the greatest reduction relative to control levels (60) (Fig. 9). Also, peritoneal MØ collected from 24-hour TNF-α–pretreated rats exhibit a marked reduction of the membrane content of the Gi$_{\alpha 3}$ subunit compared to control MØ, whereas the G$_{\alpha 1}$, Gi$_{\alpha 2}$, and Gβ subunits were not significantly affected (58). Since tolerance induced by either TNF-α or LPS produces similar differential changes in peritoneal MØ mediator production, these changes may, in part, be a consequence of altered signal transduction via specific Gi proteins.

THP-1 cells also exhibit reduced LPS-induced TxB$_2$ production upon preexposure to lower concentrations of LPS. However, unlike rat peritoneal MØ, neither the G-protein content determined by immunoblots nor the GTPγS-binding capacity were altered in LPS-desensitized THP-1 cells in response to LPS (48). The results suggest that in THP-1 cells LPS-induced tolerance is not associated with changes in the content of the Gi

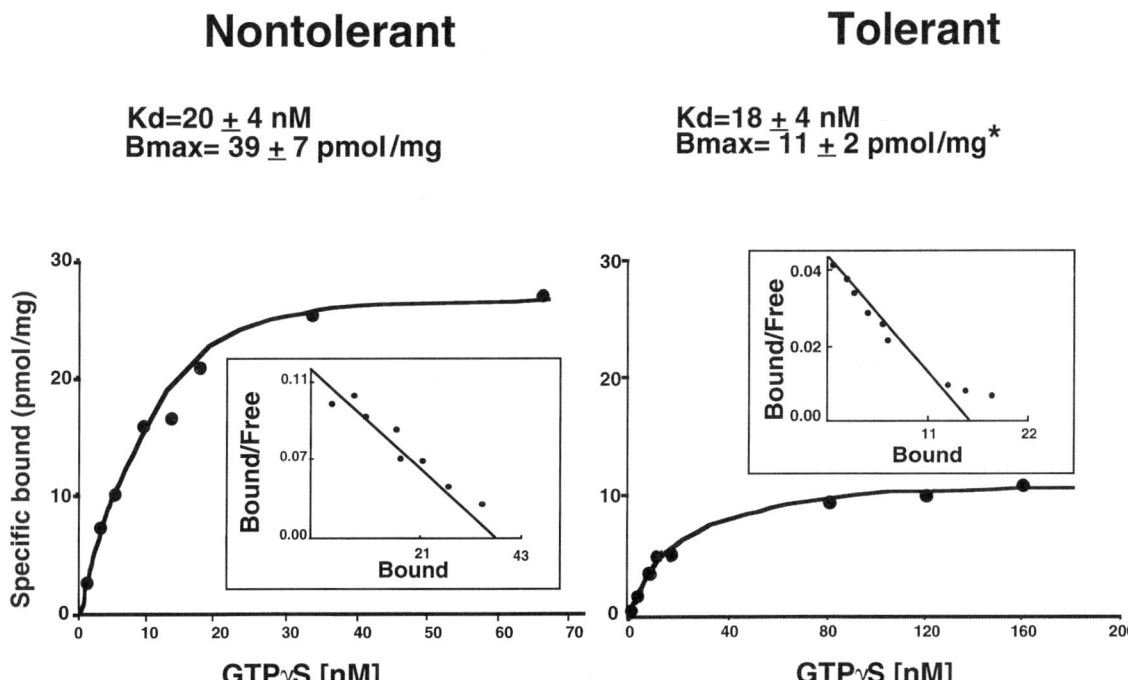

Fig. 8 Representative equilibrium-binding data for GTPγS to control (A) and tolerant (B) peritoneal MØ membranes for five different experiments. Membranes from each group were incubated with various concentrations of cold and radiolabeled GTPγS and the bound radioactivity was determined. The graph in the insert represents the Scatchard analysis of equilibrium binding data. The B$_{max}$ values for the control and tolerant groups were 38.5 and 14.5 pmol/mg protein and the Kd values were 13 and 20 nM, respectively. *$p < 0.05$ versus nontolerant.

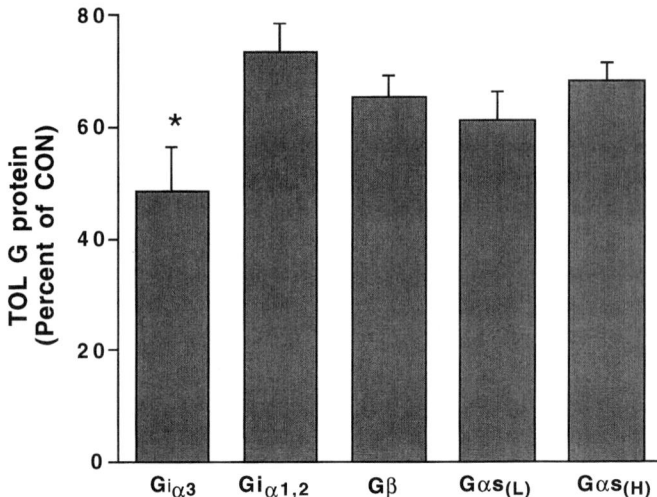

Fig. 9 Densitometric analysis of the membrane G protein levels of rat tolerant MØ over the control MØ. The membranes from control and tolerant rat MØ were analyzed by immunoblotting with antibody specific for G-protein subunits. The intensity of the signal was quantitated by densitometric analysis. Data are presented as means ± SE for three different experiments (*$p < 0.05$).

class of proteins observed with MØ from tolerant rats. These differences may be as a result of species differences or due to in vitro versus in vivo tolerance (whereby in vivo tolerance could be a more complex phenomenon involving additional paracrine factors induced by LPS). Although G-protein content is not diminished in desensitized THP-1 cells, there nevertheless appears to be a profound defect in early signaling in the tolerant THP-1 cells (49). LPS tolerance is associated with decreased MAPK activation (ERK) but does not impair MAPK activation by the protein kinase C activator phorbol myristic acid. Thus, LPS tolerance appears to selectively alter signal transduction upstream of ERK perhaps at the level of or coupled to Gi proteins (49).

Zhang and Morrison (61), using a thioglycolate-elicited murine MØ system, demonstrated a response to a stimulating concentration of LPS that is dependent on the tolerance-inducing (primary) LPS concentration. They observed that at lower primary LPS concentrations the LPS stimulation of NO decreased, whereas TNF-α production was increased. At higher primary LPS concentrations the opposite results were observed. They referred to this concentration-dependent phenomenon as "MØ reprogramming." They further showed these reprogramming events are not PTx sensitive suggesting that they are not mediated via Gi protein signaling. Thus, despite the fact that Gi proteins may be involved in some LPS activation events of MØ and monocytes, signaling pathways mediating LPS tolerance may not be Gi protein coupled (62).

FUTURE DIRECTIONS

While much progress has been made with regard to understanding LPS-mediated cellular signal-transduction mechanisms, there is still much to learn (31). Many studies suggest the involvement of the Gi subclass of G proteins in LPS-mediated cellular activation. However, the involvement of specific receptors or localization of G proteins in the complex LPS-mediated cellular signaling events are not clear. It is indeed possible that LPS receptors are not directly coupled to G proteins, but may be indirectly modulated by cytokine autocrine regulation. We and others have shown that the LPS-stimulated production of specific mediators takes place through a PTx-sensitive pathway. However, PTx has limitations in its application as an experimental tool. Indeed it has been suggested that the B oligomer subunit of PTx and LPS binds to different sites on a LPS-binding protein (P73) in murine splenocytes that recognize both LPS and PTx (63). Similar binding characteristics between PTx and LPS with a membrane 70 kDa protein were noted in Jurkat cells (64). Thus, PTx may alter LPS binding to such receptors by steric hindrance. Also, importantly PTx does not distinguish between the various isoforms of Gi. Thus, it remains to be determined if LPS receptors are coupled to a single subclass or interact with other G-protein isoforms. Other, more refined experimental approaches are required to distinguish the involvement of the isoforms. It is likewise uncertain which pathways are activated by specific Gi isoforms. For example, $Gi_{\alpha 3}$ has been shown to stimulate phospholipase C–induced increases in intracellular calcium and phospholipase A_2–mediated increases in arachidonic acid metabolism and to inhibit adenylyl cyclase (9). In LPS tolerance both the membrane GTPase activity, GTP-binding sites, and G-protein content are reduced in rat peritoneal MØ, which may contribute to LPS resistance and the cross-tolerance phenomenon. The reduction in the amount of G proteins may be due to changes in transcription or stability of the G-protein message. Further, there could be translational or posttranslational control mechanisms. The G-protein subunits can be posttranslationally modified by addition of palmitate or myristate (65). Palmitoylation is a reversible process, and the degree of palmitoylation of the Gα subunit determines its affinity to the membrane or to the βγ to form active trimeric structures (66,67). Myristoylation facilitates the association of the α subunit with βγ subunits (68). Also in U937 cells, LPS has been shown to induce the activation of protein kinase C and phosphorylation of $Gi_{\alpha 2}$ proteins (53). Such covalent modification of G proteins may have important functional consequences on G protein function. It has been shown that GM-CSF, which can modulate LPS responses, can change membrane $Gi_{\alpha 2}$ content in human neutrophils without altering of its message or its total $Gi_{\alpha 2}$ cellular protein (69). Membrane $Gi_{\alpha 2}$ levels are increased in LPS primed neutrophils (56), and membrane $Gi_{\alpha 2}$ and $Gi_{\alpha 3}$ levels are increased in polymorphonuclear leukocytes in response to TNF-α (70). TNF-α increases the rate of translocation of $Gi_{\alpha 2}$ in human polymorphonuclear leukocytes within a time frame of cellular activation while decreasing $Gi_{\alpha 2}$ transcript levels (71). It is suggested that stimulation of $Gi_{\alpha 2}$ translation, while promoting $Gi_{\alpha 2}$ mRNA degradation, provides a mechanism by which TNF-α stimulation produces a transient but self-limiting increase in G-protein expression. Future studies are needed to further assess the significance of transcriptional, translational, or posttranslational modifications of G proteins in cellular signaling in response to LPS.

Given all the possibilities of the control of G-protein subunit expression and the dynamic nature of the protein in cell signal-transduction processes, sorting out the role of specific G proteins in LPS-mediated signaling and altered signaling during LPS tolerance will prove to be a challenging research area.

ACKNOWLEDGMENT

This work was supported by NIH GM 27673.

REFERENCES

1. Rodbell M, Birnbaumer L, Pohl SL, Krans HMJ. The glucagon-sensitive adenyl cyclase system in plasma membranes of rat liver. J Biol Chem 1971; 246:1877–1882.
2. Pfeuffer T. GTP-binding proteins in membranes and the control of adenylate cyclase activity. J Biol Chem 1977; 252:7224–7234.
3. Gill DM, Meren R. ADP-ribosylation of membrane proteins catalyzed by cholera toxin: basis of the activation of adenylate cyclase. Proc Natl Acad Sci 1978; 75:3050–3054.
4. Moss J, Vaughn M. Mechanism of action of choleragen. J Biol Chem 1977; 252:2455–2457.
5. Hazeki O, Ui M. Modification β islet-activating protein of receptor-mediated regulation of cyclic AMP accumulation in isolated rat heart cells. J Biol Chem 1981; 256:2856–2862.
6. Miki N, Keirns JJ, Marcus FR, Freeman J, Bitensky MW. Regulation of cyclic nucleotide concentrations in photoreceptors: an ATP-dependent stimulation of cyclic nucleotide phosphodiesterase by light. Proc Natl Acad Sci 1973; 70:3820–3824.
7. Kalkbrenner F, Dippel E, Wittig B, Schultz G. Specificity of interaction between receptor and G protein: use of antisense techniques to relate G-protein subunits to function. Biochim et Biophys Acta 1996; 1314:125–139.
8. Itoh H, Toyama R, Kozasa T, Tsukamoto T, Matsuoka M, Kaziro Y. Presence of three distinct molecular species of Gi protein α subunit. J Biol Chem 1988; 263:6656–6664.
9. Gillman AG. G proteins regulation of adenylyl cyclase. Biosci Rep 1995; 15:65–97.
10. Hunt TW, Carroll RC, Peralta EG. Heterotrimeric G proteins containing $G_{\alpha i3}$ regulate multiple effector enzymes in the same cell. J Biol Chem 1994; 269:29565–29570.
11. Gettys TW, Sherriff-Carter K, Moonma J, Taylor IL, Raymond JR. Characterization and use of crude α-subunit preparations for quantitative immunoblotting of G proteins. Anal Biochem 1994; 220:82–91.
12. Carty DJ, Premont T, Ivengar R. Quantitative immunoblotting of G-protein subunits. Meth Enzymol 1991; 195:302–315.
13. Paulssen EJ, Paulssen RH, Haugen TB, Gautvik KM, Gordeladze JO. Regulation of G protein mRNA levels by thyroliberin, vasoactive intestinal peptide and somatostatin in prolactin-producing rat pituitary adenoma cells. Acta Physiol Scand 1991; 143:195–201.
14. Offermanns S, Schultz G, Rosenthal W. Identification of receptor-activated G proteins with photoreactive GTP analog [α-^{32}P] GTP azidoanilide. Meth Enzymol 1991; 195:286–301.

15. Meinkoth JL, Goldsmith PK, Spiegel AM, Feramisco JR, Burrow GM. Inhibition of thyrotropin-induced DNA synthesis in thyroid follicular cells by microinjection of an antibody to the stimulatory G protein of adenylate cyclase, G_s*. J Biol Chem 1992; 267:13239–13245.
16. Caulfield MP, Jones S, Vallis Y, Buckley NJ, Kim GD, Milligan G, Brown DA. Muscarinic M-current inhibition via $G_{\alpha q/11}$ and α-adrenoreceptor inhibition of Ca^{2+} current via $G_{\alpha o}$ in rat sympathetic neurons. J Physiol (Lond.) 1994; 477:415–422.
17. Lepretre N, Mironeau J, Armaudeau S, Tanfin Z, Harbone S, Guillon G, Ibarrondon J. Activation of Alpha-1A adrenoceptors mobilizes calcium from the intracellular stores in monocytes from rat portal vein. J Pharm Exp Ther 1994; 268;167–174.
18. LaMorte VJ, Goldsmith PK, Spiegel AM, Meinkoth JL, Feramisco JR. Inhibition of DNA synthesis in living cells by microinjection of G_{i2} antibodies. J Biol Chem 1992; 267:691–694.
19. Ting TD, Lee RH, Ho YK. The GTPase cycle: transducin. In: Dickey BF, Birmbaumer L, eds. Handbook of Experimental Pharmacology, Vol. 108: GTPases in Biology II. New York: Springer, 1993:99–117.
20. Tang WJ, Gilman AG. Adenylyl cyclase. Cell 1992; 70: 869–972.
21. Park D. Phospholipase C-B isoenzymes activated by Gaq members. In: Dickey BF, Birmbaumer L, eds. Handbook of Experimental Pharmacology, Vol. 108: GTPases in Biology II. New York: Springer, 1993: 239–249.
22. Gierschik P, Camps M. Stimulation of phospholipase C by G-protein Bg subunits. In: Dickey BF, Birmbaumer L, eds. Handbook of Experimental Pharmacology, Vol. 108: GTPases in Biology II. New York: Springer, 1993: 251–264.
23. Hescheler J, Rosenthal W, Trautwein W, Schultz G. The GTP-binding protein, G_0, regulates neuronal calcium channels. Nature 1987; 325:445–447.
24. Katada T, Ui M. Direct modification of the membrane adenylate cyclase system by islet-activating protein due to ADP-ribosylation of a membrane protein. Proc Natl Acad Sci 1982; 79:3129–3133.
25. Kugi M, Kitamura K, Cottam GL, Miller RT. Expression of $G\alpha_{i2}$ mimics several aspects of LPS priming in a murine macrophage-like cell line. J Inflamm 1995; 45:175–182.
26. Migeon JC, Thomas SL, Nathanson SM. Regulation of cAMP mediated transcription by wild type and mutated G-protein α subunits. J Biol Chem 1994; 269:29146–29152.
27. Ziegler-Heitbrock HWL. Molecular mechanism in tolerance to lipopolysaccharide. J Inflamm 1995; 45:13–26.
28. Cook JA, Wise WC, Halushka PV. Thromboxane A_2 and prostacyclin production by lipopolysaccharide-stimulated peritoneal macrophages. J Reticulo Soc 1981; 30:445–450.
29. Tracey KJ, Lowry SF, Cerami A. Cachectin/TNF-α in septic shock and septic adult respiratory distress syndrome. Am Rev Respir Dis 1988; 138:1377–1379.
30. Szabo C, Thiemermann C. Invited opinion: role of nitric oxide in hemorrhagic, traumatic, and anaphylactic shock and thermal injury. Shock 1994; 2:145–155.
31. Sweet M, Hume D. Endotoxin signal transduction in macrophages. J Leuk Biol 1996; 60:8–26.
32. Kravchenko VV, Steinemann S, Kline L, Feng L, Ulevitch RJ. Endotoxin tolerance is induced in chinese hamster ovary cell lines expressing human CD14. Shock 1996; 5(3):194–201.
33. Solomon KR, Rudd CE, Finberg RW. The association between glycosyl-phosphatidylinositol-anchored proteins and heterotrimeric G protein α subunits in lymphocytes. Proc Natl Acad Sci 1996; 93:6053–6058.
34. Coffee KA, Halushka PV, Wise WC, Cook JA. Altered responses to modulators of guanine nucleotide binding protein activity in endotoxin tolerance. Biochim Biophys Acta 1990; 1035:201–205.
35. Coffee KA, Halushka PV, Ashton SH, Tempel G, Wise WC, Cook JA. Endotoxin tolerance is associated with altered GTP-binding protein. J Appl Physiol 1992; 73: 1008–1013.
36. Geppert D, Whitehurst C, Thompson P, Beutier B. Lipopolysaccharide signals activation of tumor necrosis factor biosynthesis through the Ras/Raf-1/MEK/MAPK pathway. Molec Med 1994; 1:93–103.
37. Weinstein SL, June CH, DeFranco AL. Lipopolysaccharide-induced protein tyrosine phosphorylation in human macrophages is mediated by CD14. J Immunol 1993; 151:3829–3838.
38. Geng Y, Zhang B, Lotz M. Protein tyrosine kinase activation is required for lipopolysaccharide induction of cytokines in human blood monocytes. J Immunol 1993; 151:6692–6700.
39. Han J, Lee J, Tobias PS, Ulevitch RJ. Endotoxin induces rapid protein tyrosine phosphorylation in 70Z/3 cells expressing CD14. J Biol Chem 1993; 268:25009–25014.
40. Weinstein SL, Sanghera JS, Lemke K, DeFranco AL, Pelech SL. Bacterial lipopolysaccharide induces tyrosine phosphorylation and activation of mitogen-activated protein kinases in macrophages. J Biol Chem 1992; 267:14955–14962.
41. LaRue KEA, McCall CE. A labile transcriptional repressor modulates endotoxin tolerance. J Exp Med 1994; 180:2269–2275.
42. Faure M, Voyno-Yasenetskaya TA, Bourne HR. cAMP and beta gamma subunits of heterotrimeric G proteins stimulate the mitogen-activated protein kinase pathway in COS-7 cells. J Biol Chem 1994; 269: 7851–7854.
43. Crespo P, Xu N, Simonds WF, Gutkind JS. Ras-dependent activation of MAP kinase pathway mediated by G-protein beta gamma subunits. Nature 1994; 369:418–420.
44. Luttrell LM, Hawes BE, van Biesen T, Luttrell DK, Lansing TJ, Lefkowitz RJ. Role of c-Src tyrosine kinase in G protein-coupled receptor and G beta gamma subunit-mediated activation of mitogen-activated protein kinases. J Biol Chem 1996; 271:19443–19450.
45. Pace AM, Faure M, Bourne HR. Gi2-mediated activation of the MAP kinase cascade. Mol Biol Cell 1995; 6:1685–1695.

46. Wan Y, Kurosaki T, Huang XY. Tyrosine kinases in activation of the MAP kinase cascade by G-protein-coupled receptors. Nature 1996; 380:541–544.
47. Touhara K, Hawes BE, van Biesen T, Lefkowitz RJ. G protein beta gamma subunits stimulate phosphorylation of Shc adapter protein. Proc Natl Acad Sci 1995; 92: 9284–9287.
48. Durando M, Ashton S, Makhlouf M, Wagner RS, Halushka PV, Cook JA. Endotoxin-induced desensitization of THP-1 cells is not associated with altered G protein function. J Endo Res 1997; 4:97–103.
49. Durando M, Meir MK, Cook JA. Endotoxin activation of mitogen activated protein kinase (MAPK) in THP-1 cells; diminished activation following endotoxin stimulation. J Leuk Biol 1998; 64:259–264.
50. Zingarelli B, Chen H, Caputi AP, Halushka PV, Cook JA. Reorientation of macrophage mediator production in endotoxin tolerance. In Bacterial Endotoxins: Lipopolysaccharides from Genes to Therapy 1995; Wiley-Liss Inc., New York 529–537.
51. Jakway JP, DeFranco AL. Pertusis toxin inhibition of B cell and macrophage responses to bacterial lipopolysaccharide. Science 1986; 234:743–746.
52. Wang J, Kester M, Dunn MJ. Involvement of a pertussin toxin-sensitive G-protein-coupled phospholipase A_2 in lipopolysaccharide-stimulated prostaglandin E_2 synthesis in cultured rat mesangial cells. Biochim Biophys Acta 1988; 963:429–435.
53. Daniel-Issakani S, Spiegel AM, Strulovici B. Lipopolysaccharide response is linked to the GTP binding protein, G_{i2}, in the promonocytic cell line U937. J Biol Chem 1989; 264:20240–20247.
54. Altavilla D, Squadrito F, Canale P, Ioculano M, Campo GM, Squadrito G, Urna G, Sardella A, Caputi AP. Endotoxin tolerance impairs a pertussis-toxin-sensitive G protein regulating tumour necrosis factor release by macrophages from tumour-bearing rats. Pharmacol Res 1996; 33:203–209.
55. Zhang X, Morrison DC. Pertussis toxin-sensitive factor differentially regulates lipopolysaccharide-induced tumor necrosis factor-α and nitric oxide production in mouse peritoneal macrophages. J Immunol 1993; 150(3):1011–1018.
56. Yasui K, Becker EL, Sha'afi RI. Lipopolysaccharide and serum cause the translocation of G-protein to the membrane and prime neutrophils via CD14. Biochem Biophys Res Comm 1992; 183:1280–1286.
57. Landmann R, Knopf HP, Link S, Sansano S, Schumann R, Zimmerli W. Human monocyte CD14 is upregulated by lipopolysaccharide. Infect Immun 1996; 64:1762–1769.
58. Zingarelli B, Makhlouf M, Halushka PV, Caputi AP, Cook JA. Altered macrophage function in tumor necrosis factor α and endotoxin-induced tolerance. J Endotox Res 1995; 2:247–254.
59. Rogers TS, Halushka PV, Wise WC, Cook JA. Differential alteration of lipooxygenase and cyclooxygenase metabolism by rat peritoneal macrophages induced by endotoxin tolerance. Protaglandins 1986; 31:639–650.
60. Makhlouf M, Ashton SH, Hildebrandt J, Mehta N, Gettys TW, Halushka PV, Cook JA. Alterations in macrophage G proteins are associated with endotoxin tolerance. Biochim Biophys Acta 1996; 1312:163–168.
61. Zhang X, Morrison DC. Lipopolysaccharide-induced selective priming effects on tumor necrosis factor α and nitric oxide production in mouse peritoneal macrophages. J Exp Med 1993; 177:511–516.
62. Zhang X, Morrison DC. Lipopolysaccharide structure-function relationship in activation versus reprogramming of mouse peritoneal macrophages. J Leuk Biol 1993; 54:444–450.
63. Lei D, Morrison DC. Evidence that lipopolysaccharide and pertussis toxin bind to different domains on the 73kd receptor on murine splenocytes. Infect Immun 1993; 6:1359–1364.
64. Armstrong GD, Clark GC, Heerze LD. The 70 kilodalton pertussis toxin-binding protein in Jurkat cells. Infect Immun 1994; 62:2236–2243.
65. Wedegaertner PB, Wilson PT, Bourne HR. Lipid modifications of trimeric G proteins. J Biol Chem 1995; 270:503–506.
66. Iiri T, Backlund Jr PS, Jones TLZ, Wedegaertner PB, Bourne HR. Reciprocal regulation of $G_{s\alpha}$ by palmitate and the βγ subunit. Proc Natl Acad Sci 1996; 93: 14592–14597.
67. Grassie M, McCallum JF, Guzzi F, Magee AI, Milligan G, Parenti M. The palmitoylation status of the G-protein $G_o1\alpha$ regulates its avidity of interaction with the plasma membrane. Biochem J 1994; 302:913–920.
68. Bigay J, Faurobert E, Franco M, Chabre M. Roles of lipid modifications of transducin in their GDP-dependent association and membrane binding. Biochemistry 1994; 33:14081–14090.
69. Durstin M, McColl SR, Gomez-Cambronero J, Naccache PH, Sha'afi RI. Up-regulation of the amount of $G_{i\alpha2}$ associated with the plasma membrane in human neutrophils stimulated by granulocyte-macrophage colony-stimulating factor. Biochem J 1993; 292:183–187.
70. Klein JB, Scherzer JA, Harding G, Jacobs AA, McLeish KR. TNF-α stimulates increased plasma membrane guanine nucleotide binding protein activity in polymorphonuclear leukocytes. J Leuk Biol 1995; 57:500–506.
71. Scherzer JA, Lin Y, McLeish KR, Klein JB. TNF translationally modulates the expression of G protein α_{i2} subunits in human polymorphonuclear leukocytes. J Immunol 1997; 158:913–918.

35

Immediate Cytokine Responses to Endotoxin: Tumor Necrosis Factor-α and the Interleukin-1 Family

Charles A. Dinarello
University of Colorado Health Science Center, Denver, Colorado

EARLY CYTOKINE RESPONSE TO LIPOPOLYSACCHARIDE IN HUMANS

In 1988, together with other investigators, we published an amazing finding: the intravenous injection of a very small amount of *E. coli* lipopolysaccharide (LPS) (4 ng/kg) into healthy human subjects induced a rapid appearance in circulating tumor necrosis-α (TNF-α) within 15 minutes following the injection (1). Although experimental animals had been studied for the production of TNF activity following lipopolysaccharide (LPS), the exceedingly low dose of LPS and the specificity of the ELISA for measuring TNF-α immunoreactivity was a considerable advance. These experiments in humans also confirmed the exquisite sensitivity of humans to the cytokine-inducing properties of LPS.

The peak elevation in TNF-α occurred at 1.5 hours followed by a rapid decline. The absolute increase in TNF-α correlated with the leukopenia such that there appeared to be a relationship between the production of TNF-α and the margination of leukocytes to the endothelium. Although much of this information is well established today, in 1988 these results had solidified several years of in vitro investigation using human monocytes. Although we had developed a sensitive radioimmunoassay for detecting IL-1β production from human mononuclear cells in vitro (2), we were rather disappointed that there was hardly any IL-1β detectable in the same samples. However, we next tried different extraction methods (as is commonly done for some hormones) and reported that a chloroform extraction of plasma collected in calcium chelator anticoagulants (not heparin) was optimal for measuring IL-1β in the circulation (3). The chloroform extraction was used by Gordon Duff, who reported a highly positive correlation of plasma IL-1β with acute disease activity in patients with rheumatoid arthritis (4).

We returned to the issue of circulating IL-1β in volunteers injected with LPS, and using the chloroform extraction method, Cannon et al. reported an elevation of IL-1β 3–4 hours following LPS (5). In addition, we observed that IL-1β and TNF-α levels were elevated in patients with septic shock (5). Nevertheless, the most important and consistent observation was that the absolute rise in IL-1β levels compared to TNF-α levels following LPS were small, in the order of 3-fold. In contrast, the rise in circulating TNF-α was about 8- to 10-fold.

About one year after the first paper on TNF-α circulating after LPS in humans, Fong et al. reported that anti-TNF-α antibodies administered to baboons prevented an increase in IL-1β levels (6). This report suggested that the IL-1β response to endotoxin in vivo was TNF-α dependent. Indeed, TNF-α induces IL-1β from human blood mononuclear cells in vitro and in vivo (7). Is there no role for LPS induction of IL-1β, or is it all secondary to TNF-α induction of IL-1β? The evidence suggests that in vivo, the amount of IL-1β induced by LPS is a combination of the direct effect of LPS on IL-1β gene expression as well as a contribution of TNF-α–induced IL-1β. However, with the testing of the IL-1 receptor antagonist (IL-1Ra), there is a role for IL-1–induced TNF (8). There is also direct evidence that IL-1 induces TNF (9).

EXPRESSION OF TNF AND IL-1 IN ENDOTOXIN-STIMULATED HUMAN CELLS IN VITRO

As shown in Figure 1, there is an ordered stimulation of TNF and IL-1 gene expression. TNF is clearly a very early gene in monocytes after exposure to LPS. In most studies, human peripheral blood mononuclear cells (PBMC) are incubated with low concentrations of LPS (1–10 ng/ml) and within 10–15 minutes there is an increase in the steady-state levels of TNF-α mRNA. After 3–4 hours these are dramatically reduced. IL-1β is slower in onset compared to TNF-α. One usually does not observe a significant increase in IL-1β steady-state mRNA levels until 3 minutes, but the expression is longer and more sustained than that for TNF-α. In the THP-1 cell line, the expression of IL-1β has been studied in detail (10). Both TNF-α and IL-1β have a 3′ untranslated sequence that effects mRNA stability (11). This region is thought to be responsive to LPS effects.

Gene expression for IL-1β and TNF-α can be complicated because both genes will increase their expression simply by adherence to glass surfaces. When taken from the blood, freshly obtained cells do not express IL-1β or TNF-α (12). In addition, cytokines like IL-8 also are not expressed in freshly obtained blood (13). In a careful study of IL-1β and TNF-α gene expression in human PBMC, all endotoxin-like materials were removed using ultrafiltration of tissue culture media through a hollow fiber polyamide or polysulfone filter under pressure (14). Under these conditions, PBMC do not translate IL-1β or TNF-α mRNA into their respective proteins. Nevertheless, there are large amounts of both IL-1β and TNF-α mRNA species in adherent PBMC. The half-lives of these two mRNA species are normal (15). The elevated level of steady-state mRNA for these cytokines is due to the stimulation of the glass surfaces. However, in the absence of any exogenous microbial stimulation, translation does not occur (15). Upon exposure of these adherent cells to small amounts of LPS or IL-1 itself, there is enhanced synthesis. This is similar to a "superinduction" protocol accomplished by cycloheximide.

Is the mRNA for IL-1β or TNF-α not attaching to the ribosome? Although this is a possibility, the evidence is that adherence-stimulated TNF-α and IL-1β mRNA is attached to the ribosome but translation in the absence of LPS does not elongate and protein "falls off" before translation is completed. This was studied using recombinant C5a as a stimulant for IL-1β and TNF-α production (16). In human PBMC, C5a will stimulate the same high levels of IL-1β and TNF-α mRNA as that observed with LPS but without any translation. However, the mRNA for these cytokines is fully polyadenylated and attached to ribosomes (17).

THE IMPORTANCE OF IL-1 AND TNF PRODUCTION FOR ENDOTOXIN EFFECTS

Endotoxin stimulates the same signaling kinase pathway as does IL-1 and TNF. That is to say, one can see the same increase in gene expression and phosphorylation of kinases in cells exposed to LPS as in cells exposed to IL-1β or TNF-α. For example, the MAP 38 kinase is phosphorylated in macrophages using either LPS, IL-1β, or TNF-α (18). However, endotoxin is cleared rapidly and, without additional infection, is not produced once the appropriate antibiotic is used to treat the infection. On the other hand, IL-1β and TNF-α synthesis remain elevated for several hours (see Fig. 1). For the clinician, the patient is exposed to endotoxin for a relatively short period (the time of untreated bacteremia), whereas synthesis of IL-1β and TNF-α can persist for hours. As such, the early symptoms of endotoxin may be due to direct effects, whereas the more persistent effects are likely due to IL-1β and TNF.

Although the systemic effects of TNF-α and IL-1β have been studied in animals, there are now data on the effects and sensitivity of humans to IL-1 and TNF-α. In general, the effects are similar to either cytokine. Acute toxicities of either IL-1α or IL-1β were greater following intravenous compared to subcutaneous injection; subcutaneous injection was associated with significant local pain, erythema, and swelling (19,20). Chills and fever are observed in nearly all patients, even in the 1 ng/kg dose group (21). The febrile response increased in magnitude with increasing doses (22–26), and chills and fever were abated with indomethacin treatment (27). In patients receiving IL-1α (25,26) or IL-1β (22,23), nearly all subjects experienced significant hypotension at doses of 100 ng/kg or greater. Systolic blood pressure fell steadily and reached a nadir of 90 mmHg or less 3–5 hours after the infusion of IL-1. At doses of 300 ng/kg, most patients required intravenous pressors. By comparison, in a trial of 16 patients given IL-1β 4–32 ng/kg subcutaneously, there was only one episode of hypotension at the highest dose level (19). These results suggest that

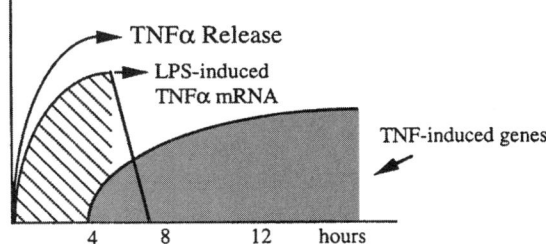
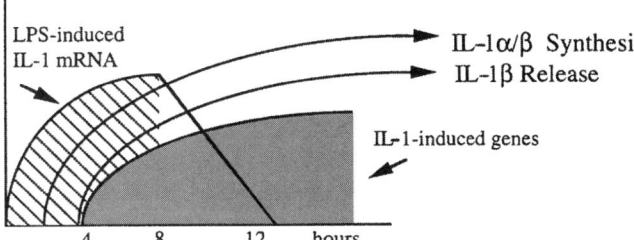

Fig. 1 Kinetics of TNF-α and IL-1 gene expression and release following endotoxin stimulation in human PBMC. TNF-α is shown in the upper graph. Early expression of mRNA for TNF-α is observed 5–10 minutes after exposure to LPS, and release of biologically active TNF-α is found in the supernatant within the first 45 minutes. mRNA levels fall rapidly after 3–4 hours. TNF-induced genes (such as IL-6) take place in the same cell (this drawing does not indicate other LPS-induced genes). In the lower graph, IL-1 gene expression is shown. Although IL-1α and IL-1β gene expression can be observed early after LPS, the synthesis of proIL-1α and proIL-1β are delayed compared to TNF-α. In addition, both IL-1α and IL-1β precursors accumulate in the cell before release can be observed in the supernatant. IL-1α in human PBMC is not commonly observed, whereas the enzymatic cleavage of proIL-1β into the mature peptide by the IL-1β converting enzyme (ICE) takes place at the level of the cell membrane and is linked to the secretion of mature IL-1β. The time scale is approximate and is influenced by the conditions of the culture. These times are based on numerous studies using freshly obtained PBMC and not adherent monocytes.

the hypotension is probably due to induction of NO, and elevated levels of serum nitrate have been measured in patients with IL-1–induced hypotension (26).

At 30–100 ng/kg of IL-1β, a sharp increase in cortisol levels were seen 2–3 hours after the injection. Similar increases were noted in patients given IL-1α. In 13 of 17 patients given IL-1β, there was a fall in serum glucose within the first hour of administration, and in 11 patients glucose fell to 70 mg/100 ml or lower (23). In addition, there were increases in ACTH and thyroid-stimulating hormone, but a decrease in testosterone (26). No changes were observed in coagulation parameters such as prothrombin time, partial thromboplastin, or fibrin-degradation products. This latter finding is to be contrasted to TNF-α infusion into healthy humans, which results in a distinct coagulopathy syndrome (28).

Not unexpectedly, IL-1 infusion into humans significantly increased circulating IL-6 levels in a dose-dependent fashion (26). At a dose of 30 ng/kg, mean IL-6 levels were 500 pg/ml 4 hours after IL-1 (baseline <50 pg/ml) and 8000 pg/ml after a dose of 300 ng/kg. In another study, infusion of 30 ng/kg of IL-1α induced elevated IL-6 levels within 2 hours (29). These elevations in IL-6 are associated with a rise in C-reactive protein and a decrease in albumin. In two studies, one with IL-1α (30) and one with IL-1β (31), a rapid increase in circulating IL-1Ra and TNF-soluble receptors (p55 and p75) were observed following a 30-minute intravenous infusion.

SYNERGISTIC ACTIONS OF IL-1 AND TNF

In considering the production of TNF-α and IL-1β in response to LPS, the synergism between these two cytokines should be considered. Of greatest importance is the synergistic effect of these two cytokines for reproducing the hemodynamic and hematologic effects of septic shock (32). In fact, there are few examples where the synergism between TNF-α and either IL-1α or IL-1β has not been demonstrated. As shown in Table 1, the synergism between IL-1 and TNF is highly consistent and a frequently reported phenomenon. In addition, the synergism between IL-1 and TNF is also observed

Table 1 Synergistic Activities of IL-1 and TNF

Hemodynamic shock and lactic acidosis in rabbits
Radioprotection
Generation of Shwartzman reaction
Luteal cell $PGF_{2\alpha}$ synthesis
PGE_2 synthesis in fibroblasts
Galactosamine-induced hepatotoxicity
Sickness behavior in mice
Circulating nitric oxide and hypoglycemia in malaria
Nerve growth factor synthesis from fibroblasts
Insulin release and beta islet cell death
Insulin resistance
Loss in lean body mass
IL-8 synthesis by mesothelial cells

in vivo, whereas the synergism between IL-1 and IL-6, IL-1 and bradykinin, or IL-1 and the various growth factors is mostly on prostanoid synthesis and primarily an in vitro finding.

The mechanism for IL-1/TNF synergism in the synthesis for PGE_2 likely involves the ability of these cytokines to release arachidonate and to stimulate cyclooxygenase type-2 (COX-2) synthesis. The mechanism for synergism may also involve receptor modulation; however, in the case of IL-1 and TNF synergism, receptors for TNF are downregulated by IL-1 (33,34). Could the synergism be explained at the level of signal transduction? Although this is an attractive hypothesis, no pathway of IL-1 or TNF signal transduction appears unique to either cytokine at the present time to account for synergism. In fact, since signal mechanisms appear similar, additive rather than synergistic effects should be observed. The true mechanism for this synergism between IL-1 and TNF remains unclear.

The signal transduction of IL-1 appears to be similar to that of TNF. Although the signaling pathway of IL-1's postreceptor binding remains unclear, several mechanisms have been proposed. These include activation of a GTP-binding protein with no associated increase in adenyl cyclase, activation of adenyl cyclase, hydrolysis of phospholipids by nonphosphatidylinositol phospholipase C, release of ceramide from sphingomyelin following activation of sphingomyelinase, and release of arachidonic acid from phospholipids via cytosolic phospholipase A_2 (PLA_2) and its activation by PLA_2-activating protein (PLAP) (reviewed in Ref. 35). Like IL-1, TNF also stimulates hydrolysis of phosphatidylcholine, release of ceramide from sphingomyelin following activation of sphingomyelinase (36), and release of arachidonic acid from phospholipids via cytosolic PLA_2 and activation of PLAP (37,38). In addition, some of the kinases that are activated by IL-1 are also activated in cells stimulated with TNF (39–43).

EXPRESSION OF VARIOUS GENES IN CELLS EXPOSED TO IL-1 AND TNF

A fundamental property of IL-1 and TNF is the ability to induce a wide variety of genes. In most cases, IL-1 and TNF induce new transcripts in cells that express these genes only during disease. There are several examples, but the most dramatic appear to be other members of the cytokine family and inducible enzymes regulating small molecular weight mediators. Mediators such as prostaglandins (PG), leukotrienes (LT), and NO require cellular enzymes to convert precursors to active molecules. IL-1 and TNF are potent inducers of these enzymes. For PGE_2 and LT synthesis, IL-1 and TNF stimulate inducible type-2 phospholipase A_2 and type-2 cyclooxygenase gene expression. The genes coding for IL-8 and other members of the chemokine family are not expressed in most cells in health, but picomolar concentrations of IL-1 and TNF trigger gene expression and protein expression in monocytes, fibroblasts, and endothelial cells. IL-1 and TNF will induce the expression of their own genes. However, it is important to note that IL-1 and TNF suppress the transcription of the genes coding for albumin and the cytochrome P450 family.

EFFECTS MEDIATED BY PROSTANOIDS

Many IL-1– and TNF-induced changes are mediated by prostaglandins, particularly PGE_2. In fact, the use of cyclooxygenase inhibitors for a variety of inflammatory conditions is often a therapeutic strategy to reduce IL-1– and TNF-induced PGE_2. Humans injected with IL-1 and TNF experience fever, headache, myalgias, and arthralgias, each of which is reduced by co-administration of cyclooxygenase inhibitors. One of the more universal activities of IL-1 and TNF is the induction of gene expression for PLA_2 and COX-2. IL-1 and TNF induce transcription of COX-2, and neither cytokine increases production of COX-1. Moreover, once triggered, COX-2 production is elevated for several hours, and large amounts of PGE_2 are produced in cells stimulated with IL-1 or TNF. It comes as no surprise that inflammation is reduced by administration of cyclooxygenase inhibitors and that many biological activities of IL-1 and TNF are actually due to increased PGE_2 production.

There appears to be selectivity in cyclooxygenase inhibitors in that some nonsteroidal anti-inflammatory agents are better inhibitors of COX-2 rather than COX-1. Similar to COX-2 induction, IL-1 and TNF preferentially stimulate new transcripts for the inducible form (type-2) of PLA_2, which cleaves the fatty acid in the number 2 position of cell membrane phospholipids. In most cases, this is arachidonic acid. The release of arachidonic acid is the rate-limiting step in the synthesis of prostaglandins and leukotrienes.

IL-18 IN RESPONSE TO LPS

In 1989, an endotoxin-induced serum activity that induced interferon-γ (IFN-γ) from mouse spleen cells was described by Nakamura and coworkers (44). This serum activity functioned not as a direct inducer of IFN-γ, but rather as a co-stimulant together with IL-2 or mitogens. An attempt to purify the activity from postendotoxin mouse serum revealed an apparently homogeneous 50–55 kDa protein (45). Since other cytokines can act as co-stimulants for IFN-γ production, the failure of neutralizing antibodies to IL-1, IL-4, IL-5, IL-6, or TNF to neutralize the serum activity suggested that it was a distinct factor. In 1995, the third report was published from the same scientists demonstrating that the endotoxin-induced co-stimulant for IFN-γ production was present in extracts of livers from mice preconditioned with *P. acnes* (46). In this model, the hepatic macrophage population (Kupffer cells) expands dramatically, and in these mice a low dose of bacterial LPS, which in nonpreconditioned mice is not lethal, becomes lethal. The factor, named IFN-γ–inducing factor (IGIF), was purified to homogeneity from 1200 g of *P. acnes*–treated mouse livers. Its molecular weight was 18–19 kDa and an N-terminal amino acid sequence was reported (46). Similar to the endotoxin-induced serum activity, IGIF purified from liver homogenates did not induce IFN-γ by itself but functioned primarily as a co-stimulant with mitogens or IL-2.

Neutralizing antibodies to mouse IGIF were shown to prevent liver damage induced by LPS in *P. acnes* preconditioned mice (47). Others had previously reported the importance of IFN-γ as a mediator of LPS-induced toxicity in preconditioned mice. For example, neutralizing anti-IFN-γ antibodies protected mice against Shwartzman-like shock (48), and galactosamine-treated mice deficient in the IFN-γ receptor were resistant to LPS-induced death (49). Hence, it was not unexpected that neutralizing antibodies to murine IGIF protected *P. acnes*–preconditioned mice against LPS-induced liver toxicity (47).

The ability of IFN-γ–inducing factor (now termed IL-18) to induce IFN-γ production is primarily in the context of a second stimulus in that it acts with IL-12, mitogens, or microbial agents to augment IFN-γ production (50–52). Alone, IL-18 does not induce IFN-γ production from T lymphocytes (47,51,53). However, in vitro LPS- and zymosan-induced IFN-γ production from murine spleen cells is strongly reduced using neutralizing antibodies to murine IL-18 (54), confirming similar findings in vivo (47) and suggesting that endogenous IL-18 is an essential for IFN-γ production by microbial agents. Because of its ability to induce TNF-α, IL-1β, and both CXC and CC chemokines (55) and because IL-18 induces Fas ligand as well as nuclear translocation of NFκB (56,57), IL-18 ranks with other pro-inflammatory cytokines as a likely contributor to systemic and local inflammation.

Consistent with stimulating TNF production (55), IL-18 upregulates Fas ligand–mediated cytotoxic activity of NK (58), T cells (59), and the myelomonocytic cell line KG-1 (60). In addition, IL-18 is a potent inducer of the human immunodeficiency virus (HIV-1) in human macrophagic cells (61) and IL-18–induced HIV-1 expression of competent virus is via a nitric oxide–dependent pathway (62). In addition to macrophagic cells, keratinocytes produce functional IL-18 following stimulation with contact sensitizers, and hence IL-18 may have a role in the inflammatory process after allergen contact (63). During endotoxin-induced liver damage in mice, neutralizing antibodies to IL-18 reduced tissue damage (47). Elevated levels of IL-18 have been found in patients with acute lymphoblastic leukemia and acute as well as chronic myeloid leukemia (64) similar to that observed for IL-1β in these patients. Also similar to IL-1β, IL-18 is expressed in the adrenal cortex after cold-induced stress and is also found constitutively in the neurohypophysis (65). IL-18 is evolving as a major pro-inflammatory cytokine with implications for a role in inflammatory and infectious diseases (see Fig. 2). On the other hand, it may also play a role in autoimmune diseases.

MOLECULAR CLONING OF IGIF

Degenerate oligonucleotides derived from amino acid sequences of IGIF purified from liver homogenates were used to clone a murine IGIF cDNA (47). Recombinant murine IGIF did not induce IFN-γ by itself but only in the presence of a mitogen or IL-2. However, most importantly, the co-induction of IFN-γ was independent of IL-12 induction of IFN-γ. After the murine form was cloned (47), the human cDNA sequence for IGIF was reported in 1996 (53). Recombinant human IGIF exhibited IGIF activity reported for the naturally occurring molecule (53). For example, human recombinant IGIF was without direct IFN-γ–inducing activity on human T cells but acted as a co-stimulant for production of IFN-γ and other T-helper cell-1 (Th1) cytokines (53). IGIF induced T cell and NK cell IFN-γ production independently of IL-12 (and vice versa) (46,47).

Hence, IGIF has been thought of as primarily an inducer for Th1 cytokine production by acting as a co-stimulant for IFN-γ production with IL-12, IL-2, microbial agents, or mitogens, particularly in NK cells (51,66,67). The cytokine also acts as a co-stimulant for

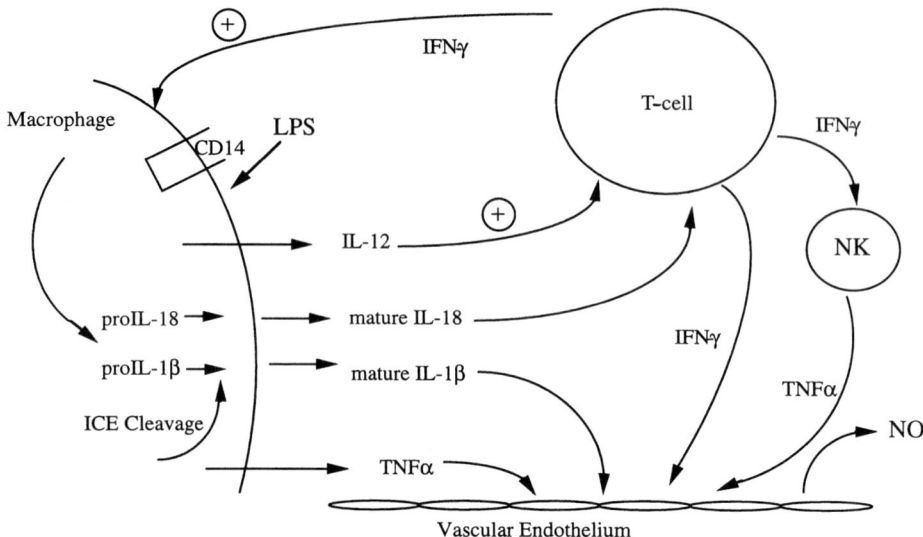

Fig. 2 Cytokine cascade following LPS stimulation of macrophages. LPS (via the CD14 receptor) triggers the synthesis of TNF-α and IL-12, which are readily released from the cell. LPS also triggers proIL-1β and proIL-18, each of which requires cleavage by ICE (caspase-1) for the release of the mature, active molecule. IL-18 targets the CD4+ T cell for IFN-γ production. IFN-γ in turn stimulates macrophage function, including increased killing of microbes, as well as increased ICE activity. T-cell production of IFN-γ also activates NK cells to release TNF-α. The endothelium is a target for each of these cytokines, and nitric oxide synthesis is stimulated by TNF-α, IL-1β, IFN-γ.

Fas ligand–mediated cytotoxicity of murine T cells and NK cell clones (58,59). In the human myelomonocytic cell line KG-1, IL-18 directly induces Fas ligand and apoptosis (60). In vivo, endogenous IGIF activity appears to account for IFN-γ production in *P. acnes*–preconditioned mice injected with LPS (46). This has been demonstrated using neutralizing antimouse IL-18 as well as the ICE-deficient mouse (see below) (68,69).

IGIF AND IL-1β AND THE NAMING OF IL-18

Scientists at DNAX analyzed the computer-generated protein folding pattern of murine IGIF and compared its pattern to those of other cytokines. Using a compatibility relatedness program, mature murine IGIF had the highest score with mature human IL-1β; furthermore, the IGIF amino acid sequence matched best with amino acids that form the all-β-pleated sheet folding pattern of human IL-1β (70). A high degree of alignment was present in the sequences that comprise the 12 β-sheets of the mature IL-1β structure. Using this alignment of conserved amino acids, there is a 19% positional identity of mature murine IGIF to mature human IL-1β and a 12% identity to human IL-1α. Using this same positional alignment, the identity of IL-1β to IL-1α is 23%. It was suggested that the name IGIF be changed to interleukin-1γ (IL-1γ) (70). Does IGIF bind to IL-1 type I receptors? This would be an essential criterion for assigning the name IL-1γ since the type I IL-1 receptor is the ligand receptor and an essential chain of the IL-1 receptor signaling complex for the biological activity of IL-1. In the absence of evidence that IGIF binds to the IL-1 receptor type I (71), IL-18 rather than IL-1γ is a more appropriate name (53).

IL-18 AND ICE

Similar to precursor IL-1β (proIL-1β), precursor IL-18 (proIL-18) does not contain a signal peptide required for the removal of the precursor amino acids with subsequent secretion. There are few cytokines and growth factors that do not contain signal peptides, but the precursor forms of these proteins are often fully active (proIL-1α and ciliary neurotrophic factor are active), and processing does not appear to be required for biological activity. However, for IL-1β, the activity of the precursor is exceedingly low, but after cleavage by ICE the mature form is fully active. ICE, also called caspase-1 (72), is the first member of the family of intracellular cysteine proteases cleaving at an aspartic acid residue. The N-terminal amino acid sequence of the secreted form of murine IL-18 (47) was consistent with that following cleavage after an aspartic acid residue, a typical cleavage site for ICE. In fact, this analysis alerted investigators that the cleavage of proIL-18 at the aspartic acid site would likely require ICE (70). Therefore, it was not surprising that ICE cleaved proIL-18 (after the aspartic acid 36) and resulted in the mature and active protein (68,73). In addition to caspase-1, caspase-3 (also known as CPP32) cleaves mature or proIL-18 after aspartic 71 and 76, resulting in inactive peptides (74). As reported for proIL-1β (75), proIL-18 may serve as a negative regulator of Fas-mediated cell death by acting as a sink for the enzymatic activity of caspase-1 and caspase-3. It is caspase-3, however, that is likely to play a greater role in cell death than ICE.

IL-18 REGULATION OF IFN-γ PRODUCTION

Unlike mice deficient in IL-1β, ICE-deficient mice do not exhibit circulating IFN-γ following LPS, and like IFN-γ receptor deficient mice, are resistant to LPS lethality. Compared with wild-type mice, splenocytes from ICE-deficient mice produced low IFN-γ following LPS or zymosan (54). In contrast, IFN-γ production was unimpaired in ICE-deficient mice using the mitogen conA. Since comparable results were obtained when endogenous IL-18 was blocked with a neutralizing antimurine IL-18 antibody, one concludes that IL-18 is essential for IFN-γ production in models of infection and inflammation. On the other hand, there is no apparent impairment of IFN-γ production induced by a T-cell mitogen (54). Also, in antigen-sensitized ICE-deficient mice, proliferation of lymph node cells in response to the specific antigen was not altered. Most importantly, the reduced levels of IFN-γ in ICE-deficient mice were not due to a lack of IL-12. By comparison, there is no reduction in IFN-γ production in mice deficient in IL-1β. Therefore, the ICE-deficient mice reveal that IFN-γ production can be independent of IL-12. The role of IL-12 in IFN-γ production appears to be due, in part, to the ability of IL-12 to increase IL-18 binding to T cells (76).

IS IL-18 A MEMBER OF THE IL-1 FAMILY?

The IL-1 family members are presently IL-1β, IL-1α, and IL-1Ra. They are considered a family primarily because they bind to the same receptor (the IL-1 receptor type I, IL-1RI). They also bind to the IL-1R type II decoy receptor (albeit with quite different affinities compared to the IL-1RI). The IL-1 receptor signaling complex is comprised of the IL-1 ligand binding IL-1RI and the IL-1R accessory protein (IL-1R-AcP), similar to the IL-2 receptor complex in which the p55 ligand binding IL-2Rα forms a complex with the p75 IL-2Rβ. Although the three IL-1 cytokines share less than 30% primary amino acid sequence homologies, they are structurally very similar since each is an all-β-pleated, barrel-like form. In that regard, two other cytokines have all-β-pleated forms: acidic fibroblast growth factor and IL-18. Fibroblast growth factor is not part of the IL-1 family because this growth factor does not bind to IL-1 receptors. However, recent studies suggest that IL-18 may now have to be considered part of the IL-1 family. This is based on the finding that IL-18 is the natural ligand for the IL-1 receptor family member called IL-1R–related protein (IL-1Rrp) (57).

This receptor was originally cloned using degenerate oligonucleotides from the members of the IL-1R family (77). At the time, IL-1Rrp had no known ligand, but IL-1 was certainly not its ligand. When a neutralizing antibody to a cell-surface receptor was used to purify an IL-18–binding protein (receptor), the amino acid sequence matched that of the IL-1Rrp (57). Therefore, IL-18 has started to look like a member of the IL-1 family because part of the IL-18 receptor complex is a member of the IL-1R family. The other component of the putative IL-18 receptor complex is discussed below. IL-1 receptor signaling recruits the IL-1R activating kinase (IRAK) (78) which has been linked to the activation of NFκB-inducing kinase. Indeed, IL-1 and IL-18 each activate translocation of NFκB and IL-18 recruits IRAK (66). It is presently unclear whether IL-18 recruits IRAK to the IL-1Rrp or the IL-18Rα (see below). IL-18 may also claim its relationship to IL-1β because like IL-1β it is first synthesized as a precursor lacking a signal peptide requiring ICE cleavage.

IL-18 AS A PLEIOTROPIC CYTOKINE

Is there a broader role for IL-18? Does the cytokine possess other biological activities, not only as a costimulant for Th1 cytokines, but as a direct acting proinflammatory cytokine? The ability of IL-18 to induce CC and CXC chemokines places IL-18 in a strategic role in inflammation. Using freshly obtained human peripheral blood mononuclear cells (PBMC), mature, recombinant IL-18 induced IL-8 gene expression and synthesis (55). Using IL-1 receptor antagonist (IL-1Ra) to block IL-1 activity in these cells, IL-18–induced IL-8 production was reduced by 40%. When TNF activity was inhibited using soluble receptors for TNF, IL-18–induced IL-8 production was reduced by 80% suggesting that the primary action of IL-18 was via a TNF-dependent pathway.

Although IL-8 in PBMC cultures was derived from the CD14+ cell population (and not the CD3+ cells), the source of TNF was the CD3+ and the NK cells. Therefore, the inflammatory cytokine cascade initiated by IL-18 in the mixed PBMC population starts with triggering of constitutive IL-18 receptors on lymphocytes and NK cells leading to TNF production; in turn, TNF stimulates IL-1 and IL-8 from the CD14+ cell. The induction of TNF can be observed at the level of gene expression within 2 hours following exposure to IL-18. IL-18 also stimulates the synthesis of macrophage chemoattractant protein-1 in these cultures (55). These results are consistent with the ability of IL-18 to activate the translocation of NFκB in peripheral T cells and fibroblasts (56,57,66). Table 2 compares IL-18 and IL-1β activities.

In vivo, antibodies to IL-18 inhibit induction of TNF-α mRNA in the liver following LPS injection into mice preconditioned with *P. acnes* (79). In this model, TNF-α mRNA induction in the liver is biphasic. Anti-IL-18 antibodies reduce the second (4–6 hours), but not the first (1–2 hours), peak of TNF-α mRNA, suggesting that IL-18 participates in the induction of late TNF-α production in vivo. An alternate interpretation is that late TNF-α induction in the liver requires IFN-γ, which is absent using anti-IL-18.

Consistent with the induction of TNF in these models is the observation that IL-18 induces human immunodeficiency virus-1 (HIV-1) expression. At concentrations as low as 100 pM, mature IL-18 induces p24 synthesis in a chronically infected macrophage cell line called U1 (61). The increase in p24 is an indication of production of competent virus by these cells. Similar to the observation in PBMC, HIV-1 p24 production by IL-18 is inhibited by blocking the activity of endogenously produced TNF (61). Furthermore, the induction of p24 by IL-18 is reduced by inhibition of endogenous nitric oxide synthesis (62). Again, these observations are consistent with the ability of IL-18 to activate NFκB translocation. As discussed below, one chain of the IL-18 receptor complex is the IL-1Rrp, which upon transient transfection into COS-1 cells results in NFκB translocation (57). In murine cloned T cells, IL-18 stimulation results in a macromolecular complex precipitated by an antibody to IRAK (66). In these same cells, IL-18 does not activate STAT 4, which characteristically occurs following IL-12 treatment (66). Since IRAK activates a kinase cascade leading to degradation of IκB and activation of NFκB, these findings are consistent with IL-18 induction of TNF from T cells and IL-8 from macrophages (55).

HOST-DEFENSE FUNCTIONS OF IL-18

In BALB/C mice, which are highly susceptible to infection with *Mycobacteria*, there is low expression of IL-18 and IFN-γ compared to the resistant strain DBA/2, in which high levels of IFN-γ and IL-18 are observed (80). For nearly all cytokines, including the proinflammatory cytokines, there are well-established animal models in which pretreatment affords protection from infection, increased host defense, or reduced disease severity. Similarly, IL-18 administered into mice as an adjuvant for increased tumor immunogenicity provided enhanced NK-cell activation, production of IL-2, and sequentially decreased IL-10 (81). Lotze et al. also reported increased tumor immunogenicity using pretreatment with IL-18 in a melanoma model (82). However, unexpectedly, the ability of IL-18 to afford protection was also observed in IFN-γ receptor–deficient mice. Since the protection afforded by IL-12 in this model is dependent upon IFN-γ receptor expression, it appears that IL-18 may possess host-defense functions independent of IFN-γ.

In models of infection, pretreatment or early treatment rather than treatment postinfection is the hallmark. Using a model of live cryptococcal infection, pre- and early treatment with IL-18 reduced infection severity but did not increase survival at 3–6 weeks (83). The amount of IL-18 administered to these mice

Table 2 A Comparison of IL-18 and IL-1β

Property	IL-18	IL-1β
IRAK activation	+	+
NFκB translocation	+	+
Precursor requires ICE	+	+
↑ Expression in NOD	+	+
Endothelial cell activation	+/−	+
Induction of IL-8/IL-1β/TNF-α	+	+
Activation of Th1 responses	+	+/−
Activation of Th2 responses	−	−
Induction of IFN-γ	+/−[a]	+/−[a]
Adjuvant for tumor immunity	+	+
Elevated levels in leukemias	+	+
Synergy with IL-12 for IFN-γ	+	+
Protection against infection	+	+

[a] Neither IL-18 nor IL-1 induces IFN-γ alone but rather acts as an enhancing cofactor for IFN-γ production with microbial agents, mitogens, IL-12, or IL-2.

was 100-fold higher than the amount of IL-1β needed to protect mice against *Candida* lethality (84).

LEVELS OF IL-18 IN HUMAN DISEASE

To date, there is one study measuring circulating IL-18 in patients with various leukemias (64). Using a specific ELISA for human mature IL-18, healthy subjects exhibited levels of plasma IL-18 between 50 and 150 pg/ml. In contrast, patients with acute lymphoblastic leukemia, chronic lymphocytic leukemia, acute myelogenous leukemia, and chronic myelogenous leukemia had elevated levels (200–1200 pg/ml). Patients with acute lymphoblastic leukemia and chronic myelogenous leukemia had the highest levels, although patients with acute myelogenous leukemia had the greatest number of samples above the level measured in healthy subjects (>150 pg/ml). It should be noted that these are the same leukemias in which a pathogenic role for IL-1β has been demonstrated (reviewed in Ref. 35).

CLINICAL IMPORTANCE OF THE PRO-INFLAMMATORY CYTOKINE IL-18

Preventing the activity of IL-18 is a sensible clinical strategy. The concept of anti-inflammatory cytokine therapy in humans is expanding. Reduction of activity of IL-1 or TNF in patients with rheumatoid arthritis in large placebo-controlled, double-blinded trials has been reported. The naturally occurring IL-1 receptor antagonist (IL-1Ra) dose-dependently reduced pain and joint swelling and the progression of bone erosions (85,86). Patients with rheumatoid arthritis have experienced dramatic improvements with strategies for neutralizing TNF-α (87,88). In patients with steroid-resistant Crohn's disease, similar improvements following anti-TNF therapy have been reported (89). It is anticipated that human trials of ICE inhibitors will show similar results. However, ICE inhibitors will be somewhat non-specific since ICE inhibition prevents processing of both IL-1β and IL-18. At the present time, there are three therapeutic options for specific blockade of IL-18: neutralizing anti-18 antibodies, soluble receptors to IL-18, and nonagonistic antibodies that bind either the ligand binding IL-18R or the IL-1Rrp.

SUMMARY

The paradigm of the cytokine cascade in response to an infectious challenge, particularly to endotoxins, now must include IL-18. Although TNF-α is one of the first cytokines to be released following stimulation by endotoxins, IL-18 may also be released. The ability of IL-18 to augment endotoxin-induced IFN-γ is of fundamental importance because IFN-γ increased TNF receptors and, together with TNF, IFN-γ augments nitric oxide synthesis. IFN-γ also increases ICE activity and hence there is more processing and release of IL-1β and IL-18. In mice, deficiency of IFN-γ receptors renders the animals resistant to death. However, in humans in clinical trails in burn units and for cancer treatment, IFN-γ administration has not been associated with increased mortality or worsening of the clinical picture even in patients with active infections. Therefore, the role of IFN-γ and IL-18 as a IFN-γ–inducing factor for human responses to endotoxins remains unclear, whereas in murine models of endotoxemia, IFN-γ and IL-18 clearly play a pathological role. Figure 2 illustrates the interaction of IL-18, TNF-α, and IL-1β with IFN-γ in the cytokine response to endotoxins.

ACKNOWLEDGMENT

These studies are supported by NIH Grant AI-15614.

REFERENCES

1. Michie HR, Manogue KR, Spriggs DR, et al. Detection of circulating tumor necrosis factor after endotoxin administration. N Engl J Med 1988; 318:1481–1486.
2. Endres S, Ghorbani R, Lonnemann G, van der Meer JWM, Dinarello CA. Measurement of immunoreactive interleukin-1 beta from human mononuclear cells: optimization of recovery, intrasubject consistency, and comparison with interleukin-1 alpha and tumor necrosis factor. Clin Immunol Immunopathol 1988; 49:424–438.
3. Cannon JG, van der Meer JW, Kwiatkowski D, et al. Interleukin-1 beta in human plasma: optimization of blood collection, plasma extraction, and radioimmunoassay methods. Lymphokine Res 1988; 7:457–467.
4. Eastgate JA, Symons JA, Wood NC, Grinlinton FM, di Giovine FS, Duff GW. Correlation of plasma interleukin 1 levels with disease activity in rheumatoid arthritis. Lancet 1988; 2:706–709.
5. Cannon JG, Tompkins RG, Gelfand JA, et al. Circulating interleukin-1 and tumor necrosis factor in septic shock and experimental endotoxin fever. J Infect Dis 1990; 161:79–84.
6. Fong Y, Tracey KJ, Moldawer LL, et al. Antibodies to cachectin/tumor necrosis factor reduce interleukin 1β and interleukin 6 appearance during lethal bacteremia. J Exp Med 1989; 170:1627–1633.
7. Dinarello CA, Cannon JG, Wolff SM, et al. Tumor necrosis factor (cachectin) is an endogenous pyrogen and

induces production of interleukin 1. J Exp Med 1986; 163:1433–1450.

8. Aiura K, Gelfand JA, Wakabayashi G, Burke JF, Thompson RC, Dinarello CA. Interleukin-1 (IL-1) receptor antagonist prevents *Staphylococcus epidermidis*-induced hypotension and reduces circulating levels of tumor necrosis factor and IL-1β in rabbits. Infect Immun 1993; 61:3342–3350.

9. Ikejima T, Ikusawa S, Ghezzi P, van der Meer JWM, Dinarello CA. IL-1 induces TNF in human PBMC in vitro and a circulating TNF-like activity in rabbits. J Infect Dis 1990; 162:215–221.

10. Fenton MJ, Clark BD, Collins KL, Webb AC, Rich A, Auron PE. Transcriptional regulation of the human prointerleukin 1 beta gene. J Immunol 1987; 138:3972–3979.

11. Caput D, Beutler B, Hartog K, Thayer R, Brown-Shimer S, Cerami A. Identification of a common nucleotide sequence in the 3′-untranslated region of mRNA molecules specifying inflammatory mediators. Proc Natl Acad Sci USA 1986; 83:1670–1674.

12. Mileno MD, Margolis NH, Clark BD, Dinarello CA, Burke JF, Gelfand JA. Coagulation of whole blood stimulates interleukin-1β gene expression: absence of gene transcripts in anticoagulated blood. J Infect Dis 1995; 172:308–311.

13. Shapiro L, Dinarello CA. Cytokine expression during osmotic stress. Exp Cell Res 1997; 231:354–362.

14. Schindler R, Dinarello CA. Ultrafiltration to remove endotoxins and other cytokine-inducing materials from tissue culture media and parenteral fluids. Bio Techniques 1990; 8:408–413.

15. Schindler R, Clark BD, Dinarello CA. Dissociation between interleukin-1β mRNA and protein synthesis in human peripheral blood mononuclear cells. J Biol Chem 1990; 265:10232–10237.

16. Schindler R, Gelfand JA, Dinarello CA. Recombinant C5a stimulates transcription rather than translation of IL-1 and TNF; cytokine synthesis induced by LPS, IL-1 or PMA. Blood 1990; 76:1631–1638.

17. Kaspar RL, Gehrke L. Peripheral blood mononuclear cells stimulated with C5a or lipopolysaccharide to synthesize equivalent levels of IL-1β mRNA show unequal IL-1β protein accumulation but similar polyribosome profiles. J Immunol 1994; 153:277–286.

18. Han J, Lee J-D, Bibbs L, Ulevitch RJ. A MAP kinase targeted by endotoxin and hyperosmolarity in mammalian cells. Science 1994; 265:808–811.

19. Laughlin MJ, Kirkpatrick G, Sabiston N, Peters W, Kurtzberg J. Hematopoietic recovery following high-dose combined alkylating-agent chemotherapy and autologous bone marrow support in patients in phase I clinical trials of colony stimulating factors: G-CSF, GM-CSF, IL-1, IL-2 and M-CSF. Ann Hematol 1993; 67:267–276.

20. Kitamura T, Takaku F. A preclinical and phase I clinical trial of IL-1. Exp Med 1989; 7:170–177.

21. Tewari A, Buhles WC Jr, Starnes HF Jr. Preliminary report: effects of interleukin-1 on platelet counts. Lancet 1990; 336:712–714.

22. Nemunaitis J, Appelbaum FR, Lilleby K, et al. Phase I study of recombinant interleukin-1β in patients undergoing autologous bone marrow transplantation for acute myelogenous leukemia. Blood 1994; 83:3473–3479.

23. Crown J, Jakubowski A, Kemeny N, et al. A phase I trial of recombinant human interleukin-1β alone and in combination with myelosuppressive doses of 5-fluorouracil in patients with gastrointestinal cancer. Blood 1991; 78:1420–1427.

24. Crown J, Jakubowski A, Gabrilove J. Interleukin-1: biological effects in human hematopoiesis. Leuk Lymphoma 1993; 9:433–440.

25. Smith JW, Longo D, Alford WG, et al. The effects of treatment with interleukin-1α on platelet recovery after high-dose carboplatin. N Engl J Med 1993; 328:756–761.

26. Smith JW, Urba WJ, Curti BD, et al. The toxic and hematologic effects of interleukin-1 alpha administered in a phase I trial to patients with advanced malignancies. J Clin Oncol 1992; 10:1141–1152.

27. Iizumi T, Sato S, Iiyama T, et al. Recombinant human interleukin-1 beta analogue as a regulator of hematopoiesis in patients receiving chemotherapy for urogenital cancers. Cancer 1991; 68:1520–1523.

28. van der Poll T, Bueller HR, ten Cate H, et al. Activation of coagulation after administration of tumor necrosis factor to normal subjects. N Engl J Med 1990; 322:1622–1627.

29. Tilg H, Trehu E, Atkins MB, Dinarello CA, Mier JW. Interleukin-6 (IL-6) as an anti-inflammatory cytokine: induction of circulating IL-1 receptor antagonist and soluble tumor necrosis factor receptor p55. Blood 1994; 83:113–118.

30. Tilg H, Trehu E, Shapiro L, et al. Induction of circulating soluble tumour necrosis factor receptor and interleukin 1 receptor antagonist following interleukin-1α infusion in humans. Cytokine 1994; 6:215–219.

31. Bargetzi MJ, Lantz M, Smith CG, et al. Interleukin-1 beta induces interleukin-1 receptor antagonist and tumor necrosis factor binding proteins. Cancer Res 1993; 53:4010–4013.

32. Okusawa S, Gelfand JA, Ikejima T, Connolly RJ, Dinarello CA. Interleukin 1 induces a shock-like state in rabbits. Synergism with tumor necrosis factor and the effect of cyclooxygenase inhibition. J Clin Invest 1988; 81:1162–1172.

33. Holtmann H, Wallach D. Down regulation of the receptors for tumor necrosis factor by interleukin 1 and 4 beta-phorbol-12-myristate-13-acetate. J Immunol 1987; 139:1161–1167.

34. Brakebusch C, Varfolomeev EE, Batkin M, Wallach D. Structural requirements for inducible shedding of the p55 tumor necrosis factor receptor. J Biol Chem 1994; 269:32488–32496.

35. Dinarello CA. Biological basis for interleukin-1 in disease. Blood 1996; 87:2095–2147.

36. Mathias S, Younes A, Kan C-C, Orlow I, Joseph C, Kolesnick RN. Activation of the sphingomyelin signaling pathway in intact EL4 cells and in a cell-free system by IL-1β. Science 1993; 259:519–522.

37. Gronich J, Konieczkowski M, Gelb MH, Nemenoff RA, Sedor JR. Interleukin-1α causes a rapid activation of cytosolic phospholipase A_2 by phosphorylation in rat mesangial cells. J Clin Invest 1994; 93:1224–1233.

38. Clark MA, Özgür LE, Conway TM, Dispoto J, Crooke ST, Bomalaski JS. Cloning of a phospholipase A$_2$-activating protein. Proc Natl Acad Sci USA 1991; 88:5418–5422.
39. Kracht M, Truong O, Totty NF, Shiroo M, Saklatvala J. Interleukin-1α activates two forms of p54α mitogen-activated protein kinase in rabbit liver. J Exp Med 1994; 180:2017–2027.
40. Guesdon F, Saklatvala J. Identification of a cytoplasmic protein kinase regulated by IL-1 that phosphorylates the small heat shock protein hsp27. J Immunol 1991; 147:3402–3407.
41. Guesdon F, Freshney N, Waller RJ, Rawlinson L, Saklatvala J. Interleukin 1 and tumor necrosis factor stimulate two novel protein kinases that phosphorylate the heat shock protein hsp27 and beta-casein. J Biol Chem 1993; 268:4236–4243.
42. Guesdon F, Waller RJ, Saklatvala J. Specific activation of β-casein kinase by the inflammatory cytokines interleukin-1 and tumour necrosis factor. Biochem J 1994; 304:761–768.
43. Freshney NW, Rawlinson L, Guesdon F, et al. Interleukin-1 activates a novel protein cascade that results in the phosphorylation of hsp27. Cell 1994; 78:1039–1049.
44. Nakamura K, Okamura H, Wada M, Nagata K, Tamura T. Endotoxin-induced serum factor that stimulates gamma interferon production. Infect Immun 1989; 57:590–595.
45. Nakamura K, Okamura H, Nagata K, Komatsu T, Tamura T. Purification of a factor which provides a co-stimulatory signal for gamma interferon production. Infect Immun 1993; 61:64–70.
46. Okamura H, Nagata K, Komatsu T, et al. A novel co-stimulatory factor for gamma interferon induction found in the livers of mice causes endotoxic shock. Infect Immun 1995; 63:3966–3972.
47. Okamura H, Tsutsui H, Komatsu T, et al. Cloning of a new cytokine that induces interferon-γ. Nature 1995; 378:88–91.
48. Heremans H, van Damme J, Dillen C, Dikman R, Billiau A. Interferon-γ, a mediator of lethal lipopolysaccharide-induced Shwartzman-like shock in mice. J Exp Med 1990; 171:1853–1861.
49. Car BD, Eng VM, Schnyder B, et al. Interferon γ receptor deficient mice are resistant to endotoxic shock. J Exp Med 1994; 179:1437–1444.
50. Micallef MJ, Ohtsuki T, Kohno K, et al. Interferon-γ-inducing factor enhances T helper 1 cytokine production by stimulated human T cells: synergism with interleukin-12 for interferon-γ production. Eur J Immunol 1996; 26:1647–1651.
51. Kohno K, Kataoka J, Ohtsuki T, et al. IFN-γ-inducing factor (IGIF) is a co-stimulatory factor on the activation of Th1 but not Th2 cells and exerts its effect independently of IL-12. J Immunol 1997; 158:1541–1550.
52. Yoshimoto T, Okamura H, Tagawa Y-I, Iwakura Y, Nakanishi K. Interleukin-18 together with interleukin-12 inhibits IgE production by induction of interferon-γ production from activated B cells. Proc Natl Acad Sci USA 1997; 94:3948–3953.
53. Ushio S, Namba M, Okura T, et al. Cloning of the cDNA for human IFN-γ-inducing factor, expression in Escherichia coli, and studies on the biologic activities of the protein. J Immunol 1996; 156:4274–4279.
54. Fantuzzi G, Puren AJ, Harding MW, Livingston DJ, Dinarello CA. IL-18 regulation of INF-γ production and cell proliferation as revealed in interleukin-1β converting enzyme-deficient mice. Blood 1998; 91.
55. Puren AJ, Fantuzzi G, Gu Y, Su MS-S, Dinarello CA. Interleukin-18 (IFN-γ-inducing factor) induces IL-1β and IL-8 via TNFα production from non-CD14+ human blood mononuclear cells. J Clin Invest 1998; 101:711–721.
56. Matsimoto S, Tsuji-Takayama K, Aizawa Y, et al. Interleukin-18 activates NFkB in murine T helper type I cells. Biochem Biophys Res Commun 1997; 234:454–457.
57. Torigoe K, Ushio S, Okura T, et al. Purification and characterization of the human interleukin-18 receptor. J Biol Chem 1997; 272:25737–25742.
58. Tsutsui H, Nakanishi K, Matsui K, et al. IFN-γ-inducing factor up-regulates Fas ligand-mediated cytoxtoxic activity of murine natural killer cell clones. J Immunol 1996; 157:3967–3973.
59. Dao T, Ohashi K, Kayano T, Kurimoto M, Okamura H. Interferon-γ-inducing factor, a novel cytokine, enhances Fas ligand-mediated cytotoxicity of murine T helper cells. Cells Immunol 1997; 173:230–235.
60. Ohtsuki T, Micallef MJ, Kohno K, Tanimoto T, Ikeda M, Kurimoto M. Interleukin-18b enhances Fas ligand expression and induces apoptosis in Fas-expressing human myelomonocytic KG-1 cells. Anticancer Res 1997; 17:3253–3258.
61. Shapiro L, Puren AJ, Razeghi P, Heidenreich KA, Meintzer MK, Dinarello CA. IL-18 induces HIV-1 production in U1 cells with dependence on tumor necrosis factor, IL-6 and the p38 mitogen activated protein (MAP) kinase (abstr). Cytokine 1997; 9:893.
62. Muhl H, Shapiro L, Dinarello CA. The nitric oxide synthase inhibitor L-NMMA is a potent inhibitor of IL-18 induced expression of HIV-1 in the chronically infected macrophage U1 cell (abstr). Cytokine 1997; 9:947.
63. Stoll S, Mueller G, Kurimoto M, et al. Production of IL-18 (IFN-γ-inducing factor) messenger RNA and functional protein by murine keratinocytes. J Immunol 1997; 159:298–302.
64. Taniguchi M, Nagaoka K, Kunikata T, et al. Characterization of anti-human interleukin-18 (IL-18)/interferon-gamma inducing factor (IGIF) monoclonal antibodies and their application in the measurement of human IL-18 ELISA. J Immunol Method 1997; 206:107–113.
65. Conti B, Jeong JW, Tinti C, Son JH, Joh TH. Induction of IFN-γ-inducing factor in the adrenal cortex. J Biol Chem 1997; 272:2035–2037.
66. Robinson D, Shibuya K, Mui A, et al. IGIF does not drive Th1 development but synergizes with IL-12 for interferon-γ production and activates IRAK and NFκB. Immunity 1997; 7:571–581.
67. Hunter CA, Timans J, Pisacane P, et al. Comparison of the effects of interleukin-1α, interleukin-1β and interferon-γ inducing factor on the production of interferon-γ by natural killer. Eur J Immunol 1997; 27:2787–2792.

68. Ghayur T, Banerjee S, Hugunin M, et al. Caspase-1 processes IFN-gamma-inducing factor and regulates LPS-induced IFN-gamma production. Nature 1997; 386:619–623.
69. Gu Y, Kuida K, Tsutsui H, et al. Activation of interferon-γ inducing factor mediated by interleukin-1β converting enzyme. Science 1997; 275:206–209.
70. Bazan JF, Timans JC, Kaselein RA. A newly defined interleukin-1? Nature 1996; 379:159.
71. Udagawa N, Horwood NJ, Elliot J, et al. Interleukin-18 is produced by osteoblasts and acts via granulocyte macrophage colony-stimulating factor and not via interferon-γ to inhibit osteoclast formation. J Exp Med 1997; 185:1005–1012.
72. Alnemri ES, Livingston DJ, Nicholson DW, et al. Human ICE/CED-3 protease nomenclature. Cell 1996; 87:123.
73. Gu Y, Wu J, Faucheu C, et al. Interleukin-1β converting enzyme requires oligomerization for activity of processed forms in vivo. EMBO J 1995; 14:1923–1931.
74. Akita K, Ohtsuki T, Nukada Y, et al. Involvement of caspase-1 and caspase-3 in the production and processing of mature human interleukin-18 in monocytic THP.1 cells. J Biol Chem 1997; 272:26595–26606.
75. Tatsuta T, Cheng J, Mountz JD. Intracellular IL-1β is an inhibitor of Fas-mediated apoptosis. J Immunol 1997; 157:3949–3957.
76. Ahn H-J, Maruo S, Tomura M, et al. A mechanism underlying synergy between IL-12 and IFN-γ-inducing factor in enhanced production of IFN-γ. J Immunol 1997; 159:2125–2131.
77. Parnet P, Garka KE, Bonnert TP, Dower SK, Sims JE. IL-1Rrp is a novel receptor-like molecule similar to the type I interleukin-1 receptor and its homologues T1/ST2 and IL-1R AcP. J Biol Chem 1996; 271:3967–3970.
78. Croston GE, Cao Z, Goeddel DV. NFkB activation by interleukin-1 requires an IL-1 receptor-associated protein kinase activity. J Biol Chem 1995; 270:16514–16517.
79. Tsutsui H, Matsui K, Kawada Y, et al. Interleukin-18 accounts for both tumor necrosis factor-α and FAS ligand mediated hepatotoxic pathways in endotoxin-induced liver injury in mice (abstr). Cytokine 1997; 9:964.
80. Kobayashi Y, Appella E, Yamada M, Copeland TD, Oppenheim JJ, Matsushima K. Phosphorylation of intracellular precursors of human IL-1. J Immunol 1988; 140:2279–2287.
81. Micallef MJ, Tanimoto T, Kohno K, Ikeda M, Kurimoto M. Interleukin-18 induces the sequential activation of natural killer cells and cytotoxic T lymphocytes to protect syngeneic mice from transplantation with Meth A sarcoma. Cancer Res 1997; 57:4557–4563.
82. Lotze MT, Osaki T, Peron J-M, Shurin B, Baar J, Tahra H. Cytokines which induce interferon-γ, IL-2, IL-12 and IL-18: regulation of tumor immunity (abstr). J Leuk Biol 1997:25.
83. Kawakami K, Qureshi MH, Zhang T, Okamura H, Kurimoto M, Saito A. IL-18 protects mice against pulmonary and disseminated infection with Cryptococcus neoformans by inducing IFN-γ production. J Immunol 1997; 159:5528–5534.
84. Van't Wout JW, van der Meer JWM, Barza M, Dinarello CA. Protection of neutropenic mice from lethal Candida albicans infection by recombinant interleukin 1. Eur J Immunol 1988; 18:1143–1146.
85. Bresnihan B, Lookbaugh J, Witt K, Musikic P. Treatment with recombinant human interleukin-1 receptor antagonist in rheumatoid arthritis: results of a randomized, double-blind, placebo-controlled multicenter trial (abstr). Arthrit Rheumat 1996; 39:S73.
86. Watt I, Cobby M. Recombinant human interleukin-1 receptor antagonist reduces the rate of joint erosion in rheumatoid arthritis. Arthrit Rheumat 1996; 39:S123.
87. Moreland LW, Baumgartner SW, Schiff MH, et al. Treatment of rheumatoid arthritis with a recombinant human tumor necrosis factor receptor (p75)-Fc fusion protein. N Engl J Med 1997; 337:141–147.
88. Elliott MJ, Maini RN, Feldman M, et al. Randomised double-blind comparison of chimeric monoclonal antibody to tumour necrosis factor alpha (cA2) versus placebo in rheumatoid arthritis. Lancet 1994; 344:1105–1110.
89. van Dullemen HM, van Deventer SJH, Hommes DW, et al. Treatment of Crohn's disease with anti-tumor necrosis factor chimeric monoclonal antibody (cA2). Gastroenterology 1995; 109:129–135.

36

Platelet-Activating Factor in Sepsis: An Update

Reuven Rabinovici and Fizan Abdullah
Yale University School of Medicine, New Haven, Connecticut

Guenther Mathiak
University of Cologne, Cologne, Germany

Giora Feuerstein
DuPont Pharmaceuticals, Wilmington, Delaware

INTRODUCTION

Despite advances in physiology, immunology, and molecular biology, the pathophysiology of sepsis remains obscure, and this disease continues to be a major cause of mortality in critically ill patients (1). Sepsis is defined as systemic inflammation in the presence of confirmed infection (2), and current data suggest that gram-negative sepsis (referred to as sepsis throughout this chapter) is the result of multiple inflammatory events initiated by the circulatory release of large amounts of bacterial lipopolysaccharide (LPS) endotoxin. The net result of these events ultimately determines outcome and is affected by the extent of injury (3), the presence of repeated insults (4), and genetic predisposition (5). It is within this context that data related to the potential role of PAF in the pathogenesis of sepsis will be reviewed.

More than two decades ago, Benveniste and associates observed that with the addition of antigen, rabbit basophils sensitized by IgE released histamine and a novel substance (6) they termed platelet-activating factor (PAF) capable of inducing platelet aggregation and platelet release of histamine. The factor was later identified as 1-O-alkyl-2-acetyl-sn-glycero-3-phosphorylcholine (7–9) generated by a variety of cells including platelets (10), neutrophils (11), macrophages (11), endothelial cells (11), and epithelial cells (12). Subsequent investigations characterized the diverse inflammatory and immunological actions (for review, see Ref. 13) of PAF in addition to its involvement in stress situations (14). Notably, many investigators focused on the role of PAF in the development of sepsis. The present chapter aims to review the available information regarding PAF's role in the disease processes leading to sepsis. Specifically, this review analyzes investigations that addressed the following issues: (1) Do serum and tissue PAF levels rise in experimental and clinical sepsis? (2) Does the infusion of PAF replicate features of sepsis? (3) Does inhibition of PAF attenuate sepsis? (4) Does PAF interact with other mediators of sepsis? This chapter concludes with a discussion of future research directions aimed to further define the role of PAF in the pathogenesis of sepsis and the use of PAF antagonists in human sepsis.

IS PAF PRODUCED IN SEPSIS?

Numerous animal studies in a variety of species have focused on the demonstration of elevated serum PAF levels in sepsis. In these reports (Table 1), elevated serum PAF levels have been demonstrated in rats, pigs, cats, and mice as early as 1 minute and as late as 4 hours after LPS administration. Serum PAF levels ranged between 12 pg/ml and 4.3 ng/ml in control an-

Table 1 PAF Serum Levels in Experimental Endotoxemia

	Species	Anesthesia	LPS dose/route	Peak PAF level	Control PAF level	p-value	Sampling time	Assay	Ref.
1	Rat	Y	1 mg/kg, i.v. bolus or 30 mg/kg, i.v. bolus	1.52 ng/ml 0.53 ng/ml	0.19 ng/ml 0.24 ng/ml	<0.05 <0.05	90 min 1 min	RIA	15
2	Rat	Y	25 mg/kg, i.v. bolus	20.0 pg/ml	12.4 pg/ml	<0.05	20 min	Gas/mass spectrometry	16
3	Rat	n.a.	0.5 mg/kg, i.v. bolus	8.0 ng/ml	2.1 ng/ml	<0.05	n.a.	Serotonin release	17
4	Rat	Y	20 mg/kg, i.p. bolus	13.7 ng/ml	4.3 ng/ml	<0.01	20 min	Serotonin release	18
5	Rat	Y	50 mg/kg, i.v. bolus	2.6 ng/ml	<0.1 ng/ml	<0.01	10 min	Platelet aggregation	19
6	Pig	Y	5 µg/kg, i.a. bolus	1300 pg/ml	80–400 pg/ml	<0.05	30 min	Platelet aggregation	20
7	Pig	Y	0.5 µg/kg, i.v. bolus at 0.5 and 10 hr	250 pg/ml	50 pg/ml	<0.05	60 min	RIA	21
8	Pig	Y	5 µg/kg, i.v., infusion for 1 hr followed by 2 µg/kg i.v., infusion for 3 hr	7.0 ng/ml	1.25 ng/ml	<0.05	30 min	Platelet aggregation	22
9	Cat	Y	1 mg/kg, i.v. bolus	927 pg/ml	316 pg/ml	n.a.	4 hr	RIA	23
10	Mouse	n.a.	25.7 µg/g, i.p. bolus	753.1 pg/ml	210 pg/ml	<0.05	60 min	RIA	24

i.a., Intraarterial; i.p., intraperitoneal; i.v., intravenous; n.a., not available; RIA, radioimmunoassay; Y, yes.

imals, with a peak concentration of 13.7 ng/ml following LPS stimulation.

Complementary evidence supporting PAF generation during sepsis was provided by studies in various animal models, which identified local production of PAF in duodenal, gastric, ileal, and pulmonary tissue (Table 2). Elevated concentrations of PAF were also reported in the spleen and gallbladder although no statistical significance was provided in these studies.

The number of studies investigating the generation in humans of PAF during sepsis is comparatively small (Table 3). Heretofore, statistically significant elevations of PAF levels in septic patients were reported in only two studies. In the first study (1 in Table 3), the serum PAF levels of eight patients was measured at a single time point within 24 hours after admission to the ICU. All patients received conventional therapy including crystalloid and colloid solutions, broad-spectrum antibiotics, and mechanical ventilation. PAF serum concentrations were fivefold higher in the studied group as compared to control groups consisting of normal healthy volunteers or nonseptic ventilated patients. The elevated serum PAF levels correlated with the Apache II score of these patients. In parallel to serum levels, samples of bronchoalveolar lavage fluid taken within 48 hours of ICU admission from five septic shock patients showed significant increases PAF concentrations. Furthermore, comparison between survivors ($n = 4$) and nonsurvivors ($n = 4$) showed that PAF levels were higher in the nonsurvivor group (8.8×10^{-10} M) as compared to the survivor group (3.3×10^{-10} M), although this difference was not statistically significant. In a second study (6 in Table 3), blood collected at unspecified time points from seven septic patients with positive blood cultures during the 48 hours prior to sampling showed a statistically significant elevation in PAF levels as compared to 17 healthy volunteers.

Elevated serum PAF levels in septic patients as compared to healthy volunteers (4, 5, and 7 in Table 3) were reported by three other investigators; however, no statistical analysis was provided in these studies, which were conducted in 12 adult and 8 pediatric patients.

In contrast to the above reports, other studies failed to detect elevated PAF levels in the serum of septic patients (2, 3 in Table 3). For example, in one study, despite PAF being detected in 10 of 12 patients (83%), the mean value of PAF in afflicted patients was not significantly higher than that in normal volunteers (2 in Table 3). Also, a recent study that measured PAF levels in 13 severely ill patients with septic shock at the time of presentation of symptoms, at day 2 and at day 7, reported no significant differences as compared to healthy controls (3 in Table 3).

Normal volunteers ranged had serum PAF concentration from less than 5.6 pg/ml to less than 0.5 ng/ml (Table 3). Possible explanations for these diverse values will be discussed later.

DOES PAF INFUSION PRODUCE A SEPSIS-LIKE STATE?

Initial studies aimed at defining a role for PAF in sepsis were based on the demonstration that PAF infusion produced hypotension and death in a variety of species (35–38). Only in later studies did it become apparent that the shock syndrome produced by PAF involves multiple characteristics of sepsis and septic shock (sepsis with tissue hypoperfusion despite adequate fluid resuscitation). For example, PAF was shown to compromise cardiac performance, which can be attributed to its documented cardiac effects including coronary vasoconstriction (38–40), reduced contractility (41–43), and reduced preload (37). Additionally, PAF has been shown to produce peripheral vasodilatation (39,40) and pulmonary vasoconstriction (38), both of which are typical features of the clinical sepsis profile.

PAF's ability to increase one of the hallmark indexes of human sepsis, microvascular permeability, has been widely documented. A direct agonistic effect of the mediator on endothelial cells was suggested by studies showing that PAF stimulates Ca^{2+} influx in cultured human endothelial cells (44). Indeed, PAF was later shown to dose-dependently induce cultured endothelial cells to retract, lose reciprocal contact, and exhibit increased permeability as measured by ^{125}I-albumin diffusion (45). Moreover, impairment of endothelial cells by PAF in vivo has also been observed. Following 5–10 minutes of superperfusion with the autacoid (10^{-7} M) in the mesenteric bed of anesthetized guinea pigs, the endothelial cells retract in the area corresponding to the site of application of PAF, resulting in exposure of subintimal tissue to the blood stream and substantial thrombus formation (46). Further formation of blebs and interstitial edema have also been observed accompanying the described PAF-induced changes. In parallel to these data, PAF was documented to induce hemoconcentration, plasma extravasation, and edema in rats (47), guinea pigs (48), sheep (49), and dogs (37). Locally, injection of PAF causes plasma extravasation in the guinea pig (50) and rabbit skin as well as rabbit retina (51).

Table 2 PAF Tissue Levels in Experimental Endotoxemia and Bacteremia

	Species	Anesthesia	Model	Dose	Organ	Peak PAF level	Control PAF level	p-value	Sampling time	Assay	Ref.
1	Rat	Y	LPS, i.v. bolus	25 mg/kg	Duodenum	95 pg/g	10 pg/g	<0.05	30 min	RIA	25
					Stomach	82 pg/g		<0.05	60 min		
2	Rat	Y	LPS, i.v. bolus	16 mg/kg	Ileum mucosa	15.0 ng/g	5.82 ng/g	<0.05	2.5 hr	Platelet aggregation	26
3	Rat	Y	LPS, i.p. bolus	20 mg/kg	Lung	312 ng	32.3 ng	<0.01	20 min	Serotonin release	18
4	Rat	Y	E. coli, i.p. bolus	2×10^8 CFU/rat	Spleen	6 ng/g	not detectable	n.a.	3 hr	Serotonin release	27
5	Cat	Y	LPS, i.v. bolus	1 mg/kg	Gallbladder	481 pg/mg	301 pg/ml	n.a.	4 hr	RIA	23

i.p., Intraperitoneal; i.v., intravenous; n.a., not available; RIA, radioimmunoassay; Y, yes.

Table 3 PAF Serum Levels in Human Sepsis

	Ref.	Year	Patients	Peak serum PAF level	Control PAF level	p-value	Sampling time	Assay
1	28	1994	8	28.0 pg/ml	<5.6 pg/ml	<0.01	Within 24 hr after ICU admission	Column purification followed by gas/mass spectrometry
2	29	1994	12	17.4 pg/ml	13.5 pg/ml	n.s.	n.a.	Platelet aggregation
3	30	1993	13	0.19 ng/ml (+ blood cultures, 10 pts.) 0.13 ng/ml (− blood cultures, 3 pts.)	0.25 ng/ml	n.s.	At diagnosis, day 2, day 7	Column purification followed by serotonin release
4	31	1993	5	91.8 pg/ml	35.0 pg/ml	n.a.	n.a.	Column purification followed by gas/mass spectrometry
5	32	1991	7	4.97 ng/ml (+ blood cultures, 4 pts.) Undetectable (− blood cultures, 3 pts.)	<0.5 ng/ml	n.a.	n.a.	Platelet aggregation
6	33	1989	7	1.4 ng/ml (+ blood cultures, 7 pts.)	0.17 ng/ml	<0.01	n.a.	Serotonin release
7	34	1987	8	1.48 ng/ml	Undetectable	n.a.	n.a.	Platelet aggregation

n.a., Not available; n.s., not significant; ICU, intensive care unit; pts., patients.

Gastrointestinal hemorrhage (52,53) and thrombocytopenia (54), which are examples of other sepsis-associated features, have been shown to be produced by PAF infusion in several species. Changes in the whole-body glucose metabolism that mimic those observed during endotoxemia have also been reported, i.e., systemic infusion of PAF has been shown to elevate both plasma glucose and the rate of glucose appearance in the rat (55). Also, intravenous bolus injection of PAF resulted in a reduced basal and insulin-stimulated amino acid transport in isolated rat skeletal muscle (56,57).

Most recently, PAF has been shown to interact with other pro-inflammatory mediators generated during the septic state and has been shown to be a trigger to their production as detailed ahead.

DOES PAF INHIBITION ATTENUATE CONSEQUENCES OF SEPSIS?

Studies utilizing highly specific PAF antagonists that have demonstrated significant protective effects are another line of evidence that support the involvement of PAF in the pathophysiology of sepsis (Table 4). Notably, 11 studies using a variety of structurally unrelated PAF antagonists demonstrated complete protection against LPS-induced mortality in several species. In only two of these studies was the PAF antagonist given concomitantly or after LPS administration (Table 4), while in the remaining experiments the PAF antagonists were administered prior to endotoxin stimulation. Partial protection by PAF antagonists against mortality induced by the intravenous administration of LPS or live bacteria was reported by eight investigators (Table 4). Also, 17 and 8 studies with PAF antagonists demonstrated statistically significant prevention or reduction of the hypotensive effect of LPS, respectively (Table 5). In 7 of the complete protection group and in 2 of the partial protection group, the PAF antagonist was posttreatment. In contrast, 12 studies testing the efficacy of a variety of PAF antagonists including WEB 2086 (94–98), BN 52021 (20,66,99), SRI 63-675 (100), SRI 63-441 (98,101), CV 6209 (102), and CV 3988 (98) failed to show protection against endotoxin-induced mortality or hypotension.

The efficacy of PAF antagonists has also been demonstrated in animal models of septic/endotoxemic lung injury. For example, the PAF receptor antagonists SRI 63-675 and WEB 2086 attenuated pulmonary vasoconstriction, edema, microvascular injury, and hypoxia induced by E. coli endotoxin infusion in pigs (95,103). Similarly, pretreatment with the antagonists SRI 63-441 or CV 3988 blocked the increased capillary leakage and edema in perfused lungs isolated from endotoxin-treated rats (18). Consistent with the above studies, pretreatment with WEB 2170 attenuated leukosequestration and microvascular injury in lungs from rats infused with LPS (104). Oliguric acute renal failure and gastrointestinal ulcerations are other features typically associated with sepsis. Several PAF antagonists including BN 52021 and SR 63-675 were recently shown to inhibit alterations in renal blood flow induced by LPS, thus producing significant improvement in renal function (105,106). Other antagonists such as CV-3988, RO 193704, and BN 52021 have been shown to reduce the endotoxin-induced gastrointestinal injury in the rat (84,107) while WEB 2086 has been shown to attenuate the inhibition by LPS of amino acid transport in rat skeletal muscle (57).

To the best of our knowledge, only two clinical trials have been conducted to test the efficacy of PAF antagonists in clinical sepsis (1, 2 in Table 6). In 1994, the effect of the PAF antagonist, BN 52021, on the 28-day all-cause mortality of 132 severely septic patients was evaluated in a prospective, randomized, placebo-controlled, double-blind, multicenter study (2 in Table 6). Although mortality rate for the treatment group was 42% compared to 51% in the placebo control group ($n = 130$), the difference was not statistically significant ($p = 0.17$). However, further analysis of only patients with gram-negative sepsis revealed a statistically significant 42% ($p = 0.01$) reduction of mortality with the BN 52021 treatment. Similarly, PAF inhibition also significantly ($p = 0.01$) reduced (43%) mortality rate among patients with gram-negative sepsis who presented with shock at the time of entry into the study and among patients >60 years of age (58%). Recently, a second prospective, randomized, placebo-controlled, double-blind, multicenter study testing the efficacy of PAF inhibition in sepsis was reported (1 in Table 6). In this study, 12 patients with SIRS and sepsis were given the PAF antagonist TCV-309 in addition to standard therapy. The control group consisted of 16 matching patients treated with placebo. Although no difference in the all-cause 26- and 58-day mortality rates was observed, pulmonary and hematological failure scores improved significantly in TCV-309–treated patients. The cytokine profile did not differ statistically between both groups.

The effect of PAF inhibition on healthy volunteers infused with endotoxin has also been evaluated (3 in Table 6). Five subjects received pretreatment with the PAF antagonist Ro 24-4736 18 hours prior to LPS (4 ng/kg, i.v.) administration in this double-blinded and

Table 4 PAF Antagonists That Protect Against LPS-Induced Mortality

	PAF antagonist	Antagonist dose/route	Time of administration versus LPS	Species	Anesthesia	Model	LPS/bacterial dose/route	Effect	Ref.
1	WEB 2086	10 mg/kg, p.o. bolus and 10 mg/kg, i.v. bolus	−60 min	Rat	Y	LPS	7.5 mg/kg, i.v. bolus	+	58
2	BN 52021	5 mg/kg, s.c. bolus	0 min	Rat	N	LPS	5 mg/kg, i.v. bolus	+	59
3	BN 52021	25 mg/kg, s.c. bolus	−15 min	Rat	Y	LPS	20 mg/kg, i.p. bolus	±	60
4	BN 52021	5 mg/kg, i.v. bolus	−30 min	Dog	N	LPS	1 mg/kg, i.v. bolus	+	61
5	BN 52021	5 mg/kg, i.v. bolus	−30 min and 240 min	Dog	Y	LPS	1 mg/kg, i.v. bolus	+	62
6	CV 3988	10 mg/kg, i.v. bolus and 5 mg/kg, i.v. bolus	−15 min	Rat	Y	LPS	20 mg/kg, i.p. bolus	±	18
			−5 min and +50 min						
7	CV 3988	10 mg/kg, i.v. bolus	0 min	Rat	Y	LPS	5 mg/kg, i.v. bolus	±	63
8	BN 50739	10 mg/kg, i.p. bolus	−30 min	Rat	N	LPS	14.3 mg/kg, i.v. bolus	+	64
9	BN 50739	10 mg/kg, i.p. bolus	−30 min	Rabbit	N	LPS	50 μg/kg, i.v. bolus	±	65
10	CL 184,005	5 mg/kg, i.v. bolus	0 min	Rabbit	Y	Staphylococcus epidermidis	3×10^{11} CFU/kg, i.v. bolus	±	66
11	CL 184,005	1–20 mg/kg, i.p. bolus	−120 min	Mouse	n.a.	Staphylococcus aureus	Dose n.a., i.v. bolus	±	67
12	SM 12502	119 mg/kg, i.p. bolus	5 min	Mouse	N	LPS	60 mg/kg, i.v. bolus	+	68
13	PCA 4248	10 mg/kg, i.p. bolus	−60 and 60 min	Mouse	N	LPS	20 mg/kg, i.v. bolus	±	69
14	ST 899	30 mg/kg, i.p. bolus	−60 and −10 min	Mouse	N	LPS	40 mg/kg, i.p. bolus	+	70
15	ABT-299	0.1 mg/kg, i.v. bolus	−15 min	Rat	Y	LPS	8.5 mg/kg, i.v. bolus	+	71
16	ABT-299	1 mg/kg/h, i.v. infusion and 0.3 mg/kg/h, i.v. infusion	−30–0 min	Pig	Y	LPS	0.5 μg/kg/h for 6 hr, i.v. infusion	+	72
17	CV-3988 and	6 mg/kg, i.v. bolus	0–6 hr −20 min	Rat	n.a.	LPS	7 mg/kg, i.p. bolus	±	73
18	ONO-1078 or SR 27388	150 mg/kg, p.o. bolus 250 μg/kg, i.v. bolus	−60 min −5 min	Mouse	n.a.	LPS	7.5 mg/kg, i.v. bolus	+	74
19	WEB 2086 SR 27417 or WEB 2086	500 μg/kg, i.v. bolus 0–2.5 mg/kg, i.v. bolus 500 μg/kg, i.v. bolus	−5 min or 1 hr	Mouse	n.a.	LPS	7.5 mg/kg, i.v. bolus	+	75

i.p., Intraperitoneal; i.v., intravenous; p.o., per os; s.c., subcutaneously; n.a., not available; Y, yes; N, no; + complete prevention; ±, partial prevention.

Table 5 PAF Antagonists That Protect Against LPS-Induced Hypotension

	PAF antagonist	Antagonist dose/route	Time of administration versus LPS	Species	Anesthesia	LPS dose/route	Effect	p-value	Ref.
1	WEB 2086	1 mg/kg, i.v. bolus	−30 min	Rat	n.a.	2 mg/kg, i.v. bolus	+	n.a.	76
2	WEB 2086	5 mg/kg, i.v. bolus	−20 min	Rat	Y	10 mg/kg, i.v. bolus	±	<0.05	77
3	WEB 2086	10 mg/kg, i.v. bolus	Pretreatment	Rat	Y	10 mg/kg, i.v. bolus	+	n.a.	78
4	WEB 2086	5 mg/kg, i.v. bolus	−5 min	Rat	N	5 mg/kg, i.v. bolus	+	<0.05	79
5	WEB 2086	1–10 mg/kg, p.o. or 0.1–5.0 mg/kg, i.v. bolus	−60 min −5 min	Rat	Y	15 mg/kg, i.v. bolus	+ +	<0.001	58
6	TCV 309	0.1 mg/kg, i.v. bolus	−30 min	Rat	Y	10 mg/kg, i.v. bolus	+	<0.01	80
7	TCV-309	100 µg/kg, i.v. bolus and 500 µg/kg/hr for 35 min, i.v. infusion	−35 min −35 min–0	Dog	Y	2 mg/kg, i.v. bolus	±	<0.01	81
8	TCV-309	100 µg/kg, i.v. bolus and 150 or 500 µg/kg/hr, i.v. infusion	−20 min 20–360 min	Dog	Y	1.5 mg/kg, i.v. bolus	+	<0.05	82
9	BN 52021	5 mg/kg, i.v. bolus	−30 min	Rat	N	20 mg/kg, i.v. bolus	±	<0.05	60
10	BN 52021	4 mg/kg, i.v. bolus	−5 min	Pig	Y	5 µg/kg/hr for 1 hr, i.a. infusion	+	<0.05	20
11	BN 52021	1–6 mg/kg, i.v. bolus	−15 min	Guinea pig	Y	0.3 mg/kg, i.v. bolus	±	<0.05	83
12	CV 3988	10 mg/kg, i.v. bolus and 5 mg/kg, i.v. bolus	−5 min 50 min	Rat	Y	20 mg/kg, i.p. bolus	+	<0.05	18
13	CV 3988	5 mg/kg, i.v. bolus or 0.1–5 mg/kg, i.v. bolus	−5 min 10 min	Rat	Y	15 mg/kg, i.v. bolus	+ +	<0.05	63
14	CV 3988	1 mg/kg/min, i.v. infusion	−20 to −10 min	Rat	Y	1.25 mg/kg/min, i.v. infusion for 10 min	+	<0.001	84
15	SRI 63-441	5 mg/kg, i.v. bolus	−20 min and 30 min	Rat	n.a.	0.5 mg/kg, i.v. bolus	+	<0.01	17
16	SRI 63-441	0.18 mg/kg, i.v. bolus	1 min	Rat	Y	5 mg/kg, i.v. bolus	+	n.a.	85
17	SM 12502	10–30 mg/kg, i.v. bolus or 3–10 mg/kg, i.v. bolus	−2 min 90 min	Rat	Y	1 mg/kg, i.v. bolus	+ +	<0.01	15
18	E 5880	1–100 µg/kg, i.v. bolus	−5 min or 80 min	Rat	Y	25 mg/kg, i.v. bolus	+ +	<0.05	16
19	BN 50739	10 mg/kg, i.p. bolus	−30 min	Rat	N	14.4 or 43.2 mg/kg, i.v. bolus	±	<0.05	64

20	L-652,731	500 nmol/rat, i.v. bolus and 25 nmol/min/rat for 30 min, i.v. infusion	6 min 8–30 min	Rat	50 mg/kg, i.v. bolus	+	<0.01	86
21	Kadsurenone	0.5 μmol/rat, i.v. bolus and 0.2 μmol/min/rat for 30 min, i.v. infusion	−6 min 5–35 min	Rat	50 mg/kg, i.v. bolus	±	<0.005	19
22	SR 63-072	1 mg/kg, i.v. bolus	3 min	Rat	15 mg/kg, i.v. bolus	+	n.a.	87
23	ABT-299	0.1 mg/kg, i.v. bolus	20 min	Rat	25 mg/kg, i.a. bolus	+	<0.05	71
24	ABT-299	1 mg/kg/hr, i.v. infusion and 0.3 mg/kg/hr, i.v. infusion	−30–0 min 0–6 hr	Pig	0.5 μg/kg/hr for 6 hr, i.v. infusion	±	<0.05	72
25	SRI 63-675	10 mg/kg/hr, i.v. infusion and 3 mg/kg/hr, infusion	−25–0 min 0–240 min	Pig	5 μg/kg/hr for 1 hr and 2 μg/kg/hr for 4 hr, i.v. infusion	±	<0.05	22
26	E 6123	1–10 μg/kg, i.v. bolus	80 min	Rat	15 mg/kg, i.v. bolus	+	<0.05	88
27	ONO-6240	12 mg/kg/hr, i.v. infusion and 0.1 mg/kg/min, i.v. infusion	−15–0 min and 0–5 hr	Sheep	1 μg/kg/hr for 0.5 hr, i.v. infusion	+	<0.05	89
28	SR 27417	1 and 6 mg/kg, i.v. bolus	−5 min	Guinea pig	15 mg/kg, i.v. bolus	+	n.a.	90
29	WEB2170	0.5–10 mg/kg, i.v. bolus	−60, −5 and 6 min	Rat	15 mg/kg, i.v. bolus	+	n.a.	91
	WEB 2086 STY 2108	0.1–5.0 mg/kg, i.v. bolus 0.004–0.085 mg/kg, i.v. bolus				+		
30	CV 3988 WEB 2086 SRI 63441	10^{-4} M/rat, i.v. bolus	−15 min	Rat	5 mg/kg, i.v. bolus	+	<0.05	92
31	BN 52021 WEB 2086	5 mg/kg, i.v. bolus	−5 min	Rat	20 mg/kg, i.v. bolus	±	<0.05	93

i.a., Intraarterial; i.p., intraperitoneal; i.v., intravenous; p.o, per os; n.a., not available; Y, yes; N, no; +, complete prevention; ±, partial prevention.

Table 6 Clinical Trials with PAF Antagonists in Human Sepsis and Endotoxemia

	Ref.	Year	Diagnosis	Patients	Antagonist	Dose	Route	Initiation of therapy	Effect	p-value
1	108	1996	SIRS/Sepsis	12	TCV-309	1 mg/kg BID for 7 days	i.v.	Undetermined time after diagnosis	Pulmonary failure score ↓ Hematological failure score ↓ No effect on mortality	<0.05 <0.05
2	109	1994	Sepsis	132	BN 52021	120 mg/pt. BID for 4 days	i.v.	Undetermined time after diagnosis	↓ mortality: −18% overall, −42% in gram-negative sepsis −58% in pts. >60 yrs. of age	0.17 0.01 0.04
3	110	1994	Endotoxin infusion, i.v.	5*	Ro 24-4736	10 mg	p.o.	18 hr after LPS infusion	↓ rigor score ↓ myalgia score ↓ serum cortisol (−30%) ↓ epinephrine (−48%)	<0.05 <0.05 <0.05 <0.05

i.v., Intravenous; p.o., per os; pts., patients; yrs., years; ↓, decrease; *, healthy volunteers.

placebo-controlled study. PAF antagonist–treated subjects experienced fewer symptoms including rigors at 1 hour and myalgias at 1–4 hours. This was associated with diminished peak cortisol and epinephrine secretion and almost complete inhibition of PAF-induced platelet aggregation ex vivo.

DOES PAF INTERACT WITH OTHER MEDIATORS OF SEPSIS?

Given the central role of endotoxin in the pathogenesis of sepsis, it was quite natural that the PAF-endotoxin relationship has been extensively studied. First, PAF was reported to augment the LPS-stimulated production of other inflammatory mediators, e.g., co-incubation with PAF has been shown to increase the LPS-induced NOS activity in murine macrophage and rat vascular smooth muscle cell lines (104) as well as enhance IL-1 mRNA expression in human monocytic cells (THP-1) subjected to LPS stimulation (111). Additionally, pretreatment with PAF remarkably augmented TNF-α release from LPS-stimulated rabbit alveolar macrophages (112). Further, exposure of rat alveolar macrophages to PAF resulted in increased LPS-stimulated transcription of COX-2, the inducible isoform of prostaglandin synthase that mediates IL-6 production (113).

Second, several studies suggested a synergistic or priming relationships between PAF and LPS. These included the demonstration that co-stimulation with PAF and LPS amplified by 80% the induction of tissue factor activity in heparinized blood as compared to LPS alone (114). Pretreatment with PAF has also been shown to prime the LPS-induced procoagulant activity from rabbit alveolar (112) and murine peritoneal macrophages (115).

Third, several typical responses to endotoxin stimulation are known to be attenuated by PAF inhibition, i.e., the PAF receptor antagonists WEB 2170 and RP 5227 effectively blocked the TNF-α production from rat Kupffer cells in response to LPS (116). Other PAF antagonists such as BN 50739 and E6123 reduced the LPS-stimulated nitric oxide synthase (NOS) activity in diverse cell systems (104). Furthermore, WEB 2086 attenuated LPS-induced inhibition of amino acid transport in the rat soleus and extensor digitorum longus muscle, while WEB 2170 strongly inhibited TNF production from murine peritoneal macrophages stimulated by LPS (117). In addition, the LPS-triggered increase in the median volume of platelets harvested from healthy volunteers was completely abolished by the PAF antagonist HAGEPH (118).

Fourth, enhanced PAF production as measured by the [^{3}H]serotonin release from washed rabbit platelets was observed in rat spleen lymphocytes and peritoneal cells harvested 2 hours after the intraperitoneal injection of *E. coli* (119). Interestingly, this elevated PAF production preceded the development of PAF actions in vivo including increased microvascular permeability, suggesting that hemoconcentration and volume contraction could very well be secondary to PAF production.

Fifth, recent in vitro experiments have demonstrated that LPS-binding protein (LBP) and the cell membrane receptor CD14 modulate the LPS-induced PAF production from human monocytes and mesangial and endothelial cells (120). Specifically, LBP enhances LPS-induced PAF production by glomerular mesangial cells, while soluble CD14 blocks LPS-LBP-induced PAF production in human monocytes possibly by competing with cell membrane CD14. In contrast, soluble CD14 was required for LPS-induced PAF production from HUVEC, which do not express membrane CD14.

Intensive investigation into the pathophysiology of sepsis has unveiled many inflammatory mediators, among which cytokines and nitric oxide (NO) seem to play a central role. Thus, the relationships between PAF, cytokines, and nitric oxide have been the focus of numerous investigators, and much attention has been devoted to discern the interactions between PAF and the pro-inflammatory cytokine, TNF-α. PAF and TNF-α were reported to have the ability to trigger reciprocal production in several cell systems including rat (11) and pig peritoneal macrophages (121), human peripheral blood-derived monocytes (122,123), and human and rodent lymphocytes (124). PAF and TNF-α interactions have also been demonstrated in a model of endothelial cell injury where PAF antagonists attenuated the endothelial response caused by TNF-α–primed PMNs, including oxygen radical and leukotriene production (125). Further interdependence can be demonstrated by inhibition of LPS-induced TNF-α production by isolated murine macrophages by pre- or posttreatment with a PAF antagonist (126). In vivo, pretreatment of mice with either *S. typhosa* endotoxin or TNF-α resulted in increased mortality caused by PAF, whereas the administration of both TNF-α and endotoxin alone did not affect mortality (127). In addition, endogenously produced TNF-α (primed by zymosan) increased the hypotensive response and bowel injury induced by LPS in rats (17). These responses were characterized by higher serum PAF and TNF-α levels in the zymosan-primed rats and showed marked improvement with pretreatment by a PAF antagonist (17). Data from our laboratory have also suggested that PAF–TNF-α interactions play a major role in the pathophysiology of septic shock. For example, rats pretreated with the PAF antagonist BN 50739 had lower mortality after recombinant human TNF-α infusion (64). Moreover, in the rat (64) but not in the rabbit (65), LPS-induced elevations of serum TNF-α were also attenuated by pretreatment with the PAF antagonist, suggesting that PAF in some species exerts part of its effects in endotoxemia indirectly through the induction of TNF-α synthesis or by priming macrophages to produce TNF-α in response to endotoxin. The latter possibility is supported by the elevation of serum TNF-α and the development of hypotension, hemoconcentration, leukopenia, thrombocytopenia, and lung injury following the combined administration of relatively low doses of LPS and PAF, while separate infusion of these mediators at similar doses resulted in little to no response (128). Taken together, these studies are supportive of a "priming" mechanism, and additional studies have suggested that this priming effect of PAF for LPS-induced TNF-α production is associated with severe ARDS-like lung injury and mortality when the LPS-PAF combination regimen is repeated (129). Interestingly, complement and TNF-α seem to partially mediate these events since pretreatment with sCR-1, a blocking agent of the complement system (130), or anti-TNF-α monoclonal antibody (mAb) (131) attenuated the shock-like state produced by the LPS-PAF paradigm.

Aside from TNF-α, PAF has been shown to interact with other cytokines implicated in the sepsis-induced inflammatory cascade, e.g., long-term parenteral treatment of rats with PAF dose-dependently stimulated IL-1 and interleukin 2 (IL-2) production from isolated spleen mononuclear cells (132). This effect was attenuated by very high doses of PAF or treatment with the PAF antagonist BN 52021. Other reports have demonstrated PAF's ability to significantly enhance IL-1 production from muramyl dipeptide–stimulated human monocytes (133) in addition to IL-1's parallel ability to enhance PAF production from cultured endothelial (134) and rabbit synovial (135) cells as well as mouse alveolar macrophages (136). Also, use of the PAF antagonist UK-74 has been reported to significantly reduce leukocyte migration across mesenteric microvessels harvested from IL-1–treated rats (137). Additionally, pretreatment with a highly specific PAF receptor antagonist, BN 50739, attenuated lung injury produced by systemic recombinant human IL-2 infusion (138).

PAF has also been demonstrated to interact with NO, a known mediator of sepsis. Several experiments in which NO inhibition modulated the biological effects of PAF support this notion. For example, in the hamster cheek pouch model, NOS inhibitors reduced microvascular leakage and arteriolar constriction induced by PAF (139). Of note, studies using radiolabeled albumin in conscious rats demonstrated an opposite effect, i.e., NO inhibition markedly potentiated PAF-induced albumin extravasation in large airways, liver, spleen, pancreas kidney, stomach, and duodenum (140). NO inhibition was also reported to reduce the vasodilatation of rabbit glomerular afferent arterioles (141) and rat bowel injury (142) produced by PAF in vitro and in vivo, respectively. Additional studies were successful in demonstrating NO's ability (directly or indirectly) to abrogate responses to PAF infusion in vivo. Specifically, inhalation of NO by anesthetized pigs reduced PAF-induced pulmonary hypertension (143) and the administration of S-nitroso-N-acetyl penicillamine, which spontaneously generates NO, inhibited PAF-induced gastrointestinal leakage. Finally, the PAF-NO association was further demonstrated in a study in which PAF receptor blockade prevented the hypotensive response to a combination of hypoxia and LNMA administration in the rat (144).

CRITIQUE OF DATA AND CONCLUSION

To analyze PAF's involvement in sepsis, we have evaluated the data summarized in this chapter with reference to the adequacy of (1) methodologies of PAF measurement in sepsis, (2) animal models of sepsis, (3) pharmacological approaches used in experimental studies, (4) clinical studies, and (5) studies on PAF-endotoxin interactions.

Methodology

A careful review of the accuracy, validity, and reproducibility of methods employed to measure PAF in sepsis suggest that while they appear convincing, they in fact need much improvement.

Accuracy

The majority of PAF measurements were performed by functional bioassays such as platelet-serotonin release and platelet aggregation. These studies are laborious to perform, have low specificity, and may be difficult to interpret in the presence of natural inhibitors or agonists with the same biological activity. The recently developed radioimmunoassay kits may also be deficient because of potential cross-reactivity of the antibody with other glycerophospholipids. Gas/Mass spectrometry, the only available method that allows for accurate identification of specific molecular structure, is expensive and labor intensive. Thus, it is not surprising that this superior method was used in only one experimental (16) and one clinical study (104).

Validity

It is widely assumed that elevated levels of inflammatory agents in the serum suggest their involvement in the disease process. Nevertheless, serum concentrations are likely to represent "the tip of the iceberg," which reflects a delicate balance between mediator production and clearance as well as the level of receptor occupancy at target cells. On the other hand, undetectable serum PAF levels may not refute a central role for this mediator in sepsis because subinjurious doses of PAF combined with minute doses of endotoxin have been shown to produce cardiovascular and pulmonary collapse (128–131). Thus, the importance of elevated serum PAF levels in sepsis is obscure. Additionally, PAF may be acting locally, making tissue level measurement a better index of its involvement in sepsis. Indeed, five studies measured PAF concentration in tissues (Table 2), three of which (18,25,26) demonstrated significantly elevated levels in the duodenum, stomach, ileum, and lung. In the remaining two studies, elevated PAF levels were also reported in the spleen and gallbladder (23,27), but no statistical analysis was given.

Another pitfall in the determination of PAF tissue levels is the timing of blood sampling, because PAF is promptly metabolized by acetylhydrolases present in both cells and plasma. Unfortunately, PAF was mostly measured within 90 minutes after shock induction, and no attempts were made to determine levels at later time points that are compatible with the insidious onset of clinical sepsis.

Reproducibility

PAF tissue and serum determination in sepsis varied considerably. For example, basal as well as peak values ranged from the picomolar to the nanomolar scale in both humans and animals. Thus, the "normal" value of PAF serum concentration is not yet defined, which further complicates correlation between PAF levels and disease.

Animal Models

Most experimental paradigms that attempted to implicate PAF in the pathogenesis of sepsis utilized intravenous injection of high doses of endotoxin or bacteria (Table 1). The clinical relevance of these paradigms was a subject of much criticism for the following reasons. First, serum endotoxin or bacterial levels in human sepsis are often undetectable or very low (145). Second, these protocols pose an acute and toxic infectious insult unlikely to occur in human sepsis. Third, the cardiovascular and metabolic responses in these models differ from clinical sepsis. Most notably, systemic injection of LPS or bacteria produces an hypodynamic state (119,146) in contrast to the hyperdynamic response of human sepsis (147). Fourth, all experiments used "isolated" pharmacological paradigms in which standard ancillary therapies such as intravenous fluid resuscitation and antibiotic administration were ignored. Finally, these paradigms were often conducted in anesthetized animals while the effects of most anesthetic agents on the inflammatory response to stress are well described.

It should be noted, however, that the deficiencies of the current animal models of sepsis represent a general problem in sepsis research that is not specific to PAF.

Pharmacological Approach

In the clinical setting, the treatment of sepsis is most often initiated after diagnosis has been made. Thus, PAF antagonists should be evaluated in a "therapeutic mode," i.e., after induction of sepsis. It is therefore disappointing that only 7 (15,16,71,75,82,86,87) of the 50 preclinical studies that examined the effect of PAF on endotoxin-induced mortality (Table 4) or hypotension (Table 5) gave PAF antagonists as posttreatment. Also, in these seven studies, the time interval between shock induction and treatment was very short (3–90 min), which may not replicate the usual clinical settings in which sepsis is diagnosed and treated much later (several days) when clinical symptoms erupt.

In this chapter we presented many experiments demonstrating PAF's ability to mimic features of human sepsis. While these studies may support a role for PAF in the pathogenesis of sepsis, they are deficient in many aspects. For example, continuous infusion of PAF was required to produce responses because of its short half-life. Since PAF measurements failed to show sustained production during sepsis, it is conceivable that PAF infusion does not mimic the clinical syndrome but rather represents a state of acute PAF toxicity that casts doubt on the validity of such a strategy.

Clinical Studies

Data derived from clinical studies in support of PAF's role in the development of sepsis are scarce. Only two studies provided statistically significant evidence that PAF is produced during sepsis (28,33) of which one

(33) did not contain information pertaining to sampling time and did not compare PAF values to matching nonseptic ICU control patients. Also, three other studies that demonstrated elevated PAF levels (31,32,34) did not present statistical analysis, and therefore their scientific value is limited. Further doubt regarding the generation of PAF during sepsis was introduced by two studies in which serum PAF levels were undetectable in 25 septic patients (29,30).

The only two clinical studies that tested the efficacy of PAF antagonists in sepsis (108,109) also generated disappointing data. In these double-blind, placebo-controlled studies no effect of PAF inhibition on overall mortality was observed. Only retrospective analysis of some subgroups of patients (gram-negative sepsis with shock at the entry into the study and among patients >60 years of age), which is statistically defective, showed beneficial effect of PAF blockade on outcome.

PAF-Endotoxin Interactions

This chapter provided a variety of in vitro and ex vivo experiments that point to a modulatory effect of PAF on multiple actions of endotoxin, the most prominent trigger molecule of human sepsis. Also, PAF has been shown to interact with other networks of inflammatory mediators, including cytokines and NO, which were implicated in the initiation and propagation of sepsis. These complex relationships between PAF, endotoxin, and other inflammatory mediators support the possibility that PAF inhibition at certain time points may be efficacious in clinical sepsis. However, because spatial and temporal relationships of endotoxin, cytokines, and NO in sepsis are still obscure, the significance of PAF's interactions with these mediators remains speculative.

This critical analysis along with the discrepancy between experimental data and the failure of PAF receptor antagonists to make it into clinical practice in sepsis call for several recommendations. First, to accurately determine whether PAF is generated during sepsis it should be measured by gas/mass spectrometry in combination with functional tests, determined in target organs, and sampled at multiple time points including later phases of sepsis.

Second, clinically relevant models of sepsis that display insidious onset, hyperdynamic response, morbidity and mortality as well as integrate current standard therapies (intravenous fluids and antibiotics) should be rapidly developed in conscious animals. To that end, studies must aim to define as many clinical criteria to reflect the clinical disease; as morbidity is difficult to assess in animals, mortality must be a cardinal endpoint. This issue, which probably has crippled many investigations aimed to develop novel therapies for sepsis, is undoubtedly a most fundamental one and should clearly be prioritized.

Third, preclinical studies to test the ability of PAF receptor antagonists to ameliorate consequences of sepsis should focus on the "treatment mode" in which these compounds will be administered only after overt sepsis has developed. In parallel, pretreatment experiments should be conducted in newly developed animal models that mimic immunocompromised conditions or in prophylactic surgical paradigms.

Fourth, prospective, placebo-controlled, double-blind studies with potent and highly specific PAF antagonists should be launched in more defined patient populations such as patients with gram-negative septic shock or the elderly. The cardinal endpoint in these studies should be morbidity and mortality, but specific organ outcome should be monitored as well.

Fifth, future studies to discern the nature of PAF's interactions with other inflammatory mediators should be conducted not only to better understand the pathophysiology of sepsis but also to facilitate the development of novel therapeutic modalities to combat this syndrome.

In conclusion, we believe that the research regarding the role of PAF in sepsis is flawed with inaccurate and possibly irrelevant measurements of the mediator, suboptimal animal models, irrelevant endpoints, and inappropriate focus on prophylactic rather than therapeutic needs. To improve our understanding of PAF's role in sepsis, it is essential that these deficiencies be recognized and prevented.

REFERENCES

1. Lowry SF. Sepsis and its complications: clinical definitions and therapeutic prospects. Crit Care Med 1994; 22:S1–S2.
2. American College of Chest Physicians–Society of Critical Care Medicine Consensus Conference. Definitions for and organ failure and guide-lines for the use of innovative therapies in sepsis. Crit Care Med 1989; 20:864–875.
3. Bone RC. Why sepsis trials fail. JAMA 1996; 276: 565–566.
4. Demling RH, Lalonde C, Ikegami K. Physiologic support of the septic patient. In: Deitch EA, ed. The Surgical Clinics of North America. Philadelphia: WB Saunders Company, 1994:637–658.
5. Jacob CO, Fronek Z, Lewis GD, Koo M, Hansen JA, McDevitt HO. Heritable major histocompatibility complex class II-associated differences in production of tumor necrosis factor alpha: relevance to genetic

predisposition to systemic lupus erythematosus. Proc Natl Acad Sci USA 1990; 87:1233–1237.
6. Benveniste J, Henson PM, Cochrane CG. Leukocyte-dependent histamine release from rabbit platelets: The role of IgE, basophils, and platelet-activating factor. J Exp Med 1972; 136:1356–1377.
7. Benveniste J, Tence M. Varenne P, Bidault J, Polonsky J. Semi-synthesis and proposed structure of platelet-activating factor (PAF): Paf-acether an alkyl ether analog of lysophosphatidylcholine [French]. CR Seances Acad Sci 1979; 289:1037–1040.
8. Blank ML, Snyder F, Byers WL, Brooks B, Muirhead EE. Antihypertensive activity of an alkyl ether analogue of phosphatidylcholine. Biochem Biophys Res Commun 1979; 90;523–534.
9. Demolpoulus CA, Pinckard RN, Hanahan DJ. Platelet-activating factor: evidence of 1-O-alkyl-2-acetyl-sn-glyceryl-3-phosphorylcholine as the active component (a new class of lipid chemical mediators). J Biol Chem 1979; 254:9355–9358.
10. Henson PM. Release of vasoactive amines from rabbit platelets induced by sensitized mononuclear leokocytes and antigen. J Exp Med 1979; 131:287–306.
11. Camussi G, Bussolino F, Salvidio G, Baglioni C. Tumor necrosis factor/cachectin stimulates peritoneal macrophages, polymorphonuclear neutrophils and vascular endothelial cells to synthesize and release platelet-activating factor. J Exp Med 1987; 166:1390–1404.
12. Salari H, Wong A. Generation of platelet-activating factor (PAF) by a human lung epithelial cell line. Eur J Pharmacol 1990; 175:253–259.
13. Braquet P, Touqui L, Shen T, Vargaftig BB. Perspectives in platelet-activating factor research. Pharmacol Rev 1987; 39:97–145.
14. Rabinovici R, Yue TL, Feuerstein G. Platelet-activating factor in cardiovascular stress situations. Lipids 1991; 26:1257–1263.
15. Natsume Y, Imanishi N, Koike H, Morooka S. Effect of platelet activating factor antagonist (+)-cis-3,5-dimethyl-2-(3-pyridyl)thiazolidin-4-one hydrochloride on endotoxin-induced hypotension and hematological parameters in rats. Arzeneimittelforschung 1994; 44:1208–1213.
16. Nagaoka J, Harada K, Kimura S. Kobayashi S, Murakami M, Yoshimura T, Yamada K, Asano O, Katayama K, Yamatsu I. Inhibitory effects of the novel platelet activating factor antagonist, 1-ethyl-1[N-(2-methoxy)benoyl-N-[(2R)-2-methoxy-3-(4-octadecylcarbamoyloxy) piperidinocarbonyloxypropyloxy] carbonyl] aminomethyl-pyridinum chloride, in several experimentally induced shock models. Arzeneimittelforschung 1991; 41:719–724.
17. Sun XM, Hsueh W, Torre-Amione G. Effects of in vivo 'priming' on endotoxin induced hypotension and tissue injury. Am J Pathol 1990; 136:949–956.
18. Chang SW, Feddersen CO, Henson PM, Voelkel NF. Platelet-activating factor mediates hemodynamic changes and lung injury in endotoxin-treated rats. J Clin Invest 1987; 79:1498–1509.
19. Doebber TW, Wu MS, Robbins JC, Ma Choy B, Chang MN, Shen TY. Platelet-activating factor (PAF) involvement in endotoxin induced hypotension in rats. Studies with PAF-receptor antagonist Kadsurenone. Biochem Biophys Res Commun 1985; 127:799–808.
20. Mozes T, Heiligers JPC, Tak CJAM, Zijlstra FJ, Ben-Efraim S, Saxena PR, Bonta IL. Platelet activating factor is one of the mediators involved in endotoxic shock in pigs. J Lipid Med 1991; 4:309–326.
21. Klosterhalfen B, Horstmann-Jungemann K, Vogel P, Flohe S, Offner F, Kirkpatrick CJ, Heinrich PC. Time course of various inflammatory mediators during recurrent endotoxemia. Biochem Pharmacol 1992; 43:2103–2109.
22. Dobrowsky RT, Voyksner RD, Olson NC. Effect of SRI 63-675 on hemodynamics and blood PAF levels during porcine endotoxemia. Am J Physiol 1991; 260:H1455–H1465.
23. Kaminski DL, Feinstein WK, Deshpande YG. The production of experimental cholecystitis by endotoxin. Prostaglandins 1994; 47:233–245.
24. Chang TW, Wu MH, Yang YJ. Blood levels of platelet-activating factor in endotoxin-sensitive and endotoxin-resistant mice during endotoxemia. J Formos Med Assoc 1992; 91:1133–1137.
25. Autore G, Cicala C, Cirino G, Maiello FM, Mascolo N, Capasso F. Essential fatty acid-deficient diet modifies PAF levels in stomach and duodenum of endotoxin-treated rats. J Lipid Mediat Cell Signal 1994; 9:145–153.
26. Defaux JP, Thonier F, Baroggi N, Etienne A, Braquet P. Involvement of platelet-activating factor in endotoxin or ischaemia-induced intestinal hyperpermeability in the rat. J Lipid Med 1993; 7:11–21.
27. Inarrea P, Gomez-Cambronero J, Pascual J, Del Carmen Ponte M, Hernando L, Sanchez-Crespo M. Synthesis of PAF-acether and blood volume changes in gram-negative sepsis. Immunopharmacology 1985; 9:45–52.
28. Sorensen J, Kald B, Tagesson C, Lindahl M. Platelet-activating factor and phospholipase A2 in patients with septic shock and trauma. Intensive Care Med 1994; 20:555–561.
29. Shinozaki K, Kawasaki T, Kambayashi J, Sakon M, Shiba E, Uemura Y, Ou M, Iwamoto N, Mori T. A new method of purification and sensitive bioassay of platelet-activating factor (PAF) in human whole blood. Life Sci 1994; 54:429–537.
30. Graham RM, Strahan ME, Norman KW, Watkins DN, Sturm MJ, Taylor RR. Platelet and plasma platelet activating factor in sepsis and myocardial infarction. J Lipid Mediat Cell Signal 1994; 9:167–182.
31. Ono S, Tamakuma S, Mochezuki H. Kinoshita M, Ohkusa Y, Aosasa S, Oda Y, Hiroshi O. Clinical and experimental studies on the role of platelet-activating factor (PAF) in the pathogenesis of septic DIC. Surg Today 1993; 23:228–233.
32. Heuer HO, Darius H, Lohmann HF, Meyer J, Schierenberg M, Treese N. Platelet-activating factor type activity in plasma from patients with septicemia and other diseases. Lipids 1991; 26:1381–1385.
33. Dicz FL, Nieto ML, Fernandez-Gallardo S, Gijion MA, Sanchez-Crespo M. Occupancy of platelet receptors for platelet-activating factor in patients with septicemia. J Clin Invest 1989; 83:1733–1740.

34. Bussolino F, Porcellini MG, Varese L, Bosia A. Intravascular release of platelet activating factor in children with sepsis. Thromb Res 1987; 48:619–620.
35. Feuerstein G, Zukowska-Grojec Z, Krausz M, Blank M, Snyder F. Cardiovascular and sympathetic effects of 1-O-hexadecyl-2-acetyl-sn-glycero-3-phosphorylcholine in conscious SHR and WYK rats. Clin Exp Hypertension 1982; A4:1335–1350.
36. Feuerstein G, Lux WE, Ezra D, Hayes E, Snyder F. Thyrotrophin releasing hormone blocks the hypotensive effect of platelet activating factor in unanesthetized guinea-pig. J Cardiovasc Pharmacol 1985; 7:335–340.
37. Bessin P, Bonnet J, Apfel D, Soulard C, Desgroux L. Acute circulating collapse caused by platelet activating factor (PAF acether) in dogs. Eur J Pharmacol 1983; 86:403–413.
38. Argiolas L, Fabi F, del Basso P. Mechanisms of pulmonary vasoconstriction and bronchoconstriction produced by PAF in the guinea-pig: role of platelets and cyclo-oxygenase metabolites. Br J Pharmacol 1995; 114:203–209.
39. Feuerstein G, Boyd LM, Ezra D, Goldstein RE. Effect of platelet activating factor on coronary circulation of the domestic pig. Am J Physiol 1983; 246:H446–H471.
40. Sybertz EJ, Watkins RW, Baum T, Pula K, Rivelli M. Cardiac, coronary and peripheral vascular effects of acetyl ether phosphoryl choline in the anaesthetized dog. J Pharmacol Exp Ther 1984; 232:156–162.
41. Levi R, Burke JA, Guo ZG, Hattori Y, Hoppens CM. Acetyl glyceryl ether phosphorylcholine (AGEPC): a putative mediator of cardiac anaphylaxis. Circ Res 1984; 54:117–124.
42. Camussi G, Aglietta M, Malavasi F, Tetta C, Piacibelo W. The release of platelet activating Factor from human endothelial cells in culture. J Immunol 1983; 131:2397–2403.
43. Benveniste J, Boullet C, Brink C, Labat C. The actions of PAF acether (platelet activating factor) on guinea pig isolated heart preparation. Br J Pharmacol 1983; 80:81–83.
44. Bussolino F, Aglietta M, Sanavino F, Stacchini A, Lauri D, Camussi G. Alkyl-ether phosphoglycerides influence calcium fluxes into human endothelial cells. J Immunol 1985; 135:2748–2753.
45. Bussolino F, Camussi G, Aglietta M, Braquet P, Bosia A, Pescarmona G, Sanavio F, d'Urso N, Marchisio PC. Human endothelial cells are a target for platelet activating factor. J Immunol 1987; 139:2439–2446.
46. Bourgain RH, Maes L, Braquet P, Andries R, Toqui L, Braquet M. The effect of 1-O-alkyl-2-acetyl-sn-glycero-3-phosphocholine (PAF-acether) on the arterial wall. Prostaglandins 1985; 30:185–197.
47. Sanchez-Crespo M, Alonso F, Inarrea P, Alvarez V, Egido J. Vascular actions of synthetic PAF-acether (a synthetic platelet activating factor in the rat), evidence for a platelet independent mechanism. Immunopharmacology 1982; 4:173–185.
48. Hwang SB, Lam MH, Lee CL, Shen TY. Releases of platelet activating factor and its involvement in the first phase of carrageenin-induced rat foot edema. Eur J Pharmacol 1986; 120:33–41.
49. Cox CP, Mojarad M, Attiah A. Platelet activating factor (PAF) increases pulmonary vascular permeability in awake sheep. Am Rev. Respir Dis 1984; 129(suppl):A334.
50. Morley J, Page CP, Paul W. Inflammatory actions of platelet activating factor (PAF-acether) in guinea-pig skin. Br J Pharmacol 1983; 80:503–509.
51. Braquet P, Vidal RF, Braquet M, Hamard H, Vargaftig BB. Involvement of leukotrienes and PAF-acether in the increased microvascular permeability of the rabbit retina. Agent Action 1984; 15:82.
52. Gonzalez-Crussi F, Hsueh W. Experimental models of ischemic bowel necrosis. The role of platelet activating factor and endotoxin. Am J Pathol 1983; 112:127–135.
53. Rosam AC, Wallace JL. Whittle BJ. Potent ulcerogenic actions of platelet activating factor on the stomach. Nature 1983; 319:54–56.
54. Smallbone BW, Taylor NE, McDonald JW. Effect of L-652,731, a platelet-activating factor (PAF) receptor antagonist, on PAF- and complement-induced pulmonary hypertension in sheep. J Pharmacol Exp Ther 1987; 242:1035–1040.
55. Lang HL, Dobrescu C, Hargrove DM, Bagby GJ, Spitzer JJ. Platelet-activating factor-induced increases in glucose kinetics. Am J Physiol 1988; 254:E193–E200.
56. Boruff JS, Karlstad MD. Effect of Platelet-Activating Factor on basal and insulin-mediated system A amino acid transport in rat soleus muscle. Circ Shock 1993; 40:75–80.
57. Buripakdi D, Karlstad MD. The role of platelet-activating factor in alterations of system A amino transport in rat soleus and extensor digitorum longus muscle during endotoxic shock. Shock 1994; 2:53–59.
58. Casals-Stenzel J. Protective effect of WEB 2086, a novel antagonist of platelet activating factor, in endotoxin shock. Eur J Pharmacol 1987; 135:117–122.
59. Etienne A, Hecquet F, Soulard C, Spinnewyn B, Clostre F, Braquet P. In vivo inhibition of plasma protein leakage and Salmonella enteritidis-induced mortality in the rat by a specific PAF-acether antagonist: BN 52021. Agent Action 1985; 17:368–370.
60. Fletcher JR, Disimone AG, Earnest MA. Platelet activating factor receptor antagonist improves survival and attenuates eicosanoid release in severe endotoxemia. Ann Surg 1990; 211:312–316.
61. Moore JM, Earnest MA, Disimone AG, Abumrad NN, Fletcher JR. A PAF receptor antagonist BN 52021, attenuates thromboxane release and improves survival in lethal canine endotomemia. Circ Shock 1991; 35:53–59.
62. Fletcher J, Earnest M, Moore J, Disimone A, Abumrad N. Platelet activating factor (PAF) antagonist improves survival and attenuates eicosanoid release in lethal endotoxemia. Circ Shock 1989; 27:359.
63. Terashita ZI, Kawamura M, Takatani M, Tsushima S, Yoshimi I, Nishikawa K. Beneficial effects of TCV-309, a novel potent and selective Platelet Activating Factor antagonist in endotoxin and anaphylactic shock in rodents. J Pharmacol Exp Ther 1992; 260:748–755.
64. Rabinovici R, Yue T-L, Farhat M, Smith III EF, Slivjak M, Feuerstein G. Platelet activating factor (PAF)

and tumor necrosis factor-α (TNF-α) interactions in endotoxemic shock: studies with BN 50739, a novel PAF antagonist. J Pharmacol Exp Ther 1990; 255: 256–263.
65. Yue TL, Farhat M, Rabinovici R, Perera PY, Vogel S, Feuerstein G. Protective effect of BN 50739, a new platelet-activating factor antagonist, in endotoxin-treated rabbits. J Pharmacol Exp Ther 1990; 254: 976–981.
66. Dirkes K, Harris BH, Connolly RJ, Schwaitzberg SD, Dinarello CA, Gelfand JA. Platelet activating factor-antagonist improves survival in experimental staphylococcal septicemia. J Pediatr Surg 1994; 29:1055–1058.
67. DeJoy SQ, Jeyaseelan R, Torley LW, Pickett WC, Wissner A, Wick MM, Oronsky AL, Kerwar S. Effect of CL 184,005, a PAF-antagonist in murine model of *Staphylococcus aureus*-induced gram-positive sepsis. J Infect Dis 1994; 169:150–156.
68. Imanishi N, Murakami-Uchida M, Koike H, Natsume Y, Morooka S. Biological effects of the new platelet activating factor receptor antagonist (+)-cis-3,5-dimethyl-2-(3-pyridyl)thiazolidin-4 one hydrochloride. Drug Res 1994; 44:317–322.
69. Fernandez-Gallardo S, Ortega MDP, Piergo JG, De Casanjuana MF, Sunkel C, Sanchez Crespo M. Pharmacological action of PCA 4248, a new platelet activating factor receptor antagonist: in vivo studies. J Pharmacol Exp Ther 1990; 255:34–39.
70. Ruggiero V, Chiapparino C, Manganello S, Pacello L, Foresta P, Martelli EA. Beneficial effects of a novel platelet-activating factor receptor antagonist, ST 899, on endotoxin-induced shock in mice. Shock 1994; 2: 275–280.
71. Albert DH, Luo G, Magoc JT, Tapang P, Holms JH, Davidsen SK, Summers JB, Carter GW. ABT-299, a novel PAF antagonist, attenuates multiple effects of endotoxemia in conscious rats. Shock 1996; 6:112–117.
72. Kruse-Elliott KT, Albert DH, Summers JB, Carter GW, Zimmerman JJ, Grossman JE. Attenuation of endotoxin-induced pathophysiology by a new potent PAF receptor antagonist. Shock 1996; 5:265–273.
73. Yoshikawa D, Fumio G. Effect of platelet-activating factor antagonist and leukotriene antagonist on endotoxin shock in rats. Circ Shock 1992; 38:29–33.
74. Herbert JM. Fraisse L, Bachy A, Valette G, Savi P, Laplace MC, Lassalle J, Roche B, Lale A, Keane PE, Maffrand JP. Biochemical and pharmacological properties of SR 27388, a dual antioxidant and PAF receptor antagonist. J Lip Med 1993; 8:31–51.
75. Herbert JM, Lespy L, Maffrand JP. Protective effect of SR 27417, a novel PAF antagonist, on lethal anaphylactic and endotoxin-induced shock in mice. Eur J Pharmacol 1991; 205:271–276.
76. Huang L, Tan X, Crawford SE, Hsueh W. Platelet-activating factor and endotoxin induce tumor necrosis factor gene expression in rat intestine and liver. Immunology 1994; 83:65–69.
77. Szabo C, Chin-Chen W, Mitchell JA, Gross SS, Thiemermann C, Vane JR. Platelet-activating factor contributes to the induction of nitric oxide synthase by bacterial lipopolysaccharide. Circ Res 1993; 73:991–999.
78. Damas J. Involvement of platelet activating factor in the hypotensive response to zymosan in rats. J Lipid Med 1991; 3:333–344.
79. Qi M, Jones SB. Contribution of platelet-activating factor to hemodynamic and sympathetic responses to bacterial endotoxin in conscious rats. Circ Shock 1990; 32:153–163.
80. Ueneo A, Ishida H, Oh-ishi S. Comparative study of endotoxin-induced hypotension in kininogen-deficient rats with that in normal rats. Br J Pharm 1995; 114: 1250–1256.
81. Yamanaka S, Iwao H, Yukimura T, Kim S, Miura K. Effect of the platelet activating factor antagonist, TCV-309, and the cyclo-oxygenase inhibitor, ibuprofen, on the hemodynamic changes in canine experimental endotoxic shock. Br J Pharm 1993; 110: 1501–1507.
82. Kawamura M, Kitayoshi T, Terashita Z, Fujiwara S, Takatani M, Nishikawa K. Effects of TCV-309, a novel PAF antagonist, on circulatory shock and hematological abnormality induced by endotoxin in dogs. J Lipid Mediat Cell Signal 1994; 9:255–265.
83. Adnot S, Lefort J, Braquet P, Vargafting BB. Interference of the PAF-acether antagonist BN 52021 with endotoxin-induced hypotension in the guinea-pig. Prostaglandins 1986; 32:791–802.
84. Wallace JL, Whittle BJR. Prevention of endotoxin-induced gastrointestinal damage by CV-3988, an antagonist of platelet activating factor. Eur J Pharm 1986; 124:209–210.
85. Handley DA, Van Valen RG, Tomesch JC, Melden MK, Jaffe JM. Ballard FH, Saunders RN. Biological properties of the antagonist SRI 63-441 in the PAF and endotoxin models of hypotension in the rat and dog. Immunopharmacology 1987; 13:125–132.
86. Wu MS, Biftu T, Doebber TW. Inhibition of the platelet activating factor induced in vivo responses in rats by trans-2,5-(3,4,5-trmethoxyphenyl) tetrahydrofuran (L-652,721), a PAF receptor antagonist. J Biol Chem 1986; 260:15639–15645.
87. Handley DA, Lee ML. Saunders RN. Evidence for a direct effect on vascular permeability of platelet activating factor induced hemoconcentration in the guinea pig. Thromb Haemost 1983; 54:756–759.
88. Sakuma Y, Shirato M, Nagaoka J, Obiashi H, Tsunoda H, Katayama S, Ono H, Katayama K. Pharmacological activities of a novel thienodiazepine derivative as a platelet activating factor antagonist. Drug Res 1991; 41:1255–1259.
89. Toyufuko T, Kubo K, Kobayashi T, Kusama S. Effects of ONO-6240, a platelet activating factor antagonist, on endotoxin shock in unanesthetized sheep. Prostaglandins 1986; 31:271–281.
90. Bernat A, Herbert JM. Salel V, Lespy L, Maffrand JP. Protective effect of SR 27417, a novel antagonist, on PAF- or endotoxin-induced hypotension in the rat and the guinea-pig. J Lipid Med 1992; 51:41–48.
91. Muacevic G, Heuer HO. Platelet-activating factor antagonists in experimental shock. Arzeneimittelforschung 1992; 42:1001–1004.

92. Salari H, Demos M, Wong A. Comparative hemodynamics and cardiovascular effects of endotoxin and platelet activating factor in rats. Circ Shock 1990; 32:189–207.
93. Lagente V, Fortes ZB, Garcia-Leme J, Vargaftig BB. PAF-acether end endotoxin display similar effects on rat mesenteric microvessels: inhibition by specific antagonists. J Pharmacol Exp Ther 1988; 247:254–261.
94. Siebeck M, Kohl J, Enders S, Spannagl M, Machleidt W. Delayed treatment with platelet activating factor receptor antagonist WEB 2086 attenuates pulmonary dysfunction in porcine endotoxin shock. J Trauma 1994; 37:745–751.
95. Siebeck M, Weipert J, Keser C, Kohl J, Spannagl M, Machleidt W, Schweiberer L. A triazolodiazepine platelet activating factor receptor antagonist (WEB 2086) reduces pulmonary dysfunction during endotoxin shock in swine. J Trauma 1991; 31:942–950.
96. Bouvier Ch, Oguz Guc M, Furman BL. Parratt JR. Platelet activating factor impairs pressor responses to noradrenaline in the anaesthetized rat but does not mediate endotoxin-induced hyporeactivity. Circ Shock 1994; 42:14–19.
97. Qi M, Jones SB. Contribution of platelet activating factor to hemodynamic and sympathetic responses to bacterial endotoxin in conscious rats. Circ Shock 1990; 32:153–163.
98. Salari H, Demos M, Wong A. Comparative hemodynamics and cardiovascular effects of endotoxin and platelet-activating factor in rat. Circ Shock 1990; 32:189–207.
99. Mulder MF, van Lambalgen AA, van Kraats AA, Scheffer PG, Bouman AA, van den BOS GC, Thijs LG. Systemic and regional hemodynamic changes during endotoxin or platelet activating factor (PAF)-induced shock in rats. Circ Shock 1993; 41:221–229.
100. Gans ROB, DiPirro JC, Ueda Y, Niesen N, Lee KH, Brentjens JR. The effect of SRI 63-675, a competitive platelet-activating factor receptor-antagonist, in the generalized Shwartzman reaction. J Lipid Mediat 1994; 10:229–242.
101. Carrick JB, Deem Moris D, Moore JN. Administration of a receptor antagonist for platelet-activating factor during equine endotoxaemia. Equine Vet J 1993; 25:152–157.
102. Inoue Y, Kohno S, Miyazaki T, Yamaguchi K. Effect of a platelet activating factor antagonist and antithrombin III on septicemia and endotoxemia in rats. J Exp Med 1991; 163:175–185.
103. Olsen NC, Joyce PB, Fleisher LN. Role of platelet-activating factor and eicosanoids during endotoxin-induced lung injuries in pigs. J Am Physiol 1990; 258:H1674–H1686.
104. Anderson BO, Pogetti RS, Shanley PF, Bensard DD, Piman JM, Nelson DW, Whitman GJR, Banerjee A, Harken AH. Primed neutrophils injure rat lung through a platelet-activating factor-depending mechanism. J Surg Res 1991; 50:510–514.
105. Ha B, Tolins JP, Vercellotti G. The role of platelet-activating factor in endotoxemic acute renal failure in the male rat. Kid Int 1988; 33:358.
106. Tolins jP, Vercellotti GM, Wilkowske B. Platelet-activating factor mediates endotoxin induced acute renal insufficiency in rats. J Lab Clin Med 1989; 113:316–324.
107. Wallace JL, Steel G, Whittle BJR. Evidence for platelet-activating factor as a mediator of endotoxin-induced gastrointestinal damage in the rat. Effects of three platelet-activating factor antagonists. Gastroenterology 1987; 93:765–773.
108. Froon AMF, Greve JWM. Buurman WA, Van Der Linden CJ, Langemeijer HJM, Ulrich C, Bouregois M. Treatment with the platelet-activating factor antagonist TCV-309 in patients with severe systemic inflammatory response syndrome: a prospective, multi-center, double-blind, randomized phase II trial. Shock 1996; 5:313–319.
109. Dhainaut JFA, Tenaillon A, Le Tulzo Y, Schlemmer B, Solet JP, Wolff M, Holzapfel L, Zeni F., Dreyfuss D, Mira JP, De Vathaire F, Guinot P. Platelet-activating factor receptor antagonist BN 52021 in the treatment of severe sepsis: a randomized, double-blind, placebo-controlled, multicenter clinical trial. Crit Care Med 1994; 22:1720–1728.
110. Thompson WA, Coyle S, Van Zee K, Oldenburg H, Trousdale R, Rogy M, Felsen D, Moldawer L, Lowry SF. The metabolic effects of platelet-activating factor in endotoxemic man. Arch Surg 1994; 129:72–79.
111. Barthelson RA, Valone FH. Platelet-activating factor stimulates expression of IL-1 beta mRNA in THP-1 cells. Lipids 1991; 26:257–261.
112. Maier RV, Hahnel GB, Fletcher JR. Platelet-activating factor augments tumor necrosis factor and procoagulant activity. J Surg Res 1992; 52:258–264.
113. Thivierge M, Rola-Plezczynski. Up-regulation of inducible cyclooxygenase gene expression by platelet-activating factor in activated rat alveolar macrophages. J Immun 1995; 154:6593–6599.
114. Osterud B. Platelet-activating factor enhancement of lipopolysaccharide-induced tissue factor activity in monocytes: requirement of platelets and granulocytes. J Leuk Biol 1992; 51:462–465.
115. Kucey DS, Kubicki EI, Rotstein OD. Platelet-activating factor primes endotoxin-stimulated macrophage procoagulant activity. J Surg Res 1991; 50:436–441.
116. Zhang F, Decker K. Platelet-activating factor antagonists suppress the generation of tumor necrosis factor-alpha and superoxide induced by lipopolysaccharide or phorbol ester in rat liver macrophages. Eur Cytokine Netw 1994; 5:311–317.
117. Engelberts, Von Assmuth EJU, Van der Linden CJ, Burman WA. The interrelation between TNF, IL-6, and PAF secretion induced by LPS in an in vivo and in vitro murine model. Lymphokine Cytokine Res 1991; 10:127–131.
118. Nystrom ML, Barradas A, Jeremy JY, Mikhailidis DP. Platelet shape change in whole blood: differential effects of endotoxin. Thromb Haemost 1994; 71:646–650.
119. Brackett DJ, Lerner MR, Wilson MF. Dimethyl sulfoxide antagonizes hypotensive, metabolic and pathologic responses induced by endotoxin. Circ Shock 1991; 33:156–163.
120. Camussi G, Mariano F, Biancone L, DeMArtino A, Bussolati B, Montrucchio G, Tobias PS. Lipopolysac-

charide binding protein and CD14 modulate the synthesis of platelet-activating factor by human monocytes and mesangial cells stimulated with lipopolysaccharide. J Immun 1995; 155:316–324.
121. Hayashi H, Kudo I, Kato T, Inoue K. Platelet-activating factor and diseases. In Saito K, Hanahan DJ, eds. Tokyo: Int Med Publishers, 1989:51–68.
122. Bonavida B, Mencia-Huerta JM, Braquet P. Effects of platelet activating factor on peripheral blood monocytes: induction and priming for TNF secretion. J Lipid Mediat 1990; 2:S65–76.
123. Bonavida B, Mencia Huerta JM, Braquet P. Effect of platelet activating factor on monocyte activation and production of tumor necrosis factor. Int Arch Allergy Appl Immunol 1989; 88:157–160.
124. Rola-Pleszczyski M, Bosse J, Bissonnette E, Dubois C. PAF-acether enhances the production of tumor necrosis factor by human and rodent lymphocytes and macrophages. Prostaglandins 1988; 5:802.
125. Braquet P, Hosford D, Koltz P, Guilbaud J, Paubert-Braquet M. Effect of platelet activating factor on tumor necrosis factor-induced superoxide generation from human neutrophils. Possible involvement of G proteins. Lipids 1991; 26:1071–1075.
126. Floch A, Bousseau A, Hetier E, Floch F, Bost P-E, Cavero I. RP 55778, a PAF receptor antagonist, prevents and reverses LPS-induced hemoconcentration and TNF release. J Lipid Mediators 1989; 1:349–360.
127. Heuer H. Effect of a new and specific PAF-antagonist, WEB 2086, on PAF and tumor necrosis factor induced changes in mortality and intestinal transit velocity. Prostaglandins 1988; 35:814.
128. Rabinovici R, Esser KM, Lysko PG, Yue TL, Griswold DE, Hillegass LM. Bugelski PJ, Hallenbeck JM, Feuerstein G. Priming by PAF of endotoxin induced lung injury and cardiovascular shock. Circ Res 1991; 69:12–25.
129. Rabinovici R, Bugelski P, Esser KM, Hillegass M, Vernick J, Feuerstein G. ARDS-like lung injury produced by endotoxin in platelet-activating factor-primed rats. J Appl Physiol 1993; 74:1791–1802.
130. Rabinovici R, Yeh GC, Hillegass LM. Griswold DE, DiMartino M, Vernick J, Fong KLL, Feuerstein G. Role of complement/platelet-activating factor-induced lung injury. J Immunol 1992; 5:1744–1750.
131. Rabinovici R, Bugelski PJ, Esser KM. Hillegass LM. Griswold DE, Vernick J, Feuerstein G. Tumor necrosis factor-alpha mediates endotoxin-induced lung injury in platelet-activating factor-primed rates. J Pharmacol Exp Ther 1993; 3:1550–1557.
132. Pignol B, Henane S, Sorlin B, Rola-Pleszczynski M, Mencia-Huerta JM, Braquet P. Effect of long term treatment with platelet-activating factor on IL-1 and IL-2 production by rat spleen cells. J Immunol 1990; 145:980–984.
133. Poubelle PE, Gingras D, Demers C, Dubois C, Harbour D, Grassi J, Rola-Pleszczynski M. Platelet-activating factor (PAF-acether) enhances the recombinant production of tumor necrosis factor-alpha and interleukin-1 by subsets of human monocytes. Immunology 1991; 72:181–187.
134. Bussolino F, Arese M, Silvestro L, Soldi R, Benfenati E, Sanavio F, Aglietta M, Bosia A, Camussi G. Involvement of a serine protease in synthesis of platelet-activating factor by endothelial cells stimulated by tumor necrosis factor-alpha or interleukin-1 alpha. Eur J Immunol 1994; 24:3131–3139.
135. Gutierrez S, Palacios I, Egido J, Zarco P, Miguelez R, Gonzalez E, Herrero-Beaumont G. IL-1 beta and IL-6 stimulate the production of platelet-activating factor (PAF) by cultured rabbit synovial cells. Clin Exp Immunol 1995; 99:364–368.
136. Warren JS. Relationship between interleukin-1 beta and platelet-activating factor in the pathogenesis of acute immune complex alveolitis in the rat. Am J Pathol 1992; 141:551–560.
137. Nourshargh S, Larkin SW, Das A, Williams TJ. Interleukin-1-induced leukocyte extravasation across rat mesenteric microvessels is mediated by platelet-activating factor. Blood 1995; 85:2553–2558.
138. Rabinovici R, Sofronski M, Renz JF, Hillegas LM, Esser KM. Vernick J, Feuerstein G. Platelet-activating factor mediates interleukin-2-induced lung injury in the rat. J Clin Invest 1992; 89:1669–1673.
139. Ramirez MM, Quardt SM, Kim D, Oshiro H, Minnicozzi M, Duran WN. Platelet activating factor modulates microvascular permeability through nitric oxide synthesis. Microvasc Res 1995; 50:223–234.
140. Filep JG, Foldes-Filep E. Modulation by nitric oxide of platelet-activating-factor-induced albumin extravasation in the conscious rat. Br J Pharmacol 1993; 110:1347–1352.
141. Juncos LA, Ren YL, Arima S, Ito S. Vasodilator and constrictor actions of platelet-activating factor in the isolated microperfused afferent arteriole of the rabbit kidney. Role of endothelium-derived relaxing factor/nitric oxide and cyclooxygenase products. J Clin Invest 1993; 91:1374–1379.
142. MacKendrick W, Caplan M, Hsueh W. Endogenous nitric oxide protects against platelet-activating factor-induced bowel injury in the rat. Pediatr Res 1993; 34:222–228.
143. Albertini M, Clement MG. In pigs, inhaled nitric oxide (NO) counterbalances PAF-induced pulmonary hypertension. Prostaglandins Leuokot Essent Fatty Acids 1994; 51:357–362.
144. Caplan MS, Hedlund E, Hill N, MacKendrick W. The role of endogenous nitric oxide and platelet-activating factor in hypoxia-induced intestinal injury in rats. Gastroenterology 1994; 106:346–352.
145. Postel J, Schloerb PR, Furtado D. Pathophysiologic alterations during bacterial infusions for the study of bacteremic shock. Surgery 1977; 141:683–689.
146. Horton JW. Hemodynamic and regional blood flow responses of neonatal and adult dogs to endotoxin challenge. Circ Shock 1993; 41:26–34.
147. Parker MM, Shelhamer JH, Bachrach SL. Profound but reversible myocardial depression in patients with septic shock. Ann Intern Med 1987; 100:483–490.

37

Interleukin-10 and Other Suppressive Cytokines

Arnaud Marchant and Michel Goldman
Erasme Hospital, Free University of Brussels, Brussels, Belgium

Tom van der Poll
University of Amsterdam, Amsterdam, The Netherlands

INTRODUCTION

Most of the toxic effects of lipopolysaccharide (LPS, endotoxin) are related to the release of proinflammatory cytokines such as tumor necrosis factor (TNF), interleukin-1 (IL-1), IL-12, and interferon-γ (IFN-γ) by immune cells, leading to tissue damage and organ failure. These proinflammatory properties of LPS are described in detail in other chapters of this book. Recently, it has become apparent that together with the release of proinflammatory mediators, LPS also triggers the production of antiinflammatory cytokines controlling immune cells activation and reducing LPS toxicity. This group includes a growing number of cytokines, the best characterized being interleukin-10 (IL-10). In this chapter, we will first summarize the data showing that IL-10 is a potent monocyte/macrophage deactivator controlling LPS toxicity in a variety of experimental settings. We will then discuss the relevance of these findings to the situation of human septic shock. Finally, we will describe other members of the group of antiinflammatory cytokines suppressing the responses to endotoxin.

INTERLEUKIN-10

Interleukin-10 Molecular Characteristics and Gene Expression

Interleukin-10 (IL-10) is a 18 kDa protein containing two intramolecular disulfide bonds. It is acid-labile and appears in soluble form as a homodimer. The gene encoding IL-10 is located on chromosome 1 and contains several noncoding sequences, which are thought to control its transcription and the stability of the corresponding mRNA (1). Cloning of mouse and human IL-10 revealed extensive homology between both molecules, which share common sequences with an open-reading frame product of the Epstein-Barr virus, now considered as the viral form of IL-10 (1,2).

IL-10 is produced by a variety of cell types including monocytes, macrophages, and T and B lymphocytes (3–6). At the level of monocyte/macrophage, the production of IL-10 is induced by microbial products and pathogens including LPS, intracellular parasites, fungi, and viruses (4,7–9).

A number of monocyte/macrophage-derived cytokines can also promote the production of IL-10 by acting directly at the level of either the monocyte/macrophage or the T lymphocyte. TNF has been shown to stimulate the production of IL-10 by human monocytes, an effect that could be dependent on NF-κB activation (1,10–12). The role of TNF in the induction of IL-10 production by endotoxin is controversial. In some experiments, the release of IL-10 during experimental endotoxemia was shown to be TNF-dependent, but this has not been a constant finding (13–15). The reason for this difference may be related to the cell source of IL-10. Indeed, TNF appears to be involved in the production of IL-10 by LPS-stimulated human monocytes (10,16) but not by activated mouse macrophages (14). As will be discussed below, TNF and IL-10 production can be differentially regulated during experimental endotoxemia indicating that TNF can promote but is not required for the production of IL-10 induced by LPS. Other monocyte/macrophage-derived cytokines stimulating the production of IL-10 by monocyte/macrophage include IL-1, transforming growth factor-β (TGF-β) and IFN-α (17–19). After LPS stimulation, monocyte/macrophage cell lines also produce cytokines that will promote IL-10 production by T lymphocytes, including IL-6, IL-12, and IFN-α (11,19–21).

The production of IL-10 by the monocyte/macrophage can be suppressed by cytokines. IFN-γ was shown to be a potent inhibitor of IL-10 production, and IL-10 controls its own gene expression in human monocytes (22,23). Two other deactivating cytokines that will be described below, IL-4 and IL-13, inhibit IL-10 production by LPS-activated monocytes, although this effect appears to be highly dependent on experimental conditions. Pre- or co-incubation of monocytes in the presence of IL-4 or IL-13 generates opposite results (4,24,25).

Several reports have emphasized the important role of cytoplasmic cAMP levels in the regulation of IL-10 gene expression. Indeed, cAMP-inducing agents including PGE_2, pentoxifylline, and epinephrine promote IL-10 synthesis by macrophages (16,26). The use of specific inhibitors has shown that the cyclic nucleotide phosphodiesterase type IV participates in the regulation of IL-10 production (26). The ability of cAMP-inducing agents to stimulate the production of IL-10 contrasts with their inhibitory effect on TNF production and indicates that IL-10 and TNF production is differentially regulated at the level of the macrophage. Other inhibitors of TNF synthesis that have been shown to promote LPS-induced IL-10 production either in vivo or in vitro at the level of the macrophage include steroids, chlorpromazine, sigma ligand, and thalidomide (14,27–29). The role of the upregulation of IL-10 production in the antiinflammatory properties of these compounds remains to be established.

Effects of Interleukin-10 on Endotoxin Responses In Vitro

IL-10 Is a Monocyte/Macrophage Deactivating Cytokine

IL-10 was first described as a cytokine produced by murine Th2 lymphocytes that inhibits cytokine production by Th1 cells (30). The effects of IL-10 on T lymphocytes will not be discussed here and have been described in detail elsewhere (31). One of the major biological properties of IL-10 is to deactivate monocytes and macrophages. In vitro, IL-10 is a very potent inhibitor of cytokine production (TNF, IL-1α and IL-1β, IL-6, IL-8, IL-12, and GM-CSF) by LPS-activated monocytes and macrophages by acting both at the transcriptional and posttranscriptional levels (4,5,23,32–34). Interestingly, IL-10 does not inhibit the production of all monocyte-derived cytokines as it stimulates the release of MCP-1 and IL-15 (35,36). Further studies are needed to establish the biological consequences of these differential regulations.

In addition to its effects on cytokines, IL-10 also modulates a number of other activities induced by endotoxin at the level of the monocyte/macrophage. IL-10 inhibits the tissue factor–dependent procoagulant activity induced by LPS on human monocytes (37). IL-10 also inhibits the release of free radicals and nitric oxide and thereby suppress the antimicrobial properties of macrophages against various pathogens (38). IL-10 is an inhibitor of the synthesis of arachidonic acid products and cyclooxygenase type 2 by LPS-stimulated monocytes (39). Finally, IL-10 also suppress the synthesis of metalloproteinases by macrophages, whereas it stimulates the release of the metalloproteinase tissue inhibitor type 1 (40).

As mentioned above, endotoxin as well as a number of cytokines produced by LPS-activated monocyte/macrophage cell lines induce the synthesis of IL-10. In vitro neutralization of endogenous IL-10 was shown to potentiate the release of proinflammatory cytokines by LPS-activated monocytes (4). These data indicate that the production of endogenous IL-10 represents an autoregulatory mechanism controlling monocyte/macrophage activation by endotoxin.

Effects of IL-10 on Neutrophils

In addition to its effects on monocytes and macrophages, IL-10 also affects the function of neutrophils. IL-10 inhibits the production of inflammatory cytokines including TNF, IL-1α and IL-1β, IL-8, and IL-12 by LPS-activated neutrophils (41).

IL-10 and the Release of Anti-Inflammatory Mediators

Interestingly, while decreasing the proinflammatory properties of monocytes/macrophages and granulocytes, IL-10 promotes the release of antiinflammatory mediators by these cells. Indeed, IL-10 does not affect the production of the antiinflammatory cytokine TGF-β by LPS-activated monocytes and stimulates the release of a natural antagonist of IL-1, IL-1Ra, by monocytes and granulocytes (4,42). Some data suggest that IL-10 also decreases the expression of membrane TNF receptor by inducing the release of its soluble form, but this issue appears to be controversial (43,44).

IL-10 Receptor and Intracellular Signaling

IL-10 recognizes a 90–110 kD membrane receptor that has been cloned (45). The human IL-10 receptor shows 60% homology with the amino acid sequence of its murine homolog. As for other cytokines, binding of IL-10 to its receptor activates the Jak tyrosine kinase pathway. At the level of monocytes and T lymphocytes, IL-10 stimulates the phosphorylation of Jak1 and Tyk2 and activates the "signal transducers and activators of transcription" STAT1 and STAT3 (46). This pattern of activation shares similarities with the one triggered by IFN-α, which stimulates Jak1, Tyk2, STAT-1, STAT-2, and STAT-3 and is probably involved in the activation of specific genes such as type 1 Fc receptor for IgG by IL-10 (47).

In contrast, the mechanisms involved in the monocyte/macrophage deactivating properties of IL-10 are not yet clear. IL-10 was shown to interfere with very early events of monocyte and neutrophil activation by LPS, inhibiting the phosphorylation of tyrosine kinases like p53/56$_{lyn}$ (48,49). More recently, IL-10 was shown to activate p70S6 and phosphatidylinositol 3-kinases in monocytes, but this effect does not appear to be involved in the deactivating properties of the cytokine (50).

Conclusion

Taken together, these data indicate that IL-10 has very potent anti-inflammatory properties in vitro, suppressing the production of a large number of proinflammatory mediators and promoting the release of other anti-inflammatory mediators. It is interesting to note that IL-10 does not inhibit the production of all mediators involved in an inflammatory response. IL-10 increases the release of some chemokines by monocytes/macrophages and endothelial cells (35,51) and could thereby promote cell recruitment at the site of inflammation (52).

Effects of Interleukin-10 on LPS Responses In Vivo

LPS-Induced Cytokine Release and LPS Toxicity In Vivo

The macrophage-deactivating properties of IL-10 led several investigators to examine whether this cytokine is able to modulate the inflammatory response induced by LPS in vivo. Administration of pharmacological doses of recombinant IL-10 (rIL-10) was shown to markedly suppress the release of TNF and IFN-γ during murine endotoxemia (53–55). In parallel with the inhibition of proinflammatory cytokines, rIL-10 also prevented LPS-induced lethality (53,54). Interestingly, rIL-10 did not influence the serum levels of bioactive IL-6 in this model, suggesting that other cell types such as endothelial cells could be resistant to IL-10 and play an important role in the production of specific cytokines like IL-6 (51,56).

The anti-inflammatory properties of rIL-10 observed in mice were also found in models of endotoxemia in primates and human volunteers (57–61). In baboons, rIL-10 administration markedly inhibits the production of TNF and IL-12 and significantly affects the release of IL-8 (57). In this model, rIL-10 injection led to a significant reduction of IL-6 release. In parallel with the decreased production of inflammatory cytokines, rIL-10 also diminishes the degranulation of neutrophils, an effect that is probably dependent on the inhibition of TNF production (57). In healthy humans, recombinant human IL-10, given as a single dose of 25 μg/kg directly prior to endotoxin, reduced the rise in body temperature and in plasma TNF, IL-6, IL-8, and IL-1Ra concentrations. Endotoxin-induced granulocyte accumulation in the lungs, as determined by dynamic granuloscintigrams, was prevented by IL-10 treatment, while granulocyte degranulation was blunted (60). Further, in contrast to what has been observed in baboons, IL-10 also inhibited the activation of the fibrinolytic and the coagulation systems (57,61). In humans, adverse effects of rIL-10 injection included mild-to-moderate flu-like syndrome at the highest doses administered (59).

Cytokine Release and Lethality During Experimental Endotoxemia

Following LPS activation, monocytes produce IL-10 in vitro (4,14). We observed that LPS induces the rapid release of IL-10 in vivo in mice, in primates, and in human volunteers (13,14,55,62). Neutralization of this endogenously produced IL-10 resulted in an increased production of both TNF and IFN-γ after LPS injection (55,63). Together with the increased production of inflammatory cytokines, anti-IL-10–treated mice displayed an increased lethality, an effect that was shown to be dependent on the increased production of TNF (56,63). Similar data have been observed in mice chronically injected from birth with anti-IL-10 mAb and in IL-10 knock-out mice (IL-10ko) in which the

IL-10 gene has been specifically inactivated (64). IL-10ko mice were also shown to be extremely sensitive to the Shwartzman reaction induced by LPS (64). Interestingly, Standiford et al. observed that the protective role of endogenous IL-10 is also related to the control of MIP-2 production (63). Taken together, these data indicate that the endogenous production of IL-10 represents an important autoregulatory mechanism controlling the production of inflammatory cytokines and the toxicity of endotoxin in vivo.

Production and Role of Interleukin-10 During Severe Bacterial Infection

The protective role played by endogenous IL-10 during endotoxin shock in mice led a number of groups to evaluate the production of IL-10 in patients with severe bacterial infection (62,65–69). It was found that IL-10 is produced by patients with sepsis with and without septic shock, higher IL-10 plasma levels being found in patients with shock (65). We observed that patients with shock of nonseptic origin also produce IL-10, indicating that circulatory shock by itself could contribute to the release of IL-10 during septic shock (66). Similar levels of IL-10 were measured in patients infected with gram-positive and gram-negative organisms (65). Two reports have shown that meningococcal septic shock is associated with the transient release of very large amounts of IL-10 (67,69). Evidence indicates that the IL-10 protein detected in the plasma of patients with septic shock is bioactive. First, van Deuren et al. showed that monocytes from patients with meningococcal infection have a defect in LPS-induced TNF production, whereas the release of IL-1Ra is increased as compared to healthy volunteers (70). As IL-10 differentially regulates TNF and IL-1Ra production by LPS-stimulated monocytes, it could play a role in monocyte abnormalities found during meningococcal infection. Second, Brandtzaeg et al. demonstrated that IL-10 accounts for most of the monocyte-deactivating properties of the plasma obtained from patients with meningococcal septic shock (71). We did not observe any influence of ex vivo IL-10 neutralization on the deactivation state of monocytes obtained from septic shock patients (66). These data suggest that either IL-10–suppressive signals were already given to monocytes in vivo and that they cannot be inhibited by anti-IL-10–neutralizing antibodies ex vivo or that other mechanisms are involved in the monocyte deactivation (66). These possibilities will be discussed below in the discussion of the role of IL-10 in the induction of LPS tolerance.

In some studies, the plasma levels of IL-10 correlated with the levels of inflammatory mediators such as TNF and with the severity of shock (66,69). Various factors could explain this observation. First, in patients with disseminated infection, a large load of bacterial products would trigger the release of large amounts of both TNF and IL-10 by monocytes and macrophages. Second, TNF has been shown to stimulate the production of IL-10 by monocytes in vitro and in vivo in primates (10,13). Finally, the circulatory shock itself triggers the production of IL-10 independently of any infection (66). Experimental data obtained in animals indicate that in the absence of IL-10, the magnitude of the inflammatory response would be higher (55,63,64). In models of endotoxin shock, a stronger antiinflammatory response was shown to be deleterious (55,63,64).

Endotoxin shock is considered to be a good model of fulminant meningococcal septic shock in which the release of large amounts of bacterial products in the circulation triggers the rapid, massive, and transient production of cytokines. In the majority of patients with septic shock of nonmeningococcal origin, an infection that is primarily localized in an organ progressively leads to a systemic inflammatory response characterized by the presence of moderate amounts of cytokines for prolonged periods of time.

Several animal models have been proposed that are considered to be more representative of this clinical situation. One of these models is the peritonitis induced by the ligation and puncture of the cecum in mice. We observed that IL-10 is produced during septic peritonitis in mice and that this endogenously produced IL-10 controls the release of inflammatory cytokines and protects mice from lethality (72). Recent data reported by Walley et al. indicate that in this model the balance between IL-10 and TNF production is related to the severity of sepsis, the highest IL-10 levels being found in mice with the less severe peritonitis (73).

In contrast, in mice with septic pneumonia caused by infection with *Streptococcus pneumoniae* or *Klebsiella*, the production of endogenous IL-10 is responsible for a decreased survival (74,75). This reduced survival is probably the consequence of the suppression of TNF production that leads to poor control of bacterial proliferation. It appears, therefore, that endogenous IL-10 is a potent antiinflammatory cytokine controlling the production of inflammatory cytokines by monomacrophages when these cells are activated with endotoxin or other bacterial products. This effect will protect the host against excessive inflammatory reactions but can decrease his ability to control bacterial proliferation. The balance between these two phenomena will determine whether endogenously produced IL-10 plays a protective or a deleterious role in patients with septic shock. Döcke et al. recently suggested that treatment with recombinant IFN-γ could be beneficial to patients with septic shock by improving the functions of their monocytes (76).

Role of Interleukin-10 in the Induction of LPS Tolerance

Monocyte deactivation observed in patients with septic shock is analogous to a phenomenon called LPS tolerance. Because another chapter of this book describes LPS tolerance in detail, we will only discuss the potential role of IL-10 in the induction of the phenomenon. LPS tolerance is a refractory state that follows the initial phase of monocyte/macrophage activation and is characterized by a reduced ability of the cells to produce TNF in response to a secondary LPS challenge. The production of TNF in LPS-tolerant macrophages is suppressed at multiple levels in the TNF gene expression pathway including transcription and translation (77–79).

As IL-10 affects TNF gene expression at multiple levels as well, it is conceivable that this cytokine plays a role in the induction of LPS tolerance. LPS tolerance can be induced in vivo in experimental animals or in human volunteers and in vitro at the level of isolated monocyte/macrophage. Randow et al. reported that the LPS tolerant monocytes in vitro display a defect in IL-10 production, but that IL-10 is required, in conjunction with TGF-β, in the induction of the tolerance phenomenon (80). Opposite results were reported by Frankenberger et al. (81). They showed that a monocytic cell line rendered tolerant to LPS produces increased amounts of IL-10 but that the induction of the state of tolerance in this experimental system does not require IL-10. Taken together, these data suggest that LPS tolerance in vitro can be induced in the absence of IL-10 but that this cytokine could play a role in some experimental conditions.

The role of IL-10 in the induction of LPS tolerance in vivo has been studied by Berg et al. (64). These authors reported that LPS tolerance in vivo is not dependent on IL-10 as it can be induced in IL-10ko mice. The induction of LPS tolerance in vivo in mice has been shown to involve the production of glucocorticoids (82).

OTHER SUPPRESSIVE CYTOKINES

Interleukin-6

IL-6 is a cytokine produced by a variety of cell types including monocytes/macrophages, neutrophils, and endothelial cells in response to LPS stimulation. The role of IL-6 in endotoxin-induced inflammation appears to be limited. IL-6 is a potent inducer of the acute phase reaction. Recently, IL-6 was shown to play a central role in the activation of the coagulation cascade during experimental endotoxemia (83). Several reports indicate that IL-6 also has anti-inflammatory properties. In vitro, IL-6 inhibits TNF and IL-1 release by LPS-activated blood mononuclear cells (84). Data published more recently indicated that IL-6 could influence monocyte/macrophage functions by stimulating the production of suppressive cytokines, including IL-4 and IL-10, by T lymphocytes (11,85).

These anti-inflammatory properties of IL-6 observed in vitro are also apparent in vivo as IL-6 inhibits the production of TNF and IL-1 during murine endotoxemia (86). IL-6 gene-deficient (IL-6ko) mice have similar mortality rates after systemic endotoxin challenge when compared with wild-type mice (87), and inflammatory changes elicited by endotoxin, such as corticosterone release, hypoglycemia, anorexia and weight loss, were also not altered in IL-6–deficient mice (88). IL-6ko mice have elevated TNF concentrations after injection of endotoxin, confirming the inhibitory effect of IL-6 on endotoxin-induced TNF production (88). IL-6 may be a more important mediator during infection with live bacteria than anticipated from endotoxin studies: IL-6ko mice are more susceptible to infection with *Listeria monocytogenes*, *S. pneumoniae*, and *Escherichia coli* (87,89–91).

IL-4 and IL-13

IL-4 and IL-13 are structurally and functionally related cytokines produced by T lymphocytes, which share a number of biological activities including the inhibition of cytokine production by blood mononuclear cells (24). This effect appears to be more complex than in the case of IL-10 and depends upon the experimental conditions. Indeed, TNF production is suppressed when monocytes are coincubated in the presence of LPS and IL-4 or IL-13 but is increased when cells are preincubated in the presence of the cytokines (24,25). A similar phenomenon has been described for IL-10; IL-13 and IL-4 decrease IL-10 production or increase IL-10 production by mouse macrophages after coincubation or preincubation (25). Data reported by Muchamuel et al. indicate that IL-13 inhibits TNF production and protects mice from LPS-induced lethality (92). The role of IL-4 and IL-13 in the biology of LPS and during severe bacterial infections is not yet established. Several authors have failed to demonstrate increased levels of IL-4 or IL-13 in the circulation of LPS-challenged human volunteers or in patients with septic shock, but these data do not exclude local production of the cytokines in affected organs (62,71).

TGF-β

TGF-β is a cytokine produced by a variety of cell types including LPS-activated monocytes and macrophages. Because TGF-β is an inhibitor of TNF and IL-1β production, it is considered to be another autoregulatory mechanism controlling monocyte/macrophage activation by endotoxin (93). The effect of TGF-β on macrophages appears to be complex, because this cytokine stimulates the production of TNF and IL-1 by resting cells (93). The anti-inflammatory properties of TGF-β could be enhanced by its ability to increase the production of IL-10 (18). As mentioned earlier, TGF-β could play a role in the induction of LPS tolerance in vitro (80). The role of TGF-β in the biology of LPS in vivo and during several bacterial infections is not yet well characterized.

IL-11

IL-11 is an hemopoietic cytokine sharing the gp130 receptor subunit with IL-6. Recently, Trepicchio et al. reported data showing that IL-11 inhibits the production of TNF and IL-1β by mouse macrophages (94). This effect observed in vitro was also apparent in vivo as they showed that IL-11 inhibits the production of these cytokines during experimental endotoxemia in mice. The role of IL-11 in the biology of LPS in vivo and during several bacterial infections has not yet been studied.

Secretory Leukocyte Protease Inhibitor

The secretory leukocyte protease inhibitor (SLPI) is not considered to be a cytokine but is known as an epithelial inhibitor of leukocyte serine proteases. Data recently reported by Jin et al. indicate that SLPI is also a potent inhibitor of LPS-induced NF-κB activation and TNF production by macrophages (95). They showed that macrophages from mice genetically resistant to LPS produce high amounts of SLPI and that transfection of SLPI into normal macrophages suppressed their ability to respond to LPS. Further studies are needed to establish the role of SLPI in the biology of endotoxin.

CONCLUDING REMARKS

Activation of the immune system by bacterial products, including endotoxin, is a very controlled phenomenon in which both inflammatory and anti-inflammatory cytokines are involved. The production of large amounts of inflammatory cytokines, of which TNF is the prototype, can lead to tissue damage, organ failure, and death. In order to prevent these excessive inflammatory responses, the immune system develops protective mechanisms controlling its own activation. Among these mechanisms, the production of anti-inflammatory cytokines such as IL-10 plays a central role. In the absence of IL-10, a dose of endotoxin that would otherwise be safe becomes lethal. If the anti-inflammatory response is clearly beneficial during a reaction that is mainly inflammatory, the situation can be very different when the host has to control an invasive bacterial infection. In this situation, the production of inflammatory cytokines like TNF is central for host defenses and its suppression by anti-inflammatory mediators can be deleterious.

ACKNOWLEDGMENTS

This work has been supported by grants from the Fonds de la Recherche Scientifique Médicale (Belgium), the Biotech Programme of the European Commission (contract No. BI02-CT92-0316), the Communauté Francaise de Belgique (Action de Recherche Concertée), and the Fondation Docteur André Loicq. Tom van der Poll is a fellow of the Royal Dutch Academy of Arts and Sciences.

REFERENCES

1. Kim JM, Brannan CI, Copeland NG, Jenkins NA, Khan TA, Moore KW. Structure of the mouse IL-10 gene and chromosomal localization of the mouse and human genes. J Immunol 1992; 148:3618–3623.
2. Moore KW, Vieira P, Fiorentino DF, Trounstine ML, Khan TA, Mosmann TR. Homology of cytokine synthesis inhibitory factor (IL-10) to the Epstein-Barr virus gene BCRFI [published erratum appears in Science 1990 Oct 26;250(4980):494]. Science 1990; 248:1230–1234.

3. Yssel H, De Waal Malefyt R, Roncarolo MG, et al. IL-10 is produced by subsets of human CD4+ T cell clones and peripheral blood T cells. J Immunol 1992; 149:2378–2384.
4. De Waal Malefyt R, Abrams J, Bennett B, Figdor CG, de Vries JE. Interleukin 10(IL-10) inhibits cytokine synthesis by human monocytes: an autoregulatory role of IL-10 produced by monocytes. J Exp Med 1991; 174:1209–1220.
5. Fiorentino DF, Zlotnik A, Mosmann TR, Howard M, O'Garra A. IL-10 inhibits cytokine production by activated macrophages. J Immunol 1991; 147:3815–3822.
6. O'Garra A, Stapleton G, Dhar V, et al. Production of cytokines by mouse B cells: B lymphomas and normal B cells produce interleukin 10. Int Immunol 1990; 2:821–832.
7. Sieling PA, Abrams JS, Yamamura M, et al. Immunosuppressive roles for IL-10 and IL-4 in human infection. In vitro modulation of T cell responses in leprosy. J Immunol 1993; 150:5501–5510.
8. Vecchiarelli A, Retini C, Monari C, Tascini C, Bistoni F, Kozel TR. Purified capsular polysaccharide of Cryptococcus neoformans induces interleukin-10 secretion by human monocytes. Infect Immun 1996; 64:2846–2849.
9. Borghi P, Fantuzzi L, Varano B, et al. Induction of interleukin-10 by human immunodeficiency virus type 1 and its gp120 protein in human monocytes macrophages. J Virol 1995; 69:1284–1287.
10. Wanidworanun C, Strober W. Predominant role of tumor necrosis factor-alpha in human monocyte IL-10 synthesis. J Immunol 1993; 151:6853–6861.
11. Daftarian PM, Kumar A, Kryworuchko M, Diazmitoma F. IL-10 production is enhanced in human T cells by IL-12 and IL-6 and in monocytes by tumor necrosis factor-alpha. J Immunol 1996; 157:12–20.
12. Mori N, Gill PS, Mougdil T, Murakami S, Eto S, Prager D. Interleukin-10 gene expression in adult T-cell leukemia. Blood 1996; 88:1035–1045.
13. van der Poll T, Jansen J, Levi M, ten Cate H, ten Cate JW, van Deventer SJ. Regulation of interleukin 10 release by tumor necrosis factor in humans and chimpanzees. J Exp Med 1994; 180:1985–1988.
14. Marchant A, Amraoui Z, Gueydan C, et al. Methylprednisolone differentially regulates IL-10 and tumor necrosis factor (TNF) production during murine endotoxaemia. Clin Exp Immunol 1996; 106:91–96.
15. Barsig J, Küsters S, Vogt K, Tiegs G, Wendel A. Lipopolysaccharide-induced interleukin-10 in mice: role of endogenous tumor necrosis factor-α. Eur J Immunol 1995; 25:2888–2893.
16. van der Poll T, Coyle SM, Barbosa K, Braxton CC, Lowry SF. Epinephrine inhibits tumor necrosis factor-alpha and potentiates interleukin 10 production during human endotoxemia. J Clin Invest 1996; 97:713–719.
17. Tilg H, Atkins MB, Dinarello CA, Mier JW. Induction of circulating interleukin 10 by interleukin 1 and interleukin 2, but not interleukin 6 immunotherapy. Cytokine 1995; 7:734–739.
18. Maeda H, Kuwahara H, Ichimura Y, Ohtsuki M, Kurakata S, Shiraishi A. TGF-beta enhances macrophage ability to produce IL-10 in normal and tumor-bearing mice. J Immunol 1995; 155:4926–4932.
19. Aman MJ, Tretter T, Eisenbeis I, et al. Interferon-alpha stimulates the production of interleukin-10 in activated CD4(+) T cells and monocytes. Blood 1996; 87:4731–4736.
20. Meyaard L, Hovenkamp E, Otto SA, Miedema F. IL-12-induced IL-10 production by human T cells as a negative feedback for IL-12-induced immune responses. J Immunol 1996; 156:2776–2782.
21. Schandene L, Del Prete G, Cogan E, et al. Recombinant interferon-alpha selectively inhibits the production of interleukin-5 by human CD4(+) T cells. J Clin Invest 1996; 97:309–315.
22. Chomarat P, Rissoan MC, Banchereau J, Miossec P. Interferon gamma inhibits interleukin 10 production by monocytes. J Exp Med 1993; 177:523–527.
23. Brown CY, Lagnado CA, Vadas MA, Goodall GJ. Differential regulation of the stability of cytokine mRNAs in lipopolysaccharide-activated blood monocytes in response to interleukin-10. J Biol Chem 1996; 271:20108–20112.
24. D'Andrea A, Ma X, Aste-Amezaga M, Paganin C, Trinchieri G. Stimulatory and inhibitory effects of interleukin (IL)-4 and IL-13 on the production of cytokines by human peripheral blood mononuclear cells: priming for IL-12 and tumor necrosis factor production. J Exp Med 1995; 181:537–546.
25. Kambayashi T, Jacob CO, Strassmann G. IL-4 and IL-13 modulate IL-10 release in endotoxin-stimulated murine peritoneal mononuclear phagocytes. Cell Immunol 1996; 171:153–158.
26. Kambayashi T, Jacob CO, Zhou D, Mazurek N, Fong M, Strassmann G. Cyclic nucleotide phosphodiesterase type IV participates in the regulation of IL-10 and in the subsequent inhibition of TNF-alpha and IL-6 release by endotoxin-stimulated macrophages. J Immunol 1995; 155:4909–4916.
27. Mengozzi M, Fantuzzi G, Faggioni R, et al. Chlorpromazine specifically inhibits peripheral and brain TNF production, and up-regulates IL-10 production, in mice. Immunology 1994; 82:207–210.
28. Bourrie B, Bouaboula M, Benoit JM, et al. Enhancement of endotoxin-induced interleukin-10 production by SR 31747A, a sigma ligand. Eur J Immunol 1995; 25:2882–2887.
29. Corral LG, Muller GW, Moreira AL, et al. Selection of novel analogs of thalidomide with enhanced tumor necrosis factor alpha inhibitory activity. Mol Med 1996; 2:506–515.
30. Fiorentino DF, Bond MW, Mosmann TR. Two types of mouse T helper cell. IV. Th2 clones secrete a factor that inhibits cytokine production by Th1 clones. J Exp Med 1989; 170:2081–2095.
31. Moore KW, O'Garra A, De Waal Malefyt R, Vieira P, Mosmann TR. Interleukin-10. Annu Rev Immunol 1993; 11:165–190.
32. Bogdan C, Paik J, Vodovotz Y, Nathan C. Contrasting mechanisms for suppression of macrophage cytokine

release by transforming growth factor-beta and interleukin-10. J Biol Chem 1992; 267:23301–23308.
33. D'Andrea A, Aste Amezaga M, Valiante NM, Ma X, Kubin M, Trinchieri G. Interleukin 10 (IL-10) inhibits human lymphocyte interferon gamma-production by suppressing natural killer cell stimulatory factor/IL-12 synthesis in accessory cells. J Exp Med 1993; 178: 1041–1048.
34. Wang P, Wu P, Siegel MI, Egan RW, Billah MM. Interleukin (IL)-10 inhibits nuclear factor kappa B (NF kappa B) activation in human monocytes. IL-10 and IL-4 suppress cytokine synthesis by different mechanisms. J Biol Chem 1995; 270:9558–9563.
35. Yano S, Yanagawa H, Nishioka Y, Mukaida N, Matsushima K, Sone S. T helper 2 cytokines differently regulate monocyte chemoattractant protein-1 production by human peripheral blood monocytes and alveolar macrophages. J Immunol 1996; 157:2660–2665.
36. Doherty TM, Seder RA, Sher A. Induction and regulation of IL-15 expression in murine macrophages. J Immunol 1996; 156:735–741.
37. Pradier O, Gerard C, Delvaux A, et al. Interleukin-10 inhibits the induction of monocyte procoagulant activity by bacterial lipopolysaccharide. Eur J Immunol 1993; 23:2700–2703.
38. Gazzinelli RT, Oswald IP, James SL, Sher A. IL-10 inhibits parasite killing and nitrogen oxide production by IFN-γ activated macrophages. J Immunol 1992; 148:1792–1796.
39. Endo T, Ogushi F, Sone S. LPS-dependent cyclooxygenase-2 induction in human monocytes is down-regulated by IL-13, but not by IFN-gamma. J Immunol 1996; 156:2240–2246.
40. Lacraz S, Nicod LP, Chicheportiche R, Welgus HG, Dayer JM. IL-10 inhibits metalloproteinase and stimulates TIMP-1 production in human mononuclear phagocytes. J Clin Invest 1995; 96:2304–2310.
41. Cassatella MA, Meda L, Gasperini S, D'Andrea A, Ma X, Trinchieri G. Interleukin-12 production by human polymorphonuclear leukocytes. Eur J Immunol 1995; 25:1–5.
42. Cassatella MA, Meda L, Gasperini S, Calzetti F, Bonora S. Interleukin 10 (IL-10) upregulates IL-1 receptor antagonist production from lipopolysaccharide-stimulated human polymorphonuclear leukocytes by delaying mRNA degradation. J Exp Med 1994; 179:1695–1699.
43. Joyce DA, Gibbons DP, Green P, Steer JH, Feldmann M, Brennan FM. Two inhibitors of pro-inflammatory cytokine release, interleukin-10 and interleukin-4, have contrasting effects on release of soluble p75 tumor necrosis factor receptor by cultured monocytes. Eur J Immunol 1994; 24:2699–2705.
44. Hart PH, Hunt EK, Bonder CS, Watson CJ, Finlay-Jones JJ. Regulation of surface and soluble TNF receptor expression on human monocytes and synovial fluid macrophages by IL-4 and IL-10. J Immunol 1996; 157: 3672–3680.
45. Liu Y, Wei SH, Ho AS, De Waal Malefyt R, Moore KW. Expression cloning and characterization of a human IL-10 receptor. J Immunol 1994; 152:1821–1829.
46. Finbloom DS, Winestock KD. IL-10 induces the tyrosine phosphorylation of tyk2 and jak1 and the differential assembly of STAT1 and STAT3 complexes in human T cells and monocytes. J Immunol 1995; 155: 1079–1090.
47. Lehmann J, Seegert D, Strehlow I, Schindler C, Lohmann Matthes ML, Decker T. IL-10-induced factors belonging to the p91 family of proteins bind to IFN-gamma-responsive promoter elements. J Immunol 1994; 153:165–172.
48. Geng Y, Gulbins E, Altman A, Lotz M. Monocyte deactivation by interleukin 10 via inhibition of tyrosine kinase activity and the Ras signaling pathway. Proc Natl Acad Sci USA 1994; 91:8602–8606.
49. Gasperini S, Donini M, Dusi S, Cassatella MA. Interleukin-10 decreases tyrosine phosphorylation of discrete lipopolysaccharide-induced phosphoproteins in human granulocytes. Biochem Biophys Res Commun 1995; 209:87–94.
50. Crawley JB, Williams LM, Mander T, Brennan FM, Foxwell BMJ. Interleukin-10 stimulation of phosphatidylinositol 3-kinase and p70 S6 kinase is required for the proliferative but not the antiinflammatory effects of the cytokine. J Biol Chem 1996; 271:16357–16362.
51. Mantovani A, Busolino F, Introna M. Cytokine regulation of endothelial cell function: from molecular level to the bedside. Immunol Today 1997; 18:231–240.
52. Wogensen L, Lee MS, Sarvetnick N. Production of interleukin-10 by islet cells accelerates immune-mediated destruction of beta-cells in nonobese diabetic mice. J Exp Med 1994; 179:1379–1384.
53. Gérard C, Bruyns C, Marchant A, et al. Interleukin 10 reduces the release of tumor necrosis factor and prevents lethality in experimental endotoxemia. J Exp Med 1993; 177:547–550.
54. Howard M, Muchamuel T, Andrade S, Menon S. Interleukin 10 protects mice from lethal endotoxemia. J Exp Med 1993; 177:1205–1208.
55. Marchant A, Bruyns C, Vandenabeele P, et al. Interleukin-10 controls interferon-gamma and tumor necrosis factor production during experimental endotoxemia. Eur J Immunol 1994; 24:1167–1171.
56. Marchant A, Vincent JL, Goldman M. Interleukin-10 as a protective cytokine produced during sepsis. In: Morrison DC, Ryan JL, eds. Novel Therapeutic Strategies in the Treatment of Sepsis. New York: Marcel Dekker, 1996:301–311.
57. van der Poll T, Jansen PM, Montegut WJ, et al. Effects of IL-10 on systemic inflammatory responses during sublethal primate endotoxemia. J Immunol 1997; 158: 1971–1975.
58. Chernoff AE, Granowitz EV, Shapiro L, et al. A randomized, controlled trial of IL-10 in humans. Inhibition of inflammatory cytokine production and immune responses. J Immunol 1995; 154:5492–5499.
59. Huhn RD, Radwanski E, O'Connell SM, et al. Pharmacokinetics and immunomodulatory properties of intravenously administered recombinant human interleukin-10 in healthy volunteers. Blood 1996; 87:699–705.
60. Pajkrt D, Camoglio L, Tiel-van Buul MCM, et al. Attenuation of proinflammatory response by recombinant human IL-10 in human endotoxemia; the effect of timing of rhIL-10 administration. J Immunol 1997; 158: 3971–3977.

61. Pajkrt D, van der Poll, Levi M, et al. Interleukin-10 inhibits activation of coagulation and fibrinolysis during human endotoxemia. Blood 1997; 89:2701–2705.
62. van der Poll T, de Waal Malefyt R, Coyle SM, Lowry SF. Antiinflammatory cytokine responses during clinical sepsis and experimental endotoxemia: sequential measurements of plasma soluble interleukin (IL)-1 receptor type II, IL-10, and IL-13. J Infect Dis 1997; 175:118–122.
63. Standiford TJ, Strieter RM, Lukacs NW, Kunkel SL. Neutralization of IL-10 increases lethality in endotoxemia. Cooperative effects of macrophage inflammatory protein-2 and tumor necrosis factor. J Immunol 1995; 155:2222–2229.
64. Berg DJ, Kuhn R, Rajewsky K. Interleukin-10 is a central regulator of the response to LPS in murine models of endotoxin shock and the Shwartzman reaction but not in endotoxin tolerance. J Clin Invest 1995; 96:2339–2347.
65. Marchant A, Deviere J, Byl B, De Groote D, Vincent JL, Goldman M. Interleukin-10 production during septicaemia. Lancet 1994; 343:707–708.
66. Marchant A, Alegre ML, Hakim A, et al. Clinical and biological significance of interleukin-10 plasma levels in patients with septic shock. J Clin Immunol 1995; 15:265–272.
67. Derkx B, Marchant A, Goldman M, Bijlmer R, van Deventer S. High levels of interleukin-10 during the initial phase of fulminant meningococcal septic shock. J Infect Dis 1995; 171:229–232.
68. Gomez Jimenez J, Martin MC, Sauri R, et al. Interleukin-10 and the monocyte/macrophage-induced inflammatory response in septic shock. J Infect Dis 1995; 171:472–475.
69. Lehmann AK, Halstensen A, Sornes S, Rokke O, Waage A. High levels of interleukin 10 in serum are associated with fatality in meningococcal disease. Infect Immun 1995; 63:2109–2112.
70. van Deuren M, van der Ven Jongekrijg J, Demacker PN, et al. Differential expression of proinflammatory cytokines and their inhibitors during the course of meningococcal infections. J Infect Dis 1994; 169:157–161.
71. Brandtzaeg P, Osnes L, Ovstebo R, Joo GB, Westvik AB, Kierulf P. Net inflammatory capacity of human septic shock plasma evaluated by a monocyte-based target cell assay: identification of interleukin-10 as a major functional deactivator of human monocytes. J Exp Med 1996; 184:51–60.
72. van der Poll T, Marchant A, Buurman W, et al. Endogenous interleukin 10 protects mice from death during septic peritonitis. J Immunol 1995; 155:5397–5401.
73. Walley KR, Lukacs, Standiford TJ, Strieter RM, Kunkel SL. Balance of inflammatory cytokines related to severity and mortality of urine sepsis. Infect Immun 1996; 64:4733–4738.
74. van der Poll T, Marchant A, Keogh CV, Goldman M, Lowry SF. Interleukin-10 impairs host defense in murine pneumococcal pneumonia. J Infect Dis 1996; 174:994–1000.
75. Greenberger MJ, Strieter RM, Kunkel SL, Danforth JM, Goodman RE, Standiford TJ. Neutralization of IL-10 increases survival in a murine model of *Klebsiella pneumonia*. J Immunol 1995; 155:722–729.
76. Döcke WD, Randow F, Syrbe U, et al. Monocyte deactivation in septic patients: restoration by IFN-γ treatment. Nature Med 1997; 3:678–681.
77. Virca GD, Kim SY, Glaser KB, Ulevitch RJ. Lipopolysaccharide induces hyporesponsiveness to its own action in RAW 264.7 cells. J Biol Chem 1989; 264:21951–21956.
78. Ziegler Heitbrock HW, Wedel A, Schraut W, et al. Tolerance to lipopolysaccharide involves mobilization of nuclear factor kappa B with predominance of p50 homodimers. J Biol Chem 1994; 269:17001–17004.
79. Marchant A, Gueydan C, Houzet L, et al. Defective translation of TNF mRNA in LPS-tolerant macrophages. J Inflamm 1996; 46:114–123.
80. Randow F, Syrbe U, Meisel C, et al. Mechanism of endotoxin desensitization: involvement of interleukin 10 and transforming growth factor beta. J Exp Med 1995; 181:1887–92.
81. Frankenberger M, Pechumer H, Ziegler-Heitbrock HW. Interleukin-10 is upregulated in LPS tolerance. J Inflamm 1995; 45:56–63.
82. Evans GF, Zuckerman SH. Glucocorticoid-dependent and -independent mechanisms involved in lipopolysaccharide tolerance. Eur J Immunol 1991; 21:1973–1979.
83. van der Poll T, Levi M, Hack CE, et al. Elimination of interleukin 6 attenuates coagulation activation in experimental endotoxemia in chimpanzees. J Exp Med 1994; 179:1253–1259.
84. Schindler R, Mancilla J, Endres S, Ghorbani R, Clark SC, Dinarello CA. Correlations and interactions in the production of interleukin-6 (IL-6), IL-1, and tumor necrosis factor (TNF) in human blood mononuclear cells: IL-6 suppresses IL-1 and TNF. Blood 1990; 75:40–47.
85. Rincon M, Anguita J, Nakamura T, Fikrig E, Flavell RA. Interleukin (IL)-6 directs the differentiation of IL-4 producing CD4(+) T cells. J Exp Med 1997; 185:461–469.
86. Aderka D, Le JM, Vilcek J. IL-6 inhibits lipopolysaccharide-induced tumor necrosis factor production in cultured human monocytes, U937 cells, and in mice. J Immunol 1989; 143:3517–3523.
87. Dalrymple SA, Slattery R, Aud DM, Krishna M, Lucian LA, Murray R. Interleukin-6 is required for a protective immune response to systemic *Escherichia coli* infection. Infect Immun 1996; 64:3231–3235.
88. Fattori E, Cappelletti M, Costa P, et al. Defective inflammatory response in interleukin-6-deficient mice. J Exp Med 1994; 180:1243–1250.
89. Kopf M, Baumann H, Freer G, et al. Impaired immune and acute-phase responses in interleukin-6 deficient mice. Nature 1994; 368:339–342.
90. Dalrymple SA, Lucian LA, Slattery R, et al. Interleukin-6-deficient mice are highly susceptible to *Listeria monocytogenes* infection: correlation with inefficient neutrophilia. Infect Immun 1995; 63:2262–2268.
91. van der Poll T, Keogh CV, Guirao X, Buurman WA, Kopf M, Lowry SF. Interleukin 6 gene deficient mice are more susceptible to pneumococcal pneumonia. J Infect Dis. 1997; 176:439–444.

92. Muchamuel T, Menon S, Pisacane P, Howard MC, Cockayne DA. IL-13 protects mice from lipopolysaccharide-induced lethal endotoxemia. Correlation with down-modulation of TNF-α, IFN-γ, and IL-12 production. J Immunol 1997; 158:2898–2903.
93. Bogdan C, Nathan C. Modulation of macrophage function by transforming growth factor beta, interleukin-4, and interleukin-10. Ann New York Acad Sc 1993; 685:713–739.
94. Trepicchio WL, Bozza M, Pedneault G, Dorner AJ. Recombinant human IL-11 attenuates the inflammatory response through down-regulation of proinflammatory cytokine release and nitric oxide production. J Immunol 1996; 157:3627–3634.
95. Jin FY, Nathan C, Radzioch D, Ding A. Secretory leukocyte protease inhibitor: a macrophage product induced by and antagonistic to bacterial lipopolysaccharide. Cell 1997; 88:417–426.

38

Nitric Oxide as a Signaling Molecule in the Systemic Inflammatory Response to LPS

Michael J. Parmely
University of Kansas Medical Center, Kansas City, Kansas

INTRODUCTION

Inflammatory responses to endotoxic lipopolysaccharides (LPS) are characterized by the expression of a set of genes whose products include cytokines, enzymes, and cell adhesion molecules that interact in a complicated fashion to produce the syndrome of sepsis. In turn, the transcriptional activation of these specific pro-inflammatory genes is controlled by many of the secondary mediators of sepsis. Important examples are the reactive oxygen species (ROS) that regulate gene transcription through the induction of reductive/oxidative (redox) stress in cells. The study of a number of different biological systems has revealed that gene expression in both prokaryotes and eukaryotes is significantly affected by the redox state of cells, with agents that alter redox balance serving as signals for initiating or suppressing gene transcription. This chapter reviews some of the current models for redox control of gene expression, with special emphasis on nitric oxide (NO) as a signaling molecule. Because unusually large quantities of NO are produced in response to endotoxic LPS, NO is a potentially important molecule for regulating redox-dependent gene expression and inflammatory responses to endotoxin challenge. Evidence will be reviewed indicating that the process by which NO regulates biological responses often involves the S-nitrosylation of protein sensors as a critical intermediate step. This suggests a link between NO production, the posttranslational modification of proteins and altered gene expression that may prove to have substantial implications for pathogenic host responses to LPS and other conditions in which NO production is a significant feature.

ORIGIN OF NITRIC OXIDE IN ENDOTOXEMIA

Nitric oxide is a labile, diffusible, relatively hydrophobic molecule produced by perhaps all mammalian cells that serves a number of functions related to cell signaling, homeostatic regulation, and host defense. Large amounts of NO are produced in endotoxin-challenged animals in vivo and by a variety of cells exposed to bacterial LPS in vitro. This finding has led to an intense investigation of both the source of NO in endotoxemia and the potential role of reactive nitrogen species in the pathophysiological responses to endotoxic LPS.

The biosynthesis of NO is catalyzed by one of the three homologous isoforms of NO synthases (NOS) that are widely distributed among mammalian cells. Each enzyme contains heme-binding and P-450 reductase domains and catalyzes the formation of NO from molecular oxygen and the guanido nitrogen of L-arginine by two sequential oxygenase/reductase reactions (Fig. 1). Enzymatic activity requires the cofactors calmodulin, the flavins FMN and FAD, tetrahydrobiopterin, and heme. Reduced NADPH serves as a cosubstrate at each enzymatic step, and the final products, NO and L-citrulline, are produced in equal molar quantities. The three NOS isoforms have endured a variety

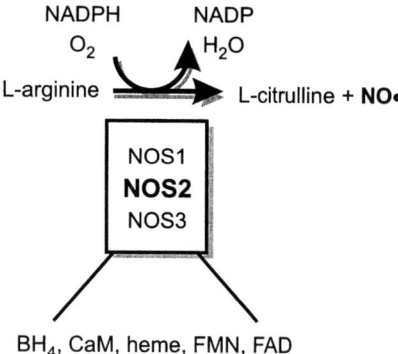

Fig. 1 Nitric oxide biosynthesis.

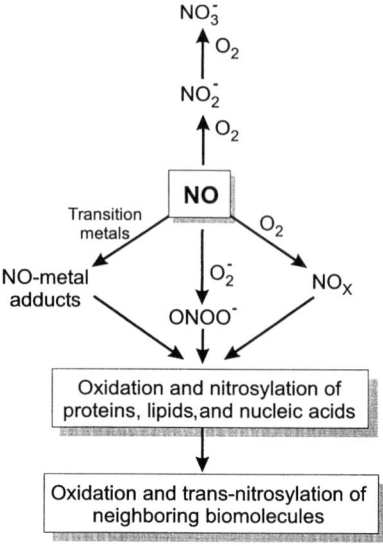

Fig. 2 Important reactions involving nitric oxide in biological systems.

of names based either on the cells in which they were initially described, their calcium dependence, the origin of their cloned genes, or their inducibility by cytokines or other inflammatory stimuli. Two isoforms are constitutively expressed and are now referred to as NOS1 (neuronal or nNOS) and NOS3 (endothelial or eNOS). The expression of the third synthase (NOS2, inducible NOS, or iNOS) differs from the constitutive enzymes both in the level of its induction and the duration of its expression (1). For these reasons, NOS2 is thought to mediate "high-output" NO synthesis (2) and probably accounts for the vast majority of nitrite/nitrate produced in endotoxin-challenged laboratory animals and, perhaps, septic patients. Several other attributes distinguish the NOS isoforms. In contrast to NOS1 and NOS3, the expression of NOS2 is regulated primarily at a transcriptional level (2,3), although changes in the stability of NOS2 mRNA or the rate of protein synthesis provide additional mechanisms for regulating its expression (4,5). While all three NOS enzymes bind calmodulin, NOS2 shows avid calmodulin binding and is fully activated at the basal calcium concentrations found in most cells (2). NOS1 and NOS3 require significantly elevated intracellular calcium for their activation.

NITRIC OXIDE IN BIOLOGICAL SYSTEMS

In aqueous solutions containing oxygen, the nitric oxide radical (NO·) decomposes to either nitrite (NO_2^-) or nitrate (NO_3^-), depending on the presence of additional oxidizing agents, like oxyhemoglobin or oxymyoglobin (6) (Fig. 2). Thus, the terminal oxidation of NO to nitrate is favored in close proximity to blood vessels. Alternatively, interactions can occur between NO and either molecular oxygen (O_2), superoxide (O_2^-), or metals that lead to the formation of several nitrosating species (i.e., NO_x, $ONOO^-$, and metal-NO adducts, respectively). These can transfer an NO^+ species to an appropriate nucleophile of a target protein, lipid, or nucleic acid (7,8). For example, in nitrosylation reactions between NO_x and protein thiol groups, an S-nitrosothiol (RSNO) is formed. This reaction is thought to proceed in vivo not only by the formation of reactive nitrogen intermediates (e.g., N_2O_3 or N_2O_4), but also through the attack on metals (e.g., Fe or Cu) to form metal-nitrosyl adducts, reactions between NO and the thiyl radical (RS·), or through the action of peroxynitrite ($ONOO^-$) (7,8). The resulting thionitrite group can be relatively stable at neutral pH (9) or may further alter protein structure by either the *trans*-nitrosation of reactive vicinal groups or the formation of intramolecular disulfide bonds (10,11). S-Nitrosylated proteins (e.g., nitroso-albumin or nitroso-glutathione) can also donate NO^+ groups to other target macromolecules (7,11,12).

Proteins with thiol-containing residues (e.g., cysteines) at or near their active sites (e.g., glyceraldehyde-3-phosphate dehydrogenase) comprise some of the more sensitive targets of NO modification (13,14). Metalloproteins, especially those containing heme-metal complexes (e.g., guanylate cyclase) or iron-sulfur clusters (e.g., aconitase) (15,16), are particularly sensitive to reactions with NO_x. Attack of metal-sulfur clusters not only leads to the modification of protein function but the release of transition metals (17) that catalyze Fenton-like reactions to form even more toxic radicals (e.g., OH·). It is important to appreciate that nucleophilic attack of proteins by NO does not always lead to protein inactivation. The activation of the enzyme guanylate cyclase by the binding of NO to its heme iron (18,19) is a model for NO-mediated cell signaling. Similarly, the constitutively active cellular enzyme aconitase is converted to an RNA-binding regulatory protein following NO-induced disassembly of its core iron-sulfur cluster (16).

Some of the better characterized effects of protein modification by NO include changes in the activities of membrane channels, transcription factors, cellular enzymes, and even antioxidants that would normally protect the cell from oxidative stress (7) (Table 1). A wide range of physiological effects results from these reactions initiated by NO. These include cellular toxicity, neurotransmission, the maintenance of blood pressure, and signal transduction for de novo gene expression (20). Identifying the particular NO-derived chemical species (i.e., NO_2, N_2O_3, and N_2O_4, $ONOO^-$ or NO-metal adducts) that mediates any one of these diverse biological responses is complicated by the intracellular concentration and life span of the reactive species, the redox environment in which it is produced, and the biochemical nature of its macromolecular target. Furthermore, these reactions all compete with one another (Fig. 2). Oxyhemoglobin, oxymyoglobin, ascorbate, and reduced glutathione constitute major scavenging systems for NO, and their wide distribution within tissues contributes to the relatively short half-life of the molecule (21).

Wink et al. (21) have concluded that very little NO_x formation occurs within cells at NO concentrations below 5 μM due to the more favorable rates of NO consumption by mitochondrial respiration. However, other important reactions do occur at these lower intracellular NO concentrations (22). These probably include the direct attack of heme proteins (e.g., guanylate cyclase activation) and reactions in which NO serves as an antioxidant to prevent either lipid peroxidation by ROS (23,24), transition metal toxicity (25), or alkyl peroxide-mediated cell damage (22). The direct effects of NO are undoubtedly also important in gram-negative sepsis, as demonstrated by the fact that LPS-induced NO production activates guanylate cyclase (26), which mediates vasodilation and inhibits platelet aggregation (18,19), two important physiological responses to endotoxin.

PATHOPHYSIOLOGICAL ROLE OF NITRIC OXIDE IN GRAM-NEGATIVE SEPSIS

Sepsis is a response to infection in which an attack on the microvascular endothelium plays a key pathogenic role. Blood vessel injury results in micro-circulatory ischemia, reduced systemic vascular resistance, hypotension, impaired organ perfusion, and derangements in oxygen delivery. Acute respiratory distress syndrome, myocardial depression, splanchnic ischemia, liver and kidney toxicity, and altered cerebral blood flow can follow and represent the critical multiorgan manifestations of these initial hemodynamic changes. The evidence that NO participates in these pathological changes has recently been reviewed (27–29). Two findings derived from studying patients undergoing sepsis provide important support for the conclusion that NO mediates vascular dysfunction in this disease in human beings. First, the oxidized products of NO, nitrite and nitrate, are elevated in sepsis (30–32), and the levels of these metabolites correlate with changes in two important hemodynamic parameters, mean arterial pressure, and systemic vascular resistance. Second, a number of clinical studies have shown that substrate analog inhibitors of NO synthases, when administered in sufficient quantities to reduce circulating nitrites and nitrates, partially correct the abnormal hemodynamic parameters of septic patients (33,34).

While these observations have justifiably led to the conclusion that NO production is a toxic event in sepsis, more recent evidence suggests that this radical also

Table 1 Selected Biological Effects of Nitric Oxide

Effect	Examples
Interactions with cellular antioxidants	Oxyhemoglobin, glutathione, ascorbate
Scavenging ROS	Peroxynitrite formation, scavenging of H_2O_2
Iron sequestration in iron-nitrosyl complexes	Reduced iron-dependent toxicity
Enzyme activation	Guanylate cyclase
Enzyme inactivation	Aconitase, γ-glutamylcysteine synthase, glutathione peroxidase, NADPH oxidase
Activation of RNA-binding regulatory proteins	IRP-1,2
Inactivation of DNA-binding transcription factors	Fos/Jun, NF-κB/Rel

serves as an important signal for regulating other aspects of the systemic inflammatory response. Endotoxemia and sepsis are accompanied by oxidative stress and redox imbalance. To the extent that changes in cellular redox status result in modified gene expression, the synthesis of high levels of NO would be expected to have far-reaching effects on the production of other inflammatory mediators and the course of disease.

REDOX STRESS IN GRAM-NEGATIVE SEPSIS

The term *redox stress* can be defined as a change in the normal redox conditions of the cell that leads to altered cell function. This condition often results from exposure to radicals, including a number of reactive oxygen species (ROS). In the latter case, the term *oxidative stress* has more often been used, because the biological effects are typically attributed to the oxidation of biomolecules.

The detection of lipid peroxidation products in the tissues of septic patients (33–40) suggests that ROS production occurs in response to endotoxic LPS in vivo and that radicals are important "secondary mediators" in sepsis (41). Several groups have also observed an acute depletion of tissue glutathione and other endogenous antioxidants during endotoxemia (42–47) that can lead to an oxidative intracellular environment and the exacerbation of LPS-induced tissue damage (48). Conversely, exogenous antioxidants can lessen the toxic effects of LPS in vivo (49–54) and attenuate gram-negative sepsis (55–57), although the therapeutic efficacy of these agents appears to vary widely among experimental systems (49,56,58,59).

Nitric Oxide and Cellular Redox Status

Although an important effect of ROS in sepsis is to alter the redox balance of cells, reduction products of molecular oxygen are not the only cause of redox stress. Indeed, several characteristics of NO production in endotoxemia and sepsis indicate that reactive nitrogen species also contribute significantly to this condition (Table 2).

The expression of the NOS2 isoform signals the potential for producing unusually high levels of NO for extended periods of time. This is an important consideration, because NO at concentrations above 10 μM has a number of biochemical effects (e.g., nitrosylation of thiol groups) that are not readily seen at lower NO concentrations (21).

Table 2 Origins of Redox Imbalance Associated with Nitric Oxide Production

Generation of nitrosylating species
Peroxynitrite formation
Glutathione depletion or glutathione redox cycle imbalance
Reduced levels of ascorbate and α-tocopherol
Generation of NO donors (e.g., nitroso-glutathione)
Release of transition metals that catalyze Fenton-like reactions

The expression of "high output" NO concomitant with O_2^- production can lead to the formation of $ONOO^-$, a strong oxidant (60). Whether this regularly occurs at a high rate in LPS-stimulated NOS2-expressing cells is not clear, given that LPS is not known as a potent inducer of respiratory bursts, and NO and O_2^- do not usually reach high concentrations at the same times following cell activation with LPS. However, a recent report (61) indicates that LPS-induced NOS2 can catalyze both the production of NO and O_2^- in a mouse macrophage-like cell line. Essential for the generation of O_2^- by NOS2 was the depletion of cellular L-arginine, a condition that would be expected to occur in activated macrophages after extended periods of time (1).

The ability of NO to diffuse between cells suggests that it may have significant paracrine effects, and this property should be NO concentration-dependent (62).

NO_x has a high affinity for the thiol groups of reduced glutathione (GSH) and can rapidly deplete this important cellular antioxidant system. Notwithstanding any consideration of the direct toxicity caused by NO itself, a significant change in the ability of the glutathione system to scavenge ROS would have important consequences for the cell.

These properties of the high output NO pathway suggest that unique biological responses can occur in LPS-activated cells that would alter the overall redox environment.

Nitric Oxide and Cellular Antioxidant Systems

The interaction between NO and endogenous cellular antioxidants is well documented and, as just mentioned, includes the S-nitrosylation of reduced glutathione (GSH) (12,63,64). Because GSH is not taken up by cells from their extracellular environments, the rate at which they reduce nitroso-glutathione back to GSH or resynthesize GSH de novo substantially determines the likelihood of NO-induced redox changes and toxicity (63). For this reason it is noteworthy that NO has also been reported to affect two enzymes that are essential for maintaining an intact glutathione redox cycle (65). The first is γ-glutamylcysteine synthase, a rate-limiting enzyme in glutathione biosynthesis (65), whose activity is inhibited by exposure to NO or S-nitroso compounds, including nitroso-glutathione (66). The second activity is that of the glutathione peroxidases (GPx), a group of enzymes that catalyze the scavenging of oxygen radicals, hydrogen peroxide, and alkyl hydroperoxides by GSH. Glutathione concentrations in rat macrophages are slightly below the K_m of macrophage GPx (67), indicating that small variations in either substrate concentration or enzyme activity should have significant effects on redox balance. Asahi et al. (68) reported that the NO donor S-nitro-N-acetyl-DL-penicillamine (SNAP) inhibited the GPx activity of RAW 264.7 mouse macrophage-like cells. This would be expected to inhibit the ability of the cell to detoxify ROS even at normal intracellular GSH concentrations.

Reactions between NO and glutathione not only deplete GSH but generate additional reactive species (e.g., GS· and GSO_3H) and lower the cellular levels of the antioxidants ascorbic acid and α-tocopherol, which are normally maintained by GSH (Table 2). In contrast, interactions between NO and GSH may also have a positive effect in that they reduce NO reactions with O_2^- and metal-sulfur clusters that would otherwise lead to the formation of toxic intermediates. However, this scavenging effect of GSH would occur only as long as favorable GSH:NO ratios were maintained within the cell. Finally, nitroso-glutathione is relatively stable and may provide a long-term source of reactive nitrosothiols (7). Indeed, nitrosoglutathione has been shown to mimic the vasodilatory effects of the NO donors nitroglycerine and nitroprusside (9,69).

Nitric Oxide as an Antioxidant

Nitric oxide not only acts as a weak oxidizing agent, but can provide important antioxidant functions for cells as well. It can scavenge O_2^- radical (24) and reduce O_2^- production by directly inhibiting components of the NADPH oxidase complex (70). The iron-dependent oxidation of lipids initiated by Fenton-like chemistry is inhibited by NO gas and nitroso-myoglobin (25), indicating that NO-iron complexes can prevent peroxide-initiated cell injury. Wink et al. (22,23) have shown that NO derived from NONOates (compounds with the formula $R_1R_2N[N(O)NO]^-$) prevented H_2O_2 or alkyl hydroperoxide-mediated toxicity of fibroblasts in vitro by a mechanism that appeared to involve ROS scavenging. Thus, although the production of high NO levels has historically been taken as a sign of toxicity, the capacity of NO to function as an antioxidant cannot be ignored and may explain some of the unexpected side effects associated with the therapeutic use of NOS inhibitors for the treatment of sepsis.

NITRIC OXIDE-INDUCED REDOX STRESS AND GENE EXPRESSION

Considerable evidence indicates that changes in cellular redox balance can substantially affect the expression of many host pro-inflammatory genes and may, for certain genes, be required for the activation of their transcription (71). Among the first well-characterized effects was the inhibition of transcription dependent on the Fos/Jun family of transcription factors. These are inducible dimeric protein complexes encoded by the c-fos and c-jun proto-oncogenes (or closely related genes) that bind to AP-1 DNA motifs and activate gene transcription (72). The activation of these AP-1–binding proteins appears to involve both the phosphorylation/dephosphorylation of preformed Fos/Jun dimers within the nucleus and the transcription of their immediate early genes (71). There is evidence that both the de novo synthesis and the DNA-binding activity of Fos/Jun dimers is redox sensitive (73–75). For example, treating nucleoproteins from activated cells with sulfhydryl modifying or oxidizing agents destroys their binding activity by attacking two cysteine residues at conserved positions within the Fos and Jun proteins that are flanked by basic amino acids (74,75). These findings have suggested an important mechanism for regulating gene expression in which the DNA-binding properties of transactivating proteins are regulated by the redox condition of key amino acid side chain groups.

Nitric Oxide and Transcriptional Activation

Influence of Nitric Oxide–induced Redox Stress on Cellular Iron Metabolism

Only recently has it become clear that NO can also regulate gene expression by a similar mechanism. This is illustrated by its effects on iron metabolism in mammalian cells. The iron-regulated biosynthesis of the transferrin receptor and ferritin, two proteins that control iron uptake and storage, is dependent on consensus iron-response elements (IRE) located in the 3′ or 5′ untranslated regions of their mRNAs. Two mammalian iron regulatory proteins, IRP-1 and IRP-2, that exist in all vertebrate tissues bind to the IREs and have been shown to be responsive to oxidative stress (16). For example, H_2O_2 can induce IRP-1 binding to the IRE of ferritin mRNA, which inhibits its translation efficiency and reduces intracellular ferritin levels (76). Similar high-affinity binding of IRP-1 to the IREs in the 3′ untranslated region of transferrin receptor mRNA reduces nucleolytic degradation of this message (77–79). Together, these two effects of H_2O_2 increase iron uptake and decrease the ability of the cell to store intracellular iron in a bound form, thus increasing the intracellular pool of free iron.

The binding affinity of IRP-1 for these mRNA motifs is regulated by its conformation, which is maintained by a 4Fe-4S cluster in the protein core. ApoIRP-1 shows high-affinity binding to IREs, whereas 4Fe-4S-IRP-1 acts as a cytosolic aconitase, but does not bind the mRNA response elements. This conversion from an active metabolic enzyme to an mRNA-binding regulatory protein can also be induced by NO (8,16). In mouse macrophages activated by LPS and IFN-γ, there is a strong correlation between NOS2 expression, NO production, IRP-1 activation, and the inhibition of ferritin biosynthesis (76,80). It is not known precisely how NO activates IRP-1, although it is postulated that a direct attack on its iron-sulfur cluster followed by a conformational rearrangement (79–81) may occur, similar to that which explains the activation of soluble guanylate cyclase by NO (18). Alternatively, exposure to NO may favor the apoprotein conformation by sequestering iron from the cellular pool in the form of iron-nitrosyl complexes (76,82).

Two additional properties distinguish NO-induced cellular iron regulation in this model. First, endogenously produced NO can affect IRP-regulated iron metabolism (79). Second, NO-induced IRP activation differs from that induced by either iron starvation or H_2O_2 in terms of the specific mRNA targets affected, the kinetics of the reactions, and the involvement of cellular phosphatases in the two signaling pathways (79,83). Therefore, the effect of NO on IRPs that are important in iron metabolism serves as a model for the control of gene expression by the redox-dependent posttranslational modification of a sensing metallo-protein.

Effect of Nitrosative Stress on Antioxidant Gene Expression in Prokaryotes

The activation of IRPs in eurokaryotic cells by H_2O_2 and NO illustrates a dedicated regulatory system in which a stress signal initiates a specific set of related cellular responses, in this case the uptake, storage, and metabolism of iron. A similar sensor has been described in *E. coli* and *S. typhimurium* that regulates cellular antioxidant responses to H_2O_2 or NO. This stress response regulon encodes the OxyR protein, a transcriptional activator that initiates the expression of a number of antioxidant genes, including catalase, glutathione reductase, and alkyl hydroperoxide reductase. Each is known to mediate hydroperoxide detoxification. In this fashion, OxyR serves as a sensing unit for a protective cellular response. Activation of OxyR by hydroperoxides is thought to result from a conformational change brought about by the oxidation of a key cysteine residue. Recent evidence indicates that the nitrosothiol S-nitroso-cysteine can also activate OxyR by S-nitrosylation. Hausladen et al. (84) have coined the term "nitrosative stress" to describe conditions that deplete intracellular thiols and activate OxyR in prokaryotic cells. Thus, S-nitrosylation of this cellular protein provides the molecular basis for activating antioxidant genes under conditions of NO-induced stress. Not surprisingly, the outcome of exposure to NO is critically dependent on the antioxidant status of the cell. Mutant bacteria defective in glutathione synthesis can be induced by nitrosothiols to express OxyR-dependent genes under conditions in which the isogenic wild-type strain does not respond (84).

Effect of Nitric Oxide on Activation of NF-κB/Rel and AP-1 Transcription Factors

Members of the NF-κB/Rel family of transcription factors have an important role in the transcription of many pro-inflammatory genes activated by LPS. In unstimulated cells, preformed inactive NF-κB/Rel dimers are found in the cytoplasm complexed with proteins of the IκB inhibitor family. With cell activation, IκB is phosphorylated, ubiquitinated, and degraded, releasing NF-κB to translocate to the nucleus, where it binds to κB DNA enhancer motifs (85–87). In LPS-activated mouse macrophages, transactivating NF-κB/Rel complexes most often consist of p50-p65 heterodimers (88–90), and these have been shown to bind to and transactivate κB elements from a number of LPS-responsive macrophage genes (e.g., *TNF-α*, *IL-1β*, *IL-6*, and *NOS2*) (88–95).

The activation of NF-κB appears to require careful regulation of the redox conditions of the cell, with both oxidative and reductive conditions playing important roles. For example, NF-κB can be activated in human Jurkat T cells by oxidant stress [e.g., H_2O_2 (96)], and exogenous antioxidants (e.g., *N*-acetylcysteine or αlipoic acid) have been shown to inhibit NF-κB activation induced by TNF-α or IL-1β (96–98). Conversely, oxidative conditions can also *inhibit* the DNA-binding and transactivating potential of NF-κB dimers (73,99). Once oxidized, NF-κB/Rel proteins can be reactivated by reduction with 2-mercaptoethanol or dithiothreitol (73,99,100). This reflects the presence of cysteine residues in the DNA-binding sites of these proteins whose sulfhydryl groups are redox-regulated (73). The cysteine within a conserved RxxRxRxxC motif that is found in all NF-κB/Rel proteins appears to be particularly susceptible to oxidative inactivation (99). There is also evidence that oxidative stress can affect IκBα degradation (101,102), although the exact mechanism of this effect has not been defined.

Nitrosylation of NF-κB peptides appears to provide a similar mechanism for redox control of gene expression. Nitric oxide donors have been shown to partially inhibit the expression of κB-dependent reporter genes in cytokine-activated vascular endothelial and smooth muscle cells (103,105). Similarly, NOS inhibitors augmented κB-dependent reporter gene transcription as well as nucleoprotein κB-binding assessed by gel mobility shift assays (103,105). Cyclic GMP analogs were without effects in these systems, suggesting that NO has the potential to inhibit pro-inflammatory gene expression by a cGMP-independent effect on NF-κB. In contrast to these findings, Lander et al. (106,107) reported that NO donors and NO gas stimulated modest NF-κB activation in resting human blood mononuclear cells by a process that appeared to be G protein dependent (107). This discrepancy may relate to the relative intracellular concentrations of NO achieved in these two experimental systems, differences in the antioxidant status of the various cell types, or the use of activated versus resting cells. Overall, it would appear that both G protein–dependent and –independent pathways are triggered by NO, which can influence NF-κB activation.

Two mechanisms have been proposed to explain the effects of NO on NF-κB–dependent transcriptional activation. The first presumably involves an effect on IκBα phosphorylation and/or proteolytic degradation. Thus, Peng et al. (108) reported that the NO donors sodium nitroprusside or S-nitroso-glutathione inhibited TNF-α–induced NF-κB activation in endothelial cells, but not the DNA-binding activity of nucleoproteins prepared from previously activated cells. They also showed that nitrosoglutathione blocked IκBα degradation induced by TNF-α and induced the expression of IκBα mRNA in unstimulated cells. It is not known whether these effects were due to changes in cellular kinase or phosphatase activity (102,106) or simply the ability of nitrosoglutathione to scavenge superoxide.

The second mechanism by which NO inhibits NF-κB is the oxidation of target sulfhydryl groups in the N-terminal Rel homology domains of NF-κB/Rel proteins, sequences that are necessary for DNA-binding activity (73,99). Treating LPS or cytokine-activated astroglial cells with the NO donor spermine NONOate has been shown to inhibit NF-κB activation (109,110). Direct evidence for S-nitrosylation of NF-κB peptides has been provided by Matthews et al. (111), who showed that the DNA-binding activity of recombinant p50-p65 heterodimers was markedly inhibited by incubation with SNAP in the presence of reducing agents (e.g., dithiothreitol). Treatment with SNAP did not inhibit the κB-binding activity of a mutant p50 protein, in which a serine was substituted for the normal cysteine at position 62. Mass spectrometry revealed that NO gas caused the S-nitrosylation of a synthetic NF-κB p50 peptide at cysteine-62.

Of interest, NOS inhibitors induce elevated levels of nuclear NF-κB/Rel complexes in murine astroglial and endothelial cells (103,110). This may indicate that in certain tissues endogenous NO maintains NF-κB/Rel peptides in an oxidized state, thereby assuring that many pro-inflammatory genes are silent in resting cells. Macrophage activation by LPS also appears to be followed by an NO-dependent autocrine feedback inhibition of NOS2 expression (112,113). It should be noted here that NO is a diffusible product (62,114) that can potentially regulate gene expression in adjacent cells that do not themselves synthesize the radical.

An unexpected finding was reported by Togashi et al. (110), who showed that the putative radical scavenger pyrrolidine dithiocarbamate (PDTC) *activated* NFκB in resting astroglial cells. The authors attributed this effect to the ability of the thiocarbamate to form iron chelates (115) that scavenged NO. This finding illustrates some of the complexities inherent in the use of PDTC as a probe for κB-dependent transcriptional activation. Besides radical scavenging, thiocarbamates can affect cellular redox status by a variety of mechanisms, including the promotion of metal uptake by cells, intracellular chelation of free metals, carbamoylation of proteins, and direct inhibitory effects on the glutathione redox cycle (100,115,116). Thus, in many respects the "antioxidant" PDTC should be considered a pro-oxidant. Moreover, the effects of PDTC are not specific to NF-κB, as the activation of AP-1 has also been shown to be PDTC-sensitive (117).

NO donors can either promote (118,119) or inhibit (120,121) AP-1 activation and AP-1–dependent gene expression. The latter effect appears to occur both at the level of DNA binding and the expression of c-fos mRNA (120). However, a direct comparison between the inhibitory effects of NO on NF-κB and AP-1 activation in endothelial cells suggests that the NF-κB response is far more NO-sensitive in this cell type (103). Similarly, Das et al. (122) reported that various thiol-modifying agents (e.g., dithiothreitol and diamide) modulated NF-κB activation in a lung adenocarcinoma cell line, but had inconsistent effects on the activation of AP-1. In neither of these cases was direct evidence for S-nitrosylation of Fos/Jun family proteins reported. In still other cell types, changes in cellular redox status often have opposite effects on the activation of NF-κB and AP-1, even within the same cell (101,123,124). For example, in human T cells redox stress associated with severe glutathione depletion caused a reduction in NF-κB DNA-binding and transactivating activity but enhanced these activities dependent upon AP-1 (123). Dithiothreitol more efficiently reversed oxidative damage to NF-κB than to AP-1. This may reflect the relative redox potentials of amino acids within the DNA-binding domains of NF-κB/Rel versus Fos/Jun proteins. The finding that AP-1 and NF-κB are not coordinately regulated by redox conditions within cells suggests that the effects of both ROS and NO on the expression of a given gene that contains both elements would be difficult to predict.

Effect of Nitric Oxide on Expression of LPS-Induced Genes

A number of pro-inflammatory genes, whose expression is induced by LPS, are subject to redox-dependent regulation. These include the genes coding for the cytokines IL-1, IL-6, TNF-α, and GM-CSF (98,125–130), the chemokines IL-8 and MCP-1 (117,131–133), the cell adhesion molecules ICAM-1 and V-CAM-1 (104), and enzymes involved in mediator biosynthesis (e.g., COX2 and NOS2) (109,134,135) (Table 3). Treatment of LPS-activated cells not only with ROS or antioxidants, but also with NO donors, NO scavengers, or NOS inhibitors (109,113,117,136) has been shown to affect the expression of mRNA for inflammatory mediators in vitro, suggesting that changes in the redox environment by either ROS or NO primarily affect the transcription of these genes.

Based on the results of these studies, several additional generalizations can be made about NO-induced effects. First, NO differentially alters the expression of pro-inflammatory genes (113,117,136). For example, in LPS-activated rat alveolar macrophages, N^G-monomethyl-L-arginine enhanced IL-1β and IL-6, but not TNF-α, production (136). Likewise, the NO donor diethylamine di-NO inhibited NOS2, but not TNF-α, mRNA expression in LPS-stimulated ANA-1 mouse macrophage-like cells (113). Second, NO is a potent inhibitor of NOS2 expression. This effect is manifest at the level of NOS2 mRNA expression (109,113) and may provide a mechanism of feedback control of NO production in addition to the well-characterized inhibitory effects NO has on NOS enzymatic activity (112,137–139). Third, changes in gene expression induced by redox stress are accompanied by similar effects on NF-κB and/or AP-1 activation (98,105,109,130,133,135). For example, Colasanti et al. (135) showed that the NO donor sodium nitroprusside inhibited both the expression of NOS2 mRNA and the activation of NF-κB in LPS-treated microglial cells. While similar findings have been reported by others (105,109,130), data that would link these two effects in a causal manner are lacking.

Indeed, there are a number of reasons to believe that the ability of NO to inhibit the expression of *NOS2* (and perhaps other genes) in many LPS-activated cells is not due primarily to its inhibition of NF-κB. Nitric oxide does not appear to inhibit TNF-α mRNA expression in LPS-stimulated mouse macrophages (113), a response that is highly κB-dependent (91–93). As noted above, NO can also affect AP-1–dependent transcriptional activation (118–121), indicating that its effects on gene expression are likely to be complex. Despite these reservations, it seems safe to conclude that NO alters responses to LPS, affects genes that encode important mediators of gram-negative sepsis, and may act, in part, by mechanisms similar to those described for ROS-dependent gene regulation.

Redox stress that leads to the altered expression of pro-inflammatory genes is often accompanied by changes in intracellular thiol content and/or the activities of thiol-modifying enzymes, like those of the glutathione redox cycle (63,64,100). Glutathione (γ-glutamylcysteinylglycine) is the most plentiful non-protein thiol associated with protection against redox stress and is maintained in a highly reduced state within cells (i.e., GSH > GSSG). Recent evidence suggests

Table 3 Pro-Inflammatory Genes Whose Expression Is Differentially Regulated by Nitric Oxide

IL-1	IL-8	ICAM-1	COX-2
IL-6	MCP-1	VCAM-1	NOS2
TNF-α			
GM-CSF			

that glutathione disulfide (GSSG) may mediate one or more of the redox-sensitive steps in the activation of NF-κB (100,123) and AP-1 (75). Elevated intracellular levels of GSSG were associated with enhanced NF-κB activation in T cells as measured both by gel shift assays and reporter gene transcription experiments (123). Oxidative activation of NF-κB in T cells by H_2O_2 was not seen if the cells were first depleted of glutathione by treatment with buthionine sulfoximine, a specific inhibitor of glutathione biosynthesis (140). Even higher levels of intracellular GSSG coincided with the inhibition of NF-κB in T cells. Likewise, Frame et al. (75) showed that GSSG greatly inhibited the formation of AP-1/DNA complexes in vitro. These findings have been interpreted as indicating that a certain intracellular concentration of GSSG may be required for NF-κB activation, while excess quantities may lead to a loss of its DNA-binding activity by oxidative damage. Given the efficiency with which NO nitrosylates GSH, a mechanism in which nitroso-gluthathione or glutathione disulfide serves as an intermediate sensing protein may similarly explain several aspects of NO-induced redox stress.

SYNTHESIS: THE MULTIPLE LIVES OF NITRIC OXIDE

The cellular redox changes that accompany gram-negative sepsis are now recognized as important events that determine not only cell survival, but the nature of cellular responses to LPS challenge. Both reduction products of oxygen and NO-derived reactive nitrogen species are produced by LPS-activated cells and elicit similar adaptive responses. Nitric oxide has been shown to S-nitrosylate transcription factors in vitro, especially those of the NF-κB and Fos/Jun families, and alter gene expression. There is no reason to believe that other NO-sensitive signaling molecules within LPS pathways will not be described. However, we currently lack a detailed understanding of the significance of these findings. For example, studies that would causally link the (in)activation of NF-κB by NO with the impaired expression of specific genes are needed. What is the molecular basis for defining a redox-sensitive, LPS-activated response? What are the endogenous effects of NO production, and how do they relate to studies with NO donors, which are less than perfect models of NO production and reactivity in vivo (60)? What are the roles of cellular antioxidant systems in regulating responses to LPS, and which are important in preventing or correcting NO-induced redox stress? Does glutathione also serve as a signaling sensor that transmits an oxidative signal in the NO cascade?

The release of a freely diffusible mediator like NO in response to LPS in vivo has significant implications for future therapies of gram-negative sepsis. While we tend to consider first the negative consequences of redox stress, a certain level of cellular redox imbalance may be beneficial for controlling inflammatory mediator production. In this sense, reducing NO production in sepsis may have consequences we cannot presently predict.

ACKNOWLEDGMENTS

The author thanks Dr. Fred Samson for the stimulating and thoughtful discussions that contributed to this chapter. Financial support for this work was derived, in part, from a grant from the American Heart Association.

REFERENCES

1. Vodovitz Y, Kwon NS, Pospischil M, Manning J, Paik J, Nathan C. Inactivation of nitric oxide synthase after prolonged incubation of mouse macrophages with IFN-γ and bacterial lipopolysaccharide. J Immunol 1994; 152:4110–4118.
2. MacMicking J, Xie Q-W, Nathan C. Nitric oxide and macrophage function. Ann Rev Immunol 1997; 15: 323–350.
3. Murphy WJ. Transcriptional regulation of the genes encoding nitric oxide synthase. In: Laskin J, Laskin DL, eds. Cellular and Molecular Biology of Nitric Oxide. New York: Marcel Dekker; 1999:1–55.
4. Vodovitz Y, Bogdan C, Paik J, Xie Q-W, Nathan C. Mechanisms of suppression of macrophage nitric oxide release by transforming growth factor-β. J Exp Med 1993; 178:605–613.
5. Park SK, Murphy S. Nitric oxide synthase type II mRNA stability is translation- and transcription-dependent. J Neurochem 1996; 67:1766–1769.
6. Ignarro LJ, Fukuto JM, Griscavage JM, Rogers NE, Byrns RE. Oxidation of nitric oxide in aqueous solution to nitrite but not nitrate: comparison with enzymatically formed nitric oxide from L-arginine. Proc Natl Acad Sci USA. 1993; 90:8103–8107.
7. Stamler JS. S-Nitrosothiols and the bioregulatory actions of nitrogen oxides through reactions with thiol groups. Curr Top Microbiol Immunol 1995; 196: 19–36.
8. Stamler JS. Redox signaling: nitrosylation and related target interactions of nitric oxide. Cell 1994; 78: 931–936.
9. Keaney JF, Simon DI, Stamler JS, Jaraki O, Scharfstein J, Vita JA, Loscalzo J. NO forms an adduct with serum albumin that has endothelial-derived re-

laxing factor-like properties. J Clin Invest 1993; 91:1582–1589.
10. Gopalakrishna R, Chen ZH, Gundimeda U. Nitric oxide and nitric oxide-generating agents induce a reversible inactivation of protein kinase C activity and phorbol ester binding. J Biol Chem 1993; 268:27180–27185.
11. Stamler JS, Simon DI, Jaraki O, Osborne JA, Francis S, Mullins M, Singel D, Loscalzo J. S-Nitrosylation of tissue-type plasminogen activator confers vasodilatory and antiplatelet properties on the enzyme. Proc Natl Acad Sci USA 1992; 89:8087–8091.
12. Clancy RM, Levartovsky D, Leszczynska-Piziak J, Yegudin J, Abramson SB. Nitric oxide reacts with intracellular glutathione and activates the hexose monophosphate shunt in human neutrophils: evidence for S-nitrosoglutathione as a bioactive intermediary. Proc Natl Acad Sci USA 1994; 91:3680–3684.
13. Dimmeler S, Brune B. Nitric oxide preferentially stimulates auto-ADP-ribosylation of glyceraldehyde-3-phosphate dehydrogenase compared to alcohol or lactate dehydrogenase. FEBS Lett 1993; 315:21–24.
14. Molina y Vedia L, McDonald B, Reep B, Brune B, Di Silvio M, Billiar TR and Lapentina EG. Nitric oxide-induced S-nitrosylation of glyceraldehyde-3-phosphate dehydrogenase inhibits enzymatic activity and increases endogenous ADP-ribosylation. J Biol Chem 1992; 267:24929–24932.
15. Hentze MW. Iron-sulfur clusters and oxidative stress responses. Trends Biol Sci 1996; 21:282–283.
16. Hentze MW, Kuhn LC. Molecular control of vertebrate iron metabolism: mRNA-based regulatory circuits operated by iron, nitric oxide, and oxidative stress. Proc Natl Acad Sci USA 1996; 93:8175–8182.
17. Reif DW, Simmons RD. Nitric oxide mediates iron release from ferritin. Arch Biochem Biophys 1990; 283:537–541.
18. Ignarro LJ. Haem-dependent activation of guanylate cyclase and cyclic GMP formation by endogenous nitric oxide: a unique transduction mechanism for transcellular signaling. Pharm Toxicol 1990; 67:1–7.
19. Traylor TJ, Sharma GS. Why NO? Biochemistry 1992; 31:2847–2849.
20. Davies MG, Fulton GJ, Hagen P-O. Clinical biology of nitric oxide. Br J Surg 1995; 82:1598–1610.
21. Wink DA, Hanbauer I, Grisham MB, Laval F, Nims RW, Laval J, Cook J, Pacelli R, Liebmann J, Krishna M, Ford PC, Mitchell JB. Chemical biology of nitric oxide: regulation and protective and toxic mechanisms. Curr Topics Cell Reg 1996; 34:159–187.
22. Wink DA, Cook JA, Krishna MC, Hanbauer I, DeGraff W, Gamsom J, Mitchell JB. Nitric oxide protects against alkyl peroxide-mediated cytotoxicity: further insights into the role nitric oxide plays in oxidative stress. Arch Biochem Biophys 1995; 319:402–407.
23. Wink DA, Hanbauer I, Krishna MC, DeGraff W, Gamson J, Mitchell JB. Nitric oxide protects against cellular damage and cytotoxicity from reactive oxygen species. Proc Natl Acad Sci USA 1993; 90:9813–9817.
24. Wink DA, Hanbauer I, Laval F, Cook JA, Krishna MC, Mitchell JB. Nitric oxide protects against the cytotoxic effects of reactive oxygen species. Ann NY Acad Sci 1994; 738:265–278.
25. Kanner J, Harel S, Granit R. Nitric oxide as an antioxidant. Arch Biochem Biophys 1991; 289:130–136.
26. Moncada S, Palmer RMJ, Higgs EA. Nitric oxide: physiology, pathophysiology, and pharmacology. Pharm Rev 1991; 43:109–142.
27. Evans TJ, Cohen J. Mediators: nitric oxide and other toxic oxygen species. Curr Top Microbiol Immunol 1996; 216:189–207.
28. Cobb JP, Danner RL. Nitric oxide and septic shock. J Am Med Assoc 1996; 275:1192–1196.
29. Wolf TA, Dasta JF. Use of nitric oxide synthase inhibitors as a novel treatment for septic shock. Ann Pharmacother 1995; 29:36–46.
30. Ochoa JB, Udekwu AO, Billiar TR, Curran RD, Cerra FB, Simmons RL, Peitzman AB. Nitrogen oxide levels in patients after trauma and during sepsis. Ann Surg 1991; 2143:621–626.
31. Evans TJ, Carpenter A, Kinderman H, Cohen J. Evidence of increased nitric oxide production in patients with the sepsis syndrome. Circ Shock 1993; 41:77–81.
32. Petros A, Bennett D, Vallance P. Effect of nitric oxide synthase inhibitors on hypotension in patients with septic shock. Lancet 1991; 338:1557–1558.
33. Petros Lamb G, Leone A, Moncada S, Bennett D, Vallance P. Effects of a nitric oxide synthase inhibitor in humans with septic shock. Cardiovasc Res 1994; 28:34–39.
34. Lorente JA, Landin L. De Pablo R, Renes E, Liste D. L-Arginine pathway in the sepsis syndrome. Crit Care Med 1993; 21:1287–1295.
35. Wong C, Flynn J, Demling RH. Role of oxygen radicals in endotoxin-induced lung injury. Arch Surg 1984; 119:77–82.
36. Yoshikawa T. Oxy radicals in endotoxin shock. Meth Enzymol 1990; 186:660–665.
37. Peavy DL, Fairchild EJ, II. Evidence for lipid peroxidation in endotoxin-poisioned mice. Infect Immun 1986; 52:613–616.
38. Demling RH, Lalonde C, Jin L-J, Ryan P, Fox R. Endotoxemia causes increased lung tissue lipid peroxidation in unanesthetized sheep. J Appl Phys 1986; 60:2094–2100.
39. Taylor DE, Ghio AJ, Piantadose CA. Reactive oxygen species produced by liver mitochondria of rats in sepsis. Arch Biochim Biophys 1995; 316:70–76.
40. Kokk T, Talvik R, Zilmer M, Kutt E, Talvik T. Markers of oxidative stress before and after exhange transfusion for neonatal sepsis. Acta Paediatr 1996; 85:1244–1246.
41. Redl H, Gasser H, Schlag G. Involvement of oxygen radicals in shock related cell injury. Br Med Bul 1993; 49:556–565.
42. Keller GA, Barke R, Harty JT, Humphrey E, Simmons RL. Decreased hepatic glutathione levels in septic shock. Arch Surg 1985; 120:941–945.
43. Pacht ER, Timerman AP, Lykens MG, Merola AJ. Deficiency of alveolar fluid glutathione in patients with sepsis and ARDS. Chest 1991; 100:1397–1403.
44. Portoles MT, Tatala M, Anton A, Pagani R. Hepatic response to the oxidative stress induced by E. coli en-

dotoxin: glutathione as an index of the acute phase during the endotoxic shock. Mol Cell Biochem 1996; 159:115–121.
45. Minamiyama Y, Takamura S, Koyama K. Yu H, Miyamoto M, Inoue M. Dynamic aspects of glutathione and nitric oxide metabolism in endotoxemic rats. Am J Physiol 1996; 271:G575–581.
46. Dhaunsi GS, Singh I, Hanevold CD. Peroxisomal participation in the cellular response to the oxidative stress of endotoxin. Mol Cell Biochem 1993; 126:25–35.
47. Maulik N, Watanabe M, Engelman D, Engelman RM, Kagan VE, Kisin E, Tyurin V, Cordis GA, Das DK. Myocardial adaptation to ischemia by oxidative stress induced by endotoxin. Am J Physiol 1995; 269:C907–916.
48. Lee KJ, Andrejuk T, Dziuban SW, Goldfarb RD. Deleterious effects of buthionine sulfoximine on cardiac function during continuous endotoxemia. Proc Soc Exp Biol Med 1995; 209:178–184.
49. McKechnie K, Furman BL, Parratt JR. Modification by oxygen free radical scavengers of the metabolic and cardiovascular effects of endotoxin infusion in conscious rats. Circ Shock 1986, 19:429–439.
50. Kumimoto F, Morita T, Ogawa R, Fujita T. Inhibition of lipid peroxidation improves survival rate of endotoxemic rats. Circ Shock 1987; 21:15–22.
51. Sugino K, Dohi K, Yamada K, Kawasaki T. Changes in the levels of endogenous antioxidants in the liver of mice with experimental endotoxemia and the protective effects of the antioxidants. Surgery 1989; 105:200–206.
52. Schneider J, Friderichs E, Heintze K, Flohe L. Effects of recombinant human superoxide dismutase on increased lung vascular permeability and respiratory disorder in endotoxemic rats. Circ Shock 1990; 30:97–106.
53. Mendez C, Garcia I, Maier RV. Antioxidants attenuate endotoxin-induced activation of alveolar macrophages. Surgery 1995; 118:412–420.
54. Blackwell TS, Blackwell TR, Holden EP, Christman BW, Christman JW. In vivo antioxidant treatment suppresses nuclear factor-κB activation and neutrophilic lung inflammation. J Immunol 1996; 157:1630–1637.
55. Warner BW, Hasselgren P-O, Fisher JE. Effect of allopurinol and superoxide dismutase on survival in rats with sepsis. Curr Surg 1986; 43:292–293.
56. Henderson A, Hayes P. Acetylcysteine as a cytoprotective antioxidant in patients with severe sepsis. Ann Parmacother 1994; 28:1086–1088.
57. Villa P, Ghezzi P. Effect of N-acetyl-L-cysteine on sepsis in mice. Eur J Pharm 1995; 292:341–344.
58. Broner CW, Shenep JL, Stidham GL, Stokes CD, Fairchlough D, Schonbaum GR, Rehg JE, Hilder WK. Effect of antioxidants in experimental E. coli septicemia. Circ Shock 1989; 29:77–92.
59. Groeneveld ABJ, den Hollander W, Straub J, Nauta JP, Thijs LG. Effects of NAC and terbutaline treatment on hemodynamics and regional albumin extravasation in porcine septic shock. Circ Shock 1990; 30:185–205.
60. Beckman JS. Biochemistry of nitric oxide and peroxynitrate. In: Kubes P. ed. Nitric Oxide: A Modulator of Cell-Cell Interactions in the Microcirculation. Austin: Landes Co. 1995:1–18.
61. Xia Y, Zweier JL. Superoxide and peroxynitrite generation from inducible nitric oxide synthase in macrophages. Proc Natl Acad Sci USA 1997; 94:6954–6958.
62. Lancaster JR. Simulation of diffusion and reaction of endogenously produced nitric oxide. Proc Natl Acad Sci 1994; 91:8137–8141.
63. Luperchio S, Tamir S, Tannenbaum SR. NO-induced oxidative stress and glutathione metabolism in rodent and human cells. Free Rad Biol Med 1996; 21:513–519.
64. Bolanos JP, Heales SJR, Peuchen S, Barker JE, Land JM, Clark JB. Nitric oxide-mediated miochondrial damage: a potential neuroprotective role for glutathione. Free Rad Biol Med 1996; 21:995–1001.
65. Deneke SM, Fanburg BL. Regulation of cellular glutathione. Am J Phys 1989; 257:L163–173.
66. Han J, Stamler J, Griffith OW. Inhibition of γ-glutamylcysteine synthetase by nitric oxide donors. FASEB J 1994; 8:A1288.
67. Forman JH, Skelton DC. Protection of alveolar macrophages from hyperoxia by γ-glutamyl transpeptidase. Am J Physiol 1991; 259:L102–107.
68. Asahi M, Fujii J, Suzuki K, Seo HG, Kuzuya T, Hori M, Tada M, Fujii S, Taniguchi N. Inactivation of glutathione peroxidase by nitric oxide. J Biol Chem 1995; 270:21035–21039.
69. Ignarro LJ, Lippton H, Edwards JC, Baricos WH, Hyman AL, Kadowitz PH, Gruetter CA. Mechanism of vascular smooth muscle relaxation by organic nitrates, nitrites, nitroprusside and nitric oxide: evidence for the involvement of S-nitrosothiols as active intermediates. J Pharm Exp Ther 1981; 218:739–749.
70. Clancy Rm, Leszczynska-Piziak J, Abramson SB. Nitric oxide, an endothelial cell relaxation factor, inhibits neutrophil superoxide anion production via a direct action on the NADPH oxidase. J Clin Invest 1992; 90:1116–1121.
71. Sun Y, Oberly LW. Redox regulation of transcriptional activators. Free Rad Biol Med 1996; 21:335–348.
72. Curan T, Franza BR. Fos and Jun: The AP-1 connection. Cell 1988; 55:395–397.
73. Tolendano MB, Leonard WJ. Modulation of transcription factor NF-κB binding activity by oxidation-reduction in vitro. Proc Natl Acad Sci USA 1991; 88:4328–4332.
74. Abate C, Patel L, Rauscher FJ, III, Curran T. Redox regulation of Fos and Jun DNA-binding activity in vitro. Science 1990; 249:1157–1161.
75. Frame MC, Wilkie NM, Darling AJ, Chudleigh A, Pintzas A, Lang JC, Gillespie DAG. Regulation of AP-1/DNA complex formation in vitro. Oncogene 1991; 6:205–209.
76. Weiss G, Goossen G, Doppler W, Fuchs D, Pantopoulos K, Wernere-Felmayer G, Wachter H, Hentze MW. Translational regulation via iron-responsive elements by nitric oxide/NO-synthase pathway. EMBO J 1993; 12:3651–3657.
77. Casey JL, Hentze MW, Koeller DM, Caughman SW, Rouault TA, Klausner RD, Harford JB. Iron-respon-

sive elements: Regulatory RNA sequences that control mRNA levels and translation. Science 1988; 240: 924–928.
78. Mullner EW, Kuhn LC. A stem-loop in the 3' untranslated region mediates iron-dependent regulation of transferrin receptor mRNA stability in the cytoplasm. Cell 1988; 53:815–825.
79. Pantopoulos K. Hentze MW. Nitric oxide signaling to iron-regulatory protein: direct control of ferritin mRNA translation and transferrin receptor mRNA stability in transfected fibroblasts. Proc Natl Acad Sci USA 1995; 92:1267–1271.
80. Drapier J-C, Hirling H, Wietzerbin J, Kaldy P, Kuhn LC. Biosynthesis of nitric oxide activates iron regulatory factor in macrophages. EMBO J 1993; 12: 3643–3649.
81. Hentze MW. Iron-sulfur clusters and oxidant stress response. Trends Biol Sci 1996; 21:282–283.
82. Drapier J-C, Pellat C, Henry Y. Generation of EPR-detectable nitrosyl-iron complexes in tumor target cells cocultured with activated macrophages. J Biol Chem 1991; 266:10162–10167.
83. Pantopoulos K. Weiss G, Hentze MW. Nitric oxide and oxidative stress (H_2O_2) control mammalian iron metabolism by different pathways. Mol Cell Biol 1996; 16:3781–3788.
84. Hausladen A, Privalle CT, Keng T, DeAngelo J, Stamler JS. Nitrosative stress: activation of the transcription factor OxyR. Cell 1996; 86:719–729.
85. Cordle SR, Donald R, Read MA, Hawiger J. Lipopolysaccharide induces phosphorylation of MAD3 and activation of c-Rel and related NF-κB proteins in human monocytic THP-1 cells. J Biol Chem 1993; 268: 11803–11810.
86. Traenckner EB-M, Wilk S, Baeuerle PA. A proteasome inhibitor prevents activation of NF-κB and stabilizes a newly phosphorylated form of IκBα that is still bound to NF-κB. EMBO J 1994; 13:5433–5441.
87. Henkel T, Machleidt T, Alkalay I, Kronke M, Ben-Neriah Y, Baeuerle PA. Rapid proteolysis of IκBα is necessary for activation of transcription factor NF-κB. Nature 1993; 365:182–185.
88. Ohmori Y, Tebo J, Nedospasov S, Hamilton TA. κB binding activity in a murine macrophage-like cell line. Sequence-specific differences in κB binding and transcriptional activation functions. J Biol Chem 1994; 269:17684–17690.
89. Xie QW, Kashiwabara Y, Nathan C. Role of transcription factor NF-κB in induction of nitric oxide synthase. J Biol Chem 1994; 269:4705–4708.
90. Murphy W, Muroi Y, Zhang X, Suzuki T, Russell SW. Both basal and enhancer κB elements are required for full induction of the mouse inducible nitric oxide synthase gene. J Endotoxin Res 1996; 3:381–393.
91. Shakhov AN, Collart MA, Vassalli P, Nedospasov SA, Jongeneel CV. κB type enhancers are involved in lipopolysaccharide-mediated transcriptional activation of the tumor necrosis factor α gene in primary macrophages. J Exp Med 1990; 171:35–47.
92. Collart MA, Baeuerle P, Vassalli P. Regulation of tumor necrosis factor alpha transcription in macrophages: involvement of four κB-like motifs and of constitutive and inducible forms of NF-κB. Mol Cell Biol 1990; 10:1498–1506.
93. Drouet C, Shakhov AN, Jongeneel CV. Enhancers and transcription factors controlling the inducibility of the tumor necrosis factor-α promoter in primary macrophages. J Immunol 1991; 147:1694–1700.
94. Hiscott J, Marois J, Garoufalis J, D'Addario M, Roulston A, Kwan I, Pepin N, LaCoste J, Nguyen H, Bensi N, Fenton M. Characterization of a functional NF-κB site in the human interleukin 1β promoter. Mol Cell Biol 1993; 13:6231–6240.
95. Dendorfer U, Oettgen P, Libermann TA. Multiple regulatory elements in the interleukin-6 gene mediate induction by prostaglandin, cyclic AMP and lipopolysaccharide. Mol Cell Biol 1994; 14:4443–4454.
96. Schreck R, Rieben P, Baeuerle PA. Reactive oxygen intermediates as apparently widely used messengers in the activation of the NF-κB transcription factor and HIV-1. EMBO J 1991; 10:2247–2258.
97. Staal FJT, Roederer M, Herzenberg LA, Herzenberg LA. Intracellular thiols regulate activation of nuclear factor κB and transcription of human immunodeficiency virus. Proc Natl Acad Sci USA 1990; 87: 9943–9947.
98. Shibanuma M, Kuroki T, Nose K. Inhibition by N-acetyl-L-cysteine of IL-6 mRNA induction and activation of NFκB by TNF-α in a mouse fibroblastic cell line, Balb/3T3. FEBS Lett 1994; 353:62–66.
99. Kumar S, Rabson AB, Gelinas C. The RxxRxRxxC motif conserved in all Rel/κB proteins is essential for DNA binding activity and redox regulation of the v-Rel oncoprotein. Mol Cell Biol 1992; 12:3094–3106.
100. Brennan P, O'Neill LAJ. 2-Mercaptoethanol restores the ability of nuclear factor κB (NF-κB) to bind DNA in nuclear extracts from interleukin 1-treated cells incubated with pyrollidine dithiocarbamate (PDTC). Evidence for the oxidation of glutathione in the mechanism of inhibition of NF-κB by PDTC. Biochem J 1996; 320:975–981.
101. Meyer M, Schreck R, Baeuerle PA. H_2O_2 and antioxidants have opposite effects on activation of NF-κB and AP-1 in intact cells: AP-1 as secondary antioxidant-responsive factor. EMBO J, 1993; 12:2005–2012.
102. Kretz-Remy C, Mehlen P, Mirault M-E, Arrigo A-P. Inhibition of IκBα phosphorylation and degradation and subsequent NF-κB activation by glutathione peroxidase overexpression. J Cell Biol 1996; 133: 1083–1093.
103. Peng HB, Rajavashisth TB, Libby P, Liao JK. Nitric oxide inhibits macrophage-colony stimulating factor gene transcription in vascular endothelial cells. J Biol Chem 1995; 270:17050–17055.
104. Shin WS, Hong YH, Peng HB, De Caterina R, Libby P, Liao JK. Nitric oxide attenuates vascular smooth muscle cell activation by interferon-γ. The role of constitutive NF-κB activity. J Biol Chem 1996; 271: 11317–11324.
105. Khan BV, Harrison DG, Olbrych MT, Alexander RW, Medford RM. Nitric oxide regulates vascular cell adhesion molecule 1 gene expression and redox-sensitive transcriptional events in human vascular endothe-

lial cells. Proc Natl Acad Sci USA 1996; 93:9114–9119.
106. Lander HM, Sehajpal P, Levine DM, Novogrodsky A. Activation of human peripheral blood mononuclear cells by nitric oxide-generating compounds. J Immunol 1993; 150:1509–1516.
107. Lander HM, Ogiste JS, Pearce SFA, Levi R, Novogrodsky A. Nitric oxide-stimulated guanine nucleotide exchange on p21^{ras}. J Biol Chem 1995; 270:7017–7020.
108. Peng H-B, Libby P, Liao JK. Induction and stabilization of IκBα by nitric oxide mediates inhibition of NF-κB. J Biol Chem 1995; 270:14214–14219.
109. Park KK, Lin HL, Murphy S. Nitric oxide regulates nitric oxide synthase-2 gene expression by inhibiting NF-κB binding to DNA. Biochem J 1997; 322:609–613.
110. Togashi H, Sasaki M, Frohman E, Taira E, Ratan, RR, Dawson TM, Dawson VL. Neruonal (type I) nitric oxide synthase regulates nuclear factor κB activity and immunologic (type II) nitric oxide synthase expression. Proc Natl Acad Sci USA 1997; 94:2676–2680.
111. Matthews JR, Botting CH, Panico M, Morris HR, Hay RT. Inhibition of NF-κB DNA binding by nitric oxide. Nuc Acid Res 1996; 24:2236–2242.
112. Assreuy J, Cunha FQ, Liew FY, Moncada S. Feedback inhibition of nitric oxide synthase activity by nitric oxide. Br J Pharm 1993; 108:833–837.
113. Sheffler LA, Wink DA, Melillo G, Cox GW. Exogenous nitric oxide regulates IFN-γ plug lipopolysaccharide-induced nitric oxide synthase expression in mouse macrophages. J Immunol 1995; 155:886–894.
114. Gally JA, Montague R, Reeke GN, Edelman GM. The NO hypothesis: possible effects of a short-lived, rapidly diffusible signal in the development and function of the nervous system. Proc Natl Acad Sci USA 1990; 87:3547–3551.
115. Sarte B, Stanford J, LaPrice WJ, Uhrich DL, Lockhart TE, Gelerinter E, Duffy NV. Nitrosylbis(diorganodithiocarbamato)iron complexes. Effect of organic substituents. Inorgan Chem 1978; 17:3361–3365.
116. Becker K. Schirmer RH. 1,3-Bis(1-chloroethyl)-1-nitrosourea as thiol-carbamoylating agent in biological systems. Meth Enzymol 1995; 251:173–188.
117. Munoz C, Pascual-Salcedo D, Castellanos MC, Alfranca A, Aragones J, Vara A, Redondo JM, de Landazuri MO. Pyrrolidine dithiocarbamate inhibits the production of interleukin-6, interleukin-8, and granulocyte-macrophage colony-stimulating factor by human endothelial cells in response to inflammatory mediators: modulation of NF-κB and AP-1 transcription factor activity. Blood 1996; 88:3482–3490.
118. Pilz RB, Suhasini M, Idriss S, Mienkoth JL, Boss GR. Nitric oxide and cGMP analogs activate transcription from AP-1-responsive promoters in mammalian cells. FASEB J 1995; 9:552–558.
119. Peunova N, Enikolopov G. Amplification of calcium-induced gene transcription by nitric oxide in neuronal cells. Nature 1993; 364:450–453.
120. Tabuchi A, Sano K, Oh E, Tsuchiya T, Tsuda M. Modulation of AP-1 activity by nitric oxide (NO) in vitro: NO-medicated modulation of AP-1. FEBS Lett 1994; 351:123–127.
121. Tabuchi A, Oh E, Taoka A, Sakurai H, Tsuchiya T, Tsuda M. Rapid attenuation of AP-1 transcriptional factors associated with nitric oxide (NO)-mediated neuronal cell death. J Biol Chem 1996; 271:31061–31067.
122. DAS KC, Lewis-Molock Y, White CW. Thiol modulation of TNFα and IL-1 induced MnSOD gene expression and activation of NF-κB. Mol Cell Biochem 1995; 148:45–57.
123. Galter D, Mihm S, Droge W. Distinct effects of glutathione disulfide on the nuclear transcription factors κB and the activator protein-1. Eur J Biochem 1994; 221:639–648.
124. Tran-Thi T-A, Decker K, Baeuerle PA. Differential activation of transcription factors NF-κB and AP-1 in rat liver macrophages. Hepatology 1995; 22:613–619.
125. Ghezzi P, Dinarello CA, Bianchi M, Rosandich ME, Repine JE, White CW. Hypoxia increases production of interleukin-1 and tumor necrosis factor by human mononuclear cells. Cytokine 1991; 3:189–194.
126. Neuschwander-Tetri BA, Bellezzo JM, Britton RS, Bacon BR, Fox ES. Thiol regulation of endotoxin-induced release of tumor necrosis factor α from isolated rat Kupffer cells. Biochem J 1996; 320:1005–1010.
127. Florquin S, Amraoui Z, Dobois C, Decuyper J, Goldman M. The protective role of endogenously synthesized nitric oxide in Staphylococcal enterotoxin B-induced shock in mice. J Exp Med 1994; 180:1153–1158.
128. Eigler A, Moeller J, Endres S. Exogenous and endogenous nitric oxide attenuates tumor necrosis factor synthesis in the nurine macrophage cell line RAW 264.7. J Immunol 1995; 154:4048–4054.
129. Munoz C, Castellanos MC, Alfranca A, Vara A, Esteban MA, Redondo JM, de Landazuri MO. Transcriptional up-regulation of intracellular adhesion molecule-1 on human endothelial cells by the antioxidant pyrrolidine dithiocarbamate involves the activation of activating protein-1. J Immunol 1996; 157:3587–3597.
130. Peng H-B, Rajavashisth TB, Libby P, Liao JK. Nitric oxide inhibits macrophage-colony stimulating factor gene transcription in vascular endothelial cells. J Biol Chem 1995; 270:17050–17055.
131. Satriano JA, Shuldiner M, Hora K, Xing Y, Sahn Z, Schlondorff D. Oxygen radicals as second messengers for expression of the monocyte chemotactic protein, JE/MCP-1, and the monocyte colony-stimulating factor, CSF-1, in response to tumor necrosis factor-α and immunoglobulin G. Evidence for involvement of reduced nicotinamide adeninedinucleotide phosphate (NADPH)-dependent oxidase. J Clin Invest 1993; 92:1564–1571.
132. DeForge LE, Preston AM, Takeuchi E, Kenney J, Boxer LA, Remmick DG. Regulation of interleukin 8 gene expression by oxidant stress. J Biol Chem 1993; 268:25568–25576.
133. Andrew PJ, Harant H, Lindley IJD. Nitric oxide regulates IL-8 expression in melanoma cells at the tran-

scriptional level. Biochem Biophys Res Comm 1995; 214:949–956.
134. Feng L, Xia Y, Garcia GE, Hwang D, Wilson CG. Involvement of reactive oxygen intermediates in cyclooxygenase-2 expression induced by interleukin-1, tumor necrosis factor-α, and lipopolysaccharide. J Clin Invest 1995; 95:1669–1675.
135. Colasanti M, Persichini T, Menegazzi M, Mariotto S, Giordano E, Caldarera CM, Sogos V, Laura GM, Suzuki H. Induction of nitric oxide synthase mRNA expression. Suppression by exogenous nitric oxide. J Biol Chem 1995; 270:26731–26733.
136. Persoons JHA, Schornagel K, Tilders FFH, de Vente J, Berkenbosch F, Kraal G. Alveolar macrophages autoregulate IL-1 and IL-6 production by endogenous nitric oxide. Am J Respir Cell Mol Biol 1996; 14: 272–278.
137. Patel JM, Zhang J, Block ER. Nitric oxide-induced inhibition of lung endothelial cell nitric oxide synthase via interaction with allosteric thiols: role of thioredoxin in regulation of catalytic activity. Am J Respir Cell Mol Biol 1996; 15:410–419.
138. Rengasamy A and Johns RA. Regulation of nitric oxide synthase by nitric oxide. Mol Pharmacol 1993; 44: 124–128.
139. Griscavage JM, Rogers NE, Sherman MP, Ignarro LJ. Inducible nitric oxide synthase from a rat alveolar macrophage cell line in inhibited by nitric oxide. J Immunol 1993; 151:6329–6337.
140. Griffith OW, Meister A. Potent and specific inhibition of glutathione synthesis by buthionine sulfoximine (S-n-butyl homocysteine sulfoximine). J Biol Chem 1979; 254:7558–7560.

39

Studies of the Primate Inflammatory Hemostatic Axis and Its Response to Inflammatory Mediators

Fletcher B. Taylor, Jr.
Oklahoma Medical Research Foundation, Oklahoma City, Oklahoma

INTRODUCTION

The response of the hemostatic system of the baboon to IV infusion of LD_{100} *E. coli* ranges from a capillary leak syndrome and shock with a relatively minor hemostatic response, to a massive consumptive coagulopathy and hemorrhage, to a thrombotic coagulopathy and irreversible microvascular thrombosis (1). Two major questions have arisen from our experience with these responses over the last 10 years. First, under what circumstance does malfunction of the hemostatic system (consumptive/thrombotic) play a role in the lethal chain of events leading to death versus simply serving as a marker of the primate response to *E. coli*? Second, in cases where this system is a link in the lethal chain of events, does it contribute by obstructing blood flow, by amplifying or perpetuating the inflammatory response, or both?

Before these questions can be addressed, those factors that control hemostatic activity (mediators, regulators) must be examined. This includes examination of the role of *modulators* (hormonal, metabolic), which while not directly involved in establishing this balance can shift it in favor of one of the three responses described above. In this chapter we describe the baboon hemostatic system as an integrated system. We describe how this system responds to hemostatic and inflammatory stress and how the hemostatic and inflammatory systems influence each other's responses to *E. coli* organisms and toxins.

THE FOUR FUNCTIONAL DOMAINS OF THE HEMOSTATIC SYSTEM

Figure 1A shows the four functional domains of the hemostatic system. The anticoagulant and coagulant domains acting through the vascular endothelium and platelets, respectively, compete to control *clot formation*. The fibrinolytic and antifibrinolytic domains compete in the same manner to control *clot removal*. All four domains respond to thrombin through its action on the vascular endothelium, platelets, and fibrinogen (2). The key to understanding this integrated system is that there is a hierarchy of responses to thrombin. The vascular endothelium and its associated anticoagulant/fibrinolytic domains are the most responsive, platelets are next, and fibrinogen is the last and least responsive. Under normal circumstances the small amounts of thrombin that are generated "spontaneously" in the microvasculature bind to receptors on the vascular endothelium such as thrombomodulin (TM) (3,4). These interactions lead to the generation of activated protein C (APC) (3) and the release of tissue plasminogen activator (t-PA) (5,6). The activated protein C inhibits further production of thrombin by inactivating factors Va and VIIIa in a negative feedback (3,4). The tissue plasminogen activator lyses any fibrin (fdp) that might have formed by activating plasminogen to plasmin (7).

Figure 1B shows that the coagulant and antifibrinolytic domains come into play in circumstances where the endothelium either is absent (laceration of a vessel)

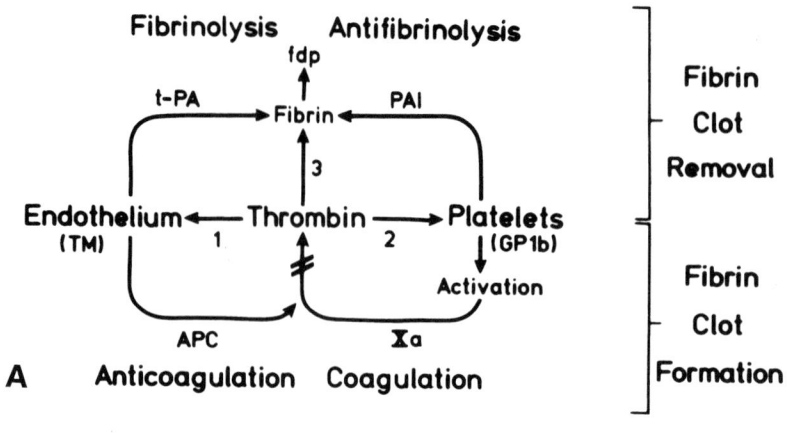

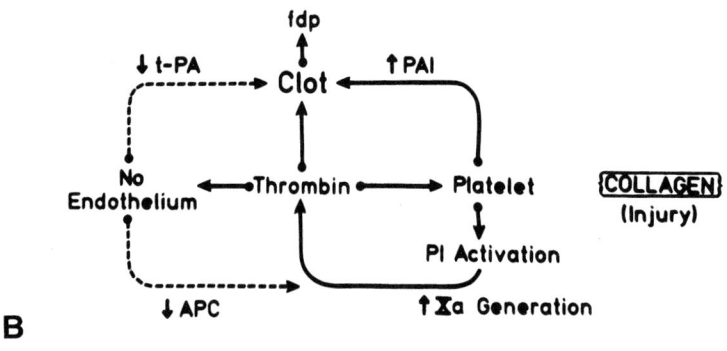

Fig. 1 (A) The four-domain construct of the unperturbed hemostatic system: The hemostatic system can be viewed as consisting of four functional domains: coagulation-anticoagulation and fibrinolysis-antifibrinolysis, all of which are regulated by thrombin. The coagulant-anticoagulant domains regulate clot formation, and the fibrinolytic-antifibrinolytic domains regulate clot removal. The numbers 1, 2, and 3 refer to the sensitivity to thrombin of thrombomodulin, platelets, and fibrinogen in that order. The other mechanistic aspects of this figure are reviewed in the text. (B) The effect of removal of the vascular endothelium as in a laceration: The removal of vascular endothelium as would occur when one lacerates or cuts oneself does two things. First, it *removes* the endothelium the source of anticoagulant-fibrinolytic activity, which allows a clot to form and seal the breech in the vascular wall. Second, it exposes the blood to tissue factor and collagen, which promote adherence and aggregation of platelets and initiate clot formation.

or is rendered dysfunctional by disorders of metabolism (e.g., diabetes) (8) or by inflammatory mediators (cytokines, endotoxins) (9–12). Absence of endothelium exposes blood platelets to collagen, initiating platelet aggregation, the assembly of clotting factors (Xa) (13), and the release from these platelets of plasminogen activator inhibitor (PAI) (14–16). Induction of dysfunctional endothelium or activation of monocyte/macrophages by inflammatory mediators results in expression of tissue factor (11,17), which also initiates both the procoagulant and antifibrinolytic events described above.

In summary, this four-quadrant system is designed, on the one hand, to keep the blood fluid and, on the other hand, to seal leaks in the vessel wall. Pathologic thrombosis occurs only when *both* the elements that preserve fluidity (the anticoagulant/fibrinolytic quadrants) and those that seal leaks (the coagulant/antifibrinolytic quadrants) are rendered dysfunctional by abnormal metabolic or inflammatory conditions. The balance between these four functional domains of the baboon hemostatic system will be reviewed next. This will be followed by studies of the effect of inflammatory mediators on this balance.

HEMOSTATIC STRESS: RESPONSES OF THE HEMOSTATIC SYSTEM TO INFUSION OF CLOTTING FACTOR Xa PLUS PHOSPHOLIPID VESICLES

IV infusion of factor Xa plus phospholipid vesicles (Xa/PCPS) produces a short burst of a very high concentrations of thrombin (18). Figure 2A shows that the generation of this large amount of thrombin in the baboon has no adverse effect because most of the thrombin produced binds preferentially to endothelial cell receptors including thrombomodulin (TM) (3,4). This thrombin-thrombomodulin complex activates equally large amounts of protein C to activated protein C (APC), which in turn inactivates clotting cofactors Va and VIIa in a negative feedback (3,4). As a result both platelet activation and fibrin formation is very limited. Also, relatively low concentrations of tissue plasminogen activator (t-PA) are produced, and there is very limited fibrinolytic activity as reflected by the production of low concentrations fibrin degradation products (fdp) (18).

In contrast, Figure 2B shows what happens when factor Xa PCPS is infused together with an antibody to protein C (18). In the absence of regulation by the anticoagulant domain, factor Xa PCPS mediates production of even larger concentrations of thrombin leading to a full blown consumptive coagulopathy (DIC) in which platelets are activated and sequestered and fibrin is formed depleting the baboon of circulating platelets and fibrinogen. This figure also illustrates that a very large concentration of t-PA is released from the endothelium by thrombin. The resulting intense and very rapid fibrinolytic response prevents the consumptive coagulopathy from becoming a catastrophic thrombotic coagulopathy.

Finally, Figure 2C shows what happens when factor XaPCPS is infused together with antibodies to both protein C *and* tissue plasminogen activator (unpublished observation). In the absence of both an anticoagulant and fibrinolytic regulator response, factor XaPCPS mediates a microvascular thrombotic coagulopathy which is lethal (MVT). It should be noted that thrombotic disorders of the hemostatic system, such as hemolytic uremic syndrome or thrombocytopenic thrombotic purpura (19) can exhibit an extensive microvascular thrombotic coagulopathy without exhibiting a consumptive coagulopathy. The reason for this is that even though fibrin is being deposited in the microvasculature, the turnover of fibrin is not sufficiently rapid to exceed the capacity of the liver to produce more fibrinogen. Except in unusual circumstances increased activity of *both* the coagulant and fibrinolytic domains is required to produce consumptive coagulopathy such as is shown in Figure 2B (18).

Before leaving this section covering the response of *regulatory* and *mediator* components of the hemostatic system to factor XaPCPS, the topic of *modulators* of this system should be introduced. Modulators are agents that alter the resting status of the host such that its normal response to the same stimulus is changed. Thus, depending on the condition of the host at the time of challenge, the host could exhibit responses to a stimulus such as XaPCPS that vary from a consumptive to a thrombotic coagulopathic response or both. A modulator does not mediate or regulate the hemostatic response. It operates within physiological "norms" either in the form of a physiological cycle (estrogen, cortisol, insulin, etc.) or as a part of a compensated host response to stress [heat shock proteins, macrophage inhibitory factor (MIF), interferon-γ (IFN-γ), or interleukin 6 (IL-6), etc.]. We believe that this is an extremely important distinction because it affords some insight into why two hosts, which appear to be normal and similar to each other (e.g., baboons from the same troop), sometimes react differently to the same stimulus (e.g., capillary leak and shock versus a consumptive coagulopathy and hemorrhage versus a thrombotic coagulopathy and renal failure.).

We believe that modulators may "tilt" the four-quadrant hemostatic playing field to favor one or another of its functional domains either by altering the balance between mediator or regulator components or the sensitivity of target tissues to the action of these components. Figure 2D shows the effect of IL-6 (40 μg) injected i.m. daily for 6–10 days on the response to XaPCPS (20). The controlled response shown in Figure 2A is converted into a microvascular thrombotic coagulopathic (MVT) response shown in Figure 2D. Although IL-6 has other effects, this change is presumably due to increases in the concentrations of both fibrinogen and platelets, shifting the balance in favor of the coagulant (↑fibrinogen) and antifibrinolytic (↑platelet PAI-1) domains. IL-6 may exert other effects on this system. These include effects on the vascular endothelium (dysfunctional) and elements of the anticoagulant and fibrinolytic domains, which operate from the endothelium such as tissue plasminogen activator (21), components of the protein C system (protein S) (22–24), tissue factor pathway inhibitor (TFPI) (25,26) antithrombin (AT-III) (27) and thrombin-activated fibrinolytic inhibitor (TAFI) (28).

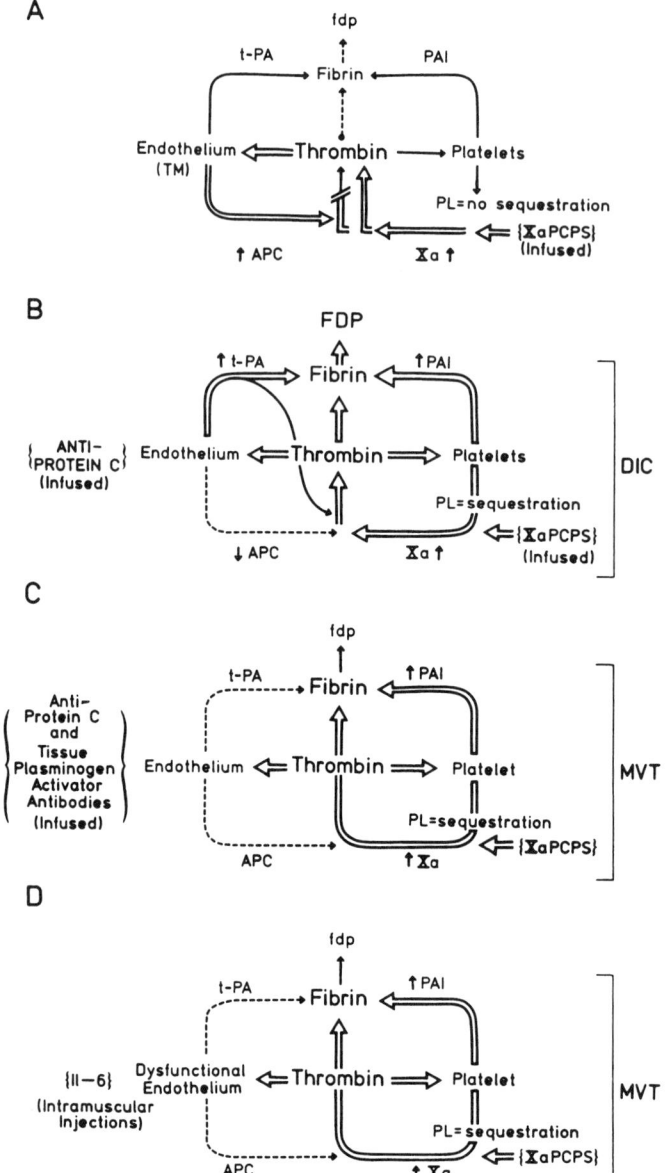

Fig. 2 (A) The effect of infusion of fXaPCPS on the normal hemostatic system: The infusion of fXaPCPS puts the normal hemostatic system under a procoagulant stress with which it is able to cope through an equally effective anticoagulant response. Even though there is a short burst of thrombin activity, it is immediately shut off by the generation of activated protein C through the assembly of the thrombin/thrombomodulin/protein C complex on the vascular endothelium. (B) The effect of infusion of fXaPCPS plus anti–protein C antibody on the normal hemostatic system: The infusion of both fXaPCPS and antibody to protein C inhibits the anticoagulant regulatory activity and allows thrombin generation to run unopposed. This leads to increased fibrin formation, which, however, is immediately reversed by the fibrinolytic regulatory activity (DIC). This emanates from the release of tissue plasminogen activator (t-PA) induced by thrombin from the vascular endothelium, which in this case is still intact and functioning normally. (C) The effect of infusion of fXaPCPS plus anti–protein C and anti–tissue plasminogen activator antibodies: The infusion of these two antibodies together with fXaPCPS removes the regulatory activity of both the anticoagulant and antifibrinolytic domains. In this case there is no fibrinolytic back-up activity to rescue the host from a massive microvascular thrombosis (MVT). This is driven by thrombin and facilitated even further by the thrombin-induced release of plasminogen activator inhibitor from platelets and endothelium. In this case the endothelium is still intact and functioning normally but is prevented from performing its regulatory functions by the two antibodies. (D) The effect of infusion of fXaPCPS following multiple IM injections of *interleukin-6* (*IL-6*): Infusion of fXaPCPS following injections of IL-6 also produces a lethal microvascular thrombosis (MVT). The reasons for this, however, are not as clear as for the microvascular thrombosis model shown above. IL-6 does promote increases in fibrinogen and platelet concentrations, which favor a more intense procoagulant response. In addition, however, IL-6 can also induce the development of a dysfunctional endothelium and a reduction of the fibrinolytic response to infusion of fXaPCPS.

INFLAMMATORY STRESS I: RESPONSES OF THE HEMOSTATIC SYSTEM TO INFUSION OF TUMOR NECROSIS FACTOR PLUS ANTIPROTEIN C PLUS PHOSPHOLIPID VESICLES

More obvious examples of *nonphysiological* modulators of the hemostatic system include members of the cytokine family such as TNF. Under what circumstances might tumor necrosis factor shift the balance of the four quadrant hemostatic system in favor of a thrombotic and/or consumptive coagulopathy? Tumor necrosis factor can induce the vascular endothelium and monocyte/macrophages to express tissue factor as shown by both in vitro (11,12) and in vivo (29) studies. Tissue factor in turn initiates activation of the coagulant domain of the hemostatic system to produce activated factor X (Xa) and thrombin (13).

Table 1 shows, however, that infusion of tumor necrosis factor alone through a catheter advanced through the superficial femoral vein of the baboon does *not* lead to development of a venous thrombus, even though it produces a transient leukopenia, rise in temperature, and sympathomimetic response (30). If however, tumor necrosis factor *plus* antibody to protein C is infused, a large venous thrombus will develop slowly from the clot in the ligated stump over a period of 48–72 hours until it occupies the venous tributaries including the inferior vena cava (30). Finally, Table 1 shows that co-infusion of active site–inhibited factor VIIa (fVIIai) prevents this thrombus from developing (31). Thus, the three conditions involving abnormalities of *flow*, vessel *wall*, and *blood* described by Virchow as being necessary for development of venous thrombosis are met (32,33).

First, there is a *flow* abnormality with the site of initial clot formation provided by the ligated stump of the superficial femoral vein opening into the common femoral vein. Second, there is the expression of tissue factor induced by tumor necrosis factor by monocytes and the endothelial *wall* of the veins draining the extremity. Third, there is the hypercoagulability of the *blood* induced by antibody to protein C. All three of these conditions are necessary to shift the balance of the four-quadrant hemostatic system in favor of the growth of a pathological venous thrombus from a clot at the site of venous ligation. It is important to emphasize that inflammatory stress can contribute to meeting

Table 1 Deep Vein Thrombotic Model: Establishing Conditions and Controlling Variables in the Baboon Model of Deep Vein Thrombosis

Agent and study conditions	Thrombin-antithrombin (TAT) complexes	Fibrinogen consumption	Thrombus
Catheter/ligation superficial femoral vein only	−	−	−
Catheter/ligation + TNF (100 µg/kg)	+	−	−
Catheter/ligation + anti-PC (2 mg/kg)	+	−	−
Catheter/ligation + anti-PC (2 mg/kg) + TNF (100 µg/kg)	+	−	+
Catheter/ligation + anti-PC (2 mg/kg) + TNF (100 µg/kg) + DEGR VIIa (280 µg × 5)[a]	−	−	−

In addition to cannulation and ligation, tumor necrosis factor (TNF) and anti–protein C antibody (anti-PC) are necessary for development of deep vein thrombosis. Details of procedures are as reported by Taylor et al. (31). Briefly, the femoral artery and vein are cannulated under anesthesia. The TNF and anti-protein C are infused as a bolus through a catheter in the cephalic vein at T − 10 and T − 5 minutes, respectively. The animals are observed (vital sign, blood CBC, chemistries, ultrasound, and [125I] fibrinogen scan) for 6 hours and again at 24 or 48 hours. The animals are sacrificed, autopsy is performed, and tissue samples from the area of thrombosis are harvested for light- and electron-microscopic examination and immunostaining. The thrombus is then dissected free, dried, and weighed.
[a]DEGR VII: Factor VIIa inactivated with dansyl-glu-gly-arg chloromethyl ketone.

these conditions. This combination of factors demonstrates the principle that both upregulation of one of the effector quadrants (coagulant) and downregulation of one of the regulator quadrants (anticoagulant) of the hemostatic system must occur simultaneously in order for an *uncompensated* coagulopathic response to develop.

Under what conditions can this relatively localized thrombotic coagulopathy be converted into a systemic consumptive coagulopathy? Figure 3A shows the combination of a flow abnormality, the induction of a dysfunctional endothelium locally with an upregulation of tissue factor expression (TF) by tumor necrosis factor, and the downregulation of protein C anticoagulant activity by antibody to protein C are sufficient to promote growth of a thrombus, however, they are not sufficient to produce a *systemic* coagulopathy (30). Since phospholipid vesicles (PCPS) amplify the effects of infused *exogenous* factor Xa (Fig. 2), co-infusion of these vesicles with tumor necrosis factor and antibody to protein C should provide the additional lipid surface needed to amplify the activity of the *endogenous* factor Xa and produce a systemic consumptive coagulopathy. Figure 3B illustrates that this is the case and that this response (30) is similar to the one shown in Figure 2C in which exogenous fXa (instead of TNF-induced endogenous

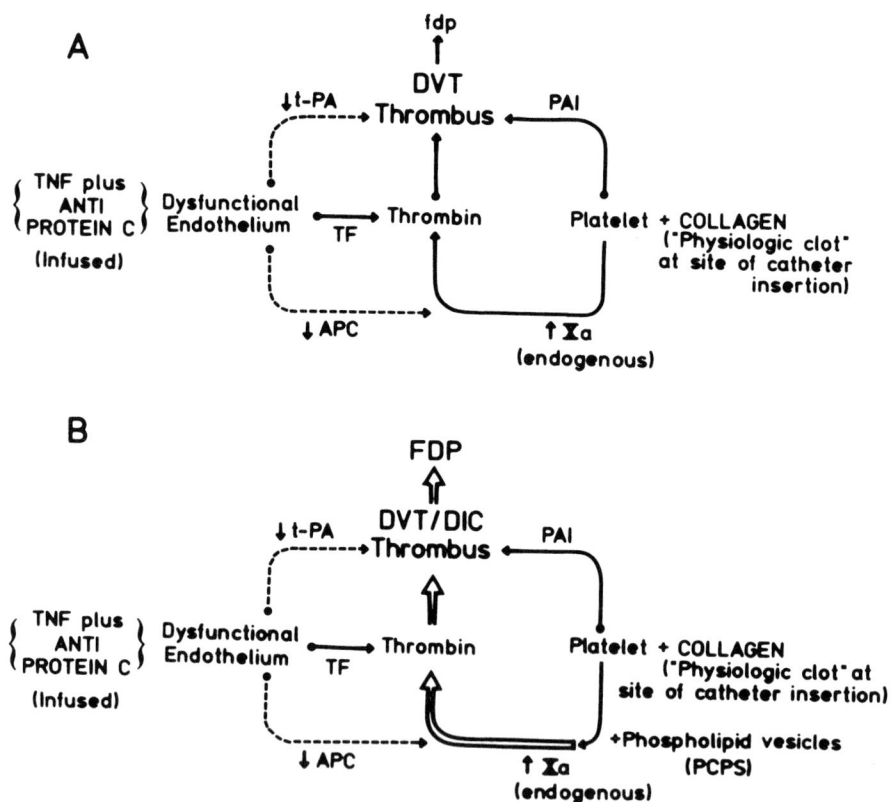

Fig. 3 (A) Effect of placement of a venous catheter and infusion of tumor necrosis factor (TNF) plus anti–protein C antibody: The venous catheter, which is removed followed by ligation of the superficial femoral vein, produces a flow abnormality and a venous stump in which a venous clot can form and *grow* into a large pathological thrombus. This requires, however, the coinfusion of tumor necrosis factor (TNF) and anti–protein C antibody. The combination of a dysfunctional endothelium and monocytes expressing tissue factor (TF), a reduced protein C activity, and the above described flow abnormality is sufficient to produce a deep vein thrombus (DVT). (B) Effect of coinfusion of phospholipid vesicles (PCPS) into the model of deep vein thrombosis described above. Coinfusion of phospholipid vesicles (PCPS) together with tumor necrosis factor (TNF) and anti–protein C antibody into this model of deep vein thrombosis provides additional phospholipid surface area. This additional surface area is sufficient to concentrate the *endogenous* factor Xa produced in the vicinity of the deep vein thrombus and convert a local thrombotic (DVT) into a systemic consumptive (DIC) coagulopathy.

fXa) was co-infused with the antibody to protein C (18). This amplification by infused phospholipid vesicles can be likened to the production of showers of microvesicles by complement (C5b-9) from platelet and cell membranes such as occurs in Dengue fever (34) and certain malignancies (35).

INFLAMMATORY STRESS II: SYSTEMIC CONSUMPTIVE COAGULOPATHIC RESPONSE OF THE HEMOSTATIC SYSTEM TO INFUSION OF *E. COLI* ORGANISMS

Descriptive Studies

Although the systemic consumptive coagulopathic (DIC) response to LD_{100} *E. coli* is more complex than has been shown in the reconstruction studies described above, it still can be rationalized using the four-quadrant construct. First, however, the clinical and clinical chemical responses over time will be described and divided into *stages* (36,37). Second, these events, which have been divided into stages, will be described in anatomical terms in which the blood elements that interact with the structures of the vessel wall are described (36,37).

Figure 4A shows the sequence of clinical and clinical chemical responses to *E. coli* over time and divides the time course into four stages (I–IV). Figure 4B pictures the structural elements involved that produce each of these four stages (1–4). Figure 4A shows that Stage I begins with the infusion of *E. coli* and ends 2 hours later when the white cell count has reached its nadir and the inflammatory mediators (TNF, t-PA, elastase) have reached their peak. During this period the *E. coli* organisms are phagocytosed by neutrophils and fixed macrophages of the lungs, lever, and spleen. Figure 4B shows that during Stage I the activated macrophages release the inflammatory mediators (TNF, interleukins) and the activated neutrophils marginate and adhere to the capillary and postcapillary endothelium releasing proteolytic enzymes. The neutrophils release elastase, and the perturbed endothelium in turn releases tissue plasminogen activator. Figure 4A shows that all this occurs during this 2-hour interval of Stage I. These events are reflected by a sharp fall in white cell count, the appearance in plasma of TNF and elastase from the activated macrophages and neutrophils, and the appearance of tissue plasminogen activator from the endothelium, which is bearing the brunt of this aberrant inflammatory response. This first stage is termed the inflammatory stage.

Figure 4A shows that Stage II begins at T = +2 hours with a fall in fibrinogen concentration just after a transitory fall in blood pressure and ends at T = +6 hours when the fall in fibrinogen concentration reaches its nadir. Figure 4B shows that Stage I is followed by Stage II in which there is expression of tissue factor (TF) not only by activated monocyte/macrophages but also by the dysfunctional microvascular endothelium of certain vascular beds. This activates the clotting cascade to produce a massive consumptive coagulopathy, or DIC, on the luminal surface of the endothelium. This second stage is termed the coagulopathic stage. These first two stages coincide with hyperdynamic or warm shock and with the clearance of *E. coli* organisms from the circulation at T = +6 hours.

Figure 4A shows that Stage III begins at T = +6 hours with a rise of the plasma concentrations of markers of cell injury (e.g., SGPT), which are accompanied by a continuing steady decline in blood pressure and platelet count. Figure 4B shows that in this process the combined inflammatory and coagulopathic events of Stage I and II drives the microvascular endothelium deeper into a dysfunctional state leading to abnormal function as reflected by a more severe DIC (luminal surface) and by capillary leak (abluminal surface). This results in intra- and extracellular edema of parenchymal tissues and a transition from reversible to the irreversible cell injury of Stage IV beginning at approximately 10–12 hours. These two later stages are termed cell injury (reversible) and cell degeneration (irreversible). They coincide with the hypodynamic or cold shock phase of the baboon response to LD_{100} *E. coli* (Fig. 4A).

Mechanism Studies

Unlike the baboon models of the hemostatic (XaPCPS) or inflammatory stress (TNF) described earlier, which were reconstructed component by component, analysis of the mechanism of the response to the whole *E. coli* organisms required intervention with specific antibodies against those mediators and regulator factors thought to be most important. A key question in these studies was whether or not the coagulopathy seen in this model was a link in a chain of lethal events or an epiphenoma. We concluded that it was an epiphenoma for the following reasons. First, microvascular thrombosis, while it occurred, was *not* irreversible and was not uniformly found at autopsy (38). Second, complete inhibition of tissue thrombosis, hemorrhage, and infarction with a delayed infusion of tissue factor pathway inhibitor *did not* protect against LD_{100} *E. coli* (39).

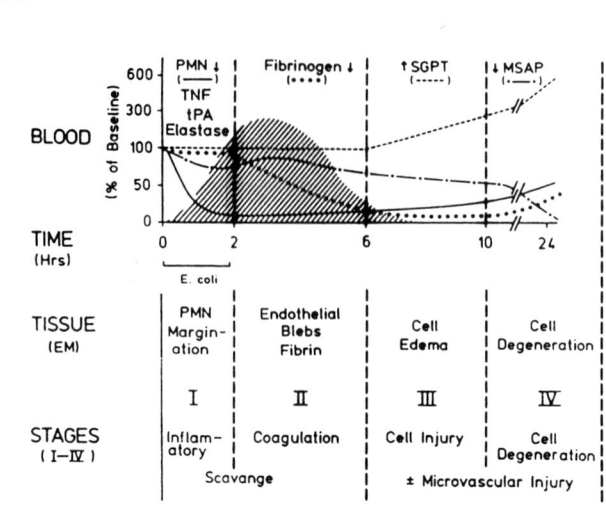

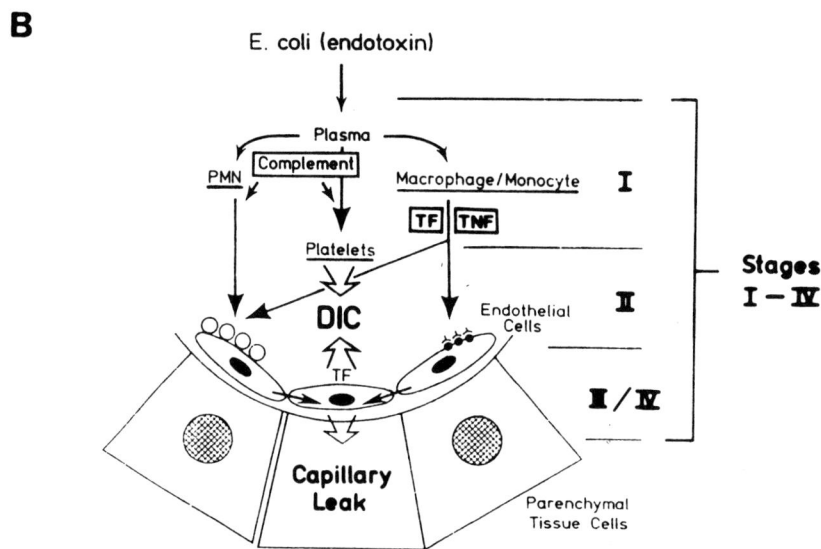

Fig. 4 (A) Disseminated intravascular coagulation (DIC) model: description of the sequence of events and stages (I to IV) following infusion of LD_{100} *E. coli*: The sequence of events in this model can be divided into four stages: inflammatory stage 1 (0–2 hr), coagulopathic (DIC) stage 2 (2–6 hr), and cell injury and cell degeneration, stages 3 and 4 (6–10 and 10+ hr, respectively). The hatched area depicts the concentration of *E. coli* organisms in the blood. They reach a maximum concentration of approximately 1×10^7 CFU/ml blood at 2 hr, after which they are cleared by 8 hr. Stage I is dominated by the fall in leukocyte count and the appearance of tumor necrosis factor, elastase, and tissue plasminogen activator in plasma, all of which reach peak concentrations by T + 2 hr. Stage II is dominated by the fall in fibrinogen concentration and rise in fibrin degradation products. Stages III and IV are dominated by a rise in concentration of markers of cell injury (SGPT) and a steadily falling platelet count and blood pressure (MSAP). Animals die between 16 and 30 hr. (B) Disseminated intravascular coagulant (DIC) model of baboon response to LD_{100} *E. coli*: *E. coli* activates complement components, neutrophils, and macrophage/monocytes which release mediators. These mediators perturb the vascular endothelium to initiate and amplify activation of coagulant factors (tissue factor, TF) to produce a massive systemic disseminated intravascular coagulation (DIC) response. Finally, in response to these explosive events occurring within the vasculature, the adjacent parenchymal tissue on the abluminal side of the endothelium accumulates fluid (capillary leak) and ultimately undergoes degeneration.

Third, complete inhibition of both the consumptive and thrombotic coagulopathic responses to LD_{100} *E. coli* with active site inhibited factor Xa also did not protect (40). Conversely, antibody to tumor necrosis factor, while it protected, did *not* inhibit the consumptive coagulopathy (41). This suggests that *microvascular thrombosis*, though an excellent marker of the severity of the response, is not a link in the lethal chain of events.

On the other hand, inhibition of the coagulopathic response with an antibody to tissue factor (42), active site inhibited factor VIIa (43), or tissue factor pathway inhibitor given early (25,26) protected at least 60–70 % of the animals against LD_{100} *E. coli*. This suggests another role, perhaps a pro-inflammatory role, for these more proximal coagulation factors, particularly the tissue factor/VIIa complex. It is now known that the active site inhibited fVIIa inhibits calcium flux across cell membranes (44,45) and that both active site inhibited factor VIIa (43) and tissue factor pathway inhibitor (25,26) attenuate the appearance of IL-6 and IL-8 in plasma following LD_{100} *E. coli* infusion. These results suggest that activation of proximal components of the hemostatic system can amplify and/or perpetuate an inflammatory response once it is initiated.

In contrast to the above intervention studies against *mediators* in which antibodies were used to protect against lethal *E. coli*, antibodies against *regulators* were used in the sublethal *E. coli* model to see if it could be converted into a lethal model. Intervention with an antibody that prevents the activation of protein C but does not inhibit activated protein C converted a sublethal into a lethal response to sublethal *E. coli* (46). This was reversed by co-infusion of activated protein C (46). Essentially the same results were obtained co-infusing an antibody to tissue factor pathway inhibitor with endotoxin into rabbits (47). Conversely, supplementation of these two naturally occurring regulators by infusing them in to the LD_{100} *E. coli* model protected 60–70% of the animals (25,46). As stated earlier, this was accompanied by attenuation of IL-6 and IL-8 release into plasma. Supplementation with AT-III, which inhibits proximal coagulant factors IX, X as well as thrombin, also protected, although infusion of large amounts of antithrombin were required before the infusion of *E. coli* (27). The rationale that argues for an anti-inflammatory role of agents that inhibit the proximal coagulation factors in the sublethal *E. coli* model also appears to hold in the LD_{100} *E. coli* studies. It is important to add that these regulatory agents also share a close association with the microvascular endothelium through their direct association with endothelial cell receptors and surface structures (48). These include tissue factor pathway inhibitor binding to glycosaminoglycans (49,50), protein C to thrombomodulin (3,4), activated protein C to endothelial protein C receptor (51), factor VIIa to tissue factor (52,53), and antithrombin to glycosaminoglycans (54). An additional rationale, therefore, explaining the relative efficacy of the inhibitors of proximal coagulation factors versus inhibitors such as active site inhibited factor Xa and hirudin may be that they protect the surface of the microvascular endothelium from injury by elements of an uncontrolled, aberrant, inflammatory response to *E. coli*.

Modulators are as important in this *E. coli* model as in the fXaPCPS hemostatic stress model. As stated in the introduction, the baboon response to LD_{100} *E. coli* can range from capillary leak, shock, and death in less than 6 hours to microvascular thrombosis, renal failure, and death in 5 days. Animals in estrus receiving sublethal *E. coli* develop massive microvascular thrombosis. Animals pretreated with 40 μg/kg IL-6 IM for 6–10 days do likewise. Other factors that are elevated in chronic subclinical inflammatory states include macrophage inhibitory factory (MIF) and interferon-γ (IFN-γ), both of which block the cellular responsiveness to cortisol (55). The response to *E. coli* of animals made deficient in or unresponsive to cortisol is a rapid cardiovascular collapse with capillary leak (56). Thus, the prechallenge status of the host as determined by its hormonal, metabolic, and underlying immune status can in turn determine the type of clinical response to *E. coli*. The same variability of responses appears to apply to patients. This complicates intervention strategy even when the type of the organism is known. Thus, in absence of a polyvalent therapy, we believe that an understanding of the underlying conditions that predispose to one or the other type of response is needed for effective prophylaxis and/or intervention.

Review and Integration (Four-Quadrant Constructs)

The preceding description of the clinical events and mechanisms of some of the events following *E. coli* can be viewed conceptually using the four-quadrant construct of the hemostatic system presented previously. Figure 5A and B compares the response to fXaPCPS plus antibody to protein C with that to LD_{100} *E. coli*. The two illustrations are similar, except that the animal experiencing a pure hemostatic stress (A) lives, while the animal experiencing a combined inflammatory/hemostatic (DIC) stress (B) dies.

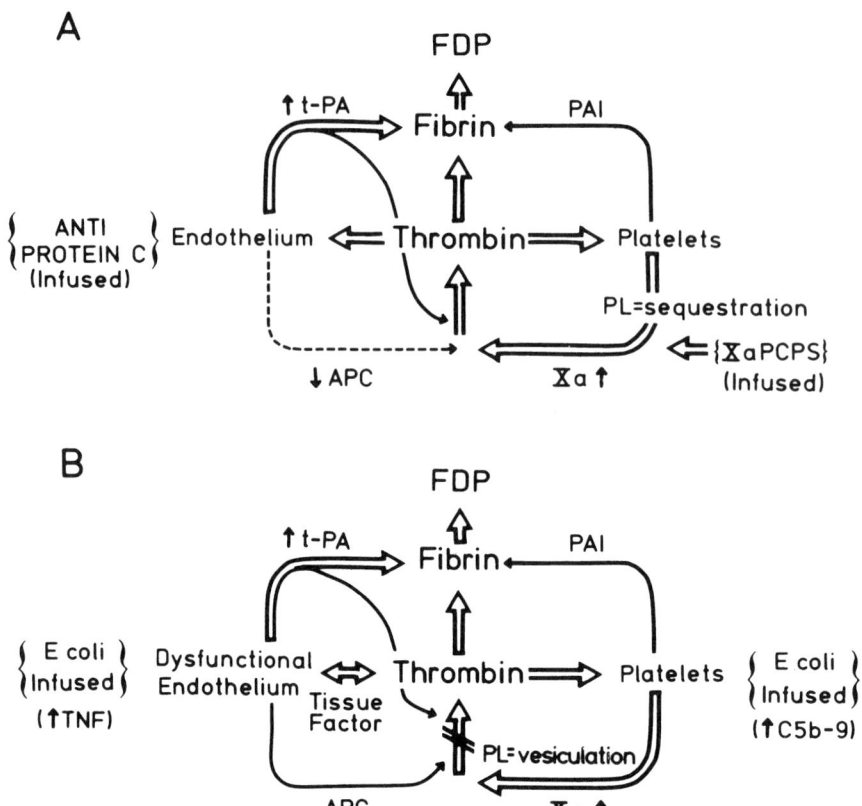

Fig. 5 Disseminated intravascular coagulant (DIC) response to infusion of factor XaPC/PS plus anti-protein C antibody (A) and to LD_{100} *E. coli* (B): The inflammatory effect of LD_{100} *E. coli* mimics the consumptive DIC response to factor XaPC/PS. There is an intense upregulation of coagulant (tissue factor) (Xa) activity and downregulation of anticoagulant (APC) activity followed by an explosive generation of thrombin and fibrinolytic (t-PA) activity such as is seen following infusion of factor XaPC/PS. The only differences are the (1) the host in responding to LD_{100} *E. coli* uses endogenous factors to reproduce the effects of factor XaPC/PS and anti-protein C antibody, and (2) the inflammatory response to LD_{100} *E. coli*, in addition to triggering the DIC response, triggers a succession of self-propagating inflammatory responses (i.e., free radical injury) that are irreversible and ultimately kill the host.

INFLAMMATORY STRESS III: LOCAL THROMBOTIC COAGULOPATHIC RESPONSE OF THE HEMOSTATIC SYSTEM TO INFUSION OF *E. COLI* 0157:H7, SHIGA-LIKE TOXIN

Descriptive Studies

Although the localized thrombotic coagulopathic response to *E. coli* 0157:H7 or Shiga-like toxin also is complex, it also can be rationalized using the four-quadrant construct described above. First, however, the clinical and clinical chemical responses over time will be described and divided into stages. Second, these events or stages will be described in anatomical terms in which the elements that interact to produce this disorder are described.

Figure 6 shows the sequence of clinical and clinical chemical responses to Shiga-like toxin over time and divides the time course into three stages (I–III). Figure 7 pictures the structural elements that are involved and that produce these responses. Figure 6 shows that stage I begins with the bolus infusion of Shiga-like toxin (0.1 μg/kg) and ends 12 hours later when the first signs of renal failure begin (57). During this period there is *no* significant change in temperature and white count and only a modest rise in fibrinogen concentration and fibrin degradation products. During this period the Shiga-like toxin (β subunit) binds (Kd = 4.6×10^{-8} M) to glycolipid receptors (globotriasyl ceramide, Gb_3) (58), which are located on proximal tubular and lower gastrointestinal mucosal epithelium (58). SLT-1 is a specific N glycosidase that inhibits protein synthesis by

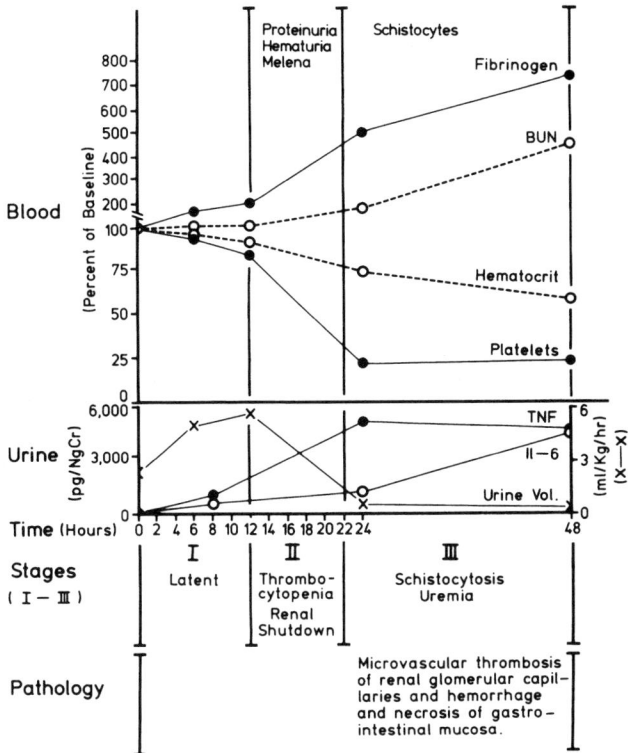

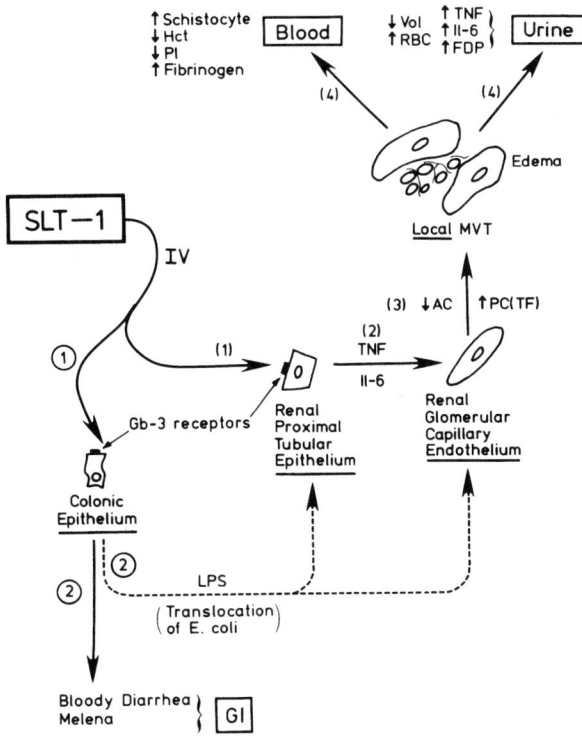

Fig. 6 The Response to Shiga-like toxin: Stage I following a bolus infusion 12 hours is termed the latent stage. During this T − 0 to T + 12 hour interval there is only a mild inflammatory response. Stage II, which runs from T + 12 to T + 24 hours is termed the thrombocytopenic/renal failure stage. During this interval the platelet count and urine volume fall and the urine TNF/IL-6 and plasma fibrinogen concentrations rise. It is not until stage III that the hematocrit begins to fall significantly and the creatinine concentration and schistocyte counts rise. This stage is termed the schistocytosis/uremic stage and runs from T + 24 hours until the time of death. Proteinuria and hematuria are first observed at T + 12 hours, and melena is first observed at T + 24 hours.

Fig. 7 Possible mechanism by which SLT-1 produces a microvascular thrombosis. The toxin first binds to constitutively expressed Gb_3 receptors on the epithelium of the proximal tubules of the kidney and of the gastrointestinal mucosa. The SLT-1 is directly toxic to these cells. In addition, however, the host renal parenchymal cells (epithelium; mesangial cells) produce cytokines (TNF, IL-6), which sensitize the adjacent glomerular capillary endothelium to effects of the toxin (induces Gb_3 synthesis). This endothelium becomes dysfunctional and expresses tissue factor. This dysfunctional endothelium together with platelets combine to produce microvascular thrombosis of the renal glomerular capillaries. There is surprisingly little inflammatory cellular activity in the renal tissues, and it is not clear to what degree these thrombotic lesions contribute to renal failure in this model. In contrast to the kidney, there is considerable local inflammatory activity and hemorrhage in the gastrointestinal mucosa. It is through this route that we postulate that a true systemic gram-negative sepsis could develop through translocation of bacteria from the gut to the general circulation.

depurinating a single adenine residue from 28_smRNA (59). Figure 7 shows that cytokines (TNF, IL-6) can be produced by renal parenchymal cells adjacent to glomerular capillary endothelium (60,61). These might include the proximal tubular epithelial cells themselves, which can produce cytokines on exposure to endotoxin (61), or possibly renal mesangial cells. Whatever the local source of cytokines, it is established that the glomerular capillary endothelium can be sensitized to SLT-1 by exposure cytokines and/or endotoxin and hence rendered dysfunctional (62). This suggests that

the organelle disintegration of these cells carrying the Gb3 receptors plus the microvascular thrombosis of adjacent glomerular capillaries, which do not carry Gb3 receptors, is a product of the combined direct toxic and host mediator effects of SLT-1, respectively. This first stage is called the latent stage.

Figure 6 shows that stage II begins at T + 12 hours with a sharp fall in platelet count accompanied by a decline in hematocrit, urine volume together with proteinuria, hematuria, and melena. This stage ends at T = 24 hours. In the process, even though the local epithelial cell injury now is accompanied by dysfunction of the endothelial cells of adjacent capillaries, and even though platelet-rich thrombi are producing a local MVT (Fig. 7), the only systemic indication of a thrombotic coagulopathy during stage II is the fall in platelet count. There is no fall in fibrinogen concentration, which instead continues to rise. Finally, Figure 6 shows that the concentration of both tumor necrosis factor (TNF) and interleukin-6 (IL-6) in the urine peak during this stage, while neither of these cytokines appear in the blood. This suggests, as discussed above, that the initial renal dysfunction is local, involving the proximal tubular epithelium through the *direct* effects of the toxin and the glomerular capillary endothelium *indirectly* through a host response (i.e., priming by cytokines from adjacent cells). This stage is termed the thrombocytopenic stage.

Figure 6 shows that stage III begins at T + 24 hours with the appearance if fragmented red blood corpuscles (schistocytes) reflecting the breakup of these corpuscles passing through the partially obstructed vessels. This also is accompanied by an increasing anion gap and a rising blood urea nitrogen (BUN) and creatinine. Although TNF and IL-6 appear in quantity in the urine during stages II and III, at no time are cytokines detected in the blood (detection limit ≥ 5 ng). There are also no significant changes in the white cell count or the appearance of thrombin-antithrombin complexes. The animals die in 48–72 hours with microvascular thrombosis, hemorrhage, and necrosis of renal and mucosal epithelial structures. There is no influx of neutrophils. Finally, Figure 7 shows a hypothetical breakdown of the mucosal barrier of the gastrointestinal tract followed by translocation of *E. coli* organisms into the general circulation in which case this localized *thrombotic* coagulopathy is converted into a systemic *consumptive* coagulopathy.

Mechanism Studies

Studies of the fXaPCPS model demonstrated that co-infusion of antibodies to both protein C and tissue plasminogen activator was necessary to produce a thrombotic coagulopathy. The above description of the thrombotic microangiopathy induced by Shiga-like toxin suggests that platelets that contain plasminogen inhibitor (PAI-1) may play a dominant role not unlike that thought to be associated with the hemolytic uremic syndrome (HUS) (63) and thrombotic thrombocytopenic purpura (TTP) (64) seen in clinical practice. Because of the localized characteristics of this disorder produced by Shiga-like toxin, intervention studies were not attempted. Reconstitution studies, however, using the native acute-phase protein C4BP to neutralize the protein S cofactor of the protein C anticoagulant system were done. Infusion of C4b binding protein alone produced a slow decline in platelet count, which was prevented by saturating the C4b binding protein with protein S before infusion (22,23). Thus, the decline in platelet count was clearly due to failure of a *physiological* function of the protein C system. Infusion of C4b binding *plus* sublethal *E. coli* was sufficient to produce a thrombotic coagulopathy in which these dysfunctional platelets played a dominant role (23,24). This conclusion was supported by the observation that this response could be prevented by the co-infusion of antibody to platelet glycoprotein IIb-IIIa (65).

Review and Integration (Four-Quadrant Construct)

The preceding description of clinical events and mechanism of some of these events following Shiga-like toxin can be viewed conceptually using the four-quadrant construct. Figure 8 compares the effect of infusing fXaPCPS plus antibodies to protein C and tissue plasminogen activator (Fig. 8A) with the effect of infusing Shiga-like toxin (Fig. 8B). The effects on the hemostatic system of Shiga-like toxin are similar to those of factor XaPCPS *plus* the above two antibodies except that TNF and IL-6 produced by the renal parenchymal cells (60,61) on the abluminal side of the capillary endothelium substitute for the two antibodies by producing a dysfunctional endothelium. The combined effects of activated platelets coming into contact with this dysfunctional capillary endothelium produces platelet-rich microvascular thrombi. The plasminogen activator inhibitor released locally by these platelets prevents dissolution of the thrombi. This produces an irreversible microvascular thrombotic coagulopathy with hemorrhage and infarction in both the kidney and GI tract. This is in contrast to the systemic consumptive coagulopathy following *E. coli* in which a fully operational fibrinolytic system lyses the microvascular thrombi. These local thrombotic events, however, can become systemic consumptive events if there is sufficient breakdown of the mucosal barrier of the gastrointestinal tract to permit translocation of *E. coli* organisms into the general circulation.

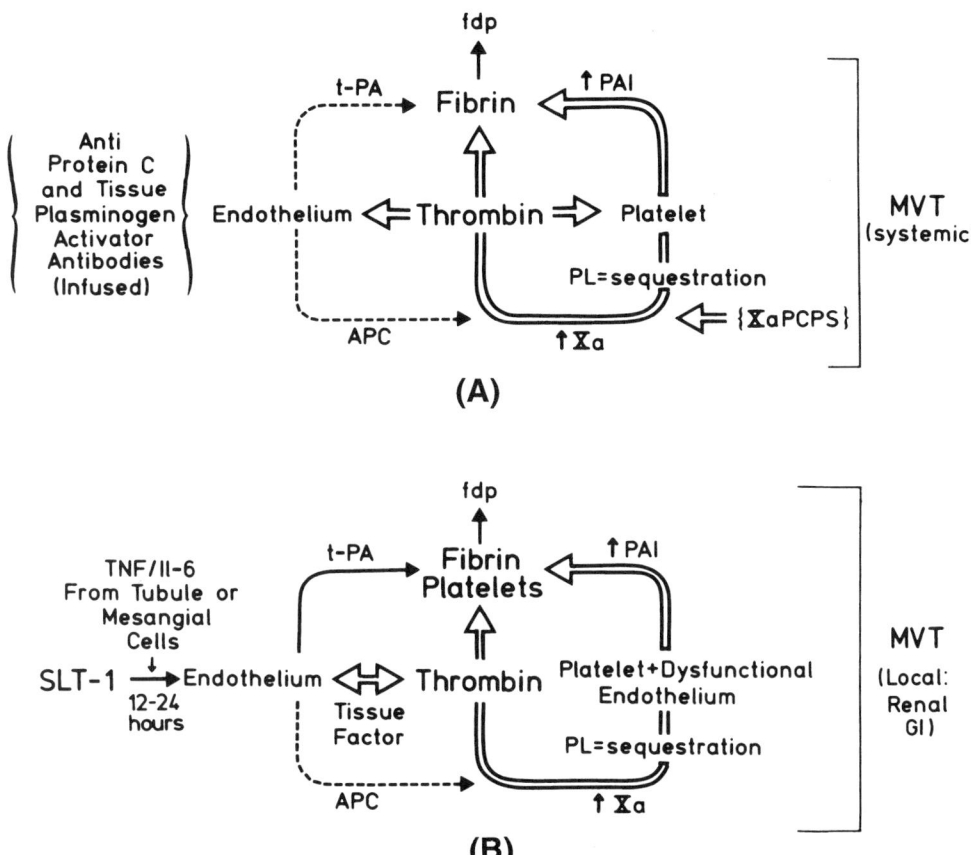

Fig. 8 (A) The effect of infusion of fXaPCPS plus anti–protein C and anti–tissue plasminogen activator antibodies: The infusion of antibodies together with fXaPCPS removes the regulatory activity of both the anticoagulant and anti fibrinolytic domains. In this case there is no fibrinolytic back-up activity to rescue the host from a massive microvascular thrombosis (MVT). This is driven by thrombin and facilitated even further by release of plasminogen activator inhibitor from platelets and endothelium induced by thrombin. In this case the endothelium is still intact and functioning normally, but is prevented from doing so by the two antibodies. (B) The effect of infusion of Shiga-like toxin: The bolus infusion of Shiga-like toxin (SLT-1) also attenuates the regulatory activity of both the anticoagulant and fibrinolytic domains arising from the glomerular capillary endothelium of the kidney. In contrast to the rapid systemic microvascular thrombosis (MVT) produced by XaPCPS and the two antibodies described above, the microvascular thrombosis produced by SLT-1 requires 12–24 hours and is restricted to the renal glomerular capillaries. This is because the action of SLT-1 is restricted to those tissues which bear the appropriate Gb_3 receptors. First, SLT-1 binds to receptors on the proximal tubular epithelium. Second, adjacent parenchymal cells produce cytokines. Third, these cytokines sensitize the glomerular capillary endothelium to SLT-1. Fourth, these dysfunctional endothelial cells produce tissue factor and interact with platelets to produce renal microvascular thrombosis.

PERSPECTIVES ON THE ROLE OF HEMOSTATIC FACTORS UNDER VARIOUS CONDITIONS BASED ON STUDIES OF THE BABOON MODEL OF SEPSIS

The response of the hemostatic system to *E. coli* can be a part of a chain of lethal events or an epiphenomenon depending on the status of the host and on the extent to which proximal (tissue factor/VIIs) and terminal (thrombin, fibrin) factors of the coagulation cascade come into play.

In circumstances where the onset of sepsis is rapid and overwhelming, such as in septic abortion, the combination of estrogen/progesterone effects plus the easy access of bacteria to the general circulation through an exposed endometrial microvasculature favor an explosive consumptive coagulopathy (DIC) and thrombosis. This DIC in itself, irrespective of associated inflammatory events, can be lethal because of hemorrhage on a massive scale.

In other conditions that are acute but are not influenced by hormonal factors, activation of the hemostatic system including consumption of terminal factors such as fibrinogen and development of thrombosis may not be critical in the chain of lethal events. This may be true even in cases where the classic picture of hemorrhagic adrenal glands is found at autopsy. Here the events driven by inflammatory mediators predominate over those involving the coagulation factors, as evidenced by a massive capillary leak and shock. In these cases there is undoubtedly expression and activation of proximal clotting factors, which may amplify the original inflammatory signal in a positive feedback. Receptors serving both the hemostatic mediator (tissue factor) and regulator (endothelial protein C receptor) arms of the hemostatic system can play such a role by altering the calcium flux in the course of forming complexes with their ligands. The intensity of these events involving inflammatory and proximal clotting factors and the efficiency of the fibrinolytic response are such that oc-

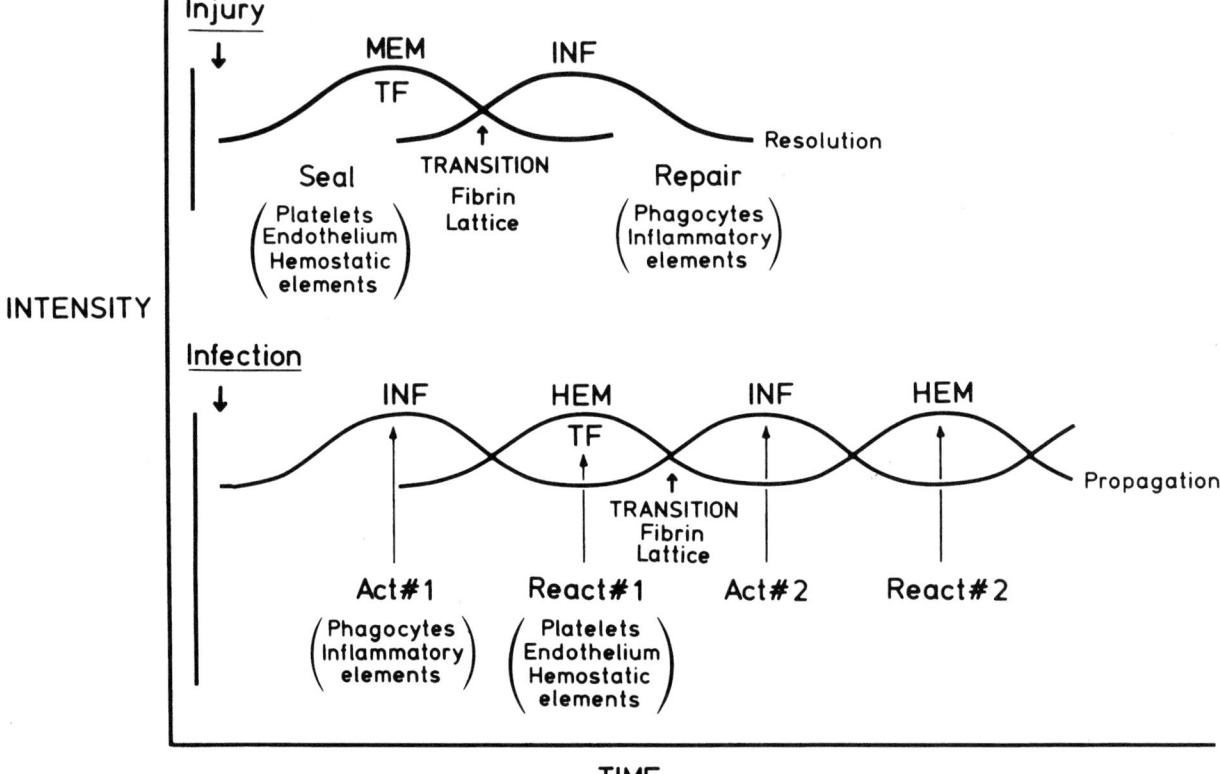

Fig. 9 Wound healing versus deep vein thrombosis—a comparison of physiological and pathophysiological interactions between hemostatic and inflammatory elements: Following an open laceration of a vessel wall the hemostatic elements beginning with tissue factor seal the wound and lay down a fibrin network in a transition phase. This sets the stage for a "coordinated" influx of inflammatory elements including phagocytes, which release cytokines, growth factors, etc. This leads to repair and resolution.

Following a closed hypoxic/mechanical injury of a vessel wall (valve leaflet) the inflammatory elements beginning with phagocytes initiate an inflammatory response (Act #1). The hemostatic elements including tissue factor in this case respond and lay down a fibrin network (React #1). In this case, however, this network is out of place, growing to occlude a vein instead of growing to seal a wound. As a result the inflammatory elements including phagocytes bind to the fibrin network (CD11c/CD18 to RGD sequences of fibrin). In this second cycle cytokines (TNF) are released by the adherent monocytes (Act #2). This leads to activation of neighboring cells and a second round of tissue factor expression and laying down of more fibrin (React #2). This constitutes a series of events in which the asynchronous interactions between these two systems feed on themselves to perpetuate the inflammatory response.

In this case the inflammatory elements initiate the hemostatic rather than following the hemostatic process. Tissue factor and fibrin operating out of sequence serve to propagate an aberrant sequence of inflammatory/hemostatic responses (propagation) rather than coordinating a normal inflammatory response leading to resolution. As a result, venous thrombi at autopsy often appear to be laminated with red layers of fibrin/red blood corpuscles alternating with gray layers of leukocytes.

clusive microvascular thrombosis may not play as important a role in these *acute* cases of cardiovascular collapse, capillary leak, and shock as they do in the subacute and chronic cases described below.

In circumstances where the patients survive the risks of a massive consumptive coagulopathy (DIC) or capillary leak and shock, the hemostatic system again can function in completely different ways depending on the status of the host. In contrast to the acute conditions, the consequences of irreversible fibrin deposition in subacute or chronic conditions become as important as the expression of tissue factor. Intravascular intra-alveolar deposition of fibrin including other acute phase proteins (e.g., C-reactive protein) can perpetuate as well as amplify inflammatory activity by numerous mechanisms. These include (1) facilitation of neutrophil aggregation by platelets via P-selectin (66), (2) recruitment and fixation of C3 component of complement by C-reactive protein/phospholipase A_2 complexes (67), and (3) recruitment and binding of phagocytes (neutrophils, monocytes) to the fibrin matrix through binding to the RGD sequences of fibrin via their CD11c/CD18 receptors (68). In these three ways fibrin, platelets and other fixed hemostatic components can substitute for bacteria in pulmonary alveolae (hyaline membrane) and in spaces lined by mesothelium (joints, pericardial, peritoneal, pleural spaces) to perpetuate that inflammatory activity originally initiated by foreign agents (bacteria, etc.). Figure 9 is an attempt at illustrating how the hemostatic system (HEM) reacting out of sequence with the inflammatory system (INF) could assist in perpetuating the inflammatory response. Thus, in conditions such as systemic inflammatory response syndrome (SIRS) in which the cycling of monocytes and macrophages between the tolerant and sensitized state is thought to be an important aspect of this disorder (69), an alternating feedback between inflammatory and hemostatic components might contribute to this phenomena creating the conditions that perpetuate this disorder long after the original pathogenic organisms have been cleared.

ACKNOWLEDGMENTS

I wish to acknowledge the valuable assistance and discussions I have had with Drs. C. Esmon, L. Hinshaw, and J. Morrissey. I also wish to acknowledge the technical assistance of ACK Chang, D. Carey, and G. Peer, and the work of P. Antkowiak, D. Irish, and Terry Young in preparation of figures and typescript. Supported in part by National Institutes of Health (Bethesda, MD) Grant #GM37704.

REFERENCES

1. Taylor FB, Kosanke S, Emerson T, et al. Retrospective description and experimental reconstitution of three different responses of the baboon to lethal *E. coli*. Shock 1994; 42:92–103.
2. Taylor FB Jr. Baboon models of three variants of hemostatic response to inflammatory stimuli. In: Brigham KL, ed. Endotoxin and the Lungs. New York: Marcel Dekker, 1994:171.
3. Esmon CT. The roles of protein C and thrombomodulin in the regulation of blood coagulation. J Biol Chem 1989; 264:4743.
4. Esmon CT, Fukudome K. Mather T, Bode W, Esmon NL, Regan LM, Stearns-Kurosawa DJ, Kurosawa S. Inflammation—the protein C pathway. 3rd International Symposium on Fibrogen-Hemostasis, Inflammation and Cardiovascular Disease, Ulm, Germany, May 3-4, 1996.
5. Hanss M, Collen D. Secretion of tissue-type plasminogen activator and plasminogen activator inhibitor by cultured human endothelial cells: modulation by thrombin, endotoxin, and histamine. J Lab Clin Med 1987; 109:97.
6. Loskutoff DJ, Edgington TS. Synthesis of a fibrinolytic activator and inhibitor of endothelial cells. Proc Natl Acad Sci USA 1977; 74:3903.
7. Collen D. On the regulation and control of fibrinolysis. Thromb Haemost 1980; 43:77–89.
8. Tesfamariam B. Free radicals in diabetic endothelial cell dysfunction. Free Radicals Biol Med 1994; 16: 383–391.
9. Read MA, Meyrick BO. Effects of endotoxin on lung endothelium. In: KL Brigham, ed. Endotoxin and the Lungs. New York: Marcel Dekker, 1994:83.
10. Old LJ. Tumor necrosis factor. Sci Am 1988; 258:59.
11. Nawroth PP, Stern DM. Modulation of endothelial cell hemostatic properties by tumor necrosis factor. J Exp Med 1986; 163:740.
12. Pohlman TH, Stannes KA, Beatley PG, Ochs DD, Harlan JM. An endothelial cell surface factor(s) induced in vitro by lipopolysaccharide, interleukin 1, tumor necrosis factor-α increases neutrophil adherence by a CD2-18-dependent mechanism. J Immunol 1986; 136:4548.
13. Davis EW, Fujikawa K. Kisiel W. The coagulation cascade. Initiation, maintenance, and regulation. Biochemistry 1991; 30:10366.
14. Booth NA, Anderson JA, Bennett B. Platelet release proteins which inhibit plasminogen activators. J Clin Pathol 1985; 38:825.
15. Erickson LA, Ginsberg MG, Loskutoff DJ. Detection and partial characterization of an inhibitor of plasminogen activator in human platelets. J Clin Invest 1984; 74:1465.
16. Sprengos ED, Kluft C. Plasminogen activator inhibitors. Blood. 1987; 69:381–387.

17. Drake TA, Cheng J, Chang ACK, Taylor FB, Jr. Expression of tissue factor, thrombomodulin and E-selectin in baboons with lethal *D. coli* sepsis. Am J Pathol 1983; 142:1458.
18. Taylor FB Jr., Hoogendorn H, Chang ACK, Peer G, Nesheim ME, Catlett R, Stump DC, Giles AR. Anticoagulant and fibrinolytic activities are promoted, not retarded, in vivo after thrombin generation in the presence of a monoclonal antibody that inhibits activation of protein C. Blood 1991; 79:1720–1728.
19. George JN, El-Harake J. Thrombocytopenia due to enhanced platelet destruction by nonimmunologic mechanisms. In: Beutler E, Lichtman M, Coller B, Kipps TJ, eds. Williams Hematology, 5th ed. New York: McGraw-Hill, 1996:1290.
20. Taylor FB Jr., Chang AK, Peer G, Li A, Burstein S. IL-6 Amplifies the coagulopathic response of the baboon to E. coli. 15th Congress of the International Society on Thrombosis and Haemostasis, Jerusalem, Israel, June 11-16, 1995.
21. Kruithof EKO, Mestries JC, Gascon MP, Ythier A. The coagulation and fibrinolytic responses of baboons after in vivo thrombin generation-effect of interleukin 6. Thromb Haemost 1997; 77(5):905–910.
22. Taylor F, Chang ACK, Ferrell G, Mather T, Catlett R, Blick K. Esmon CT. C4b-binding protein exacerbates the host response to *E. coli*. Blood 1987; 78(15):918.
23. Taylor FB Jr., Dahlback B, Esmon CT, Chang ACK, Lockhart MS, Hatanaka K. The role of free protein S and C4bBP protein in regulation of the coagulant response to *E. coli*. Blood 1995; 86:2642.
24. Dahlback B. Interaction between vitamin K-dependent protein S and the complement protein, C4b-binding protein. A link between coagulation and the complement system. Semin Thromb Hemost 1984; 10:139.
25. Creasey AA, Chang ACK, Fiegen L, Wun TC, Taylor FB, Jr., Hinshaw LB. Tissue factor pathway inhibitor (TFPI) reduces mortality from *Escherichia coli* septic shock. J Clin Invest 1993; 91(6):2850–2860.
26. Carr C, Bild GS, Chang ACK, Peer GT, Palmier MO, Frazier RB, Gustafson ME, Wun TC, Creasey AA, Hinshaw LB, Taylor FB Jr., Galluppi GR. Recombinant *E. coli*-derived tissue factor pathway inhibitor reduces coagulopathic and lethal effects in the baboon gram-negative model of septic shock. Circ Shock 1995; 44:126.
27. Taylor FB Jr., Emerson TE Jr., Jordan R, Chang ACK, Blick KE. Antithrombin-III prevents the lethal effects of Escherichia coli infusion in baboons. Circ Shock 1988; 26:227.
28. Bajzar L, Taylor F, Tracy PB. A baboon model can be used to assess the physiologic function of TAFI (abstr). Thrombosis Haemostasis 1997; (suppl):596.
29. Anguo L, Chang ACK, Peer GT, Hinshaw LB, Taylor Jr. FB. Comparison of the capacity of rhTNF-α and Escherichia coli to induce procoagulant activity by baboon mononuclear cells in vivo and in vitro. Shock 1996; 5:274–279.
30. Taylor FB Jr., He SE, Chang ACK, Box J, Ferrell G, Lee D, Lockhart M, Peer G, Esmon CT. Infusion of phospholipid vesicles amplifies the local thrombotic response to TNF and anti-protein C into a consumptive response. Thromb Haemost 1996; 75(4):578–584.
31. Taylor FB, Chang AKC, Peer G, Ezban M, Hedner U. Active site inhibited factor VIIa inhibits deep vein thrombosis in the primate. Blood. In press.
32. Virchow R. Phlagose und Thrombose in Gefassystein. In: Virchow R, ed. Gesammelte Abhandlungen für Wissenschaftlichen Medizin. Frankfurt: von Meidingen Sohn, 1856:458–636.
33. Slack SM, Cui Y, Turitto VT. The effects of flow on blood coagulation and thrombosis. Thromb Haemost 1993; 1:129–133.
34. Bokisch VA, Top FH Jr., Russell PK, Dixon FJ, Muller-Eberhard HJ. The potential pathogenic role of complement in Dengue hemorrhagic shock syndrome. N Engl J Med 1973; 289:996–1003.
35. Dvorak HF, Van De Water L, Bitzer AM, Dvorak AM, Anderson D, Harvey VS, Bach R, Davis GL, De Wolf W, Carvalho ACA. Procoagulant activity associated with plasma membrane vesicles shed by cultured tumor cells. Cancer Res 1983; 43:4434–4443.
36. Taylor FB Jr., Esmon CT, Hinshaw LB. Summary of staging, mechanisms, and intervention studies in the baboon model of E. coli sepsis. J Trauma 1991; 30(suppl):S-197.
37. Taylor FB Jr. Studies of the natural history and mechanism of the primate (baboon) response to *Escherichia coli*. In: JL Ryan, DC Morrison, eds. Bacterial Endotoxic Lipopolysaccharides. Vol. 2. Immunopharmacology and Pathophysiology. Boca Raton, FL: CRC Press, 1991:239.
38. Voss BL, DeBault LE, Blick KE, Chang ACK, Steiers DL, Hinshaw LB, Taylor FB. Sequential renal alterations in septic shock in the primate. Circ Shock 1991; 33:142.
39. Randolf MM, White GL, Kosanke SK, Bild G, Carr C, Galluppi G, Hinshaw LB, Taylor FB Jr. Complete inhibition of tissue thrombosis and hemorrhage by ALA-TFPI does not account for its protection against E. coli: a comparative study of treated and untreated non-surviving baboons challenged with LD_{100} *E. coli*. Thromb Haemost 1998; 79:1048–1053.
40. Taylor FB Jr, Chang ACK, Mather T, Blick K, Catlett R, Esmon CT. DEGR-factor Xa blocks disseminated intravascular coagulation initiated by *E. coli* without preventing shock or organ damage. Blood 1991; 78:364.
41. Hinshaw LB, Tekamp-Olsen P, Chang ACK, et al. Survival of primates in LD_{100} septic shock folowing therapy and antibody to tumor necrosis factor (TNF). Circ Shock 1990; 30:279.
42. Taylor FB Jr., Chang ACK, Ruf W, Morrissey JH, Hinshaw LB, Blick KE, Edgington TS. Lethal *E. coli* septic shock is prevented by blocking tissue factor with monoclonal antibody. Circ Shock 1991; 33:127.
43. Taylor FB Jr. Role of tissue factor and factor VIIa in the coagulant and inflammatory response to LD_{100} *Escherichia coli* in the baboon. Haemostasis 1996; 26(suppl 1):83–91.
44. Rottingen JA, Enden T, Camerer E, Iversen JG, Prydz H. Binding of human factor VIIa to tissue factor induces cytosolic Ca^{2+} signals in J82 cells, transfected COS-1 cells, Madin-Darby canine kidney cells and in human endothelial cells induced to synthesize tissue factor. J Biol Chem 1995; 270:4650.

45. Camerer E, Rottingen JA, Iversen JG, Prydz H. Coagulation factors VII and X induce CA^{2+} oscillations in Madin-Darby canine kidney cells only when proteolytically active. J Biol Chem 1996; 271:29034.
46. Taylor FB Jr., Chang A, Esmon CT, D'Angelo A, Vigano-D'Angelo S, Blick KE. Protein C prevents the coagulopathic and lethal effects of *Escherichia coli* infusion in the baboon. J Clin Invest 1987; 79:918.
47. Sandset PM, Warm-Cramer BJ, Rao LV, Maki SL, Rapaport SI. Immunodepletion of extrinsic pathway inhibitor sensitizes rabbits to endotoxin-induced intravascular coagulation. Blood 1991; 78:1496–1502.
48. Taylor FB Jr. Role of tissue factor in the coagulant and inflammatory response to LD_{100} *E. coli* sepsis and in the early diagnosis of DIC in the baboon. In: Muller-Berghaus T, et al., eds. DIC: Pathogenesis, Diagnosis and Therapy of Disseminated Intravascular Fibrin Formation. New York: Elsevier Science, 1993:19–32.
49. Broze GJ Jr. Tissue factor pathway inhibitor and current concept of blood coagulation. Blood Coagul Fibrinol 1995; 6:S7–S13.
50. Rapaport SI. The extrinsic pathway inhibitor: a regulator of tissue factor-dependent blood coagulation. Thromb Haemost 1991; 66(1):6–15.
51. Kurosawa S, Stearns-Kurosawa DJ, Hidari N, Esmon CT. Identification of functional endothelial Protein C receptor in human plasma (abstr). J Clin Invest 1997; 100:411.
52. Davie EW. Biochemical and molecular aspects of the coagulation cascade. Thromb Haemost 1995; 74:1–6.
53. Bom VJJ, Bertina RM. The contributions of Ca^{2+}, phospholipids, and tissue factor apoprotein to the activation of human blood coagulation factor X by activated factor VII. Biochem J 1990; 265:327–336.
54. Marcum IA, Rosenberg RD. Anticoagulantly active heparin-like molecules from vascular tissue. Biochemistry 1984; 23:1730.
55. Calandra T, Bucala R. Macrophage migration inhibitory factor: A counterregulator of glucocorticoid action and critical mediator of septic shock. J Inflamm 1996; 47:39–51.
56. Hinshaw LB, Beller BK, Chang ACK, Murray CK, Flournoy DJ, Passey RB, Archer LT. Corticosteroid/Antibiotic treatment of adrenalectomized dogs challenged with lethal *E. coli*. Circ Shock 1985; 16:265–277.
57. Taylor FB Jr., Siegler R, Chang ACK, Li A, Tesh V, Pyscher T, DeBault L. Characterization of the baboon responses to SHIGA-Like toxin: Descriptive study of new toxic hemolytic uremic responses to SLT-1 (abstr). Shock 1997; 7(suppl 1):59.
58. Tesh VL, Burris JA, Owens JW, Gordon VM, Wadolkowski EA, O'Brien AD, Samuel JE. Comparison of the relative toxicities of Shiga-like toxins Type I and Type II for mice. Infect Immun 1993; 61:3392–3402.
59. Tesh VL, Samuel JE, Burris JA, Owens JW, Taylor FB Jr., Siegler RL. Quantitation and localization of shiga toxin/shiga-like toxin-binding glycolipid receptors in human and baboon tissues. In: Karmali MA, Goglio AG, eds. Recent Advances in Verocytotoxin-Producing *Escherichia coli* Infections. Amsterdam: Elsevier Science, 1994:189–192.
60. Eckhardt JV, von Asmuth EJU, Dentener MA, Ceska M, Buurman WA. IL-6, IL-8 and TNF production by cytokine and lipopolysaccharide-stimulated human renal cortical epithelial cells in vitro. Eur Cytokine Netw 1994; 5:301–310.
61. Yard BA, Daha MR, Kooymans-Couthino M, Bruijn JA, Paape ME, Schrama E, van Es LA, van der Woode BJ. IL-1α stimulated TNFα production by cultured human proximal tubular epithelial cells. Kidney Int 1992; 42:383–389.
62. Louise CB, Obrig TG. Shiga toxin-associated hemolytic uremic syndrome: Combined cytotoxic effects of Shiga toxin, interleukin-1β and tumor necrosis factor-α on human vascular endothelial cells in vitro. Infect Immun 1991; 59:4173–4179.
63. Remuzzi G. HUS and TTP: Variable expression of a single entity. Kidney Int 1986; 323:292–308.
64. Richardson SE, Karmali MA, Becker LE, Smith CR. The histopathology of the hemolytic uremic syndrome (HUS) associated with verocytotoxin-producing *Escherichia coli* infections. Hum Pathol 1988; 19:1102–1108.
65. Taylor FB Jr., Coller BS, Chang ACK, Peer G, Jordan R, Engellener W, Esmon CT. 7E3 (ab')$_2$, a monoclonal antibody to the platelet GPIIb/IIIa receptor, protects against microangiopathic hemolytic anemia and microvascular thrombotic renal failure in baboons treated with C4b binding protein and a sublethal infusion of *Escherichia coli*. Blood 1997; 89(11):4078–4084.
66. McEver RP. Leukocyte interactions mediated by selectins. Thromb Haemost 1991; 66:80–87.
67. Hack CE, Wolbink GJ, Schalwijk C, Speijer, Hermens TW, Bosch H. A role for secretory phospholipase A$_2$ and C-reactive protein in removal of injured cells. Immunol Today 1997; 18:111–114.
68. Taylor FB Jr., Lockhart MS, Wakefield T, Greenfield L, Boersma Y, Hack E, Chang ACK, Peer G, Li A, He S, Blick KE, Jansen P, Esmon CT. Further characterization of the baboon model of deep vein thrombophlibitis: reconstitution studies defining optimum conditions and inflammatory mediators. Am J Pathol.
69. Rangel-Frausto MS, Pittet D, Costigan M, Hwang T, Davis CS, Wenzel RP. The natural history of the systemic inflammatory response syndrome (SIRS). JAMA 1995; 273(2):117–123.

40

Biological Functions of Lipopolysaccharide Antibodies

Matthew Pollack
Uniformed Services University of the Health Sciences, F. Edward Hébert School of Medicine, Bethesda, Maryland

LIPOPOLYSACCHARIDE STRUCTURE AND FUNCTION

Lipopolysaccharides (LPS), or endotoxins, are biologically active structural components of the outer cell membrane of all gram-negative bacteria. These heterogeneous macromolecules protect bacteria from immunological defense mechanisms of the infected host and mediate a variety of proinflammatory and toxic functions causally related to sepsis and septic shock (1).

The complete LPS macromolecule is comprised of lipid A and core oligosaccharide, both relatively conserved structures in phylogenetic terms, together with the hypervariable O polysaccharide (O side chain). The LPS O side chain is produced by smooth bacterial phenotypes such as *Escherichia*, *Salmonella*, and *Pseudomonas*, while LPS or lipooligosaccharides (LOS) from naturally rough phenotypes of many species of *Acinetobacter*, *Bordetella*, *Campylobacter*, *Bacteroides*, *Neisseria*, *Haemophilus*, and *Chlamydia* lack O side chains (2).

A majority of LPS-related proinflammatory and toxic functions originate in the lipid A moiety, although some of these biological activities may be modified by structural elements of the contiguous core oligosaccharide (3). The repeating oligosaccharide subunit composition of the LPS O side chain and its exposed position on the bacterial cell surface also appear to contribute to the virulence role of bacterial LPS as well as its immunogenicity.

Because LPS is produced in abundance on the external surface of gram-negative bacteria and is also released into the surrounding environment as a natural by-product of bacterial growth and cell division, it is a prime target for antibodies with antibacterial and endotoxin-neutralizing potential.

SPECIFICITY AND FUNCTION OF LPS ANTIBODIES

All three structural domains of LPS are antigenic and immunogenic in the mammalian host, as evidenced by LPS subcomponent-specific antibody responses following natural exposure to gram-negative bacteria or immunization with LPS-containing vaccines (4–6). In general, however, the better exposed O side chain of smooth (S-form) LPS and outer core of rough (R-form) LPS induce brisker antibody responses than do more cryptic structural elements in the inner core or lipid A regions of the LPS macromolecule. Moreover, antibodies to conserved epitopes in the lipid A and inner core regions of LPS tend to exhibit greater cross-reactivity than antibodies specific for more heterogeneous portions of the outer core or hypervariable O side chain. There are frequent exceptions to this generalization, however, as illustrated by monoclonal antibodies (mAbs) specific for non–cross-reactive epitopes on rough mutant *E. coli* J5 LPS (7) and the coexistence of distinct genus-specific and cross-reactive epitopes on rough phenotype *Chlamydia* LPS (8,9). It is clear, accordingly, that the cross-reactivity of antibodies specific for epitopes in the core/lipid A region of bacterial endotoxin is often restricted by phylogenetic differences in covalent core structure (10,11) as well as by epitope concealment due to overlying carbohydrate structures (7).

The whole-cell reactivity of antibodies specific for different structural elements of LPS corresponds in general with the specificity of these antibodies for isolated LPS (12). The O side chains of smooth LPS and more distal portions of phenotypically rough LPS are readily recognized by antibodies on the bacterial cell surface. The inner core and lipid A moieties, in contrast, are less susceptible to antibody attack on the bacterial surface (12,13). Nevertheless, lipid A– and core-specific antibodies sometimes display better reactivity with intact bacteria than with isolated LPS (14), although the whole-cell reactivity of LPS antibodies may be restricted to specific bacterial subpopulations in relation to variations in growth conditions or growth phase (12).

Overall, gram-negative infections induce more vigorous antibody responses to homologous, strain-specific epitopes on the O polysaccharide and outer core of S-form and R-form LPS, respectively, than to deeper LPS substructures, although immunization with rough LPS may induce a cross-reactive antibody response as well (15).

ENDOTOXIN-NEUTRALIZING ACTIVITY OF LPS ANTIBODIES

There is ongoing debate concerning the ability of anti-LPS antibodies to neutralize endotoxin, i.e., to inhibit in vitro or in vivo toxic or proinflammatory activities of the LPS macromolecule and its bioactive lipid A moiety (16). A further debate involves the relationship between antibody specificity for different LPS substructures and endotoxin-neutralizing activity. Still further controversy concerns the cross-reactivity of lipid A– or core-specific antibodies and whether antibodies with putative cross-reactivity are able to cross-neutralize or cross-protect animals against challenge with heterologous LPS.

It is clear that selected anti-LPS antibodies are capable of inhibiting the LPS-induced in vitro production of proinflammatory cytokines such as tumor necrosis factor-α (TNF-α) and interleukin 6 (IL-6) (17–22). Similarly, some anti-LPS antibodies block LPS-induced cytokine release in vivo and protect LPS challenged animals against endotoxin-induced death (23,24). One view holds that antibodies specific for lipid A but not other, less biologically active portions of the LPS macromolecule are capable of neutralizing endotoxin. Recent evidence suggests, however, that antibodies reactive with the core oligosaccharide and O side chain may effectively inhibit in vitro and in vivo biological activities of endotoxin (22,25). Physicochemical factors that appear to influence the endotoxin-neutralizing activity of LPS antibodies include LPS concentration and phenotype; antibody specificity, isotype, and affinity; and serum conditions. Cross-reactive LPS antibodies that recognize phylogenetically conserved epitopes in the core/lipid A region have been identified, and some of these antibodies are capable of cross-neutralizing heterologous LPS and cross-protecting experimental animals against homologous and heterologous LPS challenges and associated mortality (19,26).

The identification of membrane-bound CD14 (mCD14) as part of a multicomponent LPS receptor functionally linked to the initiation of intracellular events associated with LPS-induced leukocyte activation has facilitated our understanding of at least one mechanism of antibody-mediated LPS neutralization (27,28). This mechanism resides in the ability of LPS antibodies to block the cellular uptake of LPS through CD14 receptors on host target cells, such as human monocytes, which correlates with the capacity for modulating LPS-induced cytokine release, an important proinflammatory pathway. This phenomenon has been documented in the case of both core- and O side chain–specific mAbs and relates to both rough and smooth LPS phenotypes (21).

The hypothesis that anti-LPS antibodies neutralize endotoxin by blocking cellular uptake through mCD14 has been further tested in CD14-transfected Chinese hamster ovary (CHO-CD14) fibroblasts, cells that express no known LPS recognition elements other than mCD14 (22). LPS core- and O side chain–specific mAbs inhibit mCD14-mediated LPS uptake by these cells and downregulate LPS-induced nuclear factor-κB (NF-κB) translocation from cytoplasm to cell nucleus, an intracellular event associated with cell activation (22).

The findings raise the important question as to why core- or O side chain–specific mAbs, which do not react with the lipid A component of the LPS macromolecule recognized by mCD14, are nonetheless capable of blocking CD14-mediated LPS uptake and LPS-induced TNF-α secretion and NF-κB translocation. A possible explanation is that although the mAbs do not block the lipid A moiety per se, they effectively do so by producing a conformational change in the LPS macromolecule or by favoring the formation of LPS aggregates that are less easily recognized by mCD14 receptors than are more disaggregated forms of LPS (21,22).

ANTI- AND PROINFLAMMATORY PROPERTIES OF LPS ANTIBODIES— DUAL ROLE IN ADAPTIVE AND INNATE IMMUNITY

In the presence of fresh serum, a rich source of LPS-binding protein (LBP) and serum complement, LPS mAbs may prevent the proinflammatory uptake of LPS by monocytes via mCD14 while simultaneously enhancing proinflammatory complement-mediated uptake through membrane-bound complement receptor 1 (CR1) (21,22). These observations underscore the capacity of LPS mAbs to exert differential effects on discrete cellular pathways of LPS recognition, uptake, and function.

It follows from monocyte and CHO-CD14 cell data (22) that in the presence of heat-inactivated serum, LPS mAbs are capable of blocking proinflammatory, mCD14-mediated LPS uptake. In the presence of fresh serum, however, this inhibitory effect may be offset by the simultaneous enhancement of uptake mediated through complement and membrane-bound CR1, in the case of human monocytes, or through other as yet unidentified serum ligands and cell receptors, in the case of CHO-CD14 cells, with proinflammatory consequences including NF-κB activation and cytokine release (21,22).

LPS antibodies may thus contribute to the maintenance of homeostasis by exerting both anti-inflammatory and proinflammatory influences on LPS-target cell interactions in conjunction with discrete serum ligands and cell receptors, of which LBP and mCD14, respectively, are prime examples.

The recognition of LPS by mCD14 on host target cells provides a nonimmunological mechanism, part of "innate immunity" (29,30), for initiating inflammatory responses to LPS. This nonimmunological, proinflammatory pathway is blocked by LPS antibodies. Concurrently, however, LPS antibodies may stimulate inflammatory responses to LPS through their complement-activating and opsonic properties. Thus, LPS antibodies, a product of adaptive immunity, play a pivotal role in regulating both nonimmune (innate) and immune (adaptive) responses to LPS (22).

ENDOTOXIN CLEARANCE AND TRANSPORT FUNCTIONS OF LPS ANTIBODIES

Gram-negative bacteria are cleared efficiently from the bloodstream by the reticuloendothelial system in concert with opsonic LPS antibodies and complement (31,32). O side chain–specific antibodies are particularly active with respect to this in vivo function (32,33). Likewise, LPS antibodies promote the clearance of circulating endotoxin through enhanced uptake by the liver, lung, and spleen (34–36).

The binding of LPS to high-density lipoproteins (HDL) is a major determinant of plasma half-life. Unbound LPS leaves the bloodstream quite rapidly, often in association with formed blood elements such as platelets and erythrocytes. HDL-bound LPS, on the other hand, exits the circulation more slowly and is taken by tissues that utilize HDL cholesterol. HDL binding requires that LPS be in a disaggregated state. LPS antibodies prevent LPS-HDL binding in part by blocking LPS disaggregation. IgG antibodies to homologous LPS opsonize LPS and LPS-HDL complexes and increase their uptake by tissues such as liver and spleen, which are rich in phagocytic cells (16,36). Recent evidence from a canine septic shock model (25) suggests that O side chain–specific mAbs are more efficient than core-specific mAbs in lowering levels of circulating endotoxin. A reasonable conclusion is that antibodies to homologous LPS have a major impact on the handling of plasma endotoxin in the immune host, while HDL-mediated clearance mechanisms play a more critical role in the disposition of LPS in the nonimmune host (36).

The infected host rids itself of foreign antigens through the formation of soluble immune complexes, their opsonization by complement, and their elimination from the circulation and from local sites of infection. A great deal has been learned in recent years concerning the role of antibody and complement in this process, the cellular receptors that recognize these ligands, and the cells involved in this immunological transport and disposal system (37–39).

Despite limited information regarding the formation and clearance of immune complexes comprised of LPS and LPS antibodies, it must be assumed that the formation of such complexes and their opsonization by complement play a critical role in the disposal of endotoxin by the infected host. Indeed, chronic *Pseudomonas aeruginosa* lower respiratory tract infection complicating cystic fibrosis is a stunning example of gram-negative bacterial disease in part due to a failure of the infected host to properly clear potentially pathogenic LPS- and LPS antibody–containing immune complexes. Precipitation of such immune complexes in the infected lung tissue of cystic fibrosis patients is thought to contribute to the chronic inflammatory lung lesions found in these patients (40–43).

The human lipid A–specific mAb, HA-IA, has been shown in vitro to augment the serum complement-dependent immune adherence of LPS to human erythrocytes (RBC) and polymorphonuclear leukocytes (PMN) in a dose-dependent manner (44). The binding of immune complexes comprised of HA-1A, LPS, and C' to RBC and PMN occurs via the classical complement pathway, and this binding takes place through membrane-bound CR1 receptors. It has been hypothesized that mAb HA-IA is capable of reducing the bioavailability of endotoxin in vivo by mediating its binding to and possible clearance by human RBC, acting in conjunction with the reticuloendothelial system or via direct internalization by PMN (44).

Subsequent work by others (45) demonstrates that the anti–lipid A mAb, E5, binds to rough gram-negative bacteria, fixes C3, and facilitates binding of bacterial immune complexes to both human RBC and monocytes. On the basis of these findings it was postulated that mAb E5 may enhance bacterial clearance by (1) facilitating direct complement fixation, (2) augmenting the binding of opsonized bacteria to cells of the mononuclear phagocyte system, and (3) enabling bacteria to bind to erythrocyte CRI, allowing "safe carriage" in the circulation to the fixed macrophages of the liver and spleen (45).

ANTIBACTERIAL ACTIVITY OF LPS ANTIBODIES

Most pathogenic gram-negative bacteria are resistant to the direct bactericidal effects of fresh, complement-containing human serum (46,47). The outer membrane–associated LPS of many such serum-resistant bacteria, while a target for direct complement fixation, appears to divert complement activation away from susceptible sites on the bacterial cell surface where subsequently formed terminal membrane attack complexes (C5b-C9) would otherwise produce lytic damage to the bacterial outer membrane. LPS-specific antibodies, in contrast, are capable of activating complement by the alternative pathway through the covalent interaction of nascent C3b with the non-Fc portion of bound antibody. This results in the redirection of subsequently formed membrane attack complexes to "lethal" sites on the bacterial cell surface associated with complement-induced bacteriolysis. Thus, while LPS protects serum-resistant bacteria from complement-mediated lysis, LPS-reactive antibodies circumvent this natural resistance to complement-mediated killing by binding complement components in a functionally active form at critical sites on the bacterial cell surface (16,48).

Another mechanism of bacterial killing by LPS antibodies is through opsonophagocytosis. LPS antibodies act as opsonins, both alone and in conjunction with complement. This activity may be directed toward live bacteria, resulting in their opsonophagocytic killing by host phagocytes (47), or toward nonviable bacteria, outer membrane fragments, or free LPS, resulting in the uptake and clearance of these various bacterial components by the reticuloendothelial system (35,36).

The constant region of IgG antibodies is recognized by Fc receptors on polymorphonuclear leukocytes and macrophages (49). There are no corresponding receptors for IgM antibodies on these phagocytic cells. However, anti-LPS antibodies of both the IgG and IgM classes activate complement on the surfaces of gram-negative bacteria by the classical or alternative complement pathways. These antibodies facilitate deposition of C3b and iC3b on the bacterial cell surface in a stable and functionally active form (50) recognized by the complement receptors, CR1 and CR3, on phagocytic cell membranes. This leads to the cellular adherence of opsonized bacteria, possible ingestion, and intracellular killing (51).

Antibodies and complement exhibit overlapping yet distinct opsonic functions (49,51). Recognition of surface-bound IgG by Fc receptors on phagocytic cells results in weak bacterial adherence, ingestion, and metabolic events associated with intracellular killing. Ligation of complement receptors by C3b and iC3b, on the other hand, produces more efficient adherence to phagocytes than that mediated by Fc receptors, but does not result in phagocytosis unless the complement receptors are first activated by an independent signal (52–55). Ligated complement receptors, even then, do not activate oxidative metabolism or microbicidal functions. LPS antibodies and serum complement thus serve complementary attachment and phagocytosis-promoting opsonic functions (51), resulting in collaborative mechanisms of opsonophagocytic killing.

Phenotypically rough gram-negative pathogens like *Haemophilus influenzae* type b and *Neisseria meningitidis* are killed by complement-dependent bacteriolysis in the presence of LPS-specific antibodies. Smooth bacterial phenotypes, however, are often not susceptible to antibody- and complement-mediated bactericidal mechanisms in the absence of phagocytic cells. Serum-resistant *P. aeruginosa* blood isolates, for example, are subject to opsonophagocytic killing by PMN in the presence of LPS immunotype-specific antibodies and heat-labile serum factors but are not susceptible to killing by humoral factors alone (47). This pattern of serotype (or immunotype)-specific immunity based upon opsonic antibodies to the O-polysaccharide of smooth LPS acting in concert with professional phagocytes is broadly representative (56). Opsonic antibodies to various *P. aeruginosa* immunotypes are found in normal human serum, for example, and increase in titer in response to infection or immunization with LPS O-polysaccharide–containing vaccines (57). "Natural" antibodies to *P. aeruginosa* LPS belong mainly to the IgM class, while those arising in response to infection or immunization exhibit IgG isotypes. In both cases, opsonic activity is mostly complement-dependent, although serotype-specific IgG may mediate opsonophagocytic killing at high concentrations in the absence of heat-labile serum factors (47).

PROTECTIVE ROLE OF LPS ANTIBODIES

The ubiquitousness of LPS antibodies in human serum is directly related to the frequency with which gram-negative bacteria and LPS are found in the human environment, both internal and external. These antibodies are directed toward all three structural domains of the LPS macromolecule. They arise as a consequence of bacterial colonization of mucosal surfaces, as in the case of the endogenous microflora of the gastrointestinal tract, or from clinical or subclinical gram-negative infections, especially when these infections are associated with an invasive component. LPS antibodies may also be induced artificially through immunization with vaccines comprised of intact bacteria, isolated LPS, or LPS part-structures.

Several studies have documented the protective role of LPS antibodies in gram-negative sepsis by associating high preexisting serum titers of these antibodies with subsequent survival (56,58–60). Such protection included both serotype-specific protection conferred by O side chain–specific antibodies and phylogenetic cross-protection afforded by core-reactive antibodies.

Similarly, many experimental studies in animals (24,25,61–64) have demonstrated the protection afforded by LPS antibodies against infectious challenge, including instances in which such antibodies were actively induced by LPS-containing vaccines or passively provided in the form of immune serum, isolated polyclonal antibodies, or mAbs. The majority of such studies have been performed in small rodent models in which survival was the sole endpoint. Some of these studies have documented the therapeutic superiority of type- or strain-specific antibodies directed against exposed epitopes on the LPS O side chain or outer core compared with cross-reactive antibodies recognizing phylogenetically conserved yet less well-exposed structural elements of the core oligosaccharide or lipid A moieties.

A particularly revealing study was carried out in a canine septic shock model in which isotype-matched core- and O side chain–specific mAbs were compared for in vivo efficacy against *E. coli* sepsis resulting from intraperitoneal implantation of an infected fibrin clot (25,33). Both mAbs exerted favorable effects on survival in this canine model, but by apparently distinct mechanisms. The O side chain–specific mAb increased survival, independent of simultaneously administered antibiotics, lowered levels of bacteremia and circulating endotoxin, but had no effect on sepsis-induced decrements in mean arterial pressure. In contrast, the core-reactive mAb produced a favorable effect on survival, but only in the presence of an antibiotic. The core-reactive mAb did not influence levels of bacteremia or endotoxemia, but did produce a sustained elevation of mean arterial pressure. These study results emphasize the fact that LPS antibodies that recognize distinct epitopes on the endotoxin macromolecule may mediate very different in vivo functions. In this case, the O side chain–specific antibody mediated bacterial killing and LPS clearance, while the core-specific antibody appeared to block endotoxin-induced circulatory collapse. Other lessons emphasized by this experiment include possible differences among animal models with respect to the pathogenic and protective roles of LPS and LPS antibodies, respectively; the importance of documenting physiological functions other than survival when assessing the in vivo protective activities of LPS antibodies; and the likelihood of synergistic interactions between particular LPS antibodies and antibiotics (16).

A comprehensive review of the preparation and testing of LPS vaccines and LPS antibodies in connection with preclinical or clinical evaluations is beyond the scope of this chapter. However, there is ample evidence that selected LPS antibodies are capable of mediating protective immunity, which is either serotype-specific (i.e., O side chain– or outer core–related) or core-specific (i.e., inner core– or lipid A–related). Nevertheless, large-scale clinical evaluations of lipid A–specific mAbs in patients with gram-negative sepsis have yielded contradictory data (65–68), perhaps in part because of the heterogeneity of the patient populations studied and in part because of flaws in study design (69–71). Despite the current lack of definitive data, however, LPS-based cross-protective immunity still holds considerable promise as a basis for the immunoprophylaxis or immunotherapy of gram-negative sepsis.

MOLECULAR MIMICRY OF HOST STRUCTURES BY LPS, POLYREACTIVE LPS ANTIBODIES, AND AUTOIMMUNITY

A surprisingly high number of B lymphocytes from normal subjects produce "polyreactive" or "natural" antibodies that bind to self as well as to exogenous antigens, including LPS (72). A discrete B-cell subset, CD5+, is responsible for much of the production of these polyreactive autoantibodies. CD5+ lymphocytes represent a major component of the normal human B-cell repertoire, correspond to mouse Ly-1 B cells, and produce mainly low-affinity IgM antibodies. A major role for polyreactive antibodies produced by CD5+ B lymphocytes is likely that of a first line of defense against invading microorganisms based upon enhanced phagocytosis and complement-mediated lysis antecedent to a later developing specific immune response (72). Polyreactive IgM antibodies may be unable to eliminate foreign invaders completely, however, because of their relatively poor fit for surface antigens on microorganisms and low affinity. This process is likely completed by high-affinity, monoreactive IgG antibodies that appear as a result of antigen-driven somatic mutation ("antibody maturation") (73).

Two polyreactive human mAbs (one of them a progenitor of the HA-1A antibody studied in large-scale clinical trials) have been reported to bind to the lipid A component of LPS, a ligand on human B lymphocytes, and the i antigen on human erythrocytes derived from cord blood (74). The common structure recognized by these mAbs is an acyl-substituted disaccharide. The antibodies share the V_H4-21 gene associated with the "autoimmune repertoire," and it has been postulated that these antibodies play an important role in overall host immunity against invading microorganisms. It has been further hypothesized that antibody reactivity with the lipid A domain of bacterial LPS or other cross-reactive epitopes could lead to the selective expansion of B lymphocytes bearing the corresponding binding site, and, paradoxically, that such interactions might ultimately prevent excessive stimulation of the immune system (74).

More recently, a polyreactive, human, lipid A–specific mAb has been discovered that cross-reacts with *P. aeruginosa* core- and O side chain–specific polysaccharides as well as certain human autoantigens (75). This mAb confers protection against mouse infections caused by multiple immunotypes of *P. aeruginosa*, further documenting the polyreactivity of certain LPS antibodies and the cross-protective immunity some such antibodies are capable of mediating (75).

Polyreactive LPS antibodies may play a more sinister pathogenic role in human autoimmune disease based upon their cross-reactivity with host tissues containing molecular structures mimiced by bacterial LPS (76). For example, the core oligosaccharides of certain *Campylobacter jejuni* LPS serotypes mimic the molecular structure of the human GM1 ganglioside. Clinical isolates of *C. jejuni* that bear these serotypes and produce LPS structures closely resembling human ganglioside are strongly associated with development of the neuromuscular disorder, Guillain-Barré syndrome (GBS). Antibodies that cross-react with *C. jejuni* LPS and GM1 ganglioside are considered to play an important role in the pathogenesis of GBS (77). Moreover, GM1 antibodies inducible through immunization of experimental animals with *C. jejuni* LPS (78) are similar to those recovered from patients with *C. jejuni* enteritis (77,79).

The repeating O-specific oligosaccharide subunits of *Helicobacter pylori* LPS closely resemble the LewisX and LewisY human blood group antigens (76,80). Lewis blood group antigens are expressed in normal human gastric mucosa and human gastric carcinoma. It has been postulated that host-like structures, in this case LPS molecules closely resembling Lewis antigens expressed on the cell surface of *H. pylori* bacteria, might camouflage the organism as it invades the gastric mucosa, allowing the bacteria to somehow circumvent normal host-defense mechanisms. Over the ensuing course of infection of the gastric mucosa by *H. pylori*, seropositivity against the organism has been found to correlate with the generation of autoantibodies against the antral gastric mucosa. Moreover, antibodies reacting with human gastric mucosa have been demonstrated in mice immunized with *H. pylori*. Thus, progressive infection of the gastric mucosa accompanied by an immune response to LPS antigens on the bacterial cell surface bearing a close resemblance to mucosal Lewis antigens of the infected host might explain the observed production of autoantibodies and their implication in the development of chronic gastritis. Supporting this hypothesis is the finding of antibodies against LewisX and LewisY antigens in *H. pylori*–infected patients (76).

CONCLUSION

LPS antibodies are characterized by diverse specificities and biological functions. This diversity embraces antibody reactivity with all three functional domains of the LPS macromolecule, including both highly specific and cross-reactive epitopes shared among many different bacterial and nonbacterial species, including humans. LPS antibodies mediate complex biological functions including endotoxin neutralization, transport, and clearance; either inhibition or enhancement of LPS-induced inflammation; complement-dependent antibacterial activities; and in vivo protection against gram-negative infection. LPS antibodies serve as vehicles of adaptive immunity and modulators of innate immunity to endotoxin. Moreover, polyreactive LPS antibodies contribute to early, "natural" immunity, but also participate in the immunopathogenesis of autoimmunity. Cross-protective LPS antibodies continue to hold promise for the immunoprophylaxis and therapy of gram-negative bacterial disease, but this potential is to some extent clouded by inconclusive or contradictory clinical data.

REFERENCES

1. Hancock REW. Bacterial outer membranes: evolving concepts. Specific structures provide gram-negative bacteria with several unique advantages. ASM News 1991; 57:175–182.
2. Rietschel ET, Brade L, Holst O, Kulshin VA, Lindner B, Moran AP, Schade UF, Zähringer U, Brade H. Molecular structure of bacterial endotoxin in relation to bioactivity. In: Nowotny A, Spitzer JJ, Ziegler EJ, eds. Cellular and Molecular Aspects of Endotoxin Reactions. Amsterdam: Elsevier Science Publishers, 1990: 15–32.
3. Lüderitz T, Brandenburg K, Seydel U, Roth A, Galanos C, Rietschel ET. Structural and physicochemical requirements of endotoxins for the activation of arachidonic acid metabolism in mouse peritoneal macrophages in vitro. Eur J Biochem 1989; 179:11–16.
4. Galanos C, Lüderitz O, Westphal O. Preparation and properties of antisera against the lipid-A component of bacterial lipopolysaccharides. Eur J Biochem 1971; 24: 116–122.
5. Lüderitz O, Staub AM, Westphal O. Immunochemistry of O and R antigens of *Salmonella* and related Enterobacteriaceae. Bacteriol Rev 1996; 30:192–255.
6. Pollack M, Oishi K, Chia J, Evans ME, Guelde G, Koles, NL. Specificity and function of monoclonal antibodies reactive with discrete structural elements of bacterial lipopolysaccharide. Adv Exp Med Biol 1990; 256:331–340.
7. Pollack M, Chia JKS, Koles NL, Miller M, Guelde G. Specificity and cross-reactivity of monoclonal antibodies reactive with the core and lipid A regions of bacterial lipopolysaccharide. J Infect Dis 1989; 159:168–188.
8. Brade L, Nurimenen M, Mäkelä PH, Brade H. Antigenic properties of *Chlamydia trachomatis* lipopolysaccharide. Infect Immun 1985; 48:569–572.
9. Holst O, Brade L, Kosma P, Brade H. Structure, serological specificity, and synthesis of artificial glycoconjugates representing the genus-specific lipopolysaccharide epitope of *Chlamydia* spp. J Bacteriol 1991; 173: 1862–1866.
10. Holst O, Brade H. Chemical structure of the core region of lipopolysaccharides. In: Morrison DC, Ryan JL, eds. Bacterial Endotoxic Lipopolysaccharides. Vol. 1. Molecular Biochemistry and Cellular Biology. Boca Raton, FL: CRC Press, 1992:135–170.
11. Jansson E, Lindberg AA, Lindberg B, Wollin R. Structural studies on the hexose region of the core in lipopolysaccharides from enterobacteriaceae. Eur J Biochem 1981; 115:571–577.
12. Evans ME, Pollack M, Hardegen NJ, Koles NL, Guelde G, Chia JK. Fluorescence-activated cell sorter analysis of binding by lipopolysaccharide-specific monoclonal antibodies to gram-negative bacteria. J Infect Dis 1990; 162:148–155.
13. Gigliotti F, Shenep JL. Failure of monoclonal antibodies to core glycolipid to bind intact smooth strains of *Escherichia coli*. J Infect Dis 1985; 151:1005–1011.
14. Pollack M, Raubitschek A, Larrick J. Human monoclonal antibodies that recognize conserved epitopes in the core-lipid A region of lipopolysaccharides. J Clin Invest 1987; 79:1421–1430.
15. Cross AS, Sidberry H, Sadoff JC. The human antibody response during natural bacteremic infection with gram-negative bacilli against lipopolysaccharide core determinants. J Infect Dis 1989; 160:225–236.
16. Pollack M. Specificity and function of lipopolysaccharide antibodies. In: Ryan JL, Morrison DC, eds. Bacterial Endotoxic Lipopolysaccharides. Vol II. Immunopharmacology and Pathophysiology. Boca Raton, FL: CRC Press, 1992:347–374.
17. Battafarano RJ, Burd RS, Cody CS, Kellogg TA, Raymond CS, Ratz CA, Dunn DL. Anti-lipopolysaccharide monoclonal antibodies inhibit macrophage TNF messenger RNA synthesis in vitro. J Surg Res 1993; 54: 342–348.
18. Burd RS, Battafarano RJ, Cody CS, Farber MS, Dunn DL. Anti-endotoxin monoclonal antibodies inhibit secretion of tumor necrosis factor-α by two distinct mechanisms. Ann Surg 1993; 218:250–261.
19. Di Padova FE, Mikol V, Barclay GR, Poxton IR, Brade H, Rietschel ET. Anti-lipopolysaccharide core antibodies. Prog Clin Biol Res 1994; 388:85–94.
20. Fang IS, Wisniewski MA, Huntenburg CC, Knight LS, Bubbers JE, Schneidkraut MJ. Inhibition of lipopolysaccharide-associated endotoxin activities in vitro and in vivo by the human anti-lipid A monoclonal antibody SdJ5-1.17.15. Infect Immun 1993; 61:3873–3878.
21. Pollack M, Espinoza AM, Guelde G, Koles NL, Wahl LM, Ohl CA. Lipopolysaccharide (LPS)-specific monoclonal antibodies regulate LPS uptake and LPS-induced tumor necrosis factor-α responses by human monocytes. J Infect Dis 1995; 172:794–804.
22. Pollack M, Ohl CA, Golenbock DT, Di Padova F, Wahl LM, Koles NL, Guelde G, Monks BG. Dual effects of LPS antibodies on cellular uptake of LPS and LPS-induced proinflammatory functions. J Immunol 1997; 159:3519–3530.

23. Battafarano RJ, Burd RS, Kurrelmeyer KM, Ratz CA, Dunn DL. Inhibition of splenic macrophage tumor necrosis factor-α secretion in vivo by antilipopolysaccharide monoclonal antibodies. Arch Surg 1994; 129:179–186.
24. Baumgartner JD, Heumann D, Gerain J, Weinbreck P, Grau GE, Glauser MP. Association between protective efficacy of anti-lipopolysaccharide (LPS) antibodies and suppression of LPS-induced tumor necrosis factor-α and interleukin 6. J Exp Med 1990; 171:889–896.
25. Hoffmann WD, Pollack M, Banks SM, Koev LA, Solomon MA, Danner RL, Koles N, Guelde G, Yatsiv I, Mouginis T, Elin RJ, Hosseini JM, Bacher J, Porter JC, Natanson C. Distinct functional activities in canine septic shock of monoclonal antibodies specific for the O polysaccharide and core regions of *Escherichia coli* lipopolysaccharide. J Infect Dis 1994; 169:553–561.
26. Bailat S, Heumann D, Le Roy D, Baumgartner JD, Rietschel ET, Glauser MP, Padova FD. Similarities and disparities between core-specific and O-side-chain-specific antilipopolysaccharide monoclonal antibodies in models of endotoxemia and bacteremia in mice. Infect Immun 1997; 65:811–814.
27. Kitchens RL, Munford RS. Enzymatically deacylated lipopolysaccharide (LPS) can antagonize LPS at multiple sites in the LPS recognition pathway. J Biol Chem 1995; 270:9904–9910.
28. Ulevitch RJ, Tobias PS. Recognition of endotoxin by cells leading to transmembrane signaling. Curr Opin Immunol 1994; 6:125–130.
29. Fearon DT, Locksley RM. The instructive role of innate immunity in the acquired immune response. Science 1996; 272:50–54.
30. Wright SD. CD14 and innate recognition of bacteria. J Immunol 1995; 155:6–8.
31. Benacerraf B, Sebestyen MM, Schlossman S. A quantitative study of the kinetics of blood clearance of p^{32}-labelled *Escherichia coli* and staphylococci by the reticulo-endothelial system. J Exp Med 1959; 110:27–48.
32. Pelkonen S, Pluschke G. Roles of spleen and liver in the clearance of *Escherichia coli* K1 bacteraemia in infant rats. Microb Pathog 1989; 6:93–102.
33. Hoffman WD, Pollack M, Banks SM, Koev L, Solomon MA, Danner RL, Koles N, Guelde G, Yatsiv I, Mouginis T, Elin RJ, Hosseini JM, Porter JC, Natanson C. A controlled trial of IgG monoclonal anti-endotoxin antibodies in a canine model of human septic shock. Clin Res 1991; 39:164A.
34. Freudenberg M, Galanos C. The metabolic fate of endotoxins. Prog Clin Biol Res 1988; 272:63–75.
35. Goodman JS, Rogers DE, Koenig MG. The hepatic uptake of bacterial endotoxin. I. The influence of humoral and cellular factors. Proc Soc Exp Biol Med 1969; 132:372–375.
36. Munford RS, Dietschy JM. Effects of specific antibodies, hormones, and lipoproteins on bacterial lipopolysaccharides injected into the rat. J Infect Dis 1985; 152:177–184.
37. Fearon DT. Complement, C receptors, and immune complex disease. Hosp Pract 1988; 23:63–72.
38. Frank MM, Lawley TJ, Hamburger MI, Brown EJ. Immunoglobulin G Fc receptor-mediated clearance in autoimmune diseases. Ann Intern Med 1983; 98:206–218.
39. Schifferli JA, Ng YC, Peters DK. The role of complement and its receptor in the elimination of immune complexes. N Engl J Med 1986; 315:488–495.
40. Berdischewsky M, Pollack M, Young LS, Chia D, Osher AB, Barnett EV. Circulating immune complexes in cystic fibrosis. Pediatr Res 1980; 14:830–833.
41. Moss RB, Hsu Y-P, Lewiston N, Curd JG, Milgrom H, Hart S, Dyer B, Larrick, JW. Association of systemic immune complexes, complement activation, and antibodies to *Pseudomonas aeruginosa* lipopolysaccharide and exotoxin A with mortality in cystic fibrosis. Am Rev Respir Dis 1986; 133:648–652.
42. Wisnieski JJ, Todd EW, Fuller RK, Jones PK, Dearvorn DG, Boat TF, Naff GB. Immune complexes and complement abnormalities in patients with cystic fibrosis. Increased mortality associated with circulating immune complexes and decreased function of the alternative complement pathway. Am Rev Respir Dis 1985; 132:770–776.
43. Hornick DB, Fick RB. The IgG subclass composition of immune complexes and the pathogenesis of cystic fibrosis lung disease. Chest 1987; 92:79S.
44. Krieger JI, Fletcher RC, Siegel SA, Fearon DT, Neblock DS, Boutin RH, Taylor RP, Daddona PE. Human anti-endotoxin antibody HA-1A mediates complement-dependent binding of Escherichia coli J5 lipopolysaccharide to complement receptor type 1 of human erythrocytes and neutrophils. J Infect Dis 1993; 167:865–875.
45. Seelen MA, Athanassiou P, Lynn WA, Norsworthy P, Walport MJ, Cohen J. The anti-lipid A monoclonal antibody E5 binds to rough gram-negative bacteria, fixes C3, and facilitates binding of bacterial immune complexes to both erythrocytes and monocytes. Immunology 1995; 84:653–661.
46. Schoolnik GK, Buchanan TM, Holmes KK. Gonococci causing disseminated gonococcal infection are resistant to the bactericidal action of normal human sera. J Clin Invest 1976; 58:1163–1173.
47. Young LS, Armstrong D. Human immunity to Pseudomonas aeruginosa. I. In-vitro interaction of bacteria, polymorphonuclear leukocytes, and serum factors. J Infect Dis 1977; 126:257–276.
48. Oishi K, Koles NL, Guelde G, Pollack M. Antibacterial and protective properties of monoclonal antibodies reactive with *Escherichia coli* O111:B4 lipopolysaccharide: relation to antibody isotype and complement-fixing activity. J Infect Dis 1992; 165:34–45.
49. Unkeless JC, Wright SD. Phagocytic cells: Fc-gamma and complement receptors. In: Gallin JI, Goldstein IM, Snyderman R, eds. Inflammation: Basic Principles and Clinical Correlates. New York: Raven Press, Ltd., 1988:343–362.
50. Bjornson AB, Magnafichi PI, Schreiber RD, Bjornson HS. Opsonization of bacteroides by the alternative complement pathway reconstructed from isolated plasma proteins. J Exp Med 1987; 164:777–798.
51. Silverstein SC, Greenberg S, di Virgilio F, Steinberg TH. Phagocytosis. In: Paul WE, ed. Fundamental Immunology. New York: Raven Press, Ltd., 1989:703–720.

52. Ehlenberger AG, Nussenzweig V. The role of membrane receptors for C3b and C3d in phagocytosis. J Exp Med 1977; 145:357–371.
53. Mantovani B. Different roles of IgG and complement receptors in phagocytosis by polymorphonuclear leukocytes. J Immunol 1975; 115:15–17.
54. Newman SL, Johnston, Jr. RB. Role of binding through C3b and IgG in polymorphonuclear neutrophil function: studies with trypsin-generated C3b. J Immunol 1979; 123:1839–1846.
55. Scribner DJ, Fahrney D. Neutrophil receptors for IgG and complement: their roles in the attachment and ingestion phases of phagocytosis. J Immunol 1976; 116:892–897.
56. Pollack M, Young LS. Protective activity of antibodies to exotoxin A and lipopolysaccharide at the onset of *Pseudomonas aeruginosa* septicemia in man. J Clin Invest 1979; 63:276–286.
57. Young LS. Human immunity to *Pseudomonas aeruginosa*. II. Relationship between heat-stable opsonins and type-specific lipopolysaccharides. J Infect Dis 1972; 126:277–287.
58. McCabe WR, Kreger BE, Johns M. Type-specific and cross-reactive antibodies in gram-negative bacteremia. N Engl J Med 1972; 287:261–267.
59. Pollack M, Huang AI, Prescott RK, Young LS, Hunter KW. Enhanced survival in *Pseudomonas aeruginosa* septicemia associated with high levels of circulating antibody to *Escherichia coli* endotoxin core. J Clin Invest 1983; 72:1874–1881.
60. Zinner SH, McCabe WR. Effects of IgM and IgG antibody in patients with bacteremia due to gram-negative bacilli. J Infect Dis 1976; 133:37–45.
61. Greisman SE, DuBuy JB, Woodward CL. Experimental gram-negative bacterial sepsis: prevention of mortality not preventable by antibiotics alone. Infect Immun 1979; 25:538–557.
62. Dunn DL, Bogard, Jr. WC, Cerra FB. Efficacy of type-specific and cross-reactive murine monoclonal antibodies directed against endotoxin during experimental sepsis. Surgery 1985; 98:283–290.
63. Pennington JE, Menkes E. Type-specific vs. cross-protective vaccination for gram-negative bacterial pneumonia. J Infect Dis 1981; 144:599–603.
64. Priest BP, Brinson DN, Schroeder DA, Dunn DL. Treatment of experimental gram-negative bacterial sepsis with murine monoclonal antibodies directed against lipopolysaccharide. Surgery 1989; 106:147–155.
65. Ziegler EJ, Fisher CJ, Sprung CL, Straube RC, Sadoff JC, Foulke JC, Wortel CH, Fink MP, Dellinger RP, Teng NNH, Allen IE, Berger HJ, Knatterud GL, Lobuglio AF, Smith CR, and the HA-1A Study Group. Treatment of gram-negative bacteremia and septic shock with HA-1A human monoclonal antibody against endotoxin. N Engl J Med 1991; 324:429–436.
66. McCloskey RV, Straube RC, Sanders C, Smith SM, Smith CR, and the CHESS Trial Study Group. Treatment of septic shock with human monoclonal antibody HA-1A. A randomized, double-blind, placebo-controlled trial. Ann Intern Med 1994; 121:1–5.
67. Greenman RL, Schein RMH, Martin MA, Wenzel RP, MacIntyre NR, Emmanuel G, Chmel H, Kohler RB, McCarthy M, Plouffe J, Russel JA, and the XOMA Sepsis Study Group. A controlled clinical trial of E5 murine monoclonal IgM antibody to endotoxin in the treatment of gram-negative sepsis. J Am Med Assoc 1991; 266:1097–1102.
68. Bone RC, Balk RA, Fein AM, Perl TM, Wenzel RP, Reines HD, Quenzer RW, Iberti TJ, Macintyre N, Schein RMH, and the E5 Sepsis Study Group. A second large controlled clinical study of E5, a monoclonal antibody to endotoxin: results of a prospective, multicenter, randomized controlled trial. Crit Care Med 1995; 23:994–1005.
69. Cross AS. Antiendotoxin antibodies: a dead end? Ann Intern Med 1994; 121:58–60.
70. Cohen J, Heumann D, Glauser M. Do monoclonal antibodies and anticytokines still have a future in infectious diseases? Am J Med 1995; 99:6A-45S–6A-52S.
71. Verhoef J, Hustinx W, Frasa H, Hoepelman A. Issues in the adjunct therapy of severe sepsis. J Antimicrob Chemother 1996; 38:167–182.
72. Casali P, Notkins A. Probing the human B-cell repertoire with EBV: polyreactive antibodies and CD5+ B lymphocytes 1,2. Ann Rev Immunol 1989; 7:513–535.
73. Watts R, Daisenberg D. Autoantibodies and antibacterial antibodies: from both sides now. Ann Rheum Dis 1990; 49:961–965.
74. Bhat NM, Bieber MM, Chapman CJ, Stevenson FK, Teng NNH. Human antilipid A monoclonal antibodies bind to human B cells and the i antigen on cored red blood cells. J Immunol 1993; 151:5011–5021.
75. Yokota S, Ohtsuka H, Kohzuki T, Noguchi H. A polyreactive human anti-lipid A monoclonal antibody having cross reactivity to polysaccharide portions of *Pseudomonas aeruginosa* lipopolysaccharides. FEMS Immunol Med Microbiol 1996; 14:31–38.
76. Moran A, Prendergast M, Appelmelk B. Molecular mimicry of host structures by bacterial lipopolysaccharides and its contribution to disease. FEMS Immunol Med Microbiol 1996; 16:105–115.
77. Jacobs BC, Endtz H, Meche V, Hazenberg MP, Klerk MA. Humoral immune response against *Campylobacter jejuni* lipopolysaccharides in Guillain-Barré and Miller Fisher syndrome. J Neuroimmunol 1997; 79:62–68.
78. Ritter G, Fortunato S, Cohen L, Noguchi Y, Bernard E, Stockert E, Old L. Induction of antibodies reactive with GM2 Ganglioside after immunization with lipopolysaccharides from *Campylobacter jejuni*. Int J Cancer 1996; 66:184–190.
79. Wirguin W, Briani C, Suturkova-Milosevic L, Fisher T, Della-Latta P, Chalif P, Latov N. Induction of anti-GM1 ganglioside antibodies by *Campylobacter jejuni* lipopolysaccharides. J Neuroimmunol 1997; 78:138–142.
80. Amano K, Hayashi S, Kubota T, Fujii N, Yokota S. Reactivities of Lewis antigen monoclonal antibodies with the lipopolysaccharides of *Helicobacter pylori* strains isolated from patients with gastroduodenal diseases in Japan. Clin Diagn Lab Immunol 1997; 4:540–544.

41

Specificity and Neutralizing Properties of Cross-Reactive Anti-Core LPS Monoclonal Antibodies

Franco E. Di Padova
Novartis Pharma Ltd, Basel, Switzerland

Didier Heumann and Michel Pierre Glauser
Centre Hospitalier Universitaire Vaudois, Lausanne, Switzerland

Ernst T. Rietschel
Research Center Borstel, Borstel, Germany

PRODUCTION OF CROSS-REACTIVE ANTI-CORE LIPOPOLYSACCHARIDE ANTIBODIES

Lipopolysaccharides (LPS) are major antigenic and toxic molecules expressed by gram-negative bacteria that contribute to the pathogenesis of many diseases. Attempts to neutralize their toxic activities have been numerous, but until now no significant improvement in the treatment of patients with sepsis has been achieved. Nonetheless, polyclonal and monoclonal antibodies (mAbs) directed against LPS still remain suitable options.

LPS are structurally complex, amphipathic, microheterogeneous macromolecules consisting of three regions: the O-specific polysaccharide chain, the core oligosaccharide, and lipid A. These regions are genetically, biochemically, and antigenically distinct. However, the core region and lipid A retain common structural features (1,2). The core consists of an inner and an outer region. The inner core is structurally conserved among *Escherichia coli*, *Salmonella enterica*, and *Shigella* species (3), while the outer core determines the five different *E. coli* chemotypes (R1, R2, R3, R4, and K-12) and the core of *Salmonella* spp. (1,3,4). Additional microvariability is due to nonstoichiometric substitutions with phosphate and ethanolamine groups (4). Similarly shared structures are also found in LPS of *Pseudomonas aeruginosa* serotypes (5) and *Klebsiella* (6–8). The lipid A component is the most conserved region and carries the endotoxic principle (1,2).

Many attempts have been made to develop neutralizing antibodies (Abs) to the common inner core region of LPS using the UDP-gal-4-epimerase-deficient J5 mutant of *E. coli* O111:B4 (RcP+ chemotype) or *Salmonella minnesota* R595 (Re chemotype) (9–16). The concept of cross-protection was supported by studies on the chemical structure of the LPS core, which demonstrated the presence of shared structural regions among Enterobacteriaceae (1–3) and from the finding that Abs to core LPS antigens are detected in serum of healthy subjects (17–19). Moreover, experimental studies of passive immunotherapy with antisera from rabbits immunized with *E. coli* J5 and *S. minnesota* R595 rough mutants (20–22) and clinical studies, showing that human *E. coli* J5 antiserum increased the survival of patients with gram-negative bacteremia and septic shock in surgical patients at high risk of infection (23), suggested the existence of cross-reactive Abs and their potential protective role. However, direct experimental proof of the presence of LPS-neutralizing Abs after *E. coli* J5 or *S. minnesota* R595 vaccination was lacking, the epitopes involved in cross-reactivity were not identified, and a potential contamination with LPS or the presence of Abs to LPS O side chains or other bacterial structures were not always controlled or excluded. Moreover, cross-reactive anti–core LPS or anti–lipid A Abs were not detected in J5 antisera (24). In conclusion, many contradictory and confusing results were generated (24–26) as recently discussed in detail by Greisman and Johnston (27). More recently it was also suggested that the protection offered by J5 antisera in humans could be explained by Abs to conserved outer membrane proteins (28).

In addition, none of the many anti-core LPS mAbs generated in mice against the J5 mutant of *E. coli* O111:B4 or *S. minnesota* R595 showed a wide cross-reactivity for different LPS (9–15). They had mainly a narrow specificity, did not bind to S-form LPS, recognized the terminal oligosaccharides, and their Ab-binding site was masked by addition of more distal sugars of the core. Only a few exhibited some limited cross-reactivity and only occasionally neutralized LPS (16). In addition, the HA-1A mAb that was selected after immunization of a human volunteer with *E. coli* J5 vaccine turned out not to be cross-protective in animals (25) or specific for lipid A. Actually, it recognizes a carbohydrate epitope, which is expressed on isolated and purified lipid A, on the i Ag present on cord red blood cells, on human B lymphocytes, and on certain autoantigens (29,30). Therefore, HA-1A has the characteristics of a polyreactive, cold agglutinin that utilizes the VH4.21 gene segment in germline configuration (29).

In conclusion, all these negative findings support the belief that cross-reactive anti–core LPS Abs do not exist. On the other hand, the approach of using bacteria expressing only LPS partial core structure as the immunizing agent may not be valid. The tridimensional conformation of LPS complete core structures is lost in LPS partial core structure (31,32). This may explain why Abs generated against LPS partial core structures do not recognize LPS with a complete core. It is well known that in smooth strains of Enterobacteriaceae only LPS with a complete core structure with or without O antigens are found (33). Therefore, we began to use bacteria with a complete LPS core structure in our immunization schemes. This novel approach led to the generation of several cross-reactive and cross-neutralizing mAbs in different strains of mice (34,35) and clearly demonstrated that protective and cross-reactive Abs can be generated. Using this approach, we also generated mAbs specific for a single core type such as *E. coli* R1, R2, and R3, respectively (36).

Therefore it can be concluded that the approach to using bacteria with partial core structures to stimulate the production of cross-reactive Abs is unsatisfactory in animals and probably also in humans. These findings are relevant also for the development of novel strategies useful for the induction of an immune response in patients at high risk of infection and might open new areas for the development of bacterial vaccines.

SPECIFICITY AND BINDING TO ISOLATED AND PURIFIED LPS

Five anti–core LPS mAbs of different subclasses that are cross-reactive for all known *E. coli* and *Salmonella* core types have been described (34,35). They recognize the five known core types of *E. coli* (R1-R4, K-12) and react with *E. coli* J5 (RcP+ core) but not with *E. coli* F515 (Re) or *E. coli* free lipid A. H5 13-23 (IgG3k) and H5 415-6 (IgG2ak) show a lower level of reactivity with *E. coli* R3. Additional differences are observed in the binding pattern to rough core structures of *Salmonella*. All the mAbs recognize *S. minnesota* R60 (Ra) and, with the exception of H1 61-2 (IgG1k), they bind *S. minnesota* R345 (Rb2). mAbs WN1 222-5 (IgG2ak) and WN1 58-9 (IgG2bk) show some residual binding to more defective core structures such as *S. minnesota* RcP− and Rd1P−. H5 13-23 and H5 415-6 show a lower degree of reactivity with R-form LPS of *S. typhimurium* and with other *Salmonella* strains. In summary, WN1 222-5 and WN1 58-9 seem to recognize deeper core regions in comparison to H5 13-23, H5 415-6, and H1 61-2. The finding that none of the mAbs shows reactivity with Re structures from *E. coli* or *Salmonella* and with free mono- and bis-phosphoryl lipid A suggests that the epitopes must be localized in the inner outer core region. However, the fine specificity of the five mAbs remains to be elucidated. In fact, the lack of homogeneous and synthetic LPS core structures, the microvariability of the core structures isolated from bacteria, and the tendency of LPS to aggregate and form complex three-dimensional structures (37) make it difficult to determine whether the mAbs recognize overlapping or complementary epitopes. Actually, this is a critical point as anti-core Abs, in contrast to the anti–O side chain Abs, can recognize only one epitope per LPS molecule. The possibility to define more than one cross-reactive epitope on the same molecule might favor the search for combinations of different anti-core mAbs.

The cross-reactivity of the mAbs was confirmed using different experimental conditions and procedures, including enzyme-linked immunosorbent assay (ELISA) on purified LPS, ELISA on high-density lipoprotein (HDL)–LPS complexes, competitive ELISA, passive hemolysis assay and sodium deoxycholate-polyacrylamide gel electrophoresis (DOC-PAGE) and Western blotting of LPS. Different methods have been used to confirm binding of the mAbs to LPS to avoid the criticism that coating of LPS onto microtiter plates might alter its conformation, facilitating the expression of the epitope. In particular, in passive hemolysis assays, the presentation of LPS to the environment is more similar to that observed in the bacterial membrane as the lipid A portion is intercalated in the plasma membrane (38). Moreover, in passive hemolysis it is possible to verify the capacity of the Abs to activate complement. Binding of the Abs to LPS in the presence of HDL has also been analyzed since HDL are probably involved in binding of LPS in serum (39–42).

As shown in Table 1, WN1 222-5 binds to smooth LPS in ELISA in both the absence or the presence of HDL. Moreover, as shown in Western blotting, this core epitope remains accessible to the mAbs in S-form LPS, indicating that it is not masked by additional oligosaccharides in the outer core and/or by the O-specific chain (34,35). In conclusion all these data confirm the existence of common epitopes in the core LPS which remain available to the mAbs when the LPS is free or complexed to HDL or inserted in a plasma membrane. Moreover, the binding to LPS of these mAbs fulfills all the requirements of a high-affinity interaction (Table 2) as it shows low background binding, significant binding at low Ab and Ag concentrations, and specific and reproducible activity in more than one assay system. Therefore, our findings confirm at the immunochemical level the similarity in the inner-core region of LPS predicted by genetic and chemical analysis. Moreover, conformational similarities in the outer core (hexose region) of *Salmonella* and of *E. coli* R1-R4 and K-12 LPS have been predicted by semi-empirical calculations (43). These results are in line with our data, which suggest that the outer core may contribute to the Ab-binding surface.

More recently, WN1 222-5 has been chimerized into a human IgG1k Ab (SDZ 219-800), which expresses specificities and biological activities identical to those of WN1 222-5 (44). These data confirm that chimerization is a reliable way to maintain the biological properties of murine mAbs.

As all these experiments were performed with isolated and purified LPS preparations, one might wonder whether anti–core LPS mAbs would bind to LPS in a native form, i.e., once released in vivo from a bacterial membrane. The fact that WN1 222-5 is able to capture different LPS in the plasma of bacteremic animals is a proof that this mAb binds also to "native" LPS and is further indication that this epitope is also available to the immune system of animals infected with bacteria expressing smooth LPS (45).

BINDING TO BACTERIA

A critical point is whether anti–core LPS Abs are able to recognize LPS on intact bacteria. WN1 222-5 was therefore tested in ELISA against a large collection of heat-killed bacteria, isolated from clinical samples (34). This mAb reacts strongly with all blood, urinary, and fecal isolates of *E. coli* and *Salmonella*, with some *Citrobacter*, and weakly with some *Enterobacter* and some *Klebsiella*, showing that the epitope is widely

Table 1 Comparison of Coating with Free LPS and HDL-LPS Complexes for Measuring Binding of WN1 222-5 to S-Form LPS

LPS	LPS	HDL-LPS
E. coli O4	1.16	>2.00
E. coli O6	1.14	0.79
E. coli O8	0.81	0.40
E. coli O111	0.53	0.46
E. coli O127	0.86	0.60
S. minnesota	0.62	0.61
S. typhimurium	0.78	1.07

Values are shown as optical densities (O.D.) obtained under these conditions: WN1 222-5, 500 ng/ml; dilution of conjugated antibody, 1:1000; incubation of peroxidase substrate, 6 min. Results are expressed after subtraction of background values. LPS, lipopolysaccharide; HDL, high-density lipoprotein. For additional details, see Ref. 64.

Table 2 Affinity Calculations of WN1 222-5 Using the Biacore System

Coated antigen	WN1 222-5 (10^{-9} M)[a]
BSA-*E. coli* R1 (P-) LPS[b]	1.92
BSA-*E. coli* R2 (P-) LPS[b]	1.99
BSA-*E. coli* R3 (P-) LPS[b]	1.11
BSA-*E. coli* R4 (P-) LPS[c]	1.24

[a]WN1 222-5 was titrated among 0.5 and 8 mg/ml or among 0.625 and 10 mg/ml using twofold dilutions.
[b]Coating: 50 μg/ml.
[c]Coating: 25 μg/ml.

distributed among Enterobacteriaceae and is accessible in heat-killed bacteria. It is interesting to note that the pattern of cross-reactivity of WN1 222-5 for the Enterobacteriaceae appears to reflect the evolutionary relationship among these families, since *E. coli*, *Salmonella*, and *Citrobacter* are the most closely related groups (46).

However, heat-killed bacteria are not the same as dead bacteria in the plasma during a septic episode. In addition, the contribution of dead bacteria to clinical situations such as sepsis is not clear. A more relevant question is whether such an Ab could bind to its epitope located in the inner-outer core region of LPS in a living bacterium where the O-polysaccharide, the tight conformation of the bacterial membrane, or a capsule may inhibit its accessibility. The binding of WN1 222-5 to *E. coli* O18:K1 whole bacteria has therefore been analyzed by ELISA and by fluorescence-activated cell sorter. When living bacteria are coated onto microtiter plates, WN1 222-5 shows binding and its binding is enhanced after antibiotic treatment with sub-MIC concentrations of ampicillin and temocillin, but not of ciprofloxacin or gentamicin (47). However, even if these data support the role of certain antibiotics in enhancing the binding of Abs to bacteria and would suggest the presence and accessibility of the epitope in bacteria, in such an assay it is not possible to differentiate between dead and living bacteria and to conclude that the Ab will help the opsonization of living bacteria.

A more convincing answer is offered by flow cytometry on bacteria, as this technique allows differentiation of dead from living bacteria. In these experimental conditions WN1 222-5 shows no or minimal binding to *E. coli* O18:K1 (unpublished data). Binding can be significantly enhanced by treatment with sub-MIC concentrations of ceftazidime, i.e., in the absence of bacterial death. On the contrary, binding of WN1 222-5 to live *E. coli* O111:B4 is detected even in the absence of antibiotics. In ELISA, binding of WN1 222-5 to purified LPS of *E. coli* O111:B4 and *E. coli* O18 is comparable even if these strains have different core structures (36). These data support the notion that the location of the epitope to which WN1 222-5 binds might not be readily accessible in certain bacterial strains, but may be exposed in others. It is a matter of speculation whether the accessibility to the core of WN1 222-5 might correlate with the pathogenicity of the strains. The lack of binding of WN1 222-5 to certain bacteria suggests that it will not invariably activate complement, function as an opsonin, and cause bacterial lysis. The mAb probably will only facilitate removal of dead bacteria. Whether removal of LPS or dead bacteria is clinically relevant in septic shock remains to be determined.

NEUTRALIZATION OF LPS IN VITRO

The endotoxin-neutralizing properties of WN1 222-5 have been evaluated in vitro and in vivo. For all these studies the greatest care had to be taken to ensure that the Ab preparations and all other reagents were not contaminated with LPS. Under several experimental conditions, the mAb is able to block LAL activity and LPS-induced release of IL-6 and TNF-α by freshly explanted mouse peritoneal cells. Monokines such as TNF-α and IL-6 represent critical pathophysiological mediators in the initiation of gram-negative sepsis and septic shock (48). Moreover, a strict correlation between Ab binding and in vitro neutralizing activity was observed. Different LPS preparations including both S-form and R-form LPS were used. Structure-activity relationship studies have shown that lipid A is the active component in the LAL assay (49), and it is mainly responsible for the induction of monokines by host cells (50,51). As WN1 222-5 does not bind to free lipid A, its inhibitory activity in LPS-induced LAL coagulation and monokine release could be explained by some form of steric hindrance, different aggregation of supramolecular LPS structures, or a modification of the conformation of the LPS molecule after Ab binding.

Recently Pollack et al. (52) analyzed the mechanism by which WN1 222-5 regulates LPS uptake by human peripheral blood mononuclear cells and by CD14 transfected CHO cells. This study confirms the ability of WN1 222-5 to inhibit the recognition and uptake of S-form and R-form LPS by target cells expressing membrane-associated CD14 (mCD14). The inhibition is observed in the presence or the absence of LPS-binding protein (LBP), suggesting that anti-LPS Abs might inhibit both mCD14-dependent and -independent LPS uptake. However, this latter mechanism of uptake might become more evident at the relatively high LPS concentrations used. In this study, the anti-LPS mAbs downregulate the proinflammatory uptake of LPS by human peripheral blood mononuclear cells through mCD14, while enhancing complement-dependent opsonic uptake via membrane associated CR1. These Abs might therefore modify how LPS is seen by host monocytes, and they might exert complex modulating effects on LPS-related inflammatory and immune functions. Moreover, the inhibitory activity of anti-LPS mAbs could also be affected by other proteins able to bind

LPS such as soluble CD14, HDL, or bactericidal/permeability-increasing protein (BPI).

On the other hand, this study shows that WN1 222-5 is less efficient than anti–O side chain Abs in activating complement (52). This finding supports other data, which suggest that anti–core LPS Ab might be more endotoxin neutralizing than opsonic. This may be linked to the fact that the mAb recognizes an epitope that is closer to the lipid A region and to the fact that only one epitope is present per LPS molecule. It is also remarkable that the inhibition by anti-LPS mAbs of LPS uptake by mCD14-bearing target cells does not necessarily result from direct Ab interaction with a putative CD14 binding site on the biologically active lipid A moiety. Rather, Ab binding to other portions of the LPS macromolecule, namely the core oligosaccharide or O-polysaccharide, may result in different LPS arrangements that render LPS-containing immune complexes less recognizable by mCD14 than uncomplexed LPS. It seems likely that anti-LPS Abs interfere with the formation of the ternary complex of LPS, LBP, and mCD14, thus preventing LPS attachment to the cell surface, LPS internalization, and generation of LPS-induced proinflammatory signals (52).

NEUTRALIZATION OF LPS IN VIVO

The ability of WN1 222-5 to suppress endotoxic activity in vivo has been analyzed in animal models exploring important pathological effects of LPS. For these assays to be meaningful, special precautions were taken to exclude contamination of WN1 222-5 samples by LPS (53). All mAb preparations were LAL-negative and nonpyrogenic in rabbits. Moreover, heat treatment of the preparation abolished protection, and WN1 222-5 was not active against LPS to which it did not bind, indicating that induction of cross-tolerance was not responsible for the neutralizing properties of the Ab. WN1 222-5 blocks, in a dose-dependent fashion, endotoxin-induced pyrogenicity in rabbits and lethality in D-galactosamine–sensitized mice (34). In D-galactosamine–sensitized mice, survival was associated with a suppression of plasma TNF-α levels (54) (Figs. 1, 2). The inhibition of TNF-α in these mice correlates with the fact that, in this model, lethality is mediated by TNF-α (55).

As these models are dependent on the administration of purified LPS, it was relevant to analyze additional models in which LPS toxic effects are not caused

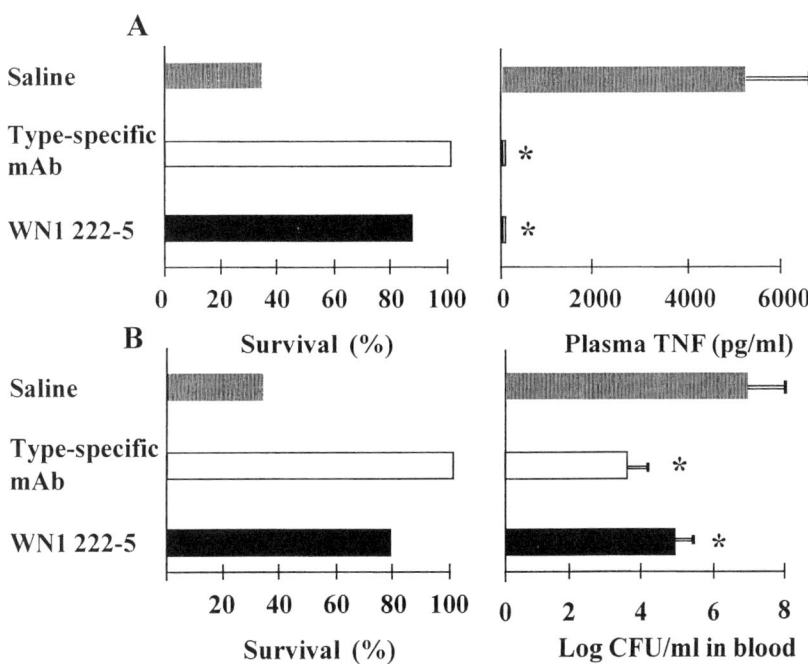

Fig. 1 (A) LPS model. Mice were challenged with 50 ng/mouse of LPS of E. coli O111 together with 15 mg D-galactosamine. Peak plasma TNF levels were measured by bioassay 90 minutes after LPS challenge. (B) Bacterial model. Mice received 10^9 CFU of E. coli O111 intravenously. Bacterial counts were done in blood 5 hours after challenge. In both models, antibodies were given 1 hour after challenge. *$p < 0.001$ compared to saline group.

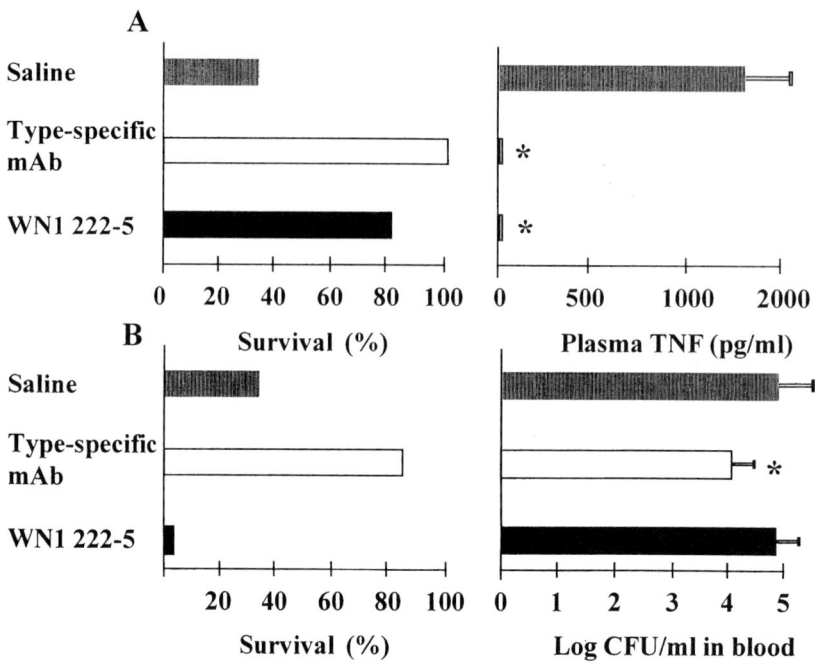

Fig. 2 (A) LPS model. Mice were challenged with 100 ng/mouse of LPS of *E. coli* O18 together with 15 mg D-galactosamine. Peak plasma TNF levels were measured by bioassay 90 minutes after LPS challenge. (B) Bacterial model. Mice received 10^9 CFU of *E. coli* O18 intravenously. Bacterial counts were done in blood 5 hours after challenge. In both models, antibodies were given 1 hour after challenge. *$p < 0.001$ compared to saline group.

through a direct administration of the purified endotoxin. Gut ischemia induces disruption of the intestinal mucosal barrier, allowing translocation of bacteria and endotoxin into the blood, which may trigger a systemic inflammatory response, lung injury, and lethality. It was, therefore, evaluated whether WN1 222-5 reduces the mortality rate in rats and protects from the lung injury following ischemia reperfusion injury to the gut (56). From this study it appears that WN1 222-5 is protective, and this finding clearly suggests that endotoxin derived from enteric bacteria might play an important role in the pathogenesis of lung injury. Moreover, WN1 222-5 almost completely neutralized the plasma endotoxin concentrations observed in control animals. Similar conclusions have been reached by showing that selective gut decontamination can reduce the generation of LPS, TNF, and the severity of lung damage that often follows ischemia and reperfusion of the intestine in rats (57).

The study of Bahrami et al. (56) is relevant in that it shows that the LPS-neutralizing properties of anti-core LPS mAbs are not limited to purified and isolated LPS but extend to naturally occurring LPS, confirming previous data (45). It also suggests that Enterobacteriaceae present in the gastrointestinal tract are the main contributors of pathogenic endotoxin, at least in rats, confirming a previous report (58). Moreover, animals treated with WN1 222-5 had a statistically lower combined incidence of translocating bacteria in the intestinal wall of jejunum, ileum, and colon. Whether this effect is due to a direct bactericidal effect of WN1 222-5 or to an indirect effect due to lower endotoxin levels and reduced damage to the gut and bacterial translocation is not established. Moreover, no data are available about the intriguing possibilities that WN1 222-5 might be able to diffuse into the lumen of the gastrointestinal tract and trap LPS there or that it might affect gut permeability.

The fact that WN1 222-5 exhibits neutralizing activity not only in vitro but also in vivo is further evidence of the accessibility of the core-located epitope on the LPS molecule in vivo. The protective effect of WN1 222-5 might be due to direct neutralization of LPS with prevention of monokine induction (TNF and IL-6) through mCD14 and to clearance of LPS-antibody complexes through the Fc receptor or the complement receptor or some other mechanism, as shown in vitro (52).

BACTERIAL MODELS

The potential protective effect of WN1 222-5 has been also analyzed in models of sepsis caused by gram-negative organisms. These are models in which the relative contribution of bacterial load and toxicity from LPS and/or other bacterial products is not clearly established (59).

WN1 222-5 in a model of intravenous injection of bacteria was able to protect mice and to reduce blood bacterial counts against *E. coli* O111 bacteria (54) (Fig. 1) but not against an inoculum of *E. coli* O18 (Fig. 2). In this acute lethality model, large bacterial inocula are needed to induce lethality, TNF-α levels do not directly correlate with mortality, and the model is highly dependent on effective reduction of bacterial counts. The fact that WN1 222-5 is able to reduce the bacterial load of some bacteria even if less efficiently than an anti–O side chain mAb, confirms that this mAb binds to *E. coli* O111:B4 and retains the ability to activate complement even if less efficiently than anti–O side chain Ab (52).

Lower inocula are used in murine peritonitis models in which bacteria are injected intraperitoneally in conjunction with mucin-hemoglobin. This model can be set up with or without antibiotics, and the readout is again lethality. In this model, TNF-α levels are lower than those observed after intravenous injection of bacteria. Unlike anti–O side chain Abs, WN1 222-5 was not protective in these models (54). These discrepant results in vivo are difficult to explain. However, differences in tissue distribution and in growth conditions of the bacteria and lower accessibility of the relevant epitope in bacteria growing in the peritoneal cavity may explain the different level of protection afforded by WN1 222-5 in these different models against the same strain. A lower number of mAb molecules per bacterium must be responsible for not achieving that critical level needed to activate efficiently complement-associated mechanisms of clearance.

The different potency between WN1 222-5 and the anti–O side chain mAb can be more easily explained by differing capacity to activate complement, mAb isotype variations, and accessibility of the recognized epitopes. The restricted number of epitopes per LPS molecules (multiple vs. single) and the deeper location of the core epitope in the bacterial membrane further limit anti-core mAbs activity. It is apparent that acute lethality models in animals can be used to differentiate the function and mechanism of action of anti-core and anti–O specific mAbs.

CONCLUSIONS AND PERSPECTIVES

In conclusion, by using a novel strategy for immunization, we have been able to generate cross-reactive and endotoxin neutralizing mAbs. The ability of these mAbs to neutralize endotoxin has been established in vitro and in vivo in different animal models. On the other hand, their capacity to show an antibacterial effect remains controversial. Most data in vitro and in vivo show that the place of WN1 222-5 as an antibacterial agent is limited. This is probably not due to an intrinsic inability of the Ab to activate complement because WN1 222-5 is active in passive hemolysis assays were LPS is directly loaded onto the red blood cell membrane. It seems more related to a lower accessibility of the recognized epitope on the viable bacterium.

Clearly a recurrent major question is the potential use of such a mAb in the clinic and therefore the potential relative contribution of its endotoxin-neutralizing and antibacterial activities. One might argue that potent antibiotics are routinely employed in sepsis to eradicate bacterial growth, but that they have not dramatically improved the outcome. In fact they might transiently aggravate the clinical picture by enhancing the release of toxic bacterial products, supporting the view that an antiendotoxic principle could be more useful than another antibacterial approach. On the other hand, the relative contribution of endotoxin versus other bacterial products as well as the relevance of endotoxin originating from Enterobacteriaceae present in the gut have not been clarified (60). Endotoxemia has been often identified in patients with gram-positive and fungal infection and even in the absence of a focus of infection, and therefore a potential gut origin has been suggested (61). However, selective decontamination of the human gastrointestinal tract to reduce bacterial load has generally not been successful (62). Whether gut decontamination in humans comes too late is also not clear. In most animal experiments, selective decontamination of the gastrointestinal tract is effective when performed before or at the time of the noxious event (63).

The finding of the existence of cross-reactive and neutralizing anti-core mAbs makes it possible to address other scientifically interesting questions. For example, attempts to generate anti-core mAbs able to cross-react with and neutralize an even broader spectrum of LPSs including those of *Pseudomonas* or *Klebsiella* have become feasible. In addition, it is now feasible to test and compare the neutralizing and/or antibacterial activity of different anti-core mAbs including those recognizing one *E. coli* chemotype. These Abs recognize an epitope that is present only once per LPS molecule but is closer to the surface of the bacterium. Therefore it will be tested whether they have advantages in terms of antibacterial activity. Moreover, IgM class Abs can be engineered to offer an improvement in complement activation. All this will lead to the definition of a mixture of anti-core Abs with enhanced antiendotoxin and antibacterial properties with potentiality for the prevention or treatment of endotoxin-related diseases.

REFERENCES

1. Rietschel ET, Kirikae T, Schade FU, Mamat U, Schmidt G, Loppnow H, Ulmer AJ, Zähringer U, Seydel U, Di Padova F, Schreier M, Brade H. Bacterial endotoxin: molecular relationships of structure to activity and function. FASEB J 1994; 8:217–225.
2. Rietschel E, Brade H, Holst O, Brade L, Müller-Loennies S, Mamat U, Zähringer U, Beckmann F, Seydel U, Brandenburg K, Ulmer AJ, Mattern T, Heine H, Schletter J, Loppnow H, Schönbeck U, Flad HD, Hauschildt S, Schade UF, Di Padova F, Kusumoto S, Schumann RR. Bacterial endotoxin: chemical constitution, biological recognition, host response and immunological detoxification. Curr Top Microbiol Immunol 1996; 216:39–81.
3. Holst O, Brade H. Chemical structure of the core region of lipopolysaccharides. In: Morrison DC, Ryan JL, eds. Bacterial Endotoxic Lipopolysaccharides. Vol I. Molecular Biochemistry and Cellular Biology. Boca Raton, FL: CRC Press, 1992:135–170.
4. Jansson PE, Lindberg AA, Lindberg B, Wollin R. Structural studies on the hexose region of the core in lipopolysaccharides from Enterobacteriaceae. Eur J Biochem 1981; 115:571–577.
5. Beckmann F, Jaeger KE, Zähringer U. 7-O-Carbamoyl-L-glycero-D-manno-heptose: a new core constituent in lipopolysaccharide of *Pseudomonas aeruginosa*. Carbohydr Res 1995; 267:C3–C7.
6. Süsskind M, Müller-Loennies S, Nimmich W, Brade H, Holst O. Structural investigation on the carbohydrate backbone of the lipopolysaccharide from *Klebsiella pneumoniae* rough mutant R20/O1. Carbohydr Res 1995; 269:C1–C7.
7. Süsskind M, Brade L, Brade H, Holst O. Identification of a novel heptoglycan of α1→2-linked D-glycero-D-manno-heptopyranose. Chemical and antigenic structure of lipopolysaccharides from *Klebsiella pneumoniae* ssp. *pneumoniae* rough strain R20 (O1:K20). J Biol Chem 1998; 273:7006–7017.
8. Trautmann M, Vogt K, Hammack C, Cross AS. A murine monoclonal antibody defines a unique epitope shared by *Klebsiella* lipopolysaccharides. Infect Immun 1994; 62:1282–1288.
9. Appelmelk BJ, Verweij-Van Vught AM, Maaskant JJ, Schouten WF, DeJonge AJR, Thijs LG, Mac Laren DM. Production and characterization of mouse monoclonal antibodies reacting with the lipopolysaccharide core region of gram negative bacilli. J Med Microbiol 1988; 26:107–114.
10. Aydintung MK, Inzana TJ, Letonja T, Davis WC, Corbeil LB. Cross-reactivity of monoclonal antibodies to *Escherichia coli* J5 with heterologous gram-negative bacteria and extracted lipopolysaccharides. J Infect Dis 1989; 160:846–857.
11. Gigliotti F, Shenep JL. Failure of monoclonal antibodies to core glycolipid to bind to intact smooth strains of *Escherichia coli*. J Infect Dis 1985; 151:1005–1011.
12. Miner KM, Manyak CL, Williams E, Jackson J, Jewell M, Gammon MT, Ehrenfreund C, Hayes E, Callahan III LT, Zweerink H, Sigal NH. Characterization of murine monoclonal antibodies to *Escherichia coli* J5. Infect Immun 1986; 52:56–62.
13. Mutharia LM, Crockford G, Bogard WC, Hancock REW. Monoclonal antibodies specific for *Escherichia coli* J5 lipopolysaccharide: cross-reaction with gram-negative bacterial species. Infect Immun 1984; 45:631–636.
14. Nelles MJ, Niswander CA. Mouse monoclonal antibodies reactive with J5 lipopolysaccharide exhibit extensive serological cross-reactivity with a variety of gram-negative bacteria. Infect Immun 1984; 46:677–681.
15. Pollack M, Chia JKS, Koles NL, Miller M, Guelde G. Specificity and cross-reactivity of monoclonal antibodies reactive with the core and lipid A regions of bacterial lipopolysaccharides. J Infect Dis 1989; 159:168–188.
16. Salles MF, Mandine E, Zalisz R, Guenounou M, Smets P. Protective effects of murine monoclonal antibodies in experimental septicemia: *E. coli* antibodies protect against different serotypes of *E. coli*. J Infect Dis 1989; 159:641–647.
17. Barclay GR, Scott BB. Serological relationships between *Escherichia coli* and *Salmonella* smooth- and rough-mutant lipopolysaccharides as revealed by enzyme-linked immunosorbent assay for human immunoglobulin G antiendotoxin antibodies. Infect Immun 1987; 55:2706–2714.
18. Hamilton-Davies C, Barclay GR, Murphy WG, Machin SJ, Webb AR. Passive immunisation with IgG endotoxin core antibody hyperimmune fresh frozen plasma. Vox Sang 1996; 71:165–169.
19. Law BJ, Marks MI. Age-related prevalence of human serum IgG and IgM antibody to the core glycolipid of *Escherichia coli* strain J5, as measured by ELISA. J Infect Dis 1985; 151:988–994.
20. Braude AI, Douglas H. Passive Immunization against the local Shwartzman reaction. J Immunol 1972; 108:505–512.
21. Chedid L, Parant M, Parant F, Boyer F. A proposed mechanism for natural immunity to enterobacterial pathogens. J Immunol 1968; 100:292–301.
22. McCabe WR. Immunization with R mutants of *Salmonella minnesota*. I: Protection against challenge with heterologous gram-negative bacilli. J Immunol 1972; 108:601–610.
23. Ziegler EJ, McCutchan JA, Fierer J, Glauser MP, Sadoff JC, Douglas H, Braude AI. Treatment of gram-

negative bacteremia and shock with human antiserum to a mutant *Escherichia coli*. N Engl J Med 1982; 307: 1225–1230.
24. Baumgartner JD, Heumann D, Calandra T, Glauser MP. Antibodies to lipopolysaccharides after immunization of humans with the rough mutant *Escherichia coli* J5. J Infect Dis 1991; 163:769–772.
25. Baumgartner JD, Heumann D, Gerain J, Weinbreck P, Grau GE, Glauser MP. Association between protective efficacy of anti-lipopolysaccharide (LPS) antibodies and suppression of LPS-induced tumor necrosis factor alpha and interleukin 6: comparison of O side chain-specific antibodies with core LPS antibodies. J Exp Med 1990; 171:889–896.
26. Greisman SE, Johnston CA. Failure of antisera to J5 and R595 rough mutants to reduce endotoxemic lethality. J Infect Dis 1988; 157:54–64.
27. Greisman SE, Johnston CA. Evidence against the hypothesis that antibodies to the inner core of lipopolysaccharides in antisera raised by immunization with enterobacterial deep-rough mutants confer broad spectrum protection during gram-negative bacterial sepsis. J Endotox Res 1997; 4:123–153.
28. Hellman J, Zanzot E, Loiselle PM, Amato SF, Black KM, Ge Y, Kurnick JT, Warren HS. Antiserum against *Escherichia coli* J5 contains antibodies reactive with outer membrane proteins of heterologous gram-negative bacteria. J Infect Dis 1997; 176:1260–1268.
29. Bhat NM, Bieber MM, Chapman CJ, Stevenson FK, Teng NN. Human antilipid A monoclonal antibodies bind to human B cells and the i antigen on cord red blood cells. J Immunol 1993; 151:5011–5021.
30. Bieber MM, Bhat NM, Teng NNH. Anti-endotoxin human monoclonal antibody A6H4C5 (HA-1A) utilizes the VH4.21 gene. Clin Infect Dis 1995; 21(suppl 2): 186–189.
31. Brandenburg K, Seydel U. Conformational analysis of endotoxins from Salmonella minnesota with varying sugar chain lengths by FT-IR spectroscopy. In: Merlin JC, Turrell S, Huvenne JP, eds. 6th European Conference on Spectrosc Biol Mol. Dordrecht, Kluwer, 1995: 393–394.
32. Kastowsky M, Gutberlet T, Bradaczek H. Molecular modeling of the three-dimensional structure and conformational flexibility of bacterial lipopolysaccharide. J Bacteriol 1992; 174:4798–4806.
33. Mäkelä PH, Stocker BAD. Genetics of lipopolysaccharide. In: Rietschel ET, ed. Handbook of Endotoxin: Vol 1: Chemistry of Endotoxin. Amsterdam: Elsevier Science Publishers BV, 1984:59–137.
34. Di Padova FE, Brade H, Barclay GR, Poxton IR, Liehl E, Schuetze E, Kocker HP, Ramsay G, Schreier MH, McClelland DBL, Rietschel ET. A broadly cross-protective monoclonal antibody binding to *Escherichia coli* and *Salmonella* lipopolysaccharides. Infect Immun 1993; 61:3863–3872.
35. Di Padova F, Gram H, Barclay R, Poxton I, Liehl E, Rietschel ET. Monoclonal antibodies to endotoxin core as a new approach in endotoxemia therapy. In: Morrison DC, Ryan JL, eds. Novel Therapeutic Strategies in the Treatment of Sepsis. New York: Marcel Dekker, 1995:13–31.
36. Gibb AP, Barclay GR, Poxton IR, Di Padova F. Frequencies of lipopolysaccharide core types among clinical isolates of *Escherichia coli* defined with monoclonal antibodies. J Infect Dis 1992; 166:1051–1057.
37. Brandenburg K, Seydel U, Schromm AB, Loppnow H, Koch MHJ, Rietschel ET. Conformation of lipid A, the endotoxic center of bacterial lipopolysaccharide. J Endotox Res 1996; 3:173–178.
38. Galanos C, Luderitz O, Westphal O. Preparation and properties of antisera against the lipid A component of bacterial lipopolysaccharides. Eur J Biochem 1971; 24: 116–122.
39. Baumberger C, Ulevitch RJ, Dayer JM. Modulation of endotoxic activity of lipopolysaccharide by high-density lipoprotein. Pathobiology 1991; 59:378–383.
40. Massamiri T, Tobias PS, Curtiss LK. Structural determinants for the interaction of lipopolysaccharide binding protein with purified high density lipoproteins: role of apolipoprotein A-I. J Lipid Res 1997; 38: 516–525.
41. Munford RS, Hall CL, Dietschy JM. Binding of *Salmonella typhimurium* lipopolysaccharides to rat high-density lipoproteins. Infect Immun 1981; 34:835–843.
42. Freudenberg MA, Bog-Hansen TC, Back U, Galanos C. Interaction of lipopolysaccharides with plasma high-density lipoprotein in rats. Infect Immun 1980; 28:373–380.
43. Jansson PE, Wollin R, Bruse GW, Lindberg AA. The conformation of core oligosaccharides from *Escherichia coli* and *Salmonella typhimurium* lipopolysaccharides as predicted by semi-empirical calculations. J Mol Recogn 1989; 2:25–36.
44. Di Padova F, Mikol V, Barclay GR, Poxton IR, Brade H, Rietschel ET. Anti-lipopolysaccharide antibodies. Prog Clin Biol Res 1994; 388:85–94.
45. Saxen H, Vuopio-Varkila J, Luk J, Lindberg A, Lang A, Di Padova F, Cryz SJ, Mertsola J, McCracken GH, Hansen EJ. Detection of enterobacterial lipopolysaccharides and experimental endotoxemia by means of an immunolimulus assay using both serotype-specific and cross-reactive antibodies. J Infect Dis 1993; 168:393–399.
46. Lawrence JG, Ochman H, Hartl DL. Molecular and evolutionary relationship among enteric bacteria. J General Microbiol 1991; 137:1911–1921.
47. Edmond DM, Poxton IR, Di Padova F. The accessibility of cross-reactive anti-lipopolysaccharide-core monoclonal antibodies to *Escherichia coli* grown in sub-MICs of temocillin and other antibiotics. J Antimicrob Chemother 1993; 31:673–680.
48. Beutler B, Cerami A. The endogenous mediator of endotoxic shock. Clin Res 1987; 35:192–197.
49. Takada H, Kotani S, Tanaka S, Ogawa T, Takahashi I, Tsujimoto M, Komuro T, Shiba T, Kusumoto S, Kusunose N, Hasegawa A, Kiso M. Structural requirements of lipid A species in activation of clotting enzymes from the horseshoe crab, and the human complement cascade. Eur J Biochem 1988; 175:573–580.
50. Loppnow H, Brade H, Durrbaum I, Dinarello CA, Kusumoto S, Rietschel ET, Flad HD. IL-1 induction-capacity of defined lipopolysaccharide partial structures. J Immunol 1989; 142:3229–3238.

51. Zähringer U, Lindner B, Rietschel ET. Molecular structure of lipid A, the endotoxic center of bacterial lipopolysaccharides. Adv Carbohydr Chem Biochem 1994; 50:211–276.
52. Pollack M, Ohl CA, Golenbock DT, Di Padova F, Wahl LM, Koles NL, Guelde G, Monks BG. Dual effects of lipopolysaccharide (LPS) antibodies on cellular uptake of LPS and LPS-induced proinflammatory functions. J Immunol 1997; 159:3519–3530.
53. Chong KT, Huston M. Implications of endotoxin contamination in the evaluation of antibodies to lipopolysaccharides in a murine model of gram-negative sepsis. J Infect Dis 1987; 156:713–719.
54. Bailat S, Heumann D, Le Roy D, Baumgartner JD, Rietschel ET, Glauser MP, Di Padova F. Similarities and disparities between core-specific and O-side-chain specific antilipopolysaccharide monoclonal antibodies in models of endotoxemia and bacteremia in mice. Infect Immun 1997; 65:811–814.
55. Lehmann V, Freudenberg MA, Galanos C. Lethal toxicity of lipopolysaccharide and tumor necrosis factor in normal and D-galactosamine-treated mice. J Exp Med 1987; 165:657–663.
56. Bahrami S, Yao YM, Leichtfried G, Redl H, Schlag G, Di Padova FE. Monoclonal antibody to endotoxin attenuates hemorrhage-induced lung injury and mortality in rats. Crit Care Med 1997; 25:1030–1036
57. Sorkine P, Szold O, Halpern P, Gutman M, Greemland M, Rudick V, Goldman G. Gut decontamination reduces bowel ischemia-induced lung injury in rats. Chest 1997; 112:491–495.
58. Yao YM, Yu Y, Sheng ZY, Tian HM, Wang YP, Lu LR, Yu Y. Role of gut-derived endotoxaemia and bacterial translocation in rats after thermal injury: effects of selective decontamination of the digestive tract. Burns 1995; 21:580–585.
59. Hoffman WD, Danner RL, Quezado ZMN, Banks SM, Elin RJ, Hosseini JM, Natanson C. Role of endotoxemia in cardiovascular disfunction and lethality: virulent and nonvirulent *Escherichia coli* challenges in a canine model of septic shock. Infect Immun 1996; 64:406–412.
60. Nieuwenhuijzen GA, Deitch EA, Goris RJ. Infection, the gut and the development of the multiple organ dysfunction syndrome. Eur J Surg 1996; 162:259–273.
61. Danner RL, Elin RJ, Hosseini JM, Wesley RA, Reilly JM, Parillo JE. Endotoxemia in human septic shock. Chest 1991; 99:169–175.
62. Lingnau W, Berger J, Javorsky F, Lejeune P, Mutz N, Benzer H. Selective intestinal decontamination in multiple trauma patients: prospective, controlled trial. J Trauma 1997; 42:687–694.
63. Yao YM, Lu LR, Yu Y, Liang HP, Chen JS, Shi ZG, Zhou BT, Sheng ZY. Influence of selective decontamination of the digestive tract on cell-mediated immune function and bacteria/endotoxin translocation in thermally injured rats. J Trauma 1997; 42:1073–1079.
64. Heumann D, Baumgartner JD, Jacot-Guillarmod H, Glauser MP. Antibodies to core lipopolysaccharide determinants: Absence of cross-reactivity with heterologous lipopolysaccharides. J Infect Dis 1991; 163:762–768.

42

Effects of Lipopolysaccharide on T Cells

Masayasu Nakano and Teruo Kirikae
Jichi Medical School, Tochigi-ken, Japan

Toshimasa Nitta
Ohu University School of Dentistry, Koriyama, Japan

INTRODUCTION

Lipopolysaccharide (LPS) endotoxin has a variety of biological effects on numerous types of mammalian cells, including immunocompetent cells (1). The effects on B-lymphocyte and macrophage function have been extensively studied. B lymphocytes are activated by LPS to proliferate and differentiate into antibody-secreting plasma cells in the absence of specific antigens—polyclonal B-cell activation. The macrophages activated by LPS produce a variety of active substances such as tumor necrosis factor (TNF), interleukin (IL)-1, IL-6, procoagulant activators, eicosanoids, and reactive oxygen and nitrogen intermediates. However, in contrast to the well-studied B-lymphocyte and macrophage response to LPS, little is known concerning the effect of biological activities of LPS on T-lymphocyte function. Mitogenic proliferative response and cytokine production can hardly be seen in the purified T-cell population in response to LPS, although LPS binds to T lymphocytes (2). The majority of T cells are refractory to LPS, but some T cells seem to respond to LPS in certain conditions.

PROLIFERATIVE RESPONSE OF T CELLS IN RESPONSE TO LPS

Specific or Nonspecific Effect

In 1983, Vogel et al. (3) demonstrated that a cloned murine IL-2–dependent cytotoxic T-cell line proliferated in response to LPS. The results suggest that some murine T-cell populations can respond directly to LPS. They also showed that highly purified murine splenic T-cell populations also had the ability to respond mitogenically to LPS. Approximately 3% of T cells in the purified splenic T-cell population prepared from LPS-responsive C3H/He mice could respond to LPS. However, no T cells capable of responding to LPS in the spleen cells obtained from LPS-hyposensitive C3H/HeJ mice retained the ability to respond to various antigens. Therefore, the results suggested that the stimulatory effect of LPS to C3H/He splenic T cells might depend on its endotoxic activity rather than the antigenicity.

A 1983 report by Milner et al. (4) suggested a possibility of T-cell response specific to LPS. They demonstrated that the T cells prepared from lymph nodes of in vivo LPS-sensitized mice and purified through nylon wool column increased their [^{3}H]-thymidine uptake in vitro by stimulation with LPS, while no response to LPS was observed in the nonsensitized T cells. They also showed that the proliferative response of LPS-sensitized T cells prepared from LPS-nonresponder C57BL/10ScN mice to LPS were apparently lower than that from LPS responder C57 BL/6J mice, but higher than that from unsensitized mice. Therefore, they concluded that LPS interacted with antigen-specific murine T cells and stimulated them to proliferate.

Recently Tough et al. (5) showed that under in vivo conditions, LPS also induced strong stimulation of T cells. As manifested by CD69 upregulation, LPS injection stimulated both CD4 and CD8$^+$ T cells, and, at high doses, stimulated naive (CD44^{lo}) T cells as well as memory (CD44^{hi}) T cells. However, in terms of cell division, the response of T cells after LPS injection was limited to the CD44^{hi} subset of CD8$^+$ cells. Based on studies with LPS-nonresponder and gene-knockout mice, they concluded that LPS-induced proliferation of CD44^{hi} CD8$^+$ cells operates via an indirect pathway involving LPS stimulation of antigen-presenting cells and release of type I (α,β) interferon.

Mattern et al. (6–8) reported that LPS as well as its hydrophobic part, lipid A, were capable of inducing proliferation of human T cells in the presence of monocytes. The LPS-induced proliferation was strongly dependent on direct cell-to-cell contact, and the interaction of the cell surface molecules CD28 and/or CTLA-4 on T cells with their ligands CD80 and/or CD86 on monocytes was especially important. These ligands were differentially regulated on monocytes of LPS responders and LPS nonresponders. There was a clear correlation of CD80 expression on monocytes after LPS stimulation and their capacity to support T-cell proliferation. The interaction between T cells and monocytes was not MHC-restricted, suggesting that LPS did not act like an antigen, and the proliferative T cells were Th1 type helper T cells, since the proliferative cells expressed mRNA for the Th1 cell–derived cytokines interferon (IFN)-γ and IL-2, but not for the Th2 type of helper T-cell–derived cytokines IL-4, IL-5, and IL-10.

Costimulation with LPS and Lectin or Antigen

LPS enhances nonspecifically the proliferation of murine T cells stimulated with lectin. Murine thymic T cells are unresponsive to LPS stimulation. However, when cultured with minimally mitogenic levels of T-cell mitogens in combination with LPS, the cells demonstrate levels of DNA synthesis de novo equal to or greater than those induced by the T-cell mitogen alone. Forbes et al. (9) demonstrated that the ordinarily non-susceptible thymus cell population was triggered to proliferate by LPS in the presence of concomitant T-cell activation initiated by low concentration of concanavalin A (ConA) or by submitogenic amount of alloantigens. We (10) also demonstrated that when the hydrocortisone-resistant thymic T cells were cultured in the presence of LPS and ConA together, [^{3}H]-thymidine uptake of the cells was enhanced synergistically in comparison with those cultured with either of the mitogens alone. The synergistic effect on thymic T cells was dependent on Ia-positive accessory cells such as macrophages and was mainly due to the action of the enhanced production of IL-2, which was released by the T cells in response to ConA and LPS.

Synergistic effect of antigen and LPS on T-cell proliferation was also reported. Mita et al. (11) showed that the enhanced T-cell proliferation occurred when lipid A, the active principle of endotoxin, was given to mice immunized with sheep red blood cells. Bismuth et al. (12) demonstrated that LPS could markedly potentiate the thymus-dependent antigen-induced specific proliferation of in vitro cultured helper T-cell lines. IL-1 was not involved in the proliferation. The responses were obtained with LPS from different bacterial sources and was partially reproduced with lipid A but not with the polysaccharide fraction of LPS. These results suggested that LPS/lipid A enhanced nonspecifically antigen-stimulated T-cell proliferation.

Proliferative Response of T Cells Bearing $\gamma\delta$–T-Cell Receptor

The T-cell receptor (TCR) is a molecular complex consisting of CD3 and its associated molecules (13,14). Proliferation of T cells after antigenic stimulation requires the help of antigen-presenting cells such as macrophages. Processed antigen in macrophages appears on the cell membranes bound with MHC class II molecules and stimulates TCR on the cell membrane of T cells. The stimuli recognized by TCR induce T-cell proliferation. TCR$^+$ cells bear either $\alpha\beta$-TCR or $\gamma\delta$-TCR chains (15). We found that LPS induced proliferative response of the highly purified T cells prepared from protease peptone-induced peritoneal exudate cells (PEC), but not those from spleen cells and thymocytes, in LPS-responsive C3H/HeN mice (Fig. 1) (16). Antigenic stimulus and accessory cells were not required in the response, and the response was abolished when the PEC-T cells were pretreated with complement and anti–$\gamma\delta$-TCR antibody, but not anti–$\alpha\beta$-TCR antibody (Fig. 2), indicating that the cells responding to LPS were apparently the cells with $\gamma\delta$-TCR. T cells expressing the $\gamma\delta$ form of the TCR are relatively abundant in the gut-associated lymphocyte system. T cells with $\gamma\delta$-TCR comprise approximately 1.5% of the nonadherent PEC (17) and less than 1% of adult thymus in mice (18). In the spleens of the mice, T cells with $\alpha\beta$-TCR are quite dominant, but $\gamma\delta$-TCR–positive T cells are few.

Furthermore, we found that although the thymic T cells did not show any increase in [^{3}H]-thymidine uptake in response to LPS or anti–$\gamma\delta$-TCR antibody, obvious uptake did occur when the cells were stimulated

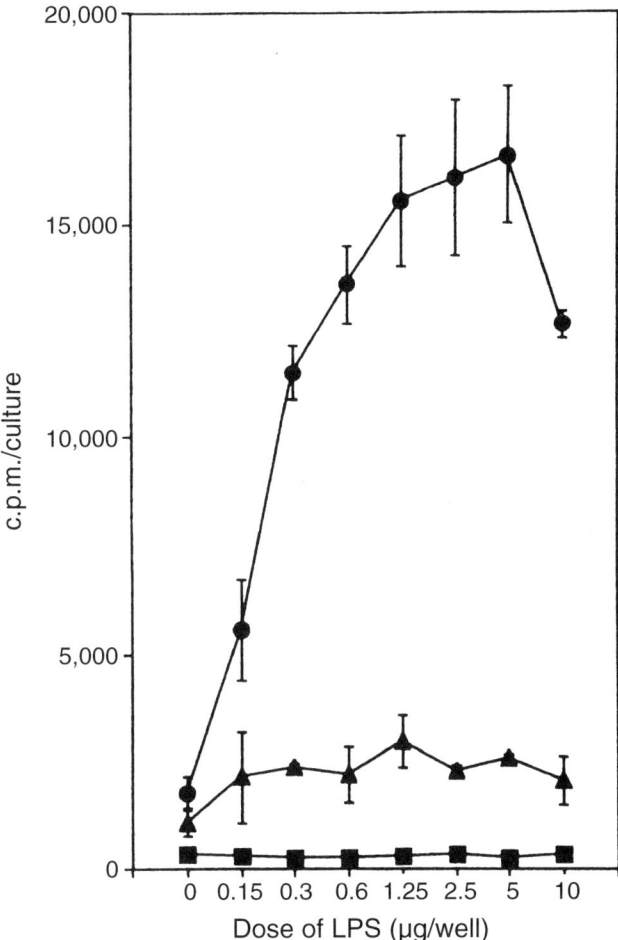

Fig. 1 Mitogenic responses of T cells purified from the peptone-induced PEC (filled circles), spleen cells (filled triangles), and thymus cells (filled squares) of C3H/HeN mice. Purified T cells from PEC and spleen cells were prepared to be passed through the nylon fiber column, treated with anti-LR-1 antibody and anti-murine macrophage antibody plus complement (C), and passed through Sephadex G-10 column. Purified T cells from thymus cells were prepared by treatment of anti-Ia^k antibody and anti-murine macrophage antibody plus C. These purified T cells were cultured in the presence (or absence) of LPS for 48 hours. Doses of LPS are indicated on the abscissa. [^{3}H]-TdR was added to the cultures 20 hours before the end of the culture period. Each symbol represents the mean value of triplicate cultures and the SD. (From Ref. 16.)

with LPS and anti–γδ-TCR antibody, but not anti–αβ-TCR antibody (Fig. 3) (16). IL-1 did not substitute for LPS in the response. Previous treatment of the cells with anti–γδ-TCR antibody and complement abrogates the response. Therefore, these findings indicate that when thymic T cells with γδ-TCR get a signal via γδ-TCR and another mitogenic signal induced by LPS, the cell proliferation occurs. It is known that stimulation of TCR is not enough to cause the proliferative response of T cells; normal T-cell activation requires delivery to the cell of both the TCR signals and a second signal, a protein in the B7 family of cell surface molecules, generated by accessory cells (13,14,19,20). Accessory cells and cytokines produced by accessory cells can provide the second signal (21–23). Thymic T cells bearing γδ-TCR may proliferate in response to LPS (the second signal) when their TCR is stimulated with the antibody (or specific antigen) (the first signal). In the case of PEC T-cell proliferation (Fig. 1), peptone used as a PEC inducer and/or various antigens derived from digested food or bacteria in intestinal flora may provide the first signal to T-cell populations bearing γδ-TCR, and LPS may act as the second signal. How LPS functions as the second signal to γδ-TCR–positive cells is unclear

INDUCTION OF T-CELL CYTOKINE BY LPS

IFN-γ is produced by activated CD4$^+$T helper 1 (Th1) cells, CD8$^+$T cells, or natural killer (NK) cells with the help of accessory cells such as macrophages and monocytes, and the IFN-γ produced is involved in inflammatory responses caused by LPS (24). Although T cells and NK cells are thought not to be traditional targets of LPS (25), it was reported that LPS induces IFN-γ production in vivo (26,27) and in vitro (28–31). These reports suggested that cytokines produced by the LPS-activated accessory cells and interaction between the accessory cells and IFN-γ–producing cells are necessary to induce IFN-γ production rather than direct action of LPS to IFN-γ–producing cells.

Recently, Tennenberg and Weller (32) showed that in the presence of endothelial cells as well as monocytes, LPS could activate human peripheral blood CD3$^+$T cells for IFN-γ secretion. IL-2 was not involved in the T-cell activation. Therefore, they presumed that the interaction of adhesion molecules and the costimulatory signals provided during the adhesion of T cells to endothelial cells might be a necessary factor for LPS-induced T-cell activation and subsequent IFN-γ secretion, and the produced IFN-γ might be involved in the pathogenesis of endotoxin-induced acute microvascular organ injury.

The heterodimer cytokine IL-12 is produced by phagocytic cells and B cells in response to LPS and bacteria. IL-12 is capable not only of augmenting the cytotoxic activity of T cells and NK cells and regulat-

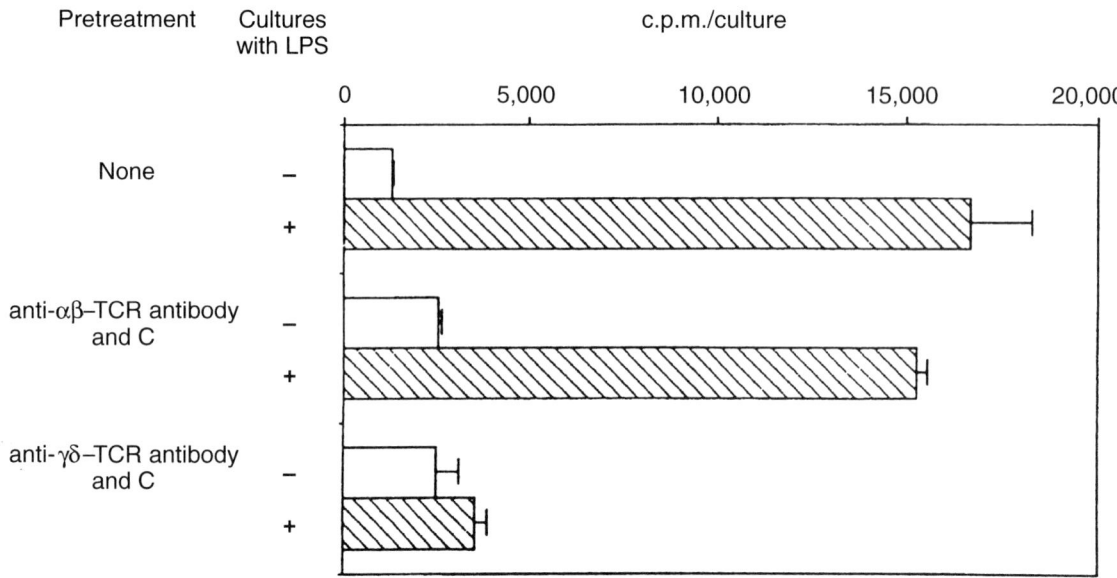

Fig. 2 Abrogation of LPS-induced [^{3}H]-TdR uptake among the PEC-T cells pretreated with anti-γδTCR antibody and complement (C). The PEC-T cells that had been treated previously with anti-γδTCR or anti-αβTCR antibody plus C, or not treated (control), were cultured in the presence or absence of LPS (10 μg/well) for 48 hours. [^{3}H]-TdR was added to the cultures 20 hours before the end of the culture period. Each bar represents the mean value of triplicate cultures and the SD. (From Ref. 16.)

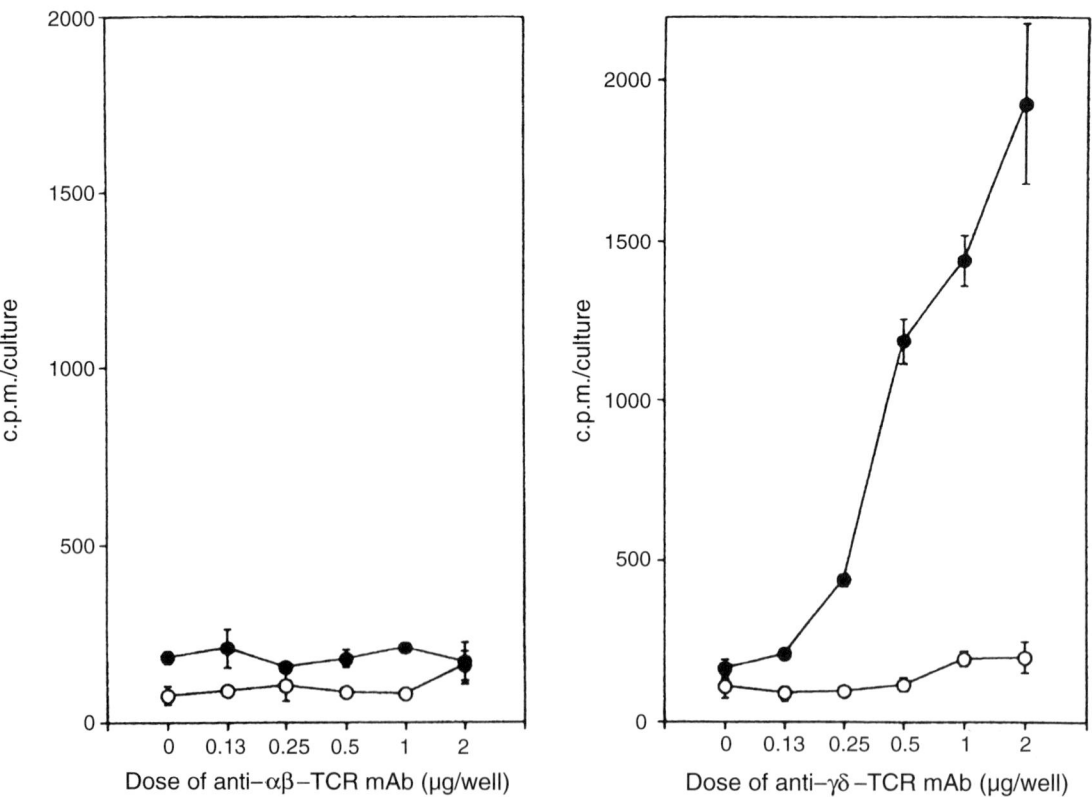

Fig. 3 [^{3}H]-TdR uptake among the thymic T cells that were co-stimulated with LPS and anti-TCR antibody. The thymic T cells were cultured in the presence of (A) anti-αβTCR antibody or (B) anti-γδTCR antibody with (filled circles) or without (open circle) LPS (10 μg/culture) for 48 hours. [^{3}H]-TdR was added to the cultures 20 hours before the end of the culture period. Each symbol represents the mean value of triplicate cultures and the SD. (From Ref. 16.)

ing IFN-γ production, but also of promoting the development of CD4⁺TH1 cells (33). LPS-induced IL-12 may activate Th1 cells to produce IFN-γ.

A novel cytokine for IFN-γ induction has been found. Okamura et al. (34,35) reported the cloning of an IFN-γ–inducing factor (IGIF) or IL-18 that augments T-cell proliferation and NK activity in spleen cells. IL-18 was found in the livers of mice treated with *Propionibacterium acnes* and subsequently challenged with LPS. IL-18, a protein with 157 amino acids, is detected in Kupffer cells and macrophages. It shares structural features with the IL-1 family of proteins and functional properties of IL-12. Like IL-12, it is a potent inducer of IFN-γ from T cells and NK cells. IL-18 synthesis and LPS-induced IFN-γ production are regulated by interleukin-1β–converting enzyme (ICE, recently termed caspase-1), which converts proIL-18 to mature IL-18, and caspase-1–deficient (ICE$^{-/-}$) mice are defective in LPS-induced IFN-γ production (36,37). ICE$^{-/-}$ mice have less IL-1α, TNF-α, and IL-6 and are resistant to septic shock induced by LPS (38).

PARTICIPATION OF T CELLS IN ADJUVANT EFFECT OF LPS

For antibody production against certain antigens (thymus-dependent, or TD, antigen), but not thymus-independent (TI) antigen, three type of cells—B cells for antibody secretion, helper T cells, and antigen-presenting cells (macrophages and others)—are necessary. LPS and lipid A was found to profoundly modulate TD antigen–induced antibody production in vivo as well as in vitro (39). The antigen-dependent immune adjuvant activity of LPS is quite prominent. It is obviously a synergistic effect of endotoxin and antigen, and not an additive effect of both of these immunostimulants, since the magnitude of the response cannot be explained by the simple sum of the polyclonal B-cell activation by LPS and the response to antigen alone. The adjuvant action of LPS/lipid A is influenced by the administration of LPS relative to antigen. When LPS is administered at the same time as antigen or shortly after the injection of antigen, the antibody response to the antigen is augmented. However, if LPS or lipid A is given prior to antigen, antibody production is suppressed (39–41). Thus, the effect of LPS on the antigen-dependent response is immunoregulatory, and the timing of administration between antigen and LPS influences the direction of regulation toward the positive or negative response.

T cells are necessary to induce the adjuvant effect of LPS on the antibody response to TD antigen (39–43). Although evidence for the direct action of LPS or lipid A on T cells is obscure, LPS can support the proliferative response of T cells after stimulation with antigen (or ConA) via mediators produced by LPS-stimulated macrophages. When stimulated with LPS or lipid A, the macrophages increase their phagocytic or pinocytic capability of engulfing antigens and releasing immunopotentiating factors such as IL-1, IL-6, IL-8, IL-10, IL-12, and TNF. These cytokines enhance antibody synthesis and secretion by B cells and help the development and cytokine secretion by helper T cells. T helper 1 (Th1) and T helper 2 (Th2) subsets have been implicated in the regulation of many immune responses (44). Th1 cells produce IL-2 and IFN-γ. Some B-cell help can be provided by Th1 but at higher Th1 cell numbers, this can become suppression (45,46). Th2 cells produce IL-4, IL-5, IL-6, IL-9, IL-10, and IL-13, and these Th2 cytokines encourage antibody production. The macrophages can also produce prostaglandin E_2, which inhibits antibody secretion by B cells (47). LPS or lipid A acts on many kinds of cells, and the LPS-stimulated cells produce a variety of the immunoactive substances. The adjuvant effect (and the suppressive effect) of LPS should be the consequence of the interactions among these activated cells and/or the substances involved in immune networks or cascades, although the precise mechanism of T-cell involvement has not been clarified yet.

INDUCTION OF CYTOTOXIC T CELLS BY LPS

LPS has a potent antitumor activity, which is thought to be mainly due to the action of various mediators, including TNF and NO, which are released from LPS-stimulated phagocytic cells (1). LPS-activated NK cells also release the mediators and participate in tumor cell destruction (48). Although direct activation of cytotoxic CD8⁺T cells by LPS is not known, IL-12 derived from LPS-activated macrophages may augment the cytotoxic effect of CD8⁺T cells and NK cells (49–52). Recently, Takahasi et al. (53) reported that systemic administration of LPS as well as synthetic lipid A induced IL-12 production by Kupffer cells (one of the macrophage population) in the liver of mice and that the IL-12 produced induced the increase of cytotoxic NK1.1⁺ αβ T cells with intermediate TCR expression (NK1⁺TCR^{int}). The NK1⁺TCR^{int} cells showed strong MCH-unrestricted cytotoxicity against NK-sensitive and NK-resistant target cells.

EFFECT OF LIPID A ON INDUCTION OF MEMORY CELLS

Administration of LPS with antigen to mice can alter the proliferative response of memory T cells to antigen restimulation (54). The highly purified T-cell populations were prepared from the PEC of mice that had been primed with horse red blood cells (HRBC) as an antigen and/or synthetic lipid A 2 weeks in advance. The T cells [T(HRBC + lipid A) cells] primed in vivo with antigen and lipid A together proliferated in vitro in response to HRBC or anti–αβ-TCR antibody in the absence of macrophages. However, the T cells [T(HRBC) cells] primed with antigen alone did not proliferate in response to HRBC and anti–αβ-TCR antibody, although they can proliferate if macrophages were supplemented. CD45 is a large (180–220 kDa) cell surface molecule that is expressed by all leukocytes, including all T lymphocytes. Signal transduction by TCR requires the co-expression of CD45, whose cytoplasmic domain has tyrosine phosphatase activity (55). T cells that lack CD45 cannot respond to antigen, even though they express normal levels of TCR. CD45 protein can exist in several distinct isoforms, and the isoforms are expressed differently on naive and memory T cells (56,57). The constituents of CD45 isoforms on the memory T cells primed with antigen and lipid A seem to be altered from those on naive T cells or the T cells primed with antigen, although it is not resolved whether lipid A acts directly or indirectly on the naive T cells that are destined to become memory cells. Previous treatment of T(HRBC + lipid A) cells with anti-CD45 (180 kDa) antibody plus complement abolished the response to antigen (HRBC) and anti–αβ-TCR antibody, while the same treatment of T(HRBC) cells did not abolish the response in the presence of macrophages. The previous treatment of T(HRBC) cells or T(HRBC + lipid A) cells with anti-CD45 (220 kDa, a different CD45 isoform from the 180 kDa) antibody plus complement eliminated their proliferative response. These findings indicate that the T memory cells that were generated by priming with antigen and lipid A become capable of responding to antigen or anti-TCR antibody without help of macrophages, and a change in CD45 isoform distribution occurs on the surface membranes of T(HRBC + lipid A) cells such that they become serologically distinct from T(HRBC) cells. Although highly purified T(HRBC + lipid A) cells can proliferate in response to HRBC or anti–αβ-TCR antibody in the absence of macrophages, the response is abrogated by treatment of anti-NK1 antibody and complement (unpublished observation). Therefore, a trace of NK cells in the T(HRBC + lipid A) cell population are required to support the response.

EFFECTS OF LPS ON T-CELL DEATH

Many studies as described above indicate that LPS activates T cells and their function. Although LPS is a toxic substance, LPS-activated T cells seem to strengthen host-defense systems. LPS is known to confer both beneficial and harmful effects to its hosts. It may induce T-cell injury and death. Zhang et al. (58) showed that in vivo administration of LPS to mice could induce apoptosis (programmed cell death) of thymic T cells. However, obvious DNA fragmentation (apoptosis) by LPS was not observed on the cultured thymic T cells, indicating that the apoptosis in vivo was an indirect effect of LPS. Since the apoptosis was not seen in adrenalectomized mice and was inhibited by injection with anti–TNF-α antibody, the participation of adrenal hormones and TNF-α was suggested. In contrast, Vella et al. (59) reported that LPS very effectively prevents T-cell death driven by superantigen.

CONCLUSIONS

Much knowledge has been accumulated on the effects of LPS on T-cell function. Even though LPS may act directly on some T-cell populations, the majority of the effects of LPS on T cells seem to be derived indirectly from the interaction between T cells and LPS-activated accessory cells such as macrophages and from biologically active mediators released from the LPS-activated accessory cells and cytokine-activated T cells. Once a small population of resting T cells is activated directly or indirectly by LPS, activation of other T cells may occur without the help of accessory cells and MHC restriction (60). LPS-induced T-cell activation may participate in various inflammatory and immune responses.

REFERENCES

1. Morrison DC, Danner RL, Dinarello CA, et al. Bacterial endotoxins and pathogenesis of Gram-negative infections: current status and future direction. J Endotoxin Res 1994; 1:71–83.

2. Lei M-G, Stimpson SA, Morrison DC. Specific endotoxic lipopolysaccharide-binding receptors on murine splenocytes. III. Binding specificity and characterization. J Immunol 1991; 147:1925–1932.
3. Vogel SN, Hilfiker ML, Caulfield MJ. Endotoxin-induced T lymphocyte proliferation. J Immunol 1983; 130:1774–1779.
4. Milner ECB, Rudbach JA, Voneschen KB. Cellular responses to bacterial lipopolysaccharide: T cells recognize LPS determinants. Scand J Immunol 1983; 18:21–28.
5. Tough DF, Sun S, Sprent J. T cell stimulation in vivo by lipopolysaccharide (LPS). J Exp Med 1997; 185:2089–2094.
6. Mattern T, Thanhäuser A, Reiling N, et al. Endotoxin and lipid A stimulate proliferation of human T cells in the presence of autologous monocytes. J Immunol 1994; 153:2996–3005.
7. Mattern T, Flad H-D, Ulmer AJ. Stimulation of human T lymphocytes by lipopolysaccharide (LPS) in the presence of autologous and heterologous monocytes. In: Levin MP, Yokochi T, Nakano M, eds. Endotoxin and Sepsis: Molecular Mechanisms of Pathogenesis, Host Resistance, and Therapy. New York: John Wiley & Sons, Inc., 1998:243–254.
8. Mattern T, Brade L, Flad H-D, Rietschel ET, Ulmer AJ. Stimulation of human T-lymphocytes by lipopolysaccharide is MHC-unrestricted, but strongly dependent on B7-interactions. J Immunol 1997; 159.
9. Forbes JT, Nakao Y, Smith RT. T mitogens trigger LPS responsiveness in mouse thymus cells. J Immunol 1973; 114:1004–1007.
10. Nitta T, Konno-Ejiri H, Nemoto K, Okumura S, Ozawa A, Nakano M. The role of interleukins in the synergistic effect of B-cell mitogens with concanavalin A on hydrocortisone-resistant thymic cell population. Immunology 1986; 59:209–216.
11. Mita A, Ohta H, Mita T. Induction of specific T cell proliferation by lipid A in mice immunized with sheep red blood cells. J Immunol 1982; 1982:1709–1711.
12. Bismuth G, Duphot M, Theze J. LPS and specific T cell responses: interleukin 1 (IL-1)-independent amplication of antigen-specific T helper (TH) cell proliferation. J Immunol 1985; 134:1415–1421.
13. Rudd CE, Janssen O, Cai Y-C, da Silva AJ, Raab M, Prasad KV. Two-step TCRζ/CD3-CD4 and CD28 signaling in T cells: SH2/SH3 domains, protein-tyrosine and lipid kinases. Immunol Today 1994; 15:225–234.
14. Paul WE, Seder RA. Lymphocyte responses and cytokines. Cell 1994; 76:241–251.
15. Rocha B, Vassalli P, Guy-Grand D. The extrathymic T-cell development pathway. Immunol Today 1992; 13:449–454.
16. Nitta T, Imai H, Ogasawara Y, Nakano M. Mitogenicity of bacterial lipopolysaccharide on the T lymphocyte population bearing the γδ T cell receptor. J Endotoxin Res 1994; 1:101–107
17. Emoto M, Danbara H, Yoshikai Y. Induction of γδ T cells in murine salmonellosis by an avirulent but not by a virulent strain of Salmonella cholerasuis. J Exp Med 1992; 176:362–372.
18. Cron RQ, Koning F, Maloy WL, Pardoll D, Coligan JE, Bluestone JE. Peripheral murine CD3$^+$,CD4$^-$,CD8$^-$ T lymphocytes express novel T cell receptor structures. J Immunol 1988; 141.
19. Schwartz RH. Costimulation of T lymphocytes: the role of CD28, CTLA-4, and B7/BB1 in interleukin-2 production and immunotherapy. Cell 1992; 71:1065–1068.
20. Azuma M, Cayabyab M, Buck D, Phillips JH, Lanier LL. CD28 interaction with B7 costimulates primary allogenic proliferative responses and cytotoxicity mediated by small, resting T lymphocytes. J Exp Med 1992; 175:353–360.
21. Sperline AI, Sinsley PS, Barrett TA, Bluestone JA. CD28-mediated costimulation is necessary for the activation of T cell receptor-γδ+T lymphocytes. J Immunol 1993; 151:6043–6050.
22. Harding F, McArthur JG, Gross JA, Raulet DH, Allison P. CD28-mediatedsignalling costimulates murine T cells and prevents induction of anergy in T-cell clones. Nature 1992; 356:607–609.
23. Bretscher P. The two-signal model of lymphocyte activation twenty-one years later. Immunol Today 1992; 13:74–76.
24. Heremans H, Damme JV, Dillen C, Dijkmans R, Billiau A. Interferon γ is a mediator of lethal lipopolysaccharide-induced Shwartzman-like shock reactions in mice. J Exp Med 1990; 171:1853–1869.
25. Heremans H, Dillen C, Billiau A. Role of IFN-γ producing NK cells in the lethal endotoxin-induced generalized Shwartzman reacton in mice. J Interferon Res 1991; 11:187–195.
26. Okamura H, Kawaguchi K, Shoji K, Kawade Y. High-level induction of gamma interferon with various mitogens in mice pretreated with Propionibacterium acnes. Infect Immun 1989; 38:440–443.
27. Wada M, Okamura H, Nagata K, Shimoyama T, Kawade Y. Cellular mechanisms in vivo production of gamma interferon induced by lipopolysaccharide in mice infected with Mycobacterium bovis BCG. J Interferon Res 1985; 5:431–443.
28. Okamura H, Wada M, Nagata K, Tamura T, Shoji K. Induction of murine gamma interferon production by lipopolysaccharide and interleukin 2 in Propionibacterium acnes-induced peritoneal exudate cells. Infect Immun 1987; 55:335–341.
29. Matsumura H, Nakano M. Endotoxin-induced interferon γ production in culture cells derived from BCG-infected C3H/HeJ mice. J Immunol 1988; 140:494–500.
30. Le J, Lin J-X, Henriksen-DeStefano D, Vilcek J. Bacterial lipopolysaccharide-induced interferon-γ production: roles of interleukin 1 and interleukin 2. J Immunol 1986; 136:4525–4530.
31. Blanchard DK, Djew JY, Klein TW, Friedman H, Stewart II WE. Interferon-γ induction by lipopolysaccharide: dependence on interleukin 2 and macrophages. J Immunol 1986; 136:963–970.
32. Tennenberg SD, Weller JJ. Endotoxin activates T cell interferon-γ secretion in the presence of endothelium. J Surg Res 1996; 63:73–76.
33. Trinchieri G, Scott P. The role of interleukin 12 in the immune response, disease and therapy. Immunol Today 1994; 15:460–463.
34. Okamura H, Nagata K, Komatsu T, et al. A novel costimulatory factor for gamma interferon induction

found in the livers of mice causes endotoxic shock. Infect Immun 1995; 63:3966–3972.
35. Okamura H, Tsutsui H, Komatsu T, et al. Cloning of a new cytokine that induces IFN-γ production by T cells. Nature 1995; 378:88–91.
36. Gu Y, Kuida K, Tsutsui H, et al. Activation of interferon-γ inducing factor mediated by interleukin-1β converting enzyme. Science 1997; 275:206–209.
37. Ghayur T, Banerjee S, Hugunin M, et al. Caspase-1 processes IFN-γ-inducing factor and regulates LPS-induced IFN-γ production. Nature 1997; 386:619–623.
38. Li P, Allen H, Banerjee S, et al. Mice deficient in IL-1β-converting enzyme are defective in production of mature IL-1β and resistant to endotoxic shock. Cell 1995; 80:401–411.
39. Nakano M, Matsuura M. Lipid A. In: Stewart-Tull DES, ed. The Theory and Practical Application of Adjuvant. Chichester, England: Wiley, 1995:315–335.
40. Nakano M, Uchiyama T. In vitro adjuvant effect of endotoxin. Review and experiment. In: Nowotny A, ed. Beneficial Effects of Endotoxin. New York: Plenum, 1983:255–272.
41. Uchiyama T, Jacobs M. Modulation of immune response by bacterial lipopolysaccharide (LPS): cellular basis of stimulatory and inhibitory effects of LPS on the in vitro IgM antibody response to a T-dependent antigen. J Immunol 1978; 121:2347–2351.
42. Armerding D, Katz DH. Activation of T and B lymphocytes in vitro. I. Regulatory influence of bacterial lipopolysaccharide (LPS) on specific T-cell helper function. J Exp Med 1974; 139:24–43.
43. Hamaoka T, Katz DH. Cellular site of action of various adjuvants in antibody response to hapten carrier conjugates. J Immunol 1973; 111.
44. Rosmann TR, Sad S. The expanding universe of T-cell subsets: Th1, Th2 and more. Immunol Today 1996; 17:138–146.
45. Del Prete GF, De Carli M, Ricci M, Romagnani S. Helper activity for immunoglobulin synthesis of T helper type 1 (TH1) and Th2 human T cell clones: the help of Th1 clones is limited by their cytolytic capacity. J Exp Med 1991; 174:809–813.
46. Coffman RL, Seymour BW, Lebman DA, et al. The role of helper T cell products in mouse B cell differentiation and isotype regulation. Immunol Rev 1988; 102:5–18.
47. Saito-Taki T, Nakano M. Suppression of lipopolysaccharide-induced polyclonal B cell activation of murine spleen with heat-aggregated murine immunoglobulin G. J Immunol 1983; 130:2022–2026.
48. Conti P, Dempsey RA, Reale M, et al. Activation of human natural killer cells by lipopolysaccharide and generation of IL-1 alpha, beta tumor neclosis factor and interleukin-6: effect of IL-1 receptor antagonist. Immunology 1991; 73:450–456.
49. Schoenhaut DS, Chua AO, Wolitzky AG, et al. Cloning and expression of murine IL-12. J Immunol 1992; 148:3433–3440.
50. Wolf SF, Temple PA, Kobayashi M, et al. Cloning of cDNA for natural killer cell stimulatory factor, a heterodimeric cytokine with multiple biologic effects on T and natural killer cells. J Immunol 1991; 146:3074–3081.
51. Nastala CL, Edinton HD, McKinney TG, et al. Recombinant IL-12 administration induces tumor regression in association with IFN-γ production. J Immunol 1994; 153:1697–1706.
52. Brunda MJ, Luistro L, Warrier RR, et al. Antitumor and antimetastatic activity of interleukin 12 against murine tumors. J Exp Med 1993; 178:1223–1230.
53. Takahasi M, Ogasawara K. Takeda K, et al. LPS induces NK1.1$^+$ αβT cells with potent cytotoxicity in the liver of mice via production of IL-12 from Kupffer cells. J Immunol 1996; 156:2436–2442.
54. Nitta T, Imai H, Okamoto A, Matsuura M, Nakano M. In vivo priming of mice with antigen and lipid A bestows proliferative ability on peritoneal exudate T cells in response to antigen or anti-αβ TCR antibody in the absence of macrophages in vitro. J Endotoxin Res 1996; 3:103–110.
55. Trowgridge LS, Thomas ML. CD45: an emerging role as a protein tyrosine phosphatase required for lymphocyte activation and development. Annu Rev Immunol 1994; 12:85–116.
56. Mackay CR. T-cell memory: the connection between function, phenotype and migration pathways. Immunol Today 1991; 12:189–192.
57. Cerottini J-C, MacDonald HR. The cellular basis of T-cell memory. Ann Rev Immunol 1989; 7:77–89.
58. Zhang Y-H, Takahashi K, Jiang G-Z, Kawai M, Fukada M, Yokochi T. In vivo induction of apoptosis (programmed cell death) in mouse thymus by administration of lipopolysaccharide. Infect Immun 1993; 61:5044–5048.
59. Vella AT, McCormack JE, Linsley PS, Kappler JW, Marrack P. Lipopolysaccharide interferes with the induction of peripheral T cell death. Immunity 1995; 2:261–270.
60. Bouchonnet F, Lecossier D, Bellocq A, Hamy I, Hance AJ. Activation of T cells by previously activated T cells. HLA-unrestricted alternative pathway that modifies their proliferative potential. J Immunol 1994; 153:1921–1935.

43

Immunological Properties of Microbial Outer Membrane Proteins and Their Effects as Modulators of LPS Immunobiology

Kathryn Nixdorff, Dagmar Schilling, and Waltraud Ruiner
Darmstadt University of Technology, Darmstadt, Germany

INTRODUCTION

The outer membrane of gram-negative bacteria contains several components that can exert a variety of effects on cells of the immune system, and they are therefore highly relevant substances immunologically. Lipopolysaccharide (LPS) represents a classic agent in this category. A prime target of LPS action is the macrophage, but it can react with various other types of cells including B lymphocytes. When purified LPS interacts with B cells, a specific immune response is elicited that is predominantly IgM in character, with little or no memory production, especially when LPS is completely free of contaminating proteins (1,2). LPS is also a mitogen for murine B lymphocytes, and it can induce these cells in an antigen-nonspecific manner to differentiate and secrete antibodies (3). In the native outer membrane, LPS is tightly bound to some of the major proteins in this structure (4), and the tight bond becomes particularly apparent when attempts are made to purify the separate components. Purification is often only successful by the use of such drastic methods as hot phenol-water extraction in the presence of detergents (5), and even then it is difficult to be certain that all contaminating quantities of LPS have been removed from some of the proteins. Outer membrane proteins are generally very active substances immunologically. In addition to their immunogenic properties, some outer membrane proteins are, like LPS, mitogens for murine B lymphocytes (6–9). Furthermore, bacterial lipoprotein and its synthetic analogs can function as immunoadjuvants (10) and can also activate macrophages to produce tumor necrosis factor (TNF-α) (11).

It is therefore not unreasonable to assume that the immune system of a host has to deal not only with single components but also with complexes of immunologically active substances from bacteria. Considering this possibility, we have been interested in determining the immunomodulating activities of defined complexes of outer membrane components of gram-negative bacteria, particularly the effects that proteins have on the interaction of LPS with cells of the immune system. The aim of the present chapter is to briefly summarize some pertinent immunological properties of outer membrane proteins and then discuss different ways in which these proteins can differentially modulate LPS immunobiology.

IMMUNOLOGICAL PROPERTIES OF OUTER MEMBRANE PROTEINS

Mitogenic Activities of Outer Membrane Proteins

Mitogenic Activity of Endotoxin Protein

Some of the first reports concerning possible mitogenic effects of outer membrane proteins on lymphocytes came from studies using LPS extracted by aqueous butanol (12) or trichloroacetic acid (13), which also extracts protein associated with LPS in the outer membrane. It was subsequently shown that the mitogenic activity of these preparations on spleen cells of LPS hyporesponsive C3H/HeJ mice (14) could be attributed to the protein fraction, which could be removed by phenol-water extraction (12,13,15). This fraction was called endotoxin protein or lipid A–associated proteins (LAP), which was shown to be a mixture composed predominantly of several different major outer membrane proteins whose patterns depended on the organism from which it was extracted (16). Significantly, LAP was found to be mitogenic for human peripheral blood lymphocytes, which are refractory to stimulation by LPS and purified protein derivative of tuberculin, another murine B-cell mitogen and polyclonal activator (14,17). Although the mechanism of activation of B lymphocytes and in particular C3H/HeJ B cells by LAP has not been elucidated, more recent evidence suggests that protein kinase C plays a prominent role (18). New models for investigating the C3H/HeJ defect have been constructed in the hope that they will be able to shed some light on the nature of the product and function of the Lps^n gene (19,20). For a further discussion, the reader is referred to Chapter 49.

Mitogenic Activities of Single Outer Membrane Proteins

The lipoprotein from *Escherichia coli* with an apparent molecular mass of 7.2 kDa (Braun lipoprotein) has three fatty acids bound to glycerylcysteine in the N-terminal region of the polypeptide chain (21). As such, this protein represented a prime candidate for interaction with cells, and it was subsequently shown that the lipoprotein was indeed a very effective murine B lymphocyte mitogen when compared to the activity of LPS and purified protein derivative of tuberculin (6). The lipoprotein lost its mitogenic activity when the ester-linked fatty acids were removed by alkaline hydrolysis. Furthermore, it was not mitogenic for C3H/Tif mouse thymus cells but could stimulate B cells from LPS hyporesponsive C3H/HeJ mice. In the same study (6), it was shown that the lipoprotein could effect the differentiation of B cells to polyclonal IgM secretion.

Several outer membrane proteins that are not lipoproteins have also been characterized as B-cell mitogens. OmpF and OmpC of *E. coli* are prominent porins that exist as trimers in the outer membrane, forming nonspecific diffusion pores for small hydrophilic molecules (22). OmpA is a heat-modifiable protein in that its apparent molecular weight is increased by boiling in the presence of sodium dodecyl sulfate, and it has a protease-sensitive stretch of amino acids that is available for cleavage even when the protein is contained in cell walls of the bacterium (22). Furthermore, a single molecule of the protein can apparently form a hydrophilic diffusion pore in the outer membrane, although the rate of diffusion is about two times slower than that measured with the trimolecular porins OmpF and OmpC (23). Bessler and Henning (7) showed that a mixture of OmpF and OmpC from *E. coli* K12 had strong mitogenic activity for murine B lymphocytes from normal and from C3H/HeJ mice. OmpA was also mitogenic for these cells, but its activity was relatively weak in comparison.

OmpA from *Proteus mirabilis* was also shown to be a mitogen for murine B lymphocytes (9). In that study, LPS was rigorously removed from the isolated protein by hot phenol-water extraction in the presence of detergent followed by gel filtration, and the protein was free of detectable amounts (<0.0075%) of LPS according to gas chromatographic analysis of 3-hydroxytetradecanoic acid content. As in the case of LPS and the other outer membrane protein mitogens, OmpA from *P. mirabilis* did not react in vitro with T lymphocytes. It should be noted that even though the OmpA proteins from *E. coli* and *P. mirabilis* are very similar, there are some notable structural differences. Like the *E. coli* protein, the apparent molecular weight of OmpA from *P. mirabilis* is modified after boiling the protein in the presence of sodium dodecyl sulfate. Furthermore, OmpA from *P. mirabilis* has a similar stretch of amino acids sensitive to trypsin. While the amino acid sequences of OmpA proteins from *P. mirabilis* and *E. coli* show a similarity index of 92%, the identity index of 74% indicates definite differences in the structure of the two proteins (9). In this regard, OmpA from *P. mirabilis* has an apparent molecular mass of 39 kDa, while that of the *E. coli* protein is 33 kDa.

Differences can also be seen in the reactivities of the two OmpA proteins with monoclonal antibodies (mAb) to OmpA from *P. mirabilis* (Fig. 1). The three mAb depicted in Figure 1 were from hybridomas that had been subcloned three times, and these antibodies did not react detectably with LPS or other outer membrane proteins from *P. mirabilis* in enzyme-linked immunosorbent assay (ELISA) and Western blot analyses (24). The mAbs 2.14.1 and 2.31.1 react primarily with intact OmpA from *P. mirabilis* (Fig. 1, lanes 1 and 5), while mAb 2.18.1 also reacts strongly with fragments of OmpA contained in the protein preparation from this organism (Fig. 1, lane 3). mAb 2.14.1 and mAb 2.31.1 reacted just as strongly with *E. coli* OmpA (Fig. 1, lanes 2 and 6) as they did with the homologous *P. mirabilis* protein (Fig. 1, lanes 1 and 5) in Western blot

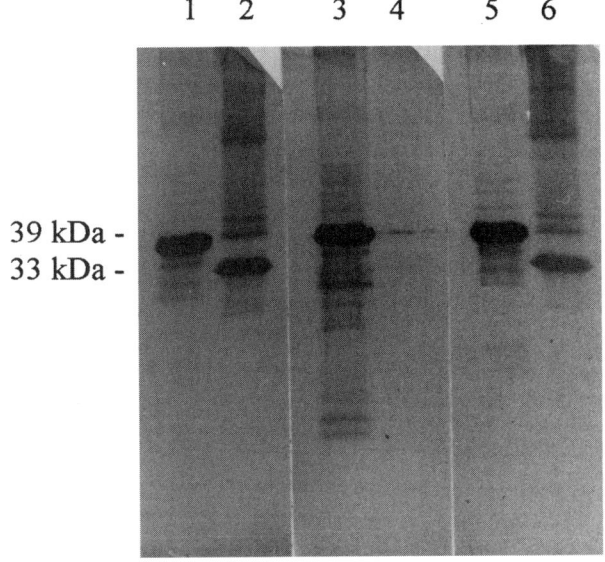

Fig. 1 Western blot analyses of reactions of mAbs to OmpA from *P. mirabilis* with OmpA proteins from *P. mirabilis* and from *E. coli*. The OmpA proteins were applied to the gel in amounts of 15 μg each per lane. Lanes 1, 3, and 5, OmpA from *P. mirabilis* (39 kDa); lanes 2, 4, and 6, OmpA from *E. coli* (33 kDa). Electrophoresis and blotting onto nitrocellulose paper were carried out as described in Ref. 24. The paper was cut into strips and reacted with mAb 2.14.1 (lanes 1 and 2), mAb 2.18.1 (lanes 2 and 3), and mAb 2.31.1 (lanes 4 and 5). Reactions were detected with a secondary goat antibody to mouse IgG conjugated with peroxidase.

analyses. However, mAb 2.18.1 showed no visible reaction at all with the *E. coli* protein (Fig. 1, lane 4). Thus, variant structural or physicochemical properties might contribute to differences in the activities of OmpA proteins from different organisms in interactions with cells of the immune system. One property that seems to contribute to the mitogenic and adjuvant activities of outer membrane proteins is the ability to aggregate (see below), which may be decidedly different for the various proteins.

The outer membrane proteins OprF, OprH, and OprI from *Pseudomonas aeruginosa* represent additional B-lymphocyte mitogens (8). All were shown to exert very strong mitogenic effects on B cells from C3H/HeJ mice and to be nonreactive with thymocytes. OprH is a lipoprotein with a molecular mass of approximately 17 kDa, while OprI is similar to the *E. coli* lipoprotein in molecular mass, but apparently has a different amino acid and lipid composition (8). OprF is a porin having 33% identity and 55% similarity to OmpA of *E. coli* in the carboxy-terminal half of the molecule (25).

T-Cell–Dependent Immune Responses to Outer Membrane Proteins

Although the majority of the outer membrane proteins tested act as B-cell mitogens and induce polyclonal IgM responses in vitro, they apparently elicit T-cell–dependent, predominantly IgG responses typical of proteins in vivo, which has been clearly documented for OmpA from *P. mirabilis* (2) and OprF from *P. aeruginosa* (26). In addition to acting as strong immunogens, outer membrane proteins have been reported in several studies to be very effective adjuvants in modulating the responses to other antigens. In this regard, it was recognized early that endotoxin protein was an excellent adjuvant for the immunization of mice with procholeragenoid (27).

The extensive studies of Wolfgang Bessler and co-workers have documented the adjuvant effect of the *E. coli* 7.2 kDa lipoprotein and its analogs on immune responses to a variety of antigens. Synthetic lipopentapeptides containing an amino acid sequence identical to the N-terminus of native lipoprotein but having a different fatty acid composition consisting solely of palmitic acid have been investigated most thoroughly (10,28). The tripalmitoyl pentapeptide (P_3CSSNA) was shown to have biological activities similar to the native lipoprotein and enhanced both IgM and IgG in vitro responses to trinitrophenyl (TNP)-SRBC considerably, although IgG levels were predominantly augmented (28). Furthermore, synthetic influenza virus nucleoprotein oligopeptides covalently linked to the lipotripeptide P_3CSS were able to induce viral peptide-specific cytotoxic T lymphocytes in vivo (29). Lipopeptide-antigen conjugates have also been used to generate HIV-specific antibodies (30).

The adjuvant effect of outer membrane proteins has also been applied in attempts to produce more effective vaccines against chronic pulmonary infection with *P. aeruginosa*. OprF and OprI of *P. aeruginosa* are outer membrane proteins that are conserved in all 17 known serogroups of this microorganism (31) and represent good vaccine candidates. In a rat model, immunization with purified OprF enhanced the ability of the animals to clear a challenge inoculum of *P. aeruginosa* from the lungs and significantly reduced the incidence and severity of pulmonary lesions (26). However, it has been indicated that immunization with the single OprF and OprI proteins from *P. aeruginosa* was not as protective as immunization with LPS-based vaccines (31). On the other hand, OprI-OprF fusion proteins seem to offer more promise (31), and improved methods of expressing the fusion proteins have been developed (32).

Another approach has been to insert foreign antigenic determinants into the OprF molecule through linker insertion mutagenesis (33). Also, antigens inserted into the third or fourth loop of OmpA of *E. coli* and expressed in *Salmonella typhimurium* have been constructed. The strategy here is to deliver antigens to the gut-associated lymphoid tissue to induce secretory, humoral, and cellular responses (34).

Activation of Macrophages by Outer Membrane Proteins

Outer membrane proteins have been used to elucidate signaling pathways in macrophages. Particular attention has been given to the ability of lipid A–associated proteins to stimulate LPS-hyporesponsive C3H/HeJ mouse macrophages. The strategy is to define differences in signaling between LPS and LAP in order to learn more about regulation of LPS responsiveness. Investigations by Hogan and Vogel (35,36) have shown that LAP can provide a "second signal" in the activation of C3H/HeJ macrophages to a fully tumoricidal state, in response to a "first signal" or priming by recombinant interferon-gamma (rIFN-γ). In contrast, LPS purified to contain no protein was unable to induce tumoricidal activity in rIFN-γ–primed C3H/HeJ macrophages. Indeed, LAP could induce tumoricidal activity against P815 mastocytoma cells in rIFN-γ–primed macrophages from C3H/HeJ mice comparable to the activity induced by the LAP in rIFN-γ–primed macrophages from normal mice (35,36). On the other hand, LAP induced much less TNF-α in IFN-γ–primed C3H/HeJ macrophages than in IFN-γ–primed macrophages from normal mice (36). Dong et al. (37) showed that C3H/HeJ macrophages could respond to the synthetic lipopeptide CGP 31362 (an analog of a fragment of the 7.2 kDa lipoprotein from gram-negative bacteria) to produce TNF-α and tyrosine phosphorylation of the proteins p39, p41, and p45. Manthey et al. (5) have clearly shown that C3H/HeJ murine macrophages are not responsive to LPS (even rough form) when protein has rigorously been removed by phenol-water extraction in the presence of detergent. Interestingly, it was found that responses to LAP in C3H/HeJ macrophages were augmented by the presence of LPS. Comparing LPS and LAP activation of macrophages, these investigators could differentiate early gene expression into two distinct pathways based on induction of a set of six early genes. Wild-type C3H/OuJ mouse macrophages could be induced by LPS to express all six genes encoding TNF-α, IL-1β, type II p75 TNF receptor (TNFR-2), IP-10, D3, and D8. In addition, tyrosine phosphorylation of 41, 44, and 47 kDa proteins was induced. In contrast, LAP induced a tyrosine phosphorylation pattern similar to the one induced by LPS, but it activated only the genes encoding TNF-α, TNFR-2, and IL-1β and not those encoding IP-10, D3, or D8.

Galdiero et al. (38) have also established that human monocytes can be activated with porins from the outer membrane of *S. typhimurium* to produce cytokines TNF-α, IL-1α, and IL-6. This activity was attributed to the proteins and not to the very small amounts of LPS contaminating the preparations, because the preparations were still active after addition of polymyxin B.

Significance of Outer Membrane Proteins in Infection with Gram-Negative Bacteria

Porin preparations from a rough mutant of *S. typhimurium* were good immunogens in rabbits and mice, producing high titers of antibodies to porins and to LPS and protecting mice significantly from infection by challenge (39). Furthermore, O-antigenic oligosaccharides coupled covalently to *S. typhimurium* porins could effectively protect mice against infection with the virulent homologous organism (40).

E. coli is the most common gram-negative bacterium causing neonatal meningitis, and it has recently been shown that OmpA of *E. coli* contributes to invasion of brain microvascular endothelial cells. In this respect, OmpA antibodies could significantly inhibit invasion of brain microvascular endothelial cells by a cerebrospinal fluid isolate of *E. coli* (41).

Proteosome Character of Outer Membrane Proteins

As discussed above, several outer membrane proteins are mitogenic for B lymphocytes in vitro and induce predominantly T-cell–dependent IgG responses in vivo. In addition, some have proved to be strong adjuvants for the immune responses to various antigens. Proteins that exhibit these properties have been termed proteosomes (42,43), a name coined to describe the multimolecular structures first observed with purified preparations of meningococcal outer membrane proteins. These preparations were used to increase the efficacy of polysaccharide vaccines for protection against group B *Neisseria meningitidis* infection (44). One property that seems to contribute to the strong adjuvant activity of proteosomes is their ability to aggregate and form complexes with antigens.

DIFFERENTIAL MODULATION OF THE RESPONSES TO LPS BY OUTER MEMBRANE PROTEINS

Modulation of LPS Immunobiology by LAP and by Proteosomes

Outer membrane proteins have also been shown to be very effective modulators of the immune responses to LPS. Again, this was first seen with LPS extracts containing endotoxin-associated proteins in studies investigating the antibody plaque-forming cell responses in mice to LPS from *Salmonella enteritidis* (1). In these studies, mice immunized with LPS extracted by the phenol-water method responded with an antibody-producing cell response that was primarily IgM in nature on days 10 and 20 after immunization. In contrast, the responses of mice immunized with LPS extracted by the trichloroacetic acid method, which contains proteins from the outer membrane, were initially IgM in character, but by day 20 a significant number of IgG plaque-forming cells was produced. After secondary immunization on day 21, IgG antibody-producing cells were detected in mice immunized with LPS extracted either with phenol-water or with trichloroacetic acid, but the responses of mice immunized with the trichloroacetic acid LPS extract were higher and the ratio of IgG to IgM antibody-producing cells was greater. Treatment of the trichloroacetic acid extracts with proteolytic enzymes or extraction with phenol-water greatly reduced both the strength and the IgG character of the responses to the extracts.

Killion and Morrison (45) also used lipid A–associated proteins in complex with LPS to investigate the determinants of immunity to salmonellosis in C3H/HeJ mice in vivo. Immunization of these mice with LPS alone afforded the animals 40% survival when challenged with the homologous strain. Accordingly, immunity was shown to correlate with the production of LPS-specific antibodies 21–28 days after immunization. Protection was increased considerably when LAP-LPS complexes were used for immunization. Consistent with a modulating effect of LAP on LPS responses, immunization of mice with LAP-LPS complexes enhanced IgM and IgG responses specific for LPS. IgG responses specific for two proteins in the LAP preparation were also detected, suggesting again that the responses to these proteins are mainly T-cell dependent. In the same investigation, experiments were carried out using the transfer of antiserum and/or immune spleen cells to naive mice in order to assess the relative contributions of humoral and cellular components to immunity in this system. Neither antiserum nor immune spleen cells alone were sufficient to confer immunity to naive C3H/HeJ mice, but a combination of both was effective.

Recently, Orr et al (46,47) used *Shigella flexneri* 2a or *Shigella sonnei* LPS hydrophobically complexed with group B type 2b *N. meningitidis* proteosomes as candidate vaccines. The immunogenicity of these complexes was measured in mice and guinea pigs. Complex formation between LPS and the proteosomes was demonstrated by size exclusion chromatography. It was also shown that the level of complex formation between LPS and the proteosomes was important for immunogenicity. Immunization with these complexes was effective by either the oral or the intranasal route, and the antibody responses elicited were type specific for LPS. Both the oral and nasal routes of administration induced comparable levels of IgG and IgA antibodies in the serum of mice. Comparable levels of IgA antibodies in intestines were induced by both routes, but the intranasal route of administration was more effective for induction of IgA in the lungs of mice. The results of Sereny tests in guinea pigs to evaluate the protective capacity of the vaccines showed that complexes delivered by both oral and intranasal routes could protect in vivo against mucosal infection with homologous bacteria (46). This demonstrated that LPS-proteosome vaccines are potent mucosal immunogens. Addition of cholera toxin B subunit to the LPS-proteosome preparations to enhance targeting to mucosal epithelium augmented the immune responses when immunization with the LPS-proteosome complex was suboptimal (47).

Modulation of LPS Immunobiology by Isolated Outer Membrane Proteins

Modulation of Antibody Responses to LPS

Our interest in the modulating effects of proteins on the responses to LPS stemmed from the consideration that LPS is tightly complexed to phospholipids and proteins in the native outer membrane structure, and the question naturally arose as to what effects these components might have on the quality and the quantity of the responses to the single substances. We addressed this question by studying the modulation of immune responses in a model system consisting of complexes produced from defined amounts of isolated, purified components of the outer membrane of *P. mirabilis*, which served as modulators (LPS, phospholipids, proteins) and as immunogens (LPS, proteins). The bacterium *P. mirabilis* was chosen for these investigations because we had been carrying out structure-function studies on the outer membrane of this organism and had gained some experience in the isolation and purification of the different components of this system.

Antibody-producing cell responses to the various components of the outer membrane of *P. mirabilis*, mixed together in different combinations, were measured in mice. Complexes were formed by mixing the purified components in aqueous buffer by sonication. The different combinations tested included the responses to LPS modulated by bacterial membrane phospholipids (48), the responses to LPS modulated by the outer membrane proteins OmpA, lipoprotein, and a 36 kDa porin protein (49,50), as well as the immune responses to OmpA modulated by phospholipids and LPS (2). For these studies LPS was extracted from whole cells by the phenol-water method and was reextracted to remove all traces of protein measured by amino acid analysis. The OmpA protein of *P. mirabilis* was extracted with 1% sodium deoxycholate, purified by gel filtration, reextracted with phenol-water in the presence of detergent and purified again by gel filtration. It contained less than 0.0075% LPS as determined by gas chromatographic analysis of 3-hydroxytetradecanoic acid (9,49).

The most dramatic effects in this system were obtained with the modulation of the immune responses to LPS by phospholipids and the OmpA (39 kDa) protein. In both cases, a strong enhancement of the LPS-specific antibody-producing cell responses was obtained, and an alteration in the antibody class after secondary stimulus from mainly IgM to predominantly IgG was observed (Table 1), similar to the modulating effects of endotoxin-associated protein reported by Hepper et al. (1). The data for the peaks of the primary (day 4) and the secondary (day 19) responses are presented in the table, along with measurements made on day 14 just before secondary stimulation. Lipoprotein from *P. mirabilis* also enhanced LPS responses, but the effect was somewhat weaker than that afforded by OmpA on a weight basis. The primary IgM responses were rather nonspecific, but the secondary IgG responses were shown to be strictly specific for the type of LPS used (49). Interestingly, very characteristic adjuvant effects on par-

Table 1 IgM and IgG Subclass Responses in Mice to LPS Alone and in Combination with Phospholipids, OmpA, and Lipoprotein from *P. mirabilis*

Immunogen[a]	Antibody type	PFC per 10^6 spleen cells on day[b]			IgG subclass (% total IgG on day 19)	Adjuvant factor on day 19[c]
		4	14	19		
LPS	IgM	32 ± 4	7 ± 1	10 ± 1		1
	IgG1	0	3 ± 1	17 ± 3	40	1
	IgG2	0	1 ± 1	19 ± 5	44	1
	IgG3	0	2 ± 1	7 ± 1	16	1
LPS plus phospholipids	IgM	55 ± 9	16 ± 2	26 ± 15		2
	IgG1	0	69 ± 22	1,793 ± 507	82	105
	IgG2	0	5 ± 2	248 ± 12	11	13
	IgG3	0	7 ± 1	147 ± 21	7	21
LPS plus OmpA	IgM	141 ± 5	5 ± 1	31 ± 5		3
	IgG1	0	38 ± 3	337 ± 75	25	20
	IgG2	0	59 ± 1	818 ± 133	62	43
	IgG3	0	3 ± 3	168 ± 49	13	24
LPS plus lipoprotein	IgM	49 ± 2	5 ± 1	12 ± 1		1
	IgG1	n.d.	n.d.	369 ± 64	59	22
	IgG2	n.d.	n.d.	145 ± 21	23	8
	IgG3	n.d.	n.d.	113 ± 24	18	16

[a]Dosages per injection were 25 μg of LPS, 300 μg of phospholipids, and 12.5 μg of proteins. Mice received a primary injection on day 0 and a secondary injection on day 14.
[b]Responses were measured against *P. mirabilis* LPS coupled to SRBC. Values represent the geometric means ± standard errors of the means of the numbers of plaque-forming cells (PFC) from three to five separate experiments.
[c]Adjuvant factor was calculated by dividing the PFC obtained with LPS alone as immunogen (taken as factor of 1) into the PFC of the respective Ig isotype obtained with LPS plus OmpA as immunogen.
n.d., Not determined.
Source: Ref. 49.

ticular IgG subclasses for the different modulators were seen. In this regard, phospholipids and lipoprotein enhanced IgG1 responses to the greatest extent, while the OmpA protein exerted its greatest adjuvant effect on IgG2 responses. It should be noted that LPS also had a strong adjuvant effect on the responses to OmpA (2). In this case, secondary IgM responses to the protein were increased 5-fold, while IgG responses were enhanced 10-fold when OmpA was mixed with LPS.

At a later point, when the appropriate reagents were available, the experiments with OmpA as modulator were repeated and the differentiation of IgG2 responses in IgG2a and IgG2b was made (50). At the same time, we used a special plaque counter, which allowed visualization of even very small plaques. The results presented in Table 2 clearly indicate that OmpA enhanced IgG2a LPS-specific responses to the greatest extent, although it must be stressed that all IgG responses were augmented to some degree. Subsequent investigations of secondary responses to LPS-OmpA complexes in vitro showed that the IgG responses specific for LPS were heavily dependent on the presence of T cells at the time of secondary stimulation (50). Treatment of spleen cells with anti-Thy 1 plus complement before secondary stimulation in vitro dramatically reduced all IgG subclass responses elicited by the LPS-protein complex, compared to the same spleen cell population treated with complement alone. Spleen cells stimulated with LPS alone were unaffected by treatment with antibody and complement.

It was further shown that LPS and the modulators had to be mixed with each other before administration to mice in order to achieve the enhancing effect on the responses to LPS (49). If the substances were injected simultaneously but separately, there was little augmentation of LPS responses. Complex formation between LPS and OmpA from *P. mirabilis* after sonication of the two isolated components together was investigated by centrifugation of the substances in sucrose density gradients (24). The results showed that the main portion of LPS was found in the top half of the gradient, but it shifted together with OmpA to fractions of greater density toward the bottom half of the gradient after mixing with the protein. At the same time, the results indicated that an entire array of complexes of very different densities was distributed throughout the bottom half of the gradient, suggesting that complexes of different sizes were formed. As mentioned above, complex formation between LPS from *Shigella* and proteosomes from *N. meningitidis* was very important for the induction of strong IgG and IgA responses to LPS in mice and guinea pigs (46).

Because it appeared as if complex formation between LPS and modulators was necessary for the enhancing effect on the IgG responses in our system, experiments were performed to test whether the

Table 2 Comparison of Secondary In Vivo Responses to LPS Alone and to a Mixture of LPS and OmpA from *P. mirabilis*

Immunogen[a]	Antibody type	PFC per 10^6 spleen cells on day 18[b]	% IgG subclass of total IgG	Adjuvant factor[c]
LPS	IgM	704 ± 31		1
	IgG1	92 ± 10	37	1
	IgG2a	11 ± 5	4	1
	IgG2b	133 ± 38	54	1
	IgG3	10 ± 4	4	1
LPS plus OmpA	IgM	1183 ± 69		2
	IgG1	915 ± 14	22	10
	IgG2a	1132 ± 86	27	103
	IgG2b	1286 ± 138	31	10
	IgG3	841 ± 54	20	84

[a]Dosages per injection were 25 μg LPS and 12.5 μg OmpA. Mice received a primary injection on day 0 and a secondary injection on day 14.
[b]Responses were measured against *P. mirabilis* LPS coupled to SRBC. Values represent the geometric means ± standard errors of the means of the numbers of plaque-forming cells (PFC) from three separate experiments.
[c]Adjuvant factor was calculated by dividing the PFC obtained with LPS alone as immunogen (taken as factor of 1) into the PFC of the respective Ig isotype obtained with LPS plus OmpA as immunogen.
Source: Ref. 50.

modulator was acting as a carrier in the classic immunological sense (51). In this case, a strong secondary IgG response to LPS would occur only if the same modulator were employed for the primary and the secondary injections (52), which would imply that modulator-specific T cells are involved. The results of these studies (51) showed that a strong secondary IgG response to LPS occurred regardless of which adjuvant was used for the first and the second stimulus, and this suggested a lack of a classical carrier effect for the modulators. A significant increase in the IgG responses was also observed when mice were given a primary injection of LPS-phospholipids and a secondary injection of LPS alone. In contrast, even the most effective modulator could not induce a secondary IgG response when the mice were first given a primary injection of LPS alone. It was interesting in this respect that the primary injection determined the quality of the responses. That is, when OmpA was the LPS modulator for the primary injection, predominantly IgG2 responses were induced. On the other hand, when lipoprotein or phospholipids were administered with LPS for the primary stimulus, mainly IgG1 responses were augmented. In general, the results suggested that T cells were involved in the secondary IgG responses to LPS, but their antigenic specificity did not play a decisive role. Instead, they were probably providing a nonspecific signal that determined the Ig character of the responses.

The elucidation of the role of T-helper (Th) cell subsets and their cytokines in the regulation of immunoglobulin isotype production (reviewed in Refs. 53–55) has shed some light on our findings. Work with murine Th cell clones has provided evidence for the existence of at least two different Th cell subsets. Th1 cells preferentially produce IL-2, IFN-γ, and lymphotoxin (LT)-α, while Th2 cells produce IL-4, IL-5, IL-6, IL-10, and IL-13. Th1 clones can help B cells to produce IgG2a, provided the concentration of IFN-γ is optimal. On the other hand, Th2 clones help B cells in the production of IgM and especially IgG1, IgA, and IgE (54). Different cytokines apparently influence the development of the two Th subsets (56,57). In this regard, IL-12 directs development into the Th1 subset, while IL-4 directs development into the Th2 subset. Later on in the process, IFN-γ augments Th1 cell development, while IL-10 and IL-4 support Th2 cell development.

Apparently, different antigens can preferentially induce one or the other Th cell subset. Differential effects on immune responses due to the participation of different Th cells have been demonstrated for responses to the parasitic helminth *Schistosoma mansoni*, which were characterized by downregulation of Th1 cytokine production accompanied by induction of Th2 responses (58). Also, the adjuvant effect of cholera toxin on the responses to tetanus toxoid in mice led to the induction of predominantly Th2 cells (59). In that report, quantitative increases in IL-4, but not IFN-γ mRNA from tetanus toxoid-specific $CD4^+$ T cells, could be correlated with serum IgG1 and IgE as well as with mucosal IgA responses. In contrast, recombinant *Salmonella* expressing fragment C of tetanus toxoid administered orally to mice elicited predominantly Th1-type responses. At the same time, Th2-type cells producing IL-10 and macrophages producing IL-6 apparently supported IgA responses in both mucosal and systemic tissues (60). Thus, production of particular isotypes of immunoglobulin can be regulated by the participation of different Th cell subsets and their cytokines in B-cell responses in vivo. We are presently investigating these relationships in our model system.

Modulation of the Responses of Macrophages to LPS

The OmpA protein from *P. mirabilis* has several properties of proteosomes. It readily forms aggregates with LPS when the two are mixed together (24), it is a mitogen for B lymphocytes in vitro (9), and it is a potent adjuvant for the IgG responses to LPS in vivo (49). It has been speculated that a part of the excellent adjuvant effect of proteosomes is the promotion of enhanced uptake and processing of antigens by antigen-presenting cells (61). Because the participation of T cells in immune responses is regulated to a great extent by antigen-presenting cells, and because macrophages, which can function as antigen-presenting cells, represent the main target of LPS action, we were interested in investigating possible immunomodulating effects of OmpA on LPS-macrophage interaction. In particular, we wanted to examine whether the protein could enhance the uptake of LPS by macrophages and if the activation of macrophages to produce cytokines could be affected by complexing LPS with OmpA.

Enhancement of the Uptake of LPS in Macrophages by OmpA from P. mirabilis

Monoclonal antibodies to LPS and to OmpA of *P. mirabilis* were generated and used to measure the uptake of LPS and OmpA in macrophages derived from the bone marrow of C57BL/6 mice (24). It was important that the antibodies were monoclonal in order to clearly distinguish between LPS and OmpA in the assay. Uptake was measured by a modified ELISA in a microtiter culture system. Cocktails of four mAb specific for LPS and three specific for OmpA from *P. mirabilis* were used to detect these substances; differentiation between antigen on the cell surface and that which had been internalized could be made in the assay (see below). Also, the fates of LPS and OmpA after uptake by macrophages could be monitored differentially by using the respective mAb cocktails.

In general, the results of this study (24) showed that OmpA from *P. mirabilis* could indeed enhance uptake of LPS by macrophages, as illustrated by the data presented in Figure 2. At given times after application of LPS, macrophages were fixed and made permeable for the penetration of antibodies. Uptake of LPS applied in a concentration of 10 μg/ml (closed circles) was just detectable in this system. In contrast, the rate of uptake of 0.5 μg LPS per ml contained in a sample of OmpA applied to the macrophage culture at a concentration of 5 μg/ml (open triangles) was much more rapid. Because uptake of LPS was enhanced in the sample that contained a total concentration of LPS and OmpA (0.5 μg LPS plus 5 μg OmpA per ml) that was less than the concentration of the sample containing LPS alone (10 μg LPS per ml), clearly more effective uptake was not dependent on the quantity but rather on the composition of the sample. Control cultures (open squares) stimulated with mixtures of LPS and OmpA and measured with mouse IgG instead of the mAb showed no significant uptake. The kinetics of uptake of LPS in our study were very similar to those measured using radioactive LPS for detection in macrophage-like cell lines (62) and thioglycollate-elicited peritoneal mouse mac-

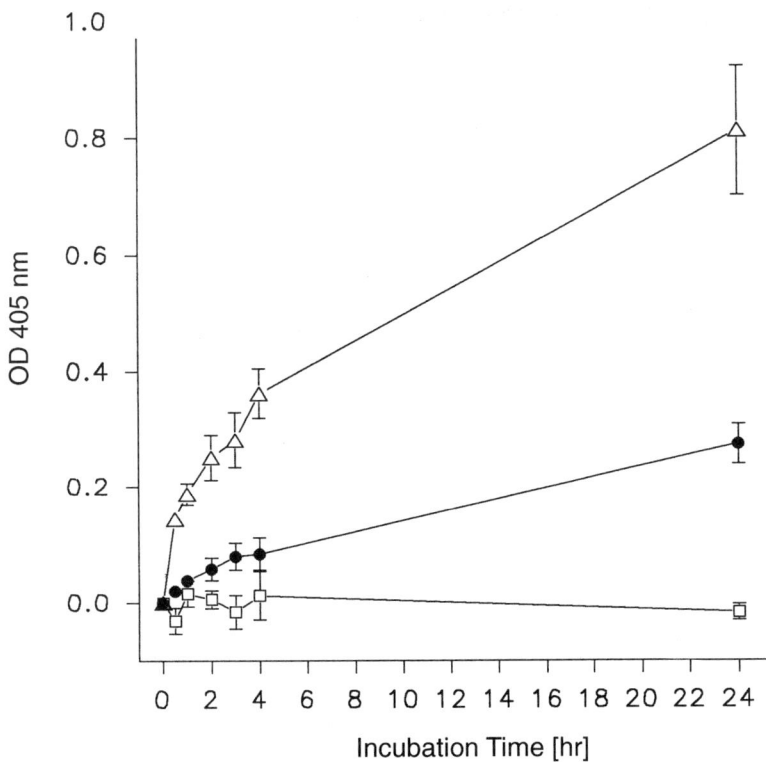

Fig. 2 Uptake of LPS and LPS-OmpA complexes by macrophages. Macrophages derived from the bone marrow of C57BL/6 mice (8 × 10⁴ cells per well) were incubated with LPS or LPS-OmpA complexes for 30 minutes and 1, 2, 3, 4, and 24 hours. The cells were treated thereafter with a mixture of methanol and acetone to make them permeable to the antibodies. LPS was detected with a cocktail of equal concentrations (10 μg/ml) of mAb 1.2.1, 2.6.2, 2.29.1, and 3.2.2 specific for LPS from *P. mirabilis* (24). Reactions were visualized with a secondary goat antibody to mouse IgG, conjugated with alkaline phosphatase. After addition of substrate, reactions were measured in an ELISA reader. LPS was applied in a concentration of 10 μg/ml (●). A sample of OmpA containing 10% LPS was applied in a concentration of 5.0 μg protein per ml plus 0.5 μg LPS per ml (△). Control cultures stimulated with the LPS-OmpA mixture were treated with mouse IgG instead of the mAb (□). Data represent the means ± standard errors of values taken from two to three separate experiments. OD 405 nm, optical density at 405 nm. (From Ref. 24.)

rophages (63), which demonstrates the validity of measurements in our system.

Different combinations and amounts of LPS and OmpA were tested in this study with the result that LPS was always taken up more rapidly when OmpA was included. Differentiation was made between the amount of LPS on the surface of macrophages and that taken up internally (Fig. 3). In this case, one portion of the cultures was permeabilized with acetone and methanol to measure total uptake (solid triangles), while another portion was left untreated to measure LPS on the surface (Fig. 3, open circles). The difference between these measurements was taken as the amount of LPS internalized (dashed line). A rather large concentration of the complex (50 μg each of LPS and OmpA per ml) was used in order to be able to detect the small amounts on the surface. The results indicated that the greater proportion of the LPS was found intracellularly within 30 minutes to 4 hours after addition to the macrophage cultures. A good portion could still be detected intracellularly even after 24 hours. The kinetics of uptake of OmpA contained in the complexes, measured with the mAb to the protein, were very similar to those measured with the LPS mAb (data not shown). Also, uptake of the LPS-OmpA complex was more efficient than that of OmpA alone, although the effect of LPS on OmpA was not as pronounced as the effect of OmpA on LPS. However, detection of OmpA in complexes after uptake in macrophages was more difficult than detection of LPS, even though the OmpA mAb reacted very strongly with the protein in ELISA and Western blot assays.

The results of pulse-chase experiments (pulsing for 30 minutes and washing to remove nonbound sub-

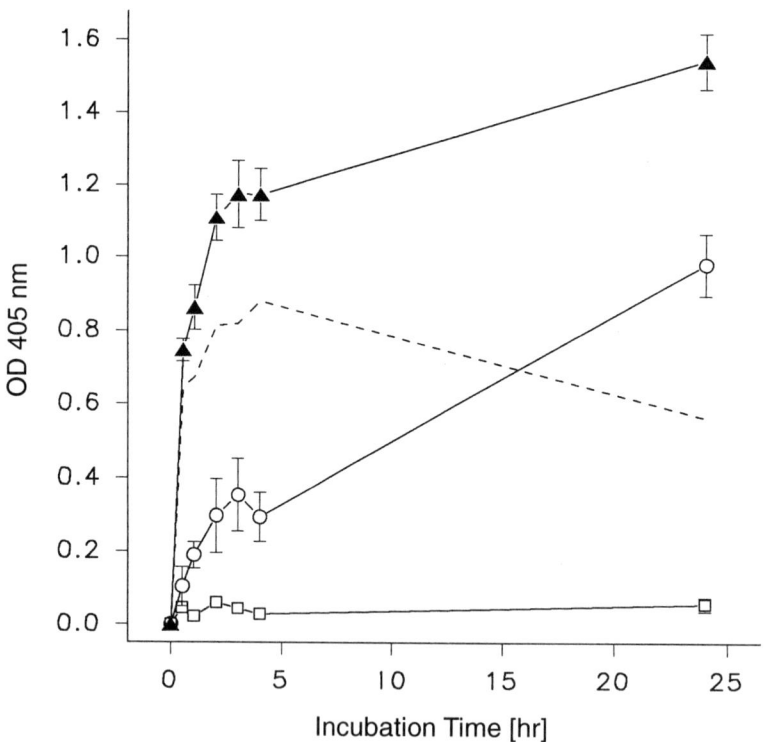

Fig. 3 Distribution of LPS intracellularly and on the surface of macrophages incubated with a mixture of LPS and OmpA from *P. mirabilis*. A mixture of LPS and OmpA, each at a concentration of 50 μg/ml, was added to macrophage cultures (8 × 10^4 cells per well). After the given incubation periods, a portion of the cultures was treated with a mixture of methanol and acetone to fix and permeabilize the cells (▲), while another portion was left untreated to measure LPS on the cell surface (○). Subtraction of the values for nonpermeabilized cells from those of permeabilized cells represents LPS intracellularly(- - -). Control cultures (□) stimulated with the LPS-OmpA mixture were treated with mouse IgG instead of the mAb. LPS was detected with a cocktail of equal concentrations (10 μg/ml) of mAb 1.2.1, 2.6.2, 2.29.1, and 3.2.2 specific for LPS from *P. mirabilis* (24). Reactions were visualized with a secondary goat antibody to mouse IgG, conjugated with alkaline phosphatase. After addition of substrate, reactions were measured in an ELISA reader. Data represent the means ± standard errors of values taken from three separate experiments. OD 405 nm, optical density at 405 nm. (From Ref. 24.)

stances) measuring LPS and OmpA on the surface of macrophages (24) seemed to indicate that the largest amounts of LPS and OmpA measured on the surface appeared within the first 30 minutes after the pulse. The levels of both components then progressively decreased on the surface for the next 3 hours, which indicated internalization. Thereafter, LPS rose to an increased level on the surface during the 24-hour observation period. In contrast, OmpA did not reappear on the surface in a form detectable by the mAb.

Modulation of LPS-Induced Cytokine Production in Macrophages by OmpA from P. mirabilis. Macrophages derived from the bone marrow of C57BL/6 mice were used to test the effects of OmpA from *P. mirabilis* on the interaction of LPS with macrophages. In the course of these studies (64) it was observed that the OmpA protein exerted very differential modulating effects on the TNF-α and IL-1 responses of macrophages to LPS. The effect on the IL-1 responses is illustrated by the data presented in Figure 4 (closed triangles). The results showed that IL-1 production, measured in an IL-2–dependent IL-1 assay, was inhibited in a dose-dependent manner by the protein when it was mixed with LPS by sonication before addition to the macrophage cultures. Interestingly, bovine serum albumin was just as effective as the OmpA protein in inhibiting the IL-1 responses. On the other hand, OmpA from *P. mirabilis* substantially enhanced the cytotoxic activity of culture supernatants directed against L929 target cells (Fig. 4, closed circles). As little as 0.1 μg/ml OmpA mixed with 10 μg/ml LPS gave detectable augmentation over the activity of cultures stimulated with LPS alone (840 ± 60 U/

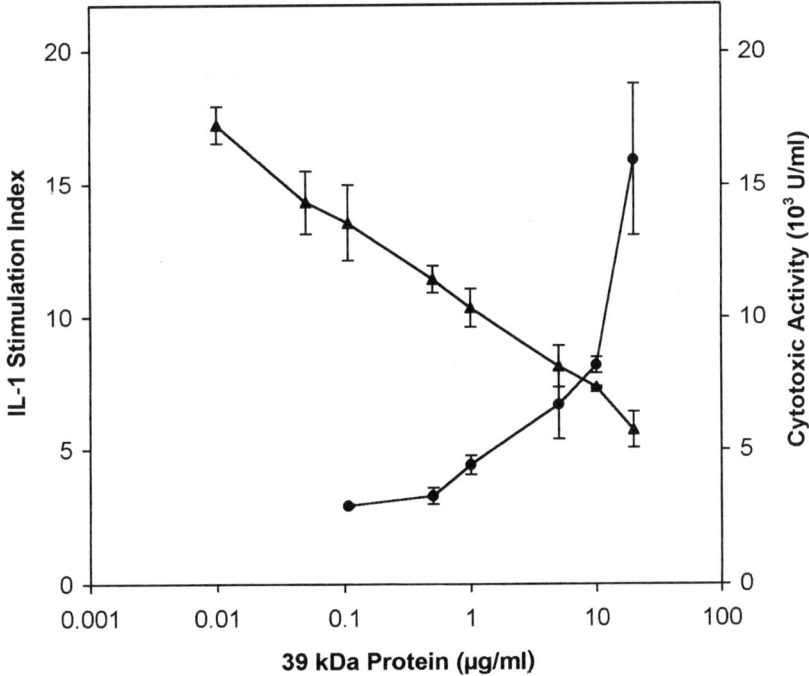

Fig. 4 Inhibition of IL-1 production and enhancement of cytotoxicity of LPS-stimulated murine macrophages by mixing LPS with OmpA (39 kDa protein) from *P. mirabilis*. Macrophages (1 × 10^6/ml) derived from the bone marrow of C57BL/6 mice were stimulated for 24 hours with 10 μg/ml LPS and various concentrations of OmpA. IL-1 (▲) production was measured in an IL-2–dependent IL-1 assay. The stimulation index was calculated by setting cpm (incorporation of [^{3}H]TdR) of unstimulated cultures (208 ± 22 cpm) equal to 1. The stimulation index value for IL-1 production induced by 10 μg/ml LPS alone was 17.96 ± 1.48. Cytotoxicity of culture supernatants (●) was tested using L929 target cells and measuring surviving cells photometrically after crystal violet staining. Cytotoxic activity (U/ml) represents the reciprocal dilution of supernatant giving 50% specific cytolysis, per ml, calculated from the relative OD of treated cultures, compared with control cultures having an OD at 570 nm of 1.22 ± 0.05. Cytotoxic activity induced with 10 μg/ml LPS alone was 838 ± 60 U/ml and that induced with 20 μg/ml OmpA alone was 214 ± 44. Data represent the means ± standard errors of values taken from two separate experiments. (From Ref. 64.)

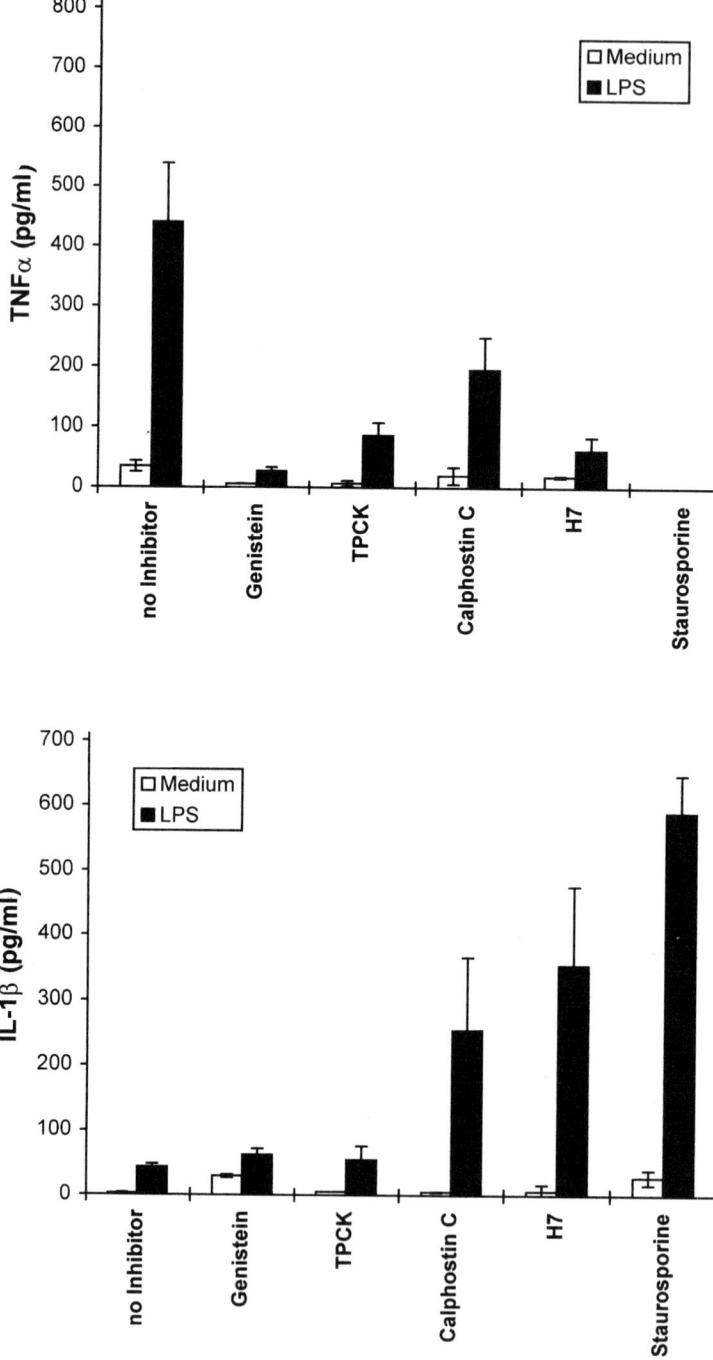

Fig. 5 Effects of inhibitors on secretion of TNF-α (top) and IL-1β (bottom) in culture supernatants of macrophages (8×10^5 cells per well) derived from the bone marrow of C57BL/6 mice after 24-hour stimulation with 1 μg/ml LPS from *P. mirabilis* in 24-well culture dishes. Inhibitors were added 30 minutes before stimulation. Concentrations of inhibitors were 148 μM genistein, 25 μM *N*-tosyl-L-phenylalanine chloromethyl ketone (TPCK), 200 nM calphostin C, 20 μM 1-(5-isoquinolinylsulfonyl)-2-methylpiperazine (H7) dihydrochloride, and 50 nM staurosporine. Control cells received no LPS stimulus (medium). TNF-α and IL-1β were measured using sandwich ELISA kits according to instructions provided by the supplier (Amersham Buchler GmbH & Co KG, Braunschweig, Germany). Data represent the means ± standard errors of the means of values taken from two to four separate experiments. (From Ref. 65.)

ml), whereas 20 µg/ml protein produced a 15-fold enhancement. The cytotoxic activity was due to TNF, because the cytotoxicity of macrophage supernatants could be reduced 80–90% by a hamster mAb to mouse TNF-α/β. In contrast to its activity in inhibiting IL-1 production, bovine serum albumin had no effects on the LPS induction of cytotoxic activity in macrophages.

From the results of these studies, it seemed as if the OmpA protein was exerting positive effects on TNF-α production but negative effects on IL-1 production by macrophages in response to LPS. Our most recent investigations applying inhibitors of intracellular signaling to modulate LPS activation of macrophages have provided indication of a negative regulatory mechanism that is apparently involved in the production of IL-1β (65). Data from these studies in Figure 5 show that the protein kinase inhibitor staurosporine produced a strong increase in IL-1β levels (Fig. 5, bottom) in supernatants of macrophages stimulated with LPS. At the same time, calphostin C and 1-(5-isoquinolinylsulfonyl)-2-methylpiperazine (H7) dihydrochloride, specific inhibitors of protein kinase C (PKC), also caused enhancement of IL-1β secretion, but the effect was not as pronounced as that of staurosporine. In contrast, all three substances caused strong inhibition of TNF-α production (Fig. 5, top). Other inhibitors such as the tyrosine kinase inhibitor genistein and the serine protease inhibitor N-tosyl-L-phenylalanine chloromethyl ketone (TPCK) effectively reduced the amounts of TNF-α produced, but had little effect on secretion or production of IL-1β. The inhibitors did not cause a reduction in viability or integrity of the cells in the concentrations applied, as determined by trypan blue dye exclusion and lactate deyhydrogenase tests. Our studies further showed that staurosporine promotes enhanced production of IL-1β in the cellular fraction of cultures, the greater portion of which is not secreted. At the same time, staurosporine effects an increase in either the levels or the stability of IL-1β–specific transcripts (65). A similar enhancing effect of staurosporine on IL-6 production in RAW 264.7 macrophages has been reported by Tremblay et al. (66), who attributed this effect to a negative regulatory mechanism dependent on PKC. Whether PKC is involved in a negative regulation of IL-1β production in our system will have to be rigorously examined. It is not known at this point if the inhibiting effects of the OmpA protein on the LPS-induced IL-1β production in macrophages are connected to the negative regulation of the cytokine that is apparently relieved by staurosporine. Studies addressing these issues are in progress.

REFERENCES

1. Hepper KP, Garman RD, Lyons MF, Teresa GW. Plaque-forming cell response in BALB/c mice to two preparations of LPS extracted from *Salmonella enteritidis*. J Immunol 1979; 122:1290–1293.
2. Karch H, Nixdorff K. Antibody-producing cell responses to an isolated outer membrane protein and to complexes of this antigen with lipopolysaccharide or with vesicles of phospholipids from *Proteus mirabilis*. Infect Immun 1981; 31:862–867.
3. Andersson J, Melchers F, Galanos C, Lüderitz O. The mitogenic effect of lipopolysaccharide on bone marrow-derived mouse lymphocytes. Lipid A as the mitogenic part of the molecule. J Exp Med 1973; 137:943–953.
4. Freudenberg MA, Meier-Dieter U, Staehelin T, Galanos C. Analysis of LPS released from *Salmonella abortus equi* in human serum. Microb Pathog 1991; 10:93–104.
5. Manthey CL, Perera P-Y, Henricson BE, Hamilton TA, Qureshi N, Vogel SN. Endotoxin-induced early gene expression in C3H/HeJ (*LPS^d*) macrophages. J Immunol 1994; 153:2653–2663.
6. Melchers F, Braun V, Galanos C. The lipoprotein of the outer membrane of *Escherichia coli*: a B lymphocyte mitogen. J Exp Med 1975; 142:473–482.
7. Bessler WG, Henning U. Protein I and protein II* from the outer membrane of *Escherichia coli* are mouse B-lymphocyte mitogens. Z Immunitätsforsch 1979; 155:387–398.
8. Chen Y-HU, Hancock REW, Mishell RI. Mitogenic effects of purified outer membrane proteins from *Pseudomonas aeruginosa*. Infect Immun 1980; 28:178–184.
9. Korn A, Kroll H-P, Berger H-P, Kahler A, Heßler R, Brauberger J, Müller K-P, Nixdorff K. The 39-kilodalton outer membrane protein of *Proteus mirabilis* is an OmpA protein and mitogen for murine B lymphocytes. Infect Immun 1993; 61:4915–4918.
10. Bessler WG, Baier W, Esche Uvd, Hoffmann P, Heinevetter L, Wiesmüller K-H, Jung G. Bacterial lipopeptides constitute efficient novel immunogens and adjuvants in parenteral and oral immunization. Behring Inst Mitt 1997; 98:390–399.
11. Hoffmann P, Heinle S, Schade UF, Loppnow H, Ulmer AJ, Flad HD, Jung G, Bessler WG. Stimulation of human and murine adherent cells by bacterial lipoprotein and synthetic lipopeptide analogues. Immunobiology 1988; 177:158–170.
12. Morrison DC, Betz SJ, Jacobs DM. Isolation of a lipid A bound polypeptide responsible for "LPS-initiated" mitogenesis of C3H/HeJ spleen cells. J Exp Med 1976; 144:840–846.
13. Sultzer BM, Goodman GW. Endotoxin protein: a B-cell mitogen and polyclonal activator of C3H/HeJ lymphocytes. J Exp Med 1976; 144:821–827.
14. Sultzer BM, Nilsson BS. PPD-tuberculin—a B cell mitogen. Nature New Biol 1972; 240:198–200.
15. Skidmore BJ, Morrison DC, Chiller JM, Weigle WO. Immunologic properties of bacterial lipopolysaccharide (LPS). II. The unresponsiveness of C3H/HeJ mouse spleen cells to LPS-induced mitogenesis is dependent

on the method used to extract LPS. J Exp Med 1975; 142:1488–1508.
16. Goodman GW, Sultzer BM. Characterization of the chemical and physical properties of a novel B-lymphocyte activator, endotoxin protein. Infect Immun 1979; 24:685–696.
17. Goodman GW, Sultzer BM. Further studies on the activation of lymphocytes by endotoxin protein. J Immunol 1979; 122:1329–1334.
18. Bandekar JR, Castagna R, Sultzer BM. Roles of protein kinase C and G proteins in activation of murine resting B lymphocytes by endotoxin-associated protein. Infect Immun 1992; 60:231–236.
19. Vogel SN, Wax JS, Perera PY, Padlan C, Potter M, Mock BA. Construction of a BALB/c congenic mouse. C.C3H-Lps^d, that expresses the Lps^d allele: analysis of chromosome 4 markers surrounding the Lps gene. Infect Immun 1994; 62:4454–4459.
20. Kang AD, Wong PMC, Chen H, Castagna R, Chung S-W, Sultzer BM. Restoration of lipopolysaccharide-mediated B-cell response after expression of a cDNA encoding a GTP-binding protein. Infect Immun 1996; 64:4612–4617.
21. Hantke K, Braun V. Covalent binding of lipid to protein. Diglyceride and amide-linked fatty acid at the N-terminal end of the murein-lipoprotein of the *Escherichia coli* outer membrane. Eur J Biochem 1973; 34:284–296.
22. Nikaido H, Vaara M. Molecular basis of bacterial outer membrane permeability. Microbiol Rev 1985; 49:1–32.
23. Sugawara E, Nikaido H. Pore-forming activity of OmpA protein of *Escherichia coli*. J Biol Chem 1992; 267:2507–2511.
24. Korn A, Rajabi Z, Wassum B, Ruiner W, Nixdorff K. Enhancement of uptake of lipopolysaccharide in macrophages by the major outer membrane protein OmpA of gram-negative bacteria. Infect Immun 1995; 63:2697–2705.
25. Finnen RL, Martin NL, Siehnel RJ, Woodruff WA, Rosok M, Hancock REW. Analysis of the *Pseudomonas aeruginosa* major outer membrane protein OprF by use of truncated OprF derivatives and monoclonal antibodies. J Bacteriol 1992; 174:4977–4985.
26. Gilleland HE Jr, Gilleland LB, Matthews-Greer JM. Outer membrane protein F preparation of *Pseudomonas aeruginosa* as a vaccine against chronic pulmonary infection with heterologous immunotype strains in a rat model. Infect Immun 1988; 56:1017–1022.
27. Sultzer BM, Craig JP, Castagna R. Endotoxin associated proteins and their polyclonal and adjuvant activities. In: Szentivanyi A, Friedman H, Nowotny A, eds. Immunobiology and Immunopharmacology of Bacterial Endotoxins. New York: Plenum Publishing Corporation, 1986:435–447.
28. Lex A, Wiesmüller K-H, Jung G, Bessler WG. A synthetic analogue of *Escherichia coli* lipoprotein, tripalmitoyl pentapeptide, constitutes a potent immune adjuvant. J Immunol 1986; 137:2676–2681.
29. Deres K, Schild H, Wiesmüller K-H, Jung G, Rammensee H-G. In vivo priming of virus-specific cytotoxic T lymphocytes with synthetic lipopeptide vaccine. Nature 1989; 342:561–564.
30. Loleit M. Ihlenfeld H-G, Brünjes J, Jung G, Müller B, Hoffmann P, Bessler WG, Pierres M, Haas G. Synthetic peptides coupled to the lipotripeptide P_3CSS induce in vivo B and T_{helper} responses to HIV-1 reverse transcriptase. Immunobiol 1996; 195:61–76.
31. von Specht B-U, Knapp B, Muth G, Bröker M, Hungerer K-D, Diehl K-D, Massarrat K. Seemann A, Domdey H. Protection of immunocompromised mice against lethal infection with *Pseudomonas aeruginosa* by active or passive immunization with recombinant *P. aeruginosa* outer membrane protein F and outer membrane protein I fusion proteins. Infect Immun 1995; 63:1855–1862.
32. Gabelsberger J, Knapp B, Bauersachs S, Lenz U, von Specht B-U, Domdey H. A hybrid outer membrane protein antigen for vaccination against *Pseudomonas aeruginosa*. Behring Inst Mitt 1997; 98:302–314.
33. Hancock REW, Wong R. Potential of protein OprF of *Pseudomonas* in bivalent vaccines. Behring Inst Mitt 1997; 98:283–290.
34. Schorr J, Knapp B, Hundt E, Küpper HA, Amann E. Surface expression of malarial antigens in *Salmonella typhimurium*: induction of serum antibody response upon oral vaccination of mice. Vaccine 1991; 9:675–681.
35. Hogan MM, Vogel SN. Lipid A-associated proteins provide an alternative ''second signal'' in the activation of recombinant interferon-γ-primed, C3H/HeJ macrophages to a fully tumoricidal state. J Immunol 1987; 139:3697–3702.
36. Hogan MM, Vogel SN. Production of tumor necrosis factor by rIFN-γ-primed C3H/HeJ (Lps^d) macrophages requires the presence of lipid A-associated proteins. J Immunol 1988; 141:4196–4202.
37. Dong Z, Qi X, Fidler IJ. Tyrosine phosphorylation of mitogen-activated protein kinases is necessary for activation of murine macrophages by natural and synthetic bacterial products. J Exp Med 1993; 177:1071–1077.
38. Galdiero F, Cipollaro de L'ero G, Benedetto N, Galdiero M, Tufano MA. Release of cytokines induced by *Salmonella typhimurium* porins. Infect Immun 1993; 61:155–161.
39. Kuusi N, Nurminen M, Saxen H, Valtonen M, Mäkelä PH. Immunization with major outer membrane proteins in experimental salmonellosis of mice. Infect Immun 1979; 25:857–862.
40. Svenson SB, Nurminen M, Lindenberg AA. Artificial *Salmonella* vaccines: O-antigenic oligosaccharide-protein conjugates induce protection against infection with *Salmonella typhimurium*. Infect Immun 1979; 25:863–872.
41. Prasadarao NV, Wass CA, Weiser JN, Stins MF, Huang S-H, Kim KS. Outer membrane protein A of *Escherichia coli* contributes to invasion of brain microvascular endothelial cells. Infect Immun 1996; 64:146–153.
42. Lowell GH, Smith LF, Seid RC, Zollinger WD. Peptides bound to proteosomes via hydrophobic feet become highly immunogenic without adjuvants. J Exp Med 1988; 167:658–663.
43. Lowell GH, Ballou WR, Smith LF, Wirtz RA, Zollinger WD, Hockmeyer WT. Proteosome-lipopeptide vac-

cines: enhancement of immunogenicity for malaria CS peptides. Science 1988; 240:800–802.
44. Frasch CE, Peppler MS. Protection against group B *Neisseria meningitidis* disease: preparation of soluble protein and protein-polysaccharide immunogens. Infect Immun 1982; 37:271–280.
45. Killion JW, Morrison DC. Determinants of immunity to murine salmonellosis: studies involving immunization with lipopolysaccharide-lipid A-associated protein complexes in C3H/HeJ mice. FEMS Microbiol Immunol 1988; 47:41–54.
46. Orr N, Rubin G, Cohen D, Arnon R, Lowell GH. Immunogenicity and efficacy of oral or intranasal *Shigella flexneri* 2a and *Shigella sonnei* proteosome-lipopolysaccharide vaccines in animal models. Infect Immun 1993; 61:2390–2395.
47. Orr N, Arnon R, Rubin G, Cohen D, Bercovier H, Lowell GH. Enhancement of anti-*Shigella* lipopolysaccharide (LPS) response by addition of the cholera toxin B subunit to oral and intranasal proteosome-*Shigella flexneri* 2a LPS vaccines. Infect Immun 1994; 62:5198–5200.
48. Ruttkowski E, Nixdorff K. Qualitative and quantitative changes in the antibody-producing cell response to lipopolysaccharide induced after incorporation of the antigen into bacterial membrane phospholipid vesicles. J Immunol 1980; 124:2548–2551.
49. Karch H, Gmeiner J, Nixdorff K. Alteration of the immunoglobulin G subclass responses in mice to lipopolysaccharide: effects of nonbacterial proteins and bacterial membrane phospholipids or outer membrane proteins of *Proteus mirabilis*. Infect Immun 1983; 40:157–165.
50. Nixdorff K, Weber G, Kaniecki K, Ruiner W, Schell S. Bacterial protein-LPS complexes and immunomodulation. In: Friedman H, Klein TW, Yamaguchi H, eds. Microbial Infections. Role of Biological Response Modifiers. New York: Plenum Publishing Corporation, 1992:49–61.
51. Karch H, Nixdorff K. Modulation of the IgG subclass responses to lipopolysaccharide by bacterial membrane components: differential adjuvant effects produced by primary and secondary stimulation. J Immunol 1983; 131:6–8.
52. Mitchison NA. The carrier effect in the secondary response to hapten-protein conjugates. II. Cellular cooperation. Eur J Immunol 1971; 1:18–27.
53. Mosmann TR, Coffman RL. TH1 and TH2 cells: different patterns of lymphokine secretion lead to different functional properties. Annu Rev Immunol 1989; 7:145–173.
54. Street NE and Mosmann TR. Functional diversity of T lymphocytes due to secretion of different cytokine patterns. Fed Am Soc Exp Biol J 1991; 5:171–177
55. VanCott JL, Kweon M, Fujihashi K. Yamamoto M, Marinaro M, Kiyono H, McGhee JR. Helper T subsets and cytokines for mucosal immunity and tolerance. Behring Inst Mitt 1997; 98:44–52.
56. Seder RA, Paul WE. Acquisition of lymphokine-producing phenotype by CD4$^+$ T cells. Annu Rev Immunol 1994; 12:635–673.
57. Trinchieri G. Interleukin-12: a proinflammatory cytokine with immunoregulatory functions that bridge innate resistance and antigen-specific adaptive immunity. Annu Rev Immunol 1995; 13:251–276.
58. Pearce EJ, Caspar P, Grzych J-M, Lewis FA, Sher A. Downregulation of Th1 cytokine production accompanies induction of Th2 responses by a parasitic helminth, *Shistosoma mansoni*. J Exp Med 1991; 173:159–166.
59. Marinaro M, Staats HF, Hiroi T, Jackson RJ, Coste M, Boyaka PN, Okahashi N, Yamamoto M, Kiyono H, Bluethmann H, Fujihashi K, McGhee JR. Mucosal adjuvant effect of cholera toxin in mice results from induction of T helper 2 (Th2) cells and IL-4. J Immunol 1995; 155:4621–4629.
60. VanCott JL, Staats HF, Pascual DW, Roberts M, Chatfield SN, Yamamoto M, Coste M, Carter PB, Kiyono H, McGhee JR. Regulation of mucosal and systemic antibody responses by T helper cell subsets, macrophages, and derived cytokines following oral immunization with live recombinant *Salmonella*. J Immunol 1996; 156:1504–1514.
61. Lowell GH. Proteosomes, hydrophobic anchors, iscoms, and liposomes for improved presentation of peptide and protein vaccines. In: Woodrow GC, Levine MM, eds. New Generation Vaccines. New York: Marcel Dekker, 1990:141–160.
62. Hampton RY, Raetz CRH. Macrophage catabolism of lipid A is regulated by endotoxin stimulation. J Biol Chem 1991; 266:19499–19509.
63. Munford RS, Hall CL. Uptake and deacylation of bacterial lipopolysaccharides by macrophages from normal and endotoxin-hyporesponsive mice. Infect Immun 1985; 48:464–473.
64. Weber G, Link F, Ferber E, Munder PG, Zeitter D, Bartlett RR, Nixdorff K. Differential modulation of the effects of lipopolysaccharide on macrophages by a major outer membrane protein of *Proteus mirabilis*. J Immunol 1993; 151:415–424.
65. Schilling D, Brauburger J, Ruiner W, Nixdorff K. Modulation of interleukin 1β production in macrophages stimulated with lipopolysaccharide by the protein kinase inhibitor staurosporine. J Endotoxin Res 1997; 4:251–260.
66. Tremblay P, Houde M, Arbour N, Rochefort D, Masure S, Mandeville R, Opdenakker G, Oth D. Differential effects of PKC inhibitors on gelatinase B and interleukin 6 production in the mouse macrophage. Cytokine 1995; 7:130–136.

44

Opsonization of *Actinobacillus actinomycetemcomitans* by LPS-Directed IgG Antibodies in Sera of Juvenile Periodontitis Patients

Mark E. Wilson
University of Medicine and Dentistry of New Jersey, Newark, New Jersey

INTRODUCTION

Periodontitis, commonly known as gum disease, is an inflammatory process that leads to the progressive loss of periodontal ligament cells and alveolar bone that provide support for the teeth. Bacteria resident in dental plaque are considered to be the primary etiologic agents of periodontal disease. A number of distinct forms of periodontitis are currently recognized, differing in microbial etiology, rate of progression, and response to periodontal therapy (1). While many adults experience some degree of periodontitis ("adult periodontitis") after the third decade of life, certain forms of periodontitis, termed early-onset periodontitis (EOP), become manifest at an earlier age. EOP develops during childhood or adolescence and is thought to be associated with defects in host defense, especially involving production or function of cirulating polymorphonuclear neutrophils.

Localized juvenile periodontitis (LJP) is a form of EOP, which has its onset during the circumpubertal period of life and is characterized by rapid loss of alveolar bone localized to the permanent first molar and incisor teeth. LJP is a relatively uncommon form of periodontal disease. Based on the results of a recent U.S. survey of over 11,000 adolescents 14–17 years of age, the prevalence of LJP has been estimated to be 0.53% (2).

Significantly, however, the results of this survey also indicated that African Americans are nearly 15 times more likely to develop LJP than Caucasians, a pattern that has been noted in other studies (3,4). There is a tendency for cases of LJP to cluster within families, suggesting that susceptibility to this form of EOP may be inherited (5).

Microbiological studies of LJP patients have revealed that the subgingival microflora of these patients differ from that of periodontally healthy individuals and patients with adult periodontitis. *Actinobacillus actinomycetemcomitans*, a capnophilic gram-negative coccobacillus that is closely related to the oral haemophili (6–8), has been identified as a prominent member of the subgingival microflora of these patients. *A. actinomycetemcomitans* is present, albeit in relatively low numbers, in the subgingival microflora of approximately 20% of periodontally healthy juveniles and adults and a similar percentage of patients with adult periodontitis (7). In contrast, more than 95% of LJP subjects examined harbor this organism in high numbers. Further evidence for an association between *A. actinomycetemcomitans* and LJP derives from studies showing that elimination of this organism from subgingival plaque following antibiotic therapy correlates with resolution of periodontal lesions in these patients (9). Moreover, recurrence of disease activity is accompanied by the reappearance of *A. actinomycetemcomitans* in periodontal lesions of LJP patients. Several studies have provided evidence that *A. actinomycetemcomitans* may be transmitted among members of families with LJP (10,11).

SEROLOGY OF
A. actinomycetemcomitans INFECTION

Early serological studies, performed using rabbit polyclonal antisera in combination with indirect immunofluorescence, revealed the existence of at least three distinct serotypes among oral isolates of *A. actinomycetemcomitans* (12). A reexamination of the serological characteristics of *A. actinomycetemcomitans* employing monoclonal antibodies provided evidence for two additional serotypes (d and e), as well as a number of nontypable strains (13). Zambon and coworkers examined the distribution of *A. actinomycetemcomitans* serotypes among periodontally healthy subjects as well as patients with either adult periodontitis or LJP (12). Serotype c strains were determined to be poorly represented in the oral cavity, comprising ≤10% of all *A. actinomycetemcomitans* isolates recovered from healthy subjects and LJP patients. Serotypes a and b were recovered at approximately equal frequency (50% serotype a, 43% serotype b) from subgingival plaque samples of healthy subjects. A similar distribution of these two serotypes was noted for adult periodontitis patients. In contrast, the prevalence of serotype b was found to be roughly twofold greater than that of serotype a (62% vs. 28%) among plaque samples of LJP patients. The prevalence of other typable and nontypable strains of *A. actinomycetemcomitans* in the oral cavity of healthy and periodontally diseased (particularly LJP) subjects remains unclear. Hence, current evidence suggests that *A. actinomycetemcomitans* serotype b is strongly associated with the development of LJP.

HOST RESPONSE TO
A. actinomycetemcomitans SEROTYPE b

Immunological studies have also yielded evidence to support an association between *A. actinomycetemcomitans* (particularly serotype b) and LJP. Utilizing an enzyme-linked immunosorbent assay (ELISA), in conjunction with a formalinized serotype b strain (Y4, ATCC 43718) of *A. actinomycetemcomitans*, Ebersole and coworkers (14) determined that sera of LJP patients often contain significantly higher titers of IgG, IgA, and IgM antibodies to this species than sera of healthy subjects or patients with other clinical forms of periodontitis. Similar results have been reported by other groups (15,16). Subsequent studies examined the antigenic specificity of serum IgG antibodies reactive toward *A. actinomycetemcomitans* serotype b in this patient group. Such studies revealed that LJP sera contain IgG antibodies that recognize a diverse array of *A. actinomycetemcomitans* antigens including outer membrane proteins (17,18), leukotoxin (19,20), and lipopolysaccharide (21,22). However, a serotype b-specific antigen appears to induce the most exuberant IgG (as well as IgA and IgM) antibody response to *A. actinomycetemcomitans* serotype b in these patients (23-25). Several reports have referred to this serotype b antigen as the "immunodominant" antigen of *A. actinomycetemcomitans*. Given the apparent immunological significance of the serotype b antigen, a number of groups sought to define its structure.

CHARACTERISTICS OF THE SEROTYPE b
ANTIGEN AND ITS RELATIONSHIP TO LPS

Initial efforts to define the nature of the serotype b antigen revealed that this molecule exhibits resistance to heat treatment (100°C, 45 min) and to papain digestion, properties consistent with those of a carbohydrate (23). Based upon immunoreactivity of whole *A. actinomycetemcomitans* with serotype-specific monoclonal antibodies (26,27), as well as the loss of such reactivity in serotype b antigen-defective mutants generated by insertion of transposon Tn916 (28), the serotype b antigen appears to be exposed on the bacterial surface. Amano and coworkers isolated a serotype-specific polysaccharide from an autoclaved extract of *A. actinomycetemcomitans* strain Y4 (serotype b) by chromatography on diethylaminoethyl-Sephadex G-25 and Sephacryl S-300 (29). Based upon compositional and structural analysis of this fraction, these investigators proposed that the serotype b antigen is a linear polymer composed of disaccharide units containing L-rhamnose and D-fucose.

Phenol-extracted LPS from *A. actinomycetemcomitans* strain Y4 has been found to contain the neutral sugars rhamnose and fucose in approximately equal amounts (22,30), as is the case for the serotype b antigen described by Amano and coworkers (29). This prompted us to examine the relationship between the serotype b antigen and the polysaccharide (particularly the O-antigen) of LPS derived from this serotype. Initial efforts were directed to demonstrating that the serotype b antigen is associated with the LPS molecule. To this end, we isolated a high molecular weight (HMW) polysaccharide from *A actinomycetemcomitans* serotype b by hot aqueous phenol extraction, which was further resolved by gel permeation chromatography in an LPS-disaggregating buffer (31). As shown by double immunodiffusion analysis, the HMW polysaccharide reacted with rabbit polyclonal antiserum to *A. actinomycetemcomitans* serotype b, but not with antisera specific for other serotypes of this species (Fig. 1A). Moreover, immunoblot analysis (Fig. 1B) of this fraction employing serum from a high-titer LJP subject produced a diffuse band, or "smear" pattern, which is characteristic for the serotype b antigen (23). Compositional analysis revealed that the HMW polysaccharide contained rhamnose and fucose in equal proportion. Also present in this fraction were L-glycero-D-manno-heptose and 3-hydroxytetradecanoic acid, constituents commonly found in LPS, including LPS extracted from *A. actinomycetemcomitans* (32,33). Indeed, hydrolysis of the HMW polysaccharide in 2% acetic acid (100°C, 2 hr) yielded an insoluble precipitate whose structure was found to be identical to that reported for lipid A from this strain (34–36). These findings indicated that the serotype b antigen of *A. actinomycetemcomitans* is associated with the polysaccharide region of LPS. Similar conclusions were reached in an independent study by Page and coworkers (37).

We subsequently sought to determine whether the core polysaccharide or O-polysaccharide (O-PS) defines serological specificity for *A. actinomycetemcomitans*. Accordingly, we performed structural analyses of the O-PS and core polysaccharides of LPS of *A.*

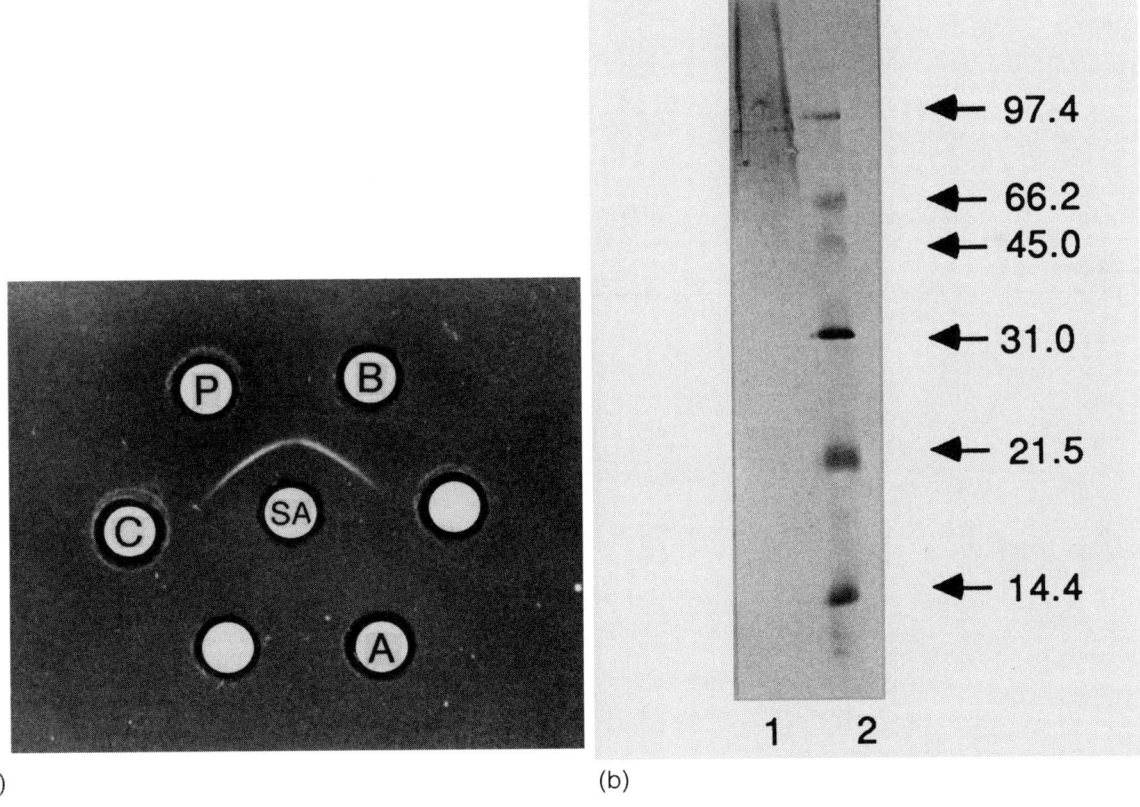

Fig. 1 Immunoreactivity of *A. actinomycetemcomitans* serotype b LPS (SA) with serotype-specific rabbit polyclonal antisera and with high-titer LJP serum (P). Double immunodiffusion analysis (panel A) indicated that LPS from a representative serotype b strain (Y4) formed a precipitin band with serotype b–specific rabbit antiserum, but not with antisera to serotypes a or c. Also note that serotype b–specific antiserum and LJP serum produced a line of identity following their reaction with serotype b LPS. As shown by immunoblot analysis (panel B), the reaction between serotype b LPS and high-titer LJP serum yielded a "smear" pattern characteristic of the serotype b antigen. (From Ref. 31.)

actinomycetemcomitans serotypes a–e by means of methylation, periodate oxidation, and one- and two-dimensional ^{1}H- and ^{13}C-NMR methods (35,36). The O-PS of each *A. actinomycetemcomitans* serotype was found to be structurally distinct (see Table 1). The O-PS of serotype b was identified as a polymer of a repeating trisaccharide unit composed of L-rhamnose, D-fucose, and *N*-acetyl-D-galactosamine (35), with D-GalNAc being linked to the O-3 positions of L-Rha. Structures of the O-PS of serotypes a and c were found to be identical to those of the serotype a– and c–specific polysaccharides described previously by Shibuya and coworkers (38). The core oligosaccharides of all five serotypes had the same glycose compositions and yielded identical specific optical rotations as well as ^{1}H-NMR and ^{13}C-NMR spectra, indicating that this structure is conserved among different serotypes. Given the conserved nature of the core oligosaccharide and the unique structures of the O-PS of different *A. actinomycetemcomitans* serotypes, it appears evident that the O-PS of LPS defines serological specificity for this species.

LPS-REACTIVE IgG ANTIBODIES IN LJP SERA: SPECIFICITY AND SUBCLASS CHARACTERISTICS

Sera of LJP patients often contain markedly elevated IgG antibody titers to *A. actinomycetemcomitans* serotype b LPS. As determined by ELISA employing HMW serotype b LPS as test antigen, IgG antibody titers in LJP sera were determined to be >20-fold greater than in sera of a periodontally healthy, race-matched control group (geometric mean titers: 2573 vs. 111; $n = 35$) (31). In order to determine if these IgG antibodies recognize determinants in the polysaccharide moiety, core glycolipid, or both, we performed inhibition ELISA experiments. Serum from an LJP subject with an elevated titer of IgG antibody to serotype b LPS was incubated with variable concentrations of lipid A or lipid-free oligosaccharide (obtained by mild acid hydrolysis of serotype b LPS) and subsequently reacted with polystyrene microtiter plates coated with whole serotype b LPS. Incubation of LJP serum with lipid A resulted in a modest reduction in serum IgG

Table 1 O Polysaccharide Structures of *A. actinomycetemcomitans* Serotypes a–e

Serotype	Strain(s) used	O-Polysaccharide structure[a]
a	ATCC 29523	→3)-α-D-6dTal*p*-(1→2)-α-D-6dTal*p*(1→ 2 ↑ Ac
b	ATCC 43718 JP2	→3)-α-D-Fuc*p*-(1→2)-α-L-Rha*p*-(→ 3 ↑ 1 β-D-Gal*p*NAc
c	SUNY 67	→3)-α-L-6dTal*p*-(1→2)-α-L-6dTal*p*(1→ 4 ↑ Ac
d	OMZ 542	→3)-β-D-Glc*p*-(1→4)-β-D-Man*p*-(1→4)-α-D-Man*p*(1→ 3 ↑ 1 α-L-Rha*p*
e	OMZ 534	→4)-α-D-Glc*p*NAc-(1→3)-α-L-Rha*p*-(1→

[a]From Perry et al. (35,36).
Abbreviations used: Rha, rhamnose; Fuc, Fucose; Glc, glucose; Gal, galactose; Tal, talose; Man, mannose; NAc, N-acetyl.

titer to LPS, whereas incubation with lipid-free oligosaccharide produced a marked reduction (87%) in IgG titer. These results suggest that LJP sera contain IgG antibodies that recognize both the core glycolipid and polysaccharide regions of serotype b LPS, although antibodies to the latter predominate.

Human serum IgG is comprised of four distinct subclasses (IgG1–IgG4), which are defined by subtle structural differences in the constant region of the γ heavy chain (39). These four subclasses differ with respect to their relative proportion in normal serum, their biological properties, and the nature of the antigens that induce their production. One key biological property of IgG entails the capacity to activate the complement cascade, the products of which promote (1) phagocyte motility, (2) membranolytic attack on susceptible gram-negative organisms, (3) increased vascular permeability, and (4) opsonization. Antibodies of the IgG1 and IgG3 subclasses are potent activators of the classical pathway of complement, whereas IgG2 is comparatively weak. IgG4 antibodies appear to lack detectable complement-fixing activity. Another important function of IgG involves binding to membrane receptors for the Fc domain of IgG (termed FcγR). Mononuclear and polymorphonuclear phagocytes display FcγR, which exhibit preferential binding of IgG1 and IgG3 over IgG2 and IgG4 (40). Such findings indicate that the subclass distribution of IgG antibodies produced in response to a particular antigen may influence the capacity of those antibodies to support host defense, at least as regards complement fixation and FcγR binding.

The results of analysis of the IgG subclass profiles of patients following natural infection or vaccination suggest that protein and carbohydrate antigens evoke distinct IgG subclass responses (41,42). In general, protein antigens preferentially induce production of IgG1 and IgG3 antibodies. IgG4 antibodies may also appear following prolonged protein antigen exposure. On the other hand, polysaccharide antigens, including bacterial lipopolysaccharides, mainly induce IgG2 antibodies, although significant amounts of IgG1 antibody may also be produced.

Serospecific determinants in the O-PS of *A. actinomycetemcomitans* serotype b have been shown to represent a major target for IgG antibodies in sera of LJP patients. We sought to examine the subclass distribution of IgG antibodies to serotype b LPS in sera of these patients. Anti-LPS IgG antibodies were quantified in an ELISA employing human IgG subclass–restricted murine monoclonal antibodies. Optical density readings were subsequently converted to gravimetric units via heterologous interpolation of dose-response curves

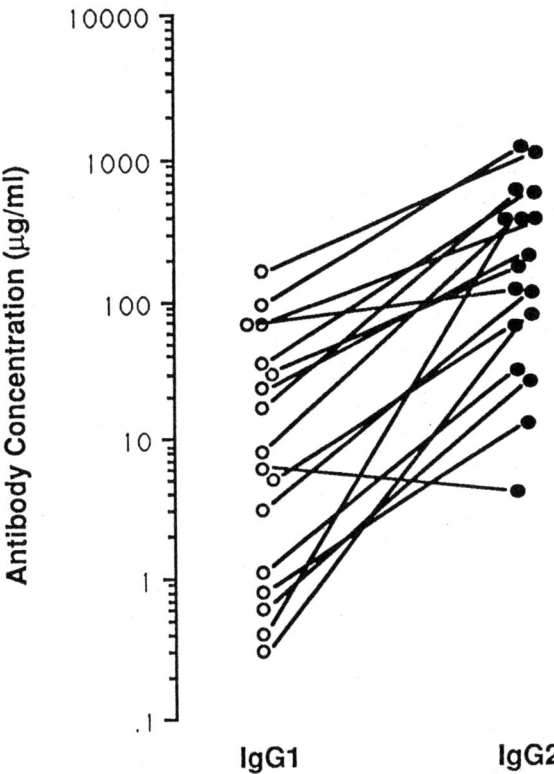

Fig. 2 Concentrations of IgG1 and IgG2 subclass antibodies reactive toward *A. actinomycetemcomitans* serotype b LPS in sera of high-responder LJP patients. IgG subclass responses were determined using an enzyme-linked immunosorbent assay, in conjuction with human IgG subclass-restricted monoclonal antibodies. Optical density readings were converted to gravimetric units by heterologous interpolation of a reference curve generated using human-mouse chimeric antibodies specific for the hapten 4-hydroxy-3-nitrophenylacetyl. (From Ref. 43.)

generated using human-mouse chimeric antibodies containing murine variable regions that define specificity for the hapten 4-hydroxy-3-nitrophenylacetyl and human constant regions for IgG1, IgG2, IgG3, or IgG4 (43). The subclass distribution of IgG antibodies to serotype b LPS was examined in sera of 17 high-titer and 6 low-titer LJP patients. Among high-titer patients, the geometric mean IgG2 concentration was more than 17-fold greater than the corresponding mean IgG1 antibody concentration (136.54 vs. 7.78 μg/ml), with 16/17 sera (94%) exhibiting a predominant IgG2 response to LPS (Fig. 2). Concentrations of both IgG1 and IgG2 antibodies to serotype b LPS were significantly elevated in high-titer LJP sera as compared with low-titer sera. These results indicated that *A. actinomycetemcom-*

itans serotype b LPS-reactive IgG antibodies in sera of LJP patients are principally of the IgG2 subclass, a pattern observed with other polysaccharide antigens.

PHAGOCYTOSIS AND KILLING OF *A. actinomycetemcomitans* BY HUMAN NEUTROPHILS

Circulating neutrophils play an important role in host defense against infections caused by extracellular bacteria and are also thought to protect the periodontium against infections by gram-negative bacteria present in dental plaque, including *A. actinomycetemcomitans*. Indeed, we found that *A. actinomycetemcomitans*, which is resistant to complement-mediated killing, is highly susceptible to phagocytosis and subsequent intracellular killing by human neutrophils (44). Intracellular killing was most efficient under normoxic conditions, but was also demonstrable under anaerobic conditions, indicating that both oxidative and nonoxidative bactericidal mechanisms were involved (45). The organism was found to be particularly sensitive to the bactericidal properties of the myeloperoxidase–hydrogen peroxide–halide antimicrobial system of the neutrophil (46).

Coating of bacterial surfaces with IgG antibodies can lead to enhanced phagocytosis and killing of many species of bacteria through a process termed opsonization and often requires the presence of complement. We examined the opsonic requirements for neutrophil killing of *A. actinomycetemcomitans*. In the absence of opsonins, minimal killing was observed during a 2-hour incubation period (47). Similarly, serum complement alone did not support killing, nor did IgG antibodies isolated from serum of a high-titer LJP patient. However, addition of both IgG and complement significantly augmented neutrophil killing (>90% loss of viability) of *A. actinomycetemcomitans*. Opsonic IgG antibody activity was detected in sera of two additional LJP patients, but was absent in serum of a fourth LJP patient as well as in nonimmune serum of periodontally healthy subjects. These results indicate that *A. actinomycetemcomitans*-specific IgG antibodies play an important role in supporting phagocytosis and killing of this organism by circulating neutrophils.

Opsonic Activity of IgG2 Antibodies in LJP Sera

As noted above, serospecific determinants in the O-polysaccharide of serotype b LPS are a major target of IgG antibodies in sera of LJP patients. Moreover, these LPS-directed antibodies belong chiefly to the IgG2 subclass. Given the apparent predominance of this subclass in the IgG response of LJP patients to *A. actinomycetemcomitans*, we performed studies designed to evaluate the ability of IgG2 antibodies to support phagocytosis and killing of this organism by human neutrophils.

Phagocyte recognition of IgG-opsonized targets is mediated by membrane receptors for the Fc region of the IgG molecule (FcγR). Three main classes of FcγR are presently recognized (48,49), differing in their cellular distribution as well as their affinity and specificity for human IgG subclass antibodies. FcγRI (CD64) is primarily expressed on mononuclear phagocytes, but can be induced on neutrophils through exposure to interferon gamma or granulocyte colony-stimulating factor. This receptor exhibits high affinity (Ka, 10^8–10^9 M^{-1}) for monomeric human IgG1 and IgG3 and moderate affinity for human IgG4. Affinity of FcγRI for IgG2 is at least two orders of magnitude lower than for IgG1 and IgG3. FcγRII (CD32) and FcγRIII (CD16) are low-affinity receptors (Ka, 1–3×10^7 M^{-1}), which bind to IgG in complexed (or polymeric), but not monomeric, form. FcγRIII, the most abundant class of IgG Fc receptors expressed on mononuclear and polymorphonuclear phagocytes, only binds human IgG1 and IgG3.

FcγRII, which is expressed on virtually all leukocytes, also binds IgG1 and IgG3 preferentially but can, under certain conditions (see below), bind human IgG2. FcγRII is a receptor family consisting of six different isoforms encoded by three genes (designated FcγRIIA, IIB, and IIC) found on human chromosome 1. The FcγRIIA gene, the product of which is expressed on neutrophils and monocytes, exhibits a biallelic polymorphism that influences binding of human IgG2. In particular, the FcγRIIa allotype containing a histidine residue at position 131 of the second extracellular domain binds IgG2-coated erythrocytes and bacteria efficiently, whereas the allotype bearing an arginine residue at this same position does not (50–52).

Inasmuch as the H131 allotype of FcγRIIA is the only member of the FcγR family known to bind IgG2 efficiently, we employed neutrophils from donors who were homozygous for the H131 allele as effectors in an assay designed to evaluate the opsonic properties of IgG2 antibodies in sera of LJP patients (53). IgG2 (greater than 99% subclass-restricted) was prepared from sera of periodontally healthy subjects and from a high-titer LJP patient by affinity chromatography using murine monoclonal antihuman IgG2. IgG2 from serum of the LJP patient exhibited concentration-dependent opsonic activity toward *A. actinomycetemcomitans* serotype b when employed in conjunction with a complement source (Fig. 3). On the other hand, IgG2 from sera of periodontally healthy individuals lacked perceptible opsonic activity. These results indicated that IgG2-mediated opsonization of *A. actinomycetemcomitans* requires the presence of specific antibody.

In vitro expression of the opsonic activity of IgG2 antibodies toward *A. actinomycetemcomitans* was dependent upon the presence of heat-labile (complement) opsonins. Thus, IgG2 exhibited minimal opsonic activity during a 2-hour incubation period in the presence of heat-inactivated (56°C, 30 min) hypogammaglobulinemic human serum (Fig. 4). Incubation of *A. actinomycetemcomitans* with IgG2 (50 µg/ml) in the presence of native hypogammaglobulinemic serum, in contrast, resulted in greater than 90% loss of bacterial viability during the same period. Complement alone (i.e., native hypogammaglobulinemic serum) did not support killing of *A. actinomycetemcomitans* by neutrophils. These results indicate that the cooperative interaction between FcγRIIa and complement receptors (particularly CD11b/CD18 [CR3]) may be required for optimal phagocytosis and killing of *A. actinomycetemcomitans*.

The aforementioned studies indicated that IgG2 antibodies in sera of LJP patients are capable of supporting opsonophagocytosis of *A. actinomycetemcomitans*, at least when used in conjunction with phagocytes expressing the H131 allotype of FcγRIIa. We next examined the extent to which the H131/R131 polymorphism associated with this receptor may influence the expression of IgG2-dependent opsonization of *A. actinomycetemcomitans*. Utilizing a fluorochrome phagocytosis assay, we compared ingestion and intracellular killing of IgG2-opsonized *A. actinomycetemcomitans* by neutrophils from donors who were homozygous for either the H131 or R131 allele of FcγRIIA (54). Neutrophils expressing the H131 allotype of FcγRIIa exhibited significantly greater phagocytosis and killing of IgG2-coated *A. actinomycetemcomitans* than neutrophils expressing the R131 allotype, both in the presence and absence of complement. In the presence of both IgG2 and complement, neutrophils from H131-homozygous individuals ingested nearly twice as many bacteria as neutrophils from R131 homozygotes. Thus, the allotype of FcγRIIa

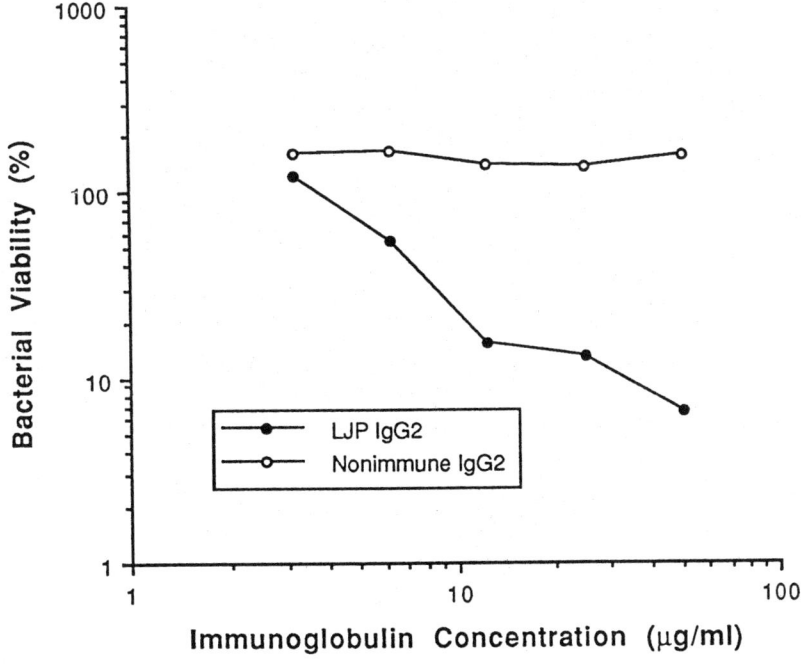

Fig. 3 Opsonic activities of affinity-purified IgG2 prepared from immune (LJP) and nonimmune (periodontally healthy) sera. *A. actinomycetemcomitans* serotype b was incubated with immune or nonimmune IgG2 in the presence of 5% (v/v) hypogammaglobulinemic serum and neutrophils expressing the H131 allele of FcγRIIA for 90 minutes at 37°C. Viability is expressed as a percentage of the starting inoculum. (From Ref. 53.)

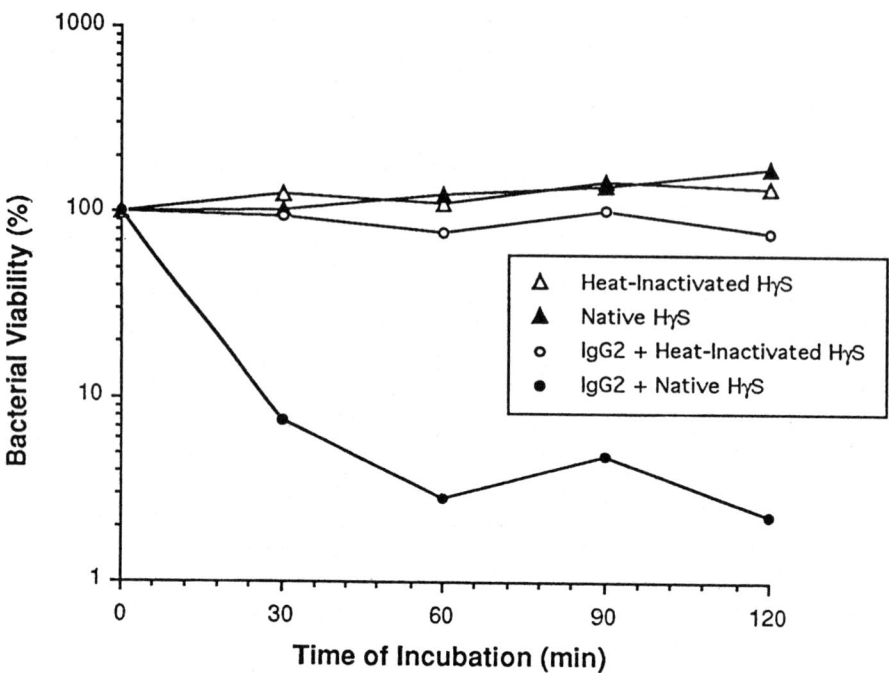

Fig. 4 Kinetics of neutrophil killing of *A. actinomycetemcomitans* serotype b in the presence of different opsonins. *A. actinomycetemcomitans* was incubated for specified intervals at 37°C with neutrophils (bearing the H131 allele of FcγRIIA) and native or heat-inactivated complement (HγS) alone or in combination with affinity-purified IgG2 from high-titer LJP serum. Bacterial viability is expressed as a percentage of the starting inoculum. (From Ref. 53.)

expressed on circulating neutrophils significantly influences the extent to which IgG2 antibodies in sera of LJP patients are capable of supporting phagocytosis and killing of *A. actinomycetemcomitans*.

A. actinomycetemcomitans O-PS as a Target for Opsonic Antibody

Our previous studies revealed that serospecific determinants in the O-PS of *A. actinomycetemcomitans* serotype b LPS are an important target for IgG antibodies in sera of LJP patients. Moreover, these LPS-directed antibodies were found to belong primarily to the IgG2 subclass. Given that IgG2 antibodies in LJP sera are opsonically active and react strongly with O-PS, we reasoned that O-PS may constitute an important target for opsonic IgG antibody. To test this hypothesis, we isolated O-PS specific IgG antibodies from high-titer LJP serum by chromatography on an affinity column constructed by coupling O-PS from serotype b LPS to CH Sepharose via a diaminopropane spacer (55). The IgG fraction of LJP serum applied to this column contained elevated levels of antibody to both LPS and at least two of the principal outer membrane proteins of *A. actinomycetemcomitans*. Specificity of the affinity-purified anti–O-PS antibodies for serotype b LPS was confirmed by double immunodiffusion analysis. These antibodies lacked detectable reactivity toward *A. actinomycetemcomitans* outer membrane proteins, as determined by immunoblot analysis. The affinity-purified anti–O-PS IgG was found to be enriched in content of IgG2. As determined using an IgG subclass-specific ELISA, the anti–O-PS contained 66.2% IgG2, compared with 37.0% IgG2 in the total IgG fraction from which these antibodies were isolated. On the other hand, content of IgG1 was nearly 50% lower in the affinity-purified fraction (26.5%) than in the total IgG fraction (50.6%). These results were consistent with our previous observation that IgG2 is the predominant subclass of IgG antibodies reactive toward *A. actinomycetemcomitans* serotype b LPS.

We examined the ability of the affinity-purified O-PS IgG, as well as IgG depleted of LPS-reactive antibody (but containing antibodies to outer membrane proteins), to support phagocytosis and killing of *A. actinomycetemcomitans* serotype b. Inasmuch as the anti–O-PS IgG contained significant amounts of IgG2 antibody, we once again employed neutrophils expressing the H131 allele of FcγRIIA as effectors. As depicted in Figure 5, O-PS–specific IgG antibodies were

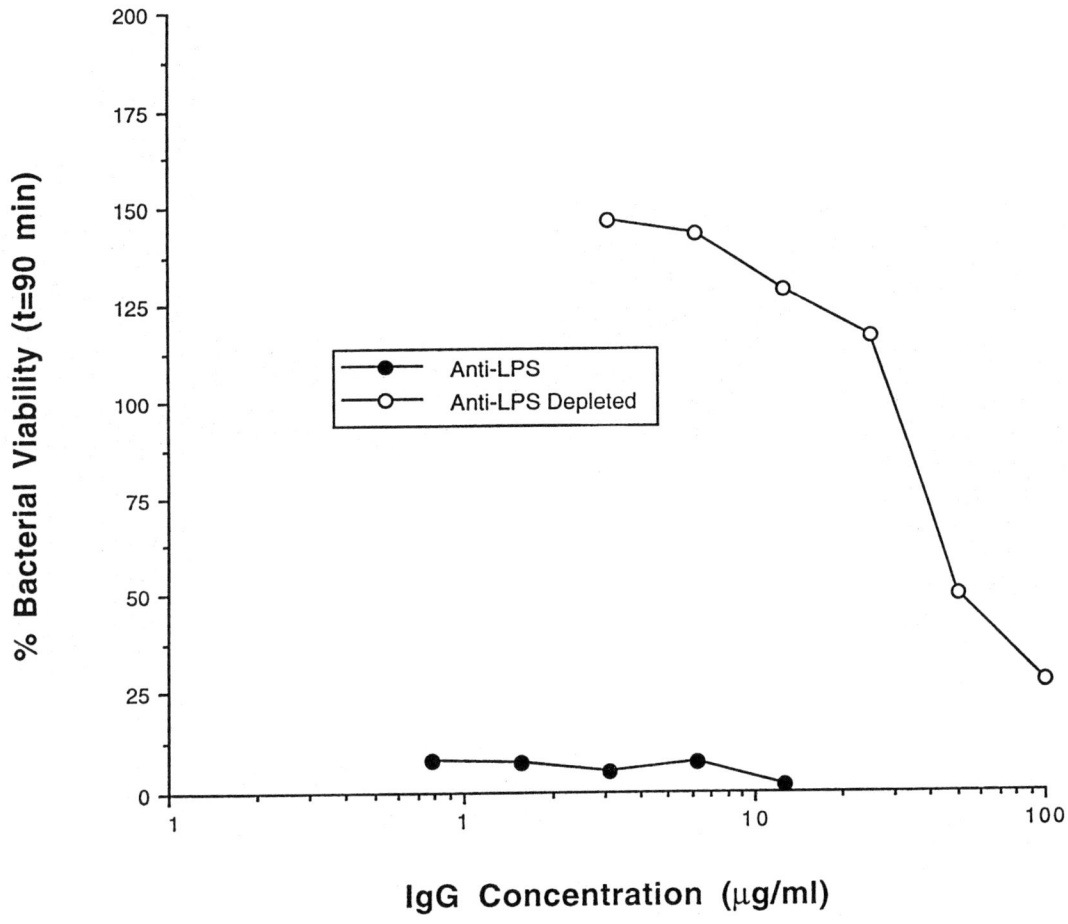

Fig. 5 Opsonic activities of anti-serotype b LPS IgG and IgG depleted of LPS-reactive antibody. Anti-LPS IgG was prepared from high-titer LJP serum by affinity chromatography on CH-Sepharose coupled to the O-PS of *A. actinomycetemcomitans* serotype b via a diaminopropane spacer. IgG depleted of LPS-reactive antibody was collected in the unbound fraction. *A. actinomycetemcomitans* Y4 (serotype b) was incubated with the indicated concentration of antibody, 5% (v/v) hypogammaglobulinemic serum, and neutrophils expressing the H131 allele of FcγRIIA for 90 minutes at 37°C. Viability is expressed as a percentage of the starting inoculum. (From Ref. 55.)

highly effective in promoting killing of *A. actinomycetemcomitans*, whereas IgG depleted of O-PS–reactive antibody was several orders of magnitude less active. These results indicate that the O-PS moiety of serotype b LPS is a major target for opsonic antibodies in LJP sera. Monoclonal IgG antibodies specific for the serotype b antigen are also opsonic for *A. actinomycetemcomitans* in the presence of complement (56).

Contribution of LPS-Directed IgG Antibodies to Host Defense Against *A. actinomycetemcomitans*

Ranney and coworkers observed that juvenile periodontitis patients whose sera contain precipitating IgG antibodies against *A. actinomycetemcomitans* have fewer teeth exhibiting attachment loss (57). Moreover, titers of serum IgG antibodies against *A. actinomycetemcomitans* serotype b were found to correlate inversely with attachment loss (58). These studies yielded indirect evidence favoring the hypothesis that IgG antibodies against *A. actinomycetemcomitans* are protective, but did not provide insight as to the nature of the antigen(s) recognized by these IgG antibodies, nor upon the mechanism whereby they might exert their protective effect. Subsequent studies revealed that IgG antibodies in sera of LJP patients play a critical role in promoting phagocytosis and killing of *A. actinomycetemcomitans* serotype b (47) and that the O-PS is a key target for such opsonic IgG antibodies (55). If the process of opsonization is important in the clearance of *A. actinomycetemcomitans* from host tissues and opsonic IgG antibodies are chiefly directed to the O-PS of serotype b LPS, then periodontitis patients whose sera contain high levels of IgG antibody against *A. actinomycetemcomitans* serotype b LPS should exhibit a lesser degree of periodontal disease activity than subjects with low levels of anti-LPS antibody. Recent studies indicate that this is, in fact, the case. Hence, generalized early-onset periodontitis patients with high levels of serum IgG antibody against *A. actinomycetemcomitans* serotype b LPS were characterized by significantly lower levels of attachment loss than patients with low to moderate anti-LPS antibody levels (59). These results suggest that the O-PS of serotype b represents an important target for IgG-mediated immunity against *A. actinomydetemcomitans*.

Future Studies

Serospecific determinants in the O-PS of *A. actinomycetemcomitans* serotype b represent a major target for IgG antibodies in sera of juvenile periodontitis patients. Moreover, these O-PS–directed antibodies have been shown to promote phagocytosis and killing of *A. actinomycetemcomitans* by human neutrophils. Opsonically active anti–O-PS IgG antibodies are enriched in content of IgG2, but also contain antibodies of the IgG1 subclass. Although we have demonstrated that affinity-purified IgG2 antibodies derived from sera of LJP patients exhibit opsonic activity, the antigenic specificity of these opsonic IgG2 antibodies was not defined. Hence, we are currently preparing IgG1 and IgG2 subclass-restricted fractions of O-PS–specific IgG and will evaluate the opsonic activity of each fraction. Moreover, we will examine the influence of genetic polymorphisms of FcγR upon the expression of such opsonic activity, as well as define FcγRIIA (and FcγRIIIB) genotype in patients with juvenile periodontitis. Finally, we are conducting longitudinal studies designed to ascertain when IgG antibodies against *A. actinomycetemcomitans* serotype b O-PS arise during the course of periodontal infection and how the emergence of such antibody influences the progression of clinical disease.

REFERENCES

1. Caton J. Periodontal diagnosis and diagnostic aids. Proceedings of the World Workshop in Clinical Periodontics. Chicago: The American Academy of Periodontology, 1989:I-1122.
2. Löe H, Brown LJ. Early onset periodontitis in the United States of America. J Periodontol 1991; 62:608–616.
3. Melvin W, Sandifer J, Gray J. The prevalence and sex ratio of juvenile periodontitis in a young racially mixed population. J Periodontol 1991; 62:330–334.
4. Neely AL. Prevalence of juvenile periodontitis in a circumpubertal population. J Clin Periodontol 1992; 19:367–372.
5. Schenkein HA. Genetics of early-onset periodontal diseases. In: Genco R, Hamada S, Lehner T, McGhee J, Mergenhagen S, eds. Molecular Pathogenesis of Periodontal Disease. Washington, DC: American Society for Microbiology, 1994:373–386.
6. Slots J, Reynolds HS, Genco RJ. Actinobacillus actinomycetemcomitans in human periodontal disease: a cross-sectional microbiological investigation. Infect Immun 1980; 29:1013–1020.
7. Zambon JJ. *Actinobacillus actinomycetemcomitans* in human periodontal disease. J Clin Periodontol 1985; 12:1–20.
8. Zambon JJ, Christersson LA, Slots J. *Actinobacillus actinomuycetemcomitans* in human periodontal disease: prevalence in patient groups and distribution of biotypes and serotypes within families. J Periodontol 1983; 54:707–711.
9. Slots J, Rosling B. Suppression of the periodontopathic microflora in localized juvenile periodontitis by systemic tetracycline. J Clin Periodontol 1983; 10:465–486.
10. Gunsolley JC, Ranney RR, Zambon JJ, Burmeister JA, Schenkein HA. *Actinobacillus actinomycetemcomitans* in families afflicted with periodontitis. J Periodontol 1990; 61:643–648.
11. Alaluusua S, Asikainen S, Lai C-H. Intrafamilial transmission of *Actinobacillus actinomycetemcomitans*. J Periodontol 1991; 62:207–210.
12. Zambon JJ, Slots J, Genco RJ. Serology of oral *Actinobacillus actinomycetemcomitans* and serotype distribution in human periodontal disease. Infect Immun 1983; 41:19–27.
13. Gmür R, McNabb H, van Steenbergen TJM, Baehni P, Mombelli A, van Winkelhoff AJ, Guggenheim B. Seroclassification of hitherto "non-typeable" *Actinobacillus actinomycetemcomitans* strains: evidence for a new serotype e. Oral Microbiol Immunol 1993; 8:116–120.
14. Ebersole JL, Taubman MA, Smith DJ, Genco RJ, Frey DE. Human immune responses to oral micro-organisms. I. Association of localized juvenile periodontitis (LJP) with serum antibody responses to *Actinobacillus actinomycetemcomitans*. Clin Exp Immunol 1982; 47:43–52.
15. Listgarten MA, Lai CH, Evian CI. Comparative antibody titers to *Actinobacillus actinomycetemcomitans* in juvenile periodontitis, chronic periodontitis, and periodontally healthy subjects. J Clin Periodontol 1981; 8:155–164.
16. Genco RJ, Zambon JJ, Murray PA. Serum and gingival fluid antibodies as adjuncts in the diagnosis of *Actinobacillus actinomycetemcomitans*-associated periodontal disease. J Periodontol 1985; 56(suppl):41–50.
17. Watanabe H, Marsh PD, Ivanyi L. Antigens of *Actinobacillus actinomycetemcomitans* identified by

immunoblotting with sera from patients with localized juvenile periodontitis and generalized severe periodontitis. Archs Oral Biol 1989; 34:649–656.
18. Wilson ME, IgG antibody response of localized juvenile periodontitis patients to the 29 kilodalton outer membrane protein of *Actinobacillus actinomycetemcomitans*. J Periodontol 1991; 62:211–218.
19. McArthur WP, Tsai C-C, Baehni PC, Genco RJ, Taichman NS. Leukotoxic effects of *Actinobacillus actinomycetemcomitans*. Modulation by serum components. J Periodontal Res 1981; 16:159–170.
20. Tsai C-C, McArthur WP, Baehni PC, Evian C, Genco RJ, Taichman NS. Serum neutralizing activity against *Actinobacillus actinomycetemcomitans* leukotoxin in juvenile periodontitis. J Clin Periodontol 1981; 8:338–348.
21. Farida R, Wilson M, Ivanyi L. Serum IgG antibodies to lipopolysaccharides in various forms of periodontal disease in man. Archs Oral Biol 1986; 11:711–715.
22. Sims TJ, Moncla BJ, Darveau RP, Page RC. Antigens of *Actinobacillus actinomycetemcomitans* recognized by patients with juvenile periodontitis and periodontally normal subjects. Infect Immun 1991; 59:913–924.
23. Califano JV, Schenkein HA, Tew JG. Immunodominant antigen of *Actinobacillus actinomycetemcomitans* Y4 in high-responder patients. Infect Immun 1989; 57:1582–1589.
24. Califano JV, Schenkein HA, Tew JG. Immunodominant antigens of *Actinobacillus actinomycetemcomitans* serotype b in early-onset periodontitis patients. Oral Microbiol Immunol 1992; 7:65–70.
25. Lu H, Califano JV, Schenkein HA, Tew JG. Immunoglobulin class and subclass distribution of antibodies reactive with the immunodominant antigen of *Actinobacillus actinomycetemcomitans* serotype b. Infect Immun 1993; 61:2400–2407.
26. McArthur WP, Stroup S, Wilson L. Detection and serotyping of *Actinobacillus actinomycetemcomitans* isolates on nitrocellulose ppaper blots with monoclonal antibodies. J Clin Periodontol 1986; 13:684–691.
27. Place DA, Scidmore NC, McArthur WP. Monoclonal antibodies to *Actinobacillus actinomycetemcomitans*. Infect Immun 1988; 56:1394–1398.
28. Saito S, Takamatsu N, Okahashi N, Matsunoshita N, Inoue M, Takehara T, Koga T. Construction of mutants of *Actinobacillus actinomycetemcomitans* defective in serotype b-specific polysaccharide antigen by insertion of transposon Tn916. J Gen Microbiol 1992; 138:1203–1209.
29. Amano K, Nishihara T, Shibuya N, Noguchi T, Koga T. Immunochemical and structural characterization of a serotype-specific polysaccharide antigen from *Actinobacillus actinomycetemcomitans* Y4 (serotype b). Infect Immun 1989; 57:2942–2946.
30. Kiley P, Holt SC. Characterization of the lipopolysaccharide from *Actinobacillus actinomycetemcomitans* Y4 and N27. Infect Immun 1980; 30:862–873.
31. Wilson ME, Schifferle RE. Evidence that the serotype b antigenic determinant of *Actinobacillus actinomycetemcomitans* Y4 resides in the polysaccharide moiety of lipopolysaccharide. Infect Immun 1991; 59:1544–1551.
32. Brondz I. Determination of acids in whole lipopolysaccharide and free lipid A from *Actinobacillus actinomycetemcomitans* and *Haemophilus aphrophilus*. J Chromatogr 1984; 308:19–29.
33. Brondz I, Olsen I. Chemical differences in lipopolysaccharides from *Actinobacillus* (*Haemophilus*) *actinomycetemcomitans* and *Haemophilus aphrophilus*: clues to differences in periodontopathogenic potential and taxonomic distinction. Infect Immun 1989; 57:3106–3109.
34. Masoud H, Weintraub ST, Wang R, Cotter R, Holt SC. Investigation of the structure of lipid A frm *Actinobacillus actinomycetemcomitans* strain Y4 and human clinical isolate PO 1021-7. Eur J Biochem 1991; 200:775–779.
35. Perry MB, MacLean LL, Gmür R, Wilson ME. Characterization of the O-polysaccharide structure of lipopolysaccharide from *Actinobacillus actinomycetemcomitans* serotype b. Infect Immun 1996; 64:1215–1219.
36. Perry MB, MacLean LL, Brisson J-R, Wilson ME. Structures of the antigenic O-polysaccharides of lipopolysaccharides produced by *Actinobacillus actinomycetemcomitans* serotypes a, c, d and e. Eur J Biochem 1996; 242:682–688.
37. Page RC, Sims TJ, Engel LD, Moncla BJ, Bainbridge B, Stray J, Darveau RP. The immunodominant outer membrane antigen of *Actinobacillus actinomycetemcomitans* is located in the serotype-specific high-molecular-mass carbohydrate moiety of lipopolysaccharide. Infect Immun 1991; 59:3451–3462.
38. Shibuya N, Amano K, Azuma J, Nishihara T, Kitamura Y, Noguchi T, Koga T. 6-Deoxy-D-talan and 6-deoxy-L-talan. Novel serotype-specific polysaccharide antigens from *Actinobacillus actinomycetemcomitans*. J Biol Chem 1991; 266:16318–16323.
39. Smith TF. IgG subclasses. Adv Pediatr 1992; 39:101–126.
40. Anderson CL, Looney RJ. Human leukocyte IgG Fc receptors. Immunol Today 1986; 7:264–266.
41. Hammarström I, Smith CI. IgG subclasses in bacterial infections. Monogr Allergy 1986; 19:122–133.
42. Hammarström I, Smith CI. IgG subclass changes in response to vaccination. Monogr Allergy 1986; 19:241–252.
43. Wilson ME and Hamilton RG. Immunoglobulin G subclass response of localized juvenile periodontitis patients to *Actinobacillus actinomycetemcomitans* Y4 lipopolysaccharide. Infect Immun 1992; 60:1806–1812.
44. Wilson M, Genco R. The role of antibody, complement and neutrophils in host defense against *Actinobacillus actinomycetemcomitans*. In: van Oss CJ, ed. Immunological Investigations. Vol. 18. Immunology and Immunopathology of the Alimentary Canal. New York: Marcel Dekker, 1989:187–209.
45. Miyasaki KT, Wilson ME, Brunetti AJ, Genco RJ. Oxidative and nonoxidative killing of *Actinobacillus actinomycetemcomitans* by human neutrophils. Infect Immun 1986; 53:154–160.
46. Miyasaki KT, Wilson ME, Genco RJ. Killing of *Actinobacillus actinomycetemcomitans* by the human neutrophil myeloperoxidase-hydrogen peroxide-chloride system. Infect Immun 1986; 53:161–165.

47. Baker PJ, Wilson M. Opsonic IgG antibody against *Actinobacillus acinomycetemcomitans* in localized juvenile periodontitis. Oral Microbiol Immunol 1989; 4:98–105.
48. van de Winkel JGJ, Anderson CL. Biology of human immunoglobulin G Fc receptors. J Leukocyte Biol 1991; 49:511–524.
49. van de Winkel JGJ, Capel PJA. Human IgG Fc receptor heterogeneity: molecular aspects and clinical implications. Immunol Today 1993; 14:215–221.
50. Salmon JE, Edberg JC, Brogle NL, Kimberly RP. Allelic polymorphisms of human Fcγ receptor IIA and Fcγ receptor IIIB: independent mechanisms for differences in human phagocyte function. J Clin Invest 1992; 89:1274–1281.
51. Bredius RGM, de Vries CEE, Troelstra A, van Alphen L, Weening RS, van de Winkel JGJ, Out TA. Phagocytosis of *Staphylococcus aureus* and *Haemophilus influenzae* type b opsonized with polyclonal human IgG1 and IgG2 antibodies. Functional hFcγRIIa polymorphism to IgG2. J Immunol 1993; 151:1463–1472.
52. Sanders LAM, Feldman RG, Voorhorst-Ogink MM, de Haas M, Rijkers GT, Capel PJA, Zegers BJM, van de Winkel JGJ. Human immunoglobulin G (IgG) Fc receptor IIA (CD32) polymorphism and IgG2-mediated bacterial phagocytosis by neutrophils. Infect Immun 1995; 63:73–81.
53. Wilson ME, Bronson PM, Hamilton RG. Immunoglobulin G2 antibodies promote neutrophil killing of *Actinobacillus actinomycetemcomitans*. Infect Immun 1995; 63:1070–1075.
54. Wilson ME, Kalmar JR. FcγRIIa (CD32): a potential marker defining susceptibility to localized juvenile periodontitis. J Periodontol 1996; 67:323–331.
55. Wilson ME, Bronson PM. Opsonization of *Actinobacillus actinomycetemcomitans* by IgG antibodies to the O polysaccharide of lipopolysaccharide. Infect Immun 1997; 65:4690–4695.
56. Yamaguchi N, Kawasaki M, Yamashita Y, Nakashima K. Koga T. Role of the capsular polysaccharide-like serotype-specific antigen in resistance of *Actinobacillus actinomycetemcomitans* to phagocytosis by human polymorphonuclear leukocytes. Infect Immun 1995; 63:4589–4594.
57. Ranney RR, Yanni NR, Burmeister JA, Tew JG. Relationship between attachment loss and precipitating serum antibody to *Actinobacillus actinomycetemcomitans* in adolescents and young adults having severe periodontal destruction. J Periodontol 1982; 53:1–7.
58. Gunsolley JC, Burmeister JA, Tew JG, Best AM, Ranney RR. Relationship of serum antibody to attachment level patterns in young adults with juvenile periodontitis or generalized secvere periodontitis. J Periodontol 1987; 58:314–320.
59. Califano JV, Gunsolley JC, Nakashima K, Schenkein HA, Wilson ME, Tew JG. Influence of anti-*Actinobacillus actinomycetemcomitans* Y4 (serotype b) lipopolysaccharide on severity of generalized early-onset periodontitis. Infect Immun 1996; 64:3908–3910.

45

Apoptotic Cell Death in Response to Lipopolysaccharide

Takashi Yokochi
Aichi Medical University, Nagakute, Aichi, Japan

INTRODUCTION

Cell death can generally proceed via necrosis or apoptosis (programmed cell death) (1–4). Necrosis is characterized by the formation of tubular lesions (pores) in the plasma membrane. On the other hand, apoptosis causes cell death in a way that differs morphologically and biochemically from necrosis. The common core mechanism of apoptosis is DNA fragmentation and morphological lesions, such as condensation and fragmentation of nucleus and cytoplasm.

Recently two common forms of cell death, necrosis and apoptosis, have been extensively characterized. Necrosis refers to the morphology most often seen when cells die from severe and sudden injury, such as ischemia, sustained hypothermia, or physical or chemical trauma (1). In necrosis there are early changes in mitochondrial shape and function, and the cell rapidly becomes unable to maintain homeostasis. The plasma membrane may be the major site of damage. Its contents are spilled into the surrounding tissue space and provoke an inflammatory response. On the other hand, apoptosis is a form of cell death in which the process is more subtle. It is often equated with programmed cell death. Apoptosis seems to be the most common morphology when cell death is physiologically determined or acceptable. Three main sets of events have been characterized in the dying cells (2). Morphologically, cytoplasmic condensation, condensation of nuclear material, and, at a later stage, massive cellular fragmentation has been reported. The most extensively studied biochemical event in apoptosis involves the fate of nuclear DNA. Synchronously with the compaction of chromatin observed monophonically, a striking fragmentation of the nuclear DNA into units of 180–200 base pairs corresponding to the size of a nucleosome has been suggested. Metabolically the synthesis of macromolecules is required in the process of apoptosis. Therefore, this type of cell death can often be inhibited by RNA synthesis inhibitors or protein synthesis inhibitors. The cell that dies must synthesize certain macromolecules in order to die. Although apoptosis was originally defined on the basis of its morphology, biochemical studies are unraveling molecular events in apoptotic cell death. Now its pathological or medical significance should be also elucidated.

Bacterial lipopolysaccharide (LPS) is present on the outer membranes of all gram-negative bacteria and causes the systemic inflammatory response syndrome (SIRS) and septic shock, which finally develop to multiorgan failure, accompanied by extensive cell death. Cell death in LPS-associated diseases, such as endotoxic shock, septic shock, and multiorgan failure, has not been fully characterized. The participation of apoptotic cell death in LPS-induced cell death should be elucidated. Further, the exact mechanism of apoptotic cell death in response to LPS should be clarified. The capacity to control apoptosis in response to LPS may have major medical consequences in the future. In this chapter I review our recent studies on apoptotic cell death in in vivo response to LPS.

APOPTOSIS IN RESPONSE TO LPS

LPS-Induced Apoptosis in Normal Mice

Apoptosis of Thymocytes

Previously we reported that the administration of LPS into normal mice induced marked apoptotic cell death in lymphoid organs. Especially the injection of LPS into mice induced DNA fragmentation in the thymus (5). LPS (100 μg) was injected intraperitoneally (i.p.) into normal mice. Fragmented DNA in the thymus was confirmed by agarose gel electrophoresis and laser flow cytometry. The DNA fragmentation of thymocytes was predominantly detected in LPS-injected young mice. The DNA fragmentation in the thymus was roughly dependent on injected dose of LPS and reached the peak level approximately 18 hours after the injection. Based on the fact that DNA fragmentation is one of the typical characteristics in apoptotic cell death, it was strongly suggested that LPS could induce apoptosis in the thymus in vivo. The addition of LPS into in vitro cultures of thymocytes did not cause DNA fragmentation, suggesting that LPS was unable to induce the apoptosis of thymocytes directly. The injection of LPS induced no significant DNA fragmentation of thymocytes in adrenalectomized mice. The injection of anti-tumor necrosis factor-alpha (TNF-α) antibody together with LPS significantly inhibited the appearance of DNA fragmentation in the thymus. It was, therefore, suggested that LPS-induced apoptosis of thymocytes occurring in the thymus might be mainly mediated by the adrenal hormones (5). Further, it was likely that TNF-α might participate in thymocyte apoptosis (5). Localization of thymocytes undergoing apoptosis by administration of LPS was studied using the in situ specific labeling of fragmented DNA (6). This method clearly stained the nuclei of immature thymocytes present at the cortex of the thymus (7). It was suggested that the cortex in the thymus is where the LPS-induced apoptosis occurs. Further, cell-surface marker analysis with laser flow cytometry was performed. The administration of LPS into normal mice mainly induced the apoptosis of CD4+8+ thymocytes (8). The mechanism of LPS-induced thymocyte apoptosis was studied in detail. The injection of anti-TNF-α antibody, or RU38486, a glucocorticoid receptor antagonist, together with LPS into normal mice definitely inhibited LPS-induced apoptosis of thymocytes. Addition of sera obtained one hour after the injection of LPS into in vitro cultures of thymocytes caused thymocyte apoptosis. The thymocyte apoptosis induced by the sera was prevented by the addition of either anti-TNF-α antibody or RU38486 into in vitro cultures. Further, it became clear that recombinant TNF-α and hydrocortisone collaborated in induction of the in vitro thymocyte apoptosis. Thus, the in vivo phenomenon of LPS-induced apoptosis of thymocytes could be reproducible by the in vitro experimental system using TNF-α, hydrocortisone, and their inhibitors. Therefore, it was strongly suggested that both TNF-α and glucocorticoid present in the circulation of mice injected with LPS might participate and collaborate as effector molecules in LPS-induced apoptosis of thymocytes. Recently we have found that interferon-gamma (IFN-γ) regulates LPS-induced thymocyte apoptosis in vivo (9). The simultaneous injection of anti-IFN-γ antibody and LPS into mice completely prevented thymocyte apoptosis. On the other hand, pretreatment of mice with recombinant IFN-γ markedly augmented LPS-induced thymoyte apoptosis. Thymocyte apoptosis augmented by IFN-γ occurred in the thymic cortex, and target cells undergoing apoptosis were CD4+8+ immature thymocytes. It seemed that IFN-γ amplified LPS-induced thymocyte apoptosis quantitatively. However, IFN-γ itself did not induce thymocyte apoptosis directly in vivo and in vitro. IFN-γ exhibited no synergistic action with effector molecules, such as TNF-α and glucocorticoid in vitro. Further, it was shown that IFN-γ did not enhance the susceptibility of thymocytes to apoptosis. Pretreatment of mice with IFN-γ significantly augmented serum TNF-α level and serum cortisol level in response to LPS (9). Therefore, we suggested that IFN-γ might augment LPS-induced thymocyte apoptosis through elevating serum TNF-α and cortisol levels.

Apoptosis of B Cells

Mouse B cells seemed to undergo apoptosis in response to LPS. LPS was administered into sheep red blood cell (SRBC)–primed mice, and the effect of LPS on SRBC-specific memory cells was investigated (10). Spleen cells from SRBC-primed mice that were injected with LPS exhibited much lower in vitro secondary plaque-forming cell (PFC) responses to SRBC than those from untreated SRBC-primed mice. The in vitro anti-SRBC response of the spleen cells to LPS was also reduced. The combination experiments of B cells and T cells fractionated from SRBC-primed mice which were injected with or without LPS demonstrated that the reduction of immune responses to SRBC after administration of LPS was caused by the defect of SRBC-specific B memory cells, but not T memory cells. B-cell-type rosette-forming cells (RFC) for SRBC (SRBC-specific B memory cells) markedly decreased after injection of LPS, while PFC as antibody-forming cells did not increase subsequently. Therefore, the reduction of RFC was not due to their differentiation of RFC into PFC. The lymphoid follicles in the spleens from mice injected with LPS were stained positively by the specific labeling of fragmented DNA. A large part of Ig$^+$ spleen cells sorted from SRBC-primed mice which were injected with LPS were also stained positively with the in situ specific labeling of fragmented DNA. The injection of hydrocortisone into SRBC-primed mice induced similar reduction of B memory cells. It was, therefore, suggested that LPS might induce apoptosis of B memory cells and regulate B cell memory in antigen-nonspecific manner (10). It is unclear whether or not naive B cells undergo apoptosis in response to LPS. Possibly, LPS may activate naive B cells and cause them to proliferate.

Recently, we have shown that an i.p. administration of LPS into normal mice induced marked reduction of CD5+ B cells in the peritoneal cavity (11). CD5+ B cells express the low level of CD5 antigen (12), which was originally demonstrated as a T-lymphocyte antigen. In adults, CD5+ B cells predominate in peritoneal and pleural cavities (13,14) and in the lamina propria of the gut (15) but account for only a few percent of spleen B cells and are rare in lymph nodes, thymus, and peripheral blood (12). In contrast to conventional B cells, CD5+ B cells characteristically possess self-replenishing capacity (16). The reduction of CD5+ B cells was not induced by an intravenous, subcutaneous, or oral administration of LPS, suggesting the requirement of direct interaction of LPS and CD5+ B cells for apoptosis. The reduction continued for about 10 days after the injection and recovered to the normal state about 14 days after the injection. A significant number of peritoneal cells from mice injected with LPS 1 day before were stained positively with the specific labeling of fragmented DNA. The reduction of peritoneal CD5+ B cells might be caused by their apoptotic cell death. The antibodies produced by CD5+ B cells are characterized by broad specificity and low affinity (17) and react against multivalent antigens of bacterial cell wall components (18,19), such as LPS (20). It was, therefore, studied how the apoptosis of CD5+ B cells by LPS affected the antibody production to LPS. The injection of LPS did not cause the production of antibody to LPS. On the other hand, the i.p. injection of heat-killed gram-negative bacteria did not induce the reduction of peritoneal CD5+ B cells and elicited the definite production of antibody to LPS. Therefore, LPS-induced apoptosis of CD5+ B cells seemed to be closely related to the failure in the production of anti-LPS antibody in i.p. administration of LPS because CD5+ B cells are known to produce anti-LPS antibody (20). In addition, the apoptosis of CD5+ B cells could not be induced by the i.p. administration of either hydrocortisone or TNF-α.

Apoptosis of Other Cell Types

The administration of LPS to normal mice induced the apoptosis of bone marrow cells as well as thymocytes and B cells. The apoptosis of bone marrow cells was supported by morphology of fragmented nuclei and the specific labeling of fragmented DNA, although fragmented DNA in the bone marrow was undetectable with agarose electrophoresis. It was assumed that the apoptosis of bone marrow cells would be due to the release of glucocorticoid by LPS because immature bone marrow cells are susceptible to glucocorticoid. The i.p. injection of LPS induced the reduction of T cells present in the peritoneal cavity. Both αβ T cells and γδ T cells disappeared 2 days after the injection. It is still unclear whether the reduction of those T cells is due to apoptosis or not.

LPS-Induced Apoptosis in D-Galactosamine–sensitized Mice as an Experimental Endotoxic Shock Model

Apoptosis of the Liver

It is of particular interest whether cell death of hepatocytes in clinical endotoxic shock is due to apoptosis or necrosis because necrosis of the liver was one characteristic of clinical endotoxic shock. In this section we describe the details of hepatocyte apoptosis in an experimental endotoxic shock model. When LPS (100 μg) was administered to normal mice, apoptotic cell death was hardly detected in the liver, kidney, heart, and lung. We utilized D-galactosamine (D-GalN)–sensitized mice as an endotoxic shock model. The sensitization with D-GalN extremely increases the sensitivity of animals to LPS and augments the lethal activity of LPS (21,22). A major feature of the D-GalN model is now clearly recognized to be an increased sensitivity to TNF-mediated effects (22). LPS-induced lethality in the system is characterized by liver failure, accompanied by severe hepatic injuries. The lethal action of LPS on D-GalN–sensitized mice is usually thought of as an experimental model for clinical endotoxic shock or septic shock (22). The characterization of LPS-induced apoptosis in D-GalN–sensitized mice might be also useful for understanding the role of apoptotic cell death in clinical endotoxin shock or septic shock. Mice were injected i.p. with D-GalN and LPS 3, 7, or 16 hours before, and apoptotic cells in various organs were followed by the in situ specific labeling of fragmented DNA (23). In the livers of mice injected with D-GalN and LPS, a number of nuclei of hepatocytes were stained 7 hours after the injection. Many fewer hepatocytes stained positively 3 and 24 hours after the injection. There were no positively stained nuclei in the livers of mice injected with either D-GalN or LPS. The apoptotic cell death of hepatocytes was detectable only in the livers from mice injected with D-GalN and LPS. Fragmented DNA detected by agarose gel electrophoresis was also found in the livers of mice injected with LPS and D-GalN, while no significant fragmented DNA was detected in the livers from mice injected with either D-GalN or LPS. Similar results were reported by the other group (24). Mice were injected i.p. with D-GalN and LPS, and blood was taken from those mice 1 hour later. This serum (1 ml) was injected i.p. together with D-GalN. The liver was inspected 7 hours after injection of the sera by the in situ specific labeling of fragmented DNA. The sera significantly induced hepatocyte apoptosis in the liver, whereas normal control sera did not. DNA fragmentation in the livers of those mice was also confirmed by agarose gel electrophoresis. It was suggested that a humoral factor appearing 1 hour after the injection of LPS might participate in induction of hepatocyte apoptosis. It has been reported that TNF-α is released into the serum 1 hour after the injection of LPS (22), and that LPS-induced hepatic injuries in D-GalN-sensitized mice is mediated by TNF-α. Therefore, we studied whether or not anti–TNF-α antibody inhibited LPS-induced apoptosis of the livers in D-GalN–sensitized mice. Mice were injected i.p. with D-GalN, LPS, and anti–TNF-α antibody, and apoptotic cells were inspected 7 hours after injection with the in situ specific labeling of fragmented DNA. Anti–TNF-α antibody completely inhibited the induction of LPS-induced apoptosis in the liver, whereas anti–IFN-γ or IL-1 antibody did not protect it. The participation of TNF-α in LPS-induced hepatocyte apoptosis of D-GalN–sensitized mice was suggested. In addition, the sera that were used for the transfer experiment contained TNF-α at a concentration of >5000 pg/ml. From the above experiment the possibility was raised that TNF-α might play a crucial role in LPS-induced apoptosis of hepatocytes. It was therefore examined whether administration of TNF-α caused the apoptosis in the livers of D-GalN–sensitized mice as well as LPS. Mice were injected i.p. with TNF-α (50 ng) and D-GalN. Administration of TNF-α into D-GalN–sensitized mice definitely caused apoptotic cells in the liver. Fragmented DNA was also found in the livers of mice injected with TNF-α and D-GalN. Further, it has been reported that Fas antigen plays an important role in induction of apoptosis in the liver (25). It was studied whether Fas antigen might play a role in the induction of hepatic apoptosis in D-GalN–sensitized mice by using Fas antigen–negative MRL/*lpr* mice (26). MRL/*lpr* mice were injected i.p. with LPS and D-GalN. A large number of positively stained cells were diffusely present in the liver. Fragmented DNA was also detectable by agarose gel electrophoresis. Therefore, Fas antigen was not involved in apoptosis of hepatocytes in response to LPS and D-GalN.

Apoptosis of the Kidney

Apoptotic cells were detected in the medullary pyramid of the kidney by the administration of LPS into D-GalN-sensitized mice (23). The details of apoptosis of the kidney in D-GalN-sensitized mice was not fully clarified.

Apoptosis of the Lymphoid Organs

LPS induced the apoptosis of the thymus, spleen and lymph nodes in D-GalN-sensitized mice. The apoptotic cells were localized diffusely in those organs (23).

LPS-Induced Apoptosis in Generalized Shwartzman Reaction as an Experimental Disseminated Intravascular Coagulation Model

Apoptosis of Vascular Endothelial Cells

Recently we reported the apoptosis of vascular endothelial cells in the generalized Shwartzman reaction (GSR) (27). GSR was induced in mice by two consecutive injections of LPS (27). The optimal dose of LPS (5μg) was injected intradermally into the footpads of mice as a preparative injection for the priming of GSR. Twenty-four hours later, a provocative injection of LPS (300 μg) was administered intravenously. More than 80% of the mice were dead 12 hours after the provocative injection of LPS. GSR is usually thought as an experimental model for clinical disseminated intravascular coagulation (DIC) model. Participation of apoptotic cell death in GSR was examined. Vascular endothelial cells in various organs, such as the livers, kidneys, and lungs, of mice were stained positively by the in situ specific labeling of fragmented DNA. Histologically fragmented nuclei of vascular endothelial cells were also seen. It was, therefore, suggested that apoptotic cell death might participate in the development of vascular endothelial cell damage in GSR. Simultaneous administration of anti–IFN-γ antibody and LPS completely blocked apoptosis of vascular endothelial cells. Priming with recombinant IFN-γ in replacement of LPS could produce apoptosis of vascular endothelial cells. It was suggested that IFN-γ might play a critical role in the sensitization of endothelial cells for LPS-induced apoptosis. However, the effector molecule that induces apoptosis of vascular endothelial cells in the provocative injection of LPS is still unidentified. In fact, the administration of anti–TNF-α antibody and LPS in the provocative injection partially inhibited the apoptosis of vascular endothelial cells. We could not exclude the participation of other unknown molecules in the apoptosis of vascular endothelial cells. In the mechanism of apoptosis of vascular endothelial cells the participation of adhesion molecules in systemic vascular injuries of GSR was studied (28). Intercellular adhesion molecule-1 (ICAM-1) was expressed on vascular endothelial cells in GSR-induced mice. On the other hand, LPS did not affect the expression of vascular cell adhesion molecule (VCAM-1). The preparative injection of LPS induced ICAM-1 expression on vascular endothelial cells, and the provocative injection of LPS for GSR augmented it further. The simultaneous administration of anti–IFN-γ antibody in the preparative injection of LPS completely inhibited ICAM-1 expression on vascular endothelial cells. The injection of recombinant IFN-γ in replacement of LPS resulted in ICAM-1 expression. The administration of anti–ICAM-1 antibody together with the provocative injection of LPS significantly blocked the apoptosis of vascular endothelial cells in GSR-induced mice. It was suggested that ICAM-1 expression on vascular endothelial cells might be involved in their apoptosis in GSR and that it might be regulated by IFN-γ.

Apoptosis of Renal Tubular Cells

Renal tubular cells in kidneys of GSR-induced mice were stained positively by the in situ specific labeling of fragmented DNA. The positive staining of renal tubules was focal. It was suggested that apoptotic cell death might also participate in the development of acute tubular necrosis in GSR. Simultaneous administration of anti–IFN-γ antibody in the injection of LPS significantly blocked apoptosis of renal tubular cells. Priming with recombinant IFN-γ in replacement of LPS could produce apoptosis of tubular cells as well as vascular endothelial cells. IFN-γ might play a critical role in the sensitization of renal tubular cells for LPS-induced apoptosis. The provocative injection of TNF-α in replacement of LPS produced marked apoptosis of tubular cells in LPS-primed mice. The intensity of the apoptosis induced by TNF-α in replacement of LPS was more marked. Therefore, effector molecules for apoptosis of tubular cells might be TNF-α. This was supported by the finding that priming of mice

with LPS resulted in the expression of TNF receptors on tubular cells. ICAM-1 was also expressed on renal tubular cells in GSR-induced mice. However, the role of ICAM-1 in the induction of apoptosis of tubular cells was still unclear.

DISCUSSION

A series of studies in our laboratory demonstrated that apoptotic cell death may be involved in LPS-induced cell death. We could not find any participation of necrosis in cell death in in vivo response to LPS. However, the participation of necrosis cannot be excluded because it is difficult to discriminate between apoptosis and necrosis. It is well known that necrosis of the liver and renal tubules is a characteristic of endotoxic shock, DIC, and septic shock. Recently extensive studies on cell death have unraveled apoptotic cell death. Based on new evidence and our findings, it is strongly suggested that LPS-induced tissue injury is mainly due to apoptotic cell death, but not necrotic cell death.

LPS leads to the apoptosis of various cell types. Cell types that undergo apoptosis in response to LPS are clearly dependent on the host condition. LPS-induced apoptosis considerably varies in normal, D-GalN–sensitized and GSR-induced mice. In fact, the administration of LPS into normal mice induces the apoptosis of lymphoid organs, such as the thymus, spleen, and bone marrow. Target cells are mainly lymphocytes, especially thymocytes and B cells. Glucocortocoids and TNF-α are suggested to be effector molecules. In D-GalN–sensitized mice used as an endotoxic shock model, the injection of LPS exclusively results in the apoptosis of hepatocyte, which can be reproduced by administration of TNF-α. Furthermore, LPS-induced apoptosis in GSR-induced mice as an experimental DIC model is characterized by the apoptosis of vascular endothelial cells and renal tubular cells. Several molecules seem to be involved in the apoptosis of vascular endothelial cells and renal tubular cells. The participation of cytokines and adhesion molecules has been suggested in this system. ICAM-1 molecules are thought to play a critical role in apoptosis in experimental DIC models. From a series of studies on LPS-induced apoptosis, it is supported that LPS may induce apoptosis of various cell types through different mechanisms. In fact, there are a number of experimental systems for characterizing the pathophysiology of LPS. LPS-induced cell death in an in vitro culture system may be different from that in an in vivo experimental system. Necrosis mediated by nitric oxide seems to be dominant in LPS-induced in vitro cytotoxicity of peritoneal exudate macrophages (unpublished). Thus, the participation of apoptosis in LPS-induced cell death must be characterized in each experimental system in vivo and in vitro, respectively.

Clinically endotoxin causes SIRS and septic shock, which finally develop to multiorgan disturbance and failure, accompanied by extensive cell death. However, the nature of cell death in clinical endotoxin-associated diseases has not been fully characterized. The cell death in LPS-induced tissue injuries has been clinically referred to as hepatocyte necrosis or renal tubular necrosis. Recent studies support that the injury of hepatocytes and vascular endothelial cells may be responsible for apoptotic cell death, but not for necrotic cell death. The characterization of LPS-induced apoptosis might also be useful for understanding the pathophysiology of clinical endotoxin shock or septic shock. The capacity to control apoptosis in response to LPS might have major medical consequences in the future.

ACKNOWLEDGMENT

This work was supported by a grant from the Ministry of Education, Science and Culture of Japan.

REFERENCES

1. Cohen JJ. Programmed cell death in the immune system. Adv Immunol 1991; 50:55–85.
2. Goldstein P, Ojcius DM, Young JDE. Cell death mechanisms and the immune response. Immunol Rev 1991; 121:29–65.
3. Kerr JF, Wyllie AH, Currie AR. Apoptosis: a basic biological phenomenon with wide ranging implications in tissue kinetics. Br J Cancer 1972; 26:239–247.
4. Wyllie AH, Kerr JFR, Currie AR. Cell death: the significance of apoptosis. Int Rev Cytol 1980; 68:251–306.
5. Zhang YH, Takahashi K, Jiang GZ, Kawai M, Fukada M, Yokochi T. In vivo induction of apoptosis (programmed cell death) in mouse thymus by administration of lipopolysaccharide. Infect Immun 1993; 61:5044–5048.
6. Gavrieli Y, Sherman Y, Ben-Sasson SA. Identification of programmed cell death in situ via specific labeling of nuclear DNA fragmentation. J Cell Biol 1992; 119:493–501.
7. Zhang SM, Morikawa A, Takahashi K, Jiang GZ, Kato Y, Sugiyama T, Kawai M, Fukada M, Yokochi T. Localization of apoptosis (programmed cell death) in mice by administration of lipopolysaccharide. Microbiol Immunol 1994; 38:669–671.
8. Kato Y., Morikawa A, Sugiyama T, Koide N, Jiang G-Z, Takahashi K, Yokochi T. Role of tumor necrosis factor-a and glucocorticoid on lipopolysaccharide (LPS)-induced apoptosis of thymocytes. FEMS Immunol Med Microbiol 1995; 12:195–204.
9. Kato Y, Morikawa A, Sugiyama T, Koide N, Jiang G-Z, Lwin T, Yoshida T, Yokochi T. Augmentation of lipopolysaccharide-induced thymocyte apoptosis by interferon-γ. Cell Immunol. 1997; 177:103–108.
10. Yokochi T, Kato Y, Sugiyama T, Koide N, Morikawa A, Jiang G-Z, Kawai M, Yoshida T, Fukada M, Takahashi K. Lipopolysaccharide induces apoptotic cell death of B memory cells and regulates B cell memory in antigenic nonspecific manner. FEMS Immunol Med Microbiol 1996; 15:1–8.
11. Paeng N, Kido N, Kato Y, Sugiyama T, Koide N, Naruse M, Jiang G-Z, Lwin T, Yoshida T, Yokochi T. Marked reduction of mouse peritoneal CD5$^+$B cells by intraperitoneal administration of lipopolysaccharide. Infect Immun 1997; 65:122–126.
12. Hayakawa K, Hardy RR, Parks DR, Herzenberg LA. The "Ly-1 B" cell subpopulation in normal immunodefective, and autoimmune mice. J Exp Med 1983; 157:202–218.
13. Hayakawa K, Hardy RR, Herzenberg LA. Peritoneal Ly-1 B cells: Genetic control, autoantibody production, increased lambda light chain expression. Eur J Immunol 1986; 16:450–456.
14. Marcos MA, Huetz F, Pereira P, Andreu JL, Martinez AC, Coutinho A. Further evidence for coelomic-associated B lymphocytes. Eur J Immunol 1989; 19:2031–2035.
15. Murakami M, Tsubata TT, Shinkura R, Nishitani S, Okamoto M, Yoshioka H, Usui T, Miyawaki S, Honjo T. Oral administration of lipopolysaccharides activates B-1 cells in the peritoneal cavity and the lamina propria of the gut and induces autoimmune symptoms in an autoantibody transgenic mouse. J Exp Med 1994; 180:111–121.
16. Kantor AB. The development and repertoire of B-1 cells (CD5 B cells). Immunol Today 1991; 12:389–391.
17. Murakami M, Honjo T. Involvement of B-1 cells in mucosal immunity and autoimmunity. Immunol Today 1995; 16:534–539.
18. Allison AC, Nawata Y. Cytokines mediating the proliferation and differentiation of B-1 lymphocytes and their role in ontogeny and phylogeny. Ann NY Acad Sci 1992; 651:200–219.
19. Sidman CL, Shultz, LD, Hardy RR, Hayakawa CK, Herzenberg LA. Production of immunoglobulin isotypes by Ly-1$^+$ B cells in viable motheaten and normal mice. Science 1986; 232:1423–1425.
20. Su SD, Ward MM, Apicella MA, Ward RE. The primary B cell response to the O/core region of bacterial lipopolysaccharide is restricted to the Ly-1 lineage. J Immunol 1991; 146:327–331.
21. Galanos C, Freudenberg MA, Reutter W. Galactosamine-induced sensitization to the lethal effects of endotoxin. Proc Natl Acad Sci USA 1979; 76:5939–5943.
22. Galanos C, Freudenberg MA, Katschinski T, Salomao R, Mossmann H, Kumazawa Y. Tumor necrosis factor and host response to endotoxin. In: Ryan JL, Morrison DC, eds. Bacterial Endotoxic Lipopolysaccharide. Vol. 2. Boca Raton, FL: CRC Press, 1992:75–104.

23. Morikawa A, Sugiyama T, Kato Y, Koide N, Jiang G-Z, Takashashi K, Tamada Y, Yokochi T. Apoptotic cell death in the response of D-galactosamine-sensitized mice to lipopolysaccharide as an experimental endotoxic shock mode. Infect Immun 1996; 64:734–738.
24. Leist M, Gantner F, Bohlinger I, Tieg G, Germann PG, Wendel A. Tumor necrosis factor-induced hepatocyte apoptosis precedes liver failure in experimental murine shock models. Am J Pathol 1995; 146:1220–1234.
25. Ogasawara J, Watanabe-Fukunaga R, Adachi M, Matsuzawa A, Kasugai T, Kitamura Y, Itoh N, Suda T, Nagata S. Lethal effect of the anti-Fas antibody in mice. Nature 1993; 364:806–809.
26. Adachi M, Watanabe-Fukunaga R, Nagata S. Aberrant transcription caused by the insertion of an early transposable element in an intron of the Fas antigen gene of *lpr* mice. Proc Natl Acad Sci USA 1993; 90:1756–1760.
27. Koide N, Abe K, Narita K, Kato Y, Sugiyama T, Jiang G-Z, Yokochi T. Apoptotic cell death of vascular endothelial cells and renal tubular cells in the generalized Shwartzman reaction. FEMS Immunol Med Microbiol 1996; 16:205–211.
28. Koide N, Abe K, Narita K, Kato Y, Sugiyama T, Yoshida T, Yokochi T. Expression of intercellular adhesion molecule-1 (ICAM-1) on vascular endothelial cells and renal tubular cells in the generalized Shwartzman reaction as an experimental disseminated intravascular coagulation model. FEMS Immunol Med Microbiol 1997; 18:67–74.

46

Nontoxic RsDPLA As a Potent Antagonist of Toxic Lipopolysaccharide

Nilofer Qureshi
William S. Middleton Memorial Veterans Hospital, and University of Wisconsin–Madison, Madison, Wisconsin

Bruce W. Jarvis
Promega Corporation, Madison, Wisconsin

Kuni Takayama
William S. Middleton Memorial Veterans Hospital, Madison, Wisconsin

TOXIC LIPOPOLYSACCHARIDE, ENDOTOXIC SHOCK, AND ANTAGONISM BY RsDPLA

Lipopolysaccharide (LPS) is an amphipathic glycolipid found on the outer surface of the outer membrane of gram-negative bacteria (1,2). Isolated LPS has been shown to have a wide range of immunological and pathophysiological effects (3,4). The LPS of the Enterobacteriaceae consists of three structural regions: the O-antigen (strain-specific repeating polysaccharide; confers serologic specificity), the core oligosaccharide (consisting primarily of heptose and 2-keto-3-deoxyoctonate [Kdo]; relatively conserved region), and lipid A (lipophilic and highly conserved region). We established the structure of the lipid A moiety of LPS from *Salmonella* spp. in 1983 (5).

Sepsis afflicts approximately 400,000 Americans per year with an associated mortality of approximately 35–50%. One of the most important initiating agents in gram-negative sepsis is thought to be LPS. For the past 3 years, we have studied how toxic LPS interacts with cells that cause the inflammatory process (macrophages, monocytes, and neutrophils) resulting in death of animals due to gram-negative septic shock. In this regard, we have studied the properties of a special nontoxic pentaacyl diphosphoryl lipid A derived from the LPS of *Rhodobacter sphaeroides* (RsDPLA). We have found that RsDPLA is a powerful antagonist of toxic LPS in both in vivo and in vitro experimental conditions. We have concluded that it has the potential to be an effective therapeutic agent in blocking the initiation of gram-negative sepsis. As a competitive inhibitor of activation of immune cells by toxic LPS, RsDPLA has become a useful tool to study LPS-induced signal transduction.

In this chapter we will discuss recent studies on the antagonistic properties of RsDPLA, other nontoxic LPS and their lipid A derivatives, as well as synthetic lipid A analogs. Earlier studies have been reviewed by Qureshi et al. (6).

ANTAGONISTS OF TOXIC LPS

A list of known antagonists of toxic LPS ig given in Table 1. These compounds have been studied for the

Table 1 Nontoxic LPS and Lipid A That Are Antagonists of Toxic LPS

Deacylated LPS[a]	Strong antagonist in human cells (7)
RcapLPS[b]	Strong antagonist in human cells (8)
Lipid X[c]	Weak antagonist (9)
Lipid IV$_A$[d]	Strong antagonist in human cells but not in murine cells (10,11)
RsDPLA[e]	Strong antagonist in both human and murine cells (6,11,12)

[a]RcLPS treated with acyloxyacyl hydrolase to remove the two acyloxy-linked normal fatty acids in the lipid A moiety. The lipid A moiety is identical to lipid IV$_A$.
[b]LPS of *Rhodobacter capsulatus*.
[c]Monosaccharide lipid A precursor.
[d]Incomplete lipid A obtained from the temperature-sensitive mutant of *Salmonella typhimurium*. It is also referred to as lipid Ia.
[e]Pentaacyl diphosphoryl lipid A derived from the LPS of *R. sphaeroides*.

past decade, and some have proven to be more effective than others. Deacylated LPS represents RcLPS that has been treated with acyloxyacyl hydrolase to yield a product with the lipid A moiety that is identical to lipid IV$_A$ (7). It was shown to have good antagonistic activity. RcapLPS is similar to nontoxic LPS from *R. sphaeroides* in both structure and biological activity (8). Lipid X appears to lack the structural features necessary for strong antagonistic activity (9). Although lipid IV$_A$ is inactive in human cells and can function as a LPS antagonist in human cells, it can directly activate murine cells (10,11). RsDPLA is a potent antagonist of toxic LPS in both human and murine cells (6,11,12). It is by far the most extensively studied LPS antagonist.

Two synthetic monosaccharide lipid A analogs based on the structure of lipid X were synthesized and tested for antagonistic activity. SDZ 880.431 was a weak antagonist in that it was only threefold more potent than lipid X (13). SDZ MRL 953 was more effective in that it induced tolerance to LPS and blocked the induction of TNF-α and IL-6 by toxic LPS in human subjects (14). E5531 is a synthetic lipid A based on the structure of the lipid A moiety of LPS of *R. sphaeroides* and *R. capsulatus* (15). It was shown to be a potent antagonist of toxic LPS. Like the RsDPLA, E5531 has the critical structural features necessary for strong antagonistic activity. It is a pentaacyl diphosphoryl lipid A that contains relatively short hydroxy fatty acids (OHC$_{10}$).

We have previously shown the RsDPLA is an LPS antagonist that is effective in both human and murine cell types (6,16). It blocked the binding of LPS to the putative LPS receptor, thus not allowing the induction of TNF-α, IL-1β, tumor necrosis receptor type 2 (TNFR-2), interferon-γ–inducible protein-10 (IP-10), D-3, and D-8 in macrophages by LPS. RsDPLA inhibited the LPS-induced protein tyrosine phosphorylation in macrophages. It also blocked the induction of TNF-α, IL-1β, and IL-6 by LPS in vitro (macrophages) and in vivo (mice). RsDPLA protected normal and galactosamine-sensitized mice against the lethal effects of LPS (6,16).

CHEMICAL STRUCTURE OF AN LPS AGONIST (ReLPS) AND ANTAGONIST (RsDPLA)

We use the highly purified and well-characterized deep rough chemotype LPS from *Escherichia coli* D31m4 (hexaacyl diphosphoryl lipid A, ReLPS) as the highly toxic agonist (17–20) (Fig. 1A). For the antagonist we use the nontoxic RsDPLA (Fig. 1B). Our preparations of RsDPLA are also highly purified and well characterized. The complete structure was established by soft ion mass spectrometry and NMR (19). This structure was reconfirmed by further and more extensive mass spectral analyses (20) after our reported structure was questioned by Christ et al. (21).

USE OF RsDPLA TO STUDY THE SIGNALING CASCADE

Background

It is generally accepted that the first step in the LPS activation of responding cells (leading to the induction of gene expression and ultimately to the release of cytokines) is the binding of this ligand to a specific membrane receptor(s) (22,23). CD14 is a cell surface 55 kDa glycosyl phosphatidyl inositol–linked protein on macrophages, monocytes, and neutrophils (24,25). Other LPS-responding cell types (e.g., lymphocytes, endothelial cells, and fibroblasts) express very low levels or are devoid of CD14. CD14 is the receptor protein that recognizes and binds LPS originating from the complexes formed between monomeric LPS and circulating LPS-binding protein (LBP). Presently it is thought that CD14 first forms the LPS-CD14 complex and transfers the LPS to as-yet-unidentified physiological LPS receptor(s) to initiate signal transduction. Coprecipitation of membrane-bound CD14 with src protein kinase activities suggests that CD14 is closely associated with a signaling protein tyrosine kinase (26).

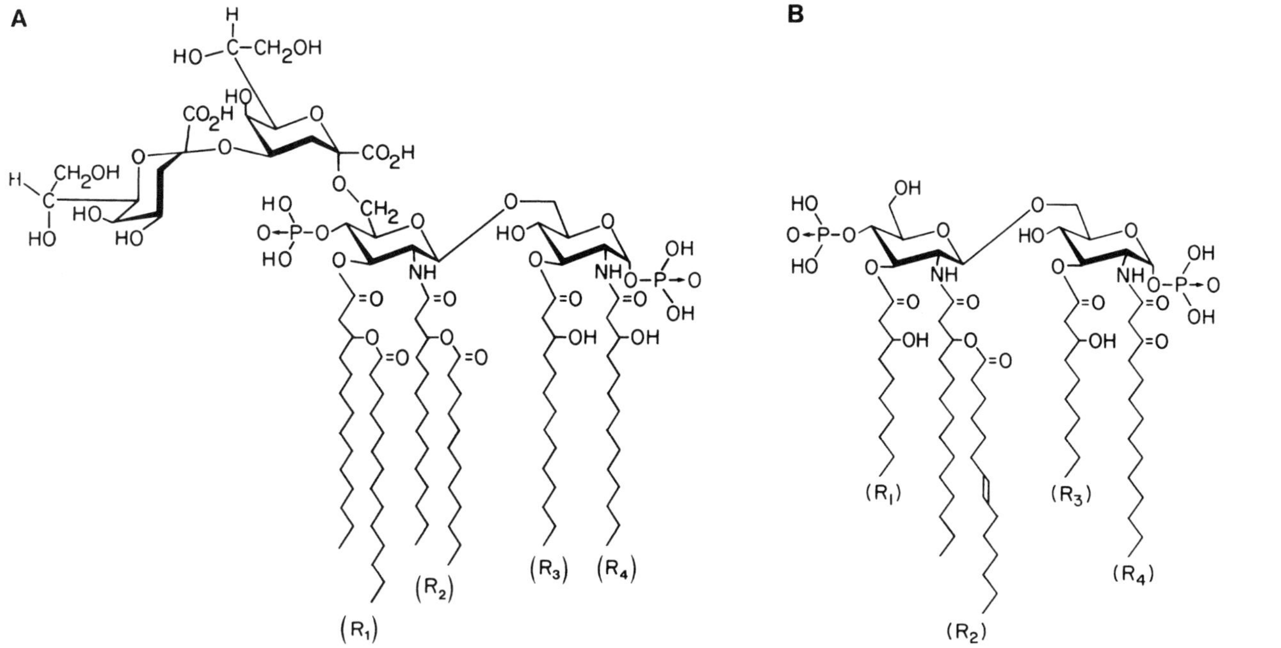

Fig. 1 Structures of ReLPS from *E. coli* D31m4 (A) and RsDPLA derived from the LPS of *R. sphaeroides* ATCC 17023 (B).

Over the last decade or so, many cellular proteins have been reported to bind LPS and lipid A (27–33). However, with the exception of CD14, these proteins have not been characterized.

CD14-Dependent Pathway for the Activation of Immune Cells

We showed that a 1:1 molar ratio of RsDPLA:[^{125}I]LPS was sufficient to block [^{125}I]LPS binding to the murine macrophage-like cell line J774.1 (34). RsDPLA did not induce TNF-α or IL-6 release by the J774.1 cells. However, RsDPLA inhibited the induction of these two cytokines by ReLPS from *Salmonella minnesota* R595. Maximal inhibition occurred when the ReLPS to RsDPLA mass ratio was 1:10. The binding study showed that the RsDPLA strongly inhibited the binding of [^{125}I]ASD-ReLPS [a 2-(*p*-azidosalicylamido)ethyl-1,3′-dithioproprionate derivative] to the cells, and that this binding inhibition was about 16-fold greater than that by the unlabeled ReLPS on a mass basis. These results showed that RsDPLA is competing with ReLPS for binding sites on the cell surface, including the physiological receptors linked to signal transduction in the J774.1 cells. Membrane-bound CD14 is thought to mediate this binding to the cell.

It is well documented that LBP and membrane-associated CD14 are involved in the activation of macrophages by LPS (23–27). Both LBP and CD14 form complexes with LPS. In an in vitro experiment we analyzed the complex formation of [^{3}H]ReLPS with LBP and soluble CD14 (sCD14) by native gel electrophoresis and gel filtration mode of high-performance liquid chromatography (HPLC) (a newly developed method) (35). RsDPLA was tested as a potential antagonist for the complex formation. By HPLC analysis, the formation of the [^{3}H]ReLPS-LBP and [^{3}H]ReLPS-sCD14 complexes were observed. RsDPLA inhibited the formation of these complexes at a 1:1 ratio of RsDPLA:ReLPS. The formation of the complex between [^{3}H]ReLPS and the truncated form of sCD14 (sCD15$_{1-152}$) was also shown, and this was inhibited by RsDPLA. These results suggested that RsDPLA blocks the activation of macrophages by LPS at the very beginning of the process. It effectively competes with the toxic LPS for the binding sites on both LBP and CD14 and prevents the formation of the complexes. HPLC analysis showed that only LBP and not sCD14 was bound to the aggregated forms of ReLPS. Soluble CD14 appears to bind only monomeric ReLPS. There was no evidence for the formation of an ReLPS-LBP-sCD14 ternary complex. This newly developed HPLC method should be useful in the purification and analysis of the various LPS-protein complexes (35).

CD14-Independent Pathway for the Activation of Immune Cells

The murine bone marrow stromal ST2 cells were used since these cells respond to LPS but do not express CD14 mRNA (36). These cells expressed IL-6 mRNA and induced IL-6 in response to LPS (at 1–10 ng/ml). RsDPLA inhibited the LPS-induced expression of IL-6 mRNA in a dose-dependent manner. These results suggested that LPS and RsDPLA are recognized by the same receptor(s) on the ST2 cells and that this receptor(s) functions independently of membrane-bound CD14 or sCD14.

LPS-Induced Gene Expression and Protein Tyrosine Phosphorylation in Macrophages

One of the earliest biochemical consequences of LPS binding to macrophages is the activation of the kinase cascades, which are thought to be essential for gene expression. Within 15 seconds, autophosphorylation of 41(lyn), 42, and 45 kDa proteins are detectable. The 42 and 45 kDa proteins have been identified as members of the ERK (extracellular-signal–regulated kinase) family of kinases, which is one of three known groups of mammalian mitogen-activated protein kinases (MAPKs) (37,38). The other two groups are p38 (p38 kinase) and JNK (c-Jun NH2 terminal kinase). The JNK family includes the stress-activated protein (SAPK). LPS strongly activates the ERK 1 and 2, p28, and the JNK (39,40). In contrast, peptidoglycan strongly activates the ERK 2, moderately activates JNK, and only weakly activates p38 kinase (41).

RsDPLA also competitively inhibited the expression of a panel of six LPS-inducible "early" genes and protein tyrosine phosphorylation when RsDPLA was present at a concentration of 1 µg/ml in the murine macrophage cultures. At higher concentrations of RsDPLA and LPS, inhibition of LPS-inducible protein tyrosine phosphorylation was blocked in a noncompetitive fashion (i.e., the inhibition of RsDPLA could not be overridden by increasing concentrations of LPS). The LPS-inducible expression of select "early" genes (IP-10) was also blocked noncompetitively, whereas the induction of other genes (TNF-α) was enhanced approximately fourfold (42). These results suggest a positive relationship between IP-10 gene expression and tyrosine phosphorylation and a negative relationship between TNF-α gene expression and tyrosine phosphorylation. Noncompetitive inhibition of LPS-inducible protein tyrosine phosphorylation by RsDPLA suggests that at high concentrations, RsDPLA induces an uncoupling between putative LPS receptors and effector molecules that leads to protein tyrosine phosphorylation (43).

In addition, we tested different analogs of RsDPLA and found that the pentaacylated RsDPLA is the most effective analog for blocking LPS-induced protein tyrosine phosphorylation in murine macrophages. The tetraacylated RsDPLA was not as effective in this assay. Inhibition of TNF-α secretion and induction of in vitro tolerance induced by toxic LPS or monophosphoryl lipid A (MPLA) of *S. minnesota* LPS were blocked most efficiently with RsDPLA and somewhat less efficiently by pentaacyl *R. sphaeroides* MPLA (43,44) suggesting that the phosphoryl group at the 1-position contributes significantly to the interaction of this analog with the putative LPS signaling receptor. RsDPLA showed no protein tyrosine phosphorylation activity in endothelial cells (45).

Transcription Factor NF-κB

During the inflammatory response to LPS, the NF-κB–responsive elements are required for the function of many cytokine promoters (46). Prior to LPS stimulation, the predominant NF-κB heterodimer is composed of 50 and 65 kDA subunits complexed to the inhibitory IκB. After LPS activation of murine macrophages, IκB dissociates and p50/p65 complex translocates to the nucleus (47–49) as shown in Figure 2. This ReLPS-inducible activation of NF-κB is blocked by RsDPLA (50). Evidence from this laboratory supports the hypothesis that the induction of corticosteroids is partly responsible for the blocking mechanisms (51). Recent evidence suggests that corticosteroids induce IκB protein synthesis and promote the reassociation of NF-κB and its inhibitor IκB in the presence of LPS or TNF-α (46,52). This would prevent the translocation of NF-κB from the cytoplasm to the nucleus.

LPS-Induced κ Light Chain Expression in 70Z/3 Cells

We found that the induction of immunoglobulin kappa (κ) light chain expression in the pre-B cell line, 70Z/3 cells, by toxic LPS is effectively blocked by RsDPLA. Induction of κ expression by LPS is dependent on at least two transcription factors, Oct-2 and NF-κB. RsDPLA alone activated NF-κB binding activity within

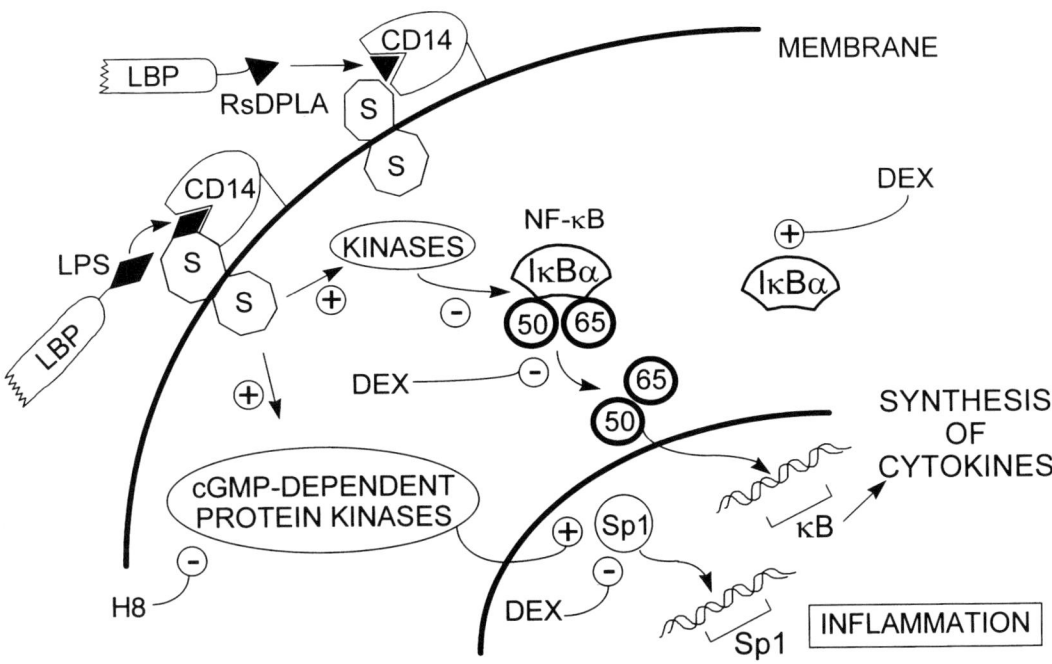

Fig. 2 A simplified scheme showing how RsDPLA might function as an antagonist of toxic LPS in macrophages. In this experimental situation, both toxic LPS and RsDPLA are present in suspension in the culture medium along with macrophages and serum (containing LBP). LBP dissociates the lipid aggregates to the monomers and form complexes. These complexes then transfer the monomeric lipids to membrane-bound CD14, which are thought to be associated with a (as yet, unidentified) transmembrane protein. CD14-LPS and the CD14-RsDPLA complex formation is shown in this figure. These complexes associate with another transmembrane protein (a signaling protein, S) to initiate signal transduction. Only the interaction of CD14-LPS complex with S leads to a strong signal. This results in the activation of kinases and transcription factors (including Sp1 and NF-κB).

30 minutes of treatment, but the activation was not sustained beyond 9 hours as observed with LPS. The NF-κB complexes activated by both RsDPLA and LPS were composed predominantly of the p50-RelA (p65) heterodimer. In addition, RsDPLA antagonized the activation of Oct-2 mRNA by toxic LPS. These results suggest that the physiological receptors on B cells transmit quantitatively different signals depending on the nature of the ligand that binds to them; furthermore, the fatty acyl groups of LPS play an important role in activating signal transduction (53).

Accumulation of p50/p50 Homodimers in 70Z/3 Cells

Upon activation of 70Z/3 cells by toxic LPS, the NF-κB complex dissociates from the IκB subunit (α and β) and migrates from the cytoplasm into the nucleus. Recently, we studied the effect of RsDPLA and toxic ReLPS on other NF-κB complexes such as p50/c-rel, p65/c-rel, and p50/p50. The accumulation of the p50/p50 complexes have been associated with tolerance to toxic LPS (54). ReLPS induced a sustained increase in both p50/c-rel and p50/p50 complexes. In contrast, RsDPLA showed some increase in p50/c-rel but not accumulation of p50/p50 homodimers, consistent with the finding that RsDPLA does not induce these homodimers in macrophages (50) and fails to induce tolerance in macrophages. We have also studied the effect of ReLPS and RsDPLA on the degradation of IκB-α and IκB-β in 70Z/3 cells. We found that ReLPS and RsDPLA are relatively slow to initiate degradation of IκB and that maximal translocation of NF-κB/rel complexes into the nucleus compartment tool about 60 minutes. In contrast, stimulation with phorbol myristic acid was far more rapid, reaching a peak at 30 minutes (O. Lawrence, N. Qureshi, and C. H. Sibley, unpublished results).

LPS-Induced Activation of NF-κB in Macrophages

The interaction of toxic LPS with macrophages results in the induction of a cascade of cytokines. NF-κB has been shown to be important for the induction of many LPS-inducible cytokine genes including TNF-α, IL-1β, and IL-6. Like LPS, taxol (an antitumor agent) is able to stimulate translocation of the p50/p65 heterodimers of NF-κB. This stimulation was observed only in macrophages derived from the LPS-responsive mice and not from the LPS-hyporesponsive mice. As was observed for gene expression and tyrosine phosphorylation of MAP kinases, RsDPLA inhibited both the LPS- and taxol-induced NF-κB activation in macrophages (50). This suggests that the two agonists share a common signaling element blocked by RsDPLA. In contrast, RsDPLA alone failed to initiate NF-κB translocation in murine macrophages.

Transcription Factor Sp1 and cGMP-Dependent Protein Kinase

Background

Like NF-κB, Sp1 is in another family of transcription factors that is activated by LPS during the inflammatory response (55). It is constitutively expressed and present in a wide range of cell types, where it binds to a GC-rich consensus sequence present in many cellular and viral promoters (56–60). Sp1 contains three zinc fingers that mediate DNA binding and four domains that mediate transcriptional activation (61). After binding to the GC box within the promoters of many cytokine genes, Sp1 becomes phosphorylated at multiple sites by double-stranded DNA-dependent kinase (61).

Sp1 is an important transcription factor involved in the regulation of expression of membrane CD14 (62). CD14 upregulation in the process of monocyte-specific differentiation induced by vitamin D_3 occurs mainly at the level of gene transcription. This induction requires new protein synthesis. Using stable transfection of the monocytoid U937 cell line with a series of deletion mutants of the CD14 5′ upstream sequence coupled to a reporter gene construct, Zhang et al. (62) showed that bp-128 to -70 is the critical region for the induction of CD14 expression. Moreover Sp1-binding sites are present in the promoter regions of a number of LPS-inducible genes (62–67), although promoter analysis demonstrating the requirement for Sp1 in the induction of these genes by LPS has not been demonstrated. In addition to mediating tissue-specific expression, Sp1 plays a critical role in this induction. Sp1 binding sites are also present in multiple monocytic specific genes such as CD11b (64) and the M-CSF receptor promoter (65). Sp1 plays a critical role in tissue-specific promoter activity of monocyte genes.

The formation of cGMP from GTP is catalyzed by an enzyme called guanylate cyclase. The cellular effects of cGMP appear to be mediated by several types of cGMP receptor proteins. The best characterized are the cGMP-dependent protein kinases, which are a class of closely related enzymes. Cyclic GMP-dependent protein kinases are serine/threonine protein kinases that belong to the very large protein kinase family. There are two catalytic site inhibitors of the cGMP-dependent protein kinase: the isoquinoline H-8 and KT5823 (68). These inhibitors selectively inhibit purified cGMP-dependent protein kinase with a Ki of 0.48 μM and 0.234 μM, respectively. These compounds also inhibit cAMP-dependent protein kinase but with much higher Ki of 1.2 μM and >10 μM, respectively. Cyclic GMP plays a major role in pathological situations, which range from endotoxic shock to various types of cardiovascular disorders, hypertension, and atherosclerosis (69).

LPS-Induced Activation of Sp1 Is Coupled to cGMP-Dependent Protein Kinase in RAW 264.7 Cells

In addition to NF-κB, LPS has been shown to activate Sp1, which is phosphorylated after binding to DNA enhancer elements. The promoters for a number of cytokine genes, including TNF-α and IL-1β, contain consensus sequence binding sites for Sp1 and NF-κB, among others, suggesting that these transcription factors may also regulate LPS-induced cytokine production. In human monocytes, there is a functional Sp1 binding site in the promoter region for the CD14 gene. We have found three agents that inhibit the ReLPS-initiated Sp1 activity in RAW 264.7 macrophages: RsDPLA, H-8 (a cGMP-dependent protein kinase inhibitor used at discriminating concentrations), and dexamethasone (a synthetic glucocorticoid and a potent regulator of gene activation), as shown in Figure 2. This work was the first report of inhibition by these three agents involving the LPS-induced binding of Sp1 and suggests that the activation of Sp1 is coupled to the activation of cGMP-dependent protein kinase (55).

Theoretical Basis for Inhibition of Signaling by RsDPLA

A question can be raised as to how RsDPLA blocks signal transduction in the activation of macrophages, monocytes, and neutrophils. As shown in Figure 2, this inhibition can occur at two levels. At the initial level, both toxic LPS and RsDPLA compete for the binding site on LBP. In the process of binding, LBP-LPS and LBP-RsDPLA complexes are formed. These complexes then compete (at the second level of binding) for the site on the membrane-bound CD14, which might be associated with a transmembrane protein "S." This results in the formation of CD14-LPS and CD14-RsDPLA complexes. Beyond this point very little is known (it is a "black box") until we get to the activation of specific components of the signaling cascade as discussed above. Thus, we offer our speculative views on this part of the pathway.

These two complexes could interact with another transmembrane protein (this would be the signaling protein) and cause a dimerization of these proteins. We suggest that when the CD14-LPS complex associates with the signaling protein (this would constitute the third level of competitive binding), a strong signal is generated, which initiates the activation of the cell. This signaling protein could be one or more of the three classes of transmembrane receptors described by Bernard (70). We suggest that when CD14-RsDPLA associates with the signaling protein, only a weak signal is generated, which is not adequate to activate the cell. A similar mechanism was proposed by Davis et al. (71) for the activation of the ciliary neurotrophic factor receptor complex. An alternative mechanism might be proposed where CD14 transfers the toxic LPS directly to the receptor (the signaling unit) to initiate signal transduction.

EFFECT OF HORMONES ON SEPTIC SHOCK

Corticosteroids

We have found that endogenous corticosterone may be beneficial in that its induction can protect animals from septic shock. We have established that prior administration of RsDPLA (30 min before LPS challenge) to mice blocks LPS-induced IL-1β and TNF-α in these animals (72,73). Consistent with its effect in reducing serum TNF-α, RsDPLA pretreatment protects galactosamine-sensitized mice from the lethal ReLPS challenge. We have also shown that RsDPLA induces corticosterone which reaches a maximum within 15 minutes after administration (to be discussed). This endogenous corticosterone could then block the synthesis of TNF-α and IL-1β at the pretranslational step. This protective effect of RsDPLA was essentially eliminated when mice were pretreated with the glucocorticoid receptor antagonist RU 486 (to be discussed). These results show that endogenous glucocorticoids modulate the endotoxic effects of LPS by inhibiting the synthesis of inflammatory cytokines and show the importance of the macrophage-neuroendocrine axis in the protection of animals from septic shock (73).

RU 486 Eliminates the Protective Effect of RsDPLA in LPS Lethality Tests

We examined the effect of RU 486 (a glucocorticoid receptor antagonist) on the protective effect of RsDPLA in galactosamine-sensitized mice challenged with toxic ReLPS. In this experiment, ReLPS challenge (0.5 μg, $n = 6$), caused 86% lethality, whereas pretreatment with RsDPLA (50 μg) 2 hours prior to challenge reduced the lethality to 20% ($n = 6$). When the mice were pretreated with both Ru 486 (0.5 mg) and RsDPLA, the protective effect of RsDPLA was essentially eliminated (93% lethality in 72 hr). We also followed the TNF-α levels of these treated mice and found that ReLPS caused an increase to 2543 ± 201 pg/ml. Pretreatment of RsDPLA reduced the TNF-α level to 516 ± 147 pg/ml. When RU 486 was included in the pretreatment, the TNF-α level rose to 1558 ± 46 pg/ml. These results are consistent with the lethality data where high serum TNF-α levels correlated well with high mortality (73).

RsDPLA PREVENTS HYPOGLYCEMIA INDUCED BY TOXIC LPS IN MICE

We have shown that RsDPLA blocks the induction of TNF-α by toxic LPS and induces a corticosteroid response. We wanted to determine if RsDPLA pretreatment would prevent the hypoglycemic response induced in mice by toxic LPS. ReLPS challenge of mice reduced the level of glucose from 112 to 50 mg/dl. RsDPLA treatment alone had no effect on the blood glucose level of mice. Moreover, when the mice were first pretreated with RsDPLA and then challenged 2 hours later with ReLPS, the blood glucose level remained essentially normal. These results showed that RsDPLA pretreatment can prevent the development of hypoglycemia in mice challenged with toxic LPS (16).

RsDPLA DOES NOT INDUCE IL-10 IN MICE

It has been suggested that early endotoxin tolerance in animals is due to the induction of IL-10 (74). We examined the ability of RsDPLA to induce IL-10 in mice. In the same experiment, we followed the induction of TNF-α and corticosterone as controls. Toxic ReLPS induced IL-10 (1032 pg/ml), TNF-α (2917 pg/ml), and corticosterone (1100 ng/ml). However, RsDPLA induced only a minimal level of IL-10 (102 pg/ml), a baseline level of TNF-α, and 989 ng/ml of corticosterone. Thus, unlike tolerance induced by LPS, the anti-inflammatory activity of RsDPLA is not likely to be mediated by IL-10 (16).

NATURAL CORTICOSTEROID PROTECTS MICE AGAINST LPS-INDUCED SHOCK

We wanted to demonstrate that treatment of mice with natural corticosteroid would protect the animals from toxic LPS challenge. In this experiment, ReLPS challenge alone (0.5 μg) resulted in 25% survival of galactosamine-sensitized mice. When the mice were first pretreated with natural corticosterone (2 mg) and then challenged with ReLPS, 92% of the mice survived. Corticosterone at a lower dosage of 200 μg afforded no such protection (16).

Endogenous corticosteroids must play a very important role since adrenalectomized animals (75) and RU 486–treated animals are very susceptible to septic shock. All of the experiments mentioned above suggest that RsDPLA protects the animal by two separate mechanisms: (1) it acts as an LPS antagonist by blocking receptor sites on cells, and (2) it induces corticosteroids. Although the experiments with RU 486 clearly demonstrate the importance of the endogenous corticosterone in protecting the animals from shock, the relative importance of the two mechanisms is presently difficult to determine.

INDUCTION OF CORTICOSTEROIDS IN MICE BY RsDPLA

Since we had established that RsDPLA clearly induces corticosteroids in mice (73), we sought to examine the time course of this induction as well as the induction of the related ACTH. Toxic ReLPS gave a biphasic rise in the level of ACTH. There was an initial modest rise at 15 minutes (100 pg/ml) and a large rise at 120–180 minutes (230 pg/ml). The second large rise is probably due to the induction of cytokines. RsDPLA also showed the initial modest rise (170 pg/ml) at 30 min), but there was no corresponding second rise at later time intervals (16).

The time course of induction of corticosteroid by toxic ReLPS showed a gradual increase from a basal level to a maximum at 180 minutes (1350 ng/ml). RsDPLA caused maximum induction of corticosteroid after only 15 minutes (1000 ng/ml), and this was maintained for 120 minutes after which it decreased sharply. The results of corticosteroid induction by ReLPS are similar to those reported by Rivier et al. (76). The initial modest rise in both ACTH and corticosterone must be due to the direct action of ReLPS and RsDPLA and not indirectly through the cytokines. This conclusion is based on the fact that IL-1β is induced much later, about 90–120 minutes after an LPS challenge. We also discovered that even though TNF-α and other cytokine levels are dramatically lower in the RsDPLA-pretreated mice (which are subsequently challenged with toxic LPS), their corticosterone levels are high. The early corticosterone induction by RsDPLA does not appear to be dependent on cytokine release in mice (16).

INFLAMMATORY RESPONSES TO TOXIC LPS AND ANTI-INFLAMMATORY RESPONSES TO NONTOXIC RsDPLA IN THE MURINE MODEL

Based on our present knowledge, we propose a simplified scheme to explain how LPS activates the immune and neuroendocrine systems (Fig. 3). Cell wall fragments containing LPS are shed from the surface of gram-negative bacteria undergoing autolysis. These particles interact with various proteins in the blood, including LBP (Fig. 2). The LBP-monomeric LPS complexes are formed. These complexes then transfer the LPS to sCD14 and membrane-CD14 to form the CD14-monomeric LPS complexes. These latter complexes interact with the signaling LPS receptors on the cell surface or in the plasma membrane proper of macrophages, monocytes, and neutrophils to initiate signal transduction. The nature of these signaling receptors has not yet been determined.

The very early responses to LPS are an influx of Ca^{2+} into the cells, activation of protein tyrosine kinases, and induction of certain transcription factors (including Sp1 and NF-κB). The macrophage/monocyte cell lines then produce cytokines, including TNF-α, IL-1β, IL-6, IL-12, and IL-8. Macrophages also induce the formation of arachidonic acid metabolites—leukotrienes, prostaglandins, and thromboxanes. Other products—superoxide anion, hydrogen peroxide, and nitric oxide—are formed. T cells and NK cells are also activated by IL-12 to produce IFN-γ, which further activates the macrophages. TNF-α, IL-1β, and chemotactic factors stimulate the neutrophils to produce lysosomal enzymes, superoxide anion, and hydrogen peroxide. This cascade of events is central to the inflammatory response that usually leads to the elimination of the bacterial and

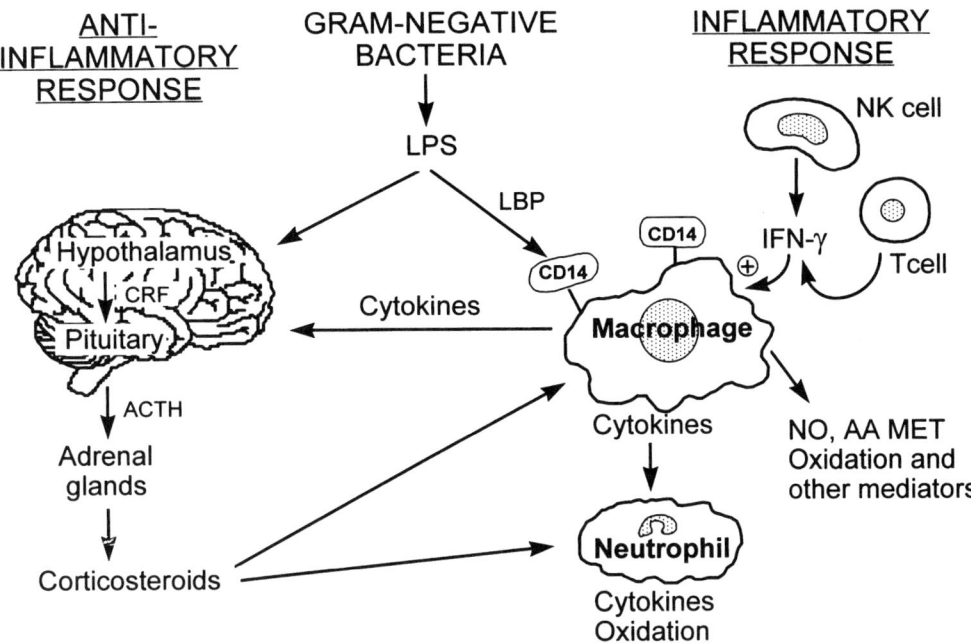

Fig. 3 Mechanism of inflammatory and anti-inflammatory responses of humans to gram-negative bacteria. The neuroendocrine system (left side) controls the anti-inflammatory responses, whereas the macrophage/neutrophil/lymphocyte system (right side) controls the inflammatory responses. The primary component of the neuroendocrine system is the corticosteroids, which can be induced by RsDPLA. RsDPLA can block the initiation of the inflammatory responses by toxic LPS. LPS, lipopolysaccharide; LBP, LPS-binding protein; NO, nitric oxide; AA MET, arachidonic acid metabolism; IFN-γ, interferon-gamma; NK cell, natural killer cells; CRF, corticotropin-releasing factor; ACTH, adrenocorticotrophic hormone.

viral infections. However, excess inflammatory response can lead to shock.

LPS also activates the production of IL-10, ACTH, and corticosteroids as a part of the anti-inflammatory response. Inflammatory mediators, including IFN-γ, histamine released from mast cells, serotonin, and leukotrienes, also induce corticosteroids. This negative regulatory loops has an essential function in preventing excess activation of the acute phase response. Whereas toxic LPS induces both inflammatory and anti-inflammatory responses, RsDPLA at high doses only induces an anti-inflammatory response, namely the induction of corticosteroids. Corticosteroids normalizes the system by reestablishing an equilibrium between the inflammatory and anti-inflammatory responses. Any defect in the biosynthesis of corticosteroids or the inflammatory responses would lead to shock in the animal model system. It is clear that the macrophage-neuroendocrine axis must be considered in explaining the mechanism of gram-negative septic shock. Thus, a basic concept has emerged from our studies and states that the induction of endogenous corticosteroids can control the detrimental effect of toxic LPS by inhibiting the synthesis of the inflammatory cytokines (6). Similarly Walley et al. (77) have shown that the balance of inflammatory mediators is related to the severity and mortality of murine sepsis.

FUTURE PERSPECTIVES

The last decade has brought much new knowledge about the structure and antagonistic properties of *Rhodobacter* lipid A. RsDPLA is not only an indispensable tool for studying LPS-induced signal transduction, but is also able to prevent LPS-induced shock in animals. RsDPLA has been shown to block all LPS-induced biological activities tested so far. RsDPLA competes for the LPS binding site on LBP. This RsDPLA-LBP complex then competes with the LPS-LBP complex at the second level of binding for the membrane-bound CD14. RsDPLA may also compete at the third level of binding at the LPS signaling protein level. However, such experiments cannot be performed until the identity of this signaling protein is revealed. Studies on the identification of this protein are in progress in several laboratories. The other practical aspect of this study would be to develop a combination of RsDPLA and other drugs to treat gram-negative septic shock. These studies are also in progress.

ACKNOWLEDGMENTS

This review and related work were supported in part by the Medical Research Service of the Department of Veterans Affairs and by National Institutes of Health grant GM-50870 (NQ).

REFERENCES

1. Mühlradt PF, Golecki JR. Asymmetrical distribution and artificial reorientation of lipopolysaccharide in the outer membrane bilayer of *Salmonella typhimurium*. Eur J Biochem 1975; 51:343–352.
2. Funahara Y, Nikaido H. Asymmetrical localization of lipopolysaccharides on the outer membrane of *Salmonella typhimurium*. J Bacteriol 1980; 141:1463–1465.
3. Morrison DC, Ulevitch RJ. The effects of bacterial endotoxins on host mediation systems. Am J Pathol 1978; 93:527–617.
4. Beutler B, Cerami A. Cachectin: more than a tumor necrosis factor. N Engl J Med 1987; 316:379–385.
5. Qureshi N, Takayama K. Structure and function of lipid A. In: Iglewski BH, Clark VL, eds. The Bacteria. Vol. XI. Molecular Basis of Bacterial Pathogenesis. New York: Academic Press, 1990:319–338.
6. Qureshi N, Takayama K, Hofman J, et al. In: Morrison DC, Ryan JL, eds. Novel Therapeutic Strategies in the Treatment of Sepsis: Diphosphoryl Lipid A from *Rhodobacter sphaeroides*: A novel lipopolysaccharide Antagonist. New York: Marcel Dekker, 1996:111–131.
7. Riedo FX, Munford RS, Campbell WB, et al. Deacylated lipopolysaccharide inhibits plasminogen activator inhibitor-1, prostacyclin, and prostaglandin E2 induction by lipopolysaccharide but not by tumor necrosis factor-alpha. J Immunol 1990; 144:3506–3512.
8. Loppnow H, Libby P, Freundenberg M, et al. Cytokine induction by lipopolysaccharide (LPS) corresponds to lethal toxicity and is inhibited by nontoxic *Rhodobacter capsulatus* LPS. Infect Immun 1990; 58:3743–3750.
9. Proctor RA, Will JA, Burhop KE, Raetz CRH. Protection of mice against lethal endotoxemia by a lipid A precursor. Infect Immun 1986; 52:905–907.
10. Loppnow H, Brade H, Durrbaum I, et al. Interleukin 1 induction capacity of defined lipopolysaccharide partial structures. J Immunol 1989; 142:3229–3238.
11. Golenbock DT, Hampton RY, Qureshi N, et al. Lipid A-like molecules that antagonize the effects of endotoxins on human monocytes. J Biol Chem 1991; 266:19490–19498.
12. Takayama K, Qureshi N, Beutler B, Kirkland TN. Diphosphoryl lipid A obtained from *Rhodopseudomonas sphaeroides* ATCC 17023 blocks induction of cachectin in macrophage by lipopolysaccharide. Infect Immun 1989; 57:1336–1338.
13. Stuetz PL, Aschauer H, Hildebrandt J, et al. In: Nowotny A, Spitzer JJ, Ziegler EJ, eds. Cellular and Molecular Aspects of Endotoxin Reactions. Chemical Synthesis of Endotoxin Analogues and Some Structure Activity Relationships. Belle Mead, NJ: Excerpta Medica, Inc., 1991:129–144.
14. Stern A, Engelhardt R, Foster C, et al. In Levin J, Alving CR, Munford RS, Redl H, eds. Bacterial Endotoxins: Lipopolysaccharides from Genes to Therapy: SDZ MRL 953, a Lipid A Analog as Selective Cytokine Inducer. New York: Wiley-Liss, 1995:549–565.
15. Christ WJ, Asano O, Robidoux ALC, et al. E5531, a pure endotoxin antagonist of high potency. Science 1995; 268:80–83.
16. Qureshi N, Jarvis B, Takayama K, et al. In: Levin J, Pollack M, Yokichi T, Nakano M, eds. Endotoxin and Sepsis: Molecular Mechanisms of Pathogenesis, Host Resistance, and Therapy: Natural and Synthetic LPS and Lipid A Analogs or Partial Structures That Antagonize or Induce Tolerance to LPS. New York: John Wiley, 1998:289–300.
17. Qureshi N, Takayama K, Mascagni P, et al. Complete structural determination of lipopolysaccharides obtained from deep rough mutant of *Escherichia coli*: purification by high performance liquid chromatography and direct analysis by plasma desorption mass spectrometry. J Biol Chem 1988; 263:11971–11976.
18. Qureshi N, Honovich JP, Hara H, et al. Location of fatty acids in lipid A obtained from lipopolysaccharides of Rhodopseudomonas sphaeroides ATCC 17023. J Biol Chem 1988; 263:5502–5504.
19. Qureshi N, Takayama K, Meyer KC, et al. Chemical reduction of 3-oxo and unsaturated groups in fatty acids of diphosphoryl lipid A from the lipopoylsaccharide of *Rhodopseudomonas sphaeroides*: comparison of biological properties before and after reduction. J Biol Chem 1991; 266:6532–6538.
20. Kaltashov IA, Doroshenko V, Cotter RJ, et al. Confirmation of the structure of lipid A derived from the lipopolysaccharide of *Rhodobacter sphaeroides* by a combination of MALDI, LSIMS, and Tandem mass spectrometry. Anal Chem 1997; 69:2317–2322.
21. Christ WJ, McGuinness PD, Asano O, et al. Total synthesis of the proposed structure of Rhodobacter sphaeroides lipid A resulting in the synthesis of new potent lipopolysaccharide antagonists. J Am Chem Soc 1994; 116:3637–3638.
22. Morrison DC, Ryan JL. Endotoxin and disease mechanisms. Ann Rev Med 1987; 38:417–432.
23. Wright SD. Multiple receptors for endotoxin. Curr Opin Immunol 1988; 3:83–90.
24. Schumann RR, Leong SR, Flaggs GW, et al. Structure and function of lipopolysaccharide-binding protein. Science 1990; 249:1429–1431.
25. Wright SD, Ramos RA, Tobias PS, et al. CD14, a receptor for complexes of lipopolysaccharides (LPS) and LPS binding protein. Science 1990; 249:1431–1433.
26. Ziegler-Heitbrock HWL, Ulevitch RJ. CD14: cell surface receptor and differentiation marker. Immunol Today 1993; 14:121–125.
27. Tobias PS, Soldau K, Kline L, et al. Cross-linking of lipopolysaccharide (LPS) to CD14 on THP-1 cells mediated by LPS-binding protein. J Immunol 1993; 150:3011–3021.

28. Kirkland TN, Virca GD, Kuus-Reichel T, et al. Identification of lipopolysaccharide-binding proteins in 70Z/3 cells by photoaffinity cross-linking. J Biol Chem 1990; 265:9520–9525.
29. Lei MG, Morrison DC. Specific endotoxic lipopolysaccharide-binding receptors on murine splenocytes. I. Detection of lipopolysaccharide-binding sites on splenocytes and splenocyte subpopulations. J Immunol 1991; 141:996–1005.
30. Vita N, Lefort S, Sozzani P, et al. Detection and biochemical characteristics of the receptor for complexes of soluble CD14 and bacterial lipopolysaccharide. J Immunol 1997; 3457–3462.
31. Hara-Kuge S, Amano F, Nishijima M, Akamatsu Y. Isolation of a lipopolysaccharide (LPS)-resistant mutant with defective LPS binding of cultured macrophage-like cells. J Biol Chem 1990; 265:6606–6610.
32. Hampton RY, Golenbock DT, Raetz CRH. Lipid A binding sites in membranes of macrophage tumor cells. J Biol Chem 1988; 263:14802–14807.
33. Schletter J, Brade H, Brade L, et al. Binding of lipopolysaccharide (LPS) to an 80-kilodalton membrane protein of human cells is mediated by soluble CD14 and LPS-binding protein. Infect Immun 1995; 63:2576–2580.
34. Kirikae T, Schade FU, Kirikae F, et al. Diphosphoryl lipid A derived from the lipopolysaccharide (LPS) of *Rhodobacter sphaeroides* ATCC 17023 is a competitive LPS inhibitor in the murine macrophage-like J774.1 cells. FEMS Immunol Med Microbiol 1994; 9:237–243.
35. Jarvis B, Lichenstein H, Qureshi N. Diphosphoryl lipid A from *Rhodobacter sphaeroides* inhibits the complexes that form in vitro between lipopolysaccharide (LPS)-binding protein, LBP, soluble CD14 and spectral-pure lipopolysaccharide. Infect Immun 1997; 65:3011–3016.
36. Kirikae T, Kirikae F, Tominaga K, et al. *Rhodobacter sphaeroides* diphosphoryl lipid A inhibits interleukin-6 production in CD14-negative murine marrow stromal cells stimulated with lipolysaccharide or Paclitaxel (taxol). J Endotoxin Res 1997; 4:115–122.
37. Weinstein SL, Gold MR, DeFranco AL. Bacterial lipopolysaccharide stimulate tyrosine phosphorylation in macrophages. Proc Natl Acad Sci USA 1991; 88:4148–4152.
38. Kyriakis JM, Avruch J. Sounding the alarm: protein kinase cascades activated by stress and inflammation. J Biol Chem 1996; 271:24313–24316.
39. Sanghera JS, Weinstein SL, Aluwalia M, et al. Activation of multiple proline-directed kinases by bacterial lipopolysaccharide in murine macrophages. J Immunol 1996; 156:4456–4465.
40. Nick JA, Avdi NJ, Gerwins P, et al. Activation of a p38 mitogen-activated protein kinase by lipopolysaccharide. J Immunol 1996; 156:4867–4875.
41. Dziarski R. Differential activation of extracellular signal-regulated kinase (ERK) 1, ERK2, p38, and c-Jun NH2-terminal kinase mitogen-activated protein kinases by bacterial peptidoglycan. J Infect Dis 1996; 174:777–785.
42. Manthey CL, Perera PY, Qureshi N, et al. Inhibition of lipopolysaccharide-induced macrophage gene expression by *Rhodobacter sphaeroides* lipid A and SDZ 880.431. Infect Immun 1993; 61:3518–3526.
43. Vogel SN, Manthey CL, Perera PY, et al. Dissection of LPS-induced signaling pathways in murine macrophages using LPS analogs, LPS mimetics, and agents unrelated to LPS. Prog Clin Biol Res 1995; 392:421–431.
44. Henricson BE, Perera PY, Qureshi N, et al. *Rhodopseudomonas sphaeroides* lipid A derivative block in vitro induction of tumor necrosis factor and endotoxin tolerance by smooth lipopolysaccharide and monophosphoryl lipid A. Infect Immun 1992; 60:4285–4290.
45. Arditi M, Zhou J, Torres M, et al. Lipopolysaccharide stimulates the tyrosine phosphorylation of mitogen-activated protein kinases p44, p42, and p41 in vascular endothelial cells in a soluble CD14-dependent manner. Role of protein tyrosine phosphorylation in lipopolysaccharide-induced stimulation of endothelial cells. J Immunol 1995; 155:3994–4003.
46. Scheinman RI, Cogswell PC, Lofquist AK, Baldwin AS. Role of transcriptional activation of IκBα in mediation of immunosupression by glucocorticoids. Science 1995; 270:283–286.
47. Sen R, Baltimore D. Multiple nuclear factors interact with the immunoglobulin enhancer sequences. Cell 1986; 46:705–716.
48. Rooney JW, Emery DW, Sibley CH. 1.3E2, a variant of the B lymphoma 70Z/3 defective in activation of NF-κB and OTF-2. Immunogenetics 1990; 31:73–78.
49. Zabel U, Bauerle P. Purified IκB can rapidly dissociate the complex of the NF-κB transcription factor with its cognate DNA. Cell 1990; 61:255–265.
50. Perera PY, Qureshi N, Vogel S. Paclitaxel (Taxol)-induced NF-kappa-B translocation in murine macrophages: Infect Immun 1996; 64:878–884.
51. Qureshi N, Takayama K, Hofman J, Zuckerman SH. In: Levin J, Alving CR, Munford RS, Stutz PL, eds. Bacterial Endotoxin: Recognition and Effector Mechanisms: Diphosphoryl Lipid A Obtained from Nontoxic Lipopolysaccharide (LPS) of *Rhodobacter sphaeroides* Is an LPS Antagonist and an Inducer of Corticosteroids. Amsterdam: Elsevier Press, 1993:361–371.
52. Marx J. How the glucocorticoids suppress immunity. Science 1995; 270:232–233.
53. Lawrence O, Rachie N, Qureshi N, et al. Diphosphoryl lipid A from *Rhodobacter sphaeroides* transiently activates NF-kB but inhibits lipopolysaccharide (LPS) induction of kappa immunoglobulin and Oct-2 in the B cell lymphoma 70Z/3. Infect Immun 1995; 63:1040–1046.
54. Ziegler-Heitbrock HW, Wedel A, Schraut W, et al. Tolerance to lipopolysaccharide involves mobilization of nuclear factor kappa B with predominance of p50 homodimers. J Biol Chem 1994; 269:17001–17004.
55. Jarvis B, Qureshi N. Inhibition of lipopolysaccharide-induced transcription factor Sp1 binding by spectrally pure diphosphoryl lipid A from *Rhodobacter sphaeroides*, protein kinase inhibitor H-8, and dexamethasone. Infect Immun 1997; 65:1640–1643.
56. Kadonaga JT, Jones KA, Tjian R. Promoter-specific activation of RNA polymerase II transcription by Sp1. TIBS 1986; 11:20–23.

57. Hagen G, Muller S, Beato M, Suske G. Sp1-mediated transcriptional activation is repressed by Sp3. EMBO 1994; 13:3843–3851.
58. Dennig J, Hagen G, Beato M, Suske G. Members of the Sp transcription factor family control transcription from the uteroglobin promoter. J Biol Chem 1995; 270:12737–12744.
59. Briggs MR, Kadonaga JT, Bell SP, Tjian R. Purification and biochemical characterization of the promoter-specific transcription factor, Sp1. Science 1986; 234:47–52.
60. Kriwachi RW, Schultz SC, Steitz TA, Caradonna JP. Sequence-specific recognition of DNA by zinc-finger peptides derived from the transcription factor Sp1. Proc Natl Acad Sci USA 1992; 89:9759–9763.
61. Jackson SP, MacDonald JJ, Lees-Miller S, Tjian R. GC box binding induces phosphorylation of Sp1 by a DNA-dependent protein kinase. Cell 1990; 63:155–165.
62. Zhang D-E, Hetherington CJ, Tan S, et al. Sp1 is a critical factor for the monocytic specific expression of CD14. J Biol Chem 1994; 269:11425–11434.
63. Ebert Sn, Wong DL. Differential activation of rat phenylethanolamine N-methyltransferase gene by Sp1 and Egr-1. J Biol Chem 1995; 270:17299–17305.
64. Pahl HL, Scheibe RJ, Zhang D-E, et al. The proto-oncogene PU.1 regulates expression of the myeloid-specific CD11b promoter. J Biol Chem 1993; 268:5014–5020.
65. Zhang D-E, Hetherington CJ, Chen H-M, Tenen DG. The macrophage transcription factor PU.1 directs tissue-specific expression of the macrophage cology-stimulating factor receptor. Mol Cell Biol 1994; 14:373–381.
66. Xu J, Thompson KL, Shephard LG, et al. T3 receptor suppression of Sp1-dependent transcription from the epidermal growth factor receptor promoter via overlapping DNA-binding sites. J Biol Chem 1993; 268:16065–16073.
67. Desai-Yajnik V, Samuels HH. The NF-κB and Sp1 motifs of the immunodeficiency virus type 1 long terminal repeat function as novel thyroid hormone response elements. Mol Cell Biol 1993; 13:5057–5069.
68. Lincoln TM. Cyclic GMP receptor proteins. In: Cyclic GMP: Biochemistry, Physiology and Pathophysiology. Austin: R.G. Landes Company, 1994:57–82.
69. Lincoln TM. Pathophysiological roles for cyclic GMP. In: Cyclic GMP: Biochemistry, Physiology and Pathophysiology. Austin: R.G. Landes Company, 1994:151–164.
70. Bernard EA. Protein structures in receptor classification. Ann NY Acad Sci 1977; 812:14–28.
71. Davis S, Aldrich TH, Stahl N, et al. LIFRβ gp130 as heterodimerizing signal tranducers of the tripartite CNTF receptor. Science 1993; 260:1805–1808.
72. Qureshi N, Takayama K, Kurtz R. Diphosphoryl lipid A obtained from the nontoxic lipopolysaccharide of *Rhodopseudomonas sphaeroides* is an endotoxin antagonist in mice. Infect Immun 1991; 59:441–444.
73. Zuckerman SH, Qureshi N. In vivo inhibition of LPS induced lethality and TNF synthesis by *Rhodobacter sphaeroides* diphosphoryl lipid A is dependent on corticosterone induction. Infect Immun 1992; 60:2581–2587.
74. Gerard C, Bruyns C, Merchant A. Interleukin-10 reduces the release of tumor necroses factor and prevents lethality in experimental endotoxemia. J Exp Med 1993; 177:547–550.
75. Bertini R, Bianchi M, Ghezzi P. Adrenalectomy sensitizes mice to the lethal effects of interleukin 1 and tumor necrosis factor. J Exp Med 1988; 167:1708–1712.
76. Rivier C, Chizzonite R, Vale W. In the mouse, the activation of the hypothalamic-pituitary-adrenal axis by a lipopolysaccharide (endotoxin) is mediated through interleukin-1. Endocrinology 1989; 125:2800–2805.
77. Walley ER, Lukacs NW, Standiford TJ, et al. Balance in inflammatory cytokines related to severity and mortality of murine sepsis. Infect Immun 1996; 64:4733–4738.

47

Synthetic Endotoxin Antagonists

Daniel P. Rossignol and Lynn D. Hawkins
Eisai Research Institute, Andover, Massachusetts

William J. Christ
Eisai Merrimack Valley Laboratories, Inc., Andover, Massachusetts

Seiichi Kobayashi and Tsutomu Kawata
Eisai Company, Ltd., Tsukuba, Ibaraki, Japan

Melvyn Lynn
Eisai Inc., Teaneck, New Jersey

Isao Yamatsu and Yoshito Kishi
Eisai Company, Ltd., Bunkyo-ku, Tokyo, Japan

INTRODUCTION: WHY ANTAGONIZE ENDOTOXIN?

Bacterial infection can be life-threatening, requiring that a potential host organism be highly diligent in maintaining its antimicrobial "state of awareness." Endotoxin or lipopolysaccharide (LPS) is an integral component of the outer membrane of gram-negative bacteria. Insofar as LPS is a unique noneukaryotic molecule common to all gram-negative pathogenic bacteria, it can be argued that the presence of LPS would qualify as an excellent surveillance signal for infection by the host. Triggered by this signal, a rapid and robust antibacterial challenge can be part of a healthy immune response that includes a complex cascade of cytokines and other cellular mediators. These secretory products can be directly bactericidal or can recruit other inflammatory cells to eradicate the infecting organism. Unfortunately, higher doses of endotoxin or the continued presence of endotoxin in the blood due to persistent infection or even fragments of lysed bacteria can result in a pathophysiological overreaction resulting in the release of toxic quantities of cytokines and other cellular mediators (Fig. 1). This deleterious overreaction can result in the septic inflammatory response syndrome (SIRS) and include the life-threatening symptoms of vascular fluid leak, tissue damage, hypotension, shock, and organ failure.

A variety of approaches has been taken to alleviate the morbidity and mortality of patients due to SIRS and septic shock. These approaches have focused on different points in the cascade of events that lead to severe sepsis. These simplified steps, labeled in Figure 1, include (1) blocking initial LPS-signaling events by preventing the generation of cell-surface signals, (2) blocking the intracellular signals induced by endotoxin or the synthesis of cytokines and other cellular mediators induced by endotoxin, (3) inhibiting the release of these cytokines and cellular mediators, or (4) inhibiting further "downstream" events triggered by cytokine release.

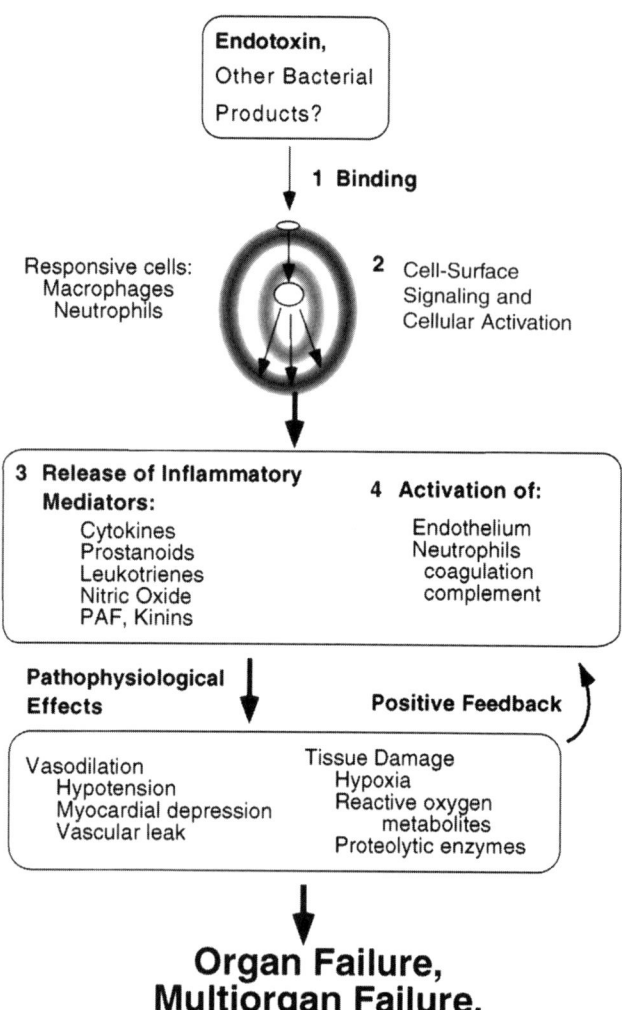

Fig. 1 Sequence of events leading to sepsis and septic shock. Endotoxin from gram-negative bacteria activates CD14+ cells, inducing release of cytokines and other cellular mediators. Once these inflammatory mediators reach "toxic" levels, other cells such as endothelial cells become activated, further enhancing the cascade of release of more cytokines and other inflammatory mediators, augmenting inflammation. The resultant response can set up a "positive feedback loop" leading to organ failure, multiorgan failure, and death.

Blocking Events Downstream of LPS Binding

Most attempts at treating sepsis and septic shock have been targeted towards the later stages of endotoxin response (steps 2 through 4 in Fig. 1). To this end, blocking the damage caused by cytokines from cells responding to LPS has been approached by desensitizing cells towards LPS (2,3) by the use of "attenuated" synthetic LPS-like agonists (3–8). Blocking cytokine synthesis/release has been attempted through interference with transduction of the cell-surface signal triggered by LPS (9–16). In addition, neutralization of released cytokines has been attempted by passive immunization [e.g., anti–TNF-α antibodies (17–20)], neutralization with soluble receptor [soluble TNF-α receptor (22)], and blocking the cytokine cell-surface receptors (e.g., IL-1 receptor) with a specific receptor antagonist (23–25).

The majority of these approaches have demonstrated efficacy in both in vitro and animal models; however, to date, none has proven to be effective at treating human sepsis. While it is unlikely that a simple explanation can account for all of these failures, it is likely that the approach of blocking only a single cytokine may be inadequate due to the large diversity and quantity of cytokines released by activated cells. The choice of cytokine targeted for antagonism was based on the likelihood that it is one of the more toxic cytokines. For instance, administration of TNF-α into humans demonstrates pathologies similar to that of endotoxin administration (26,27), and TNF-α administration can be lethal in animal models (28). However, it is possible that blocking a single cytokine may be insufficient to protect against the cascade of inflammatory mediators released during sepsis.

With this in mind, it seems likely that a more promising strategy for blocking events leading to sepsis and SIRS may be through antagonizing the interaction of LPS with its cell-surface "receptor(s)" (step 1, Fig. 1). This would block cellular activation, thereby inhibiting the release of *all* of the downstream cytokines and inflammatory cellular mediators induced by LPS.

Antagonizing LPS at Its Cell-Surface "Receptor"

Several varied approaches have focused on inhibiting the initial event of LPS interaction with the cell. This has traditionally been performed by the use of various antibodies directed against LPS or parts of the LPS molecule with the hope that this interaction would enhance LPS clearance or neutralize the ability of LPS to activate cells (29,30).

Bacterial LPS is a large and complex molecule consisting of a polysaccharide (O-antigen) region, a core region containing, in part, 3-deoxy-D-*manno*-oct-2-ulosonic acid (KDO), and the "lipid A" region (described in more detail below). Lipid A can be derived by acid hydrolysis of LPS and has been identified as the toxicophore of LPS containing the elements necessary for the toxic activity of the entire LPS molecule (31).

Neutralizing polyclonal antibodies directed against the entire LPS molecule (in general the O-antigen region) would tend to be specific for the species from which the LPS was derived. This would limit its usefulness to countering infection by only a narrow range of bacterial serotypes. In contrast, antibodies directed against the more conserved lipid A region have proven in vitro to work against a broader range of LPS. Unfortunately, all of these antibodies have shown varying degrees of activity both in vitro and in vivo, and have been ineffective in the clinic (for a review of these studies, see Ref. 32). More recently, it has been shown that antigenicity of chemically derived lipid A may be focused at the newly generated C-6' hydroxyl group, which is produced by acid hydrolysis of LPS. This would make antibodies to lipid A reactive only at lipid A, leaving the remaining polysaccharide region of LPS nonreactive (33).

The central role that lipid A plays in activating cells makes it clear that blocking the cellular "receptor" for lipid A may halt the chain of events leading to systemic inflammatory response and septic shock. A pharmacological receptor antagonist is often best derived from modification of the parent agonistic molecule. For this reason, we have modified lipid A from a variety of bacterial species as lipid A (and LPS) antagonists. Lipid A from *E. coli* is the toxic principle of the LPS molecule, first isolated by Westphal and Luderitz (34), and found to have lethal toxicity, pyrogenicity, as well as other inflammatory properties. The common features among the isolated toxic lipid As such as that from *E. coli* (see Fig. 3) include a polyacylated (usually hexa) β-1,6-disaccharide of D-glucosamine as well as phosphates at the anomeric (O-1) and O-4' positions (see Fig. 5).

Studies toward elucidating the structure-activity relationships of substructures of *E. coli* lipid A have provided valuable insight into the significance of several portions of the molecule (34–38). It is of key importance that naturally derived lipid A from some bacteria, for example, *R. capsulatus* and *R. sphaeroides*, are both nontoxic (nonagonistic) and can potently antagonize the agonistic activity of *E. coli* lipid A (40,41). While these compounds may provide interesting structural leads, it is projected that difficulties involved in obtaining sufficient amounts of homogeneous material with pharmaceutically acceptable purity and stability would severely limit the use of bacterially derived compounds as therapeutic endotoxin antagonists.

This review focuses on attempts to develop a pharmaceutically acceptable antagonist that would interact with the putative LPS receptor on the cell surface and block the action of LPS without demonstrating any "LPS-like" cellular activation on its own.

STRUCTURE/ACTIVITY OF SYNTHETIC LPS ANTAGONISTS

Types of Synthetic Endotoxin Antagonists

Synthetic endotoxin antagonists are structurally based on natural lipid A and their biosynthetic precursors. These antagonists can be classified into either monosaccharides or disaccharides. This chapter discusses the evolution of two potent antagonists based on the proposed structures of the lipid As of both *R. capsulatus* and *R. sphaeroides*. We note that this report is not a historical representation of how the synthetic strategy evolved, but a discussion of the structure-activity relationships of antagonistic activity in order of structural and synthetic complexity.

Monosaccharide Antagonists

Lipid X (Figure 2) a monosaccharide biosynthetic precursor of lipid A, was first isolated from a variant of *E. coli* and identified through the pioneering work of Raetz and colleagues (44,45). Lipid X antagonizes LPS activity in vitro (46) and has been reported to protect mice and sheep from LPS-induced lethality (47,48), although lipid X has been shown to be ineffective in vivo in a canine sepsis model (49). Efforts to develop a monosaccharide-based endotoxin antagonist by us and others (50) have met with limited success, yielding compounds with low antagonistic potency. In our hands, many monosaccharide antagonists such as ERI-1 (Fig. 2) demonstrated good activity in in vitro assays utilizing cultured cells or primary cultures of monocytes containing low concentrations of plasma. However, when the compounds were tested in vivo in the mouse endotoxin challenge models or in vitro in the presence of higher concentrations of plasma, such as in whole blood assays, the activity of the monosaccharides was attenuated.

Workers at Sankyo recently reported (42,43) two similar tri-acyl monosaccharides containing an anomeric carboxylic acid and O-4-phosphate (Sankyo-1 and -2, Fig. 2) that were shown to be potent inhibitors of LPS-induced TNF-α production in monocytes at 5 and 17 nM, respectively. The use of an anomeric carboxyl group prevents the formation of undesirable,

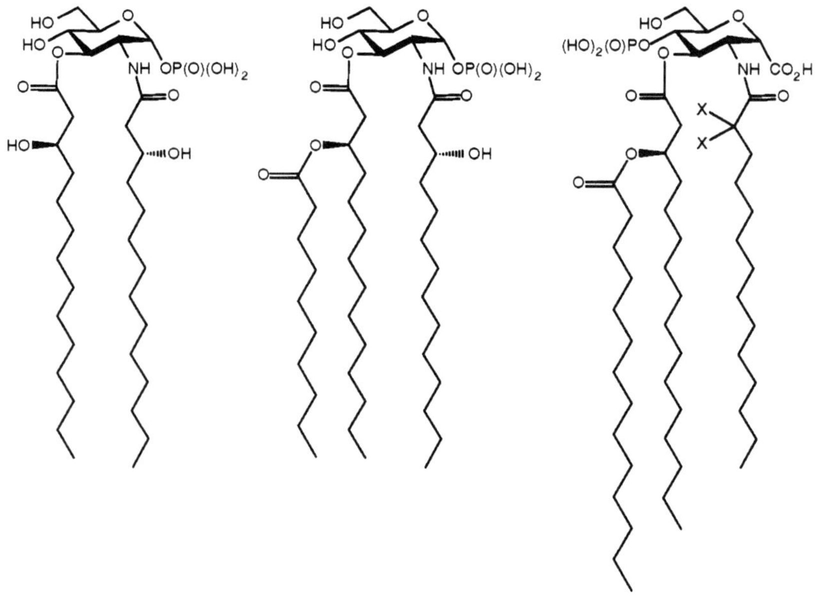

Fig. 2 Structures of monosaccharide antagonists. Lipid X is a monosaccharide precursor of lipid A, ERI-1 is the synthetic trisubstituted "right-half" of lipid A, and Sankyo-1 and -2 are previously described (42,43). For convenience and clarity, compounds are given ERI numbers if not previously disclosed, or B or E numbers were retained if previously described in literature reports.

agonistic disaccharides as described by Stütz and coworkers at Sandoz (51,52). The in vivo activities of these potential antagonistic monosaccharides have not been reported.

Disaccharide Antagonists Based on R. capsulatus Lipid A

The potential utility of disaccharide endotoxin antagonists was verified by reports describing the lipid A from *R. capsulatus* [RcLA (40)] and *R. sphaeroides* [RsLA (41,53,54)]. As shown in Figure 3, the reported proposed structures of the nontoxic lipid A from *R. capsulatus* and *R. sphaeroides* differ from toxic *E. coli* lipid A in the number of acyl chains on the B-ring of the disaccharide, the difference in length of the individual acyl chains, and the presence of unsaturation on the acyloxyacyl moiety on RcLA and RsLA. These antagonists also possess two β-ketoacyl chains at the N-2 and N-2' positions on RcLA and one β-ketoacyl chain at the N-2 position on RsLA.

Both in vitro and in vivo, *R. capsulatus* lipid A possesses antiendotoxin activity with a potency desirable of pharmaceutical antagonists (41). Synthetic RcLA (sRcLA; Table 1) appears to retain antagonistic activity ascribed to the bacterially derived molecule. Although the natural product can be prepared on a laboratory scale, our experience demonstrated that RcLA could not be readily synthesized and purified to pharmaceutical standards on the scale necessary for development (unpublished observations). In addition, during chemical synthesis, it was shown that the material did not possess a minimum shelf life necessary for pharmaceutical development. Upon standing for extended periods or after treatment with acid or base, side products are generated that are weak agonists in murine systems. For example, deacylation at the O-3 or O-3' position of the disaccharide generated weak agonists (ERI-2 and ERI-3 in Fig. 4). Additional side products included the elimination of the β-acyloxy chain at the O-3' position (ERI-4) along with the O-3 deacylation of the eliminated by-product (ERA-5) (see Fig. 4 for biological activities).

Based on these observations, we stabilized RcLA-like molecules with O-3 and O-3' ester bonds by conversion to ethers. Shiozaki et al. (57) have shown that stabilization of *E. coli* lipid A with the generation of O-3 and O-3' alkyl chains produced an analog that loses endotoxin-like activity. Although Shiozaki's results seemed to pose a drawback to this approach, conversion from esters to ethers appeared to be necessary

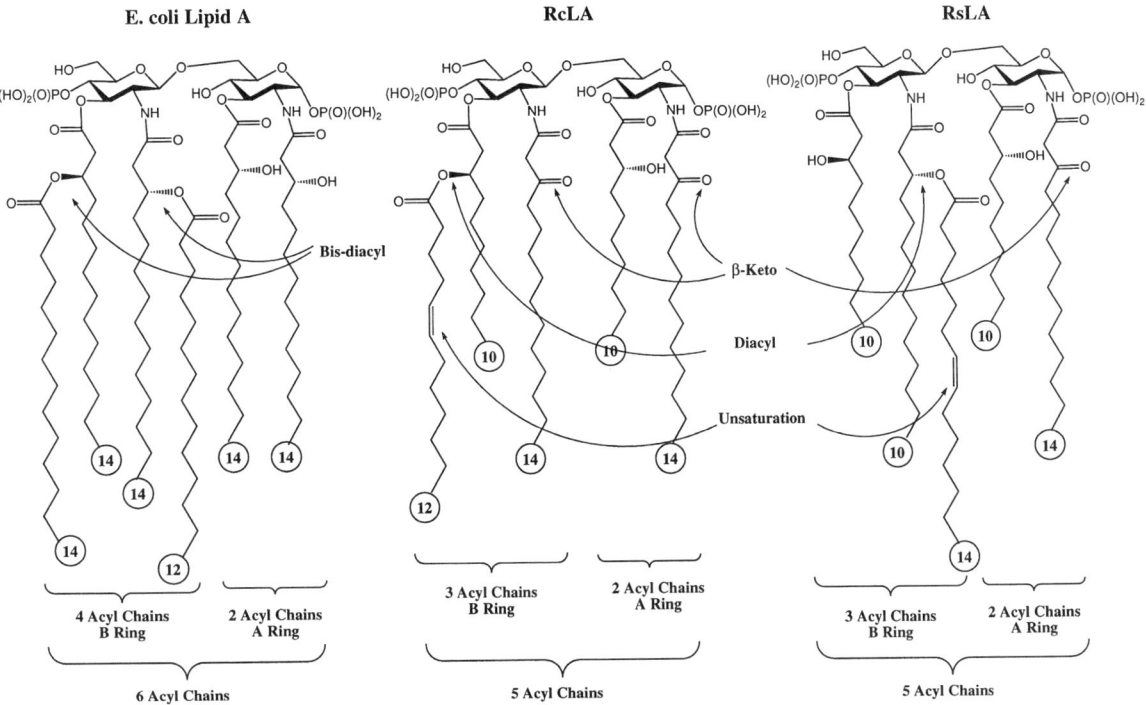

Fig. 3 Comparison of structures of natural agonistic and antagonistic lipid A from *E. coli*, *R. sphaeroides* proposed structure, and *R. capsulatus* proposed structure. The structural elucidation of bacterially derived RsLA and RcLA has been described (53,62) but the (R/S) configurations of the O-3 and O-3′ substituents of the acyl chains and the configuration of the unsaturated chain were not assigned in the natural product.

for stabilizing a potential disaccharide antagonist. Fortunately, however, this conversion of RcLA to ERI-6 (Fig. 5 and Table 1), yielded a potent LPS antagonist both in vitro and in mice. During the synthesis of this molecule, small amounts of impurities arose that appeared to be generated by the nucleophilic attack of the free C-6′ hydroxy group toward the adjacent O-4′ phosphate, rendering purification of the final product difficult. Since natural lipid A are connected to the core region and O-antigen portion of the LPS molecule at the O-6′ position, an alteration or "capping" of this position seemed to be a logical modification of ERI-6. Capping the C-6′ hydroxyl group with a methyl group not only eliminated the formation of impurities but blocked possible formation of cyclic phosphate. This final molecule, E5531 (Fig. 5 and Table 1), is a fully stabilized endotoxin antagonist with no detectable agonistic activities. Lack of toxicity and mutagenicity, as well as availability of stable formulation, established E5531 as a candidate for clinical development. E5531 is currently in clinical evaluation for the treatment of sepsis and septic shock, as previously described (58,59).

Additional E5531 Analogs

Concurrent with the clinical development of E5531, additional studies were undertaken to determine the structural requirements for a second-generation compound with structure simplification a major goal. To understand the structural requirements for antagonistic activity, we synthesized further analogs of E5531 and studied their antagonistic and agonistic activity in human whole blood and, in some cases, in vitro in murine model systems or in vivo in BCG-primed mice (Table 1). Structural modifications of E5531 provided molecules that, compared to E5531, varied in their ability to antagonize the activity of LPS. It should be noted that in some cases modifications that rendered the disaccharide less soluble or insoluble, such as removal of either phosphate functionality, resulted in compounds that appeared to be inactive (data not shown) but may have been too insoluble to assay at sufficiently high concentrations.

Physiologically, modification of the O-4 position of the A-ring of E5531 (Table 1 and Fig. 5) altered activity, sometimes dramatically. The alkylation of this po-

Table 1 Activity of E5531 and Analogs

Compound	Modification to E5531	Antagonistic activity in human whole blood (fold of E5531)[a]	Mouse agonism (% of LPS activity)[b]
sRcLA	O-3 and O-3′ acyl chains	1–2	<1[c]
E5531	O-3 and O-3′ alkyl chains	1	0.06
ERI-6	C-6′ Hydroxy on B-Ring	1	<0.2[d]
ERI-7	O-4 Methyl on A-Ring	2	NT[e]
ERI-8	O-4 Acetate on A-Ring	0.1–0.2	NT
ERI-9	C-4 Galacto configuration	0.06	NT
ERI-10	C-1 Hydroxy on A-Ring	NT	NT
ERI-11	C-4′ Hydroxy on B-Ring	<0.01	NT
ERI-12	N-2 β-(R)-Hydroxy-C14 acyl chain	0.5	0.3
ERI-13	N-2 β-(S)-Hydroxy-C14 acyl chain	1	0.2
ERI-14	N-2′ β-(R)-Hydroxy-C14 acyl chain	2.5	<0.2
ERI-15	N-2′ β-(S)-Hydroxy-C14 acyl chain	2	<0.15
ERI-16	O-3 γ-(R)-Methoxy-C10 alkyl chain	0.5	NT
ERI-17	O-3 γ-(R)-Acetoxy-C10 alkyl chain	0.1–0.4	NT
E5531	O-3 γ-[(R)-Hydroxy]-O-3′ γ-[(R)-acyloxy]-	1	0.06
ERI-18	O-3 γ-[(S)-Hydroxy]-O-3′ γ-[(R)-acyloxy]-	5.1	NT
ERI-19	O-3 γ-[(R)-Hydroxy]-O-3′ γ-[(S)-acyloxy]-	1.8	NT
ERI-20	O-3 γ-[(S)-Hydroxy]-O-3′ γ-[(S)-acyloxy]-	2.6	NT
ERI-21	O-3′ γ-[(R)-Hydroxy]-C10 alkyl chain	<0.06	<0.1
ERI-22	O-3′ γ-[(R)-(E)-Δ^5-C12 Acyloxy]-C10 alkyl chain	0.65	NT
ERI-23	O-3′ γ-[(R)-(Z,Z)-Δ^5, Δ^7-C12 Acyloxy]-C10 alkyl chain	0.94	NT
ERI-24	O-3′ γ-[(R)-C12 Acyloxy]-C10 alkyl chain	0.5–2	NT
B975	O-3′ γ-[(R)-(Z)-Δ^5-C12 Alkyloxy]-C10 alkyl chain	0.75–1	<0.2
E5531	N-2 Acyl chain = 14 carbons, O-3 alkyl chain = 10 carbons, O-2′ acyl chain = 14 carbons, and O-3′ alkyl chain = 10 carbons	1	0.06
ERI-25	O-3 Alkyl chain is 2 carbons shorter	0.54	NT
ERI-26	O-3 Alkyl chain is 4 carbons longer	1.5	NT
ERI-27	O-3 Alkyl chain and O-3′ alkyl chains are 2 carbons shorter	0.013	NT
B464	O-3 Alkyl chain and O-3′ alkyl chains are one carbon longer	1	<0.2[d]
ERI-28	O-3 Alkyl chain and O-3′ alkyl chains are 4 carbons longer	0.19	70–95
ERI-29	N-2 Acyl chain and N-2′ acyl chain are 2 carbons shorter	0.62	NT
ERI-30	N-2 Acyl chain and N-2′ acyl chain are 6 carbons shorter	<0.01	NT
ERI-31	N-2 Acyl chain and N-2′ acyl chain are 2 carbons longer	0.11	NT
ERI-32	N-2 Acyl chain, O-3 alkyl chain, O-2′ acyl chain and O-3′ alkyl chain are each 2 carbons shorter	<0.01	0.1
LPS		N/A	100
Lipid A		N/A	34
Vehicle only		N/A	0.1

[a]Synthesis and preparation and assay of all compounds was as described (55,56,58,59). Antagonistic activity was tested by measurement of TNF-α by ELISA assay after 3-hour incubation of various concentrations of the indicated compound in human whole blood along with LPS (10 ng/ml). These results can be compared to the IC$_{50}$ for E5531 found in the same assay (generally between 3 and 15 nM). Human agonistic activity was tested by measurement of TNF-α by ELISA assay after 3-hour incubation in human whole blood at 100 and 10 μM final concentration and was not detectable (<15 pg/ml) for all compounds.

[b]Compounds (100 μg/mouse) or LPS (200 ng/mouse) or lipid A (200 ng/mouse) were injected i.v. into BCG-primed male C57BL/6 mice mice and blood assayed for TNF-α by ELISA after 90 minutes (55,59). In these assays, plasma levels of TNF-α varied between 150 to 450 ng/ml in response to LPS.

[c]Tested after injection of 30 μg.

[d]Tested in vitro using γ-interferon primed mouse macrophage cells as described (54,55).

[e]NT = Not tested.

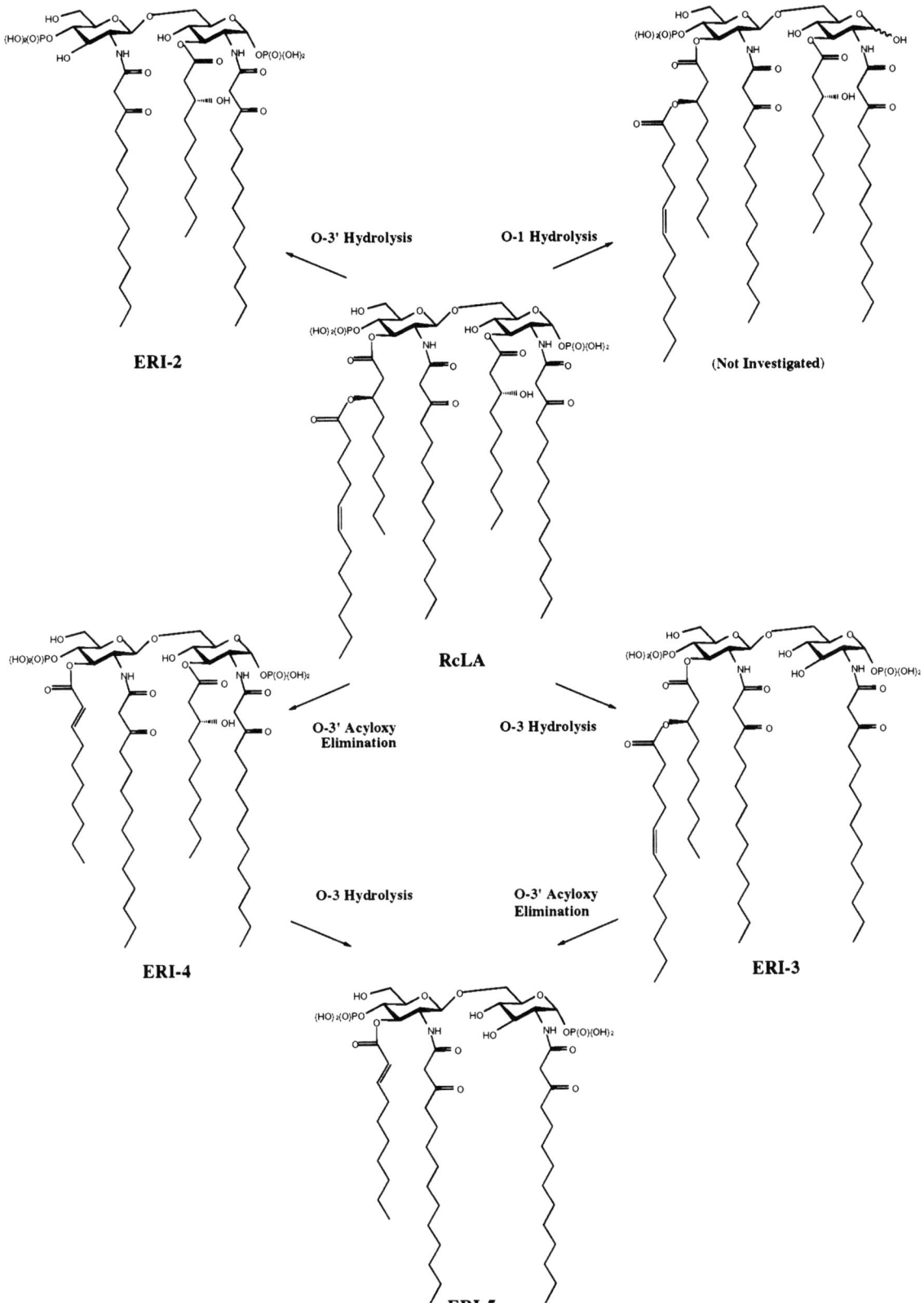

Fig. 4 Potential degradation products of RcLA. RcLA posesses hydrolytically sensitive structures at the O-1, O-3, and O-3′ acyl positions. Antagonism of LPS-mediated activation of human monocytes was assayed as described in Rose et al. (54); sRcLA (synthetic material of the proposed structure of RsLA) has an IC$_{50}$ of 5.9 ± 2.4 nM. ERI-3 was 12% as potent an antagonist as sRcLA, whereas ERI-4, ERI-2, and ERI-5 possessed less than 1% of the activity of sRcLA.

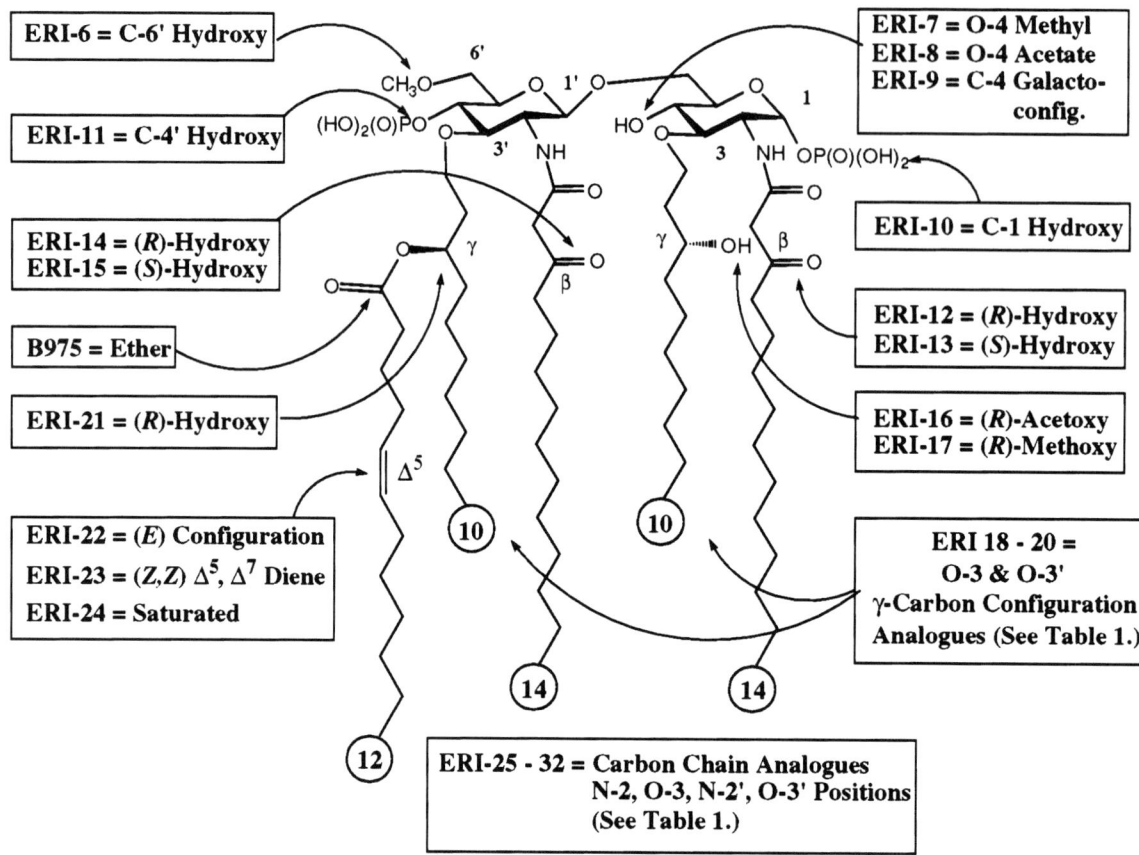

Fig. 5 E5531 and analogues. E5531 is a stabilized RcLA-like molecule with O-3 and O-3' ester bonds converted to ether-linked alkyl chains. In addition, the C-6' hydroxy of RcLA is converted to methoxy. Synthetic analogues of this molecule and their biological activities are described in Table 1.

sition with a methyl group (ERI-7) provided a molecule with an approximately twofold enhancement in antagonistic activity. Interestingly, acetylation of the O-4 position (ERI-8) decreased the biological activity by 80–90% in comparison to E5531. The decrease in activity from this last alteration is proposed to be due to the steric bulk of this modification, thus causing a drastic conformational change in the molecule, and not due to the electronics at the C-4 position. The lack of a C-4 H-bond donor in ERI-7 may also play a role in its enhanced activity. In order to investigate the above prediction, synthesis of the C-4 deoxy analog of E5531 was undertaken. Unfortunately, the synthesis of this analog was found to provide unstable intermediates that could not be elaborated to the final targeted compound. Changing the configuration at the C-4 from *gluco* to *galacto* (ERI-9) results in a fivefold loss in activity. Thus, the position of the C-4 oxygen plays a significant role in the biological activity of these molecules.

The role of phosphate functionality on the O-1 position of the A-ring and the O-4' position on the B-ring toward antagonistic activity was also evaluated. Unfortunately, the O-1 dephosphorylated molecule ERI-10 was insoluble in the assay media and could not be tested. However, the O-4' dephosphorylated molecule ERI-11 was evaluated and found to have no measureable antagonistic activity. These results demonstrate the necessity of the phosphate functionalities on the parent compound.

Replacement of the β-ketone at the N-2 position with either a β-(R) hydroxy (ERI-12) or a β-(S) hydroxy (ERI-13) provided analogs that retained the activity of E5531. On the other hand, the replacement of the β-keto functionality at the N-2' position with either a β-(R) hydroxy (ERI-14) or a β-(S) hydroxy group (ERI-15) yielded compounds slightly more active than E5531.

Modification of γ-hydroxy on the O-3 alkyl side chain to a methoxy group (ERI-16) or acetoxy group

(ERI-17) generated molecules with only 10–50% of the activity of E5531. Thus, we can predict that any addition of bulk to the O-3 alkyl chain will decrease the biological activity of E5531. The stereochemistry of the hydroxy functionality on the γ-carbon of the O-3 alkyl chain and the acyloxy functionality on the γ-carbon of the O-3′ alkyl chain of E5531 are both of the *R*-configuration. Surprisingly, changing the configuration of the γ-carbon of the O-3 alkyl chain to the *S*-configuration (ERI-18) enhanced the antagonistic activity approximately fivefold. The change of the configuration of the γ-carbon of the O-3′ alkyl chain to the *S*-form, ERI-19, also slightly enhanced activity. Likewise, changing the stereochemistry of both γ-carbons on O-3 and O-3′ alkyl chains to the *S*-configuration, ERI-20, also enhances the activity approximately threefold. Although these exciting results demonstrated the potential for analogs with increased potency, the increases in potency were not significant enough to override our primary goal of finding a simplified structure for a second-generation lipid A antagonist.

Deletion of the γ-acyl group on the O-3′ alkyl chain analog forming an O-3′ γ-(*R*)-hydroxyalkyl chain (ERI-21) resulted in a significant loss in activity (>94% loss compared to E5531). However, minor structural changes to the γ-acyl chain had only weak effects on biological activity. Conversion of the Δ^5-(*Z*)-unsaturation (*cis* double bond) on the γ-acyl chain to a (*E*)-configuration (*trans* double bond; ERI-22), substitution with a diene containing *cis* double bonds at the 11- and 13-positions (ERI-23), or substitution with a saturated chain (ERI-24) yielded compounds of about the same activity as E5531. Conversion of the γ-acyl chain linkage to an γ-alkyl linkage (B975) retained the activity of the parent molecule, although this analog was less water soluble.

Finally, variations in the length of the O-3 and O-3′ alkyl chains and the N-2 and N-2′ acyl chains were investigated. The E5531 analogs with the O-3 alkyl chain length shortened to 8 carbons (ERI-25) or lengthened to 14 carbons (ERI-26) did not substantially alter the antagonistic activity. On the other hand, when both the O-3 and O-3′ alkyl chain lengths are shortened to 8 carbons (ERI-27), activity is reduced by almost two orders of magnitude. In contrast, a molecule possessing these O-3 and O-3′ alkyl chain lengthened by one carbon (i.e., 11 carbons long; B464) retains the activity of the parent molecule. Going one step further by lengthening the chains to 14 carbons (ERI-28) results in a significant loss of antagonistic activity along with a substantial increase in agonistic activity.

Lengthening the N-2 and N-2′ acyl chains to 12 carbons (ERI-29) decreased the antagonistic activity slightly. But shortening them to 8 carbons (ERI-30) essentially inactivated the molecule. Extending both acyl chains by 2 carbons or 16 carbons total (ERI-31) not only significantly decreases antagonistic activity but also increases agonistic activity. Shortening all of the N- and O-linked chains on E5531 by two carbons (ERI-32) generates a molecule that loses nearly all activity in human blood and gains some agonistic activity in mice.

While several of these analogs were more active than E5531, none possessed a suitable combination of increased potency and structural simplicity to meet our criteria for further development. For this reason, we investigated the modification of the tetra-acylated *E. coli* lipid A and penta-acylated *R. sphaeroides* lipid A.

Disaccharide Antagonists Based on *E. coli* Lipid A

The biosynthetic precursor of LPS, lipid IV$_A$ (Fig. 6), is a diphosphorylated disaccharide containing four saturated β-hydroxyacyl substituents (60). Lipid IV$_A$ is a potent LPS antagonist in human in vitro assays but is an agonist in murine systems (61). Since the structure of lipid IV$_A$ presents a synthetically simpler disaccharide, we began the investigation into the feasibility of using this biosynthetic intermediate as a lead structure. Synthetic analogs of lipid IV$_A$ were generated in order to investigate the structural constraints of the antagonistic activity of lipid IV$_A$, to eliminate the agonistic activity of this molecule, and to determine the potential for stabilization of the compound for pharmaceutical development. To this end, modifications of the disaccharide were made by substituting an alkyl chain at the O-3 position (ERI-33) and with two alkyl chains at O-3 and O-3′ positions (B1060). Table 2 provides the biological results of these synthetic analogs. As found for the stabilized sRcLA, ERI-6 in Table 1, and contrary to the finding of Shiozaki's group with the ether-liked lipid A analog, the ether-linked analogs of lipid IV$_A$ retained some, albeit weak, antagonistic activity. Unfortunately, agonistic activity in the murine system was also retained.

R. capsulatus lipid A (RcLA) (62) contains β-ketoacyl chains attached to the N-2 and N-2′ positions. Modification of B1060 by the replacement of the N-2 and N-2′ acyl chains with β-ketoacyl chains provided ERI-34. This alteration proved to have a positive impact on the antagonistic and especially the agonistic activity. Unfortunately, even the weak agonistic activity

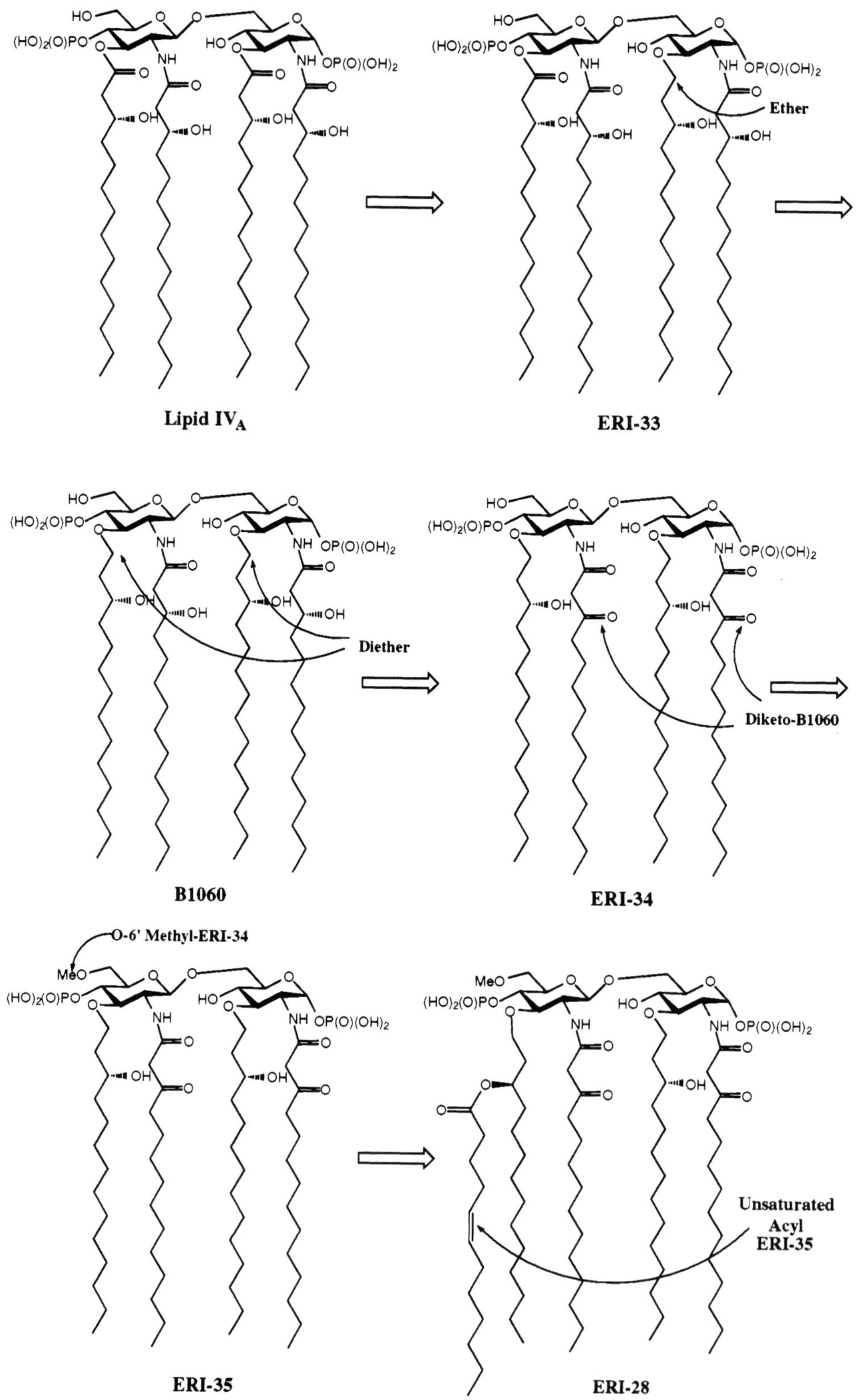

Fig. 6 Structures of lipid IV$_A$ and analogues. ERI-33 is the O-3 γ-(R)-hydroxy ether-linked analogue of lipid IV$_A$. B1060 is the O-3 γ-(R)-hydroxy ether and O-3′ γ-(R)-hydroxy ether-linked (diether) analogue of lipid IV$_A$. ERI-34 is the N-2 β-ketoacyl chain and N-2′ β-ketoacyl chain analogue of B1060. ERI-35 is the O-6′ methyl analogue of ERI-34. ERI-28 is the O-3′ (R)-γ-[(Z)(C12-Δ5-acyloxy)] alkyl chain analogue of ERI-35.

Table 2 Agonistic and Antagonistic Activity of Lipid IV$_A$ Analogs

Analog	Antagonistic activity in human whole blood (nM)[a]	Mouse agonism (% of LPS activity)[b]
ERI-33	10–30 nM	145
B1060	41–60 nM	60
ERI-34	83 nM	<1
ERI-35	Inactive	1
ERI-28	23 nM	70–95

[a]Compounds were added to freshly drawn human whole blood along with LPS (10 ng/ml) as described in the legend to Table 1. After 3-hour incubation, plasma was obtained by centrifugation and stored at −80°C until the ELISA assay for TNF-α was performed. Results are described as activity relative to E5531 tested in the same assays (generally between 3 and 15 nM). Human agonistic activity was tested by measurement of TNF-α by ELISA assay after 3-hour incubation in human whole blood at 100 and 10 μM final concentration and was not detectable (<15 pg/ml) for all compounds.

[b]All compounds (300 μg/mouse) or LPS (200 ng/mouse) were injected i.v. into BCG-primed mice and blood assayed for TNF-α by ELISA after 90 minutes. In these assays, plasma levels of TNF-α varied between 150 to 450 ng/ml in response to LPS. When tested in vitro for induction of nitric oxide synthesis at 1 μM in γ-interferon primed mouse macrophage cells as described (54), ERI-28, ERI-34, ERI-35, and B1060 induced a 7- to 11-fold stimulation of NO release over that of 5 U/ml γ-interferon alone. This is 50–90% as agonistic as 10 ng/ml LPS (12.5 stimulation of NO release).

of ERI-34 was too significant to pursue this analog as a structural lead. In addition, the chemical stability of this compound was still questionable. Potential synthetic problems such as cyclization of the O-4' phosphate during synthesis was circumvented by "capping" the O-6' position of ERI-34 with a methyl group. This alteration generated compound ERI-35, which was, surprisingly, a relatively inactive molecule.

As seen in Figure 3, RcLA also contains a unsaturated side chain acylated on the β-hydroxy group of the O-3' acyl chain not found on lipid IV$_A$. Addition of this functionality to ERI-35 provided the stabilized analog of both lipid IV$_A$ and RcLA, ERI-28. Although this alteration enhanced antagonistic activity, agonistic activity of the molecule was increased significantly. Furthermore, this addition of another chain into the molecule failed to meet the goal of providing a more simple endotoxin antagonist lacking agonistic activity. As with lipid IV$_A$, the biological activity of these molecules strongly depends on the assay system used. While none of these compounds demonstrated agonistic activity in human whole blood, all demonstrated LPS-like agonistic activity in murine systems as shown in Table 2. The relevance of this agonistic activity is discussed in the biology section below.

Disaccharide Antagonists Based on R. sphaeroides Lipid A

The lipid A from *R. sphaeroides* (RsLA) also inspired interest as it possesses potent antagonistic activity in both human and murine systems. Material of the proposed structure for *R. sphaeroides* lipid A (sRsLA) has been prepared (63). Comparison of this synthetic material to the natural product using ^{1}H- and ^{13}C-NMR and HPLC co-injection experiments showed bacterially derived, commercially produced RsLA to be not identical to its synthetic proposed structure. And while the synthetic material of the proposed structure of RsLA was as potent as bacterially derived material at antagonizing LPS in human monocytes, human whole blood, and murine macrophages, it lacked the weak agonistic activity present in the bacterially derived materials (54). Some unsaturation appears to be important; saturating the acyloxy double bond on ERI-36 decreases activity more than 80%. However, conversion of the *cis* double bond to the *trans* or E-configuration in ERI-37 (reported as 2'-*trans* LA in Ref. 54) or substitution with a C-12 Δ^5 (ERI-38) did not alter activity. Deletion of this side chain (ERI-39) dramatically decreases antagonistic activity and imparts mouse agonistic activity (Table 3). This observation leads us to speculate that contamination of bacterially derived RsLA by this compound may be responsible for the weak agonistic activity previously described (54).

As outlined above for the RcLA-to-E5531 transformations, attempts at developing stabilized materials based on the structure of RsLA were similarly focused towards obtaining a pure, stable material with high potency (see Table 3). The O-3 and O-3' acyl chains were again converted to alkyl chains as well as alkylation of the O-6' position with a methyl group.

In this series, we were also able to "collapse" the unsaturated acyloxy chain located on the β-position of the N-2' acyl chain onto the N-2' acyl chain. This transformation provided a single acyl chain 18 carbons in length containing the (Z)-unsaturation at approximately the correct position "measured out" from the disaccharide core. In addition, we have found that removal of the hydroxy group located at the γ-position of the O-3 alkyl chain increased activity and made this new molecule more synthetically accessible. Finally, alkylation of the γ-hydroxy group on the O-3' alkyl chain with a methyl group optimized this series for

Table 3 Antagonists Based on the Structure of RsLA

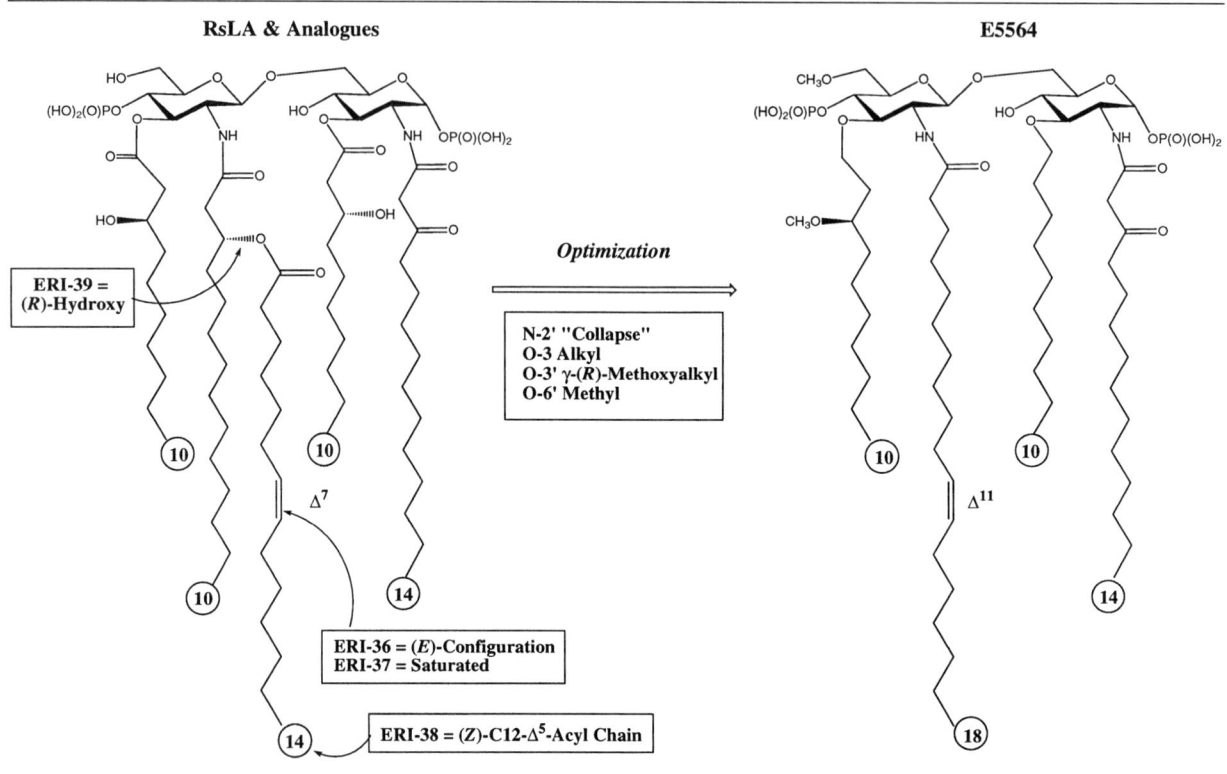

Analog	Modification to RsLA	Antagonistic activity (fold of E5531)[a]	Mouse agonism (% of LPS activity)[b]
sRsLA	Proposed structure (Fig. 4)	7	0.5–1.7
ERI-36	N-2′ β-Saturated-acyloxy chain on RsLA	0.17	NT
ERI-37	N-2′ β-(E)-Δ^7-Acyloxy chain on RsLA	8	3.4
ERI-38	N-2′ β-cis-Δ^5-C12 Acyloxy chain on RsLA	8	NT
ERI-39	N-2′ β-(R)-Hydroxy-(deacyl) RsLA	<0.03	50
E5564	(As shown)	10	<0.1

[a]Compounds were added to freshly drawn human whole blood along with LPS (10 ng/ml) as described in the legend to Table 1. Human agonistic activity was tested by measurement of TNF-α by ELISA assay after three hours incubation in human whole blood at 100 and 10 μM final concentration and was not detectable (<15 pg/ml) for all compounds.
[b]Tested at 100 μM and compared to LPS (10 ng/ml in vitro using γ-interferon primed mouse macrophage cells as described in the legend to Table 1 and in reference 54. In addition, E5564 demonstrated no measureable agonistic activity after iv. injection (100 μg/mouse) into BCG-primed mice and blood assayed for TNF-α as described in the legend to Table 1.
NT = Not tested.

both biological activity and ease of synthesis. The resultant molecule, E5564, demonstrates improved potency over that of E5531. This molecule is stable, more readily synthesized, and has no detectable agonistic activities. This molecule is presently being developed at Eisai as a second-generation antisepsis drug. Clinical evaluation will begin shortly.

Biology

In Vitro Analysis of LPS Antagonists

As sepsis is a consequence of a bloodborne bacterial infection, in vitro analysis of whole blood provides a unique opportunity to test drugs in a mileu in which they must actually work. Thus, LPS stimulation of human whole blood using release of TNF-α as a marker provides a relevant and convenient assay system to test potential antisepsis therapeutics. Furthermore, inhibition of release of TNF-α as an accurate measure of cellular activation is validated by co-analysis of release of other cytokines in this assay system. For both E5531 and E5564, inhibition of TNF-α release correlates with suppression of release of all LPS-induced cytokines and cellular mediators measured, including TNF-α, IL-1β, IL-6, IL-8, and IL-10. In these assays, the IC_{50} for inhibition of release of all of these cytokines was generally 15 nM or less.

Activity of E5531 and E5564 is not greatly affected by high serum concentrations, indicating that their activity is neither stimulated nor inhibited by serum components such as LBP or BPI. The requirement that drugs work in whole blood argues that compounds immediately inactivated by serum would clearly eliminate them as possible therapeutics. During primary screening, antagonistic activity of most analogs was also measured in primary cultures of human monocytes/macrophages cultured in 1% human serum. For our lead compounds, IC_{50}s were in the range of 0.1–0.5 nM, with complete inhibition of TNF-α release observed at 10 nM or more. Comparison of these IC_{50}s to those obtained in whole blood (<20 nM) indicates that dramatic increases in serum concentration have little effect on the activity of these two compounds. This is not the case for all compounds, where, as mentioned above for monosaccharide antagonists such as lipid X, addition of serum greatly reduces antagonistic activity (data not shown).

Analysis of LPS-Like Agonistic Activity in Mice

It is now well established that a dichotomy exists when lipid A–like compounds are tested in human and rodent systems for agonism and antagonism. As described above and in previous reports (54,55,58,59,64), several compounds such as lipid IV_A, B1060, ERI-5, ERI-28, and ERI-33 demonstrate antiendotoxin activity without exhibiting agonistic activity in human cells and yet are highly agonistic both in vitro and in mice. With this in mind, one may question the usefulness of the rodent in vitro and in vivo models for testing compounds for human therapeutic application. Does the establishment of the criteria that our lead molecules be nonagonistic in murine systems lead us to discard potential therapeutics?

We believe that the use of these animal model systems can be supported by several observations. First, it would be difficult and expensive to replace these rodent species for primary in vivo screens in advanced models for drug activity as well as relevant models for toxicity studies. In addition, we have observed that cells from rat blood can be activated (release of TNF-α) by high concentrations of molecules that are mouse agonists and human antagonists (data not shown). This observation is supported by a report that Ulmer et al. (64) found that agonism in mice correlates with agonistic activity in other animals such as rabbits, guinea pigs, and dogs, indicating that mouse agonism may predict agonistic activity in other nonhuman species. For this reason, we believe that assessment of drug agonistic activity in the mouse system can be used as a sensitive and reliable method for screening out species-specific agonism-based toxicity that would complicate animal model assessments needed for drug development. In conclusion, we believe that our criteria that a molecule be nonagonistic in murine systems in order to be selected for development appears to be well justified.

One highly sensitive assay for mouse agonism in vitro is the measurement of release of nitric oxide and TNF-α in murine macrophages (RAW cells), especially after sensitization (priming) with murine γ-interferon (IFN-γ) (54,65). Compounds such as ERI-35, ERI-5, ERI-3, and B1060 that are in vivo mouse agonists (described above), as well as certain preparations of bacterially derived RsLA (54) stimulated release of NO up to 11-fold when added along with γ-interferon (see Tables 2 and 3). E5531 and E5564 are devoid of agonistic activity in this system when used at concentrations up to 100 μM. This lack of agonistic activity argues that it is not likely that molecules such as E5531 and E5564 would demonstrate agonistic activity or agonism-based toxicity (especially in repeated-dose studies) in preclinical animal model studies.

In Vivo Evaluation of LPS Antagonistic Activity

In vivo, LPS activates more than bloodborne cells. Besides circulating cells, the presence of fixed macrophages and other responsive cells in the liver and other organs may complicate antagonism of LPS in vivo. One hour after intravenous injection of 100 ng of *E. coli* LPS into a BCG-primed mouse, plasma TNF-α levels rise from ~200 (basal) to ~150,000 pg/ml. This response is approximately 20 times greater than that seen when whole blood is challenged with 100 ng/ml LPS (TNF-α increases from ~50 to ≤3000 pg/ml). This result argues strongly that noncirculating or fixed macrophages and/or other cells are major contributors to the overall in vivo endotoxin response.

In order to determine if E5531 possesses in vivo agonistic activity, LPS or E5531 were injected i.v. into the tail vein of BCG-primed mice and induction of TNF-α, IL-1α, and IFN-γ were measured over time. As shown in Figure 7, intravenous injection of 200 ng LPS induces first TNF-α at ~1 hour, IL-1α after 2–3 hours, then IFN-γ after 3–4 hours. E5531 alone induces less than 5% of the IL-1α and less than 1% of the TNF-α or IFN-γ induced by 200 ng of LPS in BCG-primed mice. This result indicates that E5531 possesses no measurable agonistic activity of its own, even when multiple cytokines are assayed over a broad time course of in vivo response.

Antagonistic activity was measured after co-injection of E5531 with either 100 or 200 ng of LPS. As shown in Figure 8, activity of E5531 is dependent on dose of agonist as well as dose of antagonist. In this case, the IC_{50} for E5531 was roughly 33 μg or less, but greater doses of antagonist were required for complete block of the higher endotoxin challenge.

In another in vivo model testing the effect of LPS antagonist against live bacterial infection, co-administration of E5531 and antibiotic blocks septic shock due to infection by viable *E. coli* (58). These studies have also demonstrated that in vivo administration of antibiotic triggers an increase in plasma endotoxin levels.

Finally, another in vivo observation provides some insight into the mechanism of drug activity. Bacterially derived RsLA has been proposed to indirectly antagonize LPS by inducing glucocorticoids, which suppresses in vivo LPS response (66). However, it is unlikely that E5531 works through this mechanism. In mice, levels of plasma corticosterone assayed 90 minutes after injection of 200 ng of LPS increased from 51.7 ± 14.8 (preinjection) to 145.0 ± 29.8 ng/ml ($p < 0.05$; $n = 4$ for each treatment). Similar treatment with 30–300 μg E5531 resulted in no significant increases in plasma corticosterone; plasma levels changed from 51.7 ± 14.8 ng/ml (preinjection) to 41 ± 10 ng/ml (30 μg E5531), 66.8 ± 14 ng/ml (100 μg E5531), and 42 ± 11 ng/ml (300 μg E5531) at 90 minutes with no dose-response correlation observed ($p > 0.05$ for all changes). In addition, no rapid time-dependent induction of corticosterone was measurable at time points as

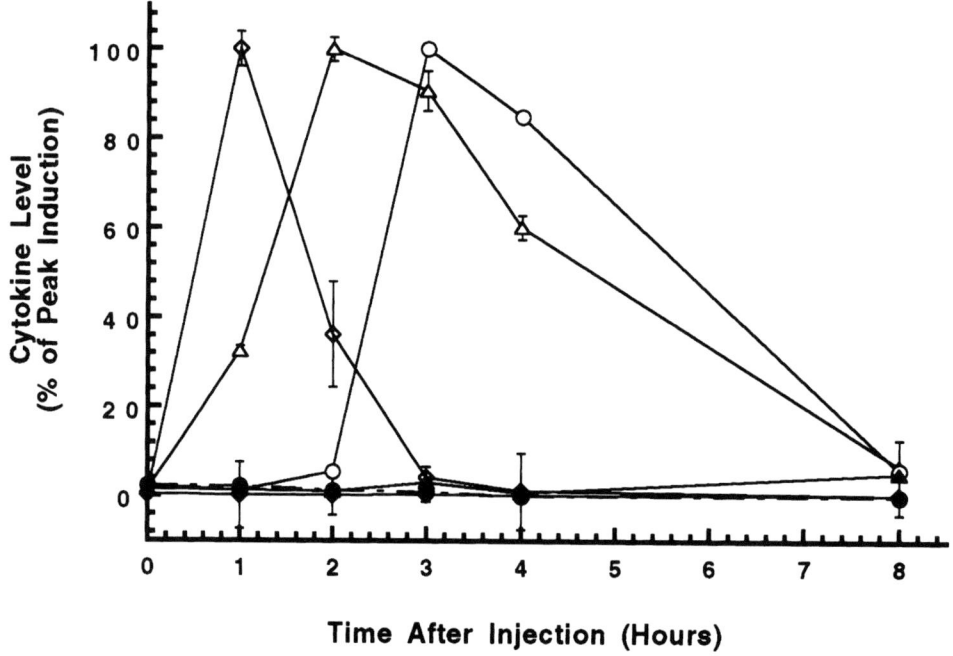

Fig. 7 In vivo time course for cytokine release after intravenous injection of LPS or E5531. Mice were primed by IV tail vein injection of live BCG 11 days before challenge with 200 ng LPS (open symbols) or 100 μg E5531 (closed symbols). At the indicated times, blood was obtained (3 animals per time point), plasma prepared, and stored at −80°C until the cytokine assays could be performed. TNF-α (◇,◆), IL-1α (△,▲), and γ-interferon (○,●) were measured by ELISA assays using Genzyme assay kits for TNF-α and IL-1α and Biosource (γ-interferon) kits. Peak plasma levels were 320,000 ± 24,363 pg/ml (TNF-α), 1008 ± 119 pg/ml (IL-1α) and 1253 ± 56 pg/ml (γ-interferon). For E5531 only, peak plasma levels (above basal) were 108 ± 224 pg/ml (TNF-α), 37 ± 9 pg/ml (IL-1α), and 0 ± 0 pg/ml (γ-interferon).

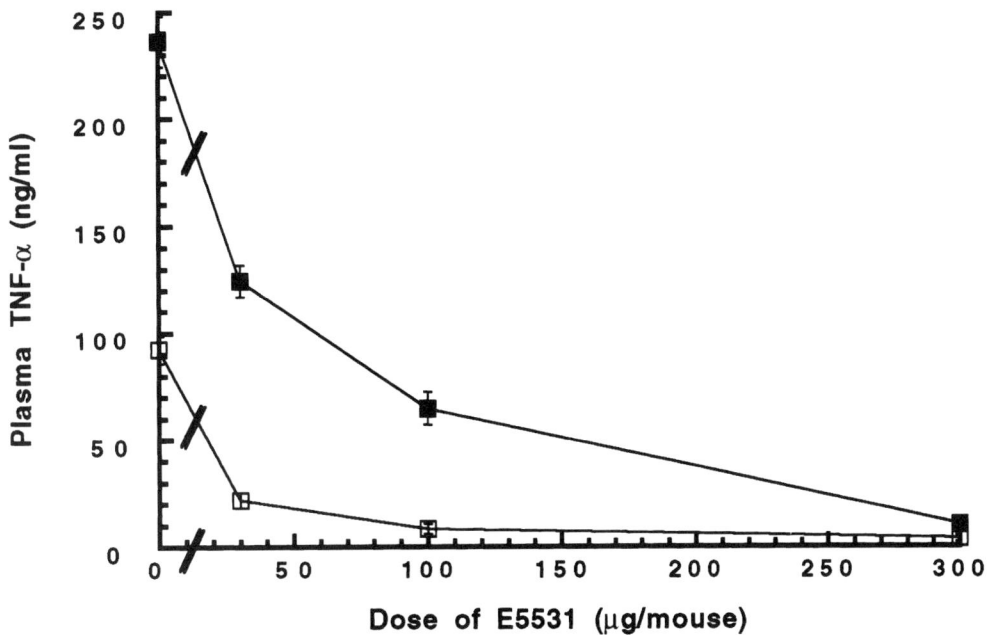

Fig. 8 In vivo inhibition of LPS response by E5531; Effect of LPS dose. Male C57BL/6 mice were primed by IV tail vein injection of live BCG 14 days before challenge with 100 ng LPS (□) or 200 ng LPS (■) plus the indicated amount of E5531 as previously described (55,58). After 90 min., blood was obtained (3 animals per time point), plasma prepared, and stored at −80°C until the ELISA assay for TNF-α could be performed.

early as 20 minutes after administration of E5531 (data not shown). These results argue that suppression of LPS response by E5531 is not related to the more indirect mechanism of corticosterone induction.

Human In Vivo Antiendotoxin Activity

E5531 has been administered intravenously in single doses ranging up to 1 mg to normal male volunteers in a double-blind, placebo-controlled, randomized phase 1 clinical study. None of these subjects treated with drug experienced any drug-associated signs, symptoms, or laboratory abnormalities, including those associated with endotoxin agonistic activity.

In a double-blind, placebo-controlled, randomized study using a human endotoxin challenge model, subjects who received intravenous endotoxin (4 ng/kg) and concurrent placebo experienced altered vital signs (fever, tachycardia, decreased blood pressure), endotoxin-induced symptoms (headache, nausea, myalgia, chills, photophobia), neutropenia and lymphopenia followed by neutrophilia, increased cardiac output, decreased systemic vascular resistance (SVR), and increased concentrations of TNF-α, IL-6, and C-reactive protein. When administered concurrently with 4 ng/kg endotoxin, intravenous E5531 blocked or significantly ameliorated all of these effects in a dose-related manner. Other agents that have been evaluated in this model either possess LPS-agonistic activity and/or lack the ability to completely inhibit the LPS response in this model (Table 4). E5531 is currently being evaluated in a phase 2 clinical study in patients with severe sepsis.

CONCLUSIONS

E5531 and E5564 are chemically and biologically optimized lipid A–like disaccharides that potently antagonize the toxic actions of LPS. Intravenous administration of E5531 or E5564 to BCG-primed mice blocks lethality induced by LPS or *E. coli* infection, apparently by directly antagonizing LPS and not indirectly by inducing glucocorticoids or other cytokines. In an endotoxin-challenge model in humans, E5531 has demonstrated a broad spectrum of activity that is superior to that of any other reported molecule aimed at the treatment of sepsis and septic shock, suggesting that this class of compounds may be clinically useful in the treatment of these and other endotoxin-related human diseases.

Table 4 Comparison of Drugs and Biologics in the Human Endotoxemia Model

			Significant blockade of LPS-induced effects or changes						
Drug	LPS dose	Agonist activity	Symptoms	Vital signs	Pro-inflammatory cytokines (>90% inhibition)	C-reactive protein	Neutrophilia/ lymphopenia	Cardiac output/ SVR	Ref.
E5531	4 ng/kg[a]	−	+	+	+	+	+	+	67
MPLA	20 EU/kg	+	+/−	+/−	−	−	−	nd	68
IL-1ra	3 ng/kg	−	−	−	−	−	+/−	nd	69,70
IL-10	Ex vivo	−/+(WBC)	nd	nd	+	nd	nd	nd	71
rBPI	40 EU/kg	−	−	−	−	nd	−	nd	72
TNFR:Fc	4 ng/kg	−	−	−	+/−	−	+/−	−	73
rHDL	4 ng/kg	−	−	−	+	nd	+	nd	74
Steroids	4 ng/kg	−	+/−	+/−	−	nd	nd	nd	75
Pentoxyfylline	4 ng/kg	−	−	−	−	nd	−	nd	76
Ibuprofen	4 ng/kg	+(TNF-α, IL-8)	+	+	−	−	−	nd	76–78

nd, Not done.
[a] 4 ng/kg = 40 EU/kg

ACKNOWLEDGMENTS

This project has involved an enormous amount of both chemical and biological effort by the following people: Karen Ackermann, Osamu Asano, Marylou Baker, Alex Bridges, John Bristol, Jacquelin Budrow, Gloria Dubuc, AnaMaria Garcia, Wendy Gavin, Ieharu Hishinuma, Koichi Katayama, Akifumi Kimura, Arthur Lee, Michael Lewis, Laura McGuigan, Pamela McGuinness, Maureen Mullarkey, Michel Perez, Andrea Robidoux, Jeffrey Rose, Emiko Sato, Dean Stamos, and Yuan Wang.

The authors also wish to thank Dr. J. P. Carter for his critical review of this manuscript.

REFERENCES

1. Pugin J, Ulevitch RJ, Tobias PS. Tumor necrosis factor-alpha and interleukin-1 beta mediate human endothelial cell activation in blood at low endotoxin concentrations. J Inflamm 1995; 45:49–55.
2. Mathison JC, Virca GD, Wolfson E, Tobias PS, Glaser K, Ulevich RJ. Adaptation to bacterial lipopolysaccharide-induced tumor necrosis factor production in rabbit macrophages. J Clin Invest 1990; 85:1108–1118.
3. Mathison JC, Wolfson E, Ulevich RJ. Participation of tumor necrosis factor in the mediation of gram negative bacterial lipopolysaccharide-induced injury in rabbits. J Clin Invest 1988; 81:1925–1937.
4. Matsura M, Nakano M. Mechanisms to induce tolerance by synthetic monosaccharide lipid A analogues against LPS lethality in mice. Prog Clin Biol Res 1995; 392:539–548.
5. Kiso M, Tanaka S, Fujuita M, Fujishima Y, Ogawa Y, Ishida H, Hasegawa A. Synthesis of the optically active 4-O-phosphono-D-glucosamine derivatives related to the nonreducing sugar subunit of bacterial lipid A. Carbohydrate Research 1987; 162:127–140.
6. Kumazawa Y, Nakatsuka M, Takimoto H, Furuya T, Naguma T, Yamamoto A, Homma Y, Inada K, Yoshida M, Kiso M. Hasegawa A. Importance of fatty acid substituents of chemically synthesized lipid A-subunit analogs in the expression of immunopharmacological activity. Infect Immun 1988; 56:149–155.
7. Pedron T, Girard R, Eustache J, Bulusu ARCM, Macher I, Radzyner-Vyplel H, Stutz PL, Chaby R. New synthetic analogs of lipid A as lipopolysaccharide agonists or antagonists of B lymphocyte activation. Int Immunol 1992; 4:533–540.
8. Sato K, Yoo YC, Fukushima A, Saiki I, Takahashi TA, Fujihara M, Tono-oka S, Azuma I. A novel synthetic lipid A analog with low endotoxicity, DT-5461, prevents lethal endotoxemia. Infect Immun 1995; 63:2859–2966.
9. Bulusu M, Hildebrandt J, Lam C, Liehl E, Loibner H, Macher I, Scholz D, Schutze E, Stütz P, Vyplel H, Unger F. Enzymatic synthesis of analogs of bacterial lipid A and design of biologically active LPS-antagonists and -mimetics. Pure Appl Chem 1994; 66:2171–2174.
10. Rice GC, Brown PA, Nelson RJ, Bianco JA, Singer JW, Bursten S. Protection from Endotoxic shock in mice by pharmacologic inhibition of phosphatidic acid. Proc Natl Acad Sci USA 1994; 91:3857–3861.
11. Zabel P, Schade FU, Schlaak M. Modulating effects of pentoxyfylline on cytokine release syndromes. In: Levin J, Alving CR, Munford RS, Stütz PL, eds. Bacterial Endotoxin: Recognition and Effector Mecha-

nisms. Amsterdam: Elsevier Science Publishers, 1993: 413–421.
12. Peppel K, Beutler B. Biological properties of a recombinant tumor necrosis factor inhibitor. In: Levin J, Alving CR, Munford RS, Stütz PL, eds. Bacterial Endotoxin: Recognition and Effector Mechansims. Amsterdam: Elsevier Science Publishers, 1993:447–454.
13. Cohen PS, Nakshatri H, Dennis J, Caragine T, Bianchi M, Cerami A, Tracey KJ. CNI-1493 inhibits monocyte/macrophage tumor necrosis factor by suppression of translation efficiency. Proc Natl Acad Sci USA 1996; 93:3967–3971.
14. Lee JC, Laydon JT, McDonnell PC, Gallagher TF, Kumar S, Green D, McNulty D, Blumenthal MJ, Heys JR, Landvatter SW, Strickler JE, McLaughlan MM, Siemans IR, Fisher SM, Livi GP, White JR, Adams JL, Young PR. A protein kinase involved in the regulation of inflammatory cytokine biosynthesis. Nature 1994; 372:739–746.
15. Bianchi M, Ulrich P, Bloom O, Meistrell M, Zimmerman GA, Schmidtmayerova H, Bukrinsky M, Donnelley T, Bucala R, Sherry B, Manogue KR, Tortolani AJ, Cerami A, Tracey KJ. An inhibitor of macrophage arginine transport and nitric oxide production (CNI-1493) prevents acute inflammation and endotoxin lethality. Mol Med 1995; 1:254–266.
16. Bianchi M, Bloom O, Raabe T, Cohen PS, Chesney J, Sherry B, Schmidtmayerova H, Calandra T, Zhang X, Bukrinsky M, Ulrich P. Cerami A, Tracey KJ. Suppression of proinflammatory cytokines in monocytes by a tetravalent guanylhydrazone. J Exp Med 1996; 183: 927–936.
17. McGeehan GM, Becherer JD, Bast RC Jr., Boyer CM, Champion B, Connolly KM, Conway JG, Furdon P, Karp S, Kidao S, McElroy AB, Nichols J, Pryzxansky KM, Schoenen F, Sekut L, Truesdale A, Verghese M, Warner J, Ways JP. Regulation of tumor necrosis factor-alpha processing by a metalloproteinase inhibitor. Nature 1994; 370:558–561.
18. Galloway CJ, Madanat MS, Mitra G. Monoclonal anti-tumor necrosis factor (TNF) antibodies protect mouse and human cells from TNF cytotoxicity. J Immunol Methods 1991; 140:37–43.
19. Pauli U, Bertoni G, Duerr M, Peterhans E. A bioassay for the detection of tumor necrosis factor from eight different species: evaluation of neutralization rates of a monoclonal antibody against human TNF-alpha. J Immunol Methods 1994; 171:263–265.
20. Beutler B, Milsark IW, Cerami AC. Passive immunization against cachectin/tumor necrosis factor protects mice from lethal effects of endotoxin. Science 1985; 229:869–871.
21. Reinhart K, Wiegand-Lohnert C, Grimminger F, Kaul M, Withington S, Treacher D, Eckart J, Willatts S, Bouza C, Krausch D, Stockenhuber F, Eiselstein J, Daum L, Kempeni J. Assessment of the safety and efficacy of the monoclonal anti-tumor necrosis factor antibody-fragment, MAK 195F, in patients with sepsis and septic shock: a multicenter, randomized, placebo-controlled, dose-ranging study [see comments] [published erratum appears in Crit Care Med 1996; 24: 1608] Crit Care Med 1996; 24:733–742.
22. Hale KK, Smith CG, Baker SL, Vanderslice RW, Squires CH, Gleason TM, Tucker KK, Kohno T, Russell DA. Multifunctional regulation of the biological effects of TNF-alpha by the soluble type I and type II TNF receptors. Cytokine 1995; 7:26–38.
23. Dinarello CA, Gelfand JA, Wolff SM. Anticytokine strategies in the treatment of the systemic inflammatory response syndrome. J Am Med Assoc 1993; 269: 1829–1835.
24. Redmond HP, Chavin KD, Bromberg JS, Daly JM. Inhibition of macrophage-activating cytokines is beneficial in the acute septic response. Ann Surg 1991; 214: 502–509.
25. Natanson C, Hoffman WD, Suffredini AF, Eichacker PQ, Danner RL. Selected treatment strategies for septic shock based on proposed mechanisms of pathogenesis. Ann In Med 1994; 120:771–783.
26. Sherman ML, Spriggs DR, Arthur KA, Imamura K, Frei ED, Kufe DW. Recombinant human tumor necrosis factor administered as a five-day continuous infusion in cancer patients: phase I toxicity and effects on lipid metabolism. J Clin Oncol 1988; 6:344–350.
27. Michie HR, Spriggs DR, Manogue KR, Sherman ML, Revhaug A, O'Dwyer ST, Arthur K, Dinarello CA, Cerami A, Wolff SM, et al. Tumor necrosis factor and endotoxin induce similar metabolic responses in human beings. Surgery 1988; 104:280–286.
28. Cerami A. Inflammatory cytokines. Clin Immunol Immunopathol 1992; 62:S3–S10.
29. Young LS, Gascon R, Alam S, Bermudez LEM. Monoclonal antibodies for treatment of gram negative infections. Rev Infect Dis 1989; 11(suppl 7):S1562–S1571.
30. Ziegler EJ, Fisher CJ, Sprung CL, Straube RC, Sadoff JC, Foulke GE, Wortel CE, Fink MP. Dellinger RP, Teng NNH, Allen IE, Berger HJ, Knatterud GL, LoBuglio AF, Smith CR, HA-1A sepsis group. Treatment of gram-negative bacteremia and septic shock with HA-1A human monoclonal antibody against endotoxin. N Engl J Med 1991; 324:429–436.
31. Galanos C, Luderitz O, Rietschel ET, Westphal O, Brade H, Brade L, Freudenberg M, Schade U, Imoto M, Yoshimura H, et al. Synthetic and natural *Escherichia coli* free lipid A express identical endotoxic activities. Eur J Biochem 1985; 148:1–5.
32. Freeman BD, Natanson C. Clinical trials in sepsis and septic shock in 1994 and 1995. Curr Opin Crit Care 1995; 1:349–357.
33. Brade L, Engel R, Christ WJ, Rietschel ET. A nonsubstituted primary hydroxyl group in position 6' of free lipid A is required for binding of lipid A monoclonal antibodies. Infect Immun 1997; 65:3961–3965.
34. Westphal O, Luderitz O. Chemische Erforschung von Lipopolysacchariden gramnegativer Bakerien. Angew Chem 1954; 66:407.
35. Rietchel ET, Kirikae T, Schade FU, Mamat U, Schmidt G, Loppnow H, Ulmer AJ, Zähringer U, Seydel U, Padova FD, Schreier M, Brade H. Bacterial endotoxin: molecular relationships of structure to activity and function. FASEB J 1994; 8:217–225.
36. Brandenburg K, Seydel U, Schromm AB, Lpponow H, Koch MHJ, Rietschel ET. Conformation of lipid A, the endotoxic center of bacterial lipopolysaccharide. J Endotoxin Res 1996; 3:173–178.

37. Wang Y, Hollingsworth RI. An NMR spectroscopy and molecular mechanics study of the molecular basis for the supramolecular structure of lipopolysaccharides. Biochemistry 1996; 35:5647–5654.
38. Bulusu MARC, Waldstätten P, Hildbrandt J, Schütze E, Schulz G. Acyclic analogues of lipid A: synthesis and biological activities. J Med Chem 1992; 35:3463–3469.
39. Tanamoto K. Predominant role of the substitutents on the hydroxyl groups of 3-hydroxy fatty acids of non-reducing glucosamine in lipid A for the endotoxic and antagonistic activity. FEBS Lett 1994; 351:325–329.
40. Loppnow H, Libby P, Freudenberg M, Krauss JH, Weckesser J, Mayer H. Cytokine induction by lipopolysaccharide (LPS) corresponds to lethal toxicity and is inhibited by non-toxic *Rhodobacter capsulatus* LPS. Infect Immun 1990; 58:3743–3750.
41. Takayama K, Qureshi N, Beutler B, Kirkland TN. Diphosphoryl lipid A from *Rhodopseudomonas sphaeroides* ATCC 17023 blocks induction of cachectin in macrophages by lipopolysaccharide. Infect Immun 1989; 57:1336–1338.
42. Shiozaki M, Mochizuki T, Wakabayashi T, Kurakata S, Tatsuta T, Nishijima M. Synthesis of 2,6-anhydro-3-deoxy-5-*O*-[(*R*)-3-(tetradecanoylosy)tetradecanoyl]-D-glycero-D-ido-heptonic acid as a new potent endotoxin antagonist and its dimeric analogue. Tetrahedron Lett 1996; 37:7271–7274.
43. Shiozaki M, Deguchi N, Macindoe WM, Arai M, Miyazaki H, Mochizuki T, Tatsuta T, Ogawa J, Maeda H, Kurakata S. Syntheses of 1-*O*-carboxyalkyl GLA-60 analogues. Carbohydrate Res 1996; 283:27–51.
44. Ray BL, Painter G, Raetz CRH. The biosynthesis of gram-negative endotoxin: formation of lipid A disaccharides from monosaccharide precursors in extracts of *Escherichia coli*. J Biol Chem 1984; 259; 4852–4859.
45. Bulawa CE, Raetz CRH. The biosynthesis of gram-negative endotoxin: identification and function of UDP-2,3-diacylglucosamine in *Escherichia coli*. J Biol Chem 1984; 259:4846–4851.
46. Danner RL, Joiner KA, Parrillo JE. Inhibition of endotoxin-induced priming of human neutrophils by lipid X and 3-aza-lipid X. J Clin Invest 1987; 80:605–612.
47. Proctor RA, Will JA, Burhop KE, Raetz CRH. Protection of mice against lethal endotoxemia by a lipid A precursor. Infect Immun 1986; 52:905–907.
48. Golenbock DT, Will JA, Raetz CRH, Proctor RA. Lipid X ameliorates pulmonary hypertension and protects sheep from death due to endotoxin. Infect Immun 1987; 55:2471–2476.
49. Danner RL, Doerfler ME, Eichacker PQ, Reilly JM, Ratica D, Wilson J, MacVittie TJ, Stütz P, Parrillo JE, Natanson C. A therapeutic trial of lipid X in a canine model of septic shock. J Infect Dis 1993; 167:378–384.
50. Danner RL, Van Dervort AL, Doerfler ME, Stuetz P, Parrillo JE. Antiendotoxin activity of lipid A analogues: requirements of the chemical structure. Pharm Res 1990; 7:260–263.
51. Aschauer H, Grob A, Hildebrandt J, Schuetze E, Stuetz P. Highly purified lipid X is devoid of immunostimulatory activity. Isolation and characterization of immunostimulating contaminants in a batch of synthetic lipid X. J Biol Chem 1990; 265:9159–9164.
52. Lam C, Hildebrandt J, Schutze E, Rosenwirth B, Proctor RA, Liehl E, Stütz P. Immunostimulatory, but not antiendotoxin, activity of lipid X is due to small amounts of contaminating N,O-acylated disaccharide-1-phosphate: in vitro and in vivo reevaluation of the biological activity of synthetic lipid X. Infect Immun 1991; 59:2351–2358.
53. Qureshi N, Honovich JP, Hara H, Cotter RJ, Takayama K. Location of fatty acids in lipid A obtained from lipopolysaccharide of *Rhodopseudomonas sphaeroides* ATCC 17023. J Biol Chem 1988; 263:5502–5504.
54. Rose JR, Christ WJ, Bristol JR, Kawata T, Rossignol DP. Agonistic and antagonistic activities of bacterially-derived *Rhodobacter sphaeroides* lipid A; comparison to synthetic material of the proposed structure and analogs. Infect Immun 1994; 63:833–839.
55. Kawata T, Bristol JR, Rose JR, Rossignol DP, Christ WJ, Asano O, Dubuc GR, Gavin WE, Hawkins LD, Kishi Y, McGuinness PD, Mullarkey MA, Perez M, Robidoux ALC, Wang Y, Kobayashi S, Kimura A, Katayama K, Yamatsu I. Anti-endotoxin activity of a novel synthetic lipid A analog. In: Levin J, Alving CR, Munford RS, Redl H, eds. Bacterial Endotoxins: Lipopolysaccharides from Genes to Therapy. Proceedings of the 3rd International Endotoxin Society. New York: Wiley-Liss, 1995:499–509.
56. Christ WJ, Kawata T, Hawkins LD, Asano O, Kobayashi S, Rossignol DP. Anti-endotoxin compounds and related molecules and methods. U.S. patent No. 5,530,113, issued June 25, 1996.
57. Shiozaki M, Kobayashi Y, Arai M, Ishida N, Hiraoka T, Nishijima M, Kuge S, Otsuka T, Akamatsu Y, Synthesis of a 3-ether analogue of lipid A. Carbohydrate Res 1991; 222:69–82.
58. Christ WJ, Asano O, Robidoux ALC, Perez M, Wang Y, Dubuc GR, Gavin WE, Hawkins LD, McGuinness PD, Mullarkey MA, Lewis MD, Kishi Y, Kawata T, Bristol JR, Rose JR, Rossignol DP, Kobayashi S, Hishinuma I, Kimura A, Asakawa N, Katayama K, Yamatsu I. E5531, a pure endotoxin antagonist of high potency. Science 1994; 268:80–83.
59. Kawata T, Bristol JR, Rose JR, Rossignol DP, Christ WJ, Asano O, Dubuc GR, Gavin WE, Hawkins LD, Lewis, MD, McGuinness PD, Mullarkey MA, Perez M, Robidoux ALC, Wang Y, Kishi Y, Kobayashi S, Kimura A, Hishinima I, Katayama K, Yamatsu I. Specific lipid A analog which exhibits exclusive antagonism of endotoxin. In: Morrison DC, Ryan JL, eds. Novel Therapeutic Strategies in the Treatment of Sepsis. New York: Marcel Dekker, 1995:171–186.
60. Raetz CRH. Biochemistry of endotoxins. Ann Rev Biochem 1990; 59:129–170.
61. Golenbock DT, Hampton RY, Qureshi N, Takayama K, Raetz CRH. Lipid A-like molecules that antagonize the effects of endotoxins on human monocytes. J Biol Chem 1991; 266:19490–19498.
62. Krauss JH, Seydel U, Weckesser J, Mayer H. Structural analysis of the nontoxic lipid A of *Rhodobacter capsulatus* 37b4. Eur J Biochem 1989; 180:519–526.
63. Christ WJ, McGuinness PD, Asano O, Wang Y, Mullarkey MA, Perez M, Hawkins LD, Dubuc GR, Robidoux AL. Total synthesis of the proposed structure of

Rhodobacter sphaeroides lipid A resulting in the synthesis of new potent lipopolysaccharide antagonists. J Am Chem Soc 1994; 116:3637–3638.

64. Ulmer AJ, Heine H, Feist W. Kusumoto S, Kusama T, Brade H, Schade U. Biological activity of synthetic phosphonooxyethyl analogs of lipid A. Infect Immun 1992; 60:3309–3314.

65. Lorsbach RB, Murphy WJ, Lowenstein CJ, Synder SH, Russell SW. Expression of the nitric oxide synthase gene in mouse macrophages activated for tumor cell killing: molecular basis for the synergy between interferon-γ and lipopolysaccharide. J Biol Chem 1993; 268:1908–1913.

66. Zuckerman SH, Qureshi N. In vivo inhibition of lipopolysaccharide-induced lethality and tumor necrosis factor synthesis by *Rhodobacter sphaeroides* diphosphoryl lipid A is dependent on corticosterone induction. Infect Immun 1992; 60:2581–2587.

67. Lynn M, Rogers SL, Parrillo JE, Bunnell E, Neumann A, Habet K, Friedhoff LT. Dose-response of E5531, a lipid A antagonist in the supression of response to endotoxin in normal volunteers. Shock (suppl 1) 1995;3:68.

68. Astiz ME, Rackow EC, Still JG, Howell ST, Cato A, Von Eschen KB, Ulrich JT, Rudbach JA, McMahon G, Vargas R, Stern W. Pretreatment of normal humans with monophosphoryl lipid A induces tolerance to endotoxin: a prospective, double-blind, randomized, controlled trial. Crit Care Med 1995; 23:9–17.

69. Van Zee KJ, Coyle SM, Calvano SE, Oldenburg HS, Stiles DM, Pribble J, Catalano M, Moldawer LL, Lowry SF. Influence of IL-1 receptor blockage on the human response to endotoxemia. J Immunol 1995; 154:1499–1507.

70. Granowitz EV, Porat R, Mier JW, Orencole SF, Callahan MV, Cannon JG, Lynch EA, Ye K, Poutsiaka DD, Vannier E, Shapiro L, Pribble JP, Dtiles DM, Catalano MA, Wolff SM, Dinarello CA. Hematologic and immunomodulatory effects of an interleukin-1 receptor antagonist coinfusion during low-dose endotoxemia in healthy humans. Blood 1993; 82:2985–2990.

71. Chernoff AE, Granowitz EV, Shapiro L, Vannier E, Lonnemann G, Angel JB, Kennedy JS, Rabson AR, Wolff SM, Dinarello CA. A randomized controlled trial of IL-10 in humans. Inhibition of inflammatory cytokine production and immune responses. J Immunol 1995; 154:5492–5499.

72. von der Möhlen MAM, Kimmings AN, Wedel NI, Mevissen ML, Jansen J, Friedmann N, Lorenz TJ, Nelson BJ, White ML, Bauer R, Hack CE, Eerenberg AJM, van Deventer SJH. Inhibition of endotoxin-induced cytokine release and neutrophil activation in humans by use of recombinant bactericidal/permeability-increasing protein. J Infect Dis 1995; 172:144–151.

73. Suffredini AF, Reda D, Banks SM, Tropea M, Agosti JM, Miller R. Effects of recombinant dimeric TNF receptor on human inflammatory responses following intravenous endotoxin administration. J Immunol 1995; 155:5038–5045.

74. Pajkrt D, Doran JE, Koster F, Lerch PG, Arnet B, van der Poll T, ten Cate JW, van Deventer SJ. Antiinflammatory effects of reconstituted high-density lipoprotein during human endotoxemia. J Exp Med 1996; 184:1601–1608.

75. Santos AA, Scheltinga MR, Lynch E, Brown EF, Lawton P, Chambers E, Browning J, Dinarello CA, Wolff SM, Wilmore DW. Elaboration of interleukin 1-receptor antagonist is not attenuated by glucosteroids after endotoxemia. Arch Surg 1993; 128:138–144.

76. Martich GD, Danner RL, Ceska M, Suffredini AF. Detection of interleukin 8 and tumor necrosis factor in normal humans after intravenous endotoxin: the effect of antiinflammatory agents. J Exp Med 1991; 173:1021–1024.

77. Michie HR, Manogue KR, Spriggs DR, Revhaug A, O'Dwyer S, Dinarello CA, Cerami A, Wolff SM, Wilmore DW. Detection of circulating tumor necrosis factor after endotoxin administration. N Engl J Med 1988; 318:1481–1486.

78. Revhaug A, Michie HR, Manson JM, Watters JM, Dinarello CA, Wolff SM, Wilmore DW. Inhibition of cyclo-oxygenase attenuates the metabolic response to endotoxin in humans. Arch Surg 1988; 123:162–170.

48

Bacteria-Induced Hypersensitivity to Endotoxin

Marina A. Freudenberg, Thomas Merlin, Andreas Sing, and Chris Galanos
Max Planck Institut für Immunbiologie, Freiburg im Breisgau, Germany

Reinaldo Salomao
Escola Paulista de Medicina, São Paulo, Brazil

INTRODUCTION

Administration of lipopolysaccharide (endotoxin, LPS) in experimental animals and human volunteers leads to the generation of numerous biological activities that can be both harmful and beneficial for the host (1). In general, toxic effects are observed with higher doses of LPS, the magnitude of which depends on the susceptibility of the host. Susceptibility to LPS varies enormously among the different mammalian species. For all practical purposes it may be said that some experimental animals, such as mice and rats, are relatively LPS resistant; others, like rabbits and swine, are relatively sensitive. Interestingly, humans probably represent the most LPS-susceptible species known so far.

Mice are the most frequently used experimental animals in biological studies on endotoxins. Investigations in these animals revealed only moderate differences in susceptibility to LPS among the different inbred and outbred strains. In the authors' laboratory C57BL/10ScSn mice were reproducibly found to be two to three times more susceptible to the lethal effects of LPS than C57BL/6 mice and four to five times more susceptible than C3H/HeN or BalbC mice. Such moderate differences in LPS sensitivity exist throughout the spectrum of inbred and outbred strains used in endotoxin research.

In mice, sensitivity to LPS is determined by a locus on chromosome 4 designated the *Lps* gene (2–4). A mutation of this gene results in unresponsiveness towards the biological action of LPS. Mice carrying this mutation are resistant to all LPS effects. For example, in the D-galactosamine lethality model, the 50% lethal dose of purified LPS for LPS-resistant mice is of the order of one million times higher than that for LPS-sensitive mice (5). So far three mutant mouse strains—C3H/HeJ (6,7), C57BL/10 ScCr (8), and C57BL/10ScN (9)—have been identified. In addition to the above mutants that occurred naturally, a fourth LPS nonresponder strain BalbC/e was produced in two laboratories independently by introducing the defective *Lps* gene of C3H/HeJ strain into BalbC mice (10,11). Mice carrying the mutation are designated Lps^d (for defective response), those carrying the normal gene, Lps^n (for normal response) mice. The product of the *Lps* gene is still unknown. However, it has been shown that the defect concerns an early event in LPS–target cell interaction.

The susceptibility of animals to LPS is also dependent on nongenetic factors. This has been demonstrated repeatedly by showing that the susceptibility of experimental animals to LPS can be altered under a variety of conditions, depending on which a reduced (hyporeactivity or tolerance) or an enhanced susceptibility (hyperreactivity or hypersensitivity) can be obtained (12–15). Therefore, it may be said that endotoxin possesses no toxicity in absolute terms and that its biological activity is determined solely by the reactivity of the host, which is dependent on intrinsic and extrinsic factors.

LPS-HYPERSENSITIVITY

As seen in Table 1, LPS hypersensitivity may be established under vastly different experimental and natural conditions. These include treatment with hepatotoxic agents such as D-galactosamine (16), live (infection) or killed bacteria (17–22), and isolated bacterial components (23).

The development of LPS hypersensitivity has also been observed during a number of viral, protozoal, and fungal infections (24–28) and during growth of certain malignant tumors (29). Further relevant examples of conditions leading to LPS hypersensitivity are elevation of environmental temperature (30) and glucocorticoid depletion (31). Some of the sensitizing conditions mentioned above represent pathological processes and are of clinical interest. The hypersensitivity induced by gram-negative microorganisms is of particular importance, because these bacteria not only produce LPS but also sensitize the infected host to its toxic activity.

BACTERIA-INDUCED HYPERSENSITIVITY

Infection of mice with a number of gram-negative or gram-positive bacteria, such as *Salmonella typhimurium* (21), *Coxiella burnetii* (20), *Propionibacterium acnes* (19,22), Calmette-Guérin bacillus (BCG) (18), leads to the development of hypersensitivity to the toxic effects of LPS. LPS hypersensitivity may be induced by lethal and sublethal infections, and its time course as well as duration depend on the infecting microorganisms. Thus, the hypersensitivity induced by a lethal infection with *S. typhimurium* becomes detectable on day 2 or 3 after inoculation and usually increases up to the time the animals succumb to the lethal effects of infection (32). In the case of sublethal infection with *S. typhimurium*, hypersensitivity is maximum on days 7–8, decreasing thereafter and returning to preinfection value several weeks later (21). For the induction of LPS hypersensitivity, infection with live bacteria can often be replaced by appropriate amounts of killed bacteria. The development of hypersensitivity induced by killed bacteria such as *P. acnes* or *C. burnetii* (20,22) is usually maximum 7–10 days after bacterial treatment. However, its duration may vary depending on the bacteria used. Thus, the *P. acnes*–induced hypersensitivity persists for approximately 3 weeks, but that induced by *C. burnetii* lasts for several months. With the elimination of the bacteria from the organism, hypersensitivity finally decreases and disappears. The LPS hypersensitivity induced by bacteria is expressed both as an enhanced susceptibility to the lethal activity of LPS and as enhanced production of cytokines, such as TNF-α,

Table 1 Induction of Hypersensitivity to Endotoxin

Treatment	Sensitization factor
Microbial Infections	
Gram-negative	
Salmonella	100 to >1000
E. coli	
Klebsiella	
Coxiella burnetii	
Gram-positive	
Propionibacterium acnes	
BCG	
Bacterial products	
Proteins	50
MDP	100
Parasitic infections	
Malaria (*P. chabaudi chabaudi*)	>100
Growing tumor	
Lewis lung carcinoma	>10,000
EMT6 sarcoma	200
Hepatotoxic agents	
D-Galactosamine	100,000
α-Amanitine	>1000
Other agents	
Carbon tetrachloride	>1000
Lead acetate	>1000
Actinomycin-D	>10,000
Hyperthermia	
Environmental temperature 30–33°C	>1000
Glucocorticoid deficiency	
Adrenalectomy	>1000
Hypophysectomy	>1000

upon LPS treatment of the animals. Both activities may be used as parameters of hypersensitivity measurement.

ENHANCED LETHAL EFFECTS OF LPS

At the peak of sensitization with most bacteria, the lethal activity of LPS for mice increases approximately by a factor of 100–1000. All Lps^n mouse strains investigated so far (>20) develop a marked hypersensitivity to LPS when treated with sensitizing bacteria. Sensitization was also found to proceed in the Lps^d C3H/HeJ and BalbC (BalbC/1) mouse strains (15,20,33). These mice, which when healthy tolerate more than 2 mg of LPS without apparent symptoms of illness, succumb to submilligram amounts when pretreated with *P. acnes* (Table 2) or other sensitizing bacteria. It should be noted, however, that the susceptibility of the above two Lps^d mouse strains after

Table 2 Changes in Sensitivity of Lps^n and Lps^d Mice to the Lethal Effects of LPS After Pretreatment with *P. acnes*

Mouse strain	Approx. LD$_{50}$ (μg LPS) in mice	
	Control	*P. acnes*–treated
Lps^n		
C3H/HeN	400	0.2
BalbC	300	2
C57BL/10ScSn	100	0.1
Lps^d		
C3H/HeJ	>2000	100
BalbC/1	>2000	150
C57BL/10ScCr	>2000	>2000

sensitization with bacteria is still lower than that of likewise sensitized Lps^n mice (Table 2). In fact, the LPS susceptibility of sensitized Lps^d mice is comparable to that of unsensitized Lps^n mice. Interestingly, sensitization by *P. acnes* does not proceed in the Lps^d C57BL/10ScCr mice (22,34).

In general, in bacteria-sensitized mice, the onset of illness and lethality after LPS challenge occurs faster than in unsensitized controls, suggesting that the mechanisms underlying the development of LPS shock in control and infected animals are not identical.

TNF-α OVERPRODUCTION

Following treatment with bacteria, sensitized mice exhibit an enhanced production of pro-inflammatory cytokines, such as TNF-α, IL-6, and/or IFN-γ, in response to LPS (13,18,19,35, and unpublished data). Previous studies have clearly shown that in certain lethality models TNF-α is a prominent mediator of LPS toxicity (36–39). Therefore, the enhanced inducibility of TNF-α in sensitized mice may be regarded as an important factor contributing to the enhanced toxic effects of LPS. Lps^n mice infected 4 days earlier with *S. typhimurium* or treated 7 days earlier with killed *P. acnes* may produce 100–1000 times higher TNF-α levels upon LPS challenge than controls. An example of the characteristic TNF-α hyperresponse of bacteria-sensitized mice is shown in Figure 1 (left), in which *P. acnes* was used as the sensitizing bacterial agent for C57BL/ScSn and BalbC mice.

As mentioned above, Lps^d C3H/HeJ and BalbC/1 mice also become partially susceptible to the lethal effects of LPS following treatment with bacteria. In the case of BalbC/1 mice, the acquisition of LPS susceptibility is further indicated by their ability to exhibit a TNF-α response when treated with LPS (Fig. 1, right). This response, although much lower than that of Lps^n mice, is nevertheless highly significant because in the absence of sensitization, TNF-α responses to LPS are not detectable in BalbC/1 mice. In contrast, bacteria-treated C57BL/10 ScCr mice, which remain resistant to the lethal effects of LPS, also fail to produce TNF-α in response to a LPS challenge (Fig. 1, right).

HYPERSENSITIVITY TO THE LETHAL EFFECTS OF TNF-α

Considering the toxic properties of TNF-α, an overproduction of this mediator, such as the one induced by LPS in bacteria-sensitized mice, would represent a potential hazard for the infected host. This hazardous situation is aggravated further by the fact that mice sensitized by bacteria not only overproduce TNF-α, but are also hypersensitive to the lethal activity of this mediator (13). Therefore the enhanced production of TNF-α and its higher toxicity are the main reasons for the enhanced susceptibility of bacteria-treated mice to the lethal effects of LPS. Interestingly, C57BL/10 ScCr mice treated with certain live or killed microorganisms frequently exhibit an enhanced susceptibility to the lethal activity of TNF-α, although at the same time they remain resistant to LPS. The enhanced susceptibility to TNF-α in these mice is seen after infection with *S. typhimurium* or *Plasmodium chabaudi chabaudi*, whereas following treatment with killed *P. acnes* susceptibility to TNF-α does not increase (unpublished data).

IFN-γ, A KEY MEDIATOR OF THE BACTERIA-INDUCED HYPERSENSITIVITY TO LPS

As shown in the aforegoing sections, treatment of Lps^d C3H/HeJ or BalbC/1 mice with sensitizing bacteria results in a change of LPS phenotype from resistant to partially sensitive. Further, a similar treatment of C57BL/10ScCr has no effect on their LPS-resistant phenotype (22,34), indicating that the mechanisms underlying the development of LPS hypersensitivity in these mice are defective. A clue as to the possible reason for the absence of sensitization in C57BL/10ScCr mice by bacteria was obtained when the cytokines produced by these mice in response to different agents were investigated. These studies revealed an impaired IFN-γ production in response to different gram-nega-

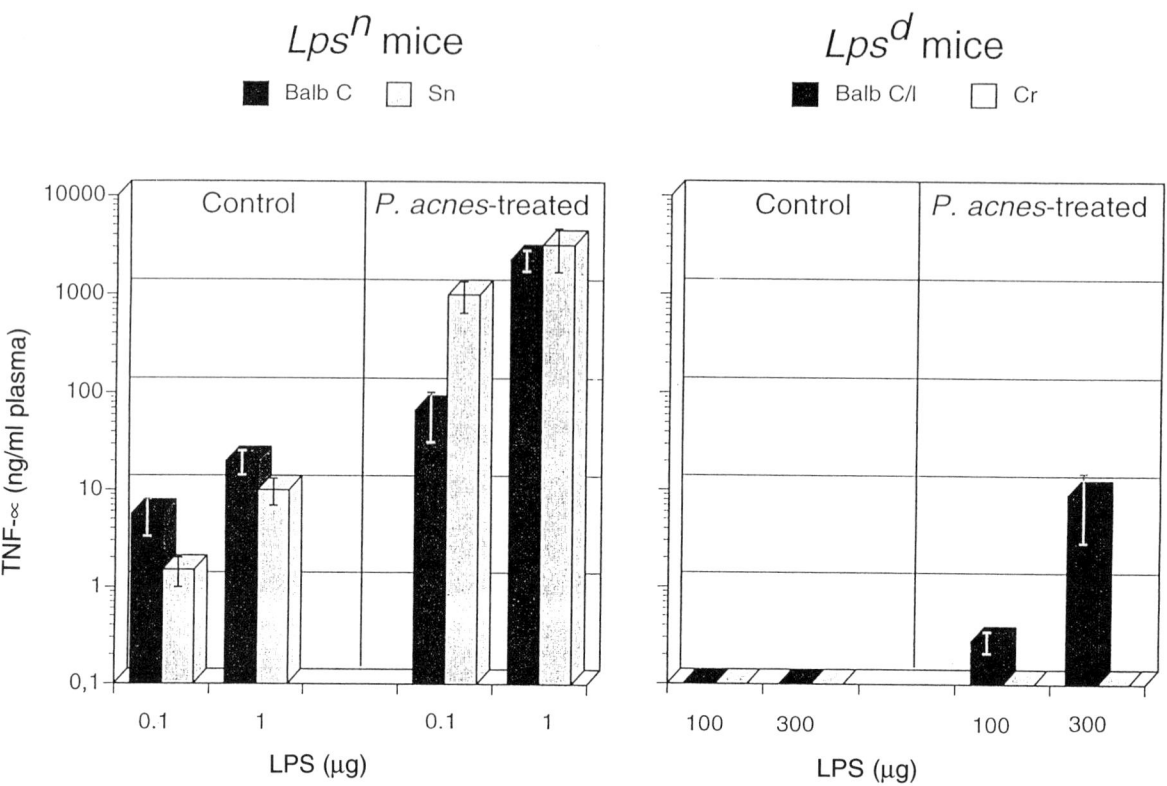

Fig. 1 Effect of *P. acnes* treatment on LPS-induced TNF-α production of Lps^n and Lps^d mice. Mice received killed *P. acnes* (25 μg/g b.w., i.v.) or remained untreated. Seven days later all mice were challenged with LPS of *S. abortus equi* i.v., and TNF-α present in plasma collected 1 hour after challenge was measured by bioassay. The vertical lines represent SD of mean values.

tive and gram-positive bacteria and certain parasites in vivo and in vitro (Fig. 2) (34,40,41). IFN-γ is a pro-inflammatory cytokine, which, among other functions, also acts as a macrophage-activating factor, increasing the ability of these cells to produce endogenous mediators, like TNF-α and IL-1 in response to LPS (42–44). Also Lps^d C3H/HeJ macrophages, after treatment with IFN-γ in vitro, acquire the ability to produce TNF-α in response to LPS (45). Therefore, an involvement of IFN-γ in the sensitization towards LPS by bacteria seemed likely, and the defective IFN-γ production of C57BL/10ScCr mice would explain the absence of LPS sensitization in these animals. Further support for the role of IFN-γ in the development of LPS hypersensitivity was obtained by showing that administration of exogenous IFN-γ conferred the ability to C57BL/10ScCr mice to respond to LPS with a measurable TNF-α production and rendered them susceptible to the lethal activity of LPS after D-galactosamine sensitization (22). More direct evidence for the key role of IFN-γ in bacteria-induced sensitization was obtained by showing that administration of specific anti–IFN-γ antibodies in *P. acnes*–treated Lps^n C57BL/10ScSn mice inhibited in a dose-dependent manner the development of hypersensitivity to LPS (22). The antibodies inhibited both the enhanced TNF-α production and the enhanced lethal toxicity induced by LPS. Final evidence for the essential role of IFN-γ in bacteria-induced sensitization was obtained when IFN-γR–deficient mice became available. IFN-γR–deficient mice, in contrast to wild-type animals, do not develop hypersensitivity to LPS when treated with *P. acnes* (Fig. 3), BCG, or *S. typhimurium* (46,47, and unpublished data).

Using IFN-γR–deficient mice it was further demonstrated that not only enhanced susceptibility to LPS, but also enhanced susceptibility to the lethal effects of TNF are mediated by IFN-γ. Treatment of IFN-γR–deficient mice with *P. acnes* or *S. typhimurium* did not alter their susceptibility to human rTNF-α, while a similar treatment of wild-type mice rendered them highly susceptible to the lethal effects of this mediator (46; unpublished data).

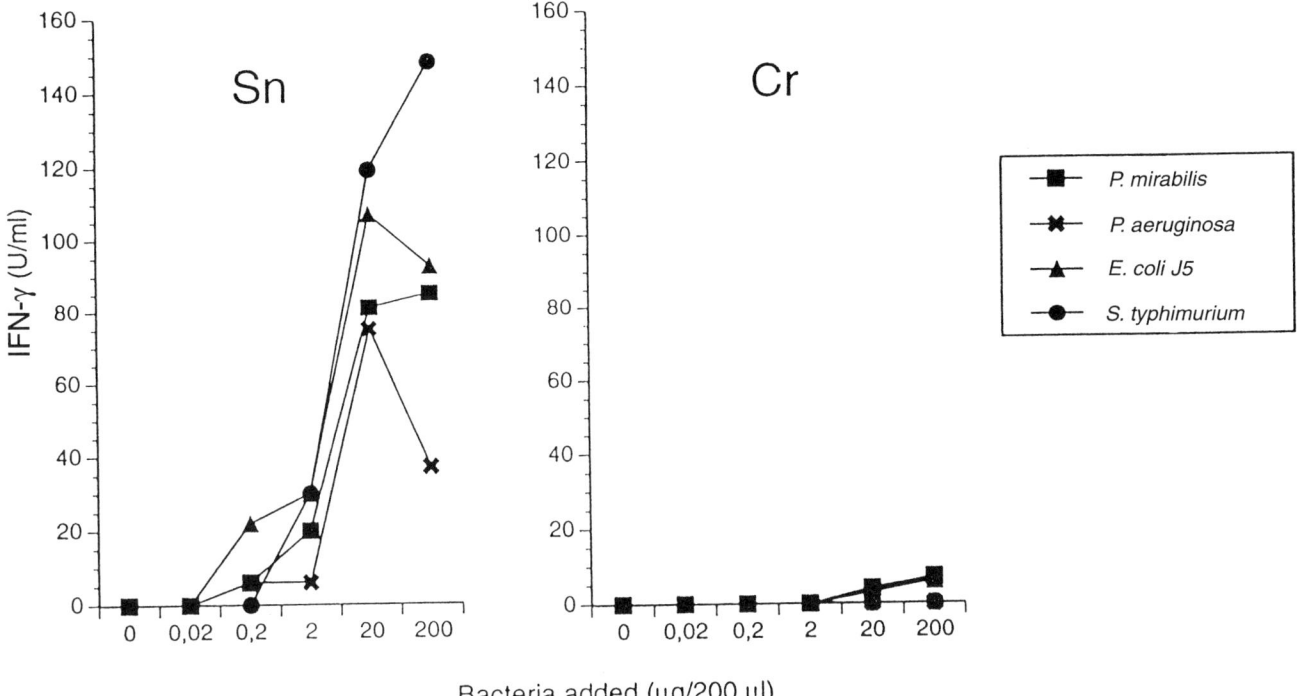

Fig. 2 Impaired IFN-γ production by splenocytes of C57BL/10ScCr mice stimulated with different gram-negative bacteria. The IFN-γ response of C57BL/10ScSn (Sn) and C57BL/10ScCr (Cr) splenocytes (10^7/ml DMEM) to different killed bacteria was measured in 24-hour culture supernatants by a specific ELISA.

ANALYSIS OF THE IMPAIRED IFN-γ RESPONSE TO MICROORGANISMS EXHIBITED BY C57BL/10ScCr MICE

C57BL/10ScCr mice exhibit, in addition to their LPS nonresponsiveness, a heavily impaired IFN-γ response to several live and killed gram-negative and gram-positive bacteria (22,34,40) (Fig. 3) and to parasites such as *Plasmodium chabaudi chabaudi* and *Leishmania major* (34,41). This impaired IFN-γ response is not due to a general defect of IFN-γ production, since C57BL/10ScCr splenocytes in culture produce reasonable levels of IFN-γ when stimulated with the T-cell mitogen ConA or with monoclonal CD3 antibodies (34,40,41). At present, the IFN-γ defect of C57BL/10ScCr mice would seem to concern responses induced by microbial agents. An investigation in which gram-negative bacteria were used as stimulus revealed that the IFN-γ defect is related to an impaired function of accessory cells (macrophages, dendritic cells), while the IFN-γ–producing cells (T and natural killer cells) are normal (40). This was shown in experiments in which splenocytes of defective C57BL/10ScCr and normal C57BL/10ScSn mice were reconstituted with macrophages of each other. C57BL/10ScCr spleen cell cultures in which adherent cells were replaced by C57BL/10ScSn macrophages attained the property of exhibiting normal IFN-γ responses to gram-negative bacteria, while C57BL/10ScSn spleen cell cultures in which adherent cells were replaced by C57BL/10ScCr macrophages lost the ability to display IFN-γ responses to these bacteria. An essential function of accessory cells is the production of cytokines such as IL-1, IL-12, IL-18, and TNF-α, which participate in the induction of IFN-γ. In an in vitro study the cytokine response of C57BL/10ScCr macrophages to different gram-negative and gram-positive bacteria was measured and compared to that of macrophages of the related Lps^n C57BL/10ScSn mice (5,40, and unpublished data). The study revealed that the macrophages of both strains produced TNF-α, IL-1α, IL-1β, IL-6, IL-10, and IL-12. However, IFN-β was produced only by C57BL/10ScSn macrophages. Similarly, following treatment with *S. typhimurium*, in vivo expression of IFN-β mRNA was observed in the

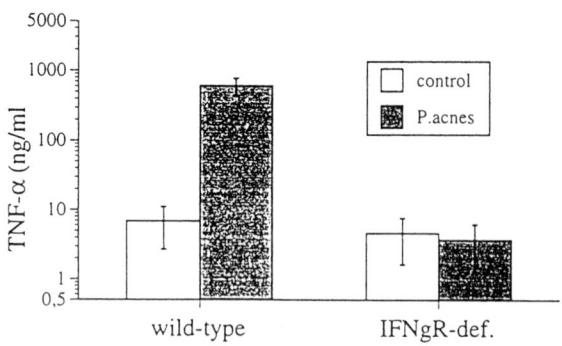

Fig. 3 Absence of LPS hypersensitivity in *P. acnes*–treated IFN-γR–deficient mice. IFN-γR–deficient and wild-type (129/Sv/Ev) mice received killed *P. acnes* (25 μg/g b.w., i.v.) or remained untreated. 7 days later all mice were challenged with the indicated amounts of LPS of *S. abortus equi* i.v. TNF-α in plasma collected 1 hour after challenge was measured by a bioassay. The vertical lines represent SD of mean values. Lethality was scored up to 72 hours after challenge.

spleen of C57BL/10ScSn mice but absent from the spleen of C57BL/10ScCr mice. Thus, unlike C57BL/10ScSn mice, C57BL/10ScCr mice are unable to produce IFN-β in response to gram-negative bacteria. For this reason the possible function of IFN-β as a cofactor in the induction of IFN-γ was investigated (40). It was shown that C57BL/10ScCr splenocytes supplemented with supernatants of activated C57BL/10ScSn macrophages acquired the ability to produce IFN-γ in response to gram-negative bacteria. This helper activity of macrophage supernatants was completely inhibitable by antibodies to murine IFN-β. Moreover, IFN-β antibodies partially inhibited the bacteria-induced IFN-γ production by C57BL/10ScSn splenocytes (Fig. 4). Finally, the addition of either murine rIFN-β or rIFN-α directly to C57BL/10ScCr splenocyte culture conferred the ability of producing IFN-γ after stimulation of the cells with gram-negative bacteria (Fig. 5). These results provide evidence that IFN-β is an important cofactor in the induction of IFN-γ by gram-negative bacteria.

The induction of IFN-β by gram-negative bacteria is a function of the LPS component and thus expressed only in Lps^n mice. Therefore, the inability of C57BL/10ScCr mice to produce IFN-β does not represent an additional defect in these mice. It is simply related to their LPS nonresponsiveness. Consequently, IFN-β responses to gram-negative bacteria are also absent from Balb/1 mice (unpublished data). The property of inducing IFN-β in mice seems to be confined to gram-negative bacteria. A large number of killed and a limited number of live gram-positive bacteria investigated failed to induce IFN-βmRNA in either Lps^n or Lps^d mice in vivo and in vitro (unpublished data). This shows that in all probability gram-positive bacteria possess no alternative component to LPS that is capable of inducing IFN-β. Interestingly, the experiments with C57BL/10ScCr splenocytes revealed that while rIFN-β or rIFN-α makes possible the induction of IFN-γ by gram-negative bacteria, they do not do so in the case of gram-positive bacteria (Fig. 4). Consequently, IFN-γ responses to gram-positive bacteria are always IFN-β independent. On the other hand, the induction of IFN-γ via the pathway that utilizes IFN-β as accessory factor remains the sole property of gram-negative microorganisms expressed in Lps^n but not in Lps^d mice. Therefore, the IFN-γ induced by gram-negative bacteria in Lps^d BalbC/1 mice is IFN-β independent. On the other hand, it can be safely assumed that the IFN-γ stimulated by gram-negative bacteria in Lps^n mice results from a contribution of both pathways of IFN-γ induction. In complete contrast, both pathways of induction are defective in C57BL/10ScCr mice. Thus, C57BL/10ScCr mice exhibit, apart from their LPS defect, additional defects that distinguish them from Lps^d Balb/1 mice.

A comparison of the cytokine responses induced by bacteria in BalbC/1 and C57BL/10ScCr mice revealed that apart from IFN-γ, which is induced only in BalbC/1 mice, no further differences between the two strains are apparent; this aroused the suspicion that the defect may involve accessory factors.

DEFECTIVE INDUCTION OF IFN-γ BY IL-12 IN C57BL/10ScCr MICE

Of the different cytokines thought to be possible cofactors of IFN-γ induction, IL-12 has been documented in recent literature to be involved in the induction of IFN-γ in T and NK cells (48). After its formation, IL-

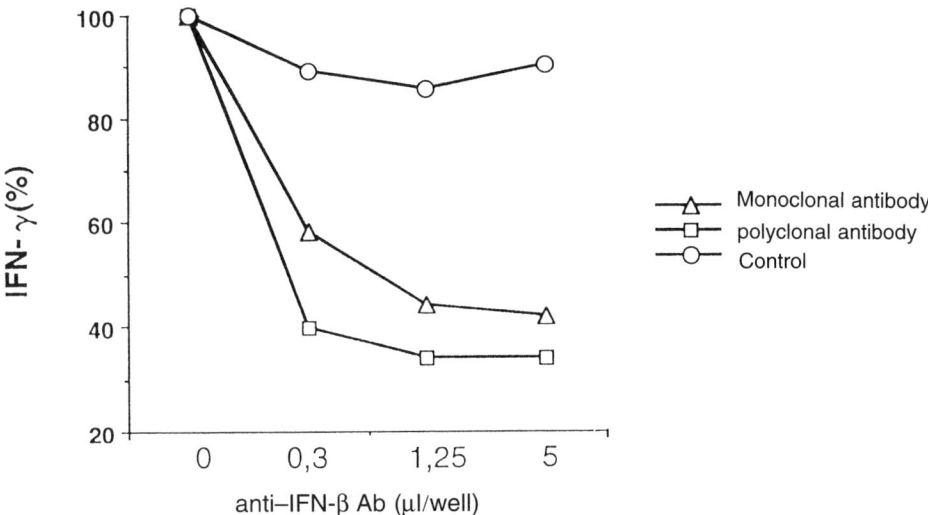

Fig. 4 Inhibition of the IFN-γ response of C57BL/10ScSn splenocytes to *S. typhimurium* by anti IFN-β. IFN-γ production by splenocytes ($2 \times 10^6/0.2$ ml DMEM) supplemented with the indicated amounts of monoclonal (10 mg/ml) or polyclonal anti-IFN-β, or control antibodies, and stimulated with killed *S. typhimurium* (20 μg) was measured after 24 hours of culture by a specific ELISA.

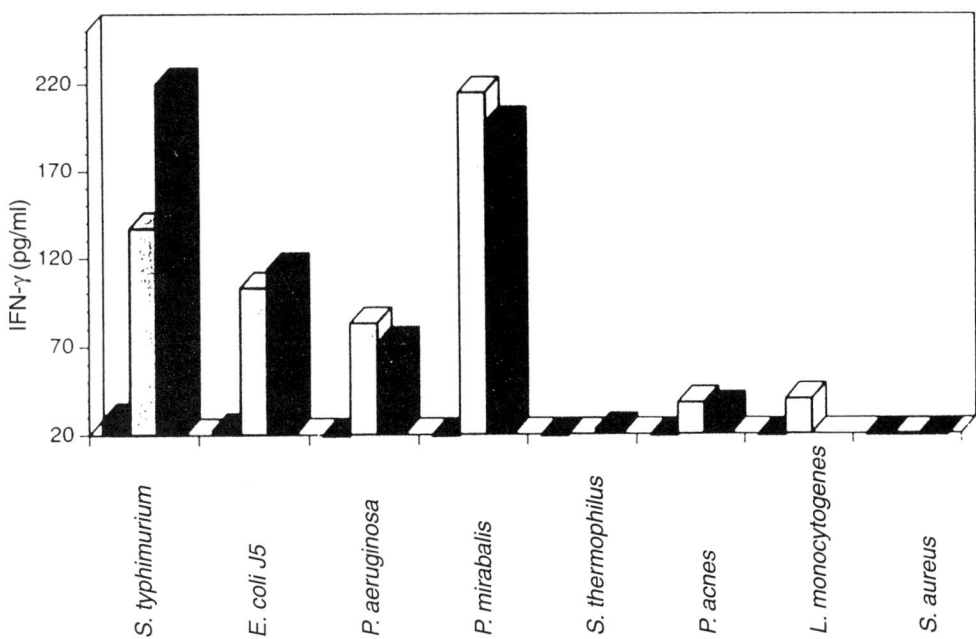

Fig. 5 IFN-γ production by C57BL/10ScCr splenocytes supplemented with rIFN-β or rIFN-α and stimulated with different gram-negative and gram-positive bacteria. Splenocytes ($2 \times 10^6/0.2$ ml DMEM) were supplemented with murine rIFN-β or rIFN-α (10,000 U) and stimulated with the different bacteria indicated (20 μg) for 24 hours. IFN-γ was estimated in culture supernatants by a specific ELISA. First column of each group: control splenocytes (no supplement); second column: splenocytes supplemented with rIFN-β; third column: splenocytes supplemented with rIFN-α.

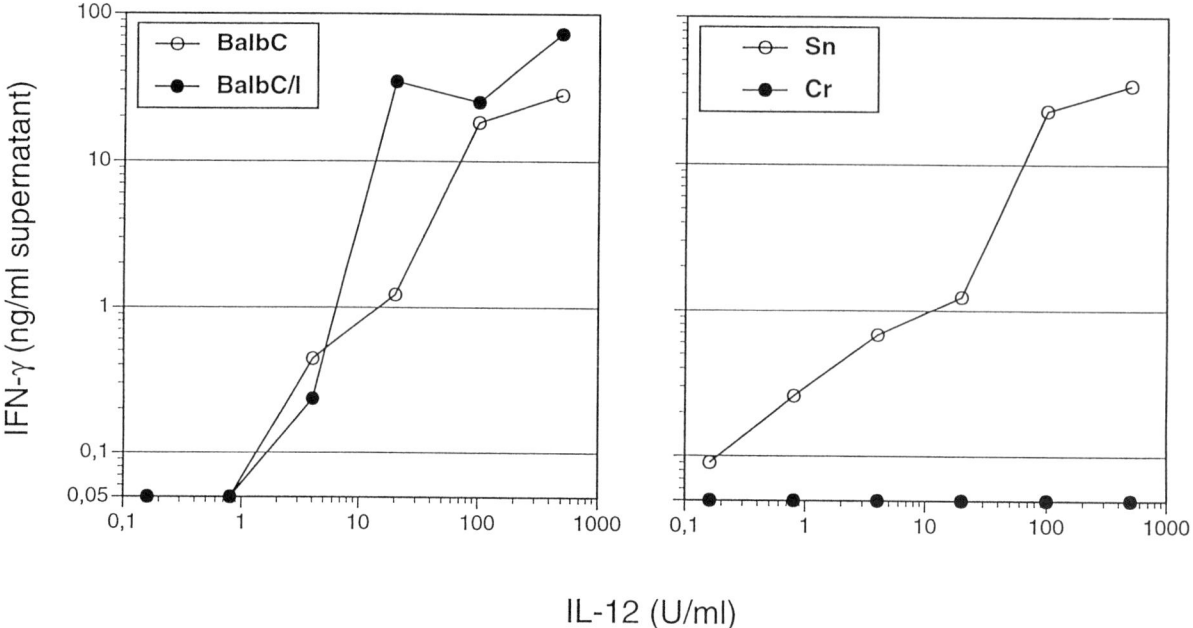

Fig. 6 Absence of IFN-γ response to rIL-12 in Lps^d C57BL/10 ScCr splenocytes. The IFN-γ response of splenocytes (10^7/ml DMEM) cultured for 24 hours with different amounts of murine rIL-12 was measured by a specific ELISA.

12 induces IFN-γ directly and therefore mediates the induction of IFN-γ by different exogenous stimuli, including LPS and bacteria. Since, as mentioned in the previous chapter, the production of IL-12 in C57BL/10ScCr mice is normal, we investigated if their IFN-γ defect might be related to an unresponsiveness to this cytokine. For this reason we compared the effect of exogenous IL-12 on the induction of IFN-γ in splenocytes of C57BL/10ScCr and other mouse strains (unpublished data). As shown in Figure 6, splenocytes of C57BL/10ScSn, BalbC, and BalbC/1 produce high levels of IFN-γ in response to murine rIL-12, while those of C57BL/10ScCr produce no IFN-γ at all. This finding indicates that the IL-12 responsiveness is defective in C57BL/10ScCr mice. The involvement of IL-12 in the development of LPS hypersensitivity was indicated by experiments in which BCG-primed mice treated with neutralizing anti–IL-12 antibodies were completely protected against a 5 × LD_{50} of LPS (49). A defect in IL-12 responsiveness would therefore explain the impaired IFN-γ production and the resultant absence of sensitization in bacteria-treated C57BL/10ScCr mice. It is not known at present whether the IL-12 unresponsiveness of C57BL/10ScCr mice involves the IL-12 receptor, signal transduction, or is due to other reasons. These questions are currently being studied.

REQUIREMENT FOR LPS-BINDING PROTEIN IN HYPERRESPONSE OF P. acnes–SENSITIZED MICE TO LPS

The mechanisms of the in vivo LPS effects are complex and not yet completely understood. The biological activities of LPS are initiated after interaction of the LPS with humoral and cellular targets. Macrophages are important effector cells of LPS action (50,51). Earlier studies showed that interaction of LPS with macrophages and their subsequent activation in vitro can proceed in the complete absence of plasma components (52–54). In vivo, however, the interaction is modulated by a number of LPS-binding proteins. One of these is high-density lipoprotein (HDL), which functions as a carrier protein for circulating LPS, modulating its uptake by the phagocytic system (55). In recent years two new proteins, CD14 in soluble and membrane-bound form and LPS-binding protein (LBP), an acute phase protein, were identified as being involved in LPS binding and signaling (56–58). The two proteins are interdependent for optimal triggering function, LBP binding LPS and transferring it to CD14. It has been shown further that the binding to LBP enhances the activity of LPS in vitro. A role for LBP in vivo was indicated by experiments in which administration of LBP anti-

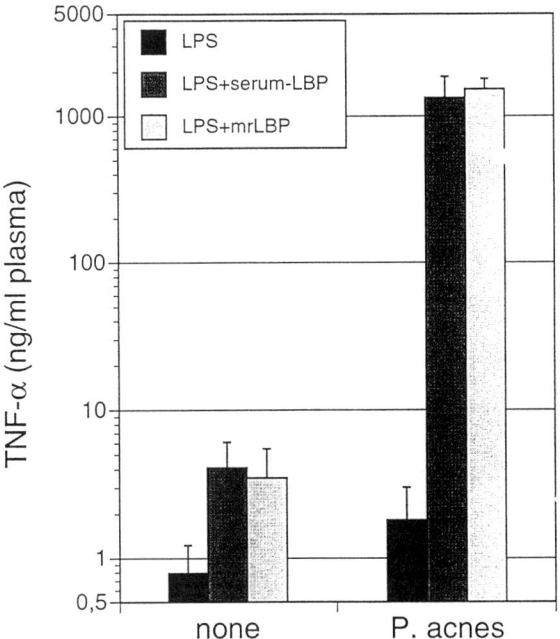

Fig. 7 Effect of exogenous LBP on the TNF-α response to LPS of unsensitized and *P. acnes*–sensitized LBP-/- mice. Mice received killed *P. acnes* (625 μg, i.v.) or remained untreated. Seven days later all mice were challenged with LPS of *S. abortus equi* (0.5 μg), or a mixture of LPS (0.5 μg) with either serum-LBP (1.8 μg) or murine rLBP (0.8 μg) i.v. TNF-α was measured in plasma collected 1 hour after challenge by a bioassay. All weights of injected materials given apply to 25 g body weight. The vertical lines represent SD of mean values.

bodies reduced the TNF-α levels and lethality by LPS in D-galactosamine–sensitized mice (59,60). To assess fully the importance of LBP in the development of LPS effects, LBP-deficient mice were generated (61). These mice were used in experiments in which the effect of LBP in LPS triggering was compared in unsensitized and *P. acnes*–sensitized mice (62).

Unsensitized LBP-/- (129 Sv × BalbC) mice exhibited a similar susceptibility to the lethal effects of LPS as wild-type 129 Sv and BablC mice. The approximate LD$_{50}$ in all three strains was between 300 and 400 μg of *S. abortus equi* LPS (unpublished data), showing that the induction of lethality by LPS in unsensitized mice, which requires relatively high amounts of LPS, proceeds via LBP-independent mechanisms. The unsensitized LBP-/- and wild-type mice were also compared in their ability to produce TNF-α in response to LPS (62). LBP-/- mice produced somewhat lower amounts of TNF-α than BalbC and marginally lower than 129 Sv mice when challenged with a relatively low dose of *S. abortus equi* LPS (0.1–1.0 μg) LPS. Treatment of wild-type mice with *P. acnes* enhanced their TNF-α response to LPS (0.5 μg) 75- to 200-fold compared to untreated controls. A similar treatment of LBP-/- mice enhanced their TNF-α production only by a factor of 4 (Fig. 7). The characteristic TNF-α hyper-response was absent, suggesting that the enhanced reaction to LPS in sensitized mice is expressed only in the presence of LBP. Evidence for this was obtained by showing that administration of exogenous LBP restored fully the inducibility of a TNF-α hyperresponse by LPS in *P. acnes*–sensitized LBP-/- mice (Fig. 7). Their response to 0.5 μg LPS was 1000-fold higher than that of sensitized controls without LBP. A similar LBP treatment of unsensitized LBP-/- mice increased the TNF-α response by a factor of only 5. In similar experiments different R-form LPS preparations and free lipid A were also shown to require LBP in inducing TNF-α hyperresponses in sensitized mice (unpublished data). It could be shown that all LPS preparations and lipid A behaved like S-form LPS, i.e., inducing low TNF-α responses in sensitized LBP-/- mice that could be enhanced considerably by exogenous LBP. In analogous experiments IFN-γ–sensitized LBP-/- mice also exhibited TNF-α overproduction in response to LPS of *S. abortus equi* only in the presence of exogenous LBP (62). The results show that while low TNF-α responses to LPS can proceed in the absence of LBP, the characteristic hyper-response to LPS seen in animals sensitized by *P. acnes* is LBP dependent.

ROLE OF CD14 IN SUSCEPTIBILITY TO LPS

LBP enhances responses to LPS by transferring LPS to another binding protein, CD14 (56–58,63,64). Two forms of CD14 have been described: membrane-bound (m)CD14 and soluble (s)CD14. Membrane-bound CD14 is expressed predominantly on monocytes and macrophages, the primary target cells of LPS action, and was shown to be involved in the triggering of these cells by LPS. sCD14 is present in significant amounts in plasma of humans and rabbits. Its function, however, is not completely understood yet. In in vitro experiments it was demonstrated that LPS/sCD14 complexes can activate cells lacking mCD14. Interestingly, unlike humans and rabbits, normal unstimulated mice exhibit only a very low expression of either form of CD14. The plasma levels of sCD14 and CD14 mRNA in the liver are below detection limits and in other organs very low (11,66). Remarkably, the majority of liver macrophages in unstimulated mice are CD14 negative

(65). The low expression of CD14 in the liver is of special interest because liver is the main organ of LPS clearance (55). Considering that the primary function of this organ is to remove LPS from circulation and render it harmless, rather than being itself activated by LPS, this low expression in the liver may represent a mouse-specific protective mechanism against bacterial LPS. Interestingly, the expression of CD14 in Lps^n and Lps^d mice is similar (11).

In contrast to the CD14 expression, the levels of plasma LBP in mice (Lps^n and Lps^d) do not differ from those observed in humans. The relatively low contribution of LBP to LPS effects in healthy, unsensitized mice might be therefore a result of a low CD14 expression in this animal species. Also the large differences in susceptibility to LPS that exist between mice (low susceptibility) and humans (high susceptibility) might, among other reasons, be related to the quantitative differences in CD14 expression. Interestingly, very recent studies revealed that, in contrast to unsensitized animals, *P. acnes*– or *S. typhimurium*–sensitized mice exhibit detectable levels of plasma sCD14 and a high expression of CD14 mRNA in the liver (unpublished data). In this connection, it has been reported that transgenic mice expressing human CD14 were three to four times more sensitive to LPS than the nontransgenic controls (67). This increase in sensitivity is still very low when compared to the true hypersensitivity seen in bacteria-sensitized mice or to the high LPS sensitivity of human beings. The low efficiency of the human transgene in mice might however, be explained by species differences. The significance of CD14 upregulation for the enhanced LPS responsiveness of sensitized mice, is currently under study.

EFFECTS OF GRAM-NEGATIVE AND GRAM-POSITIVE BACTERIA IN LPS-HYPERSENSITIVE MICE

Gram-negative bacteria possess in addition to LPS a number of components that express biological activity, similar to that of LPS. DNA and lipoprotein are two examples of bacterial components expressing mitogenic activity for B cells and inducing cytokine production and subsequent lethal shock in D-galactosamine–sensitized mice (68–73). Conclusive evidence for the participation of non-LPS components in the induction of TNF-α and of lethal shock by gram-negative bacteria was obtained in experiments using Lps^d mice (5). Lps^d mice sensitized by D-galactosamine, while completely resistant to the lethal effects of purified LPS, succumb to an injection of appropriate amounts (30–100 μg) of whole killed gram-negative bacteria. Gram-positive bacteria also possess components with LPS-like activities. The best known examples of such components are DNA (72) and superantigens, mostly exotoxins, evidently inducers of toxic shock in humans and experimental animals (74–76). In addition, whole heat-killed gram-positive bacteria, such as *Staphylococcus aureus* and *Mycobacterium phlei* were shown to induce TNF-α and thereby to be lethal for D-galactosamine–sensitized Lps^n and Lps^d mice (5).

The question therefore arises whether the hypersensitivity that develops in mice treated with live or killed bacteria is restricted towards the LPS component or whether other biologically active bacterial components exhibit enhanced activity in the sensitized mice. This question was investigated recently in Lps^n BalbC and Lps^d BalbC/1 mice sensitized by *S. typhimurium* infection (unpublished data). Sensitized and control mice were challenged with killed gram-negative bacteria or with killed gram-positive *S. aureus* and their TNF-α responses determined. Mice of both strains exhibited a TNF-α hyperresponse to killed *S. typhimurium* 4 days after infection (Fig. 8). Similar hyperresponses were observed in these mice also towards other gram-negative bacteria, including *Pseudomonas aeruginosa*, *Proteus mirabilis*, *E. coli* O5 and *S. minnesota* R595 (not shown in Fig. 8). Compared to the strongly enhanced TNF-α response induced by gram-negative bacteria, in infected BalbC and BalbC/1 mice, *S. aureus* induced only a moderate increase in the TNF-α response. A similar situation was also found after sensitization of BalbC and BalbC/1 mice with killed *P. acnes*. Both mouse strains exhibited a strong hypersensitivity towards gram-negative bacteria and only a moderate sensitization towards gram-positive bacteria. These data indicate that sensitization by bacteria increases the susceptibility of mice not only to LPS but also to other bacterial components, but to a very different extent.

Since bacteria-induced hypersensitivity to LPS is mediated by IFN-γ, we addressed the question whether the hypersensitivity to bacteria might also be mediated by IFN-γ. For this reason the experiments with BalbC and BalbC/1 mice described above were repeated in the absence of IFN-γ activity (unpublished data). To this purpose IFN-γR–deficient mice, which are normal LPS responders, and C57BL/10ScCr mice, which are both LPS nonresponders and impaired in their IFN-γ production, were used. Both mouse strains were infected with *S. typhimurium* and 4 days later challenged with killed *S. typhimurium* or *S. aureus* or LPS

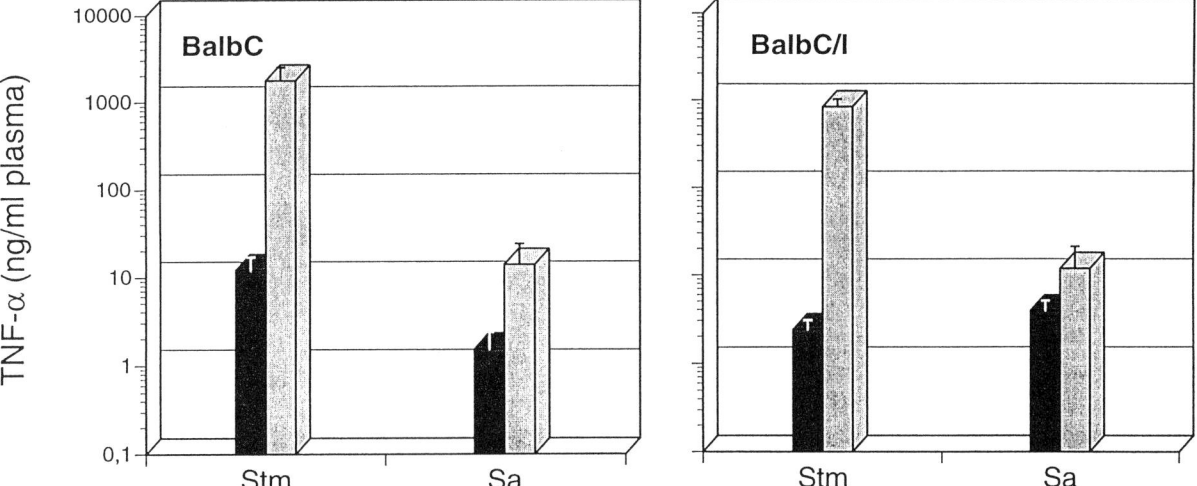

Fig. 8 Enhanced TNF-α production induced by killed bacteria in *S. typhimurium*–infected mice. Mice were infected with 300 CFU of *S. typhimurium* (light bars) or remained untreated (dark bars). Four days later, all mice were challenged with 15 μg/g b.w. of killed *S. typhimurium* (Stm) or *Staphylococcus aureus* (Sa) i.v. TNF-α was measured in plasma collected 1 hour after challenge by a bioassay. The vertical lines represent SD of mean values.

(Fig. 9). As expected, infection of these mice did not alter their LPS sensitivity. However, it did enhance their sensitivity to both bacteria used for challenge to a comparable, albeit only a moderate extent. Results comparable in all respects were also obtained when instead of infection with *S. typhimurium*, treatment with killed *P. acnes* was used for sensitization of BalbC, BalbC/1, IFN-γR–deficient, and C57BL/10ScCr mice (unpublished data).

The following clear results emerged from these studies: infection with *S. typhimurium* or treatment with killed *P. acnes* sensitizes mice moderately towards the TNF-α–inducing activity of killed gram-positive bacteria. The development of the sensitization is IFN-γ

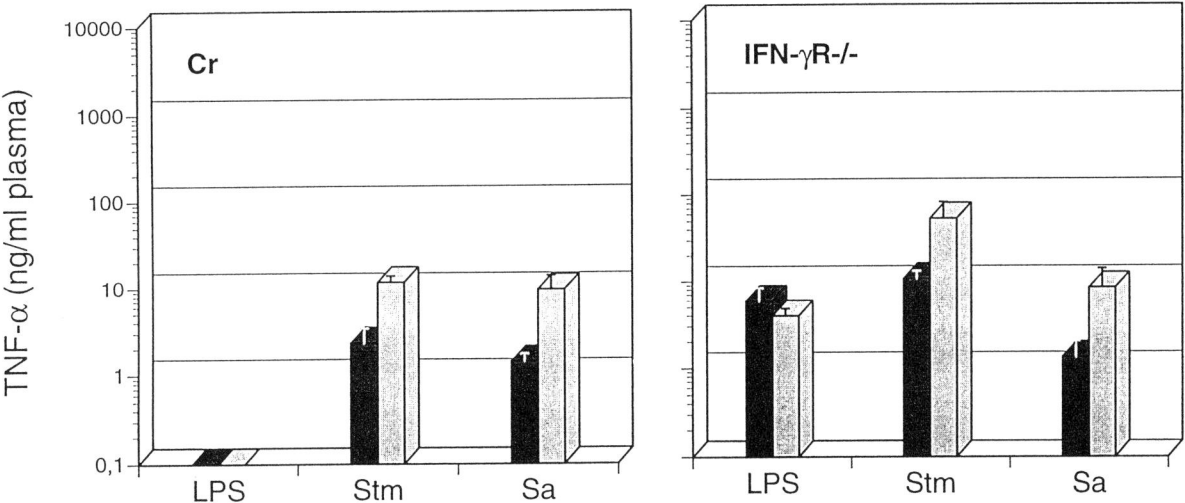

Fig. 9 Production of TNF-α induced by LPS and killed bacteria in *S. typhimurium*–infected C57BL/10ScCr and IFN-γR–deficient mice. Mice of the above strains were infected with 300 CFU and 15000 CFU respectively of *S. typhimurium* (light bars) or remained untreated (dark bars). Four days later, the mice were challenged i.v. with LPS or killed bacteria. The dose of LPS for C57BL/10ScCr and IFNγR-/- mice was 100 and 10 μg, respectively. The dose of killed *S. typhimurium* (Stm) or *Staphylococcus aureus* (Sa) was 15 μg/g b.w. for all animals. TNF-α was measured in plasma collected 1 hour after challenge by a bioassay. The vertical lines represent SD of mean values.

independent. This is evident from the results showing that all sensitized mice (BalbC, BalbC/1, IFN-γR–deficient, and C57BL/10ScCr) exhibit an almost identical enhancement of their TNF-α responses to *S. aureus*. Further, two distinct mechanisms, a potent, IFN-γ dependent, and a weaker, IFN-γ independent one, participate in the sensitization to gram-negative bacteria. Sensitization to purified LPS or to the LPS component of gram-negative bacteria is fully IFN-γ dependent.

Regarding sensitization to non-LPS component(s) of gram-negative bacteria, the evidence so far suggests an IFN-γ–independent mechanism being involved in its induction. It cannot be excluded, however, that an additional, IFN-γ–dependent pathway of sensitization towards these components may exist.

BENEFITS AND HAZARDS OF LPS HYPERSENSITIVITY

The data reviewed in this chapter show that, during infection, two types of hypersensitivity to bacterial components can develop: a strong hypersensitivity mediated by IFN-γ and directed towards purified LPS and gram-negative bacteria, and a weaker, IFN-γ-independent one, directed towards non-LPS components of gram-negative and gram-positive bacteria. In addition, during infection a IFN-γ–mediated hypersensitivity to TNF-α develops. The hypersensitivity to TNF-α, like LPS hypersensitivity, is mediated by IFN-γ (46). Therefore, animals made hypersensitive to TNF-α appear to be hypersensitive to any agent that is capable of inducing this cytokine. The biological consequences of IFN-γ–mediated hypersensitivity become manifest during infections not only with LPS-containing gram-negative bacteria, but with all microorganisms that can induce TNF-α.

The most important aspect of LPS hypersensitivity is probably the ability of the host to sense minute amounts of LPS and respond to these with the production of cytokines, such as TNF-α and IFN-γ. Since these cytokines play a crucial role in antimicrobial defense (77,78), the development of LPS hypersensitivity may be considered a meaningful reaction of the host to infection. It enables the immune system of the infected organism to react to infection at a time when the numbers of microorganisms are low and can be efficiently dealt with. Alternatively, since hypersensitivity, once established, may last for many weeks, upon reinfection during this time the hypersensitive organism can sense and react against invading bacteria that would otherwise, because of their low numbers, evade the defense mechanisms of the host.

Failure of the organism to eliminate at an early stage the invading pathogen, despite an activated antimicrobial defense, would result in the multiplication of the pathogen in the hypersensitive host. In this case, the presence of risk of hypersensitivity would be potentially hazardous, because it lowers the threshold of development of endotoxin shock.

NOTE IN PROOF

During the last stages of the preparation of this chapter, the *Lps* gene was identified and cloned (79). The product of the *Lps* gene is Tlr4, a member of the Toll-like receptor protein family. In the Lps^d C3H/HeJ mice the mutation corresponds to a single nucleotide substitution (C-A) resulting in a exchange of proline by histidine at position 712 of the polypeptide chain. In C57BL/10 ScCr mice the Tlr4 mRNA could not be detected, indicating that a null mutation is present.

REFERENCES

1. Morrison DC, Ulevitch RJ. The effects of bacterial endotoxins on host mediation systems. Am J Pathol 1978; 93:527–617.
2. Watson J, Riblet R. Genetic control of responses to bacterial lipopolysaccharides in mice. I. Evidence for a single gene that influences mitogenic and immunogenic response to lipopolysaccharides. J Exp Med 1974; 140: 1147–1161.
3. Watson J, Kelly K, Largen M, Taylor BA. The genetic mapping of a defective LPS response gene in C3H/HeJ mice. J Immunol 1978; 120:422–424.
4. Vogel SN. The Lps gene: insights into the genetic and molecular basis of LPS responsiveness and macrophage differentiation. In: Beutler B, ed. Tumor Necrosis Factors. The Molecules and Their Emerging Role in Medicine. New York: Raven Press, 1992:485–513.
5. Freudenberg MA, Galanos C. Tumor necrosis factor alpha mediates lethal activity of killed gram-negative and gram-positive bacteria in D-galactosamine-treated mice. Infect Immun 1991; 59:2110–2115.
6. Heppner G, Weiss DW. High susceptibility of strain A mice to endotoxin and endotoxin-red blood cell mixtures. J Bacteriol 1965; 90:696–703.
7. Sultzer BM. Genetic control of leucocyte responses to endotoxin. Nature 1968; 219:1253–1254.
8. Coutinho A, Meo T. Genetic basis for unresponsiveness to lipopolysaccharide in C57BL/10 Cr mice. Immunogenetics 1978; 7:17–24.
9. Vogel SN, Hansen CT, Rosenstreich DL. Characterization of a congenitally LPS-resistant, athymic mouse strain. J Immunol 1979; 122:619–622.

10. Vogel SN, Wax JS, Perera PY, Padlan C, Potter M, Mock BA. Construction of a BALB/c congenic mouse, C.C3H-Lps^d, that expresses the Lps^d allele: analysis of chromosome 4 markers surrounding the Lps gene. Infect Immun 1994; 62:4454–4459.
11. Takakuwa T, Knopf H-P, Sing A, Carsetti R, Galanos C, Freudenberg MA. Induction of CD14 expression in Lps^n, Lps^d and tumor necrosis factor receptor-deficient mice. Eur J Immunol 1996; 26:2686–2692.
12. Chedid L, Parant M. Role of hypersensitivity and tolerance in reactions to endotoxin. In: Kadis S, Weinbaum G, Ajl SJ, eds. Microbial Toxins, Bacterial Endotoxins. New York: Academic Press, 1971:415–459.
13. Galanos C, Freudenberg MA, Katschinski T, Salômao R, Mossmann H, Kumazawa Y. Tumor necrosis factor and host response to endotoxin. In: Ryan JL, Morrison DC, eds. Bacterial Endotoxin Lipopolysaccharides, Immunopharmacology and Pathophysiology. Boca Raton, FL: CRC Press, 1992:75–104.
14. Cavaillon J-M. The nonspecific nature of endotoxin tolerance. Trends Microbiol 1995; 3:320–324.
15. Freudenberg MA, Salômao R, Sing A, Mitov I, Galanos C. Reconciling the concepts of endotoxin sensitization and tolerance. In: Levin J, Pollack M, Yokichi T, Nakano M, eds. Endotoxin and Sepsis: Molecular Mechanisms of Pathogenesis, Host Resistance, and Therapy. Nagoya: Proceedings of the 4th Conference of the International Endotoxin Society, 1998 (in press).
16. Galanos C, Freudenberg MA, Reutter W. Galactosamine-induced sensitization to the lethal effects of endotoxin. Proc Natl Acad Sci USA 1979; 76:5939–5943.
17. Abernathy RS, Bradley GM, Spink WW. Increased susceptibility of mice with brucellosis to bacterial endotoxins. J Immunol 1958; 81:271–275.
18. Suter E, Ullman GE, Hoffmann RG. Sensitivity of mice to endotoxin after vaccination with BCG (bacillus Calmette-Guérin). Proc Soc Exp Biol Med 1958; 99:167–169.
19. Halpern BN, Prévot A-R, Biozzi G, Stifel C, Mouton D, Morard JC, Bouthillier Y, Decreusefond C. Stimulation de l'activité phagocytaire du système réticuloendothélial provoquée par Corynebacterium parvum. RES J Reticuloendothel Soc 1964; 1:77–96.
20. Schramek S, Kazar J, Sekeyova Z, Freudenberg MA, Galanos C. Induction of hyperreactivity to endotoxin in mice by Coxiella burnetii. Infect Immun 1984; 45:713–717.
21. Matsuura M, Galanos C. Induction of hypersensitivity to endotoxin and tumor necrosis factor by sublethal infection with Salmonella typhimurium. Infect Immun 1990; 58:935–937.
22. Katschinski T, Galanos C, Coumbos A, Freudenberg MA. Gamma interferon mediates Propionibacterium acnes-induced hypersensitivity to lipopolysaccharide in mice. Infect Immun 1992; 60:1994–2001.
23. Takada H, Galanos C. Enhancement of endotoxin lethality and generation of anaphylactoid reactions by lipopolysaccharides in muranyl-dipeptide-treated mice. Infect Immun 1987; 409–413.
24. Barlow JL. Hyperreactivity to endotoxin in mice infected with lymphocytic choriomeningitis virus. In: Landy M, Braun W, eds. Bacterial Endotoxins. New Brunswick, NJ: 1964:448–454.
25. Loose LD, Trejo R, Di Luzio NR. Impaired endotoxin detoxification as a factor in enhanced endotoxin sensitivity of malaria-infected mice. Proc Soc Exp Biol Med 1971; 137:794–797.
26. Clark IA. Does endotoxin cause both the disease and parasite death in acute malaria and babesiosis? Lancet 1978; 2:75–77.
27. Akagawa G, Abe S, Yamaguchi H. Mortality of candida albicans-infected mice is facilitated by superinfection of Escherichia coli or administration of its lipopolysaccharide. J Infect Dis 1995; 171:1539–1544.
28. Nansen A, Pravsgaard Christensen J, Marker O, Randrup Thomsen A. Sensitization to lipopolysaccharide in mice with asymptomatic viral infection: role of T cell-dependent production of interferon-γ. J Infect Dis 1997; 176:151–157.
29. Bartholeyns J, Freudenberg MA, Galanos C. Growing tumors induce hypersensitivity to endotoxin and tumor necrosis factor. Infect Immun 1987; 55:2230–2233.
30. Kass EH, Atwood RP, Porter PJ. Observations on the locus of lethal action of bacterial endotoxin. In: Landy M, Braun W, eds. Bacterial Endotoxins. New Brunswick, NJ: 1964:596–601.
31. Parant M, Le Contel C, Parant F, Chedid L. Influence of endogenous glucocorticoid on endotoxin-induced production of circulating TNF-α. Lymphok Cytok Res 1991; 10:265–271.
32. Galanos C, Freudenberg MA, Krajewska D, Takada H, Georgiev G, Bartholeyns J. Hypersensitivity to endotoxin. EOS-J Immunol Immunopharmacol 1986; 6:78–80.
33. Vogel SN, Moore RN, Sipe JD, Rosenstreich DL. BCG-induced enhancement of endotoxin sensitivity in C3H/HeJ mice. I. In vivo studies. J Immunol 1980; 124:2004–2009.
34. Freudenberg MA, Kumazawa Y, Meding S, Langhorne J, Galanos C. Gamma interferon production in endotoxin-responder and -nonresponder mice during infection. Infect Immun 1991; 59:3484–3491.
35. Old L. Tumor necrosis factor. In: Bonavida B, Granger G, eds. Tumor Necrosis Factor: Structure, Mechanisms of Action, Role in Disease and Therapy. Basel: Karger, 1990:1–30.
36. Beutler B, Milsark IW, Cerami A. Passive immunization against cachectin/tumor necrosis factor protects mice from lethal effect of endotoxin. Science 1985; 229:869–871.
37. Tracey KJ, Fong Y, Hesse DG, Manogue KR, Lee AT, Kuo GC, Lowry SF, Cerami A. Anticachectin/TNF monoclonal antibodies prevent septic shock during lethal bacteraemia. Nature 1987; 330:662–664.
38. Galanos C, Freudenberg MA, Coumbos A, Matsuura M, Lehmann V, Bartholeyns J. Induction of lethality and tolerance by endotoxin are mediated by macrophages through tumor necrosis factor. In: Bonavida B, Gifford GE, Kirchner H, Old LJ, eds. Tumor Necrosis Factor/Cachectin and Related Cytokines. Basel: Karger, 1988:114–127.
39. Rothe H, Lesslauer W. Lötscher H, Lang Y, Koebel P, Köntgen F, Althage A, Zinkernagel R, Steinmetz M, Bluethmann H. Mice lacking the tumour necrosis factor receptor 1 are resistant to TNF-mediated toxicity but

highly susceptible to infection by *Listeria monocytogenes*. Nature 1993; 364:798–802.
40. Yaegashi Y, Nielsen P, Sing A, Galanos C, Freudenberg MA. Interferon β, a cofactor in the interferon γ production induced by gram-negative bacteria in mice. J Exp Med 1995; 181:953–960.
41. Müller I, Freudenberg M, Kropf P, Kiderlen AF, Galanos C. Leishmania major infection in C57BL/10 mice differing at the Lps locus: a new non-healing phenotype. Med Microbiol Immunol 1997; 186:75–81.
42. Collart MA, Belin D, Vassalli JD, De Kossodo S, Vassalli P. γ-Interferon enhances macrophage transcription of the tumor necrosis factor-cachectin, interleukin-1, and urokinase genes, which are controlled by short-lived repressors. J Exp Med 1986; 164:2113–2118.
43. Gifford GE, Lohmann-Matthes ML. Gamma-interferon priming of mouse and human macrophages for induction of tumor necrosis factor by bacterial lipopolysaccharides. J Natl Cancer Inst 1987; 78:121–124.
44. Danis VA, Franic GM, Rathjen DA, Brooks TM. Effects of granulocyte macrophage colony stimulating factor (GM-CSF), IL-2, interferon-gamma (IFN-γ), tumor necrosis factor-alpha (TNF-α) and IL-6 on the production of immunoreactive IL-1 and TNF-α by human macrophages. J Clin Exp Immunol 1991; 85:143–150.
45. Beutler B, Tkacenko V, Milsark I, Krochin N, Cerami A. Effect of γ-interferon on cachectin expression by mononuclear phagocytes. Reversal of the lps^d (endotoxin resistance) phenotype. J Exp Med 1986; 164:1791–1796.
46. Freudenberg MA, Kopf M, Galanos C. Lipopolysaccharide-sensitivity of interferon-γ receptor deficient mice. J Endotox Res 1996; 3:291–295.
47. Kamijo R, Le J, Shapiro D. Mice that lack the interferon-γ receptor have profoundly altered responses to infection with Bacillus Calmette-Guérin and subsequent challenge with lipopolysaccharide. J Exp Med 1993; 178:1435–1440.
48. Adorini L. IL-12. Chemical Immunology. Vol. 68. Basel: Karger, 1997.
49. Wysocka M, Kubin M, Viera LQ, Ozmen L, Garotta G, Scott P, Trinchieri G. Interleukin-12 is required for interferon-γ production and lethality in lipopolysaccharide-induced shock in mice. Eur J Immunol 1995; 25:672–676.
50. Rosenstreich DL, Vogel SN. Central role of macrophages in the host response to endotoxin. In: Schlessinger D, ed. Microbiology. Washington, DC: American Society for Microbiology, 1980:11–15.
51. Freudenberg MA, Keppler D, Galanos C. Requirement for lipopolysaccharide-responsive macrophages in galactosamine-induced sensitization to endotoxin. Infect Immun 1986; 51:891–895.
52. Freudenberg MA, Galanos C. Induction of tolerance to lipopolysaccharide (LPS)-D-galactosamine lethality by pretreatment with LPS is mediated by macrophages. Infect Immun 1988; 56:1352–1357.
53. Heumann D, Gallay P, Barras C. Control of lipopolysaccharide (LPS) binding and LPS-induced tumor necrosis factor secretion in human peripheral blood monocytes. J Immunol 1992; 148:3505–3512.
54. Amura CR, Chen L-C, Hirohashi N, Lei M-G, Morrison DC. Two functionally independent pathways for lipopolysaccharide-dependent activation of mouse peritoneal macrophages. J Immunol 1997; 159:5079–5083.
55. Freudenberg MA, Galanos C. Metabolism of LPS in vivo. In: Ryan JL, Morrison DC, eds. Bacterial Endotoxic Lipopolysaccharides, Immunpharmacology and Pathophysiology. Boca Raton, FL: CRC Press, 1992:275–294.
56. Schumann RR, Leong SR, Flaggs GW. Structure and function of lipopolysaccharide binding protein. Science 1990; 249:1429–1431.
57. Wright SD, Ramos RA, Tobias PS, Ulevitch RJ, Mathison JC. CD14, a receptor for complexes of lipopolysaccharide (LPS) and LPS binding protein. Science 1990; 249:1431–1433.
58. Ulevitch RJ, Tobias PS. Receptor-mediated mechanisms of cell stimulation by bacterial endotoxin. Annu Rev Immunol 1995; 13:437–457.
59. Gallay P, Heumann D, Le Roy D, Barras C, Glauser M-P. Lipopolysaccharide-binding protein as a major plasma protein responsible for endotoxemic shock. Proc Natl Acad Sci USA 1993; 90:9935–9938.
60. Gallay P, Heumann D, Le Roy D, Barras C, Glauser M-P. Mode of action of anti-lipopolysaccharide-binding protein antibodies for prevention of endotoxemic shock in mice. Proc Natl Acad Sci USA 1994; 91:7922–7926.
61. Jack RS, Fan X, Bernheiden M, Rune G. Lipopolysaccharide-binding protein is required in vivo to combat a gram negative bacterial infection. Nature 1997; 386:742–745.
62. Freudenberg MA, Gumenscheimer M, Jack R, Merlin T, Schütt C, Galanos C. A strict requirement for LBP in the TNFα hyper-response of *Propionibacterium acnes*-sensitized mice to LPS. J Endotox Res 1997; 4:357–361.
63. Tobias PS, Soldau K, Gegner JA, Mintz D, Ulevitch RJ. Lipopolysaccharide binding protein-mediated complexation of lipopolysaccharide with soluble CD14. J Biol Chem 1995; 270:10482-10488.
64. Yu B, Wright SD. Catalytic properties of lipopolysaccharide (LPS) binding protein transfer of LPS to soluble CD14. J Biol Chem 1996; 211:4100–4105.
65. Matsuura K, Ishida T, Setoguchi M, Higuchi Y, Akizuki S, Yamamoto S. Upregulation of mouse CD14 expression in Kupffer cells by lipopolysaccharide. J Exp Med 1994; 179:1671–1676.
66. Fearns C, Kravchenko VV, Ulevitch RJ, Loskutoff DJ. Murine CD14 gene expression in vivo: extramyeloid synthesis and regulation by lipopolysaccharide. J Exp Med 1995; 181:857–866.
67. Ferrero E, Jiao D, Tsuberi BZ. Transgenic mice expressing human CD14 are hypersensitive to lipopolysaccharide. Proc Natl Acad Sci USA 1993; 90:2380–2384.
68. Melchers F, Braun V, Galanos C. The lipoprotein of the outer membrane of Escherichia coli: a B-lymphocyte mitogen. J Exp Med 1975; 142:473–482.
69. Hoffmann P, Heinle S, Schade UF, Loppnow H, Ulmer AJ, Flad H-D, Jung G, Bessler WG. Stimulation of human and murine adherent cells by bacterial lipoprotein and synthetic lipopeptide analogues. Immunobiology 1988; 177:158–170.
70. Krieg AM, Yi A-K, Matson S, Waldschmidt TJ, Bishop GA, Teasdale R, Koretzky GA, Klinman DM. CpG mo-

tifs in bacterial DNA trigger direct B-cell activation. Nature 1995; 374:546–549.
71. Klinman DM, Yi A-K, Beaucage SL, Conover J, Krieg AM. CpG motifs present in bacterial DNA rapidly induce lymphocytes to secrete interleukin 6, interleukin 12, and interferon γ. Proc Natl Acad Sci USA 1996; 93:2879–2883.
72. Sparwasser T, Miethke T, Lipford G, Borschert K, Häcker H, Heeg K, Wagner H. Bacterial DNA causes septic shock. Nature 1997; 386:336–337.
73. Sparwasser T, Miethke T, Lipford G, Erdmann A, Häcker H, Heeg K, Wagner H. Macrophages sense pathogens via DNA motifs: induction of tumor necrosis factor-α-mediated shock. Eur J Immunol 1997; 27: 1671–1679.
74. Miethke T, Duschek K, Wahl C, Heeg K, Wagner H. Pathogenesis of the toxic shock syndrome: T cell mediated lethal shock caused by the superantigen TSST-1. Eur J Immunol 1993; 23:1494–1500.
75. Miethke T, Wahl C, Regele D, Gaus H, Heeg K, Wagner H. Superantigen mediated shock: A cytokine release syndrome. Immunobiology 1993; 189:270–284.
76. Blank C, Luz A, Bendigs S, Erdmann A, Wagner H, Heeg K. Superantigen and endotoxin synergize in the induction of lethal shock. Eur J Immunol 1997; 27:835–833.
77. Huang S, Hendricks W, Althage A, Hemmi S, Bluethmann H, Kamijo R, Vilcek J, Zinkernagel RM, Aguet M. Immune response in mice that lack the interferon-γ receptor. Science 1993; 259:1742–1745.
78. Rothe J, Lesslauer W, Lötscher H, Lang Y, Koebel P, Köntgen F, Althage A, Zinkernagel R, Steinmetz M, Bluethmann H. Mice lacking the tumor necrosis receptor 1 are resistant to TNF-mediated toxicity but highly susceptible to infection by *Listeria monocytogenes*. Nature 1993; 364:798–802.
79. Poltorak A, He X, Smirnova I, Liu MY, Huffel CV, Du X, Birdwell D, Alejos E, Silva M, Galanos C, Freudenberg M, Ricciardi-Castagnoli P, Layton B, Beutler B. Defective LPS signaling in C3H/HeJ and C57BL/10ScCr mice: mutations in Tlr4 gene. Science 1998; 282:2085–2088.

49

Genetic Control of Endotoxin Responsiveness: The *Lps* Gene Revisited

Stefanie N. Vogel and Nayantara Bhat
Uniformed Services University of the Health Sciences, Bethesda, Maryland

Danielle Malo and Salman T. Qureshi
McGill Centre for the Study of Host Resistance, Montreal General Hospital, Montreal, Quebec, Canada

INTRODUCTION

Recently Kuhns et al. (1) described a young patient with recurrent bacterial infections who was highly refractory to lipopolysaccharide (LPS) both in vivo and in vitro. Based on the failure to produce various cytokines in response to LPS, in the face of normal expression of CD14 and other cell surface markers and normal responsiveness to other stimuli, the authors concluded that the LPS hyporesponsiveness exhibited by this patient was the result of a mutation very early in the LPS signal transduction pathway. This report represents the first description of a human condition that is phenotypically very similar to that of inbred mouse strains that possess a defective allele at the *Lps* locus. In addition to their profound refractoriness to LPS, both in vivo and in vitro, mice that bear a mutation within the *Lps* gene exhibit greatly increased susceptibility to bacterial infections as well (reviewed in Ref. 2). In spite of the fact that a number of knockout mice have been recently demonstrated to exhibit mitigated responses to LPS (3), a mutation within *Lps* leads to the most profound state of endotoxin hyporesponsiveness described to date.

BACKGROUND

The genetic basis for the profound state of hyporesponsiveness to the immunostimulatory and pathophysiological effects of LPS exhibited by the C3H/HeJ mouse strain was established using classical breeding experiments more than 20 years ago (reviewed in Ref. 2). Following the initial characterization of the C3H/HeJ strain as LPS hyporesponsive approximately 30 years ago (4,5), a large number of genetic analyses were carried out to identify the gene(s) responsible for this trait. Analysis of F1, F2, and backcross hybrids from crosses between LPS-responsive and LPS-hyporesponsive mice showed that the defective LPS responses of C3H/HeJ mice in vivo and in vitro (e.g., B-cell mitogenic response and macrophage sensitivity to LPS as measured by the production of prostaglandin E_2 and macrophage cytotoxic activity) were controlled by a single autosomal gene (6–8). Two alleles were assigned to the *Lps* gene in inbred mouse strains: the LPS low responder Lps^d (LPS defective) and the LPS responder Lps^n (normal) alleles (7).

Lps was mapped to mouse mid-chromosome 4 by its linkage to *Mup1* (major urinary protein 1) using the

BXH recombinant inbred strains (RIS) (9–11). Combined mapping data in two additional panels of backcross animals, (C3H/HeJ × C57BL6J)F1 × C3H/HeJ and (C3H/HeJ × C57BL/6y-Ps)F1 × C3H/HeJ, position Lps with respect to two markers on chromosome 4. The early combined recombinational data placed Lps between $Mup1$ proximally and Ps (polysyndactyly) distally at 6 ± 2 cM and 13 ± 7 cM, respectively (7), with inferred linkage to b, the brown coat color locus (a more detailed discussion of the physical mapping of the Lps gene appears below). Hyporesponsiveness to LPS has been characterized independently in two additional mouse strains (C57BL/10ScCR and its progenitor strain, C57BL/10ScN; 12–14) and has also been mapped to the same chromosomal location on mouse chromosome 4 (13). F1 progeny derived from C3H/HeJ and C57BL/10ScCR crosses fail to respond to LPS, suggesting that these two strains of mice possess noncomplementary mutations within the same genetic locus on mouse chromosome 4.

One controversial issue that remains regarding the inheritance of LPS responsiveness in F1 progeny of Lps^d and Lps^n parental mice is the biological significance of a dominantly inherited response pattern (i.e., the F1 response is comparable to the response of the Lps^n parental) versus a codominant pattern of responsiveness (i.e., F1 progeny exhibit a response that is intermediate relative to the parental responses). Both patterns of F1 responses have been reported in the literature and even at times within the same study. Discrepancies in the literature were first directly addressed by McGhee et al. (15), where issues of LPS dosage, the potential contribution of background genes unrelated to Lps, and assay sensitivity are discussed. Nonetheless, in many experimental systems, the response of $Lps^d × Lps^n$ F1 progeny is clearly intermediate. As early as 1975, Coutinho et al. (16) offered several potential mechanisms to explain the meaning of an "intermediate" response pattern. They proposed that either (1) half the cells can be activated by LPS and the other half lack a "triggering receptor" (an hypothesis analogous to allelic exclusion) or (2) all cells can be activated, but at doses tested only half the cells achieve the threshold for activation. They dismissed the latter possibility because the F1 responses in their study were intermediate, even at very high concentrations of LPS. Perhaps the biological significance of an "intermediate" response should be revisited in the context of what we now know about intracellular signaling in response to LPS. Nevertheless, the pleiotropic nature of the Lps defect is compatible with either a dominant or codominant phenotype depending on the contribution of Lps to the particular phenotype being measured. Identification of the Lps gene will facilitate the precise determination of its mode of inheritance.

In addition to the well-characterized C3H/HeJ, C57BL/10ScN, and C57BL/10ScCR Lps^d inbred mouse strains (reviewed in Ref. 2) and the recombinant inbred BXH strains that have been classified as Lps^d (9–11), two newly characterized Lps^d strains have been reported. Specifically, the C.C3H-Lps^d strain (17) is congenic for a segment of chromosome 4 (estimated to be ~5.5 centimorgans, based on analysis of chromosome 4 markers surrounding b and Lps loci) and contains the Lps^d allele on a BALB/cAnPt genetic background. These mice have been extensively inbred (>N20) and are now available from Jackson Laboratories (Bar Harbor, ME). In addition, the BXH recombinant inbred strain, BXH11/Ty, was recently reclassified as endotoxin hyporesponsive (18), unlike the original classification for this strain (9). Based on glucose utilization and nitric oxide production by LPS-stimulated macrophages and LPS-induced spleen cell mitogenesis, the authors suggest that this discrepancy might be explained if this strain had been classified prior to its stabilization or if it mutated to a hyporesponsive phenotype within the past 20 years. Unfortunately, complementation studies with C3H/HeJ mice were not performed in this recent report. Nonetheless, these two strains represent potentially important genetic resources for the analysis of genetic markers in this region and for examining functional effects of Lps^d expression on a BALB/c background.

An important cautionary point with regard to the Lps^d phenotype is that the defect appears to be highly selective for protein-free LPS preparations. Most LPS preparations, particularly commercial ones, are contaminated with significant levels of endotoxin-associated proteins, which can be largely eliminated through repurification (19). These endotoxin-associated proteins are highly bioactive in and of themselves and when purified (i.e., rendered free of contaminating LPS) are equipotent in cells derived from Lps^n and Lps^d strains; like LPS, they exhibit synergistic bioactivity in the presence of IFN-γ (20). Hence, when considering papers that claim LPS responsiveness in Lps^d mice or in cells derived from Lps^d strains, it is necessary to evaluate the source and purity of the LPS used in the study. A corollary to this caveat can be found in studies in which C3H/HeJ mice or their cells respond differentially to a non-LPS stimulant. In such cases, sufficient information should be provided to established that low levels of contaminating endotoxin are not contributing to the observed response pattern (such as reagent testing with the *Limulus* amebocyte lysate assay). A good example of these issues can be found in a recent publication by Dunn and Chuluyan (21). Administration of a commercial preparation of LPS to C3H/HeN and C3H/HeJ mice led to the intriguing finding that the C3H/HeJ strain, as expected, failed to mount a normal catecholamine and pituitary-adrenal response, whereas induction of indolaminergic responses by this LPS preparation were similar in the two strains. In the absence of information on the purity of this LPS preparation or confirmation using a second, more purified LPS preparation, it is difficult to interpret these findings. Certainly, the more interesting possibility is that the induction of tryptophan and the downstream serotonin catabolite 5-HIAA represents a signaling pathway that is not disrupted by the expression of the Lps^d allele in vivo. However, in the absence of this critical information, the reader is left to wonder if contaminating moieties within the LPS preparation are able to elicit the indolaminergic pathway independently of LPS.

As mentioned above, many studies have compared the in vivo responses of Lps^n and Lps^d mice to LPS and have found the C3H/HeJ strain to be, at best, LPS hyporesponsive (reviewed in Ref. 2). Evidence also suggests that the LPS defect is expressed in virtually all cells derived from Lps^d strains. However, two intriguing reports suggest that the phenotypic expression of LPS hyporesponsiveness may not be uniform throughout all organs of the mouse. Ryan and Vermeulen (22) have shown that alveolar macrophages (AM) derived from C3H/HeJ mice respond to LPS by producing TNF-α, albeit at higher concentrations than required for AM derived from LPS-normoresponsive C3HeB/FeJ mice. Under the same conditions, neither resident nor thioglycolate-induced peritoneal macrophages from C3H/HeJ mice were LPS responsive. This apparent circumvention of the Lps^d defect is incomplete, however, since exposure to LPS at a dose sufficient to induce TNF secretion failed to result in the induction of IL-1β mRNA in C3H/HeJ AM. The authors suggest that C3H/HeJ AM may employ an alternate signaling mechanism by which they can produce TNF in response to high concentrations of LPS, which is not expressed in peritoneal macrophages and is apparently distinct from the signaling pathway(s) required to induce IL-1β gene expression. Similarly, Nill et al. recently compared the pulmonary responses of C3H/HeN and C3H/HeJ mice using semi-quantitative RT-PCR and in situ immunostaining procedures to examine the induction of TNF-α, IL-1α, and IL-10 gene expression and protein, respectively (23). Intraperitoneal injection of 5 mg/kg LPS (stated to contain 0.13 mg LPS-associated protein) resulted in the induction of significant levels of TNF-α mRNA from lung homogenates, even in the C3H/HeJ mice; IL-1α mRNA levels were much lower in C3H/HeJ than C3H/HeN mice, with IL-10 mRNA exhibiting the most striking difference between the two strains. At the level of protein synthesis, immunoreactive cytokines were associated with bronchial epithelial cells, infiltrating neutrophils, alveolar macrophages, and type II pneumocytes with significantly less staining and a diminished number of infiltrating neutrophils in C3H/HeJ lung tissue. These data support the hypothesis put forth by Ryan and Vermeulen that in the C3H/HeJ strain, the lung represents a unique site with respect to LPS responsiveness in the C3H/HeJ strain that appears to be somewhat permissive for the induction of TNF-α.

The *Lps* gene is arguably the most extensively studied of the thousands of genes involved in the host response to endotoxin, however, its identity remains obscure despite aggressive identification efforts using a variety of functional and molecular approaches. To date, the accrued data from many laboratories support the hypothesis that the binding of LPS to cells of Lps^d mice appears to be normal (reviewed in Ref. 2). Like Lps^n macrophages, C3H/HeJ macrophages demonstrated an LPS-binding protein (LBP)–dependent enhancement of LPS binding, supporting the notion that the defect involves a step distal to LPS binding (24). In fact, recent studies even indicate that the very earliest (within seconds) intracellular signaling events can be detected in LPS-stimulated C3H/HeJ macrophages (25,26) (discussed below) followed by what appears to be a global block in downstream signaling. Thus, the defect may reflect a critical protein in the LPS signaling pathway that is disabled or, alternatively, a gene that has been mutated such that the gene product acts as a dominant negative repressor of normal LPS responsiveness. In reviewing the older literature with the specific goal of "rethinking" the nature of the LPS defect, it was striking that among the many papers published on this mouse strain, five actually suggested that the defect may mediate a "suppressor" phenotype (27–31). The most recent of these studies demonstrates an inhibitory effect of high concentrations of LPS (>10 μg/ml) on mitogen-driven proliferation of C3H/HeJ B or T cells or of prostaglandin E_2 release in C3H/HeJ macrophages. In the case of mitogen-induced B-cell proliferation, LPS was found to inhibit progression of cell cycle from G1 to S, with a concurrent inhibition of uptake of [^{3}H]uridine.

Nonetheless, in the past 5 years, previous findings associated with the *Lps* gene have been extended and new phenotypes recently discovered that suggest potential functions for this gene product in LPS-induced events. The authors regret not being able to include all citations; the sheer number of publications on this topic precludes this possibility. Therefore, in the remainder of this chapter, we will focus on the major findings that have been reported within the past 5 years and direct readers to several earlier, extensive reviews on this subject (2,32,33).

Lps^d MACROPHAGES EXHIBIT DEFECTIVE INTERFERON PRODUCTION: POSSIBLE ROLE IN INCREASED SUSCEPTIBILITY TO INFECTION

Previous studies demonstrated that Lps^d macrophages exhibit a reduced capacity to produce IFN, as evidenced by increased susceptibility to vesicular stomatitis virus infection in vitro or a more rapid decay in the capacity to transfer antiviral activity in tissue culture supernatants (reviewed in Ref. 2). Although Lps is clearly distinct from Ifa and Ifb loci (11,34,35), the differential expression of IFN may provide insights into the increased susceptibility phenotype of Lps^d mice to a variety of pathogens. Using nuclear run-on transcription assays, resident peritoneal macrophages from Lps^d mice were shown to transcribe low levels of IFN-β mRNA, even though they could not transfer an antiviral state (36). Treatment of Lps^n macrophages with cycloheximide resulted in a marked accumulation of IFN-β mRNA, in contrast to the lack of IFN-β mRNA accumulation in C3H/HeJ macrophages. Stabilization of IFN-β mRNA in the LPS-responsive macrophages was correlated with an increase in the accumulation of cytoplasmic factors that interact with the AU-rich sequences in the 3′ untranslated region of IFN-β mRNA. The authors concluded that C3H/HeJ macrophages exhibit an impaired capacity to stabilize IFN-β mRNA, thus resulting in a low expression of antiviral activity (37). In subsequent studies, semi-quantitative reverse transcription-polymerase chain reaction (RT-PCR) was utilized to demonstrate much lower basal expression of not only IFN-β, but also IFN-α and IFN-γ steady-state mRNA in C3H/HeJ peritoneal exudate macrophages when compared to C3H/OuJ macrophages (38,39), with the most profound difference being in levels of IFN-γ mRNA. Basal mRNA levels of IFN regulatory factor (IRF)-1, a well-characterized, IFN-inducible nuclear transactivating factor, were approximately 15-fold greater in LPS-responsive macrophages. In contrast, C3H/HeJ macrophage steady-state mRNA levels for IRF-2, an IFN-inducible transrepressive nuclear binding protein that is highly homologous to IRF-1, was approximately 18-fold greater than in LPS-responsive macrophages. Thus, the balance between the IRFs may represent yet an additional pathway in the complex response to LPS. Finally, C3H/HeJ macrophages failed to be "primed" by LPS to produce augmented levels of IFN bioactivity when subsequently stimulated by poly I:C, in contrast to the normal response of these cells to exogenous IFN-α or IFN-γ as the priming agent (39). Taken collectively, these data suggest the possibility that LPS-induced IFNs may act in an autocrine fashion to provide a replenishable source of "primed" macrophages that are functionally more responsive to "triggering" signals in the form of bacterial or viral challenge. Thus, the failure of environmental LPS to stimulate basal IFN levels in C3H/HeJ macrophages may well underlie the increased susceptibility to infection exhibited by this strain.

In this regard, recent studies by Cross et al. (40) provide experimental support for the hypothesis that the failure of C3H/HeJ macrophages to be activated by LPS in vivo underlies their inability to control gram-negative infection. Briefly, these investigators infected C3H/HeN (Lps^n) and C3H/HeJ mice with an extraintestinally invasive strain of $E.\ coli$ that possesses both a complete LPS and K1 capsule. Although both strains cleared the $E.\ coli$ from the circulation within 4 hours of intravenous injection (C3H/HeN mice were somewhat more efficient), electron microscopy of liver sections revealed a marked difference in the ability of C3H/HeN versus C3H/HeJ strains to control bacterial replication within the phagolysosomes of Kupffer cells. Activation of C3H/HeJ macrophages in vitro with various cytokines (e.g., IFN-γ, M-CSF, IL-1 or TNF) or activation of macrophages in vivo by administration of BCG resulted in greatly augmented killing of the $E.\ coli$ and increased survival, respectively. Thus, the authors concluded that an LPS-initiated program of cytokine-mediated activation is lacking in C3H/HeJ mice, accounting for their hypersusceptibility to gram-negative infection. These data are also consistent with the previous findings that BCG infection of C3H/HeJ mice largely normalizes their in vivo responsiveness to LPS, accompanied by augmented levels of cytokines and acute phase reactants (41).

NOVEL PHENOTYPES ASSOCIATED WITH Lps

Hyporesponsiveness to Soluble Glucan

In a study carried out by Gallin et al. (42), the actions of soluble and particulate glucan (β-1,2-D-polyglucose) preparations derived from *Saccharomyces cerevisiae* were compared in C3H/HeN and C3H/HeJ resident peritoneal macrophages. Macrophages possess β-1,3-glucan receptors and are one of the primary targets for glucan. Both particulate and soluble glucan preparations were found to contain <1 ng LPS/10 mg glucan as assessed in a *Limulus* amebocyte lysate assay. Although both particulate and soluble glucans inhibited the ability of macrophages from both strains to ingest zymosan, only particulate glucan inhibited uptake of IgG opsonized erythrocytes and latex beads. Particulate glucan was also as effective as zymosan as an inducer of superoxide and IL-1 release in macrophages derived from C3H/HeJ and C3H/HeN macrophages. In contrast, soluble glucan was effective at enhancing PMA-induced superoxide or IL-1 in C3H/HeN, but not in C3H/HeJ macrophages. The authors concluded that while C3H/HeJ macrophages exhibit normal responsiveness to particulate glucan, they are hyporesponsive to soluble glucan. Although glucans derived from *S. cerevisiae* have been recognized for decades as potent immunomodulators, the molecular mechanisms for their actions have not been well delineated. However, many of glucan's actions on the reticuloendothelial system are highly reminiscent of those elicited by LPS, and it is possible that soluble glucan stimulates an intracellular signaling pathway shared by LPS that is defective in C3H/HeJ macrophages.

Defective Platelet-Activating Factor Responsiveness

Platelet-activating factor (PAF) is a potent LPS-inducible, proinflammatory mediator that has been implicated in endotoxicity and sepsis (reviewed in Ref. 43). One of the more interesting phenotypes recently associated with the C3H/HeJ mouse strain is its resistance to PAF administered in vivo. Sun et al. (44) reported that Lps^n (C3H/OuJ or C3H/HeN) mice responded to PAF administered via the carotid artery with shock, hemoconcentration, complement activation, intestinal hyperperfusion and necrosis, and death. In contrast, Lps^d C3H/HeJ mice were highly refractory to PAF, exhibiting only mild hypotension, slight hemoconcentration, minimal intestinal injury, and no complement activation or mortality relative to Lps^n mice. In addition, LPS-responsive mice exhibited endogenous PAF production and intestinal phospholipase A_2 activation, whereas neither was induced in the C3H/HeJ mice in response to PAF administration. These findings suggest that the C3H/HeJ defect may lie within the PAF signaling pathway. Although the PAF receptor gene (*Ptafr*) maps to mouse chromosome 4 [~62 cM from the centromere (45)], it is distal to Lps (~30 cM from the centromere). More recently, Shinji et al. (26) found that Lps^n macrophages respond to LPS to release two peaks of inositol 1,4,5-triphosphate (IP3): the first IP3 peak, which occurs within 30 seconds of LPS stimulation, was blocked by the presence of a PAF receptor antagonist. The presence of this inhibitor, however, failed to alter TNF secretion and mRNA production. The second peak of IP3 generation, which occurs within minutes of LPS stimulation, was associated with phosphorylation on tyrosine of a 140 kDa protein, which is possibly phospholipase C-γ_2 and whose presence was correlated with TNF secretion and mRNA induction. In C3H/HeJ macrophages, LPS stimulates the first peak of IP3 release, but not the second. Thus, appropriate signal transduction is observed within the first 30 seconds of LPS stimulation of C3H/HeJ macrophages; however, the failure to proceed to the next peak of IP3 generation indicates that the Lps^d defect influences one or more upstream signaling events that are detectable within minutes of LPS stimulation. Although the obvious interpretation of the finding of Shinji et al. (25) is that PAF mediates the very early generation of IP3 after LPS signaling, it is also possible that the LPS signaling shares with the PAF receptor a critical signaling element that is sequestered in the presence of the PAF receptor antagonist. The work of Henricson et al. (25) also supports the notion that in C3H/HeJ macrophages, the very earliest of LPS signals are transmitted: in Lps^n C3H/OuJ macrophages, stimulation with either LPS or the LPS-mimetic Taxol™ results in an initial depression of Lyn autophosphorylation that is detectable within 15 seconds of stimulation, followed within minutes by autophosphorylation of both p53 and p56 Lyn species. In contrast, Lps^d macrophages responded to LPS and Taxol with the initial decrease in activity but failed to exhibit the subsequent increase in autophosphorylation, a situation that can be mimicked in Lps^n C3H/OuJ macrophages by the presence of tyrosine phosphatase inhibitors. A further discussion of these findings will be presented in a later section.

Defective Taxol Responsiveness

Taxol is a diterpenoid derived from the bark of the Pacific yew tree that bears no apparent structural homology to LPS and has as its only known molecular target β-tubulin (reviewed in Refs. 46 and 47). In cells, Taxol binds with high affinity to β-tubulin in the context of polymerized microtubules, resulting in highly stabilized microtubules (48–51), and thus it blocks mitosis. Aihao Ding and colleagues (52) first reported that Taxol induced TNF secretion and caused rapid involution of TNF receptors on murine macrophages, two activities shared by LPS. More interestingly, they found that Taxol failed to stimulate Lps^d macrophages (53). The observation of Taxol-induced TNF secretion in Lps^n macrophages was confirmed and extended to show that Taxol induced a large panel of LPS-inducible, immediate early genes in C3H/OuJ, but not C3H/HeJ macrophages. Both LPS and Taxol induced tyrosine phosphorylation of the MAP kinases (54–57), Lyn kinase autophosphorylation (25), and NF-κB translocation (58) in Lps^n, but not Lps^d macrophages. Yet, in spite of their failure to respond to Taxol as an LPS mimetic, Taxol-treated C3H/HeJ macrophages form microtubule bundles that are indistinguishable from those observed in LPS-responsive macrophages (54). This finding, coupled with a lack of correlation among various Taxol analogs to induce microtubule binding and their ability to act as LPS mimetics (47,59), suggests that the LPS mimetic effects of Taxol are distinct from mechanisms related to blocking microtubule depolymerization. The capacity of LPS structural antagonists to block Taxol-induced LPS mimetic effects further supports the notion that these two agents engage a common signaling apparatus that is defective in the C3H/HeJ strain. Recent studies in which a photoaffinity Taxol analog was used to cross-link proteins in membranes derived from C3H/HeJ versus C3H/OuJ mice failed to reveal any significant differences in Taxol-binding proteins (60).

Defective Ceramide Responsiveness

Ceramide is a lipid second messenger that is principally generated by the hydrolysis of membrane sphingomyelin by sphingomyelinases. Because of its lipid nature, the interaction of ceramide with specific signaling pathways is strictly membrane associated (reviewed in Ref. 61). Ceramide-activated intracellular signaling involves a cytosolic ceramide-activated protein phosphatase (62) and a membrane-associated, ceramide-activated protein (CAP) kinase. CAP kinase has been postulated to initiate TNF-α–induced signaling by phosphorylating and activating Raf-1 and was recently demonstrated to be identical to "kinase suppressor of Ras" (KSR) (63). On the basis of structural similarities between lipid A and ceramide, Joseph et al. (64) postulated that LPS might activate the ceramide-activated kinase directly, in the absence of sphingomyelinase activity, by mimicking the second-messenger function of ceramide.

Subsequently, Barber et al. (65) demonstrated that exogenous sphingomyelinase, like LPS, induced TNF secretion in C3H/OuJ macrophages but, unlike LPS, failed to induce detectable IFN bioactivity. In addition, cell-permeable ceramide analogs were demonstrated to induce a subset of LPS-inducible genes, suggesting that ceramide is a partial LPS mimetic in that it may engage only certain signaling pathways that are activated by LPS. C3H/HeJ macrophages not only failed to respond to LPS, but also failed to secrete cytokines or to express LPS-inducible genes when stimulated with either cell-permeable analogs of ceramide or exogenous sphingomyelinase (66). These findings suggest that a common critical molecule, encoded by the Lps gene, regulates both ceramide and LPS signaling pathways. In an extension of the comparison of LPS- and ceramide-inducible activities in Lps^n versus Lps^d macrophages, Lakics and Vogel (67) recently demonstrated that both LPS and cell-permeable ceramide analogs induce macrophage cytotoxicity in C3H/OuJ macrophages in the absence or presence of IFN-γ. However, C3H/HeJ macrophages were completely refractory to LPS-mediated cytotoxicity (up to 50 μg/ml LPS plus 50 U/ml IFN-γ) and hyporesponsive to 25 μM C2 ceramide plus IFN-γ. In contrast to LPS, however, higher concentrations of C2 ceramide plus IFN-γ overcame this defect, eliciting comparable cytotoxicity in C3H/HeJ and C3H/OuJ macrophages.

A related phenotype was recently reported by Thiéblemont and Wright (68). In this study, both C3H/HeJ macrophages and those derived from a second Lps^d strain, C57BL/10ScN, were compared to normal macrophages for the abilities to bind and endocytose fluorescent LPS and ceramide derivatives. Although initial binding of fluorescent LPS or ceramide was quantitatively comparable to that observed in Lps^n macrophages, the normal pattern of movement of the labeled derivatives from the cell perimeter, with ultimate consolidation in the perinuclear region, was not observed with subsequent incubation of LPS-hyporesponsive macrophages. Interestingly, the failure of C3H/HeJ macrophages to consolidate the fluorescent ceramide analog, but not LPS, in the perinuclear region of the macrophage appears to be a kinetic difference since a longer incubation results in a "normal" consolidation pattern. The authors suggest that a defect in vesicular transport of LPS and ceramide may result in the failure of Lps^d macrophages to respond to these two mediators. However, it is not clear from these studies if a LPS- or ceramide-induced signal is necessary for the observed movement of these substances to the perinuclear location or if the translocation of the LPS or ceramide to the perinuclear region initiates intracellular signaling that leads to gene expression.

Others have persued the hypothesis that structural differences in the membranes of Lps^n and Lps^d cells may account for differential responsiveness to LPS stimulation. Printen et al. (69) studied the influence of LPS stimulation on the lateral mobility of lipid probes in the B-cell membrane. Both fluorescent lipid analogs—TRITC-LPS and 3,3′-dioctadecylindocarbocyanine iodide (DiI)—showed decreased mobility in LPS-stimulated normal B cells but not in C3H/HeJ B cells. Although the mechanism responsible for the observed difference was not identified, the authors concluded that simple perturbation of the lipid bilayer by LPS was unlikely since most LPS molecules interact nonspecifically with B-cell membranes via lipid A acyl side chain insertion into the membrane (69). Moreover, equivalent amounts of LPS bind to Lps^n and Lps^d cells (reviewed in Ref. 2). Previous studies have indicated that the composition of various gangliosides in the membranes of Lps^n and Lps^d B cells differ, and in macrophages altered accessibility of certain macrophage gangliosides was observed after stimulation with LPS (reviewed in Refs. 2 and 70). More recently, Macala and Yohe (71) further characterized ganglioside expression in C3H/HeN and C3H/HeJ resident and thioglycolate-elicited macrophages unstimulated and after stimulation with E. coli. They concluded that altered macrophage ganglioside accessibility appears sometimes as a consequence, but not as a cause of C3H/HeJ hyporesponsiveness. Since ceramide is a critical structural component of gangliosides, and since gangliosides have been shown in several earlier studies to inhibit LPS-induced responses both in vitro and in vivo (72,73), it is tempting to speculate that there may be some association between the ceramide hyporesponsiveness and ganglioside architecture in these cells. Finally, it has also been speculated that LPS hyporesponsiveness may be associated with ganglioside architecture (71).

Finally, Haimovitz-Friedman et al. have shown that a lethal dose of LPS elicits endothelial apoptosis induced by TNF-α which is dependent upon the action of acid sphingomyelinase (74). Both lethality and endothelial apoptosis could be blocked by concurrent administration of basic fibroblast growth factor (bFGF), a cytokine previously demonstrated to protect endothelial cells from radiation-induced apoptosis (75). A number of years ago, Neta and colleagues (76; R. Neta, personal communication) observed that sublethal doses of LPS, IL-1, and TNF-α could induce radioresistance in C3H/HeN mice, but not in C3H/HeJ mice. Since both IL-1 and TNF-α are known to induce the sphingomyelin pathway, the possibility exists that this phenotype may be related to diminished ceramide sensitivity in the C3H/HeJ strain.

Overproduction of Secretory Leukocyte Protease Inhibitor

Recently, the technique of differential display analysis was employed to identify gene products that are differentially expressed by Lps^n and Lps^d macrophages. Using this approach to compare mRNA species generated from HeNC2 (derived from Lps^n, C3H/HeN mice) and LPS-hyporesponsive, GG2EE (derived from Lps^d, C3H/HeJ mice) macrophage cell lines, Jin et al. (77) found differential expression of the gene that encodes secretory leukocyte protease inhibitor (SLPI). In macrophage cell lines that responded strongly to LPS, SLPI mRNA and protein were underexpressed as compared to levels observed in the GG2EE macrophages. Interestingly, the J774.1 macrophage cell line, derived from the Lps^n BALB/c mouse strain, also exhibited overexpression of SLPI mRNA and protein. This fact alone suggested that SLPI was unlikely to be the product of the Lps gene, a finding that was confirmed by the report that the gene that encodes SLPI does not map to chromosome 4 (77). Nevertheless, when the SLPI gene is overexpressed in the fully LPS-responsive HeNC2 macrophage cell line, it appears to be a potent inhibitor of LPS-induced nitrite and TNF-α release, as well as NF-κB translocation.

It is important to put the association of SLPI overexpression with diminished LPS-responsiveness in the context of previous observations. Ku et al. (78) previously reported that serine esterase inhibitors, such as diisopropylfluoro-phosphate, blocked LPS-induced mitogenesis in normally LPS-responsive B cells. Thus, it is possible that SLPI, a serine esterase inhibitor, acts analogously in macrophages. Consistent with the findings of Ku et al., Kuus-Reichel and Ulevitch (79) later demonstrated that treatment of C3H/HeJ B cells with the serine protease trypsin partially restored their capacity to respond to LPS as a mitogenic stimulus. Interestingly, Jin et al. have suggested an alternate hypothesis for SLPI's antagonism of LPS-induced effects, i.e., it may act like plasminogen activator inhibitor-1, another serine protease inhibitor, which has been shown to disrupt the interaction of cellular integrins with the extracellular matrix (80,81). Since the β$_2$ integrins CD11c/CD18 and CD11b/CD18 have been demonstrated in certain systems to serve as LPS signaling molecules (reviewed in Refs. 82–84), it is possible that overproduction of SLPI could result in a modification of CD11b/CD18 expression on macrophages, resulting in a mitigated LPS response phenotype.

Differential Expression of Specific Intracellular Signaling Molecules in Lps^n versus Lps^d Membranes

Using a modification of the procedure of Liu et al. (85), Bhat and colleagues (86) prepared membranes from thioglycollate-elicited C3H/OuJ and C3H/HeJ macrophages. Membrane proteins were separated by polyacrylamide gel electrophoresis and then transferred to Immobilon-P membranes for Western blot analysis. Analysis for the presence of a variety of structural proteins (e.g., β-tubulin, actin, STIP-1), cell surface proteins (e.g., CD14, CD18, I-A^k), membrane-associated signaling molecules [e.g., ceramide activated protein kinase (KSR-1), G proteins (Gsα, Gαi2, Gβ), chaperone proteins (e.g., HSP-70 and HSP-90), and intracellular signaling molecules (e.g., Lyn, Hck, Fgr, Ras, and c-Raf] (Fig. 1) failed to reveal any significant difference in the electrophoretic mobility or intensity of staining between the two strains. However, two important signaling molecules were differentially detected in C3H/HeJ versus C3H/OuJ macrophage membrane preparations: Vav and Rsk (Fig. 1). In both cases, C3H/HeJ membranes exhibited significantly more immunoreactive Vav and Rsk (also referred to as $p90^{rsk}$) than C3H/OuJ membranes. It is important to recognize that these membranes were prepared from macrophages that were not exposed to exogenous LPS, suggesting that these differences are observed with basal expression of the gene products of the Lps^n versus Lps^d alleles.

Vav was recently shown to be phosphorylated upon LPS stimulation (87), suggesting that it is one of the many kinases activated by LPS signaling. *Vav* is a proto-oncogene, and its gene product, Vav, has been suggested to play a role in the signal transduction pathways of hematopoietic cells (reviewed in Ref. 88). We have found that Vav becomes phosphorylated in LPS-treated macrophages earlier than the MAP kinases (J. Blanco, unpublished observation). Since early activation of Rsk kinase activity by phosphorylation is co-

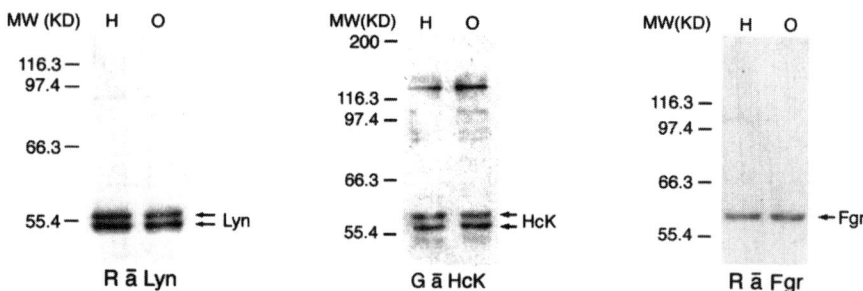

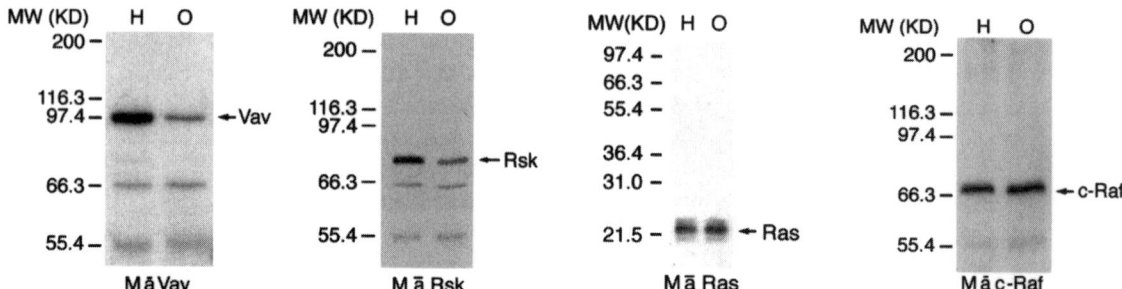

Fig. 1 C3H/HeJ (H) and C3H/OuJ (O) membrane preparations were subjected to denaturing SDS-PAGE and were analyzed by Western blot analysis for the indicated proteins. The letters under each blot indicate the species in which the antibodies were prepared (e.g., R, rabbit; G, goat; M, mouse).

ordinate with the activation of MAP kinases, it has been suggested that Rsk acts to phosphorylate DNA-binding proteins and cytoplasmic proteins (reviewed in Ref. 89). *Vav* does not map to chromosome 4 (90), and another closely related gene, *Vav2*, was recently found to map close to the telomer of the long arm of human chromosome 9, far distal to the predicted location of the human *Lps* homolog (91). *Rsk*, also referred to as p90[rsk] or *Rps6ka1* [ribosomal protein S6 kinase II alpha 1 (92)] has not been mapped in mouse; however, a human sequence called ribosomal protein S6 kinase 2 (93), which is 98% homologous to *Rps6ka1*, has been mapped to human chromosome 1 in the region of the PAF receptor, i.e., homologous to distal mouse chromosome 4 (94). Thus, the finding of increased presence of these two kinases in the membranes of C3H/HeJ macrophages may suggest that their localization is affected by the Lps^d mutation.

EXPRESSION CLONING OF THE *Lps* GENE

A novel strategy for isolating the *Lps* gene was developed, based upon acquisition of LPS responsiveness in C3H/HeJ macrophages after transfection with cDNA clones from C3H/HeN (Lps^n) macrophages (C. R. Maliszewski, M. Schoenborn, and S. N. Vogel, unpublished observations). Transfected C3H/HeJ macrophages were plated in chamber slides, cultured for 2 days, and then LPS-stimulated. A cDNA clone conferring LPS sensitivity (i.e., complementing the Lps^d defect) should permit the C3H/HeJ macrophages to respond to LPS by producing TNF-α. To detect TNF-α–producing cells, slide chambers were incubated with ^{125}I-anti-murine TNF-α monoclonal antibody, washed, and then dipped in photographic emulsion. After 3 days in the dark, slides were developed, fixed, and scanned microscopically for the rare cell laden with dark grains. This method was verified through various means, including demonstration that a single Lps^n macrophage could be detected against a background of 1000 Lps^d macrophages. A highly enriched cDNA library generated from C3H/HeN (Lps^n) macrophage mRNA in the pDC303 plasmid (a mammalian expression vector that contains a CMV-based promoter cassette) was screened using this method. This vector had been used successfully to express specific genes in transfected bone marrow–derived macrophages. Unfortunately, screening 500,000 cDNA clones (from pools of 500 clones) in transfected rGM-CSF–derived C3H/HeJ bone marrow macrophages failed to yield any convincingly positive clones.

Nevertheless, a recent report used a similar functional screening strategy to identify a gene that could reconstitute LPS responsiveness in primary C3H/HeJ B cells. In their study, Kang et al. (95) prepared a cDNA library from LPS-stimulated C3H/OuJ spleen cells that had been depleted of T cells and erythrocytes. The plasmid library was constructed using the Okayama-Berg expression vector, which was introduced by electroporation into B-cell–enriched, C3H/HeJ spleen cells. The functional "readout" in this assay is the ability of LPS-stimulated transfectants to form anti-sheep erythrocyte plaque-forming cells (PFC) in vitro. Screening of six sublibraries, each consisting of only $1-2 \times 10^4$ independent clones, was carried out with progressive reduction of the number of independent clones from a single positive sublibrary until a single plasmid species (pCD-LPS) was identified for its ability to increase the LPS-stimulated plaque-forming response of the C3H/HeJ B cells. Introduction of the pCD-LPS plasmid into splenic C3H/HeJ B cells significantly increased the number of plaques that formed in the presence of LPS (from 12 to 37 PFC), with a control plasmid resulting in only 6 PFC. Although electroporation of the control vector reduced the number of LPS-responsive PFCs in the C3H/OuJ B cell cultures by approximately two thirds, in the presence of pCD-LPS, the number of PFCs was increased from 25 to 84. Analysis of the cDNA sequence demonstrated the clone to be identical to the gene that encodes Ran/TC4 GTPase (96). The latter possesses highly conserved regions among diverse species and is a member of a unique family of GTP-binding proteins that localize in the nucleus. In their discussion, Kang et al. indicate that when four independent clones of Ran/TC4 derived from C3H/HeJ splenic B cells were sequenced, all showed a single base substitution from T to C at position 870, within the 3′-untranslated region. In addition, the authors state that by Northern blot analysis, no significant difference in Ran/TC4 mRNA levels was observed between C3H/HeJ and C3H/OuJ mRNA preparations, suggesting that the mutation would result in a posttranscriptional modification. They postulate that this Ran GTPase may play a role in nuclear transport, cell cycle progression, or cell differentiation and suggest that the apparent point mutation in the C3H/HeJ B cells may result in an alteration in the intracellular distribution of Ran/TC4.

Although Kang et al. also indicate in their discussion that gene encoding Ran/TC4 maps to chromosome 4 in the mouse, the fine mapping of this gene was not presented. More recent studies indicate that *Ran* does not map to the subchromosomal region surrounding *Lps* (S. Qureshi and D. Malo, unpublished data). It is possible that the signal detected by in situ hybridization represents a cross-hybridization signal since *Ran* is part of a gene family that contains ≥12 highly homologous members; two of these are known to be located on chromosome 4, proximal to *Lps*.

Given the plethora of literature that suggests that LPS signaling is defective at a very early point in the signaling pathway, it is difficult to understand how a nuclear GTPase such as Ran could modulate these critical early events in the signaling pathway. Specifically, two recent reports indicate that some of the earliest measurable responses to LPS are preserved in LPS-stimulated C3H/HeJ macrophages: Henricson et al. (25) and Shinji et al. (26) found that modulation of Lyn kinase activity and IP3 release, respectively, occurred in both C3H/OuJ and C3H/HeJ macrophages within seconds of LPS stimulation, followed by a curtailment of the normal LPS response pattern in the C3H/HeJ strain. In another study, Shinji et al. (97), demonstrated that two minutes after LPS stimulation, Lps^n macrophages exhibit peak activation and translocation of protein kinase C-β, while C3H/HeJ macrophages fail to exhibit this response, despite the fact that they respond vigorously to phorbol myristic acetate. This identifies a block in LPS signaling that normally occurs between 40 seconds and 2 minutes after LPS stimulation.

Despite these criticisms, the hypothesis that a defective cellular G protein could account for LPS hyporesponsiveness remains plausible, given previous studies that have implicated G proteins as critical components in LPS signaling (reviewed in Ref. 98). In studies by Zhang and Morrison (99), treatment of Lps^n macrophages with pertussis toxin (a selective G-protein inhibitor) was found to augment LPS-induced TNF-α while inhibiting LPS-dependent nitric oxide production. This treatment did not correct the LPS hyporesponsiveness of C3H/HeJ macrophages, however, suggesting that their defect is unlikely to be at the level of this pertussis toxin–sensitive factor. It is also interesting to note that in macrophages derived from endotoxin-tolerized rats, significantly reduced levels of membrane GTPase activity coupled with a significant diminution in the expression of Gi3α, but not of other G proteins, has been reported (100). Moreover, the failure of endotoxin-tolerized rat macrophages to respond poorly to G protein agonists, such as GTP-γ-S and sodium fluoride, was correlated with a significant decrease in GTP-binding capacity (101). Finally, it is also possible that a defect in a critical G protein associated with the PAF receptor underlies the reported failure of C3H/HeJ mice to respond to PAF (see above), since the PAF receptor is known to be G protein coupled (reviewed in Ref. 43). In slices of brain cortex and hippocampus, PAF-induced inhibition of K$^+$-stimulated [^{3}H]acetyl choline release was found to be mediated by a pertussis toxin–sensitive, Gα$i_{1/2}$ protein (102); however, no apparent differences in electrophoretic mobility in Gα$i_{2/3}$ have been observed in C3H/OuJ and C3H/HeJ macrophage membranes (86).

POSITIONAL CLONING OF THE *Lps* GENE

A great deal of effort has been made to elucidate the genetic basis of the *Lps* defect by studying the normal pathway of LPS-induced cellular activation; however, the complexity of the signaling pathways involved and the pleiotropic cellular responses to LPS have made it difficult to pinpoint the specific defect of the C3H/HeJ mouse mutant. Therefore, an effort to identify the *Lps* gene by positional cloning has been initiated by several groups. Application of this approach relies upon the localization of the gene and narrowing down of the candidate region to the smallest possible interval by genetic linkage studies. This is followed by the construction of a physical map of the region and the development of a clone contig map that encompasses the candidate region. The area is then examined for the presence of transcription units, with the aim of identifying a gene for the presence of mutational defects. Although difficult, the advent of novel technologies and genetic resources have improved the feasibility of this type of approach.

As presented in the introduction of this chapter, the genetic basis for the profound state of hyporesponsiveness to LPS exhibited by the C3H/HeJ mouse strain was established using BXH recombinant inbred strains and in classical breeding experiments between LPS-responsive and Lps^d parental strains. By 1980, *Lps* had been mapped to mouse mid-chromosome 4, between *Mup1* proximally and *Ps* distally at 6 ± 2 cM and 13 ± 7 cM, respectively (7). Studies carried out nearly a decade later using BXH mice demonstrated a clear separation of *Lps* from the gene cluster than encodes interferon-α (*Ifa*), in spite of a number of phenotypes that supported the possibility that *Lps* might be a member of the *Ifa* gene family (11). Subsequently, a congenic strain (C.C3H-Lps^d) was made by transferring the Lps^d allele from C3H/HeJ to an Lps^n inbred strain (BALB/cPt) through a series of backcrosses and selection (17). This strain was confirmed in functional studies to provide a model of LPS hyporesponsiveness equivalent to the parental C3H/HeJ strain. The congenic fragment was defined by the DNA markers *D4Mit56* on the proximal side and *Jun* (Jun oncogene) distally and estimated to be ≥5.5 cM in size. This congenic strain was developed to refine the *Lps* candidate interval and to define DNA markers to be used in isolation of the *Lps* gene. More recently, Peiffer-Schneider published the mapping of three microsatellite markers near the *Lps* locus using 96 progeny from a cross between C3H/HeJ and *Mus musculus castaneus* (35). Two of these markers (*D4Mit25* and *D4Mit324*) co-segregate with *Lps*.

Getting from the initial localization to the identification of a gene can be extremely labor intensive and necessitate the creation of genetic resources including a large number of animals segregating the phenotype under study, as well as clones and markers within the region. However, this task has been facilitated by the development of publicly available genetic and physical maps of the mouse genome. In more recent linkage studies involving 1345 progeny from two different intraspecific backcrosses and 50 DNA markers (cDNA, microsatellite, and microclone probes), the *Lps* interval was reduced to a 1.1 cM interval, flanked proximally by a large cluster of markers, including four known genes (*Cd301*, CD30 antigen ligand; *Tnc*, tenascin C; *Lv*, aminolevulinate dehydratase; and *Ambp*, alpha-1-microglobulin/bikunin precursor) and distally by two microsatellite markers (*D4Mit7* and *D4Mit178*) (34) (Fig. 2). For phenotypic analysis, all recombinant progeny were typed for mitogenic response to LPS in vitro, employing splenocyte cell cultures. Based on the hypothesis that the response to LPS may be linked with the ability to survive gram-negative infection (103), a subset of the recombinant progeny (325(DBA/2J × C3H/HeJ)F1 × C3H/HeJ) was tested by successive phenotyping of LPS responsiveness and *Salmonella* susceptibility as measured by bacterial growth in the liver. Inheritance between LPS responsiveness and susceptibility to infection with *Salmonella typhimurium* was perfectly concordant (34), further supporting the hypothesis that *Lps* is a *Salmonella* susceptibility locus.

To refine the genetic interval harboring the *Lps* gene further, new informative DNA probes needed to be developed. DNA clones that overlap the minimal interval defining the *Lps* region were first isolated by initiating a bidirectional chromosome walk in genomic DNA libraries (P1, BAC, and YAC) using the closest markers flanking *Lps* as entry probes. These DNA clones were then used to generate new informative probes for linkage and physical mapping to further narrow and orient the *Lps* interval and to provide the starting material to search the region for transcription units. Probes closely linked to *Lps* by combined mapping data in interspecific crosses and not informative in intraspecific crosses

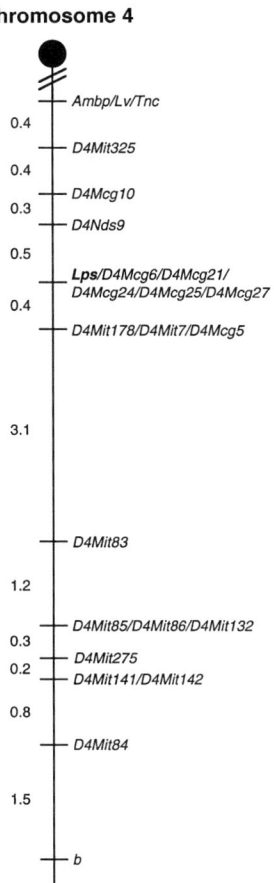

Fig. 2 A partial genetic linkage map of mouse chromosome 4 in the vicinity of the *Lps* locus. The order of the mapped loci was determined by pedigree analysis. Calculated recombination distances in centimorgans (cM) are shown to the left of the chromosome. The centromere is represented by a black circle. Gene symbols: *Ambp*, alpha-1-microglobulin/bikunin precursor; *Lv*, aminolevulinate dehydatase; *Tnc*, tenascin; *b*, brown. *D4Mcg* markers are from S. Qureshi et al. (unpublished).

were transformed to microsatellites. Forty-five new DNA probes, all located within the 1.1 cM interval carrying the *Lps* gene have been derived from cloned DNA. These probes were shown to be very useful to refine the chromosomal localization of *Lps* to a region less than 1 cM in size and to identify several markers that do not recombine with *Lps* (D4Mcg6, *D4Mcg21*, *D4Mcg24*, *D4Mcg25*, and *D4Mcg27*) (S. Qureshi and D. Malo, unpublished data).

This region of mouse chromosome 4 that harbors *Lps* is homologous to human chromosome 9 (104). The mouse-human homology is in three segments, two regions of synteny with human chromosome 9p are separated by a region syntenic with 9q (91,105). A syntenic breakpoint exists in the vicinity of *Lps*, indicating that the human homolog may be located near either 9p22-pter or 9q32-9q34.

The formal demonstration that a candidate gene and *Lps* are allelic may necessitate the conversion of the defective LPS phenotype to the normal phenotype by transfer of specific cloned DNA sequences. This can be achieved either in vitro by transfection of cells derived from the nonresponder C3H/HeJ strain or in vivo in transgenic animals. Another alternative is the creation and phenotypic analysis of a null allele at the *Lps* locus by homologous recombination. Once the *Lps* gene is cloned, a similarity search will be performed to identify orthologues and family members in other species. This analysis may provide important clues toward the elucidation of the *Lps* gene product function and insight into the cellular responses to LPS.

SUMMARY

The purpose of this chapter was to review what is known about the C3H/HeJ mouse strain, with particular emphasis on the findings that have unfolded over the past 5 years. Often it is difficult to distinguish between a phenotype that stems from a primary defect that may affect many intracellular signaling pathways from one that is the indirect result of the mutation (i.e., an "epiphenomenon"). What is clear from the data that have accumulated, particularly the data published in the past 5 years, is that Lps^d mice exhibit a variety of phenotypes that are not obviously related to their LPS hyporesponsiveness.

However, the C3H/HeJ strain has now been recognized as LPS hyporesponsive for nearly 30 years, yet unambiguous identification of the gene that maps to *Lps* has yet to be presented. This seems strange, given the number of excellent people and clever approaches that have been employed to tackle this problem. It is also fair to say that many more people than the literature suggests have worked on this problem over the past three decades. Perhaps this is telling us that the genetics of this system is controlled in a more complex fashion than has been suggested by classical patterns of inheritance. All the conventional data on the inheritance of LPS responsiveness/hyporesponsiveness would indicate that this gene is inherited in a simple Mendelian fashion; however, functional studies that suggest the possibility of "suppressor" phenotype, or that *Lps* may be acted on by another gene that controls its expression to give an apparent "responder" and "nonresponder" phenotype, should be considered. Hopefully, these questions will be resolved as novel approaches are taken to identify the *Lps* gene and its true function.

NOTE IN PROOF

Since this chapter was submitted, the molecular defect underlying LPS has been identified as the murine homologue of the human Toll-like receptor 4 (*Tlr4*) (107, 108). The C3H/HeJ strain has a missense mutation that predicts a Pro712His substitution in the intracytoplasmic region of the TLR4 protein, while the C57BL/10ScN and C57BL/ScCR mice have the gene deleted. TLR4 is a member of a family of transmembrane proteins that share significant homology with the IL-1 receptor type I and the IL-18 receptor, which have been shown to activate NF-κB translocation upon interaction with ligand (109). Moreover, LPS-induced signaling via human TLR2 has been demonstrated in transfected embryonic kidney fibroblasts (110, 111). At this time, the mechanism by which the C3H/HeJ defect in *Tlr4* blocks LPS signaling has not been fully elucidated. However, F1 progeny of C3H/HeJ mice and hemizygous mice that bear a copy of chromosome 4 in which *Lps*/*Tlr4* is deleted (e.g., $Lps^d \times Lps^0$ F1 mice) exhibit an intermediate TNF response to LPS challenge that is equivalent to that elicited in F1 progeny form C3H/HeJ $\times Lps^n$ crosses (112). The latter suggests that (i) the C3H/HeJ defect exerts a dominant negative effect on LPS signaling leading to TNF production, and (ii) an intact *Tlr4* gene is apparently not required for LPS signaling. Future experiments will now focus on the molecular interactions between TLR4 and other components within an undoubtedly intricate LPS signaling complex.

ACKNOWLEDGMENTS

The authors wish to express their sincere thanks to the many people from whom we sought advice during the preparation of this chapter. Special gratitude is expressed to Jorge Blanco, Giora Feuerstein, Charles Maliszewski, Pin-Yu Perera, and John L. Ryan for their insightful comments and suggestions. This work was supported by NIH grant AI-18797 (S.N.V.) and by MRC (Medical Research Council of Canada) grant MT-12038 (D.M.). D.M. is a scholar from Fonds de la recherche en santé du Québec, and S.Q. is a MRC clinician scientist.

REFERENCES

1. Kuhns DB, Long Priel DA, Gallin JI. Endotoxin and IL-1 hyporesponsiveness in a patient with recurrent bacterial infections. J Immunol 1997; 158:3959–3964.
2. Vogel SN. The Lps gene. Insights into the genetic and molecular basis of LPS responsiveness and macrophage differentiation. In: Beutler B, ed. Tumor Necrosis Factors: The Molecules and Their Emerging Role in Medicine. New York: Raven Press, Ltd. 1992: 485–513.
3. Gutierrez-Ramos JC, Bluethmann H. Molecules and mechanisms operating in septic shock: lessons from knockout mice. Immunol Today 1997; 19:329–334.
4. Heppner G, Weiss DW. High susceptibility of strain A mice to endotoxin and endotoxin-red blood cell mixtures. J Bacteriol 1965; 90:696–703.
5. Sultzer BM. Endotoxin-induced resistance to a staphylococcal infection: cellular and humoral responses compared in two mouse strains. J Infect Dis 1968; 118:340–348.
6. Watson J, Riblet R. Genetic control of responses to bacterial lipopolysaccharides in mice. I. Evidence for a single gene that influences mitogenic and immunogenic responses to lipopolysaccharides. J Exp Med 1974; 140:1147–1161.
7. Watson J, Kelly K, Largen M, Taylor BA. The genetic mapping of a defective LPS response gene in C3H/HeJ. J Immunol 1978; 120:422–424.
8. Rosenstreich DL, Vogel SN, Jacques AR, Wahl LM, Oppenheim JJ. Macrophage sensitivity to endotoxin: genetic control by a single codominant gene. J Immunol 1978; 121:1664–1670.
9. Watson J, Riblet R, Taylor BA. The response of recombinant inbred strains of mice to bacterial lipolysaccharides. J Immunol 1977; 118:2088–2093.
10. Watson J, Largen M, McAdam KPWJ. Genetic control of endotoxic responses in mice. J Exp Med 1978; 147:39–49.
11. Fultz MJ, Vogel SN. The physical separation of Lps and Ifa loci in BXH recombinant inbred mice. J Immunol 1989; 143:3001–3006.
12. McAdam KPWJ, Ryan JL. C57BL/10CR mice: nonresponders to activation by the lipid A moiety of bacterial lipopolysaccharide. J Immunol 1978; 120:249–253.
13. Coutinho A, Meo T. Genetic basis for unresponsiveness to lipopolysaccharide in C57BL/10Cr mice. Immunogenetics 1978; 7:17–24.
14. Vogel SN, Hansen CT, Rosenstreich DL. Characterization of a congenitally LPS-resistant, athymic mouse strain. J Immunol 1979; 122:619–622.
15. McGhee JR, Michalek SM, Moore RN, Mergenhagen SE, Rosenstreich DL. Genetic control of in vivo sensitivity to lipopolysaccharide: evidence for codominant inheritance. J Immunol 1979; 122:2052–2058.
16. Coutinho A, Möller G, Gronowicz E. Genetic control of B-cell responses. IV. Inheritance of the unresponsiveness to lipopolysaccharides. J Exp Med 1975; 142:253–258.
17. Vogel SN, Wax JS, Perera P-Y, Padlan C, Potter M, Mock BA. Construction of a BALB/c congenic mouse, C.C3H-Lps^d, that expresses the Lps^d allele: analysis of chromosome 4 markers surrounding the Lps gene. Infect Immun 1994; 62:4454–4459.
18. Bergeron AJ, Yohe HC. Reclassification of the BXH11/Ty mouse as an endotoxin-hyporesponsive strain. J Endotoxin Res 1996; 3:165–168.
19. Manthey CL, Vogel SN. Elimination of trace endotoxin protein from rough chemotype LPS. J Endotoxin Res 1994; 1:84–91.
20. Manthey CL, Perera P-Y, Henricson BE, Hamilton TA, Qureshi N, Vogel SN. Endotoxin-induced early gene expression in C3H/HeJ (Lps^d) macrophages. J Immunol 1994; 153:2653–2663.
21. Dunn AJ, Chuluyan HE. Endotoxin elicits normal tryptophan and indolamine responses but impaired catecholamine and pituitary-adrenal responses in endotoxin-resistant mice. Life Sciences 1994; 54:847–853.
22. Ryan LK, Vermeulen MW. Alveolar macrophages from C3H/HeJ mice show sensitivity to endotoxin. Am J Respir Cell Mol Biol (United States) 1995; 12:540–546.
23. Nill MR, Oberyszyn TM, Ross MS, Oberyszyn AS, Robertson RM. Temporal sequence of pulmonary cytokine gene expression in response to endotoxin in C3H/HeN endotoxin-sensitive and C3H/HeJ endotoxin-resistant mice. J Leuk Biol 1995; 58:563–574.
24. Corradin SB, Mauel J, Gallay P, Heumann D, Ulevitch RJ, Tobias PS. Enhancement of murine macrophage binding of and response to bacterial lipopolysaccharide (LPS) by LPS binding protein. J Leukoc Biol 1992; 52:363–368.
25. Henricson BE, Carboni JM, Burkhardt AL, Vogel SN. LPS and Taxol activate lyn kinase autophosphorylation in Lps^n, but not in Lps^d, macrophages. Mol Med 1995; 1:428–435.
26. Shinji H, Akagawa KS, Tusji M, Maeda M, Yamada R, Matsuura K, Yamamoto S, Yoshida T. Lipopolysaccharide-induced biphasic inositol 1,4,5-trisphosphate response and tryosine phosphorylation of 140-kilodalton protein in mouse peritoneal macrophages. J Immunol 1997; 158:1370–1376.
27. Skidmore BJ, Chiller JM, Morrison DC, Weigle WO. Immunologic properties of bacterial lipopolysaccharide (LPS): correlation between the mitogenic, adjuvant, and immunogenic activities. J Immunol 1975; 114:770–775.
28. Glode LM, Scher I, Osborne B, Rosenstreich DL. Cellular mechanism of endotoxin unresponsiveness in C3H/HeJ mice. J Immunol 1976; 116:454–461.
29. Haas GP, Johnson AG, Nowotny A. Suppression of the immune response in C3H/HeJ mice by protein-free lipopolysaccharide. J Exp Med 1978; 148:1081–1086.
30. Sultzer BM, Castagna R. Inhibition of activated nonresponder C3H/HeJ lymphocytes by lipopolysaccharide endotoxin. Infect Immun 1988; 56:3040–3045.
31. Sultzer BM, Bandekar JR, Castagna, Abu-Lawi K, Sadeghian M, Norin AJ. Suppression of C3H/HeJ cell activation by lipopolysaccharide endotoxin. Infect Immun 1992; 60:3533–3538.
32. Scibienski RJ. Defects in murine responsiveness to bacterial lipopolysaccharide: the C3H/HeJ and

C57BL/10ScCr strains. In: Gershwin ME, Merchant B, eds. Immunological Defects in Laboratory Animals. Vol. 2. New York: Plenum Press, 1981:241–258.

33. Sultzer BM, Castagna R, Bandekar J, Wong P. Lipopolysaccharide nonresponder cells: the C3H/HeJ defect. Immunobiology 1993; 187:257–271.

34. Qureshi S, Larivère L, Sebastiani G, Clermont S, Skamene E, Gros P, Malo D. A high-resolution map in the chromosomal region surrounding the *Lps* locus. Genomics 1996; 31:283–294.

35. Peiffer-Schneider S, Schutte BC, Murray JC, Frees KL, Williamson K, Leysens NJ, Schwartz DA. Exclusion of *Ifa* and *Ifb* as the *Lps* gene and mapping of three markers near the *Lps* locus. Mamm Genome 1997; 8:785–786.

36. DiMarzio P, Gessani S, Locardi C, Borghi P, Baglioni C, Belardelli F. Effects of different biological response modifiers on interferon expression in bacterial lipopolysaccharide (LPS)-responsive and LPS-hyporesponsive mouse peritoneal macrophages. J Gen Virol 1990; 71:2585–2591.

37. Gessani S, Dieffenbach CW, Conti L, DiMarzio P, Wilson KO, Belardelli F. Selective alteration of the turnover of interferon beta mRNA in peritoneal macrophages from LPS-hyporesponsive mice and its role in the defective expression of spontaneous interferon. Virology 1993; 193:507–509.

38. Barber SA, Fultz MJ, Salkowski CA, Vogel SN. Differential expression of interferon regulatory factor 1 (IRF-1), IRF-2, and interferon consensus sequence binding protein genes in lipopolysaccharide (LPS)-responsive and LPS-hyporesponsive macrophages. Infect Immun 1995; 63:601–608.

39. Fultz MJ, Vogel SN. Autoregulation by interferons provides an endogenous 'priming' signal for LPS-responsive macrophages. J Endotoxin Res 1995; 2:77–84.

40. Cross A, Asher L, Seguin M, Yuan L, Kelly N, Hammack C, Sadoff J, Gemski P. The importance of a lipopolysaccharide-initiated, cytokine-mediated host defense mechanism in mice against extraintestinally invasive *Escherichia coli*. J Clin Invest 1995; 96:676–686.

41. Vogel SN, Moore RN, Sipe JD, Rosenstreich DL. BCG-induced enhancement of endotoxin sensitivity in C3H/HeJ mice. I. In vivo studies. J Immunol 1980; 124:2004–2009.

42. Gallin EK, Green SW, Patchen ML. Comparative effects of particulate and soluble glucan on macrophages of C3H/HeN and C3H/HeJ mice. Int J Immunopharmacol 1992; 14:173–183.

43. Kuijpers TW, van der Poll T. The role of PAF in endotoxin-related disease. In: Brade H, Opal SM, Vogel SN, Morrison DC, eds. Endotoxin in Health and Disease. New York: Marcel Dekker, 1999:449–462.

44. Sun X, Caplan MS, Liu Y, Hsueh MD. Endotoxin-resistant mice are protected from PAF-induced bowel injury and death. Role of TNF, complement activation, and endogenous PAF production. Dig Dis Sci 1995; 40:495–502.

45. Ishii S, Matsuda Y, Nakamura M, Waga I, Kume K, Izumi T, Shimizu T. A murine platelet-activating factor receptor gene: cloning, chromosomal localization and up-regulation of expression by lipopolysaccharide in peritoneal resident macrophages. Biochem J 1996; 314:671–678.

46. Manthey CL, Vogel SN. Taxol: a promising endotoxin research tool. J Endotoxin Res 1994; 1:189–198.

47. Vogel SN, Carboni JM, Manthey CL. Paclitaxel, a mimetic of bacterial lipopolysaccharide (LPS) in murine macrophages (Chapter 12). In: Chen TT, Ojima I, Vyas DM, eds. Taxane Anticancer Agents Basic Science and Current Status. Washington, DC: American Chemical Society, 1995:162–172.

48. Carboni JM, Farina V, Rao S, Hauck SI, Horwitz SB, Ringel I. Synthesis of a photoaffinity analog of Taxol as an approach to identify the Taxol binding site on microtubules. J Med Chem 1993; 36:513–515.

49. Diaz JF, Andreu JM. Assembly of purified GDP-tubulin into microtubules induced by taxol and taxotere: reversibility, ligand stoichiometry, and competition. Biochemistry 1993; 32:2747–2755.

50. Rao S, Horwitz SB, Ringer I. Direct photoaffinity labeling of tubulin with taxol. J Natl Cancer Inst 1992; 84:785–788.

51. Rao S, Krauss NE, Heerding JM, Swindell CS, Ringel I, Orr GA, Horwitz SB. 3′-(p-Azidobenzamido) taxol photolabels the N-terminal 31 amino acids of β-tubulin. J Biol Chem 1994; 269:3132–3134.

52. Ding AH, Porteu F, Sanchez E, Nathan CF. Down-regulation of tumor necrosis factor receptors on macrophages and endothelial cells by microtubule depolymerizing agents. J Exp Med 1990; 171:715–727.

53. Ding A, Porteu F, Sanchez E, Nathan CF. Shared actions of endotoxin and taxol on TNF receptors and TNF release. Science 1990; 248:370–372.

54. Manthey CL, Brandes ME, Perera P-Y, Vogel SN. Taxol increases steady-state levels of LPS-inducible genes and protein-tyrosine phosphorylation in murine macrophages. J Immunol 1992; 149:2459–2465.

55. Manthey CL, Qureshi N, Stütz PL, Vogel SN. Lipopolysaccharide antagonists block Taxol-induced signaling in murine macrophages. J Exp Med 1993; 178:695–702.

56. Carboni JM, Singh C, Tepper MA. Taxol and lipopolysaccharide activation of a murine macrophage cell line and induction of similar tyrosine phosphoproteins. Monogr Natl Cancer Inst 1993; 15:95–101.

57. Ding A, Sanchez E, Nathan C. Taxol shares the ability of bacterial lipopolysaccharide to induce tyrosine phosphorylation of microtubule-associated protein kinase. J Immunol 1993; 151:5596–5602.

58. Perera P-Y, Qureshi N, Vogel SN. Paclitaxel (Taxol)-induced NF-kB translocation in murine macrophages. Infect Immun 1996; 64:878–884.

59. Kirikae T, Ojima I, Kirikae F, Ma Z, Kuduk SD, Slater JC, Takeuchi CS, Bounaud P-Y, Nakano M. Structural requirements of taxoids for nitric oxide and tumor necrosis factor producion by murine macrophages. Biochem Biophys Res Comm 1996; 227:227–235.

60. Vogel SN, Perera P-Y, Detore GR, Bhat N, Carboni JM, Haziot A, Goyert SM. CD14 dependent and independent signaling pathways in murine macrophages from normal and CD14 "knockout" (CD14KO) mice stimulated with LPS or Taxol. Prog Clin Biol Res 1998; 397:137–146.

61. Krönke M. The mode of ceramide action: the alkyl chain protrusion model. Cytokine Growth Factor Rev 1997; 8:103–107.
62. Dobrowsky RY, Hannun YA. Ceramide-activated protein phosphatase: partial purification and relationship to protein phosphatase 2A. Adv Lipid Res 1993; 178:91–104.
63. Zhang Y, Yao B, Delikat S, Bayoumy S, Lin XH, Basu S, McGinley M, Chan-Hui PY, Lichenstein H, Kolesnick R. Kinase suppressor of Ras is ceramide-induced protein kinase. Cell 1997; 89:63–72.
64. Joseph CK, Wright SD, Bornmann WG, Randolph JR, Kumar ER, Bittman R, Liu J, Kolesnick RN. Bacterial lipopolysaccharide has structural similarity to ceramide and stimulates ceramide-activated protein kinase in myeloid cells. J Biol Chem 1994; 269:17606–17610.
65. Barber SA, Detore G, McNally R, Vogel SN. Stimulation of the ceramide pathway partially mimics lipopolysaccharide-induced responses in murine peritoneal macrophages. Infect Immun 1996; 64:3397–3400.
66. Barber SA, Perera P-Y, Vogel SN. Defective ceramide response in C3H/HeJ (Lps^d) macrophages. J Immunol 1995; 155:2303–2305.
67. Lakics V, Vogel SN. LPS and ceramide use divergent signaling pathways to induce cell death in cultured peritoneal exudate macrophages. (submitted).
68. Thiéblemont N, Wright SD. Mice genetically hyporesponsive to lipopolysaccharide (LPS) exhibit a defect in endocytic uptake of LPS and ceramide. J Exp Med 1997; 185:2095–2100.
69. Printen JA, Woodard SL, Herman JR, Roess DA, Barisas BG. Membrane changes in lipopolysaccharide-stimulated murine B lymphocytes associated with cell activation. Biochim Biophys Acta 1993; 1148:91–96.
70. Yohe H, Brown L, Ryan JL. Plasma membrane gangliosides as potential binding sites for bacterial endotoxins. In: Morrison DC, Ryan JL, eds. Bacterial Endotoxin Lipopolysaccharides. Vol. I: Molecular Biochemistry and Cellular Biology. Boca Raton, FL: CRC Press, 1992:269–283.
71. Macala LJ, Yohe HC. In situ accessibility of murine macrophage gangliosides. Glycobiology 1995; 5:67–75.
72. Berenson CS, Ryan JL. Murine peritoneal macrophage gangliosides inhibit lymphocyte proliferation. J Leukoc Biol 1991; 50:393–401.
73. Mond J, Witherspoon K, Yu RK, Perera P-Y, Vogel SN. Inhibition of LPS-mediated cell activation in vitro and in vivo by gangliosides. Circ Shock 1995; 45:57–62.
74. Haimovitz-Friedman A, Cordon-Cardo C, Bayoumy S, Garzotto M, McLoughlin M, Gallily R, Edwards CK, Schuchman EH, Fuks Z, Kolesnick R. Lipopolysaccharide induces disseminated endothelial apoptosis requiring ceramide generation. J Exp Med 1997; 186:1831–1841.
75. Fuks Z, Persaud RS, Alfieri A, McLoughlin M, Ehleiter D, Schwartz JL, Seddon AP, Cordon-Cardo C, Haimovitz-Friedman A. Basic fibroblast growth factor protects endothelial cells against radiation-induced programmed cell death in vitro and in vivo. Cancer Res 1994; 54:2582–2590.
76. Neta R, Oppenheim JJ, Douches SD. Interdependence of the radioprotective effects of human recombinant Interleukin 1α, Tumor Necrosis Factorα, Granulocyte Colony-Stimulating Factor, and murine recombinant Granulocyte-Macrophage Colony-Stimulating Factor. J Immunol 1988; 140:108–111.
77. Jin FY, Nathan C, Radzioch D, Ding A. Secretory leukocyte protease inhibitor: a macrophage product induced by and antagonistic to bacterial lipopolysaccharide. Cell 1997; 88:417–426.
78. KuGSB, Quigley JP, Sultzer BM. The inhibition of the mitogenic stimulation of B lymphocytes by a serine protease inhibitor: commitment to proliferation correlates with an enhanced expression of a cell-associated arginine-specific serine enzyme. J Immunol 1983; 131:2494–2499.
79. Kuus-Reichel K and Ulevitch RJ. Partial restoration of the lipopolysaccharide-induced proliferative response in splenic B cells from C3H/HeJ mice. J Immunol 1986; 137:472–477.
80. Deng G, Curriden SA, Wang S, Rosenberg S, Loskutoff DJ. Is plasminogen activator inhibitor-1 the molecular switch that governs urokinase receptor-mediated cell adhesion and release? J Cell Biol 1996; 134:1563–1571.
81. Steffanson S, Lawrence DA. The serpin PAI-1 inhibits cell migration by blocking integrin $\alpha_v\beta_3$ binding to vitronectin. Nature 1996; 383:441–443.
82. Petty HR, Todd RF. Integrins as promiscuous signal transduction devices. Immunol Today 1996; 17:209–212.
83. Ingalls RR, Golenbock DT. CD11c/CD18, a transmembrane signaling receptor for lipopolysaccharide. J Exp Med 1995; 181:1473–1479.
84. Ingalls RR, Arnaout MA, Golenbock DT. Outside-in signaling by lipopolysaccharide through a tailless integrin. J Immunol 1997; 159:433–438.
85. Liu J, Mathias S, Yang Z, Kolesnick RN. 1994. Renaturation and tumor necrosis factor-α stimulation of a 97-kDa ceramide-activated protein kinase. J Biol Chem 1994; 269:3047–3052.
86. Bhat N, Blanco J, Vogel SN. (In preparation.)
87. English BK, Orlicek SL, Mei Z, Meals EA. Bacterial LPS and IFN-γ trigger the tyrosine phosphorylation of *vav* in macrophages: evidence for involvement of the *hck* tyrosine kinase. J Leukoc Biol 1997; 62:859–864.
88. Katzav S. Annotation: *Vav*: A molecule for all haemopoiesis? Br J Haematol 1992; 81:141–144.
89. Blenis J. Signal transduction via the MAP kinases: proceed at your own RSK. Proc Natl Acad Sci USA 1993; 90:5889–5892.
90. Copeland NG, Jankins NA, Gilbert DJ, Eppig, JT, Maltais LJ, Miller JC, Dietrich WF, Weaver A, Lincoln SE, Steen RG, et al. A genetic linkage map of the mouse: current applications and future prospects. Science 1993; 262:57–66.
91. Pilz A, Woodward K, Povey S, Abbott C. Comparative mapping of 50 human chromosome 9 loci in the laboratory mouse. Genomics 1995; 25:139–149.

92. Alcorta DA, Crews CM, Sweet LJ, Bankston L, Jones SW, Erikson RL. Sequence and expression of chicken and mouse rsk: homologs of *Xenopus laevis* ribosomal S6 kinase. Mol Cell Biol 1989; 9:3850–3859.
93. Moller DE, Xia CH, Tang W, Zhu AX, Jakubowski M. Human rsk isoforms: cloning and characterization of tissue-specific expression. Am J Physiol 1994; 226 (2 Pt 1):C351–C359.
94. The Human Transcription Map. http://www.ncbi.nlm.nih.gov/Science96.
95. Kang AD, Wong PMC, Chen H, Castagna R, Chung S-W, Sultzer BM. Restoration of lipopolysaccharide-mediated B-cell response after expression of a cDNA encoding a GTP-binding protein. Infect Immun 1996; 64:4612–4617.
96. Coutavas EE, Hsieh CM, Ren M, Drivas GT, Rush MG, D'Eustachio PD. Tissue-specific expression of Ran isoforms in the mouse. Mamm Genome 1994; 5:623–628.
97. Shinji H, Akagawa KS, Yoshida T. LPS induces selective translocation of protein kinase C-β in LPS-responsive mouse macrophages, but not in LPS-nonresponsive mouse macrophages. J Immunol 1994; 153:5760–5771.
98. Fernando LP, Makhlouf MA, Cook JA. Multi-functional G proteins; implication in endotoxin cellular responses. In: Morrison DC, Brade H, Opal SM, Vogel SN, eds. Endotoxin in Health and Disease. New York: Marcel Dekker, 1999:537–547.
99. Zhang X, Morrison DC. Pertussis toxin-sensitive factor differentially regulates lipopolysaccharide-induced tumor necrosis factor-α and nitric oxide production in mouse peritoneal macrophages. J Immunol 1993; 150:1011–1018.
100. Zingarelli B, Makhlouf M, Halushka PB, Caputi AP, Cook JA. Altered macrophage function in tumor necrosis factor α and endotoxin-induced tolerance. J Endotoxin Res 1995; 2:247–254.
101. Makhlouf M, Ashton SH, Hildebrandt J, Mehta N, Gettys TW, Halushka PB, Cook JA. Alterations in macrophage G proteins are associated with endotoxin tolerance. Biochim Biophys Acta 1996; 13121:163–168.
102. Wang H-Y, Yue T-L, Feuerstein G, Friedman E. Platelet-activating factor: diminished acetylcholine release from rat brain slices is mediated by a Gi protein. J Neurochemistry 1994; 63:1720–1725.
103. O'Brien AD, Rosenstreich DL, Scher I, Campbell GH, MacDermott RP, et al. Genetic control of susceptibility to *Salmonella typhimurium* infection in mice: Role of the *Lps* gene. J Immunol 1980; 124:20–24.
104. Mock BA, Neumann PE, Fiedorek FT Jr. Mouse chromosome 4. Mamm Genome 1997; 7:S60–S79.
105. Kwiatkowski DJ, Armour J, Bale AE, Fountain JW, Goudie D, et al. Report on the Second International Workshop on Human Chromosome 9. Cytogenet Cell Genet 1993; 64:94–106.
107. Poltorak A, He X, Smirnova I, Liu M-Y, Van Huffel C, Du X, Birdwell D, Alejos E, Silva M, Galanos C, Freudenberg M, Ricciardi-Castagnoli P, Layton B, Beutler B. Defective LPS signaling in C3H/HeJ and C57BL/10ScCr mice: mutations in *tlr4* gene. Science 1998; 282:2085–2088.
108. Qureshi ST, Larivière L, Leveque G, Clermont S, Moore KJ, Gros P, Malo D. Endotoxin-tolerant mice have mutations in Toll-like receptor 4 (Tlr4). J Exp Med 1999; 189:615–625.
109. O'Neill LAJ, Greene C. Signal transduction pathways activated by the IL-1 receptor family: ancient signaling machinery in mammals, insects, and plants. J Leuk Biol 1998; 63:650–657.
110. Yang R-B, Mark MR, Gray A, Huang A, Xie MH, Zhang M, Goddard A, Wood WI, Gurney AL, Godowski PJ. Toll-like receptor-2 mediates lipopolysaccharide-induced cellular signalling. Nature 1998; 295:284–288.
111. Kirschning CJ, Wesche H, Ayers TM, Rothe M. Human Toll-like receptor 2 confers responsiveness to bacterial lipopolysaccharide. J Exp Med 1998; 188:2091–2097.
112. Vogel SN, Johnson D, Perera P-Y, Medvedev A, Larivière L, Gureshi ST, Malo D. Functional characterization of the effect of the C3H/HeJ defect in mice that lack an Lps^n gene: *In vivo* evidence for a dominant negative mutation. J Immunol 1999. In press.

50

Endotoxin Tolerance

F. Ulrich Schade, Regina Flach, Sascha Flohé, Matthias Majetschak,
Ernst Kreuzfelder, Emilio Domínguez-Fernández, Jochen Börgermann, Martin Reuter,
and Udo Obertacke
University Hospital of Essen, Essen, Germany

THE PHENOMENON

During the last decade, great progress has been achieved in the understanding of the pathophysiology of severe infections such as sepsis. In particular, multiple mediators of the inflammatory host response have been identified, and based on this knowledge therapeutic strategies were developed to counteract the detrimental consequences of the host response to bacterial components, in particular, endotoxins. In connection with this, the phenomenon of endotoxin tolerance has attracted substantial attention, since it provides a host with increased resistance to endotoxin and to possibly unrelated assaults such as hemorrhagic shock. The elucidation of the molecular mechanisms of endotoxin tolerance may, therefore, assist in finding pharmacological tools for as yet unmanageable clinical situations.

The first to recognize endotoxin tolerance were physicians who used bacterial vaccines (heat-killed gram-negative bacilli) therapeutically to treat patients with infectious diseases by fever therapy. The repetitive use of such vaccines made it necessary to infuse increasing quantities to maintain elevated temperatures. Since some patients required as much as 250 ml of typhoid vaccine in a single day (1), endotoxin tolerance in conjunction with fever therapy was considered an undesired effect. The view of endotoxin tolerance experienced a redirection, when during studies on the antitumor properties of endotoxin it was recognized that endotoxin tolerance is not restricted to the attenuation of pyrogenicity. It was shown in 1944 that a preparation of *Serratia marcescens* administered daily to mice in progressively greater doses resulted in the survival of the mice challenged with 16 times the LD_{50} for untreated animals (2). Lethal tolerance was found to develop within one day, was most pronounced on day 3, and slowly disappeared over the subsequent 2 weeks (3).

PARADIGMS OF ENDOTOXIN TOLERANCE

Centanni was the first investigator to demonstrate that repeated application of a purified pyrogenic preparation from bacterial culture filtrates resulted in acquired resistance to its own pyrogenic activity (4). He suggested that the observed reduction in fever was due to a cellular mechanism and not based on a serological immune response (5). This view was supported by the findings of others that resistance to the pyrogenic fraction prepared from *Salmonella typhosa* developing in humans did not correlate with circulating antibody titers to this pyrogen (6). The former results were confirmed in a series of studies performed in rabbits injected daily with endotoxin. The animals showed an apparent lack of O-specificity, and serum transferability and pyrogenic tolerance disappeared rapidly when the daily injections were stopped (7–9). In particular, the finding that tolerance was accompanied by enhanced blood clearance of endotoxin led to the proposition that pyrogenic tolerance was based upon enhanced uptake of circulating endotoxin by the reticuloendothelial system (RES). The classical definition of endotoxin tolerance emerged from a series of studies conducted in the laboratories of Greisman and others (reviewed in Ref. 12). They found that in both humans and rabbits after a single infusion or bolus injection of endotoxin, the refractory state became evident within hours and was specific for endotoxins as a class, provided that the quantities of endotoxin used for tolerance induction were not too high. This tolerance could be elicited with endotoxins lacking O-specific polysaccharide side chains and showed no serotype specificity. Since it was nearly absent after 48 hours it was termed *early phase* tolerance. The early state of tolerance is relative and can be overcome by increasing the amount of endotoxin used for challenge. When the pyrogenic reaction was tested at later times after the first treatment, the *late phase* of tolerance became apparent. This late phase was manifested at 74 hours and increased over the next several days. Moreover, it was specific for he homologous endotoxin used for the initial injection. Since serum fractions prepared from rabbits previously immunized daily with smooth gram-negative bacteria or endotoxins prevented fever in normal rabbits (11), it was assumed that the late phase of endotoxin tolerance is primarily based on the formation of antibodies against the O-specific chain of LPS. From these findings it was concluded that the phenomenon of endotoxin tolerance includes at least two distinct mechanisms: an *early, transient,* refractory state that embraces different serotypes of lipopolysaccharides and is *specific for endotoxins as a class*. Apart from that, antibody formation against the O-chain antigens of LPS constitutes the late phase tolerance. These two stages of tolerance could be differentiated by varying the application of the endotoxin. While single injections or continuous infusions would rapidly induce pyrogenic unresponsiveness (early phase), after several daily injections tolerance would become increasingly dependent on the formation of antibodies to endotoxin. During a daily treatment with repetitive injections of endotoxin, however, both phases are coexisting. The experimental and historical background leading to the above definition of endotoxin tolerance has been expertly summarized in detail in a recent review (12). Since that time, new insight has been gained into the molecular mechanisms underlying the inflammatory response in general and endotoxin effects in particular. In this chapter we attempt to review several aspects of this new information with regard to cellular and molecular components that have been recognized to contribute to the early phase of endotoxin tolerance.

THE ROLE OF ENDOGENOUS FACTORS IN ENDOTOXIN TOLERANCE

Cytokines

When endotoxin tolerance is induced in experimental animals or humans, the first treatment with endotoxin will provoke cytokine synthesis. This reaction was found to be practically absent upon subsequent treatment with endotoxin in rats (13), as well as in rabbits (14) and humans (15). The extent of the attenuation is not the same for all cytokines formed. While TNF and IL-1 were suppressed throughout the duration of endotoxin tolerance, formation of colony-stimulating factor (CSF) recovered to a certain degree (15), and IL-6 synthesis was restored completely (13). A major conclusion of these findings was that the reduced capacity to form mediators of endotoxin action constitutes a main portion of endotoxin tolerance.

Once it was generally accepted that cytokines are important for the elaboration of endotoxin effects, it was a logical step to assume an involvement of endogenous cytokines in the development of endotoxin tolerance. The first cytokine to be tested for its ability to induce pyrogenic tolerance was endogenous pyrogen (EP); however, repeated injections of a crude preparation of EP did not lead to pyrogenic tolerance (16). This was later confirmed in a study attempting to induce tolerance with human recombinant IL-1 (17). When, however, IL-1 was combined with human recombinant TNF, it was observed that these factors acted synergistically in generating a state of early endotoxin tolerance. This was assessed by a decreased production of CSF in response to LPS administered 3 days later. The doses of IL-1/TNF used in this study were chosen so as to induce levels of circulating CSF in vivo comparable to those induced by a tolerance-inducing LPS injection. Coherent with a role of IL-1 in the development of endotoxin tolerance was the finding that administration of an IL-1 receptor antagonist to mice prevented early endotoxin tolerance (18).

Soon after TNF was available as a recombinant agent, attempts were undertaken to exploit its antineoplastic effect therapeutically (19). During these efforts it was noted that repetitive injections of human recombinant TNF in experimental animals resulted in anorectic tachyphylaxis (20) and tolerance to the lethal effect of TNF in rats (21). TNF tolerance developed within 3 days, was dependent on the dose of TNF used, was present 2–4 days after pretreatment, and dissipated by 2 weeks. It was excluded that the TNF-tolerant state was due to increased clearance or degradation or dependent on antibodies, since human TNF was used (21). Of particular interest was the finding that cross-tolerance exists between TNF and LPS, since pretreatment with LPS protected against TNF challenge with a lethal dose and vice versa. The notion, therefore, that the major characteristic of LPS tolerance consists in a diminished cytokine (TNF) synthesis by macrophages needs to be extended, since obviously LPS tolerance in addition protects against the action of TNF.

The role of TNF-α in LPS tolerance has also been studied in D-galactosamine (D-GalN)–sensitized mice (reviewed in Ref. 22). Treatment with the hepatotoxin D-GalN represents an experimental model of extreme hypersensitivity to endotoxin (22,23). A single dose of D-GalN administered i.p. increases the sensitivity of mice to endotoxin by a factor of up to 100,000 (23). Since in parallel to the sensitizing effect of D-GalN, depletion of uridine triphosphate (UTP) in hepatocytes was noted and application of UTP abolished the effect of D-GalN completely, the mechanism of D-GalN sensitization is thought to consist of the inability of the liver to form macromolecules in response to LPS. Pretreatment of mice with LPS prior to endotoxin challenge in combination with D-GalN was found to provide substantial protection within a time frame between 60 minutes and 48 hours after the first LPS application (22). The protective efficacy of a pretreatment with 0.01 μg/mouse in this model is such that a dose of LPS, representing an LD_{100} in nontolerant D-GalN–treated mice (0.1 μg/mouse), results in 100% survival (23). Although the challenge doses necessary to cause lethality in endotoxin-tolerant D-GalN mice are magnitudes (~1000-fold) lower than in normal mice (not sensitized by D-GalN), within the model development of tolerance can be demonstrated clearly. Differences between D-GalN–sensitized and normal mice exist, as far as the chronological development of early tolerance is concerned. Whereas in D-GalN mice tolerance evolves within hours, in normal mice a stage of increased sensitivity precedes low responsiveness acquired after 48 hours (24).

It was found that for the induction of lethal tolerance in the D-GalN model. TNF-α can replace LPS as a pretreatment stimulus (22). Since mice made tolerant to LPS turned out to be resistant to TNF shock, the above conclusion holds true that diminished TNF synthesis in LPS tolerance does not fully account for endotoxin tolerance. Another interesting effect was noted when mice were rendered tolerant with TNF in the D-GalN model: the mice were protected against lethal challenge with LPS; however, TNF was produced in the LPS-challenged animals in quantities comparable to those in nontolerant mice. Again, this argues for a protective mechanism in LPS tolerance with regard to TNF activities in addition to attenuated cytokine formation.

Besides D-GalN, actinomycin D (ActD) sensitizes mice to a high degree with regard to the lethal effects of IL-1 and TNF (25). The sensitizing effects of both ActD and D-GalN were counteracted by pretreating the animals with tiny amounts of either IL-1 or TNF (25,26), but not of IL-6 (26). Contaminating LPS as a cause for the observed protection by induction of tolerance was excluded. The desensitizing effects of TNF and IL-1 extended to protection against LPS shock in BCG-sensitized mice (25). It should be noted that in all experiments carried out in sensitized mouse models [BCG, ActD (25), D-GalN (22,26)], the time frame was different (desensitization with IL-1, TNF, or LPS within hours) from the models used without sensitizing agents (tolerance within days). Furthermore, IL-1 alone was not able to induce tolerance in normal animals (17,27).

The mechanisms of desensitization to IL-1 or TNF and tolerance to LPS are poorly understood at present. It has been suggested that the desensitizing effects in the models mentioned above originated from the liver (22,25,26). This notion is based on the assumption that sensitization by D-GalN, ActD, or partial hepatectomy (28) originates from the destruction of liver function, and pretreatment with either LPS, IL-1, or TNF somehow could protect hepatic performance. It should, however, be kept in mind that this protection is only partial and does not reverse sensitization. This should briefly be delineated for D-GalN (22). The LD_{50} of LPS in normal, LPS-responsive mice as described by Galanos and colleagues (22) is 270 μg/mouse. The LD_{50} of the same LPS in combination with D-GalN is 0.0068 μg/mouse, which is a factor of ~40,000. Pretreatment of the mice with LPS resulted in 100% protection against a dose of 0.1 μg/mouse, which caused 100% lethality in nonpretreated mice. However, a dose of 1–10 μg/mouse would also be lethal for LPS-tolerant mice. On the other hand, restoration of normal liver function by supplementation of the mice with UTP eliminates D-GalN sensitization completely (23). This illustrates that under the conditions of D-GalN treatment, endotoxin tolerance develops. It is, nevertheless, only partially capable of overcoming D-GalN sensitization, as it is achieved by restoration of the liver function with UTP. The role of protecting factors of the liver in the development of tolerance, therefore, still seems hypothetical.

Glucocorticoids

It was observed as early as 1954 that cortisone protects mice against the lethal action of endotoxin (29). Corticosteroids have also been found to inhibit the synthesis of TNF (22,30), IL-1 (31), and IL-6 (26). On the other hand, abrogation of endogenous cortisone production sensitizes mice to endotoxins (32) and to IL-1 and TNF (33) to a considerable degree. It was further noted that application of endotoxins (34) and IL-1 (35) results in the release of endogenous corticosteroids, which could be responsible for desensitization and development of tolerance encountered by such a treatment. This question was addressed in several experimental models, which shall be briefly reviewed here.

D-GalN–Sensitized Mice

In mice sensitized with D-GalN, it has been shown that glucocorticoids protect against the lethal activity of LPS (22). Since it was noted that TNF formation in response to LPS is inhibited in these mice, the protection by glucocorticoids was thought to be due to suppression of TNF formation. When TNF/D-GalN, however, was applied to the mice, no protective effects of dexamethasone were observed (22,26). Induction of endotoxin tolerance in D-GalN mice was achieved by pretreatment with LPS 90 minutes prior to challenge with LPS/D-GalN (22). The LPS-tolerant mice were equally resistant to challenge with TNF/D-GalN, and such protection could not be achieved by dexamethasone. This suggests that endotoxin tolerance in D-GalN mice involves mechanisms other than endogenous corticoids. These observations are in agreement with the findings of Libert and colleagues (36), who reported that dexamethasone, adrenocorticotropic hormone, and IL-6 were ineffective in protecting mice against a lethal challenge with TNF/D-GalN. They made the additional observation that TNF and IL-1 had protective effects when given before the challenge event, to a greater extent than LPS. Of notice, it was shown in this report that both IL-6 and IL-1 induced cortisone in the mice to the same degree; however, IL-1 was highly protective, while IL-6 did not protect. From these results it was concluded that the protection achieved with the different compounds was independent of endogenous corticosteroids. In variance with these results, Evans and Zuckerman reported that they failed to induce tolerance to LPS in D-GalN mice (37). The reason may be a major difference in the experimental protocols used. While Galanos et al. (22) and Libert et al. (9926) used LPS or TNF pretreatment with nonlethal doses of LPS 12–24 hours prior to the challenge with LPS/D-GalN, in the Evans study (37) LPS/D-GalN at a dose exceeding an LD_{50} was used as pretreatment and a challenge with the same dose was applied 6 hours later. Consequently, in the latter study mortality was a frequent problem at 20 hours postinjection. It may be doubtful, thus, whether a dose of LPS far exceeding the LD_{50} is appropriate to induce endotoxin tolerance within 6 hours. In contrast, the proper dosage of LPS pretreatment as well as TNF or IL-1 will induce desensitization, including cross-protection between LPS, TNF, and IL-1. The finding that corticosteroids were not able to counteract TNF lethality, while LPS pretreatment was highly protective, strongly argues against an involvement of corticosteroids in the development of endotoxin tolerance.

Adrenalectomized Mice

Adrenalectomy has been found to represent another way to dramatically increase sensitivity to endotoxin (36). Thus, Chedid and colleagues reported in 1965 that in adrenalectomized adult mice the LD_{50} of *S. enteritidis* LPS was 0.03 µg, while in normal control mice an LD_{50} of 200 µg was determined (reviewed in Ref. 32). When mice were rendered tolerant by injections of sublethal doses of endotoxin 24 hours before adrenalectomy and then were challenged 24 hours after surgery, the LD_{50} was 5 µg compared with an LD_{50} of 0.03 µg for controls without endotoxin pretreatment. The authors concluded that "tolerance to the second injection of endotoxin can be evidenced in the absence of corticoids" (32). In order to evaluate a possible contribution of glucocorticoids to the development of tolerance, incited by the first LPS treatment, the endotoxin pretreatment was given briefly after adrenalectomy. Challenge with 0.1 µg LPS given 24 hours postoperatively resulted in high mortality in nontolerant animals (25/29 mice died), whereas no deaths occurred in tolerant mice ($n = 16$). It was concluded that a state of increased resistance can be established in the absence of the adrenal gland. As in the other models of hypersensitivity, hence, tolerance to the lethal effects of endotoxin can be induced in adrenalectomized mice. This tolerance induction takes place at a level of increased sensitivity when compared to normal animals. It was noted by the same authors (32) that exogenous cortisone protected both adrenalectomized and normal mice. This was the case for a single injection. If, however, cortisone was repeatedly injected (4 or 8 doses daily) into normal mice, they became highly sensitive to LPS ($LD_{50} = 0.5$ μg compared to 200 μg in normal mice). It is of great interest that under the conditions of cortisone-induced hypersensitivity, endotoxin tolerance could be established by treatment with LPS after the last cortisone application. An LD_{50} of 0.5 μg in nontolerant mice could be increased to an LD_{50} to 50 μg in tolerant mice (32). In summary, these results suggest that in a model of decreased ability to form glucocorticoids (adrenalectomy) and in a model of attenuated effectiveness for corticosteroids (repeated cortisone application), tolerance to endotoxin can develop.

In a more recent study (37) it was found that adrenalectomized mice did not develop tolerance as a consequence of a single LPS treatment, as measured by TNF serum levels after a second LPS application 20 hours after the first injection. The contrasting results may have methodological reasons. Pretreatment with LPS for tolerance induction was carried out with 0.1 or 0.2 μg per mouse. Since the authors state "in the adrenalectomized . . . model there was a decrease in the LD_{50} from approximately 100 μg to 0.1 μg," the pretreatment dose thus corresponded to approximately an LD_{50}, which does not represent a dose usually applied for induction of endotoxin tolerance. Instead, sublethal quantities of LPS are applied to animals in order to induce tolerance. The reason for the failure to induce endotoxin tolerance in this study may, therefore, consist in an inappropriate dose of LPS used for induction of tolerance.

Glucocorticoid Antagonists

In order to study the influence of endogenous glucocorticoids on the development of desensitization by IL-1, a pharmacological approach to prevent glucocorticoid activity by using the antagonist RU 38486 was chosen (39). It was reported there that the synthetic corticoid receptor blocker did not interfere with the protection exerted by IL-1 against the lethal effects of TNF. On the other hand, it was observed by the same authors that RU38486 sensitizes normal mice to a high degree towards the lethal activity of TNF in mice (40). In analogy to these findings, it was noticed that RU38486 greatly enhances the sensitivity of normal mice for the lethal toxicity of LPS (41). Although the question whether tolerance or desensitization could develop was not specifically addressed, these experiments suggested that the sensitization to LPS or TNF observed in RU38486-treated mice was due to the absence of endogenous glucocorticoids. It could, therefore, be suspected that in analogy to adrenalectomized mice, endotoxin tolerance may also develop in RU38486-sensitized mice.

Indirect evidence that endotoxin tolerance can be achieved in the presence of RU38486 was reported by Lazar et al. (42). These authors showed that in the murine cecal ligation and puncture model with a 20-gauge needle, 20% of the animals survived 24 hours after puncture. In contrast, in animals that had been rendered tolerant to endotoxin prior to induction of peritonitis, survival was 90% after 5 days. A single intravenous injection of RU38486 concurrent with puncture carried out in these experiments with a larger 21-gauge needle reduced survival to 15% compared to the controls without RU38486 (71% survival). In parallel, in endotoxin-tolerant mice the antiglucocorticoid lowered survival to 35% compared to the control group of 90%. Hence, protection against the lethal consequences of cecal puncture was provided by endotoxin tolerance in the presence of RU38486 when compared to the nontolerant group receiving RU38486. Although the lethal effects of endotoxin were not examined in this study, it is suggested that endotoxin tolerance is evident in mice in whom endogenous glucocorticoids are antagonized. These results are similar to mice in whom glucocorticoid synthesis was eliminated by adrenalectomy. From all of these findings it is concluded that glucocorticoids are not of major relevance in the genesis of endotoxin tolerance.

IL-10

IL-10 has been identified as an inactivator of macrophage functions (43,44). It was found to protect mice from endotoxin shock (45), and its involvement in the desensitization of human monocytes in vitro (46) by LPS has been proposed. Since in vitro desensitized monomac 6 cells produced IL-10 in response to LPS stimulation, while TNF and IL-1 were downregulated (47), IL-10 seemed a likely candidate as a mediator of endotoxin tolerance. A strong argument against such a function for IL-10 was found, however, when mice whose capacity to produce IL-10 was eliminated by disruption of the IL-10 gene developed endotoxin tolerance. These mice were, on the other hand, more sensitive to the lethal effects of endotoxin as assessed in a model of endotoxin shock and the generalized Shwartzman reaction (48). Disruption of the IL-10 gene represents, therefore, another model in which the sensitivity to LPS is markedly enhanced; yet, within the framework of endotoxin hypersensitivity, the mechanism of endotoxin tolerance remains operative. IL-10, however, does not seem to be of foremost relevance for its development. In agreement with these data, it was recently found that IL-10 levels in endotoxin-tolerant mice challenged with endotoxin were strongly reduced compared to IL-10 levels in normal mice (49). Accordingly, when mononuclear cells were isolated from endotoxin-tolerant human volunteers and stimulated with LPS in vitro, IL-10 synthesis was very low compared to IL-10 levels elaborated by MNC, isolated from the individuals before tolerance was induced (50).

An Unidentified Inhibitor of TNF Synthesis in Endotoxin Tolerance

In 1960, Freedman reported that pyrogenic tolerance in rabbits could be passively transferred with the serum of the tolerant animals (51,52). Since no difference was found between the protecting activity between sera from donors receiving LPS, either as daily injections over several weeks or only on 6 days, it was concluded that antibodies against LPS were not relevant for the observed protective properties of the sera (51). It should be emphasized that the doses used for tolerance induction were relatively high (6 injections increasing from 2.5 to 10 μg) compared to those used in other classical investigations [single injection of about 1 μg (12)]. The characterization of the factors responsible for the transfer of pyrogenic tolerance was, however, not further pursued. Recently, in our laboratory some findings were made which are in line with those formerly made by Freedman. Mice were made tolerant by a single, yet relatively high amount of LPS (80 μg/mouse), challenged with LPS (160 μg/mouse) on day 4, and serum was prepared within 30 minutes. Incubation of this serum with endotoxin-stimulated murine whole blood resulted in a marked suppression of TNF synthesis. This serum-inhibitory activity was not present in tolerant animals unchallenged on day 4 or in normal mice after LPS challenge (49). It was not identical with IL-10, and it suppressed the synthesis of TNF and not its action. A similar activity is formed by peritoneal macrophages isolated from endotoxin-tolerant mice (134). At present the identity of this factor is unknown.

CELLULAR COMPONENTS OF ENDOTOXIN TOLERANCE

Reticuloendothelial System

The finding that early pyrogenic tolerance is independent of antibodies and could not be transferred with serum led to the proposition that "nonimmunological" mechanisms were involved. In particular the discovery that tolerance was accompanied by enhanced blood clearance of endotoxin (7,9) suggested that cellular elements of the reticuloendothelial system (RES) determined endotoxin tolerance. It was proposed that pyrogenic tolerance was achieved by enhanced uptake of circulating endotoxin by the RES, which prevented more susceptible tissues from having contact with LPS. Major support for this hypothesis came from the finding that animals rendered tolerant by daily injections of endotoxin could be resensitized by treating the animals with thorotrast (thorium dioxide), a substance that causes a "blockade" of the reticuloendothelial system (7,9). This agent, therefore, seemed to "abolish" pyrogenic tolerance by suspending the diversion of LPS from the RES. For many years, thus, the increased phagocytic capacity of the RES was considered an essential element of endotoxin tolerance.

Subsequent and more detailed studies, however, demonstrated that the phagocytic properties of the RES are of minor relevance for endotoxin tolerance. It was shown that thorotrast, besides blockading the RES, markedly increases the pyrogenic response of both tolerant and normal animals (53,54). Obviously, thorotrast sensitizes rabbits in general to the pyrogenic action of LPS, however, the differences between normal animals and those made tolerant persist within the framework of pyrogenic hyperreactivity caused by thorotrast. Further studies in humans revealed that during development of pyrogenic tolerance, no enhanced phagocytic activity of the RES could be determined (55).

Low Responsiveness of Mononuclear Phagocytes Isolated from Endotoxin-Tolerant Hosts

Considerable progress in the understanding of endotoxin tolerance developed in parallel with the presently accepted concept that endotoxins do not act directly to cause their toxic effects but rather activate host cells to produce endogenous mediators. The first of these endogenous mediators to be recognized was "endogenous pyrogen" (EP) (16), which was later identified as interleukin-1 (56). While it was originally suggested that polymorphonuclear cells were the major source of EP (57), it became clear that EP was a macrophage secretory product (58–60). The first study to lend direct proof for altered macrophage functions in endotoxin-tolerant animals was published in 1968 (61). Kupffer cells isolated from the livers of rabbits made tolerant by injection of typhoid vaccine on 7–14 consecutive days were incubated with endotoxin for 18 hours in parallel with blood leukocytes and lung macrophages. Production of EP was determined by injection of the respective supernatants into normal rabbits and quantitation of the pyrogenic response. In comparison to the corresponding cells from nontolerant rabbits, Kupffer cells from tolerant donors were refractory to EP synthesis in vitro. On the other hand, blood leukocytes or lung macrophages did not differ in their EP synthetic capacity between tolerant and normal rabbits.

Another highly interesting observation was reported in this paper. When isolated blood leukocytes, lung macrophages, spleen cells, or liver cell suspensions from normal rabbits were incubated with endotoxin in the presence of serum from tolerant rabbits, the production of EP by spleen and liver cells was considerably inhibited in comparison to incubation with normal rabbit serum. This suggests that the serum of endotoxin-tolerant rabbits contains an inhibitor of EP production. In the meantime, it was shown with more refined methods for the isolation of cells and quantitation of IL-1 that blood mononuclear cells isolated from human volunteers who were tolerant to LPS are less responsive to endotoxin activation in vitro than nontolerant individuals (62).

Low responsiveness of isolated macrophages to endotoxin ex vivo is not restricted to the synthesis of EP/IL-1. Thus, it was shown that peritoneal macrophages from mice harvested on day 4 after a prior treatment of the mice with a single dose of LPS (10 μg/animal) showed a considerably reduced capacity to produce prostaglanin E_2 and/or $F_{2\alpha}$ (63,64). Studies in rats demonstrated that induction of endotoxin tolerance in the animals altered arachidonic acid turnover (65) and its metabolism (66) in isolated peritoneal macrophages.

The importance of cytokines, in particular tumor necrosis factor α, as a mediator of host defense and inflammatory responses is well documented (67). A key role has been attributed to TNF in models of shock induced by endotoxin and gram-negative bacteria (68,69). The finding that TNF formation in rabbits and rats pretreated with LPS was considerably blocked upon challenge with LPS, therefore, provided new insight in the mechanism of endotoxin tolerance (13,14,70). The assumption that mononuclear phagocytes have a central function in endotoxin tolerance was suggested by the finding that macrophages from tolerant rats in vitro were greatly impeded in their ability to produce TNF (70,71).

Zuckermann and Evans studied TNF and IL-1 mRNA levels in macrophages isolated from endotoxin-tolerant mice after LPS challenge in vivo (72). It was determined that mRNA for both products was substantially lower in macrophages isolated from mice that received a pretreatment of LPS 20 hours plus an LPS challenge 1 hour before preparation of the cells compared to macrophages from mice receiving only the challenge dose. Consistent with the reduction in TNF and IL-1 message in the macrophages from tolerant animals, a reduced amount of NF-κB was determined in gel-shift assays in these cells, whereas NF-κB activation was high in cells from normal, LPS-challenged mice. In macrophages from endotoxin-tolerant rats it was found that guanine nucleotide–binding protein activity was significantly decreased compared to their normal counterparts (73).

Although the functional receptor for LPS has not yet been characterized (89), one obvious mechanism for this refractoriness would be downregulation of its receptor. However, controversial data about binding of LPS to cells have been published. While it was found to be downregulated in monocytes of endotoxin-tolerant humans compared to nontolerant volunteers (90), no correlation was found in mouse macrophages between binding of LPS and the state of tolerance (91). CD 14, which has been identified as an important LPS-binding protein on macrophages also is not downregulated in tolerance (92). It can be concluded from the above findings that endotoxin tolerance is related to a decreased capacity of a tolerant host's mononuclear phagocytes to elaborate proinflammatory mediators such as TNF, IL-1, and prostaglandins.

Macrophages in the Development of Endotoxin Tolerance

Unambiguous evidence for the key function of macrophages in the development of endotoxin tolerance was provided by Galanos and Freudenberg in a series of convincing experiments. These authors found that sensitization with D-GalN toward the lethal effects of endotoxin was not possible in C3H/HeJ mice (74). When, however, these mice were supplemented with bone marrow–derived macrophages (BMM) from the responsive strain C3H/HeN at a cell number of 2×10^7 in addition to D-GalN, these mice became sensitive to LPS. When the transfer of C3H/HeN BMM cells was carried out in the absence of D-GalN and the recipient C3H/HeJ mice were treated with minute amounts of LPS, the typical early endotoxin tolerance could be induced (75). This was tested by challenging these mice with C3H/HeN BMMs together with a challenge dose of LPS. The time between pretreatment and challenge was 24 hours. The mice treated this way were protected against a dose that caused 100% mortality in mice receiving only the challenge treatment (BMMs and LPS). Furthermore, when the pretreatment was carried out with cells alone in the absence of the eliciting LPS, no protective effects upon challenge was observed. These data indicate that LPS-responsive macrophages are prerequisite for both the development of endotoxin tolerance and the elaboration of toxic effects of endotoxin. Examination of the phenotype of macrophages in endotoxin-tolerant mice revealed that the precursor pool of these cells in the bone marrow is enlarged (76).

Nonmacrophage Cells

In several experimental models the potential contribution of cells other than macrophages to the development of endotoxin tolerance has been examined. Granulocytes have been suspected to be of great importance for the pyrogenic response itself and also for pyrogenic tolerance. This was mainly based on the report by Beeson (57) that granulocytes were the major source of EP in response to endotoxin. This view was challenged, however, by the finding that patients with agranulocytosis developed a normal febrile response to typhoid vaccine, which disappeared upon repeated injections, i.e., refractoriness developed in the complete absence of polymorphonuclear neutrophils (77). A role of spleen cells could likewise be ruled out by the use of splenectomized rabbits (78) and mice (37,79). In all studies, early endotoxin tolerance readily developed. Furthermore, early-phase endotoxin tolerance could be induced in athymic nude mice and B-cell–deficient (*xid*) mice (79). It is, therefore, suggested that lymphoid cell subsets or the spleen do not contribute significantly to the induction of early endotoxin tolerance.

In Vitro Desensitization of Macrophages to Endotoxin

Detailed studies of possible mechanisms underlying the downregulation of cytokine synthesis by macrophages observed in isolated cells of tolerant animals or humans were initiated by the finding that isolated macrophages treated ex vivo with tiny amounts of endotoxin acquire a low responsiveness towards later endotoxin challenge. To differentiate this in vitro induced hyporesponsiveness to endotoxin from the classical tolerance that occurs after repeated injections of LPS in vivo, it will be further refered to as "desensitization." The phenomenon of desensitization was first observed in the production of colony-stimulating factor (80). Furthermore, it was shown that a first treatment with endotoxin will result in decreased synthesis of TNF in rabbit peritoneal macrophages (81), mouse peritoneal macrophages (82–84), human monocytes (46), RAW 264.7 cells (85), and monomac 6 cells (86). In monomac 6 cells desensitization with LPS was found to extend to IL-1 and IL-6 (87). In murine macrophages, however, IL-1β mRNA was not suppressed by prior desensitization (84). Desensitization of murine macrophages inhibited the induction of the inducible NO-synthase (88).

Desensitization can be induced with LPS at doses that are 1000-fold less than required to induce TNF production (81). In rabbit macrophages desensitized with LPS in vitro, a decrease of TNF mRNA as compared to nondesensitized cells was determined after a second dose of LPS. Furthermore, an accelerated mRNA half-life in the desensitized cells could be excluded (81). Two characteristics of desensitized macrophages suggest that this effect may be closely related to endotoxin tolerance. The first is that LPS-desensitized cells have an attenuated response to LPS, however, they readily elaborate TNF message and TNF activity in response to heat-killed *Staphylococcus aureus* (81). The second characteristic is that desensitization can be overcome by challenge of the cells with high doses of LPS. The primary LPS dose causes a marked reduction in the amount of TNF mRNA induced by challenge with 0.001 μg/ml LPS. In contrast, challenge with 0.1 μg/ml results in TNF mRNA induction and TNF release (81). Based on the findings of another study (82), posttranscriptional regulation was suggested to be involved in desensitization of macrophages by LPS in vitro. The authors of the latter publication report that mouse peritoneal macrophages desensitized in vitro with LPS showed high levels of TNF mRNA, while only low quantities of TNF were secreted into the supernatants upon challenge with LPS.

In monomac 6 cells, Ziegler-Heitbrock and colleagues have studied desensitization by LPS and its effects on the mobilization of the nuclear factor NF-κB (93). These authors showed that monomac 6 cells precultured with LPS for 2 days can still mobilize NF-κB efficiently after LPS restimulation. Typically, NF-κB consists of p50/p65 heterodimers. In LPS-desensitized monomac 6 cells, however, the same authors showed that there is a predominance of p50p50 homodimers (94). P50p50 homodimers differ from p50p65 heterodimers in that they occupy the respective DNA-binding sites without transactivation, thereby serving to prevent transcription of TNF. Consistent with the predominance of the p50p50 homodimers in LPS-desensitized monomac 6 cells, an increased p105 mRNA that codes for a p50 precursor was found (94). It is not clear at present whether this mechanism is specific for in vitro desensitized cells or the monomac 6 cell line since in peritoneal macrophages prepared from endotoxin tolerant mice 1 hour postchallenge with LPS, no NF-κB activation was observed on a gel-shift assay, while abundant activation was seen in macrophages prepared from normal mice 1 hour postchallenge (72). In a study of LPS desensitization of human monocytes, it was observed that IL-1 expression was decreased (95). Remarkably, this repression could be overcome by treating desensitized monocytes with cycloheximide. The authors concluded that a protein with a short half-life time, depending on continuous protein synthesis is repressing IL-1β expression. It was suggested that IκBα might be involved in the observed effect (95).

Prostaglandins have been implicated in the desensitizing effects of endotoxins in vitro because LPS-stimulated PGE_2 production was found to be upregulated in desensitized human monocytes (96). PGE_2 (97) and PGI_2 (98) have been found to suppress TNF production in macrophage cultures, which argues in favor of the idea of an involvement of PGs in endotoxin desensitization in macrophages. A major role of these arachidonic acid metabolites seems, however, to be unlikely, since they do not prevent IL-1 or IL-6 (99,100) production, which are downregulated in desensitized macrophages. This finding in LPS-desensitized macrophages is furthermore contrasted by results obtained from macrophages isolated from mice and rats rendered tolerant by LPS in vivo showing substantially decreased synthesis of prostanoids upon LPS stimulation in vitro (63,66).

Interleukin-10 possesses considerable anti-inflammatory properties. It is of interest here that in vitro desensitization of monomac 6 cells resulted in increased production of IL-10 (47). IL-10 has been implicated in the process of in vitro desensitization since antibodies against IL-10 prevented the downregulation of TNF synthesis in cultured human monocytes by preincubation with LPS (46). On the other hand, the same authors reported that LPS desensitization of human monocytes resulted in impaired IL-10 formation of the cells when restimulated with LPS (46).

In contrast to macrophages isolated from endotoxin-tolerant rats, in which altered G-protein function has been shown (73), pretreatment of the human monocyte-like cell line U 937 with endotoxin in vitro did not interfere with the function of these signal proteins (101).

In summary, the data presented in this section show that there is good evidence to conclude that macrophages have a key role in the development of endotoxin tolerance. As shown in Table 1, the tolerance-inducing/desensitizing effects of endotoxin have been studied using different experimental approaches, including macrophages of tolerant hosts and in vitro LPS-desensitized cells. Although there are some properties in common between the different systems, e.g., a decreased capacity to form TNF, other parameters differ considerably. For example, desensitization of monomac 6 cells increases IL-10 synthesis, while it is blocked in desensitized human monocytes and monocytes or macrophages from endotoxin-tolerant humans or mice. Such differences are equally seen for NO synthesis: in macrophages from endotoxin-tolerant rats NO synthesis is increased, while desensitized monocytes show decreased NO production. Of interest, there may be organ differences, since in lungs of endotoxin-tolerant mice, NO synthesis was found to be decreased, while it was increased in the serum of tolerant rats challenged with LPS. Care should, therefore, be taken in extrapolating properties determined in isolated cells or cell lines to endotoxin tolerance in general.

SPECIFICITY OF ENDOTOXIN TOLERANCE

One of the paradigms of early endotoxin tolerance as defined by Greisman (12) is its specificity for lipopolysaccharides as a class. This specificity holds true for pyrogenic tolerance to endotoxin, when certain quantities of the tolerizing LPS are not exceeded. As Greisman pointed out: "The early-phase pyrogenic tolerance is specific for endotoxins as a class. No tolerance occurs to other pyrogens, such as staphylococcal enterotoxin, influennza virus, or old tuberculin specifically sensitized animals (provided massive doses of endotoxin are not given)." Therefore, the specificity of endotoxin tolerance seems to have its limitations when high amounts of endotoxins are used for induction of tolerance. In this section the issue of specificity of endotoxin tolerance is addressed in experimental models other than pyrogenicity.

Nonspecific Increase of the Resistance to Infection

In parallel to inducing endotoxin tolerance, the capacity of endotoxins and other bacterial products to influence nonspecifically host responsiveness to infectious microbes has been recognized for many years (reviewed

Table 1 Properties of In Vitro Desensitized Macrophage-like Cell Lines, Macrophages (Mø), and Macrophages Isolated from Endotoxin-Tolerant Hosts

Parameter	Cell line in vitro desensitized[a]	Isolated Mø in vitro desensitized[b]	Mø isolated from LPS tolerant host[c]	In vivo LPS tolerant[d]
Decreased	TNF (85,86)	TNF (81)	TNF (70,71)	TNF (13,14,70)
	IL-1 (131)		IL-1 (61,62)	
	IL-6 (131)		CSF (50)	
		IL-10 (46)	IL-10 (50)	IL-10 (49)
		NO synthesis (88)	GTP-binding protein (73)	NO synthesis
		Leishmanicidal	NFκB (72)	(rat lung) (116)
		activity (88)	Prostanoids (63,132)	
Unaltered	GTP-binding protein (THP1) (101)	IL-1 (84)		
			NO synthesis (133)	NO metabolites in serum (133)
Increased	IL-10 (monomac6) (47)	Prostanoids (84)		IL-6 (13)
	p50 of NFκB (monomac6) (94)		Bactericidal activity (102)	Resistance to bacterial infection (102)

[a] Cells were pretreated with low amounts of LPS in vitro for several hours and were restimulated with LPS in a second incubation.
[b] Macrophages were isolated from a normal host, pretreated with low amounts of LPS in vitro for several hours, and restimulated with LPS in a second incubation.
[c] Macrophages were isolated from an endotoxin-tolerant host and stimulated in vitro with LPS.
[d] Parameter was determined in serum or tissue of an endotoxin-tolerant host.

in Ref. 102). In numerous studies it has been shown that pretreatment with low amounts of endotoxin 1–4 days before challenge with gram-negative bacteria as single or repeated doses will afford significant protection to mice. It was further found that endotoxins will equally affect the pathogenicity of a wide variety of infectious agents in addition to gram-negative and gram-positive bacteria, including parasites, fungi, and viruses. The mononuclear phagocyte system has been proposed as the main target for endotoxin action. It should, therefore, be considered that endotoxin tolerance and nonspecific protection of a host against infections represent two sides of the same coin. The separation of these two phenomena may experimentally be possible by the exact titration of LPS; however, a uniform mechanism may underlie both effects. It is assumed, therefore, that induction of endotoxin tolerance, e.g., by repeated injections of endotoxin, will coincidently generate an increased anti-infectious potential represented by activated macrophages. Of note, when these macrophages are isolated and stimulated in vitro, they will produce cytokines at a lower level. Nevertheless, these cells are assumed to be in an activated state.

Effects on Tumor Necrosis or Regression

A vast literature exists on the effects of endotoxins on tumor growth in vivo and in vitro (reviewed in Ref. 102). A typical protocol to examine the antitumor effects of endotoxin shows many similarities to those often used to induce tolerance. As one example, Grohsman et al. (103) established that injection of endotoxin into C57bl/10Sr mice 24 hours before challenge with tumor (TA3-Ha) significantly reduced the development to fatal tumor as compared to tumor development in untreated mice. Daily injections of endotoxin from 3 days before tumor challenge to the day of tumor application yielded optimal results. In retrospect it can be assumed that in an attempt to increase the antitumor potential in these mice, a state of endotoxin tolerance was induced.

Hemorrhagic Shock, Ischemia/Reperfusion

The notion that endotoxin tolerance provides protection against assaults that are not primarily imposed on experimental animals by endotoxin can also be concluded from studies on hemorrhagic shock. Protection against shock achieved by repeated injections of LPS has long been recognized (104). In more recent studies on the possible contribution of neutrophils to the beneficial effects of tolerance in hemorrhage (105), it was found that adhesion of circulating PMNs to nylon fibers in vitro and the number of PMNs adhering to the endothelium in the mesentery in vivo was significantly lower in tolerant than in normal rats. It was concluded that LPS pretreatment produces a reduction in the activated circulating PMNs and in the degree of PMN adhesion to endothelium with subsequent improvement of survival after hemorrhagic shock. It should be taken into consideration, however, that endotoxin translocated from the gut may contribute to hemorrhagic shock. This has been suggested by findings presented by Yao et al. (106). They reported that bactericidal/permeability-increasing protein (BPI) increased survival of rats after hemorrhagic shock from 37.5% in controls to 68.8% in the treatment group. Since BPI belongs to a family of endogenous LPS-binding proteins, which neutralizes LPS in vivo and in vitro by binding to lipid A with high specificity (107), it was suggested that endotoxin is involved in the pathogenesis of hemorrhagic shock.

A key mediator of the multiple organ failure following hemorrhagic shock seems to be TNF, since the application of anti-TNF antibodies attenuated the hemorrhage-related pathophysical alterations and improved survival after hemorrhage (108). BPI treatment, however, failed to decrease TNF plasma concentrations in the aforementioned study (106). The reason for this could consist, as the authors discussed, "of the initial TNF formation by the LPS-stimulated, gut-associated macrophages before the entry of LPS into the circulation and its inactivation by BPI. In addition, there also are several lines of evidence suggesting that changes in organ blood flow induced by hemorrhage could lead to regional hypoxia, which also could contribute directly or indirectly to the induction of TNF." Independent of the nature of the TNF-inducing agent, we would conclude, nevertheless, that endotoxin tolerance by an unknown mechanism prevents the interaction of granulocytes with endothelial cells. This appears to be key to the protection by endotoxin tolerance against multiorgan failure occurring as a consequence of hemorrhagic shock.

Ischemia/reperfusion represents another experimental entity of potential clinical relevance in which endotoxin tolerance proved to be protective. Colletti and colleagues found that reperfusion of the liver following a 90-minute period of ischemia of one liver lobe in the rat resulted in massive increases in lung permeability (109). Since reperfusion was accompanied by increased TNF levels in the circulation and the pulmonary capillary leakage could be prevented by treating the animals with an antiserum against TNF, it was suggested that TNF formation caused by the hepatic reperfusion injury was responsible for lung edema. Application of LPS on 4 consecutive days prevented the increase in lung permeability completely (110). Surprisingly, in the tolerant animals, increased amounts of TNF were formed in response to ischemia/reperfusion of the liver. Although the authors did not provide an explanation for the higher TNF levels in tolerant rats, this finding is in agreement with the notion that endotoxin tolerance includes mechanisms that interfere with the activity of proinflammatory cytokines such as TNF.

In several studies, the effect of endotoxin tolerance on myocardial ischemia/reperfusion was examined (111,112) and found to limit ischemia/reperfusion injury in isolated rat hearts. In addition, it limited ventricular arrhythmias (113) and infarct size (113,114) after coronary occlusion in rats. Since PMNs appear to be key participants in the development of injury in the heart by reperfusion following coronary occlusion (115), it was suggested that reduction of PMN activation and adhesion to endothelium in addition to downregulation of cytokine formation associated with tolerance may account for the observed effects (114).

It has been suggested that a major effect of tolerance with regard to its myocardial protection consists in a diminished induction of NO synthase (116) caused by endogenous corticosteroids. This was mainly based on the finding that the corticosteroid antagonist RU 38486 counteracted the suppression of NO synthase. In this study, however, the possibility that RU 38486 may lead to hypersensitivity to endotoxin has not been considered. Furthermore, evidence has been published that NO itself may be beneficial in myocardial ischemia (117,118). Moreover, the beneficial effects of endotoxin on infarct size and on ventricular arrhythmias were markedly attenuated by the prior administration of dexamethasone (113), which would also argue against a participation of corticosteroids in myocardial protection by endotoxin tolerance.

The studies reviewed in this section provide good evidence that endotoxin tolerance serves to protect against infection, tumor resistance, hemorrhagic shock, and myocardial ischemia/reperfusion injury. Although it is unclear whether the protective effects exerted by endotoxin tolerance in these experimental models are based on the same mechanisms as its protection against endotoxin pyrogenicity, the notion seems to be justified that endotoxin tolerance affects systems beyond the specificity towards endotoxin as a class. A key to its broad base of actions may be the interference with the interaction between leukocytes and endothelial cells. The advantageous effects of endotoxin tolerance include those of great clinical relevance, and the elucidation of the underlying detailed mechanisms deserves future attention.

Possible Involvement of Endotoxin Tolerance in Posttraumatic and Septic Disease

The assumption that endotoxin tolerance extends its effects to infectious diseases, thus providing protection against endotoxin-unrelated assaults, gives rise to some questions. Is it possible that infections with gram-negative bacteria induce a state similar to endotoxin tolerance, and is endotoxin tolerance responsible for immunosuppression observed in septic or polytraumatized patients? While there is good experimental evidence that endotoxin tolerance induced by repeated treatment with LPS generates protection to bacterial infection, little information is available as to whether infections cause low responsiveness to LPS. It has, however, been found that patients recovered from typhoid or paratyphoid fever (119) and from malaria (120) show a significantly reduced pyrogenic response to injections with LPS compared to that of normal volunteers. Therefore, in patients who have overcome an infection, low responsiveness to endotoxin, i.e., tolerance, may have developed.

On the other hand, it has been documented in infected volunteers (121, 122) that LPS treatment during the active infection resulted in increased pyrogenic responses. Analogous to the human situation, infection of experimental animals highly increased their sensitivity to the lethal activity of LPS applied during active infection (22). That infection per se would not induce tolerance can be inferred from the finding that in both infected humans and infected animals, endotoxin tolerance could be induced by treating them with small amounts of LPS (22,121,122). Obviously it was not invoked by the infection alone. At present, therefore, very little experimental evidence seems to support the notion that infections and in particular sepsis lead to the development of endotoxin tolerance.

It has been found that whole blood or monocytes isolated from septic (123–125) or polytraumatized patients (36,126–128) show reduced synthesis of cytokines, when stimulated with LPS ex vivo in comparison to monocytes from normal persons. When patients recovered, the capacity to form cytokines was found to occur to a normal extent. It has been suggested that this phenomenon observed at the cellular level represented the development of endotoxin tolerance in these patients (36,124,129). Before this conclusion can be drawn, however, some objections to this assumption should be considered.

1. In polytraumatized patients, diminished cytokine synthesis in response to LPS ex vivo is detectable as early as 1 hour posttrauma (128). The development of endotoxin tolerance in the human requires a longer time (15,62).
2. In serum of polytraumatized patients, a circulating inhibitor for the synthesis of TNF was determined, whose appearance chronologically correlated with the diminished TNF synthesis in patients' blood (128). This inhibitor was neither specific for LPS-induced TNF synthesis nor identical with soluble TNF receptors or cortisone. Mononuclear cells isolated from patients' blood with diminished TNF synthesis were stimulated with LPS in the presence of normal human serum and produced TNF to a similar extent as MNC isolated from healthy donors (135). It occurs that this inhibitor is responsible for the suppression of cytokine synthesis in the blood of traumatized patients. In contrast, in cultures of isolated MNC from endotoxin-tolerant humans in the presence of normal human serum, the response to endotoxin ex vivo was found to be downregulated with regard to TNF, IL-1, and IL-10 synthesis (50,62). Thus, MNC of polytraumatized patients and MNC from tolerant individuals differ considerably in their capacity to synthesize cytokines.
3. Monocytes from septic patients in most cases are markedly reduced in the expression of HLA-DR on their surface (130). This shows that an important functional parameter related to antigen presentation in the cells of septic patients is impaired. In monocytes of endotoxin-tolerant humans, no change in HLA-DR expression could be detected as compared to pretolerant monocytes (50). In peritoneal macrophages from endotoxin-tolerant mice, the phagocytic capacity was increased compared to normal cells (63), suggesting that these cells, unlike the monocytes of septic patients, showed enhanced functional capabilities.
4. Finally, one general aspect of endotoxin tolerance argues against the notion of polytraumatized or septic patients as being endotoxin tolerant. As

pointed out above, the induction of endotoxin tolerance increases the resistance to bacterial and viral infections. This is obviously not the case in severely injured patients who are in general referred to as immunodepressed.

These objections against an implication of endotoxin tolerance in sepsis or polytrauma do not exclude that similar mechanisms may underlie the observed diminished cytokine production. Septic or polytraumatized patients and endotoxin-tolerant individuals may express closely related regulatory factors. It should, however, be kept in mind that endotoxin tolerance embraces a multitude of functional changes in the host besides downregulation of the synthesis of several cytokines. Care should be taken, therefore, in implying it in situations resembling endotoxin tolerance in certain aspects.

CONCLUDING REMARKS

Endotoxin tolerance by its classical definition can be induced by application of endotoxins in small doses and renders protection to a host against subsequent challenge with LPS. In addition to the protection against LPS, it increases in a host resistance against unrelated assaults such as reperfusion injury or infection. Macrophages have been recognized as a central component for the development of endotoxin tolerance. On the other hand, macrophages represent the major source for factors transmitting deleterious endotoxic activities. The molecular mechanisms transforming macrophages from a mediator element for endotoxin toxicity to promoters of tolerance are at present unknown. The attenuation of the synthesis of inflammatory cytokines by macrophage appears to be a substantial contribution to endotoxin tolerance. Some evidence exists that macrophages from tolerant hosts do not become unresponsive to LPS, they rather alter their spectrum of products and synthesize factors inhibiting the synthesis and possibly the action of proinflammatory mediators such as TNF or IL-1. The identification of such factors is supposed to provide new insight into the nature of endotoxin tolerance.

ACKNOWLEDGMENTS

The studies carried out in our laboratories described here were supported by Grants of the Deutsche Forschungsgemeinschaft (Schm 74/13-1 and 13-2) and by the Fond der Chemischen Industrie (F.U.S.).

REFERENCES

1. Heyman A. The treatment of neurosyphilis by continuous infusion of typhoid vaccine. Vener Dis Inform 1945; 26:51–57.
2. Shear MJ, Perrault A. Chemical treatment of tumors. IX. Reactions of mice with primary subcutaneous tumors to injections of a hemorrhage-producing bacterial polysaccharide. J Natl Cancer Inst 1944; 4:461–476.
3. Wharton DRA, Creech HJ. Further studies of the immunological properties of polysaccharides from *Serratia marcescens* (*Bacillus prodigiosus*). II. Nature of the antigenic action and the antibody response in mice. J Immunol 1949; 62:135–153.
4. Centanni E. Untersuchungen über das Infections fieber—das Fiebergift der Bacterien. Dtsch Med Wochenschr 1894; 20:148–150.
5. Centanni E. Immunitätserscheinungen im experimentellen Fieber mit besonderer Berücksichtigung des pyrogenen Stoffes aus Typhusbakterien. Klin Wochenschr 1942; 21:664–669.
6. Favorite GO, Morgan HR. Effects produced by the intravenous injection in man of a toxic antigenic material derived from *Eberthella thyphosa*: clinical, hematological, chemical and serological studies. J Clin Invest 1942; 21:589–599.
7. Beeson PB. Development of tolerance to typhoid bacterial pyrogen and its abolition by reticulo-endothelial blockade. Proc Soc Exp Biol Med 1946; 61:248–250.
8. Beeson PB. Tolerance to bacterial pyrogens. I. Factors influencing its development. J Exp Med 1947; 86:29–38.
9. Beeson PB. Tolerance to bacterial pyrogens. II. Role of the reticulo-endothelial system. J Exp Med 1947; 86:39–44.
10. Greisman SE, Hornick RB. Mechanisms of endotoxin tolerance with special reference to man. J Inf Dis 1973; 128:S265–S276.
11. Greisman SE, Young EJ, Carozza FA, Jr. Mechanisms of endotoxin tolerance. V. Specificity of the early and late phases of pyrogenic tolerance. J Immunol 1969; 103:1223–1236.
12. Johnston CA, Greisman SE. Mechanism of endotoxin tolerance. In: Hinshaw LB, ed. Handbook of Endotoxin. Vol 2. Pathophysiology of Endotoxin. New York: Elsevier, 1985:359–391.
13. Flohé S, Heinrich PC, Schneider J, Wendel A, Flohe L. Time course of IL-6 and TNF alpha release during endotoxin-induced endotoxin tolerance in rats. Biochem Pharmacol 1991; 41:1607–1614.
14. Mathison JC, Wolfson E, Ulevitch RJ. Participation of tumor necrosis factor in the mediation of gram negative bacterial lipopolysaccharide-induced injury in rabbits. J Clin Invest 1988; 81:1925–1937.
15. Mackensen A, Galanos C, Wehr U, Engelhardt R. Endotoxin tolerance: regulation of cytokine production and cellular changes in response to endotoxin application in cancer patients. Eur Cytokine Netw 1992; 3:571–579.
16. Atkins E. Pathogenesis of fever. Physiol Rev 1960; 40:580–646.

17. Vogel SN, Jaufman EN, Tate MD, Neta R. Recombinant interleuken-1α and recombinant tumor necrosis factor α synergize in vivo to induce early endotoxin tolerance and associated hematopoietic changes. Infect Immun 1988; 56:2650–2657.
18. Henricson BE, Neta R, Vogel SN. An interleukin-1 receptor antagonist blocks lipopolysaccharide-induced colony-stimulating factor production and early endotoxin tolerance. Infect Immun 1991; 59:1188–1191.
19. Chapman PB, Lester TJ, Cooper ES, et al. Clinical pharmacology o recombinant human tumor necrosis factor in patients with advanced cancer. J Clin Oncol 1987; 5:1942–1965.
20. Cerami A, Beutler B. The role of cachectin/TNF in endotoxic shock and cachexia. Immunol Today 1988; 9:28–31.
21. Fraker DL, Stovroff MC, Merino MJ, Norton JA. Tolerance to tumor necrosis factor in rats and the relationship to endotoxin tolerance and toxicity. J Exp Med 1988; 168:95–105.
22. Galanos C, Freudenberg M, Katschinski T, Salomoa R, Mossmann H, Kumazawa Y. Tumor necrosis factor and host response to endotoxin. In: Ryal JL, Morrison DC, eds. Bacterial Endotoxic Lipopolysaccharides. Boca Raton, FL: CRC Press, 1992:75–104.
23. Galanos C, Freudenberg MA, Reutter W. Galactoseamine-induced sensitization to the lethal effects of endotoxin. Proc Natl Acad Sci USA 1979; 76:5939–5943.
24. Greer GG, Rietschel ET. Lipid A-induced tolerance and hyperreactivity to hypothermia in mice. Infect Immun 1978; 19:357–368.
25. Wallach D, Holtmann H, Engelmann H, Nophar Y. Sensitization and desensitization to lethal effects of tumor necrosis factor and IL-1. J Immunol 1988; 140:2994–2999.
26. Libert C, Van Bladel S, Brouckaert P, Shaw A, Fiers W. Involvement of the liver, but not of IL-6, in IL-1-induced desensitization to the lethal effects of tumor necrosis factor. J Immunol 1991; 146:2625–2632.
27. Takahashi N, Brouckaert P, Fiers W. Induction of tolerance allows separation of lethal and antitumor activities of tumor necrosis factor in mice. Cancer Res 1991; 51:2366–2372.
28. Fukushima H, Ikeuchi J, Tohkin M, Matsubara T, Harada M. Lethal shock in partially hepatectomized rats administered tumor necrosis serum. Circ Shock 1988; 26:1–7.
29. Geller P, Merril ER, Jawetz E. Effects of cortisone and antibiotics on the lethal action of endotoxins in mice. Proc Soc Exp Biol Med 1954; 86:716–725.
30. Beutler B, Krochin N, Milsark IW, Luedke C, Cerami A. Control of cachectin (tumor necrosis factor) synthesis: mechanism of endotoxin resistance. Science 1986; 232:977–980.
31. Snyder DS, Unanue ER. Corticosteroids inhibit murine macrophage Ia expression and interleukin 1 production. J Immunol 1982; 129:1803.
32. Chedid L, Parant M, Boyer F, Skarnes RC. Nonspecific host responses in tolerance to the lethal effect of endotoxins. In: Landy M, Braun W, eds. Bacterial Endotoxin. New Brunswick, NJ: Rutgers University Press, 1964:500–516.
33. Bertini R, Bianchi M, Ghezzi P. Adrenalectomy sensitizes mice to the lethal effects of interleukin 1 and tumor necrosis factor. J Exp Med 1988; 167:1708–1712.
34. Zuckerman SH, Qureshi N. In vivo inhibition of lipopolysaccharide-induced lethality and tumor necrosis factor synthesis by Rhodobacter sphaeroides diphosphoryl lipid A is dependent on cortisone induction. Infect Immun 1992; 60:2581–2587.
35. Besedovsky H, Del Rey A, Sorkin E, Dinarello CA. Immunoregulatory feedback between interleukin 1 and glucocorticoid hormones. Science 1986; 233:652–655.
36. Keel M, Schregenberger N, Steckholzer U, et al. Endotoxin tolerance after severe injury and its regulatory mechanisms. J Trauma 1996; 41:430–438.
37. Evans GF, Zuckerman SH, Glucocorticoid-dependent and -independent mechanisms involved in lipopolysaccharide tolerance. Eur J Biochem 1991; 21:1973–1979.
38. Chedid L, Skarnes RC, Parant M. Characterization of a Cr 51-labeled endotoxin and its identification in plasma and urine perenteral administration. J Exp Med 1963; 117:561–571.
39. Libert C, Everaerdt B, Takahashi N, Brouckaert P, Fiers W. Interleukin-1-induced protection against the lethal activity of tumor necrosis factor is mediated by the liver. 2nd International Congress on the Immune Consequences of Trauma, Shock and Sepsis, 1991.
40. Brouckaert P, Everaerdt B, Fiers W. The glucocorticoid antagonist RU38486 mimics interleukin-1 in its sensitization to the lethal and interleukin-6-inducing properties of tumor necrosis factor. Eur J Immunol 1992; 22:981–986.
41. Lazar G, Agarwal MK. The influence of a novel glucocorticoid antagonist on endotoxin lethality in mice strains. Biochem Med Metab Biol 1986; 36:70–74.
42. Lazar G, Jr. Lazar G, Agarwal MK. Modification of septic shock in mice by the antiglucocorticoid RU 38486. Circ Shock 1992; 36:180–184.
43. Bogdan C, Vodovotz Y, Nathan C. Macrophage deactivation by interleukin 10. J Exp Med 1991; 174:1549–1555.
44. De Waal Malefyt R, Abrams J, Bennet B, Figdor CG, De Vries JE. Interleukin 10 (IL-10) inhibits cytokine synthesis by human monocytes: an autoregulating role of IL10 produced by monocytes. J Exp Med 1991; 174:1209–1220.
45. Howard M, Muchamuel T, Andrade S, Menon S. Interleukin 10 protects mice from lethal endotoxinemia. J Exp Med 1993; 177:1205–1208.
46. Randow F, Syrbe U, Meisel C, et al. Mechanism of endotoxin desensitization: involvement of interleukin 10 and transforming growth factor beta. J Exp Med 1995; 181:1887–1892.
47. Frankenberger M, Pechumer H, Ziegler-Heitbrock HWL. Interleukin 10 is upregulated in LPS tolerance. J Inflamm 1995; 45:56–63.
48. Berg DJ, Kühn R, Rajewski K, et al. Interleukin-10 is a central regulator of the response to LPS in murine models of endotoxin shock and the Shwartzman reaction but not endotoxin tolerance. J Clin Invest 1995; 96:2339–2347.

49. Schade FU, Schlegel J, Hofmann K, Brade H, Flach R. Endotoxin-tolerant mice produce an inhibitor of tumor necrosis factor-synthesis. J Endotox Res 1996; 3:455–462.
50. Flach R, Flohé S, Laschinski M, Hofmann K, Kreuzfelder E, Schade FU. Interleukin 10 is downregulated in mononuclear cells from endotoxin tolerant humans. J Endotox Res 1997.
51. Freedman HH. Passive transfer of tolerance to pyrogenicity of bacterial endotoxin. J Exp Med 1960; 111:453–463.
52. Freedman HH. Further studies on passive transfer of tolerance to pyrogenicity of bacterial endotoxin. J Exp Med 1960; 112:619–634.
53. Greisman SE, Carozza FA Jr, Hills JD. Mechanism of endotoxin tolerance. I. Relationship between tolerance and reticuloendothelial system phagocytic activity in the rabbit. J Exp Med 1963; 117:663–674.
54. Wolff SM, Mulholland JH, Ward SB. Quantitative aspects of the pyrogenic response of rabbits to endotoxin. J Lab Clin Med 1965; 65:268–275.
55. Greisman SE, Wagner HN, Jr, Iio M, Hornick RB. Mechanisms of endotoxin tolerance. II. Relationship between endotoxin and reticuloendothelial system phagocytic activity in man. J Exp Med 1964; 119:241–264.
56. Dinarello CA, Cannon JG, Wolff SM. New concepts on the pathogenesis of fever. Rev Infect Dis 1988; 10:168–189.
57. Beeson PB. Temperature elevating effect of a substance obtained from polymorphoneclear leukocytes (abstr). J Clin Invest 1948; 27:524.
58. Atkins E, Bodel PT, Francis L. Release of an endogenous pyrogen in vitro from rabbit mononuclear cells. J Exp Med 1967; 126:357–384.
59. Hahn HH, Char DC, Postel WB, Wood WB Jr. Studies on the pathogenesis of fever XV. The production of endogenous pyrogen by peritoneal macrophages. J Exp Med 1967; 126:385–394.
60. Bodel P. Atkins E. Release of endogenous pyrogen by human monocytes. N Engl J Med 1967; 276:1002–1008.
61. Dinarello CA, Bodel PT, Atkins E. The role of the liver in the production of fever and in pyrogenic tolerance. Trans Assoc Am Phys 1968; 81:334–344.
62. Granowitz EV, Reuven P, Mier JW, et al. Intravenous endotoxin suppresses the cytokine response of peripheral blood mononuclear cells of healthy volunteers. J Immunol 1993; 151:1637–1645.
63. Schade FU, Rietschel ET. Differences in prostaglandin production of peritoneal macrophages from normal, endotoxin tolerant and hyperreactive mice. In: Eaker D, Wadström T, eds. Natural Toxins. New York: Pergamon Press, 1980:271–277.
64. Rietschel ET, Schade FU, Jensen M, Wollenweber HW, Lüderitz O, Greisman SE. Bacterial endotoxins: chemical structure, biological activity and role in septicaemia. Scand J Infect Dis 1982; 31(suppl):8–21.
65. Rogers TS, Halushka PV, Wise WC, Cook JA. Arachidonic acid turnover in peritoneal macrophages is altered in endotoxin-tolerant rats. Biochim Biophys Acta 1989; 1001:169–175.
66. Wise WC, Cook JA, Halushka PV. Arachidonic acid metabolism in endotoxin tolerance. Adv Shock Res 1983; 10:131–142.
67. Grunfeld C, Palladino MA, Jr. Tumor necrosis factor: immunologic, antitumor, metabolic, and cardiovascular activities. Adv Intern Med 1990; 35:45–72.
68. Beutler B, Milsark IW, Cerami AC. Passive immunization against cachectin/tumor necrosis factor protects mice from lethal effect of endotoxin. Science 1985; 229:869–871.
69. Tracey KJ, Fong Y, Hesse DG, et al. Anti-cachectin/TNF monoclonal antibodies prevent septic shock during lethal bacteremia. Nature 1987; 330:662–664.
70. Haslberger AT, Sayers T, Reiter H, Chung J, Schütze E. Reduced release of TNF and PCA from macrophages of tolerant mice. Circ Shock 1988; 26:185–192.
71. Moore JN, Cook JA, Morris DD, Halushka PV, Wise WC. Endotoxin-induced procoagulant activity, eicosanoid synthesis, and tumor necrosis factor production by rat peritoneal macrophages: effect of endotoxin tolerance and glucan. Circ Shock 1990; 31:281–295.
72. Zuckerman SH, Evans GF. Endotoxin tolerance: in vivo regulation of tumor necrosis factor and interleukin-1 synthesis is at the transcriptional level. Cell Immunol 1992; 140:513–519.
73. Coffee KA, Haluska PV, Wise WC, Cook JA. Altered responses to modulators of guanine nucleotide binding protein activity in endotoxin tolerance. Biochim Biophys Acta 1990; 1035:201–205.
74. Freudenberg MA, Keppler D, Galanos C. Requirement for lipopolysaccharide responsive macrophages in galactosamine-induced sensitization to endotoxin. Infect Immun 1986; 51:891–895.
75. Freudenberg MA, Galanos C. Induction of tolerance to lipopolysaccharide (LPS)-D-galactosamine lethality by pretreatment with LPS is mediated by macrophages. Infect Immun 1988; 56:1352–1357.
76. Madonna GS, Vogel SN. Early endotoxin tolerance is associated with alterations in bone marrow-derived macrophage precursor pools. J Immunol 1985; 135:3763–3771.
77. Page AR, Good RA. Studies on cyclic neutropenia. Am J Dis Child 1957; 94:623–661.
78. Greisman SE, Young EJ, Workman JB, Ollodart RM, Hornick RB. Mechanisms of endotoxin tolerance. The role of the spleen. J Clin Invest 1975; 56:1597–1607.
79. Madonna GS, Vogel SN. Induction of early-phase endotoxin tolerance in athymic (nude) mice, B-cell-deficient (xid) mice and splenectomized mice. Infect Immun 1986; 53:707–710.
80. Sullivan R, Gans PJ, McCarroll LA. The synthesis and secretion of granulocyte-monocyte colony-stimulating activity (CSA) by isolated monocytes: kinetics of the response to bacterial endotoxin. J Immunol 1983; 130:800–807.
81. Mathison JC, Virca GD, Wolfson E, Tobias PS, Glaser K, Ulevitch RJ. Adaptation to bacterial lipopolysaccharide controls lipopolysaccharide-induced tumor necrosis factor production in rabbit macrophages. J Clin Invest 1990; 85:1108–1118.
82. Zuckerman SH, Evans GF, Synder YM, Roeder WD. Endotoxin-macrophage interaction: post-translational

regulation of tumor necrosis factor expression. J Immunol 1989; 143:1223–1227.
83. Fahmi H, Chaby R. Selective refractoriness of macrophages to endotoxin-induced production of tumor necrosis factor, elicited by an autocrine mechanism. J Leukoc Biol 1993; 53:45–52.
84. Takasuka N, Tokunaga T, Akagawa KS. Preexposure of macrophages to low doses of lipopolysaccharide inhibits the expression of tumor necrosis factor-alpha mRNA but not of IL-1β mRNA. J Immunol 1991; 146:3824–3830.
85. Virca GD, Kim SY, Glaser KB, Ulevitch RJ. Lipopolysaccharide induces hyporesponsiveness to its own action in RAW 264.7 cells. J Biol Chem 1989; 264: 21951–21956.
86. Haas JG, Thiel C, Blömer K, Weiss EH, Riethmüller G, Ziegler-Heitbrock HWL. Downregulation of tumor necrosis factor expression in human mono-mac-6 cell line by lipopolysaccharide. J Leuk Biol 1989; 46:11–14.
87. Ziegler-Heitbrock HWL, Blumenstein M, Kafferlein E, et al. In vitro desensitization to lipopolysaccharide suppresses tumour necrosis factor, interleukin 1 and interleukin 6 gene expression in a similar fashion. Immunology 1992; 75:264–268.
88. Severn A, Xu D, Doyle J, et al. Pre-exposure of murine macrophages to lipopolysaccharide inhibits the induction of nitric oxide synthase and reduces leishmanicidal activity. Eur J Immunol 1993; 23:1711–1714.
89. Heumann D, Glauser MP. Pathogenesis of sepsis. Sci Am Sci Med 1994; 1:28–37.
90. Larsen NE, Sullivan R. Interaction between endotoxin and human monocytes: characteristics of the binding of ^{3}H-labeled lipopolysaccharide and ^{51}Cr-labeled lipid A before and after the induction of endotoxin tolerance. Proc Natl Acad Sci USA 1984; 81:3491–3495.
91. Fahmi H, Chaby R. Desensitization of macrophages to endotoxin effects is not correlated with a downregulation of lipopolysaccharide-binding sites. Cell Immunol 1993; 150:219–229.
92. Mathison JC, Wolfson E, Steinemann S, Tobias P, Ulevitch RJ. Lipopolysaccharide (LPS) recognition in macrophages. J Clin Invest 1993; 92:2053–2059.
93. Haas JG, Bäuerle PA, Riethmüller G, Ziegler-Heitbrock HWL. Molecular mechanism in down-regulation of tumor necrosis factor expression. Proc Natl Acad Sci USA 1990; 87:9563–9567.
94. Ziegler-Heitbrock HWL, Wedel A, Schraut W, et al. Tolerance to lipopolysaccharide involves mobilization of nuclear factor κB with predominance of p50 homodimer. J Biol Chem 1994; 269:17001–17004.
95. LaRue KEA, McCall CE. A labile transcriptional repressor modulates endotoxin tolerance. J Exp Med 1994; 180:2269–2275.
96. Matic M, Simon SR. Tumor necrosis factor release from lipopolysaccharide-stimulated human monocytes: lipopolysaccharide tolerance in vitro. Cytokine 1991; 3:576–583.
97. Kunkel SL, Spengler M, May MA, Spengler R, Larrick J, Remick DG. Prostaglandin E_2 regulates macrophage-derived tumor necrosis factor gene expression. J Biol Chem 1988; 263:5380–5384.
98. Schade FU, Schudt C. The specific type III and IV phosphodiesterase inhibitor zardaverine suppresses formation of tumor necrosis factor by macrophages. Eur J Pharmacol 1993; 230:9–14.
99. Scales WE, Chensue SW, Otterness I, Kunkel SL. Regulation of monokine gene expression: prostaglandin E_2 suppresses tumor necrosis factor but not interleukin 1α or b-mRNA and cell-associated bioactivity. J Leuk Biol 1989; 45:416–421.
100. Bailly S, Ferrua B, Fay M, Gougerot-Pocidalo MA. Differential regulation of IL-6, IL-1α, IL-1β and TNF-α production in LPS-stimulated human monocytes: role of cyclic AMP. Cytokine 1990; 2:205–210.
101. Durando M, Ashton SH, Makhlouf MA, Simmons-Wagner R, Halushka PV, Cook JA. Endotoxin-induced desensitization of THP-1 cells is not associated with altered G protein binding or content. J Endotox Res 1997; 4:97–103.
102. Morrison DC, Ryan JL. Bacterial endotoxins and host immune responses. Adv Immunol 1979; 28:293–450.
103. Grohsman J, Nowotny A. The immune recognition of TA3 tumors, its facilitation by endotoxin, and abrogation by ascites fluid. J Immunol 1972; 109:1090–1095.
104. Zweifach BW. Contribution of the reticuloendothelial system to the development of tolerance to experimental shock. Ann NY Acad Sci 1960; 88:203–212.
105. Barroso-Aranda J, Chavez-Chavez R, Mathison JC, Suematsu M, Schmid-Schönbein GW. Circulating neutrophil kinetics during tolerance in hemorrhagic shock using bacterial lipopolysaccharide. Am J Physiol 1994; 266:H415–421.
106. Yao YM, Bahrami S, Leichtfried G, Redl H, Schlag G. Pathogenesis of hemorrhage-induced bacteria/endotoxin translocation in rats. Effects of recombinant bactericidal permeability-increasing protein. Ann Surg 1995; 221:398–405.
107. Marra MN, Wilde CG, Griffith JE, et al. Bactericidal/Permeability-increasing protein of neutrophils. J Immunol 1990; 144:662–668.
108. Bahrami S, Yao YM, Leichtfried G, et al. Efficacy of monoclonal antibody (MAB) to tumor necrosis factor (TNF) against hemorrhage-induced mortality in rats. Intens Care Med 1994; 20(suppl 1):S61.
109. Colletti LM, Burtch GD, Remick DG, et al. The production of tumor necrosis factor alpha and the development of a pulmonary capillary injury following hepatic ischemia/reperfusion. Transplantation 1990; 49: 268–272.
110. Colletti LM, Remick DG, Campbell DA, Jr. LPS pretreatment protects from hepatic ischemia/reperfusion. J Surg Res 1994; 57:337–343.
111. Brown JM, Grosso MA, Terada LS, et al. Endotoxin pretreatment increases endogenous myocardial catalase activity and decreases ischemia reperfusion injury of isolated rat hearts. Proc Natl Acad Sci USA 1989; 86:2516–2520.
112. Brown JM, Anderson BO, Repine JE, et al. Neutrophils contribute to TNF induced myocardial tolerance to ischemia. J Mol Cell Cardiol 1992; 24:485–495.
113. Song W, Furman BL, Parrat JR. Delayed protection against ischemia-induced ventricular arrhythmias and

infarct size limitation by the prior administration of *Escherichia coli* endotoxin. Br J Pharmacol 1996; 118:2157–2163.
114. Eising GF, Mao L, Schmid-Schönbein GW, Engler RL, Ross J. Effects of induced tolerance to bacterial lipopolysaccharide on myocardial infarct size in rats. Cardiovasc Res 1996; 31:73–81.
115. Engler RL, Dahlgren MD, Morris D, Petersen M, Schmid-Schönbein GW. Roel of leukocytes in response to acute myocardial ischemia and reflow in dogs. Am J Physiol 1986; 251:H314–322.
116. Szabo C, Thiemermann C, Wu CC, Perretti M, Vane JR. Attenuation of the induction of nitric oxide synthese by endogenous glucocorticoids accounts for endotoxin tolerance in vivo. Proc Natl Acad Sci USA 1994; 91:271–275.
117. Nakanishi KJ, Vinten-Johansen J, Lefer DJ, et al. Intracoronary L-arginine during reperfusion improves endothelial function and reduces infarct size. Am J Physiol 1992; 263:H1650–1658.
118. Martorana PA, Kettenbach B, Bohn H, Schonafinger K, Henning R. Antiischemic effects of pirsidomine, a new nitric oxide donor. Eur J Pharmacol 1994; 257:267–273.
119. Neva FA, Morgan AR. Tolerance to the action of endotoxins of enteric bacilli in patients convalescent from typhoid and paratyphoid fevers. J Lab Clin Med 1950; 35:911–922.
120. Heyman A, Beeson PB. Influence on the various disease states upon febrile response to intravenous injection of typhoid bacterial pyrogen, with particular reference to malaria and cirrhosis of liver. J Lab Clin Med 1949; 34:1400–1403.
121. Greisman SE, Hornick RB, Woodward TE. The role of endotoxin during typhoid fever ad tularemia in man. III. Hyperreactivity to endotoxin during infection. J Clin Invest 1964; 43:1747–1757.
122. Greisman SE, Hornick RB, Wagner HN, Jr, Woodward WE, Woodward TE. The role of endotoxin during typhoid fever and tularemia in man IV. The integrity of the endotoxin tolerance mechanism during infection. J Clin Invest 1969; 48:613–629.
123. Munoz C, Carlet J, Fitting C, Misset B, Bleriot JP, Cavaillon JM. Dysregulation of in vitro cytokine production by monocytes during sepsis. J Clin Invest 1991; 88:1747–1754.
124. McCall CE, Grosso-Wilmoth LM, LaRue K, Guzman RN, Cousart SL. Tolerance to endotoxin-induced expression of the interleukin-1β gene in blood neutrophils of humans with the sepsis syndrome. J Clin Invest 1993; 91:853–861.
125. Ertel W, Kremer JP, Kenney J, et al. Downregulation of proinflammatory cytokine release in whole blood from septic patients. Blood 1995; 85:1341–1347.
126. Fabian TC, Croce MA, Fabian MJ, et al. Reduced tumor necrosis factor production in endotoxin-spiked whole blood after trauma. Surgery 1995; 118:63–72.
127. Majetschak M, Flach R, Heukamp T, et al. A circulating inhibitor of cytokine synthesis in the initial phase of polytrauma. Shock 1996; 5:17–18.
128. Majetschak M, Flach R, Heukamp T, et al. Regulation of whole blood tumor necrosis factor production upon endotoxin treatment after severe blunt trauma. J Trauma 1997.
129. Cavaillon JM. The nonspecific nature of endotoxin tolerance. Trends Microbiol 1995; 3:320–324.
130. Döcke WD, Syrbe U, Meinecke A, et al. Improvement of monocyte function—a new therapeutic approach? In: Reinhart K, Eyrich K, Sprung C, eds. Sepsis, Current Perspectives in Pathophysiology and Therapy. Berlin: Springer, 1994:473–500.
131. Ziegler-Heitbrock HWL, Blumenstein M, Käfferlein E, et al. In vitro desensitization to lipopolysaccharide suppresses tumor necrosis factor, interleukin 1 and interleukin 6 gene expression in a similar fashion. Immunology 1992; 75:264–268.
132. Rogers TS, Halushka PV, Wise WC, Cook JA. Differential alteration of lipoxygenase and cyclooxygenase metabolism by rat peritoneal macrophages induced by endotoxin tolerance. Prostaglandins 1986; 31:639–650.
133. Zingarelli B, Halushka PV, Caputi AP, Cook JA. Increased nitric oxide synthesis during the development of endotoxin tolerance. Shock 1995; 3:102–108.
134. Flach R, Schade FU. Peritoneal macrophages from tolerant mice produce an inhibitor of tumor necrosis factor α synthesis and protect against endotoxin shock. J Endotox Res 1997; 4:241–250.
135. Majetschak M, Flach R, Henkamp T, Jennissen V, Obertacke U, Neudeck F, Schmit-Neuerburg KP, Schade FU. Regulation of whole blood TNF production upon endotoxin stimulation after severe blunt trauma. J Trauma 1997; 43:880–887.

51

Glucocorticoid Control of Endotoxin Responses

Richard Silverstein, Donald C. Johnson, and Mari Norimatsu
University of Kansas Medical Center, Kansas City, Kansas

GLUCOCORTICOID AND DEFENSE

Endogenous Glucocorticoids

Introduction

The observation that either adrenalectomy or hypophysectomy will markedly sensitize mice to the lethal effects of endotoxin, coupled with the observation that glucocorticoid administration to animals can confer a measure of protection against endotoxin, particularly when given with a pretreatment period, has long been used in support of the concept that glucocorticoids are important to host defense against endotoxin. This subject has been reviewed in detail (1) and will be summarized in this chapter.

The adrenalectomized or hypophysectomized mouse model has recently been utilized within the framework of endogenous versus exogenous glucocorticoid protection against lipopolysaccharide (LPS), with particular focus on the ability of glucocorticoid to act as an anti-TNF agent. The steroid receptor antagonist RU 486 (Roussel) and related compounds, although not limited exclusively to glucocorticosteroid antagonism, also serve as a chemical alternative to adrenalectomy or hypophysectomy. These studies have been used to investigate the endocrine system's role in natural host defense and defense derived from early endotoxin tolerance.

The Adrenalectomized or Hypophysectomized Mouse Model

It has been shown (2) that subcutaneous administration of naturally occurring glucocorticoid and/or mineralocorticoid, namely corticosterone and deoxycorticosterone, even in large doses and for 10 days—and in contrast to the synthetic glucocorticoid dexamethasone (3)—is unable to protect adrenalectomized mice from doses of LPS as low as 20 μg, which is a full order of magnitude less than the LD_{50} for LPS in intact mice. To underscore the potential importance of such differences between natural steroid and dexamethasone, it should be noted that dexamethasone protects adrenalectomized mice in the D-galactosamine model, even though endogenous glucocorticoids again appear ineffective in defense. This latter observation may be inferred from the fact that, in the D-galactosamine model, adrenalectomized mice are no more sensitive to LPS lethality than are D-galactosamine–treated mice with intact adrenals (3).

It is well established that glucocorticoid treatment can effectively downregulate the release of TNF from the LPS-stimulated macrophage, as previously reviewed (1). It could be hypothesized that deficient endogenous glucocorticoid production following adrenalectomy might bring about an even higher LPS-derived TNF response than would otherwise occur in normal mice. Parant et al. (4) have indeed shown that peak serum TNF levels in LPS-challenged adrenalectomized mice, even at one-third the LPS dose, are fivefold higher than among corresponding normal mice. In both groups, TNF levels peaked between 1.5 and 2 hours into the challenge. It is important to note that it is not simply the fact that the mice per se were hypersensitive to LPS that elicited the increase in TNF. The same authors demonstrated that treatment with D-galactosamine did not influence TNF levels, even though such mice were even more sensitive to the lethal effects of the LPS. In an independent study, it was shown that marked increases in TNF were also evident in mice that had previously been hypophysectomized (5).

RU 486

Studies with RU 486 with intact mice have tended to reproduce the above findings with adrenalectomized or hypophysectomized mice, underscoring the likelihood of glucocorticoid involvement. First, LPS lethal dose-response profiles for adrenalectomized versus RU 486–treated mice are similar to each other and quite different from the lethal LPS dose response in normal mice (2). Second, as in adrenalectomized or hypophysectomized mice, RU 486 mediates an elevation in circulating TNF levels in LPS-challenged mice. In that regard, McCallum et al. (6) not only reported an enhanced TNF response, but also noted that the RU 486 treatment abolished the early hyperglycemic effect that precedes classic LPS-driven hypoglycemia. The same laboratory had earlier reported a similar observation with adrenalectomized mice (7). Lazar et al. (8) also reported enhanced production of TNF in LPS-challenged mice following RU 38486 treatment, as evident in serum, liver, spleen. RU 38486 also increased TNF cytotoxicity, as assessed with actinomycin D–sensitized cultured fibroblasts.

Lazar et al. (9) found that, in the cecal ligation and puncture model, RU 38486 administered intravenously, at the same time as puncture, resulted in sensitization, reducing survival from ensuing septic peritonitis from 75 to 15%. RU 486 also reduced survival in mice that had been previously injected with LPS to induce early endotoxin tolerance (see below). Hawes et al. (10) showed protection by corticosterone against RU 486 sensitization using a single jugular catheter to infuse various agents into Lewis rats. RU 486 pretreatment significantly increased lethal sensitivity to LPS, whereas concomitant pretreatment with corticosterone via the same catheter reversed the effect while attenuating circulating TNF and IL-6 levels. It is conceivable, based upon these and other studies (2), that the mode of glucocorticoid administration, as well as time of administration relative to that of loss or attenuation of endogenous glucocorticoid capability, may prove critical. With regard to the latter, concern with respect to the diurnal circadian rhythm associated with glucocorticoid levels as an important experimental parameter in addressing septic shock has recently been reemphasized (11,12).

Glucocorticoids in Early Endotoxin Tolerance

Introduction

It is well known that small amounts of endotoxin can induce tolerance toward a subsequent larger amount of endotoxin even if the intervening time frame is sufficiently short so as not to allow antibody formation. This phenomenon, commonly termed early endotoxin tolerance, is not an example of endogenous host defense as the LPS is exogenously introduced. Nevertheless, it may be conceived as beginning during any endotoxin challenge, even if its impact may not necessarily become manifest in the course of sepsis. It would therefore seem important in the conceptual development of therapeutic intervention strategies against LPS, and, by extrapolation, against gram-negative sepsis, to understand the properties and underlying mechanisms associated with such tolerance. As previously reviewed (1), it is well recognized that LPS challenge elicits an increase in serum glucocorticoid, and thus it is not unreasonable to expect that early endotoxin tolerance may, at least in part, be glucocorticoid mediated.

The Adrenalectomized or Hypophysectomized Mouse Model

Evans and Zuckerman (13) found that when intact mice are injected with 10 μg of LPS followed by a second 10 μg injection of LPS 20 hours later, the second injection, unlike the first, does not lead to a second increase in serum TNF, indicative of tolerance. By contrast, when the same experiment is conducted with adrenalectomized mice, not only does the second TNF peak appear, but it is at least as large as the first, indicating that adrenalectomy had the effect of abrogating tolerance. Moreover, when adrenalectomized mice were pretreated with dexamethasone, tolerance was restored in a dose-dependent manner. It was also observed in this study that a relatively low-dose D-galactosamine model (6 mg/mouse) also had the effect of abrogating tolerance by the same criterion, even though circulating glucocorticoid levels were unchanged by the D-galactosamine treatment. The authors concluded, therefore, that there were both glucocorticoid-dependent and -independent mechanisms involved.

In this regard, our laboratory has tested whether or not adrenalectomy or hypophysectomy abrogates early endotoxin tolerance in the murine D-galactosamine lethality model (20 mg/mouse) (14). When 0.02 μg LPS given to intact mice 90 minutes prior to 0.10 μg LPS plus D-galactosamine, early endotoxin tolerance could be demonstrated and sustained. When, however, the

Table 1 Effects of Hypophysectomy or Adrenalectomy in D-Galactosamine–Treated Mice

Experiment	Surgical treatment	Pretreatment (−1.5 hr)	Challenge D-Gal N model	Lethality (deaths/total) 8 hr	24 hr
1	Sham	Vehicle	0.1 μg LPS	4/5*	5/5*
	Sham	0.02 μg LPS	0.1 μg LPS	0/6	0/6
	HYPOX	Vehicle	0.1 μg LPS	6/6*	6/6
	HYPOX	0.02 μg LPS	0.1 μg LPS	0/6	5/6
	HYPOX	0.02 μg LPS	PBS only	0/6	0/6
2	ADX	Vehicle	0.1 μg LPS	6/12*	12/12
	ADX	0.02 μg LPS	0.1 μg LPS	1/8	7/8
	ADX	0.02 μg LPS	PBS only	0/8	1/8
3	ADX	Vehicle	0.1 μg LPS	8/12*	11/12
	ADX	0.02 μg LPS	0.1 μg LPS	0/9	7/9
	ADX	0.02 μg LPS	PBS only	0/9	0/9

Mice were CF1 females, 13–15 weeks, from Charles River Labs. D-GalN, administered at a dose of 20 mg/mouse, was from Sigma Chemical Co. and was given at the time of challenge with LPS, *Salmonella abortus equi*, in freshly prepared PBS. All injections were i.p.
*$p < 0.05$ compared with each corresponding LPS-pretreated (−1.5 hr) group.

same experiment was conducted with either adrenalectomized or hypophysectomized mice, early endotoxin tolerance could not be sustained (Table 1). These data are consistent with a measure of glucocorticoid involvement. Nevertheless, lethality was significantly delayed in these mice (Table 1), suggesting a glucocorticoid-independent component to the tolerance as well. Thus, using a rather different approach than in the above study, with differences in lethality as the criterion for tolerance, a qualitatively similar conclusion emerged. Glucocorticoid appears to be implicated in early endotoxin tolerance, but not as its sole mediator.

RU 486

There has long been an interest in LPS analogs as vehicles to induce tolerance. Qureshi et al. (15) examined the nontoxic lipid A from *Rhodobacter sphaeroides* (RsDPLA) within the framework of glucocorticoid mediation. In a particularly interesting finding, it has been shown that the RsDPLA-stimulated increase in serum ACTH and corticosterone was lower in overall magnitude but occurred much earlier than the corresponding increases elicited by the toxic lipid A analog SDZ MRL 953 or by LPS itself. RU 486 was shown to completely abolish tolerance indicating the importance of this early rise in glucocorticoid, which occurs within 5 minutes after injection and is sustained for 2 hours (15). It would seem reasonable that timing, and especially very early timing, may be particularly important in manifestations of early endotoxin tolerance.

Exogenous Glucocorticoid Administration

Introduction

Much of the clinical thrust in attacking sepsis has focused on attacking two components, LPS and TNF. Thus, if glucocorticoid is to be considered as one component of any multifaceted approach to address bacterial sepsis, it may prove critical to systematically consider the LPS-TNF relationship, i.e., within different experimental systems. There is ample evidence that exogenous and endogenous glucocorticoid have the capacity to downregulate LPS-derived production of TNF. This has been reported to occur at the transcriptional and, to a greater extent, the posttranscriptional levels (17).

LPS and TNF

As previously reviewed (18), the D-galactosamine model, as developed by Galanos and coworkers, has been shown to acutely sensitize to the lethal effects of TNF and, as a consequence of this sensitivity, to the lethal effects of LPS. Thus, in a model of bacterial sepsis that involves LPS and/or TNF, among additional components contributing to lethality, treatment with a sensitizing dose of D-galactosamine may be expected, other things being equal, to expand the importance of TNF-mediated lethal events. Correspondingly, D-galactosamine sensitization might conceivably enhance the capacity of glucocorticoid (through downregulation of TNF production) to protect were LPS the source of the TNF contribution. If, however, LPS was not the source of TNF-mediated lethality, then it is not at all clear why glucocorticoid should necessarily protect in the D-galactosamine model. Indeed, Galanos and Freudenberg (19) have investigated this point and found that dexamethasone given at the time of challenge will protect against the lethal effects of challenge from endotoxin, but not from direct challenge with TNF!

We have confirmed that dexamethasone given at the time of challenge protects against LPS (3) but not against TNF (Table 2); in addition, dexamethasone given under the same conditions will protect adrenalectomized mice against TNF (Table 2; Figure 1) (R. Silverstein, D. C. Johnson, and D. C. Morrison, unpublished results). Consistent with this last finding, Broukaert et al. (20) showed that RU 38486 also sensitizes to lethal challenge from TNF. Thus, it remains possible that in bacterial sepsis models, dexamethasone might conceivably protect—independently—against both LPS and TNF. In the case of the D-galactosamine model, however, dexamethasone may be expected to protect against LPS but not directly against TNF challenge.

Experimental Bacterial Sepsis

For the reasons outlined earlier, dexamethasone might be expected to protect against LPS and, by extension, against gram-negative bacterial sepsis, especially in the D-galactosamine model. Likewise, it would not be expected to protect against sepsis from gram-positive bacteria, which lack LPS. Our laboratory (21) recently tested this concept using *Staphylococcus aureus* and *Escherichia coli* as prototype gram-positive and gram-negative bacteria, respectively. Dexamethasone at a dose of 100 µg given at the time of challenge was unable to protect against acute challenge from the live gram-positive bacteria in both normal and D-galactosamine–treated mouse models. Moreover, D-galactosamine elicited little or no sensitization to the challenge, consistent with a lesser importance of TNF-mediated lethality in live *S. aureus* bacterial challenge models.

By contrast, D-galactosamine sensitized to gram-negative lethal infection by 4 orders of magnitude. This degree of sensitization is comparable to that seen with LPS purified from the same bacterial strain of *E. coli*. In addition, dexamethasone, again at a dose of 100 µg given at the time of challenge, was able to protect against the gram-negative lethal bacterial infection in both normal and D-galactosamine–treated mice. In the latter model, the LD_{50} was raised by 2 orders of magnitude.

In a follow-up study, bacterial isolates prepared from each of 19 clinical blood, urine, and laboratory specimens were injected into both normal and D-galactosamine–treated mice, and dexamethasone protection was assessed. As above, differences in the ability of D-galactosamine to sensitize became evident in relationship to gram-positive versus gram-negative

Table 2 Effect of Dexamethasone on TNF Lethality in C3H/HeJ Mice

		Lethality (deaths/total)	
	TNF (µg)	Saline	Dexamethasone, 400 µg
D-Galactosamine model	0.3	3/6	1/4
	0.6	3/4	3/4
	1.2	5/6	4/4
Adrenalectomy model			
Expt. 1	3.0	2/7	1/6
	12.0	5/8	0/6*
Expt. 2	15.0	5/7	0/7*

Mice were females, 8 weeks, from Jackson Labs. D-GalN, administered at a dose of 20 mg/mouse, was from Sigma Chemical Co. and was given at the time of challenge. TNF was generously provided by Philips Petroleum Co. (Bartlesville, OK). Dexamethasone was given in saline immediately preceding challenge.
*$p < 0.05$.

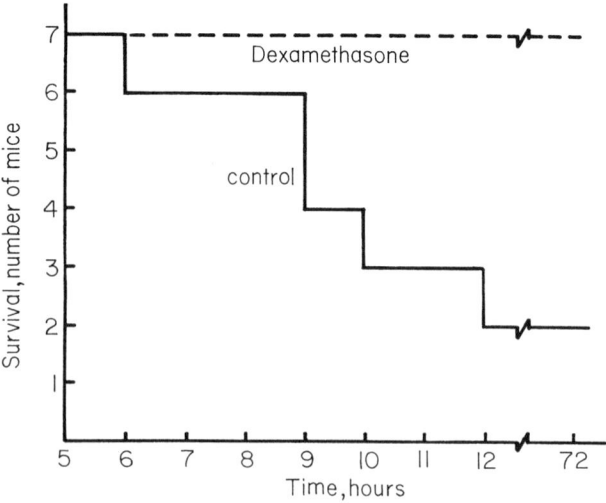

Fig. 1 Dexamethasone protection against TNF in adrenalectomized mice. Data correspond to Table 1, Experiment 1. CF1 adrenalectomized mice were given 400 µg dexamethasone at the time of challenge with 15 µg TNF (seven mice per group). For conditions of adrenalectomy, see Ref. 3.

infection. Moreover, the extent of dexamethasone protection was found to be directly proportional to the extent of D-galactosamine sensitization (22); dexamethasone remained (with one exception) ineffective against gram-positive infection.

Antibiotics

Clinical trials with glucocorticoid have failed to significantly reduce mortality from sepsis, as previously reviewed (23). These trials, conducted over several years, did include antibiotic treatment, but their application was not, in general, rigorously controlled. In particular, differences in the ability of antibiotics to mediate endotoxin release was not a consideration in the outcome analysis. More recently, Bucklin and Morrison (24) have shown that when sensitivity to endotoxin is amplified by D-galactosamine, imipenem (a carbapenem with minimal endotoxin release potential in vitro) is more effective than ceftazidime (a cephalosporin with marked endotoxin release effects) following an *E. coli* bacterial challenge.

The phenomenon of LPS release as a consequence of gram-negative bacterial killing is reviewed in Chapter 5 of this volume. This phenomenon may be an important consideration in assessing glucocorticoid effectiveness in protection against bacterial sepsis. In that regard, we have compared the ability of dexamethasone to protect against gram-negative infection when imipenem (lower LPS release) and ceftazidime (higher LPS release) are employed at otherwise comparable protective efficacy (Norimatsu M, Morrison DC, Silverstein R, unpublished results). As shown in Table 3, dexamethasone is more effective at protection when coupled with ceftazidime than with imipenem in this defined lethality model. It would seem reasonable, therefore, to conclude that (1) as LPS becomes increasingly important as a lethal component to sepsis, the ability of dexamethasone to confer protection will correspondingly rise, and (2) differences in antibiotic-mediated LPS release may prove important within that setting.

Conclusion

Studies with several mouse-sensitization models (adrenalectomy, hypophysectomy, RU 486, and D-galactosamine) support the concept that endogenous glucocorticoids are vitally important to general host defense against endotoxin-mediated lethality. Recent studies have also consistently demonstrated that glucocorticoids are important as an essential contributing factor to early endotoxin tolerance.

Table 3 Extent of Dexamethasone Protection Against Live *Escherichia coli* Lethality in D-Galactosamine-Treated Mice[a]

Protecting agent[b]	Extent of protection[c]	Theoretical[d]
Ceftazidime	30-fold	
plus DEX	7900-fold	5400-fold
Imipenem	30-fold	
plus DEX	1700-fold	5400-fold

[a]Mice were CF1 females, 9–11 weeks, from Charles River Labs. D-GalN, administered at a dose of 20 mg/mouse, was from Sigma Chemical Co. and was given at the time of challenge with *E. coli* 0111:B4, from List Labs, mid-log phase.
[b]Ceftazidime, from Glaxo, and imipenem-cilastatin, Merck Chemical Co., were administered as previously described (25) and given at doses of 1000 and 500 μg, respectively, to allow for comparable extents of protection. 100 μg dexamethasone was given in saline immediately preceding challenge.
[c]Based on LD_{50} determinations, using the method of Reed and Muench (26). Minimum of 8 mice per datum.
[d]Calculated assuming that antibiotic and dexamethasone protection are mutually independent. Dexamethasone (DEX) protection in the absence of antibiotic was determined to be 180-fold, calculated based on the method of Reed and Muench (26; see footnote immediately above).

In addition, the D-galactosamine model, because of its remarkable capacity to hypersensitize mice to TNF and consequently to amplify that component's contribution to lethal endotoxin and gram-negative septic shock, has been applied to test the effectiveness of dexamethasone under such extreme conditions. In the D-galactosamine model, dexamethasone, given at the time of challenge, has been shown to protect against endotoxin and gram-negative, but not gram-positive, bacterial sepsis. Further, the degree of that protection may be expected, and confirmed experimentally, to be influenced by the nature of accompanying antibiotic treatment.

Extrapolation of dexamethasone protection from LPS-mediated lethality in D-galactosamine–sensitized mice to other systems must remain a major concern. Nevertheless, the demonstration of protection against live experimental gram-negative bacterial challenge and, in particular, following antibiotic treatment provides a measure of anticipation that as more is learned of the mechanistic details associated with sepsis pathophysiology, glucocorticoids may yet be reevaluated in future therapeutic regimens. In that regard, glucocorticoids exert a number of important effects both metabolically and immunologically, some having been uncovered only within the past few years. Such regulatory effects may well impact on endotoxicity, with a complexity just now starting to be defined. It is to this exciting subject that attention is now directed.

GLUCOCORTICOID MECHANISMS OF ACTION

Arachidonic Acid Metabolism

Phospholipase A_2

The well-recognized ability of glucocorticoids to inhibit phospholipase A_2 (PLA_2) activity, the first step of arachidonic acid metabolism from phospholipid, may conceivably be important in the regulation of both the cyclooxygenase and lipoxygenase pathways. Serum PLA_2 Group II activity has been shown to continue to rise for at least 8 hours in a rat model of endotoxin shock. mRNA levels of PLA_2 increase in several tissues, including lung and thymus, but not in liver or kidney. Dexamethasone, given 30 minutes prior to endotoxin challenge, completely suppressed the increase in PLA_2 mRNA (27). It is noteworthy that the increase in circulating PLA_2 activity occurs despite a corresponding increase in circulating endogenous corticosteroid levels that also accompanies endotoxin shock. The hypercortisonemia is, moreover, not without clinical impact, as suggested by studies of the acute-phase response in human volunteers (28).

With respect to the cyclooxygenase pathway that follows PLA_2 action, there continues to be controversy as to the beneficial and/or detrimental effects associated with prostaglandin action and, in particular, of PGE_2 (1,29). There is considerable interest as to whether glucocorticoid may modulate prostaglandin synthesis not only at the PLA_2 level, but also, more directly, at the level of cyclooxygenase action.

Cyclooxygenase

Masferrer et al. (30–32) recently reported a cyclooxygenase (COX-2) in murine peritoneal macrophages, which, unlike the constitutive isoform (COX-1), is selectively induced two- to threefold in adrenalectomized mice, and an additional fourfold following stimulation by LPS. Dexamethasone, given at the time of challenge and under conditions such that it protects against LPS lethality, blocked the induction while not affecting COX-1 levels.

It is conceivable that differences in the modulation of COX-1 and COX-2 may, in part, contribute to the complexity associated with effects of the non-steroidal antiinflammatory drug (NSAID) indomethacin. Thus, it has been observed that indomethacin inhibits the "early" (1 hr) release of prostaglandins following LPS stimulation of mice that appears to be associated with COX-1, whereas dexamethasone inhibits the "delayed" release (2 hr) associated with COX-2 inhibition (33,34). In that regard, it had previously been reported (1) that indomethacin can protect mice against the lethal effects of an acute challenge from TNF. Nevertheless, as shown recently in a murine model of live *Pseudomonas aeruginosa* infection, indomethacin increased lethality in this experimental sepsis model and did so with an accompanying increase in TNF expression which peaked 2–4 hours after bacterial challenge (35).

It is interesting that dexamethasone inhibits the induction of both COX-2 and inducible nitric oxide synthase (iNOS). To what extent these two inducible enzymes are linked to each other in modulating endotoxicity is not year clear. It has, however, been shown that, in studies with LPS-stimulated RAW cells, the well-recognized iNOS inhibitors N^G-monomethyl-L-arginine or aminoguanidine also blocked production of PGE_2 in response to LPS (36,37).

Evidence was summarized earlier in this section that if indomethacin, alternatively, is to be protective or harmful against bacterial, endotoxin, or direct cytokine challenge, the outcome may depend, in part, on the degree of acuteness associated with the particular challenge. In that regard it has also been reported that when TNF is administered to mice not as a single lethal bolus injection, but as a series of injections, under such circumstances indomethacin, rather than being protective, has the effect of blocking tolerance (38).

In human volunteers pretreated with another NSAID, ibuprofen, the administration of endotoxin intravenously resulted in an increase in serum TNF that was heightened by the ibuprofen treatment (39). Finally, IL-1 has been shown to induce COX-2 mRNA in human rheumatoid synovium microvessel endothelial cells, while dexamethasone was able to block this effect (40).

Macrophage-Derived Glucocorticoid Antagonism: Macrophage Migration Inhibitory Factor

Berry and coworkers initially identified one or more partially purified protein factors from the LPS-stimulated macrophage that was shown to antagonize aspects of glucocorticoid function (reviewed in Ref. 1). Studies of this "glucocorticoid antagonizing factor" revealed that it affected a variety of activities such as antagonism of glucocorticoid induction of phosphoenolpyruvate carboxykinase within the hepatocyte.

More recently, Calandra and Bucala et al. (41–45) identified and characterized another protein that is released from the LPS-stimulated macrophage and antagonizes glucocorticoid action at the cellular level. Remarkably, this factor is itself released in response to physiological concentrations of glucocorticoid, and hence it has been termed a "counterregulator" of glucocorticoid action. This protein, also derived from the anterior pituitary, had in fact been studied for two decades in another context and termed macrophage migration inhibitory factor (MIF).

MIF injection was found to enhance LPS lethality in mice. Correspondingly, anti-MIF antibody was able to protect against LPS lethality even in the absence of added MIF (45). MIF can be detected in the serum at 2 hours following LPS challenge (44). Perhaps of importance, it is not as yet clear which tissues or cells are the prime source of this early increase in circulating MIF. In that regard, LPS stimulated RAW 264.7 macrophages to release MIF which was first detected 6–9 hours after LPS stimulation, and peaked 3–6 hours later (44).

MIF was found to potentiate LPS lethality in mice even in the presence of otherwise protective doses of dexamethasone (1.25 mg/kg, 25 μg, or 6.7×10^{-8} moles). Correspondingly, MIF was shown to override glucocorticoid inhibition of cytokine release from LPS-stimulated human monocytes. Dexamethasone was unable to block the LPS-potentiating effect of MIF. Furthermore, as demonstrated in RAW 264.7 macrophages, dexamethasone actually induced the secretion of MIF. This became manifest upon incubation of the macrophages with either dexamethasone or hydrocortisone at concentrations of 10^{-12}–10^{-10} M. MIF release was not evident at lower—or higher—glucocorticoid concentrations, with the profile reported as bell-shaped. Finally, in what may prove to be seminal experiments toward our understanding of differential aspects in glucocorticoid control of endotoxin, MIF was first detectable in the culture medium 3 hours after stimulation of the macrophages with glucocorticoid (43).

Glucocorticoid Receptor and Postreceptor Regulation

LPS, IFN-γ, and Glucocorticoid Receptor Expression

It is well recognized that one of the body's responses to LPS is to increase circulating levels of glucocorticoids, while another is the release of factors, such as MIF, that antagonize glucocorticoid action. Further levels of complexity occur at the cell surface where other macrophage products may also enhance glucocorticoid responsiveness of cells. Vogel and coworkers have been studying the balance of parameters that modulate macrophage responsiveness in a number of different respects (46,47), including the effects of IFN-γ and LPS on glucocorticoid receptor properties and number (48,49).

IFN-γ, which is well recognized as a potential contributor to the lethal effects of LPS (50), was found to significantly increase both the number and affinity of glucocorticoid receptors in RAW 264.7 macrophages, with a doubling by 24 hours and a maximum increase of fourfold observed at 36 hours. Inherent affinity of these macrophages was also enhanced twofold by the IFN-γ treatment. Isolated peritoneal macrophages also showed increased numbers of glucocorticoid receptors following IFN-γ stimulation in vitro. Similar studies with LPS-stimulated macrophages also showed an increase in receptor number. Stimulation of peritoneal macrophages with 10 ng/mg LPS showed a significant increase in glucocorticoid receptor number by 4 hours, was maximal at 12 hours, and remained higher through 48 hours (49).

Heat Shock

IFN-γ induces the expression of FcγI, II, and IIIα receptors for IgG on macrophages and as part of the immune response. Glucocorticoids modulate these three inductions differentially and in opposing directions. Interestingly, heat shock mimics the complexity of that response (51). Ubiquitin-mediated proteolysis is associated with heat shock, and it has also been shown that ubiquitin-mediated muscle proteolysis may be associated with sepsis (52). Further, RU 486 has been shown to inhibit sepsis-induced muscle proteolysis. To what extent glucocorticoid modulatory effects that are also inherently associated with heat shock contribute to the control of endotoxin remains to be established.

Regulation of NF-κB via IκBα

Nuclear factor (NF)-κB is important to the expression of a number of cytokines intimately involved in the manifestations of endotoxin action, including TNF (see Chapter 34). It therefore becomes of considerable interest to know whether glucocorticoids can downregulate that response, and by what mechanism(s). It is well established that NF-κB is maintained in the cell cytosol bound to one or more inhibitory proteins (IκB).

When IκB is phosphorylated and/or degraded, NF-κB is then allowed to migrate to the nucleus to function.

It has been shown that glucocorticoid can inhibit NF-κB translocation to the nucleus through increased transcription and subsequent synthesis of one of these inhibitory proteins, specifically IκBα (53,54). It was further shown that direct interaction of the glucocorticoid receptor with NF-κB is not a requirement for this glucocorticoid-mediated inhibitory effect. While these studies indicate an IκBα-mediated mechanism with monocytes and T lymphocytes, glucocorticoids also mediate repression of NF-κB activity in endothelial cells, but by a mechanism that does not involve IκBα synthesis (55).

One of the important pathological manifests in the vascular system that is associated with endotoxin is leukocyte-endothelial cell adhesion and its associated properties, including leukocyte rolling (56). Recently, it has been shown that dexamethasone inhibits the expression of E-selectin (ELAM-1) on endothelial cells. This glucocorticoid action is also mediated by inhibition of NF-κB nuclear translocation (57). ELAM-1 is an important regulator of the adhesive activity of CD11/CD18 molecules on neutrophils (58).

Mechanisms in Early Endotoxin Tolerance

Vane and coworkers (59,60) studied endotoxin tolerance in Wistar rats by monitoring endotoxin-induced hypotensive properties, i.e., decrease in mean arterial pressure and hyporeactivity to norepinephrine ex vivo. They provided evidence that glucocorticoid-dependent early endotoxin tolerance includes an ability to downregulate iNOS synthase via lipocortin 1. Importantly, N^G-nitro-L-arginine methyl ester partially reversed norepinephrine hyporeactivity. Further, RU 486 had the effect of reversing early endotoxin tolerance in rats already tolerant and, as monitored by the above pathophysiological parameters, as well as iNOS induction. As the authors note, pretreatment of rats (61) or mice (62) with anti-TNF antibody or IL-1 receptor antagonist (63) also downregulates iNOS synthase.

A recent report indicates possible involvement of NF-κB in early endotoxin tolerance at the level of IκB kinase (64). This enzyme is responsible for the hyperphosphorylation of IκB that precedes its degradation and subsequent translocation of the released NF-κB to the nucleus. This kinase was specifically absent in mice in which early endotoxin tolerance was manifest. Moreover, dexamethasone did not inhibit its induction. The latter observation would seem to indicate that NF-κB participation in early endotoxin tolerance does not of itself require glucocorticoid mediation. There is also evidence that a component of early endotoxin tolerance is NF-κB independent. Haas et al. (65) showed with human Mono-Mac-6 cells that PGE_2 and phorbol 12-myristate 13-acetate can synergize to mimic early endotoxin tolerance in terms of LPS-stimulated TNF release. Even in these desensitized cells, however, NF-κB was still induced by LPS in the nucleus.

Conclusion

The discovery of COX-2 led to the recognition that glucocorticoids have the capacity to modulate cyclooxygenase directly as well as indirectly. It is also clear that cyclooxygenase inhibitors must be further subclassified in terms of COX-1 and COX-2 modulation. Indomethacin, for example, modulates both isoforms, whereas the direct effects of dexamethasone appear to be specific to COX-2. These differences may be expected to be reflected in terms of the time course of events following COX-2 induction and in relation to others mediators in the endotoxin cascade, particularly iNOS. With respect to the importance of timing, in general, it is noteworthy that glucocorticoid elevation appeared as early as 5 minutes following treatment with RsDPLA, and Remick et al. (66) have observed that TNF mRNA appears in peritoneal cells as early as 30 minutes following challenge of mice with LPS. Doses as high as 4 mg/kg of dexamethasone are no longer effective at TNF downregulation if treatment is delayed by only 20 minutes following the LPS challenge.

The "rediscovery" of MIF and its relationship to glucocorticoid should more strictly be considered under the umbrella of endotoxin control of glucocorticoid, rather than the reverse. With respect to the systematic multifaceted targeting of sepsis, control of MIF may prove particularly helpful, inasmuch as harnessing the actions of MIF may potentiate endogenous regulatory substances such as glucocorticoids.

Studies with RU 486 have shown that antagonism of glucocorticoid receptor action can be tantamount to blocking their pituitary/adrenal source, as by hypophysectomy or adrenalectomy. Conversely, the now-recognized ability of endotoxin and IFN-γ to enhance glucocorticoid receptor number and binding raises the intriguing possibility that such enhancement can prove beneficial physiologically. However, it is conceivable that enhancement of endogenous glucocorticoid activity may also prove counterproductive if it leads in turn to undesirably increased MIF activity.

As the properties of glucocorticoids in relation to endotoxin continue to be clarified with respect to their inherent complexity, the same may also prove true for glucocorticoids in relation to experimental sepsis. Within the D-galactosamine model of acutely enhanced sensitivity to LPS and TNF, dexamethasone, given at the time of challenge, has significantly different effects depending on the identity of the invading bacteria, with strong correlation between gram-negative (protective action) versus gram-positive (ineffective) groups. These studies underscore the potential importance of early identification of the invading bacteria before immunomodulatory treatment strategies are initiated. These studies also indicate that pretreatment may not be an absolute in order for synthetic glucocorticoids to prove beneficial. In that regard, we have shown in a normal mouse model that administration of 400 μg of dexamethasone as late as 2 hours after challenge with LPS was still protective against its lethal effects (9/20 vs. 1/20 deaths; $p < 0.005$).

Finally, it has been said that in research it is more important to ask the right questions than to accumulate "answers." The past 5 years have perhaps been unusual in that a number of very cogent and different studies have raised a number of provocative, if complex, questions with respect to glucocorticoid control of endotoxin—and the reverse.

ACKNOWLEDGMENT

The author wishes to acknowledge support from an unrestricted grant provided by Merck & Co to David C. Morrison, PI.

REFERENCES

1. Silverstein R. The endocrine response to endotoxin. In: Ryan JL, Morrison DC, eds. Bacterial Endotoxic Lipopolysaccharides, Vol II. Boca Raton, FL: CRC Press, 1992:295–309.
2. Silverstein R, Hannah P, Johnson DC. Natural adrenocorticoids do not restore resistance to endotoxin in the adrenalectomized mouse. Circ Shock 1993; 41:162–165.
3. Gonzalez JC, Johnson DC, Morrison DC, Freudenberg MA, Galanos C, Silverstein R. Endogenous and exogenous glucocorticoids have different roles in modulating endotoxin lethality in D-galactosamine-sensitized mice. Infect Immun 1993; 61:970–974.
4. Parant M, Le Contel C, Parant F, Chedid L. Influence of endogenous glucocorticoid on endotoxin-induced production of circulating TNF-α. Lymphokine Cytokine Res 1991; 10:265–271.
5. Johnson DC, Freudenberg MA, Jia F, Gonzalez JC, Galanos C, Morrison DC, and Silverstein R. Contribution of tumor necrosis factor-α and glucocorticoid in hydrazine sulfate-mediated protection against endotoxin lethality. Circ Shock 1994; 43:1–8.
6. McCallum RE, Lloyd SS, Hyde SR. Enhanced TNF response in endotoxic mice after treatment with the glucocorticoid antagonist RU 486. In: Nowotny A, Spitzer JJ, Ziegler EJ, eds. Cellular and Molecular Aspects of Endotoxin Reactions. Amsterdam: Elsevier, 1990:493–500.
7. McCallum RE, Seale TW, Stith RD. Influence of endotoxin treatment on dexamethasone induction of hepatic phosphoenolpyruvate carboxykinase. Infect Immun 1983; 39:213–219.
8. Lazar G, Jr, Duda E, Lazar G. Effect of RU 38486 on TNF production and toxicity. FEBS Lett 1992; 2:137–140.
9. Lazar G, Jr., Lazar G, Agarwal MK. Modification of septic shock in mice by the antiglucocorticoid RU 38486. Circ Shock 1992; 36:180–184.
10. Hawes AS, Rock CS, Keogh CV, Lowry SF, Calvano SE. In vivo effects of the antiglucocorticoid RU 486 on glucocorticoid and cytokine responses to *Escherichia coli* endotoxin. Infect Immun 1992; 60:2641–2647.
11. Hrushesky WJM, Wood PA. Circadian time structure of septic shock; timing is everything [letter]. J Infect Dis 1997; 175:1283.
12. Pollmächer T, Mullington J, Holsboer F, Galanos C. Circadian time structure of septic shock; timing is everything [reply]. J Infect Dis 1997; 175:1284.
13. Evans GF, Zuckerman SH. Glucocorticoid-dependent and -independent mechanisms involved in lipopolysaccharide tolerance. Eur J Immunol 1991; 21:1973–1979.
14. Silverstein R, Johnson DC, Galanos C, Morrison DC, unpublished results.
15. Qureshi N, Jarvis B, Takayama K, Sattar N, Hofman J, Stütz P. Natural and synthetic analogs or partial structures that antagonize or induce tolerance to LPS. 4th International Endotoxin Society Conference, Nagoya, Japan, Oct. 22–25, 1996.
17. Han J, Brown T, Beutler B. Endotoxin-responsive sequences control cachectin/tumor necrosis factor biosynthesis at the translational level. J Exp Med 1990; 171:465–475.
18. Galanos C, Freudenberg MA, Katschinski T, Salomao R, Mossmann H, Kumazawa Y. Tumor necrosis factor and host response to endotoxin. In: Ryan JL, Morrison DC, eds. Bacterial Endotoxic Lipopolysaccharides, Vol II. Boca Raton, FL: CRC Press, 1992:75–104.
19. Galanos C, Freudenberg MA. Tumor necrosis factor mediates endotoxin shock: the protective effects of antibodies and cortisone. In: Bonavida B, Granger G, eds. Tumor Necrosis Factor: Structure, Mechanism of Action, Role in Disease and Therapy. Basel: Karger, 1990: 187–193.
20. Broukaert P, Ameloot P, Cauwells P, Everaerdt B, Libert C, Takahashi N, Van Molle W, Fiers W. The glucocorticoid antagonist RU38486 mimics interleukin-1 in its sensitization to the lethal and interleukin-6-inducing properties of tumor necrosis factor. Eur J Immunol 1992; 22:981–986.
21. Silverstein R, Norimatsu M, Morrison DC. Fundamental differences during gram-positive versus gram-negative sepsis become apparent during bacterial challenge

22. Silverstein R, Xue Q, Horvat RT, Lee CY, Luchi MJ, Sau S, Worley PA, Morrison DC. Different role for TNF-α in experimental sepsis caused by gram-positive vs. gram-negative clinical isolates. 37th Annual Interscience Conference on Antimicrobial Agents and Chemotherapy, Toronto, Sept 28–30, 1997.
23. Zeni F, Freeman B, Natanson C. Anti-inflammatory therapies to treat sepsis and septic shock: a reassessment. Crit Care Med 1997; 25:1095–1100.
24. Bucklin SE, Morrison DC. Differences in therapeutic efficacy among cell wall-active antibiotics in a mouse model of gram-negative sepsis. J Infect Dis 1995; 172:1519–1527.
26. Reed LJ, Muench HA. A simple method of estimating fifty percent endpoints. Am J Hygiene 1938; 27:493–497.
27. Nakano T, Arita H. Enhanced expression of group II phospholipase A_2 gene in the tissues of endotoxin shock rats and its suppression by glucocorticoid. FEBS Lett 1990; 273:23–26.
28. Rock CS, Coyle SM, Keogh V, Lazarus DD, Hawes AS, Leskiw M, Moldawer LL, Stein TP, Lowry SF. Influence of hypercortisolemia on the acute-phase response to endotoxin in humans. Surgery 1992; 112:467–474.
29. Schade UF, Engel R, Jakobs D, Kirikae T. Endotoxin-induced oxidative metabolism of unsaturated fatty acids. In: Ryan JL, Morrison DC, eds. Bacterial Endotoxin Lipopolysaccharides, Vol II. Boca Raton, FL: CRC Press, 1992:27–55.
30. Masferrer JL, Seibert K, Zweifel B, Needleman P. Endogenous glucocorticoids regulate an inducible cyclooxygenase enzyme. Proc Natl Acad Sci USA 1992; 89:3917–3921.
31. Mansferrer JL, Zweifel BS, Seibert K, Needleman P. Selective regulation of cellular cyclooxygenase by dexamethasone and endotoxin in mice. J Clin Invest 1990; 86:1375–1379.
32. Masferrer JL, Reddy ST, Zweifel BS, Seibert K. Needleman P, Gilbert RS, Herschman HR. In vivo glucocorticoids regulate cyclooxygenase-2 but not cyclooxygenase-1 in peritoneal macrophages. J Pharm Exp Ther 1994; 270:1340–1344.
33. Salvemini DS, Settle SL, Masferrer JL, Seibert K, Currie MG, Needleman P. Regulation of prostaglandin production by nitric oxide, an in vivo analysis. Br J Pharmacol 1995; 114:1171–1178.
34. Masferrer JL, Zweifel BS, Manning PT, Hauser SD, Leahy KM, Smoth WG, Isakson PC, Seibert K. Selective inhibition of inducible cyclooxygenase 2 in vivo is antiinflammatory and nonulcerogenic. Proc Natl Acad Sci USA 1994; 91:3228–3232.
35. Campanile F, Giampietri A, Grohmann U, Belladonna ML, Firoetti MC, Puccetti P. Evidence for tumor necrosis factor α as a mediator of the toxicity of a cyclooxygenase inhibitor in gram-negative sepsis. Eur J Pharmacol 1996; 307:191–199.
36. Salvemini D, Misko TP, Masferrer JL, Seibert K, Currie MG, Needleman P. Nitric oxide activates cyclooxygenase enzymes. Proc Natl Acad Sci 1993; 90:7240–7244.
37. Salvemini D, Manning PT, Zweifel BS, Seibert K, Connor J, Currie MG, Needleman P, Masferrer JL. Dual inhibition of nitric oxide and prostaglandin production contributes to the antiinflammatory properties of nitric oxide synthase inhibitors. J Clin Invest 1995; 96:301–308.
38. Takahashi N, Brickyard P, Fiers W. Cyclooxygenase inhibitors prevent the induction of tolerance to the toxic effects of tumor necrosis factor. J Immunotherapy 1993; 14:16–21.
39. Spinas GA, Bloesch D, Keller U, Zimmerli W, Cammisuli S. Pretreatment with ibuprofen augments circulating tumor necrosis factor-α, interleukin-6, and elastase during acute endotoxinemia. J Infect Dis 1991; 163:89–95.
40. Szczepanski A, Moatter T, Carley WW, Gerritsen ME. Induction of cyclooxygenase II in human synovial microvessel endothelial cells by interleukin-1. Arthritis Rheumatism 1994; 37:495–503.
41. Calandra T, Bucala R. Macrophage migration inhibitory factor (MIF): a glucocorticoid counter-regulator within the immune system. Crit Rev Immunol 1997; 17:77–88.
42. Bucala R. MIF rediscovered:cytokine, pituitary hormone, and glucocorticoid-induced regulator of the immune response. FASEB J 1996; 10:1607–1613.
43. Calandra T, Bernhagen J, Metz CN, Spiegel LA, Bacher M, Donnelly T, Cerami A, Bucala R. MIF as a glucocorticoid-induced modulator of cytokine production. Nature 1995; 377:68–71.
44. Calandra T, Bernhagen J, Mitchell RA, Bucala R. The macrophage is an important and previously unrecognized source of macrophage migration inhibitory factor. J Exp Med 1994; 179:1895–1901.
45. Bernhagen J, Calandra T, Mitchell RA, Martin SB, Tracey KJ, Voelter W, Manogue KR, Cerami A, Bucala R. MIF is a pituitary-derived cytokine that potentiates endotoxaemia. Nature 1993; 365:756–759.
46. Sivo J, Salkowski CA, Politis AD, Vogel SN. Differential regulation of LPS-induced IL-1 and IL-1 receptor antagonist mRNA by IFNα and IFNγ in murine peritoneal macrophages. J Endotox Res 1994; 1:30–37.
47. Sivo J, Politis AD, Vogel SN. Differential effects of interferon-γ and glucocorticoids on FcγR gene expression in murine macrophages.
48. Salkowski CA, Vogel SN. IFN-γ mediates increased glucocorticoid receptor expression in murine macrophages. J Immunol 1992; 148:2770–2777.
49. Salkowski CA, Vogel SN. Lipopolysaccharides increase glucocorticoid receptor expression in murine macrophages. J Immunol 1992; 149:4041–4047.
50. Bucklin SE, Russell SW, Morrison DC. Augmentation of anti-cytokine immunotherapy by combining neutralizing monoclonal antibodies to interferon-γ and the interferon-γ receptor: protection in endotoxin shock. J Endotox Res 1994; 1:45–51.
51. Sivo J, Harmon JM, Vogel SN. Heat shock mimics glucocorticoid effects on IFN-γ-induced FCR1 and Ia messenger RNA expression in mouse peritoneal macrophages. J Immunol 1996; 156:3450–3454.
52. Tiao G, Fagan J, Roegner V, Lieberman M, Wang J-J, Fischer JE, Hasselgren P-O. Energy-ubiquitin-depen-

dent muscle proteolysis during sepsis in rats is regulated by glucocorticoids. J Clin Invest 1996; 97:339–348.
53. Scheinman RI, Cogswell PC, Lofquist AK, Baldwin, Jr AS. Role of transcriptional activation of IκBα in mediation of immunosuppression by glucocorticoids. Science 1995; 270:283–286.
54. Auphan N, DiDonato J, Rosette C, Helmberg A, Karin M. Immunosuppression by glucocorticoids: inhibition of NF-κB activity through induction of IκB synthesis. Science 1995; 270:286–290.
55. Brostjan C, Anrather J, Csizmadia V, Stroka D, Soares M, Bach FH, Winkler H. Glucocorticoid-mediated repression of NFκB activity in endothelial cells does not involve IκB synthesis. J Biol Chem 1996; 271:19612–19616.
56. Harris NR, Russell JM, Granger DN. Mediators of endotoxin-induced leukocyte adhesion in mesenteric postcapillary venules. Circ Shock 1994; 43:155–160.
57. Brostjan C, Anrather J, Csizmadia V, Natarajan G, Winkler H. Glucocorticoids inhibit E-selectin expression by targeting NF-κB and not ATF/c-Jun. J Immunol 1997; 158:3836–3844.
58. Lo SK, Lee S, Ramos RA, Lobb R, Rosa M, Chi-Rosso G, Wright SD. Endothelial-leukocyte adhesion molecule 1 stimulates the adhesive activity of leukocyte integrin CR3 (CD11b/CD18, Mac-1, $\alpha_M\beta_2$) on human neutrophils. J Exp Med 1991; 173:1493–1500.
59. Szabó C, Thiemermann C, Wu C-C, Perretti M, Vane JR. Attenuation of the induction of nitric oxide synthase by endogenous glucocorticoids accounts for endotoxin tolerance in vivo. Proc Natl Acad Sci USA 1994; 91:271–275.
60. Wu C-C, Croxtall JD, Perretti M, Bryant CE, Thiemermann C, Flower RJ, Vane JR. Lipocortin 1 mediates the inhibition by dexamethasone of the induction by endotoxin of nitric oxide synthase in the rat. Proc Natl Acad Sci USA 1995; 92:3473–3477.
61. Thiemermann C, Wu C-C, Szabó C, Perretti M, Vane JR. Role of tumour necrosis factor in the induction of nitric oxide synthase in a rat model of endotoxic shock. Br J Pharmacol 1993; 110:177–182.
62. Green SJ, Chen TY, Crawford RM, Nacy CA, Morrison DC, Meltzer MS. Cytotoxic activity and production of toxic nitrogen oxides by macrophages treated with IFN-γ and monoclonal antibodies against the 73-kDa lipopolysaccharide receptor. J Immunol 1992; 149:2069–2075.
63. Szabó C, Wu C-C, Gross SS, Thiemermann C, Vane JR. Interleukin-1 contributes to the induction of nitric oxide synthase by endotoxin in vivo. Eur J Pharmacol 1993;
64. Kohler NG, Joly A. The involvement of an LPS inducible IκB kinase in endotoxin tolerance. Biochem Biophys Res Commun 1997; 232:602–607.
65. Haas JG, Baeuerle PA, Riethmüller G, Ziegler-Heitbrock HWL. Molecular mechanisms in down-regulation of tumor necrosis factor expression. Proc Natl Acad Sci USA 1990; 87:9563–9567.
66. Remick DG, Strieter RM, Lynch III JP, Nguyen D, Eskandari M, Kunkel SL. In vivo dynamics of murine tumor necrosis factor-α gene expression. Lab Invest 1989; 60:766–771.

52

Gene Knockout Technology and the Host Response to Endotoxin: Role of CD14 and Other Inflammatory Mediators

Sanna M. Goyert
North Shore University Hospital, Manhasset, New York

The response to endotoxin and/or gram-negative bacteria is a complicated cascade of events mediated by a plethora of molecules including cell surface receptors, cytokines, chemokines, adhesion molecules, and transcription factors regulating the expression of these molecules. The precise role of each of these molecules and their relative importance in various biological phenomena such as septic shock and inflammation is poorly understood. However, the recent application of gene "knockout" technology has led to the production of animals specifically lacking one or more of these genes; this now allows detailed, in vivo studies of the individual function and role of these genes in inflammation and shock. Among the genes that have been specifically "knocked out" to determine their role in inflammation and the response to endotoxin is CD14, a major receptor for lipopolysaccharide (LPS). This review attempts to summarize the results of studies in various knockout mice on the response to endotoxin and gram-negative bacteria, with a specific focus on CD14.

ROLE OF CD14 IN THE RESPONSE TO LPS

CD14 is a cell-surface glycoprotein anchored to the membrane by a covalently linked lipid moiety known as GPI (glycosylphosphatidylinositol). CD14 is highly expressed on monocytes/macrophages and weakly expressed by neutrophils (1,2). Soluble forms of CD14 (sCD14) can be found in serum at a concentration of approximately 2 μg/ml. They are produced by a poorly understood mechanism, which results in removal of the membrane-anchoring GPI moiety and subsequent release of CD14 from the cell surface (2–4). In addition, a small amount of sCD14 lacking the GPI anchor is naturally secreted by cells (5). The membrane form of CD14 serves as a receptor for LPS. Stimulation of monocytes/macrophages and neutrophils with very low levels of LPS (<1 ng/ml) in the presence of serum [providing a source of LBP (LPS-binding protein)] results in their activation, which is often measured by the release of cytokines or other pro-inflammatory mediators (4,6,7). This activation can be inhibited by monoclonal antibodies to CD14 (4,6,7).

Role of CD14 Expression in the Lethal Response to LPS

We have produced mice that lack expression of the CD14 gene using the technique of homologous recombination to determine the relative importance of CD14 in LPS-induced lethality, a classic murine model of endotoxin shock. Our studies reveal that mice deficient in CD14 display no obvious gross response to LPS (8); mice injected with a dose of LPS which is 100% lethal for a normal mouse (20 mg/kg) do not die and show none of the classic symptoms of response [ruffled fur (fever), eye exudate, weakness and reduced activity, labored breathing]. Furthermore, mice injected with 10 times this lethal dose (200 mg/kg) also do not die (Fig. 1), although they do display some minor symptoms of response; eye exudate can be observed in some of the

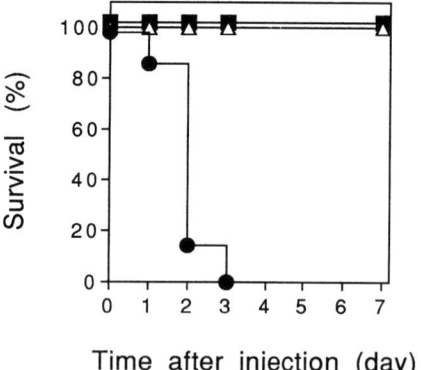

Fig. 1 Response of CD14-deficient and control mice to LPS. Mice were injected (i.p.) with a lethal (20 mg/kg) dose or 10 times a lethal dose (100 mg/kg) of LPS (*S. minnesota*, wild type) and monitored for survival for 7 days. Solid circle, control mice injected with 20 mg/kg; open triangle, CD14-deficient mice injected with 20 mg/kg; solid square, CD14-deficient mice injected with 200 mg/kg. (From Ref. 8.)

absolutely required for the lethal response to endotoxin and that no other LPS receptors can fulfill this requirement in the absence of CD14. These studies also show that CD14 expression is required for the production of high levels of circulating cytokines in response to LPS. Furthermore, the presence of large amounts of circulating IL-6 in CD14-deficient mice suggests the existence of another receptor for LPS; however, this receptor very likely has a much lower affinity for LPS than CD14 since large amounts of LPS are required to induce the secretion of IL-6. This receptor does not appear to be expressed by peritoneal macrophages since CD14-deficient macrophages stimulated in vitro with large amounts of LPS do not secrete measurable levels of IL-6 (Fig. 3).

D-Galactosamine–Sensitized CD14-Deficient Mice and LPS-Induced Lethality

Sensitization of mice with D-galactosamine causes them to be hypersensitive to the effects of LPS and very low levels of TNF-α (9–12). In this model, very low levels of LPS will induce a shock-like death. To determine whether CD14-deficient mice are also resistant to the effects of LPS after sensitization with D-galactosamine, CD14-deficient and control mice were injected with D-galactosamine (0.6 g/kg) and 2 μg LPS (*Salmonella minnesota*, wild type). With this dose, all CD14-deficient mice survive while all control mice die, indicating that lack of CD14 confers resistance to LPS in the D-galactosamine model of endotoxin shock (Fig. 4).

mice, but they do not display weakness, reduced activity, or labored breathing. Analyses of serum cytokine levels show that although control mice injected with a lethal dose of LPS have high levels of circulating TNF after 1 hour, CD14-deficient mice have little or no circulating TNF-α (Fig. 2a) with either a lethal dose of LPS or 10 times a lethal dose. Similarly, CD14-deficient mice have little or no circulating IL-6 when injected with a lethal dose of LPS; however, injection with 10 times the lethal dose results in circulating IL-6 levels, which are approximately 50% those of normal controls (Fig. 2b). These studies show that CD14 is

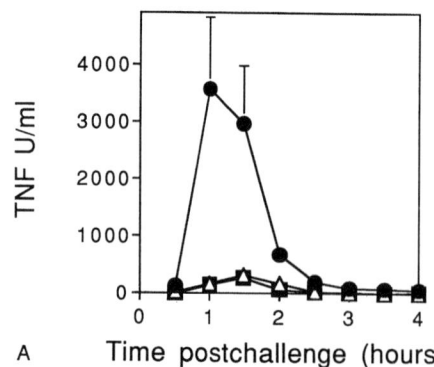
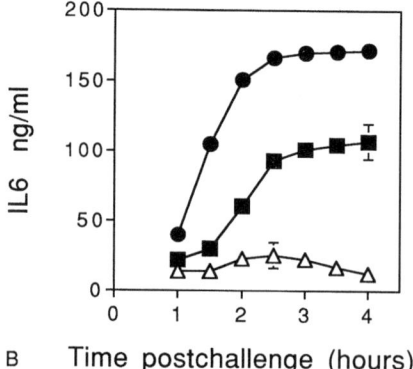

Fig. 2 Cytokines (IL-6 and TNF-α) produced by CD14-deficient and control mice after injection of a lethal dose of LPS. Mice were injected (i.p.) with a lethal dose of LPS (20 or 200 mg/kg) and bled from the tail vein at various times. Plasma concentrations of TNF-α and IL-6 were determined by bioassay and ELISA, respectively. Closed circle, control mice injected with 20 mg/kg; open triangle, CD14-deficient mice injected with 20 mg/kg; solid square, CD14-deficient mice injected with 200 mg/kg. (From Ref. 8.)

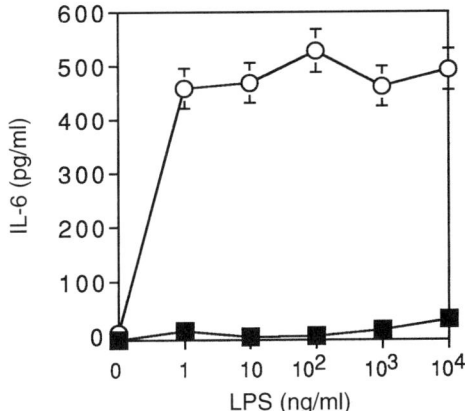

Fig. 3 Secretion of IL-6 by peripheral blood mononuclear cells (PBMC) from CD14-deficient and control mice after stimulation with LPS. PBMC were stimulated with increasing concentrations of LPS for 3 hours in the presence of 1% autologous mouse serum, and the cell-free supernatant was assayed for IL-6 by ELISA. Open circle, control mice; solid square, CD14-deficient mice. (From Ref. 8.)

ROLE OF CD14 IN THE RESPONSE TO GRAM-NEGATIVE BACTERIA

CD14-Deficient Mice in a Lethal Challenge of *E. coli* 0111

To examine the role of CD14 in the lethal response to a gram-negative bacterium, CD14-deficient and control mice were injected with *E. coli* 0111 and were monitored for gross symptoms, mortality, and cytokine responses (8). CD14-deficient mice are completely resis-

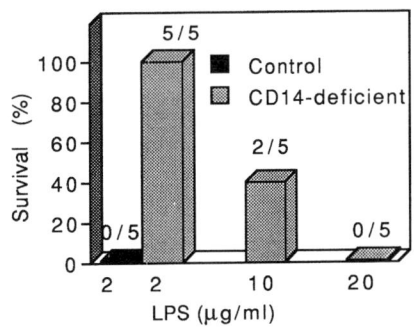

Fig. 4 Response of CD14-deficient and control mice to LPS following sensitization with D-galactosamine. Mice were injected (i.p.) with D-galactosamine (0.6 mg/kg) and LPS (*Salmonella minnesota*, wild type) and monitored for 7 days. All deaths occurred within 24 hours. (From Ref. 8.)

tant to a lethal dose of *E. coli* 0111 (5×10^6 CFU) injected i.p. or i.v. (Fig. 5A). Not only do they not die, but they also fail to show any gross symptoms of response. Furthermore, they produce little or no circulating TNF-α or IL-6 in response to this dose of *E. coli* (Fig. 5B,C). In order to get a 100% lethal response in CD14-deficient mice, the dose of bacteria must be raised fivefold to 3×10^7 CFU. With this dose, control animals are very sick by 4 hours and die within 24 hours, while CD14-deficient mice do not display any symptoms for 12–16 hours, after which they begin to appear sick and die after more than 36 hours. However, even at this high dose of bacteria, CD14-deficient mice produce 10-fold less TNF-α and 48-fold less IL-6 than control mice (Fig. 5B,C). These studies demonstrate

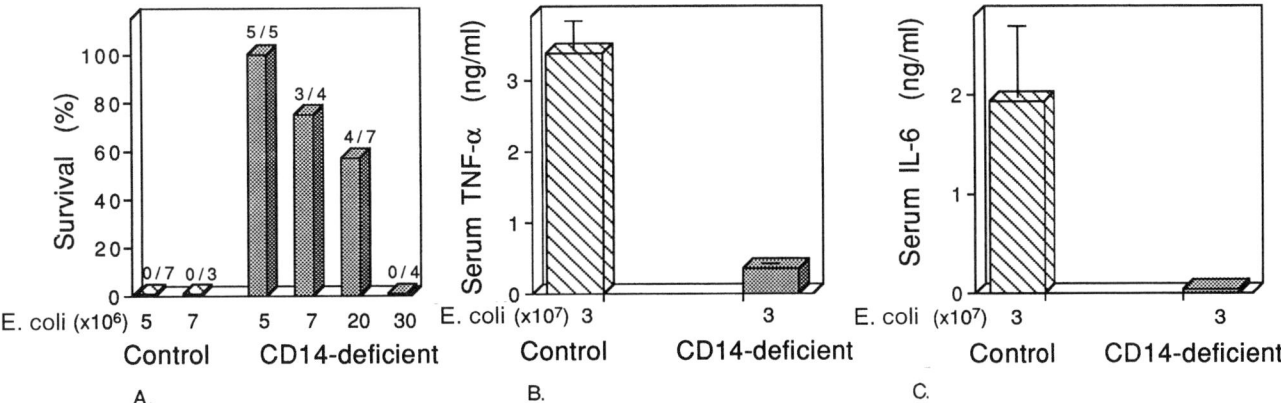

Fig. 5 Responses of CD14-deficient and control mice to *E. coli* 0111. (A) Lethality induced after infection with gram-negative bacteria. Mice were injected (i.p.) with the indicated dose of *E. coli* and monitored for 3 weeks for lethality. All deaths occurred within 48 hours of infection. Numbers above the bars indicate survivors/total number of mice injected. (B,C) Production of cytokines by CD14-deficient and control mice after infection. Mice were bled 7 hours after infection, and plasma levels of TNF-α and IL-6 were measured as described above. Results are represented as ±SEM. (From Ref. 8.)

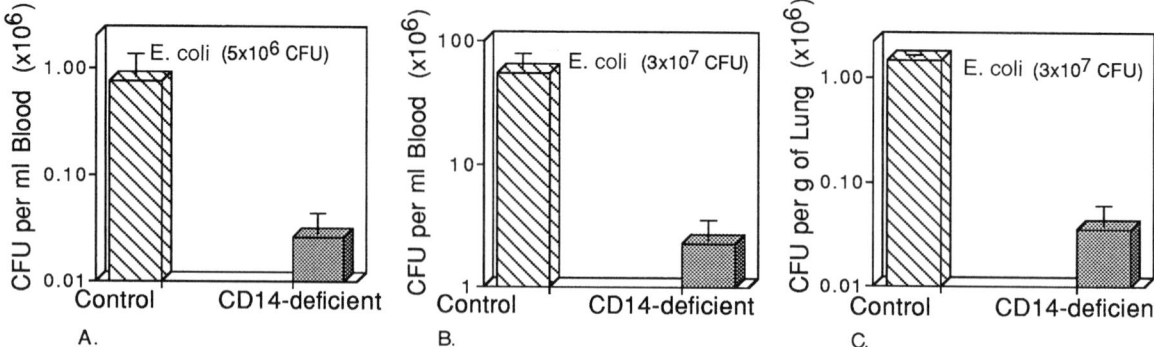

Fig. 6 Number of live *E. coli* recovered from blood and lung following infection. Seven hours after mice (3 per group) were infected (i.p.) with the indicated dose of *E. coli*, their blood and lungs were removed and processed for determination of the number of live bacteria. Results are represented as SEM. (From Ref. 8.)

that although CD14 is required for the lethal response to *E. coli*, at very high doses of bacteria other factors can lead to lethality.

Lack of CD14 and Reduced Bacterial Dissemination

To examine the effects of CD14 deficiency on bacterial spread, blood and tissue samples were taken 7 hours after injection with live *E. coli* 0111 and were processed for the enumeration of recoverable live bacteria (8). Surprisingly, CD14-deficient mice injected with the dose of bacteria that is 100% lethal for control mice (5×10^6 CFU) or the high dose that is lethal for the CD14-deficient mice (3×10^7 CFU) show a significantly reduced number of bacteria in their blood and organs as compared to control mice (Fig. 6). CD14-deficient mice injected with the lethal dose for control mice (5×10^6 CFU) have 27-fold fewer bacteria in their blood than control mice; when injected with a dose that is lethal for both groups of mice (3×10^7 CFU), CD14-deficient mice have 35-fold fewer bacteria in their blood and 24-fold fewer bacteria in their lungs than control mice. These studies suggest that lack of CD14 confers resistance to bacterial spread and promotes bacterial clearance. The molecular mechanisms responsible for reduced bacterial dissemination in CD14-deficient animals are currently under investigation.

EVIDENCE FOR OTHER RECEPTORS FOR LPS

Peritoneal Macrophages and LPS Lethality

To examine whether peritoneal macrophages (PEM) might have functional receptors for LPS other than CD14, PEM were stimulated with increasing concentrations of LPS. It was observed that only at very high concentrations of LPS (≥ 1 μg/ml), 1000-fold more than the amount of LPS required for the secretion of TNF-α from control macrophages (Fig. 7), did CD14-deficient PEM secrete TNF-α. The ability of this dose of LPS to elicit a TNF-α response from CD14-deficient PEM was confirmed by measuring the amount of TNF-α mRNA transcribed in response to LPS (13); similar results were obtained for IL-1β. The ability of CD14-deficient macrophages to express TNF-α and IL-1β when stimulated with very high levels of LPS suggests that there may be a very low-affinity receptor for LPS on macrophages. Studies by others (14–16) have suggested that CD14 is not able to transmit a signal alone and thus may be associated with one or more additional

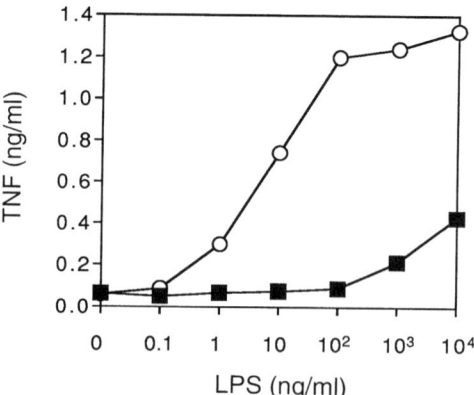

Fig. 7 Secretion of TNF-α by PBMC from CD14-deficient and control mice after stimulation with LPS. PBMC were stimulated as described in Figure 3, and TNF-α was measured by bioassay. Open circle, control mice; solid square, CD14-deficient mice. (From Ref. 8.)

proteins required for signal transduction. One possibility is that this low-affinity receptor is the signal transducer and that CD14 is required for a high-affinity response. In any event, our studies clearly show that this low-affinity receptor does not play a role in LPS-induced lethality in the absence of CD14.

A Low-Affinity Receptor for an LPS:sCD14 Complex

We and others have shown that soluble CD14 [sCD14, found in serum (3,4,17)] can bind to LPS and that LPS:sCD14 complexes can stimulate CD14-negative cells, such as endothelial cells, via an as yet unknown receptor pathway. Binding to this receptor results in the secretion of cytokines and the upregulation of various cell surface molecules, including adhesion molecules (18,19). To determine whether macrophages deficient in membrane CD14 are similarly able to respond to a complex of LPS:CD14, the response of CD14-deficient macrophages to a mixture of LPS and human recombinant sCD14 (rsCD14) was measured. Cells (CD14-deficient and control) were exposed to increasing amounts of LPS in the presence or absence of human recombinant CD14, and the TNF-α response was measured. The presence of sCD14 in combination with LPS results in TNF-α secretion by CD14-deficient PEM at a dose of LPS (100 ng/ml) that is one log less than when soluble CD14 is absent (Fig. 8). The ability of CD14-deficient macrophages to respond to LPS:sCD14 complexes was confirmed by analysis of TNF-α and IL-1β gene expression after stimulating CD14-deficient and control macrophages with LPS in the presence of fetal bovine serum as a source of soluble CD14 (13). These studies suggest that the receptor for LPS:sCD14 on CD14-deficient macrophages is also of low affinity, since at least 100 times more LPS is required for a response in CD14-deficient macrophages compared to control macrophages, which respond predominantly via the membrane CD14 pathway. These studies do not address a possible relationship between the low-affinity LPS:sCD14 receptor and the low-affinity receptor for LPS alone expressed by CD14-deficient macrophages; they could be the same or they could be distinct.

Non-CD14 Receptor in LPS Induction of Acute Phase Proteins

In addition to the symptoms characteristic of endotoxin shock, which are produced in response to LPS (cytokine secretion, fever, eye exudate, weakness, labored breathing, death), injection of LPS also stimulates the

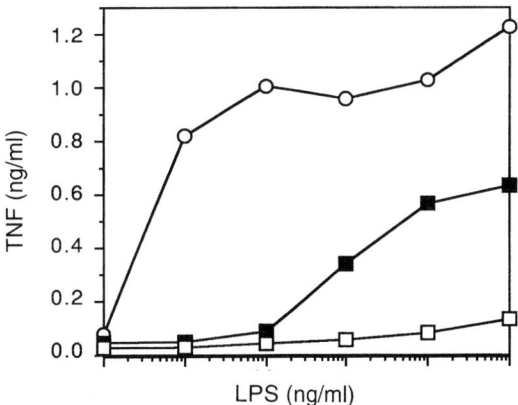

Fig. 8 Secretion of TNF-α by PBMC from CD14-deficient and control mice after stimulation with complexes of LPS:sCD14. PBMC were stimulated as above with various doses of LPS mixed with rsCD14 (20 μg/ml) in the presence of autologous serum (1%). TNF-α was measured by bioassay. Open circle, control; solid square, CD14-deficient plus rsCD14; open square, CD14-deficient. (From Ref. 8.)

production of acute phase proteins (APP) by hepatocytes (20). Since cytokines alone can stimulate the production of APP, it has commonly been assumed that LPS stimulation of APP production occurs via the CD14 pathway, where cytokine production is induced. To determine whether CD14 does indeed play a role in the induction of APP by LPS, the ability of CD14-deficient mice to produce APP in response to LPS was examined. CD14-deficient and control mice were injected with various doses of LPS, and the production of several representative APP such as serum amyloid A (SAA), fibrinogen, ceruloplasmin, and LBP (21) was measured. These studies showed that CD14-deficient mice produce APP at levels equivalent to control mice (Table 1), indicating that CD14 expression is not required for the induction of APP by LPS. In contrast, C3H/HeJ mice, which have a normal CD14 molecule (22; X.-Y. Lin and S.M. Goyert, unpublished observa-

Table 1 Production of Acute Phase Proteins by Mice Injected with LPS

Mouse	SAA	Fibrinogen	Ceruloplasmin
Control	8[a]	5	2.3
CD14-KO	7	5	3.6
C3H/HeN	8	ND	ND
C3H/HeJ	0	ND	ND

[a]-Fold increase over basal levels.
Source: Ref. 21.

tions) but are hyporesponsive to LPS due to a mutation (Lps^d) on an unknown gene on chromosome 4, are unable to produce APP. Thus, the gene (Lps) that is required for the cytokine response to LPS via the CD14 pathway also participates in the non-CD14 receptor pathway required for the induction of APP by LPS.

RESPONSES OF OTHER GENE KNOCKOUT MICE TO LPS AND BACTERIA

Responses of Other Gene Knockout Mice to High Doses of LPS

As described above, CD14 plays a preeminent role in the lethal response to LPS and live *E. coli* 0111. The production of mice lacking other specific genes encoding molecules that play a role in inflammation is beginning to reveal key aspects of their individual functions. Table 2 is a compilation of studies in gene knockout mice of several molecules involved in the inflammatory response directed towards understanding their role in the response to LPS and/or bacteria. These studies show that of all the molecules listed, only animals lacking CD14, ICE, LDLR, or ICAM-1 show a strong resistance to the lethal response to LPS in the absence of any sensitizing agent. In addition, mice deficient in GMCSF or α_2-macroglobulin show partial resistance to the effects of endotoxin. The resistance of CD14-deficient mice to high doses of LPS (200 mg/kg) correlates with very low levels of TNF-α, IL-1β, and IL-6 production as compared to controls. In contrast, mice deficient in TNF-α or its receptor are not resistant to high doses of LPS. Similarly, mice deficient in IL-1β alone or in the IL-1 type 1 receptor have a normal lethal response to a high dose of LPS (35–37). These results suggest that elimination of a single cytokine is not sufficient to confer resistance to high doses of LPS; however, reduction in the levels of multiple cytokines (as in the CD14-deficient mice) will confer resistance to LPS.

Although deletion of cytokine genes (or their receptors) does not protect against endotoxin-induced death, such deletions may nevertheless have subtle consequences on some facets of the response to endotoxin. This is illustrated with the IL-1R1 knockout mice. One of the signs of severe shock is damage to endothelial cells (EC) and detachment of EC from the internal elastic lamina (IEL). Although IL-1R1–deficient mice exhibit a normal lethal response to a high dose of endotoxin [50 mg/kg (37)], they surprisingly show a normal intact adhesion of the endothelial layer to the IEL under these conditions (54); this is in sharp contrast to the endothelial layer of control mice, which is damaged and completely detached from the IEL. Taken together, these two studies suggest that a lethal response can occur in the absence of endothelial cell damage; however, since these studies were performed in two different laboratories and since the effects on mortality were not specifically measured in the latter study, this conclusion may be premature. Nevertheless, these studies indicate that IL-1 activity may play an important role in LPS-induced endothelial cell damage.

In view of the sensitivity of mice deficient in the IL-1 type 1 receptor, which transmits a signal for both IL-1α and IL-1β molecules, to high levels of LPS, the apparent resistance of ICE (IL-1β–converting enzyme)–deficient mice to LPS is puzzling. ICE-deficient mice lack mature forms of IL-1β and have extremely low levels of IL-1α (38). However, recent studies have shown that ICE can also process the precursor of IFN-γ to its mature form (58), indicating that ICE activity is not restricted to IL-1 and may in fact control the expression of additional molecules not yet identified. Such additional activities of ICE may have profound effects on the response to LPS and may possibly explain the resistance of ICE-deficient mice to LPS.

As in the ICE-deficient mice described above, mice that are deficient in IL-1ra (a protein that inhibits the binding of IL-1 to its receptor) produce low levels of IL-1α and IL-1β after endotoxin challenge (high dose); however, unlike the ICE-deficient mice, this reduction in IL-1 is accompanied by an increased mortality in comparison to controls (39). These studies suggest that the heightened sensitivity to IL-1 in mice deficient in IL-1ra can more than compensate for the reduced levels of IL-1 seen in these mice.

LDLR-deficient mice are also resistant to the lethal effects of LPS. This resistance is associated with low TNF-α and IL-1α levels and increased levels of LDL, previously shown to neutralize LPS (46).

Resistance of mice deficient in ICAM-1 to the lethal effects of LPS is thought to be associated with a lack of cell-cell interaction. ICAM-1 (intercellular adhesion molecule 1) is an adhesion molecule expressed by leukocytes and vascular endothelium; it is a ligand for Mac-1 and LFA-1, members of the β_2 subfamily of integrins expressed on leukocytes (44). ICAM-1 mediates at least one third of the binding of lymphocytes and monocytes to resting endothelial cells based on in vitro studies; stimulation of endothelial cells with LPS results in upregulation of ICAM-1 expression and an increase in the binding of leukocytes. Accordingly, ICAM-1–deficient mice have high blood leukocyte counts, with two- to sixfold more blood neutrophils and two- to threefold more blood lymphocytes than control mice. Administration of LPS results in a strong reduction in the number of neutrophils in the blood of both control and ICAM-1-deficient mice 2 hours after administration; however, 6 hours after administration of LPS, the ICAM-1–deficient animals show a major increase in the number of blood neutrophils (at least fivefold higher), which continues to increase to at least ninefold higher than controls by 24 hours. This increase of blood neutrophils is mirrored by a strongly reduced emigration of neutrophils to the liver in ICAM-1-deficient mice by 24 hours and an increase in neutrophils in the sinusoids (44). These studies show that the resistance to LPS seen in ICAM-1–deficient mice is not due to reduced levels of cytokines, but to a decrease in neutrophil transmigration.

In addition to the knockout models described above, which display strong resistance to a high dose of LPS, mice deficient in GM-CSF show partial resistance to endotoxin. GM-CSF stimulates the proliferation and differentiation of granulocytes and monocytes and can also stimulate or enhance various functional activities of mature myeloid cells. One of these functions is the ability of GM-CSF to prime monocytes/macrophages to LPS, resulting in increased cytokine responses. Mice deficient in GM-CSF challenged with LPS at a concentration of 5 mg/kg (a nonlethal dose) exhibit reduced symptoms (fever, weight loss) compared to controls and a reduction in circulating IL-6, IL-1, and IFN-γ; in contrast, serum TNF-α levels peak similarly to control mice but decrease more slowly, resulting in slightly prolonged TNF-α expression (53). GM-CSF–deficient mice injected with a high dose of LPS (25 mg/kg) show normal symptoms but a partial resistance to lethality (58% survival) as compared to controls (0% survival). These results confirm previous in vitro studies, suggesting that GM-CSF plays a role in LPS-induced shock independent of TNF-α (53).

Mice deficient in α_2-macroglobulin are also partially protected against the effects of endotoxin; they are nearly completely resistant (90% survival) to a high dose of LPS (28.3 mg/kg), which results in only 20% survival of control animals. However, much of this resistance is overcome when fivefold more LPS is administered, resulting in 100% mortality for controls and 80% mortality for the deficient mice (51). Resistance to the effects of LPS in the α_2-M–deficient mice may in part be due to lack of neutralization of TGF-β by α_2-M, normally present in serum, which allows excess TGF-β to suppress the production of nitric oxide as well as other processes that potentiate responses to LPS (52). Additionally, as noted by the authors, these experiments were done in mice with a mixed genetic background; thus, the true sensitivity of these mice to LPS needs to be confirmed in mice with a defined genetic background.

The importance of genetic background applies to all studies on the effects of gene ablation, since a mixed background can give anomalous results. This technical issue is important since most knockout mice are produced by mutating embryonic stem cells obtained from 129/Sv mice and implanting into blastocysts of C57BL6 mice giving rise to chimeras. Chimeras are, in turn, bred with C57BL6 to determine germline transmission, and mice carrying the mutation (heterozygous F_1s) are intercrossed to produce homozygous mutants or heterozygous controls. Thus, although the knockout mice and the control mice have similar percentages of the 129 background, each mouse is unique with respect to the derivation (129 or B6) of each background gene. These background differences of the resulting homozygous mutant mice and their littermate controls may lead to spurious results, since some inbred strains, such as C57BL6, are more sensitive to LPS than others, such as 129/Sv. Similarly, comparison of the response of homozygous KO mice to that of pure C57BL6 or 129/Sv alone may result in incorrect interpretations of the effects of genes on function. An F_1 produced by crossing C57BL6 and 129/Sv is a better control, although still not ideal. Thus, the most reliable data come from mice in which the mutation has been placed onto one specific genetic background by backcrossing to an inbred strain more than 7 times (99.2% pure) or 10 times (99.9% pure), with the ultimate being 20 times (the number of crosses defining a congenic strain); however, it takes a minimum of 3.5 years to obtain 20 backcrosses, making this an unattractive alternative.

Responses of Gene Knockout Mice to LPS After Sensitization with D-Galactosamine, *P. acnes*, BCG, or Carrageenan

Mice deficient in CD14 as well as several other genes (LBP, TNF-α, TNFR1) are resistant to a lethal dose of LPS following administration of D-galactosamine (D-gal), an agent that sensitizes mice to the effects of low levels of TNF-α (Table 2). Thus, it is not surprising that mice deficient for TNF-α or the TNFRp55 are resistant to LPS-induced lethality in this model. ICE-deficient mice and IL-6–deficient mice were not tested in this model, but IL-1β–deficient mice, which make normal levels of TNF-α, show a normal sensitivity to LPS in the D-gal model. The LBP-deficient mice are resistant to LPS-induced lethality in the D-gal model, and

Table 2 Responses of Gene Knockout Mice to LPS and Bacteria

Knockout target	Function	LPS (mg/kg)	Symptoms	Mortality	Serum TNF-α	Serum IL-1α	Serum IL-1β	Serum IL-6	Mortality induced by LPS after sensitization	LPS-induction of APP	Bacterial challenge	Ref.
CD14	LPS receptor	20	None	Resistant	Low	ND	Low	Low	Resistant (D-gal)	Normal	Resistant to *E. coli* 0111, reduced dissemination	1
		200	Low	Resistant	Low		Low	Low				
SR-A	Scavenger receptor Type A; binds polyanionic ligands; detoxifies LPS; mediates uptake of modified LDL	25	ND	Higher than controls	Higher than controls	ND	ND	Higher than controls	More sensitive than controls (BCG)	ND	ND	23
LBP	LPS-binding protein; delivers LPS to CD14 and HDL	0.1	ND	ND	Normal	ND	ND	ND	Resistant (D-gal)	ND	More sensitive to *S. typhimurium*	24,25
IL-6	Proinflammatory cytokine	1	Normal	ND	Increased	Normal	ND	0	ND	Nearly normal	More sensitive to *Listeria*	26–28
TNFR p55	High-affinity receptor for TNF	2.5	ND	Normal	Normal	ND	ND	Low	Resistant (D-gal)	ND	Highly sensitive to *Listeria*	29–31
		20	ND	ND	High	ND	ND	ND				
		35	ND	Normal	ND	ND	ND	ND				
		50	ND	Normal	ND	ND	ND	ND				
		60	ND	Normal	ND	ND	ND	ND				
TNFR p75	Low-affinity receptor for LPS	35	ND	Normal	Increased	ND	ND	ND	Normal (D-gal)	ND	Nearly normal response to *Listeria*	31,32
		50			Increased							
TNFR p55/p75	Double knockout	20	ND	Normal	Very high	ND	ND	ND	Resistant (D-gal)	Normal	Highly sensitive to *Listeria*	31
TNF-α	Proinflammatory cytokine	5	ND	ND	0	ND	Normal	Normal	Resistant (D-gal)	Normal	Highly sensitive to *Listeria* and *Candida albicans*	33,34
		50	ND	ND	0	ND	Normal	ND				
		60	Normal	Normal		ND	Normal	ND				
IL-1β	Proinflammatory cytokine	High dose	Normal	Normal	Normal	Normal	0	Normal	Normal (D-gal)	Normal	Normal response to *Listeria*	35,36
IL-1R (type I)	Signaling receptor for IL-1α and IL-1β	50	ND	Normal	ND	ND	ND	ND	Normal (D-gal)	Normal	Normal response to *Listeria*	37

Gene	Function	(dose)							Infection	Refs
IL-1R/TNFR p55 or p75	Double knockout of receptors for IL-1 and TNF	50	ND	Normal	ND	ND	ND	ND	ND	31
ICE	Converts pro-IL-1β to mature forms	40	Reduced	Resistant	Moderately decreased	Very low or negative	No mature forms	Moderately decreased	ND	ND	38
IL-1ra	Interleukin-1 receptor antagonist	10	ND	Increased	Normal	Reduced	Reduced	Reduced	ND	More resistant to Listeria	39
IFN-γR	Receptor for interferon-gamma	0.05	ND	ND	Normal	ND	ND	ND	Normal?/ resistant? (D-gal); resistant (BCG and P. acnes)	More sensitive to Listeria	40–43
		5	Reduced	ND	Reduced	ND	ND	ND			
		10	ND	Normal	ND	ND	ND	ND			
		25	Normal	ND	ND	ND	ND	ND			
		40	ND	Normal	ND	ND	ND	ND			
ICAM-1	Adhesion molecule	40	Normal	Resistant	Normal	Normal	ND	ND	Normal (D-gal)	ND	44,45
LDLR	Receptor for low-density lipoprotein	50	ND	Resistant	Low	Low	Normal	ND	ND	K. pneumoniae: slightly delayed death and reduced mortality	46
iNOS	Catalyzes oxidation of L-arginine for production of NO	12.5	ND	Normal	ND	ND	ND	ND	Normal (P. acnes)	More sensitive to Listeria	47,48
		25	ND	Normal	ND	ND	ND	ND			
5LX	5-Lipoxygenase; converts aracidonate into leukotrienes	ND	ND	ND	ND	ND	ND	ND	Normal (carrageenan)	More sensitive to K. pneumoniae	49,50
α2M	Murine α2-macroglobulin (MAM); member of family of proteinase inhibitors	28.3	ND	Resistant	ND	ND	ND	ND	ND	ND	51,52
		141.5	ND	Sensitive	ND	ND	ND	ND			
GM-CSF	Granulocyte-monocyte-colony-stimulating factor	5	Reduced	ND	Normal	Reduced	Reduced	Reduced	ND	ND	53
		25	Normal	Partial resistance	ND	ND	ND	ND	ND	ND	

Normal, same as controls; ND, not described; KO, knockout (genetically deficient); D-gal, D-galactosamine (LPS agent); BCG, bacillus Calmette-Guérin.
aSensitizing agent.

this resistance is associated with very low levels of TNF-α (10 pg/ml vs. 1.1 µg/ml for controls) (24). However, Wurfel et al. (25), who also produced LBP-deficient mice, observed normal levels of TNF-α in vivo after injection with high concentrations of LPS alone; this result contradicts those described by Jack et al. (24) using the D-gal model. However, when Wurfel et al. (25) stimulated LBP-deficient mice with low concentrations of LPS alone, lower levels of TNF-α were produced by the LBP-deficient mice than controls (25; D. Golenbock, personal communication), similar to the results found by Jack et al. using the D-gal model, where very small amounts of LPS were used. Thus, LBP deficiency resulted in a decreased TNF-α secretion at low concentrations of LPS; however, at high concentrations of LPS, normal levels of TNF-α appear to be produced.

In addition, LBP-deficient mice show reduced responsiveness to LPS after pretreatment with *P. acnes*, an agent that sensitizes mice to the effects of IFN-γ (55) and results in a strong increase in TNF-α production after LPS administration; control mice pretreated with *P. acnes* show a 75- to 200-fold increase in TNF-α in response to low levels of LPS, whereas LBP-deficient mice display only a fourfold increase (56). Administration of recombinant LBP together with LPS to LBP-deficient mice restored the ability to develop a strong TNF-α response following *P. acnes* sensitization (56). These results show that the increased TNF-α production occurring after *P. acnes* sensitization requires expression of LBP. The effects on survival were not reported in this study.

The results of LPS-induced lethality in IFN-γR–deficient mice sensitized with D-gal are controversial; Car et al. (42) found them to be resistant to the effects of LPS under these conditions, suggesting that IFN-γ (and its receptor) plays a role in this response, whereas Freudenberg et al. (43) found IFN-γR–deficient mice gave a normal response under similar conditions, suggesting that this response is independent of IFN-γ. The latter group did show, however, that IFN-γR–deficient mice are at least 1000 times less sensitive to LPS than normal mice when sensitized with *P. acnes*, confirming a role for IFN-γ in the response to LPS after *P. acnes* sensitization. IFN-γR–deficient mice are also resistant to low levels of LPS after sensitization with another bacterial agent, BCG (41); this resistance is accompanied by a strong reduction in TNF-α, IL-1α, and IL-6, indicating that IFN-γ also plays an important role in BCG-sensitization of LPS responses. In contrast, IFN-γR–deficient mice give a normal lethal response to a high dose of LPS administered alone (43).

ICAM-1–deficient mice are resistant to the effects of a high dose of LPS alone due the above-mentioned deficiency in neutrophil transmigration, but they show a normal sensitivity to LPS administered following D-gal sensitization (45). This result correlates with a normal production of TNF-α by ICAM-1–deficient mice.

Type A scavenger receptor–deficient mice are more sensitive than control mice to both high doses of LPS alone and low doses of LPS in BCG-sensitized mice; these responses are associated with very high levels of circulating TNF-α. Although further studies may be required to confirm this result in animals with a pure genetic background, these studies suggest a critical role for SR-A in the clearance of, and response to, LPS.

Mice deficient in 5-lipoxygenase (5LX) and sensitized to the effects of LPS by pretreatment with carrageenan (57) show no difference in the lethal response to LPS when compared to controls (49); however, 5LX-deficient mice are resistant to lethality induced by administration of a dose of platelet-activating factor, which causes 75% of control animals to die of anaphylactic shock within 15–25 minutes, although 5LX-deficient mice do exhibit a similar listless behavior for the first 20 minutes before recovering (49).

Production of Acute Phase Proteins by Gene Knockout Mice After Stimulation with LPS

All of the deficient mice tested for LPS induction of acute phase proteins showed a normal response, including a double knockout for the IL-1 receptor and the TNF1 receptor (Table 2). This is consistent with the normal APP response in CD14-deficient mice, which fail to produce any significant levels of cytokines after exposure to LPS. Thus, the critical molecules regulating this portion of the inflammatory response remain to be identified.

Responses of Gene Knockout Mice to Bacterial Challenge

As described earlier, mice deficient for CD14 are 100% resistant to a lethal dose of the extracellular bacterium *E. coli* 0111; this resistance is accompanied by a strong reduction of cytokine production (TNF-α, II-1β, and IL-6) and significantly decreased bacterial dissemination. These studies indicate that lack of CD14 confers a strong resistance to infection with *E. coli* 0111 after either intraperitoneal or intravenous administration (8).

In contrast to the results obtained with CD14-deficient mice, LBP-deficient animals are more sensitive to the intracellular gram-negative pathogen *Salmonella ty-*

phimurium than control animals. This sensitivity is associated with a high bacterial load compared to normal animals and a reduced ability of LBP-deficient macrophages to produce TNF-α. The LDL-receptor-deficient mice show a slight resistance to infection with *Klebsiella pneumoniae* (normally an encapsulated gram-negative strain) and exhibit delayed death and a slightly lower mortality. They also show increased plasma cholesterol levels, similar to the levels seen in the response of LDLR-deficient mice to LPS and at least three times higher than the level in control mice.

In contrast, mice deficient in 5-lipoxygenase are more sensitive to the lethal effects of intratracheal infection with *K. pneumoniae* (50). Although no defect in neutrophil recruitment to the lung is detected, their alveolar macrophages are defective in phagocytosis and microbial killing, indicating a critical role for 5LX in this model of infection.

The IL-6–, TNFRp55-, TNF-α–, iNOS-, and IFN-γR–deficient mice all display a high sensitivity to the gram-positive, intracellular pathogen *Listeria monocytogenes* while the TNFRp75-, IL-1R–, and IL-1β–deficient mice show a normal response. In general, sensitivity to *Listeria* infection correlates with low TNF-α expression, a cytokine known to confer resistance to *Listeria*. However, IL-6–deficient mice, which are able to produce high levels of TNF-α in response to a high dose of LPS, are sensitive to *Listeria*. It has been speculated that a defect in the activation of T cells in IL-6–deficient mice may account for their enhanced sensitivity to *Listeria* (27).

CONCLUSION

A great deal of information concerning the role of various molecules in the response to LPS and bacteria has been gained with gene knockout mice; however, many questions remain, including the role of these and other molecules in the response to different types of gram-negative bacteria (i.e., encapsulated, nonencapsulated, those that are invasive in tissue, those causing meningitis, etc.) and in different infection models, including acute as well as chronic infection. Such studies will aid in further elucidating the role of such molecules in inflammation and the potential for new therapies based on inhibiting or enhancing their function.

ACKNOWLEDGMENTS

Some of the studies described here were supported by the National Institutes of Health Grants AI23859 to S.M.G.

REFERENCES

1. Goyert SM, Ferrero E, Rettig WJ, Yenamandra AK, Obata F, Le Beau MM. The CD14 monocyte differentiation antigen maps to a region encoding growth factors and receptors. Science 1988; 239:497–500.
2. Haziot A, Chen S, Ferrero E, Low MG, Silber R, Goyert SM. The monocyte differentiation antigen, CD14, is anchored to the cell membrane by a phosphatidylinositol linkage. J Immunol 1988; 141:547–552.
3. Bazil V, Strominger JL. Shedding as a mechanism of down-modulation of CD14 on stimulated human monocytes. J Immunol 1991; 147:1567–1574.
4. Haziot A, Tsuberi B, Goyert SM. Neutrophil CD14: biochemical properties and role in the secretion of TNF-α in response to LPS. J Immunol 1993; 150:5556–5565.
5. Bufler P, Stiegler G, Schuchmann M, Hess S, Krueger C, Stelter F, Eckerskorn C, Schutt C, Engelmann H. Soluble lipopolysaccharide receptor (CD14) is released via two different mechanisms from human monocytes and CD14 transfectants. Eur J Immunol 1995; 25:604–610.
6. Rietschel ET, Brade H, Holst O, Brade L, Muller-Loennies S, Mamat U, Zahringer U, Beckmann F, Seydel U, Brandenburg K, Ulmer AJ, Mattern T, Heine H, Schletter J, Loppnow H, Schonbeck U, Flad HD, Hauschildt S, Schade UF, Di Padova F, Kusumoto S, Schumann RR. Bacterial endotoxin: chemical constitution, biological recognition, host response, and immunological detoxification. Curr Top Microbiol Immunol 1996; 216:38–81.
7. Ulevitch RJ, Tobias PS. Receptor-dependent mechanisms of cell stimulation by bacterial endotoxin. Ann Rev Immunol 1995; 13:437–457.
8. Haziot A, Ferrero E, Kontgen F, Hijiya N, Yamamoto S, Silver J, Stewart CL, Goyert SM. Resistance to endotoxin shock and reduced dissemination of Gram-negative bacteria in CD14-deficient mice. Immunity 1996; 4:407–414.
9. Galanos C, Freudenberg MA, Reutter W. Galactosamine-induced sensitization to the lethal effects of endotoxin. Proc Natl Acad Sci USA 1979; 76:5939–5943.
10. Decker K, Keppler D. Galactosamine hepatitis: key role of the nucleotide deficiency period in the pathogenesis of cell injury and cell death. Rev Physiol Biochem Pharmacol 1974; 71:77–106.
11. Galanos C, Freudenberg MA. Mechanisms of endotoxin shock and endotoxin hypersensitivity. Immunobiology 1993; 187:346–356.
12. Lehmann V, Freudenberg MA, Galanos C. Lethal toxicity of lipopolysaccharide and tumor necrosis factor in normal and D-galactosamine-treated mice. J Exp Med 1987; 165:657–663.
13. Perera PY, Vogel SN, Detore GR, Haziot A, Goyert SM. CD14-dependent and CD14-independent signaling pathways in murine macrophages from normal and CD14 knockout mice stimulated with lipopolysaccharide or taxol. J Immunol 1997; 158:4422–4429.
14. DeLude RL, Savedra R, Zhao H, Thieringer R, Yamamoto S, Fenton MJ, Golenbock DT. CD14 enhances

cellular responses to endotoxin without imparting ligand-specific recognition. Proc Natl Acad Sci USA 1995; 92:9288–9292.
15. Kitchens RL, Ulevitch RJ, Munford RS. Lipopolysaccharide (LPS) partial structures inhibit responses to LPS in a human macrophage cell line without inhibiting LPS uptake by a CD14-mediated pathway. J Exp Med 1992; 176:485–494.
16. Lee JD, Kravchenko V, Kirkland TN, Han J, Mackman N, Moriarty A, Leturcq D, Tobias PS, Ulevitch RJ. Glycosyl-phosphatidylinositol-anchored or integral membrane forms of CD14 mediate identical cellular responses to endotoxin. Proc Natl Acad Sci USA 1993; 90:9930–9934.
17. Grunwald U, Krüger C, Westermann J, Lukowsky A, Ehlers M, Schütt C. An enzyme-linked immunosorbent assay for the quantification of solubilized CD14 in biological fluids. J Immunol Methods 1992; 155:225–232.
18. Haziot A, Rong GW, Silver J, Goyert SM. Recombinant soluble CD14 mediates the activation of endothelial cells by lipopolysaccharide. J Immunol 1993; 151:1500–1507.
19. Haziot A, Katz I, Rong GW, Lin XY, Silver J, Goyert SM. Evidence that the receptor for soluble CD14:LPS complexes may not be the putative signal-transducing molecule associated with membrane-bound CD14. Scand J Immun 1997; 46:242–245.
20. Mackiewicz A, Kushner I, Baumann H, eds. Acute Phase Proteins: Molecular Biology, Biochemistry, and Clinical Applications. Boca Raton, FL: CRC Press, 1993:686.
21. Haziot A, Lin XY, Zhang F, Goyert S. The induction of acute phase proteins by lipopolysaccharide uses a novel pathway that is CD14-independent. J Immunol 1998; 160:2570–2572.
22. Takakuwa T, Knopf HP, Sing A, Carsetti R, Galanos C, Freudenberg MA. Induction of CD14 expression in Lpsn, Lpsd and tumor necrosis factor receptor-deficient mice. Eur J Immunol 1996; 26:2686–2692.
23. Haworth R, Platt N, Keshav S, Hughes D, Darley E, Suzuki H, Kurihara H, Kodama T, Gordon S. The macrophage scavenger receptor type A is expressed by activated macrophages and protects the host against lethal endotoxic shock. J Exp Med 1997; 186:1431–1439.
24. Jack RS, Fan X, Bernheiden M, Rune G, Ehlers M, Weber A, Kirsch G, Mental R, Furll B, Freudenberg M, Schmitz G, Stelter F, Schütt C. Lipopolysaccharide-binding protein is required to combat a murine gram-negative bacterial infection. Nature 1997; 389:742–745.
25. Wurfel MM, Monks BG, Ingalls RR, Dedrick RL, Delude R, Zhou D, Lamping N, Schumann RR, Thieringer R, Fenton MJ, Wright SD, Golenbock D. Targeted deletion of the lipopolysaccharide (LPS)-binding protein gene leads to profound suppression of LPS responses ex vivo, whereas in vivo responses remain intact. J Exp Med 1997; 186:2051–2056.
26. Fattori E, Cappelletti M, Costa P, Sellitto C, Cantoni L, Carelli M, Faggioni R, Fantuzzi G, Ghezzi P, Poli VJ. Defective inflammatory response in interleukin 6-deficient mice. Exp Med 1994; 180:1243–1250.
27. Kopf M, Baumann H, Freer G, Freudenberg M, Lamers M, Kishimoto T, Zinkernagel R, Bluethmann H, Kohler G. Impaired immune and acute-phase responses in interleukin-6-deficient mice. Nature 1994; 368:339–342.
28. Chai Z, Gatti S, Toniatti C, Poli V, Bartfai T. Interleukin (IL)-6 gene expression in the central nervous system is necessary for fever response to lipopolysaccharide or IL-1 beta: a study on IL-6-deficient mice. J Exp Med 1996; 183:311–316.
29. Pfeffer K, Matsuyama T, Kundig TM, Wakeham A, Kishihara K, Shahinian A, Wiegmann K, Ohashi PS, Kronke M, Mak TW. Mice deficient for the 55 kd tumor necrosis factor receptor are resistant to endotoxic shock, yet succumb to *L. monocytogenes* infection. Cell 1993; 73:457–467.
30. Rothe J, Lesslauer W, Lotscher H, Lang Y, Koebel P, Kontgen F, Althage A, Zinkernagel R, Steinmetz M, Bluethmann H. Mice lacking the tumour necrosis factor receptor 1 are resistant to TNF-mediated toxicity but highly susceptible to infection by *Listeria monocytogenes*. Nature 1993; 364:798–802.
31. Peschon JJ, Torrance DS, Stocking KL, Glaccum MB, Otten C, Willis CR, Charrier K, Morrissey PJ, Ware CB, Mohler KM. TNF receptor-deficient mice reveal divergent roles for p55 and p75 in several models of inflammation. J Immunol 1998; 160:943–952.
32. Erickson SL, de Sauvage FJ, Kikly K, Carver More K, Pitts-Meek S, Gillett N, Sheehan KC, Schreiber RD, Goeddel DV, Moore MW. Decreased sensitivity to tumour-necrosis factor but normal T-cell development in TNF receptor-2-deficient mice. Nature 1994; 372:560–563.
33. Marino MW, Dunn A, Grail D, Inglese M, Noguchi Y, Richards E, Jungbluth A, Wada H, Moore M, Williamson B, Basu S, Old LJ. Characterization of tumor necrosis factor-deficient mice. Proc Natl Acad Sci USA 1997; 94:8093–8098.
34. Pasparakis M, Alexopoulou L, Episkopou V, Kollias G. Immune and inflammatory responses in TNF alpha-deficient mice: a critical requirement for TNF alpha in the formation of primary B cell follicles, follicular dendritic cell networks and germinal centers, and in the maturation of the humoral immune response. J Exp Med 1996; 184:1397–1411.
35. Zheng H, Fletcher D, Kozak W, Jiang M, Hofmann K, Conn C, Soszynski D, Grabiec C, Trumbauer M, Shaw A, Kostura M, Stevens R, Rosen H, North R, Chen H, Tocci M, Kluger M, Van der Ploeg LHT. Resistance to fever induction and impaired acute-phase response in interleukin-1β-deficient mice. Immunity 1995; 3:9–19.
36. Fantuzzi G, Dinarello CA. The inflammatory response in interleukin-1 beta-deficient mice: comparison with other cytokine-related knock-out mice. J Leukoc Biol 1996; 59:489–493.
37. Glaccum MB, Stocking KL, Charrier K, Smith JL, Willis CR, Maliszewski C, Livingston D, Peschon JJ, Morrissey PJ. Phenotypic and functional characterization of mice that lack the type I receptor for IL-1. J Immunol 1997; 159:3364–3371.
38. Li P, Allen H, Banerjee S, Franklin S, Herzog L, Johnston C, McDowell J, Paskind M, Rodman L, Salfeld J, Towne E, Tracey D, Wardwell S, Wei FY, Wong W,

Kamen R, Seshadri T. Mice deficient in IL1β-converting enzyme are defective in production of mature IL-1β and resistant to endotoxic shock. Cell 1995; 80:401–411.

39. Hirsch H, Irikura VM, Paul SM, Hirsch D. Functions of interleukin 1 receptor antagonist in gene knockout and overproducing mice. Proc Natl Acad Sci 1996; 93:11008–11013.

40. Huang S, Hendriks W, Althage A, Hemmi S, Bluethmann H, Kamijo R, Vilcek J, Zinkernagel RM, Aguet M. Immune response in mice that lack the interferon-gamma receptor. Science 1993; 259:1742–1745.

41. Kamijo R, Le J, Shapiro D, Havell EA, Huang S, Aguet M, Bosland M, Vilcek J. Mice that lack the interferon-gamma receptor have profoundly altered responses to infection with bacillus Calmette-Guerin and subsequent challenge with lipopolysaccharide. J Exp Med 1993; 178:1435–1440.

42. Car BD, Eng VM, Schnyder B, Ozmen L, Huang S, Gallay P, Heumann D, Aguet M, Ryffel B. Interferon-gamma receptor deficient mice are resistant to endotoxic shock. J Exp Med 1994; 179:1437–1444.

43. Freudenberg MA, Kumazawa Y, Meding S, Langhorne J, Galanos C. Gamma interferon production in endotoxin-responder and nonresponder mice during infection. Infect Immun 1991; 59:3484–3491.

44. Xu H, Gonzalo JA, St Pierre Y, Williams IR, Kupper TS, Cotran RS, Springer TA, Gutierrez-Ramos JC. Leukocytosis and resistance to septic shock in intercellular adhesion molecule 1-deficient mice. J Exp Med 1994; 180:95–109.

45. Sligh JE Jr, Ballantyne CM, Rich SS, Hawkins HK, Smith CW, Bradley A, Beaudet AL. Inflammatory and immune responses are impaired in mice deficient in intercellular adhesion molecule 1. Proc Natl Acad Sci USA 1993; 90:8529–8533.

46. Netea MG, Demacker PN, Kullberg BJ, Boerman OC, Verschueren I, Stalenhoef AF, van der Meer JW. Low-density lipoprotein receptor-deficient mice are protected against lethal endotoxemia and severe gram-negative infections. J Clin Invest 1996; 97:1366–1372.

47. Laubach VE, Shesely EG, Smithies O, Sherman PA. Mice lacking inducible nitric oxide synthase are not resistant to lipopolysaccharide-induced death. Proc Natl Acad Sci USA 1995; 92:10688–10692.

48. MacMicking JD, Nathan C, Hom G, Chartrain N, Fletcher DS, Trumbauer M, Stevens K, Xie QW, Sokol K, Hutchinson N, et al. Altered responses to bacterial infection and endotoxic shock in mice lacking inducible nitric oxide synthase. Cell 1995; 81:641–650.

49. Chen XS, Sheller JR, Johnson EN, Funk CD. Role of leukotrienes revealed by targeted disruption of the 5-lipoxygenase gene. Nature 1994; 372:179–182.

50. Bailie MB, Standiford TJ, Laichalk LL, Coffey MJ, Strieter R, Peters-Golden M. Leukotriene-deficient mice manifest enhanced lethality from *Klebsiella pneumoniae* in association with decreased alveolar macrophage phagocytic and bactericidal activities. J Immunol 1996; 157:5221–5224.

51. Umans L, Serneels L, Overbergh L, Lorent K, Van Leuven F, Van den Berghe H. Targeted inactivation of the mouse alpha 2-macroglobulin gene. J Biol Chem 1995; 270:19778–19785.

52. Webb DJ, Wen J, Lysiak JJ, Umans L, Van Leuven F, Gonias SL. Murine α-macroglobulins demonstrate divergent activities as neutralizers of transforming growth factor-β and as inducers of nitric oxide synthesis. J Biol Chem 1996; 271:24983–24988.

53. Basu S, Dunn AR, Marino MW, Savoia H, Hodgson G, Lieschke GJ, Cebon J. Increased tolerance to endotoxin by granulocyte-macrophage colony-stimulating factor-deficient mice. J Immunol 1997; 159:1412–1417.

54. Sutton ET, Norman JG, Newton CA, Hellermann GR, Richards IS. Endothelial structural integrity is maintained during endotoxic shock in an interleukin-1 type 1 receptor knockout mouse. Shock 1997; 7:105–110.

55. Katechinski T, Galanos G, Coumbos A, Freudenberg MA. Gamma interferon mediates *Propionibacterium acnes*-induced hypersensitivity to lipopolysaccharide in mice. Infect Immun 1992; 60:1994–2001.

56. Freudenberg MA, Gumenscheimer M, Jack R, Merlin T, Schutt C, Galanos C. A strict requirement for LBP in the TNF hyper-response of *Propionibacterium acnes*-sensitized mice to LPS. J Endotoxin Res 1997; 4:357–361.

57. Ogata M, Yoshida S, Kamochi M, Shigematsu A, Mizuguchi Y. Enhancement of lipopolysaccharide-induced tumor necrosis factor production in mice by carregeenan pretreatment. Infect Immun 1991; 59:679–683.

58. Gu Y, Kuida K, Tsutsui H, Ku G, Hsiao K, Fleming MA, Hayashi N, Higashino K, Okamura H, Nakanishi K, Kurimoto M, Tanimoto T, Flavell R, Sato V, Harding MW, Livingston DJ, Su MS-S. Activation of interferon-γ inducing factor mediated by interleukin-1β converting enzyme. Science 1997; 275:206–209.

53

Endotoxemia in Primate Models

Heinz Redl, Günther Schlag, and Soheyl Bahrami
Ludwig Boltzmann Institute for Experimental and Clinical Traumatology, Vienna, Austria

INTRODUCTION

Complications that occur during sepsis are ultimately the consequence of an overreaction of host defense mechanisms to bacterial insult. The reaction is the so-called generalized inflammatory response, which is an essential reaction of the host to both sepsis and trauma. When overactivated the generalized inflammatory response can act against the host in a self-destructive manner. While the initial inflammatory response can be initiated both by trauma and sepsis, the perpetuation of the response is usually associated with an often not detectable bacterial component. One of the central principles by which bacteria induce the inflammatory response is via bacterial toxins, of which endotoxin from gram-negative bacteria is the most prominent. In order to interfere with an overreacting inflammatory response in patients, interventions at different stages in the inflammatory response have been studied. Such intervention studies are not only performed to investigate therapeutic efficacy of treatment strategies. They are also performed to learn about the pathophysiological role of certain mediators (1) and to test new diagnostic procedures in a defined trauma or sepsis setting (2,3). For such investigations, animal models (together with in vitro and clinical studies) are used to study the pathophysiology and the possible choices of pharmacological interventions in trauma and sepsis. However, when we use animal models, which typically represent human disease only to a limited extent, we should be aware of their inherent limitations.

Limitations of Models

It is beyond the scope of this chapter to cover all facets of animal model use in medical research, therefore, only a few limited examples are outlined. For example, canine models of hemorrhagic shock cannot be used to study the gut (translocation), since a species-specific pooling of splanchnic blood with concomitant bleeding in the intestine is present (4). Studies using porcine and ovine models should be interpreted with care when pulmonary artery pressure (PAP) response or reticuloendothelial system (RES) clearance is measured. In contrast to humans, RES activity in these species is mediated by intravascular macrophages that are especially prevalent in the pulmonary region (5). The special RES location results in a PAP response, for example, at endotoxin challenge (6), that is not seen in humans (7), baboons (1), or chimpanzees (8). Moreover, porcine and ovine granulocytes should not be considered representative of their human counterparts because elastase release and activity differ in the former due to prevalent cytosolic inhibitors (9,10).

To avoid many limitations inherent in other animal models, the use of nonhuman primates has gained recent popularity. The main reasons are comparability with human physiology, as well as cross-reactivity with human therapeutic and diagnostic reagents. The cross-reactivity allows, on the one hand, testing of new species-specific therapies such as antihuman antibodies and, on the other hand, monitoring with available human analytical procedures. Furthermore, marker molecules such as neopterin, a marker of macrophage activation, are only found in primates (11).

Types of Primate Models

Two major groups of animals are used in primate research—baboons (several subspecies) and chimpanzees. The use of *Macaca* monkeys has also been described (12,13). While baboons are used both for sublethal and lethal experiments, chimpanzees are only used (under rather mild conditions) in sublethal experiments, similar to human volunteers.

Baboon models were pioneered by Hinshaw's group (14). These models have been further developed (1,15). The baboon offers all the advantages of a large animal, it is comparable in nearly all physiological and immunological aspects to humans, and it is available for acute and chronic studies (including survival). The majority of chimpanzee studies were performed by the group of van der Poll et al. (8).

SOURCES OF ENDOTOXEMIA

The physical form in which endotoxin is presented to plasma and cells determines how it is handled. For example, when the liver is presented with live bacteria, endotoxin can be localized in Kupffer cells, while the endotoxin purified from the same bacteria and infused into animals is localized in hepatocytes (16). This difference in localization may have important implications for the pathophysiological processes elicited and consequently for the evaluation of treatment strategies.

Bacterial toxins can arise from infectious sites such as abscesses or from nosocomial infections, from bacteria after killing with antibiotics, and from the gut due to translocation, sometimes referred to as occult "undrained" abscess or multiorgan failure (17). Thus, because of differences in endotoxin action based upon the source and physical form, it is important to separate the primate models associated with endotoxemia into different categories.

Role of Lipopolysaccharide Binding Protein and CD14 in Baboon Endotoxemia

There is now a consensus that endotoxin stimulation of human cells such as monocytes occurs via a reaction cascade, which involves lipopolysaccharide (LPS) binding to serum proteins and CD14. In particular, lipopolysaccharide-binding protein (LBP) interacts with the lipid A component of LPS and facilitates its delivery to membrane-bound (or soluble) CD14. The components of this LPS-binding cascade have also been studied in primates. LBP has been both isolated as well as measured from baboon (18,19). LBP was found to increase after both trauma and sepsis in an acute phase reaction with increases in plasma levels similar to C-reactive protein. Baboon LBP was isolated from septic animals with procedures previously described (20), and it was found that it has chromatographic and electrophoretic properties similar to those of its human equivalent. Its similarity is further supported by data from N-terminal sequence analysis.

The occurrence and potential role of CD14 in nonhuman primates was studied both with FACS analysis as well as in vitro peripheral blood cell incubation (19). In addition nonhuman primates (cynomolgus monkeys) were used to study the efficacy of an anti-CD14 therapeutic approach using monoclonal antibodies against CD14 (21). Animals were pretreated with one of two different inhibitory anti-CD14 (mAbs), then challenged with intravenous endotoxin (375 µg/kg/hr) for 8 hours. The anti-CD14 treatment regimens were successful in preventing profound hypotension, reducing plasma cytokine levels, and inhibiting the alteration in lung epithelial permeability that occurred in placebo-treated animals. In vitro studies revealed that stimulation of whole blood specimens with endotoxin in the presence of anti-human CD14 antibodies such as 3C10 and MEM 18 could block the LPS-mediated activation and subsequent cytokine formation of peripheral blood cells (19). This is similar to previous reports in humans.

Sensitivity to Endotoxin

Endotoxin sensitivity is dependent on the species (Table 1) and the type of endotoxin used. Concerning sensitivity, only the chimpanzee compares to humans (8), while other nonhuman primates such as the baboon are rather insensitive and compare with the rat. Sensitivities of other species also differ. The rabbit is used for its higher sensitivity to endotoxin and therefore stronger tumor necrosis factor (TNF) response than mice or rats. Sheep are sensitive to very small doses of endotoxin/bacteria and demonstrate cardiopulmonary changes that resemble the human sepsis syndrome (6,22) except for pulmonary hypertension, which is specific for this model. Pigs are also sensitive to LPS, however, as with sheep, the pulmonary clearance of endotoxin and/or bacteria is due to intravascular macrophages, which induce pulmonary hypertension (mainly via thromboxane release).

Table 1 Examples of Endotoxin Doses Used in Different Species

Species	Dose (μg/kg)	LPS type	Application	Partly lethal	Ref.
Rat	15000	*E. coli* O 26:B6	Bolus	Yes	120
Baboon	1500	*E. coli* O 26:B6	10 min	Yes	58
Pig	5 μg/kg/hr	Salm.abort.equii	Hours	Yes	121
Sheep	0.3–2	*E. coli*	Bolus	Yes	6
Rabbit	1	*E. coli*	30 min	Yes	122
Chimpanzee	0.004	*E. coli* (EC-5)	Bolus	No	8
Human	0.004	*E. coli* (EC-5)	Bolus	No	7

SOURCES OF ENDOTOXEMIA

Trauma-Associated Bacteremia/Endotoxemia (Endogenous Bacteremia/Endotoxemia)

The concept that a breakdown of the intestinal mucosa, a vital component of body's defenses against endogenous infection, may lead to the access of intraluminal pathogens into the body, contributing to morbidity and mortality of systemic sepsis, was developed by Fine in the 1950s (23). Reports of experimental studies, mainly in rodents, that different etiologies may lead to the entry of gut-associated bacteria into the lymphatic systems led Meakins and Marshall to suggest that "the gut is the motor of multiple organ failure" (24), a hypothesis that is still a matter of dispute (25,26). In general, impaired gut barrier permeability might be caused either during the shock period by decreased intestinal blood flow and reduced oxygen delivery (27), during reperfusion resulting in a stage of increased intestinal blood flow, or even at a later stage again by reduced flow (28). In this respect, infections with gut-associated bacteria but with no infectious focus at autopsy have been reported in patients undergoing intraoperative stress, victims of trauma, and patients developing multiple organ failure (MOF) syndrome (24,29). As diagnostic approaches often fail to document the phenomenon of translocation (the access of intraluminal bacteria/endotoxin into the systemic/lymphatic circulation) in patients, data obtained from primate studies performed under standardized and controlled conditions may provide useful information. It should help us better understand the entity of bacterial/endotoxin translocation and its possible role in the development of systemic sepsis/endotoxemia and the related pathophysiological alterations. In baboons subjected to a hemorrhagic traumatic shock (HTS) consisting of bleeding the animals to maintain a mean arterial pressure (MAP) of 40 mmHg for 3 hours, femur fracture, and soft tissue trauma followed by a 3-hour resuscitation period translocation of bacteria into the small vessels of the colon was observed (30). Positive blood cultures were noted in HTS baboons even during the shock and/or reperfusion period (30). At the end of a 6-hour acute experiment in baboons the highest bacterial counts were noted in the mesenteric lymph nodes, followed by the liver and spleen (30). Elevated plasma LPS levels were found in HTS baboons at the end of a 3-hour shock period and 1 hour after the beginning of reinfusion. Similarly, in a subchronic model in baboons consisting of oxygen debt controlled hemorrhage together with an infusion of zymosan-activated plasma (with activated complement components), the highest plasma endotoxin levels were detected at the end of the 3- to 4-hour shock period and at 1 hour after the start of reinfusion (31). The results obtained in subhuman primate models are in agreement with reports of bacterial translocation, mainly in rodents (32–34), and provide further evidence that the phenomenon of translocation may not be limited to lower animals only.

Bacterial Challenge Models

Live *E. coli* i.v.—Single Challenge

The i.v. infusion of bacteria is the most common form of sepsis/septic shock induction in primates. Other than a few studies with gram-positive bacteria (e.g., Ref. 35), in most cases live *E. coli* is used. *E. coli* infusion resulted in an accompanying endotoxemia during and shortly after bacteremia. Nearly all primate studies were performed using serum-sensitive strains (e.g., Ref. 36), which was criticized by Cross and Opal (37), who recommend the use of non–serum-sensitive strains (e.g., 018K1H7). In preliminary studies the non–serum-sensitive bacteria were more aggressive, about equivalent to one log step higher serum-sensitive bacteria, but otherwise similar including the induction of an equivalent endotoxemia (G. Schlag et al., unpublished observations).

Although the strain of bacteria was often the same or similar in many live bacteria/endotoxemia primate studies, considerable differences were seen in the associated host response. The response was found to be dependent on the dose used, e.g., 10^8–10^9/kg (1), 10^{10}/kg (36,38,39), and 10^{11}/kg (40,41). Probably the most important further difference between models is the amount of fluid infused, with only basal amounts in the model of Hinshaw/Taylor (36) but aggressive rescuscitation in the setting by Schlag et al. (titration according to wedge pressure) (1) or Moldawer's group (39). Fluid resuscitation is essential to achieve a normo- to hyperdynamic response (1). Nonaggressive fluid resuscitation leads to a hypo-dynamic pattern, which a former author responding to his own work called "Another Unacceptable Model of Primate Septic Shock" (42) (using *E. coli* injected into the gallbladder of rhesus monkeys).

Bacteria infusion leads to an accompanying endotoxemia in the range of one to several hundred ng/ml and is routinely monitored, e.g., in the model of Schlag et al. (1), where endotoxemia is generally found only up to 4 hours (after a 2-hour infusion of *E. coli*).

Part of the problem of highly variable LPS measurements may result from the type of analysis used. Although the limulus (LAL) assay is the standard method, variable techniques of plasma inactivation (e.g., heat), types of LAL, and particularly measurement methods—endpoint vs. kinetic (43)—make a lot of a difference. Due to a sharp increase and wide concentration range, a kinetic method (without necessary matrix dilution artifacts) is the method of choice.

A baboon model with elements that are similar to those observed in human diffuse intravascular coagulation, thrombotic thrombocytopenic purpura, and hemolytic-uremic syndrome has been developed (44). The infusion of C4b-binding protein (C4bBP) before infusion of a sublethal dose of *E. coli* produced a consumptive coagulopathy characterized by marked reductions in platelets and plasma fibrinogen, microangiopathic hemolytic anemia, and microvascular thrombosis-induced renal failure (44). The C4bBP infusion decreased the protein S levels, thus inhibiting the protein C anticoagulant pathway and predisposing the animals to thrombotic complications (45). This most likely accounts for the exaggregated response to the sublethal *E. coli* infusion. In this model protection was observed using a antibody to the platelet receptor GPIIb/IIIa (46).

Live *E. coli*—Multiple Challenge Model

In order to avoid several problems of the single-challenge model (high bacterial dose, partly hypodynamic response, overwhelming endotoxemia), Schlag et al. introduced a "multiple challenge" model, where live *E. coli* are administered three times every 24 hours (1). In contrast to a repeated endotoxin infusion, bacteria (with accompanying endotoxemia) did not induce the well-known tolerance phenomenon with attenuation of the TNF response. Multiple bacterial challenges induce repeated secretion of TNF and other cytokines every 24 hours (with only a slight trend to lower levels). The beauty of this model, which was set up to simulate recurrent endotoxemia/bacteremia in septic patients, is that it allows true late ("posttreatment") therapeutic investigations, which were performed for anti-TNF antibodies (47).

A specific "multiple approach" was designed by Welty-Wolf et al. (48), who showed that exposure to heat-killed bacteria modifies pathological and hemodynamic consequences of sepsis with live *E. coli*. Animals who were primed before the onset of sepsis developed more lung injury as measured by qualitative and quantitative pathology. The authors used the term priming to describe the findings in this study, though they encountered both increased and decreased injury responses in different organ systems in these animals. The authors used whole heat-killed bacteria rather than LPS because endotoxin in whole bacteria may be processed somewhat differently by the immune system (48). Unfortunately no data on the degree of endotoxemia are available.

LPS i.v.

The use of primate models that use infusion of chemically prepared endotoxin offers three logistic advantages over the use of (live) bacteria: no need for bacterial cultures, easy dosage, and easy storage. The following disadvantages must also be considered: different physicochemical status (and action), fast clearance, and lack of other bacterial factors such as peptidoglycans, which could synergize with LPS in the host. An anesthetized endotoxemic baboon model has been developed by infusing 2 mg *E. coli* endotoxin/kg i.v. over 1 hour that is characterized by early (2–6 hr) cardiovascular and metabolic derangement, late organ dysfunction (24–48 hr), subclinical DIC, and a 30%

mortality between 48 and 72 hours (49). These results are in contrast to most other endotoxin models in baboons, where with large challenge doses up to 30 mg/kg (12) early circulatory shock, multiple organ failure, severe DIC, complement consumption (50), and deaths occurred within hours (51–53). Several years ago an LD_{70} model over 72 hours was used to study several therapeutic interventions by Fletcher's group (54–56), and in 1973 a lead-sensitized model (using 70 μg LPS/kg) was reported (57). A model of endotoxic shock in baboons with 1.5 mg/kg *E. coli* endotoxin as a 10-minute intravenous infusion was used to study soluble adherence molecules. Soluble E-selectin was released in vivo after application of endotoxin and reached a peak level after 24 hours (58). Most recently a paper was published describing a sublethal baboon model using 0.5 mg/kg, which was used to determine the effects of IL-10 on LPS-induced inflammatory responses (59).

In addition to baboons, chimpanzees have been used in endotoxin infusion studies to test anti-TNF mAb (8) in a manner similar to human volunteers studies. The low dose of endotoxin chosen (4 ng/kg—similar to humans) has been shown to induce reproducible systemic inflammatory responses. Endotoxin elicited activation of the coagulation and the fibrinolytic system, leukocytosis, neutrophilia, and lymphopenia. Following endotoxemia, a TNF-dependent bronchoalveolar activation of coagulation and depression of fibrinolysis as pathogenetic mechanism of adult respiratory distress syndrome (ARDS) was seen (60). Severe clinical signs of sepsis and symptoms are minimal. Chimpanzee studies have shown as an advantage high cross-reactivity with human reagents and LPS sensitivity as found in humans. A disadvantage is that no severe sepsis models are available, only short-term experiments under anesthesia can be conducted, and no organ dysfunction or survival studies are possible.

ACTIONS OF ENDOTOXIN-PATHOPHYSIOLOGICAL RESPONSES OF PRIMATES

Cardiovascular System

The cardiovascular response to low-dose endotoxin infusion in chimpanzees is similar to those found in human volunteers (7). An increase in heart rate, cardiac output, but no or minor changes in pulmonary artery pressures are observed, which is in contrast to species like sheep or pigs (2). With severe endotoxemia, such as seen in live *E. coli* sepsis in baboons, there can also be an initial drop in cardiac output, followed by a hyperdynamic response if adequately resuscitated (1,61) or a continuing hypodynamic pattern if not adequately resuscitated (36). In a time period beyond about 12 hours, the consequence of previous endotoxemia is a declining mean arterial pressure from increased nitric oxide (NO) production and its actions on vascular smooth muscle cells. The increased NO is mainly the product of the inducible NO-synthase (iNOS), which is induced by endotoxin. Increased iNOS activity is supposed to be the underlying mechanism of increased plasma levels of NO_2/NO_3 (a breakdown product of NO) after about 8 hours postendotoxemia in baboons (62). The NO-mediated decline in MAP with a decreased peripheral resistance (SVR) (if normo/hyperdynamic) leads to microcirculatory derangement, cell damage, and organ dysfunction. Countermeasures have been taken in a recent baboon trial, where an inhibitor of NOS (546C88 N-mono-methylarginine) infused during the phase of vasoplegia (12–48 hours) could lower NO_2/NO_3 plasma levels and attenuate the drop in SVR and organ damage. As a consequence a significantly improved survival rate was found after 6 days (63). Again an important species difference was found. The increase in NO_2/NO_3 in baboons was similar to humans, while it is more pronounced in rodents.

In addition to endogenous vasodilators such as NO, there are also endotoxin-inducible vasoconstrictors such as endothelin. Endothelin and its precursor big-endothelin (big ET-1) are endothelium-derived vasoactive substances that play an important role in regulating the vascular tone. LPS induces the release of big ET-1 in a subhuman primate model, and tumor necrosis factor is an important mediator of big ET-1 release (64).

Cytokine Induction

As has been demonstrated in different species, but probably most convincingly in primates (due to available therapeutic and diagnostic reagents) endotoxin (either liberated from bacteria or isolated LPS) induces "secondary" cellular responses via "primarily" released cytokines, in particular TNF (and IL-1). There are many reactions induced (for review, see Ref. 65). The induction and release of a number of so-called secondary cytokines are partly dependent on TNF or IL-1, in particular IL-8 (66), IL-6 (66,67), and IL-1 (66,67).

Results indicate that TNF, in part, also mediates the induction of IL-10 in endotoxemia, resulting in an autoregulatory feedback loop (68), which was not evident in live *E. coli* bacteremia (69). IL-6 was found to be an important mediator of low-grade endotoxin-induced coagulation.

Due to the multifunctional properties of TNF, blockade by either antihuman TNF antibody (47) or TNF receptor fusion protein (41) gave protection in baboon bacteremia-associated endotoxemia. The effect of anti-TNF antibody was found to be dose dependent (70). In chimpanzees, treatment with the anti-TNF-α antibody completely neutralized endotoxin-induced TNF-α activity. The release of soluble TNF receptors was strongly (80–90%) inhibited in the presence of the neutralizing antibody (71).

Endotoxemia leads to shedding of TNF receptor molecules (TNFR) and thus also to higher plasma levels of soluble TNFR (72). In parallel studies with live *E. coli* (and accompanying endotoxin release), the application of anti-TNF antibodies attenuated shedding of TNF receptors (73), but much less than in pure endotoxemia (74). Similar effects were seen when IL-1 activity was eliminated by IL-1 receptor antagonist (IL-1ra), which attenuated the rise in sTNFR-I (75). Furthermore, in septic baboons, a significant reduction of the formation of thrombin, t-PA, PAI was observed in the group receiving IL-1ra. Thus, IL-1 is similar to TNF as it contributes to activation of various other mediator systems in severe sepsis in nonhuman primates (76).

The induction of cytokines by endotoxin is influenced by several factors such as catecholamines, especially epinephrine (studied in chimpanzees) (77), corticosteroids (78), or platelet activating factor (PAF). The specific and potent PAF antagonist TCV-109 reduced the endotoxin-induced rise in levels of cytokines such as TNF, IL-6, and IL-8. TCV-309 also reduced the appearance of soluble TNFR (79).

Various drugs that block TNF production by LPS-activated macrophages in vitro have been used to mitigate LPS toxicity in vivo. Drugs that act at the transcriptional level, such as pentoxifylline (80) or derivatives thereof such as lisofylline (81) and HWA138, are found to protect against endotoxin shock (82) but are not fully protective in baboon sepsis (G. Schlag et al., unpublished data). Chimpanzees were infused with pentoxyfylline shortly before low-dose endotoxin administration (83). Pentoxyfylline markedly inhibited increases in the levels of TNF and IL-6, as well as the effects on coagulation and fibrinolysis. The discrepancy between the effect seen in the baboon *E. coli* model and the LPS chimpanzee model is probably due to the large differences in the level of endotoxemia. Thus, high endotoxin levels (ng/ml) in the baboon *E. coli* model LPS appears to induce responses additional to cytokine induction (which is attenuated by pentoxyfylline). Such differences in response to LPS are similar to previous results in the rat, where HWA 138 prevented cytokine formation only in (sensitized) rats receiving low doses (μg) of LPS, but not in (normal) rats receiving high endotoxin doses (84).

Endothelial Activation

Endotoxin causes endothelial activation with induction of cytokine production, procoagulatory as well as vasopressor/vasodilator responses, and an increase in adhesion for leukocytes. Cell adhesion molecules, which are responsible for the adhesion and later for the migration of leukocytes into the perivascular space, play a critical role in the development of many inflammatory conditions. One interesting molecule is E-selectin. E-selectin is an endothelial cell–specific adhesion molecule that appears to only be expressed in response to inflammatory stimuli (85). In vivo, E-selectin induction has been seen after local injection of TNF-α in baboon skin (86) and after infusion of live *E. coli* in baboons (87) as well as endotoxin infusion in cynomolgus monkey (88). Using skin biopsies it was possible to establish a kinetic expression pattern; expression was not prominent at 4–6 hours and not detectable after 10–24 hours, while serum levels of soluble E-selectin were maximal after 24 hours of endotoxic shock conditions. While E-selectin can therefore serve as a marker for endothelial activation, it cannot be determined directly in vivo except by invasive biopsy techniques. Alternatively, the serum level of the soluble form (sE-selectin) can be determined. sE-selectin was released in vivo after application of endotoxin (1.5 mg/kg) in baboons and reached a peak level after 24 hours (58). In baboons with hemorrhagic shock much lower levels (about one third) of sE-selectin were found. Lower sE-selectin indicates a lower endothelial activation after experimental hemorrhagic traumatic shock (with endotoxin translocation from the gut) as compared to endotoxic shock, probably due to much lower endotoxin levels in traumatic shock (58).

Baboons were also used in sepsis studies with antiadherence therapy approaches such as anti-E-selectin (unpublished results) or anti-L-selectin (unpublished). Anti-αICAM-1 did not prevent acute lung injury in *E. coli* sepsis, and septic animals given anti-αICAM-1 have worsened metabolic parameters and decreased survival (89). A similar observation was made with an anti-CD18 antibody treatment (90).

PMN/Macrophage Activation

It is a common observation in nonhuman primates that endotoxin (or bacteria) infusion leads to an initial phase of neutropenia followed by neutrophilia, which is seen in baboon typically at 12–48 hours (1,91). The endotoxin-induced PMN activation also leads to a degranulation response with release of elastase [in baboons (91), in chimpanzees (79)] and lactoferrin similar to the human situation, which cannot be detected in some other species. The elastase release is also partially dependent on endotoxin-induced mediators such as TNF in *E. coli*–challenged baboons, as demonstrated with anti-TNF antibodies (66,92). Such dependence could only be observed if the *E. coli* dose was moderate, e.g., 10^8–10^9 CFU/kg (1), but not with 10^{10} CFU/kg (36).

The release of neutrophil elastase was also significantly attenuated in IL-1a–treated animals (76). In chimpanzees injected with *E. coli* endotoxin (4 ng/kg) neutrophil degranulation, as measured by the plasma concentration of elastase–α_1-antitrypsin complexes, was only slightly reduced by one anti-TNF monoclonal antibody (8), but another antibody (MAK 195F) significantly abrogated neutrophil degranulation as measured by plasma concentration of lactoferrin (71). The antibody MAK 195F is interesting in that it is one of the rare examples of anti-human antibodies that cross-react with chimpanzees but not with baboons.

While PAF is important in endotoxin-induced cytokine production, endotoxin-induced neutrophil degranulation, as monitored by elastase (79), was not effected by the PAF antagonist TCV-309.

Activated leukocytes in context with activated endothelium lead to leukostasis of organs, tissue damage, and organ failure. The central event is the interaction between adherence molecules, which blockade [e.g., P-selectin (93)] might lead to an attenuation. In septic baboons such attenuation was not found for an anti-CD18 (90) or anti L-selectin (unpublished) approach, while reduction of PMN activation was found with anti-TNF therapy (attenuated leukostasis of the lung) (unpublished).

There is consensus that monocytes/macrophages are central in the inflammatory response to endotoxin, being the primary source of cytokines and many other mediators. Since many other cells contribute to the cytokine production, cytokines are not specific for macrophage activation. In primates there is, however, a specific marker of macrophage activation via an endotoxin–IFN-γ pathway, by which neopterin, a low molecular weight substance of the pteridine system, is produced. Neopterin in reasonable amounts can only be found in primates in response to endotoxemia in a dose-dependent fashion (62).

Coagulation Induction

There is clear evidence that the tissue factor–dependent pathway of blood coagulation is central in the initial endotoxin-induced activation of coagulation. Tissue factor (TF), expressed on the surface of activated monocytes and endothelial cells, forms cell surface complexes with free circulating factors VII and VIIa (94). Administration of the 12D10 Fab fragment of a factor VII/VIIa-neutralizing murine monoclonal antibody immediately preceding the endotoxin bolus injection effectively blocked the endotoxin-induced activation of coagulation (94). Endotoxin-induced generation of thrombin was completely prevented by the administration of tissue factor–neutralizing antibodies or the naturally occurring inhibitor tissue factor pathway inhibitor (TFPI), which was successfully used in baboons (95) and chimpanzee (83). New data indicate that a lethal outcome in severe sepsis is associated with an increased induction of Fas-mediated apoptosis and that TFPI interferes with this induction (96). Pixley and coworkers (97) demonstrated that infusion of a monoclonal antibody that blocks factor XII activity can partially prevent the fall in blood pressure and significantly prolong survival in septic baboons.

Anti-IL-6 markedly attenuated endotoxin-induced activation of coagulation, monitored with the plasma levels of the prothrombin fragment F1+2 and thrombin–antithrombin III complex, whereas activation of fibrinolysis, determined with the plasma concentrations of plasmin–α_2-antiplasmin complexes, remained unaltered in low-grade endotoxemia in chimpanzee (98).

Studies on the inflammatory-coagulant axis in the baboon response to *E. coli* (and endotoxin) revealed regulatory roles of proteins C, S, C4bBP, and of inhibitors of tissue factor (99). There is an additional effect on anticoagulant systems. Inhibition of protein S activity (possibly by one of the forms of C4b-binding proteins) might be one of the factors contributing to microvascular thrombotic disorder after live *E. coli* in baboons, such as the hemolytic-uremic syndrome (45).

Fibrinolysis Induction

Besides the endotoxin-induced activation of coagulation, an important role for endotoxin in the activation of the fibrinolytic system has been found (100), however, the fibrinolytic response triggered by endotoxin is not dependent on the generation of thrombin (83). Infusion of endotoxin induced a rapid increase in plasminogen activator activity and tissue-type plasminogen activator antigen levels and subsequent plasmin generation, reaching peak levels 2 hours after endotoxin administration. Plasminogen activator inhibitor 1 levels remained constant for the first 2 hours, after which time a steep increase was observed. Plasminogen activator activity and plasmin generation decreased simultaneously with the rise in plasminogen activator inhibitor 1 levels. Fibrinolytic activity remained suppressed during the remainder of the study owing to sustained increased levels of plasminogen activator inhibitor 1 (100). The administration of anti-TNF antibodies blocked the fibrinolytic response induced by infused LPS (100), while scavenging TNF in live *E. coli* with accompanying endotoxemia only partly reduced the fibrinolytic response (101). Also the use of a 55 kDa tumor necrosis factor receptor-immunoglobulin fusion protein attenuated activation of coagulation, but not of fibrinolysis, during lethal bacteremia in baboons (102).

THERAPEUTIC INTERVENTIONS

Interventions in the inflammatory response can be performed at several levels: (1) block the inducer (e.g., by anti-LPS); (2) interference with induction; (3) block at the intermediate mediator level (e.g., by anti-TNF); (4) block at the final mediator level (e.g., by anti-PMN elastase); and (5) block at the target (e.g., by membrane stabilization or enhanced antioxidant defense). Since a thorough discussion of the many possible approaches is beyond the scope of this chapter, the use of nonhuman primates in intervention studies at levels 2–5 is summarized in Table 2 (and were partly referred to in the Pathophysiological Response section), while the few studies at level 1 (with scavenging of the inducers) are described below.

Among the possibilities of intervention, binding and inactivating bacterial toxins is the most attractive because it occurs at the top of the inflammatory network such that it avoids downstream multiple mediator induction and does not directly interfere with the immune system, minimizing unpredictable side effects and blockade of favorable host responses. Due to the physiological similarity with humans as well as the cross-reactivity with human reagents, primates are (and should be) used as ultimate models before going into clinical trials. For this review of studies in primate models, the concept was used to separate according to the source of endotoxin: (1) translocation from the endogenous sources, (2) endotoxin from bacterial infections, and (3) exogenously applied endotoxin.

Trauma Studies—Translocation of Endotoxin/Bacteria from the Gut

Among the several possibilities to interfere therapeutically with the phenomenon of endotoxin/bacterial translocation in trauma and hemorrhage (for review, see Ref. 103), therapy with antibiotic (104) or scavenging of endotoxin has been shown to be beneficial. As an example a murine monoclonal antibody, which is widely cross-protective against LPS from members of Enterobacteriaceae (105), not only protected the animals but significantly reduced the incidence of bacterial translocation in a rat model of prolonged hemorrhagic shock (106). In the same rat model as an alternative, endotoxin (translocated during or after hemorrhagic shock) was blocked with bactericidal/permeability increasing protein (BPI), which led to a favorable outcome (107). Based upon the successful rat results, studies in nonhuman primates are currently in progress, but no final results are available.

Sepsis Studies (Endotoxin from Bacterial Infection)

Antibody

To block bacterial toxins with antibodies is and has long been an attractive approach to deal with the deleterious sequelae of sepsis. Cross-reactivity among different endotoxins (or a mixture of specific antibodies) is a precondition, which was attempted with anti-core antibodies (e.g., Ref. 105) or anti-idiotype antibodies to a core LPS-epitope (108). Despite successful rodent studies to our knowledge, none of these drugs were thoroughly tested in nonhuman primates except for HA-1A. However, the experiments in a baboon live *E. coli* model were performed parallel or after the clinical trial and revealed no reduction in (LAL-measured) endotoxin levels by HA-1A administration (109). Such testing beforehand might have avoided negative trials in humans (as with HA-1A or E5).

Bacterial/Permeability-Increasing Protein

In contrast to anti-endotoxin antibodies, another LPS-neutralizing approach (with additional bactericidal activity)—the infusion of exogenous bacterial/permeability-increasing protein (BPI)—has been tested in nonhuman primates. BPI is a cationic protein (55 kDa)

Table 2 Examples of Therapeutic Interventions in Nonhuman Primate Endotoxemia

Agent	Model	Ref.
Scavenging inducers		
rBPI$_{21/55}$	Baboon sepsis	115,116
anti-Lipid A	Baboon sepsis	109
Interference of induction		
anti-CD14	Cynomologous sepsis	21
Pentoxyfylline	Baboon sepsis	G. Schlag et al., unpublished
Pentoxyfylline	Chimpanzee LPS	83
Corticosteroid	Baboon sepsis	78
Epinephrine	Chimpanzee LPS	77
PAF-antagonist	Chimpanzee LPS	79
IL-10	Baboon LPS	59
Intermediate mediator level		
anti-TNF	Baboon sepsis	36,40,47,102,123
anti-TNF	Chimpanzee LPS	8
IL-1ra	Chimpanzee LPS	39,76
anti-IL6	Chimpanzee LPS	98
Downstream mediators		
TFPI	Baboon sepsis	95
anti-TF	Chimpanzee LPS	83
anti-FVII/VIIa	Baboon sepsis	94
anti-FXII	Baboon sepsis	97
Protein C	Baboon sepsis	99
AT III	Baboon sepsis	124
anti-ICAM	Baboon sepsis	89
anti-CD18	Baboon sepsis	90
anti-L-Selectin	Baboon sepsis	G. Schlag et al., unpublished
anti-E-Selectin	Baboon sepsis	G. Schlag et al., unpublished
NO-Synthase Inhib.	Baboon sepsis	63
Target		
Corticosteroids	Baboon sepsis	78

naturally found in azurophilic granules of polymorphonuclear leukocytes (PMN) (110,111), which is also released during PMN activation and degranulation. Both a 55 kDa as well as a 23/21 kDa recombinant NH$_2$-terminal fragment of BPI (BPI$_{21}$) were developed (112,113), which retain the bactericidal and endotoxin-neutralizing properties.

Administration of *E. coli* endotoxin (4 ng/kg) elicited subsequent increases in the plasma concentrations of endogenous BPI. The rise in BPI levels remained unaltered also if TNF was scavenged (114). The endogenously released BPI is, however, not sufficient for endotoxin scavenging in sepsis, therefore recombinant BPI (either full length—55 kDa—or fragments thereof—BPI$_{21}$ or fusion proteins) was used in live *E. coli* sepsis (109). BPI-treated baboons demonstrated significantly lower circulating LPS-limulus amebocyte lysate (LAL) activity compared with the control animals, but this reduction in LPS-LAL activity was not associated with improved survival (115). TNF-α and IL-6 levels likewise were attenuated, and sTNFR I was significantly reduced in the BPI group.

In contrast to the results with BPI-55, the administration of BPI-21 was found to be highly protective in another *E. coli* baboon model (116). Intravenous BPI$_{21}$ attenuated sepsis-related organ failure and increased survival significantly. Microcirculation and organ function was improved. Bacteremia was significantly reduced in the BPI group at 2 hours after start of *E. coli* infusion, while circulating LPS was not affected. The in vivo formation of TNF was significantly suppressed by the BPI$_{21}$ treatment regimen (116). In a further acute baboon experiment, animals receiving BPI-55 or a genetically engineered variant of BPI (NCY103) revealed decreased blood levels of LPS (115). Leukocytopenia and granulocytopenia were significantly lessened (115).

Exogenous Endotoxin

Not surprisingly, only a few studies have been performed with endotoxin infusion and endotoxin scavenging except for studies with high-density lipoproteins, which were done in "human" primates (117). Other LPS-binding compounds such as cationic antibacterial proteins (e.g., CAP-18 with LPS-binding and LPS-neutralizing activity) (118) have not been used so far in primate models. Interference with the endotoxin response by LPS antagonists (119) has not been tested in primate models to our knowledge.

CONCLUSION

We believe that nonhuman primates offer the best opportunity to study endotoxemia-related pathophysiology in trauma and sepsis. Therapeutic measures for sepsis in these animals react with high similarity to human models in many aspects. Of course not every diagnostic or therapeutic approach can be tested in primates, but if there are successful preclinical studies in other species, final experiments in nonhuman primates should be obligatory before clinical testing.

ACKNOWLEDGMENTS

We are indebted to M. Grossauer and C. Wilfing for the help in preparation of this paper as well as Dr. McCulloch for language revisions. Part of the described studies were supported by Lorenz Böhler Fond.

REFERENCES

1. Schlag G, Redl H, Davies J, van Vuuren CJJ, Smuts P. Live *Escherichia coli* sepsis models in baboons. In: Schlag G, Redl H, eds. Pathophysiology of Shock, Sepsis, and Organ Failure. Berlin: Springer-Verlag, 1993:1076–1107.
2. Redl H, Schlag G, Bahrami S, Yao Y-M. Animal models as the basis of pharmacologic interventions in trauma and sepsis. World J Surg 1996; 20:406–410.
3. Redl H, Schlag G. Biochemical/immunological monitoring in the posttraumatic course. In: Risberg B, editor. Trauma Care—An Update. Göteborg: Pharmacia and Upjohn Sverige AB, 1996:80–87.
4. Selkurt EE, Rothe CF. Pressure gradients in the splanchnic bed of the monkey during hemorrhagic shock. Proc Soc Exp Biol Med 1962; 111:57.
5. Dehring DJ. Sheep and pigs as animal models of bacteremia. In: Schlag G, Redl H, eds. Pathophysiology of Shock, Sepsis, and Organ Failure. Berlin: Springer-Verlag, 1993:1060–1075.
6. Traber DL, Traber LD, Redl H, Schlag G. Models of endotoxemia in sheep. In: Schlag G, Redl H, eds. Pathophysiology of Shock, Sepsis, and Organ Failure. Berlin: Springer-Verlag, 1993:1031–1047.
7. Suffredini AF, Fromm RE, Parker MM, Brenner M, Kovacs JA, Wesley RA, et al. The cardiovascular response of normal humans to the administration of endotoxin. N Engl J Med 1989; 321:280–287.
8. Van der Poll T, Levi M, Van Deventer SJH, Ten Cate H, Haagmans BL, Biemond BJ, et al. Differential effects of anti-tumor necrosis factor monoclonal antibodies on systemic inflammatory responses in experimental endotoxemia in chimpanzees. Blood 1994; 83:446–451.
9. Junger WG, Hallström S, Liu FC, Redl H, Schlag G. The enzymatic and release characteristics of sheep neutrophil elastase: a comparison with human neutrophil elastase. Biol Chem Hoppe Seyler 1992; 373:691–698.
10. Geiger R, Junk A, Jochum M. Leukocyte elastase-inhibitor complexes in porcine blood. II. Isolation and characterization of porcine leukocyte elastase. J Clin Chem Clin Biochem 1985; 23:821–823.
11. Strohmaier W, Werner ER, Wachter H, Redl H, Schlag G. Pteridine and nitrite/nitrate formation in experimental septic and traumatic shock. Shock 1996; 6:254–258.
12. Vaughn DL, Gunter CA, Stookey JL. Endotoxin shock in primates. Surg Gynecol Obstet 1968; 126:1309–1317.
13. Vaughn DL, Peterson E. Pathophysiology of endotoxin shock in primates and the effect of various therapeutic agents. Obstet Gynecol 1969; 34:271–276.
14. Hinshaw LB. Application of animal shock models to the human. Circ Shock 1985; 17:205–212.
15. Hawes AS, Fischer E, Marano MA, VanZee KJ, Rock CS, Lowry SF, et al. Comparison of peripheral blood leukocyte kinetics after live Escherichia coli endotoxin or interleukin 1 alpha administration. Studies using a novel interleukin 1 receptor antagonist. Ann Surg 1993; 218:79–90.
16. Ge Y, Ezzell RM, Tompkins RG, Warren HS. Cellular distribution of endotoxin after injection of chemically purified lipopolysaccharide differs from that after injection of live bacteria. J Infect Dis 1994; 169:95–104.
17. Marshall JC, Christou NV, Meakins JL. The gastrointestinal tract. The undrained abscess of multiple organ failure. Ann Surg 1993; 218:111–119.
18. Natmessnig B, Redl H, Schlag G, Hatlen L, Tobias PS. Baboon lipopolysaccharide binding protein: isolation, initial characterisation, and concentration in experimental sepsis. Intensive Care Med 1994; 20, Suppl 1:S12.
19. Natmessnig BE. Induction of cytokines by endotoxin —comparison of non-human and human primates

with special emphasis on the CD14/LBP system. Thesis Vienna Technical University, 1996.
20. Schumann RR, Leong SR, Flaggs GW, Gray PW, Wright SD, Mathison JC, et al. Structure and function of lipopolysaccharide binding protein. Science 1990; 249:1429–1431.
21. Leturcq D, Moriarty AM, Talbott G, Winn RK, Martin TR, Ulevitch RJ. Antibodies against CD14 protect primates from endotoxin induced shock. J Clin Invest 1996; 98:1533–1538.
22. Krösl P, Pretorius J, Redl H, Schlag G. Myocardial function in septic sheep. Shock 1994; 1:325–334.
23. Fine J, Frank E, Rutenburg S, Schweinberg F. The bacterial factor in traumatic shock. N Engl J Med 1959; 260:214–220.
24. Carrico CJ, Meakins JL, Marshall JC, Fry D, Maier RV. Multiple-organ-failure syndrome. The gastrointestinal tract: the "motor" of MOF. Arch Surg 1986; 121:196–208.
25. Moore F, Poggetti R, McAnena O, Peterson V, Abernathy C, Parsons P. Gut bacterial translocation via the portal vein: a clinical perspective with major torso trauma. J Trauma 1991; 31:629–638.
26. Peitzman A, Udekwu A, Ochoa J, Smith S. Bacterial translocation in trauma patients. J Trauma 1991; 31:1083–1087.
27. Haglund U. Hypoxic damage. In: Schlag G, Redl H, eds. Pathophysiology of Shock, Sepsis, and Organ Failure. Berlin: Springer-Verlag, 1993:314–321.
28. Groeneveld AB, Kester ADM, Nauta JJP, Thijs LG. Relation of arterial blood lactate to oxygen delivery and hemodynamic variables in human shock states. Circ Shock 1987; 22:35–53.
29. Garrison RN, Fry DE, Berberich S, Polk HC. Enterococcal bacteremia: clinical implications and determinants of death. Ann Surg 1982; 196:43–47.
30. Schlag G, Redl H, Dinges HP, Davies J, Radmore K. Bacterial translocation in a baboon model of hypovolemic-traumatic shock. In: Schlag G, Redl H, Siegel JH, Traber DL, eds. Shock, Sepsis, and Organ Failure. Second Wiggers Bernard Conference. New York: Springer-Verlag, 1991:53–83.
31. Schlag G, Redl H, Davies J, Van Vuuren CJ, Smuts P. Bacterial translocation during traumatic shock in baboons. In: Schlag G, Redl H, eds. Pathophysiology of Shock, Sepsis, and Organ Failure. Berlin: Springer-Verlag, 1993:279–291.
32. Deitch EA, Baker T, Berg R, Ma L. Hemorrhagic shock promotes the systemic translocation of bacteria from the gut. J Trauma 1987; 27:815.
33. Jiang JX, Bahrami S, Leichtfried G, Redl H, Oehlinger W, Schlag G. Kinetics of endotoxin and tumor necrosis factor appearance in portal and systemic circulation following hemorrhagic shock in rats. Ann Surg 1995; 221:100–106.
34. Rush BFJ, Redan JA, Flanagan JJ, Heneghan JB, Hsieh J, Murphy TF, et al. Does the bacteremia observed in hemorrhagic shock have clinical significance? A study in germfree animals. Ann Surg 1989; 210:342–347.
35. Simon GL, Gelfand JA, Connolly RA, O'Donnel TF, Gorbach SL. Experimental bacteroides fragilis bacteremia in a primate model: evidence that bacteroides fragilis does not promote the septic shock syndrome. J Trauma 1985; 25:1156–1162.
36. Hinshaw LB, Tekamp Olson P, Chang ACK, Lee PA, Taylor FB, Murray CK, et al. Survival of primates in LD100 septic shock following therapy with antibody to tumor necrosis factor. Circ Shock 1985; 12:273–284.
37. Cross AS, Opal SM. Endotoxin's role in gram-negative bacterial infection. Curr Opin Infect Dis 1995; 8:156–163.
38. Wilson MF, Brackett DJ, Hinshaw LB, Tompkins P, Archer LT, Benjamin BA. Vasopressin release during sepsis and septic shock in baboons and dogs. Surg Gynecol Obstet 1981; 153:869–872.
39. Fischer E, Marano MA, Van Zee KJ, Rock CS, Hawes AS, Thompson WA, et al. Interleukin-1 receptor blockade improves survival and hemodynamic performance in Escherichia coli septic shock, but fails to alter host responses to sublethal endotoxemia. J Clin Invest 1992; 89:1551–1557.
40. Tracey KJ, Fong Y, Hesse DG, Manogue KR, Lee AT, Kuo GC, et al. Anti cachectin/TNF monoclonal antibodies prevent septic shock during lethal bacteremia. Nature 1987; 330:662–664.
41. Van Zee KJ, Kohno T, Fischer E, Rock CS, Moldawer LL, Lowry SF. Tumor necrosis factor soluble receptors circulate during experimental and clinical inflammation and can protect against excessive tumor necrosis factor alpha in vitro and in vivo. Proc Natl Acad Sci U S A 1992; 89:4845–4849.
42. Shatney CH, Read G. Another unacceptable model of primate septic shock. Adv Shock Res 1981; 6:1–13.
43. Bahrami S, Leichtfried G, Redl H, Schlag G. A kinetic chromogenic method to determine endotoxin on microplates. Eur Clin Lab News 1991; Oct. 8–9.
44. Taylor FB, Jr., Chang AK, Ferrell G, Mather T, Catlett R, Blick K, et al. C4b-binding protein exacerbates the host response to Escherichia coli. Blood 1991; 78:357–363.
45. Taylor FB, Jr., Dahlback B, Chang AC, Lockhart MS, Hatanaka K, Peer G, et al. Role of free protein S and C4b binding protein in regulating the coagulant response to Escherichia coli. Blood 1995; 86:2642–2652.
46. Taylor FB, Coller BS, Chang ACK, Peer G, Jordan R, Engellener W, et al. 7E3 f(ab')2, a monoclonal antibody to the platelet GPIIb/IIIa receptor, protects against microangiopathic hemolytic anemia and microvascular thrombotic renal failure in baboons treated with C4b binding protein and a sublethal infusion of Escherichia coli. Blood 1997; 89:4078–4084.
47. Schlag G, Redl H, Davies J, Haller I. Anti-tumor necrosis factor antibody treatment of recurrent bacteremia in a baboon model. Shock 1994; 2:10–18.
48. Welty-Wolf KE, Carraway MS, Huang Y-CT, Simonson SG, Kantrow SP, Piantadosi CA. Bacterial priming increases lung injury in gram negative sepsis. In press.
49. Lindsey DC, Emerson TE, Jr., Thompson TE, John AE, Duerr ML, Valdez CM, et al. Characterization of an endotoxemic baboon model of metabolic and organ dysfunction. Circ Shock 1991; 34:298–310.

50. Coalson JJ, Benjamin B, Archer LT, Beller B, Gilliam CL, Taylor FB, et al. Prolonged shock in the baboon subjected to infusion of *E. coli* endotoxin. Circ Shock 1978; 5:423–437.
51. Selmyer JP, Reynolds DG, Swan KG. Renal blood flow during endotoxin shock in the subhuman primate. Surg Gynecol Obstet 1973; 137:3–6.
52. Hinshaw LB, Emerson TE, Jr., Reins DA. Cardiovascular responses of the primate in endotoxin shock. Am J Physiol 1966; 210:335–340.
53. Hinshaw LB, Shanbour LL, Greenfield LJ, Coalson JJ. Mechanism of decreased venous return in sub-human primate administered endotoxin. Arch Surg 1970; 100:600–606.
54. Fletcher JR, Ramwell PW. Indomethacin treatment following baboon endotoxin shock improves survival. Adv Shock Res 1980; 4:103–111.
55. Fletcher JR, Ramwell PW. Lidocaine treatment following baboon endotoxin shock improves survival. Adv Shock Res 1979; 2:219–232.
56. Fletcher JR, Ramwell PW, Harris RH. Thromboxane, prostacyclin, and hemodynamic events in primate endotoxin. Adv Shock Res 1981; 5:143–148.
57. Holper K, Trejo RA, Brettschneider L, DiLuzio NR. Enhancement of endotoxin shock in the lead sensitized subhuman primate. Surg Gynecol Obstet 1973; 136:593–601.
58. Kneidinger R, Bahrami S, Redl H, Schlag G, Robinson M. Comparison of endothelial activation during endotoxic and posttraumatic conditions by serum analysis of soluble E-selectin in nonhuman primates. J Lab Clin Med 1996; 128:515–519.
59. Van der Poll T, Jansen PM, Montegut WJ, Braxton CC, Calvano SE, Stackpole S, et al. Effect of IL-10 on systemic inflammatory responses during sublethal primate endotoxemia. J Immunol 1997; 158:1971–1975.
60. Levi M, Van der Poll T, Biemond BJ, Ten Cate H, Kuipers B, ten Cate JW. TNF-dependent bronchoalveolar activation of coagulation and depression of fibrinolysis as pathogenetic mechanism of ARDS in experimental endotoxemia in chimpanzees. Eur Cytokine Netw 1996; 7:263.
61. Schlag G, Redl H, Hallström S, Radmore K, Davies J. Hyperdynamic sepsis in baboons. I. Aspects of hemodynamics. Circ Shock 1991; 34:311–318.
62. Strohmaier W, Werner ER, Redl H, Wachter H, Schlag G. Plasma nitrate and pteridine levels in experimental bacteremia in baboons. Pteridines 1995; 6:8–11.
63. Schlag G, Redl H, Gasser H, Davies J, Rees D, Grover R. Delayed treatment with the NO-synthase inhibitor 546C88 in a baboon model of septic shock. Am J Respir Crit Care Med 1997; 155:A263.
64. Redl H, Schlag G, Bahrami S, Kargl R, Hartter W, Woloszczuk W, et al. Big-endothelin release in baboon bacteria is partially TNF dependent. J Lab Clin Med 1994; 124:796–801.
65. Redl H, Schlag G. TNF in Sepsis. CD-Rom Vienna: European Shock Society, 1996.
66. Redl H, Schlag G, Ceska M, Davies J, Buurman WA. Interleukin-8 release in baboon septicemia is partially dependent on tumor necrosis factor. J Infect Dis 1993; 167:1464–1466.
67. Fong Y, Tracey KJ, Moldawer LL, Hesse DG, Manogue KB, Kenney JS, et al. Antibodies to cachectin tumor necrosis factor reduce interleukin 1a and interleukin 6 appearance during lethal bacteremia. J Exp Med 1989; 170:1627–1633.
68. Van der Poll T, Jansen J, Levi M, Ten Cate H, ten Cate JW, van Deventer SJ. Regulation of interleukin 10 release by tumor necrosis factor in humans and chimpanzees. J Exp Med 1994; 180:1985–1988.
69. Junger WG, Hoyt DB, Redl H, Liu FC, Loomis WH, Davies J, et al. Tumor necrosis factor antibody treatment of septic baboons reduces the production of sustained T-cell suppressive factors. Shock 1995; 3:173–178.
70. Hinshaw LB, Emerson TE, Jr., Chang AC, Duerr M, Peer G, Fournel M. Study of septic shock in the nonhuman primate: relationship of pathophysiological response to therapy with anti-TNF antibody. Circ Shock 1994; 44:221–229.
71. Van der Poll T, Levi M, Ten Cate H, Jansen J, Biemond BJ, Haagmans BL, et al. Effect of postponed treatment with an anti-tumor necrosis factor (TNF) F(ab′)2 fragment on endotoxin-induced cytokine and neutrophil responses in chimpanzees. Clin Exp Immunol 1995; 100:21–25.
72. Redl H, Schlag G, Adolf GR, Natmessnig B, Davies J. Tumor necrosis factor (TNF)-dependent shedding of the p55 TNF receptor in a baboon model of bacteremia. Infect Immun 1995; 63:297–300.
73. Redl H, Schlag G, Paul E, Bahrami S, Buurman WA, Strieter RM, et al. Endogenous modulators of TNF and IL-1 response are under partial control of TNF in baboon bacteremia. Am J Physiol 1996; 271:R1193–R1198.
74. Jansen J, Van der Poll T, Levi M, Ten Cate H, Gallati H, ten Cate JW, et al. Inhibition of the release of soluble tumor necrosis factor receptors in experimental endotoxemia by an anti-tumor necrosis factor-alpha antibody. J Clin Immunol 1995; 15:45–50.
75. Van der Poll T, Fischer E, Coyle SM, Van Zee KJ, Pribble JP, Stiles DM, et al. Interleukin-1 contributes to increased concentrations of soluble tumor necrosis factor receptor type I in sepsis. J Infect Dis 1995; 172:577–580.
76. Jansen PM, Boermeester MA, Fischer E, de Jong IW, Van der Poll T, Moldawer LL, et al. Contribution of interleukin-1 to activation of coagulation and fibrinolysis, neutrophil degranulation, and the release of secretory-type phospholipase A2 in sepsis: studies in nonhuman primates after interleukin-1 alpha administration and during lethal bacteremia. Blood 1995; 86:1027–1034.
77. Van der Poll T, Levi M, Dentener M, Jansen PM, Coyle SM, Braxton CC, et al. Epinephrine exerts anticoagulant effects during human endotoxemia. J Exp Med 1997; 185:1143–1148.
78. Hinshaw LB, Archer LT, Beller-Todd BK, Coalson JJ, Flournoy DJ, Passey R, et al. Survival of primates in LD100 septic shock following steroid antibiotic therapy. J Surg Res 1980; 28:151–170.
79. Kuipers B, Van der Poll T, Levi M, van Deventer SJ, Ten Cate H, Imai Y, et al. Platelet activating factor

antagonist TCV-309 attenuates the induction of the cytokine network in experimental endotoxemia in chimpanzees. J Immunol 1994; 152:2438–2446.
80. Strieter RM, Remick DG, Ward PA, Spenger RN, Lynch JP, Larrick J, et al. Cellular and molecular regulation of tumor necrosis factor alpha production by pentoxifylline. Biochem Biophys Res Commun 1988; 155:1230–1236.
81. Rice GC, Rosen J, Weeks R, Michnick J, Bursten S, Bianco JA, et al. CT-1501R selectively inhibits induced inflammatory monokines in human whole blood ex vivo. Shock 1994; 1:254–266.
82. Bahrami S, Yu Y-H, Redl H, Schlag G. Acute lung injury by endotoxin-induced mediators: prevention by HWA138, a new xanthine derivative. J Lab Clin Med 1995; 125:487–492.
83. Levi M, Ten Cate H, Bauer KA, Van der Poll T, Edgington TS, Büller HR, et al. Inhibition of endotoxin induced activation of coagulation and fibrinolysis by pentoxifylline of by a monoclonal anti tissue factor antibody in chimpanzees. J Clin Invest 1994; 93:114–120.
84. Bahrami S, Redl H, Buurman WA, Schlag G. Influence of the xanthine derivate HWA138 on endotoxin-related coagulation disturbances: effect in non-sensitized vs D-galactosamine sensitized rats. Thromb Haemost 1992; 68:418–423.
85. Leeuwenberg JFM, Smeets EF, Neefjes JJ, Shaffer MA, Cinek T, Jeunhomme TMAA, et al. E-selectin and intracellular adhesion molecule 1 are released by activated human endothelial cells in vitro. Immunology 1992; 77:543–549.
86. Munro JM, Pober JS, Cotran RS. Tumor necrosis factor and interferon gamma induce distinct patterns of endothelial activation and associated leukocyte accumulation in skin of papio anubis. Am J Pathol 1989; 135:1–13.
87. Redl H, Dinges HP, Buurman WA, Van der Linden CJ, Pober JS, Cotran RS, et al. Expression of endothelial leukocyte adhesion molecule-1 in septic but not traumatic/hypovolemic shock in the baboon. Am J Pathol 1991; 139:461–466.
88. Engelberts I, Samyo SK, Leeuwenberg JFM, Van der Linden CJ, Buurman WA. A role for ELAM-1 in the pathogenesis of MOF during septic shock. Ann Surg 1992; 53:136–144.
89. Welty-Wolf KE, Carraway MS, Huang YC, Simonson SG, Kantrow SP, Que LG, et al. Antibody to intercellular adhesion molecule-1 does not decrease lung injury in primates with Escherichia coli sepsis. Am J Respir Crit Care Med 1997; 155:A263.
90. Redl H, Schlag G, Davies J, Robinson M. Detrimental effects of the application of anti-CD18 antibodies in baboon live E. coli sepsis. Circ Shock 1993; Suppl. 2:33.
91. Redl H, Schlag G, Bahrami S, Schade U, Ceska M, Stütz P. Plasma neutrophil-activating peptide-1/interleukin-8 and neutrophil elastase in a primate bacteremia model. J Infect Dis 1991; 164:383–388.
92. Redl H, Schlag G, Schiesser A, Davies J. Thrombomodulin release in baboon sepsis: its dependence on the dose of Escherichia coli and the presence of tumor necrosis factor. J Infect Dis 1995; 171:1522–1527.
93. Coughlan AF, Hau H, Dunlop LC, Berndt MC, Hanhock WW. P-selectin and platelet-activating factor mediate initial endotoxin-induced neutropenia. J Exp Med 1994; 179:329–334.
94. Biemond BJ, Levi M, Ten Cate H, Soule HR, Morris LD, Foster DL, et al. Complete inhibition of endotoxin-induced coagulation activation in chimpanzees with a monoclonal Fab fragment against factor VII/VIIa. Thromb Haemost 1995; 73:223–230.
95. Randolph M, Bild G, Carr C, Galluppi G, Hinshaw L, Taylor F, et al. Protective effect of TFPI in gram negative sepsis (abstr). Shock 1995; 1:30.
96. Jansen PM, van Lopik T, Lubbers Y, de Jong IW, Brouwer M, Chang ACK, et al. The coagulant-inflammatory axis in the baboon response to E. coli: effects of tissue factor pathway inhibitor on hemostatic balance, the cytokine network and the release of the apoptosis marker sFas. Thesis, University of Amsterdam, The Netherlands, 1997.
97. Pixley RA, De La Cadena R, Page JD, Kaufman N, Wyshock EG, Chang A, et al. The contact system contributes to hypotension but not disseminated intravascular coagulation in lethal bacteremia. In vivo use of a monoclonal anti-factor XII antibody to block contact activation in baboons. J Clin Invest 1993; 91:61–68.
98. Van der Poll T, Levi M, Hack CE, Ten Cate H, van Deventer SJ, Eerenberg AJ, et al. Elimination of interleukin 6 attenuates coagulation activation in experimental endotoxemia in chimpanzees. J Exp Med 1994; 179:1253–1259.
99. Taylor FB, Jr. Studies on the inflammatory-coagulant axis in the baboon response to E. coli: regulatory roles of proteins C, S, C4bBP and of inhibitors of tissue factor. Prog Clin Biol Res 1994; 388:175–194.
100. Biemond BJ, Levi M, Ten Cate H, Van der Poll T, Buller HR, Hack CE, et al. Plasminogen activator and plasminogen activator inhibitor I release during experimental endotoxaemia in chimpanzees: effect of interventions in the cytokine and coagulation cascade. Clin Sci Colch 1995; 88:587–594.
101. Redl H, Schlag G, Bahrami S, Davies J, Stevens S, Foulkes R, et al. Influence of TNF antibody treatment on mediator release in baboon bacteremia/endotoxemia. In: Leven J, Alving CR, Munford RS, Stütz PL, eds. Bacterial Endotoxin: Recognition and Effector Mechanisms. Proceedings of the 2nd Congress of the International Endotoxin Society, Vienna, August 17–20, 1992. Amsterdam: Elsevier Science Publisher B.V., 1993:433–441.
102. Van der Poll T, Jansen PM, Van Zee KJ, Hack CE, Oldenburg HA, Loetscher H, et al. Pretreatment with a 55-kDa tumor necrosis factor receptor-immunoglobulin fusion protein attenuated activation of coagulation, but not of fibrinolysis, during lethal bacteremia in baboons. J Infect Dis 1997; 176:296–299.
103. Redl H, Bahrami S, Schlag G. Possible therapeutic approaches to deal with bacterial/endotoxin translocation. In: Vincent JL, ed. Yearbook of Intensive Care and Emergency Medicine. Berlin: Springer-Verlag, 1995:693–702.
104. Gathiram P, Wells MT, Raidoo D, Brock-Utne JG, Gaffin SL. Changes in lipopolysaccharide concentra-

tions in hepatic portal and systemic arterial plasma during intestinal ischemia in monkeys. Circ Shock 1989; 27:103–109.
105. Di Padova FE, Brade H, Barclay GR, Poxton IR, Liehl E, Schuetze E, et al. A broadly cross-protective monoclonal antibody binding to *Escherichia coli* and *Salmonella* lipopolysaccharides. Infect Immun 1993; 61:3863–3872.
106. Bahrami S, Yao YM, Leichtfried G, Redl H, Schlag G, Di Padova FE. Monoclonal antibody to endotoxin attenuates hemorrhage-induced lung injury and mortality in rats. Crit Care Med 1997; 25:1030–1036.
107. Yao YM, Bahrami S, Leichtfried G, Redl H, Schlag G. Pathogenesis of hemorrhage-induced bacteria-endotoxin translocation in rats: effects of recombinant bactericidal-increasing protein ($rBPI_{21}$). Ann Surg 1995; 221:398–405.
108. Field SK, Morrison DC. A hamster-anti-idiotype monoclonal antibody, mimicking the inner-core region of Gram-negative bacterial lipopolysaccharide (LPS), stimulates LPS inner-core-specific serum antibodies in hamsters. J Endotox Res 1994; 1:120–130.
109. Rogy MA, Moldawer LL, Oldenburg HSA, Thompson WA, Montegut WJ, Stackpole S, et al. Anti-endotoxin therapy in primate bacteremia with HA1A and BPI. Ann Surg 1994; 220:77–85.
110. Weiss J, Elsbach P, Olsson I, Odeberg H. Purification and characterization of a potent bactericidal and membrane active protein from the granules of human polymorphonuclear leukocytes. J Biol Chem 1978; 253:2664–2672.
111. Weiss J, Franson R, Beckerdite S, Schmeidler K, Elsbach P. Partial characterization and purification of a rabbit granulocyte factor that increases permeability of *Escherichia coli*. J Clin Invest 1975; 55:33–42.
112. Ooi CE, Weiss J, Doerfler ME, Elsbach P. Endotoxin-neutralizing properties of the 25 kD N-terminal fragment and a newly isolated 30 kD C-terminal fragment of the 55–60 kD bactericidal/permeability-increasing protein of human neutrophils. J Exp Med 1991; 174:649–655.
113. Weiss J, Elsbach P, Shu C, Castillo J, Grinna L, Horwitz A, et al. Human bactericidal/permeability-increasing protein and a recombinant NH2-terminal fragment cause killing of serum-resistant gram-negative bacteria in whole blood and inhibit tumor necrosis factor release induced by the bacteria. J Clin Invest 1992; 90:1122–1130.
114. von der Mohlen MA, Van der Poll T, Jansen J, Levi M. Release of bactericidal/permeability-increasing protein in experimental endotoxemia and clinical sepsis. Role of tumor necrosis factor. J Immunol 1996; 156:4969–4973.
115. Rogy MA, Oldenburg HS, Calvano SE, Montegut WJ, Stackpole SA, Van Zee KJ, et al. The role of bactericidal/permeability-increasing protein in the treatment of primate bacteremia and septic shock. J Clin Immunol 1994; 14:120–133.
116. Schlag G, Redl H, Davies J. Protective effect of bactericidal/permeability increasing protein (rBPI21) on sepsis induced organ failure in non human primates. Shock 1995; 3:64.
117. Pajkrt D, Doran JE, Koster F, Lerch PG, Arnet B, Van der Poll T, et al. Antiinflammatory effects of reconstituted high-density lipoprotein during human endotoxemia. J Exp Med 1996; 184:1601–1608.
118. Larrick JW, Hirata M, Shimomoura Y, Yoshida M, Zheng H, Zhong J, et al. Rabbit CAP18 derived peptides inhibit gram negative and gram positive bacteria. In: Levin J, Van Deventer SJH, Van der Poll T, Sturk A, eds. Bacterial Endotoxins: Basic Science to Anti-Sepsis Strategies. New York: Wiley-Liss, 1994:125–135.
119. Christ WJ, Asano O, Robidoux ALC, Perez M, Wang Y, Dubuc GR, et al. E5531, a pure endotoxin antagonist of high potency. Science 1995; 268:80–83.
120. Bahrami S, Redl H, Leichtfried G, Yu Y, Schlag G. Similar cytokine but different coagulation responses to lipopolysaccharide in D-galactosamine-sensitized versus nonsensitized rats. Infect Immun 1994; 62:99–105.
121. Holzer K, Thiel M, Moritz S, Kreimeier U, Messmer K. Expression of adhesion molecules on circulating PMN during hyperdynamic endotoxemia. J Appl Physiol 1996; 81:341–348.
122. Mileski WJ, Winn RK, Harlan JM, Rice CL. Sensitivity to endotoxin in rabbits is increased after hemorrhagic shock. J Appl Physiol 1992; 73:1146–1149.
123. VanZee KJ, Moldawer LL, Oldenburg HSA, Thompson WA, Stackpole SA, Montegut WJ, et al. Protection against lethal *Escherichia coli* bacteremia in baboons (papio anubis) by pretreatment with a 55-kDa TNF receptor (CD120a)-Ig fusion protein, Ro-45-2081. J Immunol 1996; 156:2221–2230.
124. Taylor FB, Jr., Esmon CT, Hinshaw LB. Summary of staging mechanism and intervention studies in the baboon model of *E. coli* sepsis. J Trauma 1990; 30:S197–S203.

54

The Value of Animal Models in Endotoxin Research

Steven M. Opal
Brown University School of Medicine, Providence, and Memorial Hospital of Rhode Island, Pawtucket, Rhode Island

INTRODUCTION

To understand the multitude of endotoxin-mediated effects on the entire organism, it is essential that animal models be utilized in endotoxin research. The integration of the diverse array of complex interactions between endotoxin and multiple organ systems necessitates investigation in controlled laboratory conditions with animal models. The dynamic relationship between endotoxin and host binding proteins, the distribution and clearance of endotoxin, and the global host response to endotoxin-mediated pathophysiological effects can only be studied in living organisms (1).

Animal models provide an opportunity to carefully dissect these components and to manipulate these effects with a variety of genetic, immunological, and pharmacological strategies. Animal models provide information about endotoxin-induced neuro-endocrine, immunological, metabolic, hematologic, and cardiovascular responses that cannot be duplicated in tissue culture systems or the most sophisticated computer modeling programs (1–3). The principal challenge in animal experimentation is to derive the most relevant information for a more complete understanding of endotoxin's interactions with the human host. Substantial differences exist between various animal models and between laboratory findings in animals and human responses to endotoxin. These differences must be considered in the design of animal experiments if the research results are to be extrapolated to human medicine (4–7).

DIFFERENCES IN ANIMAL MODELS USED IN ENDOTOXIN RESEARCH

Mouse Models

Animal species differ remarkably with respect to their susceptibility to bacterial endotoxin. Traditionally, small animals have been widely utilized in endotoxin research. Mice have numerous advantages in endotoxin research because the genetics and immunology of mice have been extensively studied (8). The knowledge of murine genetics approaches the level of sophistication of human genomics (9). The immune system of mice is arguably as well known as that of humans. Mouse-derived immune reagents, cytokines, monoclonal antibodies, and recombinant proteins are widely available.

Early investigations focused on mice with specific immune dysfunctions [e.g., athymic nude mice (10), B-cell–deficient CBA/N mice (11), and severe combined immunodeficiency (SCID) mice (12)] to evaluate the cellular and humoral immune responses involved in the mammalian endotoxin response. It is evident that T-cell and B-cell products (and NK cells) are not critical to the initial host response to endotoxin. SCID mice (treated with anti-asialo-GM1 antibody to remove NK cells) remain susceptible to endotoxin-induced lethality and generate early cytokine profiles comparable to that observed in BALB/c mice with normal immune responses (13).

The discovery of a spontaneous mutation in a mouse strain led to the identification of the endotoxin-resistant

C3H/HeJ mouse strain, and this finding has greatly facilitated the understanding of the genetic control of endotoxin responsiveness (14–16). Moreover, the hypersusceptibility of this mouse strain to gram-negative infections, in contrast to fully endotoxin-sensitive syngeneic strains, points out a fundamental discrepancy between endotoxicity models and bacterial infection models (discussed below).

The relatively recent availability of both transgenic mice and gene knockout mice have greatly facilitated the study of endotoxin activities in this animal species. Transgenic mice allow for the detailed investigation of a single selected gene from humans or other animal species, which has been inserted into the genome of mice. This allows for the evaluation of the specific transgene product in comparison with isogenic, wild-type mice. The ability to isolate specific genes in this fashion can be supplemented by targeted deletion of the selected gene of interest by gene knockout techniques (16–18).

The availability of gene knockout mice has allowed for functional distinctions between lymphotoxin-α (19,20), TNF-α (21), and type I (p55) and type II (p75) TNF receptors (22–24). Using this technique, it is now clear that the type I receptor is primarily responsible for the endotoxin-induced cytotoxicity of TNF, whereas the type II receptor is primarily involved in TNF-induced cellular proliferation. It is possible that membrane-bound TNF on the cell surface of effector cells may interact extensively with type II TNF receptors as well.

The availability of gene knockout mice has been instrumental in deciphering the roles of IL-1β ligand (25), IL-1 receptor antagonist (26), IL-1 converting enzyme, and the type I (24) and type II IL-1 receptors. The central roles of CD14 (27) and nitric oxide (28,29) in many endotoxin-mediated pathophysiological responses have been confirmed with specific gene knockout mice. Transgenic mice expressing elevated levels of apolipoprotein A-1 and high-density lipoprotein (HDL) are protected from endotoxin-induced lethality (30). This attests to the physiological relevance of HDL and other lipoproteins as endotoxin-binding substances in the blood stream. This serves as a therapeutic rationale for the administration of HDL in endotoxemic states (31).

Another advantage of murine studies in endotoxin research is the fact that these animals are widely available to research laboratories at relatively modest expense. Compared with larger animals, mice are generally quite easy to house and maintain at a research facility. This allows for the study of large numbers of animals using mortality endpoints with sufficient numbers for detailed statistical analysis.

Unfortunately, the mouse is highly resistant to bacterial endotoxin. This is in marked contrast to humans, who are highly susceptible to endotoxin. This necessitates the administration of extremely high doses of endotoxin in order to observe a physiological effect. Alternatively, these animals may be rendered endotoxin susceptible by a variety of priming methods such as adrenalectomy, administration of bacillus Calmette-Guérin (BCG) or D-galactosamine, or other sensitizing methods (32).

D-Galactosamine is a convenient method of sensitizing animals to endotoxin. It creates a metabolic block in uridine synthesis, which induces reversible hepatoxicity and potentiates endotoxin sensitivity by at least three orders of magnitude (33). Endotoxin-sensitizing agents may reduce the intrinsic resistance of mice to endotoxin, but they also add an additional undesirable study variable. It is not clear that artificial sensitization of these animals duplicates the physiological response seen in endotoxin-sensitive species such as humans.

Rat Models

Rats are also used extensively in endotoxin research, but they suffer from the same major problem of intrinsic resistance to endotoxin as observed in mice. These animals are used extensively in the study of intestinal translocation of endotoxin (34), and rat studies were instrumental in the discovery of the role of platelet-activating factor (PAF) in the mammalian response to endotoxin (35).

The cardiovascular physiology of rats and mice differs considerably from human beings. This limits the value of hemodynamic information derived from the study of these animals (36,37). Nonetheless, it is possible to obtain some hemodynamic information from the study of rats, whereas this is exceedingly difficult, if not impossible, in mice. Additionally, rats have the advantage of being of sufficient size that they can tolerate multiple blood samples over the course of a single experiment. This is often not the case in experiments with mice. Furthermore, the gastrointestinal microflora of rats is quite similar to that observed in human beings with an extensive population of facultatively anaerobic, gram-negative, fermentative bacteria, and enteric anaerobic organisms (38). Transgenic rats are increasingly available to research laboratories, which should greatly facilitate the use of this animal species in endotoxin research (39).

Rabbit Models

Another species widely employed in endotoxin research is the rabbit. Rabbits are remarkably susceptible to endotoxin activity, which is similar to human physiology (37,40). Classic investigations into the origin of fever, the role of endogenous pyrogens, and the generalized and the dermal Shwartzman reactions were first conducted in rabbits. The animal size and disposition makes it possible to obtain multiple blood samples and follow the course of endotoxin-mediated events over prolonged time periods. However, reagents and assays for rabbit immune reactions are not widely available. Moreover, rabbits have primarily a gram-positive microbial flora in their alimentary tract. This limits the value of rabbits in the study of translocation of gut bacteria and other effects of the hepatosplenic circulation in endotoxin research.

Dog Models

Dogs are increasingly difficult to study in that there are ethical and emotional problems with studying an animal species that is widely prized as a trusted, domestic pet. Furthermore, dogs are relatively endotoxin-resistant and have a cardiovascular physiology that exhibits some important differences from that of humans. Dogs typically develop a low cardiac output state after an endotoxin challenge (37,38). A high cardiac output state characterizes the usual human response to endotoxemia.

Nonetheless, a great deal has been learned from endotoxin research in dogs. Very small doses of endotoxin infused continuously in dogs produces a hemodynamic response that more closely resembles the hemodynamics of human sepsis (41). These animals are easy to handle and can be extensively instrumented under laboratory conditions that approximate the clinical care of septic patients. A wealth of hemodynamic, metabolic, and immunological data has been derived from this type of animal research (38,40–42).

Large Animal Models in Endotoxin Research

Sheep and Pig Models

These animals make excellent models for endotoxin research but suffer from inherent difficulties related to their care, expense, and handling. They are relatively endotoxin sensitive and have a cardiovascular physiology that is remarkably similar to humans in many respects. They are of sufficient size and constitution that fluid resuscitation, antimicrobial agents, and hemodynamic support can be provided in a manner that approaches the clinical care of septic patients (38).

However, both porcine and ovine models may be complicated by rapid and at times rather striking elevation in pulmonary artery pressures upon exposure to endotoxin. Pulmonary hypertension may occur in human sepsis, but it is an unusual occurrence; yet, it is routinely observed in sheep and pigs. This hemodynamic effect upon the pulmonary circulation complicates the interpretation of endotoxin effects in these animals in relationship to human physiology. It is feasible, with some difficulty, to instrument these animals and obtain detailed hemodynamic, metabolic, and immunological information from these animals. The large size of sheep and pigs may be problematic in the early development of new therapeutic agents if reagent quantities are limited. Ovine and porcine models are often employed after a new agent has shown promise in small animal models.

Horses in Endotoxin Research

Horses in many respects would be the ideal animal model for endotoxin research. Horses are equally or perhaps more susceptible to endotoxin-induced pathophysiological changes than human beings. Horses can be intensively monitored in veterinary laboratories, and multiple blood samples can be obtained in these animals. Nonetheless, emotional and ethical imperatives preclude the use of horses for this type of research. Horses are highly prized animals with a long and rich history with humans and are very expensive to purchase and to maintain in a laboratory setting (36). No one would seriously consider the use of horses in mortality endpoint experiments in endotoxin research; however, veterinary experience with endotoxin-mediated disorders in horses may provide useful insights into endotoxin effects in human diseases.

Nonhuman Primate Models

Nonhuman primates, especially baboons, have been used extensively in the past in endotoxin research. While chimpanzees appear to be as exquisitely susceptible to endotoxin as humans, baboons and rhesus monkeys are relatively endotoxin resistant. This limits the applicability of research in these nonhuman primates since large quantities of endotoxin or bacterial suspensions are necessary to observe pathophysiological effects. The major advantage of primate research is that the immunology of primates is sufficiently close to humans that human antibodies and other reagents are essentially interchangeable. Species specificity of the cytokines, cytokine receptors, recombinant proteins, and antibodies limits the use of human molecules in animal research. This problem is largely obviated by the study of nonhuman primates. Furthermore, primates are susceptible to many specific infectious diseases to which humans remain susceptible (37,38).

Despite the numerous desirable attributes of research in nonhuman primates, there are major disadvantages, which limit the availability of these animals in endotoxin research. Animals are exceedingly expensive and very difficult to handle. Some animal species are becoming rare in the natural environment, and it is not possible to do endotoxin research on endangered animal species. The use of genetically engineered small mammals such as the SCID-hu mice (43) (severe combined immunodeficient mice reconstituted with human bone marrow elements) may suffice as an animal model that will respond in a similar manner to human immune elements. This does not eliminate the value of primate research but may serve to minimize the need for heavy reliance on primate research in the future. A summary of important attributes of animal models in endotoxin research is found in Table 1.

ENDOTOXIN CHALLENGE MODELS VERSUS GRAM-NEGATIVE INFECTION MODELS

Standard endotoxin challenge models usually employ a single bolus injection via the intraperitoneal or intravenous route in previously healthy animals. The pathophysiological events that follow are monitored by a number of readout systems (i.e., lethality, hypotension, fever cytokine release, development of organ failure, etc). A variety of potential therapeutic interventions may then be studied in an attempt to modulate the effects of endotoxin administration. Injection of a standard dose of laboratory-prepared gram-negative bacteria for animal experiments are also used widely for the study of endotoxin-mediated pathophysiological events. These types of observations have provided investigators with a wealth of information about the actions attributable to endotoxin itself.

However, it should be noted that while the results of such experiments are of considerable value, they do not replicate the complex and dynamic molecular interactions that occur when a virulent microbial pathogen invades a vertebrate host (32,44,45). These experimental models represent intoxication models, which differ significantly from actual infection models from invasive, replicating, gram-negative bacterial organisms.

In order to initiate septic shock, a gram-negative bacterium must evade a formidable array of innate and adaptive host defense mechanisms and proliferate within the host. The organism must first colonize the host; penetrate the integument; evade complement, immunoglobulin, and phagocytic defenses; proliferate at a rate greater than host clearance mechanisms; release adequate levels of endotoxin and other injurious microbial products; and induce a sustained and pathological host immune response (1,32). This is a progressive and dynamic interaction within the host where a complex network of inflammatory mediators are activated

Table 1 Attributes of Common Animal Models in Endotoxin Research

Animal	Level of LPS sensitivity	Favorable attributes	Unfavorable attributes
Mice	Highly resistant (+)	Inexpensive; immune reagents, assays, and genetically engineered mice available	Hemodynamic measures and multiple blood samples unavailable
Rats	Highly resistant (+)	Inexpensive; multiple blood samples and hemodynamic measures feasible	Immune and cardiovascular physiology differs from humans
Rabbits	Sensitive (++++)	Endotoxin sensitivity similar to humans; multiple blood samples feasible	Alimentary tract microbial flora is primarily gram-positive
Dogs	Resistant (++)	Hemodynamic monitoring feasible; easy to handle	Ethical problems; low cardiac output response to endotoxin
Sheep/Pigs	Sensitive (+++)	Cardiovascular effects to LPS similar to humans; enteric microflora gram-negative	Difficult to handle; expensive; pulmonary hypertension to LPS
Horses	Very sensitive (++++)	Intensive monitoring available	Ethical problems; expensive
Primates	Resistant (+)	Immunology similar to humans	Ethical problems; extreme expense

which may upregulate, downregulate, augment, inhibit, or synergize with other components of the host immune response. The end result (septic shock) is the product of a multitude of molecular events in which endotoxin-mediated pathophysiological effects are only one element of a complex and intricate host-pathogen interaction (1,2,32,45,46).

During actual infections with gram-negative infections, endotoxin is released from the organism in variable quantities during growth and death of the bacteria from exposure to complement-mediated bacteriolysis, interactions with phagocytic cells, and the action of antimicrobial agents (47,48). Very little endotoxin is found in the circulation as free, monomeric LPS molecules. Endotoxin is often released as a complex with other cell wall fragments or as a micellar complex with multiple other hydrophobic LPS molecules. Ge and coworkers (49) have shown that the tissue distribution of endotoxin differs when it is delivered to an animal in the form of a chemically defined material or as a live bacterial challenge. This is particularly striking in the liver where purified endotoxin injections localize to hepatocytes while endotoxin from viable bacterial injections is localized to Kupffer cells.

Endotoxin interacts with multiple endotoxin binding proteins within the host including LPS-binding protein (LBP), bactericidal/permeability-inducing protein (BPI), albumin, lipoproteins, and endotoxin receptor molecules such as CR3 receptors, soluble and membrane-bound CD14, and LPS scavenger receptors. The host response is primarily dependent upon the relative concentrations and availability of these host proteins (50). Interactions with BPI and HDL will effectively neutralize endotoxin activity. In contrast, interactions with LBP facilitates the delivery of endotoxin to CD14-bearing effector cells, which results in a prompt response to minute quantities of endotoxin. The levels of these endotoxin-binding moieties vary depending upon the physical location within the body (e.g., abscess cavities, peritoneum, blood), the availability of neutrophils, and the physiological state of the host.

The levels of LBP increase over time as part of the hepatic acute phase response, while HDL levels fall in response to acute stress. BPI levels are orders of magnitude higher than LBP in abscess cavities, yet the opposite is true in the bloodstream. Thus the response to endotoxin release from pathogenic microorganisms varies as a consequence of the relative concentrations of these endotoxin-binding proteins in specific tissue sites (51). This is a dynamic variable over the course of an actual infection, and it cannot be accounted for in a single endotoxin challenge experiment (50). Moreover, the cytokine response to endotoxin is highly variable and depends upon the state of endotoxin sensitivity of the host cells over the course of an infection.

The expression of proinflammatory cytokines is markedly dependent upon the genetic makeup of the host, prior exposure to endotoxin, the stress hormone response, and the levels of anti-inflammatory cytokines such as IL-10 and IL-4(1,4,6,52,53). The phenomenon of endotoxin tolerance may develop over the course of an invasive gram-negative infection, which alters responsiveness at the transcriptional level to subsequent endotoxin exposure. The topic of endotoxin tolerance is considered in detail in another chapter in this volume. These complex and time-dependent interactions are not accessible in endotoxin challenge experiments (5,32). This is potentially important when the results of animal experiments are analyzed and interpreted in a clinical context (1–8,54).

The differences between endotoxin intoxication models and animal systems with actual gram-negative bacterial infections are made apparent when cytokine inhibitors are studied. C3H/HeJ mice have a transcriptional and translational block in cytokine synthesis upon exposure to endotoxin, which renders the animal resistant to the lethal effects of large doses of endotoxin. However, when these animals are exposed to systemic infection from a virulent strain of *E. coli* O18:K1, they are more susceptible to lethal infection than endotoxin-sensitive strains of C3H/HeN mice. If the C3H/HeJ mice are given low doses of either IL-1α, TNF-α, or both simultaneously, they tolerate the bacterial challenge nearly as well as the C3H/HeN mice (55).

Administration of proinflammatory cytokines may benefit animals in certain experimental models (56). Blockade of proinflammatory cytokines may prove to be detrimental to experimental animals in the presence of systemic infection (57–59). Mice rendered insensitive to IL-1 or TNF by receptor knockout experiments are susceptible to lethal infection from *Listeria* or *E. coli* challenge (23,24).

The differential effects of cytokine blockade in animal models is best illustrated in more complex systems such as the cecal ligation and puncture (CLP) model of intra-abdominal sepsis. There is ample evidence that proinflammatory cytokine blockade is either ineffective or detrimental in models of infectious peritonitis (60,61), while the same inhibitors are beneficial in endotoxin-challenge models. Similar findings are found in animal models in which the anti-inflammatory cytokine IL-10 is administered therapeutically or removed by antibodies (62,63). Interleukin-10 will protect animals from endotoxin challenge (62), yet the same cytokine may be deleterious in the presence of an invasive gram-negative infection where a coordinated host inflammatory response is essential for survival (63).

Combination therapy with inhibitors to both TNF and IL-1 improves the outcome of endotoxin-challenged rodents when compared to single inhibitor treatments (50). However, combination therapy with TNF and IL-1 inhibitors worsens the outcome from live bacterial challenges in immunocompromised rats, while single inhibitor treatments benefit these animals (65). These studies serve to indicate that the complexities of endotoxin-mediated effects in the process of actual gram-negative bacterial infection cannot be analyzed fully in simple endotoxin-challenge animal experiments.

THE PROBLEM OF DOSE-RESPONSE IN ANIMAL MODELS OF SEPSIS

When evaluating the therapeutic efficacy of a given new sepsis agent, the quantitative aspects of outcome measures need to be considered in selection of appropriate preclinical models of sepsis. In such studies, one or more treatment protocols are compared with untreated or placebo-treated animals to provide data that attempt to prove that the treatment protocol being evaluated is effective. Often such data take the form of standard Kaplan-Meier survival curves where cumulative percent survival is plotted as a function of time following challenge and/or treatment. This sort of data generally depicts relatively high levels of survival in the treatment group in comparison to relatively high mortality in the untreated control group.

While such findings may well provide statistically valid data to support the claim of clinical efficacy of an experimental treatment, it should be remembered that in many animal models of endotoxin lethality, the actual dose-response lethality profiles manifest very steep slopes at the inflection point. What this means is that a relatively modest increase in the dose can dramatically affect lethality. Often, simply doubling or tripling the dose of endotoxin challenge can alter the lethality function from 0% to virtually 100%. As a consequence, were a given therapeutic agent capable of only a 50% shift in the dose-response profile of lethality, this might well be reflected in a 100% reduction to 0% mortality on a Kaplan-Meier survival curve.

Such concepts should, therefore, be kept in mind when evaluating the potential therapeutic efficacy of new antisepsis drugs. In this respect, it might be preferable to perform complete preclinical studies in which the effect of various treatment doses of the experimental agent is tested against a range of doses of endotoxin challenges. The generation of actual dose-response curves would provide a more comprehensive analysis of the preclinical efficacy of the antisepsis drug under investigation (1,32).

SUMMARY AND CONCLUSION

Lack of full appreciation for the magnitude and multitude of endotoxin-mediated effects in the context of an actual human gram-negative infection may have contributed to inconsistent and largely unsuccessful clinical trial results in human sepsis (1,6,8,54). Overly optimistic predictions of efficacy from preclinical work may be expected because animal studies, by design, are weighted in favor of favorable treatment effects by the experimental agent (1,5,6). Potential pitfalls in the interpretation of animal studies in relationship to predicted efficacy in human medicine are numerous and are listed in Table 2. These problems may not be appreciated by clinicians, investigators, investors, or the public. The value and limitations of animal models in endotoxin research must be understood in the future development of new therapeutic strategies in human sepsis.

Table 2 Differences Between Animal Models and Human Sepsis

Factor	Animal models	Human sepsis
Genetic characteristics	Known and well defined	Highly variable
Age, weight, nutritional status	Defined and tightly controlled	Highly variable
Comorbidities, underlying disease	None or single lesion	Highly variable
Type of challenge	Often an intoxication model	Progressive infection
Infecting microorganism(s)	Strain, dose, route of infection known	Usually highly variable
Onset of septic insult	Often known precisely	Usually unknown
Cause of lethality	Septic insult itself	Underlying disease ± sepsis

REFERENCES

1. Opal SM. Lessons learned from clinical trials in sepsis. J Endotoxin Res 1995; 2:221–226.
2. Hayashi M, Gillam IC, Bondy G, et al. Molecular mechanisms of sepsis: molecular biology of the cell. J Crit Care 1995; 10:82–95.
3. Dinarello CA, Gelfand JA, Wolff SM. Anti-cytokine strategies in the treatment of systemic inflammatory response syndrome. JAMA 1993; 269:1829–1835.
4. Piper RD, Cook DJ, Bone RC, Sibbald WJ. Introducing critical appraisal to studies of animal models investigating novel therapies in sepsis. Crit Care Med 1996; 24:2059–2070.
5. Baumgartner JD. Immunotherapy with antibodies to core lipopolysaccharide: a critical appraisal. Infect Dis Clin North Am 1991; 5:915–927.
6. Natanson C, Hoffman WD, Suffredini AF, Eichacker PQ, Danner RL. Selected treatment strategies for septic shock based upon proposed mechanisms of pathogenesis. Ann Intern Med 1994; 220:771–783.
7. Cohen J. TNF and anti-TNF in sepsis. Experimental basis of evaluation of new therapeutic approaches to sepsis. Crit Care Med 1992; 16:308–314.
8. Dellinger RP, Zimmerman J, Opal SM, et al. From the bench to the bedside: the future of sepsis research. Executive summary of an American College of Chest Physicians. National Institute of Allergy and Infectious Diseases, and National Heart, Lung, and Blood Institute Workshop. Chest 1997; 111:744–753.
9. Pasparakis M, Kollias G. Production of cytokine transgenic and knockout mice. In: Balkwill FR, ed. Cytokines: A Practical Approach. 2d ed. Oxford: IRL Press, 1995:297–325.
10. Vogel SN, Hilfiker ML, Caulfield MJ. Endotoxin-induced T lymphocyte proliferation. J Immunol 1983; 130:1774–1779.
11. Rosenstreich DL, Vogel SN, Jacques A, et al. Differential endotoxin sensitivity of lymphocytes and macrophages from mice with an X-linked defect in B cell maturation. J Immunol 1978; 121:684–690.
12. Schuler W, Weeler IJ, Schuler A, et al. Rearrangement of antigen receptor genes is defective in mice with severe immunodeficiency. Cell 1986; 46:963–972.
13. Falk LA, McNally R, Perera PY, et al. LPS-inducible responses in severe combined immunodeficiency (SCID) mice. J Endotoxin Res 1995; 2:273–280.
14. Watson J, Riblet R, Taylor BA. The response of recombinant inbred strains of mice to bacterial lipopolysaccharides. J Immunol 1977; 118:2088–2093.
15. Watson J, Largen M, McAdam KPWJ. Genetic control of endotoxin responses in mice. J Exp Med 1978; 147:39–49.
16. Vogel SN, Wax JS, Perera PY, et al. Construction of a BALB/c congenic mouse, C.C3H-Lps^d that expresses the Lps^d allele: analysis of chromosome 4 markers surrounding the Lps gene. Infect Immun 1994; 62:4454–4459.
17. Yeung RS, Penninger J, Mak TW. Genetically modified animals and immunodeficiency. Curr Opin Immunol 1993; 5:585–594.
18. Franz WM, Mueller OJ, Hartong R, et al. Transgenic animal models: new avenues in cardiovascular physiology. J Mol Med 1997; 75:115–129.
19. Matsumoto M, Mariathasam S, Nahm MH, et al. Role of lymphotoxin and the type 1 TNF receptor in formation of germinal centers. Science 1996; 271:1289–1291.
20. Amiot F, Belkaid Y, Labastard M, et al. Abnormal organization of the splenic marginal zone and the correlated leukocytosis in lymphotoxin-α and TNFα in double deficient mice. Eur Cytokine Netw 1996; 7:733–739.
21. Pasparakis M, Alexopoulou L, Episkopou V, Kollias G. Immune and inflammatory responses in TNF-α-deficient mice: a critical requirement for TNFα in the formation of primary B cell follicles, follicular dendritic cell networks and germinal centers, and the maturation of the humoral immune response. J Exp Med 1996; 184:1397–1411.
22. Neumann B, Machleidt T, Lifka A, et al. Critical role of 55-kilodalton TNF receptor in TNF-induced adhesion molecule expression in leukocyte organ infiltration. J Immunol 1996; 156:1587–1593.
23. Rothe J, Lesslauer W, Lotscher H, et al. Mice lacking the tumor necrosis factor receptor 1 are resistant to TNF-mediated toxicity but highly susceptible to infection with Listeria monocytogenes. Nature (Lond) 1993; 364:798–802.
24. Acton RD, Dahlberg PS, Uknis ME, et al. Differential sensitivity to Escherichia coli infection in mice lacking tumor necrosis factor p55 or interleukin-1 P80 receptors. Arch Surg 1996; 131:1216–1221.
25. Fantuzzi G, Dinarello CA. The inflammatory response in interleukin-1 beta-deficient mice: comparison with other cytokine-related knockout mice. J Leukoc Biol 1996; 59:489–493.
26. Hirsch E, Irikura VM, Paul SM, Hirsh D. Functions of interleukin-1 receptor antagonist in gene knockout and overproducing mice. Proc Natl Acad Sci USA 1996; 93:11008–11013.
27. Haziot A, Ferrero E, Kontgen F, et al. Resistance to endotoxin shock in reduced dissemination of gram-negative bacteremia in CD14-deficient mice. Immunity 1996; 4:407–414.
28. Gross SS, Kilbourn RG, Griffith OW. NO in septic shock: good, bad, or ugly? Lessons learned from iNOS knockouts. Trends Microbiol 1996; 4:47–49.
29. Wei X-Q, Charles IG, Smith A, et al. Altered immune responses in mice lacking inducible nitric oxide synthase. Nature 1995; 375:408–411.
30. Levine DM, Parker TS, Donelly TM, et al. In vivo protection against endotoxin by plasma high density lipoprotein. Proc Natl Acad Sci USA 1993; 90:12040–12044.
31. Pajkart D, Doran JE, Koster F, et al. Anti-inflammatory effects of reconstituted high-density lipoprotein during human endotoxemia. J Exp Med 1996; 184:1601–1608.
32. Cross AS, Opal SM, Sadoff JD, Gemski P. Choice of bacteria in animal models of sepsis. Infect Immun 1993; 62:2741–2744.
33. Galanos C, Freudenberg MA, Reutter R. Galactosa-

mine-induced to the lethal effects of endotoxin. Proc Natl Acad Sci USA 1979; 76:5939–5943.
34. Alverdy J, Chi HS, Sheldon GF. The effect of parenteral nutrition on gastrointestinal immunity. Ann Surg 1985; 302:681–684.
35. Chang SW, Fedderson CQ, Henson PM, Voelkel NF. Platelet activating factor mediates hemodynamic changes and lung injury in endotoxin-treated rats. J Clin Invest 1987; 79:1498–1509.
36. Adams R. Animal models that simulate sepsis in humans. Perspectives Shock Res 1988; 299:241–242.
37. Zweifach BW. Aspects of comparative physiology of laboratory animals relative to the problem of experimental shock. Fed Proc 1961; suppl 9:18–29.
38. Fink MP, Heard SO. Laboratory models of sepsis and septic shock. J Surg Res 1990; 49:186–196.
39. Sharreau B, Tesson L, Soulillou JP, et al. Transgenesis in rats: technical aspects and models. Transgenic Res 1996; 5:223–234.
40. Wihterman KA, Baue AE, Chaudry IH. Sepsis and septic shock—a review of laboratory models and a proposal. J Surg Res 1980; 29:189–201.
41. D'Orio V, Wahlen C, Rodriguez L-M, et al. A comparison of *Escherichia coli* endotoxin single bolus injection with low-dose endotoxin infusion on pulmonary and systemic vascular changes. Circ Shock 1987; 21:207–211.
42. Hussian SNA, Roussoa C. Distribution of respiratory muscle and organ blood flow during endotoxic shock in dogs. J Appl Physiol 1985; 59:1802–1807.
43. McCune JM. Development and applications of the SCID-hu mouse model. Semin Immunol 1996; 8:187–196.
44. Bagby GJ, Plessala JK, Wilson LA, et al. Divergent efficacy of antibody to tumor necrosis factor-alpha in intravascular and peritonitis models of sepsis. J Infect Dis 1991; 163:83–88.
45. Cross AS, Asher L, Seguin M, et al. The importance of a lipopolysaccharide-initiated, cytokine-mediated host defense mechanism in mice against extra-intestinally invasive *Escherichia coli*. J Clin Invest 1995; 96:676–686.
46. Heumann D, LeRoy D, Glauser MP. Contribution of TNF and of LPS in endotoxemic shock in mice. A reappraisal. J Endotoxin Res 1996; 3:87–92.
47. Mertsola J, Rimilo O, Mustafa MM, et al. Release of endotoxin after antibiotic treatment for gram-negative bacterial meningitis. Pediatr Infect Dis J 1989; 8:904–906.
48. Jackson JJ, Kropp H. Beta-lactam antibiotic-induced release of free endotoxin: in vitro comparison of penicillin-binding protein (PBP-2) to -specific imipenem in PBP-3-specific ceftazidime. J Infect Dis 1992; 165:1033–1041.
49. Ge Y, Ezzell RM, Templeins RG, Warren HS. Cellular distribution of endotoxin differs from that after injection of live bacteria. J Infect Dis 1994; 169:95–104.
50. Horn DL, Opal SM, LoMastro E. Antibiotics, endotoxin and cytokine release, a complex and dynamic process. Scand J Infect Dis 1996; 101:9–13.
51. Opal SM, Marra MN, McKelligan B, et al. Relative concentrations of endogenous endotoxin binding proteins in infected body fluids. Lancet 1994; 344:429–431.
52. Galanos C, Freudenberg MA. Mechanisms of endotoxin shock and endotoxin hypersensitivity. Immunobiology 1993; 189:346–356.
53. Bone RC. Toward a theory regarding the pathogenesis of the systemic inflammatory response syndrome: What we do and do not know about cytokine regulation. Crit Care Med 1996; 24:163–172.
54. Zeni F, Freeman B, Natanson C. Anti-inflammatory therapies to treat sepsis and septic shock: a re-appraisal. Crit Care Med 1997; 25:1095–1100.
55. Cross AS, Sadoff JC, Kelly N, Burton E, Gemski P. Pretreatment with recombinant murine tumor necrosis factor α/cachectin in murine interleukin-1 α protects mice from lethal bacterial infection. J Exp Med 1989; 169:2021–2027.
56. Malangoni MA, Livingston DH, Sonnfeld G, Polk HC. Interferon γ and tumor necrosis factor α use in gram-negative infection after shock. Arch Surg 1990; 125:444–446.
57. Mancilla J, Garcia P, Dinarello CA. Interleukin-1 receptor antagonist can either reduce or enhance the lethality of *Klebsiella pneumoniae* sepsis in newborn rats. Infect Immun 1993; 51:926–932.
58. Havell EA. Evidence that TNF has an important role in antibacterial resistance. J Immunol 1989; 143:2894–2899.
59. Nakane A, Minagawa T, Kato K. Endogenous tumor necrosis factor cachectin is essential to host resistance against *Listeria monocytogenes*. Infect Immun 1988; 56:2563–2568.
60. Eskandari MK, Bolgos G, Miller C, et al. Anti-tumor necrosis factor antibody therapy fails to prevent lethality after cecal ligation and puncture or endotoxemia. J Immunol 1992; 148:2724–2730.
61. Bagby GJ, Plessala KJ, Wilson LA, et al. Divergent efficacy of antibody to tumor necrosis factor-α in intravascular and peritonitis models of sepsis. J Infect Dis 1991; 163:83–88.
62. Howard M, Muchamuel T, Andrade S, Memon S. Interleukin-10 protects mice from lethal endotoxemia. J Exp Med 1993; 177:1205–1208.
63. Greenberger MJ, Strieter RM, Kunkel SL, et al. Neutralization of IL-10 increases survival in a murine model of Klebsiella pneumonia. J Immunol 1995; 155:722–729.
64. Russel DA, Tucker KK, Khinookoswong N, Thompson RC, Cohno T. Combined inhibition of interleukin-1 and tumor necrosis factor in rodent endotoxemia improves survival in organ function. J Infect Dis 1995; 171:1528–1538.
65. Opal SM, Cross AS, Jhung JW, et al. Potential hazards of combination immunotherapy in the treatment of experimental sepsis. J Infect Dis 1996; 273:1415–1421.

55

Pathophysiological Responses to Endotoxin in Humans

Anthony F. Suffredini and Naomi P. O'Grady
National Institutes of Health, Bethesda, Maryland

INTRODUCTION

Bacterial products including endotoxin have been administered to humans over the past century as therapeutic, diagnostic, and experimental agents (1–5). During the last decade, endotoxin has uniquely contributed to the explosion of knowledge associated with the biology of inflammation by providing an in vivo model of human inflammatory responses including phagocytic and endothelial cell activation, leukocyte kinetics, stress hormone responses, hemostasis, cytokine biology, and defects in endotoxin signaling (4,5). Using molecules that enhance or neutralize specific components of the inflammatory response, the model has elucidated some of the complex regulatory mechanisms that comprise the acute phase response. Further, while demonstrating a proof of purpose for some novel therapeutic agents for sepsis, the model has also revealed some of the intricate and unexpected interactions that result from modulating inflammatory mediators. These observations are relevant to understanding the factors that initiate and regulate inflammation and provide a rationale for the development of new therapies for sepsis and septic shock.

The acute phase response is the early immediate host response to infection or tissue injury and results in fever, leukocytosis, changes in vascular permeability, and altered metabolic responses in various organs (6). These reactions constitute the basic components of innate or natural immunity (7). Endotoxin administration initiates the acute phase response and clinically models the systemic inflammatory response syndrome (e.g., fever, tachycardia, tachypnea, and leukocytosis), a prodrome to the development of sepsis and septic shock (8).

The majority of human endotoxin studies performed during the last decade have administered intravenous U.S. Standard reference endotoxin (*E. coli* O:113) using a dose of 20–40 endotoxin units/kg body weight (2–4 ng/kg) (9). Other endotoxin preparations described in recent human studies include endotoxin derived from *Salmonella abortus equi* (dose range 0.2–1 ng/kg) (3,10). Investigations evaluating pulmonary responses to inhaled endotoxin have used intact-killed *Enterobacter agglomerans* or its extracted endotoxin, as well as *E. coli* endotoxin (11,12).

While the human basis, reproducibility, and availability of biological assays make this a useful model to study inflammation, it is important to consider some limitations of the model. The participants in these studies are usually healthy subjects who receive a single noninfectious, short-lived inflammatory stimulus with a low toxicity. This results in normal, well-compensated inflammatory responses. As such, it is not a model of infection, shock, or multiple organ failure but provides a means to study the initial host responses to a bacterial component. The magnitude of these changes are qualitatively similar to the inflammatory response that develops during sepsis (13). The dose-related nature of these inflammatory responses is apparent in a report of shock and organ failure after self-administration of 1 mg of *Salmonella minnesota* endotoxin (14).

SYSTEMIC RESPONSES TO INTRAVENOUS ENDOTOXIN

Following intravenous administration, endotoxin is cleared from the blood within 15–30 minutes (15,16). Cells are activated and inflammatory mediators initiate responses that culminate in fever, leukocytosis, and changes in vascular tone. Coordinately, multiple control mechanisms are activated that limit and ultimately terminate the inflammatory response. Within one hour of the administration of intravenous endotoxin, subjects may experience symptoms including chills, rigors, headache, myalgias, arthralgias, and nausea (13,17). These symptoms resolve within 3–5 hours. A monophasic rise in core temperature starts by one hour with maximum increases of 1–1.75°C above baseline after 3–4 hours and returns to normal by 8–12 hours postendotoxin (13,17,18).

Alterations in Leukocytes, Platelets, and Endothelial Cells

Circulating leukocyte counts fall by 50–75% within one hour of the administration of endotoxin, and this likely represents margination in postcapillary venules (19). Following the leukopenia, a leukocytosis develops composed of mature and immature neutrophils. This response is maximum by 8–12 hours and returns to baseline by 24 hours (17,20). In parallel to the rise in neutrophils, the percentage and number of peripheral blood monocytes and lymphocytes fall. Peripheral cells expressing CD3 (T cell), CD4 (helper T cell), and CD56 (natural killer cells) surface antigens, in general, decrease in number (20–22). The percentage of CD14-positive peripheral blood mononuclear cells initially falls at 3 hours and then increases at 6 hours (21). Platelet number decreases 20–30% by 3–5 hours after endotoxin administration and returns to a normal range within 24 hours (23).

Accompanying the leukocytosis, several markers of cell activation are detected on cell surfaces or in the blood. Neutrophil cell adhesion molecules (CD11b/CD18) and membrane-associated bactericidal permeability increasing protein (BPI) are upregulated while L-selectin expression is downregulated (24,25). Levels of soluble TNF receptors (sTNFR) (type I and II or p55 and p75) rise in the blood, and their surface expression on neutrophils and monocytes is downregulated (26,27). Neutrophil granule constituents including lactoferrin, metalloproteinases, BPI, and elastase are released into the circulation (25,28–31). Cell-derived activated proteases are detected in the blood complexed with their respective antiproteases (i.e., elastase-α_1-antitrypsin) (28).

Endothelial cells are activated and release E-selectin, von Willebrand factor antigen (vWF), and two major components of the fibrinolytic pathway: tissue plasminogen activator (tPA) and its inhibitor, plasminogen activator inhibitor (PAI-1) (15,28,30). The former activates plasminogen to form plasmin, which is detected in the circulation bound to its protease inhibitor (plasmin-α_2-plasmin inhibitor complexes) (15,28).

Humoral Mediators

Humoral defense mechanisms are rapidly activated after endotoxin administration. The contact system composed of Factors XII, XI, prekallikrein, and high molecular weight kininogen amplifies several host inflammatory responses including the initiation of coagulation via the intrinsic pathway, activation of plasminogen and neutrophils, as well as the generation of bradykinin, a potent vasodilator. Within 2 hours of endotoxin administration, functional prekallikrein and Factor XI levels fall accompanied by a rise in α_2-macroglobulin–kallikrein complexes (32).

Simultaneous to the activation of the kallikrein-kinin and fibrinolytic systems, activation of the common coagulation pathway occurs manifested by increased blood levels of prothrombin F1+2 fragments and thrombin-antithrombin III complexes (15). No increase in complement components (C3a, C5a) occurs after endotoxin administration, suggesting that complement is not critical to these initial responses to endotoxin (15).

Endogenous Mediators of Inflammation

Cytokines serve as important links to amplify the inflammatory response initiated by endotoxin. Major proinflammatory cytokines detected within the first 2 hours postendotoxin include TNF-α and low levels of IL-1β (33–35). These cytokines cause the release of other inflammatory mediators, activate various cells, and initiate anti-inflammatory responses that downregulate the acute response to endotoxin.

IL-6 appears to be an important link to acute phase protein synthesis, downregulation of cytokine production, and the activation of coagulation (36–38). Other anti-inflammatory cytokines that are detected within the first 3 hours postendotoxin include IL-1 receptor antagonist and IL-10 (39). Levels of interferon-γ, TNF-β, IL-1α, IL-2, IL-4, IL-13, transforming growth factor-β, granulocyte-macrophage colony stimulating factor, and leukemia inhibitory factor do not rise significantly in the circulation during the acute phase response to endotoxin in humans (34,40,41). Alpha (CXC) chemokines such as IL-8 and GRO-α and -β (C-C) chemokines such as MIP-1 and MCP-1 rise acutely after endotoxin administration (30,42,43). Granulocyte-colony stimulating factor (G-CSF), a growth factor for hematopoietic cells, is detected in the blood after 3 hours (44).

Other inflammatory mediators detected after endotoxin administration include secretory phospholipase A_2 (45), procalcitonin (46), and neopterin (47). Urinary metabolites of prostacyclin, thromboxane A_2, and nitric oxide are detected within 24 hours post–endotoxin administration (48,49). Finally increased amounts of exhaled nitric oxide can be detected in expired gases after endotoxin administration (49).

Levels of stress hormones rise in the blood including adrenal corticotrophic hormone (ACTH), α-melanocyte–stimulating hormone, arginine vasopression, cortisol, epinephrine, and norepinephrine (17,33,50–52). The diurnal variation in endogenous cortisol may alter the host response to endotoxin. For example, the temperature response is greater when subjects are challenged with endotoxin in the evening compared with the morning (53). Thus, a variety of mediators from diverse pathways are released acutely, emphasizing the wide breadth of inflammatory responses elicited by a single dose of endotoxin.

ORGAN-SPECIFIC INFLAMMATORY RESPONSES FOLLOWING INTRAVENOUS ENDOTOXIN

Hepatic synthetic function is rapidly upregulated in response to endotoxin. Acute phase proteins synthesized in the liver include lipopolysaccharide-binding protein, serum amyloid A, C-reactive protein, and fibrinogen and are detected within 12–24 hours following endotoxin administration (17,25,54,55). These proteins have major effects on mediating inflammatory and antiinflammatory responses to endotoxin (38). Cytochrome P450–mediated drug metabolism is inhibited after 24 hours and results in decreased metabolism of several drugs including hexobarbital, theophylline, and antipyrine (56).

Several metabolic substrates undergo processing during the acute response to endotoxin. Resting energy consumption is increased and is associated with mild hyperglycemia, hypoaminoacidemia, and increased levels of free fatty acids (57). The limb musculature increases lactate and free fatty acid output, and glucose uptake is increased. Splanchnic blood flow increases and is associated with an increase in the uptake of oxygen, lactate, amino acids, and free fatty acids and an increase in glucose output (57). In addition to the marked changed in splanchnic metabolic activity and substrate utilization, splanchnic levels of TNF were twofold greater than levels in the peripheral blood, suggesting that the splanchnic organs serve as a significant source of this cytokine (57).

The cardiovascular response to endotoxin is characterized by an elevated cardiac index and heart rate, decreased mean arterial pressure and systemic vascular resistance, and reversible depression of left ventricular ejection fraction (13). Measurements derived from pulmonary artery catheterization and radionuclide cineangiograms show that myocardial function, compared to normal subjects, is impaired within 6 hours of endotoxin administration and these changes are reversible by 24 hours (13). This pattern of hemodynamic change is qualitatively similar to the hemodynamic response observed in clinical sepsis (13).

Despite activation of a wide variety of circulating mediators and cells, the lung appears to be relatively protected from the major inflammatory effects of a single intravenous bolus of endotoxin. Radiolabeled neutrophils accumulate in the lung, suggesting that this is a major site of margination of neutrophils during the systemic response to endotoxin (58). However, no alveolar neutrophil influx occurs after intravenous endotoxin administration. Bronchoalveolar lavage performed at the time of neutrophil margination (1 hour), during the subsequent leukocytosis (6 hours), or at the time of resolution of the systemic inflammatory response (24 and 48 hours) did not demonstrate increases or changes in the lavage cell counts or cell differentials (44,59). Increased permeability to small inhaled radiolabeled molecules (technitium-labeled diethylenetriamine pentacetate, ^{99m}Tc-DTPA) occurred within the first 3 hours of endotoxin administration. However, there was no increase in permeability to larger proteins such as albumin or total protein (59).

In addition to the mild changes in lung permeability, alveolar macrophages obtained after i.v. endotoxin challenge and then stimulated with endotoxin in vitro are primed to produce enhanced amounts of IL-1 and prostaglandin E_2 (60). The mediators that account for this priming effect after systemic endotoxin are unknown. Despite high levels of circulating cytokines, there was no increase in lavage cytokine protein (e.g., TNF-α, IL-1, G-CSF, IL-6, IL-8) or messenger RNA (e.g., IL-1α, IL-1β, TNF-α, IL-6, IL-8) (44). Thus, the lung is relatively spared from the high levels of inflammatory mediators that are released after endotoxin administration.

Changes in gut permeability occur after endotoxin administration. The absorption of oral mannitol and lactulose, low molecular weight sugars, is increased after endotoxin challenge (61).

Central nervous system function is altered, and these effects depend in part on diurnal rhythms. When endotoxin is administered in the morning or evening, rapid eye movement (REM) sleep is suppressed, whereas non-REM sleep is suppressed only when endotoxin is given in the evening (10,62). In summary, systemic inflammation after endotoxin administration results in a variety of organ-specific effects including changes in neurological function and altered vascular, alveolar epithelial, and gut epithelial permeability.

PULMONARY RESPONSES TO DIRECT ENDOTOXIN EXPOSURE

Endotoxin has been implicated in the pathogenesis of gram-negative bacterial pneumonia as well as the development of acute and chronic lung inflammation due to occupational dust exposure (i.e., grain dust, compost, cotton bract) (63,64). Inhalation of (20–300 μg) intact *E. agglomerans* or endotoxin extracted from the same bacteria strain resulted in fever, chest tightness, increased bronchial reactivity, and mild decreases in pulmonary function (forced expiratory volume in one second and diffusion of carbon monoxide) (11). In contrast to the systemic endotoxin administration, inhaled endotoxin was associated with an increase in bronchoalveolar lavage neutrophils and increased lavage cytokine levels (TNF, IL-1, IL-1ra, IL-6, IL-8). Peripheral blood leukocytes and levels of C-reactive protein rose after pulmonary challenge with endotoxin (12,65).

Similar local pulmonary responses were observed after the instillation of *E. coli* endotoxin (2–4 ng/kg) into segmental airway bronchi. After 6 hours, total protein and cell numbers increased with a significant rise in the numbers of neutrophils. This was associated with an increase in acute inflammatory mediators in the lavage from the endotoxin challenged segment (i.e., TNF, IL-6, IL-8). By 24 hours after endotoxin administration, these mediators and the increased cell number were returning towards baseline values (66,67). Blood leukocytes and C-reactive protein increased after local pulmonary challenge with endotoxin, while only small rises in blood cytokines were detected (i.e., TNF, IL-6), suggesting that local and systemic factors influence the recruitment of neutrophils to the lung (66).

MODULATING INFLAMMATORY RESPONSES TO ENDOTOXIN: BLOCKING ENDOTOXIN-INITIATED INFLAMMATORY RESPONSES

One of the important applications of the endotoxin model in humans is the evaluation of interventions that alter inflammatory responses. These studies have provided insight into the interactions and complexity of acute inflammatory responses (Tables 1,2).

Endotoxin Inhibitors

Several molecules that directly antagonize the initiation of endotoxin-induced inflammation have been studied in humans. Recombinant bactericidal/permeability-increasing protein ($rBPI_{23}$) binds to the lipid A moiety of endotoxin and neutralizes the inflammatory effects of endotoxin (16). When given 3 minutes prior to endotoxin, $rBPI_{23}$ decreased endotoxin levels and blunted and leukopenia and secondary leukocytosis. Increases in cytokine (TNF, IL-6, IL-8, IL-10), sTNFR, and neutrophil lactoferrin and elastase levels were attenuated (16). The fibrinolytic and procoagulant responses were significantly reduced (23). In addition, increases in heart rate and cardiac index were blunted (68). However, despite these anti-inflammatory effects, fever, symptoms, decreases in platelet number, and mean arterial pressure were not altered (16). This may have been due to an insufficient dose of $rBPI_{23}$ or possibly the activity of the $rBPI_{23}$-endotoxin complex.

E5531 is an endotoxin analog antagonist derived from *Rhodobacter capsulatus* lipid A. When administered prior to endotoxin, E5531 decreased or improved fever, symptoms, cytokine levels (TNF, IL-6), leukocytosis, and C-reactive protein levels (69). The cardiovascular response (heart rate, cardiac index, mean arterial pressure, systemic vascular resistance, fractional shortening, and peak systolic pressure to end systolic volume ratios) evaluated by serial echocardiographic measurements was improved compared to subjects given placebo (70). Thus, $rBPI_{23}$ and E5531 blocked the interaction of endotoxin with cell receptors and were associated with improvement of several components of the host inflammatory response.

In the circulation, endotoxin binds to lipopolysaccharide-binding protein, which can transfer the molecule to a signaling receptor (e.g., membrane or soluble CD14) or to lipoproteins that neutralize its effects (71). Reconstituted human high-density lipoprotein (rHDL) given to human subjects reduced endotoxin-associated symptoms, cytokines (TNF, IL-6, IL-8, IL-1ra), sTNFR, as well as blunted the leukocytosis and expression of neutrophil CD11b/CD18 (72). Activation of fibrinolysis and coagulation was reduced (73). However, fever was not diminished and increases in IL-10 and PAI-1 levels were not attenuated (72). Administration of rHDL alone downregulated CD14 expression on monocytes, suggesting that rHDL may inhibit endotoxin effects by both binding and neutralizing endotoxin and downregulating its cell receptor (72).

Not all lipoprotein compounds, however, have antiinflammatory effects. Fat emulsions composed of triglycerides and lecithin are used routinely for parenteral nutrition. When incubated with whole blood and endotoxin, fat emulsions inhibit cytokine production (74). However, when administered 2 hours prior to endotoxin challenge, fat emulsion did not alter the fever, symptoms, increased heart rate, leukocytosis, or increased levels of TNF and sTNFR (74). Instead, the fat emulsion potentiated IL-6 and IL-8 levels and enhanced neutrophil degranulation (74). Further, a coagulopathic state characterized by increased levels of prothrombin F1+2 fragments, thrombin–antithrombin III complexes, and PAI-1 developed, while fibrinolysis was not altered (75). Thus, fat emulsions may potentiate inflammatory responses to endotoxin.

Tolerance to endotoxin inflammatory responses develops rapidly after endotoxin administration. Peripheral blood mononuclear cells taken from subjects given intravenous endotoxin have depressed cytokine production (TNF, IL-1, IL-6, and IL-8) when stimulated ex vivo with either LPS, IL-1β, or toxic shock syndrome toxin-1 (21). Some of these immunosuppressant effects may be due to the effects of cyclooxygenase products. When peripheral blood mononuclear cells are cultured ex vivo after systemic endotoxin administration, proliferation in response to phytohemagglutinin antigen (PHA), and production of IL-1 and IL-2 are depressed (22). When subjects were pretreated with ibuprofen, a cyclooxygenase inhibitor, responsiveness to PHA was restored to normal and less IL-2 suppression was observed (22).

Tolerance to endotoxin can be elicited in vivo by continuous infusion of endotoxin to volunteers (76) or by the use of weak but immunostimulating lipid A analogs (77). When given alone (20 μg/kg) to volunteers, it resulted in mild to moderate symptoms with increases in temperature, leukocytosis, heart rate, and cytokine levels (TNF, IL-6, IL-8, IL-1ra) (77). When given 24 hours prior to reference endotoxin administration (20 EU/kg), fever, tachycardia, cytokine levels (TNF, IL-6, IL-8) were reduced, whereas few or no effects were observed on the endotoxin-associated leukocytosis or levels of cortisol, norepinephrine, or epinephrine (77).

Endogenous Mediators That Alter Host Responses to Endotoxin

Host responses to endotoxin can be altered by prior exposure to clinically relevant interventions including parenteral nutrition or infusions of cortisol or epinephrine. These interventions alter the physiological responsiveness to endotoxin and may resemble the altered state of patients who have had stress responses (e.g., postsurgical or traumatic injury) prior to exposure to endotoxin.

Subjects given intravenous endotoxin after 7 days of parenteral nutrition had increased temperature and heart rate responses and enhanced levels of TNF (both in peripheral blood and hepatic vein samples), glucagon, epinephrine, and C-reactive protein (78). Metabolic responses were also altered and showed increases in extremity efflux of lactate and amino acids without the hyperglycemia and increased extremity uptake of glucose observed in enterally fed control subjects given a defined formula diet (78). The exact mechanism that accounts for these enhanced inflammatory responses is unknown. It is conceivable that intestinal mucosal atrophy or other factors associated with nutritional support contributed to these findings. A second study based on parenteral and enteral nutrition using historical controls did not observe accentuated responses to endotoxin (79).

Glucocorticoids serve as an important regulatory component of inflammatory responses. The interaction of glucocorticoids with the acute phase response is complex and varies by dose and time interval from steroid exposure. When hydrocortisone was given in stress doses (100 or 200 mg i.v.) prior to administration of endotoxin, symptoms and increases in temperature were attenuated and cytokine levels (TNF, IL-6, and IL-8) were blunted (80). Increases in heart rate and IL-1ra were not altered by bolus steroid administration (80). Administration of hydrocortisone (infused at 3 μg/kg/min) for 6 hours prior to and 6 hours after endotoxin administration blocked the development of symptoms, fever, increased heart rate, and prevented rises in TNF, epinephrine, and C-reactive protein (81). However, increases in IL-6 levels and resting energy expenditure were not diminished.

When endotoxin administration occurred after a 6-hour infusion of steroids and a 6-hour steroid-free interval, clinical responses (symptoms, heart rate, resting energy expenditure, and levels of cortisol, epinephrine, and C-reactive protein) were similar to endotoxin alone despite absent TNF, decreased soluble TNF receptor levels, and enhanced IL-6 levels (81,82). If a longer

Table 1 Modulation of Endotoxin-Mediated Inflammation in Humans by Anti-Inflammatory Agents, Stress Hormones, or Endotoxin Antagonists

	LPS alone	Cortisol: bolus or 6-hr infusion pre- and post-LPS[a]	Cortisol infusion for 6 hr, then 6 hr interval prior to LPS[b]	Cortisol infusion for 6 hr, then 24 or 144 hr interval pre-LPS[b]	Epinephrine infusion 3 or 24 hr pre-LPS[c]	rBPI$_{23}$[d]	E5531[e]	rHDL[f]	Triglyceride-rich fat emulsion pre-LPS[g]	Ibuprofen[h]
Symptoms	↑	→	no Δ		no Δ	no Δ	→	→	no Δ	→
Temperature	↑	→	no Δ		no Δ	no Δ	→	no Δ	no Δ	no Δ or →
Heart rate	↑	↓ or no Δ	no Δ	no Δ		→	→	no Δ	no Δ	→
Blood pressure	→					no Δ				no Δ
Respiratory rate	↑					no Δ				no Δ
Cardiac index	↑						blunts →			no Δ
LV ejection fraction	→				←					
Leukocyte number (1 hr)	↑				→	blunts →	→	blunts →	no Δ	no Δ
Leukocyte number (after 1 hr)	←				no Δ	→ →	→	→	no Δ	←
Neutrophil number	←		→ → →	← ←		→			←	←
Neutrophil degranulation[i]			←							
TNF-α	↕[j]	→	→ → →	← ←		→ →	→	→ → →	no Δ	←
Soluble TNFR[k]	←		←							
IL-1ra	←	↓ or no Δ			no Δ				no Δ	
IL-6	←	↓ or no Δ			no Δ				no Δ	

Parameter							
IL-8	↑	→					
IL-10	↑						
C-reactive protein	↑	→	no Δ				
Fibrinolysis[l]	↑			→			
Fibrinolysis inhibition[m]	↑		↑	→	→		
Coagulation[n]	↑		no Δ	→	no Δ		
Kallikrein activation[o]	→	→		→	→		
E-selectin	↑	→	↑		no Δ	↑	
Epinephrine	↑		no Δ		no Δ	no Δ	no Δ
Cortisol	↑		no Δ		→	←	no Δ

LPS, endotoxin; rBBI$_{23}$, recombinant bactericidal/permeability-increasing protein; E5531, lipid A antagonist derived from *Rhodobacter capsulatus*; rHDL, reconstituted high-density lipoprotein.

[a]From Refs. 80, 81.
[b]From Refs. 81, 82.
[c]From Ref. 83.
[d]From Ref. 16.
[e]From Refs. 69, 70.
[f]From Refs. 72, 73.
[g]From Refs. 74, 75.
[h]From Refs. 42, 44, 51, 94, 95.
[i]Increased blood levels of lactoferrin or elastase.
[j]TNF bioactivity or TNF antigen.
[k]sTNFR-soluble TNF receptors type I (p55) and type II (p75).
[l]Increased blood levels of tissue plasminogen activator or plasmin–α_2-plasmin inhibitor complexes.
[m]Increased levels of plasminogen activator inhibitor (PAI-1).
[n]Increased blook levels of prothrombin F1 + 2 and or thrombin–anti-thrombin III complexes.
[o]Decreased blood prekallikrein activity, increased kallikrein–α_2-macroglobulin complexes.

Table 2 Modulation of Endotoxin-Mediated Inflammation in Humans by Cytokines or Cytokine Antagonists

					Compared to LPS alone				
	LPS alone	IL-1ra pre-LPS[a]	IL-1R1 pre-LPS[b]	TNFR:Fc pre-LPS[c]	IL-10 pre-LPS[d]	IL-10 post-LPS[d]	G-CSF 2 hr prior[e]	G-CSF 12 hr prior[f]	G-CSF 24 hr prior[g]
Symptoms	↑	↓ or no Δ	↓	no Δ	no Δ	no Δ	↑	no Δ	no Δ
Temperature	↑	no Δ	no Δ	no Δ delayed rise	↓	no Δ	no Δ	↑	no Δ
Heart rate	↑	no Δ	no Δ	no Δ	no Δ	no Δ	↑	no Δ	no Δ
Blood pressure	↓	no Δ	no Δ	no Δ	no Δ	no Δ	no Δ		no Δ
Respiratory rate	↑	no Δ	no Δ	no Δ					
Cardiac index	↑		no Δ	no Δ					
LV ejection fraction	↓		no Δ	no Δ					
Leukocyte number (1 hr)	↓		no Δ	blunts ↓	no Δ	no Δ	increased ↓	increased ↓	increased ↓
Leukocyte number (after 1 hr)	↑	no Δ	no Δ	no Δ	blunts ↑	no Δ	↑	↓	↑
Neutrophil number	↑	↓	no Δ	↓	↓	no Δ	↑		↑
Neutrophil degranulation[h]	↑	no Δ	no Δ	↓	↓	no Δ	↑		↑
TNF-α	↑[i]	no Δ	↑	↓[j]	↓	no Δ	↑	↑	no Δ
Soluble TNFR[k]	↑	no Δ		↓			↑	↑	↑
IL-1β	↑ or no Δ	no Δ	↓	↓					
IL-1ra	↑		↓	↓	↓	no Δ	↑	↑	no Δ
IL-6	↑	no Δ	no Δ	↓	↓	no Δ	↑	no Δ	no Δ
IL-8	↑	no Δ	↑	↓	↓	no Δ	↑		↓
IL-10	↑			no Δ	↑	↑			
C-reactive protein	↑	no Δ	↑	no Δ					
Fibrinolysis[l]	↑			↓	↓	no Δ			
Fibrinolysis inhibition[m]	↑			↓	↓	↓			
Coagulation[n]	↑			no Δ	↓	↓			
Kallikrein activation[o]	↓			delayed					
E-selectin	↑			↓	↓				
Epinephrine	↑	no Δ		↓					
Cortisol	↑	no Δ		↓	↓	↓			

LPS, endotoxin; IL-1ra, interleukin-1 receptor antagonist; IL-1R1, soluble interleukin-1 type 1 receptor; TNFR:Fc, soluble dimeric TNF receptor (p80); IL-10, interleukin 10; G-CSF, granulocyte-colony stimulating factor.
[a]From Refs. 86, 87.
[b]From Ref. 35.
[c]From Refs. 30, 88–90.
[d]From Refs. 91, 92.
[e]From Ref. 58.
[f]From Ref. 93.
[g]From Ref. 58.
[h]Increased blood levels of lactoferrin or elastase.
[i]TNF bioactivity of TNF antigen.
[j]TNF bioactivity decreased, TNF antigen increased.
[k]sTNFR-soluble TNF receptors type I (p55) and type II (p75).
[l]Increased blood levels of tissue plasminogen activator or plasmin–α_2-plasmin inhibitor complexes.
[m]Increased levels of plasminogen activator inhibitor (PAI-1).
[n]Increased blood levels of prothrombin F1 + 2 and or thrombin–anti-thrombin III complexes.
[o]Decreased blood prekallikrein activity, increased kallikrein–α_2-macroglobulin complexes.

time interval (24 or 144 hours) occurs after the 6-hour infusion of steroids, clinical responses were similar to endotoxin alone but peak levels of TNF and IL-6 levels were three- to fourfold higher than subjects given LPS alone (81). The mechanism that accounts for these enhanced cytokine responses is unknown. These studies demonstrate that elevations in glucocorticoid levels have important early and late effects on host inflammatory responses and demonstrate the dissociation of blood levels of cytokines from clinical responses associated with endotoxin.

Epinephrine can alter host responsiveness to endotoxin by increasing cellular levels of cyclic adenosine monophosphate (cAMP). Normal volunteers given a continuous infusion of epinephrine (30 ng/kg/min for 3 or 24 hours prior to endotoxin infusion and 6 hours postendotoxin) had significant decreases in TNF and increases in IL-10 levels, while IL-6 and IL-8 levels were similar to those in control subjects (83). Similarly, the rise in blood levels of sTNFR and expression of monocyte (but no neutrophil) TNF receptors was blunted by the epinephrine infusion (84). Increases in the levels of soluble E-selectin were attenuated. Fibrinolysis was enhanced and coagulation diminished after the epinephrine infusions (85). Despite these broad effects, changes in symptoms and increases in temperature were unaltered by the epinephrine infusion (83).

Inhibition of the Effects of Secondary Cytokines Induced by Endotoxin

Blood levels of IL-1β are low or undetectable in many human studies of acute endotoxemia (34,57). However, levels of cell-associated IL-1β were at least 10-fold higher than blood levels, suggesting that IL-1β may contribute to local inflammatory responses (35). Infusion of recombinant IL-1ra did not significantly alter the magnitude or the intensity of the inflammatory response to endotoxin (86,87). The effects of IL-1ra were limited to decreasing symptoms or decreasing the magnitude of the neutrophilia and peripheral blood mononuclear cell proliferation (86,87). IL-1ra infusion had no effects on fever, respiratory rate, heart rate, blood pressure, cytokine levels (IL-1β, IL-6, IL-8, G-CSF), sTNFR, C-reactive protein, serum amyloid A, stress hormones, or factor VIII levels (86,87).

Recombinant soluble type I IL-1 receptor is an alternative means of blocking IL-1–mediated responses. When given prior to endotoxin administration, the incidence of chills was less and blood levels of IL-1β were decreased (35). However, IL-1R1 has a high affinity for both IL-1β and IL-1ra, and following endotoxin and IL-1 receptor administration, levels of IL-1ra decreased compared to controls. This resulted in a loss of IL-1ra anti-inflammatory effects and enhanced levels of cell-associated IL-1β, and blood TNF, IL-8, and C-reactive protein levels (35). IL-1R1 did not attenuate the leukopenia, leukocytosis, platelet decrease, lactoferrin levels, or cardiovascular responses (35). Thus, results from studies inhibiting IL-1 effects with either IL-1ra and type I soluble IL-1 receptor suggest that circulating IL-1 contributes only to a limited degree to endotoxin responses in humans.

Tumor necrosis factor is considered to play an essential role in directing the magnitude and intensity of the inflammatory response after endotoxin. A recombinant fusion protein composed of two p75 TNF receptors linked to the Fc fragment of IgG1 (TNFR:Fc) is a potent inhibitor of TNF bioactivity. When given to volunteers prior to endotoxin, blood TNF bioactivity was completely inhibited.

TNF inhibition blocked the leukopenic response, and while the magnitude of the leukocytosis was similar to controls, the percentage of neutrophils was diminished overall (30). Markers of neutrophil and endothelial cell activation were decreased and stress hormone responses were blunted after TNFR:Fc treatment (30). Some secondary cytokine responses were diminished (IL-1β, IL-1ra, IL-8, G-CSF, GRO-α), while increases in other cytokines (MIP-1α, IL-10) were not affected (30). Monocyte TNFR expression was blunted with TNFR:Fc, while neutrophil TNFR or type II IL-1 receptor expression was not reduced (88). Kallikrein activation was delayed by several hours but was similar in magnitude to the control subjects (89). Notably, fibrinolysis but not coagulation was inhibited by TNFR:Fc. These data demonstrate that an imbalance favoring coagulation and fibrin deposition may be potentiated by giving TNFR:Fc prior to endotoxin administration (30,89,90).

Despite inhibition of circulating TNF by low- and high-dose TNFR:Fc, symptoms were unaltered, fever was similar in magnitude but peaked one hour later, and systemic hemodynamic responses (cardiac index, heart rate, mean arterial pressure, left ventricular ejection fraction) were similar to those of control subjects (30). High-dose TNFR:Fc suppressed inflammatory responses (e.g., cytokine, stress hormones, and kallikrein activation) less than low-dose TNFR:Fc (30). This may be due to effects of the TNFR:Fc complex or extravascular sources of TNF. These results emphasize the contributions of circulating TNF to an inflammatory response that is composed of many overlapping pathways, which are influenced but not completely dependent on the presence of a single mediator.

IL-10 is an important counterregulatory cytokine with potent anti-inflammatory effects. When given to subjects prior to the administration of endotoxin, IL-10 reduced temperature, cytokine levels (TNF, IL-6, IL-8, IL-1ra), sTNFR, cortisol, and the magnitude and differential of the leukocytosis (91). Neutrophil elastase and endothelial cell activation was blunted and accumulation of labeled neutrophils in the lungs was diminished (91). Fibrinolysis and coagulation activation were inhibited (92). Thus, IL-10 had potent generalized anti-inflammatory effects when given prior to endotoxin.

In contrast, when IL-10 was given one hour after the administration of endotoxin, symptoms, fever, heart rate, and blood pressure were similar to controls (91). The leukocyte response was unaltered, and cytokine (TNF, IL-6, IL-8) responses were not diminished (91). Cortisol levels, however, were decreased (91). Late treatment with IL-10 suppressed the activation of coagulation and the release of PAI-1, an inhibitor of fibrinolysis (92). However, the fibrinolytic response was similar to control subjects (92). These studies revealed a previously unappreciated effect of IL-10 on hemostasis that occurs independent of fibrinolysis and emphasize the importance of timing in anti-inflammatory effects of IL-10.

Effects of a Proinflammatory Cytokine, G-CSF, on Inflammatory Responses After Endotoxin Administration

Granulocyte colony-stimulating factor (G-CSF), an important growth factor and activator of neutrophils, has been proposed as a means to enhance host responses during infection. When G-CSF is given 12 hours prior to endotoxin (*S. abortus equii*) administration, neutrophil numbers rose approximately sevenfold above control values at baseline (93). Subsequent endotoxin-associated increases in temperature were enhanced, while the severity of symptoms and increases in heart rate were unchanged (93). Levels of TNF, sTNFR, and IL-1ra were greater than the control responses, while IL-6 levels were similar (93).

A similar priming effect on inflammatory responses occurred when G-CSF was administered 2 hours prior to endotoxin (*E. coli*) (58). Symptoms and heart rate responses were increased, while changes in temperature and mean arterial pressure were unaltered. Cytokine levels (TNF, IL-6, IL-8, IL-1ra), sTNFR, leukopenia, leukocytosis, and neutrophil degranulation were greater than control responses. Thus, G-CSF enhanced the release of inflammatory mediators when given within 2 hours of endotoxin administration (58).

However, when G-CSF is given 24 hours prior to endotoxin, symptoms, fever, and decreased mean arterial pressure were unchanged, while heart rate was increased compared to control responses (58). Cytokine (TNF, IL-6, IL-1ra) responses were similar to control responses while IL-8 levels were decreased. Levels of sTNFR, leukopenia, leukocytosis, and neutrophil degranulation were enhanced (58). With both the 2- and 24-hour G-CSF treatment regimens, accumulation of labeled neutrophils in the lung was decreased (58). While G-CSF had no effect of expression of CD11b or CD18, L-selectin surface expression was diminished with both treatment regimens (58). Thus, both IL-10 and G-CSF given to decrease or enhance host inflammatory responses, respectively, have different effects when given before or after an inflammatory stimulus. These studies have important implications for understanding the clinical use of these agents in sepsis.

Pharmacological Agents That Alter the Inflammatory Responses Associated with Endotoxin

Ibuprofen is a cyclooxygenase inhibitor that alters many of the associated LPS responses. When given prior to endotoxin administration, ibuprofen significantly decreased the magnitude of symptoms, fever, respiratory frequency, and resting energy expenditure but did not change the cardiovascular response (heart rate, mean arterial pressure, cardiac output, left ventricular ejection fraction) (51,94). Leukocyte responses were not altered but neutrophil elastase release was enhanced (95). Stress hormone responses were diminished (51). Cyclooxygenase products are important inhibitors of cytokine production, and blocking prostaglandin production in subjects given endotoxin was associated with enhanced levels of TNF and IL-6 (42,44,95). However, the enhanced blood cytokine levels were not associated with any symptoms or increased alterations in cardiovascular responses (94). These data demonstrate that the cardiovascular response is not dependent on fever and that prostaglandins have important effects on cytokine responses in vivo.

Pentoxifylline is a phosphodiesterase inhibitor that enhances cAMP levels. Subjects pretreated with i.v. pentoxifylline have decreased levels of TNF but similar temperature responses and symptoms (96). IL-6 and cortisol levels and leukocyte responses were similar to control subjects (96). Oral pentoxifylline had no effects on symptoms, temperature, cardiovascular, or cytokine responses associated with endotoxin administration (42).

The contribution of platelet-activating factor (PAF) to the acute response to endotoxin appears has been evaluated with a specific PAF-receptor antagonist (Ro 24-4736). When Ro 24-4736 is given 18 hours prior to endotoxin administration, symptoms and stress hormone responses were decreased (97). PAF-induced platelet aggregation was inhibited. However, the PAF receptor antagonist did not alter changes in temperature, heart rate, blood pressure, resting energy expenditure, TNF, sTNFR55, or IL-6 (97).

CONCLUSIONS

These studies provide some insight into the complexity and difficulty in modulating early host responses to endotoxin. The dose and timing of interventions intended to modulate the inflammatory response are critical variables in altering the responses initiated by endotoxin. For example, the effects of corticosteroids, IL-10, and G-CSF vary considerably with the their administration before or after endotoxin challenge.

In addition, the hierarchy of responses are only partially dependent on a single mediators such as TNF. Blocking TNF or IL-1 did not change the fever or cardiovascular response associated with endotoxemia (30,35). Further, these studies have shown the dissociation of blood cytokine levels from clinical responses associated with endotoxin. Inhibition of endotoxin-associated fever and symptoms by ibuprofen did not alter the cardiovascular response and was associated with enhanced levels of proinflammatory cytokines (42,94).

These studies provide some insights into the difficulties of using mediator-specific therapies on clinical sepsis and septic shock (98). The acute phase host response to endotoxin is a well-conserved inflammatory process composed of interrelated pathways with multiple overlapping activities. In parallel to advances in the biology of inflammation, the observations associated with endotoxin administration to humans will continue to provide unique information regarding the initiation, control, and interrelationships of acute inflammatory pathways.

REFERENCES

1. Nauts HC. Bacteria and cancer—antagonisms and benefits. Cancer Surv 1989; 8:713–723.
2. Stamm VH, Eichenberger E. Thromboemboliebehandlung mit Pyrexal. Geburtshilfe Frauenheilkd 1958; 18: 451–461.
3. Engelhardt R, Mackensen A, Galanos C. Phase I trial of intravenously administered endotoxin (*Salmonella abortus equi*) in cancer patients. Cancer Res 1991; 51: 2524–2530.
4. Kuhns DB, Long Priel DA, Gallin JI. Endotoxin and IL-1 hyporesponsiveness in a patient with recurrent bacterial infections. J Immunol 1997; 158:3959–3964.
5. Martich GD, Boujoukos AJ, Suffredini AF. Response of man to endotoxin. Immunobiology 1993; 187:403–416.
6. Baumann H, Gauldie J. The acute phase response. Immunol Today 1994; 15:74–80.
7. Fearon DT, Locksley RM. The instructive role of innate immunity in the acquired immune response. Science 1996; 272:50–53.
8. Bone RC, Balk RA, Cerra FB, et al. Definitions for sepsis and organ failure and guidelines for the use of innovative therapies in sepsis. The ACCP/SCCM Consensus Conference Committee. American College of Chest Physicians/Society of Critical Care Medicine. Chest 1992; 101:1644–1655.
9. Hochstein HD, Mills DF, Outschoorn AS, Rastogi SC. The processing and collaborative assay of a reference endotoxin. J Biol Stand 1983; 11:251–260.
10. Pollmacher T, Schreiber W, Gudewill S, et al. Influence of endotoxin on nocturnal sleep in humans. Am J Physiol 1993; 264:R1077–R1083.
11. Rylander R, Bake B, Fischer JJ, Helander IM. Pulmonary function and symptoms after inhalation of endotoxin. Am Rev Respir Dis 1989; 140:981–986.
12. Clapp WD, Becker S, Quay J, et al. Grain dust-induced airflow obstruction and inflammation of the lower respiratory tract. Am J Respir Crit Care Med 1994; 150; 611–617.
13. Suffredini AF, Fromm RE, Parker MM, et al. The cardiovascular response of normal humans to the administration of endotoxin. N Engl J Med 1989; 321:280–287.
14. Taveira da Silva AM, Kaulbach HC, Chuidian FS, Lambert DR, Suffredini AF, Danner RL. Brief report: shock and multiple-organ dysfunction after self-administration of *Salmonella* endotoxin. N Engl J Med 1993; 328: 1457–1460.
15. Van Deventer SJH, Buller HR, Ten Cate JW, Aarden LA, Hack CE, Sturk A. Experimental endotoxemia in humans: analysis of cytokine release and coagulation, fibrinolytic, and complement pathways. Blood 1990; 76:2520–2526.
16. von der Mohlen MA, Kimmings AN, Wedel NI, et al. Inhibition of endotoxin-induced cytokine release and neutrophil activation in humans by use of recombinant bactericidal/permeability-increasing protein. J Infect Dis 1995; 172:144–151.
17. Elin RJ, Wolff SM, McAdam KPWJ, et al. Properties of reference *Escherichia coli* endotoxin and its phthalylated derivative in humans. J Infect Dis 1981; 144: 329–336.
18. Hochstein HD, Fitzgerald EA, McMahon FG, Vargas R. Properties of US Standard Endotoxin (EC-5) in human male volunteers. J Endotoxin Res 1994; 1:52–56.

19. Malech HL. Phagocytic cells: egress from marrow and diapedesis. In: Gallin JL, Goldstein IM, Synderman R, eds. Inflammation: Basic Principles and Clinical Correlates. New York: Raven Press, Ltd., 1988:297–308.
20. Richardson RP, Rhyne CD, Fong Y, et al. Peripheral blood leukocyte kinetics following in vivo lipopolysaccharide (LPS) administration to normal human subjects. Influence of elicited hormones and cytokines. Ann Surg 1989; 210:239–245.
21. Granowitz EV, Porat R, Mier JW, et al. Intravenous endotoxin suppresses the cytokine response of peripheral blood mononuclear cells of healthy humans. J Immunol 1993; 151:1637–1645.
22. Rodrick ML, Moss NM, Grbic JT, et al. Effects of in vivo endotoxin infusions on in vitro cellular immune responses in humans. J Clin Immunol 1992; 12:440–450.
23. von der Mohlen MA, van Deventer SJ, Levi M, et al. Inhibition of endotoxin-induced activation of the coagulation and fibrinolytic pathways using a recombinant endotoxin-binding protein (rBPI23). Blood 1995; 85:3437–3443.
24. Kuhns DB, Alvord WG, Gallin JI. Increased circulating cytokines, cytokine antagonists, and E-selectin after intravenous administration of endotoxin in humans. J Infect Dis 1995; 171:145–152.
25. Calvano SE, Thompson WA, Marra MN, et al. Changes in polymorphonuclear leukocyte surface and plasma bactericidal/permeability-increasing protein and plasma lipopolysaccharide binding protein during endotoxemia or sepsis. Arch Surg 1994; 129:220–226.
26. Spinas GA, Keller U, Brockhaus M. Release of soluble receptors for tumor necrosis factor (TNF) in relation to circulating TNF during experimental endotoxinemia. J Clin Invest 1992; 90:533–536.
27. Van der Poll T, Calvano SE, Kumar A, et al. Endotoxin induces downregulation of tumor necrosis factor receptors on circulating monocytes and granulocytes in humans. Blood 1995; 86:2754–2759.
28. Suffredini AF, Harpel PC, Parrillo JE. The promotion and subsequent inhibition of plasminogen following the intravenous endotoxin administration to normal humans. N Engl J Med 1989; 320:1165.
29. Pugin J, Widmer M-Cl, Kossodo S, Liang C-M, Preas HL, Suffredini AF. Endotoxin and proinflammatory mediators induce a rapid secretion of gelatinase B (MMP-9) by neutrophils in whole human blood in vitro and in vivo. Am J Resp Cell Molecular Biol 1999. In press.
30. Suffredini AF, Reda D, Banks SM, Tropea M, Agosti JM, Miller R. Effects of recombinant dimeric TNF receptor on human inflammatory responses following intravenous endotoxin administration. J Immunol 1995; 155:5038–5045.
31. von der Mohlen MA, Van der Poll T, Jansen J, Levi M, van Deventer SJ. Release of bactericidal/permeability-increasing protein in experimental endotoxemia and clinical sepsis. Role of tumor necrosis factor. J Immunol 1996; 156:4969–4973.
32. DeLa Cadena RA, Suffredini AF, Page JD, Kaufman N, Parrillo JE, Colman RW. Activation of the kallikrein-kinin system after endotoxin administration to normal human volunteers. Blood 1993; 81:3313–3317.
33. Michie HR, Manogue KR, Spriggs DR, et al. Detection of circulating tumor necrosis factor after endotoxin administration. N Engl J Med 1988; 318:1481–1486.
34. Cannon JC, Tompkins RG, Gelfand JA, et al. Circulating interleukin-1 and tumor necrosis factor in septic shock and experimental endotoxin fever. J Infect Dis 1990; 161:79–84.
35. Preas HL, Reda D, Tropea M, et al. Effects of recombinant soluble type I IL-1 receptor on human inflammatory responses to endotoxin. Blood 1996; 88:2465–2472.
36. Fong Y, Moldawer LL, Marano M, et al. Endotoxemia elicits increased circulating beta 2-IFN/IL-6 in man. J Immunol 1989; 142;2321–2324.
37. Van der Poll T, Levi M, Hack CE, et al. Elimination of interleukin 6 attenuates coagulation activation in experimental endotoxemia in chimpanzees. J Exp Med 1994; 179:1253–1259.
38. Tilg H, Dinarello CA, Mier JW. IL-6 and APPs: antiinflammatory and immunosuppressive mediators. Immunol Today 1997; 18:428–432.
39. Granowitz EV, Santos AA, Poutsiaka DD, et al. Production of interleukin-1 receptor antagonist during experimental endotoxaemia. Lancet 1991; 338:1423–1424.
40. Granowitz EV, Porat R, Orencole SF, et al. Granulocyte-macrophage colony-stimulating factor synthesis during experimental endotoxemia in humans. J Infect Dis 1992; 166:1204.
41. Van der Poll T, de Waal Malefyt R, Coyle SM, Lowry SF. Antiinflammatory cytokine responses during clinical sepsis and experimental endotoxemia: sequential measurements of plasma soluble interleukin (IL)-1 receptor type II, IL-10, and IL-13. J Infect Dis 1997; 175:118–122.
42. Martich GD, Danner RL, Ceska M, Suffredini AF. Detection of interleukin 8 and tumor necrosis factor in normal humans after intravenous endotoxin: the effect of antiinflammatory agents. J Exp Med 1991; 173:1021–1024.
43. Sylvester I, Suffredini AF, Boujoukos AJ, et al. Neutrophil attractant protein-1 and monocyte chemoattractant protein-1 in human serum. Effects of intravenous lipopolysaccharide on free attractants, specific IgG autoantibodies and immune complexes. J Immunol 1993; 151:3292–3298.
44. Boujoukos AJ, Martich GD, Supinski E, Suffredini AF. Compartmentalization of the acute cytokine response in humans after endotoxin administration. J Appl Physiol 1993; 74:3027–3033.
45. Pruzanski W, Wilmore DW, Suffredini AF, et al. Hyperphospholipasemia A2 in human volunteers challenged with intravenous endotoxin. Inflammation 1992; 16:561–569.
46. Dandona P, Nix D, Wilson MF, et al. Procalcitonin increase after endotoxin injection in normal subjects. J Clin Endocrinol Metab 1994; 79:1605–1608.
47. Bloom JN, Suffredini AF, Parrillo JE, Palestine AC. Serum neopterin levels following intravenous endotoxin administration to normal humans. Immunobiology 1990; 181:317–323.
48. McAdam BF, Fitzgerald GA. Enzymatic regulation of the prostaglandin response in a human model of inflam-

49. Vandivier RW, Eidsath A, Banks SM, et al. Downregulation of nitric oxide production by ibuprofen in human volunteers. J Pharmacol Exp Ther 1999. In press.
50. Catania A, Suffredini AF, Lipton JM. Endotoxin causes α-MSH release in normal subjects. Neuroimmunomodulation 1995; 2:248–253.
51. Revhaug A, Michie HR, Manson JM, et al. Inhibition of cyclooxygenase attenuates the metabolic response to endotoxin in humans. Arch Surg 1988; 123:162–170.
52. Michie HR, Majzoub JA, ODwyer ST, Revhaug A, Wilmore DW. Both cyclooxygenase-dependent and cyclooxygenase-independent pathways mediate the neuroendocrine response in humans. Surgery 1990; 108:254–259.
53. Pollmacher T, Mullington J, Korth C, et al. Diurnal variations in the human host response to endotoxin. J Infect Dis 1996; 174:1040–1045.
54. Michie HR, Spriggs DR, Manogue KR, et al. Tumor necrosis factor and endotoxin induce similar metabolic responses in human beings. Surgery 1988; 104:280–286.
55. Kuipers B, Rock CS, Coyle SM, et al. Influence of hypercortisolemia on the acute-phase protein response to endotoxin in humans. Surgery 1992; 112:467–474.
56. Shedlofsky SI, Israel BC, McClain CJ, Hill DB, Blouin RA. Endotoxin administration to humans inhibits hepatic cytochrome P450-mediated drug metabolism. J Clin Invest 1994; 94:2209–2214.
57. Fong YM, Marano MA, Moldawer LL, et al. The acute splanchnic and peripheral tissue metabolic response to endotoxin in humans. J Clin Invest 1990; 85:1895–1904.
58. Pajkrt D, Manten A, Van der Poll T, et al. Modulation of cytokine release and neutrophil function by granulocyte colony-stimulating factor during endotoxemia in humans. Blood 1997; 90:1415–1424.
59. Suffredini AF, Shelhamer JH, Neumann RD, Brenner M, Baltaro RJ, Parrillo JE. Pulmonary and oxygen transport effects of intravenously administered endotoxin in normal humans. Am Rev Respir Dis 1992; 145:1398–1403.
60. Smith PH, Suffredini AF, Allen JB, Wahl LM, Parrillo JE, Wahl SM. Endotoxin administration to humans primes alveolar macrophages for increased production of inflammatory mediators. J Clin Immunol 1994; 14:141–148.
61. O'Dwyer ST, Michie HR, Ziegler TR, Revhaug A, Smith RJ, Wilmore DW. A single dose of endotoxin increases intestinal permeability in healthy humans. Arch Surg 1988; 123:1459–1464.
62. Korth C, Mullington J, Schreiber W, Pollmacher T. Influence of endotoxin on daytime sleep in humans. Infect Immun 1996; 64:1110–1115.
63. Dehoux MS, Boutten A, Ostinelli J, et al. Compartmentalized cytokine production within the human lung in unilateral pneumonia. Am J Respir Crit Care Med 1994; 150:710–716.
64. Castellan RM, Olenchock SA, Kinsley KB, Hankinson JL. Inhaled endotoxin and decreased spirometric values. An exposure-response relation for cotton dust. N Engl J Med 1987; 317:605–610.
65. Michel O, Duchateau J, Plat G, et al. Blood inflammatory response to inhaled endotoxin in normal subjects. Clin Exp Allergy 1995; 25:73–79.
66. Preas HL, Tropea M, Reda D, et al. Bronchial instillation of endotoxin in normal humans results in local and systemic inflammatory responses (abstr). Am J Respir Crit Care Med 1995; 153:A442.
67. Preas HL, Tropea M, Reda D, O'Grady NP, Culpepper S, Suffredini AF. Bronchial instillation of endotoxin in humans alters concentrations of α and β chemokines (abstr). Am J Respir Crit Care Med 1995; 153:A442.
68. de Winter RJ, von der Mohlen MA, van Lieshout H, et al. Recombinant endotoxin-binding protein (rBPI23) attenuates endotoxin-induced circulatory changes in humans. J Inflamm 1995; 45:193–206.
69. Bunnell E, Lynn M, Parrillo JE, Habet K, Friedhoff LT, Rogers SL. Effect of E5531 on systemic responses to endotoxin in healthy volunteers (abstr). Crit Care Med 1995; 23 (suppl):A147.
70. Bunnell E, Neumann A, Lynn M, et al. E5531, an endotoxin antagonist, blocks the hyperdynamic and depressant cardiovascular effects of endotoxin in healthy subjects (abstr). Crit Care Med 1995; 23 (suppl):A151.
71. Ulevitch RJ, Tobias PS. Receptor-dependent mechanisms of cell stimulation by bacterial endotoxin. Ann Rev Immunol 1995; 13:437–457.
72. Pajkrt D, Doran JE, Koster F, et al. Antiinflammatory effects of reconstituted high-density lipoprotein during human endotoxemia. J Exp Med 1996; 184:1601–1608.
73. Pajkrt D, Lerch PG, Van der Poll T, et al. Differential effects of reconstituted high-density lipoprotein on coagulation, fibrinolysis and platelet activation during human endotoxemia. Thromb Haemost 1997; 77:303–307.
74. Van der Poll T, Braxton CC, Coyle SM, et al. Effect of hypertriglyceridemia on endotoxin responsiveness in humans. Infect Immun 1995; 63:3396–3400.
75. Van der Poll T, Coyle SM, Levi M, et al. Fat emulsion infusion potentiates coagulation activation during human endotoxemia. Thromb Haemost 1996; 75:83–86.
76. Johnston CA, Greisman SE. Mechanisms of endotoxin tolerance. In: Hinshaw LB, ed. Handbook of Endotoxin, Vol. 2: Pathophysiology of Endotoxin. New York: Elsevier Science Publishers B.V., 1985:359–401.
77. Astiz ME, Rackow EC, Still JG, et al. Pretreeatment of normal humans with monophosphoryl lipid A induces tolerance to endotoxin: a prospective, double-blind, randomized, controlled trial. Crit Care Med 1995; 23:9–17.
78. Fong Y, Marano MA, Barber A, et al. Total parenteral nutrition and bowel rest modify the metabolic response to endotoxin in humans. Ann Surg 1989; 210(4):449–459.
79. Santos AA, Rodrick ML, Jacobs DO, et al. Does the route of feeding modify the inflammatory response? Ann Surg 1994; 220:155–631.
80. Santos AA, Scheltinga MR, Lynch E, et al. Elaboration of interleukin 1-receptor antagonist is not attenuated by glucocorticoids after endotoxemia. Arch Surg 1993; 128:138–144.

81. Barber AE, Coyle SM, Marano MA, et al. Glucocorticoid therapy alters hormonal and cytokine responses to endotoxin in man. J Immunol 1993; 150:1999–2006.
82. Barber AE, Coyle SM, Fischer E, et al. Influence of hypercortisolemia on soluble tumor necrosis factor receptor II and interleukin-1 receptor antagonist responses to endotoxin in human beings. Surgery 1995; 118:406–410.
83. Van der Poll T, Coyle SM, Barbosa K, Braxton CC, Lowry SF. Epinephrine inhibits tumor necrosis factor-alpha and potentiates interleukin 10 production during human endotoxemia. J Clin Invest 1996; 97:713–719.
84. Van der Poll T, Calvano SE, Kumar A, Coyle SM, Lowry SF. Epinephrine attenuates down-regulation of monocyte tumor necrosis factor receptors during human endotoxemia. J Leukocyte Biol 1997; 61:156–160.
85. Van der Poll T, Levi M, Dentener M, et al. Epinephrine exerts anticoagulant effects during human endotoxemia. J Exp Med 1997; 185:1143–1148.
86. Van Zee KJ, Coyle SM, Calvano SE, et al. Influence of IL-1 receptor blockade on the human response to endotoxemia. J Immunol 1995; 154:1499–1507.
87. Granowitz EV, Porat R, Mier JW, et al. Hematologic and immunomodulatory effects of an interleukin-1 receptor antagonist coinfusion during low-dose endotoxemia in healthy humans. Blood 1993; 82:2985–2990.
88. Van der Poll T, Coyle SM, Kumar A, Barbosa K, Agosti JM, Lowry SF. Down-regulation of surface receptors for TNF and IL-1 on circulating monocytes and granulocytes during human endotoxemia: effect of neutralization of endotoxin-induced TNF activity by infusion of a recombinant dimeric TNF receptor. J Immunol 1997; 158:1490–1497.
89. DeLa Cadena RA, Majluf-Cruz A, Stadnicki A, et al. Recombinant tumor necrosis factor receptor (TNFR:Fc) alters endotoxin-induced activation of the kinin, fibrinolytic, and coagulation systems in normal human subjects. Thromb Haemost 1998; 80:114–118.
90. Van der Poll T, Coyle SM, Levi M, et al. Effect of a recombinant dimeric tumor necrosis factor receptor on inflammatory responses to intravenous endotoxin in normal humans. Blood 1997; 89:3727–3734.
91. Pajkrt D, Camoglio L, Tiel-van Buul MCM, et al. Attenuation of proinflammatory response by recombinant human IL-10 in human endotoxemia. J Immunol 1997; 158:3971–3977.
92. Pajkrt D, Van der Poll T, Levi M, et al. Interleukin-10 inhibits activation of coagulation and fibrinolysis during human endotoxemia. Blood 1997; 89:2701–2705.
93. Pollmacher T, Korth C, Mullington J, et al. Effects of granulocyte colony-stimulating factor on plasma cytokine and cytokine receptor levels and on the in vivo host response to endotoxin in healthy men. Blood 1996; 87:900–905.
94. Martich GD, Parker MM, Cunnion RE, Suffredini AF. Effects of ibuprofen and pentoxifylline on the cardiovascular response of normal humans to endotoxin. J Appl Physiol 1992; 73:925–931.
95. Spinas GA, Bloesch D, Keller U, Zimmerli W, Cammisuli S. Pretreatment with ibuprofen augments circulating tumor necrosis factor-alpha, interleukin-6, and elastase during acute endotoxemia. J Infect Dis 1991; 163;89–95.
96. Zabel P, Schonharting MM, Wolter DT, Schade UF. Oxpentifylline in endotoxaemia. Lancet 1989; 2:1474–1477.
97. Thompson WA, Coyle S, Van Zee K, et al. The metabolic effects of platelet-activating factor antagonism in endotoxemic man. Arch Surg 1994; 129:72–79.
98. Workshop Writing Committee. From the bench to the bedside: the future of sepsis research. Executive summary of an American College of Chest Physicians, National Institute of Allergy and Infectious Disease, and National Heart, Lung, and Blood Institute Workshop. Chest 1997; 111:744–753.

56

Endotoxin Detection in Body Fluids: Chemical Versus Bioassay Methodology

Thomas J. Novitsky
Associates of Cape Cod, Inc., Falmouth, Massachusetts

INTRODUCTION

The first practical assay for endotoxin (as "pyrogen") was a rabbit bioassay based on the work of Seibert (1). Over a period spanning her entire research career, Florence Seibert elucidated the cause of "injection" and "protein" fevers, proposed methods to produce nonpyrogenic intravenous solutions, suggested a standard "pyrogen," and provided a model for the first compendial test for pyrogen (2–5). The current compendial pyrogen test consists of injecting a solution into the marginal ear vein of a rabbit under standardized conditions and measuring its rectal temperature over a period of time (6). The pyrogen test has been used with only slight changes since it was first described in Volume XII of the *Pharmacopoeia of the United States of America* in 1942 (7). Because the chemical nature of pyrogen as endotoxin was only beginning to be elucidated by the time the pyrogen test was published, no alternative (i.e., chemical) test could even be considered. The early purification work of Boivin et al. (8) and Westphal et al. (9) paved the way for characterization of the lipopolysaccharide component of endotoxin to the point where several unique components were identified. Two of these in particular, 2-keto-3-deoxy-octonic acid (or 3-deoxy-D-manno-2-octulosonic acid, KDO) and β-OH fatty acids, could be used as a chemical test for endotoxin (10,11).

In 1964, Frederik Bang and his colleague Jack Levin described an in vitro response to endotoxin with blood cells from the North American horseshoe crab, *Limulus polyphemus* (12). This reaction later became the Limulus amebocyte lysate (LAL) test for endotoxin. The LAL test (as the bacterial endotoxins test) was also the first (and so far only) assay to qualify as a substitute for the compendial pyrogen test (13). In essence, of course, the LAL reaction may be considered as a biological response to endotoxin. Given the almost ubiquitous ability of endotoxin to elicit biological responses, it should come as no surprise that other well-characterized and controlled responses should therefore have potential to serve as assays for LPS as well. A summary of the various endotoxic responses that have the potential to be used as assays for LPS are listed in Table 1.

Finally, compounds that bind endotoxin should have assay potential. Polymyxin B and its derivatives are a well-known example of an endotoxin-binding peptide, and these agents have been employed in a number of assays (14). Recently, however, a number of endotoxin-binding proteins have been defined and characterized. Some of these, in particular the endotoxin-neutralizing protein (ENP), tachyplesins and polyphemisins (from the horseshoe crab), and bactericidal permeability protein (BPI) have already been described as having utility in assays for LPS and it is likely that additional LPS-binding reagents will continue to be described (15–17).

Purified endotoxin can now be analyzed chemically in any number of ways, e.g., gel electrophoresis, nuclear magnetic resonance, electron spin resonance, gas chromatography–mass spectrometry, etc. (18–22), and likewise biologically, e.g., LD_{50}, Shwartzman reaction,

Table 1 Some Biological Effects of Endotoxin

Pyrogenicity*	Abortion
Clotting in Limulus*	Shwartzman phenomenon
Endogenous pyrogen release*	Immunomodulation
Immunogenicity*	Induction of cytokines*
Lethality*	Activation of complement
Leukopenia	Induction of tolerance
Radiation protection	Shock

*Has been used or has potential as endotoxin assay.

etc. Purified endotoxins could also be used to elicit an antigenic response in a number of animal species (23). Homologous serum or purified antibodies from immunized animals could then be used to detect and quantify homologous and cross-reacting endotoxin species. Alternatively, using immunological reagents specific for LPS constituents sensitive ELISA or RIA tests can be employed (24,25). For the purpose of this chapter, however, only those methods designed for the detection and quantification of endotoxin in an impure state, i.e., mixed with other chemical components, either synthetic or naturally occurring, will be considered. The ultimate goal, of course, is to be able to accurately and precisely measure endotoxin in body fluids in order to better understand its clinical implications.

CHEMICAL ASSAYS

In the context of this review, chemical assays for endotoxin are those defined as analytical procedures that recognize and quantify either the entire endotoxin molecule or any part of the endotoxin molecule through a chemical reaction or signal, including binding to a specific ligand. It can be argued that the Limulus amebocyte lysate test, which relies on the activation of the clotting cascade isolated from the blood of the horseshoe crab and has been shown to correlate quite well with a number of physiological responses to endotoxin, is a bioassay. LAL will, however, be treated here as a chemical test, since endotoxin reacts with an LAL enzyme (factor C) and does not require a living cell or cellular system in order to elicit a dose-dependent response.

Limulus Amebocyte Lysate

Direct Test

The LAL test is perhaps the most widespread assay for endotoxin currently in use in the world. Its utility for the detection of endotoxin in human blood has been extensively reviewed (26–28). It has also been used to detect endotoxin in other biological samples of clinical significance, including spinal fluid and urine as an aid in the diagnosis of gram-negative meningitis and bacteriuria, respectively, and for a variety of other diseases (29). It should be noted that although the LAL test serves primarily as a test for endotoxin, it has also been used as a rapid bacterial test. The best example of this is the bacteriuria application. In blood, however, it is doubtful that circulating intact bacteria contribute significantly to a positive LAL test. The exception to this may be cases of meningococcal or yersinia bacteremias.

The LAL reagent, a crude extract of the blood cells (amebocytes) of the horseshoe crab, consists of a mixture of enzymes, proteins, salts, and other soluble biological materials including a natural (protein) substrate. The assay as originally described was primarily a qualitative endpoint type of endotoxin determination, referred to as the gel-clot test. More recent modifications of the assay incorporate a spectrophotometer and measure turbidity or color if a synthetic peptide substrate that contains a chromophore is added to the LAL reagent (Fig. 1). Thus, the LAL assay is an enzymatic cascade and the consequent formation of a gel (gel-clot assay) and/or increases of turbidity (turbidimetric assay), or color (chromogenic assay) can be quantitatively measured as a function of endotoxin added. Although all these assay variations have been used successfully to measure endotoxin, the chromogenic assay has enjoyed the most widespread use in clinical research due to its highly reproducible quantitative nature, sensitivity, and micro plate format.

Although the LAL test has enjoyed widespread use, it has not been considered a totally reliable diagnostic test for endotoxemia. There are also significant arguments on optional methodologies for sample preparation of clinical samples prior to carrying out the LAL assay. It has been recognized since Levin's initial clinical studies with LAL (26,30) that inhibitors exist in blood that can mask the ability of the LAL assay to detect endotoxin. The removal or inactivation of these inhibitors would be expected to increase the sensitivity of the assay, with the result that the diagnostic utility of the LAL should be improved. Subsequent refinements of the LAL test, therefore, focused on increasing the sensitivity of the LAL assay per se, removing or inactivating the inhibitor(s), and making the assay more robust, i.e., less easy to contaminate (also includes sample collection), quicker, and less sensitive to interference.

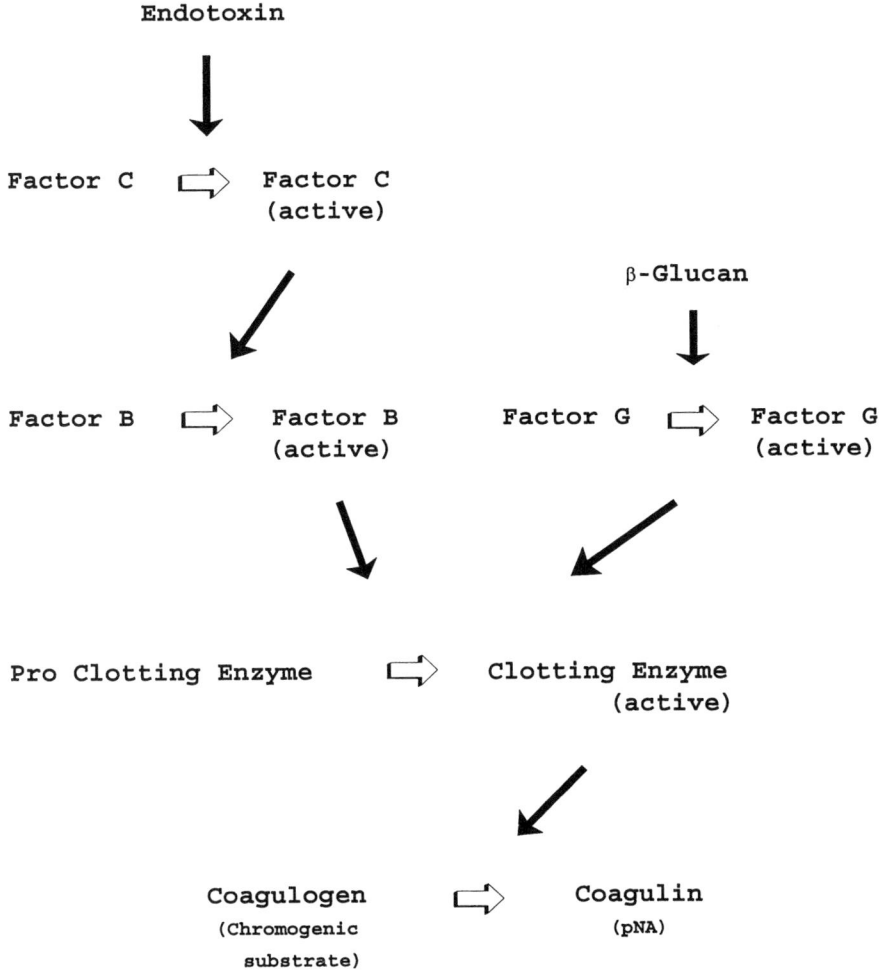

Fig. 1 Biochemistry of the LAL cascade showing endotoxin and glucan pathways.

The chromogenic methodologies currently in use are capable of achieving sensitivities of approximately 0.005 EU/ml using *E. coli* endotoxin as a standard. Although a commercially available turbidimetric LAL reagent is available with a sensitivity of 0.001 EU/ml and published techniques describe more sensitive modifications, the available evidence would support the conclusions that the practical working limit of the test at 0.001–0.005 EU/ml has been reached. Without special precautions in sample collection and test performance (beyond the already stringent recommended precautions), an increase in sensitivity much beyond this value is not likely to be possible. Such modifications would, however, render the test impractical in the routine clinical laboratory environment. Further recent data also suggest that clinically healthy patients can have measurable levels of circulating endotoxin (26).

A level of 0.1 EU/ml has been established in several studies, while a recent study employing a large number of controls found a background of 0.45 EU/ml (31). These levels are well within the sensitivity of the existing commercially available reagent.

On the basis of the above-cited studies, it has been assumed that all the circulating endotoxin was detected by the LAL test. Although the methodology for removing or inactivating inhibitors from the sample varies depending upon the specifics of the assay, two major approaches have been used. These include the heat/dilution method and the chemical treatment method. The objectives of both methods are (1) to release endotoxin bound to any blood components that might interfere with its ability to activate the clotting cascade, e.g., high-density lipoprotein; (2) to inactivate any components present in the biological samples that may directly interfere positively (enhancement) or negatively (inhibition) with the LAL reagent, e.g., serine proteases color; and (3) to adjust the concentration of endotoxin and other factors, e.g., divalent cations, for optimal reaction with LAL. Dilution/heating is the easiest method, although there is currently not full agreement on the optimal dilution factor or temperature. An additional problem occurs when heat-treated plasma or other proteins, e.g., albumin, is used as a sample in the turbidimetric assay (32). This problem manifests itself as an increase in turbidity (vs. the unheated control). Thus, a heated plasma sample with a known quantity of added endotoxin will appear to have more endotoxin than was added. Although the exact reason(s) for this phenomenon is unknown, it is likely due to a nonspecific increase in turbidity caused by an interaction between denatured (plasma) protein and coagulin. Increasing dilution (greater than 10-fold) will usually result in a more accurate result, however, this is accompanied by a loss of assay sensitivity. The extent of the nonspecific increase also varies with the sample (patient), necessitating use of an internal control. These problems do not seem as pronounced with either a chemical extraction or the chromogenic assay. Dilution will also decrease the sensitivity of the chromogenic assay, although this does not appear to be a drawback since dilution is limited to 10-fold or less. Chemical treatment, while easier to control, requires rather precise attention to the preparation of the extraction solutions. These consist of either acids and/or bases coupled with detergents and buffers (26). The addition of sample to the chemical extractant also results in some dilution, however, as with the heat/dilution method, this is not considered a drawback. Problems with this approach are that not only must the reagents used be prepared strictly in an endotoxin-free environment, but their stability must be taken into consideration. Thus, extraction "kits" are, in general, not commercially available. One trial involving a kit (SepTest) was successful in demonstrating sensitive endotoxin detection and practicality for potential use in a clinical laboratory, but was not commercialized because the endotoxin detected in septic patients (enrolled in the trial to evaluate this kit) did not show a strong correlation with any of the usual symptoms associated with sepsis (27). One, therefore, should not confuse the lack of clinical utility of endotoxin in the blood with the ability of LAL to detect circulating endotoxin.

Finally, sample collection or the nature of the sample itself needs to be considered. In the literature, the most widely used clinical sample used for detection of endotoxin has been plasma. More recently, platelet-rich plasma (PRP) has been used, since it has been shown that endotoxin can readily bind to platelets (26). In the SepTest trial and a few other studies, whole blood was used in an attempt to detect all circulating endotoxin, free, bound, or even internalized in cells. From these and other studies it is clear that similar endotoxin levels are found regardless of whether plasma, PRP, or whole blood is used. Therefore, it may be advisable to use clinical samples based upon whichever methodology requires the least amount of handling. This would eliminate chance contamination as well as allow more rapid results to be obtained. In the SepTest trial, whole blood was used, resulting in a time from sample collection to final result of approximately 2 hours. Because of volume differences between whole blood and its corresponding plasma, however, the endotoxin content on a volume basis was slightly higher for plasma than for blood. Since the values were highly correlated, either blood or plasma could be used with equal conclusions. In virtually all blood assays for the detection of endotoxin, anticoagulants have been employed to prevent blood from coagulating. In the literature, the anticoagulant used most often in studies involving LAL has been heparin. Unfortunately, heparin from animal sources is often contaminated with endotoxin. This fact necessitates thorough screening of all containers used for blood/plasma collection. Further, it has been shown that heparin can interfere directly with the LAL test (33). To resolve the former problem, a commercially available, endotoxin-free heparin-containing blood collection tube is now available (endo tube ET). Alternatively, some studies have used EDTA-containing collection tubes. These tubes are usually endotoxin-free and the EDTA concentration present does not interfere with LAL. It should also be noted that if heparin is used to irrigate in-dwelling catheters and blood is collected from these for the LAL test, an additional heparin control may be required with the test. In the SepTest trial, however, comparison between samples obtained from patients with in-dwelling catheters versus those from whom blood was obtained by fresh sticks showed no difference with respect to endotoxin contamination.

Combination Test

A number of assays have been developed that employ an initial endotoxin capture step followed by a quantitative LAL assay. This strategy would be expected to eliminate sample interference but would also strongly depend on the ability of the capture agent to bind all the endotoxin present in the sample. Also the capture agent must be one that does not inactivate the biological activity of the bound endotoxin. This is not a trivial feat since endotoxin can be strongly bound to a number of components in blood/plasma, e.g., high-density lipoprotein (HDL) and LPS-binding protein (LBP), which have substantial affinity for endotoxin.

An early attempt at endotoxin capture was described by Harris and Feinstein (30). The procedure employed by these investigators involved small polymer beads, which adsorbed endotoxin from plasma. After washing, the beads were placed in a solution of LAL. It is interesting to note that the adsorption of endotoxin to the bead did not appear to inhibit its ability to activate LAL. This assay utilized a rather insensitive gel-clot LAL technology. Perhaps because of this and the semiquantitative nature of the method, it has not been widely employed.

Another assay involves capture with immobilized histidine followed by a chromogenic LAL test (34). This method was reported to be more sensitive than any of the direct assay methods. This type of method would eliminate the need to remove or inactivate inhibitors present in blood or plasma. Histidine, a nonspecific and weak endotoxin-binding substance, is unlikely to remove endotoxin under all physiological conditions and from all the substances in blood/plasma that have been recognized to bind endotoxin.

The LAL assay has also been coupled with specific antibodies as the capture agents as will be discussed below. It should be noted that, although coupling with specific capture reagents may increase the specificity of the LAL test, the additional steps required increase the assay time and make contamination more likely. In addition, there is little evidence to indicate that circulating endotoxin is entirely derived from the infectious organism. Therefore, apart from research applications, where the species of endotoxin needs to be known or differentiated, there does not seem to be a practical clinical utility for specificity. Immunoassays, on the other hand, are usually intended to identify the infecting bacteria, not the species of endotoxin.

Immunoassays

Direct

Endotoxin is, of course, an excellent antigen, usually eliciting the rapid formation of high titers of several classes of antibodies (23). However, with few exceptions, endotoxin is used in assays for detecting endotoxin-specific antibodies rather than the reverse. One assay developed using monoclonal antibodies to *Haemophilus influenzae* b was found to be highly specific but not sufficiently sensitive to be of clinical value (35). A similar conclusion would be expected with assays directed to LPS from other organisms. An exception may be in situations where the clinical profile suggests only a limited number of possibilities for the infectious agent. Thus, in a patient with suspected spinal meningitis, immunoassays directed to the LPS from *Neisseria meningitidis*, *H. influenzae*, or *E. coli* may be clinically useful. It should be noted that the LAL assay has also been used successfully for rapid detection of gram-negative meningitis albeit not species specific (36).

Capture Assay

One method for increasing the sensitivity to endotoxin while retaining specificity uses antibodies to capture endotoxin and the LAL assay to measure (amplify) the captured material. A number of these assays have been described and patented (37–40). This type of assay works, since most antibodies raised to endotoxin react with the O-antigen or core and do not neutralize biological activity, i.e., they do not inhibit the LAL assay. These assays, although more sensitive than direct immunoassays, suffer from the same drawbacks, i.e., they are only useful in situations where the infectious agent is limited to a single or at most a few bacterial species.

β-Hydroxy Myristic Acid

Since β-hydroxy acids are a unique component of the lipid A component of LPS (11), their detection could serve as an indication of LPS presence and quantity. As with most chemical assays, however, LPS must first be isolated. All the criteria for preventing sample contamination must be employed. In addition, for β-hydroxy acids the LPS must also be hydrolyzed and fatty acid derivatives made. Currently this method is relatively insensitive in comparison with the LAL assay. It also is rather time consuming and expensive, and thus not suited for routine analyses. It may, however, be useful for the confirmation of LPS presence determined by other (e.g., LAL) test means, especially when sufficient sample is available (11,41). With improvements in technology, automated sample preparation and more widespread availability of equipment, it is likely this method may see more widespread use. This method and others based on the chemical components of LPS would continue to be of value if absolute quantitation of LPS were needed to be determined (in contrast to biological activity).

KDO

Like β-hydroxy myristic acid, KDO is a unique component of LPS. Several methods have been described to measure KDO including both chemical assays via thiobarboturic acid as well as immunoassays (10,42,43). The same advantages and disadvantages of chemical assays that were discussed earlier with respect to β-hydroxy myristic acid also apply to KDO.

Polymyxin B

Polymyxin B and derivatives thereof are a class of antibiotics with well-known high binding affinity for LPS (14). Not surprisingly, this property has led to their use in the development of endotoxin detection assays (44,45). One interesting assay for endotoxin in blood cleverly uses red blood cells present in the sample as the indicator (hemagglutination) (44–46). This assay employs a chemical conjugate of a monoclonal antibody or a lectin specific for red blood cells, which in themselves do not cause coagulation, and PMB or a derivative thereof. When the test reagent (SimpliRED Endotoxin Test) is mixed with a whole blood sample, the antibody-PMB conjugate coats the red blood cells. In the presence of endotoxin, binding to the PMB portion of the complex will cause visible hemagglutination. In a recent study (46), SimpliRED was compared to LAL and clinical variables associated with sepsis. The SimpliRED test had the advantage of being rapid (2 minutes), which would potentially be of value for "bedside" analysis of blood samples for endotoxin. Sensitivity of the test was 1.2 EU/ml. In this comparative set of studies, the LAL method employed a chromogenic LAL on heat/diluted PRP samples collected in heparin tubes. Although the sensitivity of the LAL reagent was 0.005 EU/ml, the +/− level chosen for the assay was 5 EU/ml. Of all clinical outcomes studied, the SimpliRED test showed a positive correlation for patients with multiorgan dysfunction, while the LAL did not. These results are somewhat surprising in that SimpliRED and LAL should both be measuring relative endotoxin levels. It is possible, as the authors suggest, that the heat-extraction procedure or the physical state of circulating endotoxin caused a discrepancy between the two assays. It is more likely that +/− levels selected (1.2 vs. 5) and the collection/test control had more influence on these results. Additional studies are needed here, but the goal of a rapid, simple assay is commendable.

Endotoxin-Binding Proteins/Peptides

A number of endotoxin-binding proteins/peptides have been described in the literature. These have been isolated from a number of sources, but the ones that seem most amenable for use in an endotoxin assay are probably those from the horseshoe crab and some of the mammalian proteins. Those from the horseshoe crab include tachyplesins, polyphemisins, limulus anti-LPS factor (LALF), and endotoxin-neutralizing protein (ENP) (15,17). The ENP isolated from the horseshoe crab has both a high affinity for endotoxin and a fluorescent signature (15). In the current assay with ENP, the fluorescent signal for tryptophan (and, to a lesser extent, tyrosine) is quenched in the presence of endotoxin in a stoichiometric relationship. Although the sensitivity of this assay is only in the nanogram range, the high specificity and affinity of ENP for LPS make it an ideal candidate for assay development. For example, ENP could possibly be used as a capture assay in an ELISA-type format. Bactericidal/permeability-increasing protein (BPI) is a human neutrophil granular protein, which binds to the lipid A portion of LPS with high affinity. BPI has many similarities to ENP and has been used in a capture assay similar to the antiendotoxin monoclonal antibodies (16).

BIOASSAYS

Pyrogen Test

The compendial pyrogen test has been used for over 50 years as a detection system for endotoxin contamination in pharmaceuticals (6). Generally, although the rabbit has been extensively used as an experimental animal model for endotoxin research, it is not considered a "diagnostic" test for endotoxin per se. Basically the rabbit pyrogen test involves injecting a solution into the marginal ear vein of a rabbit and recording the increases in rectal temperature over a period of 3 hours. Attempts have been made to make the measurement of fever in animals more quantitative, e.g., the fever index test (47), however, animal assay systems involving endotoxin continue to vary widely from laboratory to laboratory. Coupled with the relatively large expense of maintaining suitably controlled animal colonies, it is doubtful that the routine use of animals for the quantitation of endotoxin will be practical in the future. In contrast, animals will by necessity continue to be used extensively as model systems for testing sepsis and antiendotoxin therapies. In this regard, therefore, it is useful to use standard endotoxin assays such as the pyrogen test as a control with these animal models.

Cell Culture Assays

A number of assays that rely on the ability of endotoxin to stimulate release of cytokines have been described. These tests have employed a number of cell types, ranging from freshly harvested biological samples (48–51) to immortalized cell lines (52). These tests are known by a number of different titles including macrophage stimulation (48), human leukocyte stimulation (49), peripheral blood monocyte test (50,51), and human monocytoid cell test (52). The argument in favor of these tests is that they may be more representative of the actual biological manifestations of endotoxin activity than the chemical tests or even the Limulus amebocyte lysate test. This may also be true of bioassays in general. Although the reproducibility of these assays is still somewhat questionable, development of sensitive immortal cell lines (vs. freshly collected monocytes) should serve to improve the assay. One obvious drawback, of course, is that nonendotoxin substances are also active in these assays.

An interesting assay involving the blood cell neutrophil has been recently described that is purported to allow detection of small amounts of endotoxin/anti-endotoxin antibody. This assay relies on the opsonin-triggered oxidant production in neutrophils present in whole blood when stimulated by antibody-antigen complexes (31). The assay measures the light emitted following the oxidation of luminol present in the cell culture incubation mixture using a luminometer or scintillation counter and is dependent upon activation through the CR1 and/or CR3 receptor. It is predicated upon the fact that endotoxin itself is not capable of stimulating the oxiditive burst in human neutrophils (53). The addition of complement-fixing antibody, however, and a source of complement allows neutrophils present in a blood sample to become activated through the NADPH oxidase pathway. The assay is run by preincubating equal volumes of whole blood containing a suspected source of endotoxin and purified antibody to the endotoxin, followed by the addition of luminol. After an additional incubation, the sample is transferred to a chemiluminometer and a volume of human complement opsonized zymosan was added. Standard curves based on the light emission versus endotoxin concentration could then be constructed. A background control consisting of whole blood minus the anti-endotoxin antibody and a positive control consisting of a maximally reactive dose of endotoxin and antibody were included. This assay has the advantage of using small volumes of whole blood (which also serves as its own control). The assay also compared favorably with a chromogenic LAL test.

Animal Lethality Assays

Although used as a measure of endotoxin, the relatively large amounts required for lethality, even in sensitized animals, precludes the use of this type of bioassay except in research applications (54). A number of other animal bioassays have been reported in the literature, most of which rely on the measurement of endotoxin-induced changes. One such example is the measurement of acute phase proteins (55). It is possible that some of these tests can be standardized to the point where they may be useful in research settings.

SUMMARY

Due to the numerous biological responses to endotoxin and the complexity of the molecule, it is not surprising that quite a few assays for endotoxin have been described in the literature. It is quite obvious, however, that only a few of these, namely the LAL assay (and variations thereof) and the cell culture assays, lend themselves to widespread use. The question still remaining, however, is how suitable these tests are for assessing the bioactivity of endotoxin. Is it sufficient to know the total amount (as a chemical entity) of endotoxin present or its bioactivity (however that may be defined)? Of course the methods need much more testing before we can attempt to answer these questions. It seems, however, that a sensitive, accurate, and precise chemical test is required before the bioactivity issue can be resolved. Whether the LAL assay can be standardized to a point that meets these criteria (the LAL assay does have a bioactivity issue) remains to be seen.

REFERENCES

1. Seibert FB. Pyrogens from an historical viewpoint. Transfusion 1963; 3:245–249.
2. Seibert FB. Fever producing substance in some distilled waters. Am J Physiol 1923; 67:90–104.
3. Seibert FB, Mendel LB. Protein fevers. Am J Physiol 1923; 67:105–123.

4. Seibert FB. The cause of many febrile reactions following intravenous injections. I. Am J Physiol 1925; 71:621–651.
5. Welch H, Calvery HO, McClosky WT, Price CW. Method of preparation and test for bacterial pyrogen. J Am Pharm Assoc (Sci ed) 1943; 32:65–69.
6. Pyrogen test. Pharmacopoeia of the United States of America. Vol. XXIII. Taunton, MA: Rand McNally, 1995:1718–1719.
7. Pharmacopoeia of the United States of America. Vol. XII. Easton, PA: Mack Printing Co, 1942.
8. Boivin A, Mesrobeanu I, Mesrobeanu L. Technique pour la préparation des polysaccharides microbiens spécifiques. Soci Biol Buchar 1933; 113:490–492.
9. Westphal O, Lüderitz O, Bister F. Über die Extraktion von Bakterien mit Phenol/Wasser. F Z Naturforsch 1952; 7B:148–155.
10. Unger FM. The chemistry and biological significance of 3-deoxy-D-manno-octulosonic acid (Kdo). Adv Carbohydr Chem Biochem 1981; 38:323–387.
11. Maitra SK, Nachum R, Pearson FC. Establishment of beta-hydroxy fatty acids as chemical marker molecules for bacterial endotoxin by gas chromatography-mass spectrometry. Appl Environ Microbiol 1986; 52:510–514.
12. Levin J, Bang FB. The role of endotoxin in the extracellular coagulation of Limulus blood. Bull Johns Hopkins Hospital 1964; 115:265–274.
13. Bacterial endotoxins test. Pharmacopoeia of the United States of America. Vol. XXIII. Taunton, MA: Rand McNally, 1995:1696–1697.
14. Harvey W, Wilson M. Endotoxin assay. UK patent GB 2197470 (December 21, 1988).
15. Wainwright NR, Miller RJ, Paus E, et al. Endotoxin binding and neutralizing activity by a protein from *Limulus polyphemus*. In: Cellular and Molecular Aspects of Endotoxin Reactions. Amsterdam: Elsevier Science Publishers BV, 1990:315–325.
16. Capodici C, Chen S, Sidorczyk Z, et al. Effect of lipopolysaccharide (LPS) chain length on interactions of bactericidal/permeability-increasing protein and its bioactive 23-kilodalton NH_2-terminal fragment with isolated LPS and intact *Proteus mirabilis* and *Escherichia coli*. Infect Immun 1994; 62:259–265.
17. Nakajima H, Yamamoto H. Beta-glucans detection reagents and methods of detecting beta glucans. U.S. patent 5,571,683 (November 5, 1996).
18. Tsai C-M, Frasch CE. A sensitive silver stain for detecting lipopolysaccharides in polyacrylamide gels. Anal Biochem 1982; 119:115–119.
19. Victorov AV, Medvedeva NV, Gladkaya EM, et al. Composition and structure of lipopolysaccharide-human plasma low density lipoprotein complex. Analytical ultracentrifugation, ^{31}P-NMR, ESR and fluorescence spectroscopy studies. Biochim Biophys Acta 1989; 984:119–127.
20. Baltzer LH, Mattsby-Baltzer I. Heterogeneity of lipid A: structure determination by ^{13}C and ^{31}P NMR of lipid A fractions from lipopolysaccharide of *Escherichia coli* 0111. Biochemistry 1986; 25:3570–3575.
21. Gamian A, Katzenellenbogen E, Romanowska E, et al. Lipopolysaccharide core region of *Hafnia alvei*: structure elucidation using chemical methods, gas chromatography-mass spectrometry, and NMR spectroscopy. Carbohydr Res 1995; 266:221–228.
22. Strain SM, Armitage IM. Selective detection of 3-deoxymannoctulosonic acid in intact lipopolysaccharides by spin-echo ^{13}C NMR. J Biol Chem 1985; 260:12974–12977.
23. Neter E. Immunogenicity of endotoxin. In: Beneficial Effects of Endotoxins. New York: Plenum Press, 1983:91–109.
24. Fink PC, Galanos C. Determination of anti-lipid A and lipid A by enzyme immunoassay. Immunobiology, 1981; 158:380–390.
25. Gutowski JA, Jacobs DM. A solid phase radioimmunoassay for bacterial lipopolysaccharide. Immunol Commun 1979; 8:347–364.
26. Novitsky TJ. Limulus amebocyte lysate (LAL) detection of endotoxin in human blood. J Endotox Res 1994; 1:253–263.
27. Hurley JC. Concordance of endotoxemia with gram-negative bacteremia in patients with gram-negative sepsis: a meta-analysis. J Clin Microbiol 1994; 32:2120–2127.
28. Hurley JC. Reappraisal with meta-analysis of bacteremia, endotoxemia, and mortality in gram-negative sepsis. J Clin Microbiol 1995; 33:1278–1282.
29. Prior RB. Clinical Applications of the Limulus Amoebocyte Lysate Test. Boca Raton, FL: CRC Press, Inc., 1990.
30. Harris NS, Feinstein R. In vitro process for detecting endotoxin in a biological fluid. U.S. patent 3,944,391 (March 16, 1976).
31. Foster DM, Doig GS, Romaschin AD, et al. Reliable accurate detection of endotoxemia in critically ill patients using a point of care assay (abstr). Proceedings from the IBC Sixth Sepsis Meeting, Cambridge, MA, June 24 and 25, 1996.
32. Novitsky TJ, Roslansky PF. Quantification of endotoxin inhibition in serum and plasma using a turbidimetric LAL assay. In: Bacterial Endotoxins: Structure, Biomedical Significance, and Detection with the Limulus Amebocyte Lysate Test. New York: Alan R. Liss, Inc., 1985:181–193.
33. Sullivan JD, Watson SW. Inhibitory effect of heparin on the limulus test for endotoxin. J Clin Microbiol 1975, 2:151.
34. Prior RB. Comparative evaluation of a more sensitive immobilized/histidine Limulus amoebocyte lysate (LAL) assay for the detection of low concentrations of native bacterial endotoxins in human plasma. Abstracts of American Society for Microbiology, May 25, 1994.
35. Mertsola J, Munford RS, Ramilo O, et al. Specific detection of *Haemophilus influenzae* type b lipooligosaccharide by immunoassay. J Clin Microbiol 1990; 28:2700–2706.
36. Nachum R. Detection of gram-negative bacterial meningitis. In: Clinical Applications of the Limulus Amebocyte Lysate Test. Boca Raton, FL: CRC Press, Inc., 1990:67–80.
37. Young LS. Monoclonal antibodies specific for lipid-A determinants of gram negative bacteria. U.S. patent 4,918,163 (April 17, 1990).

38. Hansen EJ, Munford RS, Mertsola J. Method for detection of gram-negative bacterial lipopolysaccharides in biological fluids. U.S. patent 5,198,339 (March 30, 1993).
39. Hansen EJ, Munford RS, Mertsola J. Method for detection of gram-negative bacterial lipopolysaccharides in biological fluids. U.S. patent 5,356,778 (October 18, 1994).
40. Wood DM, Parent JB, Gazzano-Santoro H, et al. Reactivity of monoclonal antibody E5® with endotoxin. I. Binding to lipid A and rough lipopolysaccharides. Circ Shock 1992; 38:55–62.
41. Maitra SK, Schotz MC, Yoshikawa TT, Guze LB. Determination of lipid A and endotoxin in serum by mass spectroscopy. Proc Natl Acad Sci USA 1978; 75:3993–3997.
42. Karkhanis YD, Zeltner JY, Jackson JJ, Carlo DJ. A new and improved microassay to determine 2-keto-3-deoxyoctonate in lipopolysaccharide of gram-negative bacteria. Anal Biochem 1978; 85:595–601.
43. Pedron T, Girard R, Kosma P, Chaby R. Preparation and binding specificity of a monoclonal antibody recognizing 3-deoxy-D-manno-2-octulosonic acid (Kdo) in lipopolysaccharides of Re chemotype. Hybridoma 1992; 11:765–777.
44. Rylatt D, Wilson K, Kemp BE, et al. A rapid test for endotoxin in whole blood. In: Bacterial Endotoxins: Lipopolysaccharides from Genes to Therapy. New York: Wiley-Liss, Inc., 1995:273–284.
45. Smith NL. A field test for endotoxin in the horse. Vet Rev 1993; 13:433–440.
46. Kollef MH, Eisenberg PR. A rapid qualitative assay to detect circulating endotoxin can predict the development of multiorgan dysfunction. Chest 1997; 112(1): 173–180.
47. Wolff SM, Mulholland JH, Ward SB. Quantitative aspects of the pyrogenic response of rabbits to endotoxin. J Lab Clin Med 1965; 65:268–275.
48. Doe WF, Yang ST, Morrison DC, et al. Macrophage stimulation by bacterial lipopolysaccharides. II. Evidence for differentiation signals delivered by lipid A and by a protein rich fraction of lipopolysaccharide. J Exp Med 1978; 148:557–568.
49. Dinarello CA, O'Connor JV, LoPreste G, Swift RL. Human leukocytic pyrogen test for detection of pyrogenic material in growth hormone produced by recombinant *Escherichia coli*. J Clin Microbiol 1984; 20:323–329.
50. Duff GW, Atkins E. The detection of endotoxin by in vitro production of endogenous pyrogen: comparison with limulus amebocyte lysate gelation. J Immunol Meth 1982; 52:323–331.
51. Poole S, Thorpe R, Meager A, et al. Detection of pyrogen by cytokine release. Lancet 1988; 8577:130.
52. Eperon S, Jungi TW. The use of human monocytoid lines as indicators of endotoxin. J Immunol Meth 1996; 194:121–129.
53. Aida Y, Pabst MJ. Neutrophil responses to lipopolysaccharide: effect of adherence on triggering and priming of the respiratory burst. J Immunol 1991; 146(4):1271–1276.
54. Pieroni RE, Broderick EJ, Bundeally A, Levine L. A simple method for the quantitation of submicrogram amounts of bacterial endotoxin. Proc Soc Exp Med 1970; 133:790–794.
55. Poole S, Gaines Das RE, Baltz M, Pepys MB. Detection of endotoxin in mice by measurement of endotoxin-induced changes in plasma concentrations of zinc and of the acute-phase protein serum amyloid P-component. J Pharm Pharmacol 1986; 38:807–810.

57

The Relevance of Endotoxin Detection in Sepsis

James Hurley
Ballarat Base Hospital, Ballarat, Victoria, Australia

Jack Levin
University of California School of Medicine, and the Department of Veterans Affairs Medical Center, San Francisco, California

INTRODUCTION

Gram-negative bacterial infections account for as many as a half of documented infections in patients with sepsis, especially those who develop the serious sequelae of sepsis, multiple organ dysfunction (1). Endotoxin, a cell wall component of gram-negative bacteria, is known to activate the cytokine pathways that lead to the release of pro-inflammatory mediators, which result in a wide range of pathophysiological effects (2). Furthermore, endotoxin is known to be released by antibiotic therapy (3). Hence, endotoxin is of particular interest from two perspectives: as a possible marker of infection with gram-negative bacteria and as a putative mediator of the complications of sepsis.

ENDOTOXIN DETECTION AND QUANTIFICATION: METHODOLOGICAL CONSIDERATIONS

Methods to detect and quantify levels of endotoxin in the plasma of patients with sepsis have been described for over 30 years (4,5). The most sensitive assay described is the Limulus amebocyte lysate (LAL) assay. Increased understanding of the complex biological properties of both endotoxin and the LAL reagent has been achieved as a result of numerous investigations during the last 20 years. Several modifications to the assay have been made in an attempt to allow the quantitative measurement of endotoxin and its application to blood samples (6,7).

As an assay for endotoxin, the LAL assay has several desirable properties, which include its sensitivity, specificity, and potential for adaptation to a quantitative format. In a range of body fluids other than blood, the detection of endotoxin with the LAL test has been studied as a marker of the presence of gram-negative bacteria. Numerous published studies of body fluids such as cerebrospinal fluid (CSF) and urine have found that endotoxin detection with the LAL test can be used as a reliable marker of gram-negative infection (8).

When the LAL assay is used to detect endotoxin in blood, two obstacles are encountered: the complex and poorly understood inhibitory factors present in blood, and the low levels of endotoxin in blood, which are often at the limit of test detection. Figure 1 illustrates the complex interaction between components of blood, endotoxin, and LAL. Endotoxin interacts with several components of plasma, including bile salts, proteins, and lipoproteins. These interactions may lead to disaggregation, inhibition, or formation of complexes between endotoxin and plasma components. It remains to be determined as to how much the concentrations of interfering factors, such as the endotoxin-binding proteins lipopolysaccharide-binding protein (LBP) and bactericidal/permeability-increasing protein (BPI), vary during the development of sepsis (9). Variation in interference could have substantial implications for quantitative assays for endotoxin.

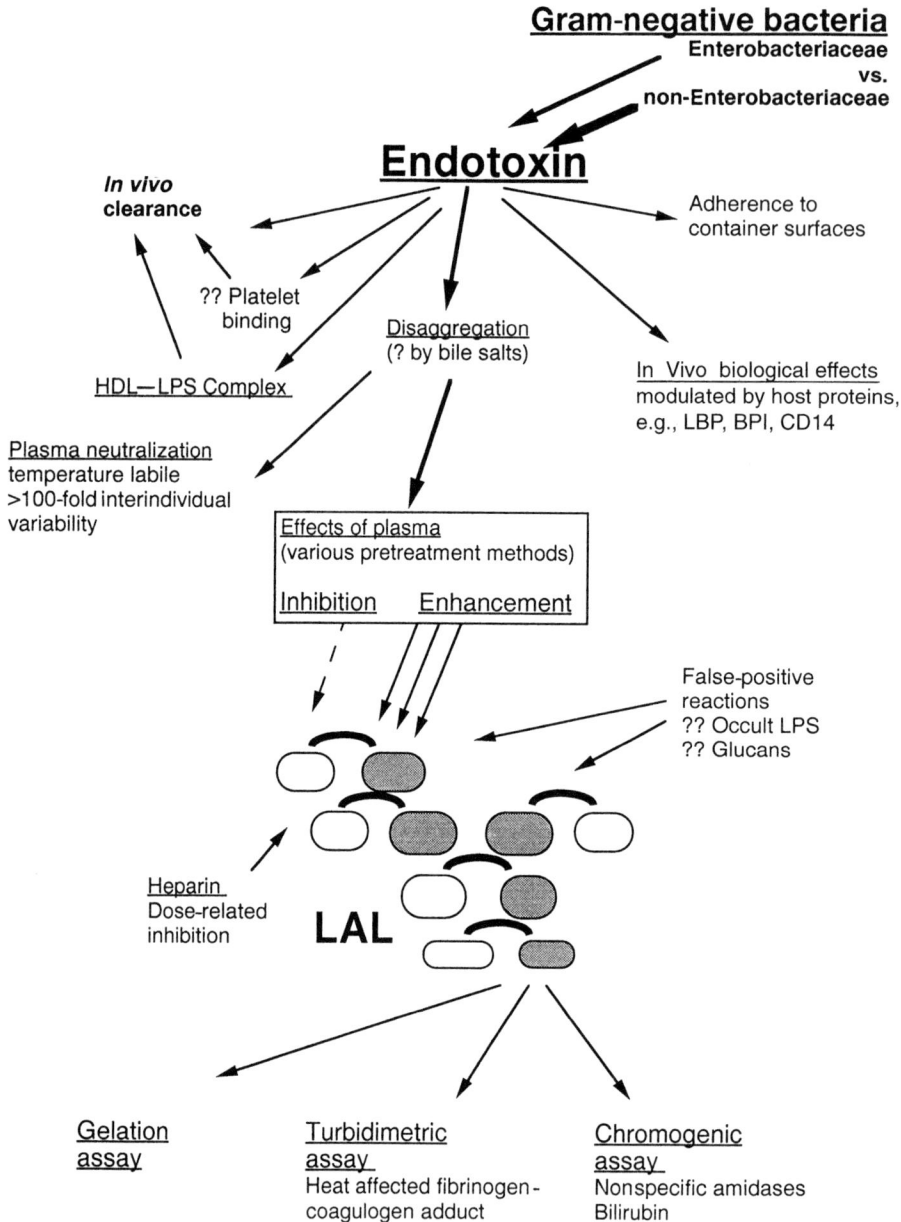

Fig. 1 Schematic overview of the factors that interfere with the detection of endotoxemia by the LAL assay, which is represented as the cascade in the center of the figure. "Occult LPS" refers to positive assay results from endotoxin contamination in reagents or containers; LPS, lipopolysaccharide; HDL, high-density lipoprotein; BPI, bactericidal-permeability-increasing protein; LBP, LPS-binding protein. Heat affected fibrinogen adduct as described in Ref. 68. (From Ref. 6.)

Methodological considerations and optimal techniques for the application of the LAL assay to blood samples are described in detail elsewhere (10). In brief, blood for endotoxin assay should be collected into sterile, pyrogen-free specimen containers that contain pyrogen-free heparin at a final concentration of 20 units per ml of blood. The container should immediately be placed on ice for separation of plasma in a refrigerated centrifuge. Either the inhibitors of endotoxin in plasma should be immediately inactivated, or the sample should be stored at below −20°C for later inactivation. The method of choice for the removal of the inhibitory factors in plasma is the method of dilution and heating, since it is the most simple and efficient (10). A sample of pretreated plasma spiked with exogenously added endotoxin (spiked sample) should be assayed in parallel to validate the inactivation procedure.

There are several suitable methods of LAL assay, and the method chosen depends on the desired level of sensitivity of the assay and whether a qualitative or quantitative result is required. For a quantitative assay, standardization of the optimal reaction conditions requires particular attention. Moreover, the endotoxin standard curve against which unknown samples are compared should be made up in treated normal pooled plasma (11).

While the quantification of endotoxin can be achieved by endpoint methods, the limitations of this technique, in contrast to a kinetic assay, should be recognized (11). Because of the properties of the LAL cascade, a standard curve derived from endpoint methodology provides usually no more than a single 10-fold range in concentration. For concentrations outside a 10-fold range, a variation in the reaction period or assay conditions will be required in an endpoint assay.

There have been hundreds of publications and studies on the use of the LAL test to detect endotoxemia in patient groups in which there was an increased risk of sepsis (e.g., neutropenia, burns, or various pediatric conditions) (6). Despite this extensive experience, the LAL test has yet to gain widespread regulatory approval for the detection of endotoxemia in clinical practice. Currently licensed kits specify that they are not approved for that purpose. Japan is one country where the LAL test has been approved for clinical use.

CONCORDANCE OF ENDOTOXEMIA WITH GRAM-NEGATIVE BACTEREMIA

Because the LAL assay is so highly sensitive, there has been no alternative assay against which to validate the results of plasma testing. Even in expert hands, the plasma of as many as 6% of healthy volunteers may yield a positive LAL test (12). The question thus arises as to how else validation can be accomplished. Gram-negative bacteremia has often been used as a standard for comparison with the result of the assay for endotoxin in plasma (13). Gram-negative bacteremia has many desirable qualities as a gold standard. It is objective, commonly reported, and always clinically significant. Gram-negative bacteremia is useful for validation in animal models of sepsis. In a particular model, the level of endotoxin broadly correlates with bacterial counts determined by quantitative blood cultures and outcome, although the exact correlation will vary for different bacterial agents (14,15). In fluids other than blood (CSF, urine), the presence (and concentration of gram-negative bacteria serves as a convenient reference standard to compare with the result of the endotoxin assay (8).

However, the use of gram-negative bacteremia as a basis for comparison with endotoxemia in clinical studies is problematic and not necessarily appropriate. The LAL assay is not an assay for gram-negative bacteria. Moreover, endotoxemia and gram-negative bacteremia are not necessarily interdependent phenomena. Endotoxemia can arise from an infected tissue site without concomitant bacteremia. From the earliest studies with LAL (16,17) and even earlier studies using the rabbit bioassay (4,5), it has been apparent that there is a poor concordance between gram-negative bacteremia and endotoxemia. Endotoxemia is detected in approximately half or less of those with gram-negative bacteremia, and, similarly, gram-negative bacteremia is detected in approximately half of those with endotoxemia (13). Pertinently, microbial antigens may be found in the circulation in the absence of bacteremia in gram-positive bacterial infections as well (18).

This apparent lack of concordance between endotoxemia and gram-negative bacteremia has prompted numerous attempts to improve the sensitivity of the LAL assay (7). A hitherto unrecognized confounding factor in the use of gram-negative bacteria in blood as the reference standard for endotoxemia is that the type of organism causing bacteremia may be at least as important as the limit of sensitivity of the assay used (13). From a compilation of data derived from published comparisons of endotoxemia and gram-negative bacteremia, the concordance is less common with the Enterobacteriaceae than with non-Enterobacteriaceae organisms (Table 1). The rate at which endotoxemia accompanies bacteremia varies from as low as 41% and 46% for *E. coli* and *Enterobacter* species to as much as 63% and 74% for *Pseudomonas aeruginosa* and *Neisseria meningitidis*, respectively. It is curious that endotoxemia is found at lower frequency in *E. coli* bacteremia than in *P. aeruginosa* bacteremia. The clinical experience appears to be the converse of experimental findings (13,14). In an experimental canine model of septic shock, bacteremia with an *E. coli* was associated with a 10-fold higher concentration of endotoxin than bacteremia with *P. aeruginosa* (14).

Discordance between the LAL test and blood culture results may occur for various technical reasons. In many studies, it is unclear when blood was taken for endotoxin detection and culture in relation to the initiation of antibiotic therapy. The duration of endotox-

Table 1 Frequency of Endotoxemia with GN Bacteremias

	Any[a]		Nonfatal		Fatal	
	Present	Absent	Present	Absent	Present	Absent
Enterobacteriaceae and *Anaerobes*[b]	171 (46%)	198 (54%)	52[c] (43%)	69 (57%)	33[c] (57%)	25 (43%)
Non-Enterobacteriaceae[d]	93 (69%)	42 (31%)	38[e] (64%)	21 (36%)	31[e] (91%)	3 (9%)

[a]Any outcome, whether specified as fatal or nonfatal or outcome not specified.
[b]Includes *E. coli, Klebsiella,* spp., *Enterobacter* spp., *Proteus* spp., *Serratia* spp., *Salmonella* spp., *Yersinia pestis,* enteric GN rods, and anaerobes.
[c]Fatal versus nonfatal; Chi square 3.1; $p = 0.092$; DF = 1.
[d]Includes *Pseudomonas* spp., *Neisseria meningitidis, Neisseria gonorrhoeae, Xanthomonas* spp., *Haemophilus influenzae, Acinetobacter* spp., *Aeromonas* spp., *Flavobacterium meningosepticum.*
[e]Fatal versus nonfatal; Chi square 8.1; $p = 0.005$; DF = 1.
Source: Modified with permission from Ref. 13 with additional data from Refs. 12,19–22, including data provided as personal communication (12,19,21,22).

emia following the initiation of antibiotic therapy is related to the type of infection. Endotoxin may be cleared within 12 hours for typhoid bacteremia (23), 24 hours for urosepsis (21), or 36 hours for meningococcemia (24,25). In other clinical situations endotoxin may be detectable for 48–60 hours with sepsis (12,19), 5 days in neutropenic patients (26), 10 days with plague (27), and up to 20 days with leptospirosis (28). The rapidity with which blood samples are obtained when sepsis is suspected and the number of samples obtained may also affect the likelihood of endotoxin detection (29). In addition, false-negative and false-positive tests can result as a consequence of inappropriate collection procedures. False-positive results for endotoxin resulting from contamination may occur as often as 9% of the time even with expert collection (30).

The sensitivity level of the assay and the method of plasma preparation are two variables under the control of the investigator. There has been no direct comparison of the two types of LAL assay—the original gelation version and the recently developed more sensitive chromogenic version—in patients with suspected gram-negative bacteremia. In a meta-analysis of 45 studies that provided data on the concordance between gram-negative bacteremia and endotoxemia, results were similar with the two types of assay (13).

The method of plasma treatment prior to endotoxin measurement may be of particular importance (7,10). In a recent study of 69 bacteremic patients, the majority of whom (56%) were neutropenic, endotoxemia was detectable in only three (11%) of the 27 patients with gram-negative bacteremia (20). These results could be criticized for the apparent failure to control for interfering factors in plasma. By contrast, other investigations with neutropenic patients (26,31,32), with plasma samples that were treated by dilution and heating, endotoxemia was detectable in the majority of episodes of gram-negative bacteremia.

A second important limitation in the use of bacteremia as a basis for evaluation of the clinical significance of endotoxemia is that gram-negative bacteremia is itself a relatively weak predictor of clinical outcome (33,34). Gram-negative bacteremia is predictive only in studies large enough to enable a stratification of the patients into different categories of illness, age groups, types of pathogens, and grades of bacteremia. Among a group of patients with sepsis, patients with gram-negative bacteremia could not be prospectively identified by the use of simple clinical criteria (35). Even in the context of controlled animal models of gram-negative sepsis (14,15), the relationship between endotoxemia, gram-negative bacteremia, and the outcome of gram-negative sepsis is complex.

ENDOTOXEMIA: QUANTITATIVE RESULTS

It is common practice to report the level of endotoxin from a quantitative assay by weight (e.g., pg/ml) or potency (e.g., EU/ml). The equivalence between weight and potency is typically in the range of 60–600 pg per endotoxin unit in the LAL assay. Even allowing for such variations in potency, it remains unclear how to interpret a quantitative result. The technical difficulties in performing and standardizing a quantitative LAL assay have already been mentioned. Even in a controlled animal model, it is difficult to equate an exogenous dose of endotoxin with an administered dose of bacteria, whether as a killed or as a live bacterial challenge, for potency in induction of pathophysiological effects or cytokine pathway activation (36). Moreover, endotoxemia can be detected in the absence of sepsis, for example, in patients undergoing surgery requiring routine cardiopulmonary bypass (37), at levels comparable to the levels in patients with the complications of sepsis. In these settings, the endotoxin is presumed to arise from the gastrointestinal tract.

There is little correlation between levels of endotoxemia and numbers of gram-negative bacteria detected in quantitative blood cultures where this has been studied (25,38,39). In patients with sepsis syndrome, there is no difference between levels of endotoxemia in patients with gram-negative bacteremia versus patients with nonbacteremic gram-negative infections (19,40) or in those with other types of infection (29,40). Furthermore, endotoxin levels in patients with gram-negative bacteremia cannot be accounted for on the basis of the numbers of bacteria present in the blood stream, typically less than 10 colony-forming units per ml (CFU/ml) in adults. Moreover, the variable rate of detection of endotoxemia in association with different types of gram-negative bacteremia (Table 1) is especially apparent among those bacteremias where the patient outcome was specified as fatal. Endotoxemia is significantly more frequent when the outcome was fatal than when the outcome was nonfatal for bacteremias with non-Enterobacteriaceae, but not so in the case of bacteremias with Enterobacteriaceae.

PROGNOSTIC SIGNIFICANCE OF ENDOTOXEMIA

The complications of gram-negative sepsis, such as the acute respiratory distress syndrome (ARDS), are generally believed to be mediated by endotoxin. These complications are associated with greatly increased mortality and are often refractory to conventional treatment including appropriate therapy for gram-negative sepsis. On the basis of extensive animal model data (41), organ failure complications such as ARDS are suspected to be at least in part mediated by endotoxin. Organ failure is associated with endotoxemia in some (25,29,42) but not all (30) clinical studies. Hence it would be expected that the detection of endotoxin in patient groups at risk of these complications would have great prognostic value.

The conclusions derived from studies of endotoxemia as an indicator of the severity and prognosis of gram-negative sepsis have been highly variable. In some reports, there is a direct quantitative correlation between the level of endotoxemia and frequency of adverse events and mortality. Such correlations have been found in patients with meningococcemia (25), plague (38), and leptospirosis (28).

By contrast, this correlation with outcome has not been seen in a variety of investigations of patients with sepsis (Table 2). In this patient group, the organ failure complications are known to be high and yet the levels of endotoxemia are poorly predictive of outcome. One investigation found the LAL test predictive of outcome (mortality), although only in the subgroup with bacteremia (29). In another investigation, the levels were predictive if they were incorporated into a LPS-cytokine score that also included levels of TNF-α, IL-1β, and IL-6 (40). By contrast, another investigation in septic patients found no correlation between the level of endotoxin and mortality or maximum daily organ failure scores. Endotoxin levels were not a significant predictor of outcome even when incorporated into a logistic regression analysis with circulating cytokines and cytokine inhibitors (22).

Overall, the conclusion of numerous recent studies in sepsis is that the plasma endotoxin level did not predict a fatal outcome (12,19,22,29,30,40). The correlation with concurrent markers of severity such as ARDS or acute renal failure has been variable, being found in some (19,29,42) but not in other studies (22,30). Interestingly, in two studies where a correlation with concurrent markers of severity and endotoxin levels was found, this correlation was especially apparent in the subgroup with bacteremia (19,29).

To reconcile the disparate findings, the patient population under study and, more important, the range in patient prognoses are variables that deserve much closer scrutiny. It should be recognized that the clinical utility of a diagnostic test is determined not only by the properties of the test itself, but also by the epidemiology of the disease in the setting in which the test is utilized. Diagnostic tests may have excellent operating characteristics, yet provide unhelpful or even misleading information if they are applied inappropriately in patient populations with a very low or a very high frequency of the disease of interest (45).

Clinical investigations of the relevance of endotoxin detection in sepsis have approached the problem in one of two ways. One group of studies examined the possible value of the detection of endotoxin in blood as a marker of gram-negative infection in patients with sus-

Table 2 Endotoxin Detection in Studies of Patients Meeting the Criteria for Sepsis ("Mediator Studies")

	Number of patients (%)		Endotoxin detection			
Total	With GN bacteremia	With endotoxemia	Level (EU/ml	pg/ml)	Mortality (%)	Ref.
1052	136 (13)	NS	NS	NS	56	33
97	18 (19)	87 (89)	2.6[a]	—	46	40
100	19 (19)	43 (43)	4.4	440[b]	24	29
20	6 (30)	10 (50)	NS	NS	25	43
33[c]	NS	20 (61)	—	120[d]	30	44
146	12 (8)	96 (66)[e]	0.36	23[d]	49	22
93	33 (35)	44 (47)[f]	—	60[b]	53	19
346	49 (14)[g]	119 (33)[g]	1.07	NS	34[g]	12
82	32 (39)	27 (33)	NS	NS	48	30

GN = Gram-negative, NS = not stated
EU = endotoxin unit.
[a]Mean
[b]Mean peak
[c]Placebo recipients only
[d]Median
[e]At study entry
[f]On day 1
[g]Range between eight centers; numbers of patients, 19–82; percent with endotoxemia, 16–44%; percent mortality, 17–49%.

pected sepsis. In general, these "marker" studies have compared the finding of endotoxemia with the results of blood cultures. Some have provided data on mortality rates, although this was usually not a primary objective of these types of studies. The patient inclusion criteria used in these marker studies have usually been broad, e.g., "patients with suspected sepsis." These patients may have had some characteristic in common, such as neutropenia or urological sepsis, or they may all have had an infection with a single type of gram-negative bacteria.

The second group of studies appeared more recently. The focus in this group has been exclusively patients who meet rigorously defined criteria of sepsis. The primary aim of these "mediator" studies has been to examine the relationship between the presence of the presumed mediator, endotoxin, in blood and outcome in terms of complications and mortality. Because of the use of widely accepted criteria of severe illness as the basis for inclusion in each study, these patients groups were nearly all found in an intensive care unit setting. By contrast, in the marker studies, which lacked rigorous severity of illness inclusion criteria, the patient populations, prognosis, and settings were diverse.

Endotoxemia as Marker Studies

There are several reasons why the results of marker studies might differ from each other and from mediator studies, as described below. The pathogenesis of endotoxemia may vary in different patient groups. For example, in neutropenic patients increased gut permeability may be a mechanism for endotoxemia as well as the ingress of gram-negative bacteremia. In patients with leptospirosis (28), endotoxemia is associated with renal dysfunction. This association may be a reflection of the specific localization of this infection in the kidney. Impaired hepatic clearance of endotoxin has been postulated as a mechanism for endotoxemia in patients with typhoid fever (46).

Even within a study, the frequency of detection and level of endotoxin in blood may vary depending on factors such as the age group, the grade of injury or the phase of infection. In a study of children with shigellosis, endotoxemia was found more commonly in younger children (<2 years). In this study, the complications, such as hemolysis and renal failure, were more common in children with endotoxemia than in those without (47). A second study in neonates with suspected sepsis also found the relevance of endotoxin detection to be age dependent, being more useful beyond the first week of life (48). In different studies of patients with burn injuries, endotoxemia has been found to correlate with the size of the burn injury (49,50) and colonization of the burn with gram-negative bacteria (51), as well as with nonsurvival (49,52).

Endotoxemia as Mediator Studies

The sepsis criteria are based on clinically determined features including vital signs such as pulse and blood pressure together with a limited number of biochemical indicators of systemic toxicity or impaired peripheral perfusion. They were derived in an attempt to enable the rapid identification of patients at high risk for developing the complications of sepsis for the purpose of their inclusion into studies of new antisepsis agents (53). These criteria are widely accepted with occasional minor modifications. Less than half of patients with severe sepsis have positive blood cultures. Others develop the syndrome in the setting of severe soft tissue, pulmonary, or intra-abdominal infections without detectable bacteremia, particularly if it is suppressed by antibiotic therapy (1). Some patients have systemic inflammatory response syndrome related to a noninfectious cause.

It should be recognized that patients who meet these criteria are a highly selected subpopulation of those with suspected sepsis, even within a teaching hospital setting. For example, in a study where the criteria were applied prospectively to hospitalized patients having blood cultures (1509 episodes among 1200 patients), only 34 episodes (3% of patients) occurred in patients meeting the criteria of sepsis. Moreover, only 12 of 298 episodes of gram-negative infections and only 5 of 40 episodes of gram-negative bacteremia occurred in patients meeting the sepsis criteria (54).

Despite satisfying the defined clinical criteria of sepsis, the patient groups in the mediator studies are by no means uniform. Among eight published series, the mean patient age varied between 46 and 63 years, the frequency of gram-negative bacteremia ranged between 13 and 40%, the proportion with endotoxemia was between 33 and 89%, and the mortality rate ranged between 24 and 56% (Table 2). This variability in key population characteristics presumably reflects intrinsic differences in factors such as the underlying conditions of patients, assay methodologies, and study designs.

Meta-Analysis: An Attempt to Resolve the Conflict

Meta-analysis is a statistical technique by which the key data of several individual studies can be summarized. Moreover, it provides an objective test of heterogeneity, which is a measure of how disparate the results of individual studies are from each other. Two meta-analyses were recently published that used this technique to evaluate and summarize the published experience with the Limulus lysate assay in patients with suspected gram-negative sepsis. The first meta-analysis applied the technique to examine the concordance between endotoxemia and gram-negative bacteremia (13). The second meta-analysis examined the relationship between these two factors on the one hand and the fatality risk on the other (55).

Figure 2 presents a revision of the second meta-analysis using the same statistical methods. In this revision, the studies have been stratified into two groups: mediator studies, which selected patients for inclusion using the sepsis criteria, and marker studies, which selected patients without using these criteria. The two groups are analyzed separately.

A total of 14 studies (1436 patients) were identified from which patient survival data could be obtained in relation to the detection of gram-negative bacteremia and the detection of endotoxemia by the LAL assay (Tables 3 and 4). Note that the marker studies were usually limited to specific patient groups: pediatric age group (58), postsurgery (57), hematological (20,31) or meningococcal sepsis (25). By contrast, the patients in six of the seven mediator studies were stated to be in an intensive care unit setting. One investigation was published as an abstract only (56). For one study (30), in which patients were randomized to receive antiendotoxin therapy or placebo, only the placebo recipients were included in this analysis. Figure 2 includes five studies—four mediator and one marker—that were published since the previous meta-analysis. Also, small and incomplete reports in the previous meta-analysis have been omitted for simplicity in the revision.

The three calculated odds ratios (OR) for death derived from the four groups of each study—group 1 (patients with both gram-negative bacteremia and endotoxemia); group 2 (patients with gram-negative bacteremia without endotoxemia); group 3 (patients with endotoxemia without gram-negative bacteremia); and group 4 (patients with neither gram-negative bacteremia nor endotoxemia)—are displayed in Figure 2 for the two strata of studies. It can be seen that many of

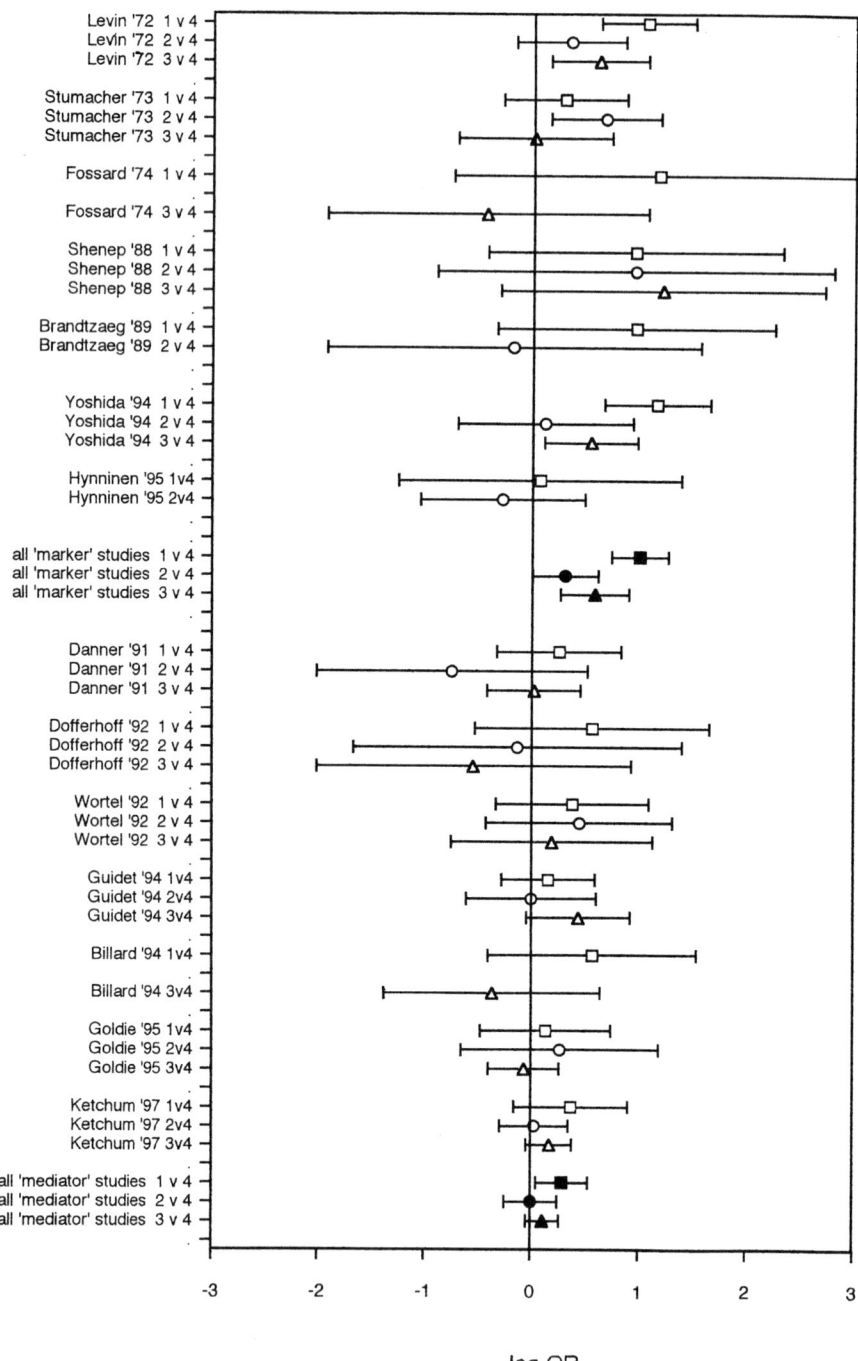

Fig. 2 Fatality risk as study-specific (open symbols) and summary (closed symbols) odds ratios, with 95% confidence intervals, for patients with gram-negative bacteremia and endotoxemia (Group 1, square), patients with isolated gram-negative bacteremia (Group 2, circle), and patients with isolated endotoxemia (Group 3, triangle), respectively, versus patients with these factors absent (Group 4). (See also Tables 3 and 4.) They are grouped separately based on whether patients had been included on the basis of sepsis criteria (mediator studies) or not (marker studies). Seven studies (127 patients) included in the previous meta-analysis (55) have been omitted for simplicity (e.g., studies in which the data were incomplete or studies with fewer than 15 patients). Recalculation with data from these seven studies included does not alter the results (data not shown). Chi-square tests for heterogeneity are presented in Table 5.

Table 3 Endotoxin Detection in Marker and Mediator Studies

Patient characteristics		
Entry criteria, hospital unit or setting	Age—mean or range (yr)	Ref.
Studies without sepsis syndrome criteria (marker studies)		
Suspected GN sepsis, NS	NS	16
Suspected bacteremia, NS	NS	39
Clinical infection, postsurgery	44–86	57
Suspected GN sepsis, pediatric	<1–20	58
Documented meningococcal disease, NS	<1–70	25
Febrile and neutropenic, hematology	NS	31
Suspected bacteremia, hematology	48	20
Studies with sepsis syndrome criteria (mediator studies)		
Clinically defined septic shock, ICU	46	29
Clinically defined sepsis, ICU	67	43
Suspected GN sepsis, multicenter trial[a]	57.6	30
Clinically defined sepsis syndrome, ICU	52.7	19
Septic shock, ICU	64.2	56
Sepsis syndrome, ICU	63	22
Sepsis syndrome, 78% of patients in ICU	59	12

NS = Not stated or not specified; ICU = intensive care unit; GN = gram-negative.
Source: Modified from Ref. 55.

the 95% confidence intervals for odds ratios from individual studies were broad, reflecting the small numbers in individual groups.

A summary result derived through meta-analysis must be interpreted with caution if the studies in the analysis are heterogeneous with respect to either their data or the patient populations studied. Indeed, the meta-analysis technique may produce erroneous conclusions if heterogeneity among study results is ignored (59). The heterogeneity among the data of individual studies can be tested for statistically, although this test is relatively insensitive at the conventional 0.05 breakpoint. In attempting to derive odds ratios incorporating all 14 studies, the tests for heterogeneity were notable with two of the three odds ratios (Table 5), albeit at the 0.10 level of significance for one and 0.01 for the other. This degree of heterogeneity precludes derivation of meaningful summary odds ratios for all 14 studies combined. While summary results have been derived from the seven marker studies, these should be interpreted with caution given the possible heterogeneity (at the 0.10 level) that is associated with one of these three odds ratios together with the known lack of homogeneity in the patient groups of these studies where rigorous inclusion criteria were not used.

The respective summary odds ratios appear to be more conservative, i.e., lower, in the case of the mediator studies than for the marker studies. This is a consequence of the low mortality in the group 4 patients in the marker studies. The mortality rate in group 4 patients in all marker studies is below 17%, whereas in all mediator studies' the rate for group 4 patients is at least 17% and usually between 25 and 50%. Moreover, the mortality rates in the groups other than group 4 in all 14 studies are typically not less than 20%: indeed, only three such groups with more than 10 patients had a mortality rate below 20%.

The data from a study with 473 patients (60) could not be included in the quantitative analysis because a different measure of outcome, rate of onset of sepsis rather than mortality rate, had been used in this study. This investigation is a marker study because the sepsis criteria were not used for patient selection. Thirty-one (7%) of 473 patients were found to have endotoxemia, 53 (11%) were found to have gram-negative bacteremia, and 17 (4%) had both. The results of this investigation appear to be consistent with those of the other marker studies in the meta-analysis as the rate of onset of sepsis for the group 1 patients was considerably higher than for any of the other three groups. Only 19 (4%) of the patients in this study developed sepsis, of whom 13 were in group 1.

As in the previous meta-analysis (55), this revised meta-analysis incorporating data from the 14 published studies shows that the outcome of gram-negative sepsis varies in different patient groups depending on the presence or absence of endotoxin and gram-negative bacteria in blood. Whether among marker studies or mediator studies, mortality is more frequent in patients with both endotoxemia and gram-negative bacteremia when compared with patients with neither factor detected. For patient groups with either isolated endotoxemia or isolated gram-negative bacteremia, an increased mortality is seen only in the case of marker studies, not in the case of mediator studies. Among the mediator studies, the mortality is not increased for either isolated endotoxemia or isolated gram-negative bacteremia. Moreover, the detection of endotoxin in

Table 4 Frequencies of Fatal Events

Survival census day	Mortality rates (percent)[a]				Ref.
	Group 1 GNB + Endotoxin +	Group 2 + −	Group 3 − +	Group 4 − −	
Studies without sepsis criteria (marker studies)					
14	14/20 (70)	4/14 (29)	7/16 (44)	27/168 (16)	16
NS	6/28 (21)	15/37 (41)	2/18 (11)	4/34 (12)	39
NS	2/2 (100)	−/− (−)	2/22 (9)	0/1 (0)	57
5	2/9 (22)	0/1 (0)	1/3 (33)	0/13 (0)	58
5	9/24 (38)	0/11 (0)	−/− (−)	0/7 (0)	25
NS	14/21 (67)	1/9 (11)	11/35 (31)	8/71 (11)	31
14	0/3 (0)	1/24 (4)	−/− (−)	10/96 (10)	20
Studies with sepsis syndrome criteria (mediator studies)					
14	4/11 (36)	0/8 (0)	8/32 (25)	12/49 (25)	29
21	2/4 (50)	0/2 (0)	0/6 (0)	1/6 (17)	43
28	6/8 (75)	4/5 (80)	2/3 (67)	13/25 (52)	30[b]
28	13/24 (54)	4/9 (44)	14/20 (70)	18/40 (45)	19
NS	5/6 (83)	−/− (−)	1/4 (25)	4/8 (50)	56
30	5/9 (56)	2/3 (67)	36/83 (43)	18/38 (47)	22
28	5/10 (50)	12/39 (31)	42/109 (39)	59/198 (30)	12

[a]GNB = gram-negative bacteria. Group 1 = GNB+, Endotoxin+; Group 2 = GNB+, Endotoxin−; Group 3 = GNB−, Endotoxin+; Group 4 = GNB−, Endotoxin−. Data are deaths per total number of patients in group (percent).
[b]Data for placebo recipients only.
Source: Modified from Ref. 55.

isolation had no prognostic value in any single mediator study or in the summary result. This is an unexpected finding, given the presumption that endotoxin is a critical mediator of gram-negative sepsis.

The reasons for the disparate conclusions from the different studies are likely to be complex. The proportion of non-Enterobacteriaceae among the gram-negative bacteremia isolates varies considerably and this warrants closer examination. For example, the study most critical of the value of the LAL assay as a clinical test investigated a patient group in which this proportion was only 9% (39). By contrast, the proportion of non-Enterobacteriaceae was 60% (31) and 100% (25) in studies with conclusions more favorable to the clinical application of the LAL test. In five mediator studies from which data were ascertainable, the proportion

Table 5 Chi-Square Tests for Heterogeneity (and degrees of freedom) Associated with Mortality Odds Ratios for Studies in Figure 2[a]

Group comparisons	Marker studies	Mediator studies	All studies
1 vs. 4	12.7 (6)[b]	1.6 (6)	28.9 (12)[c]
2 vs. 4	5.8 (5)	4.5 (5)	12.0 (10)
3 vs. 4	5.4 (4)	5.6 (6)	17.9 (10)[b]

[a]The chi-square statistic for heterogeneity is not very sensitive, and to allow for this, heterogeneity has been defined at the 0.10 level of significance rather than the usual 0.05 level as applied to the null hypothesis.
[b]$p < 0.10$.
[c]$p < 0.01$.

of non-Enterobacteriaceae among the gram-negative bacteremia isolates ranged between 10 and 33% (12,19,22,29,43).

Future investigations should give close attention to the types of gram-negative isolates. The frequency of endotoxemia is much greater in patients with non-Enterobacteriaceae bacteremias who had a fatal outcome; a negative LAL test in this group of patients implies survival (Table 1). Although gram-negative bacteremia is generally regarded as a weak indicator of prognosis (33–35), this is not uniformly the case for different gram-negative bacterial agents. Bacteremia with *P. aeruginosa* is a powerful independent predictor of poor prognosis (61). Hence it is worth considering whether the detection of endotoxemia may be an indicator of the pathogenic potential of specific gram-negative bacteria found in the blood stream. For example, could the detection of endotoxin indicate other pathogenic properties of specific gram-negative bacteria in patients with sepsis (62)? Indeed, an aphorism applied in the context of a review of experimental studies, "For the proper study of sepsis, study the bacteria that cause sepsis" (36), can equally be cited in the context of clinical studies.

ENDOTOXEMIA AND SEPSIS THERAPY

The detection of endotoxemia could be used as a basis for guiding therapy from three perspectives: endotoxemia as a therapeutic target per se, endotoxemia as a marker of gram-negative sepsis, or clearance of endotoxemia as an indicator of response to therapy.

There have been several therapies designed to target endotoxin and the endotoxin-activated cytokine networks, including monoclonal antibodies specific for the core region of the lipopolysaccharide molecule. However, efforts to reproduce the promising effects seen in subgroups of early trials on an intention-to-treat basis have been unsuccessful (63). Failures in numerous large clinical trials have in turn prompted a reevaluation of the current hypothesis regarding the pathogenesis of sepsis (64,65) and the role of endotoxin (66), and the place of these novel forms of therapy remains uncertain.

The response to therapy in relation to the presence of endotoxemia was examined in a subset of one of the double-blind antiendotoxin monoclonal antibody trials (30). Among the patients who were endotoxemic at study entry, there was a significant reduction in the mortality of monoclonal antibody recipients (5 of 16 HA-1A vs. 8 of 11 controls; $p = 0.034$) (30). However, endotoxemia remained detectable at 24 hours after therapy in a similar proportion of patients receiving the monoclonal antibody (4 of 12) as in placebo recipients (4 of 9; $p = 0.673$).

There is inconsistent information on the response to therapy in relation to clearance of endotoxemia in patients with sepsis. In one study of 30 patients with urosepsis randomized to either imipenem or ceftazidime therapy, those who received ceftazidime had a delayed defervescence at 24 hours. This was thought to correlate with prolonged detection of endotoxin in the plasma of the ceftazidime-treated patients (21). However, only seven patients had detectable endotoxemia in this study.

In two observational studies (26,67) significant decreases in the level of endotoxin occurred in patients who ultimately survived, but levels remained elevated in the nonsurvivors. In one study of 17 neutropenic patients with sepsis receiving adjuvant therapy with a polyclonal antibody (26), endotoxin levels declined within 18 hours of treatment in 10 survivors but remained elevated in the 7 nonsurvivors (26). A similar pattern was observed in the second study of 31 intensive care unit patients treated for septic shock (67). However, others have not found that persistently elevated levels of endotoxemia are associated with lethality (19,29). In one study, mortality rates were no different between 5 patients with increasing levels of endotoxin versus 12 patients with decreasing levels of endotoxin (19). In 10 of 12 patients with endotoxemia, this was cleared to undetectable levels before a fatal outcome was observed (29).

The detection of endotoxemia using the LAL test is most likely to be helpful in patient groups that are not limited to sepsis and in populations that have a high proportion of non-Enterobacteriaceae among the gram-negative bacteremias. Potential groups include patients with neutropenia or burns, where *Pseudomonas* infections are frequent, or epidemic outbreaks of meningococcemia. The LAL test could serve as a basis for selection of antibiotic therapy in these settings and warrants evaluation in prospective studies.

CONCLUSIONS

The relevance of endotoxin detection in patients with sepsis is problematic, and seemingly disparate findings have been reported in the published clinical studies. The published studies have been designed in attempts to address two different questions: whether endotoxin is a marker of gram-negative sepsis and whether it is a mediator of the complications of sepsis. The presence or absence of gram-negative bacteremia has been commonly used as a convenient standard for comparison with the detection of endotoxemia, but the results are clearly not congruous, especially when the bacterial isolate is a member of the Enterobacteriaceae.

Using the statistical techniques of meta-analysis, the mortality risk among different subgroups from 14 studies with available data were compared. The 14 studies could be stratified into 7 marker studies and 7 mediator studies on the basis of whether sepsis criteria were used as the basis for patient inclusion. In both types of studies, the presence of endotoxemia appears to have the most prognostic significance when it is detected in the presence of gram-negative bacteremia. This is particularly apparent among the marker studies.

The LAL test has promise as a prognostic marker in patient groups that are not limited to those with sepsis syndrome, since endotoxemia clearly has predictive value in most marker-type studies. The test may be particularly applicable when there is a high proportion of non-Enterobacteriaceae among the gram-negative bacteremias. In the mediator studies, the detection in blood of either endotoxin alone or gram-negative bacteria alone has no additional prognostic significance over that associated with the defining criteria of severe illness.

In general, the clinical significance of endotoxemia is not as might be expected from the experience derived from studying experimental gram-negative sepsis in animal models. The apparently conflicting conclusions in clinical studies are probably a consequence of heterogeneity among the different patient groups, the range in underlying disease states, and the bacterial isolates in the various investigations. The clinical significance of endotoxemia varies depending on the presence and type of gram-negative bacteria in blood cultures. Comparisons of quantitative results obtained with different reagents, different assay runs, different types of gram-negative infection, and different patient groups are complex and difficult. Although LPS levels decline after effective therapy, it remains to be determined whether clinical improvement can be accelerated by therapies designed to antagonize endotoxin.

ACKNOWLEDGMENTS

The provision of unpublished data from Drs. Guidet (19), Prins (21), Danner (29), Yoshida (31), Ketchum (12), and Goldie and Fearon (22) is gratefully acknowledged.

REFERENCES

1. Young LS. Sepsis syndrome. In: Mandell GL, Bennet JE, Dolin R, eds. Principles and Practice of Infectious Diseases. 4th ed. New York: Churchill Livingstone, 1995:690–705.
2. Morrison DC, Ryan JL. Endotoxins and disease mechanisms. Ann Rev Med 1987; 38:417–432.
3. Hurley JC. Antibiotic induced release of endotoxin: a reappraisal. Clin Infect Dis 1992; 15:840–854.
4. Porter PJ, Spievack AR, Kass EH. Endotoxin like activity of serum from patients with severe localized infections. N Engl J Med 1964; 271:445–447.
5. Douglas GW, Beller FK, Debrovner CH. The demonstration of endotoxin in the circulating blood of patients with septic abortion. Am J Obstet Gynecol 1963; 87:780–788.
6. Hurley JC. Endotoxemia: methods of detection and clinical correlates. Clin Microbiol Rev 1995; 8:268–292.
7. Novitsky TJ. Limulus amebocyte lysate (LAL) detection of endotoxin in human blood. J Endotoxin Res 1994; 1:253–263.
8. Jorgensen JH. Clinical applications of the Limulus amebocyte lysate test. In: Proctor RA, ed. Clinical Aspects of Endotoxic Shock: Handbook of Endotoxin. Vol 4. Amsterdam: Elsevier, 1986:127–160.
9. Opal SM, Palardy JE, Marra MN, Fisher CJ Jr, McKelligon BM, Scott RW. Relative concentrations of endotoxin-binding proteins in body fluids during infection. Lancet 1994; 344:429–431.
10. Roth RI, Levin FC, Levin J. Optimization of detection of bacterial endotoxin in plasma with the Limulus test. J Lab Clin Med 1990; 116:153–161.
11. Hurley JC, Tosolini FA, Louis WJ. Quantitative Limulus lysate assay for endotoxin and the effects of plasma. J Clin Pathol 1991; 44:849–854.
12. Ketchum PA, Parsonnet J, Stotts LS, et al. Utilization of a chromogenic Limulus amebocyte lysate blood assay in a multi-center study of sepsis. J Endotoxin Res 1997; 4:9–16.

13. Hurley JC. Concordance of endotoxaemia with gram-negative bacteraemia in patients with gram-negative sepsis: a meta-analysis. J Clin Microbiol 1994; 32: 2120–2127.
14. Danner RL, Natanson C, Elin RJ, Hosseini JM, Banks S, Mac Vittie TJ, Parillo JE. Pseudomonas aeruginosa compared with *Escherichia coli* produces less endotoxemia but more cardiovascular dysfunction and mortality in a canine model of septic shock. Chest 1990; 98: 1480–1487.
15. Natanson C, Danner RL, Elin RJ, et al. Role of endotoxemia in cardiovascular dysfunction and mortality. *Escherichia coli* and *Staphylococcus aureus* challenges in a canine model of human septic shock. J Clin Invest 1989; 83:243–251.
16. Levin J, Poore TE, Young NS, et al. Gram-negative sepsis: detection of endotoxemia with the Limulus test. With studies of associated changes in blood coagulation, serum lipids and complement. Ann Intern Med 1972; 76:1–7.
17. Levin J, Poore TE, Zauber NP, Oser RS. Detection of endotoxin in the blood of patients with sepsis due to gram-negative bacteria. N Engl J Med 1970; 283:1313–1316.
18. Kenney GE, Foy HM. Detection and quantitation of circulating polysaccharide in pneumococcal pneumonia by immunoelectroosmosis (counterelectrophoresis) and rocket electrophoresis. In: Schlessinger D, ed. Microbiology—1975. Washington, DC: American Society for Microbiology, 1975:97–102.
19. Guidet B, Barakett V, Vassal T, Petit JC, Offenstadt G. Endotoxemia and bacteremia in patients with sepsis syndrome in the intensive care unit. Chest 1994; 106: 1194–1201.
20. Hynninen M, Valtonen M, Vaara M, et al. Plasma endotoxin and cytokine levels in neutropenic and non-neutropenic bacteremic patients. Eur J Clin Microbiol Infect Dis 1995; 14:1039–1045.
21. Prins JM, van Agtmael MA, Kuijper EJ, van Deventer SJH, Speelman P. Antibiotic induced endotoxin release in patients with gram-negative urosepsis: a double blind study comparing imipenem and ceftazidime. J Infect Dis 1995; 172:886–891.
22. Goldie AS, Fearon KCH, Ross JA, et al. Natural cytokine antagonists and endogenous anti-endotoxin core antibodies in sepsis syndrome JAMA 1995; 274:172–177.
23. Magliulo E, Scevalo D, Fumarola D, Vaccaro R, Bertotto A, Burberi S. Clinical experience in detecting endotoxemia with the Limulus test in typhoid fever and other *Salmonella* infections. Infection 1976; 4:21–24.
24. Gardlund B, Sjolin J, Nilsson A, Roll M, Wickerts CJ, Wretlind B. Plasma levels of cytokines in primary septic shock in humans; correlation with disease severity. J Infect Dis 1995; 172:296–301.
25. Brandtzaeg P, Kierulf P, Gaustad P, et al. Plasma endotoxin as a predictor of multiple organ failure and death in systemic meningococcal disease. J Infect Dis 1989; 159:195–204.
26. Behre G, Schedel I, Nentwig B, Wormann B, Essink M, Hiddemann W. Endotoxin concentration in neutropenic patients with suspected gram-negative sepsis: correlation with clinical outcome and determination of anti-endotoxin core antibodies during therapy with polyclonal immunoglobulin M enriched immunoglobulins. Antimicrob Agent Chemother 1992; 36:2139–2146.
27. Butler T, Levin J, Cu DQ, Walker RI. Bubonic plague: detection of endotoxemia with the Limulus test. Ann Intern Med 1973; 79:642–646.
28. Watt G, Padre LP, Tuazon M, Calubaquib C. Limulus lysate positivity and Herxheimer-like reactions in leptospirosis: a placebo-controlled study. J Infect Dis 1990; 162:564–567.
29. Danner RL, Elin RJ, Hosseini JM, et al. Endotoxemia in human septic shock. Chest 1991; 99:169–175.
30. Wortel CH, von der Mohlen MAM, van Deventer SJH, et al. Effectiveness of a human monoclonal anti-endotoxin antibody (HA-1A) in gram-negative sepsis: relationship to endotoxin and cytokine levels. J Infect Dis 1992; 166:1367–1374.
31. Yoshida M, Obayashi T, Tamura H, et al. Diagnostic and prognostic significance of plasma endotoxin determination in febrile patients with haematological malignancies. Eur J Cancer 1994, 30A:145–147.
32. Rintala E, Pulkki K, Mertsola J, Nevalainen T, Nikoskelainen J. Endotoxin, interleukin-6 and phospholipase-A2 as markers of sepsis in patients with haematological malignancies. Scand J Infect Dis 1995; 27:39–43.
33. Brun-Buisson C, Doyon F, Carlet J, et al. Incidence, risk factors and outcome of severe sepsis and septic shock in adults. A multicenter prospective study in intensive care units. JAMA 1995; 274:968–974.
34. Sprung CL, Peduzzi PN, Shatney CH, et al. Impact of encephalopathy on mortality in the sepsis syndrome. Crit Care Med 1990; 18:801–806.
35. Peduzzi P, Shatney C, Sheagren J, Sprung C, and The Veterans Affairs Systemic Sepsis Cooperative Study Group. Predictors of bacteremia and gram-negative bacteremia in patients with sepsis. Arch Intern Med 1992; 152:529–535.
36. Cross AS, Opal SM, Sadoff JC, Gemski P. Choice of bacteria in animal models of sepsis. Infect Immun 1993; 61:2741–2747.
37. Riddington DW, Venkatesh B, Boivin CM, et al. Intestinal permeability, gastric intramucosal pH, and systemic endotoxemia in patients undergoing cardiopulmonary bypass. JAMA 1996; 275:1007–1012.
38. Butler T, Levin J, Linh NN, Chau DM, Adickman M, Arnold K. Yersinia pestis infection in Vietnam. II. Quantitative blood cultures and detection of endotoxin in the cerebrospinal fluid of patients with meningitis. J Infect Dis 1976; 133:493–499.
39. Stumacher RJ, Kovnat MJ, McCabe WR. Limitations of the usefulness of the limulus assay for endotoxin. N Engl J Med 1973; 288:1261–1264.
40. Casey LC, Balk RA, Bone RC. Plasma cytokine and endotoxin levels correlate with survival in patients with the sepsis syndrome. Ann Intern Med 1993; 119:771–778.
41. Martin MA, Silverman HJ. Gram-negative sepsis and the adult respiratory distress syndrome. Clin Infect Dis 1992; 14:1213–1228.
42. Parsons PE, Worthen GS, Moore EE, Tate RM, Henson PM. The association of circulating endotoxin with the

43. Dofferhoff ASM, Bom VJJ, de Vries-Hospers HG, et al. Patterns of cytokines, plasma endotoxin, plasminogen activator inhibitor, and acute phase proteins during the treatment of severe sepsis in humans. Crit Care Med 1992; 20:185–192.
44. Fisher CJ Jr, Agosti JM, Opal SM, et al. Treatment of septic shock with the tumor necrosis factor receptor: Fc fusion protein. N Engl J Med 1996; 334:1697–1702.
45. Perkins MD, Mirrett S, Reller LB. Rapid bacterial antigen detection is not clinically useful. J Clin Microbiol 1995; 33:1486–1491.
46. Adinolfi LE, Utili R, Gaeta GB, Perna P, Ruggiero G. Presence of endotoxemia and its relationship to liver dysfunction in patients with thyphoid fever. Infection 1987; 15:359–362.
47. Koster F, Levin J, Walker L, et al. Hemolytic-uremic syndrome after shigellosis: relation to endotoxemia and circulating immune complexes. N Engl J Med 1978; 298:927–933.
48. Scheifele DW, Melton P, Whitchelo V. Evaluation of the limulus test for endotoxemia in neonates with suspected sepsis. J Pediatr 1981; 98:899–903.
49. Endo S, Inada K, Kikuchi M, et al. Are plasma endotoxin levels related to burn size and prognosis? Burns 1992; 18:486–489.
50. Winchurch RA, Thupari JN, Munster AM. Endotoxemia in burn patients: level of circulating endotoxins are related to burn size. Surgery 1987; 102:808–812.
51. Jones RJ, Roe EA. Measurement of endotoxins with the Limulus test in burned patients. J Hyg Camb 1979; 83:151–156.
52. Dobke MK, Simoni J, Ninnemann JL, Garrett J, Harnar TJ. Endotoxemia after burn injury: effect of early excision on circulating endotoxin levels. J Burn Care Rehabil 1989; 10:107–111.
53. American College of Chest Physicians/Society of Critical Care Medicine Consensus Committee. Definitions for sepsis and organ failures and guidelines for the use of innovative therapies in sepsis. Chest 1992; 101:1658–1662.
54. Bates DW, Lee TH. Projected impact of monoclonal anti-endotoxin antibody therapy. Arch Intern Med 1994; 154:1241–1249.
55. Hurley JC. Bacteremia, endotoxemia and mortality in gram-negative sepsis: a reappraisal with meta-analysis. J Clin Microbiol 1995; 33:1278–1282.
56. Billard J-L, Berthier-Berrada S, Page Y, et al. Endotoxemia in human septic shock: relation to gastric intramucosal pH (abstr). Crit Care Med 1994; 22.
57. Fossard DP, Kakkar VV. The Limulus test in experimental and clinical endotoxemia. Br J Surg 1974; 61:798–804.
58. Shenep JL, Flynn PM, Barrett FF, Stidham GL, Westenkirchner DF. Serial quantitation of endotoxemia and bacteremia during therapy for gram-negative bacterial sepsis. J Infect Dis 1988; 157:565–568.
59. Hurley JC. Meta-analysis and the investigation of anti-infective therapies. Exp Opin Invest Drugs 1997; 6:159–167.
60. van Deventer SJH, Buller HR, ten Cate JW, Sturk A, Pauw W. Endotoxaemia: an early predictor of septicaemia in febrile patients. Lancet 1988; 1:605–608.
61. Miller PJ, Wenzel RP. Etiologic organisms as independent predictors of death and morbidity associated with blood stream infections. J Infect Dis 1987; 156:471–477.
62. Hurley JC. Reappraisal of the role of endotoxin in the sepsis syndrome. Lancet 1993; 341:1133–1135.
63. McCloskey RV, Straube RC, Sanders C, Smith SM, Smith CR, and the CHESS Trial Study Group. Treatment of septic shock with human monoclonal antibody HA-1A. A randomized, double-blind, placebo-controlled trial. Ann Intern Med 1994; 121:1–5.
64. Bone R. Why sepsis trials fail. JAMA 1996; 276:565–566.
65. Warren HS. Strategies for the treatment of sepsis. N Engl J Med 1997; 336:952–953.
66. Hurley JC. Endotoxemia and novel therapies for the treatment of sepsis. Exp Opin Invest Drugs 1995; 4:163–174.
67. McCartney AC, Banks JG, Clements GB, Sleigh JD, Tehrani M, Ledingham IMcA. Endotoxemia in septic shock: clinical and post mortem correlations. Intensive Care Med 1983; 9:117–122.
68. Nitsche D, Kriewitz M, Seifert J, Hammelmann H. Investigations on an essential plasma factor disturbing the photometric determination of endotoxin in plasma samples with the Limulus test. Prog Clin Biol Res 1987; 231:331–341.

58

The Role of Gut-Derived Endotoxin in the Pathogenesis of Multiple Organ Dysfunction

Mitchell P. Fink
Beth Israel Deaconess Medical Center and Harvard Medical School, Boston, Massachusetts

Michael G. Mythen
University College of London Hospital, London, England

INTRODUCTION

The pathological mechanisms underlying the development of multiple organ dysfunction in critically ill patients remain poorly understood. Nevertheless, the most widely held view is that poorly controlled systemic activation of the inflammatory response somehow leads to cellular dysfunction and/or destruction at locations distant from the initiating site(s) of injury or infection. Although a large body of data supports the idea that the systemic inflammatory response syndrome (SIRS) and the multiple organ dysfunction syndrome (MODS) are causally related, the SIRS/MODS model is insufficient by itself to explain some common clinical observations. For example, most patients subjected to a major pro-inflammatory stress, such as a major operation or cardiopulmonary bypass (using the heart-lung machine), enjoy an uneventful recovery. However, a substantial number suffer some sort of complication involving distant organs, such as confusion or dyspnea or oliguria. An unfortunate few develop severe organ dysfunction, necessitating a prolonged stay in an intensive care unit (ICU). Of course, the likelihood of developing MODS depends to a certain extent on the magnitude of the initial injury. But because the magnitude of the initial injury or other stressful event fails to accurately predict the risk of MODS, it seems probable that one or more other, possibly host-related, factors is also involved. One of these additional factors may be the leakage of bacteria or microbial products, notably lipopolysaccharide (LPS), from the lumen of the gut into the systemic compartment, leading to initiation or amplification of a deleterious disseminated inflammatory response.

MICROBES AND MICROBIAL PRODUCTS IN THE GUT

The gastrointestinal (GI) tract represents a large endogenous reservoir of microbes and microbial products. The proximal GI tract (stomach and duodenum) is normally sterile or nearly so. However, in the ileum, bacterial counts range from 10^3 to 10^7 organisms per milliliter of luminal contents (1). In the colon, there typically are 10^{10}–10^{12} organisms per gram of feces (1). Under normal conditions, the indigenous intestinal microflora constitutes a very stable ecosystem, typically consisting of more than 500 species and strains living in an intimate relationship to the epithelium (2). Normally, the vast majority of the organisms in the intestine and colon are obligate anaerobes, including *Bacteroides*, *Bifidobacteria*, and *Clostridia* spp., as well as peptostreptococci and peptococci (1). In addition, most healthy humans also are colonized by facultatively anaerobic enteric organisms, particularly *Escherichia coli*, and approximately 20–40% of healthy humans carry yeasts in the GI tract (2). During critical illness, the normal ecology of the gut is altered, and colonization of the GI tract by potentially pathogenic bacteria and yeasts has been documented in a number of studies (3–5).

In addition to microbes, the GI tract also contains large quantities of microbial products shed or secreted by viable organisms or derived from dead bacteria and yeasts. LPS is present in large quantities in the human gut (6). In addition to LPS, however, the gut contains high concentrations of other pro-inflammatory products of microbial origin, including various N-formylmethionyl oligopeptides (7) and peptidoglycan-polysaccharide polymers (8).

GUT MICROBES AND MICROBIAL PRODUCTS IN CRITICAL ILLNESS

Since the GI tract contains huge numbers of microbes and large quantities of potentially toxic microbial products, it is attractive to hypothesize that leakage of bacteria or yeast or microbial toxins from the gut into the systemic compartment might contribute to systemic illness, even in the absence of a gross disruption of mural integrity. Indeed, versions of this concept have been appearing in the biomedical literature for many years. As early as 1903, the Russian biologist Metchnikoff suggested that the systemic absorption of microorganisms or their products from the GI tract could lead to mortality (9). Subsequently, in 1923, the great Harvard physiologist Walter B. Cannon suggested that a toxic factor originating from the gut was responsible for the development irreversibility in cases of prolonged, profound shock (10). Later still, a surgeon, Jacob Fine, and his colleagues developed the hypothesis that endotoxin derived from intraluminal gram-negative bacteria is the gut-derived toxic factor responsible for irreversibility in shock (11). The work by Fine and colleagues, however, was repudiated by Zweifach et al. (12,13) and others (14,15), who showed that mortality rates after hemorrhage were similar in conventional animals and animals raised under gnotobiotic ("germ-free") conditions. Still, the notion that gut-derived bacteria and/or toxins play a crucial role in the pathogenesis of MODS refused to die, and, indeed, this concept regained considerable currency in the 1980s, largely as a result of the work of Border et al. (16), Deitch et al. (17), Wilmore (18), and Carrico et al. (19).

In its current incarnation, the idea that the gut-derived microbes and/or their products somehow facilitate the development of MODS proposes that the barrier function of the GI tract is deranged in patients with trauma, burns, hemorrhagic shock, sepsis, or other forms of critical illnesses. One potential consequence of deranged barrier function is bloodstream invasion by gut-derived pathogens, leading to "primary" bacteremias or fungemias or even metastatic infections. Another hypothetical consequence of gut barrier dysfunction is predicated upon the recognition that in the portal venous system the gut is proximal to the liver. Thus, if mucosal barrier function is deranged due to critical illness, gut-derived microbes or microbial products can leak into the portal circulation, leading ultimately to the activation of Kuppfer cells and/or hepatocytes in the liver with the release of various inflammatory mediators (cytokines, nitric oxide, eicosanoids, platelet-activating factor) implicated in the pathogenesis of MODS. A third hypothetical consequence of gut barrier dysfunction is prompted by the recognition that the GI tract itself is richly endowed with immunocytes and inflammatory cells and thus might become an important source of inflammatory mediators patients with critical illness (20–22).

Microbial Translocation

The movement of bacteria from the intestinal lumen to mensenteric lymph nodes (MLNs), other organs, or blood is a process referred to as *microbial translocation*. In experimental studies using animals, translocation usually is quantitated by enumerating viable colony-forming units in MLNs, liver, spleen, lung, and blood. More recently, the extent of translocation also has been assessed by using radioactively labeled bacteria to estimate the total load of bacteria and bacterial products (i.e., viable organisms plus nonviable organisms plus bacterial fragments) traversing the intestinal barrier (23,24). Data obtained in this way suggest that simply culturing MLNs substantially underestimates the extent of translocation in experimental animals, because most microbes breaching the epithelial barrier are killed (24). Therefore, it seems probable that the increases in translocation have been observed in numerous experimental animal models, reflect the combined effects of increased transmucosal penetration by microbes *and* decreased killing.

In order to move from the lumen to the lamina propria, translocating microbes must breach the barrier imposed by the continuous epithelial lining of the intestine. Many species and strains of bacteria, such as *E. coli* and *Salmonella typhimurium*, appear to translocate by moving through enterocytes, rather than by disrupting the junctional complexes between adjacent cells to traverse the epithelium via a paracellular route (25,26).

However, it is possible that certain strains or species of bacteria pass through intercellular spaces (27). Moreover, if the epithelium is denuded or ulcerated, then microbes can easily gain access to the lamina propria.

Intestinal Epithelial Hyperpermeability

The normal intestinal epithelium exhibits a phenomenon called "selective permeability" (28), which enables the gut to absorb small molecules and ions (e.g., water, Na^+, Cl^-, amino acids) but effectively precludes the transepithelial passage by means of the paracellular route of potentially toxic or pro-inflammatory molecules with a Stokes radius greater than about 11.5 Å (29). Abnormal increases in intestinal permeability to larger hydrophilic substances, however, can occur in a variety of pathological conditions.

In clinical studies, intestinal permeability typically is measured by monitoring the renal excretion of various marker substances after introduction into the gastrointestinal tract by oral ingestion or administration through a feeding tube. In such studies, two different nonmetabolizable probes are commonly employed. One probe (e.g., mannitol) is a relatively small molecule, which permeates moderately well through even normal mucosa. The second probe (e.g., lactulose) is a larger molecule, which permeates the normal mucosa to only a minimal extent. By monitoring the differential recovery of the two probes in the urine, it is possible to assess intestinal permeability while factoring out confounding effects related to changes in intestinal motility or renal function. However, it is not known whether increased intestinal permeability to a relatively small molecule like lactulose is associated with increased permeability to larger molecules of biological interest, such as LPS (30). Thus, the pathophysiological significance of intestinal epithelial hyperpermeability remains to be elucidated.

In experimental studies, a variety of perturbations have been shown to increase intestinal epithelial permeability. These perturbations include oxidant stress (31,32); cellular hypoxia (33); metabolic inhibition (34); in vitro or in vivo exposure to acidic conditions (35–37); in vitro exposure of enterocytic monolayers to nitric oxide (NO·) or various NO· donors (36,38,39), certain cytokines (40–44), or bacterial toxins (45–49); total parenteral nutrition (50–52); and induction (in experimental animals or human volunteers) of hemorrhagic shock (53), thermal injury (54), mesenteric ischemia (55–58), or acute endotoxemia (59–63). The mechanisms responsible for pathological changes in epithelial permeability are not well understood, but many studies suggest that alterations in the actin-based cytoskeleton (33,35,36,38,64,65) and depletion of intracellular ATP stores (33,35,36,38,66) may play important roles.

THE CASE IN FAVOR OF THE GUT/MODS HYPOTHESIS

The notion that the gut is somehow a pivotal organ in critical illness is tenable largely because of the existence of an extensive, albeit entirely circumstantial, body of evidence. In experimental animals, derangements in gut barrier dysfunction, manifested by increased bacterial translocation and/or increased intestinal mucosal permeability, can be induced by a wide variety of pathological insults, including hemorrhage (67), trauma (68), thermal injury (69,70), total parenteral nutrition (71–73), bowel ischemia and reperfusion (57,74,75), pancreatitis (76,77), obstructive jaundice (78), major liver resection (79), endotoxemia (59,60,63,80–83), radiation therapy (84), cytotoxic chemotherapy (85), and sterile inflammation (20). Furthermore, in experimental animals, endotoxin can be detected in the circulation after a variety of stresses, such as superior mesenteric arterial ischemia and reperfusion (86), heat stroke (87), or arterial hypoxemia (88). Furthermore, administration of a polyvalent antiserum directed against a number of O-specific LPS determinants has been shown to improve survival in rabbits subjected to mesenteric ischemia and reperfusion (89), cats subjected to hemorrhagic shock (90), mice subjected to a lethal dose of whole body x-rays (91), and monkeys with experimental heat stroke (92).

Pathological increases in intestinal permeability also have been documented in humans. For example, O'Dwyer et al. showed that gut mucosal permeability to lactulose increases in volunteers injected with a small dose of LPS (61). Other investigators have documented the development of gut mucosal hyperpermeability in a number of other acute clinical conditions, including cardiopulmonary bypass (93–97), major vascular surgery (98), polysystem trauma (99–101), thermal injury (102,103), and sepsis (104). Bacteremia, presumably secondary to translocation across the gut mucosal barrier, has been reported to occur in trauma victims with hemorrhagic shock (105). Translocation to MLNs is commonly observed in patients with bowel obstruction (106,107) and occasionally in patients undergoing operations for a wide variety of causes (108). A number of other laboratories also have found detectable levels of circulating endotoxin during or after cardiopulmonary bypass (97,109–112) and, in some instances, have correlated the degree of endotoxemia with the development of subsequent complications (97,113). Interestingly, the presence of high titers of naturally occurring anti-endotoxin antibodies predicts a lower incidence of fever and other complications in patients undergoing open heart surgery (113). These findings support the view that circulating, presumably gut-derived, endotoxins are detectable in a number of clinical situations and possibly contribute to the development of complications.

Another set of observations that tend to support the idea that the gut is a pivotal organ in critically ill patients comes from several well-designed and carefully controlled trials showing that septic complications are diminished by providing early enteral nutrition to trauma patients (114,115). In addition, colonization of the proximal GI tract with certain pathogens, particularly *Pseudomonas aeruginosa* and coagulase-negative staphylococci, is associated with subsequent development of ICU-acquired infections due to these same organisms (4,116).

The frequent occurrence of gastrointestinal mucosal acidosis in critically ill patients is another observation that lends support to the idea that gut barrier dysfunction is important in the pathogenesis of MODS. Gastrointestinal mucosal pH, commonly referred to as pHi, can be estimated by tonometrically measuring the partial pressure of carbon dioxide (CO_2) in the mucosa (117). In spite of a number of theoretical and actual flaws in the tonometric technique, clinical and experimental studies have shown that mucosal acidosis is an excellent indicator of mucosal hypoperfusion (117–121). In many studies, the development of low gastric pHi has been shown to predict the development of MODS and/or death in critically ill patients (122–127). Indeed, to date, all published studies that have examined the relationship between pHi and outcome following accidental trauma or major surgery have concluded that the presence of persistent mucosal acidosis is a significant correlate of mortality. Moreover, in those studies that have compared tonometry to other routinely used monitors of circulatory and pulmonary function, measurement of gastric pHi has been shown to be a better predictor of morbidity and mortality than such other commonly measured parameters as blood pressure, cardiac output, or urine production (123,124,128). In several prospective clinical trials, resuscitation guided by measurements of gastric pHi has been shown to be associated with fewer complications or deaths than is resuscitation guided by more conventional indices of circulatory function (128–132).

The beneficial effects of selective digestive decontamination (SDD) is yet another bit of circumstantial evidence in support of the concept that gut-derived microbes or microbial products contribute to the development of MODS in critically ill patients. SDD refers to a clinical regimen wherein a combination of topical, enteric nonabsorbable, and parenteral antibiotics are used in a prophylactic fashion to reduce the number of potentially pathogenic microorganisms colonizing the GI tract. The available data suggest that the use of SDD has little impact on survival in critically ill patients (133,134). Nevertheless, the results from a large number of clinical trials provide conclusive evidence that SDD significantly decreases the incidence of nosocomial pneumonia in patients requiring care in an ICU (133). Moreover, in a prospective randomized clinical trial of SDD in patients undergoing cardiac surgery, use of the decontamination regimen was associated with lower circulating levels of LPS and the inflammatory cytokine IL-6 (110).

Probably the most persuasive evidence to date in support of the gut/MODS hypothesis derives from studies on the relationship between the presence of naturally occurring antibodies to endotoxin and the development of postoperative complications and/or death in high-risk patients. In an early study of 86 patients subjected to cardiopulmonary bypass, Freeman and Gould reported that the incidence of fever and infections was lower in the subset of 30 patients with detectable anti-endotoxin antibodies than in the remaining patients without detectable antibodies against endotoxin (113). Subsequently, Barclay described a simple ELISA assay system that permits detection and quantification of naturally occurring ("EndoCAb") antibodies directed against highly conserved epitopes located in the core region of the LPS molecule (135). Using this ELISA, Barclay tested samples from more than 1000 blood donors in Scotland and found a large (greater than 300-fold) variation in naturally occurring EndoCAb levels in the normal population (135). Prompted by this observation, Goldie et al. measured EndoCAb levels in 146 critically ill patients and showed that the mortality rate was inversely related to circulating anti-endotoxin antibody levels (136). Similarly, in a study of 26 patients undergoing various high-risk surgical procedures, Mythen et al. found a lower incidence of various markers of systemic inflammation, such as activation of the intrinsic clotting cascade and neutrophil degranulation, in subjects with high preoperative EndoCAb levels (137). And, in a study of 58 patients undergoing valvular heart surgery, Hamilton-Davies and colleagues found that patients with low preoperative EndoCAb levels had a higher risk of postoperative morbidity (138).

Finally, in a prospective blinded cohort study of more than 300 patients undergoing cardiac surgery at Duke University Medical Center, Bennett-Guerrero et al. measured preoperative EndoCAb as well as total IgG and IgM levels (139). In this carefully designed study, a large number of preoperative factors known to be associated with a poor outcome were recorded and used to assign a risk score using a previously validated scoring system. Even when other preoperative risk factors were simultaneously considered, low circulating levels of IgM EndoCAb antibodies were shown to be a significant risk factor for the development of major complications after heart surgery. Although the presence of high levels of EndoCAb antibodies might just be a marker for better immunological and/or physiological reserves, it seems more likely that the presence of preexisting antibodies to core region epitopes confers protection against endotoxemia, presumably secondary to leakage from the gut, during and after cardiopulmonary bypass.

THE CASE AGAINST THE GUT/MODS HYPOTHESIS

Although Rush et al. reported a high incidence of bacteremia and endotoxemia in blood samples obtained from patients with hemorrhagic shock, these findings are open to criticism because most of the isolates in this study were gram-positive, not enteric gram-negative bacilli as one would expect based upon findings obtained in rodent models of hemorrhagic shock (105). Winchurch et al. showed that endotoxin is detectable in plasma samples obtained from burn patients (140), but also found that prophylactic treatment with an endotoxin-neutralizing antibiotic (polymyxin B), while effectively lowering concentrations of endotoxin, fails to decreased mortality or the levels of IL-6 in plasma (141). Thus, it may be that the presence of endotoxemia in burn patients is an epiphenomenon rather than a primary factor leading to organ dysfunction.

Moore et al. (142) were unable to detect elevated endotoxin levels in portal venous or systemic blood samples obtained over the first 5 days of hospitalization from 20 victims of major abdominal trauma, even though 60% of the patients were in shock at the time of presentation and six of the subjects developed major organ dysfunction. Although 9 (2%) of 424 systemic and portal venous blood cultures were positive, many of the isolates were nonenteric organisms and probably were contaminants. Peitzman et al. (143) obtained biopsies of mesenteric lymph nodes from 25 trauma victims. Although 40% of the patients had one or more major complications, all of the lymph node cultures were sterile. As a control for technique, lymph node biopsies also were obtained from patients with primary gastrointestinal diseases, and in three of four cases cultures were positive enteric organisms. Taken together, these two studies cast doubt on the idea that translocation of gut-derived bacteria or endotoxin plays a crucial role in the development of serious complications in victims of major trauma.

CONCLUSIONS AND FUTURE DIRECTIONS

From the preceding, it should be clear that it remains to be established that leakage of endotoxin, or other microbial products, from the gut is important in the pathogenesis of MODS in patients with critical illnesses. While this hypothesis certainly is attractive on theoretical grounds, the gut might play a key role in critical illness in other ways, for example, by releasing cardiodepressant substances (144) or acting as "priming bed" for circulating polymorphonuclear leukocytes (145,146). There are two obvious ways to test the idea that gut-derived endotoxin is important in the pathogenesis of MODS in humans. First, prophylactic treatment with various effective LPS-neutralizing agents, such as polymyxin B-dextran conjugate (147) or bactericidal/permeability-increasing protein (BPI) (148), might be shown to protect against the development of complications, particularly MODS, in high-risk patients. Second, clinical studies could be performed to determine whether efforts to prophylactically immunize patients against highly conserved epitopes on the LPS molecule can protect against the development of complications, particularly MODS, in high-risk patients.

REFERENCES

1. Simon RH, Gorbach SL. Intestinal flora in health and disease. Gastroenterology 1984; 86:174–193.
2. Stoutenbeek CP, van Saene HKF. Infection prevention in intensive care by selective decontamination of the digestive tract. J Crit Care 1990; 5:137–156.
3. Aerdts SJA, Claesner HAL, van Dalen R, Van Lier HJJ, Vollaard EJ, Festen J. Prevention of bacterial colonization of the respiratory tract and stomach of mechanically ventilated patients by a novel regimen of selective decontamination in combination with initial systemic cefotaxime. J Antimicrob Chemother 1990; 26(suppl. A):59–76.

4. Marshall JC, Christou NV, Meakins JL. The gastrointestinal tract. The "undrained abscess" of multiple organ failure. Ann Surg 1993; 218:111–119.
5. Garvey BM, McCambley JA, Tuxen DV. Effects of gastric alkalinization on bacterial colonization in critically ill patients. Crit Care Med 1989; 17:211–216.
6. van Deventer SJM, tenCate JW, Tytgat GNJ. Intestinal endotoxemia: clinical significance. Gastroenterology 1988; 94:823–831.
7. Chadwick VS, Mellor DM, Myers DB, Selden AC, Keshavarzian A, Broom MF, et al. Production of peptides inducing chemotaxis and lysozomal enzyme release in human neutrophils by intestinal bacteria in vitro and in vivo. Scand J Gastroenterol 1988; 23:121–128.
8. Kool J, Ruseler-van Embden JG, van Lieshout LM, de Visser H, Gerrits-Boeye MY, van den Berg WB, et al. Induction of arthritis by soluble peptidoglycan-polysaccharide complexes produced by human intestinal flora. Arthritis Rheum 1991; 34:1611–1616.
9. Metchnikoff E. The nature of man. In: Mitchell PC, ed. Studies in Opportunistic Philosophy. New York: G.P. Putnam's Sons, 1908:309.
10. Cannon WB. Traumatic Shock. New York: Appleton, 1923.
11. Fine J, Frank ED, Rutenberg SH, Schweinburg FB. The bacterial factor in traumatic shock. N Engl J Med 1959; 260:214–216.
12. Nadler AL, Zweifach BW. Pathogenesis of experimental shock. II. Absence of endotoxic activity in blood of rabbits subjected to graded hemorrhage. J Exp Med 1961; 114:195–204.
13. Zweifach BW, Gordon HA, Wagner M, Reyniers JA. Irreversible hemorrhagic shock in germfree rats. J Exp Med 1958; 107:437–450.
14. Heneghan JB, Hemorrhagic shock in unanesthetized gnotobiotic rats. In: Miyakawa M, Luckey TD, eds. Advances in Germfree Research and Gnotobiology. Cleveland: CRC Press, 1968:165–171.
15. McNulty WP, Jr., Linares R. Hemorrhagic shock of germfree rats. Am J Physiol 1960; 198:141–144.
16. Border JR, Hasset J, LaDuca J, Seibel R, Steinberg S, Mills B, et al. The gut origin septic states in blunt multiple trauma (ISS = 40) in the ICU. Ann Surg 1987; 206:427–448.
17. Deitch EA, Maejima K, Berg R. Effect of oral antibiotics and bacterial overgrowth on the translocation of the GI tract microflora in burned rats. J Trauma 1985; 25:385–391.
18. Wilmore DW, Smith RJ, O'Dwyer ST, Jacobs DO, Ziegler TR, Wang X-D. The gut: a central organ after surgical stress. Surgery 1988; 104:917–923.
19. Carrico CJ, Meakins JL, Marshall JC, Fry D, Maier RV. Multiple-organ-failure syndrome. Arch Surg 1985; 121:196–208.
20. Nieuwenhuijzen GAP, Haskel Y, Lu Q, Berg RD, van Rooijen N, Goris JA, et al. Macrophage elimination increases bacterial translocation and gut-origin septicemia but attenuates symptoms and mortality rate in a model of systemic inflammation. Ann Surg 1993; 218:791–799.
21. Deitch EA, Xu D, Franko L, Ayala A, Chaudry IH. Evidence favoring the role of the gut as a cytokine-generating organ in rats subjected to hemorrhagic shock. Shock 1994; 1:141–145.
22. Mainous MR, Ertel W, Chaudry IH, Deitch EA. The gut: a cytokine-generating organ in systemic inflammation? Shock 1995; 4:193–199.
23. Gianotti L, Alexander JW, Pyles T, Fukushima R, Babcock GR. Prostaglandin E_1 analogues misoprostol and enisoprost decrease microbial translocation and modulate the immune response. Circ Shock 1993; 40:243–249.
24. Fukushima R, Gianotti L, Alexander JW, Pyles T. The degree of bacterial translocation is a determinant factor for mortality after burn injury and is improved by prostaglandin analogs. Ann Surg 1992; 216:438–445.
25. Alexander JW, Boyce ST, Babcock GF, Gianotti L, Peck MD, Dunn DL, et al. The process of microbial translocation. Ann Surg 1990; 212:496–510.
26. Takeuch A. Electron microscope studies of experimental *Salmonella* infection. Am J Pathol 1967; 59:109–136.
27. Wells CL, Jechorek RP, Omsted SB, Erlandsen SL. Effect of LPS on epithelial integrity and bacterial uptake in polarized human enterocyte-like cell line Caco-2. Circ Shock 1993; 40:276–288.
28. Hollander D. Crohn's disease—a permeability disorder of the tight junction? Gut 1988; 29:1621–1624.
29. Madara JL. Loosening tight junctions: Lessons from the intestine. J Clin Invest 1989; 83:1089–1094.
30. Fink MP. Interpreting dual-sugar absorption studies in critically ill patients: what are the implications of apparent increases in intestinal permeability to hydrophilic solutes? Intensive Med 1997; 23:489–492.
31. Otamiri T. Oxygen radicals, lipid peroxidation, and neutrophil infiltration after small-intestinal ischemia and reperfusion. Surgery 1989; 105:593–597.
32. Baker RD, Baker SS, LaRosa K. Polarized Caco-2 cells. Effect of reactive oxygen metabolites on enterocyte barrier function. Digest Dis Sci 1995; 40:510–518.
33. Unno N, Menconi MJ, Salzman AL, Smith M, Hagen S, Ge Y, et al. Hyperpermeability and ATP depletion induced by chronic hypoxia or glycolytic inhibition in Caco-2$_{BBe}$ monolayers. Am J Physiol 1996; 270:G1010–G1021.
34. Matthews JB, Smith JA, Tally KJ, Menconi MJ, Nguyen H, Fink MP. Chemical hypoxia increases junctional permeability and activates electrogenic ion transport in human intestinal epithelial monolayers. Surgery 1994; 116:150–158.
35. Menconi MJ, Salzman AL, Unno N, Ezzell RM, Casey DM, Brown DA, et al. Acidosis induces hyperpermeability in Caco-2$_{BBe}$ cultured intestinal epithelial monolayers. Am J Physiol 1997; 272:G1007–G1021.
36. Unno N, Menconi MJ, Smith M, Aguirre DG, Fink MP. Nitric Oxide-induced derangements in the permeability barrier of cultured intestinal epithelial monolayers: effects of low extracellular pH. Am J Physiol 1997; 272:G923–G934.
37. Gonzalez PK, Doctorow SR, Malfroy B, Fink MP. Role of oxidant stress and iron delocalization in acidosis-induced intestinal epithelial hyperpermeability. Shock 1997; 8:108–114.

38. Salzman AL, Menconi MJ, Unno N, Ezzell RM, Casey DM, Gonzalez PK, et al. Nitric oxide dilates tight junctions and depletes ATP in cultured Caco-2BBe intestinal epithelial monolayers. Am J Physiol 1995; 268:G361–G373.
39. Kennedy M, Denenberg AG, Szabó C, Salzman AL. Poly(ADP-ribose) synthetase activation mediates increased intestinal epithelial permeability induced by peroxynitrite in Caco-2BBe cells. Gastroenterology 1998; 114:510–518.
40. Madara JL, Stafford J. Interferon-y directly affects barrier function of cultured intestinal epithelial monolayers. J Clin Invest 1989; 83:724–727.
41. Adams RB, Planchon SM, Roche JK. IFN-y modulation of epithelial barrier function: time course, reversibility, and site of cytokine binding. J Immunol 1993; 150:2356–2363.
42. Colgan SP, Resnick MB, Parkos CA, Delp-Archer C, McGuirk D, Bacarra AE, et al. IL-4 directly modulates function of a model human intestinal epithelium. J Immunol 1994; 153:2122–2129.
43. Unno N, Menconi MJ, Smith M, Fink MP. Nitric oxide mediates interferon-gamma-induced hyperpermeability in cultured human intestinal epithelial monolayers. Crit Care Med 1995; 23:1170–1176.
44. Colgan SP, Parkos CA, Matthews JB, D'andrea L, Awtrey CS, Lichtman AH, et al. Interferon-y induces a cell surface phenotype switch on T84 intestinal epithelial cells. Am J Physiol 1994; 267:C402–C410.
45. Moore R, Pothoulakis C, LaMont JT, Carlson S, Madara JL. C. difficile toxin A increases intestinal permeability and induces Cl⁻ secretion. Am J Physiol 1990; 259:G165–G172.
46. Lycke N, Karlsson U, Sjolander A, Magnusson KE. The adjuvant action of cholera toxin is associated with an increased intestinal permeability for luminal antigens. Scand J Immunol 1991; 33:691–698.
47. von Ritter C, Sekizuka E, Grisham MB, Granger DN. The chemotactic peptide, N-formyl-methionyl-leucyl-phenylalanine, increased mucosal permeability in the distal ileum of the rat. Gastroenterology 1988; 95:651–656.
48. Wells CL, van de Westerlo EM, Jechorek RP, Feltis BA, Wilkins TD. Bacteroides fragilis enterotoxin modulates epithelial permeability and bacterial internalization by HT-29 enterocytes. Gastroenterology 1996; 110:1429–1437.
49. Fasano A, Baudry B, Pumplin DW, Wasserman SS, Tall BD, Ketley JM, et al. Vibrio cholerae produces a second enterotoxin, which affects intestinal tight junctions. Proc Natl Acad Sci 1991; 88:5242–5246.
50. Purandare S, Offenbartl K, Westrom B, Bengmark S. Increased gut permeability to fluorescein isothiocyanate-dextran after total parenteral nutrition in the rat. J Gastroenterol 1989; 24:678–682.
51. Purandare S, Offenbartl K, Westrom B, Bengmark S. Increased gut permeability to fluorescein isothiocyanate-dextran after total parenteral nutrition in the rat. Scand J Gastroenterol 1989; 24:678–682.
52. Li J, Langkamp-Henken B, Suzuki K, Stahlgren LH. Glutamine prevents parenteral nutrition-induced increases in intestinal permeability. J Parent Enter Nutr 1994; 18:303–307.
53. Russell DH, Barreto JC, Klemm K, Miller TA. Hemorrhagic shock increases gut macromolecular permeability in the rat. Shock 1995; 4:50–55.
54. Carter EA, Tompkins RG, Schiffrin E, Burke JF. Cutaneous thermal injury alters macromolecular permeability of rat small intestine. Surgery 1990; 107:335–341.
55. Horton JW. Alterations in intestinal permeability and blood flow in a new model of mesenteric ischemia. Circ Shock 1992; 36:134–139.
56. Bulkley GB, Kvietys PR, Parks DA, Perry MA, Granger DN. Relationship of blood flow and oxygen consumption to ischemic injury in the canine small intestine. Gastroenterology 1985; 89:852–857.
57. Otamiri T, Sjodahl R, Tagesson C. An experimental model for studying reversible intestinal ischemia. Acta Chir Scand 1987; 153:51–56.
58. Schlichting E, Grotmol T, Kahler H, Naess O, Steinbakk M, Lyberg T. Alterations in mucosal morphology and permeability, but no bacterial translocation takes place after intestinal ischemia and early reperfusion in pigs. Shock 1995; 3:116–124.
59. Fink MP, Antonsson JB, Wang H, Rothschild HR. Increased intestinal permeability in endotoxic pigs: mesenteric hypoperfusion as an etiologic factor. Arch Surg 1991; 126:211–218.
60. Ciancio MJ, Vitiritti L, Khar A, Chang EB. Endotoxin-induced alterations in rat colonic water and electrolyte transport. Gastroenterology 1992; 103:1431–1443.
61. O'Dwyer ST, Michie HR, Ziegler TR, Revhaug A, Smith RJ, Wilmore DW. A single dose of endotoxin increases intestinal permeability in healthy humans. Arch Surg 1988; 123:1459–1464.
62. Chen K, Inoue M, Okada A. Expression of inducible nitric oxide synthase mRNA in rat digestive tissues after endotoxin and its role in intestinal mucosal injury. Biochem Biophys Res Commun 1996; 224:703–708.
63. Unno N, Wang H, Menconi MJ, Tytgat SHAJ, Larkin V, Smith M, et al. Inhibition of inducible nitric oxide synthase ameliorates lipopolysaccharide-induced gut mucosal barrier dysfunction in rats. Gastroenterology 1997; 113:1246–1257.
64. Madara JL, Barenberg D, Carlson S. Effects of cytochalasin D on occluding junctions of intestinal absorptive cells: further evidence that the cytoskeleton may influence paracellular permeability and junctional charge selectivity. J Cell Biol 1986; 102:2125–2136.
65. Madara JL, Moore R, Carlson S. Alteration of intestinal tight junction structure and permeability by cytoskeletal contraction. Am J Physiol 1987; 253:C854–C861.
66. Unno N, Menconi MJ, Smith M, Hagen SJ, Brown DA, Aguirre DE, et al. Acidic conditions ameliorate both ATP depletion and the development of hyperpermeability in cultured Caco-2 enterocytic monolayers subjected to metabolic inhibition. Surgery 1997; 121:668–680.
67. Deitch EA, Bridges W, Ma L, Berg R, Specian RD, Granger DN. Hemorrhagic shock-induced bacterial translocation: the role of neutrophils and hydroxyl radicals. J Trauma 1990; 30:942–952.

68. Deitch EA, Winterton J, Berg R. Effect of starvation, malnutrition, and trauma on the gastrointestinal tract flora and bacterial translocation. Arch Surg 1987; 122: 1019–1024.
69. Jones WGI, Minei JP, Barber AE, Fahey TJI, Shires GTI, Shires GT. Splanchnic vasoconstriction and bacterial translocation after thermal injury. Am J Physiol 1991; 261:H1190–H1196.
70. Alexander JW, Gianotti L, Pyles T, Carey MA, Babcock GF. Distribution and survival of *Escherichia coli* translocating from the intestine after thermal injury. Ann Surg 1991; 213:558–567.
71. Alverdy JC, Moss GS. Total parenteral nutrition promotes bacterial translocation from the gut. Surgery 1988; 104:185–190.
72. Helton WS, Garcia R. Oral prostaglandin E_2 prevents gut atrophy during intravenous feeding but not bacterial translocation. Arch Surg 1993; 128:178–184.
73. Deitch EA, Xu D, Naruhn MB, Deitch DC, Lu Q, Marino AA. Elemental diet and IV-TPN-induced bacterial translocation is associated with loss of intestinal mucosal barrier function against bacteria. Ann Surg 1995; 221:299–307.
74. Sheng ZY, Dong DL, Wang XH. Bacterial translocation and multiple system organ failure in bowel ischemia and reperfusion. J Trauma 1992; 32:148–153.
75. Salzman AL, Wollert PS, Wang H, Menconi MJ, Youssef ME, Compton CC, et al. Intraluminal oxygenation ameliorates ischemia/reperfusion-induced gut mucosal hyperpermeability in pigs. Circ Shock 1993; 40:37–46.
76. Ryan CM, Schmidt J, Lewandrowski K, Compton CC, Rattner DW, Warshaw AL. Gut macromolecular permeability in pancreatitis correlates with severity of disease in rats. Gastroenterology 1993; 104:890–895.
77. Ryan CM, Schmidt J, Lewandrowski K, Compton CC, Rattner DW, Warshaw AL, et al. Gut macromolecular permeability in pancreatitis correlates with severity of disease in rats. Gastroenterology 1993; 104:890–895.
78. Deitch EA, Sittig K, Li M, Berg R, Specian RD. Obstructive jaundice promotes bacterial translocation from the gut. Am J Surg 1990; 159:79–84.
79. Wang XD, Parsson H, Andersson R, Soltesz V, Johansson K, Bengmark S. Bacterial translocation, intestinal ultrastructure and cell membrane permeability early after major liver resection in the rat. Br J Surg 1994; 81:579–584.
80. Deitch EA, Berg R, Specian R. Endotoxin promotes the translocation of bacteria from the gut. Arch Surg 1987; 122:185–190.
81. Sorrells DL, Friend C, Koltuksuz U, Courcoulas A, Boyle P, Garrett M, et al. Inhibition of nitric oxide with aminoguanidine reduces bacterial translocation after endotoxin challenge in vivo. Arch Surg 1996; 131:1155–1163.
82. Mishima S, Xu D, Lu Q, Deitch EA. Bacterial translocation is inhibited in inducible nitric oxide synthase knockout mice after endotoxin challenge but not in a model of bacterial overgrowth. Arch Surg 1997; 132: 1190–1195.
83. Deitch EA, Specian RD, Berg RD, Endotoxin-induced bacterial translocation and mucosal permeability: role of xanthine oxidase, complement activation, and macrophage products. Crit Care Med 1991; 19:785–791.
84. Souba WW, Klimberg VS, Hautamaki RD, Mendenhall WH, Bova FC, Howard RJ, et al. Oral glutamine reduces bacterial translocation following abdominal radiation. J Surg Res 1990; 48:1–5.
85. Zaloga GP, Roberts P, Black KW, Prielipp R. Gut bacterial translocation/dissemination explains the increased mortality produced by parenteral nutrition following methotrexate. Circ Shock 1993; 39:263–268.
86. Gathiram P, Wells MT, Brock-Utne JG, Gaffin SL. Oral administrated non-absorbable antibiotics prevent endotoxemia in primates following intestinal ischemia. J Surg Res 1988; 45:187–193.
87. Gathiram P, Gaffin SL, Brock-Utne JG, Wells MT. Time course of endotoxemia and cardiovascular changes in heat-stressed primates. Aviation Space Environ Med 1987; 58:1071–1075.
88. Gaffin SL, Brock-Utne JG, Zanotti A, Wells MT. Hypoxia-induced endotoxemia in primates: role of reticuloendothelial system function and anti-lipopolysaccharide plasma. Aviation Space Environ Med 1986; 57:1044–1049.
89. Zanotti AM, Gaffin SL. Prophylaxis of superior mesenteric artery occlusion shock in rabbits by antilipopolysaccharide (anti-LPS) antibodies. J Surg Res 1985; 38:113–115.
90. Gaffin SL, Grinberg Z, Abraham C, Birkhan J, Shechter Y. Protection against hemorrhagic shock in the cat by human plasma containing endotoxin-specific antibodies. J Surg Res 1981; 31:18–21.
91. Gaffin SL, Wells M, Jordan JP. Anti-lipopolysaccharide toxin therapy for whole body X-irradiation overdose. Br J Radiol 1985; 58:881–884.
92. Gathiram P, Wells MT, Brock-Utne JG, Gaffin SL. Antilipopolysaccharide improves survival in primates subjected to heat stroke. Circ Shock 1987; 23:157–164.
93. Ohir SK, Bjarnason I, Pathi V, Somasundaram S, Bowels CT, Keogh BE, et al. Cardiopulmonary bypass impairs small intestinal transport and increases gut permeability. Ann Thorac Surg 1993; 55:1080–1086.
94. Sinclair DG, Haslam PL, Quinlan GJ, Pepper JR, Evans TW. The effect of cardiopulmonary bypass on intestinal and pulmonary endothelial permeability. Chest 1995; 108:718–724.
95. Riddington DW, Venkatesh B, Boivin CM, Bonser RS, Elliott TSJ, Marshall T, et al. Intestinal permeability, gastric intramucosal pH, and systemic endotoxemia in patients undergoing cardiopulmonary bypass. J Amer Med Assoc 1996; 275:1007–1012.
96. Sinclair DG, Houldsworth PE, Keogh B, Pepper JR, Evans TW. Gastrointestinal permeability following cardiopulmonary bypass: a randomised study comparing the effects of dopamine and dopexamine. Intensive Care Med 1997; 23:510–516.
97. Oudemans-van Straaten HM, Jansen PG, Hoek FJ, van Deventer SJ, Sturk A, Stoutenbeek CP, et al. Intestinal permeability, circulating endotoxin, and postoperative systemic responses in cardiac surgery patients. J Cardiovasc Vasc Anesth 1996; 10:187–194.

98. Roumen RMH, van der Vliet JA, Wevers RA, Goris RJA. Intestinal permeability is increased after major vascular surgery. J Vasc Surg 1993; 17;734–737.
99. Roumen RMH, Hedriks T, Wevers RA, Goris JA. Intestinal permeability after severe trauma and hemorrhagic shock is increased without relation to septic complications. Arch Surg 1993; 128:453–457.
100. Langkamp-Henken B, Donovan TB, Pate LM, Maull CD, Kudsk KA. Increased intestinal permeability following blunt and penetrating trauma. Crit Care Med 1995; 23:660–664.
101. Pape H-C, Dwenger A, Regel G, Auf'm'Kolck M, Gollub F, Wisner D, et al. Increased gut permeability after multiple trauma. Br J Surg 1994; 81:850–852.
102. Ryan CM, Yarmush ML, Burke JF, Tompkins RG. Increased gut permeability early after burns correlates with the extent of burn injury. Crit Care Med 1992; 20:1508–1512.
103. Deitch EA. Intestinal permeability is increased in burn patients shortly after injury. Surgery 1990; 107:411–416.
104. Ziegler TR, Smith RJ, O'Dwyer ST, Demling RH, Wilmore DW. Increased intestinal permeability associated with infection in burn patients. Arch Surg 1988; 123:1313–1319.
105. Rush BF, Sori AJ, Murphy TF, Smith S, Flanagan JJJ, Machiedo GW. Endotoxemia and bacteremia during hemorrhagic shock. Ann Surg 1988; 207:549–554.
106. Sagar PM, Macfie J, Sedman P, May J, Mancey-Jones B, Johnstone D. Intestinal obstruction promotes gut translocation of bacteria. Dis Colon Rectum 1995; 38:640–644.
107. Deitch EA. Simple intestinal obstruction causes bacterial translocation in man. Arch Surg 1989; 124:699–701.
108. Sedman PC, Macfie J, Sagar P, Mitchell CJ, May J, Mancey-Jones B, et al. The prevalence of gut translocation in humans. Gastroenterology 1994; 107:643–649.
109. Jansen NJ, van Oeveren W, Gu YJ, van Vliet MH, Eijsman L, Wildevuur CR. Endotoxin release and tumor necrosis factor formation during cardiopulmonary bypass. Ann Thorac Surg 1992; 54:744–747.
110. Martinez-Pellus AE, Merino P, Bru M, Conejero R, Seller G, Munoz C, et al. Can selective digestive decontamination avoid the endotoxemia and cytokine activation promoted by cardiopulmonary bypass. Crit Care Med 1993; 21:1684–1691.
111. Rocke DA, Gaffin SL, Wells MT, Koen Y, Brock-Utne JG. Endotoxemia associated with cardiopulmonary bypass. J Thorac Cardiovasc Surg 1987; 93:832–837.
112. Bowles CT, Ohri SK, Klangsuk N, Keogh BE, Yacoub MH, Taylor KM. Endotoxemia detected during cardiopulmonary bypass with a modified Limulus amoebocyte lysate assay. Perfusion 1995; 10:219–228.
113. Freeman R, Gould FK. Prevention of fever and gram negative infection after open heart surgery by antiendotoxin. Thorax 1985; 40:846–848.
114. Moore FA, Moore EE, Jones TN, McCroskey BL, Peterson VM. TEN versus TPN following major abdominal trauma-reduced septic morbidity. J Trauma 1989; 29:916–923.
115. Kudsk KA, Croce MA, Fabian TC, Minard G, Tolley EA, Poret HA, et al. Enteral versus parenteral feeding: effects on septic morbidity after blunt and penetrating abdominal trauma. Ann Surg 1992; 215:503–513.
116. Marshall JC, Christou NV, Horn R, Meakins JL. The microbiology of multiple organ failure: the proximal gastrointestinal tract as an occult reservoir of pathogens. Arch Surg 1988; 123:309–315.
117. Antonsson JB, Boyle CC, Kruithoff KL, Wang H, Sacristan E, Rothschild HR, et al. Validation of tonometric measurement of gut intramural pH during endotoxemia and mesenteric occlusion in pigs. Am J Physiol 1990; 259:G519–G523.
118. Fink MP, Kaups KL, Wang H, Rothschild HR. Maintenance of superior mesenteric arterial perfusion prevents intestinal mucosal permeability in endotoxic pigs. Surgery 1991; 110:154–161.
119. Salzman AL, Strong KE, Wang H, Wollert PS, VanderMeer T, Fink MP. Intraluminal "balloonless" air tonometry: a new method for determination of gastrointestinal mucosal PCO_2. Crit Care Med 1994; 22:126–134.
120. Antonsson JB, Engstrom L, Rasmussen I, Wollert S, Haglund U. Changes in gut intramucosal pH and gut oxygen extraction ratio in a porcine model of peritonitis and hemorrhage. Crit Care Med 1996; 23:1872–1881.
121. Grum CM, Fiddian-Green RG, Pittenger GL, Grant BJB, Rothman ED, Dantzker DR. Adequacy of tissue oxygenation in intact dog intestine. J Appl Physiol 1984; 56:1064–1069.
122. Doglio GR, Pusajo JF, Egurrola MA, Bonfigli GC, Parra C, Vetere L, et al. Gastric mucosal pH as a prognostic index of mortality on critically ill patients. Crit Care Med 1991; 19:1037–1040.
123. Maynard N, Bihari D, Beale R, Smithies M, Baldock G, Mason R, et al. Assessment of splanchnic oxygenation by gastric tonometry in patients with acute circulatory failure. J Am Med Assoc 1993; 270:1203–1210.
124. Marik PE. Gastric intramucosal pH: a better predictor of multiorgan dysfunction syndrome and death than oxygen-derived variables in patients with sepsis. Chest 1993; 104:225–229.
125. Bjork M, Hedberg B. Early detection of major complications after abdominal aortic surgery: predictive value of sigmoid colon and gastric intramucosal pH monitoring. Br J Surg 1994; 81:25–30.
126. Frenette L, Doblar DD, Singer D, Cox J, Ronderos J, Poplawski S, et al. Gastric intramural pH as indicator of early graft viability in orthotopic liver transplantation. Transplantation 1994; 58:292–297.
127. Mythen MG, Webb AR. Intra-operative gut mucosal hypoperfusion is associated with increased post-operative complications and cost. Intensive Care Med 1994; 20:99–104.
128. Chang MC, Cheatham ML, Nelson LD, Rutherford EJ, Morris JA, Jr. Gastric tonometry supplements information provided by systemic indicators of oxygen transport. J Trauma 1994; 37:488–494.
129. Gutierrez G, Palizas F, Doglio G, Wainsztein N, Gallesio A, Pacin J, et al. Gastric intramucosal pH as a

therapeutic index of tissue oxygenation in critically ill patients. Lancet 1992; 339:195–199.
130. Ivatury RR, Simon RJ, Havriliak D, Garcia C, Greenbarg J, Stahl WM. Gastric mucosal pH and oxygen delivery and oxygen consumption indices in the assessment of adequacy of resuscitation after trauma: a prospective randomized study. J Trauma 1995; 39:128–136.
131. Mythen MG, Webb AR. Perioperative plasma volume expansion reduces the incidence of gut mucosal hypoperfusion during cardiac surgery. Arch Surg 1995; 130:423–429.
132. Hamilton-Davies C, Mythen MG, Salmon JB, Jacobson D, Shukla A, Webb AR. Comparison of commonly used clinical indicators of hypovolaemia with gastrointestinal tonometry. Intensive Care Med 1997; 23:276–281.
133. Kollef MH. The role of selective digestive tract decontamination on mortality and respiratory tract infections: a meta-analysis. Chest 1994; 105:1101–1108.
134. Selective Decontamination of the Digestive Tract Trialists' Collaborative Group. Meta-analysis of randomised controlled trials of selective decontamination of digestive tract. Br Med J 1993; 307:525–532.
135. Barclay GR. Endogenous endotoxin-core antibody (EndoCAb) as a marker of endotoxin exposure and a prognostic indictor: a review. Prog Clin Biol Res 1995; 392:263–272.
136. Goldie AS, Fearon KC, Ross JA, Barclay GR, Jackson RE, Grant IS, et al. Natural cytokine antagonists and endogenous antiendotoxin core antibodies in sepsis syndrome. The Sepsis Intervention Group. J Am Med Assoc 1995; 274:172–177.
137. Mythen MG, Barclay GR, Purdy G, Hamilton-Davies C, Mackie IJ, Webb AR, et al. The role of endotoxin immunity, neutrophil degranulation and contact activation in the pathogenesis of post-operative organ dysfunction. Blood Coagul Fibrinolysis 1993; 4:999–1005.
138. Hamilton-Davies C, Barclay GR, Cardigan RA, McDonald SJ, Pudy G, Machin SJ, et al. Relationship between perioperative endotoxin immune status, gut perfusion, and outcome from cardiac valve replacement surgery. Chest 1997; 112:1189–1196.
139. Bennett-Guerrero E, Ayuso L, Hamilton-Davies C, White WD, Barclay GR, Smith PK, et al. Relationship of preoperative antiendotoxin core antibodies and adverse outcomes following cardiac surgery. J Am Med Assoc 1997; 277:646–650.
140. Winchurch RA, Thupari JN, Munster AM. Endotoxemia in burn patients: levels of circulating endotoxin are related to burn size. Surgery 1987; 102:808–812.
141. Munster AM, Smith-Meek M, Dickerson C, Winchurch RA. Translocation. Incidental phenomenon or true pathology? Ann Surg 1993; 218:321–327.
142. Moore FA, Moore EE, Poggetti R, McAnena OJ, Peterson VM, Abernathy CM, et al. Gut bacterial translocation via the portal vein: a clinical perspective with major torso trauma. J Trauma 1991; 31:629–638.
143. Peitzman AB, Udekwu AO, Ochoa J, Smith S. Bacterial translocation in trauma patients. J Trauma 1991; 31:1083–1087.
144. Lundgen O, Haglund U, Isaksson O, Abe T. Effects on myocardial contractility of blood-borne material released from the feline small intestine in simulated shock. Circ Res 1976; 38:307–315.
145. Kim FJ, Moore EE, Moore FA, Biffl WL, Fontes B, Banerjee A. Reperfused gut elaborates PAF that chemoattracts and rimes neutrophils. J Surg Res 1995; 58:636–640.
146. Biffl WL, Moore EE. Splanchnic ischaemia/reperfusion and multiple organ failure. Br J Anaesthesia 1996; 77:59–70.
147. Camerota AJ, Lögdberg L, Lake P, Larkin VA, Fink MP. Delayed therapy with a polymyxin B-dextran conjugate (PMX-622) improves survival in rabbits with Gram-negative peritonitis. J Endotoxin Res 1997; 4:285–292.
148. Koyama S, Shibamoto T, Ammons WS, Saeki Y. $rBPI_{23}$ attenuates endotoxin-induced cardiovascular depression in awake rabbits. Shock 1995; 4:74–78.

59

Therapeutic Approaches Targeting Endotoxin-Derived Mediators

Jean-Daniel Baumgartner
Hôpital de Morges, Morges, Switzerland

Didier Heumann and Michel Pierre Glauser
Centre Hospitalier Universitaire Vaudois, Lausanne, Switzerland

For years septic shock has been described as a complication of gram-negative bacteremia, so that initial efforts at devising therapeutic approaches have focused essentially on the development of antibodies directed against endotoxin (LPS), the major toxic component of gram-negative bacteria. Recent advances in molecular biology and mostly immunology have greatly contributed to our understanding of the pathophysiology of septic shock initiated not only by LPS but also by other microbial products. Infections begin when bacteria penetrate host barriers such as skin and mucosa, sometimes overwhelming host defenses and releasing toxic bacterial products that activate plasma factors (complement and clotting molecules) as well as cells of the immune system (monocytes, PMN, endothelium). The result is that the host's inflammatory response contributes substantially to the development of shock and organ failure. Accordingly, many novel approaches for the treatment of septic shock have been developed in addition to anti-LPS antibodies, aimed at blocking the cytokine pro-inflammatory response.

Invading bacteria and bacterial products trigger within the host the release of a complex array of mediators, including cytokines. Cytokines are believed to be autocrine and paracrine (cell-to-cell) molecules that act locally at their site of production to control the host response to invading organisms. In fact, by influencing coagulation and leukocyte transmigration and by activating professional phagocytes, cytokines assist the host in containing a local infection. However, in shock, this process is out of control, and it is generally admitted that an overproduction of cytokines generates a systemic activation, which affects vascular permeability and resistance, cardiac function, and induces many metabolic derangements so that tissue necrosis, leading eventually to multiorgan failure and death, may result.

PATHOGENESIS OF SEPSIS: THE CONVENTIONAL HYPOTHESIS

The concept that multiple organ failure is related to an uncontrolled systemic inflammatory state originated mainly from models of LPS infusions in animals. Central to the process was the discovery that macrophages and PMN activated by LPS release numerous mediators by the interaction of LPS with the CD14 receptor, a process influenced by LPS-binding protein (LBP) (1). In fact, the sole macrophage activation with its subsequent cytokine release or the injection of recombinant cytokines can produce experimentally in animals a syndrome indistinguishable from the response to severe infections.

The synthesis of pro-inflammatory molecules, especially tumor necrosis factor (TNF) and interleukin (IL)-1 directly induces endothelial damage and facilitates the upregulation of adhesion molecules on endothelium to promote accumulation of activated PMN. This leads in turn to further cytokine production and discharge of toxic molecules from PMN resulting in endothelial necrosis and vascular permeability.

The pro-inflammatory cascade contains several counterregulatory systems, including anti-inflammatory cytokines (IL-10, IL-4, IL-13, IL-15, TGF-β, G-CSF), antagonists (IL-1ra) or soluble TNF and IL-1 receptors, and glucocorticoids. The balance between these two arms of the inflammatory response determines the net response of the host. The balance is disrupted in severe shock with an increased contribution of the pro-inflammatory response.

The current strategy to attenuate the detrimental effect of cytokines in human sepsis is essentially based on observations coming from animal studies. These studies have permitted hypotheses that now must be validated in clinical studies either planned or currently underway. The review of animal models suggests that conclusions drawn from observations in animal models may have somewhat limited applicability to the human sepsis/septic shock syndrome, thus stressing the need for the careful performance and analysis of the results of clinical trials.

ANIMAL MODELS

TNF Blockade in Animal Models

Among pro-inflammatory cytokines, TNF stands as the most toxic molecule, and the pioneering work of Beutler and colleagues has opened a new avenue in the limited repertoire of therapies for septic shock, since this work was the first demonstration that anticytokine therapies might be a useful approach to treat the disease (2). Since this work, many attempts have been made to confirm the role played by TNF in animal models. Models have been parenteral injections of LPS or of gram-negative bacteria. In a few studies, gram-positive bacteria were also evaluated.

Anti-TNF Antibodies and Parenteral (or i.p.) LPS Challenge

Animals were protected against an i.p. or i.v. LPS challenge in most experiments, provided the anti-TNF antibodies were given before or during infusion of LPS (2–7). Failure to protect the animals by simultaneous injection of LPS and anti-TNF antibodies was reported only once (8). A poor increase in survival was observed in another report (9). With regard to delaying TNF blockade with anti-TNF antibodies after LPS challenge, it is inefficient in endotoxic models. Only one observation of relative success has been reported (2), while most failures have probably not been reported.

Anti-TNF Antibodies and Parenteral Bacterial Challenge

Following i.v. challenge of *Escherichia coli* (10–13), anti-TNF antibody administration was shown to be protective up to 30 minutes after the infusion of bacteria in various animal species. A delayed administration given 2 hours after challenge did not increase survival in mice (12), whereas in a study conducted in baboons, a modest protective effect of the anti-TNF antibody was observed when given up to 4 hours after bacterial challenge (50% protection over 100% death in controls) (14). Upon challenge with gram-negative species other than *E. coli*, treatment with anti-TNF antibodies partially protected neutropenic rats from an oral challenge with *Pseudomonas aeruginosa* (15) but did not protect mice from *Klebsiella* infections (12). Finally, after i.v. challenge with gram-positive organisms, one study reported protection (16), the other did not (17).

Taken together, the studies of anti-TNF antibody administration after parenteral live bacterial challenge have shown that protection was not uniformly present and that early treatment was generally associated with a better protection. Information on delayed therapy for many models is lacking, but it is generally admitted that delayed treatment, such as after i.v. LPS challenge, is less efficient than early treatment or treatment given before bacterial challenge.

TNF Blockade by Means of Soluble TNF Receptors

Another approach for TNF blockade is based on soluble TNF receptors. Two distinct forms of TNF receptor (p55 form and p75) are shed from cell surfaces during inflammatory reactions. The TNF-soluble receptor shedding is believed to be a regulatory mechanism that prevents circulating TNF to interact with cellular TNF receptors, thus attenuating the effects of excessive TNF production. Fusion proteins have been developed in which each of the two receptors was linked with hinge regions of the human IgG. This dimeric construction extends the half-life of the molecule in vivo and increases the affinity for TNF. Four variants have been constructed using IgG1 or IgG3 as the partner: sTNFR-IgG1 p75, sTNFR-IgG3 p75, sTNFR-IgG1 p55, and sTNFR-IgG3 p55.

These variants have been investigated in animal studies. All published experiments but one were conducted in mice, with most models being parenteral LPS injections. With respect to the sTNFR p55 construct, protection was very efficient when the construct was given prophylactically, and a variable effect was observed depending on the Ig fusion partner (9,18–21). Similarly, the sTNFR p55 construct appeared effective in a model of i.v. bacterial challenge (21). With respect to the sTNFR75, the IgG1 construct was found protective in a model of bolus LPS injection (22). After a lethal challenge with live *E. coli*, lethality was delayed without influencing overall death (21). The sTNFR-IgG3 p75 construct was not protective in parenteral LPS injection model (20) and was not tested after live bacterial challenge.

Taken together, these experiments showed a promising effect of the sTNFR 55 constructs in limiting TNF toxicity, with a superior efficacy of the construct over conventional anti-TNF antibody in one study (9). In some models, treatment with the sTNFR 55 could be delayed several hours, but this was not true for all models. In contrast, the potential use of the sTNFR 75 constructs appeared more limited. This difference in efficacy might be related to different functions and TNF-binding affinities of the two receptors.

Blockade of IL-1 by Means of IL-1 Receptor Antagonist

Together with TNF, IL-1 is a potent cytokine with broad proinflammatory responses. IL-1 synergizes with TNF, and both cytokines share an impressive number of functions (for review, see Ref. 23). Experimentally, IL-1 showed great promise as a therapeutic tool in animal models; an alternative strategy to TNF blockade was blocking of IL-1 by means of the IL-1 receptor antagonist (IL-1ra). Indeed, the IL-1ra inhibits IL-1 by competing with IL-1 for cell receptor sites (23). Experimentally, IL-1ra has been shown to improve various hemodynamic parameters in models of parenteral LPS or bacterial challenges and to improve survival (24–28). Due to the short half-life of IL-1ra, it has to be given in multiple repeated injections or in continuous infusion to be efficacious. Importantly, one report suggested a potential deleterious effect of IL-1 blockade. While a single injection of IL-1ra reduced the lethality of *Klebsiella pneumoniae* sepsis in newborn rabbits, a second injection 24 hours after challenge enhanced lethality (29).

Cytokine Blockade in Models of Focal Tissue Infections and in Models of Intracellular Infections

While TNF and IL-1 blockade have demonstrated efficacy in increasing survival following parenteral LPS or bacterial challenges, several studies have shown no effect or even a deleterious effect in models of local infections with gram-negative bacteria. In murine or rat models of peritonitis, including cecal/ligature puncture (CLP) (5,6,8,30,31), anti-TNF antibodies failed to afford protection, even when given prophylactically. Only one study reported a protective effect in a rat model of peritonitis caused by *Neisseria meningitidis* (32). One study reported a transient early survival in rat pups pretreated with anti-TNF antibodies and infected i.p. with group B streptococci (33). However, by 96 hours this protection was no longer apparent. No study reported the effect of delayed treatment with anti-TNF antibodies in these models. The effect of IL-1 blockade has not been reported in peritonitis models or in the CLP model.

While pro-inflammatory cytokines (TNF, IL-1, but also IL-12 and IFN-γ) play a detrimental role in many types of extracellular infections, their role is the opposite in intracellular infections. It has been well demonstrated in models of intracellular infections that blocking these cytokines was associated with increased lethality by preventing the normal process of macrophage activation leading to bacterial killing and/or phagocytosis. This was observed in mice treated with blocking antibodies to TNF, with IL-1 receptor antagonist, or in mice with a disrupted TNF or IFN-γ receptor after challenge with intracellular bacterial such as *Listeria monocytogenes*, *Salmonella typhimurium*, or *Chlamydia trachomatis* (34–40).

Comparison of Parenteral and Tissue Models of Infection

When evaluating the potential efficacy of anticytokine strategies, a clear distinction emerges between models of parenteral LPS or live bacterial challenge (extracellular vs. intracellular) and models of focal infection.

On the one hand, the parenteral models of LPS challenge are to be considered as models of intoxication rather than true models of infections. With regard to live bacterial parenteral challenge models, inocula used in most studies ranged from between 10^9 and 10^{12} CFU/kg. Both the parenteral LPS challenge as well as the high bacterial numbers induce extremely high concentrations of cytokines (usually 20–150 ng/ml of TNF). Such high concentrations of TNF are probably toxic for the host, as suggested by the observations after parenteral challenge, which have shown that prophylactic TNF blockade protected from death. The TNF response peaked 1 and 2 hours following LPS or bacterial challenge and was followed by a rapid decline despite the persistence of circulating LPS or of live bacteria. This rapid decline might provide an explanation for the failure of delayed TNF blockade after bacterial challenge, when TNF is no longer present in the circulation.

On the other hand, models of tissue infection in which bacterial inocula are lower (usually less than 10^6 CFU) are perhaps more like true infectious models, because bacteria have to multiply to invade tissues and eventually the bloodstream. The time course of cytokine production is different than after parenteral challenge. In mice challenged i.p. with *E. coli*, TNF levels are at least 10 times lower than after parenteral challenge (usually less than 1 ng/ml) and, in strong contrast to parenteral challenge models, are sustained during the whole observation period (6). In fact, cytokine profiles in these models of tissue infection are very reminiscent of profiles observed in patients with shock. Indeed, TNF levels in these patients are of low magnitude (usually less than 500 pg/ml) and sustained (up to 10 days after onset of shock) (41).

Given the fact that TNF blockade failed in the models of focal infection or even worsened the prognosis, this might be interpreted as a suggestion that TNF does not play a toxic role in these conditions but rather helps fight infection. Although there is no clear and formal evidence of extracellular infection, this is certainly the case for intracellular infections, for which pro-inflammatory cytokines are required to control infection. Alternatively, one could also postulate that TNF blockade might not be efficacious in tissues due to the low penetration of blocking antibodies, allowing TNF to play a detrimental role locally. Indeed, while in parenteral models it was generally well documented that the anti-TNF antibodies used were administered in sufficiently large doses so as to neutralize circulating TNF levels, similar observations were usually not made in models of focal infections.

CLINICAL STUDIES

In the early 1980s, Ziegler and colleagues showed that the administration of plasma enriched with polyclonal antibodies to the core structure of LPS was associated with improved survival among patients with septic shock (42). This pioneering work paved the way to the multiple modern immunotherapies. The failure of the first multicentric clinical studies aimed at blocking endotoxin by means of anti–lipid A antibodies have generated enormous controversy (43). Ten years ago, methylprednisolone was studied in sepsis patients as a nonspecific way of downregulating the inflammatory process. Two large multicenter trials were negative (44,45). The reasons for this lack of success were unclear, however, because a moderate protective effect could have been hindered by some other biological effects of corticosteroids. Soon thereafter, the cytokine network was recognized as a major regulation mechanism of inflammation. IL-1 and TNF were found to be in the midst of the cascade of mediators leading to organ system dysfunction and death in sepsis. Attempts were then made to block these inflammatory cytokines in patients with sepsis.

Clinical Trials of TNF Blockade

A great number of cellular effects of TNF are relevant to the pathophysiology of septic shock. In particular, TNF is a potent activatory of neutrophils and endothelium, increasing adhesion of neutrophils to the endothelium and inducing procoagulant activity in endothelial cells. The compelling evidence that TNF seemed to mediate many effects induced by endotoxin suggested that it might also be responsible for shock and coagulopathy. Since this proved to be the case in many animal models, clinicians embarked upon large clinical trials aimed at blocking TNF-α which appears to be the major toxic molecule of the TNF family. The first studies performed with antibodies to TNF-α were followed by studies conducted with TNF receptor constructs.

Treatment of Sepsis Syndrome with Murine
IgG1 Anti-TNF-α Monoclonal Antibodies
(Norasept I Study, Bayer)

In this phase II/III study, the murine IgG1 anti-TNF-α mAb was administered to patients with sepsis syndrome with or without shock. Nine hundred and ninety-four patients were randomized into three groups: placebo, 7.5 mg/kg, or 15 mg/kg of a single dose of mAb (46). The three groups were similar at randomization. The 28-day mortality was the primary endpoint of the study. There were no statistically significant differences for patients with or without shock (Table 1). However, 3-day mortality was significantly reduced in the 478 patients with septic shock. The patients without shock were not protected. The study was planned originally for 1200 patients, but interim analysis showed that, based on the observed differences in mortality rates, the power would not be sufficient to show an overall difference for all patients. This study was then stopped and replaced by the Norasept II study, which focused on the subgroup that seemed to benefit from the mAb, i.e., the group of patients with septic shock.

Table 1 28-Day Mortality Rates Observed During the Norasept I Study of the Bayer Murine Anti-TNF Monoclonal Antibody

Variable	No. deaths/total (%)			p value	
	15 mg/kg	7.5 mg/kg	Placebo	15 mg/kg	7.5 mg/kg
All patients	101/323 (31.3)	95/322 (29.5)	108/326 (33.1)	0.61	0.33
Patients with shock	61/162 (37.7)	59/156 (37.8)	73/160 (45.6)	0.15	0.20
Patients without shock	40/161 (24.8)	36/166 (21.7)	35/166 (21.1)	0.37	0.98

Source: Adapted from Ref. 46.

Treatment of Sepsis Syndrome with Murine IgG1 Anti-TNF-α Monoclonal Antibodies (Intersept Study, Bayer)

This three-arm study used the same mAb as in the Norasept studies, but with different dosages (3 mg/kg and 15 mg/kg) and with centers both in the United States and Europe. Otherwise, the study was similar (47). After the interim analysis of the Norasept I study was completed, the arm without shock was stopped. Overall, 564 patients were included, with 420 patients in shock. The 28-day mortality for the patients with septic shock was 44.6% in the 15 mg/kg group, 36.7% in the 3 mg/kg group, and 42.9% in the placebo group (not statistically significant). The duration of shock among survivors, which was a secondary objective of the study, was significantly shorter among the mAb recipients than among the placebo recipients (3.7 and 3.9 days vs. 7.0 days, respectively; $p < 0.01$). This study thus suggested some protective activity for the mAb, but not of the magnitude that had been expected when the study was initiated.

Treatment of Septic Shock with Murine IgG1 Anti-TNF-α Monoclonal Antibodies (Norasept II Study, Bayer)

This study was recently completed. The first results were presented at the 1997 ICC meeting in Sydney, Australia. A total of 1916 patients with septic shock were randomized between placebo and anti-TNF mAb (7.5 mg/kg). The 28-day mortality for all patients was the primary endpoint. No differences in disease severity between both groups were detectable at study entry. The 28-day mortality rates were 42.8% and 40.3%, respectively, a nonsignificant different. In patients with any organ failure, there was a trend for protection, since the mortality rates at 7 days were reduced from 39.7% in the placebo group to 20.8% in anti-TNF mAb recipients ($p = 0.023$). The difference was no longer significant at 28 days (mortality rates 45.4% vs. 42.6%, respectively). Therefore, this study failed to confirm that anti-TNF mAb could decrease the 28-day mortality rate in patients with septic shock. However, it should be stressed that only a third of patients died directly from septic shock, the underlying diseases or new medical problems significantly contributing to death in the majority of patients.

Treatment of Severe Sepsis with Murine IgG3 F(ab')2 Anti-TNF-α Fragments (MAK 105F, Knoll)

In a Phase II study, three dosages of MAK 195F were tested in 122 patients. Nine doses of 0.1, 0.3, or 1.0 mg/kg MAK 195F or placebo were administered during 3 days after randomization (48). The 28-day mortality was not different among the four groups (Table 2). However, in a retrospective stratification, it was found that the 37 patients with IL-6 serum levels ≥1000 pg/ml at randomization appeared to benefit from the mAb. The confirmatory study was prematurely stopped, but results have not been made available.

Treatment of Septic Shock with p75 TNF Receptor-IgG1 Fc Fragment Fusion Protein (TNFR2:Fc, Immunex)

In this phase II study, 141 patients with septic shock were administered a single dose of placebo or one of three doses of TNFR2:Fc (0.15, 0.45, or 1.5 mg/kg) (49). Septic shock was defined as a sepsis syndrome with hypotension, with or without organ dysfunction. The 28-day mortality was 30% in the placebo group,

Table 2 28-Day Mortality Rates Observed During the Study of the Knoll Murine Anti-TNF F(ab')2 Fragments MAK 195F

Variable	No. deaths/total (%)			
	Placebo	0.1 mg/kg	0.3 mg/kg	1.0 mg/kg
All patients	12/29 (41.4)	19/34 (55.9)	14/30 (46.7)	11/29 (37.9)
Patients with IL-6 ≥ 1000 pg/ml	4/5 (80)	9/11 (81.8)	6/10 (60)	4/11 (36.4)
Patients with IL-6 < 1000 pg/ml	7/23 (30.4)	9/22 (40.9)	6/18 (33.3)	7/17 (41.2)

Source Adapted from Ref. 48.

30% in the 0.15 mg/kg group, 48% in the 0.45 mg/kg group, and 53% in the 1.5 mg/kg group. The dose-response relationship was statistically significant ($p = 0.02$). The higher mortality rate with higher doses of TNFR2:Fc was not attributable to a measurable imbalance of the severity of illnesses at randomization. This study suggested therefore that TNFR2:Fc could worsen the outcome of patients with sepsis syndrome with hypotension.

Treatment of Sepsis Syndrome With or Without Septic Shock with p55 TNF Receptor-IgG1 Fc Fragment (TNFR1:Fc, Roche)

The purpose of this phase II study was to investigate the safety and efficacy of three dosages of TNFR1:Fc (0.083, 0.042, and 0.008 mg/kg) compared to a placebo in patients with severe sepsis, stratified according to the presence or not of a refractory shock (50). During an interim analysis of 201 patients, the arm with the lowest TNFR1:Fc dose was discontinued because a statistically nonsignificant trend toward increased mortality was observed. The study was thus completed with three arms and stopped after 498 patients had been enrolled. The 28-day mortality in the overall patients as well as in the patients with refractory shock was similar in the three study groups, but the 247 patients with severe sepsis without refractory shock appeared to benefit from the highest dose of the fusion protein ($p = 0.07$) (Table 3). In a prospectively planned logistic regression analysis using predicted mortality and serum IL-6 levels as covariates, this difference was statistically significant ($p = 0.01$). In contrast to what the Knoll study suggested, the protection was not limited to patients with high IL-6 serum levels. The differences from the Norasept I study might be attributable to different randomization criteria. Indeed, the definition of shock in the Roche study was different from that in the Norasept I study, and many patients classified as severe sepsis patients in the Roche study would have been classified as patients with shock according to the criteria of the Norasept I study. A phase III study of TNFR1:Fc is now running.

Differences Between the Anti-TNF Studies

The effects of the various anti-TNF agents do appear somewhat contradictory. This might be explained by the fact that the agents were different and that different criteria were sometimes applied to classify the patients. The murine mAbs, especially the F(ab')2 fragments, might have too short an inhibitory effect. The p75 and the p55 soluble TNF receptors are not equivalent in their ability to inhibit TNF. While both receptors in

Table 3 28-Day Mortality Rates Observed During the Study of the Roche TNF Receptor p55:Fc Fusion Protein

	No. deaths/total (%)				
	0.083 mg/kg	0.042 mg/kg	0.008 mg/kg[a]	Placebo	p[b]
All patients	52/159 (33)	53/145 (37)	31/54 (57)	54/140 (39)	0.20
Refractory shock	32/72 (44)	23/63 (37)	18/31 (68)	26/62 (42)	0.70
Severe sepsis	20/87 (22)	30/82 (37)	13/23 (57)	28/78 (36)	0.07

[a]Group halted after an interim analysis at 201 patients.
[b]The 0.008 mg/kg group was not included in the final statistical analysis.
Source: Adapted from Ref. 50.

equilibrium bind TNF with high affinity, there is a difference in binding kinetics, the exchange rate of TNF bound to the p75 receptor being more than 65-fold faster than that of the p55 receptor (21). The fast TNF exchange of the p57 receptor resulted in persistant bioactive TNF serum concentrations and death in animals treated with the p75 fusion protein, whereas treatment with the p55 fusion molecule completely neutralized TNF and protected mice in a mouse model of gram-negative sepsis (21). A similar explanation may account for the results of the positive Roche trial and of negative Immunex trial, based on TNFR1:Fc (p55) and TNFR2:Fc (p75), respectively. Finally, the results of small phase II studies should not be overinterpreted, because many observed differences might be due to chance when the number of patients is so small for such heterogeneous diseases. We must wait for the results of the Roche phase III study to know whether anti-TNF blockade should be further pursued in sepsis patients.

Clinical Trials of IL-1 Receptor Antagonist

The biological actions of IL-1 can be blocked by infusion of the naturally occurring IL-1ra. Recombinant human IL-1ra (IL-1ra) was first studied in a phase II trial in 99 patients with sepsis syndrome or septic shock (51). Patients received an intravenous loading dose of either IL-1ra (100 mg) or placebo, followed by a 72-hour intravenous infusion of either one of three doses of IL-1ra (17, 67, or 133 mg/hr) or placebo. The 28-day mortality rate was 44% in placebo recipients. An apparent dose-dependent survival benefit was found in patients receiving IL-1ra, mortality rates being 32, 25 and 16% in patients receiving 17, 67, and 133 mg/hr, respectively.

A larger phase III trial was then initiated. A total of 893 patients with sepsis syndrome received an intravenous loading dose of 100 mg of IL-1ra or placebo followed by a continuous 72-hour intravenous infusion of IL-1ra (1.0 or 2.0 mg/kg/hr) or placebo (52). The two primary efficacy analyses specified a priori for this trail were the survival time among all patients or among patients with shock at study entry. There was not a significant increase in survival time for IL-1ra recipients in both categories (Table 4). Results from secondary analyses suggested an increase in survival time among patients with dysfunction of one or more organs and among patients with a predicted risk of mortality of 24% or greater (Table 4). This phase III study could therefore not reproduce the promising results of the phase II study.

A confirmatory phase III was initiated in patients with severe sepsis or septic shock (53). This trial also failed to demonstrate a statistically significant reduction in mortality when compared with standard therapy. Furthermore, the mortality rate did not significantly differ between subgroups defined in the first phase III trial, subgroups for which a benefit of IL-1ra treatment was found retrospectively (Table 5).

Why Did So Many Clinical Trials Fail Despite Promising Preclinical Experimental Data?

At the time this chapter was written, with the exception of the phase III study of the Roche TNFR1:Fc, which is still underway, all the major clinical studies of immunomodulators in sepsis have failed (corticosteroids, anti–lipid A, PAF antagonist, IL-1ra, anti-TNF). The discrepancies between the results observed in animal models and in clinical trials warrant some discussion.

Differences Between Animal Models and Clinical Trials

Animal models are designed to investigate specific modes of action of molecules and drugs. They are not designed to truly resemble clinical septic shock in hu-

Table 4 28-Day Mortality Rates Observed During the Synergen Phase III IL-1ra Study

	No. deaths/total (%)				
	n	1 mg/kg	2 mg/kg	Placebo	p
All patients	893	31	29	34	0.22
Patients with shock	713	31	31	36	0.23
Documented infections	718	28	29	34	0.05
Dysfunction ≥ 1 organ	563	40	33	43	0.009
Predicted risk of mortality ≥ 24%	580	38	35	45	0.005

Source: Adapted from Ref. 52.

Table 5 28-Day Mortality Rates Observed During the Confirmatory Synergen Phase III IL-1ra Study

		No. deaths/total (%)		
	n	100 mg IL-1ra/patient	Placebo	p
All patients	696	33	36	0.36
Patients with shock	255	28	37	0.071
Documented infections	599	32	38	0.12
Dysfunction ≥ 1 organ	246	46	46	NS
Predicted risk of mortality ≥ 24%	464	42	42	NS

Source: Adapted from Ref. 53.

mans. For instance, animal models can be designed to demonstrate that TNF can have a lethal effect in certain circumstances. TNF blockade improves survival in these models. However, these models do not demonstrate that TNF is responsible for lethality in septic shock in humans. There are major differences between animal models and clinical septic shock.

1. In animal models, the cascade of events ending with the death of the animals follows a predictable time course. The initial stimulus is usually given as a single and defined dose, via the same route, in animals genetically defined, and maintained in the same experimental conditions. Therefore, the events are perfectly synchronized from the stimulation of inflammatory mediators to the counterregulatory mechanisms that rapidly ensue. The resulting cytokine production follows a predictable time course, with usually a single transient burst of each cytokine, and experimental protocols to block one cytokine cascade are relatively straightforward. In many cases the course of the experiment is hours or a few days. In contrast, the sequence of events leading to septic shock in humans is much more complex and asynchronous, and most of the time the evolution of illness is much longer than in animal models. Therefore, immunological interventions have to face a very complex situation due to a mixture of upregulatory and downregulatory mechanisms.

2. The rationale for clinical trials was essentially based on observations performed in animals receiving parenteral single injections of LPS or bacteria, in which cytokine blockade was beneficial, not on observations performed in animals with focal infections, in which such therapy was usually not efficient.

3. Cytokine blockade was efficient in animal models only when performed prophylactically or very early after challenge, which is obviously not feasible in humans. Interventions were not successful when applied later.

Problems in the Design of Clinical Trials

Several problems raised by the design of clinical trials are related to the criteria used for the selection of patients (broad recruitment vs. narrow selection of well-defined infections), for the selection of the endpoints (mortality vs. subrogate markers, intent-to-treat analysis vs. death directly due to septic shock, 28-day mortality vs. early mortality), and for the existence of confounding factors (decision not to resuscitate, adequacy of surgical and medical treatment, presence of underlying disease). The purpose of this chapter is not to discuss all these points, but only to make a few comments.

All clinical trials described here were aimed at improving the deleterious systemic manifestations of infection. The selection of patients was basically based on the criteria of the "sepsis syndrome." The concept of sepsis syndrome relies on the belief that the pathophysiology of sepsis and organ dysfunction is similar whatever the cause of infection may be, because the clinical systemic manifestations of infections are rather uniform. The problem is that the systemic manifestations of infections that can be readily recorded are highly nonspecific. According to the concensus conference of the American College of Chest Physicians and the Society of Critical Care Medicine in 1992 (54), "sepsis" is defined as alterations of two or more of the four following variables in a patient with a focus of infection: body temperature, pulse rate, respiratory rate, and leukocyte count. "Severe sepsis" is defined as sepsis associated with at least one organ dysfunction. The

Table 6 Microbiological Documentation of Infections in the Studies of TNF Blockade

	n	% Infections				
		Unknown	Pure G−	Pure G+	Others	Bacteremic
Intersept	420	23	29	24	42	38
Norasept I	994	27	39	36	24	39
MAK 195F	122	45	21	9	25	14
p55-IgG1	498	39	22	21	11	34
p75-IgG1[a]	141	18	43	47	12	39

[a]The number of infections with more than one organism was not reported.

weakness of this concept is obvious: these variables are nonspecific and may easily be altered by conditions unrelated to infections in ICU patients. Indeed, in the various clinical studies infection was defined on clinical grounds. From 18 to 45% of infections were of unknown origin, and only 14–39% of the patients were bacteremic in the anti-TNF studies (Table 6). Therefore, the existence of a focus of infection may not be the cause, or the sole cause, of the observed alterations. In addition, there were significant differences among the entry criteria in the various clinical trials (Table 7), making comparisons between the studies difficult. Se-

Table 7 Comparison of the Entry Criteria of the Anti-TNF Studies

Criteria used for the definition of:	NORASEPT I NORASEPT II INTERSEPT	MAK 195F	p75-IgG1	p55-IgG1
Sepsis	Fever or hypothermia, tachycardia, tachypnea	Fever or hypothermia, tachycardia, tachypnea	Fever or hypothermia, tachycardia, tachypnea	Fever or hypothermia, tachycardia, tachypnea, leucocytosis
Severe sepsis (organ dysfunction or hypoperfusion abnormality)	Altered mental status, hypoxemia, elevated plasma lactate or metabolic acidosis, oliguria, coagulopathy	Altered mental status, hypoxemia, elevated plasma lactate or metabolic acidosis, oliguria, coagulopathy thrombocytopenia, elevated cardiac index with decreased systemic vascular resistance		Hypoxemia, elevated plasma lactate or metabolic acidosis, oliguria, coagulopathy, hypotension
Study group 1	*Nonshock*: all 3 criteria for sepsis + ≥1 criteria for severe sepsis	*Severe sepsis*: all 3 criteria for sepsis + ≥2 criteria for severe sepsis or hypotension	*Shock*: all 3 criteria for sepsis + hypotension	*Severe sepsis with or without early shock*: ≥3 criteria for sepsis + ≥2 criteria for severe sepsis
Study Group II	*Shock*: all 3 criteria for sepsis + ≥1 criteria for severe sepsis + hypotension	*None*	*None*	*Refractory shock*: ≥3 criteria for sepsis + ≥1 criteria for severe sepsis + refractory hypotension

verity scoring systems such as APACHE III or SAPS II are used to predict mortality at randomization and to assess that the groups are well balanced. However, they rely also on physiological variables that are not specific for infections, and this is true also for the models modified for sepsis patients. Although their value for predicting mortality rates has been well established, none of these scores is able to estimate the very impact of infection on the predicted mortality rates. In other words, for a given predicted mortality, the contribution of infection is variable from one patient to the other. Two groups of patients with similar predicted mortality rates are likely to have similar observed overall mortality rates, but the contribution of infection to these mortality rates may be different for the two groups. This finding might explain why several agents that were promising in phase II trials dramatically failed in phase III trials. In phase III trials, the number of participating centers is usually greater, and investigators are prompted to randomize as many patients as possible. This might lead to a decrease in the proportion of patients in whom infection is actually the main cause of the so-called sepsis syndrome. Thus, the impact on the overall mortality of an antisepsis agent might seem to be lower than in a previous phase II trial.

It is obvious that the sepsis syndrome encompasses a broad range of diseases, which goes from the fulminant meningococcal purpura to necrotizing fasciitis, peritonitis, nosocomial pneumonia, and catheter-related bacteremia, for instance. It may not be appropriate to group all these different pathologies together, in particular when considering that many of these infections have naturally different outcomes. In addition, the critical role of the underlying condition may not reflect on the APACHE or SAPS scores. The adequacy of medical or surgical treatment is also a critical factor that is often unrecognized. Obviously, the randomized, double-blinded design of modern trials is aimed at minimizing imbalances between groups, but it is well known that significant imbalances may nevertheless occur. Therefore, the very factors that affect the severity of infections should be measured, and neither the concept of sepsis syndrome nor the APACHE or SAPS severity scoring systems will do it.

The concept of sepsis syndrome has become very popular as a way of randomizing patients because it allows one to include a substantial number of ICU patients in clinical trials. Given the failures of the recent clinical trials based on the sepsis syndrome concept, it might be appropriate to narrow down in new planned trials the definition of infection and sepsis.

CONCLUSION

There is one major problem in trying to delineate the best potential interventions of cytokine blockade for the treatment of septic shock—the uncertainties about which experimental model mimics best the series of events that lead to septic shock in patients. Clinical trials based on anticytokine strategies were essentially based on the lessons learned from models of parenteral injections of LPS or bacteria, not from models of focal infections. However, most cases of septic shock in humans are due to focal infections. Much is still to be done before we understand the cascade of events leading to septic shock in humans. Although the clinical studies performed so far have not shown TNF or IL-1 blockade to be effective in patients with sepsis or septic shock, we do not yet really know whether these agents have any protective efficacy or not because of the selection of patients based on the nonspecific criteria of the sepsis syndrome.

REFERENCES

1. Ulevitch RJ, Tobias PS. Recognition of endotoxin by cells leading to transmembrane signaling. Curr Biol 1994; 6:125–130.
2. Beutler B, Milsark IW, Cerami A. Passive immunization against cachectin/tumor necrosis factor protects mice from lethal effect of endotoxin. Science 1985; 229:869–871.
3. Mathison JC, Wolfson E, Ulevitch RJ. Participation of tumor necrosis factor in the mediation of bacterial lipopolysaccharide-induced injury in rabbits. J Clin Invest 1988; 81:1925–1937.
4. Fiedler VB, Loof I, Sander E, et al. Monoclonal antibody to tumor necrosis factor-α prevents lethal endotoxin sepsis in adult rhesus monkeys. J Lab Clin Med 1991; 120:574–588.
5. Bagby GJ, Plessala KJ, Wilson LA, et al. Divergent efficacy of antibody to tumor necrosis factor-α in intravascular and peritonitis models of sepsis. J Infect Dis 1991; 163:83–88.
6. Zanetti G, Heumann D, Gerain J, et al. Cytokine production after intravenous or peritoneal Gram negative bacterial challenge in mice. Comparative protective efficacy of antibodies to TNFα and to LPS. J Immunol 1992; 148:1890–1897.
7. Emerson TE, Lindsey DC, Jesmok GJ, et al. Efficacy of monoclonal antibody against tumor necrosis factor alpha in an endotoxemic baboon model. Circ Shock 1992; 38:75–84.
8. Eskandari MK, Bolgos G, Miller C, et al. Anti-tumor necrosis factor antibody therapy fails to prevent lethality after cecil ligation and puncture or endotoxemia. J Immunol 1992; 148:2724–2730.

9. Jin H, Yang R, Marsters SA, et al. Protection against rat endotoxic shock by p55 tumor necrosis factor (TNF) receptor immunoadhesin: comparison with anti-TNF monoclonal antibody. J Infect Dis 1994; 170:1323–1326.
10. Tracey KJ, Fong Y, Hesse DG, et al. Anti-cachectin/TNF monoclonal antibodies prevent septic shock during lethal bacteraemia. Nature 1987; 330:662–664.
11. Hinshaw LB, Tekamp-Olson P, Chang ACK, et al. Survival of primates in LD100 septic shock following therapy with antibody to tumor necrosis factor (TNF alpha). Circ Shock 1990; 30:279–292.
12. Silva AT, Bayston KF, Cohen J: Prophylactic and therapeutic effects of a monoclonal antibody to tumor necrosis factor-α in experimental gram-negative shock. J Infect Dis 1990: 162:421–427.
13. Jesmok G, Lindsey C, Duerr M, et al. Efficacy of monoclonal antibody against human recombinant tumor necrosis factor in *E. coli*-challenged swine. Am J Pathol 1992; 141:1197–1207.
14. Schlag G, Redl H, Davies J, et al. Anti-tumor necrosis factor antibody treatment of recurrent bacteremia in a baboon model. Shock 1994; 2:10–18.
15. Opal SM, Cross AS, Sadoff JC, et al. Efficacy of antilipopolysaccharide and anti-tumor necrosis factor monoclonal antibodies in a neutropenic rat model of *Pseudomonas* sepsis. J Clin Invest 1991; 88:885–890.
16. Hinshaw LB, Emerson TEJ, Taylor FB, et al. Lethal staphylococcus aureus-induced shock in primates: prevention of death with anti-TNF antibody. J Trauma 1992; 33:568–573.
17. Martin RA, Silva AT, Cohen J. Effect of anti-TNF-α treatment in an antibiotic treated murine model of shock due to *Streptococcus pyogenes*. FEMs Microbiol Lett 1993; 110:175–178.
18. Lesslauer W, Tabuchi H, Gentz R, et al. Recombinant soluble tumor necrosis factor receptor proteins protect mice from lipopolysaccharide-induced lethality. Eur J Immunol 1991; 21:2883–2886.
19. Ashkenazi A, Marsters SA, Capon DJ, et al. Protection against endotoxin shock by a tumor necrosis factor receptor immunoadhesin. Proc Natl Acad Sci USA 1991; 88:10535–10539.
20. Loetscher H, Angehrn P, Schlaeger EJ, et al. Efficacy of a chimeric TNFR-IgG fusion protein to inhibit TNF activity in animal models of septic shock. In: Levin J, Alving CR, Munford RS, et al., eds. Bacterial Endotoxin: Recognition and Effector Mechanisms. Amsterdam: B.V., Elsevier Science, 1993:455–462.
21. Evans TJ, Moyes D, Carpenter A, et al. Protective effect of 55- but not 75-kD soluble tumor necrosis factor receptor-immunoglobulin G fusion proteins in an animal model of gram-negative sepsis. J Exp Med 1994; 180:2173–2179.
22. Mohler KM, Torrance DS, Smith CA, et al. Soluble tumor necrosis factor (TNF) receptors are effective therapeutic agents in lethal endotoxemia and function simultaneously as both TNF carriers and TNF antagonists. J Immunol 1993; 151:1548–1561.
23. Dinarello CA. Biologic basis for interleukin-1 in disease. Blood 1996; 87:2095–2147.
24. Fischer E, Marano MA, Van Zee KJ, et al. Interleukin-1 receptor blockade improves survival and hemodynamic performance in *Escherichia coli* septic shock, but fails to alter host responses to sublethal endotoxemia. J Clin Invest 1992; 89:1551–1557.
25. Wakabayashi G, Gelfand JA, Burke JF, et al. A specific receptor antagonist for interleukin 1 prevents *Escherichia coli*-induced shock in rabbits. FASEB J 1991; 5:338–343.
26. Alexander HR, Doherty GM, Buresh CM, et al. A recombinant human receptor antagonist to interleukin 1 improves survival after lethal endotoxemia in mice. J Exp Med 1991; 173:1029–1032.
27. Ohlsson K, Björk P, Bergenfeldt M, et al. Interleukin-1 receptor antagonist reduces mortality from endotoxin shock. Nature 1990; 348:550–552.
28. Aiura K, Gelfand JA, Burke JF, et al. Interleukin-1 (IL-1) receptor antagonist prevents *Staphylococcus epidermidis*-induced hypotension and reduces circulating levels of tumor necrosis factor and IL-1 β in rabbits. Infect Immun 1993; 61:3342–3350.
29. Mancilla J, Garcia P, Dinarello CA. The interleukin-1 receptor antagonist can either reduce or enhance the lethality of *Klebsiella pneumoniae* sepsis in newborn rats. Infect Immun 1993; 61:926–932.
30. Echtenacher B, Falk W, Männel DN, et al. Requirement of endogenous tumor necrosis factor/cachectin for recovery from experimental peritonitis. J Immunol 1990; 145:3762–3766.
31. Evans GF, Snyder YM, Butler LD, et al. Differential expression of interleukin-1 and tumor necrosis factor in murine septic shock models. Circ Shock 1989; 29:279–290.
32. Nassif X, Mathison JC, Wolfson E, et al. Tumour necrosis factor alpha antibody protects against lethal meningococcaemia. Mol Microbiol 1992; 6:591–597.
33. Teti G, Mancuso G, Tomasello F. Cytokine appearance and effects of anti-tumor necrosis factor alpha antibodies in a neonatal rat model of group B streptococcal infection. Infect Immun 1993; 61:227–235.
34. Nakane A, Minagawa T, Kato K. Endogenous tumor necrosis factor (cachectin) is essential to host resistance against Listeria monocytogenes infection. Infect Immun 1988; 56:2563–2569.
35. Havell EA. Evidence that tumor necrosis factor has an important role in antibacterial resistance. J Immunol 1989; 143:2894–2899.
36. Havell EA, Moldawer LL, Helfgott D, et al. Type I IL-1 receptor blockade exacerbates murine listeriosis. J Immunol 1992; 148:1486–1492.
37. Nauciel C, Espinasse-Maes F. Role of gamma interferon and tumor necrosis factor alpha in resistance to *Salmonella typhimurium* infection. Infect Immun 1992; 60:450–454.
38. Williams DM, Magee DM, Bonewald LF, et al. A role in vivo for tumor necrosis factor alpha in host defense against *Chlamydia trachomatis*. Infect Immun 1990; 58:1572–1576.
39. Rothe J, Lesslauer W, Lötscher H, et al. Mice lacking the tumour necrosis factor receptor 1 are resistant to TNF-mediated toxicity but highly susceptible to infec-

tion by *Listeria monocytogenes*. Nature 1993; 364:798–802.
40. Locksley RM. Interleukin 12 in host defense against microbial pathogens. Proc Natl Acad Sci USA 1993; 90:5879–5880.
41. Calandra T, Baumgartner JD, Grau GE, et al. Prognostic values of tumor necrosis factor/cachectin, interleukin-1, α-interferon and γ-interferon in the serum of patients with septic shock. J Infect Dis 1990; 161:982–987.
42. Ziegler EJ, McCutchan JA, Fierer J, et al. Treatment of gram-negative bacteremia and shock with human antiserum to a mutant *Escherichia coli*. N Engl J Med 1982; 307:1225–1230.
43. Fink MP. Adoptive immunotherapy of gram negative sepsis: use of monoclonal antibodies to lipopolysaccharide. Crit Care Med 1993; 21:S32–S39.
44. The Veterans Administration Systemic Sepsis Cooperative Study Group. Effect of high-dose glucocorticoid therapy on mortality in patients with clinical signs of system sepsis. N Engl J Med 1987; 317:659–665.
45. Bone RC, Fisher CJ, Clemmer TP, et al. A controlled clinical trial of high-dose methylprednisolone in the treatment of severe sepsis and septic shock. N Engl J Med 1987; 317:653–658.
46. Abraham E, Wunderick R, Sliverman H, et al. Efficacy and safety of monoclonal antibody to human tumor necrosis factor-α in patients with sepsis syndrome. A randomized, controlled, double-blind, multicenter clinical trial. JAMA 1995; 273:934–941.
47. Cohen J, Carlet J, The INTERSEPT Study Group. INTERSEPT: an international, multicenter, placebo-controlled trial of monoclonal antibody to human tumor necrosis factor-α in patients with sepsis. Crit Care Med 1996; 24:1431–1440.
48. Reinhart K, Wiegand-Löhnert C, Grimminger F, et al. Assessment of the safety and efficacy of the monoclonal anti-tumor necrosis factor antibody-fragment, MAK 195F, in patients with sepsis and septic shock: a multicenter, randomized, placebo-controlled, dose-ranging study. Crit Care Med 1996; 24:733–742.
49. Fisher CJ, Agosti JM, Opal SM, et al. Treatment of septic shock with the tumor necrosis factor receptor: Fc fusion protein. N Engl J Med 1996; 334:1697–1702.
50. Abraham E, Glauser MP, Butler T, et al. p55 tumor necrosis factor receptor fusion protein in the treatment of patients with severe sepsis and septic shock. A randomized controlled multicenter trial. JAMA 1997; 277:1531–1538.
51. Fisher CJ, Slotman GJ, Opal SM, et al. Initial evaluation of human recombinant interleukin-1 receptor antagonist in the treatment of sepsis syndrome: a randomized, open-label, placebo-controlled multicenter trial. Crit Care Med 1994; 22:12–21.
52. Fisher CJ, Dhainaut JF, Opal SM, et al. Recombinant human interleukin 1 receptor antagonist in the treatment of patients with sepsis syndrome. Results from a randomized, double-blind, placebo-controlled trial. JAMA 1994; 271:1836–1843.
53. Opal SM, Fisher CJ, Dhainaut JF, et al. Confirmatory interleukin-1 receptor antagonist trial in severe sepsis: a phase III, randomized, double-blind, placebo-controlled, multicenter trial. Crit Care Med 1997; 25:1115–1124.
54. American College of Chest Physicians, Society of Critical Care Medicine Consensus Conference: definitions for sepsis and organ failure and guide-lines for the use of innovative therapies in sepsis. Chest 1992; 101:1644–1655.

60

Human Responses to Endotoxin: Role of the Genetic Background

Frank Stüber
Bonn University, Bonn, Germany

INTRODUCTION

Despite rapid development of new antibiotic agents, severe infection remains a major cause of death even in modern intensive care units. Endotoxin is an important mediator of microbial toxicity. Pathophysiological mechanisms of endotoxin binding to protein (lipopolysaccharide-binding protein) and receptor (CD14) has well been documented (1). Human responses to gram-negative infection show a high degree of interindividual variation (2). Comparable amounts of infectious units of microbial organisms induce a wide range of disease human populations. Endotoxin, a constituent of the outer membrane of gram-negative bacteria, is a major inducer of inflammatory responses in humans. Exogenous infection by gram-negative organisms and the vast endogenous source of gram-negative bacteria within the human gastrointestinal tract contribute to the total endotoxin pool that is able to challenge the human immune system (3).

The immune response to endotoxin involves a complex pattern of primary, secondary, and tertiary humoral and cellular responses. In this view, the role of the genetic background in responses to endotoxin is determined by genetic variabilities of endogenous mediators, which constitute the pathways of endotoxin-induced pathophysiology.

Primary responses to endotoxin challenge are mediated by proinflammatory cytokines such as tumor necrosis factor (TNF) and interleukin-1 (IL-1) (4). Secondary proinflammatory mediators like interleukin-6 (IL-6) and interleukin-8 (IL-8) are induced by TNF and IL-1 (4). Tertiary mediators comprise factors of different, even noncytokine origin such as proteases, coagulation factors, kinins, eicosanoids, nitric oxide, and others, which take part in the distal part of mediator cascades (5).

Recent evidence suggests not only that proinflammatory mechanisms contribute to organ failure and death induced by gram-negative sepsis but that anti-inflammatory mediators have important effects on the host's immune system as well (6). Anti-inflammatory mediators induce a state of immunosuppression in sepsis, which has been named "immunoparalysis" (7). This state of decreased immunoreactivity is accompanied by high levels of anti-inflammatory cytokines such as interleukin-10 (IL-10) and interleukin-1 receptor antagonist (IL-1ra) (8). Symptoms of immunosuppression comprise a decreased number of circulating monocytes expressing surface HLA class II molecules and impaired ex vivo responses of macrophages and lymphocytes to lipopolysaccharide (LPS) (9).

Pro- and anti-inflammatory responses at the same time contribute to the outcome of gram-negative sepsis, which represents the ultimate endotoxin challenge in humans. Therefore, all genes encoding proteins involved in inflammatory responses are candidate genes to determine the human genetic background responsible for interindividual differences in systemic inflammatory responses to infection.

Genetically determined capacity of cytokine production and release, heat-shock protein expression, nitric oxide synthase activity, or gene polymorphisms of coagulation factors and many other genes involved in inflammation may contribute to a wide range of clinical manifestations in inflammatory disease states. A patient with peritonitis, for example, may present without symptoms of sepsis and recover within days or may suffer from fulminant septic shock resulting in death within hours.

What is the benefit of having information on the genetic background of an individual's inflammatory response to endotoxin challenge? Besides the interest of basic science concerning role and interaction of mediators, there are several very practical and clinical considerations: Which group of patients carries the greatest risk of developing severe sepsis and multiple organ dysfunction in the situation of an endotoxin challenge? Is it possible to identify a high-risk group for nonsurvival? Will certain patients benefit more than others from antimediator strategies because of their genetic determination to high cytokine release in response to endotoxin?

Studies of the structure and function of single genes that obviously contribute to a disease have been conducted particularly in the field of cytokine research. Members of the cytokine family like TNF are highly conserved in evolution (10). Few genomic polymorphisms are known, whereas the functional relevance of most genomic variations remains obscure.

One goal in determining the role of the genetic background in responses to endotoxin is to identify genomic markers suitable for clinical use and risk stratification of patients. Another goal is to understand the influence of genomic variation on gene regulation and protein expression.

GENETIC BACKGROUND OF PROINFLAMMATORY CYTOKINES

Primary proinflammatory cytokines like TNF and IL-1 induce secondary pro- and anti-inflammatory mediators like IL-6 and IL-10. They have been shown to contribute substantially to the host's primary response to infection. Both, TNF and IL-1 are capable of inducing the same symptoms and the same severity of septic shock and organ dysfunction as endotoxin in experimental settings as well as in humans (11). Genetic variations in the TNF and IL-1 genes are of major interest concerning genetically determined differences in the response to endotoxin.

Tumor Necrosis Factor

TNF is considered one of the most important mediators of endotoxin-induced effects. Interindividual differences of TNF release have been described (12,13).

The TNF locus consists of three functional genes. TNF is positioned between lymphotoxin α (LTα) in the upstream direction and lymphotoxin β (LTβ) in the downstream direction (Fig. 1). Genomic polymorphisms within in the TNF locus have been under intense investigation. Genetic variation within the TNF locus is rare as the TNF gene is well conserved throughout evolution (10). The coding region, in particular, is highly conserved.

The main interest has been focused on the genomic variations of the TNF locus depicted in Figure 1. Two allele polymorphisms defined by restriction enzymes (NcoI, AspHI) or single base changes (−308, −238) as well as multiallelic microsatellites (TNFa–e) have

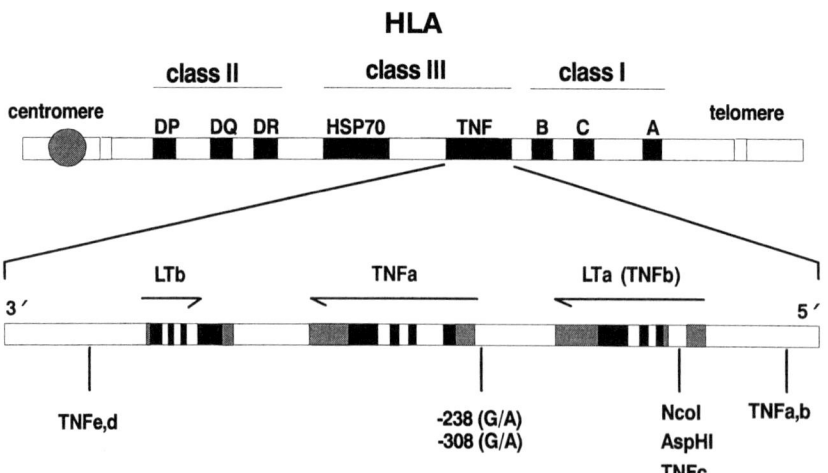

Fig. 1 Genomic polymorphisms located within the highly conserved TNF locus (chromosome 6, short arm).

been investigated in experimental studies and also in various diseases in which TNF has a pathogenic role. Functional importance for the regulation of the TNF gene has been suggested for two polymorphisms within the TNF promoter region. Single base changes have been detected at position −308 and position −238 (14,15).

A G-to-A transition at position −308 has been associated with susceptibility to cerebral malaria (16). The rare allele TNF2 (A at position −308) was thought to be linked to high TNF promoter activity (16). Autoimmune diseases like diabetes mellitus or lupus erythematosus did not show differences of allele frequencies or genotype distribution between patients and controls (17,18). In addition, patients with severe sepsis and a high proportion of gram-negative infection also did not display altered allele frequencies concerning the biallelic promoter polymorphism (position −308) (19). Analysis of the TNF promoter by means of reporter gene constructs revealed contradictory results. The first report found that there was a functional importance of the −308 G to A transition (16). Two papers could not confirm differences of the TNF promoter activity in relation to the −308 polymorphism (19,20). A recent paper reports on a possible influence on TNF promoter activity by the −308 G to A transition in a B cell line (21). Genotyping of this polymorphism in patients with severe sepsis does not contribute to risk assessment (Fig. 2). The −308 polymorphism is not a marker for susceptibility to or outcome of severe sepsis caused by gram-negative infection (19).

In contrast to genomic variations located in the promoter region, intronic polymorphisms are more difficult to associate with a possible functional relevance. Two biallelic polymorphisms located within intron two of LTα have been studied in autoimmune disease (22,23). One polymorphism is characterized by the absence or presence of a NcoI restriction site. First reports demonstrated genomic blots revealing characteristic 5.5 or 10.5 kb bands after genomic NcoI digest, which hybridize to TNF specific probes (24). These bands correspond to the presence and absence, respectively, of a NcoI restriction site within intron two of LTα.

The allele TNFB2 of this NcoI polymorphism (10.5 kb band) has been shown to be associated with high TNFα release ex vivo (25). One study showed no differences between genotypes in ex vivo TNF induction (13), while another study suggests an increased LTα response in TNFB2 homozygotes (25). The question of which genotype is clearly associated with a high proinflammatory response in the clinical situation of severe gram-negative infection and severe sepsis cannot yet be answered by ex vivo studies. Our own results show increased TNF mRNA levels induced by LPS ex vivo in whole blood drawn from healthy volunteers typed TNFB2 homozygous (data not shown), while TNFB2 homozygous patients with severe sepsis display high initial TNF plasma values (Fig. 3).

Different conditions of cell culture and cytokine induction contribute to differing results. In addition, the genomic NcoI polymorphism within intron two of the LTα gene may represent a genomic marker without evidence for its functional importance in gene regulation. This genomic marker may coincide with as-yet-undetected genomic variations, which are responsible for genetic determination of a high proinflammatory response to infection.

Several studies in chronic inflammatory autoimmune diseases suggest an association between TNFB2 and the incidence or severity and outcome of the disease (22,23,26). Studies in acute inflammatory diseases like severe sepsis in patients on surgical intensive care units showed a correlation between TNFB2 homozygosity and mortality (Fig. 4). TNFB2 homozygotes display a relative risk of death from severe sepsis of 2.9 when compared to corresponding genotypes.

Interleukin-1

Besides TNF, IL-1 is another potent proinflammatory cytokine released by macrophages in response to endotoxin. IL-1 is capable of inducing the symptoms of septic shock and organ failure in animal models and is regarded as a primary mediator of the systemic inflammatory response. Antagonizing IL-1 in endotoxin-chal-

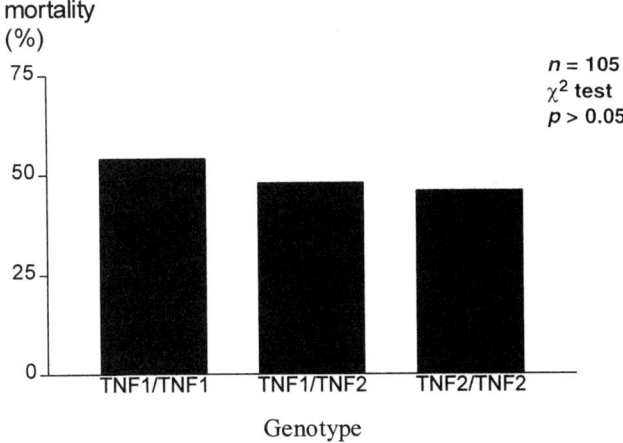

Fig. 2 The genotypes of the −308 TNF promoter polymorphismus are not associated with outcome in severe sepsis.

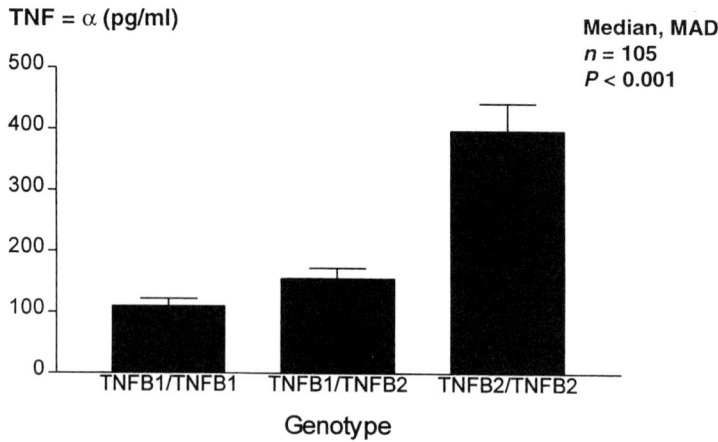

Fig. 3 TNFB2 homozygotes show high TNF levels within the first 96 hours after diagnosis of severe sepsis.

lenged animals including primates abrogates the lethal effects of endotoxin (27). IL-1β, the secreted cytokine, is regarded to be more important as the membrane-bound protein IL-1α. A biallelic TaqI polymorphism has been described within the coding region (exon 5) of IL-1β (28,29). Despite the finding that a homozygous TaqI genotype correlates with high IL-1β secretion (28), genotyping of patients with severe sepsis did not reveal any association with incidence or outcome of the disease (data not shown).

Interleukin-6

IL-6 is a secondary mediator with important immunological functions like the enhancement of B-lymphocyte proliferation. It is released by macrophages and endothelial cells. Although direct toxic effects mediated by IL-6 in severe sepsis and septic shock have not yet been demonstrated, the proinflammatory activity is obvious. The signaling of IL-6 through the IL-6 receptor also exerts some anti-inflammatory effects mediated by binding to the GP130 receptor.

IL-6 gains importance as a new clinical parameter for monitoring the inflammatory activity in the course of acute inflammatory diseases. A study testing the anti-TNF monoclonal antibody approach in the treatment of severe sepsis used IL-6 levels as a criterion of hyperinflammation to randomize patients (30).

Genomic polymorphisms of the IL-6 gene have been described in the 3'-flanking region (31). In addition, two single base changes have been reported for a MspI and a BcgII restriction site (32,33). Functional studies of these genomic variations concerning influences on gene transcription or mRNA stability do not exist. An-

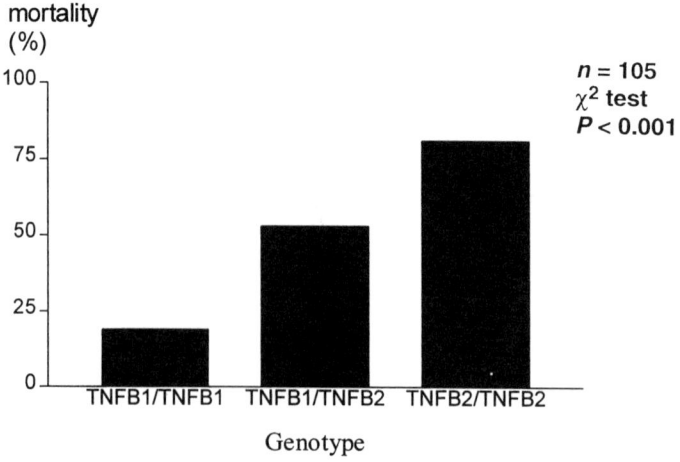

Fig. 4 TNFB2 homozygotes have a poor prognosis in severe sepsis.

other study of allele frequencies and genotype distributions of the BcgII biallelic polymorphism in patients with rheumatoid arthritis does not reveal differences when compared to normal controls (26). For the second biallelic polymorphism characterized by the presence or absence of MspI site, the functional relevance of this polymorphism is unknown.

Currently, there are no data available in the field of acute systemic inflammatory disease. Ex vivo data of the relationship of genomic variations with the quantity of IL-6 release do not exist, nor have associations of the IL-6 polymorphisms with the incidence or outcome of severe sepsis been studied. Therefore, the contribution of IL-6 gene polymorphisms to the genetic background of responses to endotoxin remains obscure.

GENETIC BACKGROUND OF ANTI-INFLAMMATORY CYTOKINES

Interleukin-1 Receptor Antagonist

Proinflammatory mediators comprise the hyperinflammatory side of gram-negative infection and endotoxin challenges. At the same time, anti-inflammatory mediators are induced by proinflammatory cytokines and function to counterbalance the overshoot of inflammatory activity. This physiological process of limiting the extent of inflammation by release of anti-inflammatory proteins may escape the physiological boundaries of local and systemic concentrations of these mediators. Proteins like IL-4, IL-10, IL-11, or IL-13 or IL-1ra contribute to a very powerful downregulation of soluble and cellular proinflammatory activities. This downregulation results in decreased expression of class II molecules in antigen presenting cells as well as reduced ex vivo responses of immuncompetent cells to inflammatory stimuli. This state of immunosuppression has also been termed "immunoparalysis" (7). It results in a state of deactivation and diminished capacity to eliminate microbial pathogens. A new term for this anti-inflammatory state has recently been created: the compensatory anti-inflammatory response syndrome (CARS) (34). Outcome of patients with severe sepsis is not only influenced by hyperinflammation with progressive organ dysfunction but may also be affected by immunosuppression and lack of restoration of immune function. In this view, innate interindividual differences in the release of anti-inflammatory mediators contribute to the human response to endotoxin to a similar extent as proinflammatory responses.

A genomic polymorphism of the anti-inflammatory cytokine IL-1ra is located within intron two and consists of variable numbers of a tandem repeat (VNTR) of a 86 bp motif. This 86-base-pair motif contains at least three known binding sites for DNA-binding proteins (35). Ex vivo experiments suggest higher IL-1ra responses combined with alleles containing low numbers of the 86 bp repeat. Ex vivo studies also demonstrate a higher level of IL-1ra protein expression and protein release of A2 homozygous individuals compared to heterozygotes following stimulation with lipopolysaccharide (36). In LPS-stimulated whole blood cultures, A2 homozygotes express higher levels of IL-1ra mRNA (Fig. 5) and protein as well (Fig. 6). The

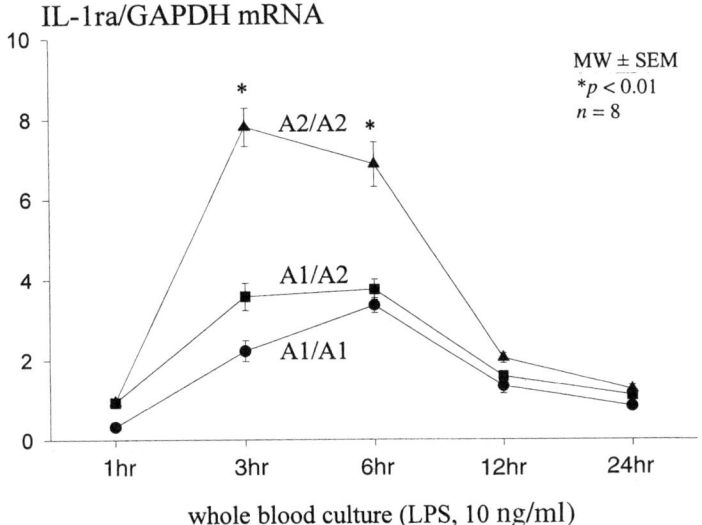

Fig. 5 IL-1ra A2 homozygotes display high amounts of inducible IL-1ra mRNA ex vivo.

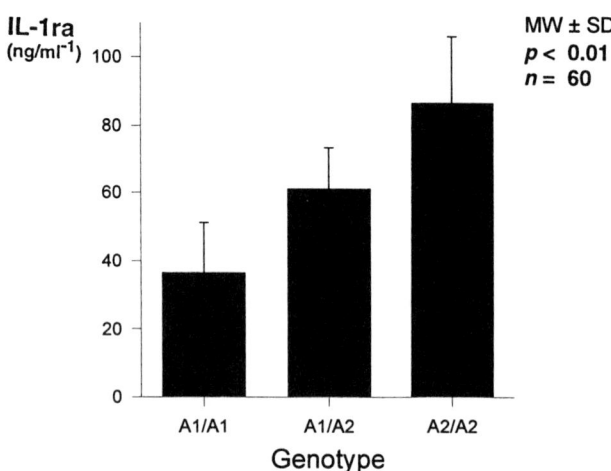

Fig. 6 IL-1ra A2 homozygotes secrete high amounts of inducible IL-1ra ex vivo.

allele A2 has been associated with an increased incidence of autoimmune diseases like lupus erythematosus and insulin-dependent diabetes mellitus (37,38). In acute systemic inflammation, there is no difference between surviving or nonsurviving patients with severe sepsis. This finding is in contrast to the results concerning the biallelic NcoI polymorphism within intron one of LTα: homozygotes for the TNFB2 genotype revealed a high mortality when compared to heterozygotes and TNFB1 homozygotes. The overall group of patients with severe sepsis did not show an increase in the TNFB2 allele frequency. For the IL-1ra polymorphism, however, an increase of the allele A2 in the patients with severe sepsis was detected (Fig. 7). Patients carrying the TNFB2 homozygous and A2 homozygous haplotype did not survive in this study. Analysis of a larger group of patients has to prove whether these two alleles as a haplotype contribute to mortality. The expectation of a balanced immune response in patients with a high TNF and high IL-1ra secretion is not fulfilled in this observation. Consequently, the hypothesis that a high proinflammatory response to LPS induces a similar overshooting anti-inflammatory mediator release (i.e., IL-1ra) and subsequent severe immunosuppression with poor prognosis has to be considered.

Interleukin-10

Recent work suggested that the anti-inflammatory cytokine IL-10 contributes significantly to the counter-regulation of the proinflammatory response evoked by LPS in human sepsis (8). In a murine model of peritonitis, therapeutic intervention using IL-10 attenuated the rise of proinflammatory serum cytokines.

The genomic structure of the IL-10 gene reveals nucleotide variations in the regulatory promoter region of the gene. Biallelic polymorphisms (RsaI and MaeIII restriction sites) as well as dinucleotide repeats with 16 different alleles have been described. Associations between IL-10 genotype and the individual's capacity of IL-10 secretion have been demonstrated. Innate low IL-10 secretion was correlated to a high rejection rate in organ transplant recipients. Another study reported a correlation between certain IL-10 microsatellite alleles and autoantibody production in lupus erythematosus. Data on secretory capacity of IL-10 in dependence of genomic variations of the IL-10 gene are still rare. Also, no data are available on allele frequencies and genotype distribution in sepsis. The importance of the IL-10 molecule in regulating inflammation deserves further investigation of the genetic background of IL-10 expression in human sepsis.

NONCYTOKINE GENES

Major efforts have been made in cytokine research to disclose the mechanisms of endotoxin-mediated effects. In addition, there are a number of noncytokine mediators that contribute significantly to the toxicity of endotoxin. Interest has been focused upon nitric oxide (NO) and the pathways controlling NO release (44). Eicosanoids take part in the organism's response to endotoxin. Complex coagulation cascades are activated and may lead to disseminated intravascular coagulation followed by impaired organ perfusion and organ dysfunction. Potential benefit and protection against cellular stress induced by endotoxin and its mediators is provided by intracellular heat-shock proteins. These proteins are capable of playing role as molecular chaperones facilitating protein synthesis and folding as well as restructuring partially denatured proteins (45).

Nitric Oxide and NO-Synthase

Nitric oxide contributes to the state of hypotension in severe sepsis and septic shock. Endotoxin and proinflammatory mediators like TNF and IL-1 activate an inducible form of NO-synthase (iNOS2) in macrophages and the vascular endothelium (46). In macrophages, NO is a potent radical and contributes to killing of invading pathogens. Excessive release of NO leads

Healthy blood donors (n = 123)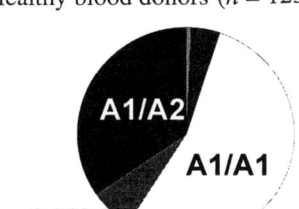

Patients with severe sepsis (n = 105)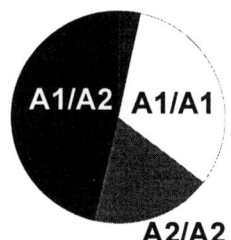

Fig. 7 Patients with severe sepsis show an increased frequency of the allele IL-1ra A2 because of an increased frequency of A1/A2 heterozygotes and A2/A2 homozygotes.

to severe hypotension in septic shock. Pharmacological antagonism of NO synthesis reverses hypotension in animal models of septic shock as well as in patients (47).

Constitutive and inducible isoforms of NOS exist, which are regulated differently. Some constitutively expressed forms are tissue specific (brain, neuron, vessels), while inducible forms are expressed in endothelial cells and macrophages. A single base change has been discovered in the promoter region of the human endothelial NO-synthase (NOS3). This polymorphism has been associated with the smoking-dependent risk of coronary artery disease (48). Still there is a lack of studies examining NOS3 promoter activities and NOS3 expression depending on allele and genotype. Associations of NOS polymorphisms and incidence or outcome of septic shock have not been reported yet.

Coagulation: Tissue Factor, Coagulation Factors

Failure of the otherwise well-balanced system of coagulation and anticoagulation is a very common feature of endotoxin-induced physiological disturbances (49). Disseminated intravascular coagulation (DIC) contributes to stasis of the microvascular blood flow. Endotoxin activates coagulation cascades by activation of coagulation factors (50). Tissue factor is a very central factor in the regulation of clotting mechanisms. Interindividual differences in susceptibility to DIC may be based on genomic variations in the genes encoding single coagulation or anticoagulation factors. Genomic polymorphisms are known for factor VII, tissue factor, and other clotting factors (51). Susceptibility to DIC in relation to these genomic variations has not been studied in patients with severe sepsis or septic shock so far.

Heat-Shock Protein Genes

Heat-shock proteins represent a physiological response to cellular stress, which is highly conserved in evolution. Expression of heat-shock proteins as molecular chaperones facilitates intracellular folding of synthesized proteins (45). Inhibition of the heat-shock response exacerbates septic shock in animal models. On the other hand, induction of heat-shock proteins prior to lethal endotoxin challenges reduces the lethality of septic shock (45).

A well-studied member of the heat-shock protein family is HSP 70. Two genomic polymorphisms have been described within the HSP 70-2 and the HSP 70 HOM gene. The first is an inducible form of HSP, while the latter is expressed constitutively. No study testing the relation between allele and HSP 70-HOM expression has been performed. The HSP 70-2 polymorphism has been related to variable expression of HSP 70-2 mRNA (52). The functional relevance of these HSP polymorphisms still remains to be elucidated.

Genotyping of patients with severe sepsis has not shown any significant differences in allele frequencies or genotype distribution compared to healthy controls (data not shown). A significant linkage exists between HSP 70-2 2 homozygotes and TNFB2 homozygotes. In this view, the allele HSP 70-2 2 may be included in an extended haplotype comprised of alleles linked to each other and defining individuals at risk for developing or even dying from severe sepsis when challenged by gram-negative infection and endotoxemia.

CONCLUSION

It is very important to elucidate the human responses to endotoxin in a genetic context. Currently, there is a lack of understanding the regulation of systemic inflammatory responses, a lack of diagnostic tools identifying individuals at risk of developing or dying from severe sepsis, and a lack of knowledge about which patients to treat with immunomodulatory agents. Evaluation of this genetic background just has started, and much work remains to be completed. Genomic markers will need to be studied in the context of extended haplotypes, which will characterize an individual patient's genetic background of inflammatory responses in the future.

REFERENCES

1. Schumann RR, Lamping N, Kirschning C, et al. Lipopolysaccharide binding protein: its role and therapeutical potential in inflammation and sepsis. Biochem Soc Trans 1994; 22:80–82.
2. Sasse KC, Nauenberg E, Long A, et al. Long-term survival after intensive care unit admission with sepsis. Crit Care Med 1995; 23:1040–1047.
3. Edmiston CE, Jr, Condon RE. Bacterial translocation. Surg Gynecol Obstet 1991; 173:73–83.
4. Blackwell TS, Christman JW. Sepsis and cytokines: current status. Br J Anaesth 1996; 77:110–117.
5. Bernard GR, Wheeler AP, Russell JA, et al. The effects of ibuprofen on the physiology and survival of patients with sepsis. The Ibuprofen in Sepsis Study Group. N Engl J Med 1997; 336:912–918.
6. van der Poll T, de Waal Malefyt R, Coyle SM, et al. Antiinflammatory cytokine responses during clinical sepsis and experimental endotoxemia: sequential measurements of plasma soluble interleukin (IL)-1 receptor type II, IL-10, and IL-13. J Infect Dis 1997; 175:118–122.
7. Volk HD, Reinke P, Krausch D, et al. Monocyte deactivation—rationale for a new therapeutic strategy in sepsis. Intensive Care Med 1996; 22(suppl 4):S474-81:S474–81.
8. Fisher CJ, Jr., Slotman GJ, Opal SM, et al. Initial evaluation of human recombinant interleukin-1 receptor antagonist in the treatment of sepsis syndrome: a randomized, open-label, placebo-controlled multicenter trial. The IL-1RA Sepsis Syndrome Study Group. Crit Care Med 1994; 22:12–21.
9. Docke WD, Randow F, Syrbe U, et al. Monocyte deactivation in septic patients: restoration by IFN-gamma treatment. Nat Med 1997; 3:678–681.
10. Nedwin GE, Naylor SL, Sakaguchi AY, et al. Human lymphotoxin and tumor necrosis factor genes: structure, homology and chromosomal localization. Nucleic Acids Res 1985; 13(17):6361–6373.
11. Weinberg JR, Boyle P, Meager A, et al. Lipopolysaccharide, tumor necrosis factor, and interleukin-1 interact to cause hypotension. J Lab Clin Med 1992; 120:205–211.
12. Westendorp RG, Langermans JA, Huizinga TW, et al. Genetic influence on cytokine production and fatal meningococcal disease. Lancet 1997; 349:170–173.
13. Whichelow CE, Hitman GA, Raafat I, et al. The effect of TNF*B gene polymorphism on TNF-alpha and -beta secretion levels in patients with insulin-dependent diabetes mellitus and healthy controls. Eur J Immunogenet 1996; 23:425–435.
14. Wilson AG, di Giovine FS, Blakemore AI, et al. Single base polymorphism in the human tumour necrosis factor alpha (TNF alpha) gene detectable by NcoI restriction of PCR product. Hum Mol Genet 1992; 1:353.
15. Brinkman BM, Huizinga TW, Kurban SS, et al. Tumour necrosis factor alpha gene polymorphisms in rheumatoid arthritis: association with susceptibility to, or severity of, disease? Br J Rheumatol 1997; 36:516–521.
16. McGuire W, Hill AV, Allsopp CE, et al. Variation in the TNF-alpha promoter region associated with susceptibility to cerebral malaria. Nature 1994; 371:508–510.
17. Pociot F, Wilson AG, Nerup J, et al. No independent association between a tumor necrosis factor-alpha promoter region polymorphism and insulin-dependent diabetes mellitus. Eur J Immunol 1993; 23:3050–3053.
18. Wilson AG, Gordon C, di Giovine FS, et al. A genetic association between systemic lupus erythematosus and tumor necrosis factor alpha. Eur J Immunol 1994; 24:191–195.
19. Stuber F, Udalova IA, Book M, et al. −308 tumor necrosis factor (TNF) polymorphism is not associated with survival in severe sepsis and is unrelated to lipopolysaccharide inducibility of the human TNF promoter. J Inflamm 1995; 46:42–50.
20. Brinkman BM, Zuijdeest D, Kaijzel EL, et al. Relevance of the tumor necrosis factor alpha (TNF alpha) −308 promoter polymorphism in TNF alpha gene regulation. J Inflamm 1995; 46:32–41.
21. Wilson AG, Symons JA, McDowell TL, et al. Effects of a polymorphism in the human tumor necrosis factor alpha promoter on transcriptional activation. Proc Natl Acad Sci USA 1997; 94:3195–3199.
22. Pociot F, Molvig J, Wogensen L, et al. A tumour necrosis factor beta gene polymorphism in relation to monokine secretion and insulin-dependent diabetes mellitus. Scand J Immunol 1991; 33:37–49.
23. Bettinotti MP, Hartung K, Deicher H, et al. Polymorphism of the tumor necrosis factor beta gene in systemic lupus erythematosus: TNFB-MHC haplotypes. Immunogenetics 1993; 37:449–454.
24. Badenhoop K, Schwarz G, Trowsdale J, et al. TNF-alpha gene polymorphisms in type 1 (insulin-dependent) diabetes mellitus. Diabetologia 1989; 32:445–448.
25. Pociot F, Briant L, Jongeneel CV, et al. Association of tumor necrosis factor (TNF) and class II major histocompatibility complex alleles with the secretion of TNF-alpha and TNF-beta by human mononuclear cells: a possible link to insulin-dependent diabetes mellitus. Eur J Immunol 1993; 23:224–231.
26. Vinasco J, Beraun Y, Nieto A, et al. Polymorphism at the TNF loci in rheumatoid arthritis. Tissue Antigens 1997; 49:74–78.
27. Boermeester MA, Van Leeuwen PA, Coyle SM, et al. Interleukin-1 blockade attenuates mediator release and dysregulation of the hemostatic mechanism during human sepsis. Arch Surg 1995; 130:739–748.

28. Pociot F, Molvig J, Wogensen L, et al. A TaqI polymorphism in the human interleukin-1 beta (IL-1 beta) gene correlates with IL-1 beta secretion in vitro. Eur J Clin Invest 1992; 22:396–402.
29. Guasch JF, Bertina RM, Reitsma PH. Five novel intragenic dimorphisms in the human interleukin-1 genes combine to high informativity. Cytokine 1996; 8:598–602.
30. Reinhart K, Wiegand-Lohnert C, Grimminger F, et al. Assessment of the safety and efficacy of the monoclonal anti-tumor necrosis factor antibody-fragment, MAK 195F, in patients with sepsis and septic shock: a multicenter, randomized, placebo-controlled, dose-ranging study. Crit Care Med 1996; 24:733–742.
31. Bowcock AM, Ray A, Erlich H, et al. Rapid detection and sequencing of alleles in the 3′ flanking region of the interleukin-6 gene. Nucleic Acids Res 1989; 17:6855–6864.
32. Fugger L, Morling N, Bendtzen K, et al. BglII polymorphism in the human interleukin 6 (IL 6) gene. Nucleic Acids Res 1989; 17:7548.
33. Fugger L, Morling N, Bendtzen K, et al. MspI polymorphism in the human interleukin 6 (IL 6) gene. Nucleic Acids Res 1989; 17:4419.
34. Bone RC. Sir Isaac Newton, sepsis, SIRS, and CARS. Crit Care Med 1996; 24:1125–1128.
35. Tarlow JK, Blakemore AI, Lennard A, et al. Polymorphism in human IL-1 receptor antagonist gene intron 2 is caused by variable numbers of an 86-bp tandem repeat. Hum Genet 1993; 91:403–404.
36. Danis VA, Millington M, Hyland VJ, et al. Cytokine production by normal human monocytes: inter-subject variation and relationship to an IL-1 receptor antagonist (IL-1Ra) gene polymorphism. Clin Exp Immunol 1995; 99:303–310.
37. Blakemore AI, Tarlow JK, Cork MJ, et al. Interleukin-1 receptor antagonist gene polymorphism as a disease severity factor in systemic lupus erythematosus. Arthritis Rheum 1994; 37:1380–1385.
38. Metcalfe KA, Hitman GA, Pociot F, et al. An association between type 1 diabetes and the interleukin-1 receptor type 1 gene. The DiMe Study Group. Childhood Diabetes in Finland. Hum Immunol 1996; 51:41–48.
39. Rongione AJ, Kusske AM, Ashley SW, et al. Interleukin-10 prevents early cytokine release in severe intraabdominal infection and sepsis. J Surg Res 1997; 70(2):107–112.
40. Eskdale J, Kube D, Tesch H, et al. Mapping of the human IL10 gene and further characterization of the 5′ flanking sequence. Immunogenetics 1997; 46(2):120–128.
41. Eskdale J, Gallagher G. A polymorphic dinucleotide repeat in the human IL-10 promoter. Immunogenetics 1995; 42(5):444–445.
42. Turner D, Grant SC, Yonan N, et al. Cytokine gene polymorphism and heart transplant rejection. Transplantation 1997; 64(5):776–779.
43. Eskdale J, Wordsworth P, Bowman S, et al. Association between polymorphisms at the human IL-10 locus and systemic lupus erythematosus. Tissue Antigens 1997; 49(6):635–639.
44. Fink MP, Payen D. The role of nitric oxide in sepsis and ARDS: synopsis of a roundtable conference held in Brussels on 18–20 March 1995. Intensive Care Med 1996; 22:158–165.
45. Buchman TG. Manipulation of stress gene expression: a novel therapy for the treatment of sepsis? Crit Care Med 1994; 22:901–903.
46. Spitzer JA. Cytokine stimulation of nitric oxide formation and differential regulation in hepatocytes and nonparenchymal cells of endotoxemic rats. Hepatology 1994; 19:217–228.
47. Pfeilschifter J, Eberhardt W, Hummel R, et al. Therapeutic strategies for the inhibition of inducible nitric oxide synthase—potential for a novel class of anti-inflammatory agents. Cell Biol Int 1996; 20:51–58.
48. Meier J, Affeldt M, Opitz C, et al. A common base change in the promoter region of the human endothelial NO- synthase (NOS3) gene. Hum Mutat 1996; 8:394.
49. Levi M, van der Poll T, ten Cate H, et al. The cytokine-mediated imbalance between coagulant and anticoagulant mechanisms in sepsis and endotoxaemia. Eur J Clin Invest 1997; 27:3–9.
50. Deitcher SR, Eisenberg PR. Elevated concentrations of cross-linked fibrin degradation products in plasma. An early marker of gram-negative bacteremia. Chest 1993; 103:1107–1112.
51. Takamiya O. Genetic polymorphism (Arg353 → Gln) in coagulation factor VII gene and factor VII levels (coagulant activity, antigen and binding ability to tissue factor) in 101 healthy Japanese. Scand J Clin Lab Invest 1995; 55:211–215.
52. Pociot F, Ronningen KS, Nerup J. Polymorphic analysis of the human MHC-linked heat shock protein 70 (HSP70-2) and HSP70-Hom genes in insulin-dependent diabetes mellitus (IDDM). Scand J Immunol 1993; 38:491–495.

61

Endotoxin, Antibiotics, and Inflammation in Gram-Negative Infections

Jan M. Prins
Academic Medical Center, Amsterdam, The Netherlands

INTRODUCTION

Although appropriate antimicrobial treatment is essential in the treatment of severe gram-negative infections, effective chemotherapy alone is not sufficient to ensure a successful clinical outcome. The underlying condition of the patient is important, as was demonstrated by Kreger et al. (1). In a large series of patients with gram-negative bacteremia, all treated with appropriate antibiotics, mortality increased from 11% if no underlying disease was present to 29% if the patient had a rapidly fatal underlying disease. But even if no underlying disease is present, antibiotic treatment can sometimes lead to a worsening of symptoms. It was already known in the early days of antibiotic treatment that fatal vasomotor collapse may occur after a loading dose of chloramphenicol for typhoid fever (2). This was explained by the rapid lysis of a large number of bacteria. Treatment of brucellosis also often results in a temporary deterioration of clinical symptoms, and the same is true for the treatment of secondary syphilis, the so-called Jarisch-Herxheimer reactions. These reactions suggest an antibiotic-mediated generation of one or more microbial factors, leading to sustained or even increased systemic inflammation.

As discussed in other chapters of this book, endotoxin (or lipopolysaccharide, LPS) is a constituent of the bacterial cell wall of gram-negative bacteria, and in particular the lipid A component of LPS is known to be the major bacterial stimulus of the inflammatory reaction occurring in gram-negative infections. The presentation of endotoxin in an accessible form is important for detection in the Limulus assay, as well as for induction of cytokine production by monocytes. In viable bacteria 98% of the endotoxin remains bound to the bacteria (3), and it has been demonstrated that bound endotoxin is 20 times less active in the Limulus assay than shed endotoxin (4). Therefore, destruction of the bacteria by antibiotics might increase the bioavailability of the endotoxin, thereby enhancing its deleterious effects. In this chapter the current knowledge of the mechanisms of antibiotic-induced endotoxin release during the treatment of gram-negative infections is reviewed, as well as its possible clinical implications.

SPONTANEOUS VERSUS ANTIBIOTIC-INDUCED ENDOTOXIN RELEASE

During bacterial growth in culture medium, endotoxin is continuously shed. This phenomenon has been observed in, among others, *Escherichia coli* and other Enterobacteriaceae (4,5), in *Pseudomonas aeruginosa* (5,6), and in *Neisseria meningitidis* (7,8). During normal growth, until the stationary phase most endotoxin remains cell-bound (3,9–11), and a stable ratio of bacterial counts and (total and shed) endotoxin concentration is usually observed in vitro and in vivo (3,9,10,12,13).

Endotoxin release by *Escherichia coli* during exposure to antibiotics in vitro has been demonstrated in many studies (3,13–17) (see Table 1), and the same observations have been done with other Enterobacteriaceae (5,18), *Haemophilus influenzae* type b (19), and *Pseudomonas aeruginosa* (20). In these experiments, a proportional increase in both bacterial count and endotoxin level was observed in all untreated cultures. In all cultures exposed to an effective antibiotic, a decrease in bacterial count occurred after antibiotic treatment, with an increase in total (unfiltered) or free (filtrable) endotoxin concentrations.

Free endotoxin levels are usually defined as the levels measured after passing culture samples through a filter (in general a 0.45-μm filter). In this manner a separation is made between bacterial cell-bound endotoxin (which will not pass through the filter) and nonbacterial cell-bound or free endotoxin. In the studies summarized in Table 1, the increase in the amount of total endotoxin was usually largest in the untreated cultures, whereas the increase in amount of free endotoxin was often larger after antibiotic treatment as compared to no treatment. Whether total or free endotoxin is the most important determinant of the host response remains a matter of controversy. The time course of endotoxin release does not always parallel the bacterial killing curve. In two of the in vitro studies, most bacterial killing preceded the peak endotoxin levels (19,20), but in another study the increase in endotoxin concentration leveled off after 2 hours, whereas bacterial counts were still decreasing (13).

At least three mechanisms have been demonstrated

Table 1 Endotoxin Release During Killing of *E. coli* In Vitro

Ref	Antibiotic	Concentration (mg/liter)	Duration of exposure (hr)	$\log_{10}$ change from baseline		Bacterial titer
				Endotoxin		
				Total	Free	
15	Control		3	+0.43		+0.6
	Ampicillin	50(8×MIC)	3	+0.90		−1.7
	Streptomycin	50(8×MIC)	3	+0.93		−1.8
	Tetracycline	50(8×MIC)	3	+0.36		−1.1
	Polymyxin B	6(7.5×MIC)	3	+0.26		−1.8
16	Control		2		+0.69	+1.0
	Enoxacin	3.7(23×MIC)	2		+1.05	−2.9
	Ofloxacin	5.3(265×MIC)	2		+1.05	−4.0
	Pefloxacin	4.3(54×MIC)	2		+0.92	−3.4
	Norfloxacin	1.5(37.5×MIC)	2		+0.92	−3.4
	Ciprofloxacin	2.3(115×MIC)	2		+0.69	−3.4
14	Control		2		+0.85	+1.0
	Gentamicin	8(133×MIC)	2		+0.40	−3.6
	Ciprofloxacin	5(250×MIC)	2		+1.04	−4.0
3	Control		4	+2.3	+1.3	+2.3
	Tobramycin	8(16×MIC)	4	+0.7	+1.2	−3.4
	Cefuroxime	75(37.5×MIC)	4	+1.3	+2.1	−2.0
	Ceftazidime	100(400×MIC)	4	+0.6	+1.3	−3.0
	Aztreonam	100(800×MIC)	4	+1.7	+2.3	−2.0
	Imipenem	100(800×MIC)	4	+0.2	+0.9	−3.0
13	Control		6	+3.33		+4.0
	Gentamicin	4(16×MIC)	6	+0.51		−4.9
	Amoxycillin	64(16×MIC)	6	+0.34		−3.9
	Ciprofloxacin	0.128(16×MIC)	6	+1.63		−3.8
17	Control		4	+1.99		+2.1
	Ceftazidime	2.5(50×MIC)	4	+1.10		−2.0
	Imipenem	4.0(50×MIC)	4	−0.07		−2.0
	Gentamicin	12(50×MIC)	4	−0.31		−2.0
	Ciprofloxacin	0.2(50×MIC)	4	+0.34		−2.0

Source: Adapted from Ref. 64.

to contribute to the increase in amount of endotoxin after exposure to antibiotics (17):

1. Accumulation of bacterial biomass following antibiotic treatment, for instance, because of filament formation in case of certain β-lactams (see below)
2. An increase in the accessibility of endotoxin that remains bound to the bacteria, which was demonstrated even when sub-MIC levels of antibiotics were used (21)
3. Release of free endotoxin

It is conceivable that during antimicrobial treatment other microbial constituents are also released, contributing to inflammation. For instance, during treatment of *H. influenzae* type b meningitis, peptidoglycans released from the bacterial cell wall into the CSF also contribute to meningeal inflammation (22), and during in vitro killing of *Staphylococcus epidermidis* TNF-inducing bacterial products like soluble peptidoglycan and teichoic acid are released into the culture medium (23). During killing of *Klebsiella pneumoniae* with ceftazidime or ciprofloxacin, the production and release of capsular polysaccharide increases (24). However, the predominant proinflammatory constituent released during antibiotic killing of gram-negative rods appears to be endotoxin. This is supported by several lines of evidence. First, the monocyte-stimulating activity present in culture supernatants was reduced at least 10-fold in the presence of the endotoxin-neutralizing agent polymyxin B (25), and polymyxin B also inhibited TNF generation during antibiotic killing of bacteria in a whole blood stimulation model (17,26). Second, fractionation of culture supernatants by velocity sedimentation on sucrose gradients or by isopycnic density gradient ultracentrifugation on CsCl gradients demonstrated that the vast majority of proinflammatory activity for human monocytes cofractionated with the LPS component (25). These data all suggest that LPS is the predominant proinflammatory mediator in the supernatants of antibiotic-treated gram-negative bacteria.

In a number of studies the biological activity of the released endotoxin was investigated by measuring TNF-α release from mononuclear cells or human whole blood ex vivo exposed to filtrates of antibiotic-killed *E. coli* (27) or *H. influenzae* type b (26). Whereas low levels of TNF were produced in response to filtrates of viable bacteria, exposure to filtrates of antibiotic-treated cultures produced a significant increase in TNF levels. This suggests that the free endotoxin released during antibiotic killing of *E. coli* or *H. influenzae* type b is biologically active. The contribution of the increase in bacterial biomass or the increase in the accessibility of bacteria-bound endotoxin to the overall increase in amount of biologically active endotoxin after antibiotic treatment is more difficult to study. Nevertheless, it is clear that effective antibiotic treatment reduces bacterial numbers, but also results in an increase in the amount of bioactive endotoxin.

ANIMAL STUDIES

Antibiotic-induced endotoxin release has been studied in various models of experimental infection in animals. In these studies a decrease in bacterial counts concurrent with an increase in endotoxin level was demonstrated in antibiotic-treated animals. This was demonstrated in experimental sepsis with *E. coli* (10,11,28,29) or *H. influenzae* (19) and in meningitis with *E. coli* (30,31) or *H. influenzae* type b (32,33). In three studies on *E. coli* sepsis in rats or dogs, no differences in endotoxin levels between control and antibiotic-treated animals were observed (34–36). However, in these studies the endotoxin:CFU ratio increased in the antibiotic-treated animals due to a sharp decrease in bacterial counts.

More important than endotoxin levels are changes in parameters of infection after the start of antibiotic treatment. In a number of these animal studies the higher endotoxin levels in the antibiotic-treated animals were associated with a deterioration of hemodynamic parameters (28,29) or, in the case of meningitis, more brain edema (31) and higher leukocyte counts and protein and lactate levels in CSF (30,32,33). Interestingly, the co-administration of the endotoxin-neutralizing agent rBPI$_{21}$ prevented not only the increase in free endotoxin levels in plasma, but also the deterioration of hemodynamic parameters (29).

These adverse effects are most pronounced early after the start of treatment: after a few hours, (free) endotoxin and TNF fall to low levels, while they continue to rise in untreated animals (30,32,33). In summary, the animal data are in agreement with the data from the in vitro studies. Several studies have demonstrated that endotoxin release is parallelled by a deterioration in the parameters of infection.

CLINICAL STUDIES

It was already known in the early days of antibiotic treatment that fatal vasomotor collapse could occur after a loading dose of chloramphenicol for typhoid fever (2). These and other adverse clinical effects associated with destruction of the typhoid bacilli were explained by an overwhelming release of endotoxin due to the antibiotic killing of the bacteria. However, this hypothesis could not be proven, as endotoxin measurements in patients were not possible at that time.

In two recent studies endotoxin levels were serially assessed in patients with suspected gram-negative sepsis before and after institution of antibiotic treatment (3,37). In the first study, in the majority of patients with blood cultures positive for gram-negative microorganisms a 2- to 50-fold increase in free endotoxin level followed administration of antibiotics (37). In a comparable study, a 2- to 15-fold increase in total endotoxin level after the start of treatment was found in 3 out of 10 patients with detectable endotoxemia on admission (3), with a concomitant decrease in blood pressure and rise in serum lactate levels. Both studies also reported that endotoxin shifted from the cell-bound to the free form after antibiotic treatment. In another study children with *H. influenzae* meningitis treated with ceftriaxone were followed (9). In 8 patients in whom a second lumbar puncture was performed, a sharp decrease in bacterial numbers was noted, with a decrease in the total endotoxin level and a shift to and increase in the free endotoxin level. The increase in free endotoxin level strongly correlated both with the initial CSF bacterial concentration as well as with the decline in bacterial count after treatment (Fig. 1). The increase in the free endotoxin level was parallelled by an increase in mean CSF lactate and LDH levels and a decrease in CSF glucose level, which suggested an enhanced inflammatory response. Antibiotic-induced endotoxin release was finally studied in chronically bacteriuric patients (12). These patients received a low dose of various antibiotics, which resulted in an average 0.93 $\log_{10}$ decrease in CFU, with an average 0.59 $\log_{10}$ increase in (total) endotoxin levels in urine. Adverse clinical effects were not reported. In nontreated patients, the bacterial counts and endotoxin levels remained unchanged during the study period. These observations in humans suggest that treatment with antibiotics may cause systemic or local endotoxin liberation. Whether this effect is related to an adverse clinical outcome remains unproven.

ANTIBIOTIC POTENTIAL FOR ENDOTOXIN LIBERATION

The in vitro, animal, and clinical studies discussed support the hypothesis that endotoxin is liberated during antibiotic treatment of gram-negative infections. However, despite this potential to cause endotoxin release, appropriate antibiotic treatment is known to reduce mortality in sepsis (1). Therefore, the critical question is whether equally effective antibiotics have a differential effect on endotoxin release and clinical outcome.

This question has been studied most extensively in vitro (see Table 1). In the first place, it is of interest to compare bactericidal and bacteriostatic antibiotics. In two studies bactericidal and (effective) bacteriostatic antibiotics were compared (15,19). In the first study ampicillin and streptomycin caused more killing of *E. coli* and more endotoxin release than tetracycline (15), and in *H. influenzae* ampicillin and chloramphenicol treatment were comparable with regard to both bacterial killing and endotoxin release (19). However, many arguments against the use of these bacteriostatic antibiotics in the treatment of severe gram-negative infections are more important than a possible lower amount of endotoxin released during treatment.

In a number of studies bactericidal antibiotics were compared during the in vitro killing of *E. coli* (3,5,13,14,17,18,38). In these studies treatment with aminoglycosides or imipenem resulted in less endotoxin release than treatment with quinolones, whereas the highest levels of endotoxin were recorded after treatment with cephalosporins and aztreonam. The bactericidal potential of the regimens were in all cases comparable. Addition of tobramycin to cefuroxime resulted in a significant decrease in endotoxin liberation in comparison to cefuroxime monotherapy (3,39). This decrease was also found when amikacin was added to ampicillin or cefotaxime monotherapy (40). In addition to differences in absolute amount of endotoxin release, the time pattern of endotoxin release can differ between

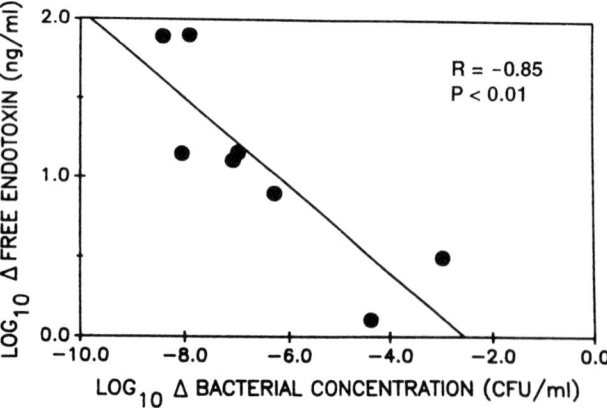

Fig. 1 Correlation between the change in CSF bacterial concentration and the change in CSF free endotoxin levels 2–6 hours after ceftriaxone in eight patients. (From Ref. 9.)

antibiotics. For instance, imipenem shows a fast killing (38,41,42) and an early peak in endotoxin release (38,41,42), whereas ceftazidime or ceftriaxone show a slower killing and a later peak in endotoxin release. Also in *P. aeruginosa* ceftazidime releases up to 10-fold more endotoxin than imipenem (20,38,42–44).

Scanning electron microscopy revealed large differences in bacterial morphology following treatment with various antibiotics (17). At a concentration of 50 times the MIC, treatment with ceftazidime resulted in the formation of long (≤0.1 mm) filaments. Treatment with ciprofloxacin also resulted in elongation of the bacteria, although it was less conspicuous than that after treatment with ceftazidime. In contrast, treatment with imipenem resulted in conversion into smaller, somewhat round bacteria, and gentamicin induced no remarkable morphological changes during treatment. These morphological differences were also reported in a number of other studies (3,13,18,20). In several animal studies it was proven that this filament formation due to ceftazidime, ceftriaxone, or aztreonam treatment can also occur in vivo (35,45,46).

In order to study how these differences would affect the (human) host response, the TNF-α release by isolated mononuclear cells or whole blood ex vivo or the IL-6 production by human umbilical vein endothelial cells in response to (filtrates of) antibiotic-treated cultures was measured (17,26,27,39,41) (Fig. 2). In all of these studies cultures treated with cephalosporins or aztreonam induced more TNF-α or IL-6 as compared to cultures treated with aminoglycosides or imipenem. Treatment with ciprofloxacin gave intermediate levels. High concentrations of cephalosporins resulted in less TNF production than low concentrations (39). Imipenem also induced significantly lower levels of nitric oxide production by isolated mouse peritoneal cells than ceftazidime (43). LPS appears to be the predominant proinflammatory mediator in the supernatants of the antibiotic-treated bacteria (25) as the addition of polymyxin B or BPI (which bind endotoxin) inhibited TNF generation (17,26,41). Differences in cytokine levels are the result of differences in the amount of endotoxin generated by antibiotics during bacterial killing.

BACKGROUND OF ENDOTOXIN RELEASE

Significant differences have been demonstrated between antibiotics with respect to their endotoxin-liberating potential. These differences are related to the modes of antibacterial activity.

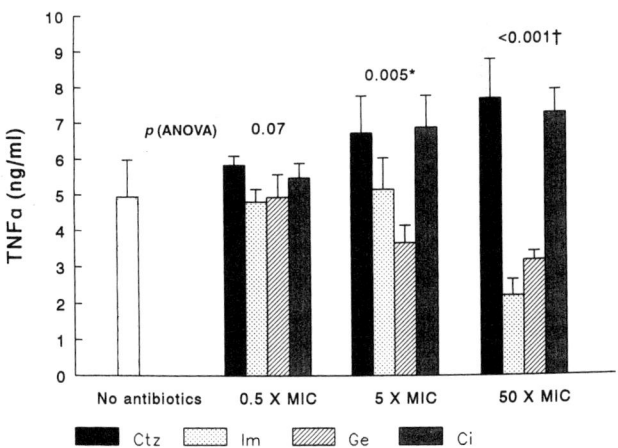

Fig. 2 TNF-α level after 4 hours of incubation of *E. coli* O4:K2 in whole blood. Ceftazidime (Ctz), imipenem (Im), gentamicin (Ge), and ciprofloxacin (Ci) were added in a final concentration of 0.5, 5, or 50 times the MIC. Treatment groups were compared by one-way ANOVA. Multiple comparisons between groups were performed by the Newman-Keuls test. *$p < 0.05$ (Newman-Keuls): Ctz and Ci vs. Ge. †$p < 0.05$ (Newman-Keuls): Ctz and Ci vs. Im and Ge. (From Ref. 17.)

Penicillin-binding proteins (PBPs) are the primary biochemical targets of β-lactam antibiotics in bacteria. These PBPs catalyze terminal stages in the assembly of the peptidoglycan network of the bacterial cell wall (47). Whereas the older penicillins (penicillin G) aspecifically bind to all of these PBPs, the newer β-lactams often specifically bind to only one or two of the PBPs. Treatment of Enterobacteriaceae with β-lactam antibiotics that have a high selective affinity for PBP 1a and especially 1b causes rapid and extensive killing of the bacteria, with degradation of cell wall material and cellular lysis. Antibiotics with selective affinity for PBP 2 cause conversion of the bacilli to round-shaped cells (also called spheroplasts). Inhibitors of PBP 3 cause selective inhibition of bacterial septation, which leads to the formation of long filaments, but initially only limited bactericidal activity and lysis takes place (47).

Examples of antibiotics selective for PBP 1 (of *E. coli*) are cefazolin, cefoxitin, cephalothin, cephaloridine, and cefsulodin (47–50). PBP 2–selective antibiotics include mecillinam, clavulanic acid, and imipenem. PBP 3–selective agents are piperacillin, mezlocillin, moxalactam, and aztreonam and, at low concentrations, cephalexin, cefuroxime, cefotaxime, and ceftazidime. At higher concentrations these cephalosporins also have affinities for PBP 1a (49,50). Antibiotics that target PBP 1a and 3 induce filamentous changes as well as killing without lysis (51,52), depending on the concentration used, which may explain the different effects of low and high concentrations of cephalosporins (39,42,51).

The higher amount of endotoxin released in PBP 3–specific β-lactam antibiotics presumably results from the increased cell mass and the ongoing production of LPS during continued bacterial growth due to filamentation. The lower endotoxin release resulting from treatment with PBP 1– or 2–specific antibiotics may be explained by their rapid bacterial action, which prevents an increase in total cell mass (3,13,17,20).

The effects of treatment with ciprofloxacin and other quinolones are very similar to the effects of PBP 3–specific antibiotics, i.e., filamentation with an increase in (nonviable) bacterial biomass and an important increase in endotoxin production (13,17,18). Aminoglycoside treatment results in bacterial growth arrest and loss of viability in the absence of lysis (13,17,18). It is likely that the low amount of endotoxin bioactivity induced by the aminoglycosides is partially explained by their ability to bind endotoxin (53,54).

DIFFERENCES IN ANTIBIOTICS IN ENDOTOXIN RELEASE: ANIMAL AND CLINICAL STUDIES

Results from animal studies that addressed the overall outcome and the cytokine levels following treatment with a low-endotoxin–releasing as compared to treatment with a high-endotoxin–releasing antibiotic are conflicting. In experimental *E. coli* bacteremia in rabbits, treatment with moxalactam and gentamicin resulted in equal killing of bacteria, but there was a 7- to 20-fold higher release of free endotoxin 1–4 hours after the start of moxalactam treatment as compared to gentamicin treatment. This was not associated with differences in survival (11). In *E. coli* meningitis in rabbits, differences between antibiotic-induced CSF endotoxin levels 2 hours after the start of treatment did not result in differences in CSF white blood cell count or TNF-α levels (30). In a comparable study, cefotaxime treatment induced equal bacterial killing, but up to 10-fold more endotoxin was released as compared to chloramphenicol treatment after 3 hours of therapy, and significantly more brain edema was observed in the cefotaxime-treated animals (31).

In rabbits with an experimental *K. pneumoniae* endocarditis treatment with ceftriaxone resulted in significant higher serum TNF levels as compared to treatment with imipenem or gentamicin (46). Differential endotoxin and TNF generation was also demonstrated by Opal et al. in a rat model of *E. coli* or *P. aeruginosa* sepsis (45). They demonstrated higher endotoxin and TNF levels after ceftazidime than after imipenem treatment, especially during the first 4 hours of treatment. However, differences between these antibiotics in endotoxin release were not found with *K. pneumoniae* as the challenge organism.

Finally, Morrison et al. (55) used a mouse model of gram-negative sepsis in which CF-1 mice were rendered hypersusceptible to the effects of endotoxin by treatment with D-galactosamine. Although imipenem and ceftazidime showed an equivalent MIC for the challenge strain of *E. coli*, imipenem offered a significantly better protective efficacy from lethality than ceftazidime. As the administration of an anti-LPS antibody significantly improved survival as compared to antibiotic treatment alone, LPS apparently participates as mediator of mortality in this model. Since the bactericidal potential was comparable, the difference between the two antibiotics in protective efficacy was attributed to differences in amount of endotoxin liberated from the bacteria during antibiotic activity. Comparable results were demonstrated for *P. aeruginosa* sepsis (55).

In only two clinical studies the endotoxin-liberating abilities of antibiotics were compared. No differences were found in the potential of ticarcillin, cephalothin, ceftazidime, gentamicin, or ciprofloxacin to induce endotoxin-release in the urine of chronically bacteriuric patients (12). However, these patients had no clinically apparent infection, and bacterial growth rate was therefore probably lower than is the case during serious infections.

We compared imipenem, which has a low endotoxin-liberating potential in vitro, with ceftazidime, which has a high endotoxin-liberating potential, in the treatment of gram-negative urosepsis in humans (56). Fifteen patients in each group completed 72 hours of therapy. The two groups were well balanced with regard to all relevant baseline variables. Baseline plasma endotoxin levels were above the detection limit in 3 of 15 imipenem-treated patients and in 4 of 15 ceftazidime-treated patients. Two of the four endotoxemic patients on ceftazidime had a rise in blood endotoxin level after 4 hours of treatment, whereas this level decreased in all endotoxemic imipenem-treated patients. The endotoxin level in the urine decreased (nonsignificantly) less after 4 hours of treatment with ceftazidime. In these patients a slower defervescence was also noticed as well as a 10–40% increase in serum and urine cytokine levels 4 hours after the start of treatment as compared to no increase in cytokine levels in the imipenem-treated patients ($p > 0.05$). There were no relevant differences between the two antibiotics in cytokine levels later in treatment (Fig. 3).

Endotoxin release during antibiotic killing in vitro was assessed for all microorganisms isolated in these patients. The mean endotoxin concentrations after 4 hours of incubation were 10-fold higher after incubation with ceftazidime as compared to the level obtained with imipenem (both in a concentration of 50 times the MIC). This in vitro difference between the two antibiotics was observed for all bacterial species (44).

Several factors may explain why the 10-fold difference found in endotoxin release in vitro did not result in a comparable difference in cytokine levels in patients. First, at higher concentrations ceftazidime also binds penicillin-binding protein 1a, which prevents filament formation (50). Indeed, the differences in endotoxin release in vitro were small in the clinical isolates with the highest MIC, and therefore the highest concentration ceftazidime added, as compared to the differences in the isolates exposed to lower ceftazidime concentrations. As ceftazidime is excreted through the kidney, high concentrations may be achieved in urine. Second, even in in vitro experiments, endotoxin concentration and TNF-α release are correlated in a semi-logarithmic fashion (57), and a 10-fold increase in endotoxin concentration results in only a doubling of TNF-α levels. Finally, in septic conditions, the production of proinflammatory cytokines has been shown to be downregulated (58,59). As a consequence, even considerable increases in endotoxin levels may lead to only modest corresponding increases in cytokine levels.

Nevertheless, the results of this study indicated that differences between antibiotics in endotoxin release may indeed influence the speed of defervescence and may affect the inflammatory response in sepsis early after the start of treatment.

ANTIBIOTIC-INDUCED ENDOTOXIN RELEASE IN MENINGOCOCCAL INFECTIONS

In patients with severe meningococcal infections endotoxin levels are considerably higher than the levels measured in septic shock caused by gram-negative rods (60), so it is conceivable that especially during this infection any additional endotoxin release due to the antibiotic treatment might be harmful. In two in vitro studies in which meningococci were exposed to penicillin in vitro for 1–2 hours, an increase in filtrable endotoxin was measured (8,61), whereas the total endotoxin level was comparable to the level in untreated cultures (61). This increase in free endotoxin level did not occur after treatment with chloramphenicol (8,61). After 6 hours of treatment, free endotoxin levels were comparable for penicillin, chloramphenicol, or untreated cultures (61). The question is how this influences the human host response. We therefore measured bacterial killing and endotoxin levels in culture medium and cytokine production in whole blood ex vivo during antibiotic killing of meningococci (62). We observed that bacterial killing in vitro was more efficient with penicillin or ceftriaxone than with chloramphenicol. Free endotoxin levels were significantly lower after exposure to antibiotics as compared with no treatment, and endotoxin levels were lowest after exposure to chlorampheniol. In three out of four meningococcal strains studied, exposure to antibiotics resulted in considerably lower cytokine levels as compared to no treatment, and TNF-α levels were significantly lower after exposure to penicillin or ceftriaxone than after chloramphenicol treatment. In the fourth strain cytokine levels were comparable for untreated and treated cultures. We therefore concluded that the efficacy of bacterial killing was a more important determinant of cytokine production than the level of free endotoxin. In only one human study were total endotoxin levels serially assessed after initiation of antibiotic treatment (60). A decrease in total endotoxin levels was observed in all patients, although in this study data were collected from the surviving patients only.

In conclusion, fear of endotoxin release caused by effective antibiotics is not justified in the treatment of severe meningococcal infections. On the contrary, effective bacterial killing must be the main goal of therapy in order to reduce the production of proinflammatory cytokines.

CLINICAL SIGNIFICANCE OF ANTIBIOTIC-INDUCED ENDOTOXIN RELEASE

In conclusion, many in vitro, animal, and clinical studies have shown that the endotoxin concentration in blood, urine, and CSF may increase following antibiotic treatment of gram-negative infections. In several animal and clinical studies, endotoxin release paralleled deterioration of parameters of illness. Differences between antibiotics in this respect have clearly been demonstrated in vitro. In general, imipenem and the aminoglycosides have only a modest endotoxin-liberating ability, whereas in most studies quinolones, and espe-

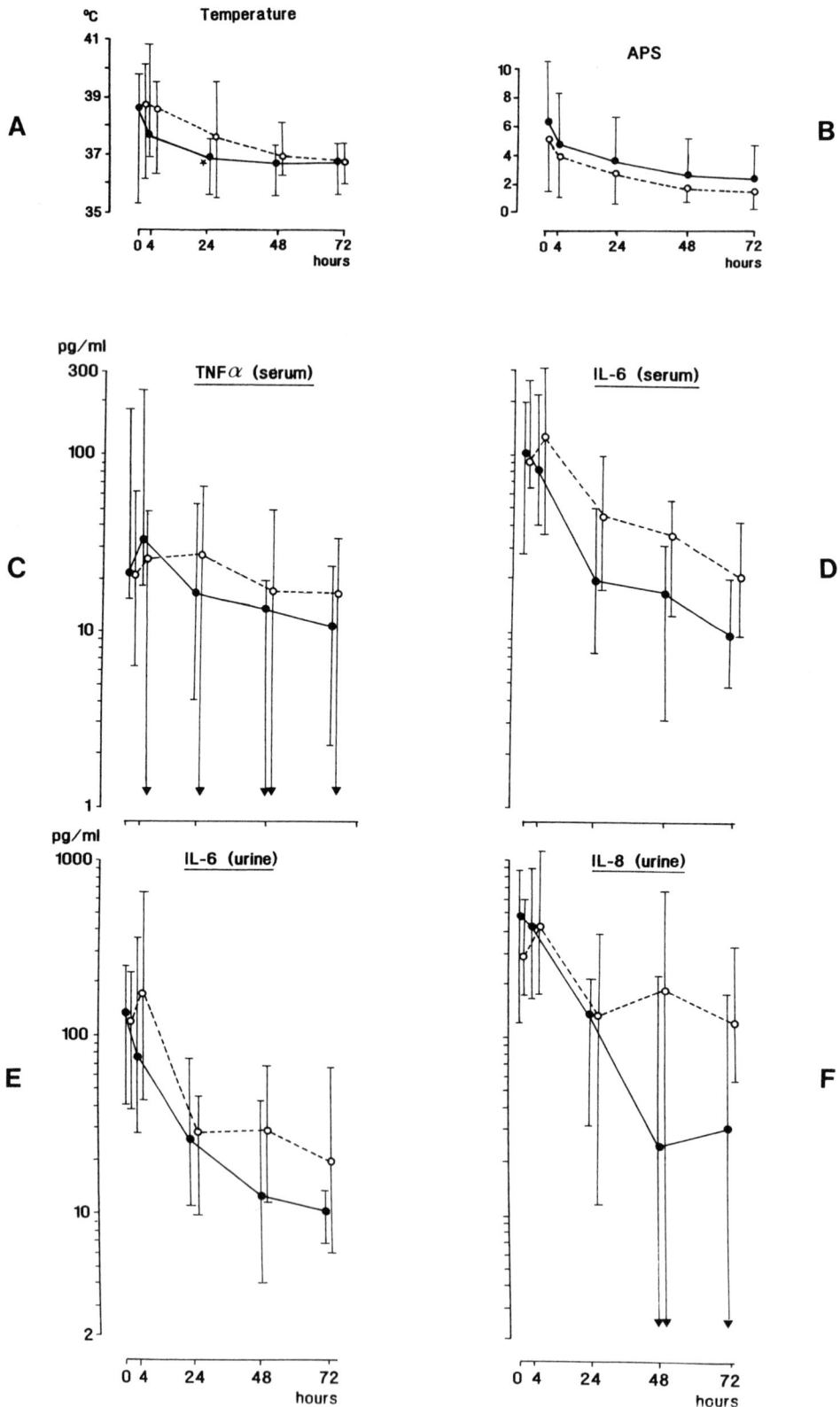

Fig. 3 Body temperature (median; range) (A), acute physiology score (APS) (mean ± SD) (B), and levels (median; 25–75th percentile) of TNF-α and IL-6 in serum (C,D) and IL-6 and IL-8 in urine (E,F) during treatment of gram-negative urosepsis. (●–●–●) Imipenem; (○–○–○) ceftazidime. Difference between the two arms (ANOVA for repeated measures): $p > 0.05$ in all cases. * $p = 0.02$ (Mann-Whitney test). (From Ref. 56.)

cially cephalosporins, cause relatively large quantities of endotoxin to be released. These differences are related to the different modes of activity of these antibiotics. In a number of animal studies, these differences between antibiotics in endotoxin release resulted in differences in morbidity and/or mortality. In the only study in which antibiotics were compared in this respect in a group of septic patients, differences between antibiotics in endotoxin release resulted in differences, albeit small, in the inflammatory response during treatment.

Should the endotoxin-liberating potential of a given antibiotic be a major determinant in the choice of the antibiotic in case of a serious gram-negative infection? It is clear that the clinical studies discussed do not justify that conclusion. In the mentioned comparative study in patients with urosepsis, all patients treated had a favorable outcome. However, urosepsis is a relatively mild form of sepsis, and therefore recovery was expected in all patients provided that an adequate antibiotic treatment was given. It is conceivable that in more severe forms of septic shock, any additional endotoxin release and/or cytokine production following antibiotic treatment might influence survival.

Conclusive clinical studies that will answer the question whether differences in antibiotic-induced endotoxin release are clinically relevant are difficult to design (63). A major problem is the definition of relevant endpoints. Application of the LAL assay to detect antibiotic-induced endotoxin release may not always be appropriate. It does not distinguish "free" from membrane-bound (and perhaps less biologically active) endotoxin, and endotoxin neutralized by HDL will also be detected in the LAL assay. Whether circulating proinflammatory cytokines would serve as appropriate endpoints is also questionable. Cytokines may be released intermittently, and many cytokines have a very short half-life. In addition, the interpatient variation in cytokine levels is usually markedly, making statistical comparison of groups difficult. Clinical endpoints like time to defervescence and hemodynamic parameters are probably more appropriate.

Another important difficulty in the design of such studies is the fact that mixed gram-positive and gram-negative infections are common in critically ill patients. Therefore, pending the results of blood cultures, monotherapy with imipenem would be possible, but if ceftazidime or quinolones were given, additional coverage for gram-positive organisms and anaerobes would be necessary. This could substantially influence endotoxin release by these antibiotics. Second, these very ill patients are usually treated in the intensive care unit, and therefore secondary infections like ventilator-associated pneumonia and catheter sepsis are likely to occur. These infections make evaluation of the effects of the antibiotic treatment of the primary infection very difficult. Therefore, mainly studies including patients with mild forms of sepsis, like urinary tract infections, are feasible at the moment, and the importance of differences in endotoxin-liberating potential has to be established in animal studies and these mild forms of sepsis in humans. As the number of these studies is still limited, and the results not unequivocal, a final conclusion about the clinical relevance of antibiotic-induced endotoxin release cannot be drawn yet. Given the potential clinical consequences, the number of these studies is likely to expand during the coming years.

REFERENCES

1. Kreger BE, Craven DE, McCabe WR. Gram-negative bacteremia. IV. Re-evaluation of clinical features and treatment in 612 patients. Am J Med 1980; 68:344–355.
2. Patel JC, Banker DD, Modi CJ. Chloramphenicol in typhoid fever. A preliminary report of clinical trial in 6 cases. Br Med J 1949; ii:908–909.
3. Dofferhoff ASM, Nijland JH, de Vries-Hospers HG, Mulder POM, Weits J, Bom VJJ. Effects of different types and combinations of antimicrobial agents on endotoxin release from gram-negative bacteria: an in-vitro and in-vivo study. Scand J Infect Dis 1991; 23:745–754.
4. Mattsby-Baltzer I, Lindgren K, Lindholm B, Edebo L. Endotoxin shedding by enterobacteria: free and cell-bound endotoxin differ in Limulus activity. Infect Immun 1991; 59:689–695.
5. Eng RHK, Smith SM, Fan-Havard P, Ogbara T. Effect of antibiotics on endotoxin release from gram-negative bacteria. Diagn Microbiol Infect Dis 1993; 16:185–189.
6. Cadieux JE, Kuzio J, Milazzo FH, Kropinski AM. Spontaneous release of lipopolysaccharide by *Pseudomonas aeruginosa*. J Bacteriol 1983; 155:817–825.
7. Andersen BM, Solberg O, Bryn K, et al. Endotoxin liberation from *Neisseria meningitidis* isolated from carriers and clinical cases. Scand J Infect Dis 1987; 19:409–419.
8. Mellado MC, Rodríguez-Contreras R, Mariscal A, Luna JD, Delgado Rodríguez M, Galvez-Vargas R. Effect of penicillin and chloramphenicol on the growth and endotoxin release by *N. meningitidis*. Epidemiol Infect 1991; 106:283–288.
9. Arditi M, Ables L, Yogev R. Cerebrospinal fluid endotoxin levels in children with *H. influenzae* meningitis before and after administration of intravenous ceftriaxone. J Infect Dis 1989; 160:1005–1011.
10. Shenep JL, Mogan KA. Kinetics of endotoxin release during antibiotic therapy for experimental gram-negative bacterial sepsis. J Infect Dis 1984; 150:380–388.

11. Shenep JL, Barton RP, Mogan KA. Role of antibiotic class in the rate of liberation of endotoxin during therapy for experimental gram-negative bacterial sepsis. J Infect Dis 1985; 151:1012–1018.
12. Hurley JC, Louis WJ, Tosolini FA, Carlin JB. Antibiotic-induced release of endotoxin in chronically bacteriuric patients. Antimicrob Agents Chemother 1991; 35: 2388–2394.
13. Van den Berg C, De Neeling AJ, Schot CS, Hustinx WNM, Wemer J, De Wildt DJ. Delayed antibiotic-induced lysis of *Escherichia coli* in vitro is correlated with enhancement of LPS release. Scand J Infect Dis 1992; 24:619–627.
14. Cohen J, McConnell JS. Release of endotoxin from bacteria exposed to ciprofloxacin and its prevention with polymyxin B. Eur J Clin Microbiol 1986; 5:13–17.
15. Goto H, Nakamura S. Liberation of endotoxin from *Escherichia coli* by addition of antibiotics. Jpn J Exp Med 1980; 50:35–43.
16. McConnell JS, Cohen J. Release of endotoxin from *Escherichia coli* by quinolones. J Antimicrob Chemother 1986; 18:765–773.
17. Prins JM, Kuijper EJ, Mevissen MLCM, Speelman P, van Deventer SJH. Release of tumor necrosis factor alpha and interleukin 6 during antibiotic killing of *Escherichia coli* in whole blood: influence of antibiotic class, antibiotic concentration and presence of septic serum. Infect Immun 1995; 63:2236–2242.
18. Crosby HA, Bion JF, Penn CW, Elliott TSJ. Antibiotic-induced release of endotoxin from bacteria in vitro. J Med Microbiol 1994; 40:23–30.
19. Walterspiel JW, Kaplan SL, Mason Jr EO. Protective effect of subinhibitory polymyxin B alone and in combination with ampicillin for overwhelming *Haemophilus influenzae* type B infection in the infant rat: evidence for in vivo and in vitro release of free endotoxin after ampicillin treatment. Pediatr Res 1986; 20:237–241.
20. Jackson JJ, Kropp H. β-Lactam antibiotic-induced release of free endotoxin: in vitro comparison of penicillin-binding protein (PBP) 2-specific imipenem and PBP 3-specific ceftazidime. J Infect Dis 1991; 165:1033–1041.
21. Nelson D, Delahooke TES, Poxton IR. Influence of subinhibitory levels of antibiotics on expression of *Escherichia coli* lipopolysaccharide and binding of anti-lipopolysaccharide monoclonal antibodies. J Med Microbiol 1993; 39:100–106.
22. Burroughs M, Cabellos C, Prasad S, Tuomanen E. Bacterial components and the pathophysiology of injury to the blood-brain barrier: Does cell wall add to the effects of endotoxin in gram-negative meningitis? J Infect Dis 1992; 165(suppl 1):S82–S85.
23. Mattson E, van Dijk H, Verhoef J, Norrby R, Rollof J. Supernatants from *Staphylococcus epidermidis* grown in the presence of different antibiotics induce differential release of tumor necrosis factor alpha from human monocytes. Infect Immun 1996; 64:4351–4355.
24. Held TK, Adamczik C, Trautmann M, Cross AS. Effects of MICs and sub-MICs of antibiotics on production of capsular polysaccharide of *Klebsiella pneumoniae*. Antimicrob Agents Chemother 1995; 39:1093–1096.
25. Leeson MC, Fujihara Y, Morrison DC. Evidence for lipopolysaccharide as the predominant proinflammatory mediator in supernatants of antibiotic-treated bacteria. Infect Immun 1994; 62:4975–4980.
26. Arditi M, Kabat W, Yogev R. Antibiotic-induced bacterial killing stimulates tumor necrosis factor-α release in whole blood. J Infect Dis 1993; 167:240–244.
27. Simon DM, Koenig G, Trenholme GM. Differences in release of tumor necrosis factor from THP-1 cells stimulated by filtrates of antibiotic-killed *Escherichia coli*. J Infect Dis 1991; 164:800–802.
28. Røkke O, Revhaug A, Østerud B, Giercksky KE. Increased plasma levels of endotoxin and corresponding changes in circulatory performance in a porcine sepsis model: the effect of antibiotic administration. Progr Clin Biol Res 1988; 272:247–262.
29. Lin Y, Leach WJ, Ammons WS. Synergistic effect of a recombinant N-terminal fragment of bactericidal/permeability-increasing protein and cefamandole in treatment of rabbit gram-negative sepsis. Antimicrob Agents Chemother 1996; 40:65–69.
30. Friedland IR, Jafari H, Ehrett S, et al. Comparison of endotoxin release by different antimicrobial agents and the effect on inflammation in experimental *Escherichia coli* meningitis. J Infect Dis 1993; 168:657–662.
31. Täuber MG, Shibl AM, Hackbarth CJ, Larrick JW, Sande MA. Antibiotic therapy, endotoxin concentration in cerebrospinal fluid, and brain edema in experimental *Escherichia coli* meningitis in rabbits. J Infect Dis 1987; 156:456–462.
32. Mertsola J, Ramilo O, Mustafa MM, Sáez-Llorens X, Hansen EJ, McCracken Jr GH. Release of endotoxin after antibiotic treatment of gram-negative meningitis. Pediatr Infect Dis J 1989; 8:904–906.
33. Mustafa MM, Ramilo O, Mertsola J, et al. Modulation of inflammation and cachectin activity in relation to treatment of experimental *Hemophilus influenzae* type b meningitis. J Infect Dis 1989; 160:818–825.
34. Almdahl SM, Østerud B. Effect of antibiotics on gram-negative sepsis in the rat. Acta Chir Scand 1987; 153: 283–286.
35. Dofferhoff ASM, Potthoff H, Bom VJJ, et al. The release of endotoxin, TNF and IL-6 during the antibiotic treatment of experimental gram-negative sepsis. J Endotox Res 1995; 2:37–44.
36. Natanson C, Danner RL, Reilly JM, et al. Antibiotics versus cardiovascular support in a canine model of human septic shock. Am J Physiol 1990; 259:H1440–1447.
37. Shenep JL, Flynn PM, Barrett FF, Stidham GL, Westenkirchner DF. Serial quantitation of endotoxemia and bacteremia during therapy for gram-negative bacterial sepsis. J Infect Dis 1988; 157:565–568.
38. Lamp KC, Rybak MJ, McGrath BJ, Summers KK. Influence of antibiotic and E5 monoclonal immunoglobulin M interactions on endotoxin release from *Escherichia coli* and *Pseudomonas aeruginosa*. Antimicrob Agents Chemother 1996; 40:247–252.

39. Dofferhoff ASM, Esselink MT, de Vries-Hospers HG, et al. The release of endotoxin from antibiotic-treated *Escherichia coli* and the production of tumour necrosis factor by human monocytes. J Antimicrob Chemother 1993; 31:373–384.
40. Bingen E, Goury V, Bennani H, Lambert-Zechovsky N, Aujard Y, Darbord JC. Bactericidal activity of β-lactams and amikacin against *Haemophilus influenzae*: effect on endotoxin release. J Antimicrob Chemother 1992; 30:165–172.
41. Arditi M, Zhou J. Differential antibiotic-induced endotoxin release and interleukin-6 production by human umbilical vein endothelial cells (HUVECs): amplification of the response by coincubation of HUVECs and blood cells. J Infect Dis 1997; 175:1255–1258.
42. Dofferhoff ASM, Buys J. The influence of antibiotic-induced filament formation on the release of endotoxin from gram-negative bacteria. J Endotox Res 1996; 3:187–194.
43. Yokochi T, Kusumi A, Kido N, et al. Differential release of smooth-type lipopolysaccharide from *Pseudomonas aeruginosa* treated with carbapenem antibiotics and its relation to production of tumor necrosis factor alpha and nitric oxide. Antimicrob Agents Chemother 1996; 40:2410–2412.
44. Prins JM. Antibiotic induced release of endotoxin—clinical data and human studies. J Endotox Res 1996; 3:269–273.
45. Opal SM, Horn DL, Palardy JE, et al. The in vivo significance of antibiotic-induced endotoxin release in experimental gram-negative sepsis. J Endotox Res 1996; 3:245–252.
46. Mohler J, Fantin B, Mainardi JL, Carbon C. Influence of antimicrobial therapy on kinetics of tumor necrosis factor levels in experimental endocarditis caused by *Klebsiella pneumoniae*. Antimicrob Agents Chemother 1994; 38:1017–1022.
47. Livermore DM, Williams JD. β-Lactams: mode of action and mechanisms of bacterial resistance. In: Lorian V. Antibiotics in Laboratory Medicine. 4th ed. Baltimore: Williams and Wilkins, 1996:502–578.
48. Tomasz A. Penicillin-binding proteins and the antibacterial effectiveness of β-lactam antibiotics. Rev Infect Dis 1986; 8(suppl 3):S260–S278.
49. Neu HC. Penicillin-binding proteins and role of amdinocillin in causing bacterial cell death. Am J Med 1983; 75(suppl 2A):9–20.
50. Neu HC. Relation of structural properties of beta-lactam antibiotics to antibacterial activity. Am J Med 1985; 79(suppl 2A):2–13.
51. Hanberger H, Nilsson LE, Kihlström E, Maller R. Postantibiotic effect of β-lactam antibiotics on *Escherichia coli* evaluated by bioluminescence assay of bacterial ATP. Antimicrob Agents Chemother 1990; 34:102–106.
52. Tuomanen E, Gilbert K, Tomasz A. Modulation of bacteriolysis by cooperative effects of penicillin-binding proteins 1a and 3 in *Escherichia coli*. Antimicrob Agents Chemother 1986; 30:659–663.
53. Artenstein AW, Cross AS. Inhibition of endotoxin reactivity by aminoglycosides. J Antimicrob Chemother 1989; 24:826–828.
54. Focà A, Matera G, Iannello D, Berlinghieri MC, Liberto MC. Aminoglycosides modify the in vitro metachromatic reaction and murine generalized Shwartzman phenomenon induced by *Salmonella minnesota* R595 lipopolysaccharide. Antimicrob Agents Chemother 1991; 35:2161–2164.
55. Bucklin SE, Morrison DC. Differences in therapeutic efficacy among cell wall-active antibiotics in a mouse model of gram-negative sepsis. J Infect Dis 1995; 172:1519–1527.
56. Prins JM, van Agtmael MA, Kuijper EJ, van Deventer SJH, Speelman P. Antibiotic-induced endotoxin release in patients with gram-negative urosepsis: a double-blind study comparing imipenem and ceftazidime. J Infect Dis 1995; 172:886–891.
57. Bruin KF, von der Möhlen MAM, Derkx BHF, Jansen J, ten Cate JW, van Deventer SJH. Characterization of the endotoxin-induced TNF release in whole blood and peripheral blood mononuclear cells. In: Bruin KF, ed. Endotoxin Responsiveness in Humans. Ph.D. thesis, University of Amsterdam, 1994:31–42.
58. Munoz C, Carlet J, Fitting C, Misset B, Blériot JP, Cavaillon JM. Dysregulation of in vitro cytokine production by monocytes during sepsis. J Clin Invest 1991; 88:1747–1754.
59. Mengozzi M, Ghezzi P. Cytokine down-regulation in endotoxin tolerance. Eur Cytokine Netw 1993; 4:89–98.
60. Brandtzaeg P, Kierulf P, Gaustad P, Skulberg A, Bruun JN, Halvorsen S, Sørensen E. Plasma endotoxin as a predictor of multiple organ failure and death in systemic meningococcal disease. J Infect Dis 1989; 159:195–204.
61. Andersen BM, Solberg O. The endotoxin-liberating effect of antibiotics on meningococci in vitro. Acta Path Microbiol Scand Sect B. 1980; 88:231–236.
62. Prins JM, Speelman P, Kuijper EJ, Dankert J, van Deventer SJH. No increase in endotoxin release during antibiotic killing of meningococci. J Antimicrob Chemother 1997; 39:13–18.
63. Workshop summary. Assessment of antibiotic-mediated endotoxin release. J Endotox Res 1996; 3:275–279.
64. Prins JM, et al. Clinical relevance of antibiotic-induced endotoxin release. Antimicrob Agents Chemother 1994; 38:1211–1218.

62

Lipopolysaccharide from Oral Bacteria: Role in Innate Host Defense and Chronic Inflammatory Disease

Brian W. Bainbridge and Richard P. Darveau
University of Washington, School of Dentistry, Seattle, Washington

INTRODUCTION

Dental plaque is a microbial biofilm that consists of more than 300 bacterial species (1). Biofilms are defined as "matrix-enclosed bacterial populations adherent to each other and/or to surfaces or interfaces" (2). They are notoriously resistant to host-defense mechanisms (3). Perhaps the single most significant ramification of biofilm formation on the tooth surface is that bacteria continuously release cell surface components into the oral cavity and gingival sulcus (Fig. 1). It is these components that the host uses to govern its response to bacterial infection in close physical proximity to the highly vascularized periodontal tissue surrounding the tooth root surfaces. It is now clear that the host employs a mechanism of low-level inflammation in clinically healthy tissue to "arm" itself against potential tissue infection. The well-characterized cellular infiltrate observed in healthy tissue (4) is consistent with the more recently described demonstration of IL-8 (5) and E selectin (6–8) expression in clinically normal tissue (Fig. 2). It is likely that antigens shed from the dental plaque biofilm are involved in the expression of these mediators. However, little is known about what bacterial antigens are involved in the maintenance of clinically healthy tissue.

In contrast, there is a clear association between an increase in the number of gram-negative bacteria that can be isolated from dental plaque samples and disease. The total number of bacteria that can be isolated from clinically healthy sites is usually between 10^2–10^3 bacteria, of which approximately 15% are gram-negative. Gingivitis, a disease characterized by an increase in the redness and swelling of the gingiva surrounding the tooth root surface (and an analogous increase in the number and types of inflammatory cells), is associated with an increased microbial load (10^4–10^5 organisms), of which 15–50% are gram-negative bacteria. Periodontitis, a chronic inflammatory disease, which results in the loss of connective tissue and the alveolar bone that surrounds and supports the tooth root, has a further increase in microbial load (10^5–10^8 organisms) and a clear association with specific gram-negative bacteria (3). A recent World Workshop on Periodontal Disease (9) has identified *Porphyromonas gingivalis*, *Bacteroides forsythus*, and *Actinobacillus actinomycetemcomitans*, three gram-negative bacteria, as agents of disease, as opposed to merely being associated with disease, in recognition of the important role of these bacteria in the induction of periodontitis.

Since it is well known that lipopolysaccharide (LPS) is an inflammatory mediator, it is of great importance to understand the potential contribution that this molecule may play in both the maintenance of healthy tissue as well as the development of gingivitis and periodontitis. LPS has been shown to permeate gingival tissues (10–13) and may thus potentially interact with all cell types in the gingiva and initiate a wide spectrum of beneficial or detrimental processes. The challenge of the host is to sort through the numerous LPS types

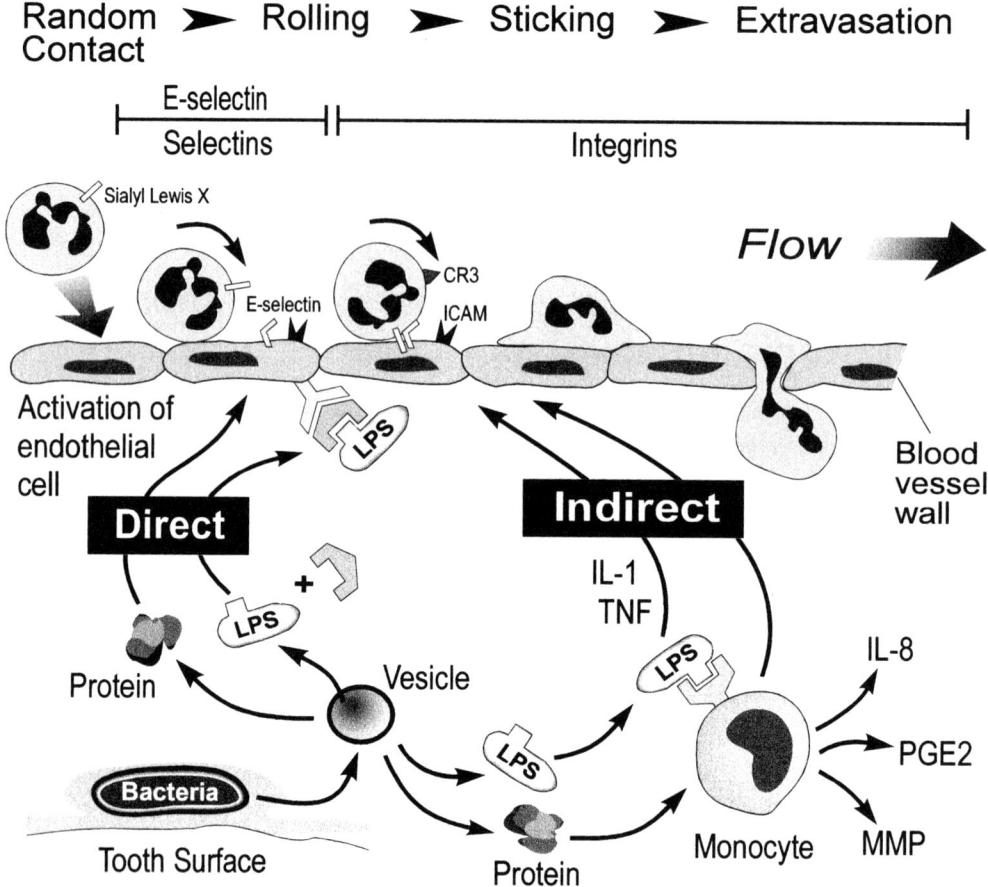

Fig. 1 Bacterial shedding and the role of CD 14 in the host response. Bacterial shedding of both LPS and protein provides the major form of communication between dental plaque bacteria and the host. Vesicles released from the dental plaque surface interact with both myeloid (monocytes) and nonmyeloid (endothelial) cells. Direct activation of the endothelium by LPS can occur as well as indirect activation by bacteria first interacting with monocytes, which release cytokines. The net result of these interactions is the exit of leukocytes from the bloodstream to surrounding tissues to remove the bacterial stimulus. However, the recalcitrant nature of dental plaque on the tooth surface prevents complete removal and contributes to a constant stimulation by bacterial antigens. ICAM: Intracellular adhesion molecule; PGE_2: prostaglandin E_2, CR3: CD11/CD18-β2 integrin; ᘯ: membrane or soluble CD14; LPS: lipopolysaccharide; TNF: tumor necrosis factor; MMP: matrix metalloproteinase. (From Ref. 2.)

being presented and mount an appropriate response. The role of LPS in the regulation of the expression of inflammatory mediators is currently not known, but based upon its action as an inflammatory mediator and the compositional changes associated with disease, it is likely to be a key component in the transition from oral health to disease. For the purpose of this review, discussions of changes in oral health to disease will be limited to those changes that occur in the development of gingivitis and periodontitis. Both of these disease states are recognized as inflammatory disorders associated with the bacterial composition of the dental plaque biofilm.

ORAL LPS COMPOSITION AND STRUCTURE

The best characterized LPS is from the family Enterobacteriacea such as *Escherichia coli*. The predominant molecular species present in these LPS have a highly conserved lipid A consisting of a β-(1,6)-D-glucosamine (1,4′) phosphate disaccharide, with 3-hydroxytetradecanoic acid substituted at positions 2,3,2′, and 3′ and dodecanoic and tetradecanoic acids in either linkage on the 2′ and 3′ hydroxy fatty acids, respectively (Fig. 3a). The lipid A is linked to an inner core containing L-glycero-D-manno-heptose and keto-deoxyoctulosonic acid (KDO) and a less highly conserved

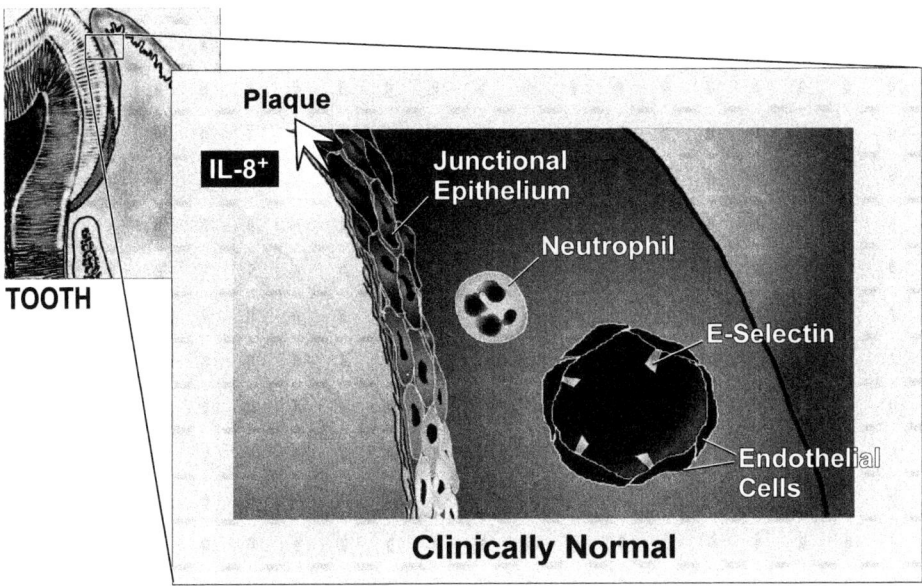

Fig. 2 Innate host defense status in clinically normal tissue. Recently it has been demonstrated that clinically healthy tissue displays low-level expression of select inflammatory mediators. The expression of E-selectin on the vascular endothelium, for example, is believed to facilitate leukocyte exit from the bloodstream into surrounding tissue, where they remove bacteria. A gradient of IL-8 expression (indicated by the shades of gray) exists in normal tissue to guide leukocytes to the site of bacterial colonization. Recent evidence (Darveau RP et al., unpublished) suggests that the biofilm of gingival health may provide the stimulus for expression of these mediators suggesting a commensal relationship between the host and these bacteria. (From Ref. 3.)

outer core composed primarily of sugars such as glucose, galactose, mannose, and glucosamine. An O-polysaccharide, which can vary greatly in composition and structure, even within a species, may be linked to the outer core. LPS isolated from any gram-negative organism, however, is composed of multiple molecular species, differing not only in the presence or quantity of attached O-polysaccharide but also in the core and lipid A composition. The LPS core may contain additional sugars or substituted sugars. Individual species may contain lipid A with altered acylation, altered number of phosphates, or substituted phosphate (14).

Oral LPS with Short, Long, or Intermediate-Length Fatty Acids

The LPS of oral bacterial species have diverse chemical compositions (see Table 1) indicating a great diversity in structure. The bacteria may be divided based on fatty acid chain length into three groups: medium-length fatty acid (14C) LPS similar to *E. coli*, long-length fatty acid (17C) LPS similar to *Bacteroides fragilis*, and short fatty acid (12–13C) LPS similar to *Pseudomonas aeruginosa*.

LPS obtained from *A. actinomycetemcomitans*, *Haemophilus parainfluenzae*, *Fusobacterium nucleatum*, and *Campylobacter rectus* contain readily detectable 3-hydroxy-tetradecanoate (3-OHC14), KDO, and L-glycero-D-manno-heptose, similar to enterobacterial LPS. The presence of D-glycero-D-manno-heptose and phosphorylated KDO in *A. actinomycetemcomitans* and the presence of 3-hydroxy-hexadecanoate in *C. rectus* and *F. nucleatum* in addition to the more usual 14-carbon component, however, point out potentially significant differences.

The LPS from *P. gingivalis*, *B. forsythus*, and *Prevotella intermedia* contain 3-OH-15-methyl-hexadecanoic acid instead of 3-OH-tetradecanoic acid as the major hydroxy fatty acid, suggestive of the LPS of *B. fragilis* (22). The presence of heptose and KDO in *P. gingivalis* LPS has been a topic of debate. Recently, however, phosphorylated KDO has been found to be present both by chemical methodologies employing dephosphorylation and by mass spectral analysis of the inner core region (27–29). Whether heptose exists in the inner core as a substituted, not easily detected derivative remains to be determined, although Johne et al. (29) reported its presence in one out of three

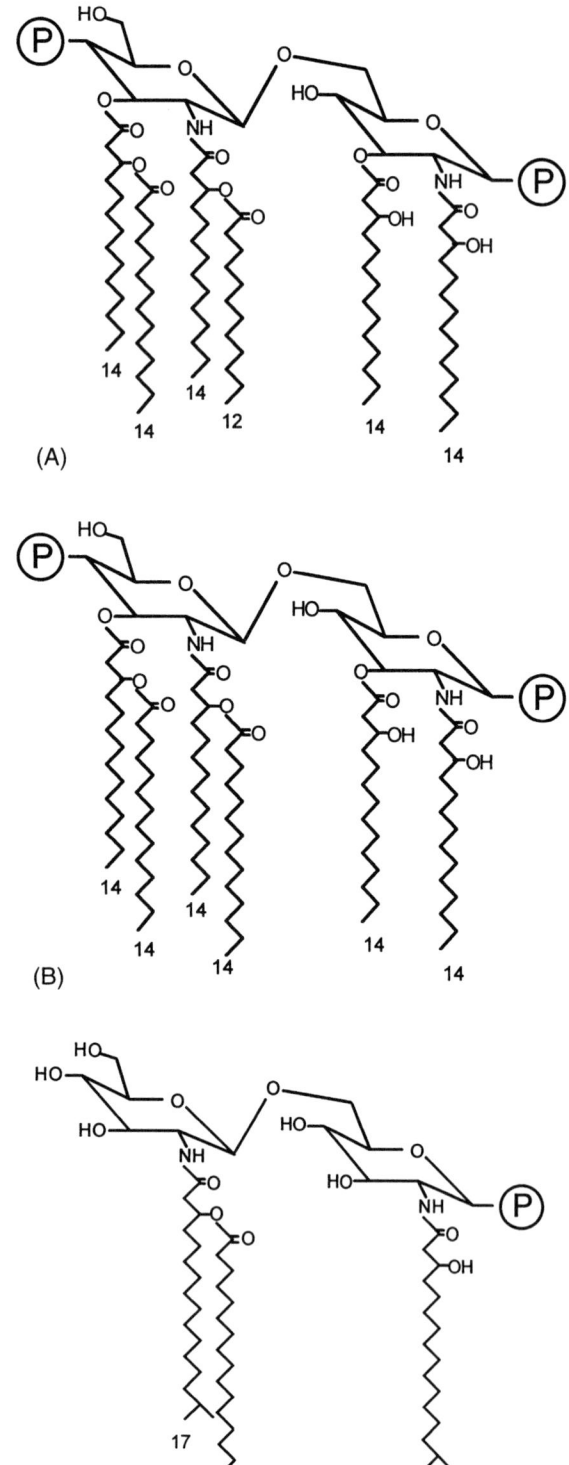

Fig. 3 Chemical structure of three well-characterized lipid A molecules from (a) *E. coli* (14), (b) *A. actinomycetemcomitans* (15), and (c) *P. gingivalis* (25).

strains examined using gas chromatographic analysis of triflouroacetyl derivatives. *Capnocytophaga sputigena* is also known to contain 3-hydroxy-15-methyl-hexadecanoic acid (32), but it is not known if the LPS is structurally related to *Bacteroides* type LPS (33).

Other oral bacterial have LPS with fatty acid chain lengths less than are found in enterobacteria. *Eikenella corrodens* has an LPS containing 3-hydroxy-dodecanoic acid and dodecanoic acid (35). *Centipeda periodontia*, *Selenomonas sputigena*, and *Veillonella* spp. have an LPS based primarily on 3-hydroxy-tridecanoic acid and 13-carbon, nonhydroxylated fatty acids (34,36,37). All three LPS were found to contain KDO and heptose as determined by conventional methodologies (34,36,37). The oral spirochete *Treponema denticola* has been reported to synthesize hydroxy fatty acids with chain lengths of 12–13 carbons, suggesting that this organism may have a short-chain fatty acid LPS (38), although the existence of treponemal LPS has not been firmly established.

Structure of Lipid A Obtained from Oral Bacteria

LPS from oral bacteria have the same general lipid A structure as the better known enterobacterial type, consisting of a β-(1,6)-glucosamine disaccharide substituted with hydroxylated and nonhydroxylated fatty acids in addition to phosphate (14). Although complete lipid A structures are known for only a few oral species, notably *P. gingivalis* (Fig. 3c) (25,26) and *A. actinomycetemcomitans* (Fig. 3b) (15), it is clear that there is great diversity. *A actinomycetemcomitans*, an organism associated with localized juvenile periodontitis, has a lipid A structure (Fig. 3b) very similar to that of *E. coli*, differing only in the replacement of the dodecanoate with tetradecanoate on the 2'-linked hydroxy-tetradecanoic acid resulting in a structure identical to that reported previously for *Haemophilus influenzae* (15). Studies of the isolated lipid A of *P. gingivalis* LPS reveal a structure similar to the lipid A of *B. fragilis* (22) with a small number of long-chain fatty acids linked to a β-(1,6)-D-glucosamine disaccharide, phosphorylated only at the 1-carbon position (Fig. 3c). Ogawa (25) found that the major species present in the LPS of strain 381 contained only three fatty acids: two amide-linked 3-hydroxy-15-methyl hexadecanoic acids and one hexadecanoic acid in an ether linkage on the 2'-hydroxy fatty acid. Kumada et al. (26) found that the major species present in strain SU63 contained an additional fatty acid, either 3-hydroxy-13-methyl-tetradecanoic acid in the 3'-position or 3-hydroxy-hexadecanoic acid in the 3-

Table 1 Lipid A and Core Composition and Select Biological Activity of LPS from Various Oral Bacteria

Organism	Heptose	KDO	#P/LA	Short	Medium	Long	Other	Shwartzman	E-selectin	Ref.
E. coli[a]	+	+	2	–	H14	–	–	+	+	14,39
A. actinomycetemcomitans	+	+	2	–	H14	–	DD-Hep	+	+	15,16,17,35
C. rectus	+	+	2	–	H14,H16	–	–	+		18
F. nucleatum	+	+	2	–	H14,H16	–	–	+		19,35
H. parainfluenzae	+	+	2	–	H14	–	–	+		20
L. buccalis	+			–	–	–	–	+		21
B. fragilis[a]		+	1	–	–	H17i	–	–	–	22,40
P. gingivalis	+/–	+	1	–	–	H17i	P-KDO	–	–	23–29,35
B. forsythus				–	–	H17i	–	–	–	30
B. loescheii				–	–	H17i,H16	–	–		29,35
P. intermedia	+/–	+		–	–	H17i	P-KDO	–	–	29,31,35
C. sputigena	+	+		–	–	H17i,H15i	–	–		32,33
S. sputigena	+	+		H13	–	–	GalN in LA	+		34
E. corrodens				H12	–	–	–	+		35
C. periodonti	+	+		H13	–	–	–	+		36
Veillonella spp.	+	+		H13	–	–	–	+		37
T. denticola ?				H12,H13	–	–	–	–		38

[a]Nonoral species.
DD-Hep = D-glycero-D-mannoheptose; P-KDO = phosphorylated KDO.
(+) = Component has been reported; (–) = absence of component has been reported.
Blank boxes indicate the component or activity has not been reported.

position. It is not clear whether the reported structures are different due to interstrain differences.

The Polysaccharide of Oral LPS

The polysaccharide portions of the LPS molecules of oral bacteria are not as well characterized as those of enterobacteria. Although detailed structural information on the core region is not available, it is apparent from compositional analysis that the core of many oral LPS vary from enterobacteria as well as from one another. This is particularly apparent from the presence of unusual components such as D-glycero-D-mannoheptose in species such as *A. actinomycetemcomitans* and *C. rectus* and phosphorylated KDO in *P. gingivalis* and *P. intermedia* (15,18,27,29). From the compositional data given in Table 2, it appears that the common constituent sugars in the oral LPS examined are rhamnose, fucose, glucose, galactose, mannose, glucosamine, galactosamine, KDO, and heptose. This small number of component sugars suggests that the strains examined have O-polysaccharide chains consisting of these sugars as well or, alternatively, are lacking in O-polysaccharide repeating units. *C. rectus* strain 33238, for example, is known to have a repeating unit consisting exclusively of rhamnose (18). LPS from 12 selected strains of *P. gingivalis* all contained the sugars rhamnose, mannose, glucose, galactose, glucosamine, and galactosamine, with one strain also containing fucose, suggesting that there is not the high degree of variability in the polysaccharide portion of *P. gingivalis* LPS that there is in many enterobacterial species (23,24,28,29). *Selenomonas sputigena* is known to have a repeating unit consisting of galactose and glucosamine (34). *F. nucleatum* strain JCM 8532, on the other hand, was found to have an O-polysaccharide consisting of the less common D-quinovosamine (19). And although the most pathogenic serotype of *A. actinomycetemcomitans* was found to have an O-polysaccharide having the repeating structure (-3-α-D-Fuc-(1→2)-α-L-Rha-(3→1)-β-D-GalNAc), other serotypes contained repeating structures of deoxy-L-talose or deoxy-D-talose (16,17). It therefore appears that in general there maybe less diversity in the O side chain composition found in LPS obtained from oral species. The significance of this observation is not fully understood.

FACTORS INFLUENCING COMPARISONS OF LPS BIOLOGICAL ACTIVITY

Lipopolysaccharide exhibits a wide variety of biological activities in vitro and affects a number of characteristic responses in vivo (14,41–43). However, studies to examine the relative potency of isolated LPS can be affected by the source of LPS and the assay conditions employed to determined biological activity. The observed activity of a purified LPS from a particular strain can depend upon the method of isolation used, the type and quantity of contaminating substances, and the physical state of the LPS (44). Additionally, some investigators have used LPS prepared in a manner that may select a population out of the total cellular LPS, which may, in turn have an altered biological activity (45–47).

The most obvious factor affecting the makeup of an LPS preparation is the bacterial material from which the LPS is derived. Two laboratory strains of the same organism may yield LPS products that vary in the presence and character of an O-antigen, the composition of the core polysaccharide, or the structure of the associated lipid A, factors which have been reported to affect activity (4,16,17,42,48). Changes in LPS structure for a given strain may be influenced by culture conditions selected. Substitutions of phosphate groups and modifications of acylations on lipid A structure have been reported to be influenced by divalent cations with resulting alterations in biological activity (49). Medium composition was reported to affect both the LPS profile and the serum susceptibility of 12 *Bacteroides* strains (50). LPS from *P. gingivalis* grown in medium with altered heme concentration has been reported to have altered activity in addition to a change in chemical identity (51).

Since LPS is a heterogeneous mixture of multiple molecular species, the structure and activity of the preparation can be influenced by the method used to isolate the LPS. Extraction by the phenol/chloroform/petroleum ether method results in a product deficient in O-polysaccharide containing molecules, extraction by phenol/water results in a preparation heavy in long-chain molecules, and the cold magnesium chloride/ethanol precipitation technique results in a preparation containing a mix of short- and long-chain molecules more representative of the LPS in the nonfractionated bacterium (46). Some methods have been reported to result in small amounts of degradation (46), an effect that should be minimized during the isolation and kept in mind when considering the activity of the preparation. All methods produce LPS isolates, which contain varying amounts of contaminating substances including proteins, polysaccharides, and procedure-related detergents, all of which have been reported to have the potential to influence the level of activation observed (42,52,53). In order to remove these contaminants,

Table 2 Carbohydrate Constituents of LPS from Oral Bacteria

Organism	rha	fuc	glc	gal	man	glcN	galN	KDO	Heptose	Others	Strains[a]	Ref.
A. actinomycetemcomitans	+	+	+	+		+	+	+	+	DD-Hep	Y4	15,16,17
D. rectus	+		+		+	+		+	+		33238	18
F. nucleatum			+	+		+		+	+	Quinovosamine	JCM8532	19
L. buccalis										Deoxy sugars	19616	21
P. gingivalis	+		+	+	+	+	+	+	+/−	P-KDO	12 strains	23–29
B. forsythus											None	
B. loescheii	+	+	+	+	+	+	+				15930	29
P. intermedia	+	+	+	+	+	+	+			Unknown sugar	2 strains	29,31
S. sputigena				+	+	+	+	+	+	Hexosamine	3 strains	32,33
S. sputigena			+	+	+			+	+		33150	34
E. corrodens						+				Hexose	1073	35
C. periodonti								+	+	Hexosamine	35019	36
Veillonella spp.	+		+	+		+		+	+		4 strains	37
T. denticola ?											None	

[a]Since the carbohydrate components may be different depending upon the bacterial strain, we have listed the numbers of strains examined for each strain.
DD-Hep = D-glycero-D-mannoheptose; P-KDO = phosphorylated KDO.
(+) = Component has been reported; (−) = absence of component has been reported.
Blank box indicates component has not been reported.

some workers have resorted to column chromatography (24,48,54), or cesium chloride density centrifugation (55). These methods may result in the fractionation of the original LPS, and activation studies using such preparations should be interpreted accordingly. In order to remove bound organic cations and produce a more uniformly soluble preparation, some investigators have converted the isolated LPS to uniform sodium or triethylamine salts. Salt form and aggregation state of LPS have long been known to affect biological activity (42,44), however, and therefore only appropriate comparisons should be made.

Serum proteins such as CD14 and lipopolysaccharide-binding protein (LBP) are known to greatly affect LPS activation of cells. Activation of myeloid cell types (monocytes, neutrophils) occurs primarily through LBP-catalyzed binding of LPS to membrane-bound CD14 while activation of nonmyeloid cell types (endothelial, fibroblast, epithelial) occurs through the LBP catalyzed binding of LPS to soluble serum CD14, which subsequently binds to a putative receptor on the target cell surface (56,57). Although not all LPS bind to CD14 at the same rate in the presence of LBP (58), it appears that discrimination of specific LPS structures occurs after the binding of LPS to CD14 (59). Reports of LPS activity in in vitro systems have employed different concentrations of serum or no serum, serum from various species, or heat-inactivated serum, all of which are known to affect the level of observed activation (55–57,60,61).

Further, the animal cells used to compare activation dictate the magnitude and specificity of the response. Response depends on the cell type (monocyte or endothelial) and species of cell examined (mouse or human) or the use of cell lines versus primary cells. Cells isolated from different donors or different biopsies as well as cells of a particular clone or passage may yield varied results. Level of activation may depend on the chosen indicator of activity or the time after stimulation that the activity is assessed (14,56,59,62–66). LPS activation of cells should be compared only in like systems with cautious extrapolation to in vivo processes.

STIMULATION OF INFLAMMATORY MEDIATORS BY ORAL LPS

The biological activity of LPS is thought to be primarily the result of the lipid A portion of the molecule (14,42). Studies of chemically synthesized analogs to enterobacterial lipid A have demonstrated that lipid A can effect all the classic endotoxic activities and have given considerable structure-function information. Evidence suggests that the structure required for the full range of biological activities is highly conserved. Removal or addition of a single fatty acid residue or removal or modification of either phosphate group results in significant alterations in biological properties such as pyrogenicity, Schwartzman reaction, and chick embryo lethality. Other activities, however, such as immunoadjuvanticity, lethality in galactosamine-sensitized mice, or IL-1 induction in monocytes appear to have less strict structural requirements, with activity being maintained for structures with reduced acylations or those lacking either phosphate group. The structural requirements for activation of human cells, however, is not identical to that for murine cells since the presence of acyloxyacyl fatty acid is not required for activation of murine cells indicating caution is required in extrapolating results obtained in animal systems to humans (14).

Relationship of Altered Activity of Structurally Different Oral LPS to the Activity of E. coli Lipid A Structures

The structurally different LPS of oral bacteria have altered biological activity in some or all assay types, similar to the alteration in activity observed for synthetic analogs. Thus, *P. gingivalis* LPS has lower potency than *E. coli* type LPS in some or all activities due to the absence of 4'-phosphate and reduced number of fatty acids and acyloxyacyl groups on its lipid A while other LPS such as that from *A. actinomycetemcomitans* or *Campylobacter rectus* have activity similar to *E. coli* LPS. To obtain a positive Schwartzman reaction, for example, it is known that a lipid A structure containing at least one acyloxyacyl group in addition to a biphosphorylated glucosamine disaccharide is required (14). Many oral LPS such as *A. actinomycetemcomitans*, *C. rectus*, *E. corrodens*, and *F. nucleatum* are capable of inducing a positive Schwartzman reaction and meet or appear to meet these structural requirements (Table 1, Fig. 3b). LPS from *P. gingivalis* gives a negative reaction consistent with its lipid A structure, which lacks 4'-phosphate and has reduced acylation (Fig. 3c). Furthermore, since it is known that adhesion molecule expression is required for a positive reaction (67), one would predict that Shwartzman-positive LPS would also have the ability to stimulate upregulation of adhesion molecules in vitro. Consistent with this idea, it has been demonstrated that in contrast to *E. coli* LPS, a potent stimulator of E-selectin, the Schwartzman negative LPS of *P. gingivalis* and *B. forsythus* had little or no ability to stimulate this adhesion molecule (68).

Interactions of Oral LPS with Epithelial Cells

Although epithelial cells are presumably the first type of cell to contact LPS in the gingiva, surprisingly little is known about how these cells respond to lipopolysaccharide obtained from oral bacteria. Sugiyama et al. (69) reported that primary human epithelial cells produced IL-1 activity in response to *E. coli* LPS but did not examine LPS from any oral bacteria. On the other hand, a recent study found that primary gingival epithelial cells did not produce IL-8 in response to LPS from either *E. coli* or *P. gingivalis* even though the same cells responded to whole cells from *E. coli* or several oral strains of bacteria excluding *P. gingivalis* (70).

Interactions of Oral LPS with Fibroblast Cells

Fibroblasts are the most thoroughly studied cell type due to the ease with which they may be obtained as well as the potential role they play in tissue degradation. Some investigators have found that LPS from *P. gingivalis* can stimulate release of cytokines (IL-8, MCP-1) with equal or greater potency than LPS from *E. coli*, while others have found that *P. gingivalis* LPS is unable to stimulate this cell type (63,65,71–73). Watanabe et al. (73) reported that the monocyte chemotactic protein (MCP-1) response to *P. gingivalis* LPS was similar to that to *E. coli* LPS in the presence of human serum, but higher in the absence of serum. Fibroblast cultures, however, are known to be heterogeneous at both the cellular level and the population level. Koka et al. (65) observed that *P. gingivalis* LPS–induced inflammatory mediator release by human fibroblast cells depended upon (1) the subject from whom the cells were isolated, (2) the source of the biopsy used to prepare the cells (gingival fibroblasts or periodontal ligament fibroblasts), and (3) the mediator examined (IL-1, IL-6, or PGE_2).

Further, Ogawa et al. (63) reported that inflamed gingival fibroblast cells from patients with chronic periodontitis had a reduced IL-8 response to *P. gingivalis* LPS compared to fibroblasts from periodontally normal individuals, although the response to LPS from *E. coli* was the same regardless of disease state. A short preexposure to *P. gingivalis* LPS was found to potentiate the response of normal fibroblasts, while a longer exposure significantly reduced the response, suggesting that tolerance may play a role in the hyporesponsiveness ob-

served in patients. Similarly, it has been reported that periodontal ligament fibroblast responsiveness to LPS may be altered in vitro by the addition of TNF-α to the culture medium (74). These results suggest that the apparently conflicting results obtained by various labs may represent an inherent variability in the responsiveness of fibroblast populations to respond to LPS. Moreover, fibroblasts in diseased periodontium may have a different innate level of response to LPS and the fibroblasts may acquire a different level of responsiveness as a result of the disease.

Interactions of Oral LPS with Monocytes

Monocytes are capable of responding to exceedingly low concentrations of LPS by producing a wide array of inflammatory mediators and indirectly activating many other cell types. As in the case of fibroblasts, reports have not been in complete agreement about both the magnitude and the specificity of the response to LPS of various types. For example, some studies have suggested that *P. gingivalis* LPS has an ability to stimulate cytokine production comparable to that of *E. coli*, while others suggest a much lower potency (55,64,68,75–79). Interestingly, studies utilizing purified *P. gingivalis* lipid A or the corresponding synthetic structure have indicated that *P. gingivalis* lipid A has low activity in stimulating some cytokines (IL-1, TNF) but a comparable activity in stimulating others (IL-6, IL-1ra) (64,75). Such a differential response indicates that inflammatory disease progression may depend on the type as well as the quantity of mediators produced by the host in response to particular bacteria. Ogawa and Uchida (64) further showed that the IL-1β response to *P. gingivalis* lipid A, but not *E. coli* lipid A could be blocked by an inhibitor of calmodulin kinase, indicating that the differential response may be determined by different intracellular signaling pathways rather than differential activation of a common pathway.

Interactions of Oral LPS with Neutrophils

Although monocytes are the highest producers of inflammatory mediators in inflamed gingival tissue, neutrophils have been shown to release cytokines in response to LPS from periodontal bacteria (62). The higher numbers of neutrophils present at the inflamed site indicates that neutrophil-derived mediators may be important in the disease process. Recently, *P. gingivalis* and *Capnocytophaga ochracea* LPS were shown to have lower ability to stimulate IL-1, IL-8, and TNF-α release than LPS from either *E. coli* or the oral bacteria *A. actinomycetemcomitans* and *F. nucleatum* but a similar ability to stimulate release of IL-1 receptor antagonist (IL-1ra) (62). Likewise, it has been reported that higher concentrations of *P. gingivalis* LPS are required to prime neutrophils for superoxide release than either *E. coli* or *A. actinomycetemcomitans* LPS (80). Interestingly, it was found that there was no biological IL-1 activity in *P. gingivalis* LPS–stimulated neutrophil culture supernatants despite the presence of immunologically detectable IL-1, apparently due to the inhibitory effect of IL-1ra (62). This result suggests there may be an important role for anti-inflammatory mediators like IL-1ra in the type and degree of inflammation that results from the presence of a specific organism.

Role of LPS Carbohydrate in Immune Cell Stimulation

Although lipid A is commonly thought to be the biologically active portion of the LPS molecule, the polysaccharide itself can play an important role in the interaction of the bacterium with the host (14,81,82). O-polysaccharide variability in oral microbes as well as the presence of unusual core constituents (phosphorylated KDO, D-glycero-D-mannoheptose) may result in LPS with pathologically relevant biological effects.

The LPS polysaccharides of oral bacteria have been reported to have a role in monocyte activation (83,84) and bone resorption in rats (85), although these effects may be due to changes in solubility or solution structure rather than a specific interaction of the polysaccharide with a cellular receptor such as CD14. Nature of the O-polysaccharide determines not only the serotype of the organism, but may also determine serum susceptibility (14) and systemic clearance by the reticular endothelial system (82) and may play a role in such processes as invasiveness and cell-to-cell spread (86).

LPS AND DESTRUCTIVE INFLAMMATORY PROCESSES

LPS is present on the tooth surface in amounts sufficient to activate cells (41), and the LPS is able to permeate the gingival tissues (12,87). Cells of the gingival tissue may respond by the production of various inflammatory mediators (68,69,73,77,80). Normally such mediators are part of a beneficial host response to a bacterial challenge and lead to suppression of the bacterial presence. However, in inflammatory diseases, these mediators have a destructive role (Fig. 4). Each cell type may respond with a different array of mediators or a different level of mediator (41,62,64,68,88).

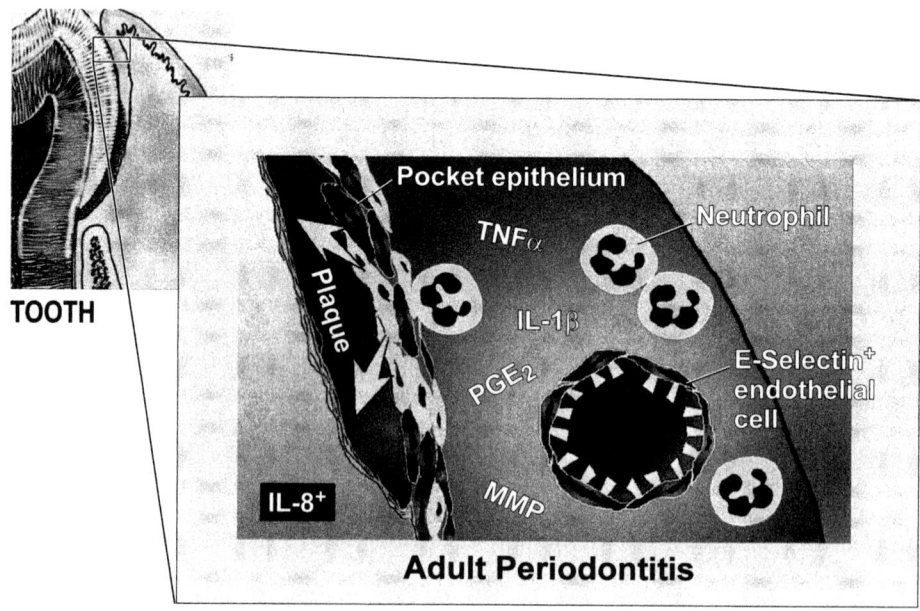

Fig. 4 Innate host-defense status in adult periodontitis. In adult periodontitis, the molecular mediators of inflammation that are expressed in clinically healthy tissue are expressed at higher levels and new mediators are present. The gradient of IL-8 expression found in healthy tissue is disrupted and a pocket epithelium forms. PGE_2: Prostaglandin E_2, TNF: tumor necrosis factor; MMP: matrix metalloproteinase. (From Ref. 3.)

Cell types may require different quantities of LPS to activate or may have different responses to LPS from different species (62,64,68,77,88).

Cells of the gingiva may be stimulated to produce inflammatory mediators either by the direct effect of the LPS or indirectly because the produced mediator may stimulate a second cell type. Both types of stimulation may occur with either myeloid or nonmyeloid cell types (88). For example, LPS may stimulate a macrophage (myeloid cell type) to secrete IL-1β, and this cytokine may then stimulate nonmyeloid cell types (endothelial, epithelial, fibroblast) to secrete further mediators. How various cell types respond, either directly or indirectly, to bacterial products such as LPS is important in determining the net inflammatory response.

Mediators such as TNF-α, IL-1β, prostaglandin E_2 (PGE_2), and matrix metalloproteinases (MMP) are known to be induced by LPS in vitro (55,77–79,88) and are found in elevated concentrations in periodontal tissues and fluid during active periodontal inflammation relative to health (89–91). Levels decrease following successful treatment (91,92). Administration of anti-inflammatory drugs that reduce levels of these mediators can suppress bone and tissue destruction (93).

Perhaps the best studied mediator of periodontal inflammation is IL-1-β. IL-1 induces many pro-inflammatory effects including proliferation of T cells, neutrophil influx and degranulation, the release of cytokines from fibroblasts, the release of PGE_2 from monocytes and fibroblasts, the release of MMP from a variety of cell types, and in vitro bone resorption (41,91,94). IL-1 release is stimulated by IL-1 (autocrine). Interferon-γ suppresses autocrine but enhances LPS-mediated stimulation (94). Human monocytes, neutrophils, and fibroblasts have been shown to secrete IL-1 in response to LPS from oral pathogens (41,43,62,64,76–78). Elevated IL-1 levels have been reported in the gingival tissue of periodontitis patients and in their gingival crevicular fluid at biologically significant concentrations (91). Crevicular fluid levels were reported to decrease to near control levels after periodontal treatment (91).

Monocytes are the main source of TNF-α in response to LPS. TNF-α stimulates bone resorption although it is less potent than IL-1 (95). It also increases vascular permeability, stimulates recruitment and degranulation of PMN, and induces responses in various cell types including the release of PGE_2 by fibroblasts and IL-1 by monocytes (41). TNF-α is known to be produced by monocytes in response to a wide array of LPS derived from oral bacteria, indicating a role for this cytokine in periodontal inflammation (43,55,77).

Prostaglandin E_2, a bioactive lipid derived from arachidonic acid in the cell membrane, has been strongly implicated as an important mediator in periodontal tissue destruction. Gingival crevicular PGE_2 levels are elevated in disease, increase with severity of disease and correlate with disease progression and remission (92). Nonsteroidal anti-inflammatory drugs, which block prostaglandin synthesis, can arrest tissue destruction (93). LPS from oral bacteria have been shown to directly stimulate the release of PGE_2 from monocytes (41,79,88) and indirectly in fibroblasts via IL-1 (88).

The family of enzymes collectively referred to as the matrix metalloproteinases (MMP), consists of at least nine distinct Zn^{2+} endopeptidases (96). MMP participate in destruction of connective tissue and perhaps in bone resorption as well (97). Oral LPS may stimulate macrophages to secrete MMP, or LPS-stimulated IL-1 or TNF may result in MMP production by nonmyeloid cells such as fibroblast and endothelial cells (88,96,97).

It is believed that LPS-induced inflammatory mediator production leads to destruction of periodontal tissues and bone loss. LPS from oral bacteria are able to directly stimulate resorption of bone (41) and are potent stimulators of IL-1B, PGE_2, and TNF-α, all potent inducers of bone resorption (77–79,94,97,98). Metalloproteinases degrade soft tissues and may be involved in bone destruction as well (97). Further, LPS-induced mediators amplify the local inflammation by stimulating other cells to release mediators and by upregulating adhesion molecules and chemotactic peptides (62,68,73,77) to recruit additional inflammatory cells.

LPS MODULATES THE HOST RESPONSE

Dental plaque bacteria not normally associated with disease-shed antigens, including LPS, lipoteichoic acids, and proteins that penetrate the gingival tissues. These bacterial components appear to interact with various cell types inducing inflammatory mediator production. This results in a low-level inflammatory state, which does not progress to disease, but rather serves to keep the host protected from further invasion of dental plaque bacteria. For example, Tonetti et al. (5) has shown that normal gingival tissue displays an mRNA IL-8 gradient (see Fig. 2), the highest levels being found in tissue adjacent to plaque bacteria. This is consistent with earlier histological observations demonstrating a cellular infiltrate in clinically healthy tissue (4). LPS from plaque organisms with short (*E. corrodens*) or intermediate-length fatty acyl chains (*F. nu-*

cleatum) and biphosphorylated lipid A are potent stimulators of these mediators and shed LPS from such organisms may play a role in forming and maintaining this gradient. Thus bacterial release of LPS that stimulates host-defense cells may have a beneficial effect on the innate host system providing a component of the innate host surveillance system associated with clinical health.

In contrast, *P. gingivalis* may play a role in the disruption of innate host surveillance by interfering with the ability of the host to position sufficient leukocytes in close proximity to bacterial colonization. *P. gingivalis* LPS blocks E-selectin expression in response to LPS from other bacteria, an adhesion molecule necessary for efficient leukocyte exit from the bloodstream (68). In addition, after *P. gingivalis* invasion of gingival epithelial cells, the cells no longer secrete IL-8 in response to other oral bacteria (70). *P. gingivalis* LPS also has been shown to have reduced cytokine-inducing activity both in human monocytes in vitro and in a murine inflammation model in vivo (62,64,68,79,99). The ability to block both endothelial and epithelial cell response to other bacteria combined with poor activation of myeloid cells is consistent with an innate host-impairment strategy of pathogenesis. In the absence of an effective innate host defense leukocyte barrier, bacterial numbers of multiple species could increase dramatically leading to more bacterial antigen release. An increase in the bacterial population surrounding the tooth root surface accompanied by the failure of the host to remove them is consistent with the etiology of periodontitis. Bacterial stimulation without adequate homing to bacterial infection could account for the accumulation of activated inflammatory cells in diseased tissues rather than at the site of bacterial colonization. These activated cells are responsible for the release of numerous cytokines and other inflammatory mediators amplifying the host response resulting in tissue destruction instead of bacterial removal (Fig. 4).

It is also possible that some pathogens with high biological activity LPS, for example, *A. actinomycetemcomitans*, may alter the structure of their LPS under certain conditions that occur within the host. Predicted changes would involve removal of select fatty acids rendering the LPS undetectable by components of the innate host defense system. Some organisms are known to alter the structure of their LPS in response to growth conditions resulting in LPS with reduced biological activity (49,51). *P. gingivalis*, for example, is known to respond to environmental heme concentration with an altered LPS, which has been shown to have an altered ability to prime neutrophils (51). These organisms may regulate the level of certain inflammatory mediators via structural changes in their LPS, evading innate host defense and creating more favorable conditions for their persistence in tissue.

SUMMARY

Dental plaque is a biofilm composed of numerous bacterial species. The bacteria release LPS with varied molecular structures and biological activities, and the LPS permeates the gingival tissues. Periodontal health has characteristic microflora that release LPS and other components leading to low-level activation of the innate host defense, suppressing but not eliminating the bacterial presence and maintaining an innate host inflammation surveillance system. Initiation of disease involves an alteration in the bacterial composition of dental plaque, resulting in the presentation of an altered LPS repertoire. The diverse LPS from oral pathogens have varied capacity to stimulate a biological responses and may promote disease both by the stimulation of destructive inflammatory processes and by the modification of the innate host response.

REFERENCES

1. Moore W, Moore L. The bacteria of periodontal diseases. Periodontology 2000 1994; 5:66–77.
2. Costerton JW, Lewandowski Z, Debeer D, Korber D, James G. Biofilms, the customized microniche. J Bacteriol 1994; 176:2137–2142.
3. Darveau, RP, Tanner A, Page RC. The microbial challenge in periodontitis. Periodontology 2000 1997; 14: 12–32.
4. Page RC, Schroeder HE. Current status of the host response in chronic marginal periodontitis. J Periodontol 1981; 52:477–491.
5. Tonetti MS, Imboden M, Gerber L, Lang N. Localized expression of mRNA for phagocyte-specific chemotactic cytokines in human periodontal infections. Infect Immun 1994; 62:4005–4014.
6. Gemmell E, Walsh L, Savage N. Adhesion molecule expression in chronic inflammatory periodontal disease tissue. J Periodontal Res 1994; 29:46–53.
7. Moughal NA, Adonogianaki E, Thornhill M, Kinane D. Endothelial cell leukocyte adhesion molecule-1 (ELAM-1) and intercellular adhesion molecule-1 (ICAM-1) expression in gingival tissue during health and experimentally-induced gingivitis. J Periodont Res 1992; 27:623–630.
8. Nylander K, Danielsen B, Fejerskov O, Dabelsteen E. Expression of the endothelial leukocyte adhesion molecule-1 (ELAM-1) on endothelial cells in experimental gingivitis in humans. J Periodontol 1993; 64:355–357.
9. Consensus Report. Periodontal diseases: pathogenesis and microbial factors. Ann Periodont 1996; 1:926–932.

10. Hamada S, Takada H, Ogawa T, Fujiwara T, Mihara J. Characterization and immunobiologic activities of lipopolysaccharides from periodontal bacteria. Adv Dent Res 1990; 2:284–291.
11. McCoy SA, Creamer HR, Kawanami M, Adams D. The concentration of lipopolysaccharide on individual root surfaces at varying times following in vivo root planing. J Periodontol 1987; 58:393–399.
12. Moore J, Wilson M, Kieser J. The distribution of bacterial lipopolysaccharide in relation to periodontally involved root surfaces. J Clin Periodontol 1986; 13:748–751.
13. Wilson M, Moore J, Kieser J. Identity of Limulus amoebocyte lysate-active root surface materials from periodontally involved teeth. J Clin Periodontol 1986; 13:743–747.
14. Morrison DC, Ryan JL, eds. Bacterial Endotoxic Lipopolysaccharides. Vol. 1. Boca Raton: CRC Press, 1992.
15. Masoud H, Weintraub S, Wang R, Cotter R, Holt S. Investigation of the structure of lipid A from *Actinobacillus actinomycetemcomitans* strain Y4 and human clinical isolate PO 1021-7. Eur J Biochem 1991; 200:775–779.
16. Perry MB, MacLean L, Gmur R, Wilson M. Characterization of the O-polysaccharide structure of lipopolysaccharide from *Actinobacillus actinomycetemcomitans* serotype b. Infect Immun 1996; 64(4):1215–1219.
17. Perry MB, MacLean L, Brisson J, Wilson M. Structures of the antigenic O-polysaccharides of lipopolysaccharides produced by *Actinobacillus actinomycetemcomitans* serotypes a, c, d and e. Eur J Biochem 1996; 242(3):682–688.
18. Kumada H, Watanabe K, Umemoto T, Kato K, Kono S, Hisatsune K. Chemical and biological properties of lipopolysaccharide, lipid A and degraded polysaccharide from *Wolinella recta* ATCC 33238. J Gen Microbiol 1989; 135(4):1017–1025.
19. Onoue S, Niwa M, Isshiki Y, Kawahara K. Extraction and characterization of the smooth-type lipopolysaccharide from *Fusobacterium nucleatum* JCM 8532 and its biological activities. Microbiol Immunol 1996; 40(5):323–331.
20. Tuyau J, Sims W. Aspects of the pathogenicity of some oral and other hemophilus. J Med Microbiol 1983; 16(4):467–475.
21. Knox K, Parker R. Isolation of a phenol soluble endotoxin from *Leptotrichia buccalis*. Archs Oral Biol 1973; 18:85–93.
22. Weintraub A, Zahringer U, Wollenweber H, Seydel S, Rietschel E. Structural characterization of the lipid A component of *Bacteroides fragilis* strain NCTC 9343 lipopolysaccharide. Eur J Biochem 1989; 183:425–431.
23. Bramanti T, Wong G, Weintraub S, Holt S. Chemical characterization and biologic properties of lipopolysaccharide from *Bacteroides gingivalis* strains W50, W83, and ATCC 33277. Oral Microbiol Immunol 1989; 4(4):183–192.
24. Schifferle R, Reddy M, Zambon J, Genco R, Levine M. Characterization of a polysaccharide antigen from *Bacteroides gingivalis*. J Immunol 1989; 143(9):3035–3042.
25. Ogawa T. Chemical structure of lipid A from *Porphyromonas gingivalis* lipopolysaccharide. FEBS 332:197–201.
26. Kumada H, Haishima Y, Umemoto T, Tanamoto K. Structural study on the free lipid A isolated from lipopolysaccharide of *Porphyromonas gingivalis*. J Bacteriol 1995; 177:2098–2106.
27. Kumada H, Kondo S, Umemoto T, Hisatsune K. Chemical structure of the 2-keto-3-deoxyoctonate region of lipopolysaccharide isolated from *Porphyromonas gingivalis*. FEMS Microbiol Lett 1997; 108(1):75–79.
28. Fujiwara T, Ogawa T, Sobue S, Hamada S. Chemical, immunobiological and antigenic characterizations of lipopolysaccharides from *Bacteroides gingivalis* strains. J Gen Microbiol 1990; 136:319–326.
29. Johne B, Olsen I, Bryn K. Fatty acids and sugars in lipopolysaccharides from *Bacteroides intermedius*, *Bacteroides gingivalis* and *Bacteroides loescheii*. Oral Microbiol Immunol 1983; 3:22–27.
30. Gersdorf H, Meissner A, Pelz K, Krekeler K, Gobel U. Identification of *Bacteroides forsythus* in subgingival plaque from patients with advanced periodontitis. J Clin Microbiol 1993; 1 31(4):941–946.
31. Johne B, Bryn K. Chemical composition and biological properties of a lipopolysaccharide from *Bacteroides intermedius*. Acta Pathol Microbiol Immunol Sacand B 1986; 94(4):265–271.
32. Dees SB, Karr DE, Hollis D, Moss CW. Cellular fatty acids of Capnocytophaga species. J Clin Microbiol 1982; 18:779–783.
33. Poirier T, Mishell R, Trummel C, Holt S. Biological and chemical comparison of butanol and phenol water extracted lipopolysaccharide from *Capnocytophaga sputigena*. J Periodontal Res 1983; 18:541–557.
34. Kumada H, Watanabe K, Nakamu A, Haishima Y, Kondo S, Hisatsune K, Umemoto Y. Chemical and biological properties of lipopolysaccharide from *Selenomonas sputigena* ATCC 33150. Oral Microbiol Immunol 1997; 12(3):162–167.
35. Mashimo J, Michiko Y, Ikeuchi K, Hata S. Fatty acid composition and Schwartzman activity of lipopolysaccharides from oral bacteria. Microbiol Immunol 1985; 29(5):395–403.
36. Kokeguchi S, Tsutsui O, Kato K, Matsumura T Isolation and characterization of lipopolysaccharide from *Centipeda periodontii*. Oral Microbiol Immunol 1990; 5(2):108–112.
37. Hewett M, Knox K. Biochemical studies on lipopolysaccharides of Veillonella. Eur J Biochem 1971; 19:169–175.
38. Dahle U, Tronstad L, Olsen L. 3-hydroxy fatty acids in a lipopolysaccharide-like material from *Treponema denticola* strain FM. Endodont Dent Traumatol 1997; 12:202–205.
39. Ogawa T. Immunobiological properties of chemically defined lipid A from lipopolysaccharide of *Porphyromonas gingivalis*. Eur J Biochem 1994; 219:737–742.
40. Sveen K. The capacity of lipopolysaccharides from bacteroides, Fusobacterium and Veillonella to produce skin inflammation and the local and generalized Schwartz-

man reaction in rabbits. J Periodont Res 1977; 12:340–350.
41. Wilson M, Biological activities of lipopolysaccharides from oral bacteria and their relevance to the pathogenesis of chronic periodontitis. Sci Prog 1995; 78:19–34.
42. Galanos C, Luderitz O, Rietschel E, Westphal O. Newer aspects of the chemistry and biology of bacterial lipopolysaccharides. In: Goodwin T, ed. Biochemistry of Lipids. Baltimore: University Park Press, 1977:239–335.
43. Hamada S, Takada H, Ogawa T, Fujiwara T, Mihara J. Lipopolysaccharides of oral anaerobes associated with chronic inflammation: chemical and immunomodulating properties. Intern Rev Immunol 1990; 6:247–261.
44. Wilson M. Biological activities of lipopolysaccharide and endotoxin. In: Biology of the Species *Porphyromonas gingivalis*. Shah HN, ed. Boca Raton, FL: CRC Press, 1993:171–197.
45. Delahooke DM, Garclay GR, Poxton IR. A re-appraisal of the biological activity of bacteroides LPS. J Med Microbiol 1995; 42:102–112.
46. Darveau RP, Hancock REW. Procedure for isolation of bacterial lipopolysaccharides form both smooth and rough *Pseudomonas aeruginosa* and *Salmonella typhimurium* strains. J Bacteriol 1983; 155:831–838.
47. Millar SJ, Goldstein EG, Levine MJ, Hausmann E. Modulation of bone metabolism by two chemically distinct lipopolysaccharide fractions from *Bacteroides gingivalis*. Infect Immun 1986; 51:302–306.
48. Fischer W. Purification and fractionation of lipopolysaccharide from gram-negative bacteria by hydrophobic interaction chromatography. Eur J Biochem 1990; 194:655–661.
49. Lin G, Lim KB, Gunn JS, Bainbridge BW, Darveau RP, Hackett M, Miller SI. Regulation of lipid A modifications by *Salmonella typhimurium* virulence genes phoP-phoQ. Science 1997; 276:250–253.
50. Allan E, Poxton IR. The influence of growth medium on serum sensitivity of *Bacteroides* species. J Med Microbiol. 1994; 41:45–50.
51. Champagne CM, Holt SC, Van Dyke TE, Gordon BJ, Shapira L. Lipopolysaccharide isolated from *Porphyromonas gingivalis* grown in hemin-limited chemostat conditions has a reduced capacity for human neutrophil priming. Oral Microbiol Immunol 1996; 5:319–325.
52. Wilson M, Reddi K, Henderson B. Cytokine-inducing components of periodontopathogenic bacteria. J Periodont Res 1996; 31:393–407.
53. Ribi E, Anacker RL, Brown R, Haskins WT. Reaction of endotoxin and surfactants: I.Physical and biological properties of endotoxin treated with sodium deoxycholate. J Bacteriol 1966; 92:1493–1509.
54. Takada H, Iki K, Sakuta T, Sugiyama A, Sawamura S, Hamada S. Lipopolysaccharides of oral black pigmented bacteria and periodontal diseases. In: Levin J, Alving CR, Munford RS, Redl H, eds. Bacterial Endotoxins: Lipopolysaccharides from Genes to Therapy. New York: Wiley-Liss, 1995:59–68.
55. Shapira L, Takashiba S, Amar S, VanDyke TE. *Porphyromonas gingivalis* lipopolysaccharide stimulation of human monocytes: dependence on serum and CD14 receptor. Oral Microbiol Immunol 1994; 9:112–117.
56. Pugin J, Schurer-maly C-C, Leturcq D, Moriary A, Ulevitch RJ, Tobias PS. Lipopolysaccharide activation of human endothelial and epithelial cells is mediated by lipopolysaccharide-binding protein and soluble CD14. Proc Natl Acad Sci 1993; 90:2744–2748.
57. Wright SD, Ramos RA, Tobias PS, Ulevitch RJ, Mathison JC. CD14, a receptor for complexes of lipopolysaccharide and LPS binding protein. Science 1990; 249:1431–1433.
58. Cunningham MD, Seachord C, Ratcliffe K, Bainbridge B, Aruffo A, Darveau RP. *Helicobacter pylori* and *Porphyromonas gingivalis* lipopolysaccharides are poorly transferred to recombinant soluble CD14. Infect Immun 1996; 64:3601–3608.
59. Delude RL, Savedra R, Zhao H, Thieringer R, Yamamoto S, Fenton MJ, Golenbock DT. CD14 enhances cellular responses to endotoxin without imparting ligand specific recognition. Proc Natl Acad Sci 1995; 92:9288–9292.
60. Grunwald U, Kruger C, Schutt C. Endotoxin neutralizing capacity of soluble CD14 is a highly conserved specific function. Circ Shock 1993; 39:220–225.
61. M'esz'aros K, Aberle S, White M, Parent JB. Immunoreactivity and bioactivity of lipopolysaccharide-binding protein in normal and heat-inactivated sera. Infect Immun 1995; 63:363–365.
62. Yoshimura A, Hara Y, Kaneko T, Kato T. Secretion of IL-1 beta, TNF-alpha, IL-8 and IL-1ra by human polymorphonuclear leukocytes in response to lipopolysaccharides from periodontopathic bacteria. J Periodont Res 1997; 32:279–286.
63. Ogawa T, Ozakai A, Shimauchi H, Uchida H. Hyporesponsiveness of inflamed human gingival fibroblasts from patients with chronic periodontal diseases against cell surface components of *Porphyromonas gingivalis*. FEMS Immunol Med Microbiol 1997; 18:17–30.
64. Ogawa T, Uchida H. Differential induction of IL-1β and IL-6 production by the nontoxic lipid A from *Porphyromonas gingivalis* in comparison with synthetic *Escherichia coli* lipid A in human peripheral blood mononuclear cells. FEMS Immunol Med Microbiol 1996; 14:1–13.
65. Koka S, Maze C, Reinhardt R, Dyer J. Variability of inflammatory mediator production by human periodontal firoblasts stimulated with bacterial lipopolysaccharide. In Vitro Cell Dev Biol 1996: 32:528–530.
66. Delahooke DM, Barclay GR, Poxton IR. Tumor necrosis factor induction by an aqueous phenol extracted lipopolysaccharide complex from Bacteroides species. Infect Immun 1995; 63:840–846.
67. Argenbright LW, Barton RW. Interactions of leukocyte integrins with intercellular adhesion molecule 1 in the production of inflammatory vascular injury in vivo. J Clin Invest 1992; 89:259–272.
68. Darveau RP, Cunningham MD, Bailey T, Seachord C, Ratcliffe K, Bainbridge B, Dietsch M, Page RC, Aruffo A. Ability of bacteria associated with chronic inflammatory disease to stimulate E-selectin expression and promote neutrophil adhesion. Infect Immun 1995; 63:1311–1317.
69. Sugiyama A, Arakaki R, Ohnishi T, Arakaki N, Daikuhara Y, Takada H. Lipoteichoic acid and Interleukin

1 stimulate synergistically production of hepatocyte growth factor (scatter factor) in human gingival fibroblasts in culture. Infect Immun 1996; 64:1426–1431.
70. Darveau RP, Belton CM, Reife RA, Lamont RJ. Local chemokine paralysis: a novel pathogenic mechanism for *Porphyromonas gingivalis*. Infect. Immun. 1998; 66:1660–1665.
71. Reiffe R. Unpublished.
72. Yamaji Y, Kubota T, Sasaguri K, Sato S, Suzuki Y, Kumada H, Umemoto T. Inflammatory cytokine gene expression in human periodontal ligament fibroblast stimulated with bacterial lipopolysaccharides. Infect Immun 1995; 63:3576–3581.
73. Watanabe A, Takeshita A, Kitano S, Hanazawa S. CD14 mediated signal pathway of *Porphyromonas gingivalis* lipopolysaccharide in human gingival fibroblasts. Infect Immun 1997; 64:4488–4494.
74. Quintero JC, Piesco NP, Langkamp HH, Bowen L, Agarwal S. LPS responsiveness in periodontal ligament cells is regulated by tumor necrosis factor. J Dent Res 1995; 74:1802–1811.
75. Tanamoto K, Azumi S, Haishima Y, Kumada H, Umemoto T. Endotoxic properties of free lipid A from *Porphyromonas gingivalis*. Microbiology 1997; 143:63–71.
76. Lindemann RA, Economou JS. *Actinobacillus actinomycetemcomitans* and *Bacteroides gingivalis* activate human peripheral monocytes to produce interlukin-1 and tumor necrosis factor. J Periodontol 1988; 59:728–730.
77. Agarwal S, Piesco NP, Johns LP, Riccelli AE. Differential expression of IL-1β TNF-α IL-6, and IL-8 in human monocytes in response to lipopolysaccharides from different microbes. J Dent Res 1995; 74:1057–1065.
78. Roberts FA, Richardson GJ, Michalek SM. Effects of *Porphyromonas gingivalis* and *Escherichia coli* lipopolysaccharides on mononuclear phagocytes. Infect Immun 1997; 65:3248–3254.
79. Bainbridge BW, Page RC, Darveau RP. Immunization of *Macaca fascicularis* with *Porphyromonas gingivalis* elicits an ability to neutralize LPS induced production of PGE_2 by mononuclear cells. Infect Immun 1997; 65: 4801–4805.
80. Aida Y, Kukita T, Takada H, Maeda K, Pabst MJ. Lipopolysaccharides from periodontal pathogens prime neutrophils for enhanced respiratory burst: differential effect of a synthetic lipid A precursor IVA (LA-14-PP). J Periodont Res 1995; 30:116–123.
81. Shimizu T, Akiyama S, Masuzawa T, Yanagihara Y, Nakamoto S, Achiwa K. Biological activities of chemically synthesized 2-keto-3-deoxyoctonic acid-(alpha 2–6)-D-glucosamine analogs of lipid A. Infect Immun 1987; 55:2287–2289.
82. Ohno A, Isii Y, Tateda K, Matumoto T, Miyazaki S, Yokota S, Yamaguchi K. Role of LPS length in clearance rate of bacteria from the bloodstream in mice. Microbiology 1995; 141:2749–2756.
83. Gillespie MJ. The *Campylobacter rectus* lipopolysaccharide core is essential for maximum prostaglandin E2 elicitation in mouse macrophages. Anaerobe 1996; 2: 313–320.
84. Nishihara T, Koga T, Hamada S. Suppression of murine macrophage interleukin-1 release by the polysaccharide portion of *Haemophilus actinomycetemcomitans* lipopolysaccharide. Infect Immun 1988; 56:619–625.
85. Sveen K, Skaug N. Bone resorption stimulated by lipopolysaccharides from Bacteroides, Fusobacterium and Veillonella and by the lipid A and the polysaccharide part of Fusobacterium lipopolysaccharide. Scand J Dent Res 1980; 88:535–542.
86. Hong M, Payne SM. Effect of mutations in *Shigella flexneri* chromosomal and plasmid encoded lipopolysaccharide genes on invasion and serum resistance. Mol Microbiol 1997; 24:779–791.
87. Schwartz J, Stinson FL, Parker RB. The passage of tritiated bacterial endotoxin across intact gingival crevicular epithelium. J Periodontol 1972; 43:270–276.
88. Heath JK, Atkinson SJ, Hembry RM, Reynolds JJ, Meikle MC. Bacterial antigens induce collagenase and prostaglandin E_2 synthesis in human gingival fibroblasts through a primary effect on circulating mononuclear cells. Infect Immun 1987; 55:2148–2154.
89. Golub LM, Sorsa T, Lee HM, Ciancio S, Sorbi D, Ramamurthy NS, Gruber B, Salo T, Knottinen YT. Doxycycline inhibits neutrophil type matrix metalloproteinases in human adult periodontitis gingiva. M Clin Periodontal 1995; 22:100–109.
90. Goodson JM, Dewhirst FE, Brunetti A. Prostaglandin E_2 levels and human periodontal disease. Prostaglandinis 1974; 6:81–85.
91. Masada MP, Persson R, Kenney JS, Lee SW, Page RC, Allison AC. Measurement of interleukin 1α and 1β in gingival crevicular fluid: implications for the pathogenesis of periodontal disease. J Periodont Res 1990; 25: 156–163.
92. Offenbacher S, Odle BM, Van Dyde TE. The use of crevicular fluid protaglandin E2 levels as a predictor of periodontal attachment loss. J Periodont Res 1986; 21: 101–112.
93. Offenbacher S, Braswell LD, Loos AS, Johnson HG, Hall CM, McClure H, Orkin JL, Strobert EA, Green MD, Odle BM. The effects of flurbiprofen on the progression of periodontitis in *Macaca mulatta*. J Periodont Res 1987; 22:473–481.
94. Page RC. The role of inflammatory mediators in the pathogenesis of periodontal disease. J Periodont Res 1991; 26:230–242.
95. Mundy GR. Local factors in bone remodeling. Recent Prog Horm Res 1989; 45:507–531.
96. Reynolds JJ, Meikle MC. Mechanisms of connective tissue matrix destruction in periodontitis. Periodontology 2000 1997; 14:144–157.
97. Birkedal-Hansen H. Role of cytokines and inflammatory mediators in tissue destruction. J Periodont Res 1993; 28:500–510.
98. Van Dyke TE, Lester MA, Shapira L. The role of the host response in periodontal disease progression: implications for future treatment strategies. J Periodontal 1993; 64:792–806.
99. Reife RA, Shapiro RA, Bamber BA, Berry KK, Mick GE, Darveau RP. *Porphyromonas gingivalis* lipopolysaccharide is poorly recognized by molecular components of innate host defense in a mouse model of early inflammation. Infect Immun 1995; 63:4686–4694.

63
Endotoxin and Cancer

Minghuang Zhang and Kevin J. Tracey
North Shore University Hospital and The Picower Institute for Medical Research, Manhasset, New York

HISTORY

A combination of gram-positive, heat-killed streptococcus plus gram-negative, heat-killed *Seiratia marcescens* is commonly referred to as Coley's toxin, named after the New York surgeon William B. Coley, who developed this pioneer attempt for cancer immunotherapy in 1892 (1,2). Although Coley's toxin initiated an important experimental strategy for cancer therapy and showed some promising results in clinical treatment of cancer, it lacked specificity to tumors and ultimately proved quite toxic. Subsequent efforts focused on separating the therapeutic from toxic effects, which led to the isolation of endotoxin (lipopolysaccharide, or LPS) in 1943 (3). Mechanistic studies of LPS-mediated antitumor effects revealed that LPS caused tumor regression by stimulating the production of a host-derived serum cytotoxic factor. This factor was isolated and named tumor necrosis factor (TNF) for its putative role in mediating LPS-induced tumor regression (4). It soon became clear that macrophages are the primary cell source of TNF. The implication of TNF and macrophages in LPS-induced tumor regression has shifted much of the interest in endotoxin and cancer to the study of mediators such as TNF. Unfortunately, in addition to mediating tumor necrosis, TNF also mediates the systemic toxicity of LPS (5). Included here is an abridged summary of the antineoplastic activities of LPS.

ANTITUMOR ACTIVITY OF LPS

Animal Tumor Models

LPS has been extensively studied in tumor models for more than 50 years. It has become clear that the tumoricidal action of endotoxin is complex and that the outcome of endotoxin treatment of tumor-bearing hosts depends on a number of variables including animal species, tumor type, dose, dosing route, and time of application in relation to the stage of tumor development.

Gratia and Linz observed in guinea pigs (6), and Shwartzman and Michailovsky in mice (7), that small doses of endotoxin induced hemorrhagic necrosis of solid tumors within 24 hours after i.v. or i.p. injection. These investigators suggested that tumor necrosis was not a direct effect of LPS, because the doses administered were below the quantity needed to mediate direct cytotoxicity (8). LPS exerts little cytotoxicity directly on tumor cells in vitro, whereas it is a potent stimulator to endogenous antineoplastic mediators. In vivo parental injection of bacterial endotoxin results in extensive hemorrhagic necrosis of many solid murine tumors, but complete regression is observed only in some (9). Sarcomas such as SA1 and Meth A are exquisitely sensitive to LPS, and LPS treatment causes hemorrhagic necrosis and subsequent regression of these tumors (10–12). In many other tumors, however, LPS causes vascular breakdown and extensive central necrosis but not complete regression of surrounding tumor cells, suggesting that central hemorrhagic necrosis and complete tumor regression induced by LPS in sarcomas are independent events. Whereas hemorrhagic necrosis is mediated by the cytotoxicity of LPS to tumor vasculature, complete tumor regression is immunologically mediated (12,13). Thus, maximal therapeutic responses to LPS consist of both significant intratumor central necrosis, which reduces tumor size, and an immune response, which destroys the surrounding peripheral tumor cells (13).

The antineoplastic activities of LPS have been severely limited by systemic toxicity, because near-lethal doses of LPS are required for tumor regression (14,15). Recent studies have been performed in an effort to develop less toxic LPS analogs that retain antitumor activity. Lipid A is a hydrophobic component of LPS that mediates both beneficial and detrimental immunopharmacological activities of LPS (16,17). A number of lipid A analogs have been synthesized (18–21), some of which exhibit potent antitumor activity with less toxicity. In animal tumor models, some of these analogs exhibit potent antitumor activity against various tumors with a 1000-fold decrease in systemic toxicity (18).

Clinical Treatment of Cancer

The clinical development of LPS for the treatment of cancer has been widely studied. The maximal tolerated dose (MTD) of LPS and side effects of intravenously administered LPS have been defined (MTD II ≈ 5 ng/kg body weight) (22,23).

Tumor response to LPS in human has been studied in two clinical trials (24,25). The first trial included 23 colorectal cancer patients and 14 non–small-cell lung cancer patients (24). Intravenous LPS (4.0 ng/kg body weight) was administered once every 2 weeks for 4 injections with cotreatment of ibuprofen (800 mg). No significant tumor response was observed in non–small-cell lung cancer patients, but one colorectal cancer patient experienced a complete remission and two had partial remission (13% response rate). It was suggested that the limited antitumor effect of LPS in this clinical trial may in part be due to the development of LPS tolerance occurring after repeated LPS applications (26). Factors such as interferon-gamma (IFN-γ) have been reported to prevent the development of tolerance (27) and enhance the tumoricidal activity of LPS (28) in animal models. Unfortunately, IFN-γ also significantly increases the cytotoxicity of TNF (29), so that it remains uncertain whether the efficacy of LPS would be significantly improved by cotreatment with IFN-γ.

In the second clinical trial (25) five patients with advanced cancers that failed to respond to conventional chemotherapy were investigated. The patient population included one patient with uterine cervical cancer, three patients with ovarian cancer, and one patient with malignant meningioma of brain. Previous studies (30,31) showed that intradermal administration of LPS had better antitumor effects and significantly less systemic cytotoxicity as compared to intravenous dosing in animal. In this clinical trial, significant tumor response rates and reduced toxicity were also observed when LPS with cyclophosphamide was intradermally administered. The MTD was >1800 ng/kg, and three (two ovarian cancer and one meningioma) of five evaluable tumors had a significant response. These results suggest that a combination of intradermal LPS with standard chemotherapeutic agents may prove useful for the treatment of cancer, but a large-scale, randomized prospective trial is needed prior to reaching a conclusion.

LPS IN CANCER RISK

It has been postulated that environmental exposures that stimulate host-defense mechanisms might decrease the risk of cancer. Endotoxin is a potent activator of host defense, and it has even been suggested that LPS may be necessary to maintain a viable immune system (32). Theoretically, therefore, LPS could prevent the development of cancer through triggering immune responses.

Epidemiological studies have provided evidence that environmental endotoxin may play an important role in preventing cancer development. For instance, cotton textile workers are chronically exposed to airborne endotoxin (up to 5 μg^{-3}) (33,34) and have an unusually low death rate from cancer (35,36). The standardized mortality ratios (SMRs) for cancer in cotton textile workers are consistently lower than the SMRs for all causes. Moreover, death rates from cancer are consistently lower in workers heavily exposed to cotton dust (37) as well as in workers exposed to cotton dust for longer period of times (38). Another interesting observation is that the SMRs for respiratory cancer are even lower than the SMRs for all cancer, possibly due to the direct contact of respiratory system with airborne endotoxin. The low mortality of lung cancer was significantly reduced in cotton textile workers (39) (SMR for lung cancer was 27.3 as compared to 59.6 for all cancers and 88.4 for all causes). In another study (40), the SMR for lung cancer was 54.8 (CI 54–102) as compared to 69.6 for all tumors and 79.3 for all causes.

A decreased incidence of lung cancer is not only found in cotton textile workers but also in other populations with high levels of exposure to bacterial endotoxin, including farmers (41), workers at municipal waste incinerators (42), coal miners (43), and people working with livestock feed (44). Proportional mortality ratio (PMR) is lower for occupations involving exposure to grain and wood dust, which is high in LPS, but the PMR in groups exposed to plywood dust (which does not contain endotoxin because of special treatments in plywood manufacturing) is higher than normal (36).

The potential importance of endotoxin in reducing cancer risk has also been demonstrated in animal experiments, because lower tumor frequency has been observed in animals exposed to endotoxin or cotton dust by inhalation (45). Treatment with *Bordetella pertusis* prior to tumor development also decreases the incidence of spontaneous neoplasm in the AKR mouse (46). When considered together, these studies suggest that LPS stimulation of host defense may be beneficial in prevent tumor formation, although other environmental factors may also be involved in this effect.

TUMORICIDAL MECHANISMS OF LPS

Through evolution, mammals have developed extremely sensitive mechanisms for recognizing endotoxin (47). LPS bioactivity is dependent upon activation of host-defense systems, resulting in inflammation, enhanced immune responses, clotting, and acute phase reaction (48). These biological responses are the mechanistic basis for endotoxin-induced physiological and pathological effects, including tumoricidal activity. It is well documented that LPS acts on numerous cellular functions through the process of cell activation and damage. It has direct effects on most immune cells including macrophages, PMNs, and B cells (49). These cells are activated, or primed, by endotoxin and contribute to LPS-mediated tumor regression. It is generally believed that the macrophage/monocyte system plays a primary role in endotoxin-induced hemorrhagic tumor necrosis. For complete tumor regression, however, other cells, particularly T cells, are also required.

Macrophage Mediation

The concept that macrophages play a role in host defense against foreign intruders was first elaborated by Metchnikoff (50). It has subsequently become clear that macrophages play an essential role in regulating immune and inflammatory responses through phagocytosis, antigen presentation, and cytokine secretion. Mononuclear cell infiltration is a common histological observation in cancers, and clinically the presence of mononuclear cell infiltrates is associated with a better prognosis as compared to similar tumor types lacking infiltrates (51,52). Experimental studies have demonstrated that activation of tumor-infiltrating macrophages can modulate tumor regression (53,54).

LPS induces many changes in macrophages, including increased cytokine secretion, enhanced tumoricidal activity, and phenotype alteration (55). LPS-induced tumor regression and hemorrhagic necrosis requires macrophage activation and the release of macrophage-derived mediators. The interactions between LPS-induced mediators (Table 1) represent a network that controls the expression of macrophage-mediated antitumor activity. Among these mediators, the cytokines TNF and IL-1, reactive oxygen intermediate NO, and arachidonic acid intermediate PGE_2 have been shown to be deeply involved in the mediation of macrophage antitumor activity.

The main effects of these factors in the expression of LPS-mediated antitumor activity are briefly summarized.

1. Macrophage cytokines TNF and IL-1 are primary mediators of LPS-induced tumoricidal responses,

Table 1 Partial List of Macrophage-Derived Mediators

Proteins	Cytokines
	Tumor necrosis factor (TNF)
	Interleukin 1 (IL-1)
	Chemokines
	MIPs
	MCPs
	IL-8
	Coagulation factors
	Enzymes
	Complement
Reactive oxygen intermediates	Nitric oxide (NO)
	Hydrogen peroxide (H_2O_2)
	Superoxide anion radical (O_2^-)
Lipids	Arachidonic acid intermediates
	Prostaglandins
	Leukotrienes
	Thromboxane
	PAF (secreted) PLA_2

and each is cytotoxic or cytostatic to tumor cells (56–58). TNF and IL-1 act synergistically or additively in killing tumor cells (56,59).

2. Macrophage-derived NO is also a potent cytotoxin, which is directly cytotoxic to tumor cells (58) and induces tumor cell apoptosis (60). NO is also a secondary mediator of many cytokine effects, and NO release can be further stimulated by TNF and IL-1 as part of a feedforward signaling (61). The role of NO in mediating the antitumor activity of human macrophage has been debated, since human macrophages release only low levels of NO as compared to murine macrophages, which release high levels of NO (62,63).

3. PGE_2 inhibits tumor cell growth and enhances IL-1 inhibition of tumor cell growth in some instances (64,65). However, the increased production of PGE_2 from macrophages is inversely correlated with antitumor activity in other cases (66). These divergent effects may be in part because high levels of PGE_2 can suppress macrophage activation and decrease cytokine production (67–69).

In addition to secreted factors, direct macrophage–tumor cell contact participates in the antitumor activity of LPS-activated macrophages. For example, LPS-activated macrophages express a membrane-associated form of TNF, which is cytotoxic to target cells through cell-to-cell contact (70). Morphological study of macrophage–tumor cell interaction indicates that lysosomal organelles can be transferred from cytotoxic macrophages directly into the cytoplasm of adjacent target tumor cells (71). Electron microscopic observations reveal that the direct destruction of target tumor cells by activated macrophages is a nonphagocytic, lytic process (71,72) and that firm physical binding of macrophages to tumor cells is required to activate this pathway (73).

T-Cell Mediation

Although T cells are not stimulated directly by LPS, several observations suggest that T cells participate in LPS-induced tumor regression in vivo. Only immunogenic tumors undergo complete regression in response to LPS (11,74), possibly because T-cell activation is required to eliminate tumor cells surrounding the necrotic core. In these cases, tumor regression depends on the acquisition by the host of a sufficient number of tumor-sensitized CD4+ T cells at the time that endotoxin is given (75). Complete tumor regression fails to occur in TXB mice that are incapable of generating concomitant immunity because they have been made T-cell–deficient by thymectomy or gamma-radiation, although tumor hemorrhagic necrosis can be induced in these mice (11,74). Tumors in TXB mice can be primed for LPS-induced regression by i.v. infusion of T cells from concomitantly immune donors bearing the same LPS-susceptible tumors (13,75). The donor T cells that prime the recipient tumors for LPS-induced regression have been shown to be the Ly-1+2- T cells, since anti-Ly-1 antibody but not anti-Ly-2 antibody inhibits this effect (13,75), suggesting that the antitumor activity of LPS is optimized by Ly-1+2- tumor-sensitized T cells

In vivo activation of T cells by LPS facilitates the development of long-lived immunity to tumor cell implantation, mediated by host of T cells that can be passively transferred to induce an immunity in normal receipts (13,75). The activation of T cells by LPS requires antigen-presenting cells (APC). LPS induces the potentiation of antigen-specific proliferative response of T-helper (TH) cells, and the effect of LPS on specific TH cell responses is dependent upon the interaction between TH lymphocytes and APC through antigen-specific recognition (76). T-cell–endothelial interactions may also play a crucial role in promoting T-cell activation during LPS-induced inflammation. This interaction is important in T-cell activation for secretion of IFN-γ (77), which can increase the cytotoxic effects of other macrophage-derived cytokines. The presence of T lymphocytes has been shown to be required for optimal production of various antitumor factors from LPS-stimulated monocytes such as procoagulant tissue factor (TF) (78). Nontoxic MPL (monophosphoryl lipid A) significantly reduces the expression of suppressor T-cell (Ts) activity (79), suggesting that some of the adjuvant effects of LPS in tumor regression may be attributed to eliminating the inhibitory effects of Ts.

B-Cell Mediation

Although LPS has a strong direct stimulatory activity on B cells (80), the exact role of B cell activation in LPS-mediated tumor regression has not been well studied. Exposure to LPS accelerates the phenotypic maturation of pre-B and immature B lymphocytes (81). LPS also induces the proliferation of a selective subpopulation of mature B cells, which differentiate into antibody-secreting plasma cells (82,83). It has been suggested that a membrane receptor is involved in B-lymphocyte activation by LPS, because insertion of a membrane fraction from LPS-responsive B cells into LPS-hyporesponsive C3H/HeJ murine B cells confers increased LPS sensitivity (84). Cytotoxic antibody responses have been implicated in the mediation of LPS-induced tumor regression and decreased incidence of spontaneous leukemia in AKR mice (46).

Endothelial Cell Mediation

LPS exerts both direct and indirect effects on endothelial cells that contribute to the antineoplastic activity of LPS. LPS is cytotoxic to human endothelium in the presence of protein synthesis inhibitors (85). The direct action of LPS on endothelial cells is mediated by soluble CD14 (86,87), and anti-CD14 antibodies reduce the response of endothelial cells to LPS (88). LPS can also activate endothelial cells indirectly through the effects of macrophage-derived cytokines, which upregulate endothelial activation antigens, procoagulant activity, and cytotoxicity (89–92). TNF is a predominant mediator of these endothelial effects. It promotes coagulation (92), inhibits angiogenesis (93–95), and induces structural damage of endothelial cells (95–100). LPS-induced endothelial injury damages the barrier function of the endothelial tissue in blood vessels and produces increased permeability to fluids and macromolecules. All these direct and indirect effects of LPS on endothelial cells occupy an important role in the mechanisms of LPS-induced tumor hemorrhagic necrosis, a characteristic that is similar to a local Shwartzman reaction (101).

A number of other vasoactive agents are induced by LPS, including histamine (102), serotonin (102,103), catecholamines (104), adrenaline (105), and bradykinin (106). LPS also stimulates vasodilatation directly by enhancing the release of endothelium-derived relaxing factor (EDRF or NO) (107). In addition, LPS-induced endothelial activation stimulates adherence of PMN, which are required for the development of tumor necrosis.

TUMOR NECROSIS FACTOR

A significant body of evidence suggests that TNF is a major mediator of LPS-induced tumor regression. TNF was identified functionally as a serum factor produced in endotoxin-treated mice and was shown to cause hemorrhagic necrosis of established sarcomas (4). The purification of TNF from macrophages and the expression of rTNF enabled the confirmation that TNF is the mediator of LPS-induced tumor regression (106). Endotoxin-induced tumor regression occurs in association with the intratumoral synthesis of TNF (13). TNF treatment causes tumor hemorrhagic necrosis, and anti-TNF antibodies prevent LPS-induced tumor regression (13).

Antiproliferative Effect of TNF In Vitro

Early studies suggested that TNF is selectively cytotoxic to tumor cells and transformed cell lines (4), with 16 out of 34 tumor cell lines exhibiting sensitivity to TNF as compared to none of the normal human cell lines from colon, lung, and fetal skin. Subsequent studies, however, revealed that TNF is also cytotoxic to certain types of normal cells including endothelial cells and smooth muscle cells (95,108), adipocytes (109), fibroblasts (110), keratinocytes (111), and hematopoietic progenitor cells (112). The sensitivity of cells to TNF has been shown to be associated with certain cellular phenotypic characteristics. For instance, cell communication through gap junction (113) or other cell contact (114) may diminish TNF sensitivity. TNF receptor expression also influences sensitivity (115), and the presence of growth factors can modulate the antiproliferative effect of TNF (116). In addition, cytokines and oncogenes induced by TNF itself (116,117) have been implicated in conferring protection against TNF cytotoxicity.

The mechanisms of TNF-induced cytotoxicity are attributed to the development of apoptosis or necrosis (118). Although the mechanism for TNF-induced necrotic cell death remains unclear, several molecular targets and events have been linked to TNF-induced apoptosis. Early studies implicate proteases in TNF cytotoxicity since TNF-induced apoptosis is inhibited by a variety of protease inhibitors (119,120).

Recently, it became clear that TNF induces cell apoptosis by activating a family of "caspase" proteases via the TNF receptor 1 (TNFR1)–mediated signal pathway (Fig. 1). Following receptor aggregation triggered by TNF binding, activated TNFR1 recruits an adaptor molecule termed TNFR1-associated death domain protein (TRADD) (121). Once TRADD binds to the TNFR1 death domain in the presence of TNF, it causes recruitment of both Fas-associating protein with death domain (FADD; MORT1) and receptor interacting protein (RIP) through the C-terminal death domain and recruitment of TNF receptor-associated factor 2 (TRAF2) through N-terminal TRAF-interacting domain (122). These two TNFR1-TRADD signaling cascades appear to bifurcate at TRADD, initiating two different

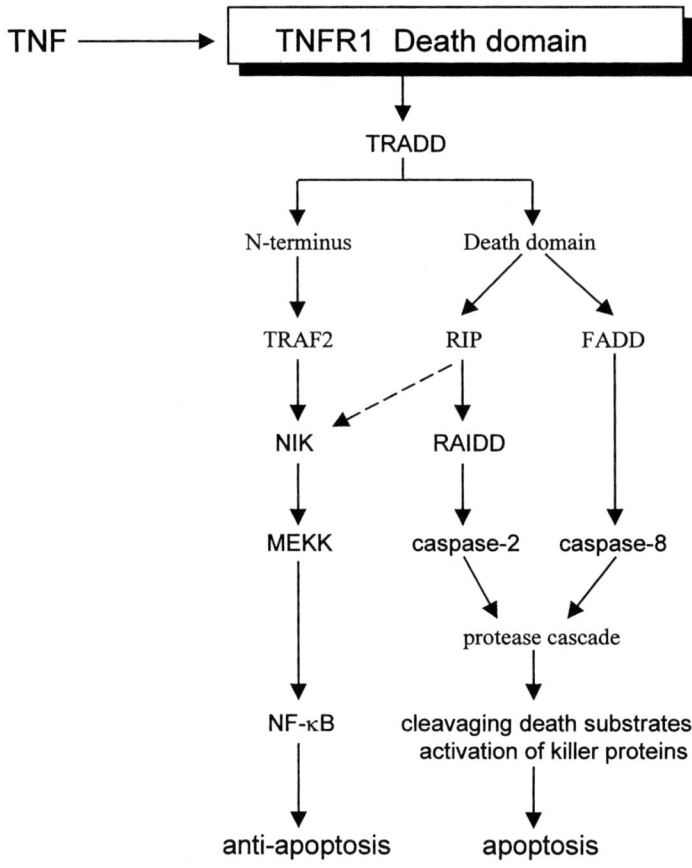

Fig. 1 TNF-triggered apoptotic and anti-apoptotic pathways through the death domain of TNF receptor 1.

pathways responsible for apoptosis induction and apoptosis protection, respectively.

A protease termed caspase-8 (cysteine aspase-8; FLICE for FADD-like ICE; MACH for MORT1-associated CED-3 homolog) has been found to bind FADD, and overexpression of caspase-8 causes apoptosis (123,124), possibly by cleaving "death substrates" such as lamin, actin, and polymerase or by activating proteins that induce DNA fragmentation and kill the cells. Another death domain–containing adaptor termed RIP-associated Ich-1/CED-3 homologous protein with death domain (RAIDD) has recently been identified as an RIP-binding protein (125). RAIDD binds RIP through its death domain and recruits caspase-2 (Ich-1) to RIP to activate a protease cascade leading to apoptosis via an alternate pathway. A TRAF2-NF-κB–mediated anti-apoptotic pathway is also initiated following TNF binding to its target cells (126). Thus, the ultimate effect of TNF on target cells is the result of a balance between TNF-triggered apoptotic and anti-apoptotic activities.

Tumoricidal Effect of TNF In Vivo

Treatment with TNF produces variable effects depending on the tumor models, but it rarely results in complete regression of the tumor tissue surrounding the central hemorrhagic necrosis core. This is quite similar to the effects of endotoxin therapy. Like LPS, the sensitivity of tumors to TNF is dependent upon tumor immunogenicity and vasculature. The indirect effects of TNF on tumor vasculature and host-specific T-cell–mediated immunity, and not its direct antiproliferative actions, predominately account for the mechanisms of TNF antitumor activity.

The importance of TNF in tumor vasculature in TNF-induced tumor regression is strongly supported by the key observation that TNF selectively eradicates vascularized tumors but is much less effective in killing avascular implants (100,127–129). The marked effects of TNF on tumor vasculature have been observed in both in vitro and in vivo tumor models (114,130). TNF is a stimulator of angiogenesis (131) but can also be an angiogenesis inhibitor (93–95). The local concentration of TNF is critical in this respect because low doses of TNF induce angiogenesis, whereas high doses inhibit angiogenesis. TNF induces rearrangements of actin filaments with structural damage to the endothelial cells and loss of tight junctions that cause capillary leakage syndrome as plasma proteins and water leak into the tissues (95,97–100). TNF increases the procoagulant activity of endothelial cells by enhancing the expression of tissue factor and suppressing cofactor activity (thrombomodulin) for the anticoagulant protein C (89–92). This results in the development of clotted capillaries, which reduce perfusion of tumors and cause tumor necrosis.

Meth A–bearing mice are specifically resistant to tumor rechallenge after TNF treatment, indicating that TNF treatment can induce the development of specific immunity (132). In the SA1 tumor model, complete tumor regression subsequent to tumor necrosis only occurs when tumors grow in immunocompetent mice (130) and only in mice with highly immunogenic tumors (133). These results indicate that antitumor immunity is critical for complete tumor regression induced by TNF. In addition, TNF has a wide range of effects on other immune cells (134), and some of these cells, including tumoricidal macrophages (135), cytotoxic T cells (136), NK cells (137), and neutrophils (138), also play an indispensable role in TNF-mediated antitumor activity.

Clinical Application

The therapeutic application of TNF as a single and systemic antitumor agent is generally limited by the toxic side effects at tumoricidal doses (130,134). A number of strategies have been proposed to avoid the limitations imposed by TNF toxicity. High doses of TNF have been given intra-arterially directly into the tumor compartment (139). It may be possible to identify tumors sensitive to a nontoxic level of TNF. Nontoxic doses of TNF have been given in combination with other antitumor agents. In this case, while synergistic antitumor effects have been reported (140), enhanced side effects can also occur (29,141). Much effort has been focused on redesigning a TNF molecule with maximum efficacy as an antitumor agent but reduced side effects. Finally, chemical modification of TNF with polyethylene glycol (PEG) may selectively increase antitumor potency and reduce systemic toxicity (142).

CONCLUSIONS

There is a long history of the use of LPS as an experimental therapy of cancer. Advances in this area have revealed that complete tumor regression by LPS or TNF requires both nonspecific inflammatory response and specific immunity. Investigation of LPS-mediated tumor responses will continue to advance our undersanding of the complexicity of host-derived responses to invasive diseases. It is hoped that a better understanding of these mechanisms will lead to improved therapies for patients with sensitive tumors such as sarcoma, lymphoma, or other tumors of mesodermal origin, which have been demonstrated in clinical and experimental studies to be the most sensitive to LPS and TNF. The major limitation of this approach is the toxicity of high-dose LPS or TNF. For future clinical application, better LPS analogs and recombinant TNF with more therapeutic activity and less toxicity remain to be identified. In addition, it may be possible to reduce the side effects of TNF by co-administration of cytoprotective compounds. For instance, we recently showed that CNI-1493 (a low molecular inhibitor of TNF translation) protects against IL-2 toxicity without affecting antitumor efficacy (143). It is plausible that a similar approach may prevent TNF side effects in cancer patients.

REFERENCES

1. Coley WB. The treatment of malignant tumors by repeated inoculations of erysipelas. Am J Med Sci 1893; 105:487–511.
2. Starnes CO. Coley's toxins in perspective. Nature 1992; 357:11–12.
3. Shear MJ, Turner FC. Chemical treatment of tumors: isolation of hemorrhagic-producing fraction from *Serratia marcescens* (*Bacillus prodigiosus*) culture filtrate. J Natl Cancer Inst 1943; 4:81.
4. Carswell EA, Old LJ, Kassel RL, et al. An endotoxin-induced serum factor that causes necrosis of tumors. Proc Natl Acad Sci USA 1975; 72:3666–3670.
5. Tracey KJ, Beutler B, Lowry SF, et al. Shock and tissue injury induced by recombinant human cachectin. Science 1986; 234:470–474.
6. Gratia A, Linz R. Le phenomene de Shwartzman dans la sarcome du cobaye. CR Soc Biol 1931; 108:427.
7. Schwartzman G, Michailovsky N. Phenomenon of local skin reactivity to bacterial filtrates in the treatment of mouse sarcoma 180. Proc Soc Exp Biol Med 1932; 29:737.
8. Shapiro CJ. The effect of a toxic carbohydrate complex from *S. enteritidis* on transplantable rat tumors in tissue culture. Am J Hyg 1940; 31:14.

9. Shear MJ. Chemical treatment of tumors. IX. Reactions of mice with primary subcutaneous tumors to injection of hemorrhagic-producing bacterial polysaccharide. J Natl Cancer Inst 1944; 4:461.
10. Berendt MJ, North RJ, Kirstein DP. The immunological basis of endotoxin-induced tumor regression. requirement for a pre-existing state of concomitant antitumor immunity. J Exp Med 1978; 148:1560–1569.
11. Berendt MJ, North RJ, Kirstein DP. The immunological basis of endotoxin-induced tumor regression. requirement for T-cell-mediated immunity. J Exp Med 1978; 148:1550–1559.
12. Freudenberg N, Joh K, Westphal O, et al. Haemorrhagic tumour necrosis following endotoxin administration. I. Communication: morphological investigation on endotoxin-induced necrosis of the methylcholanthrene (meth A) tumour in the mouse. Virchows Arch A Pathol Anat Histopathol 1984; 403:377–389.
13. North RJ, Havell EA. The antitumor function of tumor necrosis factor (TNF) II. analysis of the role of endogenous TNF in endotoxin-induced hemorrhagic necrosis and regression of an established sarcoma. J Exp Med 1988; 167:1086–1099.
14. Berendt MJ, Saluk PH. Tumor inhibition in mice by lipopolysaccharide-induced peritoneal cells and an induced soluble factor. Infect Immun 1976; 14:965–969.
15. Silverstein R, Christoffersen CA, Morrison DC. Modulation of endotoxin lethality in mice by hydrazine sulfate. Infect Immun 1989; 57:2072–2078.
16. Morrison DC, Ryan JL. Bacterial endotoxin and host immune response. Adv Immunol 1983; 28:293.
17. Somlyo B, Csanky E, Shi XM, et al. Molecular requirements of endotoxin (ET) actions: changes in the immune adjuvant, TNF liberating and toxic properties of endotoxin during alkaline hydrolysis. Int J Immunopharmacol 1992; 14:131.
18. Yang D, Satoh M, Ueda H, et al. Activation of tumor-infiltrating macrophages by a synthetic lipid a analog (ono-4007) and its implication in antitumor effects. Cancer Immunol Immunother 1994; 38:287–293.
19. Kusama T, Soga T, Shioya E, et al. Synthesis and antitumor activity of lipid a analogs having a phosphonooxyethyl group with alpha- or beta-configuration at position 1. Chem Pharm Bull (Tokyo) 1990; 38:3366–3372.
20. Johnson AG, Tomai MA, Chen YF, et al. A comparison of the immunomodulating properties of two forms of monophosphoryl lipid A analogues. J Immunother 1991; 10:398–404.
21. Tanamoto KI, Abe C, Homma Y, et al. A compound possessing antitumor and interferon-inducing activities derived from the common antigen (oep) of *Pseudomonas aeruginosa*. J Biochem (Tokyo) 1978; 83:711–718.
22. Engelhardt R, Mackensen A, Galanos C. Phase I trial of intravenously administered endotoxin (*Salmonella abortus equi*) in cancer patients. Cancer Res 1991; 51:2524–2530.
23. Engelhardt R, Mackensen A, Galanos C, et al. Biological response to intravenously administered endotoxin in patients with advanced cancer. J Biol Response to intravenously administered endotoxin in patients with advanced cancer. J Biol Response Modif 1990; 9:480.
24. Engelhardt R, Otto F, Mackensen A, et al. Endotoxin (*Salmonella abortus equi*) in cancer patients. clinical and immunological findings. Prog Clin Biol Res 1995; 392:253–261.
25. Goto S, Sakai S, Kera J, et al. Intradermal administration of lipopolysaccharide in the treatment of human cancer. Cancer Immunol Immunother 1996; 42:255–261.
26. Mackensen A, Galanos C, Wehr U, et al. Endotoxin tolerance: regulation of cytokine production and cellular changes in response to endotoxin application in cancer patients. Eur Cytokine Netw 1992; 3:571–579.
27. Mackensen A, Galanos C, Engelhardt R. Modulating activity of interferon-gamma on endotoxin-induced cytokine production in cancer patients. Blood 1991; 78:3254–3258.
28. Lorence RM, Edwards CK, 3d, Walter RJ, et al. In vivo effects of recombinant interferon-gamma: augmentation of endotoxin-induced necrosis of tumors and priming of macrophages for tumor necrosis factor-alpha production. Cancer Lett 1990; 53:223–229.
29. Schiller JH, Witt PL, Storer B, et al. Clinical and biologic effects of combination therapy with gamma-interferon and tumor necrosis factor. Cancer 1992; 69:562–571.
30. Inagawa H, Chiba Y, Nishizawa T, et al. Intradermally administered lipopolysaccharide stimulates TNF-driven inflammation systemically to cure implanted tumor. Eur Cytokine Netw 1994; 5:273.
31. Chiba Y, Inagawa H, Okutomi T, et al. Antitumor effect with complete regression by lipopolysaccharide (LPSp) purified from *Pantoea agglomerans* by intradermal administration. Eur Cytokine Netw 1994; 5:231.
32. Rietschel ET, Brade H. Bacterial endotoxins. Sci Am 1992; 267:54–61.
33. Cavagna G, Foa V, Vigliani SC. Effects in man and rabbits of inhalation of cotton dust or extracts or purified endotoxin. Br J Ind Med 1969; 26:314–321.
34. Cinkotai FF, Lockwood MG, Rylander R. Airborne bacteria and the prevalence of byssinotic symptoms in cotton mills. Am Ind Hyg Assoc 1977; 38:554–559.
35. Enterline PE, Sykora JL, Keleti G, et al. Endotoxins, cotton dust, and cancer. Lancet 1985; 2:934–935.
36. Rylander R. Environmental exposures with decreased risks for lung cancer? Int J Epidemiol 1990; 19(suppl 1):S67–72.
37. Merchant JA, Ortmeyer C. Mortality of employees of two cotton mills in North Carolina. Chest 1981; 79:6s–11s.
38. Daum SM, Seidman H, Heimann H, et al. Mortality experience of a cohort of cotton textile workers. In: Final Progress Report on Contract No. HSM 99-72-71 (NIOSH), March 1, 1975.
39. Enterline PE. Mortality among asbestos products workers in the United States. NY Acad Sci 1965; 132:156–165.
40. Henderson V, Enterline PE. An unusual mortality experience in cotton textile workers. JOM 1973; 15:717–719.

41. Olsen JH, Dragsted L, Autrup H. Cancer risk and occupational exposure to aflatoxins in Denmark. Br J Cancer 1988; 58:392–396.
42. Rapiti E, Sperati A, Fano V, et al. Mortality among workers at municipal waste incinerators in Rome: a retrospective cohort study. Am J Ind Med 1997; 31: 659–661.
43. Kjuus H, Skjaerven R, Langard S, et al. A case referent study of lung cancer, occupational exposures and smoking. I comparison of title-based and exposure-based occupational information. Scand J Work Environ Health 1986; 12:193–202.
44. Vella V, Ng T. Analysis of occupational mortality in England and Wales, Scotland, Norway and Finland. La Medicina Del Lavora 1986; 77:162–171.
45. Lange JH. A review of epidemiological evidence of anti-cancer properties of dust. In: Wakelyn PJ, Jacobs RR, eds. Proceedings from the Twelfth Cotton Dust Conference, 1988: 124–127.
46. Sinkovics JG, Ahearn MJ, Shirato E, et al. Viral leukemogenesis in immunologically and hematologically altered mice. J Reticuloendothel Soc 1970; 8:474–492.
47. Le Grand EK. Endotoxin as an alarm signal of bacterial invasion: current evidence and implications. J Am Vet Med Assoc 1990; 197:454–456.
48. Dinarello CA. Interleukin-1. Rev Infect Dis 1984; 6: 51–95.
49. Chaby R, Girard R. Interaction of lipopolysaccharides with cells of immunological interest. Eur Cytokine Netw 1993; 4:399–414.
50. Metchnikoff E. Limmununite dans les Maladies Infectieuses. Masson, Paris, 1901.
51. Flamm J. Die Bedeutung der lokalen tumorassoziierten Infiltration mononuklearer Leukozyten für persistierende Tumorformationen nach transurethraler Resektion (TUR) von Harnblasenkarzinomen. Z Urol Nephrol 1985; 78:65–70.
52. Pedersen L, Zedeler K, Holck S, et al. Medullary carcinoma of the breast. prevalence and prognostic importance of classical risk factors in breast cancer. Eur J Cancer 1995; 31A:2289–2295.
53. Singh RK, Fidler IJ. Synergism between human recombinant monocyte chemotactic and activating factor and lipopolysaccharide for activation of antitumor properties in human blood monocytes. Lymphokine Cytokine Res 1993; 12:285–291.
54. Xu ZL, Bucana CD, Fidler IJ. In vitro activation of murine kupffer cells by lymphokines or endotoxins to lyse syngeneic tumor cells. Am J Pathol 1984; 117: 372–379.
55. Drysdale BE, Agarwal S, Shin HS. Macrophage-mediated tumoricidal activity: mechanisms of activation and cytotoxicity. Prog Allergy 1988; 40:111–161.
56. Ichinose Y, Tsao JY, Fidler IJ. Destruction of tumor cells by monokines released from activated human blood monocytes: evidence for parallel and additive effects of IL-1 and TNF. Cancer Immunol Immunother 1988; 27:7–12.
57. Onozaki K, Matsushima K, Aggarwal BB, et al. Human interleukin 1 is a cytocidal factor for several tumor cell lines. J Immunol 1985; 135:3962–3968.
58. Ben-Efraim S, Tak C, Fieren MJ, et al. Activity of human peritoneal macrophages against a human tumor: role of tumor necrosis factor-alpha, PGE2 and nitrite, in vitro studies. Immunol Lett 1993; 37:27–33.
59. Ruggiero V, Baglioni C. Synergistic anti-proliferative activity of interleukin 1 and tumor necrosis factor. J Immunol 1987; 138:661–663.
60. Xie K, Dong Z, Fidler IJ. Activation of nitric oxide synthase gene for inhibition of cancer metastasis. J Leukoc Biol 1996; 59:797–803.
61. Kuemmerle JF. Synergistic regulation of NOS II expression by IL-1 beta and TNF-alpha in cultured rat colonic smooth muscle cells. Am J Physiol 1998; 274: G178–85.
62. Takema M, Inaba K, Uno K, et al. Effect of L-arginine on the retention of macrophage tumoricidal activity. J Immunol 1991; 146:1928–1933.
63. Werner ER, Werner-Felmayer G, Fuchs D, et al. Biochemistry and function of pteridine synthesis in human and murine macrophages. Pathobiology 1991; 59: 276–279.
64. Elliott GR, Tak C, Bonta IL. Prostaglandin E2 enhances, and leukotriene c4 inhibits, interleukin-1 inhibition of Wehi-3b cell growth. Cancer Immunol Immunother 1989; 28:74–76.
65. Santoro MG, Philpott GW, Jaffe BM. Inhibition of tumour growth in vivo and in vitro by prostaglandin E. Nature 1976; 263:777–779.
66. Bonta IL, Ben-Efraim S. Involvement of inflammatory mediators in macrophage antitumor activity. J Leukoc Biol 1993; 54:613–626.
67. Heidenreich S, Gong JH, Schmidt A, et al. Macrophage activtion by granulocyte/macrophage colony-stimulating factor. priming for enhanced release of tumor necrosis factor-alpha and prostaglandin E2. J Immunol 1989; 143:1198–1205.
68. Renz H, Gong JH, Schmidt A, et al. Release of tumor necrosis factor-alpha from macrophages. enhancement and suppression are dose-dependently regulated by prostaglandin E2 and cyclic nucleotides. J Immunol 1988; 141:2388–2393.
69. Bonta IL, Ben-Efraim S, Tak C, et al. Involvement of prostaglandins and cytokines in antitumor cytostatic activity of human peritoneal macrophages. Adv Prostaglandin Thromboxane Leukot Res 1991; 21B:879–882.
70. Kriegler M, Perez C, De Fay K, et al. A novel form of TNF/cachectin is a cell surface cytotoxic transmembrane protein: ramifications for the complex physiology of TNF. Cell 1988; 53:45–53.
71. Bucana C, Hoyer LC, Hobbs B, et al. Morphological evidence for the translocation of lysosomal organelles from cytotoxic macrophages into the cytoplasm of tumor target cells. Cancer Res 1976; 36:4444–4458.
72. Meltzer MS, Tucker RW, Breuer AC. Interaction of BCG-activated macrophages with neoplastic and noneoplastic cell lines in vitro: cinemicrographic analysis. Cell Immunol 1975; 17:30–42.
73. Marino PA, Adams DO. Interaction of Bacillus calmette–guerin-activated macrophages and neoplastic cells in vitro. I. conditions of binding and its selectivity. Cell Immunol 1980; 54:11–25.

74. North RJ. The therapeutic significance of concomitant antitumor immunity. II. passive transfer of concomitant immunity with ly-1+2- T cells primes established tumors in T cell-deficient receipients for endotoxin-induced regression. Cancer Immunol Immunother 1984; 18:75–79.
75. Digiacomo A, North RJ. Subtherapeutic numbers of tumour-sensitized, L3T4+, ly 1+2-T cells are needed for endotoxin to cause regression of an established immunogenic tumour. Immunology 1987; 60:367–373.
76. Bismuth G, Duphot M, Theze J. LPS and specific T cell responses: interleukin 1 (IL 1)-independent amplification of antigen-specific T helper (TH) cell proliferation. J Immunol 1985; 134:1415–1421.
77. Tennenberg SD, Weller JJ. Endotoxin activates T cell interferon-gamma secretion in the presence of endothelium. J Surg Res 1996; 63:73–76.
78. Edwards RL, Rickles FR. The role of human T cells (and T cell products) for monocyte tissue factor generation. J Immunol 1980; 125:606–609.
79. Baker PJ, Taylor CE, Stashak PW, et al. Inactivation of suppressor T cell activity by the nontoxic lipopolysaccharide of Rhodopseudomonas sphaeroides. Infect Immun 1990; 58:2862–2868.
80. Chiller JM, Skidmore BJ, Morrison DC, et al. Relationship of the structure of bacterial lipopolysaccharides to its function in mitogenesis and adjuvanticity. Proc Natl Acad Sci U S A 1973; 70:2129–2133.
81. Paige CJ, Kincade PW, Ralph P. Murine B cell leukemia line with inducible surface immunoglobulin expression. J Immunol 1978; 121:641–647.
82. Morrison DC, Ryan JL. Bacterial endotoxins and host immune responses. Adv Immunol 1979; 28:293–450.
83. Hammerling U, Chin AF, Abbott J. Ontogeny of murine B lymphocytes: sequence of B-cell differentiation from surface-immunoglobulin-negative precursors to plasma cells. Proc Natl Acad Sci USA 1976; 73:2008–2012.
84. Jakobovits A, Sharon N, Zan-Bar I. Acquisition of mitogenic responsiveness by nonresponding lymphocytes upon insertion of appropriate membrane components. J Exp Med 1982; 156:1274–1279.
85. Pohlman TH, Munford RS, Harlan JM. Deacylated lipopolysaccharide inhibits neutrophil adherence to endothelium induced by lipopolysaccharide in vitro. J Exp Med 1987; 165:1393–1402.
86. Pugin J, Schurer-Maly CC, Leturcq D, et al. Lipopolysaccharide activation of human endothelial and epithelial cells is mediated by lipopolysaccharide-binding protein and soluble CD14. Proc Natl Acad Sci USA 1993; 90:2744–2748.
87. Haziot A, Rong GW, Silver J, et al. Recombinant soluble CD14 mediates the activation of endothelial cells by lipopolysaccharide. J Immunol 1993; 151:1500–1507.
88. von Asmuth EJ, Dentener MA, Bazil V, et al. Anti-CD14 antibodies reduce responses of cultured human endothelial cells to endotoxin. Immunology 1993; 80:78–83.
89. Bauer KA, ten Cate H, Barzegar S, et al. Tumor necrosis factor infusions have a procoagulant effect on the hemostatic mechanism of humans. Blood 1989; 74:165–172.
90. Nawroth PP, Bank I, Handley D, et al. Tumor necrosis factor/cachectin interacts with endothelial cell receptors to induce release of interleukin 1. J Exp Med 1986; 163:1363–1375.
91. Nawroth PP, Stern DM. Modulation of endothelial cell hemostatic properties by tumor necrosis factor. J Exp Med 1986; 163:740–745.
92. Bevilacqua MP, Pober JS, Majeau GR, et al. Recombinant tumor necrosis factor induces procoagulant activity in cultured human vascular endothelium: characterization and comparison with the actions of interleukin 1. Proc Natl Acad Sci USA 1986; 83:4533–4537.
93. Sato N, Fukuda K, Nariuchi H, et al. Tumor necrosis factor inhibiting angiogenesis in vitro. J Natl Cancer Inst 1987; 79:1383–1391.
94. Mano-Hirano Y, Sato N, Sawasaki Y, et al. Inhibition of tumor-induced migration of bovine capillary endothelial cells by mouse and rabbit tumor necrosis factor. J Natl Cancer Inst 1987; 78:115–120.
95. Sato N, Goto T, Haranaka K, et al. Actions of tumor necrosis factor on cultured vascular endothelial cells: morphologic modulation, growth inhibition, and cytotoxicity. J Natl Cancer Inst 1986; 76:1113–1121.
96. Stolpen AH, Golan DE, Pober JS. Tumor necrosis factor and immune interferon act in concert to slow the latral diffusion of proteins and lipids in human endothelial cell membranes. J Cell Biol 1988; 107:781–789.
97. Stolpen AH, Guinan EC, Fiers W, et al. Recombinant tumor necrosis factor and immune interferon act singly and in combination to reorganize human vascular endothelial cell monolayers. Am J Pathol 1986; 123:16–24.
98. Damle NK, Doyle LV. IL-2-activated human killer lymphocytes but not their secreted products mediate increase in albumin flux across cultured ednothelial monolayers. implications for vascular leak syndrome. J Immunol 1989; 142:2660–2669.
99. Kreil EA, Greene E, Fitzgibbon C, et al. Effects of recombinant human tumor necrosis factor alpha, lymphotoxin, and Escherichia coli lipopolysaccharide on hemodynamics, lung microvascular permeability, and eicosanoid synthesis in anesthetized sheep. Circ Res 1989; 65:502–514.
100. Brett J, Gerlach H, Nawroth P, et al. Tumor necrosis factor/cachectin increases permeability of endothelial cell monolayers by a mechanism involving regulatory G proteins. J Exp Med 1989; 169:1977–1991.
101. Movat HZ, Cybulsky MI, Colditz IG, et al. Acute inflammation in gram-negative infection: endotoxin, interleukin 1, tumor necrosis factor, and neutrophils. Fed Proc 1987; 46:97–104.
102. Davis RB, Balley WL, Hanson NP. Modification of serotonin and histamine release after E. coli endotoxin administration. Amer J Physiol 1963; 205:560–566.
103. David RB, Meeker WRj, Balley WL. Serotonin release by bacterial endotoxin. Proc Soc Exp Biol NY 1961; 108:774–776.

104. Genberg JC, Lillehei RC, Moran WH, et al. Effect of endotoxin on plasma catecholamines and serum serotonin. Proc Soc Exp Biol NY 1959; 102:335–337.
105. Omas L, Zweifach BW, Benacerraf B. Mechanisms in the production of tissue damage and shock by endotoxins. Trans Ass Am Phys 1957; 70:54.
106. Palladino MA, Jr., Patton JS, Figari IS, et al. Possible relationships between in vivo antitumour activity and toxicity of tumour necrosis factor-alpha. Ciba Found Symp 1987; 131:21–38.
107. Walter R, Schaffner A, Schoedon G. Differential regulation of constitutive and inducible nitric oxide production by inflammatory stimuli in murine endothelial cells. Biochem Biophys Res Commun 1994; 202:450–455.
108. Schuger L, Varani J, Marks RM, et al. Cytotoxicity of tumor necrosis factor-alpha for human umbilical vein endothelial cells. Lab Invest 1989; 61:62–68.
109. Kawakami M, Watanabe N, Ogawa H, et al. Cachectin/tnf kills or inhibits the differentiation of 3T3-L1 cells according to developmental stage. J Cell Physiol 1989; 138:1–7.
110. Palombella VJ, Vilcek J. Mitogenic and cytotoxic actions of tumor necrosis factor in Balb/c 3T3 cells. role of phospholipase activation. J Biol Chem 1989; 264:18128–18136.
111. Pillai S, Bikle DD, Eessalu TE, et al. Binding and biological effects of tumor necrosis factor alpha on cultured human neonatal foreskin keratinocytes. J Clin Invest 1989; 83:816–821.
112. Broxmeyer HE, Williams DE, Lu L, et al. The suppressive influences of human tumor necrosis factors on bone marrow hematopoietic progenitor cells from normal donors and patients with leukemia: synergism of tumor necrosis factor and interferon-gamma. J Immunol 1986; 136:4487–4495.
113. Fletcher WH, Shiu WW, Ishida TA, et al. Resistance to the cytolytic action of lymphotoxin and tumor necrosis factor coincides with the presence of gap junctions uniting target cells. J Immunol 1987; 139:956–962.
114. Gerlach H, Lieberman H, Bach R, et al. Enhanced responsiveness of endothelium in the growing/motile state to tumor necrosis factor/cachectin [published erratum appears in J Exp Med 1989 nov 1; 170(5):1793]. J Exp Med 1989; 170:913–931.
115. Johnson SE, Baglioni C. Tumor necrosis factor receptors and cytocidal activity are down-regulated by activators of protein kinase C. J Biol Chem 1988; 263:5686–5692.
116. Sugarman BJ, Lewis GD, Eessalu TE, et al. Effects of growth factors on the antiproliferative activity of tumor necrosis factors. Cancer Res 1987; 47:780–786.
117. Kohase M, Zhang YH, Lin JX, et al. Interleukin-1 can inhibit interferon-beta synthesis and its antiviral action: comparison with tumor necrosis factor. J Interferon Res 1988; 8:559–570.
118. Laster SM, Mackenzie JM, Jr. Bleb formation and f-actin distribution during mitosis and tumor necrosis factor-induced apoptosis. Microsc Res Tech 1996; 34:272–280.
119. Baglioni C, Ruggiero V, Latham K, et al. Cytocidal activity of tumour necrosis factor: protection by protease inhibitors. Ciba Found Symp 1987; 131:52–63.
120. Ruggiero V, Johnson SE, Baglioni C. Protection from tumor necrosis factor cytotoxicity by protease inhibitors. Cell Immunol 1987; 107:317–325.
121. Hsu H, Xiong J, Goeddel DV. The TNF receptor 1-associated protein TRADD signals cell death and NF-kappa b activation. Cell 1995; 81:495–504.
122. Hsu H, Shu HB, Pan MG, et al. TRADD-TRAF2 and TRADD-FADD interactions define two distinct TNF receptor 1 signal transduction pathways. Cell 1996; 84:299–308.
123. Muzio M, Chinnaiyan AM, Kischkel FC, et al. Flice, a novel FADD-homologous ice/ced-3-like protease, is recruited to the CD95 (Fas/Apo-1) death–inducing signaling complex. Cell 1996; 85:817–827.
124. Alnemri ES, Livingston DJ, Nicholson DW, et al. Human ice/ced-3 protease nomenclature [letter]. Cell 1996; 87:171.
125. Duan H, Dixit VM. RAIDD is a new 'death' adaptor molecule. Nature 1997; 385:86–89.
126. Malinin NL, Boldin MP, Kovalenko AV, et al. MAP3k-related kinase involved in NF-kappab induction by TNF, CD95 and IL-1. Nature 1997; 385:540–544.
127. Isaka T, Yoshimine T, Maruno M, et al. Morphological effects of tumor necrosis factor-alpha on the blood vessels in rat experimental brain tumors. Neurol Med Chir (Tokyo) 1996; 36:423–427.
128. Thomson AM. Neuroscience. more than just frequency detectors? [comment]. Science 1997; 275:179–180.
129. Naredi PL, Lindner PG, Holmberg SB, et al. The effects of tumour necrosis factor alpha on the vascular bed and blood flow in an experimental rat hepatoma. Int J Cancer 1993; 54:645–649.
130. Havell EA, Fiers W, North RJ. The antitumor function of tumor necrosis factor (TNF), I. therapeutic action of TNF against an established murine sarcoma is indirect, immunologically dependent, and limited by severe toxicity. J Exp Med 1988; 167:1067–1085.
131. Frater-Schroder M, Risau W, Hallmann R, et al. Tumor necrosis factor type alpha, a potent inhibitor of endothelial cell growth in vitro, is angiogenic in vivo. Proc Natl Acad Sci U S A 1987; 84:5277–5281.
132. Palladino MA, Jr., Shalaby MR, Kramer SM, et al. Characterization of the antitumor activities of human tumor necrosis factor-alpha and the comparison with other cytokines: induction of tumor-specific immunity. J Immunol 1987; 138:4023–4032.
133. Asher A, Mule JJ, Reichert CM, et al. Studies on the anti-tumor efficacy of systemically administered recombinant tumor necrosis factor against several murine tumors in vivo. J Immunol 1987; 138:963–974.
134. Zhang M, Tracey KJ. Tumor necrosis factor. In: Thomson AW, ed. The Cytokine Handbook. 3rd ed. Academic Press, 1998:515–547.
135. Talmadge JE, Tribble HR, Pennington RW, et al. Immunomodulatory and immunotherapeutic properties of recombinant gamma-interferon and recombinant tumor necrosis factor in mice. Cancer Res 1987; 47:2563–2570.

136. Nakano K, Okugawa K, Furuichi H, et al. Augmentation of the generation of cytotoxic T lymphocytes against syngeneic tumor cells by recombinant human tumor necrosis factor. Cell Immunol 1989; 120:154–164.
137. Ostensen ME, Thiele DL, Lipsky PE. Tumor necrosis factor-alpha enhances cytolytic activity of human natural killer cells. J Immunol 1987; 138:4185–4191.
138. Shalaby MR, Aggarwal BB, Rinderknecht E, et al. Activation of human polymorphonuclear neutrophil functions by interferon-gamma and tumor necrosis factors. J Immunol 1985; 135:2069–2073.
139. Ohkawa S, Wright KC, Mahajan H, et al. Hepatic arterial infusion of human recombinant tumor necrosis factor-alpha. An experimental study in dogs. Cancer 1989; 63:2096–2102.
140. Zimmerman RJ, Chan A, Leadon SA. Oxidative damage in murine tumor cells treated in vitro by recombinant human tumor necrosis factor. Cancer Res 1989; 49:1644–1648.
141. Yang SC, Fry KD, Grimm EA, et al. Phenotype and cytolytic activity of mouse tumor-bearer splenocytes and tumor-infiltrating lymphocytes from K-1735 melanoma metastases following anti-CD3, interleukin-2, and tumor necrosis factor-alpha combination immunotherapy. J Immunother 1991; 10:326–335.
142. Tsutsumi Y, Kihira T, Tsunoda S, et al. Molecular design of hybrid tumour necrosis factor alpha with polyethylene glycol increases its anti-tumour potency. Br J Cancer 1995; 71:963–968.
143. Kemeny MP, Botchkina GI, Ochani M, et al. The tetravalent guanylhydrazone CNI-1493 blocks the toxic effects of interleukin-2 without diminishing anti-tumor efficacy. Proc Natl Acad Sci USA 1998; 95:4561–4566.

64

The Future of Endotoxin Research

Helmut Brade
Research Center Borstel, Borstel, Germany

Steven M. Opal
Brown University School of Medicine, Providence, and Memorial Hospital of Rhode Island, Pawtucket, Rhode Island

Stefanie N. Vogel
Uniformed Services University of the Health Sciences, Bethesda, Maryland

David C. Morrison*
Saint Luke's Hospital and University of Missouri–Kansas City, Kansas City, Missouri

These are exciting times indeed in the field of molecular biology in general and in endotoxin research in particular. The pace of discovery is steadily increasing and the opportunity for significant advances in the understanding of human biology has never been brighter than it is today. From a variety of microorganisms, the entire genomes have been sequenced and verified (e.g., *Escherichia coli*, *Haemophilus influenzae*). By the close of the twentieth century the genomes of virtually all of the recognized, pathogenic bacterial organisms that cause disease in humans are likely to have been sequenced. Further, the entire genetic sequence of the lower eukaryote *Saccharomyces cerevisiae* has been completed, and the human genome sequencing project is expected to be completed early in the next decade. When one considers the level of sophistication in basic biology and genetics that existed at the beginning of this century, progress in the last 100 years has been truly remarkable.

Comparative genomics will provide new insights into the molecular pathogenesis of disease and provide new targets for therapeutic intervention. The availability of the molecular "blueprints" for each invasive microbial organism will give forthcoming investigators an enormous advantage over bacterial pathogens. The mechanisms by which pathogens evade host defenses and cause disease can now be characterized in precise molecular detail. Promising new vaccine targets, novel chemotherapeutic agents, and improved methods to avoid microbial injury should lead to substantial advances for the health of humankind in the near future. There is, therefore, considerable reason for optimism if scientists and policymakers proceed wisely in an organized, ethical, and logical manner.

Transgenic animals, advanced bioengineering, combinatorial chemistry, designer monoclonal antibodies, recombinant molecules, and computer innovations will be exploited to unravel previously unsolvable mysteries in microbiology and human medicine. It is increasingly clear that biological systems are exquisitely complex, with an extremely large, yet finite number of variables that affect outcome. Advances in basic immunology and genetics continue to provide new evidence that patterns exist even in seemingly chaotic biological systems. This may well provide opportunities to modify, if not precisely regulate, the dynamic interactions between micro-

---
*Formerly at The University of Kansas Medical Center, Kansas City, Kansas.

bial pathogens and vertebrate hosts. It would appear that endotoxin research, in particular, would benefit from innovations in molecular biology that have been developed in the latter half of the twentieth century.

There are many pressing research priorities for future endotoxin research. Some of these include (1) a fundamental molecular understanding of the handling of endotoxin in the extracellular space, at the cell membrane, and in the intracellular space; (2) the molecular basis of endotoxin tolerance; (3) the genetic substrate(s) for the variability of endotoxin effects; (4) the mechanisms governing molecular interactions between endotoxin and other microbial toxins, mediators, and superantigens derived from other microbial pathogens; (5) the importance of antimicrobial agents in modifying the presentation of endotoxin to the host immune system; (6) structure-function relationships and the nature of the immune response to endotoxin; and (7) the variables that principally determine the host response to the presence of endotoxin over the course of a miocrobial infection.

It is abundantly clear that there is an immediate need for improved methods to modulate endotoxin activity in the presence of generalized infections from gram-negative infections. The optimal methods to derive maximal benefit from a large number of promising experimental anti-endotoxin strategies are of critical importance to current clinical research efforts. The role of endotoxin in local inflammatory states in vasculitis, periodontal disease, neoplastic diseases, immune-mediated disorders, and other pathological states needs to be defined more clearly. The physiological and pathological consequences of translocation of gut-derived endotoxin need to be clarified in careful clinical investigations.

We fully anticipate that the answers to many of these research priorities will be characterized before the next edition of *Endotoxin in Health and Disease* is written. Major advances in the field of endotoxin research are likely to occur in the next few years. We expect that the readers of this volume will lead the way and contribute to a greater understanding of the importance of endotoxin in basic biology and human medicine. The editors wish to express our deep appreciation to each of the contributors to this text for their thoughtful presentations. This work is intended to provide a conceptual basis for future investigations into the nature of endotoxin-related disorders. We sincerely hope that we have succeeded in presenting the current state of knowledge in this ever-expanding area of basic and applied clinical research. We look forward to the wealth of new knowledge that will be uncovered in the years to come.

Index

γδ-TCR-positive T cells, 644–645
ε^{34}, 29
β-agglutinogens, 11
β-glycosyltransferase, 21
1-O-alkyl-2-acetyl-sn-glycero-3-phosphoryl-choline (see PAF)
1-O-alkyl-2-acetyl-*sn*-glyceryl-3-phosphocholine, 449
2,3-Diaminohexuronic acid, 164
^{31}P-NMR, 105
3-Deoxy-D-manno-2-octulosonic acid, 831, 835–836 (see also 2-Keto-3-deoxy-octonic acid, KDO)
3-Deoxy-D-manno-oct-2-ulosonic acid, 233
3-Deoxy-D-manno-oct-2-ulopyranosonic acid (KDO), 115–116
3-Deoxy-lyxo-heptulosaric acid, 172
3-Hydroxy fatty acids:
    resolution of, 244
    preparation of, 243
3-Hydroxytetradecanoic acid, content in LPS, 652, 656
4-aminoarabinose, 35
6-Deoxy-altrose, 172
6-Deoxy-D-gulose, 162
6-Deoxy-L-talose, 169, 172
70Z/3 cells, 690

A-band LPS, 332, 338
AbcA, 342–343
ABC-transporter dependent, 334, 336–337, 338, 341–342, 343, 347
ABC-transporter, 334, 342–343, 351
    in *Aeromonas salmonicida*, 343
    ATP-binding component, 343
    complementation, 343, 350
    dependent, 334, 336–337, 338, 341–342, 343, 374
    mechanism of action, 343
    protein structure, 342–343
Acetobacter, 168

Acetylated LDL, 437, 439
Acetyltransferase, 338–339, 346, 347
Acidosis, mucosal, 858
Acinetobacter, 168, 272
ACL (see Undecaprenol phosphate)
acrA, 36
Actinobacillus actinomycetemcomitans, 667
Actinobacillus, 169
Acute phase response, 817
Acute renal failure, 845, 846
Acute respiratory distress syndrome (ARDS), 449, 458, 845
Acyloxyacyl hydrolase (AOAH), 530
Adhesion molecule:
    selectins, 455, 566
    integrin receptors, 455, 456
Adjuvant effect, 647
Adrenalectomy, 769
Aeromonas salmonicida, 343
    IS elements, 350
Aeromonas, 263, 271
Agglutinins, 10
Aggregate:
    critical concentration, 196, 206
    structure, 196, 206–208
Akt, 478
Albumin, human serum, 413–422
    binding sites, 414
    conformational change, 415
    dansylsarcosine, 415
    fluorescent probes, 415
    and lipid A binding, 414–416
    and lipopolysaccharide binding, 416–417
    scatchard analysis, 415
Albumoses, 7
Alcaligenes, 168
Alternative complement cascade, 331

Altruonic acid, 159
Amides, 156, 158, 160, 162, 165
Amino acids, unusual, 162
Anabaena, 338
Anaphylatoxins, 79
Angiogenesis, 929–921
Animal models:
   cecal ligation and puncture, 813
   D-galactosamine, 810
   differences between species, 812
   differences with human sepsis, 814
   dose-response effects with endotoxin, 814
   endotoxin challenge, 812–813
   gene knockout animals, research in, 812–813
   infection challenge, 812–813
   intraperitoneal bacterial challenges, 866–867
   intraperitoneal injections of LPS, 866–867
   intravenous bacterial challenges, 866–867
   intravenous injections of LPS, 866–867
   transgenic animals, research in, 810
   species characteristics and differences, 812
      dogs (canine), 811
      horses (equine), 811
      mice (murine), 809–810
      pigs (porcine), 811
      primates, nonhuman, 811–812
      rabbits (lapine), 811
      rats, 810
      sheep (ovine), 811
   transgenic animals, research in, 810
   value of, 809–814
Anomeric configuration, 331–332
Antagonist of lipid A, 243, 252, 248
Antibacterial drugs, 195
Antibiotic class induced endotoxin release, 69
   other factors, influence on, 72
      choice, dose, 72
      interval between treatment, 72
      methods of administration, 72, 73
      number of treatments, 72, 73
      pharmacokinetics, 72
      route of administration, 72
      timing of treatment, 72, 73
   specificity of PBP binding affinity, 69
      bacterial morphology associated with, 70, 71, 72
      bacterial species influence on, 70
Antibiotics, 773
Antibiotics and endotoxin release, 887–897
Antibodies, anti-endotoxin, 858, 859
Antibody-producing cells, 655 (*see also* plaque-forming cells)
Anticoagulant factor C, 921

Antigen presenting cell (APC), 918
Antigen-carrier lipid (*see* Undecaprenol phosphate)
Antigenes glycido-lipidiques, 10
Antigenic diversity, 348–350
Anti-inflammatory agents, 497
Anti-lipopolysaccharide (LPS), 802
Antioxidants, 594, 596–597
Antitermination, 348
Antitoxins, 7
AP-1, 597–598
Apoptosis, 502, 648, 679
   of B cells, 680
   basic fibroblast growth factor and, 505
   of bone marrow cells, 681
   death domain complexes in, 503, 505
   endothelial, 741
   of hepatocytes, 681
   macrophage cytotoxicity, 740
   of renal tubular cells, 683
   of T cells, 681
   of thymocytes, 679
   of vascular endothelial cells, 682
Aqueous butanol and LPS extraction, 651
Arachidonic acid, 449, 451
Asymetric membrane, 197–204
Asymmetric and symmetric distribution of fatty acids, 107
ATP-binding cassette transporter (*see* ABC-transporter)
ATP-binding motif, 342–343
Atypical codon use, 349
Azurocidin, 34

B cell activation, 918
B lymphocytes, stimulation by LPS, 651
Baboons, 796
Bacillosamine, 164
Bacitracin, 339, 348
Bacteremia, 832
Bacteria cell wall components, 463–467
Bacteria:
   dissemination of, 784, 790, 791
   E. coli, 783
   K. pneumoniae, 791
   listeria monocytogenes, 791
   S. typhimurium, 790, 791
Bacterial endotoxin (*see* Lipopolysaccharide)
Bactericidal/permeability-increasing protein (BPI), 34, 203–204, 369–374, 380, 468, 802
   function, 369–373
      antibacterial, 369–370
      anti-LPS, 369–371
      animal, 371

Index

[Bactericidal/permeability-increasing protein]
    human hemorrhagic trauma, 371–372
    human meningococcemia, 371–372
    human opsonization/phagocytosis, 371
  localization, 369
    granules, 369
    inflammatory fluid, 371
    other, 369
    polymorphonuclear leukocytes, 369
  LPS-binding proteins, 370–371
    BPI-complexes with LPS, 370–371
    CAP18, 372
    LBP-complexes with LPS, 370–371
  production, 371
  structure, 370–371
    fragments, 370–371
    x-ray, 371
  synergy, 371–373
    p15s, 371
    complement, 372–373
    defensins, 372
    PLA2, 373
Bacteriophage (*see also* Phage):
  ε^{34}, 347
  SF6, 347
Bacteriuria, 832
Bacteroides, 263, 855
Basal Core, 13
Bayer junctions, 345
B-band LPS, 332
BCG, sensitization to shock with, 790
B-hydroxymyristic acid, 831, 835–836
Bifidobacteria spp., 855
Biological activity of LPS:
  factors influencing comparisons of, 904–905
  role of LPS carbohydrate in, 908
  of structurally different oral LPS, 906, 910
Biosynthesis of lipid A, 234, 243, 250, 254
Biosynthetic precursor Ia, 243
Black lipid membranes (BLM), 198–200, 203
Blood group antigen-like, 162
Blood group specificity, 156
Bone marrow derived macrophages, 658–663
Bordetella pertussis, 271
BPI, 34, 203–204
Brain microvascular endothelial cells, role of OmopA in invasion by *E. coli*, 654
Braun lipoprotein:
  as adjuvant, 653
  mitogen for C3H/HeJ cells, 652
Brucella sp., 35
Burkholderia, 166, 272

c19:0 (11,12-methylene-octadecanoic acid), 95
C1q:
  binding requirements, 80
  role in opsinization and bacterial killing, 86–87
  structure of, 80
C3:
  role in triggering the oxidative burst, 86
  thioester bond, 77–78
C3 convertase:
  generation by classical pathway, 77
  generation by alternative pathway, 77–78
C3H/HeJ macrophages, 919
  priming with IFNγ, 654
  responses to endotoxin protein, 652
  responses to LAP-LPS complexes, 655
  tumoricidal activity of, 654
C3H/HeJ, 919
C5 convertase, generation of, 78
C9, polymerization of, 81–82
CAC, 206
Cancer (*see also* Tumor):
  immmunotherapy, 915
  risk, 916
  treatment, 915–916
CAP37, 34
CAP57, 34
Capillary leakage syndrome, 921
CAPK, 499
  autophosphorylation of, 499
  LPS activation of, 501
  Raf-1 activation by, 499
CAPP, subunits of, 499
Capsular polysaccharides (*see also* K-antigens):
  streptococcal, 334
CarbBank, 155
Cardio-pulmonary bypass, 845
Carrageenan, sensitization to shock with, 790
Carrier effect, 658
Caryophyllose, 166
Caryose, 166
Caspase protease, 919–920
Caspase-I, 647
Caspases, 502–503
Cationic detergents, 34
Caveolae, 526
CD14, 197, 212–213, 455, 457, 463–468, 521, 727, 796, 906, 908, 919
  and anti-LPS monoclonal antibodies, 636
  dependent pathway, 689
  independent pathway, 690
  membrane CD14 (mCD14), 521, 781
  mice deficient in, 781–785

[CD14]
  monoclonal antibody, 463, 464, 466
  on monocytes/macrophages, 781
  mutants, 464
  on neutrophils, 781
  pattern recognition receptor, 467
  soluble CD14 (sCD14), 383, 463–468, 522, 781, 785
CD18, 801
CD32 polymorphism, 672
CD45, 648
CD5+ B cell, 681
Cecropin A, 34
Ceftazidime and endotoxin release, 890–893
Cell activation, model of, 213
Cellular sources of complement proteins, 78–79
Cellulose, 334
Cellulose synthase, 344
Centrifugal partition chromatography (CPC) of lipid A, 246
Ceramide activated protein kinase (*see* CAPK)
Ceramide activated protein phosphatase (*see* CAPP)
  Ceramide, 473, 497, 740–741
    apoptosis and, 502–503
    endothelial cell responses to, 500
    endocytosis of, 740
    JNK/SAPP cascade and, 499
    lipid A similarity, 500–501
    NF-κB activation by, 499
    neutrophil responses to, 499
    PKC-ζ activation by, 499
    Raf-1 activation by, 499
    response to, 740–741
    structure of, 501
    synthesis of, 498
Cerebral spinal fluid, 841, 843
Chain length distribution, 339, 341, 343
Chemokines:
  interleukin-8, 451, 453, 457
  monocyte-chemotactic protein, 454, 456
Chimpanzees, 796
Chitin, 334
Chitin synthase, 344
Chlamydia, 229:
  crystal structure of Kdo disaccharide, 260
  synthesis of genus-specific LPS epitope, 259, 261, 263
CHO-cells, transfected, 466
Cholera, 164
  and induction of T-helper cell subsets, 658
  toxin, B subunit, 655

Chromobacterium, 35, 169
Chronic infection, 232
Circadian rhythm, 770
$cis\Delta^{11}$-18:1 (octadec-11-Z-enoid acid), 95
Citrobacter, 156, 269
  and anti-LPS monoclonal antibodies, 635
Cld, 339, 341
Clinical studies of endotoxemia:
  "marker" studies, 846–847
  "mediator" studies, 847–849
Clinical trials:
  therapy in, 868–874
  problems in design of, 871–874
Clonal groups, 350
Cloning, 743–747
  expression cloning, 743–744
    Ran/TC4 GTPase, 743–744
  positional cloning, 744–747
    Tlr4, 747
Clostridia spp., 855
Coated pits, clathrin-coated, 523
Coccobacteria septica, 6
ColE1, 350–351
Coley's toxin, 12, 915
Comamonas testosteroni, 35
Common basal core, 14
Complement activation, 201
  activating surfaces, 77–78, 81
  alternative pathway:
    antibody-dependent activation of, 78
    antibody-independent activation of, 77–78, 81, 82
    composition of, 77
  biological consequences of, 79
  classical pathway:
    antibody-dependent activation of, 77
    antibody-independent activation of, 77, 79–80, 82
  correlation with lethality, 84–85, 86
  correlation with lipopolysaccharide-induced shock, 85
  general formula of, 77
  LPS structural requirements for, 82–83
  role of MBP-MASP in, 80–81
Complement:
  component C9, 201–202
  pores, 201
  receptors (CRs), 79, 86–87
Complement-mediated bacterial killing (*see also* Membrane attack complex):
  effect of altering lipopolysaccharide (LPS) biosynthesis, 85–86

Complement-mediated lipopolysaccharide (LPS) clearance:
  differences in clearance of purified LPS *vs.* intact bacteria, 85
  effect of altering O-antigen synthesis, 85–86
  effect of anti-LPS antibodies, 86
  role of erythrocytes, 86
Complement-mediated lipopolysaccharide (LPS) release:
  association with serum proteins, 83
  complement "releasable" fraction, 83
  EDTA "releasable" fraction, 83
  role in meningococcal meningitis, 83–84
Compound, 406, 464, 466
Computerized analysis, 156, 160
Conformation, 160, 209–211
Contagion, 5
Core biosynthesis, 348
Core oligosaccharide:
  genes:
    gmhA, 307, 310–311
    gmhB/C, 307, 310–311
    gmhD, 307, 310–311, 317, 321–322
    kdsA, 307, 308–309
    rfaH, 307, 322–323
    waaA, 307–309, 317, 323–324
    waaC, 307, 312, 315, 317
    waaD, 307, 315, 317–319
    waaF, 307, 312, 315, 317
    waaG, 307, 315–317
    waaI, 307, 315, 317–318
    waaJ, 307, 315, 317, 318
    waaK, 307, 315, 317, 319–320
    waaL, 307, 315, 317, 320–321
    waaM, 307
    waaN, 307
    waaO, 307, 315, 317–318
    waaP, 307–308, 313–314
    waaQ, 307–308, 312
    waaR, 307, 315, 317, 319
    waaS, 307, 315, 317, 323–324
    waaT, 307, 315, 317, 319
    waaU, 307, 315, 317, 319–320
    waaV, 307, 315, 317–318
    waaW, 307, 315, 317, 320
    waaX, 307, 315, 317–318
    waaY, 307–308, 314, 317
    waaZ, 307, 317, 323–324
    wabA, 307, 317, 323–324
  regulation, 321–323
  structures, 306, 308, 315
Core region, 115–154
  Acinetobacter, 140–141
    baumannii, 140–141
    haemolyticus, 140–141
  Aeromonas hydrophila, 125, 128–129
    salmonicida, 125, 129
  Bacteroides fragilis, 140–143
  Bordetella pertussis, 139, 140
  Bradyrhizobium japonicum, 140–142
  Burkholderia cepacia, 132–133, 135
  Campylobacter jejuni, 133, 136–137
    coli, 133, 137
    lari, 133, 137–138
  Chlamydia trachomatis, 140, 143
    psittaci, 140, 143
    pneumoniae, 140
  Citrobacter freundii, 117, 121
  Coxiella burnetii, 133, 139
  Erwinia carotovora, 117, 121
  Escherichia coli, 117–119
  Haemophilus influenzae, 130, 132
    ducreyi, 130–132
  Hafnia alvei, 117, 120
  Helicobacter pylori, 133, 138
    mustelae, 133, 138
  Klebsiella pneumoniae, 121–122, 124
  Legionella pneumophila, 140, 143
  Moraxella catarrhalis, 140–141
  Neisseria gonorrhoeae, 125–126
    meningitidis, 125–126
  Ochrobacterium anthropi, 140, 143
  Pasteurella haemolytica, 131–132
  Phenylobacterium immobile, 139–140
  Proteus mirabilis, 122, 124
  Pseudomonas aeruginosa, 132, 134–135
    fluorescens, 132, 135
    maltophila, 140, 143
  Rhizobium etli, 140, 142
    leguminosarum, 140, 142
    meliloti, 140, 142
  Rhodobacter sphaeroides, 140, 143
    capsulatus, 140, 143
  Rhodocyclus tenuis, 139–140
    gelatinosus, 139–140
  Salmonella enterica, 116, 118
  Shigella, 117, 120
  Vibrio cholerae, 125, 127
    parahaemolyticus, 125, 127–128
    salmonicida, 125, 128
    ordalii, 125, 128
  Yersinia, 123–125

Core structures, 156
Corticosteroids, 693–694
Coxiella burneti, 348, 720
CPS, 155
Cps3S, 344
Critical aggregate concentration (CAC), 206
Critical micelle concentration (CMC), 223
Cubic structure, 206–209
Cyanobacteria, 338
Cyclooxygenase, 774
Cytokines, 694–695, 799
   levels and prognosis, 845
   induction of release, 464–467
   production in macrophages:
      modulation by calphostin C, 663
      modulation by 1-(5-isoquinolinylsulfonyl)-2methylpiperazine (H7), 663
      modulation by *N*-tosyl-L-phenylalanine chloromethyl ketone (TPCK), 663
      modulation by OmpA, 661, 663
      modulation by staurosporine, 662
   role in regulation of complement genes, 87–88
   role in shock:
      in mice deficient in, 786–790
      TNF, 782–784, 787, 790, 791
      IL-6, 782, 783, 787
      IL-1, 784, 787
      TGFβ, 787
      anti-TNF reagents, 865–874
      anti-IL-1 reagents, 866, 867
Cytotoxic activity of macrophages:
   induction by LPS, 661, 663
Cytotoxic effect, 647–648

Dansylpolymyxin B, 417
Deacylation (deacylated LPS), 531
Decontamination, selective digestive, 858
Deep core LPS mutants, 34
Defensin, 34
Dendritic cells, 532
Dental plaque biofilm and shed LPS, 899, 909
Dephosphorylation, 532
D-Galactan I, 336–338, 347, 350
D-galactosamine, 720, 753, 769:
   sensitization to shock with, 787, 790
   shock model of, 782
D-*Glycero*-D-*manno*-heptose, 115–116
D-*Glycero*-D-*talo*-oct-2-ulopyranosonic acid (Ko), 115–116
D-G*lycero*-D-*talo*-octulosonic acid (Kdo), 272
DIC, 682
Dipole potential, 198

Direct and indirect pathways of activation, 909
DMDS (dimethyldisulfide), 95
DNA group, 168
Drugs:
   antibacterial, 195
   polycationic, 202–203
Duodenal ulcers, 166

E. coli (*see* Bacteria and *Escherichia coli*)
Early gene activation in macrophages, 654
ECA, 332, 335–336
   biosynthesis locus, 345
Edema, 449, 456, 458
EDTA, 34
Elastase, 801
Electron microscope, 522
Electrospray ionization mass spectrometry (ESI-MS), 100–101
Endoantigen, 10
Endosomes, 523
Endothelial activation, 800
Endothelial cell, 463, 466, 919
Endothelin, 799
Endotoxemia:
   clearance with antibiotic therapy, 844, 851
   clinical studies of, 845–850
      heterogeneity, 849–851
      as a marker of sepsis, 846–847
      as a mediator of sepsis, 846–847
      meta analysis of, 847–850
   levels in burn injury, 847
   levels in neonates, 846–847
   prognostic significance, 845–850, 852
   quantitative levels, 844–845
Endotoxic conformation, 22, 212
Endotoxic principle, 93
Endotoxic shock, 682
Endotoxicity, 166
Endotoxin (*see* LPS)
Endotoxin administration to humans, 817
   arachidonate metabolites, 819
      prostacyclin, urinary metabolite, 819
      thromboxane A2, urinary metabolite, 819
   acute phase proteins, 819, 823–824
      C-reactive protein, 819, 823–824
      fibrinogen, 819
      lipopolysaccharide binding protein, 819
      serum amyloid A, 819
   cardiovascular response, 819
      cardiac index, 819, 822, 824
      heart rate, 819, 822, 824
      left ventricular ejection fraction, 819, 822, 824

[Endotoxin administration to humans]
    mean arterial pressure, 819, 822, 824
    systemic vascular resistance, 819
  chemokines, 819, 823–824
    GRO-(, 819
    IL-8, 819, 823–824
    MCP-1, 819
    MIP-1, 819
  coagulation activation, 818, 823–824
    prothrombin F1+2 fragments, 818, 823–824
    thrombin-antithrombin III complexes, 818, 823–824
  complement activation, 818
  cytochrome P450-mediated drug metabolism inhibition, 819
  endotoxin clearance, 818
  endothelial cell activation, 818
    E-selectin, 818, 823–824
    von Willebrand factor antigen, 818
  fibrinolysis, 818, 823–824
    plasmin-($_2$-plamin inhibitor complexes, 818, 823–824
    plasminogen activator inhibitor (PAI-1), 818, 823–824
    tissue plasminogen activator (tPA), 818, 823–824
  granulocyte-colony stimulating factor (G-CSF), 819, 823–824
  gut permeability, 820
  inhaled, 820
    Enterobacter agglomerans, 817, 820
  instilled, 820
    Escherichia coli O:113, 820
  intravenous, 817–827
    Escherichia coli O:113, 817
    Salmonella abortus equi, 817
    salmonella minnesota, 817
    U.S. Standard reference endotoxin, 818
  kallikrein-kinin contact system activation, 818
    α_2-macroglobulin-kallikrein, 818
    Factor XII, 818
    Factor XI, 828
    prekallikrein, 818
    high molecular weight kininogen, 818
  markers of cell activation:
    adhesion molecules, 818, 822, 824
    bactericidal permeability increasing protein, 818
    L-selectin, 818
    soluble TNF receptors, 818, 822, 824
  metabolic response, 819
    free fatty acids, 819
    hyperglycemia, 819
    hypoaminoacidemia, 819

[Endotoxin administration to humans]
  neopterin, 819
  neutrophil granule constituents, 818, 822, 824
    bactericidal permeability increasing protein, 818
    elastase, 818
    elastase-α1-antitrypsin, 818
    lactoferrin, 818, 822, 824
    metalloproteinases, 818
  nitric oxide, exhaled, 819
  phospholipase A2, 819
  procalcitonin, 819
  proinflammatory cytokines, 818, 823–824
    IL-1β, 818
    IL-6, 818, 823, 824
    IL-10, 818, 823, 824
    TNF-α, 818, 822, 824
  pulmonary function following inhalation, 820
  pulmonary responses, 819
    alveolar macrophages, 819
    bronchoalveolar lavage, cell constituents, cytokines, 819
    lung permeability, technitium labeled diethylenetriamine pentacetate, (^{99m}Tc-DTPA), 819
  rapid eye movement (REM) sleep suppression, 820
  resting energy consumption, 819
  splanchnic blood flow, 819
  stress hormones, 819, 823–824
    α-melanocyte stimulating hormone, 819
    adrenal corticotrophic hormone (ACTH), 819
    arginine vasopression, 819
    cortisol, 819, 823–824
    epinephrine, 819, 823–824
    norepinephrine, 819
  systemic inflammatory response associated with:
    fever, 818, 822, 824
    leukocytosis, 818, 822, 824
    leukopenia, 818, 822, 824
    lymphopenia, 818
    platelet number, 818
    monocytopenia, 818
Endotoxin antagonism, 23
Endotoxin protein as adjuvant, 653 (*see also* Lipid A-associated protein)
Endotoxin release, antibiotic-induced, 887–897
  animal data, 889, 892
  clinical studies, 889–890, 892–893
  differences between antibiotics in, 890–893
  in vitro, 887–889
  mechanisms of, 889
  in meningococci, 893
Endotoxin release, spontaneous, 887

Endotoxin shock, sensitization to:
  BCG, 790
  carrageenan, 790
  D-galactosamine, 782, 787, 790
  P. acnes, 790
Endotoxin tolerance, 23, 540–545:
  lethality, 751, 752, 753
  early phase, 752
  endogenous factors, 752
  pyrogenicity, 752, 757
  TNFα, 753
  glucocorticoids, 754
  adrenalectomy, 754
  IL-10, 755, 759
  mononuclear phagocytes, 756
  NF-κB, 757, 758
  CD14, 757
  bone marrow-derived macrophages, 757
  desensitization, 758
  specificity, 759
  resistance to infection, 759
  tumor necrosis, 760
  hemorrhagic shock, 760
  ischemia/reperfusion, 760
  myocardial protection, 761
  immunosuppression, 762
Endotoxin, 8, 915 (*see also* LPS)
  animal lethality assays for, 837
  binding protein, 390–391
  bioassays for, 836–837
  biological activities, 389–390, 393–399
    lethality in mice, 396–397
    Limulus amebocyte lysate activation, 393–394
    platelet adherence, 395–396
    tissue factor production, 389, 394–395
    tumor necrosis factor production, 397–398
  blood distribution, 392
  cardiovascular system, 799
  cell culture assays for, 837
  chemical assays for, 832–836
  disaggregation by hemoglobin, 391–394
  ELISA for, 832
  and hemoglobin clearance, 397
  and hemoglobin oxidation, 391–392
  and hypoglycemia, 397
  infusion, 798
  immunoassays for, 835
  interaction with PAF, 571, 574
  intravascular clearance, 391–392
  in vivo clearance, 391–392
  lipopolysaccharide component of, 831
  and oxygen affinity, 391

[Endotoxin]
  potency equivalence, 844
  and reticuloendothelial cell system clearance, 397
  sensitivity, 796
Endotoxin-associated proteins:
  contamination of LPS preparations with, 736–737
Endotoxin-binding proteins/peptides, 836
  bactericidal/permeability-increasing protein (BPI), 836
  endotoxin neutralizing protein (ENP), 831–836
  Limulus anti-LPS factor (LALF), 836
  polyphemisin, 831, 836
  tachyplesin, 831, 836
Enteritis, 165
Enterobacter, and anti-LPS monoclonal antibodies, 635
Enterobacteriaceae, synthesis of inner core, 265, 266
Enterobacterial common antigen, 286–287, 332
envA, 34
Enzymatic preparation of lipid A analogs, 254
Enzymatic resolution of 3-hydroxy fatty acids, 243
Enzymes, 4
Erythromycin, 33
Escherichia coli K-12, synthesis of outer core units, 271
Escherichia coli, synthesis of *Re* LPS, 263–265
Escherichia coli, 31–36, 156, 633, 855, 856
  IS elements, 347
  JUMPstart, 348
  J5, 633
  K-12, 346
  O7, 350
  O8, 337
  O9, 337–338
  O16 antigen, 337
  O111, 347
    O-polysaccharide biosynthesis cluster, 346
  O18:K1, 636, 639
  O111:B4, 636, 639
  SØ874, 346
  and anti-LPS monoclonal antibodies, 634, 635, 639
  and Braun lipoprotein, 652
  chemotypes, 633
  diarrhoeagenic *E. coli*, 156
  enteroaggregative *E. coli*, 156
  enterohemorrhagic *E. coli*, 156
  enteroinvasive *E. coli*, 156
  enteropathogenic *E. coli*, 156
  enterotoxigenic *E. coli*, 156
  and porins OmpA, OmpC, and OmpF, 652
E-selectin, 799, 800, 899, 906, 910
Eubacterium sabbureum, 158
Evasion of host defense, 910

Index

Exotoxin, 7
ExoU, family of glycosyltransferases, 344
Export 40:
    ABC-transport-dependent, 342–343
    synthase-dependent, 344
    to the cell surface, 344, 351
        model, 344–345
        Tol A, 344–345
    Wzx-mediated, 339
Extracellular Signal-Regulated Kinase (ERK), 498

Factor C, 832
Factor *T*, 19
Fas antigen, 682
Fas, 502
Fast atom bombardment mass spectrometry (FAB-MS), 100
Fatty acid chain length in oral LPS, 901–902, 909
Fatty acids:
    3-acyloxy, 243
    3-hydroxy, 243
    3-keto, 248
    unsaturated, 248
FcγR, 672
Fever, 6
Fibroblast Growth Factor (FGF), 505
firA, 34
Flippase (*see* Wzx)
Fluidity:
    bacteria, 206
    LPS, 200
Form variation:
    S. enterica,
        sv. enteritidis, 332, 334
        sv, typhimurium, 332
Fragmented DNA, 680
Free lipid A, 17
Frigorigenine, 6
Fumonisin B1, 498
Furanose, 331–332
Fusidic acid, 33

G proteins, 537–541, 543, 544, 545
G+C content, 349
Gangliosides, 741
Gastritis, gastric ulcers, 166
GCL (*see* Undecaprenol phosphate)
Gel phase, 205–206
Genetic analysis, 735–738, 743–747
    BXH recombinant inbred strains, 736, 745
        BXH11/Ty, 736
    co-dominant phenotype, 736

[Genetic analysis]
    dominant negative mutation, 737, 747
    hyporesponsiveness:
        to ceramide, 740–741
        to glucan, 738–739
        to LPS, 735–747
        to platelet activating factor, 739, 744
        to taxol, 739–740
    intermediate phenotype (*see* co-dominant phenotype)
GlcNAc-pryophosphorylundecaprenol, 286
Glc*p*N3N, 94
Glucan (*see* hyporesponsiveness)
Glucocorticoid antagonizing factor, 774
Glucosylation, 347
Glutathione, 594
    nitric oxide effects on, 595, 598–599
Glycerol-3-phosphate dehydrogenase, 293
Glycolipid A, 16
Glycolipids, 16
Glycopolymers, 258, 259, 265, 269
Glycospingolipid (GSL), 196–201
Glycosyl phosphate, preparation of, 245, 246, 248
Glycosyl transferases, 332, 343, 344, 350
    initiating, 335–336
    non-initiating, 335–336
    nonprocessive, 343
    post-initiation, 338
    specificity, 332, 334, 351
Glycosylation shifts, 159
Glycosylation, 347 (*see also* Glucosylation)
Glycosyl-carrier lipid (*see* Undecaprenol phosphate)
Glycosylphosphatidylinositol anchor (*see* GPI anchor)
Glycosylphosphitidylinositol, 523
gmhD, 346
gnd, 349
GPI anchor, 781
G-protein-coupled receptor:
    alternative splicing, 451, 452
    desensitization, 451, 452
    family, 451, 453
G-proteins, 464
Gram negative bacteremia:
    as a clinical predictor of outcome, 844
    concordance with endotoxemia, 843–844
    Enterobacteriaceae, 843–844
    non-*Enterobacteriaceae*, 843–844, 850
Gram negative bacteria and outer membrane components, 651
Gram negative, 9
Granulocytes (neutrophils), 449, 451, 453, 454, 456, 458

GTPγS^{35}, 541, 543
Guillian-Barr, 165
Gut, 797, 802

Haemophilus influenzae LPS, 39
  gal genes, 44
  molecular mimicry, 40
  monoclonal antibodies, 40, 42–47
  phase variation, 40–48, 50–51
    lex1, 44
    lex2, 45, 47
    lgtC, 45–47
    lic1, 41–44, 47, 50–51
    lic2, 43–44, 47, 50–51
    lic3, 43–45
  structure, 46, 48
    digalactoside, 40–41, 44, 46–47
  substituents, 40, 42, 46
    phosphorylcholine, 42, 47
  synthesis of heptose oligosaccharide units, 270, 271
  synthesis of Kdo phosphates of *H. influenzae* I-69 Rd−/b+, 263
  tetranucleotide repeats, 42–48, 50–51
    deduced amino acid sequence, 50
    flanking sequence, 51
    serotype specificity, 47, 50
  virulence determinant, 41, 47, 51
Haemophilus somnus LPS, tetranucleotide repeats, 42, 47
Hafnia, 158
HasA, 344
  family of glycosyltransferases, 344
Heat shock, 775
Helicobacter pylori, 166
Hemoglobin:
  native, 389–391, 393–398
  αα-crosslinked, 390–398
  ββ-crosslinked, 390, 397
  co-infused with endotoxin, 391–392, 396–398
    synergistic toxicity, 392–396–398
  complexes with endotoxin, 390–391, 394
  complement activation, 396
  endothelial cell injury, 389
  hemolysis, 393
  oxidation, enhancement by endotoxin, 391
  oxygen affinity, effect of endotoxin, 391
  resuscitation from hemorrhage, 389, 398–399
  and tissue factor, 389, 394–395
Hemorrhagic necrosis, 915–921
Hemostasis, 605
  four functional domains in, 606

Hemostatic stress, 607
  coagulopathy in, 607
  fXa on, 608
  phospholipid on, 610
Hepatocyte necrosis, 684
Heptose, 169, 900–904
Heterogeneity, 332
Heteronuclear correlated $^{13}C,^{1}H$-NMR spectroscopy, 105
Heteropolymers, 332
High density lipoprotein, 842
Homopolymers, 158, 166, 332
Host immune responses, 331
Housekeeping genes, 346
H-repeat, 347
HT1080 cells, 466
htrB mutant, 102
htrB, 34
Hydrocortisone, 680
Hydrophobic antibiotics, 32–36
Hydrophobicity, 335–336
Hypersensitivity, 719
Hypoglycemia, 693
Hypophysectomy, 769
Hypothermia, 6

I-κB translocation, 464
IκBα protein, 509–515
  autoregulation of, 514
  proteolysis of, 512–514
    by ubiquitin-proteasome system, 514
ICAM-1, 683
ICAM-1, 800
IcsA, 331
IFN, 736–738, 745
  interferon regulatory factors (IRF), 738
IFN-α, 725
IFN-β, 723, 724
IFN-γ-inducing factor (IGIF), 647
IFN-γ, 466, 646–647, 721, 916
IgG subclass response, 672
IL-1, 466, 917–918
  and IL-1ra, 907–909
  therapy with receptor antagonist, 865, 871
IL-1β-converting enzyme (ICE), 647
IL-10, 693
IL-12, 466, 646–647
IL-18, 466
IL-2, 468
IL-6, 466, 467, 907
IL-6, and anti-LPS monoclonal antibodies, 636
IL-8, 466, 899, 907, 909–910
Interleukin (*see* IL)

Index

Imipenem and endotoxin release, 890–893
Immunity, 916, 920–921
Immunogenic tumor, 920–921
Inaba, 349
Indomethacin, 774
Induction:
 coagulation, 801
 fibrinolysis, 801
Infection, 735, 738
 gram negative, 738
 susceptibility to, 735, 738, 745
Inflammation, 917
Inflammatory disease, 899
 modulation of the host response in, 909–910
 role for LPS in destructive processes in, 908–909
Inflammatory stress, 609, 611, 614
 coagulopathy in, 609, 611, 614
 catheter on, 609
 E. coli on, 611
 fXa on, 610
 mechanism of, 612, 613, 616
 phospholipid on, 610
 shiga toxin on, 615
 tumor necrosis factor on, 609
Initiation, 334–335
 in *Anabaena*, 338
 in non-enteric bacteria, 338
 in *P. aeruginosa*, 338
 in *Vibrio cholerae* O139, 338
Innate host defense:
 role in health, 899, 901, 909–910
 role in inflammatory disease, 908–909
Inner core:
 heptose
  biosynthesis, 307, 309–312
  biosynthetic pathway, 310
  phosphorylation, 308, 310, 312–314
  transfer, 312–313
 Kdo
  biosynthesis, 307
  transfer, 307–309
iNOS, 799
Interferon (*see* IFN)
Internalization, 521
Intracellular signaling, 737, 739, 744, 737
Invaginations, 522
Inverted hexagonal structure H_{II}, 206–209
Ionic states of ReLPS, 224
IS elements, 349
 in *Aeromonas salmonicida*, 349–350
 in *E. coli*, 349–350
 in *Vibrio cholerae*, 349

IS*1*, 349–350
IS*1358*, 349–350
Ischemia (-reperfusion), 456, 458
Iso-muramic acid, 162

JUMPstart, 347–348
 involvement of RfaH, 348

K. pneumoniae, 337–338, 347 (*see also* Bacteria)
 JUMPstart, 348
 O-PS structure, 332
 O1, 350
 O8, 350
K-antigens, 332
KDO, 93, 233 (*see also* 3-deoxy-D-*manno*-oct-2-ulosonic acid)
 analogs, 263
 glycosyl donors, 258, 259, 263, 267
 phosphate, 263
 phosphorylated, 900–904, 908
 synthesis of oligosaccharides, 258–265
Kinase suppressor of Ras (KSR), 499
Kinases, 690–692
Klebsiella, 162
Klebsiella spp., and anti-LPS monoclonal antibodies, 633
K_{LPS}, 332
Knockout mice, A-SMase, 504–505
Kupffer cells, 440

Lactyl ethers, 162
Lamellar structure, 206–209
Laser desorption mass spectrometry (LD-MS), 100
Lateral gene transfer, 349–350
LBP, 197, 212–213
 as acute phase protein, 785
 mice deficient in, 787, 790, 791
 role in shock, 781
LDL receptor (LDLR), 523
Leaching, 172
Lectin, 465, 568
Legionella, 169
Leishmania major, 723
Leptospirosis, 846
Leukotrienes, 449, 451, 453
Lewis determinants, 166
L-*glycero*-D-*manno*-heptose, 115–116
L-*glycero*-D-*manno*-heptose:
 glycosyl donors, 265, 267, 269, 271
 oligosaccharide synthesis, 265–267, 269–271
 phosphate, 267, 270, 271
 synthesis of enterobacterial heptose-Kdo-lipid A region, 265, 266

Ligase, 344
Ligation, 334, 343
Lilpoarabinomannan, 465–467
Limulus amebocyte lysate (LAL) assay, 793, 841
    assays used in combination with, 834–835
    biochemistry of, 833
    chromogenic, 832–833, 835
    clinical application, 841–852
    clinical utility of, 832
    effect of plasma, 842
    end point methods, 843
    false positive reactions, 843, 844
    gel-clot, 832, 835
    inhibitory factors in blood, 841–842
    interference with, 834
    kinetic methods, 843
    sensitivity of, 832–833
    SepTest kit, 834
    turbidimetric, 832–834
Limulus amebocyte lysate, 389–394 (*see also* Endotoxin)
Limulus amebocyte, 23
Lipase, 243
Lipid A, 13, 16, 17, 403–410, 633, 916
    acyloxyacyl substituents, 295–297
        accA, 296
        accB, 296
        lpxk, 293, 296
        msbA, 296
        orfE, 296
        waaM (*htrb*), 295–296
    aggregate structure, 206–208
    of *Aquifex aeolocus*, 299
    binding site of, 408, 404
    biosynthesis of, 283–304
    CAC, 207
    chemical synthesis of, 243–256
    CMP-Kdo synthetase, 293–294
        kdsB, 294
    conformation, 197, 206–208
    detoxification of, 404
    of *Escherichia coli*, 243, 245
    Kdo:
        kdsA, 286
        linkage to lipid A, 284
        synthesis of, 293–295
    Kdo transferase, 294–295
        gseA, 294
        kdtA, (*waaA*), 294
        of *Chlamydia pneumoniae*, 295
        of *Chlamydia psittici*, 295
        of *Chlamydia trachomatis*, 294

[Lipid A]
    of *Haemophilus influenzae*, 294–295
        substrate specificity, 294
    lipid IV_A, 286, 293–297
    lipid IV_B, 296–297
    lipid X, 286, 292–293
        lpsH, 293
    lipid Y, 286, 296
    micelles of, 403
    molecular geometry, 211
    of *Neisseria meningitidis*, 291
    permealizer of the blood-brain barrier, 406
    phase diagram, 208
    phase transition, 205
    polar substituents, 297–298
        4-aminoarabinose, 297–298
        phoP regulatory system, 297, 299
        phosphoethanolamine, 297
        effect of polymyxin on, 297–298
    precursor Ia, 464–466
    of *Pseudomonas aeruginosa*, 296
    pyrogenic activity of, 406
    of *Rhodobacter sphaeroides*, 248, 249
    size of peptides binding to, 405, 410
    somnogenic activity of, 406
    structural analogs of, 250–254 (*see also* Structural analogs)
    structure of, 284–285
    supramolecular architecture of, 403
    synthetic, 207, 210
    tetraacyldisaccharide 1-P, 292, 293, 296
        4'-kinase, 293
        lpxB, 293
        lpxk, 296
        pgsA, 293
        pgsB, 293
    UDP-*N*-acetyl-D-glucosamine, 286–287
    UDP-*N*-acetylglucosamine acyltransferase, 287–291
        lpxA, 287, 289, 291
        substrate specificity, 287, 289, 291
    UDP-2,3-diacylglucosamine, 292–293
        lpxD, 287, 291–293
        synthesis of, 292
        substrate for tetraacyldisaccharide synthesis, 293
    UDP-3-*O*-monoacylglucosamine *N*-acyltransferase, 292
        lpxD, 292
        firA, 292
        ssc gene of *Salmonella enterica* serovar *typhimurium*, 292
Lipid A analog, 916

Lipid A associated protein (LAP):
  differentiation of signaling pathways in macrophages, 654
  induction of TNFα in macrophages, 654
  as mitogen for human and murine B cells, 652
  as polyclonal activator of B cells, 652
  as second signal in activation of C3H/HeJ mouse macrophages, 654
Lipid A backbone, 18, 94, 107
Lipid A conformation, 106
Lipid A core, as anchor:
  in *E. coli* K-antigen, 332
  in ECA, 332
  in *Pseudomonas aeruginosa*, 332
  in *Serratia marcescens*, 332
  in T1 and T2 antigens, 332
Lipid A mutants, 34
Lipid A of oral bacteria:
  bioactivity of, 906
  structure of, 902
Lipid A synthase, 254
Lipid B, 93
Lipid bilayer, 196
Lipid carrier (*see Undecaprenol phosphate*)
Lipid IV$_A$, 243
    lipid IV$_A$, 226
    hexaacyl bisphosphoryl lipid A (*E. coli*), 226
    pentaacyl bisphosphoryl lipid A (*R. sphaeroides*), 226
Lipid IVA, 464, 466
Lipid transfer protein, 213
Lipid X, 18
Lipocarbohydrate, 10
Lipooligosaccharide, 15, 331
Lipopeptide CGP 31362:
  induction of TNFα in macrophages, 654
  induction of tyrosine phosphorylation of proteins in macrophages, 654
Lipopeptide, 465
Lipopolysaccharide (LPS), 12, 13, 403–410, 439, 463–468, 497, 633 (*see also* Endotoxin, LPS, Shock):
  acute phase response to, 785, 790
  aggregates, 522
  aggregate structure, 208–209
  amphipathic structure of, 403
  analog, 916
  antagonists, 464, 466, 467
  antineoplastic activity, 916
  association with lipoproteins, 382
      reconstituted-high-density lipoprotein, endotoxin-neutralizing effects of, 384

[Lipopolysaccharide]
    in animal models, 384, 385
    in human endotoxemia, 385
  biological activity of, 403
  binding of, 463–468
  binding and neutralization of, 381
    high-density lipoprotein (HDL), 382
  binding protein, 781 (*see also* Lipopolysaccharide binding protein)
  capsule, *Vibrio cholerae* O139, 345
  endothelial cell apoptosis and, 504
  endotoxin-binding proteins, 385
    CD14, downregulation of, 385
  FITC-labelled, 466, 467
  fluidity, 196, 200
  gut-derived, 855–859
  heterogeneity, 40
  hyporesponsive, 919
  induced hemorrhagic necrosis, 915–919
  induced tumor regression, 915–919
  inducible genes, 501
  induction of chemokine proteins by, 407
  induction of proinflammatory cytokines by, 407
  inflammatory activity of, 407
  internalization, 467
  interaction mechanisms, 212–213
  interaction with cell membranes, 197
  interaction with complement, 201–202
  interaction with drugs, 202–204
  interaction with porins, 199–200
  lethality in mice and anti-LPS monoclonal antibodies, 637
  lipid A moiety of, 501
  macrophage responses to, 501
  molecular area, 198
  molecular shape, 196
  and monoclonal antibodies, 633
  monomers, 522
  Neisseria meningitidis Group a and B, 410
  of oral bacteria:
    composition, 900–902
    effects on endothelial cells, epithelial cells, fibroblasts, monocytes and neutrophils, 907
    strain differences in, 901, 903–904
    structure, 902–904
  phase variation, 39–51, 348
  physicochemical parameters in relation to bioactivity, 211–212
  pyrogenicity in rabbits and anti-LPS monoclonal antibodies, 637
  receptors, 463, 466, 784, 785
  rough (R-) and smooth (S-) chemotypes, 403

[Lipopolysaccharide]
and septic shock, 497
S-form, model, 210
tissue distribution of, 382
toxicity, 915–916
toxicity in target organs, 408
treatment of cancer, 916
Lipopolysaccharide binding protein (LBP), 197, 212–213, 463, 466–468, 521, 726, 796
and anti-LPS monoclonal antibodies, 636
pK of, 419
Lipoprotein:
association with lipopolysaccharides, 382
high-density lipoprotein (HDL), 379
metabolism of, 379
and modulation of LPS antibody responses, 656
low-density lipoprotein (LDL), 379
lipid composition of, 379, 380
reconstituted high-density lipoprotein (R-HDL), 384
endotoxin-neutralizing effects of, 384
transporters of, 380
cholesteryl ester transfer protein (CETP), 380
lecithin-cholesterol acyltransferase (LCAT), 380
lipopolysaccharide-binding protein (LBP), 380
very-low-density lipoprotein (VLDL), 379
Lipoteichoic acid, 465, 467
Lipotripeptide (P3CSS) as adjuvant, 653
Liquid crystalline phase, 205–206
Liquid secondary ion mass spectrometry (LSI-MS), 100
Listeria monocytogenes (see Bacteria)
L-Perosamine, 165
LPS (see Endotoxin, Lipopolysaccharide, Shock)
LPS antagonists:
deacylated LPS, 688
E5531, 688
lipid IVA, 688
lipid X, 688
R. capsulata LPS, 688
RsDPLA, 687–696
SDZ MRL 953, 688
LPS gene, role of:
in acute phase response, 785, 786
in shock, 785, 786
Lps gene:
Lps^n allele, 735–747
Lps^d allele, 735–747
mapping of, 735, 744–745, 747
LPS hyporesponsive mouse strains:
C57BL/10ScCR, 736, 747
C57BL/10ScN, 736, 747
BXH11/Ty, 736

[LPS gene]
C.C3H-Lps^d, 736, 745
C3H/HeJ, 735–747
LPS responsiveness, 735–747
suppressor phenotype, 737, 746
LPS polysaccharide of oral bacteria:
composition of, 904
role in bioactivity, 908
LPS, response to:
in mice deficient in cytokine receptors, 786–790
in mice deficient in cytokines, 786–790
LPSd, 719
LPS^n gene, 652
LPSn, 719
LPS-OmpA complexes and modulation of LPS antibody responses, 657, 658
lpxA, 34, 36
lpxC, 34
lpxD, 34
L-selectin, 800
Lysogenic phase conversion, 346, 347
and serum resistance, 347
in Acetobacter methanoliticus, 349
in Klebsiella, 347
in Pseudomonas aeruginosa, 347
in Salmonella enterica, 346–347
in Shigella flexneri, 347

Macaca monkeys, 796
Models:
bacterial challenge, 797
live E. coli-multiple challenge, 798
Macrophage, 437, 449, 456, 457, 458, 463, 467, 521, 538–545 (see also monocytes/macrophages)
activation, 917
alveolar macrophages, 737
chemokine, 917
cytokine, 917–918
endocytosis, 740
mediated antitumor activity, 917–918
membranes, 742–743
cell surface proteins, 742
chaperone proteins, 742
signaling molecules: intracellular, membrane-associated, Rsk, Vav, 742–743
migration inhibitory factor (MIF), 775
tumor cell interaction, 918
"priming" of, 738
stimulation by LPS, 651
structural proteins of, 742
uptake of LPS by, 658–661
Western blot analysis of, 742

Macropinocytosis, 527
Malaria, 5
MALDI-TOF mass spectrometry, 158
manB, 350
manC, 350
Mannan, 166
Mannose binding protein (MBP):
  binding to C1q, 80
  MBP-associated serine protease, 80
  role in opsinization, 86
MAP kinase, 464, 474
Mass spectrometry, 95
Matrix metalloproteinases (MMP), 909
Matrix-assisted laser desorption/ionization mass spectrometry (MALDI-MS), 100, 101, 102
MEK, 475
Membrane:
  adhesion sites, 344–345 (*see also* Bayer junctions)
  function, 199
  fusion proteins, 344–345
  potential, 197–199
Membrane attack complex (MAC), 201, 331
  composition of, 77
  generation of, 78, 81
  mechanism of bacteriolysis, 81
  role in phagocyte intracellular killing, 87
Memory T cells, 648
Meningitis, gram-negative, 832
Meningococcemia, 844, 845
Microsporon septicum, 6
Mitogen activated protein kinase (MAPK), 539, 544
Mitogen, 651
MO-calculations, 106
Modality, 339, 341
Modulating inflammatory responses to endotoxin, 820–827
  E5531 (*Rhodobacter capsulatus* lipid A), 820, 822–823
  endotoxin tolerance, 821
  epinephrine, continuous infusion of, 822–823, 825
  fat (triglycerides and lecithin) emulsions, 821, 822–823
  glucocorticoids, 821, 822–823
  granulocyte-colony stimulating factor, administration of, 824, 826
  hydrocortisone stress doses, bolus, 821, 822–823
  hydrocortisone stress doses, infusion, 821, 822–823
  ibuprofen, 822–823, 826
  IL-1ra, recombinant, 824–825
  IL-1 receptor soluble type I, recombinant, 824–825
  IL-10, recombinant, 824, 826
  monophosphorylated lipid A, 821

[Modulating inflammatory responses to endotoxin]
  parenteral nutrition, 821
  pentoxifylline, 826–827
  platelet activating factor (PAF)-receptor antagonist (Ro 24-4736), 827
  recombinant bactericidal/permeability-increasing protein (rBPI$_{23}$), 820, 822–823
  reconstituted human high density lipoprotein (rHDL), 821, 822–823
  TNF receptor fusion protein (TNFR:Fx) recombinant dimeric, 824–825
Molecular conformation, 209–211
Monoclonal antibodies, 234
  anti-core lipopolysaccharide, 633
    and bacterial models in mice, 639
    binding of, 634
    and complement activation, 637
    and IL-6, 636
    and ischemia reperfusion injury in rats, 638
    and lethality in mice, 637
    LPS neutralization of, 636
    and passive hemolysis, 635
    and pyrogenicity in rabbits, 637
    specificity of, 634
    and TNF, 636, 637, 639
  HA-1A, 634
  to LPS from *Proteus mirabilis*, 658
  to OmpA from *Proteus mirabilis*, 652, 658
  SDZ 219-800, 635
  WN1 225-5, 634
Monocyte chemotactic protein-1 (MCP-1), 907
Monocytes, 463–467, 521 (*see also* monocytes/macrophages)
Monocytes/macrophages, CD14 on, 781
Monomeric ReLPS, 224
Mononuclear cell infiltration, 917
Montal-Mueller technique, 199
Moraxella catarrhalis:
  synthesis of inner core units, 272, 274
  synthesis of outer core units, 272, 274
MS/MS-analysis, 100
msbB mutant, 108
msbB, 34
Multimodal distributions, 341
Multiple organ failure (MOF), 797
Muramyl dipeptide (MDP), 464
Mycobacterium phlei, 728

Necrosis, 679
Neisseria gonorrhoeae LPS, 39
  digalactoside, 40, 49
  lacto-N-neotetraose, 40, 49–50

[*Neisseria gonorrhoeae*]
    monoclonal antibodies, 48, 50–51
    phase variation, 48–50
        homopolymeric tract, 49–51
        lgt locus, 49–50
    virulence, 39–40
Neisseria meningitidis LPS, 39
    digalactoside, 40
    lacto-N-neotetraose, 40, 49–50
    molecular mimicry, 40
    monoclonal antibodies, 41, 48–49
    phase variation, 48–50
        homopolymeric tract, 49–51
        lgt locus, 49–50
    sialic acid, 40, 50
    virulence, 39–40, 48, 50
Neisseria meningitidis, 843–844:
    artificial antigen of *N. meningitidis* group B, 269
    peptide conjugate of *N. meningitidis* immunotype, 6, 267–269
    synthesis of inner core units from *N. meningitidis* immunotypes L1-L9, 267, 268
Neoglycoproteins, 258, 263, 269
Neopterin, 801
Neutropenia, 844, 851–852
Neutrophils, 521, 671
    CD14 on, 781
NFκB, 509–517, 596–598
    activation mechanisms of, 510–515
        role of IκBα protein kinases in, 510–512
        role of IκBα proteolysis in, 512–514
    binding affinity of, 509–510
    biological significance of, 515
    in *iNOS* gene activation, 515–516
    in LPS tolerance, 516
    inhibition (down-regulation) of, 512, 514, 515–517
    properties of, 509
    translocation, 464, 466
N-Formyl-methionine-leucine-phenylalanine, 453, 454
Nitric oxide synthase, 479, 591
    isoforms of, 592
    feedback inhibition of, 596
Nitric oxide (NO), 541, 543, 544, 799, 857, 917–918
    as an antioxidant, 595
    as a cellular signal, 591
    biochemistry of, 592–593
    biological effects of, 593, Table 1
    biosynthesis of, 591–592
    nitrosative stress caused by, 596
    pathophysiological role in sepsis, 593
Nitrogen fixation, 172
NMR (nuclear magnetic resonance), 96, 103

NodX, 347
Non-carbohydrate substitutions, 332, 338–339, 347
Noninitiating transferases, 335–336
Nonprocessive β-glycosyltransferases, 344
Non-reducing terminus, 334, 341–342
Nonulosonic acid, 158, 162, 164, 169
Nuclear Overhauser effect (NOE), 106
NusG, 348

O unit transporter, 351
oac, 347
O-acetyl transferase, 347
O-acetylation, 339, 347
oafA, 347
O-antigen 2 (*see also* O-polysaccharide):
    evasion of host immune responses, 331
    serotype specificity, 331
O-antigen, 11, 13
    capsule, 349
    complex, 10
O-antigenic oligosaccharides, coupled to *S. typhimurium* porins, and protection of mice, 654
O-chain length, 339, 341, 343, 347
Ochrobactrum, 172
O-deacylated lipid A, 104
Ogawa, 349
O-hapten, 339
OmpA:
    and enhancement of uptake of LPS by macrophages, 658-651
    modulation of LPS antibody responses, 656
    modulation of LPS-induced cytokines, 661
    modulation of LPS stimulation of macrophages, 658
    proteosome character of, 658
omsA, 34
Operon polarity suppressor, 348
O-polysaccharide, 346–347
    ABC-transport-dependent, 341–342
        O chain length distribution, 343
    assembly:
        ABC transport-dependent, 334
        synthase-dependent, 334
        Wzy-dependent, 334, 339
    biosynthesis cluster, 345–346
        Enterobacteriaceae, 345
        G+C, 349
        in *E. coli* O111, 346
        non-enteric, 346
        operon structure, 346
        in *P. aeruginosa*, 346
        in *Salmonella enterica* O54, 347, 351
        in *Shigella boydii*, 346

[O-polysaccharide]
- in *Shigella dysenteriae*, 346
- in *Shigella flexneri*, 346
- in *Shigella sonnei*, 346
- transcriptional regulation, 348
- export to the cell surface, 351
- growth conditions, 347
- heterogeneity, 332
- host immune responses, 331
- *Klebsiella pneumoniae*, 332
- initiation reactions, 335–338
- ligation, 344
- multimodal distributions, 341
- polymerase specificity, 351
- polymerization, 339
    - ABC-transport-dependent, 341–342
- repeating unit structure, 331–332
- SDS-PAGE, 332
- *Serratia plymuthica*, 332
- synthase-dependent, 344
- transporter, ABC, 342–343
- Wzy-dependent, 343
    - chain length distribution, 339, 341

ops, 347, 348
Opsonization, 86–87, 671
O-repeat unit, 3, 17, 29, 30
O-specific chain, 15
O-specific side chain, 13, 14
otnA, 345
otnB, 345
Outer core assembly, 314–321
Outer membrane, 31–36
- function, 199
- permeability barrier, 32–36
- proteins, 651
    - and activation of macrophages, 654
    - OmpA, OmpC, OmpF, nitrogenic activities of, 652

Oxidative stress (*see* Stress, redox)
Oxidized LDL, 437, 439

P. acnes, sensitization to shock with, 787
P15s, 372:
- cathelicidins, 372
- localization, 372
    - polymorphonuclear leukocytes, 372
    - granules, 372
    - inflammatory fluids, 371–372
- function, 372
    - anti-bacterial, 372
    - anti-LPS, 372
- structure, disulfides, 372

[P15s]
- synergy, 371–372
    - BPI, 371–372
    - other, 372

p70 S6 kinase, 478
PAF:
- antagonists inhibit sepsis, 566–571
- cellular production of, 561
- critique of measurement assays, 572–573
- discovery of, 561
- efficacy of antagonists, 573
- elevation in human sepsis, 563, 573–574
- infusion mimics sepsis, 563–566
- interaction with endotoxin, 574
- interaction with other mediators of sepsis, 571–572
- production in animal models, 561–563
- synergistic relationship with endotoxin, 571

Partition column chromatography of lipid A, 246
Pattern recognition receptor, 467
Pectinatus, 172
Penicillin-binding proteins (PBPs), 891–892
Pentoxifylline, 800
Peptides, 403–410
- accumulation in tissues of, 409
- affinity of binding, 403
- algorithmn for prediction of, 405
- amphipathicity, 404
- antagonists of calmodulin, 408
- antibiotic activity of, 405
- binding stoichiometry of, 403, 408
- binding to lipid A, 403, 405, 410
- binding, interference by serum, 408
- biodegradability of, 409
- cationic, 403
- competitive inhibition, 404
- complex formation with lipid A, 404
- composition of, 403
- cyclic, 403
- D-amino acid analogs, 408
- detoxification mechanism of, 408
- detoxification of LPS-based vaccines by, 409
- diffusion to organ of, 409
- half-life time of, 408
- immunogenic activity of, 409
- inhibition of TNF by, 408
- inhibitors of phosphodiesterases, 408
- inhibitors of phosphokinases, 408
- LPS removal by, 409
- minimal inhibitory concentration (MIC) of, 407
- NMR spectra of, 405
- optimal sequence of, 405, 410
- organ-specific TNF inhibition by, 409

[Peptides]
    peptide-activated solid phase matrixes, 409
    predicted in the BPI structure, 406
    predicted in the CD14 structure, 406
    predicted in the LALF structure, 406
    predicted in the LBP structure, 406
    predicted in the LEBP-PI structure, 406
    primary structure of, 404
    prophylactic activity, 408
    SAEP, 404
    safety of, 409
    secondary structure of, 404
    selectivity of, 405
    size comparison to protein epitopes, 410
    synergism with antibiotics, 405, 407
    synthetic, 403
    therapeutic activity, 408
    tertiary structure of, 405
    treatment of endotoxemia by, 409
    treatment of sepsis by, 409
Peptidoglycan, 463–465, 467
    soluble, 464, 465, 467
    FITC-labelled, 464, 467
    binding of, 464, 467
    internalization, 467
Periodontitis, 899, 909–910
    juvenile localized, 667
Permeability, intestinal epithelial, 857
Pertussis toxin, 537–540, 544
Phage, 339, 347 (*see also* Bacteriophage)
Phage ε, 15, 29
Phage conversion (*see* Lysogenic phage conversion)
Phagocytosis, 522
Phase variation, 39–51
    mechanism, 41, 50
Phase:
    diagram, 208–209
    states, 205–206
    transition, 200, 205, 206
Phenol-water extraction method, 12, 172, 651
PhoP-PhoQ, 35
Phosphate substituents, 95
Phosphate-linked substituents, 96
Phosphatidylinositol 3-kinase, 478
Phospholipase A_2, 373, 449, 452, 458
    localization, 373
        circulation, 373
        inflammatory fluids, 373
    function, 373
        anti-bacterial-(gram-negative bacteria), 373
        anti-bacterial-(gram-positive bacteria), 373
        inflammation, 373

[Phospholipase A_2]
    structure, 373
    synergy, 373
        BPI, 373
        p15s, 373
        complement, 373
Phospholipase A_3, 774
Phospholipid leaflet, 198
Phospholipids, 449, 454, 455, 468
Phytopathogen, 166
Pig, 797
PKB, 478
PKC-ζ, 499
Planar lipid bilayer, 198–199
Plaque-forming cells (PFC), 655–657
Plasma:
    effects on Limulus amebocyte lysate assay, 842
    pre-treatment methods, 842
    desorption mass spectrometry (PD-MS), 101
Plasmid-encoded biosynthesis determinants, 346–347
Plasmid encoding, 162
Plasmodium chabaudi chabaudi, 723
Platelet activating factor (PAF), 739, 744, 801
    acetyl-hydrolase, 456, 459
    antagonist, 800
    biosynthesis, 449, 458
    membrane-bound, 455, 456
    receptor:
        antagonists, 456, 458, 459
        cloning, 451
        expression, 454, 458
        signaling, 452
    release, 449
Plesiomonas, 165
pmrA, 34–35
pmrB, 35
Polyacrylamide copolymers, 258, 259, 263, 265, 267, 269
Polycationic drugs, 202–203
Polylysines, 34
Polymerase, specificity, 351
Polymerization, 334, 339
    SDS-PAGE analysis, 332
    vectorial, 334
Polymorphonuclear neutrophils (PMN), 801
Polymyxin B, 34, 202–203, 403–408, 417, 831, 836
    (*see also* SimpliRED endotoxin test)
    affinity of binding, 403
    binding stoichiometry of, 403, 408
    binding to lipid A, 403
    biodegradability of, 404
    composition of, 403
    proteolytic stability of, 404

Polypeptides, 405–410
  agonists of opioid receptors, 407
  binding to LPS, 405
  cecropins, 405
  defensins, 405
  defensins-related peptides, 405
  dynorphin A, 407
  LPS recognition by, 410
  magainins, 405
  melittins, 405
  natural antibiotics, 405
  natural sequences, 405
  neurotactin, 407
  nociceptin or orphanin FQ, 407
  opioid-related peptides, 406, 407
  permealizer of the outer membrane, 405
  rabbit cationic proteins, 405
  tachyplesin, 405
  tubulin, 406
Porins, 199–201, 345
Postsource decay (PSD), 102
Primary fatty acids, 94
Primary toxicity, 19
Processive β-glycosyltransferases, 344
Proinflammatory cytokines, antibody-mediated inhibition of, 866, 868–869
Propionibacterium acnes, 720
Prostaglandin (PGE$_2$), 917–918
Prostaglandin E2, 907, 909
Protamine, 34
Protein kinase C, role in activation of C3H/HeJ B cells, 652
Protein T, 10
Proteins, 405
  binding to LPS, 405
  BPI, 405
  CD14, 405
  LALF, 405
  LBP, 405
  LEBP-PI, 405
  receptors of LPS, 405
Proteobacteria, 172
Proteosomes:
  complexes with LPS from Shigella flexneri 2a or Shigella connei, 655, 657
  modulation of LPS antibody responses, 655
  role as adjuvants, 654
Proteus mirabilis 35
  OmpA of, 652, 653, 656–661
    antibody-producing cell responses to, 656
    outer membrane components of, 656
Proteus, 159

Proton-carbon heteronuclear multiple-quantum coherence NMR spectroscopy (^{1}H,^{13}C-HMQC), 105
Providencia, 162
Pseudomonas aeruginosa, 332, 338, 346, 347, 843–844, 851–852
  and anti-LPS monoclonal antibodies, 633
  O5, 338
  Wzz, 346
  and OprI-OprF fusion protein, 653
  and outer membrane proteins OprF, OprH, OprI as adjuvants, 653
Pseudomonas cepacia, 35
Pseudomonas, 166
Purification of synthetic lipid A:
  by centrifugal partition chromatography (CPC), 246
  by partition column chromatography, 246
Putrefaction, 4
Pyranose, 332
Pyretogenine, 6
Pyrogen, 831
  chemical nature of, 831
  compendial test, 831, 836–837
  rabbit bioassay for, 831, 836–837
Pyrogenic material, 5
Pyrotoxina bacterica, 8, 9
Pyrrolidine dithiocarbamate, 597

Rabbit bio-assay, 843
Rabbit, 797
Radiation, 499
  ionizing, 499
  SMase and, 499
Radio-labeled lipid A analog, 251
Rat, 797
RAW 264.7 macrophages, and IL-6 production, modulation by staurosporine, 663
Reactive oxygen species, 591, 594
Reconstituted outer membrane, 198–199
Reducing terminus, 334
Regulation, 347
Re-lipopolysaccharide:
  synthesis of 1-dephospho LPS, 263, 265
  synthesis of Kdo units, 258
Renal tubular necrosis, 684
Repeat unit, fidelity, 332, 334, 341–342
Repeating units, 14
Repressor, 347
Resolution of 3-hydroxy fatty acids, 244
rfa, 334
rfaD, 346
rfaH, 348
rfb, 334

rfbE, 349
rfbK, 350
rfbT, 349
Rhizobium, 172
Rifampin, 33
RsDPLA, 771
   mechanism of action of, 688–696
   structure of, 688–689
RU 486, 769
RU38486, 680

Salmonella, 162, 633
   abortus equi, 724
   and anti-LPS monoclonal antibodies, 634
   enterica serovar *minnesota* R345, synthesis of core pentasaccharide, 258, 260, 261
   enterica serovar *minnesota* R4, 267
   enterica serovar *minnesota* R7, 267
   enterica serovar *minnesota* R*e* 595, 258
   enteritidis LPS, antibody-producing cell responses to, 655
   minnesota R595, 633
   synthesis of heptose region, 269–271
   synthesis of outer core units, 272, 273
   synthesis of R*a* core structures, 272, 273
   typhimurium, 31–36, 710, 856 (*see also* Bacteria)
      as carrier of antigens, 654
      porins and activation of macrophages, 654
Scavenger receptor, 437
Schistosoma mansoni, and induction of T-helper cell subsets, 658
Secondary fatty acids, 94
Secretory leukocyte protease inhibitor (SLPI), 741–742
Sepsin, 5
Sepsis, 23, 383, 449, 456, 458, 497
   and endotoxemia, 389
   lipid metabolism, 383
   mortality from, 398–399
   definition of, 497
   pathogenesis of, 497
   syndrome, 845–847
      epidemiology, 846
      and organ failure, 845
Sepsis, animal models of:
   attenuation by PAF antagonists, 566–571
   clinical relevance of, 573
   PAF production in, 561–563
Sepsis, human, 561
   elevation of PAF in, 573–574
Septic shock, 497
   ceramide in, 503–504

[Septic shock]
   immunotherapy of, 868–874
   pathogenesis of, 865–866
   TNF-α in, 497
Sereny tests in guinea pigs, 655
Serotype b antigen, 668
Serum complement, 673
Serum resistance, mechanisms of, 81–82
Shc, 475
Sheep, 797
Shigella:
   flexneri serotype, 6, 271
   sonnei, 271
   spp., and anti-LPS monoclonal antibodies, 633
Shigellosis, 846
Shock, endotoxin:
   in mice deficient in cytokine receptors, 786–791
   in mice deficient in cytokines, 786–791
   role of cytokine receptors in, 786–791
   role of cytokines in, 782, 783, 786–791
   symptoms of, 781, 783
Shock, in mice deficient in:
   α2-macroglobulin, 787
   5-lipoxygenase, 790, 791
   GM-CSF, 787
   ICAM-a, 786, 790
   ICE, 786
   IFNγR, 790
   IL-1, 786
   IL-1R, 786
   IL-6, 791
   I-NOS, 791
   LBP, 787
   LDLR, 786
   TNFR, 787
Shock, traumatic, 797
Shwartzman reaction, 682, 831, 906, 919
Signal transduction, 464, 466, 467
   protein, 197, 212
   smooth LPS, 222
Smooth muscle cells, 463
Solubility of ReLPS, 224
Soluble TNF receptors, therapy with, 867, 869–870
Sphingolipids, 465
Sphingomonas paucimobilis, 35
Sphingomyelin pathway, 497–498
   inflammation and, 500
   receptors linked to, 498
   sphingomyelinases in, 498
Sphingomyelin, 497
Sphingomyelinase (SMase), 498 (*see also* Ceramide)
   acid, 498

[Sphingomyelinase]
alkaline, 498
Nieman-Pick disease and, 498, 502
neutral, 498
receptors linked to, 498
zinc stimulation of, 500
Spontaneous neoplasm, 917
ssc, 34
S-specific side chain, 14
Staphylococcus aureus, 728
Staurosporine:
effects on cytokine production in macrophages, 663
modulation of IL-6 production in RAW 264.7 macrophages, 663
Stereochemistry of 3-hydroxylated fatty acids, 97
Stress, redox:
role of nitric oxide in, 594 (Table 2)
in sepsis, 594–595
Structural analogs of lipid A:
monosaccharide, 250
disaccharides, 250–254
carboxylated, 253
phosphonooxyethyl, 251
tritium-labeled, 251
Structural analysis, 233
Supramolecular structure, 206–209
Syk, 474
Syndrome:
multiple organ dysfunction (MODS), 855, 856, 858, 859
systemic inflammatory response (SIRS), 855
Synthesis of lipid A, 243–256
Systemic inflammatory response syndrome and organ failure, 845

T cell:
activation, 918
deficient mice, 918
dependent antibody responses, 653, 657
receptor (TCR), 644
response to LPS, 643–650
suppressor T cell, 918
Tachyplesin, 34
Taxol, 739–740
Tetanus toxoid, C fragment of, 658
and induction of T-helper cell subsets, 658
T-helper cell subsets:
and cytokine production, 658
induction of, 658
Therapeutic interventions, 802
THP-1 cells, 466, 522, 539, 540, 544
Thrombomodulin, 921

Thromboxane B_2, 539, 540, 544
Tissue factor, 389, 394–395 (*see also* endotoxin)
pathway inhibitor (TFPI, 801
TLR4, 719, 747
TNF-α (*see* Tumor necrosis factor alpha)
TNF (*see* Tumor necrosis factor)
Toxalbumin, 7
Toxic carbohydrate, 10
Toxic component *T*, 10
Toxin, 6
Toxophore, 19
Transcription:
factors, 596–598
NF-κB, 690
Sp1, 692
LPS-induced, 598–599
nitric oxide effects on, 595–598
Transferrin, 523
Translocation, 797, 802
microbial, 856, 857, 859
Trichloroacetic acid, and LPS extraction, 651
Tripalmitoyl pentapeptide (P3CSSNA), enhancement of trinitrophenyl (TNP)-SRBC responses, 653
Tris, 34
Tritium-labeled lipid A analog, 252
Tumor necrosis factor alpha (TNF-α), 443, 497, 499–500, 541, 544, 907–908
overproduction, 721
production by macrophages, 651, 661, 663
Tumor necrosis factor (TNF), 397, 398, 456, 457, 458, 464, 466, 473, 799 (*see also* Endotoxin)
antibody, 800, 919
and anti-LPS monoclonal antibodies, 636, 637, 639
antiproliferative effect, 919–921
cytotoxicity, 919
induced apoptosis, 919–920
induced necrosis, 919
procoagulant activity, 919
receptor, 498, 800, 919, 920
toxicity, 915, 921
Tumor:
immunogenicity, 920
infiltrating macrophage, 917
necrosis, 915–921
regression, 915–921
vasculature, 920
Two dimensional nuclear Overhauser effect spectroscopy (NOESY), 106
Two-dimensional (2D) ^{1}H-NMR spectroscopy, 103
Typhoid fever, 846
Typhotoxin, 6
Tyrosine kinase, 464, 466

U373 cells, 465–467
Undecaprenylphosphate, 286
Unsaturated fatty acids, 97
Urine, 841, 843
Urosepsis, 851

Vancomycin, 33
Vascular leakage, 451, 458
Vasoactive agent, 919
Vasodilatation, 919

Vibrio, 263
  vibrio ordalii, 272
Voltage-clamp, 199

waaN mutant, 108
Warfarin, 415
wecA, 286

Z,E(*cis/trans*)-configuration of fatty acids, 97
Zymoid, 6